Abbreviations Dictionary

Eighth Edition

Ralph De Sola

★ Abbreviations ★ Acronyms ★ Airlines and Airports
★ Appellations ★ Astronomical Terminology ★ Bafflegab
Divulged (euphemisms explained) ★ Birthstones ★ British and
Irish County Abbreviations ★ Canadian Provinces ★ Chemical
Elements ★ Citizen's-Band Call Signs ★ Computer Jargon
★ Contractions ★ Corrections Facilities ★ Criminalistic Terms
★ Data Processing ★ Diacritical Marks ★ Dysphemistic
Place-Names ★ Earthquake Data ★ Eponyms ★ Fishing Ports
★ Geographical Equivalents ★ Government Agencies ★ Greek
Alphabet ★ Historical, Musical and Mythological Data
★ International Conversions ★ International Vehicle License
Letters ★ Inventions and Inventors ★ Medical Terms ★ Nations
★ Nicknames ★ Numbered Abbreviations ★ Numeration
★ Phobias ★ Ports of the World ★ Railroads ★ Roman Numerals
★ Russian Alphabet ★ Short Cuts ★ Short Forms ★ Signs &
Symbols ★ Slang ★ Steamship Lines ★ Superlatives ★ Weather
Symbols (Beaufort Scale) ★ Wedding Anniversaries ★ Winds of
the World ★ Zip Coding ★ Zodiac

CRC Press
Boca Raton   Ann Arbor   London   Tokyo

Library of Congress Cataloging-in-Publication Data

De Sola, Ralph, 1908–
    Abbreviations dictionary / Ralph De Sola. — 8th ed.
      p.   cm.
    ISBN 0-8493-4247-3
    1. Abbreviations, English—Dictionaries. 2. Signs and symbols—
Dictionaries. 3. Acronyms. I. Title.
PE1693.D4   1992
423′.1—dc20   91-26324
                                   CIP

Direct all inquiries to CRC Press, Inc., 2000 Corporate Blvd., N.W., Boca Raton, Florida, 33431.

© 1992 by CRC Press, Inc.

International Standard Book Number 0-8493-4247-3

Library of Congress Card Number 91-26324

Printed in the United States of America   1 2 3 4 5 6 7 8 9 0

This is the short and the long of it.

—*Shakespeare*

Contents

Preface

Contemporary conversation and printed communication continue to be filled with undefined abbreviations, acronyms, appellations, contractions, geographical equivalents, initialisms, nicknames, and a host of specialized terms occupying more than 25 percent of the mass of words we hear or see in print. And anyone from another discipline, industry, profession, or occupation is almost completely baffled by such talk and writing.

This expanded and revised eighth edition of the *Abbreviations Dictionary* contains many items not found in other reference works: bell-code signals, Canadian provinces, Mexican states, nations of the world, zip-coded automatic-processing abbreviations, plus a host of criminal, medical, and military terms.

New items of interest have been collected from newspapers and other periodicals published in many parts of the world. The author's summertime trips aboard freighters produced entries from all parts of the world. Some appear in his *Worldwide What & Where,* a geographical glossary and traveler's guide, while others are duplicated, in part, in his *Crime Dictionary.* The underground and the underworld have not been overlooked. The Law Enforcement Assistance Administration was most helpful. The compiler's ongoing effort to create order out of abbreviatorial acronymical chaos continues. Extensive and intensive listening, looking, and reading reveal new short forms emerging daily. And the publisher's staff is forever plagued by the author's steady flow of so-called last-minute entries deserving of inclusion. Every effort is made to keep this reference up to date.

Readers and reference librarians are again solicited to direct new or revised findings to the authors.

Bureaucratically buttressed government creates a host of new short forms. Only old timers seem to recall the attempts of President Hoover to consolidate and streamline government, although at least eight presidents since then have pledged themselves and their administration to stop creating more agencies, more bureaus, more commissions, more committees, and more governmental adornments laden with special abbreviations and acronyms. The end is not in sight. This expanded international eighth edition contains more abbreviations, acronyms, and other short forms than ever before.

RALPH DE SOLA

Acknowledgments

Arthur EE Ivory of Christchurch, New Zealand traded much material found in previous editions of our *Abbreviated Dictionary* for new items in his *Pacific Index of Abbreviations and Acronyms in Common Use in the Pacific Basin.* His splendid compendium was published in 1982 by Whitcoulls.

William A Reid of Union Postal Universel in Bern, Switzerland provided excellent up-to-date information about touring club abbreviations the world over.

MA Brossard, Secretary General of Interpol in Saint-Cloud, France, was most helpful and sent very useful material about prisons.

Dr. Irmaisabel Lovera De-Sola of Caracas, Venezuela furnished most of the Venezuelan entries and others from other Latin American lands. She also put us in touch with Venezuela's best-known criminologist—Dr Helio Gómez Grillo and he also gave us entries.

Book lover and journalist Dean A Stahl of San Diego submitted more than two-hundred-and-fifty new entries plus some more from Karen Kerchelich, a well-known San Diego researcher-writer. Someday one or the other might prove eligible to take on the task of continuing editions of this *Abbreviations Dictionary.*

Expatriate New Yorkers John and Fay Silverstein continued to supply vital information as did ex-New Yorker Lorraine Sherkin of Toronto, Ontario and her daughter Kari.

Another Canadian source was David Allen of Halifax, Nova Scotia.

The largest contribution in the field of education came from James C Palmer who compiled and edited the *ERIC Dictionary of Educational Acronyms, Abbreviations and Initialisms* published by the ERIC Clearinghouse for Junior Colleges in 1981.

Lots of liturgical abbreviations were sent by Brother John-Charles of the Society of Saint Francis American Province in Mount Sinai, New York. Another contributor was Mary Bucher, the senior writer of the Battelle Columbus Laboratories in Columbus, Ohio.

Maritime contributors include Chief Officer Aage I Helde of the *Salvador* of the Ivaran Lines, Chief Officer Andrew Milligan of the Blue Star Line's *California Star,* and Chief Petty Officer Martin Parr of the United States Navy.

Very recently deceased contributors include my pen pal of some 40 years standing: Julius B Kaiser of New York and Hollywood, and Harold Q Driscoll of San Diego. Often they would contribute ideas and items on an almost weekly basis.

San Diegans who helped include a former student: Anthony Marquez; Harold Cary and his son Dr David Cary; Dr Ira Levine; Dr Warren Kessler; Dr Rodrigo Muñoz.

The *List of Acronyms* compiled by Linda Blocki, technical editor and optics librarian of the University of Daton Research Institute at Kirtland Air Force Base in New Mexico, was most welcome. Another contributor was Professor PA Doyle of Williston Park, New York.

San Diego's many reference librarians were most helpful and include the names of Keith Anderson, Alyce and Michael Archuleta, Marianne Avila, Eileen Boyle, Elizabeth Byrne, Christina Clifford, Lucy van Donck and her cousin Laura Gulotta, Debbie Gray, Sally Hamburg, Susanna Hardy, Jean Hughes, Matt Katka, Evelyn Roy Kooperman, Curt Lang, Jean Lowerison, Anna M Martinez, Sharon Nelson, Jeanne C Newhouse, Angela Patterson, Margaret Queen, Margo Sasse, Jim Shaff, Don Silva, Lyn Slomowitz, Barbara Tuthill, John Vanderby, Vere Wolf, and many others whose names appear in previous editions of the *Abbreviations Dictionary*.

Dr Ronda De Sola Chervin's enthusiasm for the task of selecting and defining euphemisms was contagious and helped flesh out the Bafflegab Divulged addendum suggested by Lucy van Donck. We thank both of them and all the foregoing including those mentioned in previous editions, and that indefatigable booster, Dan Pezze of Lakewood, New Jersey.

The entire staff at Elsevier continued to aid in the editing, marketing, and production of this and previous editions. Special thanks go to Caryl P Dreiblatt, Edmée Froment, Louise Calabro Gruendel, Marion Hess, Ethel G Langlois, Linda Leopold, Dr AW Kenneth Metzner, Phyllis Oehler, and Toni Ann Scaramuzzo of Elsevier, and to freelancers Charles Beaulein, Paul Duchesne, Chris Schreiber, and Maria Schreiber.

Maura Grant, working with Unitron Graphics, also deserves special thanks as this eighth edition of the *Abbreviations Dictionary* goes to press.

The largest contributor of new short forms was my good neighbor—Julian Gotkiewicz. Almost daily he would bring some periodicals loaded with short forms.

Many who helped include Vanessa Browne in England, Mell Carey in Nepal, Jane Conrad in Colorado, the late Paul B Hoeber in New York, Samuel Lubin in Tel-Aviv, Chris Mahan of Metropolitan Opera fame, Marge Mignacca in North Syracuse, Alice Picard in Staten Island, Daniel Russell in South Mission Hills, Dr Walter Teutsch in Point Loma, and Ralph de Toledano in Washington DC.

CRC's staff, who made this expanded edition possible, included Joel Claypool, executive editor, and Sandy Pearlman, director of production services. Special thanks go to Camilla Ayers, copy editor, and Todd Solomon and Anthony de Luna of Innodata Processing Corporation.

Introduction

Definition of Terms

abbreviations abridged contractions such as acdt: accident; AEC: Atomic Energy Commission; NASA: National Aeronautics and Space Administration.

acronyms words formed from letters in a series of related words such as ABLE: Activity Balance Line Evaluation; AGREE: Advisory Group on Reliability of Electronic Equipment; DYNAMO: Dynamic Action Management Operations.

anonyms attempts of authors to enjoy anonymity while maintaining their identity by such devices as the capitalized diphthong AE standing for Aeon, pen name of George William Russell.

contractions words shortened by dropping nonpronounced letters; omitted letter(s) which are indicated by apostrophes as in can't: can not; li'l: little; doesn't: does not; let's: let us.

eponyms designations derived from family names, nicknames or names of places or persons, e.g., Hapsburg dynasty, *Eroica* symphony, Paris of America (Montreal), Raynaud's disease.

geographical equivalents entries such as Far East: countries and islands of East Asia or the Pacific—eastern Siberia, China, Japan, Taiwan, Korea, Indochina, the Philippines, the Malay Peninsula.

initials FDR: Franklin Delano Roosevelt; HST: Harry S Truman; JFK: John Fitzgerald Kennedy; LBJ: Lyndon Baines Johnson; initials of all American Presidents are included as well as initials of other noted personalities.

nicknames Al: Alfred; Bea: Beatrice; Hal: Harold; Ike: Dwight David Eisenhower; Isaac.

place name pseudonyms see *Cannery Row, Main Street, Middletown, Red Gap, Spoon River, Tortilla Flat, Winesburg, Yoknapatawpha, Zenith* entries.

short forms amps: amperes; Olds: Oldsmobile; pots.: potentiometers.

signs $ & ¢—dollars and cents.

slang shortcuts B-girl: bar girl; C-note: $100 bill; !-G: $1000.

symbols Al: aluminum; Pt: platinum; Rx: prescription; recipe.

toponym splace names convicts use when telling you were they were imprisoned.

Editors–Teachers–Writers

Editors, teachers, and writers will perform a splendid service for readers if they insist that abbreviations and acronyms be defined the first time they are used. The old argument, "everyone knows what that stands for," no longer is true. Many abbreviations and acronyms stand for several different things.

The style of writing abbreviations and acronyms requires the attention of editors, teachers, and writers so that paragraphs do not become cluttered with unexplained capital-letter combinations. Technical literature will become almost impossible to read if the permissive trend continues wherein all abbreviations and acronyms appear in solid capital letters and without benefit of preliminary definition.

Throughout this *Abbreviations Dictionary* an attempt is made to follow the rules of English grammar. Capital letters are reserved for proper nouns. Lowercase letters are used for common nouns. However, when custom has become so strong that correctly written short forms are not recognized quickly, their more common equivalents are added parenthetically; icbm (ICBM): intercontinental ballistic missile.

Explanations

If readers and researchers did not continue to find themselves engulfed and ensnared in the modern abracadabra of abbreviations and acronyms, in the bewildering bafflegab and gobbledygook of computerese, corporationese, initialese, officialese, pentagonese, politicalese, and technicalese, there would be no need to provide this new international eighth edition of the *Abbreviations Dictionary*.

Because so many creators of abbreviations and coiners of acronyms fail to define their shortcuts the first time they use them, and because so many who use them also fail to define these things, it becomes increasingly difficult to understand what people are saying or writing when their sayings and writings are filled with abbreviations, acronyms, and anonyms, contractions, initials, nicknames, pseudonyms, short forms, signs, slang short-cuts, and symbols created for their own convenience, without regard for their ability to create communicative and easily understood statements.

Daily speech, newspapers, magazines, books, signs along the airways, highways, railways, and waterways reveal the universal use of these short-cuts to communication and the growing tendency to use and devise more and more of them. This appears to be done in response to the rapid development of technological civilization. Witness the confusion compounded when someone without a knowledge of Spanish turns on the C tap in a shower bath in Acapulco, Buenos Aires, or Madrid. Hot water streams out instead of cold. North is N in most languages of western civilization, but west can be W or O or even V.

Abbreviations of every sort cover contemporary civilization like an ever-deepening snowdrift, concealing the main features of the landscape, leaving the beholder mystified and perplexed by the overwhelming obscurity imposed by these letter and number combinations. Usually these short-cuts to communication are created without reference to the niceties of typography, the requirements of official and logical regulations, or even the rules of grammar. Most appear without definitions. More and more appear each year. And more and more duplicate already existing abbreviations. The letter *a,* for example, stands for more than 25 different things, capital *A* stands for more than 30 different things, and so it goes through the alphabet, with many varied combinations of letters and numbers, signs, and symbols.

Arrangement

Everything in this book is arranged in alphabetical and numerical order. For entries containing the same letter, lowercase precedes capital (aa, AA); roman precedes italics (AWA, *AWA*); unpunctuated precedes punctuated (BAE, B.A.E.). An Arabic numeral precedes its Roman equivalent (3, III).

The following connectives are ignored in the alphabetical arrangement: & (ampersand), and, by, in, of, or, + (plus), the, to. All other articles, particles, prepositions, and the like (between, de, del, di) are treated alphabetically. For example, U of P is alphabetized as UP; *U de ST* appears as if it were UdeST.

A dollar sign ($) is treated as if it were a lowercase "d," the pound sign (£) like a lowercase "P," and a mu (μ) like a lowercase "m."

In the case of a parenthetical plural ending, the parenthesis will be ignored [e.g., paren(s) is treated as parens].

Golden Rule

"When in doubt, spell it out," insisted Ralph Bayless when he was chief engineer of all General Dynamics engineering organizations of Convair. He urged all to define abbreviations the first time they were used.

If, for example, the Gulf Missile Range is being described, and the term *GMR* will be used again and again, the text should begin something like this:

> The Gulf Missile Range (GMR) affords facilities for national defense and space exploration. GMR personnel are active in all phases of aerospace research, development, and engineering. GMR headquarters are in Mobile

Common sense rules about abbreviations are most often ignored. Therefore it is necessary to repeat that short words like Maine, Ohio, Samoa, etc. should not be abbreviated, although their unofficial abbreviations exist and are shown in this book. Similarly it is best to avoid the truncation of words spelling other words when abbreviated: catl: catalog; king.: kingdom; man.: management.

Because this is a reference dictionary there are many duplications. Many items are included so it will not be necessary for readers to try to guess what the abbreviations are intended to mean. Many unauthorized abbreviations are included for the same reason—to help readers find their way through the alphabet soup.

Capitalization

Capitalization of abbreviations, according to Department of Defense Military Standard 12-B (Mil-Std 12-B), must follow the rules of Egnlish grammar. All proper nouns are capitalized. All common nouns are lowercased. Units of weight, measure, and velocity, such as lb, kg, in., cc, mm, rpm, and the like, appear in lowercase to avoid confusion with other letter combinations they resemble.

Many military establishments and officers use full capitals for everything because message machines are provided only with capital letters. That is why many engineering drawings supplied the armed forces contain all abbreviations in capital letters. It is also true that many draftsmen are afraid small letters will fill in, especially *a*'s, *b*'s, *e*'s, *g*'s, *o*'s, and the like. Therefore they also like to use capital letters. In text, however, 1500 RPM presents a typographical blob, as compared to the more sophisticated 1500 rpm.

At first loran was LORAN. As people became more used to it, it became Loran. Today it is loran. The same is true of other combinations. The trend is to capitalize only those letters standing for proper nouns, running all common nouns in lowercase. Nevertheless, for the sake of readers and researchers, some incorrectly rendered abbreviations appear in this book. Many people have a marked tendency to capitalize everything they think is important. If this tendency is unchecked, confusion follows. All abbreviations and acronyms look alike. So follow the commonsense rules of good grammar and correct usage.

Chemical element symbols, however, have the first letter capitalized: Au (gold), Zn (zinc), etc. The second letter of a chemical symbol always appears in lowercase.

Exceptions

The singular, plural, and tense of the word abbreviated do not alter the abbreviation except in a few instances, such as fig.: figure; figs.: figures; no.: number; nos: numbers; p: page; pp: pages; S: Saint; SS: Saints.

However, readers should be aware that the International (*SI*) System of Measurements calls for the abolition of all pluralized abbreviations. Hence in. stands for inch or inches, lb for pound or pounds, oz for ounce or ounces. This system will probably gain widespread approval.

Documentary abbreviations are rendered as follows: FARs (Failure Analysis Reports), or IRs (Inspector's Reports) or RARs (Reliability Action Reports). In the singular they appear as FAR, IR, RAR.

Italics

Items from Latin and other non-English languages, as well as titles of books and periodicals, are usually set in italic type. Many physical symbols are also set in italics to differentiate them from other letter combinations.

Punctuation

Short forms are devised to save time and space and to overcome the necessity of repeating long words and phrases. All punctuation is avoided in modern practice unless the form is taken from Latin or there is some conventional use demanding punctuation, as in the case of academic degrees and a few governmental designations. U.S.A. is the country; USA is the army. D.C. is the District of Columbia; DC is direct current when used as a noun. Cash on delivery is not cod but c.o.d. Similarly, fig., figs., and no. require periods to keep readers from thinking they may be words instead of abbreviations for figure, figures, and number. Again, when in doubt, spell it out.

Capitalization and Punctuation Trends

American as well as British and Canadian publishers appear to be following the trend to capitalize only those letters normally capitalized: proper nouns and important words in titles. They reserve lowercase letters for abbreviations consisting of adjectives and common nouns. This obviates the chaos brought about by those who capitalized all the letters in every abbreviation and then compounded their error by placing unnecessary full stops or periods after every letter, as was the custom in bygone times.

Most periods are dropped because it is generally realized that the purpose of all abbreviation is the thoroughgoing promotion of brevity. More than a decade ago, when Rudolf Flesch compiled one of his many useful books, *How To Be Brief—An Index to Simple Writing,* he stated:

> To save even more space, leave out abbreviation periods whenever you can. The British omit them regularly...*Mr, Mrs, Dr, St* (Saint), *Thos, Chas, jr.* Periods are often left out after standard abbreviations like *US, UN, FCC, PTA*...following the pattern of most telephone books (e.g., *plmbg & heatg supls, atty, flrst, acctnts, svce, rl est*).

Signs and Symbols

Frequently used signs and symbols appear in the back of this dictionary. Many are found on typewriters and computer keyboards (&: ampersand—the *and* sign; *: asterisk; ¢: cent; $: dollar; %: percent).

Symbols include the chemical elements (Al: aluminum; Au: gold—from the Latin *aurum;* C: carbon; Sn: tin—from the Latin *stannum*). All are listed in the alphabetical section without special definition to indicate they are not abbreviations but symbols. The chemical elements are also grouped together in the back of this dictionary.

Airlines use two-letter symbols for convenience in baggage handling, ticketing, and scheduling operations. Thus American Airlines is AA, Delta Air Lines is DL, National Airlines is NA, Pan American World Airways is PA, United Air Lines is UA. These two-letter designations are listed in a separate section at the back of the book as well as alphabetically along with other multiletter airline abbreviations.

Railroads and steamship lines are included both in the alphabetical section and in their own sections at the end of the book. Naval craft are designated by many arbitrary symbols. All available are given in the alphabetical section.

Abbreviations Dictionary

Eighth Edition

a abbreviation; absent; absolute; acceleration in feet per second; account; acre; acronym; adjective; adult; aerial; afternoon; altitude; altitude intercept; amateur; ampere; annealing; anthracite; arc; are (unit of metric land); area; argent; at; atmosphere; attendance; audit(or); automatic; available; aviation; aviator; axis; azure; distance from leading edge to aerodynamic center

a' all (contraction); minute angle; a prime

a" double prime; second (angle)

a am, an, an der (German—on the, at the); angle of attack; *annus* (Italian—year) (Latin—year); *antes* (Spanish before); *arteria* (Latin—artery); attenuation constant (symbol); autonomous consumption (macro-economic symbol); (Italian—and)

A absolute; absolute temperature; academy; accumulator (computerese); acid; acoustic source; actual weight of an aircraft; address (computer symbol); adenine; adulterer; adulteress (branded on the foreheads of all convicted of this crime in early New England)—also known as the scarlet letter because branding caused bleeding; aircraft; airman; Alaska Steamship Company; Alcoa Steamship Company; Alfa; ambassador; America; American; Americanization; Americanize; Amos, The Book of; ampere;

amphibian; Anchor Line; anode; anterior; April; argon; Army; artillery; aspect ratio; astragal; Atlantic; atomic weight; attack; August; Austria (auto plaque); chemical activity; first van der Waals constant; Fraunhofer line due to oxygen; linear acceleration; mean sound absorption coefficient; total acidity

Å angstrom unit

A *abajo* (Spanish—down); *abasso* (Italian—down); *absolvo* (Latin—I absolve, I acquit); *alas* (Finnish—down); *albus* (Latin—white); *Alp(en)* (German—Alp(s)); *Alpe(s)* (French—Alp(s)); *Alpi* (Italian—Alps); *Alt* (German—old); *Alteza* (Spanish—Highness); *aprobado* (Spanish—approved)—passed an examination; arrival; *arrivare* (Italian—arrival); arrive; *arrivé* (French—arrival); *Atlantic Reporter; auf* (German—up); *Aulus* (Latin—Aulus Gellius)—2nd-century author noted for his *Noctes Atticae* about languages and literature as well as natural history; *aus* (German—out); *avbeta* (Swedish—departure); mountain meadow(s)

Å *aas* (Dano-Norwegian—hills)

a (A) analysis

A+ A-plus; A-positive

A− A-minus; A-negative

A-1 air personnel officer; angstrom unit; excellent; first class; first rate; *Lloyd's Register* symbol indicating first rate equipment; personnel

section of an air force staff; skyraider single-engine general-purpose attack aircraft flown from aircraft carriers; top quality; tops; very best

A-I (motion pictures) for general patronage

A1c Airman, first class

A/1C Airman First Class

A-1 Skyraider Douglas single-engine attack aircraft (formerly AD)

A-2 air intelligence officer; almost A-1 in quality; intelligence section of an air force staff; just short of the best

A-II (motion pictures) for adults and adolescents only

A₂ aortic second sound; Asian influenza virus

A/2C Airman Second Class

A²C² see AACC

A²ᵈ *Atlantic Reporter second series*

A-3 air operations and training officer; operations and training section of an air force staff; Skywarrior twin-engine turbojet tactical all-weather attack aircraft operating from aircraft carriers; training and operations

A-III (motion pictures) for adults only

A/3C Airman Third Class

A-3 Skywarrior Douglas carrier-based twin-engine jet reconnaissance and light bombing plane (formerly A3D)—USAF B-66 Destroyer

A-4 air material and supply officer; material and supply section of an air force staff; Sky-hawk single-engine turbojet attack aircraft operating from aircraft carriers; supply and materiel

A-IV (motion pictures) for adults with reservations

A-5 planning; supersonic twin-engine turbojet all-weather attack aircraft operating from aircraft carriers

A-6 communications

A-6A Intruder twin-engine turbojet long-range carrier-based low-altitude attack aircraft

A-6 Intruder Grumman carrier-based twin-engine jet low-level attack bomber (formerly A2F)

A-7 Corsair II Ling-Temco-Vought carrier-based single-engine jet light-attack bomber

A-32 Lansen (Swedish—A-32 Lance)—Saab single-seat single-engine jet fighter-interceptor

A-37 radar-homing or television-guided air-to-surface missile

A-60 Saab twin-engine two-place jet trainer-utility aircraft also called the Saab 105

A-68 protein found in the brains of Alzheimer's victims

A-106 Agusta antisubmarine-warfare single-engine single-seat helicopter

A-109 Agusta high-performance eight-seat twin-engine helicopter

aa acetic acid; achievement age; acting appointment; adjectives; alveolar-arterial; always afloat; aminoacetone; *ander andere* (German—among others); approximate absolute; armature accelerator; arteries; ascending aorta; atomic absorption; author's alteration; equal parts

aa (AA) achievement age; anti-aircraft; ascorbic acid

a-a air-to-air

a/a antiaircraft

a & a abbreviations and acronyms; additions and amendments; aid and attendance

aa *arterias* (Latin—arteries); (Hawaiian—block lava)—pronounced *ah-ah*

AA absolute alcohol; absolute altitude; achievement age; Addicts Anonymous; Administrative Assistant; Adoption Agency; Aerolineas Argentinas (Argentine Airlines); Affirmative Action; Airman Ap-

prentice; Alcoholics Anonymous; Aluminum (Company of) America; American Airlines; American Association; Ann Arbor (railroad); Ansett Airways; antiaircraft; Appropriate Authority; arithmetic average; Arlington Annex; Asian-African; Athletic Association; author's alteration(s); Automobile Association; Aviation Annex; *Aviatsionnaya Armiya* (Russian—Air Army)

A/A Agnostics/Atheists

A.A. Associate in Accounting; Associate in Arts

AA *Air Almanac; Astronautica Acta* (Journal of the International Astronautical Federation)

aaa abdominal aortic aneurism; acquired aplastic anemia; acute anxiety attack; amalgam; androgenic anabolic agent

aa & a armor, armament, and ammunition

Aaa Alaska (government style is to spell it out)

AAA Agricultural Adjustment Administration; Agricultural Aircraft Association; Alaska; All American Aviation; Allegheny Airlines (3-letter coding); Allied Artists of America; American Academy of Advertising; American Academy of Allergy; American Accordionists Association; American Accounting Association; American Airship Association; American Antarctic Association; American Anthropological Association; American Arbitration Association; American Association of Anatomists; American Astronomers Association; American Australian Association; American Automobile Association; antiaircraft artillery; Antique Airplane Association; Appraisers Association of America; Archives of American Art; Area Agency on Aging; Argentine Anti-communist Alliance; Armenian Assembly of America; Army Audit Agency; Associated Agents of America; Association of Attenders and Alumni (Hague Academy of International Law); Association of Average Adjusters

AAA (AFL-CIO) Actors and Artistes of America

A.A.A. Amateur Athletic Association (British)

AAAA American Association for the Advancement of Atheism; American Association of Advertising Agencies; Army Aviation Association of America; Associated Actors and Artists of America; Association of Accredited Advertising Agencies (Singapore); Australian Advertising Advisory Authority; Australian Association of Advertising Agencies

AAAANZ Association of Accredited Advertising Agencies of New Zealand

AAAB American Association of Architectural Bibliographers

AAAC American Association for the Advancement of Criminology; Antiaircraft Artillery Command; Australian Army Aviation Corps

AAACE American Association of Agricultural College Editors

AAACE *Alianza Apostolica y Anti-Comunista de España* (Spanish—Apostolic and Anti-Communist Alliance of Spain)—also known as the Triple-A

AAA-CPA American Association of Attorney-Certified Public Accountants

AAAD American Athletic Association for the Deaf

AAAE American Association of Airport Executives; Australian Association of Adult Education

AAAEE American Afro-Asian Educational Exchange

A.A. Ag. Associate of Arts in Agriculture

AAAH American Association for the Advancement of the Humanities

AAAI Affiliated Advertising Agencies International; American Academy of Allergy and Immunology

AAAID Arab Authority for Agricultural Investment and Development

AAAIMH American Association for the Abolition of Involuntary Mental Hospitalization

AAAIP Advanced Army Aircraft Instrumentation Pro-

gram
AAAIS Antiaircraft Artillery Information Service; Antiaircraft Artillery Intelligence Service
AAAIWA Automobile, Aerospace, and Agricultural Implement Workers of America
aaal abolish all abortion laws
AAAL American Academy of Arts and Letters
AAALAC American Association for Accreditation of Laboratory Animal Care
AAAM American Association of Aircraft Manufacturers; American Association for Automotive Medicine
AAAN American Academy of Applied Nutrition
AAAOC Antiaircraft Artillery Operation Center
AAAR American Association for Aerosol Research; Association for the Advancement of Aging Research
AAARC Antiaircraft Artillery Reception Center
AAARG American Atheist Addiction Recovery Groups
AAAS American Academy of Arts and Sciences; American Academy of Asian Studies; American Association for the Advancement of Science; Australian Association for the Advancement of Science
A.A.A.S. Associate in Arts and Science
AAASA Association for the Advancement of Agricultural Sciences in Africa; Australian and Allied All Services Association
AAASS American Association for the Advancement of Slavic Studies
AAASUSS Association of Administrative Assistants and Secretaries to United States Senators
AAAUS Association of Average Adjusters of the United States
AAB Aircraft Accident Board; American Air Brake; American Association of Bioanalysts; Army Air Base; Army Artillery Board; Association of Applied Biologists
AABB American Association of Blood Banks
AABC American Amateur Baseball Congress; American Association of Bible Colleges; Association for the

Advancement of Blind Children
AABD Aid to the Aged, Blind, or Disabled
AABEVM Association of American Boards of Examiners in Veterinary Medicine
AABGA American Association of Botanical Gardens and Arboretums
AABH Australian Association for Better Hearing
AABI American Association of Bicycle Importers; Antilles Air Boats Incorporated
AABL Associated Australian Banks of London; Australian Associated Banks in London
aaBm analytical anatomy by the Braille method
AABM Association of American Battery Manufacturers; Australian Association of British Manufacturers
AABNCP Advanced Airborne Command Post
AABP Australian Association of Business Publications
AABPDF Allied Association of Bleachers, Printers, Dyers, and Finishers
aabshil aircraft anti-collision-beacon-system high-intensity light(ing)
AABT Association for the Advancement of Behavior Therapy
AABTM American Association of Baggage Traffic Managers
aaby as amended by
aac automatic aperture control; average annual cost
AAC Aeronautical Advisory Council; Aeronautical Approach Chart; Aircraft Armament Change; Alaskan Air Command; All-American Canal (serving California and Baja California); Alumnae Advisory Center; American Academy of Criminalistics; American Alpine Club; American Alumni Council; American Archery Council; American Association of Criminology; American Atheist Center; American Cement Corporation (stock exchange symbol); Anglo-American Corporation; Antiaircraft Command; Army Air Corps; Association of American Choruses; Association of American Colleges; Australian Agricultural Company;

Australian Air Corps; Australian Association of Chiropractors; Automotive Advertisers Council
AAC *Associação Academica de Coimbra* (Portuguese—Coimbra Academic Association); *Auto Avia Contruzione* (Italian automaker: Enzo Ferrari's first company)
A.A.C. *anno ante Christum* (Latin—year before Christ)
AACA Antique Automobile Club of America; Automotive Air Conditioning Association
AACAP Association of American Colleges Arts Program
AACB Aeronautics and Astronautics Coordination Board; Australian Association of Clinical Biochemists
AACBC American Association of College Baseball Coaches
AACBP American Academy of Crown and Bridge Prosthodontics
aacc all-attitude control capability; automatic approach control complex
AACC American Association of Cereal Chemists; American Association of Clinical Chemists; American Association for Contamination Control; American Association of Credit Counselors; American Automatic Control Council; Association for the Aid of Crippled Children; Australian-American Chamber of Commerce
A.A.C.C.A. Associate of the Association of Certified and Corporate Accountants
AACCLA Association of American Chambers of Commerce in Latin America
AACCP American Association of Colleges for Chiropody-Podiatry
AACD American Association for Counseling and Development
AACDP American Association of Chairmen of Departments of Psychiatry
AACE Airborne Alternate Command Echelon (NATO); American Association of Cost Engineers
AACFO American Association of Correctional Facility Officers
AACFT Army Aircraft
AACHS German geographical

place-name equivalent of Aixla-Chapelle on the Belgian-Dutch borders of Germany

AACHS Afro-American Cultural and Historical Society

AACI American Association for Conservation Information; Association of Americans and Canadians in Israel

AACJC American Association of Community and Junior Colleges

AACM American Academy of Compensation Medicine

AACO Advanced and Applied Concepts Office (USA); American Association of Certified Orthoptists; Assault Airlift Control Office(r)

AACOBS Australian Advisory Council on Bibliographical Services

AACOMS Army Area Communications System

AACP American Academy for Cerebral Palsy; American Academy for Child Psychiatry; American Academy of Clinical Psychiatrists; American Association of Colleges of Podiatry; American Association of Commercial Publications; American Association of Convention Planners; American Association of Correctional Psychologists

AACPP Association of Asbestos Cement Pipe Producers

AACPR American Association for Cleft Palate Rehabilitation

AACR American Association for Cancer Research

AACR Anglo-American Cataloguing Rules

AACRAO American Association of Collegiate Registrars and Admissions Officers

AACS Airborne Astrographic Camera System; Airways and Air Communications Service; Army Airways Communications System; Australian Amateur Cine Society

AACSA Anglo-American Corporation of South Africa

AACSB American Association of Collegiate Schools of Business

AACSL American Association for the Comparative Study of Law

AACSM Airways and Air Communications Service Manual

AACT American Association

of Commodity Traders; Armenian Assembly Charitable Trust

AACTE American Association of Colleges for Teacher Education

AACTP American Association of Correctional Training Personnel

AACUBO American Association of College and University Business Officers

aad active acoustic(al) device

aad (AAD) alloxazine adenine dinucleotide

AAD Aircraft Assignment Directive; American Academy of Dentists; American Academy of Dermatology; Army Air Defense

AADA Advanced Air Depot Area; American Academy of Dramatic Arts; American Association of Deaf Athletes; Army Air Defense Area; Australian Automobile Dealers Association

AADC Army Air Defense Command(er)

AADCCS Army Air Defense Control and Coordination System

AADCP Army Air Defense Command Post

AADE American Association of Dental Editors; American Association of Dental Examiners

AA de L Academia Argentina de Letras (Spanish—Argentine Academy of Letters)

AADGB American Association of District Governing Boards

AA Dip Architectural Association Diploma

AADIS Army Air Defense Information Service

AADLA Art and Antique Dealers League of America

AADM American Academy of Dental Medicine

AADMS Advanced Academic Degree Management System

AADN American Association of Doctors' Nurses

AADOO Army Air Defense Operations Office(r)

AADP American Academy of Denture Prosthetics

AADPA American Academy of Dental Practice Administration

AADS Advanced Army Defense System; American Association of Dental Schools; American Association of

Dermatology and Syphilology; Army Air Defense System

aae (AAE) above airport elevation; acute allergic encephalitis; average annual earnings

AAE American Association of Endodontists; American Association of Engineers; Army Aviation Engineers; Asia Australia Express

AAEA American Agricultural Editors Association

AAEC Association of American Editorial Cartoonists; Australian-American Engineering Corporation; Australian Army Educational Corps; Australian Atomic Energy Commission

AAEDC American Agricultural Economics Documentation Center (USDA)

A.Ae.E. Associate in Aeronautical Engineering

AAEE American Academy of Environmental Engineers; American Association of Economic Entomologists; American Association of Electromyography and Electrodiagnosis

AAEF Australian-American Education Foundation

AAEFA Army Aviation Engineering Flight Activity

AAEH Association to Advance Ethical Hypnosis

AAEKNE American Association of Elementary-Kindergarten-Nursery Educators

AAELSS Active-Arm External-Load Stabilization System

AAEOCJ American Association of Ex-Offenders in Criminal Justice

AAEP American Association of Equine Practitioners

AAES Advanced Aircraft Electrical System; Australian Agricultural Economics Society; Australian Army Education Service

AAESA Alabama Association of Elementary School Administrators

AAESDA Association of Architects, Engineers, Surveyors, and Draughtsmen of Australia

AAESPH American Association for the Education of the Severely/Profoundly Handicapped

AAEW Atlantic Airborne Early Warning

aaf (AAF) acetylaminofluorine; ascorbic acid factor

a-a-f acetic-alcohol-formalin (fixing fluid)

AAF American Advertising Federation; American Air Filter (company); American Architectural Foundation; American Astronautical Federation; Army Air Field; Army and Air Force; Army Air Forces

AAFA Architectural Aluminium Fabricators Association; Australian Amateur Fencing Association

A.A.F.A. Associate in Arts in Fine Arts

A.A. Fair Erle Stanley Gardner

AAFB Auxiliary Air Force Base

aafc (AAFC) antiaircraft fire control

AAFC Air Accounting and Finance Center; Army Air Forces Center; Army Air Force Classification Center; Association of Advertising Film Companies

AAFCE Allied Air Force, Central Europe

AAFCO Association of American Feed Control Officials; Association of American Fertilizer Control Officials

AAFCWF Army and Air Force Central Welfare Fund; Army and Air Force Civilian Welfare Fund

AAFE Advanced Applications Flight Experiment; American Association of Feed Exporters

AAFEA Australian Airline Flight Engineers Association

AAFEC Army Air Forces Engineering Command

AAFEMPS Army and Air Force Exchange and Motion Picture Service

AAFES Army and Air Force Exchange Service

AAFFA Australian Air Freight Forwarding Association

AAFH Academy of American Franciscan History

AAFIS Army Air Forces Intelligence School

AAFM American Association of Feed Microscopists

AAFMC Army Air Forces Materiel Center

AAFMPS Army and Air Force Motion Picture Service

AAFNE Allied Air Force, Northern Europe

AAFNS Army Air Forces Navigation School

AAFOIC Army Air Forces Officer in Charge

AAFP American Academy of Family Physicians

AAFPS Army and Air Force Pilot School; Army and Air Force Postal Service

AAFRS American Academy of Facial, Plastic, and Reconstructive Surgery

AAFS American Association of Foot Specialists; American Academy of Forensic Sciences

AAFSE Allied Air Force, Southern Europe

AAFSS Advanced Aerial Fire Support System

AAFSW Association of American Foreign Service Women

AAFTS Army Air Forces Technical School

AAFU All-African Farmers' Union

AAFWB Army and Air Force Wage Board

AAG Air Adjutant General; Association of American Geographers; Australian Association of Gerontology

AAGC American Association of Gifted Children

AAGFO American Academy of Gold Foil Operators

AAGL American Association of Gynecological Laparoscopists

AAGp Aeromedical Airlift Group (USAF)

AAGP American Academy of General Practice

A. Agr. Associate in Agriculture

AAGR Air-to-Air Gunnery Range

A.Agri. Associate in Agriculture

AAGS All-American Gladiolus Selections

AAGUS American Association of Genito-Urinary Surgeons

aagw (AAGW) air-to-air guided weapon(s)

aah (AAH) anti-armor helicopter

AAH Alcoholism Awareness Hour (tv); American Academy of Homiletics; Australian Auxiliary Hospital

aaha awaiting action of higher authority

AAHA American Animal Hospital Association; American Association of Handwriting Analysts; American Association of Homes for the Aging; American Association of Hospital Accountants

AAHC American Academy of Humor Columnists; American Association of Hospital Consultants

AAHD American Academy of the History of Dentistry

AAHDC American Association of Hospital Dental Chiefs

AAHE American Association for Higher Education; American Association of Housing Educators

A.A.H.E. Associate in Arts in Home Economics

AAHM American Association for the History of Medicine; Association of Architectural Hardware Manufacturers

Aahp Army artificial heart pump

AAHP American Association for Hospital Planning; American Association of Hospital Podiatrists; American Association for Humanistic Psychology

AAHPA American Association of Hospital Purchasing Agents

AAHPER American Association for Health, Physical Education, and Recreation

AAHPERD American Alliance for Health, Physical Education, Recreation, and Dance

AAHPhA American Animal Health Pharmaceutical Association

AAHPS Australian Association for the History and Philosophy of Science

AAHQ Allied Air Headquarters

AAHRA Asia and Australia Hotel and Restaurant Association

AAHS American Aviation Historical Society

AAHSLD Association of Academic Health Sciences Library Directors

aai air-to-air identification; angle-of-approach indicator; azimuth angle increment

AAI African-American Institute; Afro-American Institute; Agricultural Ammonia Insti-

tute; Akron Art Institute; Alfred Adler Institute; Allied Armies in Italy (World War II); American Association of Immunologists
A.A.I. Associate of the Chartered Auctioneers' and Estate Agents' Institute
AAIA Association of American Indian Affairs
A.A.I.A. Associate of the Association of International Accountants
AAIAL American Academy and Institute of Arts and Letters
AAIAN Association for the Advancement of Instruction about Alcohol and Narcotics
AAIB American Association of Instructors of the Blind
AAIC Allied Air Intelligence Center; Australian Advertising Industry Council
AAICD American Association of Imported Car Dealers
AAICJ American Association for the International Commission of Jurists
AAICU Alabama Association of Independent Colleges and Universities
AAID American Academy of Implant Dentistry; American Academy of Implant Dentures; American Association of Industrial Dentists
AAIE American Association of Industrial Editors; American Association of Industrial Engineers
AAII American Association of Individual Investors; Association for the Advancement of Invention and Innovation
AAIM American Association of Industrial Management
AAIMS An Analytical Information Management System
AAIN American Association of Industrial Nurses
AAIP Academic Administration Internship Program
AAIPS American Association of Industrial Physicians and Surgeons
AAIS Associate of the Australian Institute of Secretaries
AAIT American Association of Inhalation Therapists; Associate of the Australian Institute of Travel
AAJ American Association for Justice; Arab Airways, Jerusalem; Axel Axelson Johnson (Johnson Line)

AAJ Australian Anthropological Journal
AAJA Afro-Asian Journalists' Association; Asian-American Journalists Association
aajc automatic antijam circuit
AAJC American Association of Junior Colleges
AAJE American Association for Jewish Education
AAJR American Academy for Jewish Research
AAJS American Association for Jesuit Scientists
AAJSA American Association of Journalism School Administrators
AAK Alfred A Knopf
aal above aerodrome level; anterior axillary line
AAL Aid Association for Lutherans; American Airlines; Ames Aeronautical Laboratory; Arctic Aeromedical Laboratory; Association of Assistant Librarians; Australian Air League
AALA Afro-American Liberation Army; American Auto Laundry Association; American Automotive Leasing Association; Asian-American Librarians Association
AALAPSO Afro-Asian-Latin-American People's Solidarity Organization
AALAS American Association of Laboratory Animal Science
AALASO Afro-Asian Latin-American Students' Organization
AALC African-American Labor Center (AFL-CIO)
AALD Australian Army Legal Department
AALE Associate of Arts in Law Enforcement
AALL American Association of Law Libraries
aalmg (AALMG) antiaircraft light machine gun
AALPA Association of Auctioneers and Landed Property Agents
AALPP American Association for Legal and Political Philosophy
AALR American Association for Leisure and Recreation
AALS American Association of Language Specialists; Association of American Law Schools; Association of American Library Schools
AAMI American Association

of Library Trustees
AALU Association for Advanced Life Underwriting
aam (AAM) air-to-air missile
AAM Academy of Ancient Music; Acoustics Analysis Memo; American Association of Microbiology; American Association of Museums; Australian Air Mission
A-AM Afro-American Museum
AAMA American Academy of Medical Administrators; American Apparel Manufacturers Association; American Architechtural Manufacturers Association; American Association of Medical Assistants; Architectural Aluminum Manufacturers Association
A-A MAN Afro-American Men Against Narcotics
AAMBP Association of American Medical Book Publishers
AAMC American Association of Marriage Counselors; American Association of Medical Clinics; American Association of Medico-Legal Consultants; Army Air Materiel Command; American Association of American Medical Colleges; Australian Army Medical Corps
AAMCA Army Advanced Materiel Concepts Agency (USA)
AAMCH American Association for Maternal and Child Health
AAMD American Association on Mental Deficiency; Association of Art Museum Directors
aame (AAME) acetylarginine methyl ester
AAMES American Association for Middle East Studies
AAMF American Association of Music Festivals
AAMFT American Association for Marriage and Family Therapy
aamg (AAMG) antiaircraft machine gun
AAMGA American Association of Managing General Agents
AAMHPC American Association of Mental Health Professionals in Corrections
AAMI American Association of Machinery Importers; American Athletic Motiva-

tion Institute; Association for the Advancement of Medical Instrumentation; Association of Allergists for Mycological Investigation
AAMIH American Association for Maternal and Infant Health
AAML Arctic Aeromedical Laboratory
AAMMC American Association of Medical Milk Commissioners
AAMOA Afro-American Music Opportunities Association
AAMP American Academy of Maxillofacial Prosthetics
AAMR American Academy on Mental Retardation
AAMRL American Association of Medical Record Librarians
AAMS American Air Mail Society
AAMSW American Association of Medical Social Workers
AAMU Army Advanced Marksmanship Unit
A.A. Mus. Associate in Arts in Music
AAMVA American Association of Motor Vehicle Administrators
AAMW Association of Advertising Men and Women
AAMWS Australian Army Medical Women's Service
aan (**AAN**) aminoacetonitrile; assignment action number(s)
AAN Advance Acquisition Notification; American Academy of Neurology; American Academy of Nutrition; American Association of Neuropathologists; American Association of Nurserymen
A.A.N. Associate in Arts in Nursing
AANA American Association of Nurse Anesthetists; Australian Association of National Advertisers
AANNT American Association of Nephrology Nurses and Technicians
AANR American Association of Newspaper Representatives
AANS American Academy of Neurological Surgery; American Association of Neurological Surgery; Australian Army Nursing Service
AANSW Archives Authority of New South Wales

aao amino-acid oxidase
aaO *am angeführten Ort* (German—in the place cited); *an anderen Orten* (German—elsewhere, in the place cited)
AAO Academy of Applied Osteopathy; Aircraft Approach Overlay (zone); American Academy of Ophthalmology; American Academy of Optometry; American Association of Orthodontists; Anglo-Australian Observatory
AAO Abastumanskaya Astrofizicheskaya Observatoriya (Russian—Abastumani Astrophysical Observatory)
AAOA Ambulance Association of America
AAOC American Association of Osteopathic Colleges; Antiaircraft Operations Center; Australian Army Ordnance Corps
AAOD Army Aviation Operating Detachment
AAODC American Association of Oilwell Drilling Contractors
AAOG American Association of Obstetricians and Gynecologists
AAOGAS American Association of Obstetricians, Gynecologists, and Abdominal Surgeons
AAOM American Academy of Occupational Medicine; American Academy of Oral Medicine
AAOME American Association of Osteopathic Medical Examiners
AAOMS American Association of Oral and Maxillofacial Surgeons
AAONMS Ancient Arabic Order of Nobles of the Mystic Shrine
AAO & O American Academy of Ophthalmology and Otolaryngology
AAOP American Academy of Oral Pathology; Antiaircraft Observation Post
AAOPB American Association of Pathologists and Bacteriologists
AAOPS American Association of Oral and Plastic Surgeons
AAOR American Academy of Oral Roentgenology
AAOS American Academy of Orthopaedic Surgery
aap advise if able to proceed; air at atmosphere pressure

AAP Academy of American Poets; Advance Acquisition Plan(ning); Affirmative Action Program; Allied Administrative Publication; American Academy of Pediatrics; American Academy of Periodontology; American Atheist Press; Angina Awareness Program; Association for the Advancement of Psychoanalysis; Association for the Advancement of Psychology; Association for the Advancement of Psychotherapy; Association of American Physicians; Association of American Publishers; Association of Applied Psychoanalysis; Australian Associated Press
A-A P Afro-American Police
AAP Allied Army Procedures (or *Publications*)
AAPA American Amateur Press Association; American Association of Physical Anthropologists; American Association of Port Authorities
A-A PA Anglo-American Press Association
AAPB American Association of Pathologists and Bacteriologists
AAPBS Australian Association of Permanent Building Societies
AAPC Advertising Agency Production Club; All-African Peoples' Conference; Australian Aluminium Production Commission
AAPCC American Association of Poison Control Centers; American Association of Psychiatric Clinics for Children
AAPCM Association of American Playing Card Manufacturers
AAPCO Association of American Pesticide Control Officials
AAPD American Academy of Physiologic Dentistry
AAPE American Academy of Physical Education; American Association of Philatelic Exhibitors
AAPF Academy of American Poets Fellowship
AAPG American Association of Petroleum Geologists
AAPH American Association for Partial Hospitalization; American Association of Pro-

fessional Hypnologists
AAPHD American Association of Public Health Dentists
AAPHI Associate of the Association of Public Health Inspectors
AAPHP American Association of Public Health Physicians
AAPHR American Association of Physicians for Human Rights
AAPICU American Association of Presidents of Independent Colleges and Universities
AAPIU Allied Aerial Photographic Interpretation Unit
AAPL Afro-American Policemen's League; American Academy of Psychiatry and the Law; American Artists Professional League; American Association of Petroleum Landmen
AAPLE American Academy for Professional Law Enforcement; American Association for Professional Law Enforcement
aapm amphiapomict
AAPMR American Academy of Physical Medicine and Rehabilitation
AAPO All-African Peoples' Organization
AAPOR American Association for Public Opinion Research
AAPP American Association of Police Polygraphists; Association of Amusement Park Proprietors; Australian Association of Psychology and Philosophy
AAPPP American Association of Planned Parenthood Physicians
AAP/PSP Association of American Publishers—Professional and Scholarly Publishing Division
AAPRA All-African People's Revolutionary Army
AAPRCO American Association of Private Railroad Car Owners
AAPRM American Association of Passenger Rate Men
AAPRP All-African People's Revolutionary Party
AAPS American Association of Phonetic Sciences; American Association of Plastic Surgeons; American Association for the Promotion of Science; Association for Ambulatory Pediatric Services; As-

sociation of American Physicians and Surgeons
AAPSC American Association of Psychiatric Services for Children
AAPSD Alternative Automotive Power System Division (EPA)
AAPSE American Association of Professors in Sanitary Engineering
AAPSS American Academy of Political and Social Sciences
AAPSW Associate of the Association of Psychiatric Social Workers
AAPT American Association of Physics Teachers; Association of Asphalt Paving Technologists
AAPTO American Association of Passenger Traffic Officers
AAPTSR Australian Association for Predetermined Time Standards and Research
AAPY American Association of Professors of Yiddish
AA & QMG Assistant Adjutant and Quartermaster General
aar after action report; against all risks; average annual rainfall
aar (AAR) antigen-antiglobulin reaction
Aar Aarhus; Australia antigen radioimmunoassay
AAR Aircraft Accident Record; Aircraft Accident Report; American Academy in Rome; American Academy of Religion; Army Area Representative; Association of American Railroads; Australian Associated Resources; Automotive Affiliated Representatives
AARA Australian Association of Reprographic Arts
aa rating average-audience rating (percentage of tv-equipped homes viewing the average minute of a national telecast)
AARB Australian Road Research Board
AARC Ann Arbor Railroad Company; Association for the Advancement of Released Convicts; Australian Aeronautical Research Committee; Australian Applied Research Centre; Australian Automobile Racing Club
AARCO Afro-Asian Rural Construction Organisation

AARD American Academy of Restorative Dentistry
AARDCO Association of American Railroad Dining Car Officers
AARDS Australian Advertising Rate and Data Service
AARE Australian Association for Research in Education
AARF Australian Accounting Research Foundation
aarg aargang (Dano-Norwegian or Swedish—yearbook)
Aarh Aarhus
AARL Advanced Applications and Research Laboratory; Australian Academic and Research Libraries
aarp annual advance retainer pay
AARP American Association of Retired Persons; Ancient American Rocket Pioneers
AARPS Air-Augmented Rocket-Propulsion System
AARR Ann Arbor Railroad
AARRC Army Aircraft Requirements Review Committee
AARRO Afro-Asian Rural Reconstruction Organization
AARS All-America Rose Selections (award); American Association of Railroad Superintendents; American Association of Railway Surgeons; Army Aircraft Repair Ship; Army Amateur Radio System
AART American Association for Rehabilitation Therapy; American Association of Retired Teachers
AARTA American Association of Railroad Ticket Agents
aarv aerial armored reconnaissance vehicle
AARWBA American Auto Racing Writers and Broadcasters Association
aas adjusted air speed; advanced antenna system; aortic arch syndrome
aa's author's alterations
AAs Alcoholics Anonymous members; American Atheists; Asian Americans; author's alterations
AAS Aberdeen Art Society; Academy of Applied Science; Aircraft Airworthiness Section; All-America Selections; American Amaryllis Society; American Antiquarian Society; American Astronautical Society; American

Astronomical Society; Army Air Service; Army Attache System; Arnold Air Society; Association for Asian Studies; Australian Academy of Science; Australian Acoustic Society; Australian Air Services; Australian Art Society; Australian Association of Surgeons; Automatic Adjusted Suspension

A.A.S. Academiae Americanae Socius (Latin—Fellow of the American Academy of Arts and Sciences)

AASA American Association of School Administrators; Associate of the Australian Society of Accountants

aa-sat advanced anti-satellite

AASB Alabama Association of School Boards; American Association of Small Business

AASC Acupuncture Association of Southern California; Aerospace Applications Studies Committee (NATO); Allied Air Support Command; American Association of Specialized Colleges; Australian Accounting Standards Committee; Australian Army Service Corps

aascm awaiting action summary court martial

AASCO Association of American Seed Control Officials

AASCU American Association of State Colleges and Universities

aasd antiaircraft self-destroying

AASD American Association of Social Directories

AASDJ American Association of Schools and Departments of Journalism

AASE American Academy of Sanitary Engineers; American Association of Special Educators; Associated Australian Stock Exchanges; Association for Applied Solar Energy

AASEC American Association of Sex Educators and Counselors

AASECT American Association of Sex Educators, Counselors, and Therapists

AASF Advanced Air Striking Force

AASFE American Association of Sunday and Feature Editors

AASG Association of American State Geologists

AAS & GP American Association of Soap and Glycerin Producers

AASH American Association for the Study of the Headache

AASHO American Association of State Highway Officials

AASHTO American Association of State Highway and Transportation Officials

AASI Advertising Agency Service Interchange; American Academy for Scientific Interrogation; American Association for Scientific Interrogation

A'asia(n) Australasia(n)

aasir advanced atmospheric sounder and imaging radiometer

aasl antiaircraft searchlight

AASL American Antiquarian Society Library; American Association of School Librarians; American Association of State Librarians

A & ASL American & Australian Steamship Line

AASLH American Association for State and Local History

AASM Abigail Adams Smith Museum, New York City; Association of American Steel Manufacturers

AASMB Australian Association of Stud Merino Breeders

AASND American Association for the Study of Neoplastic Diseases

AASO Association of American Ship Owners

AASP American Association for Social Psychiatry

AASPA American Association of School Personnel Administrators

AASPB American Association of State Psychology Boards

AASPRC American Association of Sheriff's Posses and Riding Clubs

aasr airport and airways surveillance radar

AASR Abhazian Autonomous Soviet Socialist Republic; Adjarian Autonomous Soviet Socialist Republic

AASRC American Association of Small Research Companies

ASSRI Arctic and Antarctic Scientific Research Institute

ASSRM Ancient and Accepted Scottish Rite Masons

AASS Afro-American Students

Society; American Association for Social Security; *Americanae Antiquarianae Societatis Socius* (Latin—Associate of the American Antiquarian Society)

AASSP Arkansas Association of Secondary School Principals

AAST American Association for the Surgery of Trauma

AASTA Antiaircraft Station

AASTC Associate in Architecture—Sydney Technical College

AASTD Association for the Advancement of the Science and Technology of Documentation

AASU Afro-American Student Union

AASW Australian Association of Scientific Workers; Australian Association of Social Workers

AASWA American Association for the Study of World Affairs

AASWI American Aid Society for the West Indies

aat acute abdominal tympany; after acid treatment; auditory attending task

aat (AAT) alpha-1 antitrypsin

AAT Achievement Anxiety Test; Anglo-Australian Telescope (in NSW); Auditory Apperception Test; Australian Antarctic Territory

A-AT Anglo-Australian Telescope

AATA American Association of Teachers of Arabic; Anglo-American Tourist Association

AATB Advanced Amphibious Training Base

AATC Anti-Aircraft Training Center; Army Aviation Test Command

AATCC American Association of Textile Chemists and Colorists

AATCLC American Association of Teachers of Chinese Language and Culture

AATCO Army Air Traffic Coordinating Office

AATD Australian Association of Teachers of the Deaf

AATE American Association of Teachers of Esperanto; Association of Australian Teachers of English

AATEA American Association of Teacher Educators in Agriculture

AATEFL Australian Associa-

tion for the Teaching of English as a Foreign Language
AATF American Association of Teachers of French
AATG American Association of Teachers of German
AATH American Association of Teaching Hospitals
AATI American Association of Teachers of Italian
AATM American Academy of Tropical Medicine
AATNU *Administration de l'assistance technique des Nations Unies* (French—United Nations Technical Assistance Administration)
AATOE American Association of Theatre Organ Enthusiasts
AATP American Academy of Tuberculosis Physicians
AATPA American Association of Traveling Passenger Agents
AATPS Australian Association of Temporary Personnel Services
AATRACEN Anti-Aircraft Training Center
AATRIS Army Air Traffic Regulation and Identification System
AATS American Academy of Teachers of Singing; American Association of Theological Schools; American Association for Thoracic Surgery
AATSEEL American Association of Teachers of Slavic and Eastern European Languages
AATT American Association for Textile Technology
AATTA Arab Association of Tourism and Travel Agents
AAT & TC Anti-Aircraft Training and Test Center
AATU Association of Air Transport Unions
AATUF All-African Trade Union Federation
AAU Administrative Area Unit; Al-Azhar University; Amateur Athletic Union; Associated Aviation Underwriters; Association of American Universities; Association of Atlantic Universities (Canada); Australian Athletics Union
AAUA Amateur Athletics Union of Australia; American Association of University Administrators
AAUBO Association of Atlantic University Business Officers (Canada)
AAUCG Americans Against Union Control of Government

AAUCS Australian-Asian Universities Cooperation Scheme
AAUN American Association for the United Nations; Australian Association for the United Nations
AAUP American Association of University Presses; American Association of University Professors
AAUQ Associate in Accountancy—University of Queensland
AAUTA Australian Association of University Teachers of Accounting
AAUTI American Association of University Teachers of Insurance
AAUUS Amateur Athletic Union of the U.S.
AAUW American Association of University Women
aav airborne assault vehicle
AAV Antiaircraft Volunteer
AAVA American Association of Veterinary Anatomists; Australian Automatic Vending Association
AAVB American Association of Veterinary Bacteriologists
AAVC Australian Army Veterinary Corps
AAVCS Automatic Aircraft Vectoring Control System
aavd automatic alternative voice/data
AAVIM American Associations for Vocational Instructional Materials
AAVMC Association of American Veterinary Medical Colleges
Aavn Army aviation
AAVN American Association of Veterinary Nutritionists
AAVP American Association of Veterinary Pathologists
AAVRO American Association of Vital Records and Organizations
AAVS American Anti-Vivisection Society
AAVSO American Association of Variable Star Observers
AAW Advertising Association of the West; American Atheist Women; Anti-Air Warfare
AAWA American Automatic Weapons Association
AAWB American Association of Workers for the Blind
AAWC Australian Advisory War Council
AAWD Association of American Women Dentists
AAWEXINPT Antiair Warfare

Exercises in Port
AAWg Aeromedical Airlift Wing (USAF)
AAWM American Association of Waterbed Manufacturers
AAWO Afro-Asian Workers' Organization
AAWPI Association of American Wood Pulp Importers
AAWs Anti-Armor Weapons
AAWS American Association of Wardens and Superintendents
AAWU Amateur Athletic Western Union
AAXICO American Air Export and Import Company
AAYM American Association of Youth Museums
AAZK American Association of Zoo Keepers
AAZM *Aqueducto y Alcantarillado de la Zona Metropolitana* (Spanish—Aqueducts and Watercourses of the Metropolitan Area)
AAZN American Association for Zoological Nomenclature
AAZPA American Association of Zoological Parks and Aquariums
ab abnormal; abortion; about; abscess; adapter booster; afterburner; airbrake; alcian blue; ambient brine; anchor bolt; antibody; asbestos body; asthmatic bronchitis; at bats; axiobuccal
a/b acid-base (ratio)
a/b (A/B) airborne
a & b applejack and benedictine; assault and battery
ab *abril* (Spanish—April); (Latin prefix—away from, off); (Persian—water)
aB *auf Bestellung* (German—on order)
Ab Abner; abnormal; Abraham; alabamine
Ab *Abade* (Portuguese—abbot; fat man)
AB able-bodied seaman; Accessories Bulletin; Aid to the Blind; Air Base; Alberta; Arnold Bernstein (steamship line); Aryan Brotherhood; Assembly Bill; Atlantic Beach
A-B Allen-Bradley; Ambrose Bierce; Anton Bruckner
A.B. *artium baccalaureus* (Latin—Bachelor of Arts)
A/B Aid to the Blind; Airman Basic
A & B Antigua & Barbuda
AB *Analecta Biblica*
A/B *Aktiebolag* (Swedish—lim-

ited company)
AB-47 Agusta-Bell three-place utility helicopter
AB-204 Agusta-Bell gunship twin-engine helicopter
AB-205 Agusta-Bell ten-place troop-transport helicopter
AB-206 Agusta-Bell five-seat turbine-powered helicopter
aba antibacterial activity
ab-a abampere
ABA Aaron Burr Association; American Badminton Association; American Bakers Association; American Bandmasters Association; American Bankers Association; American Bar Association; American Bell Association; American Berkshire Association; American Billiard Association; American Bison Association; American Booksellers Association; American Bowhunters Association; American Brazilian Association; American Buddhist Association; Annual Budget Authorization; Australian Badminton Association; Australian Bankers Association; Australian Booksellers Association; Australian Boomerang Association; Australian Bowhunters Association; Australian Bridge Association; Ayrshire Breeders Association
ABAA Antiquarian Booksellers Association of America
ab ab. *ab absurdo* (Latin—to the absurd)
abac a basic coursewriter
ABAC Abraham Baldwin Agricultural College
Abaco Great and Little Abaco islands in the Bahamas north of New Providence Island
abact abacterial
ABACUS Air Battle Analysis Center Utility System; Autonetics Business and Control United Systems
-abad (Hindi—city or town)
ABAD Air Battle Analysis Division
ab aet. *ab aeterno* (Latin—until eternity)
ABAFA Association of British Adoption and Fostering Agencies
ABAG Association of Bay Area Governments (San Francisco)
A-bahn *Autobahn* (German—superhighway)
ABAI American Bell Associa-

tion International; American Boiler and Affiliated Industries
ABAJ Antiquarian Booksellers Association of Japan
ABAJ *American Bar Association Journal*
ABAK *Association di Biblioteka i Archivo di Korsow* (Papiamento—Association of Libraries and Archives of Curaçao)
abamp absolute ampere (10 amperes)
aband abandoned
abandt abandonment
ABAO *Asociación Bilbaina de Amigos de la Opera* (Spanish—Bilbaoan Association of Friends of the Opera)
abap antibody against panel
ABARE Australian Bureau of Agriculture and Resource Economics
ABAS American Board of Abdominal Surgeons; Australian Buying Advisory Service
abat abattoir
a batt *a battuta* (Italian—by the beat)—musical term
ABATU Advance Base Air Task Unit; Advance Base Aviation Training Unit
ABAUSA Amateur Basketball Association of the United States of America
abb ablating blunt body
abb *abbassamento* (Italian—abatement, decline, diminution, fall of temperature, lowering, subsiding); *abbonamento* (Italian—subscription); *abbuono* (Italian—allowance, bonus, discount)
Abb Abbess; Abbey; Abbot
Abb *Abbildung* (German—illustration)
Abb. *abbas* (Latin—abbot)
Ab of B Archbishop(ric) of Bremen
ABB Akron & Barberton Belt (railroad); Australian Barley Board
ABBA American Blind Bowling Association; American Board of Bio-Analysis; American Brahman Breeders Association; American Bread & Breakfast Association; Australian Brahma Breeders Association
ABBARS American Bread & Breakfast Associated Reservation Services
ABBB Association of Better Business Bureaus

Abb^e *Abbaye* (French—abbey)—monastery
ABBF Association of Bronze and Brass Founders
Abbild *Abbildungen* (German—illustrations)
ABBIM Association of Brass and Bronze Ingot Manufacturers
ABBMM Association of British Brush Machinery Manufacturers
A.B.B.O. Associate of the British Ballet Organisation
Abbot Vickers self-propelled fortress including 105mm gun, turret-mounted howitzer, and 7.62 machine gun
Abbotsford British Columbia's Matsqui Institution (for narcotic addicts) at Abbotsford; a town west of Wausau, Wisconsin
Abbot(t) Abbotson
Abbott & Costello Bud Abbott and Lou Costello
abbr abbreviate; abbreviated; abbreviation
ABBRA American Boat Builders and Repairers Association
abbrev *abbreviatura* (Italian—abbreviation)
abbrevia abbreviations
abbreviaz *abbreviazione* (Italian—abbreviation)
abbrevio abbreviomania(c) (al) (ly)
abbrev(s) abbreviation(s)
ABBS Australian Bibliography and Bibliographical Services
abc abecedarium (alphabet primer); acid-balance control; aconite, belladonna, chloroform; advanced base camp; advanced biomedical capsule (ABC); alphabet; atomic, biological, chemical (ABC); alum, blood, charcoal; automatic bass compensation; automatic brightness control; axiobuccocervical
abc (ABC) advance-booking charter; alarms by carrier (panic-button device alerting fire and police stations)
AbC American-born Chinese
Ab of C Archbishop(ric) of Cologne
ABC Aberrant Behavior Center; A Better Chance (scholarship program for the poor); Abridged Building Classification; Advanced Booking Charters; Aerated Bread Company; Air Bridge to Canada; Alcohol Beverage Control;

Alliance for Better Chicago (public schools); American Ballet Company; American Baptist Churches; American Bar Center; American Book (prices) Current; American Bowling Congress; American Brass Company; American, British, Canadian; American Broadcasting Company; American Business Conference; Animal Birth Control; Argentina, Brazil, Chile; Asahi Broadcasting Company; Asian Banking Council; Assisi Bird Campaign; Associated Bottlers Company; Atanasoff-Berry Computer; atomic, biological, chemical (warfare); Audit Bureau of Circulation; Australian Band Council; Australian Bankruptcy Cases; Australian Baseball Council; Australian Bowling Council; Australian Bridge Council; Australian Broadcasting Commission; Australian Broadcasting Corporation; automatic bandwidth control; automation of bibliography through computerization; Automotive Boosters Clubs

AB & C Atlanta, Birmingham and Coast (railroad)

ABC Academia Brasileira de Ciencias (Portuguese—Brazilian Academy of Sciences); Spanish newspaper

ABC³ Airborne Battlefield Command and Control Center

ABCA American Business Communications Association; American-British-Canadian-Australian; Antique Bottle Collectors Association; Army Bureau of Current Affairs; Australian Bushmen's Carnival Association

ABCAL Associated British Cables (NZ)

ABC-ASP American-British-Canadian—Army Standardization Program

abcb air-blast circuit breaker

ABCB American Bottlers of Carbonated Beverages; Australian Broadcasting Control Board

ABCC Association of British Chambers of Commerce; Atomic Bomb Casualty Commission

A-B CC Arab–British Chamber of Commerce

ABCCC Airborne Battlefield

Command and Control Center

ABC—Clio American Bibliographical Center—Clio Press (Santa Barbara, California)

ABCCTC Advanced Base Combat Communication Training Center

abcd above and beyond the call of duty; airway (opened), breathing (restored), circulation (restored), definitive (therapy); atomic, biological, chemical, and damage (control); awaiting bad conduct discharge

ABCD Accelerated Business Collection and Delivery (of mail); Action for Boston Community Development; Advanced Base Construction Depot; America, Britain, China, Dutch East Indies (ABCD Powers during World War II); American Society of Bookplate Collectors and Designers

ABCDCAL Alcoholic Beverage Control Department—California

ABCE Adult Basic and Continuing Education

ABC-fm Australian Broadcasting Commission—frequency modulation (system)

ABCI Australian Bureau of Criminal Intelligence

ABCL American Birth Control League

ABC Line Antwerp Bulk Carriers (container line)

ABCM Associate of Bandsmen's College of Music; Association for the Bedouin Culture Museum;

ABC-tv Australian Broadcasting Commission—television

abd all but dissertation

Abduction Abduction from the Seraglio (Mozart's comic opera *Entführung aus dem Serail*)

ABE American Ballet Ensemble

ABECOR Associated Banks of Europe Corporation

abel air-breathing electric laser

Abel Abelard

Abes Abyssinia(n)

ABES Association for Broadcast Engineering Standards

abess abessinisch (German—Abyssinian)

ABEU Australian Bank Employees Union

abf absolute bloody final (beverage)

ABF American Bar Foundation; Australian Bridge Federation

abfab absolutely fabulous

ABFP American Board of Family Practice; American Board of Forensic Psychology

ABGC Australian Banana Growers Council

Abh Abhandlungen (German—transactions)

abi abstracted business information

ABI Associazíone Bibliotecari Italiani (Italian—Association of Italian Librarians)

ABIC Adaptive Behavior Inventory for Children

abi/inform abstracted business information/information needs

ab init. ab initio (Latin—from the beginning)

ABINZ Australian Banking Institute of New Zealand

ABIP Australian Books in Print

ABISA Association of Burglary Insurance Surveyors—Australia

ABJ Association of Broadcasting Journalists

Abk Abkürzung (German—abbreviation)

Abkürz Abkürzung (German—abbreviation)

abl ablative

abl abril (Spanish—April)

ABLA American Blind Lawyers Association; American Business Law Association

ABLE Ability Based on Long Experience; Advocates for Border Law Enforcement

ablon abalone

abm (ABM) anti-ballistic missile

ABM Adventist Board of Missions; Australian Board of Missions (Anglican)

ABM Asociación de Banqueros de México (Spanish—Mexican Bankers Association)

ABMJ American Board of Missions to the Jews

ABMR Australian Board of Mineral Resources

ABM Treaty Anti-Ballistic Missile Treaty

ABN Anti-Bolshevik Bloc of Nations

ABNC American Bank Note Company

abndn(d) abandon(ed)

ABNY Association for a Better New York

ABO Admiralty Berthing Officer (UK)

ABOA Australian Bank Officials Association
abol abolish(able); abolisher; abolishment; abolition(ary); abolitionism; abolitionist
abortus aborted fetus
abortus pill abortion-inducing pill, developed in France as Mifepristone, generic name: RU-486
abp absolute boiling point
ABPP American Board of Professional Psychology
ABPS American Board of Plastic Surgery
ABQ Annapolis Brass Quintet
ABR Army Ballistic Research
Abra Abraham
ABRACADABRA Abbreviations and Related Acronyms Associated with Defense, and Radioelectronics (a Raytheon publication)
Abram Abraham
ABRC Advisory Board for the Research Councils
abrev abreviatura (Portuguese/Spanish—abbreviation)
abrév abréviation (French—abbreviation)
ABRS Australian Biological Resources Study
abs absent; absorb(ent)
abs (ABS) acrylonitrile-butadiene-styrene
ABs Aryan Brotherhood members
Abs Absatz (German—paragraph)
ABS Anti-lock Braking System; Association of Banks in Singapore; Australian Ballet Society; Australian Book Society; Australian Bureau of Statistics
abs art abstract art
Abschn Abschnitt (German—chapter or paragraph)
ABSEL Association for Business Stimulation and Experiential Learning
Absen Absender (German—sender)
abse. re. absente reo (Latin—defendant absent)
abs exp abstract expressionism
ABSO Auxiliary Business Service Organization
abso(l) absolute; absolutely
absol absolument (French—absolutely)
ABS-PS Adaptive Behavior Scale—Public School (version)
ABSTI Advisory Board on Scientific and Technical Information (Canadian)
abstr abstract
abt about
abt (ABT) after beginning of time
Abt Abteilung (German—division or part)
ABT American Ballet Theater; American Board of Trade; Australian Broadcasting Tribunal
ABTA American Bridge Teachers Association; Australian-British Trade Association
ABTR Association for Brain Tumor Research
ABU American Board of Urology; Asian Broadcasting Union
abul abulia; abulic
ABWA American Business Women's Association
ABWUA Amateur Boxing and Wrestling Union of Australia
ABY AB Byers (stock-exchange symbol)
ac absolute ceiling; accelerator; acetyl; acetyl-choline; acoustical(ly); adrenal cortex; aerodynamic center; air condition(ed); air conditioning; air conduction; air cool; air-cooled; alternating current; anodal closure; anticorrosive; antiphlogistic corticoid; area controller; arithmetic computation; asbestos cement; atriocarotid; auriculocarotid; auxiliary console; axiocervical; azimuth comparator
ac (AC) average cost
a-c alternating-current
a/c account; account current; air conditioning; aircraft
a & c addenda and corrigenda; arts and crafts
ac (Latin prefix—to, toward)
a.c. ante cibos (Latin—before meals)
a/c ao cuidado de (Portuguese—in care of)
a C avanti Cristo (Italian—before Christ)
Ac actinium; altocumulus
AC Adelbert College; Adelphi College; Aden Colony; Administration of Correction (Puerto Rico); Adrian College; aerodynamic center (symbol); Air Canada; Alabama College; Albion College; Albright College; Allegheny College; Alliance College; Alma College; alternating current; Alverno College; Amarillo College; Ambulance Corps; Amherst College; Anderson College; Andrew College; Annhurst College; anodal contraction or closure; Antioch College; Appeal Cases; Aquinas College; Arcadia College; Area Code; Arithmetic Computation (test); Arkansas College; Armstrong College; Asbury College; Ashland College; Assumption College; Athens College; Athletic Club; Augusta College; Augustana College; Aurora College; Austin College; Australia Council; Australian Companion (Companion of the Order of Australia); Australian Corps; Averett College; Azusa College
A-C Allis-Chalmers
A/C Air Commodore; aircraft; Aviation Cadet
AC Ação Catolica (Portuguese), *Acción Católica* (Spanish), *Action Catholique* (French), *Azione Cattolica* (Italian—Catholic Action); *Appellation Contrôlée* (French—brand name control of fine wines); *Atlanta Constitution*
A.C. année courante (French—current year); *Año Cristo* (Spanish—Year of Our Lord)—A.D.
A & C Antony and Cleopatra
AC-47 DC-3 Douglas 21-passenger transport also called C-47 Dakota or Skytrain
AC-119 Fairchild-Hiller armed gunship complete with Vulcan cannons and 7.2mm miniguns (C-119 Flying Boxcar conversion)
AC-130 Lockheed armed gunship similar to AC-119 but with more guns
aca adenocarcinoma; anterior cerebral artery; azimuth control amplifier
ac a acetic acid
Aca Acapulco (inhabitants—Acapulqueños)
ACA Acapulco, Mexico (airport); Adult Children of Alcoholics; Aero Club of America; Aircraft Castings Association; Alberta College of Art; American Camping Association; American Canoe Association; American Carnivals Association; American Casting Association; American Cat Association; American Cemetery Association; American Chiro-

practic Association; American Civic Association; American College of Allergists; American College of Anesthesiologists; American College of Apothecaries; American Communications Association; American Composers Alliance; American Congregational Association; American Correctional Association; American Cryptogram Association; American Crystallographic Association; Americans for Constitutional Action; Anti-Corruption Agency (Singapore); Arms Control Agency; Arms Control Association; Arts Council of America; Arts Council of Australia; Assembly Constitutional Amendment; Associated Chiropodists of America; Association of Correctional Administrators

A.C.A. Associate of the Institute of Chartered Accountants (of England and Wales)

ACAA Agricultural Conservation and Adjustment Administration

ACAAE Australian Council on Awards in Advanced Education

ACAAI Air Cargo Agents Association of India

ACAB Air-Conditioning Advisory Bureau; Army Contract Adjustment Board

ACAC Allied Container Advisory Committee; Association of College Admission Counsellors; Australian Conciliation and Arbitration Commission; Australian Corporate Affairs Commission

ACACA Army Command and Administration Communication Agency

acad academic; academician; academy

Acad Acadia; Academy

Acad Academia (Latin—academy)

ACAD American Conference of Academic Deans

Acad aper Academy of Motion Picture Arts and Sciences aperture (of sound films)

Acad B-A Académie des Beaux-Arts (French—Academy of Fine Arts)

Academic Academic Press

Acad Fran Académie Française (French Academy)

ACADI Association des Cadres

Dirigeants de l'Industrie (French—Association of Industrial Executives)

Acadia national park occupying Mount Desert Island, half of Isle au Haut, and Schoodic Point on the Maine coast; old name for French-speaking Canada and Nova Scotia

Acadia(n) Novia Scotia(n); native Louisianians of French origin

Acad Ins B-L Académie des Inscriptions et Belles-Lettres (French—Academy of Inscriptions and Literature)

Acad mask Academy of Motion Picture Arts and Sciences mask (enclosing the aperture area of sound films)

Acad Med Academy of Medicine

Acad Mgmt Academy of Management

Acad Mus Academy of Music

Acad Pr Academic Press

Acad Pr Ark Academic Press of Arkansas

Acad Sci Académie des Sciences (French—Academy of Science)

Acad Sin Academia Sinica (Chinese Academy of Science)

Acad St Cec Academia di Santa Cecilia, Rome

Acad Therapy Academic Therapy Publications

Acad U Acadia University

ACAE American Council for the Arts in Education; Australian Commission on Advanced Education

AC & AE Association of Chemical and Allied Employees

acaf automatic circuit assurance feature

ACAF Amphibious Corps, Atlantic Fleet; Australian Citizen Air Force

ACAM Asian Centre for Agricultural Machinery; Australian Confederation of Apparel Manufacturers

ACAN Action Committee Against Narcotics; Army Command and Administrative Network

ACAnes American College of Anesthetists

a. cant. after cant frames

Acanth Acanthocephala

acanthite silver sulfide

Acap Acapulco

ACAP American Council on

Alcohol Problems; Annapurna Conservation Area Project; Army Contract Appeals Panel

ACAPA American Concrete Agricultural Pipe Association

Acap gold Acapulco gold (high-grade marijuana of the type grown near the Mexican resort of Acapulco)

a capp a cappella (Italian—in chapel style, without musical accompaniment)

Acapulco Acapulco de Juárez (Mexico's leading seaside resort)

ACAR Australian Coal Association Research

ACARD Advisory Council for Applied Research and Development

ACAs Arms Control Associations

ACAS Aboriginal Children's Advancement Society; Advisory, Conciliation, and Arbitration Service; Airborne Collision Avoidance System; Association of Casualty Accounts and Statisticians

AC/AS Assistant Chief of Air Staff

ACAS Asociación Civil Amigos de la Salud (Spanish—Friends of Health Civil Association)

ACASP Australian Commonwealth Association of Simplified Practice

ACAST Advisory Committee on Applications of Science and Technology (UNESCO)

acata acatalectic(al)

acb air circuit breaker; asbestos cement board

acb (ACB) aortocoronary saphenous vein bypass; arterialized capillary blood

ACB Advertising Checking Bureau; Airman Classification Battery; Army Classification Battery; *Association Canadienne des Bibliothèques* (Canadian Library Association); Association of Customers' Brokers; Association of the Customs Bar; Australian Cricket Board

ACB Association Canadienne des Bibliothèques (French—Canadian Library Association)

ACBA Academy of Comic Book Artists

ACB of A Associated Credit Bureaus of America

ACBB American Council for Better Broadcasts

ACBCC Australia-China Business Cooperation Committee
ACBFC Academy of Comic-Book Fans and Collectors
ACBL American Commercial Barge Line; American Contract Bridge League; Australian Contract Bridge League
acbm atomic cesium beam maser
ACBM Associated Corset and Brassiere Manufacturers; Aviation Chief Boatswain's Mate
ACBNZ Associated Credit Bureau of New Zealand
ACBO Association of Chief Business Officials
ACBs American Conference of Bishops; Associated Credit Bureaus
ACBS Accrediting Commission for Business Schools
ACBWS Automatic Chemical Biological Warning System
acc accept; accident(al) (ly); accommodate(d); accommodation; accompaniment; according; account(ed); accounting; accusative; altocumulus castellatus (clouds); alveolar cell carcinoma; anodal closing contraction; astronomical great circle course (ACC); automatic chroma circuit (tv)
acc (ACC) accumulator
Acc Lucius Accius (Latin—name of a tragic poet)
ACC Abilene Christian College; Academy of Canadian Cinema; Accra, Ghana (airport); accumulator; Adirondack Community College; Administrative Committee on Coordination; Air Center Commander; Air Control Center; Air Coordinating Committee; Allied Control Commission; Allied Control Council; Ambulatory Care Center; American Cat Council; American College of Cardiology; American Concert Choir; American Conference of Cantors; American-Chilean Council; American Craftsmen's Council; Anglican Consultative Council; Army Chemical Center; Army Cooperation Command; Assistant Chief Constable; Association of Choral Conductors; Auburn Community College; Australian Chamber of Commerce; Australian Chiropody

Council; Australian Computer Conference; Australian Croquet Council
A-C-C Appleton-Century-Crofts
ACCA Aeronautical Chamber of Commerce of America; American Clinical and Climatological Association; American College of Clinic Administrators; American Correctional Chaplains Association; American Cotton Cooperative Association; Art Collectors Club of America; Associated Chambers of Commerce of Australia
Accad Accadèmia (Italian—academy)
Accad Ball Accadèmia di Ballo (Italian—dancing academy)
Accad di Mil Accadèmici di Milano (Italian—Academicians of Milan)
Accad Mus Nap Orch Cam Accadèmia Musicale Napoletana Orchestra de Camera (Italian—Neapolitan Musical Academy Chamber Orchestra)
Accad Nav Accadèmia Navale (Italian—naval academy)
Accad Sta Cec Accadèmia di Santa Cecilia (Italian—Academy of Santa Cecilia), musical academy and orchestra in Rome
ACCAP Autocoder-to-Cobol Conversion Aid Program
ACCAs American Correctional Chaplains Association members
acc & aud accountant and auditor
ACCC Ad Hoc Committee for Competitive Communications; Advisory Council on College Chemistry; Alternate Command and Control Center; American Council of Christian Churches; Association of Canadian Community Colleges; Association of Community Cancer Centers
ACCCA Association of California Community College Administrators
ACCCE Association of Consulting Chemists and Chemical Engineers
ACCCF American Concert Choir and Choral Foundation
Acc Chem Res Accounts of Chemical Research
ACCCI Associated Chinese Chambers of Commerce and Industry (Singapore); Austra-

lia-China Chamber of Commerce and Industry
AC & CCI American Coke and Coal Chemicals Institute
accd accelerated construction completion date
acc dec acceptable deception(s)—half truth(s)
acce acceptance
ACCE American Chamber of Commerce Executives; American Council for Construction Education
accel accelerate; accelerate(d); accelerating; acceleration
accel accelerando (Italian—accelerating)
ACCELS Automated-Circuit Card-Etching Layout System
accepon acceptation (French—acceptance)
access accessory
ACCESS American College of Cardiology Extended Study Services; Architects Central Constructional Engineering Surveying Service; Association of Community Colleges for Excellence in Systems and Services; Automated Catalog of Computer Equipment and Software Systems (USA); Automatic Computer-Controlled Electronic Scanning System
A.C.C.E.S.S. A Cooperative Community Educational School System
ACCF American Committee for Cultural Freedom; American Council for Capital Formation; Association of Community College Facilities
ACCFA Agricultural Credit Cooperative Finance Administration
AcCh acetylcholine
ACCH Association for the Care of Children in Hospitals
ACCHAN Allied Command Channel (NATO)
acci accidental injury
ACCI American Cottage Cheese Institute
accid accident(al)
ACCION Americans for Community Cooperation in Other Nations
accis accismus
accl anodal closure clonus
ACCL American Council of Commercial Laboratories
ACCM American College of Clinic Managers
ACCM P-H Appleton-Century-Crofts Medical (imprint of) Prentice-Hall

ACCN Associated Court and Commercial Newspapers

acco accompagnamento (Italian—accompaniment)

ACCO Associate of the Canadian College of Organists; Association of Child Care Officers

AcCoA acetyl coenzyme A

accom accommodation; accompaniment

accom ad lib accompaniment ad libitum

accom oblto accompaniment obligato

accomp accompaniment; accomplish

ACCOMP Academic Computer Group

ACCOR Australian Coal Corporation

ACCORD Action Coalition to Create Opportunities for Retirement with Dignity

ACCORDS Acoustic Correlation and Detection System

Accountemps company providing temporary financial, banking, EDP, credit, bookkeeping and accounting services

ACCP American College of Chest Physicians

ACCP Asociación de Camaras de Comercio del Perú (Spanish—Peru Association of Chambers of Commerce)

ACCR American Council on Chiropractic Roentgenography

ACCRA Abortion and Contraception Counselling and Research Association; American Chamber of Commerce Researchers Association; Australian Chart and Code for Rural Accounting; Australian Committee for Coding Rural Accounts

accrd int accrued interest

accred accredited

accres accrescendo or *accresciménto* (Italian—augmented or increasing)

ACCS Automated Calibration Control System

A.C.C.S. Associate of the Chartered Corporation of Secretaries

ACCSA Allied Communications and Computer Security Agency

acct account; accountant; accounting

ACCT Association of Community College Trustees

acctd accented

acctncy accountancy

ACCTU All-Union Central Council of Trade Unions (USSR)

accu automatic combustion-control unmanned

a-c cu alternating-current control unit

accum accumulate

accur accuratissime (Latin—most accurately)

accus accusative

accv (ACCV) armored cavalry cannon vehicle

accw alternating current continuous wave

accy accessory

acd absolute cardiac dullness; accord; accordion; acid-citrate-dextrose; active duty commitment; adopted child; advance delivery of correspondence; advice of duration and charges; anodal duration contraction; average daily census; axiodistocervical

acd (ACD) acid citrate dextrose; arms control through defense

ACD Administrative Commitment Document; Allied Chemical Corporation (stock exchange symbol); American Choral Directors; American College of Dentists; Australian College of Dentistry; Australian College of Dermatologists

ACD American College Dictionary

ACDA Advisory Committee on Distinction Awards; American Choral Directors Association; Arms Control and Disarmament Agency; Asian Centre for Development Administration; Aviation Combat Development Agency

a-c/d-c alternating current/direct current; slang—bisexual

ACDC Australian Counter-Disaster College

ACDCM Archbishop of Canterbury's Diploma in Church Music

ACDE American Council for Drug Education

ACDFA American College Dance Festival Association

acdl asynchronous circuit-design language

ACDM Association of Chairmen of Departments of Mechanics

ACDMS Automated Control of Document Management System

A Cdre Air Commodore

ACDS Advanced Combat Direction System; Australian College of Dental Surgeons

acdt accident

A Cdt Air Commandant

acdu active duty

acdutra active duty for training

ACDUTRA Active Duty Reserve Army

ace. acceptance checkout equipment; acetic; adrenal cortical extract; aerospace control environment; air crash equipment; alcohol-chloroform-ether (anesthetic mixture); attitude control electronics; automatic checkout equipment; automatic circuit exchange

ace. (ACE) angiotensin converting enzyme

ACE African Container Express; Allied Command, Europe; American Cinema Editors; American Coaster Enthusiasts; American Conservatory Theatre; American Council on Education; American Hard Rubber Company (trademark); Army Corps of Engineers; Association for Community Education; Association of College Entrepreneurs; Australian College of Education; Auxiliary Corps of Executives; Aviation Construction Engineers

ACEA Air Line Communication Employees Association; Association of Consulting Engineers—Australia

ACEAA Advisory Committee on Electrical Appliances and Accessories

acearts airborne countermeasures environment and radar target simulator

ACEB Association Canadienne des Ecoles Bibliothecaires (French—Canadian Association of Library Schools)

Ace Bks Ace Books

ACEC Alcoholism Counseling and Education Center; American Consulting Engineers Council; Army Communications and Electronic Command; Army Communications and Electronic Command (USA)

ACEC Ateliers de Constructions Electriques de Charleroi (French—Electrical Construction Workshops of Charleroi)—in Belgium

A.c.Ed. Associate in Commercial Education

ACED Advanced Communications Equipment Depot

ACEEE American Council for an Energy Efficent Economy

ACEF Aboriginal Child Education Fund(ing); Aboriginal Children's Education Fund; Asian Cultural Exchange Foundation; Association of Commodity Exchange Firms; Australian Council of Employers Federations

a-c-e-g (musical mnemonic—all cows eat grass)—bass clef note names of the four spaces

ACEI Association of Consulting Engineers of Ireland; Association for Childhood Education International

ACEID Asian Centre of Educational Innovation for Development

ACEJ American Council on Education for Journalism

ACEJMC Accrediting Council on Education in Journalism and Mass Communications

ACEL Air Crew Equipment Laboratory

ACELF Association Canadienne des Educateurs de Langue Française (French—Canadian Association of French Language Teachers)

ACEM Aviation Chief Electrician's Mate

a cemb a cembalo (Italian—by the harpsichord)

ACEN Assembly of Captive European Nations

ACENET Allied Command Europe Communications Network (NATO)

ACENZ Association of Consultant Engineers of New Zealand

ACEORP Automotive and Construction Equipment Overhaul and Repair Plant

ACEP American College of Emergency Physicians; American Council for Emigrés in the Professions

ACEPD Automotive and Construction Equipment Parts Depot

ACEQ Association of Consulting Engineers of Québec

ACER Australian Council for Educational Research

ACERP Advanced Communications-Electronics Requirements Plan

ACERT Advisory Council for Education of Romanies and other Travellers

aces automatic control evaluation simulator

ACES Alternative Consumer Energy Society; Americans for the Competitive Enterprise System; Area Cooperative Educational Service; Association for Counselor Education and Supervision; Australian Council for Educational Standards

ACESA Arizona Council of Engineering and Scientific Associations; Australian Commonwealth Engineering Standards Association; Australian Computer Equipment Suppliers Association

ace-s/c acceptance checkout equipment—spacecraft

ACESIA American Council for Elementary School Industrial Arts

acet acetone

ACET Advisory Committee on Electronics and Telecommunications; Australian Council for Education through Technology

ACETA Australian Commercial and Economics Teachers Association

ACE Test American Council on Education Test

acetl acetylene

acetyl-co A acetyl-coenzyme A

ACEUR Allied Command, Europe

ACEWR American Committee for European Worker's Relief

acf accessory clinical findings

acf (ACF) air-combat fighter (aircraft)

ACF Alternate Communications Facility; American Car & Foundry; American Checker Federation; American Chess Foundation; American Choral Foundation; American Culinary Federation; Anglican Charismatic Fellowship; Anti-Crime Foundation; Army College Fund; Association of Consulting Foresters; Australian Canoe Federation; Australian Chess Federation; Australian Conservation Federation

ACF Académie Canadienne Française (French-Canadian Academy); *Automobile-Club de France* (French—Automobile Club of France)

ACFA American Cat Fanciers

Association; Association of Commercial Finance Attorneys

ACFAS Association Canadienne-Française pour l'Avancement des Sciences (French—French-Canadian Association for the Advancement of Science)

ACFB Australian Canned Fruits Board

ACFC Aviation Chief Fire Controlman

ACFEA Air Carrier Flight Engineers Association

ACFEL Arctic Construction and Frost Effects Laboratory (Greenland)

ACFES Association of Canadian Faculties of Environmental Studies

acfg automatic continuous function operation

ACFHE Association of Colleges for Further and Higher Education

ACFL Atlantic Coast Football League

ACFM Association of Canadian Fire Marshals

ACFN American Committee for Flags of Necessity

ACFO American College of Foot Orthopedics

ACFOD Asian Cultural Forum for Development (Thailand)

ACFP Advisory Commission on Federal Pay

ACFR Advisory Committee of the Federal Register; Advisory Council on Federal Reports; American College of Foot Roentgenologists

ACFS American College of Foot Surgeons

ACFSA American Correctional Food Service Association

acft aircraft

ac ft acre feet; acre foot

ACFT Aircraft Flying Training

acftc aircraft carrier

ACFTU All China Federation of Trade Unions (PRC)

acg automatic caution guard; automatic control gear

acg (ACG) apex cardiogram

ac-g accelerator globulin

ACG Airborne Coordinating Group; Air Cargo Express (symbol); Airline Carriers of Goods; American College of Gastroenterology; American Council on Germany; Association for Corporate Growth

ACG An Comunn Gaidhealach (The Gaelic Society)—also

called the Highland Society
ACGA American Cranberry Growers' Association; Australian Cane Growers Association; Australian Citrus Growers Association
ACGB Arts Council of Great Britain
ACGBI Automobile Club of Great Britain and Ireland
ACGC Australian Cane Growers Council
ACGD Association for Corporate Growth and Diversification
ACGF American Child Guidance Foundation
ACGFC Associate of the City and Guilds Finsbury College
ACGI Associate of the City and Guilds Institute
ACGIH American Conference of Governmental Industrial Hygienists
ACGLA Alcoholism Council of Greater Los Angeles
ACGM Aircraft Carriers General Memorandum
ACGP Army Career Group; Australian College of General Practitioners
ACGPOMS American College of General Practitioners in Osteopathic Medicine and Surgery
ACGS Aerial Cartographic and Geodetic Squadron; American Council on German Studies
ACGSq Aerial Cartographic and Geodetic Squadron (USAF)
ach acetylcholine (Ach); actual obtained achievement; arm, chest, height
ach (Ach) (ACH) acetylcholine; adrenal cortical hormone
ach (ACH) automated clearing house
ACh acetylcholine
ACH Association for Computers and the Humanities; Australian Camp Hospital; Automated Clearing House
ACHA American Catholic Historical Association; American College Health Association; American College of Hospital Administrators
AC & HBR Algoma Central and Hudson Bay Railway
ACHCA American College of Health Care Administrators
ache. acetylcholinesterase
ACHE Alabama Commission on Higher Education
achiev achievement

ach index arm (girth), chest (depth), hip (width) index (of nutrition)
ACHNHP Appomattox Court House National Historical Park
ACHP Advisory Council for Historic Preservation
ACHPER Australian Council for Health, Physical Education, and Recreation
achr acetylcholine receptor
ACHR American Council of Human Rights
achrom achromatism
A Ch S Associate of the Society of Chiropodists
ACHS Association of College Honor Societies
ACHSA American Correctional Health Services Association
ACHTR Advisory Committee for Humid Tropics Research (UNESCO)
achvit achievement
aci airborne-controlled interception; automatic car identification
aci (ACI) adult correctional institution; anticlonus index
aci assure contre l'incendie (French—insured against fire)
ACI Air Cargo Incorporated; Air Combat Information; Air Combat Intelligence; Alliance Coopérative Internationale (International Cooperative Alliance); Alloy Casting Institute; American Carpet Institute; American Concrete Institute; American Cryogenics Incorporated; Associated Colleges of Indiana; Australian Consolidated Industries
ACI Association Cartographique Internationale (French—International Cartographic Association); *Azione Cattolica Italiana* (Italian Catholic Action)
acia asynchronous communications interface adapter
ACIA Associate of the Catering Institute of Australia; Associated Cooperage Industries of America
ACIAA Australian Commercial and Industrial Artists' Association
ACIAS American Council of Industrial Arts Supervisors
ACIASAO American Council of Industrial Arts State Association Officers
ACIATE American Council of Industrial Arts Teacher Edu-

cation
ACIB Associate of the Corporation of Insurance Brokers
acic acicular
ACIC Aeronautical Chart and Information Center; Allied Captured Intelligence Center; Australian Chemical Industry Council; Auxiliary Combat Information Center
ACICU Arkansas Council of Independent Colleges and Universities
acid acidosis; acidulated drop; hallucinogenic drug such as LSD-25
acid phos acid phosphatase
acid p'tase acid phosphatase
ACIF All Canada Insurance Federation
ACIGS Assistant Chief of the Imperial General Staff
ACII Associate of the Chartered Insurance Institute
ACIID A Critical Insight Into Israel's Dilemmas
ACIL American Council of Independent Laboratories
acim axis-crossing interval meter
ACIM American Committee on Italian Migration
ACIMS Aircraft Component Management System
ACIO Air Combat Intelligence Office(r)
acip aviation career incentive pay
ACIP Advisory Committee on Immunization Practices; American College of International Physicians
ACIPCO American Cast Iron Pipe Company
ACIR Advisory Committee on Intergovernmental Relations; Automotive Crash Injury Research
AC/IREF American Chapter—International Real Estate Federation
ACIRL Australian Coal Industries Research Laboratories
ACIS American Committee for Irish Studies; Arms Control Impact Statement
A.C.I.S. Associate of the Chartered Institute of Secretaries
acit air-cannon impact tester
ACIV Associate of the Commonwealth Institute of Valuers
ACIWLP American Committee for International Wild Life Protection
ACJ American Council for Ju-

daism

A.C.J. Associate in Criminal Justice

ACJA American Criminal Justice Association

ACJA–LAE American Criminal Justice Association–Lambda Alpha Epsilon

ACJC Assembly Criminal Justice Committee

ACJHSIS Arkansas Criminal Justice Highway Safety Information System

ACJP Airways Corporations Joint Pensions

ACJS Academy of Criminal Justice Sciences

ack acknowledge; acknowledgment

ACK accidentally killed; acknowledge; Armstrong Cork (stock exchange symbol)

ack-ack antiaircraft

Ack-Ack Aluminum Company of America (stock exchange nickname)

ackt acknowledgment

acl air-cushion landing; allowable cabin load

aCl aspiryl chloride

ACL Adjective Check List; Aeronautical Computers Laboratory (USN); American Canadian Line; American Classical League; American Cruise Line; Association of Cinema Laboratories; Association for Computational Linguistics; Atlantic Coast Line (railroad); Atlantic Container Line; Audit Command Language; Aviation Circular Letter

ACL *Automobile Club de Luxembourg* (French—Automobile Club of the Grand Duchy of Luxembourg)

ACLA American Comparative Literature Association; American Cotton Linter Association; Anti-Communist League of America; Australasian Communications Law Association

ACLAM American College of Laboratory Animal Medicine

AClant Allied Command, Atlantic

ACLC Air Cadet League of Canada; Assessment of Children's Language Comprehension

acld aircooled

ACLD Association for Children with Learning Disabilities

aclg air-cushion landing gear

ACLI American Council of Life

Insurance

ACLICS Airborne Communications Location, Identification, and Collection System

ACLM American College of Legal Medicine

ACLN American Catholic Lay Network

ACLO Association of Cooperative Library Organizations

ACLP Association of Contact Lens Practitioners

ACLR *American Criminal Law Review*

acls automatic carrier landing system

ACLS American Council of Learned Societies; Automatic Carrier Landing System

ACLU American Civil Liberties Union; American College of Life Underwriters; Atlantic Container Line Unit; Australian Civil Liberties Union

aclv accrued leave

acm active countermeasure(s); anatomy-covering material; anatomy-covering memo; asbestos-covered metal

acm **(ACM)** advanced cruise missile

a-c-m albumin-calcium-magnesium

Ac M Academy of Music, Philadelphia

ACM Air Chief Marshal; Air Commerce Manual; Air Court-Martial; American Campaign Medal; American College of Musicians; American Craft Museum; Associated Colleges of the Midwest; Association for Computing Machinery; Australian Carpet Manufacturers; Australian Consolidated Minerals; auxiliary mine layer (3-letter symbol); Aviation Chief Metalsmith

ACM *Automobile Club de Monaco* (French—Automobile Club of Monaco)

ACMA Acidproof Cement Manufacturers Association; Air Carrier Mechanics Association; Alumina Ceramic Manufacturers Association; American Certified Morticians Associations; American Circus Memorial Association; American Comedy Museum Association; American Cutlery Manufacturers Association; Associated Chambers of Manufacturers of Australia

acme. attitude control and ma-

neuvering electronics

ACME Adult Community Movement for Equality; Advisory Council on Medical Education; Association of Consulting Management Engineers; Australian Contemporary Music Ensemble

ACMET Advisory Council on Middle East Trade

ACMF Air Corps Medical Forces; Allied Central Mediterranean Forces; American Corn Millers' Federation; Australian Commonwealth Military Forces

ACMI Air Combat Maneuvering Instrumentation system; American Cotton Manufacturers Institute; American Cystoscope Makers, Incorporated

ACML Association of Canadian Map Libraries

ACMM Aviation Chief Machinist's Mate

A.C.M.M. Associate of the Conservatorium of Music—Melbourne

acmp accompany

ACMP Amateur Chamber Music Players; Assistant Commissioner of the Metropolitan Police

ACMRR Advisory Committee on Marine Resources Research (FAO)

acmru audio commercial-message repeating unit

ACMS Advanced Configuration Management System; Army Command Management System; Australian Clay Minerals Society; Automated Career Management System

ACMT Advanced Cruise-Missile Technology; American College of Medical Technologists

ACMWA Amon Carter Museum of Western Art (Fort Worth)

acn activities control number; acute conditioned necrosis (ACN); all concerned notified; assignment control number (ACN); automatic celestial navigation (ACN)

ACN American Chain & Cable (stock exchange symbol); American College of Neuropsychiatrists; American Comedy Network; American Council on NATO; Authorized Code Number

A.C.N. *Ante Christum Natum*

(Latin—before the birth of Christ)

ACNA Advisory Council on Naval Affairs; Arctic Institute of North America

ACNB Australian Commonwealth Naval Board

ACNE Action Committee for Narcotics Education; Alaskans Concerned for Neglected Environments

ACNHA American College of Nursing Home Administrators

ACNIL Azienda Comunale Navigazione Interna (Italian—City Rapid Transit Shipping Company)

ACNM American College of Nurse Midwifery; American College of Nurse-Midwives

ACNO Assistant Chief of Naval Operations

ACNOT Assistant Chief of Naval Operations—Transportation

ACNS American Council for Nationalities Service; Associated Correspondents News Service

ACNUR Alto Comisionado de las Naciones Unidas para las Refugiados (Spanish—High Commission of the United Nations for Refugees)

ACNY Advertising Club of New York

ACNYC Art Commission of New York City

aco anodal closing odor

a co a cargo (Spanish—against)

ACO Administrative Contracting Officer; Air Cargo (Leopoldville—Republic of the Congo); American Academy of Optometry; American College of Orgonomy; Australian College of Ophthalmologists; Australian College of Organists

ACOA Administrative and Clerical Officers Association; Adult Children of Alcoholics

ACOC Air Command Operations Center (NATO)

ACOCA Army Communication Operations Center Agency

acodac acoustic data capsule

ACODS Army Container-Oriented Distribution System (USA)

ACOFO American College of Foot Orthopedists

ACOFS Australian Council of Film Societies

acog (ACOG) aircraft on ground

ACOG American College of Obstetricians and Gynecologists

ACOHA American College of Osteopathic Hospital Administrators

ACOI American College of Osteopathic Internists

acol(s) acolyte(s)

acom automatic coding machine

ACOM Aviation Chief Ordnanceman

A.Comm. Associate in Commerce

A.Comm.A. Associate of the Society of Commercial Accountants

ACOMPLIS A Computerized London Information Service (GLC)

ACOMR Advisory Committee on Oceanic Meteorological Research (UN)

Acon Aconcagua (South America's highest mountain in the Andes where it rises in Argentina and towers over Valparaiso, Chile)

ACOOG American College of Osteopathic Obstetricians and Gynecologists

ACOP American College of Osteopathic Pediatricians; Association of Chief Officers of Police (England and Wales)

ACOPS Advisory Committee on Oil Pollution of the Sea

ACOR Auditory Cognition of Relationships

ACORD Advisory Council on Research and Development

ACORDE A Consortium on Restorative Dentistry Education

acorn. acronym-oriented nut

ACORN Association of Community Organizations for Reform Now; Associative Content Retrieval Network

ACOS American College of Osteopathic Surgeons

ACOSH Appalachian Center for Occupational Safety and Health; Australian Council on Smoking and Health

acous acoustical; acoustics

acous coup acoustic coupler (device allowing other electronic devices to listen to or make other sounds transmitted by an ordinary telephone)

acousid acoustic seismic intrusion detector

acoust acoustical(ly), acoust-

ic(s)

acoustint acoustical intelligence

acp acetyl-carrier protein (ACP); acid phosphatase; anodal closing picture; aspirin, caffeine, phenacetin; auxiliary control panel; azimuth change pulse

a-c-p aspirin-caffeine-phenacetin

a & cp anchors and chains proved

aCp (ACP) automatic Colt pistol

ACP Academy for Contemporary (Criminal) Problems; African, Caribbean, and Pacific countries; Agricultural Conservation Program; Air Control Point; Airline Carriers of Passengers; Allied Communications Publications; American College of Pharmacists; American College of Physicians; Anti-Comintern Pact; Area Characteristic Plan-(ning); Area Community Physician; Associated Collegiate Press; Association of Clinical Pathologists; Association of Correctional Psychologists; Australian Conservative Party; Australian Consolidated Press

ACP Automóvel Clube de Portugal (Portuguese—Automobile Club of Portugal)

ACP80(s) Air-Cargo Processing in the '80s

ACPA Affiliated Chiropodists-Podiatrists of America; American Capon Producers Association; American Cleft Palate Association; American College Personnel Association; American Concrete Paving Association; American Concrete Pipe Association; Association of Computer Professionals Australia; Association of Computer Programmers and Analysts; Australasian Corporation of Public Accountants; Australian Clay Products Association

A-CPA Asbestos-Cement Products Association

ACPAE Association of Certified Public Accounts Examiners

a/c pay accounts payable

ACPC American College of Probate Counsel; American Council of Parent Cooperatives

ACPCACP Atlantic, Caribbean, and Pacific Countries Association of Canadian Publishers

ACPD Anti-trust and Consumer Protection Division; Army Control Program Directive

ACPE American Council on Pharmaceutical Education; Association for Continuing Professional Education; Australian College of Physical Education

ACPF Amphibious Corps, Pacific Fleet

ACPFB American Committee for Protection of Foreign Born

ACPIC American Council for Private International Communications

ACPL American College of Probate Counsel

acpm attitude-control propulsion motor(s)

ACPM American College of Preventive Medicine; American Congress for Preventive Medicine

ACPMR American Congress of Physical Medicine and Rehabilitation

ACPO Association of Chief Police Officers

acpp (ACPP) adrenocorticopolypeptide

acpr acoustical paper

ACPRA American College Public Relations Association

ACPs Area Concept Papers

ACPS American Coalition of Patriotic Societies; Arab Company for Petroleum Services (OPEC)

ACPSAHMWA American Commission for the Protection and Salvage of Artistic and Historical Monuments in War Areas

acpt accept

ACPT Arizona Congress of Parents and Teachers

ACPTA Australian Commonwealth Post and Telegraph Association

acpu auxiliary computer power unit

acq acquire; acquittal

ACQ Australian Church Quarterly

Acq Libr Acquisition(s) Librarian

ACQT Aviation Cadet Qualifying Test

acquis acquisition(s)

acr abandon call and retry; acrylic; advanced capabilities radar; aerial combat recon-

naissance; airfield-controlled radar; anti-constipation regimen

acr (ACR) acre(age)

ACR Abstracts of Classified Reports; Advisory Commission on the Realm; Aircraft Control Room; Algoma Central Railway; Allied Commission on Reparations; American Academy in Rome; American College of Radiology; Area Characterization Report

AC & R American Cable and Radio (Corp)

ACR American Criminal Review

ACRA American Collegiate Retailing Association; Association of Company Registration Agents

ACRB Aero-Club Royal de Belgique (Royal Belgian Aero Club); Army Council of Review Boards

ACRC Air Compressor Research Council

acrd accrued

ACRD Australian Council for Rehabilitation of the Disabled

ACRDC Australian Code of Residential Design and Construction

ACRE Automatic Call-Recording Equipment; Automatic Checkout and Readiness Equipment

a/c rec accounts receivable

ACREC American College of Real Estate Consultants

acre ft acre foot

ACRES Airborne Communication Relay Station

ACRFAET Aircraft Crash Rescue Field Assistance and Evaluation Team (USAF)

acrg acreage

ACRI Air Conditioning and Refrigeration Institute; American Cocoa Research Institute

ACRiLIS Australian Centre for Research in Library and Information Science (Riverina College of Advanced Education)

acrim (ACRIM) active cavity radiometer irradiance

ACRIM Association for Correctional Research and Information Management

ACRL American Cities Racing League; Association of College and Research Libraries

ACRM Aviation Chief Radioman

acro acrobat(ic); acrophobe; acrophobia

acrol acrolect(ic) (al) (ly)

acron acronym

ACRONYM Allied Citizens Representing Other New York Minorities

acronymiz acronymization; acronymizing; acronymizor(s)

ACRONYMS Acceptable Contractions of Randomly Organized Names Yielding Meritorious Spontaniety

Acropolis Acropolis Hill, Athens, containing the Erectheum, the Parthenon, the Temple of Nike, the Acropolis Museum

across acrostic

ACRR American Council on Race Relations

ACRS Accelerated Cost Recovery System; Advisory Committee on Reactor Safeguards

ACRT Aviation Chief Radio Technician

ACRU Advisory Committee on Resource Use

acrw aircrew

ACRW American Council of Railroad Women

ACRY Australian Council of Rural Youth

acs alternating current synchronous; anodal closing sound; autograph card signed

acs (ACS) antireticular cytotoxic serum; attitude-control system

a-c s (ACS) alternating-current synthesizer

Ac of S Academy of Sciences (USSR); Assistant Chief of Staff

ACs appeal cases

ACS Advanced Communications System; Airline Charter Service; Alaskan Communications System; American Camellia Society; American Cancer Society; American Carnation Society; American Ceramic Society; American Cetacean Society; American Chemical Society; American College of Surgeons; American Colonization Society; American Crystal Sugar (company); Armament Control System; Army Community Service; Assistant Chief of Staff; Associated Counseling Services; Association of Clinical Scientists; Asynchronous Communications Server; Australian Cancer Society;

Australian Ceramic Society; Australian Chamber of Shipping; Australian Cinematographers Society; Australian Computer Society; Automatic Control System
AC/S Assistant Chief of Staff
A.C.S. Associate in Commercial Science
ACS *Automobile Club de Suisse* (French—Automobile Club of Switzerland)
ACSA Allied Communications Security Agency (NATO); American Cotton Shippers Association; Association of California School Administrators; Association of Collegiate Schools of Architecture; Australian Corps of Signals Association
ACSB Australian Commonwealth Shipping Board
acsc automated contingency support capability
ACSC Air Carrier Service Corporation; Air Command and Staff College; American Council on Schools and Colleges; Association of Casualty and Surety Companies; Australian Coastal Shipping Commission
A-C Scale Anti-Caucasian Scale (measuring negative attitudes towards white persons)
ACSCP Association of California State College Professors
ACS/DCI American Chemical Society/Division of Chemical Information
ACSDO Air Carrier Safety District Office(r)
ACSE Association of Consulting Structural Engineers
ACSEA Air Command—Southeast Asia; Allied Command South-East Asia
ACSEF Australian Coal and Shale Employees Federation
acsf artificial cerebrospinal fluid
ACSF Attack Carrier Striking Force
a-c sg alternating-current signal generator
ACSI Assistant Chief of Staff for Intelligence
ACSIL Admiralty Centre for Scientific Information and Liaison (United Kingdom)
ACSI-MATIC Assistant Chief of Staff—Intelligence (automatic processing system for intelligence information)

ACSL Assistant Cub Scout Leader
acsm acoustic (warfare) support measure(s)
ACSM American Congress of Surveying and Mapping
ACSMA American Cloak and Suit Manufacturers Association
ACSN Association of Collegiate Schools of Nursing
ACSO Australian Council of School Organisations
ACSOC Acoustical Society of America
ACSP Advisory Council on Scientific Policy (United Kingdom); Association of Collegiate Schools of Planning
ACSPA Australian Council of Salaried and Professional Associations
A/cs Pay Accounts Payable
acsr aluminum cable, steel reinforced
A/cs Rec Accounts Receivable
acss (ACSS) analog computer subsystem; automated color-separation system
ACSS Air Command and Staff School; American Cheviot Sheep Society; Army Chief of Support Services
ACSSAVO Association of Chief State School Audio-Visual Officers
ACSSN Association of Colleges and Secondary Schools for Negroes
ACSSO Australian Council of State School Organisations
ACSSRB Administrative Center of Social Security for Rhine Boatmen
acst acoustic; acoustical; acoustics
ACST Army Clerical Speed Test; Australian College of Speech Therapists
acst plas acoustical plaster
acst t acoustical tile
ACSW Academy of Certified Social Workers
acsyn aircraft synthesis
acsys accounting computer system software
act. acting; action; activated coagulation time; active; actor; actress; actual; actuarial; actuary; actuate; actuating; anticoagulant therapy; atropine coma therapy; azimuth control torquer
act. (ACT) advanced coronary treatment

ACT Access to Careers in Technology; Action for Children's Television; actual; Advanced Computer Techniques; Air Control Team; algebraic compiler and translator; American College Testing (program); American Conservatory Theatre; Anglican Theological College (British Columbia); Associated Community Theaters; Associated Container Transportation; Association of Christian Therapists; Association of Classroom Teachers; Australian Capital Territory; Australian College of Theology; Australian Commonwealth Territory; automatic code translation; Aviation Classification Test
a cta *a cuenta* (Spanish—on account)
ACTA Aircoach Transport Association; American Community Theatre Association; Associated Container Transportation Australia
Acta Chem Scand *Acta Chemica Scandinavia* (Latin—Scandinavian Chemistry Review)
Acta Crystallog *Acta Crystallographica* (Latin—⊃Crystallography Review)
Acta Math Acad Sci Hung *Acta Mathematica Academiae Scientiarum Hungaricae* (Latin—Journal of the Hungarian Mathematics Academy)
Acta Metall *Acta Metallurgica* (Latin—Metallurgy Review)
Acta Oto-Laryngol *Acta Oto-Laryngologica* (Latin—Otolaryngology Review)
Acta Phys *Acta Physica* (Latin—Physics Review)
Acta Phys Austriaca *Acta Physica Austriaca* (Latin—Austrian Review of Physics)
Acta Phys Pol *Acta Physica Polonica* (Latin—Polish Review of Physics)
Acta Phys Sin *Acta Physica Sinica* (Latin—Chinese Journal of Physics)
ACTAR Asian Centre for Tax Administration and Research
ACTB Aircrew Classification Test Battery
ACTC Air Commerce Type Certificate; Australian Ceramic Tile Council
A.C.T.C. Art Class Teacher's Certificate
act. ct actual count

acte anodal closure tetanus
ACTFL American Council on the Teaching of Foreign Languages
actg acting
ACTG Advance Carrier Training Group
Actg Legal Adv Acting Legal Advisor
acth (ACTH) adrenocorticotrophic hormone
ACTI Advisory Committee on Technology Innovation
act/ic active—in commission
ACTIL Australian Cotton Textile Industries Ltd
ACTION American Council To Improve Our Neighborhoods
act/is active—in service
ACTIS Auckland Commercial and Technical Information Service
activ activity
ACTIV Army Concept Team in Vietnam
Activase see TPA
ACTL American College of Trial Lawyers
ACTM Association of Cotton Textile Merchants of New York
ACTMC Army Clothing, Textile and Material Center
actn (ACTN) adrenocorticotrophin
ACT–NOW AIDS Coalition to Network, Organize, and Win
actnt accountant
ACTNZ Associated Container Transportation New Zealand
acto automatic computing transfer oscillator
ACTO Action Office(r)—USA; Advisory Council on the Treatment of Offenders
act/oc active—out of commission
actol air-cushion takeoff and landing
ACTOR Askania cine-theodolite optical-tracking range
act/os active—out of service
actp (ACTP) adrenocorticotrophic polypeptide
ACTR Air Corps Technical Report
ACTREP Activities Report
actrl acoustic trial(s)
ACTRSWD Asian Centre for Training and Research in Social Welfare and Development
Acts The Acts of the Apostles
ACTS Acoustic Control and Telemetry System; Air Corps Tactical School; Airline Computer Tracing System; Ameri-

can Catholic Theological Society; American Christian Television System; Association of Career Training Schools; Automatic Cage Transmission System; Automatic Computer Telex System
ACT-SO Afro-Academic Cultural Technological and Scientific Olympics
ACTSU Association of Computer Time-Sharing Users
ACTT Association of Cinematograph and Television Technicians; America's Christmas Train and Trucks
ACTU Association of Catholic Trade Unionists; Australian Council of Trade Unions
ACTUP AIDS Coalition To Unleash Power
actv activate
actv (ACTV) armored cavalry towing vehicle
ACTV American Coalition for Traditional Values
act. val actual value
act. wt actual weight
ACTWU Amalgamated Clothing and Textile Workers Union
ACTWUA Amalgamated Clothing and Textile Workers Union of America (formerly ACWA and TWUA)
acu acuity; aculeat; acumen; acuminate; acupressure; acupuncture; acute; address control unit; arithmetic computer; assault craft unit; automatic calling unit
ACU Abilene Christian University; Abused Child Unit (Los Angeles Police Department); American Church Union; American Congregational Union; American Conservation Union; American Cycling Union; anti-crime unit; Association of College Unions; Association of Commonwealth Universities; Australian Church Union; Autocycle Union
ACUA Association of Cambridge University Assistants; Association of College and University Auditors
ACUCAA Association of College, University, and Community Arts Administrators
ACUCM Association of College and University Concert Managers
ACUE American Committee of United Europe

ACUERI American Conservative Union Education and Research Institute
ACUG Association of Computer User Groups
ACUHO Association of College and University Housing Officers
ACUI Association of College Unions International
ACUIIS Association of Colleges and Universities for International Intercultural Studies
Acuña Ciudad Acuña (Mexican border town)
ACUNY Associated Colleges of Upper New York
ACUP Association of Canadian University Presses; Association of College and University Printers
acupunc acupuncture; acupuncturist
ACURA Association for the Coordination of University Religious Affairs
ACURL Association of Caribbean University and Research Libraries
ACUs Asian Currency Units
ACUS Administrative Conference of the United States; Atlantic Council of the United States
ACU-SACC American Conservative Union—Save Our Canal Committee
ACUSNY Association of Colleges and Universities of the State of New York
ACUTE Accountants Computer Users for Technical Exchange
acv actual cash value; air-cushion vehicle; alarm check valve
ACV air-cushion vehicle; auxiliary aircraft carrier or tender (3-letter symbol)
ACVAFS American Council of Voluntary Agencies for Foreign Service
ACVC American Council of Venture Clubs; Arms Control Verification Committee
acvd acute cardiovascular disease
ACVO American College of Veterinary Ophthalmologists
ACVP American College of Veterinary Pathologists
ACVS Australian College of Veterinary Scientists
ACVSF Air-Cooled Vault Storage Facility
acw aircraft control and warning; alternating continuous

waves; automatic car wash
ac/w acetone/water
ACW Air Control and Warning (system); Aircraftwoman; Alcoholism Center for Women; American Chain of Warehouses
AC & W Air Communications and Weather (naval group)
ACWA Amalgamated Clothing Workers of America; Australian Chinese Women's Association
A.C.W.A. Associate of the Institute of Cost and Work Accountants
, **ACWAI** Automatic Car Wash Association International
ACWB Australian Council of Wool Buyers
acwbcn action will be cancelled
ACWC Advisory Committee on Weather Control
ACWF All-China Women's Federation; American Council for World Freedom; Army Central Welfare Fund
ACWL Army Chemical Warfare Laboratory
ACWM Americans for Customary Weights and Measures
ACWO Aircraft Control and Warning Officer
ACWRON Aircraft Control and Warning Squadron
ACWRRE American Cargo War Risk Reinsurance Exchange
ACWS Aircraft Control and Warning System
AC & WS Aircraft Control and Warning Station(s)
ACWW Associated Country Women of the World
acy average crop yield
ACY Akron, Canton & Youngstown (railroad); American Cyanamid Company (stock exchange symbol); Atlantic City, New Jersey (airport)
AC & Y Akron, Canton & Youngstown (railroad)
ACYF Administration for Children, Youth, and Families; Australian Council of Young Farmers
acyl-co A coenzyme A ester (general symbol for an organic compound)
acyro acyrologia; acyrologic(al); acyrology
ad a drink; a drug (addict); active duty; addict; advertisement; advertising; aerodynamic decelerator; after drain; air dried; airdrome; area drain;

average deviation
ad (AD) address; aggregate demand; athletic director
a/d altitude/depth; analog-to-digital
a & d ascending and descending
a & d (A & D) accounting and disbursing
'ad had
ad (Latin prefix—to, toward)—as in adhesion, admixture, adopt, adoring, etc.
a d a droit (French—to the right)
a.d. *auris dexter* (Latin—right ear)
a D ausser Dienst (German—retired)
Ad Ada; Adah; Adalbert; Adam; Adams; Adán; Addington; Addis; Addison; Adela; Adelaide; Adelard; Adelardo; Adelbert; Adele; Adelina; Adeline; Adelle; Adelsteen; Adeodato; Adlai; Adna; Adolf; Adolfine; Adolfo; Adolph; Adolphe; Adolpho; Adolphus; Adriaan; Adriaen; Adrian; Adriano; Adrianus; Adrien; Adrienne; Aedh; Alzheimer's disease
AD Accession Document; Aden Airways; Air Defense; Air Depot; Air Division; Airdrome; Airframe Design (division); Airworthiness Directive; Appellate Division; Assembly District; assistant director; associate director; Astia Document; Atlantic & Danville (railroad); Australian Dame (Dame of the Order of Australia); Australian Democrats; Australian dollar (telex); *Aviatsionnaya Diviziya* (Russian—Aviation Division); destroyer tender (naval symbol)
A-D Albrecht Dürer; Antonin Dvořák
A/D Air Depot; analog-to-digital
A & D Atlantic & Danville (railroad)
AD *Acción Democratica* (Spanish—Democratic Action Party)—Venezuela's democratic movement begun by Romulo Betancourt
A.D. *Anno Domini* (Latin—in the Year of our Lord)
ad 2 vic. *ad duas vices* (Latin—for two doses, for two times)
ada action data automation; actuarial data assembly; airtight drywall approach; average

daily attendance; average deviation adjustment
ada (ADA) adenosine deaminase
ada adalah (Arabic—equity, justice); American Dental Association (logotype)
Ada (ADA) computer programming language promoted by the Department of Defense and named for Augusta Ada Byron, Lord Byron's daughter, who designed instructions for a 19th-century mechanical analytical machine
Ad of A Archduchy of Austria; Archduke of Austria
ADA Air Defense Area; American Dairy Association; American Dehydrators Association; American Dental Association; American Dermatological Association; American Diabetes Association; American Dietetic Association; Americans for Democratic Action; Assistant District Attorney; Atomic Development Authority; Australian Dental Association; Australian Department of Agriculture; Australian Design Award; Australian Draughts Association; Automatic Data Acquisition; Automobile Dealers Association
ADA# American Diabetes Association diet number
ADAA American Dental Assistants Association; Art Dealers Association of America; Australian Development Assistance Agency
ADABAS Adaptable Data Base System
adac automatic direct analog computer
ADAC Acoustic Data Analysis Center
***ADAC** Allgemeiner Deutscher Automobilclub* (German Automobile Club)
adacx automatic data acquisition and computer complex
adad air defense artillery director
ADADs Automatic Dialing and Announcing Devices
***Adag** adagio* (Italian—slowly and expressively)
***ADAGP** Association pour la diffusion des arts graphiques et plastiques* (French—Association for the Diffusion of Graphic and Plastic Arts)
ADAIS Aerodynamic Data Analysis and Integration Sys-

tem
adal action data automation language
adaline adaptive linear neuron
adam adamantine; adaptive arithmetical method; advanced data management; air deflection and modulation; area denial artillery munition; automatic distance and angle measurement
A'dam Amsterdam
ADAM Agriculture Department Automated Manpower; Alzheimer's Disease Association of Maryland; Automatic Document Abstracting Method
ADAMHA Alcohol, Drug Abuse, and Mental Health Administration
adamite basic zinc arsenate
adaml advise by airmail
adamm (ADAMM) area defense anti-missile missile
Adam's Adam's Bridge (30-mile-long island chain linking Ceylon and India); Adam's Peak (7,000-foot-high mountain, Ceylon, also called Samanaliya or Sri Padastanaya)
ADAMS Advanced Design Aluminum Metal Shelter (prefabricated ADAMS hut)
Adams Mans Adams Mansion (birthplace and home of John Adams and his son John Quincy Adams in Quincy, Massachusetts)
adandac administrative and accounting purposes
ADAOD Air Defense Artillery Operations Detachment
ADAOO Air Defense Artillery Operations Office(r)
adap adapted
ADAP Airport Development Aid Program (FAA)
ADAPCP Alcohol and Drug Abuse Prevention and Control Program
ADAPS Automatic Display and Plotting System
ADAPSO Association of Data Processing Service Organizations
adapt. adaption of automatically-programmed tools
ADAPT Alcohol and Drug Abuse Prevention Team
adapticom adaptive communication
ADAPTS Air-Deliverable Anti-Pollution Transfer System (USCG)

adar advanced development array radar; analog-to-digital-to-analog recording
ADAR Air Defense Area
adare advise date of receipt
ADARF Alcoholism and Drug Addiction Research Foundation (Ontario, Canada)
ADAS Action Data Automation System; Agricultural Development Advisory Service; Airborne Dynamic Alignment System
ADASC Auto Dismantlers Association of Southern California (often called ADA)
adash advise date of shipment
ad ast. *ad astra* (Latin—to the stars)
adat automatic data accumulation and transfer
adaval advise availability
ADAWS Action Data Automation and Weapons System
A-day assault day
adb accidental death benefit
adb *automatisk databehandling* (Dano-Norwegian—automatic data handling)
adB acceleration decibel(s)
ADB Apollo Data Bank (NASA); Asian Development Bank; Atlantic Development Board (Canada)
A.D.B. Bachelor of Domestic Arts
ADB *Australian Dictionary of Biography*
ADBA American Dog Breeders Association
ADBC American Defenders of Bataan and Corregidor
ADBM Association of Dry Battery Manufacturers
ADBPA *Association pour le Développement des Bibliothèques Publiques en Afrique* (French—Association for the Development of Public Libraries in Africa)
adc active-duty commitment; adopted child; advance delivery of correspondence; albumin, dextrose, catalase; all damn confusion; anodal duration contraction; axiodistocervical
ADC Aerophysics Development Corporation; Aerospace Defense Command; Agricultural Development Council; Aid to Dependent Children; Aide-de-Camp; Air Defense Command; Air Development Center; Air Diffusion Council; Alaska Defense Command;

American Distilling Company; American Dock Company; analog-to-digital converter; Area Dissemination Coordinator; Asian Development Centre; Australian Dairy Corporation; Australian Development Corporation; Australian Diabetic Council; Aviation Development Council
adca advanced-design composite aircraft
ADCA Australian Department of Civil Aviation
adcad airways data collection and distribution
ad cap. *ad captandum* (Latin—for pleasing, made attractive)
ADCC Air Defense Control Center
ADCG Australian Defence Co-operation Group
ADC Gen Aide-de-Camp General
ADCI American Die Casting Institute
ADCIS Association for the Development of Computer-based Instruction Systems
ADC/NORAD Air Defense Command/North American Air Defense (Command)
ADCO Alcohol and Drug Control Office(r); American Dredging Company
ADCOC Area Damage Control Center
ADCOM Administrative Command; Aerospace Defense Command
adcon advance concepts; advise or issue instructions to all concerned; analog-to-digital computer
ADCONSEN advice and consent of the Senate (of the United States)
a & d control alcohol and drug control
ADCOP Area Damage Control Party
adc's analog-to-digital converters
ADCSP Advanced Defense Communications Satellite Program
adct assisted-draft crossflow tower
ADCT Art Director's Club—Toronto; Association of District Council Treasurers
ad curtain advertisement curtain (theater)
add. addenda; addendum; address; airborne digital de-

coder; attention deficit disorder; automatic drawing device; average daily dose
ad & d accidental death and dismemberment (insurance)
add. *addendum* (Latin—addition)
ADD Abstracts of Declassified Documents; Addis Ababa, Ethiopia (airport); Administration on Developmental Disabilities; Aerospace Defense Division; Aviastiia Dalnego Deistviia (Russian—Long-Range Bombing Force)
ADDA Air Defense Defended Area
addar automatic digital data acquisition and recording
ADDAS Automatic Digital Data Assembly System
ADDC Air Defense Direction Center; Australian Dairy Development Council
Add-Can Addicts-Canada
ADDDS Automatic Direct Distance Dialing System
addee addressee
ad. def. an. *ad defectionem animi* (Latin—to the point of fainting)
ad. deliq. *ad deliquium* (Latin—to fainting)
adder. automatic-digital data-error recorder
ADDF Abu Dhabi Defense Force
ad diag admitting diagnosis
ADDIC Alcoholic and Dependency Intervention Council
addict. addiction
Addis Addis Ababa, Ethiopia
addit additional
ADDL Anti-Digit Dialing League
addm automated drafting and digitizing machine
addn addition
Addo Addo Elephant Park near Port Elizabeth, South Africa
ADDP Air Defense Defended Point
ADDR address
ADDS Alcohol and Drug Dependence Service; American Digestive Diseases Society; American Diversified Dog Society (for mutts); Apollo Document Descriptions Standards (NASA); Automatic Data Digitizing System; Automatic Direct Distance Dialing System
addsd addressed
addu additional duty

Addy Ada; Adela; Adelaide; Adelina; Adeline
ade automated drafting equipment; automatic data entry; average daily enrollment
ADE Animal Disease Eradication (Department of Agriculture division); Association of Departments of English; Association for Documentary Editing; Australian Driver Education
ADEA American Driver Education Association
ADEA *Age Discrimination in Employment Act*
ADECOS *Acción Democrática* (Spanish—Democratic Action)—Venezuelan political party
adeda advise effective date
ADEDS Advanced Electronic Display System
adee addressee
Adee Adelaide, Australia
Adeen Aberdeen (shire)
ad effect. *ad effectum* (Latin—until effective)
A de JC *Antes de Jesucristo* (Spanish—before Jesus Christ)
Adel Adelaide; Adelar(d); Adelbern; Adelbert; Adelbold; Adelfred; Adelfrid; Adelgar; Adelhart; Adelmo; Adelochorda; Adelpho; Adelquist; Adelric; Adelrik; Adelwin
ADELA Atlantic Community Development Group for Latin America
Adélie Adélie Land (French Antarctica)
adem acute disseminated encephalomyelitis
ADEMS Advanced Diagnostic Engine Monitoring System
aden augmented deflector exhaust nozzle
Aden part of the Republic of Yemen
ADEOS Advanced Earth Observation Satellite
ADEP Air Depot
ADEPO Automatic Dynamic Evaluation by Programmed Organizations
adept. a definitely empirical power of theorems; automatic data extractor and plotting table
ADEPT Agricultural and Dairy Educational Political Trust
a des *a destra* [Italian—at (to) the right]
A de S *Académie des Sciences* (French—Academy of Sciences)

ADES Automatic Digital Encoding System
ADESA *Asamblea Democrática por Sufragio Efectivo* (Spanish—Democratic Assembly for Effective Suffrage)
A de T S Alex de Tocqueville Society
ad eund. grad. *ad eundem gradum* (Latin—to the same degree)
adex advanced antisubmarine warfare exercise
adf after deducting freight; air direction finder
adf (ADF) automatic direction finder
Adf Adolf
ADF Air Defense Force; Air Development Force; Arab Deterrent Forces; Army Distaff Foundation; Asian Development Fund
ADFA Australian Defence Force Academy; Australian Department of Foreign Affairs; Australian Dried Fruits Association
adfc adiabatic film cooling
ADFC Air Defense Filter Center
ADFF Australian Dairy Farmers Federation
ADFI American Dog Feed Institute
ADFIAP Association of Development Finance Institutions of Asia and the Pacific
ad fin. *ad finem* (Latin—to the end)
ADFL Association of Departments of Foreign Languages
ADFOR Adriatic Force
ADFS American Dentists for Foreign Service; Australian Documentary Facsimile Society
ADFSC Automatic Data Field Systems Command
ADFW Assistant Director of Fortifications and Works
adg axiodistogingival
ADG degaussing vessel (3-letter symbol)
ADGA American Dairy Goat Association
ADGB Air Defence of Great Britain
A & D G C Alliance and Dublin Consumers Gas Company
adge air-defense ground environment
adgo adagio
adgo *adàgio* (Italian—gently, peacefully, slowly, softly)

ad grat. acid. *ad gratam acidatem* (Latin—to a pleasing acidity)

ad grat. gust. *ad gratum gustum* (Latin—to an agreeable taste)

ADGRU Advisory Group

adh (ADH) alcohol dehydrogenase; antidiuretic hormone (vasopressin)

ADH Academy of Dentistry for the Handicapped; Association of Dental Hospitals

ADHA American Dental Hygienists Association

ADHC Air Defense Hardware Committee (NATO); Australian Department of Housing and Construction

adhca advise this headquarters of complete action

ADHD Attention-Deficit Hyperactive Disorder

adhib *adhibeatur* (Latin—administer)

ad h.l. *ad hunc locum* (Latin—at this place)

ad hoc (Latin—for this special purpose)

ad hom. *ad hominem* (Latin—to the man)

adi acceptable daily intake; adiabat(ic); air defense intercept; air defense interceptor; alien declared intention; antidetonation injection; attitude direction indicator; automatic direction indicator

adi (ADI) area of dominant (radio or tv station) influence

ADI Acoustical Door Institute; Air Defense Interceptor; Air Distribution Institute; American Documentation Institute; Assistance Doss International

ADI *Agencia para el Desarrollo Internacional* (Spanish—Agency for International Development)—AID

ADIC American Dental Interfraternity Council

ad id. *ad idem* (Latin—both are the same, likewise)

ad ig. *ad ignorantiam* (Latin—to ignorance)

adil (ADIL) air defense identification line

ADIL *Annual Digest of International Law*

adimd advise immediately by dispatch

ad inf *ad infinitum* (Latin—to infinity)

ad init. *ad initium* (Latin—at the beginning)

adinsp administrative inspec-tion

ad int. *ad interim* (Latin—in the interim, meanwhile)

Ad Intel Cen Advanced Intelligence Center

ADIOS Automatic Digital Input-Output System

ADIP Advanced Developing Institutions Program

adipu advise whether individual may be properly used in your installation

Adirondacks Adirondack Mountains, New York

ADIS Air Defense Integrated System; Association for the Development of Instructional Systems; Automatic Data Interchange System

adit analog digital integrating translator

ADIT Alien Documentation, Identification, and Telecommunications

ADIU Armament and Disarmament Information Unit, Sussex University

ADiv Air Division

ADIZ Air Defense Identification Zone

adj adjacent; adjective; adjoint; adjust

Adj Adjutant

ADJ adjust

Adj. A. Adjunct in Arts

ADJAG Assistant Deputy Judge Advocate General

Adj Gen Adjutant General

Adjt Adjutant

adl activities of daily living; armament data line; automatic data link(ing); average decreasing line

Adl Adelaide

ADL Adelaide, Australia (airport); Admiral Corporation (stock exchange symbol); Anti-Defamation League (B'nai B'rith); Arthur D Little (corporation); Authorized Data List; Automatic Data Link(ing); Automotive Discount Leasing

ADLA Art Directors' (club) Los Angeles

Adlai Adlai Stevenson II

Adler Adler Planetarium (Chicago)

ad lib. *ad libitum* (Latin—at one's pleasure)

ADLIPS Automatic Data-Link Plotting System

adlm (ADLM) aerial-delivered land mine

ADLO Air Defense Liaison Office(r)

ad loc. *ad locum* (Latin—at this passage or place)

ADLOG Advance Logistical Command

ADLP Australian Democratic Labour Party

ADLS Air Dispatch Letter Service

ADLT Activities of Daily Living Test

ADLTDE Association of Dark Leaf Tobacco Dealers and Exporters

adm administration; administrative; administrator; admission; admit; air defense missile; atomic demolition munition; average daily membership

adm (ADM) air-launched decoy missile

Adm Admiral; Admiralty

ADM Action Description Memorandum; Advanced Diploma in Midwifery; Affiliated Dress Manufacturers; Air Defense Missile; American Drug Manufacturers (association); Archer Daniels Midland Company

ADM *American Demographics Magazine*

adma automatic drafting machine

ADMA Aircraft Distributors and Manufacturers Association; American Drug Manufacturers Association; Australian Direct Marketing Association; Australian Display Manufacturers Association; Aviation Distributors and Manufacturers Association

admad advise method and date of shipment

Ad Man Advertising Manager

admap advise by mail as soon as possible

admass advertising & mass media effect

Adm Cen Administration Center; Administrative Center; Admiralty Center

Adm Co Admiralty Court

ADMI American Dry Milk Institute

ADMIG Australian Drug and Medical Information Group

admin administration; administrative; administrator

Admin administration

AdminInstr Administrative Instructions

AdminO Administrative Order(s)

adminord administrative order

adminplan administrative plan

ADMIRAL Automatic and Dynamic Monitor with Immediate Relocation, Allocation, and Loading (system)
admire. automatic diagnostic maintenance information retrieval
ADMIRES Automatic Diagnostic Maintenance Information Retrieval System
admix administratrix
adml average daily member load
Adml Admiral; Admiralty
admn administration
admon administración (Spanish—administration)
Admor Administrador (Spanish—Administrator)
admos automatic device for mechanical order selection
ad mov. ad moveatur (Latin— let it be moved)
admr administrator
adms administrator
ADMS American Donkey and Mule Society; Assistant Director of Medical Services
ADMSC Automatic Digital Message Switching Centers (DoD)
ADMSLBN Air Defense Missile Battalion (USA)
admsn admission
ADMT Association of Dental Manufacturers and Traders
Adm Ter Administered Territories (Gaza Strip, Golan Heights, much of the Sinai, and West Bank of the Jordan as well as all of Jerusalem annexed by the Israelis in 1967)
Admty Admiralty
admx administratrix
adn (ADN) ácido desoxirribonucleico (Spanish—desoxyribo-nucleic acid)—dna (DNA)
Adn Aden
ADN Accession Designation Number; Allgemeiner Deutscher Nachrichtendienst (General German News Service); Alternative Defense Network; Ashley, Drew & Northern (railroad)
A.D.N. Associate Degree in Nursing
ADNA Assistant Director of Naval Accounts
ADNAC Air Defense of the North American Continent
ad naus. ad nauseam (Latin— boring to the point of nausea)
ADNC Air Defense National Center; Air Defense Notifica-

tion Center; Assistant Director of Naval Construction
ad neut. ad neutralizandum (Latin—until neutral)
ADNI Assistant Director of Naval Intelligence
ADNOC Abu Dhabi National Oil Company
adnok advise if not correct
'ad n't had not
ado advanced development objective; axiodistoclusal
Ado adagio (Italian—slowly and expressively)
ADO Administration Duty Officer; Air Defense Officer; Air Defense Operations; Area Dental Officer
ADOBE Atmospheric Dispersion of Beryllium (program)
ADOC Air Defense Operations Center
ADOF Assistant Director of Ordnance Factories
ADOGA American Dehydrated Onion and Garlic Association
adoit automatically-directed out-going intertoll trunk (Bell)
Adolph Adolphus
ADONIS Automatic Digital On-Line Instruments System
adop adoption
ADOPT Approach to Distributed Processing Transactions
ADOS Assistant Director of Ordnance Services
adot automatically-directed outbound trunk (Bell)
ADOT Australian Department of Transportation
adoxograph adoxographer; adoxographic(al)(ly); adoxography
adp adenosine diphosphate; advanced data processing; airborne data processor; ammonium dihydrogen phosphate; automatic data processing
ADP Academy of Denture Prosthetics; Air Defense Position; Airport Development Program; ammonium dihydrogen phosphate; Animal Disease and Parasite (Research Division—Department of Agriculture); Automatic Data Processing
ADPA American Defense Preparedness Association
ADPB Australian Dairy Produce Board
ADPC Abu Dhabi Petroleum Company; Automatic Data Processing Center
adpcm adaptive-differential pulse-code modulation

adpe automatic data processing equipment
ADPEA Australian Data Processing Employees Association
ADPESO Automatic Data Processing Equipment Selection Office (USN)
adpl average daily patient load
adplan advancement planning; advertising planning
adpo aircraft depot
ADPO Automatic Data Processing Operations
ad pond. om. ad pondus omnium (Latin—to the whole weight)
ADPR Assistant Director of Public Relations
ADPRIN Automatic Data-Processing Intelligence Network (U.S. Bureau of Customs)
ADPs Allied Defense Publications; Artillery Destruction Programs
ADPS Automatic Data Processing System(s)
ADPSD Automatic Data Processing Systems Development
ADPSO Association of Data Processing Service Organizations
ADPSR Architects, Designers, and Planners in Social Responsibility
adpt adapted; adapter
adr address; asset depreciation range
a-d r analog-to-digit recorder
adr addresse (Dano-Norwegian—address); *address* (Swedish—address)
Adr Adrian; Adriatic
Adr Adresse (German—address)
ADR Accepted Dental Remedies; adder; Aircraft Direction Room; Australian Defence Representative; Australian Design Rules (automotive); European Agreement on the International Carriage of Dangerous Goods by Road
ADR Accepted Dental Remedies; Association pour le Développement de la Recherche (French—Association for the Development of Research)
ADR 27-a Australian Design Rule 27-a (vehicle emission control)
ADRA Animal Diseases Research Association; Australian Drag Racing Association
adrac automatic digital recording and control

ADRB Army Disability Review Board; Army Discharge Review Board

ADRC Alzheimer's Disease Research Center

ADRCP Alzheimer's Disease Research Centers Program

ADRDA Alzheimer's Disease and Related Disorders Association

adrde advise reason for delay

ADRDE Air Defense Research and Development Establishment

adren adrenal; adrenalin

adrenals adrenal glands

ADRES Army Data Retrieval System

ADRI Angkatan Darat Republik Indonesia (Indonesian Army)

Adria Adriatic Sea

Adria die Adria (German—the Adriatic)—*das Adriatische Meer* (The Adriatic Sea)

Adriatic Adriatic Sea

ADRIS Automatic Dead Reckoning Instrument Systems

adrm airdrome

ADRM Analog-to-Digital Remastering of musical recordings

Adro Alejandro

ADROBN Airdrome Battalion

adrp airdrop

ADRs American Depository Receipts

ADRS Analog-to-Digital Data Recording System; Automatic Document Request Service

adrt analog data recording transcriber

adr tel adresse telegraphique (French—telegraphic address)

ads advertisements; antibody deficiency syndrome; antidiuretic substance; area, date, subject; autograph document signed; automatic door seal

AD²ᵈ Appellate Division–second series

ADS Aerial Delivery System; Air Defense Sector; Alzheimer's Disease Society; American Daffodil Society; American Dahlia Society; American Dental Service; American Dental Society; American Denture Society; American Dialect Society; Analog & Digital Systems; Association of Diesel Specialists; Automated Drafting Services

ADS Academie des Sciences (French—Academy of Science)

ADSA American Dairy Science Association; American Dental Society of Anesthesiology; Atomic Defense Support Agency

ad. saec. ad saeculum (Latin—to the century)

adsap advise as soon as possible

ADSAS Air-Derived Separation Assurance System

ad sat. ad saturandum (Latin—to saturation)

ADSATIS Australian Defence Science and Technology Information System

adsc average daily service charge (in hospitals)

ADSC Advanced Section Communication Zone; Automatic Data Service Center

ADSCAT Association of Distributors to the Self-service and Coin-operated Laundries and Allied Trades

adscom advanced shipboard communications

adsda advise earliest date

ad sec. ad sectam Latin—at suit of (legal)

AdSec Advanced Section

adshpdat advise shipping data

ADSIA Allied Data Systems Interoperability Agency

adsid air-delivered acoustic-implant seismic-intrusion detector

ADSID Air Defense Systems Integration Division

ADSL Assembly Department Shortage List

adsm (ADSM) air defense suppression missile

ADSM American Defense Service Medal

ADSMO Air Defense Systems Management Office

ADSN Accounting and Disbursing Station Number

ADSOC Administrative Support Operations Center

ADSOT Automatic Daily System Operability Test

adss analysis of digitized seismic signals

ADSS Aircraft Damage Sensing System; Australian Defense Scientific Service

ADST Atlantic Daylight Saving Time

adstadis advise status and/or disposition

adst.feb. adstante febre (Latin—when fever is present)

adstkoh advise stock on hand

adsu advanced direct support unit

ADSUP Automatic Data Systems Uniform Practice(s)

adsym automobile defog-defrost system model

adt aided tracking; any damn thing (a placebo); automatic damage template; automatic debit transfer; average daily dose

adT an demselben Tage (German—the same day)

ADT American District Telegraph; Applied Drilling Technology; Atlantic Daylight Time

ADTA American Dental Trade Association

adtam air-delivered target-activated munitions

ADTC Air Defense Technical Center; Air Defense Test Center; Armament Development Test Center (USAF)

adtech advanced decoy technology

ad tert. vic. ad tertium vicem (Latin—three times)

ADTF At-Depth Test Facility

ADTI American Dinner Theatre Institute

ADTIC Arctic, Desert, Tropic Information Center

ADTR Australian Department of Tourism and Recreation

ADTS Automatic Data and Telecommunications Service

ADTSEA American Driver Traffic Safety Education Association

adtu automatic digital test unit; auxiliary data translator unit

adu acceleration-deceleration unit; accumulation-distribution unit

ADU Aircraft Delivery Unit

adult. adulterant; adulterate; adulteration

ad us. ad usum (Latin—according to custom)

ad us. ext. ad usum externum (Latin—for external use)

adv advance; advanced; advances; advantage; adverb(ial); advertising

a/dv arterio/deep venous (injection)

adv advokat (Dano-Norwegian—attorney)

adv. adversum (Latin—adversely, against)

Adv. Adventist; Adviser

ADV advance

advac advise acceptance

ad val. *ad valorem* (Latin—according to value)
Adv Appl Mech *Advances in Applied Mechanics*
Adv At Mol Phys *Advances in Atomic and Molecular Physics*
advb adverb; adverbial
Adv Bse Advanced Base
Adv Chem Phys *Advances in Chemical Physics*
adv chgs advance charges
advdisc advance discontinuance of allotment
advec advection
advect advection(al)(ly); advective
adven adventure; adventurer
adversat adversative
advert advertising
advertique advertising antique
adverts advertisements
adv frt advance freight
Adv Intel Cen Advanced Intelligence Center
ad virus adenovirus
ADVISE Area Denial Visual Identification Security Equipment
advl adverbial
advm adaptive delta voice modulation
Adv Magn Reson *Advances in Magnetic Resonance*
adv mtr advertising matter
Advoc Advocate
advof advise this office
advon advanced echelon; advanced operations unit
Adv Phys *Advances in Physics*
adv pmt advance payment
adv poss adverse possession
Adv Quantum Chem *Advances in Quantum Chemistry*
advr advisor
ADVS Assistant Director of Veterinary Services
advst advance stoppage
advt advertise; advertisement; advertiser; advertising
advul air-defense vulnerability
adw assault (with) deadly weapon
ADW Air Defense Warning
ADWA Atlantic Deeper Waterways Association
ADWC Air Defense Weapons Center (USAF)
ADWKP Air Defense Warning Key Point
adx automatic data exchange
ADX Adams Express Company (stock exchange symbol); Air Defence Exercise (Australian)
Adyg Adygey
adz advise

ADZ Air Defense Zone
Adzh Adzhar
ae above the elbow; account executive; aircraft equipment; air escape; almost everywhere
a/e absorptivity-emissivity
a & e accident and emergency; aerospace and electronic; armaments and electronics; azimuth and elevation
ae *aetatis* (Latin—aged, at the age of)
AE Agricultural Engineering (Department of Agriculture research division); Air Explorer; Airborne Equipment (naval division); American English; ammunition ship (naval symbol); Automatic Electric
A-E Adam and Eve; Architect-Engineer; Astro-Eugenics
A.E. Aeronautical Engineer; Agricultural Engineer; Architectural Engineer; Associate in Education; Associate in Engineering
A & E Agricultural and Engineering; Architectural and Engineering; Arts and Entertainment cable television network
AE *Aeon* (pen name of George William Russell); *Aktiebolaget Atomenergi* (Swedish—Atomic Energy Corporation); *American Ephemeris; Atomnaya Energiya* (Russian—Atomic Energy); *Australian Encyclopaedia*
A & E *Adolphus and Ellis*
aea actual expenses allowable; assignment eligibility and availability
AEA Actors' Equity Association; Adult Education Association; American Economic Association; American Educational Association; American Electrical Association; American Electronics Association; American Enterprise Association; American Entrepreneurs Association; American Export Airlines; Arizona Education Association; Arkansas Education Association; Artists Equity Association; Arts, Education, and Americans; Atomic Energy Authority; Automotive Electric Association
AEAA *Asociación de Escritores y Artistas Americanos* (Spanish—Association of American Writers and Artists)
AEAF Allied Expeditionary Air Force

AEAO Airborne Emergency Actions Officer
AEARC Army Equipment Authorizations Review Center
AEAs American Entertainers Abroad
aea sol alcohol-ether-acetone solution
AEAUSA Adult Education Association of the United States of America
AEB Adult Education Board (Singapore); Area Electricity Board; Atomic Energy Bureau; Australian Egg Board
AEBIG Aslib Economics and Business Information Group
aec additional extended coverage; altitude engine control; at earliest convenience; attitude engine control
AEC Aeronautical Research Council; Agricultural Economics (division of Department of Agriculture); Aircraft Radio Corporation; Airworthiness Examination Committee; Alaska Engineering Commission (Alaska Railroad); Aluminum Extruders Council; American Engineering Council; Anglican Episcopal Church; Army Education Center; Army Educational Center; Army Educational Corps; Army Electronics Command (formerly Signal Corps); Atlantic & East Carolina (railroad); Atlas Educational Center; Atomic Energy Commission; Australian Environment Council
A & EC Atlantic & East Carolina (railroad)
AEC-A Atomic Energy Commission—Albuquerque Operations Office
AEC-AI Atomic Energy Commission—Argonne, Illinois
AEC-ANM Atomic Energy Commission—Albuquerque, New Mexico
AEC-ASC Atomic Energy Commission—Aiken, South Carolina
AECB Arms Export Control Board; Atomic Energy Control Board (Canada)
AEC-BC Atomic Energy Commission—Berkeley, California
AECC Aeromedical Evacuation Control Center
AEC-CC Atomic Energy Commission—Canoga Park, California
AECE *Asociación Española de*

Cooperación Europea (Spanish—Spanish Association for European Cooperation)

AEC-FOA Atomic Energy Commission—Fernal Office Area, Cincinnati, Ohio

AEC-HW Atomic Energy Commission—Hanford, Washington

AECI African Explosives and Chemical Industries

AECIA Australian Electronics Consumer Industry Association

AEC-II Atomic Energy Commission—Idaho Falls, Idaho

AECL Anglo-European Container Line; Atomic Energy of Canada, Limited

AEC-LN Atomic Energy Commission—Las Vegas, Nevada

AEC-LOC Atomic Energy Commission—Lockland Aircraft Reactors Operations, Cincinnati, Ohio

AECM Albert Einstein College of Medicine

AEC-NY Atomic Energy Commission—New York Operations Office

AECOM Army Electronic Command

AEC-OR Atomic Energy Commission—Oak Ridge Operations Office

AEC-OT Atomic Energy Commission—Oak Ridge, Tennessee

aecp altitude engine control panel

AECP Airman Education and Commissioning Program

AEC-PP Atomic Energy Commission—Pittsburgh, Pennsylvania

AEC-PR Atomic Energy Commission—Pittsburgh Naval Reactors Operations Office

AECPSUPC Association for the Encouragement of Correct Punctuation, Spelling, and Usage in Public Communications

AEC-RW Atomic Energy Commission—Richland, Washington

AECS Australia-Europe Container Service

AECT Association for Educational Communications and Technology

AEC-UN Atomic Energy Commission—Upton, LI, NY

aed (AED) automatic engineering design

A.Ed. Associate in Education

AED Academy for Educational Development; Associated Equipment Distributors; Association of Electronic Distributors

A.E.D. Artium Elegantium Doctor (Latin—Doctor of Fine Arts)

AEDB Apollo Engineering Development Board

AEDC Arnold Engineering Development Center

aedcm (AEDCM) advanced electrochemical depolarized concentrator module

AEDD Air Engineering Development Division

AEDE Association Européenne des Enseignants (French—European Teachers' Association)

AEDF Australian Executive Development Foundation

AEDO Aircraft Engineering District Office

AED-RCA Astro-Electronics Division-RCA

AEDS Association of Educational Data Systems; Association of Electronic Data Systems; Atomic Energy Detection System

AEDU Admiralty Experimental Diving Unit

aee absolute essential equipment; absolutely essential equipment

Ae.E. Aeronautical Engineer

AEE Alliance for Environmental Education; Association for Experimental Education; Atomic Energy Establishment

AE.E. Associate in Engineering

AEEB Association Européenne de l'Equipement de Bureau (French—European Office Equipment Association)

AEEC Airlines Electronic Engineering Committee

AEEI Arthur E E Ivory

AEEL Aeronautical Electronic and Electrical Laboratory

AEEN Agence Européenne pour l'Energie Nucléaire (French—European Agency for Atomic Energy)

AEEP Association of Environmental Engineering Professors

AEET Atomic Energy Establishment, Trombay (India)

AEEW Atomic Energy Establishment—Winfrith

AEF Advertising Educational Foundation; Aerospace Education Foundation; Aircraft Engineering Foundation; Al-

bert Einstein Foundation; Allied Expeditionary Force; American Economic Foundation; American European Foundation; American Expeditionary Force; Americans for Economic Freedom; Armenian Educational Foundation; Artists Equity Fund; Australian Expeditionary Force(s); Aviation Engineer(ing) Force

AEFC Australian-European Finance Corporation

A-effect alienation effect

AEFM Association Européenne des Festivals de Musique (French—European Association of Music Festivals)

AEFORT American-European Friends of ORT (Organization for Rehabilitation through Training)

AEFR Aurora, Elgin & Fox River (railroad)

aeg active element group(ing); air encephalogram(s)

aeg. aeger (Latin—sick)

Aeg Aegean

AEG Association of Engineering Geologists

AEG Allgemeine Elektrizitats Geseläschaft (German—General Electric Company)

Aegean Aegean Sea (arm of the Mediterranean between Greece and Turkey)

Aegeans Aegean Islands (Cyclades, Dodecanese, Sporades, etc.)

AEGIMRDA Army Engineer Geodesy, Intelligence and Mapping Research and Development Agency

Aegis U.S. Navy cruiser class

AEGIS Active Electronic Gimballess Inertial System; Aid for the Elderly in Government Institutions; Assessment of Effectiveness of Geologic Isolation Systems

AEGp Aeromedical Evacuation Group (USAF)

Aeg S Aegean Sea

AEH A(lfred) E(dward) Housman

AEH Archives of Environmental Health

AEHA Army Environmental Health Agency

AEHA Anuario Español e Hispano-Americano (Spanish and Hispanic-American Annual)

AEHL Army Environmental Health Laboratory

aei azimuth error indicator

AEI Air Express International; American Enterprise Institute; American Express Institute; American Express International; Annual Efficiency Index; Associated Electrical Industries

AEI *Association des Ecoles Internationales* (French—Association of International Schools)

AEIB Association for Education in International Business

AEIBC American Express International Banking Corporation

AEIC Association of Edison Illuminating Companies

AEIDC American Express International Development Company

AEIL American Export Isbrandtsen Lines

AEIMS Administrative Engineering Information Management System

A.E.I.O.U. *Austria Erit In Orbe Ultima* (Latin—Austria will be the world's last survivor)—ancient acrostic of House of Hapsburg

AEIP Allied Electrical Industry Publications

AEIPPR American Enterprise Institute for Public Policy Research

AEIS Association of Electronic Industries in Singapore

AEJ Association for Education in Journalism

AEJI Association of European Jute Industries

aek all-electric kitchen

ael audit error list

AEL Admiralty Engineering Laboratory (UK); Aeronautical Engine Laboratory; Aircraft Engine Laboratory; American Electronic Laboratories; American Emigrants League; American Express Line; Americanism Education League; Animal Education League; Automation Engineering Laboratory

AELC American Evangelical Lutheran Church; Association of Evangelical Lutheran Churches

AELE Americans for Effective Law Enforcement

AELE *Association Européenne de Libre-Echange* (French—European Free Trade Association)

AELR *All England Law Reports*

AEL-Rx Appalachian Educational Laboratory—Regional Exchange

AELTC All England Lawn Tennis Club

aem atomic emission monitoring

AEM Advance Engineering Memorandum; Aircraft and Engine Mechanic; American Meter Company (stock exchange symbol); Applications Explorer Mission; Association of Electronic Manufacturers; Association of Evangelical Ministers; Aviation Electrician's Mate

AEMA Australian Electrical Manufacturers Association

AEMIE *Association Européene de Médecine Interne d'Ensemble* (French—European Association for Doctors in Internal Medicine)

AEMIS Aerospace and Environmental Medicine Information System

AEMP Association of European Management Publishers

AE & MP Ambassador Extraordinary and Minister Plenipotentiary

AEMS American Engineering Model Society

AEMSA Army Electronics Material Support Agency

aen advance evaluation note

aen. *aeneus* (Latin—made of bronze or copper)

Aen. *Aeneid* (Virgil's epic poem)

A.En. Associate in English

AEN *Asahi Evening News* (Japan)

A & EN Arts & Entertainment Network

AENA All-England Netball Association

A.Eng. Associate in Engineering

AEO Air Engineer(ing) Office(r); Appeal Examining Office(r); Australian Electoral Office

AEOB Advanced Engine Overhaul Base

AEODPs Allied Explosive Ordnance Disposal Publications

AEOE Association for Environmental and Outdoor Education

Aeol Aeolian; Aeolic

Aeol Chamb Play Aeolian Chamber Players

Aeolians Aeolian Islands off

Sicily–Lipari, Stromboli, and Vulcano

Aeol Quart Lon Aeolian Quartet of London

AEOO Aeromedical Evacuation Operations Officer

aeop amend existing orders pertaining to

AEOS Ancient Egyptian Order of Sciots; Astronomical, Earth, and Ocean Sciences

aep accrued expenditure paid; average evoked potential

AEP Addo Elephant Park (South Africa); Adult Education Program; American Electric Power

AE & P Ambassador Extraordinary and Plenipotentiary

AEP *Agence Européenne de Productivite* (French—European Production Agency)

AEPC Appalachian Electric Power Company

AEPCO American Elsevier Publishing Company

AEPEM Association of Electronic Parts and Equipment Manufacturers

AEPG Army Electronic Proving Ground

AEPI American Educational Publishers Institute

aep(s) auditory-evoked potential(s)

AEPs Allied Engineering Publications; Allied Equipment Publications

AEPS Aircrew Escape Propulsion System; American Electroplaters Society

aeq age equivalent

aeq. *aequales* [Latin—equal(s)]

AEQA Alabama Environmental Quality Association

AEqPs Allied Equipment Publications

aer aldosterone excretion rate; alteration equivalent to a repair; auditory-evoked response; average evoked response

AER Abbreviated Effectiveness Report; Aeronautical Engineering Report; After Engine Room; Airman Effectiveness Report; Army Emergency Relief; Association for Education by Radio; *Association Européenne pour l'Etude du Probleme des Réfugies* (French—European Association for the Study of the Refugee Problem)

aera aeration

AERA American Educational

Research Association; American Engine Rebuilders Association; Australian Endurance Riders Association

AERB Army Education Requirements Board

AERC Association of Executive Recruiting Consultants

aercab advanced escape/rescue capability; advanced aircrew escape/rescue capability

AERDL Army Electronics Research and Development Laboratory

Aer.E. Aeronautical Engineer

AERE Atomic Energy Research Establishment

AERI Agricultural Economics Research Institution; Automotive Exhaust Research Institute

aerl aerial

AERL Aero-Elastic Research Laboratory (M.I.T.)

Aer Méx *Aero México* (Spanish—Air Mexico formerly Aeronaves de México)

AERNO Aeronautical Equipment Reference Number

aero aerographer; aeronautical; aeronautics

AERO Association of Electronic Reserve Officers

aerobatics aeronautical acrobatics

aerobee aerojet/bumblebee (naval missile)

aerob(s) aerobic exercise(s)

aerocade aerial parade; aviation parade (massed formations, stunts aloft)

Aero Commander U-4 transport aircraft

aerodyn aerodynamics

Aero E Aeronautical Engineer

Aer Of Aerological Officer

AEROFLOT Aero Flotilla (Soviet Air Lines)

aerol aerological

aeromed acromedical

aeromod aerodynamic modelling

aeromus aeronautical museum

aeron aeronautical

aéron *aéronautique* (French—aeronautical)

Aeron Aeronaut; Aeronautics

aeronaut aeronautical(ly), aeronautics

AERONAVES *Aeronaves de México* (Spanish—Airships of Mexico)

AERONORTE *Empresa de Transportes Aereos Norte do Brasil* (Spanish—North Brazil Airways)

Aero O/Y Finnair (Finnish Airlines)

Aerop *Aeropuerto* (Spanish—airport)

aeropost aerodynamic postprocessing

AEROS Aerometric and Emissions Reporting System; Artificial Earth Research and Orbiting Satellite

AEROSAT Aeronautical Communications Satellite System

aerosp aerospace

aerospace aeronautics + space

aerospacecom aerospace communication(s)

AEROTAL *Aerolineas Territoriales de Colombia* (Spanish—Territorial Airlines of Colombia)

aerotel airplane hotel (hangar)

Aerovias "Q" Aerovias Cubana (Cuban Airlines)

AERS Atlantic Estuarine Research Society

AERT Association for Education by Radio-Television

AERU Agricultural Economics Research Unit (NZ)

aes annual expectation of sales; auger electron spectroscopy

Aes *Aesop* (Greek fabulist); (Latin—bronze or copper)— used by numismatists to denote bronze or copper coins or coins of such colors

AES Aerospace Electrical Society; Agricultural Estimates (division of Department of Agriculture); Agricultural Experiment Station; Aircraft Electrical Society; Airways Engineering Society; American Electrochemical Society; American Electroencephalographic Society; American Electroplaters Society; American Entomological Society; American Epidemiological Society; American Equilibration Society; American Ethnological Society; American Eugenics Society; Apollo Extension System; Army Exchange Service; Atlantic Estuarine Society; Audio Engineering Society; Australian Educational Secretariat; Australian Entomological Society

A&ES Arson and Explosion Squad

AESBOW Association of Engineers and Scientists of the Bureau of Weapons (USN)

AESC American Engineering Standards Committee

Aescul Aesculapius (Greek god of medicine)

AESD Acoustic Environment Support Detachment (USN Office of Naval Research)

AESE Association of Earth Science Editors

AESHS Alfred E Smith High School

AESM Albert Einstein School of Medicine

AESO Aircraft Environmental Support Office (USN)

AESOP Artificial Earth Satellite Observation Program; Automated Engineering and Scientific Optimization Program (NASA)

AESq Aeromedical Evacuation Squadron (USAF)

AESQ Air Explorer Squadron

AESRS Army Equipment Status Reporting System

AESS American Ethnic Science Society

AEST Aeromedical Evacuation Support Team

AESTE Association for the Exchange of Students for Technical Experience

aesth aesthete; aesthetic; aesthetician; aesthetics

AESU Aerospace Environmental Support Unit

aet absorption-equivalent thickness

aet. *aetatis* (Latin—at or of the age of)

AET Australian Eastern Time

A.E.T. Associate in Electrical Technology; Associate in Electronic Technology

AET *Aerlinte Eireann Teoranta* (Irish Airlines)

AETA American Educational Theatre Association

AETD Aero-Electronic Technology Department (USN)

AETE Aerospace Engineering Test Establishment (Canada)

AETFAT *Association pour l'Etude Taxonomique de la Flore d'Afrique Tropicale* (French—Association for the Taxonomic Study of African Tropical Flora)

A et M *Arts et Métiers* (French—arts and crafts)

AETM Aviation Electronic Technician's Mate

AETN American Educational Television Network

AEtPs Allied Electronic Publications

AETR Advanced Engineering

Test Reactor
AETS Association for the Education of Teachers in Science
AETT Australian Elizabethan Theatre Trust
aeu accrued expenditure unpaid
AEU Amalgamated Engineering Union; American Ethical Union; Asia Electronics Union
aeuia alleluia (Italian—hallelujuh)
aev (AEV) aerothermodynamic elastic vehicle
AEV Asociación de Escritores Venezolanos (Spanish—Association of Venezuelan Writers)
AEVA Australian Equine Veterinary Association
aevac air evacuation
AEW Airborne Early Warning
AEWB Army Electronic Warfare Board (USA)
AEW & C Airborne Early Warning and Control
AEWCAP Airborne Early Warning Combat Air Patrol
AEWES Army Engineers Waterways Experiment Station
AEWHA All-England Women's Hockey Association
AEWIS Army Electronic Warfare Information System (USA)
AEWL Association of Employers of Waterside Labour (Australian)
AEWLA All-England Women's Lacrosse Association
AEWRON Airborne Early Warning Squadron
AEWS Advanced Earth Satellite Weapon System (USAF); Aircraft Early Warning System (DoD)
AEWSPS Aircraft Electronic Warfare Self-Protection System
aex automatic electronic exchange (facilitating telephony)
AExO Assistant Experimental Officer
af audiofrequency
af (AF) ale firkin; audio fidelity; autofocus
a-f anti-foam; audio-frequency
a/f *a favor* (Spanish—in favor)
af afgang (Danish—departure); *anno futuro* (Italian—next year); (Latin prefix—movement toward a central point)
a.f. ad finem (Latin—to the end)
af attofarad(s)
Af Africa; Afrikaans; African(s)

Af Académie française (French Academy); *Armée francaise* (French Army)
AF Advance Freight; Africa(n); Air Force; air freight; Anglo-French; Armored Force; Arthritis Foundation; Aviation Photographer's Mate; provision stores ship (2-letter symbol)
A-F Anglo-French
A.F. Admiral of the Fleet
A/F Air Field
A & F Agriculture and Forestry (Senate Committee)
A of F Admiral of the Fleet
AF35m auto-focus 35mm (camera)
afa azimuth follow-up amplifier
AFA Actors Fund of America; Advertising Federation of America; Advertising Federation of Australia; Aerophilatelic Federation of the Americas; Air Force Association; Alien Firearms Act; American Federation of the Arts; American Finance Association; American Forensic Association; American Forestry Association; American Foundrymens Association; American Freedom Association; Association of Federal Appraisers; Association of Flight Attendants; Australian Field Artillery; Australian Foundation for Alcoholism
AF of A. Advertising Federation of America
A.F.A. Associate in Fine Arts
AFAA Adult Film Association of America; Aerobics and Fitness Association of America; Air Force Audit Agency; Automatic Fire Alarm Association
AFAAEC Air Force Academy and Aircrew Examining Center
afac (AFAC) airborne forward air controller
AFAC Air Force Armament Center; American Fisheries Advisory Committee; Arkansas Foundation of Associated Colleges
afactplan affirmative action plan(ning)
AFADD Australian Foundation for Alcoholism and Drug Dependence
AFADO Association of Food and Drug Officials
AFAFC Air Force Accounting and Finance Center

AFAG Airforce Advisory Group
AFAIDSR American Foundation for AIDS Research
AFAIM Associate Fellow of the Australian Institute of Management
AFAITC Armed Forces Air Intelligence Training Center
AFAL Air Force Avionics Laboratory
afam airfield attack ammunition; automatic frequency-assignment model
AF & AM Ancient Free and Accepted Masons
Af-Am(s) African-American(s); Afro-American(s)
afap artillery-fired atomic projectile
AFAP Australian Federation of Airline Pilots
AFAPL Air Force Aero-Propulsion Laboratory
afar airborne fixed-array radar
AFAR Azores Fixed Acoustic Range (NATO)
AFAR Australian Foreign Affairs Record
Afars and Issas formerly French Somaliland now the Republic of Djibouti
AFAS Air Force Aid Society; Automated Frequency-Assignment System
AFAS Association française pour l'avancement des sciences (French—Association for the Advancement of Science)
AFASE Association for Applied Solar Energy
AFA-SEF Air Force Association—Space Education Foundation
AFASIC Association For All Speech-Impaired Children
AFAUD Air Force Auditor General
afb acid-fast bacillus; anti-friction bearing
afb afbeelding (Dutch—illustration)
AFB Air Force Base; American Farm Bureau; American Foundation for the Blind
AFBF American Farm Bureau Federation
AFBI Australian Fibre Box Industry
AFBMA Antifriction Bearing Manufacturers Association
AFBMD Air Force Ballistic Missile Division
AFBNM Agate Fossil Beds National Monument (Nebraska)

AFBS American and Foreign Bible Society; Ansett Flying Boat Services

AFBSD Air Force Ballistic Systems Division

afc antibody forming cells; automatic frequency control

afc (AFC) average fixed cost

AFC Air Force Cross; American Football Conference; American Forest Council; Apollo Flight Control (NASA); Area Forecast Center; Australian Film Commission; Australian Flying Corps

AFCAI Associate Fellow of the Canadian Aeronautical Institute

AFCAL Association Française de Calcul (French—French Calculus Association)

AFCBS Australian Federation of Commercial Broadcasting Stations

AFCC Air Force Communications Center; Air Force Cost Center; Australian Federal Cycling Council; Australian Federation of Construction Contractors

AFCCB Air Force Configuration Control Board

AFCCDD Air Force Command and Control Development Division

AFCCP Air Force Component Command Post

AFCD Air Force Cryptologic Depot

afce automatic flight-control equipment

AFCE Associate in Fuel Technology and Chemical Engineering

AFCEA Armed Forces Communications and Electronics Association

AFCEC Australian Federation of Civil Engineering Contractors

AFCent Allied Forces, Central Europe

afcfs advanced fighter-control-flight simulator

AFCI American Foot Care Institute

AFCL Africa Container Lines

AFCM Air Force Commendation Medal

AFCMA Aluminum Foil Container Manufacturers Association; Australian Fibreboard Container Manufacturers Association

AFCMC Air Force Contract Maintenance Center

AFCMD Air Force Contract Management Division

AFCMO Air Force Contract Management Office

AFCN American Friends of the Captive Nations

afco automatic fuel cutoff

AFCO Admiralty Fleet Confidential Order; Air Force Contracting Office(r); Australian Federation of Consumer Organisations

AF Compt Air Force Comptroller

AFCOMS Air Force Commissary Service

AFCOMSECCEN Air Force Communications Security Center

AFCON Air Force Controlled (units)

AFCOS Air Force Combat Operations Staff; Armed Forces Courier Service

AFCPMC Air Force Civilian Personnel Management Center

AFCR American Federation for Clinical Research

AFCRC Air Force Cambridge Research Center

AFCRL Air Force Cambridge Research Laboratories

AFCs Air Force Circulars

AFCS Active Federal Commissioned Service; Adaptive Flight Control System; Air Force Communications Service; Automatic Flight Control System

AFCS E&I Air Force Communications Service—Engineering and Installation

AFCSA Air Force Center for Studies and Analyses

AFCSL Air Force Communications Security Letter

AFCSM Air Force Communications Security Manual

AFCSP Air Force Communications Security Pamphlet

AFCUL Australian Federation of Credit Union Leagues

AFCW Association of Family Case Workers

AFCWF Air Force Civilian Welfare Fund

afd accelerated freeze drying; alternate full day

afd afdeling (Dano-Norwegian, Dutch—department, division, section)

AFD Air Force Depot; Association of Food Distributors; Association of Footwear Distributors; mobile floating drydock (naval symbol)

AFDA American Flag Day Association; Australian Defence Force Academy

AFDAA Air Force Data Automation Agency

AFDAP Air Force Directorate of Advanced Technology

AFDATACOM Air Force Data Communications System

AFDATASTA Air Force Data Station

AFDB African Development Bank; Air Force Decorations Board; large auxiliary floating drydock (naval symbol)

AFDC Aid for Dependent Children; Aid for Families with Dependent Children; Australian Film Development Corporation

AFDCB Armed Forces Disciplinary Control Board

AFDCMI Air Force Policy on Disclosure of Classified Military Information

AFDCS American First Day Cover Society

AFDCUF Aid to Families with Dependent Children of Unemployed Fathers

AFDC-UP Aid to Families of Dependent Children—for Unemployed Parents

AFDE American Fund for Dental Education

AFDEA American Funeral Directors and Embalmers Association

AFDH American Fund for Dental Health

AFDL small auxiliary floating drydock (naval symbol)

AFDM medium auxiliary floating drydock (naval symbol)

AFDO Air Force Duty Officer; Association of Food and Drug Officials

AFDOA Armed Forces Dental Officers Association

AFDOUS Association of Food and Drug Officials of the United States

AFDP Air Force Development Plan

AFDRB Air Force Disability Review Board; Air Force Discharge Review Board

AFDRD Air Force Director of Research and Development

AFDRQ Air Force Director of Requirements

AFDS Air Fighting Development Squadron

AFDSC Air Force Data Services Center

AFDW Air Force District of Washington
AFE Administración de Ferrocarriles del Estado (Spanish—State Railway Administration)
AFEA American Farm Economic Association; American Film Export Association
AFEB Armed Forces Epidemiological Board
AFEE Airborne Forces Experimental Establishment
AFELIS Air Force Engineering and Logistics Information System
AFEM Armed Forces Expeditionary Medal
AFEMS Air Force Equipment Management System
AFEOC Air Force Emergency Operations Center
AFEOS Air Force Electro-Optical Site
AFER Air Force Engineering Responsibility
AFERB Air Force Educational Requirements Board
AFERO Asia and the Far East Regional Office (FAO)
AFES Air Force Exchange Service; American Far Eastern Society; Armed Forces Examining Stations
AFESA Air Force Engineering and Services Agency
AFESC Air Force Engineering and Services Center
AFETR Air Force Eastern Test Range
AFEX Air Forces Europe Exchange
aff above finished floor; affairs
AFF affinity; Army Field Forces
AFFA Air Freight Forwarders Association; Angels Forever, Forever Angels (slogan of Hell's Angels motorcycle gang)
affaire affaire de coeur (French—affair of the heart)—love affair
AFFC Air Force Finance Center
affd affixed; afford(able); affordability
affd per cur affirmed by the court
AFFDL Air Force Flight Dynamics Laboratory
AFFE Air Force Far East; Allied Forces Far East; Army Forces Far East
affec affectation; affection; affective

affet affettuoso (Italian—tenderly, with pathos)
afff aqueous film-forming foam
AFFFA American Forged Fitting and Flange Association
AFFI American Frozen Food Institute
Affie Alfred
affil affiliated
affirm affirmative
AFFJ American Fund for Free Jurists
affl affluent
AFFL Agricultural Finance Federation, Limited
afflat afflatus
AFFLC Air Force Film Library Center
AFFOR Air Force Forces (joint task force element)
affores afforestation
affret affrettando (Italian—speeding the tempo)
AFFS American Federation of Film Societies
afft affidavit
AFFTC Air Force Flight Test Center; Air Force Flying Training Command
afg above finished grade; analog function generator
afg afgang (Dano-Norwegian—departure)
Afg Afghan; afghani (currency); Afghanistan; Afghans
AFG Allied Freighter Guard
AFGC American Forage and Grassland Council
AFGCM Air Force Good Conduct Medal
Afg Dem Jam Afghanistan Democrateek Jamhuriat (Pashto—Afghan Democratic Republic)
AFGE American Federation of Government Employees
Afghan Afghanistan(i)
afghani monetary unit of Afghanistan
Afghanistan Republic of Afghanistan, *Doulat i Jamhouri ye Afghánistán*
AFGIS Aerial Free Gunnery Instruction School
AFGL Air Force Geophysics Laboratory
AFGM American Federation of Grain Millers
AFGU Aerial Free Gunnery Unit
AFGW American Flint Glass Workers
AFGWC Air Force Global Weather Central
AFH Air Force Hospital; American Foundation for Ho-

meopathy; Associated Federated Hotels; Australian Field Hospital
AFHC Air Force Headquarters Command
AFHF Air Force Historical Foundation; American Foot Health Foundation
AFHQ Air Force Headquarters; Allied Forces Headquarters; Armed Forces Headquarters
AFHW American Federation of Hosiery Workers
afi amaurotic familial idiocy
AFI Air Filter Institute; Air Force Intelligence; American Film Institute; American Filter Institute; American Forest Institute; American Friends of Israel; Armed Forces Institute; Association of Federal Investigators; Atlantic Refining Company (stock exchange symbol); Australian Film Institute; Australian Foundry Institute; Australian Frontier Incorporated
AFIA Air Force Intelligence Agency; American Footwear Industries Association; American Foreign Insurance Association
AFIAS Associate Fellow of the Institute of the Aerospace Sciences
afib atrial fibrillation
afic aficionado (Spanish—admirer, devotee, fan)
AFIC Air Force Intelligence Center; Australian Fishing Industry Council
AFICCS Air Force Interim Command and Control System
AFICE Air Forces—Iceland
AFIED Armed Forces Information and Education Division
AFII American Federation of International Institutes
AFIIM Associate Fellow of the Institute of Industrial Managers
AFINE Association Française pour l'Industrie Nucleaire d'Equipement (French—French Association for the Nuclear Equipment Industry)
AFINS Airways Flight Inspector
AFIO Association of Former Intelligence Officers
AFIP Air Force Intelligence Publication; Armed Forces Information Program; Armed Forces Institute of Pathology
AFIPS American Federation of

Information Processing Societies
AFIR Air Force Installation Representative
AFIRAN Africa-Indian Ocean Region Air Navigation
AFIRO Air Force Installations Representative Officer
AFIS Air Force Intelligence Services; American Forces Information Service; Armed Forces Information School; Armed Forces Information System; Automated Field Interview System; Automated Fingerprint Identification System
AFISC Air Force Inspection and Safety Center
afism aluminum-free inorganic suspended material
AFISR Air Force Industrial Security Regulations
afit airblast fuel-injection tube
AFIT Air Force Institute of Technology
AFITAE Association Française d'Ingénieurs et Techniciens de l'Aéronautique et de l'Espace (French—French Association of Aeronautical and Aerospace Engineers and Technicians)
AFJAG Air Force Judge Advocate General
AF JINTACCS Air Force Joint Interoperability of Tactical Command and Control Systems
AFJKT Air Force Job-Knowledge Test
afk afkorten, afkorting (Dutch—abbreviation)
Afk Afrikaans
afl abstract family of languages; anti-fatty liver; atrial flutter
afl aflevering (Dutch—part)
AFL Aeroflot (Soviet Air Lines); Air Force Letter; American Federation of Labor; American Football League; Applied Fisheries Laboratory (University of Washington); Arena Football League; Association for Family Living; Australasian Federation League; Australian Fertilizers Limited
AFLA Adolescent Family Life Act; Amateur Fencers League of America; American Foreign Law Association; Asian Federation of Library Associations
AFLAT Air Force Language Aptitude Test

aflatox aflatoxin
AFLC Air Force Logistics Command
AFL-CIO American Federation of Labor and Congress of Industrial Organizations
AFLCPs Air Force Logistics Command Pamphlets
afld airfield
aflir advanced forward-looking infrared
AFLM Accredited Farm and Life Member
AFLP American Farmer Labor Party; Armed Forces Language Program
AFLRL Army Fuels and Lubricants Research Laboratory
AFLS Air Force Library Service
AFLSA Air Force Longevity Service Award
AFLSC Air Force Legal Services Center
aflt afloat
afm antifriction metal
AFM Air Force Manual; Air Force Medal; Air Force Museum; American Federation of Musicians; Associated Fur Manufacturers
AFMA American Footwear Manufacturers Association; Armed Forces Management Association
AFMA Air Force Manual of Abbreviations
AFMBE Australian Federation for Medical and Biological Engineering
AFMBT Artificial Flower Manufacturers Board of Trade
AFMDC Air Force Missile Development Center
AFME American Friends of the Middle East
AFMEA Air Force Management Engineering Agency
AFMEC African Methodist Episcopal Church
AFMed Allied Forces, Mediterranean
AFMF Air Fleet Marine Force
AFMH American Foundation for Mental Hygiene
AFMIC Air Force Materials Information Center
AFML Air Force Materials Laboratory; Armed Forces Medical Library
AFMMFO Air Force Medical Materiel Field Office
afmo afectísimo (Spanish—most affectionate)
AFMPA Armed Forces Medical Publication Agency
AFMPC Air Force Military Per-

sonnel Center
afmr antiferromagnetic resonance
AFMR American Foundation for Management Research; Armed Forces Master Records
AFMS Air Force Medical Service; American Federation of Minerological Societies; Australian Farm Management Society
AFMSC Air Force Medical Specialist Corps
AFMTC Air Force Missile Test Center
AFMVOP Air Force Motor Vehicle Operator Test
AFMW Australian Federation of Medical Women
afn active filter network
AFN *Afrique du Nord* (French North Africa); Air Force Finance Center; Alaska Federation of Natives; American Forces Network; Armed Forces Network
AF of N Alaska Federation of Natives
AFNA Accordion Federation of North America; Air Force with Navy; American Foundation for Negro Affairs
AFNB Armed Forces News Bureau
AFNC Air Force Nurse Corps
AFNE Allied Forces, Northern Europe; Americans For Nuclear Energy
AFNIL Agence Francophone pour la Numérotation Internationale du Livre (French—French Agency for the International Numbering of Books)
AFNOR Association Française de Normalisation (French—French Standards Association)
AFNorth Allied Forces, Northern Europe
AFO Accounting and Finance Office(r); Admiralty Fleet Order; Airports Field Office; Atlantic Fleet Organization
AFOAR Air Force Office for Aerospace Research
AFOAS Air Force Office of Aerospace Sciences
AFOAT Air Force Office for Atomic Energy
AFOB American Foundation for Overseas Blind
AFOC Air Force Operations Center
AFOECP Air Force Officer Education and Commissioning Program

AFOG Asian Federation of Obstetrics and Gynaecology
AFOIC Air Force Officer in Charge
AFOMS Air Force Office of Medical Support
AFOQT Air Force Officer Qualifying Test
AFORG Air Force Overseas Replacement Group
a fort a fortiori (Italian—with greater force)
AFOS Advanced Field Operations System; Automation of Field Operations and Services (NWS)
AFOSI Air Force Office of Special Investigations
AFOSP Air Force Office of Security Police
AFOSR Air Force Office of Scientific Research
AFOTEC Air Force Operation Test and Evaluation Center
AFOUA Air Force Outstanding Unit Award
afp anterior faucial pillar; automated filter photometer
afp (AFP) alphafetoprotein
AFP Agence France-Presse (successor to Havas); Air Force Pamphlet; Air Force Police; Alternate Flight Plan; American Family Publishers; American Federation of Police; Anglican Fellowship of Prayer; Annual Funding Program; Armed Forces of the Philippines; Armed Forces Police; Authority for Purchase
AF of P American Federation of Police
afpa automatic flow process analysis
AFPA Aquarama and Fairmount Park Aquarium; Australian Fire Protection Association
AFPAO Air Force Property Accountable Office(r)
AFPAV Air Force Pavement
AFPB Air Force Personnel Board
AFPC Air Force Personnel Council; Air Force Procurement Circular; American Food for Peace Council; Armed Forces Policy Council
AFPCB Armed Forces Pest Control Board
AFPD Armed Forces Police Detachment
AFPE American Foundation for Pharmaceutical Education; American Foundation for Political Education

AFPH American Federation of the Physically Handicapped
AFPI Air Force Procurement Instructions; American Forest Products Industries
AFPP Air Force Procurement Procedures
AFPPA American Federation of Poultry Producers Associations
AFPR Air Force Plant Representative
AFPRB Armed Forces Pay Review Board
AFPRO Action for Food Production; Air Force Plant Representative's Office
AFPs American Freeway Patrol cars
AFPS Armed Forces Press Service
AFPT Air Force Personnel Test
AFPTRC Air Force Personnel and Training Research Center
AFPU Air Force Postal Unit; Australian Federation of Police Unions
AFQ Association Forestière Québeçoise (French—Quebec Forestry Service)
AFQA Air Force Quality Assurance
AFQC Air Force Quality Control
AFQT Armed Forces Qualification Test
afr acceptable failure rate; airframe; air-fuel ratio; alternating frequency rejection; applicable federal rate; automatic field/format recognition; away from reactor
afr afrikansk (Dano-Norwegian—African)
Afr Africa; African; Africans; Afrikaans (South African Dutch)
A Fr Algerian franc
A-Fr Anglo-French
AFR Air Force Regulation(s); Air Force Reserve
AFR Australian Financial Review
afra average freight rate assessment
AFRA American Farm research Association; American Federation of Television and Radio Artists
AFRAeS Associate Fellow of the Royal Aeronautical Society
AFRAM Afro-American
A-frame capital-A-shaped support frame
Aframerican African + American

can
AFRASEC Afro-Asian Organization for Economic Cooperation
Afrasia Africa + Asia
AFRB Air Force Retiring Board
AFRBA Armed Forces Relief and Benefit Association
AFRBO Air Force Review Boards Office
AFRBSG Air Force Reserve Base Support Group
AFRC Air Force Records Center; Air Force Regional Civil Engineer; Armed Forces Recreation Center
AFRCC Air Force Rescue Coordination Center; Air Force Reserve Coordination Center
AFRCSTC Air Force Reserve Combat Support Training Center
afrd acute febrile respiratory disease (AFRD)
AFRD Air Force Research Division; Air Force Reserve Division; Association of Fund-Raising Directors
A-freak acid (LSD) user
AFRes Air Force Reserve
AFRESM Armed Forces Reserve Medal
AFRESNAVSQ Air Force Reserve Navigation Squadron
AFRFI American Friends of Religious Freedom in Israel
afri acute febrile respiratory illness (AFRI)
AFRI Applied Forest Research Institute
Afric Africa; African
African languages Hausa in Central and West Africa; Swahili in parts of East Africa; Yoruba and Ibo in West Africa; Rwanda in southern Central Africa; Somali in East Africa; Xhosa and Zulu in South Africa
Africs Africans
afrik afrikansk (Dano-Norwegian—African)
Afrik Afrikaans; Afrikaner State (northern Cape Province of South Africa)
Afrik Afrikaans (Dutch dialect spoken in South Africa, the language of the Boers)
afrm airframe
AFRMA American Fancy Rat and Mouse Association
Afr Nat Cnl African National Council
AFRO African Regional Office (FAO)
Afro-Am Afro-American(ese)

Afro-Amer Afro-American
Afro-American African-American
Afro-Bras Afro-Brasiliero (Afro-Brazilian)
Afro-Carib Afro-Caribbean; Afro-Caribeño
Afro(s) Afro-American(s)— Black(s), Negro(es)
AFROTC Air Force Reserve Officers Training Corps
AFRPL Air Force Rocket Propulsion Laboratory
AFRR Air Force Reserve Region
AFRRG Air Force Reserve Recovery Group
AFRRI Armed Forces Radiobiology Research Institute
afr's auditor freight receipts
AFRS Air Force Reserve Sector; Armed Forces Radio Service
AFRTS Armed Forces Radio-Television Service
AFRTVS Armed Forces Radio and Television Services
AFRVN Air Force of the Republic of Viet Nam
afs aerial fire support; aforesaid; atomic fluorescence spectroscopy
afs afsender (Danish—sender)
AFS Active Fusing System; Air Force Specialty; Air Force Station; Air Force Supply; Airline Feed System; Airways Facilities Shop; Alaska Ferry Service; American Federation of Scientists; American Feline Society; American Fern Society; American Field Service; American Fisheries Society; American Folklore Society; American Foundrymen's Society; American Fuchsia Society; Aviation Facilities Service; Azimuth Follow-up System
AFSA Air Force Sergeants Association; American Federation of School Administrators; American Flight Strips Association; American Foreign Service Association; Armed Forces Security Agency
AFSAB Air Force Science Advisory Board
AFSAS American Federation of School Administrators and Supervisors
AFSATCOMS Air Force Satellite Communications System
AFSAW Air Force Special Activities Wing
Af-Sax Afro-Saxon (black per-

son of part Anglo-Saxon parentage)
AFSB American Federation of Small Business
AFSBO American Federation of Small Business Organizations
AFSC Air Force Service Command; Air Force Specialty Code; Air Force Supply Catalog; Air Force Systems Command; American Federation of Soroptimist Clubs; American Friends Service Committee; Armed Forces Staff College
AFSCC Air Force Special Communications Center; Armed Forces Supply Control Center
AFSCF Air Force Satellite Control Facility
AFSCM Air Force Systems Command Manual
AFSCME American Federation of State, County, and Municipal Employees
afsd aforesaid
AFSE Allied Forces Southern Europe (NATO)
AFSec Air Force Section
AFSF Air Force Stock Fund
AFSIL Accommodations for Students in London
AFSINC Air Force Service Information and News Center
AFSM Association for Food Service Management
AFSMAAG Air Force Section—Military Advisory Group
AFSN Air Force Serial Number; Air Force Service Number; Air Force Stock Number
AFSNCOA Air Force Senior Noncommissioned Officers' Academy
AFSouth Allied Forces, Southern Europe
AFSPACECOM Air Force Space Command
AFSRS Australian Flying Saucer Research Society
AFSS Air Force Security Service; Air Force Service Statement
AFSSD Air Force Space Systems Division
AFSSO Air Force Special Security Office
AFSTC Air Force Space Test Center
AFSUB Army Air Forces Anti-Submarine Command
AFSWA Armed Forces Special Weapons Agency

AFSWC Air Force Special Weapons Center
AFSWP Armed Forces Special Weapons Project
aft after; afternoon; at, near, or toward the rear; automatic fine tuning
aft. (AFT) automatic fund transfer
Aft Aftenposten (Danish—Evening Post, Oslo)
AFT Air Freight Terminal; American Federation of Teachers; Annual Field Training (USA)
AFT (AFL-CIO) American Federation of Teachers
AFTA Atlantic Free Trade Area; Australian Federation of Travel Agents
AFTAC Air Force Technical Applications Center
AFTAU American Friends of Tel Aviv University
aftb afterburner
AFTB Air Force Test Base
AFTC Airborne Flight Training Command; American Fair Trade Council; American Fox Terrier Club; American Free Trade Clubs
AFTE American Federation of Technical Engineers
AFTEC Air Force Test and Evaluation Center
AFTF Air Force Task Force
AFTI Advanced Fighter Technology Integration (USAF)
AFTIA Armed Forces Technical Information Agency
AFTLI Association of Feeling Truth and Living It
AFTM American Foundation for Tropical Medicine
AFTMA American Fishing Tackle Manufacturers Association
aftn afternoon
AFTN Aeronautical Fixed Telecommunications Network
afto afecto (Spanish—affectionate, fond)
AFTO Air Force Technical Order
AFTOSB Air Force Technical Order Standardization Board
aftp additional flight-training period
AFTR American Federal Tax Reports
AFTRA American Federation of Television and Radio Artists
AFTRC Air Force Technical Training Command
afts automatic frequency tone

shift
AFTS Aeronautical Fixed Telecommunications Service; Aseptic Fluid Transfer System; Australian Film and Television School
AFTTH Air Force Technical Training Headquarters
AFTU Arizona Federation of Teacher Unions
afu all fucked up
AFU Advanced Flying Units; American Fraternal Union; Assault Fire Unit (U.S. Army)
AFU Association Fonciere Urbaine (French—Urban Land Association)
AFUA Annual Fuel Utilization Agency
AFULE Australian Federated Union of Locomotive Enginemen
AFUS Air Force of the United States; Armed Forces of the United States
AFUW Australian Federation of University Women
afv armored fighting vehicle; armored force vehicle
AFVA Air Force Visual Aid
AFvg Anglo-French variable geometry
AFVN Armed Forces Vietnam Network
AFVOA Aberdeen Fishing Vessel Owners Association
AFW Association for Family Welfare
AFWA Air Force with Army
AFWAL Air Force Wright Aeronautical Laboratories
AFWE Air Forces Western Europe (NATO)
AFWETS Air Force Weapons Effectiveness Testing System
AFWL Air Force Weapons Laboratory; Armed Forces Writers League
AFWN Air Force with Navy
AFWOFS Air Force Weather Observing and Forecasting System
AFWR Atlantic Fleet Weapons Range
AFWST Armed Forces Women's Selection Test
AFWTR Air Force Western Test Range (see WTR)
Afyon Afyonkarahisar (Turkish—Black Castle of Opium)—town in western central Turkey where much of the world's opium is grown
afz afzender (Dutch—from)
AFZ Australian Fishing Zone
ag against; agar-agar; agency;

agent; aggie; aggressive; agribusiness; agricultural; agriculture; agrobiology; agroindustrial; agrology; agronomy; albumen gland; alternate geologies; armor grating; atrial gallop; axiogingival
a-g air-to-ground; anti-gas
a/g air-to-ground; albumin-globulin ratio
a g à gauche (French—to the left)
Ag Agostino
Ag argentum (Latin—silver)
AG Adjutant General; Aeronautical Standards Group; Air Group; Allegheny Ludlum Steel (stock exchange symbol); Artists Guild; Attorney General; Auditor General; escort research vessel (naval symbol); miscellaneous auxiliary vessels (naval symbol); sonar research ship (naval symbol); technical research ship (naval symbol)
AG Aktien Gesellschaft (German—company, joint stock company); *Alberghi per la Gioventu* (Italian—Youth Hostels); *Arkansas Gazette; Asamblea General* (Spanish—General Assembly); *Astronomische Gesellschaft* (German—Astronomical Society)
aga accelerated growth area; appropriate for gestational age
a/g/a air-to-ground-to-air
AGA Abrasive Grain Association; Adjutants General Association; Alabama Gas (symbol); American Gas Association; American Gastroenterological Association; American Gastroscopic Association; American Genetic Association; American Glassware Association; American Goiter Association; American Gold Association; Australian Garrison Artillery; Australian Gas Association
AGAA Art Galleries Association of Australia
AGAAC Acuerdo General sobre Aranceles Aduaneros y Comercio (Spanish—General Accord Custom's Duties and Commerce)
AGAC American Guild of Authors and Composers
Aga cooker Aktiebolaget gas-accumulative cooker
agacs automatic ground-air-communication system

AGAFBO Atlantic and Gulf American Flag Berth Operators
AGAL Australian Government Analytical Laboratory
AGARD Advisory Group for Aeronautical Research and Development (NATO)
AGAS Australian Government Advertising Service
agate chalcedony
Agathon Agathon Press
agave. automatic gimballed antenna vectoring equipment
agb accessory gear box; any good brand
AGB Audits of Great Britain (television survey); icebreaker (3-letter symbol)
AGBAD Alexander Graham Bell Association for the Deaf
AGBI Artists' General Benevolent Institution
agbio agrobiology
AGBUC Association of Governing Boards of Universities and Colleges
agbus agribusiness; analog ground bus
agc air-ground communications; automatic gain control
AGC Adjutant General's Corps; Aerojet-General Corporation; American Grassland Council; amphibious force flagship (naval symbol); Armed Guard Center; Associated General Contractors; astronomical great circle (course); Australian Government Centre; Australian Guarantee Corporation
AGC Amgueddfa Cenedlaethol Cymru (Welsh—National Museum of Wales)
agca automatic ground control approach
AGCA Associated General Contractors of America
AGCan Auditor General of Canada
agcl automatic ground-controlled landing
AGCM Army Good Conduct Medal
AGCMWA Amon G Carter Museum of Western Art
agcol agricultural college
AGCRSP Army Gas-Cooled Reactor Systems Program
AGCSB Atlantic-Gulf Coastwise Steamship Freight Bureau
AGCSD Attorney General's Consumer Services Department
AGCT Army General Classifi-

cation Test
AGCTS Armed Guard Center Training School
Ag-Cu al silver-copper alloy (U.S. coin facing)
agcy agency
agd agreed; axial gear differential
AGD Academy of General Dentistry; Adjutant General's Department; American Gage Design; Auditor General's Department
AGD Australian Government Digest
AGDA American Gasoline Dealers Association; American Gun Dealers Association
AGDC Assistant Grand Director of Ceremonies
AGDE escort research ship (naval symbol)
Ag. Dei Agnus Dei (Latin— Lamb of God)
Ag Dept Agriculture Department
AGDS American Gage Design Standard
age. (AGE) aerospace ground equipment; automatic guidance electronics
Age The Age (Melbourne)
Ag.E. Agricultural Engineer
AGE AG Edwards; Amarillo Grain Exchange; Asian Geotechnical Engineering (Thailand)
A.G.E. Associate in General Education
AGEC Army General Equipment Command
A.G.Ed. Associate in General Education
AGED Advisory Group on Electronic Devices
AGEH hydrofoil research ship (naval symbol)
AGEHR American Guild of English Handbell Ringers
agents provocs agents provocateurs (French—secret agents)—persons hired to provoke others to commit crimes so arrest and conviction can follow
ageocp aerospace ground equipment out of commission for parts
AGEP Advisory Group on Electronic Parts
AGER environmental research ship (naval symbol)
agerd aerospace ground-equipment requirements data
AGERS Auxiliary General Electronics Research Ship(s)

AG & ES American Gas & Electric System
AGET Advisory Group on Electronic Tubes
AGF Army Ground Forces; miscellaneous command ship (naval symbol)
AGFA Aktiengesellschaft für Anilinfabrikation (German— Corporation for Aniline Manufacture)
ag.feb. *aggrediente febre* (Latin—when fever increases)
Ag and Fish Ministry of Agriculture and Fisheries
AGFRTS Air and Ground Forces Resources and Technical Staff (U.S. Army)
AGFSRS Aircraft Ground Fire Suppression and Rescue System (DoD)
agg agammaglobulinaemia(c); agglutination(ed); aggravate(ed); aggregate(d); aggregation
aggie agriculture
aggies agate playing marbles; students of agricultural colleges or schools
agglut agglutination (ed)
aggr aggregate
AGGR Air-to-Ground Gunnery Range
aggred. feb. *aggrediente febre* (Latin—while fever is developing)
aggro aggression; aggressiveness
aggs anti-gas gangrene serum
AGGS American Good Government Society
Aggy Agatha; Agnes
AGH Addison Gilbert Hospital, Gloucester, Mass.; Auditor General's Hotline; Australian General Hospital
AGHE Association for Gerontology in Higher Education
agi adjusted gross income
AGI Adjusted Gross Income; American Geographical Institute; American Geological Institute; Annual General Inspection
AGI Agenzia Giornalistica Italiana (Italian News Agency); *Associazione Guide Italiane* (Italian Girl Guides' Association)
AGIC Air-Ground Information Center
AGIFORS Airlines Group of International Federation of Operations Research Societies
agil airborne general illumination light

agile. airborne general illumination light; analytic geometry interpretative language
AGILE Autonetics General Information Learning Equipment
ag imps hnd agricultural implements hand
ag imps ot hand agricultural implements other than hand
ag'in' against
agind agroindustrial
AGIO Australian Government Insurance Office
AGIP Azienda Generale Italiana Petroli (Italian—National Italian Oil Company)
agipa adaptive ground-implemented-phased array
agit agitate(d); agitation; agitato; agitator
agit. *agitatum* (Latin—shaken)
agit. *ante sum agita ante sumendum* (Latin—shake before using)
agit. a. us. *agita ante usum* (Latin—shake before using)
agit. bene agita bene (Latin— shake well)
agit-prop agitation and propaganda
agit. vas. *agitato vase* (Latin— shaking the vessel)
agl above ground level; acute granulocytic leukemia; airborne gun laying; aminoglutethimide
Agl Argelia (Spanish—Algeria)
AGL Australian Gas Light (company); lighthouse tender (3-letter symbol)
agland(s) agricultural land(s)
AGLC Air-to-Ground Liaison Code
AGLINET Agricultural Libraries Information Network (UN)
aglm agglomerate
AGLS Association of General and Liberal Studies
AGLSP Association of Graduate Liberal Studies Programs
A-glue airplane glue
agm (AGM) air-to-ground missile
AGM American Guild of Music; Annual General Meeting (of shareholders); Australian Glass Manufacturers; missile range instrumentation ship (naval symbol)
AGM-53A North American-Rockwell Condor air-to-surface missile
AGMA American Gear Manufacturers Association; Ameri-

can Guild of Musical Artists; Athletic Goods Manufacturers Association

AGMA (AFL-CIO) American Guild of Musical Artists

AGMIS Adjutant General Management Information System

Agmk African green monkey kidney

AGMR major communications relay ship (naval symbol)

agn active galactic nuclei; acute glomerulonephritis; again; agnomen

Agn Augustín

Agñ Agaña, Guam

AGN Aerojet-General Nucleonics

Agncy Agency (postal placename abbreviation)

agnos agnostic; agnosticism

agnos *agnostos* (Greek—unknowable), origin of *agnostic* invented in 1869 by Thomas Henry Huxley

AGNS Allied General Nuclear Services

ago. atmospheric gas oil

ago *agitato* (Italian—agitated); *agosto* (Spanish—August)

Ago *agosto* (Italian/Spanish—August)

AGO Adjutant General's Office; Air Gunnery Officer; American Guild of Organists; Attorney General's Office; Attorney General's Opinion; Auditor General's Office

AGOR Auxiliary General Oceanographic Research (vessel)

agp above-ground pool; automatic guidance programming

AGP Academy of General Practice; Achievement Goals Program (to improve test scores in minority schools); Achievement Guidance Program; Adjutant General's Pool; Army Ground Pool; motor torpedo boat tender (naval symbol)

AGPA American Group Psychotherapy Association

AGPC Adjutant General Publications Center

agpe angle plate

agpi automatic ground position indicator

AGPL *Administração-Geral do Porto de Lisboa* (Portuguese—Port of Lisbon Authority)

ag prov agent provocateur

AGPS Australian Government Publishing Service

agr agree(ment); agricultural; agriculture

agr (AGR) advanced gas-cooled graphite-moderated reactor

AGRA Australian Garrison Royal Artillery

agrar agrarian; agrarianism; agrarians

a/g ratio albumin-globulin ratio

Agra U Agra University

agrbl agreeable

agrd agreed

AGRE Atlantic Gas Research Exchange

AGREE Advisory Group on Reliability of Electronic Equipment

agrep agricultural research project(s)

AGRF American Geriatric Research Foundation

agri agricultural; agriculturalist; agriculture; agriculturist

agribusiness agricultural business (large-scale farming)

agric agriculture

Agric E Agricultural Engineer

agricrime agricultural crime (theft of crops and/or equipment)

agridollars agricultural (market) dollars

agri-indus agricultural-industrial (complex)

agrimech agriculture mechanized

agripower agricultural power

AGRIS Agricultural Information System

AGRM Adjutant General—Royal Marines

agrmt agreement

agro aggravation; agrobiological; agrobiologist; agrobiology; agrologic; agrological; agronomical; agronomics; agronomist; agronomy

agro *agronomiae* (Dano-Norwegian—agronomy)

agrobio agrobiologic(al)(ly); agrobiologist; agrobiology

agrogeol agrogeology

agroind agroindustrial(ly); agroindustrialization; agroindustrialize(r); agroindustry

agron agronomy

agronome (Russian—agricultural expert)

agros agrostology

ags adrenogenital syndrome; agencies

ags *angelsächsisch* (German—Anglo-Saxon)

ags (Ags) antigens

ags (AGS) alternation gradient synchrotron

Ags Aguascalientes (inhabitants—Hidrocalidos)

AGs attorneys general

AGS Abort Guidance System; Academic Guidance Service; Aircraft General Standards; Alabama Great Southern (railroad); Allied Geographic Section; American Gem Society; American Geographical Society; American Geriatrics Society; American Goat Society; American Gynecological Society; Army General Staff; Army Guard School; Association of Graduate Schools; Australian Geographic Society; Australian Geomechanics Society; surveying ship (naval symbol)

A.G.S. Associate in General Studies

AGSA Art Gallery of South Australia

AGSI Automatic Government Source Inspection

AGSM American Gold Star Mothers; Associate of the Guildhall School of Music; Australian Graduate School of Management

AGSRO Association of Government Supervisors and Radio Officers

AGSS American Geographical and Statistical Society

agst against

agt agent; agreement

agt (AGT) antiglobulin test

AGT Art Gallery of Toronto; Association of Geology Teachers

AGTA Australian Geography Teachers Association

AGTC Airport Ground Traffic Control

AGTE Association of Group Travel Executives

AGTELIS Automated Ground Transportable Emitter Location and Identification System

AGTELS Automated Ground Tactical Emitter Location System

AGTF Alternate Geology Test Facility

agto *agosto* (Portuguese and Spanish—August)

agtt (AGTT) abnormal glucose tolerance test

agtv advanced ground transportation vehicle

AGU American Geophysical

Union
Aguacates Aguacate Mountains, Costa Rica
Aguas Aguascalientes
Agu Cur Agulhas Current
Agunalaksh (Aleut—Unalaska)—shores where the sea breaks on the Aleutian Islands
agv aniline gentian violet
AGV Automatic Guided Vehicle
AGVA American Guild of Variety Artists
AGvga Anglo-German variable-geometry aircraft
agw actual gross weight; allowable gross takeoff weight
AGWA Australian Government Workers Association
AGWAC Australian Guided Weapons and Analog Computer
AGWD *Australian Government Weekly Digest*
AGWI American Gulf and West Indies (steamship line)
agy agency
agz actual ground zero
ah abdominal hysterectomy; acetohexamide; after hatch; alter heading; amenorrhea and hirsutism; aminohippurate; antihalation; antihyaluronidase; arterial hypertension; astigmatism hypermetropic
a-h ampere-hour
a/h at home
a & h accident and health; alive and healthy
Ah ampere-hour; hyperopic astigmatism
AH Airfield Heliport; Alfred Holt's Blue Funnel Line (house flag and funnel mark); Allis Chalmers (stock exchange symbol); Animal Husbandry (division of Department of Agriculture); Army Hospital; hospital ship (naval symbol)
A-H American-Hawaiian Line; Arrow-Hart & Hegeman Electric Company
A & H Arm and Hammer (trade mark)
AH *Akademiya Nauk* (Russian—Academy of Sciences)
A.H. *Anno Hebraico* (Latin—in the Hebrew Year); *Anno Hegirae* (Latin—Year of Hegira)—Moslem
AH-1 Huey Cobra gunship military aircraft carrying machinegun pods on its stub wings, a 7.62mm minigun in its nose plus a grenade

launcher
AH-64 attack helicopter
aha acquired hemolytic anemia; all have automobiles; autoimmune hemolytic anemia
AHA Adirondack Historical Association; American Hardboard Association; American Heart Association; American Hereford Association; American Historical Association; American Hospital Association; American Hotel Association; American Humane Association; American Humanist Association; American Hypnotherapy Association; Area Health Authority; Association of Handicapped Artists; Association for Humane Abortion; Australian Hospital Association; Australian Housewives Association
ahab attacking hardened air bases
AHAM Association of Home Appliance Manufacturers
ahas (AHAS) acetohydroxy acid synthase
AHAUS Amateur Hockey Association of the U.S.
ahc acute haemorrhagic conjunctivitis
AHC Academy of Hospital Counselors; American Hardware Corporation; American Hockey Coaches; American Horticultural Council; American Hospital Corps; Army Hospital Corps; Australian Heritage Commission
ahca (AHCA) American Health Care Association
AHCEI American Histadrut Cultural Exchange Institute
AHCo Assault Helicopter Company (USAF)
ahd ahead; airhead; aired head; arteriosclerotic heart disease; atherosclerotic heart disease; audio high density; auto-immune haemolytic disease
AHD *American Heritage Dictionary*
A-H DT Alaska-Hawaii Daylight Time
ahe acute hemorrhagic encephalomyelitis
AHE Association for Higher Education
A.H.E. Associate in Home Economics
AHEA American Home Economics Association; American Hungarian Educators Association

AHEAD Army Help for Education and Development
AHEL Army Human Engineering Laboratory (USA)
AHEM Association of Hydraulic Equipment Manufacturers
AHEPA American Hellenic Educational Progressive Association
AHES American Humane Education Society
ahf anti-hemophilic factor
Ahf Argentinian hemorrhagic fever
AHF American Health Foundation; American Heritage Foundation; American Hobby Federation; American Hungarian Foundation; Associated Health Foundation
AHF *American Hospital Formulary; Azod Hind Fouj* (Indian National Army)
AHFCR Anderson Hospital for Cancer Research
AHFS American Hospital Formulary Service
ahg antihemolytic globulin; antihuman globulin
ahg (AHG) antihemophilic globulin
AHG American Housing Guild
ahh alpha-hydrazine analog of histidine; arylhydrocarbon hydroxylase
AHHA Allied Home Health Association
AHHS Alexander Hamilton High School
AHI American Health Institute; American Honey Institute; American Hospital Institute; Animal Health Insurance
AHIL Association of Hospital and Institution Libraries
AHIP Australian Health Insurance Program
AHIRS Australian Health Information and Research Service
AHIS American Hull Insurance Syndicate
ahl alcohol-induced hyperlipidemia
a.h.l. *ad hunc locum* (Latin—at this place)
AHL Alaska Historical Library; American Hockey League; Associated Humber Lines; Association for Holistic Living
ahle acute hemorrhagic leukoencephalitis
AHLMA American Home Laundry Manufacturers Association

ahls antihuman-lymphocyte serum
ahm ampere-hour meter
Ahm Ahmadabad; Arnhem
AHM *American Health Magazine*
ahma advanced hypersonic manned aircraft
AHMA American Hardware Manufacturers Association; American Hemisphere Marine Agencies; American Hotel and Motel Association
AHMC Association of Hospital Management Committees
Ahmed Mohammed Ahmed ibn-Seyyid Abdullah—the Mahdi
AHMI Appalachian Hardwood Manufacturers Incorporated
AHMPS Association of Headmistresses of Preparatory Schools
ahm(s) Asiatic homosexual male(s)
AHMS American Home Missionary Society
AHMSA *Altos Hornos de México* (Spanish—Great Ovens of Mexico)—steel mills
AHN Assistant Head Nurse
AHNA Accredited Home Newspapers of America
a-hole ass hole, anus
AHOP Assisted Home Ownership Plan
ahp acute hemorrhagic pancreatitis; air at high pressure; air horsepower; aviation horsepower
AHP American Home Products; Assistant House Physician; Association for Humanistic Psychology
AHPA American Horse Protection Association
AHPC American Heritage Publishing Company
AHPR Academy of Hospital Public Relations
ahps auxiliary hydraulic power supply
AHQ Air Headquarters; Allied Headquarters; Army Headquarters
ahr acceptable hazard rate
AHR Academy of Human Rights; Association for Health Records
AHRC Association for the Help of Retarded Children; Australian Housing Research Council; Australian Humanities Research Council
AHRGB Association of Hotels and Restaurants of Great Britain

ahs ablative heat shield
AHS Aerospace High School; American Harp Society; American Hearing Society; American Helicopter Society; American Hibiscus Society; American Home Security; American Horticultural Society; American Hospital Supply (stock exchange symbol); American Humane Society; American Hypnodontic Society; Assistant House Surgeon; Association for Humanistic Studies; Aviation High School; Aviation Historical Society
AHS *Anno Humanae Salutis* (Latin—the Year of Human Salvation)
AHSA American Hampshire Sheep Association; American Horse Shows Association; Art, Historical, and Scientific Association; Aviation Historical Society of Australia
AHSB Authority Health and Safety Branch
AHSC American Hospital Supply Company
A-H Scale Anti-Hispanic Scale (measuring negative attitudes toward persons of Latin American origin)
AHSCo Assault Helicopter Support Company (USAF)
ahse assembly, handling, and shipping equipment
AHSNZ Aviation Historical Society of New Zealand
ahsr air height-surveillance radar
AHSRC American High-Speed Railway Corporation
AHSS Association of Home Study Schools
AHSSPPE Association of Handicapped Student Service Programs in Postsecondary Education
A-H ST Alaska-Hawaii Standard Time
aht antihyaluronidase titer
AHT Animal Health Technician; Animal Health Trust; Augmented Histamine Test
ahtm(s) Asiatic heterosexual male(s)
AHTN Association of Hospital Television Networks
ahu after hangup
ahv (AHV) alternative fuel vehicle
a.h.v. ad hunc vocem (Latin—at this word)

AHV *Altos Hornos de Vizcaya* (Spanish—Great Ovens of Vizcaya)—steel mills
Ahvenanmaa (Finnish—Ahvenanmaa Islands)—called Åland by the Swedes
AHWA Association of Hospital and Welfare Administrators
AHWG Ad Hoc Working Group (USA)
ai accidentally incurred; achievement via independence; airborne intercept; anti-icing; aortic incompetence; aortic insufficiency; apical impulse; articulation index; artificial insemination; artificial intelligence; axio-incisal; azimuth indicator
a&i accident and indemnity
a & i abstracting and indexing
a. i. ad interim (Latin—in the interim)
AI Aaland Islands; Admiralty Islands; Adult Institutions (New Hampshire); Air India; Air Inspector; Air Installation(s); Airways Inspector; Alianza Interamericana (Inter-American Alliance); Allegheny International; American Institute; Amnesty International; Arctic Institute; Army Intelligence; Aspen Institute (Aspen, Colorado); Astrologers International; Avery Island
AI *Altesse impériale* (French—imperial highness)
A/I Aptitude Index
A & I Afars and Issas, Republic of Djibouti (formerly French Somaliland); agricultural and industrial (college or school or subjects); Arts and Industries
aia advise if able; anti-icing additive
AiA Accuracy in Academia
AIA Aerospace Industries Association; Allergy Information Association; Allred Interaction Analysis; American Institute of Accountants; American Institute of Aeronautics; American Institute of Architects; American Insurance Association; Antivenin Institute of America; Archeological Institute of America; Arctic Institute of America; *Association Internationale d'Allergologie;* Association of Insurance Attorneys; Australian Insurance Association; Australian Italian Association

A.I.A. Associate of the Institute of Actuaries

AIAA Aerospace Industries Association of America; American Industrial Arts Association; American Institute of Aeronautics and Astronautics

AIAC Air Industries Association of Canada

AIAD Acronyms, Initialisms, & Abbreviations Dictionary

AIADA American Imported Automobile Dealers Association

AIAE Association of Institutes of Automobile Engineers

AIAESD American International Association for Economic and Social Development (AIA)

AIAL Associate of the Institute of Arts and Letters

AIAOS Academic Instructor and Allied Officer School

AIAP Ardmore Industrial Air Park

AIArb Associate of the Institute of Arbitrators

AIAS Australian Institute of Aboriginal Studies; Australian Institute of Agriculture and Science

AIAT Attitude-Interest Analysis Test

AIAW Association for Intercollegiate Athletics for Women

aib aminoisobutyric acid

AIB Academy for International Business; Accident Investigative Branch; Accident Investigative Bureau; American Institute of Baking; American Institute of Banking; Anti-Inflation Board; Assassination Information Bureau; Australian Infantry Battalion; Australian Institute of Building

A.I.B. Associate of the Institute of Bankers

AIB Association des Industries de Belgique (Association of Belgian Industries); Associazione Italiana Biblioteche (Italian Library Association)

aiba amino-isobutyric acid

AIBA American Industrial Bankers Association

AIBA Association Internationale de Boxe Amateur (French—International Amateur Boxing Association)

AIBC Architectural Institute of British Columbia

AIBCS American Intersociety Board of Certification of Sanitarians

AIBD Associate of the Institute of British Decorators and Interior Designers

aibf advanced internally blown jet flap

aibm (AIBM) anti-intercontinental ballistic missile

AIBM Association Internationale des Bibliothèques Musicales (French—International Association of Music Libraries)

AIBP Associate of the Institute of British Photographers

AIBS American Institute of Biological Sciences

aic aminoimidazole carboxamide

aic (AIC) aircraft in commission

AIC Accelerator Information Center; Advanced Intelligence Center; Agricultural Institute of Canada; Aircraft Industries Center; Allied Intelligence Center; Allied Intelligence Committee; American Indian Center; American Institute of Chemists; American Institution of Cooperation; Ammunition Identification Code; Arab Information Center; Arab Investment Company; Army Industrial College; Army Intelligence Center; Art Information Center; Art Institute of Chicago; Artificial Illumination Centre; Australian Institute of Cartographers; Australian Institute of Criminology

AICA Australasian Institute of Cost Accountants

AICA Association Internationale des Critiques d'Art (French—International Association of Art Critics); Associazione Italiana per il Calco Automatico (Italian—Italian Association for Automatic Data Processing)

aicar amino-imidazolecarboxamide ribonucleotide

AICB Association Internationale Contre le Bruit (French—International Association Against Noise)

aicbm (AICBM) anti-intercontinental ballistic missile

aicc antibody-induced cell-mediated cytoxicity

AICC All-India Congress Committee

AICCC American Institute of Child Care Centers

AICCU Association of Independent California Colleges and Universities

AICE Agency for Information and Cultural Exchange (formerly the USIA); American Institute of Chemical Engineers; American Institute of Consulting Engineers

AI-CE Atomic International–Combustion Engineering

aicf auto-immune complement fixation

AICF America-Israel Cultural Foundation

aich automatic integrated container handling

AIChE American Institute of Chemical Engineers

AICK Associated Independent Colleges of Kansas

AICM Association of Independent Conservators of Music; Australian Institute of Credit Management

AICMA Association Internationale des Constructeurs de Matériel Aéronautique (French—International Association of Builders of Aeronautical Material)

AICMDM Association of Independent Copy Machine Dealers and Manufacturers

AICO Action Information Control Office(r)—USN; American Insulator Corporation

AICPA American Institute of Certified Public Accountants

AICQ Associazione Italiana per il Controllo della Qualità (Italian—Italian Association for Quality Control)

AICR American Institute for Cancer Research

AICRO Association of Independent Contract Research Organizations

AICS Air Induction Control System; American Institute of Ceylonese Studies; Association of Independent Colleges and Schools

A.I.C.S. Associate of the Institute of Chartered Shipbrokers

AICS Association Internationale du Cinéma Scientifique (French—International Scientific Film Association)

AICTA Associate of the Imperial College of Tropical Agriculture

AICU Association of International Colleges and Universities

AICUM Association of Independent Colleges and Univer-

sities in Massachusetts; Association of Independent Colleges and Universities of Michigan
AICUN Association of Independent Colleges and Universities of Nebraska
AICUO Association of Independent Colleges and Universities of Ohio
aicv armored infantry combat vehicle (AICV)
aid. acute infectious disease; applications interface device; artificial insemination donor; avalanche injection diode
Aid Aideen
AID Agency for International Development; Airline Interline Development; American Institute of Decorators; American Instructors of the Deaf; Arkansas Information Dissemination; Army Information Digest; Army Intelligence Department; Artificial Insemination Donor; Association for International Development
A & ID Acquisition and Improvement District
AID Acronyms and Initialisms Dictionary; Association Internationale des Documentalists et Techniciens de l'Information (French—International Association of Documentalists and Information Technicians)
aida attention-interest-desire-action (marketing formula); automatic instrumented diving assembly; automobile information data advertising
aida (AIDA) automatic intruder-detector alarm
AIDA American Indian Development Association; Associated Independent Dairies of America; Australian Industries Development Association
AIDATS Army In-flight Data Transmission System
AIDB Association of International Bond Dealers
AIDC American Industrial Development Council; Arkansas Industrial Development Commission; Asian Industrial Development Council; Association of Information and Dissemination Centers; Australian Industries Development Corporation
AIDD American Institute of Design and Drafting
aide. airborne insertion display equipment; aircraft installation

diagnostic equipment
AIDE American Institute of Driver Education; Arizona Information Dissemination for Educators
AIDE Association Internationale des Distributions d'Eau (French—International Water Supply Association)
aidecs automatic inspection device for explosive charge shell
AIDI Associazione Italiana per la Documentazione e l'Informazione (Italian—Italian Association for Documentation and Information)
AIDIA Associate of the Industrial Design Institute of Australia
AIDIS Asociación Interamericana de Ingeneria Sanitaria (Spanish—Inter-American Association of Sanitary Engineering)
AIDL Auckland Industrial Development Laboratory; Australian Industrial Development Laboratories
AIDP Advanced Institutional Development Program; Associate of the Institute of Data Processing
AIDP Association Internationale de Droit Pénal (French—International Association of Penal Law)
AIDRB Army Investigational Drug Review Board
aids acute infectious deseases
aids (AIDS) acquired immune deficiency snydrome
AIDS Abstracts Information Dissemination System; Account Identification and Description Services; acquired immune deficiency syndrome; Action Information Display System; Activity Information Data System; Administrative Information Data System; Advanced Integrated Data System; Aerospace Intelligence Data System; Aircraft Intrusion Detection System; Alabama Information Development System; American Institute for Decision Services; Automated Identification Division System; Automatic Inventory Dispatching System
AI & DSC Army Information and Data System Command
AIDT All-Inclusive Deed of Trust
AIDUS Automated Information

Directory Update System; Automated Input and Document Updating System
AIE American Institute of Esthetics; Australian Institute of Export
AIEA Agence Internationale de l'Energie Atomique (French—International Atomic Energy Agency); *Agencia Internacional de Energia Atómica* (Spanish—International Atomic Energy Agency)
AIEAL Australian Institute of Engineering Associates Ltd
AIEC Association of Iron Exporting Countries
AIECF American Indian and Eskimo Cultural Foundation
A.I.Ed. Associate in Industrial Education
AIEE American Institute of Electrical Engineers
AIEF Association Internationale des Etudes Françaises (French—International Association for French Studies)
AIEI Association of Indian Engineering Industry
aiep amount of insulin extracted from the pancreas
AIER American Institute for Economic Research
AIERI Association Internationale des Etudes et Recherches sur l'Information (French—International Association for Mass Communication Research)
AIES Accreditation and Institutional Eligibility Staff
AIESEC Association Internationale des Etudiants en Sciences Economiques et Commerciales (French—International Association of Students in Economics and Commerce)
AIEST Association Internationale d'Experts Scientifiques du Tourisme (French—International Association of Scientific Experts in Tourism)
AIF Air Intelligence Force; American Institute of France; Amphibian Imperial Forces; Army Industrial Fund; Atomic Industrial Forum; Atomic International Forum; Australian Imperial Forces; Australian Institute of Fuel
AIF Agencia Internacional de Fomento (Spanish—International Development Agency); *Agenzia Internazionale Fides* (Italian—International Faith

Agency—Vatican State News Service); *Alliance Internationale des Femmes* (French—Women's International Alliance); *Asociación Interna-cional de Fomento* (Spanish— International Development Association)— IDA

AIFA Associate of the International Faculty of Arts

AIFCS Airborne Interception Fire-Control System

AIFD Alaska Institute for Fisheries Development

AIFE Associate of the Institution of Fire Engineers

AIFLD American Institute for Free Labor Development

AIFM Association Internationale des Femmes Médecins (French—International Association of Women Doctors)

AIFR American Institute of Family Relations

AIFS American Institute for Foreign Study

AIFST Australian Institute of Food Science and Technology

AIFT American Institute for Foreign Trade; Americans for Indian Future and Traditions

AIFTA Anglo-Irish Free Trade Area

aifv (AIFV) armored infantry fighting vehicle

aig all inertial guidance; angle of inner gimbal

Aig Aiguille (French—needle, peak)

AIG Address Indicating Group; Adjutant Inspector General

AIG Association Internationale de Géodesia (French—International Geodesy Association)

AIGA American Institute of Graphic Arts

AIGAM Australian Institute of Graphic Art Management

AIGCM Associate of the Incorporated Guild of Church Musicians

AIGS Agricultural Investment Grant Scheme

AIGT Association for the Improvement of Geometrical Teaching

aih artificial insemination by husband

AIH American Institute of Homeopathy; Aspen Institute of the Humanities; Australian Institute of Horticulture

AIH Association Internationale de l'Hôtellerie (French—International Hotel Association)

aiha autoimmune hemolytic anemias

AIHA American Industrial Hygiene Association

AIHC American Industrial Health Conference

AIHE Association for Innovation in Higher Education

AIHEC American Indian Higher Education Consortium

AIHED American Institute for Human Engineering and Development

AIHR Australian Institute of Human Relations

AIHS American Indian Historical Society; American Irish Historical Society; Aspen Institute of Human Studies; *Association Internationale d'Hydrologie Scientifique* (French—International Association of Scientific Hydrology); Australian Institute of Health Surveyors

AIHSC Auto Industry Highway Safety Committee

AII Air India International; Australian Insurance Institute

AIIA Association of International Insurance Agents; Australian Institute of Incorporated Accountants; Australian Institute of International Affairs

AIIC Army Imagery Intelligence Corps; Associate of the Insurance Institute of Canada

AIID American Institute of Interior Designers

AIIDC Authorized Item Identification Data Collaborator

AIIDR Authorized Item Identification Data Receiver

AIIDS Authorized Item Identification Data Submitter

AIIE American Institute of Industrial Engineers

AIIG American International Group (financial and insurance services)

AIIMS All-India Institute of Medical Sciences

AIInfSc Associate of the Institute of Information Scientists

AIIP Australian Institute of Industrial Psychology

AIIS American Institute of Indian Studies; American Intraocular Implant Society

AIK Assistance-in-Kind (funds)

AIKD American Institute of Kitchen Dealers

ail aileron

Ail Aileen

AIL Aeronautical Instruments Laboratory; Air Intelligence Liaison; Airborne Instruments Laboratory; American Institute of Laundering; American Israeli Lighthouse; Art Institute of Light; Association of International Libraries; Australian Institute of Librarians; Aviation Instrument Laboratory

A.I.L. Associate of the Institute of Linguistics

AILA American Institute of Landscape Architects; Australian Institute of Landscape Architects

A.I.L.A. Associate of the Institute of Landscape Architects

AILA Association Internationale de Linguistique Appliquée (French—International Association of Applied Linguistics)

AILAS Automatic Instrument Landing Approach System

AILC American Indian Law Center

AILC American International Law Cases

Ailie Aileen; Alice; Alicia; Alison; Helen; Helena

AILO Air Intelligence Liaison Office(r)

AILS Advanced Integrated Landing System

AILSA American Indian Law Students Association

ailuroph ailurophile (fancier or lover of cats); ailurophobe (one who fears or hates cats)

aim active inert missile; aerotriangulation (by observation of) independent models; air intercept missile; air-isolated monolithic (circuit)

AiM Anglicans in Mission

A i M Accuracy in Media; Adventures in Movement

AIM Abstracts of Instructional Materials; Academy Introduction Mission (USCG); Accuracy In Media; Aide Inter-Monasteries; American Indian Movement; American Institute of Management; American Institute of Musicology; Army Installation Management; Association for the Integration of Management; Australian Institute of Management; Australian Institute of Metals

AIM Abstracts of Instructional Material; Airman's Informa-

tion Manual
AIM-9 Sidewinder air-to-air missile
AIM-47A Hughes air-to-air missile
aima as interest may appear
AIMA All-India Management Association
AIMACC Air Material Command Compiling (system)
AIMACO Air Materiel Command Compiler (language)
AIM Bankers Asian International Merchant Bankers (Singapore)
AIMBW American Institute of Men's and Boy's Wear
AIMC American Indian Medical Clinic; Association of Interstate Motor Carriers
aime (AIME) average indexed monthly earnings
AIME American Institute of Mechanical Engineers
AIMES Association of Interns and Medical Students
AIMF American International Music Fund; Australian Institute of Metal Finishing
AIMH Academy of International Military History
AIMILO Army/Industrial Material Information Liaison Office(s)
AIMIT Associate of the Institute of Musical Instrument Technicians
AIML All-India Muslim League
AIMLT Australian Institute of Medical Laboratory Technology
AIMM Australasian Institute of Mining and Metallurgy
AIMME American Institute of Mining and Metallurgical Engineers
AIMMPE American Institute of Mining, Metallurgical, and Petroleum Engineers
aimo air mold; audibly-instructed manufacturing operations
AIMO Audibly Instructed Manufacturing Operation
aimp air intercept missile package
AIMPA Association Internationale de Météorologie et de Physique de l'Atmosphère (French—International Association of Meteorology and Atmospheric Physics)
AIMPE Australian Institute of Marine and Power Engineers
AIMS Advanced Intercontinen-

tal Missile System; Air Traffic Control Radar Beacon/Identification Friend or Foe/Mark XII/System; American Institute for Marxist Studies; American Institute for Mathematical Statistics; American Institute for Mental Studies; American Institute of Merchant Shipping; American Institute of Musical Studies; Association of Independent Maryland Schools; Association for International Medical Study; Australian Institute of Marine Science; Australian Institute of Marine Studies; Automated Instructional Materials Services; Automatic Industrial Management System
AIMT Australian Institute of Medical Technologists
A.I.M.T.A. Associate of the Institute of Municipal Treasurers and Accountants
AIMU American Institute of Marine Underwriters
ain acute interstitial nephritis; approved item name
AIN American Institute of Nutrition; Association of Interpretive Naturalists; Australian Institute of Navigation
aina automated immunoephelometric assay
AINA American Indian Nurses Association; American Institute of Nautical Archeology; American Israel Numismatic Association; Arctic Institute of North America
A-Ind Anglo-Indian
AINDT Australian Institute for Non-Destructive Testing
AINEC All-India Newspaper Editors' Conference
AINL Association of Immigration and Naturalization Lawyers
AINS Assateague Island National Seashore (Maryland and Virginia)
AINSE Australian Institute of Nuclear Science and Engineering
ainsuf aortic insufficiency
ain't am not, are not, has not, have not, is not
AINWR Aleutian Islands National Wildlife Refuge (Alaska)
AINZ Advertising Institute of New Zealand
aio activity, interest, opinion (marketing factors); activity-interest-option (marketing fac-

tor scores)
Aio Aioi
AIO Action Information Organization; Air Installation Office; Air Intelligence Organization; Americans for Indian Opportunity; Anglican Information Office; Arecibo Ionospheric Observatory; Artillery Intelligence Officer
AIOB American Institute of Oral Biology
AIOEC Association of Iron Ore Exporting Countries
AIOPI Association of Information Officers in the Pharmaceutical Industry
aip ablative insulative plastic; accident insurance policy; acute intermittent porphyria; average intravascular pressure
aip (AIP) aldosterone-induced protein; automated imagery processing
AIP Aeronautical Information Publication; Aerovias Panama (Panamanian airline); American Independent Party; American Institute of Physics; American Institute of Planners; American Institute for Psychoanalysis; Australian Institute of Petroleum; Australian Institute of Physics
A-I-P Afghanistan-Iran-Pakistan
AIP Association Internationale de Papyrologues (French—International Association of Papyrologists); *Association Internationale de Pediatrie* (French—International Pediatric Association)
AIPA American Indian Press Association
AIPA Association Internationale de la Psychologie Adlerienne (French—International Association of Adlerian Psychology)
AIPAC American-Israeli Public Affairs Committee
AIPC All Indian Pueblo Council; *Association Internationale de Prophylaxie de la Cécité* (French—International Association for the Prevention of Blindness); *Association Internationale des Ponts et Charpentes* (French—International Association of Bridges and Scaffolds); Australian Institute of Pest Control
AIPCEE Association des Industries du Poisson de la Communauté Economique

Européenne (French—Association of Fishing Industries of the European Economic Community)

AIPCN *Association Internationale Permanente des Congrés Navigation* (French—Permanent Sailing International Association of Congresses)

AIPCR *Association Internationale Permanente des Congrés de la Route* (French—Permanent Road International Association of Congresses)

AIPE American Institute of Park Executives; American Institute of Plant Engineers; American Institute for Professional Education

AIPG American Institute of Professional Geologists

AIPHE Associate of the Institution of Public Health Engineers

AIPLU American Institute for Property and Liability Underwriters

AIPO American Institute of Public Opinion; Asian Inter-Parliamentary Organisation

AIPR American Institute of Pacific Relations; Australian Institute of Parks and Recreation

AIPs Allied Intelligence Publications; Association of Irish Priests

AIPS AIDS-induced panic syndrome; Australian Institute of Political Science; Automatic Indexing and Proofreading System

AIPS *Association Internationale pour la Prevention du Suicide* (French—International Association for the Prevention of Suicide)

AIQ Associate of the Institute of Quarrying

AI & Q Animal Inspection and Quarantine

AIQS Australian Institute of Quantity Surveyors

A.I.Q.S. Associate of the Institute of Quantity Surveyors

air. average injection rate

a-i-r artist-in-residence

AIR Action for Industrial Recycling; Air Control Products; All-India Radio; American Institute of Refrigeration; American Institute of Research; Arkansas Intermediate Reformatory; Army Institute of Research; Army Intelligence Reserve; Association for Immigration Reform; Association for Institutional Research; Australian Industrial Refractories; Australian Institute of Radiography

AIR *Amnesty International Report; Asociación Interamericana de Radiodifusión* (Spanish— Interamerican Broadcasters Association)

AIR-2A Douglas air-to-air rocket fitted with a nuclear warhead and called Genie

AIRA Air Attaché

airac (AIRAC) aeronautical information regulation and control

AIRAH Australian Institute of Refrigeration, Air Conditioning, and Heating

AIRB Alabama Inspection and Rating Bureau; Arkansas Inspection and Rating Bureau; Australian Industrial Relations Bureau

AIRBALTAP Allied Air Forces Baltic Approach (NATO)

airbm (AIRBM) anti-intermediate-range ballistic missile

AIRBR *Association Internationale du Registre des Bateaux du Rhin* (French—International Association of Rhine Ship Registry)

AIRCAL Air California

Air Can Air Canada (formerly Trans-Canada Air Lines)

aircat automated integrated radar control for air traffic

Air Cav Airmobile Cavalry

Air Cdr Air Commander

AIRCENT Allied Air Forces, Central Europe

AIRCEY Air Ceylon

Air Cmdre Air Commodore

AIRCO Air Reduction Chemical Company

Air Coal Airport Coalition (*see* AIRPORT COALITION)

AIRCOM Air Force Communication Complex

AIRCOMNET Air Communications Network

AIRCOMS Airways Communications System

aircond air condition(ed); air conditioning

AIRDEF Air Defense (NATO division)

AIRDEP Air Deputy (NATO)

AIREA American Institute of Real Estate Appraisers

AIREASTLANT (Naval) Air Forces Eastern Atlantic (NATO)

AiRepDn Aircraft Repair Division

airew airborne infrared early warning

AIRFA American Indian Religious Freedom Act

airfil air filter(s)

Air Force I Air Force One (aircraft reserved for or used by the President of the U.S.)

AIRH *Association Internationale des Recherches Hydrauliques* (French—International Association of Hydraulic Research)

air hp air horsepower

AIRI Atomic Industry Research Institute

AIRIMP ATC/IATA (*q.q.v.*) reservations interline message procedures

Air Jam Air Jamaica

AIRL Aeronautical Icing Research Laboratory

AIRLEX Air Landing Exercise

Air LO Air Liaison Officer

AIRLORDS Airlines Load Optimization Recording and Display System

Air Mad Air Madagascar

airmada airplane aramada

airmap air monitoring, analysis, and prediction

AIRMIC Association of Insurance and Risk Managers in Industry and Commerce

airmiss aircraft-in-flight collision barely missed

air/mmh acoustic intercept receiver/multimode hydrophone

AIRMOVEX Air Movement Exercise

Air NG Air National Guard

Air Niu Air Niugini (national airline of Papua New Guinea)

AIRNON Allied Air Forces in Northern Norway (NATO)

AirNorth Allied Air Forces, Northern Europe

Air NZ Air New Zealand

AIROPNET Air Operational Network

AIRPAC Air Pacific

AIRPASS Aircraft Interception Radar and Pilots Attack Sight System

airpl airplane(s)

AIRPORT COALITION fifteen San Diego organizations united to control air and noise pollution by relocating Lindbergh Field

airpt art airport art

airr (AIRR) adjusted internal rate of return

Air Res Squad Air Reserve Squadron

airs. (AIRS) advanced inertial reference sphere

AIRS Aircraft Inventory Reporting System; Airline Interline Reservations System; Automatic Image Retrieval System

AIRSONOR Allied Air Forces in Southern Norway (NATO)

AIRSouth Allied Air Forces, Southern Europe

Air-Std Air Force International Standard

airsurance air insurance

Air Svc Air Service

airtel air + hotel, airport hotel

airvan airmobile van

AIRWORK Airwork Atlantic Limited

AIRX American Industrial Radium and X-Ray Society

ais ablating inner surface; agreed industry standard; answer in sentence; average insurance set

ais (AIS or Lunik III) automatic interplanetary station

AIS Abbreviated Injury Scale; Academic Instructors School (USAF); Administrative and Information Services; Advanced Information System; Aeronautical Information Service; Air Intelligence Service; Alexander I Solzhenitsyn; American Indian Scholarships; American Indian Studies; American Interplanetary Society; American Israeli Shipping (Zim Lines); Army Intelligence School; *Association Internationale de la Savonnerie et de la Detergence* (French—International Association of Soaps and Detergents); *Association Internationale de Sociologie* (French—International Sociology Association); Association of Iron and Steel; Australian Information Service; Australian Institute of Sport; Australian Iron and Steel

AI & S Army Intelligence and Security

aisa analytical isoelectrofocusing scanning apparatus

AISA Associate of the Incorporated Secretaries Association; Australian Institute of Systems Analysis

AISA Association Internationale pour la Sécurité Aérienne (French—International Air Security Association)

AISB Artificial Intelligence and Simulation of Behavior

AISC American Institute of Steel Construction; Association of Independent Software Companies; Australian Institute of Steel Construction

AISE Association of Iron and Steel Engineers

AISE Association Internationale des Sciences Économiques (French—International Association of Economic Sciences)

AISI American Iron and Steel Institute

AISJ Association Internationale des Sciences Juridiques (French—International Association of Juridical Sciences)

AISM Association Internationale des Sociétés de Microbiologie (French—International Association of Microbiology Societies)

AISS Association Internationale de la Science du Sol (French—International Soil Science Association)

AIST Australian Institute of Science Technology

ait auto-ignition temperature

AiT Anjuman-i-Tarikh (Historical Society of Afghanistan)

AIT Agency for Instructional Television; American Institute in Taiwan; American Institute of Technology; Army Intelligence Translator; Australian Institute of Travel; Automatic Information Test

AIT Académie Internationale du Tourisme (French—International Academy of Tourism)

AITA Act Inside the Army (antiwar society); Air Industries and Transport Association; Australian Industrial Truck Association

AITC Action Information Training Center; American Institute of Timber Construction

AITC Association Internationale des Traducteurs de Conférence (French—International Association of Conference Translators)

aitd all-inclusive trust deed

AITD Australian Institute of Training and Development

AITI Aero Industries Technical Institute

AITO Association of Independent Tour Operators

AITVS Association of Independent Television Stations

aiu abort interface unit; absolute iodine uptake; advanced instrumentation unit

AIU Action for Interracial Understanding; Aero Insurance Underwriters; American International Underwriters

AIU Alliance Israelite Universelle (French—Universal Israelite Alliance)

AIUM American Institute of Ultrasound in Medicine

AIUS Australian Institute of Urban Studies

AIUSA Amnesty International in the United States of America

aiv accelerated inverse voltage

AIV Association Internationale de Volcanologie (French—International Association of Vulcanology)

AIVAF American-Israeli Vocal Arts Foundation

aiw auroral intrasonic wave

AIW Atlantic Intracoastal Waterway (Cape Cod to Florida Bay)

AIWM American Institute of Weights and Measures

Aix Aix-en-Provence

AIX Advanced Interactive Executive operating system

AIYW Association for International Youth Work

aj ankle jerk; antijamming; apple juice

aj a jini (Czech—and others)

AJ Air Jordan; Alma & Jonquieres (railroad); Americans for Justice; Andrew Jackson (7th U.S. President); Andrew Johnson (17th U.S. President); Associate Justice

A.J. Associate in Journalism

AJ American Jurisprudence; Architects Journal; l'Armée Juife (French—Jewish Army)—anti-Nazi resistance group

AJ-37 Swedish Thunderbolt multimission combat aircraft also called Viggen

AJA American Jail Association; American Jewish Archives; American Judges Association; Australian Journalists Association

A-JA Anglo-Jewish Association

AJA *American Journal of Anatomy*

AJAC *American Journal of Alzheimer's Care*

AJADD *Australian Journal of Alcoholism and Drug Dependence*

AJAG Assistant Judge Advocate General

ajai antijamming anti-interference

AJAs Americans of Japanese Ancestry

AJAS *Australian Journal of Applied Science*

AJASS African Jazz Art Society Studios

Ajax Douglas Nike-Ajax surface-to-air missile; mythological Greek hero of the Trojan Wars and title of a play by Sophocles

AJAZ American Jewish Alternatives to Zionism

AJB Arthur J(ames) Balfour

AJB *Association des Juifs de Belgique* (French—Association of the Jews of Belgium); *Australian Journal of Botany*

AJBC American Junior Bowling Congress

AJBP Association of Jewish Book Publishers

aJc *antes de Jesucristo* (Spanish—before Jesus Christ)

AJC Altus Junior College; American Jewish Committee; American Jewish Congress; Anderson Junior College; Australian Jockey Club

AJCA Association of Juvenile Compact Administrators

AJCC Alternate Joint Communications Center

AJCL Australia-Japan Container Line

AJC-RC American Jewish Committee—Records Center

AJCSA All-Japan Cotton Spinners Association

AJCU Association of Jesuit Colleges and Universities

AJCW Association of Jewish Center Workers

AJDC American Joint Distribution Committee; Asian-Japan Development Corporation

AJE Adult Jewish Education

AJEI Australia-Japan Economic Institute

AJF Asian-Japan Forum; Australia-Japan Foundation

AJHS American Jewish Historical Society; Andrew Jackson High School

AJI American Justice Institute

AJIF Australia-Japan International Finance

AJIL *American Journal of International Law*

AJIS Automated Jail Information System

AJJUST Automated Juvenile Justice System Technique

AJL Association of Jewish Libraries; Association of Junior Leagues

AJLAC American Jewish League Against Communism

AJNHS Andrew Johnson National Historic Site, Greeneville, Tennessee

ajo antijam operator

ajp alarm and jettison panel

AJP Additional-Jobs Programme (New Zealand)

AJPA American Jewish Press Association

AJR Association of Jewish Refugees

AJR *Australian Jurist Reports*

AJRC American Junior Red Cross

AJRJ Association of Japanese Residing in Japan

AJS American Judicature Society; Australia-Japan Society

AJS *American Journal of Sociology*

AJSJ American Justinian Society of Jurists

AJSR *Australian Journal of Scientific Research*

ajvd abrupt junction varactor doubler

AJY Association for Jewish Youth

AJYB *American Jewish Year Book*

ak above the knee; ass kisser

a k *alter kocker* (Yiddish colloquialism—old man)

Ak Auckland, New Zealand

AK Alaska; Alaska Coastal—Ellis Airlines; Australian Knight (Knight of the Order of Australia); cargo ship (2-letter naval designation)

AK *Avtomat Kalasnikov* (Russian—submachine gun)

AK 47 automatic rifle

aka above-the-knee amputation; also known as

Aka Akasaka (Tokyo district)

AKA American Kiteflyers Association; Associated Klans of America; Australian Karate Association; Australian Kidney Association; cargo vessel, attack (3-letter coding)

Akad *Akademie* (German—

Academy); *Akademi* (Dano-Norwegian—Academy)

Akad Nauk Akademiya Nauk (USSR Academy of Sciences)

AKAG Albright-Knox Art Gallery

ak amp above-the-knee amputation

Akan Akan National Park on Hokkaido Island, Japan

AKBS Advanced Kinematic Bombing System

AKC American Kennel Club; Associate King's College

AKCF Arthur Kill Correctional Facility

AKF Australian Koala Foundation

AKG *Astronomischen Gesellschaft Katalog* (German—Astronomical Society Catalog)

AKI Alfred Kinsey Institute; American Kynol Incorporated

akk *akkusativ* (Dano-Norwegian—accusative)

Akk Akkadian

AKL *Algemene Kunstzijde Unie* (Artist's Union); Auckland, New Zealand (airport)

AKLD Auckland

AKM Soviet military weapon capable of firing 600 rounds per minute

AKN King Salmon, Alaska (airport)

Akr Akron

Akr *Akra* (modern Greek—cape); *Akrotírion* (Modern Greek—Cape)

AKR vehicle cargo ship (naval symbol)

Akropolis (Greek—Upper City)—Acropolis Hill section of Athens

AKS general stores issue ship (3-letter symbol)

Akt *Aktiebolag* (Swedish—limited company)

AKT Akita Television (Japan)

Akt Ges *Aktiengesellschaft* (German—corporation or joint stock company)

Aktieb *Aktiebolag* (Swedish—limited company)

Akties *Aktieselskab* (Swedish—joint stock company)

Akust Zh *Akusticheskii Zhurnal* (Russian—Acoustics Journal)

AKV cargo ships and aircraft ferries (3-letter symbol)

akwic author and key word in context

al absolute limen; accommodation ladder; air lock; albumin; alcohol; alias; all lengths; annual leave; autograph letter; axiolingual

al alumnos (Spanish—pupils, students)

a l après livraison (French—after delivery)

aL assumed latitude

Al accommodation ladder; air lock; Alan; Albert; Albin; alcohol; Alden; Alex(ander); Alf; Alfred; alias; Allan; Allen; Alley; Allied; all lengths; Alton; aluminum; Alva; Alvah; Alvin; Alvina; Alyn; annual leave; autograph letter

Al. Book of Alma

AL Abraham Lincoln; Accession List(s); Acoustics Laboratory; Aeronautical Laboratory; Air Liaison; Aircraft Laboratory; Aircraft Logistics; Alabama; Allegheny Airlines; Aluminum Limited; aluminum (machine shop symbol); *América Latina* (Portuguese or Spanish—Latin America); American League; American Legion; *Angkatan Laut* (Indonesian—Naval Forces); Anglo-Latin; Annual Lease; Annual Leave; Arab League; Architectural League; Assumed Latitude; Astronomical League; Aviation Electronicsman

A-L Allegheny-Ludlum; Anglo-Latin

A/L airlift

A.L. Anno Lucis (Latin—in the Year of Light)

AL-60 Lockheed Associates Conestoga—six-seat piston-powered transport aircraft

ala alanine; alternate living arrangement; axiolabial

ala (ALA) alighting area

Ala Alabama; Alabamian; Alameda(n)

ALA Alternate Living for the Aged; Amalgamated Lithographers of America; American Landrace Association; American Landscape Architects; American Laryngological Association; American Latvian Association; American Legion Auxiliary; American Liberal Association; American Library Association; American Livestock Association; American Lung Association; Arkansas Library Association; Army Launching Area; Assembly of the Librarians of the Americas; Associa-

tion of Legal Administration; Australian Lebanese Association; Australian Legal Aid; Authors League of America

ALA (I) Amalgamated Lithographers of America (Independent)

A.L.A. Associate in Liberal Arts; Associate of the (British) Library Association

ALAA Associate of the Library Association of Australia

alaar air-launched air-recoverable rocket

A-lab acid (LSD) laboratory

alabaster calcite (onyx marble); variety of gypsum

ALABEL American Library Association Board of Education for Librarianship

alabol algorithmic and business-oriented language

ALAC Alcoholic Liquor Advisory Council (New Zealand); Artificial Limb and Appliance Centre; Australian Labor Advisory Council

ALACP American League to Abolish Capital Punishment

alad abnormal left axis deviation; aminolevulinic acid dehydrase; automatic liquid agent detector

aladdin atmospheric layer and density distribution of ions and neutrons

ALADI Asociación Latino-americano de Integración (Spanish—Latin American Integration Association)

ALAEA Australian Licensed Aircraft Engineers Association

alag axiolabiogingival

Al Ahr Al Ahram (Arabic—The Pyramids)—Cairo's daily paper

alairs advance low-altitude infrared-reconnaissance sensor

ALA-ISAD American Library Association—Information Science and Automation Division

alal axiolabiolingual

ALAL Association of Legal Aid Lawyers

ALALC Asociación Latino-americana de Libre Comercio (Spanish—Latin American Free Trade Association)

ALAM Associate of the London Academy of Music

Alan Alain; Allan; Allen

Åland (Swedish—Aland Islands)—between the Gulf of

Bothnia and the Baltic Sea separating Sweden from Finland

Alanders Aland islanders

Alands Aland Islands

alanon alcoholics' anonymous

ALAO Australian Legal Aid Office

alara as low as reasonably achievable

alarm. automatic light aircraft readiness monitor

ALARM Assessment of Language and Reading Maturity

Alas. Alaska; Alaskan

ALAS Army Library Automated Systems; Automated Literature Alerting System

A.L.A.S. Associate in Letters, Arts, and Sciences

Alas Cur Alaska Current

Alas DST Alaskan Daylight Saving Time

Alasia Australasia

Al-Ass Al-Assifa (Syrian terrorist group)

Alas ST Alaskan Standard Time (150th meridian west of Greenwich)

Alastair Alexander

alateen program for teenagers affected by someone else's drinking (*see* prealateen)

a la v a la vista (Spanish—at sight, payable upon presentation)

alb albanisch (German—Albanian); albumin; aluminum-bronze

alb (ALB) air-land battle

Alb Albania; Albanian; Albany; Albert; Alberta; Albertan; Albion; Albalasserdam

ALB American League Baseball

ALBA Aluminium Bahrain; American Lawn Bowling Association; American Leather Belting Association

Albac Albacete

Alban Albania; Albanian

Alban Albanensis (Latin—of St. Albans)

Albania People's Republic of Albania (smallest of the Balkan nations, once called Illyria by the Romans), *Republika Popullore Socialiste e Shqipipërïse*

Albanian Ports (north to south) Shengjin, Durres, San Nicolo, Vlore

albany adjustment of large blocks with *any* number of photos, points, or images, using *any* photogrammetric-measuring instrument on *any* computer

Albatross Grumman amphibian transport aircraft; name of a series of oceanographic survey ships flying the American flag; Piaggo P-166M coastal patrol aircraft built in Italy

ALBE Air League of the British Empire

Alber Alberic; Albern; Albert(ino)

Albert Albert Canal connecting Antwerp and Liege; Albert National Park in Zaire; Albert Nyasa (Albert Lake, Africa's third largest); Halbert; Halbertus

Alberta Girls Alberta Institution for Girls

albi air-launched booster intercept

Albion Correctional Albion State Institution and Western Correctional Facility, Albion, NY

ALBIS Australian Library-Based Information System

albm (**ALBM**) air-launched ballistic missile

Alb Mus Albany Museum, Grahamstown, South Africa

Albn Albanian

Albq Albuquerque

Albr Albrecht

ALBS Anti-Lock Braking System

Albt Albert

Albturist Albanian Tourism

Albuq Albuquerque

albus all bureaus (naval coding)

alc alcohol; approximate lethal concentration; avian leukosis complex; axiolinguocervical

a l c a la carte (French—on the menu)

ALC Accredited Land Consultant; Air Logistics Center (USAF); Alabama Central (railroad); American Life Convention; American Lutheran Church; Area Logistics Center; Area Logistics Command; Armament Logistics Center; Armament Logistics Command; Asian Law Collective; Associated Lutheran Charities

ALCA American Leather Chemists Association; Associated Landscape Contractors of America

ALCAC Airlines Communications Administrative Council

AlCan Alaska-Canada

ALCAN Aluminium Company of Canada

Alcanzamos Alcanzamos por fin la Victoria (Spanish—Finally we attained victory), Panama's national anthem

alcapp automatic list classification and profile production

ALCC Airborne Launch-Control Center

ALCC Asociación de Libre Comercio del Caribe (Spanish—Caribbean Free Trade Association)

Alc^de Alcalde (Spanish—justice of the peace, mayor)

alch approach-light contact height

alchem alchemy

alcid alcohol + acid

alcism alcoholism (addiction to alcohol)

ALCL Association of London Chief Librarians

alcm (**ALCM**) air-launched cruise missile

ALCM Associate of the London College of Music

ALCMs Air-Launched Cruise Missiles

ALCO Aluminum Company of Colorado; American Lava Corporation

ALCOA Aluminum Company of America

alcoh alcohol

alcohol ethyl alcohol (C_2H_5OH)

alcolic alcoholic

alcom algebraic compiler; algebraic computer

alcon all concerned

ALCOP Alternate Command Post

alcotrician alcohol + nutrition (adverse effects of alcohol on good nutrition)

Alcott Amos Bronson Alcott or his daughter Louisa May

alcr aluminum crown (dental)

ALCS Airborne Launch-Control System; Authors' Lending and Copyright Society

ALCTP Academic Library Consultant Training Program

ald a later date; acceptable limit for dispersion; aldolase

Ald Aldabra; Alderman; Aldermanic

ALDA Air Line Dispatchers Association; American Land Development Association; Australian Land Development Association

ALDCS Active Lift Distribution Control Center

aldehyde al(cohol) dehy(drogenated)—dehydrogenated

(oxidized) alcohol

aldep automated layout design program

Alder Alderic; Alderley

Alderson minimum-security Federal Reformatory for Women at Alderson, West Virginia

ALDEV African Land Development

ALDHA Appalachian Long-Distance Hikers Association

Aldm Alderman

aldo aldosterone

Aldo Teobaldo; Teobaldo Manuzio, 16th-century Venetian printer and typographer

aldp automatic language-data processing

ALDS Apollo Launch-Data System

Ale Alemania (Spanish—Germany)

ALE Association for Liberal Education

ALEA Airline Employees Association

alec algebraic components and coefficients

ALEC American Legislative Exchange Council; American Lutheran Evangelical Churches

Alec(k) Alexander

ALECS Automated Law-Enforcement Communications System

ALECSO Arab League Educational, Cultural, and Scientific Organization

alegar ale + vinegar (vinegar derived from ale)

Ale^jo Alejandro

ALEOA American Law Enforcement Officers Association

Alep (Turkish—Aleppo)—Syrian city

Ale RD República Democrática Alemana (Spanish—German Democratic Republic)—East Germany

Ale RF República Federal Alemana (Spanish—Federal German Republic)—Germany

alerfa alert phase

ALERT Affiliated League of Emergency Radio Teams; Automatic Linguistic Extraction and Retrieval Technique

ALERT II Automatic Law Enforcement Response Time

ale(s) additional living expense(s)

Ales Alessandro

ALESCO American Library

and Educational Service Company
Aleut Aleutian; Aleutian Islands
Aleut Cur Aleutian Current
Aleutians Aleutian Mountains; Aleutian islanders; Aleutian Islands
Aleut Is Aleutian Islands
A-levels advanced levels (of educational tests)
alex alexandrine (verse)
alex (ALEX) alert exercise
Alex Alexander; Alexandra; Alexandria
Alexa Alexandra
Alexanders Alexander Archipelago; Alexander cocktails; Alexander islanders; Alexander Islands of southeastern Alaska
Alex City Alexander City, Alabama
alf automatic letter facer
Alf Alfa; Alfhild; Alfonso; Alford; Alfred; Alfred(o)
ALF American Life Federation; American Life Foundation; Animal Liberation Front; Arab Liberation Front; Association of Libertarian Feminists; Australasian Labour Federation; Australian Labor Federation
Alfa letter A radio code
ALFA Anonima Lombarda Fabbrica Automobili
ALFC Aboriginal Land Fund Commission (Australia)
ALFCE Allied Land Forces in Central Europe (NATO)
Alfie Alfred
Alfo Alfonso
ALFORD Appalachian Laboratory for Occupational Respiratory Diseases
ALFS U.S. Navy's Airborne Low-Frequency Sonar program
ALFSEA Allied Land Forces—South-East Asia
ALFSH Allied Land Forces in Schleswig-Holstein (NATO)
alft airlift
alg algae; algal; algebra; algebraic; allergic; allergical; allergy; along; alongside; antilymphocyte globulin (ALG)
alg *algemeen* (Dutch—generally or universally); algerisch (German—Algerian)
Alg Algeria; Algiers
ALG Air Algérie; Algiers, Algeria (airport)
ALGC Association of Local

Government Clerks
Alge Algeciras
Algeb algebra
Alger Algernon
Algeria Democratic and Popular Republic of Algeria (North African Arab nation), *El Djemhouria El Djazairia Demokratia Echaabia* (Arabic name); *République Algérienne Démocratique et Populaire* (French name)
algett *allegretto* (Italian—brisk or jolly)
Algie Algernon
algins algae derivatives
alglyn aluminum glycinate
algol algebraically oriented language (algorithmic international language)
Algonquin Algonquin Peak in the Adirondacks; Algonquin Provincial Park, Ontario
algtto *allegretto* (Italian—brisk or jolly)
ALGU Association of Land Grant Colleges and Universities
ALGWA Australian Local Government Women's Association
Algy Algernon
alh anterior lobe hormone; anterior lobe of the hypophysis
Alh Alhambra
ALH Australian Light Horse
ALHS Abraham Lincoln High School
Alht Apollo lunar hand tool
Alhtc Apollo lunar hand tool carrier
ali. *alibi* (Latin—elsewhere)
'ali (Arabic—high)
Ali Alicante
ALI American Law Institute; American Library Institute
ALIA Royal Jordanian Airlines
Alianza *La Alianza Federal de las Mercedes* (Spanish—Federal Alliance of Mercedes)—New Mexican organization founded by Reies Lopez Tijerina to reclaim Mexican land acquired by the United States
Alic Alicante
ALIC Association of Life Insurance Counsel
alice (ALICE) automatic laundering instrument control equipment
Alice The Alice—Alice Springs, Northern Territory, Australia; Allis-Chalmers Manufacturing Company (stock exchange slang)

ALICE Artillery Line Communications Equipment
Alick Alexander
ALICS Advanced Logistics Information and Control System (USAF)
alien alienist
'Alifax (Cockney—Halifax)
align. alignment
alim (ALIM) air-launched interceptor missile
ALIMD Association of Life Insurance Medical Directors
ALIMDA Association of Life Insurance Medical Directors of America
Al Imp Reps Alert Implementation Reports
Aline Adeline
alirt (ALIRT) adaptive long-range infrared tracker
ALIS Advanced Life Information System; Automated Library Information Service; Automated Library Information System
alit automatic line insulation tester
Alitalia Italian national airlines (AZ)
ALITALIA Italian International Airline
A.Litt. Associate in Letters
ALJ Administrative Law Judge
ALJ *Australian Law Journal; Australian Library Journal*
aljak aluminum-jacketed coaxial cable
ALJC Alice Lloyd Junior College
ALJH Association of Libraries of Judaica and Hebraica (in Europe)
ALJR *Australian Law Journal Reports*
al-Jum al-Iraq *al-Jumhuriyah al Iraqiyah* (Arabic—Republic of Iraq)
al-Jum al-Jaz ad-Dim ash-Shab *al-Jumhuriya al-Jazairiya ad-Dimuqratiya ash-Shabiya* (Arabic—Democratic and Popular Republic of Algeria)
alk alkali
Alkali Soviet air-to-air radar-guided homing missile (NATO)
Alkan Valentin Alkan, Charles-Valentin Morhang
alki alcohol; homeless alcoholic
alkie(s) alcoholic(s)
alk phos alkaline phosphatase
alkums air-launched cruise missiles (ALCMs)
all. above lower limit; acute

lymphocytic leukemia; allergy

al.l. *alia lectio* (Latin—a different reading)

All Alley; Alloa; Aloha

Al-L Alsace-Lorraine

ALL Admiralty Lines Limited; Airborne Laser Laboratory; American Life Lobby against abortion

ALL Admiralty List of Lights

All 8va all'ottava (Italian—in the octave)

all'8va all'ottava (Italian—an octave higher)

ALLA Allied Long Lines Agency (NATO)

Allagash Allagash River and Allagash Wilderness Waterway, Maine

all. **(ALL)** airborne laser laboratory

allcat all critical atmospheric turbulence (programs)

alld allowed

Alld Allahabad

alleg allegation; allegoric; allegorical; allegory

Alleghenies Allegheny Mountains of Pennsylvania, Maryland, Virginia, and West Virginia

Allem Allemagne (French—Germany)

allergol allergologic(al)

allg allgemein (German—general)

allgem allgemein (German—general)

All H All Hallows (Halloween)

all hands all hands on deck (everyone)

Allie Alice; Alison

Alligators Alligator Rivers of Australia's Northern Territory (East, South, and West Alligator)

all'ingr all'ingrosso (Italian—wholesale)

Allison Allison Division, General Motors

allit alliteration; alliterative

ALLNAVSTAS All Naval Stations

allo allonym

allo (Greek prefix—other)—allele, allopathic, allopatric, allophone, alloplasm, allotetrapoid

Allo allegro (Italian—lively, quickly)

alloc allocate; allocation

allop allophone

all'ott all'ottava (Italian—an octave higher)

allow. allowance

allp audiolingual language pro-

gramming

All Quiet All Quiet on the Western Front (Lewis Milestone film, 1930)

All S All Souls College, Oxford

ALLS Apollo Lunar Logistic Support

All Saint's All Saint's Day (November 1)

allstat all-purpose statistical (package)

Alltto allegretto (Italian—lively but less so than *allegro*)

allu allude; allusion; allusively

allus allusion

alluv alluvial; alluvium

Ally Pally Alexandra Palace in North London

alm. alarm

alm almindelig (Dano-Norwegian—common, frequent, plain, simple)

Alm Almería

Alm Almirante (Spanish—Admiral); (German—mountain pasture)

ALM American Leprosy Missions

A & LM Arkansas & Louisiana-Missouri (railroad)

ALM Antilliaanse Luchtvaart Maatschappij (Dutch—Antillean Airline Company)

A.L.M. *Artium Liberalium Magister* (Latin—Master of Liberal Arts)

ALMA Aircraft Locknut Manufacturers Association; Association of Literary Magazines of America

ALMAC Association of Labor-Management Administrators and Consultants

ALMACA Association of Labor-Management-Administrators and Consultants on Alcoholism

ALMAJCOM All Major Commands

al-Mam al-Urd al-Hash al-Urd al-Mamlakah al-Urdunniyah al Urdun (Arabic—Hashemite Kingdom of Jordan)

ALMC Army Logistic Management Center

alme acetyl-lysine methyl ester

Almer Almeric

almi anterior lateral myocardial infarct

ALMIDS Army Logistics Management Integrated Data System

ALMs Amindivi, Laccadive, and Minicoy Islands off India's Malabar Coast

ALMS Analytic Language Manipulation System

ALMT Association of London Tailors

Almte Almirante (Spanish—Admiral)

aln anterior lymph node

aln **(ALN)** accounting line number

alnico aluminum, nickel, copper (magnet alloy containing iron and cobalt)

alnmt alignment

ALNP Abraham Lincoln National Park

ALNZ Air League of New Zealand

alo axiolinguoclusal

alo' alow

Alo Alonso

ALO Accreditation Liaison Officer; Agency Liaison; Air Liaison Office(r); Allied Liaison Office(r); Aloha Airlines; Amalgamated Lace Operatives; American Liaison Office(r); Army Liaison Office(r)

ALOA Amalgamated Lace Operatives of America; Amalgamated Lithographers of America; Assembly of Librarians of the Americas; Associated Locksmiths of America

aloc air lines of communication; allocation

ALOC Air Line of Communication

ALOE A Lady Of England—Charlotte Maria Tucker

alof' aloft

aloft. airborne light optical fiber technology

ALOHA Aboriginal Lands of Hawaiian Ancestry; Aloha Airlines

ALON Air Liaison Officer Net

alo'-'n'-alof' alow and aloft (everywhere aboard ship—in the lower rigging and in the upper rigging)

ALOO Albuquerque Operations Office

alor advanced lunar orbital rendezvous

alot allotment

aloteen alcoholic teenagers (rehabilitation program)

alotm allotment

ALOTS Airborne Lightweight Optical Tracking System

Alouette Aerospatiale armed helicopter made in 4-passenger and 6-passenger versions

Aloys Aloysius

alp. anterior lobe (of) pituitary;

assembly language program (data processing); autocode list processing; automated language processing
Alp Alphen; Alpine
ALP Air Liaison Party; Allied Liaison and Protocol; Ambulance Loading Post; American Labor Party; Antigua Labour Party; Australian Labor Party; Australian Liberal Party; Automated Learning Process; Automated Library Program
ALP Agence Lao Press (French—Lao Press Agency)
ALPA Air Line Pilot's Association
ALPAC Automatic Language Processing Advisory Committee (National Research Council)
alpak algebra package
ALPB American Lutheran Publicity Bureau
ALPC Army Logistics Policy Council; Australian Library Promotion Council
ALPCA Auto License Plate Collectors Association
Alpen (German—Alps)—Richard Strauss's *Alpine Symphony*
ALPGA Australian Liquefied Petroleum Gas Association
Alph Alphonse
alpha alphabetical
Alpha letter A radio code
ALPHA Action League for Physically Handicapped Advancement
alphameric alphanumeric and alphabetic-numeric
alphametic alphabet arithmetic
alphanumeric alphabetical-numerical
alpha order alphabetical order
ALPHAS Automatic Literature Processing, Handling, and Analysis System
ALPL Advanced Lunar Projects Laboratory
alpo (ALPO) Apollo lunar polar orbiter
ALPO Allen Products; Amalgamation of Left Political Organizations; Association of Lunar and Planetary Observers
ALPOWAD Alaska Power Administration
Alps Alpine Mountains of south central Europe
ALPS Accidental Launch Protection System; Advanced Linear Programming System; Automated Library Process-

ing Services; Automatic Landing Positioning System
ALPSP Association of Learned and Professional Society Publishers
Alpujarras Alpujarras Mountains of Almería and Granada in Spain
ALPURCOMS All-Purpose Communications System
ALQAS Aircraft-Landing Quality-Assessment Scheme
alr. aliter (Latin—otherwise)
ALR Australian League of Rights
ALR American Law Reports
ALRA Abortion Law Reform Association; Agricultural Labor Relations Act
ALRANZ Abortion Law Reform Association of New Zealand
ALRB Agriculture Labor Relations Board; Agriculture Labor Relations Bureau
ALRC Anti-Locust Research Center; Australian Law Reform Commission
alri airborne long-range input
ALRI Angkatan Laut Republik Indonesia (Indonesian Navy)
ALRM Aboriginal Legal Rights Movement (Australia)
ALROS American Laryngological, Rhinological, and Otological Society
a l r p de V M a los reales pies de Vuestra Majestad (Spanish—at the royal feet of Your Majesty)
ALRS Admiralty List of Radio Signals
ALRTF Army Long-Range Technological Forecast
als amyotrophic lateral sclerosis; antilymphocytic serum; autograph letter signed
als (ALS) amyotrophic lateral sclerosis (Lou Gehrig's disease); automatic line supervision
ALS Aboriginal Legal Service; Advanced Laser System; Advanced Launch System; Alton & Southern (railroad); American Littoral Society; anti-lymphocyte stimulator; Approach Light System; Area Licensing Scheme (Singapore); Australian Literary Society; Australian Literary Studies
A.L.S. Associate of the Linnean Society
ALSA American Land Sailing Association; American Law Student Association

ALSAA Americans of Lebanese-Syrian Ancestry for America
alsam (ALSAM) air-launched surface-attack missile
Alsat Alsatian
ALSC American Lumber Standards Committee; Association for Library Service to Children
ALSCP Appalachian Land Stabilization and Conservation Program
alse aviation life-support equipment
Al seg al segno (Italian—return to the sign)
alsep (ALSEP) apollo lunar surface experiments package
Alsk Alaska(n)
Also Also Sprach Zarathustra (German—Thus Spake Zarathustra)—symphonic poem by Richard Strauss
ALSO Alex Lindsay String Orchestra (New Zealand)
alsor air-launch sounding rocket
alss airline system simulator
ALSS Advanced Location Strike System; Airborne Location and Strike System; Apollo Logistics Support System
ALST Alaska Standard Time
alt academic learning time; alter(ation); altered; altering; alternative; alternator; altimeter; altitude
alt altitud; altura (Spanish—altitude, height)
Alt alternating (light)
Alt Altesse (French—Highness)
ALT Aboriginal Lands Trust; Aer Lingus (Irish Air Lines); alteration
ALT Australian Law Times
Alta Alberta
ALTA Agricultural Landlords and Tenants Act; American Land Title Association; American Library Trustee Association; Association of Local Transport Airlines
altac algebraic translator and compiler
altair (ALTAIR) ARPA (*q.v.*) long-range tracking and instrumentation radar
Altais Altai Mountains
altan alternate alerting network
altare automatic logic testing and recording equipment
Altay high mountains rising above northern edge of Gobi Desert in Central Asian por-

tion of Russia
alt-ch alternate-channel
ALTCOMIND Alternate Commander, Indian (USN)
ALTCOMLANT Alternate Commander, Atlantic (USN)
ALTCOMPAC Alternate Commander, Pacific (USN)
altd altered
alt. dieb. *alternis diebus* (Latin—alternate days)
alter. alteration; alternate
Alt F Fl alternating fixed and flashing (light)
Alt F Gp Fl alternating fixed and group flashing (light)
Alt Fl alternating flashing (light)
Alt Got Alternate Gothic
Alt Gp Occ alternating group occulting (light)
Alt Gr Fl alternating group flashing (light)
altho although
alt. hor. *alternis horis* (Latin—at alternate hours)
altm altimeter
alt. noc. *alternis noctibus* (Latin—on alternate nights)
altnr alternator
Alt Occ alternating occulting (light)
ALTPR Association of London Theatre Press Representatives
altran algebraic translator
altrec automatic life testing and recording of electronic components
altru altruism; altruist; altruistic
ALTS Advanced Lunar Transportation System; Airborne Laser Tracker System
alt set. altimeter setting
ALTUC All-India Trade Union Congress
alt udk *alt udkomne* (Dano-Norwegian—all published)
alt xyl alto xylophone
alu (ALU) arithmetic and logic unit
Alucon Aluminium Conductors (New Zealand)
alue admissible linear unbiased estimator
alum. alumna; alumnae; alumni; alumnus; hydrated potassium aluminum sulfate
alv alveolar
alv (ALV) avian leukemia virus(es)
älv (Swedish—river)
ALV Annual Lease Value
alv. adstrict. *alvo adstricto* (Latin—bowels being constipated)
ALVAO Association des Lang-

ues *Vivantes pour l'Afrique Occidentale* (French—West African Modern Languages Association)
alv. deject. *alvi dejectiones* (Latin—intestinal discharges)
Alver Alvern(on)
Alv⁰ Alvaro
alvx alveolectomy
alw allowance; arch-loop whorl
Alweg Axel Lennert Wenner-Gren (Swedish industrialist's name applied to monorailroad systems)
alwin algorithmic wiswesser notation
ALWL Army Limited War Laboratory
alwt advanced lightweight torpedo
Alx Alexandria
aly alloy
Aly Alley
Alyce Girls Alyce D. McPherson School for (delinquent) Girls at Ocala, Florida
ALYESKA Alaska Pipeline Service
alz *alzamento* (Italian—heaving, lifting, raising)
am. aircooled motor; ammeter; amplitude modulation; auditory (sequential) memory
am. (AM) air-locked module
a.m. *ante meridiem* (Latin—before noon)
a/m auto/manual
a & m agricultural and mechanical; ancient and modern; architectural and mechanical; archy and mehitabel
am *amerikansk* (Dano-Norwegian—American)
Am Amazonas; America; American; americium; myopic astigmatism (symbol)
Am. *Amós* (Spanish—The Book of Amos)
AM Academy of Management; Aeronaves de México (Mexican Airlines); Air Marshal; Air Medal; Air Ministry; Alexander Mackenzie (Canada's second Prime Minister); *Almacenes Maritimos; Alpes Maritimes* (Maritime Alps); amplitude modulation; angular momentum; Arthur Meighen (Canada's tenth and twelfth Prime Minister); Australian Member (Member of the Order of Australia); Aviation Medicine; Aviation Structural Mechanic; large minesweeper (naval symbol); metric angle (symbol)

A-M Addressograph-Multigraph; Alpes-Maritimes
A.M. Air Mail
A/M Aviation Medicine
A & M Agricultural and Mechanical; Agricultural and Mechanical College of Texas; Ancient and Modern (hymns)
A of M Academy of Music
AM *Almanaque Mundial* (Spanish—World Almanac)
A.M. *artium magister* (Latin—Master of Arts); *Ave Maria* (Latin—Hail Mary)
a/m¹ amperes per square meter
AM-3C Aeritalia-Aermacchi single-engine three-place armed-trainer aircraft
ama actual mechanical advantage; against medical advice
amᵃ *amiga* (Spanish—female friend)
AMA Academy of Model Aeronautics; Acoustical Materials Association; Aerospace Medical Association; Agricultural Marketing Administration; Air Materiel Area; Aircraft Manufacturers Association; Amarillo, Texas (airport); Amateur Trapshooting Association; Ambulance Manufacturers Association; American Machinery Association; American Management Association; American Maritime Association; American Marketing Association; American Matthay Association; American Medical Association; American Ministerial Association; American Monument Association; American Motel Association; American Motorcycle Association; American Municipal Association; Arena Managers Association; Association of Metropolitan Authorities; Australian Medical Association; Australian Meteorological Association; Automobile Manufacturers Association
A & MA Advertising and Marketing Association
AMAA Adhesives Manufacturers Association of America; Army Mutual Aid Association; Association of Medical Advertising Agencies
AMAB Army Medical Advisory Board
Am Acad Pol Soc Sci American Academy of Political and Social Science
Am Acad Rel American Acad-

emy of Religion
AMACO American Medical Assurance Company
AMACUS Automated Microfilm Aperture Card Updating System
amad aircraft-mounted accessory drive
Amad Amadeus
AMA-DE American Medical Association—Drug Evaluation(s)
Amad Quart Amadeus Quartet
AMAE American Museum of Atomic Energy; Association of Mexican-American Educators
AMAERF American Medical Association Education and Research Foundation
amal amalgam; amalgamate; amalgamation
AMAL Aero-Medical Acceleration Laboratory; American Medical Acceleration Laboratory
amalg amalgamated
amalgam mercury and silver mixture
a M (a/M) am Main (German—on the Main River)
Aman Agaf Modiin (Hebrew—Military Information Bureau) —Israel
amap advanced multiprogramming analysis
AMAR Annual Major Additions Rate
AMARC Army Materiel Acquisition Review Committee
AMARS Air Mobile Aircraft Refueling System; Automatic Message Address Routing System
AMAS American Military Assistance Staff; Automatic Message Accounting System
Am Assn Blood American Association of Blood Banks
Am Assn Coll Pharm American Association of Colleges of Pharmacy
Am Assn Comm Jr Coll American Association of Community and Junior Colleges
amat amateur
AMATC Air Materiel Armament Test Center
amatol ammonia & toluene (explosive)
A-matter advance matter (written in advance of a newspaper story)
AMATYC American Math-

ematical Association of Two-Year Colleges
AMAUS Aero Medical Association of the United States
AMAWA American Medical Association Women's Auxiliary
AMAX American Metal Climax
amb amber; ambient; ambulance
Amb Ambassador
Amb Amberes (Spanish—Antwerp)
AMB Airways Modernization Board; *Associação Médica Brasileira* (Portuguese—Brazilian Medical Association); Australian Meat Board
AMB Asociación Mundial de Boxeo (Spanish—World Boxing Association); *Association Maritime Belge* (French—Belgian Maritime Association); *amba andelsselskab med begraenset ansvar* (Dano-Norwegian—limited liability cooperative)
AMBAC American Bosch Arma Corporation; American Municipal Bond Assurance Corporation
Am Bankr Reps American Bankruptcy Reports
Am Baptist American Baptist Historical Society
AMBBA Associated Master Barbers and Beauticians of America
Amb Brdg Ambassador Bridge (Detroit—**Am Acad Pol Soc Sci**—Windsor)
Amb Col Ambassador College
ambel ambiguity eliminator
Amber Amberes (Spanish—Antwerp)
Amb Ex Ambassador Extraordinary
Amb Ex/Plen Ambassador Extraordinary and Plenipotentiary
ambi (Latin prefix—both)—ambidextrous
Am Bibl American Bibliographic Center—Clio Press
ambidex ambidextrous
ambig ambiguity; ambiguous
ambisex ambisextrous (bisexual)
ambish ambition
ambit algebraic manipulation by identity translation
ambiv ambivalence; ambivalent
Am Bk American Book Company

Am Bk Prices American Book Prices Current
ambl ambulatory
amblads advise method, bill of lading, and date shipped
Amb Lib Ambrosian Library (Milan)
Ambo Ambrose
Ambon Amboina, Indonesia
Am Booksellers American Booksellers Association
Amboys New Jersey's Perth Amboy and South Amboy
Am Brass Quin American Brass Quintet
Ambridge American Bridge (company)
AMBRL Army Medical Biomechanical Research Laboratory
ambros ambrosia
Ambrosian Ambrosian Library (Milan)
ambt ambulant
ambu ambulance
ambul ambulacral; ambulacrum; ambulance; ambulant; ambulate(d); ambulating; ambulation; ambulatorily; ambulatory
Amburgo (Italian—Hamburg)
amc arthrogryposis multiplex congentia (AMC); automatic mixture control; axiomesiodistal
amc (AMC) armed merchant cruiser
AMc coastal minesweeper (3-letter naval symbol)
AMC Aerospace Manufacturers Council; Air Mail Center; Air Materiel Command; Aircraft Manufacturers Council; Albany Medical Center; Albany Medical College; Alternate Media Center; American Maritime Cases; American Mining Congress; American Mission to the Chinese; American Motors Corporation; American Music Center; Animal Medical Center; Appalachian Mountain Club; Army Materiel Command; Army Medical Center; Army Medical Corps; Army Missile Command; Army Mobility Command; Army Munitions Command; Association of Management Consultants; automatic message counting
AMCA Air Moving and Conditioning Association; American Medical College Association; American Mosquito Control Association; Austra-

lian Management Consultants Association

AMCALMSA Army Materiel Command Automated Logistics Management Systems Agency

Am Camping American Camping Association

Am Can American Can

AMCAS American Medical College Application Service

AMC-ASC Air Materiel Command–Aeronautical Systems Center

AMCAWS Advanced Medium-Caliber Aircraft Weapon System

amcbh auxiliary machine casing bulkhead

AMC & BW Amalgamated Meat Cutters and Butcher Workmen

AMCD American Medical Center at Denver

AMCEA Advertising Media Credit Executives Association

AMCFSA Army Materiel Command Field Safety Agency

Am Chem American Chemical Society

AMCI&SA Army Materiel Command Installations and Service Agency

amcl amended clearance

AMCL African Metals Corporation Limited; Association of Metropolitan Chief Librarians

AMCLDC Army Materiel Command Logistic Data Center

AMCLSSA Army Materiel Command Logistics Systems Support Agency

amcm airborne mine countermeasures

AMCM Air Materiel Command Manual; Army Materiel Command Memorandum

AMCMFO Air Materiel Command Missile Field Office

AMCO American Manufacturing Company

AMCOA AiResearch Manufacturing Company of Arizona

AMCOM American Stock Exchange Communications

Am Con American Consul(ate)

AMCOS Aldermaston Mechanized Cataloguing and Ordering System; Australian Mechanical Copyright Owners Society

AMCPI Army Materiel Command Procurement Instruction(s)

AMCR Air Materiel Command Regulation(s); Army Materiel Command Regulation(s)

AMCRD Air Materiel Command Research and Development; Army Materiel Command Research and Development

AMCS Airborne Missile Control System; Association of Military Colleges and Schools

AMCSA Army Materiel Command Support Activity

AMCSOF Army Combat Surveillance Office

AMCST Associate of the Manchester College of Science and Technology

AMCTB Associated Motor Carriers Tariff Bureau

AMCU Australian Malaria Control Unit

am. cur. amicus curiae (Latin—a friend at court)

amd air movement designator; alpha-methyldopa; axiomesiodistal

AMD Accident Model Document; Aero-Mechanics Department (USN); Aerospace Medical Division; Air Movement Data; Army Medical Department; Atomic and Molecular Data

AMD Aerospace Material Document

AMDA Advanced for Mutual Defense Assistance; Advances for Mutual Defense Assistance; Airlines Medical Directors Association

AMDB Agricultural Machinery Development Board

AMDEA Associated Manufacturers of Domestic Electrical Appliances

Am Dec American Decisions

AMDEC Associated Manufacturers of Domestic Electric Cookers

Amdel Australian Mineral Development Laboratories

Am Dent American Dental Association

a.m. D.g. ad majorem Dei gloriam (Latin—to the greater glory of God)—also A.M.D.G.

AMDI Associazione Medici Dentisti Italiani (Italian—Association of Italian Medical Dentists)

Am Dig American Digest of Public International Law Cases

AmdlEvac aeromedical evacuation

Amdoc American Doctors (organization)

Am Doc Inst American Documentation Institute

AMDS Advanced Missions Docking Subsystem (NASA); Association of Military Dental Surgeons; Automatic Message Distribution System

amdsbsc amplitude-modulation double-sideband suppressed carrier

amdt amendment

ame angle-measuring equipment; automatic microfiche editor

AME Admiralty Mining Establishment (UK); Aero-Medical Evacuation; African Methodist Episcopal; Aviation Medical Examiners

A.M.E. Advanced Master of Education

AMEB Australian Music Examinations Board

amec aft master-events controller

AMEC Airframe Manufacturing Equipment Committee; Australian Minerals and Energy Council

amecd antimechanized

amech account mechanical (failure or malfunction)

ameda automatic-microscope electronic-data accumulator

AMedD Army Medical Department

AMEDDPAS Army Medical Department Property Accounting System

AMedP Army Medical Publication(s)

AMedS Army Medical Service

AMEE Admiralty Marine Engineering Establishment

AMEG Association for Measurement and Evaluation in Guidance

AMEIC Associate Member of the Engineering Institute of Canada

AMEL Aero Medical Equipment Laboratory

amelior amelioration

Am Elsevier American Elsevier Publishing Company

AMEM African Methodist Episcopal Mission

Am Emb American Ambassador; American Embassy

AMEME Association of Mining, Electrical, and Mechanical Engineers

AMEMIC Association of Mill and Elevator Mutual Insurance Companies

amend. amendment(s)

Am Engr *American Engineer*

Amenia Girls Amenia Center for (delinquent) Girls at Amenia, New York

amens amenities

amer *amerikansk* (Dano-Norwegian—American)

amér *américain* (French—American)

Amer America; American

AMERADC Army Mobility Equipment Research and Development Center

Amerasians American-Asians (offspring of Americans and Asians)

Amer-Eng American-English

Amer F American French

AMERICAL Americans in New Caledonia (Army division)

American Grove *New Grove Dictionary of American Music*

American Virgins U.S. Virgin Islands

Americans United Americans United for Separation of Church and State

Americas Western Hemisphere; including North, Central, and South America

America's Tropical Islands Florida Keys; Guam; Hawaiian Islands; Padre Islands, Texas; Puerto Rico; U.S. Virgin Islands

Americo-Libs Americo-Liberians (Liberian descendants of American Blacks)

Ameridish American Yiddish

Amerind American & Indian (American Indian or Eskimo)

Amer Ind American Indian

Amerindians American Indians

Ameringlish American English

Ameritech American Information Technologies

Amer Men Sci *American Men of Science*

AmerSp American Spanish

Amer Spec *American Spectator*

Amer Std American Standard

Amer St Papers *American State Papers*

Amer Trauma Soc American Trauma Society

AMeS American Meteorological Society

AMES Association of Marine Engineering Schools

Ameslan American sign language

AMETA Army Management Engineering Training Agency

AMETS Artillery Meteorological System

AMEWA Associated Manufacturers of Electric Wiring Accessories

Amex American Stock Exchange

AMEX American Express

AMEX *Agencia Mexicana de Noticias* (Mexican News Agency)

Amexco American Express Company

AMEZ African Methodist Episcopal Zion

AMEZC African Methodist Episcopal Zionist Church

amf (AMF) airmail facility

AMF AIDS Medical Foundation; Air Material Force; Air Mobile Force; American Machine and Foundry; Arab Monetary Fund; Arctic Marine Freighters; Australian Marine Force; Australian Military Forces

AMF(A) Allied Mobile Force (Air)—NATO

AmFAR American Foundation for AIDS Research

Am Feed American Feed Manufacturers Association

AMFGC Association of Midwest Fish and Game Commissioners

AMFIC Automatic Microfilm Information System

AMFIE Association of Mutual Fire Insurance Engineers

AMFIS American Microfilm Information Society; Automatic Microfilm Information System

AMF(L) Allied Mobile Force (Land)—NATO

am/fm amplitude modulation/frequency modulation

Am Friends American Friends Service Committee

amg automatic magnetic guidance; axiomesiogingival

AMG Aircraft Machine Gunner; Albertus Magnus Guild; Allied Military Government; Australian Map Grid

Am Geol American Geological Institute

Am Geophysical American Geophysical Union

AMGNY Associated Musicians of Greater New York

AMGOLD Anglo-American

Gold Investment Trust

AMGOT Allied Military Government

AMGS Acceleration Monitoring Guidance System

Am Guidance American Guidance Service

amh *amharisch* (German—Amharic); astigmatism with myopia predominating; automated medical history

Amh Amharic

AMH Australian Military Hospital

AMHA American Motor Hotel Association

AMHCI Associate Member of the Hotel and Catering Institute

Am Heart American Heart Association

Am Heritage American Heritage Publishing Co

AMHIS American Marine Hull Insurance Syndicate

Am Hist Res American History Research Associates

Am Home Prod American Home Products

AMHS Alaska Marine Highway Authority; American Material Handling Society; Australian Maritime Historical Society

AMHT Automated Multiphasic Health Testing

ami acute myocardial infarction; advanced manned interceptor; air mileage indicator; amitriptyline; axiomesioincisal

AMI Advanced Manned Interceptor; American Management Institute; American Marine Institutes; American Meat Institute; American Medical International, Inc.; American Military Institute; American Museum of Immigration; American Mushroom Institute; American Music Institute; Association of Medical Illustrators; Association for Multi-Image; Australian Marketing Institute; Australian Motor Industries

AMI *Aeronautica Militare Italiana* (Italian Air Force); *Association Montessori Internationale* (French—International Montessori Association)

AMIA American Metal Importers Association; American Mutual Insurance Alliance

AMIADB Army Member—Inter-American Defense Board

AMIAE Associate Member of

the Institute of Automobile Engineers

AMIAMA Associate Member of the Incorporated Advertising Managers Association

AMIAP Associate Member of the Institution of Analysts & Programmers

AMIC Aerospace Materials Information Center; Air Movement Information Center (NATO); Australian Mining Industry Council

AMICA Automobile Mutual Insurance Company of America

AMICE Associate Member of the Institution of Civil Engineers

AMICI Association Mondiale des Interprètes de Conférences International (French— World Association of International Conference Interpreters)

AMICO American Measuring Instrument Company

AMICOM Army Missile Command

AMIDS Advanced Multispectral Image Descriptor System; Area Manpower Instructional Development System

amigo ants, mice, gophers (electromagnetic device affecting the neurological system of such pests)

AMIGOS Americans Interested In Giving Others a Start

AMII Association of Musical Instrument Industries

AMILO Army-Industry Materiel Information Liaison Office

amilpri amilprilose hydrocholoride (rheumatoid arthritis drug)

AMIN Advertising and Marketing International Network

AMINA Association Mondiale des Inventeurs (French— World Association of Inventors)

Am Ind American Indian

Am Indus Arts American Industrial Arts Association

AMINOIL American Independent Oil (company)

aminos amino acids (protein building blocks)

Am Inst American Institute

Am Inst Disc American Institute of Discussion

AMINTAPHIL American Section, International Association for Philosophy of Law and Social Philosophy

AM International Addressograph-Multigraph International

AMIO Arab Military Industrialization Organization

AMIOP Associate Member of the Institute of Printing

AMIPA Associate Member of the Institute of Practitioners in Advertising

Amirantes Amirante Islands

AMIRS Alternative Mortgage Instruments Study

AMIS Aircraft Movement Information Section; Automated Mask Inspection System

Amistad Amistad National Recreation Area near Del Rio, Texas

AMJ Assemblée Mondiale de la Jeunesse (French)—World Assembly of Youth)

Am Jour Sci American Journal of Science

Am J Phys American Journal of Physics

Am J Psy American Journal of Psychiatry

aml acute monocytic leukemia; acute myelocytic leukemia; acute myoblastic leukemia

aml (AML) adjustable-mortgage loan; amplitude-modulated link

Aml Amlwch

Am L American Lawyer

AML Aberdeen Marine Laboratory; Admiralty Materials Laboratory; Aeromedical Laboratory; American Mail Line; Applied Mathematics Laboratory

AML-60 French four-wheeled armored car with 7.5mm machineguns and a 60mm mortar

AML-90 French four-wheeled armored car with 7.5mm machineguns and a 90mm mortar

AMLC Aerospace Medical Laboratory (USAF)

Am Lib Dir American Library Directory

Am Librarians American Librarians' Agency

Am-Lib(s) Americo-Liberian(s)

amls antimouse lymphocyte serum

AMLS American Medico-Legal Society; Master of Arts in Library Science

amm agnogenic myeloid metaplasia; ammonia; ammunition; anti-missile missile (AMM)

AMM Air Mining Mission; Amman, Jordan (airport); Anti-Missile Missile; Associated Millinery Men; *Association Medicale Mondiale* (French—World Medical Association); Aviation Machinist's Mate

AM & M Applied Mathematics and Mechanics

AMMA Adult Movies and Magazines Association; American Museum of Marine Archeology; Assistant Masters and Mistresses Association

Am Mach American Machinist

Am Malacologists American Malacologists

Am Management American Management Association

Am Map American Map Company

Am Math Soc American Mathematical Society

AMMC Aviation Materiel Management Center

Am Media American Media

Am Metal Mkt American Metal Market/Metalworking News

Am Meteorite American Meteorite Laboratory (Denver)

ammeter amperemeter (current-measuring instrument)

AMMI American Merchant Marine Institute

AMMINET Automated Mortgage Management Information Network

AMMIS Aircraft Maintenance Manpower Information System (USAF)

amml acute myelomonocytic leukemia

AMMLA American Merchant Marine Library Association

ammo ammunition

Ammo American Motors

ammobr ammunition bearer

ammon ammonia

Ammon Ammonite

ammonia water ammonium hydroxide (NH$_4$OH)

Am Motors American Motors

AMMPE American Mining, Metallurgical, and Petroleum Engineers

ammrpv advanced multimission remotely-piloted vehicle

Am Mus Mag American Museum of Magic

amn airman

amnes amnesia(c)(al)(ly)

AMNH American Museum of Natural History

amnip adaptive man-machine nonarithmetic information

processing
AmnM Airman's Medal
AMN & PA Australian Monthly Newspapers and Periodicals Association
amnswp acoustic mine-sweeping
AMNZIE Associate Member of the New Zealand Institution of Engineers
amo (AMO) air mail only; alternant molecular orbit
amo amigo (Spanish—male friend); axiomesio-occlusal
AMO Accredited Management Organization; Advance Material Order; Aircraft Material Officer; Air Ministry Order; American Medical Optics; American Motors (stock exchange symbol); Area Medical Officer
amob automatic meterological oceanographic buoy
AMOCO American Oil Company
amol acute monocytic leukemia
Amon Carter Amon Carter Museum of Western Art
AMOP Association of Mail Order Publishers
amor amorphous
AMORC Ancient Mystic Order Rosae Crusis (Rosicrucian Order)
amorph amorphous
amort. amortizable, amortization, amortize(d), amortizement, amortizing
amos antireflection-coated metal-oxide semiconductor
AMOS Acoustic, Meteorological, and Oceanographic Survey; American Monument and Outdoor Sculpture Database; Associated Migrant Opportunity Services; Automatic Meterological Observation Station
amp acid mucopolysaccharide; adenosine monophosphate (hormonal chemical); amperage; ampere; amphetamine; ampicillin; amplification; amplifier; amplitude; ampule; amputation; average mean pressure
amp (AMP) automatic multi-pattern (camera meter)
AMP Air Mail Pioneers; American Museum of Photography; Army Mine Planter; Association of Media Producers; Aurora Memorial Park (Philippines); Automated Mathematics Program; Aviation Modern-

ization Program
AMPA American Medical Publishers Association; Associate of the Master Photographers Association; Australian Magazine Publishers Association
AMPAC American Medical Political Action Committee
AMPAS Academy of Motion Picture Arts and Sciences
AMPC Automatic Message Processing Center
AMPCO Associated Missile Products Corporation; Association of Major Power Consumers of Ontario
Am Peace American Peace Society
Ampersand Ampersand Press (Princeton)
Ampersand NYC Ampersand Press (New York City)
AMPFTA American Military Precision Flying Teams Association
amph amphibian; amphibious; amphimict; amphoric
Amph Amphibia
AMPH Association of Management in Public Health
amphet amphetamine (stimulant)
amphetamine alphamethylpenethylamine
amphets amphetamines
amphi (Greek prefix—both, both sides of, two)—amphibian, amphibolite, amphibological
amphib amphibia(n); amphibious
amphibex amphibious exercise
amphig amphigoric; amphigorical; amphigorist; amphigory
Am Philatelic American Philatelic Society
Am Philos Soc American Philosophical Society
Amphoto American Photographic Book Publishing Co
amp hr ampere hour
AMPI Associated Milk Producers, Incorporated; Associated Music Publishers, Incorporated
Ampico American Piano Company
ampl a macroprogramming language; amplifier; amplitude
ampl ampliata (Italian—enlarged); *amplus* (Latin—large)
AMPOL American Petroleum
ampp advanced microprogrammable processor
ampr advanced multipurpose radar; automatic manifold pressure regulator

AMPR Aeronautical Manufactures Planning Report; Airframe Manufacturers Planning Report; Area Manpower Planning Report; Area Manpower Planning Review
amps amperes; ampules; atmospheric, magnetospheric, and plasmas in space
AMPS Accrued Military Pay System; American Metered Postage Society; Army Mine Planter Service; Army Motion Picture Service; Associated Music Publishers; Association for Media Psychology; Automatic Message Processing System
AMPSS Advanced Manned Precision Strike System
AMPT alpha methyl paratyrosine
AMPTC Arab Maritime Petroleum Transport Company
AMPTEs Active Magnetosphere Particle Explorers
AMPTP Association of Motion Picture and Television Producers
amp-turns ampere-turns
Am Public Health American Public Health Association
ampul. ampulla (Latin—ampule)
ampus (s) amputee (s)
AMQ American Medical Qualification
AMQUA American Quaternary Association
amr (AMR) automatic message routing
AMR Abnormal Mission Routine; Advanced Material Request; Airman Military Record; American Airlines (stock exchange symbol); Atlantic Missile Range; Auxiliary Machinery Room (USN)
AMR Applied Mechanics Reviews
A.M.R. Master of Arts in Research
AMRA American Medical Records Association; Army Materials Research Agency; Australian Model Railways Association
AMRAAM Advanced Medium-Range Air-to-Air Missile
AMRAC Anti-Missile Research Advisory Council
Am Radio American Radio Relay League
AMRC Advanced Metals Research Corporation; Army Mathematics Research Center; Automotive Market Research

Council
AMRCA American Miniature Racing Car Association
AMRCUS Alternative Marriage and Relationship Council of the United States
AMR & DL Air Mobility Research and Development Laboratory (USA)
Am Record American Record Collectors Exchange
Am Red American Red Cross
Am Res American Research Council
AMREX American Real Estate Exchange
AM & RF African Medical and Research Foundation
AMRINA Associate Member of the Royal Institution of Naval Architects
AMRIP Avionics Module Repair Improvement Program
Amrit Amritsar
AMRL Aerospace Medical Research Laboratories; Army Medical Research Laboratory
AMRNL Army Medical Research and Nutrition Laboratory
AMRO Amsterdam-Rotterdam (bank); Association of Medical Record Officers
amrpd applied manufacturing research and process development
AMRS Air Ministry Radio Station; American Moral Reform Society; Australian Media Research Services
ams aggravated in military service; auditory memory span; automated multiphasic screening; automatic music scan
Ams Amsterdam
AMs auxiliary motor minesweeper
AMS Acute Mountain Sickness; Administration and Management Services; Administration Management Society; Advanced Marketing Services; Aeronautical Material Specification; Agricultural Marketing Service; American Management Systems; American Marketing Service; American Mathematical Society; American Meteor Society; American Meteorological Society; American Microscopical Society; American Mineral Spirits; American Montessori Society; American Museum of Safety; American Musicological Society; Am-

sterdam, Netherlands international airport; Anglican Men's Society; Army Map Service; Army Medical Service; Association of Messenger Services; Association of Museum Stores; Australial Museum, Sydney; Australian Medical Services; Aviation Marketing Services (Australia)
AM & S Australian Mining and Smelting
AMS Acta Medica Scandinavica
amsa **(AMSA)** advanced manned strategic aircraft
AMSA American Metal Stamping Association; American Museum of Social Anthropology; Association of Metropolitan Sewerage Agencies; Australian Marine Sciences Association; Australian Medical Students Association
AMSACP Advanced Multistage Axialflow Compressor Program (NASA)
amsam anti-missile surface-to-air-missile
Am Sam American Samoa
AMSANZ Aviation Medicine Society of Australia and New Zealand
amsat amateur satellite
AMSAT Amateur Satellite
AMSC Army Medical Specialist Corps
Am Sch Athens American School of Classical Studies at Athens
Am School American Scholar
Am Sci & Eng American Science and Engineering, Inc
AMSCO American Mineral Spirits Company; American Sterilizer Company
AMSE Associate Member of the Society of Engineers
amsef anti-mine-sweeping explosive float
AMSGA Association of Manufacturers and Suppliers for the Graphic Arts
AMSH Association for Moral and Social Hygiene
AMSIR Agriculture, Marine, Scientific, and Industrial Research Ministry
amsl above mean sea level
AMSMH Association of Medical Superintendents of Mental Hospitals
AMSO Air Member for Supply and Organisation (RAF)
AMSOC American Miscellaneous Society

Am Soc Afr Cult American Society of African Culture
Am Soc HRAC Eng American Society of Heating, Refrigerating, and Air-Conditioning Engineers
Am Society Pr American Society Press
Am Soc Indxrs American Society of Indexers
Am Soc Metals American Society for Metals
Am Soc Not American Society of Notaries
Am Soc Soc American Sociological Society
Am Soc Tool and Mfg Eng American Society of Tool and Manufacturing Engineers
Am Sp American Spanish (Latin American)
AMSP Army Master Study Program
AMSPAR Association of Medical Secretaries, Practice Administrators, and Receptionists
AMSq Avionics Maintenance Squadron (USAF)
ams s autographed manuscript signed
AMSS Advanced Meterological Sounding System; Automatic Music Select System
AMSSEE Area Museum Service for South-Eastern England
AMSSFG Association of Manufacturers of Small Switch and Fuse Gear
a *mss* s autographed manuscripts signed
Amst Amsterdam
AMSTAT News American Statistical Association News
Amst Concert Amsterdam Concerto
Amster Amsterdam(mer)
AMSUS Association of Military Surgeons of the United States
Am Sym Orch American Symphony Orchestra
amt alpha-methyltyrosine; amethopterin; amount; amphetamine
amt **(AMT)** alternative minimum tax
AMT Academy of Medicine, Toronto, Canada; Advanced Manufacturing Technology; Aerial Mail Terminal; Air Mail Transfer; American Medical Technologists; Area Management Team; Astrograph Mean Time
A.M.T. Associate in Mechani-

cal Technology; Associate in Medical Technology; Master of Arts—Teaching

amta airborne moving target attack

AMTA American Massage and Therapy Association

amtank amphibious tank

AMTC Airframe Manufacturing Tooling Committee

A.M.T.C. Art Master's Teaching Certificate

AMTCL Association for Machine Translation and Computational Linguistics

AMTD Automatic Magnetic Tape Dissemination (Service)

AMTDA Agricultural Machinery and Tractor Dealers Association

Am Technical American Technical Society

Am Tech Soc American Technical Society

AMTEG Australian Metal Trades Export Group

Am Tel & Tel American Telephone and Telegraph

amtex air mass-transportation experiment

Am Theatre Assoc American Theatre Association

amti airborne moving target indicator

Amtorg Amerikanskaya Torgovlya (Russian—American Trading Company)

AMTPI Associate Member of the Town Planning Institute (UK)

amtrac amphibious tractor

Amtrak American railroad tracks, the National Railway Passenger Corporation

amtran automatic mathematical translator

amt(s) amphetamine(s)

AMTS Associate Member of the Television Society

amu air mileage unit; air mission unit; astronaut maneuvering unit; atomic mass unit

AMU Alaska Methodist University; American Malacological Union; American Marksmanship Unit; Arab Maghreb Union; Army Marksmanship Unit; Associated Midwestern Universities; Association of Marine Underwriters

AMUA Associate in Music—University of Adelaide

AMUBC Association of Marine Underwriters of British Columbia

Am U Field American Univer-

sities Field Staff

Am Univ Artforms American Universal Artforms

AMURT Anada Marga Universal Relief Team (India)

A.Mus. Associate in Music

A.Mus.A. Associate in Music—Australia

A.Mus.C. Associate in Music—Canada

A.Mus.L.C.M. Associate in Music–London College of Music

A.Mus.N.Z. Associate in Music–New Zealand

A.Mus.S.A. Associate in Music–South Africa

A.Mus.T.C.L. Associate in Music–Trinity College of Music–London

amv alfalfa-mosaic virus; avian myeloblastitis virus

AMV Association Mondiale Vétérinaire (French—World Veterinary Association)

AMVAP Associate Manufacturers of Veterinary and Agricultural Products

AMVER Atlantic Merchant Vessel Report; Automated Mutual Assistance Vessel Rescue (USCG)

AMVERS Automated Merchant Vessel Reporting System

AMVETS American Veterans

AMVM Administrative Motor Vehicle Management

amw actual measurement weight

AMW Antimissile Warfare; Association of Married Women

AMWA American Medical Women's Association; American Medical Writers' Association

AMWC Association of Workers for Maladjusted Children

Am West American West Publishing Company

AMWG Academy of Master Wine Growers

AMWM Association of Manufacturers of Woodworking Machinery

AMWSU Amalgamated Metalworkers and Shipwrights Union (Australia)

AMX-13 French light tank carrying SS-11 antitank guided missiles and a 75mm gun

AMX-30 French medium tank carrying a 105mm gun plus machineguns (antiaircraft and ground)

AMX-105 French self-propelled

105mm howitzer

AMX-155 French self-propelled 155mm howitzer

AMX-VTT French armored personnel carrier (crew of 2 plus 12 troops)

amy amytal (barbituate depressant and sedative)

Amy Amelia; Amoy, China

amyl amyl nitrate

amys amyl nitrate

an. above named; airman; annual; anode

an' and

a/n acidic and neutral

an (Greek—lacking, not, without)—anaerobic, anemic, anonymous, anorexia

an. anno (Latin—year); *ante* (Latin—before)

An Annam; Annamese

A$_n$ normal atmosphere

AN Acid Number; Advance Note; Aerodynamic Note; Air Force-Navy; Airmail Notice; Air Navigation; Air Navigator; Air Reduction (stock exchange symbol); alphanumeric; Anglo-Norman; Apalachicola Northern (railroad); Army-Navy; net laying vessel (naval symbol)

A.N. Associate in Nursing

A & N Army and Navy

AN-2 Soviet Antonov 14-passenger biplane nicknamed Colt by NATO forces

AN-12 Soviet Antonov 100-passenger cargo plane nicknamed Cub by NATO

AN-14 Soviet Antonov 6-seat transport aircraft nicknamed Clod by NATO

AN-14M Soviet Antonov 15-passenger turboprop plane

AN-22 Soviet Antonov 22 (super transport plane)

AN-26 Soviet Antonov 50-passenger transport plane nicknamed Coke by NATO

ana anesthesia; anesthesiac

ana (ANA) antinuclear antibodies

ana (Greek prefix—up or up against)—analogy, analysis, anatomy

Ana Anaheim; Anita; Anna; Annabel(la)

'Aña Agaña, Guam

ANA Administration for Native Americans; Air Force-Navy Aeronautical; All Nippon Airways; American Nature Association; American Neurological Association; American Newspaper Association; Amer-

ican Numismatic Association; American Nurses' Association; Anti-Nicaragua Association; Arab Network of America; Armenian National Army; Army-Navy Aeronautical; *Asociación Nacional Automovilista* (Spanish—National Automobile Association); Association of National Advertisers; Association of Nurse Administrators; Australian National Airways; Australian Natives Association
ANA Automotive News Almanac
ANAAS Australian and New Zealand Association for the Advancement of Science
anab anabasis
anac anachronism; anachronistic
ANACHEM Association of Analytical Chemists
anacol anacoluthon
anacom analog computer
anacreon anacreontic(s); anacreontist
anacru anacrusis
ANADIS Australian National Animal Disease Information System
ANADP Association of North American Directory Publishers
anaesth anaesthesia; anaesthetic(s); anaesthesiologist; anaesthesiology
ANAF Army, Navy, Air Force
anag anagram; anagrammatic(al) (ly); anagramist; anagrams
ANA/HEW Administration for Native Americans—HEW
ANAHL Australian National Animal Health Laboratory
ANAI Asociación Nacional de Administradores de Inmuebles (Spanish—National Association of Real Estate Administrators)
anal analogy; analysis; analytical
Anal Chem Analytical Chemistry
analg analgesic
anal psychol analytical psychology
analyst psychoanalyst
analyt analytical
Anambas Anambas Islands in the South China Sea where Indonesians permitted Vietnamese boat-people to land while awaiting international aid
anap agglutination negative, absorption positive

ANAP Asociación Nacional de los Agricultores Pequeños (Spanish—National Association of Small Farmers)
ANAPO Alianza Nacional Popular (Spanish—Popular National Alliance)—Colombia
ANARC Association of North American Radio Clubs
anarch anarchist; anarchism; anarchy
ANARE Australian National Antarctic Research Expeditions
ANAS Australian National Airlines Commission
anat anatomical; anatomist; anatomy
Anat Anatomy
anath anathema; anathematize
anatran analog translator
anav area navigation
ANB Army-Navy-British Standard
ANB Australian National Bibliography
anbs (ANBS) armed nuclear bombardment satellite
ANB & TC American National Bank and Trust Company
anc all numbers calling; ancient
Anc Ancona
ANC African National Congress; African National Council; Air Force-Navy-Civil; American News Company; Anchorage, Alaska (airport); Arlington National Cemetery; Army and Navy Civil Committee on Aircraft; Army Nurse Corps; Australian Newspaper Council
ANCA Allied Naval Communications Agency; American National Cattlemen's Association
ANCAM Association of Newspaper Classified Advertising
ANCAP Administratión Nacional de Combustibles Alcohol y Portland (Spanish—National Administration of Alcohol and Portland Fuel)
ANCAR Australian National Committee for Antarctic Research
ancc anodal closure contraction
ANCC Australian National Cattlemen's Council
ANCCAC Australian National Committee on Computation and Automatic Control
anch anchorage
Anch Anchorage, Alaska
anchor. alphanumeric character generator

Anchorage Youth McLaughlin Youth Center, Anchorage, Alaska
Anchor Your Age founded in 1915 in Anchorage
ancienn anciennement (French—formerly)
ANCIRS Automated News Clipping, Indexing, and Retrieval System
ANCLD Australian National Committee on Large Dams
Anco Ancohuma
ANCO Andersen-Collingwood (tanker service)
ANCOA Aerial Nurse Corps of America
ANCOM Andean Common Market
ANCON Asociación Nacional para la Conservación de la Naturaleza (Spanish—National Association for the Conservation of Nature)—in Panama
ancova analysis of covariance
ancr aircraft not combat ready
ANCs African National Congress members
ANCS American Numerical Control Society
ANCSA Alaska Native Claims Settlement Act
ANCUN Australian National Committee for the United Nations
ANCW Australian National Council of Women
ANCXF Allied Naval Commander Expeditionary Force
and altnordisch (German—Old North German); *andante* (Italian—of moderate speed)
And Andalucía; Andaman Islands; Andromeda
AND Army-Navy Design
AND Australian News Digest
ANDA Australian National Dance Association
andalusite aluminum silicate
Andamans Andaman islanders; Andaman Islands
Andamans and Nicobars Andaman and Nicobar Islands off Burma in the Bay of Bengal
ANDB Air Navigation Development Board
ANDC Australian National Dairy Committee
Andes Cordillera de los Andes; Los Andes; mountain chain of South America
And I Andaman Islands
Andie Andrew
Andno andantino (Italian—

slower than andante)
Ando Andorra; Andorran
Andorra Valleys of Andorra (Pyrenees principality)
Andover Hawker-Siddeley troop transport designated HS-748 and holding 40 paratroopers or 58 regular soldiers, Phillips Andover Academy
Andr Andromeda
ANDRA Agence National pour la Gestion des Dechets Radioactifs (French—National Radioactive Management Agency)
Andre Andrea
Andreaofs Andreaof Islands
Andrews twenty-dollar bills bearing the portrait of President Andrew Jackson
andro androsterone
andro (Greek prefix—old)—androgen(ic)(al)(ly)
androg androgyn (girlish male)
Andryusha (Russian nickname—Andrei)—Andrew; Andy
Ands Andreas
ANDSA Almacenes Nacionales de Deposito SA (Spanish—National Depository Warehouse Company)
andte anodal duration tetanus
Andte andante (Italian—of moderate speed)
Andy Andrew
andz anodize
ane acoustic noise environment
ANEAC Australian National Energy Advisory Committee
anec anecdotal; anecdote(s)
Aneda (Latin—Edinburgh)
ANEDA Association Nationale d'Etudes pour la Documentation Automatique (French—National Association for Automatic Documentation Studies)
ANEF American Nepal Education Foundation
ANEQ Association Nationale des Estudiants du Québec (French—National Association of Students of Québec)
ANERA American Near East Refugee Aid
anes anesthesia; anesthesiologist; anesthesiology; anesthetician; anesthetic(s)
anesth anesthetic
anesthesiol anesthesiology
an. ex anode excitation
anf anchored filament; antinuclear factor(s)
ANF Advance Notification

Form; American Nurses Foundation; Atlantic Nuclear Force (NATO); Atomic Nuclear Forum; Australian Nursing Federation
anfe (ANFE) aircraft not fully equipped
anfi automatic noise-figure indicator
ANFIA Associazione Nazionale fra le Industrie Automobilistiche (Italian—National Association of Automobile Industries)
anfo ammonium nitrate fuel oil (explosive)
ang angiogram; angle; angular
ang angaende (Danish, Norwegian, Swedish—concerning)
Ang Anchorage; Angel (phonograph records); Anglo-; Angola
ANG Air Force-Navy-Army Guided Missiles; Air National Guard; American Newspaper Guild; Australian National Gallery; Australian New Guinea
ANGAU Australian New Guinea Administrative Unit
Ang Chamb Orch Angelicum Chamber Orchestra
Angela Angelica
Angeles Port Angeles, Washington opposite Victoria, British Columbia
angew angewandte (German—applied, put into practice)
Ångfart Ångfartygas (Swedish—steamship company)
Angie Angela; Angelina; Angeline
angiol angiology
Angkor Angkor Thom (Walled City) or Angkor Vat (Temple City)—Cambodia
Angl Anglican
Angl Angleterre (French—England)
Anglic Anglican; Anglicism
ANGLICO Air and Naval Gunfire Liaison Company
Anglo- combining form meaning England
Anglo (Spanish—an English-speaking person)
Anglo-Afg Anglo-Afghan
Anglo-Afr Anglo-African
Anglo-Amer Anglo-American
Anglo-Ant Anglo-Antarctic(an); Anglo-Antillean
Anglo-Arab Anglo-Arabian
Anglo-Arg Anglo-Argentine
Anglo-Art Anglo-Arctic
Anglo-Aus Anglo-Australian; Anglo-Austrian

Anglo-Bah Anglo-Bahaman
Anglo-Barb Anglo-Barbadian
Anglo-Bas Anglo-Basque
Anglo-Bel Anglo-Belizian
Anglo-Belg Anglo-Belgian
Anglo-Bhu Anglo-Bhutanese
Anglo-Bol Anglo-Bolivian
Anglo-Bots Anglo-Botswana
Anglo-Braz Anglo-Brazilian
Anglo-Bul Anglo-Bulgarian
Anglo-Bur Anglo-Burman; Anglo-Burundian
Anglo-CA Anglo-Central American
Anglo-Cam Anglo-Cameroonian
Anglo-Can(ad) Anglo-Canadian
Anglo-Cat Anglo-Catalan
Anglo-Cath Anglo-Catholic
Anglo-Cey Anglo-Ceylonese
Anglo-Chi Anglo-Chinese
Anglo-Chil Anglo-Chilean
Anglo-Col Anglo-Colombian
Anglo-Cub Anglo-Cuban
Anglo-Cyp Anglo-Cypriot
Anglo-Czech Anglo-Czechoslovak(ian)
Anglo-Dah Anglo-Dahomean
Anglo-Dan Anglo-Danish
Anglo-Du Anglo-Dutch
Anglo-Ecu Anglo-Ecuadorean
Anglo-Egypt Anglo-Egyptian
Anglo-Epis Anglo-Episcopal(ian)
Anglo-Ethio Anglo-Ethiopian
Anglo-Fin Anglo-Finnish
Anglo-Fr Anglo-French
Anglo-Gam Anglo-Gambian
Anglo-Ger Anglo-German
Anglo-Gr Anglo-Greek
Anglo-Guy Anglo-Guyanese
Anglo-Hond Anglo-Honduran
Anglo-Hung Anglo-Hungarian
Anglo-Ice Anglo-Icelandic
Anglo-Ind Anglo-Indian
Anglo-Indo Anglo-Indonesian
Anglo-Ir Anglo-Iranian; Anglo-Iraqi; Anglo-Irish
Anglo-Isr Anglo-Israeli
Anglo-Ital Anglo-Italian
Anglo-Jam Anglo-Jamaican
Anglo-Jap Anglo-Japanese
Anglo-Jew Anglo-Jewish
Anglo-Jor Anglo-Jordanian
Anglo-Ken Anglo-Kenyan
Anglo-Kuw Anglo-Kuwaiti
Anglo-Lat Anglo-Latin
Anglo-Mal Anglo-Malawian; Anglo-Malaysian; Anglo-Maltese
Anglo-Mald Anglo-Maldivian
Anglo-Mex Anglo-Mexican
Anglo-N Anglo-Norse
Anglo-Nep Anglo-Nepalese
Anglo-Nig Anglo-Nigerian

Anglo-Nor Anglo-Norwegian
Anglo-Norm Anglo-Norman
Anglo-NZ Anglo-New Zealand
Anglo-Pak Anglo-Pakistani
Anglo-Para Anglo-Paraguayan
Anglo-Per Anglo-Persian; Anglo-Peruvian
Anglo-Pol Anglo-Polish
Anglo-Port Anglo-Portuguese
Anglo-Rho Anglo-Rhodesian
Anglo-Rom Anglo-Romanian
Anglo-Rus(s) Anglo-Russian
Anglo(s) Anglo-Saxon(s)
Anglo-SA Anglo-South African; Anglo-South American
Anglo-Sam Anglo-Samoan
Anglo-Sax Anglo-Saxon
Anglo-Scot Anglo-Scottish
Anglo-SL Anglo-Sierra Leonean
Anglo-Som Anglo-Somali
Anglo-Sov Anglo-Soviet
Anglo-Span Anglo-Spanish
Anglo-Sud Anglo-Sudanese
Anglo-Swi Anglo-Swiss
Anglo-Tanz Anglo-Tanzanian
Anglo-Tob Anglo-Tobagan
Anglo-Togo Anglo-Togolese
Anglo-Ton Anglo-Tongan
Anglo-Trin Anglo-Trinidadian
Anglo-Turk Anglo-Turkish
Anglo-Ugan Anglo-Ugandan
Anglo-Uru Anglo-Uruguayan
Anglo-Ven Anglo-Venezuelan
Anglo-W Anglo-Welsh
Anglo-Yem Anglo-Yemini
Anglo-Yugo Anglo-Yugoslav-(ian)
Anglo-Zamb Anglo-Zambian
Anglo-Swe Anglo-Swedish
***Angl-Yingl** Anglish-Yinglish* (Yiddish in American life and literature)
Angola People's Republic of Angola
***Ang Pam Akla** Ang Pambansang Aklatan* (Filipino—The National Library)—in Manila
***ang pec** angina pectoris* (Latin—strangling of the chest)—heart attack
Ang-Sax Anglo-Saxon
Ang-Sax Amer Anglo-Saxon America (Canada, United States, Bahamas, Belize, Guyana, Jamaica, Trinidad, West Indian islands, Bermuda)
ANGTS Alaska Natural Gas Transportation System
Angus Aeneas
ANGUS Acoustically-Navigated Geophysical Underwater Survey; Air National Guard of the United States
anh anhydrite; anhydrous
***Anh** Anhang* (German—appendix)

ANH Australian National Highway; Australian National Hotels
ANHA American Nursing Home Association
ANHAL Australian National Humanities and Arts Library
anhed anhedral
ANHF Australian National Heart Foundation
anhic anhydritic
ANHS Adams National Historic Site
anhyd anhydrous
ani automatic number identification
***ani** atmosphère normale internationale* (French—international normal atmosphere)
ANI Army-Navy-Industry; Australian National Industries
A & NI Andaman and Nicobar Islands
***ANI** Agência Nacional de Informaçao* (Portuguese—National Information Agency); *Agencia Nacional de Informaciones* (Spanish—National Information Agency)
ANIB Australian News and Information Bureau
***ANICA** Associazione Nazionale Industrie Cinematografiche e Affini* (Italian—National Association of Cinematographic and Related Industries)
ANICO American National Insurance Company
anil aniline
aniline phenyl amine
anim animal; animate; animism
***anim** animato* (Italian—animated)
ANIM Association of Nuclear Instrument Manufacturers
animad animadversion
ANIP Army-Navy Instrumentation Program
aniv anniversary
***aniv** aniversario* (Spanish—anniversary)
ank ankle
***ank** ankomen* (Dutch—arrival); *ankomst* (Danish—arrival); *ankunft* (German—arrival)
Ank Ankara
ANK Ankara, Turkey (airport)
***Ankerplatz** Ankerplatz der Freude* (German—Anchorage of Joy)—St Pauli's Reeperbahn section of Hamburg
anl anneal; annoyance level (aircraft noise); automatic noise limiter
ANL Argonne National Labora-

tory; Australian National Library; Australian National Line; net-laying ship (naval symbol)
A-N L Anti-Nazi League
ANLCA Alaska Native Land Claims Act
anld annealed
ANLINE Australian National Line
an. lt anchor light
anlys analysis
***anm** anmaerkning* (Danish, Norwegian, Swedish—footnote, note, remark, observation)
***Anm** Anmerkung* (German—footnote, note)
ANM Anacostia Neighborhood Museum; Australian Newsprint Mills
***ANM** Admiralty Notices to Mariners*
ANMC American National Metric Council
ANMCC Alternate National Military Command Center
ANMEF Australian Naval and Military Expeditionary Force
anmi air navigation multiple indicator
ANMI Allied Naval Maneuvering Instructions (NATO)
***ANMPE** Asociación Nacional de Municipalidades del Perú* (Spanish—National Association of Municipalities of Peru)
ANMRC Australian Numerical Meteorology Research Centre
ANMS Automated Notice to Mariners System
ann announce(ment); announcer; annual(ly); annuity; annunciator
***ann** annonce* (Dano-Norwegian—advertisement, announcement)
***ann.** anni* (Latin—years); *anno* (Latin—year)
Ann Anastasia; Angela; Angelina; Angeline; Anita; Anna; Annabelle; Annelida; Annetta; Annette; Annie; Antoinette
***Ann** Annalen* (German—annals); *Annales* (French—annals); *Annali* (Italian—annals)
ANN All Nippon News (network)
Anna Annabella; Annapolis, Maryland; Annette
ANNA Army, Navy, NASA, Air Force
ANNAF Army/Navy/NASA/Air Force
Ann Arbor Pub Ann Arbor Publishers
***Ann Chim Phys** Annales de

Chimie et de Physique (French—Annals of Chemistry and Physics)
ANNECS Automated Nikkei News Editing
Ann Fluid Dyn *Annals of Fluid Dynamics*
Ann Geophys *Annales de Geophysique* (French—Annals of Geophysics)
Ann Inst Henri Poincaré *Annales de l'Institut Henri Poincaré* (French—Annals of the Henri Poincaré Institute)
anniv. *anniversarium* (Latin—anniversary)
Ann Math *Annals of Mathematics*
Annng Annapolis graduate
annot. annotated; annotation
Ann Oto Rhino Laryngol *Annals of Otology, Rhinology, and Laryngology*
Ann Phys *Annalen der Physik* (German—Annals of Physics); *Annales de Physique* (French—Annals of Physics); *Annals of Physics* (New York)
Ann Rept Annual Report
Ann Rev Nucl Sci *Annual Review of Nuclear Science*
annu annual; annuale; annuario
annuit annuitant
Annuit Coeptis *Annuit coeptis novus ordo seculorum* (Latin—He has favored our undertakings—a new secular order of the ages is born) Motto on the reverse of the great seal of the United States
annul. annulment
Annunc Annunciation
ano above-named officer
ANO Air Navigation Office; Anti-Narcotics Office; Area Nursing Officer
anoc anodal opening contraction
ANOC Authorized Notice of Change
anod. anodize
anom anomia; anomiac; anomiacal
Anon. *anonymous* (Latin—nameless)
anop assembly no operation
ANOP Australian Nationwide Opinion Polls
ANOPP Aircraft Noise Prediction Program
anorex anorexia nervosa
anorm aircraft not operationally ready—maintenance
anor nerv anorexia nervosa (Latin) Psychological and endocrine disorder characterized

by a pathological fear of weight gain leading to faulty eating habits, malnutrition, and weight loss
anors aircraft not operationally ready—supplies
ANOS Australian Native Orchid Society
anot annotate
A.N. Other another (person)
anov analysis of variance
anova analysis of variance
anp aircraft nuclear propulsion
A-np A-norprogesterone
ANP Aberdare National Park (Kenya); Acadia National Park (Maine); Adult Nurse Practitioner; Aircraft Nuclear-propulsion Program; Akan National Park (Japan); Albert National Park (Zaire); American Nazi Party; Angkor National Park (Cambodia); Arusha National Park (Tanzania); Associated Negro Press; Australian Nationalist Party; Awash National Park (Ethiopia)
ANP Administración Nacional de Puertos (Spanish—Colombia's National Administration of Ports); *Algemeen Nederlandsch Persbureau* (Dutch—Netherlands Press Bureau)
ANPA American Newspaper Publishers Association; Australian National Publicity Association; Australian Newspaper Proprietors Association
ANPAT American Newspaper Publishers Abstracting Technique
ANPI Associazione Nazionale Partigiani d'Italia (Italian—National Association of Italian Partisans)
ANPO Aircraft Nuclear Propulsion Office
anpod antenna-positioning device
ANPP Aircraft Nuclear Propulsion Program
ANPPF Aircraft Nuclear Power Plant Facility
ANPPIA Associazione Nazionale Perseguitati Politici Italiani Antifascisti (Italian—National Association of Italian Antifascist Political Victims)
ANPRM Advanced Notices of Proposed Rule Making (FAA)
ANPs Allied Navigation Publications
ANPS American Nail Producers Society
anpt aeronautical national taper

pipe threads
ANPWS Australian National Parks and Wildlife Service
ANQUE Asociación Nacional de Quimicos de España (Spanish—National Chemical Association of Spain)
anr another
ANR American Natural Resources (formerly American Natural Gas); American Newspaper Representatives; Antwerp, Belgium (airport); Australian National Railways
ANR Asociación Nacional Republicana (Spanish—National Republican Association)—Paraguay's Colorado Party
ANRA Amistad National Recreation Area (Texas); Arbuckle National Recreation Area (Oklahoma)
anrac aids navigation radio control
ANRAO Australian National Radio Astronomy Observatory
ANRC American National Red Cross; Animal Nutrition Research Council; Australian National Railways Commission; Australian National Research Council
ANRF Australian Nomads Research Foundation
ANRPC Association of Natural Rubber Producing Countries
ANRT Association Nationale de la Recherche Technique (French—National Association of Technical Research)
ans answer; answered; answering; autograph note signed; autonomic nervous system
Ans Anselm; Anselmo
ANS Agencia Noticiosa Saporiti (Argentine press service); American Name Society; American Nuclear Society; American Numismatic Society; American Nutrition Society; Army Newspaper Service; Army News Service; Army Nursing Service; Astronomical Netherlands Satellite (first joint United States-Netherlands satellite)
ansa aminonapthosulfonic acid; automatic new structure alert
ANSA Australian National Sportfishing Association
A(N)SA American (National) Standards Association
ANSA Agenzia Nazionale Stampa Associata (Italian—

National Press Association Agency)
ansam (ANSAM) antimissile surface-to-air missile
ANSC American National Standards Committee
A-N Scale Anti-Negro Scale (measuring negative attitudes)
ANSCO Anthony and Scovill (New York camera and film manufacturer merged with AGFA to become Agfa-Ansco and more recently GAF—General Aniline and Film Corporation)
Ansel Anselm
ANSETT Ansett Airways
ANSETT-ANA Ansett Australian National Airways
ANSI American National Standards Institute; Australian National Standards Institute
ANSIC Aerospace Nuclear Safety Information Center
ANSL Australian National Standards Laboratory
ANSP Academy of Natural Sciences of Philadelphia; Australian National Socialist Party
ANSS American Nature Study Society
ANSSL Australian National Social Sciences Library
ANSSMFE Australian National Society of Soil Mechanics and Foundation Engineering
ANST Appalachian Nature Scenic Trail (Maine to Georgia)
ANSTEL Australian National Scientific and Technological Library
answ (ANSW) antinuclear submarine warfare
ant. antenna(s); anterior; anticipated; antilog; antilogarithm; antiquarian; antique; antiquities; antiquity; antonym
ant anterior (Spanish); *antología* (Spanish—anthology)
ant. antico (Italian—antique); *antiporta* (Italian—half-title)
Ant Antigua; Antillean; Antillea—West Indian Federation; Antilles; Antlia (constellation); Antwerp
ANT American National Theater; AN Tupolev (Soviet aircraft designer's initials and designation of planes he designed); Australian Northern Territory
ANT Australian National Times
ANTA American National Theater and Academy; Australian National Travel Association
antag antagonistic

Antarc Antarctic; Antarctica
Antarc O Antarctic Ocean
ant. ax line anterior axillary line
Ant & Bar Antigua and Barbuda
ANTC Australian National Television Council
Ant & Cl Antony and Cleopatra
Ant Cur Antilles Current
ant. d anterior diameter
ante (Latin prefix—before, in front of); (Latin—before)
antec annual technical conference
ANTELCO Administración Nacional de Telecomunicaciones (Spanish—Paraguayan National Telecommunication Administration)
antennafier antenna + radio-frequency amplifier
antennamitter antenna + transmitter
antennaverter antenna + converter
Antf Antofagasta
Ant f Antillean florin (guilder)
anthol anthological(ly); anthologist, anthologize, anthology
anthro anthropogeography; anthropological; anthropologist; anthropology; anthropometry; anthropomorphism; anthropophagy
anthroco anthrocosis (sickness due to coal-dust inhalation)
anthrop anthropology
anthropol anthropologist; anthropology
anthropom anthropometry
Anthroposophic Anthroposophic Press
Anthy Anthony
anti (Latin prefix—against)—anticommunist
antibio(s) antibiotic(s)
antichlor anti + chlorine; antichloristic
anticli anticlimactic(al)(ly); anticlimax; anticlinal; anticline; anticlinorium
antidis antidisestablishmentarianism
antid(s) antidote(s)
antifreeze grain or methyl alcohol (CH_3OH) mixture
Antig Antigua
Antigua West Indian island nation, officially Antigua and Barbuda including nearby Redonda
antikv antikvarisk (Dano-Norwegian—antiques)
Antillas (Spanish—Antilles) West Indies

Antillas Mayores (Spanish—Greater Antilles) Cuba, Hispaniola, Jamaica, Puerto Rico
Antillas Menores (Spanish—Lesser Antilles) Leeward and Windward Islands extending from the Virgin Islands to Aruba
Antillas Neerlandesas (Spanish—Netherlands Antilles) Aruba, Bonaire, Curacao, Saba, Sint Eustatius, and half of Sint Maarten
Antilles West India Islands excluding Bahamas
antilog antilogarithm
antimag antimagnetic
antimat antimatter
antinuke anti nuclear (energy, power, or war)
Antioch Antioch Press
antip antiparasitic; antiparticle; antipasti; antipasto; antipathetic; antipathy; antiperiodic; antipersonnel; antiperspirant; antipodal; antipode; anti-poetic; antipollution; anti-poverty; antiproton; antipsychotic; antipyretic; antipyrine
antiphon antiphonal(ly)
Antipodes Australia and New Zealand; rocky islands off Dunedin, New Zealand
antipol antipollutant; antipollution
antiporn antipornographic; antipornography
antiq antiquarian, antique; antiquities; antiquity
antiq égypt antiquité égyptienne (French—Egyptian antiquity)
antiq gr antiquité grecque (French—Grecian antiquity)
antiq hébr antiquité hébraique (French—Hebraic antiquity)
antiq rom antiquité romaine (French—Roman antiquity)
antiquar antiquarian
Antiques Antiques Publications
antisem antisemite; antisemitic; antisemitism (pseudoscientific term for Jew hatred)
antisex antisexual
antivox antivoice-operated transmission
ant. jentac. ante jentaculum (Latin—before breakfast)
Antl Antlia
Ant Lat Antique Latin
ant. ld antique laid
Anto Antofagasta
Ant° Antonio
anton antonym
Anton Antonio; Antony

Ant Ops Antarctic Operations

ant. pit. anterior pituitary

ant. prand. *ante prandiu-matin*—before dinner)

antr apparent net transfer rate

Antr Antrim

Antrims Antrim Mountains of Northern Ireland

ant(s). antonym(s)

ant. sup. spine anterior superior spine

antu alpha-naphthyl-thiourea (rat poison)

ANTU Atlantic (Line container) Unit

Antuer *Antuérpia* (Portuguese—Antwerp)

Antw *Antwerpen* (Dutch, Flemish, German—Antwerp)

ant. wo antique wove

ANU Australian National University (Canberra); St John's, Antigua (3-letter code)

ANUP Australian National University Press

anv *anvendelse* (Dano-Norwegian—application, use)

anvo accept no verbal orders

ANWA *Abstracts of New World Archeology*

ANWG Apollo Navigation Working Group (NASA)

ANWR Agassiz National Wildlife Refuge (Minnesota); Aransas NWR (Texas); Arctic Natural Wildlife Refuge in Alaska; Arrowhead NWR (North Dakota); Audubon NWR (North Dakota)

anx. annex

Anx Annex (postal place-name abbreviation)

ANX Anixter Brothers (stock-exchange symbol)

anytimeteller anytime bank teller

Anz Anzania (African name for South Africa)

ANZ Air New Zealand; Australia and New Zealand (bank)

ANZA Association of New Zealand Advertisers

ANZAAS Australia and New Zealand Association for the Advancement of Science

ANZAC Australia and New Zealand Army Corps

ANZAM Australia, New Zealand, and Malaysia (defense pact)

ANZAMRS Australia and New Zealand Association for Medieval and Renaissance Studies

ANZAR Australian and New Zealand Association of Radi-

ology

ANZASA Australian and New Zealand American Studies Association

ANZ Bank Australia and New Zealand Bank

ANZCAN Australia-New Zealand-Canada (submarine cable system)

ANZCP Australian and New Zealand College of Psychiatrists

ANZDEC Asian-New Zealand Development Consultants

ANZECO Australia and New Zealand Exploration Company

ANZECS Australia-New Zealand-Europe Container Service

ANZESC Australian and New Zealand Eastern Shipping Conference

ANZGAM Australian and New Zealand Graduates Association of Malaysia (Singapore)

ANZHES Australian and New Zealand History of Education Society

ANZIA Associate of the New Zealand Institute of Architects

ANZIC Associate of the New Zealand Institute of Chemists

ANZJS *Australian and New Zealand Journal of Sociology*

ANZLA Associate of the New Zealand Library Association

ANZPC Australia and New Zealand Passenger Conference

ANZPPA Australia and New Zealand Professional Photographers Association

ANZQR *Australia and New Zealand Bank Quarterly Review*

ANZS Africa New Zealand Service

ANZSNM Australian and New Zealand Society of Nuclear Medicine

ANZSOM Australian and New Zealand Society of Occupational Medicine

ANZSOS Australian and New Zealand Society of Oral Surgeons

ANZSTS Australia and New Zealand Society for Theological Study

ANZUK Australia, New Zealand, United Kingdom (cultural, military, and trading alliance)

ANZUS Australia, New Zealand, United States (mu-

tual security pact)

ao access opening; anodal opening; anterior oblique; anti-oxidant; aorta; aortic opening; area of operation(s); axioocclusal

ao (AO) accuracy only; account of; arresting officer

a/o (A/O) account of

A° *anno* (Latin—year)

AO Administration Office; Airdrome Office(r); American Optical (company); Arkansas & Ozarks (railroad); Australian Officer (Officer of the Order of Australia); Autonomous Oblast; Aviation Ordnanceman; fleet tanker (2-letter naval designation); Order of Australia

A/O Administrator's Office (EPA)

AO *Ahonim Ortalik* (Turkish—Anonymous Company)—joint stock company; *Avtonómnaya Oblast* (Russian—Autonomous Region)—province

A-O *Auslands-Organisation* (German—Overseas Organization)—Hitler's overseas public relations agency

aoa abort once around; at or above

AoA Administration on Aging (HEW)

AOA American Oceanology Association; American Optometric Association; American Ordnance Association; American Orthopedic Association; American Orthopsychiatric Association; American Osteopathic Association; American Overseas Airlines; American Overseas Association; Australian Optometrical Association; Australian Orthopaedic Association

AOAC Association of Official Agricultural Chemists; Association of Official Analytical Chemists

AOAD Army Ordnance Arsenal District

a O (a/O) *an der Oder* (German—on the Oder River)

AOATC Atlantic Ocean Air Traffic Control

aob alcohol on breath; angle on the bow; any other business; at or below

aob (AOB) annual operating budget

AOB Advanced Operational Base

AO-BIRMDis Army Ordnance–Birmingham District
AOBMO Army Ordnance Ballistic Missile Office (USA)
AO-BOSTDis Army Ordnance–Boston District
AOBs Antediluvian Order of Buffaloes
AOBSR Air Observer
aoc anodal opening contraction; any other color
AoC Architect of the Capitol (D.C.)
A o C Academy on Computers
AOC Air Officer Commanding; Air Operations Center; Airport Operators Council; American Optical Company (stock exchange symbol); American Orthoptic Council; Arabian Oil Company; Army Ordnance Corps; automatic output control; Aviation Officer Candidate
AOCA American Osteopathic College of Anesthesiologists
AOCCOS Australian Organisations Coordinating Committee for Overseas Students
AO-CHIDis Army Ordnance–Chicago District
AOCI Airport Operators Council International
AO C-in-C Air Officer Commander-in-Chief
aocl anodal opening clonus
AO-CLEVDis Army Ordnance–Cleveland District
aocm advanced optical countermeasures
aocm (AOCM) aircraft out of commission for maintenance
AOCO Atomic Ordnance Cataloging Office
aocp (AOCP) aircraft out of commission for parts
AOCs American Olympic Committee members; Association of Old Crows
AOCS American Oil Chemists' Society; Atlantic Outer Continental Shelf
aod arterial occlusive disease; as of date
A o D Airport of Departure
AOD Academy of Oral Dynamics; Air Officer of the Day
ao diag acridine-orange diagnosis (cancer)
AODs Ancient Order of Druids
AODS All Ordnance Destruct System
aoe airborne operational equipment; auditing order error
AoE Aerodrome of Entry
A o E Airport of Entry

AOE Association of Overseas Educators
AOEHI American Organization for the Education of the Hearing Impaired
AOEM Automotive Original Equipment Manufacturers
AOER Army Officers' Emergency Reserve
AOF *Afrique Occidentale Francaise* (French—French West Africa); Ancient Order of Foresters; Australian Olympic Federation
AOFS Active Optical Fuzzing System
aog (AOG) aircraft on ground
AOG Atlantic Oceanographic Group; Australian Oil and Gas; gasoline tanker (3-letter symbol)
AOGM Army of Occupation of Germany Medal
AOH Ancient Order of Hibernians
aoi accent on information; angle of incidence; area of interest
A o I Aims of Industry
AOIB Anglo-Oriental International Bank
AOIPS Atmospheric and Oceanic Information Processing System
aoiv automatically-operated inlet valve
ao. J.C. *après Jésus-Christ* (French—after Christ) A.D.
AOJDD Anglican Orthodox Joint Doctrinal Discussions
aok all okay; everything in good order
A OK Armee-Oberkommando (German—Army High Command)
aol absent over leave
AOL American-Oriental Lines; Atlantic Oceanography Laboratories
AO-LADis Army Ordnance–Los Angeles District
aolo advanced orbit laboratory operations
AOLP Action Organization for the Liberation of Palestine
AOM Army of Occupation Medal
A.O.M. Master of Obstetric Art
AOMA American Occupational Medical Association
AOMAA Apartment Owners and Managers Association of America
AOMC Army Ordnance Missile Command
AOMSA Army Ordnance Missile Support Agency

AOMSC Army Ordnance Missile Support Center
aonb area of outstanding natural beauty
AONB Area of Outstanding Natural Beauty
AO-NYDis Army Ordnance–New York District
aoo anodal opening odor
AOO American Oceanic Organization
AOOA Aircraft Owners and Operators Association
aop anodal opening picture; aortic-pressure pulse
AOP Apprenticeship Outreach Program; Association of Optical Practitioners; Association of Osteopathic Publications
AOPA Aircraft Owners and Pilots Association
AOPC Australian Overseas Projects Corporation
AOPE Associated Organization for Professionals in Education
AOPEC Arab Organization of Petroleum Exporting Countries
AO-PHILDis Army Ordnance–Philadelphia District
aoProf auszerordentlicher Professor (German—associate professor or special lecturer)
A Ops Air Operations
AOPs Allied Ordnance Publications
AOPU Asian Oceanic Postal Union (China, Korea, Philippines, Thailand)
aoq average outgoing quality
AOQC Australian Organisation for Quality Control
aoql average outgoing quality limit
aor album-oriented rock music (FM radio performance); angle of reflection; aorist; area of responsibility
a/or and/or
AOR Army Operational Research; Australian Oil Refining; auxiliary oil replenishment (USN)
Aorangi (Maori—Cloud Piercer)—Mount Cook, New Zealand's tallest towering to 3764 meters or 12,349 feet
AORB Aviation Operational Research Branch
AORG Army Operational Research Group (United Kingdom)
AORL Apollo Orbital Research Laboratory
AORN Association of Operating Room Nurses

AOrPA American Orthopsychiatric Association
AORS Army Operations Research Symposium
AORT Association of Operating Room Techniques
AORTF American Organization for Rehabilitation through Training Federation
aort regurg aortic regurgitation
aort sten aortic stenosis
aos acquisition of signal; add or subtract; angle of sight; anodal opening sound
AOS American Opera Society; American Ophthalmological Society; American Orchid Society; American Oriental Society; American Otological Society; Australian Overseas Smelting; Azimuth Orientation System
AOSC Association of Oilwell Servicing Contractors
A-O Scale Anti-Oriental Scale (measuring negative attitudes)
AOSE American Order of Stationary Engineers
AOSO Advanced Orbiting Solar Observatory
aosp automatic operating and scheduling program
AOSPS American Otorhinologic Society for Plastic Surgery
AOSs Ancient Order of Shepherds
AO-STLDis Army Ordnance–St Louis District
AOSTRA Alberta Oil Sands Technology and Research Authority
AOSW Association of Official Shorthand Writers
aot active on target; anodal opening tetanus
Aot Askania optical tracker
AOT Alameda-Oakland Tunnel; Association of Occupational Therapists
AOTA American Occupational Therapy Association; Australian Overseas Transport Association
AOTC Aspen Opera Theater Center; Australian Overseas Trading Corporation
Aotearoa (Maori—Long White Cloud)—New Zealand
ao technique acridine-orange technique (two-color fluorescent test for cancer)
AOTOI American Organization of Tour Operators to Israel
AotOS Admiral of the Ocean Sea (U.S. Merchant Marine

award)
AOtS American Otological Society
aou apparent oxygen utilization; azimuth orientation unit
AOU American Ornithologists' Union; Australian Ornithologists Union
AOUSC Administrative Office of the United States Courts
AOUW Ancient Order of United Workmen
aov analysis of variance; any other variety
AOVP Australian Ordnance Vehicle Park
AOW Articles of War
ap above proof; access panel; acid phosphatase; acid proof; action potential; acute proliferation; aerial port; aiming point; airplane; alkaline phosphatase; alum precipitated; aminopeptidase; angina pectoris; antepartum; anterior pituitary; anteroposterior; aortic pressure; appendectomy; appendices; appendix; arithmetic progression; armor piercing; arterial pressure; artificial pneumothorax; as prescribed; association period; attached processor; author's proof; axiopulpal; (Welsh prefix—son of)
ap (AP) advanced placement; automatic payment; average product
a/p after perpendicular; air port (porthole); angle point; authority to pay; authority to purchase; autopilot
a&p agricultural and pastoral; anterior and posterior; apogee and perigee (apex and antapex); auscultation and percussion
a$_p$ geomagnetic index
ap anno passato (Italian—last year)
ap. apud (Latin—according to)
a.p. ante prandium (Latin—before a meal)
Ap Apothecary
Ap. Apostolus (Latin—Apostle)
AP Acquisition Plan; Additional Premium; Advanced Placement; Air Police; Air Publication; Airport; Allied Publication; American Party; American President Lines; Andra Pradesh; Associated Press; Australia Party; Aviation Pilot; personnel transport (naval symbol)
A-P American Plan (includes

meals)
A/P allied papers; authority to pay
A & P Agricultural and Pastoral Society of New Zealand; Atlantic & Pacific Travel International; Great Atlantic & Pacific Tea Company
AP Acción Popular (Spanish—Popular Action); Administration Pénitentiare (French—Penitentiary Administration); Almonester y Pontalba (New Orleans: Don Andres Almonester and his daughter the Baroness de Pontalba); Arbeiderpartiet (Norwegian—Det Norske Arbeiderpartiet—The Norwegian Labor Party); Atlanska Plovidba (Russian—Atlantic Press); Australia Post; Aviapolk (Russian—Air Regiment)
A.P. a protester (French—to be protested later)
apa aldosterone-producing adenoma; aminopenicillanic acid; antipernicious anemia factor; axial pressure angle
APA Adult Parole Authority; Advertising Photographers of America; Aerovias Panamá Airways; Agricultural Publishers Association; Airline Passenger Association; American Patients Association; American Pharmaceutical Association; American Philological Association; American Philosophical Association; American Photoengravers Association; American Physiotherapy Association; American Pilots Association; American Planning Association; American Podiatry Association; American Polygraph Association; American Poultry Association; American Press Association; American Protective Association; American Psychiatric Association; American Psycho-analytical Association; American Psychological Association; American Psychosomatic Association; American Psychotherapy Association; American Pulpwood Association; Animation Producers Association; Anthracite Producers Association; Anti-Papal Association; Apache Railway; Associated Press of Australia; Association of Paroling Au-

thorities; Association of Physicians of Australasia; Association for the Prevention of Addiction (London); Australian Provincial Insurance; Automobile Protection Agency; transport attack vessel (naval symbol)
APA Austria Presse Agentur (German—Austrian Press Agency)
APAA Australian Port Authorities Association
APAAC Asian-Pacific-American Advocates of California
APAC Aerial Photographic Analysis Center
apache analog programming and checking
APACHE Accelerator for Physics and Chemistry of Heavy Metals; Application Package for Chemical Engineers
APACI Associación Peruana para el Avance de la Ciencia (Spanish—Peruvian Association for the Advancement of Science)
APACL Asian Pacific Anti-Communist League; Asian People's Anti-Communist League
apacs adaptive planning and control sequence (marketing)
APADA Australian Petroleum Agents and Distributors Association
APADS Automatic Programmer and Data System
APAE Association of Public Address Engineers
apaf antipernicious anemia factor
APAFA ASEAN Professional Association on Food and Agriculture
APAG Atlantic Policy Advisory Group; Atlantic Political Advisory Group (NATO)
APAIS Australian Public Affairs Information Service
APAL American Puerto-Rican Action League
AP & AM Adler Planetarium and Astronomical Museum
APANZ Associate of the Public Accountants of New Zealand
APAP American People for American Prisoners
APAP Asociación Peruana de Agencias de Publicidad (Spanish—Peruvian Association of Advertising Agencies)
apar apparatus
APAR Automatic Programming

and Recording
apart. apartment
a-part alpha particle(s)
apart apartheid (Afrikaans—apartness, racial segregation)—South African government policy
APAs American Polygraph Association members; Anti-Papal Association members and sympathizers
APAS Automatic Performance Analysis System
APASCO Australian Planning and Systems Company
APATS Antenna Pattern Test System (USA)
APAVIT Asociación Peruana de Agencias de Viajes y Turismo (Spanish—Peruvian Association of Travel and Tourism Agencies)
apb atrial premature beat; auricular premature beat
apb (APB) all-points bulletin
APB barracks ship, self-propelled (3-letter symbol)
APB All-Points Bulletin
APBA American Power Boat Association; Association Press Broadcasters Association
APBPA Association of Professional Ball Players of America
APBS Accredited Poultry Breeding Scheme; Australian Permanent Building Society
APBSD Advanced Post Boost System Development
APBSQ Association of Permanent Building Societies of
apc (APC) average propensity to consume
a/p c autopilot capsule
APc coastal transport (3-letter symbol)
APC Aeronautical Planning Chart; Aerospace Primus Club; Agricultural Productivity Commission; Air-Pollution Control; American Parents Committee; American Peace Corps; American Philatelic Congress; Area Positive Control; Arkansas Polytechnic College; armored personnel carrier; Army Petroleum Center; Army Policy Council; Association of Private Camps; Association of Pulp Consumers; Australian Police College; Australian Postal Commision; Avian Propagation Center
APCA Air Pollution Control Association; American Petroleum Credit Association; American Planning and Civic

Association; Anglo-Polish Catholic Association; Australian Physical Culture Association
APCB Air Pollution Control Board
apcbc armor-piercing carbide ballistic cap
apc-c aspirin, phenacetin, caffeine—with codeine
APCC Asian and Pacific Coconut Community
APCD Air Pollution Control District
APCG Association of Pacific Coast Geographers
apche automatic programmed checkout equipment
apci armor-piercing capped incendiary
apcit armor-piercing capped incendiary with tracer
APCK Association for Promoting Christian Knowledge
APCL Association of Professional Color Laboratories
apcm authorized protective connecting module
APCM Asiatic-Pacific Campaign Medal
APCO Air Pollution Control Office; Alabama Power Company; Associated Public Safety Communications Officers
apcr armor-piercing-composite rigid
APCS Air Photographic and Charting Service; Associative Processor Computer System
apct armor-piercing capped with tracer
a/p ctl autopilot control
apc virus adenoidal, pharyngeal, conjuctival virus
APCYA A Presidential Classroom for Young Americans
apd action potential duration; aiming point determination; anteroposterior diameter
APD Air Pollution Division (U.S. Dept Agriculture); Air Procurement District; high-speed troop transport (3-letter naval symbol)
APDA American Parkinson's Disease Association
APDC Agricultural Pest Destruction Council; Albany Port District Commission
Ap Del Apostolic Delegate
APDF Asian-Pacific Dental Federation
apdl algorithmic processor description language
Apdo Apartado (Spanish—post-

office box)
apds armor-piercing discarding sabot
APdS American Pediatric Society
APDSMS Advanced Point Defense Surface Missile System
APDU Association of Public Data Users
APDUSA African People's Democratic Union of Southern Africa
apdy appropriate duty
ape. adapted physical educator; aerial port of embarkation (APE); aminophylline, phenobarbital, ephedrine; anterior pituitary extract; apparent effect; automatic photomapping equipment
APE aerial port of embarkation; Amalgamated Power Engineering
A.P.E. Air Pollution Engineer
APEA Association of Professional Engineers (Australia); Australian Petroleum Exploration Association
APEC Asia–Pacific Economic Cooperation
Apeco American Photocopy Equipment Company
APELR Association for the Preservation of English Language Rights
Apennines Apennine Mountains of the Italian Peninsula
aper aperture
APER Air Pollutant Emissions Report
apers antipersonnel
APETS Australia-Papua New Guinea Education and Training Scheme
apex. advance-purchase excursion (airline fare); air-pollution exercise; assembler and process executive
APEX Advance-Purchase Excursion (Plan) Association of Professional Executive and Management Staff
apf acidproof floor; animal protein factor
APF American Progress Foundation; Anglican Pacifist Fellowship; Asia Pacific Forum; Asian Packaging Federation; Association of Pacific Fisheries; Association of Protestant Faiths; Australian Parachute Federation
APF Australian Pharmaceutical Formulary
APFA American Pipe Fittings Association; Associated Poul-

try Farmers of Australia; Association of Professional Flight Attendants
APFA Association des Professeurs Franco-Americains (French—Association of Franco-American Professors)
APFC Asia-Pacific Forestry Commission
APFF American Police and Fire Foundation
APFM Australian Postgraduate Federation in Medicine
APFO Association on Programs for Female Offenders
APFRI American Physical Fitness Research Institute
apfsds armor-piercing fin-stabilized discarding sabot
Apg Appingedam
APG Aberdeen Proving Ground; Air Proving Ground; American Pewter Guild; Army Planning Group; Army Proving Ground; Australian Proving Ground
APG Autoridad Portuaria de Guayaquil (Spanish—Port Authority of Guayaquil)
APGA Alabama Personnel and Guidance Association; American Personnel and Guidance Association; Apple and Pear Growers Association; Arizona Personnel and Guidance Association; Arkansas Personnel and Guidance Association
APG-BRL Aberdeen Proving Ground—Ballistics Research Laboratory
APGC Air Proving Ground Center
apgcu autopilot ground control unit
APG/HEL Aberdeen Proving Ground—Human Engineering Laboratory
APG/OBDC Aberdeen Proving Ground–Ordnance Bomb Disposal Center
APGOEF Air Proving Ground–Eglin, Florida
APGp Aerial Port Group
aph antepartum hemorrhage; anterior pituitary hormone (APH); aphelion
APH Access Permit Holder; transport fitted for evacuation of wounded (3-letter symbol)
A.P.H. A(lan) P(atrick) Herbert
APhA American Pharmaceutical Association
APHA American Printing History Association; American Protestant Hospital Association; American Public Health

Association; Australian Pneumatic and Hydraulic Association
APHB American Printing House for the Blind; Army Pearl Harbor Board
APHCA Animal Production and Health Commission for Asia, Far East and South West Pacific
aphet aphetic
APHI Association of Public Health Inspectors
APHIA Association for the Promotion of Humor in International Affairs
A & PHIS Animal and Plant Health Inspection Service
APhO Area Pharmaceutical Officer
Aph of Lath Aphorism of Lathem ("There is nothing edible or potable failing to find someone to take it as a sovereign remedy for some disease and upon the earnest recommendation of some eminent physician.")
aphor aphorism; aphorist(ic) (ally); aphorize
aphp antipseudomonas human plasma
aphro(s) aphrodisiac(s)
APHS Arizona Pioneer Historical Society; Australian Psychology and Hypnotherapy Association
api acceptable periodic inspection; air position indicator; armor-piercing incendiary tracer
API Academic Press, Inc; Alabama Polytechnic Institute; American Paper Institute; American Petroleum Institute; American Pipe Institute; American Potash Institute; American Press Institute; Animal Protection Institute; armor-piercing incendiary; Australian Pharmaceutical Industries; Australian Planning Institute
API Association Phonetique Internationale (French—International Phonetic Association); *Associazione Pionieri Italiani* (Italian—Italian Boy Scouts Association)
APIA Animal Protection Institute of America
APIC Apollo Parts Information Center; Army Photo Interpretation Center; Association for Practitioners in Infection Control
APICORP Arab Petroleum In-

vestments Corporation

APICP Association for the Promotion of the International Circulation of the Press

APICS American Production and Inventory Control Society

APICSC Atlantic-Pacific Interoceanic Canal Study Commission

apicult apiculture

APID Army Photo Interpretation Detachment

APIDC Andhra Pradesh Industrial Development Corporation

APIF Automated Process Information File

APIJ Australian Planning Institute Journal

A-pill abortion pill

APIM Association Professionelle Internationale des Médecins (French—International Professional Association of Physicians)

APIN Atlas Propulsion Information Notice

apipocc appropriating property in possession of (a) common carrier

APIS Army Photographic Intelligence Service; Australian Professional Interpreting Society

APIU Army Photo Interpretation Unit

apivr artificial pacemaker ventricular rhythm

APJ American Power Jet (company)

APJE Association of Philosophical Journals Editors

apl a programming language; adult performance level; aluminum-polythene laminate; automatic premium loan

apl (APL) anterior pituitary-like hormone

a/pl armorplate

Apl Appledore

APL A Programming Language (mathematically-oriented computer language developed by Kenneth E Iverson of IBM); Air Pacific Limited; Air Provost Marshal; Akron Public Library; Albany Public Library; Albuquerque Public Library; American Pioneer Line; American President Lines; Applied Physics Laboratory; Assembly Parts List; Augusta Public Library; Australian Pensioners League; barracks ship (naval symbol)

AP–LS American Psychology–

Law Society

A-PL All-Purpose Linotype

APLA American Patent Law Association; Armenian Progressive League of America; Atlantic Provinces Library Association

ap/lat anteroposterior and lateral

APLC Automated Parking Lot Control

Aplcrs Applecross

apld (APLD) automatic program locate device

APLE Association of Public Lighting Engineers

APLIC Association of Parliamentary Librarians in Canada

apll analog phased-locked loop(s)

aplns applications

APLQ Agence de Presse Libre du Québec (French—Free Press Agency of Québec)

APLS American Plant Life Society

AP-LS American Psychology-Law Society

APLU American President Line (container) Unit

apm apomict; associative principle for multiplication

apm (APM) antipersonnel missile

APM Academy of Physical Medicine; Academy of Psychosomatic Medicine; Air Power Museum; Association for Psychoanalytic Medicine; Australian Paper Manufacturers; Australian Paper Mills

apma advance payment of mileage authorized

APMA Absorbent Paper Manufacturers Association; Automatic Phonograph Manufacturers Association

APMAC A.P. Moller Associated Concerns

APMC Academy of Physchologists in Marital Counseling; Andhra Pradesh Mining Corporation; Asia and Pacific Maritime Cooperation Centre

a/p mcu autopilot monitor and control unit

APME Associated Press Managing Editors (Association)

APMF Australian Paper Manufacturers Federation

APMG Assistant Postmaster General

APMHC Association of Professional Material Handling Consultants

apmi area precipitation mea-

surement indicator

APMIS Automated Project Management Information System

APMR Association for Physical and Mental Rehabilitation

APMT Antenna Pattern Measuring Test (USA)

apn artificial pneumothorax; average peak noise

APN American Practical Navigator

APNA Asia-Pacific News Agencies

APNEC All-Pakistan Newspaper Employees Confederation

APNP Arthur's Pass National Park (South Island, New Zealand)

apo airport; apogee; apolipoprotein

APO Accountable Property Office(r); Advanced Post Office; Air Force (Army) Post Office; American Potash & Chemical (stock exchange symbol); Animal Procurement Office(r); Area Patrol Office(r); Area Petroleum Office(r); Association of Physical Oceanographers; Australian Post Office

apob airplane observation

Apoc Apocalypse; Apocrypha; Apocryphal

APOC Army Point of Contact

APOD Aerial Port of Debarkation

APOE Aerial Port of Embarkation

apol apologete; apologetic(al); apologetics; apologia; apologise; apologist(s); apologize; apology

APOLLO Article Procurement and On-Line Local Ordering

APON Association of Pediatric Oncology Nurses

APOP Australian Public Opinion Polls

apos apostrophe

APOS Advanced Polar Orbiting Satellite

apost apostasy; apostate

a post. a posteriori (Latin—derived from or relating to reasoning from observed facts)

Apostle Islands national lakeshore, Wisconsin

Apostles Apostle Islands, Lake Superior; New Testament disciples

apota automatic positioning of telemetering antenna

apotek apoteket (Danish—apothecary)—drugstore

apoth apothecaries' (weight); apothecary

A-powered atomic-powered

app apparatus; apparel; apparent; appeal; appelate; appendage; appended; appendix; apperception; appetite; appetizer; applause; applied; appointed; apprehended; apprentice; approach; appropriate; appropriation; approval; approve; approximate

App Appellate; Lucius Appuleius

App Apparat (German—apparatus); Lucius Appuleius (Roman philosopher)

App. Apostoli (Latin—Apostles)

APP Advanced Placement Program; Air Parcel Post; Algonquin Provincial Park (Ontario); *Alianza Para Progreso* (Spanish—Alliance for Progress); Army Procurement Procedure; Association of Professional Photogrammetrists; automatic priority processing

appa advise present position and altitude

APPA American Penal Press Association; American Probation and Parole Association; American Public Power Association; American Pulp and Paper Association

APPAG All Party Penal Affairs Group (Britain)

Appalachia eastern Kentucky, southeastern Ohio, eastern Tennessee, and western West Virginia; region in the Appalachian Mountains extending from Québec to northern Georgia

Appalachians Appalachian Mountains

appar apparatus

APPAUC Association of Physical Plant Administrators of Universities and Colleges

APPC Advance Procurement Planning Council; Advanced Program-to-Program Communications

appd approved

App Div Appellate Division (NY Supreme Court Reports)

APPECS Adaptive Pattern-Perceiving Electronic Computer System

appellat appellative

appi advanced planning procurement information

APPITA Australian Pulp and Paper Industries Technical Association

appl applicable; application; applied

APPL Advance Procurement Planning List(s)

applan. applanatus (Latin—flattened)

APPLE Advanced Propulsion Payload Effects (NASA); Association of Public and Private Labor Employees

Appleton Appleton-Century-Crofts

applican. applicandus (Latin—applied, to be applied)

appln application

Appl Opt Applied Optics

Appl Phys Lett Applied Physics Letters

Appl Spectrosc Applied Spectroscopy

APPM Association of Publication Production Managers

appmt appointment

appn appropriation

App^{no} Appennino (Italian—Appenines)

a/p poi autopilot positioning indicator

APPP Association of Planned Parenthood Physicians

appr approval; approve; approved

APPR Army power package reactor

appren apprentice

appro approval

approp appropriation

approx approximate(d); approximately; approximation; approximative

apps appendices; appendixes

APPS Australian Physiological and Pharmacological Society; Australian Plant Pathology Society

appt appoint; appointment

apptd appointed

App Thorn Appleton Thorn prison, Lancashire, England

APPU Australian Primary Producers Union

appx appendix

appy appendectomy

apr amoebic prevalence rate; annual percentage rate; anterior pituitary reaction; apprentice

apr (APR) aerial photographic reconnaissance

apr aprile (Italian—April)

Apr April

Apr Aprel (Russian—April)

APR Air Pictorial Service; Air Priority Raging; Airman Performance Report; American

Public Radio; Annual Percentage Rate; Annual Progress Reports; Association of Petroleum Re-Refiners; Association for Promoting Retreats; Association of Publishers' Representatives

Apra San Luis de Apra (Guam's principal port)

APRA Aircraft Resources Production Agency; American Popular Revolutionary Alliance (Peru); Australian Performing Rights Association; Australian Plastics Research Association

APRA Alianza Popular Revolucionaria Americana (Spanish—Popular American Revolutionary Alliance)—Peru's Aprista Party of Haya de la Torre

aprax apraxia(l)

APRC Army Physical Review Council

APRDC Army Polar Research and Development Center

APRE Air Procurement Region—Europe

APREF Asia Pacific Real Estate Federation

après-40 after 40 years of age

après JC après Jesus Christ (French—after the birth of Jesus Christ)

APRF Army Pulse Radiation Facility

APRFE Air Procurement Region—Far East

a pri. a priori (Latin—derived from or relating to self-evident propositions)

APri air priority

april automatically programmed remote indication logged

A/Prin Assistant Principal

Apr J-C Après Jésus-Christ (French—after Christ) A.D.

APRL American Prosthetic Research Laboratory

Aprmay April and May

aprmd appointment recommended

APRO Aerial Phenomena Research Organization; Army Personnel Research Office

apróx apróximadamente (Spanish—approximately)

AprS American Proctologic Society

APRS Association of Professional Recording Studios

aprt airport

APRTA Associated Press Radio and Television Association

aprthd *Apartheid* (Afrikaans—apartness)

APRU Asia Pacific Research Unit

aprx approximately

aps accessory power supply; adenosine phosphosulfate; autograph postcard signed; auxiliary power supply; auxiliary propulsion system

aps (APS) average propensity to save

Aps Apus

ApS *AnpartsSelskab* (Dano-Norwegian-limited company)

APS Aboriginal Protection Society; Academy of Political Science; adenosine phosphosulfate; Alternative Press Syndicate; American Metal Products (stock exchange symbol); American Pediatric Society; American Pheasant Society; American Philatelic Society; American Philosophical Society; American Physical Society; American Physiological Society; American Phytopathological Society; American Plant Selections; American Poinsettia Society; American Polar Society; American Proctologic Society; American Prosthodontic Society; American Psychosomatic Society; Army Pilot School; Army Postal Service; Association of Photo Sensitizers; Atmospheric Protection System; Australian Pig Society; Australian Psychological Society; Australian Public Service; submarine transport (naval symbol)

APS *Algerie Presse Service* (French—Algerian Press Service); *Associated Press Stylebook and Libel Manual*

APSA Aerolíneas Peruanas, South America; American Political Science Association; Associate of the Photographic Society of America; Association of Professional Scientists of Australia; Australian Pharmaceutical Sciences Association; Australian Political Studies Association; Australian Pre-School Association

A & PSA Aden and Protectorate of South Arabia

APSA *Aerolíneas Peruanas SA* (Spanish—Peruvian Airlines Corporation); *Azufrera Panamericana SA* (Spanish—Pan-

american Sulfur Corporation)

APsaA American Psychoanalytic Association

APSB Aid to the Potentially Self-supporting Blind; Australian Public Service Board

APSC Assembly Public Safety Committee (California)

APSE Abstracts of Photographic Science and Engineering

APSF Alfred P Sloan Foundation; Australian Public Service Federation

APS/HEW Administration for Public Services—HEW

APSq Aerial Port Squadron

APSS Association for the Psychophysiological Study of Sleep; Association of Professional Sleep Societies

A.P.S.T. Associate in Public Service Technology

APsychoA American Psychoanalytic Association

APsychosomS American Psychosomatic Society

APsychpthA American Psychopathological Association

apt automatic photometric telescope

apt. airborne pointer-tracker; alum-precipitated toxoid; apartment; armor-piercing with tracer; automatic picture transmission; automatically-programmed tool(s)

apt. (APT) afterpeak tank

apt *apartadero* (Spanish—platform)

APT Advanced Passenger Train; Airman Proficiency Test; Applied Performance Test(s); Automatic Photoelectric Telescope; Automatic Picture Transmission; Automotive Professional Training

APTA American Physical Therapy Association; American Pioneer Trails Association; American Platform Tennis Association; American Public Transit Association

APTC Allied Printing Trades Council

APTD Aid to the Permanently and Totally Disabled

Aptdo *apartado* (Spanish—post office box)

apte advance passenger train express (149 mph British turbine-powered train)—APTE

apth apthong (a silent letter like the *p* in pneumatic)

apti actions per time interval

APTI Association of Principals

of Technical Institutions

APTIC Air Pollution Technical Information Center

apto aluminum plastic tearoff (container cover)

Apto *apartamento* (Spanish—apartment)

apt(s) apartment(s)

APTs Advanced Passenger Trains

APTS Automatic Picture Transmission System

APTU African Postal and Telecommunications Union; Australian Postal and Telecommunications Union

apu auxiliary power unit

APU Army Postal Unit; Asian Parliamentary Union

apu-hs automatic program unit–high speed

apu-ls automatic program unit–low speed

apv automatic-patching verification

APV All-Purpose Vehicles; Avenida Presidente Vargas, Rio de Janeiro, Brazil

apw anti-personnel weapon; architectural projected window

APW Accelerated Public Works; American Prisoner of War; Apia, Western Samoa (airport)

APWA American Public Welfare Association; American Public Works Association

APWP Accelerated Public Works Program

APWSS Asian-Pacific Weed Science Society

APWU American Postal Workers Union

apx appendix; approximate(ly)

APZ Assiniboine Park Zoo

aq accomplishment quotient; achievement quotient; any quantity; aqueous

a-q aircraft quality

aq. *aqua* (Latin—water)

AQ achievement quotient; aviation fire-control technician (USAF symbol); Schreiner Aerocontractors (Hague)

AQ *Australian Quarterly*

AQA Australian Quadriplegic Association

AQAB Air Quality Advisory Board

AQAPs Allied Quality Assurance Publications

Aqar Aquarius

aq. astr. *aqua adstricta* (Latin—ice)

aq. bull. *aqua bulliens* (Latin—boiling water)

aq. cal. aqua calida (Latin— warm water)

AQCL Analytical Quality Control Laboratory

aq. com. aqua communis (Latin ⊃—ordinary water)

AQCR Air Quality Control Region (EPA)

aq. dest. aqua destillata (Latin—distilled water)

aqdm air quality display model

AQE Airman Qualifying Examination

aq. ferv. aqua fervens (Latin— hot water)

aq. fluv. aqua fluvii (Latin— river water)

aq. font. aqua fontana (Latin— spring water)

aq. fort. aqua fortis (Latin—nitric acid)

aq. frig. aqua frigida (Latin— cold water)

aqgv azimuth-quantized gated video

AQHA Australian Quarter Horse Association

Aqil Aquila

aql acceptable qualifying levels; acceptable quality level; approved quality level

Aql Aquila

aq. mar. aqua marina (Latin— sea water)

AQMD Air-Quality Management District (Southern California)

aq. ment. pip. aqua menthae piperitae (Latin—peppermint water)

AQMG Assistant Quartermaster General

AQMP Air Quality Master Plan

aqn azimuthal quantum number

aq. niv. aqua nivalis (Latin— snow water)

aq. pluv. aqua pluvialis (Latin—rain water)

aq. pur. aqua pura (Latin— pure water)

Aqr Aquarius

AQREC Army Quartermaster Research and Engineering Command

aq. regia aqua regia (Latin— royal water) hydrochloric and nitric acid

aqs additional qualifying symptoms

AQT Applicant Qualification Test

aq. tep. aqua tepida (Latin— tepid water)

aqu aqueous

aqua. aquaria; aquarium; aquatic

aquacade aquatic parade (or exhibition of diving, swimming, water sports)

aquacult aquaculture

aquar aquarium

Aquar Aquarius

aque aqueduct

Aqued Aqueduct

Aqueduct Aqueduct Books

ar achievement ratio; acid resisting; active resistance; alarm reaction; all rail; all risks; allocated reserve; analytical reagent; antireflection; aromatic; arrival; artificial respiration; aspect ratio; auditory reception

ar (AR) address register; armed robbery; average revenue

a/r all risks; armed robbery; at the rate of

a & r approved and removed; artists and repertory; assault and robbery; assembly and repair

ar arabisch (German—Arabian); *avis de reception* (French—return receipt)

a/R *am Rhein* (German—on the Rhine River)

Ar Arab; Arabia; Arabian; Arabic; Aragon; argon; Aries; aryl

Ar Arabic; *Arroyo* (Spanish— brook, creek, or rivulet)

AR Aberdeen & Rockfish (railroad); Administrative Ruling; Aerodynamic Report; Aerolineas Argentinas (Argentine Airlines); Aeronautical Radionavigation; Airman Recruit; Airship Rigger; Amendment Request; American Smelting & Refining (stock exchange symbol); Annual Report; Arkansas (zip code); Army Regulation(s); Army Reserve; repair ship (naval symbol)

A/R Antwerp-Rotterdam (range of ports)

AR Agencja Robotnicza (Polish—Workers' Press Agency); *Aller et Retour* (French— roundtrip); *American Rationalist* (freethought magazine); *Andata-Ritorno* (Italian— roundtrip);

A.R. Anno Regni (Latin—In the Year of the Reign of)

A/R Aksjerederi (Norwegian— shipping company)

ara assigned responsible agency (DoD)

ara (ARA) aerial rocket artillery

Ara Ara (a three-letter constel-

lation without an abbreviation); Arabic; Argentina

ARA Academy of Rehabilitative Audiology; Aerospace Research Association; American Radio Association; American Railway Association; American Rationalist Association; American Relief Association; American Rental Association; American Republics Area; American Revenue Association; American Rheumatism Association; Applied Research of Australia; Arcade & Attica (railroad); Area Redevelopment Administration; *Armada República Argentina* (Argentine Navy); Artists' Representatives Association; Association of Radiologists of Australasia; Association of River Authorities; Auckland Regional Authority; Australian Regular Army; Australian Retailers Association; Automatic Retailers of America

ARA (AFL-CIO) American Radio Association

A.R.A. Associate of the Royal Academy

A/R/A Antwerp-Rotterdam-Amsterdam (range of ports)

ARA Acçao Revolucionaria Armada (Portuguese—Armed Revolutionary Action)

Arab. Arabia; Arabian; Arabic

ARAB Australian Radio Advertising Bureau

Arab Emirates United Arab Emirates including the seven Trucial Sheikdoms along the southern shore of the Persian Gulf with Abu Dhabi its capital

Arab League League of Arab States (Algeria, Bahrain, Egypt, Iraq, Jordan, Kuwait, Lebanon, Libya, Morocco, Oman, Qatar, Somalia, Sudan, Syria, Tunisia, United Arab Emirates, Yemen)

ARABS Australian Racing and Breeding Stables

Arabsat Arab Communications Satellite

ARAC Aerospace Research Applications Center; Associate of the Royal Agricultural College; Australian Refugee Advisory Council

arach arachnology

Arach Arachnida

ARACI Associate of the Royal Australian Chemical Institute

arad airborne radar and doppler

ARAD Associate of the Royal Academy of Dancing

ARADCOM Army Defense Command

ARAeS Associate of the Royal Aeronautical Society

Arafat Yasir Arafat (Palestinian leader)

Arafura Arafura Sea between Australia and New Guinea

Arag Aragón

ARAgS Associate of the Royal Agricultural Society

ARAIA Associate of the Royal Australian Institute of Architects

aral automatic record analysis language

Aral Aral Sea

Aram Aramaic

ARAM Association of Railroad Advertising Managers

A.R.A.M. Associate of the Royal Academy of Music

Aramco Arabian-American Oil Company

Arans Aran Islands off County Clare, Ireland

Aransas Aransas National Wildlife Refuge near Rockport, Texas

ARAPCS Association for Research, Administration, Professional Councils, and Societies

Ararat Mount Ararat (Turkey's highest mountain and place where Noah's Ark is believed to have landed; Armenian name of the mountain the Turks call Agri Dagi on the border of Soviet Armenia)

aras ascending reticular activating system

ARAS Accept (each person you encounter), Respect (each person whatever their position), Affect (each person with the warmth of your heart), Support (each person in the place they are now); Ascending Reticular Activating System; Associate of the Royal Astronomical Society

Arava Israeli IAI-201 light transport plane

arb arbitrageur; arbitrary; arbitration; arbitrator; labor arbitrator

arb arbeid(er) [Dano-Norwegian—work(s)]; *arbejde* (Dano-Norwegian—job, labor, work)

Arb Arbroath

ARB Accident Records Bureau

(NYC Police Dept.); Air Registration Board; Air Research Bureau; Air Resources Board; Armored Rifle Battalion; Army Rearming Base; Army Retiring Board; Asian Reserve Bank; ASTIA Report Bibliography; battle damage repair ship (naval symbol)

ARBA American Railway Bridge and Building Association; American Road Builders Association; Associated Retail Bakers of America

ARBA American Reference Books Annual

arb & aw arbitration and award

ARBED Aciéries Réunies de Burbach-Eich-Dudelange

arbit arbitrageur

ARBM Association of Radio Battery Manufacturers

arbo arthropod-borne (viral diseases)

arbor. arboriculture

arbor. virus arthropod-borne virus

ARBP Associated Reinforcing Bar Producers

ARBs Air Resources Boards (pollution-control agencies)

ARBS Angular-Rate Bombing System; Associate of the Royal Society of British Sculptors

Arb Trib Arbitration Tribunal

arbtrn arbitration

arbtror arbitrator

Arbuckle Arbuckle National Recreation Area near Sulphur, Oklahoma

arb. vit. arbor vitae (Latin—tree of life) any of various evergreens with overlapping or compressed scale leaves often grown for ornament in hedges

arc Association for Retarded Citizens

arc. arcade; auto-refrigerated cascade

arc arco (Italian—bow, indicating end of *pizzicato* passages)

Arc Arachon; Arcade; Archaic; Arctic

ARC Aboriginal Research Club; Addicts Rehabilitation Center; Agricultural Relations Council; Agricultural Research Council; AIDS-Related Complex(es); Air Rescue Center; Air Reserve Center; Aircraft Radio Corporation; Airlines Recording Corporation; Airworthiness Requirements Committee; Alcoholic Rehabilitation Center; Ameri-

can Red Cross; Ames Research Center (NASA); Appalachian Regional Commission; *Armada República de Colombia* (Spanish—Colombian Navy); Asian Research Center (Harvard); Association of Rehabilitation Centers; Association of Retail Confectioners; Association for Retarded Citizens; Atlantic Research Corporation; Atomedic Research Center; automatic radio control; cable laying or repair ship (naval symbol)

ARC Association des Restauratrice-Cuisinières (French—Association of Women Restaurant-Cook Owners)

ARCA Associate of the Royal College of Art

ARCA Associate of the Royal Canadian Academy of Arts

ARCAD Association of Renault and Chrysler for Automotive Development

arcade. automatic radar-control-and-data equipment

Arc Arch Arctic Archipelago (Canadian Arctic)

ARCAS Automatic Radar Chain Acquisition System

ARCB Air Resources Control Board

arccos arccosine

arccot arccotangent

arccsc arccosecant

Arc Cur Arctic Current

arce amphibious river-crossing equipment

ARCE American Record Collectors Exchange

ARCen Air Reserve Center

arch. archaic; archipelago; architect(s); architectural; architecture

Arch archive, archivum, archiwum, archway

Arch archipiélago (Spanish—archipelago)

ARCH Articulated Computing Hierarchy

archaeol archaeologist; archaeology

Archam Archambault; prison in Archambault, Québec

Arch-Bish Archbishop

archcrit arch critic; architechtural critic(ism)

Archd Archdeacon; Archduke

Arch de Cln Archipelago de Colón

Archdn Archdeacon

Arch E Architectural Engineer

archeo archeological; archeologist; archeology

archeo (Latin prefix—beginning)—archaic, archeologist, archeology
archeol archeology
Archeol Archeology
Archeoz Archeozoic
Arches Arches National Monument, Utah
archi archival; archive; archivist
ARCHI *Asociación de Radiodifusoras de Chile* (Spanish—Association of Chilean Broadcasters)
Archie Archibald
archip archipelago
archipel (Dutch or French—archipelago)
archit architecture
Archit Architecture
Archive Archive Press
Archives Soc Hist Archives of Social History
Arch Neurol *Archives of Neurology*
Arch Rec Bks Architectural Record Books
archv archive
Archy Archibald
ARCI Addiction Research Center Inventory; American Railway Car Institute
ARCIC Anglican-Roman Catholic International Commission
Arclos Army Close support
ARCM Associate of the Royal College of Music
ARCNet Attached Resource Computer Network
ARCNS American Red Cross Nursing Services
arco *arcata* (Italian—bow) stroke of a bow (music)
Arco Arco Publishing Company
Arc O Arctic Ocean Command
ARCO Associate of the Royal College of Organists; Atlantic Richfield Company
ARCom Army Research
ARCOMET Area Commander's Meeting (NATO)
ARCON Advanced Research Consultants
ARCOPS Arctic Operations
arcos arc cosine
ARCOS Anglo-Russian Cooperative Society
ARCOV Army Combat Operations Vietnam
arcp air refueling control point
ARCR Arthritis and Rheumatism Council for Research
ARCRL Agricultural Research Council Radiobiological Laboratory
ARCs Alcoholic Rehabilitation

Centers
ARC(s) AIDS-Related Complex(es);
ARCS Achievement Rewards for College Scientists; Air Resupply and Communication Service; Australian Red Cross Society
A.R.C.S. Associate of the Royal College of Science; Associate of the Royal College of Surgeons
ARCSA Aviation Requirements for the Combat Structure of the Army
arcsec arcsecant
arcsin arcsine
ARCST Associate of the Royal College of Science and Technology
arct air refueling control time
Arct Arctic
arctan arctangent
Arctic Unicorn one-tusked narwhal of the Canadian arctic
ARCUK Architects' Registration Council of the United Kingdom
ARCUP Atlantic Region Canadian University Press (news cooperative)
ARCVS Associate of the Royal College of Veterinary Surgeons
arc/w arcweld
ard acute respiratory disease
ar & d aeronautical research and development; air research and development
Ard Ardrossan
ARD Accelerated Rehabilitated Disposition; Accelerated Rural Development; Air Reserve District; American Research and Development (corporation); Appalachian Regional Development; *Arbeitsgemeinschaft Rundfunkanstalten Deutschland* (German—German National Broadcasting); Army Renegotiation Division; Association of Research Directors; auxiliary floating dock (naval)
AR & D air research and development
arda analog recording dynamic analyzers
ARDA Advanced Reactor Development Associates; American Railway Development Association
ARDB Australian Resources Development Ban
ARDC Aberdeen Research and

Development Center; Action Resource Development Center; Agricultural Refinance and Development Corporation; Air Research and Development Command; American Racing Drivers Club
ARDCM Air Research and Development Command Manual
ARDCO Applied Research and Development Company
arddie analysis, requirements determination, design, development, implementation, and evaluation
ARDE Armament Research and Development Establishment (Ministry of Supply)
Ardennes Ardennes Forest of Belgium, France, and Luxembourg
ARDG Army Research and Development Group (USA)
ARDG(E) Army Research and Development Group (Europe)
ARDG(FE) Army Research and Development Group (Far East)
ARDIS Army Research and Development Information System
ARDM medium auxiliary repair drydock (naval symbol)
Ardnamurchan Point Ardnamurchan—Scottish headland
ardo *ardito* (Italian—bold, daring, fearless)
ard's analog recording dynamic analyzers
ARDS Aviation Research and Development Service
ARDU Analytical Research and Development Unit
ARDXC Australian Radio DX Club
are. (ARE) air reactor experiment
ARE Admiralty Research Establishment; Arab Republic of Egypt; Association for Research and Enlightenment
A.R.E. Associate in Religious Education
AREA Aerovias Ecuatorianas (Ecuadorian Airways); American Railway Engineering Association; American Recreational Equipment Association; Army Reactor Experimental Area; Association of Records Executives & Administrators; Australian Remedial Education Association
a'ready all ready
AREC Amateur Radio Emergency Corps

AREFS Air Refueling Squadron
AREI Associate of the Real Estate and Stock Institute (Australia)
ARENA Adoption Resource Exchange of North America
ARENA *Alianza Republicana Nacionalista* (Spanish—National Republican Alliance) rightist party in El Salvador; *Alianca Renovadora Nacional* (Portuguese—National Renovating Alliance)—political party in Brazil
aren't are not
ARENTS ARPA Environmental Test Satellite
Areop Areopagite; *Areopagitica* (Milton's pamphlet advocating freedom of the press); Areopagus
arestem a recording stray energy monitor
ARETO Arab Republic of Egypt Telecommunications Organization
AREUEA American Real Estate and Urban Economics Association
arf acute renal failure; acute respiratory failure; dog's bark
ARF Addiction Research Foundation; Advertising Research Foundation; African Research Foundation; Air Reserve Force(s); American Radio Forum; American Rationalist Federation; American Rehabilitation Foundation; American Retail Foundation; American Rose Foundation; Armour Research Foundation; Arthritis and Rheumatism Foundation; Australian Road Federation
ARFA Allied Radio Frequency Agency
ARFC Air Reserve Flying Center
ARFCOS Armed Forces Courier Service
ARFDC Atomic Reactor and Fuel Development Corporation
ARFL Australian Rugby Football League; Australian Rules Football League
arfo after receipt firm order
arfor area forecast
ARFPC Air Reserve Forces Policy Committee
ARFU Australian Rugby Football Union
arfx automatic riveting fixture
arg argent; argot; argument; argumentation; argumentative; argumentator (a controversialist); argus; arresting; arresting gear
arg (Arg) arginine
arg argang (Dano-Norwegian—yearbook); *argol* (Mongolian—dried camel or cattle dung fuel)
Arg Argentina; Argentinian; Argyll
Arg Argelia (Spanish—Algieria)
ARG Aerolineas Argentinas; repair ship, internal combustion engine
ARG American Record Guide
arga appliance, range, adjust (data processing)
arg. ad bac. argumentum ad baculum (Latin—argument of the club) appeal to force
arg. ad hom. argumentum ad hominem (Latin—personal attack)
arg. ad ignor. argumentum ad ignorantum (Latin—argument from ignorance)
arg. ad miser. argumentum ad misericordiam (Latin—appeal to pity)
arg. ad veri. argumentum ad vericundiam (Latin—argument based on misuse of authority)
ARGCA American Rice Growers Cooperative Association
Argel Argelia (Spanish—Algeria); *Argelía* (Portuguese—Algeria)
Argen Argentine; Argentinian
Argentina Argentine Republic, *República Argentina*
Argie(s) Argentino(s)
argmt arrangement
Argosy Argosy-Antiquarian Limited; Hawker-Siddeley four-engine turboprop transporting 54 paratroopers or 69 regular troops
ARGS American Rock Garden Society
argus advanced research on groups under stress
Argus Canadair long-range reconnaissance plane designated CL-38
ARGUS Automatic Routine Generating and Updating System
Argyll Argyllshire
arh (ARH) advanced reconnaissance helicopter
a Rh am Rhein (German—on the Rhine)
ARH heavy-hull repair ship (3-letter symbol)
ARHA Associate of the Royal Hibernian Academy
arh/ir anti-radiation homing/infrared
ARHS Associate of the Royal Horticultural Society; Australian Railway Historical Society
Ari Aries (constellation); Aristotle
ARI Acne Research Institute (Newport Beach, California); Air-Conditioning and Refrigeration Institute; Alcoholic Rehabilitation, Inc; Aluminum Research Institute; American Reciprocal Insurers; American Refractories Institute; American Russian Institute; Atherosclerosis Research Institute; Automatic Radio Information
ARIA Accounting Research International Association; Adult Reading Improvement Association; American Risk and Insurance Association
ARIANA Ariana Afghan Airlines
ARIB Asphalt Roofing Industry Bureau
ARIBA Associate of the Royal Institute of British Architects
ARIC Associate of the Royal Institute of Chemistry
ARICRSU American Russian Institute for Cultural Relations with the Soviet Union
ARICS Associate of the Royal Institution of Chartered Surveyors
ARIEL Automated Real-Time Investments Exchange Limited
ARIEM Army Research Institute of Environmental Medicine
aries astronomical radio interferometric earth surveying
ARIES Advanced Radar Information Evaluation System
arima autoregressive integrated moving average
ARINA Associate of the Royal Institution of Naval Architects
ARINC Aeronautical Radio Incorporated
ARIO Association of Retired Intelligence Officers
arip automatic rocket impact predictor
aris (ARIS) advanced range instrumentation ships
ARIS Activity-Reporting Information System; Advanced

Research Instrument System; Aerial Radio Instrument System; Aircraft Research Instrumentation System
Arist Aristotle
ARIST Annual Review of Information Science and Technology
Arista high-school honor society
Aristo Aristocles; Aristol; Aristotle
aristocat(s) aristocratic cat(s)
Aristoph Aristophanes
aristo(s) aristocrat(s)
ARISTOTLE Annual Review and Information Symposium on the Technology of Training, Learning, and Education (DoD)
arit aritmética (Portuguese or Spanish—arithmetic)
A.R.I.T. American Registered Inhalation Therapist
arith arithmetic(al)(ly); arithmetician
Arith Arithmetic
Ariz Arizona; Arizonian
Ariz Hist Found Arizona Historical Foundation
Arizona Girls Arizona Girls School at Phoenix (correctional facility)
ARJIS Automated Regional Justice Information System
Ark Arkansas; Arkansan
Ark City Arkansas City, Arkansas
Ark f Fys Arkiv for Fysik (Danish—Physics Archives)
ARKIA Israel Inland Airlines
arl acceptable reliability level; air run landing; average remaining lifetime
ARL Admiralty Research Laboratory (UK); Aeromedical Research Laboratory; Aeronautical Research Laboratory; Aerospace Research Laboratory; American Reefer Line; American Republics Line; American Roque League; Americans for Religious Liberty; Anesthesia Research Laboratories; Applied Research Laboratory (Johns Hopkins University); Association of Research Libraries; Australian Rugby League; landing craft repair ship (3-letter naval symbol)
ARLA Arab Latin American Bank
Arlanda Stockholm, Sweden's airport
ARLD Army Logistics Re-

search and Development
ARLHS Australian Railway and Locomotive Historical Society
Arlington Arlington Books (Louisville, Ky); Arlington House (New Rochelle, NY); Arlington National Cemetery in Arlington, Virginia
Arlington House Robert E Lee's home overlooking the Potomac
Arlington Hse Arlington House
ARLIS Arctic Research Laboratory Island (USN)
ARLIS/NA Art Libraries Society/North America
ArLO Army Liaison Officer
ARLO Art Reference Libraries of Ohio
Arlon Luxembourg, Belgium's provincial capital
ARLSEA Active-Retired Lighthouse Service Employees Association
arm (ARM) adjusted-rate mortgage
arm. anti-radar missile (ARM); anti-radiation missile; area radiation monitor(s); armature; arming; armor(ed)
arm. (ARM) adjustable-rate mortgage; anti-radar missile; anti-radiation missile
arm armenisch (German—Armenian)
Arm Armagh; Armenia(n)
Ar.M. Architecturae Magister (Master of Architecture)
ARM Abortion Rights Mobilization; Accredited Resident Manager; Animal Rights Movement; Aryan Resurgence Movement (neo-Nazi organization); Auditory Rehabilitation Mobile; Australian Reform Movement
A.R.M. Allergy Relief Medicine
arma armature
ARMA American Bosch Arma Corporation; American Records Management Association; Anglican Renewal Ministries Australia; Association of Records Managers and Administrators; Australian Rubber Manufacturers Association
a&r man artist and repertory man (supervising phonograph record production)
Armand Armando
ARMCM Associate of the Royal Manchester College of Music

ARMCOM Armament Command (USA)
ARMCOMSAT Arab Communications Satellite System
armd armored
armd (ARMD) age-related macular degeneration
Armen Armenia(n)
armet area forecast (given in metric system)
armgrd armed guard
ARMH Academy of Religion and Mental Health
ARMI American Rack Merchandisers Institute; American Research Merchandising Institute; Army Resources Management Institute
Armin Armindo
ARMIS Agricultural Research Management Information System
ARMIT Associate of the Royal Melbourne Institute of Technology
arml airmail
armm analysis and research methods for management
ARMM Association of Reproduction Materials Manufacturers
ARMMA American Railway Master Mechanics' Association
ARMMS Automated Reliability and Maintainability Measurement System
AR/MONP Ayers Rock/Mount Olga National Park (Northern Territory, Australia)
ARMOP Army Mortar Program
armor armorikanisch (German—Armorican) Celtic language of Brittany
Armor Armoric
armpl armorplate
armr armorer
ARMS Advanced Receiver Model System; Aerial Radiological Measuring Survey; Amateur Radio Mobile Society; Automated Records Management System
arm-saf arm-safe (switch)
armt armament
ARMU Associated Rocky Mountain Universities
army disease drug addiction
Army NG Army National Guard
Army ROTC Army Reserve Officers' Training Corps
arn (ARN) ácido ribonucleico (Spanish—ribonucleic acid)—rna (RNA)
a Rn am Rhein (German—on

the Rhine)
Arn Arnold
ARN Stockholm, Sweden (Arlanda Airport)
ArNa Army with Navy
ARNA Arab Revolution News Agency
ARNAT Asia Research News Analysis Team
arng arrange
ARNG Army National Guard
A Rn I Association of Rhodesian Industries
Arnie Arnold
ARNM Aztec Ruins National Monument
ARNMD Association for Research in Nervous and Mental Disease
Arno Arnold; Arnon; Arnot
aro after receipt of order; airborne range only
ARO Air Radio Office(r); Applied Research Objective; Army Research Office; Army Routine Order; Asian Regional Organization; Association for Research in Ophthalmology; Association of Roentgenological Organizations
AROCC Association for Research of Childhood Cancer
arod airborne range and orbit determination
ARO-FE Army Research Office—Far East
arom aromatic; artificial rupture of membranes
aromath aromatherapist; aromatherapy
AR-ONP Ayers Rock-Olgas National Park (Australia)
arp airborne radar platform; airport reference point; alternator research package; (cartoonist's symbol—dog's bark)
ARP Acreage Reduction Program; Advanced Research Project(s); Aeronautical Recommended Practice(s); Air Raid Precautions; American Registry of Pathologists; Ammunition Refilling Point; Area Redevelopment Program; Association for Realistic Philosophy; Australian Reptile Park (New South Wales); Australian Republican Party
ARP *Anti-Revolutionaire Partij* (Dutch—Anti-Revolutionary Party)
ARPA Advanced Research Projects Agency
ARPANET Advanced Research Projects Agency (computer)

Network
ARPAS Air Reserve Pay and Allowance System
ARPAT Advanced Research Projects Agency Terminal (defense system)
ARPC Air Reserve Personnel Center
arpd (ARPD) applied research planning document
arpege air-pollution episode game
ARPEL *Asistencia Recíproca Petrolera Estatal Latinoamericana* (Spanish—Latin American State Petroleum Reciprocal Assistance)—international agency
arpl a retrieval-process language
Arpo *arpeggio* (Italian—producing the tones in a chord successively rather than simultaneously)
ARPO Association of Resort Publicity Officers
ARPOA Association of Railway Professional Officers of Australia
Arprt Airport (postal placename abbreviation)
ARPS Advanced Radar Processing System; Arab Physical Society; Associate of the Royal Photographic Society; Association of Railway Preservation Societies; Australian Radiation Protection Society; Australian Royal Photographic Society
ARPSA Army Postal Service Agency
Arpt Airport
ARPT American Registry of Physical Therapists
a-r pulse apical-radial pulse
arq *arquiteto* (Spanish—architect); *arquitectura* (Spanish—architecture); *arquiteto* (Portuguese—architect); *arquitetura* (Portuguese—architecture)
ARQ automatic error correction; automatic request for repetition
Arq⁰ *Arquiteto* (Portuguese or Spanish—architect)
arquo *arquipélago* (Portuguese—archipelago)
arr airborne radio receiver; arrange(ment); arrestor; arrival; arrive; arriving
arr (ARR) arrest(ed)
arr *arrecife* (Spanish—reef)
Arr *arrondissement* (French—district)
ARR Air Regional Representa-

tive; Air Reserve Record(s); American Right to Read; Army Retail Requirements
ARRA Amateur Radio Retailers Association; Australian Rough Riders Association
ARRB Australian Road Research Board
ARRC Air Reserve Records Center; Associate of the Royal Red Cross
ARRCS Air-Raid Reporting Control Ship
ARRDO Australian Railway Research and Development Organisation
arre *arrecife* (Spanish—reef, roadbed, rocky road)
ARRES Automatic Radar-Reconnaissance Exploitation System
arrex arriving ex . . .
ARRF Australian Reading Research Federation; Automatic Recording and Reduction Facility
ARRGp Aerospace Rescue and Recovery Group (USAF)
Arri Arnold and Richter (reflex motion-picture camera)
ARRL American Radio Relay League
arr n arrival notice
arro *arroyo* (Spanish—creek, brook, stream)
arro seco *arroyo seco* (Spanish—dry creek, brook bed, streambed, or riverbed)
arrowhead symbol used to indicate direction
ARRS Aerospace Rescue and Recovery Service; Aircraft Refueling and Rearming System; American Roentgen Ray Society; American-Russian Research Society
ARRSq Aerospace Rescue and Recovery Squadron (USAF)
ARRT American Registry of Radiologic Technologists
ARRTC Aerospace Rescue and Recovery Training Center (USAF)
ARRWg Aerospace Rescue and Recovery Wing (USAF)
arry arrhythmia
ars active repeater satellite; aerospace research satellite; arsenal; asbestos roof shingles
Ars Arsenal
ARs Action Requests
ARS Advanced Record System; Aerospace Research Satellite; Agricultural Research Service; Airail Service (monorail); Air Rescue Service; American Re-

corder Society; American Records Society; American Recreation Society; American Repair Society; American Rescue Service; American Rhododendron Society; American Rocket Society; American Rose Society; Army Relief Society; salvage ship (naval symbol)
ARS Annual Report to Shareholders
ARSA Associate of the Royal School of Art
arsab arsonist sabotage; arsonist saboteur
arsabs arsonist saboteurs
ARSC Association of Recorded Sound Collections
ARSD Advanced Reentry System Deployment; salvage lifting ship (naval symbol)
ARSDEP Asian Regional Skills Development Programme
arsen arsenal
arsg arising
ARSH Associate of the Royal Society for the Promotion of Health
ARSI Associate of the Royal Sanitary Institute
arsin arc sine
Arsl Arsenal (postal place-name abbreviation)
ARSL Associate of the Royal Society of Literature
ARSM Associate of the Royal School of Mines
ARSP Aerospace Research Support Program
arspa aerial reconnaissance and surveillance penetration analysis
ARSPH Associate of the Royal Society for the Promotion of Health
arsr air route surveillance radar
ARST Aerial Reconnaissance and Security Troop; salvage craft tender (naval symbol)
ARSTAF Army Staff
ARSTRAC Army Strike Command
ARSU Alcohol Rehabilitation Services Unit (Navy Regional Medical Center in Long Beach, California)
ARSV armored reconnaissance scout vehicle (USA)
art. advanced research and technology; airborne radiation thermometer; art assembly; arterial; artery; article; articulate; articulation; artifact; artificial; artillery; artisan; artist; artistic; artistry; automatic re-

porting telephone
'art heart
art artikel (Dano-Norwegian— article)
Art Arthur; Arturo
Art Artikel (German—article)
ART Accredited Record Technician; Air Reserve Technician; American Repertory Theater; Arithmetic Reading Test; Arithmetic Reasoning Test; Aviation Radio Technician
ARTA American River Touring Association; Association of Retail Travel Agents
artac advanced reconnaissance and target acquisition capabilities
ARTADS Army Tactical Data Systems
artb (ARTB) automatic return to base
ARTC Addiction Research and Treatment Center; Addiction Research and Treatment Corporation; Air Route Traffic Control
ARTCC Air Route Traffic Control Center
Art C-Part articles of co-partnership
artcrit art critic(ism)
art deco arts décoratifs (French—decorative arts)— decorative style of the 1920s and 1930s emphasizing bold outlines, geometric forms, streamlining, and strong colors
art. dir artistic director
Art Div Artillery Division
ARTDO Asian Regional Trade Development Organisation
ARTE Admiralty Reactor Test Establishment
ARTEMIS Automatic Retrieval of Text through European Multipurpose Information Systems
Artemus Ward Charles Farrar Browne
ARTEP Army Training and Evaluation Program; Asian Regional Team for Employment Promotion
artesian artesian well (of the type originating in Artois, France)
ARTF Advanced Radiation Technology Facility; Asian Rice Trade Fund; Australian Road Transport Federation
Arth Arthropoda; Arthur; Arthurian
arthro arthron (Greek—joint)

—arthritis, arthropod
artic articulate(d); articulation
Articles Articles of Agreement between a ship's crew and its master
Artie Artemas; Artemisia; Artemus; Arthur; Artur; Arturo; Artus
artif artificer(s); artificial(ly)
Artigas José Gervasio Artigas—defender of Uruguayan independence
ARTINS Army Terrain Information System
art. insem artificial insemination
Art Kill Arthur Kill waterway between New Jersey and Staten Island, New York
art nou art nouveau
art nr artikelnummer (Dano-Norwegian—article number)
arto air run takeoff
art° articulo (Italian—article); *artículo* (Spanish—article); *artigo* (Portuguese—article)
art° artículo (Spanish—article)
ARTO Advanced Radiation Technology Office
ARTOC Army Tactical Operational Control; Army Tactical Operations Central
ARTP Army Rocket Transportation System
art. pf artist's proof
artrac advanced range testing, reporting, and control
artron(s) artificial neuron(s)
arts. articles
ARTS Advanced Radar Traffic Control System; Automatic Radar Traffic Control System
ArtSci Arts and Sciences
artsem artificial insemination
arts graph arts graphiques (French—graphic arts)
ARTSM Association of Road Sign Makers
ARTS & P Australian Radio Technical Services and Patents
artt automatic rubber tensile tester
ARTT Annual Review Travelling Team (NATO)
artu automatic range tracking unit
arty artillery
aru analog remote unit; audio response unit
Aru Aruba
ARU Air Reserve Unit; American Railway Union; Australian Railway Union
arv (ARV) advanced reentry vehicle; aeroballistic reentry

vehicle

Arv *Arvoisa* (Finnish—esteemed)

ARV aircraft engine overhaul and structural repair ship; American Revised Version

ARVA aircraft repair ship for aircraft (4-letter designation)

ARVE aircraft repair ship for engines (4-letter designation)

ARVH aircraft repair ship for helicopter (naval symbol)

ARVIA Associate of the Royal Victorian Institute of Architects

ARVN Army of the Republic of Vietnam

ARVO Association for Research in Vision and Ophthalmology

ARVSG Air Reserve Volunteer Support Group

arw attitude reaction wheel

ARW Air Raid Warden; Air Raid Warning; Air Reserve Wing (Canada)

ARWC Army War College

ARWEN Anti-Riot Weapon (shoots rubber bullets)

ARWH Air Reserve Wing Headquarters

ARWS Antiradiation Weapon System; Associate of the Royal Society of Painters in Water Colours

Aryabhata Indian spacecraft named for the fifth-century astronomer

Arz *Arzobispo* (Spanish—Archbishop)

ARZ Active Reconnaissance Zone

Arzbpo *Arzobispo* (Spanish—Archbishop)

as. airscoop; air-to-surface missile; alloy steel; antiseptic; aortic stenosis; asymmetric

a-s ascendance-submission

a/s account sales; after sight; airspeed; alongside; antisubmarine

a & s accident and sickness (insurance)

as. *aanstaande* (Dutch—next)

a.s. *auris sinistra* (Latin—left ear)

a/s. *aux soins de* (French—in care of)

As altostratus; arsenic; Asia; Asian; Asiatic; astigmatism; aunicles; Australia(n)

A$_s$ atmosphere standard

AS Abilene & Southern (railroad); Academy of Science(s); Aeronautical Standard(s); Air Service; Air Speed; Air Staff;

Air Station; Air Surveillance; Airports Service; air-to-surface missile; Alaska Airlines; Ambulatory Surgi-Center; American Samoa; Anglo-Saxon; antisubmarine; Apprentice Seaman; Army Security; Army Staff; Associated Steamships; Australian Standard(s); Auto Squad (police); submarine tender (naval symbol)

A.S. Antonius Stradivarius (initials usually accompanied by a Maltese cross, both enclosed in a double circle)

A-S Allied-Signal Corporation

A/S alongside (barge, cargo carrier, lighter)

A & S Alton & Southern (railroad); Arts and Sciences

A of S Academy of Science

AS *Anonim Sirket* (Turkish—joint stock company); *Arabia Saudita* (Spanish—Saudi Arabia); *Aviaeskadra* (Russian— air squadron)

A/S *Aksjeselskap* (Norwegian—limited company); *Aktieselkab* (Danish—joint stock company)

AS–11 Nord air-launched antitank missile

AS–12 Nord automatic-telecommand antitank missile

AS–20 Nord air-to-surface radio-controlled missile

AS–30 improved version of AS-20 with longer range and heavier warhead

AS–33 Nord air-launched inertial-guidance missile

asa acetylsalicylic acid (aspirin); antistatic additive; azimuth servo assembly

aSa anti-Soviet agitation

A-S a Adams-Stokes attack

ASA Acoustical Society of America; Actuarial Society of America; *Aerovias Sud Americana* (Spanish—South American Airways); African Studies Association; Alaska Airlines; Aluminum Siding Association; Amateur Softball Association; Amateur Swimming Association; American Scientific Affiliation; American Shorthorn Association; American Sightseeing Association; American Society for Abrasives; American Society for Aesthetics; American Society of Agronomy; American Society of Anesthesiologists; American Society of Apprais-

ers; American Society of Auctioneers; American Sociological Association; American Sociometric Association; American Softball Association; American South African Line; American Soybean Association; American Standards Association; American Statistical Association; American Stockyards Association; American Studies Association; American Sunbathing Association; American Surgical Association; Anthroposophical Society of America; Army Seal of Approval; Army Security Agency; Assistant Secretary of the Army; Associated Stenotypists of America; Association for the Study of Abortion; Association of Secretaries in Asia; Association of Southeast Asia; Atomic Security Agency; Australian Shareholders Association; Australian Sheep-breeders Association; Australian Society of Anaesthetists; Australian Society of Authors; Aviation Supply Annex

A of SA (**ASA**) Association of Southeast Asia

ASAA Amateur Softball Association of America; American Society for the Abandonment of Acronyms; Armenian Students Association of America; Asian Studies Association of Australia

ASAALH Association for the Study of Afro-American Life and History

ASAB Association for the Study of Animal Behavior

ASAC American Society for African Culture; American Society of Agricultural Consultants; Army Study Advisory Committee; Assistant Special Agent in Charge

ASAE American Society of Agricultural Engineers; American Society of Association Executives; American Society of Automotive Engineers

AS of AF Assistant Secretary of the Air Force

ASAH American Society of Association Historians

Asahi *Asahi Shimbun* (Japanese—Rising Sun Newspaper)

ASAIHL Association of Southeast Asian Institutions of Higher Learning

ASAIO American Society for Artificial Internal Organs
ASALA Armenian Secret Army for the Liberation of Armenia; Associate of the South African Library Association
asalm (ASALM) advanced strategic air-launched multi-mission missile
ASAM American Society for Abrasive Methods
ASAN Adriatica Società per Azioni di Navigazióne (Italian—Adriatic Society for Navigation)
ASAnes American Society of Anesthesiologists
ASAO Association for Social Anthropology in Oceania
asap analog system assembly pack; as soon as possible
asap (ASAP) antisubmarine attack plotter
ASAP Aircraft Synthesis Analysis Program; Airlines of South Australia Pty; Airlines of South Australia Pty; Antenna-Scatterer Analysis Program; antisubmarine attack plotter; Australian Society of Animal Production
ASAPs Alcohol Safety Action Projects
ASAPS American Society for Aesthetic Plastic Surgery; Anti-Slavery and Aborigines Protection Society
asar (ASAR) advanced surface-to-air ramjet
ASARCO American Smelting and Refining Company
ASAS Advertising Standards Authority of Singapore; American Society of Abdominal Surgery; American Society of Animal Science; Army Security Agency School; Australian Staffing Assistance Scheme
a-sat (A-Sat) anti-satellite
ASAT anti-satellite
ASATs antisatellite weapons
asatt advanced small-axial-turbine technology
ASAWS Advanced Surface-to-Air Weapon System
asb aircraft safety beacon; asbestos
asb (ASB) anxiety scale for the blind
as & b aloin, strychnine, and belladona (pills)
ASB Administration and Storage Building; Air Safety Beacon; Air Safety Board; Air Staff Board; Aircraft Safety

Beacon; American Society of Bacteriologists; Associated Student Body; Australian Savings Bonds; Australian Shipping Board
A.S.B. Associate in Science in Business; Associate in Specialized Business
ASBA Australian Sheep Breeders Association
ASBAH Association for Spina Bifida and Hydrocephalus
asb c asbestos covered
ASBC American Society of Biological Chemists
ASBC American Standard Building Code
ASBCA Armed Services Board of Contract Appeals
ASBCO American Ship Building Company
ASBD Advanced Sea-Based Deterrent Program
ASBDA American School Band Directors Association
ASBE American Society of Bakery Engineers
A.S.B.E. Associate in Science in Basic Engineering
asb & i aloin, strychnine, belladona, and ipecac
asbl assemble
ASBO Association of School Business Officials
ASBPA American Shore and Beach Preservation Association
ASBPE American Society of Business Press Editors
asc altered state of consciousness; arteriosclerosis; arteriosclerosistic; ascarid; ascaridian; ascend; ascender; ascending; ascension; ascent; ascertain; ascertainable; automatic sequence control; automatic switching center; auxiliary switch closed
as & c aerospace surveillance and control
Asc Ascidian
A.Sc. Associate in Science
ASC Adelaide Steamship Company; Aeronautical Systems Center; Air Service Command; Air Support Command; Air Support Control; Air Systems Command; Alabama State College; Alaska Steamship Company; Albany State College; All Souls College; American Samoa Commission; American Security Council; American Silk Council; American Society of Cinematographers; American So-

ciety of Criminology; American Society for Cybernetics; American Society of Cytology; Area/Site Characterizations; Arizona State College; Arkansas State College; Army Service Corps; Army Subsistence Center; Asian Socialist Conference; Associated Sandblasting Contractors; Associated Schools of Construction; Australian Schools Commission; Australian Shippers Council
A & SC Adhesive and Sealant Council
asca automatic science citation alerting; automatic subject citation alert
ASCA American School Counselor Association; American Senior Citizens Association; American Speech Correction Association; Associate of the Society of Company and Commercial Accountants; Association for Science Cooperation in Asia; Association of State Correctional Administrators
ASCAA Automobile Seat Cover Association of America
ASCAC Antisubmarine Contact Analysis Center
ASC/AIA Association of Student Chapters/American Institute of Architects
ascap at-sea calibration procedure
ASCAP American Society of Composers, Authors, and Publishers
ASCAT Antisubmarine Contact Analysis Team
ASCATS Apollo Simulation Checkout and Training System
ASCB American Society of Cell Biology
ASCC Adams State College of Colorado; Air Standardization Coordinating Committee; American Society for the Control of Cancer; Army Strategic Communications Command; Association of Senior Citizens Clubs; Australian Society of Cosmetic Chemists
ASCD Association for Supervision and Curriculum Development
ASCE American Society of Civil Engineers
ASCEA American Society of Civil Engineers and Architects
ASCEF American Security

Council Education Foundation
ASCEP Australian Society of Clinical and Experimental Pharmacologists
ASCET American Society of Certified Engineering Technicians
ASCHAL American Society of Corporate Historians, Archivists, and Librarians
ASCHE American Society of Chemical Engineers
ASCI American Society for Clinical Investigation
ASCIE American Standard Code for Information Exchange
ASCII American Standard Code for Information Interchange (pronounced *asky*)
ASCL Australia Straits Container Line
ASCLA Association of Specialized and Cooperative Library Agencies
ASCLD American Society of Crime Laboratory Directors
ASCLU American Society of Chartered Life Underwriters
ascm (ASCM) antiship capable missile
ASCM Association of Steel Conduit Manufacturers
ASCMA American Sprocket Chain Manufacturers Association
ASCN American Society of Clinical Nutrition
asco automatic sustainer cutoff
Asco Automatic Switch Company
ASCO American Society of Contemporary Ophthalmology; Arab Satellite Communications Organization
ASCom Army Service Command
ASCOPE ASEAN Council on Petroleum
ascore automatic shipboard checkout and readiness equipment
A Scot type-A Scottish influenza virus
ASCP American Society of Clinical Pathologists; American Society of Consulting Pharmacists; American Society of Consulting Planners
ASCPC American Society of Clinical Pharmacology and Chemotherapy
ASCPT American Society for Clinical Pharmacology and Therapeutics
ascr. *ascriptum* (Latin—as-

cribed to)
ASCRO Active Service Career for Reserve Officers
asc's (ASCs) altered states of consciousness
ASCs All Savers Certificates
ASCS Agricultural Stabilization and Conservation Service; American School of Classical Studies (Athens); Automatic Stabilization and Control System
ASCU Association of State Colleges and Universities
ascvd arteriosclerotic cardiovascular disease; atherosclerotic cardiovascular disease
A Sc W Association of Scientific Workers
asd aldosterone secretion defect; atrial septal defect
ASD Aeronautical Systems Division; Army Shipping Document; Artillery Spotting Division; Assistant Secretary of Defense; Association of Steel Distributors; Aviation Supply Depot
ASD (PA & E) Assistant Secretary of Defense (Program Analysis and Evaluation)
ASD *Association Suisse de Documentation* (Swiss Association of Documentation)
ASDA American Safe Deposit Association; American Seafood Distributors Association; American Stamp Dealers Association; Asbestos and Danville (railroad); Association of Structural Draftsmen of America; Atomic and Space Development Authority; Australasian Stamp Dealers Association
ASDAE Association of Seventh-Day Adventists Educators
ASD-ALA Adult Services Division—American Library Association
AsDB Asian Development Bank
ASDC Aeronomy and Space Data Center (NOAA); Automobile Safe Driving Center
asde aircraft surface detection equipment
a/s de *aux soins de* (French—in care of)
asder airfield surface-detection radar
ASDF Air Self-Defense Force (Japanese Air Force)
ASDG Aircraft Storage and Disposition Group

asdi automatic selective dissemination of information
ASDIC Anti-Submarine Detection Investigation Committee (British sonar); Armed Services Documents Intelligence Center
ASDIRS Army Study Documentation and Information Retrieval System
ASDM Apollo-Soyuz Docking Module
asdng ascending (flow chart)
asdr airport surface-detection radar
ASDR American Society of Dental Radiographers
ASDS American Society of Dental Surgeons
A/S D/S *Akties Dampskibsselskab* (Danish—steamship company, limited)
ASDT Australian Society of Dairy Technology
ASD(T) Assistant Secretary of Defense (Telecommunications)
ase airborne search equipment
ASE Amalgamated Society of Engineers; American Society of Enologists; American Steel Equipment; American Stock Exchange; Association of Science Education; Australian Society of Engineers; Australian Stock Exchange(s)
AS & E American Science and Engineering
ASEA *Allmänna Svenska Elektriska Aktiebolaget;* American Society of Engineers and Architects; Association of South-East Asia
ASEAN Association of Southeast Asian Nations (Brunei, Indonesia, Malaysia, the Philippines, Singapore, Thailand)
ASEANTA ASEAN Travel Association
ASEB Aeronautics and Space Engineering Board; Assam State Electricity Board
ASEC All Saints' Episcopal College; American Standard Elevator Code
ASECA Association for Education and Cultural Advancement (South Africa)
ASECS American Society for Eighteenth-Century Studies
ASED Aviation and Surface Effects Department
ASEE American Society for Ecological Education; American Society of Electrical Engineers; American Society for

Engineering Education; American Society for Environmental Education

ASEET Associate in Science in Electronic Engineering Technology

ASEI American Sports Education Institute

ASEM Anti-Ship Euro-Missile

ASEP American Society for Experimental Pathology; ASEAN Environment Program(s)

aseptics aseptically packaged liquids

ASESA Armed Services Electro-Standards Agency

ASESB Armed Services Explosive Safety Board

ASESS Aerospace Environment Simulation System

aset aeronautical satellite earth terminal

ASET Assistant Secretary for Energy Technology; Author System for Education and Training

ASETC Armed Services Electron Tube Committee

asew airborne and surface early warning

ASEWS Airborne and Surface Early Warning System

asf additional selection factor; amperes per square foot

a-s-f aniline-formaldehyde-sulfur

AsF America's Future

ASF Advisory Support Force; Aircraft Services Facility; Alaskan Sea Frontier; American Scandinavian Foundation; American Schizophrenia Foundation; Ammunition Storage Facility; Army Service Forces; Army Stock Fund; Association of State Foresters; Australian Soccer Federation; Australian Speleological Federation; Automotive Safety Foundation

ASFA American Steel Foundrymen's Association; Association of Superannuation Funds of Australia

ASFC All Sports Federation of China; Atlantic Salt Fish Commission (Canada)

ASFCO American Soda Fountain Company

asfe accelerometer scale factor error

ASFE American Society For Aesthetics; Association of Specialized Film Exhibitors

ASFEC Arab States Fundamen-

tal Education Center

ASFFHF Association of Science Fiction, Fantasy, and Horror Films

ASFH Albert Schweitzer Friendship House

asfip accelerometer scale factor input panel

asfir active swept-frequency interferometer radar

ASFIS Aquatic Science and Fisheries Information Service (FAO)

ASFM American Sexual Freedom Movement

ASFMRA American Society of Farm Managers and Rural Appraisers

ASFP Association of Specialized Film Producers

ASFS Australia-Soviet Friendship Society

ASFSA American School Food Service Association

asfts airborne systems functional test stand

asfx assembly fixture

asg assignment

ASG Aeronautical Standards Group (Air Force and Navy); American Saint Gobain (glass); American Society of Genetics

ASGA Advertising Specialty Guild of America

ASGB Aeronautical Society of Great Britain

ASGBI Association of Surgeons of Great Britain and Ireland

asgd assigned

ASGE American Society for Gastrointestinal Endoscopy

asgmt assignment

asgn assign; assignment

ASGp Aeronautical Standards Group (USAF)

ASGP Australian Society of General Practitioners

ASGS American Scientific Glassblowers Society

ASGW Association for Specialists in Group Work

ash. airship; armature shunt

ash. (ASH) aerial scout helicopter

Ash Ashbel; Ashburton; Ashbury; Ashdown; Asher; Asheto; Ashley; Ashman; Ashton; Ashur; Ashville; Ashvillian; Soviet infrared and radar-homing missile (NATO)

Ash Asahi Shimbun (leading Japanese newspaper)

AsH hyperopic astigmatism

ASH Action on Smoking and Health; American Society of Hematology; Ashland Oil and Refining (stock exchange symbol); Australian Society of Herpetologists; Australian Stationary Hospital

A-S-H Allen-Sherman-Hoff

A & SH Argyll and Southerland Highlanders

ASHA American School Health Association; American Social Health Association; American Social Hygiene Association; American Speech and Hearing Association

ASHACE American Society of Heating and Air-Conditioning Engineers

ASHBM Associate Scottish Hospital Bureau of Management

ASHC All-States Hobby Club

ash can ash-can school of 20th-century painters and photographers including The Eighth

ashd arteriosclerotic heart disease

ASHE American Society of Hospital Engineers; Association for the Study of Higher Education

Ashfo'd Ashford Remand Prison Center, London area

ASHG American Society of Human Genetics

ASHH American Society for the Hard of Hearing

ASHI Association for the Study of Human Infertility

ASHM Association of Scottish Hospital Matrons

Ashken Ashkenazim (Hebrew—Jews of central and northern Europe)

Ashland Youth Federal Youth Center (for delinquents) at Ashland, Kentucky

Ash Mus Ashmolean Museum

ashp airship

ASHP American Society of Hospital Pharmacists

ASHRAE American Society of Heating, Refrigerating, and Air-Conditioning Engineers

ASHS Advanced Study of Human Sexuality; American Society for Horticultural Science

ASHU Airline Stewards and Hostesses Union (New Zealand)

asi airspeed indicator; azimuthal speed indicator

ASI Advanced Scientific Instruments; Aero-Space Institute; Aerospace Studies Institute; Africa Service Institute; Air Society International; Amended Shipping Instruction(s); American Society of Indexers; American Specifications Institute; American Statistics Index; American Statistics Institute; American Swedish Institute; Asian Statistical Institute (Japan); Audience Studies, Incorporated; Australian Shipbuilding Industries; Aviation Simulations International

ASIA Airlines Staff International Association; Army Signal Intelligence Agency; Australian Scientific Industry Association; Australian Stevedoring Industry Authority

ASIAC Aerospace Structures Information and Analysis Center; Asian International Acceptances and Capital

ASIAL Australian Security Industry Association Ltd

Asian-Am(s) Asian-American(s)

ASIC Air Service Information Circular; Application Specific Integrated Circuit; Australian Standard Industry Classification

ASID American Society of Interior Designers; Australian Society of Implant Dentistry

ASIDIC Association of Scientific Information Dissemination Centers

ASIF Airlift Service Industrial Fund

ASI & H American Society of Ichthyologists and Herpetologists

ASII American Science Information Institute

ASIL American Society of International Law

ASILS Association of Student International Law Societies

ASIM American Society of Insurance Management; American Society of Internal Medicine

ASIMET *Asociacion de Industrias Metalurgicas* (Spanish—Association of Metallurgical Industries)—Chile

a sin a sinistra [Italian—at (to) the left]

ASI/NATO Advanced Study Institute/NATO

ASIO Australian Security Intelligence Organization

ASIP Army Stationing and Installation Plan

ASIRC Aquatic Sciences Information Retrieval Center (U of RI)

asis anterior superior iliac spine

ASIs American Society of Indexers

ASIS Abort-Sensing Implementation System; American Society of Industrial Security; American Society for Information Science; ammunition stores issue ship (naval designator); Association of Small Island States; Australian Security Intelligence Service

asist advanced scientific instruments symbolic translator

ASIST Alzheimer Support, Information, and Service Team

ASIWPCA Association of State and Interstate Water Pollution Control Administrators

ASJ Asiatic Society of Japan

ASJA American Society of Journalists and Authors

ASJJA Association of State Juvenile Justice Administrators

ASJMC Association of Schools of Journalism and Mass Communications

ASJSA American Society of Journalism School Administrators

ask. amplitude shift keying

Ask American standard keyboard (typewriter)

ASK Associated Students of Kansas; Association for Social Knowledge

ASKA Automatic System for Kinematic Analysis

askg asking

Askham G Askham Grange (female offender's prison in Yorkshire, England)

ASKS Automatic Station-Keeping System

ASKT American Society of Knitting Technologists

asl abandon ship ladder; above sea level

asl *altslavisch* (German—Old Slavic)

Asl (ASL) American sign language

ASL American Association of State Libraries; American Scantic Line; American Shuffleboard Leagues; American Sign Language; Anti-Saloon League; Australian Society for Limnology

A-SL Abelard-Schuman Limited

ASLA American Savings and Loan Association; American

Society of Landscape Architects; Arizona State Library Association; Australian School Library Association

ASLAB Atomic Safety and Licensing Appeal Board

ASLB Atomic Safety and Licensing Board (AEC)

AS & LB American Savings and Loan Bank

ASLBP Atomic Safety and Licensing Board Panel (NRC)

ASLE American Society of Law Enforcement; American Society of Lubrication Engineers

ASLEC Association of Street Lighting Erection Contractors

ASLEF Associated Society of Locomotive Engineers and Firemen

ASLEP Apollo Surface Lunar Experiments Package

ASLH American Society for Legal History

ASLHA American Speech-Language-Hearing Association

ASLIB Association of Special Libraries and Information Bureaus

ASLIS Association of Special Libraries and Information Services

ASLNY Art Students League of New York

aslo assembly layout

ASLO American Society of Limnology and Oceanography; Australian Scientific Liaison Office (London)

ASLP Association of Special Libraries in the Philippines

ASLR American Short Line Railroads

ASLRA American Short Line Railroad Association

aslt advanced solid logic technology; assault(ing)

aslv *assurance sur la vie* (French—life insurance)

ASLW Amalgamated Society of Leather Workers

asm air-to-surface missile; assembly; available-seat mile

AsM myopic astigmatism

ASM Air-to-Surface Missile; Alaska State Museum; American Society of Mamalogists; American Society for Metals; American Society for Microbiology; Antarctic Service Medal; Association of Systems Management

ASMA Aerospace Medical Association; American Society

of Music Arrangers; American Student Media Association

ASMAR Astilleros Maritimos (Spanish—Maritime Shipyards)—Chile

asmbl assemble (flow chart)

asmblr assembler

ASMC Army Supply and Maintenance Command (formerly Quartermaster Corps)

asmd (ASMD) anti-ship missile defense

asmd/ew antiship-missile defense/electronic warfare

ASME American Society of Magazine Editors; American Society of Mechanical Engineers; Association for the Study of Medical Education; Australian Society for Music Education

As Mem Associate Member

ASMFC Atlantic States Marine Fisheries Commission

ASMFS American Society of Maxillo-Facial Surgeons

ASMH Association for Social and Moral Hygiene

asmi airfield surface movement indication

ASMM American Supply and Machinery Manufacturers

ASMP American Society of Magazine Photographers

ASMPA Armed Services Medical Procurement Agency

ASMPE American Society of Motion Picture Engineers

asmr (ASMR) advanced short-to-median-range (twin-engine aircraft)

ASMR Australian Society for Medical Research

ASMRO Armed Services Medical Regulating Office

ASMS Advanced Surface Missile System

asmt assortment

ASMT American Society of Medical Technologists

asn average sample number

asn (ASN) asparagine (amino acid)

Asn Association

As of N Assistant Secretary of the Navy

ASN Allotment Serial Number; American Society of Naturalists; Army Serial Number; Army Service Number; Asiatic Steam Navigation; Assistant Secretary of the Navy

ASN (R & D) Assistant Secretary of the Navy (Research and Development)

ASNA Advertising Specialty National Association

asnap automatic-steerable null-antenna processor

ASNC Atlantic Steam Navigation Company

ASNDE Associate of the Society of Non-Destructive Examination

ASNE American Society of Naval Engineers; American Society of Newspaper Editors; Assistant Secretary for Nuclear Energy (DoE)

ASNLH Association for the Study of Negro Life and History

As & Ns Andamans and Nicobars (Andaman and Nicobar Islands)

ASNSW Anthropological Society of New South Wales; Art Society of New South Wales; Astronomical Society of New South Wales

ASNT American Society for Nondestructive Testing

aso arteriosclerosis obliterans; auxiliary switch open

ASO Adelaide Symphony Orchestra; Advanced Solar Observatory (in space); Aeronautica Supply Office(r); Air Signal Officer; Air Staff Officer; Air Staff Orientation; Air Surveillance Officer; Akron Symphony Orchestra; Albany Symphony Orchestra; Albuquerque Symphony Orchestra; American School of Orthodontists; American Sokol Organization; American Symphony Orchestra; Area Supply Office(r); Area Supplies Officer; Assistant Secretary's Office; Athens Symphony Orchestra; Atlanta Symphony Orchestra; Australian Security Organisation; Australian Society of Orthodontists; Aviation Supply Office(r)

ASOA American Society on Aging; Australian Shipping Officers Association

ASOC Air Support Operations Center

ASOK Angfartygas Svenska Ostasiatiska Kompaniet (Swedish—Swedish East Asiatic Steamship Line)

ASOL American Symphony Orchestra League

ASOP Atomic Standing Operation Procedures

ASOPA Australian School of Pacific Administration

ASOR American Schools of Oriental Research

ASOS American Society of Oral Surgeons

aso titer antistreptolysin titer

asp (ASP) aspartic acid

asp. affirmative self protection; ammunition supply point; aspartic acid; aspen; automatic servo plotter; automatic switching panel; automatic system procedure

a s p *accepté sous protêt* (French—accepted under protest)

Asp American selling price

ASP Amalgamated Society of Printers; American Schulzhund Products; American Society of Parasitologists; American Society of Pharmacognosy; American Society of Photogrammetry; Ammunition Supply Point; Antisubmarine Patrol; Archival Security Program; Arizona State Prison; Assistant Superintendent of Police; Association of Seattle Prostitutes; Astronomical Society of the Pacific; atmosphere-sounding projectile; Atomic Strike Plan; Australian Socialist Party; Australian Society for Parasitology; Australian Society of Prosthodontists; Automatic Schedule Procedure

A-S P Anglo-Saxon Protestant

A/S/P Aleksandr Sergeyvich

A.S.P. *accepté sans protêt* (French—accepted without protest)

ASPA Alloy Steel Producers Association; American Society for Personnel Administrators; American Society for Public Administration; Australian Sugar Producers Association

ASPAC Asia and South Pacific Area Council; Asian and Pacific Council

A-span anticipation span (eye-voice span); capital-A-shaped span

ASPAP Australian South Pacific Aid Programme

ASPB Armed Services Petroleum Board

aspc *accepté sous protêt pour acompte* (French—accepted under protest for account)

ASPC American Sheep Producers Council

ASPCA American Society for the Prevention of Cruelty to Animals

ASPCC American Society for the Prevention of Cruelty to Children

ASPERS American Society of Professional Draftsmen

aspect. acoustic short-pulse echo-classification techniques

ASPER Assembly System for Peripheral Processors; Assistant Secretary (of Labor) for Police Evaluation and Research

ASPERS Armed Services Procurement Regulations

ASPET American Society for Pharmacology and Experimental Therapeutics

ASPF Association of Specialized Film Producers; Association of Superannuation and Pension Funds

ASPFA Association of Superannuation and Provident Funds of Australia

asph asphalt; asphaltic

ASPH Australian Society of Professional Hypnotherapists

asphalt solid bitumen pitch

asphaltum mineral pitch

asphic asphaltic

asph mac asphalt macadam

asphy asphyxia

ASPI American Society for Performance Improvement

ASPIRE Associated Students Promoting Individual Rights for Everyone

ASPIRIN Automatic System for Passenger Reservation by Notation

ASPL American Society for Law Pharmacy; Associated Steamships Proprietary Limited

ASPLP American Society for Political and Legal Philosophy

ASPM American Society of Paramedics

ASPM Armed Services Procurement Manual

aspn asparagine

ASPO American Society of Planning Officials; Avionics System Project Officer

aspp alloy-steel protective plating

ASPP American Society for the Perfection of Punctuation; American Society of Picture Professionals; American Society of Plant Physiologists;

American Society of Polar Philatelists; Australian Society of Plant Physiologists

ASPPA Armed Service Petroleum Purchasing Agency

ASPPO Armed Services Procurement Planning Office

ASPQ Association Suisse pour la Promotion de la Qualite (French—Swiss Association for Quality Improvement)

ASPR American Society of Psychical Research; Armed Services Procurement Regulations; Association of South Polar Research

ASPRL Armament Systems Personnel Research Laboratory (USAF)

aspro ass prostitute (male homosexual)

ASPRS American Society of Plastic and Reconstructive Surgeons

ASPs Anglo-Saxon Protestants; Assistant Superintendents of Police

ASPS Acoustic Ship-Positioning System

ASPSPOM American Society for the Preservation of Sacred, Patriotic, and Operatic Music

ASPT American Society of Plant Taxonomists

ASPTC Army Support Center

ASQ Anthropological Society of Queensland; Anxiety-Scale Questionnaire

ASQ Administrative Science Quarterly

ASQC American Society for Quality Control

ASQDE American Society of Questioned Document Examiners

asr airport surveillance radar; air-sea rescue; answer and receive; apical sensory region; automatic send-receive; available supply rate

ASR American Society of Rocketry; American Sugar Company (stock exchange symbol); Association of Southeastern Railroads; Aviation Safety Regulation(s); submarine rescue vessel (naval symbol)

asra athwartships reference axis

asradi adaptive surface-signal recognition-and-direction indicator

ASRAPS Acoustic Sonar Range Prediction System

ASRB Australian Sales Research Bureau

asrc (ASRC) air-sea rescue craft

ASRC Air-Sea Rescue Craft; Alabama Space and Rocket Center; Atmospheric Sciences Research Center

asrd aircraft shipment readiness date

ASRE Admiralty Signal and Radar Establishment (UK); American Society of Refrigeration Engineers

ASREC American Society of Real Estate Counselors

ASRI Aluminum Smelters Research Institute

ASRL Aero-elastic and Structures Research Laboratory (M.I.T.)

ASRM American Society of Range Movement

asro (ASRO) astronomical roentgen observatory (satellite)

asroc (ASROC) antisubmarine rocket

ASRP American Society for the Republic of Panama

ASRPP American Society for Research in Psychosomatic Problems

a-s rs air-sea rescue service

ASRS Aviation Safety Reporting System

ASRT Air Support Radar Team; American Society of Radiologic Technologists

asrv angle-stop radiator valve

ASRY Arab Shipbuilding and Repair Yard (Bahrain)

ass. anterior superior spine; assurance

Ass Assyrian

ASS Accordion Symphony Society; Anglo-Swedish Society; Army Special Staff; Associated Scholastic Society; Associated Sociologists Society; Australian Security Service; Aviation Security System

A-SS Anti-Slavery Society

A.S.S. Associate in Secretarial Science; Associate in Secretarial Studies

ASSA American Society for the Study of Allergy; Army Signal Supply Agency; Astronomical Society of South Australia; Australian Society of Security Analysts

ASSArthr American Society for the Study of Arthritis

ASSASSIN Agricultural System for Storage and Subsequent Selection of Information

assassrep assassination report

A-S Scale Anti-Semitism Scale (measuring negative attitudes)

Assateague Assateague Island National Seashore linking Maryland and Virginia

assce assurance

ASSCO American Steam Ship Company

Ass Com Gen Assistant Commissary General

assd assigned

ASSE American Society of Safety Engineers; American Society of Sanitary Engineers

assem assemble

Assem God Assemblies of God

assess. analytical studies of surface effects of submerged submarines

ASSESS Airborne Science-Spacelab Experiments-Simulation System

ASSET Aerothermodynamic Elastic Structural System Environmental Tests

ASSGB Association of Ski Schools in Great Britain

ASSH American Society for Surgery of the Hand

ASSIFONTE Association de l'Industrie de la Fonte de Fromage (French—Association of the Processed Cheese Industry)

assigt assignment

assim assimilated

assist. assistant

assmt assessment

Assn Association

Assn Brain Injured New York Association for Brain Injured Children

Assn Brit Zool Association of British Zoologists

assnce assurance

Assn Clin Biochem Association of Clinical Biochemists

Assn Consumer Res Association for Consumer Research

assnd assigned

Assn Ed Comm Tech Association for Educational Communications and Technology

Assn Pr Association Press

Assn Sch Busn Association of School Business Officials of the United States and Canada

Assn Study Anim Behav Association for the Study of Animal Behaviour

Assn Supervision Association for Supervision and Curriculum Development

Assn Tchr Ed Association of Teacher Educators

Assn Under Man Association for the Understanding of Man

ASSOBANCA Associazione Bancaria Italiana (Italian— Italian Bankers' Association)

assoc associate; associated; association

Assoc Associate; Associated; Associates; Association

Assoc Bk Associated Booksellers

Assoc Coun Arts Associated Councils of the Arts

Assoc Eng Associate in Engineering

ASSOCHAM Associated Chambers of Commerce

Assoc IEE Associate of the Institution of Electrical (Electronic) Engineers

Assoc I Min E Associate of the Institute of Mining Engineers

Assoc INA Associate of the Institute of Naval Architects

Assoc ISI Associate of the Iron and Steel Institute

Assoc Met Associate of Metallurgy

Assoc Pr Associated Press

Assoc Sci Associate in Science

assoc w associated with

asson assonance

ASSP All Saints Sisters of the Poor

ASSPHR Anti-Slavery Society for the Protection of Human Rights

ASSR American Society for the Study of Religion; Armenian Soviet Socialist Republic; Autonomous Soviet Socialist Republic; Azerbaijan Soviet Socialist Republic

ASSS American Society for the Study of Sterility; Australian Society of Soil Science

asst assist; assistance; assistant

ASST American Society for Steel Treating

ASST Aziendo de Stato per i Servizi Telefonici (Italian— State Telephone Service)

Asst Att Gen Assistant Attorney General

Asst Chf Engr Assistant Chief Engineer

asstd assented; assorted

Asst Engr Assistant Engineer

Asst Pur Assistant Purser

Asst Stwd Assistant Steward

A-S Study Ascendance-Submission (reaction)

assu (ASSU) air support signal unit

ASSU American Sunday School Union

assw antistrategic submarine warfare

assy assembly

Assyr Assyria(n)

Assyr-Babyl Assyro-Babylonian

ast absolute space-time

ast (AST) advanced supersonic transport; average spring tides

Ast astigmatism; Astoria(n); Asturian; Asturias

AST Academic Salaries Tribunal; Aerial Survey Team; Air Service Training; Air Surveillance Technician; Alaska State Troopers; Alaskan Standard Time; American Radiator and Standard Sanitary (stock exchange symbol); Army Satellite Tracking; Army Specialized Training; Arts Society of Tasmania; Association for Student Training; Astronomical Society of Tasmania; Atlantic Standard Time

ASTA Aerial Surveillance and Target Acquisition; American Seed Trade Association; American Society of Travel Agents; American String Teachers Association; Army Strategy and Tactics Analysis; Australian Science Teachers Association

ASTA Allgemeiner Studentenausschuss (German—General Students Committee)

ASTAC Australian Shipping, Trading, and Chartering

ASTANO Astilleros y Talleres del Noroeste (Spanish— Dockyards and Workshops of the Northwest)

ASTAP Advanced Statistical Analysis Program

ASTAS Antiradar Surveillance and Target Acquisition System

astc (ASTC) airport surface traffic control

ASTC Appalachian State Teachers College; Arkansas State Teachers College; Aroostook State Teachers College

A.S.T.C. Associate of the Sydney Technical College

ASTD American Society of Teachers of Dancing; American Society for Training and Development; American Society of Training Directors

ASTE American Society of

Tool Engineers

astec advanced solar turboelectric concept; advanced solar turboelectric conversion

ASTEC Antisubmarine Technical Evaluation Center; Australian Science and Technology Council

ASTECNAVAIR Assistant Secretary of the Navy for Air

a sten aortic stenosis

ASTEO Association Scientifique et Technique pour l'Exploration des Oceans (French—Scientific and Technical Association for the Exploration of the Oceans)

ASTF Aeropropulsion System Test Facility; Aerospace Structures Test Facility

asth asthenopia

asti antispasticity index

asti (ASTI) antisubmarine training indicator

ASTI American School of Technical Intelligence

ASTI Applied Science and Technology Index

ASTIA Armed Services Technical Information Agency

astig astigmatic; astigmatism; astigmatizer; astigmatoscope; astigmatoscopy; astigmia; astigmometer; astigmoscope

astinator procrastinator

ASTIP Army Scientific and Technical Information Program

ASTM American Society for Testing and Materials; American Society of Tropical Medicine

ASTME American Society of Tool and Manufacturing Engineers

ASTMH American Society of Tropical Medicine and Hygiene

ASTMS Association of Scientific, Technical, and Managerial Staffs

asto antistreptolysin

as tol as tolerated (by the patient)

astor (ASTOR) antisubmarine torpedo

ASTP Apollo-Soyuz Test Project; Army Specialized Training Program

astr astronomy

ASTR American Society of Therapeutic Radiologists

astra advanced structure analyzer; advanced system for radiological assessment; automatic scheduling with time-

integrated resource allocation

ASTRAC Arizona Statistical Repetitive Analog Computer

ASTREA Air Support to Regional Enforcement Agencies (helicopter surveillance)

astrion astrionic(al)(ly); astrionics

Astrl Australia

astro astrograph(ic); astrolabe; astrology; astrometry; astronautics; astronomer; astronomical; astronomy; astrophysics

Astro Astronautics

ASTRO Air-Space Travel Research Organization

astro-ad-anon astrological adventures anonymous

astrobio astrobiological; astrobiologist; astrobiology

astrochem astrochemical(ly); astrochemist(ry)

astrochronics astrochronological relatives

astrodyn astrodynamic(al)(ly); astrodynamic(ist)

astrog astrogeological; astrogeologist; astrogeology

astrogen astrogenealogy

astrol astrology

Astrol Astrology

astromonk astronautical monkey (specimen used in biological tests)

astron astronomer; astronomic(al)(ly); astronomy

Astron Astronomy

Astron Astrophys Astronomy and Astrophysics

astronaut astronautical(ly); astronautics

Astron J Astronomical Journal

Astron Nachr Astronomische Nachrichten (German—Astronomical News)

Astron Zh Astronomicheskii Zhurnal (Russian—Astronomical Journal)

Astro Obsv Astrophysical Observatory

astrophys astrophysics

Astrophys J Astrophysical Journal

Astrophys Lett Astrophysical Letters

ASTS Alabama State Training School (for female delinquents at East Lake near Birmingham)

ASTSECNAV Assistant Secretary of the Navy

astt (ASTT) action-speed tactical trainer

ast t astronomical time

ASTT American Society of

Traffic and Transportation

A.S.T.T. Associate in Science Teacher Training

asu all screwed up

asu (ASU) administrative systems unit; acromedical staging unit

ASU American Secular Union; American Student Union; Arab Socialist Union; Arizona State University; Asuncion, Paraguay (airport); Atheist Student Union; Australian Swimming Union

ASU-57 Soviet self-propelled 57mm gun on tracked chassis

ASUA Amateur Swimming Union of the Americas

ASUC American Society of University Composers; Associated Students of the University of California

ASU Lat Am St Arizona State University Center for Latin American Studies

asupt advanced simulator for undergraduate pilot training

ASUSSR Academy of Sciences of the USSR

ASUTS American Society of Ultrasound Technical Specialists

ASUUS Amateur Skating Union of the U.S.

asv acceleration switching valve; airborne radar for detecting surface vessels; aircraft-to-surface vessel; angle stop valve

asv (ASV) automatic self-verification

a-s v anti-snake venom; arterio-superficial venous

a/sv arterio/superficial venous

ASV American Standard Version; Anthropological Society of Victoria; Astronomical Society of Victoria

ASVA Associate of the Society of Valuers and Auctioneers

ASVAB Armed Services Vocational Aptitude Battery

asveo advance space vehicle engineering operation

ASVT Applications Systems Verification Test

ASVU Army Security Vetting Unit

asw antisubmarine warfare

ASW Anti-Submarine Warfare; Association of Scientific Writers; Association of Social Workers; Australian Standard White (wheat)

ASW (LR) Antisubmarine Warning (long-range)

ASW (SR) Antisubmarine Warning (short-range)
AS & W American Steel and Wire (gage)
ASWA Anthropological Society of Western Australia
A/S WA Aviation/Space Writers Association
asw/aaw antisubmarine warfare/anti-air warfare
ASWC Antisubmarine Warfare Center (NATO)
ASWE Admiralty Surface Weapons Establishment
ASWEPS Anti-Submarine Warfare Environmental Prediction System
aswf arithmetic-series weight function(s)
ASWG American Standard Wire Gage; American Steel and Wire Gage
ASWI Antisubmarine Warfare Installations (NATO)
ASWIPT Antisubmarine Warfare In-Port Training (NATO)
ASWRC Antisubmarine Warfare Research Center (NATO)
ASWS Audubon Shrine and Wildlife Sanctuary
ASWSOW Anti-Submarine Warfare Standoff Weapon
ASWSS Antisubmarine Warfare Schoolship (USN)
ASWTDS Antisubmarine Warfare Tactical Data System
asy asylum
asym asymmetrical
async asynchronous
ASZ American Society of Zoologists
ASZD American Society for Zero Defects
at. accounting tabulating (card); airtight; asphalt; asphaltic; asphalt tile; atmosphere (technical); atomic
at. (AT) advanced technology (computer); alternative technology; appropriate technology
a t a tempo (Italian—at the speed written)
at.% atomic percent
a/t action/time; antitank; antitorpedo
a & t acceptance and transfer; assemble and test
At ampere-turn; astatine
AT Adirondack Trail; Advanced Technology; Advanced Trainer; Air Travel; antitank; Appalachian Trail; Atherton Tablelands (Queensland parks)
A T Absolute Time (compact disc)
A/T American terms
AT Antico Testamento (Italian—Old Testament)
A-T *'Alef-Tav* (Hebrew—from the first to the last letter of the alphabet)—similar to the English expression from A to Z
AT₁9 dihydrotachysterol
AT₇ hexachlorophene (disinfectant)
AT-26 Aermacchi jet-trainer ground-attack aircraft also known as Xavante
ata academic travel abroad; actual time of arrival; air-to-air; azimuthal torque amplifier
ata admission temporair (French—temporary admission)
ata (ATA) advanced tactical aircraft
ATA Advertising Typographers Association; Air Transport Association; Albanian Telegraph Agency; Amateur Trapshooting Association; American Taxicab Association; American Taxpayers Association; American Teachers Association; American Theatre Association; American Thyroid Association; American Tinnitus Association; American Title Association; American Topical Association; American Transit Association; American Translators Association; American Tree Association; American Trucking Association; American Tunaboat Association; Anatomical Transplant Association; Applied Technology Associates; Area Transportation Authority; Army Transportation Association; Asia Teachers Association; Atlantic Treaty Association; Australian Taxpayers Association; Australian Toolmakers Association; Australian Translators Association; auxiliary ocean tug (naval symbol)
A.T.A. Associate Technical Aide
ATA Agence Telegraphique Albanaise (French—Albanian News Service)
ATAA Advertising Typographers Association of America; Air Transport Association of America; Amateur Trapshooting Association of America

ata (ATA) advanced tactical aircraft
ATAC Air Transport Association of Canada; Allied Tactical Air Force; Anatomical Transplant Association of California; Army Tank Automotive Center; Army Tank and Automotive Command
ATACS Army Tactical Communications System
atad absent on temporary additional duty
ATAD Air Transport and Delivery (service)
ATAE Association of Tutors in Adult Education
ATAF Allied Tactical Air Force; American Trucking Associations Foundation
ATAFCS Airborne Target-Acquisition and Fire-Control System
ATAG Air Training Advisory Group
ATAI Air Transport Association International
ATAJ Association of Transport Advisers of Japan
ATALA Association pour l'Etude et de la Linguistique Appliquee (French—Association for the Study of Applied Linguistics)
ATAM Association for Teaching Aids in Mathematics
atan arc tangent
atar above transmitted and received
atar (ATAR) antitank aircraft rocket
ATAR Automated Travel Agents Reservation
ATARS Anti-Terrain-Avoidance Radar System
ATAs American Tinnitus Association members
ATAS Academy of Television Arts and Sciences; Air Transport Auxiliary Service
at. (AT) appropriate technology
atav atavism; atavist; atavistic(al)(ly)
atb alternate top level; asphalt tile base; at the time of bombing
atb (ATB) advanced technology bomber; stealth bomber
ATB Air Transportation Board
A & TBCB Architectural and Transportation Barriers Compliance Board
ATBI Allied Trades of the Banking Industry
atbm average time between maintenance

atbm (ATBM) advanced technology ballistic missile; antitactical ballistic missile

atbt acoustic telemetry bathythermometer

atbyropt at buyer's option

atc ablative thrust chamber; acoustic test(ing) chamber; acoustical tile ceiling; aerial tuning condenser; allergic to combat; approved type certificate; automatic temperature control; automatic tint control(tv)

atc (ATC) all-terrain cycle; automatic train control; average variable cost

atc Amsterdam Towing Company

ATC Advertising Training Center; Air Traffic Conference; Air Traffic Control; Air Training Command; Air Transport Command; Air Transportation Corps; Aircraft Technical Committee; Airport Traffic Control; Airway Traffic Control; Alcohol Treatment Center; Alpine Tourist Commission; Appalachian Trail Conference; Armament Test Center; Army Training Center; Army Transportation Corps; Associated Traffic Clubs; Associated Travel Clubs; Athletic Training Council; Australian Tariff Council; Australian Tourist Commission

ATC All Things Considered (radio program)

atca (ATCA) advanced tanker-cargo aircraft

ATCA Air Traffic Conference of America; Air Traffic Control Association; Allied Tactical Communications Agency; American Theater Critics Association

ATCAS Air-Traffic-Control Automated System

atcase aspartate transcarbamylase

ATCB Air Traffic Control Board

ATCC Air Traffic Control Center; American Type Culture Collections; Automatic Train Control Center

ATCDE Association of Teachers in Colleges and Departments of Education

atce ablative thrust chamber engine

atcen air traffic control evaluation unit

ATCF Automobile and Touring Club of Finland

atch attach; attaching; attachment

atchd attached

ATCL Associate of Trinity College of Music—London

ATCMD Atlanta Contract Management District

ATCMS Advanced Technology Cruise Missile Study

ATCMU Associated Third-Class Mail Users

ATCO Air Traffic Coordinating Office(r)

ATCOM Atoll Commander

ATCOS Atmospheric Composition Satellite

ATCRBS Air Traffic Control Radar Beacon System

atc's airtight containers; any-terrain motorcycles

ATCU Association of Texas Colleges and Universities

atd absent (on) temporary duty; actual time of departure; anthropomorphic test dummy

atd *a tak dale* (Czech—et cetera)

ATD Actual Time of Departure; Aid to the Totally Disabled; Armament Test Division; Art Teachers Diploma

atda augmented target docking adapter

ATDA American Train Dispatchers Association; Army Training Device Agency; Australian Telecommunications Development Association

atdc after top dead center (valve setting)

atdp attitudes toward disabled persons

ATDS Airborne Tactical Data System; Association of Teachers of Domestic Science; Automated Data and Telecommunications Service

ate altitude transmitting equipment; automatic test equipment

Ate Almirante (Spanish—admiral)

ATE Associated Telephone Exchanges; Association of Teacher Educators; Automatic Telephone and Electric (New Zealand company)

ATEA American Technical Education Association; American Toy Export Association; Australian Telecommunications Employees Association

ATEC Air Transport Electronics Council; Aviation Education Council

A.Tech. Associate in Technology

ATEM Aircraft Test Equipment Modification

ATEMIS Automated Traffic Engineering and Management Information System

A temp a tempo (Italian—in the speed written)

Aten Atenas (Portuguese or Spanish—Athens); *Atene* (Italian—Athens); *Athenes* (French—Athens)

ATEN Association Technique pour la production et l'utilisation de l'Energie Nucleaire (French—Technical Association for the Production and Use of Nuclear Energy)

At Energ Atomnaya Energiya (Russian—Atomic Energy)—journal

ATEP Aboriginal Teacher Education Programme

A Term Air Terminal

ATESL Association of Teachers of English as a Second Language

ATEWS Advanced Tactical Electronic Warfare System

Atex Atlantic tradewind experiment

atf accounting tabulating form; actual time of fall

ATF Acceptance Test(ing) Facility; Advanced Technical Fighter; Advanced Technology Fighter; Air Task Force; Alcohol, Tobacco, and Firearms (bureau); Alternative Test Facility; American Type Founders; Australian Teachers Federation; ocean tug (3-letter symbol)

ATFAC American Turpentine Farmers Association Cooperative

ATFCNN Allied Task Force Commander—Northern Norway (NATO)

atfi attitudes toward feminist issues scales

atfr automatic terrain-following radar

ATFS Association of Track and Field Statisticians

atg air-to-ground

ATG Accordion Teachers Guild; Army Technical Group

atgar (ATGAR) anti-tank guided air rocket

ATGSB Admission Test for

Graduate Study in Business
atgw (ATGW) antitank guided
weapon(s)
ath above the horizon; atheism;
atheist(ic); athletic
äth *äthiopisch* (German—
Ethiopian)
Ath Athens
ATH Athens, Greece (airport)
AT-H August Thyssen-Hutte
Athab Athabasca(n)
athe allotetrahydrocortisol
ath dfld atheism defiled
Athel Athel Line
Athen Athenian
atheol atheological; atheo-
logist; atheology
athodyd aerothermodynamic
duct (ramjet engine)
athsc atherosclerosis
athw athwartship
ati actual time of interception; aerial tuning inductance;
aptitude-treatment inter-
action; average total inspection
ATI Air Technical Intelligence;
American Technology Institute; American Television Institute; Ansett Transport Industries; Asbestos Technical
Institute; Asbestos Textile Institute; Australian Textile Institute
A & TI Agricultural and Technical Institute
ATI Aero Transporti Italiani
(Italian—Italian Air Freight
Line); Air Technical Index;
Azienda Tabacchi Italiani
(Italian—Italian State To-
bacco Board)
ATIC Aerospace Technical Intelligence Center; Air Techni-
cal Intelligence Center;
Antigua Tourist Information
Center; Australian Tin Infor-
mation Centre
ATIGS Advanced Tactical In-
ertial Guidance System
ATII Associate of the Taxation
Institute Incorporated
ATIL Air Target Intelligence
Liaison Program (USAF)
atiob as this is our best
ATIP Alaskan Talent, Informa-
tion, and Practices
atis automatic terminal infor-
mation service
ATIS Adirondack Trail Im-
provement Society; Air Tech-
nical Intelligence Study
ATISC Air Technical Intelli-
gence Services Command
(USAF)
ATJ Association of Teachers of

Japanese
ATJS Advanced Tactical Jam-
ming System
atk attack
a-tk anti-tank
atl analog threshold logic
Atl Atlanta; Atlantic; Australia
Atl Atlantico (Italian, Portu-
guese, or Spanish—Atlantic);
Atlantique (French—Atlan-
tic); *Australia* (Spanish—
Australia)
ATL Acoustic Test(ing) Labo-
ratory; Alexander Turnbull
Library (Wellington, NZ); As-
sociated Truck Lines; Atlanta,
Georgia (airport); Atlantic
Tankers Limited; Automatic
Totalisators Limited
ATLA Air Transport Licensing
Authority; American Theo-
logical Library Association;
American Trial Lawyers As-
sociation
Atlanta capital of Georgia;
U.S.. Penitentiary at Atlanta,
Georgia
Atlanta Youth Atlanta Youth
Development Center (for fe-
male juvenile delinquents) in
Atlanta, Georgia
ATLANTIC Atlantic Refining
Company
Atlantol Atlantologic(al)(ly);
Atlantolgist(ic)(al)(ly); Atlan-
tology
Atlas Atlas Mountains of Alge-
ria and Morocco
ATLAS Abbreviated Test Lan-
guage for Avionic Systems;
Automated Tape Label As-
signment System; Automatic,
Tabulating, Listing, and Sort-
ing System
Atlas-Agena two-stage launch
vehicle
Atlas-Centaur first American
high-energy launch vehicle
for space exploration—D-Se-
ries Atlas boosts Centaur
space vehicle
Atlas-E intercontinental ballistic missile designed to place a
thermonuclear warhead on a
9000-mile-distant target
atlas fol atlas folio—a book
about 25 inches high
Atlas icbm first American intercontinental ballistic missile
ATLB Air Transport Licensing
Board (UK)
Atl C Atlantic City
ATLD Air-Transportable Loading Dock
ATLIS Army Technical Library Improvement Studies;

Automatic-Tracking Laser-Il-
lumination System
Atl O Atlantic Ocean
Atl Pil Aut Atlantic Pilotage
Authority
Atl Sym Atlanta Symphony
atm atmosphere (normal)
atm (ATM) automatic teller
machine
at. m atomic mass
at/m ampere turns per meter
At/m ampere turns per meter
ATM Apollo Telescope Mount;
Association of Teaching Aids
in Mathematics; Associated
Tobacco Manufacturers;
Atomic Demolition Munition;
Automatic Teller Machine
*ATM Amateur Telescope Mak-
ing; Azienda Tranviaria
Municipale* (Italian—Munici-
pal Rapid Transit Board)
ATMA Adhesive Tape Manufacturers' Association
atm ab atmospheres absolute
ATMAC Air Traffic Management Automated Center
ATMC Army Transportation
Materiel Command
atm/d (ATM/D) automatic
teller machine deposit
ATMI American Textile Manufacturers Institute
atmos atmosphere; atmos-
pheric(al)(ly)
atm press atmospheric pressure
ATMS Air Traffic Management
System; Automatic Teller Machine System(s); Automatic
Transmission Measuring System
atm/w (ATM/W) automatic
teller machine cash with-
drawal
ATMX railcar used by the Department of Energy for shipping defense waste
atn actual test number; acute tubular necrosis
ATN Alabama, Tennessee and
Northern (railroad)
ATNA Australian Trained
Nurses' Association
atnav acoustic-transponder
navigation
atndt attendant
at. no. atomic number
ATNP Atherton Tablelands National Parks (Queensland)
ato according to others; assisted
takeoff; automatic train operation
ATO Academy of Teachers of
Occupations; Australian Taxation Office; ocean tug, old (3-
letter symbol)

ATOA Australian Transport Officers Association

atoll. acceptance, test, or launch language

atomdef atomic defense

atomdev atomic device

atoms. automated technical order maintenance sequence(s)

ATOM August Twenty-One Movement (Philippines)

ATOMSTATREP Atomic Status Report

aton at once

atorp antitorpedo; atomic torpedo

ATOS American Theatre Organ Society; Association of Temporary Office Services

atot actual time over target

atp actual time of penetration; array transform processor

atp (ATP) adenosine triphosphate, material found in almost all terrestrial life

atp a tout prix (French—at any price)

ATP Acceptance Test(ing) Procedure; adenosine triphosphate; Admissions Testing Program; Allied Technical Publication; Army Training Program; Association of Tennis Professionals

ATP Accorde Transports Perissable (French) European agreement on international shipment of perishable foodstuffs

atpa auxiliary turbopump assembly

ATPAS Association of Teachers of Printing and Allied Subjects

ATPase adenosine triphosphate

atpcc attitudes toward parental control of children

atpd ambient temperature and pressure–dry

ATPE Association of Teachers in Penal Establishments

ATPI American Textbook Publishers Institute

ATPM Association of Toilet Paper Manufacturers

at pres at present

atps ambient temperature and pressure–saturated with water vapor

atpu air transport pressurizing unit

atr advanced test reactor; anti-transmit-receive; transmitter-receiver

atr (ATR) after tax rate; audio-tape recorder

Atr Achilles tendon reflex

ATR Advanced Test Reactor; Association of Teachers of Russian; ocean tug, rescue (3-letter naval symbol)

ATR Admission Temporaire Roulette (French—Temporary Admission on Wheels); *Anglican Theological Review*

ATRA American Television and Radio Artists; American Tort Reform Association

atran automatic terrain recognition and navigation

atrax air-transportable communications complex

atrc anti-tracking control

ATRC Air Traffic Regulation Center

atr fib atrial fibrillation

atrid automatic target recognition, identification, and detection

atrima as their respective interests may appear

ATRIS Air Traffic Regulation Identification System (USA)

atrl atrial

atrls actual time of release

atrm after torpedo room

ATRM American Tax Reduction Movement

atro actual time of return to operation

atrop atrophy

A Tr Ps Allied Training Publications (NATO)

atrr advanced threat-reactive receiver

ATRS Australian Tape Recording Society

atrso accepts transfer as offered

atrt anti-transmit-receive tube

ats absolute temperature scale; advanced technological satellite; air-to-ship; anxiety-tension state; astronomical time switch

ATs Achievement Tests

ATS Acoustic Transmission System; Acquisition and Tracking System; Administrative Terminal System; Advanced Technological Satellite; Aeronautical Training Society; Air Tactical School; Air Traffic Services; American Theological Society; American Therapeutic Society; American Travel Service; American Trudeau Society; Anglican Truth Society; Application Technology Satellite; Army Transport Service; Association of Theological Schools; Automatic Transfer Service (bank accounts); sal-

vage tug (naval symbol)

ATSA Aero Transportes

ATSBNZ Associated Trustee Savings Banks of New Zealand

ATSC American Traffic Safety Council

ATSE Alliance of Theatrical Stage Employees

AT & SF Atchison, Topeka and Santa Fe (railway)

ATSFD Air Traffic Service Flight Services Division (FAA)

atsit automatic techniques for the selection and identification of targets

ats/jea automated test system/jet engine accessories

ATSOCC Applications Technology Satellite Operations Control Center (NASA)

ATS's Advanced Technological Satellites

AtST Atlantic Standard Time

ATSU Association of Time-Sharing Users

att attach; attempt; attorney

a t & t all tacos and tamales (American Southwestern roadside-stand short form); always talking and talking

Att Attic(a)

ATT Army Training Test

AT & T American Telephone & Telegraph

A & TT Alcohol and Tobacco Tax

atta atenta (Spanish—attentively)

ATTA Association of Travel and Tourist Agents (Singapore)

ATTC American Towing Tank Conference

atten attenuation, attenuator

ATT & F Alcohol, Tobacco Tax, and Firearms (Division of U.S. Treasury Dept)

AttGen Attorney General

Att Gen Attorney General

ATTI Association of Teachers in Technical Institutions

Attica Facility Attica Correctional Facility (for males) at Attica, New York

attn attention

atto attorney

atto atento (Spanish—attentively); 10^{-18}

attr attractive

attrd attributed

attrest(s) attitude arrest(s)—made by law-enforcement officers who dislike the attitude(s) of the person(s) ar-

rested
attrib attributive
attrit attrition
ATTS Automatic Telemetry Tracking System
atty attorney
atty & c attorney and client
Atty Gen Attorney General
AT type adenine and thymine type
atu alien tax unit; atomic time unit
Atu *Atmosphärenüberdruck* (German—atmospheric excess pressure)
ATU Alliance of Telephone Unions; Amalgamated Transit Union; Anchorage Telephone Utility; Anglo-Turkish Union; Anti-Terrorist Union; Anti-Terrorist Unit; Arab Telecommunications Union
atum antitank nonmetallic
ATURM Amphibious Training Unit—Royal Marines
atv (ATV) all-terrain vehicle
ATV Associated Tele Vision
ATV Akademiet for de Tekniske Videnskaber (Danish—Academy of Technical Sciences)
atvm attenuator thermo-element voltmeter
at. vol atomic volume
atw (ATW) antitank weapon
at/w atomic hydrogen weld
ATW American Theater Wing; Atlantic & Western (railroad)
at/wb ampere turns per weber
ATWE Association of Technical Writers and Editors
ATWg Air Transport Wing (USAF)
atws adjustable thermal wire stripper; automatic track while scanning
at. wt atomic weight
ATWU Australian Textile Workers Union
atx air taxi
at. xpl atomic explosion
ATYP Australian Theatre for Young People
A Typ I Association Typographique Internationale (French—International Typographic Association)
atyropt at your option
ATZ Air Traffic Zone
au activity unit; angstrom unit; antitotxin unit; arbitrary unit(s); author; azauridine
au aurum (Latin—gold)
a.u. aures unitas (Latin—both ears); *au usum* (Latin—according to custom)
Au angstrom unit; astronomical

unit; gold (symbol)
AU Aarhus Universitet (University of Aarhus); Air University; Alfred University; Allen University; American University; Andrews University; Army Unit; Assumption University; astronomical unit; Atheists United; Atlanta University; Auburn University
AÜ Ankara Üniversitesi (University of Ankara)
A/U advanced undersea weapons
A & U Allen & Unwin Publishers
Au¹⁹⁸ radioactive gold (symbol)
AU-23A Fairchild piston-powered stol aircraft
AUA American Unitarian Association; American Urological Association; Aruba, Netherlands West Indies (airport); Associated Unions of America; Austrian Airlines
A.U.A. Associate of the University of Adelaide
AUAF Association of University Affiliated Facilities
AUAS Academy of Underwater Arts and Sciences
aub alstublieft (Dutch—please)
Aub Aubrey
AUB American University of Beirut
AUBC Association of Universities of the British Commonwealth
AUBER Association for University Business and Economic Research
AUBTW Amalgamated Union of Building Trade Workers
Auburn Facility Auburn Correctional Facility (for males) at Auburn, New York
auc average unit cost
a.u.c. ab urbe condita (Latin—from the founding of the city, usually refers to Rome)
AUC Aberystwyth University College; American University of the Caribbean; American University Club; Australian United Corporation; Australian Universities Commission
AU of C American University of Cairo
AUCA American Unitarian Christian Association
AUCANNZUKUS Australia, Canada, New Zealand, United Kingdom, United States
AUCANUKUS Australia, Canada, United Kingdom,

United States
AUCAS Association of University Clinical Academic Staff
AUCC Association of Universities and Colleges of Canada
Auck Auckland
Aucklands Auckland Islands
AUCOA Association of United Contractors of America
AUCSRLFRVWAM All-Union Central Scientific Research Laboratory for the Restoration of Valuable Works of Art in Museums
auct auction(eer)
auct auctorum (Latin—of authors)
AUCTU All-Union Council of Trade Unions
aud audible; audit; audition; auditor; auditorium
Aud (AUD) Australian dollar(s)
Audᵃ audiencia (Spanish—court of justice, hearing)
AUDACIOUS Automatic Direct Access to Information with On-Line UDC System
audar autodyne detection and ranging
aud disb auditor disbursements
AUDDITS Automated Dynamic Digital Test System
Audel Theodore Audel
Aud Gen Auditor General
Aud Gen Nav Auditor General of the Navy
Audie Audry
auding auditory hearing, listening, and understanding
audio audiofrequency; audiogenic; audiogram; audiology; audiometer; audiometry; audiophile; audiovisual; audiovisual aids; etc.
audiol audiologist; audiology
audiom audiometer; audiometric(al)(ly); audiometrist
audiovis audiovisual; audiovisual aids
audre audio response; automatic digit recognizer
AUEC Association of University Evening Colleges
AUEW Amalgamated Union of Engineering Workers
AUF Australian Underwriters Federation
Aufdr Aufdrucke (German—imprint)
Aufl Auflage (German—edition)
AUFL Americans United For Life
AUFS American Universities Field Staff
AUFUSAF Army Unit for

United States Air Force
aug augment; augmentation; augmentative
Aug Augsburg; August; Augusta; Augustan
Augember August and September
Augie August; Augusta, Georgia; Augustine; Augustus
augm augmentation
augm augmente (French—augmented)
augra authority granted
August Augustine; Augustus
AUI Associated Universities Incorporated
auj aujourd'hui (French—today)
Auk Auckland
aul above upper limit
AUL Aberdeen University Library; Air University Library; American United for Life; American United Life (insurance)
AULC American University Language Center
auld auld lang syne (Scottish—old times fondly remembered)
AULI As You Like It
AULLA Australasian Universities Language and Literature Association
aum (AUM) air-to-underwater missile
aum aumentado (Spanish—augmented)
AUMLA Australian Universities Modern Language Association
a. u. n. abesque ulla nota (Latin—without annotation)
AUNT Alliance for Undesirable but Necessary Tasks
auntie. automatic unit for national taxation and insurance (UK)
AUO African Unity Organization; Atlantic Union Oil
AUP Andrews University Press; Australian United Press
AUPE Amalgamated Union of Public Employees (Singapore)
AUPELF Association of Wholly or Partially French Language Universities
AUPF Australian Uranium Producers Forum
AUPG American University Publishers Group
AUPHA Association of University Programs in Hospital Administration
AUPO Association of University Professors of Ophthal-

mology
AUPOSTCOM Australian Postal Commission
aur auricle; auricular; auricularis; aurum
Aur Auriga
AUR Association of University Radiologists
AURA Association of Universities for Research in Astronomy; Automated Reasoning Assistant
aureq authority is requested
aur fib auricular fibrillation
Auri Auriga
AURI Angkatan Udara Republik Indonesia (Indonesian Air Force)
auric auricular
AURISA Australian Urban and Regional Information Systems Association
aurist. auristillae (Latin—ear drops)
aurora australis (Latin—southern lights)
aurora borealis (Latin—northern lights)
Aus Austin; Austria; Austrian
AUs Area Units (New Zealand)
AUS Army of the United States; Austin, Texas (airport); Australian Union of Students
AUSA Assistant United States Attorney; Association of the United States Army; Australian Universities Sports Association
ausc auscultation
AUSCS Americans United for Separation of Church and State
Ausg Ausgabe (German—edition)
Au sh Australian serum hepatitis
AUSHC Australian High Commission
AUSIMM Australian Institute of Mining and Metallurgy
AUSLFL All-Union State Library of Foreign Literature (Moscow)
AUSS Association of University Summer Sessions
AUSSAT Australian Satellite
Aussie(s) Australian(s)
Aust Australia; Australian
Aust Alps Australian Alps of New South Wales and Victoria
Aust Cur Australian Current
Aust$ Australian dollar
austen austenitic
Austen Australian sten gen
Aust Engl Australian English

(Cockney with Australian accents)
Auster Auster-Beagle light liaison aircraft
Austin Augustina; Augustine; capital of Texas
AUSTIRAN Australian-Iran Shipping Company
Aust J Phys Australian Journal of Physics
AUSTRAFORD Ford Motors of Australia
Austrail Railways of Australia
austral unit of Argentina currency
Austral Australian
Australas Australasian
Australasian Australia, Tasmania, New Zealand, and islands of Melanesia
Australs Austral Islands of Polynesia, also called the Tubuais
Austria Republic of Austria, *Republik Österreich*
AUSTRIATOM Austrian Atomic Energy Group
Austro-Hungarian Empire Austria, Bohemia, Bosnia, Croatia, Moravia, Bukovina, Transylvania, Galicia, Hungary and part of Yugoslavia and Italy (1867–1918)
Austronesia islands of South Pacific from Madagascar in Indian Ocean to Hawaiian Islands in the Pacific
AUSUDIAP Association of U.S. University Directors of International Agricultural Programs
aut autore (Italian—author)
Aut Autriche (French—Austria)
AUT American Union Transport; Association of University Teachers
AUTA Association of University Teachers of Accounting
AUTE Association of University Teachers of Economics
AUTEC Atlantic Underwater Test Evaluation Center
AUTELCOM Australian Telecommunications Commission
auth authentic; authenticate; authenticity; author; authority; authorization; authorize(d)
Auth Authority
authab authorized abbreviation (USAF)
Auth Ver Authorized Version
AUT(I) Association of University Teachers (Ireland)
autiobio autiobiograph; autiobiographer; autobiographic(al); autobiography

autmwtr ck automatic water check

auto. automobile; automatic; automotive

auto (Latin prefix—self)— automobile (self-moving vehicle)

autobird automatic bird feeder

autocade automobile parade

AUTOCAP Automobile Consumer Action Programs(s)

autocat automatic cat feeder

autocolor automatic color (tv)

autocom automated combustor (design code)

autodidac autodidact(ic)(al)(ly)

autodin automatic digital network

autodoc automatic documentation

autodog automatic dog feeder

autog autograph

autogrom autoprompter (tape)

auto. lean automatic lean

autom automatic, automation; automobile; automotive

autom automobile (Italian— automobile); *automóvel* (Portuguese—automobile); *automóvil* (Spanish—automobile)

automag automatic-loading magnum (handgun)

automap automatic machining program

automast automatic mathematical analysis and symbolic translation

automát automática; automático (Spanish—automatic)

automatic automatic revolver

automation automation action; automatic operation

automtn automation

auton autonomous; autonomy

autonet automatic network

autop automatic pistol; autopsy

autopet automatic pet feeder

AUTOPIC Automatic Personal Identification Code

autopilot automatic pilot

autopistol automatic pistol

autoprompt automatic programming of machine tools

AUTOPROS Automated Process Planning System

AUTOPSY Automatic Operating System (IBM)

auto pts automobile parts

autoqest automatic generation of requests

auto. recl automatic reclosing

auto. rich automatic rich

autorotic(s) automobile neurotic(s)

autos automobiles; automatics

autosate automatic data systems analysis technique

autoscript automated system for composing, revising, illustrating, and typesetting

auto s & cv automatic stop-and-check valve

AUTOSERVCEN Automated Service Center

autosevocom automatic secure voice communication(s)

autospec automated specification(s)

autospot automatic system for positioning tolls

auto s & sv automatic stop and check valve

autostatis automatic statewide auto theft inquiry

AUTOSTATIS Automatic Statewide Theft Inquiry System (California)

autostrad automated system for transportation data

autosyn automatically synchronous

autotr autotransformer

autotran automatic translation

autovon automatic voice network

au tr aural training

autran automatic target-recognition analysis

AUTRANAVS Automated Transponder Navigation System

AUT(S) Association of University Teachers (Scotland)

AUT(W) Association of University Teachers (Wales)

AUU Association of Urban Universities

auv administrative use vehicle; armored utility vehicle

Au virus Australian antigen

AUVMIS Administrative Use Vehicle Management Information System (USA)

AUVS Association of Unmanned Vehicle Systems

auw airframe unit weight

auw (AUW) advanced underwater weapon(s)

AUWE Admiralty Underwater Weapons Establishment

aux auxiliary

aux (Greek—grow, increase)— *auxiliar;* (Spanish—auxiliary); auxiliary, auximones, auxocardia

aux m auxiliary machinery

AUXOPS Auxiliary Operational Members (USCG)

auxrc auxiliary recording control

av acid value; anteversion; aortic valve; arteriovenous; as-

sessed valuation; atrioventricular; auriculoventricular; average; average; aviator; avoirdupois

av (AV) audiovisual

a-v atriventricular; audio-visual

av avril (French—April)

a v a vista (Italian—at sight)

a/v (A/V) ad valorem (Latin— as valued)

Av Avenue; Aves; Avestan; Avian; Avila(n)

Av avenida (Portuguese or Spanish—avenue); *Avrum* (Yiddish—Abraham)

AV *alta voltagem* (Portuguese—high voltage); *alto voltaggio* (Italian—high voltage); *alto voltaje* (Spanish— high voltage); American viewpoint; Antonio Vivaldi; arteriovenous; audiovisual; Authorized Version; large seaplane tender (naval symbol)

AV Avtomat Kalashnikov (Russian—Kalashnikov automatic)—Soviet assault rifle

A.V. Anno Vixit (Latin—he (she) lived (a given number of) years)

AV–8B U.S. Marine Corps fighter-bomber jump-jet (capable of taking off and landing vertically)

ava arteriovenous anastomosis; azimuth versus amplitude

ava (AVA) automatic voice alarm

AVA Academy of Vocal Arts; Aerodynamische Versuchsanstalt; American Vocational Association; Asbestos Victims of America; Audio-Visual Aids; Australian Veterinary Association; Australian Volunteers Abroad; Award(s) in the Visual Arts

A-V A All-Volunteer Army

AVAC Asociación Venezolana para la Avance de la Ciencia (Spanish—Venezuelan Association for the Advancement of Science)

AVADS Autotrack Vulcan Air Defense System

av/af anteverted/anteflexed

avail. available; availability

aval availability; available

'Avana Havana

avasa abbreviated visual-approach slope indicator

AVASIS Abbreviated Visual Approach Slope Indicator System

avb avbeta (Swedish—depar-

ture)
AVB advanced aviation base ship (naval symbol)
avbl armored vehicle bridge launcher
AVCA acceleration vector control; allantoid vaginal cream; automatic volume control; average variable cost
avc (AVC) average variable cost
av C avanti Cristo (Italian—Before Christ)
AVC American Veterans Committee; Antelope Valley College; Association of Virginia Colleges; Association of Vitamin Chemists; Audio-Visual Center
AVCA Australian Volunteer Coastguard Association
AvCad Aviation Cadet
avcat aviation high-flash turbine fuel
Av Cert Aviation Certificate
avcs atrioventricular conduction system
AVCS Advanced Videocon Camera Systems; Assistant Vice Chief of Staff
avd automatic voice data; automatic voltage digitizer
avd avdeling (Dano-Norwegian—part, section)
AvD Automobil Club von Deutschland (German—German Automobile Club)
AVD Army Veterinary Department; high-speed seaplane tender (3-letter naval symbol)
Avda avenida (Spanish—avenue)
AVDA American Venereal Disease Association
a-v difference arteriovenous concentration difference
AVDO Aerospace Vehicle Distribution Office(r)
avdp avoirdupois
avdth average depth
ave automatic volume expansion
'ave have
Ave Avenue
AVE Asociación Venezolana de Ejecutivos (Spanish—Venezuelan Association of Executives)
avec amplitude vibration exciter control
AVEM Association of Vacuum Equipment Manufacturers
AVENSA Aerovias Venezolanas (Spanish—Venezuelan Airlines)
AVERA American Vocational

Education Research Association
Averroes Abul-ibn-Roshd
Aves Los Aves—Bird Islands off Venezuela, west of Curaçao in the Caribbean
avf arteriovenous fistula; azimuthally varying field
AVF All-Volunteer Force
avfr available for reassignment
avfuel aviation fuel
avg average
Avg Avgust (Russian—August)
Av Gar Avant Garde
avgas aviation gasoline
avge average
avh acute viral hepatitis
Avh Avhandlinger (Swedish—transactions)
avi adaptive user interface; airborne vehicle identification; air velocity index; aviation
AVI American Virgin Islands; *Association Universelle d'Aviculture Scientifique* (French—Universal Association of Scientific Aviculture); Audio-Visual Institute; Automatic Vehicle Identification
Aviaco Aviación y Comercio (Spanish airline)
AVIANCA Aerovias Nacionales de Colombia (National Airlines of Colombia)
AVIATECA Empresa Guatemalteca de Aviacion (Guatemalan Aviation Enterprise)
AVID Advancement Via Individual Determination (educational program); Audio-Visual Instruction Department
avigation aircraft navigation
aviob aviation observation
avionics aviation and astronautics electronics
AVIP Association of View-data Information Providers
AVIS Active Vibration Isolation System
AVISCO American Viscose Corporation
AVISPA Aerovias Interamericanas de Panamá (Spanish—Interamerican Airways of Panama)
avit (AVIT) audiovisual instruction(al) technology
av JC avant Jésus Christ (French—before Jesus Christ, B.C.)
avl average versus length
av l average length
AVL Asheville, North Carolina (airport)
AVLA Audio-Visual Language

Association
Av Labs Aviation Laboratories (USA)
AVLINE Audiovisuals On-Line (computer retrieval system)
avlm anti-vehicle land mine
avloc airborne visible-laser optical communication
AVLS Automatic Vehicle Location System
avlub aviation lubricant
avm automatic voting machine
avm (AVM) arteriovenous malformation
AVM guided-missile ship (naval symbol)
AVMA American Veterinary Medical Association
AVMF Aviatsiya Voenno Morskikh Flota (Russian—Soviet Naval Aviation)
avn atrioventricular node; aviation
Avn Avonmouth
AVN Air Vietnam
AVNA Australian Visiting Nurses Association
AVNMED Aviation Medicine (DoD)
av node arterioventricular node
AVNOJ Antifasisticko Vijece Narodnog Oslobidjenja Jugoslavije (Serbo–Croat—Anti-Fascist Council of National Liberation of Yugoslavia)
avo ampere-volt-ohm; avocado
AVO Állam Védelmi-Osztály (Hungarian—Hungarian-Secret Soviet Police); avoid verbal orders
Avog Avogadro
avoid. airfield vehicle obstacle indication device
AVOIDS Avionic Observation of Intruder Danger System
avoil aviation oil
avoir avoirdupois
avolo automatic voice link observation
avos avocados
avozvots average Australian voters
avp arginine vasopressin
AVP Assistant Vice President; seaplane tender, small (3-letter symbol); Wilkes-Barre/Scranton airport at Avoca, Penna.
AVP Aruba Volkspartie (Dutch—Aruba People's Party)
avr aortic valve replacement
AVR Army Volunteer Reserve
AVRA Audio-Visual Research Association

AVRI Animal Virus Research Institute

AVRO A.V. Roe (Ltd)

AVRO Algemeene Vereniging Radio Omroep (Dutch— General Broadcasting Association)

AVROS Algemeene Vereniging van Rubberplanters ter Oostkust van Sumatra (Dutch—General Association of Rubber Plantations of the East Coast of Sumatra)

avrp atrioventricular refractory period; audiovisual recording and presentation

AVRS Audiovisual Recording System

avs aerospace vehicle simulation; area vocation school(s)

AVS American Vacuum Society; Association for Voluntary Sterilization; aviation supply ship (naval symbol)

A-V S Anti-Vivisection Society

AVSA African Violet Society of America

AVSC Audio-Visual Support Center (USA)

AVSECOM Aviation Security Command (Philippines)

AVSL Assistant Venture Scout Leader; Association of Visual Science Librarians

avst automated visual-sensitivity test(er)

AVSYCOM Aviation Systems Command (USA)

avt audiovisual tutorial

Avt Allen vision test

AVT Adult Vocational Training; auxiliary aircraft transport (naval symbol); Aviation Medicine Technician

avta automatic vocal transaction analyzer

avtag aviation wide-cut turbine fuel

av tmp average temperature

AVTP Adult Vocational Training Program

AVTRW Association of Veterinary Teachers and Research Workers

avtur aviation turbine fuel

AVUS Automobile Versuchs and Untersuchungs Strecke (German—Automobile Test Track)

avv avvocato (Italian—advocate)—lawyer

av vales atrioventricular (heart) valves

av w average width

AVWV Antilliaans Verbond van Werknemers Vereni-

gingen (Dutch—Antillean Confederation of Workers' Unions)

AVX Avalon Bay, Catalina Island, California (airport)

aw abandoned woman; above water; acid waste; actual weight; air-to-water; anterior wall; antiwear; atomic warfare

Aw *Antwerpen* (Dutch-Antwerp)

a/w actual weight; all-water; all-weather

a & w alive and well

AW air warning; Air Work, Ltd; American Welding; Articles of War; atomic warfare; atomic weight; automatic weapons(s); distilling ship (naval symbol)

A-W Addison-Wesley Publishers

A & W Atlantic & Western (railroad)

awa absent without authority; advise when able

AWA Air Warfare Analysis; All-Weather Attack; Aluminum Wares Association; America West Airlines; American Warehousemen's Association; American Watch Association; American Waterfowl Association; American Wine Association; American Woman's Association; American Wrestling Alliance; Association of Women in Architecture; Aviation/Space Writers Association

AWA All the World's Aircraft

awac airborne warning and control

AWACS Airborne Warning and Control Systems

AWADS All-Weather Aerial Delivery System

AWAIK Abused Women's Aid in Crisis

AWAL American-West African Line

AWAM Association of West African Merchants

AWANS Aviation Weather and Notice to Airmen System

awar area-weighted average resolution

aware. advance warning equipment

AWARE Addiction Workers Alerted to Rehabilitation and Education (NYC); Association for Women's Active Return to Education

AWARS Airborne Weather and

Reconnaissance System

AWAS Acoustic Wave Analysis System; Australian Women's Army Service

AWASA Australian Women's Army Service Association

AWASM Associate of the Western Australia School of Mines

awb air waybill

AWB Agricultural Wages Board (UK); Arizona White Battalion (neo-Nazi group); Australian Wheat Board; Australian Wine Board; Australian Wool Board; Australian Wool Bureau

AWBA American World Boxing Association

AWB/CN Air Waybill or Consignment Note

AWC Air War College; American Watershed Council; American Wool Council; Anaconda Wire & Cable (stock exchange symbol); Area Wage & Classification (office); Arizona Western College; Army War College; Army Weapons Command; Australian Whaling Commission; Australian Wool Corporation

AWC Amgueddfa Werin Cymru (Welsh Folk Museum)

AWCC Australian Wine Consumer's Cooperative

AWCO Area Wage and Classification Office

awcs agency-wide coding structure

AWCS Air Weapons Control System

AWCU Association of World Colleges and Universities

awd all-wheel drive; awards

AWD Action for World Development; Air Worthiness Division

AWDA Automotive Warehouse Distributors Association

AWDCS Alternate Waste Disposal Concepts Study

awdr advanced weapon-delivery radar

awe accepted weight estimate; advise when established; average weekly earnings

AWEA American Wind Energy Association; Australian Wind Energy Association

AWEASVC Air Weather Service

AWED American Woman's Economic Development

A Weld I Associate of the

Welding Institute
AWES Army Waterways Experiment Station; Association of Western European Shipbuilders
awf awful(ly)
a wf acceptable work-load factor; adrenal weight factor
AWF African Wildlife Foundation; American Wildlife Foundation; Australian Wheatgrowers Federation
AWFS All-Weather Fighter Squadron
AWG American Wire Gage; Art Workers Guild; Australian Writers Guild
AWH Association of Western Hospitals; Australian Women's Hospital
AWHA Australian Women's Home Army
A Whitman Albert Whitman Company
AWHPS Association of White House Press Secretaries
awi anterior wall infarction
AWI All-Weather Interceptor; Animal Welfare Institute; Australian Welding Institute; Australian Wire Industry; Australian Wool Industries
AWIA American Wood Inspection Agency
AWIRA American Wax Importers and Refiners Association
AWIS Association of Women in Science
AWIU Agricultural Workers Industrial Union; Allied Workers International Union; Aluminum Workers International Union
awiy as we informed you
awk awkward
AWK Wake Island (airport)
awkm a wonderfully knowledgeable man
awl. absent with leave; administrative weight limitation; all-weather landing; artesian well lease
Awl Soviet infrared or radarguidance system (NATO)
AWL Animal Welfare League
AWLC Association of Women Launderers and Cleaners
AWLF African Wildlife Leadership Foundation
AWLNET Area Wide Library Network
AWLOGS Army Wholesale Logistic System
AWLS All-Weather Landing System

awm automatic washing machine; awaiting maintenance
AWM American War Mothers; Association of Women Mathematicians; Australian Wallcovering Manufacturers; Australian War Memorial
AWMF Andrew W Mellon Foundation
awmi anterior wall myocardial infarction
AWMPF Australian Wool and Meat Producers Federation
awn awning
AWN Activation Work Notice; Automated Weather Network
AWngSvc Air Warning Service
AWNL Australian Women's National League
AWNY Advertising Women of New York
AWO Accounting Work Order; American Waterways Operators
awol a wolf on the loose; absent without leave; absent without official leave
AWOP All-Weather Operations Panel
awp amusements with prizes
A & WP Atlanta and West Point (railroad)
AWPA Academy of Wind and Percussion Arts; American Wood Preservers Association; American Wood Products Association; Australian Women Pilots Association
AWPB American Wood Preservers Bureau
A-WPC Addison-Wesley Publishing Company
AWPL Australia-West Pacific Line
AWPs *Allied Weather Publications*
awr adaptive waveform recognition
AWR Arctic Wildlife Refuge (Alaska); Association of Western Railways; Australian Wire Rope
AWRA American Water Resources Association; Australian Welding Research Association; Australian Wool Realization Agency
AWRC Australian Water Resources Council
AWRE Atomic Weapons Research Establishment
AWRIS Army War Room Information System
AWRNCO Aircraft Warning Company (Marines)

AWRO Atomic Weapon Retrofit Order
AWRT American Women in Radio and Television
aws adjustable wire stripper
AWS Air Warning Service; Air Warning Squadron; Air Warning System(s); Air Weapon Systems; Air Weather Service; Aircraft Warning Service; Aircraft Warning System; Alston Wilkes Society; American War Standards; American Watercolor Society; American Weather Service; American Welding Society; Atlas Weapon System; Attack Warning System; Automatic Warning System; Automatic Weather Station; Aviation Weather Service
AWSA American Water Ski Association; American Woman Suffrage Association; Association of Wisconsin School Administrators
AWSF Australian Wholesale Softgoods Federation
AWSG Army Work Study Group
AWSJ *Asian Wall Street Journal*
AWSP Association of Washington School Principals
AW & ST *Aviation Week & Space Technology*
awt advanced waste treatment
AWT Aero-elastic Wind Tunnel; Associate in Wildlife Technology
AWTA American Working Terrier Association; Australian Wool Testing Authority
AWTE Association for World Travel Exchange
AWTEW *All's Well That Ends Well*
AWTI Air Weapons Training Installation
awu atomic weight unit
AWU Aluminum Workers Union; Australian Workers Union
AWWA American Water Works Association; Asian Women's Welfare Association; Australian Water and Wastewater Association
awwf all-weather wood foundation(s)
AWWU American Watch Workers Union
awx (AWX) all-weather aircraft
awy airway
ax. axiom(atic); axes; axis
AX American Air Export & Im-

port Company (stock exchange symbol)
AXAF Advanced X-ray Astrophysics Facility
axbt aircraft-expendable bathythermograph
axd auxiliary drum
axe. (AXE) automatic electronic exchange
AXF Advanced X-ray Facility
axfl axial flow
axgrad axial gradient
axidnt accident
axio axiological(ly); axiologist; axiology; axiom; axiomatic(al)(ly)
axmin(s) axminster(s)
AXO Assistant Experimental Officer
Axon Axelson (Swedish—son of Axel)
Ay Ayala
Ay Ayios (Modern Greek—Holy)
AY Allied Youth
AYA American Yachtsmen's Association
Ayat Ayatollah (Persian—Sign of God)—a religious leader among Shiite Muslims
AYC Albany Yacht Club; American Yacht Club; American Youth Congress; Arthur Young & Company; Atlantic Yacht Club; Audubon Yacht Club
AYD American Youth for Democracy
ayer (Malay—water); (Spanish—yesterday)
Ayer NW Ayer and Son
Ayers Ayers Rock National Park, in Australia's Northern Territory, features a colossal red sandstone rock—the world's largest monolith
ayf anti-yeast factor
AYF Australian Yachting Federation
AYH American Youth Hostels
AYHA Australian Youth Hostels Association
AYI Academic Year Institute (NSF)
Ayla Aylett; Aylmar; Aylmer; Aylsworth; Aylward; Aylwin
AYLC Association of Young Launderers and Cleaners
Aym Aymara
AYM Ancient York Mason; Ancient York Masonry
AYM-YWHAs Association of Young Men-Young Women's Hebrew Associations of Greater New York
AYP Alaska-Yukon Pioneers
ayr all-year 'round
Ayr Ayrshire
AYSA American Yarn Spinners Association
AYSO American Youth Soccer Organization
aytng anything
a Z aan Zee (Dutch—on sea); *auf Zeit* (German—on account, on credit)
a/z aan zee (Dutch—on sea)
Az azimuth; Azores; Aztec; Aztecan; azure
Az Azote (Greek—nitrogen)
AZ Active Zone; Alitalia (Linee Aeree Italiane); Alzheimer's disease; Arizona
A-Z Ascheim-Zondek (pregnancy test)
A to Ž from the beginning to the end; thoroughly
AZ Akademisch Ziekenhuis (Dutch—Academic Hospital)
AZA American Zionist Association
Azania South Africa's name according to African nationalists
AZAPO Azania People's Organization (militant South African blacks)
azas adjustable-zero adjustable-span
Azb Azerbaijan; Azerbaijani; Azerbaijanian

AZC American Zionist Counicl
azel azimuth elevation
Azer Azerbaidzhan (or Azerbaijan). One of 15 republics in the U.S.S.R.
AZF American Zionist Federation
azg azaguanine
AZGS Azusa Ground Station
azi azimuth
Az I Azores Islands
AZI American Zinc Institute
Azië (Dutch—Asia)
az ld azure laid (paper)
azm azimuth
azon azimuth only
Azores Azores Islands; Azores Islands in the North Atlantic far to the west of Portugal
Azorín José Martínez Ruiz
Azov Sea of Azov (landlocked body of water within the Crimean section of the USSR)
Azr Azores
azran azimuth and range
AZRI Arid Zone Research Institute
azrock asbestos rock
Azru Aztec Ruins National Monument
azs automatic zero set
azt azusa transponder
azt (AZT) azidothymidine, one of several drugs used in the treatment of AIDS
Azt Aztec; Aztecan
AZT Ascheim-Zondek Test
aztc azusa transponder coherent
Aztec Ruins Aztec Ruins National Monument, New Mexico
A-Z Test Ascheim-Zondek Test (for pregnancy)
aztran azimuth from transit
azur azauridine
azusa azimuth, speed, altitude
az wo azure wove (paper)
azy azyme (matzos, unleavened bread)

B

b baby; base; bicuspid; bituminous; black; blue; book; born; brass; breadth; bridge; bulb (camera exposure device); wing span (symbol)
b. *bis* (Latin—twice)
b span
B Bacillus; bad; *bajar* (Spanish—to descend); balboa (Panamanian currency); Baltic; bandwidth; Barber Lines; *bas* (French—down); bass; bassoon; bastard; Baume; Baume scale; bay; *Bay* (Turkish—Mister); Beatrice (Bea-trice Foods); Beech; Belgium (auto plaque); belted; Bendix; Benoist scale; unit of marijuana measurement consisting of just enough to fill a small matchbox; benzene; body; Boeing; boils at; bolivar (Venezuelan currency); boliviano (Bolivian currency); bomber; bonded; borderline; boron; Boston; bowels; Bravo—code for letter B; British; brightness (symbol); Brother; Bruning; Buddhist; Bull Lines; buoyancy; Burroughs; contrabass; flux density (symbol); Fraunhofer line caused by terrestrial oxygen
B/ balboa (Panamanian currency unit 9 $1.00 U.S.)
°B degrees Baumé
B *Baai* (Afrikaans or Dutch—bay); *Bad* (German—bay); *Bahía* (Spanish—bay); *Baía* (Portuguese—bay); *Baie* (French—bay); *Baja* (Spanish—lower); *bajar* (Spanish—to descend) *Ban* (Indo-Chinese—bay); *bas*

(French—down); *Bay* (Turkish—Mister); *Bir* (Arabic—cistern, well); *Bucht* (German—bay); *bueno* (Spanish—good); *Bukhta* (Russian—bay)
B' Ben (Hebrew—son, son of)
b 1 booster 1
B+ B-plus; B-positive
B-1 B-minus; B-negative (battery terminal); North American-Rockwell strategic supersonic bomber equivalent to the Soviet Backfire
B₁ thiamine vitamin
B-1B Rockwell International long-range supersonic swept-wing bomber
b 1 p booster 1 pitch
b 1 y booster 1 yaw
b 2 booster 2
B₂ riboflavin vitamin
B-2 Advanced Technology Fighter aircraft
B2F Boeing 320 fan jet airplane; Boeing 720 fan jet airplane
b 2 p booster 2 pitch
b 2 y booster 2 yaw
B3F Boeing 320 aircraft
b4 before
b7d buyer (has) seven days (to pay)
B7D buyer has seven days to pay
B7F Boeing 707 fan jet airplane
B8H Boosey and Hawkes
B-25 World War II light bomber called the Mitchell
B-26 modernized Douglas B-26 Invader renamed Counter Invader
B-47 Stratojet all-weather strategic medium bomber

B-52 Stratofortress all-weather intercontinental strategic heavy bomber
B-57 Canberra two-place twin-engine turbojet all-weather tactical bomber
B-58 Hustler strategic all-weather supersonic bomber
B-66 Destroyer twin-engine turbo-jet tactical all-weather light-bombardment aircraft
B 77 Bratislava 77 (viral) strain
B-707 one of a Boeing aircraft series
B-747 Boeing jumbo jet aircraft
ba baby addict; bad attitude; base line; bath(room); blind approach
b-a bare ass(ed); naked; unclothed
b/a backache; billed at; boric acid; budget authority; budget authorization; budget authorized
b.a. *balneum arenae* (Latin—sand bath)
Ba Baia (Portuguese—Bahia); barium (symbol)
BA Bank of America; Basic Airman; Bellas Artes (Fine Arts); Berkshire Athenium; Boeing (stock exchange symbol); Boston & Albany (railroad); British Academy; British Admiralty; British Airways (airline code); British Army; British Association (for the Advancement of Science); Buenos Aires; Bureau of Accounts; Bureau of Apprenticeship; Busted Aristocrat (an officer reduced to the ranks)
B-A Basses-Alpes

B.A. *Baccalaureus Artium* (Latin—Bachelor of Arts)
B/A Bank of America; British American (oil company)
B & A Bangor & Aroostook (railroad); Boston & Albany (railroad)
BA *Bayerische Landesbank* (German—Bavarian National Bank); *Biological Abstracts; Bonne Action* (French— Good Deed); *Bowker Annual; Business* Automation
baa benzoyl arginine amide; bleat of a sheep
Baa Baal; Baalam
BAA Brewers Association of America; British Acetylene Association; British Airports Authority; British Archeological Association; British Astronomical Association; Bureau of African Affairs
B.A.A. Bachelor of Applied Arts
BAAA British Association of Accountants and Auditors
BAAB British Amateur Athletic Board
BAAC Bank of Agriculture and Agricultural Cooperatives (Thailand)
BAADS Bangor Air Defense Sector
BAAF Brigade Airborne Alert Force
Baal Baalbek
BAAL Black Academy of Arts and Letters; British Association for Applied Linguistics
BAAP Bilateral Aid to Asia and Pacific (program)
BAAR Board of Aviation Accident Research
BAAS British Association for the Advancement of Science
bab (Arabic—gate, strait)
Bab Barbara; Baboo (Anglo–Indian dialect); Babylon; Babylonia; Babylonian; WS Gilbert
BAB British Airways Board; BT Babbitt (Babo cleanser)
BABA British Antiquarian Bookseller's Association
Babars Babar Islands of Indonesia
babb babbit metal
Babbie Barbara
babbitt babbitt metal (named for its inventor, Isaac Babbitt of Taunton, Massachusetts)
Babb's computer Babbage's computer (the first computer)
Babette Elizabeth
BABI Brooke Army Burn In-

stitute (San Antonio, Texas)
Babines Babine Mountains of British Columbia
bab met babbitt metal
Babo Boolean approach for bivalent optimization; BT Babbitt detergent scouring powder
babord (Norwegian—larboard) port side
Babs blind approach beacon system
Bab(s) Barbara
BABS Babbage Society; Blind-Approach Beacon System
BABT Brotherhood of Associated Book Travelers
Baby Babylon(ia); Babylonian
bac bacilli; bacillus; bacteria; bacterial; bacterial antigen complex; bacteriologist; bacteriology; blood-alcohol concentration; buccoaxiocervical
bac (BAC) binary asymmetric channel
bac *bachot* (French abbreviations for baccalaureat)— bachot also means ferryboat
Bac. Baccalaureus (Latin— Bachelor)
BAC Bendix Aviation Corporation; Black Action Committee; Blair Athol Coal; blood alcohol concentration, Boeing Airplane Company; born-again Christian; British Aircraft Corporation; British Association of Chemists; Bureau of Air Commerce; Business Advisory Council (U.S. Department of Commerce); Business Archives Council of Australia
BAC Baile Atha Cliath (Gaelic—Dublin)
BAC-145 British Jet Provost trainer aircraft
BACA British Aerospace Commercial Aircraft; Business and Consumer Affairs
BACAH British Association of Consultants in Agriculture and Horticulture
BACAIC Boeing Airplane Company Algebraic Interpretive Computing
BACAICS Boeing Airplane Company Algebraic Interpreter Coding System
BACAL Butter and Cheese Association Limited
BACAN British Association for the Control of Aircraft Noise
bacat barge aboard catamaran
BACAT Barge Canal Traffic

bac bag bactine bag
B.Acc. Bachelor of Accountancy
BACC British-American Collectors' Club
BA & CC Billiards Association and Control Council
BACCHUS Boost Alcohol Consciousness Concerning Health of University Students; British Aircraft Corporation Commercial Habitat Under the Sea
BACD Boeing Airplane Company Design
bace basic automatic checkout equipment
BACE British Association of Consulting Engineers; Bureau of Agricultural Chemistry and Engineering
BACGA British-Australian Cotton Growing Association
bach bachelor
Bach (German—brook, stream)
B.A. Chem. Bachelor of Arts in Chemistry
bach girl(s) bachelor girl(s)
bachot (French—baccalaureat, ferryboat)—an abbreviation and a definition
Bach Soc Bach Society
BACIE British Association for Commercial and Industrial Education
back. backwardation
'backs wetbacks (illegal immigrants from Mexico)
BACM British Association of Colliery Management
BACMA British Aromatic and Compound Manufacturers Association
BACNATO British Atlantic Committee of NATO
BACO British Aluminium Company
BACP Blair Athol Coal Pty
BACS Ben Asia Container Service
bact bacteria; bacteriological; bacteriologist; bacteriology; bacterium
BACT Best Available Control Technology
bacter bacteriologist
bacteriol bacteriologic(al)(ly); bacteriologist; bacteriology
bactrian bactrian camel (two-hump camel of Asia)
BACU Battle Area Control Unit
Bad Badajoz
Bad (German—Bath)—short form for more than a hun-

dred Austro-German hydro-thera-peutic resorts ranging from *Bad Abbach* to *Bad Zwis-chenahn*

BAD Bantu Administration and Development; Base Air Depot; Berlin Airlift Device; Black, Active, and Determined; British Association of Dermatology

BADA Base Air Depot Area; British Antique Dealers' Association

BADAS Binary Automatic Data Annotation System

badc binary asymmetric dependent channel

BADD Bartenders Against Drunk Driving; Bothered About Dungeons and Dragons

baddies bad ones

baddie(s) bad guy(s)—incorrigible criminal(s)

Baden Baden-Baden

Baden (German—Baths)—*Baden bei Wien* (Baden near Vienna) and Baden-Baden near Karlsruhe, Germany

BADGE Basic Air Defense Ground Environment

BADGES Base Air Defense Ground Environment System

badhouse bawdyhouse

Badian(s) Barbadian(s)

B.Admin. Bachelor of Administration

BADS British Association of Dermatology and Syphilology

bae Beacon antenna equipment

Ba e barium enema

BAe British Aerospace

BAE Board of Architectural Examiners; Bureau of Agricultural Economics; Bureau of American Ethnology

BA of E Badminton Association of England

B.A.E. Bachelor of Aeronautical Engineering; Bachelor of Agricultural Engineering; Bachelor of Architectural Engineering; Bachelor of Art Education; Bachelor of Arts in Education

BAE Buque Armada Ecuatoriana (Spanish—Ecuadorian Naval Ship)

BAEA British Actors' Equity Association

BAEC British Agricultural Export Council

B.A.Econ. Bachelor of Arts in Economics

B.A.Ed. Bachelor of Arts in Education

BAED British Airways European Division

B.Ae.E. Bachelor of Aeronautical Engineering

BAEF British-American Educational Foundation

Ba enem barium enema

BAEng. Bureau of Agricultural Engineering

baer (BAER) brainstem auditory-evoked response

baf baffle; bunker adjustment factor

ba & f budget, accounting, and finance

BAF British Air Force; Burma Air Force; Burundi Air Force

BAFCom Basic Armed Forces Communication Plan

bafflegab ambiguous expressions (*see* Bafflegab addendum)

bafgab bafflegab—*see* Bafflegab addendum

BAFL Baltic-American Freedom League

BAFM British Association of Forensic Medicine; British Association of Friends of Museums

BAFMA British and Foreign Maritime Agencies

BAFO British Air Forces of Occupation; British Army Forces Overseas

BAFS British Academy of Forensic Science

BAFSC British Association of Field and Sports Contractors

BAFSV British Armed Forces Special Vouchers

BAFTA British Academy of Film and Television Arts

BAFTM British Association of Fishing Tackle Makers

bag. bagasse; baggage; ballistic attack game; buccoaxiogingival

Bag Baghdad

B.Ag. Bachelor of Agriculture

BAG Beaverbrook Art Gallery

BAGA British Amateur Gymnastics Association

BAGBI Booksellers Association of Great Britain and Ireland

BAGDA British Advertising Gift Distributors Association

B.Ag.E. Bachelor of Agricultural Engineering

bagg buffered azide glucose glycerol

B. Agr. Bachelor of Agriculture

BAGR Bureau of Aeronautics

General Representative

B.Agr.Eco. Bachelor of Agricultural Economics

B.Agric. Bachelor of Agriculture

B.Ag.Sc. Bachelor of Agricultural Science

Bah Bahamas; Bahia; Bahrain

BAH Bahrain Island, Persian Gulf (airport); British Airways Helicopters

B-A H British-American Hospital

Baha'i (Abdul) Baha Bahai

Bahamas Commonwealth of the Bahamas off coast of Cuba (discovered by Columbus on October 12, 1492)

Bahams Bahamas

Bahia Sao Salvador de Bahia

Bah Ind Bahasa Indonesian (national language)

BAHOH British Association of the Hard of Hearing

BAHPA British Agricultural and Horticultural Plastics Association

Bahrain Bahrain Island in the Persian Gulf

Bahrains Bahrain Islands in the Persian Gulf between Qatar and Saudi Arabia

BAHS British Agricultural History Association

baht unit of currency in Thailand

Ba I Bahama Islands

BAI Bank Administration Institute; Bank of America International; Barrier Industrial Council; Book Association of Ireland; British Airports International; Bureau of Animal Industry

B.A.I. *Baccalaureus in Arte Ingeniaria* (Latin—Bachelor of Engineering)

baib beta-amino-isobutyric (acid)

BAIC Bureau of Agricultural and Industrial Chemistry

baid boolean array identifier

BAIE British Association of Industrial Editors

BAINS Basic Advanced Integrated Navigation System

bait. bacterial automated identification technique

B.A.J. Bachelor of Arts in Journalism

Baja Baja California (Spanish—Lower California)

Bajan Barbadan (inhabitant of Barbados); Barbadian English; Barbadian people; pertaining to Barbados

Baja Norte *Baja California Norte* (Spanish—Northern Baja California)—Mexican state including Ensenada, Mexicali, Tecate, and Tijuana
Baja Sur *Baja California Sur* (Spanish—Southern Baja California)—Mexican territory including Cabo San Lucas, La Paz, and Loreto
B.A.Jour. Bachelor of Arts in Journalism
Bajuns Barbadans
bak bakery
baka African pygmy
Bakery Workers Bakery and Confectionery Workers International Union of America
baktr *baktrisch* (German—Bactrian)
bal balance; balcony; baloney; blood-alcohol level; broad absorption line
bal (BAL) basic assembly language (computer programming)
Bal Baleares; Ballarat; Balthasar; Baltimore; British anti-lewisite
BAL Baltimore, Maryland (Friendship Airport); Barclays Australia Ltd; Belgian African Line; Bonanza Airlines (3-letter coding); Borneo Airways Ltd.
balance. basic and logically applied norms—civil engineering
bal. arenae *balneum arenae* (Latin—sandbath)
balast balloon astronomy
Balaton Lake Balaton, central Europe's largest lake, nicknamed the Hungarian Ocean
Balb Balboa
balboa unit of currency in Panama
balc balconette; balconied; balcony
Bald Baldwin
Baldie Archibald; Baldassare; Baldomero; Balduin; Baldur; Baldwin; Baldwina
baldie(s) bald person(s)
Baldy Baldemar; Baldram; Baldred; Baldrey; Baldric; Baldur; Baldwin
Balearics Balearic Islands of Majorca, Minorca, Ibiza, Formentera, Aire, Aucanada, Botafoch, Cabrera, Dragonera, Pinto, and El Rey
Balgol Burroughs algebraic compiler
BALH British Association for Local History

balid ballistics identification
Balkans Balkan mountains, peoples, and states in southeastern Europe (Albania, Bulgaria, Greece, Romania, Turkey, Yugoslavia)
ball. ballast
Ball Ballerup; Balliol College, Oxford
Ball Coll Balliol College—Oxford
Ballenys Balleny Islands
ballots. bibliographic automation of large library operations
ballute balloon parachute
bally ballyhoo
Ballyhouras Ballyhoura Hills of southern Ireland
bal. mar. *balneum maris* (Latin—salt-water bath, seawater bath)
Bal-Mol Ballester-Molina (45-caliber Argentine semi-automatic pistol)
balop balopticon (projector)
B Alp Basses-Alpes
balpa balance of payments; ballpark
BALPA British Airline Pilot's Association
B-alpes Basses-Alpes
bals balsam
bals. *balsamum* (Latin—balsam)
B.A.L.S. Bachelor of Arts in Library Science
BALSA Black American Law Students Association
Balt Balthasar; Baltic; Baltimore
BALTAPs Baltic Approaches (NATO)
balth balthazar (16 bottle capacity)
balthum balloon temperature and humidity
Balti Baltimore (slang)
Baltic Baltic and Mercantile Shipping Exchange (in London); Baltic Sea
Baltic(s) Estonia, Latvia, Lithuania
Baltis Baltistan (Karakoram Range between N Kashmir and NW Tibet)
Balto Baltimore
Balts Baltic peoples; Balto-Slavs (East Prussians, Estonians, Latvians, Lithuanians); Balto-Slavic-speaking peoples
Balt Sym Baltimore Symphony
Baluch Baluchistan
balun balance-to-balance (network)

balute balloon parachute
bal. vap. *balneum vapour* (Latin—steambath, vapor bath)
bam bio-activated mineral lotion; broadcasting am
Bam Bamberger
BAM Baikal-Amur-Magistral (railroad); BankAmerica Corporation (stock-exchange symbol); Basic Access Method; broadcasting AM; Brooklyn Academy of Music
B-A-M Baikal-Amur-Magistral (railway of eastern Siberia)
'Bama Alabama
BAMA British Amsterdam Maritime Agencies
bambi (BAMBI) ballistic missile bombardment interceptor
bame benzoylarginine methyl ester
BAMIRAC Ballistic Missile Radiation Analysis Center
BAMM Black Afro Militant Movement
BAMO BuAer Material Officer
bamp basic analysis and mapping program
BAMR BuAer Maintenance Representative
B.A.M.S. Bachelor of Ayurvedic Medicine and Surgery
BAMTM British Association of Machine Tool Merchants
B.A.Mus. Bachelor of Arts in Music
BAMW British Association of Meat Wholesalers
ban. best asymptotically normal
Ban Bantu; Byron Bancroft Johnson
BAN Base Activation Notice; British Association of Neurologists
BAN *Biblioteka Akademii Nauk* (Russian—Library of the Academy of Sciences)—Leningrad
Banaca *Banco Nacional de Credito Agricola* (Spanish—National Agricultural Credit Bank)—Mexico
Banace *Banco Nacional de Credito Ejidal* (Spanish—Public Land Credit National Bank)—Mexico
Banacoex *Banco Nacional de Comercio Exterior* (Spanish—National Bank of Foreign Commerce)
Banafoco *Banco Nacional de Fomento Cooperativo* (Spanish—National Bank for Co-

operative Promotion)
Banamex *Banco Nacional de México* (Spanish—National Bank of Mexico)
Bananagate Honduran scandal involving high officials bribed to lower export taxes on bananas
Banana Republics countries of Central and northern South America; Jamaica often included
Banape *Banco Nacional de la Pequeña Empresa* (Spanish—National Small Business Bank)—Mexico
BANC British Association of National Coaches
Banco *El Banco* (Spanish—The Bank)—World Bank for Reconstruction and Development
Bancomer *Banco de Comercio* (Spanish—Bank of Commerce)
Bancoop *Banco Nacional de las Cooperativas del Perú* (Spanish—National Bank of Peruvian Cooperatives)
Banc.Sup. *Bancus Superior* (Latin—Upper Bench)—King's or Queen's Bench
Band Bandung
Bandas Banda Islands of Indonesia
Bandeirante Brazilian 12-passenger transport honoring frontier pioneers, Bandeirantes
Bandelier Bandelier National Monument and cliff-dweller Indian reservation in New Mexico west of Santa Fe
Banffs Banffshire
Bang Bangalore
Bangla Bangladesh (formerly East Pakistan)
Bangladesh People's Republic of Bangladesh—formerly East Pakistan
bang(s) bombing(s); explosion(s)
banir bombing and navigation inertial reference
Banjul formerly Bathurst, The Gambia
bank. banking
Bank Bangkok
BANK International Bank for Reconstruction and Development
BankCal Bank of California
bankcy bankruptcy
Bankers Bankers Publishing (Boston); Bankers Trust (New York)

Bankhead Bankhead National Forest in northwest Alabama
BANKPAC Bankers Political Action Committee
banks. bank holidays (West Indian English)
Banks the Banks (shallow fishing banks offshore Canada—the Grand or Newfoundland Banks, or Georges Banks off New England)
banks clgs bank clearings
bankster(s) banker gangster(s) who deprive(s) you of your deposits under the cloak of bank management
Bann Bannockburn
Banobras *Banco Nacional de Obras y Servicios Publicos* (Spanish—National Bank of Public Works and Services)—Mexico
Banpeco *Banco Peruano de los Constructores* (Spanish—Peruvian Constructor's Bank)
Banrural *Banco Nacional de Credito Rural* (Spanish—National Rural Credit Bank)—Mexico
ban's bond anticipation notes
BANS Bright Alphanumeric Subsystem; British Association of Numismatic Societies
BANSA *Banco del Ahorro Nacional* (Spanish—National Savings Bank)—Mexico
BANSW Band Association of New South Wales
Bantam Bantam Books; Swedish antitank guided missile
BANTSA Bank of American National Trust and Savings Association
B.A. Nurs. Bachelor of Arts in Nursing
BANWR Bosque Apache National Wildlife Refuge (New Mexico)
BANZ Bahrain-New Zealand Trading and Storage Company
BANZARE British, Australian, New Zealand Antarctic Research Expedition
bao basal acid output
BAO Bankruptcy Annulment Order; British Association of Otolaryngologists; British-American Oil
B.A.O. Bachelor of the Art of Obstetrics; Bachelor of Arts in Oratory
BAOD British Airways Overseas Division
bao-mao basal acid output to maximal acid output (ratio)

BAOP British Atlantic Ocean Possessions (Ascension, St Helena, and Tristan da Cunha islands)
BAOR British Army on Rhine
bap baptism; baptized; base auxiliary power; beginning at a point; blood-agar plate; brachial artery pressure
bap *billets a payer* (French—bills payable)
Bap Baptist; Baptista; Baptiste
BAP Bankers Association of the Philippines; Booksellers Association of Philadelphia
B A & P Butte, Anaconda & Pacific (railroad)
BAP *Barco de la Armada Peruana* (Spanish—Ship of the Peruvian Navy)
BAPA British Airline Pilots' Association
BAP & C British Association of Print & Copyshops
BAPCO Bahrain Petroleum Company
bape baseplate
B.A.P.E. Bachelor of Arts in Physical Education
BA Phys Med British Association of Physical Medicine
BAPL Bettis Atomic Power Laboratory (AEC)
BAPM British Association of Physical Medicine
B.App.Arts Bachelor of Applied Arts
B.App.Sci. Bachelor of Applied Science
BAPS Beacon-Automated Processing System; British Association of Pediatric Surgeons; British Association of Plastic Surgeons; Bureau of Air Pollution Sciences
BAPSA Broadcast Advertising Producers Society of America
Bapt Baptist
BAPT British Association of Physical Training
baq basic allowance for quarters
BAQ Barranquilla, Colombia (airport)
BAQC Bureau of Air Quality Control
bar. baritone; barometer; barometric
bar. (BAR) buffer address register
bar *billets à recevoir* (French—bills receivable)
Bar Baroque; Baruch, Book of
Bar *Barone* (Italian—Baron)
B.Ar. Bachelor of Architecture

BAR Base Address Register (computer); Board of Airline Representatives (Singapore); Board of Anthropological Research; Broadcast Advertisers' Reports; Browning automatic rifle; Bureau of Aeronautics Representative; Bureau of Automotive Repair
BARA Bureau d'Analyse et de Recherche Appliquées (French—Bureau of Analysis and Applied Research)
Barak Israeli version of the French Mirage military aircraft
barb. barbarian; barbary ape; barbary horse; barbecue; barber; barbiturate
Barb Barbados Islands; Barbara; Barbary
barbi baseband radar bag initiator
Barbie Barbara
bar-b-q barbecue
Barbra Barbara
barbs. barbiturates
barbus barbudos (Spanish—bearded ones)
Barc Barcelona; Barclay
BARC Bay Area Reference Center; Bay Area Research Collective; British Aeronautical Research Committee
Barca Barcelona
Barc Con Cam Barcelona Conjunto Cameristico (Spanish—Barcelona Chamber Ensemble)
B.Arch. Bachelor of Architecture
B.Arch.E. Bachelor of Architectural Engineering
B.Arch. & T.P. Bachelor of Architecture and Town Planning
BARCS Battlefield Area Reconnaissance System
Barents Barents Sea in the Arctic between Norway and Russia
barg(s) bargain(s)
bari baritone; baritone saxophone
Bari Bari delle Puglie, Italy
Bariloche San Carlos de Bariloche, Argentina
Barisans Barisan Mountains of Sumatra
barite barium sulfate
Bark Barker; Barkham
Barlinnie Glasgow, Scotland's great prison
Barme Bartolome
bar mitz bar mitzvah (Hebrew—one who is respon-

sible for the Commandments)—religious ritual marking a boy's 13th birthday; similar ceremony for girls called *bas mitzvah* or *bat mitzvah*
barn. bombing and reconnaissance navigation
Barn Barnard
BARN Body Awareness Resource Network
Barna Barcelona
Barney Barnabas; Barnett; Bernard; Bernardino; silver cigarette box engraved with drawings of Barney Google and Snuffy Smith (awarded to the year's best cartoonist)
BARNS Bombing and Reconnaissance Navigation System
Barnum P(hineas) T(aylor) Barnum; 19th-century American impresario and showman who brought his circus to all parts of the United States; presented the Swedish soprano Jenny Lind, the midget General Tom Thumb, Jumbo the enormous elephant, the first hippopotamus ever shown in America; also established mo-del industrial and workers community in Bridgeport, Connecticut
BARONS Business-Accounts Reporting Operating Network System
b & arp bare and acid resisting paint
BARP British Association of Retired Persons
barq barquentine
Barq Barranquilla
barr barrister
BARR British Association of Rheumatology and Rehabilitation
Barrens Barren Grounds of northern Canada west of Hudson Bay in treeless tundra; Pine Barrens of New Jersey
Barrow Point Barrow (Alaska's most northerly point of land and settlement)
Barry Barrymore
BARS Backup Attitude-Reference System; Ballistic Analysis Research System; British Association of Residential Settlements
BARSR Biblioteca Academiei Republicii Socialiste Romania (Academic Library of the Socialist Republic of Romania)—in Bucharest

BARSTUR Barding Sands Underwater Test Range
bart bartender; barter
Bart Baronet; Bartel; Barth(el); Barthold; Bartholomew; Bartimeus; Bartlet(t); Bartley; Bartolo; Bartolomeo; Barton; Bartram
BART Bay Area Rapid Transit (San Francisco); Brooklyn Army Terminal (New York)
BARTD Bay Area Rapid Transit District
Bartenders Union Hotel and Restaurant Employees and Bartenders International Union
Bartolo Bartolomeo (Italian—Bartholomeo)
Barts Saint-Barthelemy in the French West Indies; Saint Bartholomew's Hospital in London
Barty Bartholomew
barv beach armored recovery vehicle
bas basenji; basic airspeed; basic allowance for subsistence; basilica; basophil(s); basset; beam-alignment sensor; benzyl analog of serotonin
bas (BAS) baths
Bas Basel; Basil; Basilica; Basi-licata; Bass Strait; Bastogne; Bastrop; Basuto; Basutoland
BAs Business Agents (of unions)
B.As Buenos Aires
BAS Basic Allowance for Subsistence; Behavioral Approach Scale; Brazilian-American Society; British Acoustical Society; British Antarctic Survey; British Arachnological Society
B.A.S. Bachelor of Agricultural Science; Bachelor of Applied Science
BAS Bulletin of the Atomic Scientists
Basa Baronessa (Italian—Baroness)
BASA British Architectural Students' Association; Buckeye Association of School Administrators
BASAF British and South Africa Forum
BASAM British Association of Grain, Seed Feed, and Agricultural Merchants
B. A. Sc. Bachelor of Applied Science
BASC Booth American Ship-

ping Corporation
basc b bascule bridge
basd basic active service date
base. (BASE) basic semantic element
BASE Bank-Americard Service Exchange; buildings, antennas, spans, earth formations (term used by parachutists); Business Assessment Study and Evaluation
Base 2 binary
Base 8 octal
Base 10 decimal
Base 16 hexadecimal
BASEC British Approvals Service for Electric Cables
BASEEFA British Approvals Service for Electrical Equipment in Flammable Atmospheres
BASF Badische Anilin und Soda Fabrik
bash. body acceleration given synchronously with the heartbeat
BASH Bird Aircraft Strike Hazard; Bulimia Anorexia Self Help
BASI British Association of Ski Instructors
basic. (BASIC) battle-area surveillance and integrated communications; beginner's all-purpose symbolic instruction code (computer language)
BASIC British-American Scientific International Commercial (English)
BASIC Biological Abstracts Subjects in Context
BASICO Behavior Science Corporation
basicpac basic processor and computer
basictng basic training
BASIE British Association for Commercial and Industrial Education
basil. basilect(ic)(al)(ly)
BASIL Barclays Advanced Staff Information Language
basis. (BASIS) bibliographic author of subject interactive researches
BASIS-H BASIS history
BASIS-P BASIS political science, public administration, urban studies, and international relations
BASIS-S BASIS sociology
bask baskisch (German—Basque)
Bask Baskir(ia)
Bask(er) Baskerville
BASMA Boot and Shoe Manu-

facturers' Association
bas mitz see bar mitz
BASO Base Accountable Supply Officer; Bureau of Aeronautics Shipping Order
basops base operations
baso(s) basophile(s)
baspm basic planning memorandum
BASR Bureau of Applied Social Research (Columbia University)
BASRA British Amateur Scientific Research Association
bass. bassoon
BASS Basic Analog Simulation System; Bass Anglers Sportsman Society
B.A.S.S. Bachelor of Arts in Social Science
bass con basso continuo (Italian—continuous bass)—figured bass background
Bassie Sebastian
BASSR Bashkirian Autonomous Soviet Socialist Republic; Buriat Autonomous Soviet Socialist Republic
bast. bastard; bastardization; bastardize; bastardly; bastard title; bastardy
Bast Sebastian
bas tit. bastard title (half title)
Basto Sebastiano
ba sw bell-alarm switch
BASW British Association of/for Social Workers
basys basic system
bat. battery; battle; best available technology
Bat Bartholomew; Battista
BAT Beaux Arts Trio; Blind Approach Training; Body Adjustment Test; Boeing Air Transport; Bolshoi All-azimuth Telescope; British American Tobacco; Bureau of Apprenticeship and Training
BA & T Bureau of Apprenticeship and Training
B-AT British-American Tobacco
BATAB Baker and Taylor Automated Bookordering
BATC British Amateur Television Club
bat. chg battery charger; battery charging
BATDIV battleship division
bate base activation test equipment
batea best available technology economically available
b-a test blood-alcohol test
BATF Bureau of Alcohol, To-

bacco, and Firearms (U.S. Treasury)
BATFOR battle force
bath. bathroom; best available true heading
Batham Bantam Books
bath mitz bath mitzvah
batho bathometer
bathy bathymeter; bathysphere; bathyscaphe
BATM British Admiralty Technical Mission
bat mitz see bar mitz
bato baloon-assisted takeoff
B.A.T.P. Bachelor of Arts in Town Planning
batreadcompi battle readiness and competition instructions
BATRON battleship squadron
batrop baratropic
Bat Rou Baton Rouge
Bats British-American Tobacco (stock-exchange sobriquet)
BATS Business Air Transport Service
batt batter; batteries; battery
bau basic assembly unit; British absolute unit (BTU, Btu)
Bau Bauer; Bauhaus
BAU Bangladesh Agricultural University; British Association Unit
BAUA Business Aircraft Users' Association
Baubie (Scottish—Barbara)
baud telecommunication unit measuring speed of signalling and equal to one code element or pulse per second—named for its French inventor, JME Baudot (1845–1903)
BAUS British Association of Urological Surgeons
bav bon à vue (French—good at sight);—sight draft
Bav Bavaria; Bavarian
BAV Biblioteca Apostolica Vaticana (Latin—Apostolic Vatican Library), in Rome's Vatican City
BAVA Bureau of Audio-Visual Aids (NY)
BAVE Bureau of Audio-Visual Education (Calif)
BAVIP British Association of Viewdata Information Providers
BAVTE Bureau of Adult, Vocational, and Technical Education (HEW)
baw bare aluminum wire
BAWA British Amateur Wrestling Association
BAWHA Bide-A-Wee Home

Association
BAWR Bertrand Arthur William Russell
BAWRA British Australian Wool Realization Association
BAX Burlington Air Express
bay cand dc bayonet candelabra double contact
Bayer Bayerisch (German—Bavarian)
bayr bayrisch (German—Bavarian)
BAYS British Association of Young Scientists
bazuco Colombian cocaine product
bb ball bearing; ball beaters; ball-beating contests: baseball, basketball, billiards, cricket, football, hockey, ping-pong, soccer, tennis, volleyball; bank burg-lar(y); baseline-to-baseline measure in typography; bases on balls (walks); bayonet base (lamp or socket); below bridges; bill book; blood bank; blood buffer (base); blue bloaters; both bones; both to blame; breakthrough bleeding; breast biopsy; buffer base; bungling bureaucrat; double black; pellet fired from or made for a bb gun
bb (BB) billboard (television script)
b-b black bordered; bogie-bogie (single-unit locomotive)
b/b bail bond; bottled in bond
b & b bed and board; bed and breakfast; benedictine and brandy
b or b brass or bronze (cargo)
b to b back to back
bb babord (Swedish—port side);
BB battleship; Before Bach; B'nai B'rith; Brigitte Bardot; Bureau of the Budget
BB (DCO) Barclays Bank (Dominion, Colonial and Overseas)
B B former Columbian president Belisario Betancur Cuartas
B-B Bora-Bora
B.B. Bernard Berensen; Bjørnstjerne Børnson; Boys' Brigade
B & B Brown and Bigelow
B of B Bureau of the Budget
BB Banco do Brasil (Portuguese—Bank of Brazil)
BB 62 USS *New Jersey*
bba born before arrival

BBA Big Brothers of America; born before arrival (of the midwife or doctor); British Bankers' Association
B.B.A. Bachelor of Business Administration
bbac bus-to-bus access circuit(ry)
BBAC British Balloon and Airship Club
b-ball basketball
b/bar bull bar (fender)
BBAR Broad-Band Anti-Reflection
bbb banker's blanket bond; basic boxed base; bed, breakfast, and bath; beleaguered by bullets and burglars; blood brain barrier; triple black
BBB Bach, Beethoven, Berlioz; Bach, Beethoven, Brahms; Bach, Beethoven, Bruckner; Best Berlin Broadcast; Best British Briar (pipes); Better Business Bureau
BBBC British Boxing Board of Control
BBBS Boat Builder's Benefit Society
BB & BU Bagel Boilers and Bakers Union
bbc barrels, boxes, and crates (cargo); bromobenzylcyanide (gas)
BBC Bank of British Columbia; Beautiful British Columbia; Best British Cinema; Better Breathers Club (for those with pulmonary disease); Billionaire Boys' Club; Biwako Broadcasting Corporation (Japan); British Broadcasting Corporation; Brown Boveri Corporation; Buchanan Borehole Colleries
BBC-1 British Broadcasting Corporation's first television network featuring cinema and sports
BBC-2 British Broadcasting Corporation's second television network featuring classical music
BBCC Big Bend Community College
BBCCS B'nai B'rith Career and Counseling Services
BBC dissociation Braid-Berheim-Charcot dissociation
BBC English cultured way of speaking English
BBCF British Bacon Curers' Federation
BBCL Bermuda Broadcasting

Company Limited
BBCM Bandmaster—Bandsmen's College of Music
BBCMA British Baby Carriage Manufacturers' Association
BBC Northern British Broadcasting Corporation's Northern Symphony
BBCS Browne Book-Charging System
BBC Scot British Broadcasting Corporation's Scottish Symphony
BBCSO British Broadcasting Corporation Symphony Orchestra
BBC Sym British Broadcasting Corporation Symphony
bbcw bare beryllium copper wire
bbd baby born dead; beta-binomial distribution; big bass drum; bucket-brigade device; bulletin board
bbdc before bottom dead center
BB(DCO) Barclays Bank (Dominion, Colonial, Overseas)
BBDO Batton, Barton, Durstine, and Osborne
BBE Board of Barber Examiners
BBEA Brewery and Bottling Engineers Association
bb & em bed, breakfast, and evening meal
bbf boron-based fuel
BBF Biblioteca Benjamin Franklin (Mexico City); Boilermakers, Blacksmiths, Forgers (union)
BBFC British Board of Film Censors
bbg bundle branch block
BBG Bermuda Botanical Gardens; Brooklyn Botanic Garden
BBGA British Broiler Growers' Association
bb-gun airgun shooting bb's (ball bearings)
BBHC Buffalo Bill Historical Center (Cody, Wyoming)
BBHF B'nai B'rith Hillel Foundations
BBI Barbecue Briquet Institute; Brandeis-Bardin Institute
B Bibl Bachelier en Bibliothàconomie (French—Bachelor in Library Science)
BBINA Bed and Breakfast Inns of North America
bbing bureaucratic bumbling
BBiP British Books in Print
BBIRA British Baking Indus-

tries Research Association

B Bisc Bay of Biscay

bbj ball-bearing joint

bbk breadboard kit

bbl barrel

BBL Bahia Blanca; Bangkok Bank Ltd; Barclay's Bank Limited; Big Brothers League

bbl roll barrel roller

bbls/day barrels per day

bbm big bowel movement; break-before-make

b & b magazine periodical featuring breasts and buttocks

BBMRA British Brush Manufacturers Research Association

BBNNR Braunton Burrows National Nature Reserve (England)

BBNP Big Bend National Park (Texas)

BBNR Back Bay National Refuge (Virginia)

Bbo Bilbao

B-Bomb benzedrine bomb (slang—benzedrine inhaler)

B-boy busboy; mess sergeant

bbp boxes, barrels, packages, (cargo); building block principle

BBP Beech Bottom Power Company

BBP *Boletim de Bibliografia Portuguesa* (Portuguese—Bulletin of Portuguese Bibliography)

b&b pericarditis bread-and-butter pericarditis

BBPR Bianfi, Barbiano, Peresutti, and Rogers (avant-garde Italian architects)

bbq barbecue

BBQ Brooklyn, Bronx, Queens

bbr balloon-borne radio

BBRR Brookhaven Beam Research Reactor (AEC)

BBRS Balloon-Borne Radio System

bbs ball bearings; barrels of basic sediment; box bark strips

Bbs British biscuits

BBS Barber Blue Sea (steamship line); Bermuda Biological Station; Brunei Broadcasting Service; Bulletin Board System

B.B.S. Bachelor of Business Science

BBSATRA British Boot, Shoe and Allied Trades Research Association

BBSE Board of Behavioral Science Examiners

BBSI British Boot and Shoe Institution

bbsj ball-bearing swivel joint

B & B SNC British and Burmese Steam Navigation Company

BBSR Bermuda Biological Station for Research

bbsu bid bond service undertaking

bbs & w barrels of basic sediment and water

bbt basal body temperature; bombardment

BBT Brotherhood of Book Travelers

BBTA British Bureau of Television Advertising

BB & TC Bahamas Broadcasting and Television Commission

BBU Bagel Boilers Union; Blue-Brick University (one of the highest academic traditions such as Cambridge, Harvard, Oxford, or Yale)

bbw bare brass wire

BBW B'nai B'rith Women

BBWAA Baseball Writers' Association of America

BBX Bluebird (stock-exchange symbol)

BBYO B'nai B'rith Youth Organization

bbz bearing bronze

bc back course; bad check; ballistic camera; base (shield) connection; between centers; bills for collection; binary code; binary counter; birth control; bogus check; bolometric correction; bolt circle; bone connection; born in colony; bottom (dead) center; broadcast control; budgeted cost; building center; burden center; bursa copulatrix

bc (**BC**) bio-conversion; bulk carrier

b/c bales of cotton; bills for collection; birth control; broadcast

b & c building and contents

B ͨ *Banc* (French—bank, reef or sandbank)

BC Bacone College; Baja California, Mexico; Bakersfield College; Balliol College (Oxford); Bank Clearing; Bankruptcy Court; Bard College; Barnard College; Barrington College; Barry College; Baruch College; Bates College; Battery Commander; Battle Cruiser; Beaver Col-

lege; Beckley College; Before Christ; Belgian Congo; Belhaven College; Bellarmine College; Belmont College; Beloit College; Benedict College; Bennett College; Bennington College; Berea College; Berry College; Bethany College; Bethel College; Bishop College; Blackburn College; Blinn College; Bliss Classification; Bliss College; Bloomfield College; Bluefield College; Bluffton College; Bomber Command; Borough Council; Boston College; Bourget College; Bowdoin College; Boys Club; Brandon College; Brasenose College (Oxford); Brenau College; Brentwood College; Brescia College; Brevard College; Briarcliff College; Bridgewater College; British Columbia; British Commonwealth; British Council; Brooklyn College; Bruyere College; Bryant College; Burdett College; Butler College

B-C Barber-Colman

B.C. Bachelor of Chemistry; Bachelor of Commerce; Baja California; Before Christ; British Columbia

B & C Banking and Currency (Senate Committee)

B of C Bank of Canada; Bureau of the Census

BC Baja California (Spanish—Lower California), northern section is a state whose capital is Mexicali while the southern part is a territory whose capital is La Paz; *Banco Central* (Spanish—Central Bank), Spain's largest; *Biological Conservation*

B C basso continuo (Italian—continuous bass background)

bca best cruise altitude; blood color analyzer

bca barrica (Spanish—cask, keg); *biblioteca* (Portuguese or Spanish—library)

B ͨᵃ *Boca* (Portuguese or Spanish—mouth, river mouth)

BCA Bank Credit Analyst; Battery Control Area; Billiard Congress of America; Blue Cross Association; Boys' Clubs of America; British Caledonian Airways; British Colonial Airlines; Bureau of Consular Affairs; Bu-

reau of Consumer Affairs
B/C of A British College of Aeronautics
BCAA British Columbia Automobile Association
BCAB Birth Control Advisory Bureau; British Computer Association for the Blind
bcac biology classroom activity checklist
BCAC British Conference on Automation and Computation
BCA/DoS Bureau of Consular Affairs—Department of State
BCAir British Commonwealth Air Force
BCAL British Caledonian Airways
BCALA Black Caucus of the American Library Association
BCAP Beacon Collision-Avoidance Program
BCAPT Braverman-Chevigny Auditory Projective Test
BCAR British Civil Airworthiness Requirements; British Council for Aid to Refugees
BCAS Beacon-compatible Collision-Avoidance System; British Compressed Air Society
b'cast(ing) broadcast(ing)
BCAT Birmingham College of Advanced Technology
bcb binary code box; broadcast band; button-cell battery
BCBC British Cattle Breeders' Club
bcbh boiler casing bulkhead
BC/BS Blue Cross/Blue Shield
bcc beam-coupling coefficient; blind carbon copy; body-centered cubic
BCC Battery Control Central; Berkshire Community College; Board of Crime Control (Montana); British Communications Corporation; British Council of Churches; British Crown Colony; Broadcasting Corporation of China; Bronx Community College; Bureau Central de Compensation; Bureau of Charities and Corrections (South Dakota); Burlington Community College; Bus & Coach Council
BCCA Beer Can Collectors of America; British Cyclo-Cross Association
BCCBP Biblioteca de Cataluña y Central de Bibliotecas Populares (Spanish—Library of Catalonia

and Central Public Library)—Calle Carmen, Barcelona
BCCCUS British Commonwealth Chamber of Commerce in the United States
bccd buried-channel charge-coupled device
BCCF British Cast Concrete Federation
BCCG British Cooperative Clinical Group
BCCI Bank of Credit and Commerce International
BCCLA British Columbia Civil Liberties Association
BCCO Base Consolidation Control Office(r)
BCCR Banco Central de Costa Rica (Spanish—Central Bank of Costa Rica)
BCCS British Columbia Coastal Service
bcd barrels per calendar day; binary-coded data
BCD Battelle's Columbus Division
bcdc binary coded decimal counter
BCDDP Breast Cancer Detection Demonstration Project
B-cdf B-cell differentiation factor
BCDIC Binary-Coded Decimal Interchange Code
Bce Belice (Spanish—Belize) formerly British Honduras
BCE before the Christian era
BCESL British Commonwealth Ex-Services League
BCF Battle Cruiser Fleet; Battle Cruiser Flotilla; Battle Cruiser Force
B-cgf B-cell growth factor
Bch Beach
BCHA British Columbia Health Association
bchcmbr(s) beachcomber(s)
BCHP Banco Central Hipotecario del Perú (Spanish—Central Mortgage Bank of Peru)
BCI Bat Conservation International; Birth Control Institute; Bureau of Correctional Institutions (Iowa); Bureau of Criminal Identification
BCIS Bureau of Collection and Investigation Services
BCIU Border Crime Intervention Unit
BCL Battelle's Columbus Laboratories; Bermuda Container Line; Bougainville Copper Limited
b clar basse clarinette

(French—bass clarinet)
BCLA Birth Control League of America
bcm basic control monitor; brightest cluster member
BCM Berklee College of Music; Bougainville Copper Mine (Papua, New Guinea)
BCM Banco Comercial Mexicano (Spanish—Mexican Commercial Bank)
bcme bichloromethylether
bcmg becoming
BCN British Commonwealth of Nations
BCNZ Broadcasting Corporation of New Zealand
BCP Bonus Community Plan (telephone service)
BCP Banco de Credito del Perú (Spanish—Credit Bank of Peru); *Book of Common Prayer*
BCPA British Copyright Protection Association
BCPs Black College Presidents
BCPU Border Crime Prevention Unit
BCR British Columbia Railway
BCRLRA Beverage Container Recycling and Litter Reduction Act
BCRP Banco Central de Reserva del Perú (Spanish—Central Reserve Bank of Peru)
bcs because
BCS Battle Cruiser Squadron; Black Christian Students; British Conchological Society; Budget Control System
B-C-S Bardeen-Cooper-Schrieffer
B.C.S. Bachelor of Criminal Science
BCSC British Columbia Steamship Company
BCSR Board of Certified Shorthand Reporters
BCSS Bard Cardiopulmonary Support System
BCT Bell Canyon Test
bcu big closeup
bcwp budgeted cost of work performed
bcws budgeted cost for work schedule
bcz because
bd behavior(ally) disorder(ed); board; boat deck; bond; bound; bridge deck
b & d bondage and discipline (sado-masochism)
bd bind (Dano-Norwegian—volume)

Bd board
BD Banking Department
B.D. *Bona Dea* (Latin—good goddess) Roman goddess of chastity and fertility
B/D Banker's Draft
BDA bomb damage assessment
bdc bottom dead center
BDC Berlin Documentary Centre (National Archives)
BDE Board of Dental Examiners
BDF Birth Defects Foundation
bdi both days included
BDI *Bundesverband der Deuts-chen Industrie* (Federation of German Industries)
bdl baseline demonstration laser
bdle bundle
bdm's births, deaths, marriages
BDMS Bulk Direct Mail Service
bds brands
bds *bis in die summendus* (Latin—take twice daily)
B-D Squad Bomb-Disposal Squad
bdt *bundt* (Dano-Norwegian—bundle)
bdt's back-door trots (diarrhea)
be. beam expander; bent; bilingual education; binding edge
b/e bill of entry; bill of exchange
b & e beginning and ending; breaking and entering
BE Beginning Experience (retreat program for separated divorced and widowed); Black English; Borough Engineer; British Embassy; British English
B/E bill of exchange
BE *Buque Escuela* (Spanish—Schoolship)
be4 before
Bea Beatrice
BEA Broadcast Education Association; Bureau of Economic Analysis
BEAR Bureau of Electronic and Appliance Repair
BEAR HUG *Basic Extended Acronym Human Users Guide*
Beatrix Beatrix Wilhelmina Armgard, Queen of the Netherlands, 1980–
Beau Beauford; Beaufort; Beaumont; Beauregard
beautility beauty + utility
BEBA Bureau of Economic and Business Affairs (U.S. State Department)

bec because
BECA Bureau of European and Canadian Affairs
BECC Boeing Engineering & Construction Company
Bedloe's Bedloe's Island in New York's Upper Bay, also called Liberty Island, site of the Statue of Liberty
bedoc beds occupied
bed(s). bedroom(s)
Beds Bedfordshire
Bed-Stuy Bedford-Stuyvesant
BEE Basic Earthquake Education; Basic Economic Education
B.E.E. Bachelor of Electrical Engineering
BEE *Bulletin of Environmental Education*
Beeb BBC (British Broadcasting Corporation)
beec binary error-erasure channel
Beech 99 Beechcraft seventeen-seat aircraft
Beecham Beauchamp
Beech F-33C five-place Bonanza training airplane
Beech Queen Air Beechcraft Seminole transport aircraft
beef. business-and-engineering-enriched fortran
beefalo beef cattle + buffalo (hybrid)
BEEP *Bureau Europeen del' Education Populaire* (French—European Bureau of Adult Education)
bef before; blunt end first; buffered emitter follower
Bef *Befehl* (German—command, order)
BEF Bonus Expeditionary Force; British Equestrian Federation; British Expeditionary Force
BEFA British Emigrant Families Association
befm bending form
befo' before
beg. begin; beginning
beg *begynde(lse)* [Dano-Norwegian—begin(ning)]
BEG Belgrade, Yugoslavia (airport); Bureau of Economic Geology
BEG *Bank Europaeischer Genossenschafsbanken* (German—European Cooperative Bank)
Begl *Begleitung* (German—accompaniment)
begr *begrundet* (German—established)
BEH Bureau of Education for

the Handicapped
BEHA British Export Houses Association
behav behavior; behavioral; behaviorist(ic)
Behavioral Res Behavioral Research Laboratories
BEHC Bio-Environmental Health Center
bei butanol-extractable iodine
BEI Bridgeport Engineering Institute
BEI *British Education Index*
BEIA Board of Education Inspectors' Association
Beibl *Beiblatt* (German—supplement)
beif *beifolgend* (German—sent herewith)
Beih *beihft* (German—supplement)
beil *beiliegend* (German—enclosed)
Beil *Beilage* (German—appendix, supplement)
BEIS British Egg Information Service
Beisp *Beispiel* (German—example or illustration)
Beitr *Beitrag* (German—contribution)
bek *bekendgørelse* (Dano-Norwegian—announcement)
BEK AS Beck Shoe Corp (symbol)
bel below; 10 decibels
bel (Turkish—pass)
Bel Bela; Belcher; Belden; Belem; Belen; Belfast; Belford; Belgian; Belgium; Belham; Belize; Bel-lamy; Bellanca; Belmont; Belorussia; Belos; Beltram
Bel *Bacharel* (Portuguese—Bachelor)—academic degree
BEL Belém do Pára, Brazil (airport)
BELAIR Belgian Air Staff (NATO)
Belau Republic of Belau (formerly Palau)
Belchers Belcher Islands in Hudson Bay just north of James Bay
belcrk bellcrank
Bel & Dr *Bel and the Dragon*
B.Elec. & Tel. Eng. Bachelor of Electronics and Telecommunication Engineering
Belém (Portuguese—Bethlehem)—also the Amazon River port of Belém do Pará; Lisbon suburb
bel ex *bel example* (French—fine example)—fine copy of a book, engraving, map, etc.

Belf Belfast; Belfastian(s)
belg belgisk (Dano-Norwegian—Belgian)
Belg Belgian; Belgium
Belg Belgica (Portuguese or Spanish—Belgium); *Belgio* (Italian—Belgium)
Belgium Kingdom of Belgium (lowland nation on the North Sea); *Royaume de Belgique* (French); *Koninkrijk België*
Belglais Belgian-English
Belgolux Belgium and Luxembourg
Bell Bell Aircraft; Bell System (American Telephone and Telegraph and associated companies collectively called Ma Bell or Mother Bell)
Bell 47 Sioux utility helicopter built by Bell Aircraft
Bell 204 gunship helicopter nicknamed Huey as it is designated UH-1
Bell 206 five-seat turbo-powered helicopter also called Jet Ranger or Sea Ranger
bella belladonna (drug stimulant)
Bella Arabella; Isabella
Bellas Artes Instituto Nacional de Bellas Artes (Spanish—National Institute of Fine Arts)—Mexico City
Belle Bella; Arabella; Isabella
Bellevue New York City hospital noted for its psychiatric ward
BELLMATIC Bell Laboratories Machine-Aided Technical Information Center
bells. bell-bottom pants
Bell's Cr C Bell's Crown Cases (British)
Bell Sys Tech J Bell System Technical Journal
Belmo Belmopan
BELNAV Belgian Naval Staff (NATO)
Belsen Bergen-Belsen (World War II concentration camp town in northwest Germany)
Belvac Société Belge de Vacuologie et de Vacuotechnique (French—Belgian Society for Vacuum Science and Technology)
bem (BEM) behavior engineering model
Bem Bemerkung (German—comment, note, observation)
BEM British Empire Medal
B.E.M. Bachelor of Engineering of Mines
BEMA Business Equipment Manufacturers Association

BEMB British Egg Marketing Board
BEMO Base Equipment Management Office
bems bug-eyed monsters (science-fiction jargon)
BEMS Bakery Equipment Manufacturers Society; Building Energy Management Systems
BEMSA British Eastern Merchant Shippers Association
ben. bene (Latin—good, well); *benedictio* (Latin—blessing)
Ben Benard; Benedict; Bengali; Beniah; Benito; Benjamin; Bennet(t); Benno; Beno; Be-noni; Bentley; Benvenuto
Ben Ben Hur
BEN Bureau d'Études nucleaires (French—Belgian Bureau of Nuclear Studies)
Benavides originally Ben David
Bend Bendigo
benday benday (photoengraving) process named for a 19th-century American printer, Benjamin Day
BENDEX Beneficial Data Exchange (linking Social Security Administration with state welfare agencies)
B en Dr Bachelier en Droit (French—Bachelor of Law)
bends caisson disease
bene benzine
BENECHAN Benelux Subarea Channel (NATO)
Bened Benedict; Benedictine
benef beneficiary
Benef Benefice
Ben Eil Benedenwindse Eilanden (Dutch—Leeward Islands)
Benelux economic union of Belgium, Netherlands, and Luxembourg
Bene't Benedict
benev benevolent
Beng Bengal; Bengali
B.Eng. Bachelor of Engineering
bengals bengal tigers; thick cigars
Bengis Bengalis
Bengs Bengalis
B.Eng. Sci. Bachelor of Engineering Science
B.Eng.Tech. Bachelor of Engineering Technology
Benin People's Republic of Benin (West African nation formerly called Dahomey)
Benj Benjamin

Benja Benjamin
Benjn Benjamin
Benjy Benjamin
Bennie Benjamin
bennies benzedrine stimulants
benny benzedrine
Benny Benjamin
ben sug beneficial suggestion
Bento Baruch; Benito
b & ent & pl breaking and entering and petty larceny
benz benzedrine; benzine
Benziger Benziger, Bruce, and Glencoe
BEO Borough Education Office(r)
beoc battery echelon operating control
BEOG Basic Educational Opportunity Grant
bep break-even point
BEP Bureau of Engraving and Printing
BE & P Bureau of Engraving and Printing
B.E.P. Bachelor of Engineering Physics
BEP Brevet d'Études Professionelles (French—Professional Studies Diploma)
B EpA British Epilepsy Association
BEPC Beijins Electron Positron Collider
BEPI Budget Estimates Presentation Instructions
BEPO British Experimental Pile Operation
bepoc Burrough's electrographic printer-plotter for ordnance computing
BEPQ Bureau of Entomology and Plant Quarantine
bepti bionomics, environment, plasmodium, treatment, immunity (factors in malaria epidemiology)
BEPZ Bataan Export Processing Zone (Philippines)
beq bequeath
BEQ Bachelor Enlisted Quarters; Background and Experience Questionnaire
beqd bequeathed
beqt bequest
ber buffer (flow chart)
ber berechnet (German—computed)
Ber Berber; Berlin; Berwickshire
Ber Bericht (German—report)
BER Berlin, West Germany (Tempelhof airport); Bureau of Economic Regulation
BERA Business Education Research Associated

Berb Berber
Berb *Berberia* (Spanish—Barbary Coast)
Ber Bunsenge Phys Chem *Berichte der Bunsengesellschaft für Physikalische Chemie* (German—Report of the Bunsen Society for Physical Chemistry)
BERC Biomedical Engineering Research Corporation; Black Economic Research Center
BERCO British Electric Resistance Company
BERCON Berlin Contingency (NATO)
Berdoo San Bernardino, California
berg iceberg (sometimes written 'berg)
Berg Berger; Bergin
Berg (Dutch or German—mountain)
Bergen Bergen-Belsen (site of Nazi German concentration camp); Norway seaport city; Bergen op Zoom (Dutch port); county in New Jersey; former name of Mons, Belgium
Bergman Ingmar Bergman (Swedish film-writer director); Ingrid Bergman (Swedish actress); Torbern Bergman (Swedish chemist and phy-cisist)
BERH Board of Engineers for Rivers and Harbors
Beria Lavrenty Pavlovich Beria (director of Stalin's secret service)
Bering sea linking the Arctic Ocean with the North Pacific between Alaska and the USSR; strait
Beringia Bering and Chukchi Seas area
Berk Berkeley
Berks Berkshire
Berl Berlin
Berl Tid *Berlingske Tidende* (Danish—Berling's Times)—a leading daily newspaper
Berm Bermuda Islands
Bermudas Bermuda Islands
Bern Berna; Bernal(do); Bernan; Bernard(o); Bernarr; Berner; Bernhard; Bernold
Bernh Bernhard
Bernie Bernard
Berno Bernardo
Ber Phil Berlin Philharmonic
Bert Albert; Alberta; Albertina; Bertel; Bertha; Berthel; Ber-tillon (system);

Bertin; Bertol; Berton; Bertram; Bertrand; Bertwin; Cuthbert; Delbert; Elbert; Elberta; Filbert; Gilbert; Herbert; Hilbert; Ibert; Lambert; Norbert; Philbert; Roberta; Wilbert; Zilbert
bertm berth term
Ber Tri Bermuda Triangle (North Atlantic shipwreck area extending from Bermuda to Cape Hatteras to Key West and back to Bermuda)
Berts Bertillon Measurements
BERU Building Economics Research Unit
Berw Berwick
beryl beryllium aluminum silicate
bes balanced electrolyte solution
bes *besonders* (German—especially)
Bes Bessel's functions
BES Biological Engineering Society; Bureau of Employment Security; Bureau of Environmental Statistics
B.E.S. Bachelor of Engineering Science
B es A *Bachelier des Arts* (French—Bachelor of Arts)
BESA British Engineering Standards Association
BESE Bureau of Elementary and Secondary Education
BeShT Baal Shem-Tov (Israel Ben Eliezer)
besi bus electronic-scanning indicator
B es L *Bachelier des Lettres* (French—Bachelor of Letters)
BESL British Empire Service League
BESN British Empire Steam Navigation (company)
BESRL Behavioral Science Research Laboratory (USA)
bess binary electromagnetic signature
Bess Bessemer; Mrs Harry S Truman
B es S *Bachelier es Sciences* (French—Bachelor of Science)
BESS Bank of England Statistical Summary
Bessie Bethlehem Steel (Wall Street slang); Elizabeth
Bessy Elizabeth
BESSY Bestell System (German teleordering)
best *Bestellung* (German—order); *bestyrelse* (Dano-Norwegian—board, direction)

BEST Basic Essential Skills Training; Black Efforts for Soul in Television
bet. best estimate of trajectory; between
Bet Beirut; Betsy; Elizabeth
BET Biker Enforcement Team (of law-enforcement officers); British Electric Traction
Beta Betamax (video-cassette system)
BETA Business Equipment Trade Association
Beth Bethlehem; Elizabeth
Bethlehem Bethlehem Steel Corporation
Bethnel Bethnel Green, London
Beth Steel Bethlehem Steel
betr better
betr *betreffend* (German—concerning)
Betsy Elizabeth
BETTS Bolt Extrusion Thrust Termination System
Betty Elizabeth
betw between
BEU British Empire Union
BEUC *Bureau Européen des Unions Consommateurs* (French—Bureau of European Consumer Unions)
bev bevel; beverage; billion electron volts
bev (BeV) billion electron volts
Bev Beva; Bevan; Beveridge; Beverley; Beverly; Bevis
BEV Black English Vernacular; Blake E Vance
Bevans Charles I. Bevans' compilation of *Treaties and Other International Agreements of the United States of America*
BEW Board of Economic Warfare
BEWT Bureau of East-West Trade
bex broadband exchange
bexec budget execution
BEY Beirut, Lebanon (airport)
bez *bezahlt* (German—paid); *bezuglich* (German—referring to)
Bez *Bezirk* (German—district)
bezw *bezichungsweise, beziehungsweise* (German—respectively)
bf back feed; beer firkin; before; blackface (minstrel makeup); blackface (sheep); bloody fool; board feet, board foot; boiler feed; boldface (type); bold face; both faces; boy friend; buffered;

butter fat
b-f beat-frequency
b/f black female; brought forward; brown female
b & f bell and flange
bf bassa frequenza (Italian—bass frequency); *bestemt form* (Dano-Norwegian—definite shape); *bouillon filtrate* (French—filtered bouillon)
b.f. bona fide (Latin—genuine, sincere)—in good faith; without deception; without fraud
Bf black female; Bowser factor
BF *Banque de France* (French—Bank of France); Battle Fleet; Battle Force; black female; Bowser factor; Burkina Faso; Burrhus Frederic Skinner (behaviorist)
B.F. Bachelor of Forestry
BF Beogradska Filharmonica (Serbo-Croat—Belgrade Philharmonic)
bfa basal forebrain area
BFA Black Faculty Association; British Fellmongers' Association; British Film Academy; Broadcasting Foundation of America; Bureau of Financial Assistance
B.F.A. Bachelor of Fine Arts
bfaln buffer boundary alignment
BFAP British Forces—Arabian Peninsula
BFAR Bureau of Fisheries and Aquatic Resources (Philippines)
BFB Bureau of Forensic Ballistics
BFBC British Forces Broadcasting Service
BFBPW British Federation of Business and Professional Women
BFBS British Forces Broadcasting Service; British and Foreign Bible Society
bfc benign febrile convulsion
BFCA British Federation of Commodity Associations
BFCC Brothers for Christian Community
BFCF Bremerton Freight Car Ferry
BFCS British Friesian Cattle Society
BFCSD Brewery, Flour, Cereal, Soft Drink and Distillery (Workers of America)
bfct boiler feed compound tank
bfcy beneficiary
bfd big fucking deal
BFDC Bureau of Foreign and

Domestic Commerce
bfe beam-forming electrode
BFEA Bureau of Far Eastern Affairs (U.S. Department of State)
BFEBS British Far Eastern Broadcasting Society
BFFA British Film Fund Agency
BFFC British Federation of Folk Clubs
bfg brute-force gyro
Bfg Bank für Gemeinwirtschaft (German—Bank for Municipal Management)
BFG BF Goodrich
BFHMF British Felt Hat Manufacturers' Federation
BFHS Benjamin Franklin High School
bfi beam-forming interfact
BFI British Film Institute; Business Forms Institute; Seattle, Washington (Boeing Field)
BFIA British Flour Industry Association
BFICC British Facsimile Industry Compatability Committee
bfl back focal length
BFL Barber Fern Line; Belgian Fruit Line; Blue Funnel Line (Holt's); Books For Libraries
BFLF Biblioteca della Facoltà di Lettere e Filosofia (Italian—Library of the Faculty of Letters and Philosophy)—in Florence
BFM Ballet Folklórico de México (Spanish—Folklore Ballet of Mexico)
BFMA Business Forms Management Association
BFMF British Federation of Music Festivals; British Footwear Manufacturers' Federation
BFMIRA British Food Manufacturing Industries Research Association
BFMO Base Fuels Management Officer (USAF)
BFMP British Federation of Master Printers
Bfn Bloemfontein
BFN British Forces Network
bfo beat-frequency oscillator; blood-forming organs
Bfo Buffalo
BFO Bath Festival Orchestra; Bureau of Field Operations
bfoq bona-fide occupational qualification
B.For. Bachelor of Forestry
bform budget formulation

B.For.Sci. Bachelor of Forestry Science
bfozp best-fit optic Z-plane
bfp biological false-positive (reactions); boiler feedpump
BFP British Fishing Port (registration symbols appearing on the bows of British fishing vessels and indicating their home ports)—(*see* British Fishing Port appendix)
BFPA British Film Producers Association
BFPC British Farm Produce Council
bfpdda binary floating-point digital-differential analyzer
BFPO British Field Post Office
BFPPS Bureau of Foods, Pesticides, and Product Safety (FDA)
bfpv bona fide purchaser for value
bfr biologic false reactor; blood flow rate; bone formation rate; buffer
Bfr Belgische frank (Dutch—Belgian franc)
B Fr Belgian franc
bfr(s) belgisk(e) franc(s) [Dano-Norwegian—Belgian franc(s)]
BFRS Bio-Feedback Research Society
bfR sol buffered Ringer's solution
BFS Belfast, Northern Ireland (airport); Board of Foreign Scholarships; Bureau of Family Services; Bureau of Federal Supply
B.F.S. Bachelor of Foreign Service
BFS Bundesanstalt für Flugsicherung (German—Air-Traffic Control Authority)
BFSA Black Faculty and Staff Association; British Fire Services Association
BFSS British and Foreign Sailors' Society
bft bio-feedback training
BFT Bentonite Flocculation Test
B.F.T. Bachelor of Foreign Trade
BFTA British Fur Trade Alliance
BFTC Boeing Flight Test Center
BFTS British and Foreign Temperance Society
BFUP Board of Fire Underwriters of the Pacific
BFUSA Basketball Federation of the United States of Amer-

ica

BFUW British Federation of University Women

BfV Bundesamt für Verfassungsschutz (German—Federal Office for the Protection of the Constitution)—German FBI roughly equivalent to the Special Branch in Britain

bfw boiler feedwater

bg back gear; before girls (entered the armed forces); bluish-green; buccal ganglion; buccogingival; business girl

bg (BG) background (behind tv performers); block grant

b/g bonded goods

bG bluish green

Bg Bengal; Bengalese; Bengali

Bg Berg (German—mountain); *Bogen* (German—arch or bow) musical slur

BG Benny Goodman; Birmingham Gage; British Guiana; Butchart Gardens (Victoria, British Columbia)

B-G Bach Gesellschaft; David Ben-Gurion

B & G Barton and Guestier; Bing and Grondahl; buildings and grounds

BG Bibliothèque publique et universitaire de Genève (French—Public and University Library of Geneva)

bga blue-green algae (virus)

BGA Better Government Association; British Gliding Association

BGAS Boys and Girls Aid Society

B-G b Bordet-Gengou bacillus

BGB Booksellers of Great Britain

bgc blood group class

BGC British Gas Corporation

BGCC Bowling Green College of Commerce

BGCCC Board of Governors of the California Community Colleges

Bge barrage (artificial dam)

BG & E Baltimore Gas and Electric

B.G.E. Bachelor of Geological Engineering

B.Gen.Ed. Bachelor of General Education

BGF Banana Growers' Federation; Black Guerrilla Family

BGFE Boston Grain and Flour Exchange

BGFO Bureau of Government Financial Operations

bgg booster gas generator; bo-

vine gamma globulin

BGGRA British Gelatine and Glue Research Association

bgh bovine growth hormone

bght bought

BGI Bechtel Group Inc; Bridgetown, Barbados (airport)

BGIRA British Glass Industry Research Association

B-girl bar girl

Bgk Bangkok

bgl below ground level

B.G.L. Bachelor of General Laws

BGLA Business Group for Latin America

bglb brilliant-green lactose broth

bglr burglar

bgl(s) bagel(s); beagle(s); bugle(s)

bgm background music

BGM Bethnal Green Museum; Binghamton, New York (airport)

BGMA British Gear Manufacturers' Association

bgmn baggageman

Bgn Bergen

BGN Board on Geographic Names

BGNR Barren Grounds Nature Reserve (New South Wales)

BGNY Bookbinders' Guild of New York

Bgo Bugo

bgp below-ground pool

bgr bombing and gunnery range

bgrv (BGRV) boost-glide re-entry vehicle

bgs bags

bg(s) back gear(s); bag(s)

Bgs Brightlingsea

BGS British Geriatrics Society; Brotherhood of the Good Shepherd

BGS Bundesgrenz Schutz (German-Frontier Troops)—West German NATO forces

bgsa blood-granulocyte specific activity

BGSC Belfer Graduate School of Science (Yeshiva University); Boise-Griffin Steamship Company

BGSM Bowman Gray School of Medicine

BGSU Bowling Green State University

bgt bought

Bgt Bight

Bgt Bugt (Danish—Bay)

BGT Bender Gestalt Test; British Guiana Time

BGTA Birmingham Group Training Association

BGTT Borderline Glucose Tolerance Test

B Gu British Guiana

BGU Ben Gurion University (Beersheba); Bowling Green University

BGU Biblioteca General da Universidade (Portuguese—University General Library)—in Coimbra

BgUL Bibliothèque générale de l'Université de Liège (French—General Library of the University of Liege)

bgw (BGW) battlefield guided weapon

BGW Baghdad, Iraq (airport)

bh bloody hell (British expletive); boiler house; breast height; Brinell hardness

bh bougie-heure (French—candlehour); *brysholder* or *busteholder* (Dano-Norwegian—bra or brassiere)

Bh Brinell hardness

BH Base Hospital; Bath & Hammondsport (railroad); Benjamin Harrison (23rd President U.S.); Bill of Health; Brigade Headquarters; Brinell hardness; British Honduras; Broken Hill; magnetization curve (symbol)

B/H Bill of Health; Bordeaux-Hamburg (range of ports)

B&H Bell and Howell; Breitkopf and Härtel

B of H Board of Health

BH Bonne Humeur (French—Good Humor); *Boston Herald*

B H Ben Hur

bha base helix angle

bha (BHA) butylated hydroxyanisole

BHA Boston Housing Authority; British Homeopathic Association; British Honduras Airways

bh ad broach adapter

B.H.Adm. Bachelor of Hospital Administration

BHAFRA British Hat and Allied Feltmakers' Research Association

B'ham Birmingham

BHB British Hockey Board

BHBNM Big Hole Battlefield National Monument

B Hbr boat harbor

BHBS British Honduras Broadcasting Service

bhc beaching cradle; benzene hexachloride (BHC)

BHC Barbers, Hairdressers, Cosmetologists (and Proprietors' Union); Black Hawk College; British High Commissioner; British Hovercraft Corporation; Brotherhood of the Holy Cross

BHCIUS Barbers, Hairdressers, and Cosmetologists International Union of America

BHCSA British Hospitals Contributory Scheme Association

bhd beachhead; bulkhead

BH$ British Honduras dollar

BHD Bronx House of Detention

BHDF British Hospital Doctors Federation

B.H.E. Bachelor of Home Economics

B of HE Board of Higher Education

B'head Birkenhead

BHEW Benton Harbor Engineering Works

Bhf Bahnhof (German—station)

BHF Berliner Handels und Frankfurter (bank); British Heart Foundation

bhf(s) black homosexual female(s)

bhfx broach fixture

B H & G Better Homes and Gardens

BHGMF British Hang Glider Manufacturers Federation

bhi brain-heart infusion

BHI Better Hearing Institute; British Horological Institute; *Bureau Hydrographique Internationale* (French—International Hyrdrographic Bureau)

BHI British Humanities Index

bhib beef-heart infusion broth

BHIS Burroughs Hospital Information System

BHISSA Bureau of Health Insurance, Social Security Administration

BHK type-B Hong Kong influenza virus

bhl biological half-life

BHL Borax Holdings Limited

Bhm Birmingham, England

BHM Birmingham, Alabama (airport); Bureau of Health Manpower

BHMA Bald-Headed Men of America; British Hard Metal Association

BHMC Bell & Howell/Mamiya Company

BHMH Benjamin Harrison Memorial Home (Indianapolis, Indiana)

BHMRA British Hydromechanics Research Association

bhm(s) black homosexual male(s)

BHMS Bishop's Home Mission Society

bhn bephenium hydroynaphthoate

Bhn Bremerhaven; Brinell hardness number

BHNWR Bombay Hook National Wildlife Refuge (Delaware)

B Hond British Honduras

B.Hort. Bachelor of Horticulture

B.Hort.Sci. Bachelor of Horticultural Science

bhp biological hazard potential; boiler horsepower; brake horsepower

BHP Borehole Plugging Program; Broken Hill Proprietary

bhp hr brake horsepower hour

BHPRD Bureau of Health Planning and Resource Development (HEW)

Bhpric Bishopric

BHQ Brigade Headquarters

bhr basal heart rate; biotechnology and human research

Bhr Bahrain

BHRA British Hotels and Restaurants Association; British Hydromechanics Research Association

B & HRO Biotechnology and Human Research Office (NASA)

bhs betahemolytic streptococcus

Bhs Bohus

BHS Balboa High School; Boys High School; British Home Stores; British Horse Society; Bureau of Health Services; Burlesque Historical Society; Bushwick High School

B&HS Bonhomie and Hattiesburg Southern (railroad)

B.H.Sci. Bachelor of Household Science

BHSS Bronx High School of Science

bhst bottom-hole static temperature

bht baht tical (Thai monetary unit)

bht (BHT) butylated hydroxytoluene

BHTA British Herring Trade Association

bhtm(s) black heterosexual male(s)

Bhu Bhutan

Bhutan Kingdom of Bhutan (Asian Himalayan nation); *Druk-yul* (its name in the official language called Dzonkha Bhután)

Bhv Bhavnagar

bh/vh body hematocrit/venous hemocrat (ratio)

Bhvn Bremerhaven

bhw boiling heavy water

BHW Boston Hospital for Women

BHYC Boothbay Harbor Yacht Club

B.Hyg. Bachelor of Hygiene

bi background investigation; bacteriological index; base ignition; base of prism in; biopsy; burn index; bodily injury; buffer index

bi (BI) binary

b/i battery inverter

b & i bankruptcy and insolvency; base and increment

b or i brass or iron (cargo)

Bi biot; bismuth (symbol)

B^i Bani or *Beni* (Arabic—sons of)

BI Babson Institute; background investigation; Bahama Islands; Bermuda Islands; Braniff International; British India; Brookings Institution; Bureau of Investigation; Business International; National Biscuits (stock exchange symbol)

B of I Bureau of Investigation

BI Banca d'Italia (Bank of Italy)

BIA Bicycle Institute of America; Binding and Industries of America; Board of Immigration Appeals; Braille Institute of America; Brazilian International Airlines; Building Industry Association; Bureau Issues Association; Bureau of Indian Affairs; Bureau of Insular Affairs; Bureau of Inter-American Affairs

BI & A Bureau of Intelligence and Research (US Department of State)

BIAA Bureau of Inter-American Affairs (US Department of State)

BIAC Business and Industry Advisory Committee (NATO)

BIAE British Institute of Adult Education

BIALL British and Irish Association of Law Librarians

bialy bialystok roll (holeless onion-flaked bagel)
BIAS Brooklyn Institute of Arts and Sciences
BIATA British Independent Air Transport Association
bib. bibliography; biographical inventory blank; bottled in bond
bib. (BIB) baby incendiary bomb
bib biblioteca (Italian, Latin, Portuguese, Romanian, Spanish—library); *bibliotecario* (Spanish—librarian); *biblioteka* (Albanian, Bulgarian, Macedonian, Polish, Russian, Serbo-Croatian, Slovene, Ukrainian); *bibliotek* (Dano-Norwegian or Swedish); *biblioteket* (Dano-Norwegian or Swedish); *bibliotheek* (Dutch or Flemish); *bibliotheka* (Latin); *bibliotheke* (Greek); *Bibliothek* (German); *bibliothèque* (French)
bib. *bibe* (Latin—drink)
Bib Bible; Biblical
Bib Biblica (Latin—Bible)
BIB Biennale of Illustrations Bratislava (international exhibition of children's book illustrations); Board for International Broadcasting; Bureau of International Broadcasting
BIB Berliner Institut für Betriebsführung (German—Berlin Business Management Institute)
BIBA Babson Institute of Business Administration; British Insurance Brokers' Association
Bib Amb Biblioteca Ambrosiana (Italian—Ambrosian Library)—in Milan
Bib Apo Vat Biblioteca Apostolica Vaticana (Italian—Vatican Library)
bib b biblioteksbind (Dano-Norwegian—library binding)
bibber winebibber
Bib Bod Bibliotheca Bodmeriana (Latin—Bodmer Library)—in Cologny/Geneva contains first editions of Cervantes, Dante, Goethe, Homer, Shakespeare, a Gutenberg Bible and one of the three recorded copies of Luther's *Disputio pro Declaratione Indulgentiarum* from 1517
BIBC British Isles Bowling Council

Bib Cen Biblioteca Central (Spanish—Central Library)—in Mexico City's Ciudad Universitaria
Bib Ecu Biblioteca Ecuatoriana (Spanish—Ecuadorian Library)—in Quito, also known as Padre Aurelio Espinosa Pólit
Bib Esc Biblioteca de San Lorenzo el Real de El Escorial (Spanish—Library of Royal San Lorenzo of the Escorial Palace)—monastic library within the Escorial Palace in the Guadarramas near Madrid
BIBF British and Irish Basketball Federation
bi or bin (Latin *bini*—two by two, Latin *bis*—twice)—binary, binoculars, binomial, bipolar
bibl bibliotec-; bibliotek-; bibliothec-; bibliothek; bibliothèque
biblio bibliographical imprint or note; biblioclasm (book destruction); biblioclast (book destroyer); bibliogenesis (book production); bibliognost (bibliographic expert or book expert); bibliogony (book production); bibliograph (bibliographer); bibliographer; bibliographic(al); bibliography
biblioc biblioclasm; biblioclast
bibliog bibliographer; bibliographic(al); bibliography
bibliograph bibliographer; bibliographee; bibliography
biblioklept biblioklepto-mania(c)
bibliol bibliolater (person with excessive admiration or reverence for books); bibliolatrous (characterized by bibliolatry); bibliolatry (book worship); bibliological; bibliologist; bibliology (scientific description and study of books)
bibliom bibliomancy; bibliomane (collector of books); bibliomania; bibliomaniac; bibliomanist
bibliop bibliopegic (relating to book binding); bibliopegist; bibliopegy; bibliophagist; bibliophile; bibliophilia; bibliophobe; bibliophobia; bibliopole
bibliopsy bibliopsychology (study of authors, books, and readers as well as their inter-

relationships)
bibliothec bibliotheca (bibliographer's catalog or a library); bibliothecal (belonging to the library); bibliothecar (librarian); bibliothecary
bibliother bibliotherapeutic; bibliotherapist; bibliotherapy
bibliotrain railroad car-converted into a mobile library
bibl mun bibliothèque municipale (French—city library, public library)
Bib Mus Biblioteca Musicale (Italian—Musical Library)—in Rome's Via dei Greci
Bib Nac Biblioteca Nacional (Spanish—National Library)
Bib Nar Biblioteka Narodowa (Polish—National Library)—in Warsaw
Bib Nat Bibliothèque Nationale (French—National Library—Paris)
Bib Naz Bra Biblioteca Nazionale Braidense (Italian—Braidense National Library)—in Milan
Bib Naz Cen Biblioteca Nazionale Centrale (Italian—National Central Library)
Bib Pal Biblioteca de Palacio (Spanish—Palace Library)—Madrid
BIBRA British Industrial Biological Research Association
bibs. bibliographies
Bib Soc Am Bibliographical Society of America
Bib Soc Can Bibliographical Society of Canada
Bib Sor Bibliothèque de la Sorbonne (French—Sorbonne Library)
Bib Uni Biblioteca Universitaria (Spanish—University Library)
Bic Société Bic (ballpoint pen factory founded by Baron Marcel Bich)
BIC Barrier Industrial Council; Bronx Irish Catholic; Bureau of International Commerce; *Bureau International des Containers* (French—International Bureau of Containers)
BIC Bureau International des Containers (French—International Container Bureau)
bicarb sodium bicarbonate
BICC British Insulated Callenders Cables
BICEMA British Internal Combustion Engine Manu-

facturers Association
bicept book indexing with context and entry points from text
BICERA British Internal Combustion Engine Research Association
BICERI British Internal Combustion Engine Research Institute
bichrome sodium bichromate
BICS British Institute of Cleaning Science
BICTA British Investment Casters' Technical Association
bicv biconcave
bicx biconvex
Bicycle Bicycle Thieves
bicyea best ice cream you ever ate
bicyplane bicycle-powered airplane (first cross-Channel flight achieved June 12, 1979 by the *Gossamer Albatross* designed by Paul MacCready of Pasadena, California pedalled and piloted by Bryan Allen)
bid. (BID) brought in dead
b.i.d. bis in die (Latin—twice daily)
Bid Bideford
Bld Bureau of Identification
BID Banco Interamericano de Desarrollo (Spanish—Interamerican Development Bank)
B.I.D. Bachelor of Industrial Design
B. of I.D. Bachelor of Interior Design
bidap bibliographic data processing program
Biddy Bridget; Briged; Brigid
bidec binary-to-decimal converter
BIDP Basic Institutional Development Program
BIE Bureau of Industrial Economics (DoC); *Bureau International d'Education* (French—International Bureau of Education); *Bureau International des Expositions* (French—International Bureau of Expositions)
B.I.E. Bachelor of Industrial Engineering
Bieder Biedermeier
BIEE British Institute of Electrical Engineers
bien biennial
BIEN Basic Income European Network
BIEPR Bureau of International

Economic Policy and Research
BIET British Institute of Engineering Technology
BIETA Biblioteca Interamericana de Estadistica Teórica y Aplicada (Spanish—Interamerican Library of Theoretical and Applied Statistics)
bif buyer-induced failure
BIF Bombardier's Information File; British Industries Federation
BIFN Banque Internationale pour le Financement de l'Énergie Nucléaire (French—International Bank for the Financing of Nuclear Energy)
bif O opium
BIFUS Britain, Italy, France, United States
big. best in group; bigamist; bigamy; biological isolation garment
BIG Bartoni International Gallery (Melbourne); Basic Industries Group; Beneficial Insurance Group; Better Independent Grocers; Blacks In Government
BIG Bazak Israel Guide
Big Belts Big Belt Mountains of Montana
Big Cats lions, tigers, jaguars, leopards, and pumas
big demo big demonstration
biggies big ones
big H big house (slang—penitentiary such as San Quentin or Sing Sing); hernia; heroin
Big Horns Big Horn Mountains, Wyoming
big mo big momentum
big O's obsolescence, overcrowding, overburdening omigod!
bigr bigger
bigs biological isolation garments
bigst biggest
big. unlwfl—trig awf bigamy is unlawful—trigamy is awful, explained Ogden Nash
bih benign intracranial hypertension
BIH Beth Israel Hospital
BIHA British Ice Hockey Association
bihor. bihorium (Latin—two hours)
BII Beckman Instruments Incorporated; Biosophical Institute Incorporated
BIIA British Institute of Industrial Art

BIICL British Institute of International and Comparative Law
Bij Benjamin
bijb bijbelse term (Dutch—biblical term)
bijv bijvoorbeeld (Dutch—for example)
bi k bilge keel
bike bicycle
bikers motorcyclists
biki bikini
bil bilateral; billet; billion; block input length
b-i-l brother-in-law
Bil Bilbao
BIL Billings, Montana (airport); Braille Institute Library; Brierly Investments Limited; British India Line; Bulk Items List
BILA Bureau of International Labor Affairs
bilat bilateral
Bilders Bilderbergers (now called Tri-Laterals)
bildg bill of lading
bildl bildeich (German—figuratively)
bile. balanced-inductor logical element
BILG Building Industry Libraries Group
bili bilirubin
biling bilingual(ism); bilingualist(ic)(al)(ly)
bilj biljarttern (Dutch—billiards)
bil k bilge keel
bill billede (Dano-Norwegian—illustrations)
Bill Billie; Billy; William
bill. acad billiard academy
Billie William
billion (American—a thousand million, 10^9); (British—a million million, 10^{12})
Bill of Rights first ten amendments to the *Constitution of the United States*
Billy William
bil(s) billion(s)
BILS British International Law Society; Butterworth Industrial Laws Service
bilt built
bim beginning of information marker
bi-m bi-monthly
bim bimensile (Italian—semimonthly); *bimestrale* (Italian—bimonthly); *bimestre* (Italian—two-month period)
Bim Barbadan
BIM British Institute of Management

B.I.M. Bachelor of Indian Medicine
BIM Bord Iascaigh Mhara (Gaelic—Sea Fisheries Board)
bimac bi-stable magnetic core
BIMCAM British Industrial Measuring and Control Apparatus Manufacturers Association
BIMS Business Information Management System
BIMT Bahama Islands Ministry of Tourism
bin. binary
BINA Bureau International des Normes de l'Automobile (French—International Bureau of Automobile Standards)
binac high-speed electronic digital computer
b-in-B banned in Boston (and therefore a best-seller)
BINCOS Binder Control System
bind. binding
B.Ind. Bachelor of Industry
B.Ind.Ed. Bachelor of Industrial Education
BINDT British Institute of Non-Destructive Testing
Bing Binghampton
Binj Benjamin
BINL Basic Inventory of Natural Language
BINM Buck Island National Monument, St Croix, Virgin Islands
binocam binocular and camera (combination instrument)
binocs binoculars
bins (Cockney contraction—binoculars)
BINS Barclays Integrated Network System
binsum brief intelligence summary
BINWR Blackbeard Island National Wildlife Refuge (Georgia)
BINZ Bankers Institute of New Zealand
bio biographical; biography; biological; biology
bio (Latin prefix—life)—biology, the study of life and living things
Bio Biology
BIO Base Installation Officer; Bedford Institute of Oceanography; Biological Information-Processing Organization; Broadcasting Information Office
BIOA Bureau of International

Organization Affairs (US Department of State)
bioact bioactive; bioactivity
bioastro bioastronaut(ic)(al)(ly)
bioauto bioautograph(ic)(al)(ly)
biocam binocular camera
biochem biochemical; biochemist; biochemistry
Biochem Biochemistry
biochron biochronometry
biocid biocidal; biocide
bioclean biologically clean
biocon biocontamination
biocyb biocybernetics
biodef biological defense
biodeg biodegradability; biodegradable; biodegradation; biodegrade; biodegraders; biodegrading
biodeg(s) biodegradable(s)
biodes biodestructible
biodet biodeterioration
bioelectrog bioelectrogenesis
bioelectron bioelectron(ic)(al)(ly); bioelectronics
bioeng bioengineer(ing); biological engineer(ing)
bioenv bioenvironment(al)(ly); bioenvironmentalist
bioex bioexperiment(ation)
biog biographer; biographical; biography
biogeo biogeology
biogeog biogeographer; biogeographic(al); biogeography
bioinstru bioinstrument(al)(ly); bioinstrumentation
biol biological; biologist; biology
biol biologi or *biologisk* (Dano-Norwegian—biology or biologist)
Biol Biology
Biol Abstr Biological Abstracts
BIOLWPNSYS Biological Weapon System (USA)
biomass mass of biological material
bio-mass biological mass source of ethanol and methanol from crops and trees
BIOMASS Biological Investigations of Marine Antarctic Systems and Stocks
biomath biomathematician; biomathematics
biomed biomedical; biomedicine
bionics biology + electronics
bio-org bio-organic(al)(ly)
biophys biophysical; biophysicist; biophysics
biopol(s) biopolymer(s)
bior business input-output re-

run
biore bioresearch(er)
BIOREP Biological Attack Report
bios (BIOS) biological satellite
BIOS Basic Input Output System; Biological Investigations of Space
biosat biosatellite
biosci bioscience; bioscientific; bioscientist
biosen(s) biosensor(s)
BIOSIS Biosciences Information Service of *Biological Abstracts*
biospel biospeleologist(ic)(al)(ly); biospeliology
biostat biostatistic(s)
biot biotron(ic)(al)(ly)
BIOT British Indian Ocean Territory (Aldabra, Chagos Archipelago, Des Roches, Diego Garcia, Farquhar, Mau-ritius, Seychelles)
BIOTA Biological Institute of Tropical America
biotec biotechnical(ly); biotechnological(ly); biotechnologist; biotechnology
biotel biotelemetric; biotelemetry
biotrans biotransformation; bio-transformer
biowar biological warfare
bip background interference procedure(s); bacterial intravenous protein; balanced in plane; bismuth iodoform paraffin; books in print; borough-interborough problem(s)
bip (BIP) body improvement plan
BiP Books in Print
BIP Board for International Broadcasting; Border Industrial Program; British Industrial Plastics; British Institute of Physics
BIP Banco Industrial del Perú (Spanish—Industrial Bank of Peru)
BIPAD Bureau of Independent Publishers and Distributors
bipco built-in-place components
bipd biparting doors
biphet biphetamine (drug stimulant)
bipkwele unit of currency in Equatorial Guinea
BIPL Burma Industrial Products Limited
BIPM *Bureau International des Poids et Mesures* (French—International Bu-

reau of Weights and Measures)
BIPO British Institute of Public Opinion
bipp bismuth, iodoform, paraffin paste
BIPP British Institute of Practical Psychology
BIPS British Integrated Programme Suite
BIPS Bibliographic Information Publication System
bipyr bipyramidal
biquin biquinary
bir basic incidence rate; built-in robes (closets)
Bir Birmania (Italian or Spanish—Burma); *Birmânia* (Portuguese—Burma)
BIR Board of Inland Revenue; Board of Internal Revenue; British Institute of Radiology; Bureau of Intelligence and Research; Bureau of Internal Revenue
BIRC Bio-Integral Resource Center
BIRD *Banque Internationale pour la Reconstruction et le Développement* (French—International Bank for Reconstruction and Development)
BIRDDOG Basic Investigation of Remotely Detectable Deposits of Oil and Gas
birdie battery integration and radar display equipment
BIRE British Institution of Radio Engineers
B.Ir.Eng. Bachelor of Irrigation Engineering
BIRF Brewing Industry Research Foundation
BIRF Banco Internacional de Reconstrucción y Fomento (Spanish—International Bank for Reconstruction and Development)
BIRG Basic Income Research Group (England)
birl girlish boy (transvestite)
Birm Birmingham
Birmingham notation (*see* GKD-notation)
BIRMO British Infra-Red Manufacturers' Association
BIRMPDis Birmingham Procurement District (U.S. Army)
BIRMS Battelle Interactive Resources Management System (computerized)
BIRP Beverage Industry Recycling Program
birr monetary unit of Ethiopia

BIRS Basic Indexing and Retrieval System; British Institute of Recorded Sound
birt bolt installation and removal tool
birthquake population explosion
bis best in show; bissextile
Bis Bismarck; Bissau
BIS Bank for International Settlements; Bismarck, North Dakota (airport); Board of Inspection Survey (USN); British Imperial System; British Information Service; British Interplanetary Society; Business Information System
bis in 7d. bis in septem diebus (Latin—twice in seven days, twice weekly)
BISA British International Studies Association
BISAC Book Industry System Advisory Committee
bisad business information systems analysis and design
BISAKTA British Iron, Steel, and Kindred Trades Association
BISAM Basic-Indexed Sequential-Access Method
BisArch Bismarck Archipelago
Bisc Biscayan
BISCA Building Industry Subcontractors Association
Biscayne national park south of Miami, Florida, consisting of islands leading to Key West
BISCO British Iron and Steel Corporation
bis in d. bis in dies (Latin—twice daily)
bisett bisettimanale (Italian—bi-weekly)
bisex bisexual(ism); bisexualist(ic)(al)(ly); bisexually
BISF British Iron and Steel Federation
BISFA British Industrial and Scientific Film Association
BISG Book Industry Study Group
Bish Bishop
bishaw bicycle rickshaw
Bish Mus Bishop Museum
Bishop Bishop Museum; Bishop Museum Press
BISITS British Iron and Steel Industry Translation Service (BISRA)
BISL British Information Service Library
Bismarck North Dakota's capital; Prince Otto Eduard Leopold von Bismarck-

Schönhausen—the Iron Chancellor
Bismarck Pen North Dakota Penitentiary at Bismarck
Bismarcks Bismarck Islands
BISN British India Steam Navigation (company)
bisp between ischial spines; bispinous (interspinous diameter)
BISPA British Independent Steel Producers Association
BISRA British Iron and Steel Research Associates
BISS Battlefield Identification System Study (NATO)
BISTA Bureau of International Scientific and Technological Affairs (U.S. Department of State)
Bister Bicester
bisw bisweilen (German—sometimes)
bisync binary synchronous computer
bit. binary digit; bituminous
BIT Bradford Institute of Technology; British Independent Television; *Bureau International du Travail* (French—International Labor Organization)
BITA British Industrial Truck Association
BITB Building Industry Training Board
BITC Bahamas International Trust Company
BITCH Black Intelligence Test of Cultural Homogeneity
bite. base installation test equipment; built-in test equipment
BITE Base Installation Test Equipment
bitm bituminous
BITM Birla Industrial and Technological Museum
bitn bilateral iterative network
bito burnishing tool
BITO British Institution of Training Officers
bit(s). binary digit(s)
BITTC Building Industry Technicians Training Council
bitu benzyl-thiourea
BITU Bustamante Industrial Trade Union
bitum bituminous
bitumd bituminized
bituminous soft coal
biu basic income unit
B-I U Bar-Ilan University
BIU Bureau International des Universités (French—International University Bureau)

biv bivouac
BIV *Banco Industrial de Venezuela* (Spanish—Industrial Bank of Venezuela)
bivar bivariant (function generator)
bi-w bi-weekly
BIW Bath Iron Works; Boston Insulated Wire (and Cable Company)
BIWF British Israel World Federation
BIWS Bureau of International Whaling Statistics
bix binary information exchange
biz bizarre; business
BIZ *Bank für Internationalen Zahlungsausgleich* (Bank for International Settlements)
bizad business administration
bizjet business-type jet airplane
bizmac business machine computer
bizman business man
BIZNET American Business Network (data base of Chamber of Commerce)
bj back judge (football); biceps jerk; blow job (fellatio)
b & j bone and joint
Bj Burj (Arabic—bluff, cliff, fort, tower)
BJ Benito Juarez; Byron Jackson (Borg-Warner)
B.J. Bachelor of Journalism
B & J Burke & James
B of J Bank of Japan
BJA Burlap and Jute Association; Bureau of Justice Assistance
b/Jan binding expected in January (for example)
BJC Baltimore Junior College; Bismarck Junior College; Boise Junior College; Brevard Junior College; Bureau of Juvenile Correction (Delaware)
BJCEB British Joint Communications-Electronics Board
BJCO British Joint Communications Office
BJCPA Baptist Joint Committee on Public Affairs
bjf batch-job format
BJIP Better Jobs and Incomes Program
B Jon Ben Jonson
BJOS British Journal of Occupational Safety
BJp Bence Jones protein
BJS Bureau of Justice Statistics
BJSM British Joint Services

Mission
bjt bipolar junction transistor
BJTRA British Jute Trade Research Association
BJU Bob Jones University
B.Juris. Bachelor of Jurisprudence
bk bank; below the knee; black; book; brake
Bk berkelium; Brook
B^k Bank
Bk Buku (Indonesian or Malay—hill, mountain)
B-K Blaw-Knox
BK Biblioteka Kombëtare (Albanian—National Library)—Tirana
bka below-knee amputation
BKA Bundeskriminalamt (German—Federal Criminal Ministry)
bkble bookable; bookmobile
bkbndg bookbinding
bkbndr bookbinder
bkbrd bakboord (Dutch—port) port side
bkc benzalkonium chloride (BKC)
bkcy bankruptcy
bkd blackboard
bk di brake die
bkfst breakfast
bkg banking; bookkeeping; breakage
bkgd background
bkge brokerage
bkhs blockhouse
BKII Vsesoyuenaya Kommunisticheskaya Partiya (Russian—All-Union Communist Party)
BKK Bangkok, Thailand (airport)
bklr black letter
bklt booklet
Bklyn Brooklyn
Bklyn Brdg Brooklyn Bridge
Bklyn HTF Brooklyn Homicide Task Force
Bklyn Mus Brooklyn Museum
Bklyn Phil Brooklyn Philharmonic Symphony
bkm buckram
BKM Moscow, USSR (Bykovo Airport)
bkn broken
Bkn Birkenhead
bkpg bookkeeping
bkpr bookkeeper
bkpt bankrupt
bkr baker; beaker; breaker
bks bunks; barracks; books; brakes
BKS British Kinematograph Society
bk sh bookshelves

Bks for Libs Books for Libraries
bksp backspace (flow chart)
bkt basket; bracket
Bkt Bukit (Malay—Hill, Hilly Street)
bkt(s) basket(s)
bktt below knee to toe
bkw breakwater
bkwp below-knee walking plaster (cast)
bl bank larceny; baseline; billet; bleed(ing); blood; blood loss; blue; bomb line; buccolin-gual; butt line; buttock line
b/l basic letter; bill of lading (B/L); blueline; blueprint
b & l ball and lever; business and loan
bl blad; blank (Dano-Norwegian—leaf, sheet, blank)
Bl Burkitt's lymphoma (BL)
Bl Blasinstrument (German—wind instrument; woodwind); *Blatt(er)* [German—leaf; leaves; page(s)]; *Böluk* (Turkish—company)
BL Barrister-at-Law; Basutoland; Blessed Lady; Bonanza Airlines; British Leyland; British Library (formerly British Museum Reading Room)
B-L Belgium-Luxembourg
B.L. Bachelor of Letters
B/L Bill of Lading
B & L Bausch & Lomb; Building and Loan (association or bank)
bl a blandt andet; (Dano-Norwegian—among other things); *blandt andre* (Danish—among other things)
Bla Belawan; Brasilia
BLA Bangladesh Library Association; Black Liberation Army; Board of Landscape Architects; Bombay Library Association; British Legal Association; British Library Association
B.L.A. Bachelor of Landscape Architecture; Bachelor of Liberal Arts
BL-AA Biblioteca Luis-Angel Arango (Spanish—Luis-Angel Arango Library)—Bogotá, Colombia's library named for a former bank president
BLAC British Light Aviation Center
BLACC British and Latin American Chamber of Commerce

black. blackmail

Black Black's United States Supreme Court Reports

Blackfeet Blackfoot Indians

Black Islands *see* Melanesia

Blackstairs Blackstairs Mountains of Ireland

Blackstone Sir William Blackstone's *Commentaries on the Laws of England*

blad blotting pad

blade. basic level automation of data through electronics

BLADES Bell Laboratories Automatic Design System

BLAISE British Library Automated Information Service

Blake Blakely; Blakeman; Blakeslee

Blan Blanca; Blanchard; Blanco; Bland(on)

Blanca Blanche

B.Land.Arch. Bachelor of Landscape Architecture

Blaskets Blasket Islands on Ireland's Atlantic coast

blast blastos (Greek—sprout)—blastoderm, blastodisc, blastopore, osteoblast

Blast Blastoidea

BLAST Black Legal Action for Soul in Television

Bla Sta Blackfriars Station

BLAT British Life Assurance Trust

Blatch Blatchford's United States Circuit Court Reports

BLB Boothby-Lovelace-Bulbulian (oxygen mask)

blc balance; boundary-layer control

BLC British Lighting Council

blchd bleached

blchg bleaching

bl cult. blood culture

bld blood; blood and lymphatic system; bloody; bold; boldface

BLD Burglary Larceny Division (NYPD)

bldg building

Bldg Engr Building Engineer

bldi blank die

bldr builder

BLE Brotherhood of Locomotive Engineers

B & LE Bessemer and Lake Erie (railroad)

bleap bought ledger and expenditure analysis package

BLEDCO Brooklyn Local Economic Development Corporation

Blemish Belgian & Flemish

blems blemishes (acne, blackheads, pimples)

blenno blennorrhea

BLESMA British Limbless Ex-Service Men's Association

bless. bath, laxative, enema, shampoo, and shower

bleu blind landing experimental unit

BLEU Belgium-Luxembourg Economic Union

bleve boiling-liquid expanding-vapor explosion

Blf Bluff

BLF & E Brotherhood of Locomotive Firemen and Enginemen

blg betalactoglobulin

Blg Bulgarian

BLG Burke's Landed Gentry

BLH Baldwin-Lima-Hamilton

BLHA British Linen Hire Association

BLHS Ballistic Laser Holographic System

BLI Bliss & Laughlin Industries; Buyers Laboratory Incorporated

B.L.I. Bachelor of Literary Interpretation

BLI Bank Leumi le-Israel (Bank Association of Israel)

B.Lib.S. Bachelor of Library Science

B.Lib.Sci. Bachelor of Library Science

Blick Blickensderfer portable typewriter

Blinder NATO code name for Soviet Tu-22 bomber

blip. background-limited infrared photography

BLIP Big Look Improvement Program

BLIS Bell Laboratories Interpretive System

BLISS Baby Life-Support Systems

B-lite baton-flashlight combination

B.Litt. *Baccalaureus Literarum* (Latin—Bachelor of Literature)

Blitz Blitzkreig (German—lightning war)

bliz blizzard; blizzardly; blizzardous

blk black; block; blocking; bulk

Blk Block

blkcnt block count (flow chart)

blkd bulkhead

Blk Eng Black English

blk lt black light

blk rt bulk rate

blksh blackish

blksmith blacksmith

blkstp blackstrap (molasses)

bll below lower limit

BLL Butyrka, Lefortovo, and Lubyanka (Moscow's most dreaded prisons)

BLLD British Library Lending Division (Boston Spa)

BLLRCS Bureau of Library and Learning Resources and Community Services (Office of Education)

blm bilayer lipid membrane

blm besa la mano (Spanish—a kiss to your hand)

BLM British Leather Manufacturers; British Leland Motor (corporation merging Austin, British Motor Moldings, Jaguar, Morris, Riley, Rover, Triumph, Wolseley); Bureau of Land Management (General Land Office and Grazing Service)

B.L.M. Bachelor of Land Management

BLM Bonniers Literary Magasin (Bonnier's Literary Magazine)

BLMA British Lead Manufacturers' Association

BLMC British Leyland Motor Corporation

BLMRA British Leather Manufacturers' Research Association

BLMS Book-Library-Management System

bln balloon; bronchial lymph nodes

Bln Berlin

blnk blank (flow chart)

blnkt blanket

BLNR Benton Lake National Refuge (Montana)

BLNWR Big Lake National Wildlife Refuge (Arkansas); Bitter Lake NWR (New Mexico); Buffalo Lake NWR (Texas)

BLNY Book League of New York; Booksellers League of New York

blo blower

BLOB Ban Large Office Buildings

Bloch Pub Bloch Publishing Company

block. blockade

blodi block diagram (compiler)

blokops blockade operations

bloody (Early English—By Our Lady)

Bloomies Bloomingdale's

blooper blunder and error

Blos Blossom

BLOT British Library of Tape Recordings

blou blouse
B-love being love (unselfish accepting love of another person, according to Maslow)
blp besa los pies (Spanish—a kiss to your feet)
BLP British Labor Party
BLPES British Library of Political and Economic Science (London)
bl pr blood pressure
blr boiler; breech-loading rifle; broad-line system
BLR Ballistic Research Laboratories (USA)
BLRA British Launderers' Research Association
BLRD British Library Reference Division (British Museum Library)
blrmkr boilermaker
BLROA British Laryngological, Rhinological, and Otological Association
blrp boilerplate
bls bales; barrels; binary light switch; blood sugar
BLS Boston Latin School; Brooklyn Law School; Bureau of Labor Statistics
B.L.S. Bachelor of Library Science; Bachelor of Library Service
B.L.S. Benevolenti Lectori Salutem (Latin—Salutations to the Kind Reader)
BL & SA Bank of London and South America
BLSGMA British Lampblown Scientific Glassware Manufacturers' Association
blsh bluish
blsn blowing snow
blstg pwd blasting powder
blstl billet steel
blsw barrels of load salt water
blswd barrels of load salt water per day
blt blood type; built
blt (BLT) bottom-loading transporter
b-l-t bacon, lettuce, and tomato (sandwich)
BLT Battalion Landing Team; Before Large Telescopes
Bltc Baltic
bltg belting
bltn(s) built-in(s)
blu blue
B-L u Bessey-Lowry units
Blubo Blut und Boden (German—blood and soil)
BLUCB Bancroft Library of the University of California at Berkeley
blue bullet blue-tipped bullet

color-coded to indicate its incendiary purpose
Blue Ridge Blue Ridge Mountains of the Appalachian range; national parkway in North Carolina and Virginia
Blues Blue Mountains
blunt marijuana joint rolled in a cigar
BLV British Legion Village
Blvd Boulevard
BLW Baldwin-Lima-Hamilton
BLWA British Laboratory Ware Association
Bly Blyth
BLYMSA Banco de Londres y México SA (Spanish—Bank of London and Mexico Corporation)
Blz Belize (formerly British Honduras); Belizian
blz bladzijde (Dutch—page)
Blz Belize
bm basal metabolism; basement membrane; beam; board measure; body mass; bone marrow; book of the month; bo-wel movement; buccal mass; buccomesial
bm (BM) buffer mark (flow chart); buffer modules
b/m (B/M) bill of material; black male; brown male
bm bez mista (Czech—no place of publication)
b.m. balneum maris (Latin—bath in sea water); *bene merenti* (Latin—to the well-deserving)
Bm beam; birthmark; black male; board measure; bowel movement; Burma; Burmese
BM bench mark; Boatswain's Mate; Boston & Maine (railroad); Brian Mulroney—Canada's 23rd prime minister; Brigade Major; British Museum; Brooklyn Museum; Bureau of Medicine; Bureau of Mines; Bureau of the Mint; Business Manager
B-M Bolinder-Munktell; Bristol-Myers
B.M. Bachelor of Medicine; Bachelor of Music
B & M Beaufort & Morehead (railroad); Boston & Maine (railroad)
B of M Bank of Montreal; Bishop(ric) of Münster; Bureau of Mines
BM Banca Mondiale (Italian—World Bank); *Banco de México* (Spanish—Bank of Mexico); *Banco Mundial* (Portuguese or Spanish—

World Bank); *Banque du Monde* (French—World Bank); *Beata Maria* (Latin—Blessed Mary)
BMA Baltimore Museum of Art; Bank Marketing Association; Bible Memory Association; Bicycle Manufacturers' Association; British Medical Association; British Military Authority; Stockholm, Sweden, airport (3-letter code)
BMAD Black Mothers Against Drugs
B.Mar.E. Bachelor of Marine Engineering
B.Mar.Eng. Bachelor of Marine Engineering
BMASR Bureau of Military Application of Scientific Research
bmat beginning of morning astronomical twilight
B.Math. Bachelor of Mathematics
BMB Ballistic Missile Branch (USA); British Medical Board; British Metrication Board
BMB British Medical Bulletin
B-M B Baader-Meinhof Bande (German—Baader-Meinhof Gang)—terrorist Red Army Group
BMBW Bundesministerium für Bildung und Wissenschaft (German—Ministry for Education and Science)
bmc blockhouse monitor console
BMC Ballistic Missile(s) Center; Ballistic Missiles Committee; Book-of-the-Month Club; British Mountaineering Council; Bryn Mawr College; Business Microcomputer (Technology Group)
BMCC Blue Mountain Community College
BMCS Ballistic Missile Cost Study; Bureau of Motor Carrier Safety
bmd births, marriages, deaths; bone marrow depression
BMD Ballistic Missile Defense; Bureau of Medical Devices
B-M-D Blow-Me-Down, Nova Scotia
BMDATC Ballistic Missile Defense Advanced Technology Center (USA)
BMDCA Ballistic Missile Defense Communications Agency

BMDEAR Ballistic Missile Defense Emergency Action Report

BMDITP Ballistic Missile Defense Integrated Training Plan

BMDM British Museum Department of Manuscripts

BMDMB Ballistic Missile Defense Missile Battalion (USA)

BMDMP Ballistic Missile Defense Master Plan

bmdns basic mission, design number, and series (aircraft)

BMDO Ballistic Missile Defense Operations

BMDOA Ballistic Missile Defense Operations Activity

BMDPM Ballistic Missile Defense Program Manager

BMDPO Ballistic Missile Defense Program Office(r)

bmdr bombardier

BMDSCOM Ballistic Missile Defense System Command (USA)

BMD System Ballistic Missile Defense System

bme biomedical engineering

BME Blue Mountains Expeditions; Brotherhood of Marine Engineers

B.M.E. Bachelor of Mechanical Engineering; Bachelor of Mining Engineering; Bachelor of Music Education

BMEC British Marine Equipment Council

B. Med. Bachelor of Medicine

B.M.Ed. Bachelor of Music Education

B.Med.Biol. Bachelor of Medical Biology

B.Med.Sc. Bachelor of Medical Science

BMEF British Mechanical Engineering Federation

BMEG Building Materials Export Group

BMEL Barber Middle East Line

bmep brake mean effective pressure

BM & ESA Building Materials and Equipment Southeast Asia (Singapore)

B.Met. Bachelor of Metallurgy

B.Met.E. Bachelor of Metallurgical Engineering

BMEWS Ballistic Missile Early Warning System

BMFA Boston Museum of Fine Arts

bmg business management game

B.Mgt.Eng. Bachelor of Management Engineering

BMH British Military Hospital

bmi ballistic missile interceptor (BMI)

BMI Barley and Malt Institute; Battelle Memorial Institute; Book Manufacturers Institute; Broadcast Music Incorporated; Broadway Memorial Institute

B.Mic. Bachelor of Microbiology

BMIC British Music Information Centre (London); Broadcast Music Incorporated (Canada)

BMIC Bureau of Mines Information Circular

B.Min.E. Bachelor of Mining Engineering

BMIP Basic Medical Insurance Plan

BMIS Business Management Information System

BMJ British Medical Journal

bmk birthmark; bookmark(er)

bmkr boilermaker

BML Belfast & Moosehead Lake (railroad); Bodega Marine Laboratory (University of California); Bougainville Mining Limited; British Maritime League; British Museum Library (London)

B.M.L. Bachelor of Modern Languages

B & M L Belfast & Moosehead Lake (railroad)

BMLA British Maritime Law Association

BMLG Branch and Mobile Libraries Group

BMM Belfast, Mersey and Manchester Steamships

BMMA British Mantle Manufacturers' Association

BMMA Biblioteca Municipal Mário de Andrade (Portuguese—Mario de Andrade Municipal Library)—named in honor of Brazil's musician-poet promoter of modernism

BMMFF British Man-Made Fibres Federation

BMMO Birmingham and Midland Motor Omnibus

bmn bone marrow necrosis

Bmn Bremen

BMN British Merchant Navy

BMNH British Museum (Natural History)

BMNP Bale Mountains National Park (Ethiopia); Blue Mountains National Park

(New South Wales)

BMNT beginning morning nautical twilight

bmo business machine operator

BMO Ballistic Missile Office

bmoc big man on campus

bmom base maintenance and operations model

B'mouth Bournemouth

bmp best management practices; brake mean power; buttermilk powder

BMP Bricklayers, Masons and Plasterers' (Union); Business Migration Program (Australian)

BMP Banco Minero del Perú (Spanish—Mining Bank of Peru)

BMP-76PB Soviet amphibious armored-infantry combat vehicle also designated BTRM

BMPA British Metalworking Plantmakers' Association

BMPIUA Bricklayers, Masons, and Plasterers International Union of America

bmpp benign mucous-membrane pemphigus

BMPS British Medical Protection Society; British Musicians Pension Society

BMQ Best Memory Quiz

BMQA Board of Medical Quality Assurance

bmr basal metabolic rate; bomber

BMR Basal Metabolism Rate; Bureau of Mineral Resources

BMRA British Manufacturers' Representatives' Association

BMRB British Market Research Bureau

BMRL Building Materials Research Laboratories

BMRR British Museum Reading Room

BMRS Ballistic Missile Recovery System

bms balanced magnetic switch

BMs Black Muslims; Boatswain's Mates

BMS Boston Museum of Science; British Malachological Society; British Ministry of Supply; Buffalo Museum of Science; Bureau of Medical Services; Bureau of Medicine and Surgery

B.M.S. Bachelor of Marine Science; Bachelor of Medical Science

BMSA British Medical Students' Association

BMSE Baltic Mercantile and

Shipping Exchange
BMSG British Merchant Service Guild
BMSS British and Midlands Scientific Society
BMT Basic Military Training; Boston & Maine Transportation (railroad); Brooklyn-Manhattan Transit (subway system)
B.M.T. Bachelor of Medical Technology
BMTA Boston Metropolitan Transit Authority
BMTFA British Malleable Tube Fittings Association
BMTP Bureau of Mines Technical Paper
BMTS Ballistic Missile Target System
BMTV Ballistic Missile Test Vessel
BMU British Medical Union
B. Mus. Bachelor of Music
bmv bromegrass-mosaic virus
BMVM British Military Volunteer Service
BMW Bayerische Motoren Werke (German—Bavarian Motor Works)
BMWE Brotherhood of Maintenance of Way Employees
BMWS Ballistic Missile Weapon System
bmx (BMX) bicycle motorcross
BMYC Baltimore Motor Yacht Club
bmz basement membrane zone
bn bassoon; battalion; between; billion; bloody nuisance; branchial neuritis
bn (BN) binary number (system)
bn bijvoeglijk naamwoord (Dutch—adjective)
Bn beacon (daybeacon); bearing; Benjamin
Bn Bayan (Turkish—Miss, Mrs)
B^n Bassin (French—basin, pond)
BN Braniff; Bureau of Narcotics; Burlington Northern (merger of Chicago, Burlington, and Quincy, Frisco—St Louis and San Francisco, Great Northern, Northern Pacific, Spokane, Portland, and Seattle railroads)
B-N Bloomington-Normal, Illinois
B.N. Bachelor of Nursing
B & N Barnes & Noble; Bauxite & Northern
B of N Bureau of Narcotics

BN Biblioteca Nacional (Portuguese or Spanish—National Library); *Biblioteca Nazionale* (Italian—National Library); *Bibliothèque National* (French—National Library)
bna (BNA) beta-naphthylamine
BNA Black Network Alliance; Blackwell North America; Brazil Nut Association; British Naturalists' Association; British North America; British North Atlantic; British Nursing Association; Bureau of National Affairs; Nashville, Tennessee (airport)
BNA Basle Nomina Anatomica (Basel Anatomical Nomenclature)
BNAF British North Africa Force
B'nai B'rith Benai Berith (Hebrew—Sons of the Covenant)
BNAs British Naval Attaches
BNAU Bulgarian National Agrarian Union
B.Nav. Bachelor of Navigation
BNB British National Bibliography; British North Borneo (Sabah)
BNB British National Bibliography
BNBC British National Book Centre; British North Borneo Company
BNC Biblioteca Nacional de Chile (Spanish—National Library of Chile); *Biblioteca Nacional de Colombia* (Spanish—National Library of Colombia)
B.N.C. Brasenose College (Oxford)
BNCC Bay de Noc Community College
BNCF Biblioteca Nazionale Centrale Firenze (Italian—National Central Library—Florence)
bnchbd benchboard
BNCM Bibliothèque Nationale du Conservatoire de Musique (French—National Library of the Conservatory of Music—Paris)
BNCOR British National Committee for Oceanographic Research
BNCS British Numerical Control Society
BNCSR British National Committee for Space Research (Royal Society)
BNCVE Biblioteca Nazionale Centrale Vittorio Emanuele

II (Italian—Victor Emanuel IInd Central Library)—in Rome
b/nd binding—no date available
Bnd Bend
BND Bundesnachrichtendienst (German—Federal Intelligence Service)
BNDD Bureau of Narcotics and Dangerous Drugs
bndl bundle
Bndr Bandmaster
Bndr S-L Bandmaster—Sub-Lieutenant
bndy bindery; boundary
bne but not exceeding
BNE Bank of New England; Board of National Estimates (CIA); Brisbane, Australia (airport); Buffalo Niagara Electric Corporation
BNEC British National Export Council; British Nuclear Energy Conference
BNES British Nuclear Energy Society
BNE & SAA Bureau of Near Eastern and South Asian Affairs (US Department of State)
bnf bomb nose fuse
Bnf Banff
BNF Brand Name Foundation; Braniff International Airways
BNF British National Formulary
BNFC British National Film Catalogue
BNFEX Battalion Field Exercise
BNFL British Nuclear Fuels Limited
BNFMF British Non-Ferrous Metals Federation
BNFMRA British Non-Ferrous Metals Research Association
BNFMTC British Non-Ferrous Metals Technology Centre
BNFP Barnwell Nuclear Fuel Plant
BNFSA British Non-Ferrous Smelters' Association
Bng Bangor
BNGA British Nursery Goods Association
BnG-DL Bibliothèque Nationale du Grand-Duche de Luxembourg (French—National Library of the Grand Duchy of Luxembourg)
BNGM British Naval Gunnery Mission
BNGS Bomb Navigation Guidance System
bnh burnish

BNHA Badlands Natural History Society
BNHQ Battalion Headquarters
BNHS British National Health Service
BNI Bechtel National Inc; Black Nation of Islam
BNIB British National Insurance Board
BNJ Bonn, Germany (Cologne-Bonn airport)
BNJM Biblioteca Nacional José Martí (Spanish—José Martí National Library)—Havana's library named for Cuba's apostle of independence in the late nineteenth century
Bnk bank
bnkg banking
BNL Brookhaven National Laboratory
BNL Banco Nazionale del Lavoro (Italian—National Bank of Labor); *Biblioteca Nacional de Lisboa* (Portuguese—National Library of Lisbon); *Bibliothèque Nationale du Liban* (French—National Library of Lebanon)—in Beirut
bnm bajo el nivel del mar (Spanish—below sea level)
BNM Badlands National Monument (South Dakota); *Biblio-teca Nacional de México* (Spanish—National Library of Mexico—Mexico City)
BNM Banco Nacional de México (Spanish—National Bank of Mexico); *Biblioteca Nacional de México* (Spanish—National Library of Mexico)—in Mexico City; *Biblioteca Nazionale Marciana* (Italian—Marcian National Library)—in Venice
bno barrels of new oil; bladder neck obstruction; but not over
BNO Bank of New Orleans
BNOC British National Oil Corporation; British National Opera Company
bnp (BNP) bruttonational-produkt (Dano-Norwegian—gross national product)
BNP Bahamas NP (West Indies); Bako National Park (Sarawak); (Spanish—National Bank of Panama) Banco Nacional de Panamá; Banff NP (Alberta); Belair NP (South Australia); Bontebok NP (South Africa)

BNP Banque Nationale de Paris (French—National Bank of Paris)
bnpa binasal pharyngeal airway
bnr burner
BNRDC British National Research Development Corporation
BNS Bank of Nova Scotia; Bathymetric Navigation System; British Nylon Spinners
B.N.S. Bachelor of Natural Science; Bachelor of Naval Science
B of NS Bank of Nova Scotia
B.N.Sc. Bachelor of Nursing Science
BNSM British National Socialist Movement
bnst bassoonist
Bnt Burntisland
BNTL British National Temperance League
BNU Banco Nacional Ultramarino (Portuguese—Overseas National Bank)
B Nurs Bachelor of Nursing
B-nut B-shaped nut
BNV Biblioteca Nacional de Venezuela (Spanish—National Library of Venezuela)—in Caracas
BNVE Biblioteca Nazionale Vittorio Emanuele III (Italian—Victor Emanuel III Library)—in Naples
BNW Battelle-Northwest; Bureau of Naval Weapons
BNWR Blackwater National Wildlife Refuge (Maryland); Bowdoin National Wildlife Refuge (Montana); Brigantine National Wildlife Refuge (New Jersey)
Bnx Bronx
BNX British Nuclear Export Executive
BNY Bank of New York
BNYI Brooklyn Navy Yard Incinerator
BNZ Bank of New Zealand
bnzn benzoin
bo base (of prism) out; blackout; body odor; bowel obstruction; bowels open; bucco-occlusal
bo' bore; brother
b/o back order; boiloff; brought over; budget outlay
b & o belladonna and opium
'bo hobo (vagrant)
Bo Bolivia; Bolivian
BO Baltimore & Ohio (stock exchange symbol); Base Order; black oil (bunker oil

fuel); Board of Ordnance; body odor; box office; branch office; broker's order; Bureau of Ordnance
B.O. Bachelor of Oratory
B & O Baltimore & Ohio Railroad; Bang & Olufsen
BO Boletín Oficial (Spanish—Official Bulletin)
BO-5 Messerschmidt-Bolkow-Blohm five-seat helicopter
boa. born on arrival; breakoff altitude
Boa Balboa, Panama
B o A Board of Accountancy
BOA Basic Ordering Agreement; Boat Owners Association; British Optical Association; British Orthopedic Association; British Osteopathic Association; British Overseas Airways (BOAC)
BOA (Disp) British Optical Association (Dispensing Certificate)
BOAC Better on a Camel; British Overseas Airways Corporation
BOAdicea British Overseas Airways digital information computer for electronic automation
BOADS Boston Air Defense Sector
BOAE Bureau of Occupational and Adult Education (Office of Education)
BOAFG British Order of Ancient Free Gardeners
'board aboard; all aboard; on board; starboard
Boardwalk Atlantic City, New Jersey's famous promenade
BOAS British Orphans Adoption Society
BOAT Business Operational and Administrative Training (program)
boat dk boat deck (lifeboat-boarding deck)
boatel boat + hotel (waterside hotel or motel)
boats. boatswain (bo'sun)
BOAT/US Boat Owners Association of the United States
bob back of the book; best of breed
Bob Robert
B o B Bookbuilders of Boston
BOB Bureau of the Budget
BOBA British Overseas Banks Association
Bobbie Robert
Bobbs Bobbs-Merrill
Bobby Robert(a); London policeman, named after Sir

Robert Peel who organized the London police
b-o-b cult ban-on-bathing cult
BOBMA British Oil Burner Manufacturers Association
bo-bo bo-bo-type locomotive
bobr boring bar
boc back outlet central; blowout coil; body on chassis
Boc Boccaccio
B o C Board of Cosmetology; Bureau of Correction (Pennsylvania); Bureau of Corrections (Virgin Islands)
BOC Bank of China; Brooklyn Opera Company; Burmah Oil Company
BOCA Building Officials and Code Administrators; Building Officials Conference of America
boca(s) [Spanish—gulf(s); inlet(s); mouth(s)]
BOCCI Bureau of Organized Crime and Criminal Intelligence (California)
B.Occu.Ther. Bachelor of Occupational Therapy
bocd barrels of oil per calendar day
BOCE Board of Customs and Excise
BOCES Board of Cooperative Educational Services
BoCHS Bureau of Community Health Services
BOCI *Bloque de Obreros, Campesinos e Intelectuales* (Spanish—Bloc of Workers, Farmers, and Intellectuals)
BOCM British Oil and Cake Mills
B & O—C & O Baltimore and Ohio—Chesapeake and Ohio (merged railroad)
Boc Rat Boca Raton, Florida
BOCS Board of Cooperative Services
bod beneficial occupancy date; biochemical oxygen demand; biological oxygen demand; blackout door
bod *bodega* (Spanish—wineshop); *bodoniana* (Italian—Bodoni-style type)
Bod Bodaway; Bodel; Boden; Bodil; Bodleian; Bodnar; Bodo; Bodoni
BoD Board of Directors; Bureau of Drugs
Bodl Bodleian Library
bod lang body language (communication via body movements or postures)
Bodley Bodley Head
b-o d(s) box-office dis-

aster(s)—frequently called artistic success(es)
Bod units Bodansky units
boe back outlet eccentric
BoE Bank of England
B o E Bank of England; Board of Equalization
BOE Board of Osteopathic Examiners
BOE *Boletín Oficial del Estado* (Spanish—Official State Bulletin)
Boeing 707 four-engine long-range jet transport
Boerst Boerestaat (Boer State in South Africa)
BOES Branch Ordinary Enquiry System
bof basic oxygen furnace; binary oxide film
bof *beurre, oeufs, fromages* (French—butter, eggs, cheeses)—big butter-and-egg man
BoF Bureau of Foods
B-o-F Books-on-File
B of N *Birth of a Nation* (David Wark Griffith's three-hour classic)
Bog Bogotá
BoG Board of Governors
B o G Board of Governors
BOG Bogotá, Colombia (airport); Boston Opera Group
boggan toboggan
bogh *boghandel* (Dano-Norwegian—bookstore, booktrade)
bogie unidentified aircraft
boh breakoff height
Boh Bohemia(n)
BoH Bank of Hawaii
B O'H Bernardo O'Higgins
Bohem Bohemia; Bohemian
B o HF Bureau of Home Furnishings
bohica bend over—here it comes again
BoHM Bureau of Health Manpower
BoHP&RD Bureau of Health Planning and Resources Development
BOHS British Occupational Hygiene Society
boi basis of issue; break of inspection
boi (BOI) branch output interrupt
Boi Boise
BoI Board of Investment
BOI Boise, Idaho (airport)
B & OI Bank and Office Interiors
BoIA Board of Immigration Appeals

BOIC Boarding Officer in Charge
BOIESA Bureau of Oceans and International Environmental and Scientific Affairs (US Department of State)
boil. boiling
boil.pt. boiling point
Boi Phil Boise Philharmonic
Bois (French—woods)—Bois de Boulogne park, racetrack, and recreation area of Paris
Boise St Univ Boise State University
boj booster jettison
BoJ Bank of Japan
bo juice body-odor deodorant
BOK Book-of-the-Month Club
Boko Bohner & Kohle
'boks springboks
bol bill(s) of lading; bollard(s)
bol (BOL) block output length
bol. *bolus* (Latin—large pill)
Bol Bolivia; Bolivian; boliviano
Bol *Bol'shaya* or *Bol'shoy(e)* (Russian—big)
bol-148 (also BOL-148) d-2-bromolysergic acid tartrate (hallucinogen)
Bol cols Bolivarian colors (yellow, blue, and red as in the flags of Colombia, Ecuador, and Venezuela)
BOLD Bibliographic On-Line Display (document retrieval system); Blind Outdoor Leisure Development
BOLDS Burroughs Optical Lens Docking System
bolf barge off loading facility
Bolívar monetary unit of Venezuela; Simón Bolívar (Guayaquil, Ecuador's airport)
Bolivia Republic of Bolivia (landlocked Andean nation named for Simón Bolívar who liberated it from Spain) *República Boliviana*
boliviano monetary unit of Bolivia
Bolivianos *Bolivianos—el hado es propicio* (Spanish—Bolivians, destiny is propitious) national anthem
bolo be on the lookout (for a criminal at large)
bolo(s) bolshevik(s)
Bolo(s) Bolshevik(s)
bolovac bolometric voltage and current (voltage measurement)
bols bolster
BOLSA Bank of London and South America
bolshie(s) bolshevik(s)

bolt. beam-of-light transistor

BOLT Basic Occupational Language Training; Basic Occupational Literacy Test

boltop better on lips than on paper (a kiss)

Bolv Bolivia; Bolivian

bom business office must

Bom Bombay

BoM Bureau of Mines; Bureau of the Mint

BOM Boiler Operations & Management; Bombay, India (airport)

BOMA Building Owners and Managers Association

BOMAP Barbados Oceanographic and Meteorological Analysis Project

Bomarc Boeing long-range surface-to-air missile bearing nuclear warhead

BOMARC Boeing-Michigan Research Center

BOMAS Business Opportunity and Management Advisory Service

bomb. bombardment

Bomb Bombardier

Bombay Hook Bombay Hook National Wildlife Refuge near Dover, Delaware

bombex bombing exercise

BOMC Book of the Month Club

Bom Com Bomber Command

BoMD Bureau of Medical Devices

BOMEX Barbados Oceanographic and Meteorological Experiment

bomfog brotherhood of man under the fatherhood of god

Bompo Bompensiero (Frank Bompensiero—San Diego, California's mob boss slain in 1977 when it was revealed he had provided the FBI with information about organized crime)

bomrep bombing report

BoMS Bureau of Medical Services

bomst bombsight

Bon Bonin Islands

BON Bonaire, Netherlands West Indies (airport); British Organization for Non-parents

Bon Air Girls Bon Air School (for delinquent) Girls at Bon Air, Virginia

bond. bonding

bone(s) trombone(s)

Bo'ness Borrowstounness

Boni Boniface

Boniato Santiago de Cuba's

prison noted for inhuman treatment of prisoners

Bonins Bonin Islands (Ogasawaras)

Boo Bootes

B o O Board of Optometry

B o OE Board of Osteopathic Examiners

Book Bookman

bookie(s) bookmaker(s)

bookmobile book + automobile (mobile branch library within a truck)

Bookstax Bookstax of Britain

Booklist *Booklist and Subscription Books Bulletin*

boom boomerang

boomers baby boomer generation

boonies boondocks

BOOST Broadened Opportunities for Officer Selection and Training (USN)

Boot Bootes

Boothia Boothia Peninsula in the Canadian Arctic, northernmost extension of North America

bop balance of payments; basic oxygen process(ing); bebop; best operating procedure; buy our product(s)

b-o-p balance of payments

Bop Buffalo orphan prototype (virus)

BoP Bay of Pigs (invasion); Bay of Plenty

B o P Board of Pharmacy; Bureau of Prisons (United States Department of Justice)

BoPa *Borgelige Partisaner* (Danish—Middleclass Partisans)—underground resistance against occupying German forces during World War II

BoPat Border Patrol

bopd barrels of oil per day

B o PM Board of Podiatric Medicine

bops blowout preventer stack(s)

B o PS Board of Personnel Services

bopt broken orange pekoe tea

B.Opt. Bachelor of Optometry

BOQ Bachelor Officers' Quarters; Base Officers' Quarters

bor boring; bowels open regularly

Bor Boris; Borough

BOR Board of Review; Borg-Warner (stock exchange symbol); Bureau of Outdoor Recreation

BORAD British Oxygen Re-

search and Development

Borains people of Belgium's Borinage mining district

boram block-oriented random-access memories

borax sodium tetraborate

borazon boron nitrogen compound harder than diamond; boron nitride heated and pressed with a catalyst

Bore Ro-Ro Bore Roll-on Roll-off Line

boricua(s) *borinqueño(s)* [Spanish-American slang—Puerto Rican(s)]

borino(s) *borinqueño(s)* [Spanish-American slang—Puerto Rican(s)]

BORN FREE Build Options, Reassess Norms, Free Roles through Educational Equity

boro borough

Boro Borough

Boro' Borough

Borromeans Borromean Islands in Lake Maggiore

bos basic oxygen steel

bos' bosun (boatswain)

Bos Bosphorus; Boston

Bos *Bosanski* (Serbo-Croatian—Bosnian)

BoS Bureau of Ships

BOS Boston, Massachusetts (airport); British Oil Shipping

Boschaps Boston Symphony Chamber Players

Bösend Bösendorfer

bo's'n boatswain

Bo'sn Boatswain

Bosna (Yugoslav—Bosnia-Herzegovina)

Bosnia Bosnia-Herzegovina (part of the Austro-Hungarian Empire, now a federated republic in Yugoslavia)

Bosnywash Boston-New York-Washington DC corridor

Bosox Boston Red Socks (baseball team)

BosPops Boston Pops Orchestra

BOSS Bioastronautic Orbital Space System; Boeing Operational Supervisory System; Bureau of State Security (South Africa's Secret Service)

Bost Boston

Boston Spa British Library Lending Division in Boston Spa, Wetherby, West Yorkshire

Boston Tech Boston Technical Publishers

BOSTPDis Boston Procurement District (U.S. Army)

bo'sun boatswain
Boswash Boston-to-Washington
bot balance of time (to be served by a convict); botanic; botanical; botanist; botany; bottle; bottled; bottom; bottomed; bottoming
bot (BOT) beginning of tape
b-o-t build-operate-transfer (toll roads)
bot botanik or *botanisk* (Dano-Norwegian—botany or botanist)
Bot Botany
BoT Bank of Tokyo
B o T Board of Trade (British); Board of Transport (NATO)
BOT Board of Trade; Board of Trade unit; Books On Tape
B.O.T. Bachelor of Occupational Therapy
BOTAC Board of Trade Advisory Committee
botan botanic(al)(ly); botanist; botany
BOTB British Overseas Trade Board
bot & can bottle and can
botel boat hotel
BOTEX British Office for Training Exchange
both. bombing over the horizon
botmg bottoming
BOT-ohm Board of Trade ohm
bot(s) bottle(s)
Bots Botswana
Botswana Republic of Botswana (landlocked South African country)—formerly Bechuanaland
botu botulism
BOTU Board of Trade Unit
boty bike of the year
Bou Boulogne-sur-Mer
BOU Bank Officers' Union; Boat Operating Unit; British Ornithologists' Union
boul boulevard
Boul' Mich' Boulevard St Michel in the student quarter of Paris
Boulogne Boulogne-sur-Mer (French—Boulogne by the Sea)—English Channel port
bound. boundaries; boundary
'bout about
bov best of variety; bovine; bovril; brown oil of vitriol
Bov Eil Bovenwindse Eilanden (Dutch—Windward Islands, Aruba, Bonaire, Curaçao)
Bovid Bovidae (Latin—bovines: cows, goats, oxen, other ruminants, sheep)

bovinol bovinologic(al)(ly); bovinologist; bovinology
bow. bag of water (amniotic sac); blackout window; born out of wedlock
bo & w barrels of oil and water
bowdler bowdlerize
Bowker RR Bowker Company
bowla bowlathon
BOWO Brigade Ordnance Warrant Officer
Box Post Office Box
boyc boycott (named for C C Boycott), a system of coercion brought on by not having any dealings—commercial or social—with a company, country, person, or their products or services
bozo brawny intellectual lightweight
bp back pressure; bandpass; baptized; bathroom privileges; beautiful people; bedpan; before present; behavior pattern; below proof; benzypyrene; between perpendiculars; bills payable; biotic potential; biparietal; birthplace; black pimp; blood pressure; blueprint; boiling point; broncho-plural; buccopulpal
bp (BP) back projection (tv slide-or-film background projection)
b/p baking powder; bills payable; blood pressure; blueprint
b & p bare and painted; beer and pretzels
b of p balance of payments
bp Bergstrom Paper Company; *buono per* (Italian—good for)
b.p. bonum publicum (Latin—the public good)
Bp Babinski's phenomenon; Bishop
Bp Boerenpartij (Dutch—Farmers' Party)
BP Beach Party (amphibious military operation); Beacon Press; Be Prepared (Boy Scout and Girl Guide motto); *Beschleunigter Personenzug* (German—express train); Board of Parole; Border Patrol; British Petroleum; British Pharmacopoeia; British Public; Bureau of Power; Bureau of Prisons (United States Department of Justice); Burns Philp Lines
B-P Basses-Pyrénées; Bermuda Plan (breakfast only); Lord Robert S Baden-Powell, founder of the Boy Scouts

B.P. Bachelor of Pharmacy; Bachelor of Philosophy
B/P Bills Payable
B of P Bishop(ric) of Passau; Bureau of Prisons
BP Bassposaune (German—bass trombone); *Battleship Potemkin*; *Berliner Philharmoniker* (German—Berlin Philharmonic); *Biblioteca Pubblica* (Italian—Public Library); *Biblioteca Pública* (Portuguese or Spanish—Public Library); *Boerenpartij* (Dutch—Farmers' Party); *Boite Pôstale* (French—post office box); *Box Poste* (French—mail box); *British Pharmacopoeia*
bp 120/80 lar blood pressure 120 (systolic)/80 (diastolic) left arm reclining
bpa broadband power amplifier
Bpa Bahnpostamt (German—Railway Post Office)
BPA Beach Protection Authority; Bedding Plants Australia; Biological Photographers Association; Blanket Purchasing Agreement; Bonneville Power Administration; Book Publishers Association; British Pediatric Association; British Ports Association; Broadcasters Promotion Association; Brunswick Port Authority; Bureau of Public Affairs (U.S. State Department); Bureau of Public Assistance; Bush Pilots Association (Australia); Business Press Association; Business Publications Audit (of circulation)
B.P.A. Bachelor of Professional Arts
BPA Banco Portugués do Atlántico (Portuguese—Portuguese Bank of the Atlantic)
BPAA Bowling Proprietors' Association of America
BPAC Budget Program Activity Code; Business Publications Audit of Circulation
BP-ACT Blueport Associated Container Transporation
BPA/DoS Bureau of Public Affairs—Department of State
B.Paed. Bachelor of Paediatrics
BPAGB Bicycle Polo Association of Great Britain
bpam basic partitioned access method
BPANZ Book Publishers Association of New Zealand

BPAO Branch Public Affairs Office(r)
BPAS British Pregnancy Advisory Service
BPASC Book Publishers Association of Southern California
bpay bill(s) payable
bpb bank post bills; blanket position bond; bromophenol blue
BPBD Bill Posters, Billers and Distributors (Union)
BPBF British Paper Box Federation
BPBI British Plaster Board Industries
BPBIF British Paper and Board Industry Federation
BPBIRA British Paper and Board Industry Research Association
BPBM Bernice P Bishop Museum (Honolulu)
BPBMA British Paper and Board Makers Association
bpc back-pressure control; book prices current; book and periodical circulation
BPC British Pharmaceutical Codex; British Phosphate Commission; British Printing Corporation; British Purchasing Commission; Business and Professional Code
b-p cartridge barricade-penetrating cartridge
BPCC British Printing and Communication Corporation
bpcd barrels per calendar day
BPCF British Precast Concrete Federation
BPCI Bulk Packaging and Containerization Institute
BPCR Brakes on Pedal Cycle Regulations
BPCRA British Professional Cycle Racing Association
bpctca best practicable control technology currently available
bpd barrels per day; boxes per day; bronchopulmonary dysplasia
B. Pd. Bachelor of Pedagogy
BPD Bureau of the Public Debt
bpd & a basic planning data and assumption
BPDC Berkeley Particle Data Center; Books and Periodical Development Council (Canadian)
BPDMS Basic Point-Defense Missile System
BPDP Brotherhood of Painters, Decorators, and Paperhangers

bpe bit-plane encoding; black powder express (cartridge)
BPE Bureau of Postsecondary Education (Office of Education)
B.P.E. Bachelor of Physical Education
BPE-LCA Board of Parish Education—Lutheran Church in America
B.Pet.E. Bachelor of Petroleum Engineering
bpf bottom pressure fluctuation
bpf bon pour francs (French—good for francs)
BPF British Polio Fellowship
bpg break pulse generator
BPGC Building Performance Guarantee Corporation
Bpge bearing per gyro compass
bph barrels per hour; benign prostatic hypertrophy
B.Ph. Bachelor of Philosophy
BPh British Pharmacopoeia
B.P.H. Bachelor of Public Health
B.Pharm. Bachelor of Pharmacy
B.P.H.E. Bachelor of Physical and Health Education
B.Phil. Bachelor of Philosophy
BP & HL Brown Picton and Hornby Libraries (Liverpool)
B.Phys. Bachelor of Physics
B.Phys.Ed. Bachelor of Physical Education
B.Phys.Thy. Bachelor of Physical Therapy
bpi bits per inch; bytes per inch
BPI Bernreuter Personality Inventory; British Pacific Islands; Brooklyn Polytechnic Institute; Bureau of Public Information
BPICA Bureau Permanent Internationale des Constructeurs d'Automobiles (French—Permanent International Bureau of Automobile Manufacturers)
B picture moving picture designed as a second or supporting feature in a cinema program
b-pid book-physical inventory difference
BPIF British Printing Industries Federation
BPISAE Bureau of Plant Industry, Soils, and Agricultural Engineering
BP & JC FL Birmingham Public and Jefferson County Free Library

BPKT Basic Programming Knowledge Test
bpl birthplace
Bpl Barnstaple
BPL Belfast Public Library; Binghamton Public Library; Birmingham Public Library; Boston Public Library; Brass Pounders League; Bridgeport Public Library; Brooklyn Public Library; Buffalo Public Library
BP Lib Broadcast Pioneers Library
bpm barrels per minute; beats per minute; best practical means; blows per minute
BPM Bulletin of Paleomalacology
BPMA British Photographic Manufacturers Association; British Printing Machinery Association; British Pump Manufacturers Association
BPMA/DoS Bureau of Politico-Military Affairs—Department of State
BPMD Battelle Project Management Division
BPMF British Postgraduate Medical Federation; British Pottery Manufacturers' Federation
BPMS Blood Pressure Measuring System
bpn bloody public nuisance
Bpn Balikpapan
BPNHM Banff Park Natural History Museum
BPNMA British Plain Net Manufacturers' Association
BPO Base Post Office; Base Procurement Office; Berlin Philharmonic Orchestra; Boston Pops Orchestra; British Post Office; British Postal Order; Brooklyn Philharmonia Orchestra; Brooklyn Post Office; Buffalo Philharmonic Orchestra
BPO Berliner Philharmonisches Orchester (German—Berlin Philharmonic Orchestra)
BPOE Benevolent and Protective Order of Elks
BPOEW Benevolent and Protective Order of Elks of the World
BPP Black Panther Party; Botswana People's Party
BPP Banco Popular del Perú (Spanish—Popular Bank of Peru); **British Parliamentary Papers**
BPPMA British Power Press

Manufacturers Association
BPR Bureau of Public Roads
BPR *Bloque Popular Revolucionario* (Spanish—Popular Revolutionary Block)—El Salvador; *Book Publishing Record* (periodical)
BPRA Book Publishers' Representatives' Association
bprf bulletproof
bprs brief psychiatric rating scale
bps bits per second; bytes per second
bp(s) black pimp(s)
bp's beautiful people
BPs Book Publishers (sales reports); Burns Philp steamships
B.Ps Bachelor in Psychology
BPS Balanced-Pressure System; Basic Programming System; Bayou Preservation Society; Benchmark Portability System; Border Patrol Sector; Border Patrol Station; British Police Service; British Psychological Society; Bureau of Product Safety
B_{psc} bearing per standard compass
bpsd barrels per steam day
bpsm bulk pre-sorted mail
BPsS British Psychological Society
B_{p stg c} bearing per steering compass
B.Psych. Bachelor of Psychology
bpt boiling point
bpt (BPT) bound plasma tryptophan
BPT Board of Prison Terms; British Petroleum Tanker; Bureau of Prison Terms
B.P.T. Bachelor of Physiotherapy
bpti bovine pancreatic trypsin inhibitor
bptv battleship propulsion test vehicle
bpu base production unit
BPU British Powerboating Union
BPUNP *Biblioteca Pública de la Universidad de la Plata* (Spanish—Public Library of the University of La Plata)
bpv bovine papilloma virus; bullet-proof vest
BPWA Business and Professional Women's Association
bpwr burnable poison water reactor
B-Pyr Basses-Pyrénées
bq beauty quotient; boiler qual-

ity
Bq Becquerel
BQ Bachelor's Quarters; Basic Qualification; Basically Qualified (member of USCG Aux)
B Q *Bibliothèque nationale du Québec* (French—National Library of Quebec)—Montreal
Bqa Barranquilla
BQL Bank of Queensland
BQLI Brooklyn, Queens, Long Island
BQMS Battery Quartermaster Sergeant
BQSF British Quarrying and Slag Federation
b quark bottom quark
bque barque
Bquilla Barranquilla
br bank rate; bank robber(y); berth; bill of rights; branch; bread (slang—money); breath; breeder reactor; broken; brown; builder's risk; butadiene rubber
br (BR) bedroom; bedroom steward; branch (flow chart); break (request signal)
b/r bills receivable
b & r budget and reporting
b or r bales or rolls (freight)
br *bez roku* (Czech—no date, no year)
Br Branch; Bridge; Britain; British
Br *Bachiller* (Spanish—Bachelor)—academic degree; *Bratsche* (German—viola); *Bredning* (Danish—Bay); *Brücke* (German—Bridge); *Burun* (Turkish—nose, Point)
BR Baton Rouge; bearing; branch; Brazil (auto plaque); Breeder Reactor; bridge; British; British Railways; British Resident (commissioner); British United Airways; bromine; brown (buoy); Bureau of Reclamation
B-R Bas-Rhin; Business Route
B/R Bills Receivable; Bordeaux or Rouen
B of R Bureau of Reclamation; Bureau of Rehabilitation
BR *Banco di Roma* (Italian—Bank of Rome)
B.R. *Bancus Reginae* (Latin—Queen's Bench); *Bancus Rex* (Latin—King's Bench)
BR-1150 Breguet maritime-patrol aircraft also called Atlantique
bra brassiere
Bra Beira

Bra *Brasil* (Portuguese or Spanish—Brazil)
BrA *Bibliothèque royale Albert I* (French—Albert Ist Royal Library)—Brussels library called *Koninklijke Bibliotheek Albert I* in Flemish
BRA Bankruptcy Reform Act; Bee Research Association; Boston Redevelopment Authority; Bougainville Revolutionary Army (Papua, New Guinea); British Records Association; British Robot Association; Building Renovating Association
Brab *La Brabançonne* (French—the Brabant) national anthem of Belgium
BRAC Brotherhood of Railway and Airline Clerks
brachi brachion (Greek—arm)—brachiation, brachiopod, brachium
brachycephs brachycephalics (short-skulled people)
Bra Cur Brazil Current
Brad Bradburn; Bradbury; Bra-den; Bradfield; Bradford; Bradley; Bradner; Bradshaw; Bradstreet; Brady
Bradshaw's *Bradshaw's Railway Guide*
brady bradycardia
Brady Bradenton, Florida; Brady Glacier, Alaska; Brady Lake, Ohio; Brady Mountains, Texas; Dr Brady C Hartman
braid. bidirectional reference array internally deprived
BRAINS Behavior Replication by Analog Instruction of Nervous System
Bram Abraham
Br.Am. British America safety lock invented by Joseph Bramah
Brambach Radiumbad Brambach in Saxony
Brampton Women Vanier Centre for (criminal) Women at Brampton, Ontario
BRANCHHYDRO Branch Hydrographic Office
brane bombing radar navigation equipment
Brangus 3/8 Brahman + 5/8 —Angus cattle
BRANZ Basketball Referees Association of New Zealand; Building Research Association of New Zealand
bras ballistic rocket air suppression

bra(s) brassiere(s)
Bras Brasenose College, Oxford; Brasil; Brasileiro
Bras Brasil (Portuguese or Spanish—Brazil); *Brasile* (Italian—Brazil)
BRAs Bosom-Rehabilitation Associates
BRASCFHESE Brotherhood of Railway, Airline, and Steamship Clerks, Freight Handlers, Express, and Station Employees
Bras Coll Brasenose College—Oxford
Brasilsat Brazil's satellite
b-r-a-s-s breathe, relax, aim, squeeze, shoot (marksman's acronym)
BRASS Bottom Reflecting Active Sonar System
brasses brass instruments: bugles, cornets, French horns, horns, mellophones, trombones, trumpets, tubas, wag-ner horns
brass knucks brass knuckles
BRASTACS Bradford Scientific, Technical, and Commercial Service
Brat bi-drive recreational all-terrain transporter
BRAVE Boeing Robotic Air Vehicle
Bravo letter B radio code
braz Brazil; Brazilian
Braz Brazil(ian)
Brazil Federative Republic of Brazil (South America's largest country), *Republica Federativa do Brasil*
Brazza Brazzaville
Brb Borba (Yugoslavia—Struggle)—leading newspaper in Communist-controlled Yugoslavia
BRB Benefits Review Board; British Railways Board; Builders' Registration Board
brbc bovine red blood cells
Brbds Barbados
BRBMA Ball and Roller Bearing Manufacturers Association
brbzc brass, bronze, or copper (cargo)
brc business reply card
Br.C. British Columbia
BRC Balcones Research Center (University of Texas); Base Residence Course; Bolivia Railway Company; British Research Council; Broadcast Rating Council; Brotherhood of Railway Carmen
BRCA Brotherhood of Rail-

way Carmen of America
BRCCP British Royal Commission on Capital Punishment
Brch Branch
BRCMA British Radio Cabinet Manufacturers' Association
Br Col British Columbia
BRCS British Rail Catering Service; British Red Cross Society
BRCUSC Brotherhood of Railway Carmen of the United States and Canada
Br Cwlth British Commonwealth
brd basic retirement date; board; bomb-release distance; broad
BRD Bundesrepublik Deutschland (Federal Republic of Germany)
BRDC British Racing Drivers' Club
brdcst broadcast
brdf bidirectional reflectance distribution function
BRDM Soviet amphibious reconnaissance vehicle carrying three men and antitank missiles
Brdw Broadwood
Bre Bremen; Bremerhaven
BRE Bureau of Readjustment Education
B.R.E. Bachelor of Religious Education
brec bills receivable
B/Rec Bills Receivable
breccia pyroclastic volcanic rock
Breck Breckinridge; Brecknockshire
Brecon Breconshire (Brecknockshire)
BRECSU Building Research Energy Conservation Unit
Breguet 765 Sahara flying trans-port for 145 troops
Breguet 1150 Atlantique maritime-patrol aircraft
brek breakfast
BREL British Rail Engineering Limited
'brella umbrella
Brem Bremen; Bremerhafen; Bremerhaven; Bremerton
BREMA British Radio Equipment Manufacturers Association
Brenner Brenner Pass in the Alps connecting Bolzano, Italy with Innsbruck, Austria
Brennero Brenner Pass
Brent Brentford and Chiswick
Br'er Brother

Bres Breslau
bret bretonisch (German—Breton)
Bret Brittany; Breton
brev brevet; breviary; breviate; brevier
brev breveté (French—patent); *brevetto* (Italian—patent)
brev. breviarium (Latin—abridgement or breviary)
brew. brewer; brewery; brewing
Brew Brewer; Brewster
brew'd brewed
Brewer's Brewer's Dictionary of Phrase and Fable
brf brief; briefing
BRF Bass Research Foundation; Bible Reading Fellowship; British Road Federation
BRFC British Record Fish Committee (of rod anglers)
brg bearing; brewing; bridge; brigantine
Brg Bridge
BrG British Guiana
BRG Bibliotheek van de Rijksuniversiteit te Gent (Flemish—Royal University Library of Ghent)
brghd bridgehead
brghm brougham (pronounced *broom*)
Brgo Spgs Borrego Springs
Br Gu British Guiana
BrH British Honduras
BRH Brussels, Belgium (airport); Bureau of Radiological Health
BRHL British Rail Hovercraft Limited
BrHon British Honduras
BRHS Bay Ridge High School; Betsy Ross High School
Bri Brian; Bridge; British(er)(s); Briton(s)
Br I British Isles
BRI Babson's Reports Incorporated; *Banque des Réglements Internationaux* (French—Bank of International Settlements); Biological Research Institute; Brain Research Institute; Bristol Royal Infirmary; Building Research Institute; Bureau of Rehabilitation Inc; Burlington-Rock Island (railroad)
BRI Banque des Règlements Internationaux (French—Bank for International Settlements); *Brand Rating Index*
BRIA Biological Research Institute of America; Bread Re-

search Institute of Australia
BRIC British Columbian Resource Investment Corporation
Bricklayers Union Bricklayers, Masons, and Plasterers International Union of America
BRICS British Rail Inter-City Service
BRICSHST British Rail Inter-City Service High-Speed Train
BRIDGEX Bridge Construction Exercise
brig brigantine; ship's prison
Brig Brigade; Brigadier
Brig Gen Brigadier General
Brigitte Bridget
BRIGLEX Brigade Landing Exercise
Brigton Glasgow's Bridgetown
Brilab bribery-labor (FBI investigation's code name)
brill brillante (Italian—brilliant)
BRIMEC British Mechanical Engineering Federation
BRINCO British Newfoundland Corporation
BRINDEX British Independent Oil Exploration (Companies Association)
Bris Brisbane
Brisb Brisbane
Brissie Brisbane
Brist Bristol
brit britisk (Dano-Norwegian—British)
Brit Britain; Britannia; British
Brit Encyclopaedia Britannica
Britain Great Britain (England, Scotland, and Wales)
Brit Book Centr British Book Centre
Britcoms British comedians; British comedies
Brit and For British and Foreign State Papers
Britic Briticism
Brit Info British Information Services
British pertaining to the British Commonwealth, the British Empire, or the British people (English, Scottish, and Welsh)
British Am Bks British American Books
British Bk Ctr British Book Center
British Commonwealth British Commonwealth of Nations: Great Britain and Northern Ireland; British dominions, republics, and dependencies

British Dependencies Bermuda, Cayman Islands, Falklands, Gibraltar, South Georgia, South Sandwich, South Shetlands, Turks, and Caicos Islands
British India colonial India (1757–1948)
British Virgins British Virgin Islands
Brit J Psychiat British Journal of Psychiatry
Brit J Surg British Journal of Surgery
brit met britannia metal (tin, copper, antimony alloy—sometimes bismuth, lead, and zinc)
Brit Mus British Museum
Brit Pat British Patent
Brit Phos Comm British Phosphate Commission
BritRail British Railways
Brit—Rail Hover British Railways Hovercraft
Brits British; Britons
Brit(s) British(ers); Briton(s)
Brits (Dutch—British)
Brit Sam British Samoa
BRITSHIPS British Shipbuilding Integrated Production System
Britt. Britannorum (Latin—of the Britons)
Brix Brixham; Brixton
Br J App Phys British Journal of Applied Physics
brk brick
Brk Brook
BRK The Bridge on the River Kwai (movie)
brkf breakfast
brklyr bricklayer
brkmn breakman
brks breakers
brkt bracket
brkwtr breakwater
brl bomb-release line
br/l brown line positive
BRL Babe Ruth League; Ballistic Research Laboratories; Beecham Research Laboratories; Bible Research Library; *Bibliotheek der Rijksuniversiteit te Leiden* (Dutch—Library of the Royal University in Leyden); British Research Library
BRL 1241 Beecham Research Laboratories formula 1241 (methicillan)
BRL 1341 Beecham Research Laboratories formula 1341 (penbritin)
brlg bomb radio longitudinal generator-powered

Brlp burlap
brl sys barrier ready light system
brm bedroom
BRM British Racing Motors
BRMA Board of Registration of Medical Auxiliaries; British Rubber Manufacturers' Association
BRMBR Bear River Migratory Bird Refuge (Utah)
BRMC Business Research Management Center (USAF)
BRMCA British Ready-Mixed Concrete Association
BRMF British Rainwear Manufacturers' Federation
brn brown
Brn Bahrain; Brunei; Sultanate of Brunei
BRNC Britannia Royal Naval College (Dartmouth)
brng bearing; browning; burning
BRNP Blue Ridge National Parkway
brnsh brownish; burnish
Brnx Bronx
brnz bronze; bronzing
bro broach; bronchoscopy; brother
brO brownish orange
Bro Brother
BRO Brigade Routine Order(s)
Broads The Broads (Norfolk Broads)—England's east coast holiday resort area
Broadway main north-south avenue of New York City; New York's theater district
broast(ed) broil(ed) + roast(ed)
broc brocaded
Brod Broderick; Brodie; Brody
broficon broadcast fighter control
BROILER Biopedagogical Research Organization on Intensive Learning Environment Reactions
bro-in-law brother-in-law
brok broker; brokerage
brom bromide; bromidic; bromo; bromo-seltzer
bromat bromatology (treatise on foods)
bromidrosis bromohydrosis
bromo bromidrosis; bromoform; bromo-seltzer
bromo-seltzer (bromide + seltzer)
bromot bromotology [study of smell(s)]
bronc bronco (Spanish—small half-wild horse)
bronch bronchial; bronchitis; bronchoscopic; bronchos-

copist; bronchoscopy
broncho bronchus (Greek—windpipe)—bronchi(tis)
Bronx HTF Bronx Homicide Task Force (NYPD)
Bronx Zoo New York Zoological Gardens (Bronx Park)
bronze 92% copper, 6% tin, 2% zinc
Brookings Brookings Institution
Brooklyn HTF Brooklyn Homicide Task Force (NYPD)
Brookwood Girls Brookwood Center for (delinquent) Girls at Claverack, New York
Bros brothers
brosch broschiert (German—stitched)
Brose Ambrose
brot brought
brotel brothel + hotel
BROU Banco de la República Oriental del Uruguay (Spanish—Bank of the Oriental Republic of Uruguay)
browners brown nosers
brownulated granulated brown sugar
Brown U Pr Brown University Press
Brownwood Girls State Home, Reception Center, and School for Delinquent Girls at Brownwood, Texas
brp bathroom privileges
BRP Breeder Reactor Program
BRPF Bertrand Russell Peace Foundation
brph bronchophony
brPk brownish pink
BRPL Baton Rouge Public Library
brpp basic radio propagation prediction(s)
Br Rys British Railways
brs brass
brs (BRS) break request signal
br's bedrooms
Brs Bristol
Br S Bedroom Steward
BRS Bertrand Russell Society; Bibliographic Retrieval Services (data base); Bomber Recorder System; British Road Services; British Roentgen Society; Brotherhood of Railway Signalmen; Bureau of Railroad Safety; Business Radio Service; Buyers' Research Syndicate
BRSA British Railway Staff Association
BR & SC Brotherhood of Railway and Steamship Clerks
BRSCC British Racing and

Sports Car Club
br snds breath sounds
br sounds breath sounds
brst burst
Br std British standard
brstr burster
brt bright
brt (BRT) bruttoregisterton (Dano-Norwegian—registered gross tonnage)
Brt Brest
BRT Brotherhood of Railroad Trainmen
B.R.T. Before Recorded Time
BRT Belgische Radio en Televisie (Belgian Radio and Television); *Brutto-Register-Tonnen* (German—gross register tons)
BRTA British Regional Television Association; British Road Tar Association
BRTC British Rail Travel Centre
BR & TC Bermuda Radio and Television Company
brt fwd brought forward
brtg bartering
Bru Bruce; Brunei; Bruno; Brutus
BRU Brussels, Belgium (National Airport)
BRU Bibliotheek der Rijksuniversiteit te Utrecht (Dutch—Library of the Royal University in Utrecht)
B.Ru.Eng. Bachelor of Rural Engineering
BRUFMA British Rigid Urethane Foam Manufacturers Association
Brum Brummagen (Birmingham, England's nickname)
Brum Brumaire (French—Foggy Month)—beginning October 22—second month of the French Revolutionary Calendar
Brun Brunei
brunch(eon) breakfast-lunch-(eon)
Bruns Brunswick
Brunsw Brunswick
Brun U Brunel University
B.Rur.Sci. Bachelor of Rural Science
Brus Bruselas (Spanish—Brussels); *Bruselle* (Italian—Brussels); *Brussel* (Dutch or Flemish); *Brüssel* (German—Brussels)
BRUTE British Universal Trolley Equipment
brux bruxism; bruxitic
Brux Brussels
Brux Bruxelas (Portuguese—

Brussels); *Bruxelles* (French—Brussels)
brv (BRV) ballistic reentry vehicle
BRVMA (BVA) British Radio Valve Manufacturers' Association
Brw Barrow
BRW British Relay Wireless
Brx Bronx
bry bryology
Bry Barry; Bryant
Bryce Bryce Canyon National Park, Utah; Mount Bryce, British Columbia
bryol bryology
Bryth Brythonic
brz bronze
brzg brazing
bs beam splitter; blood sugar; bluestone; bomb service; bonded single-silk (insulation); both sides; bowel sound; breath sound; bullshit
bs (BS) backspace (data-processing character); binary subtract(ion)
b's boomerangs
b/s back stamp
b/s (B/S) bill of sale
b & s beams and stringers; bell and spigot; boosters and sustainers; brandy and soda
b-S by-Sea
Bs bolivares (Venezuelan currency); bolivianos (Bolivian currency)
BS Battle Squadron; Battle Star; Bethlehem Steel; Berlin Sector; Birmingham Southern (railroad); British Standard; Broadcast Satellite; Bureau of Ships; Bureau of Standards
B-S Bedford-Stuyvesant
B.S. Bachelor of Science
B/S Bill of Sale
B & S Bank and Savill (steamship line); Brown and Sharpe; Butterfield and Swire
BS Bayerische Staatsbibliothek (German—Bavarian State Library)
bsa bismuth-sulphite agar; body surface area; bovine serum albumin; brown strain apparent
BSA Bank Stationers Association; Bibliographical Society of America; Birmingham Small Arms; Blind Service Association; Blinded Soldiers Association (Australia); Botanical Society of America; Boy Scouts Association; British School of Athens; British South Africa; Brotherhood of

St Andrew; Bruckner Society of America; Bureau of Supplies and Accounts
B.S.A. Bachelor of Agricultural Science
BSAA British South American Airways
BSA(A) British School of Archeology (Athens)
B.S.A.A. Bachelor of Science in Applied Arts
BSAC British South Africa Company; Brotherhood of Shoe and Allied Craftsmen
B.S.Adv. Bachelor of Science in Advertising
B.S.A.E. Bachelor of Science in Aeronautical Engineering; Bachelor of Science in Architectural Engineering
BSAF British Sulphate of Ammonia Federation
BSAG Bristol Social Adjustment Guides
B.S.Agr. Bachelor of Science in Agriculture
BSAM Basic Sequential Access Method
B.S.A.M. Bachelor of Suddha Ayurvedic Medicine
BSAOT Bell System American Orchestras on Tour
BSAP British South Africa Police
B.S.Arch. Bachelor of Science in Architecture
B.S.Arch. Eng. Bachelor of Science in Architectural Engineering
B.S.Art Ed. Bachelor of Science in Art Education
Bs As Buenos Aires
BSAS British Ship Adoption Society
BSAVA British Small Animals Veterinary Association
bsb body surface burned
bsb (BSB) backspace block
Bsb Brisbane
BSB Brasilia, Brazil (airport); British Satellite Broadcasting; Brotherhood of St Barnabas
BSBA British Starter Battery Association
B.S.B.A. Bachelor of Science in Business Administration
BSBC British Social Biology Council
bsbg burst and synchronous bit generator
BSBI Botanical Society of the British Isles
BSBSPA British Sugar Beet Seed Producers' Association
B.S.Bus. Bachelor of Science in Business

bsc basic; basic-message switching center; binary synchronous communication
Bsc British standard channel (steel)
B.Sc. Bachelor of Science
BSC Bank Street College; Beltsville Space Center; Bemidji State College; Bethlehem Steel Corporation; Bibliographical Society of Ca-nada; Biological Stain Commission; Biomedical Sciences Corporation; Bloomsburg State College; Bluefield State College; Booth Steamship Company; British Society of Cinematographers; British Steel Corporation; British Supply Council; Business Service Center
B.S.C. Bachelor of Science in Commerce
BSCA Bureau of Security and Consular Affairs (US Department of State)
B.Sc.Acc. Bachelor of Science in Accounting
B.Sc.Ag. & A.H. Bachelor of Science in Agriculture and Animal Husbandry
B.Sc.Agr.Bio. Bachelor of Science in Agricultural Biology
B.Sc.Agr.Eco. Bachelor of Science in Agricultural Economics
B.Sc.Agr.Eng. Bachelor of Science in Agricultural Engineering
B.Sc.Ag(ri)(c). Bachelor of Science in Agriculture
B.Sc.Arch. Bachelor of Science in Architecture
B.Sc.B.A. Bachelor of Science in Business Administration
BSCC British Society for Clinical Cytology
B.Sc.C.E. Bachelor of Science in Civil Engineering
B.Sc.Chem.E. Bachelor of Science in Chemical Engineering
B.Sc.Dent. Bachelor of Science in Dentistry
B.Sc.Dom.Sc. Bachelor of Science in Domestic Science
BSCE Bank Street College of Education
B.S.C.E. Bachelor of Science in Civil Engineering
B.S.Ch. Bachelor of Science in Chemistry
B.S.Chm. Bachelor of Science in Chemistry
B-school(s) business school(s)
bscn bit scan

B.Sc.Nurs. Bachelor of Science in Nursing
BSCO British Security Coordination Office
B.S.Comm. Bachelor of Science in Commerce
BSCorp British Steel Corporation
BSCP Brotherhood of Sleeping Car Porters; Business Service Centers Program
BSCP British Standard Code of Practice
BSCRA British Steel Castings Research Association
BSCS Biological Sciences Curriculum Study
B.Sc.S.S. Bachelor of Science in Secretarial Studies
BSCU Blue Star (Line) Container Unit
B.Sc.Vet.Sc. Bachelor of Science in Veterinary Science
bsd barrels per steaming day; beam-steering device; bit storage density; blast-suppression device; burst-slug detection
BSD Ballistic Systems Division (USAF); Bank of San Diego; Berkeley Software Distribution; British Space Development
B.S.D. Bachelor of Science in Design
BSDA British Spinners and Doublers Association
BSD Bancorp Bank of San Diego Bank Corporation
bsdc binary symmetric dependent channel
BSDC British Space Development Company; British Standard Data Code
B.S. Dent. Bachelor of Science in Dentistry
bsdg breveté sans garantie du gouvernement (French—patented without government guarantee)
B.S.D.H. Bachelor of Science in Dental Hygiene
bsdl boresight datum line
bse base support equipment; breast self-examination (cancer control)
bse (BSE) breast self-examination
BSE Base Support Equipment; Birmingham & Southeastern (railroad); Broadcast Satellite Experiment; Broadcasting Sa-tellite for Experimental Purposes; Building Service Employees (Union); Bureau of Steam Engineering

B.S.E. Bachelor of Sanitary Engineering; Bachelor of Science Education; Bachelor of Science Engineering

B & SE Birmingham & Southeastern (railroad)

B.S.Ec. Bachelor of Science in Economics

B.S.Ed. Bachelor of Science in Education

B.S.E.E. Bachelor of Science in Electrical Engineering

B.S.El.E. Bachelor of Science in Electronic Engineering

B.S.Eng. Bachelor of Science in Engineering

b's'er bullshiter

BSES British Schools Exploring Society

bsf back scatter factor; bulk shielding facilities

bsf (BSF) beta-s-fetoprotein

BSF Basic Skill Films; British Shipping Federation

B.S.F. Bachelor of Science in Forestry

BSFA British Sanitary Fireclay Association; British Steel Founders' Association; Brotherhood of St Francis of Assisi; Building Science Forum of Australia

bsfc brake specific fuel consumption

BSFC Baltic States Freedom Council

BSFF Buffer Stock Financing Facility

B.S.Fin. Bachelor of Science in Finance

BSFL British Shipping Federation Limited

B.S.For. Bachelor of Science in Forestry

B.S.F.S. Bachelor of Science in Foreign Service

BSFT Basalt Spent Fuel Test(ing)

BSF & W Bureau of Sport Fisheries and Wildlife

BSG British standard gage

B.S.G.E. Bachelor of Science in General Engineering; Bachelor of Science in Geological Engineering

B.S.Gen. Nur. Bachelor of Science in General Nursing

B.S.Geog. Bachelor of Science in Geography

B.S.Geol. Bachelor of Science in Geology

B.S.Geol.Eng. Bachelor of Science in Geological Engineering

B & S glands Bartholin and Skene's glands

bsh bushel

BSH British Society of Hypnotherapists; British Standard of Hardness

B.S.H.A. Bachelor of Science in Hospital Administration

B.S.H.E. Bachelor of Science in Home Economics

B.S.H.Eco. Bachelor of Science in Home Economics

B.S.H.Ed. Bachelor of Science in Health Education

BSHS British Society for the History of Science

bsi basic shipping instructions; bound serum iron

BSI Baker Street Irregulars; British Sailors' Institute; British Standards Institution

BSI *Business Survey Index*

BSIA Better Speech Institute of America

BSIB Boy Scouts International Bureau; British Society for International Bibliography

bsic binary-symmetric independent channel

BSIC British Ski Instruction Council

B.S.I.E. Bachelor of Science in Industrial Engineering

BSIHE British Society for International Health Education

B.S.Ind.Art Bachelor of Science in Industrial Art

B.S.Ind.Chem. Bachelor of Science in Industrial Chemistry

B.S.Ind.Ed. Bachelor of Science in Industrial Education

B.S.Ind.Eng. Bachelor of Science in Industrial Engineering

BSIP British Solomon Islands Protectorate

B.S.I.R. Bachelor of Science in Industrial Relations

BSIRA British Scientific Instrument Research Association

BSIs Baker Street Irregulars

BSIS BioScience Information Services

BSIU British Society for International Understanding

bsj balanced swivel joint; ball-and-socket joint

B.S.J. Bachelor of Science in Journalism

BSJA British Show Jumping Association

B.S.Jr. Bachelor of Science in Journalism

bsk basket(s)

Bskrvlle Baskerville

bskt basket

bsl billet split lens

bs/l bills of lading

Bsl Bislig Bay

BSL Barber Steamship Lines; Behavioral Sciences Laboratory; Black Star Line; Blue Sea Line; Blue Star Line; British Sign Language; Building Service League; Bull Steamship Lines

BSLA Bus Services Licensing Authority (Singapore)

B.S.Lab.Rel. Bachelor of Science in Labor Relations

bslb ball-and-socket lower bearing

bsln ball-and-socket lower bearing

bsl(s) bushel(s)

B.S.L.S. Bachelor of Science in Library Science; Bachelor of Science in Library Service

bsm bi-stable multivibrator; bottom sonar marker

BSM Battery Sergeant Major; Birmingham School of Music; Branch Sales Manager; Bronze Star Medal

BSM *beso sus manos* (Spanish—I kiss your hands)—respectfully yours

BSMA British Skate Makers' Association

B.S.Mar.Eng. Bachelor of Science in Marine Engineering

BSMD Business System Marketing Division

B.S.M.E. Bachelor of Science in Mechanical Engineering; Bachelor of Science in Mining Engineering; Bachelor of Science in Music Education

B.S.Med. Bachelor of Science in Medicine

B.S.Med.Rec. Bachelor of Science in Medical Records

B.S.Med.Rec.Lib. Bachelor of Science in Medical Records Librarianship

B.S.Med.Tech. Bachelor of Science in Medical Technology

B.S.Met. Bachelor of Science in Metallurgy

B.S.Met.Eng. Bachelor of Science in Metallurgical Engineering

B.S.Mgt.Sci. Bachelor of Science in Management Science

B.S.Min Bachelor of Science in Minerology; Bachelor of Science in Mining

B.S.Min.Eng. Bachelor of Science in Mining Engineering

BSMMA British Sugar Machinery Manufacturers Asso-

ciation
bsmt basement
B.S.Mus.Ed. Bachelor of Science in Music Education
bsmv barley-stripe-mosaic virus
bsn bowel sounds normal
BSN Baker School of Navigation; Broadcasting System of Niigata (Japanese)
B.S.N. Bachelor of Science in Nursing
BSN *Bayerische Staatsoper—Nationaltheater* (German—National Theater—in Munich)
bsna bowel sounds normal and active
B.S.N.A. Bachelor of Science in Nursing Administration
B.S.Nat.Hist. Bachelor of Science in Natural History
BSNDT British Society for Non-Destructive Testing
BSNH Boston Society of Natural History; Buffalo Society of Natural History
B.S.N.I.T. Bachelor of Science in Nautical Industrial Technology
B.S.Nurs. Bachelor of Science in Nursing
B.S.Nurs.Ed. Bachelor of Science in Nursing Education
bso blue stellar objects
BSO Baltimore Symphony Orchestra; Bamberg Symphony Orchestra; Birmingham Symphony Orchestra; Bombay Symphony Orchestra; Boston Symphony Orchestra; Bournemouth Symphony Orchestra; Budapest Symphony Orchestra
BSOA British Sexual Offenses Act
B.S.Occ.Ther. Bachelor of Science in Occupational Therapy
B.Soc.Sci. Bachelor of Social Science
B.Soc.St. Bachelor of Social Studies
B.Soc.Wk. Bachelor of Social Work
BSOIW Bridge, Structural and Ornamental Iron Workers
B.S.Opt. Bachelor of Science in Optometry
B.S.O.T. Bachelor of Science in Occupational Therapy
bsp bromosulphalein
Bsp British Standard pipe
BSP Bering Sea Patrol; Border Security Police (NATO); Boy Scouts of the Philippines;

British Society for Parasitology; Brotherhood of St Paul; Brunei Shell Petroleum; Bulgarian Socialist Party
B-S-P Bartlett-Snow-Pacific (foundry division)
B.S.P. Bachelor of Science in Pharmacy
BSP *Bureau de Sécurité Publique* (French—Bureau of Public Security)
BSPA Basic Slag Producers' Association; Black Students Psychological Association
B.S.P.A. Bachelor of Science in Public Administration
B.S.P.E. Bachelor of Science in Physical Education
B-Specials Belfast's special soldiers (attached to the Ulster Special Constabulary)
B.S.Per. & Pub.Rel. Bachelor of Science in Personnel and Public Relations
B.S.Pet. Bachelor of Science in Petroleum
B.S.Pet.Eng. Bachelor of Science in Petroleum Engineering
B.S.P.H. Bachelor of Science in Public Health
B.S.Phar. Bachelor of Science in Pharmacy
B.S.Pharm. Bachelor of Science in Pharmacy
B.S.P.H.N. Bachelor of Science in Public Health Nursing
B.S.Phys.Ed. Bachelor of Science in Physical Education
B.S.Phys.Edu. Bachelor of Science in Physical Education
B.S.Phys.Ther. Bachelor of Science in Physical Therapy
bspl behavioral science programming language
BSPL Blue Star Port Lines
BSPM Battlefield Systems Project Management
BSPMA British Sewage Plant Manufacturers Association
BSPP Burmese Socialist Program Party
B.S.P.T. Bachelor of Science in Physical Therapy
bsp test bromsulphalein test
B.Sp.Thy. Bachelor of Speech Therapy
bspw bare silver-plated wire
Bsq Basque
BSQ Bachelor Sergeant Quarters
bsr backspace recorder; balloon-supported rockets (rockoons); basal skin resistance; basic service rate; battle short

relay; blood sedimentation rate; blue-streak request; bore sight restricted
Bsr Basra (Busreh)
BSR British Society of Rheology
B.S.R. Bachelor of Science in Rehabilitation
BSRA British Ship Research Association
BSRC Battelle Seattle Research Center; Biological Serial Record Center
BSRD Behavioral Sciences Research Division
B.S.Rec. Bachelor of Science in Recreation
B.S.Ret. Bachelor of Science in Retailing
bsrf brain stem reticular formation
BSRI Bem Sex Role Inventory
BSRIA Building Services Research and Information Association
BSRL Boeing Scientific Research Laboratories
B.S.R.T. Bachelor of Science in Radiological Technology
bss balanced salt solution; basic shaft system; beam-steering system; black-silk suture; buffered saline solution
BSS Bibliothèque Saint-Sulpice (Montreal); Biological and Social Sciences (NSF); British Sailors Society; British Security Service; British Standard Specification; Bronze Service Star; Bureau of School Systems; Bureau of State Services
B.S.S. Bachelor of Sanitary Science; Bachelor of Science in Science; Bachelor of Secretarial Science; Bachelor of Social Science(s)
Bssa *Baronessa* (Italian—Baroness)
B.S.S.A. Bachelor of Science in Secretarial Administration
B.S.Sc. Bachelor of Sanitary Science
B.S.Sc.Eng. Bachelor of Science in Science Engineering
B.S.Sec.Ed. Bachelor of Science in Secondary Education
B.S.Sec.Sci. Bachelor of Science in Secretarial Science
BSSG Biomedical Sciences Support Grant
BSSM Blue Star Ship Management
BSSML Blue Star Ship Management Limited
BSSO British Society for the

Study of Orthodontics
B.S.Soc.Serv. Bachelor of Science in Social Service
B.S.Soc.St. Bachelor of Science in Social Studies
B.S.Soc.Wk. Bachelor of Science in Social Work
bssp broadband solid-state preamplifier
BSSP Battelle Seminars and Studies Program
BSSR Bureau of Social Science Research; Byelorussia Soviet Socialist Republic
BSSS British Society of Soil Science
B.S.S.S. Bachelor of Science in Secretarial Studies; Bachelor of Science in Social Science
B.S.S.Sc. Bachelor of Science in Social Science
B.S.Struc.Eng. Bachelor of Science in Structural Engineering
bssw bare stainless-steel wire
bst beam-steering transducer; blood serological test(ing); brief stimulus therapy
bst (BST) bovine somatotropin
b s & t blood, sweat, and tears
b/st bill of sight
BST Bering Standard Time; Blood Serological Test; British Summer Time
BSTA British Surgical Trades Association
BSTC Ball State Teachers College
bstd bastard
B.S.Text. Bachelor of Science in Textiles
bst lt blue stern light
bstm biaxial shock-test machine
bstr booster
B.S.Trans. Bachelor of Science in Transportation
bstrk bomb service truck
bstr rkt booster rocket
BSU Black Students Union; Boat Support Unit; British Standard Unit(s)
bsub ball-and-socket upper bearing
B.Sur. Bachelor of Surgery
B.Surv. Bachelor of Surveying
bsut beam-steering ultrasonic transducer
bsv Boolean simple variable
BSV Batten-Spielmyer-Vogt (syndrome)
B.S.Voc.Ag. Bachelor of Science in Vocational Education
bsw barrels of salt water
bs & w basic sediment and water

BSW Biological Society of Washington; Boot and Shoe Workers (union); Botanical Society of Washington; British Standard Whitworth
B.S.W. Bachelor of Social Work
BSWB Boy Scouts World Bureau
bswd barrels of salt water per day
BSWE Boy Scouts in Western Europe
BSWG British Standard Wire Gage
BSWIA British Steel Wire Industries Association
bt baby talk; *Bacillus thuringiensis* (biological pesticide); bathtub; bathythermograph; bedtime; bent; bitemporal; blue tetrazolium (stain); boat; boat-tail; body temperature; bombing table; bottom time; bought; brain tumor; broader term; brought; bulk transport; byssal thread
bt (BT) basic typing
b & t bacon and tomato sandwich; (cabin or room with) bath and toilet; bridges and tunnels
b of t balance of trade
Bt baronet
Bt Bukit (Indonesian or Malay—Height, Hill)
B-t Bacillus thuringiensis (biological pesticide fatal to gypsy moth caterpillars)
BT basic trainer; British Telecom/Telecommunications; Burgtheater (Vienna)
B & T Baker & Taylor
B of T Bank of Tokyo; Board of Trade
BT Berlingske Tidende (Danish—Berling's Times—Copenhagen); *Bolshoi Teatr* (Russian—Bolshoi Theater) Moscow concert hall; *Brevet Technique* (French—Technical Diploma); *Burgtheater* (German—Court Theater) Vienna opera house
BT-13 Vultee two-place basic-trainer aircraft used during World War II
bta better than average
bta (BTA) best time available (for tv broadcast)
BTA Blood Transfusion Association; Board of Tax Appeals; Border Trade Alliance; Boston Transportation Authority; Brazilian Travel Agency; Brith Trumpeldor of

America; British Tourist Authority; British Travel Association; British Tuberculosis Association
BTAM Basic Telecommunications Access Method; Basic Terminal Access Method
BTANZ British Trade Association of New Zealand
BTAO Bureau of Technical Assistance Operations (UN)
BTAP Bond Trade Analysis Program
BTASA Book Trade Association of South Africa
btb braided tube bundle; bus tie breaker
BTB Barbados Tourist Board; Belgian Tourist Bureau
BTBA Blood Transfusion Betterment Association
BTBL Braille and Talking Book Library
BTBS Book Trade Benevolent Society
btc below threshold change; beryllium thrust chamber
BTC Bankers Trust Company; Basic Training Center; Bethlehem Transportation Company; Board of Transport Commissioners; British Textile Confederation; Building Trades Council
B.T.C. Bachelor of Textile Chemistry
btca biblioteca (Spanish—library)
BTCC Bloom Township Community College; Board of Transportation Commissioners for Canada; Broome Technical Community College
B.T.C.P. Bachelor of Town and Country Planning
BTCV British Trust for Conservation Volunteers
btd bomb testing device
BTDB Bermuda Trade Development Board
btdc before top dead center
btdl basic-transient diode logic
bte battery terminal equipment; better then expected; blunt trailing edge; Boltzmann transport equation; bourdon tube element; Brayton turbo-electric engine; bulk tape eraser
bte breveté (French—patent)
BTE Board of Teacher Education; Board of Transport Economics
B.T.E. Bachelor of Textile Engineering

BTEA British Textile Employers Association

B.Tech. Bachelor in Technology

Btee Brayton turboelectric engine

BTEF Book Trade Employers' Federation

B.Tel.E. Bachelor in Telecommunications Engineering

BTEMA British Tanning Extract Manufacturers' Association

BTES Beginning-Teacher Evaluation Study

B.Text. Bachelor of Textiles

btf balance to follow; barrels of total fluid; bomb tail fuse

b/tf balance transferred

BTF British Trawlers Federation

btg ball-tooth gear; battery timing group; beacon trigger generator; burst transmission group

BTG British Technology Group; Building Trades Group

btgj ball-tooth gear joint

bth bath; bathroom; beat the hell; berth; beyond the horizon

B.Th. Bachelor of Theology

BT-H British Thompson-Houston

BTHS Brooklyn Technical High School

B t h u British thermal unit (btu, Btu, BTU)

bti bank-and-turn indicator; bridgetape isolator

bti (BTI) bacillus thuringiensis israelensis (developed for mosquito abatement)

Bti Bacillus thuringiensis israe-lensis (mosquito-control substance)

BTI Bandung Technical Institute; Benefit Tours International

BTI British Technology Index

BTIA British Tar Industries Association

BTIPR Boyce Thompson Institute for Plant Research

btj ball-tooth joint

BTJ Board of Trade Journals

btk buttock

btk l buttock line

btl beginning tape label; behind the lens (camera); bottle

BTL Bell Telephone Laboratories

BTLS Bell Telephone Laboratories System

btlv biological threshold limit value

btm bottom

btm (BTM) bromotrifluoromethane (fire extinguisher)

Btm Bottom (postal abbreviation)

BTMA British Typewriter Manufacturers Association

BTME Babcock Test of Mental Efficiency

btn button

Btn Batangas

BTN Brussels Tariff Nomenclature

bto big-time operator; bombing through overcast

bto bruto (Spanish—gross weight); *brutto* (Dano-Norwegian—bulk or gross weight); *bulto* (Spanish—bulk)

BTO Branch Transportation Office(r); British Trust for Ornithology

bto(s) big time operator(s)

BTOW Boiler Technician of the Watch (USN)

B-town Bean Town (Boston—sailor's sobriquet)

btp body temperature and pressure

BTP Bailment Test Program; British Transport Police; Bush Terminal Piers

B.T.P. Bachelor of Town Planning

btps body temperature and pressure—saturated

btr better; bus transfer

BTR Baton Rouge, Louisiana (airport); Bureau of Trade Regulation

BTR British Tax Review

BTR-40 Soviet armored personnel carrier and scout car for 10 troops including the driver

BTR-50 Soviet amphibious personnel carrier for 15 troops including the driver

BTR-60P Soviet amphibious armored personnel carrier including 12.7mm machinegun

B.T.R.A. Bachelor of Town and Regional Planning

BTRM Soviet armored-infantry combat vehicle armed with 76.2 gun and antitank missile

btrmlk buttermilk

btry battery

bts back to school; base of terminal service (USAF); Boolean time sequence

BTS Blood Transfusion Service; British Tanzania Society; British Textile Society

BTSA British Tensional Strapping Association

BTSB Bound-to-Stay-Bound Books

BTSC British Transport Staff College

BTSS Basic Time-Sharing System

BTTA British Thoracic and Tuberculosis Association

bttns battens

btu (BTU, Btu) British thermal unit

BTU Board of Trade Unit

btv basic transportation vehicle

btw between

BTW Belasting Toegevoegde Waarde (Dutch—value added tax)

BTWHS Booker T. Washington High School

btwn between

btx benzene, toluene, xylene

BTX Bildschirmtext (German—viewdata interactive videotext system)

bty battery

B-type Basedow type

bu base (of prism) up; base unit; base up; biological urge; brick unprotected; brilliant uncirculated; bromouracil; builder; burglary; bushel

Bu Bulgaria; Bulgarian; Bureau (United States Navy); butyl

Bü *Büyük* (Turkish—big)

BU Baker University; Baylor University; Bishop's University; Bloomsburg University; Board of Underwriters; Boston University; Bradley University; Brandeis University; Brown University; Brunel University; Bucknell University; Burma (symbol); Butler University

B & U Beechey and Underwood

BU Bollettino Ufficiale (Italian—Official Gazette)

BUA Belfast Urban Area; British United Airways

BuAer Bureau of Aeronautics (USN)

BUAF British United Air Ferries

BUAV British Union for the Abolition of Vivisection

Bubs Bubbles

buc buccal; buccaneer; buccinator

BUC Bangor University College

bucc buccal

BUCCS Bath University Comparative Catalogue Study
Buchar Bucharest
buck buckram
Buck House Buckingham House (Buckingham Palace—London residence of British royalty)
Buck Island Buck Island Reef National Park off the north shore of St Croix, American Virgin Islands
Bucknell U Pr Bucknell University Press
Buck Pal Buckingham Palace
Bucks Buckinghamshire
Bucks Co Hist Bucks County Historical Society
BUCOP British Union Catalogue of Periodicals
bucu burring cutter
bud. budget
Bud Buddha; Buddhism; Buddhist; Buddy; Budweiser
BUD Basic Underwater Demolition Team; Budapest, Hungary (airport)
Buda across the Danube from Pest, together known as Budapest
Bud(dy) Brother
BUDFIN Budget and Finance Division (NATO)
budgie(s) budgerigar(s)
BuDocks Bureau of Yards and Docks (USN)
Budpst Budapest
budr bromodeoxyuridine
bue built-up edge
BUE Buenos Aires, Argentina (Ezeiza airport)
Buen Buenaventura
BUET Bangladesh University of Engineering and Technology
buf buffer(ed)
Buf Buffalo (city and port)
BUF British Union of Fascists; Buffalo, New York (airport)
BUFF Big Ugly Fat Fellow (Air Force nickname for the eight-engine B-52 bomber)
Buffalo Acad Buffalo Fine Arts Academy
bufg buffing
bufno buffers, number of
Buf Phil Buffalo Philharmonic
bug. black urban ghetto
Bug Bugatti; standard-model Volkswagen (also called the Beetle)
BUG Brooklyn Union Gas (company)
BUGA-UP Billboard Utilizing Graffitists Against Unhealthy Promotions

Büg Nay Mong Bügd Nayramdakh Mongol (Khalka—Mongolian People's Republic)
BUH Bucharest, Rumania (airport)
BUIA British United Island Airways
buic (BUIC) backup interceptor control
BUIC Bureau (of Naval Personnel) Unit Identification Code (USN)
build. building
builds. builders
buisys barrier-up indicating system
Buk Bukit (Malay—Hill, Hilly Street)
Bukh Bukhta (Russian—Bay)
bul below upper limit; bulletin
Bul Bulgaria(n)
BUL Bombay University Library
BUL Bibliothèque de l'Université Laval (French—Laval University Library)—Québec
Bulg Bulgaria; Bulgarian
bulg bulgarisch (German—Bulgarian)
Bulgaria People's Republic of Bulgaria, Narodna Republika Bulgaria
bull. bulla (Latin—leaden seal, a papal pronouncement bearing such a seal)
Bull bulletin
BULL Bank Users Legislative Lookout
Bull Acad Sci Bulletin of the Academy of Sciences
Bull Am Astron Soc Bulletin of the American Astronomical Society
Bull Am Phys Soc Bulletin of the American Physical Society
Bull Astron Inst Neth Bulletin of the Astronomical Institutes of the Netherlands
Bull Chem Soc Jp Bulletin of the Chemical Society of Japan
bullet(s) bullet train(s)
bulli. bulliat (Latin—let it boil)
Bull NYZS Bulletin of the New York Zoological Society
bull(s) bulletin(s)
Bull Seismol Soc Am Bulletin of the Seismological Society of America
bullsh Australian contraction of bullshit
buloga business logistics game
BULVA Belfast and Ulster Licensed Vintner's Association
BUMA Bureau voor Muziek-

Auteursrecht (Dutch—Author's Rights Bureau)
BuMed Bureau of Medicine and Surgery
bump-and-run bump-and-run mugger-team technique (wherein two muggers run alongside the intended victim and as one knocks the victim to the sidewalk the other snatches the victim's handbag)
Bu M & S Bureau of Medicine and Surgery (USN)
B.U.M.S. Bachelor of Urani Medicine and Surgery
Bun Bunbury, Western Australia
BUN blood urea nitrogen
buna butadiene + natrium (synthethic rubber)
BUNAC British Universities North America Club
Bund German-American Volks-bund (pre-World War II alliance of Hitler's supporters in the U.S. now supplanted by the American Nazi Party); secret cells of Jewish Social Democrats in Russian Lithuania and Poland in 1897 who organized the General Union of Jewish Workers known as the Bund
Bund Deut Bundesrepublik Deutschland (German—Federal Republic of Germany)
bunsenite nickel oxide
Bunty Barbara
bunwich bun + sandwich (sandwich made in a bun)
BuOrd Bureau of Ordnance (USN)
bup backup plate; bull pup
BUP Boston University Press; British United Press
BUPA British United Provident Association
Bupers Bureau of Personnel (USN)
BuPers Bureau of Personnel (USN)
bupp backup plate perforated
Buppies black urban professionals
BuPubAff Bureau of Public Affairs
bur built-up roof(ing); bureau
Bur Burma; Burmese
BUR Burbank, California (Lockheed Airport)
Buran unmanned flight of the first Soviet space shuttle
'burbs suburbs
BURCEN Bureau of the Census

burd biplane ultralight research device

Burd suc Burdick suction

BuRec Bureau of Reclamation

Bur Eco Aff Bureau of Economic Affairs (US Department of State)

Bur Eur Aff Bureau of European Affairs (US Department of State)

burg burgess; burgomaster

Burg Burgess; Burgo; Burgos; Burgwald

Burg Burgtheater (German—Castle Theater) Vienna opera house

burger(s) hamburger(s)

burgle(d) burglarize(d)

burgrep burglary report

Bur Intl Aff Bureau of International Affairs

Burk Burke; Burkhardt

Burke's Burke's Peerage

Burk Fas Burkina Faso

burl. burlesque

Burl Burleigh; Burley; Burlingame

Burl N Burlington Northern and St Louis San Francisco (railway merger)

Burm Burmese

Burma Socialist Republic of the Union of Burma (mountainous Asian country between India and Malaysia)

Burn Burnell; Burnett; Burney

Burnaby Lower Mainland Regional Correctional Centre in British Columbia's Burnaby

buro bureau

burocrap bureaucratic excess

burp. backup rate of pitch

Bur Pub Aff Bureau of Public Affairs (US Department of State)

Burs Bursar

Bursting Bursting Day (February 18 when the sea ice bursts apart and crumbles in Ice-land's icy waters)

Burun Burundi; Burundian

Burundi Republic of Burundi (Central African land)

bus. business; omnibus

Bus autobus; Busan; business

BuSanda Bureau of Supplies and Accounts (USN)

BUSARB British-United States Amateur Rocket Bureau

busbar omnibus bar

buscrit business critic(ism)

BUSF British Universities' Sports Federation

BuS glands Bartholin's, urethral, Skene's glands

bush. bushing(s)

Bush George Herbert Walker Bush, 41st President and 43rd Vice President of the United States

BuShips Bureau of Ships (USN)

bus hrs business hours

busk(s) busker(s)

BUSM Boston University School of Medicine

Bus Mgr Business Manager

Busn Intl Business International

bust(ed) arrest(ed)—slang

Bus W Business Week

buswrec ban unsafe school-buses which regularly endanger children

but. butter; button

but. butyrum (Latin—butter)

BUT British United Traction

bute butazolidin (phenylbut-azone)

Buten Mus Buten Museum of Wedgewood

Butterick Butterick Publishing

Bu-Tyur Butyrskaya Tyurma (Russian—Butyrki Prison)—one of Moscow's major prisons

buv backscatter ultraviolet

buvs backscatter ultraviolet spectrometer

BUW Biblioteka Uniwersytecka w Warszawie (Polish—Warsaw University Library)

BuWeps Bureau of Weapons (USN)

buy. buyer; buying

buz buzzer

bv balanced voltage; bellows valve; biologic(al) value; blow valve; blood vessel; blood volume; bonnet valve; breviary; bronchovesicular

bv (BV) breakdown voltage

b/v brick veneer

bv bijvoorbeeld (Dutch—for example)

b.v. balneum vaporis (Latin—steambath, vapor bath)

Bv Benvenuto

B/v book value

BV Bureau Veritas (French ship-classification bureau)

B + V Blohm und Voss (ship-builders)

BV Bayerische Vereinsbank (German—Bavarian Union Bank); *Besloten Vennootschap* (Dutch—closed corporation, private partnership)

BV. Beata Virgo (Latin—Blessed Virgin); *bene vale* (Latin—a good farewell);

bene vixit (Latin—he lived a good life)

BV-202 Norwegian Army armored personnel carrier

BVA British Veterinary Association

B.V.A. Bachelor of Vocational Adjustment; Bachelor of Vocational Agriculture

BVAL Blackman's Volunteer Army of Liberation

bvbrf blood vessel of bronchial filament

BVC Buena Vista College

bvd beacon video digitizer; boys vear dem (New York City slang)

BVD Bradley, Vorhees & Day

BVD Binnenlandse Veiligheids-dienst (Dutch—Internal Security Service)—FBI-type organization in the Netherlands

BVDs suits of underwear (derived from BVD)

BVDT Brief Vestibular Disorientation Test

Bve Buenaventura

B.V.E. Bachelor of Vocational Education

Bventura Buenaventura

B/ventura Buenaventura, Colombia

b ver back verandah

BVES British Voluntary Euthanasia Society

B.Vet.Med. Bachelor of Veterinary Medicine

B.Vet.Sci. Bachelor of Veterinary Science

B.Vet.Sur. Bachelor of Veterinary Surgery

BVG Berliner Verkehrs-Betriebe (German—Berlin Traffic Carrier)—Berlin's transit system

bvh biventricular hypertrophy

BVH British Van Heusen

bvi blood vessel invasion

BVI Better Vision Institute; British Virgin Islands

Bville Bougainville (Papua New Guinea)

BVJ British Veterinary Journal

bvm broncho-vascular markings

B.V.M. Bachelor of Veterinary Medicine

B.V.M. Beata Virgo Maria (Latin—Blessed Virgin Mary)

BVMA British Valve Manufacturers Association

BVMGT Bender Visual-Motor Gestalt Test

B.V.M.S. Bachelor of Veteri-

nary Medicine and Surgery
BVN *Bund der Verfolgten des Nazi Regimes* (German—League of Persons Persecuted by the Nazi Regime)
BVNP Bolusan Volcano National Park (Luzon, Philippines)
bvo brominated vegetable oil
bvp beacon video processor; booster vacuum pump; boundary value problem
BVP British Visitors Program; British Volunteer Programme
BVPS Beacon Video Processing System; Booster Vacuum Pump System
bvr balanced valve regulator; black void reactor
BVR British Vehicle Registration (symbols appearing on automotive vehicle license plates)—*see* British Vehicle Registration Symbols *in appendix*; Bureau of Vocational Rehabilitation
BVRO Base Vehicle Reporting Officer
BVRR Bureau of Veterans Reemployment Rights
BVRS Breadboard Visual Reference System
BVS Best Vested Socialists; Bevier & Southern (railroad)
B-V S Brisch-Vistem System (Visican punched-cards)
B.V.S. Bachelor of Veterinary Science; Bachelor of Veterinary Surgery
B.V.Sc. Bachelor of Veterinary Science
B.V.Sc. & A.H. Bachelor of Veterinary Science and Animal Husbandry
bvt brevet; brevetted
bvv bovine vaginitis virus
bvw binary voltage weigher
bw best of winners; biological warfare (BW); birth weight; body water; body weight; both ways; braided wire (armor)
b/w black-and-white
b & w black and white; bread and water
bw *bijwoord* (Dutch—adverb); *bitte wenden* (German—please turn over)
bW blood Wassermann
BW Bendix-Westinghouse; Biological Warfare; biological weapon; Black Watch; Borg-Warner; Business Week
B-W Bendix Westinghouse Automotive Air Brake; Borg-Warner

B & W Babcock and Wilcox; Barker and Williamson; Burmeister and Wain
B of W Bishop(ric) of Würzburg
BW *Bitte Wenden* (German—please turn over); *Business Week*
bwa backward-wave amplifier; bent-wire antenna
BWA Baptist World Alliance; Baseball Writers Association; Boxing Writers Association; British West Africa; Building Waterproofers Association
BWAL Barber West African Line
Bway Broadway
BWB British Waterways Board
BWB *Bundestampt für Wehrtechnik und Beschaffung* (German—Federal Office for Military Technology and Procurement)
bwc basic weight calculator; broadband waveguide oscillator
BWC Battered Women's Coalition; British War Cabinet
BWCA Boundary Waters Canoe Area
BWCC British Weed Control Conference
bwcdi best we can do is
BWCI Beauty Without Cruelty, Incorporated
bwcp bench welder control panel
bw-cw biological warfare—chemical warfare
bwd bacillary white diarrhea; backward; barrels of water per day
BWD Baldwin Wallace College; British War Cabinet
B & WE Bristol and West of England
BWF Baha'i World Faith; Beyond War Foundation
Bwg Bowling
BWG Birmingham Wire Gage
bwh barrels of water per hour
BWH Book Week Headquarters
BWI Baltimore Washington International (airport); British West Indies
bwia better walk if able
BWIA British West Indian Airways
BWI$ British West Indian dollar
BWIP Basalt Waste Isolation Project
BWIPO Basalt Waste Isolation Project Office

BWIR British West India Regiment
BWISA British West Indies Sugar Association
BWIU Building Workers Industrial Union
bwk brickwork; bulwark
bwl belt work line
BWL Biological War Laboratory
bwlt bow light
bwm barrels of water per minute
BWM British War Medal; Broom and Whisk Makers (union)
BWMA British Woodwork Manufacturers Association
BWMB British Wool Marketing Board
BWN Basic Weather Network; Brown Company (stock-exchange symbol)
bwo backward-wave oscillator
bwoc big woman on campus
B'worth Butterworth
bwos backward-wave oscillator synchronizer
bwot backward-wave oscillator tube
bwp ballistic wind plotter
BWP Basic War Plan
bwpa backward-wave parametric amplifier
BWPA British Wood Preserving Association; British Wood Pulp Association
bwpd barrels of water per day
bwph barrels of water per hour
bwr (BWR) boiling-water reactor
BWRA British Water Research Association; British Welding Research Association
BWRC Biological Warfare Research Center
BWRWS Biological Warfare Rapid Warning System (USA)
bws beveled wood siding
BWS Bandipur Wildlife Sanctuary (India); Bank of Western Samoa; Batch Weighing System; Battered Women's Service; Battlefield Weapons System; Beaufort Wind Scale; Better World Society; Biological Weapons System; British Watercolour Society
BW & S Boyd, Weir & Sewell
BWSF British Water Ski Federation
BWSL Battlefield Weapons Systems Laboratory
bwso backward wave sweep oscillator

BWSR Bruno Walter Society Recording(s)
bwt both-way trunk
BWT Boeing Wind Tunnel
BWTA British Women's Temperance Association
BWTP Bureau of Work-Training Programs
bw-tv black-and-white television
bwu blue whale unit
bwv back-water valve
BWVA British War Veterans of America
bwvs black-and-white vertical stripes
BWW Bad Weather Watch (Coast Guard)
BWWA British Water Works Association
bx biopsy; box; electrical cable contained in flexible tubing (bx cable)
Bx Beatrix; Box (post-office box); Brix; Bronx
BX Base Exchange (USAF); Bellingham-Seattle Airways (2-letter code)
Bx-arts Beaux-arts (French—fine arts)
bx cable insulated wires within flexible tubing
bxd boxed
bxk broadband X-band klystron
bx k box keel
BXL Bakelite Xylonite Limited
Bxm Brixham
Bxng D Boxing Day (first weekday after Christmas in England, Wales, Northern Ireland, Canada, South Africa, New Zealand, and Australia when boxes of gifts are given lettercarriers, newspaper carriers, and all others who provide services)
Bx Pk Bronx Park
bxs boxes
by. billion years; brilliant yellow (litmus paper for testing alkalinity)
b-y bloody
By Buryat(ic); Byron(ic)
BY blowing spray
BYC Baltimore Yacht Club; Bayside Yacht Club; Bensonhurst Yacht Club; Beverley Yacht Club; Boston Yacht Club; Brewers Yeast Council; Bridgeport Yacht Club; Bronx Yacht Club; Buffalo Yacht Club
bydv barley yellow dwarf virus
Bye Byelorussia; Byelorussian
Byelorussia White Russia bordering on Latvia, Lithuania, and Poland
byfml by first mail
byg buying
byo bring your own
Byo Bulawayo
byob bring your own beer
byod bring your own drinks
byog bring your own girl
byp bypass
Byp Bypass
bypro(s) by-product(s)
byr(s) billion year(s)
byssin byssinosis
byt bright young things (British younger set)

byte eight adjacent binary digits that a computer processes as a unit
Byu Bayou (postal place-name abbreviation)
BYU Brigham Young University
Byz Byzantine
Byzantine Empire eastern segment of the Roman Empire
bz blank when zero; buzzer; (cartoonist's symbol—buzzing, sawing, snoring)
Bz benzene; benzodiazepine; benzoyl; Brazil; Brazilian
Bz Beobachtungszimmer (German—examining room)—hospital observation room
BZ Air Congo (Brazzaville, Congo Republic); B'nai Zion
B/Z British Zone
BZ Bild Zeitung (German—Picture Newspaper)
Bza Bizerta
BZA Board of Zoning Adjustment
bz brigade (Danish—*besaet brigade*)—occupiers claiming squatter's rights on vacant property
bzbx brazing box
Bze Belize
bzfm brazing form
bzfx brazing fixture
Bzi Benghazi
bzw beziehungsweise (German—respectively)
bzz cartoonist's symbol—buzzing; sawing; snoring
bzzz same as bzz

C

c calorie (large); candle; canine; capacity; carbon; catcher; cathode; caudal; cent; cen-tavo; center; centi (prefix); centime; centimeter; central; certified; cervical; cervix; chapter; charm; chest; child; chord length (symbol); cirrus; clearance; clonus; closure; coarse; cocaine; coefficient; cold; *colón; colones* (currency in Costa Rica and El Salvador); color; colored; common; complement; conductor; contact; contraction; control; cortex; cranial; crystal(line); cube; cubic; cubical; cycle(s); cylinder(s); cytidine; cytochrome; cytosine; heat capacity per mole (symbol); see; speed of light (symbol)

c (C) convict

c. *cibus* (Latin—meal); *circa* (Latin—about); *congius* (Latin—gallon); *cum* (Latin—with)

c/ *cargo* (Spanish—total, weight); *contra* (Spanish—against, versus)

C calculated weight (symbol); candle; capacitance; capacitor; Cape; carat; carbon; Cardinal; cargo or transport airplane; cargo vessel; carton; case; cathode; cavalry; celestial; Celsius; Celtic; Centigrade; century; cervical; chairman; Charlie—code for letter C; Chief; Christ(ian); coast; cocaine; cold; college; colored; combat aircraft; commander; compliance; computer programming language; concentration; Con-

servative; consul; control; Convair; copyright; Cosmopolitan Shipping; coulomb; council; course; Curie's constant; Fraunhofer line characteristic of hydrogen (symbol); hundredweight (symbol); molecular heat (symbol); see (popular phonetic spelling)

C. carbohydrates; cocaine; Conservative (political party)

"C" Costa Line

°C degree Celsius; degree centigrade

C *Cabo* (Spanishcape); *Cap* (Frenchcape); *centum* (Latin—one hundred); (French or Italian—highpass, pass); (Latin—Gaius)

C+ C-plus

C++ computer programming language

C- C-minus

C° *Comisario* (Spanish—Commisariat)

C₁ first class

Cᴵ bacteriologic complement

c 1° *canto primo* (Italian—first soprano part or voice)

C¹1, C¹2, C¹3, etc. complements of complements

C 1, C 2, C 3, etc. cervical nerves or vertebrae 1, 2, 3, etc.

C I, C II, C III, etc. cranial nerves I, II, III, etc.

C₁, C₂, C₃, etc. cytochromes 1, 2, 3, etc.

C₂ second class

C² D² (ARDC) Command and Control Development Division

C-3 mentally or physically defective (British equivalent of

American 4-F)

C₃ command, control, communications; third class

C³I Command, Control, Communications, and Intelligence

C.3.3. cell 3, 3rd landing, gallery C (occupied by Oscar Wilde while in *Reading Gaol* and the nom de plume he used there)

C3S College Chemistry Consultants Service

C4 Convair 440 airplane; crown quarto (7-½ x 10 inches)

C-4 military explosive

C5 Convair 580 turboprop airplane

C-5A Lockheed military cargo transport airplane

C-6 hexamethonium

C₆H₆ benzene

C8 crown octavo (5 x 7-½ inches)

c8ᵛᵃ *coll'ottava* (Italian—in octaves)

C-9 McDonnell-Douglas twin-engine jetliner designed for medical evacuation, named Nightingale to honor Florence Nightingale

C-10 decamethonium

C₁₂H₂₂O₁₁ cane sugar

C¹⁴ radioactive carbon (used in determining age of objects)

C₁₇H₂₁NO₄ cocaine (also known as blow, coke, flake, freeze, happy dust, nose candy, lady, Peruvian, white girl)—derived from the leaves of the coca plant (*Erythroxylon coca*)

C 19 ster steroids containing 19 carbon atoms

C 21 ster steroids containing

21 carbon atoms

C 33 Oscar Wilde's identification number in Reading Gaol

C-42 Brazilian Neiva four-seat utility aircraft called Regente

C-45 Beechcraft four-passenger transport plane

C-46 Curtiss-Wright World War II Commando 36-passenger transport

C-47 Douglas DC-3 Dakota or Skytrain 21-passenger air transport

C-54 Douglas DC-4 44-passenger transport called Skymaster

C-95 Brazilian 12-passenger transport aircraft named Bandeirante honoring frontier pioneers

C-118 Douglas DC-6 92-passenger transport also called Liftmaster

C-119 Fairchild-Hiller Flying Boxcar carrying 62 paratroopers or an equal weight of cargo

C-121 Lockheed Constellation or Super-Constellation transport carrying 63 or 99 passengers, respectively

C-123 Provider twin-engine assault transport

C-124 Globemaster heavy cargo four-engine transport airplane

C-130 Hercules medium-range cargo and troop transport airplane powered by four turboprop engines; Lockheed four-engine transport aircraft for military use

C-131 Convair 48-passenger military transport adapted from 24/440 commercial airliners

C-133 Cargomaster heavy four-engine turboprop cargo transport airplane

C-135 Boeing Stratofreighter military transport carrying 126 troops or equivalent cargo

C-140 Jet Star support-type transport aircraft powered by four turbojet engines

C-141 Starlifter large cargo transport airplane powered by four turbojet engines

C-212 Casa 15-seat aero medical or paratrooper transport plane made in Spain and called Aviocar

¢ cedi—monetary unit of Ghana; centmonetary unit of the United States; colon

monetary unit of Costa Rica and El Salvador

ca cable; calibrated altitude; cancer; capital asset; carbonic anhydrase; carcinoma; cardiac arrest; cathode; caudal; cen-tare; cervoaxial; chronological age; civil affairs; civil authorities; clerical aptitude; cold agglutinin; common antigen; convening authority; coronary artery; council accepted; covert action; croup associated; current assets

ca (CA) cancer; carcinoma

ca' calf; call (Scottish contraction)

c/a capital account; center angle; coated abrasive; current account

c&a (C&A) command and administration

c & a classification and audit

ca *circa* (Latin—about); *corrente alternada* (Portuguese—alternating current); *corriente alterna* (Spanish—aternating current)

cᵃ *compañia* (Spanish—company)

c a *coll'arco* (Italian—to be bowed)

Ca calcium; Canada; Canadian

Ca *Compagnia* (Italian—company)

Ca' *Casa* (Venetian—house)

Cᵃ *Cabeça* (Portuguese—head, headland); *Companhia* (Portuguese—company); *Compañía* (Spanish—company)

CA California; Canadian Army; Capital Airlines; Central America; Certificate of Airworthiness; Chargé d'Affaires; Chartered Accountant; Chemical Abstracts; Chief Accountant; Church Army; Civil Affairs; Coast Artillery; Cocaine Anonymous; Combat Aircrew; Combat Aircrewman; Commercial Agent; *Companhia de Navegação Carregadores Açoreanos* (Azore Line); Compensation Act; Comptroller of the Army; Confederate Army; Construction Authority; Construction Authorization; Consular Agent; Controlled Atmosphere; Convening Authority; Coordinating Agency; County Attorney; Court of Appeals; Cranial Academy; heavy cruiser (naval symbol)

CA (Aust) Institute of Chartered Accountants in Australia

C.A. Chartered Accountant

C & A Clemens and August Breeninkmeyer's international house of fashion

C of A College of Aeronautics; Commonwealth of Australia

CA *Centre Agricole* (French—Agricultural Center)—prison farm; *Chemical Abstracts; corriente alterna* (Spanish—alternating current)

caa caging amplifier assembly; circular aperture antenna; computer amplifier alarm; crime aboard aircraft

CAA Canadian Authors' Association; Canadian Automobile Association; Cantors Assembly of America; Caribbean Atlantic Airlines; Central African Airways; Chester Alan Arthur (21st President U.S.); Chief of Army Aviation; Civil Aeronautics Administration; Civil Aeronautics Authority; Civil Aviation Authority; Clean Air Act; Collectors of American Art; College Art Association; Colonial Athletic Association; Commercial Apiarists Association; Community Action Agencies; Community Aid Abroad; Correctional Administrators Association; Council on American Affairs; Cremation Association of America; Custom Agents Association

C.A.A. Civil Aviation Authority (United Kingdom)

CAA *Clean Air Act; Congressional Assassination Act*

CAAA Canadian Association of Advertising Agencies; College Art Association of America; Composers, Authors, and Artists of America

CAAB California Avocado Advisory Board

CAABU Council for the Advancement of Arab-British Understanding

CAAC China Aeronautics Airline Corporation (nicknamed China Airlines Always Cancels); Civil Aviation Administration of China; Customs and Allied Affairs Committee

CAADRP Civil Aircraft Airworthiness Data Recording Program (UK)

CAAE Canadian Association

for Adult Education
CAAFS Chinese Academy of Agricultural and Forestry Sciences
CAAIS Computer-Assisted Action Information System(s)
caar compressed-air-accumulator rocket
CAAR Committee Against Academic Repression
CAARC Commonwealth Advisory Aeronautical Research Council
CAAs Community Action Agencies
CAAS Ceylon Association for the Advancement of Science; Civil Aviation Authority of Singapore; Computer-Assisted Acquisition System; Connecticut Academy of Arts and Sciences
CAASE Computer-Assisted Area Source Emissions
CAAT Campaign Against Arms Trade; Canadian Academic Aptitude Test; Colleges of Applied Arts and Technology
CA Att Civil Air Attaché
CAAV Central Association of Agricultural Valuers
cab cabal; cabbage; cabin; cabinet; cable; cabochon; cabriolet; calibration; captured air bubble; cellulose acetate butyrate; taxicab
cab (CAB) cellulose acetate butyrate; coronary artery bypass
Cab Cabell; Cabot; NATO nickname for Soviet Lisunov transport plane designated Li-2
CAB Canadian Association of Broadcasters; Charles A(ustin) Beard; Circulation Audit Board; Citizens Advice Bureau; Civil Aeronautics Board; Civil Aeronautics Bulletin; Clark Air Base; Commonwealth Agricultural Bureau; Consumer Affairs Bureau; Contract Appeals Board (Veterans Administration)
CABA Charge Account Bankers Association
cabaf currency and bunker adjustment factor
cabal *cabbala* (Hebrew—something secret)—(*see also* CABAL)
CABAS City and Borough Architects Society
CABB Captured Air-Bubble

Boat (naval)
CABE California Association of Bilingual Education
CABEI Central American Bank for Economic Integration
CABEZA California and Baja California Enterprise Zone Authority
CABIC Copper and Brass Information Centre (Australia)
CABIN Campaign Against Building Industry Nationalization
CABLE Computer-Assisted Bay Area Law Enforcement (San Francisco)
cablecast broadcast by cable tv; cablecaster; cablecasting
cablecast(ing) cable television telecast(ing)
cablese cablegram language (abbreviated, telegraphic, truncated style)
cable tv community-antenna television
cablevision cable television
CABM Commonwealth of Australia Bureau of Meteorology
CABMA Canadian Association of British Manufacturers and Agencies
CABMS Chinese-oriented Antiballistic Missile System
Cabo Cabo San Lucas, Baja California; San José del Cabo, Baja California
CABO Council of American Building Officials
cabot cabotage (coastal navigation)
Ca bp calcium-binding protein
CABRA Copper and Brass Research Association
Cabral Pedro Álvarez Cabral, discoverer of Brazil
cabs. cabbages
cab(s) cabochon(s)
CABS Children's Adaptive Behavior Scale; Computer-Augmented Block System; Computerized Annotated Bibliographic System
cabtmkr cabinetmaker
CABWA Copper and Brass Warehouse Association
cac cardiac-accelerator center
Cac Caceres
CAC California Administration Code; California Advisory Council (on Vocational Education); California Aeronautics Commission; California Arts Commission; California Arts Council; Canadian Ar-

moured Corps; Central Arbitration Committee; Chief of Air Corps; City Administration Center; Civic Administration Center; Civil Administration Commission; Coast Artillery Corps; College Admissions Center; Colonial Ammunition Company; Combat Air Crew; Commander Air Center; Commission of Accreditation for Corrections; Community Administration Council; Commuter Aircraft Corporation; Consumer Advisory Council; Consumer Affairs Council; Consumers' Association of Canada; Continental Air Command; Corrective Action Commission; Corrective Action Committee; Cosmetology Accrediting Commission; County Administration Center
CAC *Comité de Acción Cultural* (Spanish—Cultural Action Committee)
CACA Canadian Agricultural Chemicals Association; Central After-Care Association
cacb compressed-air circuit breaker
CACB Council Against Cigarette Bootlegging
cacc cathodal closure contraction
CACC Civil Aviation Communications Center; Corrective Action Control Section; Council for the Accreditation of Correspondence Colleges
CA-CC Christian Anti-Communist Crusade
CACCE Council of American Chambers of Commerce in Europe
CACCI Confederation of Asian Chambers of Commerce and Industry
CACDA Combined Arms Combat Development Activity
CACE California Association for Childhood Education; Chicago Association of Consulting Engineers
CACEX *Carteira do Comercio Exterior* (Portuguese—Foreign Commerce Department)—Bank of Brazil
CACF Colombian-American Culture Foundation
Cach *Cachoeira* (Portuguese—rapids, waterfall)
cache. computer-controlled au-

tomated cargo-handling envelope

CACHE Computer Aids for Chemical Engineering Education

cachi cachivache (Spanish—broken crockery, foolish or worthless person, poor quality, pots and pans, utensils)—dialect heard around Buenos Aires also called *porteño*

CACJ California Attorneys for Criminal Justice

CACL Canadian Association of Children's Librarians

CACM Central American Common Market

Caco Cacoliche (pidgin Argentine-Spanish including many Italian words)

CACO Casualty Assistance Call Office(r)

CaCO₃ calcium carbonate (limestone)

cacoph cacaphonic; cacophony

cacp cartridge-actuated compaction press

CACS California Aqueduct Control System

CACSW Citizens' Advisory Council on the Status of Women

CACTUS Capteur Accelerometrique Capacitif Triaxial Ultra-Sensible (French—Ultra-Sensitive Triaxial Capacitive Accelerometric Detector)

CACUL Canadian Association of College and University Libraries

CACVE California Advisory Council on Vocational Education

CAC & W Continental Aircraft Control and Warning

cad. cadastral; cadaver; caddie; cadenza; cadet; cadmium; cartridge-activated device; cartridge-actuated device; cash against disbursements; cash against documents; computer-aided design; contract award date

cad. (CAD) computer-aided design

c.a.d. cash against disbursements

cad cadenza (Italian—solo passage near end of a concerto movement)

c-a-d c'est-à-dire (French—that is to say)

Cad Cadell; Cadiz; Cadmar; Cadmus; Cadogan; Cadwalader or Cadwallader (often pronounced *Calder*)

CAD California Association of the Deaf; Civil Air Defense; Claude Archille Debussy; Combat Air Division; Commission Against Discrimination; Computer-Aided Dispatch (police); Computer-Assisted Design; Consumer Affairs Department; Crown Agents Department

cada clean air dot angle

CADA Centre d'Analyse Documentaire pour l'Archéologie (French—Document Analysis Center—Archaeology)

CADAFE Compañía Anónima de Administración y Fomento Electrico (Spanish—Corporation for Electrical Administration and Development)

CADAM Computer-graphic Augmented Design and Manufacturing (registered trademark of Cadam, Inc.)

CADAN Centre d'Analyse Documentaire pour Afrique Noir (French—Document Analysis Center—Africa)

cadav cadaver(ous)

cadc central air data computer

CADC Continental Air Defense Command; Corrective Action Data Center

cad/cam computer-aided design/computer-aided manufacturing

cadco core and drum corrector

cadd computer-aided design drafting

Caddie Charlotte

Cad(dy) Cadillac

cade computer-aided design engineering; computer-aided design evaluation; computer assisted data engineering; computer-assisted data evaluation

cadet. computer-aided design experimental translator

Cadet old Russian acronym for Constitutional Democratic Party or one of its members

Cadets Constitutional Democrats (in czarist Russia)

cadf commutated antenna direction finder

CADF Central Air Defense Force; Contract Administrative Data File

cadfiss computation and data flow integrated subsystems

'Cadian(s) Acadian(s)

CADIG Coventry and District Information Group

CADIN Continental Air Defense Integration North

cadis coronary artery disease

CADIZ Canadian Air Defense Identification Zone

CADL Christian Anti-Defamation League

CADM CONUS (Continental United States) Air Defense Modernization

CADO Central Air Documents Office (USAF); Central American Development Organization; Current Actions Duty Office(r)

Ca'd'Oro Casa de Oro (Italian—House of Gold)

'cado(s) avocado(s)

CADPIN Customs Automatic Data Processing Intelligence Network (U.S. Bureau of Customs)

CADPOS Communications and Data Processing Operation System

cadre. current awareness and document retrieval for engineers

cads. cellular-absorbed-dose spectrometer

CADS Central Air Data System (USAF); Containerized Ammunition Distribution System (USA)

cadss combined analog-digital systems simulator

CADSYS Computer-Aided Design System

cadte cathodal duration tetanus

Cadwal Cadwallader

cae carrier aircraft equipment; computer-assisted electrocardiography; computer-assisted enrollment

Cae Caelum

CAE Canadian Aviation Electronics; Columbia, South Carolina (airport); Common Applications Environment; Council on Anthropology and Education

CA&E Council on Anthropology and Education

CAE Cóbrese al Entregar (Spanish—cash on delivery)

CAEA California Aviation Education Association; Chartered Auctioneers and Estate Agents

CAEAI Chartered Auctioneers and Estate Agents Institute

CAED Canadian Association of Equipment Dealers

Ca edta calcium disodium ethylene diamine tetra-acetate

Cael Caelum

CAEL Council for the Advancement of Experimental

Learning

CAEM *Conseil d'Assistance Economique Mutuelle* (French—Council for Mutual Economic Assistance)

Caer (Cornish or Welsh—fortress)—Caermarthen, Caernarvon, Caerphilly, Caerwent, and Caerwys—the Caers

CAER Centre for Applied Economic Research (New South Wales)

Caern Caernarvonshire

caerul. *caeruleus* (Latin—cerulian)—sky blue

caes compressed-air energy storage

Caes Caius Julius Ceasar

CAES Canadian Agricultural Economics Society; Connecticut Agricultural Experiment Station

caesar computerized automation by electronic system with automated reservations

CAET Corrective Action Evaluation Team

CAEU Council of Arab Economic Unity

CAEWW Carrier Airborne Early Warning Wing (USN)

caf cafeteria; caffeine; clerical, administrative, and fiscal; cost and freight; cost, assurance, and freight; currency-adjustment factor

caf *coût, assurance, fret* (French—cost, assurance, freight)

CAF Canadian Armed Forces; Central African Federation; Ceylon Air Force; Citizen Air Force; Conventional Armed Forces

CAF *Corporación Andina de Fomento* (Spanish—Andean Promotion Corporation)

CAFA Chicago Academy of Fine Arts

CAFB Canadian Air Force Base; Clark Air Force Base

CAFCINZ Campaign Against Foreign Control in New Zealand

cafd contact analog flight display

cafe (CAFE) corporate average fuel economy

cafe. computer-aided film editor; corporate average fuel economy

CAFE Conventional Armed Forces in Europe; Corporate Average Fuel Economy

CAFEA-ICC Commission on Asian and Far Eastern Affairs—International Chamber of Commerce

CAFEI Central American Fund for Economic Integration

cafetorium cafeteria-auditorium

caff caffeine

cafga computer applications for the graphic arts

CAFGU Citizens Armed Forces Geographical Unit (Philippines)

CAFI Commercial Advisory Foundation in Indonesia

CAFIC Combined Allied Forces Information Center

CAFIT Computer-Assisted Fault Isolation Test(ing)

cafm commercial air freight movement

CAFMS Continental Association of Funeral and Memorial Societies

CAFO Command Accounting and Finance Office

CAFR Comparative Annual Financial Report

C Afr Fed Central African Federation

CAFS Cartridge-Actuated Flame System

CAFSC Control Air Force Specialty Code

CAFTA Council of Australian Food Technology Associations

CAFU Civil Aviation Flying Unit

cag chronic atrophic gastritis; constant aerial glide; constant altitude glide

Cag Cagliari; Cagliostro

CAG Carrier Air Group; Civil Air Guard; Composers-Authors Guild; Computer Analysis Group; Concert Artist Guild; Corrective Action Group; heavy guided-missile cruiser (naval symbol)

CAGA California Asparagus Growers Association; Commercial and General Acceptance

CAGE Convicts' Association for a Good Environment

cagel consolidated aerospace ground equipment list

CAGEO Council of Australian Government Employee Organisations

CAGI Compressed Air and Gas Institute

CAGS Canadian Arctic Gas Study

cah congenital adrenal hyperplasia

CAH Community of All Hallows; Conzinc Asia Holdings

cahd coronary atherosclerotic heart disease

CAHOF Canadian Aviation Hall of Fame

CAHS Comprehensive Automation of the Hydrometeorological Service

CAHT Canadian Association for Humane Trapping

cai computer-aided instruction; computer-assisted instruction; confused artificial insemination

Cai Cairo; Gonville and Caius College, Cambridge

C-a I Computer-assisted Instruction

CAI Canadian Aeronautical Institute; Canadian Airlines International; Career Assessment Inventory; Computer Applications Incorporated; Confederation of Australian Industry; Configuration Audit Inspection (USA); Container Aid International; Cruelty to Animals Inspectorate; Culinary Arts Institute

CAI *Club Alpino Italiano* (Italian—Italian Alpine Club)

CAIA Customs Agents Institute of Australia

CAIB Certified Associate of the Institute of Bankers

caic computer-assisted indexing and classification

CAIC Civil Aviation Information Circular; Coalition for the Apparel Industry of California

Cai Col Gonville and Caius College—Cambridge

Caicos Caicos and Turks Islands (in British West Indies southeast of the Bahamas and north of Hispaniola)

CAIG Canadian Aircraft Insurance Group

CAIL Coal and Allied Industries Limited

CAIMAW Canadian Association of Industrial, Mechanical, and Allied Workers

CAIN CAtaloging-INdexing (National Agricultural Library data base)

CAINS Carrier Aircraft Inertial System

caint counter-air and interdiction

caiop computer analog input-output

CAIP Computer-Assisted Indexing Program (UN)

CAIQ Chamber of Automotive Industries of Queensland

CAIRA Central Automated Inventory and Referral Activity (USAF)

CAirC Caribbean Air Command

CAIRS Central Automated Inventory and Referral System (USAF); Computer-Assisted Information Retrieval System; Computer-Assisted Interactive Resources Scheduling System

CAIS Canadian Association for Information Science; Center for Advanced International Studies (Univ of Miami); Central Abstracting and Indexing Service; Connecticut Association of Independent Schools

Caith Caithness

CAITS Chemical Agent Identification Training Set

caj calked joint

CAJ Center for Administrative Justice; Confederation of ASEAN Journalists

caje consolidated antijam equipment

'cajun Acadian (native of Louisiana)

cak conical alignment kit; cube alignment kit

CAK Akron, Ohio (airport)

CAK-C Concept Assessment Kit—Conservation

cal caliber; calorie (small); computer-aided learning; computer-assisted learning; conversational algebraic language

cal calando (Italian—calming); *carbine automatique légère* (French—light automatic carbine)—*CAL*

Cal Calabar; Calabozo; Calabria; Calafat; Calahan; Calais; Calamar; Calbert; Calcutta; Caldecott; Calder; Caldwell; Cale; Caleb; Caledonia(n); Calgary; Calhoun; Caliente; California; Calixto; Calkins; Call; Callaghan; Callahan; Callao; Callcott; Callyhan; Calorie (large); Calpurnius; Calumet; Calvagh; Calvary; Calven; Calvert; Calvin; Calvus

CAL Catholic Aviation League; Center for Applied Linguistics; China Airlines; Citizens Action League;

Commonwealth Acoustic Laboratories; Computer Accounting Limited; Computer-Assisted Learning; Conference-Approved Literature (Australia); Continental Airlines; Conversational Algebraic Language; Cornell Aeronautical Laboratory; Cyprus Airways; Point Arguello (California) tracking station

CAL Comandos Armados por Liberación (Spanish—Armed Commandos for Liberation)—Puerto Rican underground militants fighting for decolonialization

cala calabozo (Spanish—cell, dungeon, jail)

CALA Chinese-American Librarians Association; Civil Aviation Licensing Act

calaham California ham (picnic ham)

calamine smithsonite (zinc carbonate)

CALANS Caribbean and Latin American News Service

CalArts California Institute of the Arts

CALAS Computer-Assisted Language Analysis System

calb computer-assisted line balancing

C_{alb} albumin clearance

calbr calibration

calc calculate(d); calculation; calculator; calculus

calc (CALC) calculate; calculator (flow chart)

Calc Calcutta

Calc Calçada (Portuguese—Street)

Calcasieu Calcasieu Lake or Calcasieu Pass in southwestern Louisiana where the lake waters flow into the Gulf of Mexico

calcd calculated

CALCOFI California Cooperative Oceanic Fishery Investigation

Calcomp California Computer Products

Calc Univ Calcutta University

cald calculated; caldera

CALDA Canadian Air Line Dispatchers Association

CALDAC California Debt Advisory Commission

CALDEA California Driver Education Association

Calder Cadwalader; Cadwallader

CALE Canadian Army Liaison

Executive

CALEA Canadian Air Line Employees Association; Commission on Accreditation for Law Enforcement Agencies

Caled Caledonia

Caled Can Caledonian Canal

Caledonia (Latin—Scotland)

Caledonian Caledonian Canal bisecting northern Scotland and connecting the Atlantic Ocean with the North Sea; pertaining to Scotland and things Scottish

calef. *calefactus* (Latin—warmed)

calen calendar; calender

CALEV Compañía Anómina Luz Electrica de Venezuela (Spanish—Electric Light of Venezuela Corporation)

Calex Calexico (California border city)

Cal Expo California Exposition (permanent show at Sacramento)

CALFAA Canadian Air Line Flight Attendants Association

calfin calendered finish

Calg Calgary

Calhan Calahan

Calhoun Calquahoun; originally Colquhoun

calib calibrate; calibration

calibn calibration

caliche calcium carbonate crust (or) dust—$CaCO_3$

Caliente Agua Caliente, Mexico; Nevada (delinquent) Girls Training Center at Caliente, Nevada; racetrack town adjacent to Tijuana, in Baja California

Cal-ID California Identification System

Calif California; Californian

CALIF California

Calif Cur California Current

Calif Hist California Historical Society

Calif Rev Pr California Review Press

Calipuerto Cali Aeropuerto (Cali, Colombia)

cal_{It} calorie (International Table calorie)

CALIT California Institute of Technology (also Caltech or CIT)

CALL Canadian Association of Law Libraries; Community Access Library Line; Community Action for Limited Learners; Composite Aeronautical Load List(ing);

Counselling at the Local Level (SBA)

callas calla lillies

Calli Callimachus of Alexandria (bibliographer-poet-scholar)

callig calligrapher; calligraphic; calligraphy

call-in call-in radio or television program soliciting audience participation; call-in telephone call advising of an anticipated absence due to illness, etc.

calm. collected algorithms for learning machines

calm calmato (Italian—calmly; tranquilly)

CALM Child Abuse Listening Mediation; Citizens Against Legalized Murder; Computer-Assisted Library Mechanization

CALMA California Marine Associates

Cal Maritime California Maritime Academy

Calmex California-Mexico

CALMS Computer Automatic Line Monitoring System

caln calculation

calo calando (Italian—softer and slower, bit by bit)

calogsim computer-assisted logistics simulation

calomel mercurous chloride (Hg_2Cl_2)

CAL/OSHA California Occupational Safety and Health Administration

CALPA Canadian Air Line Pilots Association

CALPIRG California Public Interest Research Group

Cal Poly California Polytechnic

CALRI Central Artificial Leather Research Institute

Cal Rptr California Reporter

CALS Canadian Association of Library Schools

CALSO California Transport

CalSpace California Space Institute

CalTec California Institute of Technology

CALTEX California-Texas Petroleum; Overseas Tankship Corporation

cal$_{th}$ calorie (thermochemical calorie)

CALTIP Californians Turn In Poachers (who fish with two poles or hunt out of season)

CALTRAC California Track

Caltrans California Department of Transportation

CALURA Corporation and Labour Unions Returns Act

Calv Calvin; Calvinism; Calvinist

Calvary Calvary Hill outside Jerusalem where its Aramaic name is Golgotha—Place of the Skull

Cal-VDAC California Venereal Disease Advisory Council

Calz Calzada (Spanish—boulevard, highway)

cam. camber; camouflage; circular area method; cockpit area microphone; commercial air movement; comprehensive achievement monitoring; computer-addressed memory

cam. (CAM) central address memory; checkout and automatic monitoring; computer-aided manufacturing

ca'm calm

Cam Camaguey; Cambodia; Cambodian; Camden; Camelopardalis (constellation); Cameron; Cameroons; Campbell; Campechanos; Campeche

C$_{am}$ amylase clearance

CAM Certified Administration Manager; Civil Aeronautics Manual; Civil Aviation Medicine; Composite Army-Marine; Computer-Aided Management; Computer-Aided Manufacture; Consumers Action Movement; Contract Air Mail; Contract Audit Manual; Cooperative Autoworks Malaysia; Cost-Account Manager; Course-A-Month

cama centralized automatic message accounting

CAMA Children's Aid Movement of Australia; Civil Aerospace Medical Association

camal (CAMAL) continuous air borne missle alert

C'Amalie Charlotte Amalie

CAMALS Cambridge Algebra System

camb camber; cambium; cambric linen or tea

Camb Cambrian; Cambridge

Cambod Cambodia; Cambodian

Cambodia former French Indo-Chinese colony, Peoples Republic of Kampuchea

Cambrian Cambrian Airways

Cambrians Cambrian Mountains, Wales

Cambridge UP Cambridge University Press

Cambs Cambridgeshire and Isle of Ely

CAMC Canadian Army Medical Corps; Cartoon Art Museum of California

camcorder camera and tape recorder device

CAMDA Car and Motorcycle Drivers Association

camel. common automatic manifest language; computer-assisted machine loading

CAMEO Capitol Area Motion Pictures Education Organization (D.C.)

camera. cooperating agency method for event reporting and analysis

CAMERA Committee for Accuracy in Middle East Reporting in America

Cameroon United Republic of Cameroon (equatorial African nation); *République Unie du Cameroun*

CAMESA Canadian Military Electronics Standards Agency

Cam High Camden High School; Cameron Highlanders

CAMI Civil Aeromedical Institute; Columbia Artists Management, Incorporated

camiknick camisole-knicker-bocker combination

Camillo Escamillo

CAMIS Computer-Assisted Makeup and Imaging System

Caml Camelopardus

CAML Canadian Association of Music Libraries

CAMM Canadian Association of Medical Microbiologists

CAMMIS Command Aircraft Maintenance Manpower Information System

camof camouflage

camol computer-assisted management of learning

camp. computer-assisted menu planning; cosmopolitan art—modern and personalized; cyclic adenosine monophosphate

cAMP cyclic adenosine 3',5'-monophosphate

Camp Campeche (inhabitants—Campechanos); Campion Hall, Oxford

CAMP Campaign Against Marijuana Planting; Campaign Against Moral Persecution; College-Assistance Mi-

grant Program; Computer Applications of Military Problems; Continuous Air Monitoring Program; Course-A-Month Program; Craft Attitude Monitoring Package
campan campanological; campanologist; campanology
Camp Hall Campion Hall—Oxford
Campoformido Campo Formio
CAMPS Cooperative Area Manpower Planning Systems
CAMPSA Compañia Arrendataria del Monopolio de Petroleos (Spanish—Petroleum Company)
CAMP Test Christie-Atkins-Munch-Peterson Test
CAMPUS Comprehensive Analytical Methods for Planning in University Systems
CAMQAB Consumer Affairs Medical Quality Assurance Board (California)
CAMRA Campaign for Real Ale
CAMRC Child Abuse and Maltreatment Reporting Center
cams. cybernetic anthropomorphous machines
CAMS Chinese Academy of Medical Sciences; Coastal Anti-Missile System; Communication, Advertising, and Marketing Studies (System); Confederation of Australian Motor Sports; Continuous Air-Monitoring System
CAMSI Canadian Association of Medical Students and Interns
Cam Soc Camden Society
can. canal; canalization; canalize; cancel; canceled; cancellation; canister; cannon; canon; canopy; canto; canvasback (duck)
can. (CAN) cancel character (data processing)
can canto (Italian—melody, song)
Can Caen; Canada; Canadian; Canal; Canberra; Cancer (constellation); Canfield; Cano; Canyon
Can Canal (French, Portuguese, Spanish—canal); *Canale* (Italian—canal); *Cañon* (Spanish—canyon)
Can. Cantoris (Latincantor's or preceptor's side of the choir)
CAN Canberra, Australia (airport); Citizens Against Noise; *Compagne Auxiliare*

de Navigation; Corporate Angel Network; Customs Assignment Number
CANA Canadian Army; Clergy Against Nuclear Arms
CANABRIT Canadian Navy Joint Staff in Great Britain
CANACIN Camara Nacional de la Industria (Spanish—National Chamber of Industries)—Mexico
CANACO Camara Nacional de Comercio (Spanish—National Chamber of Commerce)
CANACOR Canadian Agro-Industrial Corporation
Canad Canadian
Canada formerly the Dominion of Canada (largest nation in the western hemisphere and second largest in the world)
Canad Fr Canadian French (French Canadian)
Canadian Galápagos southern tip of Queen Charlotte Islands, British Columbia
Ca Na F Campaña Nacional Fronterizo (Spanish—National Frontier Campaign)
CANAIRDEF Canadian Air Force Defense Command
CANAIRDIV Canadian Air Force Division
CANAIRHED Canadian Air Force Headquarters
CANAIRLIFT Canadian Air Force Transport
CANAIRLON Canadian Air Force Joint Staff—London, England
CANAIRMAT Canadian Air Force Material Command
CANAIRNEW Canadian Air Force—Newfoundland
CANAIRNORWEST Canadian Air Force—Northwest, Edmonton
CANAIRPEG Canadian Air Force—Winnipeg
CANAIRTAC Canadian Air Force Tactical Command
CANAIRTRAIN Canadian Air Force Training Command
CANAIRVAN Canadian Air Force—Vancouver
CANAIRWASH Canadian Air Force Joint Staff—Washington, D.C.
Canajan Canadian speech
Canal Zone Panama Canal Zone
Can-Am Canadian-American
Can-Am Cup Canadian-

American Challenge Cup (automobile racing)
Canar Cur Canaries Current
Canaries Canary Islands in the Atlantic off southern Morocco
CANAS Canadian Naval Air Station
Canavaral Cape Canaveral, Florida also called Cape Kennedy
CANAVAT Canadian Naval Attaché
CANAVCHARGE Canadian Naval Officer in Charge
Canaveral national seashore on Florida east coast
CANAVHED Canadian Naval Headquarters
CANAVSTORES Canadian Naval Stores
CANAVUS Canadian Naval Joint Staff in United States
Canb Canberra
canc cancel; canceled; cancellation; cancelling
Canc Cancer (constellation)
Canc. Cancellarius (Latin—Chancellor)
CANCARAIRGRP Canadian Carrier Air Group
CANCEE Canadian National Committee for Earthquake Engineering
CANCIRCO Cancer International Research Cooperative
CANCOMARLANT Canadian Maritime Commander, Atlantic
CANCOMARPAC Canadian Maritime Commander, Pacific
Can Cus Canadian Customs
cand candelabra; candidate
Can$ Canadian dollar
cande command and edit (computer program)
CANDEP Canadian Naval Depot
candf cost and freight
candi cost and insurance
CANDLES Children of Auschwitz Nazis' Deadly Lab Experiment Survivors
Candn Canadian
CANDOC Canadian Electronic Document Ordering Service
cand sc candelabra screw
Candu Canadian deuterium uranium
CANDU Chelsea Against Nuclear Destruction United
CANDUR Canadian Deuterium-Uranium Reactor
Candy Candice

CANDY Cigarette Advertising Normally Directed to Youth
Canea Khania
CANEL Connecticut Advanced Nuclear Engineering Laboratory
Can-End Canton and Enderbury Islands
cane sugar saccharose or sucrose
Can F Canadian French
CANF Combined Allied Naval Forces; Cuban-American National Foundation
CANFARMS Canadian Farm Management Data System
CANFORCEHED Canadian Forces Headquarters
Can Fr Canadian French
Can I Canary Islands
CANI Committee on Non-discrimination and Integrity
Canid Canidae (Latin—dog family)
Can Imm Cen Canada Immigration Centre
canis canister
Can J Chem Canadian Journal of Chemistry
Can J Phys Canadian Journal of Physics
Can J Res Canadian Journal of Research
Can-Jud Canadian-Judeo (Canadian Jewish)
CANLANT Canadian Atlantic
Can Ltd Canadair Limited (operating unit of General Dynamics Corporation)
Can. maj. Canis Major (Latin—Greater Dog)—astronomical constellation
Can Man Cen Canadian Manpower Centre(s)
Can Met Ser Canadian Meteorological Service (EAES)
CANMET Canada's Center for Mineral and Energy Technology
Can. min. Canis minor (Latin—Lesser Dog)—astronomical constellation
canna. cannabinoid; cannabinol; cannabis
cannib cannibal(ism); cannibalistic; cannibalization; cannibalize
CANO Chief Area Nursing Officer
Canola Canadian oil, rapeseed
Can/ole Canadian on-line enquiry
Canon City Colorado State Penitentiary at Canon City
Can Op Canadian Opera Company (Toronto)

CANP Campaign Against Nuclear Power; Civil Air Notification Procedure
Can Pac Canadian Pacific
Can Pen Ser Canadian Penitentiary Service; Canadian Pension Commission
cans. canvasbacks (ducks); custom-assigned numbers
CANSAV Canadian Save the Children Fund
Can/sdi (CAN/SDI) Canadian selective dissemination of information
CANSG Civil Aviation Navigational Services Group
Can St Cannon Street (rail terminal)
Can Sym Canadian Symphony
cant canto (Italian—song; singing; voice part; instrument string tuned to the highest pitch)
cant. cannot; cantaloupe
can't can not; cannot
cant. canticum (Latin—canticle or hymn of praise)
Cant Canterbury; Canton; Cantonese
cantab cantabile (Italian—singable or songlike)
Cantab. Cantabrigiensis (Latin—of Cambridge)
Cantabrians Cantabrian Mountains of northern Spain
CANTAP Canadian Technical Awareness Programme
CANTAT Canadian Transatlantic Telephones
can. tb canine tuberculosis
cant b cantilever bridge
Cant Chin Cantonese Chinese
Can Telsat Canadian Telecommunications Satellite System
cantran cancel(led) in transmission
Can Tran Comm Canadian Transport Commission
cants cantaloupes
CANTT Cantonment Telegraph
CANTU Compañía Anónima Nacional Telefonos de Venezuela (Spanish—National Telephone Corporation of Venezuela)
Cantuar. Cantuaria or Cantuariensis (Latin—of Canterbury)
can't win mnemonic abbreviation for community property states—California, Arizona, Nevada, Texas, Wyoming, Idaho, New Mexico
canu can you
CANU Constabulary Anti-Nar-

cotics Unit (Philippines)
CANUKUS Canada—United Kingdom—United States
CANUS Canada—United States
CANUSE Canadian-United States Eastern (electric power interconnection)
CANUSPA Canada, Australia, New Zealand, and United States Parents Association
canv canvas
CANY Correctional Association of New York
Canyon de Chelly Canyon de Chelly National Monument (cliff-dweller ruins in northern Arizona)
Canyonlands Canyonlands National Park surrounding the junction of the Colorado and Green rivers in southeastern Utah
CANYPS Canadian Yellow Pages Service
canz canzone; canzonetta
CANZ Composers Association of New Zealand
cao chronic airway obstruction
CAO Canadian Association of Optometrists; Central Accounting Office(r); Chief Accounting Office(r); Chief Administrative Officer; Civil Affairs Office(r); County Administration Office(r); Crimean Astrophysical Observatory (USSR); Cultural Affairs Office(r)
caoc cathodal opening contraction
CAOC Consumers' Association of Canada
CAOGA Crown Agents for Oversea Governments and Administrations
CAOOAA Civil Air Operations Officers Association of Australia
CAORB Civil Aviation Operational Research Branch
CAORE Canadian Army Operational Research Establishment
CAOS Completely Automatic Operational System
CAOSOP Coordination of Atomic Operations—Standard Operating Procedures
CAOT Canadian Association of Occupational Therapy
cap. capacity; capital(ize); capital letter; capsule; caput; client assessment package
cap. (CAP) computer-aided production; consumer ac-

count protection
'cap handicap
cap capìtolo (Italian—chapter); *capìtulo* (Portuguese or S p a n i s h — c h a p t e r); (French—cape)
cap. capiat (Latin—take); *capsula* (Latin—capsule)
c/a/p codice di avviamento postale (Italian—mailing code)—zip coding
Cap capitol; Capricornus (constellation); captain; Captain; Caspar; Charles A. Pearce
Cap. Chapter—Number of Act of Parliament
Cap Capitán (Spanish—captain)
CAP California Assessment Program (educational); Canadian Association of Pathologists; *Certificat d'Aptitude Professionelle* (French—Certificate of Professional Aptitude); Citizens Against Pornography; Civil Addict Program; Civil Air Patrol; College of American Pathologists; Combat Air Patrol; Common Agricultural Policy; Commonwealth Association of Planners; Community Action Program; Community Advancement Program; Comprehensive Assessment Program; Computer Address Panel; Consumer Action Panel; Cooperating Accountability Project
CAPA California Association of Port Authorities; Canadian Association of Purchasing Agents; Commission on Asian and Pacific Affairs; Confederation of Asian and Pacific Accountants; Council of Australian Postgraduate Associations
capac capacity; cathodic protection
CAPAC Canadian Association of Primary Air Carriers; Composers, Authors, and Publishers Association of Canada
capal computer-and-photographic-assisted learning
CAPAR Cost-Account Problem Analysis Report
CAPC Civil Aviation Planning Committee
capche component automatic programmed checkout equipment
cap com capsule communicator

capcon(s) captured conversation(s)—electronically recorded tape of speech between two or more persons; recorded conversation
capd (CAPD) continuous ambulatory peritoneal dialysis
Cape The Cape—Good Hope, Hatteras, Horn, Verde, Aguilhas, Ann, Blanc, Blanco, Breton, Camorin, Canaveral (Kennedy), Charles, Clear, Cod, Columbia, Cornwall, Cruz, Disappointment, Fear, Finisterre, Flattery, Guardafue, Law, Leeuwin, Maisi, May, Mendocino, Muhammad, Palliser, Race, Sable, San Antonio, San Lucas, Spear, St Vincent, Wrangell, Wrath, York
CAPE California Association of Polygraph Examiners; Classification and Placement Examination; Confederation of American Public Employees; Council for American Private Education; Course and Professor Evaluation
Cape Breton Highlands Breton Highlands National Park, Cape Breton Island
Cape-Cairo Cape Town-to-Cairo Highway; Cape Town-to-Cairo Railway
Cape Cod National Seashore on coast of Massachusetts
Cape Cod National Cape Cod National Seashore conservation and recreation reservation on Cape Cod, Massachusetts
Cape Colony Cape of Good Hope Colony (South Africa)
CAPED California Association of Postsecondary Educators of the Disabled
Cape Hatteras national seashore in North Carolina coast
Cape Lookout national seashore in North Carolina
Cape Province Cape of Good Hope Province
Cape Roca Cabo da Roca, Portugal (Europe's westernmost point)
capertsim computer-assisted program evaluation review technique simulation
CAPES College Association for Public Events and Services
Cape Verde Islands Republic of Cape Verde (small island nation off Africa's westernmost tip) *Repúblíca de Cabo*

Verde
Cape Verdes Cape Verde Islands off the Senegal coast of West Africa, called Ilhas do Cabo Verde by the Portuguese
CAPEXIL Chemicals and Allied Products Export Promotion Council
CAPH California Association of Physically Handicapped
Capitol Reef Capitol Reef National Park, Utah
CAPL Canadian Association of Public Libraries; Controlled Assembly Parts List
CAPLOT Canadians Against PLO Terrorism
CAPM Computer-Aided Patient Management
cap. moll. capsula mollis (Latin—soft capsule)
CAPMS Central Agency for Public Mobilization and Statistics
Cap'n Captain
Cap^n Capitán (Spanish—captain)
capo [Italian—*capobanda*— bandmaster; *capo cameriere*—chief steward; *capo fabbrica*—factory foreman or overseer; *caporione*—ringleader; *capo stazione*—station master]
CAPO Canadian Army Post Office; Chief Administrative Pharmaceutical Office(r)
CAPOSS Capacity Planning and Operation Sequencing System
C App Chartered Appraiser
CAPPA Crusher and Portable Plant Association
capp^n capellán (Spanish—chaplain)
CAPPS Council for the Advancement of the Psychological Professions and Sciences
cap & puncless capitalization and punctuationless American author e e cummings
cap. quant. vult capiat quantum vult (Latin—allow the patient to take as much as he will)
Capr Capricornus
capri computerized advance personnel requirements and inventory
Capric Capricorn (constellation)
capris capri pants
Cap-Rouge Maison Notre-Dame de la Garde facility for juvenile delinquents at Cape-

Rouge, Québec
caps. capital letters
caps. (CAPS) computer-assisted problem solving
caps. capsule (Latin—capsule)
CAPs Community Action Programs; Consumer Action Panels
CAPS Casette Programming System; Cashiers Automatic Processing System; Clearinghouse on Counselling and Personnel Services; Coastal Aerial Photolaser Survey; Collins Adaptive Processing System; Combat Air Patrol Support; Computer-Aided Pipe Sketching System; Computer-Aided Project Study; Computer-based Aid-to-Aircraft Project Studies; Conventional Armaments Planning System; Creative Artists Public Service
capsep capsule separation
caps and lower case capital letters and lower case letters
CAPSS Computer-Assisted Public Safety System
caps and small caps upper case capital letters and small capital letters
capt caption
Capt(.) Captain
CAPT Clearinghouse for Applied Performance Testing
CAPTAINS Character and Pattern Telephone Access Information Network System
Captn Captain
Capulin Capulin Peak, Capulin Mountain National Monument, Mount Capulin—all in New Mexico
capun capital punishment
capy capacity; capybara
CAQ Craft Association of Queensland
caqa (CAQA) computer-aided quality assurance
car baggage car, boxcar, buffet car, cattle car, club car, coal car, dining car, electric car, elevated car, freight car, mail car, motor car, observation car, parlor car, passenger car, prison car, pullman car, railroad car, refrigerator car, sleeping car, steam car, street car, subway car, surface car, tourist car, trolley car
car. carat; carton; cloudtop altitude radiometer
car. (CAR) channel address register
Car Caradoc; Carberry; Car-

bury; Carel; Carew(e); Carey; Carina (constellation); Carleton; Carlow; Caroline Islands; Carter; Carvel; Carver
Car. *Carolus* (Latin—Charles)
CAR California Association of Realtors; Canadian Association of Radiologists; Central African Republic; Chief Airship Rigger; Civil Air Regulation(s); Civil Air Reserve; *Comité Agricole Régional* (French—Regional Agricultural Committee); Commonwealth Arbitration Reports; Contract Authorization Request; Corrective Action Request; US Army, Caribbean (area)
CAR *Cadena Azul de Radiodifusión* (Spanish—Blue Broadcast Chain)
cara combat air rescue aircraft
CARA California Association for Research in Astronomy; Center for Applied Research in the Apostolate; Chinese-American Restaurant Association
CARAC Civil Aviation Radio Advisory Committee
CARAL California Abortion Rights Action League
Carav Caravelle
carb carbon; carbonacious; carbonate; carburetor; carburize
CARB California Air Resources Board
carbage car-floor garbage
carbecue car + barbecue (device for melting waste out of junked automobiles)
carbo carbohydrate
carboloy carbon-cobalt-tungsten alloy
carbonado black or grayish-black industrial diamond; meat scored before grilling over charcoal
carbon dioxide carbonic acid gas
Carbonif Carboniferous
carbontet carbon tetrachloride
carbopol carboxypolymethylene
carborundum silicon carbide (SiC)
carb(s) carburetor(s)
CARBS Computer-Assisted Rationalized Building System
CARC Canadian Arctic Resources Committee
carcin (Latin prefix—cancer)—carcinogenic
Carcross Caribou Crossing

card. cardamon; cardinal
card. (CARD) compact automatic retrieval device; compact automatic retrieval display
Card Cardiganshire; Cardinal
CARD Campaign Against Racial Discrimination; Civil Aeronautics Research and Development; Compact Automatic Retrieval Device (or Display)
CARDA Continental Airborne Reconnaissance for Dammage Assessment (USAF)
cardamap cardiovascular data analysis by machine processing
CARDE Canadian Armament Research and Development Establishment
cardi (Latin prefix—heart)—cardiac arrest (heart failure)
CARDI Colorado Allergy and Respiratory Disease Institute
cardioac cardioacceleration; cardioaccelerator(y); cardioactive; cardioactivity
cardiog cardiogenesis; cardiogenetic; cardiograph(ic); cardiography
cardiol cardiolith(ic); cardiologist(ic)(al)(ly); cardiology; cardiolysin
cardiomeg cardiomegaly
cardiomyo cardiomyopathy
cardiopul cardiopulmonary
cardiov cardiovascular
cardiover cardioversion (electric-shock therapy)
CARDIV Carrier Division (naval)
Cardl *Cardenal* (Spanish—Cardinal)
CARDPACS Card Packet System
card(s). (CARDs) computer-aided research device(s)
Cards Cardinals
CARDS Combat Aircraft Recording and Data System; Computer-Assisted Recording of Distribution Systems
care. continuous aircraft reliability evaluation
Care Caretaker
CARE Citizens Association for Racial Equality; Consumer Awareness Retailer Effort; Cooperative for American Relief Everywhere; Cooperative for American Remittances to Everywhere; Cottage and Rural Enterprises
CAREIRS Conservation and Renewal Energy Inquiry and

Referral Service
CAREL Central Atlantic Regional Educational Laboratory
CARES Computer-Assisted Regional Evaluation System
CARF Canadian Amateur Radio Federation; Canadian Arthritis and Rheumatism Society; Central Altitude Reservation Facility
CARG Caribbean Ready Group (USN); Corporate Accountability Research Group (Nader's)
Cargomaster Douglas 200-passenger military transport designated C-133
cargotainer cargo container
Cari Carina
CARI Civil Aeromedical Research Institute
Carib. Caribbean
CARIBAIR Caribbean Atlantic Airlines
CARIBANK Caribbean Development Bank
Caribbean Caribbean Area (islands in or washed by the Caribbean Sea); Caribbean Sea
Caribbean countries Antigua and Barbuda, Barbados, Dominica, Dominican Republic, Grenada, Haiti, Jamaica, St Kitts and Newis, St Lucia, St Vincent and the Grenadines, Trinidad and Tobago
CARIBCOM Caribbean Command
Carib Cur Caribbean Current
Caribe (Spanish—Carib language, Caribbean Sea)
Caribous Caribou Mountains of British Columbia
CARIBSEAFRON Caribbean Sea Frontier
caric caricature; caricaturist
CARIC Contractor All-Risk Incentive Contract (USAF)
Caricom Caribbean Community Anguilla, Antigua, Barbados, Belize, Dominica, Grenada, Guyana, Jamaica, Montserrat, St Kitts-Nevis, St Lucia, St Vincent, Trinidad and Tobago
CARIFESTA Caribbean Festival of the Arts
CARIFTA Caribbean Free Trade Association
CARIH Children's Asthma Research Institute and Hospital (Denver)
CARIPLO Cassa di Risparmio

delle Provincie Lombarde (Italian—Savings Bank of the Province of Lombardy)
CARIS Computerized Agricultural Research Information System; Current Agricultural Research Information System (FAO)
carl computer-assisted reference locator
Carl Carla; Carle; Carleton; Carlisle; Carlo(s); Carlton; Carlyle
CARL Canadian Academic Research Libraries; Chatfield Applied Research Laboratories; Computer Audio Research Laboratory
Carla Carlotta; Caroline
Carl Gustaf 9mm Swedish sub-machine gun firing parabellum bullets; recoilless 84mm antitank weapon
Carm Carmathenshire; Carmichael; Carmo; Carmody
CARML County and Regional Municipality Librarians (Ontario)
carmrand civilian application of the results of military research and development
Carn Caernarvonshire
Carnegie Inst Carnegie Institution of Washington
Carnegie Tech Carnegie Institute of Technology
carni carnival
Carnics Carnic Alps
carnie(s) carnival(s); carnival workers
Caro Carolina; Caroline
Carol Carola; Carole; Carolina; Caroline; Carolyn
CAROL Caribbean Overseas Lines
Carolina Art Carolina Art Association
Carolina Pop Ctr Carolina Population Center
Carolinas North and South Carolina
Carolines Caroline Islands (Kusaie, Palau, Ponape, Truk, Yap) in the Western Pacific
carot centralized automatic recording on trunks (Bell)
carp. carpenter; carpentry; carpet(ing); computed air-release point; construction of aircraft and related procurement
Carp Carpathian; Carpentaria
CARP Campaigns Against Rising Prices; computed air-release point
CARPAS Comisión Asesora

Regional de Pesca el Atlantico Sud-Occidental (Spanish—Regional Fisheries Advisory Commission for the Southwest Atlantic)
Carpathians Carpathian Mountains between Czechoslovakia and Poland
Carpaths Carpathian Mountains
carpilf cargo pilferage
carp(s) stage carpenter(s)
Carps Carpathian Mountains
carr carriage (flow chart); carrier
Carrasco Montevideo, Uruguay's airport
Carrie Carolina; Caroline
Carrie Mac California Association of Realtors Mortgage Assistance Corporation
CARRIS Companhia Carris de Ferro de Lisboa (Portuguese—Lisbon Street Railway)
Carroll of Carrollton Charles Carroll of Carrollton, Maryland—self-identified signer of the *Declaration of Independence*
cars. community antenna relay service
CARS Canadian Arthritis and Rheumatism Society; Central American Research Station (for disease control); Community Antenna Relay Station; Community on Alcohol and Road Safety; Computer-Aided Routing System; Computer-Assisted Reliability Statistics
Carson Carson City; Nevada State Penitentiary in Carson City
Carson City Women's Women's Prison at Carson City, Nevada
CARSTRIKFOR Carrier Striking Force
cart. cartage; carton; collision-avoidance radar trainer
cart carta (Italian, Portuguese, Spanish—card, chart, document, page of music)
Cart Carter; Cartwright
CART Cargo Automation Research Team; Central Automated Replenishment Techniques; Championship Auto Racing Teams; Coalition Against Regressive Taxation; Complete Automatic Rating Technique; Complete Automatic Reliable Testing
CARTB Canadian Association

of Radio and Television Broadcasters
Carter James Earl Carter, 39th President of the United States
Carth Carthage; Carthaginian; Carthusian
cartobib cartobibliographer; cartobibliography
cartog cartographer; cartographic; cartography
cartoonitorial cartoon editorial
CARTS Computer-Automated Reserved Track System
Car Z Caribische Zee (Dutch—Caribbean Sea)
cas calibrated airspeed; casual; casualty; close air support; commence average speed
cas (CAS) cooperative applications satellite
ca's combat actions; covert actions
Cas Caracas; Casimir; Caslon; Cassiopeia (constellation); Castle
Cas ciencias (Spanish—sciences)
CAs Consumers Associations; Cooperative Associations
CAS California Academy of Sciences; Cambrian Airways (symbol); Capital Assistance Scheme; Casualty Actuarial Society; Center for Administrative Studies; Center for Auto(mobile) Safety; Change Analysis Section; Chemical Abstracts Service; Chicago Academy of Sciences; Chief of Air Staff; Children's Aid Society; Chinese Academy of Sciences; Civil Affairs Section; Civil Air Surgeon; Clean(er) Air System(s); Collision Avoidance System (aircraft); Commercial Air Service; Computer Acquisition System; Contemporary Art Society; Contract Administration Services; Cost Accounting Standards; Courier Air Services; Current Australian Serials; Current Awareness Service; Customs Agency Service
C.A.S. Certificate of Advanced Studies
ca.sa. *capias ad satisfaciendum* (Latin—writ of execution)
Casa Casablanca
CASA Campaign Against Psychiatric Abuse (in the USSR); Canadian Association of School Administrators; Canadian Automatic

Sprinkler Association; Catgut Acoustical Society of America; Citizens Against Sneakin' Aroun'; Contemporary Art Society of Australia; Court-Appointed Special Advocates; Crafts Association of South Australia
CASA Construcciones Aeronauticas, SA (Spain)
Casa Grande Casa Grande National Monument near Phoenix, Arizona
CASANZ Clean Air Society of Australia and New Zealand
CASAO Chartered Accountants Students Association of Ontario
CASB Cost-Accounting Standards Board
CASBS Center for Advanced Study in the Behavioral Sciences (Stanford)
casc computer-assisted cartography
CASC Council for the Advancement of Small Colleges
CASCADE Citizens and Scientists Concerned About Dangers to the Environment
Cascades Cascade Mountains extending from British Columbia to California
cascan casualty cancelled
CASCOMP Comprehensive Airship Sizing and Performance Computer Program
cascor casualty corrected
CASCU Cooperative Association of Suez Canal Users
CASD Center for Applied Studies in Development
casdac computer-aided ship design and construction
CASDO Computer Applications Support and Development Office (USN)
casdos computer-assisted detailing of ships
case. common-access switching equipment, computer-automated support equipment
Case Casey
CASE Coalition of Agencies Serving the Elderly; Coalition for Safe Energy; Colorado Association of School Executives; Commission on the Accreditation of Service Experiences; Committee on Academic Science and Engineering; Committee on the Atlantic Salmon Emergency; Coordinated Aerospace Supplier Evaluation; Council of Administrators of Special Educa-

tion; Council for the Advancement of Secondary Education; Council for Advancement and Support of Education; Counselling Assistance to Small Enterprises
CASEA Center for the Advanced Study of Educational Administration
CASETT Cases of Settlements and Removals
CASEX Close Air Support Exercise; Combined Aircraft Submarine Exercise
CASF Combat Alert Strike Force; Composite Air Strike Forces
CASGTC California Academy of Sciences Geology Type Collection
cash. cashier
Cash Cashlin; Cassius
CASH Catalog of Available and Standard Hardware; Citizen Action for Safer Harlems; Commission for Administrative Services in Hospitals
Casi Casimir
CASI Canadian Aeronautics and Space Institute
CASIA Chemical Abstracts Subject Index Alert
CASIG Careers Advisory Service in Industry for Girls
Casl Caslon
Ca S-L Catering Sub-Lieutenant
CASLE Commonwealth Association of Surveying and Land Economy
CASLIS Canadian Association of Special Libraries and Information Services
casm cycling air sampling monitor
CASMT Central Association of Science and Mathematics Teachers
CASNP Canadian Association in Support of Native Peoples
CASO Civil Aviation Safety Order
CASOE Computer Accounting System for Office Expenditures
casoff control and surveillance of friendly forces
CASOS Center for Advanced Study in Organization Science (U of Wisconsin)
Casp Caspar(d)
CASP Capability Support Plan; Cape Arago State Park (Oregon); Country Analysis and Strategy Paper (U.S. State Department)

Caspar Cambridge analog simulator for predicting atomic reactions

CASPER Contact Area Summary Position Estimate Report

CASPERS Computer-Automated Speech-Perception System

Caspian Caspian Sea (landlocked body of water between Iran and the USSR)

Cas Reps Casualty Reports

CASRO Council of American Survey Research Organizations

cass cassowary

Cass Casimir; Cassander; Cassandra; Casseus; Cassidy; Cassius; Casso

CASS Center for Astrophysics and Space Scientists

CASSO Command Active Sonobuoy System; Connecticut Association of Secondary Schools

CASSA Continental Army Command Automated System Support Agency (USA)

Cassi Cassiopeia

CASSI Chemical Abstracts Service Source Index

CASSIS Communication and Social Science Information Service (Canada)

CASSR Chuvash Autonomous Soviet Socialist Republic

cast. computer applications and systems technic; computer-augmented scanning technics

Cast Castel; Castile; Castilian; Castillon; Castimir; Castislav; Castle; Castor

CAST Center for Application of Sciences and Technology; Contemporary Artists Serving the Theater; Council for Agricultural Science and Technology

CAST Clearinghouse Announcements in Science and Technology

CASTE Collision-Avoidance System Technical Evaluation

CASTLE Computer-Assisted System for Theater-Level Engineering

CASTOR College Applicant Status Report

Castro San Francisco neighborhood

CASTS Canal Safe Transit System

CASW Council for the Advancement of Science Writing

cat. carburetor air temperature; catalog; catamaran; catapult; catboat; category; caterpillar tractor; caudé auxiliary telescope; clear air turbulence; compressed air tunnel; computer-assisted transcription; computerized axial tomography

cat. (CAT) choline acetyltranferase; computer-aided typesetting; computer-assisted test(ing); computerized adaptive test(ing); computerized axial test(ing); computerized axial tomography

Cat Catalán; Catalina; Catalonia; Catalonian; Cataluña; Catalunya; Catamarca; Catania; Cataño; Catarina; Caterino; Catasauqua; Catawba; Caterpillar Tractor; Catesby; Catlett

Cat Cat Duet (Rossini's comic work for two voices); Catalan (language spoken in the Spanish province of Catalonia, France and Andorra)

CAT California Achievement Test; California Advocacy for Trollops; Child's Apperception Test; Civil Air Transport; Civilian Actress Technician; Clerical Aptitude Test; Cognitive Abilities Test; College Ability Test; Colleges of Advanced Technology; Commercial Airlift Contract; Computer-Aided Training; Computer-Aided Translation; Computer-Aided Typesetting; Computer-Assisted Teaching; Consolidated Atomic Time; Container Associates Transport; Control and Assessment Team; Corrective Action Team

CAT Comisaría de Abastecimientos y Transportes (Spanish—Commisariat of Supply and Transport)

CATA Canadian Air Transport Association; Canadian Air Transportation Administration

catal catalog; catalogue

Catal Catalan; Catalonia; Cataluña

Catalina Catalina de Güines southeast of Havana, Cuba; Santa Catalina Island off Long Beach, California

catawump catawumpus (catamount, mountain lion)

catc computer-assisted test construction

CATC Commonwealth Air Transport Council; Continental (Oil), Atlantic (Refining), Tidewater (Oil), and Cities (Service) (combined in mutual drilling)

CATCALL Completely Automated Technique for Cataloging and Acquisition of Literature for Libraries

CATCC Canadian Association of Textile Chemists and Colorists; Carrier Air Traffic Control Center

CATCH Citizens Against The Concorde Here

CATCO Catalytic Construction Company

cate comprehensive automatic test equipment

CATE Current ARDC (Air Research and Development Command) Technical Efforts (program)

catec catechism; catechist(ic)-(al)(ly)

Cater Trac Caterpillar Tractor

Ca Test Calcium Test (dental)

CATF Canadian Achievement Test in French; Central American Task Force

cath cathartic; cathedral; catheter; catheterize

Cath Cathedral; Catherine; Catholic; St Catharine's College, Cambridge; St Catherine's College, Oxford

CATHAY Cathay Pacific Airways

Cathedrals Cathedral Caverns near Grant, Alabama

cath fol cathode follower

Cathie Catherine

Cath Lib Assn Catholic Library Association

Cathlibs Catholic liberals

cathol catholic; catholically; catholicly; catholicalness; catholicness; catholicate; catholice; catholicity

Cath U Pr Catholic University of America Press

Cathy Catherine

CATIB Civil Air Transport Industry Training Board

CATIE Centro Agronómico Tropical de Investigación y Enseñanza (Spanish—Tropical Agriculture Center of Investigation and Teaching)—Turrialba, Costa Rica

catk counterattack

catlg catalog(ue)

CATM Canadian Achievement Test in Mathematics

CATNIP Computer-Assisted Technique for Numerical Index Preparation

Catoctins Catoctin Mountains of Maryland and Virginia

CATOR Combined Air Transport and Operations Room

CATP Computer-Assisted Typesetting Process

catproc catalog(ue) procedure

CATRA Cutlery and Allied Trades Research Association

Catracho(s) Honduran(s)

CATRALA Car and Truck Renting and Leasing Association

CATs Civic Action Teams

CATS Canon Auto Tuning System; Certificates of Accrual on Treasury Certificates; Civil Affairs Training School (USN); Comprehensive Analytical Test System; Compute Air-Trans Systems; Computer-Assisted Test Shop; Computer-Automated Test System (AT & T)

CATSA Cooperative Air Transport System for Antarctica

CAT-scan computerized axial tomography scan that depicts body structures not shown by conventional X-ray procedures

Catskills Catskill Mountains, southeastern New York

CATSS Cataloguing Support System

catt conveyorized automatic tube tester

cattalo cattle + buffalo—hybrid

CATTCM Canadian Achievement Test in Technical and Commercial Mathematics

CAT test Computerized Axial Tomography Test

CATTS California and Texas Telecommunications System

Catty Catherine

catv cabin air temperature valve; cable television; community antenna television

catva computer-augmented total-value assessment

cau command arithmetic units

Cau Caucasian

CAU Child Abuse Unit (police department); Congress of American Unions; Consumer Affairs Union; Consumer Affairs Unit

Caucasus Caucasus Mountains between the Black Sea and the Caspian Sea

Cauc(s) Caucasian(s)

caud caudal; caudate

caud (Latin prefix—tail)—caudal appendage

CAUL Committee of Australian University Librarians

cauli cauliflower

caus causation; causative

CAUS Color Association of the United States

CAUSA *Compania Aeronautica Uruguay SA*; Confederation of Associations for the Unity of the Societies of the Americas

causat causative

'cause because

CAUSE College and University Systems Exchange; Comprehensive Assistance to Undergraduate Science Education (National Science Foundation); Counselor Advisor University Summer Education

caust caustic

caustic potash potassium hydroxide (KOH)

caustic soda sodium hydroxide (NaOH)

caut caution

CAUT Canadian Association of University Teachers

CAUTION Citizens Against Unnecessary Tax Increases and Other Nonsense (St Louis citizens)

cav cavalier; cavalry; cavitation; cavity; congenital absence of vagina; congenital adrenal virilism; continuous airworthiness visit

cav (CAV) class attendance verification; construction assistance vehicle

cav. caveat (Latin—warning, writ of suspension)

c.a.v. *curia advisare vult* (Latin—the court cares to consider)

Cav Cavanagh; Cavanaugh; Cavell; Cavendish

Cav *Cavaliere* (Italian—Knight)

cav brk cavity brick

cavd completion, arithmetic, vocabulary, directions (test)

Cav Div Cavalry Division

CAVE California Association of Vocational Educators; Catholic Audio-Visual Educators Association; Consolidated Aquanauts Vital Equipment

CAVEA Connecticut Audio-Visual Education Association

caveat code and visual entry authorization technic

cav. emp. *caveat emptor* (Latin—let the buyer beware)—also appears as *c.e.*

Cavendish Cavendish Laboratory (Cambridge University)

CAVI *Centre Audio-Visuel International* (French—International Audio-Visual Center)

caviol caviology

ca virus croup-associated virus

CAVN *Compañía Anonima Venezolana de Navegación* (Spanish—Venezuelan Steamship Line)

cavu ceiling and visibility unlimited

caw cam-action wheel; channel address word

c-a w conflict-alert warning

CAW Cables and Wireless (company); Californians Against Waste

CAWA Canadian-American Women's Association

CAWC Committee on Air and Water Conservation (American Petroleum Institute)

CAWD Canadian-American Wolf Defenders

CAWE California Association of Work Experience Educators

cawg coaxial adapter waveguide

CAWG Clean Air Working Group

CAWM College of African Wildlife Management

CAWP Center for the American Woman and Politics

CAWS Central Aural Warning System; Conflict Alert Warning System; Conservation and Wildlife Studies

CAWSPS Computer-Aided Weapon Stowage Planning System (USN)

CAWU Clerical and Administrative Workers' Union

cax community automatic exchange (telephone)

Caxton Caxton Printers, Ltd

Cay Cayenne; Cayman

Cayes Haitian seaport also called Aux Cayes or Les Cayes

Caymans Cayman Islands (Grand Cayman, Little Cayman, Cayman Brac)

cb cast brass; catch basin; cement base; center of buoyancy; chemical and biological; circuit breaker; collective bargaining; common battery;

continuous breakdown
cb (CB) container base
c-b circuit breaker
c/b caught and bowled
c & b collating and binding
c of b confirmation of balance
Cb columbium (symbol); cumulo-nimbus
Cb contre-basse (French—double bass)
CB Cape Breton (island); Caribair (airline); Caribbean-Atlantic Airlines; Carte Blanche; Cavalry Brigade; Cemetery Board; Census Bureau; Chairman of the Board (of directors); Chief Boilermaker; Children's Bureau; Church of the Brethren; citizen's band (radiofrequency); Companion of the Bath; compass bearing; confidential book; confidential bulletin; confinement to barracks; Construction Battalions (hence the nickname "seabees"); Consultants Bureau; contrabass(o) (double-bass viol); Control Branch; Counter Battery; County Borough; Cumulative Bulletin; Currency Bond; large cruiser (naval symbol); William Cullen Bryant
C-B (Sir Henry) Campbell-Bannerman
C.B. *Chirurgiae Baccalaureus* (Latin—Bachelor of Surgery); Companion of the Bath
C & B Clemens and Brenninkmeyer; Cleveland and Buffalo (steamship line)— *Seeandbee*
C of B Commonwealth of the Bahamas
CB Carte Blanche (French—white card) full discretionary power; *col basso* (Italian—with the bass); *contrabasso* (Italian—double bass)
C-B Creditanstalt-Bankverein (German—Credit Institution and Bank Association)—Austria's largest banking institution
cba chemical bond approach; chronic bronchitis with asthma; cost-benefit analysis
cba (CBA) colliding beam accelerator
CBA California Benefit Association; Canadian Bankers Association; Canadian Booksellers Association; Caribbean Atlantic Airlines; Chris-

tian Booksellers Association; Clydesdale Breeders Association; College of Business Administration; Community Broadcasters Association; Consumer Bankers Association
CBA Chemical-Biological Activities
CBAA Canadian Business Aircraft Association
CBAC Chemical-Biological Activities
cbaf cobalt-base alloy foil
CBAICP Chemical and Biological Accident and Incident Control Plan (USA)
c/bale cents per bale
cbam concerns-based adoption model
cbar counterbore arbor
CBARC California Border Area Resource Center
CBAT Central Bureau of Astronomical Telegrams
cbb commercial blanket bond
CBB Chesapeake Bay Bridge (Maryland)
CBB Centre Belge du Bois (French—Belgian Forestry Research Center)
CBBA Christian Brothers Boys Association
CBBB Council of Better Business Bureaus
CBBI Cast Bronze Bearing Institute
CBBII Council of the Brass and Bronze Ingot Industry
CBBT Chesapeake Bay Bridge-Tunnel (Maryland to Virginia)
cbc (CBC) combined blood count; complete blood count
CBC Canadian Broadcasting Corporation; Caribbean Broadcasting Company; Central Bank of China; Central Bible College; Ceylon Broadcasting Corporation; Children's Book Council; Christmas Bird Count (Audubon Society); Columbia Basin Council; Commendation for Brave Conduct; Commercial Banking Company; Commonwealth Banking Corporation; Congressional Budget Committee; Contraband Control; Corset and Brassiere Council; Cyprus Broadcasting Corporation; large tactical-command ship (naval symbol)
CBC Cadena Baja California (Spanish—Baja California

Radio Chain) Mexican border network
CBCA California Black Commission on Alcoholism
cbcc common bias—common control
CBCII California Bureau of Criminal Identification and Investigation
CB circuit common-base amplifier for junction transistors
CB Club Citizen's-Band (radio) Club
cbcm cheque book-charging method
CBCMA Carbonated Beverage Container Manufacturers Association
CBCS Commonwealth Bureau of Census and Statistics
cbct circuit board card tester
CBCT Customer-Bank Communication Terminal
cbcu counterbore cutter
cbd cash before delivery; closed bladder drainage; common bile duct
Cbd Cambodian
CBD Central Business District; Construction Battalion Detachment
CBD Commerce Business Daily
cbdn can be done
CBDNA College Band Directors National Association
CBDS Carcinogenesis Bioassay Data System
cbe cesium bombardment engine; chemical binding effect; circuit board extractor; competency-based education; compression bonding encapsulation
CBE Cheese Bureau of England; Community-Based Education; Competency-Based Education; Computer-Based Education; Conference of Biological Editors; Council for Basic Education; Council of Basic Education; Council of Biology Editors
C.B.E. Commander of the Order of the British Empire; Companion of the Order of the British Empire
CBEL Cambridge Bibliography of English Literature
CBEMA Canadian Business Equipment Manufacturers Association; Computer and Business Equipment Manufacturers Association
CBer operator or owner of a Citizen's-Band radio

CB-er(s) citizen's radio-frequency band (short-wave two-way) broadcaster(s)
cbf cerebral blood flow; coronary blood flow
CBF Children's Blood Foundation
CBFC Commonwealth Bank Finance Company
CBFCA Commander, British Forces, Caribbean
cbfm constant bandwidth frequency modulation
cbft cubic feet
cbg (CBG) corticosteroid-binding globulin; transcortin
CBG Compagnie des Bauxites de Guinée (French—Bauxite Company of Guinea)
CBH Cooperative Bulk Holding
C B & H Continent between Bordeaux and Hamburg
CBHE Connecticut Board of Higher Education
cbi complete background investigation; compound batch identification; computer-based information
CBI Cape Breton Island; Carbonated Beverage Institute; Caribbean Basin Initiative; Central Bureau of Investigation; Chesapeake Bay Institute; Chicago Bridge and Iron (company); China-Burma-India (theater of war); Coffee Brewing Institute; Confederation of British Industries; Confederation of British Industry; Council of the Building Industry; Council of Burma Industries
CB & I Chicago Bridge and Iron (company)
CBI Cumulative Book Index
CBIA California Building Industry Association
cbid counter bid
CBIS Campus-Based Information System (NSF); Communist Bloc Intelligence Service; Computer-Based Instruction(al) System
cbit (CBIT) contract bulk inclusive tour (travel plan)
cbj common bulkhead joint
CBJO Coordinating Board of Jewish Organizations
cbk checkbook
cbl cable; cement-bond log; commercial bill of lading
cb/l commercial bill of lading
c bl carte blanche (French—white card)—full power to act

CBL Configuration Breakdown List; Chesapeake Biological Laboratories
CB of L Chartered Bank of London
cbm chemical biological munitions; confidence-building measure; conventional buoy mooring; cubic meter(s)
cbm Kubikmeter (German—cubic meter)
CBM Christian Blind Mission
CBMA Carbonated Beverage Manufacturers Association
CBMC Corregidor-Bataan Memorial Commission
CBM-I Common Bahasa Malay-Indonesian
CBMIS Computer-Based Management Information System
CBMM Council of Building Materials Manufacturers
CBMPE Council of British Manufacturers of Petroleum Equipment
CBMQA California Board of Medical Quality Assurance
CBMS Conference Board of Mathematical Sciences
cbmu current bit monitor unit
CBMU Canadian Board of Marine Underwriters
CBMUA Canadian Boiler and Machinery Underwriters Association
cbn chemical, bacteriological, nuclear; courses by newspaper
C bn contrabassoon
CBN Christian Broadcasting Network; Columbia Carbon Company (stock-exchange symbol); Commonwealth Broadcasting Network
CBNE California Bureau of Narcotics Enforcement
CBNM Custer Battlefield National Monument
CBNS Commander, British Naval Staff
CBNY Chemical Bank, New York
cbo compensation by objectives
Cbo Colombo
CBO Community-Based Organizations; Conference of Baltic Oceanographers; Congressional Budget Office (U.S.A.)
CBOA Commonwealth Bank Officers Association
cboc completion bed occupancy care
CBOE Chicago Board Options Exchange
C-bomb cobalt bomb

cbore counterbore
CBOT Chicago Board of Trade
cbp ceramic beam pentode; constant boiling point
CBP Centro de Biologia Piscatória (Portuguese—Piscatorial Biological Center—Lisbon)
CBPA Connecticut Book Publishers Association
CBPBG Commonwealth Bureau of Plant Breeding and Genetics
CBPC Canadian Book Publishers' Council
CBPDC Canadian Book and Periodical Development Council
CB & PGNCS Circuit Breaker and Primary Guidance Navigation Control System
CBPI Canadian Business Periodicals Index
CBPO Consolidated Base Personnel Office
CBQ Children's Behavior Questionnaire; Civilian Bachelor Quarters
C B & Q Chicago, Burlington & Quincy (railroad)
CBQ Catholic Biblical Quarterly
cbr change board register; chemical, biological, radiological; crude birth rate
Cbr Calabar
CBR Canberra, Australia (airport); Center for Brain Research (University of Rochester)
CBRA Chemical, Biological, Radiological Agency
CBRA Consolidated Budget Reconciliation Act
CB radio citizen's band radio (26.965 to 27.405 MHz)
CBRC Crichton Behavioral Rating Scale
CBRE Chemical, Biological, and Radiological Element
CBRI Central Building Research Institute
CBRL Chemical, Biological, and Radiation Laboratories (Ottawa)
cbrn chemical, biological, radiological, and nuclear
CBRO Chemical, Biological, Radiological Officer
CBRS Canadian Bond Rating Service (Montreal); Child Behavior Rating Scale; Citizen's Band Radio Service
CBRTGW Canadian Brotherhood of Railway, Transport,

and General Workers

cbrw chemical, biological, radiological warfare

cbs chronic brain syndrome; concrete-block stucco

cb's citizen's-band transceivers

cbs concerned Black students

CBs cost-of-living benefits

CBS Canadian Biochemical Society; Central Bureau of Statistics (Jerusalem); Columbia Broadcasting System; Common Beliefs Survey; Commonwealth Bureau of Soils; Confraternity of the Blessed Sacrament; Currumbin Bird Sanctuary (Queensland); Custom Bucket Service

CBS *Centraal Bureau de Statistiek* (Dutch—Central Statistical Bureau)

cbse caboose

CBSO City of Birmingham Symphony Orchestra; City of Bournemouth Symphony Orchestra; Czechoslovak Broadcasting Symphony Orchestra

cbt cesium beam tube; computer-based training; criminal breach of trust

cbt (CBT) cognitive-behavioral therapy

CBT Chicago Board of Trade; Computer-Based Testing; Connecticut Bank and Trust (company)

CBT *Centre Belge de Traductions* (French—Belgian Translations Center)

CB & TC Connecticut Bank & Trust Company

CBTE Competency-Based Teacher Education

cbts cesium beam time standard

cbu cluster bomb unit

CBU Chicago Board of Underwriters

c/bush cents per bushel

cbv central blood volume; circulating blood volume; corrected blood volume

CB-VD citizen's-band radio-contracted venereal disease (resulting from sexual pick-ups made along byways and highways)

CBVHS Clara Barton Vocational High School

cbw chemical-biological warfare

CBX Computerized Branch Exchange

cbx's (CBXs) computerized

business exchanges (telephone service)

cby carboy

cc camp chair; canonical correlation; carbon copy (or copies); centuries; chapters; civil commotions; close control; closed cup; closing coil; cognitive complexity; color code; combustion chamber; command and control; complex conjugate; condemned cell; continuation clause; contrasting color; cubic centimeter(s); current cost; customs clearance

cc (CC) check charge; chief complaint

c/c center to center; corner/card

c & c carpets and curtains; caviar and champagne; consultation and concurrence

c of c certificate of occupancy; cost of construction

c-to-c center-to-center

c.c. corpora cardiaca (Latin—cardiac body)—heart

c/c compte courant (French); *conta corrente* (Portuguese); *conto corrente* (Italian); *cuenta corriente* (Spanish)—current account

Cc cirrocumulus

Cc. Confessores (Latin—Confessors)

CC Calvin Coolidge (30th President U.S.); Chief Constable; Circuit Court; City Center; Clare College (Cambridge); Coastal Commission; Colby College; Common Cause; Community College; County of City (Aberdeen, Dundee, Edinburgh, Glasgow); Critical Care

C-C Coca-Cola

C & C Columbia & Cowlitz (railroad); Command and Control; Computer and Communications

C-by-C Come-by-Chance, Newfoundland

C of C Conclave of Cardinals; Count(y) of Cleves

CC corriente continua (Spanish—direct current)

CC-106 Canadair version of the Britannia called Yukon

CC-109 Canadian-built medium-range transport designed by General Dynamics and known as the Cosmopolitan

CC-115 Canadian twin-engine turboprop transport called the Buffalo

cca carrier-controlled approach; cellular cellulose acetate (plastic); chromated copper arsenate; crop-condition assessment

CCA Cacchetti Council of America; California Central Airlines; California Confederation of the Arts; California Correctional Association; Camp and Cabin Association; Canadian Cat Association; Canadian Construction Association; Central City Association; Chief of Civil Affairs; Circuit Court of Appeals; Citizens for Clean Air; Citizens' Councils of America; Cleaning Contractors Association; Colorado Correctional Association; Comics Code Authority; Committee for Conventional Armaments; Commonwealth Correspondents Association; Community College Association; Community Concerts Association; Community Corrections Act; Conquest of Cancer Act; Conservative Clubs of America; Consumers Cooperative Association; Container Corporation of America; Continental Control Area; Coordinating Committee on Alcoholism; Corduroy Council of America; Corrections Corporation of America (private operator of jails in the U.S.); Crafts Council of Australia; Cruising Club of America; Current Cost Accounting

C & CA Consumer and Corporate Affairs (Canada)

CCAA Cement and Concrete Association of Australia

CCAB Canadian Circulation Audit Board

CCAC California College of Arts and Crafts

CCAD Commerce and Consumer Affairs Department

CCAE California Council for Adult Education

CCAF Commander-in-Chief—Atlantic Fleet; Community College of the Air Force

CCAHC Central Council for Agricultural and Horticultural Cooperation; Central Council for Agricultural and Horticultural Cooperatives

CCAIA California Council of the American Institute of Architects

CCAIT Community College Association for Instruction and Technology
CCAM Colby College Art Museum
ccap communication capability application program
CCAP Citizens Crusade Against Poverty
CCAPS Circuit-Card Assembly and Processing System
CCAQ Consultative Committee on Administrative Questions (UN)
CCAR Central Conference of American Rabbis
CCAs California Correctional Association members; Cruising Club of America members
CCAS Council of Colleges of Arts and Sciences
ccat conglutinating complement absorption test
CCATS Communications, Command, and Telemetry Systems
cca unit chicken-cell agglutination unit
ccb command control block; convertible circuit breaker; cubic capacity of bunkers
CCB California Canadian Bank; Civil Cooperation Bureau (South African Secret Military); command-and-control boat (naval symbol); Configuration Control Board; Criminal Courts Building
CCBD Council for Children with Behavioral Disorders
cc black conductive channel black
CCBI Central City Business Institute
CCBM Copper Cylinder and Boiler Manufacturers
CCBO Cape Clear Bird Observatory (Ireland)
CCBS California Canadian Banks
ccbv central circulating blood volume
CCBW Commission on Chemical and Biological Warfare
ccc central computer complex; command control console; computer-command control
CCC California Cheese Company; California Conservation Corps; Canadian Chamber of Commerce; Cape Cod Canal; Car Care Council; Caribbean Conservation Corporation; Caribbean Cultural

Center; Central Community Center; Central Control Commission; Central Criminal Court; Changcheng (Great Wall) Computer Corporation; Chimpanzee Crisis Concern; Chopin Cultural Center; Christ Church College; Citizens Crime Commission; Civil Construction Corps; Civilian Conservation Corps; Columbian Carbon Company; Commercial Credit Corporation; Commissioner's Coordination Council; Commodity Credit Corporation; Commonwealth Credit Corporation; Community Care Center; Consumer Credit Counselors; Copyright Clearance Center; Corning Community College; Coronary Care Certificate; Corpus Christi College; Cox Cable Communications; Crime and Correction Commission; Crime and Correction Committee; Customs Cooperation Council; Cuyahoga Community College
CC & C Command Control and Communications (USAF)
CCC Consejo de Cooperación Cultural (Spanish—Council of Cultural Cooperation)—of the Council of Europe; *Cwmni Cyfyngedic Cyhoeddus* (Welsh—Public Limited Company)
C.C.C. Constitutio Criminalis Carolina (Latin—Carolingian Criminal Code)
CCCA Classic Car Clubs of America; Conservative Christian Churches of America
CCCB Component Change Control Board (DoD)
CCCC Cape Cod Community College; Committee for the Conservation and Care of Chimpanzees; Conference on College Composition and Communication
CCCCO Chicago Coordinating Council of Community Organizations
CCCCSA California Community College Community Service Association
CCCD California Commission of College Districts
CCC-FID Central Classification Committee—Fédération Internationale de Documentation

CCC Highway Cleveland-Columbus-Cincinnati Highway
CCCI Computer Control Company, Inc
CC circuit common-collector amplifier for junction transistors
CCCJ California Council on Criminal Justice
cccl cathodal closure clonus
CCCM Central Committee for Community Medicine
CCCN Customs Cooperation Council Nomenclature
CC Co Commercial Cables Company
CCCP California Coalition for Capital Punishment; Citizens Crime Commission of Philadelphia; Council on Cooperative College Projects
CCCP (Russian transliteration—U S S R)—*Soyuz Sovetchikh Sotsialisticheckikh Respublik* (Union of Soviet Socialist Republics)
CCCPS Chicago College of Chiropody and Pedic Surgery
CCCR Center for Crime Control Research; Communications and Command Control Requirements
CCCS Concerned Citizens for Community Standards; Consumer Credit Counseling Services
CCCT California Community College Trustees
ccd charge-coupled device; computer-controlled display
ccd (CCD) charge-coupled device (photographic)
CCD Center for Curriculum Development; Central Council for the Disabled; charge-coupled device; Circuit Court Decision; Confraternity of Christian Doctrine; Cost Center Determination; Criminal Conspiracy Division
CCDA Commercial Chemical Development Association
CCDB Canadian Car Demurrage Bureau
CCDC Center City Development Corporation; Central Citizens' Defence Committee; Centre City Development Corporation; Commission on Crime, Delinquency, and Corrections (Nevada)
CCDN Central Council for District Nursing
cce carbon-chloroform extract
CCE California Cooperative Extension; *Casa de la Cul-*

tura Ecuatoriana (Spanish—House of Ecuadorian Culture); Council for a Competitive Economy
CCEA Cabinet Council on Economic Affairs
CCEBS Committee for the Collegiate Education of Black Students
CCED County Council Electoral Division
ccei composite cost-effectiveness index
CCEM Comprehensive Career Education Model
CCES Catholic Church Extension Society
CCET Carnegie Commission on Educational Television
CCETSW Central Council for Education and Training of Social Workers
CCETT *Centre Commun d'Etudes de Télévision et de Télécommunication* (French—Public Center for the Study of Television and Telecommunication)
ccf cephalin-cholesterol flocculation; compound comminuted fracture; congestive cardiac failure; chronic cardiac failure; concentrated complete fertilizer
CCF Canadian Commonwealth Federation; Citizens Council Forum; Combined Cadet Force; Common Cold Foundation; Congressional Clearinghouse on the Future; Cooperative Commonwealth Federation; Credit Commercial de France
CCFA Combined Cadet Force Association
CCFC Citizens Committee for a Free Cuba
ccfe commercial customer-furnished equipment
CCFG California Contemporary Fashion Guild
ccfm cryogenic continuous-film memory
ccfr constant current flux reset
CCG Canadian Coast Guard; Choral Conductors Guild; Control Commission of Germany
CCGB Cycling Council of Great Britain
CCGE California Council for Geographic Education
CCGEA Community College General Education Association
CCGNY Community Council

of Greater New York
CCGS Canadian Coast Guard Service; Canadian Coast Guard Ship
ccgt closed-cycle gas turbine
cch cubic capacity of holds
Cch Christchurch, New Zealand
CCH Chaminade College of Honolulu; Commercial Clearing House; Computerized Criminal Histories
C of CH Chief of Chaplains
CCHE California Coordinating Council for Higher Education; Central Council for Health Education; Coordinating Council for Higher Education
CC-HEW Clinical Center—HEW
CCHF Children's Country Holidays Fund
CCHK Crown Colony of Hong Kong
CCHPP California Council for the Humanities in Public Policy
cc/hr cubic centimeters per hour
CCHS Christopher Columbus High School; Computerized Criminal Histories System (FBI)
CCHW Citizen's Clearinghouse for Hazardous Wastes
cci chronic coronary insufficiency; circuit condition indicator; concentric coordinate incident; corrugated, cupped, or indented (cargo)
CCI Citizens Committee of Investigation (into President Kennedy's assassination); Community Concerts, Incorporated; Computer Consoles Inc; Connecticut Correctional Institution; Conservative Caucus, Inc
CCI *Central Campesina Independiente* (Spanish—Independent Peasant Central)—political party in Mexico
CCIA Consumer Credit Insurance Association
CCIAP Cooperative Committee on Interstate Air Pollution (New Jersey-New York)
ccib computerized central information bank
CCIB Chinese Commodities Inspection Bureau; Computerized Central Information Bank; Cook County Inspection Bureau
ccig cold cathode ion gage

CCIL Commander's Critical Item List (USA)
CCIM Certified Commercial Investment Member
CCINC Cabinet Committee for International Narcotic Control
ccip continuously computed impact point (USAF)
CCIPT Canadian Center for Investigation and Prevention of Torture
CCIR *Comité Consultatif International de la Radiodiffusion* (French—International Consultative Committee on Broadcasting)
CCIS Command Control Information System
CCIT California Council for International Trade
CCITT Consultative Committee in International Telephone and Telegraph
CCIW Canada Centre for Inland Waters
CCJ Center for Correctional Justice (Harvard Law School); Center for Criminal Justice (Washington, D.C.); Circuit Court Judge; Cook County Jail (Chicago); County Court Judge
CCJC Canadian Centre for Justice Statistics; Chicago City Junior College; Cook County Junior College; Custer County Junior College
CCJCA California Community and Junior College Association
CCJO Consultative Council of Jewish Organizations
cck (CCK) cholecystokinin
CCK Centre College of Kentucky
cck-pz (CCK-PZ) cholecystokinin-pancreozymin
cckw counterclockwise
ccl cancel(ed); cancellation
CCL Canadian Congress of Labour; Caribbean Cruise Lines; Commissioner for Crown Lands; Commodity Control List; Council for Civil Liberties
CCl₄ carbon tetrachloride
C-clamp C-shaped clamp
C-class Soviet nuclear-powered submarines nicknamed Charlies by NATO; capable of underwater missile launchings
CCLC Cooperative College Library Center
c clef alto clef (on the third

line); soprano clef (on the first line); tenor clef (on the fourth line)

CC List Critical Condition List

cclkws counterclockwise

CCLM Coordinating Council of Literary Magazines

ccln consignment note control label number

CCLN Council for Computerized Library Networks

CCLP Cabinet Council on Legal Policy

ccl's criminal criminal lawyers

CCLs Court of Claims

CCLS Canadian Council of Library Schools

ccm cubic centimeter(s); counter-countermeasure(s)

ccm Kubikzentimeter (German—cubic centimeter)

CCM California College of Medicine; Canadian Cycle Manufacturers; Caribbean Common Market; Certified Club Manager

CCMA Canadian Council of Management Association; Community College Media Association

ccmc coincident-current magnetic core

CCMC College-Conservatory of Music of Cincinnati; Cross-Country Motor Club

ccmd continuous-current monitoring device

CCMD Chicago Contract Management District

cc/min cubic centimeters per minute

CCMP Coalition of Concerned Medical Professionals

CCMR Central Contract Management Region

CCMS California College of Mortuary Science; Chicago Chamber Music Society; Committee on the Challenges of Modern Society (NATO)

ccmt catechol-O-methyl transferase

CCMTC Crown Cork Manufacturers' Technical Council

ccmv (CCMV) cowpea chlorotic-mottle virus

ccn coronary care nurse; coronary care nursing

CCN Cloud Condensation Nuclei; Command Control Number; Community of the Cross of Nails; *Companhia Colonial de Navegação* (Portuguese—Colonial Navigation Company); Contract Change Notice; Contract Change Notification

CCNA Canadian Community Newspapers Association

CCNDT Canadian Council for Non-Destructive Testing

CCNI Cía Chilena de Navegación Interoceanica (Spanish—Chilean Interoceanic Navigation Company)

CCNM Chaco Canyon National Monument

CCNMA California Chicano News Media Association

CCNP Callao Cave National Park (Luzon, Philippines); Carlsbad Caverns National Park (New Mexico)

CCNR Citizens Committee on Natural Resources; Consultative Committee for Nuclear Research

CCNS Cape Cod National Seashore (Massachusetts)

CCNSC Cancer Chemotherapy National Service Center

CCNV Community for Creative Non-Violence

CCNWR Cross Creeks National Wildlife Refuge (Tennessee)

CCNY Carnegie Corporation of New York; City College of the City University of New York

cco current-controlled oscillator

cc/o certificate of consignment/origin

Cco Curaçao

CCO Chicago College of Osteopathy; Clandestine Communist Organization; Commonwealth Communications Organization; Comprehensive Certificate of Origin; Crossroads Chamber Orchestra

CCOA California Correctional Officers Association; County Court Officers' Association; Crafts Council of Australia

CCOC Command Control Operations Center (USA)

CCOD Coalition to Cease Ocean Dumping

CCOFI California Cooperative Oceanic Fisheries Investigations

c conc cast concrete

CCOS Cabinet Committee on Opportunity for the Spanish Speaking

CCOU Construction Central Operations Unit

ccp control change proposal; credit card purchase

ccp conto corrente postale

(Italian—current postal account)

CCP Caribbean Conservation Program; Central Cataloging Project; Chinese Communist Party; Code of Civil Procedure; Code of Criminal Procedure; Commonwealth Centre Party; Consolidated Cryptologic Program; Cultural Center of the Philippines

ccpa cloud chamber photographic analysis

CCPC Community Crime-Prevention Centers

CCPE Canadian Council of Professional Engineers

CCPF Commander-in-Chief—Pacific Fleet

CCPG Chemical Corps Proving Ground

CCPI California Consumer Price Index

cc-pill compound-cathartic pill

CCPIT China Commission for the Promotion of International Trade

CCPL Corpus Christi Public Library

CCPO Central Civilian Personnel Office

CCPO Comité Central Permanent de l'Opium (French—Permanent Central Opium Committee)

CCPOA California Correctional Peace Officers Association

CCPOST California Commission on Peace Officer Standards and Training

CCPP Citizen Commission on Pension Policy

ccpr coherent cloud physics radar

CCPR Central Council of Physical Recreation

CCPS Certified Crime Prevention Specialist; Consultative Committee for Postal Studies; Council of Commonwealth Public Service

CCPSHE Carnegie Council on Policy Studies in Higher Education

CCPSO Council of Commonwealth Public Service Organisations

ccr closed-cycle refrigerator; combat crew; command control receiver; complex chemical reaction; computer character recognition; consumable case rocket; control circuit resistance; credit card reader; cross-channel rejection; crystal can relay; cube corner reflector

C$_{cr}$ creatinine clearance

CCR Center for Constitutional Rights; Central Commission for the Navigation of the Rhine; Commission on Civil Rights; Contract Change Request

CCRB Civilian Complaint Review Board

CCRB *Cooperatief Centraal Raiffeisen-Boerenleenbank* (Dutch—Raiffeisen's Central Cooperative Farmer's Loan Bank)—largest bank in the Netherlands

CCRDC Chemical Corps Research and Development Command

CCRE Canadian Council for Research in Education

CCRESPAC Current Cancer Research Project Analysis Center

CCRF City College Research Foundation

CCRKBA Citizens Committee for the Right to Keep and Bear Arms

CCRMA Center for Computer Research in Music and Acoustics (Stanford University)

ccros card-capacitor read-only storage

C Cr P Code of Criminal Procedure

CCR & R covenants, conditions, restrictions, and reservations

CCRs Covenants, Conditions, and Restrictions

CCRS Canadian Centre for Remote Sensing

ccrt cathodochromic cathode-ray tube

CCRT Check Collectors Round Table

ccru complete crew

CCRU Common Cold Research Unit

ccs central computer and sequencer; collective call sign; column code suppression; command, control, support (military function); computer control station(s); custom contract service(s)

cc's carcasses

cc & s central computer and sequencer

Ccs Caracas (inhabitants called Caraqueños)

CCs Community Centers; Community Colleges

CCS Canadian Cancer Society; Cape Cod System; Caracas, Venezuela (Maiquetia Airport); Casualty Clearing Station; Catholic Community Service; Center for Chinese Studies (University of California); Center for Cuban Studies; Chemical Corporation of Singapore; Chief Commissary Steward; Church of Christ, Scientist; Civil Communications Section; Combined Chiefs of Staff; Controller of Communication Services; Council of Communication Societies; Customer Conversion Statistics

CCSA Canadian Committee on Sugar Analysis; Community College Service Association; Community College Student Association

CCSB Credit Card Service Bureau

CCSC Central Connecticut State College; Central Coordinating Staff, Canada

CCSD Counter Counter-Strategic Defense

CCSEA Council of the Church of Southeast Asia

cc/sec cubic centimeters per second

ccsem computer-controlled-scanning electron microscope

ccsep cement-coated single epoxy

CCSF City College of San Francisco

CCSL Communications and Control Systems Laboratory

CCSN Center City Shelter Network

CCSO Corpus Christi Symphony Orchestra

ccsr cash-to-common-stock ratio

CCSS Charles Camille Saint Saëns; Cleveland-Cliffs Steamship (company)

CCSSO Council of Chief State School Officers

CCST Chelsea College of Science and Technology

cct cathodal closing tetanus; chocolate-coated tablet; controlled cord traction

CCT Clarkson College of Technology; Combat Control Team; Common Customs Tariff; Cumberland College of Tennessee

C & CT Chemistry and Chemical Technology

CCTA Central Computer and Telecommunications Agency; Community College Trustees Association

CCTC Chinese Cultural and Trade Center; Columbia County Teachers College

ccte cathodal closure tetanus

CCTE Canadian Council of Teachers of English

cctep cement-coated triple epoxy

CCTF California Correctional Training Facility

CC & TI Community College and Technical Institute

cctks cubic capacity of tanks

CCTP Center City Transportation Program; Coronary Care Training Project

CCTS Canaveral Council of Technical Societies; Combat Crew Training School

CCTT Cornell Critical Thinking Test

cctv closed-circuit television

CCTWg Combat Crew Training Wing (USAF)

ccu chart comparison unit; color-control unit; coronary care unit

ccu (CCU) camera-control unit (television); cardiac care unit; coronary care unit; correctional custody unit (U.S. naval vessel's brig or jail)

Ccu Calcutta

C-C u Cherry-Crandall units

CCU Calcutta, India (airport); California Conservative Union; Community College Unit; Confederation of Canadian Unions; Cooperative Care Unit; Council for Canadian Unity

CCUDA Community College Urban District Association

CCUL California Credit Union League

CCUN Collegiate Council for the United Nations

CCURR Canadian Council on Urban and Regional Research

CCUS Chamber of Commerce of the United States

c/cut crosscut

ccv closed-circuit voltage; coolant control valve

ccv (CCV) control-configured vehicle

ccw carrying a concealed weapon; channel command word; counterclockwise

CCW Caldwell College for Women; Circulation Control

Wing; Citizens Crime Watch; Combat Crew Wing
CCWA Consultative Commission on Workers Affairs
cc wr hdr canvas-covered wirerope handrail
ccws counterclockwise
ccxd computer-controlled X-ray diffractometer
ccy currency
cd caesarean delivery; candela; canine distemper; cash discount; center door; certificate of deposit; civil defense; coin dimpler; cold drawn; communicable disease; confidential document; conjugate diameter (pelvic inlet); contagious disease; convulsive disorder; convulsive dose; cord; countdown; curative dose
cd (CD) compact digital; compact disc; compact-disc player; compact-disc record(ing); companion dog; compounded daily
c-d countdown
c/d cash against documents; cigarettes per day; cigars per day
c/d (C/D) carried down (bookkeeping); certificate of deposit
c & d carpets and drapes; censorship and documents; collection and delivery
cd cadde (Turkish—street); **colla destra** (Italian—with the right hand); **corriente directa** (Spanish—direct current)
c.d. conjugata diagonalis (Latin—diagonal conjugate)—pelvic inlet diameter
Cd cadmium; caudal; coefficient of drag
C $ cordoba (Nicaraguan monetary unit)
Cd ciudad (Spanish—city)
CD Canadair turboprop airplane; Certificate of Deposit; Civil Defense; coastal defense radar (for surface-vessel detection); communicable disease; Community Development; confidential document; Control Data; **Corps Diplomatique** (French—Diplomatic Corps); Corrections Department (New Mexico); Corrections Division (Hawaii, Oregon); countdown; Customs Declaration
C.D. Chancery Division
C/D Commercial Dock, consular declaration; Customs Dec-

laration
C & D Chemist and Druggist; collection and delivery
CD Centre de Détention (French—Detention Center); **Centre Démocrate** (French—Democratic Center); **Commission du Danube** (French—Danube Commission); **Computer Design**
C & D Crime and Delinquency
cd$_{50}$ median curative dose (abolishing symptoms in 50 percent of all test cases)
Cd$_{115}$ radioactive cadmium
cda chain data address; command and data acquisition
cda (CDA) chenodeoxycholic acid
CDA California Dental Association; California Dietetic Association; Canadian Dental Association; Canadian Dietetic Association; Catholic Daughters of America; Child Development Association; **Compañía Dominicana de Aviación** (Spanish—Dominican Aviation Company); Control Data Australia; Copper Development Association
CDA Christen Democratisch Appèl (Dutch—Christian Democratic Appeal) political party
CD Act(s) Contagious Diseases Act(s)
CDAE Civil Defense Adult Education
Cd A Eng Commissioned Air Engineer
CD Aim Commissioned Airman
Cd Airn Commissioned Airman
Cdale Carriedale
CDAP Civil Defense Auxiliary Police
CDARC Chelsea Drug Addiction and Research Centre
CDAS Civil Defense Ambulance Service
cdb caliper disk brake; capacitance decode box; cast double base; central data bank; could be; current data bit
Cd B Commissioned Boatswain
CDB Caribbean Development Bank; Combat Development Branch
cdba clearance divers breathing apparatus
CDBA California Dining And Beverage Association

cdbd cardboard
CDBG Community Development Block Grant
CD/BMI Columbus Division/ Battelle Memorial Institute
Cd Bndr Commissioned Bandmaster
cdc calculated date of confinement; call direction code; career development course; command and data-handling console
CDC Cadaver Disposal Center; California Debris Commission; California Democratic Council; California Department of Corrections; Canada Development Corporation; Caribbean Defense Command(er); Centers for Disease Control; Certificate of Disposition of Classified Documents; Cesspool Detergent Chemistry; Citizens' Defense Corps; Citizens Democracy Corps; Civil Defense Coordinator; Civil Defense Council; Combat Development Command; Command Destruct Control; Commissioners of the District of Columbia; Commonwealth Development Corporation; Communicable Disease Center; Community Development Corporation; Configuration Data Control; Conservation Data Center; Control Data Corporation; Control Distribution Center; Criminal Diagnostics and Counseling
C.D.C. Commonwealth Development Corporation (formerly Colonial Development Corporation)
CDC Centro de Documentação Científica (Portuguese—Scientific Documentation Center)
CDCA chenodeoxycholic acid
cdce central data-conversion equipment
CD circuit common-drain circuit for field-effect transistors
cdcm carbon-dioxide concentration module
Cd Cmy O Commissioned Commissary Officer
Cd C O Commissioned Communications Officer
Cd Con Commissioned Constructor
CDCP Construction and Development Corporation of the Philippines
CDCR Center for Documenta-

tion and Communication Research; Control Drawing Change Request

CDCs Community Development Corporations

CDCS Civil Defense Countermeasures System; Construction Dollar Control System

CDCT *Centro de Documentación Cientifica y Téchnica* (Spanish—Center of Scientific and Technical Documentation)—Mexico City

cdd central data display; chart distribution data; coded decimal digit; color data display; command-destruct decoder; computer-directed drawing; cosmic dust detector; cratering demolition device

CDD Certificate of Disability for Discharge

cddd comprehensive dishonesty, disappearance, and destruction (insurance policy)

cddi computer-directed drawing instrument

CDDP Canadian Department of Defense Production

cde carbon dioxide economizer; contamination-decontamination experiment

cde (CDE) canine distemper encephalitis

CDE Central Document Exchange; Conference on Disarmament in Europe; Cornell-Dubilier Electronics

C.D.E. Certificate in Data Education

C de C (CDC) Canyon de Chelly

CDEE Chemical Defense Experimental Establishment

C de F *Collège de France* (French—College of France)

C de G *Ciudad de Guatemala* (Spanish—Guatemala City); *Croix de Guerre* (French— War Cross)

CDEG Chicago District Electric Generating Corporation

CDEI Control Data Education Institutes

C de J *Compañía de Jesus* (Spanish—Company of Jesus)—Society of Jesuits

cdek computer data entry keyboard

Cd El O Commissioned Electrical (Electronic) Officer

C del S *Corriere della Sera* (Evening Courier—Milan)

C de M *Ciudad de México* (Spanish—Mexico City)

C-de-N Côtes-de-Nord

Cd Eng Commissioned Engineer

CDEOS Civil Defense Emergency Operations System

CDER Center for Death Education and Research

cdf command decoder film; command decoder filter; confined detonating fuse; constant current fringes; continuous desk file

CDF California Department of Forestry; Canadian Department of Forestry; Children's Defense Fund; Civil Defence Force; Colorado Department of Forestry; Community Development Foundation; Congregation for the Doctrine of the Faith (conservative Catholics); Connecticut Department of Forestry; Council for the Defense of Freedom

CDFA California Dried Fruit Association

CDFC Commonwealth Development Finance Company

CDFGI Charles Darwin Foundation for the Galápagos Islands

CD film camouflage detection film

cdfr (CDFR) commercial demonstration fast reactor

CDFRS Charles Darwin Foundation Research Station, Academy Bay, Santa Cruz, Galápagos

CDFS Chief of Defence Force Staff

CDFSB Canadian Dairy Foods Service Bureau

cd/ft² candela per square foot

cd fwd carried forward

Cdg Cardigan; Cardiganshire

CDG Coder-Decoder Group (USA)

CDGA California Date Growers Association

CD & GB TC Chicago, Duluth and Georgian Bay Transit Company

Cd Gr Commissioned Gunner

Cd Gr O Commissioned Gunnery Officer

cdh constant differential height

CDH College Diploma in Horticulture

CDHS Comprehensive Data-Handling System

cdhv could have

cdi course deviation indicator

cdi (CDI) capacitor discharge ignition

CDI Center for Defense Information; Children's Depres-

sion Inventory; Classified Defense Information; Comprehensive Dissertation Index; Contractor Demonstration Inspection; Conventional Defense Improvement (NATO)

CDIC Canada Deposit Insurance Corporation

Cd In O Commissioned Instructor Officer

C Dip F & A Certified Diploma in Finance and Accounting

c div cum dividend

Cd J *Ciudad Juárez* (Spanish—Juarez City) (inhabitants—*Juaristas*)

CDJ California Department of Justice

CDJ *Comité de Défense des Juifs* (French—Committee of the Defense of Jews)

cdk containers (carried on) deck

c$k consumer's survival kit

cdl common display logic

Cdl Cardinal

CDL Central Dockyard Laboratory (UK); Citizens for Decency through Law; Citizens for Decent Literature; Country and Democratic League

CDLC Canadian Dental Laboratory Conference

cdm contributing to the delinquency of a minor

CDM Coalition for a Democratic Majority; Consolidated Diamond Mines (South Africa)

cd/m² candela per square meter

cdma code division multiple access

Cd M-a-A Commissioned Master-at-Arms

CDMB Civil Defense Mobilization Board

CDMMA Canadian Direct Mail Marketing Association

CDMSWA Consolidated Diamond Mines of South-West Africa

cdn (CDN) condition

CDN *Chicago Daily News*

CDNAC Canadian Daily Newspaper Advisory Council

CDNPA Canadian Daily Newspaper Publishers Association

CDNRA Coulee Dam National Recreation Area (Washington)

CDNS Chicago Daily News Service

cdnt could not

cdo (CDO) chronic drunkenness offender

Cd O Commissioned Officer

C-d'O Côte-d'Or

CDO California Disaster Office; Civil Defense Organization; Community Development Office(r)

Cd Ob Commissioned Observer

Cd O E Commissioned Ordnance Engineer

CDOGS Council for the Defence of Government Schools

Cd O O Commissioned Ordnance Officer

cdos controlled date of separation

cdo(s) commando(s)

cdp checkout data processor; communications data processor; contract definition phase

CDP Centralized Data Processing; Certified Data Plan; Critical Decision Point

C.D.P. Certificate in Data Processing

CDPA Civil Defense Preparedness Agency

cdpc central data-processing computer

CDPC California Delinquency Prevention Commission

CDPE Continental Daily Parcels Express

cd pl cadmium plate

cd player compact-disc player

cdp's comprehensive dwelling policies

cdr command-destruct receiver; composite damage risk (audiometry)

cdr (CDR) crude death rate

Cdr Commander

C d R *Casa di Risparmio* (Italian—Savings Bank)

CDR Change Design Request; Conceptual Design Report; Countdown Deviation Request; Critical Design Review(s)

CDR *Comité Defensa Revolucionario* (Spanish—Revolutionary Defense Committee)—Cuba

CDRA Canadian Drilling Research Association; Committee of Directors of Research Associations

CDRB Canadian Defense Research Board

CDRBTE Canadian Defense Research Board Telecommunication Establishment

CDRC Civil Defense Regional Commission(er)

Cdre Commodore

CDRF Canadian Dental Research Foundation

CDRI Central Drug Research Institute

cdrill center drill

Cdrngtn C Codrington College

Cd R O Commissioned Radio Officer

cdrom (CD Rom) compact-disc read-only memory

CDRS Charles Darwin Research Station

CDRSC Children's Depression Rating Scale for Classrooms

cds cards; circular date stamp; cold-drawn steel; single cotton double silk (insulation)

cds (Cd S) cadium sulfide

cd's (CDs) certificates of deposit; compact-disc players; compact-disc record(ing)s—played by a laser beam

C d S *Circolo della Stampa* (Italian—Press Club); *Codice della Strada* (Italian—Highway Traffic Code); *Consiglio di Sicurezza* (Italian—Security Council)

CDs Catholic Documents; compact discs

CDS California Dental Service; Center for Degree Studies; Civil Defence Services; Civil Defense Staff; Climatological Data Sheet; Commander, Destroyer Squadron; Community Dispute Services

Cd S B Commissioned Signals Boatswain

cdse computer-driven simulation environment

CdSh Commissioned Shipwright

CDSHA Country Day School Headmasters Association

Cd S O Commissioned Supply Officer

CDSO Commonwealth Defense Service Organization

CDSP Church Divinity School of the Pacific

CDSP *Current Digest of the Soviet Press*

CDSs Civil Disobedience Squads

CDSS British Post Office trade mark covering telecommunications and telephonic apparatus, instruments, and installations; Compressed Data Storage System; Customers' Digital Switching System

CDST Central Daylight Saving Time

cdt command-destruct transmitter; conduct; conductor

Cdt Cadet; Commandant

CDT Canadian Department of Transport; Central Daylight Time

CDT (ADA) Council on Dental Therapeutics (American Dental Association)

C.D.T. Certified Dental Technician

Cdte *Comandante* (Spanish—Commander)

CDTE Council for Distributive Teacher Education

cd tec compact-disc technology

Cdt Mid Cadet Midshipman

cdts constant-depth temperature sensor

cdu cable distribution unit; central display unit; chemical dependency unit

CDU Cable Distribution Unit; Christian Democratic Union (Germany); Civil Disobedience Unit; coastal defense (radar) unit; Computer Display Unit

CDU *Christlich-Demokratische Union* (German—Christian Democratic Union)—political party

CDUEP Civil Defense University Extension Program

cdv cadaver; *carte de visite* (visiting card, sometimes with photograph); computed dollar value; current domestic value

Cdv Commonwealth dollar value

CDV Civil Defense Volunteer(s)

cdvr cadaver

cdw charge density wave; chilled drinking water; collision damage waiver

CDW Civil Defense Warning; Collision Damage Waiver

CD & W Colonial Development and Welfare

Cd Wdr Commissioned Wardmaster

cdwe could we

Cd W O Commissioned Writer Officer

c dwr chest of drawers; chilled drinking water return

CDWS Civil Defense Wardens Service

cdwt cordwelt

cdx control differential transmitter

cdz concordant zone

Cdz Cádiz

ce carbon equivalent; career education; center of effort (naval architecture); center entrance; circular error; compression engine; constant error; consumption entry; counterespionage; critical examination; cum entitlement

ce (CE) counterespionage

c-e communications-electronics

c/e custom entry

c & e commission and exchange; customs and excise

c of e coefficient of elasticity

c.e. caveat emptor (Latin—let the buyer beware); *curvée extra* (French—special sort)—spe cial quality

Ce Ceará; cerium; Ceylon

CE Canada East; Chief Engineer; Chief Executive; Christian Endeavor; Church of England; circular error; Common Era; compass error; Corps of Engineers; cost effectiveness; Counselor of Embassy; Customer Engineer

C-E communications electronics

C.E. Chemical Engineer; Civil Engineer

C/E Chancellor of the Exchequer; Chief Engineer

C of E Church of England (Protestant Episcopal); Corps of Engineers

CE Chemical Engineering; *Comedy of Errors*

C.E. Christian Era; Civil Engineer

cea circular error average

cea (CEA) carcinoembryonic antigen

CEA Canadian Education Association; Canadian Electrical Association; Canadian Export Association; Captain's Endowment Association (police); Childbirth Education Association; Classified Employees Association; College English Association; Combined Educational Associations; Combustion Equipment Associates; Commodity Exchange Authority; Congressional Education Associates; Connecticut Educational Association; Conservation Education Association; Cooperative Education Association; Correctional Education Association; Council of Economic Advisers; Council of Engineering Associations;

County Employees Association

CEA Comisión Económica de Africa (Spanish—Economic Commission of Africa); *Commissariat à l'Energie Atomique* (French—Atomic Energy Commission)

CEAA Center for Editions of American Authors; Council of European-American Associations

CEAC Commission for European Airspace Coordination; Consulting Engineers Association of California

CEAC Commission Européenne de l'Aviation Civile (French—European Civil Aviation Commission)

CEAFU Concerned Educators Against Forced Unionism

CEAN Community Energy Action Network

CEANAR Commission on Education in Agriculture and National Resources

CEAPD Central Air Procurement District

CEARC Computer Education and Applied Research Center

CEARD Commission for Attention to Refugees and Displaced Persons (Guatemala)

CEAT Canadian English Achievement Test

ceb cryogenic expulsive bladder

Ceb Cebu

CEB Central Electricity Board; Continuing Education Books

CEB Comité Electrotechnique Belge (Belgian Electrotechnical Committee)

cebar chemical, biological, radiological warfare

CEBS Certified Employee Benefit Specialist; Church of England Boys' Society; Commonwealth Experimental Building Station

Ceb-Vis Cebu-Visayan

CEC California Energy Commission; Canadian Electrical Code; Catholic Education Council; Central Economic Committee; Ceramic Educational Council; Chief Executive Commissioner; Civil Engineer Corps; Coal Experts Committee; Commission of the European Community; Commodity Exchange Commission; Commonwealth Economic Committee; Commonwealth Edison Company;

Communications and Electronics Command; Consolidated Edison Company; Consolidated Electrodynamics Corporation; Consulting Engineers Council; Continental Entry Chart(s); Correctional Economics Center; Council for Exceptional Children

CECA Communauté Européenne du Charbon et de l'Acier (French—European Coal and Steel Community); *Comunidad Europea del Carbon y del Acero* (Spanish—European Coal and Steel Community)

CECC California Educational Computer Consortium

Cece Cecil

CECEW Catholic Education Council for England and Wales (often truncated to CEC—Catholic Education Council)

CECH Citizenship Education Clearinghouse

Cechy (Czechoslovakian—Bohemia)

CECIL Compact Electronic Components Inspection Laboratory

CE circuit common-emitter amplifier for junction transistors

CECLA Comisión Especial de Coordinación Latinoamericana (Spanish—Special Commission for Latin American Coordination)

CECR Center for Environmental Conflict Resolution; Central European Communication Region (USAF)

CECs California Ecology Corpsmen

CECS Church of England Children's Society; Communications Electronics Coordinating Section

CECS Comisión Especial de Consulta sobre Seguridad (Spanish—Special Commission for Security Consultation)

ced capacitance electronic disc; communications–electronics doctrine; computer entry device

c-e-d carbon-equivalent-difference

c & ed clothing and equipment development

Ced Ceda; Cedomil; Cedric; Cedron

CED Committee for Economic

Development; *Communauté Européenne de Defense* (French—European Defense Community); Communications–Electronics Doctrine (USAF manuals); Council for Economic Development
CED *Centro Elletronnico di Documentazione* (Italian— Electronic Documentation Center)—in Rome
CEDA California Economic Development Agency; Canadian Electrical Distributors Association
CEDA *Comisión Económica de Oeste Asiatico* (Spanish— Economic Commission of West Asia); *Confederación Española de Derechas Autonomas* (Spanish—Spanish Confederation of Autonomous Rights)—Catholic party
cedac central differential analyzer control; cooling effect detection and control
CEDAL *Centro de Estudios Democráticos de America Latina* (Spanish—Latin American Center of Democratic Studies)
CEDAM Conservation, Exploration, Diving, Archeology, Museums (organization)
CEDAR Council for Educational Development and Research
Cedar Breaks Cedar Breaks National Monument in Utah's Wasatch Mountains
CEDDA Center for Experimental Design and Data Analysis
cedi unit of currency in Ghana
CEDI *Centre Européen de Documentation et d'Information* (French—European Documentation and Information Center); *Centro Europeo de Documentación e Información* (Spanish—European Documentation and Information Center)
CEDIC Church Estates Development and Improvement Company
CEDO Centre for Educational Development Overseas (UK)
CEDPA California Educational Data Processing Association
ced's captured enemy documents
cee computer-enhanced education
CEE Center for Environmental

Education; Central Engineering Establishment; Certificate of Extended Education; Common Entrance Examination; Cultural Environment Emergency
CEE *Comunidad Económica Europea* (Spanish—European Economic Community)
CEEA *Communauté Européenne de l'Energi Atomique* (French—European Atomic Energy Community)
CEEB College Entrance Examination Board
CEEC Council for European Economic Cooperation
CEECC Consolidated-Edison Energy Control Center
Ceece Cecil
CEEED Council on Environment, Employment, Economy, and Development
ceefax see the facsimile; see the facts
CEEP *Centre Européen d'Etudes de Population* (French— European Center for Population Studies)
CEev Central European encephalitis virus
cef cellular-expansion factor; chicken-embryo fibroblasts
CEF Canadian Expeditionary Force; Children's Emergency Fund(ing); Citizens for Educational Freedom; Citizens for Energy and Freedom
C of EF Count(y) of East Friesland Country
CEFA Council for Educational Freedom in America
ceff controlled energy flow forming
CEFP Council of Educational Facility Planners
CEFT Children's Embedded Fissures Test(ing)
CEFTRI Central Food Technological Research Institute
CEG Coalition for Economic Growth
CEGB Central Electricity Generating Board
CEGGS Church of England Girls' Grammar School
CEGJA Coalition to End Grand Jury Abuse
CEGS Church of England Grammar School
CEHHS Charles Evans Hughes High School
CEHS Civilian Employee Health Service
cei contract end item
CEI Claremont Economics In-

stitute; Cleveland Electric Illuminating Company; Commission Electrotechnique Internationale (International Electrotechnical Commission); Communications-Electronics Instruction; Cost Effectiveness Index; Council of Engineering Institutions
C & EI Chicago & Eastern Illinois (railroad)
CEI *Chemical Engineering Index; Commission Electrotechnique Internationale* (French—International Electrotechnical Commission)
CEIE Commission on Excellence in Education (U.S.)
CEIF Community Employment Initiatives Fund; Council of European Industrial Federations
ceil ceiling
cein contract end-item number
CEIP Carnegie Endowment for International Peace; Communications-Electronics Implementation Plan
C-E-I-R Corporation for Economic and Industrial Research
CEIS California Education Information System; Cost and Economic Information System
ceisd customer engineering instruction system diagram
CEIWT Central Europe Inland Waterways Transport (NATO)
cej cement-enamel junction
CEJEDP Central Europe Joint Emergency Defense Plan (NATO)
CEJNSA Council of European and Japanese National Shipowners Associations
cel celluloid; cellulose
c-e-l carbon-equivalent-liquid
Cel Celeban; Celebes; Celsius
CEL Constitutional Educational League; Cryogenics Engineering Laboratory
CELAC *Comisión Económica de Latinoamérica y el Caribe* (Spanish—Economic Commission of Latin America and the Caribbean)
cel acet cellulose acetate
CELADE *Centro Latinoamericano de Demografía* (Latin American Demographic Center)
CELDS Computerized Environmental Legislative Data System
celeb celebrate; celebration;

celebrity
celebs celebrities
celest celestial
Celia Cecilia
celintrep accelerated intelligence report
cell celluloid
CELL Case Existological Laboratories Limited; Continuing Education Learning Laboratory
celli cellos (violoncellos)
'cellist(s) violoncellist(s)
cello violoncello
celnav celestial navigation
cel nitr cellulose nitrate
celo chicken embryo lethal orphan (virus)
CELOS *Centrum voor Landbouwkundig Onderzoek in Suriname* (Dutch—Center for Agricultural Research in Surinam)
cels (Cels) celsius
CELS Continuing Education for Library Staffs
cel sheet cellulose (plastic) sheet
CELSS Closed Ecological Life-Support System
celt classified entries in lateral transposition
Celt Celtic
CELT Comprehensive English Language Test(ing)
celtuce celery-lettuce (lettuce-derived vegetable whose stalks taste like celery)
cem cement; cement asbestos; cemetery; channel electron multiplier; communication-electronics and meteorological
CEM Council of European Municipalities
CEM *Confederación Evangelical Mundial* (Spanish—World Evangelical Confederation)
CEMA Canadian Electrical Manufacturers Association; Connecticut Educational Media Association; Conveyor Equipment Manufacturers Association; Council for Economic Mutual Assistance; Council for the Encouragement of Music and the Arts
CEMAA Council for Egg Marketing Authorities of Australia
cem ab cement asbestos board
cemad coherent echo modulation and detection
cemb *cembalo* (Italian—harpsichord)

CEMB Communications-Electronic-Meteorological Board (USAF)
CEMCO Continental Electronics Manufacturing Company
cemf counter-electromotive force
cem fl cement floor
c-e mix chloroform-ether mixture
CEMLA *Centro de Estudios Monetarios Latinoamericanos* (Spanish—Center of Latin American Monetary Studies)
CEMO Command Equipment Management Office
cemon customer engineering monitor(ing)
cem p cement paint
CEMPIMS Communications Electronics Meteorological Program Implementation Management System (USAF)
cem plas cement plaster
CEMR Canadian Energy, Mines, and Resources
CEMREL Central Midwestern Regional Educational Laboratory
CEMS Church of England Mens' Society
CEMT *Conférencia Europea de Ministros de Transporte* (Spanish—European Conference of Ministers of Transport)
cen center; central; centralization; centralize
Cen Cenozoic; Centaurus (constellation)
CEN Captive European Nations; Central Airlines
CEN *Comité Européen de Coordination des Normes* (French—European Committee of the Coordination of Standards)
CENA Coalition of Eastern Native Americans
CENACO *Centro Nacional de Computación* (Spanish—National Computation Center)
Cenacolo *Il Cenacolo* (Italian—Refectory, Supper Room)—Leonardo da Vinci masterpiece—*L'Ultima Cena*—The Last Supper
CENAMEC *Centro Nacional para el Mejoramiento de la Enseñanza de la Ciencia* (Spanish—National Center for the Betterment of the Teaching of Science)—Venezuelan society
CENCOMMURGN Central

Communications Region
CENCOMS Center for Communication Sciences (USA)
CENDES *Centro de Enseñanza para el Desarollo* (Spanish—Center of Learning for Development)
CENDHRRA Center for the Development of Human Resources in Rural Asia
CENDIT Centre for Development of Instructional Technology
Cen Eccl *Censura Ecclesiastica* (Latin—Ecclesiastical Censure)
CENEUR *Compañía Española de Navegación Marítima* (Spanish—Spanish Maritime Navigation Company)
CENFAM *Centro Nazionale di Fisica dell'Atmospera e Meteorologia* (Italian—National Center of Physics of the Atmosphere and Meteorology)
C Eng Chartered Engineer; Chief Engineer
CENIM *Centro Nacional de Investigaciones Metalúrgicas* (Spanish—National Center for Metallurgical Research)
cenog computerized electroneuro-ophthalmograph
cens censor; censorship
CENS China Economic News Service
censor. centrifugal solids recovery
cent. centrifugal; century
cent. *centum* (Latin—hundred)
Cent Centaurus; Century
CENTA Committee for Establishing a National Testing Authority
centac(s) central tactical report(s)
CENTACS Center for Tactical Computer Sciences (USA)
CENTAG Central European Army Group
centen centennial
Centennial(s) Coloradan(s)
centi 10^{-2}
CENTO Central Treaty Organization (Great Britain, Iran, Pakistan, Turkey)
Central Archives records of the NSDAP on microfilm in the Hoover Institution on War, Revolution, and Peace, Stanford University, California
centrex central exchange
cents. centuries
cent(s) céntimo(s); one-hun-

dredth of a peseta
CENTURY *Humanist Century* (tabloid)
ceo chick embryo origin
ceo (CEO) chief executive officer
CeO Chairman ex-Officio
CEO Chief Education Office(r); Chief Engineer's Office; Chief Executive Officer; County Employees Organization
CEOA Central European Operating Agency
CEOAS Corps of Engineers Office of Appalachian Studies (USA)
CEOED *Compact Edition of the Oxford English Dictionary*
CEOs Chief Executive Officers (conglomerate and multinational corporations)
CEOSL *Confederación Centroamericana de Organizaciones Sindicales Libres* (Spanish—Central American Confederation of Free Trade Unions)
cep circle of equal probability; circle of error probability
'cep' except
Cep Cepheus
CEP Capability Evaluation Plan; Chicano Education Project; Civil Emergency Planning (NATO); Color Evaluation Program; Concentrated Employment Program; Continuing Education Program; Council on Economic Priorities; Council on Educational Policy
CEPA Chicago Educational Publishers Association; Civil Engineering Program Applications; Consumers Education and Protective Association
CEPACS Customs Entry Processing and Cargo System
CEPAL *Comisión Económica Para América Latina* (Spanish—Economic Commission for Latin America)— UNs ECLA
CEPB Civil Emergency Planning Bureau (NATO)
CEPC City of Erie Port Commission
CEPC *Comité Européen pour les Problèmes Criminels* (French—European Committee on Crime Problems)
CEPD Career Education Planning District; Council for

Economic Planning and Development
CEPDs Communications Electronics Policy Directives (NATO)
CEPE Central Experimental and Proving Establishment; *Corporación Estatal Petrolera Ecuatoriana* (Spanish—Ecuadorian State Petroleum Corporation)
CEPEC Center for Professional Executive Career Development and Counselling
CEPEX Controlled Ecosystem Pollution Experiment
CEPG Cambridge Economic Policy Group
Ceph Cepheus
CEPH *Centre d'Etude du Polymorphisme Humain* (French—Center for the Study of Human Polymorphism)
cephal (Latin prefix—head)— cephalic
ceph floc cephalin flocculation (test)
CEPM Center for Educational Policy and Management
CEPO Central Engineering Projects Office (NATO); Corps of Engineers—Portland, Oregon
CEPR Center for Educational Policy Research
ceps civil engineering problems
CEPS Central Europe Pipeline System (NATO); Commonwealth-Edison Public Service; Cornish Engines Preservation Society
Cepsa *Compañía Española de Petróleos* (Spanish—Spanish Petroleum Company)
'cept accept; except
CEPT *Conférence Européenne des Administrations des Postes et des Télécommunications* (French—European Conference of Posts and Communications)
CEPTA Committee to End Pay Toilets in America
'cepted accepted; excepted
'cepting accepting; excepting
'ception deception; exception; perception; reception
cept(s) concept(s); precept(s)
CEQ Council on Environmental Quality (appointed by the President of the United States)
CEQA California Environmental Quality Act

ceqom combined electron quench and optical masker
cer ceramic; conditioned emotional response
c & er combustion and explosives research
CER Center for Educational Reform; Certification Evaluation Review; Combat Effectiveness Report; Community Educational Resources
CERA/ACCE Canadian Educational Researchers Association/Association Canadienne des Chercheurs en Education
ceram ceramic; ceramicist; ceramics
ceramal ceramic + alloy
CERB Coastal Engineering Research Board (USA)
cerc centralized engine-room control
CERC Coastal Engineering Research Center; Coastal Engineering Research Council
CERCA Commonwealth and Empire Radio for Civil Aviation
CERDS Charter on the Economic Rights and Duties of States
Cer.E. Ceramic Engineer
cerebro (Latin prefix— brain)—cerebral, cerebrospinal fluid
CERES Center for Research and Education in Sexuality
CERF Coastal Education and Research Foundation
CERI Center for Educational Research and Innovation; Clean Energy Research Institute (University of Miami)
CERIC Central ERIC
CERL Central Electricity Research Laboratories; Coastal Engineering Research Laboratory; Cooperative Educational Research Laboratory
CERLAL *Centro Regional para el fomento del Libro en America Latina* (Spanish— Regional Center for the Development of Books in Latin America)
cermet ceramic-metallic (powders fused to form solid nuclear fuel elements)
CERN Center for Nuclear Research
CERN *Commission Européenne pour la Recherche Nucléaire* (French—European Commission for Nuclear Research)
CEROILFOOD China Na-

tional Cereals, Oils, and Foodstuffs Import and Export Corporation

CERP Current Economic Reporting Program

CERP Centre Européen des Relations Publiques (French—European Center of Public Relations)

CERR Commonwealth Employees Redeployment and Retrenchment

CE/RRT Central Europe Railroad Transport (NATO)

cert certificate; certify

cert. *certiorari* (Latin—to certify; to be informed)—a writ from a superior court to a lower court to produce papers needed for a review of a case

CERT Communications Effectiveness Response Test; Cost-Effective Rapid Transit; Council of Energy Resources Tribes

CE/RT Central Europe Road Transport (NATO)

certif certificate(d)

cert inv certified invoice

certs certificates

cerv cervical

ces central excitatory state; compressor end seal; constant elasticity of substitution; constructive error source

CEs Council of Europe members

CES California Employment Security; Closed Ecological System; Commercial Earth Station; Committee on Earth Sciences; Commonwealth Education Scheme; Commonwealth Employment Service; Comprehensive Export Schedule; Conference on European Security; Consolidated Electronic Services; Consumer Electronics Show; Cost Effectiveness Study; Council for European Studies; Crew Escape System

CES Certificat d'Etudes Supérieures (French—Advanced Studies Certificate); *Consejo Económico y Social* (Spanish—Economic and Social Council)—UN

CESA Canadian Engineering Standards Association; Commercial Education Society of Australia; Cooperative Educational Service Agency

CESAALA Charles E Stevens American Atheist Library

and Archives (Austin, Texas)

CESAME Commonwealth Employment Service Animated Memory

CESAP Comisión Económica y Social de Asia y el Pacífico (Spanish—Economic and Social Commission of Asia and the Pacific)

CESAR Capsule Escape and Survival Applied Research

CESAR Compagnie d'Etudes des Stations Air-Route (French—Company for the Study of Airfields)

CESC Calcutta Electric Supply Corporation

cesemi computer evaluation of scanning electron microscopic image

cesi closed-entry socket insulator

CESI Council for Elementary Science International

cesk cable end-sealing kit

Cesko Soc Ceskoslovenská Socialistická (Czech—Czechoslovak Socialist Republic)

CESMM Civil Engineering Standard Method of Measurement

CESO Canadian Executive Service Overseas; Civil Engineer Support Office (USN)

CESO-W Council of Engineers and Scientists Organizations—West

c esp con espressione (Italian—with expression)

CESP Centrais Eléctricas de São Paulo

cesr conduction electron spin resonance

CESR Canadian Electronic Sales Representatives

cess assess; assessment; cessation; cession(aire); cessionary; cessment; cesspipe; cesspit; cesspool; success

cess. cesspit; cesspool; excrement

Cess Cecil

CESS Council of Engineering Society Secretaries; Crew Escape Subsystem

CESSAC Church of England Soldiers, Sailors, and Airmens Clubs

Cessna 180 6-passenger utility aircraft

Cessna 185 6-passenger utility aircraft called the Cessna 185 E Skywagon

Cessna 310 6-passenger aircraft designated U-3

Cessna FR-172 French four-place rocket launcher aircraft

CEST Career Education Study Trip(s)

Cestr Chester

Cestr. *Cestrensis* (Latin—of Chester)

cet capsule-elapsed time; controlled environmental test-(ing); corrected effective temperature; cumulative elapsed time

Cet Centus; Cetus (constellation)

CET Center for Employment Training; Central European Time; Certified Electrical Technician; Certified Electronics Technician; Common External Tariff (European Communities); Council for Educational Technology

CET Collèges d'Enseignement Technique (French—Technical Education Colleges)

CETA Chinese-English Translation Assistance; Comprehensive Employment and Training Act

CETA Centre d'études pour la Traduction (French—Center for the Study of Translation)

CETAG Centre d'études pour la Traduction, Grenoble (French—Center for the Study of Translation, Grenoble)

CETAP Centre d'études pour la Traduction, Paris (French—Center for the Study of Translation, Paris)

CETC Community and Education Center

CETC Centro de Estudios Tecnicos (Spanish—Center for Technical Studies)

CETDC China External Trade Development Council

CETEC Consolidated Engineering Technology Corporation

CETEDOC Centre de Traitement Electronique des Ducments (French—Center of Electronic Treatment of Documents)

CETEKA Ceskoslovenská Tisková Kancelár (Czechoslovakian Press Bureau)

CETEM Comprehensive Elementary Teacher Education Models

CETEX Committee on Contamination of Extra-Terrestrial Exploration (NASA)

CETF Clothing and Equip-

ment Test Facility (USA)
CETHV Council for the Education and Training of Health Visitors
ceti communications with extra-terrestrial intelligence
CETIS Centre de Traitement de l'Information Scientifique (French—Center for Processing Scientific Information)
CETME Centro de Estudios Tecnicos de Materiales Especiales (Spanish—Center of Technical Studies of Special Materials)
CETO Center for Educational Television Overseas
cet. par. ceteris paribus (Latin—other things being equal)
CETS Church of England Temperance Society; Commission on the Education of Teachers of Science; Contractor Engineering and Technical Services
ceu continuing education unit
CEU Christian Endeavor Union; Constructional Engineering Union
CEUCA Customs and Economic Union of Central Africa
CEUs Continuing Education Units
CEUSA Committee for Exports to the U.S.A.
cev cryogenic explosive valve
cevat combined environmental, vibration, acceleration, temperature
cevi contract exhibit vendor item
cew circular electric wire
CEW Church-Employed Women
cewrm communications-electronics war-readiness materiel
cex charge exchange; civil effects exercise
CEX Corn Exchange Bank (stock-exchange symbol)
Cey Ceylon; Singhalese
CEY Century Electric (stock-exchange symbol)
CEYC Church of England Youth Council
Ceyl Ceylon
Cey Rs Ceylon rupees
cf calf binding; carried forward; carrier frequency; carry forward; cement floor; center field; center of flotation; center forward; central files; central filing; centrifugal force; communication

factor; complement fixation; conception formulation; conditional freedom; continuous focusing; contract formulation; corrugated furnaces; cost and freight; counterfire; counting fingers; cubic feet; cubic foot; cystic fibrosis
c/f carried forward
c & f clearing and forwarding; cost and freight
c-to-f center-to-face
cf. confer (Latin—compare)
c.f. cantus firmus (Latin—fixed song)
Cf californium
Cf. Confessor (Latin—Confessor)
CF Cape Fear (railroad); Chaplain to the Forces; Chief of Finance; Coastal Frontier; Colorado Fuel & Iron (stock-exchange symbol); Commonwealth Fund(ing); Conservation Foundation; Consolidated Freightways; Corresponding Fellow
C/F Contract Formulation
C of F Chief of Finance
CF Carlo Felice (Italian) theater in Genoa; *Chemin de Fer* (French—Railroad); *Club de Fútbol* (Spanish—Football Club)
CF-5 Canadian version of the F-5 jet fighter
CF-86 Australian-built Canadian version of the F-86 jet fighter called Sabre
CF-100 Avro two-seat jet interceptor called the Canuck
CF-101 Canadian-built version of the F-101 jet interceptor named Voodoo
CF-104 Canadian version of the F-104 interceptor called Starfighter
cfa complement-fixing antibody; configural frequency analysis; cowl flap angle; crossed-field amplifier
cFa complete Freund's adjunct
CfA Coalitions for America
CFA Canadian Federation of Agriculture; Canadian Football Association; Canadian Forestry Association; Canadian Freight Association; Cat Fanciers' Association; Center for Astrophysics; Chartered Financial Analyst; Clearing and Forwarding Agents; Commission on Fine Arts; Commonwealth Firemen's Association; Community Facilities Administration; Con-

sumer Federation of America; Correctional Facilities Association; Council for Foreign Affairs; Country Fire Authority
CF & A Chief of Finance and Accounting (USA)
C & FA Cookery and Foods Association
CFA Colonies Française d'Afrique (French—French Colonies of Africa)
CFAA Circus Fans Association of America
c factor cleverness factor
CFAD Commander, Fleet Air Defense
CFADC Canadian Forces Air Defence Command; Controlled Fusion Atomic Data Center
cfae contractor-furnished aerospace equipment
CFAE Council for Financial Aid to Education
CFAE Centre de Formation en Aérodynamique Expérimentale (French—Training Center for Experimental Aerodynamics)
CFAF California Financial Aid Form
CFA franc unit of currency in Benin, Burkina Faso, Cameroon, Central African Republic, Chad, Comoros, Congo, Gabon, Ivory Coast, Niger, Senegal, Togo
CFAL Current Food Additives Legislation
CFAP Canadian Foundation for the Advancement of Pharmacy
cfar constant false alarm rate
CFAR Chicago for AIDS Rights
CFAT Carnegie Foundation for the Advancement of Teaching
CFAW Canadian Food and Allied Workers
cfb (CFB) circulating fluidized bed (combustion)
CFB California Farm Bureau; Canadian Forces Base; Commonwealth Forestry Bureau; Consumer Fraud Bureau; Council of Foreign Bondholders
cf black conductive furnace black
CFBS Canadian Federation of Biological Societies
CFBT Canadian Forces Base Toronto
cfc campus-free college; capil-

lary filtration coefficient; colony-forming cells; complex facility console

cfc (CFC) chlorofluorocarbon

cf & c cost, freight & commission

CFC Catholics for a Free Choice; Chrysler Financial Corporation; Citizens for a Free Cuba; Combined Federal Campaign (USA); Committee for a Free China; Consolidated Freight Classification

CFC-113 chlorofluorocarbon

CFCA Canterbury Farmers Cooperative Association

cfcb card feed circuit breaker

CFCC Canadian Forces Communications Command

CFCF Central Flow Control Facility

cfd control functional diagram; cubic feet per day

CFD Consumer Fraud Division

CFDA Catalog of Federal Domestic Assistance

CFDC Canadian Film Development Corporation

CFDT *Confederation Française et Democratique du Travail* (French—French Democratic Confederation of Labor)

CFDTS Cold-Flow Development Test System (AEC)

cfe complex features (realty); contractor-furnished equipment

CFE Canadian Forces Europe; Central Fighter Establishment; College of Further Education; Conventional Forces in Europe

CFE *Comisión Federal de Electricidad* (Spanish—Federal Electricity Commission)

CFEME Canadian Forces Environmental Medicine Establishment

cff computer forms feeder; counter flip-flop; critical flicker frequency

Cff Cardiff

CFF Cat Fanciers Federation; Commission for the Future; Compensatory Financing Facility

CFF *Chemin de Fer Fédéraux* (French—Swiss Federal Railroad)

cffc counterflow film cooling

CFFC Catholics For a Free Choice

cfg cubic feet of gas

CFG Camp Fire Girls

cfgd cubic feet of gas per day

cfgh cubic feet of gas per hour

cfgm cubic feet of gas per minute

cfh cubic feet per hour

CFH Council on Family Health

CFHO Canada-France-Hawaii Observatory

CFHQ Canadian Forces Headquarters

CFHS Canadian Federation of Humane Societies

CFHT Canada-France-Hawaii Telescope

cfi cost, freight, and insurance

CFI Canadian Film Institute; Canadian Forest Inventory; Chief Flying Instructor; Committee on Foreign Intelligence (CIA); Corporate Financial Instruction; Counselor Function Inventory; Court of First Instance

CF & I Colorado Fuel and Iron

CFI *Corporación Financiera Internacional* (Spanish—International Finance Corporation)—IFC

CFIA Cavity Foam Insulation Association; Center for Independent Action; Component Failure Impact Analysis

CFIAB Canadian Federation of Insurance Agents and Brokers

CFIP Chamber of Furniture Industries of the Philippines

CFIT Culture Fair Intelligence Test(ing)

CFJS Center For Judicial Studies

cfl context-free language

CFL Canadian Football League; Carnegie Free Library; *Chemins de Fer Luxembourgeois* (French—Luxembourg State Railways); Consolidated Fertilizers Limited; Container Fleets Limited

cflg counter flashing

cfm chlorofluoromethane; confirm; confirmation; confirmed; cubic feet per minute; cubic feet per month

CFM Canterbury Frozen Meat (New Zealand); Council of Foreign Ministers

CFMA Central Financial Management Activities

CFMC Consumer-Farmer Milk Cooperative

CFMUA Cotton Fire and Marine Underwriters Associa-

tion

CFN *Compagnie France-Navigation*

CFNI Caribbean Food and Nutrition Institute

CFNP Community Food and Nutrition Programs

cfo calling for orders; coast for orders

Cfo Channel for orders; Coast for orders

CFO Chief Fire Officer; Commonwealth Fisheries Offices; Complex Facility Operator

CFOA Chief Fire Officers Association

cfp cold frontal passage; computer forms printer; contractor-furnished property; cystic fibrosis of the pancreas

CFP Common Fisheries Policy; Consumer Fraud Protection

CFP *Colonies Française du Pacifique* (French—French Colonies of the Pacific); *Communauté Française de Pacifique* (French—French Community of the Pacific); *Compagnie Française des Petroles* (French—French Petroleum Company); *Cours du Franc Pacifique* (French—French Pacific francs)

CFPC College of Family Physicians of Canada

CFPF Central Food Preparation Facility (USA)

CFPO Center for Populations Options

CFPO *Compagnie Française des Phosphates de l'Océanie* (French—French Oceana Phosphates Company)

CFPS Central Food Preparation System (USA)

CFPTS Coalition For Peace Through Strength

cfr catastrophic failure rate; chauffeur; crash fire rescue

cfr *confronta* (Italian—compare)

CFR Center for Future Research; Code of Federal Regulations; Contact Flight Rules; Coorong Fauna Reserve (South Australia); Council on Foreign Relations

CFRC Canadian Forces Recruiting Centre

CFR engine Cooperative Fuel Research (Council) engine (for measuring quality of fuels)

cfrg carbon-fiber-reinforced glass

cfrgc carbon-fiber-reinforced glass ceramic

cfrp carbon fiber reinforced plastic

CFRPA California Fire Rescue and Paramedic Association

CFRS Central Fisheries Research Station

cfs completely-finished sets; cubic feet per second

cf's confessions of fornication (colonial-style abbreviation originating in Massachusetts and used before the American Revolution)

CFS Canadian Forestry Service; Central Federal Savings; Central Flying School; Container Freight Station; Contract Field Service

CFS *Chemins de Fer Fédéraux Suisses* (French—Swiss Federal Railways)

CFSA College Food Service Association

CFSAN Center for Food Safety and Applied Nutrition (FDA)

CFSC Canadian Forces Staff College

CFSR Commission on Financial Structure and Regulation (White House); Contract Funds Status Report

CFSTI Clearinghouse for Federal Scientific and Technical Information

cft clinical full time; complement fixation test; craft; craftsman; cubic feet; cubic foot

CFT California Federation of Teachers; Colorado Federation of Teachers; Concept Formation Test; Cooperative Field Test(ing)

CFT *Compagnie Française de Télévision* (French—French Television Company)

CFTA Cattle Food Trade Association

CFTAU Canadian Friends of Tel Aviv University

cftb controlled-flight test bed

CFTB Commonwealth Forestry and Timber Bureau

CFTC Commodity Failures Trading Commission; Commodity Futures Trading Commission; Commonwealth Fund for Technical Cooperation

c-f tests complement-fixation tests

CFTH *Compagnie Française Thomson-Houston*

cftmn craftsman

CFTR Citizens For The Republic

cfts captive firing test set(s)

CFTSD Canadian Forces Technical Services Detachment

cfu colony-forming units

CFU Central Functional Unit; Commonwealth Film Unit; Consumer Fraud Unit

CFUA Croatian Fraternal Union of America

cfv conventional friend virus

cfvd constant-frequency variable dot

CFWA Canadian Fruit Wholesalers Association

CFWI County Federation of Women's Institutes

CFWIS Central Fighter Weapons Instructor School

cfy clarify

CFZ Contiguous Fisheries Zone

cg cardiogreen; center of gravity; centigram; cerebral ganglion; choking gas (phosgene); chorionic gonadotropin; chronic glomerulonephritis; colloidal gold; corner guard

c/g coincidence guidance

c of g center of gravity

cg *Zentigram* (German—centigram)

CG Captain General; cargo glider aircraft (DoD symbol); Central of Georgia (railroad); Chaplain General; Coast Guard; Commanding General; Commissary General; Connecticut General (Life Insurance Company); Consul General; Covent Garden; guided-missile cruiser (naval symbol)

CG (ROH) Covent Garden (Royal Opera House)

C of G Central of Georgia (railway); College of Guam (Agaña)

CG *Consumer Guide; Croix de Guerre* (French—War Cross)

C G *cassa grande* (Italian—bass drum)

cga cargo (proportion of) general average; color graphics adaptor

CGA Canadian Gas Association; Coast Guard Auxiliary; Coat Guard Academy; Commonwealth General Assurance; Compressed Gas Asso-

ciation; Corcoran Gallery of Art

CGADC Commanding General, Air Defense Command

CGAIRFMLANT Commanding General, Air Fleet Marine Force, Atlantic

CGAS Coast Guard Air Station; Cornell Guggenheim Aviation Safety Center

CGB Canadian Geographic Board

CGBR Central Government Borrowing Requirement

cgc ceramic gold coating; critical grid current

CGC Chinese Grandmother Carryon (luggage); Coast Guard Cutter; Continental Grain Company

CGCARC Commanding General, Continental Army Command

CG circuit common-gate amplifier for field-effect transistors

cgd chronic granulomatous disease

cgd (CGD) cow grazing day (13.5 kg pasture matter or feed in average bale of hay)

cge carriage

CG & E Cincinnati Gas and Electric Company

CGE *Compagnie Générale d'Electricité* (French—General Electric Company)

CGEC Committee on Global Ecology Concern (U.S.— USSR)

cge fwd carriage forward

CGEL & PB Consolidated Gas, Electric Light and Power Company of Baltimore

C Gen Consul General

cge pd carriage paid; charge paid

cgf center of gravity factor; chemotaxis-generating factor; coarse-glass frit

CGF College of Great Falls

CGFA Columbus Gallery of Fine Arts; Consolidated Gold Fields of Australia

CGFMFLANT Commanding General, Fleet Marine Force, Atlantic

cgfp calcined gross fission product

CGFSA Consolidated Gold Fields of South Africa

cgg continuous grinding gage

cgh computer-generated hologram

cgh (CGH) chorionic gonadotrophic hormone

CGH São Paulo, Brazil (Congonhas Airport)

C of GH Cape of Good Hope

CGHB Cape of Good Hope Bank

CGHSB Cape of Good Hope Savings Bank

cgi computer-generated imagery; corrugated galvanized iron; cruise guide indicator

CGI Chief Ground Instructor; Chief Gunnery Instructor; City and Guilds of London Institute

CGIAR Consultative Group on International Agricultural Research

CGIC *Comisaria General de Investigación Criminal* (Spanish—Commissariat General of Criminal Investigation)—Spain's Interpol office

CGIL *Confederazione Generale Italiana del Lavoro* (Italian—Italian General Confederation of Labor)

C-girl call girl (prostitute); hundred-dollar girl

cgit compressed-gas-insulated tube

C G Jung Foun C G Jung Foundation for Analytical Psychology

cgk grid cathode capacitance

cgl center-of-gravity locator; continuous-gas laser; controlled ground landing; corrected geomagnetic latitude (CGL)

cgl (CGL) chronic granulocytic leukemia

CGL Canadian Gulf Line; Central Gulf Lines

CGL *Confederazione Generale del Lavoro* (Italian—General Confederation of Labor)

CGLAT Cassel Group Level of Aspiration Test

CGLI City and Guilds of London Institute

cg lkr cleaning gear locker

cgm centigram(s); ciliated groove to mouth

cgm (CGM) central gray matter

CGM *Compagnie Genéral Maritime*; Conspicuous Gallantry Medal

CGMA Covent Garden Market Authority

CGMIS Commanding General's Management Information System

cGMP cyclic GMP

CGMW Commission for the Geological Map of the World

cgn chronic glomerulonephritis

Cgn Cartagena, Colombia (British maritime abbreviation) (*see* Ctg)

CGN Cologne, Germany (airport); nuclear-powered guided-missile cruiser (naval symbol)

CGNM Casa Grande National Monument

cgo cargo

Cgo Chicago

CGO Committee on Government Operations; Connecticut Grand Opera

Cgo Chkr Cargo Checker

cg/oq cerebral glucose oxygen quotient

CGOT Canadian Government Office of Tourism

CGOU Coast Guard Oceanographic Unit

cgp choline glycerophosphatide; chorionic growth hormone prolactin; circulating granulocyte pool; grid plate capacitance

CGP College of General Practitioners; Comparative Guidance and Placement Program

CGP *Current Geographical Publications*

CGPM *Conférence Générale des Poids et Mesures* (French—General Conference of Weights and Measures)

CGPN Catalog of Galactic Planetary Nebulae

CGPP Comparative Guidance Placement Program

CGPS Canadian Government Purchasing System

CGPSq Cartographic and Geodetic Processing Squadron (USAF)

cgr captured gamma ray; crime on government reservation

CGRA Canadian Good Roads Association; Chinese Government Radio Administration (Taiwan)

CGRDO Coast Guard Radio

CGRLS Coast Guard Radio Liaison Station

CGRM Commandant General—Royal Marines

cgs centimeter gram second

CGS Canadian Geographical Society; Central Gulf Steamship (corporation); Chief of General Staff; Coast and Geodetic Survey; Council of Graduate Schools (in the

U.S.)

C & GS Coast and Geodetic Survey

CGSA Carriage of Goods by Sea Act; Computer Graphics Structural Analysis

CGSAC Commanding General, Strategic Air Command

CGSB Canadian Government Specifications Board

C & GSC Command and General Staff College

cgse centimeter-gram-second electrostatic

cgsfu ceramic glazed structural facing units

cgsm centimeter-gram-second-electromagnetic

CGSS Cryogenic Gas Storage System

CGSSC Columbia Gas Service Corporation

CGSTC *Centro Giovanile Scambi Turistici e Culturali* (Italian—Youth Center for Tourism and Culture)

cgsub ceramic glazed structural unit base

CGSUS Council of Graduate Schools in the United States

cgt capital gains tax(ation); chorionic gonadotropin; combustible gas tracer; gains tax(ation)

cgt (CGT) corrected geomagnetic time

CGT *Compagnie Générale Transatlantique* (French Line); *Confederación General del Trabajo* (Spanish—General Confederation of Labor); *Confederation du Travail* (French—General Confederation of Labor)

CGTA *Companie Générale de Transports Aériens* (French—Air Algeria)

CGTAC Commanding General, Tactical Air Command

CGTB Canadian Government Travel Bureau

CGTEL Coast Guard Teletype

CGTSF *Compagnie de Télégraphie San Fils* (French wireless company)

cgtt cortisone glucose tolerance test (CGTT)

cgtv (CGTV) command guidance test vehicle

cgu ceramic glazed units

CGU Canadian Geophysical Union

CGUSACE Commanding General, United States Army Corps of Engineers

CGUSACOMZEUR Commanding General, United States Army, Communications Zone, Europe

CGUSARMC Commanding General, United States Army Materiel Command

CGUSCONARC Commanding General, United States Continental Army Command

CGUSFET Commanding General, United States Forces—European Theater

cgv critical grid voltage

cgvs ciliated groove to ventral sac

CGW Chicago Great Western Railway; Coast Guard Women

Cgy Cagayan de Oro

ch case harden; chain; change; chapter; chest; chief; child; chipped; choke; choline; church; coat hook

ch (CH) critical hours (when broadcast signals can cause interference)

c/h cards per hour

c & h cocaine and heroin; cold and hot

ch chambre (French—room); *cheque* (French, Portuguese or Spanish—check)

ch. chori (Latin—choruses)

Ch Chancery; Channel; Chile; Chilean; China; Chinese; choreographer; Christchurch (New Zealand or Oxford or other); church; Clearinghouse

Ch. Chirurgiae (Latin—Surgery)

CH Camp Hospital; Carnegie Hall; Chicago Helicopter (airways); compass heading; concentration of hydrogen ions in moles per liter (symbol); Court House; Custom House; Switzerland (autoplaque)

C-H Crouse-Hinds; Cutler-Hammer

C.H. Companion of Honour

C and H California and Hawaiian Sugar Company

CH College Heights

CH Confederatio Helvetico (Latin—Swiss Confederation)

CH$_2$O formaldehyde

CH$_3$COOH acetic acid

CH-46 Boeing-Vertol twin-rotor helicopter called Sea Knight

CH-47 Boeing-Vertol helicopter called Chinook

CH-53 Sikorsky heavy-assault helicopter called Sea Stallion

CH-54 Sikorsky crane helicopter called Sky Crane or S-64

CH-113 Canadian version of Boeing-Vertol helicopter designated CH-46 and called Labrador

cha cable-harness analyzer; congenital hypoplastic anemia; cyclohexylamine

cha (CHA) cyclohexylamine

Cha Chamaeleon (constellation); Charles

CHA California Hospital Association; Catholic Health Association; Catholic Hospital Association; Chattanooga, Tennessee (airport); Chicago Helicopter Airways; Child Health Association; Community Health Association

CHABA Committee on Hearing and Bio-Acoustics (US Army)

chabak chabakano (Philippine Spanish dialect)

C-habit cocaine habit

Chaco Canyon Chaco Canyon National Monument near Bloomfield, New Mexico

chacom chain of command

chad code to handle angular data

Chad Chadburn; Republic of Chad (landlocked North African country) *République du Tchad*

CHAD Combined Health Agency Drive

CHADS Chicago Air Defense Sector

Chafarinas Chafarinas or Zafarinas Islands (in the Spanish Mediterranean off Morocco and southeast of Melilla)

CHAFB Chanute Air Force Base

chaffroc (CHAFFROC) chaff rocket

Chagos Chagos Archipelago

CHAIN California Housing, Action, and Information Network

Chair Chairman

Chairp Chairperson

chal challenge; chalumeau (ancient clarinet)

chal chaleur (French—heat, warmth)

Chald Chaldean

CHALFA Charlottesville/Albemarle Foundation for the Encouragement of Artists

chalicos chalicosis (sickness caused by metallic-dust inhalation)

chalk calcium carbonate (CaCO$_3$)

cham chamfer; champagne; champion; combustion, heat, mass

Cham Chamaeleon

chamb chamber

Chamb Chamberlain

Chamb Ency Chamber's Encyclopaedia

Chamber Chamber of Deputies

Chambers Chambers Dictionary of Science and Technology

Chambly Girls Girls' Cottage School (for delinquents) at Chambly, Québec

chammy (English slang—champagne)

champ champion(ship)

Champ Beauchamp

CHAMP Character Manipulation Procedure(s); Civilian Health and Medical Program; Community Health Air Monitoring Program

Champ Intl Champion International

champion. compatible hardware and milestone program for integrating organizational needs

Champs Champs Elysées (French—Elysian Fields)—main boulevard of Paris

CHAMPUS Civilian Health and Medical Program for the Uniformed Services

CHAMPVA Civilian Health and Medical Program of the Veterans Administration

chan channel

Chan Channel

Chanc Chancellor; Chancery

CHANCE Complete Help and Assistance Necessary for College Education

CHANCOM Channel Command (NATO)

CHANCOMTEE Channel Committee (NATO)

Chandeleurs Chandeleur Islands of Louisiana

'change exchange; produce exchange; stock exchange

'Change Royal Stock Exchange in London

Chan Isl Channel Islands (Alderney, Guernsey, Jersey, Sark)

Channel The Channel (Beagle, Bristol, English, Old Bahama, Saint George's, Santa Barbara, Ten Degree)

Channel Islands Park islands

off California
Channels Channel Islanders; Channel Islands
chans chanson (French—song)
CHANSEC Channel Committee Secretary (NATO)
Chao Phraya Krung Thep (Bangkok) river
CHAOS Committee for Halting Acronymic Obliteration of Sense; Consortium for the Hastening of the Annihilation of Organized Society
CHAOTIC Computer-and-Human-Assisted Organization of a Technical Information Center (NBS)
chap. chapter
Chap Chaplain
CHAP Certified Hospital Admissions Program; Charring Ablation Program (NASA); Child Health Assistance Program
Chap(pie) Chapin; Chapman; Chappell
chaps. chaparajos (Spanish—open backed leather overalls)
CHAPS Children Have A Potential Society; contractor-held Air Force property
Chapter 7 liquidation
Chapter 11, etc. bankruptcy—Chapter 11, etc., of the Bankruptcy Act of the U.S.
char character; characteristic; charcoal; charwoman
char (CHAR) character (data processing)
Char Charter
Char Amal Charlotte Amalie
Charbray Charolais-Brahman cattle
charc charcoal
Charl Charlottenburg
Charles University University of Prague
Charley Charles
Charlotte Corday Marie-Anne-Charlotte Corday d'Armont
Charm Charmian
char reac character reaction (sometimes simply cr)
chars characters
char(s) charwoman; charwomen
chart. charta (Latin—paper)
chart. bib. charta bibula (Latin—blotting paper)
chart. cerat. charta cerata (Latin—waxed paper)
chartul. chartula (Latin—small paper)
Char X Charing Cross (rail terminal)
chas chassis

Chas Charles
CHAS Catholic Housing Aid Society
chase. cut holes and sink 'em (navalese acronym for sinking old ammunition cases or obsolescent barges or boats)
Chase Chase Manhattan Bank
Chasn Charlestown
Chat château (French—castle)
Chat Choo-Choo Chattanooga Choo-Choo (restaurant)
chat mtg chattel mortgage
Chat(ty) Charlotte
Ch^{au} Chateau (French—castle, country mansion)
Chauc Geoffrey Chaucer
chaud chemical audit
chauf chauffeur
Chávez Jorge Chávez International Airport of Lima, Peru
chb complete heart block
Chb Cherbourg; Chiba
Ch B Chief of Bureau
Ch.B. Chirurgiae Baccalaureus (Latin—Bachelor of Surgery)
chbd chalkboard
ChBuAer Chief of the Bureau of Aeronautics
ChBuDocks Chief of the Bureau of Yards and Docks
ChBuMed Chief of the Bureau of Medicine and Surgery
ChBuOrd Chief of the Bureau of Ordnance
ChBuPers Chief of the Bureau of Naval Personnel
ChBuSanda Chief of the Bureau of Supplies and Accounts
ChBuShips Chief of the Bureau of Ships
ChBuWeps Chief of the Bureau of Weapons
chc choke coil
Ch of C Chamber of Commerce
CHC Chicago House of Correction; Christchurch, New Zealand (airport); Community Health Council; Comprehensive Health Center; Congressional Hispanic Caucus
ch cab china cabinet
CHCF Component Handling and Cleaning Facility
Chch Christchurch
Ch Ch Christ Church College, Oxford
CHCl₃ Chloroform
CHCMD Chicago Contract Management District
chd chaldron; childhood disease(s); congestive heart disease; coronary heart dis-

ease
Ch D Charles Darwin
Ch.D. Chirurgiae Doctor (Latin—Doctor of Surgery)
C-H d Chediak-Higashi disease
CHD Charles Halliwell Duell
Ch d'A Chargé d'Affaires
ch de f chemin de fer (French—railroad)
CHDF Civilian Home Defense Force
chdl computer hardware description language
chdm cyclohexanedimethanol
CHDP Child Health and Disability Prevention
che cholinesterase
che (CHE) channel end(ing)
c & he consumer and homemaking education
Che Chetverg (Russian—Thursday); Ernesto (Che) Guevara
Ch^e Chapelle (French—Chapel)
C^{he} Chaine (French—chain)
Ch E Chief Engineer
Ch.E. Chemical Engineer
CHE Chete Game Reserve; Chewore Game Reserve; Chizarira Game Reserve
C-head coke head (slang—cocaine addict)
Cheaha Cheaha Mountain or Cheaha State Park south of Anniston, Alabama
CHEAP Computerized Health Education Assessment Program
CHEAR Council on Higher Education in the American Republics
chec checked; checkered
CHEC Citizens Helping Eliminate Crime; Commonwealth Human Ecology Council
Checo Checoslovaquia (Spanish—Czechoslovakia)
Chee Chee-chee (Anglo-Indian dialect)
cheesesan cheese sandwich
cheesewich cheese sandwich
Cheka Chrezvychainaya Kommissiya po Borbe s Kontrrevolutisiei i Sabotzhem (Russian—Extraordinary Commission for Combating Counterrevolution and Sabotage)—original Soviet Secret Police founded December 20, 1917, at Lubianka Prison in Moscow (q.v.—VOT)
CHEL Cambridge History of English Literature
Chelm (ancient Jewish town in

Poland known in folklore as the Town of Fools); short form for Cheltenham

Chelon Chelonia

cheloniol cheloniologic(al)(ly); cheloniologist; cheloniology

chelons chelonians (tortoises, terrapins, turtles)

Chelt Cheltenham

chem chemical; chemist; chemistry

Chem Chemistry

chemanal chemical analysis

Chem E Chemical Engineer- (ing)

Chem.E. Chemical Engineer

Chem Econ Chemical Economic Services

Chem Ed Chemical Education Publishing Company

Chem Educ Chemical Education Publishing Co

Chem Elements Pub Chemical Elements Publishing Co

chem etch chemically etched; chemical etching

CHEMI Chemical Engineering Modular Instruction

chemly chemically

Chem & Met Eng *Chemical and Metallurgical Engineering*

chem mill chemically milled; chemical milling

chemo chemotherapist; chemotherapy

chemonuc chemonuclear

chemos chemosphere; chemospheric(al)(ly)

chemosens chemosensory

chemoster chemosterilant; chemosterilization; chemosterilize(d)

chemosurg chemosurgical(ly); chemosurgery

chemotax chemotaxonomic(al)(ly); chemotaxonomist; chemotaxonomy

chemoth chemotherapy

Chem Phys *Chemical Physics*

Chem Phys Lett *Chemical Physics Letters*

Chem Pub Chemical Publishing Company

Chem Rev *Chemical Reviews*

Chem Rubber Chemical Rubber Company

CHEMS Chemical Education Materials Study

chemsearch chemicals selected for equal, analogous, or related characters

chemsol chemical solution (for decontamination)

CHEMTREC Chemical Transportation Emergency

Center

chem war. chemical warfare

CHEN *Chail Nashim* (Hebrew-Women's Force of the Israeli Army); *chen* is the Hebrew word for grace

CHEOPS Chemical Operations System

CHEP Commonwealth Handling and Equipment Pool

Cher Cherilyn; Cherilyn Sarkisian

chert ironstone sedimentary rock

Cherv Cherville; Chervin

Ches Cheshire

Cheskey(s) Czechoslovakian(s)

chesky cherry-flavored whiskey

chester(s) (Early English—city, old fortification, town)—short form for such places as Manchester, Winchester, and even Tadcaster and Worcester

Chet Chester

chev chevron

Chev Chevalier (French—Knight)

Cheviots Cheviot Hills between England and Scotland

Chevron Standard Oil of California

Chev(y) Chevrolet

Chewko Chewing Tobacco Company

Chey Cheyenne

chf congestive heart failure; critical heart flux

Chf Chief; Crimean hemorrhagic fever

Ch F Chaplain of the Fleet

CHF Carnegie Hero Fund; Coalition for Health Funding

CHFA California Housing Finance Agency

ch-factor chutzpah factor (degree of guts or nerve)

Chf Bkr Chief Baker; Chief Bookkeeper

ChFC Chartered Financial Consultant

CHFC Carnegie Hero Fund Commission

Chf Engr Chief Engineer

Chf Libr Chief Librarian

Chf M Sgt Chief Master Sergeant

Chf Off Chief Officer

Chf Pur Chief Purser

Chf Stwd Chief Steward

Chf Surg Chief Surgeon

ch fwd charges forward

Chf Wt Ofcr Chief Warrant Officer

chg change; charge

Chg Chittagong

CHGC Committee for Hand Gun Control

chgd charged

Chgo Chicago

chgph choreographer; choreographic; choreography

chg pl change plane

chgs charges

chh cartilage-hair hypoplasia

CH&H Continent between Havre and Hamburg

chi specific magnetic susceptibility

Chi Chicago; Chichester; China; Chinese

CHI Catastrophic Health Insurance; Chicago; Crouse-Hinds (stock-exchange symbol)

Chia Chiapas

CHIA Canadian Health Insurance Association

CHIAA Crop-Hail Insurance Actuarial Association

chic cermet hybrid integrated circuit

Chic Chicago

Chicagorican Chicago Puerto Rican

Chicano (diminutive nickname for *Mexicano*)

Chich Chichester

chick. chicken

Chick Chickering

chickensand chicken sandwich

chickenwich chicken sandwich

chick(s) chicken(s)

Chico Francisco

Chicom Chinese communist

Chicos Chinese communists

Chi$ Chilean peso

Chidic Chinese dictionary

Chief Chief Engineer

CHIEF Controlled Handling of Internal Executive Functions

Chih Chihuahua (inhabitants—Chihuahuenses, chihuahua dogs characteristic of this area—*chihuahueños*)

chil children('s)

CHI-LAX Chicago—Los Angeles

Chil Cur Chilean Current

child. computer having intelligent learning and development

child. (CHILD) children having individual learning difficulties

Children's Children's Crusade (in 1212 when 90,000 children from France and Germany set out to free the Holy

Land); Children's Opera [Hansel and Gretel by Engelbert Humperdinck (1854—1921)]; Children's Hospital

Chile Republic of Chile, *República de Chile*

chilidog chile-con-carne sauced hotdog (frankfurter)

chili(es) chili pepper(s)

Chillicothe Institute Chillicothe Correctional Institute at Chillicothe, Ohio

Chillicothe School Training School for Girls at Chillicothe, Missouri

Chilterns Chiltern Hills of England

Chilton Chilton Book Company

chim chimica (Italian—chemistry); *chimie* (French—chemistry)

Chimbo Chimborazo, Ecuador; Chimbote, Peru

Chi Met Chicago Metropolitan Correctional Center

CHI-MIA Chicago—Miami

chimponaut chimpanzee astronaut (primate used in space travel experiments)

chimp(s) chimpanzee(s)

chin. chinchilla

chin chinesisch (German—Chinese)

Chin China; Chinese

Chin Chinese (world's leading language in terms of numbers)

China People's Republic of China (communist-controlled mainland); Republic of China (nationalist island of Taiwan once known as Formosa plus Matsu, Quemoy, and the Penghus or Pescadores

CHINALIGHT China National Light-Industry Products Import and Export Corporation (mainland China)

China Nac China Nacionalista (Spanish—Nationalist China)—offshore China also known as Formosa or Taiwan

China Sea(s) East China Sea and South China Sea

Chinat Chinese nationalist

CHINATEX China National Textiles Import and Export Corporation (mainland China)

CHINATIVE China National Native Produce (mainland China)

Chi Nats Chinese Nationalists

CHINATUHSU China National Trading Corporation

(mainland produce and animal by-products)

Chin J Phys Chinese Journal of Physics (Acta Physica Sinica—Wuli Xuebao)

CHINKUNG China National Trading Corporation for Light Industrial Products (mainland China)

chins. children in need of supervision

Ch Insp Chief Inspector

Chinsyn Chinese-English synthesis-oriented machine translation system

CHI-NY Chicago—New York

Chios English equivalent of Khios island in the Aegean

Chip Chipre (Portuguese or Spanish—Cyprus)

CHIP Community Housing Improvement Programme (New Zealand)

CHIPDis Chicago Procurement District (US Army)

Chipitt Chicago-to-Pittsburgh (complex of cities)

chippy fish-and-chip shop

Chippy Chipping Norton, England

chips children in need of protection and services

chips. coherent high-intensity photon source

Chips ship's carpenter

CHiPs California Highway Patrol cops

CHIPS Chemical Engineering Information Processing System; Clearing House Interbank Payment System

chir chiropody

chir chirurgia (Italian—surgery)

Chir. Doc. Chirurgiae Doctor (Latin—Doctor of Surgery)

Chiricahua Chiricahua National Monument in southeastern Arizona

Chiricahuas Chiricahua Mountains of Arizona

Chiricano(s) Panamanian(s)

chiro chirography; chiropractic; chiropractor

CHIRP Community Housing Improvement and Revitalization Program

Chis Chiapas (inhabitants—Chiapanecos)

CHI-SAN Chicago—San Diego

CHI-SEA Chicago—Seattle

CHI—SFO Chicago—San Francisco

Chish Chisholm

Chisox Chicago White Sox

(baseball team)

Chi Sym Chicago Symphony

chit chitty (Hindustani—voucher signed to cover small debts for drinks, food, tobacco, etc.)

Chita Conchita

Chi-Trib Chicago Tribune

chiv chivalry

chix chickens

Ch J Chief Justice

CHJM Carnegie Hall—Jeunesses Musicales

CHJMKHK Chung-Hua Jen-Min Kung-Ho Kuo (People's Republic of China)

chk check

chkpt checkpoint

chkr checker

chl chloroform; confinement at hard labor

Chl Chalna

CHL Central Hockey League

chlamy chlamydia (sexually-transmitted bacterial infection)

Chla Vsta Chula Vista

chlb chlorobutanol

Ch Lbr Chief Librarian

ch-lkr chiffonier-locker

Ch^{lle} Chapelle (French—Chapel)

chlor chloride; chlorination; chlorine

chloro chloroform; chlorophyll; chloroprene

chloro chlorus (Greek—green)—chlorine, chlorophyll

chloroform trichloromethane $(CHCl_3)$

chloroprene synthetic rubber (C_4H_5Cl)

chlw commercial high-level waste

chm chamber; checkmate

Chm Chairman; Chairwoman; Choirmaster; Choirmistress

Ch.M. Chirurgiae Magister (Latin—Master of Surgery)

CHM Cleveland Health Museum

C-H M Cooper-Hewitt Museum

CHMC Children's Hospital Medical Center (Boston)

CHMDDA Cooper-Hewitt Museum of Design and Decorative Arts

ch-mir chiffonier-mirror

CHMK Chung-Hua Min-Kuo (Republic of China)

chmn chairman

ChMNH Chicago Museum of Natural History

chmp chairperson

Chn Cochin

Chn China (Spanish abbreviation)

CHN College of the Holy Name; Community of the Holy Name

C-H-N carbon, hydrogen, nitrogen, oxygen, phosphorus, sulfur (compounds)

CHNAVPERS Chief, Naval Personnel

CHNAVSECMAAG Chief, Navy Section, Military Assistance Advisory Group

Chne Chaîne (French—Chain)—mountain range

chngd changed

chns chains

CHNS Cape Hatteras National Seashore (Buxton, North Carolina)

CHNSRA Cape Hatteras National Seashore Recreational Area

CHNSY Charleston Naval Shipyard (South Carolina)

CHNT Community Health Nurse Teacher/Tutor

ChNZAgCo China New Zealand Agricultural Consultants

cho (Cho) containers carried in hold

Cho Chosen (Korea)

CHO carbohydrate (generalized formula); Community Health Organization

CHOBS Chief Observer (USN)

choc chocolate

chocbar(s) chocolate bar(s)

chocmalt chocolate malted milk

choco chocolate

chocs chocolate candies; chocolate drops; chocolates

CHOD Chief of Defense

choirm choirmaster

choke choke hold (bar hold or carotid hold)

CHOKE Care How Others Keep the Environment

chol cholesterol

chol (Latin prefix—bile)—cholecyst, cholera, cholesterol(ic), cholic

chol est cholesterol esters

Cho Min Inm Kon Chosun Minchu-chui Inmin Konghwa-guk (Democratic People's Republic of Korea) North Korea

CHOMPS Canine Home Protection System

Chonos Chonos Islands

CHOP Change of Operational Control

CHOPS Chief of Operations

chor choral; choreographer; choreographist; choreography; chorus; choruses

Chord Chordata

chorégr chorégraphie (French—choreography)

C Horn Cur Cape Horn Current

chortle chuckle and snort

Chou Chou En-lai

cho/vac cholera vaccine

chovr changeover

chp child psychiatry; comprehensive health plan(ning)

Chp Chepstow

Chp Chipre (Spanish—Cyprus)

CHP California Highway Patrol; Chihuahua Pacific (railroad—Ferrocarril de Chihuahua al Pacifico)

CHPA California Highway Patrol Academy

chpae critical human performance and evaluation

ch pd charges paid

Chpn Chairperson

CHPP Cypress Hills Provincial Park (Saskatchewan)

ch ppd charges prepaid

chpx chickenpox

chq cheque

CHq Corps Headquarters

chr character; chrome; chromium; chromobacterium; chronic

c hr candle-hour

Chr Choir; Christ; Christ College, Cambridge; Christian; Church

Chr Chronicles

CHR Commission on Human Rights; Connecticut Hard Rubber (company)

CHRB California Horse Racing Board

Chr Coll Christ College—Cambridge

chrg charge

CHRG Citizens Health Research Group

CHRIE Council on Hotel, Restaurant, and Institutional Education

Chris Christian(a); Christiania; Christina; Christopher

CHRIS Cancer Hazards Ranking and Information System

Chrissie Christina; Christine

Christ. Christian; Christianity; Christmas

christie Christiania turn

Chrlstn Charleston

Chrlstn SC Charleston, South Carolina

Chrm Chairman; Chairwoman

chrom (Latin prefix—color)—chromatic, chromoplast, chromosome, chromosphere

chromite iron chromate

chromolith chromolithograph(y)

chromo(s) chromolithograph(s); chromosome(s)

chron chronogram; chronograph; chronology; chronometer; chronometry

Chron Chronicle(s)—First Book of Chronicles; Second Book of Chronicles

chrono chronologic(al)(ly); chronology; chronometer; chronometric(al)(ly)

chrono order chronological order

chro pltd chrome plated

Chrp Chairperson

Chrs Chambers; Christians; Churches

Chrys Chrysler

chrysanthemum nationalist symbol of China and Japan; symbol of the Orient Overseas Line

chrysoberyl beryllium aluminate

chrysocolla hydrous copper silicate

chrysoprase chalcedony gemstone

chs chapters; crime on the high seas

Chs Chambers; Charles; Chester

Ch of S Chamber of Shipping

C-H s Chediak-Higashi syndrome

CHS Canadian Hydrographic Service; Charleston, South Carolina (airport); Chicago Historical Society; Childrens Home Society; Citizens for Highway Safety; Community Health Service (HEW); Community of the Holy Spirit; Confederate Hammer Skins (neo-Nazi group, Dallas, Texas); Connecticut Herpetological Society; Cristobal High School; Curtis High School

CHSA Chest, Heart, and Stroke Association

ch'ship championship

Ch Skr Chief Skipper

CHSL Cleveland Health Sciences Library

CHSM China Service Medal

CHSS Children's Hypnotic Susceptibility Scale; Cooperative Health Statistics System

Ch Supt Chief Superintendent

cht cylinder head temperature

Cht Chittagong

chtg charting

CHTNP Chittagong Hill Tracts National Park (Bangladesh)

cht tanks collect, hold, transfer (raw sewage) tanks (used by naval vessels to overcome harbor pollution when in port)

chu (CHU) central heating unit

Chu Centigrade heat unit

CHU *Christelijk-Historische Unie* (Dutch-Christian Historical Union)—political party

CHUA Canadian Hail Underwriters Association

Chuck Charles

Chugach Chugach National Forest in Alaska

Chugaches Chugach Mountains of Alaska

Chukchi Chukchi Peninsula and the Chukchi Sea in the Arctic between Alaska and Siberia where the peninsula is located

Chulajuana Chula Vista-Tijuana area of southwesternmost California and northwesternmost Tijuana

CHUM *Computing and the Humanities*

Chumley (British contraction—Chalmondeley)

CHUMS Cancer Hopefuls United for Mutual Support; Care and Help for Unmarried Mothers

Chung Chungking

Chung Min *Chung-hua Min-kuo* (Mandarin Chinese—Republic of China)—offshore China

Chunnel Channel Tunnel (under the English Channel)

Chuqui Chuquicamata

Chur Churchill College, Cambridge

Churchill Sir Winston Churchill

chut cable households using tv (audience survey)

'chute parachute

ch v check valve

chw chilled water; cladding hull waste; cold-and-hot water; constant hot water

CHW Charleston, West Virginia (airport)

CH & W Canadian Health and Welfare

CHWA California Health and Welfare Agency

Chwdn Churchwarden(ess)

chx chiro-xylographic

chy chimney

C Hy Commission for Hydrology

chyd churchyard

Chy Div Chancery Division

ci cardiac index; cardiac insufficiency; cast iron; cerebral infarction; chemotherapeutic index; clinical investigator (CI); clonus index; coefficient of intelligence; colloidal iron; color index; compression ignition; contamination index; contrast index; coronary insufficiency; cost and insurance; counterintelligence; criminal informant; crystalline insulin

ci (CI) consular invoice; cooperative individual (informant)

c-i criminal-investigation

c.i. (C.I.) consular invoice

c/i carriage-to-interference (ratio); configuration interface

c/i (C/I) certificate of insurance

c & i cost and insurance; cowboys and indians

Ci cirrus; curie (unit of activity in radiation dosimetry)

Ci *cerveau isolé* (French—isolated intellect, intellectual)

CI Carnegie Institute; Cayman Islands; Channel Islands; Color Index; Combustion Institute; Communist International; Confidential Informant; Conservation International; Consumers Institute; Cranberry Institute; Curtis Institute

C.I. Lady of the Imperial Order of the Crown of India

C/I Certificate of Insurance; Consular Invoice

C & I Currier and Ives

C of I Church of Ireland (Roman Catholic)

CI *Colour Index; Comédie-Italienne* (Paris)

cia captured in action; cash in advance; child(ren) in arms; computer interface adaptor

Cia *Compagnia* (Italian—Company); *Companhia* (Portuguese—Company); *Compañía* (Spanish—Company)

CIA Caribbean International Airways; Catering Institute of Australia; Central Intelligence Agency; Commerce and Industry Association; Cook Island Airways; Correctional Industries Association; Cotton Insurance Asso-

ciation; Cowboy and Indian Alliance; Culinary Institute of America

CIA *Comité International d'Auschwitz* (French—International Auschwitz Committee); *Conseil International des Archives* (French—International Council on Archives)

CIAA College Inventory of Academic Adjustment; Coordinator Inter-American Affairs

CIAB Canadian Immigration Appeal Board

C*iac *Compania* (Spanish—company)

CIAC Canadian Independent Adjusters Conference; Career Information and Counseling (USAF)

CIAL *Communauté Internationale des Associations de la Librairie* (French—International Community of Booksellers' Associations)

CIAM *Congreso Internacional de Arquitectura Moderna* (Spanish—International Congress of Modern Architecture)

CIANY Commerce and Industry Association of New York

CIAO Congress of Italian-American Organizations

CIAP *Comité Interamericano de la Alianza para el Progreso* (Spanish—Inter-American Committee of the Alliance for Progress)—ICAP

CIAPS Customer-Integrated Automated Procurement System

CIAS California Institute of Asian Studies; Council for Inter-American Security

CIASSR Cecheno-Ingush Autonomous Soviet Socialist Republic

CIAT *Centro Interamericano de Administradores Tributarios* (Inter-American Center of Revenue Administrators); *Centro Internacional de Agricultura Tropical* (Spanish—International Center of Tropical Agriculture)

CIAU Canadian Interuniversity Athletic Union

CIAW Commission on Intercollegiate Athletics for Women

cib. *cibus* (Latin—food)

CIB California Industries for the Blind; Canadian Interna-

tional Bank; Central Intelligence Board; Commonwealth Investment Bank; Criminal Intelligence Bureau; Criminal Investigation Bureau
CIB *COBOL Information Bulletin* (USAF)
CIBA Corporation of Insurance Brokers of Australia
CIBC Canadian Imperial Bank of Commerce; Council on Interracial Books for Children
CIBG Canadian Infantry Brigade Group
cibha congenital inclusion body hemolytic anemia
cibhp closed-in-bottom hole pressure
CIBNC Cook Islands Broadcasting and Newspaper Corporation
CIBNZ Corporation of Insurance Brokers of New Zealand
CIBS Chartered Institution of Building Services
cic cardio-inhibitor center; cloud in cell; command input coupler; critical item code
Cic Marcus Tullius Cicero
CIC Caribbean Investment Corporation; Cedar Rapids & Iowa City (railroad); Center for Instructional Communications (Syracuse University); Central Inspection Commission; Central Intelligence Center; Change Identification Control; Chemical Institute of Canada; Combat Information Center; Combat Intelligence Center; Combined Intelligence Committee; *Comité International de la Conserve* (French—International Canning Committee); Command Information Center; Commander-in-Chief; Committee on Institutional Cooperation; Commonwealth Industrial Court; Commonwealth Information Centre; Community Information Centre(s); *Conseil International des Compositeurs* (French—International Council of Composers); Consumer Information Center (Pueblo, Colorado 81009); Continental Insurance Companies; Cooperative Insurance Corporation; Council of Independent Colleges; Counter-Intelligence Corps; Crime Intelligence Center; Criminal Investigation Command; Critical Issues Council; Curacao Infor-

mation Center; Customer Identification Code
CIC *Consejo Interamericano Cultural* (Spanish—Interamerican Cultural Council); *Cymdeithas yr Iaith Cymraeg* (Welsh Language Society)
CICA Canadian Institute of Chartered Accountants; Council of International Civil Aviation
CICA *Centro de Investigaciones Ciencias Agronómicas* (Spanish—Agronomic Sciences Investigation Center)
CICAR Cooperative Investigations of the Caribbean and Adjacent Regions (UNESCO)
CICAS Computer-Integrated Command-and-Attack Systems
CICB Criminal Injuries Compensation Board
CICC California Institute of Color Consulting; Criminal Injuries Compensation Commission (Hawaii)
CICCU Cambridge Inter-Collegiate Christian Union
CICESE *Centro Investigación y Educación Superior de Ensenada* (Spanish—Center of Investigation and Higher Education of Ensenada)
Cicestr. Cicestrensis (Latin—of Chichester)
CICI Composite Index of Coincident Indicators
CICJ *Comité International pour la Coopération des Journalistes* (French—International Committee for the Cooperation of Journalists)
CICMA Canadian Insurance Claims Managers Association
CICO Combat Information Center Office(r)
CICOM *Centro de Comercialización Nacional e Internacional* (Spanish—Center of National and International Marketing)
CICP Capital Investment Computer Program; Center Program(ming); Committee to Investigate Copyright Problems
CICRIS Cooperative Industrial and Commercial Reference and Information Service
CICs Community Improvement Corpsmen; Community Improvement Corpswomen

CIC's Change Information Control (numbers)
CICS Committee for Index Cards for Standards; Customer Information and Control System
CICSB Coalition of Indian-Controlled School Boards
CICS/VS Customer Information Control System/Virtual Storage
CICT Commission on International Commodity Trade
CICT *Conseil International du Cinéma et de la Télévision* (French—International Council of Cinema and Television)
cicu cardiology intensive care unit (CICU); computer-integrated converter unit; coronary intensive care unit (CICU)
CICU Commission for Independent Colleges and Universities
CICUNM Council of Independent Colleges and Universities of New Mexico
CICUP Commission for Independent Colleges and Universities of Pennsylvania
CICV Council of Independent Colleges in Virginia
CICYP *Consejo Interamericano de Comercio y Producción* (Spanish—Interamerican Council of Commerce and Production)
cid cash in drawer; charge-injection device; chick infective dose; commercial item description; cubic-inch displacement
cid (CID) cytomegalic inclusion disease
CID Center for Industrial Development; Central Institute for the Deaf; *Centre d'Information et de Documentation* (French—Center for Information and Documentation—Belgium); Change in Design; Classification of Instructional Disciplines; Commission for International Development; Council for Independent Distribution; Criminal Investigation Department (Scotland Yard); Criminal Investigation Division
CID *Colegio Interamericano de Defensa* (Spanish—Inter-American Defense College)
CIDA Canadian International

Development Agency
CIDA *Comité Interamericano de Desarollo Agricola* (Spanish—Inter-American Committee of Agricultural Development)
CIDALC *Comité International du Cinéma d'Enseignement et de la Culture* (French—International Committee of Film Education and Culture)
CIDC Cryogenic Information and Data Section; Curriculum and Instructional Development Center
cide (Latin suffix—destroy)—germicide, insecticide
CIDEM *Consejo Interamericano de Música* (Spanish—Inter-American Music Council)
CIDG Civil Indigenous Defense Group (Vietnam)
CIDH *Comisión Interamericana de Derechos Humanos* (Spanish—Inter-American Commission of Human Rights)
cidi crimping die
cidnp chemically induced dynamic nuclear polarization
CIDOC *Centro Intercultural de Documentación* (Spanish—Intercultural Documentation Center)
cids cellular immunity deficiency syndrome
CIDS Chemical Information and Data System
cidstat civil disturbance status (USA reporting activity)
cie coherent infrared energy; counter immunoelectrophoresis; customs input entry
Cie Compagnie (French—company)
CIE Center for Independent Education; Chrysler Institute of Engineering; Cleveland Institute of Electronics; Commonwealth Institute of Entomology
C.I.E. Companion of the Order of the Indian Empire
CIE Centro de Informações do Exército (Portuguese—Military Intelligence Center)—Brazil; *Comité Interamericano de Educación* (Spanish—Inter-American Committee of Education)
CIEA California Indian Education Association
CIEA *Centro Internacional de Estudios Agricolas* (Spanish—International Center of

Agricultural Studies)
CIEBM Committee on the Interplay of Engineering with Biology and Medicine
CIEC Centre International d'études Criminologiques (French—International Center of Criminological Studies)
CIECC *Consejo Interamericano para la Educación, la Ciencia, y la Cultura* (Spanish—Inter-American Council for Education, Science, and Culture)
CIEE Companion of the Institution of Electrical Engineers; Council on International Educational Exchanges
Cie Gle Transatlantique Compagnie Générale Transatlantique (French Line)
CIEM Conseil International pour l'Exploration de la Mer (French—International Commission for the Exploration of the Sea)
CIEN Comisión Interamericana de Energía Nuclear (Spanish—Inter-American Commission for Nuclear Energy)
CIENES Centro Interamericano de Enseñaza de Estadística (Spanish—Inter-American Center for the Study of Statistics)
CIENT Cambridge and Isle of Ely Naturalist Trust (England)
CIEO Catholic International Education Office
ciep counterimmunoelectrophoresis
CIEP Council on International Economic Policy
CIER Centro Interamericano de Educación Rural (Spanish—Inter-American Center of Rural Education)
CIES Comparative and International Education Society; Council for International Exchange of Scholars
CIES Consejo Interamericano Economico y Social (Spanish—Inter-American Economic and Social Council)
CIESMM Commission Internationale pour l'Exploration Scientifique de la Mer Méditerranée (French—International Commission for the Scientific Exploration of the Mediterranean Sea)
CIESPAL Centro Interna-

cional de Estudios Superiores de Periodismo para America Latina (Spanish—International Center for Advanced Studies of Journalism in Latin America)
CIET Centro Interamericano de Estudios Tributarios (Spanish—Inter-American Center of Revenue Studies)
CIETA Calcutta Import and Export Trade Association
CIETA Centre International d'Etude des Textiles Anciens (French—International Center for the Study of Ancient Textiles)
cif cash in fist; central index(ing) file; central integration facility; cost, insurance, and freight
CIF California Interscholastic Federation; Canadian Institute of Forestry; Construction Industry Foundation
CIF Commission Interaméricaine des Femmes (French—Interamerican Commission of Women); *Conseil International des Femmes* (French—International Council of Women)
CIFA Courtauld Institute of Fine Arts
cifane cost, insurance, freight, and slight difference in exchange
CIFAR Central Institute of Foreign Affairs Research
CIFAS Consortium Industriel Franco-Allemand pour Symphonie (French—Franco-German Industrial Consortium for Symphonie)—communication satellite linking systems between points in Africa, the Americas, Europe, and the Middle East
cif&c cost, insurance, freight, and commission
CIFC Council for the Investigation of Fertility Control
cifc & e cost, insurance, freight, and exchange
cifci (CIF and C & I) cost, insurance freight (plus) commission and interest
CIFE Central Index File—Europe
CIFEJ Centre International du Film pour l'Enfance et la Jeunesse (French—International Center of Films for Children and Young People)
CIFF Cannes International Film Festival; Comprehen-

sive International Freight Forwarders

cif&i cost, insurance, freight, and interest

CIFIUS Current Interagency Committee on Foreign Investment in the United States

cifLt cost, insurance, and freight, London terms

cig cigarette

CIG *Comité International de Géophysique* (French—International Geophysics Committee); Commonwealth Industrial Gases

CIGA *Compagnia Italiana dei Grandi Alberghi* (Italian—Italian Great Hotels Company)

CIGAR Common Interactive Graphics Application Routine (USA)

CIGGT Canadian Institute of Guided Ground Transportation

CIGNA Connecticut General Corporation combined with the Insurance Company of North America

CI Gov Cook Islands Government

CIGS Chief of the Imperial General Staff (Great Britain)

cigsmug cigarette smuggler

CIGTF Central Inertial Guidance Test Facility

cih carbohydrate-induced hyperglyceridemia

CIHR Clinical Institute for Human Relations

CII Chartered Insurance Institute; Coffee Information Institute; Combat Information Intelligence

CIIA Canadian Institute of International Affairs

CIIB Consumers Insurance Information Bureau

CIIC Chemical Industry Institute of Toxicology; Counter Intelligence Interrogation Center

CIIIA *Soedinennye Shtaty Ameriki* (Russian—United States of America)—U.S.A.

c-i info criminal-investigation information

CIIR Central Institute for Industrial Research

CIIS California Institute of International Studies

CIIT Chemical Industry Institute of Toxicology

cij control joint

CIJ *Consejo Interamericano de Jurisconsultos* (Spanish—

Inter-American Council of Legal Consultants)

CIJE *Current Index to Journals in Education*

CIKCU Council of Independent Kentucky Colleges and Universities

cil control interpreter language; core-image library; current-inhibit logic

Cil Cilicap

CIL Canadian Industries Limited; Center for Independent Living; Commissioner of Irish Lights; Council for Interinstitutional Leadership

C/I/L *Computer/Information/Library Sciences*

cila casualty insurance logistics automated

CILA *Centro Interamericano de Libros Académicos* (Spanish—Inter-American Center of Academic Books)

CILES Central Information Library and Editorial Section (CSIRO)

Cilla Priscilla

CILSA Chief Inspector of Land Service Ammunition

CILT Center for Information on Language and Teaching

cim capital investment model; communication-interface module(s); computer-input microfilm(ing); conductance-increase mechanism; continuous-image microfilm(ing)

CIM California Institution for Men; Canadian Institute of Management; Canadian Institute of Mining; Canadian Institute of Music; Commission for Industry and Manpower; Computer-Integrated Manufacturing; Curtis Institute of Music

C & IM Chicago & Illinois Midland (railroad)

CIM *Centro Italiano della Moda* (Italian—Italian Fashion Center); *Conseil International de la Musique* (French—International Music Council); *Consejo Internacional de Mujeres* (Spanish—International Council of Women)

CIMA Construction Industry Manufacturers Association; Coordinated Investigation of Micronesian Anthropology

Cimabue Cenni di Pepo

cimb *cimbalom* (Hungarian—dulcimer)

CIMB Construction Industry

Management Board

CIMBA Contractor Installation Make-or-Buy Authorization

CIMC California Institution for Men in Chino (state prison); Commander's Internal Management Conference

CIMC *Colegio Interamericano Médicos y Cirujanos* (Spanish—Interamerican College of Physicians and Surgeons)

cimco card image correction

CIMCO Congo International Management Corporation

CIME Council of Industry for Management Education

CI Mech E Companion of the Institution of Mechanical Engineers

CIMIC Civilian Military Cooperation

cimm constant-impedance mechanical modulation

CIMM Canadian Institute of Mining and Metallurgy

CIMMS Civilian Information Manpower Management System (USN)

CIMMYT *Centro Internacional de Mejoramiento de Maíz y Trigo* (Spanish—International Center for the Improvement of Corn and Wheat)

CIMP *Conseil International de la Musique Populaire* (French—International Folk Music Council)

CIMR Commander's Internal Management Review

cims chemical ionization mass spectrometry

CIMS Computer-Integrated Manufacturing System; Convair Integrated Management System

CIMTP *Congrès International de Médecine Tropicale et de Paludisme* (French—International Congress of Tropical Medicine and Malaria)

cimu compatibility-integration mockup

cin cervical intra-epithelial neoplasia; code identification number; component identification number; cost item number

cin (CIN) communication identification navigation

c_{in} insulin clearance

cin *cinéma* (French—motion picture theater)

Cin Cincinnati

CIN Change Incorporation Notice; Change Instrumentation

Notice; Cooperative Information Network (linking libraries by twx)

CIN Chemical Industry Notes

CINAT Cook Islands National Art Theatre

CINB&T Continental Illinois National Bank and Trust

Cinc Cincinnati

C-in-C Commander-in-Chief

CINC Commander-in-Chief

CINCAFE Commander-in-Chief, Air Forces Europe

CINCAFLANT Commander-in-Chief, Air Force Atlantic Command

CINCAFMED Commander-in-Chief, Allied Forces Mediterranean

CINCAFSTRIKE Commander-in-Chief, Air Force Strike Command

CINCAL Commander-in-Chief, Alaskan Command

CINC ATL FLT Commander-in-Chief, Atlantic Fleet

CINCEASTLANT Commander-in-Chief, Eastern Atlantic

CINCENT Commander-in-Chief, Central Europe

CINCEUR Commander-in-Chief, Europe

CINCHAN Commander-in-Chief—Channel (NATO)

CINCHF Commander-in-Chief, Home Fleet (British)

CINCHOMEFLT Commander-in-Chief, United Kingdom Home Fleet

Cinci Cincinnati

CINCIBERLANT Commander-in-Chief, Iberian Atlantic

Cincin Cincinnati

CINCLANT Commander-in-Chief, Atlantic

CINCLANTFLT Commander-in-Chief, Atlantic Fleet

CINCMEAFSA Commander-in-Chief, Middle East, Southeast Asia, Africa South of the Sahara

CINCMED Commander-in-Chief, Mediterranean

CINCMELF Commander-in-Chief, Middle-East Land Forces

CINCNELM Commander-in-Chief, U.S. Naval Forces in Europe, the Eastern Atlantic, and the Mediterranean

CINCNORAD Commander-in-Chief, North American Defense Command

CINCNORTH Commander-in-Chief, Northern Europe

CINCONAD Commander-in-

Chief, Continental Air Defense Command

CINCPAC Commander-in-Chief, Pacific

CINCPACFLT Commander-in-Chief, Pacific Fleet

CINCRDAF Commander-in-Chief, Royal Danish Air Force

CINCRDN Commander-in-Chief, Royal Danish Navy

CINCRNAF Commander-in-Chief, Royal Norwegian Air Force

CINCRNORN Commander-in-Chief, Royal Norwegian Navy

CINCSOUTH Commander-in-Chief, Southern Europe

CINCSTRIKE Commander-in-Chief, United States Strike Command

CINCUNC Commander-in-Chief, United Nations Command

CINCUSAFE Commander-in-Chief, United States Air Forces in Europe

CINCUSAFLANT Commander in Chief—United States Air Force Atlantic

CINCUSAFSTRIKE Commander-in-Chief—United States Air Force Strike

CINCWESTLANT Commander-in-Chief, Western Atlantic

Cincy Cincinnati

Cindy Cinderella; Cynthia

cine cinema; cinematography

CINECA Cooperative Investigation of the Eastern Central Atlantic

cinemactor cinema actor

cinemactress cinema actress

cinerama cinematic panorama (three-dimensional film)

CINFAC Counterinsurgency Information Analysis Center

CINFO Chief of Information

CINM Channel Islands National Monument (Southern California)

cinn cinnabar

Cinn Cincinnati

cinna cinnamon

cinnabar mercuric sulfide (H_gS)

Cinn Sym Orch Cincinnati Symphony Orchestra

CINOA Confédération Internationale des Négociants en Oeuvres d'Art (French—International Confederation of Art Dealers)

CINPDis Cincinnati Procure-

ment District (US Army)

cins child(ren) in need of supervision

CINS CENTO Institute of Nuclear Science

Cin Sym Cincinnati Symphony

CINTA Compañía Nacional del Turismo (Chilean Airline)

Cinty Cincinnati

CINVA Centro Interamericano de Vivienda y Planteamiento (Spanish—Inter-American Center of Housing and Planning)

cio central input/output (multiplexer)

CIO Church Information Office(r); *Commission Internationale d'Optique* (French—International Optical Commission); Congress of Industrial Organizations

CIOCS Communications Input-Output Control System

CIOMS Council for the International Organization of Medical Sciences

ciopw charcoal, ink, oil, pencil, and watercolor (title of a book illustrated and written by e e cummings in 1931)

CIOSL Confederación Internacional de Organizaciones Sindicales Libres (Spanish—International Confederation of Free Trade Union Organizations)

cip cast-iron pipe; cataloging in publication; certified inhalation protection; cipher

cip (CIP) capital investment program; commercially important passenger

C i P Cataloging in Publication (Library of Congress program)

CIP Canadian International Paper; Career Internship Program; Carriage/Freight and Insurance Paid (to); Cataloging-in-Publication; Citizens Involvement Project; Civilian Institution Program; Communications and Information Policy; Composite Interface Program; Consolidated Intelligence Program; Cook Islands Party; Coordinated Instrument Package; Cost Improvement Proposal

CIP Comisión Interamericana de Paz (Spanish—Inter-American Peace Commission)

CIPA Canadian Industrial Preparedness Association; Chartered Institute of Patent

Agents; Committee for Independent Political Action
CIPAC Collaborative International Pesticides Analytical Council (UK)
CIPASH Committee for an International Program in the Atmospheric Sciences and Hydrology
cipc cast-in-place concrete
CIPC Christmas Island Phosphate Commission
CIPCE *Centre d'Information et de Publicité des Chemins de Fer Européens* (French—Information and Publicity Center of the European Railways)
CIPE *Centro Interamericano para la Promoción de las Exportaciones* (Spanish—Inter-American Center for the Promotion of Exports); *Consejo Internacional de la Pelicula de Ensena; atnza* (Spanish—International Council for Educational Films)
CIPEC *Conseil Intergouvernmental des Pays Exportateurs de Cuivre* (French—Intergovernmental Council of Copper-Exporting Nations)
CIPFA Chartered Institute of Public Finance and Accountancy
ciph cipher
CIPHER Calculations of Patient and Hospital Education Resources
ciphony enciphered telephony
CIPL Canada India Pakistan Line; Commission of Inquiry into Public Libraries (Australian)
CIPL *Comité International Permanent de Linguistes* (French—Permanent International Committee of Linguists)
CIPM Council for International Progress in Management
Cipo Cipriano
CIPO *Conseil International pour la Préservation des Oiseaux* (French—International Council for the Preservation of Birds)
CIPP Cataloging-in-Publication Program (Library of Congress); Comprehensive Income and Price Policy
CIPP *Conseil Indo-Pacifique des Pêches (French—Indo-Pacific Fisheries Council)*

CIPR *Commission Internationale de Protection Contre les Radiations* (French—International Commission on Radiological Protection)
CIPRA Cast-Iron Pipe Research Association
CIPRA *Commission Internationale pour la Protection des Régions Alpines* (French—International Commission for the Protection of Alpine Regions)
CIPREC Canadian Institute of Public Real Estate Companies
CIPs Commercially-Important Persons
CIPS Canadian Information Processing Society
CIPT Canadian Institute for the Prevention of Torture
CIQ Customs, Immigration, Quarantine
cir circle; circuit; circular
cir. *circa* (Latin—about)
cIR crime on Indian Reservation
Cir Circimus; Circinis (constellation); Circle; Circus
CIR Commission on Intergovernmental Relations; Commissioner of Internal Revenue; Consumer Information Report; Cost Information Report; Court of Industrial Relations; Current Industrial Reports
CIRA Committee on International Reference Atmosphere; Conference of Industrial Research Associations
CIRA *Centro Interamericano de Reforma Agraria* (Spanish—Inter-American Center of Agrarian Reform)
CIRADS Counter-Insurgency Research and Development System
cir ant. circular antenna
cir bkr circuit breaker
circ circle; circular; circulate; circulation; circumcision; circumference; circumferential(ly); circumstance; circus
Circ Circimus; Circle; Circus
CIRC Central Information Reference and Control; Cross Interleave Reed-Solomon Code
CIRC *Centre International de Recherché sur le Cancer* (French—International Center for Cancer Research)
CIRCA Computerized Information Retrieval and Current

Awareness
circad circadic; circadian; circadianly
circal circuit analysis
CIRCALS Circle Analysis System
CIRCAS Contract Information Retrieval and Cost Accounting System
circltr circular letter
circs circumstances
circum circumcision; circumference
circum (Latin prefix—around) —circumnavigation
circum haema circumorbital haematoma (a black eye)
Circumv Stz Circumvesuviana Stazione (Neapolitan railroad station serving Herculaneum, Mt Vesuvius, and Pompeii)
Circus circular intersection (Oxford Circus, Piccadilly Circus, St Giles Circus, etc.)
circuscade circus parade
Ciren Cirencester (Sisister)
CI Rep Communist International Representative
CIRES Cooperative Institute for Research in Environmental Studies
CIRF Corn Industries Research Foundation
CIRF *Centre International d'Information et de Recherche sur la Formation Professionelle* (French—Vocational Training and Research Center)
CIRIA Construction Industry Research and Information Association
CIRIEC Canadian International Centre of Research and Information on Public and Cooperative Economy
CIRIS Completely Integrated Range-Instrumentation System (NASA)
CIRJP Commission on International Rules of Judicial Procedure
CIRM *Centro Internazionale Radio-Médico* (Spanish—International Medical Radio Center); *Comissão Interministerial de Recursos do Mar* (Portuguese—Inter ministerial Commission for the Resources of the Sea)
CIRO Consolidated Industrial Relations Office
CIRP City Improvement and Restoration Program; Cooperative Institutional Research Program
CIRVIS Communication In-

structions for Reporting Vital Intelligence Sightings (of ufo's from aircraft)

cis carcinoma in situ; cataloging in source; central inhibitory state

cis (CIS) cataloging in source

ci's conflict indicators

Cis Cecilia

CIs Current Investigations

CIS Cancer Information Service; Career Information System; Catholic Information Society; Center for International Studies (MIT); Central Industrial Secretariat; Central Instructor School; Chartered Industries of Singapore; Chartered Institute of Secretaries; Communications and Information Systems; Congressional Indexing Service; Congressional Information Service; Cost Inspection Service; Council for Inter-American Security; Cranbrook Institute of Science

CISA Canadian Industrial Safety Association; *Commission Internationale pour le Sauvetage Alpin* (French—International Commission for Alpine Rescue); Cook Islands Sports Association; Council for Independent School Aid

CISAC *Confédération Internationale des Auteurs et Compositeurs* (French—International Federation of Authors and Composers)

cisam compressed index sequential access method

CISC Commission for Investigation of Special Crimes

Cisco San Francisco

CISCO Civil Service Catering Organization; Commercial and Industrial Security Corporation (Singapore)

CISE Colleges, Institutes, and Schools of Education (Library Association)

CISF *Confédération Internationale des Sages-Femmes* (French—International Confederation of Midwives)

CISHEC Chemical Industry Safety and Health Council

CISI Command Inspection System Inspection (USAF); *Compagnie Internationale de Service et Information* (French—International Service and Information Company)

CISIR Ceylon Institute of Scientific and Industrial Research

Cisister Cirencester, England

CISL *Confederazione Italiana Sindacati Lavoratori* (Italian Confederation of Labor Syndicates)

CISLE *Centre International des Syndicalistes Libres en Exil* (French—International Center of Free Trade Unionists in Exile)

cislun cislunar; cislunarian; cislunarite

CISPES Committee in Solidarity with the People of El Salvador

CISR Center for International Systems Research; Commonwealth Inscribed Stock Registry

CISS Computerized Information Storage System

Cissie Cecilia

Cissy Cecilia

Cist Cistercian

CISTI Canada Institute of Scientific and Technical Education

CISV Children's International Summer Village

cit citation; cited; citizen(ship); citrate; compression in transit; computer interface terminal; configuration identification table(s); counterintelligence team

cit (CIT) call-in time

cit citat (Dano-Norwegian—quotation)

Cit Citadel

CIT Calcutta Improvement Trust; California Institute of Technology (Cal Tech); Carnegie Institute of Technology; Case Institute of Technology; Central Institute of Technology; Chartered Institute of Transport; Coal Industry Tribunal (Australian); Conference of Interpreter Trainers; Continental Inclusive Tour; Counterintelligence Team; Court of International Trade; Cranfield Institute of Technology

CIT (ARIA) Commission on Insurance Terminology (American Risk and Insurance Association)

CIT *Compagnia Italiana di Turismo* (Italian—Italian Travel Bureau)

cit a citric acid

CITA Commercial-Industrial-

Type Activity; Cook Islands Tourist Authority

CITAB Computer Instruction and Training Assistance for the Blind

CITARS Crop Identification Technology Assessment for Remote Sensing (NATO)

CITB Construction Industry Training Board

CITC Canadian Institute of Timber Construction; Cook Islands Trading Corporation

cite. compression ignition and turbine engine

CITE Consolidated Index of Translations into English; Coordinating Information for Texas Educators; Council of the Institute of Telecommunication Engineers; Current Information on Tapes for Engineers

CITEL *Comisión Interamericana de Telecomunicaciones* (Spanish—Inter-American Telecommunication Commission)

CITEP Canadian Indian Teacher Education Projects

CITES Convention on International Trade in Endangered Species

CITGO Cities Service Gulf Oil

cithp closed-in tubing head pressure

Citi Citibank

Citibank First National City Bank

CITIC China International Trust and Investment Corporation

Citicorp First National City Bank Corporation

CITIS Centralized Integrated Technical Information System

CITL Canadian Industrial Traffic League

CITO Charter of the International Trade Organization

cito disp. *cito dispensetur* (Latin—dispense rapidly)

CITP Civilian Industrial Technology Program

CITRAIL Centre Inter-Regional de Transit Rail-Route (French Canadian)

citricult citriculture

citrine false topaz (quartz with ferric iron)

CITS China International Travel Service

CITT Canadian Institute of Traffic and Transportation

citu (CITU) coronary intensive-care unit
Cit U City University
City The City—business and financial section of London
City Ed City Editor
CIU Coopers' International Union; Criminal Intelligence Unit (police)
CIUL Council for International Urban Liaison
CIUS Conseil International des Unions Scientifiques (French —International Council of Scientific Unions)
civ civil; civilian; civilization; civilize
CIV City Imperial Volunteers (London); Commonwealth Institute of Valuers
CIV Commission Internationale du Verre (French—International Glass Commission)
CIVA Cook Islands Visitors Association
Civ Air NM Civil Aircraft National Marking(s)
civd cold-induced vasodilation
Civ E Civil Engineering
civ eng civil engineering
Civ Eng Civil Engineer
civies civilian clothes; civilians
civiling civilingenier (Dano-Norwegian—civil engineer)
CIVIS Centro Italiano per i Viaggi d'Instruzione per Studenti (Italian—Italian Center for Students' Educational Travel)
civ svc civil service
civ svt civil servant
civvies civilian clothes; civilians
civvy civilian
ciw current instruction word
CIW California Institution for Women
CIWS Close-in Weapons System
cixa constant infusion excretory urogram
cj clip joint; conjectural; construction joint
CJ Chief Justice; Civil Jail; Court of Judiciary
C of J Collector of Junk
CJ Computer Journal
CJA Carpenters and Joiners of America
CJA Criminal Justice Abstracts
C-jam cocaine
CJB Constructors John Brown (British shipbuilders)
CJC Colby Junior College;

Community Junior College
CJC Corpus Juris Canonici (Latin—Code of Canon Law)
CJCA California Junior College Association
CJCiv Corpus Juris Civilis (Latin—Code of Civil Law)
CJCs Criminal Justice Councils
C-J disease Creutzfeldt-Jakob disease
cje corretaje (Spanish—brokerage)
CJE Citizens for Jobs and Energy
CJF Carlos J. Finley
CJFWF Council of Jewish Federations and Welfare Funds
CJI Concrete Joint Institute
CJI Comite Juridico Interamericano (Spanish—Inter-American Juridical Committee)
CJIS Criminal Justice Information System (Rhode Island)
CJLF Criminal Justice Legal Foundation
CJM Congregation of Jesus and Mary
C-joint cocaine joint
CJP Criminal Justice Publications
CJR Cecil John Rhodes
CJR Columbia Journalism Review
CJRL Criminal Justice Reference Library (Austin)
cjs cotton, jute, or sisal (cargo)
CJS Canadian Joint Staff; College of Jewish Studies
CJS Corpus Juris Secundum
CJTF Commander Joint Task Force
ck cask; certified kosher; check; coke; cork
ck ceekay (Spanish-American slang—cocaine)
Ck chalk; Creek
CK cyanogen chloride (poison gas)
C K Cape Kennedy
ckb cork base
ckbd cork board
CKC Canadian Kennel Club
CKCJP Center for Knowledge in Criminal Justice Planning
CKCL Chicago-Kent College of Law
ckd completely knocked down
CKD Certified Kitchen Designer
CKE Central Kingdom Express (see Ori Exp)
ckf cork floor
ckfm checking form

ckga checking gage
CKIC Chemical Kinetics Information Center (NBS)
ckm cents per kilometer
CKMTA Cape Kennedy Missile Test Area
cko checking operator
ck os countersink other side
ckout checkout
ckpt cockpit
cks casks; checks
ckt circuit
CKT Chung-Kuo Kung-ch'an Tang (Chinese—Chinese Communist Party)
ckt bd circuit board
ckt bkr circuit breaker
ckt cl circuit closing
ck tp check template
ck ts countersink this side
ck viv check valve
ckw clockwise
cl carload; center line; centiliter; chest and left arm (cardiology); chloride; class; clavicle; clear; clearance; climb; clinic; close; closure; conceptual level; confidence limits; corpus luteum; critical list
cl (CL) control leader (data processing)
c/l combat loss
c/l (C/L) carload lot; cash letter
cl. classis (Latin—class or collection)
Cl chlorine; chlorine gas; Cloister; Close
Cl Calle (Spanish—street)
CL Capital Airlines; chlorine; chlorine gas; Cooperative League; Critical List; light cruiser (2-letter naval symbol)
C-L Canadair Limited (Division of General Dynamics)
C/L craft loss (insurance)
C & L Canal and Lake; Coopers and Lybrand
C of L Count(y) of Lippe
CL. Clericus (Latin—cleric or clergyman)
CL-1 Computer Language One
CL-13 Canadair-built F-86 Sabre aircraft
CL-28 Canadair-built long-range reconnaissance version of the Britannia
CL-41 Canadair-built jet-trainer aircraft nicknamed Tutor
cla center line average; communication link analyzer
Cla Clare College, Cambridge
CLA California Library Association; Canadian Library Association; Canadian Lumber-

men's Association; Catholic Library Association; Chinese Librarians Association; College Language Association; Commercial Law Association; Computer Law Association; Connecticut Library Association; Conservative Library Association; Copyright Licensing Agency

C.L.A. Certified Laboratory Assistant

CLAA anti-aircraft light cruiser (4-letter naval symbol)

CLA-ACB Canadian Library Association—*l'Association Canadienne des Bibliothèques*

Clack Clackmannan(shire)

cl ad collet adapter

CLAE Council of Library Associations Executives

CLAH Conference of Latin American History

CLAIMS Class Codes Assigned Index Method Search

CLAIRA Chalk Lime and Allied Industries Research Association

clam (CLAM) chemical low-altitude missile

clam. chemical low-altitude missile

clamato clam-and-tomato juice

clamsan clam sandwich

clamwich clam sandwich

cland lit clandestine literature (underground)

cland press clandestine press

CLAO Contact Lens Association of Ophthalmologists

CLAP Citizens Lobbying Against Prostitution

clar clarification; clarify; clarinet

Clar Clarence

Clara Clarabelle; Clarissa; Clarita

Clare Clara; Clarita

Clar(en) Clarendon

clark combat launch and recovery kit

Clark William Andrews Clark Memorial Library of the University of California at Los Angeles

CLARNICO Clark, Nichols, and Coombes (confectioners)

Clarrie Clarice; Clarissa

clas classification; classify; congenital localized absence of skin

c-l-a-s crowd-lift-actuate-swing (tractor backhoe control)

CLAS Chartered Land Agents

Society; Computer Library Applications Service

CLASB Citizens League Against the Sonic Boom

CLASC *Confederación Latinoamericana de Sindicalistas Cristianos* (Spanish—Latin American Confederation of Christian Trade Unionists)

clasn classification

clasp. **(CLASP)** computer liftoff and staging program

CLASP Center for Law and Social Policy; Citizens Local Alliance for a Safer Philadelphia; Client's Lifetime Advisory Service Program; Computer Language for Aeronautics and Space Programming; Computer Launch and Separation Problem

CLASPS Coded Label Additional Security and Protection System

class. classification

Class Classical

CLASS California Library Authority for Systems and Services; Christian Leaders And Speakers Seminars; Class Action Study and Survey; Close Air-Support System; Closed-Loop Accounting for Store Sales; Computer-based Laboratory for Automated School Systems; Current Literature Alerting Search Service

class A's class-A narcotics (addictive drugs such as opium and its derivatives)

class B's class-B narcotics (almost non-addictive drugs such as codeine and nalline)

CLASSIC Classroom Interactive Computer

classif classification

CLASSMATE Computer Language to Aid and Stimulate Scientific, Mathematical, and Technical Education

class M's class-M narcotics (non-addictive drugs)

classn classification

class X's class-X narcotics (drugs containing small amounts of narcotics such as cough syrups with non-narcotic and almost non-addictive codeine)

clat communication line adapters for teletype

CLAT Confederation of Latin American Teachers

clav clavecin; clavichord; clavicle

clave autoclave; steamclave (sterilizer)

clavicemb *clavicembalo* (Italian—clavichord)

claw. clustered atomic warhead

CLAW Community Law Workshop (New Zealand)

clax claxon

Claymont Girls Woods Haven-Kruse School for (delinquent) Girls at Claymont, Delaware

Clb Caleb

CLB Church Lads' Brigade; Configuration Liaison Board

clbbb complete left bundle branch block

clbr calibration

CLBs Combat Lessons Bulletins

c & lc capital and lower case letters

CLC Canadian Labour Congress; Canners League of California; Chiriqui Land Company; Cost of Living Council; task-fleet command cruiser (naval symbol)

CLCB City of Liverpool College of Building; Committee of London Clearing Banks

CLCCS Cammel-Laird Cable-Control System

cl/cll counseling learning/community language learning

CLCMD Cleveland Contract Management District

CL & Co Cammell Laird and Company (shipbuilders)

clcs current-logic-current switching

clct collector

CLCT City of Liverpool College of Technology

cld cancelled; chronic liver disease; chronic lung disease; cleared; colored; cooled; cost laid down

cld (CLD) called (line)

CLD Central Library and Documentation

CLDAS Clinical Laboratory Data Acquisition System

cldwn cooldown

cldy cloudy

CLE Cleveland, Ohio (Hopkins Airport)

Clea Cleopatra

CLEA Canadian Library Exhibitor's Association

clean. comprehensive lake-ecosystem analyzer

CLEAN Committee for Leaving the Environment of America Natural; Commonwealth Law Enforcement Assistance Network (Pennsylvania)

CLEAPSE Consortium of Local Education Authorities for the Provision of Science Equipment

CLEAR Center for Lake Erie Area Research; Civic Leaders for Ecological Action and Responsibility; Closed-Loop Evaluation and Reporting (system); Committee to Leach the Environment of Acid Rain; County Law Enforcement Applied Regionally

Clearwaters Clearwater Mountains of Idaho

clec closed-loop ecological cycle

CLEC Citizen/Labor Energy Coalition

Clem Clemens; Clement; Clementina; Clementine

CLEMARS California Law-Enforcement Mutual-Aid Radio System

Clemte Clemente

CLENE Continuing Library Education Network and Exchange

cleo clear language for expressing orders

Cleo Cleopatra

CLEO Council on Legal Education Opportunity

cleopatra comprehensive language for elegant operating system and translator design

CLEP College-Level Education Program; College-Level Examination Program

CLEPR Council on Legal Education for Professional Responsibility

cler clerical; controlled letter contract reduction

cleric. clerical(s); clerical error; clericalism; clericality; clerically

CLES Customs Law-Enforcement Service

CLETS California Law Enforcement Telecommunications System

CLEU Coordinated Law Enforcement Unit

Cleve Cleveland

Cleve Orch Cleveland Orchestra

CLEVPDis Cleveland procurement District (US Army)

CLEW Chicago Law Enforcement Week

clf capacitive loss factor

CLF Chicano Liberation Front; Church of the Larger Fellow-

ship (Unitarian Universalist); Commonwealth Literary Fund(ing)

CLFNE Conservational Law Foundation of New England

Clfs Cliffs

clg calling; ceiling; clearing

Clg College

CLG light guided-missile cruiser (3-letter symbol)

CLGA Composers and Lyricists Guild of America

CLGES California Life Goals Evaluation Schedules

clgp (CLGP) cannon-launched guided projectile

clgsfu clear glazed structural facing units

clgsub clear glazed structural unit base

cl gt cloth gilt

CLGW Cement, Lime and Gypsum Workers (union)

Cl H Clare Hall, Cambridge

CLH *Croix de la Légion d'Honneur* (French—Cross of the Legion of Honor)

CLHU Computation Laboratory of Harvard University

cli central life interests; coin-level indicator; cost-of-living index

CLI Cost-of-Living Index

CLIA Clinical Laboratory Improvement Act; Cruise Lines International Association

C-library circulating library

CLIC Corporate Library Information Centre (Canadian)

clics computer-linked information for container shipping

Cliff Clifford; Clifton

clim climatic

CLIMAPS Climate Long-range Investigation, Mapping, and Prediction Study

climat climatological; climatologist; climatology

Climatol Climatology

CLIMPO Contract Liaison and Master Planning Office

clin clinic; clinical; clinician; clinometer

CLIN Contract Line Item Number

clin/d clinical death

clink (generic nickname—prison)—also the nickname for brothels and in London, where it originated in Clink Prison, also stands for the Southwark Fair depicted by Hogarth

clin path clinical pathology

clin proc clinical procedures

Clint Clinton

Clinton Men Clinton Correctional Facility at Dannemora, New York

Clinton Women Correctional Institution for Women at Clinton, New Jersey

clip. compiler language for information processing; confused, lacerated, incised, and punctured (wounds)

CLIP Cancel Launch in Progress (USAF); Country Logistics Improvement Program (USAF)

clips. clippings; computer launch interference problems

CLIS Central Library and Information Services; Clearinghouse for Library Information Sciences

clit clitoral; clitoridectomy; clitoris

C. Litt. Companion of Literature

clj control joint

Clj *Callejon* (Spanish—alley, blind alley, cul-de-sac, lane)

CLJ *Cambridge Law Journal*

CLJC Copiah-Lincoln Junior College

clk clerk; clock

CLK hunter-killer cruiser (naval symbol)

clkg caulking

clks clockwise

clkws clockwise

cll cholesterol lowering lipid; chronic lymphocytic leukemia; circuit load logic; community language learning

CLL Chief of Legislative Liaison

cllo *cuartillo* (Spanish—fourth of a real, pint)

Cllr Councillor

clm column; culumnar

c-lm common-law marriage

Clm Culham

CLM Canadian Liberation Movement

CLMA Cigarette Lighter Manufacturers Association; Contact Lens Manufacturers Association

CLML *Current List of Medical Literature*

CLMS Clinical Laboratory Monitoring System; Company Lightweight Mortar System

cln colon; corrective lens

Cln Colón

clnc clearance

CLNP Crater Lake National Park (Oregon)

clnr cleaner

CLNS Cape Lookout National Seashore (North Carolina)
clnt coolant
CLNTS China Lake Naval Test Station
CLNWR Crescent Lake National Wildlife Refuge (Nebraska)
clo closet; cloth; clothing; cod liver oil
Clo Callao
CLO Cali, Colombia (Calipuerto airport); Chief Liaison Officer; Citizens for Law and Order; Civic Light Opera (Los Angeles); Cornell Laboratory of Ornithology
CLOB Composite Limit Order Book
CLOC Computerized Logging and Outage Control
CLOCE Contingency Lines of Communication Europe
CLODS Computerized Logic-Oriented Design System
clog. change log (for software); computer-logic graphics
clor container loaded at owner's risk
clora closed-form ray analysis
clos closure; command to line of sight
clousy cloudy—lousy (weather)
clp control line platform; criminal law and procedure
clp (CLP) command language processor
Clp Cornell list processor (language)
CLP Cargo Loss Prevention; Carnegie Library of Pittsburgh; Country Liberal Party
CLPA Common Law Procedure Acts
cl pal cleft pallet
clpr caliper
clr center of lateral resistance; clear; clearing; cooler
clr (CLR) computer language recorder
CLR Central London Railway; Council on Library Research; Council on Library Resources
CL&R Canal, Lake, and Rail
CLR *Common Law Reports; Commonwealth Law Reports*
CLRA Consumer Law Reform Association
CLRB Canada Labour Relations Board
CLRI Council on Library Resources Incorporated
clrm classroom
clr test chloride test

CLRU Cambridge Language Research Unit
CLRV Canadian Light Rail Vehicle
cls coils
cls (CLS) close (flow chart)
CLS Certificate in Library Science; Country Library Service
CLSA Conservation Law Society of America; Contact Lens Society of Australia
CLSB California Library Services Board
CLSC Chautauqua Literary and Scientific Circle
CLSCS Cain-Levine Social Competency Scale
clsd closed
clsg closing
CLSG Contact Lens Study Group
CLSI Computer Library Services, Inc.
clsl chronic lymphosarcoma leukemia
CLSP Cape Lookout State Park (Oregon)
clsr closure
clst clarinettist
clsx close-loop support extended
clt communications line terminals
CLT Charlotte, North Carolina (airport)
CLT *Canadian Law Times*
CLTA Canadian Library Trustees Association; Canterbury Lawn Tennis Association; Chinese Language Teachers Association
C Lt-Cdr Communication Lieutenant-Commander
cltgl climatological
cltgr climatographer
cltv closed-loop television
clu central logic unit; circuit lineup
CLU Chartered Life Underwriter
CLUB Central Library of the University of Baghdad
Club Med Club Méditerranée
CLUC Central Library Union Catalog
CLUM Civil Liberties Union of Massachusetts
CLUMIS Cadastral and Land-Use Mapping Information System
clurt come let us reason together (mediator's motto)
CLUS continental limits United States
CLUSA Cooperative League

of the USA
clv clevis
Clv Cleveland
clvd calved
clvs calves
Clw Collingwood
CLW Council for a Livable World
clwg clear wire glass
clws clockwise
Cly Clyde; Clydebank
Clydebank Scotland's shipyard city on the River Clyde northwest of Glasgow
clz copper, lead, or zinc (cargo)
cm center of mass; centimeter(s); circular mil; circular muscle; contrast media; costal margin; countermortar; mechanic (symbol)
cm (CM) command module; communications multiplexor
c'm' come
c/m color modulation (tv); communications multiplexer; control and monitoring; corrected manifest current month
c&m cocaine and morphine
cm *carat métrique* (French—metric carat); *Zentimeter* (German—centimeter)
c.m. *causa mortis* (Latin—cause of death); *cras mane* (Latin—tomorrow morning)
Cm curium
Cm *Camino* (Spanish—highway)
CM absolute coefficient of pitching moments (symbol); Canadian Militia; Certificate of Merit; Certificated Master (or Mistress); Circulation Manager; Clyde-Mallory (steamship line); Command Module; Commercial Message; Common Market; Configuration Management; Corporate Member(ship); Corresponding Member; Court Martial; Cow Month (New Zealand); mine layer (naval symbol)
C-M Charente-Maritime
C.M. central meridian; *Chirurgiae Magister* (Latin—Master of Surgery)
C/M Curtis/Mathes
C of M Certificate of Merit; Count(y) of Mark
CM *Collegiate Microcomputer; Correo Marítimo* (Spanish—sea mail)
cm² square centimeter
cm³ cubic centimeter

CM4 Comet 4 jet airplane
cma civil-military affairs
Cma Camilla
C ^{ma} Cima (French or Italian—summit)
C Ma Canis major
CMA Cable Makers Australia; California Maritime Academy; California Medical Association; Canadian Manufacturers Association; Canadian Medical Association; Candle Manufacturers Association; Canterbury Manufacturers Association; Cash Management Account(ing); Casket Manufacturers Association; Central Monetary Authority; Certified Medical Assistant; Chemical Manufacturers Association; Chinese Manufacturers Association; Chocolate Manufacturers Association; Cigar Manufacturers Association; Cleveland Metal Abrasive (company); Clothespin Manufacturers Association; Colleges of Mid-America; Colorado Mining Association; Commonwealth Medical Association; Confederate Memorial Association; Continental Marketing Association; Court of Military Appeals; Crucible Manufacturers Association
CMA (C.M.A.) Certified Management Accountant
CMA Compañia Mexicana de Aviación (Spanish—Mexican Aviation Company)
CMAA Cleveland Musical Arts Association; Comics Magazine Association of America; Crane Manufacturers Association of America
CMAAC Certified Medical Assistant Administrative and Clinical
CMAAO Confederation of Medical Associations in Asia and Oceania
cmab clothing maintenance allowance, basic
CMAC Capital Military Assistance Command; Catholic Marriage Advisory Council; Computer Management and Control
CMAD Computer Manufacture and Design
cmai clothing maintenance allowance, initial
C Maj Canis Major
CMAL Clothing Monetary Allowance List; Coal Mines

Authority Limited
CMAR Can't Manage A Rifle
C/marca Cundinamarca, Colombia
CMAS Confédération Mondiale des Activités Subaquatiques (French—World Confederation of Sub-aquatic Activities); Council for Military Aircraft Standards
CMAT Canadian Mathematics Achievement Test
CMAV Coalition Mondiale pour l'Abolition de la Vivisection (French—World Coalition for the Abolition of Vivi section)
cmb carbolic methylene blue; chloromercuribenzoate
Cmb Colombo
CMB Chase Manhattan Bank; Cheese Marketing Board; Cina Motor Bus; coastal motor boat; Colombo, Ceylon (airport); Combat Maneuver Battalion(s); Compagnie Maritime Belge (Royal Belgian Lloyd Line)
CMB Consejo Mundial de Boxeo (Spanish—World Boxing Commission); cuyas manos beso (Spanish—whose hands I kiss)—very respectfully yours
CMBARMTNG Combined Arms Training
CMBI Caribbean Marine Biological Institute
CMBs Certified Mortgage Loan Brokers
cmbt combat
cmc code for magnetic characters; contact-making clock; coordinated manual control
cmc (CMC) carboxymethyl cellulose
CMc coastal mine layer (naval symbol)
CMC Canadian Marconi Company; Canadian Music Council; China Machinery Company; Commandant of the Marine Corps; Commercial Metals Company
CMCC Canadian Memorial Chiropractic College; Classified Matter Control Center
cm-cellulose carboxymethyl cellulose
cmcr continuous melting, casting, and rolling
CMCR Compagnie Maritime des Chargeurs Réunis
cmct communicate; communication
cmd command; common meter

double
CMD California Moderate Democrats; Center for Management Development; Center for Moral Democracy (NYC); Central Marine Depot; Contract Management District
cmdg commanding
Cmdr Commander
CMDR Council for Microphotography and Documentary Reproduction
Cmdre Commodore
Cmdt Commandant
cmdty commodity
cme continuing medical education
CME California Motor Express; Center for Musical Experience; Central Medical Establishment; Chicago Mercantile Exchange (formerly Chicago Butter and Egg Board); Chicago Merchandise Exchange; Courtesy Motorboat Examination (U.S. Coast Guard)
CME Conférence Mondiale de l'Energie (French—World Power Conference)
CMEA Council for Mutual Economic Assistance (also called CEMA or COMECON)
CMEC Council of Ministers of Education (Canadian)
CMERD Centre for Medical Education, Research, and Development (New South Wales)
c'mere come here
CMERI Central Mechanical Engineering Research Institute (India)
cmet coated metal
cmf calcium-and-magnesium-free; countermortar fire; cylindrical magnetic film
cmf (CMF) cyclophosphamide methotrexate 5-fluorouracil (anticarcinogen)
CMF Citizen Military Forces; Commonwealth Military Forces; Composite Medical Facility
CMFNZ Chamber Music Federation of New Zealand
CMFRI Central Marine Fisheries Research Institute
cmfsw calcium-and-magnesium-free seawater
cmg call me gov(ernor); control-moment gyroscope
CMG Computer Management Group; Corning Museum of

Glass
C.M.G. Companion of the Order of St Michael and St George
CMGH Cleveland Metropolitan General Hospital
cmh countermeasures homing
CMH Columbus, Ohio (airport); Combined Military Hospital; Congressional Medal of Honor; Council for the Mentally Handicapped; County Mental Health
cmha confidential, modified handling authorized
CMHA Canadian Mental Health Association
CMHC Central Mortgage and Housing Corporation; Community Mental Health Center(s)
CMHCA Community Mental Health Centers Act
CMHPA Cloves Memorial Hall for the Performing Arts (Indianapolis)
cmi carbohydrate metabolism index; cellular-mediated immune (response); container master information; cumulative monthly issue
cmi (CMI) computer-managed instruction
C Mi Canis Minor
CMI Can Manufacturers Institute; Career Maturity Inventory; Christian Michelson Institute (for Science and Free Thought—Bergen, Norway); Church Management Institute (Episcopal); *Comité Météorologique Internationale* (French—International Meterological Committee); Command Maintenance Inspection (US Army); *Commission Mixte Internationale* (Spanish—International Joint Commission); Commonwealth Mining Investments; Commonwealth Mycological Institute; Communications Management Inc; Consolidated Metal Industries; Cornell Medical Index
CMI Cornell Medical Index
CMIA Coal Mining Institute of America; Cultivated Mushroom Institute of America
cmid cytomegalic inclusion disease
cmif career-management individual file
CMIK Choson Minjujuui In'min Konghwaguk (North Korea)
cmil circular mil

c/min cycles per minute
C Min Canis Minor
CMIU Cigar Makers' International Union
CMJ Church Mission to the Jews; Church's Ministry among the Jews
CMJ Computer Music Journal
cml chemical; circuit micrologic; commercial; current mode logic
cml (CML) chronic myelocytic leukemia
CML Central Music Library; Colonial Mutual Life; Container Marine Lines
CML Camara Municipal de Lisboa (Portuguese—Lisbon Town Council)
CMLA Canadian Music Library Association; Central Medical Library Association; Chief Martial Law Administrator (Bangladesh); Civilian Maimed and Limbless Association
CmlC Chemical Corps
cml def chemical defense
CMLEA California Media and Library Educators Association
cmlops chemical operations
CMLS Cleveland-Marshall Law School
CM/LSCNP Cradle Mountain/Lake Saint Clair National Park (Tasmania)
CMLU Container Marine Lines (container) Unit
cmm commemorative; cubic millimeter(s); cutaneous malignant melanoma
CMM Center for Male Medicine; Central Methodist Mission; Chief Machinist's Mate (USN); Commission for Maritime Meteorology (WMO)
cmma clothing monetary maintenance allowance
CMMA Concrete Mixer Manufacturers Association
CMMBE Comissão Militar Mista Brasil-Estados Unidos (Portuguese—Joint Brazilian-American Military Commission)
cmmch combat Mach change
cmme carcinogenesis of chloromethyl-methyl ether
CMMM Chase Manhattan Money Museum (New York City)
cmmnd command(ing)
CMMP Commodity Management Master Plan

CMMS Columbia Mental Maturity Scale
cmn commission; cystic medial necrosis
CMN Common Market Nationals; Common Market Nations
CMN Common Market News; Conselho Monetario Nacional (Portuguese—National Monetary Council)—Brazil
cmn-aa cystic medial necrosis of the ascending aorta
cmnce commence
CMNH Cleveland Museum of Natural History
CMNM Capulin Mountain National Monument; Craters of the Moon National Monument
cmnr commissioner
cmo cardiac minute output; computer microfilm output
CMO Chief Medical Officer; Collateralized Mortgage Obligation; Commonwealth Medical Officer; Configuration Management Office; Contract Management Office(r)
c'mon come on
cmos (CMOS) complementary metal-oxide semiconductor
cmp corrugated metal pipe; cost of maintaining product
cmp (CMP) compare (flow chart); computation(al)
CMP Catoctin Mountain Park (Maryland); Christian Movement for Peace; Church Music Publishers; Commissioner of the Metropolitan Police; Company of Mission Priests; Competitive Medical Plan; Controlled Materials Plan; Cornell Maritime Press; Corps of Military Police
CMPC California Manpower Planning Council
CMPC Compañia Manufacturera de Papeles y Cartones (Spanish—Paper and Carton Manufacturing Company)
cmpd compound; compounded; compounding
cm pf cumulative preference; cumulative preferred (shares)
cmpl complement (flow chart)
cmpld compiled
cmplt complete
cmpltn completion
cmplx complex
Cmpn Companion
Cmpn IAP Companion of the Institution of Analysts & Programmers
cmpnt component

CMPO Calcutta Metropolitan Planning Organisation
cmps centimeters per second
cmpt component
cmptr computer
cmr cerebral metabolic rate; common-mode rejection
CMR Cape Mounted Rifles; Communications Monitoring Report; Consolidated Mail Room; Contract Management Region; Convention Merchandises Routiers; Court of Military Review
cmrO₂ cerebral metabolic rate for oxygen
CMRA Chemical Marketing Research Association
CMRB Chemicals and Minerals Requirements Board
Cmrd comrade
CMRE California Marriage-Readiness Evaluation
CMRF Childrens Medical Research Foundation
cmrg cerebral metabolic rate of glucose
CMRL Chamber of Mines and Research Laboratories
CMRN Cooperative Meteorological Rocket Network
CMRNWR Charles M. Russell National Wildlife Range (Montana)
cmro cerebral metabolic rate of oxygen
CMRO County Milk Regulations Office(r)
cmrr common mode rejection ratio
cmrs computer-managed rotation schedules
CMRs Classified Material Receipts
CMRS Countermeasures Receiver System
CMRW Coalition for the Medical Rights of Women
cms complete matched set
cm/s centimeters per second
c.m.s. *cras mane sumendus* (Latin—to be taken tomorrow morning)
CMS California Malacozoological Society; California Museum of Science; Center for Measurement Science (George Washington University); Central Monitoring Stations (Africa); Chicago Medical School; Chief Master Sergeant; Christian Medical Society; Church Missionary Society; College Music Society; *Compagnie Maritime de la Seine* (French—Seine Mari-

time Company); Computer Management System; Configuration Management System; Consumers and Marketing Service; Contemporary Music Society; Conversational Monitor System; Correctional Medical Systems; County Medical Services
CMSA Consolidated Metropolitan Statistical Area
CM & SA Canning Machinery and Supplies Association
CMSC Central Missouri State College
CMSEP Contractor Management System Evaluation Program
CMSER Commission on Marine Science, Engineering, and Resources
CMSG Canadian Merchant Service Guild
CMSgt Chief Master Sergeant
CMSI California Museum of Science and Industry
CMS & I California Museum of Science and Industry
cm/sm command module/service module
CMSN China Merchants Steam Navigation (company)
CMSTP & P Chicago, Milwaukee, St Paul and Pacific (railroad)
cmt comment; confluence of major and tributary (rivers); cut, make, and trim (clothing)
CMT California Mastitis Test; California Motor Transport; Camden Marine Terminals; Charcot-Marie-Tooth; Compulsory Military Training; Concert Memory Test; Current Medical Terminology; Current Mortuary Tables
CMT *Confédération Mondiale du Travail* (French—World Confederation of Labor)
CMTA Chinese Musical and Theatrical Association
CMTC Citizens Military Training Camp
cmt/conc cement or concrete
CMTCS Computer Management Transaction Control System
CMTCU Communications Message Traffic Control Unit
cmte committee
Cmte committee
Cmto Caminito
Cmto *Caminito* (Spanish—small street)
cmu central markup unit; chlorophenyldimethylurea; con-

crete masonry unit
CMU Central Michigan University; Church Management Institute
C-M U Carnegie-Mellon University
CMUA Commercial Motor Users Association
cmv cytomegalovirus
cmv (CMV) cytomegalovirus
cmvm contact-making (or breaking) voltmeter
CM von W Carl María von Weber
CMVPB California Motor Vehicles Pollution Board
cmw critical minimum weight
CMW Citizens for Migrant Workers
CMWA County Meat Works Association
cmy civilian man-years
cmz concordant memory zone
CMZ *Compagnie Maritime du Zaire* (French—Zaire Maritime Company)
CMZS Corresponding Member of the Zoological Society
cn canal; cannon; coordination number
cn (CN) chloroacetophenone
c/n carbon-to-nitrogen ratio; carrier-to-noise ratio
c/n (C/N) credit note
c.n. *cras nocte* (Latin—tomorrow night)
cN centinewton
Cn canon; contract number; cumulonimbus
CN absolute coefficient of yawing moments (aerodynamic symbol); Carl Nielsen; Central Airlines; Chinese Nationalists; Code Napoléon; Commercial Node (zone); Commonwealth Nations; compass north; Confederate Navy; cosine of the amplitude (mathematical symbol)
C/N Consignment Note; Cover Note
C & N communication and navigation
CN Canadian National-Grand Trunk Railways
cna code not allocated
CNA California Nurses Association; Canadian Nuclear Association; Canadian Numismatic Association; Canadian Nurses Association; Center for Naval Analyses (Franklin Institute); Central News Agency (Nationalist China); Central Northern Airways; Certified Nurses Aide;

Certified Nurses Assistant; Chemical Notation Association; Chief of Naval Air; Chief of Naval Aviation; Community Newspaper Association; Confederate Navy of America

CNA Comisión Nacional del Azúcar (Spanish—National Sugar Commission)—Mexico

CNAA Council for National Academic Awards

CNAC China National Aviation Corporation

CNACO Canada National Arts Centre Orchestra

CNADS Conference of National Armaments Directors

CNAEA California Narcotic Addict Evaluation Authority

CNAN Compagnie Navale Afrique du Nord (French—North Africa Naval Company)

CNAS Chief of Naval Air Services; Civil Navigation Aids System

CNASA Council of North Atlantic Shipping Associations

CNATra Chief of Naval Air Training

CNAV Canadian Naval Auxiliary Vessel

CNAVSTA Charleston Naval Station (South Carolina)

CNB Central Narcotic Bureau (Singapore); Crocker National Bank

CNBC Consumer News and Business Channel

Cnbr Canberra

Cnbry Canterbury

CNBS Comisión Nacional Bancaria y Seguros (Spanish—National Banking and Security Commission)—Mexico

cnc central navigation computer; computer numerical control; consecutive number control

Cnc Cancer

CNC Christopher Newport College; Counter Narcotics Center

Cncl(r) Council(or)

CNCMH Canadian National Committee for Mental Hygiene

CNCO China Navigation Company

CNCP Canadian National/Canadian Pacific (telecommunications)

cncr concurrent

cnct connect(ion)

cnd condition(ed); conduit

CND Campaign for Nuclear Disarmament; Commission on Narcotic Drugs (UN)

CND Code Names Dictionary

cndi commercial nondevelopment items

cn di combination die

CNDP Communications Network Design Procedure(s)

cnds condensate

cne chronic nervous exhaustion

CNE Canadian National Exhibition; Commission on National Elections

cnee consignee

Cnel Coronel (Spanish—Colonel)

CNEL community noise equivalent level

C'nelia Cornelia

CNEN Comisión Nacional de Energía Nuclear (Spanish—National Nuclear Energy Commission)

CNEngO Chief Naval Engineering Officer

CNEP Cable Network Engineering Program (Bell)

CNES Centre National d'Etudes Spatiales (French—National Center for Space Studies)

CNET Chief of Naval Education and Training

CNET Centre National d' Etude des Télécommunications (French—Telecommunication National Study Center)

CNEXO Centre pour d'Exploitation des Océans (French—Center for the Exploitation of the Oceans)

cnf confine

CNF Caribbean National Forest (Puerto Rico); Cleveland National Forest (near San Diego, California); Commonwealth Nurses' Federation

cng compressed natural gas

cng (CNG) compressed natural gas

CNG Connecticut Natural Gas

CNGA California Natural Gas Association

CN-gas cyanide gas

CNGB Chief, National Guard Bureau

CNGO Committee on Non-Governmental Organizations (UN)

CNGT Commonwealth New Guinea Timbers

CN-GT Canadian National

Railways-Grand Trunk Western

CNH Community Nursing Home

cnhd congenital nonspherocytic hemolytic disease

CNHI Committee for National Health Insurance

CNHM Chicago Natural History Museum (Field Museum of Natural History)

CNI Chief of Naval Information; Communications—Navigation and Identification

CNI Centro de Información Nacional (Spanish—National Information Center)—Chilean security police; *Centro Nacional de Informaciones* (Spanish—National Information Center)—Chile's intelligence service

CNIB Canadian National Institute for the Blind

CNIE Comisión Nacional de Inversión Extranjiera (Spanish—National Commission for Foreign Investment)—Mexico

CNIF Conseil National des Ingénieurs Français (French—National Council of French Engineers)

CNIN California Narcotic Information Network

CNIPA Committee of National Institutes of Patent Agents

CNJ Central of New Jersey (railroad)

cnl cancel(lation); cardiolipin natural lecithin

cnl (CNL) circuit net loss

CNL Canadian National Library (Ottawa); Commonwealth National Library (Canberra)

CNLA Canadian National Library Association; Council of National Library Associations

CNLIA Council of National Library and Information Associations

CNM Cabrillo National Monument; Chief of Naval Material; Chiricahua National Monument; Colombo National Museum; Colorado National Monument

CN-M Certified Nurse-Midwife

CNMI Commonwealth of the Northern Mariana Islands (in the western Pacific)

CNN Cable News Network; Certificated Nursery Nurse

CNN *Compagnie de Navigation Nationale* (French—National Navigation Company)
CNNI Cable News Network International
CNNR Caerlaverock National Nature Reserve (Scotland); Cairngorms National Nature Reserve (Scotland)
cno computer non-operational
*C*ⁿᵒ *Corno* (Italian—peak, summit)
CNO Chief of Naval Operations; Chief Nursing Officer
CNOA California Narcotics Officers Association
CNOAE Council of National Organizations of Adult Education
CNOBO Chief of Naval Operations Budget Office
cnop conditional no operation
cnor consignor
C-note $100 bill
CNP Canyonlands National Park (Utah); Caramoan NP (Philippines); Cleveland NP (South Australia); Colonial NP (Virginia); *Compagnie Navale des Pétroles; Compagnie de Navigation Paquet;* Corbett NP (India); Council for National Policy; Cyril Northcote Parkinson
CNP *Compañia Navegación Peruana* (Peruvian Navigation Company)
CNPA California Newspaper Publishers Association
CNPB Canadian National Parole Board
cn/pnl contractor's panel
CNPP *Centre National de Prévention et de Protection* (French—National Prevention and Protection Center)
CNPS California Native Plant Society; Central National Philharmonic Society (Beijing)
cnr carrier-to-noise ratio; composite noise rating; corner
Cnr Corner
CNR Canadian National Railway; Civil Nursing Reserve; Coleford Nature Reserve (South Africa); Council of National Representatives
CNR *Consiglio Nazionale delle Ricerche* (Italian—National Research Council)
CNRA Curecanti National Recreation Area (Colorado)
CNRS *Centre National de la Recherche Scientifique* (French—National Center for Scientific Research)
cnrt concrete
cns central nervous system; construction
cns (CNS) central nervous system
c.n.s. cras nocte sumendus (Latin—to be taken tomorrow night)
Cns Cairns
CNS Center for New Schools; Chief of the Naval Staff; Community Nursing Services; Congress of Neurological Surgeons; Copley News Service
CNS *Chubu Nippon Shimbun* (Central Japan Newspaper)
CNSA Carl Nielsen Society of America
CNSC China National Software Corporation
cnsg consolidated nuclear steam generator
cnsl console (flow chart)
Cnst Pty Constitution Party
cnstr canister
cnstrn construction
CNSWTG Commander, Naval Special Warfare Task Group
cnt celestial navigation trainer (CNT); count(er)
cnt (CNT) celestial navigation trainer
CNT Canadian National Telegraphs; Composite Negotiating Text(s)
CNT *Compañía Nacional de Teléfonos* (Spanish—National Telephone Company); *Confederación Nacional de Trabajo* (Spanish—National Confederation of Labor); *Conselho Nacional de Telecomunicação* (Portuguese—National Telecommunications Council)—government-controlled radio and television for all Brazil
CNTB Colombia National Tourist Board
CNTCA Canadian National Railway—Transcanada Airlines
cntn contain
CN Tower Canadian National Tower, Toronto, Ontario
cntr container; contribute; contribution; counter
Cntr Centaur (space vehicle)
cntrfugl centrifugal
cntrl central; control(ler)
cntrs containers
CNTU Canadian National Trade Unions; Confederation of National Trade Unions

CNUCE *Centro Nazionale Universitàrio di Calcol Elettronico* (Italian—National University Center of Electronic Calculation)
Cnut King Canute II of Denmark and England
cnv contingent negative variation
CNV Cape Canaveral, Florida (tracking station)
CNVA Committee for Non-Violent Action
cnvc conveyance
cnvr conveyor
cnvt convict
C & NW Chicago and North Western (railway)
CNWDI Critical Nuclear Weapons Design Information
CNWR Camas National Wildlife Refuge (Idaho); Chassahowitzka NWR (Florida); Chatauqua NWR (Illinois); Chincoteague NWR (Virginia); Columbia NWR (Washington)
CNX Canadian National Exposition
CNYP Central New York Power (corporation)
co carbon monoxide; cardiac output; castor oil; cervicoaxial; choir organ; cleanout; coenzyme; conscientious objector; convenience outlet; corneal opacity; criminal offense; crossover(s); cutoff; cutout
co (CO) close/open (to official correspondece)
c-o cutoff
c/o care of; carried over; cash order; complains of
co compagno (Italian—company); (Latin prefix—together)—copulation
co. compositus [Latin—compound(ed)]
Co cobalt; Colombia; Colombian; Colombiano; Columbia; Columbian; company; County
C/o complained of
*C*ᵒ *Cabeço* (Portuguese—hillock, knoll, mound)
Co cerro (Spanish—hill or peak)
CO carbon monoxide; Certificate of Origin; Cleveland Orchestra; Colonial Office; Colorado; Commanding Officer; Commissioners Office; conscientious objector; Continental Airlines (2-letter code); Controller's Office;

Copyright Office; Correctional Officer; Criminal Office; Crown Office(r)

C/O cash order; Chief Officer

C & O Chesapeake & Ohio (railroad)

C of O Count(y) of Oldenburg

co 1mo canto primo (Italian—first treble)

CO_2 carbon dioxide

Co^{60} radioactive cobalt

coa condition on admission

coa coenzyme A

CoA College of the Atlantic; Committee on Accreditation (ALA); Council of the Americas

C o A Committee on Accreditation (ALA); Court of Appeal

COA Canadian Orthopedic Association; Change Order Account; Chattanooga Opera Association; Conchologists of America; Connecticut Opera Association; Control Operating Authority (NATO); Cordova Airlines; Correctional Officers Association

CO(A) Change Order (Aircraft)

COA Comunidad Oriental Africana (Spanish—East African Community)

coac clutter-operated anti-clutter receiver

Coad Coadjutor

COADS Command and Administration System (USA)

coag coagulant; coagulate; coagulation

coag time coagulation time

Coah Coahuila (inhabitants—Coahuileños or Coahuilenses)

Coal. Coalition

coalit govt coalition government

co/alr cooper/aluminum

coam coaming; customer-owned-and-maintained equipment

coam equip customer-owned-and-maintained equipment (data processing)

CO-AMP Cost Optimization-Analysis of Maintenance Policy

coas crewman optical alignment sight

COAS Council of the Organization of American States

Coast The Coast (Pacific coast of Canada and the United States)

Coastal Eastern East-Coast-of-the-United-States English reflecting cultural influences

Coastal Express Norwegian steamship line linking ports from Bergen to Kirkenes

Coast Line Atlantic Coast Line Railroad

coax coaxial

c-o-b close of business

CoB Chief of Base (CIA)

C o B Chief of Boat (submarine)

COB Change Order Board; Command Operating Budget

COBA Correction Officers Benevolent Association

COBAS Council of Black Architectural Schools

cobble(s) cobblestone(s)

COBE Cosmic Background Explorer Satellite

COBF Cobol-F (program)

cobh carboxyhemoglobin

cobility cobol utility (program)

coblib cobol library

C & O-B & O Chesapeake and Ohio-Baltimore & Ohio (merged railroads)

COBOL common business-oriented language

cobra. (**COBRA**) coolant boiling in rod arrays

COBRA Counter Battery Radar

COBRA Computadores Brasileiros (Portuguese—Brazilian Computers)

COBRAY Cobra + Moray (anti-terrorist academy)

COBSI Committee on Biological Sciences Information

COBTU Combined Over-the-Beach Terminal Unit

coc cathodal opening clonus; cathodal opening contraction; cocaine; coccygeal; combination-type oral contraceptive

Coc Cleveland open cup

CoC Chamber of Commerce

COC Canadian Opera Company; Combat Operations Center

coca cocaina (Spanish—cocaine)

coca-colon coca-colonization; coca-colonize; coca-colonizer

C & O Canal Chesapeake and Ohio Canal

COCAST Council for Overseas Colleges of Art, Science, and Technology

cocb crossed olivochochlear bundles

cocc coccyx

cocci coccidioidomycosis

coccy coccidioidomycosis

COCESS Contractor-Operated Civil Engineer Supply Store

coch coach(es)

coch. *cochleare* (Latin—spoonful)

Coch Cochin

coch. ampl. *cochleare amplum* (Latin—tablespoonful)

COCHASE Code for Coupled-Channel Schrödinger Equations

coch. infant. *cochleare infantis* (Latin—teaspoonful)

coch. mag. *cochleare magnum* (Latin—tablespoonful)

coch. med. *cochleare medium* (Latin—dessertspoonful)

coch. parv. *cochleare parvum* (Latin—teaspoonful)

COCI Council on Consumer Information

cock. cockney (dialect of London's East End and waterfront residents born within sound of the bells of the Church of Saint Mary-le-Bow—Bow bells)

Cock Cockburn; Soviet Antonov 350-passenger plane (NATO)

cockapoo crocker spaniel-poodle mix-breed dog

cockaterr cocker-terrier mixed-breed dog

Cock Engl Cockney English (spoken by people in London's East End and traditionally born within sound of Bow bells)

cocl cathodal opening clonus

C & OC NM Chesapeake and Ohio Canal National Monument

co-co carried-on-carried-off (break-bulk cargo)

cocohol coconut-oil-extended ethanol (diesel fuel)

COCOM Coordinating Committee for Export to Communist Area(s); Coordinating Committee on Multilateral Export Controls

COCOSEER Coordinating Committee on Slavic and East European Library Services

cocp closed olivocochlear potential

cocr cylinder overflow control record

COCS Container Operating Control System

coct. *coctio* (Latin—boiling)

Co Cts County Courts

COCU Churches of Christ Uniting; Consultation on

Church Union (of Episcopalians, Methodists, Presbyterians, and others)

cod. cause of death; chemical oxygen demand; cleanout door; codeine

c-o-d cargo-on-deck

c.o.d. cash-on-delivery

Co D Costume Designer

COD coding

COD *Concise Oxford Dictionary*

CODA Came Out Decades Ago; Committee on Drugs and Alcohol

codac coordination of operating data by automatic computer

CODAC Community Organization for Drug Abuse Control

CODAF Commission on Border Development and Friendship (U.S.-Mexican)

codag combined diesel and gas (turbine machinery)

codan carrier-operated device anti-noise

CODAP Client-Oriented Data-Acquisition Process; Control Data Assembly Program

CODAS Customer-Oriented Data System

CODASYL Conference on Data Systems Languages

CODC Canadian Oceanographic Data Center

codd codices

CODE Collaborating on Drug Education; Commission on Declining Enrollments; Committee on Donor Enlistment

Code Blue (medical jargon) frequently used to mean cardiac arrest

codec coder decorder

coded. computer-oriented design of electronic devices

CODEF Chairman of Defense Committee

CODELCO *Corporación del Cobre* (Spanish—Copper Corporation)—Chile

codel(s) congressional delegation(s)

CODELS Computer Development System

Code N Code Napoléon

CODES Computer-Oriented Data Entry System

codic computer-directed communication(s)

codiphase coherent digital-phased array system

codit computer direct to telegraph

codl cash on the dotted line

cod. memb. *codex membranacius* (Latin—book printed or written on skin or vellum)

CoDoC Cooperation in Documentation and Communication

codog combined diesel or gas

CODOT Classification of Occupations and Directory of Occupational Titles (UK)

CODSIA Council of Defense Space Industries Association

coe cab over engine (truck); close of escrow (realty); crude oil equivalent

coe (COE) crossover electrophoresis

CoE Corps of Engineers

COE Commonwealth Office of Education; Corps of Engineers; Council for Occupational Education; Council on Optometric Education

CO(E) Change Order (Electronic)

COE *Conseil Aécuménique des Eglises* (French—World Council of Churches)

coea cost and operational effectiveness analysis

COEA *Consejo de la Organisación de los Estados Americanos* (Spanish—Council of the Organization of American States)

coed coeducation(al); female student

coed (COED) computer-operated electronic display

co-ed co-editor

COEDS Char Oil Energy Development Systems

COEES Central Office Equipment Engineering System (Bell)

coef coefficient

coel (Greek *koilos*—cavity, hollow)—coelenterate, coelomate, coelostat(ic)

Coel Coelenterata

COENCO Committee for Environmental Conservation

COEP Community Outreach Educational Program

COEPS Cortically-Originating Extra-Pyramidal System

COESA Committee on Extension of the Standard Atmosphere (United States)

coet (COET) crude oil equalization tax(ation)

cof cause of failure; coefficient of friction; cost of funds

CoF Chaplain of the Fleet; Chief of Finance

cofad computerized facilities design

cofc container on flat car

coff cofferdam

COFFEE Community Organization for Full-Employment Economy

c/offer counter offer

Coffs Coffs Harbour, Australia

COFHE Consortium on Financing Higher Education

COFI Committee on Fisheries (FAO)

COFIDE *Corporación Financiera de Desarrollo* (Spanish—Financial Development Corporation)

COFIPS Central Ohio Federation of Information Processing Societies

COFIS Canadian On-Line Financial Information Service

COFO Council of Federated Organizations (CORE, NAACP, SCLC, SNCC)

COFPHE Capital Outlay Fund for Public Higher Education

co/fr counter offer

COFRC Chevron Oil Field Research Company

cofron copper iron (patent medicine mixture)

COFSAF Chief of Staff, U.S. Air Force

cofx component fixture

cog center of gravity

cog. cognate

c-o-g coal to oil to gas

CoG Council of Governments

COG Change Our Gender; Change Our Goal; Council of Governments

cogag combined gas and gas

CoGARD Coast Guard

cogas coal-oil-gas

cogb certified official government business

cogent. compiler and generalized translator

cogita computerized general I.Q. test(ing)

cogn cognomen

cognit cognition(al)(ly); cognitive(ly)

cogn w cognate with

cogo coordinate geometry

COGP Commission on Government Procurement

cog/prsl cognizant personnel

cogs cost of goods sold

cogs. combat-oriented general support

COGS Continuous Orbital Guidance System

COGSA Carriage of Goods by Sea Act

cogtt cortisone-primed oral glucose tolerance test

coh cash-on-hand; coefficient of haze; coheir

COH carbohydrate (generalized formula)

COHA Council on Hemispheric Affairs

COHATA *Compagnie Haitienne des Transports Aériens* (French—Haitian Air Transport Company)

cohb carboxyhemoglobin

Co Hd coral head

COHD Copyright Office History Document

coher cohere(d); coherence; coherency; coherer; cohering; coherent(ly)

COHM Copyright Office History Monograph

coho coherent oscillator

COHO Council of Health Organization

Cohoun Colquhoun

COHSE Confederation of Health Service Employees

coi classroom observation instrument; crack-opening interferometry

COI Central Office of Information; Certificate of Origin and Interest; Coordinator of Information

COI *Comité Olimpico Internacional* (Spanish—International Olympic Committee); *Commission Oceanographique Intergouvernementale* (French—Intergovernmental Oceanographic Commission)

COIC Canadian Oceanographic Identification Center; Combined Operations Intelligence Center

CoID Council of Industrial Design

coif coiffure

COIMS Council for International Organizations of Medical Sciences

coin. coinage; counterinsurgency—anti-guerrilla warfare

coin. (COIN) complete operating information

COIN Counterinsurgency

Coin & Curr Coin and Currency Institute

coin gold 90% gold, 10% copper

coin-op coin-operated

COINS Computer and Information Sciences; Computerized Information System(s); Control in Information Systems; Cooperative Intelli-

gence Network System

coin silver 50 to 92.5% silver with balance of copper or other metals

co-intel counterintelligence

COINTELPRO Counterintelligence Program (FBI)

COIR Commission on Intergroup Relations (NYC)

Cois François

COIT Central Office of the Industrial Tribunal (UK)

COIU Congress of Independent Unions

COJ Court of Justice

COJO Conference of Jewish Organizations

Cok Cochin

coke coca drink; cocaine

Coke Coca Cola; NATO nickname for the Soviet Antonov AN-26 350-passenger transport plane

Coke Rep Coke's *English King's Bench Reports*

cokesmoke(s) cocaine smoker; cocaine smoking

col colon; colonial; colonic; colonist; colonization; colonize; colony; color; coloring; colorist; colors; column

col (COL) computer-oriented language

c-o-l (COL) cost of living

col. *colatus* (Latin—strained, as through a filter); *collum* (Latin—collar); *colon* (Latin—large intestine)

co-L co-latitude

Col Colchester; Colima; Coliseum (London); College; Cologne; Colombia(no); Colón; Colonel; Colossians; Colossus to the; Columba (constellation); Columbia(n); Coronel

COL Computer Oriented Language

cola (COLA) cost-of-living adjustment; cost-of-living allowance

cola. cost-of-living allowance

cola *colonia* (Spanish—colony)

COLA Committee on Latin America; Committee on Library Automation (ALA); Cost of Living Adjustment

COLAC Central Organization of Liaison for Application of Circuit

Col Alb College of the Albermarle

COLAs Cost-of-Living Adjustments

colat. *colatus* (Latin—strained)

col bh collision bulkhead

col C *col canto* (Italian—follow the voice)

COLC Cost of Living Council

cold. chronic obstructive lung disease; colored

Col$ Colombian peso

COLDEMAR *Compañia Colombiana de Navegación Maritima* (Spanish—Colombian Maritime Navigation Company)

colen. *colentur* (Latin—let them be strained, strain them)

Col Ency Columbia Encyclopedia

Col Ent Exam College Entrance Examination

coleop coleoptera; coleopterist

colet. *coleatur* (Latin—let it be strained, strain it)

colidar coherent light detection and ranging

Colin Nicholas

colingo compile online and go (data processing)

coll collect(or); collection; colloid(al); colloquial(ism)

Coll College; Collegiate

collab collaboration; collaborator

collabo(s) collaborator(s)

coll agc collection agency

collat collateral

collect. collection; collective; collectively

Coll Ency Colliers' Encyclopedia

Collier-Macmillan Collier-Macmillan Library Service

Collins Wm Collins Sons & Co

Collins Bay Canadian penitentiary on Collins Bay near Kingston, Ontario

Coll L Collection Letter

colloq colloquial(ism); colloquium

coll'ott *coll'ottava* (Italian—play in octaves, with the octave)

collr collector

collun. *collunarium* (Latin—nose wash)

collut. *collutorium* (Latin—mouthwash)

coll vol collective volume

Coll Wooster College of Wooster

colly colliery

collyr. *collyrium* (Latin—eyewash)

colm column

Colm Columba

COLMIS Collection Management Information System

colo colophon (printer's or publisher's device, symbol,

or trademark)
Colo Colorado; Coloradan
Colo Colossians
Colo Assoc Colorado Associated University Press
colog cologarithm
colograph color lithograph
Colom Colombia; Colombian
Colom Christovão Colom (Portuguese—Christopher Columbus)
Colombia Republic of Colombia (South American two-ocean nation), *República de Colombia*
Colombian West Indies Saint Andrew and Providence islets, seven groups of coral cays on Nicaragua's Mosquito Coast
Colombo Christoforo Colombo (Italian—Christopher Colombus)
colón unit of currency in Costa Rica and El Salvador
coloph colophon
Colorado Springs U.S. Air Force Academy at Colorado Springs, Colorado
colorectal colon rectal (area or cancer)
coloreds colored persons (South Africans of mixed blood)
Colo St U Comm Colorado State University Institute in Technical and Industrial Communications
col p color page
COLPA Commission on Law and Public Affairs
colrad collegiate research and development
COLREG Regulations Governing Collisions
COLS Communications for On-Line Systems
Col-Sgt Colour-Sergeant
colspd collapsed
Col Sym Columbia Symphony
colt computerized on-line test(ing)
COLT Council on Library Technology; Council on Library-Media Technical-Assistants
Colu Columba
Columbia School Columbia Training School at Columbia, Mississippi
Columbus House Columbus Workhouse and Women's Correctional Institution at Columbus, Ohio
com comedy; comma; command; commercial; commis-

sion; committee; common; communication(s); complement; compliment
com (COM) computer output on microfilm; computer-output microfilm(ing)
com comunicaciones (Spanish—communications)
com or *con* or *cor* (Latin *cum*—together, with)—compound, confine, correct
com. commemoratio (Latin—commemoration)
Com Coma Berenices (constellation); Comoro Islands
COM Chief Operations Manager; Council of Ministers
COMA Coke Oven Managers' Association; Council On Minority Aging
comac continuous multiple-access comparator
COMACH Confederación Maritima de Chile (Spanish—Maritime Confederation of Chile)
COMAEGEAN Commander, Aegean
COMAINT Command Maintenance
COMAIR Commercial Airways
COMAIRBALTAP Commander Baltic Air Forces (NATO)
COMAIRCENT Commander, Allied Air Forces, Central Europe
COMAIRCENTLANT Air Commander, Central Atlantic
COMAIRCHAN Maritime Air Commander, Channel
COMAIRESTLANT Air Commander, Eastern Atlantic
COMAIRLANT Commander, Air Force, Atlantic
COMAIRNON Commander, Allied Air Forces, Northern Norway
COMAIRNORLANT Air Commander, Northern Atlantic
COMAIRNORTH Commander, Allied Air Forces, Northern Europe
COMAIRSONOR Commander, Allied Air Forces, Southern Norway
COMAIRSOUTH Commander, Allied Air Forces, Southern Europe
Comalco Commonwealth Aluminum Company (Australia)
COMANSEC Computation and Analysis Section (Canadian Defense Research

Board)
COMANTDEFCOM Commander, United States Antilles Defense Command
comar computer aerial reconnaissance
COMARC Cooperative Machine Readable Cataloging
COMARRHIN Commander, Maritime Rhine
COMART Commander, Marine Air Reserve Training
comat computer-assisted training
COMAT Committee on Materials
COMATS Commander Military Air Transport Service
Com Aus Commonwealth of Australia
comb. combat; combination; combine; combing; combustion
Com Bah Commonwealth of the Bahamas
COMBALTAP Allied Command Baltic Approaches (NATO)
COMBARFORCLANT Commander, Barrier Forces, Atlantic
COMBATCRULANT Commander, Battleship-Cruiser, Atlantic Fleet
combi combination
combine. combined harvester
COMBISLANT Commander, Bay of Biscay, Atlantic
COMBLACKBASE Commander, Black Sea Defense Sector
combo combination (of musicians, or of a safe)
COMBO Combined Arts of San Diego
combo lock(s) combination lock(s)
COMBOSFORT Commander, Bosphorus Fortifications
COMBQUARFOR Combined Quarantine Force
Com Brit Comunidad Británica (Spanish—British Commonwealth of Nations)—Great Britain and former colonies
COMBRITELBE Commander, British Naval Elbe Squadron
COMBRITRHIN Commander, British Naval Rhine Squadron
combs. combinations
combu combustion
combust combustion
COMCANLANT Command-

er, Canadian Atlantic

COMCARIBSEAFRON Commander, Caribbean Sea Frontier

COMCAT Computer Output Microfilm Catalog

COMCEN Communications Center

COMCENTLANT Commander, Central Atlantic

ComCm Communications counter-measures and deception

COMCRUDESFLOT Commander Cruiser-Destroyer Flotilla

COMCRUDESPAC Commander Cruisers and Destroyers in the Pacific (USN)

COMCRULANT Commander, Cruisers, Atlantic

comd command

COMDARFORT Commander, Dardanelles Fortifications

COMDESFLOT Commander, Destroyer Flotilla

COMDEV Commonwealth Development

comdg commanding

Com Dom Commonwealth of Dominica

Comdr Commander

Comdt Commandant

COME Chief Ordnance Mechanical Engineer

comeas countermeasures

COMEASTSEAFRON Commander, Eastern Sea Frontier

COMECON Council of Mutual Economic Assistance (of communist nations)

COMED Communications Editing Unit

COMEDBASE Commander, Mediterranean Defense Sector

COMEDS Continental Meteorological Data System

COMEINDORS Composite Mechanized and Document Retrieval System

COMERMEX *Comercial Mexicano (Banco)* (Spanish—Commercial Bank of Mexico)

Com Err *Comedy of Errors*

comet. computer operated management evaluation technique

COMET Committee for Middle East Trade; Computer-Operated Management Evaluation Technique; Controllability, Observability, and Maintenance Engineering Technic

COMETS Computer-Operated Multifunction Electronic Test Station

COMEX Commodity Exchange (NY)

COMEXCO Committee for Exploitation of the Oceans

COMFAIRELM Commander, Air Fleet, Eastern Atlantic and Mediterranean

COMFAIRWINGLANT Commander, Fleet Air Wing, Atlantic

Com Fran *Comunidad Francesa* (Spanish—French Community of Nations)—France and former colonies

comfy comfortable

Com-Gen Commissary-General

COMGENEUCOM Commanding General, European Command

COMGENTHIRDAIR Commanding General, Third Air Division

COMGENUSAFE Commanding General, U.S. Air Forces, Europe

COMGENUSAREUR Commanding General, U.S. Army, Europe

COMGERNORSEA Commander, German North Sea Subarea

COMGIBLANT Commander, Atlantic Approaches Gibraltar

Com/I Commercial Invoice

COMIBERLANT Commander, Iberian Atlantic Area

COMIBOL *Corporacíon Minera de Bolivia* (Spanish—Bolivian Mining Corporation)

COMICEDEFOR Commander, United States Iceland Defense Force

COMICS Computer-Oriented Managed-Inventory Control System

COMIL Chairman of Military Committee

comin' coming

COMINCH Commander-in-Chief, United States Fleet

COMIND Commander, Indian (USN)

Cominform Communist Information Bureau

comint communications intelligence

Comintern Communist International; Cominform

Com Int Sec Committee on Internal Security (formerly

House Committee on Un-American Activities— HUAC)

Com Isl Comoro Islands

comis° *comisario* (Spanish—commissary, delegate, deputy, manager, police inspector)

comit computer operations management information training

COMITEXTIL Coordination Committee for the Textile Industry in the EEC

COMJUWATF Commander, Joint Unconventional Warfare Task Force

comkd completely knocked down

coml commercial

COMLA Commonwealth Library Association

COMLANDCENT Commander, Allied Land Forces, Central Europe

COMLANDEAST Commander, Allied Land Forces, Southeastern Europe

COMLANDMARK Commander, Allied Land Forces, Denmark

COMLANDNON Commander, Allied Land Forces, Northern Norway

COMLANDNORWAY Commander, Allied Land Forces, Norway

COMLANDSOUTH Commander, Allied Land Forces, Southern Europe

COMLANT Commander, Atlantic (USN)

COMLOGNET Combat Logistics Network

comm commerce; commercial; commission; committee; commonwealth; commune; communication; commutator

comm. *commune* (Latin—all the people, the community)

Comm community

Comm. Commodore

Comm *Commendatore* (Italian—Commander, knight)— equivalent to the British Sir

COMMAIRGIBLANT Commander, Maritime Air, Gibraltar, Atlantic

Commanders Commander Islands in the Bering Sea

Com Mat Cen Communication Materials Center (Columbia University)

Comm Bio Pest Committee for Biological Pest Control (San Ysidro, California)

COMMCEN Communications

Center
commd command(ing); commissioned
commdg commanding
Commdr Commander
Commdt Commandant
commem commemoration; commemorative
Comments Nucl Part Phys Comments on Nuclear and Particle Physics
Commerce Department of Commerce; High School of Commerce
COMMFEX Communications Field Exercise
commfu complete and utterly monumental foulup
commi communism; communist
commie commissary; communist
COMMIR Commissioner of Inland Revenue
Commiss Commissary
commn commission
commo communications
Commo Office of Communications (CIA)
commod commodity
Commons House of Commons
common salt sodium chloride (NaCl)
Commonwealth British Commonwealth of Nations: the United Kingdom, Australia, Bahamas, Bangladesh, Barbados, Botswana, Canada, Cyprus, Ghana, Grenada, Guyana, Fiji, India, Jamaica, Kenya, Lesotho, Malawi, Malaysia, Malta, Mauritius, Nauru, New Zealand, Nigeria, Sierra Leone, Singapore, Sri Lanka, Swaziland, Tanzania, The Gambia, Tonga, Trinidad and Tobago, Uganda, Western Samoa, Zambia, and their dependent territories
Commr Commissioner
commstitch communications failure detecting and switching (equipment)
commun communication
commun dis communicable disease
Commun Math Phys Communications in Mathematical Physics
Commun Pure Appl Math Communications on Pure and Applied Mathematics
commuterport(s) commuter-type airport(s)
commy commissariat; commis-

sary; communist
commz communications zone
comn common
ComNAB Commander, Naval Air Bases
COMNAVBALTAP Commander Baltic Naval Forces (NATO)
COMNAVBASE Commander, Naval Base
COMNAVBREM Commander, Bremerhaven Naval Group
COMNAVCAG Commander, Naval Forces, Central Army Group Area and Bremerhaven
COMNAVCENT Commander, Allied Naval Forces, Central Europe
COMNAVCRUITCOMINST Commander, Naval Recruiting Command Instructions
COMNAVFORCESMARIANAS Commander, Naval Forces, Marianas Islands
COMNAVFORJAPAN Commander, Naval Forces, Japan
COMNAVGERBALT Commander, German Naval Forces, Baltic
COMNAVNORCENT Commander, Northern Air Forces, Central Europe
COMNAVNORTH Commander, Allied Naval Forces, Northern Europe
COMNAVSONOR Commander, Allied Naval Forces, Southern Norway
COMNAVSOUTH Commander, Naval Forces, Southern Europe
COMNAVSUPPACT Commander, Naval Support Activity
comnd commissioned
COMNEATLANT Commander, Northeast Atlantic
COMNON Commander, Allied Forces, Northern Norway
COMNORASDEFLANT Commander, North American Anti-Submarine Defense Force, Atlantic
COMNORLANT Commander, Northern Atlantic
COMNORSEACENT Commander, North Sea Subarea, Central Europe
comnr commissioner
Como Commodore; Comodoro Rivadavia (Argentine naval hero and seaport name); Comoro

comp accompaniment; accompany; comparative; comparator; compare; comparison; compass; compensate; compensation; compensatory; compilation; compile(d); compiler; complimentary; compose(d); composition; compositor; compound(ed); comprehension; comprehensive; comptroller; rhetorician's mark meaning false comparison
comp (COMP) complainant
comp. compositus (Latin—compounded of)
COMP College of Osteopathic Medicine of the Pacific; Conceptually-Oriented Mathematics Program
comp a compressed air
compac computer-output microfilm package; computer program for automatic control
COMPAC Commander, Pacific (USN); Commonwealth Pacific Telephone Cable (linking Australia, New Zealand, and Pacific Ocean islands with the rest of the world)
COMPACS Computer-Output Microforms Program and Concept Study (USA)
compact. compatible algebraic compiler and translator; computer planning and control technique
COMPACT Computator Planning and Control Technique
compand compress to expand (radio communication term)
compar comparative
compare. computerized performance and analysis response evaluator; console for optical measurement and precise analysis of radiation from electronics
COMPASS Comprehensive Assembly System; Computerized Movement Planning and Status System
COMPATFOR Commander, Patrol Forces
COMPATFORNORLANT Commander, Patrol Forces, Northern Subarea, Atlantic
comp case compensation case
Comp Curr Comptroller of the Currency
compd compound
compdes compensator design; competitive design
COMPELS Computerized

Evaluation and Logistics System (USA)

compen compensate; compensation; compensatory

compend compendious; compendium

Compendex Computerized Engineering Index

compf composition floor

Comp Gen Comptroller General

compl complaint; complete; compilation; compiled

Compl A Lover's Complaint

COMPLEX Committee on Planetary and Lunar Exploration

complic complications

complt complainant; complaint

comp mar companionate marriage

COMPMR Commander, Pacific Missile Range

compn composition

compo compensation; component; composer; composite; composition; compositor

compool common pool; communications pool(ing)

compos components; composers; composites; compositions; compositors

compound A 11-dehydrocorticosterone

compound B corticosterone

compound E cortisone

compound F cortisol

compound S 11-deoxycortisol

Compr compressor

compreg compressed-impregnated (wood)

comprosl compound procedural scientific language

comp(s) complimentary ticket(s)

compt catecholomethyltransferase; compartment; comptroller

Compt Comptroller

Comptes Rend. Comptes rendus de l'Académie des Sciences (Proceedings of the Academy of Science)

comp time compensatory time off in lieu of overtime wages

COMPTUEX Composite Training Unit Exercise

compu computable; computability; computation(al); computer; computerization; computerize

CompuServ Computer Service (network)

Compu Sex Computer Sex (erotic messages conveyed to home-computer owners)

comput computer

computa computational

computes. computers

computime computer-computed time

Comr Commissioner

COMRAC Combat Radius Capability (DoD)

com rcm command reconnaissance

comrel community relations

COMRNDN Commander, Riverine Division (USN)

COMRNFLOT Commander, Riverine Flotilla (USN)

COMRNRON Commander, Riverine Squadron (USN)

coms communications support

COMS College of Osteopathic Medicine and Surgery (Des Moines)

comsab communist sabotage; communist saboteur

comsabs communist saboteurs

COMSAMAR Commander, Straits and Marmara Defense Sector

Comsat Communications Satellite (corporation)

comsat(s) communications satellite(s)

ComSeaFron Commander Sea Frontier (USN)

comsec communications security

COMSECONDFLT Commander, Second Fleet (USN)

COMSENEX Combined Sensor Tracking Exercise

COMSER Commission on Marine Science and Engineering Research

Com Ser Cen Community Service Center

COMSEVFLT Commander, Seventh Fleet (USN)

COMSIXFLT Commander, Sixth Fleet (USN)

comsn commission

comsoal computer method of sequencing operations for assembly lines

comstar communications satellite network

Comsteel Commonwealth Steel

comstock comstockery

COMSTRATRESCENT Commander, Strategic Reserve, Allied Land Forces, Central Europe

COMSTRIKFLTLANT Commander, Striking Fleet Atlantic (USN)

COMSTRIKFORSOUTH Commander, Naval Striking

and Forces Support, Southern Europe

COMSTS Commander Military Sea Transport Service

COMSUBEASTLANT Commander, Submarine Force, Eastern Atlantic

COMSUBLEDNOREAST Commander, Submarines, Northeast Mediterranean

COMSUBPAC Commander, Submarines, Pacific

comsy commissary

comsymp communist sympathizer

comt comptroller

comt (COMT) catechol-O-methyltransferase

COMTAC Command Tactical (USN)

COMTAFDEN Commander, Tactical Air Force, Denmark

comte committee

COMTEC Computer Micrographics Technology (group)

com tech communications technician

COMTRAC Computer-aided Traffic Control

comtran commercial translation; computer translation

COMUSAFSO Commander, United States Air Forces, Southern Command

COMUSFORAZ Commander, U.S. Forces, Azores

COMUSJAPAN Commander, U.S. Forces, Japan

COMUSKOREA Commander, U.S. Forces, Korea

COMUSMACV Commander, United States Military Assistance Command Vietnam

COMUSRHIN Commander, U.S. Rhine River Patrol

COMUSTDC Commander, U.S. Taiwan Defense Command

Com Ver Common Version (of the Bible)

com wc command weapon carrier

Comy-Gen Commissary-General

Com Z Communications Zone

con confidence (game, man); conned; conning; consolidated; control; conversation; convict

con (CON) constant (flow chart)

con (Latin prefix—together or with)—confab, conference

con. contra (Latin—against)

Con Concord(e); Connie; Conservative; Constance; Con-

suela; consul; consultation
CON Conservative; Conservative Party
con8va. *con ottava* (Italian—with octaves)
CONAC Continental Air Command
CONACS Contractor's Accounting System
CONACYT *Consejo Nacional de Ciencia y Tecnologia* (Spanish—National Council for Science and Technology)
CONAD Continental Air Defense Command
CONADE *Consejo Nacional de Desarrollo* (Spanish—National Development Council)
conaloc continuity and logic
ConArC Continental Army Command
CONASA Council of North Atlantic Shipping Associations
conc concentrate; concentration; concentric; concrete
CONCACAF *Confederación Centro Americano y Caríbean Futbal* (Spanish—Central American and Caribbean Soccer Association)
concb concrete block
conc c concrete ceiling
conc clg concrete ceiling
concd concentrated; concerned
concentr concentrate(d)
CONCEPT Computation Online of Networks of Chemical Engineering Processes
Concertg *Concertgebouworkest* (Dutch—Concert Hall Orchestra)—Amsterdam's celebrated symphony orchestra
conc f concrete floor
conc fl concrete floor
concg concentrating
conch. conchology
concis. *concisus* (Latin—cut)
concn concentration
concomp conversational computation project
Con Con Constitutional Convention
Con Cpt Constructor Captain
concr concrete
cond condenser; condition; conductivity; conductor
condeep(s) concrete deepwater structure(s)
condit conditional
condiv continental divide
condo(s) condominium(s)
condr conductor
cond ref conditioned reflex
cond resp conditioned re-

sponse
condrill concrete drill(ing)
conductimetric conductance + metric
CONE Chamber Orchestra of New England; Collectors of Numismatic Errors
CONEA Confederation of National Educational Associations
Con Ed Consolidated Edison
CONEFO Conference of New Emerging Forces
CONEG Conference of Northeastern Governors
conelrad control of electromagnetic radiation
CONESCAL *Centro Regional de Construcciones Escolares para America Latina* (Spanish—Regional Center of School Construction for Latin America)
con esp *con espressione* (Italian—with expression)
co-netic high-permeability non-shock-sensitive (alloy developed for maximum attenuation at low flux density)
conex connection(s); container export
conex (CONEX) connection(s)
Coney Coney Island
conf confer; conference; confidential
conf. *confer* (Latin—compare)
Conf Confucian; Confucius
CONF Conference Papers Index
confab confabulation; confabulate
CONFAD Concept of a Family of Army Divisions
confec. *confectio* (Latin—confection)
Conf Econ Prog Conference on Economic Progress
confed confederation
Confed Confederate
Confederacy Confederate States of America (Virginia, North and South Carolina, Georgia, Florida, Alabama, Mississippi, Louisiana, Texas, Arkansas, Tennessee—and temporarily in Kentucky and Missouri)
confer. conference
confi confidant(e); confidence; confidential
confid confidential
confid(l)(ly) confidence; confidential(ly)
confit(s) confiture(s)
confr confectioner
cong congress(ional)

cong. *congius* (Latin—gallon)
Cong Congress; congregation; Vietcong
congal *(cuarto) con gal* [Mexican-American—(room) with girl]—house of prostitution
con game confidence game; confidence trick(ery)
Cong Christ Congregational Christians
Cong Digest *Congressional Digest*
congen common specification statements generator; congenial; congenital(ly)
Con Gen Consul General
Cong Fr Congolese franc
Congl Congregational
conglom(s) conglomerate(s); conglomerateur(s); conglomerator(s)
Congo People's Republic of the Congo (formerly the French Congo in west central Africa), *République Populaire du Congo*
Cong Orat Congregation of the Oratory
Congrats congratulations
Cong Rec *Congressional Record*
Congreg Congregationalist
Cong Staff *Congressional Staff Directory*
Cong U Congregational Union (England and Wales)
CONGU Council of National Golf Unions
conics conic sections
conj conjugal; conjugate; conjunction; conjunctivitis
CONLIS Committee on National Library and Information Systems
Con Lt Constructor Lieutenant
Con Lt-Cdr Constructor Lieutenant-Commander
con man confidence man; swindler
CONMAROPS Concept of Maritime Operations
conn connection; connective; connector
Conn Connecticut; Connecticuter
CONN Connellan Airways
CONNECT Connecticut On-Line Enforcement Communication and Teleprocessing (computerized criminal file)
Connie Conrad; Constance; Consuela; Cornelia; Cornelius
Conn Turn Connecticut Turnpike
Conny Constance

co/no current operator/next operator
conobjtr conscientious objector
CONOCO Continental Oil Company
con of consisting of
conopt constrained optimization
conq conquer(ed); conquering; conqueror; conquest
Conr Conrad
conrad contour radar data
CONRAD Contraceptive Research and Development
ConRail Consolidated Rail Corporation (government-sponsored railroads including the Ann Arbor, Central Railroad of New Jersey, Erie-Lackawanna, Lehigh and Hudson River, Lehigh Valley, Penn Central, Reading)
con rod connecting rod
cons conservatory; consider; consist
con(s) confidence (games); conviction(s); convict(s)
cons. conserva (Latin—a preserve)
Cons Conservative
CON Certificate of Need
CONSAL Congress of Southeast Asian Librarians
CONSCIENCE Committee on National Student Citizenship in Every National Case of Emergency
Consc° Consejo (Spanish—Council)
con sect conic section
Cons Eng Consulting Engineer
CONSER CONversion of SERials (Council on Library Resources project)
conserv conservation; conservationist; conservatoire; conservatory
Conserv Conservatoire; Conservatory
cons. et prud. consilio et prudentia (Latin—by counsel and prudence)
Cons Gen Consul General
consgt consignment
conshelf continental shelf
conship control by ship
conshore control from shore
consid consideration
consig consignee
Con S-Lt Constructor Sub-Lieutenant
consltnt consultant
consol consolidated
Consol Consolidated Coal
consolex consolidation exercise

consols consolidated annuities
CONSORT Conversation System with On-Line Remote Terminals
consperg. consperge (Latin—dust, sprinkle)
conspic conspicuous
const constant; constitution; constitutional; construction; constructor
Const Constable; Constitution; Constructor
Const Constitution (of the United States)
constab constabulary
Constan Constantine; Constantinople (Istanbul)
Constance Lake Constance called Bodensee by the Austrians and the Germans
constit constituent(s); constitution(al)
constn constitution; construction
constocs contingency support stocks
constr construction; constructor
construct. construction(al)(ly)
Const US Constitution of the United States
consub continental-shelf submersible
consult. consultant
consumcrit consumer critic(ism)
consv conservation; conserve
cont contact; content(s); continent(al); continue(d); continuous(ly); contract(or); control(ler)
cont. contra (Latin—against); *contusus* (Latin—bruised, contused)
Cont Continent; Continental
contac continuous action
contag contagious
containerport container-ship seaport (equipped for handling and storing containers)
contam contaminant; contaminate; contamination
CONTAM Committee on National Television Audience Measurement
contax consumers and taxpayers
contb contraband
contbg contributing
cont. bon. mor. contra bonos mores (Latin—contrary to good manners)
contd contained; continued
contemp contemporary
contempo contemporary

conter. contere (Latin—rub together)
conter US conterminous United States (48 states having common boundaries)
Cont Eur Continental Europe
Cont Eur & Br I Continental Europe and British Isles
contg containing
Cont HH continental range of ports from Havre to Hamburg
cont hp continental horsepower
Conti Constantine; Constantinople
contigs contiguous states (48 abutting states that form the continental U.S.)
contig US contiguous United States (48 states that border one another)
contin continental; continuous
contin continuo (Italian—continuous); *continuetur* (Latin—let it be continued)
contin US continental United States (Alaska plus the 48 states occupying much of the North American continent)
contl continental
contr contracted; contraction; contractor
contra against; contra-indicated
contra (Latin prefix—against or opposite)—contradict, contraception
CONTRA Coalition of Non-Theist Religious Alternatives
contrail condensation trail
contralat contralateral
contran control translator
contraprop contra + propeller; contrarotating propeller
contrapun contrapuntal(ist); contrapuntist(ic)
contra(s) contraceptive(s)
Contras Contra Sandinistas (Spanish—anti-Sandinistas) Nicaraguan rebels
contr. bon. mor. contra bonos mores (Latin—contrary to good manners)
cont. rem. continuetur remedia (Latin—let the remedy be continued)
contrib contribution; contributor
contrit. contritus (Latin—broken, ground, macerated)
CONTU Commission on New Technological Uses of Copyrighted Works (Library of Congress)
contus. contusus (Latin—

bruised, contused)
cont w continuous window
CONU Contrans (container) Unit
conurb(s) conurbation(s)
Con US (CONUS) Continental United States
CONUS Intel Continental United States Intelligence (USA)
conv convalescent; convention; conventional
Conv convention, conversion
convair conveyed by air
convce conveyance
conv encl convector enclosure
convex convoy exercise
convg convergence
ConVis Convention and Visitors Bureau
Convis Bur Convention and Visitor's Bureau
convl conventional
convn convenient
convt convert(ible)
conv^{te} *conveniente* (Spanish— convenient)
CONWR Crab Orchard National Wildlife Refuge (Illinois)
Coo Coos Bay
Coo Coo *blimey* (Cockney— God blind me)
CoO Committee on Organization
CoO (COO) Chief of Outpost (CIA)
COO Chief Ordnance Officer
cooc contact with oil or other cargo
COOC Commission on Organized Crime
COOH (carboxyl group found in all organic acids)
cook. cookery
Cooks Cook Islanders; Cook Islands; Cook's Tours (Thomas Cook and Son, Ltd)
cool. coolant
Cool-Kal Coolgardie-Kalgoorlie
coon(s) coonhound(s)—contraction of racoon hounds
'coon(s) racoon(s)
coop. cooperation
co-op cooperative
Coop Cooper
COOP ED Cooperative Education
coopg cooperage
Co-op L Cooperative League
COOPLAN Continuity of Operations Plan (USN)
Coop Rep Guy Cooperative Republic of Guyana
Co-op U Co-operative Union

coorauth coordinating authority
coord coordinate; coordination; coordinator
COORS Communications Outrage Restoration Section
COOS Chemical Orbit-to-Orbit Shuttle (NASA)
cop capillary osmotic pressure; casing operating pressure; constable on patrol (origin of cop); copper; copyright; customer owned property; policeman (slang)
cop (COP) computer optimization package
cop. coefficient of performance; commencement of passage; custom of port
c-o-p change of plea
Cop Copernican; Coptic
Cop Copenhague (French, Portuguese, Spanish—Copenhagen)
COP Career Opportunity Program; Certificate of Participation; Certificate of Proficiency; City of Prineville (railroad); Coalition on Police; Combat Outpost; Commissary Operating Program; Community-Oriented Policing; Continuity of Operations Plan; Cox's Orange Pippin; Custom of the Port
Copa Copacabana
COPA Compañía Panameña de Aviación (Spanish—Panamanian Aviation Company); Council on Postsecondary Accreditation
copac continuous operation production allocation and control
COPAL Cocoa Producers' Alliance
COPANT Comisión Panamericana de Normas Tecnicas (Spanish—Panamerican Commission for Technical Standards)
COPAO Council of Philippine-American Organizations
COPAR Corrective or Preventive Action Report
COPARS Contractor-Operated Automotive Parts Store (DoD)
copd chronic obstructive pulmonary disease; coppered
COPD chronic obstructive pulmonary disease
COPDAF Continuity of Operations Plan—Department of the Air Force
cope chronic obstructive pul-

monary emphysema
COPE Champions of Private Enterprise; Committee on Political Education (AFL-CIO); Committee for Original People's Entitlement (Canadian Eskimo's claim to Canadian land); Concerned Organization of Parents to Educate; Congress on Optimum Population and Environment; Council on Population and Environment
COPEC Compañía Petrolera Chilena (Spanish—Chilean Petroleum Company)
COPEI Comité Organizador del Partido Electoral Independiente (Spanish—Organization Committee of the Independent Electoral Party)— Venezuela's Social Christian Party
Copen Copenhagen
COPES College Occupational Programs Educational System; Committee on Program Evaluation and Support; Conceptually-Oriented Program in Elementary Science
COPH Congress of Organizations of the Physically Handicapped
COPIAT Coalition to Preserve the Integrity of American Trademarks
COPICS Copyright Office Publication and Interactive Cataloging System (Library of Congress)
COPL Council of Planning Librarians
copo copolymer
COPO Council of Philatelica Organizations
copp cobaltiprotoporphyrin
Copp Copperplate
COPP Conservation Organization Protesting Pollution
copperas ferrous sulfate; green vitriol
COPPS Committee on Power Plant Siting (Nat Acad Engineering)
COPR Center for Overseas Pest Research; Critical Officer Personnel Requirement (USAF)
cops coppers; policemen (slang)
COPs Coalition on Police members
COPS California Organization of Police and Sheriffs; Chief of Operations (CIA); Committees Organized for Public

Service; Concerns of Police Survivors

Copt Coptic

coptec controller overload prediction technic

copter(s) helicopter(s)

co-ptr co-partner

copu copulate; copulation; copulatory

COPUL Council of Prairie University Libraries

copy. copyright

coq cost of quality

coq. coque (Latin—boil)

co Q coenzyme Q

coq. s.a. coque secundum artem (Latin—boil correctly)

coq. in s.a. coque in sufficiente aqua (Latin—boil in sufficient water)

coq. sim. coque simul (Latin—boil together)

cor contactor, running; corner; cornet; correction

cor corno (Italian—horn)

cor. corpus (Latin—body)

Cor Corinthians; Corona; Coronado; Coroner; Corsica; Coruña

Cor Corea (Portuguese or Spanish—Korea)

COR Commonwealth Oil Refineries

COR Comisión(es) de Orientación Revolucionaria [Spanish—Revolutionary Orientation Committee(s)]—Cuba

cora conditioned orientation reflex audiometry

Cor A Corona Australis

CORA Corporación de la Reforma Agraria (Spanish—Agrarian Reform Corporation)—Chile

coral. class-oriented ring-associated language

Coral Coral Sea; Coral Sea Island Territory beyond Australia's Barrier Reef

CORAL Coherent Optical Radar Laboratory (USAF)

Cor B Corona Borealis

cor bd corner bead

corbfus copy of reply to be furnished us

Corbin Corbin on Contracts by Arthur L Corbin

Corbu Le Corbusier (nickname of Edouard Jeanneret-Gris)

Corc Cornell computing (language)

Cor Chr Col Corpus Christi College—Cambridge

CORCO Commonwealth Oil Refining Company (Puerto Rico)

cord. computer on-line devices

cord. cordillera (Spanish—mountain range)

Cord Cordelia; Córdoba

C of Ord Chief of Ordnance

CORD Center for Occupational Research and Development; Commissioned Officer(s) Residency Deferment; Congress on Research and Dance

cordat coordinate data set

cordic coordinate rotation digital computer

Cordilleras Cordillera Mountains of the Americas

CORDIPLAN Oficina Central de Coordinación y Planificación (Spanish—Central Office of Coordination and Planning)

cordoba monetary unit of Nicaragua

cordpo correlated radar data printout

cords. corduroy pants; corduroy trousers

CORDS Civil Operations and Revolutionary Development Support

core. computed oriented reporting efficiency

CORE Competitive Operational Readiness Evaluation (Air Force); Congress of Racial Equality

corex coordinated electronic countermeasures exercise

corf classroom observational rating form

CORF Comprehensive Outpatient Rehabilitation Facility

corfam microporous artificial leather

corflu correction fluid

CORFO Corporación de Fomento (Spanish—Development Corporation); *Corporación de Fomento de la Producción* (Spanish—Production Development Corporation)—Chile

CORG Combat Operations Research Group

CORGI Confederation for Registration of Gas Installers

corin corinthian

Coriol Coriolanus

CORL Canadian Operations Research Society

CORLS Central Ontario Regional Library System

CORM Council for Optical Radiation Measurements

CORMA Corporación de la Madera (Spanish—Wood Corporation)

cormant cormorant

CORMAR Coral Reef Management and Research

Cor Mem Corresponding Member

Corn Cornelius; Cornish; Cornwall

corned-beefsan corned-beef sandwich

corned-beefwich corned-beef sandwich

Cornell Maritime Cornell Maritime Press

Cornell U Pr Cornell University Press

Corner House Central Mining and Finance Corporation (South Africa)

Corney Cornelia; Cornelius

Cornie Cornelia; Cornelio; Cornelis; Cornelisz; Corneliu; Cornelius; Cornewall; Cornwall; Cornwallis

Corning Mus Corning Museum of Glass

Corns Corn Islands in the Caribbean

coroll corollary

coron coronary

Coron Convair 990 Coronado (aircraft)

Coronados Coronado Islands *(Los Coronados)* southsouthwest of San Diego

corp (Latin prefix—body)—corporation; *corpus delicti* (the body of the crime)

Corp Corporation; Corpus Christi College, Cambridge or Oxford

Corp Coll Corpus Christi College—Oxford

Corpl Corporal

Corpn Corporation

CORPOANDES Corporación de los Andes (Spanish—Andes Corporation)

CORPORIENTE Corporación de Oriente (Spanish—Corporation of the East)—Venezuela

corppin corporeal pin (tuberculin testing)

Corpus Corpus Christi, Texas

CORPUS Corps of Reserve Priests United for Service

corr correction; correspondence; corrosion; corrugate

corr corregido (Spanish—corrected); *corriage* (French—corrected)

Corr Corriere della Sera (Italian—Daily Courier)—Milan's leading newspaper

CORRA Combined Overseas Rehabilitation Relief Appeal

corr case corrugated case

corregate correctable gate

correl correlative

corres correspondence; correspondent; corresponding

corresp corresponding

corrig corrigenda

Corr Memb Corresponding Member

corros corrosive

corrosive sublimate mercuric chloride

corr^{te} **corriente** (Spanish—current month)

corrupt. corruption

Cors Corners; Corsica; Corsican

CORS Canadian Operational Research Society

corsa (CORSA) cosmic-ray satellite

Cor Sec Corresponding Secretary

cort cortex; cortical

cort. cortex (Latin—bark)

CORT Council On Radio and Television

CORTEX Computer-based Optimization Routines and Techniques for Effective X

Cortissoz Aeropuerto Ernesto Cortissoz (Barranquilla, Colombia's airport)

CORU Coordinación de Organizaciones Revolucionarias Unidas (Spanish—Coordination of United Revolutionary Organizations)—Cuban exiles

corundolite emery

corundum aluminum oxide

Corv Corvette; Corvus

CORVA California Off-Road Vehicle Association

Cory Cornelia

cos cash-on-shipment; contactor, starting; cosine; cosmic; cosmogany; cosmography; cosmology; cosmopolitan

co's career officers

Cos Consul; Counties

Cos Kosinus (German—cosine)

CoS Chief of Staff; Chief of Station

C-o-S Clacton-on-Sea

COS Canadian Ophthalmological Society; Central Opera Service; Chamber of Shipping; Chief of Section; Colorado Springs, Colorado (airport); Conservative Opportunity Society; Czechoslovak Ocean Shipping

COS College Outline Series

cosa combat operational support aircraft

co sa come sopra (Italian—as above)

COSAD Classroom Observation System for Analyzing Depression

cosag combined steam and gas (turbine machinery)

COSAL Consolidated Shipboard Allowance List

COSAMREG Consolidation of Supply and Maintenance Regulations

COSA NOSTRA Computer-Oriented System And Newly Organized Storage-To-Retrieval Apparatus

cosar compression scanning-array radar

COSAS Congress of South African Students

COSATI Committee on Scientific and Technical Information (Federal Council for Science and Technology)

COSATU Congress of South African Trade Unions

COSBA Computer Services and Bureaus Association

COSBAL Coordinated Shore-based Allowance List

COSBO Council on Small Business Organizations

COSCO China Ocean Shipping Company

COSCOE Congress of Seniors and Coalition of Elders

COSD Council of Organizations Serving the Deaf

Cos de Mar Costa de Marfil (Spanish—Ivory Coast)

COSEBI Corporación de Servicios Bibliotecarios (Spanish—Librarian Services Corporation)—Puerto Rico

cosec cosecant

COSEC Coordinating Secretariat of National Unions of Students

cosecy company secretary

cosfad computerized safety and facility design

COSFPS Commons, Open Spaces, Footpaths Preservation Society

cosh hyperbolic cosine (symbol)

COSHTI Council for Science and Technological Information

COSI Center of Science and Industry (Columbus, Ohio); Committee on Scientific Information

Cosie Kathleen

COSINE Committee on Computer Science in Electrical Engineering Education

COSIP College Science Improvement Program

COSIRA Council for Small Industries in Rural Areas

cosis care of supplies in storage

COSLA Chief Officers of State Library Agencies

cosm cosmetic; cosmetics; cosmetologist; cosmetology

cosma computerized service for motor freight activities

COSMD Combined Operations Signals Maintenance Department (Division)

COSMEP Committee of Small Magazine Editors and Publishers

cosmetol cosmetologist(ic); cosmetology

COSMIC Computer Programmes Information Center (Univ of Georgia); Computer Software Management and Information Center

COSMIS Computer System for Medical Information Services

cosmo cosmoline; cosmopolitan

cosmog cosmogony; cosmographical; cosmography

cosmograph(s) composite photograph(s)

cosmonaut. cosmonautic(al)-(ly); cosmonautics

cosmor component open/short monitor

COSMOS Coast Survey Marine Observation Station

co so come sopra (Italian—as above)

COSPAR Committee on Space Research (International Council of Scientific Unions)

COSPEC Christian Organizations for Social, Political, and Economic Change

COSPUP Committee on Science and Public Policy (National Academy of Sciences)

cosr cutoff shear

COSR Committee on Space Research

coss. consules (Latin—consuls)

cossac cooled spectral shared-aperture concept

COSSAC Chief of Staff to the Supreme Allied Commander

cost. contaminated oil settling tank; costume

COST Congressional Office of Science and Technology;

Cost-Oriented Systems Technique

costar conversational on-line storage and retrieval

Costa Rica Republic of Costa Rica (Central American nation), *República de Costa Rica*

COSTEP Commissioned Officer Student Training and Extern Program

coster costermonger

COSTPRO Canadian Organization for the Simplification of Trade Procedures

COSTS Committee on Sane Telephone Service

COSW Citizen's Organization for a Sane World

coswap coaxial switch and alternator panel

COSWL Committee on Status of Women in Librarianship

COSY Checkout Operating System

COSYWOG Communications System Working Group

cot. card or tape reader; cathodal opening tetanus; cotangent; cotter; cotton

COT Consecutive Overseas Tour

COTA confirming telephone or message authority

COTAL *Confederación de Organizaciones Turísticas de la América Latina* (Spanish—Confederation of Tourist Organizations of Latin America)

COTAM *Commandement du Transport Aerien Militaire* (French—Military Air Transport Command)—Air Force

cotan cotangent

cotar correction tracking and ranging

CotB Commonwealth of the Bahamas

COTC Canadian Officers' Training Corps; Canadian Overseas Telecommunications Corporation

cote cathodal opening tetanus

COT & E Contractor Operation Test and Evaluation

cotfin cotton finish(ed)

cotg component tooling gage

coth hyperbolic cotangent (symbol)

COTH Council on Teaching Hospitals

cotics narcotics

cotnsd cottonseed

Coto Cotopaxi

CotP Captain of the Port

COTPAL *Comité Tecnico Permanente sobre Asuntos Laborales* (Spanish—Permanent Technical Committee for Labor Matters)

COTR Contracting Officers' Technical Representative

COTRANS Coordinated Transfer Applications System

cots. checkout test set; cottages

cot's classical organizational theories

'cot(s) apricot(s)

C o t S College of the Sea

Cotswolds Cotswold Hills of south-central England

Cott Cottesloe

COTT Central Organization for Technical Training

Cottians Cottian Alps between France and Italy

cott(s) cottage(s)

coty car of the year

cou clip-on unit; coupon

COU Coalition Unionist

couch couchant

couldn't could not

Coun Council; Councillor; Counsellor; County

Coun Biology Eds Council of Biology Editors

Coun Exc Child Council for Exceptional Children

COUP Congress of Unrepresented People

cour *courant* (French—current)

Court Courtenay; Courtland; Courtney

cous cousin

COUSA Confederation of Ontario University Staff Associations

cov concentrated oil of vitriol; covenant; cutout valve; cover

c-o v cross-over value

Cov Covell; Covenant; Coventry

COVE Citizens Opposed to the Violation of the Environment

covers. coversed sine

COVET Cooperative Venture in the Education of Teachers

covff coverings, facing, or floor (cargo)

COVINCA *Corporación Venezolana de la Industria* Naval (Spanish—Venezuelan Corporation of the Naval Industry)

cov pl coverplate

cow. chlorinated organics in wastewater; crude oil washing

COWAR Committee on Water

Research

Coward Coward, McCann and Geohegan

COWEAEX Cold-Weather Exercise (military)

cowl. cowling

Cowles Cowles Education Corporation

COWLEX Cold-Weather Landing Exercise (military)

COWPS Council on Wage and Price Stability

COWRR Committee on Water Resources Research

Cox Coxwain

cox'n coxswain (pronounced as contracted)

Cox's *Cox's Criminal Cases*

coxsec coexsecant

Coy Company

coydog(s) coyote(s) + dog(s)—mixed-breed canine(s)

COYOTE Call Off Your Old Tired Ethics

Coyte Coyte Lines

coz cousin (colloquial contraction)

Coz Cozumel Island, Mexico

cozi communication zone [indicator(s)]

cp camp; candlepower; capillary pressure; carrier packed; center of pressure; centipoise; cerebral palsy; cesspool; chemically pure; chloropurine; chloroquinine and primaquine; chronic pyelonephritis; claw plate; closing pressure; cochlear potential; code of practice; cold-punch(ed); combination product; combining power; command post (CP); compare; compound; compressed; concrete-piercing; constant pressure; cor pulmonale; creatine phosphate

cp (CP) carotid pulse; cerebral palsy; construction permit

c/p carport; change package; composition/printing; control panel

c/p (C/P) charter party

c & p carriage and packing; collated and perfect

cP centipoise; polar continental air

Cp Caucasian pimp; Chicano pimp; Chinese pimp; Compline

CP *Caminhos de ferro Portuguese* (Portuguese Railways); Canadian Pacific; Canadian Press (news agency); cerebral palsy; Characterization Plan; charter party; chemically pure; Civil Parish; Communist Party; Community Physician; Conservative Party; Constitution Party; copilot; Country Party

C-P Colgate-Palmolive

C & P Compensation and Pension

C of P Captain of the Port

CP *Centre Pénitentiaire* (French—Penitentiary Center); Crescendo Publishers

cpa claims payable abroad; closest point of approach; cost planning and appraisal; critical path analysis

c-p a cattle-prod approach (electric-shock stimulation)

CPA Canadian Pacific Airlines; Canadian Petroleum Association; Canadian Psychological Association; Canaveral Port Authority; Cathay Pacific Airways; Catholic Press Association; Certified Public Accountant; Chartered Patent Agent; Chicago Publishers Association; Civilian Production Administration; Cocoa Producers Alliance; Combat Pilots Association; Combined Pensioners Association; Commonwealth Parliamentary Association; Commonwealth Preference Area; Communist Party of Australia; Connecticut Prison Association; Consumer Protection Agency; Control of Pollution Act; Council of Professional Associations; Country Press Association; Creditors Protection Association

CPA *Community Planning Act*

CPAA *Current Physics Advance Abstracts*

CPAB California Prune Advisory Board

CPAC Center for Protection Against Corrosion; Conservative Political Action Conference; Corrosion Prevention Advisory Center

CPACS Coded Pulse Anticlutter System

CPAE Certified Public Accountant Examination

cpaf cost plus award fee

CPAG Collision Prevention Advisory Group

C_{pah} para-aminohippurate clearance

CPAI Canvas Products Association International

CP Air Canadian Pacific Air

C Pal Crystal Palace

cpam continental polar air mass

CPAM Committee of Purchasers of Aircraft Material

CPAO Country Public Affairs Office(r)

cpap continuous positive airway pressure

CPAP Committee on Pan-American Policy

CPAR Cooperative Pollution Abatement Research (Canadian)

CPARS Compact Programmed Airline Reservation System

CPAS Catholic Prisoners' Aid Society

CPAUS&C Catholic Press Association of the United States and Canada

cpaws computer-planning and aircraft-weighting scales

cpb cardiopulmonary bypass; casual payments book; cetyl pyridinium bromide; competitive protein-binding (clearance); corporation for public broadcasting

cpb (CPB) charged-particle beam(s)

cpb *cuyos pies beso* (Spanish—whose feet I kiss)

Cpb Campbelltown

CPB Casual Payments Book; Central Planning Bureau; Consumer Protection Bureau; Corporation for Public Broadcasting

CPB *Centraal Plan Bureau* (Dutch—Central Planning Bureau)

cpba competitive protein-binding analysis

cpbl capability; capable

CPBMP Committee on Purchases of Blind-Made Products

cpc card-programmed calculator; chronic passive congestion; clinicopathological conference (CPC); coated-paper copier; commerical property coverage; compound parabolic concentrator; computer-production control

CPC California Polytechnic College; Canterbury Promotion Council; Cessna Pilots Center; China Petroleum Company; China Productivity Council; Church Periodical Club; City Planning Commission; City Police Commissioner; City Projects Council; Cogswell Poly-

technical College; College Placement Council; Communist Party of China; Consumers Power Company; Creole Petroleum Corporation

CPC *Congreso Panamericano de Carreteras* (Spanish—Pan-American Highway Congress)

CPCC Central Piedmont Community College

c-p-c cycle circumspection-preemption-control (or choice) cycle

CPCG *Comité Panamericano de Ciencias Geofícicas* (Spanish—Panamerican Committee of Geophysical Sciences)

CPCGN Canadian Permanent Committee on Geographical Names (Ottawa)

CPC(M-L) Communist Party of Canada (Marxist-Leninist)

CPCN Canadian Pacific—Canadian National (telecommunications)

CPCS Canadian Pacific Consulting Services

CPCU Chartered Property and Casualty Underwriter

c-p cycle constant-pressure cycle

cpd charter pays dues; compound; contact potential difference; contagious pustular dermatitis; container-padded delivery

cpd (CPD) charter (party) pays (port) dues

CPD Committee on the Present Danger; Community Planning and Development; Consumer Protection Division; County Probation Department

CPD *Catalog of the Public Documents*

CPDA Council for Periodical Distributors Associations

cpdd command-post digital display

CPDL Canadian Patents and Developments Limited

C-P D L Christian-Patriots Defense League

cpds compounds

CPDS Computerized Preliminary Design System

cpe chlorinated polyethylene; chronic pulmonary emphysema; circular probable error; compensation, pension, and education; coupe; customer-provided equipment; cytopathic effect; cytopathogenic

effect
cpe (CPE) central programmer and evaluator
CPE Central Park East; Certificate for Proficiency in English; Certified Property Exchanger; Chief Polaris Executive (missiles); Clinical Pastoral Education; College of Physical Education; Contractor Performance Evaluation
CPEA Confederation of Professional and Executive Associations; Cooperative Program for Educational Administration
CPEC California Post-secondary Education Commission
cpe d vle coupe de ville
CPEG Contractor Performance Evaluation Group
CPEHS Consumer Protection and Environmental Health Service
c pen code pénal (French—penal code)
CPEP Contractor Performance Evaluation Plan
CPEQ Corporation of Professional Engineers of Quebec
CPERB California Public Employees Relation Board
cpf conditional peak flow; cost per flight
CPF Central Provident Fund; Church Pension Fund; Commission on Federal Paperwork; Committee to Protect the Family; Commonwealth Police Force
cpfa (CPFA) cyclopropenoid fatty acid
cpff (CPFF) cost plus fixed fee
CPFS Council for the Promotion of Field Studies
cpg controlled-pore glass; cotton piece goods
CPG College Publishers Group
CPGA California Personnel and Guidance Association; Colorado Personnel and Guidance Association; Connecticut Personnel and Guidance Association
CPGB Communist Party of Great Britain
Cpge course per gyro compass
cph cards per hour; cycles per hour
CPH Certificate of Public Health; Copenhagen, Denmark (airport); Corps of Public Health
C-PH Columbia-Presbyterian

Hospital
CPHA Canadian Public Health Association
CP & HA Canadian Port and Harbour Association
CPHC Central Pacific Hurricane Center (Honolulu)
cpi characters per inch; commercial performance index; constitutional psychopathic inferior; consumer price index; crash position indicator
CPI California Personality Inventory; California Psychological Inventory; Canadian Pacific Investments; Chemical Processing Industries; Chief Pilot Instructor; Committee on Public Information; Communist Party of India; Conference Papers Index; Consolidated Plastic Industries; Consumer Price Index
cpia close-pair interstitial atom
CPIA Chemical Propulsion Information Agency
cpiaf (CPIAF) cost-plus-incentive-award fee
cpib chlorophenoxyisobutyrate
CPIB Corrupt Practices Investigation Bureau
CPIC Canadian Police Information Centre
cpif character position in frame
cpif (CPIF) cost plus incentive fee
CPILS Correlation-Protected Integrated Landing System
CPIM Curaçaosche Petroleum Industrie Maatschappij (Dutch—Curaçao Petroleum Industry Society)
cpin crankpin
CPI-U Consumer Price Index-Urban
CPI-U-NSA Consumer Price Index—Urban—Not Seasonally Adjusted
CPI-W Consumer Price Index-revised
CPJ Communist Party of Japan (also called JCP)
CPJI Cour Permanente de Justice Internationale (French—Permanent Court of International Justice)
cpk (CPK) creatinine phosphokinase
cpkg cents per kilogram
cpl cement plaster; characters per line; common program language; complete; completion
Cpl Corporal
CPL Calgary Public Library; Canadian Pacific Limited;

Cats' Protection League; Central Public Library; Certified Parts List; Certified Products List; Charleston Public Library; Charlotte Public Library; Chattanooga Public Library; Chicago Public Library; Cincinnati Public Library; Civilian Personnel Letter; Cleveland Public Library; Clio Press Limited (Oxford); Colonial Products Laboratory; Columbus Public Library; Commercial Pilot's License; Commonwealth Parliamentary Library; Commonwealth Public Library; Coronado Public Library; Council of Planning Librarians; Council of Prison Locals; Crew Procedures Laboratory
CPLA California Palace of the Legion of Honor
cpld coupled (flow chart)
cplg coupling
cplmt complement
cplr center of pillar; coupler
CPLS Canberra Public Library Service; Certified Professional Legal Secretary
cplt copilot
cpm cards per minute; commutative principle of multiplication; condensed particulate matter; counts per minute; critical path method; cycles per minute
cpm (CPM) cost per thousand
cp/m control program/microcomputers
CPM Center for Preventive Medicine; Central Pacific Minerals; Certified Property Manager; Certified Purchasing Manager; Chief Postmaster; Colonial Police Medal (British); Communist Party of Malaya; Critical Path Method
CPMA Computer Peripheral Manufacturers Association
CPMC Columbia-Presbyterian Medical Center
CPMEEW Council for Postgraduate Medical Education in England and Wales
CPMS Civilian Personnel Management System; Computer Performance Monitoring System
cpn chronic pyelonephritis; coupon
Cpn Copenhagen
CPN Communistische Partij van Nederland (Dutch—

Netherlands Communist Party)
Cpnhgn Copenhagen
CPNP Cape Perth National Park (Western Australia)
CPNZ Communist Party of New Zealand
cpo cost proposal outline
CPO Calgary Philharmonic Orchestra; Center for Population Options; Chief Petty Officer; Chief Post Office; Civil Post Office; Civilian Personnel Office(r); Community Post Office; Community Producers Organization; Comprehensive Planning Organization; Constitutional Protection Office (Germany); County Planning Office(r); Czech Philharmonic Orchestra
CPOA California Peace Officers Association; Chief Petty Officers Association
CPOG Canadian Pacific Oil and Gas
cpp critical path plan
CPP Caltech Population Program; Canada Pension Plan; Center for Policy Process; Chemical Processing Plant; Communist Party of the Philippines; Critical Path Planning
CPP Civilian Personnel Pamphlet
CPPA Canadian Pulp and Paper Association
cppb continuous positive-pressure breathing
CPPB Canada Pension Plan Benefits; Commonwealth Prickly Pear Board
CPPCA California Probation, Parole, and Correctional Association
cppd calcium pyrophosphate dihydrate
CPPD Collaborative Program for Professional Development
CPPL Canadian Pacific Princess Lines (Vancouver-Nanaimo run)
CPPR Cassel Psychotherapy Progress Record
cpps critical path planning and scheduling
CPPS Comisión Permanente para la Explotación y Conservación de las Riquezas Maritimas del Pacífico Sur (Spanish—Permanent Commission for the Exploitation and Conservation of the Maritime Riches of the South

Pacific)
CPQ Children's Personality Questionnaire
cpr cardiopulmonary resuscitation; copper
cpr (CPR) chemical propulsion rocket
CPR Canadian Pacific Railway; Carlos Peña Romulo; Central Premonitions Registry; Cobourg Peninsula Reserve (Australian Northern Territory); Committee on Polar Research; Cost Performance Report; Council for Public Responsibility
CPRA Council for the Preservation of Rural America
CP Rail Canadian Pacific Rail
CPRE Council for the Preservation of Rural England
CPRF Cancer and Polio Research Fund
CPRG Computer Personnel Research Group
CPRI Council for the Protection of Rural Ireland
CPR-nummer Centrale Person Register nummer (Dano-Norwegian—Central Person Register number)
CPRS Council for the Protection of Rural Scotland
CPRSA Cape Peninsula Road Safety Association
CPRW Council for the Protection of Rural Wales
CPR-WBS Cost Performance Report—Work Breakdown Structure
cps characters per second; constitutional psychopathic state; coupons; creative problem solving; critical path scheduling; cycles per second
Cp(s) Caucasian pimp(s); Chicano pimp(s); Chinese pimp(s)
CP's Command Posts
CPS California Physician's Service; California Production Service; Canadian Pacific Steamships; Canadian Penitentiary Service; Catholic Pamphlet Society; Center for Population Studies (Harvard); Central Park South; Central Philharmonic Society (Beijing); Certified Professional Secretary; College Placement Service; College Press Service; Commission on Presidential Scholars; Commonwealth Public Service; Computer Processing Service(s); Condensate Pol-

ishing System; Congregational Publishing Society; Consumer Price Survey; Consumer Purchasing Service; Conversational Programming System; Current Population Survey
C.P.S. Custos Privati Sigilli (Latin—Keeper of the Privy Seal—Great Britain)
CPS Compendium of Pharmaceuticals and Specialities; Conseil Permanent de Sécurité (French—Permanent Security Council)
CPSA Canadian Political Science Association; Civil and Public Services Association (UK); Clay Pigeon Shooting Association; Commonwealth Public Service Association
CPSAA Commonwealth Public Service Artisans Associations
cpsac cycles-per-second alternating current
Cpsc course per standard compass
CPSC Consumer Product Safety Commission
CPSCU College of Physicians and Surgeons—Columbia University
cpsd cross-power spectral density
cpse counterpoise
cpsi causing pressure shut in
CPSI Council of Profit-Sharing Industries
CPSL Canadian Pacific Steamship Line
CPSLCS Complete Power Signalling Local Control System
CPSM Colonial Prison Service Medal (British)
CPSP Cove Palisade State Park (Oregon)
CPSR Calibration Procedure Status Report (Polaris); Contractor Procurement Systems Review
CPSS Certificate in Public Service Studies; Common Program Support System
Cp stg c course per steering compass
CPSU California Polytechnic State University; Combined Public Service Unions; Communist Party of the Soviet Union
cpt carpet(ed); casement-projected transom; change-parity time; chest physiotherapy; cockpit procedure trainer; continuous performance task;

counterpoint; critical path technique
cpt (CPT) California Public Television; critical path technic
Cpt Capitaine (French—Captain)
CPT Canadian Pacific Telegraphs; Cape Town, South Africa (Malan Airport); Civilian Pilot Training; Communist Party of Thailand; Continuing Performance Test(ing)
CPT *Current Physics Titles*
C.P.T. *Contador Público Titulado* (Spanish—Certified Public Accountant)
CPTB Clay Products Technical Bureau
CPTL Canadian Pacific Transport Limited
Cptn Captain
cptng mats rgs carpeting, mats, or rugs
cptr capture; carpenter; carpentry
CPTS California Public Television Stations; Coalition for Peace Through Strength; Council of Professional Technological Societies
CPTV Connecticut Public Television
cpu central processing unit
cpu (CPU) central processing unit
CPU California Pacific University; Canadian Paperworkers Union; Central Processing Unit; Commonwealth Press Union; Crime Prevention Unit
CPUBINFO Chief of Public Information Division (NATO)
CPUC California Public Utilities Commission
CPUSA Communist Party USA
cpv (CPV) cytoplasmic polyhedrosis virus
CPV Combination Pump Valve; Communist Party of Vietnam; *Compañía Peruana de Vapores* (Peruvian Steamship Line)
cpvc critical pigment volume concentration
cpvc (CPVC) chlorinated polyvinyl chloride
CPVPL Charles Patterson Van Pelt Library (University of Pennsylvania)
cpw commercial projected window

cPw polar continental air warmer than underlying surface
CPW California Press Women; Central Park West
CPWH Committee for the Preservation of the White House
cpx complex
CPX Command Post Exercise
cpy copy
CPY Communist Party of Yugoslavia
cpz chlorpromazine
CPZ Central Park Zoo
cq chloroquine quinine; circadian quotient; come quick; conceptual quotient; copy correct; copy (spelled) correctly
cq (CQ) class quotient (lowerclass, upperclass, etc.)
CQ call to quarters (radio signal meaning message following is intended for all receivers); Charge of Quarters; Conditionally Qualified
CQ *Caribbean Quarterly; Congressional Quarterly*
CQC Citizens for a Quiet City
CQCA Central Queensland Coal Associates
cqcm cryogenic quartz-crystal microbalance
CQD wireless distress signal
cqm chloroquine mustard
CQM Chief Quartermaster; Company Quartermaster
CQMS Company Quartermaster Sergeant
cqr secure anchor (British short form for a plowshareshaped single-fluke anchor)
CQR Customer Quality Representative
CQR *Church Quarterly Review*
CQs Citizens for Quieter Cities
CQS California Q-Set
CQSW Certificate of Qualification in Social Work
cqt circuit; correct
CQT College Qualification Test
CQU College Qualification Test(s)
CQUCC Commission on Quantities and Units in Clinical Chemistry
cr calculus removal; calculus removed; cardiorespiratory; carriage return; cathode ray; center; center of resistance; chest and right arm; clinical research; clot reaction; coefficient (of fat) retention; cold roll; cold-rolled; colon re-

section; complete remission; complete round; compression ratio; conditioned reflex; conditioned response; continuing resolution; cranial; created; creatinine; credit; creek; cresyl red; crew; critical; critical ratio; crown; crown-rump; cruise
cr (CR) carriage return (data processing); conditional release(parole); conditioned reflex; conditioned response; critical ratio
c-r cognitive restructuring
c/r company risk; correction requirement(s)
c & r cops and robbers
cr. *crux* (Latin—cross)
c/r *cuenta y riesgo* (Spanish—for account and risk of)
Cr chromium; Commander; creatinine; creditor
Cr *Contador* (Spanish—Bookkeeper, Cashier, Purser)
Cr. *Credo* (Latin—I believe, the creed); *Ceskoslovensky rozhlas* (Czechoslovak Radio)
CR Camping Reserve; Carriers Risk; Central Registry; *Ceskoslovenska Republika* (Czechoslovakian Republic); Change Recommendation; Characterization Report; Chief Ranger; Classified Register; Combat Ready; Commonwealth Railways (Australia); Compound Risk; Contract Requisition; cost reimbursement; Costa Rica; Costa Rican; Current Rate
C-R Crouse-Hinds; Cutler-Hammer
C/R Chicago Rawhide (manufacturing company)
C & R convoy and routing
C of R Count(y) of Ravensberg
CR *Centre de Réadtation* (French—Rehabilitation Center); *Computing Reviews; Consumer Reports*
C R *comptes rendus* (French—proceedings, report)
C.R. *Carolina Regina* (Latin—Queen Caroline); *Carolus Rex* (Latin—King Charles); *Civis Romanus* (Latin—Citizen of Rome); *Custos Rotulorum* (Latin—Roll Keeper)
cra central retinal artery
Cra *Carretera* (Spanish—highway); *Contadora* (Spanish—Bookkeeper, Cashier, Purser)

Cr A Commander at Arms; Corona Australis
CRA California Redwood Association; California Republican Assembly; Canadian Rheumatism Association; Cave Research Associates; *Centres de la Recherche Appliqué* (French—Applied Research Centers); Coal Research Association; College of Radiologists of Australia; Colorado River Aqueduct; Colorado River Authority; Community Redevelopment Agency; Community Reinvestment Act; Concentrated Rehabilitation Area; Continuing Resolution Authority; Convair Recreation Association; Conzinc Riotinto of Australia
C.R.A. Conzinc Riotinto of Australia (their periods as shown)
CRAB Central Registry at Bethesda; coastal-research amphibious buggy
CRABS Close-Range Analytical-Bundle System
crabsan crab sandwich
crabwich crab sandwich
CRAC Careers Research and Advisory Center; Community Research Action Center; high-potency cocaine
CR Acad Sci *Comptes Rendus Hebdomadaires des Seances de l'Academie des Sciences* (French—Weekly Reports of Meetings of the Academy of Sciences)
crac-coc crack-cocaine
CRAD Committee for Research into Apparatus for the Disabled; Contracted Research and Development
craf comet rendezvous asteroid flyby
CRAF Civil Reserve Air Fleet
CRAFT Commonwealth Rebate for Apprentice Full-time Training; Computerized Relative Allocation of Facilities Technic; Cycle Reporting and Fatigue Tracking
CRAG Combat Readiness Air Group
CRAGS Chemical Records and Grading System
cram. card random access memory
CRAM Contractual Requirements Recording, Analysis, and Management
cran cranial; craniology; cra-

nium
cranapple cranberry-and-apple juice
Cranch's *Cranch's United States Supreme Court Reports*
craniol craniologic(al)(ly); craniologist; craniology
craniom craniometry
cranple cranberry-apple cider or juice
cran(s) cranberries; cranberry
Cranston Juvenile (delinquent) Diagnostic Center at Cranston, Rhode Island
crap crapola
crap. completely ridiculous anthropic principle
CRAR Critical Reliability Action Request
cras coder and random access switch
CRASC Commander—Royal Army Service Corps
CRASH Center for Reproductive and Sexual Health; Citizens Rally and Appeal to Save Our Homes; Citizens to Reduce Airline Smoking Hazards; Community Resource and Self Help; Community Resources Against Street Hoodlums (Los Angeles Police Department detail)
crast. *crastinus* (Latin—of tomorrow)
'crastinator(s) procrastinator(s)
Crat Crater
Crater Lake national park in Oregon
Craters of the Moon Craters of the Moon National Monument, Idaho
C-rat(s) C-ration(s)
CRAV *Compañía Refineria de Azucar Viña del Mar* (Spanish—Viña del Mar Sugar Refining Co)
CRAW Combat Readiness Air Wing (USN)
cray(s) crayfish(es)
crb central radio bureau; curb; curbing
crb (CRB) chemical, radiological, biological (warfare)
Cr B Corona Borealis
CRB Central Reproduction Bureau; Certified Residential Broker; Change Review Board; Civilian Review Board; Commission for Relief in Belgium; Commodities Research Bureau; Cooper River Bridge (Charleston, South Carolina); County

Roads Board
crbbb complete right bundle branch block
CRBC Chinese Road and Bridge Company
cr bl credit balance
cr & br crown and bridge (dental)
CRBRP Clinch River Breeder Reactor Plant
CRBS Customer Records and Billing System
crc cavity rim cap (contraceptive device); complete round chart; cyclic redundancy check
CrC control and reporting center; Crew Chief
CRC California Rehabilitation Center; Certified Recreation Counselor; Chemical Rubber Company; Civil Rights Commission; Commonwealth Reply Coupon; Consolidated Rail Corporation; Consolidated Railroads of Cuba; Control and Reporting Center; Coordinating Research Council; Corrosion Reaction Consultant; CRC Press
CRCA Canadian Rodeo Cowboys Association
CRCC Consolidated Record Communications Center (USA)
CRCE Centaur Reliability Control Engineering
crchf crew chief
CRCNJ Central Railroad Company of New Jersey
crcp continuously reinforced concrete paving
CRCP Certificate of the Royal College of Physicians
CRCR Center for Rate-Controlled Recordings
CRCRS Civil Rights Community Relations Service
CRCS Canadian Red Cross Society; Certificate of the Royal College of Surgeons
Crct Circuit
crd chronic renal disease; chronic respiratory disease; complete reaction of degeneration
Cr$ cruzeiro (Brazilian monetary unit)
CRD Community Relations Department; Crop Research Division (USDA)
CRDHE Center for Research and Development in Higher Education
crdl cradle
CRDL Chemical Research and

Development Laboratories; Contractor Data Requirements List

crdm control-rod device mechanism

CR & DP Cooperative Research and Development Program

CRDS Clarence Ralph De Sola; Colgate-Rochester Divinity School

CRDSD Current Research and Development in Scientific Documentation

cre corrosion resistant

Cre Crescent

CRE Center for Radical Education; Commission for Racial Equality; Congress of Racial Equality; Counselor of Real Estate

CREA California Real Estate Association; Clearinghouse of Resources for Educators of Adults

C Real Ciudad Real

cream of tartar potassium acid tartrate ($KHC_4H_6O_6$)

creat creatine

CREAT Combined Resources for Editing Automated Teaching

CREATE Computational Requirements for Engineering, Simulation, Training, and Education (USAF time-sharing computer complex)

Creation Sci Creation Science Research Center

Creative Ed Creative Educational Society

crectte creciente (Spanish—crescent, growing)

cred credit; creditor

credd customer requested earlier due date

CREDO Chaplain's Religious Education Development Organization

Creek The Creek—oilfields scattered along the creeks of western Pennsylvania

CREEP Committee to Re-elect the President (Nixon)

CREF College Retirement Equities Fund

CREFAL Centro Regional de Educación Fundamental para la America Latina (Spanish—Regional Center of Fundamental Education for Latin America—United Nations organization)

CREG Cancer Research Emphasis Grants

CREI Capitol Radio Engineering Institute

crem cremation

cremains cremation remains

CREMI Credito Minero y Mercantil (Spanish—Mining and Mercantile Credit)

crem mus crematorium music (Beethoven's Marcia funebre from his *Eroica* Symphony, Berlioz's Death March from *Les Troyens*, Chopin's Funeral March, Handel's Death March from *Saul*, Mozart's Masonic Funeral Music, Rachmaninoff's *Isle of the Dead*, Richard Strauss's *Tod und Verklaerung*, Wagner's funeral music from *Siegfried*)

cremo crematorium

CREO Central Real Estate Office; Crystalline Regions Exploration Office (ONWI)

crep. crepitus (Latin—crepitation)

crepe(s) crepe(s) suzette

cres corrosion-resistant stainless steel; crescent; crescentic

cres crescendo (Italian—expanding, swelling)

Cres Crescent

CRES Center for Reproduction of Endangered Species (San Diego Zoological Society); Center for Research in Engineering Science (University of Kansas); Corrosion Resistant Stainless Steel

cresc crescendo (Italian—increasing, swelling)

Crescendo Crescendo Publishing Company

creso crescendo (Italian—increasing; swelling)

cress garden cress; watercress

CRESS Combined Reentry Effort in Small Systems; Computer Reader Enquiry Service System

crest. crew-escape and rescue techniques (USAF)

CREST Committee on Reactor Safety Technology

Cret Cretaceous

CrewTAF Crew Training Air Force

crf capital recovery factor; carrier frequency; continuous reinforcements; control relay forward; cross-reference file

crf (CRF) corticotropin-releasing factor

CRF Cancer Research Foundation; Citizens Research Foundation; Constitutional Rights Foundation

CRFA Czechoslovak Rationalist Federation of America

CRFG California Rare Fruit Growers

crfs copper reverbatory furnace slag

crf's change request forms

crg carriage

CRG Cave Research Group; Cooperative Republic of Guyana (formerly British Guiana)

cri chemical rust inhibitor; cold running intelligibility; criminal; criterion-referenced instruction

CRI Caribbean Research Institute; Coconut Research Institute; Committee for Reciprocity Information; Communications Research Institute; Composers Recordings Incorporated

CR & I Chicago River and Indiana (railroad)

CRI Croce Rossa Italiana (Italian Red Cross)

CRIB Computerized Resources Information Bank

CRIC Canon Regular of the Immaculate Conception

CRICAP Carpet and Rug Industry Consumer Action Panel

CRIEPI Central Research Institute of the Electrical Power Industry

CRIF Comité Représentatif des Israélites de France (French—Representative Committee of the Jews of France)

CRILC Canadian Research Institute of Launderers and Cleaners

CRILI Center for Research in Learning and Instruction

crim criminal; criminalism; criminalist; criminologist; criminology

crim con criminal conversation (British euphemism—adultery)

Crimea Crimean Peninsula between the Sea of Azov and the Black Sea

criminol criminologist; criminology

criminotechnol criminological technology (using electronic and photographic devices and techniques to apprehend criminals and secure evidence needed for their conviction)

criminotic criminal neurotic

crip cripple

CRI & P Chicago, Rock Island and Pacific (railroad)
CRIPA Civil Rights of Institutionalized Persons Act
crips cripples
crip(s) crippler(s)—gangster(s) noted for crippling victims
CR & IR Chicago River and Indiana (railroad)
Cris Cristina; Cristine; Cristóbal; Cristopher
CRIS Command Retrieval Information System; Conference Resource and Information Services; Current Research Information System
crisco cream received in separating cottonseed oil
CRISP Computer Resources Integrated Support Plan; Cosmic Radiation Ionization Spectrographic Program (NASA)
crit critic; critical; criticality; criticism
CRITICOMM Critical Intelligence Communications System
crits critical reactor experiments
Crk Creek; Cork
crkc crankcase
crl cross register line
CRL California Republican League; Cambridge Research Laboratory; Cardiac Research Laboratory; Center for Research Libraries; Chemical Research Laboratory; Civil Rights Law(s); County Rugby League; Crown Renewable Lease
C.R.L. Certified Record Librarian; Certified Reference Librarian
C & RL College and Research Libraries
CRLA California Rural Legal Assistance; Canadian Railway Labor Association
CRLLB Center for Research on Language and Language Behavior (Univ Mich)
crls carelessness
crm confidence rulemaking; count rate meter; counter-radar missile; cross-reacting material; crucial reaction measure(ment)
cr/m crew member
CRM Certified Records Manager; Cockpit Research Management; Combat Readiness Medal; Communications/Research/ Machines (publisher); Counter-Radar Missile

CRM Consumer Research Magazine
CRMA Cotton and Rayon Merchants Association
crmch cruise Mach change
CRMD Children with Retarded Mental Development
CRME Council for Research in Music Education
Crml Carmel
crmn crewman
crmnls criminalism; criminalist; criminalistics; criminals
crmoly chrome molybdenum
CRMP Corps of Royal Military Police
crmr continuous-reading meter relay
CRMT Community Resources Management Team (parole and probation)
CRMWD Colorado River Municipal Water District
crn crane; crown
Crn (The) Crown (The Monarchy)
CRN Customs Registered Number; Course Reference Number
CRNA Certified Registered Nurse Anesthetist
CRNL Chalk River Nuclear Laboratories (Canada)
CRNLE Center for Research in the New Literatures in English (Australian)
CRNM Capitol Reef National Monument
cr note(s) credit note(s)
CRNP Cape Range National Park (Western Australia)
CRNPTG Commission on the Review of the National Policy Toward Gambling
CRNSS Chief of the Royal Naval Scientific Service
CRNWR Cape Romain National Wildlife Refuge (South Carolina); Clarence Rhode National Wildlife Range (Alaska)
cro cathode-ray oscilloscope
Cr O chrome oxide (recording tape)
CRO Carnarvon, Australia (tracking station); Chief Recruiting Officer; Commonwealth Relations Office; Community Relations Office; Contractor's Resident Office; County Recorder's Office; Criminal Records Office
CrO₂ CrO_2 chromium dioxide (recording tape coating)
Croat. Croatia; Croatian
C Rob Christopher Robinson's

Reports of Cases Argued and Determined in the High Court of Admiralty
CROC Committee for the Rejection of Obnoxious (tv) Commercials
CROC Confederación de Revolucionarios Obreros y Campesinos (Spanish—Confederation of Revolutionary Workers and Peasants)
crock. crockery; crocks (English slang—broken-down animals or athletes)
Crockett Girls Crockett State School for (delinquent) Girls at Crockett, Texas
Croco Crocodilia
crocodiliol crocodiliologic(al)-(ly); crocodiliologist; crocodiliology
croc(s) crocodilian(s)—alligator(s), caiman(s) or cayman(s), crocodile(s), gavial(s)
Croix St Croix, American Virgin Islands
cro'jack crossjack
crom control read-only memory
Crom Cromwell
CROM Confederación Regional de Obreros Mexicanos (Spanish—Regional Confederation of Mexican Workers)
CROP Christian Rural Overseas Program; Community Response in Opposition to Poverty
cross. crossing
Cross King's Cross (Sydney, Australia's nightlife section also called The Cross)
CROSS Committee to Retain Our Segregated Schools (Arkansas); Computerized Rearrangement of Special Subjects
CROSSBOW Computerized Retrieval of Organic Structures Based on Wiswesser
'crosse lacrosse; lacrosse stick
CROWCASS Combined Registry of War Criminals and Security Suspects
Crowell Crowell Collier; Thomas Y Crowell
Crown Crown Publishers
Crozets Crozet Islands in the South Indian Ocean
crp cathode-ray tube/keyboard printer
Crp C-reactive protein
CrP creatinine phosphate
CRP Committee to Re-elect the President (Nixon's fundraising organization also

known as CREEP); Conservation Reserve Program; Control and Reporting Post; Corpus Christi, Texas (airport); Cost Reduction Program; Crime Restitution Program; Crisis Relocation Plan
CRP *Cruz Roja Peruana* (Spanish—Peruvian Red Cross)
CRPD Chicago Regional Port District
cr pl chromium plate
CRPL Central Radio Propagation Laboratory
Cr Pr Criminal Procedure
CRPR Child-Rearing Practices Report
CRPS Center for Research on Population and Security
crr constant ratio roll
CrR *Croix-Rouge* (French—Red Cross)
CRR Cost Reduction Representative
CRRA Component Release Reliability Analysis
CRRB Centaur Reliability Review Board
CRRC Costa Rica Railway Company
crrd conceptual reference repository description
CRREL Cold Regions Research and Engineering Laboratory (USA)
CRRERIS Commonwealth Regional Renewable Energy Resources Information System
CRRES Combined Release and Radiation Effects Satellite
crrl contour roller
CRRS Combat-Readiness Rating System (USAF)
crs coast radio station(s); cold-rolled steel; colon-rectal surgery; creditors; credits; crew reserve status
cr's character reactions
Crs Cristóbal, CZ
CRs counter-revolutionaries (sometimes appears as KRs)
CRS Calibration Requirements Summaries; Career Service Status (USAF); Certified Residential Specialist; Childcare Resource Service; Child Rearing Study; Coast Radio Service; Commonwealth Rehabilitation Service; Community Relations Service; Computing Research Station; Congressional Research Service; Consumer

Reservation System; Corrective and Rehabilitation Squadron (USAF)
CRS *Conseil de la Recherche Scientifique* (French—Scientific Research Council)—Quebec; *Corps Républicain de la Sécurité* (French—Republican Security Corps)—anti-riot squads
CRSA Canadian Retail Shipment Association; Cold-Rolled Sections Association; Concrete Reinforcement Steel Association; Connecticut River Salmon Association
CRSC Center for Research in Scientific Communications (Johns Hopkins)
CRSG Classification Research Study Group
CRSI Concrete Reinforcing Steel Institute
CRSM Certified Real Estate Marketer
crsp criminally receiving stolen property
CRSP Colorado River Storage Program
CRSR Center for Radiophysics and Space Research (Cornell University)
CRSS Certified Real Estate Security Sponsor; Collectors of Religion on Stamps Society; Community Refugee Settlement Scheme (Australia)
crst syndrome calcification and clinical signs of Raynaud's phenomenon, scleroderma, and telangiectasis
crt cargo-restraint transporter; cathode-ray tube; cold-rolled and tempered
Crt Court; Crater
CRT Certified Radiologic Technician; Combat Readiness Training; Criterion-Referenced Tests
cr tan lthr chrome-tanned leather
CRTC Canadian Radio-Television Commission; Cavalry Replacement Training Center
crtgc cartographer
crtkr caretaker
crtl criticality
crtn correction
crtog cartographer; cartographic; cartography
cr tp contour template
CRTPB Canadian Radio Technical Planning Board
crt's cathode-ray tubes
CRTS Commonwealth Recon-

struction Training Scheme
crtu combined receiving and transmitting unit
cru clinical research unit; combined rotating unit; crucible; cruise
Cru Crux
CRU Cecil Rhodes University; Civil Resettlement Unit; Collective Reserve Union; Crime Reduction Unit
Cru Base Cruiser Base
CRUBATFOR cruisers, battle force
CRUD Chalk River Unidentified Deposit
CRUDESLANT Cruiser-Destroyer Forces, Atlantic
CRUDESPAC Cruiser-Destroyer Forces, Pacific
CRUDIV cruiser division
CRUEL Commission on Reform of Undergraduate Education and Living (Univ Ill)
crug corrugated
cruis cruiser; cruising
CRULANT Cruiser Forces, Atlantic
CRUPAC Cruiser Forces, Pacific
cru's collective reserve units (international banking currency)
CRUSK Center for Research on Utilization of Scientific Knowledge (Univ Mich)
Crust Crustacea
crustas ice-encrusted cocktails
cruz *cruzeiro* (Brazilian currency)—also appears as *C, Cr, Cruz, Crz*
Cruz(an) St Croix Island (or person from there)—American Virgin Islands
CRUZEIRO *Serviços Aéreos Cruzeiro do Sul* (Southern Cross Air Service—Brazil)
crv central retinal vein
Crv Corvus
CRV Corvette aircraft
crvan chrome vanadium
cr. vesp. *cras vespere* (Latin—tomorrow evening)
crvf congestive right ventricular failure
CRW Clean Radwaste (System); Commission on Rural Water
CRWA Charles River Watershed Association
CRWG Computer Resources Working Group
CRWM Committee on Radioactive Waste Management (NAS-NRC)
CRWPC Canadian Radio

Wave Propagation Committee

cry. crystal(s)

CRY Citizens Redirecting Youth

cryng carrying

cryobio cryobiological(ly); cryobiologist; cryobiology

cryochem cryochemical(ly); cryochemist(ry)

cryoelectro cryoelectronic(al)-(ly); cryoelectronicist; cryoelectronics

cryogen cryogenic(al)(ly)

cryolite sodium aluminum fluoride

cryon cryonic(s)

cryosurg cryosurgeon; cryosurgic(al)(ly); cryosurgery

crypt. cryptography

crypt (Latin prefix—hidden)— cryptogram, cryptographer, cryptographic

crypta cryptanalysis; cryptanalyst

crypto cryptograph; cryptographer; cryptographic; cryptography

cryptocom cryptocommunism; cryptocommunist

cryptofasc cryptofascism; cryptofascist

cryptonet crypto-communication network

crypton(s) cryptonym (s)

cryptos cryptocommunists; cryptofascists; cryptograms

cryptozool cryptozoological(ly); cryptozoologist(s); cryptozoology

crys crystal; crystalline; crystallization; crystallize; crystallography; crystalloids

crysnet crystallographic computing network

cryst crystal; crystalline; crystallography

crystal methamphetamine

Crystal Meth Capital Crystal Methamphetamine Capital (San Diego, California)

crystd crystallized

crystn crystallization

cs caesarean section; capital stock; carbon steel; cast steel; cast stone; center section; cerebrospinal; cirrostratus; close support; cognitive style; color stabilizer; common steel (projectile); concentrated strength; conditioned stimulus; corticosteroid; crucible steel; cryptographic system; current series; current strength; cutting specification(s); cycloserine

cs (CS) central service; closeup shot (waist-up tv picture); conditioned stimulus

c/s cases; *con safos* (Spanish-American slang—impervious to attack, the same to you, you're stuck with it); cycles per second

c & s clean and sober

cs céntimos (Spanish—centimes, hundredths)—coins worth a hundredth part of any unit; *come sopra* (Italian—as above); *cours* (French—course, currency, current price); *cuartos* (Spanish—apartments, fourths)— coins worth a fourth part of any unit

c s colla sinistra (Italian—left hand); *con sordino* (Italian—with the mute)

cS centistoke(s)

Cs cesium; Chinese visitors; cirrostratus

CS Cadbury Schweppes; Call Sign; Casualty Station; Chemical Society; Chief Secretary; Chief of Staff; Civil Service; Colonial Secretary; Commonwealth Secretariat; Communications Station; Communications System; contract surgeon; Cooperative Society; Correspondence School; Credit Suisse (bank); Cryptographic System; Cultural Survival; current series; current strength; cutting specifications

C/S call signal; certificate of service; Currency Surcharge

C&S Chicago and Southern (Delta Airlines); Citizens and Southern (bank); Colorado and Southern (railroad)

C of S Chief of Staff; Chief of Service; Church of Scotland (Presbyterian)

CS Centraal Station (Dutch— Central Station); *Consejo de Seguridad* (Spanish—Security Council)—UN; Customs Service

C.S. Custos Sigilli (Latin— Keeper of the Seal)

Cs¹³⁷ radioactive cesium

CS-3A Sakura-3A (Japanese communications satellite)

CS-3B Sakura-3B (Japanese communications satellite)

CSA Canadian Standards Association; Canterbury Society of Arts; Casualty Surgeons Association; Central South Australia; Central Surgical

Association; *Ceskoslovenske Aerolinie* (Czechoslovakian Airline); Chief of Staff, Army; College of Surgeons of Australasia; Commercial Service Authorization; Common Services Agency; Commonwealth Sugar Agreement; Communication Service Authorization; Community Services Administration; Community of St Andrew; Computer Sciences of Australia; Confederate States of America; Confederate States Army; Contractor Service Action; Controlled Substances Act

C & SA Counterinsurgency and Special Activities (Joint Chiefs of Staff)

CSAA California State Automobile Association; Child Study Association of America; Committee on Space Astronomy and Astrophysics; Council of Specialized Accrediting Agencies

CSAC Cameron State Agricultural College; Citizens' Stamp Advisory Committee; Conners State Agricultural College

CSADC Canadian—South African Diamond Corporation

CSAE Canadian Society of Agricultural Engineering

CSAF Chief of Staff, United States Air Force

CSAL Central Scientific Agricultural Library (Moscow)

CSANZ Cardiac Society of Australia and New Zealand

CSAO Civil Service Association of Ontario

CSAP Canadian Society of Animal Production; Career Skills Assessment Program

csar communication satellite advanced research

CSAR Comité Secret de l'Action Révolutionnaire (French—Secret Committee of Revolutionary Action), the Cagoule active during World War II in aiding the Nazis

CSATU Congress of South African Trade Unions

CSAV Compañía Sud America de Vapores (Chilean Line)

csb calcium silicate brick; chemical stimulation (of the brain); concrete splash block

Csb Casablanca

CSB Canterbury Savings Bank; Central Statistical Board;

Christian Service Brigade; Committee for Safe Bicycling; Commonwealth Savings Bank; Copra Stabilization Board

C.S.B. Bachelor of Christian Science

CSB Centro Simón Bolívar (Spanish—Simón Bolívar Center), metropolitan management investment in Caracas, Venezuela

CSBA California School Board Association

CSBE California State Board of Education

CSBG Concerned Seniors for Better Government

CSBM Confidence and Security-Building Measures

CSBPA California Shore and Beach Preservation Association

CSBs Canada Savings Bonds

csc cartridge storage case; change schedule chart; cosecant

c & sc capital and small capital letters

CSC Canadian Shippers Council; Canadian Space Centre; Central Security Control; Central Security Council; Child Safety Council; Citizens Service Corps; Civil Service Commission; Civilian Screening Center; Colorado State College; Combat Support Company; Command and Staff College (USAF); Commonwealth Scientific Committee; Commonwealth Steel Company; Communications Satellite Corporation; Community Service Center; Computer Science Corporation; Consolidated Coal Company (stock exchange symbol); Conspicuous Service Cross; Continuous Service Certificate

CSCA Central States Corrections Association; Civil Service Clerical Association

CSCAR Citizens for Sensible Control of Acid Rain

CSCAW Catholic Study Circle for Animal Welfare

CSCC Civil Service Commission of Canada

CSCCL Center for Studies in Criminology and Criminal Law (University of Pennsylvania)

CSCD Center for Studies of Crime and Delinquency;

Community Service Center for the Disabled

CSCE Conference on Security and Cooperation in Europe

CSCFE Civil Service Council for Further Education (UK)

csch hyperbolic constant; hyperbolic cosecant

CS Ch E Canadian Society for Chemical Engineering

CS circuit common-source amplifier for field-effect transistors

CSCJ Center for Studies in Criminal Justice

CSCI Community of St Clare

cscn character scan(ning)

CScO Chief Scientific Officer

CSCP Christian Science Committee on Publications

CSCS Cost, Schedule, and Control System; Crusader S-wire Container Service

C/SCSC Cost-Schedule Control Systems Criteria

cscu countersink cutter

csd closed shelter deck(ing); constant-speed drive; controlled-slip differentials; convection suppression device(s); cortical spreading depression

CSD Civil Service(s) Department; Commonwealth Society for the Deaf; Consumer Correctional Services Department; Consumer Service(s) Division; Convair San Diego (Division of General Dynamics Corporation); Correctional Services Department; Corrective Services Department

CSD Ceskolovenske Statne Draphy (Czechoslovak State Railway)

CSDA Canadian Stamp Dealers Association

CSD-ALA Children's Services Division—American Library Association

csdc computer signal data converter

CSDE California State Department of Education; Central Servicing Development Establishment

CSDI Center for the Study of Democratic Institutions

CSDP Coordinated Ship Development Plan (USN)

CSDPH California State Department of Public Health

CSDS Chicago Sewage Disposal System

csdv closed shelter-deck vessel

cse course

cs & e crew station and escape

Cse Causse (French—limestone plateau)

CSE Calcutta Stock Exchange; Cincinnati Stock Exchange; Citizens for a Sound Economy; Certificate of Secondary Education; Committee on Special Education

CSEA California State Electronics Association; California State Employees Association; Combat System Engineering Authorization

CSEAA Civil Service Employees Association of America

csect control section; cross section

c-sect cesarian section

csed coordinated ship electronics design

CSEE Canadian Society for Electrical Engineering

csei concentrated solar-energy imitator

CSEIP Center for the Study of the Evaluation of Instructional Programs

CSEL Consolidated Support Equipment List(ing)

CSEPA Central Station Electrical Protection Association

cseq/cseqt consequences/consequent

CSERB Computers, Systems, and Electronic Research Board

CSEU Confederation of Shipbuilding and Engineering Unions

csf cerebrospinal fluid

Csf one hundred cubic feet

CSF California Scholarship Federation; Center for Southern Folklore; colony-stimulating factor; Community of St Francis; Correctional Service Federation

CSF Compagnie Générale de Télégraphie Sans Fil (French—Wireless Telegraph Company)

CSFA Canadian Scientific Film Association; Citizens Scholarship Foundation of America

CSFAC Colorado Springs Fine Arts Center

CSFE Canadian Society of Forest Engineers

CSFPA Central Station Fire Protection Association

CSFS Commonwealth Scholarship and Fellowship Scheme

CSFT Climax/Granite-Spent

Fuel Test(ing)
csf-Wr cerebrospinal fluid-Wassermann reaction
csg casing
CSG Capital Systems Group; Configuration Steering Group; Council of State Governments
CSG Centre Spatial Guyanais (French-Guiana Space Center)
CSGA Canadian Seed Growers Association; Central States Gas Corporation
CS-gas civil(ian)-security or cyanide-simulating gas also called Mace or tear gas, causes temporary blindness, burning, tearing, choking, coughing, stinging, and vomiting; used to control unruly mobs
CSGBI Cardiac Society of Great Britain and Ireland
csgn consign
csgnd consigned
c/sgnd countersigned
csgnee consignee
csgng consigning
csgnmt consignment
CSGUS Clinical Society of Genito-Urinary Surgeons
csh calcium silicate hydrate; cash
CSH Combat Support Hospital
cshaft crankshaft
CSHP Canadian Society of Hospital Pharmacists
csi contractor standard item
CSI Campus Studies Institute; Child Study Institute; Construction Specification Institute; Container Status Information; Customer Satisfaction Index
C.S.I. Companion of the Order of the Star of India
CSI Cinematique Scientifique Internationale (French—International Scientific Film Library)
CSIC Consejo Superior de Investigaciones Cientificas (Spanish—Superior Council of Scientific Investigations)
CSICC Canadian Steel Industries Construction Council
CSICOP Committee for the Scientific Investigation of Claims of the Paranormal
csid consider
csidd considered
csidl considerable
csidn consideration
CSIE Center for the Study of Information and Education

CSIES Community Sex Information and Education Service
CSigO Chief Signal Officer
csink countersink
CSIP Committee for the Scientific Investigation of the Paranormal
CSIR Council of Scientific and Industrial Research; Council for Scientific and Industrial Research (South Africa); Council of Scientific and Industrial Research (India)
CSIRA Council for Small Industries in Rural Areas
CSIRO Commonwealth Scientific and Industrial Research Organization (Australia)
CSIROLCA Commonwealth Scientific and Industrial Research Organization Laboratory Craftsmens Association
CSIRONET Commonwealth Scientific and Industrial Research Organization Computing Network
CSIROTA Commonwealth Scientific and Industrial Research Organization Technical Association
CSIS Canadian Security Intelligence Service; Center for Strategic and International Studies (Georgetown University)
CSISRS Cross-Section Information Storage and Retrieval System (AEC)
CSIT Chapin Social Insight Test
CSIVP California State Influenza Vaccine Program
CSJ Council for Soviet Jews
CSJ Christian Science Journal
CSJB Community of St John Baptist
csk cask; countersink; countersunk
CSK Cooperative Study of the Kuroshio (UNESCO)
CSK Consumer Survival Kit (public tv program)
csk hd countersunk head
csko countersink other side
csl computer simulation language; computer-sensitive language; console
csl (CSL) crane stores lighter
CSL Canada Steamship Lines; Cedar Springs Library; Chicago Short Line (railroad); Cinderella Softball League; Circle of State Librarians; Colorado State Library; Consumer Service Litigants

CSL Centre de Semi-Liberté (French—Semi-Liberty Center)—halfway house for criminals; *Conseil Supérieur du Livre* (French—Better Book Council)
CSLA Canadian School Library Association; Church and Synagogue Library Association
CSLATP Canadian Society of Landscape Architects and Town Planners
CSLB Contractor's State License Board
CSLC California State Lands Commission
CSLEA Center for the Study of Liberal Education for Adults
CSLICC Counseling Service of the Long Island Council of Churches
CSLO Canadian Scientific Liaison Office; Combined Services Liaison Office(er)
CSLP Center for Short-Lived Phenomena (Smithsonian)
cslr consular
CSLS Civil Service Legal Society (UK)
CSLT Canadian Society of Laboratory Technologists
csm cerebrospinal meningitis; combustion space monitor; command service module (CSM); corn-soya-milk (mixture)
CSM Central States Motor Freight Bureau; Chief Stipendiary Magistrate; Christian Socialist Movement; Colorado School of Mines; Command and Service Module; Commission for Synoptic Meteorology; Company Sergeant-Major; Correctional Service of Minnesota; Cosmopolitan School of Music
CSM Christian Science Monitor
csma (CSMA) carrier-sensed multiple access
CSMA Chemical Specialities Manufacturers Association
CSMA/CD Carrier Sense Multiple Access with Collision Detection
CSMC Catholic Students' Mission Crusade; Council for the Single Mother and her Child
CSMFTA Central and Southern Motor Freight Tariff Association
csmith coppersmith
CSM-LM Command Service

Module–Lunar Module (Apollo spacecraft)
CSMMG Chartered Society of Massage and Medical Gymnastics
CSMP Comprehensive School Mathematics Program; Continuous System Modeling Program
c/smp(s) counter sample(s)
CSMPS Computerized Scientific Management Planning System
CS/Ms Commander, Submarines
CSMSW Carver School of Missions and Social Work
csn colloidal suspension
CSN Canadian Switched Network; Community of the Sacred Name; *Companhia Siderurgica Nacional* (Portuguese—National Steel Company); Confederate States Navy; Contract Serial Number; Control Symbol Number
CSNAR Charles Sheldon National Antelope Refuge (Nevada)
CSNDA Center for the Studies of Narcotic and Drug Abuse (National Institute of Mental Health)
CSNH Cincinnati Society of Natural History
CSNI Committee for the Safety of Nuclear Installations
CSNMDU Center for the Study of Non-Medical Drug Use
CSNO California School Nurses Organization
CSNWR Carolina Sandhills National Wildlife Refuge (South Carolina)
cso central signoff; chained sequential operation
C^so^ *Corso* (Italian—Street)
CSO Cairo Symphony Orchestra; Canberra Symphony Orchestra; Cargo Security Office; Central Selling Organisation (diamonds sold in London); Central Statistical Office; Charlotte Symphony Orchestra; Chattanooga Symphony Orchestra; Chicago Symphony Orchestra; Chief Signal Officer; Chief Staff Officer; Chief Surgical Officer; Cincinnati Summer Opera; Cincinnati Symphony Orchestra; Clothing Supply Office(r); Columbia Symphony Orchestra; Columbus Symphony Orchestra; Com-

mand Signal Office(r); Commonwealth Scientific Office; Community Service Office(r); Community Service Organization; Community Standards Organization; Consumer Services Organization; Montevideo, Uruguay (Carrasco airport)
csocr code-sort optical-character recognition
CSOP Commission to Study the Organization of Peace (UN)
csoro conical span on receive only
CSOs Community Service Officers; Community Service Organizations
csp central switching point; concurrent spare parts; constant-speed drive
Csp Caspar; Caspean
CSP Center for the Study of the Presidency; Certified Safety Professional; Chartered Society of Physiotherapy; Connecticut State Police; Continuous Sampling Plan; Contractor Support Program; Corporation Standard Practice
C.S.P. Congregation of St Paul
CSPA California State Psychological Association; Civil Service Pensioners' Alliance; Columbia Scholastic Press Association; Council of State Planning Agencies
CSPAA Columbia Scholastic Press Advisers Association
CSPB California State Personnel Board
CSPC California State Polytechnic College
CSPCA Canadian Society for the Prevention of Cruelty to Animals
CSPCo Caledonian Steam Packet Company
C/SPCS Cost-Schedule Planning Control Specification
CSPE Columbia Storage Power Exchange
CSPI Center for Science in the Public Interest
CSPM Communications Security Publications Memorandum
cspp corrugated structural plate pipe
CSPP Community Shelter Planning Program
cspr chlorosulphonated polyethylene rubber
CSPR(s) Christian Science

Practitioner(s)
CSPS Cable-Suspended Pumping Station; Christian Science Publishing Society
csqm climax stock quartz monsonite
CSQs College Student Questionnaires
csr circumsolar radiation; compulsive security ritual; corrected sedimentation rate; corrugated steel reinforcement
C-S r Cheyne-Stokes respiration
CSR Certified Shorthand Reporter; Chartered Stenographic Reporter; Civil Service Requirement; Colonial Sugar Refining; Commonwealth Strategic Reserve
CSRA Central Savannah River Area (Planning and Development Commission)
CSRC Communication Science Research Center (Batelle Memorial Institute—Columbus, Ohio)
CSRF Childrens Surgical Research Fund
CSRG Commonwealth Special Research Grant
CSRL Center for the Study of Responsive Law
CSRO Consolidated Standing Route Order (USA)
CSRP Canadian Sprinkler Risk Pool; Cognitive Systems Research Program
CSRS Cooperative State Research Service
CSRUIDR Chemical Society Research Unit in Information Dissemination and Retrieval
css center spar station; computer systems simulator; control-stick steering
CSS Calcutta School Society; Center for Strategic Studies; Central Security Service (DoD); City Shuttle Service; Civil Service System; Clandestine Services Staff (CIA); Coded Switch System (to arm nuclear weapons); College Scholarship Service; Combat Service Support (USA); Commit Sequence Summary; Community Service Society; Computerized Shipping Service; Confederate States Ship (C.S.S.); Contractor Storage Site
C.S.S. Charles Stuart Calverley (nineteenth-century satirist whose works appear under

the initials shown)
C^{*ssa*} *Contessa* (Italian—Countess)
CSSA Cactus and Succulent Society of America; Central States Speech Association; Central Supply Support Activity
cssb compatible single sideband
CSSB Civil Service Supply Board
CSSC California Seismic Safety Commission
CSSCG Container Systems Standardization-Coordination Group
C S-S Co Cunard Steam-Ship Company
CSSD Central Sterile Supply Department
CSSDA Council of Social Science Data Archives
CSSDC Canadian Society for the Study of Diseases in Children
CSSE Canadian Society for the Study of Education
cssl continuous system simulation language
CSSL Central Sierra Snow Laboratory (Norden, California)
CSSLRP Commonwealth Secondary School Libraries Research Project
cssm compatible single-sideband modulation
CSSM Council of State Supervisors of Music
CSSO Combined State Services Organization; Consolidated Surplus Sales Office
CSSP Center for Studies of Suicide Prevention; Committee on Solar and Space Physics; Customer Standard Settlement Program
CSSR Cost Schedule Status Report
CSSRC Canadian Social Science Research Council
CSSS Canadian Soil Science Society; Council of State Science Supervisors
csst computer system science training
CSSU Crime Scene Search Unit
cst cargo ships and tankers; centistokes; channel status indicator; combined station power; convulsive shock therapy
c's/t certificates of title
CST Cat Survival Trust; Cele-

ban Standard Time; Central Standard Time; Conventional Stability Talks; Council for Science and Technology
CSta consolidating station
CSTA Canadian Society of Technical Agriculturists; Canterbury Science Teachers Association; Correspondence School Teachers Association
cs & tae combat surveillance and target acquisition equipment (DoD)
CSTAL Confederación Sindical de Trabajadores de America Latina (Spanish—Trade Union Confederation of the Workers of Latin America)
CSTC Charleston Submarine Training Center (South Carolina); Coordinating Scientific and Technical Council (UN); Coppin State Teachers College
C'sted Christiansted, St Croix
cstg casting
CSTI California Specialized Training Institute (for coping with terrorism); Chattanooga State Technical Institute
cstmr customer
cstol combined short takeoff and landing; controlled short takeoff and landing
cstr canister
CSTS Combined Systems Test Stand; Computer Science Time Sharing
cstv community-supported television
C/Stwd Chief Steward
csu catheter specimen of urine; central statistical unit; central statistical unit; circuit switching unit(s); constant-speed unit
CSU California State University; Casualty Staging Unit; Civil Service Union; Colorado State University; Combined State Unions; Connecticut State University; Crime-Suppression Unit
CSU Christlich-Soziale Union (German—Christian Social Union)—political party
CSUB Center for the Study of Urban Poverty (UCLA)
CSUC California State Universities and Colleges; California State University at Chico
CSUCA Consejo Superior Universitaria Centroamericano (Superior Council of Central American Universities)

CSUF California State University at Fresno
CSUH California State University at Humboldt
csul consult
CSULA California State University at Los Angeles
CSULB California State University at Long Beach
csuld consulted
csulg consulting
CSUN California State University at Northridge
CSUS California State University at Sacramento
CSUSA Copyright Society of the U.S.A.
CSUSB California State University at San Bernardino
CSUSD California State University at San Diego
CSUSF California State University at San Francisco
CSUSJ California State University at San Jose
CSV Community Service Volunteer
csw channel status word(ing); continuous seismic wave
CSW Certified Social Worker; Commission on the Status of Women
CSWA Chinese Seamens Welfare Association
CSWAE Commission on the Status of Women in Adult Education
CSWE Council on Social Work Education
CSWI Commission for Synoptic Weather Information
csws crew-served weapon sight
Cswy Causeway
CSX Chessie and Seaboard (railroads consolidated)
csz copper, steel, or zinc (freight)
ct cable transfer; carat; caught; cellular therapy; cent; center; center tap; central timing; ceramic tile; circuit; coated tablet; coffee table; compressed tablet; compute topography; contrast threshold; control transformer; corrective therapist; corrective therapy; court; credit; current; current transformer
ct (CT) computed tomograph(y); corrective therapist; corrective therapy
c/t conference terms
c & t classification and testing
ct. centum (Latin—hundred)
Ct celtium; Court
CT Canadian Terms; Capital

Territory (airline code); Certificate of Title; Chrysler Technologies; Cognitive Therapy; Community Transit; Connecticut; Copy Typist; Credit Transfer; Sir Charles Tupper (Canada's seventh Prime Minister)
C/T California Terms
C of T Certificate of Title; Count(y) of Tyrol
CT *Corrections Today*
CT-4 New Zealand-built Airtrainer aircraft
cta call time adjustor; cash to assume; catamenia (menstruation); cystine trypticase agar
cta (CTA) cyano-trimethyl-androsterone
cta *communiquer à toutes adresses* (French—circulate to all addresses; *cuenta* (Spanish—account)
c.t.a. *cum testamento annexo* (Latin—with the will annexed)
Ct A Control Area
CTA California Taxpayers Association; California Teachers Association; Canadian Tuberculosis Association; Caribbean Tourist Association; Chemical Toilet Association; Chicago Transit Authority; Colorado Teachers Association; Commercial Travellers Association; Container Truckers Association; Council for Technical Advancement; Covered Threads Association
cta *corr^{te}* *cuenta corriente* (Spanish—current account)
cta cte *cuenta corriente* (Spanish—current account)
CTAF Crew Training Air Force
cta/ir control area/instrument restricted
cta/iv control area/instrument visual
CTAL Container Terminals Australia Ltd
CTAL *Confederación de Trabajadores de America Latina* (Spanish—Confederation of Latin American Workers)
ctam continental tropical air mass
Ct Ap Court of Appeals
CTAU Catholic Total Abstinence Union
cta/ve control area/visual excepted
ctb cement-treated base; ce-

ramic-tile base
CTB Cable Television Bureau; California Test Bureau; Canadian Tourist Board; Commercial Traffic Bulletin; Commonwealth Telecommunications Board; Commonwealth Trading Bank; Comprehensive Test Ban; Corporation for Television Broadcasts
CTB *Centre Technique de Bois* (French—Wood Research Center)
CTBA California Toll Bridge Authority; Commonwealth Trading Bank of Australia
ctbid(s) counterbid(s)
ctbm cetyl-trimethyl-ammonium bromide
ctbore counterbore
CTBRD *Commonwealth Taxation Board of Review Decisions*
CTBS Comprehensive Tests of Basic Skills
CTBT Comprehensive Test Ban Treaty
ctc carbon tetrachloride; contact
ctc (CTC) central traffic control; central train control; chlortetracycline
CTC California Tankers Company; California Transportation Commission; Canadian Tire Corporation; Canadian Transport Commission; Canberra Technical College; Catholic Teachers College; Central Test Control; Certified Travel Consultant; Charter Travel Company; Chicago Teachers College; Chicago Technical College; Citizens Training Camp; Citizens Training Corps; Concordia Teachers College; Corn Trade Clauses; Curaçao Trading Company; Cyclists Touring Club
CTC *Congrés du Travail du Canada* (French—Canadian Congress of Labour)
CTCA Canadian Telecommunications Carriers Association; Channel and Traffic Control Agency
CTCB Contract Technical Compliance Board
ctcd contacted
ctcg contacting
Ct Cl Court of Claims
CTCL Community and Technical College Libraries
CTCOSBA Cape Town Com-

puter Services and Bureaux Association
CTCP Contract Task Change Proposal
Ct Crim Aps Court of Criminal Appeals
CTCs Community Treatment Centers (US Bureau of Prisons)
CTCSS Continuous-Tone Coded-Squelch System
ctd charge transfer device; coated; crated
c-t-d conductivity-temperature-depth
CTD Central Training Depot; Classified Telephone Directory; Convention Travel Document; Corrective Therapy Department
CTDAS Canadian Trade Document Alignment System
ctdc control track direction computer
CTDC Chemical Thermodynamics Data Center (NBS)
ctdh command and telemetry data handling
CT & DM *Canadian Transportation and Distribution Magazine*
CTDO Central Technical Documents Office (USN)
cte coefficient of thermal expansion
cte *corriente* (Spanish—current)
Cte *Comte* (French—Count)
Cte *Conte* (Italian—Count)—Earl
CTE Car Tours in Europe; *Compañía Transatlántica Espanola* (Spanish Line)
CTEB Council of Technical Examining Bodies
CTEC Chemical Transportation Emergency Center
Ctee Committee
Cten Ctenophora
Cteno Ctenocephalides (fleas)
CTES Computer Telex Exchange System (RCA)
Ctesse *Comtesse* (French—Countess)
CTETOC Council for Technical Education and Training for Overseas Countries
CT Exam Computed Tomography Examination
ctf certificate; correction to follow; cytotoxic factor
Ctf Colorado tick fever
CTF Canadian Teachers Federation; Cayman Turtle Farm; Commander Task Force
CTFA Cosmetics, Toiletry,

and Fragrance Association

CTFC Commodity Futures Trading Commission

CTFE Colleges of Technology and Further Education (subsection of the University and Research Section of the Library Association)

ctfet counterfeit

ctfm continuous-transmission frequency-modulated (sonar)

CTFT Centre Technique Forestier Tropical (French— Tropical Forest Technical Center)

ctfy certify

ctg cartage; cartridge; cutting

Ctg Cartagena, Spain (*see* Cgn)

CTG Center Theatre Group; Commander Task Group; Commercial Travellers Guild; Components Technology Group

ctge cartage; cartridge; cottage

ctgf clean tanks, gas free

CTGI Canadian Test of General Information

Cth Commonwealth

CTH Chalmers Techniska Högskola (Swedish—Chalmers Institute of Technology); Corporation of Trinity House

CTHA Chinese Tourist Hotel Association

Cthse Courthouse

cti Container Transport International (trademark)

CTI Central Technical Institute; Ceramic Tile Institute; Container Transport International; Contract Technical Instructor; Cooling Tower Institute

CTI Communication Technology Impact

CTIA Caravan Trades and Industries Association; Committee to Investigate Assassinations

CTIAC Concrete Technology Information Analysis Center (USA)

CTIC Cable Television Information Center

CTIU Container Transport International (container) Unit

CTJ Citizens for Tax Justice

ctk capacity-ton kilometer

cTk tropical continental air colder than underlying surface

CTK Ceskoslovenska Tiskova Kancelar (Czechoslovak Press Bureau)

ctl castellate; cental; central; complementary transistor logic; constructive total loss; control

ctl (CTL) checkout test(ing) language

Ctl central

CTL Certified Tool List; Cincinnati Testing Laboratories; Container Terminals Limited

CTLA California Trial Lawyers Association

ctlg catalog

ctlo constructive total loss only

ctm capacity ton mile; centrifugal turning moment; communications terminal modules

CTM Contract Termination Manual; Contractor Technical Meeting

CTM Confederación de Trabajadores de México (Spanish—Confederation of Workers of Mexico)

CTMA Collapsible Tube Manufacturers Association; Commercial Truck Maintenance Association; Country Timber Merchants Association (Australia)

ctmc communications controller; communications terminal modules

ctmdr clamptop metal drum

CTMM California Test of Mental Maturity

ctn carton; cotangent

C Tn Cape Town (British maritime contraction)

CTN Canton Island (tracking station)

ctnd contained

CTNE Compañía Telefonica Nacional de España (Spanish—National Telephone Company of Spain)

ctng containing

ctnrs containers

ctns cartons

ctn's confectioners, tobacconists, newsagents

CTNS Chicago Tribune News Service

cto cancelled to order; concerto

c^to conto (Italian—account); *cuarto* (Spanish—fourth)

CTO Central Telegraph Office; Central Treaty Organization; Chief Technical Officer; City Ticket Office; Cognizant Transportation Office; Combined Transport Operator; Container Transport Operator; Courier Transfer Officer

CTOA Commonwealth Tele-

phone Officers Association; Creative Tour Operators Association

ctocu central technical order control unit

ctofr counteroffer

ctol conventional takeoff and landing

ct ord court order

ctp central transfer point; close to profit

ctp (CTP) cytidine triphosphate

CTP Canterbury Timber Products; Columbia Television Pictures

CTP Centre de Tutelle Pénale (French—Penal Surveillance Center)—for recidivists; *Confederación de Trabajadores del Peru* (Spanish— Confederation of Workers of Peru)

CTPL Commission for Teacher Preparation and Licensing

CTPOA Commonwealth Telephone and Phonogram Officers Association

ctprt counterpart

ctpt counterpoint

CTPTA Centro Tropical de Pesquisas y Tecnologías de Alimentos (Tropical Center of Food Research and Technology)

ctptal contrapuntal

ctptst contrapuntist

ctr center; contour; controlled thermonuclear reactor; counter; cutter

Ctr Center

CTR Cash Transaction Report; Controlled Thermonuclear Reactor; Currency Transaction Report

CTRA Coal Tar Research Association

Ctr Appl Ling Center for Applied Linguistics

Ctr Appl Res Center for Applied Research in Education (New York)

Ctr Byz Center for Byzantine Studies

CTRC Caribbean Tourism Research Center

Ctr Calif Pub Center for California Public Affairs (Claremont)

Ctr Chin Stud Center for Chinese Studies (Berkeley, California)

Ctr Cont Celeb Center for Contemporary Celebration (Chicago)

Ctr Cont Poetry Center for

Contemporary Poetry (La Crosse, Wisconsin)

Ctr Info Am Center for Information on America

ctr/iv control zone/instrument visual

ctrl control (flow chart)

Ctr Land Arch Center for Landscape Architecture

Ctr Marital Sexual Center for Marital and Sexual Studies (Long Beach, Calif)

Ctr Mig Center for Migration Studies (New York)

ctrofr counteroffer

ctrofrdcl counteroffer declined

CTRP Controlled Thermonuclear Research Program

CTRP *Confederación de Trabajadores de la República de Panamá* (Spanish—Confederation of Workers of the Republic of Panama)

Ctr Pol Process Center for Policy Process (DC)

Ctr Pre-Col Center for Pre-Columbian Studies (DC)

Ctr Sci Pub Center for Science in the Public Interest (DC)

Ctr Sci Study Rel Center for the Scientific Study of Religion (Chicago)

Ctr S&SE Asian Center for South and Southeast Asian Studies (Ann Arbor, Mich)

CTRU Colonial Termite Research Unit

Ctr Urb Pol Res Center for Urban Policy Research (New Brunswick, NJ)

ctr/ve control zone/visual exempted

ctry country

cts carats; cents; computer typesetting; contralateral threshold shift (audiometry); crates

cts (CTS) communications technology satellite(s)

cts *centavos* (Spanish—cents); *centimes* (French—cents); *centimos* (Spanish—cents)

Cts courts

CTS Canadian Thoracic Society; Captive Trajectory System; Card-to-Magnetic Conversion System; Catholic Truth Society; Central Transportation System; Centralized Title Service; China Travel Service; Commercial Teachers Society; Commonwealth Teaching Service; Commonwealth Time Service; Component Test(ing) Set; Computer Test(ing) Site; Comput-

erized Type System; Consolidated Tin Smelters; Consolidated Translation Survey; Container Terminal Station; Contract Technical Services; Contractor Technical Service; Conversational Terminal System; Cosmic Top Secret; Courier Transfer Station; Custom Track Service

CTSA Catholic Theological Society of America; Coordinated Transportation Service Agency; Crucible and Tool Steel Association

CT Scan computerized tomography scan

Ct Sess Court of Sessions

ctsp contract technical services personnel

CTSS Compatible Time-Shared System

ctt capital transfer tax(ation); compressed tablet triturate

CTT Columbia Technical Translations

CTT *Correios e Telecomuniações de Portugal* (Portuguese—Postal and Telegraph Services of Portugal)

ct ta control tape

CTTB Central Trade Test Board

CTTC Central Telegraph Test Center

Cttee Committee

CTTF California Turtle and Tortoise Club

CTTP Complementary Trade Training Program

ctu centigrade thermal unit; central terminal unit; components test unit

CTU Combat Training Unit; Commander Task Unit

CTU (AFL-CIO) Commercial Telegraphers' Union

C-tube C-shaped tube

CTUS Carnegie Trust for the Universities of Scotland

CTUY Confederation of Trade Unions of Yugoslavia

ctv (CTV) cable television; color television

CTV Canadian Television; Children's Television

CTV *Confederación de Trabajadores de Venezuela* (Spanish—Confederation of Venezuelan Workers)

ctvo *centavo* (Spanish—cent)

CTVT Childrens Television Theater

CTVW Children's Television Workshop

ctw counterweight

cTw tropical continental air warmer than underlying surface

CTW Children's Television Workshop

ctwt counterweight

ctx computer telex exchange (RCA system)

Ct X Court Exhibit

Cty City; County

ctz chlorothiazide

CTZ Corps Tactical Zone

ct zone chemoreceptor trigger zone

cu cleanup; clinical unit; closeup; condition unknown; container unit (CU); contents unknown; control unit; cube; cubic; cumulus

cu (CU) credit union

c-u see you

c/u *cada uno* (Spanish—each one)

Cu Cuba; Cuban; cumulus; cuprum (Latin—copper)

Cu urea clearance

CU Cambridge University; Capital University; Carleton University; Church Union; City University; Claflin University; Clark University; Colgate University; Colorado University; Columbia University; Commercial Union; Concordia University; Consumers Union; Cooper Union; Cooperative Union; Cornell University; Creighton University; Cumberland University; Customs Union

Cu₂SO₄ copper sulfate

Cu-7 copper-constructed 7-shaped intrauterine device

cua central unit assembly; computer unit assembly

CUA Canadian Underwriters Association; Catholic University of America; Center for Urban Analysis; Council on Urban Affairs

CUAC Cambridge University Athletic Club

cuad *cuadrado* (Spanish—square)

CUAFC Cambridge University Association Football Club

CUAG Computer Users Associations Group

CUAP Catholic University of America Press

CUAS Cambridge University Agricultural Society; Cambridge University Air Squadron

CUAV Citizens United Against Violence

cub. control unit busy; cubic
cúb cúbico (Spanish—cubic)
Cu b copper band
CUB advanced unit base; Carlton and United Breweries; Citizens Utility Board; Consumers Utility Board
Cuba Republic of Cuba (largest West Indian island, *República de Cuba*
CUBANA Compañía Cubana de Aviación (Spanish—Cuba Air Company)
CUBC Cambridge University Boat Club; Cambridge University Boxing Club
cubo conduct unbecoming an officer
CUBS Congress for the Unity of Black Students
cuc chronic ulcerative colitis
CUC Canadian Unitarian Council; Canberra University College; Canterbury United Council
CUCA Carpet and Upholstery Cleaning Association
cu cap. cubic capacity
CUCC Cambridge University Cricket Club
Cuch Cuchillas (Spanish—mountain chain, range)
cu cm cubic centimeter
CUCNY Citizens Union of the City of New York
cud. congenital urinary (tract) deformities
'cuda(s) barracuda(s)
Cuddy Cuthbert
CUDN Common User Data Network
CUDS Cambridge University Dramatic Society
cue. coastal upwelling experiment; computer update equipment; configuration utilization efficiency; control unit end; correction update extension
CUE Center for Urban Education; Coastal Upwelling Experiment; Concentrated Urban Enforcement (of gun control)
CUEA Coastal Upwelling Ecosystem Analysis
CUEBS Commission on Undergraduate Education in the Biological Sciences
CUED Council for Urban Economic Development
Cuen Cuenca
CUEPACS Congress of Unions of Employees in the Public and Civil Services
CUERAC Computer-Con-

trolled Random-Access Cartridge Libraries
CUERD Committee for Upgrading Environmental Radiation Data
CUES College and University Environment Scales
CUEW Congregational Union of England and Wales
CUF Canadian Universities Foundation
CUF Companhia Uniao Fabril (Portuguese—United Manufacturing Company)—Iberian conglomerate whose company street in Barreiro is named Rua do Acido Sulfúrico (Sulfuric Acid Street)
CUFC Consortium of University Film Centers
'cuffed handcuffed
'cuffs handcuffs
cu ft cubic feet; cubic foot
cu ft min cubic feet per minute
cu ft sec cubic feet per second
cug closed-user group; cystourethrogram
CUGC Cambridge University Golf Club
CUHC Cambridge University Hockey Club
CUHK Chinese University of Hong Kong
CUIC Canadian Unemployment Insurance Commission
CUIHC California Urban Indian Health Council
cu in cubic inch
cuis cuisine (French—cookery, kitchen)
cuj. cujus (Latin—of which)
cuj. lib. cujus libet (Latin—of any you wish)
CUK São Paolo, Brazil (Combica Airport)
cukes cucumbers
CUKT Carnegie United Kingdom Trust
cul culinary
c-u-l see you later
CUL Cambridge University Libraries; Catholics United for Life; China Union Lines; Columbia University Library; Cooper Union Library; Cornell University Library
c-u later see you later
cull. cullage; cullboard; culling; cullion
CULP California Union List of Periodicals
cult. cultural; culture
CULT Chinese University Language Translation (system)

cult. anthro(s) cultural anthropologist(s); cultural anthropology
CULTC Cambridge University Lawn Tennis Club
culv culvert
cul vul(s) culture vulture(s)
cum central unit memory; cumulative
cu m cubic meter
CUM Cambridge University Mission
CUM Centro Universitario México (Spanish—Mexico University Center)
CUMA Canadian Urethane Manufacturers Association
Cumb Cumberland
Cumberland Island national seashore on coast of Georgia
Cumberlands Cumberland Caverns, Tennessee; Cumberland Islands off Queensland, Australia; Cumberland Mountains extending from Alabama to Virginia
cum/d cubic meters per day
cum d(iv) cum dividend (with dividend)
cumes cumulative audience survey (radio/tv)
cu mm cubic millimeter
Cummins Farm Cummins Prison Farm in Arkansas
CUMMM Council of Underground Mining Machinery Manufacturers
Cum Nursing Lit Cumulative Index to Nursing Literature
Cump Tecumseh
cum pref cumulative preference
CUMS Cambridge University Musical Society
cum/sec cubic meters per second
cumshaw (Chinese—grateful thanks)—quasi-legal barter system; under-the-table graft
cu mu cubic micron
CUMWA Consortium of Universities in the Metropolitan Washington Area
cun cuneiform
CUN Convent van Universiteits-bibliothecarissen in Nederland (Dutch—Association of University Librarians in the Netherlands)
CUNA Credit Union National Association
cuni cupro-nickel (coin alloy)
cu-nim cumulo-nimbus (clouds)
CUNSA Canadian University Nursing Students Association

CUNY City University of New York

CUOG Cambridge University Opera Group

cup. cupboard

CUP Cambridge University Press; Canadian University Press; Columbia University Press; Conditional Use Permit; Cornell University Press

CUPA College and University Personnel Association

CUPBEQ Canadian University Press Québec Region

CUPE Canadian Union of Public Employees

cupper cup-tie-er (athletic matches played for a trophy cup)

CUPR Catholic University of Puerto Rico

cuprite cuprous oxide

cupronic copper-nickel alloy

CUPS Consolidated Unit Personnel Section

CUPW Canadian Union of Postal Workers

cur. curiosa; curiosity; currency; current

Cur Curaçao (maritime abbreviation)

CUR Council on Undergraduate Research; Curaçao, Netherlands West Indies (Plesman Airport)

CURA Center for Urban Research and Action; Comprehensive Urban Renewal Area

CURAC Coal Utilization Research Advisory Committee

curat curative

curat. curatio (Latin—dressing, wound dressing)

CURB Campaign on the Use and Restriction of Barbiturates

cure. (C-U-R-E) care, understanding, research (organization for the welfare of drug addicts)

CURE Citizens United for Racial Equality

C-U-R-E Care, Understanding, Research (organization for the welfare of drug addicts)

CURES Computer Utilization Reporting System

CURF Citizens Union Research Foundation

CURFC Cambridge University Rugby Football Club

curio curiosa; curiosity

CURLS College, University, and Research Libraries Section (California Library Association)

CURMCO City Urban Renewal Management Corporation (NYC)

curr currency; current

curric curriculum

curt. current (Scottish—instant); curtain

Curt Curtis

Curt Quintus Curtius Rufus (Roman historian)

CURTS Common-User Radio Transmission System

curv cable-operated unmanned recovery vehicle

CURV Cable-controlled Underwater Research Vehicle

cury currently

Curzon Lord Curzon (George Nathaniel Curzon)—Viceroy and Governor General of India

cus customer

Cus Customs

CUS Cambridge Union Society; Continental United States

CUSA Conservative United Synagogue of America; Council of Unions of South Africa

cusecs cubic feet per second

Cush Cushing's Massachusetts Reports

Cus Ho Custom House

CUSIP Committee on Uniform Security Identification Procedures (for computer user protection); Committee for Uniform Securities Information

CUSIP No. CUSIP Number (assigned to every issue of a bond or stock)

cusm customs

CUSM Columbia University School of Medicine

cusmr customer

CUSO Canadian University Service Overseas

CUSP Central Unit for Scientific Photography

CUSR Canada/United States Region

CUSRPC Canada-United States Regional Planning Committee

CUSRPG Canada-United States Regional Planning Group

CUSS Continental, Union, Shell, Superior (oil companies' deep-sea oil-drilling ship)

cust custard; custodian; custody; custom(s)

Cust Ct Customs Court

custod custodian

custs custards; customers

CUSW/NAS Committee on Undersea Warfare—National Academy of Sciences

CUT California United Terminal

CUTF Commonwealth Unit Trust Fund

cutg cutting

Cuth Cuthbert; Cuthwald; Cuthwold

cuti cutis (Latin—skin)—cutaneous, cuticle, cutin

'cutor prosecutor

CUTS Computer-Utilized Turning System

Cuu Chihuahua

CUUI Center for US—USSR Initiatives

CUUS Consumers Union of the United States

cuv current use value

CUW Committee on Undersea Warfare (DoD)

CUWARFA California Urban Waterfront Area Restoration Financing Authority

Cux Cuxhaven

cu yd cubic yard

cuz cousin; because

cv caloric value; capital value; cardiovascular; carrier vehicle; cataclysmic variable; catalog value; check valve; chief valve; coefficient of variation; collection voucher; concave; constant velocity; contributing valve; convertible; culture vulture

cv caballo de vapor (Spanish), cavallo vapore (Italian), cavalo vapor (Portuguese), chevalvapeur (French)—horsepower (also appears as *CV*); curriculum vitae (Latin—biographical résumé)

c.v. conjugata vera (Latin—true conjugate)—pelvic inlet diameter; *cras vespere* (Latin—tomorrow evening); *cursus vitae* (Latin—course of life)

Cv Cove; molecular heat (symbol); specific heat at constant volume (symbol)

CV aircraft carrier (2-letter naval symbol); Central Vermont (railroad); Chula Vista; collection voucher; combat vehicle; Commercial Village (zone); Community of the Visitation; Community Volunteers; Convair

C-V Convair (Division of General Dynamics)

C/V Certificate of Value

CV *Cabo Verde* (Spanish—Cape Verde Islands); *cheval-vapeur* (French—horsepower)

CV4 Convair 440 airliner

cva cerebrovascular accident stroke); costovertebral angle

CVA attack aircraft carrier (naval symbol); Civilian Voluntary Agency; Columbia Valley Authority

CVA *Centro Venezolano América* (Spanish—Venezuelan-American Center), cultural display promoted by the U.S. Embassy

CVAA *Centre de Vulgarisation Aéro-Astronautique* (French—Center for the Popularization of Astronautics)

CVAC Consolidated Vultee Aircraft (now Convair)

cvae coordinated vocational academic education

CVALI Crime Victims Legal Advocacy Institute

CVAN nuclear-powered aircraft carrier (naval symbol)

CVAS Configuration Verification and Accounting System

cvb combined very-high-frequency band

CVB large aircraft carrier (naval symbol)

C & VB Convention and Visitors Bureau

c-v-c consonant-vowel-consonant

CVC Cauca Valley Corporation; Clinch Valley College; Consolidated Vacuum Corporation

CVCB Crime Victims Compensation Board (New York)

cvcc compound vortex-controlled combustion (Japanese automotive engine designed to reduce air pollution)

cvcm collected volatile condensable material

CVCP Committee of Vice-Chancellors and Principals

cvcr control van connecting room

cvd cardiovascular disease; cash versus documents; chemical vapor deposition; coordination of valve development; coupled vibration dissociation; current-voltage diagram

CVDE *Columbia-Viking Desk Encyclopedia*

cve (CVE) customer-vended equipment

CVE aircraft carrier, escort

(naval symbol); Council for Visual Education

C Ven Canis Venatici

CVF Caravelle fan jet airplane

CVF *Corporación Venezolano de Fomento* (Spanish—Venezuelan Promotion Corporation)

CVG Cincinnati, Ohio, airport; Corporación Venezolana de Guayana (Spanish—Venezuelan Corporation of Guyana)

cvh combined ventricular hypertrophy

cvh (CVH) compound-valve hemispherical head (cam-in-head automotive powerplant)

CVH Center for Visual History

CVHC coastal helicopter aircraft carrier (naval symbol)

CVHS Center on Violence and Human Survival (John Jay College); Chelsea Vocational High School

cvi center of visual impact; cerebrovascular insufficiency; common variable immunodeficiency

CVI Cape Verde Islands; College of the Virgin Islands

CVIA Commercial Vehicle Industry Association

C viruses Coxsackie viruses

CVIS Computerized Vocational Information System

cvk centerline vertical keel

CVL Caravelle jet airplane; small aircraft carrier (naval symbol)

CVLAI Crime Victims Legal Advocacy Institute

cvli cash value life insurance

CVM Company of Veteran Motorists

CVMA Canadian Veterinary Medical Association

cvn carbohydrate vitamin nitrogen; convene

C Vn Canis Venatici

CVN nuclear-powered aircraft carrier (naval symbol)

c.v.o. *conjugata vera obstetrica* (Latin—conjugate obstetric diameter)

CVO Chief Veterinary Officer(r)

C.V.O. Commander of the Royal Victorian Order

C/VO Certificate of Value and Origin

c voc *colla voce* (Italian—with the voice)

cvou (CVOU) cardiovascular observation unit

cvp central venous pressure; climate, vegetation, produc-

tivity

CVP *Corporación Venezolana de Petroleo* (Spanish—Venezuelan Petroleum Corporation)

cvr cardiovascular renal; cardiovascular-respiratory; cerebrovascular resistance; continuous video recorder

cvr (CVR) cockpit voice recorder; crystal video receiver

cvrd cardiovascular renal disease

cvrd hpr covered hopper (freight car)

cvs cardiovascular surgery; cardiovascular system

cvs (CVS) chorionic villus sampling

CVS antisubmarine warfare support aircraft carrier (3-letter symbol); Change Verification System

cvsd continuously variable slope-delta (modulation)

cvt chemical vapor transport; constant-voltage transformer; continuously variable transmission; controlled variable time (fuze); convertible

CVT training aircraft carrier (naval symbol)

c/vta *cuenta de venta* (Spanish—bill of sale)

Cvt Gdn Covent Garden (Royal Opera House)

cvtr charcoal viral transport medium

CVV conventional oil-powered aircraft carrier (naval symbol)

CVW attack carrier air wing (naval symbol)

CVWS Combat Vehicle Weapon System

cw call(s) waiting; canistered waste(s); cardiac work; casework(er); chemical warfare (CW); chest wall(s); child welfare; children's ward; cladding waste(s); clockwise; continuous wave; conventional wisdom; copperweld (copper-covered steel); cubic weight

c-w chronometer time minus watch time

c/w chainwheel; counterweight

c & w country and western (music)

CW Canada West; Channel Airways; chemical warfare; continuous wave

C-W Curtiss-Wright

C&W Cable and Wireless

C of W College of Wooster

(Ohio)
CW *Computer World*
CWA Civil Works Administration; Clean Water Act; Communication Workers of America; Concerned Women of America; Country Womens Association; County Water Authority; Crime Writers Association; Customer Work Authorization
CWA *Clean Water Act*
CWAA Cotton Warehouse Association of America
CWAC California Wildlife Advisory Committee; City-Wide Anti-Crime Unit (sometimes called Quacks)
CWAM Cliffs Western Australian Mining Company
cwar continuous-wave acquisition radar
cwas contractor-weighted average share
cwb clay-water-base (oil well) mud
CWB Canadian Wheat Board; Central Wages Board; Child Welfare Board
CWBS Cost Work Breakdown Structure
cwbts capillary whole blood true sugar
cw-bw chemical warfare—biological warfare
CWC Canadian Welfare Council; Central Wesleyan College; Chemical Weapons Convention
CWCA California Women's Commission on Alcoholism
CWCC Civil War Centennial Commission
c & w ck caution and warning check
CWCO China Wire and Cable Company
CWCP Combat Wing Command Post
cwd civilian war dead; clerical work data
cwe current working estimate
CWE Commonwealth Edison
C'wealth Commonwealth
CWEP Community Work Experience Program
CWET Community Work Experience Training
CWETA California Worksite Education and Training Act
CWF California Wildlife Federation; Central Wool Facility; Christian Women's Fellowship; Cornell Word Form(ation)
cwfm continuous-wave fre-

quency modulated
cwfp clean wool fibers present
CWFT Cornell Word Form Test
cwg corrugated wire glass
CWG California Writers Guild
CWGC Commonwealth War Graves Commission
cwgt counterweight
CWHSSA Contract Work Hours and Safety Standards Act
cwi cardiac work index; clear word identifier
CWI California Wine Institute; Colonial Williamsburg Incorporated; Country Women's Institute
cwik cutting with intent to kill
CWINC Central Waterways, Irrigation, and Navigation Commission
CWIS Chaim Weizmann Institute of Science
CWISS Child Welfare Information Services System
cwit concordance words in title
cwl calm waterline
CWL Catholic Women's League
CWLA Child Welfare League of America
C & W Ltd Cables and Wireless Limited
cwm (CWM) commercial waste management
CWMH Colonial War Memorial Hospital
CWMTU Cold Weather Materiel Test Unit
CWNA Canadian Weekly Newspapers Association
cwo cash with order; continuous-wave oscillator
CWO Chief Warrant Officer
CWOD California's War On Drugs
cwp childbirth with(out) pain; circulating water pump; communicating word processor; community work plan
CWP Communist Workers Party
CWPEA Childbirth Without Pain Education Association
CWPLs Childbirth Without Pain Leagues
CWPP Commercial Waste Packaging Program
CWPS Center for Women Policy Studies; Council on Wage and Price Stability
CWPU Central Water Planning Unit
cwr continuous welded rail
CWR California Western Rail-

road; Crusade for World Revival
CWRA California Water Resources Association
cwrb (CWRB) canistered waste-receiving building
CWRC Chemical Warfare Review Commission
CWRSM Case-Western Reserve School of Medicine
cws clockwise; cold-water soluble; countersunk wood screw; curve warning signal
cw & s crushed, washed, and screened
Cws chemical weapons; Cowes
CWS California Water Service; Canadian Welding Society; Canadian Wildlife Service; Chandraprabha Wildlife Sanctuary (India); Child Welfare Services; Church World Service; Clearinghouse on Women's Studies; College Work Study; College World Series; Cooperative Wholesale Society; Cunard-White Star (steamship line)
C-WS Crop-Weather Service
CWSC Canterbury Winter Sports Club; Central Washington State College
cwsfp commercial waste spent-fuel packaging
cw sig gen continuous wave signal generator
CWSP College Work-Study Program
CWSS Center for Women's Studies and Services
Cwsy Causeway
cwt centum weight; counterweight; hundredweight
CWT Central War Time; Command Word Trap; Community Work Training; Consumers for World Trade; Container Warehousing and Transportation; Cooperative Wind Tunnel
CWTC California World Trade Center
cwtd continuous-wave target detector
cwtdc continuous-wave target detection console
cwu composite weighted work unit
CWU California Western University; Chemical Workers Union; Church Women United; Culinary Workers Union
cwv continuous-wave video
CWV Catholic War Veterans
cww cruciform wing weapon

CWWC Concerned Women in the War on Crime
cwy clearway
cx cervix; chest X-ray; complex; connection; control transmitter; convex; correct copy (instruction to the printer)
cx (CX) central exchange
Cx Caxton; Caxton Printers
Cx Caixa (Portuguese—Box)—post office box; also written *cx*
CX Cathay Pacific Airways; outsize cargo-transport aircraft
Cxo Calexico
cxr carrier
cXr chest X-ray
cxs consort parallax servo
CXT Common External Tariff
cy calendar year; capacity; copy; cubic yard; currency; current year; cyanide; cyanogen; cycle; cylinder(s)
Cy City; cyanogen; Cyprus; Cyrus
CY Container Yard
cya cover your ass (protect yourself)
CYA California Youth Authority; Canadian Yachting Association; Carded Yarn Association; Catholic Youth Adoration (Society); Chicano Youth Alliance; Covenant Youth of America
cyan cyanamid; cyanic; cyanide; cyanogen; cyanotype
cyan (Latin prefix—blue)—cyanometer (device for measuring the blue in the sky)
cyanide cyanide of potassium
cyath. cyathus (Latin—cup, ladle, glass)
cyath. vin. cyathus vinarius (Latin—wineglassful)
cyb cybernetic; cyberneticist; cybernetics
CYB Canada Year Book
cyber cybernetics
cybercult cybercultural(ly); cyberculture
cyberlog cybernetic logistics
cybernat cybernated; cybernation(al)(ly)
cyborg(s) cybernetic organism(s)
cyc curb your curiosity; cyclazogine (narcotic antagonist used in curing victims of addiction); cycle; cyclorama
CYC Capital Yacht Club; Chicago Yacht Club; Cleveland Yacht Club; Colorado Youth Center; Columbia Yacht

Club; Company of Young Canadians; Corinthian Yacht Club
CYCA Clyde Yacht Clubs Association; Cruising Yacht Club of Australia
Cycl Cyclostomata
Cyclades Cyclades Islands
cyclams cyclamates
cyclaz cyclazocine
cycle bicycle; motorcycle
cyclecade bicycle parade; motorcycle parade; tricycle parade
cyclo cyclopedia; cyclopedic; cyclophosphamide; cyclopropane; cyclorama
cyclon cyclonometer
cyclon gas cyanide, chlorine, and nitrogen
cyclos (Greek—circle, ring)—bicycle, circle, cyclorama
CY/CY Container Yard to Container Yard
c yd cubic yard(s)
CYEE Central Youth Employment Executive (UK)
CYFA Club for Young Friends of Animals
cyflo cylinder overflow
Cyg Cygnus
CYHA Canadian Youth Hostels Association
cyk consider yourself kissed
cyke cyclorama
cyl cylinder; cylindrical; cylindroid
CYL Chinese Youth league
cyl l cylinder lock
cyls cylinders
cym cymbal(s)
Cym Cymraeg (Welsh); Cymric
CYMA Catholic Young Men's Association
Cymb Cymbeline
Cymr Cymric (Welsh—Wales)
CYMRU Welsh television
CYMS Catholic Young Men's Society
cyn cyanide
Cyn Canyon; Cynthia
cyni cynical; cynicism
cynol cynologic(al)(ly); cynologist; cynology
CYO Catholic Youth Organization; Civic Youth Orchestra; Community Service Corps
Cyp Cyprian; Cypriote; Cyprus
CYP Commonwealth Youth Programme; Couple Years of Protection; Cyprus Airways
Cyprus Republic of Cyprus (Mediterranean island country split between Greek and

Turkish settlers *Kypriaki Dimokratia* (Greek name for Cyprus); *Kibris Cumhuriyeti* (Turkish name)
CYRA Commission Yellowfin Regulatory Area
Cyrano Cyrano de Bergerac
cys cysteine; cystoscopy
cys (CYS) cystine (amino acid)
CYS Cheyenne, Wyoming airport
CYSA Combed Yarn Spinners Association
CYSS Community Youth Support Scheme (Australia)
cysti (Latin prefix—bladder or sac)—cystoscope (device for examining the bladder)
cysto cystoscope; cystoscopic examination
CYSYS Center for Cybernetics System Synergism
cyt cytology
cytac control of tactical aircraft
cytd calendar year to date
cyto (Latin prefix—cell)—cytology
cytoeco cytoecologic(al)(ly); cytoecologist; cytoecology
cytol cytological; cytologist; cytology
Cytol Cytology
cytomorph cytomorphologic(al)(ly); cytomorphologist; cytomorphology
cytopatho cytopathogenic(al)(ly); cytopathogenicity
cytophoto cytophotometer; cytophotometric(al)(ly); cytophotometry
cytostat cytostatic(al)(ly)
cyto syst cytochrome system
cytotech cytotechnic(al)(ly); cytotechnician; cytotechnologist; cytotechnology
cz coryza
Cz Czech; Czechoslovakia; Czechoslovakian
CZ Canal Zone; combat zone; communications zone; Consolidated Zinc
C-Z Crown-Zellerbach
C.Z. Canal Zone
CZ Ceska Zbrojovka (Czechoslovak Arms Factory)
Cza Constanza
CZA Canal Zone Authority; Coastal Zone Authority
CZAG Committee for Zero Automobile Growth
CZBA Canal Zone Biological Area
CZC Canal Zone College
CZC Canal Zone Code (legal)
CZCA Coastal Zone Conservation Act

czcs coastal-zone color scanner
czd calculated zenith distance; combat zone distance
Czech Czechoslovakia, Czechoslovakian
Czech J Phys *Czechoslovak Journal of Physics*
Czechoslovakia Czechoslovak Republic, *Ceskoslovenská Socialistická Republika*
Czech Phil Czech Philharmonic
CZF Canadian Zionist Federation

CZG Canal Zone Government
czi (CZI) crystalline zinc insulin
CZI Canal Zone Institute
CZJC Canal Zone Junior College
Cz kr Czechoslovakian kronen (monetary unit)
CZL-M Canal Zone Library-Museum (Balboa Heights)
CZm compass azimuth
CZMA Coastal Zone Management Act
Czml Cozumel

C-Zone commercial zone
CZP Chicago Zoological Park (Brookfield Park); Consolidated Zinc Proprietary
CZ Pen Canal Zone Penitentiary
czr combat zone radius
CZRs Canal Zone Regulations
CZSG Canal Zone Study Group
C-Z strain Carr-Zilber (viral) strain
c-Z-t chirp-Z-transform
czy crazy

D

d angular deformation (symbol); daily; date; daughter; day; declination; degree; dented; depth; dextrorotatory; died; differentiation; dime; dinar; diopter; divorced; dorsal; drizzling; dyne; grating space in calcite (symbol); liter (symbol); pence (symbol); penny (symbol)

d (D) demand; deposit

d' surname prefixes such as da, de, di, etc

'd (contraction—could, did, had, would)

d decimus (Latin—tenth); *der* (German—the); *denarii* (Latin—pennies); *denarius* (Latin—penny); *dexter* (Latin—right)

D December; degree of curve (symbol); Delta—code for letter D; democracy; Democrat (ic); density; Denver; department; derivation; Detroit; deuterium; diameter; dielectric flux density (symbol); Dietzgen; dioptric power (symbol); director aircraft; disaster; disaster broadcasting; dollar; dose; Douglas; down; drag (symbol); drone-control version (symbol); Dublin; Dutch; electrostatic flux density (symbol); Fraunhofer lines caused by sodium (symbol); propeller diameter (symbol)

D' surname prefixes such as Da, De, Di, Do, Du, etc.

D Dagh (Persian—*Daglar* (Turkish—mountain range); *Dagi* (Turkish—mountain range); *Dag* (Turkish—mountain); *Damas* (Portu-guese or Spanish—ladies); *damas* (Spanish—ladies); *Damen* (German—ladies); *darin* (German—in); *Darreh* (Persian—valley); *Daryaceh* (Persian—lake); *Dauer* (German—bulb type camera shutter stop); *dehors* (French—out); *départ* (French—departure); *derecha* (Spanish—right); *Deus* (Latin—God); *dexter* (Latin—right); *Don* (Spanish—Sir)—Mr in its most formal meaning; *dun* (Danish—down)

D. *Don* (Spanish—Sir)—Mr

d₁ diffusing capacity—lung

d 1/2 d dispatch money payable at one-half demurrage rate

D₁, D₂, D₃, etc. 1st dorsal vertebra, 2nd dorsal vertebra, 3rd dorsal vertebra, etc.

D₂O deuterium oxide (heavy water)

d2s & cm dressed two sides and center matched (lumber)

d2s & m dressed two sides and matched (lumber)

d2s & sm dressed two sides and standard matched (lumber)

D3 Douglas DC-3 airplane

D4 Douglas DC-4 airplane

D-5-HS dextrose 5 percent in Hartman's Solution

D-5-S dextrose 5 percent in saline (solution)

d₅w 5 percent dextrose in water

D6 Douglas DC-6 airplane

D7 Douglas DC-7 airplane

D8F Douglas D8F fan jet airplane

D8S Douglas super DC-8 fan jet airplane

D9S Douglas super DC-9 fan jet airplane

D-18 Beechcraft four-passenger transport also designated C-45

D 40 iopax (uroselectan)

D'66 Democrats 1966 (Dutch political party)

D of '98 Daughters of '98

D-150 Dimension 150 (150-degree field of vision achieved by deeply curved motion-picture screen)

da daughter; days after acceptance; defined adult; delayed action; delayed arming; density altitude; deposit account; direct action; discharge afloat; discriminant analysis; district attorney; do not answer; documents against acceptance; documents attached; doesn't answer (the telephone); double acting; double aged; drift angle

da (DA) diphenylchlorasine (deadly gas); directional antenna; dopamine

d-a direct-action (adjective)

d/a digital-to-analog

d/a (D/A) deposit account

d in a (found) dead in automobile (or) airplane

d-to-a digital-to-analog

da dansk (Dano-Norwegian—Danish); *dette år* (Dano-Norwegian—that year); *dette ar* (Norwegian—this year)

dA der Ältere (German—senior); *dette Aar* (Danish—this year)

Da Danish; Danmark

Dᵃ Doña (Spanish—lady, woman of rank)

DA Daughters of America; Defense Aid; Dental Apprentice; Department of Agriculture; Department of the Army; direct action (DA as a noun, d-a as an adjective); District Attorney; Division Artillery; does not affect; Dominion Atlantic (railroad); Dragon Airways; drift angle (symbol)

D-A Devin-Adair

D.A. Diploma in Aesthetics; Diploma in Anesthetics; Doctor of Arts

D/A digital-to-analog

D of A Defenders of Animals; Department of Agriculture

DA Dalniya Aviatsiya (Russian—Long-Range Aviation); *Dissertation Abstracts*

daa data access arrangement; direct access arrangement

DAA Danish Atlantic Association; Dental Assistants Association; Department of Aboriginal Affairs (Australian); Diploma of the Advertising Association; Direct Action Associates

DAA Défense Anti-Aérienne (French—Anti-Aircraft Defense)

D/AA Days After Acceptance; Documents Against Acceptance

DAACA Department of the Army Allocation Committee—Ammunition

DAAG Deputy Assistant Adjutant General

DAA & QMG Deputy Assistant Adjutant and Quartermaster General

dab. daily audience barometer; delayed-action bomb; dimethylaminoazobenzene

DAB Daytona Beach, Florida (airport)

DAB Deutsches Apothekerbuch (German Pharmacopoeia); *Dictionary of American Biography*

dabco diazabicyclooctane

DABPN Diplomate American Board of Psychiatry and Neurology

DABS Discrete Address Beacon System

DABSIPCS Discrete Address Beacon System with Intermittent Positive Control System

dac data acquisition and control; data assistance and control; deductible average clause;

digital-to-analog converter; digital arithmetic center; direct air cycle; dynamic amplitude control

Dac Dacca

DAC Daughters of the American Colonists; Debt Advisory Commission; Defenders of the American Constitution; Department of Adult Corrections (Alaska); Douglas Aircraft Company; Durex Abrasives Corporation

daca (DACA) diphenylaminochloroarsine

DACA Drug Abuse Control Amendments

DACAN Douglas Aircraft Company of Canada

dacbu data acquisition and control buffer unit

D.Acc. Doctor of Accountancy

DACC Dangerous Air Cargoes Committee

DACCC Defense Area Communications Control Center

DACCEUR Defense Area Communications Control Center Europe (NATO)

dachs dachsbracke (Swedish basset); dachshund

dachsaterr dachshund-terrier mixed-breed dog

dacks slacks (sport pants) made of dacron

DACL Depression Adjective Check List(s)

DACO Douglas Aircraft Corporation Overseas

DACOM Datascope Computer Output Microfilmer

dacon digital to analog converter

d/a converter device converting digital input to analog input

dacor data correction

DACOS Deputy Assistant Chief of Staff

DACOWITS Defense Advisory Committee on Women in the Services

dacr dacron (synthetic fiber)

DACRP Department of the Army Communication Resources Plan

DACs Department of the Army Civilians; Desegregation Assistance Centers

DACS Data Acquisition and Correction System

dact dactyl(ic); dactylology; dactylus; dissimilar air combat training

dacty dactylography; dactyloscopy

dactygram dactylogram (fingerprint)

dactyl dactylogic(al)(ly); dactylologist(ic); dactylology

dacum designing a curriculum

dad daddy (father); digital audio definition

dad. design-approval data; dispense as directed; double-acting door

Dad Daddy; Dadiangas

DAD Directorate of Armament Development; Double Atmospheric Density (rocket); Drop A Dime (anti-drug program)

DADEE (Dynamic-Analog Differential-Equation Equalizer

DADIT Daystrom Analog-to-Digital Integrating Translator

D.Adm. Doctor of Administration

dads (DADS) dual air density satellite

DADS Director Army Dental Service

dadsm direct-access device (for) space management

D.Ae. Doctor of Aeronautics

DAE Diploma in Advanced Engineering; Director of Army Education; Division of Adult Education

DAE Dictionary of American English

daea dimethyl aminoethyl acetate

DAEC Danish Atomic Energy Commission

DAEDARC Department of the Army Equipment Data Review Committee

D.Ae.Eng. Doctor of Aeronautical Engineering

daemon data-adaptive evaluator and monitor

DAEP Division of Atomic Energy Production

DAER Department of Aeronautical Engineering Research

D.Ae.Sc. Doctor of Aeronautical Science

daf delayed auditory feedback; described as follows; discharge afloat

DaF Delivered at Frontier

DAF Danish Air Force; Department of the Air Force; Dutch Air Force

DAF Dansk Arbejdsgiverforening (Danish Employers Confederation); *van Doorne Auto Fabriek* (Dutch—van Doorne's Auto Factory), autos and trucks

dafa data accounting flow assessment

DAFA Data Accounting Flow Assessment

dafc digital automatic frequency control

DAFCCS Department of the Air Force Command and Control System

DAFFO Dansk Forening til Fremme af Opfindelser (Danish Society for Encouraging Inventions)

daffs daffodils

daff(y) daffodil

DAFIE Directorate for Armed Forces Information and Education

dafm discard-at-failure maintenance

DAFM Department of the Army Field Manuals

DAFO Division Accounting and Finance Office

DAFS Department of Agriculture and Fisheries (Scotland); Duty Air Force Specialty

DAFSC Duty Air Force Specialty Code

DAFSO Department of the Air Force Special Order

daft. digital/analog function table

dag aquadag; decagram; dysprosium aluminum garnet

Dag Dagestan(i); Dag Hammarskjöld; Dagmar; Dagna

Dag Dagbladet (Oslo's Daily Blade)

D.Ag. Doctor of Agriculture

DAG Deputy Adjutant General; Deputy Attorney General

DAG Deutsche Angestellten-Gewerkschaft (German Employees Union)

dagc delayed automatic gain control

dagl daglig (Dano-Norwegian—daily); *dagligdags* (Dano-Norwegian—ordinary)

dagmar defining advertising goals for measured advertising results; drift-and-ground-speed-measuring radar

Dagmar Dagmar Godowsky

Dag Nyh Dagens Nyheter (Sweden's Daily News)

D.Agr. Doctor of Agriculture

D.Agr.Eng. Doctor of Agricultural Engineering

D. Agr.Sc. Doctor of Agricultural Science

DAGS Department of Accounting and General Ser-

vices

dah disordered action of the heart

Dah Dahomey

DAH disordered action of the heart

Dahlaks Dahlak Islands in the Red Sea off Eritrea, Ethiopia

dai (DAI) death from accidental injuries

Dai David

DAI Dayton Art Institute; Drug Abuse Information

DAI Dissertation Abstracts International

DAIA Department of Aboriginal and Island Affairs (Australian)

DAICS Data-Acquisition and Instrument-Control System

daigc direct-aqueous-injection gas chromatography

DAIM Data Analysis Information Memo

DAIMC Defense Advanced Inventory Management Course (USA)

DAIMS Department of the Army Integrated Material Support (USA)

DAIR Driver Aid Information and Routing (System)

DAIRE Delaware Application of Information and Research in Education

DAIS Defense Automatic Integrated Switching System

DAISY Data Acquisition and Interpretation System; Decision-Aiding Information System; Displacement-Automated Integrated System

DAISY-201 Double-Precision Automatic Interpretive System

DAJAG Deputy Assistant Judge Advocate General

Dak Dakota; Dakotan

Dakoming Dakota + Wyoming

Dakotas North and South Dakota

Dak Ter Dakota Territory

Dak Zoo Dakota Zoo (Bismarck, North Dakota)

dal decaliter

d'AL d'Amico Line

Dal Dallas; Dalmatia; Dalmatian

DAL Dallas, Texas (Love Field); Delta Air Lines; Department of Agriculture Library; *Deutsche Afrika Linien* (German Africa Line)

dala delta-amino-levulinic acid

dalapon dialphapropionic acid (herbicide)

Dalarna Swedish truncation of Dalecarlia, the lake district of folklore

Dalarna (Swedish—Dalecarlia), derived from *Dalkarl*—Man from Dalarna—Sweden's agricultural valley area

dalasi monetary unit of Gambia

DALATS Data Logging and Transmission System

DALE Drug Abuse Law Enforcement

d'Alembert Jean Le Rond d'Alembert

dalgt daylight

Dalh Dalhousie

Dall Dallas' Reports––U.S. Supreme Court

dalpo do all possible

dalr dry adiabatic lapse rate

DALRLV Department of the Army Logistics Readiness Liaison Visit(s)

DALRTF Department of the Army Long-Range Technological Forecast(ing)

dal s dal segno (Italian—from the sign)

DALS Distress Alerting and Locating System

dal seg dal segno (Italian—from the sign)

Dal Sym Orch Dallas Symphony Orchestra

dalvp delay enroute authorized chargeable as ordinary leave provided it does not interfere with reporting on date specified and provided individual has sufficient accrued leave

dam. damage; decameter; degraded amyloid; diacetyl monoxime; divided and mashed

dam. (DAM) direct-access method; down-range anti-missile (program)

Dam Damascus; Damman

DAM Damascus, Syria (airport); Dayton Art Museum; Denver Art Museum

DAM Dictionary of Abbreviations in Medicine

DAMCO Dampier Mining Company

dame. data acquisition and monitoring equipment; digital automatic-measuring equipment

Dame Kiri Dame Kiri Te Kanawa

DAMIS Department of the Army Management Information System

DAMOS Data Moving System

DAMP Down-Range Anti-

Missile Measurement Project
Dampiers Dampier Islands in the Indian Ocean off northwestern Westralia
DAMPS Data Acquisition Multiprogramming System
DAMRIP Department of the Army Management Review and Improvement Program
DAMS Defense Against Missiles System; Deputy Assistant Military Secretary
DAMT Draw-A-Man Test
DAMWO Department of the Army Modification Work Order
dan dekanewton
dän *dänisch* (German—Danish)
Dan Daniel (name); Daniel, Book of; Danish; Danmark (Denmark)
Dan Daniel
DAN Dan-Air Service
DANA Diffraction Analysis System
DANBIF *Danske Boghandleres Importrfrening* (Danish Booksellers Importation Association)
dancin' dancing
DANCOM Danube Commission (Austria, Bulgaria, Czechoslovakia, Hungary, Romania, the USSR, Yugoslavia)
Dand(ie) Andrew
Dandy Andrew
Danglish Danish-English
dang mod dangling modifier
Dani Daniel
Danl Daniel
Danl W Daniel Webster
Danm Danmark (Denmark)
Dannebrog (Danish—Danish cloth)—Denmark's flag reputed to be the oldest national symbol in western Europe
Dannemora Clinton Correctional Facility (for males) at Dannemora, New York
Dan-Nor Dano-Norwegian
Danny Daniel
DANR Department of Agriculture and Natural Resources
dans dansyl chloride (fluorescent dye)
Dansker Dane; Danish sailor
DANTES Defense Activity for Non-Traditional Education Support (USN)
DANZ Dyslexia Association of New Zealand
dao duly-authorized officer, paldao (Philippine wood)
DAO Dairy Advisory Office(r); District Accounting

Office(r); District Aviation Office(r); Division Air Office(r); Division Ammunition Office(r); Dominion Astrophysical Observatory (Victoria, British Columbia)
DAOT Director of Air Organization and Training
dap data analysis package; data automation proposal; digital audio processor; direct-agglutination pregnancy (test); do anything possible; documents against payment
dap (DAP) diaminopimelic acid; diammonium phosphate; dihydroxyacetone phosphate
d-a-p draw-a-person (psychological test)
DAP Democratic Action Party; Development Academy of the Philippines; Director of Administrative Planning; Division of Air Pollution (US Public Health)
DAPA Drug Abuse Programs of America
DAPD Directorate of Aircraft Production Development
DAP & E Diploma in Applied Parasitology and Entomology
dapi (DAPI) diamidinophenylindole
DA Plan Deposit Administration Plan
DAPM Deputy Assistant Provost Marshal
DAPMC Defense Advanced Procurement Management Course (USA)
dapon diallyl phthalate resin
Da Ponte Lorenzo Da Ponte (Mozart's librettist, born Emanuèle Conegliano)
DAPP Data Acquisition and Processing Program
D.App.Sci. Doctor of Applied Science
dapr digital automatic pattern recognition
daps downed airman power source (USN)
DAPS Direct-Access Programming System
dapsone diaminodiphenyl sulfone
dapt daptazole; direct-agglutination pregnancy test (DAPT)
DAPT Direct Latex Agglutination Pregnancy Test; Draw-a-Person Test
DAQMG Deputy Assistant Quartermaster General
dar damned average raiser

(good student whose high marks raise the grading scale); dressed all 'round
Dar Dar-es-Salaam
Dar *Dar-es-Salaam* (Arabic—There is the Peace)—capital and seaport of Tanzania
DAR Daily Activity Report; Data Automation Request; Data Automation Requirements; Daughters of the American Revolution; Dead Animal Removal; Department of Animal Regulation; Directorate of Atomic Research; Dominion Atlantic Railway
DARAS Direction and Range Acquisition System
Darb Darby(shire)
D & ARC Drug and Alcohol Resource Center
DARCEE Demonstration and Research Center for Early Education (Peabody College)
D.Arch. Doctor of Architecture
D.Arch.E. Doctor of Architectural Engineering
DARCOM Development and Readiness Command (USA)
dard data acquisition requirements document
DARD Directorate of Aircraft Research and Development
Dardan Dardanelles
Dardanelles Dardanelle Straits called Hellespont by the Greeks and Canakkale Bogazi by the Turks; link the Aegean Sea with the Sea of Marmara, the Bosporus, and the Black Sea
dare. data automatic reduction equipment; data automation research and experimentation; destination arrival research engineering; documentation automated retrieval equipment; doppler automatic reduction equipment
DARE Detroit Association for Rational Enquiry; Drug Abuse Research and Education (UCLA's neuropsychiatric institute); Drug Abuse Resistance Education; Drug Assistance, Rehabilitation, and Education; Drug Awareness Resistance Education
DARE *Dictionary of American Regional English*
daren't dare not
DARES Data Analysis and Reduction System

DARF Defense Atomic Research Facility
Darlings short form for the Darling Ranges of Westralia
darms digital alternate representation of music scores
DARPA Defense Advanced Research Projects Agency
DARR Department of the Army Regional Representative; Drawing and Assembly Release Record
Darren Darwen, England
dars differential absorption remote sensing (laser)
DARs Design Assist Reports; Development Appraisal Reports
DARS Digital Adaptive Recording System
darss diode-array rapid-scan spectrometer
dart. datagraphic automated retrieval technique(s); deployable automatic relay terminal; detection, action, and response technique(s); development advanced rate techniques; disappearing automatic retaliation target
DART Dallas Area Rapid Transit; Direct Access to Regional Transit; Disaster Assisted Radio Teams; Dublin Area Rapid Transit
Dartmouth Darmouth College at Hanover, New Hampshire; Massachusetts fishing port near New Bedford; Nova Scotia port and rail terminus across Halifax harbor from Halifax; Royal Naval College at Dartmouth near Plymouth on the English Channel
DARTS DoE Audit Report Tracking System; Dynamically-Actuated Road Transit System
Darw Darwin College, Cambridge
Darwin formerly Port Darwin, Australia
Darwin Pr Darwin Press
das data analysis station; dekastere; delivered alongside ship; dial-assistance switchboard
das (DAS) dextroamphetamine sulfate (stimulant)
da's domestic afflictions (menses)
DAs Design Assist Reports
DAS Data Acquisition System; Data Analysis System; Dean A Stahl; Defense Atomic Support Agency; Defense

Audit Service; Digital Analog Simulator; Digital Attenuator System; Director(ate) of Administrative Services; Director(ate) of Aerodrome Standards; Distributed Authorization System; Division of Apprenticeship Standards
DAS *Departamento Administrativo de Seguridad* (Spanish—Security Administration Department); *Dictionary of American Slang*
dasa dual aerospace servoamplifier
DASA Defense Atomic Support Agency
DASA-TP Defense Atomic Support Agency—Technical Publication(s)
DASC Defense Automotive Supply Center; Direct Air Support Center
D.A.Sci. Doctor of Agricultural Science
dasd direct access storage device
DASD Direct-Access Storage Device; Director of Army Staff Duties
DASDL Data and Structure Definition Language
dash. dashboard; drone antisubmarine helicopter
Dash Dashiell Hammett
DASH Delta Airlines Special Handling (of small packages); Drug Abuse Services of Hawaii
dasi diffusion of arsenic in silicon
dasm (DASM) delayed-action space missile
DASNET Data Switching Network
daso (DASO) development and shakedown operations
dasp discrimination among sound patterns; double-arm magnetic spectrometer
DASP Director(ate) of Advanced Systems Planning
dass defined antigen substrate sphere
DASS Direct Air Support Squadron (USAF)
DASSR Dagestan Autonomous Soviet Socialist Republic
DAST Division for Advanced Systems Technology
DAST *Detective-Agents-Science Fiction-Thriller*
dastard. destroyer anti-submarine transportable array detector
DASTL Defense Atomic Sup-

port Agency Technical Letters
dat date after tomorrow; dative; datum; delayed-action tablet; differential agglutination titer
dAt (DAT) dementia of the Alzheimer type
dat (DAT) differential aptitude test(ing); digital audio tape
d a t diet as tolerated
DAT Dental Aptitude Test; Development Acceptance Test (USA); Differential Aptitude Test; Docking Alignment Test (NASA)
DATA Data Acquisition and Technical Analysis; Defense Air Transportation Administration; Development and Technical Assistance (UN); Dial-a-Teacher Assistance (telephone-service program); Draughtsmen's and Allied Technicians' Association
datac digital automatic tester and classifier
DATAC Development Areas Treasury Advisory Committee
datacom data communications
datacor data correction; data correlator
datan data analysis
datanet data network
datap data transmission and processing
datar digital automatic tracking and ranging
datastor data storage
DATC Developmental and Training Center
datda (DATDA) diallytartardiamide
DATDC Data Analysis and Technic Development Center
datel data + telecommunication
datico digital automatic tape intelligence checkout
datin data inserter
DATM *Department of the Army Technical Manual*
DATO Disbursing and Transportation Office; Discover America Travel Organizations
datom data aids for training, operations, and maintenance
dator digital (data), auxiliary (storage), track (display), outputs (and) radar (display)
DATOR Data Operational Requirements Board (NATO)
datran data transmission
datrec data recording

datrix direct access to reference information

dats data accumulation/transmittal sheet

DATS Dynamic Accuracy Test System

DATSC Department of the Army Training and Support Committee

dau data adapter unit

D.Au.Eng. Doctor of Automobile Engineering

dau(s) daughter(s)

DAUS Despatch Agency of the United States

dav data above voice

dav davaerende (Dano-Norwegian—then)

Dav David

DAV Dayanana Anglo Veolic (Fiji school); Disabled American Veterans

DAVA Defense Audiovisual Agency

davc delayed automatic-volume control

Dave David

Davey David; General David C. Jones

DAVI Department of Audio-Visual Instruction (National Education Association)

davidite uranium ferric ferrous iron titanate

Davie David

da Vinci Leonardo da Vinci (Rome, Italy's airport)

Davis Mountains West Texas range of Rocky Mountain system running south through Big Bend National Park into Mexico; named after Jefferson Davis

D.Av.Med. Diploma in Aviation Medicine

DAVNO Division Aviation Office(r)

DAVRS Director of Army Veterinary and Remount Services

Davy David

DAW Directorate of Atomic Warfare

Daw al-Bah Dawait al-Bahrayn (Arabic—State of Bahrain)

DAWE Daughters Already Well-Endowed

dawid device for automatic word identification and discrimination

DAWN Drug Abuse Warning Network

Daw Qat Dawlet al-Qatar (Arabic—State of Qatar)

DAWS Director of Army Welfare Services

dax dachsund

Day Birthday; Dayton; Daytona; Natal Day; President's Birthday; President's Day (holiday)

DAY Dayton, Ohio (airport)

Day Phil Dayton Philharmonic

daysoap(s) daytime (tv) soap opera(s)

db damned bad; day book; dead body; decibel(s); dextran blue; diameter baudelocque (external pelvic conjugate diameter); diode block; disability; distobuccal; distribution board; distribution box; domestic boiler; double bass; double bayonet-base (lamp); double bed; double braid(ed); double breasted; double-biased (relay); down(ward) bound; draw bar; drop by for a few minutes; dry bulb; dynamic brake

db (DB) delayed broadcast

d/b documentary bill

d & b dead and buried

d in b (found) dead in bed

dB decibel

Db dubhium (ytterbium symbol)

DB Data Bank; Date of Birth; David Brown (tractors); Daytona Beach; Declining Balance; Defined Benefit retirement plan; Detective Bureau; Disciplinary Barracks; Dispersal Base; Dodge Brothers; Dominion Breweries

D-B Daimler-Benz

D & B Dun & Bradstreet

D of B Daughters of Bilitis

DB Danmarks Biblioteksforening (Danish Library Association); *Danske Bank* (Danish Bank); *Deutsche Bank* (German Bank); *Deutsche Bundesbahn* (German State Railways); *Dresdner Bank* (German—Dresden Bank)

D.B. Divinitatis Baccalaureus (Latin—Bachelor of Divinity)

dba design-basis accident; doing business as/at

dba (DBA) Dibenzanthracene; dihydro-dimethyl-benzopyran butyric acid

d b a doing business as

dBa decibel A

Dba Dubai

DBA Data Base Administrator; Duke Bar Association

D.B.A. Doctor of Business Administration

dbacc debit account(ing)

DBAE Discipline-Based Art Education

dbam data-base-access method(ology)

DBAP Darien Book Aid Plan

d/bar draw bar

DBAS Development Bank of American Samoa

DBAT Dating Behavior Assessment Test

dbb deals, battens, boards; detector back bias; dinner, bed, breakfast; distance between bends

db & b deals, boards, and battens

dbbal debit balance

dbbd (DBBD) dibromopolybutadiene

dbc diameter bolt circle; dry breast care; dye-binding capacity

DBC Demerara Bauxite Company; Detective Book Club; Drums and Bugle Corps

D.B.C. Doctor of Beauty Culture

DBCA Du Bois Clubs of America

dbcl dilute blood clot lysis

DBCM De Beers Consolidated Mines

DBCO Dairy Board Consulting Office(r)

dbcp (DBCP) dibromochloropropane

DBCSO DeBeers Central Selling Organisation

dbcu data bus control unit

dbd death by drugs (execution by lethal injection); doublebase diode

D Bd Distribution Board; Drug Board

dbe design-basis earthquake; design-basis event; double-bell euphonium (marching band tuba)

dbe (DBE) dibasic ester

D.B.E. Dame Commander of the Order of the British Empire

dbeats despatch payable at both ends (for) all time saved

dbed (DBED) dibenzyl-ethylene-diamine (penicillin)

D.B.Ed. Doctor of Business Education

dbelts despatch payable both ends (for) all laytime saved

dbf design-basis fire; design-basis flood

dBf decibel femtowatt; divorced Black female

DBG Division of Basic Grants
dbh diameter breast high
DBHNT Detective Bureau Hostage Negotiating Team (NYPD)
dbhp drawbar horsepower
dbi database index; development-at-birth index (DBI)
Dbi Dubai
DBib Douay Bible
D.Bi.Chem. Doctor of Biological Chemistry
D.Bi.Eng. Doctor of Biological Engineering
D.Bi.Phy. Doctor of Biological Physics
dbir double built-in (ward)-robes
D.Bi.Sc. Doctor of Biological Sciences
DBIU Dominion Board of Insurance Underwriters
DBJC Daytona Beach Junior College
dbk debark; drawback
DBK Daiichi Bussan Kaisha (Japanese steamship line); Dobeckmun (company)
dbkn debarkation
dbl double; doubler
DBL Disability Benefit Law; Displaced Business Loan (SBA)
dbl act. double acting
dblb double room with bath
dbl cnt double contract
dbl eleph fol. double elephant folio—books about 50 inches high
dblr doubler
dbls double room with shower
dbm decibels per milliwatt; diabetic management
dBm divorced Black male; decibel referred to one milliwatt
DBM Division of Biology and Medicine (Atomic Energy Commission)
D.B.M. Diploma in Business Management
DBM *Deutches Bundes Marine* (German Federal Navy)
db meter decibel meter
DBMS Data Base Management System; Director of Base Medical Services
Dbn Durban
dbo dead blackout; disassembled by owner; distobucco-occlusal; dreadful body odor
DBOS Data-Based Operating System
D-box distribution box
dbp design-basis probability; diastolic blood pressure; dis-

tobuccopulpal; drawbar pull
DBP Development Bank of the Philippines; Division of Beaches and Parks
DBP Dicionário Bibliografico Portugues (Portuguese Bibliographic Dictionary)
db part double-beaded partition
DBPO Data Buoy Project Office
dbr double book rack
D Br Defendant's Brief
DBR Division of Building Research
dbrap decibels above reference acoustic power
dbre diciembre (Spanish—December)
DBRL DeBeers Research Laboratory
dbrn data bank release notice; decibels above reference noise
DBRS Dominion Bond Rating Service (Toronto)
db rts debenture rights
dbs damn bloody soon; despeciated bovine serum; double bottoms
dbs (DBS) direct broadcast satellite
db's dirty books; dune buggies
DBS Development Bank of Singapore; Direct Broadcast Satellite; Distressed British Seaman (provided free passage home); Division of Biological Standards
dbsm decibels per square meter
dbsr double bed sitting room
dbst double bituminous surface treatment
DBST Double British Summer Time
dbt debit; design-basis tornado; dry-bulb temperature
DBT David Brown Tractors
dbtd debited
dbtfl doubtful
dbtg debiting
dbtl doubtful
dbtt ductile-brittle transmission temperature
dbtu (DBTU) dibutylthiourea
dbtw design-basis tornado and windstorm
dbuf dry-buffed (leather)
d-bug debug; debugged; debugging
dbur data bank update request
dbv decibel referred to 1 volt
DBV Deutscher Bibliotheksverband (German—Library Association); *Deutscher*

Bund für Vogelschutz (German Bird-shooters Bund)
dbw differential ballistic wind
dbx design-basis explosion
dc data collection; dead center; death cell; deck cargo; deposited carbon; deviation clause; device control; diagonal conjugate; diesel car; differential calculus; digital computer; direct credit; direct cycle; directional coupler; disorderly conduct; door closer; double cap; double certificated; double column; double contact; double crochet; double crown; down center; draft card; drawing change; drift correction; drill collar (oil well)
dc (DC) cancrizans of the duration series; debit card(s); diagonal conjugate
d-c direct-chill (casting); direct-current (adjective)
d/c deviation clause; double-column (bookkeeping)
d & c dilation and curettage
dc da capo (Italian—again); Dick Cavett
d/c dinero contante (Spanish—cash)
dC depois de Cristo (Portuguese—after Christ); *dopo Cristo* (Italian—after the birth of Christ)
DC Dana College; Dartmouth College; Davidson College; Death Certificate; decimal classification; Defiance College; Dental Corporation; Dental Corps; Department of Commerce; Deputy Chief; Deputy Commissioner; Deputy Consul; Deviation Clause; Diagnostic Center; Dickinson College; Diners Club; Dining Car; direct current (when used as a noun); District of Columbia (D.C.); District Commissioner; District Court(house); Doane College; Doctor of Chiropractic; Dominican College; Donnelly College; Dordt College; Douglas Commercial (aircraft); Downing College (Cambridge); Drury College; Duchesne College; Dumbarton College; Dyke College; D'Youville College
D-C Denver-Chicago (truck line); Dow-Corning (chemical products)
D/C drift correction
D & C Dean and Chapter; De-

troit and Cleveland (steamship line); Doctrine and Covenants

D of C Daughters of the Confederacy; Department of Commerce; Department of Communications (DoC); Department of Correction(s); District of Columbia (D.C.); Duchy (Duke) of Carinthia; Duchy (Duke) of Carniola

DC *Democrazia Christiana* (Italian—Christian Democracy)—political party; *Distrito Capital* (Spanish—Capital District)

D C *da capo* (Italian—again)

DC-1 Defense Condition-1 (war)

DC1, DC2, DC3, etc. device-control characters (data processing)

DC-2 through DC-5 Defense Condition-2 through Defense Condition-5 (stages of military alert short of war)

DC-3 Douglas 21-passenger twin-engine transport aircraft also known as the C-47, Dakota, or Skytrain

DC-4 Douglas 44-passenger four-engine transport aircraft also called C-54 or Skymaster

DC-6 Douglas 64 to 92-passenger transport also known as C-118 Liftmaster because of its cargo-carrying capacity

DC-8 Douglas DC8 jet airplane

DC-9 Douglas twin-jet short-range airplane

DC-9 Super 80 McDonnell-Douglas commercial jetliner billed as the world's quietest

DC-10 McDonnel-Douglas jumbo jetliner

dca deoxycholate citrate sugar; dollar cost averaging

Dca Dacca

DCA Dachshund Club of America; Dalmatian Club of America; Damage Control Assistant; Defence Costs Agreement (Hong Kong); Defense Communications Agency; Department of Civil Aviation; Department of Consumer Affairs; desoxycorticosterone acetate; Diamond Council of America; Diapulse Corporation of America; Digital Computers Association; Director of Civil Aviation; Disassembly Compliance and Analysis; Disc

Company of America; Distribution Contractors Association; District Court of Appeals; Division of Consumer Affairs; Document Content Architecture; Drafting Contractors Association; Drawing Change Authority; Drug Control Agency; Dynamics Corporation of America; Washington, D.C. (national airport)

DCA *Défense Contre Aéronefs* (French—anti-aircraft defense)

DCAA Defense Contract Audit Agency

DCA/A Disassembly Compliance and Analysis/Abbreviated

DCADA District of Columbia Alley Dwelling Authority

D.C.Ae. Diploma of the College of Aeronautics

DCAEUR Defense Communications Agency, Europe

DCAF Design Corrective Action Form

DCAO Deputy County Advisory Officer

DCAOC Defense Communications Agency Operations Center

d cap double foolscap (paper)

DCAP Disadvantaged Country Areas Program

DCAR Defense Contract Administration Services Region; Design Corrective Action Report; Disassembly Compliance and Analysis Report

DCAS Data Collection and Analysis System; Defense Contract Administration Services; Defense Control Administration Services; Deputy Chief of Air Staff

DCASR Defense Contract Administrative Service Region

DC-AST McDonnell Douglas Advanced Supersonic Transport

DCATA Drug, Chemical, and Allied Trades Association

D Cath *Documentation Catholique* (French—Catholic Documentation)

dcb data control block

DCB Decimal Currency Board (British)

D.C.B. Dame Commander of the Most Honourable Order of the Bath

DCBD Division for Children with Behavioral Disorders (Council for Exceptional

Children)

DCBRE Defense Chemical, Biological, and Radiation Establishment

dcc dark curtain closed; double concave; double cotton covered

dcc (DCC) decade counter code; deleted from colorectal carcinoma

DCC Damage Control Center; Day Care Center; Defense Concessions Committee; Deputy Chief Constable; Design Change Control; Disease Control Center; Dutchess Community College

DCCA Design Change Cost Analysis

DCCB Defense Center Control Building (USA)

DCCC Democratic Congressional Campaign Committee; Domestic Coal Consumers Council

DCCDCA Day Care and Child Development Council of America

d & c color drug and cosmetic color (synthetic dye)

DCCP Design Change Control Program; Directorate of Communication Components Production

DCCS Digital Command Communications System

dccu data communications control unit

dcd differential current density

DCD Daitch Crystal Dairies; Damage Control Diagram(s) (USN); Design Change Document; Directorate of Civil Disturbance

D.C.D. Diploma in Chest Diseases

DCD *Dansk Central för Dukumentation* (Danish Center for Documentation)

DCDMA Diamond Core Drill Manufacturers Association

DCDPO Directorate for Civil Disturbance Planning and Operations (USA)

dcdr decoder

dcds double cotton double silk

DCDS Digital-Control Design System

dcdt direct-current differential transformer

dce dairy cow equivalent; data conversion equipment; differential compound engine; domestic credit expansion

dce (DCE) data-communications equipment

DCE Division of Career Education; Division of Compensatory Education

D.C.E. Doctor of Civil Engineering

DCEA *Dictionary of Civil Engineering Abbreviations*

dcel direct-current electroluminescence

D.C.E.P. Diploma of Child and Educational Psychology

dcf deal-cased frame; direct centrifugal flotation; discounted cash flow

DCF Deputy Chief; Donner Canadian Foundation

dcfem dynamic crossed-field electron multiplication

dcfp dynamic cross-field photo-multiplier

dcg dancing; decigram; displacement cardiograph; dynamic cardiogram

dcg (DCG) deoxycorticosterone glucoside

dcgm decorticated groundnut meal

DCGS Deputy Chief of the General Staff

dch dicyclohexyl

D. Ch. *Doctor Chirugiae* (Latin—Doctor of Surgery)

DCH Diploma in Child Health

dcha *derecha* (Spanish—right)

DCHCL Dropsie College for Hebrew and Cognate Learning

D.Ch.E. Doctor of Chemical Engineering

dchn dicyclohexylamine nitrate

DChO Diploma in Ophthalmic Surgery

dci dichloroisoprenaline; dichloroisoproterenol; double-column inch; driving car intoxicated

DCI Department of Citizenship and Immigration; Des Moines and Central Iowa (railway); Director of Central Intelligence

DCIC Defense Ceramic Information Center

dcid decide

DCID Department of Commercial and Industrial Development

DCIGS Deputy Chief of the Imperial General Staff

DCII Defense Central Index of Information

dcisn decision

D.Civ.L. Doctor of Civil Law

DCJ Dade County Jail (Miami, Florida); Department of Criminal Justice; District Court Judge

DCJC Dawson County Junior College

dckng docking

dcl decaliter; declaration; declarative; decline

DCL Dartmouth College Library; Detroit College of Law; Deuterium of Canada, Limited; Distillers Company Limited

D.C.L. Doctor of Canon Law; Doctor of Civil Law

DCLA Deputy Chief of Staff, Logistics and Administration (NATO)

dcld declined

DCLE Department of Criminal Law Enforcement (Florida)

dclg declining

DCLI Duke of Cornwall's Light Infantry

dclrt decelerate

dcls deoxycholate citrate lactose saccharose (agar); disclose

D.Cl.Sci. Doctor of Clinical Science

dclsd disclosed

dclsg disclosing

dclsr disclosure

DCLTC Dry Cargo Loading Technical Committee (NATO)

dcltr declines transfer (offered)

DCLU Developing Countries Liaison Unit

dcm decameter; defense combat maneuvers

DCM Director of Civilian Marksmanship; Directorate of Classified Management; Distinguished Conduct Medal; District Court Martial; Dominican Campaign Medal

D.C.M. Doctor of Comparative Medicine

DCMA Defense Contract Management Association; District of Columbia Manpower Administration; Dry Color Manufacturers Association

DCMBA Dairy, Confectionery, and Mixed Business Association

D.C.M.G. Dame Commander of the Order of St Michael and St George

dcmi disclosure of classified military information

dcmps degaussing compass

dcmptr degaussing computer

DCMs Deputy Chiefs of Missions

DCMS Deputy Commissioner of Medical Services

dcmsn decommission

dcmt document

dcmu dichlorophenyldimethylurea

dcn delayed conditioned necrosis; double crown

DCN Data Change Notice; Defense Communication Network; Design Change Notice; Drawing Change Notice

DCNI Department of the Chief of Naval Information

D.Cn.L. Doctor of Canon Law

DCNO Deputy Chief of Naval Operations

DCNS Deputy Chief of Naval Staff

dco doppler cutoff; draft collection only

D_co diffusing capacity—carbon monoxide

DCO Dallas Civic Opera; Data Change Order; Deputy Chief of Staff, Operations (NATO); Director of Combat Operations; District Control(ling) Office(r); Dominion, Colonial, and Overseas (Department of Barclays Bank)

DCOBE Dame Commander—Order of the British Empire

DCOG Diploma of the College of Obstetricians and Gynecologists

d & coh daughter and co-heiress

d col double column

D. Com. Doctor of Commerce

D.Com.L. Doctor of Commercial Law

D. Comp. L. Doctor of Comparative Law

dcop displays, controls, and operation procedures

DCOR Defense Committee on Research (USAF)

DCOS Deputy Chief of Staff

dcp dental continuation pay; depot condemnation percent; development cost plan; discrete component parts

dcp (DCP) dicalcium phosphate

DCP Department of Consumer Protection; Diploma in Clinical Pathology; Disaster Control Plan; Division of Consumer Protection

dcpa (DCPA) dicylcopentenyl acrylate

DCPA Defense (Department's) Civil Preparedness Agency

DCPANDP Deputy Chief of Staff, Plans and Policy

(NATO)

DC Path Diploma of the College of Pathologists

dcpd (DCPD) dicyclopentadiene

DCPL District of Columbia Public Library

DCPO Deputy Chief of Staff, Personnel and Organization (NATO)

DCPR Defense Contractors Planning Report; Dry Cleaning Plant Registration

dcp's development concept papers

dcpta (DCPTA) dichlorophenoxy triethylamine

dcr data conversion receiver; decrease; decreasing; direct cortical response; division credit rebate

DCR Design Change Request; Design Characteristic Review; District Chief Ranger; District Court Report(s); Division of Computing Research; Drawing Change Request

DCRB Design Change Review Board

DCRE Deputy Commandant—Royal Engineers

DCRLA District of Columbia Redevelopment Land Agency

DCR (MU) Diploma—College of Radiographers (Medical Ultrasound)

DCR (NM) Diploma—College of Radiographers (Nuclear Medicine)

DCRO Dyers and Cleaners Research Organization

DCRR Drawing Change Recorder Request

DCR (R) Diploma—College of Radiography (Radiography)

DCR (T) Diploma—College of Radiographers (Radiotherapy)

dcs dorsal column stimulator; double cotton silk

DCs Douglas Commercial-type airplanes

DCS Damage Control School (USN); Data Control System; Deaf Community Services; Defense Communications System; Department of Correctional Services (Nebraska, New York); Deputy Chief of Staff; Digital Command System; Direct Coupler System; Director of Community Services; Distillers Corporation—Seagrams; Domestic

Contact Service (CIA)

DC of S Deputy Chief of Staff

D.C.S. Doctor of Christian Science; Doctor of Commercial Science

DCSA Department and Chain Store Association

DCSAB Distinguished Civilian Service Awards Board

DCSADN Defense Communication System Automatic Digital Network

DCSC Defense Construction Supply Center

DCSCD Deputy Chief of Staff for Combat Developments (NATO)

DCSCOMPT Deputy Chief of Staff, Comptroller (NATO)

DCSF Dry Caisson Storage Facility

DCSFOR Deputy Chief of Staff, Force Development (NATO)

DCSL Deputy Chief of Staff, Logistics (NATO); District Cub Scout Leader

DCSM Deputy Chief of Staff, Materiel (NATO)

DCSMIS Deputy Chief of Staff, Management Information System (NATO)

DCSO Deputy Chief Scientific Officer

DCSOI Deputy Chief of Staff for Operations and Intelligence (NATO)

dcsp digital control signal processor

DCS/P Deputy Chief of Staff for Personnel

DCSPA Deputy Chief of Staff, Personnel and Administration (NATO)

DCS/P&O Deputy Chief of Staff for Plans and Operations

DCS/P&R Deputy Chief of Staff for Programs and Resources

dc sr *da capo senza replica* (Italian—from the beginning without repeat); *da capo senza repetizione* (Italian—from the beginning play repeated parts once)

DCS/R&D Deputy Chief of Staff for Research and Development

DCSRDA Deputy Chief of Staff—Research, Development, and Acquisition (USA)

DCSRM Deputy Chief of Staff for Resource Management (NATO)

DCSROTC Deputy Chief of

Staff for Reserve Officers' Training Corps

DCS/S&L Deputy Chief of Staff for Systems and Logistics

DCST Deputy Chief of Supplies and Transport

DCSTS Deputy Chief of Staff for Training and Schools

dct depth-charge thrower; depth-control tank; distal convoluted (kidney) tubule; document(ary); documentation

DCT Department of Commerce and Trade

DCT Division de Contre-Topilleurs (French—Division of Destroyers)

DCTC District of Columbia Teachers College; Dodge County Teachers College

DCTD Diploma in Chest and Tuberculous Diseases

dctl direct-coupled transistor logic

DCTSC Defense Clothing and Textile Supply Center

dcu display and control unit; dynamic checkout unit

dcu (DCU) dichloral urea (herbicide)

dcutl direct-coupled unipolar transistor logic

dcv double cotton varnish

DCVO Dame Commander of the Royal Victorian Order; Deputy Chief Veterinary Officer

dcw dead carcass weight

DCW Detroit Chemical Works

dcwv direct-current working volts

dcx double convex

DCZ District of the Canal Zone

dd days after date; day's date; deadline date; decreased (sexual) desire; deep-drawn; deferred delivery; delayed delivery; delivered; dental discomfort; development directive; developmentally disabled; differential diagnosis; digital display; discharged dead; double draft; drunk driver(s); drydock; due date; dutch door

d-d dumb-dumb; dum-dum

d'd deceased

d/d dated; delivered at dock(s); demand draft; detergent dispersant; domicile to domicile; due date

d...d damned

d & d deaf and dumb; death

and decay; death and dying; defiled and deflowered; diarrhea and dehydration; drinking and drugging; drunk and disorderly; dungeons and dragons (game)

d & d (D & D) decontamination and decommissioning; development and demonstration

d-to-d dawn-to-dusk (daylight patrol); dusk-to-dawn (night patrol)

dd dags dato (Dano-Norwegian—days to date); *direttissimo* (Italian—express train)

dd (DD) direct deposit; double density

d.d. dono dedit (Latin—he gave as a gift)

d(D) dead; died

Dd David; Drydock

DD Deputy Director; destroyer (naval symbol); Detective District; Detective Division; Development Directive; dichloropropane dichloropropylene (insecticide); Dishonorable Discharge; E.I. du Pont de Nemours & Company (stock exchange symbol)

D.D. Doctor of Divinity

D & D Dungeons and Dragons

DD Divinitatis Doctor (Latin—Doctor of Divinity); *Doctores* (Spanish—Doctors); *Dottores* (Italian—Doctors); *Doutores* (Portuguese—Doctors)

DD-2 Second Development Decade (1971-1980)

DD-214 Department of Defense Honorable Discharge (form DD-214)

dda duty deposit account

dda (DDA) digital differential analyzer

DDA Dangerous Drug Act; Deputy Director of Administration (CIA); Diemakers and Diecutters Association; Disabled Drivers Association; Display and Decision Area; Duty Deferment Account

ddalv days delay enroute authorized chargeable as leave

DDAS Digital Data Acquisition System

DDAU Doctoral Dissertations Accepted by American Universities

ddavp (DDAVP) decamino-D arginine vasopressin

D-Day day of attack; Decimalisation Day (Feb 15, 1971

when British money was decimalized)

DDB Double Declining Balance

ddc data documentation costs; decision-difficulty checklist; direct digital control; double-deck car(s)

DDC corvette (naval symbol); Defense Documentation Center; Defensive Driving Course; Dewey Decimal Classification; Diamond Dealers Club; Digital Development Corporation

DDC Docteur en droit canonique (French—Doctor of Canon Law)

ddce digital data-conversion equipment

DDCI Deputy Director of Central Intelligence (CIA)

ddcmp digital-data communication-message protocol

ddc's deck decompression chambers

DDCs Desk and Derrick Club members (petroleum professionals)

ddd darling discipline of the decade (computer science courses in the 1975—1985 era); deadline delivery date; detail data display; digital data distributor; digital display driver; drink, drank, drunk (alcoholic's progress); dynamic dummy director

d.d. in d. de die in diem (Latin—from day to day)

DDD Department of Decentralization and Development; direct distance dialing

DDD dat, dicat, dedicat (Latin—he gives, devotes, dedicates)

ddda decimal digital differential analyzer

dd/dc diamond differential direct current

DDDIC Department of Defense Disease and Injury Code

d & dd's depraved and deprived dropouts (street people characteristic of many great cities)

DDDS Deputy Director of Dental Services

dde direct data entry; dual-displacement engine

dde (DDE) dichlorodiphenyl-dichloroethylene

DDE Designated Destroyer Escort (Canadian); dichlorodiphenyldichloroethylene (in-

secticide less toxic than DDT); Dwight David Eisenhower (34th President U.S.)

D De L Daniel De Leon

D de l'U Docteur de l'Université (French—Doctor of the University of Paris)—the Sorbonne

DDEM Dwight D. Eisenhower Museum

DDEP Defense Development Exchange Program

ddf design disclosure format; double defruit

DDF Dental Documentary Foundation

DDF Departamento del Distrito Federal (Spanish—Federal District Department)

ddg (DDG) digital display generator

DDG guided missile destroyer (naval symbol)

DDGSE Deputy Director General—Signals Equipment

DDGSR Deputy Director General of Signals Equipment

ddh diamond drill hole

DDH Digital Data Handling (system); Diploma in Dental Health

DDHA Detective Division Homicide Assault Squad

ddi depth deviation indicator; digital data indicator; discrete digital input; document disposal indicator

DDI Deputy Director, Intelligence (CIA)

dd-ing double dipping (cheating, milking the government)

ddis data display

DD & J Deacons for Defense and Justice

ddl digital data link

ddl (DDL) data definition language

DDL Data Disclosure List; Det Danske Luftfartsselskab (The Danish Airways)

ddm data demand module; difference in depth of modulation; digital drawing monitoring

DDM Diploma in Dermatological Medicine; Distributed Data Management

DDME Deputy Director of Mechanical Engineering

DDMI Deputy Director of Military Intelligence

DDMOI Deputy Director of Military Operations and Intelligence

DDMS Deputy Director of Medical Services

DDMT Deputy Director of Military Training
Ddn Dunedin, NZ
DDN Defense Data Network; nuclear-powered destroyer (naval symbol)
ddnc direct digital numerical controller
DDNI Deputy Director of Naval Intelligence
ddo despatch money payable (for) discharging only
DDO David Dunlap Observatory (Ontario); Deputy Director of Operations (CIA)
D.D.O. Diploma in Dental Orthopedics
D-dog detector dog (U.S. Customs)
DDOS Deputy Director of Ordinance Services
ddp digital data processor
ddp (DDP) distributed data processing
DDP Data Distribution Point (NATO); Declaration of Design Performance; Delivered Duty Paid; Department of Defence Production (Canadian); Deputy Director, Plans (CIA); Design Development Plan; Devalued Dollar Planning
DDPH Diploma in Dental Public Health
DDPR Deputy Director of Public Relations
DDPS Discrimination Data Processing System
ddr direct debit
DDr Doktor, Doktor (Austrian-German—person with two doctor's degrees)
DDR *Deutsche Demokratische Republik* (German Democratic Republic); radar picket destroyer (3-letter naval symbol)
D.D.R. Diploma in Diagnostic Radiology
DDRA Deputy Director—Royal Artillery
DDRD Deputy Directorate of Research and Development
DD R & D Department of Defense Research and Development
DDRE Danish Defense Research Establishment
DDR&E Defense Development Research and Engineering
DDRM Deputy Director of Repair and Maintenance
ddrr directional discontinuity ring radiator

DDRS Declassified Documents Reference System
dds diaminodiphenylsulfone; digital display scope; digital dynamics simulator
d/d's developer/demonstrators
DDS Deep-Diving System; Demos D-Scale; Department of Developmental Services; Deployable Defense System; Deputy Director of Support (CIA); Dewey Decimal System; Documentation Distribution System; Drug Delivery System
D.D.S. Doctor of Dental Science; Doctor of Dental Surgery
D.D.Sc. Doctor of Dental Science
DDSD Deputy Director of Staff Duties
DDSC Dewey Decimal System of Classification
DDSG *Donau-Dampfschiffahrts-Gesellschaft* (German Danube Steamship Travel Service)
dd & shpg dock dues and shipping
ddso diamino-diphenyl sulphoxide
DDSR Deputy Director of Scientific Research
DDST Denver Developmental Screening Test; Deputy Director of Supplies and Transport; Double Daylight Saving Time (two hours ahead)
DDS & T Deputy Director of Science and Technology (CIA)
DDSTs Denver Developmental Screening Tests
ddt deduct; digital data transceiver; digital data transmitter; digital debugging tape(s); drop dead twice (epithet); ductus deferens tumor; dynamic debugging technique
ddt (DDT) direct-decision therapy
DDT dichlorodiphenyl-trichloro-ethane (insecticide)
ddt & e design, development, test, and evaluation
DDTE Deputy Director, Test and Evaluation (NASA)
DDTF Dynamic Docking Test Facility (NASA)
ddtl dreary desk-top lunch
DDTS Dynamic Docking Test System (NASA)
DDTV Dry Diver Transport Vehicle (naval)
ddu data diagnostic unit; data

display unit; design diagnostic unit; display driver unit; distribution data unit
ddv deck drain valve
ddvp (DDVP) dimethyldichlorovinylphosphate
DDVS Deputy Director of Veterinary Services
ddw displaying a deadly weapon
DDWE&M Deputy Director of Works, Electrical and Mechanical
DDx differential diagnosis
DDY *Devlet Demiryollari* (Turkish Railways)
de deckle edge; deckle edging; deflection error; development engineering; diatomaceous earth; diesel-electric; digestive energy; direct elimination; direct entry; double end; double entry; dream elements; duration of ejection
d/e date of establishment
d & e dilation and evacuation
de *det er* (Norwegian—that is); (Latin prefix—down, from)— descent, description
DE Deere (stock exchange symbol); Delaware; Department of Education; Department of Employment; Department of the Environment; destroyer escort (naval symbol); District Engineer; Dynamite Engineer(ing)
DE *Dáil Éireann* (Irish Gaelic—Assembly of Ireland)— Ireland's House of Representatives
D.E. Doctor of Economics
D of E Department of Energy; Department of the Environment; Department of the Environment (UK)
D of E *Dictionary of Electronics*
dea (DEA) dehydroepiandrosterone
Dea Deacon
DEA Dance Educators of America; Department of External Affairs; Detectives Endowment Association; Digital Equipment Australia; Drug Enforcement Administration
deac deacon; diethylaluminum chloride
DEACONS Direct English Access and Control System
dead. (DEAD) destruction-entrusted automatic devices
Dead Horses Dead Horse Mountains between Mexico and Texas in the Big Bend

Area where it is also called Sierra del Caballo Muerto

DEADS Detroit Air Defense Sector

DEAE Division of Eligibility and Agency Evaluation (Office of Education)

DEAE-cellulose diethylaminoethyl cellulose

deal. decision evaluation and logic

dealer prep dealer preparation

DEAN Deputy Educators Against Narcotics

dear. diamonds, emeralds, amethysts, rubies

dearg. pil. deargentur pilulae (Latin—let the pills be silvered)

DEAS Delaware Educational Accountability System

Death Valley Death Valley National Monument on the border of California and Nevada

DEAUA Diesel Engineers and Users Association

deaur. pil. deaurentur pilulae (Latin—let the pills be gilded)

deb debenture; debit; debut(ante); diethylbutanediol

DEB Dental Examining Board

Debbie Deborah

Deb(by) Deborah

de Bc Honoré de Balzac

debil debilitate(d); debilitating; debilitation; debility

debk debark; debarkation

De Brücke (German—The Bridge)—pessimistic expressionism popular in Germany in the 1920s and 1930s

deb(s) debenture(s); debutante(s)

deb. spis. debita spissitudine (Latin—of the correct consistency)

deb stk debenture stock

dec decant; decanter; deceased; deciduous; decimal; decimeter; decision; declination; decompose(d); decorate; decoration; decorator; decrease(d)

dec decani (Latin—sung by the choir on the deacon's side of the church

dec. décembre (French—December); *décor* (French—decoration, stage scenery); *decubitus* (Latin—lying down)

Dec Decca; December

DEC Deductible Employee Contribution; Dental Exami-

nation Centre; Department of Environmental Conservation; Detroit Edison Company; Developmental Education Center; Digital Electric Corporation; Digital Equipment Corporation; Disaster Emergency Committee; Dominion Executive Council

deca- 10

DECA Distributive Education Club of America

decad decadence; decadency; decadent(ly)

decaf decaffeinated

decal decalcomania

DECAL detection and classification of an acoustic lens

decap(ped) decapitation(ed), behead(ed)

decasyl decasyllable; decasyllabic

decb data event control block

Decca Decca Navigation System

Deccan Deccan Plain of southern India

DECCO Defense Commercial Communications Office

decd deceased

decd est deceased estate

decel deceleration

deci 10^{-1}

decid deciduous

DECIDE Decide the problem precisely, Enumerate two groups of decision factors, Collect relevant information, Identify the best alternatives, Develop and implement a detailed plan, Evaluate the decision

decim decimeter

decis decision

decit decimal digit

decl declension

DECL Direct Energy Conversion Laboratory (NASA)

declon declaration

decm defensive electronic countermeasure

DECMD Detroit Contract Management District

decn decision; decontamination

deco direct energy conversion operation

deco (DECO) decreasing consumption of oxygen

decoct decoction

decomm decommissioning (date)

decomp decomposition

DECOMPS Decomposition Mathematical Programming System

decon decontaminate; decontamination

D. Econ. Doctor of Economics

decor decorate; decoration; decorative

decr decrease; decrement(al)(ly)

Decr Decreto (Italian, Portuguese, Spanish—Decree)

decres decrescendo (Italian—contracting, subsiding)

decrim decriminalization(al)(ly); decriminalize(r)

DECS Direct Evacuation Control System (air filtration)

decsn decision

DEC Station Digital Equipment Corporation Station

dec stories detective stories

DECTRA Decca Track and Range

decu data-exchange control unit

Decuary December and January

decub. decubitus (Latin—lying down)

DECUS Digital Equipment Computer Users Society

ded date expected delivery; dedendum; dedicate; dedicated; deduct; deducted; deduction; diesel engine driven

Ded Dedan; Dedham; Dedric(k)

D. Ed. Doctor of Education

DED Data Element Dictionary (USA)

de d. in d. de die in diem (Latin—from day to day)

dedic dedicate(d)(ly); dedicating; dedication; dedicative; dedicator(y)

dedl data element description list

dedn deduction

deduct. deduction

dee digital events recorder; discrete event evaluator

dee (DEE) diethoxyethylene

DEE Diploma in Electrical (Electronic) Engineering

dee-dee deaf and dumb

Deedee Dorothy

Dee High Doctor of Hygiene

dee jay disc jockey

deeks duck decoys

DEEP Development Economic Education Program; Diffusion of Exemplary Educational Practices; Directly-Elected European Parliament

deep 6 burial at sea; disposing of anything unwanted in at least six fathoms of water

Dee Pee Doctor of Pharmacy

Dee R doctor
dees dynamic electromagnetic environment simulator
Deeside River Dee valley around Aberdeen
deet diethyl toluamide (insecticide)
def defeated; defecate; defecation; defect; defection; defective; defector; defendant; defense; defensive; defer; deferred; deficiency; deficient; define; definite; definition; definitive; deflagrate; deflagrating; deflagration; deflect; deflecting; deflection; defoliate; defoliating; defoliation; defrost; defroster; defrosting; defunct; defunction; defunctive
déf déficit (Spanish—deficit)
def. defunctus (Latin—deceased)
def art. definite article
defcon defense condition; defensive concentration
Def Con-1 Defense Condition-1 (war)
Def Con-2 through Def Con-5 stages of military alert short of war
Def Con(s) Defense Condition(s): Def Con I—war, Def Con II—attack imminent, Def Con III—highest state of readiness for war, Def Con IV—readiness alert, etc.
defec defective
Defense Department of Defense
defi deficiency
defib defibrillate
defic deficiency; deficit
defl deflate; deflation; deflect; deflection
deflor defloration
deform deformity
DEFREPNAMA Defense Representative North Atlantic and Mediterranean
defs definitions
DEFSIP Defense Scientists Immigration Program
deft. defendant; dynamic error free transmission (DEFT)
DEFY Drug Education For Youth
deg degenerate; degeneration; degree(s)
deg (DEG) diethylene glycol
DEG guided-missile escort ship (naval symbol)
D & EG Development and Engineering Group
DEG Derechos Especiales de Giro (Spanish—Special

Drawing Rights)
de ga depth gage
de Gaulle Charles de Gaulle (Paris, France's airport)
degen degeneration
deglut. deglutiatur (Latin—let it be swallowed)
degrad(s) degradable(s)
De Graff John De Graff
degsvc degaussing service(s)
degust degustation (savoring, tasting)
de gustibus de gustibus non est disputandum (Latin—there is no disputing about tastes)
DE-H destroyer escort—hydrofoil
deha (DEHA) diethylhydroxylamine
DEHCD Department of Environment, Housing, and Community Development
DeHoCo Detroit House of Correction
DEHS Division of Emergency Health Services
dei design engineering identification; development engineering inspection; double electrically isolated
DEI Digital Electronics Incorporated; Dutch East Indies
DEIC Diver Equipment Information Center; Dutch East India Company
Deich Bib Deichmanske Bibliotek (Norwegian—Deichman's Library)—Oslo
DEIR Department of Employment and Industrial Relations
deis design engineering inspection simulation; design engineering inspection simulator
DEIS Defense Energy Information System; Draft Environmental Impact Statement
dej dento-enamel junction
Dek Dekabr (Russian—December)
deka 10
dekag dekagram
dekal decaliter
dekam decameter
Deke Deacon; Donald
dekon economic declaration (Indonesian)
del delegate; delegation; delete; deletion; deliberate; deliberation; delineate; delineated; delineation; deliver(y)
del (DEL) delete character (data processing)
del. delineavit (Latin—he or she drew it)

Del Delaware; Delawarean; Delhi; Delphinus
del acct delinquent account
delcap delay capacity
DELCO Dayton Engineering Laboratory Company
deld delivered
dele delete
deleat. deleatur (Latin—delete)
deleg delegation
del ent delete entirely
delg delivering
deli delicatessen
delib deliberate; deliberation
delic delicatamente (Italian—delicately)
deli-market delicatessen and market
DELIMCO German-Liberian Mining Company
delin delineate(d); delineating; delineation; delineative; delineator; delineatrix; delinquencies; delinquency; delinquent; delinquently; delinquents
delinq delinquent
deliq deliquescent
De L Isls De Long Islands
Dell Dell Publishing Company
Dells The Dells (the Dells of Wisconsin), scenic gorge of the Wisconsin River
Del-Mar-Va Delaware-Maryland-Virginia (Eastern Shore peninsula)
Delmarvia the Delaware-Maryland-Virginia peninsula
delmes delay message
Del Mus Nat Hist Delaware Museum of Natural History
D. Elo. Doctor of Elocution
De Longs De Long Islands in the Arctic
delphi declaiming eclectic liberalism possessively, hotly, instantaneously
delpho deliver(y) by telephone
Del Rio San Felipe del Rio
dels deliveries
DELS Direct Electrical Linkage System
delt delete; deletion
de lt deck edge light(ing)
delt. delineavit (Latin—he or she drew it)
delta. detailed labor and time analysis
Delta letter D radio code
DELTA Daily Electronic Lane Toll Audit
deltic delay line time compression
delu delusion
delv deliver
Delv Delvalle

delvd delivered
dely delivery
dem demand; dementia; democracy; democrat; democratic; demodulate; demodulator; demolish; demolition; demonstrate; demonstration; demonstrative; demote; demotion; demur; demurrage; demy; differential element movement
dem (DEM) demerol
Dem Demerera (British Guiana); democracy; Democrat; democratic; Democratic Party
DEM Department of Environmental Management
DEM Developpement-Études-Marketing (French—Marketing Studies Development)
DEMA Diesel Engine Manufacturers Association
demac deck and engine mechanic
dem adj demonstrative adjective
Demba Demarara bauxite
Dembos (Dutch truncation—'s-Hertogenbosch)
dem cap democratic capitalism
de/me decoding memory
DEME Director of Electrical (Electronic) and Mechanical Engineering
demij demijohn
demimond demimondaine; demimonde
demirep demireputation; a woman of loose morals
DEMKO Dansk Elektrische Materialkontrol (Danish Board for Approving Electrical Equipment)
demo demolition; demonstration (model)
Demo Democrat(ic)
demob demobilization; demobilize
demobed demobilized
democ democracy; democrat; democratic; democratization; democratize; democratizer
demod demodulator
demogr demographer; demographic(al); demography
demon. demonology; demonstrate; demonstration; demonstrator
demonol demonologic(al)(ly); demonologist(ic)(al)(ly); demonology; [*see* maj dem(s)]
demonstr demonstrative
demos demonstrators
demo(s) demolition(s); demonstration(s); demonstrator(s)

Demos Democrats
DEMOS Director(ate) of Estate Management Overseas
dem pro demonstrative pronoun
dem pug *dementia pugilistica* (Latin—punch drunk)
dems defensively-equipped merchant ship
Dem(s) Democrat(s)
DEMS Development Engineering Management System
demur demurrage
den denotation; dental; dentist; dentistry
den Denier (German—denier)
Den Denbighshire; Deniz; Denmark; Denver
Den Denizi (Turkish—lake, sea)
D. En. Doctor of English
DEN Denver, Colorado airport
Denali National Park in Alaska—contains Mt McKinley, North America's highest peak (20,320 ft)
denat denatured
Denb Denbighshire
dend dendrology
D en D Docteur en Droit (French—Doctor of Law)
dendro dendrometer
dendrol dendrology
D. Eng. Doctor of Engineering
D.Eng.Sc. Doctor of Engineering Science
D èn L Docteur èn Leyes (French—Doctor of Law)
D en M Docteur en Médecine (French—Doctor of Medicine)
Denmark Kingdom of Denmark (Scandinavian nation) *Kongeriget Danmark*
Denny Denis; Dennis
Dennys Dennys Lascelles
denom denomination
denot denotation; denotative-(ly); denote(ment)
dens density
dent. dental; dentist; dentistry; denture
dent. dentur (Latin—give, let it be given)
Dent JM Dent & Sons Ltd
D. Ent. Doctor of Entomology
dentac dental accounting
Dent Corps Dental Corps
Dent Hyg Dental Hygienist
Denticare Dental Care
dent. tal. dos. dentur tales doses (Latin—give of such doses)
DEO District Engineering Office; District Engineers Office; Divisional Education

Office(r); Divisional Entertainment Office(r)—British Army
DEOR Duke of Edinburgh's Own Rifles
dep depart; department; departure; dependency; dependent; depilate; depilatory; deponent; depose; deposit; depositor; depot; depotize; deputy; do everything possible
dep. depuratus (Latin—purify)
Dep Deputy
Dep Département (French—Department);*Député* (French—Deputy)
DEP Defense Electronic Products (RCA); Department of Employment and Production; Department of Environmental Protection
depa diethylene phosphoramide
DEPA Defense Electric Power Administration
depart. department; departure
Dep CFO Deputy Chief Fire Officer
dep con departmental control
depcru dependent's (daylight) cruise (USN)
dep ctf deposit certificate
Dep Dir Deputy Director
depend. dependent; dependency
depen-undepen dependably undependable
depi differential equations pseudocode interpreter
depil depilate, depilation, depilator(y)
Dep Insp Deputy Inspector
dep inst depot installed
depl depilate(d) depilation; depilator(y); deplete; depletion-(ary); deploy(ed); deployment
deplab depilatory laboratory
DEPMIS Depot Management Information System (USA)
depn dependency; dependent
DEPNAV Naval Deputy (NATO)
depo deposit
depod deposited
depog depositing
depon deponent
depor depositor
depos depositary
deposn deposition
depr depreciation; depreciative; depression
DEPRA Defense European and Pacific Redistribution Activity
DEPS Departmental Entry Pro-

cessing System; Diploma in Economics and Political Science

DEPSACLANT Deputy Supreme Allied Commander, Atlantic (NATO)

DepSO Departmental Standardization Office

dept depart; department; departure; deponent; depot; deputy

dep't (contraction—department)

Dept Ag Department of Agriculture

deptr departure

Dept State Bull Department of State Bulletin

DePU De Paul University; De Pauw University

deputn deputation

Depy Deputy

DEQ Department of Environmental Quality

der derivation; derivative; derived; dermatine; derrick(s)

der derecha (Spanish—right); *dernier* (French—last)

Der Derringer

DeR reaction of degeneration

DER Department of Environmental Resources; Development Engineering Review; Draft Environmental Report; radar picket escort ship (naval symbol)

DERAP Development Economics Research and Advisory Service

Derb(s) Derby; Derbyshire

Derby. Derbyshire

DERBY Derby Aviation

Derbys Derbyshire

DERE Dounreay Experimental Reactor Establishment

dereg deregulation

Derek Theodoric

deriv derivation; derivative

derm dermatitis; dermatology; dermatophyte

derm (Latin prefix—skin)—dermatology, epidermis

dermat dermatology

dermatol dermatologic(al)(ly); dermatologist; dermatology

Derniers Dernieres Islands

deros date eligible for return from overseas; date of estimated return from overseas service

DERR Duke of Edinburgh's Royal Regiment

Derrick Theodoric

Derry Londonderry

DERT Division Électronique, Radioélectricité et Félécom-

munications (French—Electronic, Radioelectric, and Telecommunications Division)

derv diesel-engine road vehicle

des descend(ed); descending; desert; design; designate; designation; designator; designer; desire; dessert

des (Des) diethylstilbestrol (morning-after contraceptive)

des descubrimiento (Spanish—discovery)

de S de Sola; de Solá

Des Des Moines; Desmond

Des Desierto (Spanish—desert); (German—D-flat)

De S De Sola

DES Data Exchange System; Department of Education and Science; destroyer (naval symbol); Director of Educational Services; Director of Engineering Stores; Dispersed Emergency Station; Draft Environmental Statement; Drug Education Specialist

DESAC Destroyer Sonar Analysis Center (USN)

desal desalinization; desalinize(r)

desat desaturated

DESAT Defense (Department) Small (Business) Advanced Technology (Program)

DESB Devereaux Elementary School Behavior Rating Scale

Des Base Destroyer Base

desc descend(ant)

DESC Defense Electronics Supply Center

descr description

descron description

descto descuento (Spanish—discount)

desdg descending (flow chart)

DESDIV Destroyer Division (naval)

desfex desert field exercise

desfirex desert firing exercise

desg designate; designation

desi designated hitter

DESI Division for Economic and Social Information (UN)

desid desiderata; desideratum

desider desiderative

desig designate; designer

D es L Docteur ès Lettres (French—Doctor of Literature)

DESLANT Destroyer Forces—Atlantic

desp despatch

DESP Department of Elementary School Principals

DESPAC Destroyer Forces—Pacific

despd despatched

despg despatching

despot. design performance optimization

DesRCA Designer of the Royal College of Art

DESRON destroyer squadron

dess dessiatine

d ès S Docteur ès Science (French—Doctor of Science)

D ès S Dar ès Salaam

DESS destroyer schoolship (naval symbol)

dest destination; destroy; destroyer; destruction

dest destra (Italian—right)

dest. destilla (Latin—distilled)

DEST Diplôme de l'Ecole Supérieure Technique (French—Diploma of the Technical Institute)

destdist destructive distillation

destil. destilla (Latin—distill)

destination SPPK destination Singapore, Penang, and Port Klang

destn destination

destr desires to transfer; destructor

destr fir destructive firing

desubex destroyer/submarine antisubmarine warfare exercise

DESY Deutsches Elektronen Synchrotron (German Electronic Synchrotron)

det detach; detachment; detail; detective; detector; determinant; determine; detonator; double end trimmed

det (DET) diethyltryptamine (quick-acting hallucinogen drug)

det. detur (Latin—let it be given)

Det. Detective; Detroit

DET Design Evaluation Testing; Detroit, Michigan (Detroit City Airport)

DETA Direção de Exploração dos Transportes Aéreos (Mozambique airline)

detab decision table

detab/X decision table(s)/experimental

DETAPS Decision Table Processing System

detcom(s) detected communist(s)

Det Con Detective Constable

detd determined

det. in dup. detur in duplo (Latin—give twice as much); *detur in duplo* (Latin—let

twice as much be given)
detectionary dictionary of detectives (mystery-fiction type)
determin determination
DETEST Demystify Established Standardized Tests
DETG Defense Energy Task Group
Det Insp Detective Inspector
detl detail
detm determine
DETMAHOG Deliver-the-Mail/Holy-Grail (dichotomous theory of problem protection practiced by adept bureaucracies worldwide)
detn detention
detox detoxification; detoxification center (for alcoholic and narcotic addicts)
detoxcen detoxification center
detr detector
detrins detailed routing instructions
Detroit Inst Detroit Institute of Arts
d. et s. detur et signatur (Latin—let it be given and labelled)
Det Sgt Detective Sergeant
Det Sup Detective Superintendent
Det Sym Orch Detroit Symphony Orchestra
deu data exchange unit; digital evaluation unit; display electronics unit
DEU Data Exchange Union; Drug Epidemiology Unit
DEUA Diesel Engines and Users Association
Deuce The Deuce (New York City's 42nd Street)
deuce. digital electronic universal computing engine
Deut Deuteronomy
Deut Deuteronomy
Deutschland Deutschland, Deutschland über alles (German—Germany, Germany above all)—national anthem
dev develop; developer; development; deviate; deviation; deviator
dev (DEV) duck embryo vaccine
Dev Devon; Devonian; Devonshire; Eamon De Valera
De V De Vilbiss
DEV Development Well
deva development acceptance
DEVCO Development Committee
devd device data set residence
devel developer; development

Dev-Genc Devrimci-Gencler (Turkish—Revolutionary Youth)
devil. development of integrated logistics
Devils The Devils of Loudon by Aldous Huxley
Devils Postpile Devils Postpile National Monument
Devils Tower Devils Tower National Monument on Wyoming's Belle Fourche River
Devin Devin-Adair
devlp develop
devlpd developed
devlpg developing
devlpmt development
dev^{mo} devotissimo (Italian—devotedly yours)—yours truly
Devon Devonshire
devp develop
devpt development
devs developers; devotions
DEVSIS Development of Science Information Systems
Dev-Yol Devrimoi-Yol (Turkish—Revolutionary Path)
dew. dewpoint
DEW Demineralization Water (decontamination subsystem); Distant Early Warning
dewat deactivated war trophy
dewd detailed elementary wiring diagram(s)
De Witt De Witt Clinton High School
DEWIZ Distant Early Warning Identification Zone
DEW Line Distant Early Warning Line
DEWS Diagnostic Evaluation of Writing Skills
dex dextroamphetamine tablet
dex dexter (Latin—right)
Dex Dexter
D. Ex. Doctor of Expression
dexan digital experimental airborne navigator
dexe dexedrine
dexies dexedrine tablets (stimulant drugs)
d. ex m. deus ex machina (Latin—god from a machine)—introduction of a god-like device to resolve a play or problem
dext. dexter (Latin—right)
dextrose glucose ($C_6H_{12}O_6H_2O$)
dez diethyl zinc
dez dezembro (Portuguese—December)
Dez Dezember (German—December)
Dezhda Nadezhda
df damage free; dead freight;

decontamination factor; defensive fire; defogging; degree(s) of freedom; dense film; derrick floor; diamond flap; direction finder; disposition form; double feeder; double foolscap; double fronted; draft; drinking fountain; drive fit; drop forge; drop forging; dummy fuse; dummy fuze; dunnage free
d/f defogging; direct flow; double fleece; double fronted
d & f determination and finding
d/f días fecha (Spanish—days from date)
Df Douglas fir
DF Dean of the Faculty; Defender of the Faith; Destroyer Flotilla; deuterium fluoride
D-F Dansk-Franske; deflection factor (symbol)
D of F Department of Fisheries
DF Distrito Federal (Spanish—Federal District)
D.F. Defensor Fidei (Latin—Defender of the Faith)
dfa digital fault analysis
DFA Dairy Farmers Association; Department of Foreign Affairs; Dimensional Fund Advisor; Division Freight Agent; Drop Forging Association
D.F.A. Doctor of Fine Arts
DFAC Dried Fruit Association of California
DFAR Daily Field Activity Report
dfb damage-free bulkheads; distributed feedback; distribution fuse board; dunnage-free bulkheads
dfc data format converter; discriminant function coefficient; dry-filled capsules
DFC Department Frequency Coordinator; Development Finance Corporation; Distinguished Flying Cross
dfclt difficult
dfcs digital flight-control software
DFCT Deputy Federal Commissioner of Taxation
dfcty difficulty
dfd data function diagram; defend(ed); deferred
DFD Dogs For Defense
dfdr digital flight-data recorder
DFDS Det Forende Dampskibs-Selskab (Danish United Steamship Company, Limited, Denmark)

DFDT difluoro-diphenyl trichloroethane (insecticide)
dfe derivative fighter engine; derrick floor elevation; double fish eye (buttons)
DFE Department of Further Education
dff dilutent-free formulation
dfg digital function generator; diode function generator
DFG Department of Fish and Game
dfga distributed floating-agate amplifier
DFGJPC Daniel and Florence Guggenheim Jet Propulsion Center
DFH *Danmarks Fiskeri og Havundersogelser*
dfi definite; direct-flame impingement
DFI Director(ate) of Food Investigation
DFIB Data Function Information Book
DFIC Dairy Foods Information Center
d/fing direction finding
DFISA Dairy and Food Industries Supply Association
dfitw dead flat in the water (becalmed ship)
dfiy definitely
D fl Dutch florins
DFL Daily Flight Log; Democrat Farmer-Labor; *Deutsche Forschungsanstalt für Luft und Raumfahrt* (German—German Air and Space Research Institute)
dfld defiled; deflated
DFLP Democratic Front for the Liberation of Palestine
DFLS Day Fighter Leaders' School
DFM Distinguished Flying Medal
DFMR Dazian Foundation for Medical Research
DFMS Domestic and Foreign Missionary Society
DFMSR Directorate of Flight and Missile Safety Research
dfn distance from nose
dfndt defendant
DFNWR Deer Flat National Wildlife Refuge (Idaho)
DFO Department of Fisheries and Oceans (Canadian); District Field Office(r)
d forg drop forging
dfp (DFP) diisopropyl phosphofluoridate
DFP *Detroit Free Press*
DFPA Douglas Fir Plywood Association

dfq day frequency
dfr decreasing failure rate; defrost(ing); dropped from rolls
Dfr Dounreay fast reactor
D fr Djibouti franc
DFRA Drop Forging Research Association
DFRC Dryden Flight Research Center (NASA)
dfr(d) defer(red)
DFRDBA Defence Forces Retirement and Death Benefits Authority (Australian)
dfrn differential
dfrs differs
dfs distance finding station
DFs Duty Frees (tobacco products)
DFS *Dirección Federal de Seguridad* (Spanish—Federal Security Directorate)—Mexico's famed *Federales*, the Feds or Federals
D.F.Sc. Doctor of Financial Science
DFSC Defense Fuel Supply Center
DFSM Distinguished Fire Service Medal
dfsr diffuser
dft deaerating feed tank; defendant; draft
DFT Diagnostic Function Test
dfti distance from touchdown indicator
dftmn draftsman
dft(s) draft(s)
d-f tube double-flare tube
dfu data file utility; dummy flying unit in uterus; dead fetus in uterus; dummy flying unit
dfus diffuse
DFW Dallas-Fort Worth, Texas (airport); Director of Fortifications and Works
dg dark ground; decigram(s); degenerate(d); deoxyglucose; diagnosis; diastolic gallup; digestive gland; diglyceride; disk grind; distogingival; double glass; double groove; durable gum
d/g dangerous goods; decomposed granite; directional gyroscope; displacement gyroscope
DG *Deutsche Grammophon*; Diego Garcia; Director General
DG *Déclaration de Guerre* (French—Declaration of War)
D.G. *Dei Gratia* (Latin—By the Grace of God)
dga (DGA) diglycolamine
DGA Directors Guild of Amer-

ica
DGAA Distressed Gentlefolk's Aid Association
DGAMS Director General of Army Medical Services
DGAS Double-Glazing Advisory Service
DGB *Deutscher Gewerkschaftsbund* (German Federation of Trade Unions)
DG Bank *Deutsche Genossenschaftsbank* (German Co-operative Bank)
dgbus digital ground bus
DGC Dangerous Goods Classification; Data General Corporation; Duty Group Captain
D.G.C. Diploma in Guidance and Counseling
DGCA Director General of Civil Aviation
DGCE Director General of Communications Equipment
dgd double glass doors
DGD Director Gunnery Division
DGDC Deputy Grand Director of Ceremonies
DGD & M Director General Dockyards and Maintenance
DGE Directorate General of Equipment; Division of Geothermal Energy (DoE)
DGEIS Draft Generic Environmental Impact Statement
DGG *Deutsche Grammophon Gesellschaft* (German—German Gramophone Record Company)
dgi disseminated gonococcal infection
DGI Date Growers Institute; Director General of Information; Director General of Inspection; Directorate of General Intelligence
DGI *Directorio General de Inteligencia* (Spanish—Directorate General of Intelligence)—Cuban branch of the Soviet KGB
DGIP Division of Global and Interregional Projects (UN)
DGLA dihomo-gamma linolenic acid
Dgls Douglas
Dglsh Daglish
dgm decigram
DGM Diploma in General Medicine; Director General of Manpower; Director(ate) of General Mobilization
DGMS Director General of Medical Services
DGMT Director General of

Military Training
dgmw double-gimbal momentum wheel
DGMW Director General of Military Works
Dgn Dragoon(s)
dgnast (DGNAST) design assist
dgnl diagonal
Dgo Durango
DGO Diploma in Gynecology and Obstetrics
DGP Director General of Production
DGPS Director General of Personnel Services
dgr danger(ous)(ly); door gunner
d Gr der Grosse (German—the Great)
DGR Director of Graves Registration
DGR Dirección General de Radiocomunicaciones (Spanish—General Administration of Radio Communications)—Bolivian broadcasting control
DGRR Deutsche Gesellschaft für Raketentechnik und Raumfahrt (German Society for Rocket Technique and Space Flight)
dgs double green silk
DGS Degaussing System; Diploma in General Surgery; Director General of Ships
DGSC Defense General Supply Center
DGSE Directorat Général Sécurité External (French—General Directorate of External Security); *Directorio General de Seguridad del Stado* (Spanish—Directorate General of State Security)–Nicaragua's secret police
DGSRD Director(ate) General of Scientific Research and Development
DGSS Director General Secret Service
Dgt Dumaguette
DGT Director General of Training
DGT Dirección General de Turismo (Spanish—Administration of Tourism)
DGTA Dry Goods Trade Association
DGTTT Dirección General de Transporte y Transito Terrestre (Spanish—Ministry of Communications)
DGW Director General of Weapons

dgz designated ground zero
dh dead heat; deadhead; dehydrogenase (DH); delayed hypersensitivity; designated hitter; double hung; drill hole
dh (DH) designated hitter
d & h daughter and heiress; dressed and headed
dh das heisst (German—that is to say)
Dh Moroccan dirham(s)
DH Declaration of Homestead; De Havilland (aircraft); Department of Health
D.H. Doctor of Humanities
D & H Delaware & Hudson (railroad)
D of H Degree of Honor; Degree of Honour
dha dicha (Spanish—good luck, happiness)
DHA Dhahran, Saudi Arabia (airport); Drug Houses of Australia
D & HAA Dock and Harbour Authorities Association
DHAC De Havilland Aircraft of Canada Limited
D-handle D-shaped handle
dha (DHAP) dihydroxyacetone phosphate
dhard dehaired (skins)
dhas (DHAS) dehydroepiandrosterone sulfate
dhc (DHC) dihydrochalcone
DHC Department of Housing and Construction; Detroit House of Correction
DHC-3 Canadian version of De Havilland Otter utility aircraft
DHC-6 Canadian De Havilland Twin Otter transport aircraft
DH Canada De Havilland Aircraft of Canada Limited
dhcv down-hole control valve
dhd distillate hydrosulfurization
dhdd digital high-definition display
dh di drophammer die
dhe data-handling equipment
dhea (DHEA) dehydroepiandrosterone
dheas (DHEAS) dehydroepiandrosterone sulfate
DHEW Department of Health, Education, and Welfare
DHF Dag Hammarskjöld Foundation
dhfr dihydrofolate reductase
D. Hg. Doctor of Hygiene
DHHS Department of Health and Human Services
DHI Dental Health International

DHI Deutsches Hydrographisches Institut (German—German Hydrographic Institute)
dhia dehydro-isoandrosterol (DHIA)
DHIA Dairy Herd Improvement Association
dhic dihydro-isocodeine (DHIC)
dhl distemper, hepatitis, leptospirosis (vaccine)
DHL Dag Hammarskjöld Library (UN in NYC)
D.H.L. Doctor of Hebrew Letters; Doctor of Hebrew Literature
dhllp direct high-level language processor
dhl-p distemper, hepatitis, leptospirosis—parainfluenza (vaccine)
dhlpp distemper, hepatitis, leptospirosis, parainfluenza, parvovirus (vaccine)
dhlw defense high-level waste(s)
DHM Detroit Historical Museum
dhma dehydroxymandelic acid (DHMA)
DHMPGTS Department of Her (His) Majesty's Procurator General and Treasury Solicitor
dhn dynamic hardness number
DHN Department of Hospital Nursing
dho dicho (Spanish—said)
DHO deuterium hydrogen oxide; District Health Office(r); Downhill Only (ski club)
d'Holbach Paul Henry Thiry d'Holbach
dhon dishonor(able)
D.Hor. Doctor of Horticulture
dhp developed horsepower
DHP Diplôme en Hygiène Publique (French—Diploma in Public Health)
dhpg dehydroxyphenylglycol (DHPG)
dhq mean diurnal high water inequality
DHQ Division Headquarters
dhr delayed hypersensitivity reaction(s)
DHR Division of Housing Research
Dhr de heer (Dutch—Mr)
dhrr de herrer (Dano-Norwegian—the gentlemen)
dhs dry heat sterilization
dh's deadheads (freeloaders who never buy a ticket or pay their own way)
DHS Department of Hyperten-

sion and Stress; Detroit High School; Diploma in Horticultural Science; District High School; Dublin High School

D.H.S. Doctor of Health Science(s)

DHSA Diploma—Health Service Administration

DHSC Department of Health and Social Security

dhsm dihydrostreptomycin (DHSM)

DHSS Department of Health and Social Security

dht distillate hydrotreating

dht (DHT) dihydrotestosterone

dhtv downhole television

DHUD Department of Housing and Urban Development

D.Hum.L. Doctor of Humane Letters

dhw domestic hot water; double-hung windows

DHX Dependable Hawaiian Express

D. Hy. Doctor of Hygiene

di daily inspection; de-ice; diameter; diametral; diplomatic immunity; direction indicator; display indicators; document identifier; double imperial

di (DI) diabetes insipidus; double indemnity; inversion of the duration series

di (Latin prefix—two)—dipole antenna

d i das ist (German—that is)

Di Diana; Diane; didymium; Dinorah

DI Defense Intelligence; Denizyollari Isletmesi (Turkish Maritime Lines); Department of the Interior; Departmental Instruction(s); Detective Inspector; Deterioration Index (annual rate for the deterioration of a mailing list to the point it ceases to be deliverable); Diffusion Index; direct ignition; Director of Intelligence; District Inspector; Division Instruction; Divisional Inspector; Drill Instructor (USMC)

D-I Dai-Ichi

D of I Daughters of Isabella; Declaration of Independence; Department of Insurance; Department of the Interior; Division of Intelligence

DI-5 Defense Intelligence (British agency)

dia date of initial appointment; diagram; diameter; diaphone; diathermy; due in assets

dia (Greek prefix—passing through or through)—diabetes, diagnosis, dialysis, diaphragm

DIA Defense Intelligence Agency; Defense Investigative Agency; Department of Institutions and Agencies (NJ); Design and Industries Association; Designated International Accounts; Document Interchange Architecture; Dulles International Airport (Washington, D.C.)

diab diabetic

DIAB Defense Internal Audit Board

diac di-iodothyroacetic acid (DIAC)

DIAC Defense Industry Advisory Council

diacrit diacritic(al)(ly)

di ad die adapter

diag diagnose; diagnosis; diagnostic; diagnostician; diagonal; diagram

dial. dialect; dialectical; dialectician; dialectics

DIAL Disc Interrogation and Loading (system)

dial-a-mation dial-a-cremation

dial-a-porn dial-pornography (erotic telephone service)

dialec dialectic(al)(ly); dialectician(s); dialectics; dialectologist(s); dialectological-(ly); dialectology

dialgol dialect of algol *(q.v.)*

diam diameter

DIAMANG *Companhia de Diamantes de Angola* (Portuguese—Angolan Diamond Company)

diamat dialectical materialism

diamond carbon

dian digital analog

DIAND Department of Indian Affairs and Northern Development (Canada)

diane digital-integrated attack and navigation equipment (DIANE)

DIANE Direct Information Access Network for Europe

diap. diapason (Greek—consonant harmony, octave)

diaph diaphragm

diaphor diaphoresis

DIAR Defense Intelligence Agency Regulation

dias. defense-integrated automatic switch

DIAS Drug Information and Assistance Service; Dublin Institute for Advanced Studies; Dynamic Inventory Analysis System

diast diastolic

diat diathermy

DIAT Dundee Institute of Art and Technology

diath diathermy

dib dead in bed (not physically but sexually); diameter inside bark

DIB Department of Information and Broadcasting; Department Information Bulletin

DIB *Dictionary of International Biography*

DIBA Domestic and Internal Business Administration; Dominion Investment and Banking Association

dibah (DIBAH) diisobutylaluminum hydride

dibas dibasic

dibb double-income baby boomers

DIBR Dartnell Institute of Business Research

dic data insertion converter; data item category; defense identification code; dependency and indemnity compensation; dictionary; digital integrated circuit; digital integrating computer; disseminated intravascular coagulopathy; drunk in charge; inverted cancrizans of the duration series

d & ic dependency and indemnity compensation

dic dicembre (Italian—December); diciembre (Spanish—December)

DiC diesel cargo vessel

DIC Dai Nippon Ink and Chemicals; Diplomate of the Imperial College (London); Direct Importing Company (New Zealand); Diving Information Center (USN)

DICAP Direct-Current Circuit-Analysis Program

DICASS Directional Command Active Sonobuoy System

dicautom automatic dictionary look-up

DICB Demolition Industry Conciliation Board

dice. digital intercontinental-conversion equipment; digital-interface countermeasure equipment; direct-installation coaxial equipment

DICEF Digital Communications Experimental Facility (USAF)

DIChem Diploma of Industrial Chemistry
dichlorvos dimethyldichlorovinyl phosphate (insecticide)
dicht dichterlijk (Dutch—poetic)
dick detective
Dick Richard
dickel dime and nickel
Dickie Dickman; Richard
Dickon Richard(son)
dick(s) detective(s)
Dicky Richard; Tricky Dicky
DICNAVAB Dictionary of Naval Abbreviations
dicot(s) dicotyledon(s)
dict dictated; dictation; diction; dictionary
dicta dictaphone
DICTA Diploma of the Imperial College of Tropical Agriculture
Dict Amer Slang Dictionary of American Slang
dictsort dictionary sorter
did. data item description; dead of intercurrent disease; didactic; direct in dialing
did. (DID) drum information display
Did Didot
DID Department of Industrial Development; Division of Institutional Development; Drainage and Irrigation Department
DID Daily Intelligence Digest
dida differential in-depth analysis
didac didactic(al)(ly); didacticism; didactics
didad digital data display
DIDAS Dynamic Instrumentation Data Automobile (Automotive) System
DIDC Depository Institutions Deregulation Committee
didd diddle
dident distortion identity
DIDMCA Depository Institutions Deregulation and Monetary Control Act
didn't did not
di/do data input/data output
DIDS Digital Information Display System
die. died in emergency room (DIE)
DIE Diploma in Industrial Engineering; Diploma of the Institute of Engineering; Division of International Education
DIEA Dictionary of Industrial Engineering Abbreviations
dieb. alt. diebus alternus

(Latin—on alternate days)
dieb. secund. diebus secundis (Latin—every second day)
dieb. tert. diebus tertius (Latin—every third day)
Diego (Mexican-American truncation—San Diego)— San Diego, California
die Kö Königsallee (main street of Dusseldorf)
diel dielectrics; diesel electric
di el diesel electric
DIEME Director(ate) of Inspection of Electrical (Electronic) and Mechanical Equipment
DIEPO Dieterich-Post
diet. dietary; dietetic(s); dietician
DIEX Dirección de Identificación y Extranjería (Spanish—Directorate of Identification and Immigration)
dif differ; difference; differential; diffuse(er)(s)
dif (DIF) discriminant function
DIF Defense Industrial Fund; Descriptive Item File; Discriminant Function (auditing system for income-tax returns); District Inspector(ate) of Fisheries
DIF Desarrollo Integral de la Familia (Spanish—Integral Family Development)
dif-amps differential amplifiers
difar directional frequency analysis and recording
difce difference
diff difference; differential
diff calc differential calculus
diff diag differential diagnosis
diffr diffraction
diffu diffusion
DiFr diesel fruit vessel
dift different
diftl differential
dig. digamist; digamy; digest; digestion; digestive
dig. digeratur (Latin—let it be digested)
DIG Deputy Inspector General; Discussion Interest Groups
DIGA Dynamics International Gardening Association
digas digastric
DIGEPOL Dirección General de Policías (Spanish—General Directorate of Police)— Venezuela
digger(s) gold digger(s); gold miner(s)
digi digital
Digi Digiform (business forms

typesetter)
digicom digital communications (system)
digital IC digital integrated circuit
digres digression(al)(ly); digressionary; digressive(ly); digressiveness
digrm digit/record mark(ing)
dig r-o digital readout
digs. archeological excavation; diggings
DIGs Development Import Grants
DIGS Delta Inertial Guidance System
di-H hydrogen
DIH Diploma of Industrial Health; Division of Indian Health
DIHJHU Department of International Health—Johns Hopkins University
Dij Dijon
di ji drill rig
dik drug-identification kit
DIKB Dai-Ichi Kangyo Bank
diks double income, kids
dil dilute; dissolve
dil. dilue (Latin—dilute); *dilutus* (Latin—diluted)
DIL Deliverable Items List; Director of International Logistics; Division of Insured Loans
dilat dilatation; dilate; dilation (ed)
dild diluted
dilet dilettante
dilligaf do I look like I give a fuck?
dilligas do I look like I give a shit?
diln dilution
diluc. diluculo (Latin—at daybreak)
dilut. dilutus (Latin—dilute)
dim. defense information memo; description, installation, and maintenance; digital dimmer memory; dimension; dimensional; dimension(al)-(ly); diminutive
dim dimanche (French—Sunday); *dimidius* (Latin—one half); *diminuendo* (Italian—diminishing gradually)
DIM Dialogue Inter-Monasteries; Diploma in Industrial Management
DIMA Detroit Institute of Musical Art
Dimashq (Arabic—Damascus)—capital of Syria
dimate depot-installed maintenance automatic test equip-

ment
DIMD Dorland's Illustrated Medical Dictionary
DIMDI Deutsches Institut für Medizinische Dokumentation und Information (German—German Institute for Medical Documentation and Information)
dime. dual independent map encoding
DIME Division of International Medical Education (Assn Amer Med Colleges)
DIMES Defense Integrated Management Engineering Systems
dimin diminish; diminution; diminutive
DIMIS Depot Installation Management Information System (USA)
dimn dimension
di'mon' diamond
dimorph dimorphous
dimple deuterium-moderated pile low energy
DIMS Data Information and Manufacturing System; Director International Military Staff Memo (NATO)
din. dinar(s); dining room; dinner; do it now
din dinar (Yugoslavian monetary unit)
Din Dinsdag (Dutch—Tuesday)
DIN Data Identification Number
DIN Deutsche Industrie Norm (German Industry Standard)—film rating sometimes written *din* and said to mean *das ist norm* (this is standard); *Deutsches Institut für Normung* (German Standards Institute)
Dina Dinamarca (Portuguese or Spanish—Denmark)
DINA Dirección de Inteligencia Nacional (Spanish—Directorate of National Intelligence)—Chilean secret police
dinar monetary unit of Algeria, Bahrain, Iraq, Jordan, Kuwait, Libya, South Yemen, Tunisia, Yugoslavia
diner dining car
Ding JN Darling
D.Ing. Doctor Ingeniariae (Latin—Doctor of Engineering)
dinin' dining
dinks double incomes, no kids
dino(s) dinosaur(s)

Dinosaur Dinosaur National Monument, Colorado and Utah
DINP Dunk Island National Park (Queensland)
DINS Dormant Inertial Navigation System
dio diode; direct input-output
DIO Director(ate) of Intelligence Operations; District Intelligence Office(r); Duty Intelligence Officer
diob digital input-output buffer
dioc dioceasan; diocese
diode. digital input-output display equipment
Dion Dionisio
diop di-iso-octyl phthalate (plasticizer); diopter; dioptrics
dior diorama
dios diver lockout submersible
DIOS Distributed Input-Output System
diox dioxygen
dip. digital inline pins; dipeptide; diphtheria; diphthong; diplex; diplococcus; diploma; diplomacy; diplomat; dipsomania(c); dissemination and improvement of practice; dual incline package; (slang for pickpocket); stupid person
DIP Document Improvement Program (DoD)
DIPA Diploma of the Institute of Park Administration
Dip AD Diploma in Art and Design
Dip Agr Diploma in Agriculture
Dip A Ling Diploma in Applied Linguistics
Dip AM Diploma in Applied Mechanics
Dip Amer Bd P & N Diplomate of the American Board of Psychiatry and Neurology
Dip AMS Diploma in Ayurvedic Medicine and Surgery
Dip Anth Diploma in Anthropology
Dip App Sci Diploma in Applied Science
Dip Arch Diploma in Architecture
Dip Ars Diploma in Arts
Dip Bac Diploma in Bacteriology
Dip BMS Diploma in Basic Medical Sciences
Dip CAM Diploma in Communications, Advertising, and Marketing
Dip Card Diploma in Cardiol-

ogy
Dip Com Diploma in Commerce
dipcrit diplomatic critic(ism)
Dip DP Diploma in Drawing and Painting
Dip DS Diploma in Dental Surgery
DIPEC Defense Industrial Plant Equipment Center
Dip Eco Diploma in Economics
Dip Ed Diploma in Education
Dip Eng Diploma in Engineering
Dip FA Diploma in Fine Arts
Dip For Diploma in Forestry
Dip G & O Diploma in Gynaecology and Obstetrics
Dip GT Diploma in Glass Technology
diph diphtheria
Dip HA Diploma in Hospital Administration
Dip HE Diploma in Higher Education; Diploma in Highway Engineering
diph tet diphtheria tetanus
diphth diphthong
diph tox diphtheria toxin
diph tox ap diphtheria toxin alum precipitated
Dip Hus Diploma in Husbandry
dipj distal interphalangeal joint
Dip J Diploma in Journalism
dipl diplomacy; diplomat; diplomatic
Dipl Diplom (German—Diploma)
Dip L Diploma in Languages
Dip Lib Diploma in Librarianship
Dip Lib Sci Diploma in Library Science
diplo diploma; diplomacy; diplomat; diplomatic; diplomatics; diplomatism; diplomatist
diplo diplomatico (Spanish—diplomat); *diplotienda* (Spanish—diplomat store)—special store catering only to diplomats; (Greek *diploos*—twofold)—diploid, diplomacy, diplomat(ic)
Dip ME Diploma in Mechanical Engineering
Dip MFOS Diploma in Maxial, Facial, and Oral Surgery
Dip Mgmnt Diploma of Management
Dip Micro Diploma in Microbiology
Dip Mus Edu Diploma in Musical Education
Dip NA & AC Diploma in

Numerical Analysis and Automatic Computing

Dip NS Edu Diploma in Nursery School Education

Dip NZLS Diploma of the New Zealand Library Service

Dip OL Diploma in Oriental Learning

Dip Phar Diploma in Pharmacology

Dip Phys Edu Diploma in Physical Education

Dip P & OT Diploma in Physical and Occupational Therapy

Dip Pub Adm Diploma in Public Administration

Dip RADA Diploma of the Royal Academy of Dramatic Art

Dip RSAM Diploma of the Royal Scottish Academy of Music

diprt discharge printed

dips. dipeptides; diphtheria patients; diphthongs; diplexes; diplomas; diplomats; dipsomaniacs

DIPS Dendenkosha's Information Processing System; Development Information Processing System

dipsey deep-sea lead (line for measuring depths)

Dip SMS Diploma in School Management Studies

dipso dipsomania(c); drunkard

Dip Sp Ed Diploma in Special Education

Dip SS Diploma in Social Studies

Dip SW Diploma in Social Work

Dip T Teachers Diploma

Dip T & CP Diploma in Town and Country Planning

Dip Tec Diploma in Technology

Dip TEFL Diploma in Teaching English as a Foreign Language

dipth diphthong

Dip The Diploma in Theology

Dip TP Diploma in Town Planning

dipu diputado (Spanish—deputy)

Dip VFM Diploma in Valuation and Farm Management

dIQ deviation IQ

dir direct; direction; director

dir (DIR) digital instrumentation radar

dir. directione (Latin—directions); *direxit* (Latin—directed by)

Dir Director(ate); Dirham(s)—Moroccan money

DIR Detailed Inspection Report; Diesel Inspector's Report

dir conn direct-connect

dir coup directional coupler

dircty directly

direct. directory

D.Ir.Eng. Doctor of Irrigation Engineering

Dir Gen Director General

Dir Gen Direttore Generale (Italian—General Manager)

dirham monetary unit of Morocco and United Arab Emirates

Dirk Derek; Everett McKinley Dirksen

dir max directional maximum

dir min directional minimum

diron direction

dir. prop. directione propria (Latin—with proper directions)

DIRT Department of Industrial Relations and Technology (Australian)

dis delivered into store; disability; disable(d); disciple; discipline; disconnect(ed); discontinue(d); discount(ed); disease(d); disrespect; distal; distance; distant; distribute(d); distribution

dis (Latin prefix—apart, away from)—disable, disarticulate

Dis Disney (Walt Disney); Disneyland; Disraeli (Benjamin Disraeli); Pluto

DIs Department(al) Instructions

DIS Dairy Industry Society; Defense Intelligence School; Defense Intelligence Service; Defense Investigative Service; Department of Industrial Services; Diagnostic Interview Schedule; Disney Productions (stock exchange symbol); Distribution Advisory Service; Distribution Information System; Dow Industrial Service; Drug Instruction Service; Ductile Iron Society

disab disable; disabled

disabl disability

disac digital simulator and computer

disap disapprove

Disap Disappointment (cape in Washington state, lake in western Australia, islands in French Polynesia)

disas disaster

disassy disassembly

disb disburse; disbursement

disbmt disbursement

disbt disbursement

disc. (DISC) direct-injection stratified charge (automobile engine)

disc. dimension of schooling questionnaire; discography; disconnect; discontinue; discophile; discount; discover(ed)

DISC Defense Industrial Supply Center; Distribution Stock Control System; Domestic International Sales Corporation

discd discounted

discg discounting

disch discharge; discharging

dischd discharged

dischg discharging

disc jock(s) disc jockey(s)

disco disc jockey; discotheque; discotheque music

DISCO Defense Industrial Security Clearance Office

discol discolored

discom digital selective communication(s)

discomb discombobulate(d); discombobulation

discon disconnect; disorderly conduct

DISCON Defence-Integrated Secure Communications Network (Australian)

discontd discontinued

discos discotheques

discr discriminator

discron discretion

discrtn discretion

DISCs Domestic International Sales Corporations

disct discount

discum discumbobulate(d); discumgalligumfricate(d)

DISCUS Distilled Spirits Council of the United States

discus (see DSSCS)

Discuss Faraday Soc Discussions of the Faraday Society

DISD Data and Information System Division

DISE Digital Systems Education

disemb disembark

disg disagreeable

dishon dishonest; dishonesty; dishonorable; dishonorably

disi door insulating system index

DISI Dairy Industries Society International

disid disposable seismic intrusion detector

disin disinfectant; disinfection
dis int discrete integrator
DISIP Dirección de Seguridad e Inteligencia Policiales (Spanish—Directorate of Police Security and Intelligence)
disk diskonto (Norwegian—discount)
disloc dislocation
dism dismiss; dismissal
Dismals Dismal Gardens near Phil Campbell, Alabama
dismd dismissed
dis/min disintegrations per minute
diso die shoe
disod disodium
disord disorder
DISOSS Distributed Office Support System
disp dispatch; dispensary; dispensatory; dispenser; displacement; display; disposal; disposition
disp. dispensa (Latin—dispense)
dispen dispensatories; dispensatory
displ displacement
dispr dispatcher
disr disrated
diss disassembly; dissent; dissenter; dissertation
DISS Director(ate) of Information Systems and Settlement (stock exchange)
dissd dissolved
dissec dissection
dis/sec disintegrations per second
dissed disrespectful
dissem disseminate; disseminated
dissert dissertation(s)
disson dissonance; dissonant(ly)
disspla display integrated software system and plotting language
dissyl dissyllable
dist distance; distant; distribute; distribution; distributor; district
dist. distancia (Spanish—distance); *distilla* (Latin—distill)
Dist District
distab disestablish(ment)-(tarian)(ism)
Dist Ad District Administrator
distads administrative districts
distar direct instruction
DISTAR Direct Instruction System for Teaching Arithmetic and Reading

Dist Atty District Attorney
distb distillable
distbtr distributor
Dist Ct District Court
distil distillation; distilled; distilling
Dist J District Judge
distn distillation
distng distinguish; distinguishing
Dis TP Distinction in Town Planning
distr distribute; distribution
DISTRAMS Digital Space Trajectory Measurement System
distran diagnostic fortran
distrib distribution; distributive; distributor
DISTRIPRESS Fédération Internationale des Distributeurs de Presse (French—International Federation of Wholesale Book, Newspaper, and Periodical Distributors)
Dists Districts
DISUM Daily Intelligence Summary (USAF)
disy disyllabic
dit domestic independent tour; dual input transponder
dit (DIT) diiodotyrosine
DIT Defining Issues Test; Detroit Institute of Technology; Drexel Institute of Technology; Durham Institute of Technology
DIT Deutscher Investment-Trust
DiTa diesel tanker vessel
ditar digital telemetry analog recording
DITC Disability Insurance Training Council
Ditch The Ditch, 3100-mile-long (4989-kilometer-long) Intracoastal Waterway along the Atlantic coast from Boston to Key West, as well as from the St Marks River in Florida to Brownsville, Texas
dithy dithyramb(ic)(al)(ly); dithyrambs
ditmco data information test material checkout
DITN Diabetes in the News
Dito Ernesto
diu data interface unit; digital interface unit
DIU Diversion Investigation Unit
div data in voice; digits in voice; divergence; diverse; divide; divided; dividend; divisibility; division; divisor; divorce; divorced

div divisi (Italian—divided)
Div Divide (postal abbreviation); Divine; Divinity; Division
divab digital input/voice answer back
DIVAD Divisional Air Defense
Div Arty Division Artillery
divd dividend
divde dividende (French—dividend)
Div E Division Engineer
divear diving instrumentation vehicle for environmental and acoustic research
div. en p. aeq. divide in partes aequales (Latin—divide into equal parts)
divi(s) dividend(s)
divn division
divnl divisional
div. in par. aeq. dividatur in partes aequales (Latin—divide into equal parts)
Div Jum Divehi Jumhuriya (Divehi—Republic of the Maldives)
divs dividends
divvy divide; dividend
diw dead in the water
DIW Deutsches Institut für Wirtschaftforschung (German—German Institute for Economic Research)
Dix Dixie; Fort Dix, New Jersey
diy do it yourself
diz dizionario (Italian—dictionary)
dj disc jockey; dust jacket
d J der Jüngere (German—junior); *dieses Jahres* (German—of this year)
Dj Djawa (Indonesian—Java); *Djebel* (Arabic—mount, mountain)
DJ David Jones (Australian department store chain); Department of Justice; District Judge; Divorce Judge; Don Jail (Toronto)
D-J Dow-Jones (average)
D of J Department of Justice; Dominion of Jamaica
DJ Divehi Jumhuriyya (Divehi Arabic—Republic of Maldives)—Maldive Islands
D.J. Doctor Juris (Latin—Doctor of Law)
Dja Djakarta
DJAD Department of Justice Antitrust Division
DJAG Deputy Judge Advocate General
DJCD Department of Justice

Civil Division; Department of Justice Criminal Division

DJCP Division of Justice and Crime Prevention (Virginia)

DJCRD Department of Justice Civil Rights Division

djd degenerative joint disease

dJf divorced Jewish female

Dji Djibouti

DJI Dow-Jones Industrials (average)

DJIA Dow-Jones Industrial Average

Djib Djibouti (formerly Afars and Issas Territory also known as French Somaliland)

Djibouti Republic of Djibouti (formerly French Somaliland)

Djinn Sud-Aviation two-seat helicopter built in France

DJJ Department of Juvenile Justice

Djkta Djakarta (Batavia), Java

Djl *Djalan* (Malay—road or street)

DJLNRD Department of Justice Land and Natural Resources Division

dJm divorced Jewish male

DJNR Dow Jones News Retrieval

DJO *Den Jyske Opera* (Danish National Opera)

Djokja Djokjakarta, Java, Indonesia

D.Journ. Doctor of Journalism

dj's (DJs) disc jockeys

DJs Department of Justice investigators

D.J.S. Doctor of Juridical Science

DJTA Dow Jones Transportation Average

DJTD Department of Justice Tax Division

D.Jur. Doctor of Jurisprudence

DJWWB *Dowodztwo Jednostki w Wielkiej Brytanii* (Polish—Units Command in Great Britain)

dk dark; decay; deck; didn't know; diseased kidney(s); dock; dog kidney; don't know; drop kick; duck; duct keel; dusky

d & k dining and kitchen

DK Danny Kaye

D & K Dalhoff and King

DK *Danmark* (Danish—Denmark)

DKB Dai-ichi Kangyo Bank; *Det Kongelige Bibliotek* (Danish—The Royal Library)—Copenhagen

DKC De Kalb College

dk di dinking die

dkftcol dark fast color

dkg decking; dekagram(s)

dkga (DKGA) diketogulonic acid

dkgrcol dark-ground color

dkhse deckhouse

dk hse deck house

DKI *Det Kriminalistiriske Institute* (Danish—The Criminalistic Institute)—Copenhagen

DKK *Danmark Kroner* (Danish crowns)

dkl dekaliter

dkm dekameter

dkm² square dekameter

dkm³ cubic dekameter

dkp deck passenger(s)

DKP Democratic Korea Party

DKP *Danmarks Kommunistiske Parti* (Danish Communist Party); *Deutsche Kommunistische Partei* (German Communist Party)

Dkr Dakar

DKr Danish krone(r)

DKR Dakar, Senegal (airport)

dks dekastere

DKS Deputy Keeper of the Signet; Direct Keying System

dkt docket

DKTC Door-Kewaunee Teachers College

DKW *Dampf Kraft Wagen* (German—steam powered vehicle); *Das Kleine Wunder* (German—The Little Wonder—automobile); *Deutsche Kraftfahrt Werks* (German—German Power-drive Works)

dkyd dockyard

dl data link; day letter; dead load; deadlight; deciliter; delay line; demand loan; difference limen (threshold); dog license; dollar; double acetate; drawing list; driver's license

d-l -dextro-levo

d/l data link; demand loan

Dl Daniel

DL Danger List; Delta Air Lines (2-letter symbol); Department of Labor; difference of latitude; Djakarta Lloyd; Drawing List; Drury Lane (London theater); frigate (naval symbol)

D-L Deputy-Lieutenant

D/L De Luxe

D of L Department of Labor; Department of Labour; Department of Law; Duchy (Duke) of Lancaster; Duchy

(Duke) of Lorraine; Duchy (Duke) of Luneburg

DL *Danske Lov* (Danish Law)

dla distolabial

Dla Douala

DLA Decorative Lighting Association; Defense Logistics Agency; District Licensing Authority; Divisional Land Agent (UK); Documentation, Libraries, and Archives Director(ate)

dlab disc label

dlai distolabioincisal

D.Lang. Doctor of Languages

D-L antibody Donath-Landsteiner antibody

DLAS Defence of Literature and the Arts Society

d lat difference in latitude

DLAT Defense Language Aptitude Test (USA)

dlb's dead-letter boxes

dlc direct lift control; down left center

DLC Democratic Leadership Council; Disaster Loan Corporation; Duquesne Light Company

DLCO Desert Locust Control Office

DLCO-EA Desert Locust Control Organization—East Africa

dld deadline date; delivered

dle data link escape; disseminated lupus erythematosus (DLE)

DLE Department of Law Enforcement

dlea double leg elbow amplifier

D.L.E.S. Doctor of Letters in Economic Studies

dlet delete

dletd deleted

dletg deleting

D Lett *Docteur en Lettres* (French—Doctor of Letters)

DLF Development Loan Fund(ing); Disabled Living Foundation

DLG David Lloyd George; guided-missile frigate (naval symbol)

DLG *Deutsche Landwirtschafts Gesellschaft* (German—German Agricultural Society)

DLGA Decorative Lighting Guild of America

DLGCD Department of Local Government and Community Development

DLGN nuclear-powered guided missile frigate (naval symbol)

DLH *Deutsche Lufthansa* (German airline)

DLI Defense Language Institute; Department of Labour and Industry (Australian)

DLIA Dental Laboratories Institute of America

D-library duplicating library

dlimp descriptive language for implementing macroprocessors

dlir depot-level inspection and repair

DLIS Desert Locust Information Service

D. Litt. *Doctor Litterarum* (Latin—Doctor of Letters, Doctor of Literature)

dll dial long line

DLL *Deutsche Levante-Linie* (Levant Line); Donaldson Line Limited

dllf design limit load factor

dlli dulcitol lysine lactose iron (DLLI)

dlM *des laufenden Monats* (German—this month)

DLM Daily List of Mails; Depot-Level Maintenance

DLMA Decorative Lighting Manufacturers Association; Downtown Lower Manhattan Association

dlmmjvs *domingo, lunes, martes, miércoles, jueves, viernes, sábado* (Spanish—Sunday, Monday, Tuesday, Wednesday, Thursday, Friday, Saturday)

DLNS Department of Labour and National Service (Australian)

DLNWR Des Lacs National Wildlife Refuge (North Dakota)

dlo difference in longitude; dispatch loading only; distolinguo-occlusal

D'Lo The Lord (town in Mississippi)

DLO Dead Letter Office; Difference of Longitude; District Legal Office(r)

D.L.O. Diploma in Laryngology and Otology

DLOC Division Logistical Operation Center

DLOCA Department of Law Office Consumer Affairs

d lock dial-lock

d long difference in longitude

D-love deficiency love (exploitative and possessive love of another person)

DLOY Duke of Lancaster's Own Yeomanry

dlp date of last payment; distolinguopulpal; double-large post; mean diurnal low-water inequality

DLP Democratic Labour Party (Barbados); Director of Laboratory Programs (USN)

DLPS Department of Law and Public Safety (New Jersey)

dlq deliquescent; mean diurnal low water inequality

dlr dealers; discharge, land, and reload; discharged, landed, and reshipped; dollar; double lift restow; double-lens reflex (camera)

d-l-r discharge-load-reposition (containers)

DLR District Land Registrar; Driving Licences Regulations

DLR *Distrito de la Luz Roja* (Spanish—Red Light District)

dlra door lock rotary actuator

DLRA Divorce Law Reform Association

DLRO District Labor Relations Office(r)

dlrs dollars

DLRs *Dominion Law Reports* (Canadian)

dls debt liquidation schedule; dollars

dls *dólares* (Spanish—dollars)

DLs Defence Lists

DLS Debt Liquidation Schedule; District Law Society

D.L.S. Doctor of Library Science; Doctor of Library Service

D.L.Sc. Doctor of Library Science

DLSC Defense Logistics Service Center

DLSEF Division of Library Services and Educational Facilities (U.S. Office of Education)

dls/shr dollars per share

dlt deck landing training; dry long tons

dlt (DLT) data-loop transceiver (data processing)

dlt dans le texte (French—in the text)

DLT Development Land Tax(ation); Discrimination Learning Test

D-L T Donath-Landsteiner Test

dlts deep-level transient spectroscopy

DLTS Deck Landing Training School

dlu digitizer logic unit

dlvd delivered

dlvr deliver; delivery

dlvry delivery

DLW Diesel Locomotive Works

DL & W Delaware, Lackawanna and Western (railroad)

dlwg daily weight gain

dlx deluxe

dly daily; delay; dolly

dlyd delayed

dm data management; dead meat; decimeter(s); delta modulation; demand meter; development milestone; diabetes mellitus (DM); diabetic mother; diagnostic monitor; diastolic murmur; diesel-mechanical; diphenylamine-arsine chloride (Adamsite war gas); direct monitoring; double medium; draftsman; dry matter

d/m date and month; day and month; density/moisture

d & m dressed and matched

d m *destra mano* (Italian—right hand)

d M *dieses Monats* (German—this month)

DM Deputy Master; Des Moines; Design Manual; Deutsche Mark (German mark—currency unit); Drafting Manual; Du Mont (television network); Dungeon Master; Dungeon Module; light minelayer, high-speed (naval symbol)

D.M. Doctor of Mathematics; Doctor of Medicine; Doctor of Music; Doctor of Musicology

D & M Detroit and Mackinac (railroad)

D of M Duchy (Duke) of Milan

DM *Daily Mail; Deutsche Mark* (German mark)

dm² square decimeter

dm³ cubic decimeter

dma direct memory access; direct memory asset

DMA Dance Masters of America; Defence Manufacturers Association (Australian); Defense Mapping Agency; Delicatessen Managers Association; Direct Mail Association;Direct Marketing Association; Division of Military Application; Dog Museum of America; Dominion Marine Association

DMAA Direct Mail Advertising Association

DMAAC Defense Mapping

Agency Aerospace Center

dmac dimethylacetamide (DMAC)

DMAC Des Moines Art Center

D.Ma.Eng. Doctor of Marine Engineering

DMAHC Defense Mapping Agency Hydrographic Center

DMAIAGS Defense Mapping Agency Inter-American Geodetic Survey

D-man drug-enforcement officer

DMAODS Defense Mapping Agency Office of Distribution Services

dmards disease-modifying anti-rheumatic drugs

D Mark Deutsche Mark (German mark)—currency unit

DMATC Defense Mapping Agency Topographic Center

D.Math. Doctor of Mathematics

d-max density maximum

dmb dual-mode bus

Dmb Dumbarton

dmba dimethylbenzanthracene (DMBA)

dmbc direct material balance control

DMBC Detroit Motor Boat Club

dmbl demobilization; demobilize; demobilized

dmc digital microcircuit(ry); dimethylcarbinol (DMC)—insecticide; direct manufacturing cost(s); dough moulding compound

DMC Del Mar College; Democratic Movement for Change; Developing Member Country; District Materials Center

dmcl (DMCL) device media control language

dmctc dimethylchlortetracycline (DMCTC)

DM & CW Diploma in Maternity and Child Welfare

dmd demand; diamond; disc memory drive

Dmd Duchenne's muscular dystrophy

D.M.D. *Dentariae Medicinae Doctor* (Latin—Doctor of Dental Medicine)

dmdd demanded

dmdg demanding

dme distance measuring equipment

DME Designated Medical Examiner; Director of Mechanical Engineering; Director of Medical Education

DMEA Defense Minerals Ex-

ploration Administration

DMEA Dictionary of Mechanical Engineering Abbreviations

D.Mec.E. Doctor of Mechanical Engineering

D.Mech. Doctor of Mechanics

dmed digital message entry device

D.Med. Doctor of Medicine

D.M.Ed. Doctor of Musical Education

D-men drug-enforcement officers; narcotics officers

dmet distance-measuring equipment and tacan

D.Met. Doctor of Metallurgy

DMET Director(ate) of Marine Engineering Training

D.Met.Eng. Doctor of Metallurgical Engineering

D. Meteor. Doctor of Meteorology

dmetu (DMETU) dimethylthiourea

dmf decayed, missing, or filled (teeth)

DMF Decorative Marble Federation

DMFA Direct Mail Fundraisers Association

DMFOS Diploma in Maxillo-Facial and Oral Surgery

dmg damage; damaged; damaging

DMG Defense Marketing Group

D of M-G Duchy (Duke) of Mecklenburg-Güstrow

DMGO Division(al) Machine-Gun Officer

dmh drop manhole

DMH Director of Mental Hygiene; Division of Mental Hygiene

dm/ha dry matter per hectare

DMHS Director of Medical and Health Services; Dolley Madison High School

dmi defense mechanisms inventory; deferred maintenance item

DMI Data Machines Incorporated; Data Management Inquiry; Department of Manufacturing Industry; Director(ate) of Military Intelligence

DMIAAI Diamond Manufacturers and Importers Association of America, Incorporated

DMIC Defense Metals Information Center (Batelle Memorial Institute)

D.Mi.Eng. Doctor of Mining Engineering

D.Mil.S. Doctor of Military

Science

d-min density minimum

DMIR Duluth Mesabi and Iron Range (railroad)

DMJ Diploma in Medical Jurisprudence

dml demolish; demolition

dml (DML) data manipulation language; dimyristoyl lecithin

D.M.L. Doctor of Modern Languages

d mld depth moulded

DMLS Doppler Microwave Landing System

DMLT Diploma in Medical Laboratory Technology

dmm digital multimeter

DMM Defense Market Measures; Directorate of Materiel Management

dmma (DMMA) Direct Mail/Marketing Association

dmmf dry mineral matter free

dmmp (DMMP) dimethylmethyl phosphonate

dmn dimension; dimensional

Dmn Drammen

dmna (DMNA) dimethylnitrosamine

Dmn Fst Damnation of Faust

DMNH Delaware Museum of Natural History; Denver Museum of Natural History

dmnstr demonstrator

dmo demetallized oil

DMO Deputy Medical Officer; Director of Military Operations; District Medical Officer

dmod displacement-measuring optical device

DMO & I Director of Military Operations and Intelligence

dmp difference of meridional parts; dimethylphthalate (insect repellent also abbreviated DMP)

DMP Developing Mathematical Processes; Director of Manpower Planning; Dublin Metropolitan Police

dmpa depomedroxyprogesterone (DMPA)

DMPA Dublin Master Printers' Association

DMPB Diploma in Medical Pathology and Bacteriology

dmpea (DMPEA) dimethoxyphenylethylamine

dmpi desired mean point of impact

DMPL Des Moines Public Library

DMPP Duck Mountain Provincial Park (Manitoba and Saskatchewan)

dmpr damper
DMPS Deepwater Motion Picture System
DMR Data Management Routines; Defective Material Report; Department of Main Roads; Diploma in Medical Radiology; Director of Materials Research
DMRC Deering Milliken Research Corporation
DMRD Diploma in Medical Radio-Diagnosis; Directorate of Materials Research and Development
DMRE Diploma in Medical Radiology and Electrology
DMRT Diploma in Medical Radio-Therapy
dms dermatomyositis; diacritical marking system (DMS); digital multiplex switching; drums
DMS Data Management System; Decision Making System; Defense Mapping School; Denominational Ministry Strategy; Director of Medical Services; Disk Monitoring System; Display Management System; Division of Medical Standards; Dominion Mutual Securities
D.M.S. Doctor of Medical Science
D of M-S Duchy (Duke) of Mecklenburg-Schwerin
D.M.Sc. Doctor of Medical Science
DMSC Defense Medical Supply Center
DMSDS Direct Mail Shelter Development System
DMSE Developing Models for Special Education
DMSGR Dowd's Morass State Game Reserve (Victoria, Australia)
dmsh diminish
DMSI Directorate of Management and Support of Intelligence
dmso (DMSO) dimethyl sulfoxide
DMSP Defense Meteorological Satellite Program
DMSS Data Multiplex Subsystem; Director of Medical and Sanitary Services
dmst demonstrate; demonstration
dmstn demonstration
dmstr demonstrator
dmt demountable; dimethyltryptamine—DMT (dangerous hallucinogen)

DMT Department of Motor Transport(ation); Director-(ate) of Military Training
DM & TS Department of Mines and Technical Surveys
dmu dual maneuvering unit
DMU Des Moines Union (railway)
D.Mus. Doctor of Music
D.Mus.A. Doctor of Musical Arts
D.Mus.Ed. Doctor of Musical Education
DMV Department of Motor Vehicles
D.M.V. Doctor of Veterinary Medicine
dmy dummy
DmZ demilitarized zone
DmZ (DMZ) Demilitarized Zone
dn debit note; decinem; dekanem; delta amplitude (symbol); dibucaine number; dicrotic notch; died near; down; downward
d'n damn
d/n (D/N) debit note
d & n dumb and numb (insensitivity factor)
d...n damn
d/N dextrose/nitrogen (ratio)
Dn Dale; Daniel; Dragoon(s)
D^n *Don* (Spanish—title equivalent to "Sir")
DN Department of the Navy; Division Notice
D.N. Diploma in Nursing; Diploma in Nutrition
D of N Daughters of the Nile
D.N. *Dominus Noster* (Latin—Our Lord)
dna did not attend; does not answer
Dna *Doña* (Spanish—Lady)—Mrs
DNA Defense Nuclear Agency; deoxyribonucleic acid (chromosome and gene component); Digital Network Architecture
DNA *Deutscher Normenausschusz* (German—German Committee on Standards)
DNAD Director of Naval Air Division
DNAN Department Number Assignment Notice
DNANR Department of Northern Affairs and National Resources
D.N.Arch. Doctor of Naval Architecture
dna(s) *docena(s)* [Spanish—dozen(s)]
DNase deoxyribonuclease

dnb departure from nucleate boiling; dinitrobenzene
DNB Distribution Number Bank
D.N.B. Diplomate of the National Board of Medical Examiners
DNB *Dictionary of National Biography*
dnc direct numerical control
DnC *Det Norske Creditbank* (The Norwegian Credit Bank)—also shown as *DNC*
DNC Democratic National Committee; Domestic National Committee; Director of Naval Construction
dncb dinitrochlorobenzene (DNCB)
DNCCC Defense National Communications Control Center
DNCMD Dayton Contract Management Office
dn ctl down control
dnd died a natural death
Dnd Dunedin
DND Department of National Defense; Department of National Development; Director of Navigation and Direction; Division of Narcotic Drugs (UN)
DN & D Director of Navigation and Direction
dne *douane* (French—customs)
DNE Director of Naval Equipment; Director of Nursing Education
D.N.Ed. Doctor of Nursing Education
D.N.Eng. Doctor of Naval Engineering
DNES Director of Naval Education Service
dnf did not finish
dnfb dinitrofluorobenzene
DNHW Department of National Health and Welfare (United Kingdom)
DNI Director of Naval Intelligence
DNI *Dana Normalisasi Indonesia* (Indonesian Institute of Standards)
DNIC Data Network Identification Code
DNII *Dirección Nacional de Información e Inteligencia* (Spanish—National Direction of Information and Intelligence)—Uruguay
dnj drone noise jammer
DNJ *Det Norske Justervesen* (Norwegian Bureau of Weights and Measures)

D.N.J.C. Dominus Noster Jesus Christus (Latin—Our Lord Jesus Christ)

Dnk Dunkirk

dnka did not keep appointment

dnl do not load; dynamic noise limiter

DNL Det Norske Luftfartselkap (Norwegian Airlines)

dnm data name

DNM Dinosaur National Monument

DNMS Director(ate) of Naval Medical Services; Division of Nuclear Materials Safeguards

DNO Director of Naval Ordnance; District Naval Office(r); District Nursing Officer

DNO Den Norske Opera (The Norwegian Opera)—Oslo

dnoc dinitro-orthocresol (DNOC)

d/note debit note

D-Note $500 bill

D-Notices Defense Notices

D-notice system British defense-notice system for protecting state secrets with the cooperation of the press

dnp do not publish

DNP 2, 4-dinitrophenol; Dinder National Park (Sudan)

dnpm dinitrophenyl morphine (DNPM)

D.N.P.P. Dominus Noster Papa Pontifex (Latin—Our Lord the Pope)

dnpt (DNPT) dinitrosopentamethylene tetramine

dnr does not run; do not renew; dynamic noise reduction

dnr (DNR) do not resuscitate

D/N r dextrose-to-nitrogen ratio

DNR Department of National Revenue; Department of Natural Resources; Director(ate) of Naval Recruiting

d/n ratio ratio of dextrose (glucose) to nitrogen in the urine

dns dinoyl sebacate (DNS)

Dns Downs

DNS Decimal Number System; Department of National Savings (British)

DNSA Diploma in Nursing Administration; Director of National Security Affairs

D.N.Sc. Doctor of Nursing Science

dnslp downslope

DNSS Defense Navigation Satellite System

dnt dinitrotoluene

DNT Director(ate) of Naval Training

Dntn downtown

DNTO Danish National Travel Office

dntp diethyl-nitrophenyl thiophosphate (DNTP)—insecticide

Dnus. Dominus (Latin—Lord)

DNV Det Norske Veritas (Norwegian ship classifier)

dnwind downwind

DNWR Darling National Wildlife Refuge (Florida); Delta NWR (Louisiana); Desert NWR (Nevada)

DNWS Director(ate) of Naval Weather Service(s)

do. day(s) off; defense optics; delivery order; diamine oxidase (DO); diesel oil; direct order; dissolved oxygen; ditto; dropout; dual ownership

do' door

d-o dropout

d/o delivery order

do. dictum (Latin—as before, the same); *ditto* (Italian—the same)

d:o: dito (Swedish—ditto)

d O der (die, das) Obige (German—the aforementioned)

Do Dominican; Dominican Republic; Dominican or Santo Domingan; Dornier

DO Defense Order; Department of Oceanography; Design Office; Director of Operations; Disbursing Office(r); District Office(r); Division(al) Office(r); Dominion Observatory; Dominion Office(r); Duty Officer

D.O. Doctor of Optometry; Doctor of Osteopathy

D/O Disbursing Officer

DO-27 Dornier 6-passenger utility aircraft built in West Germany and also called Skyservant

doa date of arrival; date of availability; dead on arrival; direction of approach; disposal of assets; dissolved oxygen analysis

DoA Department of Agriculture; Department of the Army (DOA)

DOA Dead on Arrival; Draft on Arrival

Doac Dubois oleic albumin complex

DOAE Defence Operational Analysis Establishment (UK)

DOAL Deutsche Ost Afrika Linie (German East Africa Line)

Doàn Quân Vietnam national anthem

DOARS Donnelley Official Airline Reservations System

dob date of birth; degree of bend(ing); diameter overbark; disbursed operating base; doctor's order book

dob (DOB) 2.5-dimethoxy-4-bromoamphetamine (hallucinogen causing blood-vessel constriction leading to possible loss of limbs)

DoB Daughters of Bilitis

DOB Date of Birth; doctor's order book; Do Our Best (Boy Scout and Girl Guide slogan)

DOB Deutsche Oper Berlin (German Opera of Berlin)

Dob(bin) Robert

Dobbs School for Girls State Training School for (delinquent) Girls at Kinston, North Carolina

'dobe adobe

dobe(s) doberman dog(s)

dobra monetary unit of São Tome and Principe

D Obst RCOG Diplomate of the Royal College of Obstetricians and Gynaecologists

doc date optimizing computer; desoxycorticosterone (DOC); died of other causes; diesel oil cement; direct operating cost; doctor; doctoral; document; documentary; documentation; drive(s) other cars

doc (DOC) desoxycorticosterone

Doc doctor

DoC Department of Commerce

D o C Department of Correction (Arkansas, Connecticut, Delaware, Indiana, Massachusetts, North Carolina, Tennessee); Department of Corrections (Arizona, California, District of Columbia, Florida, Guam, Idaho, Illinois, Kansas, Kentucky, Louisiana, Maine, Michigan, Minnesota, Mississippi, Missouri, New Jersey, Rhode Island, South Carolina, Texas, Vermont, Washington, West Virginia); Division of Corrections (Utah, Wisconsin)

D-o-C Doctors-on-Call

DOC Department of Commerce; Department of Communications; District Officer in Command; District Officer Commanding

DOC Denominazione di Origine Controllata (Italian—Place of Origin Controlled)

doca data of current appointment; deoxycorticosterone acetate

DOCA Deoxycorticosterone Acetate

doce date of current enlistment

Doc.Eng. Doctor of Engineering

docg desoxycorticosterone glucoside (DOCG)

DOCIT Directors of Central Institutes of Technology (Australian)

DOCLINE Document Delivery On-Line (computer service)

docn documentation

Doc.Pol.Sci. Doctor of Political Science

DOCS Department of Correctional Services (NY)

Doct. Doctor (Latin—Doctor)

Doctᵃ Doctora (Spanish—Doctor)—feminine

docu document(ary)

docubio documentary biographee; documentary biographer; documentary biography

docudrama documentary drama

docum document; documentary; documentation; documented

documᵗᵒ documento (Spanish—document)

DOCUS Display-Oriented Computer Usage System

dod date of death; died of disease; dust of desuetude

Dod Dodecanese

DoD Department of Defense

DOD date of death; Department of Defense; died of disease; Domestic Operations Division (CIA)

DODAS Digital Oceanographic Data Acquisition System

DoDCI Department of Defense Computer Institute

Dodd Dodd, Mead

DODD Department of Defense Directive

DoDDAC Department of Defense Damage Assessment Center

DoDDS Department of Defense Dependent Schools

Dod(dy) Dorothy; George

Dodec Dodecanese

Dodecanese Dodecanese Islanders; Dodecanese Islands

DODI Department of Defense Instruction

dodprt date of departure

Dodson's Dodson's Reports

doe. date of enlistment; dyspnea on exercise; dyspnea on exertion

DoE Department of Education (DoEd is better); Department of Energy (DoEn is better); Department of the Environment; Director(ate) of Education

DOE Department of Education (DoEd is better); Department of Energy (DoEn is better)

DoEd Department of Education

DoEn Department of Energy

DOES Disk-Oriented Engineering System

doesn't does not

dof degrees of freedom; delivery on field

DOF Defense Optics Facility; Developmental Optics Facility

dofab damned old fool about books

dofic domain-originated functional integrated circuit

DOFL Diamond Ordnance Fuze Laboratories

dog hot dog; frankfurter; sausage(s)

dog. disgruntled old graduate

Dogger Dogger Bank in the North Sea off England's east coast

dogm dogmatic; dogmatism; dogmatist

DOGMAD Dissatisfied Owners of General Motors Automotive Diesels

doh direct operating hours

Doh Doha

dohc double overhead cam; dual overhead cam

doi dead of injuries; descent orbit insertion

DoI Department of Industry; Department of the Interior (DoInt is better); Director(ate) of Information

D o I Department of Institutions (Montana); Department of the Interior; Director of Institutions (North Dakota); Division of Institutions (Oklahoma)

doin' doing

DoInt Department of the Interior

do/it digital output/input translator

DoJ Department of Justice

D Ø K Det Østasiaatiske Kompagni (Royal Danish East Asiatic Company)

Dok Akad Nauk Doklady Akademii Nauk (Russian—Proceedings of the Academy of Science)—USSR

dol dear old lady; display-oriented language; dollar

dol (DOL) dioleoyl lecithin

dol dolce (Italian—sweet); *dolor* (Latin or Spanish—pain)—the *dol* is the unit of pain; *dolore* (Italian—pain)

Dol Dolph (Adolf); dolphin; Dorothea; Dorothy

DoL Department of Labor (Do Lab is better)

D o L Department of Labor; Department of Labour

D.o.L. Doctor of Oriental Learning

D & O L Director's and Officer's Liability

Do Lab Department of Labor

DOLARS Dynamic Preferential Runway System

dolciss dolcissimo (Italian—very sweetly)

Dolf Adolph; Adolphus; Rudolph

dolichocephs dolichocephalics (long-skulled people)

Doll Dorothy

dollar monetary unit of the Bahamas, Canada, Fiji, Guyana, Jamaica, Liberia, New Zealand, Singapore, Solomon Islands, Trinidad and Tobago, the United States, and Zimbabwe

dollies dolophine pills

dolo dolophine (methadone hydrochloride used as a morphine substitute in withdrawing addicts from heroin)

Dolomites Dolomite Alps of northeastern Italy

Dolores Dolores Hidalgo, Guanajuato, Mexico

Dolph Adolph, Bardolph, Rudolph, Zardolph

DOLPHIN Dump Obsolete Laws—Prove Hypocrisy Isn't Necessary

dols dollars

dom date of marriage; digestible organic matter; dirty old man; dissolved organic matter; division-owned material(s); domestic; domicile; dominant; dominion; drawn over mandrel

dom *domenica* (Italian—Sunday); *domingo* (Portuguese or Spanish—Sunday)

Dom Domain; Domenico; Dominic; Dominican; Dominican Republic; Dominion

Dom. *Dominicus* (Latin—of the Lord, as in *Dies Dominica*—the Lord's Day)

DOM Date of Marriage; dimethoxyalpha methyl phenethylmine (psychedelic drug also called STP)

D.O.M. *Deo Optimo Maximo* (Latin—to God the Best and the Greatest)

DOMAINS Deep-Ocean Manned Instrument Station(s)

Dom Bk *Domesday Book*

Dom Can Dominion of Canada

Dom Day Dominion Day (celebrated in Canada July 1)

dom econ domestic economy (home economics)

DOMEI-KAIGI *Zen Nihon Rodo Sodomei Kumiai Kaigi* (Japanese—Confederation of Labor)

DOMES Deep-Ocean Mining Experimental Study

dom ex domestic exchange

Dom Fiji Dominion of Fiji

domi domicile

domina distribution-oriented management information analyzer

Dominica Commonwealth of Dominica (formerly a British Windward Island and the most northern of the Windwards in the Caribbean)

Dominican Republic eastern half of Hispaniola in the West Indies, *República Dominicana*

Dominion Dominion Day or Canada Day (July 1)

DOMMDA Drawing Office Material Manufacturers and Dealers Association

dom° *domingo* (Spanish—Sunday)

Dom° *Domingo* (man's name)

DOMO Dispensing Opticians Manufacturing Organization

Dom Pedro II Dom Pedro de Alcantara, emperor and president of Brazil

Dom.Proc. *Domus Procerum* (Latin—House of Lords)

Dom Rep Dominican Republic

DOMS Diploma in Ophthalmic Medicine and Surgery

domsat domestic communication satellite; domestic satellite carrier

dom sci domestic science

don' don't (do not)

don. *donec* (Latin—until)

Don Donald; Donaldo; Donegal

Don *Donderdag* (Dutch—Thursday); *Don Giovanni* (Mozart opera); *Donnerstag* (German—Thursday); (Spanish—Lord and Master, from the Latin—dominus); *Don Quixote* (fantastic variations for cello and orchestra by Richard Strauss); The Don—Mozart's two-act comic opera—*Don Giovanni*

DoN Department of the Navy

DON Diploma in Orthopaedic Nursing

Donalbane Donald Bane

Donbas Donets Basin in the Ukraine

donec alv. sol. fuerit *donec alvus soluta fuerit* (Latin—until the bowels move)

Doneg Donegal (sometimes Don)

Donets Donets Basin or Donbas of the Ukraine

dong monetary unit of Vietnam

Dong Phan Van Dong

donk donkey; donkeyback; donkeyboiler; donkey boy; donkey breakfast (sailor's straw-stuffed mattress); donkeycart; donkey crosshead; donkey engine(man); donkey house; donkeyman; donkey pump; donkey puncher; donkey sled; donkey stack; donkeywork(man)

Donnie Donald

Don Q Don Quixote

DONS Department of National Security (South Africa)

don't do not

do-nut doughnut

doo diesel oil odor

DOO Director—Office of Oceanography

doom. deep ocean optical measurement

dop dermo-optical perception; designated overhaul point; developing-out paper; dressing-out percentage

dop (DOP) diocytl phthalate

D o P Department of Prisons (Nevada)

dopa dynamic output printer analyzer

dopa (DOPA) dihydroxyphenylalanine

dopadic dope addict

dopase dopa oxidase

D. Oph. Doctor of Ophthalmology

D.Ophth. Doctor of Ophthalmology

dopl *doplene* (Czech—enlarged)

dopp ped *doppio pedale* (Italian—double pedal)—musical term

d-o psychiatrists directive-organic psychiatrists

D.Opt. Doctor of Optometry

dor date of rank; dental operating room; digital optical recording; doric; dormitory

Dor Dorado; Doric; Dorothy

DoR Department of Rehabilitation

D. Or. Doctor of Oratory

DOR Department of Offender Rehabilitation (Georgia); Director(ate) of Operational Research

dora dynamic operators research apparatus; dynamic operators response apparatus

Dora Deborah; Dorothea; Dorothy; Eudora; Theodora

DORA Defence of the Realm Act

doran Doppler range and navigation

Dord Dordogne

DORDEC Domestic Refrigerator Development Council

Doric(k) Theodoric(k)

Dorie Doris; Theodora; Theodore

Doris Doreen; Dorothea; Dorothy; Eudora; Theodora

DORIS Direct Order Recording and Invoicing System

DORL Deployable Orbital Research Laboratory

dorm(s) dormitory; dormitories

dorna deoxyribose nucleic acid

Dors Dorset; Dorsetshire

Dorset Dorsetshire

Dort Dordrecht

DORT Detroit Objective Reference Test

D Orth Diploma in Orthodontics; Diploma in Orthoptics

dos date of sale; date of separation; dosage; dose; dosimetric; dosimetry; dosiology

dos. *dosis* (Latin—dose)

Dos John Dos Passos

DoS Department of State

DOS Date of Separation; Department of State; Digital Operation System; Disk Operating System

D.O.S. Doctor of Ocular Science; Doctor of Optical Science; Doctor of Optometric Science

Dosc Dubois oleic serum complex

DOSCO Dominion Steel and Coal Corporation

Dosh Univ Doshira University

dosim dosimetry (measurement of radiation doses)

do'sn't does not

DOSS Deep-Ocean Search System

DOST Dictionary of the Older Scottish Tongue

dosv deep ocean survey vehicle

dot. deep ocean transponder; deep-ocean technology; deep-oceanic turbulence; dotation-(al); draft(s) on treasury

Dot Dorothy; Dotty

DoT Defense of the Territory; Department of Telecommunications; Department of Tourism; Department of Trade (United Kingdom); Department of Transport (Canada); Department of Transport-(ation); Department of Transportation (US); Department of the Treasury

D o T Defense of the Territory; Department of Trade; Department of Transport; Department of Transportation

DOT Deep Oil Technology (company); Department of Overseas Trade; Diploma in Occupational Therapy

DOT Dictionary of Occupational Titles

DOTE Department of the Environment

DOTIPOS Deep Ocean Test-in-Place and Observation System

DOTM Department of Ordnance, Torpedoes, and Mines

Dott Dottore (Italian—Doctor)

D o T & T Dominion of Trinidad and Tobago

Dotty Doreen; Dorothea; Dorothy; Eudora

dou (DOU) definitive observation unit

Douay Douay Version of the Bible (published at Douai, France in 1609)

double-B double-backed; double-banked; double-barreled; double-bass; double-bedded; double-benched; double-bonded; double-bottomed; double-breasted; double-brooded

Double D Doubleday

double-X doublecross; double quality; double quantity; double thickness; doubleweight; two-X; XX

doubt. doubtful

Doug Douglas(s)

Doug fir Douglas fir

dov data over voice; double oil of vitriol (sulphuric acid)

Dov Dover; Dovid

Dov Dovid (Yiddish—David)

DOV Defence of Village (educational program)

dovap Doppler velocity and position

Dover Delaware's capital named after an English Channel port; Dover Publications

dow died of wounds; dowager; dowel; dowelled

Dow Dowager

DOW Died of Wounds; Dow Chemical Company; Dow Chemicals

DoWaPO Dictionary of Word and Phrase Origins

dowb deep ocean work boat

Dow Kuw Dowlat al-Kuwait (Arabic—State of Kuwait)

Down Downing College, Cambridge

dows dowsing; dowsers

doy day of year

doz dozen

dozer bulldozer

dp damp proof(ing); dash pot (relay); data processing; deck piercing; deep penetration; deep pulse; deflection plate; departure point; dewpoint; diametral pitch; diastolic pressure; diffusion pressure; digestible protein; diphosgene (deadly gas); diprorprionate; disability pension; discriminatory power; diphosphate; displaced person; distopulpal; distribution point; donar's plasma; double paper; double plays; double pole; drip-proof; drop point; dual purpose; dump; durable press; potential difference (symbol)

dp (DP) data processing; dementia praecox

d/p delivery papers; documents against payment; door-to-port/port-to-door (delivery)

d & p developing and printing; development and printing; drain and purge

d.p. directione propria (Latin—with proper direction)

d/p días plazo (Spanish—pay days)

d. in p. divide in partes (Latin—divide)

DP by direction of the President; Dan Pezze; Democratic Party; Department of the Pacific; Detrucking Point; Dinosaur Park (Drumheller, Alberta); Director of the Port; Displaced Person

D-P Data-Phone

D.P. dementia praecox; Doctor of Pharmacy; Doctor of Podiatry

D & P Deberny and Peignot

D of P Daughters of Pennsylvania; Daughters of Pocahontas; Director of Planning; Director of Plans; Duchy (Duke) of Prussia

DP Denver Post

D.P. Domus Procerum (Latin—House of Lords)

dpa deferred payment account; diagnostic prescriptive arithmetic

dpa (DPA) diphenylamine; dipicolinic acid

dPA di Pietro Aretino

Dpa Diputada (Spanish—Deputy)—feminine

DPA Data Processing Agency; Data Protection Authority; Diabetes Press of America; Discharged Prisoners Association; Division of Performing Arts; Division of Public Affairs

D.P.A. Doctor of Public Administration

DPA Deutsche Press Agentur (German news agency); *Doulat i Padshahi ye Afghanistan* (Kingdom of Afghanistan)

d. in p. aeq. divide in partes aequales (Latin—divide into equal parts)

dpars data processing automatic record standardization

DPAS Discharged Prisoners' Aid Society

D Path Diploma in Pathology

dpb deposit passbook

DPB Department of Printed Books (British Museum Library); Domestic Purposes Benefit

dpbc double pole both connected

dpc damp-proofing course; data processing computer; data processing control; double paper single cotton

DPC Daniel Payne College; Data Processing Center; Defense Plant Corporation; Defense Procurement Center; Defense Procurement Circular; Defense Production Chief; Deputy Police Commissioner; Desert Protective Council; Displaced Persons Commission; Dissemination Policy Council; District Police Commissioner; Division Planning Corporation; Domestic Policy Council; Duke Power Company

DPCE Data Processing Customer Engineering

dpcm differential pulse-code modulation

DPCP Department of Prices and Consumer Protection (British)

dpd data project directive; diffuse pulmonary disease

DPD Data Products Division (Stromberg-Carlson); Department of Public Dispensary; Diploma in Public Dentistry

DPD *Data Processing Digest*

dpdc double paper double cotton

dp di dimple die

DPDS Defense Property Disposal Service

dp dt double pole, double throw

dpe data processing equipment; digital processing effects; digital production effects; direct plate exposure

d-p-e development-printing-enlargement

Dpe Dieppe

DPE Diploma in Physical Education; Director of Primary Education; Director of Public Education

D.P.E. Doctor of Physical Education

D.Ped. Doctor of Pedagogy

DPED Department of Planning and Economic Development

DP/ED *Data Processing for Education*

dpe service developing-printing-enlarging service

DPEWS Designed-to-Price Electronic Warfare System

dpf deferred pay fund

DPf Deutsche Pfennig (German—pfennig)

DPF Drug Policy Foundation

dpfc double pole front connected

dpft double-pedestal flat-top (desk)

dpg data processing group; deck plate girder; digital pattern generator

dpg (DPG) diphosphoglyceric acid

DPG Dugway Proving Ground

DPGA Delaware Personnel and Guidance Association

DPGs Development Planning Groups

dph diamond pyramid hardness; diphenylhydantoin (DPH)

D. Ph. *Doctor Philosophiae* (Latin—Doctor of Philosophy)

DPH Department of Public Health; Department of Public Highways; Diploma in Public Health; Domestic Packing House

D.P.H. Doctor of Public Health

D.Pharm. Doctor of Pharmacy

DPHD Diploma in Public Health Dentistry

D.Phil. Doctor of Philosophy

DPHN Diploma in Public Health Nursing

d'phone dictaphone

D.Ph.Sc. Doctor of Physical Science

D Phys Med Diploma in Physical Medicine

dpi data processing installation

DPI Department of Primary Industries; Department of Public Information; Department of Public Instruction; Disorderly Persons Investigation; Distillation Products Industries; Division of Plant Industry

DPIF *Drug Product Information File*

DPII Dairy Products Improvement Institute

dp-ing data processing; durable pressing

dpir data processing and information retrieval

dpl deferred pastoral lease; deferred payment license; diploma; diplomat; dual propellant loading; duplex

dpl (DPL) dipalmitoyl lecithin

DPL Dallas Public Library; Dayton Power and Light; Dayton Public Library; Delhi Public Library; Denver Public Library; Detroit Public Library; diplomatic corps (license plate)

DP & L Dallas Power and Light

DPL *Den Polytekniske Lae-*

ranstalt (Danish—The Polytechnic Institute)—Copenhagen

DP & LC Dundee, Perth & London (shipping) Company

dplx duplex

dpm data processing machine; disintegrations per minute; documents per minute

DPM Data Processing Manager; Deputy Prime Minister; Deputy Provost Marshal; Development Program Manual; Diploma in Psychological Medicine

D.P.M. Doctor of Podiatric Medicine

DPMA Data Processing Management Association

dpn diamond pyramid number

dpn (DPN) diphosphopyridine nucleotide

dpng deepening

dpnh (DPNH) reduced diphosphopyridine (same as nadh or NADH)

d pnl distribution panel

DPNM Devil's Postpile National Monument

dpo development planning objective

Dpo Depot (postal abbreviation)

Dpo *Diputado* (Spanish—Deputy)

DPO Dayton Philharmonic Orchestra; Discontinued Post Office; Distributing Post Office; District Pay Office(r)

dpob date and place of birth

D.Pol.Eco. Doctor of Political Economy

D.Pol.Sci. Doctor of Political Science

dpp deferred payment plan

DPP Democratic Progressive Party; Director of Public Prosecutions; Disease Prevention Program

DPPS Department of Public Printing and Stationery

dpr day press rates; double lapping of pure rubber

DPR Democratic Peoples' Republic; Differential Police Response; Director(ate) of Public Relations

DPRGR *Dewan Perwakilan Ratjat-Gotong Rojong* (Indonesian—Mutual Cooperation House of Representatives)

DPRI Disaster Prevention Research Institute

DPRK Democratic People's Republic of Korea (North Korea)

DPRS Dynamic Preferential Runway System
dps double-pole snap switch
dp's (DPs) displaced persons
dp&s data processing and software
dPs displaced Palestinians
DPs Detention Pens
DP's displaced persons
DPS Data Processing Service; Data Processing Station; Data Processing System; Defense Printing Service; Department of Public Safety; Division of Primary Standards; Domestic Policy Staff
DPSA Data Processing Supplies Association
DPSB Defense Production Supply Board (NATO)
DPSC Defense Personnel Support Center; Defense Petroleum Supply Center
DPSCS Department of Public Safety and Correctional Services (Maryland)
DPSS Data Processing Subsystem; Department of Public Social Services
dpst deposit
dp st double pole, single throw
DPsy Diploma in Psychiatry; Diploma in Psychology
D. Psych. Doctor of Psychology
D.Psy.Sci. Doctor of Psychological Science
dpt deeper-pool (pay) test (oil well); department; deponent; deposition; depth
dpt (DPT) dipropylphytamine
DPT Design Proof Test(ing); Director of Physical Therapy
Dpto Departamento (Spanish—Department)
dpt vaccines diphtheria, pertussis, tetanus vaccines
dptw double-pedestal typewriter (desk)
dpty deputy
dpu data processing unit
D.Pub.Adm. Doctor of Public Administration
dpv dry pipe valve; duty-paying value
dp/w drawbar pull/weight (ratio)
DPW Department of Public Works
DPWA Data Processing Work Assignment
DPWG Defense Planning Working Group (NATO)
DPWO District of Public Works Office
dpx duplex

dq definite quantity; deterioration quotient; direct question(s)
dqd digital quadrature detection
dqm data quality monitors
DQMG Deputy Quartermaster General
DQMS Deputy Quartermaster Sergeant
DQU Deganawidah-Quetzalcoatl University (University of California at Davis)
dqy don't quit yet
dr debit; differential rate; door; double-reduction; drachma; dram; draw; drawn; drill; drive; drum
dr (DR) data register; dead reckoning; delivery room
d/r deposit receipt
Dr debtor; doctor; Drenthe (Dutch province); Drive; drachma (Greek monetary unit)
DR Data Report; Date of Rank; Dead Reckoning; Deficiency Report; Dental Record; Dental Recruit; Design Requirements; Despatch Rider; Detailed Report; Development Report; Document Report; National Distillers and Chemical Corporation (stock exchange symbol); reaction of degeneration (symbol)
D/R date of rank; dead reckoning
DR Deutsche Reichsbahn (German State Railway)
dra dead-reckoning analyzer
dr & a data reporting and accounting
dra derecha (Spanish—right)
Dra Draco (constellation)
Dra Doctora (Spanish—woman doctor)
Drª Doctora (Spanish—doctor)—feminine form; *Doutora* (Portuguese—doctor)—feminine form
DRA Democratic Republic of Afghanistan; Division of Ratepayer Advocates
Drac Draco
DRAC Director of the Royal Armoured Corps
drachma monetary unit of Greece
dr ad drill adaptor
D^ra D^na Doctora Doña (Spanish—Madam Doctor)
Dr Ae.Sc. Doctor of Aeronautical Science
Dr Agr. Doctor of Agriculture

drai dead-reckoning analog indicator
drain. drainage
dram. drama; dramatic; dramatist
dram. (DRAM) detection radar automatic monitoring
dram. pers. dramatis personae (Latin—cast of characters)
DRAMs Dynamic Random Access Memories
dr ap dram, apothecaries'
Draper Utah State Prison at Draper
drapes draperies
Dr Arne Thomas Arne
dras derechas (Spanish—duties, fees, tariffs)
Dr Atl Gerardo Murillo
dr av dram avoirdupois
Drav Dravidian
draw. direct read after write; drawing
drb design requirements baseline
Drb Durban
DRB Defense Research Board (Canada); Discharge Review Board; Druggists' Research Bureau
DRBC Delaware River Basin Commission
dr bg drill bushing
Dr Bi.Chem. Doctor of Biological Chemistry
DRBU Dharma Realm Buddhist University
Dr Bus.Adm. Doctor of Business Administration
drc damage-risk criteria (noise-exposure limits); down right center (driving, lighting, or seating)
DRC Department of Rehabilitation and Correction (Ohio); District Recruiting Command(er); Driver Re-education Course; Drug Referral Center; Drug Rehabilitation Center; Dutch Reformed Church; Dynamics Research Corporation
Dr C Cabinet of Dr Caligari
dr canon droit canon (French—canon law)
drch drachma
Dr Chem. Doctor of Chemistry
dr ck drill chuck
DRCOG Diploma of the Royal College of Obstetricians and Gynaecologists
Dr Com. Doctor of Commerce
dr com droit commun (French—common law)
dr comm droit commercial

(French—commercial law)

dr cout droit coutumier (French—common law)

Dr D *Doctor Don* (Spanish—Sir Doctor)

DR & D Defense Research and Development

DRDO Defense Research and Development Organization

drdp detection radar data processing

DRDT Daily Record of Dysfunctional Thoughts; Division of Reactor Development and Technology (AEC)

drdto detection-radar data take-off

dre dead reckoning equipment

DRE Defense Research Establishment (Canada); Department of Real Estate (California); Director of Religious Education

DR & E Defense Research and Engineering

D.R.E. Doctor of Religious Education

DREA Defense Research Establishment, Atlantic

drec detection-radar electronic component

Dr Ec. Doctor of Economics

dred. dredging

DREE Department of Regional Economic Expansion (Canada)

drek dead reckoning

Dren Drenthe (Dutch province)

Dr Eng. Doctor of Engineering

Dr Ent. Doctor of Entomology

DREO Defense Research Establishment, Ottawa

DREP Defense Research Establishment, Pacific

Dres Doctores (Spanish—Doctors)

DRES Defense Research Establishment, Suffield

Dr es L. Docteur ès Lettres (French—Doctor of Letters)

Dr es S. Docteur ès Sciences (French—Doctor of Sciences)

DRET Defense Research Establishment, Toronto

DREV Defense Research Establishment, Valcartier

Drew Andrew; Charles E. Drew Postgraduate Medical School

drews (DREWS) direct readout equatorial satellite

drf differential reinforcement; dose reduction factor

DRF Deafness Relief Founda-

tion; Direct Relief Foundation

dr féod droit féodal (French—feudal law)

drftmn draftsman

dr fx drill fixture

drg dorsal root ganglion; drawing(s); drogue; during

DRG Detroit Rubber Group; Dickinson Robinson Group

DRGM Deutsches Reichgebrauchsmuster (German registered design)

DRGs Diagnosis-Related Groups

D & RGW Denver and Rio Grande Western (railroad)

DRH Division of Radiological Health

Dr h.c. Doctor honoris causa (Latin—honorary doctor)

dr hd drill head

Dr Hor. Doctor of Horticulture

Dr Hy. Doctor of Hygiene

dri data rate indicator; data reduction interpreter; direct reduced iron; drive

DRI Dairy Research Institute; Data Resources Inc; Defense Research Institute; Denver Research Institute; Direct Relief International; Document Retrieval Index

drib deoxyribose

DRIC Dental Research Information Center; Dispute Resolution Information Center

drid direct-readout image dissector

DRIFT Diagnostic Retrieval Information For Teachers

Drigo Rodrigo

drill. drilling

DRINC Dairy Research Incorporated

D-ring capital-D-shaped ring

Dr Ing. Doktor-Ingenieur (German—Doctor of Engineering)

drinkin' drinking

drip. digital ray and intensity projector

drir direct readout infrared

DRIS Department of Defense Retail Interservice Support Program

DRIVE Developing Resources for Instructors of Vocational Education; Digital Raster Imaging, Viewing, and Editing system

dr jg drill jig

Dr Drenthe (Dutch province)

Dr J Dr Jekyll (and Mr Hyde)

Dr J.Sc. Doctor of Judicial Science

Dr Jur. Doctor Juris (Latin—Doctor of Law)

drk dark; display request keyboard

DRK Deutsches Rotes Kreuz (German Red Cross)

Drk-Yul Druk-Yul (Dzongka—Kingdom of Bhutan)

drl data retrieval language

DRL Design Report Letter; Diamond Research Laboratory

DRLG Danish Royal Life Guards

Dr Lit. Doctor of Literature

DRLS Dispatch Rider Letter Service

drm direction of relative movement

DRM Drafting Room Manual

dr mar droit maritime (French—maritime law)

Dr Med Doktor der Medizin (German—Doctor of Medicine)

Dr Med. Doctor Medicinae (Latin—Doctor of Medicine)

Dr Mus. Doctor of Music

drn drawn

Drn Dairen; Darien

DRN Daily Reports Notice; Detroit River Navigation

drna (DRNA) deoxyribose nucleic acid

Dr Nat.Sci. Doctor of Natural Science

drnt diagnostic roentgenology

dro destructive readout

dro (DRO) differential reinforcement of other behavior; double-room occupancy

dro derecho (Spanish—custom duty, right)

DRO Disablement Resettlement Office(r)

drod delayed readout detector

DRO-LA Defense Research Office—Latin America (USA)

dromdi direct readout miss-distance indicator

'drome aerodrome; airdrome

'Drome Hippodrome

dron data reduction

dros date returned from overseas

dros derechos (Spanish—duties, fees, tariffs)

drp dead reckoning position

DRP Democratic Republican Party (Korean); *Deutsches Reichspatent* (German—patent); Diebold Research Program

DRP Deutsche Reichspartei (German Reich Party)

DRPA Delaware River Port Authority
DRPC Defense Research Policy Committee
dr pén droit pénal (French—penal law)
Dr P.H. Doctor of Public Health
Dr Phil. *Doktor der Philosophie* (German—Doctor of Philosophy)
DRPL Del Rio Public Library
Dr Pol.Sc. Doctor of Political Science(s)
DRPP Director(ate) of Research Programs and Planning
drps digital random program selector; drapes
drq discomfort relief quotient
DRR Drawing Release Record
Dr Ra.Eng. Doctor of Radio Engineering
DRRB Data Requirements Review Board (DoD)
Dr Rec. Doctor of Recreation
Dr Re.Eng. Doctor of Refrigeration Engineering
DRRI Defense Race Relations Institute (DoD)
dr rom droit romain (French—Roman law)
d-r-r-r-r-r-um snaredrum roll
drs data-reduction situation; data reduction system; digital range safety; drawers; drowsiness
DRs Development Rights; Discrepancy Reports
DRS Dairy Research Station; Data Reduction System; Data Relay Station; Debtor Reporting System; Development Reference Service; Diagnostic Research System; Diagnostic Rework Sheet(s); Document Retrieval System
DRSAM Diploma of the Royal Scottish Academy of Music
drsc direct radarscope camera
Dr Sc. Doctor of Science
Dr Sci. Doctor of Science
DRSCS Digital Range-Safety Command System
dr sh drill shell
drsmkr dressmaker
drsn drifting snow
DRSO Danish Radio Symphony Orchestra
drsr dresser
DRSS Discrepancy Report Squawk Sheet
drt data review technique; dead reckoning tracer; dead reckoning trainer

dr t dram troy
Drt Dartmouth
DRT Diagnostic Rhyme Test; Dog Rescue Team
DRTC Documentation Research and Training Center
DRTE Defense Research Telecommunications Establishment (Canada)
Dr Tech. Doctor of Technology
Dr Theol. Doctor of Theology
Dr Theol. *Doktor der Theologie* (German—Doctor of Theology)
dr tp drill template
dr trav droit du travail (French—labor law)
dru digital register unit; digital remote unit
Dru Drusila
drub digital remote unit buffer
D.Ru.Eng. Doctor of Rural Engineering
Drug Abuse Drug Abuse Council
Drug Rehab Drug Rehabilitation Program
drums kettledrums, bass drums, field or tenor drums, side or snare drums, tomtoms or bongo drums
Dr und Vrl Druck und Verlag (German—printed and published by)
Dr Uni.Par. Doctor of the University of Paris
D.Rur.Sci. Doctor of Rural Science
DRUs Directing Reporting Units (Air Force)
drv data-recovery vehicle
Drv Drive
DRV Democratic Republic of Vietnam (North Vietnam)
DRVN Democratic Republic of Vietnam
drvr driver
dr vs drill vise
drw defensive radio(logical) warfare; drawing
DRW Darwin, Australia (airport)
drwg drawing
DRWS Dirty Radwaste System
DRWW Distillery, Rectifying, Wine Workers (union)
drx drachma (Greek monetary unit)
dry. drying
dry alum aluminum and potassium sulfate
dry disco alcohol-free, drug-free, tobacco-free discotheque
Dr Z Doctor Zhivago

drzl drizzle
ds days after sight; day's sight; dead-air space; debenture stock; decanning scuttle; decimal selector; density standard; detached service; dilute strength; dioptric strength; direct support; discarding sabot; document signed; domestic service; donar's serum; double silk; double stout; double strength; double-screened; double-stitch(ed); downspout; draft stop
ds (DS) data set (data processing); duration series
d-s dead slow (ship's engine signal)
d.s. document signed
d/s dextrose in saline
d & s demand and supply; dermatology and syphilology; distribution and supply
ds destro (Italian—right)
Ds Down's syndrome; dysprosium (symbol)
Ds. *Deus* (Latin—God); *Durchführumgssatz* (German—development of a sonata)
DS Date of Service; Delphian Society; Delta Society; Dental Surgeon; Department of Sanitation; Department of State; Design Standard(s); Detached Service; Direct Support; Directing Staff; Director of Services; Distributed Services; Drill Ship; Drug Store; Durham & Southern (railroad)
D-S Deux-Sèvres; Ditlev-Simonsen Lines
D & S Durham & Southern Railway
D of S Daughters of Scotia; Department of State; Duchy (Duke) of Savoy; Duchy (Duke) of Silesia; Duchy (Duke) of Styria
DS *Danske Standardiseringsraad* (Danish Standards Institute)
D S dal segno (Italian—return to the sign:*S:*)
D/S *Dampskip* (Norwegian—steamer, steamship)
dsa data set adapter; dial service assistance; dimensionally-stabilized anode; discrete sample analyzer
dsa (DSA) digital subtraction angiography
DSA Danish Sisterhood of America; Dante Society of

America; Defense Shipping Authority; Defense Supply Agency; Defense Supply Association; Democratic Socialists of America; Dental Surgery Assistant; Department of Substance Abuse; Deputy Sheriff's Association; Design Schedule Analysis; Division Service Area; Drum Seiners Association; Duluth, South Shore and Atlantic (railroad); Duodecimal Society of America

DSAA Defense Security Assistance Agency

DSAB Dictionary of South African Biography

dsabl disable; disability

DSACEUR Deputy Supreme Allied Command, Europe

DSAHBK Defense Supply Agency Handbook

DSAM Defense Supply Agency Manual

DSANZ Direct Selling Association of New Zealand

DSAO Diplomatic Service Administration Office(r)

DSAP Data Systems Automatic Program; Data Systems Automation Program; Defense Systems Application Program

DSARC Defense Systems Acquisition Review Council

dsas dial-service-assistance switchboard

D/S A/S Dampskipaksjeselskap (Norwegian—joint stock steamship company, limited)

dsasbl disassemble

DSASO Deputy Senior Air Staff Officer

dsb double sideband; double-strength B (quality glass)

DSB Danske Stats Baner (Danish State Railways); De Sola Brothers; Defense Science Board; Drug Supervisory Body (UN)

DSBA Delaware School Boards Association

dsbg disbursing

dsbn disband

dsc downstage center; dynamic standby computer

D.Sc. Doctor of Science

DSC Defense Supply Corporation; Delaware State College; Depot Supply Center; Die Casters' Conference; Distinguished Service Cross; Document Service Center

DSC (I) Die Sinkers' Conference (International)

D.S.C. Doctor of Christian Science; Doctor of Commercial Science; Doctor of Surgical Chiropody

D & SC Defense and Space Center (Westinghouse)

DSCA Douglas Social Credit Association

dscb data set control block

DSCC Deep Space Communications Complex

D.Sc.Com. Doctor of Science in Commerce

DSCDP Delaware State Central Data Processing

D.Sc.Eco. Doctor of Science in Economics

D.Sc.Eng. Doctor of Science in Engineering

D Sch Dmitri Shostakovich (in his *Tenth Symphony* uses his initials to form a four-note theme, applying German letters D, S for Es—E-flat, C, and H—German for B natural)

D.Sch.Mus. Doctor of School Music

D.Sc.Hyg. Doctor of Science in Hygiene

D.Sc.I. Doctor of Science in Industry

D.Sc.Jur. Doctor of the Science of Jurisprudence

D.Sc.L. Doctor of the Science of Law

DSCM Diploma of the Sydney Conservatorium of Music

DSCMD Dallas Contract Management District

D.Scn. Doctor of Scientology

D. Sc. Os. Doctor of the Science of Osteopathy

DSCP Detailed Site Characterization Plan

D.Sc.Pol. Doctor of Political Science(s)

DSCR Detailed Site Characterization Report

dscs direct-set cheese starter

DSCS Defense Satellite Communication System(s)

dsd dry surgical dressing

DSD Daily Staff Digest; Defence Signals Directorate (Australian); Depressive Spectrum Disease; Director of Signals Division

DSDP Deep Sea Diving Project; Deep Sea Drilling Program; Deep Sea Drilling Project

DSDS Deep Sea Diving School (USN)

d's & d's disbelievers and doubters

dse data-storage equipment; depot support equipment; development support equipment

D.S.E. Doctor of Science in Economics

DSE Departamento de Seguridad del Estado (Spanish—Department of State Security)—Cuba

DSEA Davis Submerged Escape Apparatus; Delaware State Education Association

dsf day-second-feet (or foot)

Dsf Dusseldorf

DSF Dainippon Silk Foundation; Daughters of St Francis of Assisi; Division of Sea Fisheries

dsfc direct side force control

dsg designate; designation

DSG Deutsche Schlaf- und Speisewagen Gesellschaft (German Sleeping-and-Dining-Car Company)

dsgl desgleichen (German—ditto)

dsgn design; designed; designer

dsgnd designated

dsh domestic short hair (cat)

DSHC Defence Service Homes Corporation

dshe downstream heat exchanger

d s'horn dairy shorthorn

dsi data systems inquiry; digital speech interpolation

DSI Dairy Society International; Dalcroze Society Incorporated; Distilled Spirits Institute; Drinking Straw Institute

DSIA Diaper Service Institute of America

DSIATP Defense Sensor Interpretation and Application Training Program

DSIF Deep-Space Instrumentation Facility

dsipt dissipate

DSIR Department of Scientific and Industrial Research

DSIs Directorate of Service Intelligence members or operatives

DSIS Directorate of Scientific Information Services

D-site decoy site

dsj differential space justifier

Dsk Dvorak simplified keyboard

D Sk Daily Sketch

dsl deep scattering layer; diesel; doppler speed log

DSL Dampier Salt Ltd; Deep Scattering Layer; Defence

Standards Laboratory; Delta Steamship Lines; Dickinson School of Law; Dominican School of Law; Dominican Steamship Line

D & SL Denver and Salt Lake (railroad)

DSL *Directory of Special Libraries and Information Centers*

DSLC Defense Logistics Services Center

D-sleep desynchronized sleep; rem sleep

dsl elec diesel electric

ds lt deck surface light

dsltd dry-salted (hides)

dsm dense-staining material; dried skim milk

d & sm dressed and standard matched (lumber)

DSM Des Moines, Iowa (airport); Development Shop Memorandum; Distinguished Service Medal; District Sales Manager

DSM *Diagnostic and Statistical Manual* (of mental disorders)

DSMC Defense System Management College

dsmd dismissed

D.S.Met.Eng. Doctor of Science in Metallurgical Engineering

DSMG Designated Systems Management Group

DSM Project Development of Substitute Materials (Manhattan Engineer District secret project from 1942 to 1947, responsible for development of A-bomb)

dsn design

DSN Deep Space Network; Department of School Nurses (NEA)

DSNA Dictionary Society of North America

dsnd descend

dsndi descend immediately

dsnrv double-swivel-nose reentry vehicle

DSNWR De Soto National Wildlife Refuge (Iowa)

dso data set optimizer; deck stowage only; direct shipment order; direct shipping ore

D.So. Doctor of Sociology

DSO Dallas Symphony Orchestra; Defense Systems Operator; Denver Symphony Orchestra; Detroit Symphony Orchestra; Distinguished Service Order; District Security Office(r); District Service Office(r); District Staff Office(r); District Supply Office(r); Division Signal Officer; Duluth Symphony Orchestra

D.S.O. Doctor of the Science of Oratory

DSOC Democratic Socialist Organizing Committee

D.Soc.Sci. Doctor of Social Science

dsorg data set organization

D.So.Se Doctor of Social Service

dsp (DSP) digital signal processing

d.s.p. *decessit sine prole* (Latin—died without issue)

DSP Defense Standardization Program; Defense Support Program; Democratic Socialist Party; Detroit Steel Products; Director of Selection and Personnel; Division Standard Practice

DS & P Duell, Sloan & Pearce

DSPA Deep-Submersible Pilots Association

dspch dispatch; dispatcher

d spec(s) design specification(s)

dsph diopter spherical

dspl disposal

d.s.p.l. *decessit sine prole legitima* (Latin—died without legitimate issue)

dspln disciplinary; discipline

d.s.p.m. *decessit sine prole mascula* (Latin—died without male issue)

d.s.p.m.s. *decessit sine prole mascula superstite* (Latin—died without surviving male issue)

dspn disposition

dspo disposal; dispose; disposition

dsprsl dispersal

d.s.p.s. *decessit sine prole superstite* (Latin—died without surviving issue)

DSPS Dynamic Ship-Positioning System

d.s.p.v. *decessit sine prole virile* (Latin—died without male issue)

dsq discharged to sick quarters

D-squad death squad

dsr depolymerized scrap rubber; digit storage relay; digital stepping recorder

dsr (DSR) dynamic spatial reconstructor

ds&r data storage and retrieval; document search and retrieval

ds & r document search and retrieval

DSR Danmarks Radio (Danish radio and tv); Detroit Street Railways; Director of Scientific Research; District Sales Representative

DSRC David Sarnoff Research Center (RCA)

DSRD Director(ate) of Signals Research and Development

DSRK *Deutsche Schiffs Revision und Klassifikation* (German Ship Revision and Classification)

dsRNA double-stranded ribonucleic acid

d's & r's dailies and rushes (motion-picture film editing)

DSRS Data Storage and Retrieval System

dsrv (DSRV) deep-submergence rescue vehicle

dss developmental sentence scoring; documents signed; dry surface storage

Dss Deaconess

DSS Data Systems Services; David S(olomon) Schwab; Deaf Supportive Services; Decision Support System; Defense Supply Service; Department of Social Services; Department of Supply and Services; Director of Social Services; Director(ate) of Statistical Services

DS & S Data Systems and Statistics

D.S.S. Doctor of Sacred Scripture; Doctor of Social Science

D S S & A Duluth, South Shore & Atlantic (railroad)

DSSc Diploma in Sanitary Science

DSSC Defense Subsistence Supply Center

DSSCS Defense Special Security Communications System

DSSD Dry Surface Storage Demonstration

DSSH Department of Social Services and Housing

DSSL Delta Steamship Lines

DSSN Disbursing Station Symbol Number

DSSO Defense Surplus Sales Office; Duty Space Surveillance Officer

dssp deep-sea submergence project

DSSRG Deep Submergence System Review Group

DSSS Division of Special Schools and Services

DSSSP Division of Student Support and Special Programs (Office of Education)

DSSV Deep Submergence Search Vehicle

dst door stop; drop survival time

dst (DST) dexamethasone-suppression test

DST Daylight Saving Time; *Defense et Sécurité du Territoire* (French equivalent of FBI); Dermatology and Syphilology Technician; Desensitization Test (for allergies); Director of Supplies and Transport; Double Summer Time

DS & T Directorate of Science and Technology (CIA)

D.S.T. Doctor of Sacred Theology

d-std vehicle driver-seated vehicle

D.St.Eng. Doctor of Structural Engineering

d-stg vehicle driver-standing vehicle

dstl distill

dstn destination

DSTO Defense Sciences and Technology Organization (Australian); Divisional Sea Transport Office(r)

DSTP Director, Strategic Target Planning

dstpn dessert spoon

dstr distribution; distributor

dsu dissemination services unit; drum storage unit

DSUE *Dictionary of Slang and Unconventional English*

dsuh direct suggestion under hypnosis

dsuphtr desuperheater

D.Sur. Doctor of Surgery

D.Surg. Dental Surgeon

dsv double silk varnish

dsw door switch

DSW Department of Social Welfare

D.S.W. Doctor of Social Welfare

dsz decrement and skip on zero (calculator)

D Sz Diego Suarez

dt dead time; delirum tremens; detective; dinette; diphtheria tetanus; double throw; double time; drain tile; dual tires

dt (DT) deep tank(s)

d-t deuterium-tritium; double-throw

d/t deaths (total ratio); dictaphone typist

d of t deed of trust

dt doit (French—debit)

Dt duration tetanus

DT Daylight Time; Dental Technician; Department of Tourism; Department of Transportation; Department of the Treasury; Detroit Terminal (railroad); Director of Transport(ation); Directorate of Tests; Distance Test; Distance Test(ing); Dylan Thomas

D.T. Dental Technician; Doctor of Theology

DT Daily Telegraph (London); *Danmarks Turistrad* (Danish Tourist Board)

dta daily travel allowance; development test article; differential thermal analysis; distributing terminal assembly; double tape armored cable

DTA Defense Transportation Administration; Democratic Turnhalle Alliance (multiracial); Development Test Article; Differential Thermal Analysis; Diploma in Tropical Agriculture; *Divisão de Exploração dos Transportes Aéreos* (Portuguese—Air Transport Exploration Division)

D.T.A. Democratic-Turnhalle Alliance (of South-African oriented Namibians)

dtas diffuse thalamic activating system

DTASW Department of Torpedo and Anti-Submarine Warfare

dtbc disturbance

dtc deposit-taking company; design to cost; direct-to-consumer

DTC Day Training Center; Department of Trade and Commerce

DTC Deutscher Touring Club (German Touring Club)

DTCD Diploma in Tuberculosis and Chest Diseases

D.T.Chem. Doctor of Technical Chemistry

DTCN Direction Technique des Constructions Navales (French—Technical Direction of Naval Construction)

DTCS Digital Test Command System

dt c sk don't countersink

dtcw data transfer command word

dtd dated; direct to disc (recording system)

d.t.d. detur talis dosis (Latin—

let such a dose be given)

DTD Diploma in Tuberculosis; Director(ate) of Technical Development

DTD Decoratie voor Trouwe Dienst (Dutch—Decoration for Loyal Service)

DTDRS Direct-to-Disc Recording System

dte data terminal equipment; development; test(ing), and evaluation; diagnostic test equipment; digital television equipment; diploma test of empathy (DTE)

dte (DTE) data terminal equipment

D.Tech. Doctor of Technology

DTEE Division of Technology and Environmental Education

D.T.Eng. Doctor of Textile Engineering

D Ter Dakota Territory (before 1889)

dtf daily transaction file

DTF Deep Test(ing) Facilities; Dental Traders' Federation; Division of Training and Facilities; Domestic Traffic Federation; Domestic Textiles Federation

dtfc differential temperature-flow controller

dtfcd define the file for card

dtfcn define the file for console

dtfda define the file for direct access

dtfdi define the file for device independence

dtfdr define the file data recorder

DT & FE Department of Technical and Further Education (Australian)

dtfis define the file for indexed sequential (files)

dtfmr define the file for magnetic reader

dtfmt define the file for magnetic tape

dtfor define the file for optical reader

dtfph define the file for physical input-output multiplexer

dtfpr define the file for printer

dtfpt define the file for paper tape

dtfsd define the file for sequential direct-access storage device

dtfsr define the file for serial device file

dtg data time group; date time group(ing); display transmission generator

Dtg *Dienstag* (German—Tuesday)

dth delayed-type hypersensitivity

D.Th. Doctor of Theology

DTH Dance Theatre of Harlem; Diploma in Tropical Hygiene

D.Theol. Doctor of Theology

D ThPT Diploma in Theory and Practice of Teaching

dti dial test indicator

DTI Department of Trade and Industry (UK); Direct Trader Input

DT & I Detroit, Toledo and Ironton (railroad)

DTIC Defense Technical Information Center

d-time dream time

dt-sit drunkard

dtl detail; detailed; diode transistor logic

dtl (DTL) diode-transistor logic

DTL Detroit Testing Laboratory

dtm duration time modulation

Dtm Dortmund

DTM Diocesan Travelling Mission; Diploma in Tropical Medicine

D.T.M. Doctor of Tropical Medicine

DTMB David Taylor Model Basin

DTMBAL David Taylor Model Basin Aerodynamics Laboratory

dtmf dual-tome multifrequency (telephone)

DTMH Diplomate of Tropical Medicine and Hygiene

DTMI Dairy Training and Merchandising Institute

dt mld draft moulded

DTMO Design Test and Mission Operations; Development Test and Mission Operations

DTMS Defense Traffic Management Service

dtn detain

dtn (DTN) diphtheria toxin, normal

DTN Defense Teleprinter Network

DTN Drug Trade News

DTNM Devil's Tower National Monument

DTNSRDC David Taylor Naval Ship Research and Development Center (USN)

dto detailed test objective; dollar tradeoff; due to

dto descuento (Spanish—discount)

dtº direito (Portuguese—right)

DTO Dental Therapists of Ontario; Director(ate) of Trade and Operations; Disbursing and Transport(ation) Office(r)

DTO Dansk Teknisk Oplysningstjeneste (Danish Technical Information Service)

dtol digital test-oriented language

dtp data type punch; diphtheria; tetanus, pertussis (whooping cough)—combined vaccination

dtp (DTP) directory tape processor

DTP distal tingling on pressure

dtpb divider time pulse distributor

DTPEWS Design-to-Price Electronic Warfare System

DTPH Diploma in Tropical Public Health

dtps diffuse thalamic projection system

dtr data tape recorder; deep tendon reflexes; demand totalizing relay; double tax-(ation) relief

dtr (DTR) distribution tape reel (data processing)

d/tr documents against trust receipt

DTR Diploma in Therapeutic Radiology

DTRA Defense Technical Review Agency (USA)

DTRC David Taylor Research Center

dtrm determine

DTRP Diploma in Town and Regional Planning

dtrt deteriorate; do the right thing

Dtrt Detroit

dts dense tar surfacing

dt's deep tanks; delerium tremens; dementia tremors

DTS Data Transmission System; Defense Telephone Service; Defense Transportation System; Dynamic Test Station

D & TS Detroit and Toledo Short Line (railroad)

Dtsch Deutsch (German—German)

DTSG Data Transmission Study Group

DTSS Dartmouth Time-Sharing System

dtt diphtheria tetanus toxin; duplicate title transferred

DTT Dictionary of Technical Terms

D of TT Dominion of Trinidad and Tobago

dt/tm delayed-time/telemetry

dtu data transfer unit; data transformation unit

DTU Delft Technical University

dtur departure

dtv diver transport vehicle

DTV Deutsche Taschenbuch Verlag (German—German Pocketbook Publisher)

DTVM Diploma in Tropical Veterinary Medicine

DTVP Developmental Test of Visual Perception

DTW Dance Theater Workshop; Detroit, Michigan (Detroit Metropolitan Airport)

dtx detoxification

Dtz Dutzend (German—dozen)

DTZ Division Tactical Zone (USA)

Dtzd Dutzend (German—dozen)

du density unknown; diagnosis undetermined; died unmarried; digital unit; distribution unit; dog unit; duodenal ulcer

Du Ducal; Duchy; Duke; Dutch

Du Dutch

DU Dalhousie University; Deakin University; Denison University; diagnosis undetermined; Dillard University; Drake University; Drew University; Drexel University; Ducks Unlimited; Duke University; Duquesne University

du 26 ct du 26 mois courant (French—the 26th of this month)

dua digital uplink assembly

DUA Digitronics Users Association

DUADS Duluth Air Defense Sector

DUAH Department of Urban Affairs and Housing

dual. dynamic universal assembly language

DUAL Data Use and Access Laboratories

DUAP Dante University of America Press

dub. diameter underback; double; dubber; dubbing; dubious

dub. dubius (Latin—dubious)

Dub Dublin

DUB Dublin, Eire (airport)

DUBC Durham University Boat Club

Dubl Dublin; Dubliner

DUBS Durham University Business School
duc demonstration unity capsule
DUC Datatron Users Organization; Distinguished Unit Citation; Durban University College
D.U.C. Doctor of the University of Calgary
Duck Mountain Duck Mountain Provincial Park in western Manitoba and adjacent Saskatchewan
DUCS Deep Underground Communications System
duct. ductile
duct (Latin prefix—conduct or lead)—conductor, ductless
dud. (DUD) dependably undependable
Dud Dudley
dudat due date
DUF Democratic Unification Party (Korean); Drug Use Forecasting (drug-control program)
Duff Duffield; Duffle; Mc Duff
DUFFEL Dutch Far East Lines
Du Fl Dutch Flemish
dufus blundering person, a dip
DUH Duke University Hospital
dui driving under the influence (of alcohol and/or drugs)
DUI Driving Under the Influence (of alcohol, drugs, or narcotics)
duit2me (DUIT2ME) do it to me
Duke Marmaduke; The Duke, actor John Wayne
dukw (DUKW) code letters, pronounced *duck*, for an amphibious automotive vehicle
DUKW amphibious truck
Dul Duluth
DUL Duke University Library; Durham University Library
Dulag *Durchgangslager* (German—prisoner-of-war transit camp)
dulc. *dulcis* (Latin—sweet)
Dulles John Foster Dulles International Airport named for a former secretary of state and serving Washington, D.C.
du'log duolog (conservation wherein the conversants talk without listening to one another)
DUM Dublin University Mission(aries)

Dumb Dumbarton
DUMB Deep Underground Missile Basing; Defensive Umbrella
DUMBO seaplane used for rescue work (naval symbol)
Duke University Medical Center
Dumf Dumfries
Dumf & Gall Dumfries and Galloway
dums deep unmanned submersibles
dun. dunnage
Dun Dun Laoghaire (Dunleary); Dunbar; Duncan; Dundalk; Dundas; Dundee; Dundrennan; Dunedin; Dunellen; Dunelm; Dunfermline; Dungarvan; Dungeness; Dunglas; Dunglison; Dunlap; Dunlop; Dunmore; Dunn; Dunnachie; Dunning; Dunnsville; Dunoon; Dunscore; Dunsmuir; Dunstable; Dunstan; Dunvegan; Dunwood; Dunwoody
Dunb Dunbarton
dunc deep underwater nuclear counter
Dunc Duncan
D.Univ. Doctor of the University
Dunk Dunkerque (Dunkirk)
dunna don't know
DUNS Data Universal Numbering System
duo. duodecimo
duod duodenum
duodec duodecimo
duol duologue
dup duplicate; duplicating; duplication
DUP Diplomate of the University of Paris; Duquesne University Press
D.U.P. *Docteur de l'Université de Paris* (French—Doctor of the University of Paris)—the Sorbonne
du pa duplicating pattern
DUPA Drug Users Parent Aid
dup^{do} *duplicado* (Spanish—duplicate)
dupe. duplicate; duplicate copy
dupe. neg duplicate negative
dupes. duplicates; duplicate copies
dupl duplicate; duplication
dupli duplicate; duplicated; duplication
DUPONT EI du Pont de Nemours & Company
dur duration
dur (Latin prefix—hard)—durable

dur. duris (Latin—hard)
Dur Durango; Durban; Durham
Duraks Durak Ranges of northernmost Western Australia
duralumin durable aluminum-copper-magnesium-manganese alloy
Durant Will and Ariel Durant
DURD Department of Urban and Regional Development
dur. dolor. durante dolore (Latin—as long as the pain lasts)
Durf Durfee's Reports
durg during
durgc during climb
durgd during descent
Durh Durham
Dur Mus Durban Museum
DUS Düsseldorf, Germany (airport)
DUSA Defense Union of South Africa
DUSA Dispensatory of the United States of America
dusam dummy surface-to-air missile
DUSC Deep Underground Support Center (USAF)
dus/testing distinctness, uniformity, and stability testing
DUSW Director(ate) of the Undersurface Warfare Division
dut device under test(ing); dunnage untreated
Dut Dutch; Dutch Harbor
Dutch Leewards Aruba, Bonaire, Curaçao
Dutch provinces Drenthe, Friesland, Gelderland, Groningen, Limburg, North Brabant, North Holland, Overijssel, South Holland, Utrecht, Zeeland, plus the Netherlands Antilles
Dutch Windwards Saba, Sint Eustatius, Sint Maarten
Dutton EP Dutton & Co
Dutz Dutzend (German—dozen)
duv data under voice
duvd direct ultrasonic visualization of defects
dv dependent variable; device; dilute volume; direct vision; distemper virus; distinguished visitor; dive; double vibrations; double vision
d.v. dorsiventral
d/v declared value
d & v diarrhea and vomiting
d/v días vista (Spanish—days at sight)

DV Diploma in Venereology; Douay Version
D/V Discovery Vessel
D.V. *Deo volente* (Latin—God willing)
dva dynamic visual acuity
DVA Department of Veterans Affairs; *Distribuidora Venezolana de Azucareros* (Spanish—Venezuelan Sugar Growers Distributing Organization)
D.V.A. Doctor of Visual Aids
D-value death value (minutes needed for a lethal agent or environment to kill the population)
dvars doppler velocity altimeter radar set
dva test duration of voluntary apnoea test
DVC Diablo Valley College; Deputy Vice-Chancellor
DVCSA Delaware Valley College of Science and Agriculture
dvd direct-view device
DV & D Diploma in Venereology and Dermatology
DVDP Dry Valley Drilling Project
dve device end; digital video effect
Dve Drive
DVECC Disease Vector Ecology and Control Center
d Verf der Verfasser (German—the author)
DVES Defense Value Engineering Services
dvfr defense visual flight rules
dvg digital video generator
DvH Dietrich von Hildebrand
DVH Diploma in Veterinary Hygiene; Division for the Visually Handicapped
dvin deviation
dvl direct voice line
DVLC Driver and Vehicle Licensing Centre
DVTF Domestic Violence Task Force
dvlp development
dvm digital voltmeter
d.v.m. *decessit vita matris* (Latin—he died during his mother's lifetime)
D.V.M. Doctor of Veterinary Medicine
D.V.M.S. Doctor of Veterinary Medicine and Surgery
DvN D. Van Nostrand
DVNM Death Valley National Monument
Dvnport Devonport
Dvo Davao

DVO Divisional Veterinary Office(r)
dvom digital volt ohmmeter
dvp differential value profile; direct vision panel
d.v.p. *decessit vita patris* (Latin—he died during his father's lifetime)
DVPH Diploma in Veterinary Public Health
dvppi daylight-view plan-position indicator
dvr driver
DVR Division of Vocational Rehabilitation
D.V.R. Doctor of Veterinary Radiology
dvrg diverge
dvrsn diversion
dvs det vill säga (Swedish—that is); *det vil si* (Norwegian—that is); *det vil sige* (Danish—that is)
DVS Division of Vital Statistics
D.V.S. Doctor of Veterinary Surgery
D.V.Sc. Doctor of Veterinary Science
DVSL District Venture Scout Leader
DVSM Diploma of Veterinary State Medicine
dvst direct-view storage tube
dvt deep venous thrombosis
DVTE Division of Vocational and Technical Education
DVTI De Vry Technical Institute
dvtl dovetail
Dvwp *Deo volente*, weather permitting (God willing, weather permitting)
dw data word(ing); deadweight; delivered weight; developed width; diameter width; dishwasher; distilled water; double weight; dry wine; dumbwaiter; dust wrapper
d/w dextrose in water; dock warrant
DW Defenders of Wildlife; Department of Waters
dwa double wire armor(ed)
DWA Deadly Weapons Act; Distributive Workers of America
DWAA Dog Writers' Association of America
dwb double with bath
DWB Doctors Without Borders
dwba direct-wire burglar alarm
dwc deadweight capacity
dwcc deadweight cargo capac-

ity
DWCHS De Witt Clinton High School
DWCP Detroit-Wayne County Port
dwd died while drinking; driving while drunk; dumbwaiter door
dw di draw die
DWDL Donald W Douglas Laboratory
dweg died while eating gumbo
dwel dwelling
dwf divorced white female
dw fm draw form
dwg drawing; dwelling
DWG Diamond Walnut Growers
dwg-ho dwelling house
DWGNRA Delaware Water Gap National Recreation Area (New Jersey and Pennsylvania)
dwi driving while intoxicated
DWI Descriptive Word Index; Durable Woods Institute; Durham Wheat Institute; Dutch West Indies (Netherlands Antilles)
DWIC Disaster Welfare Inquiry Center
Dwig Dwiggins
dwim do what I mean
dwk death-wish kids (gangsters)
DWK *Deutsche Gesellschaft für Wiederaufarbeitung von Kernbrennstoffen* (German—Society for Reprocessing Nuclear Waste); *Die Wit Kommando* (Afrikaans—The White Commando)
dwl derived working level; designed waterline; displacement waterline; dowel
dwm dangerous waste material; deadweight machine(ry); divorced white male
DWM *Deutsche Waffen und Munitionsfabriken* (German—German Arms and Ordnance Factory)
dwn down
dwndfts downdrafts
dwo delta-wing orbiter
DWOP Denver War On Poverty
dwp deepwater port; dyna whirlpool
DWP Department of Water and Power
D W & P Duluth, Winnipeg & Pacific (railroad)
DWPF Defense Waste Processing Facility (Savannah River)

dwpnt dewpoint
dwr drawer
DWR Department of Water Resources; Duke of Wellington's Regiment
dws drinking-water standards; drop wood siding; double white silk
DWS Department of Water Supply
DWSG & E Department of Water Supply, Gas, and Electricity
DWSO Drainage and Water Supply Office(r)
dwt deadweight ton(nage)(s); denarius weight; double weight; pennyweight
DWT *Deutsche Gesellschaft für Wehrtechnik* (German— German Society for Defense Technology)
dw tk drinking water tank
dwuld dewooled (skins)
dwv drain, waste, and vent
dww downward
dwz *dat wil zeggen* (Dutch— that is)
dx de luxe; dextran; diagnosis; distance (radio); double cash ruled; duplex; static (symbol)
dx (DX) defense exhibit
Dx diagnosis (medical)
DX Aerotaxi (Colombia); distance radio reception or transmission; Sun Ray Mid-Continent Oil (stock exchange symbol)
dxc data exchange control
DXC Penn-Dixie Cement (stock exchange symbol)
dxd discontinued
dxda-mc ductile metals experimental diamond abrasive— metal clad
dxer (DX-er) long-distance radio receptionist
dx-ing long-distance (radio) communicating
dxm dexamethasone
dxr deep X-ray
dxrt deep X-ray therapy
dXt deep X-ray therapy
dy delivery; demy (paper); dock yard; duty; penny (nails)
Dy Dylan; dysprosium
Dy *Douay Bible* (Roman-Catholic English translation of the Latin Vulgate made at

Douay and Rheims in 1610)
D-y *Druk-yul* (Bhutanese— Bhutan)
DY De Young Memorial Museum; Druk-Yul (Kingdom of Bhutan)
DYA Department of Youth Authority (California)
dyana dynamics analyzer
dyb do your best; dynamic braking
DYB Do Your Best
dy bf hl day before holiday
DYC Detroit Yacht Club; Dominion Yeast Company
dyd dockyard
dydff dyed and fully finished (leather)
dye. dyeing
dyf damned young fool
DYF Democratic Youth Front
dy fl hl day following holiday
DYFS Division of Youth and Family Services
dyke bulldike
dykes diagonal wire cutters
dymaxion dynamic maximum
DYMM MH De Young Memorial Museum
DYMM (Malay—His Highness the Ruler or Her Highness the Ruler)
dyn dynamic; dynamics; dynamo; dynamometer; dyne
Dyn Dynasty
dyna dynamite
dynam dynamic; dynamics; dynamite; dynamo
dynamit dynamic allocation of manufacturing inventory and time
dynamo. dynamic model
DYNAMO Dynamic Action Management Operation
dynasoar dynamic soaring (space flight)
dynatac dynamic adaptive total area coverage
dyncm dyne centimeter
dynmt dynamite
dyno dynamite, undiluted drugs; dynamometer
dypso dypsomania(c)
dy r dynamic response
dys (Latin prefix—bad, difficult; painful)—dysentery, dyspepsia
DYS Department of Youth Services; Department of

Youth Services (Alabama); Division of Youth Services; Division of Youth Services (Arkansas)
dysac digitally simulated analog computer
dysen dysentery
dyslex dyslexia; dyslexic
dysm dysmenorrhea
dysp dyspepsia
dysphem dysphemistic(al)(ly); dysphemism(s)—antonym(s) for euphemism(s)
dystac dynamic storage analog computer
dystal dynamic storage allocation language
dysto dystopia(n)
dystope(s) dystopian(s)—slum dweller(s) leading a fear-filled and wretched existence
dyu do your utmost
DYW Dynamic Youth Workers
dz dizygotic; dizziness; dizzy; dozen; drizzle
dz *deppelzentner* (German— 100 kilograms); *distance zénithale* (French—zenith distance); *dozzina* (Italian— dozen)
d Z *der Zeit* (German—of the time)
Dz *Deniz* (Turkish—sea)
DZ Department of Zoology; Drop Zone
D.Z. Doctor of Zoology
DZA Drop Zone Area
DZF *Deutsche Zentrale für Fremdenverkehr* (German— German National Tourist Association)
dzg dizygotic
Dzl Delfzijl (Dutch port)
dzne *douzaine* (French— dozen)
D.Zool. Doctor of Zoology
dzt digit zero trigger
DZT *Deutsche Zentrale für Tourismus* (German Directorate for Tourism)
dz twins dizyotic (fraternal) twins
D-Zug *Durchgangszug* (German—express train, through train)
Dzun Dzungaria

E

e base for natural logarithms 2.7182818; coefficient of impact (symbol); east(ern); electron; empty; emulsifier; emulsion; error; errors; eve ning; exa(E)—10^{18} (one quintillion); longitudinal strain per unit length (symbol); numerical value of electron charge in an electron or proton (symbol)

'e he

e angle of downwash (symbol); natural logarithmic (Napierian) base; (Portuguese–and); (Spanish—and)—used when the following word begins with *i* or *hi* as in *Juana e Ignacio* or *padre e hijo*

e/ envío (Spanish—sent)

E American Export-Isbrandtsen Lines; Eagle Airways; Earth; east; eastern; eccentricity of a curve (symbol); Echo—code for letter E; Edinburgh; efficiency; einsteinium; emmetropia; engineer; engineering; England; English; Equator; equatorial; erbium; estimated weight (symbol); excellent; exempt; eye; Fraunhofer line caused by iron (symbol); instantaneous value alternating current (symbol); modulus of elasticity (symbol)

E east; Einstein unit of energy (symbol); electromotive force (symbol); (Latin—Egregius); *en* (Dutch, Portuguese, Spanish—in); Envoy Extraordinary and Minister Plenipotentiary; *est* (French or Italian—east); *este* (Portuguese or Spanish—east); *etelä*

(Finnish—south); experiment (symbol); voltage (symbol)

E¹ Lhotse I (27,890-ft adjoining peak of Mount Everest)

E1, E2, etc. East One, East Two, etc. (London postal zones)

E-2 Hawkeye airborne early-warning and fighter-control aircraft

E² Lhotse II (27,560-ft adjoining peak of Mount Everest)

E-14 Hispano Saeta twin-engine jet trainer, also designated HA-200

E 107 tribromoethanol (anesthetic)

E 605 parathion (deadly insecticide)

ea each; ends annealed; enemy aircraft; enlistment allowance

ea (EA) educational age; enrolled agent

e/a (E/A) experimental aircraft

EA East Africa(n); Eastern Air Lines; Economic Adviser; educational age; Egyptian Army; Electrical Artificer; Electronic Artificer; Electronic Associates; Environmental Action; Environmental Agency; Environmental Assessment; expectancy age; experimental aircraft

E/A Ecology Action; enemy aircraft

EA Ente Autonomo (Italian—Autonomous Corporation)

EA-6B Grumman electronic-intelligence-gathering aircraft named Intruder

eaa essential amino acid (EAA); ethylene acrylic acid (EAA)

EAA Education Amendment

Act; Electrical Appliance Association; Electronics Association of Australia; Employment Agents Association; Engineer in Aeronautics and Astronautics; Engineers and Architects Association; Equipment Approval Authority; Experimental Aircraft Association; Export Advertising Association

E.A.A. Engineer in Aeronautics and Astronautics

EAA Encyclopedia of American Associations

EAAA European Association of Advertising Agencies

EAAC East African Airways Corporation

EAAFRO East African Agriculture and Forestry Research Organization

EAAM European Association for Aquatic Mammals

EAAP European Association for Animal Production

EAB Ethnic Affairs Bureau; European American Bank; European Asian Bank

EABn Engineer Aviation Battalion

eabrd electrically-actuated band-release device

eac end around carry; erythrocyte antibody complement; estimate at completion

EAC East African Community (Kenya, Tanzania, Uganda); East Asiatic (Line) Container; East Asiatic Company; Eastern Air Command; Estate Agents Cooperative; European Atomic Commission

eaca (EACA) epsilon-amino-caproic acid

eacd eczematous allergic contact dermatitis

E & A Co Eastern and Australian Steamship Company

ea content effective-agent content

EACR Environmental Area Characterization Report

EACSO East African Common Services Organization

EACU East Asiatic (Line) Container Unit

ead equipment allowance document; error adjusted; estimated availability date; extended active duty

ead. eadem (Latin—the same)

EAD Employer Association of Detroit

EADB East African Development Bank

EADF Eastern Air Defense Force

eadi electronic attitude and direction indicator

eae experimental allergic encephalomyelitis

EAEBP European Association of Editors of Biological Periodicals

EAEC East African Economic Community; European Atomic Energy Community

EAEG European Association of Exploration Geophysicists

EAEI Ecology Action Educational Institute

EAES Environment-Atmospheric Environment Service (Canada); European Atomic Energy Society

eaf emergency action file

EAFB Eglin Air Force Base (near Pensacola, Florida)

EAFC Eastern Area Frequency Coordinator; Eastern Association of Fire Chiefs

EAFFRO East African Freshwater Fishery Research Organization

EAG Edmonton Art Gallery

EAGGF European Agricultural Guidance and Guarantee Fund

Eagle Springs Girls Samarkand Manor for females at Eagle Springs, North Carolina

EAHC East African Harbours Corporation; East African High Commission

eahf eczema, asthma, and hay fever

EAI East Asian Institute (Columbia University); Education Audit Institute

EAIC East African Industrial Council

EAID Equipment Authorization Inventory Data

EAJC Eastern Arizona Junior College

eal electromagnetic amplifying lens; estimated average life

Eal English as an additional language

EAL East Asiatic Line; Eastern Air Lines; Ethiopian Airlines

EALA East African Library Association

eam electronic accounting methods

eam (EAM) electrical accounting machine; electronic accounting machine; electronic automatic machine(ry)

EAM Eastern Atlantic and Mediterranean

EAM Ethniko Apelevtherotiko Metopo (Greek—National Liberation Front)

EAME European, African, Middle Eastern

EAMECM European-African Middle Eastern Campaign Medal

eamedpm (EAMEDPM) electric accounting machine and electronic data processing machine

EAMF European Association of Music Festivals

EAMFRO East African Marine Fisheries Research Organization

EAMP Envoy Extraordinaire and Minister Plenipotentiary

EAMPA East Anglian Master Printers' Alliance

EAMS Empire Air Mail Scheme

EAmst Elsevier Amsterdam

EAMTC European Association of Management Training Centers

EAN Emergency Action Notification

EANA Esperanto Association of North America

EANDC Edgewood Arsenal Nuclear Defense Center; European American Nuclear Data Center

EANS Emergency Action Notification System (radio broadcasting)

EANSW Electricity Authority of New South Wales

eaon except as otherwise noted

eaos expiration of active obligated service

eap engines, armament, and pyrotechnics; equivalent air pressure; eye artifact potential

eap (EAP) erythrocyte acid phosphatase

EAP Edgar Allan Poe; Emergency Action Procedure; Employee Assistance Programs; Environmental Analysis and Planning

EAP École d'Administration Pénitentiaire (French—Penitentiary Administration School)

EAPA Employment Aptitude Placement Association

EAPD Eastern Air Procurement District

EAPG Eastern Atlantic Planning Guidance (NATO)

EAPR European Association for Potato Research

'eap(s) heap(s)

EAPTC East African Posts and Telecommunications Corporation

ear electronic array radar; electronically agile radar; estimate after release

ear. electronic analog resolver

Ea-R Entartungs-Reaktion (German—degeneration reaction)

EAR East African Railways; Edwin Arlington Robinson

EARB Engineering Associates Registration Board

EARC East African Railways Corporation; Eastern Air Rescue Center

EARDHE European Association for Research and Development in Higher Education

EAR & H East African Railways and Harbours

EARI Equipment Acceptance Requirements and Inspection

Earnie Ernest; Ernestine; Ernesto

earom electrically alterable read-only memory

earp equipment anti-riot projector

EARS Electronic Airborne Reaction System; Electronically Agile Radar System; Emergency Airborne Reaction System

earssn early season

Earth Planet Sci Lett Earth and Planetary Science Letters

eas electronic article surveillance; equivalent airspeed; estimated air speed

EAs East African shilling

EAS Early American Society; Enterprise Allowance Scheme; Executive Assignment Service; Extended Area Service

EASA Electrical Apparatus Service Association; Engineers Association of South Africa

EASE Emigrant's Assured Savings Estate; European Association of Science Editors

easemt easement

EASEP Early Apollo Scientific Experiments Payload

EA sh East African shilling

easl engineering analysis and simulation language

EASS Engine Automatic Stop-and-Start System

east. easterly; eastern

East east of the Mississippi, eastern states of the U.S.; Far East

EAST Eastern Australian Standard Time

EASTAF Eastern Transport Air Force

East African Community Kenya, Tanzania, Uganda

East Bloc Albania, Bulgaria, Czechoslovakia, East Germany, Hungary, Poland, Romania, Yugoslavia

Eastcommrgn Eastern Communications Region

East L East Lothian

East Lake Girls Alabama State Training School (for female delinquents) at East Lake near Birmingham

EASTLANT Eastern Atlantic Area

East Los East Los Angeles, California

East Phil Eastern Philharmonic (North Carolina); Eastman Philharmonia

EASTROLANT Eastern Tropical Atlantic

EASTROPAC Eastern Tropical Pacific

East Sutton Park borstal for delinquent girls in Kent, England

easy. efficient assembly system; expense-account spending money

EASY Early Acquisition System (USA); Engine Analyzer System

eat. earliest arrival time (EAT); earnings after taxes; estimated arrival time (EAT);

expected approach time (EAT)

e/at. electrons per atom

EAT earliest arriving time; Economic and Technical Committee(s); Experiments in Art and Technology

EATA East Asia Travel Association

EATC Ecology and Analysis of Trace Contaminants

EATCS European Association for Theoretical Computer Science

EATRO East African Trypanosomiasis Research Organization

EATS Equipment Accuracy Test Station; Extended Area Tracking System (USN)

EATTA East Africa Tourist Travel Association

eau extended arithmetic unit

EAU Emergency Assistance Unit

EAU *Emiratos Arabes Unidos* (Spanish—United Arab Emirates)

EAVRO East African Veterinary Research Organization

eaw Electrical Association for Women; equivalent average words

EAW Electrical Association for Women

EAWP Eastern Atlantic War Plan (NATO)

EAWS East African Wildlife Society

eax electronic automatic exchange

eb electron beam; elementary body; emergency brake; engine burn; environmental buoy

e-b estate-bottled

e/b eastbound

eb *point d'ébullition* (French—boiling point)

Eb Ebba; Ebed; Eben; Ebenezer; erbium (symbol); erbium (symbol)

EB Avitour Airlines; Eesti Vabariik (Estonian Republic); Electricity Board; Easter Bunny

E-B Electric Boat (Division of General Dynamics)

E & B Ellerman and Bucknall (Ellerman Lines)

EB *Encyclopaedia Britannica; Engineering Bulletin*

eb 1 s edge bead one side (lumber)

eb 2 s edge bead two sides (lumber)

EBA Education Boards Association; Energy Business Association; English Bowling Association

EBAA Eye-Bank Association of America

EBAILL European Bureau for the Allocation of International Long Lines

ebam electron-beam-addressed memory

EBAM Electron-Beam Addressed Memory

ebar edited beyond all recognition

EBAR EB Aabys Rederi (Norwegian freight line)

EBB Elias Baseball Bureau; Elizabeth Barrett Browning

ebc enamel bonded single cotton

EBC Educational Broadcasting Corporation; European Bibliographical Center (Oxford, England)

ebcdic extended binary-coded decimal interchange code

EBCDIC Extended Binary Coded Decimal Interchange Code (pronounced *ebsidick*)

ebce experience-based career education

ebd education by discussion; effective biological dose

ebd *ebenda* (German—in the same place)

ebds enamel bonded double silk

EBEC Encyclopedia Britannica Educational Corporation

Eben Ebenezer

Eber Eberard; Eberhard; Eberhart; Ebert

EBES Electron-Beam Exposure System

ebf electronically blown flap; erythroblastosis foetalis; externally blown flap

EBF Encyclopedia Britannica Films

ebfa electron-beam fusion accelerator

ebi (EBI) emetine bismuth iodide

EBI Emerson Books, Incorporated

EBIC European Banks International Corporation (lowercase logotype appears as *ebic*)

ebicon electron-bombardment-induced conductivity

ebit earnings before interest and taxes; electron-beam ion trap

ebiv electron-beam-induced voltage

ebk embryonic bovine kidney

EBL Eastern Basketball League

ebm electronic bearing marker; expressed breast milk

EBM *Empresa Bacaladera Mexicana* (Mexican Codfishing Enterprise)

EBMC English Butter Marketing Company

EBMUD East Bay Municipal Utility District

EBNI Electricity Board for Northern Ireland

Ebnr Ebenezer

EBNY Edition Bookbinders of New York

E-boat enemy boat

Ebor. *Eboracensis* (Latin—of York); *Eboracum* (Latin—York)

ebp enamel single paper bonded

ebpa electron-beam parametric amplifier

EBQ Empire Brass Quintet

ebr electron-beam recorder; experimental breeder reactor

EBR Emu Bay Railway; Engineering Business Report

EBR-75 Panhard armored car carrying a 75mm gun

EBR-90 Panhard armored car carrying a 90mm gun

EBRA Engineer Buyers' and Representatives' Association

EBRD European Bank for Reconstruction and Development; Export Business Division (U.S. Department of Commerce)

EBRI Employee Benefit Research Institute

ebs electron-bombarded silicon; enamel single cotton

eb(s) eager beaver(s)

EBS Emergency Bed Service; Emergency Broadcast System; English Bookplate Society; Ethiopian Broadcasting Service

EBSA Estuarine and Brackish-Water Sciences Association

EBSR Eye-Bank for Sight Restoration

ebt earth-based tug (NASA); electron-beam technique; examination before trial

EBU European Broadcasting Union

ebul ebullition

EBv Epstein-Barr virus

EBV Epstein-Barr Virus

E–B V Epstein–Barr Virus

ebw exploding bridge wire

EBW Elwyn Brooks White

ebwr (EBWR) experimental boiling-water reactor

EBYC European Bureau for Youth and Childhood

ec earth closet; economics; electric(al) coding; electrolytic corrosion; electronic calculator; electronic computer; emergency capability; emulsifiable concentrate; enamel coated; enteric coated; entering complaint; error correcting; expansive classification; expiratory center; extended coverage; extension and conversion; extension course; exterior closet; extra choice (wool)

e-c ether-chloroform (mixture)

e/c estrogen-to-creatinine (ratio)

ec en cuento (Spanish—on account)

e.c. exempli causa (Latin—for example)

Ec Ecclesiastic; Ecuador; Ecuadorian

EC Earlham College; East African Airways; East Carolina (railroad); East Central; East Coast; Eastern College; Eastern Command; Edgewood College; Electricity Council; Elizabethtown College; Elmhurst College; Elmira College; Elon College; Emergency Commission(er); Emergency Coordinator; Emerson College; Emmanuel College; Energy Commission; Engineer Captain; Engineering Change; Engineering Construction; Environmental Control; Episcopal Church; Erskine College; Essex College; Established Church; Eureka College; European Communities; European Community; Evangel College; Evansville College; Executive Committee; Executive Council; Exeter College; Explorers Club

EC (followed by numbers) Enzyme Commission (numbers indicate enzyme classification)

E-C Erckmann-Chatrian (collaborators: Emile Erckmann and Alexandre Chatrian)

E & C Engineering and Construction

EC *Encyclopedia Canadiana; Era Cristiana* (Spanish—Christian Era); *Étoile du Courage* (French—Star of Courage)

EC1, EC2, etc. East Central One, East Central 2, etc. (London postal zones)

eca electronic control amplifier; electronics control assembly

ECA Earthmovers and Contractors Association; Economic Commission for Africa (UN); Economic Control Agency; Economic Cooperation Administration; Educational Communication Association; Educational and Cultural Affairs; Electrical Contractors Association; Engineering Change Analysis; Epidemiologic Catchment Area; European Confederation of Agriculture; Exchange Control Act

ECA *Empresa de Comercio Agricola* (Spanish—Agricultural Commerce Enterprise)

ECAB Early Case Assessment Bureau; Employees' Compensation Appeals Board

ECAC Eastern College Athletic Conference; Electromagnetic Compatibility Analysis Center; Extra-Curricular Activities Center

ecad error check analysis diagram

ECAFE Economic Commission for Asia and the Far East (UN)

ecal equipment calibration

ecam extended communications access method

ecan excitation, calibration, and normalization

ECAP Electronic Circuit Analysis Program; Environmental Compatability Assurance Program (USN)

ECARS Electronic Coordinatograph Readout System

ECAS Electrical Contractors Association of Scotland

ecat emission computerized axial tomography

ECB E(benezer) Cobham Brewer; Energy Conservation Board (US)

e & cb 1 s edge and center bead one side (lumber)

e & cb 2 s edge and center bead two sides (lumber)

ecbo enteric cytopathogenic bovine orphan (virus)

ecc eccentric; electrically-continuous cloth; electrodeposited composite coat(ing); emergency cardiac care;

emergency combat capability; equipment classification control; equipment configuration control; equipment control classification; error correction code; execute control cycle

ecc (ECC) electrocorticogram; exchange control copy

ecc eccetera (Italian—et cetera)

Ecc Eccellenze (Italian—Excellency)

ECC Economic Council of Canada; Educational Cultural Complex; Educational Cultural Complex; Elderly Citizens Club(s); Electronics Capital Corporation; Emergency Conservation Committee; Employees Compensation Commission; End Conscription Campaign (coalition of South African civil-rights organizations); Energy Control Center; English Conservation Center; Ethnic Communities Council; European Coordinating Committee; European Coordinating Council; European Cultural Center; European Cultural Commission

ECCA East Caribbean Currency Authority

ECCA Empresa Consolidada Cubana de Aviación (Spanish—Consolidated Cuban Aviation Enterprise)

ECCAA Executive Chefs de Cuisine Association of America

ECCC English Country Cheese Council

ECCCM Electronic Counter-Counter-Counter Measure

ECCCS Emergency Command Control Communications System

ECCDA Eastern Connecticut Clam Diggers Association

eccen eccentric; eccentrics

ECCET Engineering Casualty Control Evaluation Team (USN)

Ecc. Hom Ecce Homo (Latin—Behold the Man)

ECCI Executive Committee Communist International

eccl ecclesiastic(al)

Eccl Ecclesiastes

eccles ecclesiastic; ecclesiastical

Ecclus. Ecclesiasticus

eccm electronic counter-counter-measures

eccmo electronic counter-countermeasures operation(s); electronic counter-counter-measures operator(s)

ec^{co} eclesiástico (Spanish—clergyman, ecclesiastic, ecclesiastical, priest)

ECCP East Coast Coal Port; Engineering Concepts Curriculum Project; European Commission on Crime Problems

ECCR Engineering Calibration Cycle Request

ECCs Emitter-Coupled Circuits

ECCS Emergency Core Cooling Systems (AEC)

eccsl emitter-coupled-current steered logic

ECCT Eddy Current Conductivity Test(ing)

ECCTYC English Council of the California Two-Year Colleges

ECCU English Cross-Country Union

ecd early closing day; endocardial cushion defect; estimated completion date

ec&d electronic cover and deception

ec & d electronic components and devices

ECd East Caribbean dollar; monetary unit of Antigua and Barbuda, Grenada, St Kitts and Nevis, St Lucia, St Vincent and the Grenadines

ECD Energy Conversion Devices

EC & D Electronic Components and Devices

ecdc electrochemical diffused-collector transistor

ECDC Economic Cooperation among Developing Countries

ecdi electronic course deviation indicator

ECDIN European Chemical Data and Information Network

ecdn electrical cables down

ECDP Estimating Controlled Data Package

ECDU European Christian Democratic Union

ece eligible capital expenditure; extended coverage endorsement

ECE Early Childhood Education; Economic Commission for Europe (UN)

ECEO Economic Crime Enforcement Office (U.S. Dept. of Justice) to combat white-

collar crimes

ECES Educational and Career Exploration System

ecf extracellular fluid

ECF Edgar Cayce Foundation; Electrical Contractors Federation; Enhanced Connectivity Facilities; Episcopal Charismatic Fellowship; European Cultural Foundation; Ex-Communist Forces

ECFA Evangelical Council for Financial Accountability

ECFI Eastern Caribbean Farm Institute

ECFMG Educational Council for Foreign Medical Graduates

ECFMS Educational Council for Foreign Medical Students

ecg export credit guarantee(s)

ecg electrocardiogram; electrocardiograph(y)

ECG electrocardiogram

ECGAI Education Council of the Graphic Arts Industry

ECGB East Coast of Great Britain

ECGC Empire Cotton Growing Corporation

ECGD Export Credit Guarantee Department

ech echelon; engine compartment heater

ECH European Country Hotels

echo. enteric cytopathogenic human orphan (virus)

echo. (ECHO) electronic computing, hospital oriented

Echo letter E radio code; NATO nickname for Soviet missile-carrying nuclear-powered submarine designated E-class

ECHO Efficient Car-Handling Operations (railroad); European Commission Host Organization; Evidence for Community Health Organization; Experimental Contract Highlight Operation

ECHR European Commission on Human Rights; European Court of Human Rights

ECHS Evander Childs High School

eci extracorporeal irradiation

ECI Electronic Communications Incorporated; Extension Course Institute (Air University)

ECIAL Enseñanza de las Ciencias y de la Ingeniería en la América Latina (Spanish—Teaching of Science and Engineering in Latin

America)
ECIC Export Credits Insurance Corporation (Canada)
ECICS Export Credit Insurance Corporation of Singapore
ECIS Error-Correction Information System (NASA)
ECITO European Central Inland Transport Organization
ECIUSAF Extension Course Institute, USAF
ECJ Erie County Jail (Buffalo); European Court of Justice
eck embryonic chicken kidney
Eck(ie) Alexander; Alexandra; Alexis; Hector; Hecuba
ecl eclipse; electrocardiograph log; electronic crash locator (aircraft); extended center line
ecl (ECL) emitter-coupled logic
ecl eclairage (French—lighting)
ECL Electronic Components Laboratory; Engineering Computing Laboratory; Entertainment Corporation Limited; Equipment Component List; Europe-Canada Line
ECL Encyclopedia of Comparative Letterforms
ECLA Economic Commission for Latin America (UN)
E-class NATO designation for a Soviet class of missile-carrying nuclear-powered submarines also known as Echo
eclec eclectic; eclecticism
ecli eclipse; ecliptic
eclo emitter-coupled logic operator
ecm electric coding machine; electrochemical machining; electronic countermeasure(s); ends matched, center (lumber); extended core memory
ECM electronic countermeasures; Engineering Change Management; European Common Market
EC & M Electric Controller and Manufacturing (company)
e/cm³ electrons per cubic centimeter
ECMA Engineering College Magazines Associated; European Computer Manufacturers Association
Ecmalgol European Computer Manufacturers Association Algorithmic Language
ECMCA Eastern-Central Motor Carriers Association
ecmd electronic countermeasures display
ecme electronic countermeasures equipment
ECME Economic Commission for the Middle East (UN)
ecmex electronic countermeasures exercise
ECMF Electric Cable Makers' Federation
ECM & FS East Coast Marine and Ferry Service
ECMHP East Coast Migrant Health Project
e-c mix. ether-chloroform mixture
ecmo (ECMO) enteric cytopathogenic monkey organ (virus); extracorporeal membrane oxygenation (pulmonary therapy)
ECMR Eastern Contract Management Region
ECMRA European Chemical Market Research Association
ECMS Engine Condition Monitoring System
ECMSA Electronics Command Meteorological Support Agency (USA)
ecmtng electronic countermeasures training
ECMWF European Centre for Medium-range Weather Forecast
ECN Engineering Change Notice
ECNA East Coast of North America
ECNOS Eastern Atlantic, Channel, and North Sea (orders for ships given by NATO)
eco ecological; ecologist; ecology; economic; economist; economics; electron-coupled oscillator; exempted by commanding officer
eco (Latin prefix—home or house)—ecology, economy, ecosphere, ecosystem
ECO East Coast Overseas; Economic Corporation Organization; Effective Citizens Organization; Engineering Change Order; Environment and Conservation Organizations; Environmental Control Organization; European Coal Organization
ECOA Equal Credit Opportunity Act; Equipment Company of America
ECOCEN Economic Cooperation Center

ecocrit economic critic(ism)
ecodoom large-scale ecological destruction
ecodoomster(s) predictor(s) of large-scale ecological destruction
ecofuel ecology fuel (made from garbage and other wastes generated by man and his domestic animals)
ecog electrocorticogram
ecogeo ecogeographer; ecogeographic(al)(ly); ecogeography
ecol ecology
Ecol Ecology
ecolcrit ecological criticism; ecology critic
E coli Escherichia coli (intestinal bacillus)
Ecol Soc Am Ecological Society of America
ecom (ECOM) electronic computer-originated mail
ECOM Electronic Computer-Originated Mail (postal service); Electronics Command (USA)
ECOMINAS Empresa Colombiana de Minas (Spanish—Colombian Mining Enterprise)
econ economic; economics; economist; economy
e con. e contrario (Latin—on the contrary)
Econ Economics
Econ Jrnl Economic Journal
economan effective control of manpower
economet econometric(al)(ly); econometrician; econometrics; econometrist
Econ Rev Economic Review
ECOP Extension Committee on Organization and Policy
ECOPETROL Empresa Colombiana de Petróleos (Spanish—Colombian Petroleum Enterprise)
ecophys ecophysiologic(al)-(ly); ecophysiologist; ecophysiology
ecopow(s) economic superpower(s)—U.S.A., USSR, Japan, Germany
Eco Pty Ecology Party
ECOR Engineering Committee on Ocean Resources
ECORS Eastern Counties Operational Research Society
EcoSoc Economic and Social (Council)
ecosupow(s) economic superpower(s)—U.S.A., USSR, Japan, Germany
eco system ecological system;

economic system

ecotopia(n) ecologically ideal utopia(n)

ecou electric clip-on unit; electronic clip-on unit

ECOWAS Economic Community of West African States (Benin, Burkina Faso, Cape Verde Islands, Ivory Coast, Gambia, Ghana, Guinea, Guinea-Bissau, Liberia, Mali, Mauretania, Niger, Nigeria, Senegal, Sierra Leone, Togo)

ECP Engineering Change Proposal; Estonian Communist Party; Examiner of Commercial Practices; Executive Control Program

ECPA Evangelical Christian Publishers Association

ECPAC East County Performing Arts Center

ECPD Engineers Council for Professional Development

ecpiu electronic circuit plug-in unit

ecpnl equivalent continuous preceived noise level

ecpo enteric cytopathogenic porcine orphan (virus)

ECPO Environmental Characterization Projects Office

ecpog electrochemical potential gradient

ECPR European Consortium for Political Research

ecp(s) external casing packer(s)

ECPS European Center for Population Studies

ecpt egress cockpit procedure trainer

ECPTA European Conference of Postal and Telecommunication Administrations

ECQAC Electronic Components Quality Assurance Committee

ecr electronic cash register; energy consumption rate; error cause removal; external channels ratio

ECR Engineering Change Request; Environmental Characterization Report

ECRB Export Control Review Board

ECRC Earth Colonization Research Center; Electronic Component Reliability Center; Engineering College Research Council

ECRL Eastern Caribbean Regional Library

ecro erection counter readout

ECRO European Chemorecep-

tion Research Organization

ECRs Enemy Contact Reports

ECRS Empty Car Routing System

ecs electroconvulsive shock; emperor's clothes syndrome; ends cut square; error correction servomechanism; error correction signals; extended core storage

ECS Education Commission of the States; Electrochemical Society; Electronic Composing System; Employment Counseling Service; Engineering Change Schedule; Engineering Change Sheet; Environmental Control Systems; Episcopal Community Services; Equipment Concentration Sites; Equipment Configuration Study; Etched Circuit Society; European Communications Satellite; Experimental Communications Satellite (NASA)

ECS *Échantillons Commerciaux* (French—commercial samples)

ECSA East Coast of South America; European Communication Security Agency; Expanded Clay and Shale Association

ECSC European Coal and Steel Community

ECSCF Eastern Connecticut State College Foundation

ECS/HCS Education Career Services/Health Career Service

ECSIL Experimental Cross-Section Information Library (University of California—Livermore)

ecss extendable computer system simulator

ECST European Convention on the Suppression of Terrorism

ECSTC Elizabeth City State Teachers College

ECSU Educational Cooperative Service Unit

ect electro-convulsive therapy; engine cutoff time; enteric coated tablet

ect (ECT) electroconvulsive treatment

ECT Environmental Control Technology (DoE division); Equivalence Conversion Training; European Container Terminus

ECTA Economics and Commercial Teachers Associa-

tion; Electrical Contractors' Trading Association; Electronic Components Test(ing) Area

ectl emitter-coupled transistor logic

ecto (Latin prefix—external, outer, outside)—ectoderm

ectohorm ectohormonal; ectohormone

ectomy (Latin suffix—surgical removal)—tonsillectomy

ecu ecumania(c); ecumenism; electronic computing unit; engine compatability unit; environmental control unit; extra closeup; extreme closeup

ecu (ECU) extra closeup; extreme closeup

Ecu Ecuador; Ecuadorean; European currency unit

Ecu (ECU) European currency unit

ECU East Carolina University; Economic Crime Unit; English Church Union; European Customs Union

Ecua Ecuador; Ecuadorean

Ecuador Republic of Ecuador, *Republica del Ecuador*

ecube energy conservation using better engineering

Ecu Con Ecumenical Conference; Ecumenical Council

ecufuel eucalyptus-tree fuel

ECUK East Coast of United Kingdom

ecumen ecumenical(ism)(ist); ecumenicist; ecumenicity; ecumenics; ecumenism

ECUSA Episcopal Church of the U.S.A.

ecusat ecumenical satellite

ECUSATCOM Ecumenical Satellite Commission

ecv estimated cash value; extracellular virus

ecv (ECV) energy conservation vehicle

e & cV 1 s edge and center-V one side (lumber)

e & cV 2 s edge and center-V two sides (lumber)

ecw extracellular water

ECW Episcopal Church Women

ECWA Economic Commission for Western Asia (UN)

ECY European Conservation Year

ECYO European Community Youth Orchestra

ecz eczema(tic)

ed edge distance; edit; edited; edition; editor; editorial; educate; educated; education; educational; educator; effective dose; electronic displays; emotionally disturbed; enemy dead; error detecting; erythema dose; excused from duty; ex-dividend; existence doubtful; extra dividend; extra duty

e & d (E & D) exploration and development

ed edición (Spanish—edition); *édition* (French—edition); *edizione* (Italian—edition)

e_d price elasticity of demand

Ed Edgar; Editor; Edmond; Edmund; Edson; Edvard; Edvardson; Edward; Edwin

Ed. Editor

Ed edificio (Spanish—building)

E$ Eurodollar (American dollar deposited in Europe)

ED Consolidated Edison Company (stock exchange symbol); Eastern District; Economics Division; Education Department; Efficiency Decoration; Elder Dempster Line; Electric Dynamic; Engineering Data; Engineering Depot; Engineering Design; Engineering Draftsman; Export Declaration

E-D Electro-Dynamics (division of General Dynamics); Elsevier-Dutton

E.D. Doctor of Engineering

ed_5O median effective dose

eda early departure authorized; equipment design agent; equivalent design axles; erection digital assembly

E d A *Ejercito del Aire* (Spanish—Air Force)

EDA Economic Development Administration (Puerto Rico); Electrical Development Association; Employment Development Act; Environmental Development Administration; Environmental Development Agency

edac error detection and correction

EDAC Engineering Decision Analysis Company; Evaluation, Dissemination, Assessment Centers

E da M *Escuatrão da Morte* (Portuguese—Death Squad)—Brazilian right-wing terrorists

EDANA European Disposables and Non-Wovens Association

EDAQ Electrical Development Association of Queensland

EDARR Engineering Drawing and Assembly Release Record

edb emergency dispersal base(s); end of data block(s); ethene dibromide (EDB); extended double base

edb (EDB) ethylene dibromide (deadly pesticide)

edb elektronisk databehandling (Dano-Norwegian—electronic data handling)

Edb Edinburgh

Ed.B. Bachelor of Education

EDB Economic Development Board; Energy Development Board

edbiz educational business

EDBP Epidemiology, Demography, and Biometry Program

edc electronic digital computer; emergency digital computer; energy distribution curve; engine-driven compressor; error detection and correction; estimated date of completion; estimated date of confinement; extra dark color

EDC Eastern Defense Command; Economic Development Corporation; Educational Development Centers; Educational Development Corporation; Educational Development Council; Engineering Data Consultants; European Defense Community; Export Development Corporation

EDCC Environmental Dispute Coordination Commission

edcl electric discharge coaxial laser

edcn education

edcom editor-compiler

E/DCP Equipment/Document Change Proposal

EDCPF Environmental Data Collection and Processing Facility (USA)

EDCs Economic Development Committees

edcsa effective date of change of strength accountability

edcv enamel double cotton varnish

edcw external-device control word(ing)

edd electronic data display; estimated delivery date; expected date of delivery

edd ediderunt (Latin—published by)

edd. editiones (Latin—editions)

Ed. D. Doctor of Education

EDD Eastman Dental Dispensary; Economic Development Division (Singapore); Employment Development Department; Engineering Data Depository; Engineering Development Department; Engineering Development Design; Engineering and Development Directorate (NASA)

EDD English Dialect Dictionary

eddf error detection and decision feedback

Eddie Edgar; Edmund; Edoardo; Edouard; Edsel; Eduard; Eduardo; Edvard; Edward; Edwin; Edwina

EDDS Electronic Devices Data Service

E-D DS Elsevier-Dutton Distribution Services

Eddy Edgar; Edmund; Edward; Edwin; Edwina

ede electronic defense evaluator; emitter dip effect

ede (Latin prefix—swelling)—edema

EDE Electrical Design Engineering; Electronic Design Engineering; Elevator Design Engineering; Engineering Design Establishment

edent edentate; edentulous

EDEO Episcopal Division Ecumenical Officers

edexs education of exceptional students

EDF Environmental Defense Fund; European Development Fund; Everyman Defense Fund

Edg Edgar

EDG Export Development Group(ing)

Edgar The Edgar (bust of Edgar Allan Poe given for the best mystery novel)

EDGAR Electronic Data Gathering, Analysis, and Retrieval

EDGB Export Development Grants Board

edge. electronic data-gathering equipment

edh efficient deck hand

EDH Ego-Distonic Homosexuality

edhe experimental data-handling equipment

edi electron-diffraction instrument

EDI Economic Development Institute; Edinburgh, Scot-

land (airport); Engineering Department Instruction
EDIA Engineering Department Instruction Amendment
edic electric diesel injection control
edict. engineering document information collection technique
Edie Edith
Edim Edimburgo (Portuguese or Spanish—Edinburgh)
Edin Edinburgh
Ed-in-Ch Editor-in-Chief
edinet education instruction network
EDIP European Defence Improvement Program (NATO)
EDIS Engineering Data Information Service; Engineering Data Information System
edit. editing; edition; editor; editorial
EDIT Estate Duties Investment Trust
editar electronic digital tracking and ranging unit
EDITH Exit Drills in the Home (for fire prevention)
EDITS Electronic Data Information Technical Service; Experimental Digital Television System
edl edition de luxe; electric(al) discharge laser
EDL Elder Dempster Lines; Every-Day Life (psychological test); Executive Data Link
EDLD Employee Daily Labor Distribution
EDLNA Exotique Dancers League of North America
edm early diastolic murmur; electrical-discharge machining; electromagnetic discharge measuring; electrostatic discharge machining
Edm Edmund; St Edmund's House, Cambridge
Ed. M. Master of Education
EDM Engineering Development Model
EDM Engineering Drafting Manual (USAF)
EDMICS Engineering Data Management Information Control System
Edm & Ips St Edmundsbury and Ipswich
Edmn Edmonton
Edmo Edmonton (inhabitants—Edmontonians)
EDMS Engineering Data Microreproduction System
edn edition; electrodesiccation

Edn Edwin
EDN Engineering Department Notice
EDNA Emergency Department Nurses Association
Ednbgh Edinburgh
EDNY Eastern District, New York
edo effective diameter of objective; error demodulator output; error detector output
EDO Employee Development Officer; Engineering Duty Officer; Engineering Duty Only
edoc effective date of change
Ed Op Edmonton Opera Association
EDOPAC Enlisted Personnel Distribution Office Pacific Fleet
edp engineering data processing
edp (EDP) electronic data processing
EDP Educational Data Processing; Engineering Design Proposal
EDPAA Electronic Data-Processing Auditors Association
edpac electronic data processing air conditioning
EDPC Electronic Data-Processing Center
edp crimes electronic data-processing crimes
edpe electronic data processing equipment
ed-ped-psych-soc education-pedagogy-psychology-sociology
edpm electronic data processing machine(s)
EDPRESS Educational Press Association of America
EDPS Electronic Data Processing System
EDPT Electronic Data Processing Test
Ed & Pub Editor and Publisher
edr electrical distance recorder; electrodermal response; electronic decoy rocket; equivalent direct radiation
EDR Engineering Data Requirements; Engineering Division Regulation(s); Equipment Damage Report
EDRA Environmental Design Research Association
EDRI Electronic Distributors Research Institute
edrl effective damage risk level

EDRs European Depository Receipts
EDRS Education Document Reproductive Service; Engineering Data Retrieval System
edrt effective date of release from training
eds editors; enamel double silk; estimated date of separation
ed's endangered species
EDs Explosive Disposal specialists
E-Ds Ehlers-Danlos syndrome
EDS Electronic Data Systems; Electronic Devices Society; Engineering Data Sheet; English Dialect Society; Environmental Data Service; Episcopal Divinity School; European Demonstration Scheme
edsac electronic delayed-storage automatic computer
edsat educational television satellite (EDSAT)
EDSC Engineering Data Support Center (USAF)
Ed. Spec. Educational Specialist
edst elastic diaphragm switch technology
EDST Eastern Daylight Saving Time
edsv enamel double silk varnish
edt effective date of training
EDT Eastern Daylight Time
edta ethylene diamine tetraacetic (acid)
EDTA European Dialysis and Transplant Association
edtr experimental, developmental, test, and research
edtsr electronic dial tone speed register
edu electronic display unit; experimental diving unit
EDU European Democratic Union
educ education; educational; educator
Educ Education
Educational Film Educational Film Library Association
Educ Digest Educational Digest
educom education communication(s)
Educ Pr Educational Press; Educational Press Association of America
Educ Pub Educational Publications Services; Educational Publishers

educrat educational bureaucrat
educrit educational critic(ism)
educ(s) eductor(s)
EDUPLAN Oficina de Planeamiento Integral de Educación (Spanish—Office of Integral Planning in Education)
edutainment educational entertainment (via tv)
edutele educational television
edutherap educational therapist; educational therapy
edv end-diastolic volume
edv eau de vie (French—water of life)—cognac
edvac electronic discrete variable automatic computer
edw energy dump window
Edw Edward
Edw VII King Edward VII
Edw VIII King Edward VIII
Edward G Edward G Robinson (Emmanuel Goldberg)
eDx electrodiagnosis
ee eased edges (lumber); electric eye (camera); embryo extract; environmental education; equine encephalitis; errors excepted; exoelectron(ic)(al) emission; expiration of enlistment; extra efficiency; eye and ear
ee (EE) errors excepted
e/e electrical/electronic
e & e evacuation and evasion; evasion and escape; eye and ear
e-to-e end-to-end
'ee thee
EE Early English; Electrical Engineer(ing); Electronics Engineer(ing); Envoy Extraordinary; *Estado Español* (The Spanish State)
E.E. Electrical Engineer
EE Euer Ehrwürden (German—Your Reverence)
eea engineering evaluation article
EEA Electronic Engineering Association; Emergency Employment Act; Environmental Education Act; Ethical Education Association
EEA Electrical and Electronic Abstracts
EEAC Energy and Education Action Center
EEAIE Electrical, Electronic, and Allied Industries of Europe
eeat end-of-evening astronomical twilight
EEB Eastern Electricity Board; Educational Employees Board
EEB Enosis Ellenon Bibliotekarion (Modern Greek—Greek Library Association)
eec electronic engine control
EEC East Erie Commercial (railroad); Education Exploration Center; English Electronic Computers; European Economic Community (Belgium, Britain, Denmark, France, Germany, Greece, Ireland, Italy, Luxembourg, Netherlands, Portugal, Spain)
EECA Engineering Economic Cost Analysis
eecom electrical, environmental, and communications
EECS Electrical Engineering and Computer Service
e.e. cummings Edward Estlin Cummings
eed elastic energy density; electrical explosive device
eee eastern equine encephalitis
EEE Environmental-Ecological Education (program)
EEEP Entry Employment Experience Program
EEF Egyptian Expeditionary Force
eefi essential elements of friendly information
eeg (EEG) electroencephalogram; electroencephalograph
EEG Environmental Education Group
ee/ha ewe equivalents per hectare
EEI Edison Electric Institute; Environmental Equipment Institute; Essential Elements of Information
EEIA Electrical and Electronic Insulation Association
EEIBA Electrical and Electronic Industries Benevolent Association
EEIS Evanston Early Identification Scale
EEL Ecology and Epidemiology Laboratory; Engineering Electronic Laboratory; English Electric Limited; Evans Electroselenium Limited
eem Electronic Engineers Master (catalog)
EEMP Envoy Extraordinary and Minister Plenipotentiary
EE & MP Envoy Extraordinary and Minister Plenipotentiary
een exceptional educational needs
e'en even; evening
E Eng Early English

E-engine(s) electric engine(s)
EENT end, evening nautical twilight; eye, ear, nose, and throat
EENWR Exe Estuary National Wildlife Refuge (England)
eeo equal employment opportunity
ee & o excuses, errors, and omissions
EEO Energy Efficiency Office
EEOC Economic Employment Opportunity Committee; Equal Employment Opportunity Commission
eep electronic evaluation and procurement; electronic event programmer(s); emergency essential personnel
EEP Energy Emergency Plan(ning); Export Enhancement Program
eepnl estimated effective-perceived noise level
eeprom electrically erasable programmable read-only memory
eer energy efficiency ratio; explosive echo ranging
e'er ever
EER Engineering (Equipment) Evaluation Report; Experimental Ecological Reserves
EERC Earthquake Engineering Research Center (NSF)
EERI Earthquake Engineering Research Institute; Experience, Education, and Research Institute
EERL Electrical Engineering Research Laboratory (University of Texas)
eerom electrically erasable read-only memory
ees electronic environment simulator
EES Engineering Experiment Station; Enlisted Evaluation System; European Economic Space; European Exchange System
E E Somerville and V Martin Edith Somerville and Vivian Martin
EESS Encyclopedia of Engineering Signs and Symbols
eet estimated elapsed time
EET Eames Eye Test; Eastern European Time; Education Equivalency Test; Engineering Evaluation Test(ing)
E-et-L Eure-et-Loire
EETS Early English Text Society
EETU Electrical Electronic Telecommunication Union

EEUA Engineering Equipment Users Association

EEUU *Estados Unidos* (Spanish—United States)

EEV English Electric Valve (company)

EEVC English Electric Valve Company

eex electronic egg exchange (computer program)

eez (EEZ) eastern economic zone; exclusive economic zone

ef each face; electroflotation; elevation finder; equivalent focal length; expectant father; experimental flight; exposure factor; extra fine; extremely fine

ef (EF) electrofocus

EF Educational Foundation; Emergency Fleet; Engineering Foundation; Expeditionary Force

E & F Elders and Fyffes (steamship line)

EF *Europaiske Faellesskaber* (Danish—European Common Market)

efa essential fatty acids

EFA Empire Forestry Association; Environmental Financing Authority; Epilepsy Foundation of America; European Free Associations; Evangelical Friends Alliance

EFA *Empresa Ferrocarriles Argentinos* (Spanish—Argentine Railway Enterprise)

EFAS Enroute Flight Advisory Service

efc earth fixed coordinate; electronic feedback carburetor; engineered for color (tv); Evergreen Fir Corporation (initials)

e & fc examined and found correct

EFC Educational Facilities Center; Electronic Fabrication Center; European Forestry Commission; European Freedom Council

EFCB Emergency Financial Control Board

EFCX Evergreen Freight Car Express

efd electro fluid dynamics; excused from duty

EFDO European Film Development Office

EFDSS English Folk Dance and Song Society

efe early fuel evaporation; endocrinal fibro-elastosic; expected field emergence

EFEA European Free Exchange Area

EFEA *Empresa Ferrocarriles del Estado Argentino* (Spanish—Argentine State Railways)

EFEC Efforts From Ex-Convicts (Washington, D.C.'s parole project)

eff effect; effective; efficiency

eff *effeto* (Italian—bill, promissory note)

EFF Educational Freedom Foundation; European Furniture Federation

effcy efficiency

effect. effective; effectivity

effer efferent

Effie award for effective advertising; Euphemia

Effigy Mounds Effigy Mounds National Monument on the Mississippi in northeastern Iowa

effl efflorescent

eff wd effective wind

EFG Educational Fee Grant(s); Edward FitzGerald

EFH Eileen F Hodges

efi electronic flight instruments; electronic fuel injection

ef & i engineer, furnish, and install

EFI Educational Forces Inventory; Electronic Fuel Injection (system)

EFIB European Freight Inspection Bureau

EFIC Export Finance and Insurance Corporation

eficon electronic financial control

EFINS Enrico Fermi Institute for Nuclear Studies (Univ of Chicago)

efl effective focal length; emitter-follower logic

Efl (EFL) English as a foreign language

EFL Educational Facilities Laboratories; English as a Foreign Language

EFLA Educational Film Library Association

EFLC Engineers Foreign Language Circle

efm eight-to-fourteen modulation; electronic fuel management; electronic fuel metering

efm (EFM) electronic fetal monitor(ing)

EFM European Federalist Movement

efmamjjasond *enero, febrero,*

marzo, abril, mayo, junio, julio, agosto, septiembre, octubre, noviembre, diciembre (Spanish—January, February, March, April, May, June, July, August, September, October, November, December)

EFMCNTA Elastic Fabric Manufacturers Council of the Northern Textile Association

EFMG Electric Fuse Manufacturers Guild

EFMM Education for Mission and Ministry

EFNEP Expanded Food and Nutrition Education Program

EFNIR Exhibition/Festival for New Instrumental Resources

EFNS Educational Foundation for Nuclear Science

EFOs errors, freaks, and oddities

efp effective filtration pressure; electric(al) fuel propulsion; end of flight plan

efp (EFP) electronic field production (on-location videotape production); emergency firing panel

EFPA Educational Film Producers Association

EFPW European Federation for the Protection of Waters

efr effective filtration rate; engine firing rate

EFR Electronic Failure Report(ing)

EFRC Edwards Flight Research Center

E Fris East Frisian

efs economic farm surplus; equivalent standard fillet

EFS Edinburgh Festival Society; Emergency Feeding Service; Emergency Fire Service(s)

EFSA European Federation of Sea Anglers

EFSC European Federation of Soroptimist Clubs

EFSS Emergency Food Supply Scheme

eft earliest finish time

eft (EFT) electronic funds transfer

EFT Embedded Figures Test; Engineering Flight Test

EFTA European Free Trade Association (Austria, Finland, Iceland, Luxembourg, Norway, Sweden, Switzerland)

EFTC Electrical Fair Trading Council

eftf *efterfölger* (Dano-Norwegian—successor)

***Eftf(lg)** Efterfölgere* (Dano—Norwegian—successor)

EFTI Engineering Flight Test Instrumentation

eftm eftermiddag (Norwegian—after noon)—p.m.

efto encrypt for transmission only

EFT-POS Electronic Funds Transfer initiated at Point of Sale

EFTS Electronic Funds-Transfer System; Elementary Flying Training School

efu energetic feed unit; environmental force unit

EFU European Football Union

***EFU** Europäische Frauenunion* (German—European Women's Union)

efv equilibrium flash vaporization

EFVA Education Foundation for Visual Aids

EFWA Eastern Farmworkers Association

eg electrogalvanized; electronic guidance

e.g. exempli gratia (Latin—for example)

Eg Egypt; Egyptian

EG Electrographic (copier); Engineers Guild; Equatorial Guinea (formerly Spanish Guinea); Evaluation Group(ing); grid voltage (symbol)

ega enhanced graphics adaptor

EGA Elizabeth Garrett Anderson (hospital); Ethics in Government Act; European Golf Association

egad oh god

egad. electronegative gas detector

egads electronic ground automatic destruct sequencer (system for destroying malfunctioning missiles)

egal egalitarian(ism)

EGAT Electricity Generating Authority of Thailand

Egb Egbert

e-g-b-d-f (musical mnemonic—every good boy does fine)—treble clef note names of the five lines

egc electrogalvanized coated

egcr experimental gas-cooled reactor

EGCRNR Eilat Gulf Coral Reef Nature Reserve (Israel)

EGCS English Guernsey Cattle Society

egd electrogasdynamics

egdg electrogasdynamic generator

EGDS Equipment Group Design Specifications

ege expected grade equivalent

ege eau, gaz, electricite (French—water, gas, electricity)

***Egeo** Mar Egeo* (Spanish—Aegean Sea)

E Ger East Germany

egf epidermal growth factors

egg. electrogastrogram

EG & G Edgerton, Germeshausen & Grier

eggler egg + dealer (an egg dealer)

eggsan egg sandwich

eggwich egg sandwich

ECI Export Consignment Identifier

EGIFO Edward Grey Institute of Field Ornithology

***Egip** Egipto* (Portuguese or Spanish—Egypt)

***Egit** Egitto* (Italian—Egypt)

egl egentlig (Dano-Norwegian—actual, proper, real)

EGL Eglin, Florida (tracking station); European Guarantee Loan(s)

egm extraordinary general meeting

EGM Extraordinary General Meeting (of shareholders)

EGmc East Germanic

EGMEX Eastern Gulf of Mexico

Egmonts Egmont Islands in the Chagos Archipelago northwest of Diego Garcia

EGMRSA Edible Gelatin Manufacturers Research Society of America

EGNR Ein Gedi Nature Reserve (Israel's Dead Sea oasis)

EGO *Ankara Elektrik, Havagazi ve Otobüs Isletme Müessesesi* (Ankara Electricity, City-Gas, and Bus Traffic Department); Eccentric-Orbiting Geophysical Observatory; Educational Growth Opportunities

egomac effect of gravity on methane-air combustion

egp embezzlement of government property; exhaust gas pressure

EGP Experimental Geodetic Payload

EGPA Egyptian General Petroleum Authority

EGPC Egyptian General Petroleum Corporation

EGPS Electric Ground Power System

egr egress; exhaust gas recirculation

egr (EGR) erythrocyte glutathione reductase

EGRET Explorer Gamma-Ray-Experiment Telescope (NASA)

egrs extragalactic radio source

EGS Electronic Guidance Section; Employment Guarantee Scheme

EGSP Electronics Glossary and Symbol Panel

egt exhaust gas temperature

Egyp Egypt; Egyptian; egyptology

Egypt Arab Republic of Egypt (North African nation)

Egypt. Egyptian

egyptol egyptology

eh educationally handicapped

e/h exercise-head

e & h environment and heredity

eH oxidation-reduction potential (symbol)

EH Electra House; Ernest Hemingway; extra hazardous

***EH** Enciclopedia Hoepli* (Italian—Hoepli's Encyclopedia)

eha enroute high altitude

EHA Economic History Association; Education of the Handicapped Act; Environmental Health Association; Equipment Handover Agreement

EHB Environmental Hearing Board

ehbf extrahepatic blood flow

***EHBQ** Eerste Hulp bij Ongelukken* (Dutch—First-Aid Organization)

ehc enterohepatic circulation; enterohepatic clearance

ehc (EHC) external heart compression

E & HC Emory and Henry College

EHCD Environment, Housing, and Community Development (Australian Department of)

ehd electrohydrodynamics

ehd (EHD) epizootic hemorrhagic disease

ehec (EHEC) ethylhydroxyethylcellulose

EHES Environmental Health Engineering Services (USA)

ehf extreme high-frequency—30,000–300,000 mc

ehf (EHF) epidemic hemorrhagic fever

EHF Experimental Husbandry Farm
ehg extra high grade
EHG Edvard Hagerup Grieg
EHH Ernst Heinrich Haeckel
EHHS Erasmus Hall High School
EHI Emergency Homes, Incorporated
EHIS Emission History Information System
EHJ Emanuel Haldeman-Julius
ehl effective half life
Ehl English as a home language
EHL Eastern Hockey League; Environmental Health Laboratory (USAF)
e/h/m eggs per hen per month
EHMA Electric Hoist Manufacturers Association
EHMI Environmental Hazards Management Institutes
EHMS Engine Health Monitoring System
EHN Exploring Human Nature
EHOG European Host Operators Group
ehp effective horsepower; electric horsepower; extrahigh potency
EHP Eric Honeywood Partridge (British lexicographer)
EHPT Eddy Hot-Plate Test
ehr enhanced reflector
EHS Emergency Health Service; Environmental Health Services; Experimental Horticultural Station (UK)
ehsi electronic horizontal-situation indicator
EHSP Environment Health Safety Program
eht extra-high tension
EHTRC Emergency Highway Traffic Regulation Center
ehv extra-high voltage
EHV Empresa Hondureña de Vapores (Honduran Steamship Line)
EHVIST Ethical and Human Value Implications of Science and Technology
ehw extreme high water
EHW Environmental Health Watch
ehws electric hot water service; extreme-high-water-level spring tides
e/h/yr eggs per hen per year
ei electrical insulation; end item; engineering installation; exposure index
ei (EI) environment(al) illness

e-i electromagnetic interference; electronic interface; electronic interference; extraversion-introversion
e/i endorsement irregular
e by i execution by injection (of air or poison)
e^i elasticity of demand
Ei Eire (Irish Free State)
Ei encéphale isolé (French—isolated intellectual)
EI East Indies; Electro Institute; Embrittlement Index; Essex Institute; Eunice Institute
EI Engineering Index
eia enzyme immunoassay; equine infectious anemia
EIA East Indian Association; Electronic Industries Association; Empire Industries Association; Energy Information Administration; Engineering Institute of America; Environmental Impact Assessment; Enzyme-Immune Assay
EIA Environmental Information Abstracts
EIAC Environmental Information Analysis Center
EIAJ Electronics Industry Association of Japan
EIAR Environmental Impact Analysis Report
EIB Ernst Ingmar Bergman; European Investments Bank; Export-Import Bank
EIB Economisch Instituut voor de Bouwnijverheid (Dutch—Economics Institute of the Building Industry); **Elsevier International Bulletins**
EIBA Electrical Industries Benevolent Association
EIBs Elsevier International Bulletins
EIBUS Export-Import Bank of the United States
EIBW Export-Import Bank of Washington
eic emotional inertia concept; equipment installation and checkout
EIC East India Company; Ecology International Corporation; Education Information Center; Energy Information Center; Engineering Institute of Canada; European Investment Center
EICBL Eastern Independent Collegiate Basketball League
EICF European Investment Casters' Federation
e-i children emotionally-im-

paired children
eicm employer's inventory of critical manpower
EICR Eppley Institute for Cancer Research (Omaha)
EICS East India Company's Service
eid electron impact desorption; end item description
EID East India Dock; End Item Delivery; End Item Description; Engineering Item Description
EIDA Engineering Industries Development Agency
EIDC East Indian Defense Commission; East Indian Defense Committee (Canada)
EIDEBOEWABEW Economic Intelligence Division of the Enemy Branch of the Office of Economic Warfare Analysis of the Board of Economic Warfare
Eidg Eidgenössisch (Swiss—federal)
eid lt emergency identification light
EIDs East India Docks (London)
eie end-item equipment
EIES Electronic Information Exchange System
EIF Elderly Invalids Fund; Executive Inventory File
EIFAC European Inland Fisheries Advisory Committee (FAO)
eiff enemy identification—friend or foe
eig eigenlijk (Dutch—proper)
Eig Eigenvalue
EIG Exchange Information Group
Eight Great Eight Great Islands of Japan (largest islands of the Japanese archipelago)
Eii East Intercourse Island (Australia)
eiii Electrical Industry Information Institute
eil electron injection laser
Eil Eileen; English as an international language
Eil Eiland(en) [Afrikaans or Dutch—island(s)]
EIL Electronic Instruments Limited; Experiment in International Living
Eimac Eitel-McCullough
EIMO Electronic Interface Management Office
EIMS Engineering Installation Management System
EIN Equipment Installation

Notice
EIN *Empresa Insulana de Navegacão* (Island Navigation Line)
E-in-C Engineer-in-Chief
E Ind East Indian; East Indies
EINP Elk Island National Park (Alberta)
einschl einschliesslich (German—including)
Einw Einwohner (German—inhabitants, population)
EINZ Export Institute of New Zealand
EIO Emergency Information Office(r)
EIP Environmental Improvement Program; Experiment Implementation Plan
EIPC European Institute of Printed Circuits
eiph exercise-induced pulmonary hemorrhage
eir earned income relief (tax)
EIR East Indian Railway; Emergency Information Readiness; Engineering Information Request; Environmental Impact Report; Equipment Interchange Receipt
EIRMA European Industrial Research Management Association
eirnv extra incidence rate in non-vaccinated groups
eiro evaluation of infra-red optics
EIRs Environmental Impact Reports
EIRS Education Information Resources Service
eirv extra incidence rate in vaccinated groups
eis electrical intersection splice; electron impact spectroscopy; end interruption sequence; extended instruction set
EIS Early Implementation System; Economic Information Systems; Education Information Services; Electronic Ignition System; Electronic Inquiry System; End Item Specification; Environmental Impact Statement; Epidemic Intelligence Service (HEW); Export Intelligence Service; Exxon Information System
Eisted (*Welsh*—Eisteddfod—session, meeting for competition)
e-i student(s) emotionally-impaired student(s)
eit engineer in training

ei & t emplacement, installation, and test(ing)
EIT Electrical Information Test
EITA Electric Industrial Truck Association
EITB Engineering Industry Training Board
eitp environmental interaction theory of personality
EITS Educational and Industrial Testing Service
eiu economist intelligence unit
EIVT European Institute for Vocational Training
EIWS Engineering Installation Workload Schedule
ej elbow-jerk; expansion joint
ej ejemplo (Spanish—example)
E-J Endicott-Johnson
EJ ERIC Journal
EJA Executive Jet Aviation
EJC Edison Junior College; Engineers Joint Council; Engineers Junior College; Everett Junior College
EJCC Eastern Joint Computer Conference
eject. ejector
EJ & E R Y Elgin, Joliet & Eastern Railway
EJMA Educational Jewelry Manufacturers Association; Expansion Joint Manufacturers Association
EJN Edicott Johnson (stock exchange symbol)
ejp excitatory junction potential
EJS Engineering Job Sheet
EJT Engineering Job Ticket
EJTA Emergency Jobs Training Act
ejusd. ejusdem (Latin—of the same)
ek even keel; single enamel single cellophane (insulation symbol)
eK etter Kristi (Norwegian—after Christ)
EK Eastman Kodak
EK Eisernes Kreuz (German—Iron Cross)—military decoration; *Esperanto Klubo* (Esperanto Club)
ekc epidemic keratoconjunctivitis
EKCO EK Cole (Limited)
EKD Evangelische Kirche in Deutschland (German—Protestant Church in Germany)
Eken (Swedish slang—Stockholm)
ekg electrokardiogram (electrocardiogram); electrocardiography

EKG electrokardiogram
eKr efter Kristus (Dano-Norwegian—after Christ) A.D.
eks eksempel (Dano-Norwegian—example)
EKSC Eastern Kentucky State College
EKSD Esperanto Klubo San Diego
ekskl eksklusive (Dano-Norwegian—exclusive)
eksp expederet or *ekspedition* (Dano-Norwegian—expedite or expedition)
ekspl eksemplar (Dano-Norwegian—example or sample)
eku earliest known usage
EKU Eastern Kentucky University
ekv electron kilovolt
ekw electrical kilowatt(s)
el each layer; educational level; elastic level; elect(ed); electric(ity); electroluminescence; element(ary); elevated; elevation; elongation; extra line
el eller (Dano-Norwegian—or)
El Elbert; Elevated Railroad; Elias; Elvie; Elvira
EL Eastern League; Electrical Laboratory; Electronics Laboratory; *Empresa do Limpopo* (Limpopo Line); Engineer Lieutenant; English Leicester (sheep); Epworth League; Erie-Lackawanna (railroad); Everyman's Library; Export Licence
E-L Erie-Lackawanna (railroad)
el2 elongation in 2 inches
ELA English Language Amendment
elab elaborate(d); elaborately; elaborating; elaboration; elaborative
ELAC East Los Angeles College
e lacte. e lact (Latin—with milk)
EL AL El Al Israel Airlines
ELAM Escuela Latinoamericana de Matemáticas (Spanish—Latin American School of Mathematics)
ELAP Emergency Legal Assistance Project
ELAPR Estado Libre Asociado de Puerto Rico (Spanish—Associated Free State of Puerto Rico)—the Commonwealth of Puerto Rico's official name
elas elastic; elasticity; emergency logistical air support

ELAS *Ethnikos* *Laikos* *Apelephterotikos* *Stratos* (Greek—Hellenic Peoples' Army of Liberation)

Elasm Elasmobranchia

elasmobranchs elasmobranch fishes (cartilaginous fishes such as chimaeras, dogfishes, rays, and sharks)

El-ay Los Angeles, California

elb electronic lean burn; emergency locator beacon

Elb Egyptian pound

El of B Elector(ate) of Bavaria; Elector(ate) of Brandenburg

ELB English Language Battery

E.L.B. Bachelor of English Literature

elba emergency location beacon aircraft

El Banco (Spanish—The Bank)—World Bank for Reconstruction and Development

El'brus Mount El'brus (Europe's highest mountain in the Caucasus of the USSR)

ELBS English Language Book Society

elc extra-low carbon (electrodes)

El C El Centro

ELC Electronic Location Center

ELCA Evangelical Lutheran Church in America

El Caj El Cajon

El Cajohn El Cajon, California

El Cap El Capitan Dam; El Capitan Reservoir

elcar electric car

El Cen El Centro

ELCID Enforcement of Law Through Court Intervention and Diversion

elct electronics

elct rm electronics room

eld edge-lighted display; elder; eldest; electric load dispatcher; extra-long distance

Eldercare plan providing medical care for the elderly

el-dl electric locomotive-diesel locomotive

ELDO European Launcher Development Organization

ELDS Editorial Layout Display System

Elean Eleanor

Eleanor Mrs Anna Eleanor Roosevelt—wife of President Franklin Delano Roosevelt

elec electric; electrical; electrician; electricity; electro-;

electuary

Elec Elector; Electorate; Electra; Electricity

ELEC Election Law Enforcement Commission; European League for Economic Cooperation

ELEC/DR Electric Residential Service

Elec Engr Electrical Engineer; Electronic Engineer

elec pt electric(al) point

elecpub electronic publishing (dissemination of information via any electronic distribution means)

elect electrical

elect. election; elector; electoral; electrolyte; electrolytic

elect. *electuarium* (Latin—electuary)—confectioned drug; lollipop

elec tech electrical technician; electronic technician

elect. in. electricity installed

electn electrician

elect. pd electricity planned

ELECTRA Electrical, Electronics, and Communications Trades Association; trademark of the London Electricity Board

electraac electronic auto analysis clinic

electro electrocute; electrocution; electrotype

electrochem electrochemistry

electrocortico electrocorticograph(ic)(al)(ly); electrocorticography

electroderm electrodermal(ly)

electroenceph electroencephalography

electrogas electrogasdynamic(s)

electrogen electrogenic(al)(ly); electrogenesis

electrohyd electrohydraulic(s)

electrohydraul electrohydraulic(al)(ly)

electrol electrolysis

electrolev electronic levitation; electronically levitated

electromusic electronic music

electron. electronic(s)

Electron Lett *Electronics Letters*

electro-ocu electro-oculogram

electrophonics electrophonic instruments

electrophys electrophysics

electroret electroretinograph-(ic)(al)(ly); electroretinography

electro(s) electrotype(s)

electrosen electrosensitive;

electrosensitivity

electrostat electrostatic copy; electrostatic printing

Electrovette electric-battery-powered Chevette

electrum 50% gold, 50% silver

electy electricity

elek electric(al); electronic

Elekt *Elektrizität* (German—electricity)

elektr *elektriciteit* (Dutch—electricity)

elem element; elementary

elephantocade elephant parade

eleph fol elephant folio—books about 23 inches high

e-less e-less novel written by British author Ernest Vincent Wright in 1939 with more than 50,000 words without the letter e; Georges Perec's *La Disparition*, published in French in 1969, is also e-less

elev elevated; elevation; elevator

elex electronics; electronics exercise

elf. early lunar flare; electric-light fitting; extra low frequency; extremely low frequency

El F El Ferrol

ELF Early Lunar Flare; Eritrean Liberation Front

ELFA Electric Light Fittings Association

elfc electroluminescent ferro-electric cell

ELF-HELP Educators, Librarians, and Families—Helping Loving Prescholars

Elg Elgar; Elgin

El G El Paso Natural Gas Company

ELG European Liaison Group (USA)

elgas electricity and gas

ELGB Emergency Loan Guarantee Board

elhi elementary and high school (textbooks)

eli electricity installed; electronic line indicator

Eli Elias; Elihu; Elijah; nickname for a student or alumnus of Yale University

ELI English Language Institute; Environmental Law Institute

ELIA English Language Institute of America

ELIC Electric Lamp Industry Council

Elien. *Eliensis* (Latin—of Ely)

elig eligible

Elij Elijah
elim eliminate; eliminated; elimination
ELIM Evangelical Lutherans in Mission
elin exhibit line item number
elint electronic intelligence
elints electronic intelligence-gathering vessels
elip electrostatic latent image photography
Elis Elisabeth
elisa enzyme-linked immunosorbent assay
Elise Elizabeth
elix elixir
Eliz Elizabeth(an)
Eliza Elizabeth
Elizabeth II Elizabeth Alexandra Mary of Windsor (Queen of United Kingdom of Great Britain and Northern Ireland and Her Other Realms and Territories)
Elizabeths Elizabeth Islands; queens named Elizabeth
Elk Hills U.S. Navy's petroleum reserve at Elk Hills, California
Elk Island Elk Island National Park east of Edmonton, Alberta
ell. elbow; ellipsoid(al); elliptic(al)
ell eller (Spanish—or)
Ell English language learning
Ella Eleanor; Eleanora; Eleanore; Isabella
ELLA European Long Lines Agency
Ell Dim Elliniki Dimokratia (Greek—Hellenic Republic) - Greece
Ellen Eleanor(a)(e)
Ellerman Ellerman Lines Ltd
Ellices Ellice Islands now called Tuvalu
Ellie Alice
el lign eller lignende (Dano-Norwegian—or similar)
ellip elliptic; elliptical; elliptically
Ellis Ellis Island Immigration Examination Station, New York
ELLIS Ellis Air Lines
el lt electric light; electric lighting
Elly Eleanor
elm. element; energy-loss meter
ELM Eastern Atlantic and Mediterranean; Edgar Lee Masters
elma electromechanical aid
Elma Elizabeth Mary; Wilhelmina

ELMA Empresa Lineas Maritimas Argentinas (Argentine Lines)
Elmer R Rice Elmer Reizenstein
ELMG Engine Life Management Group (USN)
elmint electromagnetic intelligence
Elmira Men's Reception Center (for male prisoners) at Elmira, New York
ELMO Engineering and Logistics Management Office (USA)
elmobile electric automobile
ELMS Earth Limb Measurement Satellite; Experimental Library Management System
El Mus East London Museum
E Ln East London
ELN Ejército de Liberación Naciónal (Spanish—Army of National Liberation)—Bolivian and Colombian group
ELNA Esperanto League of North America
El Ng Ela Nguema (formerly San Fernando)
elngn elongation
El Niño El Niño Current (warm ocean current sweeping northward from the west coast of South America to the west coast of North America)
ELNM Edison Laboratory National Monument, West Orange, New Jersey
elo elocution; eloquence
Elo Eloheimo
ELO Electric Light Orchestra; English Language Office(r)
eloc elocution(ary); elocutionist(ic)(al)(ly)
ELOI Emergency Letter of Instruction
Eloise European large-orbiting instrumentation for solar experimentation
elong elongate; elongation
E long east longitude
eloq eloquence; eloquent(ly)
E Loth East Lothian
elox electrical spark erosion
elp electricity planned
ELP El Paso, Texas (airport)
El Pais (Spanish—The Country)—daily morning newspaper published in Madrid
elpc electroluminescence photo conductor
ELR Engineering Laboratory Report
elra electronic radar

ELRACS Electronic Reconnaissance Accessory System
elrat electrical ram air turbine
El Reno Federal Reformatory, El Reno, Oklahoma
ELRO Electronics Logistics Research Office (USA)
els extreme long shot
Els Elsinore (Helsingör); Elspeth
El of S Elector(ate) of Saxony
ELs elevated railways
ELS Escabana and Lake Superior (railroad)
Elsa Elizabeth
El Salv El Salvador (Spanish—Republic of El Salvador)
El Salvador Republic of El Salvador (Central America's smallest nation), *República de El Salvador*
elsbm exposed location single buoy mooring
ELSE European Life Science Editors
elsec electronic security
ELSEGIS Elementary and Secondary General Information System
elsets element sets
Elsev Elsevier (family of Dutch printers and publishers dating from the 16th century)—also spelled Elzevir
Elsev App Sci Elsevier Applied Science
Elsevier Sci Elsevier Scientific Publishing Co
Elsev NH Elsevier North Holland
Elsev Sci Elsevier Science Publishing Company
Elsev Seq Elsevier Sequoia
El Sgndo El Segundo
elsie electronic location of status-indicating equipment; electronic signalling and indicating equipment; emergency life-saving instant exit
Elsie Elizabeth
Elspet(h) (Scottish—Elizabeth)
ELSS Emplaced Lunar Scientific Station
elsse electronic sky screen equipment
elt electrometer; emergency locator transmitter
Elt European letter telegram
E Lt Engineer Lieutenant
ELT English Language Teaching
E Lt-Cdr Engineer Lieutenant-Commander
ELTDA English Language

Teaching Development Aid

eltec electrical technician; electronic technician

ELTI English Language Teaching Institute

ELTJ English Language Teaching Journal

ELU English Lacrosse Union

elv extra-low voltage

elw extreme low water

elwh elsewhere

El Wld Electrical World

elws extreme-low-water-level spring tides

elxr elixir

Ely easterly

Elz (*see* Elsev)

ELZ Environmental Living Zone

em electromagnetic(al)(ly); emanation; embargo; emergency maintenance; emergency mobilization; enlisted man; expanded metal

em (EM) electron microscope; electron microscopy; end of medium character (data processing)

e/m specific electronic mass

e & m endocrine and metabolism; erection and maintenance

e of m error of measurement

'em them

em eftermiddag (Dano-Norwegian—after midday)

Em Emily; Emma; Emmanuel; Emy

EM Earl Marshal; Education Manual; Electrician's Mate; electromagnetic (symbol); Engineer Manager; Engineering Memorandum; Enlisted Man (Men); Etna & Montrose (railroad); European Movement; External Memorandum

E-M Electric Machinery (company); Electro-Motive (corporation); Embden-Meyerhof (glycolitic path)

E.M. Engineering of Mines; Engineer of Mining

EM Estado-Maior (Portuguese—general staff, headquarters); *Estado Mayor* (Spanish—general staff, headquarters); *Excerpta Medica* (Elsevier logo-type)

E-M Etat-Major (French—Headquarters)

E.M. Equitum Magister (Latin—Master of Horse)

EM 1 C Electrician's Mate First Class (USN)

EM 2 C Electrician's Mate

Second Class (USN)

EM 3 C Electrician's Mate Third Class (USN)

EMA Electronics Manufacturers Association; Employment Management Association; Envelope Manufacturers Association; Environmental Media Association; European Marketing Association; European Monetary Agreement; Evaporated Milk Association; Exposition Management Association; Extended Mission Apollo

E MacD Edward MacDowell

EMAD Engine Maintenance Assembly and Disassembly

EMAIA Electrical Meter and Allied Industries Association

E-mail electronic mail

Em Ar Un Emiratos Arabes Unidos (Spanish—United Arab Emirates)

EMAS Emergency Medical Advisory Service; Emergency Message Authentication System; Employment Medical Advisory Service; Employment Medical Advisory Service (UK)

EMATS Emergency Message Automatic Transmission System

emb embankment; embargo; embark; embarkation; embassy; embroidered; embroidery; embryo; embryology

Emb Embankment; Embassy

EMB Egg Marketing Board; Energy Mobilization Board

emball emballsje (Norwegian—packing)

EMBERS Emergency Bed Request System

embgo embargo

embk embark

Embkmt Embankment

embkn embarkation

EMBL Eniwetok Marine Biological Laboratory

embo emboss(ed); embossing

EMBO European Molecular Biology Organization

embr embroider(y)

embry embryology

embryol embryology

EMBS Energy Management Bumper System

embsy embassy

emc electromagnetic capability; engineered military circuit; equilibrium moisture content

emc (EMC) electromagnetic control; encephalomyocardi-

tis

EMC Education Media Council; Einstein Medical Center; Electronic Material Change; End Mollycoddling in America; Engineering Maintenance Center; Engineer(ing) Maintenance Control; Engineering Manpower Commission; Evergreen Marine Corporation

EMCA Ethnic Minority Council of America

EMCC European Municipal Credit Community

EMCCC European Military Communications Coordinating Committee

EMCE Eastern Montana College of Education

emcee(s) master(s) of ceremonies

EMCF European Monetary Cooperation Fund

EMCMF Embarked Mine Countermeasures Force

emcon emission control

EMCS Energy Management and Control Systems

EMCU Evergreen Maritime Container Unit

emcv encephalomyocarditis virus

emd electric-motor-driven

Emd Emden

EMD Energy Management Display

E-MD Electro-Motive Division (General Motors)

EMDI Export Market Development Incentive

emdp electromotive difference of potential

EMDP Export Market Development Programme

EME Electrical and Mechanical Engineering; Electrical and Mechanical Engineer(s); Extraordinary Minister of the Eucharist

EMEA Electrical Manufacturers Export Association

EMEB East Midlands Electricity Board

EMEC Electronics Maintenance Engineering Center

EMELEC trademark of East Midlands Electricity Board

emend. emendate(d); emendating; emendation(s); emendator(s); emendatory; emender(s)

emend. emendatis (Latin—corrected, edited, emended)

emer emergency

Emer Emeritus

Emerald Triangle Humboldt,

Mendocino, and Trinity counties, California, where potent marijuana is grown

emerg emergency

emergcons emergency conditions

emerit. emeritus (Latin—retired with honor)

E.Met. Engineer of Metallurgy

E-meter electrical-resistance galvanometer

EMETF Electromagnetic Environment Test Facility (USA)

EMEU East Midlands Educational Union

emf electromotive force; erythrocyte maturing factor; every morning fix

EMF European Management Forum; European Motel Federation; Excerpta Medica Foundation

E.M.F. E(dward) M(organ) Foster

emg electromyogram; electromyography

EMG Economic Monitoring Group

EMG Estado-Maior General (Portuguese—Staff General); *Estado Mayor General* (Spanish—Staff General)

emgcy emergency

emh electrical, mechanical, and hydraulic

EMHA Electronically-Monitored Home Arrest

emi electromagnetic interference

EMI Electrical and Musical Industries; Equipment Manufacturing Incorporated; Experiences in Mathematical Ideas

EMI Edizione Musicali Italiane (Italian Musical Publications)

emia (Latin suffix—a condition of the blood)—anemia

emic emergency maternity and infant care

emid electromagnetic intrusion detector

E Midl East Midland

EMIETF Ethnic Materials Information Exchange Task Force

emig emigrant; emigration

EMILY Early Money Is Like Yeast

Emin Eminence

emip equivalent means investment period

Emirates United Arab Emirates (on the Trucial Coast of the Persian Gulf)

emis emission

EMIS Electromagnetic Intelligence System; Electronic Materials Information Services; Engineering Maintenance Information System

EMIT Engineering Management Information Technique

EMJC East Mississippi Junior College

Emjo Emmanuel Jobe

emK elektromotorische Kraft (German—electromotive force)

eml electromagnetic levitation; equal matrix languages

Eml Emily

EML Earthquake Mechanisms Laboratory; Equipment Modification List

EML Everyman's Library

em log electromagnetic log

EMLTS Electromagnetic Levitation Transportation System (wheelless railway)

emm electromagnetic measurement

Emm Emmanuel; Emmanuel College, Cambridge

emma electron microscopy and microanalysis

Emma Emma Adelheid Welhelmine Therese, Queen of the Netherlands 1879–1890; Queen Emma Bridge, Willemstad Harbor, Curaçao

Emm Coll Emmanuel College—Cambridge

Emmie Emma; Emy; Emmy

Emml Emmanuel

EM^{MO} Eminentísimo (Spanish—Most Eminent)—masculine ecclesiastical title applied to cardinals

EMMS Electronic Mail and Message System

EMMSA Envelope Makers and Manufacturing Stationers Association

E Mn E Early Modern English

EMNM El Morro National Monument

EMO Emergency Measures Organization; Emergency Services Organization; Engineering Maintainability Organization; Equipment Move Order

emol emolumentos (Portuguese or Spanish—emoluments, official fees)

EMOL Excerpta Medica On-Line

Emos Earth's mean orbital speed

emot emotion(al)

E-motor(s) electric motor(s)—submarine

EMOW Electrician's Mate of the Watch (USN)

emp electromagnetic pulses; empennage

emp. emplastrum (Latin—adhesive, a plaster)

e.m.p. ex modo prescripto (Latin—in the manner prescribed)

Emp Emperor; Empire; Empress

EMPAC Engineering Management Planning and Control

emp agcy employment agency

empath empathetic; empathy

EMPC Educational Media Producers Council

empd employed

Emperor Range mountains traversing Bougainville in the Solomon Islands

emph emphasis

emphy emphysema; emphysematous; emphyteusis; emphyteuta; emphyteutic

EMPI European Motor Products Incorporated

EMPIRE Early Manned Planetary Interplanetary Round-Trip Experiment

empl emplace; emplacement; employ; employee; employer; employment

empld employed

EMPOCOL Empresa Puertos de Colombia (Colombian Port Works)

EMPORCHI Empresa Portuaria de Chile (Spanish—Chilean Port Enterprise)

EMPPO European and Mediterranean Plant Protection Organization

Empress of India Britain's Queen Victoria of Hanover (1837–1901)

empro emergency proposal

empsked employment schedule

empsz emphasize

emp. vesic. emplastrum vesicatorium (Latin—a blistering plaster)

emq electromagnetic quiet

emr educable mentally retarded; electromagnetic resonance; electromagnetic riveting

EMR Eastern Mediterranean Region; Emerson Electric (stock exchange symbol); Engineering Master Report; Engineering Model Report; Enlisted Manning Report

EM & R Equipment Maintenance and Readiness
em-related emission-related (smog)
EMRIC Educational Media Research Information Center
EMRODA Electronic Maintenance Repair Operation Distributors Association
EMRS East Malling Research Station
ems emergency medical services; expected mean squares
ems (EMS) electronic muscle stimulation
Ems Bad Ems
EMS Econometric Society; Electronic Message System; Emergency Medical Service; Engineering Material Specification; European Mone- tary System; Export Marketing Service; Express Mail Service
EMSA Electron Microscope Society of America
EMSC Educational Media Selection Center; Electrical Manufacturers Standards Council
EMSO European Mobility Service Office (USA)
EMSP Enhanced Modular Signal Processor
EMSS Electromechanical Subsystem
EMSU Environmental Meteorological Support Unit
emt electrical metallic tubing; emergency medical technique; equivalent megatonnage
EMT Emergency Medical Technician; Evaluation Modality Test
EMTA Electro-Mechanical Trade Association
EMT-A Emergency Medical Technician—Ambulance
EMTN European Meteorological Telecommunications Network
EMT-P Emergency Medical Technician—Paramedic
emtr emitter
emu. electromagnetic unit
Emu European monetary unit
EMU Eastern Michigan University; Economic and Monetary Union
EMU Europese Monetaire et Economische Unie (Dutch—European Monetary and Economic Union)
emul emulsion
emuls. emulsio (Latin—emul-

sion)
emut electric multiple-unit train
emux electronic multiplexer
emv electromagnetic vulnerability; electron megavolt
Emy Emilia; Emily
en enema; enemy; exceptions noted
en (Greek—in, into)—encephalitis, energy, entropy, environment
En Engineer, English
EN Esquimalt and Nanaimo (railway)
EN Emissora Nacional (Portuguese—National Broadcast); *Estrada Nacional* (Portuguese or Spanish—National Highway); *Evening News*
En 1 c Engineman first class
ENA English Newspaper Association; Evening News Association
ENA Escuela Nacional de Agricultura (Spanish—National School of Agriculture); *L'Ecole Nationale d'Administration* (French—National Administration School)—France's civil-service academy
ENAB Evening Newspaper Advertising Bureau
ENAF Empresa Nacional de Fundiciones (Spanish—National Smelters Enterprise)
ENAFRI Empresa Nacional de Frigorificos (Spanish—National Freezer Enterprise)
enam enamel; enameled; enamels
ENAMI Empresa Nacional de Minería (Spanish—National Mining Enterprise)—Chile
ENAP Empresa Nacional del Petróleo (Spanish—National Petroleum Enterprise)
ENAPUPERU Empresa Nacional de Puertos del Peru (Spanish—National Enterprise of the Ports of Peru)
ENASA Empresa Nacional de Autocamiones (Spanish—National Trucking Enterprise)
ENATA Empresa Nacional de Tabaco (Spanish—National Tobacco Enterprise)
ENATEL Empresa Nacional de Telecomunicaciones (Spanish—National Telecommunication Enterprise)
ENB East New Britain; English National Board (of health visiting, midwifery, and nursing)

ENBPS Ente Nazionale per le Biblioteche Populari e Scolastiche (Italian—National Organization of Popular and Scholastic Libraries)
enc enclosed
ENC Empresa Nacional de Carbon (Spanish—National Coal Enterprise)
ENCA European Naval Communications Agency
encap encapsulate(d); encapsulation
Enc Can Encyclopedia Canadiana
ENCI Empresa Nacional de Comercializacion de Insumos (Spanish—National Enterprise for the Commercialization of Raw Materials)
Enc Jud Encyclopedia Judaica
encl enclose; enclosed; enclosure
encld enclosed
enclg enclosing
enclit enclitic
enclo enclosure
ENCO Energy Company (Humble Oil & Refining)
encom encomiast(ic); encomium(s)
ENCORE Encouragement, Normalcy, Counseling, Opportunity, Reaching out, Energies revived (YWCA program for women who have undergone breast surgery)
ENCOTEL Empresa Nacional de Correos y Telegrafos (Spanish—Post and Telegraph National Enterprise)
ENCP European Naval Communications Plan (NATO)
enct encounter
ency encyclopedia
Ency Assn Encyclopedia of Associations
Ency Brit Encyclopaedia Britannica
end. endorsement
end (Latin prefix—within)—endoderm
ENDE Empresa Nacional de Electricidad (Spanish—National Electricity Enterprise)
ENDESA Empresa Nacional de Electricidad SA (Spanish—National Electricity Enterprise Corporation)
ENDEX Environmental Data Index
endis endispiece (antonynm for frontispiece, correct term is tailpiece)
end mth end of month
endo endocrine; endocrinology

endo *endon* (Greek—within)—
endocrine, endodermic, endo-
skeletal
endocrin endocrinological; endocrinologist; endocrinology
endocrino endocrinologic(al)-
(ly); endocrinologist; endocrinology; endocrinopath(ic)-
(al)(ly); endocrinopathy; en-
docrinosis; endocrinotherapy;
endocrinous
EndocSoc Endocrine Society
endor electron nuclear double
resonance
endor(s) endorsement(s)
endow. endowment
endp endpaper(s)
ends. endpapers
ENDS Euratom Nuclear Documentation System
ENDS *Empresa Nacional de
Semillas* (Spanish—National
Seed Enterprise)
end tel *endereço telegráfico*
(Portuguese—cable address)
endv *endvidere* (Dano-Norwegian—furthermore)
endvr endeavor
endvrg endeavoring
end wk end of week
end yr end of year
ene *enero* (Spanish—January)
ENE east northeast
ENE *Escuela Nacional de
Economica* (Spanish—Na-
tional School of Economics)
ENEA European Nuclear Energy Association
ENEL *Ente Nazionale per
l'Energia Elettrica* (National
Electric-Power Company of
Italy)
enem. *enema* (Greek—injection)
ener energize
ENERGAS *Empresa Nacional
de Gas* (Spanish—National
Gas Enterprise)
energe *energicamente* (Italian—energetically)
ENEWS Effectiveness of Navy Electronic Warfare System
ENEX Engineering Export Association
ENF European Nuclear Force
en fav de *en faveur de*
(French—in favor of)
enf(d) enforce(d)
enft enforcement
eng electronic news gathering;
electronystagmogram; en-
gine; engineer(ing)
eng (ENG) electronic news
gathering (tv news reports)
eng *engelsk* (Dano-Norwe-

gian—English)
Eng Engineer; England; English
Eng *Engineering* (British periodical)
Engañol English-Spanish
ENGBCA Engineers Board of
Contract Appeals (USA)
Eng. D. Doctor of Engineering
Eng Div Engineering Division
eng dvr engine driver
eng/efp camera electronic
news-gathering/electronic
field-production camera
eng err engineering error
eng fnd engine foundation
enggmt engagement
Eng hrn English horn (a low
oboe and neither English nor
horn)
engin engineering
Eng Index *Engineering Index*
engitist engineer + scientist
engl *englisch* (German—English)
Engl England; English
Englewood Federal (delinquent) Youth Center at Englewood, Colorado
English Lit English literature
Eng Lit English Literature
Eng News-Rec *Engineering
News-Record*
eng° *engenheiro* (Portuguese—
engineer)
engr engineer
eng rm engine room
engrv engraver; engraving
Eng. Sc. D. Doctor of Engineering Science
ENGSS Engineering School-
ship (USN)
EN-H Elsevier North-Holland
ENI *Ente Nazionale Idro-
carburi* (Italian—National
Fuel Agency)
ENI *Escuela Nacional de
Ingenieria* (Spanish—National School of Engineering)
eniac electronic numerical integrator and computer
ENIC *Ente Nazionale Industrie
Cinematografiche* (Italian—
National Association of Film
Producers); *Ente Nazionale
della Cinofilia Italiana* (Ital-
ian—National Organization
of Italian Dog Lovers)
ENICO Exxon Nuclear Idaho
Company
ENIDS Ethnic Name Identifi-
cation System
Enigma Elgar's *Enigma* Varia-
tions for Orchestra with an
enigmatic program wherein
the composer dedicates its

movements to his friends de-
scribed by their initials
ENIM *Ente Nazionale dell-
'Istruzióne Media* (Italian—
National Organization for In-
termediate Instruction)
ENIT *Ente Nazionale Industrie
Turistiche* (Italian—National
Tourist Industry)
enk *enkelvoud* (Dutch—singu-
lar)
enl enlist
E n l English as a national lan-
guage
enlgd enlarged
en ml end mill
ENMU Eastern New Mexico
University
Enn Quintus Ennius (Roman
poet)
ENNWR Eastern Neck Na-
tional Wildlife Refuge
(Maryland)
eno *enero* (Spanish—January)
en° *enero* (Spanish—January)
E/no *estacionamiento no*
(Spanish—no parking)
ENO English National Opera
Enoch Pratt Enoch Pratt Free
Library
enol enology
Enos Book of Enos
Eno's Eno's Fruit Salts
ENP Egmont National Park
(North Island, New Zealand);
Etosha NP (South-West Af-
rica); Everglades NP (Flo-
rida)
ENPA *Ente Nazionale Pro-
tezione Animali* (Italian—Na-
tional Society for the Protec-
tion of Animals)—Italy
ENPI *Ente Nazionale Pre-
venzione Infortuni* (Italian—
National Institution for the
Prevention of Accidents)
ENPMA Eastern National Park
and Monument Association
enq enquire; enquiry
enr en route; equivalent noise
resistance
enr (ENR) extrathyroidal neck
radioactivity
ENR *Emissora Nacional de
Radiodifusão* (Radio Portugal)
E & NR Esquimalt and Nanaimo Railway
ENRI Electronic Navigation
Research Institute
enrpae enroute (to/from) public affairs event
enrt enroute
Ens Ensign
Ens *Ensenadas* (Spanish—in-
lets, small bays)

ENS *Empresa Naviera Santa;*
European Nuclear Society;
experimental navigation ship
ENSA Entertainments National
Service Association
Ensen Ensenada
ensi equivalent—noise side-
band input
ENSIDESA *Empresa Nacional
Siderurgica SA* (Spanish—
National Steel Works)
ENSIP Engine Structural In-
tegrity Program (USAF)
en-skids National Security In-
telligence Directives
ent ear, nose, and throat; enter;
entrance
ent ental (Dano-Norwegian—
singular)
ENT Aerolineas Argentinas
(Argentine Airlines); Ear,
Nose, and Throat; Euro–NA-
TO Training
ENTA Environmental Test
Area
entd entered
ENTE *Ente Nazionale per
l'Energia Elettrica* (Italian—
National Electric Energy Enterprise)
ENTEL *Empresa Nacional de
Telecomunicaciones* (Spanish—National Telecommuni-
cations Enterprise)
entero (Latin prefix—intestine)—enteritis
enterobact enterobacterial(ly);
enterobacteriologist; entero-
bacterium
enteropath enteropathogenic-
(al)(ly)
enterov enterovioform
*Entführung Entführung aus
dem Serail* (German—Ab-
duction from the Seraglio)—
Mozart opera
ent hall entrance hall
entl entitle
entn entertain
entom entomology
entomol entomologic(al)(ly);
entomologist; entomology
entr entrance
entspr entsprechend (German—corresponding)
Ent Sta Hall Entered at Stationers' Hall
ENTURPERU *Empresa Nacional de Tourismo del Perú*
(Spanish—National Tourist
Enterprise of Peru)
ent-vio entero-vioform (anti-
diarrhetic)
ENUF Everybody Now Undo
Foul-ups
enur enuresis

enus end user
enutech enuresis technology
(controlling bedwetting)
env envelop; envelope; environ; environment(al)(ly); envoy
Env Envoy
Env Ext Envoy Extraordinary
ENVI Envirosphere Company
environ. environment; environmental; environmental-
ism; environmentalist
ENWR Erie National Wildlife
Refuge (Pennsylvania); Eu-
faula National Wildlife Refuge (Alabama)
ENY Elsevier New York
enz enzovoort (Dutch—et
cetera)
enza influenza
En Zed(er)(s) New Zealand-
(er)(s)
eo end of operation; engine oil
e-o electro-optical; even-odd
e & o errors and omissions
e.o. ex officio (Latin—by virtue
of office)
Eo Ecuadorian escudo(s); es-
cudo(s) (Portuguese currency)
E, electric affinity (symbol)
EO Eastern Orthodox; Education Officer; Employers Organization; Engineering Order; Entertainments Office(r);
Examining Office(r); Executive Office(r); Executive Order
E & O Eastern and Oriental
eoa effective on or about; end
of address; examination,
opinion, advice (medical)
EOA Economic Oil Association; Education Officers Association; Essential Oil Association
EOARDC European Office of
the Air Research and Development Command (USAF)
eob end of block (character);
expense operating budget
EOB Executive Office Building
EOBs Explanation of Benefits
eoc electric overhead crane;
emotional-organic combination; end of card
Eoc Eocene
EOC Economic Opportunity
Commission; Educational
Opportunity Center; Elec-
tronic Operations Center; Enemy Oil Committee; Equal
Opportunities Commission;
Executive Officers Council
EOCC Engineering Operation-

al Casualty Control (USN)
EOCI Electric Overhead Crane
Institute
eocp engine out of commission
for parts
eod entry on duty; every other
day; explosive ordnance disposal
eod (EOD) explosive ordnance
device
eodad end-of-data-set address
EODAP Earth and Ocean Dynamic Applications Program
(NASA)
EODG Explosive Ordnance
Disposal Group
EODP Engineering Order Delayed for Parts
EODU Explosive Ordnance
Disposal Unit
eoe earth orbit ejection; equal
opportunity employer
e & oe errors and omissions
excepted
e and oe errors and omissions
excepted
EOE Enemy-Occupied Europe
eof end of flight
eof (EOF) end of file
EOF Earth Orbital Flight
eog effect on guarantees; electro-oculogram
EOG English Opera Group
eogb (EOGB) electro-optical
glide bomb(ing); electro-optical guided bomb(ing)
EOGs Educational Opportunity Grants
eoh end of overhaul; equipment on hand
EOH Emergency Operation
Headquarters
eohp except otherwise herein
provided
eoj (EOJ) end of job
eol effective operational
length; end of life; expres-
sion-oriented language
Eol Eolic
EOL Ex Oriente Lux (The
Light of the Orient—The
Oriental Society)
eolb end-of-line block
eolm electro-optical light modulator
eom end of month; every other
month; extra-ocular move-
ments
eom (EOM) end of message
(data processing)
EOM Employment Office
Manager(s)
EOMB Explanation of Medicare Benefits
eomi end of message incomplete

eoms end-of-message sequence
EONR European Organization for Nuclear Research
EOO Equal Opportunity Office
eooe error or omission excepted
EOOW Engineering Officer of the Watch (USN)
eop earth orbit plane; end of part; end of passage
EOP Educational Opportunity Programs; Engineering Operational Procedure; Equal Opportunity Program; Equipment Operations Procedure; Executive Office of the President; Experiment of Opportunity Program
EOPs Extended Opportunity Programs
EOPS Equal Opportunity Programs and Services; Extended Opportunity Program and Services
eoq economical ordering quantity; end of quarter
EOQC European Organization for Quality Control
eor earth orbital rendezvous; end of reel; explosive ordnance reconnaissance
eor (EOR) end of record; end of run
EOR Earth Orbit Rendezvous
EORSA Episcopalians and Others for Responsible Social Action
EORTC European Organization for Research on the Treatment of Cancer
eos eligible for overseas service; end operation suppress; end of segment
EO's Engineering Orders
EOS Earth Observation Satellite; Earth Orbiting Satellite; Earth Orbiting Shuttle (NASA); Electro-Optical System; Engine Overhaul Shop; European Orthodontic Society
eosins eosinophils
eosp economic order and stocking procedure
EOSS Earth Orbital Space Station; Engineering Operational Sequencing System
eot end of tape; end of transmission; enemy-occupied territory; engine order telegraph
EOT Eagle Ocean Transport
EOTP European Organization for Trade Promotion
eou electro-optical unit
EOU Epidemic Observation Unit

eov economic order van; end of volume
eow engine(s) over wing(s); every other week
EOx Elsevier Oxford
ep easy projection; electric primer; electrically polarized; electroplate; electroplated; electroplating; electropneumatic; endpaper(s); entrucking point; estimated position; evoked potential; exit pupil; experienced playgoer; explosion-proof; extended play (records); external publication; extreme pressure
ep (EP) extended play (45 rpm phonograph disc)
e/p endpaper
e & p exploration and production (area)
e p en passant (French—in passing)
e.p. editio princeps (Latin—first edition)
Ep. Episcopus (Latin—Bishop or overseer)
EP Eagle-Picher; *Ecole Polytechnique* (French—Polytechnic School); Engineering Personnel; Engineering Publications; entrucking point; estimated position; exceptions passed
E-P European Plan (no meals)
E & P Extraordinary and Plenipotentiary
EP Environmental Pollution
E & P Editor & Publisher
epa economic price adjustment; eicosapentanoic acid; electron probe analyzer; estimated profile analysis
EPA Eastern Psychological Association; Economic Planning Agency; Emergency Powers Act; Empire Parliamentary Association; Empire Press Agency; Employment Protection Act; Engineering Practice Amendment; Environmental Planning Authority; Environmental Protection Agency; European Productivity Agency; Evangelical Press Association; Executive Protective Agency
EPAA Educational Press Association of America; Emergency Petroleum Allocation Act; Employing Printers Association of America
EPAC Electronic Parts Advisory Committee
EPACCI Economic Planning

and Advisory Council for the Construction Industries
epam (EPAM) elementary perceiver and memorizer
epaq electronic parts of assessed quality
EPAT Every Pupil Achievement Test
epb equivalent pension benefit
EPB Electronic Planning Board; Environmental Periodicals Bibliography
epbm electroplated base metal
EPBX Electronic Private Branch Exchange
epc easy processing channel; edge-punched card; electronic program control; electroplate on copper; engine performance computer; every poor cluck
EPC Economic and Planning Council; Economic Policy Committee; Economic Policy Council; Education Products Center; Educational Publishers' Council; Environmental Policy Center; Esso Petroleum Company; European Planning Council
epca external-pressure circulatory assist
EPCA Energy Policy and Conservation Act; European Petro-Chemical Association
EPCAF El Paso Coalition Against the Fence (*see* Tortilla Curtain)
epc black easy-processing channel black
ep cells epithelial cells
epcg endoscopic pancreaticholangiography (EPCG)
EPCOT Experimental Prototype Community of Tomorrow
epcp electric plant control panel
epcrbs emergency-position communication radio beacons
EPCS Equitable Pioneers Cooperative Society
epd earliest practicable date; excess profits duty
ep & d electric power and distribution
epd en paz descanse (Spanish—may he rest in peace)
EPD Excellent Policy Duty (citation)
EPDA Education Professions Development Administration; Exhibit Producers and Designers Association
epdc economic power dispatch computer

EPDC Electric Power Development Corporation

ep disc extended-play (45 rpm) disc

epdm epidemiological; epidemiologist; epidemiology

epdm (EPDM) ethylene propylene diene monomer

epe electrical parts and equipment; electronic parts and equipment

EPE Editorial Projects for Education

EPEA Electrical Power Engineers Association

epedemiol epedemiology

EPEM Electric Parts and Equipment Manufacturers

epf exopthalmos-producing factor

EPF Employees Provident Fund; European Packaging Federation

EPF Empresa Petrolera Fiscal (Spanish—State Petroleum Enterprise)—Peru

EPFL Enoch Pratt Free Library (Baltimore)

ÉPFL École Polytechnique Fédérale de Lausanne (French—Federal Polytechnic School of Lausanne)

epg eggs per gram (parasitology); electropneumogram

EPG Economic Policy Group; Electronic Proving Ground (US Army)

EPGA Emergency Petroleum and Gas Administration

EPGS Export Programme Grants Scheme (New Zealand)

Eph Ephraim

Eph Ephesians

EPHC Eastern Pacific Hurricane Center

ephmer ephemeral; ephemerides; ephemeris

epi electronic position indicator; emotional-physiologic illness

epi (Latin prefix—after, in addition, upon)—epicardial, epidemic, epidermis, epilogue, epithelium

EPI Edwards Personality Inventory; Emergency Public Information; Environmental Policy Institute; Expanded Program on Immunization; Eysenck Personality Inventory

epic. electronic printer image construction; electron-positron intersecting complex

EPIC Early Purchase Individual Contract; Education Professional for Indian Children; El Paso Intelligence Center; Electronic Properties Information Center; Elyria Project for Innovative Curriculum; End Poverty in California; Exchange Price Indicators; Exports Payments Insurance Corporation

epicen epicenter; epicentral(ly)

Epict Epictetus

epid epidemic

EPIE Educational Products Information Exchange

EPIEI Educational Products Information Exchange Institute

epig epigastric; epigeal; epigeous; epigenesis; epigenetic; epigenic; epiglottal; epiglottic; epiglottis; epigone; epigonic; epigonism(s); epigonus; epigram; epigrammatic(al)(ly); epigrammatism; epigrammatist(s); epigrammatize; epigrammatized; epigrammatizing; epigraph(er); epigraphic(al)(ly); epigraphist(s); epigraphy; epigynous; epigyny

epil epilogue

epineph epinephrine

epingrad equal participation in the great American dream

Epiph Epiphania; Epiphany

epirb emergency position-indicating radio beacon

epirb (EPIRB) emergency position-indicating radio beacon

epis episiotomy

Epis Episcopal(ian)

Epist. Epistola (Latin—epistle or letter)

epistem epistemic(al)(ly); epistemological(ly); epistemologist(s)

epistom. epistomium (Latin—stopper)

epit epitaph; epitome

EPIT Equipment Procurement and Installation Team

epith epithelial; epithelium

epithal epithalamic; epithalamion

epivag epivaginitis

epl early programming language; extreme pressure lubricant

EPL Engineering Parts List; Erie Public Library; Evansville Public Library

EPL Ecole Polytechnique de Lausanne (French—Polytechnic School of Lausanne)

EPLF Eritrean People's Liberation Front

epm electric pedestrian mover; explosions per minute

epm en propia mano (Spanish—in good hands, the right way)

EPM Easy Pickin's Mine (Imperial County, California); Environmental Program Manager

epma electron-probe micro analysis

EPMS Engine Performance Monitoring System; Engineering Project Management System

epn effective-perceived noise

epn (EPN) ethyl paranitrophenyl

epnd effective-perceived noise decibels

epndbl effective-perceived noise decibel level

EPNG El Paso Natural Gas

epnl effective perceived noise level(s)

epns electroplated nickel silver

EPNS English Place-Name Society

epo experimental processing operation

EPO Emergency Planning Office(r); Energy Policy Office; European Patent Office

EPOC Earthquake Prediction Observation Center; Eastern Pacific Ocean Conference

EPOCA Environmental Project on Central America

EPOCS Effectual Planning for Operation of Container Systems

epon eponym(s)—designation(s) derived from proper names of families, places, or persons

epos electronic point of sale

EPOSS Environmental Protection Oil Sands System

epp end plate potential; excess personal property

epp edellä puolenpäivien (Finnish—before noon)

Epp. Episcopi (Latin—Bishops or overseers)

EPP Earth Physics Program; European Pallet Pool

EPPL El Paso Public Library

EPPO European and Mediterranean Plant Protection Organization

EPPR Engineering Procurement Proposal Request

EPPS Edwards Personal Preference Schedule; Engineering Procurement Planning

Sheet
EPQ Eysenck Personality Questionnaire
epr electronic paramagnetic resonance; engine pressure ratio; evaporator pressure regulator
epr (EPR) electric propulsion rocket
EPR Einstein-Podolsky-Rosen experiment; Engineering Power Reactor; Engineering Purchase Request; Essential Performance Requirements; External Planning Regent(s)
EPRA Early Planning for Retirement (Australia); Eastern Psychiatric Research Association
EPRC Educational Policy Research Center (Syracuse University)
EPRI Electric Power Research Institute
EPRL Electric Power Research Laboratory
eprom erasable programmable read-only memory
EPRS Engineering Proposal Requirement Specification
eps earnings per share; electric power supply; electron proton spectrometer; emergency power supply
eps (EPS) energetic particle(s) satellite(s); expanded polystyrene (insulation); extrapyramidal side effect(s)
ep's epithelial cells
EPS El Paso Southern (railroad); Emergency Power System; Emergency Pressurizing System; Emergency Procurement Service; Engineering Purchase Specification; Entry Processing Station; Escape Propulsion System
EPSA Energy Products and Services Administration
EPSA *Empresa Publica de Servicios Agropecuarios* (Spanish—Public Enterprise Agricultural Services)
epsdt early and periodic screening, diagnosis, and treatment
EPSEP *Empresa Publica de Servicios Pesqueros* (Spanish—Public Enterprise Fishing Services)
EPSIS Education Program and Studies Information Services
epsp excitatory postsynaptic potential
Eps Vle Epsom Vale

ept egress procedures trainer; ethylene-propylene terpolymer; excess profits tax; external pipe thread
EPT Early Pregnancy Test-(ing); Excess Profits Tax(ing)
EPTA Expanded Program of Technical Assistance (UN)
epte existed prior to entry
EPTG Electronic Publication Technology Group
EPTI Export Performance Taxation Incentive
epts existed prior to service
epu electrical power unit; electronic power unit; emergency power unit; entry processing unit
EPU Empire Press Union; European Payment Union
EPUL *Ecole Polytechnique de l'Université de Lausanne* (French—Polytechnic School of the University of Lausanne)
Epus *Episcopus* (Latin—Bishop)
eput events-per-unit-time
EPUY Education Program for Unemployed Youth
EPVT English Picture Vocabulary Test
epw enemy prisoner of war
epwm electroplated white metal
EPZ *Ecole Polytechnique de Zürich* (French—Polytechnic School of Zurich); Export Processing Zone
eq encephalization quotient; equal; equalization quotient; equation; equivalent; (*also see* EQ)
Eq Equator
EQ educational quotient; enthusiasm quotient; ethnic quotient
EQA Environmental Quality Act (California)
EQAA Environmental Quality Advisory Agency
EQAD Electrical (Electronic) Quality-Assurance Directorate
EQB Environmental Quality Board
EQC Environmental Quality Council
eqcc entry-query-control console
Eq Guin Equatorial Guinea
eqi environmental quality index
eqiv equivalent
eqm equal-flow manifold
eqn equation; equine

eqn prdx equine paradox (the fact there are more horses' asses than horses)
eqp equip; equipment
eqpmt equipment
eqpt equipment
eqq electric quadripole-quadripole
E.Q.R. *Eques Romanus Eques* (Latin—Roman cavalry or knights)
eqs equations
eqs (EQS) equivalents (20-foot containers)
EQSC Environmental Quality Study Council
eqt equivalent training
Eq T equation of time
eq tr equipment trust
equ equate; equation
Equ Equerry; Equuleus
Equa Equator; Equatorial
Equa C Cur Equatorial Counter-current
EQUAP Engineering Qualification Approval Program
Equa Pac Equatorial Pacific
equat equator; equatorial
Equatorial Guinea Republic of Equatorial Guinea (West African island and mainland country), *República de Guinea Ecuatorial*
Equatorial islands Borneo, Galápagos, Halmahera, Kiribati, Maldives, Moluccas, Nauru, São Tome and Principe, Sulawesi (Celebes), Sumatra
Equatorial nations Brazil, Colombia, Congo, Ecuador, Gabon, Indonesia, Kenya, Peru, Somalia, Uganda, Zaire
Equatorials Equatorial Islands in the central and South Pacific Ocean, also called the Line Islands
Equi *Equidae* (Latin—horse family)—domestic breeds, wild horses, zebras
equil equilibrium
equin equinox
equinol equinologic(al)(ly); equinologist; equinology
equip. equipment
equipt equipment
Equity Actors' Equity Association
equiv equivalent
eq & wd earthquake and war damage (insurance)
er echo ranging; electronic reconnaissance; emergency rescue; enhanced recovery (oil well); error; established reliability; external resistance

e/r editing/reviewing; en route

'er her

Er erbium; Eritrea; Eritrean

ER East Riding; East River; Edwardus Rex (King Edward); Effectiveness Report; Elizabeth Regina (Queen Elizabeth); Emergency Request; Emergency Rescue; Emergency Reserve; Emergency Room; Engine Room; Engineering Release; Engineering Report; Environmental Report; Equipment Requirement; Evaluation Report; Expense Report; Expert Rifleman; Explosives Report; Express Route; External Report

E by R English by Radio

E.R. *Elizabeth Regina* (Queen Elizabeth)

ER-200 high-speed train between Leningrad and Moscow

era. electronic reading automation; electronic ring accelerator

era. (ERA) earned run average

ERA Earthquake Risk Analysis; Economic Regulatory Administration; Economic Regulatory Agency; Electrical Research Association; Electronic Realty Associates; Electronic Representatives Association; Energy Resources of Australia; Engine Room Artificer; Engineering Research Associates; Engineering Research Association; Equal Rights Amendment; Equitable Reserve Association; Eritrean Relief Organization; European Ramblers Association; Evaporative Rate Analysis

ERA *Equal Rights Amendment*

ERAA Equipment Review and Authorization Activity

ERAI Embry-Riddle Aeronautical Institute

ERAP Economic Research and Action Project

ERAP *Entreprise de Recherches et d'Activités Petrolienes* (French—Petroleum Research Development Enterprise)

Eras Erasmus

erase. electromagnetic radiation source elimination

eraser. (ERASER) elevated radiation seeker rocket

erb economic requirement batching; electron beam recording; emergency radio beacon; enlisted record brief; epigram record bureau

'Erb Herbert

Erb *Erbitten* (German—ask for, beg for, request)

ERB Economic Research Bureau; Educational Records Bureau; Electricians Registration Board; Engineering Review Board; Engineers Registration Board; Environmental Review Board; Equipment Review Board

ERBE Earth Radiation Budget Experiment

er bh engine room bulkhead

erbm (ERBM) extended-range ballistic missile

ERBS Earth Radiation Budget Satellite

erc earnings-related compensation; en-route chart; equatorial ring current; equipment record card; expendability repair classification

ERC Economic Research Council; Economic Resources Corporation; Educational Resources Center; Electronics Research Center (NASA); Elmira Reception Center (for male prisoners in Elmira, NY); Employment Rehabilitation Center; Enlisted Reserve Corps

ERCA Educational Research Council of America

erdc equine respiratory disease complex

ERC & I Economic Reform Club and Institute

ERCO Electric Reduction Company

ercp endoscopic retrograde cholangiopancreatography

ercr electronic retina-computing reader

ERCS Emergency Rocket Communications System

erd emergency return device; equivalent residual dose

ERD Earth Resources Data; Emergency Reserve Decoration; Equipment Requirements Data

ERDA Electrical and Radio Development Association; Electronics Research and Development Agency; Energy Research and Development Administration

ERDC Earth Resources Data Center; Electronic Research and Development Command (USA)

ERDE Explosives Research and Development Establishment

ERDF European Regional Development Fund

ERDIP Experimental Research and Development Incentives Program

ERDL Engineering Research and Development Laboratory

ere expected repository environment(s)

'ere here

ERE Edison Responsive Environment

erect. erection

'Ereford(shire) [Cockney contraction—Hereford(shire)]

EREP Earth Resources Experiment Package (NASA)

erf error function

ERF Education and Research Foundation; Eye Research Foundation

ERFA European Radio-Frequency Agency

ERFAA European Radio-Frequency Allocation Agency

erfc error function complement

erg electrical resistance gage; unit of mechanical energy or work (derived from the word *energy*)

erg (ERG) erase gap

erg. electroretinogram

ERG electro-magnetic rail gun; Energy Research for the Governors; Energy Research Group

ERGOM European Research Group on Management

ergon ergonomic; ergonomical; ergonomics

ergp emergency removal gate pass

ergs (ERGS) earth geodetic satellite (USAF)

ERGS Electronic Route Guidance System

Erh Erhard

E & R: Hist Soc Evangelical and Reformed Historical Society

Eri Eridamus; Eridanus (constellation)

ERI Earthquake Research Institute (Tokyo University); Economic Research Institute; Environmental Research Institute; Erie, Pennsylvania (airport)

E.R.I. *Edwardus Rex et Imperator* (Latin—Edward, King and Emperor)

eric electronic remote and independent control; energy

rate input controller
ERIC Educational Resources Information Center (US Office of Education)
ERICA Experiment on Rapidly Intensifying Cyclones over the Atlantic
ERIC/AE Educational Resources Information Center/ Adult Education
ERIC/CAPS Educational Resources Information Center/ Clearinghouse on Counseling and Personnel Services
ERIC/CE Educational Resources Information Center/ Clearinghouse in Career Education
ERIC/CEA Educational Resources Information Center/ Clearinghouse on Educational Administration
ERIC/CEC Educational Resources Information Center/ Clearinghouse for Educational Change
ERIC/CEM Educational Resources Information Center/ Clearinghouse on Educational Management
ERIC/CHE Educational Resources Information Center/ Clearinghouse on Higher Education
ERIC/CHESS Educational Resources Information Center/Clearinghouse for Social Studies and Social Science
ERIC/CIR Educational Resources Information Center/ Clearinghouse on Information Resources
ERIC/CLL Educational Resources Information Center/ Clearinghouse on Languages and Linguistics
ERIC/CLS Educational Resources Information Center/ Clearinghouse for Library and Information Sciences
ERIC/CRESS Educational Resources Information Center/Clearinghouse on Rural Education and Small Schools
ERIC/CRIER Educational Resources Information Center/Clearinghouse on Retrieval Information and Evaluation on Reading
ERIC/CUE Educational Resources Information Center/ Clearinghouse on Urban Education
ERIC/ECE Educational Resources Information Center/ Clearinghouse on Early

Childhood Education
ERIC/HE Educational Resources Information Center/ Clearinghouse on Higher Education
ERIC/IR Educational Resources Information Center/ Clearinghouse for Information Resources
ERIC/IRCD Educational Resources Information Center/ Information Retrieval Center on the Disadvantaged
Ericofon Ericsson telephone
ERIC/RCS Educational Resources Information Center/ Clearinghouse on Reading and Communication Skills
ERIC/SMEAC Educational Resources Information Center/Clearinghouse for Science, Mathematics, and Environmental Education
ERIC/TME Educational Resources Information Center/ Clearinghouse on Tests, Measurement, and Evaluation
Erid Eridamus
Erie Erie-Lackawanna (railroad)
ERIE Eastern Regional Institute for Education
ERiEI Eastern Regional Institute for Education
Erie Phil Erie Philharmonic
erild earth rotation in lunar distances
ERIM Environmental Research Institute of Michigan
ERISA Employee Retirement Income Security Act
Erit Eritrea
ERJA E R Johnson Association
erk enroute kit
erl emergency reference level
Erl Erläuterung (German—explanatory note)
ERL Environmental Research Laboratories
erm elastic reservoir moulding; ermine
Erm European red mite
ERM exchange rate mechanism
erma electronic recording machine accounting
ERMS Educational Resource Management System
Ern Ernest; Ernst
ernic earnings-related national insurance contribution
ernie electronic random-numbering-and-indicating equipment
Ernie Ernest

ERNIE Electronic Random Number Indicator Equipment
ERO Eastman-Rochester Orchestra
eroa economic rehabilitation in occupied area(s)
eroduction(s) erotic production(s)
erom erasable read-only memory
E-room engine room
EROPA Eastern Regional Organization for Public Administration
eropt error option(s)
EROS Earth Resources Observation Satellite; Eliminate Range-Zero System; Experimental Reflector Orbital Shot (space probe)
EROSP Earth Resources Observation Systems Program
erot erotic; erotica; erotical-(ly); eroticism; eroticist; eroticization; eroticize; eroticizing; erotism(s); erotogenic(s); erotologic(al)(ly); erotologist; erotology
erotol erotologist; erotology
erp effective radiated power; electro rustproofing
ERP Easy Revolving Plan; Emerson Radio & Phonograph (stock exchange symbol); European Recovery Program
ERP *Ejército Revolucionario Popular* (Spanish—Popular Revolutionary Armed Force)—Argentina; *Ejército Revolucionario del Pueblo* (Spanish—People's Revolutionary Army)—Argentina
ERPC Eastern Railroads Presidents Conference
erpf effective renal plasma flow
ERPFI Extended-Range Floating-Point Interpretive System
ERPM East Rand Proprietary Mines
ERPSL Essential Repair Stock List
err. error; erroneous
ERR Engineering Release Record; Engineering Research Report
err & app error and appeals (legal)
errc expandability, recoverability, repairability cost
ERRDF Earth Resources Research Data Facility (NASA)
erron erroneous(ly)
ERRS Environmental Response and Referral Service

ers (ERS) environmental research satellite

ERS Earth Regeneration Society; Economic Research Service; Educational Research Service; Edwards Rocket Site; Emergency Relocation Site; Ergonomics Research Society; Experimental Research Society

ER & S Eletrolytic Refinery and Smelting (company)

ERSA Economic Research and Statistics Service

ersir earth-resources shuttle-imaging radar

E-R S O Eastman-Rochester Symphony Orchestra

ersos (ERSOS) earth-resources-survey operational satellite

ERSP Earth Resources Survey Program (NASA)

ERSR Equipment Reliability Status Report

ert electrical resistance temperature; electrical resistance thermometer; extended research telescope

ert (ERT) estrogen replacement therapy

ERT Exhibits Round Table

ERTA Economic Recovery Tax Act; Energy Research and Technology Administration

ERTC European Regional Test Center (NATO)

ERTS Earth Resources Technology Satellite; European Rapid Train System

eru emergency reaction unit

ERU English Rugby Union

erv expiratory reserve volume

ERV English Revised Version

ERVAD Engineering Release for Vendor Article Data

erw electro-resistance welding

erw (ERW) enhanced radiation weapon (neutron bomb)

erw *erweiterte* (German—enlarged, extended)

ERWS Engineering Release Work Sheet

erx electronic remote switching

ery erysipelothrixia

ER Yorks East Riding, Yorkshire

erythro (Latin prefix—red)—erythrocyte

es echo sounding; effect size; eldest son; electric starting; electrical sounding; electrostatic; enamel single silk (insulation); engine speed; engine-sized (paper); equal section; exploratory shaft

es (ES) ejection sound

e/s early shorn (sheep); en suite

es *esempio* (Italian—example)

e_s price elasticity of supply

Es einsteinium; Essen

ES East Sussex; Eastern States; Econometric Society; Educational Specialist; El Salvador; Electrochemical Society; Ellis Air Lines; Employee Suggestion; Endocrine Society; Engineering Study; Entomological Society; Environmental Studies; Espirito Santo; Ethnological Society; Etymological Society; Experiment(al) Station; Extension Service

ES *El Salvador* (Spanish abbreviation)

ESA Ecological Society of America; Economic Stabilization Agency; Economic and Statistical Analysis; Electric(al) Supplies Authority; Electrical Supply Authorities; Electrolysis Society of America; Employment Standards Administration; Engineers and Scientists of America; Entomological Society of America; Epiphyllum Society of America; European Space Agency; Euthanasia Society of America; Exceptional Service Award; Export Screw Association

ES & A English, Scottish, and Australian (Bank)

ESA *Endangered Species Act*

ESAA Electrical Supply Authorities Association; Electricity Supply Association of Australia; Emergency School Aid Act

ESAAB Energy Systems Acquisition Advisory Board

ESAB Energy Supplies Allocation Board (Canada)

ESAC Environmental Systems Applications Center

ESAEI Electric Supply Authority Engineers Institute

ESA-IRS European Space Agency—Information Retrieval Service

E Sam Eastern Samoa (American Samoa)

ESANZ Economic Society of Australia and New Zealand; Electrical Supply Authorities of New Zealand; Ergonomics Society of Australia and New Zealand

ESAP Emergency School Assistance Program

ESAPP Energy System Acquisition Project Plan

esar electronically-steered array radar

ESARS Employment Service Automated Reporting System

ESAs Eastern Socially Attractives (Ivy League graduates)

ESA System Easy, Speedy Accounting System

eSat except Saturday

ESAWC Evaluation Staff, Air War College

esb electrical stimulation (of the) brain; electric storage battery

ESB Economic Stabilization Board; Electricity Supply Board; Electric Storage Battery (company); Empire State Building

ESBA English Schools' Badminton Association

ESBBA English Schools' Basket Ball Association

esc electronic service change; electronic spark control; elongation-sensitive cell; escadrille; escalator; escape; escape character; escort; escrow; escutcheon; evanescent space charge; extended core storage

esc (ESC) escape character (data processing)

esc *escompte* (French—discount)

Esc escudo (Portuguese currency)

ESC Economic and Social Council (UN); Education Service Center; Education Systems Incorporated; Electronic Security Command; Electronic Systems Command (USN); Electronics Systems Center; Energy Security Corporation; European Shippers Council; Executive Service Corps

esca electron spectroscopy for chemical analysis

ESCA English Schools' Cricket Association; English Schools' Cycling Association

Escales (French—ports of call)—Jacques Ibert symphony

escap escapologist; escapology

ESCAP Economic and Social Commission for Asia and the Pacific (UN)

Escarp Escarpment

ESCAT Emergency Security

Control of Air Traffic
eschat eschatology
ES/CIP Employee Suggestion/
Cost Improvement Proposal
escl esclamazione (Italian—exclamation); *esclamativo* (Ital-
ian—exclamative); *esclusivo*
(Italian—exclusive)
ESCL Elias Sourasky Central
Library (Tel Aviv); Evans
Signal Corps Laboratory
ESCMA Electric Steel Conduit
Manufacturers' Association
escn electrolyte-and-steroid-
produced cardiopathy charac-
terized by necrosis
esc° escudo (Portuguese or
Spanish—coat of arms, Por-
tuguese monetary unit,
shield)
Esco Escocia (Spanish—Scot-
land); *Escócia* (Portuguese—
Scotland)
ESCO Educational, Scientific,
and Cultural Organization
(UN)
Escom Electrical Supply Commission
ESCORTDIV escort division
ESCOW Engineering and Scientific Committee on Water
(New Zealand)
escp expendable surface-cur-
rent probe
ESCP Earth Science Curriculum Project
*ESCP École Supérieure de
Commerce de Paris*
(French—Paris College of
Commerce)
escr escrow
escrit° escritura (Portuguese or
Spanish—assignment, con-
tract, deed, writ)
escrnía escribanía (Spanish—
notary's office)
escrno escribano (Spanish—
notary)
escr^no escribano (Spanish—
court clerk, notary, scribe)
escs escudos (Portuguese or
Spanish—coats of arms, Portuguese monetary units,
shields)
ESCS Economics, Statistics,
and Cooperatives Service
escudo monetary unit of Cape
Verde and Portugal
esd echo-sounding device;
electronic smoke detector;
estimated shipping date; estimated standard deviation; extended school day
esd (ESD) echo-sounding device; external symbol dictionary (data processing)

Esd (ESD) English as a second
dialect
ESD East San Diego; Elec-
tronic Systems Division
(USAF); Emergency Service
Division (NYPD)
ESDA Earth-Science Data Ac-
quisition
ESDAC European Space Data
Analysis Center (Darmstdat)
ESDAG Earth-Science Data
Acquisition Guidelines
esdp external stores data pack-
age
Esdr Esdras (The Book of
Esdras)
esE electrostatische Einheit
(German—electrostatic unit)
Ese Ensenada
ESE east southeast
ESEA Electrical Supply Engi-
neers Association; Elementa-
ry and Secondary Education
Act
ESECA Energy Supply and
Environmental Coordination
Act
ESEF Electrotyping and Ste-
reotyping Employers Federation
esf electrostatic focusing;
erythropoietic stimulating
factor
ESF Eastern Sea Frontier;
Economic Support Fund; Engineering Specification Files;
Extended Spooling Facility
esfc extended specific fuel
consumption
esfp environment-sensitive
fracture process(es)
esfswr extra-special flexible-
steel wire rope
esg electrically suspended gy-
ro(scope); electronic-sweep
generator; extended-sweep
generator
e sg e seguente (Italian—and
the following one)
Esg English standard gage
esgm electrostatically supported gyro monitor
esh equivalent solar hour(s)
ESH European Society of Haematology
ESHL Eastern Seaboard Herpetological League
eshp equivalent shaft horsepower; established standard
horsepower
esi emergency stop indicator;
equivalent spherical illumination; externally specified indexing
ESIL European Standard Inventory List (NATO)

ESIS Executive Selection In-
ventory System
Esk Eskimo
Eskie(s) Eskimo(s)
esl expected significance level
Esl English as a second lan-
guage
Esl (ESL) English as a second
language
ESL Eagle Shipping Ltd; Earth
Sciences Laboratory; Eastern
Steamship Lines; Engineer-
ing Societies Library
E S-L Engineer Sub-Lieuten-
ant
ESL Endangered Species List
ESLAB European Space Labo-
ratory (Delft)
esle engineering special labora-
tory equipment
ES/LES Equipment Section/
Loaded Equipment Section
ESLO European Satellite
Launching Organization
esm electronic support mea-
sures; electrostatic memory;
ends standard matched (lum-
ber)
ESM Eastman School of Mu-
sic; Engineering Services
Memo; Engineering Shop
Memo
ESMA Electric Sign Manufacturers Association; Electronic
Sales-Marketing Association;
Engraved Stationery Manu-
facturers Association; Episcopal Society for Ministry on
Aging
ESMC Eastern Space and Mis-
sile Center, Cape Canaveral,
Florida
ESMI Energy Studies Measurement Instrument
ESMRI Engraved Stationery
Manufacturers Research Institute
esm's electronic-support mea-
sures
esn essential
esn (ESN) educationally sub-
normal
ESN Elastic Stop Nut (corporation); Engineering Shipping
Notice; English-Speaking
Nations (NATO)
esna electrical survey net adjuster
ESNA Elastic Stop Nut Corporation of America; Empire
State Numismatic Associa-
tion
ESNE Engineering Societies of
New England
ESN-H Elsevier North-Holland
esntl essential

ESNZ Entomological Society of New Zealand

ESO Educational Services Office(r); Electronic Supply Office(r); Embarkation Staff Office(r); Emergency Services Organization; Engineering Service Order; Engineering Stop Order

ESOC European Space Operations Center

Esol English for speakers of other languages

ESOMAR European Society for Opinion Surveys and Market Research

ESOP Employees Stock Ownership Plan

esoph esophageal; esophagus

esor electronically scanned optical receiver

esot esoteric; esoterica; esoterical(ly); esotericism(s)

ESOT Employee Stock Ownership Trust

esp easy solution possible; echeloned series processor; electro-magnetic surface profiler; electro-sensitive paper; electro-sensory panel; engine sequence panel; especially; extrasensory perception

esp (ESP) electro-selective pattern (light meter system for cameras); electrosensitive programming

e & sp equipment and spare parts

esp espressivo (Italian—expressive)

Esp Esperanto; Esplanade

Esp (ESP) English for special purposes

Esp Espagne (French—Spain); *España* (Spanish—Spain); *Español* (Spanish—Spanish)

ESP East Sepik Province; Eastern State Penitentiary (Philadelphia, Pennsylvania); Elsevier Science Publishing; Emerson Select Protection; English for Special Purpose(s); Equipment Status Panel; Extrasensory Perception

ESP Ecole des Sciences Politiques (French—School of Political Science)

espa electronically steered phased array

ESPA Elementary School Principal's Association; Evening Student Personnel Association

ESPAC Elementary School Principal's Association of Connecticut

Espantuguese Spanish-Portuguese

ESPAW Elementary School Principal's Association of Washington

ESPC Elsevier Scientific Publishing Company (Amsterdam)

espec especial(ly)

Esper Esperanto

espg espionage

espi electronic speckle-pattern interferometer

Esplish Spanish-English

ESPN Entertainment and Sports Programming Network

ESPOA Electricity Supply Professional Officers Association

ESPQ Early School Personality Questionnaire

espr espressivo (Italian—expressive)

ESPR English Society for Psychical Research

espress espressivo (Italian—expressive)

ESPRI Education Service of the Plastics and Rubber Institute

esq extra-special quality

esq esquerdo (Portuguese—left)

Esq Esquire

ESQ Entomological Society of Queensland

ESQA English Slate Quarries Association

esqᵒ esquerdo (Portuguese—left)

Esqrr Esquire

ESQST Ego-Strength Q-Sort Test

esr effective signal radiated; electrical skin resistance; electron skin resonance; electronic slide rule; electronically-scanned radar; electroslag resmelting; equivalent series resistance; erythrocyte sedimentation rate

ESR Engineering Societies Library; Engineering Stop Release; Engineering Summary Report

ESRANGE European Space Research (northern rocket range)—Kiruna

esrc electronics recovery control; engine surge recovery control

ESRC European Science Research Council

ESRD End-Stage Renal Disease

ESRG Earth Sciences Review Group

ESRI Economic and Social Research Institute (Dublin)

ESRIN European Space Research Institute

ESRO European Space Research Organization

ESRP Environmental Standard Review Plans

ESRU Environmental Sciences Research Unit

ess empty solution set; essence; essences; essential; expendable sound source

ess. essentia (Latin—eseence)

eSS except Saturday and Sunday

Ess Essex

ESS Eastern Searoad Service; Educational Services Section; Electrical Standards System; Electronic Switching System; Elementary School Science; Elementary Science Study; Emplaced Scientific Station; Employment Security System; English Speaking Society; Evaluation SAGE Sector; Experimental SAGE Sector

ESS Encyclopedia of the Social Sciences

essa environmental survey satellite (weather satellite)

ESSA Environmental Science Services Administration—Central Radio Propagation Laboratory, Coast and Geodetic Survey, Weather Bureau (Department of Commerce); environmental survey satellite

Essandess Simon and Schuster

Essequibos Essequibo Islands (in the Essequibo River estuary off Guyana)

ESSEX Effects of Sub-Surface Explosions (USA)

Essie Esther

ess neg essentially negative

ESSO Esso Shipping; Standard Oil

ESSPO Electronic Support System Project Office

ess pos essentially positive

ESSR Estonian Soviet Socialist Republic

ESSS Electronic Security Surveillance System

essu electronic selective switching unit

ESSWACS Electronic Solid-State Wide-Angle Camera System

est earliest start time; elastic

surface transformation; electrolytic sewage treatment; establish; established; establishment; estate; estimate; estimated; estimation; estimator; estuary; external static pressure

est (EST) electrical stimulating treatment; electroshock therapy

est estación (Spanish—station); *estimado* (Spanish—estimated)

Est The Book of Esther; Estates (postal abbreviation); Estonia(n); Estuary

Est Estado (Spanish—State); (French—east)

EST Eastern Standard Time; Eastern Summer Time; Electrical Stimulating Treatment; English in Science and Technology; Enlistment Screening Test; Enroute Support Team; Epidemiology and Sanitation Technician; Erhard Seminars Training

ESTA Energy Systems Trade Association

estab established

Estab Establishment

Established Church Established Church of England

estab tip establecimiento tipografico (Spanish—publishing company)

estar estimated arrival

estb establish

estbl establishment

ESTEC European Space Technology Center

ESTF Exploratory Shaft Task Force; Exploratory Shaft Test Facility

estg estimating

esth esthetics

Esth Esthonia; Esthonian

Esth Esther

Esther Hester

Esthr Book of Esther

ESTI European Space Technology Institute

estm estimate

estmd estimated

estmg estimating

estmn estimation

estn estimation

estn estnisch (German—Estonian)

Estoc Estocolmo (Portuguese or Spanish—Stockholm)

ESTP Earth Science Technical Plan

ESTPP Earth Science Teacher Preparation Project

ESTRACK European Space

Satellite Tracking and Telemetry Network

Estr B Estero Bay

estriff encryptic-secure tracking-radar identification friend or foe

Ests Estates

est wt estimated weight

esu educational service unit(s); electrostatic unit

ESU Emporia State University; English-Speaking Union

E-SU English-Speaking Union

e sub excitor substance

E Suffolk East Suffolk

eSun except Sunday

ESUNA Ethiopian Students Union of North America

E Sussex East Sussex

E-SUUS English-Speaking Union of the United States

esv earth satellite vehicle; enamel single varnish (insulation code)

ESV Earth Satellite Vehicle; Experimental Safety Vehicle

ESW Ethical Society of Washington

eswl equivalent single-wheel loading

eswl (ESWL) extracorporeal shock-wave lithotripsy

Esx Essex

esy extended school year

et edge thickness; educational therapy; educational training; effective temperature; electric telegraph; electric telegraphy; electric typewriter; electrical time; electrical transcription; electronic tests; engineering test; engineering testing; equation of time

et (ET) elapsed time; electronic timing; ephemeris time; external tank; extra terrestrial

e/t (E/T) ergotamine tartrate; ergotin tartrate

e t en titre (French—in the title)

Et Ethyl; Etienne

ET East Texas (Pulp & Paper Company); Eastern Time; Educational Therapy; Electronics Technician; Employment and Training choices; English Text; English translation; Entertainment Tax; Ethiopian Airlines; European Theater (of war); Exchange Telegraph

ET Extra Terrestrial (symphonic suite by John Williams)

eta estimated time of arrival;

expect to arrive

ETA Educational Telecommunications for Alaska; Employment Training Administration; English Teachers Association; European Teachers Association; Exception Time Accounting; Express Transport Association

ETA Euzkadi ta Azkatasuna (Basque—Nation and Liberty)

Etab Etablissement (French—business establishment or factory)

ETAB Environmental Testing Advisory Board (Dow)

ETAC Environmental Technical Applications Center

et al. *et alibi* (Latin—and elsewhere); *et alia* (Latin—and others)

ETAP Expanded Technical Assistance Program

ETAQ English Teachers Association of Queensland

ETAS Escort-Towed Array System

ETASS Escort-Towed-Array Sonar System

etb early to bed; end of transmission block; estimated time of berthing

etb (ETB) end of transmission block character (data processing)

ETB Engineering Test Basis

ETBO Engineering Test Base Office

etc earth terrain camera; effluent treatment cell; electronic travel computer; electronic typing calculator; employee timecard; estimated time of completion; extraterrestrial civilization

etc. et cetera (Latin—and so forth)

e t c en tout cas (French—in any case)

ETC Electrical Technician Certificate; Electro Tech Corporation; Electronic Technician Certificate; Emergency Training Center; Engine Technical Committee; Environmental Testing Corporation; Episcopal Travel Club; European Translations Center; European Travel Commission

ETC. A Review of General Semantics (Official Organ of the International Society for General Semantics)

ETCC Eastern Tank Carrier

Conference
ETCE *Empresa Transportes Colectivos del Estado* (Spanish—State Collective Enterprise Transport)
etcg elapsed-time code generator
etcrrm electronic teleprinter cryptographic regenerative repeater mixer
etd estimated time of departure; extension trunk dialing
ETD End of Train Device; English Teaching Division
ETDS Electronic Theft Detection System
ete estimated time enroute
ete *este* (Spanish—east)
ETE Experimental Tunnelling Establishment
ETE *Escuela Technica del Ejército* (Spanish—Technical School of the Army)
ETEMA Engineering Teaching Equipment Manufacturers Association
eter estimated time enroute
etf electron-transferring flavor-protein; enhanced tactical fighter
Étg *Étang* (French—lagoon, pond)
etgm estimate to get money
eth ether; ethical; ethics; ethmoid; ethmoidal; ethnic; extraterrestial hypotheses (explaining close encounters of the third kind such as ufo's); extraterrestrial hypothesis
Eth Ethiopia; Ethiopian; Ethiopic
ETH *Eidgenössische Technische Hochschule* (German—Swiss Federal Institute of Technology)
ethanol ethyl alcohol or grain alcohol (C_2H_5OH)
Eth$ Ethiopian dollar
eth dat ethic dative
Ethel Ethelberg; Ethelberta; Ethelburg; Ethelda; Etheldrid; Ethelind; Etheljean; Ethelrede; Ethelsa; Ethelwyn
ether ethyl ether ($(CH_2H_5)_2O$)
Ethiop Ethiopia; Ethiopian
ethno ethnology
ethnoc ethnocide
ethnog ethnography
ethnograph ethnograph(er); ethnographic(al)(ly); ethnography
ethnol ethnologist(ic)(al)(ly); ethnology
ethnomus ethnomusicologist; ethnomusicology
ethnomusi ethnomusic(al)(ly);

ethnomusicologist; ethnomusicology
ethnophaul ethnophaulism (study of international slurs); ethnophaulist(ic)(al)(ly)
ethnosci ethnoscience; ethno-scientific(al)(ly); ethno-scientist(s)
etho ethylene oxide
ethog ethogram; ethographer; ethographic; ethography
ethol ethologic(al)(ly); ethologist(ic)(al)(ly); ethology
eti elapsed-time indicator; estimated time of interception
Eti *Etiopia* (Italian, Spanish—Ethiopia); *Etíopia* (Portuguese—Ethiopia)
ETI Electric Tool Institute; Electronic Technical Institute; Equipment and Tool Institute
ETIA European Tape Industry Association
ETIC English Training Information Centre (London)
etio etiocholandone
etiol etiology
ETIS-MARFO European and Technical Information Service in Machine-Readable Form
etk (ETK) erythrocyte transketolase
etkm every test known to man
etl emergency tolerance limit; ending tape label; etching by transmitted light
ETL Electrical Testing Laboratory; Electro-Technical Laboratory; Engineering Test Laboratory; Essex Terminal (railroad)
ETM Electronic Technician's Mate
ETMA English Timber Merchants Association
ETMA-A Engineering Tooling and Manufacturing Aide
ETMWG Electronic Trajectory Measurements Working Group
etn equipment table nomenclature
ETN Eastern Technical Net (USAF)
eto electric truck operator; estimated takeoff; estimated time off; ethylene oxide
ETO Energy Technology Office; European Theater of Operations; European Transport Organization; Executive Training Office(r)
etoc expected total operating cost

Et OH ethyl alcohol
etp estimated turnaround point; estimated turning point
etp (ETP) electron transfer particle
ETP Eastern Tropical Pacific; Education and Training Program; Effluent Treatment Plant; Engineering Test Plan; Evaluation Test Plan
ETPI Eastern Telecommunications Philippines Inc
et-pnl engine test panel
ETPS Empire Test Pilots School
etr effective thyroid ratio; estimated time of return; export traffic release
etr (ETR) engineering test reactor
Etr Etruscan
Etr *entrada* (Spanish—entrance)
ETR Eastern Test Range; Easy-growth Treasury Receipts; Electric Target Range; Engineering Test Reactor; Export Traffic Release; External Technical Report
etra estimated time to reach altitude
ETRC Educational Television and Radio Center; Engineering Test Reactor Critical Facility
etro estimated time of return to operation
ETRs Encrypted Traffic Reports
etry entirely
ets electronic telegraph system; estimated time of sailing; expiration term of service; expiration of time of service
Ets *Etablissements* (French—establishments)
ETS Educational Television Stations; Educational Testing Service; Electronic Telegraphic System; Engine Test Stand; Engineering Task Summary; Engineering Test Satellite
ETSA Electricity Trust of South Australia
ETSC East Tennessee State College; East Texas State College
et seq. *et sequens* (Latin—and following)
etsp entitled to severance pay
etsq electrical time superquick
ETSS Engineering Time-Sharing System; Entry Time-Sharing System; Experimental Time-Sharing System

ETSU Energy Technology Support Unit

ett early thrust termination; electromagnetic thickness tool; exercise tolerance test(ing)

ett (ETT) evasive target tank

ETT Elizabethan Theatre Trust; Explosion Tear Test-(ing)

etta electronic temperature trip and alarm

Etta Henrietta

ETTA English Table Tennis Association

ETTDC Electronics Trade and Technology Development Corporation

et to extractor tool

et tp etch template

ETTU English Table Tennis Union

etu electron tube

ETU Electrical Trades Union; Emergency Treatment Unit

ETUC European Trade Union Confederation

et ux. et uxor (Latin—and wife)

etv educational television; engine test vehicle

etv (ETV) educational television

ETV Educational Television; *Electrotechnischer Verein* (German—Electrotechnical Society); Engine Test Vehicle

etvm electrostatic transistorized voltmeter

etw empty tank weight; end-of-tape warning

etw etwas (German—something)

ETWN East Tennessee & Western North Carolina (railroad)

etx (ETX) end of text character (data processing)

etym etymologic(al)(ly); etymologist(ic)(al)(ly); etymology

E-type Jungian extrovert type

eu electron unit; emergency unit; external upset (oil well)

eu (Greek—good or well)—eubacterium, eucalyptus, euphoria

Eu entropy unit (symbol); Euler unit; Europe; European; europium; Eustace; Eustatia

EU Emory University; Evacuation Unit; Everyman's University (Tel Aviv); Experimental Unit

E-U Etats-Unis (French—United States)

EU Estados Unidos (Spanish—United States); *Europa Unie* (French—United Europe)

eua examination under anesthetic

Eua European unit of account

EUA Eastern Underwriters Association; *Estados Unidos de América* (Spanish—United States of America); *Etats-Unis Amérique* (French—United States of America)

EUA Estados Unidos da América (Portuguese—United States of America)

EUB Estados Unidos do Brasil (Spanish—United States of Brazil)

euc end-use check(ing)

EUC Euclid (railroad)

eucd emotionally unstable character disorder

Eucl Euclid

EUCLID Experimental Use Computer—London Integrated Display

EUCOM European Command

euc(s) eucalyptus tree(s)

EUDISED European Documentation and Information System for Education

euf eufemismo (Italian, Portuguese, Spanish—euphemism)

EUF European Union of Federalists

eufe eufemismo (Italian, Portuguese, Spanish—euphemism)

EUFTT European Union of Film and Television Technicians

Eug Eugene; Eugenia

eugen eugenics

Eugn eugenics

Eugº Eugenio

EUI Energy Utilization Index

EUI Enciclopedia Universal Ilustrada (Spanish—Universal Illustrated Encyclopedia)

EUL Edinburgh University Library

EUL Everyman's University Library

EUM European Mediterranean

EUM Entr'aide Universitaire Mondiale (French—World University Service); *Estados Unidos Mexicanos* (Spanish—United States of Mexico)

EUM-AFTN European-Mediterranean Aeronautical Fixed Telecommunications Network

EUMOTIV European Association for the Study of Economic, Commercial, and Industrial Motivation

EUMR Emergency Unsatisfactory Material Report

Euni Eunice

EUP Edinburgh University Press; English Universities Press

euph euphemism(s); euphemistic(al)(ly); euphemize(d); euphemizer(s); euphemizing

euphé euphémisme (French—euphemism)

Euphe der Euphemismus (German—euphemism)

euphem euphemism; euphemistic(al)

euphem euphémique (French—euphemistic); *euphémisme* (French—euphemism)

Euphie Euphemia

euphon euphonic; euphonically; euphony

eur europaeisk (Dano-Norwegian—European)

Eur Europe; European

EUR Erasmus Universiteit Rotterdam (Dutch—Erasmus University of Rotterdam)

Eurafrica Europe and Africa

Eurail European Railways

Eurailpass European tourist railroad pass

EURAS European Anodisers Association

Eurasafrica Europe, Asia, and Africa

EURASBANK European Asian Bank

Eurasia Europe and Asia; from the Caspian Sea and the Caucasus Mountains to the Ural Mountains

Eurasian(s) person(s) of European and Asian parents

Euratom six-nation atomic energy pool consisting of France, Germany, Italy, Belgium, Netherlands, and Luxembourg

EURATOM European Atomic Energy Community

Eur Ct H R European Court of Human Rights

eurex enriched uranium extraction

EURIMA European Insulation Manufacturers Association

Eurip Euripides

EURIPA European Information Providers Association

EURO European Regional Office (FAO)

Eurobonds European bonds

EUROCAE European Organization of Civil Aviation Electronics

Euro-Can(s) European-Canadian(s)

EUROCEAN European Oceanographic Association

Eurochemic European chemical processing of irradiated fuels

Euro Com European (NATO) Communications

EUROCOM European Coal Merchants Union

Eurocom(s) European communism; European communist(s)

EUROCOOP European Community of Cooperative Societies

EUROCORD European Cord, Rope, and Twine Industries

eurocrat European bureaucrat

EURODICAUTOM European Automated Dictionary

EURODIDAC European Association of Manufacturers and Distributors of Educational Materials

Euro$(s) European dollar(s)

Eurodol(s) European dollar(s)

Eurofima European Company for the Financing of Rolling Stock

Eurofinance *Union International d'Analyse Economique et Financière* (French—International Union of Economic Analysis and Finance)

EUROFINAS European Financial Houses

Eurogroup Belgium, Denmark, Germany, Greece, Italy, Luxembourg, Netherlands, Norway, Portugal, Spain, Turkey, United Kingdom

Eurolex full-text electronic legal-research network

Euro Log European (NATO) Logistics

Euro Long-Term European (NATO) Long-Term Operations

Euromart European Common Market

Euro Med European (NATO) Military Medicine

Euromissiles European-deployed medium-range nuclear missiles

EURONET European Network (data-transmission)

Europ European railway car pool

Europhot European professional photographers

Eurosac European paper sack manufacturers

Eurosat European application satellite systems

EUROSPACE European Space Study Group

eurotainer European-owned container

Euroterro European terrorism; European terrorist

EUROTEST European Association of Testing Institutions

Eurotories European Tories (conservative parties such as Britain's Conservatives and Germany's Christian Democratic Union)

Eurotox European Committee on Toxicity Hazards

Eurotunnel railway tunnel linking England and France

Eurovision European Television

EUS Eastern United States; Engineering Undergraduates Society

EUSA Eighth United States Army

EUSAFEC Eastern United States Agricultural and Food Export Council

Euseb Eusebius Pamphili

EUSIDIC European Association of Scientific Information Dissemination Centers

Eus Sta Euston Station

eutec eutectic; eutectoid

EUTELSAT European Telecommunications Satellite organization

euth euthanasia(n), euthanasic(al)(ly), euthanatize

euv energetic ultraviolet; equivalent ultraviolet; expected utility value; extreme ultraviolet

EUVE Extreme Ultraviolet Explorer satellite

euvsh equivalent ultraviolet solar hour

euw engine(s) under wing(s)

EUW European Union of Woman

Eux Euxine

Euxine Sea Black Sea

ev earned value; efficient vulcanization; electric vehicle; electron volt; enclosed and ventilated; escort vessel; evangelical; exposure value; extremely violent

ev electrón-voltio (Spanish—electron volt)—also appears as *eV; en ville* (French—local); *evangelisch* (German—Protestant)

eV electronvolt

eV eingetragener Verein (German—registered society)

Ev Evenkian; Everett

Ev Eingang vorbehalten (German—rights reserved)

Ev. Evangelium (Latin—the Gospel)

EV Elivie (Italian Heliways); English Version; Erne Valley; Everett (railroad)

eV 1 s edge-V one side (lumber)

eV 2s edge-V two sides

eva electronic velocity analyzer; ethyl-vinyl acetate; extra-vehicular activity; extra-vehicular ambulation

EVA Educational Voucher Authority; Electrical Vehicle Association; Engineer Vice Admiral

evac evacuate; evacuation

evacship evacuation ship

eval evaluate; evaluation

Evan Evangelical; Evangelist

Evans Phil Evansville (Indiana) Philharmonic

evap evaporate; evaporation; evaporator; evaporize

evapd evaporated

evaptr evaporator

evata electronic visual auditory training aid

EVC Educational Video Corporation; Electric Vehicle Council

evce evidence

evco electron vibration cutoff

EVCP Engineering Value Control Proposal

EVCS Extravehicular Communications System

evd extended voluntary departure (immigration)

EVDF Eugene V. Debs Foundation

eve evening

Eve Eveleen; Evelina; Eveline; Evelyn; Everarda; Everett; Everette; Everina; Everline

evea extravehicular engineering activities

evenin' evening

event. eventuell (German—possibly)

Ever Everest—world's highest mountain towering over the Himalayas of Nepal and Tibet

Everglades Everglades National Park in Florida

Eve Trib Evening Tribune

evf electronic viewfinder

evg evening

EVG Europäische Verteidi-

gungsgemeinschaft (German—European Defense Community)
evi evidence
EVI Extreme Value Index
evict. evaluation of intelligence-collection tasks
evid evidence
evir evirato (Italian—emasculate)—eunich
e viv. disc. e vivis discessit (Latin—departed from life)
EVL E(dward) V(errall) Lucas
evln evolution
ev-luth evangelisch-luterisch (German—Evangelical Lutheran)
evm extraneous vegetable matter
evm (EVM) earth-viewing module
evminfin everglazed minicare finish
evmu extra-vehicular material unit
evng evening
evol evolution; evolutionary; evolutionist
evop evolutionary operation
EVP Executive Vice President
evr electronic video recording
evrep event recording potential
EVRS Electronic Video Recording System
evs expected value saved
evs (EVS) extravehicular system
EvS Environmental Science
EVs electric vehicles
EVS Electronic Voice Switching (system); Electronic-optical Viewing System
evsd energy-variant sequential detection
evss extravehicular space suit
evstc (EVSTC) extravehicular suit telemetry and communications
evt educational and vocational training; effective visual transmission; equiviscous temperature; eventually; extra-value trimmed (meat)
evt eventuel (Dano-Norwegian—possible)
E v T E van Tongeren
EVT Engineering Verification-Test(ing)
EVT Europäische Vereinigung für Tierzucht (German—European Association for Animal Production)
evtl eventuell (German—eventually, perhaps, possibly)
EVV Evansville, Indiana (airport)

EVW European Voluntary Workers
EVX Electric Vehicle Experimental
evy every
evythg everything
ew each way; earthenware; effective warmth; electrically welded; electronic warfare; equivalent widths; extensive wound
ew (EW) earth watch
e/w equipped with
Ew Ewart; Ewbanke; Ewell; Ewen; Ewing
Ew Euere or *Eure* or *Eurer* (German—your)—abbreviation used in titles
EW early warning; electronic warfare; Emergency Ward; enlisted woman; enlisted women
E & W England and Wales
EWA East and West Association; East-West Airlines; Education Writers Association; Electrical Wholesalers Association
ewac electronic warfare anechoic chamber
EWACS Electronic Wide-Angle Camera System
EWAD Early Warning Air Defense
EWAS Economic Warfare Analysis Section
ewb estrogen withdrawal bleeding
ewc electric water cooler
ewc (EWC) electronic warfare coordinator
EWC East-West Center (University of Hawaii)
EWCB Electrical Workers and Contractors Board
EWCRP Early Warning Control and Reporting Post
ewd elementary wiring diagram
EWD Economic Warfare Division
ewdt early warning data transmission
ewe. electronic warfare element
ewec electromagnetic wave energy conversion
ewes electronic warfare evaluation simulator
EWES Engineering Waterways Experiment Station
ewex electronic warfare exercise
ewexipt electronic warfare exercise in port
ewf equivalent weight factor

EWF Earth, Wind, and Fire (music group); Electrical Wholesalers Federation
EWG Executive Working Group (NATO)
EWG Europäische Wirtschaftsgemeinschaft (German—European Common Market)
ewgcir early-warning ground-control-intercept radar
EWHS Eli Whitney School
ewi education with industry; entered without inspection
Ewi English winter index
EWI Earl Warren Institute
ewicb electronic-warfare interface-connection box
ewl evaporative water loss
EWL Ellerman's Wilson Line
EWLD Engineering Weekly Labor Distribution
ewma exponentially weighted moving average
EWMC Eli Whitney Metrology Center
EWO Electrical and Wireless Operators; Electronic Warfare Officer; Emergency War Order; Engineering Work Order; Essential Work Order
EWO-DS Engineering Work Order—Drawing Summary
ewops electronic warfare operations
EWOS Electronic Warfare Operational System (USAF)
EWP Emergency War Plan
EWPI Eysenck-Withers Personality Inventory
EWPs Electronic Warfare Plans
ewr early-warning radar
EWR Electrical Wiring Regulations; Engineering Work Request; Newark, New Jersey (airport)
EWRC European Weed Research Council
ews experienced workers standard
ew's edge weapons (sharp-bladed daggers, cutlasses, knives, machetes, swords, etc.)
EWS Emergency Water Supply; Emergency Welfare Service; European Wars Survey
EWSC Eastern Washington State College; Electric Water Systems Council
EWSF European Work Study Federation
ewsl equivalent single-wheel load(ing)
ewsm electronic-warfare sup-

port measures
EWT Eastern War Time (advanced time)
EWU Eastern Washington University
eww extended work week
EWWS Electronic Warfare Warning System
ex etc.; exact(ed); exacting; exactitude; exactly; examination; examine(d); examiner; examining; example; excess(ive); exclusive; exclusively; exclusivity; execute(ed); executing; exercise; exercising; experiment(al's); extra(neous)
ex (Latin prefix—out of)—excision; (Latin—from)
Ex Excelsior; Exchange; Exchequer; Exeter; Exmoor; Exmouth; Extremadura; Exuma
Ex Exodo (Spanish—The Book of Exodus); *Exodus*
EX experimental broadcasting
exacct expense account
ex af. ex affinis (Latin—of affinity)
exafs extended X-ray-absorption final structure
exag exaggerate; exaggerated; exaggeration
Ex Agt Executive Agent
exam examination; examine; examiner
examd examined
exametnet experimental meteorological sounding rocket network
examg examining
examn examination
examr examiner
exams examinations
ex aq. ex aqua (Latin—out of water)
exbedcap expanded bed capacity
Ex B/L exchange bill of lading
exc excavate; excellent; except(ion)(al)(ly); exciter
exc. excudit (Latin—he engraved it)
Exc *Excelencia* (Spanish—Excellency); Excellency
Exc Excélsior (Mexico City); *Exelencia* (Spanish—Excellency)
exca excavate; excavation
Exc^a Excelencia (Spanish—Excellency)
ex cath. ex cathedra (Latin—from the seat of authority)
Excel Excelsior
EXCEL Corporation for Excellence in Public Education;

Ex-offender Coordinated Employment Lifeline (Indiana's parole project)
exch exchange
ex champ ex-champion; former champion
Excheq exchequer
exch oper exchange operator
exchq exchequer
exchr extra charge
excl exclude; exclusion; exclusive; exclusivity
excl exclusief (Dutch—not included)
exclam exclamation; exclamatory
exclt excellent
exclu exclusive(ly); exclusivity
Exc^ma Excelentísima (Spanish—Most Excellent)—feminine
Excmo Excelentísimo (Spanish—Most Excellent)
Exc^mo Excelentísimo (Spanish—Most Excellent)—masculine
Ex Cncl Executive Council
Ex Co Executive Council
Ex Com Executive Committee
ex-con(s) ex convict(s); former convict(s)
ex cont from contract
excp except(ion)(al)(ly); execute channel program
ex cp ex coupon
excpt except(ion)(al)(ly)
Excrpt Med Excerpta Medica
excs excess
exct execution
excv exclusive
exd examined
EXDAMS Extendable Debugging and Monitoring System
ex det explosives detector
ex div ex dividend
Ex Div Experimental Division
Ex Doc Executive Document
Exe Exeter
exec execute(d); execution; executive; executive officer; executor
exec (EXEC) execute statement (data processing)
Exec Dir Executive Director
Exec Off Executive Officer
execs executives
Exec Sec Executive Secretary
exeod expects to enter on duty
exer exercise
exes expenses
Exet Coll Exeter College—Oxford
Exeter Exeter College (Oxford), Phillips Exeter Academy
exf external function

ex f extremely fine
ex fac ex factory
ex fy extra fancy
exg existing
ex ga external gage
EXGO Export Guarantee Office(r)
ex gr. exempli gratia (Latin—for example)
exh exhaust
exhib exhibit; exhibition; exhibitor
exhib. exhibeatur (Latin—let it be shown)
exhn exhibition
exh t exhaust turbine
exh v exhaust vent
ex hy extra heavy
EXIAC Explosives Information and Analysis Center (USA)
Ex-Im Export-Import Bank
EXIMBANK Export-Import Bank
ex int ex interest
exis existential; existentialism; existentialist
exist. existing
EXIT Ex-offenders In Transit (Maine's parole project)
exkl exklusiv (German—excepted, not included)
ex lib. ex libris (Latin—from the library of)
Ex^maSr^aD Excelentíssima Senhora Dona [Portuguese—Mrs (precedes full name in formal style)]
ex-mer ex-meridian
Ex^moSr Excelentíssimo Senhor [Portuguese—Mr (precedes full name in formal style of address)]
exmr examiner
ex n(ew) excluding new shares
ex-nupt(s) ex-nuptial(s)—person(s) born out of wedlock
ExO executive officer; executive order
Ex O Experimental Office(r)
EXO European X-ray Observatory
exobio exobiologic(al)(ly); exobiologist; exobiology
exocrin exocrinologic(al)(ly); exocrinologist; exocrinology
Exod Exodus
ex off. ex officio (Latin—by authority of his office)
Exon. Exonia (Latin—Exeter)
exonum exonumia(l)(ly); exonumic(al)(ly); exonumist(s)
exonym foreign-language placename such as Londres (Spanish for London)
Ex O P Executive Office of

the President
exopac exoatmospheric jettisonable control wafer
exor executor
exord exercise order
exos (EXOS) exospheric satellite
exosat (EXOSAT) European X-ray observatory satellite
exot exotic
exotheo exotheologic(al); exotheologist(s); exotheology
exp expansion; expenditure; expense; experience; experiment(al); exponential; export; Exposition; exposure; express; expulsion
exp expreso (Spanish—express)
ex p. ex parte (Latin—on one side only)
EXP Exchange of Persons (UNESCO office)
expate(s) expatriate(s)
expdivun experimental diving unit
expdn expedition
expdt expiration date
exped expedite; expedition
exper experiment; experimental
Expert Expanded Pert (program evaluation and review technique)
expi export performance taxation incentive
exp-imp export-import
expir expiratory; expiration
exp jt expansion joint
expl explain; explanation; explanatory; explosimeter; explosimetric; explosion; explosive(s)
expl exemple (French—example)
explan exercise plan
EXPLIC Export License
explo explosion; explosive
exploit. exploitation
explor exploration
Explora Exploratorium
explos explosion; explosive
expn exposition
expnd expenditure
expo expose; exposition
exp o experimental order(s)
expo expreso (Spanish—express)
Expo 67 1967 exposition in Montreal
Expo 70 1970 exposition at Tokyo
expol expanded polysterene (light-weight packing moulding)
export exportaciones (Span-

ish—exports)
expr expiration; expire
expr expressif (French—expressive)
ex-Pres ex-President
EXPRESO Expreso Aéreo Interamericano (Spanish—Interamerican Air Express)
expressway express highway
exps expenses
expt experiment
exptl experimental
expto expedite travel order
exptr exporter
expul expulsion
expur expurgate(d)
Expwy Expressway
Expy Expressway
ex-quay free on quay
exr executor
ex r ex rights
exray expendible relay
exrx executrix
exs expenses; expropriations
ex's expenses
exsec exsecant
exshi expedite shipment
ex ship delivered out of the ship
exspec exercise specification(s)
exst exempt sales tax
Ex Sta Experimental Station
ext extend; extension; exterior; external; extinguish; extinguisher; extra
ext (EXT) extraction (dental)
ext. extend (Latin—spread); *extractum* (Latin—extract)
Ext Extended; Extension
extal extra time allowance
extd extracted
ext d & cc external drug and cosmetic color
Extel Exchange Telegraph (press agency)
EXTEL Exchange Telegraph (British news agency)
extemp extemporaneous(ly)
exten extension
extend. extensus (Latin—spread)
extern external; externally
EXTERRA Extraterrestrial Research Agency (USA)
ext fl extract fluid (fluid extract)
extg extinguish(er)
extgh extinguish
exting extinguished
ext. liq. extractum liquidum (Latin—liquid extract)
extm extended telecommunications module
ex tm. ex testamento (Latin—in accord with the testament)
extn extraction

extr extract; extrude; extruded; extrusion
Extr Extremadura
extra extraordinary
extrad extradition
extradop extended range doppler
extradovap extended-range doppler velocity and position
extrap extrapolate; extrapolated; extrapolating; extrapolation; extrapolative; extrapolator
extra sess extra session (legislature)
extrd extruded
extrem extremity
Extrem Extremadura
extro extroversion; extrovert
extrx executrix
extsn extension(al)
exurb exurban; exurbanite; exurbia; exurban
ex works out of the factory (factory price exclusive of delivery charge)
exx examples; executrix
e_{xy} cross-elasticity of demand
Exy Expressway
Exz Exzellenz (German—Excellency)
eyawtkas everything you always wanted to know about sex
EYC Eastern Yacht Club; Encinal Yacht Club; European Youth Campaign
eyco estimated yearly cost of operation
EYD Ejaan Yang Disempurnakan (Indonesian—Improved Spelling System)
EYOA Economic and Youth Opportunities Agency
EYR East Yorkshire Regiment
EYS Ecumenical Youth Service
EYW Key West, Florida (airport)
ez easy; eczema; electrical zero
e-z easy
e/z equal zero
Ez Ezekiel; Ezra; The Book of Ezra
EZ Eastern Zone; Emile Zola; Extraction Zone
EZ Einelige Zwillinge (German—monozygotic twins)
EZ Duzit Easy Does It
Ezeiza Ezeiza International Airport (Buenos Aires, Argentina)
Ezek The Book of Ezekiel
Ezek Ezekiel
Ezi Ezias; Eziel; Eziongaber
EZI Electrolytic Zinc Indus-

tries

EZPERT Easy Programme Evaluation and Review Technic

Ezr Ezra

EZU Emiliano Zapata Unit (Chicano terrorists); *Europäische Zahlungsunion* (German—European Payment Union)

F

f faded; fair; family; farthing; fast; father (capitalized in religious orders); fathom; feet; female; feminine; filment; final target; fine; first class (travel); flat; focal length; fog; folio; following; following page; force; forecastle; founded; franc(s); frequency; freshwater; fuel; fugacity; full; function; latitude factor (symbol); relative humidity (symbol)

f/ relative aperture of a lens (also shown as *f:*)

f fecit (Latin—he did); *filius* (Latin—son); *forte* (Italian—loud); *fundada; fundado* (Spanish—founded); *für* (German—for)

f/ fardo(s) [Spanish—bale(s); bundle(s); package(s)]

F Fahrenheit; Fairchild; farad; Faraday; Faraday constant (symbol); Farrell Lines; fathom(s); February; Fellow; field of vision (symbol); fighter; fire; fixed; fixed broadcast; fixed broadcasting; flagship; florin; fluorine; formal(ity); formula; Foxtrot—code for letter F; France; franc(s); Fraunhofer line (caused by hydrogen); freedom; freedom, degree of (symbol); free energy (symbol); French; Friday; fuel; furlong(s); Furness Lines; Grumman; longitude factor

F. fats (dietary symbol)

°F degree Fahrenheit

F Federal Reporter; feria (Latin, Portuguese, Spanish—fair or market); *fine* (Italian—the end); *fora* (Por-

tuguese—out); *framkomst* (Swedish—ar-rival); *Frauen* (German—women); *freddo* (Italian—cold); *frio* (Portuguese, Spanish—cold); *froid* (French—cold); *fuera* (Spanish—out); *fuori* (Italian—out); (Latin—Filius)

F⁻ fluoride ion

F 1 Formula One (automobile sport)

F-1 Fury single-engine jet fighter-bomber flown from aircraft carriers

F₁ F_1 layer [lower of two atmospheric layers wherein the F region of the ionosphere splits during the day at heights varying from 90 to 150 miles (145 to 241 kilometers) above the earth's surface]; first filial generation

F₁O F_1O decimetric solar flux (symbol)

F 1C Fireman 1st Class (USN)

F1S finish one side

F 2 Formula Two (automobile sport)

F₂ F_2 layer [upper of two atmospheric layers wherein the F region of the ionosphere splits during the day at heights varying from 150 to 250 miles (241 to 402 kilometers) above the earth's surface; second filial generation

F² F^2 prostaglandin alpha (abortion-producing hormone)

F²ᵈ F^{2d} *Federal Reporter, second series*

F2S finish two sides

F-3 Demon single-engine supersonic all-weather jet fighter

F-4 Phantom II twin-engine

all-weather supersonic jet fighter-bomber

f4p fortran 4 plus

F-4U Chance-Vought single-engine fighter popular during World War II and called the Corsair

F-5 Northrup Freedom Fighter twin-jet aircraft

F-6 Skyray single-engine supersonic all-weather jet fighter

F6F Grumman single-seat piston-powered fighter aircraft named Hellcat

F-8 Crusader single-engine all-weather supersonic jet fighter

f/8 @ 1/50th camera lens aperture ratio 8 at an exposure of 1/50th of a second

F-9 single-engine single-seat jet fighter aircraft made by the Shen Yang Aircraft Production Complex of the People's Republic of China

F9F-2 Grumman Panther single-engine single-seat naval fighting aircraft

F9F-6 Grumman carrier-based transonic fighter aircraft called Cougar

F 11 fluorocarbon (concentrations and emissions)

F-11 Tiger single-engine supersonic jet fighter

f-12 freon (refrigerant)

F-13 dope; drugs; narcotics

F-14 swing-wing jet fighter aircraft nicknamed Tomcat and carried on some U.S. naval vessels

F-15 Eagle supersonic-jet fighter aircraft

F-16 high-performance low-cost air-combat fighter aircraft produced by Convair's Fort Worth Division

F-16B two-place fighter/trainer aircraft

F-18 all-weather fighter and attack airplane

F-18A McDonnell-Douglas Hornet strike fighter aircraft

F-18L McDonnell-Douglas-Northrup multirole fighter aircraft

F-27 Fokker Friendship (aircraft)

F-27M Fokker Troopship built in the Netherlands

F-28 Fokker turbojet aircraft

F-47 Republic fighting aircraft developed during World War II and called Thunderbolt

F-51 North American fighter aircraft developed during World War II and called Mustang

f/64 Group f/64 (photographers Ansel Adams, Imogen Cunningham, Edward Weston, Willard Van Dyke, and their followers)

F-80 Lockheed Shooting Star jet fighter-bomber

F-84 Republic Thunderjet fighter-bomber

F-86 North American Sabre single-engine jet fighter aircraft

F-89 Scorpion all-weather interceptor with twin turbojet engines

F-100 Super Sabre supersonic turbojet fighter

F-101 Voodoo supersonic twin-engine turbojet aircraft

F-102 Delta Dagger single-engine supersonic turbojet interceptor

F-104 Starfighter supersonic single-engine turbojet fighter

F-105 Thunderchief supersonic single-engine turbojet tactical fighter

F-106 Delta Dart supersonic single-engine turbojet interceptor aircraft

F-111 twin-engine turbojet tactical fighter-bomber all-weather interceptor aircraft (TFX)

F-111A variable-geometry supersonic fighter-bomber (TFX)

F-404 General Electric turbofan jet engine

fa family allowance; fatty acid; field activities; filterable agent; fire alarm; first aid; first attack; fluorescent antibody; folic acid; fortified aqueous; free acid; free aperture; frequency agility; friendly aircraft; fuel-air (ratio)

fa (FA) fatty acid; fluvic acid

f/a friendly aircraft; fuel-air ratio; further advances

f & a fire and allied (insurance); fore and aft

fa firma (Dano-Norwegian—company or firm); *foregående (forrige) år* (Dano-Norwegian—previous year); (Italian—fourth tone, *D* in diatonic scale, *F* in fixed-do system)

få forrige år (Dano-Norwegian—last year)

f^a factura (Spanish—invoice)

fA femtoampere

fA forrige Aar (Danish—last year)

Fa Faeroes

Fa Firma (German—firm, business)

FA Factory Act; Factory Automation; Failure Analysis; Farm Advisor; Field Allowance; Field Ambulance; Field Artillery; Final Acceptance; Financial Adviser; Fine Art(s); Fireman Apprentice; *Flota Argentina (de Navegación Fluvial)*—Argentine River Navigation Line; Football Association; Forecast Area; Foreign Affairs; Frankford Arsenal

F-A fighter-attack (aircraft)

F/A friendly aircraft

F & A Finance and Accounting; Financing and Accounting

F of A Foresters of America; Freethinkers of America

FA Forze Armate (Italian—Armed Forces); *Frontovaya Aviatsiya* (Russian—Frontal Aviation)—Soviet air force

F-A-18 McDonnell-Douglas fighter-attack aircraft named Hornet

faa field artillery airborne; formalin, acetic acid, alcohol (mixture); free of all average

FAA Federal Aviation Administration; Federal Aviation Agency; Fifth Avenue Association; Film Artists' Association; *Finska Angpartygys* (Finnish Steamship Line); Fleet Air Arm; Foreman's Association of America; Foundation for Aboriginal Affairs; Foundation for

American Agriculture; Fraternal Actuarial Association; Free Afghanistan Alliance

Faaa Papeete, Tahiti's airport

FAAA Fellow of the American Academy of Allergy

FAAAS Fellow of the American Academy of Arts and Sciences; Fellow of the American Association for the Advancement of Science

FAABMS Forward Army Anti-Ballistic Missile System

FAADS Forward Air-Defense Area System

FAAG First Advertising Agency Group

FAAI Filipinos for Affirmative Action, Inc.

FAAN First Advertising Agency Network

FAAO Federation of American Arab Organizations; Finance and Accounts Office (US Army)

FAAOS Fellow of the American Academy of Orthopaedic Surgeons

FAAP Federal Aid to Airports Program

FAAPS Fine Art, Antique, and Philatelic Squad (Scotland Yard)

faar forward area alerting radar

FAAR Feminist Alliance Against Rape

fab fable; fabric; fabricate; fabrication; fabulist; fabulous; first-aid box

fab fabrique (French—factory); *franco à bord* (French—free on board); *frei an bord* (German—free on board)

Fab Fabio; Fabius; Fabre; Fabrian; Fabrice; Fabrizio

FAB Facilities Advisory Board; Fijian Affairs Board; Fleet Air Base; *Força Aérea Brasileira* (Portuguese—Brazilian Air Force); Fourth Avenue Booksellers (NYC); Frédéric Auguste Bartholdi

FAB Força Aérea Brasileira (Portuguese—Brazilian Air Force)

FABAS Farm Amalgamations and Boundary Adjustment Schemes

fabbr fabbrica (Italian—factory)

FABI Fédération Royale des Associations Belges d'Ingénieurs (French—Royal Federation of Belgian Engineer-

ing Associations)
fabl fire alarm bell
FABMDS Field Army Ballistic Missile Defense System
FABMIDS Field Army Ballistic Missile Defense System
fabr fabricate; fabrication
Fab Soc Fabian Society
FABSS Fellow of the Architectural and Building Surveyors' Society
FABU Fleet Air Base Unit
fabx fire alarm box
fac façade; facial; facility; facsimile; factor; factory; faculty; fast as can; field accelerator; forward air cargo; forwarding agents commission
fac. factum similie (Latin—facsimile)
Fac Faculty
FAC Factor (Max, stock exchange symbol); Federal Advisory Council; Federal Aviation Commission; Financial Administrative Control; Financial Affairs Commission; Fleet Air Control; Forward Air Controller; Frequency Allocation Committee; Friday Afternoon Club (collegiate drinking group)
FACA Federal Advisory Committee Act; Fellow of the American College of Anaesthetists; Fellow of the American College of Angiology; Fellow of the Association of Chartered Accountants
FAC(A) Forward Air Controller (Airborne)
FACA Federación Argentina Cooperative Agrarias (Spanish—Argentine Agrarian Cooperative Federation)
FACAl Fellow of the American College of Allergists
FACAn Fellow of the American College of Anesthesiologists
FACB Federation of Australian Commercial Broadcasters
FACC Fellow of the American College of Cardiology; Florida Association of Community Colleges; Ford Aerospace and Communications Corporation
FACCA Fellow of the Association of Certified and Corporate Accountants
FACCC Faculty Association of the California Community Colleges
faccm fast-access charge-cou-

pled memory
facd foreign area consumer dialing
FACD Fellow of the American College of Dentistry
FACDS Fellow of the Australian College of Dental Surgeons
face. field artillery computer equipment; forced-air-cooled electronics
FACE Facilities and Communications Evaluation (USA); mnemonic for remembering the space notes of the treble clef—F, A, C, E
FACEM Federation of Associations of Colliery Equipment Manufacturers
FACES Fortran Automatic-Code-Evaluation System
facet facetious(ly)
FACFI Federal Advisory Committee on False Identification
FACFO Fellow of the American College of Foot Orthopedics
FACFP Fellow of the American College of Family Physicians
facg fast attack-craft gun
FACG Fellow of the American College of Gastroenterology
FACHA Fellow of the American College of Health Administrators; Fellow of the American College of Hospital Administrators
FACI First Article Configuration Inspection
facil facility
facile. fire and casualty insurance library edition
facl facilitate
facm fast attack-craft missile
FACMTA Federal Advisory Council on Medical Training Aids
FACO Fellow of the American College of Otolaryngology
FACOG Fellow of the American College of Obstetricians and Gynecologists
facp fast attack-craft patrol; forward air control point
FACP Fellow of the American College of Physicians
FACPM Fellow of the American College of Preventive Medicine
fac pwr ctl facility power control
fac pwr mon facility power monitor
fac pwr pnl facility power

panel
FACR Fellow of the American College of Radiology
facs facsimile(s)
fac's facilities
FACS Family and Community Services; Faxon's Automatic Claim System; Federation of American-Controlled Shipping; Fellow of the American College of Surgeons; Financial Accounting and Control System; Floating-Decimal Abstract Coding System
FACSAF Fleet Air Control and Surveillance Facility (USN)
FACSFAC Fleet Air Control and Surveillance Facility
facsim facsimile(s)
facsim(s) facsimile(s)
fact. factory; fast attack-craft torpedo; flexible automatic circuit tester; fully-automatic compiler translator
fact factura (Spanish—bill of lading, invoice)
FACT Facilitation and Coordination Therapy; Family Action Council of Texas; Fast Access Current Text Bank; Financial Accounting Control Technique; Flanagan Aptitude Classification Test; Flight Acceptance Composite Test(ing); Fully-Automatic Compiler Translator; Fully-Automatic Compiling Technique
fact^a factura (Spanish—invoice)
FACTS Facilities Administration Control and Time Schedule; Federation of Australian Commercial Television Stations; Financial Accounting and Control Techniques for Supply
facty fact filled; factory
fad. force activity designator; fracture analysis diagram; free air delivered; free air delivery
fad. (FAD) flavine adenine dinucleotide; funding authorization document
FAD Families Against Drunks; Fleet Air Defense; Food and Agricultural Department
FADA Federal Asset Disposition Association
fadac field artillery digital automatic computer
FADC Federal Alien Detention Center
FADD Fight Against Dictating

Designers
F Adm Fleet Admiral
FADM Functional Area Documentation Manager (USAF)
FADO Fellow of the Association of Dispensing Opticians
fadsid fighter-aircraft-delivered seismic intrusion detector
fadsorog false and dangerous systems of religion or government
FADT First Article Demonstration Test(ing)
fae fine-alignment equipment; forward air express; fuel air explosive
Fae Faeroese
FAE Federation of Arab Engineers; Fund for the Advancement of Education
FAE Federación de Amigos de Enseñanza (Spanish—Federation of the Friends of Teaching); *Fuerza Aérea Ecuatoriana* (Spanish—Ecuadorian Air Force)
FAEA Federation of ASEAN Economic Associations
FAECC Fellow of the Accountants and Executives Corporation of Canada
FAECT Federation of Architects, Engineers, Chemists, and Technicians
faer faerøsk (Dano-Norwegian—Faeroese)
Faer Faeroe Islands
Faeroes Faeroe Islands in the North Atlantic
fae's fuel air explosives
faeshed fuel-air-explosive-system helicopter delivered
FAETUA Fleet Airborne Electronic Training Unit, Atlantic
FAETUP Fleet Airborne Electronic Training Unit, Pacific
faf final approach fix; financial-aid form; first article flow; flyaway factory; forage acre factor; forward air freight; free at field; fuzing, arming, and firing
FAF Fafnir Bearings (stock-exchange symbol); Financial Analysts Federation; Fine Arts Foundation; French American Foundation
FAFT First Article Factory Test(s)
fag fagotto (Italian—bassoon)
FAG Failure Analysis Group; Finance and Accounting Group (USAF); Fine Arts Gallery; Finished Americans Group; Fiscal Activities Guide

Faga Fagatoa (American Samoa's seat of government facing Pago Pago harbor)
Fagatogo American Samoa's capital on Tutuila Island adjacent to Pago Pago
fagms (FAGMS) field artillery guided missiles
FAGO Fellow of the American Guild of Organists
fag(s) faggot(s)
fags fagottos (Italian—bassoons)
FAGS Federation of Astronomical and Geophysical Permanent Services; Fellow of the American Geographical Society
fagt first available government transportation
fagtrans first available government transportation
FAGU Fleet Air Gunnery Unit
fah failed to attend hearing
FAHA Finnish-American Historical Archives
fahqmt fully automatic high-quality machine translation
Fahr Fahrenheit
fai final acceptance inspection; first article inspection; frequency-azimuth intensity; fresh air intake
FAI Fairbanks Alaska (airport); *Fédération Aéronautique Internationale* (French—International Aeronautics Federation)
FAI Fédération Abolitionniste Internationale (French—International Abolitionist Federation); *Federación Anarquista Iberica* (Spanish—Iberian Anarchist Federation)
FAIA Fellow of the American Institute of Architects
FAIAS Fellow of the Australian Institute of Agricultural Science
FAIB Fédération des Associations Internationales Establies en Belgique (French—Federation of International Associations Established in Belgium)
FAIC Federation of Australian Investment Clubs; Fellow of the American Institute of Chemists
FAIEx Fellow of the Australian Institute of Export
FAIHA Fellow of the Australian Institute of Hospital Administration
FAII Fellow of the Australian Insurance Institute

FAIM Fellow of the Australian Institute of Management
FAIME Foreign Affairs Information Management Effort (Dept State)
fain. functional air index number
FAIO Field Army Issuing Office(r)
FAIP Fellow of the Australian Institute of Physics
FAIPM Fellow of the Australian Institute of Personnel Management
fair. fairing; fast-access information retrieval
FAir fleet air
FAIR Fair Access to Insurance Requirements; Federation for American Immigration Reform; Financial Assistance for Independent Rehabilitation; Fleet Air (Wing); Friends in America for Independence of Rhodesia
FAIRA Foundation for Aboriginal and Island Reserve Action (Australia)
Fairbanks Institute Northern Region Correction Institute at Fairbanks, Alaska
FAIRELM Fleet Air Eastern Atlantic and Mediterranean
FAIRS Fair and Impartial Random Selection System (military draft); Federal Aviation Information Retrieval System
fairships fleet airships
FAIS Fellow of the Amalgamated Institute of Secretaries
FAITH Fending Alone In The Home (Girl Scout program)
fak fly-away kit; freights all kinds
Fak Faktura (German—invoice)
FAK Federasie van Afrikaanse Kultuurvereniginge (Afrikaans—Federation of Afrikaans Cultural Societies)
fak-pak freight all kinds (in a box on wheels)
faks faksimile (Dano-Norwegian—facsimile)
Fakt Faktura (German—invoice)
fal fusil automatique légère (French—light automatic rifle)—*FAL*
Fal Falmouth
F a L Fathers-at-Large
FAL Frequency Allocation List; Frontier Airlines
FAL Frente Argentino de Liberación (Spanish—Argentine Liberation Front)—pro-Cu-

ban
FALA Federation of Asian Library Associations; Federation of Australian Literature and Art
Falashas (Amharic—strangers)—ones without a place
'falfa alfalfa
Falk Cur Falkland Current
Falk Isl Falkland Islands (Islas Maldivas)
Falklands Falkland Islands and Dependencies (South Georgia, South Sandwich Islands, South Shetlands)
fallex fall exercises
fall(s) waterfall(s)
Falls The Falls (any waterfall place-name such as Angel Falls, Niagara Falls, Victoria Falls, Yosemite Falls, etc.)
fallwarn fallout warning
FALN Fuerzas Armadas de Liberación Nacional (Spanish—Armed Forces of National Liberation)
FALNP Fuerzas Armadas de Liberación Nacional Puertoriqueña (Spanish—Armed Forces for Puerto Rican National Liberation)
FALS Ford Authorized Leasing System
falset falsetto
fam familiar; familiarization (of flights); family; foreign air mail; free at mill
Fam Famagusta; Family
FAM Federal Air Marshal; Football Association of Malaysia; foreign airmail; forward air mail; Free and Accepted Masons
F & AM Free and Accepted Masons
FAMA Federal Agriculture Marketing Authority; Fellow of the American Medical Association; Fire Apparatus Manufacturers Association
FAMA Fábrica Argentina de Materiales Aerospaciales (Spanish—Argentine Factory of Aerospace Materials)
FAMAS Flutter and Matrix Algebra System
FAMC Fitzsimons Army Medical Center
fame. fatty-acid methyl ester(s); financial accounting made easy
FAME Farmers Allied Meat Enterprises Cooperative; Fine Arts Magnet Program; Future American Magical Entertainers

FAMEM Federation of Associations of Mining Equipment Manufacturers
FAMEME Fellow of the Association of Mining, Electrical, and Mechanical Engineers
famex familiarization exercise
FAMHEM Federation of Associations of Materials Handling Equipment Manufacturers
FAMIS Financial and Management Information System
F-am-M Frankfurt-am-Main (Frankfurt-on-Main)
FAMOS Fleet Applications of Meteorological Observations from Satellites
FAMOUS French-American Mid-Ocean Undersea Study (of an Atlantic reef off the Azores)
Famous Potatoes State Idaho
fam per para familial periodic paralysis
fam phys family physician
fam rm family room
FAMS Fellow of the Ancient Monuments Society
FAMSF Fine Arts Museum of San Francisco
FAMU Florida A & M University
fan. fanatic (usually in sense of enthusiast); fantasia; fantasy
Fan Frances
FANA Federation of Australian Nurserymens Associations
FANAC Fabrica Nacional de Aceite (Spanish—National Oil Factory)
FANALOZA Fabrica Nacional de Loza (Spanish—National Porcelain Factory)
Faneuil Faneuil Hall meeting house in Boston's Dock Square where colonial Americans met to plot the revolution
FANK Forces Armées Nationales Khmères (French—Khmer National Armed Forces)—Cambodian armed forces
Fannie Mae Federal National Mortgage Association
Fan(ny) Frances; Francisca; Frasquita
FANPT Freeman Anxiety and Psychosomatic Test
FANS Food and Nutritional System; Fresh Air for Non-Smokers
Fanshaw Featherstonehaugh

fant fantasia; fantasy
Fant Fantasia
fantabulous fantastic + fabulous
fantac fighter analysis tactical air combat
FANU Flota Argentina de Navegación de Ultramar (Spanish—Argentine High Seas Navigation Line)
FANY First-Aid Nursing Yeomanry
FANYS First Aid Nursing Yeomanry Service
FANZAAS Fellow of the Australian and New Zealand Association for the Advancement of Science
fanzines fan + magazines
fao finish all over
FAO Farm Advisory Office(r); Field Audit Office(r); Finance and Accounts Office(r); Fleet Accountant Officer; Fleet Administration Office(r); Food and Agriculture Organization (UN); Free Albania Organization
F & AO Finance and Accounts Office (US Army)
faop full away on passage
fap final approach; fixed action pattern; floating arithmetic package
fap (FAP) familial adenomatous polyposis; final anthropic principle; fixed action pattern
fAp full American plan
FAP Failure Analysis Program; Family Assistance Plan; Family Assistance Program(ming); First Aid Post; Foreign Assistance Program; Frequency Allocation Panel
FAP Força Aérea Portuguesa (Portuguese Air Force); *Fuerza Aerea del Perú* (Spanish—Peruvian Air Force); *Fuerzas Armadas Peronistas* (Spanish—Peronist Armed Forces)—Argentine guerrilla group
FAPA Filipino-American Political Association
FAPC Food and Agriculture Planning Committee (NATO)
FAPHA Fellow of the American Public Health Association
FAPHI Fellow of the Association of Public Health Inspectors
FAPI First Article Production Inspection

FAPIG First Atomic Power Industry Group

FAPP Federation of Associations of Periodical Publishers

FAPR Federal Aviation Procurement Regulations

FAPREC Federación de Asociaciones de Padres, Representantes, y Educadores Católicos (Spanish—Federation of Associations of Fathers, Representatives, and Catholic Educators)

FAPRS Federal Assistance Programs Retrieval System

FAPS Fellow of the American Physical Society

FAPT Fellow of the Association of Photographic Technicians

faq fair average quality; free at quay

FAQ Free at Quay

faqs fair average quality of season

far. false alarm rate; farad; Faraday; faradic; farriery; farthing; finned air rocket; floor/area ratio; forward-acquisition radar

Far Faraday; Farsi (Iranian language)

FAR Failure Analysis Report; Federal Acquisition Regulations; Federal Aviation Regulations; Financial Accounts Receivable; Foundation for Australian Resources; finned air rocket; flight aptitude rating

FAR Fuerzas Armadas Rebeldes (Spanish—Rebel Armed Forces)—Guatemala; *Fuerzas Armadas Revolucionarias* (Spanish—Revolutionary Armed Forces)—Cuba

FARA Foreign Agents Registration Act

FARACS Faculty of Anaesthetists of the Royal Australian College of Surgeons

FARADA Failure Rate Data (BuWeps Program)

Farallones Farollon Islands off San Francisco

Farasans short form for the Farasan Islands of the Red Sea off Saudi Arabia

FARB Federation of Australian Radio Broadcasters

FARC Federal Addiction Research Center

FARC Fuerzas Armadas Revolucionarias de Colombia (Spanish—Armed Revolutionary Forces of Colombia)

Far East countries and islands of East Asia or the Pacific—eastern Siberia, China, Japan, Taiwan, Korea, Indochina, the Philippines, the Malay Peninsula

FARELF Far East Land Forces

faret fast reactor test

FARI Foreign Affairs Research Institute

Farl Farley

FARL Frick Art Reference Library; Lebanese Armed Revolutionary Factions

farm farmacia (Spanish—pharmacy)—drugstore

farmobile farm automobile

FARN Fuerzas Armadas de Resistencia Nacional (Spanish—Armed Forces of National Resistance)—El Salvador guerrillas

Farnes Farne Islands off England's Northumberland coast

faro. flow(ed, ing) at rate of

FARO Flare-Activated Radiobiological Observatory

Faroes Faerøerne (Faroe Islands)

FARP Fronte Antifascista e di Rinascita Populare (Italian—Antifascist Front and Popular Revival)—left-wing party

Far Pom Farther Pomerania (coastal Poland)

Farrar Farrar, Straus and Giroux

Fars Faristan

fas fetal alcohol syndrome; first and seconds; free alongside ship

FAS Facility Activity Schedule; Farm Advisory Service; Federal Agricultural Service; Federal Air Surgeon; Federation of American Scientists; Fellow of the Society of Arts; Food Advice Service; Foreign Agricultural Service; Free Alongside Ship; Frequency Assignment Subcommittee

FASA Federation of ASEAN Shipowners Associations; Fellow of the Acoustical Society of America; First Audit(or) of Sheriff's Accounts; Florida Association of School Administrators

FASAP Fellow of the Australian Society of Animal Production

FASB Financial Accounting Standards Board

FASBA Florida Association of School Business Administrators

fasc fascicule (French—part); *fasciculus* (Latin—little bundle)

FASC Federation of ASEAN Shippers Councils; Free Standing Ambulatory Surgical Center

FASCE Fellow of the American Society of Civil Engineers

fasci (Latin prefix—band)—fascia board

fascic fascicle

FASCO Forward Area Support Coordination Office(r)

fase fundamentally-analyzable simplified English

FASE Fellow of the Antiquarian Society—Edinburgh

FASEB Federation of American Societies of Experimental Biology

fasgrolia fast-growing language of initialisms and acronyms

fash (FASH) forward area support helicopter

FASH Fraternal Association of Steel Haulers

FASII Federation of Associations of Small Industries in India

FASL Florida Association of School Librarians

FASOC Forward Air Support Operations Center

FASPAC Ford Asia Pacific

faspl fair average sample

FASPM Flotte Administrative des Iles Saint Pierre et Miquelon

FASS Fine Alignment Sub-System

fast. fuel and sensor tanks; fully automatic switching teletype

fast. (FAST) facial affect scoring technique; facility for automatic sorting and testing; failure analysis by statistical technics; field data applications, systems, and technics; file analysis and selection technics; fleet-sizing analysis and sensitivity technic; flexible algebraic scientific translator; forecasting and scheduling technic; formula and statement translator; free and single tourist

FAST Factor Analysis System; Fast Answers about State

Taxes; First Atomic Ship Transport; flexible algebraic scientific translator; freight accounting system tracing; French Advances in Science and Technology (newsletter)

FASTM Freight Automated System for Traffic Management

fastnr fastener

FASWAC Food and Service Workers Association of Canada

fat fire and theft

fat. fatigue; final assembly test(ing); fixed asset transfer; free alongside terminal; free at terminal; full annual toll

FAT Family Adjustment Test; Flight Test Station; Folk Arts Theater; Fresno, California (airport)

fata fatigue test(ing) article

Fatah Harakat-Tahrir Falastin (Arabic—Palestinian terrorist underground organization)— Arabic acronyms such as this have inverted initials

fatdog fatty hotdog

fa technique fluorescent antibody technique

fatfurters fat-filled frankfurters

fath father; fathom

fath-in-law father-in-law

Father of Pasteurization Louis Pasteur

FATIS Food and Agriculture Technical Information Service

FATS Factory Acceptance Test Specification; Fast Analysis of Tape Surfaces; Fiji Air Travel Service; Firearms Training Systems

fatt fattura (Italian—invoice)

fau faucet; field action units; fixed asset utilization; forced air unit

fau (FAU) fine-alignment unit

FAU Florida Atlantic University; Friends' Ambulance Unit

F & AUA Fire and Accident Underwriters Association

FAUI Federation of Australian Underwater Instructors

FAUL Five Associated University Libraries (Binghamton, Buffalo, Cornell, Rochester, Syracuse)

Faunty Fauntleroy

FAUSA Federation of Australian University Staff Associations

FAUSST French-Anglo-U.S.

Supersonic Transport

faustite basic hydrated zinc aluminum phosphate (zinc-rich form of turquoise)

fav favor; favorable; favorite

FAVA Fixed Asset Valuation Assignment

FAVO Fleet Aviation Officer

fav's far-away visitors

FAW Fellowship of Australian Writers

FAWA Factory Assist Work Authorization; Federation of Asian Women's Associations

Fawcett Fawcett World Library

FAWCO Federation of American Women's Clubs Overseas

fawg free at wharf gate

FAWS Flight Advisory Weather Service

FAWU Fishermen and Allied Workers Union

fax facilities (tv technical equipment such as cameras, lights, microphones); facsimile transmission; facsimile(s); facts; fuel air explosion; photo facsimile transmission

Fax Faxon

FAX fixed aeronautical station

Faxon Fetherstoneaugh

fax sheet facilities sheet (tv production)

Fay Fagele; Faith; Fanny

Fayette La Fayette

FAZ Frankfurter Allgemeine Zeitung (German—Frankfurt's Universal Newspaper)

fb film bulletin; flat bar; flat bottom (rails); fog bell; foreign body; freight bill; fringe benefits; full American breakfast; full board; full-back; fully bleached

f-b full-bore (greater than 22 caliber)

f/b feedback; flock book; front to back (ratio)

f & b fire and bilge; fumigation and bath

fB female Black

f/B female Black

FB Fenian Brotherhood; Fernandina Beach; fighter bomber; Film Bulletin; Fire Brigade; Fisheries Board; Flying Boat; Forth Bridge; Free Baptist

FB-111 Convair strategic-bomber version of the F-111 with variable-geometry wings

fba fighter-bomber aircraft; fighter-bomber attack; fluo-

rescent brightening agent

FBA Farm Buildings Association; Federal Bar Association; Fellow of the British Academy; Fibre Box Association; Freshwater Biological Association; Fur Brokers Association

FBAA Fellow of the British Association of Accountants and Auditors

f'ball football

FBBDO Fibre Building Board Development Organization

FBBO Fellow of the British Ballet Organisation

fbc fallen building clause; fluidized bed combustion; fully-buffered channel; fully-buxomed charmer

FBC Family Benefit Capitalization; Federal Broadcasting Corporation; Fiji Broadcasting Commission; First Boston Corporation; Fukui Broadcasting Company (Japan)

FBCM Federation of British Carpet Manufacturers; Federation of British Cutlery Manufacturers

FBCP Fellow of the British College of Physiotherapists

FBCS Fellow of the British Computer Society; Foreground-Background Operating System

fbcw fallen building clause waiver

fbd freeboard

FB & D Ford, Bacon and Davis

FBEA Fellow of the British Esperanto Association

FBF Federal Buildings Fund; Frankfurt Book Fair

fbfm frequency feedback frequency modulation

FBFM Federation of British Film Makers

FBG Federation of British Growers

fbh fire-brigade hydrant

FBHI Fellow of the British Horological Institute

fbhp flowing bottom hole pressure (oil well)

FBHTM Federation of British Hand Tool Manufacturers

FbI foreign-born Irish

FBI Fast Boats Incorporated; Federal Bureau of Investigation; Federation of British Industries; Food Business Institute; full-blooded Irishman, Icelander, Indian, Indonesian,

Iranian, Iraqi, Israelite, Italian, or Ivory Coaster

FBIA Fellow of the Bankers' Institute of Australasia

FBIM Fellow of the British Institute of Management

FBIRA Federal Bureau of Investigation Recreation Association

FBIRE Fellow of the British Institution of Radio Engineers

FBIs Forgotten Boys of Iceland (American armed forces personnel stationed in Iceland)

FBIS Fellow of the British Interplanetary Society; Foreign Broadcast Information Service (CIA)

fbk flat back (lumber); fast buck

FBKS Fellow of the British Kinematograph Society

fbl forged billet; form block line

FBL Federal Barge Lines; Furness Bermuda Line

FBLA Future Business Leaders of America

fbm feet board measure; fleet ballistic missile; forward branch mail

FBM Fleet Ballistic Missile

FBMP Fleet Ballistic Missile Project (Polaris-Poseidon)

FBMWS Fleet Ballistic Missile Weapon System

FBN Federal Bureau of Narcotics

fbnrv fixed bent-nose reentry vehicle

fbo fixed-base operation; foreign building office

FBOA Fellow of the British Optical Association

fboe frequency band of emission

FBOU Fellow of the British Ornithologists' Union

fbp final boiling point

FBP Federal Bureau of Prisons; Federation of Podiatry Boards

FBPI Franklin Book Programs, Incorporated

FBPMC Federation of British Police Motor Clubs

FBPS Fellow of the British Psychological Society; Forest and Bird Protection Society

fbr fast burst reactor; fiber

fbr (FBR) fast breeder reactor; fast burst reactor

FBR Full Bibliographic Record(ing)

FBRAM Federation of British Rubber and Allied Manufacturers

fbrk firebrick

fbrl final bomb release line

fbro *febrero* (Spanish—February)

FBRS Farm Business Recording Scheme

fbs fasting blood sugar; fighter-bomber strike

fbs (FBS) frontal bovine serum

fb's fullbacks

FBS Fellow of the Botanic(al) Society; Fighter Bomber Squadron; Forward-Base System(s); Franco-Belgian Services; Fukuoka Broadcasting System

FBSC Fellow of the British Society of Commerce

FBSE Fellow of the Botanical Society—Edinburgh

FBSM Fellow of the Birmingham School of Music

FBTT Federal Board of Tea Tasters

f/bu flowing/buildup (oil well)

FBu Burundi Franc(s)

FBU Fire Brigades Union; Oslo, Norway (Fornebu Airport)

FBUI Federation of British Umbrella Industries

fbw full bandwidth

FBW Fighter-Bomber Wing

FBW System Fly-by-Wire System

fby future budget year

fbyracc for buyer's account

fc facilities control; file cabinet; filter center; fire clay; fire-control; first cross(ing); follow copy; foot-candle; franc; front-connected; functional code; fund code

fc (FC) field champion; fixed cost

f/c for cash; fill and check; fixed contract; flight control; foolscap; free and clear

f & c fire and casualty (insurance); full and change (tides)

fc *ferrocarril* (Spanish—railroad, railway)

f.c. *fidei commissum* (Latin—bequeathed in trust); *fieri curavit* (Latin—donor directed this be done)

Fc fractocumulus

FC Fairbury College; Farm Credit; Federal Cabinet; Federal Conference; Federal Convention; Fencing Club; Fenn College; Fighter Command; Finch College; Findlay College; fire control;

Fisheries Convention; Fitzwilliam College; Fontbonne College; Foothill College; Franconia College; Frederic Chopin; Frederick College; Free Church (Scotland); French Canada; French Canadian

F-C Franche-Comté

FC *Ferrocarril(es)* [Spanish railroad(s)]

fca frequency control and analysis; functional configuration audit

FCA Facility Change Authorization; Farm Credit Administration; Federated Confectioners Association; Federation of Canadian Artists; Federation of College Academics; Fellow (of the Institute) of Chartered Accountants; Fiji College of Agriculture; Finance Corporation of Australia; Financial Corporation of America; Fishermen's Cooperative Association; Foster Care Association; Freight Claim Agent; Freight Claim Association

FCA *Fédération Canadienne de l'Agriculture* (French—Canadian Federation of Agriculture)

FCAA Federal Clean Air Act; Florence Crittenton Association of America

FCAAA Federal Council of Australian Apiarists Association

FCAATSI Federal Council for the Advancement of Aborigines and Torres Strait Islanders

FCACS Federal Civil Agencies Communications System

FCAI Federal Chamber of Automotive Industries

f cant. forward cant frames

fcap foolscap

f/cap foolscap

FCAP Fellow of the College of American Pathologists

F-car(s) French car(s)

F Cas *Federal Cases*

FCAS Federal Council of Agricultural Societies; Fellow of the Casualty Actuarial Society

FCASA Foreign Correspondent's Association of South Africa

FCASI Fellow of the Canadian Aeronautics and Space Institute

fcb free-cutting brass

FCB Facility Clearance Board; Flight Certification Board; Foundation for Commercial Banks; Freight Container Bureau; Frequency-Coordinating Body
FCBA Federal Communications Bar Association
fcbu foreign currency banking unit
fcc face-centered cubic; facilities control console; fire-control computer; fire-control console; flat conductor cable; flight-control console; fluid catalytic cracking; fluid convection cathode; freight control computer
fcc (FCC) first-class certificate
FCC Fairbanks Correctional Center (Alaska); Farm Credit Corporation (Canada); Federal Communications Commission; Federal Council of Churches; Federal Court of Canada; First-Class Certificate; Flight Coordination Center; Florida Citrus Commission; Foreign Correspondents Club
FC of C Foundation Company of Canada
FCC Food Chemicals Codex
FCCA Federal Court Clerks Association; Fellow of the Association of Certified Accountants; Four Cylinder Club of America
fccc fire-control control sole
FCCCA Federal Council of Churches of Christ in America
FCCD Florida Council on Crime and Delinquency
fcci fuel-cladding chemical interaction
fcck fire-control check
FCCO Fellow of the Canadian College of Organists
fccp (FCCP) carbonylcyanide p-trifluoromethoxyphenylhydrazone
FCCP Fellow of the College of Chest Physicians
FCCS Federal Cost-Control Survey; Fellow of the Corporation of Certified Secretaries
FCCSS Fire-Control Control Subsystem
fccu fluid catalytic cracking unit
fcd failure-correction coding; function circuit diagram
FCDA Federal Civil Defense Administration

F & CD—IR Failure and Consumption Data—Inspector's Report
FCDNA Field-Command Defense Nuclear Agency (DoD)
FCDR Failure Cause Data Report
FCDU Foreign Currency Deposit Unit
fce food conversion efficiency
FCE Florida Citrus Exchange; Foreign Currency Exchange; French-Canadian Enterprises
FCE Fondo de Cultura Economica (Spanish—Foundation of Economic Culture)
FCECA Fishery Committee for the Eastern Central Atlantic
fcepc fire-control electrical package container; flight-control electrical package container
FCEX Fruit Growers Express
fcf front-end communications facility; functional check flight
fcg facing
FCG Foreign Clearance Guide
fcga facility gage
FCGB Forestry Committee of Great Britain
FCGI Fellow of the City and Guilds of London Institute
FCGP Fellow of the College of General Practitioners
fcgpc flight-control gyro-package container
fcgr fatigue crack growth rate
FCGS Freight Classification Guide System
FChS Fellow of the Society of Chiropodists
fci fuel-coolant interaction
FCI Federal Correctional Institution; *Federazione Calcistica Italiana* (Italian Football Association); *Federazione Ciclista Italiana* (Italian Cycling Association); *Federazione Colombotila Italiana* (Italian Carrier-pigeon Fanciers' Association); Fellow of the Clothing Institute; Fluid Controls Institute; Franklin College of Indiana
FCI Federación Cynological Internacional (Spanish—International Cynological Federation)
FCIA Fellow of the Canadian Institute of Actuaries; Fellow of the Corporation of Insurance Agents; Foreign Credit Insurance Association; Friends of Cast-Iron Archi-

tecture
FCIB Fellow of the Corporation of Insurance Brokers
FCIC Fairchild Camera and Instrument Corporation; Farm Crop Insurance Corporation; Fellow of the Chemical Institute of Canada
FCIF Flight Crew Information File
FCII Fellow of the Chartered Insurance Institute
fcim farm, construction, and industrial machinery
FCIP Federal Crime Insurance Program; Fire Company Inspection Program
FCIPA Fellow of the Chartered Institute of Patent Agents
FCIs Fast Coastal Interceptors (US Coast Guard)—fast boats; Federal Correctional Institutions
FCIS Fellow of the Chartered Institute of Secretaries; Florida Council of Independent Schools; Foreign Counterintelligence System (FBI)
FCIT Fellow of the Chartered Institute of Transport
FCIV Fellow of the Commonwealth Institute of Valuers
FCJ Foreign Criminal Jurisdiction
FCJC Flit Community Junior College
FCJS Federal Criminal Justice System
fcl freon coolant loop; front connecting loop; full container load
FCL Foundation for Christian Living
F-class NATO designation for a Soviet class of attack submarines also known as Foxtrot
f clef bass clef
f-c los fire-control line of sight
FCLS Family Colonization Loan Society
fclty facility
fcly face lying
fcm fat-corrected milk
FCM Ferrocarril Mexicano (Mexican Railway); Firestone Conservatory of Music (Akron)
FCMA Finch College Museum of Art; Fishery Conservation and Management Act
FCMI Federation of Coated Macadam Industries
FCMIE Fellow of the Colleges of Management and Indus-

trial Engineering

FCMS Fellow of the College of Medicine and Surgery

FCMSBR Federal Coal Mine Safety Board of Review

FCN Federal Catalog Number; Friendship, Commerce, and Navigation

FCNA Fellow of the College of Nursing—Australia

FCNM Ferrocarriles Nacionales de México (Spanish— National Railroads of Mexico)

fco cleanout flush with finished floor; fair copy; franking privilege; free postage

fco franco (Italian—free)

Fco Francisco; Franco

Fᶜᵒ Francisco (Spanish—Francis)

FCO Facility Change Order; Fire Control Officer; Fleet Construction Officer; Fleet Constructor Officer; Rome, Italy (Leonardo da Vinci airport, formerly Fiumicino)

F & CO Foreign and Commonwealth Office

fcos francos (Spanish—francs)

fcp final common pathway; foolscap

FCP Family Circle Publications; Fellow of the College of Preceptors; *Ferrocarril de Chihuahua al Pacifico* (Spanish—Chihuahua Pacific Railroad)

FCPA Fellow of the Canadian Psychological Association; Foreign Corrupt Practices Act

FC Path Fellow of the College of Pathology

FCPC Federal Committee on Pest Control

fc pl face plate

FCPO Fleet Chief Petty Officer

FCPS Fellow of the College of Physicians and Surgeons; France and Colonies Philatelic Society

fcr forward contactor; full cold rolled (steel sheeting)

FCR Facility Capability Report; Facility Change Request; Field Contact Report; Fire Control Room; First City Regiment; Flight Configuration Release; Flinders Chase Reserve (South Australia); Forwarders Certificate of Release

FCR Free China Review

FCRA Fellow of the College

of Radiologists of Australia; Fellow of the Corporation of Registered Accountants; Fair Credit Reporting Act

FCRLS Flight-Control Ready Light System

fcs forged carbon steel; fraction charge states; francs

fc & s free of capture and seizure (insurance)

FCS Facsimile Control System; Farm Credit System; Farmer Cooperative Service; Fellow of the Chemical Society; Financial Control System; Fire Control School; Fire Control Station; Fire Control System

F/CS Flight-Control System

fcsad free of capture, seizure, arrest or detainment (shipping insurance)

fcsb fire-control switchboard

FCSBC Ferrocarril Sonora-Baja California (Spanish— Sonora-Baja California Railroad)

FCSC Foreign Claims Settlement Commission

FCSCUS Federal Claims Settlement Commission of the United States

fcsle forecastle

fcsm fire-control system module

fc sm functional simulation

FCSP Fellow of the Chartered Society of Physiotherapy

fcsrcc free of capture, seizure, riots and civil commotion (shipping insurance)

fc & s and r & cc free of capture, seizure, riots, and civil commotion

fcst forecast

FCST Federal Council for Science and Technology (Executive Office of the President)

fcsu fire-control simulator unit

fcswbd fire-control switchboard

fct factory; filament center tap; fraction thereof; function

FCT Federal Capital Territory; Federal Commission(er) of Taxation

FCTB Fellow of the College of Teachers of the Blind

FCTC Fleet Combat Training Center (USN)

fcte fire-control test equipment

fctry factory

fcts fire-control test set; firing-circuit test set; flight-control test stand

FCTU Fiji Council of Trades Unions

fcty factory

fcu fare calculation unit; fare construction unit; fire-control unit; fuel-control unit

FCU Federal Credit Union(s); Federated Clerks Union

FCUS Federal Credit Union System

FCUSAA Federated Council of University Staff Associations of Australia

fcv fill-and-check valve

FCV Feline Calcivirus

FCW Fire-Control Workshop

FCWA Fellow of the Chartered Institute of Cost and Works Accountants

fcy fancy

fcy pks fancy packs

FCZ Ferrocarril de Coahuila y Zacatecas (Spanish—Coahuila and Zacatecas Railroad); Fishery Conservation Zone; Forward Combat Zone

fd face of drawing; faculty development; fan douche; fatal dose; field; field dependence; field discharge; finite difference; fiord; flame detector; flight deck; floor drain; focal dispatch; focal distance; forced draft; framed; free delivery; free discharge; free dispatch; freeze-dried; front of dash; full dress (formal attire or uniform); full duplex; functional description; fund

fd (Fd) ferredoxin

f/d father and daughter; free dock

f & d faced and drilled; fill and drain; findings and determination; fire and flushing; freight and demurrage

Fd Ferdinand; Fiord (Fjord)

F$ Fiji dollar

FD Federal Defender; field drum; Finance Department; Fire Department; Fleet Duties; Flight Director; Flying Dutchman (yacht); Forestry Department; Foundation for the Disabled; Free Democrat

F.D. Fidei Defensor (Latin— Defender of the Faith)

fd₅₀ median fatal dose

fda flight-direction attitude; fronto-dextra anterior; fully drawn account; functional demonstration and acceptance

FDA Fisheries Development Authority; Food and Drug Administration

FDA *Fraternidade Descendencia Americana* (Portuguese—Fraternity of American Descendants)—offspring of Confederate refugees in Brazil

FDAA Federal Disaster Assistance Administration

FDATC Flying Division, Air Training Command

fdau flight-data acquisition unit

FDAWU Food, Drinks, and Allied Workers Union

fdb field dynamic braking; forced-draft blower

FDB Fiji Development Bank

fdc fire-direction center (FDC); first-day cover (postage stamp); flight-director computer; formation density content (oil well log)

fdc *fleur de coin* (French—mint condition)

FDC Facility Design Criteria; Fire-Detection Center; Flight Data Center; Forsyth Dental Center (Harvard); Friends of Democratic Cuba; Furniture Development Council

FD & C Food, Drug, and Cosmetic (Act)

FDCC Fort Dodge Community College

F D & C-color Food, Drug, and Cosmetic (Act) color

FDCD Facility Design Criteria Document

FDCs Federal Detention Centers (Florence, Arizona and El Paso, Texas)

FDCT Franck Drawing Completion Test

fdd *franc de droits* (French—free of charge)

FDD Fondation Documentaire Dentaire (Dental Documentation Foundation)

fddc (FDDC) ferric dimethyl dithiocarbonate

fddl frequency division data link

fddlp. frequency division data link printout

fde field decelerator

FDEA Federal Drug Enforcement Administration

f/deck flat deck (truck)

F del P *Ferrocarril del Pacífico* (Spanish—Pacific Railroad)

F del S *Ferrocarril del Sureste* (Spanish—Southeast Railway—Tabasco, Campeche, Veracruz, Yucatan)

F de P *Ferrocarril de Panamá*

(Spanish—Panama Railroad)

F de PS General Francisco de Paula Santander—South American liberator assisting Bolívar

F de S *Ferrovie dello Stato* (Italian State Railways)

F de T *Fulano de Tal* (Spanish—So-and-So)

Fdez Fernández

FDF Footwear Distributors' Federation

FDFU Federation of Documentary Film Units

fdg funding

FDH Federal Detention Headquarters

FDHO Factory Department—Home Office

fdi fat depth indicator; field discharge

FDI Farm Dairy Instructor; Federal Department of Information (Malaysia); *Fédération Dentaire Internationale* (French—International Dental Federation); Fir Door Institute

FDIC Federal Deposit Insurance Corporation; Fire Department Instructor's Conference

FDIF *Fédération Démocratique Internationale des Femmes* (French—International Democratic Federation of Women)

FDIM *Federación Democrática Internacional de Mujeres* (Spanish—International Democratic Federation of Women)

FDIT Federal Daily Income Trust

FDJ *Freie Deutsche Jugend* (German—Free German Youth)

FDL Fast Deployment Logistic(s)—naval logistic(s)—naval cargo carrier(s); fleet deployment logistic ship (naval symbol); Flight Dynamics Laboratory; Foremost Defended Localities

F & DL Food and Drug Laboratory

fd ldg forced landing

FDLE Florida Department of Law Enforcement

FDLI Food and Drug Law Institute

FDLP Federal Depository Library Program

FDLS Fast Deployment Logistics Ship

fdm frequency division multi-

plexing

FDM *Forenede Dansk Motorejere* (Danish—Federation of Danish Motorists)

FDMA Fibre Drum Manufacturers Association

FDMBB Ferruccio Dante Michelangelo Benvenuto Busoni (Ferruccio Busoni—for short)

FDMHA Frederick Douglass Memorial and Historical Association

FDMS Flight Data Management System (USAF)

fdn foundation

FDN Field Designator Number

FDN *Frente Democrática Nacional* (Spanish—National Democratic Front); *Fuerza Democratica de Nicaragua* (Spanish—Nicaraguan Democratic Force)

fdnb (FDNB) fluorodinitrobenzene

Fdo Ferdinando

FDO Fighter Duty Officer; Fleet Dental Officer

F do I *Foz do Iguaçu* (Portuguese—Mouth of the Iguazu)—three miles above the gigantic Iguazu Waterfalls shared by Argentina, Brazil, and Paraguay

fdor four door (vehicle)

FDOS Floppy Disc Operating System

fdp field-developed program; foreign duty pay; forward defense post; funded delivery period

fdp (FDP) fructose 1,6-diphosphate

f/dp field despatch

FDP foreign duty pay; frontodextra posterior

FDP *Freie Demokratische Partei* (German—Free Democratic Party)

FDPA Fogg Dam Protected Area (Australian Northern Territory)

FDPC Federal Data Processing Center(s)

Fd PO Field Post Office

fdr feeder; field data recorder; functional demonstration requirement

fdr (FDR) flight data recorder

f dr fire door

Fdr Founder

FDR Franklin Delano Roosevelt—thirty-second President of the United States

FDRHS Franklin Delano Roosevelt High School

FDRL Franklin D Roosevelt Library (Hyde Park, New York)

FDRMC Franklin Delano Roosevelt Memorial Commission

FDRS Fire Department Rescue Squad; Flight Data Recording System; Flight Display Research System

fdry foundry

fds fixed disc store

FDS Fellow in Dental Surgery; fighter-director ship

FDSRCPS Glas Fellow in Dental Surgery of the Royal College of Physicians and Surgeons of Glasgow

FDSRCS Fellow in Dental Surgery of the Royal College of Surgeons

FDSRCS Edin Fellow in Dental Surgery of the Royal College of Surgeons of Edinburgh

FDSRCS Eng Fellow in Dental Surgery of the Royal College of Surgeons of England

fdt first destination transportation; fronto-dextra transverse

FDT Failure Diagnostic Team

fdte force development testing and experimentation

FDTL'O François Dominique Toussaint L'Ouverture (Haitian patriot who freed his country from Napoleon's control)

fdtn foundation

FDU Fairleigh Dickinson University

f/d vlv fill-and-drain valve

fdw feed water

fdx (FDX) full duplex (data processing)

FD-Zug *Fernschnellzug* (German—long-distance express train)

fe feather-edged (lumber); female employee; fighter escort; finite element; fire extinguisher; first edition; fisheye (buttons); flanged ends; for example; format effector; front end; further education

fe (FE) format effective character (data processing)

f/e fortnight ending

f & e facilities and equipment

Fe *ferrum* (Latin—iron)

FE Far East; Fighter Escort; Flight Engineer; Foreign Editor; Friends of the Earth

F & E Fearnley & Eger [FernVille (steamship) Lines]

F of E Friends of the Earth

FE Fonetic English (for spelling words as they sound)

Fe₂O₃ · H₂O rust

fe3dgw finite-element three-dimensional ground water (model)

Fe⁵2/³ radioactive iron

fea front-end analysis

FEA Failure Modes and Effects Analysis; Federal Energy Administration; Federal Executive Association; Federation of Employment Agencies; Fiji Electricity Authority; French Equatorial Africa

FEAA Federal Employees Appeal Authority

FEAD Fondo Especial de Asistencia para el Desarrollo (Spanish—Special Assistance Fund for Development)

FEAF Far East Air Force

Fe-Ag iron-silver blend

FEA(I) Federal Employees Association (Independent)

FE al P Ferrocarril Eléctrico al Pacífico (Spanish—Costa Rican electric railway)

FEAMIS Foreign Exchange Accounting and Management Information System

FEANI Fédération Européenne d'Associations Nationales d'Ingénieurs (French—Federation of European National Associations of Engineers)

Fearkar Farquhar

feat. frequency of every allowable term

feath feather(ed)(ing)

Feathers Featherstone

Featherstone Featherstone Prison near Wolverhampton northwest of Birmingham, England

FEAU Florida Education Association United

feb functional electronic block

feb febrero (Spanish—February)

feb. febris (Latin—fever)

Feb February

FEB Field Engineering Bulletin; Financial and Economic Board; Flying Evaluation Board

feba (FEBA) forward edge of battle area

Febarch February and March

febb febbraio (Italian—February)

FEBC Far Eastern Broadcasting Company

feb.dur. febre durante (Latin—

as long as fever lasts)

FEBF Far East Bridge Federation

feb° febrero (Spanish—February)

febr (Latin prefix—fever)—febrile

FEBs Federal Executive Boards

FEBS Federation of European Biochemical Societies

FEBTC Far East Bank and Trust Company

fec feckless; forward error correction; forward exchange control

fec foi, espérance, charité (French—faith, hope, charity)

fec. fecit (Latin—he made)

FEC Facilities Engineering Command; Faculty Exchange Center; Far East Command; Federal Election Commission; Federal Election Council; Federal Electoral Council; Federal Electric Corporation; First Edition Club; Florida East Coast (railway); Free Europe Committee; Free-standing Emergency Center

FECA Facilities Engineering and Construction Agency; Fiji Employers Consultative Association

FECB Foreign Exchange Control Board

FECEP Fédération Européenne des Constructeurs d'Equipement Pétrolier (French—European Federation of Petroleum Equipment Constructors)

FECIT Federación Española de Centros de Iniciativas y Turismo (Spanish Federation of Centers of Initiative and Tourism)

feck (Scottish abbreviation—effect, efficacy, value)

FECL Fleet Electronics Calibration Laboratory

fecm ferret electronic countermeasures

FECM Fellowship of the Elder Conservatorium of Music

feco fringes of equal chromatic order

FECONS Field Engineer Control System

fe cr ferrichrome (recording tape)

FECS Fédération Européenne des Fabricants de Céramiques Sanitaires (French—European Federation of Manufacturers of Sanitary Ceramics)

FECU Far Eastern Container Unit

FECUA Farmers Educational and Cooperative Union of America

fed. federal; federal law-enforcement officer; federal narcotics agent; federated; federation

Fed Federal; Federalist (Party); Federation; The Fed—The Federal Reserve Board

Fed Federación (Spanish—Federation)

FED Facilities Engineering Department; Field Experience Data; Fuel Element Design

FEDAPT Foundation for the Extension and Development of the American Professional Theatre

FEDC Federation of Engineering Design Consultants

FEDECAM Federación de Cameras de Comercio (Spanish—Federation of Chambers of Commerce)

FEDECAMARAS Federación Venezolana de Cameras y Asociaciones de Comercio y Producción (Spanish—Venezuelan Federation of Chambers of Commerce and Manufacturers)

FEDECAME Féderación Cafetalera de America (Spanish—Coffee-Growers' Federation of America)

FEDEX Federal Express

FEDFU Federated Engine Drivers and Firemens Union

FEDLINK Federal Library and Information Network

Fed Mal Federation of Malaya; Federation of Malay States; Malaysia

Fed Mal Sta Federated Malay States

fedn federation

fed narc federal narcotics agent

Fed Ref Federal Reformatory

Fed Reg Federal Register

Fed Rep Federal Reporter

Fed Rep Nig Federal Republic of Nigeria

Feds federal excise tax collectors; federal law-enforcement officers

Feds Federales (Spanish—federal police, federal troops)

FEDS Fixed Exchangeable Disc Store; Foreign Econom-

ic Development Service

FEDSIM Federal Computer Performance Evaluation and Simulation Center (GSA)

Fed-Spec Federal Specification(s)

Fed-Std Federal Standard

Fedya Fyodor

FEE Foundation for Economic Education; Foundation for Environmental Education

feeb feeble; feebleminded

FEEB Fleet Electronic Effectiveness Branch (USN)

Feeney Leonard Feeney

FEER Far Eastern Economic Review

fef fast-extrusion furnace

FEF Foundry Educational Foundation

FEFC Far Eastern Freight Conference

FE & FO Francis E and Freeland O Stanley of Stanley Steamer fame

FEGA Film Editors Guild of Australia

FEGLI Federal Employees Group Life Insurance

FEHB Federal Employees Health Benefit

FEHD Fair Employment and Housing Department; Far Eastern Hotel Development

fei for engineering information; for engineer's information; fluidic explosive initiator

FEI Farm Equipment Institute; Financial Executives Institute; Flight Engineers International; Free Enterprise Institute

FEIA Flight Engineers International Association

FEICRO Federation of European Industrial Cooperative Research Organizations

FEIS Fellow of the Educational Institution of Scotland; Final Environmental Impact Statement; Florida Educators Information Service

fekg fetal electrocardiogram

feks for eksempel (Dano-Norwegian—for example)

f eks for eksempel (Dano-Norwegian—for example)

fel fellow; front-end loader; front-end loading

fel (FEL) free-electron laser

Fel Felicita; Felix

FEL Financial Enterprises Limited; Food Engineering Laboratory (USA)

FELABAN Federación Latinoamericana del Caribe de

Asociaciones de Exportadores (Spanish—Latin American Federation of Caribbean Exporters Associations)

FELCRA Federal Land Consolidation and Rehabilitation Authority (Malaysian)

FELDA Federal Land Development Authority

feldspar barium, calcium, potassium, or sodium silicates (mineral mixtures such as orthoclase)

FELF Far East Land Forces

Felid Felidae (Latin—cat family)

felinol felinologic(al)(ly); felinologist; felinology

Felix Felixstowe

Fell Fellow

FELL Finland, Estonia, Latvia, and Lithuania

fella fellaheen (Arabic—tillers)—peasant farmers of Egypt, Syria, and nearby lands

FELO Far Eastern Liaison Office

f/e loader front-end loader; front-end loading

Felsto Felixstowe

felv feline complex leukemia virus(es)

FeLV Feline Leukemia Virus

fem female; feminine; femur; femoral; field-effect mode; fuel efficiency monitor

fem femininum (Dano-Norwegian—feminine)

fem. feminea (Latin—female); *femoris* (Latin—femur, thigh)

f.e.m. (fem or FEM) fuerza electromotriz (Spanish—electromotive force)

FEM Finite Element Method

FEM Fédération Européenne des Motels (French—European Motel Federation)

FEMA Farm Equipment Manufacturers Association; Federal Emergency Management Administration; Federal Emergency Management Agency; Fire Equipment Manufacturers Association; Foundry Equipment Manufacturers Association

femboy(s) feminine boy(s)

fem. ext. femur externum (Latin—external thigh)

fem gen circum female genital circumcision

FEMIC Fire Equipment Manufacturers Institute of Canada

fem.int. femur internum

(Latin—inner thigh)
femlib feminine liberation (women's liberation); feminine liberationist
femm femminile (Italian—feminine)
femo femoral
FEMSA Fire Equipment Manufacturers and Suppliers Association
FEMSACO Fabrica Electromecanica SA (Spanish—Electromechanical Factory)
fem-sem feminine seminary (woman's college)
femto 10^{-15}
FEMUSI Federación Mundial de Sindicatos de Industrias (Spanish—World Federation of Industrial Unions)
Fen Fenner; Fenwick; Fenwood
fenc fencing
fender bender fender-bending automotive vehicle accident
F/Eng Flight Engineer
Feng Yun 1 Chinese weather satellite
Fenno-Scandinavia Finland, Greenland, Iceland, Norway, Sweden, Denmark
FENSA Film Entertainment National Service Association
FENSA Fabrica Nacional de Electrodomesticos SA (Spanish—National Factory for Domestic Electric Appliances Incorporated)
Fen-Scan Fenno-Scandia; Fenno-Scandinavian
Fen St Fenchurch Street (rail terminal)
FEO Federal Energy Office; Federal Executive Office; Federation of Economic Organizations; Fleet Engineer(ing) Office(r)
FeO₂ FeO_2 ferric oxide (recording tape coating)
feov force end of volume
fep fore edges painted; front-end processing
FEP Federal Employees Program; Financial Evaluation Program
FEP Federación de Estudiantes del Perú (Spanish—Peruvian Students Federation)
FEPC Fair Employment Practices Commission; Federation of Electric Power Companies
FEPE Fédération Européenne pour la Protection des Eaux (French—European Federation for the Protection of Waters)

FEpow Far East prisoner of war
fer forward engine room
fer ferre (Latin—to bear)—fertile, fertilization, fertilize
fer. ferrum (Latin—iron)
Fer Ferdinand; Fermanagh; Ferris
FERA Federal Emergency Relief Administration; Foreign Exchange Regulation Act
Fer. Aet. Ferrea Aetas (Latin—Iron Age)—last of the four ages of the human race—the Plutonian period marked by avarice, crime, and cunning in the absence of honor, justice, or truth
FERC Federal Energy Regulatory Commission; Franco-Ethiopian Railway Company
fer con ferrule-contact
Ferd Ferdinand
Ferde Grofé Ferdinand Rudolph von Grofé
Ferdie Ferdinand
FERF Financial Executives Research Foundation
Ferg Fergus(on)
Fergie Fergus
FERIC Florida Educational Resources Information Center
FERIT Far East Regional Investigation Team
Ferm Fermanagh
fermentol fermentology
Fernando Rey Fernando Casado d'Aranbillet Velga
Fernᵈᵒ Fernando (Spanish—Ferdinand)
Fernspr Fernsprecher (German—telephone)
ferp family educational rights and privacy
FERP Far Eastern Refugee Program
FERPC Far Eastern Research and Publications Center
ferr ferrovia (Italian—railroad)
fert fertility; fertilization; fertilizer
fertd fertilized
FERTIPERU Fertilizantes del Perú (Spanish—Fertilizers of Peru)
fertz fertilizer
ferv. fervens (Latin—boiling)
Ferv Fervidor (French—Glowing Month)—synonym sometimes used for *Messidor* (see *Mess*)
fes festival(s); foil, épée, saber (fencing); fundamental electrical standards
FES Farm Employment

Scheme; Federation of Engineering Societies; Fellow of the Entomological Society; Fellow of the Ethnological Society; Final Environmental Statement; Fisheries Experiment Station; Flat Earth Society; Florida Engineering Society
FESA Fonetic English Spelling Association
FESCO Far Eastern Shipping Company
FESE Far East Stock Exchange (Hong Kong)
FESIP Fifth Estate Security Information Project
FESO Federal Employment Stabilization Office
FESPAC Far East and South Pacific
'fess confession
FESS Flywheel-Energy Storage System
'fession confession
'fessor professor
fest festival; festive; festivities; festivity
fest. festivus (Latin—festive or gay)
FEST Federation of Engineering and Shipbuilding Trades (British)
fesv feline sarcoma virus
fet field-effect transistor
FET Federal Estate Tax; Federal Excise Tax
FET Falange Española Tradicionalista (Spanish—Spanish Traditional Falange)—fascist organization
FETF Flight Engine Test Facility (National Reactor Test Station, Idaho)
fetol fetological; fetologist; fetology
fets field-effect transistors
FETS Far East Trade Service
FETU Federation of Entertainment Trade Unions
feu forty-foot equivalent container unit
FEU Federated Engineering Union
FEU Federación de Estudiantes Universitarios (Spanish—Federation of University Students)
feud. feudal; feudalism; feudalistic
fev fever(ish); forced expiratory volume
fev fevereiro (Portuguese—February); *février* (French—February)
fev 1 forced expiatory volume

in 1 second

FEVA Federal Employees Veterans Association

Fevr Fevral' (Russian—February)

FEW Federally-Employed Women

fex fleet exercise

FEX Foreign Exchange Service (telephonic)

fext far-end crosstalk

fey forever yours

Fez el Bali (Arabic—Fez the Old)—capital of Morocco

ff far field; fat-free; file finish; fixed focus; flip-flop; folded flat; following folios; form factor; form feed; fortissimo; french fried; front focal (length); front focus; fuel flow; full fashioned; full field

ff (FF) form feed character (data processing); folios

f/f face to face; fat and forward (sheep); flip-flop; full force

f & f fat and forward (stock); fire and flushing; fire-and-forget missile; fittings and fixtures; furniture and fixtures

f to f face to face; foe to foe; friend to friend

ff følgende (Dano-Norwegian—following or next); *folgende Seiten* (German—following pages); *fortissimo* (Italian—very loud)

Ff French franc(s)

Ff Fortsetzung folgt (German—to be continued)

FF Federated Farmers; Field Force(s); Field Foundation; fleet flagship (naval symbol); Ford Foundation; Foreign Friend (tourist to the People's Republic of China); Formula Ford; Freight Forwarder; Frontier Force(s)

F & F Faber & Faber

F of F field of fire; Firth of Forth

FF Faith and Freedom; Fianna Fail (Irish—Republican Party); *fratres* (Latin—brothers); *frères* (French—brothers)

ffa flexible factory automation; foreign freight agent; free for all; free of fatty acid; free foreign agency; free from alongside; for further assignment

FfA Fund for Animals

FFA Fellow of the Faculty of Actuaries; Fire Fighters Association; Foreign Freight Agent; Forum Fisheries Agency; Foundation for Foreign Affairs; Future Farmers of America; Future Fuels of America (hydrogen, methanol, etc.)

FFAC Freshwater Fisheries Advisory Council

FFACT Fiber, Fabric and Apparel Coalition for Trade

ffar folding-fin aircraft rocket; forward-fighting aircraft rocket

FFARACS Fellow of the Faculty of Anaesthetists of the Royal Australasian College of Surgeons

FFARCS Fellow of the Faculty of Anaesthetists of the Royal College of Surgeons

FFAS Fellow of the Faculty of Architects and Surveyors

ffb fat-free body

FFB Federal Financing Bank; Fellow of the Faculty of Building

ff black fine furnace black

ffbp free-fall bomb pod

ffc first flight cover; free from chlorine

ffC foreign friend of China

FFC Farmers Federation Cooperative; Federal Facilities Corporation; Federal Fire Council

FFCB Federal Farm Credit Board

ff cc ferrocarriles (Spanish—railroads)

FFCC Nales Ferrocarriles Nacionales (Spanish—Colombian National Railways)

FFCDPA Federal Field Committee for Development Planning in Alaska

FFCM Fellow of the Faculty of Community Medicine

FFCSA Florida Fresh Citrus Shippers Association

ffd focus film distance; forward floating depot; fuel failure detection; functional flow diagram

ffda fiber fineness distribution analyzer

FFDA Flying Funeral Directors of America

FFDRCS Fellow of the Faculty of Dental Surgery of the Royal College of Surgeons

FFE Fight for Free Enterprise; Fire Fighting Equipment (company)

ffex field firing exercise

fff fat, forty, and female

f,ff following folios (following pages)

fff forte fortissimo (Italian—very, very loud)

FFF Fellowship of First Fleeters; Four Freedoms Foundation; Frozen Food Foundation

ffff forte forte fortissimo (Italian—very, very, very loud)

ffft symbol for the sound of a pump spray

ffg friendly foreign government

FFG-7 guided-missile frigate

ffgt firefighter; firefighting

ffh formerly-fat housewife; formerly-fat husband

FFHC Federation of Feminist Health Centers; Freedom from Hunger Campaign

FFHMA Full-Fashioned Hosiery Manufacturers of America

FFHom Fellow of the Faculty of Homeopathy

ffi free from infection

FFI Fiji Forest Industry; Finance for Industry (Bank of England); Flanders Filters Incorporated; Freight Forwarders Institute; Frozen Food Institute; Frozen Foods Industries

FFI Forces Françaises de l'Intérieur (French—French Forces of the Interior)—underground soldiers fighting against the Germans in occupied France during World War II

ffim far-field image maximizer

ff ind fact-finding index

ffl field failure; fixed and flashing

F Fl fixed and flashing (light)

FFL Feminists for Life; *Forces Françaises Libres* (French—Free French Forces)

FFLA Federal Farm Loan Association

FFLI Frozen Food Locker Institute

ffly faithfully

ffm floating fecal material

FFM Fellowship for Freedom in Medicine; Fulton Fish Market

FFMC Federal Farm Mortgage Corporation

FFML Franklin Ferguson Memorial Library

ffn free-floating nozzle

FFN nuclear-powered frigate (naval symbol)

FFNM Fort Frederica National Monument (Georgia)

ffo furnace fuel oil

FFOPA Federal Firearms Owners Protection Act

ffp ferromagnetic fine particles; firm fixed price; fuel fabrication plant

f & fp fraud and false pretenses

FFP Forest Fires Prevention; Free Flight Plan

F & FP Force and Financial Program

ffpa free from prussic acid

FFPB Flora and Fauna Protection Board

FFPS Fellow of the Faculty of Physicians and Surgeons

F & FPS Fauna and Flora Preservation Society

FFPSG Fellow of the Faculty of Physicians and Surgeons

ffr foreign force reduction; free-flight rocket; frequency following response

FFR Fellow of the Faculty of Radiologists; Fleay's Fauna Reserve (Queensland)

FFRF Freedom From Religion Foundation

ffrr full frequency range recording

ffr(s) fransk(e) franc(s) [Dano-Norwegian—French franc(s)]

ffs fat-free solids

FFs first families

FFS Family Financial Statement; *Ferrovia Federali Svizzere* (Italian—Swiss Federal Railways); Financial Forecasting System; Free-Flying System; Fruit Frost Service

ffss full-frequency stereophonic sound

FFSS Ferrovia dello Stato (Italian—State Railways)

fft fast fourier transform; for further transfer

FFT Formation Flight Trainer (USAF); Functional Field Test(er)

FFTA Foundation of the Flexographic Technical Association

FFTB Freight Forwarders Tariff Bureau

FFTC Food and Fertilizer Technology Center

FFTF Fast Flux Test Facility

ff/tot fuel-flow totalizer

fftr firefighter

FFU Feminist Free University; Fire Fighters Union

ffv foreign fishing vessel

FFV First Families of Virginia

FFVA Florida Fruit and Vegetable Association

FFVMA Fire-Fighting Vehicle Manufacturers Association

ffw fast flood watch

F f W Foundation for Wellness

FFW Failure-Free Warranty

ffwd fast forward; full-speed forward

ffwm free-floating wave meter

Ffy Faithfully

FFY Fife and Forfar Yeomanry

FFZ Free Fire Zone (USA)

fg fencing; filter gate; fine grain(ed); fire glaze(d); fiscal guidance; flashgun; flat grain(ed); fog; friction glaze(d); frog(ged); fuel gas; full gilt; fully good

fg faubourg (French—suburb)

FG Federal Government; Field Goal; Fire Guard(s); Fitzroy Gardens

F & G Farmers and Graziers

FG Fine Gael (Irish—United Ireland Party)

fga foreign general average; free of general average

FGA Fellow of the Gemological Association; Fighter, Ground Attack; Freer Gallery of Art

FGAA Federal Government Accountants Association

FGAJ Fellow of the Guild of Agricultural Journalists

fgc facility group control; fine-grained concrete

f & gc failure and guilt complex

FGC Fish and Game Code; Friends General Conference

FGCM Field General Court Martial

fgcr (FGCR) fast gas-cooled reactor

FGCSO Florida Gulf Coast Symphony Orchestra

FGCSSWA Federation of Glass, Ceramic, and Silica Sand Workers of America

fgd flue-gas desulfurization

FGD Fish and Game Department

FGDS Fédération de la Gauche Démocrate et Socialiste (French—Federation of the Democratic and Socialist Left)

FGEX Fruit Growers Express

fgf fully good, fair

FGG-1 First-Generation Fuel-cell System

FGGE First GARP Global Experiment

fgim figures or images

FGIS Federal Grain Inspection Service

FGJS Farm Groups Joint Secretariat

FGL Federico García Lorca

f/glass fiberglass

fgm (FGM) field guided missile

FGMC Federal Government Micrographics Council

FGMD Fairchild Guided Missile Division

fgn foreign; foreigner

FGN Family Group Number(s)

FGNRA Flaming Gorge National Recreation Area (Utah and Wyoming)

FGO Fellow of the Guild of Organists; Fleet Gunnery Officer

Fg Off Flying Officer

FGP Foster Grandparent Program

FGR Franklin Game Reserve

f & g's folded-and-gathered signatures

FGS Fatstock Guarantee Scheme; Fellow of the Geological Society

FGSA Fellow of the Geological Society of America

FGSM Fellow of the Guildhall School of Music

fgt freight

FGT Federal Gift Tax

FGTO French Government Tourist Office

FGU Fuel Geoscience Unit

fh firehose; flathead; foghorn; forehatch; full hole (oil well)

fh (FH) familial hypercholesterolemia

f/h freehold(er)

f.h. fiat haustus (Latin—make a draft)

FH Fair Haven; Family History; Far Hills; Fashion Hills; Field Hospital

FH₂ dihydrofolic acid

FH₄ tetrahydrofolic acid

FH₅ Firehouse Five

FH-1100 Fairchild-Hiller observation helicopter

fha filterable hemolytic anemia

fha fecha (Spanish—date)

FHA Farmers Home Administration; Federal Highway Administration; Federal Housing Administration; Fellow of the Institute of Health Service Administrators; Finance Houses Association; Fine Hardwoods Association; Friends Historical Association; Future Homemakers of America

FHAA Field Hockey Association of America

FHAI Federal Housing Authority Insurance

FHAS Fellow of the Highland and Agricultural Society (Scotland)

FHASA Forces Hydroelectriques de l'Andorre (French—Andorra Hydroelectric Power)

fhb family hold back

FHBC Federation of Historical Bottle Clubs

FH/B USA Freedom House/Books USA

fhc firehose cabinet

FHC Freed-Hardeman College

FHCI Fellow of the Hotel and Catering Institute

FHCIMA Fellow of the Hotel and Catering Institutional Management Association

fhd first-hand distribution; fixed-head disc

fhdo fechado (Spanish—dated)

F & HE Fridays and Holidays Excepted (Moslem)

fhf (FHF) fulminant hepatic failure

FHF Federation of Hardware Factors

fhh fetal heart heard

FHI Family Health International; Fraser-Hickson Institute; Fuji-Hakone-Izu (national park on Honshu, Japan)

FHI Fédération Halterophile Internationale (French—International Weightlifting Federation)

FHIP Family Health Insurance Plan

FHKSC Fort Hays Kansas State College

FHL Friends Historical Library (Swarthmore)

FHLB Federal Home Loan Bank

FHLBB Federal Home Loan Bank Board

FHLBs Federal Home Loan Banks

FHLBS Federal Home Loan Bank System

FHLC Forest Hill Learning Centre (Toronto)

fhld freehold

FHLKH VIII The Famous History of the Life of King Henry VIII

FHLMC Federal Home Loan Mortgage Corporation

FHNWR Flint Hills National Wildlife Refuge (Kansas)

f-holes f-shaped sound holes in tops of stringed instruments such as violins, violas, cellos, double basses

fhp fractional horsepower; friction horsepower

FHP Family Health Program

FHPRP Family Housing Program Review Panel

fhr fetal heart rate; firehose rack

FHR Federal House of Representatives

fhs fetal heart sounds

FHS Fellow of the Heraldry Society; Forest History Society

fhsg family housing

fht fetal heart tone

FHT Fellowship Houses Trust

FHTA Federated Home Timber Association

fhtl first-class hotel

FHTNC Fleet Home Town News Center

FHU Foundation for Human Understanding

fhv forhenvaerende (Dano-Norwegian—former)

FHVMA Flowers and Hughes Values for Marriage Analysis

FHWA Federal Highway Administration

fhws flat-headed wood screw

FHWU Federated Hotel Workers Union

fhy fire-hydrant

fi fade in; failed item; female impersonator; field independence; field ionization; fire insurance; fixed interval; free in; fuel injection; fuel inspection; for instance

fi (FI) foreign intelligence

fi finsk (Dano-Norwegian—Finnish)

Fi Fidel; Finland; Finnie; Finnish

FI Faeroe Islands; Falkland Islands; Field Interview; Fiji Islands; Fiscal Intermediary; Fog Index; Fourth International; Franco-Iberian; Franklin Institute

F of I Fruit of Islam (Black Muslim disciplinary corps)

fia financial inventory accounting; full interest admitted

FIA Factory Insurance Association; Faculty Insurance Association; Fashion Industries of Australia; Federal Insurance Administration; Federal Intelligence Agency; Federated Ironworkers Association; Fellow of the Insti-

tute of Actuaries; Flatware Importers Association; Flight Information Area

FIA Fédération Internationale de l'Automobile (French—International Automobile Federation); *Federazión Internazionale Automobilística* (Italian—International Automobile Association); *Freedom of Information Act*

FIAB Foreign Intelligence Advisory Board

FIAB Fédération Internationale des Associations de Bibliothécaires (French—International Federation of Librarian Associations)

FIABCI Fédération Internationale des Administrateurs de Biens Conseils Immobiliers (French—International Federation of Real Estate Administrators)

FIAC Federation of International Amateur Cycling; Flanders Interaction Analysis Categories

FIAC Fédération Interaméricaine des Automobile-Clubs (French—Interamerican Federation of Automobile Clubs)

FIAF Federación Interamericana de Filatelia (Spanish—International Federation of Philately); *Fédération Internationale des Archives du Film* (French—International Federation of Film Archives)

FIAI Fédération Internationale des Associations d'Instituteurs (French—International Federation of Teachers' Associations)

FIAJ Fédération Internationale des Auberges de la Jeunesse (French—International Federation of Youth Hostels)

FIAJY Fellowship in Israel for Arab-Jewish Youth

FIAL Fellow of the Institute of Arts and Letters

FIAM Fellow of the International Academy of Management

FIAMA Fellow of the Incorporated Advertising Managers' Association

FIAMS Fellow of the Indian Academy of Medical Sciences

FIANEI Fédération Internationale d'Associations Nationales d'Élèves Ingénieurs

(French—International Federation of National Associations of Engineering Students)

FIANZ Fellow of the Institute of Actuaries of New Zealand

FIAP *Fédération Internationale de l'Art Photographique* (French—International Federation of the Photographic Art); Fellow of the Institution of Analysts & Programmers

FIAPF *Fédération Internationale des Associations de Producteurs de Films* (French—International Federation of Film Producers)

FIAR *Fabbrica Italiana Apparecchi Radio* (Italian—Italian Radio Apparatus Factory)

FIArb Fellow of the Institute of Arbitrators

fias free in and stowed

FIAS Flanders Interaction Analysis System

fiaT fix it again Tony (angry Fiat-owners' acronym)

FIAT *Fabrica Italiana Automobili, Torino* (Italian—Italian Automobile Factory—Turin)

FIAV *Fédération Internationale des Agences de Voyage* (French—International Federation of Travel Agencies)

fiawol fandom is a way of life

fib. fibro cement; fibula; free into barge; free into bond; free into bunkers

FIB Fellow of the Institute of Bankers; First Interstate Bank (formerly UCB); Fishing Industry Board; Franklin Institute of Boston

FIB *Fédération des Industries Belge* (French—Federation of Belgian Industries); *Félag Islenzkra Bifreidaeigenda* (Icelandic Automobile Association)

FIBA *Fédération Internationale de Basketball Amateur* (French—International Federation of Amateur Basketball)

FIBEX First International Biomass Experiment

FI Bio Fellow of the Institute of Biology

FIBM Fellow of the British Institute of Management

FIBOT Fair Isle Bird Observatory Trust

FIBP Fellow of the Institute of British Photographers

fibrd fiberboard

fibril fibrillation

FIBST Fellow of the Institute of British Surgical Technicians

fic fiction; freight, insurance, carriage; frequency interference control

FIC Farm Improvement Club(s); Federal Information Center(s); Federal Insurance Corporation; Federal Insurance Counsel; Fellow of the Institute of Chemistry; Flight Information Center; Forest Industries Council; Foundation for International Cooperation; Freedom of Information Committee

FIC *Federación Internacional de Carreteras* (Spanish—International Highway Federation)

FICA Federal Insurance Contributions Act; *Ferrocarriles Internacionales de Centro America* (Spanish—International Railways of Central America); Food Industries Credit Association

FICA *Foreign Intelligence Surveillance Act*

FICAP Furniture Industry Consumer Advisory Panel

FICB Federation of International Commercial Broadcasters

FICBs Federal Intermediate Credit Banks

FICCI Federation of Indian Chambers of Commerce and Industry

FICD Fellow of the International College of Dentists

FICE Fellow of the Institute of Civil Engineers

FICeram Fellow of the Institute of Ceramics

FIC/HEW Fogarty International Center—HEW

fi/ci foreign intelligence/counterintelligence

FICO Ford Instrument Company

FICOA Film Instruction Company of America

FICP Federal Information Centers Program

fic(s) aficionado(s) [Spanish—devotee(s)]

FICS Fellow of the International College of Surgeons; Fellow of the Institute of Chartered Shipbrokers

FICSA Federation of International Civil Servants Associations

fict fiction; fictitious

fict. fictilis (Latin—made of pottery)

FICWA Fellow of the Institute of Cost and Works Accountants

fid fiduciary; force identification; free induction decay

Fid Fidji (Spanish—Fiji)

FID Falkland Island Dependencies; Federation of International Documentation; Fellow of the Institute of Directors; Field Intelligence Department

FIDA Federal Industrial Development Authority

FIDA *Fondo Internacional para el Desarrollo de la Agricultura* (Spanish—International Fund for the Development of Agriculture)

fidac film input to digital automatic computer

fidal fixed-wing insecticide-dispersal apparatus, liquid (USNs defoliant spraying system)

FIDCR Federal Interagency Day Care Requirements

FIDE *Fédération Internationale des Échecs* (French—International Chess Federation)

Fidel Fidel Castro

FIDEL *Frente Izquierda de Liberación* (Spanish—Leftist Liberation Front)

FIDER Foundation for Interior Design Education Research

fidivan fiber-diameter video analyzer

fido fog investigation dispersal operation; freaks, irregulars, defects, oddities (created by minting errors); fugitive information data organizer

FIDO Facility for Integrated Data Organization; Federal Island Development Organization; Fire Incident Data Organization; Flight Dynamics Officer

FIDOR Fibre Building Board Development Organisation

FIDP Fellow of the Institute of Data Processing

FIDS Falkland Islands Dependencies Survey; Foolproof Identification System

FIE Feline Infectious Anemia

FIE *Fédération Internationale d'Escrime* (French—International Fencing Federation)

FIED Fellow of the Institution of Engineering Designers

FIEE Fellow of the Institution

of Electrical Engineers
FIEJ *Fédération Internationale des Éditeurs de Journaux et Publications* (French—International Federation of Editors of Journals and Publications)
FIEL *Fundación de Investigaciones Económicas Latinoamericanas* (Spanish—Foundation for Latin American Economic Investigations)
FIEN *Forum Italiano dell'Energia Nucleare* (Italian—Italian Nuclear Energy Forum)
FIER Foundation for Instrumentation Education and Research
FIERE Fellow of the Institute of Electronic and Radio Engineers
FIES Fellow of the Illuminating Engineering Society
fif ferric ion free
FIF First Investment Fund; Friends of Irish Freedom
fi. fa. *fieri facias* (Latin—see it done)
FIFA *Fédération Internationale de Football Associations* (French—International Federation of Football Associations)
FIFCLC *Fédération Internationale des Femmes de Carrières Libérales et Commerciales* (French—International Federation of Liberal and Commercial Career Women)
Fife Fifeshire
FIFE Fellow of the Institution of Fire Engineers
FIFE *Fédération Internationale Féline* (French—International Feline Federation)
fifo first in, first out (inventory)
FIFO Flight Inspection Field Office(r)
FIFRA Federal Insecticide, Fungicide, and Rodenticide Act
FIFSP *Fédération Internationale des Fonctionnaires Supérieurs de Police* (French—International Federation of Senior Police Officers)
FIFTU *Fédération Internationale des Femmes Diplômées des Universites* (French—International Federation of University Women)
fig. figuratively; figure
Fig. *Figur(en)* [German—fig-

ure(s)]; *Le Figaro* (Paris' oldest daily newspaper)
FIG Farmers Insurance Group
FIG *Federazione Italiana Golf* (Italian—Italian Golf Association)
FIGA Fretted Instrument Guild of America
FIGB *Federazione Italiana Gioco Bocce* (Italian—Italian Bocce Ball Association)
FIGC *Federazione Italiana Gioco Calcio* (Italian—Italian Football Federation)
FIGCM Fellow of the Incorporated Guild of Church Musicians
FIGM Fellow of the Institute of General Managers
FIGO *Fédération Internationale de Gynécologie et d'Obstétrique* (French—International Federation of Gynecology and Obstetrics)
figs (FIGS) figures shift (data processing)
fig(s). figure(s); finger-sized banana(s)
Fig(s) figure(s)
figt fully inclusive group tour
fih fat-induced hyperglycemia; free in harbor
FIH *Fédération Internationale des Hôpitaux* (French—International Federation of Hospitals)
FIHVE Fellow of the Institution of Heating and Ventilating Engineers
FII Fellow of the Imperial Institute; Foreign Investment Institute
FIIA Fellow of the Institute of Industrial Administration
FIIAL Fellow of the International Institute of Arts and Letters
FIIC Fellow of the Insurance Institute of Canada
FIICU Federation of Independent Illinois Colleges and Universities
fiigmo forget it, I've got my orders
FIIGS Federal Item Identification Guide System
FIIM Fellow of the Institute of Industrial Management
FIIN Federal Item Identification Number
FI Inf Sc Fellow of the Institute of Information Scientists
FIIP Fellow of the Institute of Incorporated Photographers
FIIT Federal Individual Income Tax

FIJ Fellow of the Institute of Journalists
FIJ *Fédération Internationale de Judo* (French—International Judo Federation); *Fédération Internationale des Journalistes* (French—International Federation of Journalists)
FIJET *Fédération Internationale des Journalists et Écrivains du Tourisme* (French—International Federation of Travel Journalists and Writers)
Fiji Dominion of Fiji (island nation in the western South Pacific)
Fijis Fiji islanders; Fiji Islands
FIJL *Fédération Internationale des Journalistes Libres* (French—International Federation of Free Journalists)
fil filament; fillet; fillister; filter; filtrate
f-i-l father-in-law
Fil Filbert; Filemón; Filiberto; Filinto; Filipevna; Filipp; Filippino; Filippo; Filley; Fillmore; Filmore; Filomena; Filpot; Filpotts
FIL Fellow of the Institute of Linguists
FILA Fellow of the Institute of Landscape Architects
Fil-Am Filipino-American
fild federal item logistics data
fildr federal item logistics data record
File *Filemón* (Spanish—Philemon)—The Epistle of St Paul to Philemon
FILE Fellow of the Institute of Legal Executives
file 13 trashcan; wastebasket
fil h fillister head
Fili *Filipinas* (Portuguese or Spanish—Philippines)
FILIM *Fédération Internationale des Langues et Litteratures Moderne* (French—International Federation of Modern Languages and Literature)
fill. filling
filo first in, last out
filos *filosofia* (Italian or Portuguese—philosophy); *filosofía* (Spanish—philosophy)
filt filter; filtrate; filtration
filt. *filtra* (Latin—filter)
FILT *Fédération Internationale de Lawn Tennis* (French—Lawn Tennis International Federation)
fim field ion microscope

Fim Finnish mark(s)
FIM Fellow of the Institute of Metallurgists; Flight Information Manual
FIM *Fédération Internationale des Musiciens* (French—International Federation of Musicians); *Fédération Internationale Motocycliste* (French—International Motorcycle Federation)
FIMA Fellow of the Institute of Municipal Administration; Food Industries of Malaysia; Forging Ingot Makers' Association; Friendly International Males' Association
FIMC Fellow of the Institute of Management Consultants
FIME Fellow of the Institute of Mechanical Engineers
FIMF *Fédération Internationale de Médecine Fisica* (French—International Federation of Physical Medicine)
FIMI Fellow of the Institute of the Motor Industry
FIMIT Fellow of the Institute of Musical Instrument Technology
FIMLT Fellow of the Institute of Medical Laboratory Technology
FIMS *Fédération Internationale de Médecine Sportive* (French—International Federation of Sporting Medicine)
FIMT Fellow of the Institute of the Motor Trade
FIMTA Fellow of the Institute of Municipal Treasurers and Accountants
fin. finance; financial; financier; finish
fin. finis (Latin—the end)
Fin Finistère; Finland; Finnic; Finnish
Fin Finnish
FIN Fellow of the Institute of Navigation
fina following items not available
FINAC Fast Interline Non-Active Automatic Control (automatic teletype service)
FINAST First National Stores
FINCANTIERI Società Finanziaria Cantleri Navali (Italian—Dockyards Finance Company)
fincl financial
FIND Friendless, Isolated, Needy, Disabled (older people)
FIND Federal Item Name Directory

fin dec final decree
fine b fine boomerang
FINEBEL France, Italy, Netherlands, Belgium, and Luxembourg (economic agreement)
fined finished
FINELETTRICA Società Finanziaria Elettrica (Italian—Electric Power Finance Company)
fines. fine particulates
fin fl finished floor
FINFO Flight Inspection National Field Office (FAA)
fing finishing
F-ing fucking (slang—copulating)
Fingal Finn Mac Cumhail (semimythical Irish fighter whose Hebrides hideaway is described in Mendelssohn's *Fingal's Cave* overture)
finif field-induced negative ion formation
Finisterre Cape Finisterre—northern Spain's westernmost cape
Finlan Finlândia (Italian or Spanish—Finland); *Finlandia* (Portuguese—Finland)
Finland Republic of Finland (north European land) *Suomen Tasavalta* (Finnish name); *Republiken Finland* (Swedish name)
FINMARE Società Finanziaria Marittima (Italian—Maritime Shipping Finance Company)
finn finnisch (German—Finnish)
Finn Finnish
FINNAIR Aero O/Y (*q.v.*, Finish Airlines)
Finnglish Finnish + English
Fi-No-Tro Finmark-Nord-Troms (fish processing)
FINS Fire Island National Seashore
Fin Sec Financial Secretary
FINSIN Finlandia Sinfonietta
FINSINDER Società Finanziaria Siderùrgica (Italian—Iron and Steel Financing Society)
F Inst Fellow of the Institute; Fellow of the Institution
F Inst F Fellow of the Institute of Fuel
F Inst P Fellow of the Institute of Physics
F Inst Pet Fellow of the Institute of Petroleum
F Inst SP Fellow of the Institute of Sewage Purification

f insulin fibrous insulin
FINTEL Financial Times Publishing Group
Fin-Ug Finno-Ugric
fio for information only; free in and out
FIO Fleet Information Office; Flight Information Office(r)
FIO Fédération Internationale d'Oléiculture (French—International Olive Growers Federation)
fio and stowed (trimmed) free in and out and stowed (trimming paid for)
Fiordland Fiordland National Park (southwest corner of New Zealand's South Island); Norway's nickname
fios free into owner's store; free in and out stowed
fiot free in and out and trimmed
fip fair in place; fi'pence (fivepence); fi'penny (fivepenny); fire insurance policy
FIP Feline Infectious Peritonitis; Fleet Introduction Program; Flight Instruction Program; Forestry Incentives Program
FIP Fédération Internationale de Philatélie (French—International Philatelic Federation); *Fédération Internationale des Phonothèque* (International Federation of Record Libraries); *Fuerze Interamericana de Paz* (Spanish—Interamerican Peace Force)
FIPA Fellow of the Institute of Practitioners in Advertising
FIPAGO Fédération Internationale des Fabricants de Papiers Gommes (French—International Federation of Manufacturers of Gummed Paper)
FIPD Fellow of the Institute of Professional Designers
FIPJF Fédération Internationale des Producteurs de Jus de Fruits (French—International Federation of Fruit Juice Producers)
FIPLV Fédération Internationale des Professeurs des Langues Vivante (French—International Federation of Professors of Living Languages)
FIPM Fédération Internationale de Psychothérapie Médicale (French—International Federation of Medical Psychotherapy)
FiPo Fire and Police (Research

Association)
FIPP *Fondation Internationale Pénale et Pénitentiare* (French—International Penal and Penitentiary Foundation); *Fédération Internationale de la Presse Périodique* (French—International Federation of the Periodical Press)
fips female iron-pipe size
FIPS Federal Information Processing Standards
FIPSE Fund for the Improvement of Postsecondary Education
FIPTP *Fédération Internationale de la Presse Technique et Periodique* (French—International Federation of the Technical and Periodical Press)
FIQ *Fédération Internationale des Quilleurs* (French—International Bowling Federation)
FIQS Fellow of the Institute of Quantity Surveyors
fir(.) financial inventory report; firkin; flight information requirement; floating-in rate(s); fuel indicator reading; future issue requirement(s)
fir. flight information region
FIR Fabrication Information Report; Field Interrogation Record; Flight Information Report
FIRA Federal Investment Review Agency; Foreign Investments Review Agency; Furniture Industry Research Association
FIRAA Fire Insurance Research and Actuarial Association
FIRB Fire Insurance Rating Bureau; Florida Inspection and Rating Bureau; Foreign Investment Review Board
FIRCE *Fiscalização e Registro de Capitais Estrangeiros Banco Central do Brasil* (Portuguese—Central Bank of Brazil's Foreign Capital Regulations)
FIRE Fellow of the Institution of Radio Engineers
FIREBRICK Federal Inter-Agency River Basin Committee
fireclay sedimentary rock containing chlorite-kaolinite with illite
fire damp methane
Fire Island National Seashore, Long Island, New

York
fires. firearms
FIRES Fire Inspection Reporting and Evaluation System
fir. ex fire extinguisher
FIRFLT First Fleet
FIRI Fellow of the Institute of the Rubber Industry; Fishing Industry Research Institute
FIRME *Fondo de Inversiones Rentables Mexicanas* (Spanish—Mexican Rental Investments Fund)
FIRREA Financial Institutions Reform, Recovery, and Enforcement Act
FIRST Fast Interactive Retrieval System Technology; Financial Information Reporting System; Foster Initial Reading Skills in Time
First Atomic City Handford, Washington
firta far infrared technical area
FIRTO Fire Insurers Research and Testing Organization
Firton Girton College, Cambridge
fis family income supplement; foam in salvage; free in store; freight, insurance, and shipping (charges)
fis *fisica* (Italian—physics)
fis *fisica* (Portuguese or Spanish—physics)
FIS Facial Identification Systems; Field Instruction System; Fighter Interceptor Squadron; Financial Information System (DoE); Flight Information Service
FIS *Fédération Internationale de Sauvetage* (French—International Life-Saving Federation)
FISA Fellow of the Incorporated Secretaries Association
FISAR Federal Institute for Snow and Avalanche Research
FISARS Fleet Information Storage and Retrieval System (USN)
FISC Federation of Infant School Clubs; Financial Industries Service Corporation
fisc irre fiscal irresponsibility
FIS countries France, Ivory Coast, Senegal
FISD *Fédération Internationale de Sténographie et de Dactylographie* (French—International Federation of Stenography and Typewriting)
FISE *Fédération Internationale Syndicale de l'Enseigne-*

ment (French—International Federation of Teachers' Unions)
fish. fishery; fishes; fishing
FISH Friends in Service Here
fishwich fish sandwich
FISIPE *Fibras Sintéticas de Portugal* (Portuguese—Synthetic Fibers of Portugal)
fisk *fiskeri* (Dano—Norwegian—fishery)
FISL Federally Insured Student Loan(s)
fiss (Latin prefix—split)—fissure
FIST Federation of Interstate Truckers; Field Intelligence Simulation Test; Fugitive Investigative Strike Team (for catching criminals at large)
fisteg fiscal integrity
FI Struct E Fellow of the Institute of Structural Engineers
fit. fabrication in transit; foreign inclusive tour; foreign independent traveler; foreign independent trip; formation interval tester (oil well); free of income tax; free in truck; freely independent traveller; fully inclusive tour
fit. (FIT) foreign independent travel (tour)
FIT Fashion Institute of Technology; Federal Income Tax; *Fédération Internationale des Traducteurs* (French—International Federation of Translators); Fellow of the Institute of Transport; Fiji Institute of Technology; Footscray Institute of Technology; Foreign Independent Tours
FITA *Fédération Internationale de Tir à l'Arc* (French—International Archery Federation)
fits. foreign individual travellers
fitw federal income tax withholding
fitwh federal income tax withholding
Fitz Fitzedward; Fitzgerald; Fitzgreen(e); Fitzhugh; Fitzjames; Fitzjohn; Fitzmaurice; Fitzrandolph; Fitzroy; Fitzsim(m)ons; Fitzwilliam(s)
Fitzbill Fitzwilliam
Fitzw Fitzwilliam College, Cambridge; Fitzwilliam Library (Cambridge)
Fitzw Coll Fitzwilliam College—Cambridge

FIU Federation of Information Users; Florida International University; Forward Interpretation Unit (US Army)

FIV Fellow of the Institute of Valuers

FIV Fondo de Inversiones de Venezuela (Spanish—Investments Fund of Venezuela)

fiva fluid inject valve actuator

FIVA Fédération Internationale des Vehicules Anciens (French—International Federation of Antique Vehicles)

fiw free in wagon

FIWC Fiji Industrial Workers Congress

fix. fixture

Fiz Fizika (Russian—physics)

Fiz Elem Chastits At Yadra Fizika Elementarnykh Chastits i Atomnogo Yadra (Russian—Journal of Particles and Nuclei)—USSR

Fiz Met Fizika Metallov i Metallovedenie (Russian—Physics of Metals and Metallography)—USSR

Fiz Nizk Temp Fizika Nizkikh Temperatur (Russian—Journal of Low-Temperature Physics)—USSR

Fiz Plazmy Fizika Plazmy (Russian—Plasma Physics)—USSR

Fiz Tekh Poluprovodn Fizika i Tekhnika Poluprovodnikov (Russian—Semiconductor Physics and Technology)—USSR

Fiz Tverd Tela Fizika Tverdogo Tela (Russian—Solid-State Physics)—USSR

fj flush joint

Fj Fjord

FJ Fiji Airways; Flying Junior

F-J Fisher-John

FJA Future Journalists of America

FJC Federal Judicial Center; Fullerton Junior College

Fjd Fjord

Fjd(s) Fiji dollar(s)

FJH Franz Josef Haydn

Fji Fiji

FJI Fellow of the Institute of Journalists

FJIC Federal Job Information Center

FJNM Fort Jefferson National Monument

fjp first job program

FJS First Jersey Securities; Fulton J Sheen

fk flat keel; fork

Fk Frank

FK Fluid Kinetics; Franz Kafka; Fujita Airways

FK Frankfurt Kassenverein (German—Frankfurt Clearinghouse)

FKBD Fort Knox Bullion Depository

FKBI Fourdrinier Kraft Board Institute

FKC Fellow of King's College

Fkd Frankford

FKI Federation of Korean Industries

FKJC Florida Keys Junior College

FKL Frauen Konzentrations-Lager (German—Women's Concentration Camp)

Fkn Franklin; Frederikshavn

fkr før Kristus (Dano-Norwegian—before Christ)

Fks Fredrikstad

FKSII Fresh Kills, Staten Island, Incinerator

FKSNS Fort Kent State Normal School

FKTU Federation of Korean Trade Unions

FKWR Florida Keys Wildlife Refuges

fl flash(ing); flash(ing) light; flight level; flood(ing); floor(ing); flourish; flow line; flow(ing); fluid loss; fluid(s); fluorescent level; flush(ing); focal length; follow(ing); footlambert; foreign language; forklift

f/l freight liner

f&l fuel and lubricants

fl flaske (Dano-Norwegian—bottle, flask); *flauti* (Italian—flute, flutes); *flauto; flores* (Latin—flowers; *florin* (Dano-Norwegian or Dutch—florin); *floruit* (Latin—he flourished)

f.l. falsa lectio (Latin—false reading)

fL foot-lambert

Fl Fall (postal abbreviation); Flemish; fluorine

Fl Fleuve (French—large river)

FL Ferdinand Laeisz; First Lady; Flag Lieutenant; Flight Lieutenant; Florida; focal length; Football League; foreign language; Frontier Airlines (2-letter code)

F.L. Franz Liszt

F for L Feminists for Life

FL Fürstentum Liechtenstein (Principality of Liechtenstein)

fla fronto-laeva anterior

f.l.a. fiat lege artis (Latin—according to the rules of art)

Fla Florida; Floridian

FLA Federal Loan Administration; Federal Loan Agency; Fellow of the Library Association; Florida; Florida East Coast Railway (symbol); Foam Laminators Association

FLA Frente de Libertação Açoriana (Portuguese—Azorian Liberation Front)

FLAA Fellow of the Library Association of Australia

flab flabby

fl abwth flush armor balanced watertight hatch

FLAC Florida Automatic Computer (USAF)

FLACCS Florida Climate and Control System

Fla Cur Florida Current

flag. flageolet

FLAI Fellow of the Library Association of Ireland

FLAIR Floating Airport

FLAIRS Fleet Locating and Information Reporting System (for police-patrol vehicles)

flak fondest love and kisses

flak Fliegerabwehrkanone (German—anti-aircraft cannon, anti-aircraft shrapnel)

flake colorful and eccentric person

flam flamländisch (German—Flemish)

FLAME Facility Laboratory for Ablative Materials Evaluation

flam(s) flamenco (songs); flaming(s); flammable(s)

flang flowchart language

FLAP Flores Assembly Program

FLAPS Flexibility Analysis of Piping Systems

flar florward-looking airborne radar

FLAS Fellow of the Land Agents Society

FLASH Foreign Fishing Vessels Licensing and Surveillance Hierarchical Information System (Canadian)

Flats Durango, Colorado's slums

flav flavor(ing)

flav. flavus (Latin—yellow)

flb flight-line bunker

FLB Federal Land Bank

FLBAs Federal Land Bank Associations

flbin floating-point binary

flbm (FLBM) fleet-launched ballistic missile

fl bp filter, band-pass

fl bs filter, band-suppression

FLC Federal Library Committee; Foundation Library Center

flcc flight-control computer

FLCM Fellow of the London College of Music

FLCO Fellow of the London College of Osteopathy

fl crs flat cars

fld failed; field; flowered; fluid

Fld Field (postal abbreviation)

FLD Friends of the Lake District

Fld Com DNA Field Command, Defense Nuclear Agency

fldec floating-point decimal

fldg folding

fldg chr folding chair(s)

fl di flare die

FL & DI Food Law and Drug Institute

fldl field length (flow chart)

fldo final limit, down

fldop field operations

fl dr fluid dram

Flds Fields

fldxt fluid extract

Fl e Flemish ell (unit of measure)

flea. flux logic element array

fleact fleet activities

flee. fast-linkage editor

FLEEC Federal Libraries' Experiment in Cooperative Cataloging

fleetex fleet exercise

Flem Flemish

fleming(s) fleming-gear hand-propelled lifeboat(s)

Flem(s) British slang for Belgian(s)

fles foreign language in elementary school

FLES Foreign Languages in Elementary Schools (linguistic teaching program)

FLETC Federal Law Enforcement Training Center

FLETRABASE Fleet Training Base (USN)

Flev Flevoland (Dutch province)

FLEWEACEN Fleet Weather Center

FLEWEAFAC Fleet Weather Facility

flex. flexible

FLEX Federal Licensing Examination

flexo flexographic

flf final limit, forward; flip flop

FLF Freedom Leadership Foundation

flg failing; flagging; flange; flashing; flooring; flying

FLG Flagship (USN)

FLGA Fellow of the Local Government Association

FLGB Federal Loan Guarantee Board

flgd flanged

flgstn flagstone

flh familial lefthandedness; final limit, hoist

fl hd flathead

flhls flashless

fl hp filter, high-pass

flia familia (Spanish—family)

flib friggin little itinerant bastard(s)

FLIC Film Library Information Council

flick(s) flicker(s), [motion picture(s)]

flicon flight control

flicr fluid-logic industrial control relay

fliden flight data entry

FLIM Flight Mechanics Internal Memorandum

Flinders Flinders Ranges of South Australia

flint variety of chalcedony

flint. (FLINT) floating interpretive language

Flint Flintshire

Flints Flintshire

flip. film library instantaneous presentation

FLIP Flexible Loan Insurance Program; Flight Information Publication; Floated Lightweight Inertial Platform; Floating Instrument Platform; Free-form Language for Image Processing

Flip(s) Filipino(s)

flir forward-look infrared

FLIRT Federal Librarians Round Table

flit functional literacy

fliv flivver

FLIWR Functional Listing and Interconnection Wiring Record

flkprt flock printed

Flks Falkland Islands

fll final limit, lower

FLL Finanglia Line Ltd; Fort Lauderdale, Florida (airport); Friends Library, London

fllar forward-looking light attack radar

fl ld floor load

Flli fratelli (Italian—brothers)

fl lp filter, low-pass

flm functional-level manager

Flm Flemish

FLM Free Library Movement

FLM Fédération Luthérienne Mondiale (French—Lutheran World Federation)

FLMI Fellow of the Life Management Institute

fl/mtr flow meter

fln fallen; following landing numbers

Fln Flensburg

FLN Frente de Liberación Nacional (Spanish—National Liberation Front); *Front de Liberation Nationale* (French—National Liberation Front)—official Algerian party

FLNC Frente de Liberación Nacional de Cuba (Spanish—National Liberation Front of Cuba)

flng falling

FLNM Fort Laramie National Monument

flo floodlight(s)

Flo Flobert; Florence; Florentz; Florian; Floris

Fl O Flight Officer

FLO Foreign Liaison Office(r)

float. floating offshore attended terminal

floatel floating motel

floc floccule; flocculent; floccus

FLOC For Love of Children

flodac fluid-operated digital-automatic computer

Fl Offr Flying Officer

FLOG Fleet Logistics Air Wing

FLOOD Fleet Observation of Oceanographic Data (USN)

flop. floating octal point

flor floriculture

flor flores (Latin—flowers); *floruit* (Latin—he flourished)

Flor Floréal (French—Flowery Month)—beginning April 20—eighth month of the French Revolutionary Calendar

Flor(a) Florence

Florence Federal Detention Headquarters at Florence, Arizona

Floribbean Floridian-Caribbean (resort area)

Florrie Flora; Florence

florsent fluorescent

floss. flossing (dental care)

Floss(ie) Florence

flot flotation; flotilla; flotsam

Flota Flota Oceanica Brasileira (Portuguese—Brazilian Oceanic Fleet)

flotel floating hotel

fl ovth flush oiltight ventilation hole

flox fluorine + liquid oxygen

Floy Florence

fl oz fluid ounce

flp fault location panel; frontolaeva posterior

FLP Free Library of Philadelphia

flpl fortran-compiled list-processing language

fl pl. flore pleno (Latin—in full bloom)

FLPMA Federal Land Policy and Management Act

fl prf flameproof

FLPS First Log Procurement Status

fl pt flashpoint; fluid pint

FLQ Front de Libération Quebecois (French—Front for the Liberation of the people of Quebec)

flr failure; final limit, reverse; flame resistant; flare(s); floor; florin; flow rate; forward-looking radar

FLR Florence, Italy Airport

FLR Federal Law Reports

FLRA Federal Labor Relations Authority

flrg flooring

flrng flash ranging

flrs flares; flowers; forward-looking radar set

fl/rt flow rate

FLRT Federal Librarians Round Table

fls floors; forward-looking sonar

Fls Falls (postal abbreviation); Flushing

FLS Fellow of the Linnaean Society; Flashing Light System

FLSA Fair Labor Standards Act

flsc flexible linear-shaped charge; flight shape charge

FLSEP Family Life and Sex Education Program

flsh flesh (side) leather

flshd fleshed (skins)

FLSO Fort Lauderdale Symphony Orchestra

FLSP Fort Lincoln State Park (North Dakota)

flst flautist; flutist

flt filter; fleet; flight; float; flotation; fork-lift truck; frontolaeva transverse

flt flertall (Dano-Norwegian—majority or plural)

Flt Flats (postal abbreviation); Fleetwood

F/Lt Flight Lieutenant

Flt Adm Fleet Admiral

fltbcst fleet broadcast (USN)

Fltcher C Fletcher College

fltck flight check

Flt Cmdr Flight Commander

fltg floating

flt ld sim flight-load simulator

Flt Lt Flight Lieutenant

Flt No. Flight Number

fltp flight template

flt/pg flight programmer

flt pln flight plan

fltr floater

flts flights

Flt Sat Com Fleet Satellite Communications System (USN)

FLTSATCOM Fleet Satellite Communications (DoD)

Flt Sgt Flight Sergeant

Flt Sgt Nav Flight Sergeant Navigator

fltstrikex full general-emergency striking force (USN)

flu fault location unit; final limit, up; first line unit; influenza

fluc fluctuant; fluctuate; fluctuating; fluctuation

FLUG Flugfelag Islands (Iceland Airways)

flummery foolish humbuggery (named after British custard made of flour or oatmeal boiled with water until almost too thick to swallow)

fluor fluor-apatite; fluorescence; fluorescent; fluorite; fluorspar; fluotaramite; synonym of flourite

fluorspar calcium fluoride (CaF_2)

flur fluorescent

fluss flüssig (German—fluid)

flv foreign leave

flw follow(s)

FLW Frank Lloyd Wright

flwd followed

flwg following

flwop forced landing without power

flx flexible

fly. flinty; flying; flyweight

FLY Flying Tiger Line

flying wing B-2 advanced technology bomber; stealth bomber

FlyTAF Flying Training Air Force

FLZO Farband-Labor Zionist Order

fm face measurement; facial measurement; facing matter; fan marker; farm; farmer; fathom; fathometer; female

white; femtometer(s); fermi; fine measurement; form; frequency modulation; from; front matter; fumigation

f-m frequency modulation

f/m feet per minute

f & m foot-and-mouth disease

fm femmes mariées (French—married women); *formiddag* (Dano-Norwegian—before noon)—a.m.; *formiddagen* (Swedish—before noon)—a.m.

f.m. fiat mistura (Latin—make a mixture)

f/M female Mexican

Fm fermium

F/m unit of permittivity

FM Fed Mart; Federated States of Micronesia; Ferrocarril Mexicano (Mexican Railroad); Field Manual; Field Marshal; Fire Marshal; Flight Mechanic; Foreign Minister; frequency modulation

F & M Franklin and Marshall College

F.M. Fraternitas Medicorum (Latin—Physicians Fraternity)

fma forward maintenance area

FMA Federal Maritime Administration; Felt Manufacturers Association; File Manufacturers Association; Financial Management Association; Fish Marketing Authority; Flour Mills of America; Food Machinery Association; Ford Motor Argentina; Forging Manufacturers Association; Fulfillment Management Association

FMACC Foreign Military Assistance Coordinating Committee

FMAI Financial Management for Administrators Institute

fman foreman

FMANA Fire Marshals Association of North America

FMAO Farm Machinery Advisory Office(r)

FMAS Foreign Marriage Advisory Service

fmb (FMB) fast missile boat

FMB Federal Maritime Board; Felix Mendelssohn Bartholdi

fmbid firm bid

FMBRA Flour Milling and Baking Research Association

FMBSA Farmers and Manufacturers Beet Sugar Association

FMC Failure Mode Center

(Reliability Laboratory); Federal Maritime Commission; Federated Motor(ing) Clubs; Federated Mountain Clubs; Federation of Mothers Clubs; Felt Manufacturers Council; Food Machinery Corporation; Ford Motor Company

FMC *Federación de Mujeres Cubanas* (Spanish—Federation of Cuban Women)

fmca forming cam

FMCA Family Motor Coach Association; Fire Mark Circle of the Americas

FM Can Ford Motor of Canada

FMCC Fulton-Montgomery Community College

F McH NM Fort McHenry National Monument

FMCL Fleet Mechanical Calibration Laboratory

FMCS Federal Mediation and Conciliation Service

fm cu form cutter

fmcw frequency-modulated continuous wave

fmd foot-and-mouth disease

FMD Federated Metals Division—American Smelting and Refining; Fisheries Management Division; Fixtures Manufacturers and Dealers; *Flota Mercante Dominicana* (Spanish—Dominican Steamship Line); Forward Metro Denver

fmdf fixed mirror-distributed focus

fm di form die

fme frequency-measuring equipment

FMEA (FEA) Failure Modes and Effects Analysis

FMECA Failure Mode, Effects, and Criticality Analysis

fmer factory mutual engineering and research

fmeva floating-point means and variance

fmf fetal movement felt; field maintenance factor

fMf (FMF) familial Mediterranean fever

FMF Fleet Marine Force; Food Manufacturers' Federation; Foreign Military Financing

FMF-A Fleet Marine Force—Atlantic

fmfb frequency-modulation feedback

FMFIC Federation of Mutual Fire Insurance Companies

FMFLANT Fleet Marine Force, Atlantic

FMF-P Fleet Marine Force—Pacific

FMFPAC Fleet Marine Forces—Pacific

fmfs fat in the moisture-free substance

fmg foreign medical graduate

FMG *Flota Mercante Grancolombiana* (Spanish—Colombian national steamship lines); franc(s) Malagasy

FMGJ Federation of Master Goldsmiths and Jewelers

fmh (FMH) fat-mobilizing hormone

FMH Friends Meeting House

FmHA Farmers Home Administration

FMHCSS Federal Mobile Home Construction and Safety Standard

FMHHS Fort McHenry Historic Shrine (Baltimore)

FMI Farmers Mutual Insurance; Fiber Materials Inc; FM Intercity (relay broadcasting); *Fonds Monétaires Internationals* (French—International Monetary Fund); Food Marketing Institute; Freight Management International

FMI *Fondo Monetario Internacional* (Spanish—International Monetary Fund)

fmicw frequency-modulated intermittent-continuous-wave (radar)

FMIG Farmers Mutual Insurance Group; Food Manufacturers' Industrial Group

FMIS Functional Management Inspection System

fmk full-mouth radiograph

Fmk Finnmark; Finnish markka (currency unit)

FML Factory Mutual Laboratories; Fermi National Laboratory (Batavia, Illinois)

FMLN *Farabundo Martí Liberación Nacional* (Spanish—Augustín Farabundo Martí National Liberation Front)—leftist rebels in El Salvador

FMLNF *Farabundo Marti National Liberation Front* (Spanish—Salvadoran Marxist guerrillas)

fmly formerly

fmly k a formerly known as

FMM Federation of Malay Manufacturers; French Military Mission

FMMA Floor Machinery Manufacturers Association

fmmd form mandrel

FMME Fund for Multinational Management Education

fmn formation

fmn (FMN) flavin mononucleotide

FMN *Fédération Motorcycliste Nationale* (French—National Motorcycling Federation); *Ferrocarril Mexicano del Norte* (Spanish—Northern Mexican Railroad)

FMNH Field Museum of Natural History

FMNM Fort Matanzas National Monument

FMO Fleet Mail Office; Fleet Medical Officer; Flight Medical Officer

FMOF First Manned Orbital Flight (NASA)

fmofr firm offer

fmp first menstrual period; functional maintenance procedure; funny-man prop

FMP Fairbanks Morse Pump; Family Medicine Program; Final Management Plan; Fourth Malaysia Plan; Frontier Mounted Police; Fuel Management Panel

FMPA Fellow of the Master Photographers' Association

FMPA *Fédération Mondiale pour la Protection des Animaux* (French—World Federation for the Protection of Animals)

FMPC Feed Materials Production Center

FMPE Federation of Master Process Engravers

FMPEC Financial Management Plan for Emergency Conditions (USA)

fm/pm phase-modulated telemetering system

fm prot fine-mesh (cover) protected

FMPS Fairbanks Morse Power Systems

fmr fair market rent; farmer; fast metabolic rate; ferromagnetic resonance; former; former(ly)

FMR Field Maintenance Reliability; Field Materials Request; Franco Maria Ricci (or his magazine *FMR*)

F-M-R Friend-Moloney-Rauscher (virus)

FMRA Fertilizer Manufacturers Research Association

FMRC Financial Management Research Center

fm rl form roll

fmrly formerly

fmrr (FMRR) financial management rate of return

FMRS Federal Mediation and Reconciliation Service; Foreign Member of the Royal Society

fms fathoms; fat-mobilizing substance; flush metal saddle; foreign military sales; free-machining steel; frequency-multiplexed subcarrier

fm's formerly-married persons

FMS Federal Mining and Smelting (company); Federated Malay States; Field Music School; Financial Management System; Flexible Manufacturing System; Floating Machine Shop; Fort Myers Southern (railroad); Frequency Monitoring System; Friends Mission Society

FMS *Fédération Mondiale des Sourds* (French—World Federation of the Deaf); Foreign Military Sales

fmsa frequency measuring spectrum analyzer

FMSA Federal Managers Support Agency; Fellow of the Mineralogical Society of America

FMSF Foreign Military Sales Financing

FMSI Friction Materials Standards Institute

FMSL Fort Monmouth Signal Laboratory

FMSM *Fédération Mondiale pour la Santé Mentale* (French—World Mental Health Federation)

fmswr flexible mild-steel wire rope

fmt flush metal threshold

fmt (FMT) format (flow chart)

FMT Factory Marriage Test; Flight Management Team (NASA)

fm to. form tool

FMTS Field Maintenance Test Station

F & MTVHS Food and Maritime Trades Vocational High School

fmu force measurement unit; freight multiple unit

FMVSS Federal Motor Vehicle Safety Standard

FMWC Federation of Medical Women of Canada

FMWS Fairbanks Morse Weighing Systems

fmx full-mouth radiography

fn fence; flatnose (projectile);

footnote; fusion

f/n freight note

fn *fête nationale* (French—national holiday)

Fn Factonimbus

F$_n$ Fibonacci number(s)

FN Flight Nurse; Fridtjof Nansen

FN *Fabrique Nationale* (French—National Factory)—Belgian arms firm's initials appearing on all its products; *Forenede Nationer* (Danish—United Nations)

FN4RM/62FAB Belgian four-wheeled armored vehicle armed with a 60mm mortar and two machineguns or a 90mm cannon

fna for necessary action

FNA following named airmen; French North Africa

FNAA Fellow of the National Association of Auctioneers

FNAF Federal Nigerian Air Force

FNAL Fermi National Accelerator Laboratory

FNB First National Bank; Food and Nutrition Board

FNBC First National Bank of Chicago

FNBP Far North Bicentennial Park (Anchorage)

fnc finance

FNC *Federación Nacional de Cafeteros* (Spanish—National Federation of Coffee Growers—Colombia); *Ferrocarriles Nacionales de Colombia* (Spanish—National Railroads of Colombia)

FNCB First National City Bank

fncg financing

fncl financial

FNCR *Ferrocarril del Norte de Costa Rica* (Spanish—Northern Railway of Costa Rica)

fnd found; foundation; foundered

FND Flinders Naval Depot (Australia)

fndd founded

fndg founding

fndn foundation

fndr founder

fndrs fenders

fndry foundry

FNDTS Fellow of the Non-Destructive Testing Society

fne fine

fnf flying needle frame

fng fuckin' new guy

fnh flashless nonhygroscopic

(gunpowder)

FNH *Ferrocarril Nacional de Honduras* (Spanish—National Railway of Honduras)

FNIC Food and Nutrition Information and Educational Materials Center

FNIE *Fédération Nationale des Industries Électriques* (French—National Federation of Electrical Industries)

FNIF Florence Nightingale International Foundation

FNIMC Florida Normal and Industrial Memorial College

fnl final

FNL Friends of the National Libraries

FNLA *Frente Nacional de Libertação de Angola* (Portuguese—Angolan National Liberation Front)

FNLO French Naval Liaison Office(r)

fnl qtr final quarter

fnly finally

fnlz finalize

FNM *Ferrocarriles Nacionales de México* (Spanish—National Railroads of Mexico); Financial Network Manager

FNMA Federal National Mortgage Association

FNN Fiji News Network; Financial News Network

FNNWR Fort Niobrara National Wildlife Refuge (Nebraska)

FNO Fleet Navigation Officer; following-named officers

FNOA following-named officers and airmen

fnp fusion point

fnp (FNP) floating nuclear-power plant

FNP Family Nurse Practitioner; Fiordland National Park (South Island, New Zealand); Fundy National Park (New Brunswick, Canada)

FNPF Fiji National Provident Fund

FNRC Federal Nuclear Regulatory Commission

FNRJ Federation Narodna Republik Jugoslavija (Yugoslavia)

fns flask-nitrogen supply

FNS Food and Nutrition Service; Frontier Nursing Service

FNSAE Fellow of the National Society of Art Education

fnsh finish

fnshd finished

fnshg finishing

fnshr finisher
FNTO Finnish National Travel Office
fnu first name unknown
FNU Forces des Nations Unies (French—United Nations Forces)
f number focal length of a lens
f-number diameter of a lens aperture in relation to its focal length
FNV Financiera Nacional de la Vivienda (Spanish—National Housing Finance)
FNWA Foreign National Weather Agency
FNWC Fleet Numerical Weather Center (USN)
FNWF Fleet Numerical Weather Facility
FNZDT Federation of New Zealand Dancing Teachers
FNZLA Fellow of the New Zealand Library Association
FNZSA Fellow of the New Zealand Society of Accountants
FNZSID Fellow of the New Zealand Society of Industrial Designers
fo faced only; fade out; fast operating; filter output; firm offer; firm order; flat oval; folio; formal offer; free out; free overside; freight on; fuel oil; full out terms; for orders
fo (FO) full organ
fo' for; four
f/o for credit of; firm offer; free overside; for orders
fᵒ folio
fᵒ firmato (Italian—signed)
f/O female Oriental
Fo Fornax
F₀ pure parental type
FO Federal Office(r); Federal Official(dom); Field Office(r); Field Operations; Field Order; Finance Office(r); Finance Officer; Fisheries Office(r); Flag Officer; Flying Officer; Foreign Office; Forward Observer
F.O. Foreign Office
F/O Flight Officer; Flying Officer
FOA Farmers Organization Authority; Football Officials Association; Foreign Operations Administration; Foresters of America; Friends of Animals
FOAC Flag Officer, Aircraft Carrier(s)
fob feet out of bed; freight on board

fob. feet out of bed; fresh off the boat; front of body (hoist); fuel on board; full of baloney
fo & b fuel oil and ballast
f.o.b. free on board; fuel on board
FoB Friends of the Bureau (FBI)
F o B Faculty of Building
FOB Federal Office Building; Forward Operating Base; Free on Board
fobcnlf free on board cars, named point, lighterage free
fobcnp free on board cars, named point
FOBFO Federation of British Fire Organisations
fobot free on board, owners trim
FOBS Fractional-Orbit Bombardment System
fobse free on board, sacks extra
fobsi free on board, sacks included
foc final operation capability; flag of convenience; focal; focus(ing); free of charge; free on car(s); free on container(s); full operational capability
f.o.c. free of charge; free on car(s); free on container(s)
FoC Father of the Chapel (printer's union)
FOC Ferrocarriles Occidentales de Cuba (Spanish—Western Railroads of Cuba); Flight Operations Center
FOCA Federation of Citizens Associations; Formula-One Constructors Association
FOCAS Ford Operating Cost Analysis System
fochr free of charge
FOCI Farrand Optical Company, Incorporated
FOCIS Financial On-Line Central Information System
FOCLA Federation of Country Local Associations
focmg forthcoming
FOCOL Federation of Coin-Operated Launderettes
FOCS Freight Operation Control System
FOCSL Fleet-Oriented Consolidated Stock List
fo'c's'le forecastle
FOCT Flag Officer Carrier Training
FOCUS Federation of Community United Services
fod fodder; foreign object dam-

age; free of damage
fod (FOD) foreign object damage
f.o.d. free of damage
FOD Flag Officer, Denmark; follow-on destroyer
fo/do fuel oil/diesel oil (consumed daily)
foe fuel oil equivalent
FOE Fraternal Order of Eagles; Friends of the Earth
fof free on field (airmail)
FOF Facts-On-File
FOFA Follow-On Forces Attack
F of F Firth of Forth
fofr firm offer
fog. flow of gold
FoG Friends of Gill
FOG Field Operations Group (US Army); Flag Officer, Germany; Florida Orange Growers
FOGA Fashion Originators Guild of America
FOGAIN Fondo de Garantia y Fomento a la Industria Mediana y Pequeña (Spanish—Fund for the Guarantee and Promotion of Medium and Small Industry)—Mexico
Fog Sig fog signal (station)
foh front of house
foi freedom of information
foi (FOI) fighter officer interceptor; follow-on interceptor
F o I Freedom of Information
FOI (station) Operations Intelligence; Fighter Officer Interceptors; Fruit of Islam (Black Nationalists)
FoIA Freedom of Information Act
FOIC Flag Officer in Charge
foil. file-oriented interpretive language
FOIR Field-of-Interest Register
fok free of knots (lacing, rope, string, twine)
f.o.k. free of knots
fol folio; folios; follow; following; follows; free-on-lorry oil and lubricants
fol. folium (Latin—leaf); *folia* (Latin—leaves)
FOL Federation of Labor; Federation of Labour (New Zealand); Foreign Office Library; Friends of the Land
fold. folding
folg folgend (German—following)
Folkes Folkestone
foll followed by
folnoaval following (items) not

available
fols folios; follows
Folsom California State Prison at Folsom
fom fat off mothers (sheep); fault of management; figure of merit
fomaj force majeure
FOMC Federal Open Market Committee
FOMCA Federation of Malaysian Consumers Association
FOMEX Fondo Nacional de las Exportaciones (Spanish—National Fund for the Promotion of Exports); *Fondo para el Fomento de las Exportaciones de Productos Manufacturados* (Spanish—Fund for the Promotion of Export of Manufactured Products)
FOMIN Fondo Nacional de Fomento Industrial (Spanish—National Fund for Industrial Promotion)
fomm functionally-oriented maintenance manual(s)
FoMoCo Ford Motor Company
fomth for one month
FON Fiber Optic Network
FONADE Fondo Nacional de Desarrollo (Spanish—National Development Fund)
FONASBA Federation of National Associations of Shipbrokers and Agents
FONATUR Fondo Nacional de Fomento al Turismo (Spanish—National Fund for Tourist Promotion)
fonc fonctionnaire (French—bureaucrat or office holder)
Fondo El Fondo (Spanish—The Fund)—International Monetary Fund—IMF
fonecon telephone conversation
FONEI Fondo Nacional de Equipamiento Industrial (Spanish—National Fund for Industrial Equipment)
fonet fonetica (Portuguese or Spanish—phonetics)
fonét fonética (Italian—phonetics)
F on F Facts-on-File
fono photograph
fonoff foreign office
Fons Alphonse; Fonseca
Fontanka Fontanka Canal linking Leninport with the main section of Leningrad and the Neva River
FONZ Friends of the National Zoo

foᵒ folio (Spanish—folio)
FOO Forward Observation Officer
foob (FOOB) firing out of the battery (artillery project)
fool's gold pyrites (copper, iron, tin, etc.)
foot(s) footnote(s)
fop. forward observation post
f/op firing/observation port
FOP feminization of poverty; Fraternal Order of Police
fopt fiber-optics photon transfer
f.o.q. free on quay
for. foreign; foreigner; forensic; forest; forester; forestry; forint (Hungarian monetary unit); free on rail; free on road
f.o.r. free on rail
for (Latin prefix—opening)—foramen
For Formosa(n); Fornax
FOR Fellowship of Reconciliation; Final Outturn Report; Flying Objects Research; Foundation for Ocean Research
forac for action
FORACS Fleet Operational Readiness Accuracy Check Site
forast formula assembler translator
FORATOM Forum Atomique Européen (French—European Atomic Forum)
for. bal forensic ballistics
FORBID Federatie van Organisaties op het gebied van Bibliotheek—Informatieen Dokumentatiewezen (Dutch—Federation of Organizations on Libraries, Information, and Documentation Services)
forbloc fortran-compiled block-oriented (simulation programme)
for. bod foreign body
forcap forward combat air patrol
for & cc free of riots and civil commotion
for'd forward
Ford Gerald R Ford—38th President of the United States, 40th Vice President; automobile manufacturer Henry Ford; English playwright John Ford; movie director John Ford; historian Paul Leicester Ford
FORD Families Opposed to Revolutionary Destruction;

Fix Or Repair Daily; Found On the Road Dead
FORDS Floating Ocean Research and Development Station
'fore before
FORE Foundation of Record Education
fore 1/4s fore-quarters (meat cuts)
foren forensic(ally); forensic medicine
fores'l foresail
FOREST Freedom Organisation for the Right to Enjoy Smoking Tobacco
FOREWAS Force and Weapon Analysis System
forf forfeit; forfeiture
forf forfattare, författarinna (Swedish—author, authoress)
förf forfatter (Dano-Norwegian—author)
forg forger; forgery; forging
f org (F Org) full organ
forint monetary unit of Hungary
fork forkortelse (Dano-Norwegian—abbreviation); *forkortning* (Swedish—abbreviation)
fork. forkortelse (Danish—abbreviation)
for. lang foreign language(s)
form. format; formation; former(ly)
form formiddag (Dano-Norwegian—morning, before midday)
forma fortran matrix analysis
formac formula manipulation compiler
formal. formaldehyde; formalin
formalin HCHO
format. fortran matrix abstraction technique(s)
for med forensic medicine
For Min Foreign Minister; Minister of Foreign Affairs
formn foreman
FORMS Federation of Rocky Mountain States
for'm'st foremast
formul formulary
forpac forecasting passengers and cargo
For Pol Foreign Policy
forr forretning (Dano-Norwegian—business or store)
for'rd forward
for.rts foreign rights
Forsch Forschung (German—research)
FORSIC Forces Intelligence Center

forsk *forskellig* (Dano-Norwegian—different, distinct, unlike)
for's'l foresail
FORSTAT Force Status and Identity Report (USAF)
fort. fortification; fortify; fortnight(ly); fortress; full-out rye terms (grain trade)
fort. *fortis* (Latin—strong)
fortel formatted teletypewriter
Fort Frederica Fort Frederica National Monument on Saint Simon's Island off Brunswick, Georgia
Fort Jeff Fort Jefferson National Monument in the Dry Tortugas in the Gulf of Mexico west-northwest of Key West
Fort Laramie Fort Laramie National Monument on the Oregon Trail in southeastern Wyoming
Fort Leavenworth U.S. Disciplinary Barracks at Fort Leavenworth, Kansas
fortly fortnightly
Fort Matanzas Fort Matanzas National Monument near St Augustine, Florida, built by the Spaniards in 1736
Fort McHenry Fort McHenry National Monument in Baltimore Harbor where the *Star Spangled Banner* was written
Fort Meade Maryland operations center of the National Security Agency
for. tox forensic toxicology
Fort Pulaski Fort Pulaski National Monument at the mouth of the Savannah River
FORTRAN formula translation computer language
For-Trans Ford Foundation Transfer Student Project
FORTRANS Formula Translating System
fortransit formula translator internal translator
Fort Riley U.S. Army Correctional Training Facility at Fort Riley, Kansas
forts *fortsaettelse* (Dano-Norwegian—continuation or sequel)
Forts *Fortsetzung* (German—continuation)
Fort Savage New York City's East Harlem police precinct
Fortschr Phys *Fortschritte der Physik* (German—Advances in Physics)
fortsim fortran simulation
Fort Sumter national monument in Charleston Harbor (South Carolina)—first shot of Civil War fell on this fort
fort. twn fortified town
Fortune Five Hundred *Fortune Magazine's* annual listing of the 500 leading corporations
Fort Union Fort Union National Monument near Santa Fe, New Mexico
Forth Worth Federal Correctional Institution at Fort Worth, Texas
forum. formula for optimizing through realtime utilization of multiprogramming
FORUM Federation Of Retired Union Members
***For Whom* **For Whom The Bell Tolls**
forwn forewoman
'forz *sforzando* (Italian—emphasized forcefully)
fos fossil; free on station; free on steamer; fuel-oxygen scrap; full of shit
fos (FOS) faint object spectography; full operational status
f.o.s. free on station; free on steamer
F-o-S Frinton-on-Sea
FOS File Organization System; Fisheries Organization Society; Fuel Oil Supply (company)
FOSATU Federation of South African Trade Unions
fosdic film optical sensing device for input to computers
FOSFA Federation of Oil Seed and Fats Association
fos fls fossil fuels (coal, natural gas, oil, etc.)
FOSG Factory Outlet Shopping Guide
FOSH Foshing (airlines)
FOSI Florida Ocean Sciences Institute
fosplan formal space-planning language
foss fear-of-success syndrome
***Fos sur Mer* **(French—Fos by the Sea)—Marseilles port
fot free of tax(ation); free on truck; frequency optimum traffic; fuel-oil transfer
f.o.t. free on truck
fot *fotographie* (Dutch—photography)—plus all derivatives
FOT Fraternal Order of Police
fot & e follow-on test(ing) and evaluation
F o t L Friends of the Library
FOTL Follow-on-to-Lance (launcher)
FOTM Friends of Old-Time Music
foto photograph(ic)
foto *fotografia* (Italian or Portuguese—photography); *fotografía* (Spanish—photography)
Foto (Jewtongo—Fort)—native nickname for Paramaribo, Surinam
fotog *fotografia* (Italian or Portuguese—photography); *fotografía* (Spanish—photography)
fotsu forward observer target survey unit
fo'ty forty
found foundation; foundling; foundry
Found Econ Educ Foundation for Economic Education
Found Phys *Foundations of Physics*
Foun Mot Dent Foundation for Motivation in Dentistry
fount fountain
Foun Than Foundation of Thanatology
FOUO For Official Use Only
Four Cs Community-Coordinated Child Care
Four H Four H (hand, head, heart, and health) Club
Four Horsemen Four Horsemen of the Apocalypse (each mounted, respectively, on a white horse symbolizing pestilence, a red horse—war, a black horse—famine, a pale horse—death); title of a novel by Vicente Blasco Ibañez—*Los cuatro jinetes del Apocalipsis*
Four Mountains Islands of the Four Mountains
FOUSA Finance Office(r), United States Army
fov field of view
fov (FOV) flyable orbital vehicle
fow first open water; free on wagon; free on warehouse; free on wharf
f.o.w. first open water; free on wagon
Foxes Fox Islands off southwestern tip of Alaska
Foxtrot letter F radio code
Foy Fowey
fp factory pass; family plan(ning); fast peak; field punishment; film pack; fine paper; fire plug; fire policy; fireplace; first performance;

first performed; first proof; fix point; fixed price; flameproof(ed); flash point; flat pad(ded); flat pattern; flat point(ed); flight pay; flight plan; floating (open) policy; flower people; focal plane; food poisoning; foot path; foot pound(s); forte piano; forward perpendicular; free piston; freezing point; fresh paragraph; frontispiece; full page; full point; full price; fully paid

fp (FP) family practitioner; flavoprotein

f/p flat pattern

f.p. fiat potio (Latin—make a potion)

FP Family Physician; Federal Parliament; *Ferrocarril del Pacífico* (Spanish—Pacific Railroad); former pupil; Franklin Pierce (14th President U.S.); Free Press

F/P Fire Policy (insurance)

FP Freiheitliche Partei (German—Freedom Party)—Austrian party

fp4c full page four colors

fpa fluorescent pen aerosol; focal plane array; free of particular average

FPA Family Planning Association; Federal Preparedness Agency; Federation of Motion Picture Producers in Asia; Flexible Packaging Association; Flying Physicians Association; Foreign Policy Association; Forest Products Association; Franklin Pierce Adams; Free Pacific Association; Freemantle Port Authority; Freethought Press Association

fpaa free from particular average, absolutely

FPAA Family Planning Association of Australia

fpaAc free of particular average, American conditions

FPAD Fund for Peaceful Atomic Development

fpaEc free of particular average, English conditions

fpaf fixed-price award fee

FPAS Federal Property and Administrative Services; Fellow of the Pakistan Academy of Sciences

FPASA Federal Property and Administrative Services Act

FPAT Family Planning Association of Tasmania

fpaucb free from particular average unless caused by (stranding, etc.)

FPB Fast Patrol Boat

FPBA Folding Paper Box Association

FPBAI Fellow of the Publishers' and Booksellers' Associations in India

FPBG fast patrol boat, guided-missile (USN)

FPBRS Fels Parent Behavior Rating Scale(s)

fpc fish protein concentrate; fixed price contract; fixed-price call; flat plate (solar) collector(s); flight progress chart; full-page color (ad); for private circulation

fp—C flash point—Celsius

FPC Facility Power Control; Family Planning Center; fast patrol craft; Federal Pacific Electric (stock exchange symbol); Federal Power Commission; Federal Prison Camp; Fiji Pine Commission; Food Packaging Council; Friends Peace Committee; Frozen Pea Council

FPCA Federal Post Card Application (for absentee ballot)

fpcc flight propulsion-control coupling

FPCC Fair Play for Cuba Committee

FPCE Fission Products Conversion and Encapsulation (AEC plant)

FPCI Federal Penal and Correctional Institutions

FPCL Fronte Paisanu Corsu di Liberazione (Italian—Corsican Peoples Liberation Front)

FPCS Fire Power Control Subsystem; Full-Page Composition System

FPD Federal Public Defender

FPD Fundación Panamericana de Desarrollo (Spanish—Pan-American Development Foundation)

FPDA Finnish Plywood Development Association; Five-Power Defence Arrangement (Malaysian)

fpdi flight path deviation indicator

FPDO Federal Public Defender Organization(s)

fpe fixed price with escalation

FPE Foundation for Personality Expression; Full Personality Expression

FPEA Fellow of the Physical Education Association

FPEB Family Planning Evaluation Branch (USPHS)

FPEBT Fire Prevention and Engineering Bureau of Texas

fpec four-pile-extended cantilever (platform)

FPED Farm Production Economics Division (USDA)

FPF French Protestant Federation

fph feet per hour (oil well drilling)

FPH Federal Pacific Hotels (Australian)

FPHA Federal Public Housing Authority

F Pharm S Fellow of the Pharmaceutical Society

fphs fallout protection in homes

F Ph S Fellow of the Philosophical Society

F Phy S Fellow of the Physical Society

fpi faded prior to interception; family pitch in; fixed price incentive

FPI Federal Prison Industries; Fellow of the Plastics Institute

FPI Fédération Prohibitionniste Internationale (French—International Prohibitionist Federation)

FPIA Family Planning International Assistance

fpif fixed-price-incentive firm

fpil full premium if lost

f. pil. fiat pilulae (Latin—make pills)

fpis fixed-price incentive successive; forward propagation ionosphere scatter

FPJMC Four-Power Joint Military Commission

fpl final protective line; fire plug; fireplace

FPL Family Protection Law; Florida Power and Light; Forest Products Laboratory

FPL Fuerzas Populares de Liberación (Spanish—Popular Forces of Liberation)—El Salvador

FPLA Fair Packaging and Labelling Act

fplce fireplace

fpm facility power monitor; feet per minute; fissions per minute; frequency pulse modulation

FPML Forest Products Marketing Laboratory

FPMR Federal Property Management Regulation(s)

FPMSA Food Processing Machinery and Supplies Association

FPMT Filter Paper Microscopic Test

FPNM Fort Pulaski National Monument

fpo fixed price open

FPO Field Post Office; Field Project Office; Fleet Post Office; Fleet Postal Organization

FPOA Federal Probation Officers Association

fpoe first port of entry

fpoh food prepared outside the home

FPOP Family Planning Organization of the Philippines

fpp facility power panel; fixed-pitch propeller; floating-point processor; forward(ing) parcel(s) post

FPP Family Planning Program; Foster Parents Plan; Foster Parents Program; Friendly Peoples Proviso

FPPB Family Planning and Population Board (Singapore)

FPPC Fair Political Practices Commission

fppe fluorescent pen-post emulsified

FPPS Flight Plan Processing System; Full-Page Phototypesetting System

fpr feet per revolution; fixed price redeterminable; flatplate radiometer; forward parcels rail

FPR Factory Problem Report; Field Personnel Record; Fishing Ports Registration

FPRC Fair Play for Rhodesia Committee

fprf fireproof

FPRI Foreign Policy Research Institute (University of Pennsylvania)

FPRL Forest Products Research Laboratory

FPRS Federal Property Resources Service; Forest Products Research Society

fps feet per second; focus projection and scanning; foot per second; foot-pound-second; frames per second

f'ps former priests

FPs Flying Physicians; Flying Psychologists

FPS Farm Placement Service; Fauna Preservation Society; Federal Protection Service; Fellow of the Pharmaceutical Society; Fellow of the Philharmonic Society; Fellow of

the Philological Society; Fellow of the Philosophical Society; Fence Protection System; Financial Planning System; Fire Protection System; Fluid Power Society

FPSA Fellow of the Photographic Society of America

FPSAA Federated Public Service Assistants Association

FPSE Federation of Public Service Employees

FPSL Fellow of the Physical Society of London

FPSO Fleet Publication Supply Office

fpsps feet per second per second

fpt female pipe thread; fill, puddle, and tamp; fixed price tenders; forepeak tank; full power trial

FPT Flight Proof Test(ing); Four Picture Test

fptm fluorescent pen-tank method

fpts forward propagation tropospheric scatter

FPTU Federation of Progressive Trade Unions

fpu field pickup unit

FPU Food Preservers Union

fpv fixed-price vendor

FPWA Federation of Professional Writers of America; Federation of Protestant Welfare Agencies

fq fiscal quarter

FQ French Quarter (New Orleans)

FQ *Faerie Queene* (Edmund Spenser allegory)

fqawt flush quick-acting watertight

fqcy frequency

FQL Food Quality Laboratory

FQO Federation of Quarry Owners

FQS Federal Quarantine Service

fque fabrique (French—factory)

fr family room; fast release (relay); father; field relay; fire retardant; fixed response; flaring and retracting; flight request; flow rate; frame; franc; frayed; frequency response; frequent; from; front; fruit

f/r fixed response; flat rack; freight release; front to rear

f & r feed and return (plumbing); force and rhythm (pulse)

fr franco (Spanish—franc);

franc(s) or *fransk* [Dano-Norwegian—franc(s) or French]

fr. folio recto (Latin—front of the sheet)

f.r. folio recto (Latin—front of the sheet)—righthand page

Fr Father; France; francium; Franco-; Franklin; Frau (German—Missus); French; Friar; Friday; Friesian(s); Frisian(s); Froude number

F/r restricted first-class (travel)

Fr Frankrijk (Dutch—France); *Frau* (German—Missus); *Fray* (Spanish—Friar); *Fredag* (Danish—Friday)

FR Facilities Request; Feather River (railroad); Federal Reformatory; Federal Register; Federal Reserve; Field Report; fighter reconnaissance (aircraft); Final Report; Fireman Recruit; flash red—enemy aircraft nearby; Fleet Reserve; Freight Release; Friden (stock exchange symbol)

F of R Fellowship of Reconciliation

FR *Federal Register*

F.R. Forum Romanum (Latin—Roman Forum)

FR-172 French-built four-place rocket-launching counterinsurgency aircraft

fra forward refueling area; functional residual air

fra factura (Spanish—invoice)

Fra Francis

Fra Francia (Spanish—France)

Fra. frater (Latin—brother, monk)

FRA Federal Railroad Administration; Fleet Reserve Association; Food Retailers Association; Footwear Research Association; Frankfurt-am-Main (airport)

frac frationator reflux analog computer

FRAC Food Research and Action Center

FRACA Failure Reporting, Analysis, and Corrective Action

FRAC Arts Foundation for Research in the Afro-American Creative Arts

FRACHE Federation of Regional Accrediting Commissions of Higher Education

FRACI Fellow of the Royal Australian Chemical Institute

FRACP Fellow of the Royal Australian College of Physicians

FRACS Fellow of the Royal Australian College of Surgeons

fract fraction; fracture

fract. dos. fracta dosi (Latin—in divided doses)

FRAD Fellow of the Royal Academy of Dancing

FRAeS Fellow of the Royal Aeronautical Society

frag fragile; fragment; fragmentary; fragmentation; fragmented

frago fragmentary order; fragmented order

FRAgS Fellow of the Royal Agricultural Societies

FRAHS Fellow of the Royal Australian Historical Society

FRAI Fellow of the Royal Anthropological Institute

FRAIA Fellow of the Royal Australian Institute of Architects

FRAIC Fellow of the Royal Architectural Institute of Canada

'fraid afraid

FRAM Fellow of the Royal Academy of Music; Fleet Rehabilitation and Maintenance (USN)

FRAME Fund for the Replacement of Animals in Medical Research

Framingham Massachusetts Correctional Institution (for female) at Framingham, Massachusetts

fran framed-structure analysis; franchise

Fran Frances; Francis; Franciscan

franc monetary unit of Andorra, Belgium, France, Luxembourg, Madagascar, Monaco, Rwanda, and Switzerland

Franc Franciscan

France French Republic

Francine Frances

Francisco I Francisco Indalécio Madero

Franco Francisco Paulino Hermenegildo Teodulo Franco-Bahamonde—Spanish dictator

Franc° Francisco (Spanish—Francis)

frangi(s) frangipani(s)

Franglais *francais + anglais* (French + English)

Fran-Jud Franco-Judeo (French Jewish)

Frank Frank; Frankford; Frankfort; Frankfurt; Frankish; Franklin

Frank Frankrike (Norwegian—France)

franklinite ferric iron and zinc crystalline compound

frank(s) frankfurter(s)

Frans Francis

FRAP Fellow of the Royal Academy of Physicians

FRAP *Frente Revolucionario de Acción Popular* (Spanish—Revolutionary Popular Action Front)—Chile

FRAPS Farm Record Analysis Pilot Scheme

Fras Francis

FRAS Fellow of the Royal Asiatic Society; Fellow of the Royal Astronomical Society

Frasca Francesca

Frasco Francisco

frat fraternity

frat fratello (Italian—brother)

FRAT Free Radical Assay Technique (heroin-morphine test)

FRATADD Foundation for Research and Treatment of Alcoholism and Drug Dependence (Australian)

frate formula for routes and technical equipment

frater fraternity brother

frats fraternities

fratting fraternizing

Frau *Die Frau Ohne Schatten* (German—The Woman Without a Shadow)—Richard Strauss opera

fraud. fraudulent

frav first available

Fraxi Pisanus Fraxi (Herbert Specer Ashbee)

FRB Federal Reserve Bank; Federal Reserve Board; Fisheries Research Board

frbb fracture of both bones; free room, board, and beverages

FRBC Fisheries Research Board of Canada

fr bel from below

FRBk Federal Reserve Bank

FRBNY Federal Reserve Bank of New York

FRBs Federal Reserve Banks

FRBS Fellow of the Royal Botanic Society; Fellow of the Royal Society of British Sculptors

frc fiber-reinforced concrete; functional residual capacity

f.r.c. free carrier

FRC Facility Review Committee; Fasteners Research Council; Federal Radiation Council; Federal Radio Commission; Federal Records Center; Federal Republic of Cameroon; Filipino Rehabilitation Commission; Flag Research Center; Flight Research Center; Foreign Relations Committee; Foreign Relations Council; Forwarder's Receipt Certificate; Fuels Research Council

FRCA Fellow of the Royal College of Art

FRC—AAP Freedom-to-Read Committee—Association of American Publishers

Fr-Can French-Canadian

FRCAT Fellow of the Royal College of Advanced Technology

fr & cc free of riots and civil commotion

FRCD Fellow of the Royal College of Dentists

frcd's floating-rate certificates of deposit

FRCGP Fellow of the Royal College of General Practitioners

FRCI Fellow of the Royal Colonial Institute

FRCM Fellow of the Royal College of Music

FRCO Fellow of the Royal College of Organists

FRCOG Fellow of the Royal College of Obstetricians and Gynaecologists

FRCP Federal Rules of Civil Procedure; Fellow of the Royal College of Physicians

FRCPath Fellow of the Royal College of Pathologists

FRCP(C) Fellow of the Royal College of Physicians of Canada

FRCPE Fellow of the Royal College of Physicians of Edinburgh

FRCPGlas Fellow of the Royal College of Physicians of Glasgow

FRCPI Fellow of the Royal College of Physicians of Ireland

FRCP Lond Fellow of the Royal College of Physicians of London

FRCPSG Fellow of the Royal College of Physicians and Surgeons of Glasgow

FRC Psych Fellow of the Royal College of Psychiatrists

FRCR Fellow of the Royal College of Radiologists

FRCrP Federal Rules of Criminal Procedure

FRCs Federal Regional Councils

FRCS Fellow of the Royal College of Surgeons

FRCSc Fellow of the Royal College of Science

FRCS(C) Fellow of the Royal College of Surgeons of Canada

FRCSE Fellow of the Royal College of Surgeons of Edinbrugh

FRCSGlas Fellow of the Royal College of Surgeons of Glasgow

FRCSI Fellow of the Royal College of Surgeons of Ireland

FRCSL Fellow of the Royal College of Surgeons of London

FRCTS Fast Reactor Core Test Facility

FRCVS Fellow of the Royal College of Veterinary Surgeons

frd formerly restricted data; friend; friendly

Frd Ford (postal abbreviation)

FRD Federal Rules Decisions

FR Dist Federal Reserve District

Frdn Friedenau

fre free energy region

fre fracture (French—invoice)

Fre Freemantle; French

Fre Freitag (German—Friday)

FRE Federal Rules of Evidence

FREB Federal Real Estate Bord

FR Econ S Fellow of the Royal Economic Society

FR Econ Soc Fellow of the Royal Economic Society

fred figure-reader electronic device

Fred Alfred; Alfredo; Freddie; Frederic; Frederick; Fredric; Fredrick; Wilfred

Freda Winifred

Fred(die) Frederica; Fredrica

Freddie Mac Federal Home Loan Mortgage Corporation

Fred(dy) Alfred; Frederick; Wilfred

Fredk Frederick

Fredk D Frederick Douglass

Fredo Alfredo

Free Freeway

FREE Florida Resources in Education Exchange

freebd freeboard

freebies free services; free things; free tickets

Freedman's Bureau Bureau of Refugees, Freedmen, and Abandoned Lands (set up after the Civil War in the United States)

Free Lib Phila Free Library of Philadelphia

freem freemason(ry)

Free-O Freemantle, Western Australia

Freep Free Press (Los Angeles underground newspaper)

freeture freedom, the wave of the future

freeway toll-free express highway

FREI Fellow of the Real Estate Institute

Freib Freiburg (Germany)

freid freidenker(ei) (German—freethinker; freethinking)—latitudinarian(ism)

FRELIMO Frente de Libertação de Moçambique (Portuguese—Mozambique Liberation Front)

FRELP Flexible Real Estate Loan Plan

Frem Fremantle

frem. voc. fremitus vocalis (Latin—vocal fremitus)

French Antilles Désirade, Guadeloupe, Les Saintes, Marie Galant, Martinique, Saint Barthélmy, and part of Saint Martin

French Can French Canadian

Frenglish frenchified English

FREntS Fellow of the Royal Entomological Society

freon tf trifluorotrichloroethane (solvent)

FREP Fleet Return Evaluation Program

freq frequency; frequent; frequentative; frequently

FrEqAfr French Equatorial Africa

freq m frequency meter

fres fire-resistant

fres frères (French—brothers)

FRES Fellow of the Royal Entomological Society

frescanar frequency scan radar

fresh. freshman; freshmen

Freud. Freudian

frev fast reverse

frf flight-readiness firing; frequency response function

fr-f french-fried (potatoes)

FRF Freedom-to-Read Foundation; Fringe Reduction Facility

FRFPS Fellow of the Royal Faculty of Physicians and Surgeons

Frf(s) French franc(s)

FRFS Fast Reaction Fighting System

Frg Forge (postal abbreviation)

FrG Federal Republic of Germany (West Germany)

Fr G Frans Guyana (Dutch—French Guiana)

FRG Facility Review Group; Federal Republic of Germany (West Germany)

FRGS Fellow of the Royal Geographical Society

frgt freight

FRHB Federation of Registered House Builders

frhgt free height

FR Hist S Fellow of the Royal Historical Society

FR Hort S Fellow of the Royal Horticultural Society

Fr hr French horn

Frhr Freiherr (German—Baron)

Fr hrn French horn

FRHS Fellow of the Royal Horticultural Society

fri feeling rough inside

Fri Friday; Friesland (Dutch province)

FRI Fellow of the Royal Institution; Fels Research Institute; Forest Research Institute; Friends of Rhodesian Independence

FRIA Fellow of the Royal Irish Academy

FRIAI Fellow of the Royal Institution of Architects of Ireland

FRIAS Fellow of the Royal Incorporation or Architects of Scotland

Frib Fribourg (Switzerland)

FRIBA Fellow of the Royal Institute of British Architects

fric frication; fricative; fricatruce; fricatrix; friction; frictional

FRIC Fellow of the Royal Institute of Chemistry

Frick Frick Collection (New York City)

FRICS Fellow of the Royal Institution of Chartered Surveyors

frict friction

fridg frigidaire (refrigerator)

fridge(s) refrigerator(s)

Friedrh Friedrichshafen

Fried Test Friedman Test (for pregnancy)

Friends Society of Friends

(Quakers)

Friends Meet Friends Meeting

fries friesisch (German—Frisian)

Fries Friesland (Dutch province); Friesic

Friesn Friesian (cattle, language, or people)

frig refrigerator

frig. frigidus (Latin—cold)

FRIGS Fellow of the Royal Imperial Geographical Society

FRIIA Fellow of the Royal Institution of International Affairs

Frim Frimaire (French—Sleety Month)—beginning November 21st—third month of the French Revolutionary Calendar

FRINA Fellow of the Royal Institution of Naval Architects

fringe. file-and-report information-processing generator

Fringlish French + English (English interlarded with French expressions and words)

fring(s) french onion ring(s)

f'r instance for instance

FRIPA Fellow of the Royal Institution of Public Administration

FRIPHH Fellow of the Royal Institute of Public Health and Hygiene

Fris Friesland; Frisia; Frisian

frisco fast-reaction integrated submarine control

Frisco (navalese—San Francisco)—no San Franciscan will use this nickname

FRISCO St Louis-San Francisco Railway

Frisco Bay (sailor's slang—San Francisco Bay)

Frisia (Latin—Friesland)—in the Netherlands

Frisians Frisian islanders or the Frisian Islands in the North Sea

Fritalux France, Italy, and Benelux nations

frits fritters

Fritz Friedrich

frjm full-range joint movement

frk fröken (Swedish—Miss)

Frk Fork (postal abbreviation); Frankfort

Frk Froken (Dano-Norwegian—Miss)

Frks Forks (postal abbreviation)

frl fractional; fuselage reference line

Frl El Ferrol

Frl Fräulein (German—Miss)

FRL Fuel Research Laboratory

FRLL Farrell Lines (container unit)

frm fiberglass-reinforced metal; fireroom; frame; framing; frequency meter

frm (FRM) former

FRM Federal Reformatory for Men

FRMA Floor Rug Manufacturers Association

FRMCM Fellow of the Royal Manchester College of Music

FR Met Soc Fellow of the Royal Meteorological Society

FRMIT Fellow of the Royal Melbourne Institute of Technology

frmn formation

frmr former

Frms Farms (postal abbreviation)

FRMS Federation of Rocky Mountain States; Fellow of the Royal Microscopical Society

FRN Federal Republic of Nigeria; Federal Reserve Note

frna foreign rations not available

FRNHS Fort Raleigh National Historic Site

FRNM Foundation for Research on the Nature of Man

FRNS Fellow of the Royal Numismatic Society

FRNSA Fellow of the Royal Navy School of Architects

Frnz Fernandez

FRNZIH Fellow of the Royal New Zealand Institute of Horticulture

'fro Afro

FRO Fellow of the Register of Osteopaths; Fire Research Organization; Friends Religious Order

FROC Federated Russian Orthodox Clubs

frof fire risk on freight

frog. free rocket over ground

FROGIE Fellowship to Resist Organized Groups Involved in Exploitation

from full range of movement

from. full range of movement

fron frontal; frontalis

FRONAPE Frota Naccional de Petroleiros (Portuguese—National Petroleum Fleet—Brazil)

front. frontispiece

FRONT BC Frontera (Fronteriza) Baja California (Spanish—Baja California Frontier)

Frontera Girls California Institution for Women at Frontera

frosh freshman; freshmen

frp fiberglass reinforced plastic; forward refueling area

frp (FRP) follicular regulatory protein (hormone)

FRP Fuel Reprocessing Plant; Fundamental Research Press

frpf fireproof

frpng fireproofing

FRPS Fellow of the Royal Photographic Society

FRPSL Fellow of the Royal Philatelic Society of London

frq frequent(ly)

FRR Facilities and Rearrangement Request

FRRA Facilities and Rearrangement Request and Authorization

frs flight reference selector; francs

frs (FRS) first readiness state

Frs Fresno; Frisian

Fr S French Somaliland (French Territory of the Afars and the Issas)

FRS Federal Reserve System; Fellow of the Royal Society; Financial Relations Society; First-Rank Symptoms (Schneiderian); Fisheries Research Society; Foundation Research Service; Freethought Reprint Series; Frequency Response Survey; Fuel Research Station

FRSA Fellow of the Royal Society of Arts

FRSAI Fellow of the Royal Society of Antiquaries of Ireland

frsc full range source code

FRSC Fellow of the Royal Society of Canada

FRSCM Fellow of the Royal School of Church Music

FRSE Fellow of the Royal Society of Edinburgh

FRSGS Fellow of the Royal Scottish Geographical Society

FRSH Fellow of the Royal Society of Health

FRSI Fellow of the Royal Sanitary Institute

FRSL Fellow of the Royal Society of Literature; Fellow of the Royal Society—London

FRSM Fellow of the Royal

Society of Medicine
FRSNA Fellow of the Royal School of Naval Architecture
FRSNZ Fellow of the Royal Society of New Zealand
Fr Som French Somaliland
FRSPS Fellow of the Royal Society of Physicians and Surgeons
FRSS Fellow of the Royal Statistical Society
FRSSA Fellow of the Royal Scottish Society of Arts
FRS(SA) Fellow of the Royal Society of South Africa
FRSSI Fellow of the Royal Statistical Society of Ireland
FRSSS Fellow of the Royal Statistical Society of Scotland
Frst Forest (postal abbreviation)
FRSTAT Fringe Software System
FRSTM & H Fellow of the Royal Society of Tropical Medicine and Hygiene
frt free return trajectory; freight; fruit
frt før vor tidregning (Dano-Norwegian—before time was reckoned)
FRT Family Relations Test
FRTC Fast-Reactor Training Center
frt/fwd freight forward
frtiso floating-point root isolation
frto flight radio telephone operator
Fr To French Togoland
FRTO Federated Road Transport Organization(s)
frt ppd freight prepaid
frtr freighter
fru fructose; fruit sugar
FRU Federal Reserve Unit; Fiji Rugby Union
fruat. frustrillatum (Latin—in small bits)
fruc. fructus (Latin—fruit)—sometimes abbreviated *fr.*
Fruc Fructidor (French—Fruitful Month)—August 18 through September 16—twelfth month of the French Revolutionary Calendar whose remaining five days were called Sansculottides and named respectively for the Virtues, Genius, Labor, Reason, and Rewards
fruct fructification, fructify, fructose, fructuous
frug frugal(ity), frugally
frugal. fortran rules used as a

general applications language
frumpie formerly-radical upwardly-mobile professional in elections
FRUS Foreign Relations of the United States
frust. frustillatim (Latin—in small portions)
fru veg fruits and/or vegetables
frv (FRV) flight-readiness vehicle
FRVIA Fellow of the Royal Victorian Institute of Architects
FRW Federal Reformatory for Women (Alderson, West Virginia)
frwd foreword, forward
FRWI Framingham Relative Weight Index
frwis frost warnings issued
frwk framework
Frwy Freeway
frx firex
Fry Ferry (postal abbreviation); Freeway (highway abbreviation)
FRYC Fall River Yacht Club
fr yr gdnce for your guidance
frz französisch (German—French)
FRZS Fellow of the Royal Zoological Society
FRZS (NSW) Fellow of the Royal Zoological Society of New South Wales
FRZS(Scot) Fellow of the Royal Zoological Society of Scotland
fs facsimile; factor of safety; far side; film strip; fin stabilized; fire station; flight service; flying status; foot second; foreign service; foresight; freight supply; front scalloped; front spar; sulfur trioxide chlorsulfonic acid (commercial short form or symbol)
fs (FS) file separator character (data processing)
f/s feet per second; first-stage
f⁵ francos (Spanish—francs)
fs faites suivre (French—please forward)
Fs fractostratus
FS Faraday Society; Feasibility Study; Federal Specification(s); Field Security; Field Service; Fighter Squadron; Financial Statement; Fire Station; Flight Sergeant; Fog Signal (Station); Foreign Service; Forest Service; Franciscan Studies; Franz Schubert; Free State; Freedom

School; freight supply (vessel); Friendly Society; Friends Society; Friendship Store(s); small freighter (naval symbol)
F-S Fenno-Shipping
F.S. Father of Sion
F/S Financial Statement
FS Filharmonisk Selskap (Norwegian—Philharmonic Orchestra); *Forente Staterna* (Swedish—United States)
fsa family separation allowance; fuel storage area
fsa (FSA) fetal sulfoglycoprotein
f.s.a. fiat secundum artem (Latin—let it be done skillfully)
FSA Farm Security Administration; Federal Security Administration; Federal Security Agency; Federal Supply Classification; Federation of South Arabia; Fellow of the Society of Antiquaries; Fellow of the Society of Arts; Field Survey Association; Finance Service—Army; Fire Support Area; Flax Spinners Association; Florida Student Association; Fraternal Scholastic Association; Free Selectors Association; Free Society Association; Freethinkers Society of America; Friendly Societies Act; Future Scientists of America
F & SA Farmers and Settlers Association (Australian)
FSAA Family Service Agency of America; Family Service Association of America; Flight Stewards Association of Australia
FSAC Freight Station Accounting Code
FSAG Fellow of the Society of Australian Genealogists
fsaga first sortie after ground alert
FSAICU Federation of State Associations of Independent Colleges and Universities
FSAL Fellow of the Society of Antiquaries of London
FSALA Fellow of the South African Library Association
f.s.a.r. fiat secundum artem regulas (Latin—let it be prepared according to the rules of the art)
FSAR Final Safety Analysis Report
FSAS Fellow of the Society of Antiquaries of Scotland
FSAScot Fellow of the Society

of Arts of Scotland

FSASM Fellow of the South African School of Mines

fsb forward space block

FSB Federal Specifications Board; Field Selection Board; Final Staging Base; Floating Supply Base

FSBA Florida School Boards Association

FSBC *Ferrocarril Sonora— Baja California* (Spanish— Sonora—Baja California Railway)

fsbl feasible

fsbly feasibility

fsbo for sale by owner

fsc foreign service credit

fsc (FSC) fast strike craft

FSC Family Services Bureau; Federal Safety Council; Federal Salary Commission; Federal Stock Catalog; Federal Stock Code; Federal Supplemental Compensation; Federal Supply Classification; Federal Supply Code; Federal Supreme Court; Fiji Sugar Corporation; Five Star Corporation; Flight Service Center; Flying Status Code; Food Standards Committee; Foreign Service Credits; Foundation for Student Communication; Foundation for the Study of Cycles; Friends Service Council; Frostburg State College

FSC *Federal Supply Catalog*

FSCA Fellow of the Society of Company and Commercial Accountants

f/scap foolscap

FSCC Federal Surplus Commodities Corporation; Fire Support Coordination Center; Food Surplus Commodities Corporation

fsce fire-support coordination element

FS Cen Flight Service Center

fscl fire-support coordination line

FSCM Federal Supply Code for Manufacturers

fscp foolscap

FSCS Fire Support Coordination Section; Flight Service Communications System

fsd flying spot digitizer; foreign sea duty; full-scale deflection; full-scale development; functional sequence diagram

fsd (FSD) focus skin distance

FSD Federal Systems Division;

Flight Service Director; Fuel Supply Depot; Sioux Falls, South Dakota (airport)

FSDC Fellow of the Society of Dyers and Colourists

fse field-support equipment; forward support element

FSE Federation of Stock Exchanges; Fellow of the Society of Engineers; Field Service Engineer

FSEA Food Service Executives Association

FSEC Florida Solar Energy Center

FSER Field Service Engineering Report

FSERI Federal Solar Energy Research Institute

FSES Federal-State Employment Service

fsf forward space file

FSF Fleet Servicing Facility; Flight Safety Foundation; Forensic Sciences Foundation

FSFA Federation of Specialized Film Associations

fsg first-stage graphitization

FSG Federal Supply Group; Fellow of the Society of Genealogists; Friends School Group

FS&G Farrar, Straus & Giroux

FSGB Foreign Service Grievance Board

FSgt Flight Sergeant

FSGT Fellow of the Society of Glass Technology

fsh (FSH) follicle-stimulating hormone

FSHM Fellow of the Society of Housing Managers

fshrf (FSHRF) follicle-stimulating hormone releasing factor

fshrh (FSHRH) follicle-stimulating hormone releasing hormone

FSHS Friendly Societies Health Services

fsh stk fish steak

FSI Federal Stock Item; Fellow of the Sanitary Institute; Fellow of the Surveyors' Institution; Foreign Service Institute; Foundation Sciences Inc; Free Sons of Israel

FSIA Fellow of the Society of Industrial Artists

FSIC Federal Savings Insurance Corporation; Foreign Service Inspection Corps (US Department of State)

FSIO Foreign Service Information Office(r)

FSIS Food Safety and Inspec-

tion Service (USDA)

FSJC Fort Smith Junior College

fsk frequency shift keying

FSK Fatigue Scales Kit

fsklf frequency shift keying low frequency

fskof for the sake of

fsl fire services levy; formal semantic language; frequency-selective limiter

FSL First Sea Lord; Folger Shakespeare Library; Food Science Laboratory (USA)

FSLA Federal Savings and Loan Association

FSLAC Federal Savings and Loan Advisory Council

FSLAs Federal Savings and Loan Associations

FSLIC Federal Savings and Loan Insurance Corporation

FSLN *Frente Sandinista de Liberación Nacional* (Spanish—Sandinista National Liberation Front)—Nicaragua

fslracc for seller's account

fsm flying-spot microscope

FSM Federated States of Micronesia; *Fédération Syndicale Mondiale* (French World Federation of Trade Unions); Fiji School of Medicine; Fort Smith, Arkansas (airport); Free Speech Movement

FSMA Friendly Societies Medical Association; Full Service Maintenance Agreement

FSMB Federation of State Medical Boards

FSMC Flora Stone Mather College

FS Method Federal Standard Method

fsmtc full-size moving target carrier

FSMWO Field Service Modification Work Order

FSN Federal Stock Number

FSNA Fellow of the Society of Naval Architects

FSNC Federal Steam Navigation Company

FSNM Fort Sumter National Monument

FSNP Fuyot Spring National Park (Philippines)

FSNWR Fish Springs National Wildlife Refuge (Utah)

FSNY Free Synagogue of New York

fso field service operation(s)

FSO Field Security Office(r); Fleet Signals Officer; Flint Symphony Orchestra; Florida

Symphony Orchestra; Flying Safety Officer; Foreign Safety Officer; Foreign Service Office(r); Fuel Supply Office(r)

FSOs Foreign Service Officers

FSOTS Foreign Service Officers Training School

fsp fiber saturation point; flat salary payroll; foreign service pay

FSP Family Survival Project; Field Security Police; Food Stamp Program; Freedom Socialist Party

FSPB Field Service Pocket Book; Forward Support Patrol Base

FSPT Federation of Societies for Paint Technology

fs&q functions, standards, and qualifications

F Sq Flying Squadron

FSQS Food Safety and Quality Service

fsr flight safety research; free of strikes and riots; full-scale repository

FSR Fellow of the Society of Radiographers; Field Service Report; Field Service Representative; Foreign Service Reserve

FSRA Federal Sewage Research Association

FSRJ *Federativna Socijalisticka Republika Jugoslavija* (Republic of Yugoslavia)

FSRS Frequency Selective Receiver System

fss finite solution set (mathematics)

FSS Federal Supply Schedule; Federal Supply Service; Fellow of the Statistical Society; Field Support System; Fire Support System; Flight Service Station; Flight Standard Service; Forward Scatter System; Friday(s), Saturday(s), Sunday(s)

FSSC Federal Standard Stock Catalog; Foreign Student Service Council

FSSCT Forer Structured Sentence Completion Test

fssd foreign service selection date

fssp fuel system supply point

FSSP Friendly Sons of Saint Patrick

FSSS Fuel Set Subsystem

fsst flying spot-scanner tube

FSSU Federated Superannuation Scheme of Universities

fsswt full-scale subsonic wind tunnel

fst forged steel; full-scale tunnel

Fst *Funkstation* (German—radio station)

FST Follow-on Soviet Tank

FSTA Food Science and Technology Abstracts

fstacoe fleet special test and checkout equipment

FSTC Farmington State Teachers College; Fayetteville State Teachers College

FS & TC Foreign Science and Technology Center (US Army)

FSTD Fellow of the Society of Typographic Designers

F'sted Frederiksted, St Croix

FSTL Future Strategic Target List

FSTMB *Federación Sindical de Trabajadores Mineros de Bolivia* (Spanish—Syndicalist Federation of Working Miners of Bolivia)

f-stop camera diaphragm setting of an f-number stop

FSTPP Foreign Service Team Preceptorship Program

fsts fuze set test set

FSTS Federal Secure Telephone Service

FSTWP Fellow of the Society of Technical Writers and Publishers

fsty firstly

fsu freak student union

FSU Family Service Unit; Florida State University; Friends of the Soviet Union

FSuH Fridays, Sundays, Holidays

F Supp *Federal Supplement*

fsv final-stage vehicle

fsv *for så vidt* (Dano-Norwegian—as far as)

FSVA Fellow of the Society of Valuers and Auctioneers

fsw final status word(ing); flexible steel wire (cable)

FSWA Federation of Sewage Works Associations

F & SWMA Fine and Specialty Wire Manufacturers Association

fswr flexible steel wire rope

fswt free-surface water tunnel

FSX Fighter-Support Experimental (advanced F-16 fighter airplane)

ft feet; firing table; fixed tannin; flat; flat top;flush threshold; foot; formal training; free of tax(ation); free trade; frequent traveller; full terms;

fumetight

f-t follow through

f/t freight ton

f & t fire and theft

ft. fiat (Latin—let it be made)

Ft Fort; forint (Hungarian currency unit)

Ft Folyoirat (Hungarian—journal, review)

FT Field Test; Flying Test; Flying Tiger Lines (2-letter coding); Functional Test(ing)

FT Financial Times (London); *Freethought Today*

ft² square feet; square foot

ft³ cubic feet; cubic foot

ft³/min cubic feet per minute

ft³/s cubic feet per second

fta failure to appear (in court); fatigue test(ing) article; film training aid; fluorescent treponemal antibody; full-throttle altitude

ftA fuck the Army (to hell with the rules)

FTA Finnish Travel Association; Free Trade Agreement (Canada–U.S.); Free Trade Area; Free Trade Association; Free Transport Association; Future Teachers of America

fta-abs fluorescent treponemal antibody absorption (test for syphilis)

FTAC Foreign Trade Arbitration Commission

FTAF Flying Training Air Force

FTAT Fluorescent Treponemal Antibody Test

ftb fails to break

FTB fleet torpedo bomber; Forestry and Timber Bureau; Franchise Tax Board; Fukui Television Broadcasting (Japan)

ftbd fit to be detained; full-term born dead

ft black fine thermal black

ftbm foot board measure

ftbrg footbridge

FTBS Free Throwers Boomerang Society

ftc fast time constant; final turn collision

ft c foot-candle

FTC Fair Trade Commission; Farmers Trading Company; Federal Telecommunications Laboratories; Federal Trade Commission; Fleet Training Center; Flight Test Center; Flying Training Command

FTCA Federal Tort Claims Act

ft. cata. *fiat cataplasma* (Latin—make a poultice)

FTCC Flight Test Coordinating Committee; French Telegraph Cable Company

ft cd foot candela

FTCD Fellow of Trinity College—Dublin

ft. cerat. *fiat ceratum* (Latin—make a wax)

ft. chart. *fiat chartulae* (Latin—let powders be made)

FTCL Fellow of Trinity College of Music—London

ft col fast color

ft. colly. *fiat collyrium* (Latin—make an eyewash)

ftcolovprt fast color overprint

ftd fails to drain; flight test(ing) direction

FTD Field Training Detachment; Florists' Telegraph Delivery; Foreign Technology Division; Fuel Testing Department

FTDA Fellow of the Theatrical Designers and Craftsmens Association

FTDC Fellow of the Society of Typographic Designers of Canada

ft di flattening die

ftdr friction-top drum

fte fracture transition elastic; full-time equivalence; full-time equivalent

FtE Free the Eagle

ftee full-time equivalency enrollment

ft. emuls. *fiat emulsio* (Latin—make an emulsion)

ft. enem. *fiat enema* (Latin—make an enema)

FTESA Foundry Trades Equipment and Supplies Association

F test Fisher Test (forestry)

ftet full-time equivalent terminals

ftf face to face

FTF *Flygtekniska Forsoksantalten* (Swedish—Aeronautical Research Institute of Sweden)

ftfet four-terminal field-effect transistor

f-t fibers fast-twitch muscle-cell fibers

ftg fitting; footing

FTG Fleet Training Group (USN); Fuji Texaco Gas

ft. garg. *fiat gargarisma* (Latin—make a gargle)

FTGSVC Fleet Training Group Services

ft hd flathead

fth(m) fathom

fthp flowing tubing head pressure (oil well)

ft/hr feet per hour

fti federal tax included; fixed time indicator; fixed time interval; frequency time indicator; frequency time intensity

FTI Facing Tile Institute; Federal Tax Included; Fellow of the Textile Institute; Functional Test(ing) Instruction(s)

FTIG Fort Indiantown Gap (USA)

FTII Fellow of the Taxation Institute Incorporated

FTIMA Federal Tobacco Inspectors Mutual Association

FT Index Financial Times Index

ft. infus. *fiat infusum* (Latin—make an infusion)

ft. injec. *fiat infectio* (Latin—make an injection)

ftir functional terminal innervation ratio

Ft-ir Fourier transform-infrared spectroscopy

ftit fan turbine inlet temperature

FTIT Fellow of the Institute of Taxation

ftk forward track kill

ftka failed to keep appointment

ftl faster than light

ft l foot -lambert

FTL Federal Telecommunications Laboratory; Flight Test Letter; Flying Tiger Line

ft lb foot pound

ft-lbf foot-pound force

Ftle Fremantle

ft. linim. *fiat linimentum* (Latin—make a liniment)

ftm flat tension mask; fractional test meal; functional testing machine(ry)

FTM Flight Test(ing) Manual; Flight Test(ing) Model; Flying Training Manual

FTM *Federación de Trabajadores de México* (Spanish—Federation of Mexican Workers)

FTMA Federation of Textile Manufacturers Associations

ft. mas. *fiat massa* (Latin—make a mass)

ft. mas. div. in pil. *fiat massa dividenda in pilulas* (Latin—make a mass and divide into pills)

ft md flattening mandrel

ft/min feet (foot) per minute

ft. mist. *fiat mistura* (Latin—make a mixture)

ftn fortification

Ftn Fountain (postal abbreviation); Freetown (maritime abbreviation)

FTN Facsimile Transmission Network

ftnd full-term normal delivery

*f*ᵗᵒ *firmato* (Italian—signed)

FTO Field Test(ing) Operations; Field Training Officer (police); Fleet Torpedo Officer; Fleet Training Officer (naval); Franciscan Third Order

FTOs Field Training Officers

ftp field terminal platform (oil well); final-turn pursuit (aircraft); folded, trimmed, and packed (books); full-time personnel (civil service)

FTP Field Test Plan; Fleet Training Publication; Flight Test Program; Functional Test(ing) Procedure

FTP *Francs Tireurs Partisans* (French—Partisan Sharpshooters)—communists in the anti-Nazi underground of France during World War II

FTPAA Film and Television Production Association of Australia

ft-pdl foot poundal

ft/pf foot-pound force

ft. pil. *fiat pilulae* (Latin—make pills)

ft/pnl fuel-tanking panel

ftpo for testing purposes only

FTPR *Federación del Trabajo de Puerto Rico* (Spanish—Federation of Labor of Puerto Rico)

FTPS Fellow of the Technical Publishing Society

ft. pulv. *fiat pulvis* (Latin—make a powder)

ftr fighter; fixed-transom; flat-tile roof; fusion test reactor

F Tr flag tower

FTR Final Technical Report; flag tower (chart and map designation); Flight Test Report; Fruehauf (stock exchange symbol); Functional Test Report; Functional Test Request

ftrac full-tracked (vehicle)

FTRF Freedom-to-Read Foundation

ftro fighter operations

FTRO Flight Test Release Order

ftrp fighter plans

fts favorite-track selection

ft/s feet (foot) per second

FTS Federal Telecommunications System; Federal Telephone System; Field Test Support; Flying Traffic Specialist; Flying Training School; Forged Tool Society; Funeral Telegraph Service; Furnishing Trades Society
ft/s² foot per second squared
ft sec foot second
ft. so. fiat solutio (Latin—make a solution)
ft. suppos. fiat suppositorium (Latin—make a suppository)
ftt field test telescope; formation tester tool (oil well); framed timber trestle; full-time temporary (civil-service employee); functional test tool
FTT Fever Therapy Technician; Five Task Test
fttp full-time temporary personnel
fttr fitter
ft & tw combination flat top and typewriter (desk)
ftu field transfer unit; fuel tanking unit
Ftu Freeman time unit
FTU Federation of Trade Unions; Field Torpedo Unit; First Training Unit
FTUC Fiji Trades Union Congress
ft. ung. fiat unguentum (Latin—make an ointment)
FTUR Flight Test Unsatisfactory Report
FTV Flight Test Vehicle; Fukushima Television; Functional Test Verification
ftw free-trade wharf
Ft W Fort Worth
Fty Factory
F-type Jungian feeling type
FTZ Foreign Trade Zone; Free Trade Zone
FTZB Foreign Trade Zones Board
fu Farmers Union; feed(ing) unit; frame unprotected (insurance classification)
f/u fine used (postage stamps)
Fu Finsen unit
F-u fuck you (underground slang—very insulting epithet)
FU Fairfield University; Farmers Union; Fisk University; Fordham University; Franklin University; *Freie Universität* (Berlin Free University); Friends University; Furman University
FUA Farm Underwriters Association

FUB Freie Universität Berlin (German—Free University, Berlin)
fubar fouled up beyond all recognition
fubb fouled up beyond belief
fuc full usable capacity
FUC Ferrocarriles Unidos de Yucatan (Spanish—United Railroads of Yucatan)
FUCA Federal Unemployment Compensation Act
fuchsite chrome mica
fucm (FUCM) full-utility cruise missile
FUDR Failure and Usage Data Report
FUE Federated Union of Employers
FUEL Fuel-Users Emergency Line
FUEN Federal Union of European Nationalities
Fuente (Spanish—fountain, source, spring)—short form for such places as Fuente-Alamo, Fuente de Cantos, Fuente-Palmera, Fuente Vaqueros, etc.
fuetap (concrete) formed under elevated temperature and pressure
fufo fly under, fly out
FUG-1966 Hungarian-built armored vehicle based on Soviet model
Führer (German—leader)—Hitler's title
FUIB Fire Underwriters Inspection Bureau
Fuji Alberto Fujimori, president of Peru; Fujinoyama, Fujisan, or Mount Fuji (Japan's highest peak)
Fujita Leonardo Fujita
Fujiyama Europeanized form for Fujisan or Mount Fuji—Japan's highest peak—3775 meters or 12,388 feet above sea level
Ful Fulcran; Fulgence; Fulgencio; Fulke; Fuller; Fullerton; Fulton; Fulvia; Fulvius
FULICO Fidelity Union Life Insurance Company
fulnm full name
fum fuming
FUM Friends United Meeting
fumi fumigant; fumigate; fumigation
fumtu fouled up more than usual
fun. funeral; funerary
FUNAI Fundaçáo Nacional do Indio (Portuguese—National Foundation of the Indian)

funamb funambulation; funambulist (tightrope or tightwire walker)
func function(al)
funct function; functional; functionally
fund. fundamental; fundamentalism; fundamentalist
fund. fundador (Spanish—founder)
FUND International Monetary Fund
fundies fundamentalists
Fundy Bay of Fundy; Fundy National Park on the north shore of the Bay of Fundy in New Brunswick, Canada
fungi. fungicide
Funk Funk & Wagnalls
FUNK Front Uni National du Kampuchea (French—Khmer National United Front)—Cambodia and Khmer forces
Funk&W Funk & Wagnalls
FUNM Fort Union National Monument
FUNNs For Your Nieces and Nephews
F-U-N trio flourine, uranium, nitrogen tests for relative dating
FUNU Force d'Urgence de Nations Unies (French—United Nations Emergency Force)
fuo fever (of) unknown origin
fup fusion point
f/up follow up
FUP Friends United Press; Furman University Press
FUP Frente Unido del Pueblo (Spanish—People's United Front)—Colombia
fuposat follow-up on supply action taken
fur. furlong; further
fur furiant (Czech—Bohemian folk dance)
FUR Follow-up Report
FURG Fundacão Universidade de Rio Grande (Portuguese—Rio Grande University Foundation)
furl. furlough
Für Liech Fürstentum Liechtenstein (German—Principality of Liechtenstein)
furlong furrow long (one eighth mile or 220 yards—201.17 meters), originally the average length of a plowman's furrow
furmr furthermore
furn furnace; furnish(es, ed, ing, ings); furniture
Furn Furnace (postal abbrevia-

tion)
furngs furnishings
furnit furniture
furn pts furniture parts
Fur Seals Fur Seal Islands (Alaska's Pribilofs)
Furtwängler Gustav Heinrich Ernst Martin Wilhelm von Furtwängler
fus far ultraviolet spectrometer; firing unit simulator; fuselage; fusing
FUS Feline Urological Syndrome
FUSA Flinders University of South Australia
FUSE Far-Ultraviolet Spectroscopic Explorer; Federation for United Science Education
FuSf Fortsetzung und Schluss folgen (German—to be concluded in the next issue)
fut future
Fut Futura
FUTA Federal Unemployment Tax Act
FUTC Fidelity Union Trust Company
futs firing unit test set
fuv far ultraviolet
FUW Farmers' Union of Wales; Federation of University Women
fv fire vent; flats vacant; flush valve; forward visibility; fuel valve; future value
f.v. *folio verso* (Latin—back of the sheet)—lefthand page
fv. folio verso (Latin—back of the sheet)
FV Falck's Flyvetjeneste (Copenhagen); fishing vessel; Fruit and Vegetable (US Department of Agriculture)
FV-432 British armored personnel carrier called Trojan
FV-1609 advanced model of the FV-432
FVA Fellow of the Valuers Association
FVB Fiji Visitors Bureau
fvc forced vital capacity
FVCQFRA Fruit and Vegetable Canning and Quick Freezing Research Association
FVDE Fighting Vehicles Design Establishment
f vd & w firearms, venereal disease, and whiskey
FVI Fellow of the Valuers' Institution
FVMMA Floor and Vaccum Machinery Manufacturers Association
FVNM Fort Vancouver National Monument

FVPA Flat Veneer Products Association
FVPRA Fruit and Vegetable Preservation Research Association
fvq full variable quality
FVR Feline Viral Rhinotracheitis
fvrbl favorable
FVRDE Fighting Vehicles Research and Development Establishment
fv's fashion victims
f. vs. fiat venaesectio (Latin—perform a venesection)
FVS Forer Vocational Survey
FVSC Fort Valley State College
fvt family vewing time
fw fire wall; fixed wing; flash welding; formula weight; fresh water; fresh weight; front wiring
f & w feed and water; feeding and watering
fw Funk & Wagnalls
f/W female White
FW Fairbanks Whitney (stock exchange symbol); Focke-Wulf; Fog Whistle; Fort Worth; Foster Wheeler
F & W Funk and Wagnalls
fwa financial working arrangement; first word address; fluorescent whitening agent
FWA Family Welfare Association; Farm Workers Association; Federal Works Agency; Forest Workers Association; French West Africa; Future Weapons Agency
FWAA Football Writers Association of America
FWAS Fort Wayne Art School
FWAT Fish and Wildlife Advisory Team (Alaskan)
fwb four-wheel brake; four-wheel braking; free-wheel bicycle; front-wheel bicycle; furnished with bed
FWB Fort Worth Belt (railroad); Free-Will Baptists
fw ball. freshwater ballast
fwc full weight contents
FWC Farmers Wholesale Cooperative; Federal Warning Center; Foster Wheeler Corporation
FW & C Furness, Withy & Company
FWCC Friends' World Committee for Consultation
fwd forward(ing); four-wheel drive; freshwater damage; freshwater draught; front-

wheel drive
F W & D Fort Worth & Denver (railroad)
FwdBL forward bomb line
fwdct fresh water drain collecting tank
fwdg forwarding
fwdr forwarder
fwe finished with engines
f-w-e finished with engine(s)
FWeldI Fellow of the Welding Institute
FWFM Federation of Wholesale Fish Merchants
fwg following
FWGE Fort Worth Grain Exchange
fwh flexible working hours; free-wheeling hubs
FWHC Feminist Women's Health Center
FWHF Federation of World Health Foundations
FWI Federation of West Indies; French West Indies
FWID Federation of Wholesale and Industrial Distributors
fwl foilborne waterline
FWL Foreign Workers Levy; Foundation for World Literacy
FWLERD Far West Laboratory for Educational Research and Development
FWO Facilities Work Order; Fleet Wireless Officer
FWOA Fort Worth Opera Association
FWONA Free World Outside North America
fwop furloughed without pay
fwp filament-wound plastic(s)
FWP Federal Writers' Project
FWPCA Federal Water Pollution Control Administration
FWPO Federal Wildlife Permit Office; Fort Wayne Philharmonic Orchestra
FWQA Federal Water Quality Administration; Federal Water Quality Association
fwr full-wave rectifier; full-wave reflector
F-W r Felix-Weil reaction
FWRC Federal Water Resources Council
FWRM Federation of Wire Rope Manufacturers
fws filter wedge spectrometer
FWS Fighter Weapons School; Fish and Wildlife Service
F & WS Fish and Wildlife Service
FWSG Farm Water Supply Grant

FWSO Fort Worth Symphony Orchestra
FWSSUSA Federation of Worker's Singing Societies of the U.S.A.
fwt fair wear and tear; featherweight
FWT Free World Trade
FWTBT Hemingway's *For Whom The Bell Tolls*
fwth flush watertight hatch
FWU Food Workers Union
FWWS Fire-Weather Warning Service
Fwy Freeway
fx extraneous (television) effects; fixed; foreign exchange; foxed; fractured; fractures; frozen section; special effects
fx for eksempel (Dano-Norwegian—for example)
Fx fracture (bone)
FX Foreign Exchange
F.X. Francis Xavier
fxd fixed; foxed
fxg fixing
fxle forecastle
fy (FY) fiscal year
Fy Ferry

FY fiscal year; Ferdinand(e) Ysabella
FY2 fuck you too (graffitic defacement)
fya first-year algebra
FYC Federal Youth Center; Florida Yacht Club
FYDP First-Year Development Program; Five-Year Defense Plan
fyg for your guidance
fyi for your information
fyig for your information and guidance
fym farmyard manure
fyou fuck you
FYP Five-Year Plan; Four-Year Plan; etc.
FYPB Five-Year Planning Base (USA)
fypi for your personal information
FYPP Five-Year Procurement Program (USA)
fyr for your reference
fys fysik or *fysisk* (Dano-Norwegian—physics, physical)
fytd fiscal year to date
FYTP Five-Year Test Program

Fyz Fyzabad
fz freeze; freezing; fuze (ordnance explosive device)
fz forzando (Italian—accented strongly)
Fz Fernández; Franz
FZ Franc Zone; Free Zone; French Zone
FZA Fellow of the Zoological Academy; Fellow of the Zoological Association
fzdz freezing drizzle
fzfg freezing fog
FZGB Federation of Zoological Gardens of Great Britain and Ireland
FZGBI Fellow of the Zoological Gardens of Great Britain and Ireland
FZIA First Zen Institute of America
fzr freezer
fzra freezing rain
FZS Fellow of the Zoological Society of Scotland
FZSL Fellow of the Zoological Society, London
FZSScot Fellow of the Zoological Society of Scotland

G

g gage; garage; gateway; gelding; gender; generator; gilbert; glucose; gold; good; grain; gram; gravitational acceleration (symbol); gravity; grease; great; green; grey; gross; ground; grunting; guide; gun; gyromagnetic ratio (symbol); Lande factor (symbol)

g (G) giga-(prefix meaning billion); glucose

g acceleration of gravity (symbol); gloom (gloomy weather symbol)

g/ *giro* (Spanish—bank check)

G conductance (symbol); control grid (symbol); Fraunhofer line caused by iron (symbol); gap; gauss; gear; general audiences (all ages admitted); German(ic); Germany; Gibbs function (free energy symbol); glider; go; God (on Masonic emblems); Golf—code for letter G; good; Goodyear; gourde (Haitian unit of currency); govern–ment (broadcasting); Grace (steamship line); Green Line; Greene Line; Greenwich; Greyhound (bus line); guanine; guineas; gulden (Netherlands guilder); gulf; Gulf Oil (stock exchange symbol); Newtonian gravitational constant (symbol); specific gravity (symbol); thousand-dollar bill

G-1 conductance (symbol)

G-7 Group of Seven (Britain, Canada, France, Germany, Italy, Japan, the United States)

G-15 Group of Fifteen (Algeria, Argentina, Brazil, Egypt, India, Indonesia, Jamaica, Malaysia, Mexico, Nigeria, Peru, Senegal, Venezuela, Yugoslavia, Zimbabwe)

G (G) government spending

G *Gade* (Danish—street); *Galica* (Latin—Gaul or Germania); *Gasse* (German—street); *Gata* (Swedish—street); *Gate* (Norwegian—street); *gawa* (Japanese—river, stream)—also shown as *kawa*; *Gebel* (Arabic—mountain); *Göl* (Turkish—lake); *Golfe* (French—gulf); *Golfo* (Italian or Spanish—gulf); *Gôlfo* (Portuguese—gulf); *Gora* (Russian—hill, mountain); *Góra* (Polish—hill, mountain); *Guba* (Russian—bay); *Gunung* (Indonesian or Malay—mountain)

G-1 Army or Marine Corps personnel section; personnel officer

G-2 military intelligence section of Army or Marine Corps; military intelligence officer

G-3 operations and training section of Army or Marine Corps; operations and training officer

G3P glyceraldehyde 3-phosphate

G-4 logistics officer or section of U.S. Army or Marine Corps; undercover anti-terrorist group within the Royal Canadian Mounted Police

G₄ dichlorophen (bactericide and fungicide)

G-5 civil affairs section of Army; civil affairs officer

G6P glucose 6–phosphate

G6PD glucose–6–phosphate dehydrogenase

G-6-pdd glucose–6–phosphate dehydrogenase deficiency

G₁₁ hexachlorophene (antibacterial agent)

G-91 Fiat-built single-engine jet fighter-bomber

ga gage; gas amplification; gastric analysis; gauge; general average; general aviation; glide angle; go ahead; greenhouse annual; ground to air; ground alert; ground attack

g/a general average; ground-to-air

g & a general and administrative

Ga gallium; Georgia; Georgian; Ghana (tribe)

Gᵃ García

GA Gage Man; Gamblers Anonymous; Garrison Adjutant; Garrison Artillery; Gemmological Association; General Accounting; General Agent; General Assembly (UN); General Assignment; Geographical Association; Geologists Association; Georgia; Georgia (railroad); Glen Alden (stock exchange symbol); Government Actuary; Government Agency; Grant Award; Graphic Arts; Gypsum Association

G-A General Atomic (Division of General Dynamics)

gaa ground-aided acquisition

GAA Gaelic Athletic Association; Gay Activists Alliance; Gemmological Association of Australia; General Aviation

Association
GAA Glossary of Aeronautical Abbreviations
GAAC Graphics Arts Advisers Council
gaafr governmental accounting, auditing, and financial reporting
GAAP Generally Accepted Accounting Principles
Ga As gallium arsenide
GAATV Gemini-Atlas-Agena Target Vehicle
gab gabardine; gabbing; gabble; gable; girth above buttress
Gab Gabon Republic (République Gabonaise); Gabriel
GAB Games and Amusement Board; General Adjustment Bureau; General Arrangements to Borrow
GABA gamma-aminobutyric acid
Gabba Wollongabba, Brisbane
Gabby Gabriel; Gabriella; Gabrielle
Gabe Gabriel
Gabl Gabriel
Gabon Gabonese Republic (West African country)
Gabr Gabriel; Gabriella; Gabrielle
gac granular-activated carbon; grilled American cheese (sandwich)
GAC General Acceptance Corporation; General Advisory Committee; General Apprenticeship Committee; General Atomic Company; General Average Certificate; Geological Association of Canada; Georgia Association of Colleges; Goodyear Aircraft Corporation; Gulfstream Aerospace Corporation; Gustavus Adolphus College
GACHAL Gush Herut Liberalim (Hebrew—Herut-Liberal Bloc)—right-wing party
g/a con general average contribution
GAD Gases Applications Development; Generalized Anxiety Disorder; Great American Desert
g/a dep general average deposit
Gadis Gaditanas (dancing girls of Cadiz who perfected abdominal dancing to a fine art representing fertility rites and child bearing)
GADNA Gdud Noar (Hebrew—Youth Corps)
GADO General Aviation District Office
gadpet graphic data presentation and edit(ing)
GADS Goose Air Defense Sector
Gae Gaelic
GAE General American English; Georgia Association of Educators
GAEC Goodyear Aircraft and Engineering Corporation; Grumman Aircraft Engineering Corporation
gael gaelisch (German—Gaelic)
Gael Gaelic
GAER Gay Alliance for Equal Rights
Gaet Gaetano
gaf General Aniline & Film Corporation (trademark)
GAF General Aniline & Film; Government Aircraft Factories
GAFB Goodfellow Air Force Base
GAFD Guild of American Funeral Directors
gaffer (motion-picture and tv slang—chief electrician)
gaffer and gammer grandfather and grandmother
GAFLAC General Accident Fire and Life Assurance Corporation
GAFTA Grain and Feed Trade Association
gag. gaging
g/a/g ground-air-ground
GAG Graphic Artists Guild
gag mm gage thickness (in millimeters)
gags glycosaminoglycans
GAHH Good American Helping Hands
gai guaranteed annual income
GAI Government Affairs Institute
gaia go and inspect aircraft
GAIA Graphic Arts Information Association
GAIF General Assembly of International Federations
Gail Abigail
GAIN Greater Avenues for Independence (from welfare)
GAIS Georgia Association of Independent Schools
GAIU Graphic Arts International Union
GAJ Guild of Agricultural Journalists
GAJC Georgia Association of Junior Colleges
GAK Garlock (stock-exchange symbol)

Gaki Akutagawa Ryunosuke
Gaku Univ Gakushuin University
gal galileo (unit of acceleration); gallon (unit of capacity)
Gal Epistle to the Galatians; Galacia(ns); Galatians; Galveston; Galway
Gal Galatians; Général (French—General
GAL Gdynia America Line; General Assembly Library (Wellington, NZ); Guggenheim Aeronautical Laboratory; Guinea Airways)
G A & L General Aircraft and Leasing (Division of General Dynamics Corporation)
GALA Gay Atheists League of America
Galap Galápagos Islands
GALAP Graphic Arts Literacy Action Program
Galápagos Galápagos Islands or Galápagos tortoise(s)
galaxy. general automatic luminosity and x y (measuring machine)
gal cap gallon capacity
GALCIT Guggenheim Aeronautical Laboratory, California Institute of Technology
Gal Col Gallaudet College
Gale Gale Research Company
galena lead sulfide
Galileo Galileo Galilei
gall. gallery
gall gallego (German—Galician)––Spanish provincial language and people
Gall Galleria (Italian—gallery or tunnel)
Gall C Gallaudet College
Gallo-Rom Gallo-Romance
Gall U Gallaudet University
Gallup Dr George Horace Gallup of Gallup Poll fame
GALOP Gay London Police Monitoring Group
gal per min gallons per minute
gals gallons
gals (GALS) generalized assembly-line simulator; geographic adjustment by least squares
gal(s) girl(s)
galt gut-associated lymphoid tissue
Galtees southern Ireland's Galty Mountains
galumphing galloping and triumphing
galv galvanic; galvanism; galvanize(d); galvanometer
Galv Galveston

galv i galvanized iron
galvnd galvannealed
galvo(s) galvanometer(s)
Galvy Galveston
Galw Galway
gam gammon (sailor's gossip, seamen's talkfest); gamut; guided-aircraft missile
gam (GAM) ground-to-air missile
gam *gamos* (Greek—marriage)—gamete(s)
Gam Gamaliel; Gambia
GAM Guest Aerovías Mexico; Guided-Aircraft Missile
GAMA Gas Appliance Manufacturers Association; General Aviation Manufacturers Association; Graphic Arts Marketing Associates
GAMAA Guitar and Accessories Manufacturers Association of America
gamba viola da gamba
Gambia Republic of The Gambia (West African coastal country)
Gambiers Gambier Islands in the South Pacific
Gamblers Anon Gamblers Anonymous
gamblin' gambling
Gamboa Penitentiary (close to the midsection of the Panama Canal)
GAMC General Agents and Managers Conference
Gamerco Gallup American Coal Company—coal-mining town near Gallup, New Mexico
GAMET Gyro Accelerometer Misalignment Erection Test
GAMIS Graphic Arts Marketing Information Service
Gamla Stan (Swedish—Old Town)—old Stockholm
gamm gimbal angle matching monitor
GAMM *Gesellschaft für Angewante Mathematik und Mechanik* (German—Association for Applied Mathematics and Mechanics)
GAMMA Guns and Magnetic Material Alarm (anti-hijacking device)
Gam Pra Bang *Gama Praj tantrï Bangladesh* (Bengali—People's Republic of Bangladesh)
GAMTA General Aviation Manufacturers and Traders Association
gan generating and analyzing networks

gan (GAN) gyrocompass automatic navigation
gan *ganado* (Spanish—cattle; livestock)
GAN Generalized Activity Network
Gandhi *Mahatma* (Hindustani—Great Souled)—Mohandas Karamchand Gandhi
Gandhi's Mahatma Gandhi's Birthday (October 2)
ganefo games of new emerging forces
gang. ganglia; ganglion
Ganges Ganges–Brahmaputra Delta, sacred waters of the Ganges, India
GANZ Gas Association of New Zealand
ganzl *gänzlich* (German—complete, entire)
gao general alert order
GAO General Accounting Office; General Administrative Order; General American Oil (company); General American Overseas (corporation); General Auditing Office; Government Accounting Office
GAO *Glavnaya Astronomicheskaya Observatoriya* (Russian—Main Astronomical Observatory)
gaof gummed all over flap
GAOR *General Assembly Official Records* (UN)
gap. gallium arsenic phosphorus; guidance autopilot
gap. (GAP) gross agricultural product
Gap The Gap—Delaware Water Gap between New Jersey and Pennsylvania on the upper Delaware River or Semangko Gap north of Kuala Lumpur in Malaysia; Pennington Gap
GAP General Assembly Program; Government Aircraft Plant; Great American Public; Great Atlantic & Pacific (Tea Company); Group for the Advancement of Psychiatry
GAP *Grupo de Amigos Personales* (Spanish—Group of Personal Friends)—Chilean President Allende's bodyguard; *Gruppo d'Azione Partigiana* (Italian—Partisan Action Group)
gapa ground-to-air pilotless aircraft
Ga-Pac Georgia-Pacific
GAPAN Guild of Air Pilots

and Air Navigators
GAPCE General Assembly of the Presbyterian Church of England
GAPEFA Graphic Arts Platemaking Employers Federation of Australia
GAPL Group Assembly Parts List
gapo giant armpit odor; gorilla armpit odor
gapp growth and profit planning)
GAPR Grant Application Request
GAPs Geographic Applications Programs
gapsfas graduate and professional students financial statement
gapt graphical automatically programmed tools
gaq general average quality; good average quality
gar garfish; garpike; gharial, gavial
gar. garage; garrison; ground avoidance radar; guided aircraft rocket
gar. (GAR) gross annual return; growth analysis and review; guided air(craft) rocket
GAR Gioacchino Antonio Rossini; Grand Army of the Republic; Guided Aircraft Rocket; Gustavus Adolphus Rex (King Gustav II of Sweden)
Gara Garamond
garade gathers, alarms, reports, displays, and evaluates
garb. garbage; green, amber, red, blue (airway priority color code)
GARB Garment and Allied Industries Requirements Board
gar-barge garbage barge
garbd garboard
garbol garbologic(al)(ly); garbologist(ic)(al)(ly); garbology (archeological study of man's discards)
garbologist garbage collector
garbz *garbanzos* (Spanish—chickpeas)
GARC Graphic Arts Research Center
G.Arch. Graduate in Architecture
Garcia Diego Garcia (Anglo-American naval base in the Chagos Archipelago or Oil Islands of the Indian Ocean)
gard gamma atomic radiation detector; garden; gardener; gardening; general address

reading device; guard
GARD Gamma Atomic Radiation Detector
gards gardenias
GA Res *General Assembly Resolution* (UN)
garg. *gargarisma* (Latin—gargle)
Garifuna Black Carib (spoken in Belize)
garioa government and relief in occupied areas
Garmo Garmo Peak (formerly Stalin Peak and the highest in the USSR)—also called Communism Peak
GARP Global Atmospheric Research Program
G.A.R.S. Gustavus Adolphus Rex Sueciae (Gustavus Adolphus King of Sweden)
gar str garboard strake
Gart Garrett
GARUDA Garuda Indonesia Airways
Gary Gareth; Garvey
gas gasoline
ga & s general average and salvage
g-a s general-adaptation syndrome
GAs Gamblers Anonymous
GAS Georgia Academy of Science; Ghana Academy of Sciences; Government of American Samoa; Grant's Acronymical Shorthand (*see* JERK); Great American Smokeout (campaign to get smokers to kick the habit); Guild of All Souls
GASAA Graphic Arts Services Association of Australia
gasahol nine parts gasoline and one part alcohol (fuel extender mixture)
GASBO Georgia Association of School Business Officials
GASC German-American Securities Corporation
GASCO General Aviation Safety Commission; General Aviation Safety Committee
gasdyn gas dynamic; gas dynamicist; gas dynamics
gaser gamma-ray laser
GAS/GR Gas Service
gasid gas-acid (indigestion)
GASL General Applied Science Laboratories
GA S&L Great American Savings & Loan
gaso gasoline
gasoff gasoline ripoff
gasohol Brazilian gasoline made from alcohol; gasoline

+ alcohol (fuel-extender fluid)
gasp. gravity-assisted space probe
Gasp Gaspar(o)
GASP Global Air Sampling Program; Greater (Washington, D.C.) Alliance to Stop Pollution (air and water); Group Against Smog and Pollution; Group Against Smokers' Pollution; Group Against Smoking Pollution
Gaspar Jasper
gasphyxiation gas + asphyxiation (death by gas)
GASS Gimbal Assembly Storage System
GASSAR General Atomic Standard Safety Analysis Report
gast gastric
Gast Gaston
gastro gastronomy
gastro (Latin prefix—relating to the stomach)—gastrointestinal
gastroc gastrocnemius
gastroenterol gastroenterology
gastrol gastrolithiasis; gastrolith(ic)(al); gastrolithograph(y); gastrologist; gastrology
gat gatling gun; ground air transmitter; gun; revolver
gat (GAT) generalized algebraic translator
g-at. gram-atom
gat *gata* (Swedish—Street); (Dano-Norwegian—channel)
GAT Georgetown Automatic Translation; Greenwich Apparent Time
GATA Graphic Arts Technical Association
gatac general assessment tridimensional analog computer
GATAP Generally Accepted Tax-Accounting Principles
GATB General Aptitude Test Battery
GATCO Guild of Air Traffic Control Officers
Gate The Gate (harbor entrance such as the Golden Gate at San Francisco, the Lion's Gate at Vancouver, the Silver Gate at San Diego)
GATE Gifted and Talented Education program; Group to Advance Total Energy (American Gas Association)
GATF Graphic Arts Technical Foundation
'gator(s) alligator(s)
gatri gamma technology re-

search irradiator
GATS Guidance Acceptance Test Set
GATT General Agreement on Tariffs and Trade
Gatti Guilio Gatti-Casazza
Gatun Girls Gatun Prison for Women and Juveniles at Gatun, Panama
g at. wt gram atomic weight
GATX General American Transportation Corporation (tank car marking)
GAU Gay Academic Union
GAUFCC General Assembly of Unitarian and Free Christian churches
Gaul. Gaulish
gav gavage(r); gavel; gavelkind; gavial; gavotte; gross annual value
g/av general average
gav(s) gavial(s); gavotte(s)
gaw guaranteed annual wage
gawam great American wife and mother
gawr gross axle weight rating
Gay Gaylord
gaylib gay liberation(ist)
Gay Lib Gay Liberation Movement
gayola homosexual payola (forced payments made by homosexual establishments to crime syndicates offering them protection)
gaz gazette; gazetteer
GAZ (Russian—*Gorki Avtomobilnii Zavod*)—Gorki Automobile Factory producing the Volga sedan-type auto
gb gall bladder; generator breaker; glide bomb; goodbye; grid bearing; gun bed
gb (GB) code name for sarin (high-toxicity warfare chemical)—$C_4H_{10}FO_2P$
g-b goof-ball (barbiturate pill)
g/b ground based
gB greenish blue
Gb gilbert
GB Gas Board; General Board; General Bronze (corporation); Georges Bizet; Girls' Brigade; Great Books; Great Britain; gunboat (naval symbol)
GB *Gran Bretaña* (Spanish—Great Britain)
gba give better address
GBAD Great Britain Allied and Dominion
gbb glossopharyngeal breathing
GBBA Glass Bottle Blowers Association

GBBCS Ground Base Beam Control System

GBC General Binding Corporation; Gibraltar Broadcasting Company; Green Belt Council of Greater London; Greenland Base Command

GB & C General Battery and Ceramic (corporation)

GB COLL George Brown College

gbd grain boundary dislocation

gb'd goofballed (underground slang—drugged)

GBDC Grand Bahama Development Company

g-b-d-f-a (musical mnemonic—good boys do fine always)—bass clef note names of the five lines (g-b-d-f-a)

GBDO Guild of British Dispensing Opticians

gbe gilt bevelled edge

G.B.E. Dame or Knight of the Grand Cross of the British Empire

GBF *Gakujitsu Bunken Fukyukai* (Japanese Society of Scientific Documentation and Information); Great Books Foundation

GBG General Baking (Stock exchange symbol); Golden Bay Group

gb gas US Army symbol for a colorless and odorless nerve gas of extreme lethality also referred to as general biological gas, goodbye gas, or gruesome business gas

GBGB Gaming Board for Great Britain

gbh girth breast height; grievous bodily harm

gbh (GBH) gamma benzene hydrochloride

GBHC Governor Bacon Health Center

gbi great bodily injury

GBI Georgia Bureau of Investigation; Grand Bahama Island (tracking station)

GB & I Great Britain and Ireland

gbiu geoballistic input unit

g/bl government bill of lading

GBL Georgian Bay Line; government bill of lading

gbm glomerular basement membrane; good bowel movement

GBMA Golf Ball Manufacturers Association

GBMC Greater Baltimore Medical Center

GBMD Global Ballistic Missile Defense

GBNE Guild of British Newspaper Editors

GBNM Glacier Bay National Monument

gbo goods in bad order

G-bomb gravitational bomb

GBOTA Greyhound Breeders, Owners, and Trainers Association

g/box gearbox

GBp Great Britain pound

GBPA Grand Bahama Port Authority

gbr give better reference; gun, bomb, rocket

gbr *gebräuchlich* (German—usual)

GBR Great Barrier Reef

GBR *Guinness Book of Records*

GBRMP Great Barrier Reef Marine Park

GBRMPA Great Barrier Reef Marine Park Authority

gbs gall-bladder series

gb's goofballs (barbiturates)

G-B s Guillain-Barré syndrome

GBS General Business Systems; George Bernard Shaw; Gifu Broadcasting System (Japanese); Guyana Broadcasting Service

GBSM Guild of Better Shoe Manufacturers

GBST Grass Block Substitution Test

GBSTC General Beadle State Teachers College

gbu (GBU) guided bomb unit

GBV Gustahlwerk Bochumer Verein (Krupp Steel)

GBW Guild of Book Workers

GB & W Green Bay & Western (railroad)

g'bye goodbye

gc gas check; gas controller; general cargo; general contract(or); geographical coordinates; gigacycle; glucocorticoid; going concern; gonorrhea case; good condition; good conduct; great circle; grid course; ground clearance; ground control(led); guidance control; gun carriage; gun control

Gc great tropic range; gyrocompass

GC Gallaudet College; Gannon College; Gas Council; Gaston College; General Command; Geneva College; Georgetown College; Gettysburg College; Girton College; Glendale College; Gliding Club; glucorticoid; Goddard College; Golf Club; Gordon College; Goshen College; Goucher College; Government Chemist; Graceland College; Grambling College; Greensboro College; Greenville College; grid course (symbol); Grinnell College; Grover Cleveland (22nd and 24th President U.S.); Guilford College; Gustave Charpentier

G.C. George Cross; gonorrhea case

G & C Gonville and Caius (Cambridge)

GC *Guardia Civil* (Spanish—Civil Guard); *Guardia Costa* (Spanish—Coast Guard)

gca group capacity assessment

gca (GCA) ground-controlled approach

GCA Girls' Clubs of America; Government Contract Committee; Green Coffee Association; Greeting Card Association; Ground Control Center

GCAA Government Corporations Athletic Association

GCAHS Guggenheim Center for Aviation Health and Safety

g cal gram calorie

G Capt Group Captain

G-car(s) German car(s)

G-Cass Gomes-Cáceres; Gomez-Cáceres

GCB Glen Canyon Bridge

G.C.B. Knight of the Grand Cross, Order of the Bath

GCBA Golf Course Builders of America

GCBS General Council of British Shipping

GCC Grand Canyon College; Ground Control Center; Gulf Coast College; Gulf Cooperation Council (six Arab nations bordering the Persian Gulf)

G & CC Gonville and Caius College—Cambridge

GCCA Graduate Careers Council of Australia

GCCC Goshen County Community College

GCCF Governing Council of the Cat Fancy (Great Britain)

GCCS Government Code and Cypher School (nicknamed Government Golf, Cheese, and Chess Society)

gcd general and complete disarmament; great circle dis-

tance; greatest common divisor
GCD Grand Coulee Dam
gce ground cooperational equipment
GCE Gas City Empire; General Certificate of Education; General College Entrance (diploma or examination)
GCEC Greater Colombo Economic Community
gcf greatest common factor
GCFI Gulf and Caribbean Fisheries Institute
gcfr gas-cooled fast reactor
GCFT Gonorrhea Complement Fixation Test
gcg gas-chamber green (institutional paint color)
GCGR Giant's Castle Game Reserve (South Africa)
GCHQ Government Communications Headquarters
gci gray cast iron; ground-controlled interception
gci (GCI) gas chromatograph intoximeter (test for drunk drivers)
GCI General Cognitive Index; Grand Canary Island (tracking station); ground-controlled interception
GCIA Granite Cutters' International Association
G.C.I.E. Knight Grand Commander of the Order of the Indian Empire
gcip guidance correction input panel
GCIS Ground Control Interception Squadron
gcitng ground-control intercept training
GCIU Graphic Communication International Union
GCJC Gulf Coast Junior College
gcl general control language
gcl (GCL) ground-controlled landing
GCL Guild of Cleaners and Launderers; Gulf Caribbean Lines
GCL *Guide to Catholic Literature*
G-class Soviet diesel-powered submarines fitted for launching missiles and nicknamed Golf by NATO
G clef treble clef
GCLH Grand Cross of the Legion of Honour
gcm gas-cut mud (oil well); greatest common measure; greatest common multiple
GCM General Court-Martial;

Gian Carlo Menotti; Good Conduct Medal; Grand Cayman, Cayman Islands (airport)
GCMA Government Contract Management Association
GcmG God calls me God
G.C.M.G. Knight Grand Cross of the Order of Saint Michael and Saint George
GCMI Glass Container Manufacturers Institute
GCMO General Court-Martial Order
gcmps gyro(scope) compass
GCMRU General Control of Mosquitoes Research Unit (India)
gcms gas chromatograph mass spectrometer
GC/MS Gas Chromotography/Mass Spectometry
GCN Greenwich Civil Noon
GCNA Guild of Carillonneurs in North America
GCNM Grand Canyon National Monument
GCNP Grand Canyon National Park (Arizona)
GCNRA Glen Canyon National Recreation Area (Arizona and Utah)
GCO Greater Coin Operators; Guidance Control Officer; Gun Control Officer
gcos general comprehensive operating supervisor
GCOS Great Canadian Oil Sands
GCPL Glasgow Corporation Public Libraries
gcr great circle route
gcr (GCR) gas-cooled graphite-moderated reactor; ground-controlled radar
GCR Geological Characterization Report; Great Central Railway
g crg gun carriage
GCRI Gilette Company Research Institute; Glasshouse Crops Research Institute
GCRO Grand Council and Register of Osteopaths
gcs gate-controlled switch; gram-centimeter-second
gc's genetic girls (real girls)
gc/s gigacycles per second
Gc/s gigacycle per second
GCS Game Conservation Society; General Computer Systems; Georgia Consumer Services; Grant and Contract Service; Group Computer Service(s)
GCSCO Göta Canal Steamship

Company
GCSE General Certificate of Secondary Education
Gc/sec gigacycles per second
G.C.S.G. Knight Grand Commander of the Order of Saint Gregory the Great
G.C.S.I. Dame or Knight Grand Commander of the Star of India
gct ground-control unit
GCT General Classification Test; Glamorgan College of Technology; Greenwich Civil Time
GCTC Green County Teachers College
gcte guidance computer test equipment
GCTS Ground Communication Tracking System
gcu generator control unit; ground control unit
GCU Glasgow Choral Union
G.C.V.O. Dame or Knight of the Grand Cross of the Victorian Order
gcw gross combination weight (of tractor and loaded trailer)
gd general duties; good; good delivery; grade; grading; granddaughter; gravimetric density; ground; ground detector; guard; guardian; gum disturbance
g-d god-damned; granddaughter
g/d gallons per day
g & d galvanized and dipped
gd *gade* (Danish—street)
Gd gadolinium
G-d God (Hebraic contraction)
G^d *Grand* (French—big, large, principal)
GD Gaol Delivery; General Discharge; General Dispensary; General Dynamics (corporation); Geologic(al) Depository (for spent fuel); George Dewey; Grand Duchy; Gudermannian or hyperbolic amplitude (symbol); Gunnery Division
G-D General Dynamics Corporation
G & D Garcia & Diaz (steamship line); Grosset & Dunlap (publisher)
GD *Globe-Democrat*
gda gun-damage assessment; gun-defended area; gunned accelerator
g'day good day (Australia)
GDA General Dynamics Ardmore
GD/A General Dynamics/As-

tronautics
GDBA Guide Dogs for the Blind Association
GDBMS Generalized Data Base Management System
gdc geocentric dust cloud
GDC General Dynamics Convair; *Gesellschaft Deutscher Chemiker* (German—Society of German Chemists)
Gd Ch Grand Chaeur (French—full choir or full organ)
GDCL General Dynamics Candair Limited
GD/Convair General Dynamics/Convair
GD/D General Dynamics/ Daingerfield
GDDQ Group Dimensions Descriptions Questionnaire
gde gilt deckle edging; gross domestic expenditure
Gde gourde (Haitian monetary unit)
G^{de} Grande (Italian, Portuguese, Spanish—big, large, principal)
GDE General Dynamics Electronics; Graduate Diploma in Extension
GD/EB General Dynamics/ Electric Boat
GDED General Dynamics Electro Dynamic
G de F Gaz de France (French—Gas of France)
g del's golden-delicious apples
GDEUT Guidance Digital Evaluation Test
GDFB Guide Dog Foundation for the Blind
GD Fort Worth General Dynamics Fort Worth
GDFW General Dynamics Fort Worth
GDGA General Dynamics General Atomic
gdh growth and development hormone
gdh (GDH) glutamate dehydrogenase
GDHS Ground Data Handling System
gdi god-damned independent (college student failing to join a fraternity or a sorority)
GDI General Dynamics International
GDIFS Gray and Ductile Iron Founders' Society
GDIS General Dynamics International Service
Gdk Gdansk (Danzig)
gdl ground dynamic laser
Gdl Guadalajara; Guadalajare-

ños (inhabitants)
GDL *Grand-Duche de Luxembourg* (French—Grand Duchy of Luxemburg); Guadalajara, Mexico (airport)
GDLC General Dynamics Liquid Carbonic
gdling good looking
GDLST General Dynamics Low-Speed Tunnel
GDM General Dynamics Manufacturing
gdml gas dynamic mixing laser
GDMO General Duty Medical Officer
GDMS General Dynamics Material Service
gdn garden
Gdn Gardener; Godown; Guardian
GDNA Gesellschaft Deutscher Naturforscher und Arzte (German—Society of German Naturalists and Physicians)
gdnce guidance
Gdnk Gdansk (Danzig)
gdnr gardener
Gdns Gardens
gdo gun direction officer
GDO Guild of Dispensing Opticians
gdop geometric dilution of precision
gdp graphic display processor; gross domestic product; guanosine diphosphate; gun director pointer
gdp (GDP) gross domestic product; guanosine diphosphate
GDP General Defense Plan; General Dynamics Pomona; Gross Domestic Product; Guanosine diphosphate
GDP(D) General Dynamics Pomona (Daingerfield)
GDPS Global Data Processing System (WMO); Government Document Publishing Service
gdr guard rail
GDR German Democratic Republic
gds goods
Gds Guards
GDS Gradual Dosage Schedule; Graphic Data System; Graphic(al) Display System; Greater Danube Society
Gdsk Gdansk (Danzig)
Gdsm Guardsman; Guardsmen
gdsob god-damned son of a bitch
GDSS General Dynamics Space Systems
gdt graphic display terminal

gd&t guidance dimensioning and tolerancing
GDTP Geologic(al) Disposal Technology Program
gdu graphic display unit
GDU Guide Dog Users
gdwnd gradient wind
Gdy Gdynia
ge gas ejection; gastroenterology; gilt edge(s); good evening; gross energy; gyroscope error
Ge German; Germanic; germanium; Germany
GE Garrison Engineer; General Election; General Electric; Great Exuma; Group Engineer
GEA Garage Equipment Association; Gravure Engravers Association; Greater East Asia
GEACS Great East Asia Co-prosperity Sphere
GE-ANPD General Electric Aircraft Nuclear Propulsion Development
gear. gearing
geb geboren (German—born); *gebunden* (German—bound)
Geb Gebergte (Afrikaans or Dutch—mountain range); *Gebirge* (German—mountains)
GEB General Education Board; Gerber Products (stock exchange symbol); Grain Elevators Board; Guiding Eyes for the Blind
gebco general bathymetric chart of the oceans
GEBECOMA Groupement Belge des Constructeurs de Matériel Aérospatial French—Belgian Group of Aerospace Builders
Gebr Gebroeders (Dutch—brothers); *Gebrüder* (German—brothers)
gec gecartonneerd (Dutch—bound in boards)
GEC General Electric Company
GECAM Gerência de Operaçôes de Cambio (Portuguese—Exchange Operations Agency)
GECC General Electric Credit Corporation
gecom general(ized) compiler
GECOMIN General Congolese Ore Company
gecref geographic reference (worldwide geographic reference system, also appears as GECREF)

ged *gedampft* (German—muted)

GED General Education Diploma; General Educational Development (testing service); General Equivalency Diploma; Geologic-(al) Exploration Department; Ground Environmental Development

Geda Goodyear electronic differential analyzer

GEDP General Educational Development Program (USA)

gedr *gedrukt* (Dutch—printed)

GEDT General Educational Development Test

GEE Generic Environmental Evaluation

GEEC General Egyptian Electricity Corporation

Geech Geechee (dialect spoken along the Ogeechee River, Georgia)

GEEIA Ground Electronics Engineering Installation Agency

geek geomagnetic electrokinetograph

Ge. Eng. Geological Engineer

geep goat + sheep (hybrid)

GEEP General Electric Electronic Evaluator

gef gonadotrophin enhancing factor

GEG Spokane, Washington (airport)

GEGAS General Electric Gas (process)

gegr *gegründet* (German—founded)

GEHP George Eastman House of Photography (Rochester)

Geh Rat *Geheimrat* (German—Privy Councillor)

GEI *Giovani Esploratori Italiani* (Italian—Italian Boy Scouts)

GEIA Generic Environmental Impact Assessment; Ground Equipment Electronics Installations Agency

GEIC Gilbert and Ellice Islands Colony

GEICO Government Employees Insurance Company

GEIS Generic Environmental Impact Statement

Geist Kon *Geistliches Konzert* (German—sacred concerto)

geistl *geistlich* (German—spiritual)

gek geomagnetic electrokinetograph

gek *gekürzt* (German—abbre-viated)

GEKTUSA Grand Encampment of the Knights Templar of the United States of America

gel gelatine; gelatinous; gelding

GEL General Electric Laboratory; Great Eastern Line

gelat gelatinous

Geld Gelderland (Dutch province)

GELISH Ground Emitter Location Identification System—High

Gell Aulus Gellius (Roman grammarian)

gel. quav. *gelatina quavis* (Latin—in some jelly)

gem. ground-effect machine; guidance evaluation missile

gem. (GEM) growing equity mortgage

Gem Gemini

GEM Gas Equipment Manufacturers; General Education Model; Growing Equity Mortgage

GEMA Gymnastic Equipment Manufacturers Association

GEMAC General Electric Measurement and Control

GEMCO Grazing Export Meat Company; Groot Eylandt Mining Company

GEMCOS Generalized Message Control System

gemi grating efficiency measurement instrument

Gemini two-man spacecraft

gemms geophysical exploration manned mobile submersible

GEMMWU General, Electrical, Mechanical, and Municipal Workers' Union

gems. growth, economy, management, and customer satisfaction

gem's ground-effect machines

GEMS Geostationary European Meteorological Satellite; Global Environmental Monitoring System; Goodyear Electronic Mapping System; Group Export Marketing Scheme

Gemy General Motors Corporation

gen gender; genealogy; genera; general; generally; generating; generator; generic; genetic(s); genital; genitive; gentian; genus

Gen General; Gennadi; Genoa; Genoese

Gen Genesis

GEN Oslo, Norway (Gardermoen Airport)

gen av general average

Gen Cls Comm General Claims Commission (U.S. and Mexico or U.S. and Panama)

Gend *Gendarme* (French—Policeman)

genda general data analysis and simulation

Gen Dyn General Dynamics Corporation

Gene Eugene; Eugenia

geneal genealogy

gen eng genetic engineer(ing)

General (Portuguese or Spanish—General)—short form for many Latin American places ranging from General Acha in Argentina to General Zuazua in Mexico

genet genetic; geneticist; genetics

Genet Janet Flanner's pen name

gen. et sp. nov. *genus et species nova* (Latin—new genus and species)

Geneva Girls Illinois State Training School for (delinquent) girls at Geneva; a similar girls training school at Geneva, Nebraska

Gen Hosp General Hospital

genic (Latin suffix—produced from, producing)—carcinogenic (cancer producing)

GENIP Geographic Education National Implementation Project

GENIRAS Generalized Information Retrieval System

genit genitive

genl general

Gen General (Spanish—General)

Gen Mgr General Manager

genn *gennaio* (Italian—January)

gen. nov. *genus novum* (Latin—new genus)

genoc genocide

gen prac general practice

gen proc general procedure

gen pub general public

genr generate; generation; generator

Gen Rel Gravit General Relativity and Gravitation

genrl general

Gensek *Generalnyi Sekretar* (Russian—Secretary General)—leader of the secretariat of the Central Commit-

tee of the Communist Party
Gen Supt General Superintendent
gent gentleman
Gent East Flanders' provincial capital in Belgium
Gen Tel & El General Telephone and Electric
gents gentlemen; gentlemen's
genvst general visiting (aboard ships of the USN inviting the public)
GENZL Geothermal Energy New Zealand Limited
geo geocentric; geochemistry; geodesy; geodetic; geodynamics; geognosy; geography; geology; geometry; geophysics; geopolitics; geostatic; geothermal (and all their derivatives)
Geo George
Geo I King George I (England, 1714–1727)
Geo II King George II (England, 1727–1760)
Geo III King George III (England, 1760–1820)
Geo IV King George IV (England, 1820–1830)
Geo V King George V (England, 1910–1936)
Geo VI King George VI (England, 1936–1952)
GEO Georgetown, Guyana (Atkinson Field)
GEOC General Estate and Orphan Chamber (trust company)
Geochim Cosmochim Acta Geochimica et Cosmochimica Acta (Latin—*Geochemical and Cosmochemical Review*)
geod geodesic; geodesist; geodesy; geodetic; geodynamic(s)
Geo Dat Pt Geodetic Datum Point (North America's geodetic datum point is the National Ocean Survey's triangulation station at Meades Ranch in Osborne County, Kansas)
Geod. E. Geodetic Engineer
geodss ground electro-optical deep-space surveillance
Geof Geoffrey; Geoffroy
Geoffrey Jeffrey
geog geographer; geographical; geography
Geog Geographic(al); Geography
geogr geografi or *geografisk* (Dano-Norwegian—geography or geographer)

geograph geographical
geohy geohygiene
GEOIS Geographic Information System
geol geologic; geological; geologist
geol geologi or *geologisk* (Dano-Norwegian—geology or geologist)
Geol Geology
Geol.E. Geological Engineer
Geol Surv Geological Survey
geom geometry
Geom Geometry
geomed geometric editor
geomorph geomorphologic(al); geomorphologist; geomorphology
geon (GEON) gyro-erected optical navigation
geoph geophysics
geophy geophysical; geophysics
geopol geopolitical; geopolitics
geor Georgian
Geordie George; Newcastle-on-Tyne, England
Georef World Geographic Reference System
Georg Georgia (USSR)
George George Herbert Walker Bush—41st President of the United States
Georges one-dollar bills bearing the portrait of President George Washington
Georgie George
Georgies one-dollar bills bearing the portrait of President George Washington
Georgy George
geos generator, earth orbital scene; geodetic earth-orbiting satellite; geodetic orbiting satellite
GEOS Geodetic Orbiting Satellite; Geodynamics Experimental Ocean Satellite
GEOSECS Geochemical Ocean Sections Study
gep gross energy product
GEPAC General Electric Programmable Automatic Comparator
Geph Gephyra
GEPI Gestioni e Partecipazioni Industriali (Italian—Industrial Management and Participation)
GEPURS General Electric General Purpose (Computer)
ger gerund; gerundial; gerundival; gerundive
Ger German; Germanic; Germany
GER Great Eastern Railway

gera geratic(al)(ly); geratologic(al)(ly); geratologist; geratology
ger grndng gerund grinding (pedagogic pedantry)
geriat geriatrics
germ. ground-effect research machine
Germ German(y)
Germ Germinal (French—Seedy Month)—beginning March 21st—seventh month of the French Revolutionary Calendar and also title of a novel by Zola
germi germicide
GERNORSEA German Naval Forces in the North Sea
Geron Geronimo
gerontol gerontology
Gerry Gerald; Gerard; Gerhard
Gersis General Electric range safety instrumentation system
gert graphical evaluation and review technique
Gert Gertie; Gertrude
Geru Gerusalemme (Italian—Jerusalem)
ges gesetzlich (German—registered)
Ges (German—G-flat); *Gesellschaft* (German—association, company, society)
GES General Engineering Service; Government Economic Service; Great Eastern Shipping
GE-S Gold Exchange—Singapore
GESAMP Group of Experts on the Scientific Aspects of Marine Pollution
gesch geschützt (German—registered)
Gesch Geschichte (German—history)
GESCO General Electric Supply Corporation
gespeg (Micmac Indian—end of the earth)—Quebec's Gaspé Peninsula
gest gas-explosive simulation technique
gest gestorben (German—dead, deceased)
Gestapo Geheime Staatspolizei (German—State Secret Police)
get. ground-elapsed time; ground-engaging tool(s)
GET Getty Oil (stock exchange symbol); Great Eastern Television (Australian); Gross Error Test(ing)
get 1/2 gastric emptying half-time

GETIS Ground Environment Technical Information System

getlo get locally

getma get from local manufacturer; purchase for local manufacturer

getol ground-effect takeoff and landing

Getty The Getty (J Paul Getty Museum in Santa Monica, California)

GEU Genetic Evaluation and Utilization

gev giga electron volt (10^9 electron volts)

gev (GEV) ground-effect vehicle

GeV billion electron volts; giga electron volt

GEVIC General Electric Variable Increment Computer

gew gram equivalent weight

gew gewoonlijk (Dutch—as a rule, generally, usually)

Gew Gewehr (German—rifle)

gez gezeichnet (German—signed)

Gez Gezira (Arabic—island)

GEZ Gosudarstvennoe knigoisdatelstvo (Russian—State Publishing House)

gf gap filler; generator field; girl friend; globular fibrous; glomerular filtrate; goldfield; government form; green feed; ground fog; growth fraction; guiltfree

g-f globular-fibrous

Gf Gottfried

GF General Fireproofing; General Foods; Georgia & Florida (railroad); Guggenheim Foundation

G F Guyane Française (French Guiana)

G & F Georgia & Florida (railroad)

gfa gas-filled ass; good fair average; good freight agent; gunfire area

GFA Game Fishing Association; Gardens For All; Glider Flying Area; Gliding Federation of Australia

GFA Générale Française (de Construction) Automobile

GFAA Game Fishing Association of Australia

g factor general factor

gfae government-furnished aerospace equipment

gfam graphics flutter analysis methods

GFC Gorilla Foundation of California

gfci ground-fault circuit interrupter

GFCM General Fisheries Council for the Mediterranean (FAO)

gfd general functional description

GFD General Freight Department

GFDL Geophysical Fluid Dynamics Laboratory

gfe government-furnished equipment

gff granolithic finish floor

GFG Good Food Guide

GFH George Frideric Handel

gfi gas-flow indicator; ground-fault interrupter

GFI General Felt Industries

GFIC Georgia Foundation for Independent Colleges

Gfk Gustafsvik

Gfl Genfle (Gävle)

GFL Glossary Function List

gfm government-furnished materiel; government-furnished missile

gfme government-furnished missile equipment

GFMVT General Foods Moisture Vapor Test

GFO General Freight Office

g-force(s) gravity force(s)

G forces acceleration forces

gfp government-furnished parts; government-furnished property

gfr gap-filled radar; glomerular filtration rate

GFR German Federal Republic; Government Facilities Request

gfrc glass-fiber reinforced cement; glass-fiber reinforced concrete

gfrp glass-fiber reinforced plastic

GFS Girls Friendly Society

gfst ground fuel start tank

gft graphical firing table

GFTU General Federation of Trade Unions

gfu glazed facing units

gfut ground fuel ullage tank

GFWC General Federation of Women's Clubs

gg gamma globulin; gas generator; gender gap; go girls; great gross

g-g ground-to-ground

gg gange (Dano-Norwegian—multiply)

Gg Georgian

GG Government Gazette; Governor General

G-G Goodrich-Gulf (chemi-

cals)

G & G Gordon and Gotch

GGA Girl Guides Association; Gulf General Atomic

GGAC Gulf General Atomic Company (formerly General Atomic division of General Dynamics)

Ggb Gorilla gorilla beringei (the mountain gorilla)

GGB Golden Gate Bridge

GGB & HD Golden Gate Bridge and Highway District

G & GBR Geologists and Geophysicists Board of Registration

ggc ground guidance computer

GGC Golden Gate College

GGCST Gleb-Goldstein Color Sorting Test

ggd great granddaughter

gge garage; generalized glandular enlargement

ggf ground gained forward

g.g.g. gummi guttae gambiae (Latin—gamboge)—cathartic

Ggg Gorilla gorilla gorilla (the lowland gorilla)

GGHNP Golden Gate Highlands National Park (South Africa)

GGI Guided Group Interaction

g gl ground-glass

ggm (GGM) ground-to-ground missile

GGNRA Golden Gate National Recreation Area (San Francisco)

Ggo Gallego

GGOC Goldovsky Grand Opera Company

g gr great gross

GGR Gambill Goose Refuge (Texas); Ground Gunnery Range

ggrs great grandson

ggs great grandson; ground gained sideways

g-g's go-go girls

Gg's Ganges gavials; Ganges gharials

GGS Ground Guidance System

GGSM Graduate of the Guildhall School of Music

ggts gravity-gradient test satellite

gh grid heading; growth hormone; guardhouse

gh (GH) growth hormone

Gh Ghana, Commonwealth of

GH General Hospital; Grosvenor House

GH Good Housekeeping

GHA Greenwich Hour Angle

GHAA Group Health Association of America

GhAF Ghanian Air Force

GHAMS Greenwich Hour Angle of the Mean Sun

Ghan Afghan Express

Ghana Republic of Ghana (West African nation), formerly the Gold Coast and British Togoland

GHANA Ghana Airways

ghar ghariyal (Hindu—gavial)—long-snouted fish-eating crocodilian

Gharb al-Gharb (Arabic—the west)—southwest Portugal, Algarve

GHATS Greenwich Hour Angle of the True Sun

GHB Good Housekeeping Bureau

ghc guidance heater control

GHC Gray Harbor College; Group Health Cooperative

GHDVHS Grace H Dodge Vocational High School

ghe ground handling equipment

G H & H Galveston, Houston & Henderson (railroad)

GHI Good Housekeeping Institute

GHMC Good Harvest Marine Company

GHMS Graduate in Homeopathic Medicine and Surgery

GhN Ghana Navy

ghost. global horizontal sounding technique

GHP Group Health Program

g/hphr gallons per horsepower hour

GHQ General Headquarters

g/hr gallons per hour

ghrf growth hormone-releasing factor

ghrh growth hormone releasing hormone

GHS Galileo High School; Girls High School

Ght Ghent

Ghub Ghubba (Arabic—bay, cove)

GHWB George Herbert Walker Bush, 41st President of the United States

ghx ground heat exchanger

GHz gigahertz (gigacycle per second)

gi galvanized iron; gastrointestinal; general issue; generation interval; gill; globulin insulin; government issue; gross income; gross inventory

g-i granuloma inguinale

Gi Giles; Guy

GI Air Guinée; American Sol-

dier (from *gi*—general issue or government issue); Garden Island; General Intelligence; Gideons International; Gimbel Brothers (stock exchange symbol); Government of India; Gunner Instructor

GI Gessellschaft für Informatik (German—Society for Data Processing)

gia grant-in-aid

GIA Garuda Indonesian Airways; Gemological Institute of America; Goodwill Industries of America; Gregorian Institute of America; Gummed Industries Association

GIAHA Gilcrease Institute of American History and Art (Tulsa)

giardiasis diarrhea due to contaminated food or water

gib guy in the back

Gib Gibraltar; Gibraltarian

GIB Gibraltar, British Crown Colony (airport); Gulf International Bank

GIBAIR Gibraltar Airways

Gib(bie) Gilbert

gibb(s) gibbon(s)

Gibfo Gibraltar for orders

GIBMED Gibraltar Mediterranean Command (NATO)

gibs guy in the back seat

gib(s) gibbon(s)—smallest of the anthropoid apes found in the forests of southeast Asia

Gibs Gibraltarians

Gibson Gibson Desert of east-central Western Australia

Gib-tv Gibraltar television

gic ground intercept control

GIC General Investment Corporation; Government Information Center; Guaranteed Income Contract; Guaranteed Investment Contract

GICA Green Island Coral Atoll (Queensland)

gi'd prepared for military-type inspection

Gid Gideon

GID General Intelligence Division

GIDAP Guidance Inertial Data Analysis Program

GIDC Georgia Information Dissemination Center

GIDEP Government-Industry Data Exchange Program

gi distress gastro-intestinal distress

gidp grounded into double plays

GIEE Graduate of the Institution of Electrical Engineers

gif (GIF) growth hormone-inhibiting factor

GIF Rio de Janeiro, Brazil (Galeo Airport)

GIFAS Groupement des Industries Françaises Aéronautiques et Spatiales (French—French Aeronautical and Aerospace Industry Association)

gift. (GIFT) gamete intrafallopian transfer

GIFT Glasgow International Freight Terminal

giga 10^9

Gig Harbor Purdy Treatment Center for Women at Gig Harbor, Washington

gigo garbage in, garbage out (computer term)

GIIS Graduate Institute of International Studies (Geneva)

GIJ Guild of Irish Journalists

Gil Gilbert; Gilchrist; Gilder; Gildo; Gilead; Giles; Gilford; Gilland; Gillespie; Gillian; Gilman; Gilmore; Gilroy

Gilberts Gilbert and Ellice Islands in the Pacific including the Line Islands and Phoenix Islands

Gill(y) Gillian

Gilo Gilberto

gim general information management; gimmick

GI Mech Eng Graduate of the Institution of Mechanical Engineers

gimic guard-ring-implanted monolithic integrated circuit

GIMLCS Generalized Information Management Language and Computer System

gimp. gimbal position(ing)

GIMPEX Guyana Import-Export

GIMRADA Geodesy, Intelligence and Mapping Research and Development Agency (US Army)

gin giugno (Italian—June)

Gin Ginebra (Spanish—Geneva)

Gina Genevieve; Virginia

ging gingival; gingivitis

ging. gingiva (Latin—gum)

gink ginkitis; ginkitological-(ly); ginkitology (science and study of elderly ginks); ginkoid(al)

Ginnie Mae nickname of the Government National Mortgage Association

Ginny Virginia

gins aborigine girls

G Inst T Graduate of the Institute of Transport

GI Nuc Eng Graduate of the Institution of Nuclear Engineers

Ginza center of downtown Tokyo

gio giovedi (Italian—Thursday)

GIO Government Information Organization; Government Insurance Office; Guild of Insurance Officials

g ion gram ion

Giorgione Giorgio Barbarelli

giorn giornaliero (Italian—daily); *giornalist* (Italian—journalist)

Giov Giovanna; Giovanni

gip gas in place (oil well); get(ting) into publication(s); get(ting) into publishing

GIP Great Indian Peninsular (railway)

Gipps Gippsland, Victoria, Australia

GIPR Great Indian Peninsula Railway

GIPSY General Information Processing System

giq giant imperial quart (of beer)

gir girder

GIR Gulf Interior Region

giraffe. graphic interface for finite elements

GIRB Georgia Inspection and Rating Bureau

gird ground-installed recording data

GIRDHS Ground Installation Reconnaissance Data Handling System

girl. generalized information retrieval language

GIRLS Generalized Information Retrieval and Listing System

Girls' Cottage Girls' Cottage School (for delinquents) at Chambly, Québec

Girls' Town correctional facility at Tecumseh, Oklahoma

giro autogiro

GIRU General Intelligence and Reconnaissance Unit (Israel's anti-terrorist commando force is GIRU 269)

gis galvanized iron sheet(ing); gastrointestinal series

gi's gastrointestinal troubles (diarrhea)

GI's enlisted men; enlisted soldiers in the US Army

GIS General Mills (stock exchange symbol); Generalized Information System; Geographic Information Systems;

Geoscience Information Society; Global Information System; Government Information Service(s)

Gisep Giuseppe

GISP Greenland Ice Sheet Program; Guided Independent Study Program

gi spasm gastrointestinal spasm

GISS Goddard Institute of Space Studies (NASA)

git guitar

git (GIT) group inclusive tour; group insurance tour (travel plan)

GIT General Information Test; Georgia Institute of Technology

Gita Bhagavad-Gita

Gitmo Guantánamo Naval Base (Guantánamo Bay, Cuba)

giu geoballistic input unit

giu giugno (Italian—June)

GIUK Greenland, Iceland, United Kingdom

Gius Giuseppe

Givhans Ferry Givhans Ferry State Park (near Charleston, South Carolina)

GIW Gulf Intracoastal Waterway

gj gigajoule (1000-million joules); grapefruit juice

GJB George Jackson Brigade (underground black extremists)

GJC Galdhøppigen Jotunheimen Climbers; Gibbs Junior College; Grand Junction Canal

GJD Grand Junior Deacon

Gjn Gijon

GJO Grand Junction Office

Gk Greek

GK Gaol Keeper

GKC Gilbert Keith Chesterton

GKD-notation Gordon-Kendall-Davison notation for chemical formulas (sometimes called Birmingham notation)

Gk I Greek Isles

GKIAE Gossurdarstveinny Komitet po Ispolzovaniyu Atomnoi Energi (Russian—State Committee for the Use of Atomic Energy)

GKN Guest, Keen, and Nettlefold

GK & N Guest, Keen & Nettleworth

gkw god knows what

gl gas lifting (oil well); general ledger; glass; glazed; gloss;

gold lease; gothic letter; ground level; gun layer; gun license

gl (GL) general liability

g/l grams per liter; guideline

Gl Glagolitic; glucinium

Gl. Gloria in excelsis Deo (Latin—Glory be to God in the highest)

GL Germanischer Lloyd's (German ship classifier); Goldstar Lines; Government Laboratory; Great Lakes (load line mark); Greek line

G.L. Graduate in Law

GL Gamle (Swedish—old); *Glacier* (French—glacier, ice field)

gla gingiovolinguo-axial

GLA gamma-linolenic acid; General Laboratory Associates; Georgia Library Association; Glasgow, Scotland (airport)

GLAA Greater London Arts Association

glab glabrous

glac glacial

GLAC Greek Library Association of Cyprus

Glacier place-name in British Columbia or Montana; Glacier Bay National Monument, Glacier Highway, or Glacier Island in Alaska, Glacier Mountain in Colorado, Glacier National Park in British Columbia and Montana, Glacier Peak in Washington

glaciol glaciologist(ic)(al)(ly); glaciolographic(al)(ly); glaciology; glaciology

Glad Gladstone; Gladwin; Gladys

GLAD Gay and Lesbian Advocates and Defenders

'Glades Florida's Everglades

glads gladiolas

Glam Glamorganshire

GLAMO Great Lakes Association of Marine Operators

Glamorgan Glamorganshire

gland. glandular

gland. glandula (Latin—gland)

glas glasnost (Russian—openness)

Glas Glasgow; Glaswegian

GLASLA Great Lakes—St Lawrence Association

glasphalt glass + asphalt (paving)

glass silicon dioxide—SiO_2

glass. glassware

glassie glass playing marble

glassteel glass + steel (skyscrapers)

Glaswegian(s) Glasgow person(s)

glau glaucous

glauberite calcium sodium sulfate

glauber's salt sodium sulfate

glauc glaucoma

Glav Red Glavnyi Redaktor (Russian—Editor-in-Chief)

glb glass block

GLB Girls' Life Brigade; Greater London Borough (City of London)

GLBA Great Lakes Booksellers' Association

glbs globes

GLBSA Greater London Building Surveyors Association

glc gas-liquid-chromatographic; global loran (navigation) chart(s)

glc (GLC) G-force-induced loss of consciousness

GLC Greater London Council; Great Lakes Carbon; Great Lakes Colleges; Great Lakes Commission

GLCA Great Lakes College Association

glcm (GLCM) ground-launched surface-to-surface cruise missile

GLCM Graduate of the London College of Music

GLCSC Gay and Lesbian Community Service Center

gld gilded; glider; gold; guilder

Gld Gelderland (Dutch province)

Gld Cst Gold Coast

GLDP Greater London Development Plan

gld pltd gold plated

gldr guilder

GLE Grand Larousse Encyclopedie (French—Great Larousse Encyclopedia)

gleep graphite low-energy experimental pile

GLEF Gay and Lesbian Emergency Fund

GLERL Great Lakes Environmental Research Laboratory

Glesca Glasgow

GLF Gay Liberation Front

GLFB Greater London Fund for the Blind

GLFC Great Lakes Fisheries Commission

Glf Mex Gulf of Mexico

Glf Str Gulf Stream

GLHA Great Lakes Harbor Association

gli glider

gli (GLI) glucagon-like immu-

noreactive factor from gastrointestinal mucosa

GLI General Time (stock exchange symbol); Great Lakes Institute (University of Toronto)

glickums ground-launched cruise missiles (GLCMs)

Glierè Reinhold Moritsevich Glierè

GLIS Greater London Information Service

glit glittering

glitch unexpected transient

glitz glitzy (Yiddish—glitter)—golden glitter

Glitz Galitzianer (Yiddish—Galician)—person of Judaic origin from Austrian or Polish Galicia

glld ground laser locator designator

GLLO Great Lakes Licensed Officer's Organization

glm graduated learning method; graduated length method

glm grand livre du mois (French—great book of the month)—best-seller

GLM Gay Liberation Movement

GLMI Great Lakes Maritime Institute

gln (GLN) glutamine (amino acid)

Gln Glen (postal abbreviation)

GLNTC Great Lakes Naval Training Center

GLO General Land Office; Goddard Launch Operations (NASA); Ground Liaison Officer(r); Gunnery Liaison Officer(r)

g LO₂ t ground liquid-oxygen tank

glob globular; globule

globecomm global communications

GLOBECOMS Global Communications System

glock glockenspiel

GLOE Gay and Lesbian Outreach to Elders

glomb glide bomb

glomex global oceanographic and meteorological experiment (GLOMEX)—1975–1980

GLOMR Global Low-Orbiting Relay satellite

GLONASS Global Navigation Satellite System

Gloria Gloria Swanson

Gloria Gloria al bravo pueblo (Spanish—Glory to the brave

people)—Venezuelan anthem

Glos Gloucestershire

gloss. glossary

gloss (Latin prefix—tongue)—glossary, glossopharyngeal

glossies slick-paper magazines

Gloster Gloucester

Glostr Glostrup

glotrac global tracking

Glou Gloucester(shire)

Gloucestr Gloucester

glow. gross liftoff weight

GLOW Gay and Lesbian—an Older Way; Gorgeous Ladies of Wrestling

glp general layout plan(ning); general letter packet

GLP Good Laboratory Practice(s); Greater London Plan

GLP Great Lakes Pilot

GLPA Great Lakes Pilotage Administration

glq greater than lot quantities

glr gas liquid ratio (oil well)

Glr Gloucester

Gls Glasgow

GLS Georgetown Law School; Graduate Library School; Greene Line Steamers (Mississippi); Gypsy Lore Society

GLSOA Great Lakes Ship Owners Association

glt gilt; greetings letter telegram; guide light

glu glutamic acid

gluco or glyco (Greek *glykys*—sweet)—glucose, glycerol, glycogen, glycoprotein

glulam(s) glue-laminated wooden beam(s)

glv globe valve

GLV Gemini Launch Vehicle

GLW Corning Glass Works (stock exchange symbol)

glwb glazed wallboard

GLWQB Great Lakes Water Quality Board (Canada-U.S.)

gly glycerine; glycerol glycogen

gly (GLY) glycine (amino acid)

Gly Gulley; Gully

glycerol glycerine—$C_3H_5(OH)_3$

glyc-pos glycerine suppositories

glykr glykrrhiza (Greek—licorice)—herbal root

glyp glyphography; glyptics; glyptography

glyph hieroglyph

glyph(s) hieroglyph(s)

GLZ General Bronze Corporation (stock exchange symbol)

gm general medicine; general mortgage; good morning; gram; gross margin; guard mail; guided missile; mutual conductance (symbol)

gm (GM) group mark (data processing); group mark(ing)

g/m gallons per minute

GM General Manager; General Medicine; General Motors; Grand Master; Guided Missile; Gunner's Mate; Gustav Mahler

G-M Geiger-Muller (detector)

G.M. George Medal; Gold Medal

GM metacentric height (symbol)

G & M Globe and Mail (Toronto)

gm² grams per square meter (paper weight)

GMA Gallery of Modern Art; Glass Manufacturers Association; Government Modification Authorization; Grocery Manufacturers of America; Grocery Manufacturers of Australia

GMAA Gold Mining Association of America

gmac gaining major air command

GMAC General Motors Acceptance Corporation

GMAD General Motors Assembly Division

GMAIC Guided Missile and Aerospace Intelligence Committee

G-man FBI law-enforcement officer also known as a special agent

GMAS Ground Munitions Analysis Study

GMAT Graduate Management Admissions Test; Greenwich Mean Astronomical Time

GMATS General Motors Air Transport Section

gm-aw gram atomic weight

gmb good merchandise brand

GMB Georg Morris Brandes (originally Cohen)

GMBE Grand Master (of the Order of the) British Empire

GmbH Gesellschaft mit beschrankter Haftung (German—incorporated, limited liability company)

gmbl gimbal

gmc gun motor carriage

Gmc Germanic

GMC General Medical Council; General Motors Corporation; George Mason College; Global Marine Corporation; Guggenheim Memorial Concerts; Guided Missile Com-

mittee; Gulf Maritime Company

GMCC Geophysical Monitoring for Climatic Change

GMCO General Mathematics Computing Option

g-m counter Geiger-Muller counter for measuring cosmic rays and radioactivity

GMCS Group Medicare Cooperative Society

gm-csf (GM-CSF) granulocyte-macrophage colony-stimulating factor

gmd (GMD) green-monkey disease

GMDRL General Motors Defense Research Laboratories

G-men FBI law-enforcement officers

g met gun-metal

GMF Glass Manufacturers Federation

GMFC General Mining and Finance Corporation

gmfp guided-missile firing panel

Gmh Grangemouth

GM-H General Motors-Holden (Australia)

GMHC Gay Men's Health Crisis

GMI General Motors Institute

GMIA Gelatin Manufacturers Institute of America

gmidg garnish moulding

gmk grand master keyed; green monkey kidney

GMK Gold Mines of Kalgoorlie

GMK Gomei Kaisha (Japanese—Mercantile Partnership)

gm/l grams per liter

GML Gold Mining Lease

gmldg garnish molding

GMMC Godden Memorial Medical Centre (Fiji)

GMNNR Glasson Moss National Nature Reserve (England)

GMNP Guadalupe Mountains National Park (Texas)

GMNZ General Motors New Zealand

Gmo Guillermo (Spanish—William)

GMO Government Medical Office(r); Guided Missile Office(r)

GM & O Gulf, Mobile & Ohio (railroad)

g mol g molecule

GMOO Guided Missile Operations Office(r)

gmp guaranteed minimum

price

gmp (GMP) guanosine monophosphate

GMP General Medical Practice; Greater Manchester Police; Green Mansion Properties

gmpa gas-metal-plasma arc

gmpg ground nautical miles per gallon

GMPI Guilford-Martin Personnel Inventory

gmq good merchantable quality

gmr ground mapping radar

GMRD Guided Missiles Range Division (Pan American World Airways)

gm rm games room

gms guidance monitor set

gms (GMS) geostationary meteorological satellite

gm & s general, medical, and surgical

Gms Grimsby

GMS General Maintenance System; General Medical Services

GMSB Guided Missile System Branch

GMSC Guangdong Manpower Service Corporation

GMSL Group Management Service Limited

GMST General Military Subjects Test

GMT General American Transportation (stock exchange symbol); Greenwich Mean Time; Greenwich Meridian Time

GMT Geo Marine Technology; Geriatric Medicine Today

GMTC General Motors Technical Center; Glutamate Manufacturers Technical Committee

GMTL Goudy Memorial Typographic Laboratory (Newhouse Communications Center—Syracuse University)

gmts guided missile test set

g-m tube geiger-müller tube

GMU George Mason University

gmv gram molecular volume

GMV Government Motor Vehicle; Government Motor Vessel

gmw gram molecular weight

GMWU General and Municipal Workers Union

gn general; glen; golden number; good night; green; guide number; guinea (21 shillings); gun

g:n glucose-nitrogen (ratio)
GN Great Northern (railroad); great novel
G.N. Graduate Nurse
G & N Gippsland and Northern
GN Gas Natural (Spanish—natural gas); Guardia Nacional (Spanish—National Guard)
GN₂ gaseous nitrogen
GN₂ s/a gaseous nitrogen storage area
g N₂ stor gaseous nitrogen storage
GNA Ghana News Agency; Great Northern Insured Annuity Corporation
GNAL Georgia Nuclear Aircraft Laboratory
GNAS Grand National Archery Society
gnat. global network of automatic telescopes
GNB Good News Bible
gnc general nuclear war
gn & c guidance, navigation, and control
GNC General Nursing Council; Good Neighbor Council(s)
gnd gross national demand; ground
gndck ground check
gnd ht xgr ground heat exchanger
gne gross national effluent
gne (GNE) gross national expenditure
gni (GNI) gross national income
gnl general
GNL Georgia Nuclear Laboratory
GNM Ghana National Museum
GNMA Government National Mortgage Association
GNN Great Northern Nekoosa
g noz grease nozzle
gnp (GNP) gross national product
g np gas, nonpersistent
GNP Geriatric Nurse Practitioner; Glacier National Park (in British Columbia and in Montana); Gombe National Park (Tanzania); Gorongoza National Park (Mozambique); gross national product
GNP & BL Great Northern Pacific & Burlington Lines
GNPC Great Northern Paper Company
gnr goods not received; gunner; gunnery

gnr (GNR) gaseous nuclear rocket
GNR Great Northern Railway
GNRA Gateway National Recreation Area (New York City's designation by the Department of the Interior)
g/n ratio glucose-nitrogen ratio
gnrh (GnRH) gonadotropin-releasing hormone
gnrl general
gnry gunnery
gns guineas
Gns Guernsey
GNS General Naval Staff
GNSRA Great North of Scotland Railway Association
GNT Grand National Teams
GNT Gesellschaft für Nukleartransporte (German—Nuclear Transport Society)
GNTC Girls' Nautical Training Corps
gnte gerente (Spanish—manager)
GNTO Greek National Tourist Organization
GNTP Georgia Narcotics Treatment Project
gnw (GNW) gross national welfare
Gny Sgt Gunnery Sergeant
go. gas operated; gear oil; growth opportunities
go' gore
g/o gear box oil
Go gadolinium; Gothic
Gᵒ Gonzalo (Spanish)
GO General Office; general order(s); George Orwell (Eric Blair); Group Officer; Gulf Oil (stock exchange symbol)
GO₂ gaseous oxygen
goa gone on arrival; gyro-(scope) output amplifier
GOA Gun Owners of America
GOAL Gay Organized Alliance for Liberation; Gun Owners Action League
goar ground-observer aircraft recognition
GOAT Give Our Animals Time (organization devoted to saving the endangered goats on California's offshore islands)
gob. gobbledygook; good ordinary brand
gob gobierno (Spanish—government)
Gob Gobernador (Spanish—Governor)
G o B Government of Belize (formerly British Honduras)
GObC Ground Observers Corps (Canada)

Gobi great desert of Central Asia in Mongolia
gobᵒ gobierno (Spanish—government)
Gobr Gobernador (Spanish—Governor)
GOBS Guardians of Better Speech
goc gas-oil content (oil well)
GOC General Officer Commanding; General Optical Council; Greek Orthodox Church; Ground Observer Corps; Gulf Oil Company
GOC in C General Officer Commanding in Chief
GO City Greater Omaha, Nebraska
goco government-owned contractor-operated
god. (GOD) government observing device
g.o.d. good old days
GODAS Graphically Oriented Design and Analysis System
GODCO Gulf Oman Oilfields Development Company
GODE Gulf Organization for the Development of Egypt (funded by Kuwait and Saudi Arabia)
GODORT Government Documents Round Table
God Save God Save the King (Queen) British national anthem
godsd godsdienst (Dutch—religion)
Godzone God's own country
goe gas, oxygen, ether (mixture); ground operating equipment
GOE General Ordination Examination
GOES Geostationary Operational Environmental Satellite
gof good old Friday; government-owned facilities
G o F Gang of Five
GOFAR Global Ocean Floor Analysis and Research
gogo government-operated government-owned
Gogol Nikolai Vasilyevich Gogol-Yanovsky
gogs goggles
goi gross operating income
Goi Goidelic
GoI Government of Indonesia
GOI Gallup Organization Incorporated
goin' going
GOIN Gosudarstvienny Okeanograficheskiy Institut (Russian—State Oceanography Institute)

gol general operating language

GOLB *Gosudarstvennaya Ordena Lenina Biblioteka* (Russian—Lenin State Library)

gold. geometric on-line definition

GOLD Grandparents Offering Love and Direction

Golda Golda Meir (Israel's first woman prime minister)

Gol de Cal *Golfo de California* (Spanish—Gulf of California)—also called Sea of Cortés or Vermillion Sea

Golden Rule *What is hateful to thee do not do unto thy neighbor.*—Hillel (30 B.C.–10 A.D.) stated in explaining essence of the Torah; *Do unto others as you would have them do unto you*

Goldie Gold; Golden; Goldilocks; Goldsborough; Goldsmith; Goldsworthy; Goldwin; Goldwyn

gold(s) gold bond(s); gold coin(s); gold medal(s)

Golf letter G radio code; Soviet G-class submarines with missile-launching capability (NATO)

Gollancz Victor Gollancz Ltd

gom (GOM) government-owned material

Gom God's own medicine (opiates)

G.O.M. Grand Old Man (sobriquet for William Ewart Gladstone)

goma general officer money allowance

GOMA Good Outdoor Manners Association

go'n' going

gon *goniff* (Yiddish—thief)

gond(s) gondola(s); railroad car(s); car(s)

Gone Gone With The Wind

gonna (American slang—going to)

GONP Gal Oya National Park (Ceylon)

GONS Gun Orientation Navigation System

Gonz Gonzàles

Goo Goole

GOO Get Oil Out (of Santa Barbara, California)

goobs going out of business sale(s)

Goochland State Industrial Farm for Women (convicts) at Goochland, Virginia

goodbye god be with you (contracted)

Good H Good Housekeeping

goodies good ones

Goodwins Goodwin Sands off Kent near the North Sea entrance to the English Channel

goof. general on-line oriented function

goof(er) stupid person

googol 10 raised to the 100th power (10^{100})

googoo good government political plank

goon *goonda* (Hindi—hired killer)

GOP Grand Old Party (Republican)

GO & P Griffith Observatory and Planetarium

gor gas/oil ratio; general operational requirement(s); gorilla

Gor Gordon; Gorham; Gorki; Gorman

GOR General Operating Room; General Operational Requirements

GORA Government Oil Refineries Administration

Gordie Gordon

Goree Unit Women's Prison at Huntsville, Texas

GORF Goddard's Optical Research Facility

g org great organ

goric paregoric (an opiate narcotic)

gorill(s) gorilla(s)

gorm gormandize(r)

gos *gosudarstvo* (Russian—state)

Gos *Gosudarstvo* (Russian—State)

GOS General Operating Specification(s); George Orwell Society; Global (weather) Observing Systems

Gos Alb *Gossamer Albatross* (see bicyplane)

GOSS Ground Operational Support System

gost government-owned special tooling

GOST Goddard Satellite Tracking

Gösta Björling Karl Gustaf Björling

got. (GOT) glutamic oxaloacetic transaminase

Got Gothenburg (Göteborg)

gotcha got you

Goten (German naval contraction—Gotenhafen)—Gdynia's name during World-War-II Nazi occupation

goth. gothic type

Goth. Gothic

Göt(h) Göteborg (Gothenburg)

gothic gothic script or gothic type

gotran go fortran

got-roy *gotong-royong* (Indonesian–cooperation, mutual aid)

Gott Gottingen

gou *gourde* (Haitian currency)

Gou Goudy

Goudy Frederic William Goudy (American printer and designer of 90 typefaces)

goulard water lead lotion

gourde monetary unit of Haiti

Gouv Gouverneur

gov government

Gov Governor

GOVA Guide to Opportunities in Volunteer Archaeology

goveclop government closest to the people

govg governing

Gov Gen Governor General

Gov Is Governor's Island

govt government

govtalk government talk

Govt Print Government Printer

GOW Grand Old Woman (Queen Victoria)

GOWA Guild of Watchmen of Australia

gox gaseous oxygen

goya & kod get off your ass and knock on doors

Goyo Gregorio

gp galley proofs; gas, persistent; general paralysis; general practice; general practitioner; general public; general purpose; geographic position; glide path; government property; grateful patient; gratitude patient; greenhouse perennial; ground pneumatic; group; guinea pig; gun pointer

g-p general purpose; graduated-payment

gp *grand prix* (French—grand prize)

g/p *giro postal* (Spanish—money order)

Gp Group

GP Gallup Poll; Gaspesian Park (Quebec); general public; Georgia-Pacific (stock exchange symbol); Giacomo Puccini; Gulf Province

G-P Georgia-Pacific (forest products); Gunier-Preston zone

GP *Generalpause* (German—general pause)—musical term

gpa grade-point average

GPA Gas Processors Association; Gay Press Association; Genealogical Publications of Australia; General Practitioners' Association; General Public Accounting; Governmental and Public Affairs; Guinnesse Peat Aviation

gpac grade point average category

gpad gallons per acre per day

gpae general-purpose aerospace equipment

GPAEVD Greater Philadelphia Alliance for the Eradication of Venereal Disease

GPAL Gold Producers Association Limited

gpate general-purpose automatic test equipment

GPATS General-Purpose Automatic Test System

gpb glossopharyngeal breathing

GPB Gosudarstvennaya Publichnaya Biblioteka (Russian—State Public Library)

gpc gallons per capita; general purpose computer; gypsum-plaster ceiling

gpc (GPC) general physical condition

GPC Genuine Parts Company; Georgia Power Company; Gulf Park College

Gp Capt Group Captain

gpcd gallons per capita per day

Gp Cmdr Group Commander

Gp Comdr Group Commander

GPCR Great Proletarian Cultural Revolution (China)

GPCT George Peabody College for Teachers

gpd gallons per day

GPDA Gypsum Plasterboard Development Association

gpdc general-purpose digital computer

GPDS General-Purpose Display System

GPDST Girls' Public Day School Trust

gpdw gypsum-plaster dry wall-(ing)

gpe good phonetic equivalents

GPE General Precision Equipment; Global Perspectives in Education; Guided Projectile Establishment

Gp. Eng. Geophysical Engineer

gperf ground passive electronic reconnaissance facility

GPES Ground Proximity Extraction System

gpete general-purpose electronic test equipment

gpf gasproof

GPF Generic Packaging Facility

Gp Fl group flashing (light)

GPFS General-Purpose Financial Statement

gpg grains per gallon

gph gallons per hour; graphite

G.Ph. Graduate in Pharmacy

GPH Game Packing House; Grand Pacific Hotel (Suva)

GPHI Guild of Public Health Inspectors

gpi general paralysis of the insane (symptom of tertiary syphilis); ground-position indicator (aviation)

gpi (GPI) general price index; glucosephosphate isomerase

GPI General Printing Ink; Gordon Personal Inventory; Government Property Inventory

gpid guidance package installation dolly

GPII Geist Picture Interest Inventory

gp int qk fl group interrupted quick flashing

gpl generalized programming language; geographic position locator; grams per liter; gypsum lath

GPL General Precision Laboratory

GPLC Guild of Professional Launderers and Cleaners

gply gingivoplasty

gpm gallons per minute; gross profit margin

gpm (GPM) graduated payment mortgage

GPM General Preventive Medicine; Grand Past Master

gpmg general-purpose machinegun

g-p mortgage graduated-payment mortgage

GPMS Gross Performance Measuring System (USAF)

GPN Graduate Practical Nurse

GPNITL Great Plains National Instructional Television Library

gpo gun position officer

GPO General Post Office; Government Printing Office; Great Plains Organization

Gp Occ group occulting (light)

gpp galley page proofs; graphic part programmer

gpp (GPP) graduated property payment(s)

GPP Gordon Personal Profile; Gross Provincial Product (Canadian)

gppl gypsum plaster

GPPT Group Personality Projective Test

GPR Glider Pilot Regiment

GPR Grand-Positif-Récit (French—great choir and swell coupled)—organ music

G.P.R. Genio Populi Romani (Latin—Genius of the Roman People)

GPRA General Practice Reform Association

gps gage pressure switch; gallons per second; general-purpose solver; ground plane simulator; guidance power supply

gp's galley proofs; guinea pigs

g-p's general practitioners (GPs)

Gps general-parents motion pictures (for youngsters only with parent's consent)

GPs Great Performances

GPS General Practitioners Society; Gibbs-Poole-Stockmeyer (algorithm); Global Positioning System; Graduated Pension Scheme; Great Persons Society; Greater Public Schools

GPSA General Practitioners Society of Australia

gpsdw general-purpose scientific document writer

gpse general-purpose simulation environment

gpss general-purpose systems simulator

GPSS General Process Simulation Studies

gpt gas power transfer; guidance position tracking; gypsum tile

gpt (GPT) glutamic pyruvic transaminase

GPT Grayson Perceptualization Test; Guild of Professional Toastmasters; Guild of Professional Translators

gpte general-purpose test(ing) equipment

gp th group therapy

gptr guidance power temperature regulator

gpu ground power unit

GPU General Postal Union

GPU Gosudarstvennoe Politicheskoe Upravlenie (Russian—State Political Administration)—secret police

GPV Gun Powder Van

GPV Gereformeerd Politiek Verbond (Dutch—Reformed Political Union)–Calvinist party

gpw gross plated weight; gypsum-plaster wall

GPW Geneva Convention Relative to Treatment of Prisoners of War

GPWS Ground Proximity Warning System

gpx generalized programming extended

GPX Greyhound Package Express

GPY Government Property Yard

GQ general quarters

gqa give quick answer; government quality assurance; grain quality analyzer

GQG *Grand Quartier Général* (French—General Headquarters)

GQNM Gran Quivira National Monument

GQR Gauss Quadrature Rule

GQS General Quarter Sessions

Gquil Guayaquil

gr gear; grab rod; grade; grain; gram; grammar; grand; grass runway; gravity; great(er); grind(er); gross; ground; group; gunner

g-r gamma ray

gr *gravida* (Latin—gravid)—pregnant

Gr Grashof number; Great (postal abbreviation); Grecian; Greece; Greek; Grove

Gr *Graben* (German—ditch, trench); Greek; *Groot* (Afrikaans—big, great); *Gross(e)* (German—big, great, vast)

GR B.F. Goodrich (stock exchange symbol); General Radio; General Reconnaissance; General Reserve; *Georgius Rex* (King George); Government Report; Grand Recorder; Grasse River (railroad); Graves Registration; Group Report; Gunnery Range

GR *Guardia Republicana* (Spanish—Republican Guard)

GRA Girls Rodeo Association; Government Reports Announcements; Governmental Research Association; Grass Roots Association; WR Grace & Company (stock exchange symbol)

gr ab grade ability

GRAB Group Rooms Availability Bank (hotel-motel convention service)

GRACE Grace Agencies; Grace Chemicals; Grace

Line; WR Grace and Company (stock exchange symbol); graphic arts composing equipment; group routing and exchange equipment (telephone)

grad gradient; grading; graduate

grad (GRAD) graduate résumé accumulation and distribution

grad. *graditim* (Latin—by degrees)

Grad IAE Graduate of the Institution of Automobile Engineers

Grad IM Graduate of the Institution of Metallurgists

Grad Inst BE Graduate of the Institution of British Engineers

Grad Inst P(hys) Graduate of the Institute of Physics

Grad Inst R(frg) Graduate of the Institute of Refrigeration

Grad IRI Graduate of the Institution of the Rubber Industry

Grad NDTS Graduate (member) of the Non-Destructive Testing Society

Grad RIC Graduate (member) of the Royal Institute of Chemistry

grad(s) gradient(s); graduate(s)

GRADS Great Falls Air Defense Sector

Grad SE Graduate of the Society of Engineers

Grad Soc Eng Graduate of the Society of Engineers

gradu gradual(ly); graduate(d); graduating

graf graphic additions to fortran

graf(s) graphic addition(s); paragraph(s)

Grahams Grahamstad or Grahamstown in South Africa

Grail Soviet shoulder-fired surface-to-air missile called SA-7 (NATO)

gral *general* (Spanish—general)

Gral *General* (Spanish—General)

gram. grammar; grammatical; gramophone

'gram cablegram; radiogram; telegram

gram (Latin suffix—record)—radiogram

Gram Grammar; Grandfather; Grandpa(pa)

Gram *Gramaphone*

gramm grammatical

Grammy National Academy of Recording Arts and Sciences award; replica of a gramophone

Grampians Grampian Hills of Scotland or the Grampian Mountains of Australia

gramp(s) grandfather

GRAMS Ground Recording and Monitoring System

gran granite; granular; granulated sugar

gran. *granulatus* (Latin—granulated)

Gran Granada; Granjon

GRAN Global Rescue Alarm Net

Grand Canyon the Grand Canyon of the Colorado in the Grand Canyon National Park in Arizona, the Grand Canyon of the Arkansas in Colorado where it is also called the Royal Gorge, the Grand Canyon of Santa Elena in the Big Bend National Park in Texas, the Grand Canyon of the Snake River in Idaho, the Grand Canyon of the Tuolumne in California, the Grand Canyon of the Yellowstone in the Yellowstone National Park in Wyoming

Grand Coulee Grand Coulee Dam; Grand Coulee Valley in eastern Washington

grando *grandioso* (Italian—grandiose)

Grand Teton Grand Teton Mountain; Grand Teton National Monument in northwestern Wyoming

Gran-Duc Lux *Grand-Duché de Luxembourg* (French—Grand Duchy of Luxembourg)

Granny Grandmother

Gran Quivira Gran Quivira National Monument in central New Mexico

grapden graphic data entry

Grapes Grapes of Wrath

graph. graphology

graph (Latin suffix—record, recording)—radiograph

grapheme written language symbol representing an oral language code

Graphics 'O' Level commercial art with some drawing or fine art

graphite carbon

graph rec graphic record(ing)

GRAPO *Grupos Antifascista Para Octubre* (Spanish—Oc-

tober 1st Antifascist Groups)—Spain
gr ar grinding arbor
GRAR Government(al) Report Authorization and Record
gras generally recognized as safe
graser gamma-ray laser
grasp. graphics-augmented structural-post processing
grat graticule
grats congratulations
GRATS Gang-Related Active Traffickers Suppression
grav gravimetric; gravitation; gravity
grazo grazioso (Italian—gracious)
grb gamma-ray burst; granolithic base
GRB Gerakan Rakjat Baru (Indonesian—New People's Movement); *Guide to Reference Books*
GRBI Gardeners' Royal Benevolent Institution
grbm (GRBM) global-range ballistic missile
Gr Br *Grande Bretagne* (French—Great Britain); Great Britain
Gr Brit Great Britain
GRBS Gardeners' Royal Benevolent Society
grc glass-reinforced cement
GRC Gale Research Company; Gerontology Research Center; Government Research Corporation; Gulf Research Corporation
GRC Gendarmarie Royale du Canada (French—Royal Gendarmarie of Canada)—Royal Canadian Mounted Police
GRCB Greyhound Racing Control Board
GRCM Graduate of the Royal College of Music
grcol ground color
Gr Cpt Group Captain
grd granddaughter; grind; ground; ground detector; guard
Grd Ground (postal abbreviation)
Gr d Greek drachmae
Gr D Grand Duchy
GRD Geophysics Research Directorate
GRDC Gulf Research and Development Company
grdl gradual(ly)
Grdn The Guardian
grd tot grand total
gre ground reconstruction

equipment
Gre Grecia (Spanish—Greece)
GRE Graduate Record Examination; Guardian Royal Exchange
Great Amer Great American Federal Savings Bank; Great American First Savings Bank
Great Communicator Ronald Reagan
Great Creole General Pierre Gustave Toutant de Beauregard
Great Dic Great Dictator
Great Dividing Great Dividing Range of Australia's New South Wales and Queensland
Great Lon Greater London
Great Sandy Great Sandy Desert of South and Western Australia
Great Smokies Great Smoky Mountains of North Carolina and Tennessee
Great Smoky Great Smoky Mountains; Great Smoky Mountains National Park in North Carolina and Tennessee
Great Vic Great Victoria Desert of Western Australia
Great-West Great-West Life and Annuity Insurance
GREB Graduate Records Examination Boards
Grec Grécia (Italian or Spanish—Greece); *Grecia* (Portuguese—Greece)
GRECC Geriatric Research, Education, and Clinical Center (Seattle)
Greece Hellenic Republic (Balkan nation whose history antedates classical antiquity), *Elliniki Dimokratia*
Greece's Principal Port Piraeus
Greek Fabulist Aesop
Greeks Greek Islands; Greek people
Green Green College (Oxford or elsewhere)
Green. Greenland
green flag all-clear signal; express; go-ahead
Greenhouse Greenhouse Effect
Greenland Arctic Greenland north of the Arctic Circle
green light all-clear signal; go-ahead signal; safety signal; starboard side of aircraft, ships, or other vessels
Green Mts Green Mountains of Vermont
Greenock Girls prison for fe-

male offenders in Greenock, Scotland
greeny conservationist or environmentalist
Grefco General Refractories
Greg Gregorian; Gregory
Greg⁰ Gregorio (Spanish—Gregory)
gr el greatest elongation
Gren Grenada
Grenada State of Grenada (West Indian island nation)
Grenadines Bequia, Cannouan, Carriacou, and Mustique islands in British Windward Islands
Grendr Grenadier
grep gets repeating patterns
Grepo Grenzpolizei (German—border-control police)
Greta Greta Garbo; Margaret
Gretchen Marguerite
Greyf Greyfriars College, Oxford
grf (GRF) growth hormone-releasing factor
gr f grass firm (on runway)
GRF Gerald Rudolph Ford—thirty-eighth President of the United States; Graphic Reproduction Federation; Grassland Research Foundation; Gravity Research Foundation
gr Fl grosse Flöte (German—full-size flute)
GRFMA Grand Rapids Furniture Market Association
grfrp graphite fiberglass-reinforced plastic
gr fx grinding fixture
grg generalized reduced gradient; gravimetric rain gage
grh (GRH) gonadotrophin-releasing hormone
GRI Gas Research Institute; General Religions International (publishing division of Humanists of South Jersey); Geothermal Resources International; Government Reports Index; Government of the Ryukyu Islands
G.R.I. Georgius Rex et Imperator (Latin—George, King and Emperor)
GRID gay-related immune deficiency
grif (GRIF) growth hormone-inhibiting factor
Grif Griffin; Griffith; Griffiths
griff griffin
GRIP Grass Roots Improvement Program
griphos general retrieval and information processor humanities-oriented studies

Gris Griswold
GRIST Grazing Incidence Solar Telescope
grit. gradual reduction in temperature; gradual reduction in tensions
grits boiled grits; hominy grits; rockahominie in Algonquian Indian
GRITS Goddard Range Instrumentation Tracking System (NASA)
griz grizzly
grizz grizzly bear
GRJC Grand Rapids Junior College
Grk Greenock
Gr-L Graeco-Latin
Gr L Gunner Lieutenant
gr lp ground lamp
Gr Lt Gunner Lieutenant
GRL Gross Regional Loss
grm gaseous radiation monitor; gram; gross rent multiplier; guidance rate measuring
grmp generalized report module program
grn green
g/r/n goods received note
Gr.N. Graduate Nurse
grnd ground
Grnd Grand
grndr grinder(s)
grnl *giornalista* (Italian—newspaperman)
GRNL Gay Rights National Lobby
Grnld Greenland
grns green skins
grnsh greenish
grnt guarantee
gro gross
Gro Grocer(y); Groningen; Grove; Guerrero
Gro. grocer
GRO Gamma Ray Observatory; Greenwich Royal Observatory
GROBDM General Register Office of Births, Deaths, and Marriages
groc grocer(y)
Groen *Groenlandia* (Italian or Spanish—Greenland); *Groenlandia* (Portuguese—Greenland)
GROIN Garbage Removal Or Income Now
Grolier Grolier Society
grom grommet
Grom Andrei Gromyko
Gron Groningen (Dutch province)
gros. *grossus* (Latin—coarse, gross)
Grose Francis Grose's *Classi-*

cal Dictionary of the Vulgar Tongue
Grosset Grosset & Dunlap
Groucho Groucho Marx (Julius Marx)
Group of Seven see G-7
Group W Westinghouse Broadcasting
Grove's Sir George Grove's *Dictionary of Music and Musicians*
GROW Gay Rights for Older Women; Group Recovery Organizations of the World
Growlers Growler Mountains of southwestern Arizona
growsy grumpy and drowsy
grp glass-reinforced plastic (fiberglass); graphite-reinforced plastic; ground relay panel; group repetition panel
Grp Group
GRP Gross Regional Profit
grp(s) group(s)
grp's gross rating points
grr growler
GRR Grand Rapids, Michigan (airport); Graphic Reproduction Request
grreg graves registration
GrReg graves registration
grs grains; grandson; grass; greens
gr s grass soft (on runway)
gr-s government rubber plus styrene (buna-S synthetic rubber)
GRs Government Regulations; Granitic Regions
GRS General Railway Signal; Graves Registration Service; Great Red Spot
GRSE Guild of Radio Service Engineers
Gr S-Lt Gunner Sub-Lieutenant
GRSM Graduate of the Royal Schools of Music (Royal Academy of Music and the Royal College of Music)
GRSP General Revenue Sharing Program
grst gross ton(s)
grsy greasy
grt gross register(ed) tonnage (tons); grout
grtee guarantee
grtg grating
grtm gross-ton mile
gr tons gross tons
grtr greater
gr tr graphite treatment
gr Tr *grosse Trommel* (German—bass drum)
Gru Grus; Gruyère
GRU *Glavnoye Razvedyvatel-*

noye Upravlenie (Russian—Intelligence Directorate of the Red Army)—(*q.v.* VOT)
grub. grubby; grubstreet
grub. (GRUB) grocery update and billing
GRULA *Grupo Latino Americano* (Spanish—Latin American Group)—UN power bloc
gr'ups grownups
Grv Grove
grvl gravel(ly)
Grwd Grunewald
gr wt gross weight
gry grocery; gross redemption yield
gs galvanized steel; gastric shield; gauss; german silver; glandular segment; glide slope; grand slalom; grandson; ground speed; guardship; guineas
gs (GS) gold standard; group separator character (data processing)
g-s grandson
g/s gallons per second
Gs force of gravity; general motion pictures (for the general public); German silver; Gomes
GS General Schedule (civil service classification system); General Secretary; General Service; General Sessions; General Staff; General Studies; General Support; Geochemical Society; Geological Survey; Gerontological Society; Gillette (stock exchange symbol); Girl Scouts; Glow Start (tractor); Government Servant; Government Service; Grand Secretary; Ground Staff; Ground Station; Guidance Station; Gunnery School; Gunnery Sergeant
GS1 Gustav Siegfried Eins (British radio station broadcasting propaganda to Germany in World War II)
G-S Gallard-Schlesinger
G & S Gilbert and Sullivan (Sir William Schwenck Gilbert, librettist, and Sir Arthur Seymour Sullivan, composer)
GS *Garda Siochana* (Gaelic—Police Force)—Ireland's police force
gsa gross soluble antigen
gsa (GSA) general (travel) sales agency; general (travel) sales agent
GSA Garden Seed Association;

Gas Service Agents; General Services Administration; Genetics Society of America; Geological Society of America; Girl Scouts of America; Gourd Society of America

G & SA Gulf and South American (steamship line)

GSABCA General Services Administration Board of Contract Appeals

GSAI General Services Administration Institute

gsap gunsight aiming point

GSAPBS General Services Administration Public Building(s) Service(s)

gsb gypsum sheathing board

gsb (GSB) go subroutine

GSB Government Savings Bank

GSBA Georgia School Boards Association

GSBAA General Service Board of Alcoholics Anonymous

gs bot glass-stoppered bottle

gsbr gravel-surface built-up roof

gsc gas-solid chromatography; geodetic spacecraft; ground speed continue; guidance systems console

GSC General Staff Corps; Genetics Society of Canada; Geological Survey of Canada; Gold Star Cable (Korean); Group Study Course; Group Switching Center

GSCBA Georgia State College of Business Administration

GSCP Generic Site Characterization Plan

GSCT Goldstein-Scherer Cube Test

GSCW General Society of Colonial Wars; Georgia State College for Women

gsd general system description; genetically significant dosage; grid sphere drag

GSD General Services Department; General Supervisor's Directive; General Supply Depot

GSDFJ Ground Self-Defense Force Japan

GSDNM Great Sand Dunes National Monument

gse (GSE) government-supplied equipment; ground-service equipment; ground-support equipment

GSE Graduate School of Education (Harvard University)

GSED Ground Support Equipment Division (USN)

GSEL Ground-Support Equipment List

GSERD Ground-Support Equipment Recommendation Data

GSES Geocentric Solar Ecliptic System (NASA)

GSE/TD General Systems Engineering/Technical Direction

gsf general scientific framework

GSF General Support Force (USAF); Government Superannuation Fund

GSFC Goddard Space Flight Center

GSFG Group of Soviet Forces in Germany

GSFLT Graduate School Foreign Language Test

GSFSR Ground Safety and Flight Safety Requirements

gsfu glazed structural facing units

GSG Grenzschutzgruppe (German—Border Protection Group)—anti-terrorist commando force

GSGB Geological Survey of Great Britain

GSGS Geographical Section—General Staff

GSGS maps General Staff, Geographical Section (British War Office) maps covering Africa, Asia, the East Indies, and Europe

gsh good study habits

GSH glutathione

gshr grand-slam home run(s)

gshv globe stop hose valve

gsi gas installed; glide scope indicator; graphic structure input; gross scheduled income; ground speed indicator

GSI General Safety Inspection; General Safety Inspector; General Service Infantry; General Steel Industries; Geological Survey of Israel; Geophysical Services International; Government Source Inspection

G & SI Gulf and Ship Island (railroad)

gsid ground-emplaced seismic intrusion detector

g sil german silver

GSIS Government Service Insurance System; Group for the Standardization of Information Services

gskt gasket

gsl guaranteed student loan

GSL Geological Society of London; Graphics Subroutine Library; Guaranteed Student League

GS & LA Guam Savings and Loan Association

gslcv globe stop lift check valve

GSLP Guaranteed Student Loan Program

gsm good sound merchantable; grams per square meter; gross sales monthly; ground-supplied material

GSM General Sales Manager; Gibson Spiral Maze; Guildhall School of Music

GSMD General Society of Mayflower Descendants; Guildhall School of Music and Drama

GSMFC Gulf States Marine Fisheries Commission

GSML General Stores Material List

GSMNP Great Smoky Mountains National Park (Tennessee and North Carolina)

GSMOL Golden State Mobile-home Owners League

GSMS Geocentric Solar Magnetospheric System (NASA); Graduate Student of the Management Society

GSNC General Steam Navigation Company

GSNWR Great Swamp National Wildlife Refuge

gso ground speed outbound

GSO General Staff Officer; Girls Service Organization; Greensboro, North Carolina (airport); Ground Safety Officer

GSOST Goldstein-Scherer Object Sorting Test

gsp gas paid for; gas planned

GSP Generalized System of Preferences; Good Service Pension

GSPA Gulfport State Port Authority

G-spot Grafenberg spot—erotic vaginal area present in some women

GSPOT Geometric Spot Analysis System

gsps guidance spare power supply

gsr galvanic skin reflex; galvanic skin response; general service reinforcement(s); ground speed return; ground surveillance radar

GSRI Gulf South Research Institute

GSRS General Support Rocket System

gsrv globe stop radiator valve

gss guidance system simulator

GSS General Service School; General Social Services; General Supply Schedule; Geo-Stationary Satellite; Gilbert and Sullivan Society; Global Surveillance System; Grumman Standard Specification

GSSF General Supply Stock Fund

GSSH Grand Street Settlement House

GSSL Genoa, Savona, Spezia, and Leghorn (ports)

GSSLNCV Genoa, Savona, Spezia, Leghorn, Naples, Civetta, and Vecchia (ports)

GSSR Georgian Soviet Socialist Republic; Ground Support System Review

GSSS Ground Support System Specification(s)

GSST Goldstein-Scheerer Stick Test

gst garter stitch (knitting); ground special tools

GST General Service Test; General Staff Target; Goods and Services Tax (Canadian); Greenwich Sidereal Time; Guamanian Standard Time

GSTC Gorham State Teachers College

gste guidance system test equipment

GSTP Generalized System of Tariff Preferences

G-string capital-G-shaped string-like genital covering worn by exotic entertainers; violin's lowest string

gsts guidance system test set

gstu guidance system test unit

gsu glazed structural units; ground-support unit

GSU General Service Unit; Georgia State University; Gulf States Utilities

gsub glazed structural unit base

gsuc ground stub-up connection

g-suit antigravity suit worn during supersonic flight

GSUSA Girl Scouts of the USA

GSUSDA Graduate School, United States Department of Agriculture

gsv globe stop valve

GSV Grumman Submersible Vehicle; Guided Space Vehicle

gsvr ground-to-surface vessel radar

gsw gunshot wound

GSW Fort Worth, Texas (Greater Southwest International Airport)

GSW 1812 General Society of the War of 1812

GS & WR Great Southern and Western Railway

gt gas turbine; gastight; gilt; gilt top; glass tube; grand total; grease trap; great; greater than; greetings telegram; gross tonnage; gross ton(s); ground transmit(ter); group technology; gun target; gut tripe

g/t gooseneck tunnel; grams per ton; granulation time; granulation tissue

gt *gate* (Norwegian—street)

gt. *gutta* (Latin—drop)

Gt Great; Greenwich time

Gt *Groot* (Afrikaans—big, large, vast)

GT General Tariff; Good Templar; Goodyear Tire & Rubber (stock exchange symbol); Grand Trunk (Pakistan road); Gran Turismo; Grand Tiler; Grupo de Transportes (Transport Group)

G/T Gas Turbine (vessel)

GT *Gamle Testamente* (Dano-Norwegian—Old Testament); *Gran Turismo* (automobile)

gta gas-tungsten arc; graphic training aid

GTA Gatt Textiles Arrangement; Geography Teachers Association; Gospel Truth Association; Government Telecommunications Agency; Graduate Teaching Assistant; Gravure Technical Association; Gun Trade Association

GTAP General Technical Assistance Program

gtaw gas turbine arc welding

Gtb Godthab

GTB Government Tourist Bureau

GTBC Guild of Teachers of Backward Children

Gt Br Great Britain

Gt Brit Great Britain

gtc gain time control; gas turbine compressor; good till cancelled

GTC Girls Training Corps; Government Training Center; Government Travel Center; Guam Territorial College; Guild of Television Camera-

men; Gulf Transport Company (railroad)

GTCs Government Training Centres (UK)

gtd geometrical theory of diffraction; guaranteed

GTDS Goddard Trajectory Determination System (NASA)

gte general total energy; gilt top edge; ground test(ing) equipment; guidance test(ing) equipment; gunner tracking evaluator

gt-e gas turbo-electric

gte *gerente* (Spanish—manager)

GT & E General Telephone and Electronics (Corporation)

GT & EA Georgia Teachers and Education Association

GTEC Georgia Institute of Technology

gtee goatee; guarantee

gtee od guaranteed overdraft

GTEIS General Telephone and Electronics Information System

GT & EL General Telephone and Electronics Laboratories

GTEP Guaranteed Training Equipment Program

gtev (GTEV) gas-turbine electric vessel

gtf glucose tolerance factor

GTF Gang Task Force (police function); General Trust Fund; Government Test(ing) Facilities; Government Test(ing) Facility; Granite Test Facility; Great Falls, Montana (airport)

gtg gas to gasoline

GTG Sappho's daddy

gth go to hell

gth (GTH) gonadotrophic hormone

gthtgr (GTHTGR) gas-turbine high-temperature gas-cooled reactor

gti general transportation importance

GTI Grand Turk Island (tracking station)

GTIL Government Technical Institute Library

GTIO German Tourist Information Office

GTL Glass Technology Laboratories

Gt Ldn Greater London

gtm good this month; gross ton(nage) mile(s)

GTM General Traffic Manager

GTMA Gauge and Tool Makers Association

Gt Man Greater Manchester

Gtmo Guantanamo Bay
GTMS Graphic Text Management System
gtn glomerulo-tubulo nephritis
GTN Government Training News
GTNP Grand Teton National Park (Wyoming)
gto gate turnoff
gto (GTO) geostationary transfer orbit
g to go to (calculator)
Gto Gunajuato
GTO Government Team of Officials
GTO Gran Turismo Omologato [hard-top type of high-performance auto certified *(omologato)* to enter Gran Turismo automobile race]
gtol graphic takeoff language; ground takeoff and landing
gtow gross takeoff weight
gtp ground test(ing) plotter
gtp (GTP) guanosine triphosphate
GTP General Test Plan
gtr gantry test rack; greater; ground test(ing) reactor
GTR Grand Trunk Railway; Gurkha Transport Regiment(als)
GT-R Grand Touring-Racing (version)
Gtr Ant Greater Antilles
gtrp general transpose
gts guidance test(ing) set
gts (GTS) geostationary technology satellite
gt's grand touring cars
g/t/s gas-turbine ship
Gts Gateshead
GTS gas turbine vessel (3-letter code); General Telephone System; General Theological Seminary; Global Telecommunications System (WMO); Greenwich Time Signal; Ground Transport System; Guinean Trawling Survey
GTSC German Territorial Southern Command (NATO)
GTSI Government Technology Services, Inc
gtss gas turbine self-contained starter
GTSTD Grid Test of Schizophrenic Thought Disorder
gtt glass transition temperature
gtt gelatin-tellurite-taurocholate
gtt. guttae (Latin—drops)
gtT gone to Texas (one jump ahead of the sheriff)
GTT Glucose Tolerance Test
gtu guidance test unit

GTU Graduate Theological Union
gtv gate valve; gravity vacuum transit
gtv (GTV) gas turbine vessel
GTV Gumma Television (Japan)
gtw good this week; gross ton(nage) weight
GTW Grand Trunk Western (railroad)
Gtwy Gateway
gty gritty
Gtz Galatz
gu gastric ulcer; genitourinary; geographically unsuitable; glycogenic unit
Gu Guinea; Gujarat; Gujarati
Gu Göteborgs Universitetsbiblioteket (Swedish—Gothenburg's University Library)
GU genito-urinary; Georgetown University; Glasgow University; Gonzaga University; Griffith University; Guam
GUA Guatemala City, Guatemala (airport)
Guad Guadeloupe
Guadal Guadalajara
Guadalupe Mountains Texan national park
Guadalupes Guadalupe Mountains of New Mexico and Texas
Guadarramas Guadarrama Mountains of central Spain *(Sierra de Guadarrama)*
Guam Pacific Island possession of United States; inhabited by Guamanians
Guam ST Guamanian Standard Time
Guana Guanajuato
'Guana Iguana Island, British Virgin Islands
Guanacastes Guanacaste Mountains of northwestern Costa Rica *(Cordillera de Guanacaste)*
Guanajay Cuban prison in Guanajay southwest of Havana in Pinar del Rio province
'guana(s) iguana(s)
GUANOMEX Guanos y Fertilizantes de México (Spanish—Guanos and Fertilizers of Mexico)
guar guarantee
Guar Guarani (Brazil)
guaraní monetary unit of Paraguay
guard. guaranteed
GUARD Government Employees United Against Discrimi-

nation
Guardian The Guardian (leading British newspaper published in London and Manchester)
guarg guaranteeing
Guat Guatemala(n)
GUATEL Empresa Guatemalteca de Telecomunicaciones (Spanish—Guatemalan Telecommunications Enterprise)
Guatemala Republic of Guatemala (Central American country), *República de Guatemala*
Guay Guayaquil
Guaya Guayaquil
gub generalized upper boundary
GUBC Guyana United Broadcasting Company (Radio Demerara)
gubernalection gubernatorial election
Guer Guerrero
GUGK Glavnoje Upravlenije Geodesii i Kartografii (Russian—Administrative Agency for Geodesy and Cartography)
GUGMS Glavnoje Upravlenije Gidrometeorologicheskoi Sluzhby (Russian—Administrative Agency of the Hydrometeorological Service)
Gug Mus Guggenheim Museum
Gui Guinea
GUI Golfing Union of Ireland
Gui-Bis Guinea-Bissau (formerly Portuguese Guinea)
Gui Cur Guinea Current
guid guidance
guide. guidance for users of integrated data equipment
guidn guidance
Gui E Guinea Ecuatorial (Spanish—Ecuatorial Guinea)
guil guilder(s)
Guil Guillaume
Guild The Newspaper Guild (American periodical)
guilder monetary unit of Netherlands, Netherlands Antilles, and Surinam
Guildhall Lib Guildhall Library (London)
Guild Prof Trans Guild of Professional Translators
Guillaume Appolinaire Guillaume Appolinaire de Kostrowitsky
Guill° Guillermo (Spanish—William)

guin guinea(s)
Guinea Republic of Guinea (West African nation), *République de Guinée*
Guinea-Bissau Republic of Guinea-Bissau (former West African colony of Portugal)
Guip Guipuzcoa
guit guitarra (Spanish—guitar); *guitarrazo* (Spanish—blow struck with a guitar); *guitar-rear* (Spanish—strumming the guitar); *guitarrista* (Spanish—guitar player); *guiterrera* or *guiterrero* (Spanish—guitar maker or guitar player)
Guj Gujarat; Gujarati
GULAG Chief Administration of Corrective Labor Camps, Prisons, Labor, and Special Settlements of the Soviet Secret Police (*q.v.* VOT)
GULC Georgetown University Law Center
Gulf Gulf of Adalia, Aden, Alaska, Alexandretta, Aqaba, Boothia, Bothnia, Cadiz, California, Cambay, Campeche, Canada, Carpentaria, Cattaro, Chihli, Chiriqui, Cutch, Darien, Eilat, Finland, Fonseca, Gabes, Genoa, Guayaquil, Guinea, Honduras, Izmir, Kotor, Kutch, Lepanto, Lions, Maine, Manaar, Maracaibo, Martaban, Mexico, Nicoya, Oman, Panama, Paria, Quarnero, Santa Catalina, Siam, Sidra, Smyrna, St Lawrence, Suez, Taranto, Tehuantepec, Tonkin, Venice; Gulf Oil; Spencer Gulf
GULF Gays United for Liberty and Freedom
Gulf Coast coastline of Texas, Louisiana, Mississippi, Alabama, and Florida
Gulf Islands national seashore along Florida and Mississippi
Gulfs Gulf Islands off Pascagoula, Mississippi
Gulf States Florida, Alabama, Mississippi, Louisiana, and Texas along the Gulf of Mexico; Iran, Iraq, Kuwait, Saudi Arabia, Bahrain, Qatar, United Arab Emirates, and Oman along the Persian Gulf
Guli Gulielma
Gull Gullah (dialect of coastal South Carolina, Georgia, and Florida)
gulp (data-processing slang—a succession of bytes)

gulp. (GULP) general utility language processor
GULP General Utility Library Program
gum. (GUM) genito-urinary malignancy
Gum Guam (container port)
GUM Guam (airport)
GUM Gosurdarstvennoe Univ-ersalny Magasin (Russian—State Universal Store)
gumbo mucilaginous alluvial mud; slang for okra or a person of Black and Cajun ancestry
Gumbo patois spoken in Louisiana
Gumbo French dialect prevailing in parts of Louisiana
gums gum trees (eucalyptus)
gun. guncotton; guncrete; gunnery; gunpowder
gun gunung (Malay—mountain)
gun dip gunboat diplomacy
gun'l gunwale
Gun Sgt Gunnery Sergeant
GUNSS Gunnery Schoolship (USN)
guo government use only
GUO Greater Union Organization
GUOOF Grand United Order of Odd Fellows
gup guppy
GUPCO Gulf Petroleum Corporation; Gulf of Suez Petroleum Company
guppie(s) gay urban professional(s)
guppy. greater underwater propulsive-powered (guppy-shaped) submarine
gups guppies
gup(s) gay urban professional(s)
GURC Gulf Universities Research Corporation
Gus August; Augustus; Gustaf; Gustave; Gustavus
GUs Guns Unlimited
GUS Globe Universal Services; Great Universal Stores; Grocers United Stores
Gus Rom Gustavo Romero
Gussies Great Universal Stores
gust gustation; gustatorily; gustatory; gustily; gustiness; gusto; gusty
Gustus Augustus
gut. gutter
Gut Gutenberg; The Gut (Valetta's redlight street on Malta)
Gutenberg Johannes Gensfleisch (German—John

Gooseflesh)—the inventor of movable type
GUTS Georgians Unwilling to Surrender
gutt. gutta (Latin—drop)
guttat. guttatim (Latin—drop by drop)
gutt. quibus. guttis quibusdam (Latin—a few drops)
guv governor
GuV Gerecht und Volkommen (German—correct and complete)
guv'nor governor
Guy Guayaquil; Guido; Guyana; Guyon
Guyana Cooperative Republic of Guyana (formerly called British Guiana or Demerara, South America's only English-speaking nation)
Guyane Guyane français (French Guiana)
Guybau Guyana Bauxite
Guy Esse Guyana Essequibo (between Guyana's Essequibo River and Venezuela's eastern border)—area claimed by Venezuela for more than a century
Guy's Guy's Hospital (London)
gv gate valve; gentian violet; government valuation; gravimetric volume; grid variation; ground visibility
gv (GV) granulosis virus
gv grande vitesse (French—fastfreight train); *gran velocidad* (Spanish—high velocity)
Gv Gustav
GV gigavolt; Giuseppe Verdi; Göta Verken (steel company); grid variation
gva general visceral afferent
GVA Geneva, Switzerland (airport); Grants by Voluntary Agencies
gvb gelatine veronal buffer
GVB Guam Visitors Bureau
GVC Grand View College
gve general visceral efferent
Gve Grove; Gustave
GVF Grazhodanskii Vozdushnyi Flot (Russian—Civil Air Fleet)
gvh graft versus host (reaction in bone marrow transplants)
gvhd graft versus host disease(s)
gvhr graft versus host reaction(s)
gvhrr geosynchronous very-high-resolution radiometer
GVI Gas Vent Institute

gvl gravel
GVL Global Van Lines
gvm generating voltmeter; gross vehicle mass
GVMDS Ground-Vehicle Mine-Dispensing System
GVN George V Novotny; Government of Vietnam
gvo gross value of output
GVP General Vice President
GVP *Gereformeerd Politiek Verbond* (Dutch—Reformed Political Union)
GVRD Greater Vancouver Regional District
GVS Government Vehicle Service
gvt government; gravity vacuum transit; gravity vacuum tube
GVT Ground Vibration Test(ing)
gvty gingivectomy
gvw gross vehicle weight
gvwr gross vehicle weight rating
gw gigawatt(s); green weight; ground wave(s); guerilla warfare
g/w gross weight(s)
GW George Washington—first President of the United States; Great Western (savings)
G-W Globe-Wernicke
G & W Gulf and Western
G + W Gulf and Western
GWA Girl Watchers of America; Golden West Airlines
GWA *Goode's World Atlas*
GWAA Golf Writers Association of America
GWB George Washington Bridge
gwc gas-water contact (oil well)
GWC George Washington Carver
GWCHS George Washington Carver High School
GWCM George Washington Carver Museum
gwcswbd gunnery weapon-control switchboard
gwe gigawatts electrical
gwen ground-wave emergency network
Gwen Gwendolyn
Gwenda Gwendolen

Gwennie Gwendolen
GWG George Washington Geist
gwh gigawatt hour
gwh/day gigawatt hours per day
GWHNWR Great White Heron National Wildlife Refuge (Florida)
GWHS George Washington High School; George Westinghouse High School
GWI Grinding Wheel Institute; Ground Water Institute
G'wich Village Greenwich Village
Gwin Gwinett
GWK Grenswisselk-Kantoren
GWMNP George Washington Memorial National Parkway
GWO General Wage Order
GWOA Guerrilla Warfare Operational Area
gwp (GWP) gross world product
GWP *Government White Paper*
GWPA *Grote Winkler Prins Atlas* (Dutch—Great Winkler Prins Atlas)
GWR General War Reserves; Great Western Railway
GWRI Ground Water Resources Institute
gws grid-wire sensor
GWS Geneva (Convention for the Amelioration of the) Wounded and Sick (in Armed Forces in the Field); George Washington School; Gir Wildlife Sanctuary (India)
gwt glazed wall tile
GWTA Gift Wrappings and Tyings Association
G W T W *Gone With The Wind*
GWU George Washington University
GWVA Great War Veterans Association
GWWD Greater Winnipeg Water District (Railway)
Gwyn Gwynedd; Gwynne
gx (GX) government exhibit
gxmtr guidance transmitter
gxt graded exercise test(ing)
gy gray; gunnery; gyro; gyrocar; gyrocompass; gyrodyne; gyroscope
gY greenish yellow

Gy gray
gya got yuh again (slang for caught you again)
GYE Guayaquil, Ecuador (airport)
gym gymnasium; gymnastics
Gym Gymnastics
GYM General Yard Master; Guyamas, Mexico (tracking station)
gymstic gymnastic(s)
gyn gynecology
gyn *gyne* (Greek—woman)— gynecologist, gynecology, misogyny, polygyny
G.Y.N. gynecologist
gynae(col) gynaecological; gynaecologist; gynaecology
gynecol gynecologic(al)(ly); gynecologist; gynecology
gyp gypsum; gypsy; cheat or swindle (slang)
GYP Guild of Young Printers
gyp bd gypsum board
gypsiol gypsiologic(al)(ly); gypsiologist(s); gypsiology
gypsum calcium sulfate $(CaSO_4 \cdot 2H_2O)$
'gyptian(s) Egyptian(s)
gyro gyrocompass; gyroplane; gyroscope
gyrocop gyrocopter
gyrocopter autogyro helicopter (rotary-wing aircraft driven forward by a conventional propeller)
gyrodyn gyrodynamic(al)(ly); gyrodynamicist; gyrodynamics
GYS Co Great Yarmouth Shipping Company
gywp gee you're wonderful, professor
gz ground zero
Gz Gomez
GZ Girozentrale Vienna
G-Z Guilford-Zimmerman test(ing)
GZ *Girozentrale Vienna* (German—Vienna Central Exchange)—Austrian international bank
GZG *Gutegemeinschaft Zinngerat* (German—Pewter Quality Society)
GZn grid azimuth
GZT Greenwich Zone Time

H

h hail; hard; hardening; hardness; hazy; hectare; hecto; height; high(er); hit(s); home; horse; hours(s); house; hundred(s); husband; hydrant; hydraulic(s); hydrodynamic head (symbol); hydrolysis; Planck's constant (symbol); Planck's element of action (symbol)
(h) per hypodermic
h altitude (symbol); atmospheric head (symbol)
H amateur broadcasting (symbol); ceiling (symbol); Fraunhofer line produced by calcium (symbol); Hamiltonian function (symbol); Hangarage; Harbor; hard; hardness; hatch; headlines; heart disease potential; heat; heater; helicopter; henry; heroin; Hill; Hindu; Hinduism; horizontal component of the earth's magnetism (symbol); hot; Hotel—code for letter H; humidity; hydrogen; hyperopia; intensity of magnetic field (symbol); latent hypermetropia (symbol); maximum altitude (symbol); McDonnel Aviation; Minneapolis-Honeywell (trademark); very hazy (symbol)
H hacienda (Spanish—customs service, treasury); *haut* (French—high); *heet* (Dutch—hot); *Herren* (German or Swedish—gentlemen); *herrer* (Norwegian—gentlemen); *het* (Norwegian—hot); *hinaus* (German—out); *hombres* (Spanish—men); *Hoyre* (Norwegian—Right)—Conservative Party
H- *Hauptstimme* (German—principal voice)—12-tone term
H¹ protium
H¹+ proton
H–1 symbol for radar air navigation system
H-2 Australian macadamia
H² deuterium (heavy hydrogen symbol)
H₂O water
H₂O₂ hydrogen peroxide
H₂S hydrogen sulfide
H₂SO₄ sulfuric acid
H₃ procaine hydrochloride (symbol)
H³ tritium
H₃BO₃ boric acid
H-13 Bell three-place helicopter named Sioux and made in Britain, Italy, and Japan
H-19 Sikorsky transport helicopter called Chickasaw or UH-19
H-23 Hiller utility helicopter used by USA and called Raven
H24 hard rolled and partially annealed (half hard)
H-34 Sikorsky troop-transport helicopter called Choctaw
H-37 Sikorsky heavy helicopter called Mojave
H-43 Kaman utility helicopter called Huskie
H-53 Sikorsky CH-53 Stallion assault helicopter
ha hahnium, unnilpentium; hardy annual; hatch(way); hectare; heir apparent; high altitude; high angle; home address; hostile aircraft; hour angle; hour aspect
ha (HA) humic acid

ha' half
h.a. *hoc anno* (Latin—in this year)
Ha hahnium (element 105); Haiti(an); Hawaii(an)
Ha (German pronunciation for B sharp)
HA Hatch Act; Hawaiian Airlines; Headquarters Administration; Heavy Artillery; Horse Artillery; Hospital Apprentice; House of Assembly; Housing Authority
H-A Hautes-Alpes
H/A Havre-Antwerp (range of ports)
HA Hardware Age
HA-200 Hispano Saeta jet trainer
HA-220 Hispano Saeta ground-attack jet fighter
haa heavy anti-aircraft; heavy antiaircraft artillery; height above airport
haa (HAA) hepatitis-associated antigen
HAA Helicopter Association of America; Hospital Activity Analysis; Hotel Accountants Association; Humanist Association of America
haaat height of (transmission) antenna above average terrain
HAAC Harper Adams Agricultural College
HAAFE Hawaiian Army and Air Force Exchange
haandb haandbog (Dano-Norwegian—handbook)
Ha'aretz (Hebrew—The Land)—Israeli newspaper
Haar Haarlem (Dutch—Harlem) provincial capital of North Holland in the Netherlands

haat height (of tv transmission antenna) above average terrain

haatc high altitude air traffic control

haaw heavy anti-tank assault weapon

hab high-altitude bombing; habitat; habitation

hab *habitantes* (Spanish-inhabitants)

Hab *Habakkuk; Habana* (Spanish—Havana); The Book of Habakkuk

HAB Hazards Analysis Board (USAF)

HAB Handels Aktie Bolag (Swedish—Limited Trading Company)

HABA Hardwood Agents and Brokers Association

hab. corp. habeas corpus (Latin—may you have the body)—prisoner's right to be brought before the court so its judge may decide on the legality of the detention

habe habeas corpus

habit. habitat (Latin—it inhabits)

habs high-altitutde bombsight

HABS Historic American Buildings Survey

habt. habeat (Latin—let him have)

hac high acceleration cockpit; high alumina cement

HAc acetic acid

HAC Hawkesbury Agricultural College; Helicopter Aircraft Command(er); Hines Administrative Center; Honourable Artillery Company; Hughes Aircraft Company

hacc high alumina cement concrete

HACC Harrisburg Area Community College

hace (HACE) high-altitude cerebral edema

hack hackney coach; hackney horse; taxicab

hack. hacking (illegally breaking into computerized systems)

Hack Hackbrett (German—cymbalom or dulcimer)

Hackworth *Hackworth's Digest of International Law*

hacls (HACLS) harpoon-type aircraft command and launch subsystem missile

HACTL Hong Kong Air Cargo Terminal Limited

HACU Hansa (Line) Container Unit

had. head acceleration device; heat-actuated device (thermostat); hereinafter described

Had Hadley

H/A or D Havre-Antwerp or Dieppe (grain trade)

H^{ada} *Hacienda* (Spanish—estate, farm, ranch)

HADA Hawaiian Defense Area

hadbn had been

HADC Holloman Air Development Center

HADES Hypersonic Air Data Entry System

HADIS Huddersfield and District Information Service

HADIZ Hawaiian Air Defense Identification Zone

hadn't had not

HADR Hughes Air Defense Radar

hads hypersonic air data sensor

hads. hypersonic air data sensor

hae hereditary angioedema

ha'e (Gaelic contraction—have)

Haeck Ernst Heinrich Haeckel; Haeckelian; Haeckelism

HAECO Hong Kong Aircraft Engineering Company

HAER Historic American Engineering Record

haes high-altitude-effects simulation

haf high-abrasion furnace; high-altitude fluorescence

HAF Hebrew Arts Foundation; Helicopter Assault Force; Hellenic Armed Forces; Helms Athletic Foundation; Helvetia-America Federation

HAFB Homestead Air Force Base (Florida)

haf black high-abrasive furnace black

HAFMED Headquarters—Allied Forces Mediterranean

HAFO Home Accounting and Finance Office (USAF)

HAFRA Hat and Allied Feltmakers Research Association

HAFSE Headquarters, Armed Forces, Southern Europe

HAFTB Holloman Air Force Test Base

Hag The Book of Haggai; The Hague

Hag Haggai

HAG Hardware Analysis Group

HAGB Helicopter Association of Great Britain

hagio hagiogracies, hagio-

gracy, hagiographer, hagriographic(al)(ly), hagiographist, hagiography, hagiolater, hagiolatrous, hagiologic(al)(ly), hagiologies, hagiologist(ic)-(al)(ly), hagiology, hagiolotry, hagioscope, hagioscopic

hagiol hagiology

Hague The Hague

HAI Hospital Audiences Incorporated

HAIA Hearing Aid Industry Association

haic hetero-atom in context

haid hand-emplaced acoustic intrusion detector

HAIL Hague Academy of International Law

H&A Ins Health and Accident Insurance

hair. high-accuracy instrumentation radar

hairdrsr hairdresser

hairies long-haired hippies

HAISS High-Altitude Infrared Sensor System

hait haitisch (German—Haitian)

Hait Haitian

Haiti Republic of Haiti (French-speaking West Indian nation occupying western half of Hispaniola), *République d'Haiti*

HAJ Hanover, Germany (airport)

Hak Hakka; Hakodate

HAKASH Hayl Kashish (Hebrew—Army of Elders)—Israel's senior-citizen corps

Haken Hakenkreuz (German—hooked cross); swastika

Hak Soc Hakluyt Society

hal halogen(ic); handicapped assistance loan; helmet audio link(age)

Hal Halawa; Halbert; Halcott; Halden; Halensee; Halex; Halford; Halfrid; Haliburton; Hallam; Halleck; Hallett; Halogen; Halsey; Halsom; Halstead; Halvar; Halworth; Harold; Harry

HAL Hamburg-Amerika Linie (Hamburg-America Line); Hamburg-Atlantic Line; Hardboards Australia Limited; Hawaiian Airlines; Holland America Line

Halawa Halawa Jail at Aiea on Oahu, Hawaii

HALDIS Halifax and District Information Service

Haleakala Haleakala National Park and Haleakala Volcano

on the Hawaiian island of Maui

half-g half-gallon(s)

Hali Halifax

halite rock salt (sodium chloride)

Halle a/S Halle an der Saale (German—Halle on the Salle River)

Halle Bach Wilhelm Friedemann Bach

Hallowell Girls Stevens School for female juvenile delinquents at Hallowell, Louisiana

hallu hallucinant; hallucinate; hallucination; hallucinogen; hallucinogenic

halluc hallucination

hallus hallucinations; hallucinogens

halo. high-altitude large optics; high-altitude low opening

Hal Orch Hallé Orchestra (Manchester)

HALS Harwell Automated Loans System

HALT Help Abolish Legal Tyranny; High-Altitude Laser Targeting; Houston Anti-Litter Team

haltata high-and-low-temperature-accuracy testing apparatus

Halterm Halifax Container Terminal

halv hamster leukemia virus

Halv Halvøy (Dano-Norwegian—peninsula)

ham. hardware-associated memory

Ham Hamal; Haman; Hamblin; Hamburg; Hamed; Hamilton; Hamitic; Hamlet; Hamlin; Hamlyn; Hammerfest; Hamnet; Hamon

HAM Hamburg, Germany (airport)

HA & M Hymns Ancient and Modern

ham ham and eggs

Hamb Hamburg

Haml Hamlet, Prince of Denmark

hamlet ham omelet

hamletom ham, lettuce and tomato (sandwich)

Hamm Hammerfest, Norway

hamma' hammer

hammer and sickle communist symbol; the crossing of the proletarian hammer and the agrarian sickle appears on the flags of the Congo and the USSR

Hammersleys short form for the Hammersley Mountains of Western Australia

ham 'n' eggsan ham-and-egg sandwich

ham 'n' eggwich ham-and-egg sandwich

Hamp Hampton Roads; Lionel Hampton

Hamptons Bridgehampton, Easthampton, Southhampton, and Westhampton, Long Island, New York; collective short form for all Hamptons such as Bridgehampton, East Hampton, Hampton Bays, Southampton, West Hampton, and West Hampton Beach—all at the eastern end of Long Island, New York, plus the original English estates and homestead placenames such as Hampton, Hampton Bishop, Hampton Court Palace, Hampton Heath, Hampton in Arden, Hampton Lovett, Hampton Lucy, Hampton Poyle, Hampton Wick, and Northampton as well as the great port of Southampton, and adjacent Southampton Water plus all other Hamptons

hams. hour angle of the mean sun

hamsan ham sandwich

hamt human-aided machine translation

HAMTC Hanford Atomic Metal Trades Council

hamwich ham sandwich

han' hand

Han Handel; Handel Society; Hanover(ian)

hand. handling

HAND Hawaii Association for National Defence

Handb Phys Handbuch der Physik (German—Handbook of Physics)

hande hydrofoil analysis and design

Handl Handlingar (Swedish—transactions)

HANDS High-Altitude Nuclear-Detection Studies

hane high-altitude nuclear effects; high-altitude nuclear explosion

HANES Health and Nutrition Examination Survey

HANG Hawaii Air National Guard

hanki handkerchief

Hanover Girls Jane Porter Barrett School for (delinquent) Girls at Hanover, Virginia

Hans Johann(es)

HANS High-Altitude Navigation System

han't has not; have not

Hants Hampshire

Hanuk Hanukkah (Hebrew—Feast of Candle Lights)

HANZ Hotel Association of New Zealand

hao hardware action officer; high-altitude observation

HAO High Altitude Observatory; Horticultural Advisory Office(r)

haoa hight angle of attack

hap happening; heading axis perturbation

hap (HAP) high-altitude platform

HAP Home Attendant Program

HAPAG Hamburg-American Line

HAPAG Hamburg-Amerikanische Packetfahrt Aktien Gesellschaft (German—Hamburg-American Packet Company)

hapd happened

hapdar hardpoint demonstration array radar

hapdec hard point decoy

HAPE High-Altitude Pulmonary Edema

ha'penny halfpenny

ha'p'orth half-pennyworth

happ high air pollution potential

haps happenings

hap's housing assistance payments

har harbor; harmonic

Har Harbin; Harbor; Harbour; Harold; Harwich

HAR Harrisburg, Pennsylvania (airport)

HARAO Hartford Aircraft Reactor Area Office

Harare, Zimbabwe formerly Salisbury, Rhodesia

harb harbor

Harbison Girls Harbison Correctional Institution for Women at Irmo, South Carolina

Harbrace Harcourt Brace Jovanovich

HarBraceJ Harcourt Brace Jovanovich

HARB Historic Architectural Review Board

HARC Human Affairs Research Center

harcft harbor craft

Harcourt Harcourt Brace Jovanovich

hard. hardware
Hard Hardangerfelen (Norwegian—Hardanger fiddle)
Harden (British contraction—Harwarden)
hard porn hard-core pornography
hardtack ship's biscuits
Hardwick Girls Georgia Rehabilitation Center for Women at Hardwick
hare. high-altitude ramjet engine
Hare Soviet Mi-1 utility helicopter
HAREP Harbour Repairs
HARES High-Altitude Radiation Environment Study
Har Hakarmel (Hebrew—Mount Carmel)
HARIS High-Altitude Radiation Instrument System
harm. harmonic; harmony
harm. (HARM) high-speed anti-radiation missile; hypervelocity anti-radiation missile
HARM high-speed anti-radiation missile; Humans Against Rape and Molestation
harn harness
harn lthr harness leather
harp. harpoon; harpsichord; harpsichordist; heater above reheat point; heating, air conditioning, refrigeration, plumbing; high-altitude relay point; high-altitude research probe
Harp Halpern's anti-radar point
HARP Helmlich-Armstrong-Rieveschi-Patrick (aerospace heart pump); Honeywell Acoustic Research Program
Harp Baz Harper's Bazaar
Harper Harper & Row
harps. harpsichord
Har-Row Harper and Row
Harry Harold; Henry
hart. (HART) hyper-velocity anti-aircraft rocket tactical
Hart Hartford
HART Halt All Racist Tours; Highway Advisory Radio Tactical; Honolulu Area Rapid Transit
Hartran Hartwell Atlas fortran
Hart Sym Orch Hartford Symphony Orchestra
HARU Harrison Line (container) Unit
harv (HARV) high-altitude research vehicle
Harv Harvard; Harvey
Harw Harwarden *(Harden)*
HARYOU Harlem Youth Opportunities Unlimited
has. high-altitude sample
Has Haselhorst
HAs Hispanic Americans; Housing Assistants
HAS Health Advisory Service; Helicopter Air Service; Hellenic Affiliation Scale; Hospital Adjustment Scale; Hospital Administrative Services; Housing Alternatives for Seniors
HASAWA Health and Safety at Work Act
HASC House (of Representatives) Armed Services Committee
HASCO Haitian-American Sugar Company
HASD Humanist Association of San Diego
hash. hashish
Hashbury Haight-Ashbury (district of San Francisco)
Hasid Hasidim (Hebrew—godly pious people)
HASL Health and Safety Laboratory (Atomic Energy Commission)
hasn't has not
hasp. hardware-assisted software polling; high-altitude sampling program; high-altitude space platform
HASP Hawaiian Armed Services Police; Houston Automatic Simulator of Peripherals
haspa high-altitude superpressure-powered aerostat
hasr high-altitude sounding rocket
hast high-altitude supersonic target
Hastings Hastings House; Hastings-on-Hudson
Ha strain Harris (viral) strain
hasvr high-altitude space-velocity radar
hat. height above touchdown; high-altitude temperature
HATA Hong Kong Association of Travel Agents
hato handling tool
hatoff highest astronomical tide of the foreseeable future
hatom highest astronomical tide of the month
hatoy highest astronomical tide of the year
HATRA Hosiery and Allied Trades Research Association
hatrack. hurricane and typhoon tracking
HATREMS Hazardous and Trace Emissions System
HATRICS Hampshire Technical Research Industrial and Commercial Service
hats. hour angle of the true sun
HATS Helicopter Advanced Tactical System
Hatteras Cape Hatteras, the Cape Hatteras National Seashore Recreational Area, Hatteras Inlet, Hatteras Island, the village of Hatteras—all part of North Carolina's Outer Banks
Hau Hausa
Hauptw Hauptwerk (German—great or chief work)
haust. haustus (Latin—a draught)
haut hautboy (oboe)
hav haversine
hAv hepatitis A virus
Hav Havre
HAV Havana, Cuba (airport)
HAVEN Help Addicts Voluntarily End Narcotics
haven't have not
havoc. histogram average ogive calculator
Havre Havre de Grace, Maryland; Le Havre (de Grace), France
haw highly active waste (radioactive); hour angle west
haw. (HAW) heavy anti-tank assault weapon
Haw Hawaii; Hawaiian
HAW Kauai, Hawaii (tracking station)
HAWA Hammond Ambassador World Atlas
Hawaii Volcanoes Hawaiian national park
hawb house air waybill
HAWC Help for Abused Women and Children, Salem, Mass.
HAWE Honorary Association for Women in Education
HAWEIT Hamburg-Wechsler Intelligence Test
hawk. (HAWK) homing-all-the-way kill (missile)
Hawks Nest Hawks Nest State Park, West Virginia
Haw'n Hawaiian
Hawna Hawaiiana
Hawn Isl Hawaiian Islands
Haw Tel Hawaii Telephone (company)
Hawthorn Hawthorn Books; Missouri state flower
hax hrir/apt interface (high-resolution infrared radiometer/automatic picture transmission)
Hay Hayle

HAYP Hire-A-Youth Program
haystaq have you stored answers to questions?
haz hazard; hazardous
hb half breadth; halfback; half-bound; half bow; handbook; hard black; hardback (book); hardy biennial; heavy barrel; heavy bombardment; heavy bombing; hemoglobin; herringbone; high band; hollow bar; homing beacon; horizontal bands; horizontal bombing; hose bib; human being
h/b handbook
Hb hemoglobin; Herbarium
H^b deuterium (heavy hydrogen symbol)
Hb Hoboe (German—oboe)
HB Hawke's Bay; Hawthorn Books; Hector Berlioz; High Bridge; House (of Representatives) Bill
H-B Huebner-Bleistein (process)
H & B Humboldt and Bonpland
HB Hindi Bharat (Hindustani—Republic of India)
Hba Habana (Spanish—Havana)
HBA Hispanic Bankers Association; Hoist Builders Association; Hollywood Bowl Association; Home Builders Account; Honest Ballot Association; Hospital Benefit Association; Housing Builders Association
h'back hatchback
h/back hardback
h B ag hepatitis B antigen
H-bar capital-H-shaped bar
HBAVS Human Betterment Association for Voluntary Sterilization
hbb hollow-bored bar
hbc high breaking capacity
HBC Hokkaido Broadcasting Company; House Budget Committee; Hudson's Bay Company
HbCO carbon monoxide hemoglobin
hbd hardboard; has been drinking; headboard; herein-before described
hbd (HBD) hydroxybutyrate dehydrogenase
HBD Harbor Board; Harbour Board
hbe hard-boiled egg(s)
H-beam capital H-shaped beam
hbf hepatic blood flow
Hbf fetal hemoglobin

Hbf Hauptbahnhof (German—depot, main station)
HBF Hospital Benefit Fund
Hbg Hamburg; Harrisburg; Helsingborg (Hälsingborg)
HBG Henry B(arbosa) Gonzalez; Hongkong Bank Group; Huntington Botanical Gardens
HBJ Harcourt Brace Jovanovich
hbk halfback; hardback (book); hatchback; hollow back (lumber); hollowback
Hbk Hoboken
HB & K Humboldt, Bonpland, and Kunth (botanists)
hblv (HBLV) human B-cell lymphotropic virus
hbm hard bowel movement
HBM His (Her) Britannic Majesty
hbn hazard beacon
HBNNR Hickling Broad National Nature Reserve (England)
HBNWR Holla Bend National Wildlife Refuge (Arkansas)
Hbo Hoboken
HBO Home Box Office
h/board hardboard; headboard
HBOG Hudson's Bay Oil and Gas
HBOI Harbor Branch Oceanographic Institution, Fort Pierce, Florida
H-bomb hydrogen bomb
HBO S oxyhemoglobin
hbp high blood pressure; hit by pitcher (baseball)
HBPS Home Building Plan Service
hbr has been reviewed
Hbr Harbor
HBR Hudson Bay Railway
HBR Harvard Business Review
hbr acw has been reviewed and concurred with
hbs (HBS) hulking building syndrome
hb's halfbacks
Hbs sickle-cell hemoglobin
HBS Harvard Business School; Hawaiian Botanical Society; Hope Botanic Gardens
Hbt Hobart
HB & T Houston Belt and Terminal (railroad)
hbt's human-breast tumors
hBv hepatitis B virus
H&BV Houston and Brazos Valley (railroad)
hbw highspeed black-and-white (photography)
Hbwr Halden boiling heavy water reactor

hby hereby
hc habitual criminal; hand control; hard copy; heating cabinet; hexachlorethane; high carbon; high-capacity; highly commended; hydrocarbons
hc (HC) hard copy
h/c hard cover(ed); held covered
h & c heroin + cocaine; hot and cold (running water)
h.c. hac nocte (Latin—tonight); honoris causa (Latin—out of respect for); hors commerce (French—not for sale, privately printed)
Hc computed altitude; Hermitian conjugate; Huntington's chorea
HC Hagerstown College; Hague Convention; Hamilton College; Hamline College; Hanover College; Harding College; Harpur College; Hartford College; Hartnell College; Hartwick College; Hastings College; Haverford College; Health Certificate; Heidelberg College; Helicopter Council; Hendrix College; Hershey College; Hertford College; Hesston College; High Commission(er); Higher Certificate; Highway Code; Hillsdale College; Hiram College; Hood College; Hope College; Hospital Consult(ation); Hospital Corps; House of Commons; House of Correction(s); Housing Commission; Housing Corporation; Howard College; Humphreys College; Hunter College; Huntingdon College; Huntington College; Huron College; Hussan College; Hutchinson College; hydrocarbon(s)
H C Holy Communion
H-C Harbison-Carborundum
H.C. High Commission
H of C House of Commons; House of Correction
HC Hartford Courant; haute-contre (French—high tenor)
HC-54 Douglas C-54 modified for search-and-rescue missions
hca held by civil authorities
HCA High Conductivity Association; Hobby Clubs of America; Hospital Corporation of America; Hotel Corporation of America; Hunting-Clan Air Transport
HC(A) Helicopter Coordinator

(Airborne)
HCAAS Homeless Children's Aid and Adoption Society
hcap handicap
H-caps heroin capsules
hcb hard-core base; hard-covered book; heating and cooling of buildings; hollow concrete block(s)
HCB House of Commons Bill
hcc hydraulic cement concrete
hcc (HCC) 25-hydroxycholecalciferol (vitamin D^3 metabolite)
HCC Hebrew Culture Council; Holyoke Community College
HCCA Health Care Consumers Association
HCCJ Harvard Center for Criminal Justice
hcd high current density
hcd (HCD) human chorionic gonadotropin
HCD Housing and Community Development
HCDCS Harmonized Commodity Description and Coding System
HC Deb *House of Commons Debates*
hce human-caused error
HCEEP Handicapped Children's Early Education Project
hcef henceforth
HC & ES Hull Chemical and Engineering Society
hcex high-speed color exterior
hcf haemolytic complement fixation; hardened compacted fibers; height-correction factor; high carbohydrate fiber (diet); high cycle fatigue; highest common factor; hundred cubic feet
hcf (HCF) high-carbon ferrochrome
HCF Health Care Financing; Honorary Chaplain to the Forces; Hospital Contribution Fund(ing); Hungarian Cultural Foundation
HCFA Health Care Financing Administration
hcfc (HCFC) halogenated chloro fluorocarbon
hcg horizontal location of center of gravity; human chorionic gonadotropin pregnancy test
hch (HCH) hexachlorocyclohexane (insecticide)
HCH Herbert Clark Hoover (31st President U.S.)
HCHI Hand Chain Hoist Institute
HCHP Harvard Community

Health Plan
HCI Handgun Control Inc; Hotel and Catering Institute
HCIL Hague Conference on International Law
HCIMA Hotel and Catering Institutional Management Association
HCIS House Committee on Internal Security
HCITB Hotel and Catering Industry Training Board
HCJ High Court of Justice
HCJC Howard County Junior College
hcl high cost of living; horizontal center line
h cl hanging closet
HCl hydrochloric acid (muriatic acid)
HCL Hod Carriers, Building and Common Laborers
H-class Soviet missile-launching nuclear-powered submarines called Hotel by NATO
HCM Her (His) Catholic Majesty
HCM *Ho Chi Minh* (Chinese—He Who Shines)
HCMC Ho Chi Minh City
hcmm heat-capacity map mission (NASA)
HCMPA Home Counties Master Printers' Alliance
hcmr heat-capacity mapping radiometer
HCMT Ho Chi Minh Trail
HCMW Hatters, Cap and Millinery Workers (union)
hcn hydrocyanic acid
HCn hydrocyanic acid
HCN House (of Representatives) Committee on Narcotics
HCNZ Housing Corporation of New Zealand
hco hydrogenated coconut oil
HCO Harvard College Observatory; Headquarters Catalog Office
HCO$_3$ bicarbonate ion
hcp handicap; hexachlorophene; hexagonal close-packed; high card point; humidity control(ler's) panel
HCP Honors Cooperative Program
HCP *House of Commons Proceedings*
HCPNI Hardware Cloth and Poultry Netting Institute
HCPT Historic Churches Preservation Trust
hcptr helicopter
HCR High Chief Ranger
HCRAO Hat Creek Radio As-

tronomy Observatory (University of California)
HCRC Honeywell Corporation Research Center
hcrit hematocrit
HCRS Heritage Conservation and Recreation Service
hcrw hot and cold running water
hcs high-carbon steel
hcs (HCS) human chorionic somatomammotropin; hydrological communications satellite
hc's hard cover books
HCs Hebrews converted to Roman Catholicism; hydrocarbons
HCS Hallé Concerts Society; Harvey Cushing Society; Home Civil Service; Hydromechanical Control System
HC & S Hawaiian Commercial and Sugar (company)
HCSA Hospital Consultants and Specialists Association; House (of Representatives) Committee on Space and Astronautics
hcsht high-carbon steel heat treated
HCSI Health Correspondence Schools International
hct hematocrit
H Ct High Court
HCT High Commission Territories; Huddersfield College of Technology
HCTBA Hotel and Catering Trades Benevolent Association
hcu homing comparator unit; hydraulic cycling unit
HCUA Honeywell Computer Users Association
HCV Housing Commission of Victoria
HCVA Historic Commercial Vehicle Association
HCVC Historic Commercial Vehicle Club
hcvd hypertensive cardiovascular disease
hd half day; hand; hard-drawn; head; hearing distance; heavy duty; high density; hogshead; horse-drawn; hourly difference; hundred; hurricane deck
hd (HD) half duplex (data processing); harmonic distortion (audio); harmonic definition (video)
h-d heavy-duty; high-density
h/d holddown
h.d. *hora decubitus* (Latin—at

bedtime)
Hd Head
Hd (HD) Huntington's disease
Hd Hochdruck (German—high pressure)
HD Hansen's Disease (leprosy); Harbor Defense; Harbor Drive; Historic District; Historical Division; Home Defense; Honorable Discharge; Hoover Dam
H.D. Hilda Doolittle
H/D Havre-Dunkirk (range of ports)
H & D Hurter & Driffield (photo emulsion speed)
hda high-duty alloy(s); horizontal danger angle; hydroxydopamine
HDA High Duty Alloys
hdatz high-density air traffic zone
HDB Housing Development Board
hdbk handbook
hdbn had been
hdc half double crochet; high-density cotton; holder in due course
HDC Helicopter Direction Center; Housing Development Corporation
HD Clinic Hansen's Disease Clinic (for lepers)
hd cr hard chromium
hdd head-down display; heavy-duty detergent
HDD Higher Dental Diploma
hddr high-density digital recording
HDDS High-Density Data System
HDE Higher Diploma in Education
h de c hidratos de carbono (Spanish—carbohydrates)
hded heavy-duty enzyme detergent
hdg heading
HDGA Hot Dip Galvanizers Association
HDH Hawker de Haviland
hdhc high-density hydrocarbon(s)
HDHD Hawaiian District Harbors Division
hdhl high-density helicopter landing (USA)
HDHQ Hostility and Direction of Hostility Questionnaire
HDI Human Development Index; Humane Development Institute
hdip hazardous-duty incentive pay
H Dip E Higher Diploma in

Education
H disease Hart's disease
hdk husbands don't know
h dk hurricane deck
hdkf handkerchief
hdl handle; hardware description language
hdl (HDL) high-density cholesterol; high-density lipoproteins
HDL Harry Diamond Laboratory (US Army Diamond Ordnance Fuze Laboratory); Hydrologic Data Laboratory (USDA)
HDLC High-level Data Link Control
hdlg handling
hdlr handler
hdls headless
hdlw hearing distance, watch at left ear
hdm high-duty metal
hdmi high-density multichip interconnect
HDML Harbor Defense Motor Launch
hdmr high-density moderated reactor
hdn harden
hdn (HDN) hemolytic disease of the newborn
H Doc House Document
hdp (HDP) hexose diphosphate
hdpe high-density polyethylene
hdpg half deck plate girder
hdqrs headquarters
hdr handrail; header
hd & r human development and relationships
HDRA Heavy-Duty Representatives Association
HDRI Hannah Dairy Research Institute
HDRSS High-Data-Rate Storage System(s)
hdrw hearing distance, watch at right ear
hds heat detectors; holidays; hundreds; hydrodesulfurization
Hds Holidays (of Obligation)
HDS Hospital Discharge Survey; Human Development Services
Hd Schm Head Schoolmaster
hdsp hardship
hdst high-density shock tube
HDST Hawaiian Daylight Saving Time
HDT Henry David Thoreau
hdta high-density-traffic airport
HDTI Human Development

Training Institute
hdtm heavy-duty target mechanism
HDTMA Heavy-Duty Truck Manufacturers Association
HDTS Harbor Drive Test Site (Convair Ramp)
hdtv (HDTV) high-definition television
hdu hemodialysis unit
h/duty heavy duty
hdv heavy-duty vehicle; high dollar value
hdw hardware
Hdwbch Handwörterbuch (German—pocket dictionary)
hdw c hardware cloth (wire screen)
hdwd hardwood
hdwe hardware
hd whl hand wheel
hdx (HDX) half duplex (data processing)
he. hammerless ejector; heat engine; heavy enamel; height of eye; high explosive; horizontal equivalent; hub end; human enteric; hydrogen embrittlement
h&e hemotoxylin and eosin; heredity and environment
h.e. hic est (Latin—this is)
He Hebraic; Hebrew; helium; Hertz
He. Book of Helaman
HE Her Eminence; Her Excellency; high explosive; His Eminence; His Excellency; Hollis & Eastern (railroad); Human Engineering; Hydraulics Engineer(ing)
H.E. His Eminence; His Excellency
HE Human Events
HEA Higher Education Act; Home Economics Association; Horticultural Education Association
heaa high-explosive anti-aircraft (shell)
HEAA Home Economics Association of Australia
head(s). headache(s)
HEADS-UP Health Care Delivery Simulator for Urban Populations
heaf heavy end aviation fuel
HEAF Higher Education Assistance Foundation
heafs high-explosive antitank fin-stabilized
HEAL Health Education Assistance Loans; Home Environment Aid for Living; Human Ecology Action League
HEALT Helicopter Employ-

ment and Assault Landing Table

HEAO High-Energy Astronomical Observatory

heap. high-explosive armorpiercing (shell)

HEAR Hospital Emergency Administrative Radio

HEARS Higher Education Administration Referral Services

HEARU Higher Education Advisory and Research Unit

heat. heat escape lessening posture; heating; high-explosive anti-tank (projectile)

HEATH Higher Education and the Handicapped

Heathrow airport near London

heavy water deuterium oxide

Heb Epistle of Paul the Apostle to the Hebrews: Hebraic; Hebrew

Heb Hebrew (classical language)

HEBA Home Extension Building Association

hebc heavy enamel bonded single cotton

hebd hebdomadal (weekly)

hebdo hebdomad(al)(ly); hebdomadaries; hebdomadary

hebdom. hebdomas (Latin—week)

hebdomag hebdomadal magazine—weekly magazine

hebdp heavy enamel bonded double paper

hebds heavy enamel bonded double silk

Heb Ety Hebretasebawit Etyopia (Amharic—Socialist Ethiopia)—formerly Abyssinia

hebr hebräisch (German—Hebraic)

Hebr Hebraic; Hebrew; Hebrides

Hebrides Hebrides Islands off Scotland

hec heavy-enamel single-cotton (insulation)

Hec Hasselblad electric camera; Hector; Hecuba; Hollerith electronic computer

HEC Hawaii Electric Company; Health Education Council; Hydro-Electric Commission

HEC Hautes Études Commerciales (French—High Commercial Studies) France's leading business school; *Heure de l'Europe Central* (French—Central European Time)

HECC Higher Education Coordinating Council (St Louis library network)

he cls b heating coils in bunkers

he cls ct heating coils in cargo tanks

HECO Hawaiian Electric Company; Hydro-Electric Commission of Ontario

he comp helium compressor

hect hectare; hectoliter

Hect Hector

HECT Hydro-Electricity Commission of Tasmania

hecticity hectic activity

hecto 10^2

hectog hectogram

hectol hectoliter

hectom hectometer

hector. heated experimental carbon thermal oscillator reactor (HECTOR)

hed horizontal electric dipole

hed (HED) high-energy detector

he'd he had; he would

HED Haupt-Einheits Dosis (German—unit skin dose)—X-rays

HEDCO Hawaii Economic Development Corporation

HEDCOM Headquarters Command

HEDL Hanford Engineering Development Laboratory

hed('s) hearing-ear dog(s)

HEDS Hall-Effect Distribution System

hed sked headline schedule

hedsv heavy-enamel double-silk varnish (insulation)

Hedy Hedvig; Hedwig

HEE High Express Emotion

HEEA Home Economics Education Association

HEED Health and Education (department or ministry)

heei high-energy electronic ignition

heent head, ears, eyes, nose, throat

HEEP Highway Engineering Exchange Program

HEERA Higher Education Employer-Employee Relations Act

hef heifer; high-energy fuel

HEF High-Energy Fuel; Hospital Employees Federation

HEFA Higher Education Facilities Act

HEFC Higher Education Facilities Commission

heg heavy-enamel single-glass (insulation)

HEGIS Higher Education

General Information Survey; Higher Education General Information System

HEH Her (His) Exalted Highness

HEHF Hanford Environmental Health Foundation (AEC)

HEHL Henry E Huntington Library

hei high-explosive incendiary; holographic exposure index

HEI Heat Exchange Institute; Hotel Enterprises Incorporated

HEI H/F Eimiskipafelag Islands (Icelandic Steamship Company)

HEIAC Hydraulic Engineering Information Analysis Center (USA)

HEIAS Human Engineering Information and Analysis Service (Tufts U)

HEIC Honourable East India Company

HEICN Honourable East India Company Navy

HEICS Honourable East India Company Service

Heidel Heidelberg

Heidelberg Ruprecht-Karl-Universität pouplarly called the University of Heidelberg

Hein Heinersdorf

heip high-explosive incendiary plug

heir app heir apparent

heir pres heir presumptive

heisd high-explosive incendiary self-destroying

heit high-explosive incendiary with tracer

heitdisd high-explosive incendiary tracer dark ignition self-destroying

heitsd high-explosive incendiary tracer self-destroying

hek heavy-enamel single-cellophane (insulation)

hek (HEK) human embryo kidney

hel helicopter

hel (HEL) hen's egg-white lysozyme; high-energy laser (beams); human embryonic lung

Hel Helen; Helena; Helsinki (Helsingfors); Helvetia (Switzerland)

Hel Hellas (Norwegian—Greece)

HEL Hartford Electric Light; Helsinki, Finland (airport); Human Engineering Laboratories (USA); Hydraulic Engineering Laboratory

HeLa Helen Lake (tumor cells)
HELCIS Helicopter Command Instrumentation System
HELCO Hilo Electric Company
heli helicopter; heliport
helio heliochrome; heliodon; heliodor; helioelectric; helioengraving; heliogram; heliograph; heliogravure; heliology; heliostat; heliotherapy; heliotrope; heliotype
helipad helicopter landing pad
he'll he will
Hell Hellerup
HELL Higher Education Learning Laboratory
Hellen Hellenic; Hellenism; Hellenistic
hellfire. (HELLFIRE) helicopter-launched fire-and-forget missile
helminthol helminthology
helo helicopter; heliport
Heloise Heloise Fulbert (abbess-scholar best remembered for her love of Abelard—a monk living in a nearby monastery he founded in 1112)
helosid helicopter-delivered seismic intrusion detector
help. high-energy-level pneumatic automobile bumpers
HELP Helicopter Electronic Landing Path; Help Elderly Locate Positions; Help Establish Lasting Peace; High School Education Law Project; Highway Emergency Locating Plan; Home Environment and Living Program; Home Equity Living Plan; Homophile Effort for Legal Protection
HELPR Handbook of Electronic Parts Reliability
hel rec health record
Hels Helsingborg, Sweden; Helsingør, Denmark; Helsinki, Finland
Hel San *Helsingin Sanomat* (Helsinki's News)
Helv Helvetia; Helvetica
Helv Chim Acta *Helvetica Chimica Acta* (Latin—Swiss Review of Chemistry)
Helv Phys Acta *Helvetica Physica Acta* (Latin—Swiss Review of Physics)
hem. hemoglobin; hemorrhage; hemorrhoid
hem. (HEM) hybrid electromagnetic wave
HEM Ernest Hemingway
hema (Latin prefix—blood)—

hematology
HEMA Heavy Engineering Manufacturers Association
hematol hematolith(ic)(al)(ly); hematologist; hematology; hematolymphangioma; hematolysis; hematolytic
HEMF Handling Equipment Maintenance Facility (USN)
hemi (Latin prefix—half)— hemisphere
hemi engine hemispherical combustion chamber engine
he. missile high-energy missile
hemlaw (HEMLAW) helicopter-mounted laser weapon
hemloc heliborne-emitter-location countermeasures
hemo (Latin—blood)—hemoglobin, hemophilia, hemorrhage, hemostat
hemolysis hemocytolysis
Hem Soc Hemlock Society
Hen Henrietta; Henry; Soviet Ka-15 light-utility helicopter
Hen V *King Henry V*
Hen VIII *King Henry VIII*
HENA Home Economics and Needlework Association
Hence Henderson
H'english Limey English
HENILAS Helicopter Night Landing System
Henk Hendrik
Henri (French—Henry)— Henri Vieuxtemps
Henriqz Henriquez
Henry Henry Ford Commercial College; Patrick Henry Commercial College; Patrick Henry High School; Patrick Henry State Junior College
Henry B Henry B Gonzalez of San Antonio, Texas
Hen wlad Hen Wlad fy Nhadau (Welsh—Land of my Fathers) national anthem of Wales
HEOC Honeywell Electro-Optics Center
HEOP Higher Education Opportunity Program
HEOr *Heure de l'Europe Oriental* (French—Eastern European Time)
heos (HEOS) high eccentric orbiting satellite
hep high-energy phosphate; high-explosive plastic
Hep Hepburn; Hepple; Hepworth
HEP Have Error-free Product; High School Equivalency Program(s); Higher Education Panel
HEPAC High-Energy Physics

Advisory Council
hepaf high-efficiency particulate air filter
HEPALIS Higher Education Policy and Administration Library and Information Service
hepat high-explosive plastic antitank
hepat (Latin prefix—liver)— hepatitis
HEPC Hydro-Electric Power Commission
HEPCAT Helicopter Pilot Control and Training
HEPCC Heavy Electrical Plant Consultative Council
hepdnp high-explosive point-detonating nose plug
hepl high-energy pulse laser
HEPL High Energy Physics Laboratory
HEPP Hoffman Evaluation Program and Procedure
her. heraldry
her. (HER) high-energy rotor (helicopter)
her. heres (Latin—heir)
her-2 human egf-receptor-related receptor
Her Hercules (constellation); Hereford; Hereford(shire)
HER *Harvard Educational Review*
hera high explosive rocket assisted
Hera (Greek—Juno)—goddess of the heavens
HERA Heavy Engineering Research Association; Housewives for ERA
HERALD Highly-Enriched Reactor—Aldermaston
herb. herbarium
Herb Herbert
Herc Hercules (constellation)
HERC Humber Estuarial Research Committee
Herdez (Spanish contraction— Hernandez)
herd° herdeiro (Portuguese— heir)
herdr herdruk(ken) [Dutch— reprint(s)]
HERE Hotel Employees and Restaurant Employees; Human Endurance Range Extender
hered heredity
hereds herederos (Spanish— heirs)
Heref Herefordshire
Herefs Herefordshire
here's here is
Here & Worcs Hereford and Worcester

herf high-energy rate forging
herfs high-energy-rate forging systems
HERI Higher Education Research Institute
herj high explosive ramjet
herm hermetically
hermes heavy element and radioactive material electromagnetic separator (HERMES)
Hermes (Greek—Mercury)—the messenger
HERMES Helicopter Energy and Rotor Management System
hero heroin
hero. hazards of electromagnetic radiation to ordnance; hot experimental reactor of 0 (zero power)—also appears as HERO
hero heroína (Spanish—heroin)
HERO Hazard of Electromagnetic Radiation of Ordnance; Heath Educational Robot; Historical Evaluation and Research Organization; Home Economics Research Organization
Herod. Herodotus
Herois Herois do mar (Portuguese—Heroes of the Sea)—Portugal's anthem
herp herpetologist; herpetology
HERP Human Exposure Rodent Potency
HERP Bulletin of the New York Herpetological Society
herpetol herpetologic(al)(ly); herpetologist; herpetology
HERPOCO Hercules Powder Company
herps herpetological books, papers, specimens; herpetologists
Herr Kaleun Herr Kapitänleutnant (German—Mr Captain Lieutenant)—U-boat commander
HERS Higher Education Resource Services; Home Economics Reading Service
Hersch Herschel
herst herstellung (German—manufacture)
Hertf Hertford College, Oxford
HERTIS Hertfordshire County Council Technical Information Service
Herts Hertfordshire
HERU Higher Education Research Unit

Hervey Allen William Hervey Allen
hes heavy enamel single silk (insulation)
he's he has; he is
Hes Hesba; Hesione, Hesper(ian)(s); Hesketh; Hesperides; Hesperus; Hessels; Hessian(s); Hessin; Hester; Hesther
HES Hawaiian Entomological Society; Health Economic Service; Health Examination Survey; History of Economics Society
HESCA Health Sciences Communication Association
hesd high-explosive self-destroying
hesh high-explosive squash head
HESIS Hazard Evaluation System and Information Service
hess human-engineering systems simulator
hest heavy-end aviation fuel emergency service tanks
h'est highest
HEST High-Explosive Simulation Test
he. stor helium storage
hesv heavy-enamel single-silk varnish (insulation)
het heavy equipment transporter
Hetch Hetchy Hetch Hetchy Dam; Hetch Hetchy Lake (both in Yosemite National Park)
hetdi high-explosive tracer dark ignition
hetero (Latin prefix—different, other)—heterosexual
heterocl heteroclite
heterog heterogeneous
heterosex heterosexual(ity); heterosexuals
HETS High-Energy Telescope System; Hyper-Environmental Test System
Hetty Hester
heu highly enriched uranium; hydroelectric units
heur heuristic (problem solution by trial and error)
Heure L'Heure Espagnole (French—The Spanish Hour)—Ravel opera
hev health and environment; heavy
HEVAC Heating, Ventilating, and Air Conditioning Manufacturers Association
hevr heavier
Hew Heward; Hewett; Hewitt; Hewlett; Hewson; Hugh;

Hugo
HEW Health, Education, and Welfare (US department); Housing, Education, and Welfare (Philippines)
HEWPR Health, Education, and Welfare (department) Procurement Regulations
hex hexachord; hexagon(al); uranium hexafluoride
hex (HEX) hexadecimal
hexa hexamethylene tetramine
hexag hexagon(al)
hex hd hexagonal head
Hez Hezekiah
hf hageman factor; half; hard finish(ed); hard firm; height finding; high frequency (3000 to 30,000 kc); hold fire; home freezer; hook fast; horse and foot (cavalry and infantry); hot finished; hyper filtration; hyperfocal
h/f held for
Hf hafnium
HF Handwriting Foundation; Home Fleet; Home Forces; hydrofluoric acid; hydrogen fluoride
H of F Hall of Fame
H/F Hlutaffelagid (Icelandic—limited company)
hfa hard factory automation
Hfa Haifa
HFA Headquarters Field Army; Holiday Fun Association; Hollywood Film Archive; Hourly Faculty Association
HFAA Holstein-Friesian Association of America
HFARA Honorary Foreign Associate of the Royal Academy
hf bd half-bound
hf bd cf half bound in calfskin (calf leather back and corners)
hf bd cl half bound in cloth (cloth back and corners or cloth sides)
hf bd mor half bound in morocco (morocco leather back and corners)
HFBLB Hokkaido Farmland Bride Liaison Bureau
hfbr high flux beam reactor
hfc hard-filled capsules; high-frequency current
HFC Household Finance Corporation; Human Freedom Center
HFCC Henry Ford Community College
hf cf half-calf
hf cl half-cloth (binding)

hfcs high-fructose corn sweetener; high-fructose corn syrup
HFCT Hawaii Federation of College Teachers
Hfd Hereford
hf-df high-frequency direction finder
hfe human factors (in) electronics; human factors engineering
hff horizontal falling film
HFFF Hungarian Freedom Fighters Federation
hfg heavy free gas
HFGA Hall of Fame for Great Americans
hfh half-hard (steel)
HfH Habitats for Humanity
hfi hydraulic fluid index
HFIA Heat and Frost Insulators and Asbestos Workers Union
hfim high-frequency instruments and measurements
hfir high flux isotope reactor
HFL Human Factors Laboratory (NBS)
hfm hold for money
HFM Henry Ford Museum
hfmd hand-foot-and-mouth disease
hfmf home-furnish monolithic floor
hf mor half-morocco
hfmr high-fat milk replacer
hfo heavy fuel oil; high-frequency oscillator; hole full of oil
Hford Hereford
HFORL Human Factors Operations Research Laboratory
Hfors Helsingfors
hfp hostile fire pay
h&f pool heated and filtered (swimming) pool
HFPS Home Fallout Protection Survey
hfr heifer; held for release; high-frequency range; high-frequency recombination; hold for release
HFR (Sir Edward) Hallstrom Faunal Reserve (New South Wales)
HFRA Honorary Fellow of the Royal Academy
h-f radar height-finder radar
HFRB Hawaii Fire Rating Bureau
Hfrz Halbfranzband (German—halfbound in calf)
hfs high-fructose syrup(s); hot-finished seamless; hyperfine structure
Hfs Helsinki (Helsingfors)

HFS Human Factors Society
HFSSA Historical Firearms Society of South Africa
hfssb high-frequency single sideband
hft hot flow test(ing)
hft hefte (Dano-Norwegian—part, issue)
Hft Heft (German—part)
HFT Hawaii Federation of Teachers; Heavy Fire Team; Human Factors Team
HFTS Human Factors Trade Studies (USU)
hfupr hourly fetal urine production rate
hfw hole full of water
Hfx Halifax
hg hand generator; hectogram; heliogram; high grade; hydrostatic gage; hypobranchial gland
h & g harden and grind
Hg *hydrargyrum* (Latin—mercury)
H^g Hegység (Hungarian—mountain, mountainous)
HG Haute-Garonne; Her (His) Grace; H(erbert) G(eorge) (Wells); High German; Home Guard; Horse Guards
H-G Haute-Garonne
HG House & Garden
hga high gain antenna
HGA Heptagonal Games Association, Hobby Guild of America; Holological Guild of Australia; Hop Growers of America; Hotel Greeters of America; Hungarian Gypsy Association
h-galv hot-galvanize
hgb hemoglobin
HGB Handelsgesetzbuch (German—Commercial Law Code)
HGCA Home-Grown Cereals Authority
HgCl$_2$ bichloride of mercury; mercuric chloride
HGD Hourglass Device
HGDH Her (His) Grand Ducal Highness
hge heavy gold electroplate; hogshead
HGEA Hawaii Government Employees Association
hgf (HGF) hyperglycemic-glucogenolytic factor
HGF Human Growth Foundation
HGFA Hang Gliding Federation of Australia
hg ga height gage
HGH human growth hormone
HGHCA Hotels, Guest

Houses, and Caterers' Association
HGI Henry George Institute
HGJP Henry George Justice Party
Hglds Highlands
HGM Human Gene Mapping workshop
HGMM Hereditary Grand Master Mason
hgo hepatic glucose output
Hgo Hidalgo
HGOA Houston Grand Opera Association
HGOAA Hobby Greenhouse Owners Association of America
hgor high gas-oil ratio
HGP Humbug Gulch Press
hgps high-grade plow steel
hg pt hard-gloss paint
hgr hangar; hanger
HGR Hluhluwe *(shlooshlooway)* Game Reserve (northern Zululand)
hgs hangars; hangers
Hgs Haugesund
HGS Hydrological Growing Season
hgsw horn gap switch
hgt height; hogget
HGTAC Home Grown Timber Advisory Committee
HGTB Haiti Government Tourist Bureau
hgts heights; hoggets
H^{gts} Heights
hgv heavy goods vehicle
Hgw Hanford (basalt) ground water
HGW Herbert George Wells
Hgy Highway
Hgz Hoogezand
hh half-hard; handhole; hands high (horse's height); heavily hinged; heavy hydrogen
h/h half height; hard of hearing; house-to-house (search or transport)
h to h heel-to-heel
hh hojas (Spanish—leaves)
hH heavy hydrogen
HH double-hard (pencils); Harry Hansen; Helen Hunt Jackson; Her (His) Highness; His Holiness; Howard Hanson; Hugo Humdinger; Huntington Hartford
H/H Havre-Hamburg (range of ports)
H & H Handy & Harman; Holland & Holland
HH Herren (German—Gentlemen)
HH-52 Sikorsky 12-passenger helicopter

HH-53 Sikorsky Sea Stallion CH-53 assault helicopter
hha half-hardy annual
hhb half-hardy biennial
HHBS Hereford Herd Book Society
hhcc higher-harmonic circulation control
hhd hogshead
HH. D. *Humanitatis Doctor* (Latin—Doctor of Humanities)
hhdws heavy handy deadweight scrap
hhf household furniture
HHFA Housing and Home Finance Agency
HHFTH Happy Horsemanship for the Handicapped (foundation)
hhg household goods
hhh triple hard
HHH Hubert Horatio Humphrey; triple-hard (pencils)
HHHC Hunt the Hunters Hunt Club (Amory Foundation funded)
HHHIPA Hubert H Humphrey Institute of Public Affairs
Hhhs Hincherton hayfever helmets
HHI Hellenic Hydrobiological Institute; Highland Home Industries; Hyundai Heavy Industries
H-hinge capital-H-shaped hinge
HHK Honor Hong Kong
HHKA Husband's Housemaid's Knee Association
hhld household
HHMS His Hellenic Majesty's Ship
hhmu hand-held maneuvering unit
HHNSR Hudson Highlands National Scenic Riverway
H-hour hostile operations commencement hour
hhp half-hardy perennial; hydraulic horsepower
HHPL Herbert Hoover Presidential Library
HHR Health and Human Resources
HHRA Heartland Human Relations Association
HHRC Health and Human Resource Center
H & HRR Harlem and Hudson River Railroad
HHS Haaren High School; Hawaiian Humane Society; Health and Human Services; Hunter High School
hhsd holographic horizontal

situation display
HHSP Highland Hammock State Park (Florida)
hht (HHT) high-temperature helium turbine
HHT Horn-Hellersberg Test
hhtg house(hold) heating
hhtv hand-held thermal viewer
HHUMC Hadassah-Hebrew University Medical Center
hhw (HHW) household hazardous waste
HHW higher high water
HHWI higher high water interval
hi contracted form of "hail"; high; high intensity; horizontal interval; humidity index
hi (HI) hyperglycemic index
h & i harassing and interdictory (artillery fire)
h.i. *hic iacet* (Latin—here lies)—also appears on tombstones as H.I.
Hi Hering illusion; High (postal abbreviation); Hindi; Hiram
H^i Hasi (Arabic—waterhole)—also appears as *Hasy*
HI Hammersley Iron; Hampton Institute; Handwriting Institute; Harris Intertype; Hat Institute; Hawaii; Hawaiian Islands; Heat Index; Henrik Ibsen; Holiday Inns; Hoover Institution; Hudson Institute; Humidity Index; Hydraulic Institute
hia hold in abeyance
HIA Handkerchief Industry Association; Hobby Industry Association; Home Improvement Association; Homeopathic Institute of Australia; Horological Institute of America; Hospital Industries Association; Housing Improvement Association; Housing Industry Association; Hungarian Imperial Association
HIAA Health Insurance Association of America; Health Insurance Association of Australia
HIAB Hydrauliska Industri AB (Swedish—Hydraulic Industry Company)
hiac high acuity
hi-ac high accuracy
HIAD Handbook of Instructions for Airplane Designers
HIAG Hilfsorganisation auf Gengenseitigkeit (German—Mutual Aid Organization)
HIAGSED Handbook of In-

struction for Aircraft Ground Support Equipment Designers
HIAP Health Insurance Advocacy Program
HIAS Hebrew Sheltering and Immigrant Aid Society
HIAVED Handbook of Instructions for Aerospace Vehicle Equipment Design
Hib *Haemophilus influenzae* type B; Hibernia (Ireland); Hibernian (Irish)
HIB Herring Industry Board
Hibbd *Halbband* (German—half binding)
hibex high-acceleration booster experiment
HIBR Huxley Institute for Biosocial Research
HIBT Howard Ink-Blot Test
hic head injury criterion; hearing-impaired children; hot isostatic compaction; hybrid integrated circuit; hydrologist in charge
HIC Heart Information Center; Herring Industries Council
HICAP Health Insurance Advocacy Program
hicapcom high-capacity communications
hicat high-altitude clear-air turbulence
hic jac hic jacet (Latin—here lies)
hiclass hierarchical classification
Hi Com High Command; High Commission; High Commissioner
HICS Hardened Intersite Cable System
hid. hallucinations, illusions, and delusions; headache, insomnia, depression (syndrome); high-intensity discharge (lamps)
Hid Hidalgo
hidal helicopter insecticide-dispersal apparatus, liquid
hidalgo hijo de algo (Spanish—son of someone)
Hidalgo Miguel Hidalgo y Costilla (Padre Hidalgo)
HIDB Highlands and Islands Development Board (Scotland)
hidvl high-intensity-discharge vapor lamp
hi-E high efficiency
HIE Hibernation Information Exchange; Histrionic Instruction Education
hier hieroglyphics
Hier. *Hierosolma* (Latin—

Jerusalem)
HIES Hadassah Israel Education Services
HIF Health Information Foundation
hifar high-flux Australian reactor (HIFAR)
hifc hog intrinsic factor concentrate
hi-fi high-fidelity
Hi Fi High Fidelity and Musical America
hiflex high flexibility
HIFNY Hospitality Industry Foundation of New York
hifo highest in, first out
hifor high-level forecast
hig hermetically sealed integrating gyroscope; higgler
Hig Higgins; Higginson
HIG Hartford Insurance Group; Hawaii Institute of Geophysics
HIGED Handbook of Instruction for Ground Equipment Designers
higher 3-Rs remedial reading, remedial writing, remedial arithmetic
Highlands Highlands of the Hudson; Highlands of the Navesink close to where Henry Hudson first landed in 1609; Highlands of Scotland
highpro high protein (diet)
high-Q high quality
high-Z high-impedance
High Sierras higher Sierra Nevada Mountains of California
High Tatras high Tatra Mountains of Czechoslovakia's Carpathians
high tech high technology
HIH Her (His) Imperial Highness
HII Health Industries Institute; Health Insurance Institute
hijack hijacked; hijacker; hijacking
hik hiking
hil high intensity lighting; high lift
Hil Hilary
hila health insurance logistics automated
hilac heavy-ion linear accelerator
hilat high-latitude satellite
HILC Hampshire Inter-Library Center (Amherst, Mount Holyoke, and Smith colleges)
Hilda Hildegarde
Hill The Hill (Capitol Hill in Washington, D.C.)
hilla (HILLA) high-input low-labor agriculture

hi-lo high-low
Hil-Vis Hiligaynon-Visayan
him. high impact; horizontal impulse
HIM Her (His) Imperial Majesty
Himal Himalaya (Sanskrit—Abode of Snow)
Himalayas Himalaya Mountains between India and Tibet
himat high-maneuverable advanced-fighter technology
hi mi high mileage
HIMM Hallberg Index of Male Menopause
HIMS Heavy Interdiction Missile System
Hin Hindi
Hinck Hinckley
hind hindustanisch (German—Hindustani)
Hind Hindi; Hindu; Hindustani
Hindenburg General Paul Ludwig Hans Anton von Benackendorf und von Hindenburg
hinga anhinga (fish-eating wading bird)
Hinglish Hindi + English (English interlarded with Hindi expressions and words)
hinil high noise-immunity logic
HINP Hundred Islands National Park (Philippines)
Hint Hinton; Hinton Test (for syphilis)
HINWR Hawaiian Islands National Wildlife Refuge
hio hypoiodite
hiomt (HIOMT) hydroxyindole-O-methyltransferase
H-ion hydrogen ion
HI-OVIS Highly Interactive Optical Video Information System
hip. hierarchical information processor; high-impact pressure; hot isostatic pressure; humanizing, individualizing, and personalizing
hip. (HIP) historically-informed performance
Hip Hippolyte; Soviet Mi-8 transport helicopter used by Afghanistan, Cuba, Czechoslovakia, East Germany, Poland, and the United Arab Republic (Egypt)
HIP Health Insurance Plan; Help for Incontinent People; Hoover Institution Press; Houston's Informed Parents
hipar high-power acquisition radar
HIPCO Hunt International Pe-

troleum Company
hipdom hippiedom
HIPERNAS High-Performance Navigation System
hipi high-performance intercept(ion)
hipo hierarchy plus input process output
hipoe high-pressure oceanographic equipment
hipot high potential
Hipp Hippocrates
hippo(s) hippopotamus(es)
hi pres high pressure
hips. hippies
hiptoc high-power testing of optical components
hi-q high iq (IQ)
hir hydrostatic impact rocket
HIR Harbour Improvement Rate; Heron Island Resort (Queensland); Honiara (Solomon Islands airport)
hiran high-precision shoran
HIRB Health Insurance Registration Board
HIRC Housing Industry Research Committee
HIRE Help Through Industry Retraining and Employment; Hooking Is Real Employment
hirel high reliability
HIRI Hawaiian Independent Refinery Incorporated
hirl high-intensity runway lights
Hirohito Emperor Hirohito Showa (Japan's 124th emperor in direct lineage)
Hiroshige Ando Hiroshige (19th-century Japanese landscape painter)
HIRS High-Impulse Retrorocket System; Holographic Information Retrieval System
HIRS/smrd High-Impulse Retrorocket System/spin-motor rotation detector
Hirt Aulus Hirtius (Roman historian)
his. (HIS) histidine (amino acid); history
h.i.s. hic iacet sepultus (Latin—here lies buried)—also appears as h.i.s.
Hi-S Hi-Standard (firearms)
HIS Health Interview Survey; Horticultural Improvement Scheme; Hospital Information System; Human Intrusion Studies
HISA Hawaii International Services Agency; Headquarters and Installation Support Activity (USA)

HISAM Hierarchical Indexed Sequential Access Method
His-Am(s) Hispanic American(s)
HISC House Internal Security Committee (formerly House Un-American Activities Committee—HUAC)
his'n his own
Hisp Hispaniola
HISPA History of Sport and Physical Education (association)
Hispan Hispanic
Hispano Hispanoamericano (Spanish American); Hispano-Suiza (automobile)
HISSG Hospital Information Systems Sharing Group
hist historical; history
hist historie or *historisk* (Dano-Norwegian—history or historian)
Hist historic(al); History
Hist Abs Historical Abstracts
Hi Stan 22 plus High Standard .22-caliber automatic plus silencer (assassin's special)
Hist Dist Historic District
histn historian
histo histoplasmosis
histo (Latin prefix—tissue or web)—histology
histocrit historical critic(ism)
histol histologic(al)(ly); histologist; histology
HISU Hoover Institution on War, Revolution, and Peace at Stanford University
hit. high-intensity tutoring; homing intercept(ion) technology
hi-T high torque
Hit Holtzman inkblot technique
HIT Health Inca Tea (with coca leaves); Health Indication Test; Hitachi Innovative Technology; Hong Kong International Terminals
Hitac Hitachi computer
Hitch Hitchborn(e); Hitchcock
hi-tec(h) high technology
hi-temp high temperature
Hitler Adole Schickphlgruber
Hit Pom Hither Pomerania (coastal Germany)
Hitt Hittite
HIU Hypnosis Investigation Unit
HIUS Hispanic Institute of the United States
HIUS Historisches Institut der Universität Salzburg (German—Historical Institute of the University of Salzburg)

HIUV Historisches Institute der Universität Vienna (German—Historical Institute of the University of Vienna)
hiv human immunodeficiency virus
hiv hiver (French—winter)
HIV Human Immunodeficiency Virus
hi-vision high-definition television
hivos high-vacuum orbital simulator
hi wat high water
Hiwi Hilfsfreiwilliger (German—auxiliary volunteer); *Hilfswillige* (German—volunteers)
HIWRP The Hoover Institution on War, Revolution and Peace
h & j hyphenation and justification
HJ *Hitler Jugend* (German—Hitler Youth); Honest John (short-range unguided missile); Howard Johnson
H. J. hic jacet (Latin—here lies)
HJBS Hashemite Jordan Broadcasting Service
HJC Hershey Junior College
HJD Heliocentric Julian Date; Hospital for Joint Diseases
hjed heliocentric Julian ephemeris date
HJPA Holmes Junge Protected Area (Australian Northern Territory)
H J Res House Joint Resolution
H.J.S. hic jacet sepultus (Latin—here lies buried)
h-k hand to knee
Hk Hakenkreuz (German—hooked cross) swastika
HK Heckler and Kock (firearms); Hong Kong
HK Handelskammer (German—Chamber of Commerce); *Helsingin Kaupunginorkesteri* (Finnish—Helsinki City Symphony Orchestra)
HKA Hong Kong Airways
HKCEC Hong Kong Catholic Education Council
hk cells human kidney cells
HKCL Hong Kong Container Line
Hkd Hakodate
HK$ Hong Kong dollar
HKDR Hong Kong Depository Receipt
HKECIC Hong Kong Export Credit Insurance Corporation
HKEL Hong Kong Export Lines

hkf handkerchief
HKFE Hong Kong Futures Exchange
H Kg Hong Kong
HKG Hong Kong, British Crown Colony (airport)
HKGMA Hosiery and Knit Goods Manufacturers Association
HKH Hans (Hendes) Kongelige Højhed [Dano-Norwegian—His (Her) Royal Highness]
HKI Helen Keller International
HKIL Hong Kong Islands Line
HKJ Hashemite Kingdom of Jordan
HKL Halldor Kilyan Laxness
HKLA Hong Kong Library Association
hkm high-velocity kill mechanism
h-k m (H-K M) hunter-killer missile
HKMA Hong Kong Management Association
H'Kong Hong Kong
HKP Hong Kong Polytechnic
HKPO Hong Kong Philharmonic Orchestra
HK & S Hong Kong and Shanghai Bank
HKSE Hong Kong Stock Exchange
HKTA Hong Kong Tourist Association
HKTC Hong Kong Training Council
HKTDC Hong Kong Trade Development Council
HKU Hong Kong University
hkups hookups
HK virus Hong-Kong type of influenza virus
HKX Hong Kong Express (container service)
hky hand knitting yarn(s)
hl hand lantern; hard labor; heavy lift; hectoliter; high level; high lift; hinge line; holiday
h-l highest and lowest (quotations)
h/l heading line; high or low
h&l door hinge resembling ligature of capital H and capital L
hl heilig (German—holy)
h.l. hoc loco (Latin—in this place)
HL Haute-Loire; Herpetologists League; Home Lines; Homestead Lease; Honours List; House of Lords; Hy-

gienic Laboratories; Hygienic Laboratory
H-L Haute-Loire
H & L Harbour and Light Department
H of L House of Lords
HL House of Lords Cases
hla (HLA) histocompatibility antigens; homologous leucocytic antibodies; human leucocytic antigen
HLA Hawaii Library Association; Human Life Amendment
HL&AG Henry E Huntington Library and Art Gallery
HLAHWG High-Level Ad-Hoc Working Group (NATO)
H-land Headland
hlb hydrophile-lipophile balance
HLB Hotel Licensing Board
HLBB Home Loan Bank Board
hlc health locus of control
HLC Hapag-Lloyd Container (steamship line); Hospital Library Council (Dublin)
HLCAS House of Lords Cases
HLCU Hapag-Lloyd Container Unit
hld held; hold; holder
HLD Harold Handley Page (aircraft)
hldg holding
hl di hole die
HLDI Highway Loss Data Institute
hlds holdings
hldw high-level defense waste(s)
hlem horizontal-loop electromagnetic method
HLF Human Life Foundation
hlg halogen
HLG High-Level Group
hlge handling equipment
hlgp heavy-lift general purpose
HLH Haroldson Lafayette Hunt
HLHS Heavy-Lift Helicopter System
HLI Highland Light Infantry
HLIC Housing Loans Insurance Corporation
hll high-level language
HLL Hellenic Lines Limited
hllw high-level liquid waste(s)
hl IX high-level language X
HLM Henry Louis Mencken
HLM Habitations à loyer modéré (French—moderately-priced housing)
HLMR Hunter-Leggitt Military Reservation

HLNP Hattah Lakes National Park (Victoria, Australia)
h/l number hydrophile/lipophile number
hlnw high-level nuclear waste(s)
HLNWR Havasu Lake National Wildlife Refuge (California); Hutton Lake National Wildlife Refuge (Wyoming)
hlo horizontal lockout
H Lords House of Lords
hlp (HLP) hyperlipidemia
hlpr helper
HLPR Howard League for Penal Reform
hlr heart-lung resuscitation
HLRS Homosexual Law Reform Society
hls heavy liquid separation; heavy logistics support; hills; holes
hl S heilige Schrift (German—holy scripture)
Hls Hills (postal abbreviation)
HLS Harvard Law School; Heavy Logistics Support; High-Level Scheduler
hl sa hole saw
hlse high-level single-ended
HLSS Harry Lundeberg School of Seamanship
HLSUA Honeywell Large Systems Users Association
hlsw high-level solidified waste(s)
hlt halt; halter
HLT Holborn Law Tutors
HLTF High-Level Task Force
hlth prof health professions
hlttl high-level transistor-translator logic
hltw high-level transuranic waste(s)
Hlu Honolulu
hlv herpes-like virus
HLVA Hospital Lady Visitors Association
hlw higher low water; high-level waste
Hlw Halbleinwand (German—half-bound cloth)
HLW higher low water
HLWI higher low water interval
HLW-ICB High-Level Waste-Interface Control Board
HLWIP High-Level Waste Immobilization Program
hlwn highest low-water neap tides
HLWRP Hoover Library on War, Revolution, and Peace (Stanford University)
Hlzbl Holzbläser (German—

woodwinds)
hm hallmark; harmonic mean; heavy metal; hectometer; hollow metal
hm? how many?; how much?
h & m hit and miss; hull and machinery
h.m. hoc mense (Latin—in this month)
Hm Haymarket (London theater); manifest hypermetropia
HM Harbour Master; Haute-Marne; Head Master; Head Mistress; Her (His) Majesty; Herman Melville; Home Missions; Houghton Mifflin
H-M Haute-Marne
HM Heure de Moscou (French—Moscow Time)
hm² square hectometer
hm³ cubic hectometer
Hma Hiroshima
HMA Head Masters Association; Her (His) Majesty's Airship; Hoist Manufacturers Association; Home Manufacturers Association
H & MA Hotel and Motel Association
HMAA Horse and Mule Association of America
HMAC Her (His) Majesty's Aircraft Carrier
HMAI Handbook of Middle-American Indians
HMARC Houston Metropolitan Archives and Research Center
HMAS Her (His) Majesty's Australian Ship
HMAV Her (His) Majesty's Armed Vessel
hmb homatropine methyl bromide (HMB)
HMB Home Mission Board; Hops Marketing Board
HMBDV Her (His) Majesty's Boom Defence Vessel
HMBI Her (His) Majesty's Borstal Institution
HMBP Heavy Machine Building Plant
hmc heavy media cyclone; heroin-morphine-cocaine (mixture); howitzer motor carriage
hmc (HMC) hydroxymethyl cystosine
HMC Harvey Mudd College; Her (His) Majesty's Customs; Hospital Management Committee
hmcc housewife/mother career concept
HMCG Her (His) Majesty's Coastguard

HMC & H Hahnemann Medical College and Hospital

HMCIF Her (His) Majesty's Chief Inspector of Factories

HMCN Her (His) Majesty's Canadian Navy

HM Comm Historical Manuscripts Commission

HMCS Her (His) Majesty's Canadian Ship

HMCSC Her (His) Majesty's Civil Service Commissioners

HMCyS Her (His) Majesty's Ceylonese Ship

hmd hollow metal door; humid; hydraulic mean depth

hmd (HMD) hyaline membrane disease

HMD Her (His) Majesty's Destroyer

HMDBA Hollow Metal Door and Buck Association

hmde hanging-mercury-drop electrode

hmdf hollow metal door and frame

hmdi (HMDI) hexamethylene diisocyanate

HMDS Hazardous Material Data System

hmf hollow metal frame

HMF Her (His) Majesty's Forces

hmf black high-modulus furnace black

HMFI Her (His) Majesty's Factory Inspectorate

hmg heavy machine gun; high modulus graphite

hmg (HMG) human menopausal gonadotrophin

HMG heavy machine gun; Her (His) Majesty's Government

HMHS Her (His) Majesty's Hospital Ship; Horace Mann High School

hmi heavy maintenance interval

HMI Hahn-Meitner Institut; Her (His) Majesty's Inspector; Hughes Medical Institute

HMI *Himpunan Mahasiswa Islam* (Indonesian—Islamic Students Society)

HMIA Haitian Migrant Interdiction Operation

HMIC Her (His) Majesty's Inspectorate of Constabulary

HMIS Her (His) Majesty's Indian Ship; Her (His) Majesty's Inspector of Schools

HMIT Her (His) Majesty's Inspector of Taxes

HMK His Majesty the King

HML Harper Memorial Library (University of Chi-

cago); Horace Mann—Lincoln Institute

HMLR Her (His) Majesty's Land Registry

hmlt hamlet

HMM Her (His) Majesty's Minister; Hyundai Merchant Marine

hmma (HMMA) 4-hydroxy-3-methodxy-mandelic acid

HMML Her (His) Majesty's Motor Launch

HMMS Her (His) Majesty's Motor Mine Sweeper

HMMWV High-Mobility Multi-Purpose Wheeled Vehicle (successor to the Jeep)

HMNAO Her (His) Majesty's Nautical Almanac Office

HMNAR Hart Mountain National Antelope Refuge (Oregon)

hmnts humanities—the arts, language, literature, music, philosophy

HMNZS Her (His) Majesty's New Zealand Ship

hmo heart minute output

HMO Health Maintenance Organization; Hospital Medical Office(r)

HMOCS Her (His) Majesty's Overseas Civil Service

hmo's (HMOs) health maintenance organizations

HMOW Her (His) Majesty's Office of Works

hmp handmade paper

hmp (HMP) hexose monophosphate

HMP Her (His) Majesty's Penitentiary; Her (His) Majesty's Prison

H.M.P. *hoc monumentum posuit* (Latin—he erected this monument)

HMPMA Historical Motion Picture Milestones Association

HMQ Her Majesty the Queen

HMRC Heineman Medical Research Center

HMRCS Her (His) Majesty's Royal Canadian Ship

H & M RR Hudson & Manhattan Railroad (Hudson Tubes)

HMRT Her (His) Majesty's Rescue Tug

hms hours, minutes, seconds

HMs Her (His) Majesty's

HMS Harvard Medical School; Her (His) Majesty's Service, Ship, or Steamer; Home Mission Society

H & MS Headquarters and

Maintenance Squadron

HMS *Hotel and Motel Systems*

HMSA Hawaii Medical Service Association

HMSAS Her Majesty's Special Air Service

HMSG Hirshhorn Museum and Sculpture Garden

HMSO Her (His) Majesty's Stationery Office

HMSS Hospital Management Systems Society

hmstd homestead

HMT Her (His) Majesty's Transport; Her (His) Majesty's Trawler; Her (His) Majesty's Treasury; Her (His) Majesty's Tug

HMTA Hotel-Motel Association

hmu (HMU) hydroxymethyl uracil

HMV His Master's Voice (phonograph records)

hmw high molecular weight

hmwp high molecular weight polyethylene

hmy too little

h.n. *hac nocte* (Latin—tonight)

Hn Herman(n); Horn

Hⁿ Horn

HN Head Nurse; Hoff und Nationaltheater (Munich)

Hna Habana

HNBI Hellenic National Broadcasting Institute

hnc hypothalamic-neurohypophysical complex

HNC Harbors and Navigation Code; High National Council; Higher National Certificate; Human Nature Cooperative; Human Nature Council; Human Nutrition Center; Human Nutrition Council

Hnd *The Hindu* (Madras)

HND Haneda-Tokyo (airport); Higher National Diploma

hndbk handbook

hnddiprt hand-discharge printed

hndflshd hand-fleshed (skins)

hndlg handling

hndlg/shpng handling & shipping charges

hndlr handler

hndovprt hand overprint(ed)

hndprt hand print(ed)

HNEI Hawaii Natural Energy Institute

HNF Home Nursing Foundation

hn fm hand form

HNG Hawaii National Guard; Houston Natural Gas

Hnl Honolulu

HNL Honolulu, Hawaii (airport)
hnml hindumeal
HNMS High NATO Military Structure
HNNNR Herma Ness National Nature Reserve (Scotland)
Hno Hanover
HNO₂ nitrous acid HNO_2 nitrous acid
HNO₃ nitric acid HNO_3 nitric acid
Hnos Hermanos (Spanish—brothers)
hnp high needle position
HNP Haleakala National Park (Maui, Hawaii)
HNP Herstigte Nasionale Partij (Afrikaans—Reformed National Party)—South African segregationists
hnr handwritten numerical recognition; hiss noise reduction
HNRC Human Nutrition Research Center
hnrna (HNRNA) heterogeneous nuclear ribonucleic acid
hnrs honors
hn(s) horn(s)
HNT Hof und Nationaltheater (German—Court and National Theater)—Munich
HNTLA Hiskey-Nebraska Test of Learning Aptitude
HNVS Hughes Night Vision System
HNWR Hagerman National Wildlife Refuge (Texas); Horicon National Wildlife Refuge (Wisconsin)
ho held over; hoist; hold over; holds; hostilities only; house
h & o hook and oil (damage to cargo)
'ho' whore
Ho Ho Chi Minh; holmium; Honduran; Honduras; Hondureño; House (of Representatives)
HO Head Office; Health Office; Home Office (British); Hydrographic Office (USN)
HO Handelsorganisation (German—trade organization)
hoa hands off—automatic
HOA Homeowners A (insurance policy); Home Owners Association
hoax (Contraction—hocus pocus)
hob. height of burst; horizontal oscillating barrel; human observation(al) blunder
Hob Anthony van Hoboken (Dutch chronologist-enumerator of Haydn's music);

Hobart, Tasmania; Hoboken (Belgian seaport near Antwerp, place near Waycross, Georgia, port city in New Jersey)
HOB Homeowners B (insurance policy); House Office Building
Hoban Holborn
hobe honeycomb before expansion
hobgob(s) hobgoblin(s)
Hob-Job Hobson-Jobson (similar-sounding words to those of other languages with some or complete loss of meaning, e.g., white rhino really the Dutch *weid rhino*—a wide-mouthed rhinoceros and really not white)
hobo. (HOBO) homing bomb
Hobo Hoboken
Hobohemia Hobo bohemia (skid-row areas such as Manhattan's Bowery)
HOBOS Homing Bombing System
HOBS High Orbital Bombardment System
Hobt Hobart
hoc heavy organic chemical(s); held on charge
hoc (HOC) hydrofoil ocean combatant
HoC House of Commons
HOC Homeowners C (insurance policy); Hydrology Overview Committee
hock. Hockheimer (Rhine wine)
HOCPC Home Office Crime Prevention Center
H.O.C.S. Hostem Occidit, Civem Servavit (Latin—A foe he slew, a citizen he saved)—inscription found on Roman civic crowns
hoc vesp. hoc vespere (Latin—this evening)
hod. hyperbaric oxygen drenching
HoD Head of Department
HOD Hoffer-Osmond Diagnostic Test
HOD Test Hoffer, Osmond, and Desmond Test (for schizophrenia)
hoe. holographic optical element
hof home owner's fee
HoF Hall of Fame
Hoff Hoffman; Hoffmann; Hofman reflex
Hoffmann Ernst Theodor Amadeus Hoffmann
hofin hostile fire indicator

HOFSL Home Office Forensic Science Laboratory (London)
HOG heavy-ordnance gunship
HO-gage ⁵/₈-inch track gauge (model railroads) $^5/_8$-inch
ho & gem heavy oil and gas-cut mud
hoh hard of hearing
HOHI Home Ownership and Home Improvement (loan program)
HOI Headquarters Operating Instruction
HOI Handbook of Operating Instructions
hoj home on jamming
HoJo Howard Johnson
H o K House of Keys
hoke hokum
Hokusai Katsushika Hokusai (19th-century Japanese engraver-illustrator-teacher)
hol holiday; hollow; holly
Hol Holland; Hollander
Hol Holanda (Portuguese or Spanish—Holland)
HoL House of Lords
HOLC Home Owners Loan Corporation
hold. holding
holidaze alcohol-or-drug-induced daze characterized by incidence of over-the-holidays accidents and fatalities
holl hollandais (French—Dutch); *hollandsk* (Dano-Norwegian—Dutch)
Holl Holland; Hollander
holland hope our love lives and never dies
Holländer Der Fliegende Holländer (German—The Flying Dutchman)—Wagner opera
hollands hollands gin (juniper flavored)—also called dutch gin as it originated in the Netherlands
Holland's Holland & Holland (British gunmakers)
Hollyw'd Hollywood
Holmes Sherlock Holmes
holo holograph
holo (Latin prefix—entire or whole)—holocaust
holog hologram; holograph(ic)al)(ly); holography
hol-ry whole rye
hols holidays
holsum wholesome
Holt Holt, Rinehart & Winston
HOLUA Home Office Life Underwriters Association
holupk holiday upkeep
Hol Via Holborn Viaduct (rail terminal)
Holw Hollow

hom homonym
hom homing; hominy
Hom Homer
Hom Homerton College, Cambridge
Hom. *Homilia* (Latin—homily, sermon)
HOMA Houston Oil and Minerals
Home Home Office (England and Wales)
HOME Home Opportunities Made Equal; Home Ownership Made Easy; Homemakers Organized for More Employment
home ec home economics
homeo homeopath; homeopathic; homeopathy
homeo (Latin—same or similar)—homeostasis, homogeneous, homogenized, homologous
Homer. Homeric
HOMES mnemonic for remembering the five Great Lakes—Huron, Ontario, Michigan, Erie, Superior
Hominoid Hominoidea [primate superfamily including the great apes (chimpanzees, gorillas, orangutans), the gibbons, and humans]
homo homeopath; homeopathic; homeopathy; homosexual; homosexuality
homo (Latin prefix—man)—homosexual
homoeo homoeopath(ic); homoeopathy
homolat homolateral
Homo ludens creative, festive, leisure-loving, playful person
homomilk homogenized milk
homop homophobia (anti-homosexual hysteria)
homosex homosexual; homosexuality
homrep homicide report
hon honey; honor; honorable; honorarium; honorary; honored
Hon Honduran; Honduras; Hondureño; Honolulu; Honorable
Hon'ble Honourable
Hon Consul Honorary Consul
hond honored
Hond Honduran; Honduras
Hong Kongese person(s) of British and Chinese descent
Hono Honolulu
hons honors
Hon Sec Honorary Secretary
HOO House Officer Observer
hood. hoodlum

'hood neighborhood
Hook Hook Point, Ireland; Hooker; The Hook—Hook of Holland *(Hoek van Holland)*; Hooky Nail; Sandy Hook, New Jersey
Hoover Hoover Dam southeast of Las Vegas, Nevada; Hoover Institution (Stanford University)
hop. high oxygen pressure; holding procedures
Hop Hopkin; Hopkins; Hopkinson; Hopwood
HOP Hong Kong Outline Plan; Hydrographic Office Publication(s)
HOPE Harbingers of Productive English; Health Opportunity for People Everywhere; Homeownership and Opportunity for People Everywhere
HOPE Help Organize Peace Everywhere
HOPEG Hotel and Public Building Equipment Group
HOPES High-Oxygen-Pulping Enclosed System
Hopkins Mt Hopkins Observatory, Amado, Arizona
hoppers grasshoppers
HOQ Hysteroid-Obsessoid Questionnaire
hor home of record; horizon; horizontal; horology
Hor Horace; Horatio
Hor Horologium (constellation)
HoR House of Representatives
H-O-R Hoover-Owens-Rentschler (engines)
Horace Quintus Horatius Flaccus
hora decub. hora decubitus (Latin—at bedtime)
hora interm. hora intermedius (Latin—at the intermediate hours)
hora som. hora somni (Latin—at bedtime)
HO & RC Humble Oil and Refining Company
HoReCa Hotel, Restaurant, and Cafe Keepers
horen horizontal enlarger
horiz horizontal
horn french horn in the British Isles, Canada, or the United States)
Horn The Horn (Cape Horn—southernmost South America); Hornblower
horo horoscope
Horo Horologium (constellation)
horol horology

Hor Q Horatius Quintus Flaccus (Roman poet)
HORSA Hut Operation Raising School-leaving Age
horse. (HORSE) hydrofoil-operated rocket submarine
Horsemonger Horsemonger Lane Gaol (notorious London prison in the early 1800's)
hort horticulture
Hort Horticulture
horti horticultural; horticulturalist; horticultural
hortic horticultural; horticulture; horticulturist
HORU Home Office Research Unit
hor. un. spatio horae unius spatio (Latin—at the end of an hour)
hos human operator simulator
'ho's whores
Hos The Book of Hosea
Hos Hosea
H-o-S Holland-on-Sea
HOS Hawaiian Orchid Society
HOSA Home Owner Services Administration
HOSC History of Science Cases
hose. hosiery
Hosp Hospital
hosp ins hospital insurance
hot. human old tuberculin
Hot high-subsonic optically-guided tube-launched Franco-German antitank missile
HOT Hamilton-Oshawa-Toronto (industrial complex); Hot Springs, Arkansas (airport)
ho/ta hold tank(s)
HOTAC Hotel Accommodation (London hotel service)
hot arts hot art dealers (trafficking in stolen works of art)
HOT Car Hands Off This Car (antitheft program)
HOTLIPS Honorary Order of Trumpeters Living in Possible Sin
HOTOL Horizontal Takeoff and Landing (aircraft)
HOTS Higher-Order Thinking Skills
Hotz Hotzenplatz (also known as Osoblaha by its Czechoslovakian citizens)
Hou Houston
HOU Houston, Texas (airport)
Houghton Houghton Mifflin
Hounds Houndsditch
Hous Houston
House House of Representatives in the United States;

House of Commons in England; London's Stock Exchange; Oxford University's Christ College
House of D (Women's) House of Detention (NYC)
Hou Sym Orch Houston Symphony Orchestra
houv houvere (Finnish—charity)
hov high-occupancy vehicle
Hovensweep Hovensweep National Monument in southwestern Colorado and southeastern Utah
how. howitzer
How Howard (U.S. Supreme Court Reports)
HoW Happiness of Womanhood
HOW Home-Owners' Warranty
Howard U Pr Howard University Press
howtar howitzer-mortar
HOW-TO Housing Operation with Training Opportunity (OEO)
Hox Hoxie
Hoxford(shire) (Cockney—Oxfordshire)
hp half pay; hardy perennial; high pass; high performance; high potency; high potential; high power; high pressure; hire purchase; holiday pay; hollowpoint; horizontal parallax; horizontally polarized; horsepower; hot press(ed); hybrid perpetual; hyperbolic; hypoid(al)
h/p house to pier
h & p history and physical (examination); hydraulic and pneumatic
HP Haute-Pyrénées; Highway Patrol; historic park; House Physician; Houses of Parliament
H-P Handley-Page; Haute-Pyrénées; Hewlett-Packard
HP Homeopathic Pharmacopoeia
hpa high power amplifier; horn-parabola antenna; hydraulic pneumatic area
HPA History, Physical, Admit; Hospital Physicists Association
HPAAS High-Performance Aerial Attack System
hpac hydropress accessor
HPAC Hawaii Performing Arts Company
HPAL Holland Pan-American Line

hpb hinged plotting board
HPBA Housing Patrolmen's Benevolent Association
hpc history of the present complaint
HPC Hercules Powder Company; Highland Park College
hpc black hard-processing channel black
HPCC High-Performance Control Center
hpchd harpsichord
hpchdst harpsichordist
HPCL Hindustan Petroleum Corporation Limited
HP Club Homing Pigeon Club
hp cyl high-pressure cylinder
hpd high-performance diesel; hydraulic pump discharge
H-P d Hough-Powell digitizer
HPD Hawaii Police Department; Housing Preservation and Development
HPDC High-Pressure Data Center (NBS)
HPDF High-Performance Demonstration Facility
hpdo high-performance diesel oil
hpe high-power effect(s)
hper health, physical education, and recreation
HPERB Hawaii Public Employment Relations Board
hpew high-power(ed) early warning
hpf highest possible frequency; high-powered field; hydropress form
HPF Horace Plunkett Foundation
hpfm hydropress form
hpg (HPG) human pituitary gonadotrophin
HPGA Hawaii Personnel and Guidance Association
hp Ge high-purity Germanium
H_pge heading per gyro compass
hp hd high-pressure high-density
hp hr horsepower hour
hpi high-power illuminator; history of present illness; homing-position indicator
HPI Handicap Problems Inventory; Heifer Project International; Hydrocarbon Processing Industry
hpl high(est) point level; human parotid lysozyme; human placental lactogen
Hpl Hartlepool
HPL Halifax Public Library; Hamilton Public Library; Hartford Public Library; Houston Pipe Line; Houston

Public Library
hplc high-pressure-liquid chromatography
hpll hybrid phase-locked loop
hplr hinge pillar
hpls hopeless
hpm high polymer molecular; human potential movement
H-P m Harding-Passey melanoma
HPM Human Potential Movement
HPMA Hardwood Plywood Manufacturers Association
hpmv high-pressure mercury vapor
hpn home parental nutrition; horsepower nominal
hpns high-pressure nervous syndrome
hpo high-pressure oxygenation
HPO Hamilton Philharmonic Orchestra; Helsinki Philharmonic Orchestra; Highway Post Office
HPOL High-Power Optical Laboratory
H-pole H-shaped telegraph or telephone pole
hpox high-pressure oxygen
hpp hydraulic pneumatic panel
HPP Harvard Project Physics; Hawker Pacific Proprietary
HPPA Horses and Ponies Protection Association
HPPB Historic Pensacola Preservation Board
h-p plan hire-purchase plan (British equivalent of American installment-plan purchasing)
HPPP High-Priority Production Program
hpr high penetration resistant; high-power(ed) radar; hopper; hot particle rolling
HPR House of Pacific Relations
HPRF Hypersonic Propulsion Research Facility
HPRP High-Performance Reporting Post; High-Power(ed) Radar Post
hps high primary sequence; high protein supplement; high-pressure sodium; high-pressure steam; hot-pressed sheet
HPS Harlem Preparatory School; Health Physics Society; High Protestant Society
H_psc heading per standard compass
hpsn hot-pressed silicon nitride
hpst harpist
H_pstgc heading per steering

compass
hpt high point; high-pressure test
HPTA High-Power Test Area; Hire Purchase Trade Association
hptn hypertension
Hptw *Hauptwerk* (German—great work)
hpu hot-plate unit; hydraulic pumping unit
hpv high-passage virus; human-powered vehicle
HPVA Human-Powered Vehicle Association
hpv-de high-passage virus (grown in) duck embryo
hpv-dk high-passage virus (grown in) dog kidney
hpw high-power window
hq headquarters
hq (HQ) high quality
h.q. *hoc quaere* (Latin—see this)
Hq Headquarters
H-Q Hydro-Quebec
HQASC Headquarters Air-Support Command (NATO)
HQBA Headquarters Base Area
hqc hydroxyquinoline citrate
HQCC Headquarters Coastal Command (UK)
HQ Comdt Headquarters Commandant
HQ COMD USAF Headquarters Command, USAF
HQD Harold Q(uinten) Driscoll
HQDA Headquarters, Department of the Army
hqdt handling qualities during tracking
HQDTMS Headquarters, Defense Traffic Management Service
HQEARC Headquarters, Equipment Authorization Review Center (USA)
HQFC Headquarters, Fighter Command (NATO)
HQFT Humanist Quest For Truth
HQMC Headquarters—Marine Corps
hq's (Hq's) headquarters
HQSC Headquarters, Signal Command (UK)
HQSTC Headquarters, Strike Command (UK)
HQT Humanist Quest for Truth
HQTC Headquarters, Transport Command (UK)
Hqtrs Headquarters
HQ USAF Headquarters,

USAF
hr hairspace; handling room; handrail; heat resisting; height range; high resilient; hinge remnant; home run; homing relay; hook rail; hoserack; hospital recruit; hot rolled; hour; human relations; relative humidity (symbol)
h(r) hail and rain (meteorological symbol)
h/r heart rate
hr *herr* (Swedish—Sir)—Mr
Hr *Herr* (Danish or German—Mr, Sir); *Horner* (German—brass instruments; horns)
HR Highland Regiment; Hillside Review (overlay zone); Hospital Recruit; House of Representatives; House Resolution; International Harvester (stock exchange symbol)
H-R Haut-Rhin; Hertzsprung–Russell
H & R Harper & Row; Harrington & Richardson; Herweg & Romine
HR *Hauptrhythmus* (German—outstanding rhythm)—12-tone term; *Hellenic Register* (Greek ship-classification book); *House* (of Representatives) *Resolution*
hra housing review account
hra (HRA) hypersonic research airplane
hRA hardness Rockwell A (scale)
Hra Herra (Finnish—Mister)
HRA Hardware Retailers Association; Health Resources Administration; Historical Records of Australia; Human Resources Administration; Hunters' Rights Association; Hypnotic Research Association
HRA *Historical Records of Australia*
H & RA Homeowners and Renters Assistance
HRAA Hypnotic Research Association of Australia
HRAF Human Relations Area File
HRAG Helena Rubinstein Art Gallery
hRB hardness Rockwell B (scale)
HRB Highway Research Board; Highway Research Bureau; Housing and Redevelopment Board
hrbr harbor
hrbr vu harbor view

hrc high rupturing capacity
hRC hardness Rockwell C (scale)
HRC Herpes Resource Center; Holy Roman Church; Horse Racing Commission; Humacao Regional College; Human Relations Commission; Human Resources Center; Human Rights Commission (OAS); Humanities Research Council
HRCC Humanities Research Council of Canada
HRCF Human Rights Campaign Fund(ing)
hrd hard; high roughage diet
HRD Hertzprung-Russell Diagram; Human Resources Development
HRDA Human Resources Development Agency
HRDF Human Resource Development Foundation
HRDI Hospital Reserve Disaster Inventory
HRDL Hudson River Day Line
hrdly hardly
hrdwd hardwood
hrdwr hardware
hre hypersonic research engine
hre (HRE) high-resolution electrocardiography
HRE Holy Roman Empire
HREBU Hotel and Restaurant Employees and Bartenders Union
H reflex Hoffmann reflex (of the tibial nerve)
H Rept House Report
HRes House Resolution (US House of Representatives)
HRET Hospital Research and Educational Trust
HREU Hotel and Restaurant Employees Union
hrf high rate of fire
Hrf *Harfe* (German—harp)
HRF Hat Research Foundation
HRFA Hudson River Fishermen's Association
hrf's health-related facilities
HRG Halford, Robins, and Godfrey
HRGs Health Research Groups
hrh high resistance hold
HRH His (Her) Royal Highness
hri height-range indicator
hri (HRI) high-resolution images
HRI Hotel Reservations International; Human Relations Inventory
HRIP Highway Research in

Progress
H.R.I.P. *hic requiescit in pace* (Latin—here rests in peace)
hrir high-resolution infrared radiometer
hrirs high-resolution infrared-radiation sounder
HRIS Highway Research Information Service; Human Resource Information System
hrl horizontal reference line
Hrl Harlingen
HRL Hughes Research Laboratories; Human Resources Laboratory
Hrm Herman
HRMA Hampton Roads Maritime Association
Hr Ms Haar Majesteits Schip (Dutch—Her Majesty's Ship)
Hrn Herren (German—gentlemen); *Horner* (German—brass instruments; horns)
HRNTWT High-Reynolds-Number Transonic Wind Tunnel
HRO Housing Referral Office (USAF)
hrp horizontal radiation pattern
HRP Hampton Roads Ports; Human Reliability Program; Huntsville Research Park
HRPA Hudson River Pilots Association
HRPP Human Rights Protection Party (Samoa)
hrr higher reduced rate (taxation)
HRRA Human Resources Research Organization
HRRC Human Resources Research Center
HRRL Human Resources Research Laboratory
HRRO Human Resources Research Office
hrs high-resolution spectrograph; high-resolution spectrometer; hot-rolled steel; hours
HRS Hamilton Rating Scale; Health and Rehabilitation Services; Health Resources Statistics; Human Resource System; Hydraulics Research Station; Hydrostatic Research System
HRSA Honorary Member of the Royal Scottish Academy
hrsg herausgegeben (German—edited or published)
Hrsg Herausgeber (German—editor)
hrsi high-temperature reusable-surface insulation
HRSRS Hartbeestehoek Radio

Space Research Station
hrt high-resolution track(er)
HRT Honolulu Rapid Transit; Hormone Replacement Therapy
hrts high risk test site
hrtwd heartwood
Hrtz Ha'aretz (Hebrew—The Land)—Israeli newspaper
HRU Hydrological Research Unit
hrv hypersonic research vehicle
HRVC Hudson River Valley Commission
HR & W Holt, Rinehart & Winston
HRWMC House of Representatives Ways and Means Committee
Hry Henry
HRYC Halifax River Yacht Club; Hampton Roads Yacht Club
HRZ Hertz Corporation (stock exchange symbol)
hs half strength; hardstand; head suppression; heating surface; hide substance; high-speed; hinged seat; horizontal shear; horizontal stripe(s); hot stuff; hypersonic
hs (HS) hardened site
h/s hand stamp(ing)
h.s. *hic situs* (Latin—laid here); *hoc sensu* (Latin—in this sense)
Hs Henriques
Hs Handschrift (German—manuscript)
HS Hakluyt Society; Haute-Saône; Haute-Savoie; Hawker Siddeley; High School; historic site; Home Secretary; Home Service; Hospital Ship; House Surgeon; Hunterian Society; hydrofoil ship (naval symbol)
H-S Haute-Saône; Haute-Savoie
H & S Health and Safety (Code); Home & School
HS Hauptsatz (German—principal theme in a sonata)
H.S. *hic sepultus* or *hic situs* (Latin—here lies buried)
HS-30 German armored-personnel carrier
HS-125 Hawker-Siddeley Dominie jet transport
HS-748 Hawker Siddeley support aircraft; Hawker-Siddeley troop transport carrying 40 paratroopers or 50 regular soldiers
hsa human serum albumin; hy-

personic aircraft (HSA)
HSA Hawker Siddeley Aviation; Health Service Area; Health Services Administration; Health Systems Agency; Herb Society of America; Hispanic Society of America; Holly Society of America; Home Servicemens Association; Hospital Savings Association; Hunt Saboteurs Association
HSAA Health Sciences Advancement Award
HSAC House (of Representatives) Science and Astronautics Committee
HSA & D High School of Art and Design
HSAL Hispanic Society of America Library (NYC)
hsb human sexual behavior
HSBC Hong Kong and Shanghai Banking Corporation
hsbr high-speed bombing radar
hsc (HSC) engine high-swirl combustion engine
HSC Health and Safety Code
H-S-C Hand-Schüller-Christian (disease)
HSCA Health Sciences Communication Association
HSCC Historical Society of Southern California
H Sch High School
Hschonhsn Hohenschonhausen
HS-Co A reduced coenzyme A
hscp high-speed card punch
hscr high-speed card reader
hsct high-speed compound terminal
HSCTB Heavy and Specialized Carriers Tariff Bureau
hsctt high-speed-card teletypewriter terminal
hsd hard-site defense; high-speed diesel (oil)
HSD Hawker Siddeley Dynamics; Health Services Department
hsda high-speed data acquisition
HSDE High-School Driver Education
HSDG Hamburg-Sudamerika Dampfschiffahrts Gesellschaft (Columbus Line)
HSDM Harvard School of Dental Medicine
hse house
hse (HSE) syndrome hemorrhagic shock and encephalopathy syndrome
Hse House (postal abbreviation)

HSE Health and Safety Executive

H.S.E. *hic sepultus est* or *hic situs est* (Latin—here lies buried)

HSERF High-Score Educational Research Foundation

H & SF Heart and Stroke Foundation

HSFI High School of Fashion Industries

hsg housing

Hsg Helsingör (Elsinore)

HSG Hawker Siddeley Group

hsgt high-speed ground transport

HSGTC High-Speed Ground Test Center

HSGTP High-Speed Ground Transportation Program

HSH Her (His) Serene Highness

h & s hole hellhole and smellhole

hsi heat-stress index; horizontal situation indicator

HSI Health Services Inc; Home and School Institute; Hotel Systems International; Humanist Science Incorporated; Human Suffering Index

HSIS Highway Safety Information Service

hsk housekeeper; housekeeping (flow chart); housekept

HSK Honorary Surgeon to the King

hskpg housekeeping

hskpr housekeeper

hsl herpes simplex labialis (HSL); hytran simulation language

HSL Hawaii State Library; Huguenot Society of London

hsla high-strength low-alloy (steel)

HSLA Home and School Library Association

HS Lab Health Service Laboratory (USA)

hslkfin high silky finish

hsltd hand salted

HSLWI Helical Spring Lock Washer Institute

hsm high-speed memory

hsm (HSM) holosystolic murmur

HSM Her (His) Serene Majesty; Historical Society of Montana

HSMA Hotel Sales Management Association

HSMB Hydronautics Ship Model Basin

HSMHA Health Services and Mental Health Association

HSN Home Shopping Network

HSNP Hot Springs National Park

HSNR Huleh Swamp Nature Reserve (Israel)

HSNY Handel Society of New York

hso high specific outlet

HSO Haifa Symphony Orchestra; Hamburg Symphony Orchestra; Hartford Symphony Orchestra; Hitachi Symphony Orchestra; Honolulu Symphony Orchestra; Houston Symphony Orchestra

HSORS High-Seas Oil-Recovery System

hsp high-speed printer

h of sp hybrid of species

H-S p Henoch-Schönlein purpura

HSP Historic State Park; Historical Society of Pennsylvania; Hospital Surgical Plan

HSP Haute Société Protestant (French—High Protestant Society)

HSPA Hawaiian Sugar Planters' Association; High School of the Performing Arts

h-span hydrolized starch-polyacrylonitrile copolymer graft(ing)

HSPG Hansard Society of Parliamentary Government

HSPH Harvard School of Public Health

HSPQ High-School Personality Questionnaire

hsptp high-speed paper-tape punch

hsptr high-speed paper-tape reader

HSQ Historical Society of Queensland; Honorary Surgeon to the Queen

hsr high-speed reader; high-speed rewind

hsr (HSR) high-speed rail(road); high-speed railway

HSR Health Service Region (USA)

hsrc high-speed rail concept

HSRC Health Sciences Resource Center (Canadian)

HSRI Health Systems Research Institute; Highway Safety Research Institute

hsro high-speed repetitive operation

hss high-speed steel

hss (HSS) hydrological-sensing satellite

H-S s Hallervorden-Spatz syndrome

HSS History of Science Society; Hungarian State Symphony

HSSA History of Science Society of America

HSSO Hungarian State Symphony Orchestra

hsss high-speed stainless steel; high-strength stainless steel

hsst (HSST) high-speed surface transport (vehicle floats above its track on a magnetic cushion)

HSSTS High-Speed Surface Transport System

hst highest spring tide; high-speed train; hoist; hypersonic transport

Hst (HST) Hubble space telescope

H St Hugo Stinnes (steamship line)

HST Harry S Truman—thirty-third President of the United States; Hawaiian Standard Time; hypersonic transport

HSTA Hawaii State Teachers Association

HSTC Henderson State Teachers College

h/stead homestead(er)(s)

HSTI Hartford State Technical Institute

HSTL Harry S Truman Library

Hstn Houston

HSTRU Hydraulic System Test and Repair Unit

hsts horizontal stabilizer trim setting(s)

HSTS House Subcommittee on Traffic Safety

HSU Hardin-Simmons University

H substance histamine-like capillary vasodilator

H-substance histamine-like substance

HSUL Haile Selassie University Libraries (Addis Ababa, Ethiopia)

HSUNA Humanist Student Union of North America

HSUS Humane Society of the United States

hsv heat-suppression valve

hsv (HSV) herpes simplex virus

HSV Huntsville, Alabama (airport)

hswf housewife

HSWP Hungarian Socialist Workers Party

HSWT High-Speed Wind Tunnel

hszd hermetically-sealed zener diode

ht half title; halftime; halftone; hard top; heat; heat treat; heat treatment; heat-treated; heavy formex; heavy tank; heavy traffic; height; height telling; high temperature; high tensile; high tension; high tide; high treason; hired transport; hollow tile; hybrid tea (rose); hydrotherapy; hypertropia; hypodermic tablet

ht (HT) horizontal tabulation (data processing)

h & t harden(ed) and temper(ed); hospitalization and treatment; hospitalize and treat

h.t. hoc tempore (Latin—at this time); hoc titulo (Latin—under this title)

Ht total hypermetropia

H^t Haut (French—high, upper)

HT Hawaiian Telephone; Hawaiian Territory; Hawaiian Theater; Hawaiian Time; Height Technician; Horsed Transport; Hospital Train

HT-2 trichothecense toxin used in biochemical warfare

hta heavier than air

HTA Hardcourt Tennis Association; Harris Tweed Association; Horticultural Trades Association

Htal Hospital (Spanish—hospital)

htb hautboy (oboe); high-tension battery

HTB Highway Tariff Bureau; Horserace Totalisator Board

htc head(ing) to come; headline to come; heat transfer coefficient; hydraulic temperature control

htd heated

HTD Hospital for Tropical Diseases

htd pl heated pool

htd rm heated room

H^te Haute (French—high, upper)

Hte-Gar Haute-Garonne

Hte-L Haute-Loire

Hte-M Haute-Marne

H^ter Hinter (German—behind, rear)

Hte-Sao Haute-Saône

Hte-Sav Haute-Savoie

Htes-Pyr Hautes-Pyrénées

htfc high-temperature fuel cell

ht fx heat treat fixture

htg heating

htgr (HTGR) high-temperature gas-cooled reactor

Htg & Vent Heating & Ventilating

hth (HTH) holiday travel hostility

hthp high temperature high pressure

hti high-temperature isotope

HTI Hand Tools Institute; High Twelve International

htk headline to come

htl hearing threshold level; heavy-traffic licence; high threshold logic

Htl Hotel

htls high torque low speed

htlv human T-cell leukemia virus

HTLV-III human T-cell lymphotropic virus

htm heat transfer medium; high-temperature metallography

HTMC High Temperature Materials Corporation

Htn Hamilton, Bermuda

hto high-temperature oxidation; horizontal takeoff

htofore heretofore

htol (HTOL) horizontal-takeoff-and-landing

h top hardtop

HTOT High-Temperature Operating Test

H-Town Hartford, Connecticut

htp high-test peroxide

h-t p half-title page

h-t-p house-tree-person (psychological drawing test)

HTP House-Tree-Person (test); Humor Test of Personality

htr heater

htr (HTR) high-temperature reactor

HTR Highway Traffic Regulations(s)

HTR Harvard Theological Review

htrac half-track

htrb high-temperature reverse bias

HTRDA High-Temperature Reactor Development Associates

htres heat resistant

H Trin Holy Trinity

hts half-time survey; heights; high-tensile steel

Hts Heights

HTS Huntington, West Virginia (airport)

HTSA Highway Traffic Safety Administration

htst high-temperature short-time (pasteurization)

htt (HTT) heavy tactical transport

ht tr heat treat

htu heat transfer unit

htv (HTV) hypersonic test vehicle

HTV Hiroshima Television

htvt heating and ventilating

htw high-temperature water

HT&W Hoosac Tunnel & Wilmington (railroad)

ht wkt hit wicket

htxgr heat exchanger

hu hyperemia unit

Hu Hungarian; Hungary

HU Haifa University; Harvard University; Hebrew University; Howard University

HUA Highway Users Association; Housing and Urban Affairs

HUAC House Un-American Activities Committee

Huascán Huascarán (Peru's highest mountain)

HUB Humboldt Universität zu Berlin (German—Humboldt University of Berlin)—library on Berlin's Clara-Zetkin Strasse

hubba hubbas (crack-type cocaine)

Hubble Hubble Space Telescope

HUC Hebrew Union College

HUCIA Harvard University Center for International Affairs

HUCJIR Hebrew Union College Jewish Institute of Religion

hucks huckleberries

Huck(y) Huckleberry Finn

hucr highest useful compression ratio

hud head-up display

Hud Huddleston; Hudson

HUD Hong Kong United Dockyard; Housing and Urban Development

HUDC Housing and Urban Development Corporation

hud-eu head-up display electronic unit

Hud Inst Hudson Institute

HUDPR Housing and Urban Development Procurement Regulations

hudson hudson seal (imitation seal made of dyed muskrat fur)

Hudson Angus Hudson, the butler in *Upstairs, Downstairs*; river rising in the Adirondacks and flowing to New York Bay

Hudson Girls New York School for (delinquent) Girls

at Hudson
Hudson Internat Legis Hudson's International Legislation
Hudson World Court Hudson's World Court Reports
Hud Val Phil Hudson Valley Philharmonic
hudwac head-up-display weapon-aiming computer
Huel Huelva
Hueneme Port Hueneme, California
Hues Huesca
huff-duff high-frequency direction finder
HUFSM Highway Users Federation for Safety and Mobility
Huggin Hugh; Hugo
HUGHES Hughes Aircraft Company
hugo highly unusual geophysical operations
HUGO Human Genome Organization
Hu H Hughes Hall, Cambridge
HUJ Hebrew University of Jerusalem
huk (HUK) hunter-killer
HUKFORLANT Hunter-Killer Forces—Atlantic (USN)
HUKFORPAC Hunter-Killer Forces—Pacific (USN)
huks (HUKS) hunter-killer submarine(s)—USN
Huks Hukbong Mapgapalayang Bayan (Philippine Communist Armed Forces)
Hul Hulbert; Huldreich; Hulton
HUL Harvard University Library; Helsinki University Library; Hokkaido University Library
Hull in England, Kingston-upon-Hull
Hully Hulbert
HULTIS Hull Technical Interloan Scheme
hum human; humane; humanism; humanities
hum. humaniora (Latin—humanities)—also appears as *H.U.M.*
Hum Humbert; Hummel; Humphrey; Humphreys; Humphry
Huma L'Humanité (French—communist daily paper)
human eng human engineering
HUMARIS Human Materials Resources Information System
Humb *humberside*

HUMBLE Humble Oil (Company)
Hum Con Humanist Conference
humer humerus
humi humidity
humint human intelligence
hummer. high-mobility multi-purpose wheeled vehicle
hummer(s) humming bird(s)
Hummon Herman
Humph Humphrey
HUMRRO Human Resources Research Office
hums humanitarian reasons
hun hundred
Hun Hungarian; Hungary
Hun Hungria (Portuguese—Hungary); *Hungría* (Spanish—Hungary); *Hun's NY Supreme Court Reports*
hund hundred
Hung Hungaria; Hungarian; Hungarica; Hungary
Hungary Hungarian People's Republic (central European nation), *Magyar Népköztársaság*
Hunnen Die Hunnenschlacht (German—The Battle of the Huns)—Liszt's Symphonic Poem No. 11
Hunt Hunter; Huntington; Huntley; Huntly
hunth hundred thousand
Huntington Huntington Library, Art Gallery, and Botanical Gardens at San Marino, California
Hunts Huntingdonshire
Huntsville Girls Goree Unit Women's Prison at Huntsville, Texas
HUP Harvard University Press; Hospital of the University of Pennsylvania
HUPAS Hofstra University Pro Arte Symphony
hur hurricane
hur (HUR) homes using radio
Hurb Hurban (Hebrew—Holocaust)
hurcn hurricane
hurevac hurricane evacuation
HURRAH Help Us Reach and Rehabilitate America's Handicapped (HEW program)
hus hemolytic uremic syndrome (HUS)
Hus Johannes Hus von Husinetz
HUSAT Human Sciences and Advanced Technology
husb husbandry
Huss Jan Huss known in German as Johannes von Husi-

netz—his death by being burned at the stake for heresy led to the Hussite War from 1419 to 1434
hustle. helium-underwater speech-translating equipment
hut. (HUT) homes using television
hutch. humidity-temperature charts
Hutch Hutcheson; Hutchings; Hutchins; Hutchinson; Hutchison
hutv home(s) using television
Hux Huxley
hv heavy; high velocity; high voltage
h-v high-voltage
h & v heating and ventilating
h.v. hoc verbum (Latin—this word)
HV Hardness Vickers (symbol); Health Visitor; Hospital Visit
H-V Haute-Volta (French—Upper Volta)
hva homovanillic
Hva Huelva
HVA Health Visitors' Association
hvac heating, ventilating, and air conditioning; high-voltage alternating current
hvap hyper-velocity armor-piercing
hvar (HVAR) high-velocity aircraft rocket
HVB Hawaii Visitors Bureau
hvbn have been
hvc hardened voice circuit
hv & c heating, ventilating, and cooling
HVCA Heating and Ventilating Contractors' Association
HVCC Hudson Valley Community College
hvd high-velocity detonation; hypertensive vascular disease
hvdc high-voltage direct current
hvdn have done
hve home video entertainment
HVEC High Voltage Engineering Corporation
hvem high-voltage transmission electron microscopy
hvf high viscosity fuel
hvg having
hvgo heavy-vacuum gas oil
hvh herpesvirus hominis
hvh (HVH) herpesvirus hominis (type 1 transmitted by mouth and marked by cold sores and fever blisters; type 2 transmitted venereally and characterized by genital le-

sions)
Hvh Herpesvirus hominus (Latin—herpes simplex virus)
H v H Hoek van Holland (Dutch—Hook of Holland)
hvhmd holographic visor-helmet mounted display
hvi high viscosity index
HVI Hartman Value Inventory; Home Ventilating Institute
H'ville Huntsville, Alabama
hvJ hemagluttinating virus of Japan
H v K Herbert von Karajan
hvl half-value layer
HVL Hanseatic Vaasa Line; Heitor Villa-Lobos
HVMC Harbor View Medical Center
Hvn Haven
HVNP Hawaii Volcanoes National Park
HVO Hawaiian Volcano Observatory
HVOT Hooper Visual Organization Test
hvp high-value package
HVP Hudson Vitamin Products
HVPO Hudson Valley Philharmonic Orchestra
hvps high-voltage power supply
hvpve high-voltage photovoltaic effect
hvr high-vacuum rectifier
HVRA Hawaiian Volcano Research Association
hvrap (HVRAP) hyper-velocity rocket-assisted projectile
hvsa high-voltage slow activity
hvss horizontal volute spring suspension
hvtp high-velocity target-practice
HVWS Hebrew Veterans of the War with Spain
hvy heavy
hw headwaiter; headwind; heartworm; herewith; high water; hot water
h/w husband and wife
Hw Hauptwerk (German—great work)
HW Helsinki Watch; high water
H-W Harbison-Walker (refractories)
H & W Harland and Wolff (Belfast shipbuilders); Hereford and Worcester
Hway Highway
Hwb Handwörterbuch (German—pocket dictionary)
hwc hot-water circulating

HWC Heriot-Watt College
hwctr heavy-water components test reactor
hwd hardwood
H'w'd Hollywood
HWDYKY How Well Do You Know Yourself? (psychological test)
hwf & c high water full and change
hwgcr (HWGCR) heavy-water-moderated gas-cooled reactor
hwi high water interval
HWI Helical Washer Institute
hwl high-water line
HWL Henry Wadsworth Longfellow; Hutchison Whampoa Limited
hwLB high water London Bridge
hwlwr (HWLWR) heavy-water-moderated boiling light-water-cooled reactor
hwm high-water mark; high-wet modulus
HWM Hiram Walker Museum (Windsor, Ontario)
HWMC House Ways and Means Committee
HWMD Hazardous Waste Management Division (EPA)
hwmnt high-water mark neap tide
hwmont high-water mark ordinary neap tide
hwmost high-water mark ordinary spring tide
hwmst high-water mark spring tide
hwnt high-water neap tide
HWO Homosexual World Organization
hwocr (HWOCR) heavy-water (moderated) organic-cooled reactor
hwont high-water ordinary neap tide
Hwood Hollywood
hwos high-water ordinary springs
hwost high-water ordinary spring tides
hwq high-water quadrature; tropic high-water inequality
hwr (HWR) heavy water reactor (AEC)
hws hot-water soluble; hot-water system
HWS Hazardous Waste Service; Hurricane Warning Service
H & WSC Hobart and William Smith Colleges
HWSS Hot Water Service System

hwst high-water spring tide
HWT Herald and Weekly Times
H-W U Heriot-Watt University
hwvr however
HWW Hochschule für Welthandel, Wien (School for World Trade—Vienna)
hwwc hand wash with care
Hwy Highway
hx hexode; history
hx (HX) headroom extension
Hx history (medical case)
Hxd Hardinxveld
hXr head X-ray
hy heavy; henry; high yield; hundred yards; hydrant
Hy Henry; Highway; Hiram; Hyman
H^y Highway
H^y Hasy (Arabic—waterhole)—also appears as *Hasi*
HY Helsingin Yliopisto (University of Helsinki)
Hya Hydra (constellation)
hyb hybrid
hyball hydraulic ball
HYC Harlem Yacht Club; Hartford Yacht Club; Haverhill Yacht Club
hycol hybrid computer link
hycon hydraulic control
hycotran hybrid computer translator
hyd hydrate; hydraulic(s); hydrostatics
Hyd Hyderabad; Hydrus (sometimes abbreviated Hyi for the genitive Hydri)
hydac hybrid digital-analog computer
hydapt hybrid digital-analog pulse time
h-y dash hundred-yard dash
hydel hydroelectric(al)
hyd/pnu hydraulic/pneumatic
hydr hydrographer
hydrarg. hydrargyrum (Latin—mercury)
HYDRAS Hydrographic Digital Positioning and Depth Recording System
hydraul hydraulic(s)
hydraweld hydraulic-drawn welded (steel tubing)
hydro hydrodynamic group of hydrodynamics (slang); hydroelectric; hydroelectrical; hydrographic; hydrology; hydrostatic
hydro (Latin prefix—water)—hydrodynamics
HYDRO Hydrographic Office
hydrodyn hydrodynamics
hydroelec hydroelectric
hydrog hydrography

HYDROIND Hydrography of the Indian Ocean
hydrol hydrology
HYDROLANT Hydrography of the Atlantic Ocean
hydrom hydromechanics
hydromag hydromagnetic(s)
hydromagnetics magnetohydrodynamics
HYDROPAC Hydrography of the Pacific Ocean
hydros hydrostatics
hydrot hydrotherapy
hydrox hydroxyline
HYDRSS High Data Rate Storage System (NASA)
hydt hydrant
hydx hydroxide(s)
hyf (HYF) hydrofoil
HYF Hong Kong and Yaumati Ferry
hyfes hypersonic flight environmental simulator
hyg hygiene; hygienic; hygroscopic
hygas hydrogen gasification
hygst hygienist
Hyk Helsingin yliopiston kirjasto (Finnish—Helsinki University Library)
hyl (Hyl) hydroxylysine
hyla hybrid language assembler
HYMA Hebrew Young Men's Association
hymnol hymnologist; hymnology

hynmm have you not made a mistake
hyp hyperbola; hyperbolic; hyphen; hyphenate; hyphenation; hypochondria(c); hypothesis; hypothetical
hyp (Hyp) 4-hydroxyproline
Hyp Hypolite
HYP Harvard, Yale, and Princeton
hype high performance (bullet); hyperbole; hypertension; hypodermic (underground slang—person who injects drugs with a hypodermic syringe)
hyper hypercritical
hyper (Latin prefix—above, beyond, excessive)—hypercritical
hyperb hyperbole; hyperbolic
hyperdip hyperdiploid(al); hyper diploidy
hyperdop hyperbolic doppler
hypersex hypersexual(ity)
hypert hypertape (flow chart); hypertension
hypn hypertension
hypno hypnotism
hypnot hypnotic; hypnotism; hypnotist
hypo hypochondria; hypochondriac; hypochondriacal; hypodermic (injection or needle); hyposulfite of soda (sodium thiosulfate—NaS_2O_3 $5H_2O$)

hypo (Latin prefix—below or under)—hypodermic
hypodip hypodiploid(al); hypodiploidy
hypoth hypothesis
hypro hydroxyproline
hys hysteria; hysteric; hysterical; hysterics
HYSAS Hydrofluidic Stability Augmentation System
hyst hysteresis; hysteria
hystad hydrofoil stabilizing device
hyster hysterectomy
hysterec hysterectomic (sterilization); hysterectomy (removal of the uterus)
HYSTU Hydrofoil Special Trials Unit (USN)
HYSURCH Hydrographic Survey and Charting System
hytemco high-temperature coefficient nickel-iron alloy
hy tr heat treat
hyv's high-yielding varieties (of grain)
hz haze; heritability zone; herpes zoster
hz (Hz) hertz (one cycle per second); hertzian
Hz Henriquez; hertz (cycles per second)
Hzbl Holzbläser (German—woodwind instruments or players)
Hzk Hezekiah
hzy hazy

I

i angle of incidence (symbol); incisor; indigo; infant(icide); instantaneous current (symbol); interceptor; interest; intransitive; isotopic fine structure (symbol); moment of photographic plate (symbol); optically inactive (symbol); rate of interest (symbol); Van't Hoff factor (symbol); vapor pressure constant (symbol)

i (I) inversion (12-tone matrix)

i' in

i Illinois Central symbol; Imperial Savings

i. id (Latin—that)

I acoustic intensity (symbol); candlepower or intensity of luminosity (symbol); conduction current (symbol); convection current (symbol); Ido (artificial language); in; inclination; Independent; India—code for letter I; Indian; industrial broadcasting; inertia; infantry; Inspector; Institute; Institution; Instructor; Intelligence; iodine; ionic strength (symbol); Ireland; Irish; Island; Isthmian Line; Italian Line; Italy—auto plaque; izzard

I *(I)* investment (macroeconomics symbol)

I Ile (French—Island, Isle); *Imperator* (Latin—Emperor); *Imperatrix* (Latin—Empress); *Imperium* (Latin—Empire); *in* (German or Italian—in); *inde* (Danish—in); *Infidelis* (Latin—infidel or unbeliever); *Isle* (French—island); *itä* (Finnish—east); *izquierda* (Spanish—left)

I-l luminous intensity (symbol)

I¹2⁸ radioactive iodine

I¹3⁰ radioactive iodine

I¹3¹ radioactive iodine

I²L integrated-injection logic

I³ Illinois Innovators and Inventors

ia immediately available; impedance angle; indicated altitude; infra-audible; initial allowance; initial appearance; international angstrom; intra-arterial; intra-articular

i & a indexing and abstracting; integration and assembly

i.a. in absentia (Latin—in the absence of)

i A im Auftrage (German—by order, for, under instruction)

Ia Ingegerda

IA Incorporated Accountant; Indian Army; Industrial Arts; Infected Area; Inspection Administration; Institute of Actuaries; Instructional Aide; Internal Affairs; International Angstrom; Iowa; Iraqi Airways; Irrigation Area; Isaac Asimov

I/A Insurance Auditor; Isle of Anglesey

I of A Inspector of Anatomy; Institute of Accountants; Institute of Acoustics; Instructor of Artillery

IA International Atlas (Rand McNally)

IAA Independent Airlines Association; Indian Association of America; Inspector Army Aircraft; Insurance Accountants Association; Interment Association of America; International Academy of Astronautics; International Acetylene Association; International Advertising Association; International Apple Association; International Association of Allergology; Intimate Apparel Associates; Inventors Association of Australia

IAA Instituto do Acuar e do Alcool (Portuguese—Sugar and Alcohol Institute); *International Aerospace Abstracts*

IAAA Institute of Air Age Activities; International Airforwarders and Agents Association

IAAAA Intercollegiate Association of Amateur Athletes of America

IAAB Inter-American Association of Broadcasters

IAABA International Association of Aircraft Brokers and Agents

IAAC International Agriculture Aviation Center; International Antarctic Analysis Center

IAACC Inter-Allied Aeronautical Control Commission

IAAE Institution of Automotive and Aeronautical Engineers

IAAER International Association for the Advancement of Educational Research

IAAF International Amateur Athletic Federation

IAAFA Inter-American Air Force Academy

IAAHU International Association of Accident and Health Underwriters

IAAI International Airports Authority of India; International Association of Arson Investigators

IAALD International Association of Agricultural Librarians and Documentalists

IAAM International Association of Auditorium Managers; International Association of Automotive Modelers

IAANZ Institute of Actuaries of Australia and New Zealand

IAAO Interlochen Arts Academy Orchestra; International Association of Assessing Officers

IAAOPA International Association of Aircraft Owners and Pilots Associations

IAAP International Association of Applied Psychology

IAAPEA International Associations Against Painful Experiments on Animals

IAAS Incorporated Association of Architects and Surveyors; Institute of Advanced Arab Studies; International Association of Agricultural Students

IAASE Inter-American Association of Sanitary Engineering

IAASS International Association of Applied Social Science

iab increasing assurance benefits

IAB Industrial Advisers to the Blind; Industrial Arbitration Board; Industry Advisory Board; Inter-American Bank; International Air Bahama; International Association of Bureaucrats

IABA Inter-American Bar Association; International Association of Aircraft Brokers and Agents

IABBE International Association for Better Basic Education

IABC International Association of Business Communicators

IABG International Association of Botanic Gardens

IAB-ICSU International Abstracting Board—International Council of Scientific Unions

IABLA Inter-American Bank for Latin America; Inter-American Bibliographical and Library Association

IABO International Associa-

tion of Biological Oceanography

IABPAI International Association of Blue Print & Allied Industries

IABPC International Association of Book Publishing Consultants

IABSE International Association for Bridge and Structural Engineering

IABSIW International Association of Bridge and Structural Iron Workers

IABTI International Association of Bomb Technicians and Investigators

iac innovative academic courses; integrating assembly contractor; integration, assembly, checkout; interview after combat

IAC Ibrahim Ali Commission (Malaysia); Indian Airlines Corporation; Industrial Arbitration Court; Industry Advisory Commission; Industry Assistance Commission; Information Analysis Center; Institute of Amateur Cinematographers; Insurance Advertising Conference; Intelligence Advisory Committee (CIA); Intermediate Air Command; International Anticounterfeiting Coalition; Interview After Combat; Irish Air Corps

IACA Independent Air Carriers Association; Inter-American College Association

IACAC Inter-American Commercial Arbitration Commission

IACB Indian Arts and Crafts Board; International Advisory Committee on Bibliography (UNESCO); International Association of Convention Bureaus

IACC International Anti-Counterfeiting Coalition; Italy-America Chamber of Commerce

IACC *Instituto Argentino de Control de la Calidad* (Spanish—Argentine Institute for Quality Control)

IACCD Inter-American Confederation of Continental Defense

IACCI International Association of Credit Card Investigators

IACCP Inter-American Council of Commerce and Produc-

tion

IACD International Association of Clothing Designers

IACDLA International Advisory Committee on Documentation, Libraries, and Archives (UNESCO)

IACE International Air Cadet Exchange

IACES International Air Cushion Engineering Society

IACHR Inter-American Commission on Human Rights

IACI Inter-American Childrens Institute; Irish-American Cultural Institute

IACID Inter-American Center for Integral Development

IACM International Association of Circulation Managers; International Association of Concert Managers

IACOMS International Advisory Committee on Marine Sciences (FAO)

IACP International Association of Chiefs of Police

IACP & AP International Association for Child Psychiatry and Allied Professions

IACRL Italian-American Civil Rights League

IACS International Annealed Copper Standard; International Association of Cooking Schools; International Association of Counseling Services; Irish-Australian Cultural Society

IACT Illinois Association of Classroom Teachers; Indiana Association of Cities and Towns

IACTE Iowa Association of Colleges for Teacher Education

IACUSD International Association of College and University Security Directors

IACVB International Association of Convention and Visitor Bureaus

IACW Inter-American Commission for Women

iad initiation area discriminator; installation, assembly, or detail; integrated automatic documentation

IAD Dulles International Airport (Washington, DC); Integrated Area Development; Internal Affairs Department; Internal Affairs Division (LAPD and NYPD); International Agricultural Distribution; International Astrophysical Decade—1965–1975

I-A D Inter-American Dialogue

IADB Inter-American Defense Board; Inter-American Development Bank

IADC Inter-American Defense College; International Association of Dredging Companies

IADF Inter-American Association for Democracy and Freedom

IADIS Irish Association for Documentation and Information Services

iadl (IADL) instrumental activities of daily living

IADL International Association of Democratic Lawyers; Italian-American Defense League

IADO Iranian Agriculture Development Organization

IADPC Inter-Agency Data Processing Committee

IADR International Association for Dental Research

IADS Integrated Air Defense System; International Association of Dental Students; International Association of Department Stores

iadt initial active duty training

iae in any event; integral absolute error

IAE Institute of Army Education; Institute of Automobile Engineers; Institution of Automobile Engineers; International Animal Exchange

IAE Institut Atomnoi Energii (Russian—Atomic Energy Institute)

IAeA Institution of Aeronautical Engineers

IAEA Inter-American Education Association; International Association for Educational Assessment; International Atomic Energy Agency

IAEC Israel Atomic Energy Commission

IAECOSOC Inter-American Economic and Social Council

IAEE International Association of Earthquake Engineers

IAEI International Association of Electrical Inspectors

IAEL International Association of Electrical Leagues

IAES International Association of Electrotypers and Stereotypers

IAESP Indiana Association of Elementary School Principals; Iowa Association of Elementary School Principals

IAESTE International Association for the Exchange of Students for Technical Experience

IAET In-Flight Aeromedical Evacuation Team

IAEVG International Association for Educational and Vocational Guidance

IAEWP International Association of Educators for World Peace

iaf immobilizing accelerating factor; interview after flight

IAF Industrial Areas Foundation; Inter-American Foundation; International Abolitionist Federation (for abolition of prostitution); International Association of Firefighters; International Astronautical Federation; Israeli Air Force

I-AF Inter-American Foundation

IAFAE Inter-American Federation for Adult Education

IAFC International Association of Fire Chiefs

iafd intentionally administered fatal dose(s)

IAFD International Association of Food Distribution

IAFE International Association of Fairs and Expositions

IAFF Institute of Australian Flora and Fauna; International Association of Fire Fighters

iafi infantile amaurotic family idiocy

IAFMM International Association of Fish Meal Manufacturers

IAFP International Association for Financial Planning

IAFV Infantry Armed Fighting Vehicle

IAFWNO Inter-American Federation of Working Newspapermen's Organizations

IAG Institute of Australian Geographers; Interagency Advisory Group; International Association of Geodesy; International Association of Gerontology

IAGA International Association of Geomagnetism and Aeronomy

IAGB & I Ileostomy Association of Great Britain and Ireland

iagc instantaneous automatic gain control

IAGC International Association for Geochemistry and Cosmochemistry

IAGFCC International Association of Game, Fish, and Conservation Commissioners

IAGLP International Association of Great Lakes Ports

IAGM International Association of Garment Manufacturers

I Agr E. Institution of Agricultural Engineers

IAGS Inter-American Geodetic Survey

IAH Houston International Airport (Texas); Inter-American Highway; International Asian Highways; International Association of Hydrology

IAHA Inter-American Hotel Association

IAHF International Aerospace Hall of Fame

IAHIC International Association of Home Improvement Councils

IAHM International Association of Head Masters

IAHP Institutes for the Achievement of Human Potential; International Association of Horticultural Producers

IAHR International Association for Hydraulic Research

IAHS International Association of Hydrological Sciences

IAI Icelandic Airlines Incorporated; International African Institute; International Association for Identification

IAI-201 Israeli Arava light transport plane

IAIA Institute of American Indian Arts

IAIAS Inter-American Institute of Agricultural Sciences

IAICM International Association of Ice Cream Manufacturers

IAIE Inter-American Institute of Ecology

IAII Inter-American Indian Institute

IAIs Israeli Aircraft Industries

IAIS Industrial Aerodynamics Information Service (UK)

ial initial; initial appearance; initialism; instrument approach and landing; interlaminar adhesive layer; international algebraic language

IAL Icelandic Airlines; Imperial Airways Limited; International Aeradio Limited; International Algebraic Language; International Arbitration League; International Association of Limnology; Irish Academy of Letters

IAL Icelandic Airlines-Loftleider

IALA International Association of Lighthouse Authorities

IALC International Association of Lions Clubs; International Association of Lyceum Clubs

IALL International Association of Law Libraries

i allg im allgemeinen (German—generally, in general)

IALP International Association of Logopedics and Phoniatrics

IALS International Association of Legal Science

iam interactive algebraic manipulation

IAM Indian-Artifact Magazine; Institute of Appliance Manufacturers; Institute of Aviation Medicine; International Academy of Medicine; International Association of Machinists; International Association of Meteorology

IAMA International Abstaining Motorists Association

IAMAM International Association of Museums of Arms and Military History

IAMAP International Association of Meteorology and Atmospheric Physics

IAMAT International Association for Medical Assistance to Travelers

IAMAW International Association of Machinists and Aerospace Workers

IAMB International Association of Microbiologists

IAMC Indian Army Medical Corps; Institute for Advancement of Medical Communication; Inter-American Music Council

IAMCA International Association of Milk Control Agencies

IAMCL International Association of Metropolitan City Libraries

IAMCR International Association for Mass Communication Research

IAMFE International Association on Mechanization of Field Experiments

IAMFS International Association of Milk and Food Sanitarians

IAML International Association of Music Libraries

IAMLT International Association of Medical Laboratory Technologists

IAMM International Association of Master Mariners; International Association of Medical Museums

IAMO Inter-American Municipal Organization

IAMP Inter-Agency Motor Pool

IAMPO International Association of Mechanical and Plumbing Officials

IAMPTH International Association of Master Penmen and Teachers of Handwriting

IAMR Institute of Arctic Mineral Resources

IAMS International Association of Microbiological Societies; International Association of Municipal Statisticians

IAMSO Inter-African and Malagasy States Organization

IAMTCT Institute of Advanced Machine Tool and Control Technology

IAMTF Inter-Agency Maritime Task Force

IAMWF Inter-American Mine Workers Federation

Ian (Gaelic—John)

IAN Instituto Agrario Nacional (Spanish—National Agrarian Institute)

IANA Inter-African News Agency

IANAP Interagency Noise Abatement Program

IANC International Airline Navigators Council

IA & ND Indian Affairs and Northern Development (Canada)

IANE Institute of Advanced Nursing Education

IANEC Inter-American Nuclear Energy Commission

Ian F Ian Fleming

IANSA Industria Azucarera Nacional SA (Spanish—National Sugar Industry Corporation)

IANSW Institute of Architects of New South Wales

iao intermittent aortic occlusion

IAO Incorporated Association of Organists

IAOC Indian Army Ordnance Corps

IAOL International Association of Orientalist Libraries

IAOR International Abstracts in Operations Research

IAOS International Association of Oral Surgeons; Irish Agricultural Organization Society

IAOT International Association of Organ Teachers

iap interceptor aim point(s)

IAP Institute of Agricultural Parasitology; Institute of Australian Photographers; Institute of Australian Photography; Institution of Analysts & Programmers; International Academy of Pathology; International Academy of Proctology

IAPA Industrial Accident Prevention Association; Inter-American Parliamentary Association; Inter-American Parliamentary Organization; Inter-American Police Academy; Inter-American Press Association; International Airline Passengers Association; International Association of Police Artists

IAPB International Association for the Prevention of Blindness

IAPC Institute for the Advancement of Philosophy for Children; International Association of Political Consultants; International Association for Public Cleansing

IAPCO International Association of Professional Congress Organizers

IAPCU Iowa Association of Private Colleges and Universities

IAPESGW International Association of Physical Education and Sports for Girls and Women

IAPG Interagency Advanced Power Group; International Association of Physical Geography

IAPH International Association of Paper Historians; International Association of Ports and Harbors

IAPHA International Association of Port and Harbor Authorities

IAPHC International Association of Printing House Craftsmen

IAPI Institute of American Poultry Industries; *Instituto Argentino de Producción Industrial* (Spanish—Argentine Industrial Production Institute)

IAPIP International Association for the Protection of Industrial Property

IAPL International Association of Penal Law

IAPM International Academy of Preventive Medicine; International Association of Progressive Montessorians

IAPMO International Association of Plumbing & Mechanical Officials

IAPN International Association of Professional Numismatists

IAPO International Association of Physical Oceanography

IAPP International Association of Police Professors

IAPPW International Association of Pupil Personnel Workers

IAPR Indian Air Patrol Reserve

iaps inductosyn angle position simulator

IAPs Industry Application Programs

IAPS Incorporated Association of Preparatory Schools; International Affiliation of Planning Societies; International Association for the Properties of Steam

IAPSC Inter-African Phytosanitary Commission

IAPSO International Association of Physical Sciences of the Oceans

IAPT International Association for Plant Taxonomy

IAPTA International Allied Printing Trades Association

IAPW International Association of Personnel Women

IAQ Independent Activities Questionnaire

IAQC International Association of Quality Circles

IAQR Indian Association for Quality and Reliability

iar intersection of air routes

IAR Institute for Air Research

IARA Inter-Allied Reparations Agency

I Arb Institute of Arbitrators

IARC Indian Agricultural Research Council; International Agency for Research on Cancer

IARD Information Analysis and Retrieval Division (American Institute of Physics)

IARF International Association for Liberal Christianity and Religious Freedom

IARI Indian Agricultural Research Institute; Industrial Advertising Research Institute

IARIGAI International Association of Research Institutes for the Graphic Arts Industry

IARIW International Association for Research into Income and Wealth

IARP Indian Association for Radiation Protection; Inflation Accounting Research Project

IARQ Intellectual Achievement Responsibility Questionnaire

IARS International Anesthesia Research Society

IARSS Illinois Association of Regional Superintendents of Schools

IARU International Amateur Radio Union

ias immediate access storage; indicated airspeed; instrument approach system

ia's infant addicts

IAS Indian Administrative Service; Industrial Arbitration Service; Infrared Astronomy Satellite; Institute for Advanced Study; Institute of the Aeronautical Sciences; Institute of Aerospace Sciences; Institute of American Strategy; Institute of Andean Studies; Instrument Approach System; Intelligent Authoring Systems; International Accountants Society; International Association of Siderographers; International Aviation Service

IASA Idaho Association of School Administrators; Illinois Association of School Administrators; Insurance Accounting and Statistical Association; International Air Safety Association; International Association of Sound Archives; Iowa Association of School Administrators

IASB Iowa Association of School Boards

IASBO Indiana Association of School Business Officials; Iowa Association of School Business Officials

IASC Indian Army Service Corps; Inter-American Safety Council; International Association of Seed Crushers

IASCH Institute for Advanced Studies in Contemporary History (formerly Wiener Library)

iasd interatrial septal defect

IASDI Inter-American Social Development Institute

IASG Inflation Accounting Steering Group

IASH International Association of Scientific Hydrology

IASI Inter-American Statistical Institute

IASL Illinois Association of School Librarians; International Association for the Study of the Liver; International Association of School Librarians; Irish Association of School Librarians

IASLIC Indian Association of Special Libraries and Information Centers

iasor ice and snow on runway

IASP International Association for Social Progress; International Association for Suicide Prevention; International Association of Scholarly Publishers

IASPEI International Association of Seismology and Physics of the Earth's Interior

IASPO International Association of Senior Police Officers

IASPS International Association for Statistics in Physical Sciences

IASS Insurance Accounting and Statistical Society; International Association for Shell Structures; International Association of Soil Science

IASSS International Association for Shell and Spatial Structures

IASSW International Association of Schools of Social Work

IASTE International Association for the Study of Traditional Environments

iasy international active sun years

iat inside air temperature

IAT Individual Acceptance Test(ing); Institute for Applied Technology; Institute of Atomic Physics (Peking); International Academy of Tourism

IATA International Air Transport Association

iatc inlet air temperature control

IATC International Association of Tool Craftsmen

iatd is amended to delete

IATE Illinois Association of Teachers of English; International Association for Television Editors; International Association for Temperance Education

IATL International Association of Theological Libraries

IATM International Association for Testing Materials

IATME International Association of Terrestrial Magnetism and Electricity

IATP Individual Aircraft Tracking Program

iatr is amended to read

IATSE International Alliance of Theatrical Stage Employees

IATTC Inter-American Tropical Tuna Commission

IATUL International Association of Technical University Libraries

iau intrusion alarm unit

IAU International Association of Universities; International Astronomical Union

IAUPE International Association of University Professors of English

IAUPL International Association of University Professors and Lecturers

IAUPPR Inter-American University Press of Puerto Rico

IAUPR Inter-American University of Puerto Rico

IAUR Institute of Art and Urban Resources

IAV International Association of Volcanology

IAVA Industrial Audio-Visual Association

iavc instantaneous automatic volume control

IAVCB International Association of Visitors and Convention Bureaus

IAVCEI International Association of Volcanology and Chemistry of the Earth's Interior

IAVE Interaction Analysis for Vocational Educators

IAVFH International Association of Veterinary Food Hygienists

IAVG International Association for Vocational Guidance

IAVRS International Audio-visual Resource Service (UNESCO)

IAVTC International Audio-Visual Technical Center

iaw in accordance with

IAW International Alliance of Women

IAWA International Association of Wood Anatomists

IAWL International Association for Water Law

IAWMC International Association of Workers for Maladjusted Children

IAWP International Association of Women Police

IAWPC International Association on Water Pollution Research and Control

IAWPR International Association on Water Pollution Research

IAWS Intercollegiate Association for Women Students; Irish Agricultural Wholesale Society

IAWWW *International Authors and Writers Who's Who*

IAZ Inner Artillery Zone

ib in bond; illegal behavior; inbound; incendiary bomb; inclusion body; index of body build; infectious bronchitis; inner bottom; instruction book; instructional brochure; invoice book(let); inward bound

i & b improvements and betterments

ib. *ibidem* (Latin—in the same place)

i b im besonderen (German—in particular)

Ib Ibadan

IB *Iberia Líneas Aéreas de España* (Spanish—Iberian Airlines of Spain); Imperial Bank; Imperial Beach; incendiary bomb; Infantry Battalion; Information Bulletin; Information Bureau; Intelligence Branch; International Baccalaureate; International Bank(ing); international broadcast(ing)

I of B Institute of Bankers; Institute of Biology

IB *Istanbul Bankasi* (Turkish—Istanbul Bank)

IBA Independent Bankers Association; Independent Bar Association; Independent Broadcasting Association; Independent Broadcasting Authority (United Kingdom); Institute for Bioenergetic Analysis; Institute of British Architects; International Banana Association; International Bar Association; International Bauxite Association; International Briqueting Association; Investing Builders Association; Investment Bankers Association

IBAA Investment Bankers Association of America; Italian Baptist Association of America

IBAE Institution of British Agricultural Engineers

IBAHP Inter-African Bureau for Animal Health and Protection

IBAM Institute of Business Administration and Management

IBAP Intervention Board of Agricultural Produce

I-bar capital-I-shaped metal bar

IBAR Inter-African Bureau of Animal Resources

IBAS Indonesian Business Association of Singapore

IBAU Institute of British-American Understanding

ibb intentional bases on balls (baseball)

IBB Illinois Inspection Bureau; Institute of British Bakers; International Bowling Board; International Brotherhood of Bookbinders

IBBD *Instituto Brasileiro de Bibliografia e Documentação* (Portuguese—Brazilian Institute of Bibliography and Documentation)

IBBISBBFH International Brotherhood of Boilermakers, Iron Ship Builders, Blacksmiths, Forgers, and Helpers

ibbm iron body bronze (or brass) mounted

IBBY International Board on Books for Young People

ibc (**IBC**) intermediate bulk carrier; intermediate bulk container

IBC Insurance Bureau of Canada; International Biographical Centre; International Broadcasting Corporation; Iwate Broadcasting Company

(Japan)
IBC *Instituto Brasileiro do Café* (Portuguese—Brazilian Coffee Institute)
IBCA Institute of Burial and Cremation Administration
IBCs Institutional Biosafety Committees
IBCS Integrated Battlefield Control System (USA)
IBCUSCAN International Boundary Commission, United States and Canada
ibd interest-bearing debentures; interest-bearing deposit
IBD Inflammatory Bowel Disease; Institute of British Decorators; International Bank of Detroit
ibda indirect bomb damage assessment
ibe integrity basis earthquake; inventory by exception
IBE Institute of British Engineers; International Bureau of Education
I-beam capital-I-shaped metal beam
IBEC International Bank for Economic Cooperation; International Basic Economy Corporation
IBECC *Instituto Brasileiro de Educação Ciencia e Cultura* (Portuguese—Brazilian Institute of Science Education and Culture)
IBEG International Book Export Group
iben incendiary bomb with explosive nose
Iber Iberia(n); Iberic(a)(n); Iberville
IBERIA *Líneas Aéreas de España* (Spanish—Iberian Airlines of Spain)
IBERLANT Iberian Atlantic
ibes integrated building and equipment scheduling
IBES Illinois Bureau of Employment Security
IBEW International Brotherhood of Electrical Workers
ibf internally blown flap; international bond fund
IBF Institute of Banking and Finance; Institute of British Foundrymen; International Boxing Federation
IBFD International Bureau of Fiscal Documentation
IBFI International Business Forms Industries
IBFMP International Bureau of the Federations of Master Printers

IBFO International Brotherhood of Firemen and Oilers
IBFs International Banking Facilities
ibg inter-block gap
IBG Institute of British Geographers
IBHA Insulation, Building, and Hardwood Association
IBhd initial beachhead
IBHE Illinois Board of Higher Education
ibi invoice book, inward
IBI Illinois Bureau of Investigation; Indiana Bureau of Investigation; Insulation Board Institute
IBI *Instituto Bancario Italiano* (Italian—Italian Banking Institute)
ibid. international bibliographical description
ibid. *ibidem* (Latin—in the same place)
IBiol Institute of Biology
IBIS International Book Information Service
IBJ Industrial Bank of Japan
IBK Institute of Bookkeepers
IBK *Institut für Bauen mit Kunststoffen* (German—Institute for Building with Plastics)
ibkr icebreaker
IBL Institute of British Launderers; Irish Biscuits Limited
ibm induced bowel movement
ibm (IBM) intercontinental ballistic missile
IBM Institute for Burn Medicine; International Business Machines
IBM *Industrias Biologicas Mexicana* (Spanish—Mexican Biological Industries)
IBMA Independent Battery Manufacturers Association
IBM JRD IBM Journal of Research and Development
IBMR International Bureau for Mechanical Reproduction
ibn identification beacon
IBN *Institut Belge de Normalisation* (French—Belgian Standards Institute)
ibnr incurred but not reported
ibo invoice book, outward
IBO International Baccalaureate Office
IBOB International Brotherhood of Old Bastards
ibol integrated business-oriented language
ibop (IBOP) international balance of payments

IBOP International Brotherhood of Operative Potters; International Business Opportunity Program
ibp initial boiling point
IBP Institute of British Photographers; International Biological Program; Iowa Beef Processors
IBPAT International Brotherhood of Painters and Allied Trades (U.S. and Canada)
IBPGR International Board for Plant Genetic Resources
IBPI International Bureau for Protection and Investigation
IBPOEW Improved Benevolent and Protective Order of Elks of the World
ibp's imperial belch pills (nonfattening diet-reduction capsules)—users enjoy flavored belch while avoiding preparation and cleanup
ibr information-bearing radiation; integral boiling reactor
ibr (IBR) infectious bovine rhinotracheitis
IBR Institute of Behavioral Research; Institute of Biosocial Research
IBRA International Bible Reading Association
IBRD International Bank for Reconstruction and Development (World Bank)
ibrl initial bomb-release line
IBRM Institute of Boiler and Radiator Manufacturers
IBRMR Institute for Basic Research on Mental Retardation
IBRO International Bank Research Organization; International Brain Research Organization; International Brewers' Research Organization
ibs inflatable boat, small; international belt skimmer (for cleaning oil slicks)
ibs (IBS) ionospheric beacon satellite; irritable bowel syndrome
IBS Ibaraki Broadcasting System; Indian Boy Scouts; Institute of Basic Standards; Institute of Biblical Studies; Intercollegiate Broadcasting System; International Bach Society; International Bank of Singapore; Irritable Bowel Syndrome; Israel Broadcasting Service
IBSA Inanimate Bird-Shooting Association; International Barber Schools Association; International Bible Student

Association
IBSC Iranian-British Shipping Company
IB Scot Institute of Bankers in Scotland
IBSGR Isiolo Buffalo Spring Game Reserve (Kenya)
ibsr individual base stock requirements; individual battle shooting range
IBSS Imperial Bureau of Soil Science
IBST Institute of British Surgical Technicians
IBSTP International Bureau for the Suppression of Traffic in Persons
ibt in-barrel time; initial boiling-point temperature
IBT Industrial Bio-Test (laboratories); International Brotherhood of Teamsters
IBTA International Baton Twirlers Association
IBTCWH International Brotherhood of Teamsters, Chauffeurs, Warehousemen, and Helpers
IBTCWHA International Brotherhood of Teamsters, Chauffeurs, Warehousemen, and Helpers of America
ib test inkblot test (Rorschach test)
IBTS International Bicycle Touring Society
IBTTA International Bridge, Tunnel, and Turnpike Association
ibu imperial bushel
IBU Inland Boatmen's Union; International Broadcasting Union
ibv infectious bronchitis vaccine
***IBVL** Instituut voor Bewaring van Landbowprodukten* (Dutch—Institute for Storing and Processing Agricultural Products)
ibw information bandwidth
IBW International Boiler Works
IBWCUSMEX International Boundary and Water Commission, United States and Mexico
IBWM International Bureau of Weights and Measures
IBWS International Bureau of Whaling Statistics
ibx (IBX) intermediate branch exchange
IBY International Book Year (1972)
Ibz Ibiza

ic in calf; in charge of; in command; ice crystals; index correction; informal communication; inspected and condemned; inspiratory capacity; inspiratory center; instruction counter; instrument correction; integrated circuit; interference control; interior communication; intermediate language; internal classification; internal combustion; internal communication; internal connection; international control; interstitial cells; interstitial cystitis; intracerebral; intracutaneous
ic. *icon* (Latin—figure, woodcut)
i-c integrated circuit
i/c in charge; in command; ice cream; intercom (intercommunication via interphone)
i&c inspected and condemned
i & c installation and checkout; installation and construction; instrumentation and control
***i.c.** inter cibos* (Latin—between meals)
Ic Iceland; Icelander; Icelandic
IC Idaho College; Identity Card; Ignatius College; Illinois Central (railroad); Illinois College; Immaculata College; Industrial Court; Information Center; Information Circular; integrated circuit; Integrating Contractor; Intelligence Corps; Interchemical Corporation; Intercity; International Control; Iola College; Iona College; Iron Curtain; Itaska College; Itawamba College; Item Code; Ithaca College
I-C Indo-China; Indo-Chine; Indo-Chinese; Iran-Contra scandal
I.C. *Iesus Christus* (Latin—Jesus Christ); Institute of Charity (Rosminian)
I & C Ictinus and Callicrates (designers of the Parthenon)
IC 4-A Intercollegiate Amateur Athletic Association of America
ica ignition control additive; Imperial Corporation of America; Institute of Contemporary Arts; instrument compressed air
ICA Illinois Correctional Association; Imperial Corporation of America; Imperial Savings Association; Indiana Correctional Association; In-

dustrial Communication Association; Industrial Coordination Act; Institute of Chartered Accountants; Institute of Contemporary Arts; Institute of Criminal Anthropology; Insurance Council of Australia; Intermuseum Conservation Association; International Cartographic Association; International Chefs' Association; International Chiropractors Association; International Christian Aid; International Claims Association; International Coffee Agreement; International Commercial Arbitration; International Commodity Agreement; International Communication(s) Agency; International Communication Association; International Cooperative Administration; International Cooperative Alliance; International Cooperative Association; International Council on Archives; Iowa Corrections Association
I of CA Institute of Chartered Accountants
***ICA** Ingenieros Civiles Asociados* (Spanish—Associated Civil Engineers)
icaa integrated cost-accounting application(s)
ICAA Institute of Chartered Accountants of Australia; International Council on Alcohol and Addictions; Invalid Children's Aid Association; Investment Counsel Association of America
ICAAAA Intercollegiate Association of Amateur Athletes of America
ICAB International Council Against Bullfighting
ICAC Independent Commission Against Corruption (Hong Kong)
icad integrated control and display
icade interactive computer-aided design evaluation
icadg interactive computer-aided design and graphics
ICADS Integrated Control and Display System
ICAE International Commission on Agricultural Engineering; International Council for Adult Education
ICAESD International Center for African Economic and Social Documentation

ICAEW Institute of Chartered Accountants in England and Wales
ICAF Industrial College of the Armed Forces; International Committee on Aeronautical Fatigue
ICAFI International Commission on Agriculture and Food Industries
ICAI Institute of Chartered Accountants in Ireland; International Commission of Agricultural Industries
ICAITI *Instituto Centroamericano de Investigacion y Tecnologia Industrial* (Spanish—Central American Institute of Industrial Investigation and Technology)
icam integrated computer-aided manufacturing
ICAM Institute of Corn and Agricultural Merchants; Internation Civil Aircraft Marking(s)
I Can Information Canada
ICAN Insurance Consumer Action Network; International Commission for Air Navigation
ICAO International Civil Airlines Operation; International Civil Aviation Organization
ICAP Institute of Certified Ambulance Personnel; Integrated Criminal Apprehension Program (computerized system); Inter-American Committee of the Alliance for Progress; International Civil Aviation Policy
ICAP *Instituto Cubano de Amistad con los Pueblos* (Spanish—Cuban Institute for Friendship with Peoples)
ICAPR Interdepartmental Committee on Air Pollution Research (UK)
ICAPS Integrated Carrier Acoustic Prediction System (for aircraft carriers)
ICAR Indian Council of Agricultural Research; International Committee Against Racism
ICARMO International Council of the Architects of Historical Monuments
I-car(s) Italian car(s)
Icarus *International Journal of the Solar System*
icas intermittent commercial and amateur service
ICAS Institute for Chartered Accountants in Scotland; Interdepartmental Committee for Atmospheric Sciences; Intermittent Commercial and Amateur Service; International Council of the Aeronautical Sciences; International Council of Aerospace Sciences
ICASALS International Center for Arid and Semi-Arid Land Studies
ICASB Iowa Council of Area School Boards
ICATS Intermediate-Capacity Automated Telecommunications System (USAF)
icav intracavity
icb interlocking concrete block; international competitive bid(ding)
ICB Indian Coffee Board; Industrial and Commercial Bank(ing); Institute of Collective Bargaining; Institute of Comparative Biology; Interface Control Board; International City Bank; International Container Bureau
ICBA International Community of Booksellers Associations
ICBC Insurance Corporation of British Columbia; International Commercial Bank of China
ICBIF Inner-City Business Improvement Forum
icbm (ICBM) intercontinental ballistic missile
ICBO International Conference of Building Officials; Interracial Council for Business Opportunities
icbp intracellular binding proteins
ICBP International Council for Bird Preservation
ICBR Institute for Child Behavior Research
ICBS Interconnected Business System; International Call-Boy Service (for women)
icbt intercontinental ballistic transport
icc improved contemporary comparison; institute cargo clauses; integrated circuit computer; international catalog card (3 x 5 inches or 7.5 x 12.5 centimeters); intracompany correspondence
ic & c invoice cost and charges
ICC Indian Claims Commission; Instrumentation Control Center; International Chamber of Commerce; International Control Commission; International Correspondence Course(s); Interstate Commerce Commission
I.C.C. Isthmian Canal Commission
icca initial cash clothing allowance
ICCA Infants' and Children's Coat Association; International Congress and Convention Association; International Consumer Credit Association; International Corrugated Case Association
ICCAD International Center for Computer-Aided Design
ICCAT International Commission for the Conservation of Atlantic Tunas
ICCB Illinois Community College Board
ICCC International Center for Comparative Criminology
iccd intensified charge-coupled device; internal coordination control drawing
ICCD Information Center on Crime and Delinquency
ICCF International Correspondence Chess Federation
ICCI Inter-Continental Computing Incorporated
ICCO International Carpet Classification Organization
ICCP International Conference on Cataloguing Principles
ICCR Indian Council for Cultural Relations; Interfaith Center on Corporate Responsibility; International Charge Card Registry
ICCS International Center of Criminological Studies
ICCSL International Commission of the Cape Spartel Light
ICCTA Illinois Community College Trustees Association
IC & CY Inns of Court and City Yeomanry
icd immune complex disease; interface control document; interface control drawing; investment certificate of deposit (ICD)
ic & d installation, checkout, and demonstration
ICD Industrial Cooperation Division; Industry Cooperation Division; Inland Clearance Depot; Inland Container Depot; Institute for the Crippled and Disabled; International College of Dentists; International Cooperative Distribu-

tors

ICD International Classification of Diseases

ICDA International Classification of Diseases, Adapted for Use in the United States

icdh (ICDH) isocitric dehydrogenase

ICDO International Civil Defense Organization

ICDRG International Contact Dermatitis Research Group

icd's investment certificates of deposit

ICDS Information Collection Dissemination System

ice addictive synthetic stimulant, crystal methamphetamine

ice. in-car entertainment; increased combat effectiveness; input-checking equipment; instruction-curriculum environment; internal combustion engine; inventory control effectiveness

ice. (ICE) crystallized smokable derivative of methamphine

Ice Iceland; Icelander; Icelandic

ICE Institution of Chemical Engineers; Institution of Civil Engineers; *Instituto Costarricense de Electricidad* (Spanish—Costa Rican Electric Institute); Instruction-Course Evaluation; Instructor Course Evaluation; International Cometary Explorer; International Cultural Exchange

ICEC Illinois Citizens Education Council

ICECAP Infrared Chemistry Experiments—Coordinated Auroral Program (DoD)

ICED International Council for Educational Development

ICEED International Center for Energy and Economic Development

ICEF International Children's Emergency Fund; International Council for Educational Films

ICEG Insulated Conductors' Export Group

ICEI International Combustion Engine Institute

Icel Icelandic

ICEL International Committee on English in the Liturgy; International Council of Environmental Law

Iceland Republic of Iceland (island nation in the North

Atlantic) *Lydveldio Island*

ICEM International Commission for European Migration

icepack. individual career exploration pack(age)

ICEPS International Center for Economic Policy Studies

ICER Information Centre of the European Railways

I Ceram Institute of Ceramics

ICEs International Customs Examinations

ICES Instructor and Course Evaluation System; Integrated Civil Engineering Systems; International Council for the Exploration of the Sea

ICESC Industry Crew Escape Systems Committee

ICET Institute for the Certification of Engineering Technicians; International Center of Economy and Technology; International Council on Education for Teaching

ICETEX *Instituto Colombiano de Especialización Tecnica en el Exterior* (Spanish—Colombian Institute of Technical Specialization Abroad)

ICETT Industrial Council for Educational and Training Technology

ICEWATER Inter-Agency Committee on Water Resources

icf inertial confinement fusion; intermediate care facilities; intracellular fluid

ICF Ice Cream Federation; Independent Cat Federation; *Ingénieur Civil de France* (French—Civil Engineer of France); Inter-bureau Citation of Funds; International Canoe Federation; International Cynological Federation; Iowa College Foundation

ICFA Independent College Funds of America; International Chicken Flying Association; International Cystic Fibrosis Association

ICFC Industrial and Commercial Finance Corporation

icff intercommunication flip-flop

ICFNC Independent College Fund of North Carolina

ICFNJ Independent College Fund of New Jersey

ICFP Institute of Certified Financial Planners

ICFPW International Confederation of Former Prisoners of War

ICFR Intercollegiate Conference of Faculty Representatives (Big Ten)

ICFTU International Confederation of Free Trade Unions

ICFU International Council on the Future of the University

icg icing

ICG Industries Consultative Group; Interface Coordination Group; International Commission on Glass; International Congress of Genetics; Interviewers Classification Guide; Iowa Corn Growers

ICGB International Cargo Gear Bureau

ICGE International Center of Genetic Epistemology

ICGS Icelandic Coast Guard Service; International Call-Girl Service (Amsterdam)

ich ichthyology; in-calf heifer

ich (ICH) infectious canine hepatitis

Ich Ichabod

ICHAM Institute of Cooking and Heating Appliance Manufacturers

ICHCA International Cargo Handling Coordination Association

IChemE Institute of Chemical Engineers

ICHEO Inter-University Council for Higher Education Overseas

ichnol ichnolite; ichnologist; ichnology

ICHPER International Council for Health, Physical Education, and Recreation

ichs ichthyologists

I/C Hs Iran-Contra Hearings

ICHS International Committee of Historical Sciences

ichth ichthyology

ichthyol ichthyologic(al)(ly); ichthyologist; ichthyology

ICI Imperial Chemical Industries; Institution of Chemistry in Ireland; International Commission on Illumination; Interpersonal Communication Inventory; Investment Casting Institute; Investment Company Institute; Investment Costing Institute

ICIA Interagency Committee on International Athletics; International Credit Insurance Association; International Crop Improvement Association

ICIANZ Imperial Chemical

Industries of Australia and New Zealand

ICIAP Interagency Committee on International Aviation Policy

ICIC International Copyrights Information Center

ICID International Commission on Irrigation and Drainage

ICIE International Council of Industrial Editors

ICIECA Interagency Council on International Educational and Cultural Affairs

ICIMP Interagency Committee for International Meteorological Programs

ICIP International Conference on Information Processing

ICIPE International Center for Insect Physiology and Ecology

ICIREPAT International Cooperation in Information Retrieval among Examining Patent Offices

ICIS International Cargo Information System

ICITA International Cooperative Investigation of the Tropical Atlantic

ICITAP International Criminal Investigative Training Assistance Program

ICITO Interim Commission for the International Trade Organization

ICJ Institute of Creative Judaism; Institute of Criminal Justice; International Commission of Jurists; International Court of Justice

ICJC Institute of Criminal Justice and Criminology

ICJP Irish Commission for Justice and Peace

ICJW International Council of Jewish Women

icky sticky

icl incoming correspondence log(ging); input capacitorless (circuitry or hi-fidelity)

ICL *Institut de Chimie de Lyon* (French—Lyon Industry of Chemistry); Institute for Continued Learning; International Computers Limited; International Confederation of Labor; International Containers Limited; Israel Chemicals Limited

ICLA International Committee on Laboratory Animals

ICLD International Center for Law in Development

Iclnd Iceland

ICLP Institute of Criminal Law and Procedure (Georgetown University)

I & CLQ International and Comparative Law Quarterly

ICLs Income Contingent Loan

ICLS Irish Central Library for Students

icm increased capability missile; intercostal margin; interference control monitor

icm (ICM) increased capability missile

ICM Increased Capability Missile; Indian Campaign Medal; Industrial and Construction Machines; Institute of Computer Management; International Confederation of Midwives

ICMA Independent Cable Makers Association; International Circulation Managers Association; International City Manager's Association

ICMF Indian Cotton Mills Federation

ICMPH International Center of Medical and Psychological Hypnosis

icmps induction compass

ICMR Indian Council of Medical Research

ICMREF Interagency Committee on Marine Science, Research, Engineering, and Facilities

ICMs Instructional Curriculum Maps

ICMS International Commission on Mushroom Science

ICMUA International Commission on the Meteorology of the Upper Atmosphere

ICN Instrumentation and Calibration Network; International Chemical and Nuclear (corporation); International Council of Nurses

I.C.N. In Christi Nomine (Latin—in Christ's name)

ICNA Infection Control Nurses Association

ICNAF International Commission for the Northwest Atlantic Fisheries

ICND International Commission on Narcotic Drugs

ICNT Informal Composite Negotiated Text

ICNV International Committee on Nomenclature of Viruses

ico in case of; iconology

ICO Immediate Commanding Officer; Immediate Control-

ling Office(r); Instrumentation Control Office(r); Interagency Committee on Oceanography; International Coffee Organization; International Commission for Optics; International Congress of Ophthalmology; Islamic Conference Association; Israel Chamber Orchestra

ICOA International Castor Oil Association

ICOD International Centre for Ocean Development

ICOGRADA International Council of Graphic Design Associations

ICOM International Council of Museums

ICOMIA International Council of Marine Industries Associations

ICOMOS International Council on Monuments and Sites

icon. iconic; iconoclasm; iconoclast; iconography

iconoc iconoclast (breaker of political or religious images)

ICONS Information Center on Nuclear Standards; Isotopes of Carbon, Oxygen, Nitrogen, and Sulfur (AEC)

ICOO Iraqi Company for Oil Operations

icop imported crude oil processing

ICOPA International Conference of Police Associations

icor incremental capital-output ratio

ICOR Intergovernmental Conference on Oceanic Research (UNESCO)

I Corr Tech Institution of Corrosion Technology

ICOS International Committee of Onomastic Sciences

ICOT Institute of Coastal Oceanography and Tides; Institute for New Generation Computer Technology

icp inventory control point

ICP Industry Cooperative Program; Infection-Control Procedure; *Institut de Chimie de Paris* (Chemical Institute of Paris); International Center of Photography; International Commerce Promoters; International Council of Psychologists

ICPA International Commission for the Prevention of Alcoholism; International Conference of Police Associations

ICPC International Criminal Police Commission (Interpol)

ICPCCP Illinois Council of Public Community College Presidents

ICPHS International Council for Philosophical and Humanistic Studies

ICPI Insurance Crime Prevention Institute

i/c/pm/m incisors, canines, premolars, molars (dentition formula, e.g., i 4/4 means 4 upper and 4 lower incisors, c 2/2 means 2 upper and 2 lower canines, etc.)

icpo intercompany payment orders

ICPO International Criminal Police Organization (Interpol)

ICPP Idaho Chemical Processing Plant (AEC)

ICPRB Interstate Commission on the Potomac River Basin

ICPS International Congress of Photographic Science; International Credit Protection Services

ICPSR Inter-university Consortium for Political and Social Research

ICQ Invested Capital Questionnaire

icr increase; increment; instrumentation control rack; ion cyclotron resonance

icr (ICR) iron-core reactor

ICR Independent Congo Republic; Institute of Cancer Research; Institute for Cooperative Research; International Council for Reprography

ICRA International Copper Research Association

ICRC International Committee of the Red Cross

ICRDB International Cancer Research Data Bank

ICRF Imperial Cancer Research Fund

ICRH Institute for Computer Research in the Humanities (NYU)

ICRI Illinois Committee for Responsible Investment; International Crop Research Institute

icrm intercontinental reconnaissance missile (ICRM)

ICRM Institute of Certified Records Managers

ICRO International Cell Research Organization

ICRP International Commission on Radiation Protection; International Commission on Radiological Protection

ICRSC International Council for Research in the Sociology of Cooperation

ICRT Individualized Criterion-Referenced Test(ing)

ICRU International Commission on Radiological Units and Measurements

ICRUM International Commission on Radiological Units and Measurements

ics installment credit selling; intercostal space; interim contractor support

ic's immediate constituents; integrated circuits

ic's (ICs) integrated circuits

ICS Imperial College of Science and Technology; Indian Chemical Society; Indian Civil Service; Information Centers Service; Inner Continental Shelf; Institute of Chartered Shipbrokers; Institute of Chartered Shipbuilders; Institute for Contemporary Studies; Institution of Computer Sciences; Integrated Command System; Integrated Container Service; Intelligence Community Staff; Interagency Communications System; Inter-Communications System; International Cardiovascular Society; International Chamber of Shipping; International Chamber of Shipping; International Clarinet Society; International College of Surgeons; International Container Service; International Contract Specialists; International Correspondence Schools; International Telephone and Telegraph Communications System; Interway Container Service

ICSA Institute of Chartered Secretaries and Administrators; International Council of Shopping Centers

ICSAC International Confederation of Societies of Authors and Composers

ICSB International Center of School Building

ICSC Independent Colleges of Southern California; Interoceanic Canal Study Commission

ICSDW International Council of Social Democratic Women

icse intermediate current stability experiment

ICSE International Committee for Sexual Equality

ICSEAF International Commission for the Southeast Atlantic Fisheries

ICSEM International Center of Studies on Early Music

ICSEMS International Commission for the Scientific Exploration of the Mediterranean Sea

icsh (ICSH) interstitial cell-stimulating hormone

ICSH International Committee for Standarization in Haematology

ICSCHM International Commission for a History of the Scientific and Cultural Development of Mankind

ICSI International Conference on Scientific Information

ICSID International Center for the Settlement of Investment Disputes; International Council of Societies of Industrial Design

ICSLS International Convention for Safety of Life at Sea

icsm instant corn-soya milk

ICSOM International Conference of Symphony and Opera Musicians

ICSP International Council of Societies of Pathology

ICSPE International Council of Sport and Physical Recreation

ICSPRO International Calcium Silicate Products Research Organization

icss intracranial self stimulation

ICSS International Council for the Social Studies

ICSSD International Committee for Social Sciences Documentation

ICSST Institute of Child Study Security Test

ICST Imperial College of Science and Technology; Institute for Computer Sciences and Technology

ICS & T Imperial College of Science and Technology

ICSTA International Cooperative Study of the Tropical Atlantic

ICSTS Intermediate Combined System Test Stand

ICSU Integrated Container Services Unit; International Council of Scientific Unions;

Interway Container Service (container) Unit
ICSW Interdepartmental Committee on the Status of Women
icswbd interior communications switchboard
ict icterus; identity conversion training; inflammation of connective tissue; insulin coma therapy (ICT)
ic/t integrated computer/telemetry
ICT Inter-City Train; International Computers and Tabulators; Wichita, Kansas (airport)
ICT *International Critical Tables*
ICTA Imperial College of Tropical Agriculture; International Center for the Typographic Arts; International Council of Travel Agents
ICTB International Customs Tariffs Bureau
icte integration and calibration test(ing) equipment
ICTF International Cocoa Trade Federation
ICTMM International Congresses of Tropical Medicine and Malaria
ICTN Industry Center for Trade Negotiations
ICTP International Center for Theoretical Physics
ICTR International Center of Theatre Research
ICTS Intermediate-Capacity Transit System
Ictus. *Iurisconsultus* (Latin—attorney, counsellor-at-law)
ic tv integrated-circuit television
icu indicator console unit; intensive care unit (medical)
icu (ICU) intensive care unit
Icu I see you
ICU International Code Use; International Cultural Understanding; International Cultural University
ICUA Institute for College and University Administrators
ICUF Independent Colleges and Universities of Florida
ICUI Independent Colleges and Universities of Indiana
ICUM Independent Colleges and Universities of Missouri
ICUMSA International Commission for Uniform Methods of Sugar Analysis
ICUS inside continental United States

ICUT Independent Colleges and Universities of Texas
icv improved capital value; intracellular virus
ICVA International Council of Voluntary Agencies
icw in connection with; interrupted continuous wave; intracellular water
ICW India-China Wing (World War II); Institute of Child Welfare; Inter-American Commission of Women; International Chemical Workers; International Commission on Whaling; International Council of Women
ICWA Indian Council of World Affairs; Institute of Cost and Works Accountants; Institute of Current World Affairs; International Coil Winding Association
ICWG International Cooperative Women's Guild
ICWL International Creative Writers League
ICWM International Committee on Weights and Measures
ICWP International Council of Women Psychologists
ICWU International Chemical Workers Union
ICX International Cultural Exchange
ICY International Cooperation Year (1965)
ICZ Intertropical Convergence Zone
ICZN International Code of Zoological Nomenclature; International Commission on Zoological Nomenclature
id idea; identification; immediate delivery; import duty; independent development; independent distributor; induced draft; industrial design; infectious disease; infective dose; inside diameter; instructional development; intercept director; interest deductible; intradermal; island; islander; item description
id (ID) import duty; instruction decoder
i & d incision and drainage
id. *idem* (Latin—the same)
Id Iraqi dinar (monetary unit of Iraq)
I'd I could; I had; I should; I would
ID Idaho; Import Declaration; Institute of Distribution; Insurance Department; Intelligence Department; Interior

(US department); Interior Department; Iraqi dinar (currency unit)
id$_{50}$ median infective dose
ida integrated digital avionics
ida (IDA) iminodiacetic acid
Ida Idah; Idaho; Idalah; Idalia; Idalina; Idaline
IDA Industrial Development Agency; Industrial Development Authority; Industrial Diamond Association; Institute for Defense Analyses; Institute for Design Analysis; Institute of Directors in Australia; Intercollegiate Dramatic Association; International Development Assistance; International Development Association; International Discotheque Association; International Dredging Association; Irish Dental Association
IDA *Import Duty Act*
IDAA Industrial Diamond Association of America; International Doctors in Alcoholics Anonymous
idac interim digital-analog converter
id. ac *idem ac* (Latin—the same as)
IDAC Import Duties Advisory Committee
IDAI Industrial Development Authority of Ireland
IDA Ireland Industrial Development Authority of Ireland
IDAS Information Display Automatic System
IDASF Institute for a Democratic Alternative for South Africa
idast interpolated data and speech transmission
idb illicit diamond buyer; illicit diamond buying; integrated data base; intercept during burning
IDB Industrial Development Board; Inter-American Development Bank; Internal Drainage Board; Islamic Development Bank(ing); Israel Diamond Building (Ramat Gan)
IDBP Industrial Development Bank of Pakistan
IDBT Industrial Development Bank of Turkey
idc interest during construction
IDC Imperial Defense College; Industrial Development Commission; Industrial Development Corporation; Industry

Development Commission; Intelligence Documentation Center; Intercontinental Dynamics Corporation; Interdepartmental Committee; Inter-departmental Communication; Inter-Departmental Correspondence; International Danube Commission; Iowa Development Commission

IDC Internationale Dokumentationgesellschaft für Chemie (German—International Chemical Documentation Society)

IDCA Industrial Design Council of Australia; International Development Corporation Agency

id card identification card

ID-card identification card

IDCAS Industrial Development Center for Arab States

idcf indirect command file

IDCF Industrial Development Completion Form

IDC(orp) International Disposal Corporation

IDCSP Initial Defense Communications Satellite Program

IDCSS Initial Defense Communications Satellite System

i-d curve intensity-duration curve

idd identified; industrial diamond drill; interface designation drawing

idd (IDD) insulin-dependent diabetes; international direct dialing

i/d/d illicit diamond dealer

IDD International Direct Dialing; Island Development Department

IDD Industrielle Designere Danmark (Danish—Denmark Industrial Design)

IDDD International Direct Distance Dialing

IDDD International Demographic Data Directory

IDDS International Dairy Development Scheme; International Digital Data Service

ide industry-developed equipment

IDE International Development Enterprises; Industrial Development Executive; Institute of Diesel Engineers; Israel Desalination Engineering

IDEA Illinois Drug Education Alliance; Institute for the Development of Educational

Activities; Institute of Diesel Engineers of Australia; Instructional Development and Effectiveness Assessment; International Dance-Exercise Association; International Downtown Executives Association; International Drug Enforcement Association

IDEAS Integrated Design and Analysis System

I de C Islas del Cisne (Spanish—Swan Islands)

IDEC Industrial and Domestic Equipment Corporation; Interior Design Educators Council; International Drug Enforcement Conference

IDECC Interstate Distributive Education Curriculum Consortium

ideea information and data exchange experimental activities

idef intercept during exo-atmospheric fall

I de F Institut de France (French—Institute of France)

iden identification; identify; identity

ident identification; identify; identity

IDEP Interagency Data Exchange Program; Interservice Data Exchange Program

IDET Institute for Development, Employment, and Training

idex initial defense experiment

idf intermediate distribution frame; international distress frequency

idf (IDF) integrated data file; interceptor day fighter

IDF International Dairy Federation; International Democratic Fellowship; International Dental Federation; International Diabetes Federation; Israeli Defense Force (Israeli Secret Service or Zahal)

IDFC Indo-Pacific Fisheries Council

IDFF Internationale Demokratische Frauenfederation (German—Women's International Democratic Federation)

idfm induced directional fm

idg integrated-drive generator

idgprt indigo print

ID grinding internal grinding

idh (IDH) isocitrate dehydrogenase

id he. index head

IDHEC Institut des Hautes Etudes Cinématographiques, Paris (French—Paris Institute of Higher Cinematographic Studies)

IDHS Intelligence Data Handling System

idi improved data interchange; inter-division invoice

IDI Industrial Designers' Institute

IDI Institut de Droit International (French—International Law Institute); *Instituto Venezolano de Derecho Imobiliario* (Spanish—Venezuelan Institute of Withholding Law)

IDIA Industrial Design Institute of Australia

IDIB Industrial Diamond Information Bureau

IDIL Institute for the Development of Indiana Law

IDIMS Interactive Digital Image Manipulation System

idio (Latin prefix—distinct, self, separate)—idiograph, idiomatic, idiot

idiot. instrumentation digital online transcriber

IDIU Interdivisional Information Unit; Interdivisional Intelligence Unit

idl (IDL) intermediate-density lipoproteins

IDL Instrumentation Development Laboratories; International Date Line; Ira D Levine; New York, New York (Kennedy International Airport—Idlewild)

IDLE Idaho Department of Law Enforcement

IDLIS International Desert Locust Information Service

id lt identification light

idm illicit diamond mining; integrated direct metering

IDM Institute of Defense Management (Indian)

IDMA Indian Drug Manufacturers Association; International Dancing Masters Association; Isaac Delgado Museum of Art

IDMS Integrated Database Management System

I.D.N. In Dei Nomine (Latin—in God's name)

idne inertial-doppler navigation equipment

IDNL Indiana Dunes National Lakeshore (Indiana)

ido industrial diesel oil

IDO Intelligence Division Of-

fice; Interim Development Order; Interim Development Ordnance; International Disarmament Organization
idoc inner diameter of outer conductor
IDOC Illinois Department of Corrections
IDOE International Decade of Ocean Exploration—1970–1980
idon. vehic. idoneo vehiculo (Latin—in a suitable vehicle)
IDOT Illinois Department of Transportation
idp information data processing; input data processing; input data processor; integrated data processing
idp (IDP) inosine diphosphate
IDP Immediate Decision Plan; Independent Development Project; Industrial Development Bank; Institute of Data Processing; Integrated Data Processing; International Driving Permit
IDPE Incorporated Data Processing Executives
IDPS Incremental Differential Pressure System
idr intercept during reentry
idr idraulica (Italian—hydraulics)
IDR Incremental Design Review; Infantry Drill Regulations; Institute for Desert Research; Institute for Dream Research
IDRC International Development Research Centre; International Drycleaning Research Committee
IDRDS International Directory of Research and Development Scientists
ids illicit diamond smuggling; inadvertent destruct; input data strobe; integrated data store; interdiction and strike; intermediate drum storage
IDs Instructural Designs
IDS Inertial Doppler System; Institute of Dental Surgery; Instructional Dimensions Study; Interior Design Society; Interior Designers Society; International Development Services; International Development Strategy; International Documents Service; Investigative Dermatological Society; Investors Diversified Services
IDSA Indian Dairy Science Association; Industrial De-

signers Society of America
IDSC Information Dissemination Service Center
IDSCS Initial Defense Satellite Communication System
IDSO International Diamond Security Organization
idt identification disposition tag(ging); inter-division transfer
idt in de text (Dutch—in the text)
IDT Industrial Design Technology; Industrial Detergents Trade; Instrument Definition Team
IDTA International Dance Teachers Association
IDTS Instrumentation Data Transmission System
idu interactive data-base utilities; intermittent drive unit; iododeoxyuridine
IDU idoxuridine; International Democratic Union; International Dendrology Union
idur intercept during unpowered rise
i Durchshn im Durchschnitt (German—on an average)
IDV International Distillers and Vinters
IDX Index to Dental Literature
IDZ Internationales Design Zentrum (German—International Design Center)
ie index error; initial equipment; inside edge; ion exchange
i-e internal-external
i/e ingress/egress
i & e identification and exposition (lines)
i.e. id est (Latin—that is); *inside english* (journal of the English Council of California Two-Year Colleges)
IE Indo-European; Industrial Engineering; Industrial Espionage; Information and Education; Institute of Education; Institute of Electronics; Institute of Engineers; Institution of Electronics; Institution of Engineers
I-E Indo-European; Internal-External Scale
I.E. Industrial Engineer
I & E Information and Education; Inspection and Enforcement
I of E Institute of Export
IE Immunitäts Einheit (German—immunizing unit)
iea intravascular erythrocyte aggregation

IEA Idaho Education Association; Illinois Education Association; Index of Economic Activity; Industrial Environmental Association; Institute of Economic Affairs; Institute of Engineers—Australia; International Association for the Evaluation of Educational Achievement; International Economic Association; International Energy Agency; International Entrepreneurs Association; International Epidemiological Association; International Study of Educational Achievement
IEA Indian Education Act (1972)
IEAF Imperial Ethiopian Air Force
IEAS Institute of East Asian Studies
IEB Institute of Economic Botany; International Energy Bank; International Environmental Bureau (of the Non-Ferrous Metals Industry); International Executive Board
iec injection electrode catheter; integrated electronic control; intra-epithelial carcinoma
iec (IEC) inherent explosion clause
IEC Industry-Education Council; *Institut d'Etudes Centrafricaines* (French—Institute of Central African Studies); International Education Center; International Electrochemical Commission; International Electrotechnical Commission
IECE Institute of Electronic and Communication Engineers
IECEJ Institute of Electronic and Communication Engineers of Japan
IECI Institute for Esperanto in Commerce and Industry
iecm internal electronic counter-measures
IECO International Engineering Company
IECOK International Economic Consultative Organization for Korea
IECP Intermediate Engineering Change Proposal
iec's integrated electronic components
ied improvised explosive device; individual effective dose
IED Institute for Educational

Development; Institution of Engineering Designers; Integrated Electronics Division (USA Electronics Command)
IEDC International Energy Development Corporation
IEDR Integrated Electric-Drive Propulsion
ied's income equalization deposits
IEDs Intermediate Educational Districts
iee inner enamel epithelium
IEE Institute of Electrical Engineers; Institute of Electronic Engineering; Institute for Environmental Education; Institute of Environmental Engineers; Institution of Electrical Engineers
Ieee I expect everything eventually
IEEE Institute of Electrical and Electronics Engineers
IEEE J Quantum Electron IEEE Journal of Quantum Electronics
IEEE Trans Antennas Propag IEEE Transactions on Antennas and Propagation
IEEE Trans Electron Devices IEEE Transactions on Electronic Devices
IEEE Trans Inf Theory IEEE Transactions on Information Theory
IEEE Trans Instrum Meas IEEE Transactions on Instrumentation and Measurement
IEEE Trans Magn IEEE Transactions on Magnetics
IEEE Trans Microwave Theory Tech IEEE Transactions on Microwave Theory and Techniques
IEEE Trans Nucl Sci IEEE Transactions on Nuclear Science
IEEE Trans Sonics Ultrason IEEE Transactions on Sonics and Ultrasonics
ieef ion-integrated evaporation filter
IEEIE Institution of Electrical & Electronics Incorporated Engineers
IEEJ Institute of Electrical Engineers of Japan
IEES International Educational Exchange System
IEETE Institution of Electrical and Electronics Technician Engineers
ief integrated electronic flash(ing)
IEF Indian Expeditionary

Force; International Ecumenical Fellowship; International Exhibitions Foundation; International Eye Foundation
IEG Information Exchange Group
IEHA International Economic History Association
iei indeterminate engineering items
IEI Industrial Education Institute; Industrial Engineering Institute; Institute of Electrical Inspectors; Institution of Engineering Inspection; Institution of Engineers of Ireland; Iran Electronics Industries
IEIC Iowa Educational Information Center
IEKV International Eisenbahn-Kongress-Vereinigung (German—International Railway Congress Association)
IEL Industrial Engineering Limited; Industrial Equity Limited; Institute for Educational Leadership
iem iemand (Dutch—a man, somebody, someone)
IEMA Iowa Educational Media Association
IEMCAP Intrasystem Electromagnetic Compatability Analysis Program (USAF)
IEME Inspectorate of Electrical and Mechanical Engineering
IEMS Institute of Experimental Medicine and Surgery
IEN Imperial Ethiopian Navy
IEO Instituto Español de Oceanografía (Spanish—Spanish Oceanographic Institute)
IEOM Institute of Environment and Offshore Medicine
ieop immunoelectro-osmophoresis
iep iso-electric point
IEP Individual Education Plan; Individual Educational Program; *Institut d'Etudes Politiques* (French—Institute of Political Studies); Institute of Earth Physics; Institute of Experimental Psychology
IEPA International Economic Policy Association
IEPG Independent European Program Group
IEPs Individualized Education Programs
ieq index of environmental quality
ier installation enhancement

release
Ier (Dutch—Irishman)
IER Industrial Equipment Reserve; Institute for Econometric Research; Institute of Educational Research; Institute of Engineering Research; Institute of Environmental Research; Interim Engineering Report
IERC Indian Education Resources Center (Bureau of Indian Affairs); International Electronic Research Corporation
IERE Institution of Electronic and Radio Engineers
Ierl Ierland (Dutch—Ireland)
IERT Institute for Education by Radio-Television
IER Test Institute of Educational Research Test (intelligence)
ie & s institutional, environmental, and safety
IES Illuminating Engineering Society; Independent Educational Services; Indian Educational Service; Industrial Electronic Systems; Information Exchange Service; Institute of Environmental Sciences; Institute of European Studies; Institution of Engineers and Shipbuilders; Intensive Employment Services
IE-S Institution of Engineers—Singapore
IESA Illuminating Engineering Societies of Australia
IESC International Executive Service Corps
IESS Institution of Engineers and Shipbuilders in Scotland
iet interest equalization tax
IET Initial Engine Test; Institute of Educational Technology
IETG International Energy Technology Group
I-et-L Indre-et-Loire
IETRA Interhuman Embryonic Transfer and Restablilization Agency
I-et-V Ille-et-Vilaine
IEWS Integrated Electronic Warfare System
if. ice fog; immunofluorescence; information feedback; infrastructure; instrument flight; intermediate frequency; internal flush(ing); interstitial fluid; intrinsic factor
if. (IF) interferon; internal focusing
i-f in-flight; intermediate fre-

quency

i & f inverter and cathode follower

if iflge (Danish—according to)

i.f. ipse fecit (Latin—he did it himself)

If Ifni; Sidi Ifni (Spanish West Africa)

IF grid current (symbol)

I-F Isotta-Fraschini

ifa instrumental fuel assembly; integrated file adapter

I f A Institutt for Atomenergi (Norwegian—Atomic Energy Institute)

IFA Industrial Forestry Association; Industry Film Association; Institute of Foresters of Australia; Intercollegiate Fencing Association; International Federation of Actors; International Federation on Aging; International Fertility Association; International Fiscal Association; International Football Association; International Footprints Association; International Franchise Association

IFA Institut Fiziki Atmosfery (Russian—Atmospheric Physics Institute)

IFABC International Federation of Audit Bureaus of Circulations

IFAC International Family Association of Canada; International Federation of Automatic Control

IFAD International Fund for Agricultural Development

IFALPA International Federation of Air-Line Pilots' Associations

IFAN *Institut Français d'Afrique Noire* (French—French Institute of North Africa, Dakar, Ivory Coast)

IFAP Industrial Foundation for Accident Prevention; International Federation of Agricultural Producers

IFAPA International Federation of Airline Pilots Association

IFAR International Foundation for Art Research

IFAS Institute for American Strategy; International Federation of Aquarium Societies

IFATCA International Federation of Air Traffic Controllers Associations

IFATCC International Federation of Associations of Textile Chemists and Colourists

IFATE International Federation of Airworthiness Technology and Engineering

IFAW International Fund for Animal Welfare

IFAWPCA International Federation of Asian and Western Pacific Contractors Associations

ifb invitation for bid(s)

IFB International Federation of the Blind; Invitation for Bid(s)

IFBA International Fire Buff Association

IFBB International Federation of Bodybuilders

IFBPW International Federation of Business and Professional Women

IFBS Interrupted Feedback System

IFBWW International Federation of Building and Woodworkers

ifc independent fire control; inflight collision; inner front cover; institute freight clause(s); integrated fire control

IFC Inland Fisheries Commission; Intellectual Freedom Committee; International Finance Corporation; International Fisheries Commission; International Freighting Corporation

IFC-ALA Intellectual Freedom Committee—American Library Association

IFCATI International Federation of Cotton and Allied Textile Industries

IFCC International Federation of Camping and Caravanning; International Federation of Clinical Chemistry

IFCCTE International Federation of Commercial, Clerical, and Technical Employees

IFCJ International Federation of Catholic Journalists

IFCL International Fixed Calendar League

IFCO Interreligious Foundation for Community Organization

ifcr interface control register

IFCS Improved Fire-Control System; International Federation of Computer Sciences

IFCU International Federation of Catholic Universities

i.f.d. in flagrante delicto (Latin—in the heat of the evil deed)—caught in the act

IFD International Federation of Documentation

IFDA Institutional Food Distributors of America

IFDP Institute for Food and Development Programs

IFE Industrial Foundation on Education; Institution of Fire Engineers

IFEBP International Foundation of Employee Benefit Plans

IFEBS Integrated Foreign Exchange and Banking System

IFEE Institute for Free Enterprise Education

IFEMS International Federation of Electron Microscope Societies

IFEP Instituttet for Elektronikmateriels Palideliged (Danish—Electronic Materials Reliability Institute)

IFEW Inter-American Federation of Entertainment Workers

iff invert fluid fill(ing)

iff (IFF) identification friend or foe

if & f intermediate flush and fill

IFF Institute of Freight Forwarders; Institute for the Future; Intermediate Financing Facility; International Flavors and Fragrances (corporation)

IFFA International Federation of Film Archives; International Frozen Food Association

IFFC Integrated Fire and Flight Control

IFFCO Indian Farmers Fertilizer Cooperative

IFFF Internationale Frauenlige für Frieden und Freiheit (German—International Women's League for Peace and Freedom)

IFFJ Independent Federation of Free Journalists

IFFJP International Federation of Fruit Juice Producers

IFFNM Internationale Ferienkurse für Neue Musik (German—International Vacation Courses for New Music)—held in Darmstadt

IFFPA International Federation of Film Producers Associations

IFFS International Federation of Film Societies

IFFTU International Federation of Free Teachers'

Unions

ifg instrument flight guide

IFGA International Federations of Grocers' Associations

IFGO International Federation of Gynecology and Obstetrics

ifh in-flight helium

IFHE International Federation of Home Economics

IFHP International Federation for Housing And Planning

IFHTM International Federation for the Heat Treatment of Materials

IFI Industrial Fasteners Institute; Institutional Functioning Inventory

IFI Instituto de Fomento Industrial (Spanish—Institute for Industrial Promotion)

IFIA International Federation of Ironmongers' and Iron Merchants' Associations; International Fence Industry Association

IFIAS International Federation of Institutes for Advanced Studies

IFIF International Foundation for Internal Freedom (hallucinogenic experimenter's society found by Richard Alpert and Timothy Leary)

IFIP Iguazu Falls International Park (shared by Argentina, Brazil, and Paraguay)—Argentinians spell it Iguazu, Brazilians—Igauçu, Paraguayans—Iguassu; International Federation of Information Processing

IFIPS International Federation of Information Processing Societies

I Fire E Institution of Fire Engineers

IFIS Integrated Flight Instrument System

IFJ International Federation of Journalists

IfK Institut für Kriminologie (German—Institute of Criminology); *Institut for Kriminologi* (Norwegian—Institute of Criminology)

IFKM Internationale Föderation für Kurzschrift und Maschinenschreiben (German—International Federation of Shorthand and Typewriting)

IfL Institut für Landeskunde (German—Geographical Institute)—at Bad Godesberg

IFL Imperial Fascist League;

Institute of Fluorescent Lighting; International Federation of Labor; International Friendship League

IFLA International Federation of Landscape Architects; International Federation of Library Associations

iflet interim focal-length optical tracker

IFLFF Internationale Frauenliga für Frieden und Freiheit (German—International Women's League for Peace and Freedom)

IFLWU International Fur and Leather Workers Union

ifm intermediate frame memory

IFM Industrial Facility Manager; Institute for Forensic Medicine

IFMA International Federation of Margarine Associations; International Foodservice Manufacturers Association

IFMBE International Federation for Medical and Biological Engineering

IFMC International Folk Music Council

IFME International Federation of Medical Electronics; International Federation of Municipal Engineers

IFMEO International Fish Meal Exporters Organization

if/mf intermediate frequency/ medium frequency

IFMI Irish Federation of Marine Industries

IFMP International Federation of Medical Psychotherapy

IFMS Integrated Financial Management System

IFMSA International Federation of Medical Students Associations

ifn information

IFN Institut Français de Navigation (French—French Institute of Navigation)

IFNAFSS International Feminist Network Against Female Sexual Slavery

IFNB Idaho First National Bank

IFNE International Federation for Narcotic Education

if nec if necessary

ifo in favor of; in front of; identified flying object; intermediate fuel oil

IFOFSAG International Fellowship of Former Scouts and Guides

IFOP Institut Français d'Opinion Publique (French— French Institute of Public Opinion)

IFOR International Fellowship of Reconciliation

IFORS International Federation of Operational Research Societies

IFOSA International Federation of Stationers' Associations

ifov instantaneous field of view; instrument field of view

ifp in-flight performance; in-flight printer; international fixed public broadcast band

IFP Imperial and Foreign Post; Institute of Fluid Power; International Federation of Purchasing

IFP Institut Français du Pétrole (French—French Petroleum Institute)

IFPA Industrial Film Producers Association; Institute for Foreign Policy Analysis

IFPAAW International Federation of Plantation, Agricultural, and Allied Workers

IFPCW International Federation of Petroleum and Chemical Workers

IFPI International Federation of the Phonographic Industry

ifpm in-flight performance monitor

IFPM International Federation of Physical Medicine

IFPMA International Federation of Pharmaceutical Manufacturers Associations

IFPMM International Federation of Purchasing and Materials Management

IFPO International Freelance Photographers Organization

IFPP Imperial and Foreign Parcel Post

IFPRA International Federation of Park and Recreation Administrators

IFPS Institute for Foreign Policy Studies

IFPTO Internation Federation of Popular Travel Organizations

IFPTS Intertype Fototronic Photographic Typesetting System

IFPW International Federation of Petroleum Workers

ifr infrared; inflight refueling

ifr (IFR) internal function register

i-f-r image-to-frame ratio
IFr Internationaler Frauenrat
(German—International
Council of Women)
IFR Institute for Food Research; Instrument Flight Rules
IFRA International Foundation for Research in the Field of Advertising
IFRB International Frequency Registration Board
IFRC International Fusion Research Council
IFREMER Institut Français de Recherches pour l'Exploitation des Mers (French Research Institute for the Exploitation of the Seas)
IFRF International Flame Research Foundation
IFRT Institute for Fitness Research and Training; Intellectual Freedom Round Table
ifru interference rejection unit
IFRU Institut Français du Royaume-Uni (French Institute of the United Kingdom)
IFS Indian Forest Service; Instrument Flight System; International Federation of Surveyors; International Foundation for Science; International Freight Services; Irish Free State
IFS International Financial Statistics
IFSA International Federation of Sound Archives
IFSDA International Federation of Stamp Dealers' Associations
IFSDP International Federation of the Socialist and Democratic Press
IFSEA International Federation of Scientific Editors Associations
IFSEM International Federation of Societies for Electron Microscopy
IFSF Independent Fuel Storage Facility; Irradiated-Fuels Storage Facility
IFSIT In-Flight Safety Inhibit Test
IFSMA International Federation of Ship Master Associations
IFSO In-Flight Safety Officer
IFSP International Federation of Societies of Philosophy
IFSPO International Federation of Senior Police Officers
IFSPS International Federation of Students in Political Sciences

ences
ifss infinite solution set
IFSS Instrumentation Flight Safety System; International Fertilizer Supply Scheme; International Flight Service Station(s)
IFSSO Irish Free State Stationery Office
IFST Institute of Food Science and Technology; International Federation of Shorthand and Typing
IFSTA International Fire Service Training Association
IFSW International Federation of Social Workers
ift inflight text
IFT Indiana Federation of Teachers; Institute of Food Technologists; International Federation of Translators; International Foundation for Telemetering; International Frequency Tables
IFT Institut für Tieflagerung (German—Institute for Geologic Disposal)
IFTA International Federation of Travel Agencies
IFTC International Film and Television Council
iftcd in-flight thrust-calculation deck
IFTF Inter-Faith Task Force
IFTPP International Federation of the Technical and Periodical Press
iftr in-flight thrust reverser
IFTR International Federation for Theatre Research
IFUW International Federation of University Women
ifv (IFV) infantry fighting vehicle
IFVME Inspectorate of Fighting Vehicles and Mechanical Equipment
IFWL International Federation of Women Lawyers
IFZ Industrial Free Zone
ig ignition; immunoglobulin; inertial guidance
ig (Ig) (IG) immunoglobulin
i/g in ground
IG Illustrators Guild; Indo-Germanic; Inspector General
IG Interessengemeinschaft (German—pool, trust)
iga integrating gyro(scope) accelerometer
IGA Independent Grocers' Alliance; Integrated Grant Administration; International Geneva Association; International Geographical Associa-

tion; International Glaucoma Association; International Golf Association; International Graduate Achievement
IGAEA International Graphic Arts Education Association
i gal imperial gallon
IGAM Internationale Gesellschaft für Allgemeinmedizin (German—International Society of General Medicine)
IGAS International General Aviation Society; International Graphic Arts Society
I Gas Eng Institution of Gas Engineers
IGB International Gravimetric Bureau
IGB International Geophysics Bulletin
igc intellectually gifted children
IGC Institute for Graphic Communication; Intergovernmental Copyright Committee; International Geophysical Cooperation
IGCA Industrial Gas Cleaning Association; International Garden Centre Association; Israel Government Corporation Authority
IGCC Inter-Governmental Copyright Committee
igce independent government cost estimate
IGCI Industrial Gas Cleaning Institute
IGCM Incorporated Guild of Church Musicians
i/g/d illicit gold dealer
IGD Inspector General's Department
ig det ignition detector
IGDS Iodine Generating and Dispensing System
ige in ground effect; individually guided education; instrumentation ground equipment
IgE Immunoglobin E
IGE International General Electric
IGF International Grieg Festival
igf-1 (IGF-1) insulin-like growth factor 1
IGFA International Game Fish Association
I.G. Farben Interessengemeinschaft der Farbenindustrie (German—German Dye Trust)
ig. fat. ignis fatuus (Latin—foolish fire)—will-o'-the-wisp; marsh gas
igfet insulated gate field-effect

transistor

igg ill-gotten gains

IGH Incorporated Guild of Hairdressers

IGI Institutional Goals Inventory

IGI 1 Grandi Interpreti (Italian—The Great Interpreters)—of classical music

IGIA Interagency Group on International Aviation

IGIS International Guild for Infant Survival

igl information grouping logic

igl *iglesia* (Spanish—church)

iglᵃ *iglesia* (Spanish—church)

igla *iglesia* (Spanish—church)

IGM Internationale Gesellschaft für Moorforschung (German—Society for Research on Moors)

ign ignite; ignition

ign. ignotus (Latin—unknown)

Ign Ignacio; Ignatius; Ignatz; Ignazio

IGN International Great Northern (railroad)

Ignᵒ Ignacio (Spanish—Ignatius)

IGO Independent Garage Owners; Intergovernmental Organization

igor injection gas-oil ratio; intercept ground optical recorder

igortt intercept ground optical recorder tracking telescope

IGOSS Integrated Global Ocean Station System

igp inside gravel pack (oil well)

IGP Industrial Government Party; Inspector General of Police

igpm imperial gallons per mile; imperial gallons per minute

igpp (IGPP) interactive graphics packaging program

IGPP Institute of Geophysics and Planetary Physics (UCLA)

igr. igitur (Latin—therefore)

igrf international geomagnetic reference field

IGROF Internationale Rorschach Gesellschaft (German—International Rorschach Society)

IGRS Irish Geneological Research Society

igs (IGS) interactive graphics system

IGS Imperial General Staff; Inert Gas System; Inertial Guidance System; Infogram

Service; Institute of General Semantics; Institute of Geological Sciences; International Geranium Society; International Graphoanalysis Society

igse interim ground-support equipment

IGSEAP Inertial Guidance System Error Analysis Program

IGSESS International Graduate School for English-Speaking Students

IGSS Instituto Guatemalteco de Seguridad (Spanish—Guatemalan Social Security Institute)

IGST Intergovernmental Committee on Science and Technology

igt (IGT) interactive graphics terminal

IGT Institute of Gas Technology

IGTO India Government Tourist Office; Israel Government Tourist Office; Italian Government Tourist Office

IGU International Gas Union; International Geographical Union

igv inlet guide vane

IGWF International Garment Workers Federation

IGWT In God We Trust

IGWU International Garment Workers Union

IGWUA International Glove Workers Union of America

IGY International Geophysical Year (July 1957 through December 1958)

ih inside height

ih (IH) infectious hepatitis

i.h. iacet hic (Latin—here lies)

IH International Harvester

I of H Institute of Hydrology

IH International Humanism

iha (IHA) idiopathic hyperal–dosteronism

IHA Institute of Hospital Administrators; International Hahnemannian Association; International Hotel Association; International House Association

IHAB International Horticultural Advisory Bureau

IHAR Institute for Human-Animal Relationships

IHAS Integrated Helicopter Avionics System

IHATIS International Hide and Allied Trades Improvement Society

IHB International Hydrograph-

ic Bureau (Monaco)

IHBR Indiana Harbor Belt Railroad

ihc interstate highway capability

IHC Intercontinental Hotels Corporation; International Help for Children

IHCA International Hebrew Christian Alliance

IHCC International Harvester Credit Company

IHCD International Holocaust Commemoration Day

ihd (IHD) ischemic heart disease

IHD Institute of High Fidelity; Institute of Human Development; International Health Division (Rockefeller Institute for Medical Research); International Hydrological Decade (1965—1974)

IHDS Interstate Highway and Defense System

IHE Institute of Highway Engineers; Institute of Home Economics; Institutions of Higher Education

I-head capital-I-shaped head (gasoline engine)

IHEU International Humanist and Ethical Union

ihf interesting historic figure

IHF Independent Health Food; Industrial Hygiene Foundation; Institute of High Fidelity; International Hockey Federation; International Hospital Federation

IHFA Industrial Hygiene Foundation of America

IHFAS Integrated High-Frequency Antenna System

ihff inhibit halt flip-flop

IHHA International Halfway House Association

IHI Ishikawajima-Harima Heavy Industries

IHK Internationale Handelskammer (German—International Chamber of Commerce)

IHL International Homeopathic League

IHM Institute of Hotel Marketing; Institute of Housing Managers

I.H.M. Immaculate Heart of Mary

iho in-house operation

IHOP International House of Pancakes

IHOU Institute of Home Office Underwriters

ihp indicated horsepower;

ischemic heart disease
IHP Integrated Humanities Program
IHPA Imported Hardwood Plywood Association
ihph indicated horsepower hour
ihp/hr indicated horsepower hour
IHQ International Headquarters
IHR Institute of Historical Research; Institute of Human Relations
IHRA Independent Human Rights Association
IHRB International Hockey Rules Board
ihrd international rubber hardness degree(s)
IHRLA International Human Rights Law Group
ihs independent hemopathic syndrome; integrated heat sink; intellectually handicapped society
i.h.s. a variant of I.H.S.
IHS Christian symbol and monogram for Jesus; Immigration Historical Society; Indian Health Service; Infrared Homing System; Institute for Humane Studies; Institute of Hypertension Studies; International Horn Society; Interstate Highway System; Irish Hospitals Sweepstakes; Ivory Hunters Society
I.H.S. Iesus Hominum Salvator (Latin—Jesus Savior of Men); *In Hoc Signo* (Latin—In This Sign), Jesus
ihsa iodinated human serum albumin
IHSA Institute of Health Service Administrators; Italian Historical Society of America
ihsbr improved high-speed bombing radar
ihss idiopathic hypertrophic subaortic stenosis (IHSS)
IHSS Integrated Hydrographic Survey System
IHT Institute of Handicraft Teachers
IHT International Herald Tribune
IHU Irish Hockey Union; Interservice Hovercraft Unit
ihv intravenous hyperalimentation
IHVE Institute of Heating and Ventilating Engineers
ihx intermediate heat exchanger

IHY International Historical Year
IHYC Indian Harbor Yacht Club; Indian Harbour Yacht Club
ii illegal immigrant; individualized instruction; ingot iron; initial issue; injectivity index (oil well); interest included; inventory and inspection
i & i intercourse and intoxication (aspects of r & r); introduce and interview
II Ikebana International; Instituto Interamericano (Interamerican Institute); Irish Institute; Islamic Institute
I/I Inventory and Inspection (Report)
I & I instruction and inspection
II Instituto Interamericano (Spanish—Interamerican Institute)
iia if incorrect advise; inertial instrument assembly; inner-inch adjustment; integrated irradiance analyzer
IIA Aerlinte Eireann (3-letter symbol for Irish Airlines); Incinerator Institute of America; Information Industry Association; Institute of Industrial Arts; Institute of Internal Auditors; Insurance Institute of America; International Information Administration; Invention Industry Association
IIAA Independent Insurance Agents Association
IIAC Industrial Injuries Advisory Council
IIAF Imperial Iranian Air Force
IIAG Interbureau Insurance Advisory Group
IIAL International Institute of Arts and Letters
IIAPCO Independent Indonesian-American Petroleum Company
IIAS International Institute of Administrative Services
IIASA International Institute of Applied Systems Analysis
IIASR Israel Institute of Applied Social Research
IIB International Investment Bank; Internordic Investment Bank
IIB Institut International de Bibliographie (French—International Institute of Bibliography); *Institut Internationale des Brevets* (French—International Patent Institute);

International Institute of Bankers
IIC Independent Insurance Conference; Insurance Institute of Canada; International Institute of Communications; International Institute for the Conservation of Historic and Artistic Works; International Inter-City (train)
IICA Indians Into Communications Association; Institute of Instrumentation and Control—Australia
IICA Instituto Interamericano de Ciencias Agrícolas (Spanish—Inter-American Institute of Agricultural Sciences)
I-ICB Isolation-Interface Control Board
IICLRR International Institute for Children's Literature and Reading Research
iid impact ionization diode; infrared intrusion detection; interior intrusion device
IID Internal Investigation Division
IID Institut International de Documentation (French—International Documentation Institute)
IIDA Irish Industrial Development Authority
IIDS Interior Intruder Detection System
IIDS Instituto Interamericano de Desarrolo Social (Spanish—Interamerican Social Development Institute)
IIE Institute of Industrial Engineers; Institute for International Economics; Institute for International Education; International Institute of Embryology
IIE Instituto Interamericano de Estadistica (Spanish—Inter-American Institute of Statistics)
IIEA International Institute for Environmental Affairs
IIEC Inter-Industry Emission Control (program)
IIEG Interest Inventory for Elementary Grades
IIEL Institut Internationale d'Études Ligures (French—International Institute for Ligurian Studies)
IIEP International Institute of Educational Planning
IIET Inspection Instructions for Electron Tubes
IIF Institute of International Finance; *Institut Internation-*

al du Froid (French—International Institute of Refrigeration)

IIFA International Institute of Films on Art

IIFT Indian Institute of Foreign Trade

IIGF Imperial Iranian Ground Forces

IIGH *Instituto Interamericano de Geografía e Historia* (Spanish—Interamerican Institute of Geography and History)

IIHCEHV International Institute of Health Care, Ethics, and Human Values

IIHF International Ice Hockey Federation

IIHS Insurance Institute for Highway Safety

III Insurance Information Institute; Inter-American Indian Institute; International Institute of Interpreters (UN); International Isostatic Institute

III *Instituto Indigenista Interamericano* (Spanish—Inter-American Indigenist Institute); *International Intertrade Index*

IIIC International Irrigation Information Center (Israeli)

IIIRI Illinois Institute of Technology Research Institute

IIJR Illinois Institute of Juvenile Research

iil integrated injection logic

IIL Intelligence International Limited

IILC *International Instituut voor Landaanwinning en Cultuurtechniek* (Dutch—International Institute of Land Reclamation and Cultivation)

IILRI International Institute for Land Reclamation and Improvement

IILS International Institute for Labour Studies

IIM Indian Institute of Management

IIME Institute of International Medical Education

IIMS Intensive Item Management System

IIMSD International Institute for Music Studies and Documentation

IIMT International Institute for the Management of Technology

IIN Item Identification Number

IIN *Instituto Interamericano del Niño* (Spanish—Inter-

American Children's Institute); *Instituto Italiano di Navigazione* (Italian—Italian Institute of Navigation)

IInfSc Institute of Information Scientists

IINZ Insurance Institute of New Zealand

IIOE International Indian Ocean Expedition

IIOOF International Independent Order of Odd Fellows

IIOS International Indian Ocean Survey

iip index of industrial production; individualized instructional planning

IIP Indian Imperial Police; Institute International de la Presse (International Institute of the Press); International Ice Patrol; International Institute of Peace; International Institute of Philosophy

IIP *Institute International de la Presse* (French—International Institute of the Press)

IIPER International Institution of Production Engineering Research

IIPs Individualized Instruction(al) Program(s)

iir imaging infra-red; isobutylene isoprene rubber

IIR International Institute of Refrigeration

IIRA International Industrial Relations Association

IIRE International Institute for Resource Economics

IIRM Irish Immigration Reform Movement

iirs instrumentation inertial reference set

IIRS Institute for Industrial Research and Standards (Erie)

ii's illegal immigrants

IIs Iberial Inquisitions in Spain and Portugal; Immigration Inspectors

IIS Indian Institute of Science; Industrial Inquiry Service; Institute of Information Science; Institute of Information Scientists; Insurance Institute of Singapore; Integrated Instrument System; Interactive Instructional System

IIS *Institut International de la Soudre* (French—International Institute of Welding); *Institut International de la Statistique* (French—International Institute of Statistics); *Internationales Institut der*

Sparkassen (German—International Institute of Savings Banks)

IIS & EE International Institute of Seismology and Earthquake Engineering

IISG *Internationaal Instituut voor Sociale Geschiedenis* (Dutch—International Institute of Social History)—in Amsterdam

IISL International Institute of Space Law

IISL *Istituto Internazionale di Studi Liguri* (Italian—International Institute for Ligurian Studies)

IISO Institution of Industrial Safety Officers

IISR International Institute for Submarine Research

IISRP International Institute of Synthetic Rubber Producers

IISS International Institute of Strategic Studies

IISWM International Institute of Iron and Steel Wire Manufacturers

iit independent inclusive tour; individual inclusive tour

IIT Illinois Institute of Technology; Indian Institute of Technology; Israel Institute of Technology

IIT *Institut International du Théâtre* (French—International Institute of the Theater)

IITB Indian Institute of Technology—Bombay

IITM Indian Institute of Technology—Madras

IITRAN Illinois Institute of Technology Translators

IITYWYBAD? If I tell you will you buy a drink?

IIW International Institute of Welding

iiwfm if it weren't for me

iiwfy if it weren't for you

iJ *im Jahre* (German—in the year)

IJ IJssel; Institute of Journalists; Irish Jurist

I of J Institute of Jamaica

IJ *Internationale Jugendbibliothek* (German—International Youth Library)—Munich

IJA Institute of Jewish Affairs; Institute of Judicial Administration; International Judiciary Association

IJA *International Journal of the Addictions*

IJC International Joint Commission (Canada—U.S.); Itawamba Junior College

IJE Institute for Journalism Education

IJF International Judo Federation

I-J FC Iselin-Jefferson Financial Company

IJIAP International Juridical Institute for Animal Protection

IJISID Imperial Japanese Institute for the Study of Infectious Diseases

IJK *Internationale Juristen-Kommission* (German—International Jurist Commission)

IJLP Islamic Jihad for the Liberation of Palestine

IJMS *Israel Journal of Medical Sciences*

IJO International Juridical Organization (for developing countries of the third world)

IJOA International Juvenile Officers Association

ijp inhibitory junction potential

IJPPR Institute for Jewish Policy Planning and Research

IJR Institute for Juvenile Research

IJS Institute of Jazz Studies; Institute of Jesuit Sources; Institute of Jewish Studies

IJszee (Dutch—Ice Sea)—Arctic Ocean, Polar Sea

IJVA International Journal of Verbal Aggression *(Maledicta)*

ik inner keel

ik *ikke* (Danish—not)

Ik Ichabod

IK *Immune Korper* (German—immune bodies)

IKAR *Internationale Kommission für Alpines Rettungswesen* (German—International Commission for Alpine Rescue)

IKB Isambard Kingdom Brunel

ike iconoscope; ikebana; ikebanism; ikebanist(ic)

Ike USS *Eisenhower* (nuclear-powered supercarrier)

IKG *Internationale Kommission für Glas* (German—International Commission for Glass)

IKI *Internationale Kali-Institut* (German—International Potash Institute)

IKN *Internationale Kommission für Numismatik* (German—International Numismatic Commission)

I-K-P In-Ko-Pah (Park or Mountains in California)

IKPK *International Kriminal-Polizei-Kommission* (German—International Criminal Police Commission)

I kr Icelandic krona (monetary unit)

ikrd inverse kinetics rod drop

IKRK *Internationale Komitee vom Roten Kreuz* (German—International Committee of the Red Cross)

ik unit infusoria killing unit

IKV *Internationaler Krankenhausverband* (German—International Hospital Federation)

IKV-91 Hagglund and Soner tank destroyer made in Sweden

il illustrate; illustrated; illustration; illustrator; including loading; incoming letter; inside layer; inside left; inside length; instrument landing; interline; interlinear; interlinearly; interpretative language

il (IL) interleukin

Il illinium

Il *Illiad*

IL Identification List(ing); Illinois; Import License; Incres Line; Independent Laboratory; Instruction Leaflet; International Logistics; Inter-ocean Line; Israel (auto plaque)

I/L Import License

I & L Installations and Logistics

I of L Institute of Linguists

IL *Institut Littéraire*

Il-12 Soviet Ilyushin transport called Coach by NATO

Il-14 Soviet Ilyushin transport called Crate by NATO

Il-18 Soviet Ilyushin transport called Coot by NATO

Il-28 Soviet Ilyushin jet bomber called Beagle by NATO

Il-38 Soviet Ilyushin transport called May by NATO

IL-62 Ilyushin 62 aircraft

ila insurance logistics automated

ila (ILA) instrument landing approach

ILA Illinois Library Association; Independent Literary Agent; Indian(a) Library Association; Indonesian Library Association; Institute of Landscape Architects; International Laundry Association; International Law Asso-

ciation; International Leprosy Association; International Linguistic Association; International Longshoremen's Association; Iowa Library Association; Iranian Library Association; Iraq Library Association; Israel Library Association

ILAA Independent Literary Agents Association; International Legal Aid Association

ILAAS Integrated Light Aircraft Avionics System; Integrated Light Attack Avionics System; International League Against Anti-Semitism

ILAB International League of Antiquarian Booksellers

ILAFA *Instituto Latinamericano del Fierro y del Acero* (Spanish—Latin American Institute of Iron and Steel)

ILAMA International Lifesaving Appliance Manufacturers Association

ILAP Individualized Language Arts Program

ILAR Institute of Laboratory Animal Resources

ilas interrelated logic accumulating scanner

ILAS Institute of Latin American Studies; Instrument Low-Approach System

ilb inshore lifeboat

ilc irrevocable letter of credit

ilc (ILC) instruction length code

ILC Independent Learning Center; Individualized Learning Center; International Law Commission (UN)

ILCA International Livestock Centre for Africa

ILCNY I Love a Clean New York

ILCOP International Liaison Committee of Organizations for Peace (UN)

ILCW Inter-Lutheran Commission on Worship

ild instructional logic diagram

ILD International Labor Defense

Ildef° Ildefonso (Spanish)

ildf integrated logistic data file

ildt item logistics data transmittal

ile individual learning expectations; isoleucine

ile (ILE) isoleucine (amino acid)

Il' *Illustre* (Spanish—illustrious)

ILE Institution of Locomotive

Engineers
ILEA Inner London Education Authority
ILEI Index of Leading Economic Indicators
ILEI *Internacia Ligo de Esperantistaj Instruistoj* (International League of Esperanto Instructors)
ILERA International League of Esperantist Radio Amateurs
ILESA International Law Enforcement Stress Association
ileu isoleucine
ilf inductive loss factor
Ilf Ilya Arnoldovich Feisliber
ILF International Landworkers Federation
ILFI International Labor Film Institute
ILFO International Logistics Field Office (USA)
ILGA Institute of Local Government Administration
ILGPNWU International Leather Goods, Plastics, and Novelty Workers Union
ILGWU International Ladies' Garment Workers' Union
ILH Imperial Light Horse; International League of Honolulu; Interscholastic League of Honolulu
ILHR International League of Human Rights
ILI Indiana Limestone Institute; Institute of Life Insurance; Inter-African Labor Institute; International Language Institute
ILIA Indiana Limestone Institute of America
ILIAS Inforonics Library Automation Services
ILIC International Library Information Center
ill. illusion; illusionary; illusionist; illustrate; illustrated; illustration; illustrator
ill. *illustrissimus* (Latin—most illustrious)
Ill Illinois; Illinoisan
I'll I shall; I will
ILL Institute of Languages and Linguistics; Institute of Lifetime Learning; Inter-Library Loan; Interstate Loan Library
ILLA Irish Ladies Lacrosse Association
ILLC Inner London Library Committee
illegals illegal aliens
illegit illegitimate
ILLIAC Illinois Automatic Computer

ILLINET Illinois Library Information Network
illit illiterate
ILLRI Industrial Lift and Loading Ramp Institute
Ill St Hist Lib Illinois State Historical Library
Ill St Hist Soc Illinois State Historical Society
illu illustrate
illud illustrated
illum illuminant; illuminate; illumination
illun illustration
illus illustrated; illustration; illustrator
illw intermediate-level liquid waste
ilm insulin-like material
ILM International Literary Management
ILMA Incandescent Lamp Manufacturers Association
Ilmo *Illustrissimo* (Italian—Most Illustrious)
Il^mo *Illustrísimo* (Spanish—Most Illustrious)
ILMP *International Literary Market Place*
ILN *Illustrated London News*
ilo in lieu of
Ilo Iloilo
I lo iodine lotion
ILO International Labour Office (UN); International Labor Organization
ILOA Industrial Life Officers Association
I Loco E Institution of Locomotive Engineers
I Loco Eng Institution of Locomotive Engineers
Ilopango San Salvador, El Salvador's airport
iloue in lieu of until exhausted
ilp instant linear programming
ILP Independent Labour Party; Israel Labor Party
ILPA Independent Labor Press Association
ILPES *Instituto Latinoamericano de Planificación Económica y Social* (Spanish—Latin American Institute for Economic and Social Planning)
ILPH International League for the Protection of Horses
ILQ *International Law Quarterly*
ilr (ILR) independent local radio
ILR Institute of Library Research; International Luggage Registry
ILR *Instituut voor Landbow-*

techniek en Rationalisatie (Dutch—Institute for Agricultural and Planning Technics); *International Law Reports*
ILRA International Log Rolling Association
ILRAD International Laboratory for Research into Animal Diseases
ILRC Indian Law Resources Center
ILREC International League for the Rational Education of Children
ILRI Indian Lac Research Institute
ILRM International League for the Rights of Man
ILS Incorporated Law Society; Industrial Locomotive Society; Instrument Landing System; Integrated Logistic Support; International Latitude Service; International Lunar Society
ILSA Insured Locksmiths and Safemen of America
ilsam international language for servicing and maintenance
ILSC International Learning Systems Corporation
ILSMT Integrated Logistic Support Management Team
ILSP Integrated Logistic Support Plan(ning)
ILSR Institute for Law and Social Research; Institute for Local Self-Reliance
ilsw interrupt-level status word
ilt interferometric landmark tracker; in lieu thereof
ilt (ILT) infectious laryngotracheitis
ILT Illinois Terminal (railroad)
ILTF International Lawn Tennis Federation
ILTS Institute of Low Temperature Science; Integration Level Test Series
iltw intermediate-level transuranic waste
ILU Institute of Life Insurance; Institute of London Underwriters
ilv induced leukemia virus(es)
ilw intermediate-level wastes
ILWU International Longshoremen's and Warehousemen's Union
ILZ Illinois Zinc (company)
ILZRO International Lead Zinc Research Organization
im immature; imperial mea-

sure; impulse modulation; infectious mononucleosis; inner marker; installation maintenance; installment mortgage; intensity modulation; intermodulation; intramuscular

im (IM) inland marine (insurance); interceptor missile

i&m improvement and modernization

'im (American slang—him)

im in dem (German—in the); *imeni* (Russian—in the name of); (Latin—in)—imprint(ing)

Im Imperial

I'm I am

IM impulse modulation; Industrial Management; Institute of Metallurgists; Institute of Metals; intermediate modulation; Inventory Manager

I of M Institute of Medicine

IM Index Medicus

ima ideal mechanical advantage

I^{ma} prima (Italian—first)

IMA Ignition Manufacturers Institute; Indian Military Academy; Individual Medical Account; Indonesian Mining Association; Industrial Marketing Association; Industrial Medical Association; Institute for Mediterranean Affairs; Institute of Municipal Administration; *Instituto Mobiliare Italiano* (Italian—Italian Security Institute), International Management Association; International Mineralogical Association; Islamic Mission of America

IMAA Indochinese Mutual Assistance Association

imac integrated microwave amplifier converter

IMAC International Metals and Commodities

imag imaginary

IMAGE Instruction in Motivation Achievement and General Education

IMAGI Index Measuring Accurate Growth of Inflation

IMAJ International Management Association of Japan

IMAR Inner Mongolia Autonomous Region (of the People's Republic of China)

IMarE Institute of Marine Engineers

IMARPE Instituto de Mar del Perú (Spanish—Sea Institute of Peru)

IMARS Institutional Manage-

ment for Accountability and Renewal System

IMAS Integrated Management Accounting System; International Marine and Shipping Conference

IMAU International Movement for Atlantic Union

IMAU Instituto Municipal de Aseo Urbano (Spanish—Municipal Institute of Urban Sanitation)

IMAWU International Molders and Allied Workers Union

imb interaction of man and the biosphere

IMB Institute of Marine Biochemistry; Institute of Marine Biology; International Maritime Bureau

IMB Internationaler Metallarbeiterbund (German—International Metalworkers Federation)

ImbarsbIdbib I may be a rotten sod but I don't believe in bullshit (early 19th century British slang)

IMBC Independent and Multi-Cultural Broadcasting Corporation

IMBE Institute for Minority Business Education

IMBLMS Integrated Medical and Behavioral Laboratory Measurement System (NASA)

IMBO Institutt for Marin Biologi (Norwegian—Institute for Marine Biology)

imc image motion compensation; instrument meteorological condition; intermediate metal(lic) conduit

IMC Industrial Management Center; Infant Mortality Commission; Institute of Management Consultants; Institute of Measurement Control; Instructional Materials Center; Instructional Media Center; International Maritime Committee; International Medical Corps; International Meteorological Committee; International Minerals & Chemical; International Mining Corporation; International Missionary Council; International Monetary Conference; International Music Council; Iran Meat Corporation

IMC Instituto Mexicano del Café (Spanish—Mexican Coffee Institute)

IMCA Institute of Management Consultants in Australia

imcc item management control code

IMCC Integrated Mission Control Center

IMCE Instituto Mexicano de Comercio Exterior (Spanish—Mexican Institute of Foreign Commerce)

IMCEA International Military Club Executives Association

IMCI Interracial Music Council, Incorporated

imco improved combustion

IMCO Inter-Governmental Maritime Consultative Organization

IMCOS International Meteorological Consultant Service

IMCOV Iron Mines Company of Venezuela

IMCS Interactive Manufacturing Control System

IMD Indian Medical Department; Inventory Management Division

IMDA Indirect Missile Damage Assessment

IMDC Internal Message Distribution Center

IMDGC International Maritime Dangerous Goods Code

IMDS International Microform Distribution Service

imdt immediate(ly)

imdtty it's my duty to tell you

ime (IME) international magnetospheric explorer

IME Indo-Malaysian Engineering; Institute of Makers of Explosives; Institute of Marine Engineers; Institution of Mechanical Engineers

I&ME Indiana and Michigan Electric Company

I of ME Institution of Mining Engineers

I Mech E Institution of Mechanical Engineers

IMEG International Management and Engineering Group

IMEO Interim Maintenance Engineering Order

imep indicated mean effective pressure

IMER Institute for Marine Environmental Research

I Met Institute of Metals

IMET International Military Education and Training

imf intermediate fuel

imf (IMF) integrated maintenance facility

IMF International Metalworkers Federation; International

Monetary Fund; International Motorcycle Federation; Interstate Motor Freight (stock exchange symbol); Israel Music Foundation
IMFC Investment and Merchant Finance Corporation
IM FI International Mineral Fiber Institute
IMFJC International Metalworkers Federation Japan Council
im/fm intensity modulated/frequency modulated
imfrad integrated multiple-frequency radar
IMF/SDR International Monetary Fund—Special Drawing Rights
imfu immense military fuckup
img informational media guarantee
IMG International Marxist Group
IMGP Internal Medicine Group Practice
IMH Institute of Materials Handling
imhe international management in higher education
IMHT Institute for Material Handling Teachers
imi improved manned interceptor
IMI Ignition Manufacturers Institute; Imperial Metal Industries; Imperial Mycological Institute; International Masonry Institute; Irish Management Institute; Israel Military Industries; Israeli Military Intelligence
IMI Instituto Mobiliare Italiano (Italian)—Italian Assets Institution)—credit bank
IMIA International Marketing Institute of Australia
IMIB Inland Marine Insurance Bureau
imid inadvertent missile ignition detection
imieo initial mass in earth orbit
IMIMI Industrial Mineral Insulation Manufacturers Institute
IMINCO Iran Marine International Oil Company
IMinE Institute of Mining Engineers
IMINOCO Iranian Marine International Oil Company
imit imitate; imitation
IMIT Institute of Musical Instrument Technicians
imitac image input to automatic computers

imit lea imitation leather
iml inside mold layer; inside mold line
Iml Imanuel
IML International Music League; Irradiated Materials Laboratory; Island Merchants Limited (Cook Islands)
IMLS Institute of Medical Laboratory Sciences
IMLT Institute of Medical Laboratory Technology
imm immediate; immune; immunization; immunologist; immunology; impairing a minor's morals; impairing the morals of a minor
Imm Immingham
IMM Institute of Mining and Metallurgy; Integrated Maintenance Management; International Mercantile Marine; International Monetary Market (Chicago)
immac inventory management and material control
immat immature; immaturity
immed immediate
immedly immediately
immie immitation marble; low-grade playing marble
immig immigrant; immigration
immob immobilization; immobilize
IMMS International Material Management Society
IMMT Integrated Maintenance Management Team
IMMTS Indian Mercantile Marine Training Ship
immun immunity; immunization
immunol immunologist; immunology; imunologic(al)(ly)
immy immediately
imn indicated mach number
IMNS Imperial Military Nursing Service
imo imitation (slang short form); immobilized
Imo Imogen(e)
IMO Integrated Marketing Organization; Inter-American Municipal Organization; International Maritime Organization; International Meteorological Organization (World Meteorological Organization)
imos inadvertent modification of the stratosphere
imp. impedance; imperative; imperfect; imperial; implement; implementation; import; imprint; improve; improvement

imp (IMP) inflatable micrometeoroid paraglide; International Match Point
imp. (IMP) indeterminate mass particle; inertial measuring platform; international match point
imp. impotentia (Latin—impotent); *imprenta* (Spanish—printing office, printing press); *imprimatur* (Latin—let it be printed); *imprimé* (French—printed); *imprimis* (Latin—especially, particularly)
Imp Imperator (Latin—supreme ruler)
Imp. Imperator (Latin—Emperor); *Imperatrix* (Latin—Empress)
IMP Instrumented Mobile Platform (oceanographic drone boat); International Monitoring Probe (space instrument); Interplanetary Monitoring Platform (space vehicle)
IMP Instituto Mexicano del Petroleo (Spanish—Mexican Petroleum Institute)
imp. 8 imperial octavo ($7^1/2$ × 11-inch or 19 × 28-cm book size)
IMPA International Master Printers Association; International Museum Photographers Association; International Myopia Prevention Association
impact. implementation planning and control technique
IMPACT Improving Public Awareness of Concepts of Telecommunications; Interdisciplinary Model Programs in the Arts for Children and Teachers
Imp B Imperial Beach
impce importance
imper imperative
imperat imperative
imperf imperfect; imperforate
Imperial Imperial Savings Association
impers impersonal
imp-exp import-export
impf imperfect
impg importing; impregnate
imp. gal imperial gallon
IMPI International Microwave Power Institute
impig impignorate; impignorated; impignorating; impignoration
impl imperial; implement
implic import license

import **importaciones** (Spanish—imports)
imposs impossible
impr improvement
impr impresión; imprenta (Spanish—edition, printing office)
impracl impracticable
impreg impregnate(d); impregnation
IMPRESS Inter-disciplinary Machine Processing for Research and Education in the Social Sciences
imprim. imprimatur (Latin—let it be printed)
Impr Nat Imprimerie Nationale (French—National Printing Office of France)
improp improper(ly)
improv improvement
imps. interplanetary measurement probes
Imps Imperial Tobacco Company
IMPS Inpatient Multidimensional Psychiatric Scale; Institute of Management Public Speaking
Imp Sav Imperial Savings
impt important
imptd imported
imptr importer
Imptypco Imperial Typewriter Company (also appears as ITC)
impv imperative
impvt improvement
impx impaction
imqc imported merchandise quantity control
imr internal mold release
imr (IMR) infant mortality rate
IMR Individual Medical Report; Institute of Marine Resources; Institute of Masonry Research; Institute for Materials Research; Institute for Medical Research; Institute for Mortuary Research; Institute for Motivational Research; Institute for Muscle Research; International Medical Research
IMRA Industrial Marketing Research Association
IMRADS Information Management, Retrieval, and Dissemination System
imran international marine radio aids to navigation
IMRC International Marine Radio Company
IMRL Individual Material Readiness List(ing)

IMRO Inspection Minor Rework Order; Interior Macedonian Revolutionary Organization
IMRS Inpatient Multidimensional Rating Scale
ims industrial methylated spirit(s); inertial measurement set
im's intramuscular injections
IMS Index Management System; Indian Medical Service; Individualized Mathematics System; Industrial Management Society; Industrial Mathematics Society; Information Management System; Institute of Management Sciences; Institute of Marine Science; Institute of Mathematical Statistics; Institute of Museum Services; Institute on Man and Science; International Magnetic System; International Magnetosphere Study; International Military Staff; International Musicological Society; International Mythological Society
IMSA International Management Systems Association; International Motor Sports Association; International Municipal Signal Association
IMSC International Maritime Satellite Corporation; International Military Sports Council
IMSCOM International Military Staff Communication (NATO)
IMS/HEW Institute of Museum Services—HEW
IMSL Independent Measurement Standards Laboratory; International Mathematical and Statistical Library
IMSM Institute of Marketing and Sales Management
IMSO Institute of Municipal Safety Officers
IMSR Isle of Man Steam Railway
imss integrated manned-system simulator
IMSS Integrated Manned Systems Simulator; International Museum of Surgical Science
IMSS Instituto Mexicano del Seguro Social (Spanish—Mexican Social Security Institute)
imt independent model triangulation
IMT International Military Tribunal (Nuremberg)

IMTA Imported Meat Trade Association; Institute of Municipal Treasurers and Accountants
IMTC Instituto Municipal de Transporte Colectiva (Spanish—Municipal Institute of Collective Transport)—metropolitan bus system
IMTD Inspectors of the Military Training Directorate
IM Tech Institute of Metallurgists Technician
IMTFE International Military Tribunal for the Far East
IMTP Industrial Mobilization Training Program
imu inertial measurement unit
imu (IMU) inertial measurement unit
IMU International Mailers Union; International Maritime Union; International Mathematical Union
IMUA Inland Marine Underwriters Association
I Mun E Institution of Municipal Engineers
imusc intramuscular
imv imperative; improve; intermittent mandatory ventilation
IMVS Institute of Medical and Veterinary Science
imw international map of the world
IMW Institute of Masters of Wine
imwxprt imitation wax prints
IMX Inquiry Message Exchange
Im Yem Imamate of Yemen
IMZ Internationales Musikzentrum (German—International Music Center)
in. inch(es); interest
in. (In) inulin
i/n item number
In India; Indian; indium; Indus; Instructor
In Indre (Norwegian—inner, interior, inside)
IN Indiana; Indian Navy; Institute of Neurobiology (Göteborg); Interested Negroes
I & N Immigration and Naturalization
I of N Institute of Navigation
in.² square inch(es)
in.³ cubic inch(es)
ina international normal atmosphere
INA Indian National Army; Inspector Naval Aircraft; Institute of National Affairs; Institute of Nautical Archaeology; Institution of Naval Ar-

chitects; Insurance Company of North America; Iraqi News Agency; Israeli News Agency

inabi inability

inacdutra inactive duty training

INACESA Industria Nacional de Cemento SA (Spanish—National Cement Industry Company)

INACH Instituto Antártico Chileno (Spanish—Chilean Antarctic Institute)

inactv inactivate; inactivation; inactive

INAEA International Newspaper Advertising Executives Association

InAF Indian Air Force

inah (INAH) isonicotinic acid hydrazide

INAH Instituto Nacional de Antropología e Historia (Spanish—National Institute of Anthropology and History)—Mexico

INALPRE Instituto Nacional de Preinversión (Spanish—National Investment Control Institute)—Bolivia

INAM Istituto Nazionale Assicurazione Malattie (Italian—National Health Insurance Board)

INAME International Newspaper Advertising and Marketing Executives

inanim inanimate; inanimative

INANTIC Instituto Nacional de Normas Tecnicas Industriales y Certificación (Spanish—National Institute of Technical Standards)

INAO Institut Nationale des Appellations d'Origine (French—National Institute of Authentic Names)

inappbl inapplicable

INAR Institute of Northern Agricultural Research

in'ards innards

INAS Inertial Navigation and Attack System(s)

inaud inaudible

inaug inaugurate; inaguration

inaug diss inaugural dissertation (thesis for doctor's degree)

in bal. in ballast

inbd inboard

inbu internal navigation battery unit

INBUCON International Business Consultants

inc in cloud; inclosure; in-

clude; income; increase; incumbent

Inc Inchon; Incorporated

In C Instructor Captain

INC Indian National Congress; Industrial National Corporation; International Narcotics Control; International Nickel Company; International Numismatic Commission; Island Navigation Company (tankers)

INC Instituto Nacional de Cultura (Spanish—National Institute of Culture)—Lima, Peru

inca inventory control and analysis

Inca Incahuasi

INCA Information Council of the Americas; International Newspaper Color Association; Inventory Control and Analysis

incair including air

incalz incalzando (Italian—increasing dynamics and tone)

incan incandescent

incap incapacitant; incapacitating

INCAP Instituto de Nutrición de Centroamerica y Panamá (Spanish—Institute of Nutrition of Central America and Panama)

incaps incapacitating agents

incb inclusion body

INCB International Narcotics Control Board

incct incorrect

inccty incorrectly

incd incendiary; incident

incdt incident

ince insurance

INCE Institute of Noise Control Engineering

incfmy inconformity

inch. inchoative; integrated chopper

In-Ch Indo-China

incho inchoate

inchoat inchoative

incid incidence; incident; incidental

incid mus incidental music

INCING International Copyright Information Center

INCIRS International Communication Information Retrieval System

incl incline; inclose; inclosure; include; including; inclusive

incl inclusivement (French—inclusively)

incld included

incln inclusion

inclntr inclinator

inclr intercooler

inclsv inclusive

INCMD Indianapolis Contract Management District

INCO International Nickel Company

incog incognito

INCOLSA Indiana Cooperative Library Services Authority

INCOMAG International Communication Agency

INCOMEX International Computer Exhibition

INCOMEX Instituto Colombiano de Comercio Exterior (Spanish—Colombian Institute of Overseas Commerce)

incomp incomplete

incompat incompatible; incompatibility

incompl incomplete

incor incorrect

INCORA Instituto Colombiano de Reforma Agraria (Spanish—Colombian Institute of Agrarian Reform)

incorp incorporated

Incorp Incorporated; Incorporation

incorr incorrect

inco(s) incorrigible(s)

incpt intercept

incr increase; increased; increasing; increasingly; increment; incremental

INCRA International Copper Research Association

INCREF International Children's Rescue Fund

incrim incriminate; incrimination; incriminatory

INCSR International Narcotics Control Strategy Report

incumb incumbent

incun incunabula

incur. incurable

ind independent; index; indicate; indicative; indicator; indirect; indigo; indorse; indorsement; industrial; industry

ind (IND) investigational new drug

in d. in diem (Latin—daily)

Ind India; Indian; Indiana; Indianapolis; Indianian; Indo-; Indonesian; Indus (constellation); Industries; Industry

Ind Indian; *Indiano* (Italian—Indian, Indian Ocean); *Indico* (Portuguese or Spanish—Indian, Indian Ocean); *Indien* (French—Indian, Indian Ocean)

IND India (auto plaque); Indianapolis, Indiana (airport)
INDA International Non-wovens and Disposables Association
indac industrial data acquisition and control
INDALUM Industria del Aluminio (Spanish—Aluminum Industry)
INDASAT Indian Scientific Satellite
INDAX Interactive Data Exchange
Ind Day Independence Day
Ind Dem Independent Democrat
Ind. E. Industrial Engineer
indecl indeclinable
INDECO Industrial Development Corporation; International Development and Construction Corporation
indef indefinite
indef art. indefinite article
indefops indefinite operations
indem indemnify; indemnity
inden indenture; indentured; indenturing
Ind Eng Industrial Engineer(ing)
Ind Eng Chem Industrial and Engineering Chemistry
Ind & Eng Chem Industrial and Engineering Chemistry
indep independent
INDEP Industria Nacional del Plomo (Spanish—National Lead Institute)
Ind-et-L Indre-et-Loire
Index Index Librorum Prohibitorum (Latin—Index of Forbidden Books); *Index on Censorship*
indi indicate; indication
India letter I radio code; Republic of India (Asian nation); *Bharat* (India's name in Hindi)
Indiana Girls Indiana Girls School (for juvenile delinquents at Indianapolis)
Indianap Indianapolis
indic indicative; indicator
indic indicateur (French—informer)
indicolite blue tourmaline
indies independents
Indies East Indies; West Indies
Ind. Imp. Indiae Imperator (Latin—Emperor of India)
Indira Indira Ghandi (India's first woman prime minister)
INDITECNOR Instituto Nacional de Investigaciones Tecnologicas y Normaliza-

ción (Spanish—National Institute for Technical and Standardization Investigation)
indiv individual
indivl individual
indiv psychol individual psychology
Ind L Independent Liberal
Ind Lab Independent Labor (party)
indm indemnity
Ind Med Index Medicus
Ind Mgr Industrial Manager
indn indication (flow chart)
Indo Indonesia; Indonesian
Ind O Indian Ocean
Indo-Afr Indo-African
Indo-Amer Indo-American
Indo-Austral Indo-Australasian
indoc indoctrinate; indoctrination
Indoc Indochina; Indochinese
Indo-Chi Indo-China; Indo-Chinese
indocin indomethacine
indocum indocumentado (Spanish—undocumented) person without identification papers
Indo-Eur Indo-European
Indo-Ger Indo-German(ic)
Indo-Mal Indo-Malayan
Indon Indonesia(n)
Indon Indonesian (modified Malay language)
Indonesia Republic of Indonesia (Asian island nation), *Republik Indonesia*
Indonesias Indonesian Islands
Indo-Pak India-Pakistan; Indo-Pakistan(i)
INDOSUEZ Banque de l'Indochine et de Suez (French—Bank of Indochina and Suez)
Indpls Indianapolis
indpol industrial pollutant; industrial pollution
ind quest indirect question
indr indicator (flow chart)
indre indenture
ind reg induction regulator
Ind Rep Independent Republican
Ind Sym Indianapolis Symphony
indt indent
Ind Ter Indian Territory (now Oklahoma)
indtr indentor
induc inductance; induction
INDUGAS Industria de Gas (Spanish—Gas Industry)
Ind U Pr Indiana University Press

indus industrial; industry
indust industrial; industrialization; industrialize; industrialized; industry
indvl individual
Indy Indianapolis; Indianapolis Speedway
Indy-style Indianapolis-style
inec inverted emulsifiable concentrate
ined. ineditus (Latin—unpublished)
INED Institute for New Enterprise Development
ineffv ineffective
inefvy ineffectively
inel inelastic
INEL Idaho National Engineering Laboratory (ERDA)
INEOA International Narcotic Enforcement Officers Association
Iness Inverness-shire
in ex. in extenso (Latin—at length)
inf infant(ile); infantry; infect-(ious); inferior; infinitive; infinity; influence; information; intervertebral foramina
inf (INF) interceptor night fighter; intermediate-range nuclear force
inf. infra (Latin—below, beneath); *infunde* (Latin—pour into)
Inf Infirmary
Inf Inférieur (French—lower, nether)
INF Intermediate Nuclear Force; Intermediate-Range Nuclear Forces; International Naturist Federation; International Nudist Federation
INFA Institut pour l' Etude du Fascisme (French—Institute for the Study of Fascism)
INFAMA International Fair Promotion and Marketing
INFANTS Iroquois Night Fighter And Night Tracker System
infarc infarction
Inf Bat Infantry Battalion
infce international nuclear-fuel-cycle evaluation
Inf Div Infantry Division
infe inferior
INFE International Newspaper Financial Executives
infect. infection; infectious
infib infibulate; infibulation
infin infinitive
infirm. infirmary
infl inflammable; inflorescence; influence(d)
in-fl in-flight

influ influence; influential
infm information
infmry infirmary
INFN *Istituto Nazionale di Fisica Nucleare* (Italian—National Institute of Nuclear Physics)—Italy
info inform; information
INFO International Fortean Organization
Info Can Information Canada
infol (INFOL) information-oriented language
infoline information line (telephone)
INFONAC *Instituto de Fomento Nacional* (Spanish—Institute for National Production)
inforem inventory forecasting and replenishment module(s)
INFORFILM International Information Film Service
Informbureau Communist Information Bureau (Cominform)
Informburo (Soviet) Information Bureau
Informex *Informaciones Mexicanas* (Mexican Information Service)
INFORS International Federation of Engineers
INFORSA *Industrias Forestal SA* (Spanish—Forest Industries Corporation)
INFOTERM International Information Center for Terminology (UNESCO)
info theory information theory
infr inferior
infra below
infra (Latin prefix—beneath)—infraspinal, infrastructure
infra dig. *infra dignitatem* (Latin—beneath one's dignity, undignified)
infral information retrieval atuomatic language
infraptum. *infrascriptum* (Latin—written below)
Infrared Phys *Infrared Physics*
infric. *infricetur* (Latin—let it be rubbed in)
infross information requirements of the social sciences
inft infant
infus infusible
infx inspection fixture
ing inguinal
ing *ingégnere* (Italian—engineer); *ingegneria* (Italian—engineering);*i ngeniør* (Dano-Norwegian—engineer)
Ing Ingmar

Ing *Ingénieur* (French—engineer); *Ingenieur* (German—engineer)
inga inspection gage
INGA Interstate Natural Gas Association
INGAA Interstate Natural Gas Association of America
INGEOMINAS *Instituto de Investigaciones Geologico Mineras* (Spanish—Institute of Geological Sources Research)
Ingg *Inggeris* (Malay—English)
Ingl *Inghilterra* (Italian—England); *Inglaterra* (Portuguese or Spanish—England)
Ingm Berg Ingmar Bergman
INGO International Non-Governmental Organization
ingred(s) ingrediient(s)
Ingria Ingermanland
Ingrid Ingrid Bergman
inh (INH) isonicotinic hydrazide
Inh *Inhaber* (German—proprietor)
INH *Instituto Nacional de Hipódromos* (Spanish—National Institute of Racetracks)
inhab inhabitant(s)
inhal inhalation
in. Hg inch of mercury
inhib inhibition; inhibitory
INHP Independence National Historical Park
INHS Indian Naval Hospital Ship
INI Indianapolis Newspapers Incorporated; Industrial Nurses Institute; *Institut National De l'Industrie* (French—National Institute of Industry)
INI *International Nursing Index*
INIA *Instituto Nacional de Investigaciones Agricolas* (Spanish—National Institute of Agricultural Research)
INIBP *Instituto Nacional de Investigaciones Biológico-Pesqueras* (Spanish—National Institute of Fish Biology Research)
INIC *Instituto Nacional de Investigaciones Cientifica* (Spanish—National Institute for Scientific Investigation)
INIF *Instituto Nacional de Investigaciones Forestales* (Spanish—National Institute for Forestry Research)
in./in. inch per inch
in init. *in initio* (Latin—in the beginning)

INIP *Instituto Nacional de Investigaciones Pecuarias* (Spanish—National Institute for Cattle Research)
INIS International Nuclear Information System
init initial
initv initiative
inj inject; injection; injections; injure; injury
inj. *injectio* (Latin—inject, injection)
inj. enema *injiciatur enema* (Latin—inject an enema)
inj. hyp. *injectio hypodermica* (Latin—hypodermic injection)
inj mldg injection moulding
inkl *inklusiv* (German—inclusive)
inl initial
inl *inlichtingen* (Dutch—information)
In L Instructor Lieutenant
INL Independent Newspapers Limited
INLA International Nuclear Law Association; Irish National Liberation Army
inlaw (INLAW) infantry laser weapon
in.-lb inch-pound
In L-Cdr Instructor Lieutenant-Commander
in lim. *in limine* (Latin—at the outset)
in litt. *in litteris* (Latin—in correspondence)
in loc. *in loco* (Latin—in the place)
in. loc. cit. *in loco citato* (Latin—in the place cited)
Inlt Inlet (postal abbreviation)
inly initially
INM Institute of Naval Medicine; International Narcotics Matter; Irish National Museum
inmarsat international marine satellite
INMARSATORG International Maritime Satellite Organization
INMAS *Intensifikasi Massnal* (Indonesian—Mass Intensification)
INMED Indians into Medicine
in mem. *in memoriam* (Latin—in memory of)
In Mem *In Memoriam*—Sir Arthur Sullivan's Overture in C
i n mi international nautical mile(s)
INMM Institute of Nuclear Materials Management

inn. inning
Inn. Innoshima
INN Instituto Nacional de Normalización (Spanish—National Institute of Standards); *Instituto Nacional de Nutrición* (Spanish—National Institute of Nutrition)
innerv innervated; innervation
Innis Inniskilling
INNOTECH Institute for Educational Innovation and Technology
inns. innings
INO Inspectorate of Naval Ordnance
inoc inoculation; inoculate
INOC Iraq National Oil Company
INOCO Indonesian Nippon Oil Corporation
inop inoperative
inorg inorganic
Inorg Chem Inorganic Chemistry
INOS Instituto Nacional de Obras Sanitarias (Spanish—National Institute of Sanitation—Venezuela)
in-out input-output
inp inert nitrogen protection
INP Inyanga National Park (Rhodesia)
INPA International Newspaper Promotion Association
INPA Instituto Nacional por Pesquisar au Amazonas (Portuguese—National Institute for Research on Amazons)
in partibus in partibus infidelium (Latin—in the region of the unbelievers)
INPFC International North Pacific Fisheries Commission
inph interphone
in p. inf. in partibus infidelium (Latin—in the region of the unbelievers)
INPO Institute of Nuclear Power Operations
INPOLSE International Police Services (CIA)
inpr in progress
in pr. in principio (Latin—in the first place)
Inprecorr International Press Correspondence
in prep in preparation
in pro in proportion
inprons information processing in the central nervous system
inps if not previously sold
INPS Istituto Nazionale di Previdenza Sociale (Italian—National Institute of Social

Security)
in pulm. in pulmento (Latin—in gruel)
inq inquiry
Inq Inquisidor (Spanish—inquisitor, investigator)
INQUA International Association on Quaternary Research
inr impact noise rating; impact noise ratio; intelligence and research
in'r inner
i-n r interference-to-noise ratio
INR Institut National de la Radio (French—National Radio Institute); Institute of Natural Resources; Intelligence and Research
INRA Instituto Nacional de la Reforma Agraria (Spanish—National Institute of Agrarian Reform)
in ref in reference (to)
in req information requested
In Res Indian Reservation
I.N.R.I. Iesus Nazarenus Rex Iudaeorum (Latin—Jesus of Nazareth, King of the Jews)
INRIA Institut National de Recherche Informatique et Automatique (French—National Institute of Informative and Automatic Research)
INRO International Natural Rubber Organization
ins inches; inscribe(d); inscription; inspector; insular; insulate(d); insulation; insurance; insure(d)
ins (INS) inertial navigation system
in's in his
in./s inch(s) per second
ins in das (German—into the)
in s in situ (Latin—in the original place)
Ins Insecta; Inverness
INS Immigration and Naturalization Service; Indian Naval Ship; Inertial Navigation System; Institute of Naval Studies; Institute of Nuclear Sciences; Institute of Nutritional Sciences; Integrated Navigation System; International News Service
I & NS Immigration and Naturalization Service
INSA International Shipowners Association
INSA Industria Nacional de Neumaticos SA (Spanish—National Tire Industry Corporation)
InsACS Interstate Airway Communication Station

Ins Agt Insurance Agent
INSAIR Inspector of Naval Aircraft
INSAT Indian National Satellite
insav interim shipyward availability
INSCAIRS Instrumentation Calibration Incident Repair Service
insce insurance
INSCO Intercontinental Shipping Corporation
inscr inscribed; inscription
INSDC Indian National Scientific Documentation Center
INSDOC Indian National Scientific Documentation Center
insd val insured value
INS & E Institute of Nuclear Science and Engineering
INSEA International Society for Education Through Art
in./sec inches per second
insecti insecticide(s)
INSEL International Nickel Southern Exploration Limited
INSENG Inspector of Naval Engineering Material
insep inseparable
Ins Gen Inspector General
insh inspection shell
insig insignificant
insinuendo insinuate + innuendo
INSIS Inter-Institutional Integrated Services Information System
INSJ Institute for Nuclear Study—Japan
insl insulate; insulation
INSMACH Inspector of Naval Machinery
INSMAT Inspector of Naval Material
INSNAVMAT Inspector of Navigational Material (USN)
insol insoluble
insolv insolvent
insoly insolubility
INSORA Instituto de Organización y Administración de Empresas (Spanish—Institute for the Organization and Administration of Enterprises)
INSORD Inspector of Naval Ordnance
insp inspect; inspected; inspection; inspector; inspiration; inspire; inspired
Insp Inspector
in-spec within specifications
INSPECC Information Services in Physics, Electrotechnology, Computers, and Control

INSPECT Infrared System for Printed-Circuit Testing; Inquiry into Pollution and Environmental Conservation; Integrated Nationwide System for Processing Entries from Customs Terminals
INSPEL International Journal of Special Libraries
INSPETRES Inspector of Petroleum Resources
Insp Gen Inspector General
inspir. inspiretur (Latin—let it be inspired)
INSPIRE Institute for Public Interest Representation
Inspr Inspector
INSRADMET Inspector of Radio Materials
insrnc insurance
INSRP Interagency Nuclear Safety Review Panel
inst install; installation; installment; instant; instantaneous; institution; instruct; instruction; instructor; instrument; instrumented; institute
Inst Institute; Institution
INSTAAR Institute of Arctic and Alpine Research
INSTAB Information Service on Toxicity and Biodegradability
insta-cam instant camera (tv)
instar inertialess scanning, tracking, and ranging
INSTARS Information Storage and Retrieval System
Inst CE Institute of Civil Engineers
Inst Ceram Institution of Ceramics
inst ctl instrumentation control
instd instead
Inst Dirs Institute of Directors
Inst EE Institute of Electrical Engineers
Inst F Institute of Fuel
Inst Gas Eng Institute of Gas Engineers
Inst Gen Sem Institute of General Semantics
Inst HE Institute of Highway Engineers
Inst Int Educ Institute of International Education
instln installation
instm instrument; instrumentation; instrumented
Inst ME Institute of Mechanical Engineers
Inst Mediaeval Mus Institute of Mediaeval Music
Inst Met Institute of Metals
Inst Mod Lang Institute of

Modern Languages
instn institution(al)
INSTN Institut National des Sciences et Techniques Nucléaires (French—National Institute of Science and Nuclear Techniques)
instns instructions
Inst P Institute of Physics
Inst Pat Institute of Patentees
Inst Pckg Institute of Packing
Inst Pet Institute of Petroleum
Inst Plan & Res Institute for Planning and Research
instpn instrument panel
Inst P S Institute of Purchasing and Supply
instr instruct; instruction; instructor; instrument(s)
instru instrumentation
instruct. instruction; instructor
Instru Soc Am Instrument Society of America
Inst W Institute of Welding
Inst WE Institute of Water Engineers
insuf insufficient
insul insulation
insur insurance
insurd insured
INSURV Board of Inspection and Survey
in sync in synchronization; perfectly synchronized
int intake; integer; integral; interest; interior; interjection; internal; international; intersection
int (INT) initial (flow chart)
int intérêt (French—interest)
Int International
INt Isaac Newton telescope
INT Air Inter *(Lignes Aériennes Intérieures)*; Interpool
INTA International New Thought Alliance
INTA Instituto Nacional de Tecnologia Agropecuaria (Spanish—National Institute of Agricultural Technology)
INTACS Integrated Tactical Communications Systems
INTACT Infants Need to Avoid Circumcision Trauma
INTAF Internal Affairs (ministry)
int. al. inter alia (Latin—among other things)
INTAL Industria Nacional de Tejidos de Alambre (Spanish—National Wire Netting Industry)
INTAMEL International Association of Metropolitan City Libraries
INTASGRO Interallied Tacti-

cal Study Group (NATO)
int. cib. inter cibos (Latin—between meals)
intcl intercoastal
intcol intelligence collecting; intelligence collection
int comb. internal combustion
Int Com Illum International Commission on Illumination
intcp intercept; interception; interceptor
int dec interior decorator
Int Doc Serv International Documents Service (Columbia University)
INTECOM International Council for Technical Communication
Integ Ed Assoc Integrated Education Associates
intel intelligence
intelpost (INTELPOST) international post (computerized postal service)
intelsat international telecommunications satellite
Intelsat International Telecommunications Satellite Organization
INTELSAT International Telecommunications Satellite Organization
inteltng intelligence training
INTEM Instituto Interamericano de Educacion Musical (Spanish—Inter-American Institute of Musical Education)
Intend Intendente (Spanish—manager, police commissioner, provincial governor, superintendent, supervisor)
intens intensive
inter intercalation; interest; intermediate; interrogation
inter (Latin prefix—among or between)—interborough, international
Interarmco International Armament Corporation
INTERASMA Association Internationale d'Asthmologie (French—International Association for the Study of Asthma)
Interavia World Review of Aviation and Astronautics
Interchem Interchemical Corporation
intercom intercommunication system
interd interested
INTERDATA Interdata Computers
interdict. intelligence detection and interdiction countermea-

sures
interdisc interdisciplinary
INTER-EXPERT International Association of Experts
INTEREXPO International Expositions
interf interference
INTERFILM International Church Film Center
INTERFLORA International Florists (telegraphic service)
interg interesting
Interior US Department of the Interior
interj interjection
INTERMARC International Machine-Readable Catalog
Intermex International Mexican Bank
InterMilPol International Military Police (NATO)
intern. internal
internat international; internationalism; internationalist
International Date Line 180° longitude
INTERNOISE International Conference on Noise Control Engineering
interp interpolation
Interpace International Pipe and Ceramics
Interpen/IAB Intercontinental Penetration Force/International Anti-communist Brigade
INTERPHOTO Fédération Internationale des Négociants en Photo et Cinéma (French—International Federation of Photograph and Cinema Merchants)
Interpol International Criminal Police Commission
INTERPOL International Criminal Police Organization (Saint-Cloud, France)
interr interrogative
interrog interrogation; interrogative
INTERSTENO International Federation of Short Hand and Typewriting Stenographers
INTERTANKO International Association of Independent Tanker Owners
Intertel International Television
INTERTELL International Intelligence Legion
intertwangled intertwined + wangled
inter/w intersection with
intest intestinal; intestine
INTEXT International Textbook Company
intfc interference

intg interrogate; interrogator
inth intrathecal
Int Harv International Harvester
inti monetary unit of Peru
intif intifadah (Arabic—uprising)—Palestinian uprising against Israel
intip integrated information processing
INTIPS Integrated Information Processing System
Int J Mag International Journal of Magnetism
Int J Quantum Chem International Jounral of Quantum Chemistry
Int J Theor Phys International Journal of Theoretical Physics
intl international
intl comb. internal combustion
Intl Ctr Envir International Center for Environmental Research
Intl Film Bur International Film Bureau
Intl Law Reps International Law Reports
Intl Legal Mats International Legal Materials
Intl Review International Review Service
Intl Univs Pr International Universities Press
intmed intermediate
int med (Int Med) internal medicine
intmt intermittent
int. noct. inter noctem (Latin—during the night)
intns intransit
INTO Irish National Teachers' Organization
intops interdiction operations
Intourist Soviet Tourist Office
intox intoxicant; intoxicate; intoxicated; intoxication
Int Pap International Paper
intpr interpret; interpretation; interpreter
int qk fl interrupted quick-flashing light
intr intransitive; intruder; intrusion
intra (Latin prefix—inside or within)—intracranial, intrastate
INTRACO International Trading Company
intran input translator
intrans intransitive
in trans in transit
in trans. in transitu (Latin—in transit)
intransit intransitive

Int Rep Intelligence Report
Int Rev Internal Revenue
intrex information transfer complex
intrmt interment
intro introduce; introduced; introducing; introduction; introductory; introversion; introvert
introd introduction
introd introduzione (Italian—introduction)
intropta. introscripta (Latin—written within)
intro(s) introduction(s)
intrp interrupt(ion)
intrpt interpret(ation); interrupt(ion)
intrvlmtr intervalometer
intsf intensification; intensify
int sig interval signal
int std d international standard depth
Int Sum Intelligence Summary
INTU Interpool (container unit)
INTUC Indian National Trades Union Congress
intv independent television
intvlmtr intervalometer
intvw intravenous
Int Wildlife International Wildlife
inu internal navigation unit
INU Nauru Island (airport)
I Nuc E Institute of Nuclear Engineering
INUIDS Instituto de las Naciones Unidas para la Investigación del Desarrol Social (Spanish—United Nations Institute for the Investigation of Social Development)
inurn inurnment
InUS inside the United States
in ut. in utero (Latin—within the uterus)
inv invent; inventor; inventory; inversion; invert; inverter; investment; invoice
inv. invenit (Latin—he devised it)
Inv Inverness
INV Instituto Nacional de la Vivienda (Spanish—National Housing Institute)
inval invalid(ate)
invc invoice
invcd invoiced
invert. invertebrate
inves investigate; investigation; investigator
investig investigate; investigation; investigator
invest(s) investigation(s)
inv. et del. invenit et delineavit

(Latin—devised and drawn)
invic. invictus (Latin—unconquerable)—title of a poem by William Ernest Henley—*Invictus*
in vit. in vitro (Latin—within glass, within a test tube or other laboratory glass vessel)
in viv. in vivo (Latin—within a living body)
invol involuntary
invos in vivo optical spectroscopy
invt inventory
invtn invitation
invtrx inventrix
INWATS Inward Wide Area Telephone Service
INWR Imperial National Wildlife Refuge (Arizona); Iroquois National Wildlife Refuge (New York)
INX Inexco Oil (stock-exchange symbol)
INZP Index to New Zealand Periodicals
io ion engine; intraocular
io (IO) inverted original (12-tone)
i/o image/orthicon; in and/or over; inboard-out-board (motorboat engine); input/output; instead of
i & o input and output
Io ionium
IO India Office; Information Officer; Inspecting Office(r); Inspection Office(r); Intelligence Office(r); Intercept Office(r); Irish Office; Issuing Office(r)
I/O Inspection Order; Investigating Officer
ioa instrument-operating assembly; instrumentation-operating area
IOA Institutional Overlay zone; Intelligence Oversight Act; International Omega Association
IOAM Institute of Appliance Manufacturers
IOAT International Organization Against Trachoma
ioau input/output access unit
iob input/output buffer; internal operating budget
I o B Institute of Bakers; Institute of Bankers; Institute of Bookkeepers; Institute of Brewers; Institute of Builders
IOB Institute of Brewing; Intelligence Oversight Board (CIA)
IOBB Independent Order of B'nai B'rith

IOBC Indian Ocean Biological Center; International Organization for Biological Control of Noxious Animals and Plants
IOBI Institute of Bankers in Ireland
IOBS Institute of Bankers in Scotland
iobyte input/output byte
ioc initial operational capability; in our culture
i-o c input-output channel(s)
I o C Index on Censorship
IOC Indian Ocean Commission (Madagascar, Mauritius, and the Seychelles); Institute of Chemistry; Intergovernmental Oceanographic Commission; International Olympic Committee; Interstate Oil Compact
IOCA Interstate Oil Compounders Association
IOCC Interstate Oil Compact Commission
IOCI Interstate Organized Crime Index
ioco industry-owned contractor operator
iocs interoffice comment sheet
IOCS Input-Output Control System
IOCU International Office of Consumers Unions; International Organization of Consumer Unions
IOCV International Organization of Citrus Virologists
IOD Imperial Order of the Dragon
IODE Imperial Order of Daughters of the Empire
I o E International Office of Education; Isle of Ely
IOE Institute of Education; International Office of Epizootics; International Organization of Employers
ioem invert oil emulsion mud (oil well)
I o F Institute of Fuel
IOF Independent Order of Foresters; International Oceanographic Foundation; International Olympic Federation; International Olympic Foundation
iofb I only fire blanks (sterile male); intraocular foreign body
IOFC Indian Ocean Fishery Commission
IOFI International Organiza-

tion of the Flavor Industry
IOFSI Independent Order of the Free Sons of Israel
ioga industry-organized government-approved
IOGP Independent Oil and Gas Producers
IOGT International Order of Good Templars
ioh item(s) on hand
IOH Institute of Heraldry
ioi internal operating instruction
IOI Industrial Oxygen Incorporated; International Ocean Institute; Israel Office of Information
I o J Institute of Journalists
IOJ International Organization of Journalists
IOJD International Order of Job's Daughters
iol intraocular lens
I o L Institute of Librarians
IOL India Office Library (London); Interoffice Letter
iol('s) interocular lens(es)
IO Ltd Imperial Oil Limited
iom input/output multiplexer
IoM Isle of Man
IOM Institute for Organization Management; Institute of Medicine; Institute of Metallurgists; Institute of Metals
I.O.M. *Iovi Optimo Maximo* (Latin—Jove the Greatest Superlative)
IOMC International Organization for Medical Cooperation
IOME Institute of Marine Engineers
i/o media input-output media
IOMM & P International Organization of Masters, Mates and Pilots
IOMR International Offshore Multihull Rule
IOM SPC Isle of Man Steam Packet Company
IOMTR International Office for Motor Trades and Repairs
iomux input/output multiplexer
Ion Ionic
ION (pseudynymic initials— George Jacob Holyoake); Institute of Navigation
IONDS Integrated Operational Nuclear Detection System
Ioanians Ionian Islands
iont in order not to
IOO Inspecting Ordnance Officer
IOOC International Olive Oil Council; Iranian Oil Operating Companies; Irish Organization of Celts

IOOF Independent Order of Odd Fellows

IOOTS International Organization of Old Testament Scholars

iop input-output processor; intraocular power; irrespective of percentage

i & op in-and-out processing

IoP Institute of Poverty; Isle of Palms; Isle of Pines

I o P Institute of Packaging; Institute of Petroleum; Institute of Physics; Institute of Plumbing; Institute of Printing

IOP Institute of Petroleum; Integrated Obstacle Plan; International Organization of Paleobotany; Iranian Oil Participants; Irish Organization of Papists

IOPAB International Organization for Pure and Applied Biophysics

IOPC Interagency Oil Policy Committee

IOPK Independent Order of Panamanian Kangaroos

IOP & LOA Independent Oil Producers and Land Owners Association

IOPS Input/Output Processing System

IOQ Institute of Quarrying

ior input/output register; item on request

I o R Institute of Roofing

IOR Independent Order of Rechabites (Quaker abstainers); International Ocean Rule; International Offshore Rules

IOR *Istituto per le Opere de Religione* (Italian—Institute for the Operation of Religion)—the Vatican Bank

IORD International Organization for Rural Development

IORM Improved Order of Red Men

IORS International Orders' Research Society

IoS Isles of Scilly; Isles of Shoals; Isles of the Sea

IOS Inspection Operation Sheet; Institute of Oceanographic Science; Institute of Oceanographic Services; International Organization for Standardization; Investors Overseas Services

IOSA Incorporated Oil Seed Association; Irish Offshore Services Association

IOSHD International Organization for the Study of Human Development

IOSM Independent Order of the Sons of Malta

IOSOT International Organization for the Study of the Old Testament

IoT Institute of Transport; Isle of Thanet

iota. [inbound-outbound traffic analysis; information overload testing aid

IOTA Institute of the Americas; Institute of Traffic Administration

IOTC International Originating Toll Center

IOT & E Interim Operational Test and Evaluation

Iotthy I'm only trying to help you

IOTTSG *International Oil Tanker and Terminal Safety Guide*

iou immediate operation use; industrial operations unit; inevitability of the unpredictable

I.O.U. I owe you

IO UBC Institute of Oceanography—University of British Columbia

I.O.U.s (plural of I.O.U.)

IOUSP *Instituto Oceanográfico da Universidade de São Paulo* (Portuguese—Oceanographic Institute of the University of São Paulo)

IOV *Instituto Oceanográfico de Valparaíso* (Spanish—Oceanographic Institute of Valparaíso)

IOVPT *Internationale Organisation für Vakuum-Physik und Technik* (German—International Organization for Vacuum Science and Technology)

IOVST International Organization for Vacuum Science and Technology

iow in other words

IoW Isle of Wight

I o W Isle of Wight

IOW Institute of Welding

iox instructional objectives exchange

IOY Iron Ore Year

IOZV *Internationale Organisation für Zivilverteidung* (German—International Organization for Civil Defense)

ip identification point; impact predictor; improvement purchase; incentive pay; india paper; induced polarization; industrial photographer; in-

dustrial photography; information provider; initial phrase; initial point; injured person; innings pitched; input primary; installment plan-(ning); integer programming; intermediate pressure; iron pipe; plate current (symbol)

ip (IP) imposter phenomenon; insurance payment

i/p input

i & p indexed and paged

iP *in Preussen* (German—in Prussia)

Ip Ipanema

I£ Israeli pound

IP Imperial Preference; *Institut Pasteur* (French—Pasteur Institute); Institute of Petroleum; Instructor Pilot; Insular Police; Interpool (container unit); *Isla de Pinos* (Spanish—Isle of Pines); plate current (symbol)

I-P Indian-Pacific (transcontinental train linking Perth, Australia with Sydney)

I & P Island and Peninsular (development bank)

I & P *Izvestia* and *Pravda* (Russian—*News* and *Truth*)

ipa including particular average; initial perceptual alphabet; intermediate power amplifier; internal power amplifier; international phonetic alphabet (IPA)

ipa (IPA) isopropanol; isopropyl alcohol

IPA Illinois Principals Association; Independent Petroleum Association; Independent Practice Association; Independent Publishers Association; Independent Publishers of Australia; Individual Practice Association; Institute for Physics of the Atmosphere; Institute of Propaganda Analysis; Institute of Public Administration; Institute of Public Affairs; International Police Association; International Peace Academy; International Pediatric Association; International Phonetic Alphabet; International Phonetic Association; International Platform Association; International Police Academy; International Police Archives (Manchester Central Library); International Police Association; International Press Association; International Psychoanalytical Association; International Publishers Association; (*see* TALA); Investment Partnership Association

IPA Information Please Almanac; International Pharmaceutical Abstracts

IPAA Independent Petroleum Association of America; Institute of Patent Attorneys of Australia; International Plan of Action on Aging (UN); International Prisoners Aid Association

ipac isopropyl acetate

IPAC Independent Petroleum Association of Canada; Iranian Pan-American Oil Company

IPACK International Packaging Material Suppliers Association

IPACS Integrated-Power/Attitude-Control System

IPAD Integrated Program for Aerospace Vehicle Design

IPAI Information Processing Association of Israel; International Primary Aluminum Institute

IPAR Institute of Personality Assessment and Research

IPARA International Publishers Advertising Representatives Association

IPARS International Programmed Airline Reservation System

IPAT Institute for Personality and Ability Testing

ipb illustrated parts breakdown

IPB International Peace Bureau; Islamic Party of Britain

ip & be initial program and budget estimate

ipbm (IPBM) interplanetary ballistic missile

IPBMM International Permanent Bureau of Motor Manufacturers

ipc industrial process control; isopropyl carbanilate

IPC Illinois Power Company; Industrial Process Control; Industrial Property Committee; Institute of Paper Chemistry; Institute of Pastoral Care; Institute of Printed Circuits; Integrated Programme for Commodities; Intelligence Priorities Committee (CIA); Inter-African Phytosanitary Commission; International Pacific Corporation; International Packings Corporation; International Paper

Chemists; International Petroleum Company; International Polar Commission; International Poplar Commission; Iraq Petroleum Company; Isopropyl Carbanilate

IPCA Industrial Pest Control Association

IPCAIL International Pacific Corporation Australian Investments Limited

IPCC Intergovernmental Panel on Climate Change (UN)

ipce independent parametric cost estimate

IPCEA Insulated Power Cable Engineers Association

IPCI International Potato Chip Institute

IPCL Institut du Pétrole des Carburants et Lubrifiants (French—Petroleum Institute for Motor Fuel and Lubricants)

IPCO International Paper Company

IPCPA Institute of Private Clinical Psychologists of Australia

IPCR Institute of Physical and Chemical Research

IPCS Institution of Professional Civil Servants; Integrated Propulsion Control System; International Peace Corps Secretariat

IPCU Intensive Psychiatric Care Unit

ip cyl intermediate-pressure cylinder

ipd individual package delivery; insertion phase delay

IPD Institute for Professional Development; Institute of Professional Designers

IPD In Praesentia Dominorum (Latin—In the Presence of the Lords)

IPDA International Periodical Distributor's Association

IPDC International Program for the Development of Communication

I pd cash I paid cash

ipe industrial plant equipment; interpret parity error

IPE Institution of Plant Engineers; Institute of Production Engineers

IPE Instituto de Providência do Estado (Portuguese—State Loan Institute); *International Petroleum Encyclopedia*

IPEC International Petroleum Exploration Company; Inter-

state Parcel Express Company

ipecac ipecacuanha

IPEU International Photo Engravers' Union

ipf initial production facilities

IPF Irish Printing Federation

IPFA Institute of Public Finance Accountants

IPFC Indo-Pacific Fisheries Council

ipfm impact form; integral pulse frequency modulation

ipg immediate participation guarantee (insurance plan)

IPG Income Property Group; Independent Publishers' Group; Information Policy Group

IPGA Illinois Personnel and Guidance Association; Iowa Personnel and Guidance Association

IPGCU International Printing and Graphic Communications Union

IPGH Instituto Panamericano de Geografía e Historia (Spanish—Pan-American Institute of Geography and History)

iph impressions per hour; inches per hour; interphalangeal

IPHC International Pacific Halibut Commission

IPHE Institute of Public Health Engineers

ipi individually planned instruction; interior point intermodal

i.p.i. in partibus infidelium (Latin—in the region of unbelievers)

IPI Industrial Production Index; Institute of Poultry Industries; International Patent Institute; International Press Institute

IPI Intelligence Publications Index (DIA)

IPICS Initial Production and Information Control System

IPIECA International Petroleum Industry Environmental Conservation Association

IPIP Information Processing Improvement Program

IPIR Initial Photographic Interpretation Report (USAF); Institute for Public Interest Representation; Integrated Personnel Information Report(s)

IPISD Interservice Procedures for Instructional Systems De-

velopment
IPKF Indian Peace-Keeping Force
IPKI Ikatan Pendukung Kemerdekaan Indonesia (Indonesian Malay—Upholders of Indonesia's Independence)
ipl information program loading; initial program load(er)
ipl (IPL) information processing language
IPL Illustrated Parts List; Integrated Parts List; Italian Pacific Line
I Plant Eng Institution of Plant Engineers
IPLE Institute for Political/Legal Education
IPLGY Institute for the Protection of Lesbian and Gay Youth
ip log(ging) induced polarization log(ging)
ipm impulses per minute; inches per minute; inches per month; incidental phase modulation; interruptions per minute; inventory policy model
ipm (IPM) integrated pest management
IPM Institute of Personnel Management; Institute of Police Management; Institute for Police Management; Integrated Post Management
IPMA International Personnel Management Association
ipmin inches per minute
IPMP Industrial Plant Modernization Program
ipms impact predictor monitor set
IPMS International Polar Motion Service
ipn inspection progress notification
ipo initial public offer(ing)(s)
ipo (IPO) input processing output
IPO Installation Production Order; International Projects Office (NATO); Israel Philharmonic Orchestra
IPO Instituut voor Perceptie Onderzoek (Dutch—Institute for Perception Research)
ipod initial phase of ocean drilling
IPOD Interstate Project on Dissemination
IPOEE Institution of Post Office Electrical Engineers
IPOH Instituto Panamericano de Geografía e Historia (Spanish—Panamerican Insti-

tute of Geography and History)
ipo's intellectual property owners
IPOT Imperial Philharmonic Orchestra of Tokyo
ipp imaging photopolarimeter; impact prediction point; india paper proof(s); intrapleural pressure
Ipp Ippolito
IPP Immediate Past President; Islamic People's Party; Ivan Petrovich Pavlov
ippa inspection, palpitation, percussion, auscultation
IPPA International Planned Parenthood Association
IPPAU International Printing Pressmen and Assistants' Union
ippb intermittent positive-pressure breathing
ippb/i intermittent positive pressure breathing/inspiratory
IPPF International Penal and Penitentiary Foundation
IPPJ Institute of Plasma Physics—Japan
IPPL International Primate Protection League
IPPNW International Physicians for the Prevention of Nuclear War
IPPP Industrial Property Policy Program
IPPPE Institute on Public Policy and Private Enterprise
ippr intermittent positive pressure respiration
IPPTA Indian Pulp and Paper Technical Association
IPPTT Internationale du Personnel des Postes, Télégraphes et Téléphones (French—International Postal, Telegraph, and Telephone Personnel)
ippv intermittent positive pressure ventilation
ipq intimacy potential quotient
IPQ International Petroleum Quarterly
ipr inches per revolution; inflow performance relationship (oil well); interpersonal process recall
IPR Individual Pay Record; Industrial Public Relations; Institute of Pacific Relations; Institute for Philosophical Research; Institute for Public Representation
IPR International Public Relations
IPRA International Profes-

sional Rodeo Association; International Public Relations Association
IPRC Institute of Puerto Rican Culture
IPRE Incorporated Practitioners in Radio and Electronics
IPRO International Patent Research Office
I Prod Eng Institute of Production Engineers
IPRR Integrated Personnel Requirement Report
ips inches per second; interceptor pilot simulator; interruptions per second; iron pipe size
Ips Ipswich
IPS Incremental Purchasing System; Indian Police Service; Industrial Planning Specification; Industrial Promotion Service; Inertial Positioning System; Information Processing Society; Institute of Pacific Studies; Institute for Policy Studies; Institute of Population Studies (Japan); Institute of Private Secretaries; Institute of Public Safety; Institute of Public Supplies; Integrated Publishing System; Intensive Probation Supervision; International Phenomenological Society; International Pipe Standard; Interpretive Programming System; Introductory Physical Science; Ionospheric Prediction Service; Israel Prison Service
IPS Instituto Poligrafico dello Stato (Italian—State Printing and Stationery office)
IPSA Independent Passenger Steamship Association; Independent Pool Service Association; Independent Postal System of America; International Political Science Association
IPSB Institute of Psycho-Structural Balancing
IPSC International Pacific Salmon Committee
IPSCE Inventory of Psychic and Somatic Complaints in the Elderly
IPSF International Pharmacy Students Federation; International Piano Symphony Foundation
IPSFC International Pacific Salmon Fisheries Commission
IPSJ Information Processing

Society of Japan
IPSM Institute of Purchasing and Supply Management
ipsp inhibitory postsynaptic potential
IPSP Industrial Personnel Security Program
IPSS International Packet Switching Service; International Pilot Study of Schizophrenia
IPSSB International Processing Systems Standards Board
ipt indexed, paged, titled; internal pipe thread
IPT Initial Production Test (USA); Institute of Photographic Technology; International Planning Team (NATO)
IPT Instituto Panameño de Turismo (Spanish—Panamanian Institute of Tourism)
IPTA International Patent and Trademark Association
IPTCCS Integrated Pipeline Transportation and Coal-Cleaning System
ipth (IPTH) immunoreactive parathyroid hormone
ipto independent power takeoff
IPTPA International Professional Tennis Players' Association
ipts international practical temperature scale
IPTS Improved Programmer Test Section; International Practical Temperature Scale
IPTT Internationale du Personnel des Postes, Télégraphes et Téléphones (French—International Postal, Telegraph, and Telephone Personnel)
ipu input preparation unit
IPU Institute for Public Understanding; International Paleontological Union; Inter-Parliamentary Union
ipv inactivated poliomyelitis vaccine; infectious pustular vaginitis; infectious pustular vulvovaginitis
ipv in plaats van (Dutch—in place of)
IPW interrogation prisoner of war
ipy inches penetration per year; inches per year
IPY International Polar Year
IPZ Investment Promotion Zone
IPZE Istituto Poligrafico dello Stato (Italian—State Poligraphic Institute)—issues paper

money and stamps
i.q. idem quod (Latin—the same as)
Iq Iraq
IQ Import Quota; intelligence quotient
I.Q. I Quit (smoking)
I of Q Institute of Quarrying
IQA Institute of Quality Assurance
IQCA Irish Quality Control Association
IQCT Institute for Quality Control Training
i.q.e.d. id quod erat demonstrandum (Latin—that which was to be proved)
iqf instant quick frozen
IQHL Institute for Quality in Human Life
IQI Instructional Quality Inventory
I Qk interrupted quick flashing (light); interrupted quick (light)
iqmf image-quality merit function
iqrp interactive query and report processor
iq & s iron, quinine, and strychnine
IQS Institute of Quality Surveyors; Institute of Quantity Surveyors
IQSA Institute of Quantity Surveyors of Australia
IQSY International Quiet Sun Year (1964–1965)
Iqu Iquique
ir ice on runway; incidence rate; information retrieval; infrared; inland revenue; inside radius; inside right; instantaneous relay; instrument rating; instrument reading; insulation resistance; intelligence resource(s); internal resistance; interrogator-responder
ir (IR) inflation rate
i-r infra-red
i/r interchangeability and replaceability
i & r information and retrieval; intelligence and reconnaissance; interchangeability and replaceability
ir irisch (German—Irish)
i R im Ruhestand (German—in retirement)
Ir Iran; Irania; Ireland; iridium; Irish
Ir Iers (Dutch)—Ireland, Irish-(man) (woman)
IR Index Register; Indian Reservation; Indian Reserve; Industrial Relations; Informal

Report; Information Request; Inspection Rejection; Inspector's Report; Institutional Research; Intelligence Report; Internal Revenue; Invention Report; Investigation Record
I-R Ingersoll-Rand
I & R Initiative and Referendum; Intelligence and Reconnaissance
ira independent retirement account (IRA)
ira (IRA) immunoregulatory alpha globulin
Ira Dr Ira D Levine; Iraq
IRA Indian Rights Association; Individual Retirement Account; Individual Retirement Arrangement; Institute of Registered Architects; Intercollegiate Rowing Association; International Racquetball Association; International Reading Association; International Recreation Association; Iranian Airways; Irish Republican Army; Israel Railway Administration
IRAA Independent Refiners Association of America
IRAB Institute for Research in Animal Behavior
irac mnemonic abbreviation helpful in making legal or logical presentations; letters stand for issue, rule, application, and conclusion
IRAC Indochina Refugee Assistance Program; Industrial Relations Advisory Committee; Intelligence Resources Advisory Committee; Interdepartmental Radio Advisory Committee; Interfraternity Research and Administrative Council
iracq instrumentation radar acquisition
irad independent research and development
IRAD Institute for Research on Animal Diseases
iraf interferogram requirements and analysis funnel
iran inspect and repair as necessary
Iran formerly Persia; Imperial Government of Iran (ancient Asian nation), *Keshvaré Shahanshahiyé Iran*
IRANAIR Iran National Airlines
Iran(ian) Persia(n)
IRANOR Instituto Nacional de Racionalización y Normali-

zación (Spanish—National Institute of Rationalization and Standards)
IRAP Indochinese Refugees Assistance Program
Iraq Republic of Iraq (Persian Gulf country formerly called Mesopotamia), *al Jumhouriya al 'Iraqia*
Iraquia Iraq (Mesopotamia)
IR/AR Inspector's Report/Action Request
iras (IRAS) infrared astronomical satellite; infrared-measuring astronomical satellite
IRAs Individual Retirement Accounts; Irish Republican Army members or supporters and sympathizers
IRASA International Radio Air Safety Association
IRASE Institute of Refrigeration and Air Conditioning Service Engineers
iraser infrared amplification by stimulated emission of radiation
irb (IRB) in-shore rescue boat
IRB Indiana Rating Bureau; Industrial Relations Bureau; Industrial Review Board; Insurance Rating Board; International Resources Bank; Irish Republican Brotherhood
IRBDC Insurance Rating Bureau of the District of Columbia
IRBEL Indexed References to Biomedical Engineering Literature
irbm (IRBM) intermediate range ballistic missile
irc infrared countermeasures; international reply coupon; item responsibility code
IRC Immigration Restriction Council; Indonesian Red Cross; Industrial Recreation Council; Industrial Relations Center; Industrial Relations Committee; Industrial Relations Council; Industrial Reorganization Corporation; Inebriate Reception Center; Information Resources Center; Institutional Research Council; Instructional Resources Center; Internal Revenue Code; International Railways of Central America (stock exchange symbol); International Rainwear Council; International Red Cross; International Relations Committee; International Relief

Committee; International Rescue Committee; International Research Council; International Resistance Company; International Rice Commission
IRCA Immigration Reform and Control Act; International Railway Congress Association; International Railways of Central America
IRCAM Institut de Recherche et de Coordination Acoustique-Musique (French—Institute of Research and Acoustic-Music Coordination)
IRCAR International Reference Center for Abroation Research
irccd infrared charge-coupled device
IRCD Information Retrieval Center on the Disadvantaged
i-r charts infrared correlation charts
ircm infrared countermeasures
IRCO Industrial Rustproof Company
IRCP International Commission on Radiological Protection
IRCs Inebriate Reception Centers (where nonviolent abusers of alcohol and other drugs accept coffee and counseling in lieu of being jailed)—pilot facility in San Diego, California
IRCS International Research Communications System
ird (IRD) internal research and development; international resource development
ir & d international research and development
IRD Institute on Religion and Democracy (anti-communist); Institute of Reading Development; *Instituto Rubén Darío*; International Resource Development
IR & D International Research and Development
IRDA Industrial Research and Development Authority
IRDC International Research Development Centre (Canadian)
ir & dg industrial research and development grants
IRDI Indian Resources Development and Internship
irdm illuminated runway distance marker
IRDN Illinois Resource and

Dissemination Network
IRDOE Institute for Research and Development in Occupational Education
irdome infrared dome
irds idiopathic respiratory-distress syndrome
irdu infrared detection unit
Ire Ireland
IRE Institute of Radio Engineers; Institute for Responsive Education; Investigative Reporters and Editors
IREC International Real Estate Corporation (Singapore); Irrigation Research and Extension Commission
IREE Institute of Radio and Electronic Engineers
IREF International Real Estate Federation
Ireland Irish Republic (North Atlantic island nation), *Eire*
IREM Institute of Real Estate Management
irene indicating random electronic numbering equipment
Ir Eng Irish English
IREQ Institute of Research—Québec
irer infrared extra rapid
IREX International Research and Exchanges (New York-based board arranging scholarly interchanges between the United States and the USSR and other
irf instrument reliability factor; interrogation repetition frequency
IRF International Road Federation; International Rowing Federation
IRFAA International Rescue and First Aid Association
IRFB Internationaler Regenmantelfabrikantenverband (German—International Rain-wear Fabric Council)
IRFC Ingersoll-Rand Finance Corporation
IRFM Industrias Reunidas Francisco Matarazzo (Spanish—Francisco Matarazzo's Reunited Industries)
IRFU Iriah Rugby Football Union
irg interrecord gap
IRG Interdepartmental Regional Group
IRG Internationale des Résistans à la Guerre (French—War Resisters' International)
Ir Gael Irish Gaelic
IRGDLP International Research Group on Drug Legis-

lation and Programs (Geneva, Switzerland)

irgl immunoreactive glucagon

IRGRD International Research Group on Refuse Disposal

IRH Internationalen Roten Hilfe (German—International Red Aid)—Red Fighting Fund

irha injured as a result of hostile action

irhd international rubber hardness degrees

iri immunoreactive insulin

Iri Irina

IRI Industrial Reconstruction Institute; Industrial Research Institute; Institute of the Rubber Industry; Islamic Republic of Iran

IRIA Infrared Information and Analysis

IRIC Inter-Regional Insurance Conference

IRICA Industrial Research Institute for Central America

iricbm (IRICBM) intermediate-range intercontinental ballistic missile

irid iridescent

IRIG Inter-Range Instrumentation Group

IRIR Interchangeability Replaceability Information Report

iris. infrared interferometer spectrometer

IRIS Industrial Research and Information Service; Infrared Information Symposia; Integrated Reconaissance Intelligence System

Irish Irish Gaelic

Irish FP Irish Fishing Port (registration symbols displayed on the bows of fishing vessels)

Irish VR Irish Vehicle Registration (symbols on automotive vehicle licenses)

IRJC Indian River Junior College

irl information retrieval language

Irl Ireland; *Irlanda* (Italian, Portuguese, Spanish—Ireland)

Irl Irlande (French—Ireland)

IRL Illustrations Requirements List

IRLA Independent Research Library Association

Irland Irlandais (French—Irish)

IRLC Illinois Regional Library Council

IRLCS International Red Locust Control Service

IRLI Immigration Reform Law Institute

IRLS Interrogation Recording Location System

irm infrared measurement; innate release mechanism; intermediate range monitor

IRM Improved Risk Mutuals; Information Resource Management; Islamic Republic of Mauritania

IRM (NYU) Institute of Rehabilitation Medicine (New York University)

irma information revision and manuscript assembly; integrated revenue and marketing applications

IRMA Individual Reverse Mortgage Account

IRMMH Institute of Research into Mental and Multiple Handicaps

IRMP Intermountain Regional Medical Program

IRMPC Industrial Raw Materials Planning Committee (NATO)

IRMRA Indian Rubber Manufacturers Research Association

IRMS Imperial Russian Musical Society; Information Retrieval and Management System

IRN Independent Radio News

IRNA Islamic Republic News Agency

IRNP Isle Royale National Park (Michigan)

iro in rear of

IRO Industrial Recycling Organization; Industrial Relations Office(r); Information Resource Office(r); Inland Revenue Office(r); Internal Revenue Office(r); International Refugee Organization; International Relief Organization

IRO-ALA International Relations Office—American Library Association

IROC International Race of Champions (autos)

irod instantaneous readout detector

iron. ironic(al)

Iron Curtain barrier between eastern and western Europe (1946–1989)

iron pyrites sulfide of iron

iron rust iron oxide

IROPCO Iranian Offshore Petroleum Company

iros ipsilateral routing of signal

IROS Increase Reliability of Operational Systems

irp initial receiving point

ir£ Irish pound

IRP Individualized Reading Program; Information Resources Press

IRPA International Radiation Protection Association

irpm individual risk premium modification

IRPS Institute of Reconstructive Plastic Surgery (NYU); International Religious Press Service (Vatican City)

irr infrared rays; infrared reflectance; internal rate of return; irredeemable; irregular-(ity)

irr (IRR) integral rocket ramjet; internal rate of return

IRR Individual Ready Reserves; Institute of Race Relations

IRRA Industrial Relations Research Association

IRRC Investor Responsibility Research Center

irrd international road research documentation

IRRD International Road Research Documentation

IRRDB International Rubber Research and Development Board

irreg irregular

irres irrespective

irrev irrevocable

irrg irregular

irrgty irregularity

irrgy irregularly

IRRI International Rice Research Institute

irrig irrigation

IRRN Illinois Research and Reference Center (libraries)

IRRP Icefield Ranges Research Project

irr/ssm (IRR/SSM) integral rocket ramjet/surface-to-surface missile

IRRT International Relations Round Table

irs incremental range summary; independent rear suspension

IRs Inspector's Reports

IRs Irish Reports

IRS Indian Register of Shipping; Industrial Rubber Sales; Ineligible Reserve Section; Information Retrieval System; Internal Revenue Service; International Re-

cruiting Service; International Referral System; International Rorschach Society; Irrigation Research Station
I & RS Information and Research Services
IRS-1A Indian-built remote-sensing satellite
IRSA Industrial Relations Society of Australia
IRSE Institution of Railway Signal Engineers; International Reactor Safety Evaluation
IRSF Inland Revenue Staff Federation
IRSG International Rubber Study Group
IRSID *Institut des Recherches de la Sidérurgie Française* (French—French Steel Research Institute)
IRSNB *Institut Royal des Sciences Naturelles de Belgique* (French—Royal Belgian Institute of Natural Sciences)
IRSP Irish Republican Socialist Party
IRSS Instrumentation Range Safety System
irt infrared temperature; infrared tracker; intermediate range technology; interrogator responder transponder
IRT Institute for Rapid Transit; Institute of Reprographic Technology; Interborough Rapid Transit (subway system)
IRT *Industria Radio y Televisión* (Spanish—Radio and Television Industry)
IRTA Illinois Retired Teachers Association
IRTAC International Round Table for the Advancement of Counseling
IRT Corp Instrumentation/Research/Technology Corporation
IRTE Institute of Road Transport Engineers
IRTP Integrated Reliability Test Program
irts infrared target seeker
IRTS International Radio and Television Society
IRTU International Railway Temperance Union
IRTWG Inter-Range Telemetering Working Group
iru industrial rehabilitation unit; inertial reference unit; infrared unit; international radium unit; international rat unit

IRU International Road Transport Union
irupt interrupt
iruptd interrupted
iruptn interruption
iruptng interrupting
irv inspiratory reserve volume
Irv Irvin; Irvine; Irving; Irwin
Irve Irving
Irving Irving Trust Company; Sir Henry Irving; Washington Irving
Irw Irwin
IRW Iowa Reformatory for Women
IRWC International Registry of World Citizens
irwl interchangeability/replaceability working list
is his
is. ingot sheet; integrally stiffened; intercoastal space; interim storage; internal shield; international status; interval signal; island; isle
i&s installation and service; investigation and suspension; iron and steel
i & s inspection and security; inspection and survey; interchangeability and substitutability
Is Islam; Islamic; Island; Isle; Israel; Israeli
Is Israeli pound
Is *Isaías* (Spanish—Isaiah)
I(s) *isla(s)* Spanish—island(s)
*I*ˢ *Îles* (French—islands); *Ilhas* (Portuguese—islands); *Islas* (Spanish—islands)
IS Identification Section (NYPD); Igor Stravinsky; Indian Summer (freeboard marking); Information Service; Instruction(s) to Ship; Irish Society
I of S Institute of Sound; Isle of Skye
IS 201 Intermediate School 201 (for example)
isa international standard atmosphere
Isa Isaiah, The Book of the Prophet
Isa *Indonesia* (Spanish—Indonesia); *Isaiah*
ISA Incest Survivors' Anonymous; Independent Showmen of America; Institute of Strategic Affairs; Institution of Surveyors—Australia; Instrument Society of America; Insulating Siding Association; Intelligence Support Activity (US Army); Intermediate Service Agency; Internal Security Act; International

Schools Association; International Scientific Affairs; International Seabed Authority; International Security Affairs; International Security Assistance; International Sign Association; International Silk Association; International Society of Appraisers; International Sociological Association; International Standards Association; International Student Association; International Sugar Agreement; Israel Space Agency
ISA *Information Science Abstracts; Irregular Serials and Annuals*
ISAA Institute of Shops Acts Administration
Isab Isabella
ISAB Institute for the Study of Animal Behavior
ISAB *Internationaler Studenbund* (German—International Student Society)
ISABPS Integrated Submarine Automated Broadcast Processing System
ISAC International Security Affairs Committee
ISACP Italian Society of Authors, Composers, and Publishers
ISAD Information Science and Automation Division (ALA)
ISADPM International Society for the Abolition of Data-Processing Machines
ISAE *Internacia Scienca Asocio Esperantista* (International Esperantist Scientific Association)
IsAF Israeli Air Force
isaf black intermediate superabrasive furnace black
ISAFP Intelligence Service of the Philippine Armed Forces
ISAGA International Simulation and Gaming Association
ISAHM International Society for Animal and Human Mycology
ISALPA Incorporated Society of Auctioneers and Landed Property Agents
ISAM Independent School Association of Massachusetts; Indexed Sequential Access Method; Institute for Studies in American Music; Integrated Switching and Multiplexing
ISAP Institute for the Study of Animal Problems
ISAPC Incorporated Society of

Authors, Playwrights, and Composers
ISAPM International Shipowners Association of Peninsular Malaysia
isar information storage and retrieval
ISAR International Society for Astrological Research
isarc installation shipping and receiving capability
ISAS Industrial Sales and Service; Institute of Social and Administrative Studies; Institute of Space and Aeronautical Science; Isotopic Source Assay System
ISAT International Society of Analytical Trilogy
ISAW International Society of Aviation Writers
isb independent sideband; intermediate sideband
ISB International Society of Biometeorology
ISBA Incorporated Society of British Advertisers; Indiana School Boards Association
ISBB International Society for Bioclimatology and Biometeorology
ISBD International Standard Book Description
ISBD(M) International Standard Bibliographic Description for Monographic Publications
ISBD(NBM) International Standard Bibliographic Description for Non-Book Materials
ISBD(S) International Standard Bibliographic Description for Serial Publications
ISBE International Society for Business Education
ISBN International Standard Book Number
ISBNA International Standard Book Numbering Agency
ISBOA Idaho School Business Officials Association
ISBP International Society for Biochemical Pharmacology
isbr interior salt basin region
ISBS Icelandic State Broadcasting Service; International Scholarly Book Services
isc intermediate slack compensation; interstate commerce; intrasite cabling; item status code; item status coding; in such case
ISC Icelandic Steamship Company; Idaho State College; Imperial Service College;

Imperial Staff College; Indian Staff Corps; Indiana State College; Indoor Sports Club; Industrial Security Commission; Institute for the Study of Conflict; Inter-American Society of Cardiology; International Salt Company; International Science Center; International Sericultural Commission; International Society of Cardiology; International Softball Congress; International Statistical Classification; International Sugar Council; International Supreme Council (World Masons); Interseas Shipping Corporation; Interservice Sports Council; Interstate Sanitation Commission
IS&C International Systems and Controls
ISCA International Sailing Craft Association; International Senior Citizens Association
iscan inertialess steerable communication antenna
ISCB International Society for Cell Biology
ISCC Inter-Society Color Council
ISCDD International Scheme for the Coordination of Dairy Development
ISCE International Society for Christian Endeavor
ISCED International Standard Classification of Education
ISCEH International Society for Clinical and Experimental Hypnosis
ISCERG International Society for Clinical Electroretinography
ISCET International Society of Certified Electronics Technicians
ISCII International Standard Code for Information Interchange
ISCM International Society for Contemporary Music
ISCO Independent Schools Careers Organization; International Standard Classification of Occupations
ISCOR Iron and Steel Industrial Corporation (South Africa)
ISCOS Institute for Security and Cooperation in Outer Space
ISCP International Society of Clinical Pathology

ISCPET Illinois Statewide Curriculum Study Center in the Preparation of Secondary School English Teachers
ISCRP International Society of City and Regional Planners
ISCS Information Service Computer System; Intermediate Science Curriculum Study
ISCTP International Study Commission for Traffic Police
isd induction system deposit; inhibited sexual desire; installation start date; instructional systems development; integrated symbolic debugger
isd (ISD) inhibited sexual desire (asexuality); international standard depth
ISD Independent School District; Information Systems Design; Institute for the Study of Diplomacy, Georgetown University; Instructional Systems Design; Instructional Systems Development; Intermediate School District; Internal Security Department; Internal Security Division (U.S. Dept of Justice); International Subscriber Dialing
ISDAIC International Staff Disaster Assistance Information Coordinator (NATO)
ISDD Institute for the Study of Drug Dependence (London, England)
ISDI International Social Development Institute
ISDN Integrated Services Digital Network
ISDO International Staff Duty Officer (NATO)
ISDRA International Sled Dog Racing Association
ISDS Inadvertent Separation Destruct System; International Serials Data System; International Sheep Dog Society
ISDSI Insulated Steel Door Systems Institute
ISDS/IC International Center of the International Series Data System (UNESCO)
ise integral square error
ISE Indian Service of Engineers; Institute for Services to Education; Institute for Sex Education; Institute of Social Ethics; Institute of Space Engineering; Institution of Sanitary Engineers; Institution of Structural Engi-

neers; International Stock Exchange; Irish School of Ecumenics
ISE Instituto de Seguros del Estado (Spanish—Institute of State Insurance)
ISEA Industrial Safety Equipment Association; Iowa State Education Association
ISEAS Institute of Southeast Asian Studies
ISEE International Sun-Earth Explorer (NASA/ESRO)
ISEEP Infrared-Sensitive Element Evaluation Program
ISEF International Science and Engineering Fair
ISELS Institute of Society, Ethics, and Life Sciences
ISEP Instructional Scientific Equipment Program; International Summer Education Project; International Society for Educational Planners; Interservice Experiments Program
isepc installation specification; insulation specification
iseq input sequence check(ing)
ISES International Ship Electric Service Association; International Society of Explosives Specialists; International Solar Energy Society
ISETU International Secretariat of Entertainment Trade Unions
ISEU International Stereotypers' and Electrotypers' Union
isf intermittent storage flood-(ing); interstitial fluid
ISF Industrial Space Facility; Intermediate-Scale Facilities; International Science Foundation; International Shipping Federation; International Society for Fat Research; International Softball Federation; International Solidarity Fund
ISFA Institute of Shipping and Forwarding Agents; Intercoastal Steamship Freight Association; International Scientific Film Association
ISFL International Scientific Film Library
ISFMS Indexed Sequential File Management System
ISFR Institute for the Study of Fatigue and Reliability
ISFSC International Society of Free Space Colonizers
isfsi independent spent-fuel storage installation
ISFSI International Society of Fire Service Instructors

isg imperial standard gallon
ISGE International Society of Gastroenterology
ISGM Isabella Stewart Gardner Museum
ISGS International Society for General Semantics
ISGW International Society of Girl Watchers
Ish Isham; Ishbel; Ishmael
ISH International Society of Hematology
ISHAM International Society for Human and Animal Mycology
ISHI Institute for the Study of Human Issues
ISHL Illinois Social Hygiene League
ISHR International Society for Human Rights
ISHS International Society for Horticultural Science
isi industrial standard item; internally specified index(ing); interstimulus interval
ISI Indian Standards Institution; Information Sciences Institute (USC); Institute for Scientific Information; Intercollegiate Society of Individualists; Intercollegiate Studies Institute; International Statistical Institute; Iron and Steel Institute
ISIB Inter-Services Ionospheric Bureau
isic (ISIC) immediate superior in command
ISIC International Standard Industrial Classification; International Student Identity Card
ISIM International Society of Internal Medicine
isip inertial system indication position
ISIP Iron and Steel Industry Profile Service
ISIR International Society for Invertebrate Reproduction
isirta I'm sorry, I'll read that again (BBC comedy)
isis (ISIS) ionospheric studies
ISIS Independent Schools Information Service; Individualized Science Instructional System; Instant Sales Indicator System; Institute of Scrap Iron and Steel; Integral Spar Inspection System; Integrated Scientific Information Service; Integrated Set of Information Systems; Integrated Ship Instrumentation System; Integrated Statistical Infor-

mation Service; International Satellites for Ionospheric Studies; International Science Information Service; International Shipping Information Services; International Species Identification System; Investigative Support Information System (FBI)
ISIT Institute for Studies in International Terrorism (SUNY)
ISIYM International Society of Industrial Yarn Manufacturers
ISJP International Society for Japanese Philately
isk insert storage key
ISK Isambard Kingdom Brunel
ISK Internationale Seidenbau Kommission (German—International Sericulture Commission)
ISKC International Society for Krishna Consciousness
iskr identification, station keeping, and rendezvous
isl intermediate-stage letter; island
isl (ISL) intersatellite link
isl islandsk (Dano-Norwegian—Icelandic); *isländisch* (German—Icelandic)
Isl island (Norwegian—Iceland); *Islanda* (Italian—Iceland); *Islandia* (Spanish—Iceland); *Islândia* (Portuguese—Iceland)
ISL Iceland Steamship Company; Interseas Shipping Lines; Iranian Shipping Lines; Irish Shipping Limited
I S-L Instructor Sub-Lieutenant
ISLA International Survey Library Association
Isle Royale Isle Royale National Park, Michigan
ISLFD Incorporated Society of London Fashion Designers
ISLIC Israel Society of Special Libraries and Information Centers
isl of Lan islands of Langerhans
ISLLSS International Society for Labor Law and Social Security
Islm Rep Pak Islamic Republic of Pakistan
isln isolation
ISLRS Inactive Status List Reserve Section
isl's initial stock lists; islands
isls L islands of Langerhans
Islw Indian spring low water

ISLWF International Shoe and Leather Workers Federation

ism industrial, scientific, medical wave length; interpretive structural modeling; interstellar matter

ism (Latin suffix—condition or state)—capitalism, communism, rheumatism

ISM Imperial Service Medal; Incorporated Society of Musicians; Industrial Sugar Mills; Institute of Sales and Marketing; Institute of Sports Medicine; Institute of Supervisory Management; International Society for Musicology

ISMA International Superphosphate Manufacturers Association

ISME International Society of Musical Education

ISMEC Information Service in Mechanical Engineering

ISMH International Society of Medical Hydrology

ISMI Institute for the Study of Mental Images

ISMLS Interim Standard Microwave Landing System

ISMR Independent Snowmobile Medical Research (organization)

ISMRC Inter-Services Metallurgical Research Council

ISMS Inherently-Safe Mining System(s)

ISMUN International Student Movement for the United Nations

Is N (Sir) Isaac Newton

ISN International Society for Neurochemistry

ISNAC Inactive Ships Navy Custody

ISNP International Society of Naturopathic Physicians

isn't is not

iso isolate; isolation; isolator; isotope; isotopic; in spite of

iso (ISO) infrared space observatory

iso (Latin suffix—equal or like)—isometric, isotonic

ISO Imperial Service Order; Indianapolis Symphony Orchestra; Individual System Operation; Information Service(s) Office(r); Information Systems Office (Library of Congress); Insurance Service(s) Office(r); International Science Organization; International Standardization Organization; International

Standards Organization; International Sugar Organization; Irish Symphony Orchestra

ISO-4 international code for the abbreviation of periodical titles

ISO-833 international list of periodical title word abbreviations

isobu isobutyl

is/oc individual system/organization cost

ISOC Internal Security Operation Command

isochr isochronal

ISODOC International Center for Standards in Information and Documentation

isogone isogonal line

isogons isogonic lines

isol isolate; isolation

Isol Isolation; Isolator

isoln isolate; isolated; isolation

isolr isolationer

isom isometric(s)

ISOMATA Idlewild School of Music and the Arts; Idyllwild School of Music and Art

isomorph isomorphic(al)(ly); isomorphism

ison isolation network

ISONET International Standards Organization Network

ISOO Information Security Oversight Office

isordil isorbide dinitrate

ISORID International Information System on Research in Documentation

ISORT Interdisciplinary Student-Originated Research Training (NSF)

ISOS International Ship Operating Services

isot isotropic

iso wd isolation ward

isp intraspinal

Isp specific impulse (symbol)

ISP Idaho State Penitentiary; Imperial Smelting Process; Index of Social Position; Indiana State Police; Industrial Security Program; Institute of Social Psychiatry; Institute of Store Planners; Integrated Support Plan(ning); Interamerican Society of Psychology

ISP *Internationale des Services Publics* (French—Public Services International)

ISPA International Screen Publicity Association; International Small Printers' Association; International Soci-

ety for the Protection of Animals; International Sporting Press Association; International Squash Players Association

ISPC International Statistical Program Center (AID)

ISPE Institute and Society of Practitioners in Electrolysis

ISPEMA Industrial Safety Personal Equipment Manufacturers' Association

ISPHS International Society of Phonetic Sciences

ISPM International Solar Polar Mission; International Staff Planners Message (NATO)

ISPMEMO International Staff Planners Memo (NATO)

ISPO Instrumentation Ships' Project Office; International Society for Prosthetics and Orthotics; International Statistical Programs Office

ISPP Inter-Services Plastic Panel

isps international standard paper sizes

ISPS International Society of Phonetic Sciences

ISPT Initial Satisfactory Performance Test(ing)

isq in status quo

ISQA Israeli Society for Quality Assurance

isr information storage and retrieval

isr *Israel* (Portuguese or Spanish—Israel); *Israele* (Italian—Israel)

Isr Israel; Israel

Isr *Israel* (Portuguese or Spanish—Israel); *Israele* (Italian—Israel)

ISR Indian State Railways; Institute for Sex Research; Institute for Social Research; Institute of Surgical Research; International Sanitary Regulations; International Society of Radiology

IS & R Information Storage and Retrieval (system)

I & SR Institutional and Staff Relations

ISRAD Institute for Social Research and Development

Israel State of Israel (Middle Eastern country), *Medinat Israel* known as Judea Palestine before the Christian era

ISRB Idaho Surveying and Rating Bureau; Inter-Services Research Bureau

ISRC International Synthetic Rubber Company

ISRD International Society for the Rehabilitation of the Disabled
ISRF International Squash Rackets Federation
ISRI Israeli Shipping Research Institute
ISRR Institute of Social and Religious Research
ISRRT International Society of Radiographers and Radiological Technicians
ISRSA International Synthetic Rubber Safety Association
isru information search and retrieval unit
ISRU International Scientific Radio Union
iss ideal solidus structures; ionscattering spectroscopy; issue
iss (ISS) immune system suspected; ionospheric sounding satellite
ISS Industry Standard Specifications; Inertial Sensor System; Information Service Specialist; Inspection Surveillance Sheet; Institute for Socioeconomic Studies; Institute of Space Sciences; Institute of Space Studies; Institutional Staff Services; Integrated Start System; International School Service; International Seismological Summary; International Shoe Company (Stock Exchange Symbol); International Social Service; International Students Society; International Sunshine Society; Israeli Secret Service
ISSA International Social Security Association
ISSAS Interactive Structural Sizing and Analysis System
ISSB Inter-Services Security Board
ISSC Institute for the Study of Social Conflict; International Social Science Council
ISSCA Institute of Steel Service Centres of Australia
ISSCAAP International Standard Statistical Classification of Aquatic Animals and Plants
ISSCB International Society for Sandwich Construction and Bonding
ISSCO Integrated Software Systems Corporation
ISSCT International Society of Sugar Cane Technologists
ISSE Inter-Sun-Earth Explorer
ISSL Initial Spares Support List
ISSLIC Israel Societies of Special Libraries and Information Centers
ISSMFE International Society of Soil Mechanics and Foundation Engineering
ISSMIS Integrated Support Services Management Information System
ISSMS Integrated Support Services Management System (USA)
ISSN International Standard Serial Number
ISSOE Instructional Support System for Occupational Education
ISSOL International Society for the Study of the Origin of Life
issr information storage, selection, and retrieval
ISSS International Society for the Study of Symbols; International Society of Soil Science
ISSSC International Society for the Suppression of Savage Customs
ISSSTE *Instituto de Seguridad y Servicios Sociales de los Trabajadores* (Spanish—Workers' Institute of Social Security and Social Services)
ISST International Society of Skilled Trades
IS Standards International Safety Standards
ist insulin shock therapy; interstellar travel
is't is it
ist *istituto* (Italian—institute)
Ist Istanbul
IST Indian Standard Time; Industrial Steel and Tube (consolidated); In-Service Training; Institute of Science and Technology (University of Michigan); Institute on Strategic Trade (Washington, D.C.); Institute for the Study of Terrorism; International Society of Toxicology; Istanbul, Turkey (airport)
IST *International Steam Table*
IS & T *International Science and Technology*
ISTA Indiana State Teachers Association; Industrial Science and Technology Agency; International Seed Testing Association
Istan Istanbul
istar information storage translation and reproduction
ISTAT *Istituto Centrale di Statistica* (Italian—Central Institute of Statistics)
ISTB Interstate Tariff Bureau
ISTC Interdepartmental Screw Thread Committee; International Shade Tree Conference; Iron and Steel Trades Confederation
ISTD Imperial Society of Teachers of Dancing; Institute for the Study and Treatment of Delinquency; International Society of Tropical Dermatology
ISTEA Iron and Steel Trades Employers' Association
ISTEM Inter-Seminary Theological Education for Ministry
isth isthmian; isthmus
Isth Isthmiam; Isthmus
ISTI Iowa State Technical Institute
istim interchange of scientific and technical information in machine language
ISTM International Society for Testing Materials
ISTO Italian State Tourist Office
istom interstate transportation of obscene matter
I Struct E Institute of Structural Engineers
istse integral square time square error
ISU Idaho State University; International Seamen's Union; International Shooting Union; International Skating Union; International Society of Urology; International Space University; Iowa Southern Utilities; Iowa State University; Italian Service Unit; Southern Iowa Railway (railroad coding)
I-sub inhibitor substance
ISUM Intelligence Summary
ISUP Iowa State University Press
ISUST Iowa State University of Science and Technology
ISV Institute for the Study of Violence (Brandeis U); International Scientific Vocabulary
ISVA Incorporated Society of Valuers and Auctioneers
ISVR Institute of Sound and Vibration Research
ISVS International Secretariat for Volunteer Service
isw interstitial water
ISW Institute for Solid Wastes

ISWA International Science Writers Association; International Solid Wastes and Public Cleansing Association
ISWG Imperial Standard Wire Gauge
ISWM Institute of Solid Wastes Management
ISWNE International Society of Weekly Newspaper Editors
isy intrasynovial
isz increment and skip on zero
it Intermediate Technology; slang term for sex appeal
it. information theory; inspection tag; internal thread; international tolerance; inventory transfer; item; itemization(s); itemize(d); in transit
it. (IT) information technology; internal tank; intertuberous; transposed inversion (12-tone)
i/t intensity duration
it *italienisch* (German—Italian); *italiensk* (Dano-Norwegian—Italian); *item* (Spanish—item)
i.t. *in transitu* (Latin—in transit)
i/t *in transitu* (Latin—in transit)
It Italian; Italy
It *Italia* (Italian, Portuguese, Spanish—Italy); Italian
IT Idaho Territory; Immunity Test; Imperial Territory; Imperial Typewriter; Income Tax; Indian Territory; Inner Temple; Institute of Technology; International Telephone and Telegraph (Wall Street slang)
ita initial teaching alphabet; inner transport area; interface test adapter(s); international teaching alphabet; international telegraph alphabet
Ita *Italia* (Italian, Portuguese, Spanish—Italy)
ITA Income Tax Act; Independent Teachers Association; Independent Television Authority; Industrial Training Act; Industrial Truck Association; Industry and Trade Administration; *Institut du Transport Aérien* (French—Air Transport Institute); International Tea Agreement; International Temperance Association; International Tin Agreement; International Touring Alliance; International Trade Administration; International Twins Association
ITAA International Transactional Analysis Association
ITAB Industry Technical Advisory Board
ITAC Interconnect Association of Canada
ITACS Integrated Tactical Air Control System
itae integrated time and absolute error
ITAI Institute of Technical Authors and Illustrators
ital italic; italicize; italics
ital *italiensk* (Dano-Norwegian—Italian)
Ital Italia; Italian; Italy
ITAL *Information Technologies and Libraries*
Italia *Italia Società di Navigazione* (Italian Line); (Italian, Latin, Spanish—Italy)
italy I trust and love you
Italy Italian Republic (European nation whose civilization antedates the Roman Empire), *Repubblica Italiana*
ITAM *Instituto Tecnologico Autonomo de México* (Spanish—Technological Institute of Mexico)
ITAP Interim Track Analysis Program
itar interstate and foreign travel (or transportation) in aid of racketeering enterprises
ITAR International Traffic in Arms Regulations; Interstate and Foreign Travel (or Transportation) in Aid of Racketeering Enterprises
ITASS Interim Towed-Array Surveillance System
Itavia Italian Aviation (domestic airline)
itax italics
itb (ITB) integrated tug barge
ITB Industrial Training Board; Integrated Tug Barge; International Theft Bureau; International Time Bureau; Invitation to Bid; Irish Tourist Board; Irish Tourist Bureau
ITB *Internationaler Turnerbund* (German—International Gymnastic Federation)
itbh internal broach
ITBL Integrated Transportation Bill of Lading
IT & BL Island Tug & Barge, Ltd.
ITBS Iowa Test of Basic Skills
itc installation time and cost;

investment tax credit
itc (ITC) Intertropical Confluence
ITC Illinois Terminal Company (railroad); Imperial Tobacco Company; Inclusive Tour Charter; Industrial Training Council; Infantry Training Center; International Tea Council; International Tin Council; International Toast-mistress Clubs; International Trade Commission; International Traders Clubs; International Training Center; Island Trading Company
IT & C Industry, Trade, and Commerce (Canada)
ITCA Independent Television Companies' Association; Interamerican Technical Council of Archives; Intercollegiate Tennis Coaches Association; International Typographic Composition Association
ITCA *Instituto Tecnologico Centroamericano* (Spanish—Central American Technical Institute)
itcan inspect, test, and correct as necessary
ITCC International Technical Cooperation Center
itcm integrated tactical counter-measures
ITCP Integrated Test and Check-out Procedures
ITCRM Infantry Training Center—Royal Marines
ITCV Inter-Tropical Convergence Zone
it'd it had; it would
ITD *International Telephone Directory*
itda indirect target damage assessment
ITDC Indian Tourist Development Corporation
ITDP Institute for Transportation and Development Policy
ITE Institute of Telecommunication Engineers; Institute of Terrestrial Ecology; Institute of Traffic Engineers
ITEC International Thoroughbred Exposition and Conference
ITED Iowa Tests of Educational Development
ITEP Indian Teacher Education Project; Integrated Test-Evaluation Program
ITER International Thermonuclear Experimental Reactor
itf inland transit floater (insur-

ance)
ITF Integration Task Force; International Television Federation; International Tennis Federation; International Trade Federation
ITF *Institut Textile de France* (French—Textile Institute of France); *Internationale Transportarbeiter Föderation* (German—International Transportworkers Federation)
ITFCA International Track and Field Coaches Association
ITFCS Institute for Twenty-First Century Studies
itfs instructional television fixed service
ITFS Instructional Television Fixed Service; International Television Fixed Service
ITG International Trumpet Guild
itga internal gage
ITGWF International Textile and Garment Workers Federation
ithp increased take-home pay
iti intertial interval
ITI Inagua Transports Incorporated; Integrated Task Indices; International Technical Institute; International Technical Institute of Flight Engineers; International Theatre Institute; International Thrift Institute
ITIB Iceland Tourist Information Bureau
ITIC International Tsunami Information Center
ITIM Interchurch Trade and Industry Mission
itin itinerary
itis (Latin suffix—inflammation)—bronchitis, meningitis
Iti(s) Italian(s)
ITITA *Instituto Veterinario de Investigaciones Tropicales y de la Altura* (Spanish—Veterinarian Institute of Tropical and High-Altitude Research)
itk *intetkøn* (Dano-Norwegian—neuter)
itl integrate-transfer-launch
Itl Italian
it'll it will
itlx italics
itm inch trim moment; information transfer module
ITM Institute of Travel Managers; Institute of Tropical Medicine
ITMA Institute of Trade Mark Agents; International

Twelve-Meter Association
ITMA *It's That Man Again* (Tommy Handley's most popular World-War-II BBC series)
ITMF International Textile Manufacturers Federation
ITMRC International Travel Market Research Council
ITN International Television Network
ITN *Independent Television News*
ITNA Independent Television News Association
ITNS Integrated Tactical Navigation System (USN)
ITO India Tourist Office; Interim Technical Order; International Terminal Operators; International Trade Organization (UN); Invitational Travel Orders
ITOA Independent Taxi Owners Association
ITOFCA Industrial Trailer-on-Flatcar Associates
ITOL International Thomson Organisation, Ltd
itom interstate transportation of obscene matter
itp (ITP) idiopathic thrombocytopenic purpura; immune thrombocytopenic purpura; inosine triphosphate
ITPA Illinois Test of Psycholinguistic Abilities
ITPP Institute of Technical Publicity and Publications
ITPS Income Tax Payers' Society
itq's in-text questions
itr incremental tape recorder; integrated test requirement(s)
ITR Indiana Toll Road
ITRA International Truck Restorers Association
ITRC International Terrorist Research Center (El Paso, Texas); International Tin Research Council
ITRP Institute of Transportation and Regional Planning
ITRU Industrial Training Research Unit
its (ITS) invitation to send (data processing)
it's it has; it is
Its Italians
ITS Idaho Test Station; Institute for Transportation Studies; Integrated Trajectory System; Interactive Training System; Intermarket Trading System; International Technogeographical Society; In-

ternational Thespian Society; International Tracing Service; International Trade Secretariat; International Transportation Service
itsa interstate transportation of stolen aircraft
ITSA Institute for Telecommunication Sciences and Aeronomy
itsb interstate transportation of strikebreakers
itsc interstate transportation of stolen cattle
ITSC International Telecommunications Satellite Consortium
itse integral time square error
itsmv interstate transportation of a stolen motor vehicle
itsp interstate transportation of stolen property
ITSS Integrated Tactical Surveillance System
itt instant-touch tuning
ITT Institute of Textile Technology; Insulin Tolerance Test
IT & T International Telephone and Telegraph
ITTA International Table Tennis Association
ITTC International Television Trading Corporation
ITTCS International Telephone and Telegraph Communications System
ITTE Institute of Transportation and Traffic Engineering
ITTF International Table Tennis Federation
Itti *Imar Arab Ittih d al-Imarat al-Arabiyah* (Arabic—United Arab Emirates)
ITTP Indian Teacher Training Program
ITTTA International Technical Tropical Timber Association
itu (ITU) intensive therapy unit; intrauterine transfusion
ITU Income Tax Unit; International Telecommunications Union; International Typographical Union
ITUA Industrial Trades Union of America
ITURM International Typographical Union Ruling Machine
itv instructional television; internal television
ITV Independent Television
ITVA Instructional Television Authority
ITVN Independent Television News

ITVs Independent Television programs broadcast from offshore Great Britain
ITVS International Television Service
itw initial training wing
ITW Illinois Tool Works
ITWF International Transport Workers Federation
itx inclusive tour excursion(s)
I-type Jungian introvert type
Itz Itzik
Itz Itzik (Yiddish—Isaac)
ITZC Intertropical Convergence Zone
ITZEL Irgun Tzvai Le'umi (Hebrew—National Military Organization)
ITZN International Trust for Zoological Nomenclature
iu immunizing unit(s); indicator unit; internal upset (oil well); international unit(s)
i of u inevitability of the unpredictable
IU Indiana University; Indianapolis Union (railroad); international unit; International Utilities
IÜ Istanbul Üniversitesi (Universityo) (University of Istanbul)
IUA International Union of Architects
IUAA International Union Against Alcoholism; International Union of Advertisers' Association; International Union of Alpine Associations
IUAES International Union of Anthropological and Ethnological Sciences
IUAI International Union of Aviation Insurers
IUAIWA International Union of Allied Industrial Workers of America
IUAJ International Union of Agricultural Journalists
IUAO International Union for Applied Ornithology
IUAPPA International Union of Air Pollution Prevention Associations
IUAT International Union Against Tuberculosis
IUB International Union of Biochemistry; Interstate Underwriters Board
IUBS International Union of Biological Sciences
IUC International Union of Chemistry; International Union for Conservation
IUCc International Union of Crystallography

iucd intrauterine contraceptive device
IUCL Istanbul University Central Library
IUCN International Union for Conservation of Nature and Natural Resources
IUCNNR International Union for Conservation of Nature and Natural Resources
IUCr International Union of Crystallography
IUCSTP Inter-Union Commission on Solar-Terrestrial Physics
IUCSTR Inter-Union Commission on Solar and Terrestrial Relationships
IUCW International Union for Child Welfare
iud intrauterine device; intrauterine diaphragm
Iud Iudicum (Spanish—Epistle of St Paul to the Hebrews)—Book of the Jews
IUD Institute for Urban Development
iudr idoxuridine
IUDTPNAPUSCAN International Union of Dolls, Toys, Playthings, Novelties, and Allied Products of the United States and Canada
IUDZG International Union of Directors of Zoological Gardens
IUE International Ultraviolet Explorer (space vehicle); International Union of Electrical Workers; International Union for Electroheat
IUEC International Union of Elevator Constructors
IUEF International University Exchange Fund
IUER & MW International Union of Electrical, Radio & Machine Workers
IUES International Ultraviolet Explorer Satellite
IUF International Union of Food and Allied Workers' Associations
IUFA International Union of Family Organizations
IUFDT International Union of Food, Drink, and Tobacco Workers' Association
IUFOST International Union of Food Science and Technology
IUFRO International Union of Forest Research Organizations
IUGG International Union of Geodesy and Geophysics

IUGS International Union of Geological Sciences; International Union of Geological Services
IUHA Industrial Unit Heater Association
IUHE International Union for Health Education
IUHPS International Union of the History and Philosophy of Science
IUHS International Union of the History of Science
IUIS International Union of Immunological Societies
IUL (Bloomington); Ibadan University Library (Nigeria); Indiana University Library
IULA International Union of Local Authorities
IULIA International Union of Life Insurance Agents
iumbb if unworkable, make best bid
IUMC Indiana University Medical Center
IUMI International Union of Marine Insurance
IUMK Istanbul Üniversitesi Merkez Kütüphanesi (Turkish—Istanbul University Central Library)
IUMM & SW International Union of Mine, Mill and Smelter Workers
IUMP International Union of the Medical Press
IUMSA International Union for Moral and Social Action
IUMSWA Industrial Union of Marine and Shipbuilding Workers of America
IUNS International Union of Nutritional Sciences
IUOE International Union of Operating Engineers
IUOPA International Union of Practitioners in Advertising
IUOPAB International Union of Pure and Applied Biophysics
IUOT Indiana University Opera Theater
IUOTO International Union of Official Travel Organizations
IUOW Industrial Union of Workers
iup intrauterine pregnancy
IUP Indiana University Press; International University Press; Irish Universities Press; Israel Universities Press
IUPA International Union of Police Associations; International Union of Practitioners in Advertising

IUPAC International Union of Pure and Applied Chemistry

IUPAP International Union of Pure and Applied Physics

IUPLAW International Union for the Protection of Literary and Artistic Works

IUPM International Union for Protecting Public Morality

IUPN International Union for the Protection of Nature

iups installed user programs

IUPS International Union of Physiological Sciences

IUPW International Union of Petroleum Workers

IUR International Union of Railways

I U Res Ctr Indiana University Research Center (for the Language Sciences)

IURN *Institut Unifé de Recherche Nucléaires* (French—Unified Institute of Nuclear Research)

ius inertial upperstage; interim upperstage

IUs international units

IUS Institute of Urban Studies; International Union of Students; International Urban Society; International Urban Studies

IUSA Institute of the U.S.A. (Soviet office charged with analyzing American news and political statements concerning the USSR)—Russian counterpart of American kremlinologists

IUSM Indiana University School of Music

IUSP International Union of Scientific Psychology

IUSS Institute of United States Studies

IUSSI International Union for the Study of Social Insects

IUSSP International Union for the Scientific Study of Population

IUT *Instituts Universitaires de Technologie* (French—University Institutes of Technology)

IUTAM International Union of Theoretical and Applied Mechanics

iutip if unworkable, telegraph idea of price

IUUAAAIWA International Union of United Automobile, Aerospace, and Agricultural Implement Workers of America

IUUCLG International Union,

United Cement, Lime & Gypsum Workers

IUVD International Union Against Venereal Diseases

IUVDT International Union against the Venereal Diseases and the Treponematoses

IUVSTA International Union for Vacuum Science Techniques and Applications

IUW Industrial Union of Workers

IUWCC Inshore Undersea Warfare Control Center (USN)

IUWWML International Union of Wood, Wire, and Metal Lathers

iv increased value; initial velocity; intravenous(ly); intravertebral; inverted vertical (engine); invoice value; ivory

i/v increased value; instrument/visual

i.v. in verbo (Latin—under the word)

i V in Vertretung (German—as a substitute, by proxy)

IV Imperial Valley; initial visit; Ivan; Ivy

IV Islas Virgenes (Spanish—Virgin Islands)

iva inspection visual aid

Iva Godiva

IVA Independent Voters Association; International Volleyball Association

IVAAP International Veterinary Association for Animal Production

ivala integrated visual approach and landing aids

Ivaran Ivar Anton Christensen's steamship company

IVBF International Volley-Ball Federation

IVBH Internationale Vereinigung für Brückenbau und Hochbau (German—International Association for Bridge and Structural Engineering)

ivc inferior vena cava; intermittent vertical chambers

IVC Imperial Valley College

ivcd intraventricular conduction defect

ivc's inner van connectors

Iv Cst Ivory Coast

ivd interpolated voice data; intervertebral disk

IVDP Identification, Validation, and Dissemination Process

ivds independent variable depth sonar

Ive Ivan; Iven

I've I have

IVE Institute of Vitreous Enamellers

IVECO Industrial Vehicles Corporation

I've had it (popular American contraction—I have had enough of it)

I-vets Iceland veterans

ivf intravenous fluid; in-vitro fertilization

IVF Innocent Victims Fund

IVFZ International Veterinary Federation of Zootechnics

IVGMMA International Violin, Guitar Makers, and Musicians Association

IVI Independent Voters of Illinois

IVIC Instituto Venezolano de Investigaciones Científicas (Spanish—Venezuelan Institute of Scientific Investigations)

IVIS International Visitors Information Service

ivjc intervertebral joint complex

IVJH Internationale Vereinigung für Jugendhilfe (German—International Union for Child Welfare)

IVK Institutet för Vaxtforskning och Kyllagring (Swedish—Institute for Foodstuff Research and Refrigeration—Sweden)

IVL Ivaran Lines

IVL Internationale Vereinigung für Theoretische und angewandte Limnologie (German—International Association of Theoretical and Applied Limnology)

IVLA International Visual Literacy Association

IVLU Ivaran Lines (container) Unit

IVMB Internationale Vereinigung der Musikbibliotheken (German—International Association of Music Libraries)

ivmu inertial velocity measurement unit

ivo in view of

Ivory Coast Republic of the Ivory Coast (West African nation)

ivox (IVOX) intravascular oxygenator

ivp initial vapor pressure; inspected variety purity (certified seeds); intravenous pyelogram

IVP Instituto Venezolano de la Petroquimica (Spanish—Venezuelan Petrochemical Institute)

ivr instrumented visual range

IVR International Vehicle Registration (symbols displayed on automotive license plates)

IVR Internationale Vereinigung des Rheinschiffsregisters (German—International Association of Rhine Ships Registers)

ivs intraventricular septum

iv's intravenous feedings; intravenous injections

IVS Integrated Versaplot Software; International Voluntary Service

IVS Instituto Venezolano de los Seguros Sociales (Spanish—Venezuelan Institute of Social Security)

ivsd interventricular septal defect

ivsi instantaneous vertical speed indicator

IVSS Internationale Vereinigung für Soziale Sicherheit (German—International Association for Social Security)

IVSU International Veterinary Students Union

ivt intravehicular transfer; intravenous transfusion

ivu intravenous urography

ivu (IVU) imposta valore aggiunto (Italian—value-added tax)

IVU International Vegetarian Union

IVU Instituto de Vivienda Urbana (Spanish—Institute of Urban Housing)

ivvs instantaneous vertical-velocity sensor

iw index word; indirect waste; individually wrapped; inside width; instruction word(ing); isotopic weight; ivory woodpecker

iw (IW) index word

i/w interchangeable with; in work

iW innere Weite (German—inside diameter)

IW Aero Trasporti Italiani (2-letter coding, Italian Air Transport)

IWA Inland Waterways Association; Institute of World Affairs; Insurance Workers of America; International Wheat Agreement; International Woodworkers of America

IWA International Wheat Agreement

IWAHMA Industrial Warm Air Heater Manufacturers Association

IWAIU Industrial Workers of America International Union

IWBP Integration with British Party (Gibraltar)

iwc in which case

IWC Inland Waterways Corporation; International Watch Company; International Whaling Commission; International Wheat Council

IWCA International World Calender Association

IWCC International Wrought Copper Council

IWCCA Inland Waterways Common Carriers Association

IWCI Industrial Wire Cloth Institute

IWCS Integrated Wideband Communications System

IWCT International War Crimes Tribunal

IWD International Waterways and Docks; International Women's Day (March 12)

IWE Institution of Water Engineers

IWG Imperial Wire Gauge; Interface Working Group; International Working Group (NATO)

IWGC Imperial War Graves Commission

IWGM Intergovernmental Working Group on Monitoring (or Surveillance)—UN

IWGMP Intergovernmental Working Group on Marine Pollution (UN)

IWHS Institute of Works and Highway Superintendents

IWI Inventors' Workshop International

iwistk issue while in stock

IWIU Insurance Workers International Union

iwl insensible water loss

IWL Institute of World Leadership

IWLA Izaak Walton League of America

IWM Imperial War Museum; Institute of Works Managers

IWMA International Working Men's Association

iwmi inferior wall myocardial infarction

IWML Imperial War Museum Library

Iwo Iwo Jima (Sulfur Island)

IWO Institute of World Order; International Wine Office;

International Workers Order

IWP Indicative World Plan (FAO)

IWPA International Word Processing Association

IWPC Institute of Water Pollution Control

IWPO International Word Processing Organizations

IWPPA Independent Waste Paper Processers Association

IWPS Institute of War and Peace Studies (Columbia)

IWRI Informal Word Recognition Inventory; International Wildfowl Research Institute

IWRMA Independent Wire Rope Manufacturers Association

IWS Inland Waterway Service; Institute of Wood Science; Intelligent Warning System; International Wool Secretariat

IWSA International Water Supply Association

IWSB Insect Wire Screening Bureau

IWSc Institute of Wood Science

IWSC Irrigation and Water Supply Commission

IWSG International Wool Study Group

IWSP Institute of Work Study Practitioners

IWST Integrated Weapon System Training

IWT Indus Water Treaty; Inland Water Transport; Institute of Women Today; International Working Team (NATO)

IWT Industriewerke Transportsystem (cargo container system)

IWTA Inland Water Transport Authority

IWTD Inland Water Transport Department

IWTO International Wool Textile Organization

iwu illegal wearing of uniform

IWU Illinois Wesleyan University; Insurance Workers Union

IWV Internationale Warenhaus Vereinigung (German—International Department Store Association)

IWVA International War Veterans Alliance

iwvmts interim water velocity meter test set

iww inland waterway

IWW Industrial Workers of

the World; Intracoastal Waterway

IWWP *International Who's Who in Poetry*

IWY International Women's Year (1975–1984)

IX unclassified vessel (2-letter naval code)

I.X. *Iesous Christos* (Greek—Jesus Christ)

ixc interexchange

IXSS unclassified miscellaneous submarine (letter symbol)

Ixta Ixtaccihuatl

iy ionized yeast

IY Imperial Yeomanry; International Petroleum (stock exchange symbol)

IYB *International Year Book*

IYC Inland Yacht Club; International Year of the Child

IYDP International Year of Disabled Persons (1981)

IYEO Institute of Youth Employment Officers

IYF International Youth Federation

IYHF International Youth Hostel Federation

IYL International Youth Library

IYRU International Yacht Racing Union

iyswim if you see what I mean

i y v ida y vuelta (Spanish—round trip)

I y v ida y vuelta (Spanish—round trip)

iz izzard; zed

Iz Izar; Izar: Izmir (Smyrna)

IZ Institute of Zoology; Israel Zangwill

IZ *I Zingari* (Italian—The Gypsies)

izd *izdanie* (Russian—edition)

Izd *izdatl'* (Russian—publisher)

IZD *Internationaler Zivildienst* (German—International Voluntary Service)

izdat *izdatel* (Russian—publisher)

IZL *Irgun Z'vai Leumi* (Hebrew—National Army Organization)

izq^a *izquierda* (Spanish—left)

izq^o *izquierdo* (Spanish—left)

izs insulin zinc suspension

IZTO Interzonal Trade Office (NATO)

Izv *Izvestia* (Russian—news)—official newspaper of the Presidium of the Supreme Soviet—published in Moscow

Iz: Wa: Izaak Walton, *The Complete Angler*, used colons after his initials

Izzie Isador; Isadora; Isadore; Isidro; Isodoro; Ysidra

Izzy (*see* Izzie)

J

j inner quantum number (symbol); jack; joint (marijuana); joist(ing); junior; junk(ie); square of minus 1 (symbol); unit vector in y direction (symbol)

j journal; *jour(nal)* (French—day, newspaper)

j. *juris* (Latin—of law); *jus* (Latin—law)

J action variable (symbol); advance ratio (symbol); electric current density (symbol); gram-equivalent weight (symbol); heat transfer factor (symbol); Jacob; Jacobean; Jacobian; Jaeger; Jaen; Jamaica; Jamaican; January; Japan; Japanese; Jesuit; jet; Jew; Jewish; joint; joule; Judaic; Judaism; Juliet—code for letter J; Julliard; July; Junction; junction devices; June; North American Aviation (symbol); polar movement of inertia (symbol); radiant intensity (symbol)

J *Jabal* (Arabic or Persia—mountain, mountain range); *Jebel* (Arabic—mountain, mountains); *Jejunium* (Latin—fast, hunger); *Jibal* (Arabic—mountain range); *Jogi* (Estonian—river); *Jøkel* (Norwegian—glacier); *Joki* (Finnish—river); *Jökull* (Icelandic—glacier); *Journal* (French—journal)

J-1 personnel section of joint military staff

J-2 intelligence section of joint military staff

J-3 operations and training section of joint military staff

J-4 logistics section of joint military staff

J-4 fuel jet-engine fuel (derived from coal oil or kerosene)

J-5 Plans and Policy (Joint Chiefs of Staff)

J-6 Communications, Electronics (Joint Chiefs of Staff)

J-32 Saab Lansen jet interceptor aircraft

J-35 Saab double-delta-wing supersonic fighter or fighter-bomber built in Sweden and named Draken (Dragon)

J-37 Swedish Thunderbolt or Viggen jet fighter aircraft

ja jack; jack adapter; jetavator assembly; job analysis; joke awful

ja (JA) jump address

j/a (J/A) joint account

j & a junk and abandon; junked and abandoned

Ja Jacob; Jacque(s); James; Japan; Japanese

JA Jamaica(n); Japan Association; Jewish Agency; John Adams (2nd President of U.S.); Judge Advocate; Junior Achievement; Justice of Appeal

JAA Japan Aeronautic Association; Japan Asia Airways; Joint Airways Association

JAAA Japan Amateur Athletic Association

JAAB Joint Airlift Allocations Board

JAAF Joint Army-Air Force

JAAFU Joint Anglo-American Foulup; Joint Anglo-American Fuckup

JAALD Japanese Association of Agricultural Librarians and Documentalists

JAAOC Joint Anticraft Operation Center (NATO)

JAAR *Journal of the American Academy of Religion*

jaarg *jaargang* (Dutch—annual volume)

JAARS Jungle Aviation and Radio Service

JAAS Jewish Academy of Arts and Sciences

Jab Jabal; Jabalpur; Jabez; Jabneel

JAB J Ashleigh Burke; Joint Amphibious Board; Junior Advisory Board

JABA Jefferson Area Board for Aging

JABC Japan Audit Bureau of Circulations

jac jet aircraft coating

Jac Jacobean; Jacobite; Jacobus; Jacumba

Jac. Book of Jacob

Jac. *Jacobus* (Latin—James)

JAC Japan Advisory Committee; Joint Advisory Committee; Joint Apprenticeship Council

JAC *Journal of Applied Chemistry*

JACA Japan Air Cleaning Association

JACC Joint Admissions Centre for Colleges (Australia); Journalism Association of Community Colleges

JACCC Japanese-American Cultural and Community Center; Joint Air Control and Coordination Center (USAF)

Jace Jason

jack jackass

Jack Jackson; Jacob; John

JACKPOT Joint Airborne Communications Center and

Command Post
jack(s) jackass(es)—male don-
key(s)—*see* hinny
JACL Japanese-American Cit-
izens League
*JACM Journal of the Associa-
tion for Computing Machin-
ery*
JACOB Junior Achievement
Corporation of Business
Jacq Jacques: Jacquin
Jacq's loom Jacquard's loom
*JACS Journal of the American
Chemical Society*
JACT Joint Association of
Classical Teachers
Jad Jadavpur, India
JAD Julian astronomical day
JADA Japan Automobile Deal-
ers Association
*JADA Journal of the American
Dental Association*
JADB Joint Air Defense Board
JADE Japanese Air Defense
Environment
jadeite sodium aluminum sili-
cate
JADF Japan Air Defense
Force
jaditbhkycc just a drop in the
basket helps keep your city
clean (anti-litter-civic-respon-
sibility campaign)
JADPU Joint Automatic Data
Processing Unit (shared by
the Home Office and the
Metropolitan Police of Lon-
don)
*Jadroplov Jadranska Slobodna
Plovida* (Yugoslav Great
Lakes Line)
J Adv Judge Advocate
J Adv Gen Judge Advocate
General
JAE Joint Atomic Exercise
(NATO)
JAEC Japan Atomic Energy
Commission; Joint Atomic
Energy Committee (U.S.
Congress)
JAEIA Japan Atomic Energy
Industrial Association
JAEIC Joint Atomic Energy
Intelligence Committee
JAEIP Japan Atomic Energy
Insurance Pool
JAERI Japan Atomic Energy
Research Institute
JAES Japan Atomic Energy
Society
JAF Japan Automobile Federa-
tion; Jordanian Air Force;
Judge Advocate of the Fleet
JAFC Japan Atomic Fuel Cor-
poration
jaff chaff and electronic jam-

ming
Jaffna Jaffnapatam
JAFPUB Joint Armed Forces
Publication
jag. jaguar; jaguarundi
Jag Jaguar
JAG James Abram Garfield
(20th President U.S.); Judge
Advocate General
JAG-A Judge Advocate Gen-
eral—Army
JAGC Judge Advocate Gen-
eral's Corps
JAGD Judge Advocate Gen-
eral's Department
JAG-N Judge Advocate Gen-
eral—Navy
jag(s) jaguarundi(s)
JAGS Judge Advocate Gen-
eral's School
JAH John Adams House
Jahrb Jahrbuch (German—
yearbook)
Jahrg Jahrgang (German—an-
nual publication, vintage of
the year); year's growth
jai juvenile amaurotic idiocy
JAIEG Joint Atomic Informa-
tion Exchange Group
JAIF Japan Atomic Industrial
Forum
JAIL Justice Against Identifi-
cation Laws
JAIMS Japan-America Insti-
tute of Management Science
JAIS Japan Aircraft Industry
Society
jak jackfruit (durian)
Jak Jakarta (Batavia)
Jal Jalisco (inhabitants nick-
named tapatios as they excel
in dancing the tapatio jarabe)
Jal Jalan (Malay—Lane,
Road, Street)
JAL Japan Air Lines; Jet Ap-
proach and Landing Chart
*JAL Journal of Academic Lib-
rarianship*
JALMA Japan Leprosy Mis-
sion for Asia
JALTOS Japan Air Lines
Computerized Air Cargo Ter-
minal System
jam. jamming; job analysis
memo
Jam Jamaica
JAM James A. Michener; Joint
Action for Mission; Joslyn
Art Museum; Sir John Alex-
ander Macdonald (Canada's
first and third Prime Minis-
ter)
JAMA Japan Automobile Ma-
nufacturers Association
*JAMA Journal of the Ameri-
can Medical Association*

JAMAG Joint American Mili-
tary Advisory Group
Jamaica English-speaking
West Indian island nation
*Jam Arab Lib Sha Ish al-
Jamahiriyah al-Arabiya al-
Libya al-Shabiya al-Ishtira-
kiya* (Arabic—Socialist Peo-
ple's Libyan Arab Republic)
*Jam Arab Sour al-Jamhouriya
al Arabia as-Souria* (Ara-
bic—Syrian Arab Republic)
JAMC Japan Aircraft Manu-
facturers Corporation
*J Am Ceram Soc Journal of
the American Ceramic Soci-
ety*
*J Am Chem Soc Journal of the
American Chemical Society*
*Jam Dim Som Jamhuriyadda
Dimugradiga Somaliya* (Ara-
bic—Somali Democratic Re-
public)
jamex jamming exercise
*J Am Geriatrics Soc Journal
of the American Geriatrics
Society*
JaMi Jacksonville-Miami
(metropolitan area including
Fort Lauderdale, Hollywood,
Tampa, and St Petersburg)—
also called Metro or Metro
Area
Jamie James
*J Am Inst Electr Eng Journal
of the American Institute of
Electrical Engineers*
Jam Ken Jamhuri ya Kenya
(Swahili—Republic of Ken-
ya)
JAMMAT Joint Military Mis-
sion for Aid to Turkey
jammies pyjamas
jamocha java & mocha (prison
argot—coffee)
jams pajamas
jamsan jam sandwich
Jamsat Japan radio amateur
satellite
JAMSTEC Japan Marine Sci-
ence and Technology Center
jamtrac jammers tracked by
azimuth crossings
JAMTS Japan Association of
Motor Trade and Service
jamwich jam sandwich
*Jam Mwu Tan Jamhuri ya
Mwungano wa Tanzania*
(Swahili—United Republic
of Tanzania)
Jam Sud Jamhuryat as-Sudan
(Arabic—Republic of the Sudan)
jan janitor; janitorial
Jan Janice; Jansen; Janson;
January; John

JAN Jackson, Mississippi (airport); Joint Army-Navy

JANAF Joint Army-Navy Air Force

JANAIR Joint Army-Navy Aircraft Instrument Research

JANAP Joint Army-Navy-Air Force Publication

JANAST Joint Army-Navy-Air Force Sea Transport

JANBMC Joint Army-Navy Ballistic Missile Committee

JanFeb January and February

JANFU Joint Army-Navy Foulup; Joint Army-Navy Fuckup

JANIS Joint Army-Navy Intelligence Surveys

jan mer jangan merokok (Malay—no smoking)

Janna Johanna

Jans Janson

JANS Jet Aircraft Noise Survey; Joint Army-Navy Specification

JANSPEC Joint Army-Navy Specification

JANSRP Jet Aircraft Noise Survey Research Program

JANSTD Joint Army-Navy Standard

janv janvier (French—January)

jap japanned

jap japanisch (German—Japanese); *japansk* (Dano-Norwegian—Japanese)

Jap Japan; Japanese; Jasper

Jap Japanese language

JA£ Jamaican pound

JAP Joint Acceptance Plan(ning); Joint Apprenticeship Program

J-AP Jewish-American Prince(ss)

JAPAC Japan Atomic Power Company; Joint Air Photo Center

Japan Asia's most productive country—*Nippon* or *Nihon*

JAPATIC Japan Patent Information Center

JAPC Joint Air Photo Center

JAPCO Japan Atomic Power Company

Jap Cur Japan Current

Japdic Japanese dictionary

Japex Japan Petroleum Exploitation Company

JAPEX Japan Express

JAPIA Japan Auto Parts Industries Association

Japlish Japanese & English

J App Crystallogr Journal of Applied Crystallography

J Appl Phys Journal of Applied Physics

Jap Soc Japan Society

jar. jargon(ish)

Jar. Book of Jarom

JARC Joint Air Reconnaissance Center (NATO)

Jardines' Jardine, Matheson & Company

JARE Japanese Antarctic Research Expedition

jarg jargon; jargonese; jargonist; jargonistic; jargonize

Jarg Soc Jargon Society

JARI Japan Automotive Research Institute

JARL Japan Amateur Radio League

JARO Johore Area Rehabilitation Organization

JARS Journalization and Recovery System

JARTS Japan Railway Technical Service

Jas James; Jason

JAS Jamaica Agricultural Society; Japan Agricultural Standards; Japan Association of Shipbuilders; Japan Astronautical Society; Japanese Amateur (radio) Satellite; Jewish Agricultural Society; Jordanian Agricultural Society

J-A S Japan-Australia Society

JASA Journal of the Acoustical Society of America

JASC Japan-Asia Sea Cable

JASDF Japan Air Self-Defense Force

JASG Joint Advanced Study Group

JASI Joint Asian Surgical Industries

JASIN Joint Air-Sea Interaction (oceanographic experiment)

JASIS Journal of the American Society for Information Science

jasp jasper; jasperoid

Jasp Jasper

Jasper Jasper National Park

Jaspr Jasper

JASRAC Japanese Society of Rights of Authors and Composers

jastop jet-assisted stop

jasu jet aircraft starting unit

JAT Jugoslovenski Aero-Transport (Yugoslav Airlines)

JATC Joint Apprenticeship Training Committee

JATCA Joinery and Timber Construction Association

JATCC Joint Aviation Telecommunications Coordina-

tion Committee

JATCRU Joint Air Traffic Control Radar Unit

JATMA Japan Automobile Tire Manufacturers Association

J Atmos Sci Journal of Atmospheric Sciences

J Atmos Terr Phys Journal of Atmospheric and Terrestrial Physics

jato jet-assisted takeoff

JATS Joint Air Transportation Service

J Audio Eng Soc Journal of the Audio Engineering Society

jaund jaundice

JAUNT Jefferson Area United Transportation

Jav Java; Javanese

Java Djawa

JAVA Jamaica Association of Villas and Apartments

JAVHS Jane Addams Vocational High School

Ja, vi elsker Ja, vi elsker dette landet (Norwegian—Yes, we love this land)—national anthem

JAWA Jane's All the World Aircraft

JAWS Japan Animal Welfare Society; Jet Advance Warning System

Jax Jacksonville, Florida

JAX Jacksonville, Florida (airport)

jaycee (JC) Junior Chamber of Commerce

jazzercise jazz-induced exercise

jb jet black; jet bomb (JB); job blank (form); joint board; junction box

Jb Jacob

Jb Jahrbuch (German—annual, yearbook)

JB Jacksonville Beach; James Buchanan (15th President U.S.); Jodrell Bank; John Bull (British empire personified); Joint Board; Stetson hat (after its original maker—JB Stetson)

J-B Jacques Barzun; Jean-Baptiste; Johannes Brahms

J.B. *Jurum Baccalaureus* (Latin—Bachelor of Laws)

JB Jerusalem Bible

JBa Jacques Barzun

JBA Japan Binoculars Association; Junior Bluejackets of America

JBAA Journal of the British Archeological Association

JBAKC John Brown Anti-Klan Committee

J-bar capital-J-shaped bar (as used in ski tow lifts)

JBC Jamaica Broadcasting Corporation; Japan Broadcasting Corporation (*q.v.* NHK); Jewish Book Council

JB & C John Brown and Company (shipbuilders)

JBCA Jewish Book Council of America

JBCNS Joint Board of Clinical Nursing Studies (England and Wales)

JB & Co John Brown and Company (shipbuilders)

JBCSA Joint British Committee for Stress Analysis

jbd jet blast deflector

JBe Japanese B encephalitis

Jber Jahresbericht (German—annual report)

JBES Jodrell Bank Experimental Station (Cheshire, England)

JBG Jewish Board of Guardians

JBHS John Bartram High School

JBIA Jewish Braille Institute of America

J-bird jailbird, convict

JBL James B Lansing (sound equipment)

JBL Journal of Biblical Literature; Journal of Business Law

JBMA John Burroughs Memorial Association

JBMMA Japanese Business Machine Makers Association

J-boat large yacht, often 76 feet or longer; small racing boat sailed by youngsters

J-bolt capital-J-shaped bolt

J-box J-shaped bleaching box; junction box

JBP Jewel Bearing Program; John Boynton Priestley

JBPA Japan Book Publishers Association

JBPI Japanese Bicycle Promotion Institute

JBPS Jamaica Banana Producers Steamship

JBS Japan British Society; John Birch Society

JBSW Joseph Bulova School of Watchmaking

jbt jail(ing) before trial

JBT Jewelers Board of Trade

JBUSDC Joint Brazil-United States Defense Commission

JBUSMC Joint Brazil-US Military Commission

JBYC Jamaica Bay Yacht Club

jc joint compound

Jc Junction

JC Brother John Charles SSF; Jackson College; Jacksonville College; Jamestown College; Jefferson City; Jefferson College; Jersey City; Jesus College (Cambridge); Jet Club; Job Corps; Jockey Club; Johannesburg Consolidated; Johnstown College; Joliet College; Judson College; Juniata College; Junior Chamber (of Commerce, members called *Jaycees*); Juvenile Corps; Juvenile Court

J.C. Jesus Christ; Julius Caesar

JC Jewish Chronicle

J-C *Jésus-Christ* (French—Jesus Christ)

J.C. *Juris Consultus* (Latin—Juris Consult)

JCA Jewelry Crafts Association; Johore Consumers Association; Joint Commission on Accreditation (of colleges and universities); Joint Communication Activity; Joint Communications Agency; Joint Construction Agency; Junior College of Albany

JCAA Japanese Civil Aviation Authority

JCAB Japan Civil Aviation Bureau

JCAE Joint Committee on Atomic Energy

JCAH Joint Committee on Accreditation of Hospitals

JCAM Joint Commission on Atomic Masses

JCAP Joint Conventional Ammunition Panel (DoD)

JCAR Joint Commission of Applied Radioactivity

J-car(s) Japanese car(s)

JCB Japan California Bank; Japan Credit Bank; Japan Credit Bureau; Joint Consultative Board (NATO)

J.C.B. *Juris Canoni Baccalaureus* (Latin—Bachelor of Canon Law); *Juris Civilis Baccalaureus* (Latin—Bachelor of Civil Law)

JCBC Junior College of Broward County

JCBL John Carter Brown Library (of Americana)—Brown University, Providence, Rhode Island

JCBSF Joint Commission for Black Sea Fisheries

JCC Jamestown Community College; Japan Cotton Center; Jefferson Community College; Jesus Christ Christian Aryan Nations Church (neo-Nazi organization); Jewish Community Center; Job Corps Center; John C Calhoun; John Caldwell Calhoun; Joint Communications Center; Joint Communications Committee; Junior Chamber of Commerce

JC of C Junior Chamber of Commerce

JCCA Joint Conex Control Agency

JCCI Japan Chamber of Commerce and Industry

JCCRG Joint Command Control Requirements Group

JCCSO Jewish Community Center Symphony Orchestra

JCD Journal of Crime and Delinquency

J.C.D. *Juris Canonici Doctor* (Latin—Doctor of Canon Law); *Juris Civilis Doctor* (Latin—Doctor of Civil Law)

JCE Johannesburg College of Education; Junior Certificate Examination

JCE Journal of Chemical Education

JCEC Joint Communication Electronics Committee

JCED Japan Committee for Economic Development

JCEE Joint Council on Economic Education

JCEG Joint Communications Electronics Group

JCENS Joint Communication Electronic Nomenclature System

JCET Joint Council on Educational Telecommunications

JCFA Japan Chemical Fibres Association

J Chem Phys Journal of Chemical Physics

J Chem Soc Journal of the Chemical Society

J Chim Phys Journal de Chimie Physique (French—Journal of Chemical Physics)

JCI Joint Communications Instruction; Junior Chamber International

JCIA Japan Camera Industry Association; Japan Chemical Industry Association

JCIC Johannesburg Consolidated Investment Company

JCIEABJ Joint Commission for the Investigation of the Effects of the Atomic Bomb

in Japan
JCII Japan Camera Inspection Institute
JCJC Jasper County Junior College; Jefferson County Junior College
Jck Jacksonville
jcl job-control language
Jcl Johnny come lately
JCL Job Control Language; John Crerar Library
J.C.L. *Juris Canonici Licentiatus* (Latin—Licentiate in Canon Law)
JCLA Joint Council of Language Associations
J-class Soviet diesel-powered missile-launching submarines nicknamed Juliet
JCLCPS *Journal of Criminal Law, Criminology, and Police Science*
JCLS Junior College Libraries Section
jcm jettison control module
JCM Joint Committee on Microcards
JCMC Joint Conference on Medical Conventions
JCN Job Change Notice
JCNAFF Joint Canadian Navy-Army-Air Force
JCNM Jewel Cave National Monument
JCO José Clemente Orozco
JCOA Jazz Composers Orchestra Association
JCOC Joint Combat Operations Center; Joint Command Operations Center
J Comput Phys *Journal of Computational Physics*
JCOSA Joint Chiefs of Staff in Australia
jcp jettison control panel; jungle canopy penetration
JCP Japan Communist Party; J.C. Penney; Joint Committee on Printing (Congress); Junior Collegiate Players; Justice of the Common Pleas
JCPCI Junior College of Packer Collegiate Institute
JCPI Japan Cotton Promotion Institute
JCR Junior Common Room
JCR *Journal of Coastal Research*
JC3RCP Joint Committee of the three Royal Colleges of Physicians (Edinburgh, Glasgow, London)
JCRFD Joint Commission for Regulation of Fishing on the Danube
JCRR Joint Commission on

Rural Reconstruction
J Cryst Growth *Journal of Crystal Growth*
JCs Job Corpsmen
JCS Jewish Community Center(s); Joint Chiefs of Staff; Joint Commonwealth Societies
JCS-ACA Joint Chiefs of Staff—Automatic Conference Arranger
JCS-IDTN Joint Chiefs of Staff—Interim Data Transmission Network
JCSMR John Curtin School of Medical Research
JCS-PUBS Joint Chiefs of Staff—Publications
JCSRE Joint Chiefs of Staff Representative, Europe (NATO)
JCSUK Jersey Cattle Society of the United Kingdom
jct junction
Jct Junction (postal abbreviation)
JCT Joint Committee on Taxation
JCTC Japanese Cultural and Trade Center; Juneau County Teachers College
jctn junction
jct pt junction point
JCU James Cook University; John Carroll University
JCUDI Japan Computer Usage Development Institute
JCUNQ James Cook University of North Queensland
J-curve shape formed on a graph when a country's currency drops in value and the trade balance worsens before it gets better
JCUS Joint Center for Urban Studies (MIT and Harvard); Judicial Conference of the United States
JCW JC Williamson
JCWA Japan Child Welfare Association
JCWI Joint Council for the Welfare of Immigrants
jd joined; joint dictionary; junior debutante; jury duty; juvenile delinquency; juvenile deliquent
jd *jemand* (German—someone, somebody)
Jd Jordanian dinar (monetary unit of Jordan)
JD Julian day; Junior Deacon; Junior Dean; Justice Department
J.D. Doctor of Jurisprudence; *Juris* or *Jurum Doctor*

(Latin—Doctor of Law or Laws)
JDA Japan Defense Agency; Japan Domestic Airline; Jefferson Davis Association
J-day Judas Day (Wednesday before Good Friday when Judas is believed to have betrayed Jesus)
JDB Japan Development Bank
JDC Joint Distribution Committee; Juvenile Delinquency Control; Juvenile Detention Center
JDCC Juneau-Douglas Community College
J/deg joule per degree
JDHS Jefferson Davis High School
JDI Juvenile Delinquency Index
JDIC Justice Data Interface Controller
jdl job description language
JDL Jewish Defense League
JDP John Dos Passos
JDPA Japan Dairy Products Association
JDR Japanese Depository Receipts
jds job data sheet
jd's juvenile delinquents
JDS Job Diagnosis Survey; John Dewey Society; Joint Defense Staff (NATO); Justice Data System
J.D.S. Doctor of Juridical Science
JDSCS Joint Defence Space Communications Station (Nurrangar, Australia)
JDSFA Japan Self-Defense Forces Academy
JDSRF Jim Dandy's Still and Refreshment Factory (Australian definition for the Joint Defense Space Research Facility near Alice Springs)
je job estimate
jé *jésus* (French—paper of super-royal size)
jea joint export agent
JEA Jesuit Educational Association; Joint Engineering Agency; Journalism Education Association
jebm jet engine base maintenance
JEC Jardine Engineering Corporation; Joint Economic Committee (Congress)
JECC Japan Electronic Computer Company
JECMA Japan Export Clothing Makers Association
JECMOS Joint Electronic

Countermeasures Operation Section (NATO)
Jed Jedediah
J Ed Journal of Education
JED Japan Engineering Development
JEDEC Joint Electron Device Engineering Council
JEDPE Joint Emergency Defense Plan, Europe (NATO)
JEDS Japanese Expeditions to the Deep Sea
JEE Japan Electronics Engineering
jeep (from GP meaning general purpose) 4-wheel-drive quarter-ton utility vehicle
JEEP Joint Emergency Evacuation Plan
jeepney Filipino-built jitney bus
Jef(f) Geoffrey; Geoffroy; Jefferson; Jeffery; Jeffry
Jeff City Jefferson City, Missouri
Jeff D Jefferson Davis
Jefferson's Thomas Jefferson's Birthday (April 13)
jefm (JEFM) jet engine field maintenance
JEFR Japan Experimental Fast Reactor
JEG John Edward Gray; Joint Exploratory Group (NATO)
Jeho Jehosaphat
JEI Japan Electronics Industry
JEIA Japanese Electronic Industries Association
JEIDA Japan Electronic Industry Development Association
jeim jet engine intermediate maintenance
JEIPAC Japan Electronic Information Processing Automatic Computer
JEJ Japan Economic Journal
jejun jejunectomy; jejunitis; jejunostomy
J Electrochem Soc Journal of the Electrochemical Society
jem jet engine modulation
Jem Jemima
JEMC Joint Engineering Management Conference
JEN Junta de la Energia Nuclear (Atomic Energy Board)
JEN Journal of Emergency Nursing
Jen Jih Jen-min Jih-pao (People's Daily)—published in Peking by Communist Party of China
jentac. jentaculum (Latin—breakfast)
JEOCN Joint European Operations Communications Net-

work
JEOL Japan Electron Optics Laboratory
JEPI Junior Eysenck Personality Inventory
JEPIA Japan Electronic Parts Industry Association
JEPO Jet Engine Project Office
JEPOSS Javelin Experimental Protection Oil Sands System
JEPS Joint Exercise Planning Staff (NATO)
Jer Jersey
Jer. Jeremiah, The Book of the Prophet
Jer Jeremiah; Jeroesjalaim (Dutch—Jerusalem)
JER Japan Economic Review
JERC Japan Economic Research Center; Joint Electronic Research Committee
Jere Jeremiah; Jerry
JERI Japan Economic Research Institute; Joint Economic Research Institute
jerk ineffectual fool
JERK Journalist's Easy Road to Knowledge (routed through an impasse of acronyms such as GAS—Grant's Acronymical Shorthand)
jerky beef jerky; buccan; charqui; jerked beef
jerob jeroboam (4-bottle capacity)
Jeron⁰ Jerónimo (Spanish—Jerome)
JERS Japanese Ergonomics Research Society
Jes Jessica; Jesus; Jesus College, Cambridge
JES James Ewing Society; Japan Electroplating Society; Japan Engineering Standards; John Ericsson Society
JES Journal of Ecumenical Studies
JESA Japanese Engineering Standards Association
Jes Coll Jesus College—Cambridge
jes min just a minute
Jess Jessica
jes sec just a second
Jesus Jesus College, Oxford
jet black lignite; jet-engine aircraft
jet. jetsam
JET Joint European Tours; Joint European Transport
JETCO Jamaican Export Trading Company
JETDS Joint Electronics Type Designation System
JETEC Joint Electron Tube

Engineering Council
jet fag jet flight fatigue
Jeth Jethro
jetma jet mechanic
jet-p jet-propelled; jet propulsion
JETP Journal of Experimental and Theoretical Physics (Academy of Sciences, USSR)
JETRO Japan Exterior Trade Research Organization; Japan External Trade Organization
JETS Joint Enroute Terminal Systems (air traffic); Junior Engineers Technical Society
jett jettison
jeu jeudi (French—Thursday)
Jev Japanese encephalitis virus
JEVA Japan Electric Vehicle Association
Jew. Jewish
Jewel The Jewel (La Jolla, California); Jewel Cave National Monument near Custer in southwestern South Dakota
Jewtong Jewtongo (Surinam dialect spoken by former slaves who picked it up from their masters who spoke Dutch, English, Portuguese, or Spanish within their hearing in the belief they would not understand)
JEZ Johannes Enschede en Zonen
JEZ Journal of Experimental Zoology
jf distant fog (meterological symbol)
j/f jigs and fixtures; journal folio
JF Jewish Federation; Joint Force
JFACT Joint Flight-Acceptance Composite Test
JFAI Joint Formal Acceptance Inspection (NATO)
jfb jet flying belt
jfc Japan Food Company
JFC Japan Film Center
JFCC Japanese Federation of Culture Collections of Microorganisms
JFCS Jewish Family and Child Service
JFEA Japan Federation of Employer's Associations
jfet junction field-effect transistor
JFK John F Kennedy international airport, New York City; John Fitzgerald Kennedy—thirty-fifth President of the United States

JFKCAS John F Kennedy College of Arts and Sciences (Trinidad)

JFKCPA John F Kennedy Center for the Performing Arts

JFKMF John F Kennedy Memorial Forest (near Jerusalem, Israel)

JFKMH John F Kennedy Memorial Highway (Baltimore, Maryland to Wilmington, Delaware)

JFKML John F Kennedy Memorial Library

JFKSC John F Kennedy Space Center

JFKYCC John F Kennedy Youth Correctional Center

jfl joint frequency list

J Fluid Mech Journal of Fluid Mechanics

JFM Jeunesses Fédéralistes Mondiales (French—Young World Federalists)

JFMAMJJASOND January, February, March, April, May, June, July, August, September, October, November, December

JFMIP Joint Financial Management Improvement Program

JFNP John Forrest National Park (Western Australia)

JFO San Francisco, California (heliport)

jfp joint frequency panel

JFP Jobs For Progress

JFPS Japan Fire Prevention Society

jfr jevnfr (Dano-Norwegian—compare)

JFR Joint Fiction Reserve

JFRC James Forrestal Research Center

JFRCA Japanese Fisheries Resources Conservation Association

JFRO Joint Fire Research Organisation (UK)

jfs jet fuel starter

JFS Japan Fishery Society; Jewish Family Service

JFS Jane's Fighting Ships

JFSOC Junior Foreign Service Officers Club

JFTC Joint Fur Trade Committee

JFU Jersey Farmers' Union

JFV Jobs For Veterans

JG junior grade

jga juxtaglomerular apparatus

JGA Japan Golf Association

JGC Japan Gas Chemical; Japan Gasoline Company

JGD John George Diefenbaker (Canada's seventeenth Prime Minister)—also known as Dief the Chief

jg di joggle die

JGE Journal of General Education

J Geophys Res Journal of Geophysical Research

J-girl joy girl (prostitute)

jgn junction gate number

JGNP Japanese Gross National Product

JGR Jaldapara Game Reserve (India); Jamaica Government Railway

JGS Joint General Staff (NATO)

JGSA John G Shedd Aquarium

JGSDF Japan Ground Self-Defense Force

jg sm joggle shims

JGTC Junior Girls Training Corps

JGW Junior Grand Warden

JGWTC Jungle and Guerrilla Warfare Training Center (USA)

jh juvenile hormone

Jh Jahresheft (German—yearly publication)

J & H Jack & Heintz

JH Jugendherberge (German—youth hostel)

jha job hazard analysis

JHA John Howard Association

JHAH John Howard Association of Hawaii

JHAI John Herron Art Institute

Jhb Johannesburg

JHC John Hancock Center

Jhd Japanese haakon dahl (log measure of 11.7 cubic feet being 100 Jhd)

JHDA Junior Hospital Doctors Association

Jhdf Japanese haakon-dahl feet

JHE Journal of Higher Education

JHH Johns Hopkins Hospital

JHI Jacob Hiatt Institute; Jesuit Historical Society; Jewish Historical Society

jhj jail-house juvenile (delinquent)

JHL John Harvard Library

JHMI Johns Hopkins Medical Institutions

JHMO Junior Hospital Medical Officer

JHO Jam Handy Organization; Japan Hydrographic Office

JHOS Johns Hopkins Oceanographic Studies

JHP Jackson Hole Preserve; Johns Hopkins Press

JHS John Howard Society; Judaic Heritage Society; Junior High School

J.H.S. Jesus Hominum Salvator (Latin—Jesus Savior of Men)

JHU Johns Hopkins University

JHUL Johns Hopkins University Library

JHUP Johns Hopkins University Press

JHUSHPH Johns Hopkins University School of Hygiene and Public Health

JHUSM Johns Hopkins University School of Medicine

JHVH Jehovah [transliteration of Hebrew tetragrammaton Yhwh, Yahwah, or Jahvah, used by Hebrew tribes in 3rd century BCE because they thought "Jehovah" was too sacred to pronounce]

JHWH (*see* JHVH)

ji jet interaction; junction isolation; junction isolator

JI Aerovias Sudamericanos (symbol)

JI Japan Interpreter

JIA Japanese Interchange Association

JIAS Jewish Immigration Aid Society

jib. job-information block

Jib Jibouti

JIB Jack-in-the-Box; Japan International Bank; Joint Intelligence Bureau

JIBA Japan Institute of Business Administration

JIBC Japan International Biological Program

JIBICO Japan International Bank and Investment Company

jíb(s) jíbaro(s) [Spanish—peasant farmer(s)]—Puerto Rican(s)

Jibuti Djibouti

jic jet-induced circulation; jet-induced combustion

JIC Joint Industrial Council; Joint Industry Council; Joint Intelligence Center; Joint Intelligence Committee

JICA Japan International Cooperation Agency; Joint Intelligence Collecting Agency

JICST Japan Information Center of Science and Technology

JICTAR Joint Industry Committee for Television Advertising Research

JID Journal of Infectious Diseases; Junta Interamericana de Defensa (Spanish—Inter-American Defense Board)
JIDA Japan Industrial Designers Association
JIDC Jamaica Industrial Development Corporation
JIDPO Japan Industrial Design Promotion Organization
JIE Junior Institution of Engineers
JIEA Japan Industrial Explosives Association
JIFA Japanese Institute of Foreign Affairs
JIFE *Junta Internacional de Fiscalización de Estupefacientes* (Spanish—International Council for the Investigation of Narcotics)
JIG Joint Intelligence Group
JIIST Japan Institute for International Studies and Training
JILA Japanese Institute of Landscape Architects; Joint Institute for Laboratory Astrophysics
Jill Jillian
Jim James
JIM Jakarta Informal Meetings; Japan Institute of Metals; Junior Index of Motivation
JIMA Japan Industrial Management Association
JIMA *Journal of the Israel Medical Association*
Jimmu Jimmu Tenno—first emperor of Japan who began his reign in 660 BCE
Jimtown Jamestown (California, New York, or North Dakota)
jimson weed *Datura stramonium*
J Inorg Nucl Chem *Journal of Inorganic and Nuclear Chemistry*
JINR Joint Institute for Nuclear Research
jins juveniles in need of supervision
JIO Joint Intelligence Organization
JIOA Joint Intelligence Objectives Agency
JIP Joint Installation Plan(ning)
JIPS Japanese Information Processing Service
JIR Jewish Institute of Religion; Job Improvement Request
JIRA Japan Industrial Robot Association

JIRP Juneau Icefield Research Project
jirv jet-interaction reentry vehicle
JIS Jail Inspection Service; Jamaica Information Service; Japan Industrial Standard; Jewish Information Society; Joint Intelligence Staff
JISA Japan Industrial Safety Association
JISC Japanese Industrial Standards Committee
JISEA Japan Iron and Steel Exporters Association
JISF Japan Iron and Steel Federation
JISP Jack Island State Park (Florida)
jit jitney bus
JIT Job Instruction Training
jizz *jizzim* (slang—semen)
jj jaw jerk
JJ Judges, Justices
J-J Jean-Jacques
J & J Johnson & Johnson
J-J *Jen-min Jih-pao* (Chinese—people's daily communist-controlled Peking newspaper)
JJA John James Audubon
JJC Juvenile Justice Center (Los Angeles); Juvenile Justice Clearinghouse
JJCA Sir John Joseph Caldwell Abbott (Canada's fourth Prime Minister)
JJCCJ John Jay College of Criminal Justice
JJHL John Jay Hopkins Laboratory for Pure and Applied Science (General Atomic Division of General Dynamics Corporation)
JJHS John Jay High School
jj's joyless jobs (pimping and prostituting)
Jj's Jewish jokes
JJS James Joyce Society
JJS *Journal of Jewish Studies*
JJSC Juvenile Justice Standards Committee
JJSS Jean-Jacques Servan-Schreiber
J-J S-S Jean-Jacques Servan-Schreiber
jk just kidding
JK Jack Kerouac; Sun World (airline code)
J/°K joule(s) per degree Kelvin (unit of entropy)
J & K Jammu and Kashmir (University)
Jka Jakarta
JKC Jane Kathryn Conrad; Japan Kennel Club

jkg joules per kilogram
JKG John Kenneth Galbraith (economist and diplomat)
J/kg°K joule(s) per kilogram degree Kelvin
JKP James Knox Polk (11th President U.S.)
JKS Julius Kayser (stock-exchange symbol)
jkt jacket
Jkt Jakarta
JKT Jakarta, Indonesia (airport); Job Knowledge Test
jl just looking (pseudo customer)
Jl Joel
JL J Lauritzen (steamship line); Japan Airlines (2-letter code); Johnson Line; Jones and Laughlin; Joseph Lewis
Jla Julia
JLA Jamaica Library Association; Japan Library Association; Jewish Librarians Association; Jordan Library Association
JLB Jewish Lads' Brigade; John Logie Baird (tv's inventor)
JLC Japan Logistical Command; Jewish Labor Committee; Joint Logistics Command(ers)
JLCU Johnson Line Container Unit
Jlem Jerusalem
JLMIC Japan Light Machinery Information Center
Jln *Jalan* (Malay—Lane, Road, Street)
JLOIC Joint Logistics, Operations, Intelligence Center (NATO)
J Low Temp Phys *Journal of Low Temperature Physics*
JLP Jamaica Labour Party
JLPPG Joint Logistics and Personnel Policy Guidance
jlr(s) jeweler(s)
JLRSS Joint Long-Range Strategic Study
jls jewels
JLS Jail Library Service (California State Library); Junior Literary Society
Jlt Juliet
J Lumin *Journal of Luminescence*
JM James Madison (4th President U.S.); James Monroe (5th President U.S.); Japan Mail; Jardine Matheson; Jewish Museum; José Martí
J-M Johns-Manville
J-M *Jiyu-Minshuto* (Japanese—Liberal Democratic

Party)
JMA joule per meter (impact strength) squared
JMA Japan Management Association; Japan Medical Association; Japan Meterological Agency; Jewish Music Alliance
J Macromol Sci Journal of Macromolecular Science
J Math Phys Journal of Mathematical Physics (published in New York City); *Journal of Mathematics and Physics* (published in Cambridge, Mass)
JMB J(ames) M(atthew) Barrie
JMBA Journal of the Marine Biological Association
JMC Japan Metals and Chemicals; Japan Monopoly Corporation; Jefferson Medical College; Jerusalem Music Centre; Joint Maritime Commission; Joint Maritime Congress
JMCC Joint Mobile Communications Center (NATO)
JMD M(alaby) Dent
JMDC Japan Machinery Design Center
JMD & S Joseph Malaby Dent and Sons
J Mech Phys Solids Journal of the Mechanics and Physics of Solids
jmed jungle message encoder decoder
JMF Jewish Music Forum; Juilliard Musical Foundation
JMHS James Madison High School; James Monroe High School; John Muir High School
JMI Japan Machinery and Metal Inspection; John Muir Institute
JMIA Japan Mining Industry Association
JMIF Japan Motor Industrial Federation
JMJ Jesus, Mary, and Joseph
JMMA Japan Materials Management Association
JMMC James Madison Memorial Commission
JMMF James Monroe Memorial Foundation
JMMII Japan Machinery and Metal Inspection Institute
J Mol Spectro Journal of Molecular Spectroscopy
JMP Jen Men Piao (Chinese—People's Bank Dollar)
jmpr jumper
JMPTC Joint Military Packag-

ing Training Center
JMRMA John and Mable Ringling Museum of Art
JMRP Joint Meteorological Radio Propagation
JMRT Junior Members Round Table
JMS James Madison Society; Japan Medical Society; Johannesburg Musical Society
JMSA Japanese Maritime Safety Association
JMSDF Japanese Maritime Self-Defense Force
JMSO Joint Meetings of Seafarers' Organization
jmt (JMT) jointly-managed trust
JMT Job Methods Training
JMTBA Japan Machine Tool Builders Association
JMTR Japan Material Testing Reactor
JMUSDC Joint Mexico-United States Defense Commission
jn join; junction
j-n jet navigation
Jn John
Jn Juan (Spanish—John)
JNA Jordan News Agency
JNA Jena Nomina Anatomica; Jugoslovenska Narodna Armija (Serbo-Croat—Yugoslav People's Army)
JNB Johannesburg, South Africa (airport)
Jnc Junction
JNC Joint Negotiating Committee
JNCA Junior Naval Cadets of America
jnd joined; just noticeable difference
JND Juvenile Narcotics Division
JNDC Jamaica National Dance Company
JNDNWR JnN. (Ding) Darling National Wildlife Refuge (Florida)
jne ja niin edespäin (Finnish—and so on)
JNEC Jamaican National Export Corporation
JNF Japan Nuclear Fuel (company); Jewish National Fund
JNI Journal of the Nautical Institute
JNIP Jamaican National Investment Promotion
Jnl Journal
JNL Japanese National Laboratory
jnls journals
jnlst journalist
JNM Journal of Nuclear Medi-

cine
JNN Japan News Network
jnnd just not noticeable difference
Jno John
JNODC Japanese National Oceanographic Data Center
J Non-Cryst Solids Journal of Non-Crystalline Solids
JNP Jasper National Park (Alberta)
JNPGC Japan Nuclear Power Generation Corporation
jnr joiner; junior
Jnr Jesurun
JNR Japanese National Railways
jns just noticeable shift
Jns Johannes
JNS Japan Nuclear Society; Jet Noise Survey
JNSDA Japan Nuclear Ship Development Agency
jnt joint; junction; juncture
JNT John Napier Turner—Canada's 17th Prime Minister if counted by name and 22nd if counted by terms in office
JNTA Japan National Tourist Association
JNTO Japan National Tourist Office; Japan National Tourist Organization
jnt stk joint stock
JNU Juneau, Alaska (airport)
J Nucl Energy Journal of Nuclear Energy
J Nucl Mater Journal of Nuclear Materials
JNUL Jewish National and University Library (Jerusalem)
JNV Junta Nacional do Vinho (Portuguese—National Wine Board)
jnwpu joint numerical weather-prediction unit
JNZ Jewelers of New Zealand
jo journalist
Jo Joel; Joseph; Josephine
JO Job Order; Juilliard Orchestra; Jupiter Orbiter
JO Journal Officiel (French—Official Journal); *Justie Ombudsman* (Swedish—representative of justice)
Joa Joachim
JOA Joint Operating Agreement
joalh joalharia (Portuguese—jewelry store); *joalheiro* (Portuguese—jeweler); *joalheria* (Portuguese—jewelry store)
Jo Bapt John the Baptist

joblib job library
jo block(s) johannson block(s)
jobman job management
JOBS Job Opportunities in the Business Sector
Jo'burg Johannesburg
joc jocose; jocular
JOC Japan Olympic Committee; Joint Operations Center; Joint Opposition Council
jock jockey; jockstrap
jock(s) jock strap(s)—nickname for physical education student(s)
Jock(s) Scot(s)
joco jocose
JOCV Japan Overseas Cooperation Volunteers (Peace Corps)
jod joint occupancy date
JODC Japanese Oceanographic Data Center
Jo Div John the Divine
Joe Joel; Joseph; Josephine
JOE Juvenile Opportunities Extension
Joe C Joe Clark (Canada's 16th Prime Minister)
JOERA Japan Optical Engineering Research Association
Jo Evang John the Evangelist
joey baby kangaroo
J-off jack off (underground slang—masturbate)
jog. joggle
JOG Joint Operations Group; Junior Ocean Group (*jay-oh-gees*—smallest sailing cruisers); Junior Offshore Group
Jogja Jogjakarta
Jogjakarta Djokjakarta
Joh St John's College, Cambridge
Joh *Johann(es)* (German—Hans, John)
Johan Johannesburg
John The Gospel According to John; St John, American Virgin Islands; St John, New Brunswick, Canada
John B John B Stetson (hat)
John D. John D Rockefeller, Sr.
JOHNNIAC John von Newman's Integrator and Automatic Compiler
Johns H Johns Hopkins University
Joh Seb Bach Johann Sebastian Bach
JOI Joint Oceanographics Institution
JOIDES Joint Oceanographic Institutions for Deep Earth Sampling
JOIDESP Joint Oceanographic

Institutions Deep Earth Sampling Program
join. joinery
JOIN Job Orientation in Neighborhoods; Jobs Or Income Now
Joint American Jewish Joint Distribution Committee
Joio Norman Dello Joio
JOIS Japan On-Line Information System
JOK Oakland, California (heliport)
jol job organization language
JOLA *Journal of Library Automation*
JoLoPo José Lopez Portillo
Jolyon Joseph Lyons
JOM Job-Oriented Manual
JOM *Johnson O'Malley Act*
Jom-Isl-Irân *Jomhori-e-Islami-e-Irân* (Farsi—Islamic Republic of Iran)
JOMO Junta of Militant Organizations (Black Nationalists)
Jon. The Book of Jonah
Jona Jonathan
Jonathan Jonathan David
JONS *Juntas de Ofensiva Nacional Sindicalista* (Spanish—United National Syndicalist Offensive)—fascist anti-syndicalists
JONSDAP Joint North Sea Data Acquisition Program
JONSIS Joint North Sea Information Systems
JONSWAP Joint North Sea Wave Project
JOOD Junior Officer of the Deck
JOOM Junior Observers Of Meteorology
JOP Joint Operating Plan(ning); Joint Operations Procedure
JOPCN Job Order Program Control Number (USA)
JOPM Joint Occupancy Plan Memo; Joint Operation Procedure Memo
JOPR Joint Operation Procedure Report
JOPS Joint Operating Study
JOPS *Journal of the Patent Office Society*
J Opt Soc Am *Journal of the Optical Society of America*
JOR Jet Operations Requirements
Jord Jordan
Jord *Jordânia* (Portuguese—Jordan); *Jordania* (Spanish—Jordan)
Jordan Hashemite Kingdom of Jordan (Middle East country

formerly called Transjordania), *Al Mamlaka al Urduniya al Hashemiya*
JORG Joint Oceanographic Research Group
Jos Joseph; Joshua; Josiah; Jossie
JOS Junior Ordinary Seaman
Josa Josepha; Josephine
José Ferrer José Vicente Ferrer de Otero y Cintrón
Josh. The Book of Joshua
Josh Joshua
Joshua Tree Joshua Tree National Monument north of the Salton Sea in southern California
JOSS JOHNNIAC Open-Shop System
Josy Joseph
jot. jump-oriented terminal; junction optimization technique
JOT Joint Observer Team
JOTS Job-Oriented Training Standards
JOUAM Junior Order of United American Mechanics
jour journal; journalese; journalism; journalist; journalistic; journey
journ journal; journalese; journalism; journalist; journalistic; journey
Jove Jupiter or Zeus
JOVE Job Placement on the Job Training Vocational Education Educational Assistance; Jupiter Orbiting Vehicle for Exploration
JOVIAL Jules' Own Version of IAL (International Algebraic Language)
jp jet penetration; jet pilot; jet power; jet propulsion; junior partner; precipitation in sight but not at weather station reporting (symbol)
j & p joists and planks
Jp Japan(ese)
JP Japan Press (news agency); Jaya Prakash Narayan, Prime Minister of India; Jet Pilot; Justice of the Peace
J.P. Jayaprakash Narayan; J Pierpont Morgan
JP *Jerusalem Post*
JP-4 jet propellant 4
jpa jack panel assembly
JPA Jamaica Press Association; Japan Petroleum Association; Japan Procurement Agency; Joint Passover Association; Joint Powers Agreement
JPB Joint Planning Board;

Joint Production Board; Joint Purchasing Board; Judah Philip Benjamin
JPBHS Judah P Benjamin High School
jpbs jettison pushbutton switch
JPC Jan Pieterszoon Coen; Japan Productivity Center; Jet Propulsion Center; Joint Planning Center; Joint Planning Council; Joint Production Council; Joint Publishers Committee; Judge of the Prize Court
JPCAC Joint Production, Consultative, and Advisory Committee
JPCC Joint Petroleum Coordination Center (NATO)
JPCRSP John Pennekamp Coral Reef State Park (Florida)
JPDC Japan Petroleum Development Corporation
JPDR Japan Power Demonstration Reactor
JPF Jewish Peace Fellowship
j-p fuel jet-propulsion fuel
JPG J Peter Grace; Job Proficiency Guide; Joint Planning Group (NATO)
JPGA Japan Professional Golf Association
JPGM J Paul Getty Museum
J Phys Journal de Physique (French—*Journal of Physics*)
J Phys Chem Journal of Physical Chemistry
J Phys Chem Ref Data Journal of Physical and Chemical Reference Data
J Phys Radium Journal de Physique et le Radium (French—Journal of Physics and Radium)
J Phys Soc Jpn Journal of the Physical Society of Japan
JPI Joint Packaging Instruction
JPIA Japan Plastics Industry Association
JPJ John Paul Jones
JPL Jacksonville Public Library; Java Pacific Line; Jet Propulsion Laboratory (California Institute of Technology); Job Parts List
J Plasma Phys Journal of Plasma Physics
Jpn Japan(ese)
Jpn J Appl Phys Japanese Journal of Applied Physics
Jpn J Phys Japanese Journal of Physics
Jpns Japanese; Japan's
JPO Japan Patent Office; Joint Petroleum Office; Joint Pro-

gram Office; Joint Project Offices; Junior Police Officer
J Pol Eco Journal of Political Economy
J Polymer Sci Journal of Polymer Science
jpp jälkeen puolenpäiven (Finnish—afternoon, P.M.)
JPPS Japan Pearl Promoting Society
JPPSOWA Joint Personal Property Shipping Office (Washington, D.C.)
JPR Joint Procurement Regulation(s)
J Prob Judge of Probate
JPRS Joint Publications Research Service
JPRST (guo) Joint Publications Research Service Translations (government use only)
JPs Jesuit priests
JPS Jet Propulsion Systems; Jewish Publication Society; Johannesburg Philharmonic Society; Joint Planning Staff; Juvenile Probation Services
J-P S Jean-Paul Sartre
JPSA Jewish Publication Society of America
JPSA Journal of Police Science and Administration
JPSO Jamaica Philharmonic Symphony Orchestra
jpt jet pipe temperature
JPT Journal of Petroleum Technology
JPTDS Joint Photographic Type Designation System
jpto jet-propelled take-off
JPV Japan Peace Volunteers
jpw job processing word
JP-X jet-propellant rocket fuel
JPz4-5 West German tank-destroyer tracked vehicle
jq job questionnaire
JQ Japan Quarterly; Journalism Quarterly
JQA John Quincy Adams (6th President U.S.)
JQAH John Quincy Adams House
J Quant Spectros Radiat Transfer Journal of Quantitative Spectroscopy and Radiative Transfer
jr jinx ratio
Jr Journal; Junior
JR Joint Resolution
JR Journal of Religion; Jugoslav Register (of shipping)
J.R. Jacobus Rex (Latin—King James)
jra junior rheumatoid arthritis
JRA Japan Racing Associa-

tion; Japan Ryokan Association; Japanese Red Army (terrorist group)
JRAI Journal of the Royal Anthropological Institute
Jr Asst Pur Junior Assistant Purser
JRATA Joint Research and Test Activity
JRB New York, New York (Wall Street Heliport)
JRC Jamaica Railway Corporation; Japan Red Cross; Joint Rivers Commission; Junior Red Cross
JRCA Junior Ruritan Clubs of America
JRCD Journal of Research in Crime and Delinquency (published semi-annually by NCCD)
jrci jamming radar coverage indicator
JRCS Jet Reaction Control System
JRD Riverside, California (heliport, 3-letter code)
JRDB Joint Research and Development Board
JRDC Japan Research and Development Corporation
JREA Japanese Railway Engineering Association
J Res Nat Bur Stand Journal of Research of the National Bureau of Standards
JRF Job Request Form; Judicial Research Foundation
jrg jaargang (Dutch—year)
jr gr junior grade
Jr HS Junior High School
JRHS Julia Richman High School
jri jail release information
JRI Japan Research Institute
JRIA Japan Radioisotope Association; Japan Rocket Industry Association; Japan Rubber Industry Association
JRMO Junior Resident Medical Officer
JRN Japan Radio Network
Jro Jerome
JROTC Junior Reserve Officers' Training Corps
JRPG Joint Radar Planning Group
JRR Japan Research Reactor
JRRC Joint Regional Reconnaissance Center (NATO)
J.R.R. Tolkien John Ronald Reuel Tolkien
JRS Jerusalem, Jordan (airport)
JRSMA Japan Rolling Stock Manufacturers Association

JRSWG Joint Reentry System Working Group

JRT Jaguar Rover Triumph Inc; Job Relations Training

JRTUR Jugoslovenska Radio-Televisija Udruzenja Radiostancia (Yugoslav Association of Radio and Television Stations)

Jrw Jarrow-on-Tyne

J's joints (of marijuana)

J/s jamming-to-signal ration

Js Jesuits

JS Al-Jamhourya as-Souriya (Syria); Jan Sibelius; Japan Society; Jet Study; Johnson Society; Judeo-Spanish; Judgement Summons; Judicial Separation; Junior Sailor

J-S Judeo-Spanish

JS-2 Soviet heavy tank of World War II vintage

JS-3 Soviet post WW II heavy tank

JSA Jewelers Security Alliance; Journeymen Stone Cutters Association; Junior Statesmen of America

jsact jetstream anti-countermeasure trainer

JSACT Joint Strategic Air Control Team

JSAE Japan Society of Automotive Engineers

JSAP Japan Society of Applied Physics

JSB Jewish Society for the Blind; Jewish Statistical Bureau; Johann Sebastian Bach

JSBs Joint Stock Banks

JSC Jackson State College; Japan Science Council; Johnson Space Center (NASA); Joint Staff Council; Joint Standing Committee; Joint Stock Company; Justice Statistics Clearinghouse

JS-C Jesus College—Cambridge (also appears as JCC, J.C.C., and Jes Coll or Jes. Coll.)

JSCA Journeyman Stone Cutters Association

JSCC Japan Securities Clearing Corporation

J.Sc.D. Doctor of Juristic Science

J-school journalism school

J Sci Instrum Journal of Scientific Instruments

JSCM Joint Service Commendation Medal

JSCP Joint Strategic Capabilities Plan

JSCR Job Schedule Change Request

JSCS Joint Strategic Connectivity Staff

JS & CS Jewish Family and Child Services

J.S.D. Jurum Scientiae Doctor (Latin—Doctor of the Science of Laws)

JSDA Japan Self-Defense Agency

JSDFA Japan Self-Defense Forces Academy

JSDFs Japan Self-Defense Forces

JSDT Sir John Sparrow David Thompson (Canada's fifth Prime Minister)

JSDTI John S Donaldson Technical Institute (Trinidad)

JSE Johannesburg Stock Exchange

JSEA Japan Ship Exporters Association

JSEE Japanese Society for Engineering Education

JSEM Japan Society of Electrical Discharge Machining; Japan Society for Electron Microscopy

JSESPO Joint Surface Effect Ships Program

Jsey Jersey

JSF Japan Scholarship Foundation; Jewish Student Federation; Junior Statesman Foundation

JSFC Japanese-Soviet Fisheries Commission

JSGMF John Simon Guggenheim Memorial Foundation

JSGMRAM Joint Study Group for Material Resource Allocation Methodology

jsi job satisfaction inventory

JSI Japanese Studies Institute

JSIA Japan Software Industry Association

JSIF Japan Spinners Inspecting Foundation

JSIIDS Joint-Services Interior-Intruder Detection System

JSL Jurong Shipyard Limited

JSLB Joint Stock Land Bank(s)

JSLE Japan Society of Lubrication Engineers

JSLS Joint Services Liaison Staff

JSM Joint Staff Mission; Juilliard School of Music

JSMA Joint Sealers Manufacturers Association

JSMB Joint Sealift Movements Board

JSMDA Japan Ship Machinery Development Association

JSME Japan Society of Mechanical Engineers

JSMEA Japan Ship Machinery Export Association

J-smoke (underground slang—marijuana cigarette)

JSNP Japan Satellite News Pool

JSO Jackson Symphony Orchestra; Jacksonville Symphony Orchestra; Joint Services Organization; Judgement Summons Order

J Soc Arts Journal of the Society of Arts

JSOP Joint Strategic Objectives Plan

J Sound Vib Journal of Sound and Vibration

JSP Japan Socialist Party

JSP Jadranska Slobodena Plovida (Yugoslavian Shipping Line)

JSPA Japan Screen Printing Association

JSPB Joint Staff Pension Board (UN)

JSPC Joint Strategic Plans Committee

J Speech Hear Disorders Journal of Speech and Hearing Disorders

J Speech Hear Res Journal of Speech and Hearing Research

jspf jet shots per foot

JSPF Joint Staff Pension Fund (UN)

JSPG Joint Strategic Plans Group

JSPS Japan Society for the Promotion of Science; Japan Sword Preservation Society

JSQC Japan Society for Quality Control

JSQS Japan Shipbuilding Quality Standard

jsrt joint short range technology

JSS Johnson Scan Star (Johnson, East Asiatic, and Blue Star lines); Joint Services Standard

JSSA Japan Science Student Awards

JSSC Joint Services Staff College; Joint Strategic Service Committee

JST Japan Standard Time; Javanese Standard Time; Job Safety Training

J Stat Phys Journal of Statistical Physics

JSTC Japan-Singapore Training Center

J-stick joystick (underground slang—marijuana cigarette)

JSTPB Joint Strategic Target Planning Board
JSTPS Joint Strategic Target Planning Staff
JSU Jewish Student Union
JSU-122 Soviet 122mm assault-gun howitzer (SU-122)
JSU-152 Soviet assault-gun howitzer (SU-152)
J-S unit Junkerman-Schoeller unit (of thyrotrophin)
JSW Japan Steel Works
JSWPB Joint Special Weapons Publications Board
JSY Jersey Airlines
jt joint; joint tenancy; junction
JT Air Oregon (2-letter code); Jamaica Air Service (symbol); John Tyler (10th President U.S.); joint tenancy; Juvenile Templar
JT *Japan Times* (Japan's oldest English newspaper); *John Thomas* (British slang—penis)
JTA Jewish Telegraphic Agency (news service)
JTAC Joint Technical Advisory Committee
JTAD Joint Tactical Aids Detachment
jt agt joint agent
jt auth joint author
jtb joint bar
JTB Jamaica Tourist Board; Japan Travel Bureau; Jute Trade Board
JTBI Japan Travel Bureau International
JTC Japan Tobacco Corporation; Joint Technical Committee; Joint Telecommunications Committee; Junior Training Corps
JTCGALNNO Joint Technical Coordinating Group for Air-Launched Non-Nuclear Ordnance (DoD)
JTCGAS Joint Technical Co-ordinating Group for Aircraft Survivability (DoD)
jt comp joint compiler
jtda joint track data storage
jtde joint technology demonstration engine
J-teacher journalism teacher
Jt Ed Joint Editor
JTES Japan Techno-Economics Society
JTF Joint Task Forces
JTF-4 Joint Task Force 4
JTFOA Joint Task Force Operating Area
Jth. Apocryphal Book of Judith
jthh justice the helping hand

JTI *Journal of the Textile Institute; Jydsk Teknologisk Institut* (Danish—Jutland Technological Institute)
JTIDS Joint Tactical Information Distribution System (USAF and USN)
JTII Japan Telescopes Inspection Institute
jtly jointly
JTM&H *Journal of Tropical Medicine and Hygiene*
jtms jamb-template machine screws
JTNM Joshua Tree National Monument
jto jump takeoff
JTO Jordan Tourist Office
J-town Juarez
JTPA Job Training Partnership Act
JTPT Job Task Performance Test
jt r joint rate
JTR Joint Termination Regulation; Joint Travel Regulation; Jordan Travel Research
JTRC Joint Theater Reconnaissance Committee (NATO)
JTRE Joint Tsunami Research Effort
JTRU Joint-Services Tropical Research Unit
JTS Jewish Theological Seminary; Job Training Standards
JTS *Journal of Theological Studies*
JTSA Jewish Theological Seminary of America
JTSG Joint Trials Subgroup (NATO)
jtst jet stream
jt stk joint stock
JTTA Japan Table Tennis Association
jt ten joint tenancy; joint tenant
jt ten. joint tenant(s)
JTWC Joint Typhoon Warning Center
jtwros joint tenant with right of survivorship
J-type Jungian judging type
ju jackup (oil well); joint use
Ju June; Junkers
JU Jacksonville University; Jadavpore University
JU *Jeunesse Universelle* (French—World Youth)
JU-52 German Junkers transport developed before World War II and used by many airlines
juana marijuana
Juana *Juana la Loca* (Span-

ish—Crazy Jane)—nickname of the demented and lisping daughter of Ferdinand and Isabella; when queen of Castile in 1504 her courtiers flattered her by lisping in the manner still called Castilian; title of an opera by Gian Carlo Menotti—*Juana la Loca*
Juan Carlos Juan Carlos de Bourbon—chief of state and king of Spain
Juan Fernández Islas Juan Fernández (South Pacific islands Robinson Crusoe, Santa Clara, and Alejandro Selkirk)
Juárez Ciudad Juárez (formerly El Paso del Norte)
juco junior college
jucund. jucunde (Latin—pleasantly)
jud judgment; judicial; judo
Jud Judah; Judaic; Judaism; Judean; Judson
J.U.D. *Juris Utriusque Doctor* (Latin—Doctor of Civil and Canon Law)
Jud-Alg Judeo-Algerian (Algerian Jewish)
Jud-Amer Judeo-American (American Jewish)
Jud-Arg Judeo-Argentinian (Argentine Jewish)
Jud-Ash Judeo-Ashkenazic (Ashkenazic Jewish) or the oriental branch of Yiddish-speaking Jews of eastern Europe (*see* Jud-Sep)
Jud-Aus Judeo-Austrian (Austrian Jewish)
Jud-Aust Judeo-Australian (Australian Jewish)
Jud-Bel Judeo-Belgian (Belgian Jewish)
Jud-Bol Judeo-Bolivian (Bolivian Jewish)
Jud-Bra Judeo-Brazilian (Brazilian Jewish)
Jud-Bul Judeo-Bulgarian (Bulgarian Jewish)
Jud-Can Judeo-Canadian (Canadian Jewish)
Jud-Chi Judeo-Chilean (Chilean Jewish)
Jud-Chr Judeo-Christian (biblical and historic connection between Jews and Christians)
JUDCLA *Juventud Demócrata Cristiana Latino-Americana* (Spanish—Latin American Christian Democratic Youth)
Jud-Col Judeo-Colombian (Colombian Jewish)
Jud-CR Judeo-Costa Rican

(Costa Rican Jewish)
judcrit judicial critic(ism)
Jud-Cub Judeo-Cuban (Cuban Jewish)
Jud-Cur Judeo-Curaçoan (Curaçoan Jewish)
Jud-Czech Judeo-Czechoslovakian (Czechoslovakian Jewish)
Jud-Dan Judeo-Danish (Danish Jewish)
Jud-Dut Judeo-Dutch (Dutch Jewish)
Jude The General Epistle of Jude
Jud-Ecu Judeo-Ecuadorean (Ecuadorean Jewish)
Jud-Egy Judeo-Egyptian (Egyptian Jewish)
Jud-Eng Judeo-English (English Jewish)
Judes Judesmo (Ladino)
Jud-Eth Judeo-Ethiopian (Ethiopian Jewish)
Jud-Fin Judeo-Finnish (Finnish Jewish)
Jud-Fre Judeo-French (French Jewish)
Judg. The Book of Judges
Judg Judges
Judge Adv Gen Judge Advocate General
Jud-Ger Judeo-German (German Jewish)—Yiddish dialect retaining many German words although written in Hebrew
Jud-Gib Judeo-Gibraltarian (Gibraltarian Jewish)
Jud-Gre Judeo-Grecian (Grecian Jewish)
judgt judgment
Jud-Guat Judeo-Guatemalan (Guatemalan Jewish)
Jud-His Judeo-Hispanic (Hispanic Jewish)—Portuguese-Spanish Jewish
Jud-HK Judeo-Hong Kongese (Hong Kongese Jewish)
Jud-Hung Judeo-Hungarian (Hungarian Jewish)
Jud-Ind Judeo-Indian (Indian Jewish)
Jud-Ire Judeo-Irish (Irish Jewish)
Jud-Irn Judeo-Iranian (Iranian Jewish)
Jud-Isr Judeo-Israeli (Israeli Jewish)
Jud-Itl Judeo-Italian (Italian Jewish)
Jud-Jam Judeo-Jamaican (Jamaican Jewish)
Jud-Jap Judeo-Japanese (Japanese Jewish)
Jud-Jor Judeo-Jordanian (Jor-

danian Jewish)
Jud-Lad Judeo-Ladino (Ladino Jewish)—Ladino-speaking Jews who settled in Muslim lands around the Mediterranean following the expulsion of the Jews from Portugal and Spain; Ladino combines medieval Castilian with Arabic, Hebrew, Turkish, and other elements local to places where they settled
Jud-Leb Judeo-Lebanese (Lebanese Jewish)
Jud-Mex Judeo-Mexican (Mexican Jewish)
Jud-Mor Judeo-Moresque (Moorish Jewish) also called Judeo-Moroccan (Moroccan Jewish)—Ladino-speaking Jews whose ancestors came to Morocco and other parts of northwest Africa when expelled from Spain by the Inquisition
Jud-Nor Judeo-Norwegian (Norwegian Jewish)
Jud-NZ Judeo-New Zealand (New Zealand Jewish)
Jud-Pan Judeo-Panamanian (Panamanian Jewish)
Jud-Par Judeo-Paraguayan (Paraguayan Jewish)
Jud-Per Judeo-Peruvian (Peruvian Jewish)
Jud-Pol Judeo-Polish (Polish Jewish)
Jud-Port Judeo-Portuguese (Portuguese Jewish)
Jud-Rho Judeo-Rhodesian (Rhodesian Jewish)
Jud-Rom Judeo-Romanian (Romanian Jewish)
Jud-Rus Judeo-Russian (Russian Jewish)
Jud-SAf Judeo-South African (South African Jewish)
Jud-Scot Judeo-Scottish (Scottish Jewish)
Jud-Sep Judeo-Sephardic (Sephardic Jewish)—Portuguese-Spanish Jewish or the occidental branch of European Jews who settled in Portugal and Spain before expulsion by the Inquisition (*see* Jud-Ash)
Jud-Sin Judeo-Singaporan (Singaporan Jewish)
Jud-Slav Judeo-Slavic (Slavic Jewish)
Jud-Span Judeo-Spanish (Spanish Jewish)—also called Ladino or Spanish Yiddish, a dialect composed of medieval Spanish plus

Arabic, Hebrew, and Turkish terms; Ladino is heard around the Mediterranean from Morocco to the Balkans, Greece, and Turkey; Ladino is written in Hebrew characters
Jud-Sur Judeo-Surinamer (Surinamer Jewish)
Jud-Swe Judeo-Swedish (Swedish Jewish)
Jud-Swiss Judeo-Swiss (Swiss Jewish)
Jud-Syr Judeo-Syrian (Syrian Jewish)
Jud-Tun Judeo-Tunisian (Tunisian Jewish)
Jud-Tur Judeo-Turkish (Turkish Jewish)—Ladino-speaking Jews who settled in the Turkish Empire after their expulsion from Spain and Portugal by the Holy Inquisition
Jud-Uru Judeo-Uruguayan (Uruguayan Jewish)
Jud-Ven Judeo-Venezuelan (Venezuelan Jewish)
Jud-Yem Judeo-Yemenite (Yemenite Jewish)
Jud-Yug Judeo-Yugoslavian (Yugoslavian Jewish)
jue jueves (Spanish—Thursday)
Juec Jueces (Spanish—Judges)
juev jueves (Spanish—Thursday)
Jug Jugoslavia (Yugoslavia)
JUG Joint Users Group
JUGC Jugolinija Container
Jugolinija Yugoslav Line
Jugoslav(ia)(n) Yugoslav(ia)(n)
Jugoslavija Yugoslavia
Jug(s) Jugoslavia(n)(s)
juil juillet (French—July)
jul julho (Portuguese—July); *julio* (Spanish—July)
Jul July
Jul Caes Julius Caesar
Juliana Juliana Louise Emma Marie, Queen of the Netherlands, 1948–1980
Julians Julian Alps (northwestern Yugoslavia)
Juliet J-class Soviet submarines (diesel-powered and missile-launching); letter J radio code
Julⁿ Julián (Spanish—Julius)
Julust July and August
Jum Arab Yam al-Jumhuriyat al-Arabiyah al-Yamaniyah (Arabic—Yemen Arab Republic)—North Yemen, merged with Yemen on May

23, 1990
Jum Dji Jumhoūrīyya Djibouti (Arabic—Republic of Djibouti)
JUMIP Juror Utilization and Management Incentive Program
Jum 'Iraqia al Jumhouriya al 'Iraqia (Arabic—Republic of Iraq)
Jum Lub al-Jumhouriya al-Lubnaniya (Arabic—Republic of Lebanon)
Jum Misr Jumhoūrīyya Misr al-Arabiya (Arabic—Arab Republic of Egypt)
JUMPS Joint Uniform Military Pay System
Jum Qum Itt Islam Jumhurīyat al-Qumur al-Itthadiyah al-Islamiyah (Swahili Arabic— Federal Islamic Republic of the Comoros)
Jum Tunis al-Jumhuriyah at-Tunisiyah (Arabic—Republic of Tunisia)
Jum Yam Dim Jumhuriyat al-Yaman ad-Dimuqratiyah (Arabic—People's Democratic Republic of Yemen)— South Yemen merged with North Yemen on May 23, 1990
jun juniore (Italian—junior); *junio* (Spanish—June)
Jun June; Juneau; Junior
Juⁿ Julián (Spanish—Julius)
Junc Junction
Junct Junction
jun part. junior partner
Junuly June and July
Jup Jupiter
jur juridical
jur juridisch (Dutch—juridical); *juridisk* or *jurist* (Dano-Norwegian—legal or lawyer)
Jur Jurassic
Jur Juridisch (German—juridical)
Juras Jura Mountains between France and Switzerland
Jur.D. Juris Doctor Latin— Doctor of Law
jurimet(s) jurimetrician(s); jurimetric(s)
juris jurisdiction
JURIS Justice Retrieval and Inquiry System (U.S. Department of Justice); Juvenile Referral Information System
jurisd jurisdiction
jurisp jurisprudence
jus justice(s)
jus' just
Jus Justin
jusc. jusculum (Latin—broth)

JUSCIMPC Joint United States-Canada Industrial Mobilization Planning Committee (NATO)
JUSE Japanese Union of Scientists and Engineers
J-U.S. FC Japan-United States Friendship Commission
JUSMAG Joint United States Military Advisory Group; Joint United States Military Aid Group to Greece
JUSMAP Joint United States Military Advisory and Planning Group
JUSMG Joint United States Military Group
JUSMMAT Joint United States Military Mission for Aid to Turkey
JUSPAO Joint United States Public Affairs Office
juss jussive
Juss Jussieu
just. justification
Just Justinian
Justice Department of Justice; Hall of Justice; United States Department of Justice
Justin Justin cowboy boots (made by Joe Justin in Fort Worth, Texas)
JUSTIS Japan-United States Textile Information Service
juv juvenile
Juv Juvenal
juve juvenile
juve delinq juvenile delinquent
juve gang juvenile gang
juven juvenile; juvenilization; juvenilized; juvenilizing
juvie juvenile delinquent; juvenile hall; juvenile law-enforcement officer
JUWTFA Joint Unconventional Warfare Task Force—Atlantic
JUWTFP Joint Unconventional Warfare Task Force—Pacific
jux juxtapose; juxtaposition
jv japanese vellum; joint venture; jugular vein; jugular venous
Jv Java; Javanese
JV Jules Verne; Junior Varsity
JVA Jordan Valley Authority
J Vac Sci Technol Journal of Vacuum Science and Technology
JVC Japan Victor Company
jvp japanese vellum proofs; jugular venous pulse
jvp (JVP) jugular venous pulse
JVS Jewish Vocational Service; Joint Vocational School

jw jacket water; jugwell (hydrocarbon storage well); junior wolf (a young philanderer)
JW Jehovah's Witnesses
JWA Japan Whaling Association
jwac jacket water aftercooled
J-walk jaywalk (cross streets against traffic lights at any part of the street except the pedestrian crossing)
J-walker jaywalker
JWB Jewish Welfare Board; Joint Wages Board; Joint Welfare Board
jwc junction wire connector
JWC Joint Working Committee
JWCA Japan Watch and Clock Association
JWDS Japan Work Design Society
JWEF Joinery and Woodwork Employers Federation
JWGA Joint War Games Agency
JWI Jack Winter (stock-exchange symbol)
JWJ James Weldon Johnson
JWJL JW Jagger Library (Cape Town)
jwl jewel; jeweler
JWL Johnston Warren Lines
jwlr jeweler
jwlry jewelry
j & wo jettison and washing overboard
JWO Jardine Waugh Organisation
JWPAC Joint Waste Paper Advisory Council
JWPT Jersey Wildlife Preservation Trust
JWR Joint War Room
JWR Jane's World Railways
JWs Jehovah's Witnesses
JWS Japan Welding Society
JWT J Walter Thompson (advertising agency)
JWTC Jungle Warfare Training Center
JWU Jewelry Workers' Union
JWV Jewish War Veterans (of the United States)
JW von G Johann Wolfgang von Goethe
JX Bougainville Air Service (2-letter code)
J.X. Jesus Christ
Jy Jenny; July; Jury
JY British United Channel Islands Airways (2-letter coding)
JYL Jugolinja-Yugoslav Line
Jyll Jylland (Danish—Jutland)

Jylland (Danish—Jutland)
JZP Jersey Zoological Park

JZS Jersey Zoological Society
JZS *Jugoslovenski Zavod za*

Standardizacija (Jugoslavian Standards Institution)

K

k Boltzman constant; carat (karat); cathode or vacuum tube; coefficient of alienation; compressibility factor; cumulus (symbol); force constant; keel; killed; kilo; kilo-(gram)(s); knot(s); kweer (homosexual); reaction velocity constant; reproduction factor; thermal conductivity; torsion constant; unit vector in Z-direction

k (K) unit of computer memory capacity = 1000 (or 1024 in binary system of bytes, characters, or words)

k kontra (German—against, an octave lower); units of capital (microeconomics)

K 1024 storage bytes; capacity (symbol); centuple calorie (symbol); curvature (symbol); equilibrium constant (symbol); Fraunhofer line produced in part by calcium (symbol); hip; kaiser; Karman constant (symbol); Kawasaki Line; kelvin; Kelvin; Kerr constant; Kidde Fire Protection; kilobyte (symbol denoting 1024 units of stored matter); Kilo—code word for letter K; kilohm(s); kilometer(s); King(dom); Kiwanis International; Knabe; Köchel, cataloger of Mozart's music; kopec(s); kosher; krone; kroner; luminous efficiency (symbol); modulus of cubic compressibility (symbol); pilotless aircraft (symbol); potassium (kalium); proportionality constant (symbol); radius of gyration (symbol); strike-out (base-

ball); tanker (naval symbol)

°K degree(s) Kelvin

K kade (Dutch—embankment, quay); Kadenz (German—cadenza); kald (Norwegian—cold); kall (Swedish—cold); kalt (German—cold); koel (Dutch—cold); köld (Danish—cold); Koln (German—Cologne); krinda (Danish—women); kvinne (Norwegian women); kvinnor (Swedish—women); kylmä (Finnish—cold)

K^2 Mount Godwin Austen, Kashmir (28,250-ft mountain, second highest in the world)

K^5 Kunlun Mountain known on the Chinese-Kashmir border as Muztagh

K-9 Corps Canine Corps (staffed by police dogs)

k9p dog piss; urine produced by coyotes, dogs, foxes, hyenas, jackals, wolves, and other canines

K-12 kindergarten through 12th grade

K-61 Soviet amphibious-assault vehicle

K98k German carbine (World War II)

ka cathode(s); kiloampere(s)

k/a ketogenic to antiketogenic (diet ratio)

Ka auroral absorption index (symbol)

Ka Komppania (Finnish—company)

KA Kapok Association; Karhumaki Airlines (Finland)

K-A King-Armstrong (units)

K of A King(dom) of Aragon

Ka-15 Soviet light-utility helicopter nicknamed Hen

Ka-18 Soviet utility-transport helicopter nicknamed Hog

Ka-20/Ka-25k Soviet helicopters built for military or commercial use with Ka-20 nicknamed Harp and Ka-25-k nicknamed Hormone

KA-25 Soviet armed helicopter called Hormone by NATO

kaa keep-alive anode

KAA Kwikasair (Australasia)

kaad kerosene, alcohol, acetic acid, dioxane (insect larva killer)

Kab Kabel; Kabul

KAB Keep America Beautiful

KAC Kuwait Airways Corporation

KACC Kaiser Aluminum Chemical Corporation; Kansas Association of Community Colleges

KACF Korean American Cultural Foundation

KACIA Korean-American Commerce and Industry Association

KADA Kemubu Agricultural Development Authority; Kemuta Agricultural Development Authority

Kadet(s) (see KD)

Kae Katherine

KAESP Kansas Association of Elementary School Principals

kaf kaffir

KAF Kenya Air Force

KAFB Kirtland Air Force Base

kaffir kaffir bean (African cowpea); kaffir beer; kaffir bread (Encephalartos fruit); kaffir (cat reputedly the ancestor of the common domestic cat); kaffir crane (black-plume gray crane of southern

Africa); kaffir piano (southern African marimba); kaffir plum (edible fruit from southern Africa also called kaffir date or kaffir date plum)
Kafka Franz Kafka
Kagan Kaganovich
KAH Kahului Railroad
KAIIN third word of Sen Nihon Kaiin Kumiai—the All Japan Seamen's Union
Kaimanawas Kaimanawa Mountains of New Zealand's North Island
Kajiwara Takuma Kajiwara
KAK Kungliga Automobil Klubben (Swedish—Royal Automobile Club)
kakis kakistocracy (government by the worst men)
kal kalamein
kal. kalendae (Latin—calends, the first day of the month)
Kal¹ Kalana, Kalmar; Kalgoorlie
Kal Kalendae (Latin—first day of the Roman month)—day when debts were paid; *Kalium* (Latin—potassium)
KAL Korean Air Lines
Ka Lae Hawaii's southernmost point also called South Cape or South Point
Kalahari Kalahari Desert or Kalahari National Park in South Africa
kald kalamein door
Kaleun Kapitänleutnant (German—Commander)
Kali Kalimantan (Borneo)
Kal Nun Kalaallit Nunaat (Greenlandic—Greenland)
Kam Kampong (Malay—Village); *Kampuchea* (Khmer—Kampuchean People's Republic)—formerly Cambodia
KAM Kimball Art Museum (Fort Worth)
Kamarans Kamaran Islands in the Red Sea
kamb kamboganisch (German—Cambodian)
Kamk keyed alike and master keyed
kamp known as male prostitute
Kamp Kampuchea (Cambodia)
Kampuchea Democratic Kampuchea (formerly Cambodia)
Kan Kansas; Kanpur
Kan Kanal (German—canal); *Kanaal* (Afrikaans or Dutch —canal)
Kanal Der Kanal (German—The Channel)—The English

Channel
Kanchen Kanchenjunga; (28,146-foot-high mountain in the Himalayas, third highest in the world)
Kangeans Kangean islanders or Kangean Islands in the Java Sea north of Bali
kang(s) kangaroo(s)
Kano Eitoku Kano (late 16th-century Japanese painter)
Kans Kansas; Kansan
k antigen capsular antigen
KANU Kenya African National Union (party)
KANUPP Karachi Nuclear Power Plant
kao kaolin
Kao Kaohsiung
KAO Kuiper Airborne Observatory
kaocon kaopectate concentrate
Kaoh Kaohsiung
kaolin aluminum silicate $(Al_2O_3 \cdot 2SiO_2 \cdot 2H_2O)$; kaolinite
kaos killing as an organized sport
kap knowledge, attitude, practice
kap kapitel (Dano-Norwegian—capital); *kapitel* (Swedish—chapter)
Kap (German—cape); *Kapital* [German—capital (money)]; *Kapitel* (Danish and German —chapter)
KAP Chinese Ministry of Public Security—external counterintelligence and internal secret police force of the People's Republic of China
KAPG Kluwer Academic Publishers Group
KAPL Knolls Atomic Power Laboratory
Kapo Kamaradschafts Polizei (German—police fellowship)—organization of nonpolitical prisoners in prison camps
Kar Karachi; Karafuto
Kar Karabiner (German—carbine)—short rifle
K/Ar potassium-argon dating
KAR King's African Rifles
KARAI Karhumaki Airways (Finland)
Karakorums Karakorum Mountains of Kashmir
Karawankens Karawanken Alps between Austria and Yugoslavia
Karel Karelia; Karelian
Kar Fin Karelian Finland (now part of the USSR)

Karimunjawas Karimunjawa Islands off the north coast of Djawa or Java
Karimuns Karimun Islands between Singapore and Sumatra
Kas Kansas
KAS Kentucky Academy of Science; Kroeber Anthropological Society
KASA Kentucky Association of School Administrators
KASC Knowledge Availability Systems Center
Kash Kashmir
KASSR Kalmyk Autonomous Soviet Socialist Republic; Karelian Autonomous Soviet Socialist Republic; Komi Autonomous Soviet Socialist Republic
kast kastilianisch (German—Castilian)
kat katalog or *katolsk* (Dano-Norwegian—catalog or Catholic); *katalonisch* (German—Catalonian)
Kat Katmandu; Katowice
Kat Katar (Spanish—Quatar)
KAT Kenosha Auto Transport
kath katholisch (German—catholic)
Kath Katherine
Kath Katholik (German—Catholic)
Kathy Katharine; Kathleen; Kathryn
Katie Catherine; Katherine
Katmai Alaska's Katmai National Monument or its Valley of Ten Thousand Smokes or its Katmai Volcano creating the foregoing smoky valley in the Aleutian Peninsula
Kats Katangese
Katteg Kattegat (North Sea between Jutland peninsula of Denmark and west coast of Sweden)
KATUSA Korean (soldier) attached to (the) United States Army
Katy Missouri-Kansas-Texas Railroad
Katyn Katyn Forest massacre of 15,000 Polish officers and other prisoners of the Soviet Union, 1940
Katzen (German—Cats)— highest mountain in the Odenwald—Katzenbuckel or the village of Katzenellenbogen (Cats' Elbows) with its ancestral castle once inhabited by the counts and countesses Katzenellenbogen

Katzet Konzentrationslager (German—concentration camp)

Kauf Kaufman

kauk kaukasisch (German— Caucasian)

K-A units King-Armstrong units

Kawa Kawasaki

Kawarthas Kawartha Lakes of southeastern Ontario

kay knockout (*kayo*—spelled abbreviation of ko); okay (truncated slang)

Kay Catherine

Kay-Cee Kansas City

Kaz Kazak(stan)

Kazak Kazakhstan(i)

Kazoo Kalamazoo

kb kilobit(s); kilobyte (1024 characters); kitchen and bathroom; kite balloon; knee brace

k & b kitchen and bathroom

Kb Kontrabass (German—double bass)

KB Knight Bachelor; Koninkrijk Belgie (Flemish—Kingdom of Belgium)

K.B. King's Bench; Knight of the Order of the Bath

K of B King(dom) of Bavaria

KB Kongelige Bibliotek (Danish—Royal Library)—in Copenhagen; *Koninklijke Bibliotheek* (Dutch—Royal Library)—in The Hague; *Koninkrijk Belgie* (Flemish—Kingdom of Belgium); *Kungliga Biblioteket* (Swedish—Royal Library)—Stockholm

kba killed by air

KBAI Koninklijke Bibliotheek Albert I (Flemish—Albert Ist Royal Library)—see *BrA*

K-band 10,900–36,000 mc

kbar kilobar(s); 1 kbar equals approx 14,500 lbs per square inch

KBART Kings Bay Army Terminal

KBASSR Kabardino-Balkar Autonomous Soviet Socialist Republic

KBC King's Bench Court; Kyushu Asahi Broadcasting

KBD King's Bench Division

kbe keyboard encoder; keyboard entry; knotted both ends

K.B.E. Knight Commander of the Order of the British Empire

kbh killed by helicopter

Kbh København (Dano-Norwegian—Copenhagen)

Kbhvn København (Copenhagen)

KBI Keyboard Immortals (record label); Klan Bureau of Investigation (Ku Klux Klan)

KBIM Kongres Buruh Islamic Merdeka (Indonesian—Islamic Trade Union Congress)

K-bit unit of computer storage capacity equal to 1024 bytes

KBL Kabul, Afghanistan (airport)

KBL Kilusang Bagong Lipunan (Filipino—Philippines New Society Movement)

kbm keyboard monitor

KBNWR Klamath Basin National Wildlife Refuges (California and Oregon)

K Bon Klein Bonaire (Netherlands Antilles)

KBP Koala Bear Park (Adelaide)

kbps kilo bits per second

kbs kilobits per second

KBS Kinki Broadcasting System (Japan); Korean Broadcasting System

KBS Kärnbränslesakerhet (Swedish—Nuclear Fuel Safety Project)

KBSI Kongres Buruh Seluruh Indonesia (Indonesian—Indonesian Trade Union Congress)

KB & TS Kuwait Broadcasting and Television Service

kbtu kilo British thermal unit (1,000 btu's)

kbv kauri-butanol value

KBW Klan Border Watch (along the Mexican border)

kc kilocycle(s); koruna (Czechoslovakian monetary unit)

Kc Kyle classification (social sciences)

KC Kalamazoo College; Kansas City; Keble College, Oxford; Kendall College; Kennedy Center; Kennel Club; Kenyon College; Keuka College; Keystone College; Keystone Shipping Company (flag code); Kilgore College; King College; King's College; Kirksville College (of osteopathy and surgery); Knox College; Knoxville College

K.C. King's Counsel; Knight Commander

K of C Knights of Columbus

KC-10A advanced tanker-cargo aircraft

KC-50 tactical aerial tanker for refueling aircraft in flight

KC-97 Stratofreighter strategic tanker-freighter equipped for inflight refueling

KC-130 Lockheed Hercules tanker aircraft

KC-135 Stratotanker multipurpose aerial tanker-transport

KCA Kitchen Cabinet Association

KCA Keesings Contemporary Archives

kcal kilocalorie(s)

kcas knots calibrated air speed

k/cb keel combined with centerboard

KCB Kenya Commercial Bank

K.C.B. Knight Commander of the Order of the Bath

KCBT Kansas City Board of Trade

kcc kathodic closure contraction; keyboard common contact

KCC Kellogg Community College; Kenai Community College; Kennedy Cultural Center; Kern County College; Ketchikan Community College; Kingsborough Community College; King's College, Cambridge

KC & C Kembla Coal and Coke

KCCD Kentucky Council on Crime and Delinquency

KCCE Keystone Center for Continuing Education

KCCI Korean Chamber of Commerce and Industry

kcd kilocandelas

KCDMA Kiln, Cooler, and Dryer Manufacturers Association

kcf thousand cubic feet

KCFF Korean Cultural and Freedom Foundation

Kch Kuching

KCH King's College Hospital

K.C.H.S. Knight Commander of the Order of the Holy Sepulchre

kCi kiloCurie(s)

KCI Key Club International

KCIA Korean Central Intelligence Agency

K.C.I.E. Knight Commander of the Indian Empire

KCl potassium chloride

KCL Kai Curry-Lindahl; King's College, London; Kirchoff's Current Law

KCLA Known Coal-Leasing Area(s)

KCLY Kent and County of London Yeomanry

KCM Kansas City Museum

kcmG kindly call me God

K.C.M.G. Knight Commander of the Order of Saint Michael and Saint George

KCM & O Kansas City, Mexico & Orient (railroad)

kcmx keyset central multiplexer

KCNA Korean Central News Agency

KCNP Kings Canyon National Park (California); Ku-ring-gai Chase National Park (New South Wales)

KCNS King's College, Nova Scotia

KCOBE Knight Commander—Order of the British Empire

KCP Key Curriculum Project

KCPA Kaolin Clay Producers Association; Kennedy Center for the Performing Arts

KCPL Kansas City Public Library

KCPO Kansas City Philharmonic Orchestra

kcps kilocycles per second

KCR Kowloon-Canton Railway

kcs Czechoslovakian koruna(s); kilocycles per second

kc/s kilocycles per second

KCS Kansas City Southern (railroad)

KCS Kansas City Star

KCSC Kansas Cosmophere and Space Center

kc/sec kilocycles per second

K.C.S.G. Knight Commander of Saint Gregory the Great

KCSI Knight Commander of the Star of India

KCSO Kansas City Symphony Orchestra

KCT Kansas City Terminal (railroad)

kcte kathodic closure tetanus

K Cur Klein Curaçao (Netherlands Antilles)

KCVO Knight Commander of the Victorian Order

kd key depression(s); killed; kiln dried; knocked down; known distance; pilotless aerial target (code)

Kd Konrad; Kuwait dinar(s)

KD Kidderpore Docks (Calcutta); *Kongeriget Danmark* (Kingdom of Denmark)

K of D King(dom) of Denmark

KD Kampuchea Democratique (French—Democratic Kampuchea)—formerly Cambodia; *Konstitutsionnodemokra-*

ticheskaya partiya (Russian—Constitutional Democratic Party)—party of the KDs or Kadets liquidated by the Bolsheviks under Lenin

KDA Kongelik Dansk Aeroklub (Danish—Royal Danish Aero Club)

KDAK Kongelig Dansk Automobile Klub (Danish—Royal Danish Automobile Club)

K Dan Vidensk Selsk, Mat-Fys Medd Kongelige Danske Videnskabernes Selskab, Matematiske-Fysiske Meddelelser (Danish—Royal Scientific Society, Mathematical and Physical Announcements)

KDAR Klein-Drohne Anti-Radar device

K-day basic date for introduction of convoy system or lane; carrier aircraft assault day

KDB Korea Development Bank

kdcl knocked down in carload lots

KDD Kokusai Denshin Denwa (Japan's Overseas Radio and Cable System)

kdf knocked-down flat

K d F Kraft durch Freude (German—Strength through Joy)—Nazi holiday association

KDG King's Dragoon Guards

KDHNM Kill Devil Hill National Memorial

KDI Kwaliteitsdienst voor de Industrie Stichting (Dutch—Industrial Quality Control Society)

kdlcl knocked down in less than carload lots

kdly kindly

kdm kingdom

KDM Kongelige Danske Marine (Danish—Royal Danish Navy)

K d N Koninkrijk der Nederlanden (Dutch—Kingdom of the Netherlands)

Kdo Kasado

K-do Kamarado (Esperanto—comrade)

KDP potassium dihydrogen phosphate

KDs Kadets

kdv kiln-dried veneer

ke kinetic energy

K_e exchangeable body potassium

KE Kaiser Engineers

K-E Krafft-Ebing

K + E Keuffel & Esser

K of E King(dom) of England; Knights of Equity

KEA Kentucky Education Association; Kiwifruit Exporters Association

Keams Keams Canyon, Arizona, Hopi Indian Reservation headquarters

keas knots equivalent airspeed; knots estimated airspeed

Keb Coll Keble College—Oxford

KEBK Korea Exchange Bank

Keble Keble College, Oxford

Kech Kechua (Quechua)

KECO Korea Electric Company

KEDDS Kansas Education Dissemination/Diffusion System

Kee Keelung

Keel Keeling

KEEP Kentucky Environmental Education Program

KEF Keflavik Airport, Iceland

KEHF King Edward's Hospital Fund

Kel Kiel (British maritime abbreviation)

KELP Kindergarten Evaluation of Learning Potential

KELTS Key English Language Teaching Scheme

kem *kemisk* (Dano-Norwegian—chemical)

KEMA Kitchen Equipment Manufacturers Association

KEMA Keuring van Electrotechnische Materialen (Dutch—Testing Institute for Electrochemical Materials)

Kemp Richard M Kemp, American naturalist; Kemp's sea turtle, sometimes known as Kemp's Bastard thought to be a cross between Green and Loggerhead sea turtles

Ken Kendal(l); Kendrick(k); Kenelm; Kenilworth; Kenley; Kenna(rd); Kennedy; Kennet; Kenneth; Kennit; Kenny; Kenric(k); Kensell; Kensington; Kent(on); Kentuckian; Kentucky; Kenward; Kenwood; Kenya; Kenyan; Kenyon

Ken Kenia (Spanish—Kenya)

Kenai Kenai Fjords National Park, Alaska

Kennedy John F Kennedy (his brothers and others); John F Kennedy International Airport (New York)

Kens Kensington

Kent Kentucky

Kentuck Kentucky
Kenyatta Jomo Kenyatta (Kamau Ngengi)
Keogh Keogh Retirement Plan
kep key-entry processing
kep' kept
Kep Kepulauan (Indonesian or Malay—archipelago)
KEPCO Kyushu Electric Power Company
KEPZ Kaohsiung Export Processing Zone
Kerala southwestern India including most of the Malabar and Travancore coasts
kerat keratometric(al)(ly); keratometry
Kerguelens Kerguelen Islands in the subantarctic South Indian Ocean
kerk kerkelijke term (Dutch—ecclesiastical term)
Kermadecs Kermadec Islands
kern kernan
kero kerosene
KESCO Kowloon Electricity Supply Company
keto ketonaemia; ketogenic; ketone; ketonuria; ketoses; ketosis
ketol ketone alcohol (compound)
kev kilo electron volt; 1,000 electron volts
keV kiloelectronvolt(s)
Kev Kelvin; Kevin
KEVs King's Empire Veterans
kew (KEW) kinetic energy weapon
Kew Gar Kew Gardens
Kew Obs Kew Observatory
key. keep extending yourself
keyboard computer input device, to input computer data; keyboard instruments: celestas, clavichords, concertinas, harpsichords, organs, piano accordions, pianolas, pianos, virginals
keyper key personnel; keywords permuted
Keys the Keys (the Florida Keys)
kf kitchen facilities; koff
KF Kaiser-Frazer; Kellogg Foundation; Kent Foundation; Kidney Foundation; *Kooperative Forbunded* (Federation of Cooperatives—Sweden); Kresge Foundation
K of F King(dom) of France
KF kleine Flöte (German—small flute or piccolo); *Konservative Folkeparti* (Danish—Conservative Party); *Kontrafagott* (German—dou-

ble bassoon); *Kooperative Forbunded* (Swedish—Federation of Cooperatives)
KFA Kenya Farmers Association; Krishnamurti Foundation of America
KFASSR Karelo-Finnish Autonomous Soviet Socialist Republic (formerly the Karelia of Finland)
Kfc Kentucky fried chicken
KFC Kentucky Fried Chicken; Kropp Forge Company
KFEA Korean Federation of Education Associations
kff keep from freezing
KFH Kaiser Foundation Hospitals
KFL Kenya Federation of Labour
kfm kaufmännisch (German—commercial)
Kfm Kaufmann (German—merchant)
KFNP Kaieteur Falls National Park (Guyana)
kfo killing federal officer
KFP Kristelig Folkeparti (Norwegian—Christian People's Party)
KFPC Kansas Foundation for Private Colleges
K-F s Klippel-Feil syndrome
KFSR Karakul Fur Sheep Registry
KFT Kansai Fishing Tackle
KFUK Kristelig Forening for Unge Kvinder (Danish—Young Women's Christian Association)
KFUM Kristelig Forening for Unge Maend (Danish—Young Men's Christian Association)
Kfz Kraftfahrzeug (German—motor vehicle)
kg keg; kilogram; known gambler
/kg per kilogram
kG kilogauss
Kg Kirghiz(ian)
Kg Kampong (Malay—village); *Kompong* (Indo-Chinese—landing place, riverside)
KG Kelly Girl
K-G Kanematsu-Gosho Ltd.
K.G. Knight of the Order of the Garter
K of G King(dom) of Granada
KG Kommanditgesellschaft (German—limited partnership)
KG5 King George V School
KGA Kitchen Guild of America

KGB Komitet Gossudarrstvennoi Bezopastnosti (Russian—Committee of State Security, Soviet Secret Police)
KGBW Kewaunee, Green Bay, and Western (railroad)
KGC Knights of the Golden Circle
K.G.C. Knight of the Grand Cross
kg cal kilogram calorie
kg-cal kilogram calorie
K.G.C.B. Knight of the Grand Cross of the Bath
kg/cm kilograms per centimeter
kg cum kilograms per cubic meter
kgf kilogram-force
Kgf Kriegsgefangener (German—prisoner of war)
KGFS King George's Fund for Sailors
kg/hl kilograms per hectoliter
kg/hr kilograms per hour
KGIS Kuder General Interest Survey
KGJT King George's Jubilee Trust
KGK Kabushiki Goshi Kaisha (Japanese—joint stock limited partnership of members with unlimited liability and shareholders with limited liability)
kgl kongelig (Dano-Norwegian—royal)
Kgl Königlich (German—royal)
kgm kilogram meter
kg/m² kilograms per square meter
kg/m³ kilograms per cubic meter
kg/ms kilograms per meter second
Kgn Kingston, Jamaica
KGNP Kalahari Gemsbok National Park (South Africa); Katherine Gorge National Park (Australian Northern Territory)
kgps kilograms per second
kgra known geothermal resource area
kgs kegs; kilograms
kg/s kilograms per second
KGS Kate Greenaway Society; Kigezi Gorilla Sanctuary (Uganda); Korean Geological Survey
kgs/ha kilograms per hectare
KG St J Knight of Grace of the Order of Saint John of Jerusalem
kg U kilogram of uranium

KGV Knight of Gustavus Vasa
KGVDs King George V Docks (London)
KGWS Keoladeo Ghana Wildlife Sanctuary (India)
kh keyhole
kH kilohertz
Kh Khmer (Cambodia)
Kh Khawr (Arabic—creek, inlet, ravine, water-course)
KH King's Hussars; Knut Hamsun
K-H Kelsey-Hayes
K of H King(dom) of Hungary
KH Karen Hayesod (Hebrew— United Israel Appeal); *Kjøbenhavns Handelsbank* (Danish—Copenhagen's Commercial Bank); *Kupat Holim* (Hebrew—Health Insurance Fund)
KH-4 Kawasaki all-purpose helicopter similar to the Bell 47
KH-11 American-made intelligence-gathering satellite
KH IV King Henry IV
KH VI King Henry VI
kha killed by hostile action
Khar Kharkov
Khazar Bahr-ul-Khazar (Arabic—Caspian Sea)
KHC Karen Horney Clinic
KHDS King's Honorary Dental Surgeon
Khi Karachi
KHI Karachi, Pakistan (airport)
Khingans Kinghan Mountains of northeast China
KHL Koninklijke Hollandsche Lloyd (Dutch—Royal Holland Lloyd)
KHM King's Harbour Master
Khn Knoop hardness number
KHNS King's Honorary Nursing Sister
khp kilohorsepower (hour)
KHP King's Honorary Physician
KHPC Karen Horney Psychoanalytic Clinic
Khr Khrebet (Russian—mountain range)
KHRI Kresge Hearing Research Institute
KHS Kennedy High School; King's Honorary Surgeon
khz (kHz) kilocycle(s)/second; kilohertz, formerly kilo cycle(s) per second
ki kilo; kitchen
KI Kathleen Investments; Kiwanis International; potassium iodide
K-I Kaiser-Illin

K of I King(dom) of Ireland; King(dom) of Italy
KI Kol Israel (Hebrew— Voice of Israel)—broadcasting service; *Kommunisticheskii Internatsional* (Russian —Communist International); *Komunisticna Internacijonala* (Yugoslav—Communist International)
kia (KIA) killed in action
Kia Kligler iron agar
kias knots indicated airspeed
KIB Kansas Inspection Bureau; Kentucky Inspection Bureau
Kib Cum Kibris Cumhuriyeti (Turkish—Republic of Cyprus)
Kibo Mount Kibo (Africa's highest peak also called Kilimanjaro)
KICF Kentucky Independent College Foundation
kid. kidney
KID Key Industry Duty
kidult kid adult (older person who enjoys juvenile entertainment)
kidvid children's television program; children's tv or video programs
kidzines magazines for children
Kiev Kiev-class 40,000-ton Soviet aircraft carrier
Kifis Kollsman integrated flight instrument system
K-i-H Kaiser-i-Hind (Emperor of India medal)
KIICC *Kommunisticheskaya Partiya Sovetskogo Soyuza* (Russian—Communist Party of the Soviet Union)
Kikdl Krokodil
kikú (Japanese—chrysanthemum)—symbol of Japan's highest order for men
kil (Dutch—channel, estuary, strait)—as in Arthur Kill, French Kill, and Kill van Kull along the shores of New York's Staten Island
kild kilderkin(s)
Kild Kildare
Kili Kilimanjaro
Kilk Kilkenny
Kill van Kill van Kull (waterway between Bayonne, New Jersey and Port Richmond, Staten Island, New York)
kilo Kilogram; 10^3
Kilo letter K radio code
kilobrick(s) kilo-weight brick(s) of marijuana measuring about 2½ x 5 x 12 inches

(64 x 127 x 300 millimeters)
kilohm kilo-ohm
kilovar kilovolt-ampere (reactive)
Kim Kimball; Kimballton; Kimberley; Kimberly; Kimble; Kimbolton; Kimborough; Kimbrough; Kimiwan; Kimmell; Kimmins; Kimmswick; Kimsquit
K i M Knudsen i Marken
Ki-mi-ga yo wa (Japanese— May thy peaceful reign last long)—national anthem
kin kinesisk (Dano-Norwegian—Chinese)
Kin Frank McKinney Hubbard; Kingston, Ontario (maritime contraction)
KIN Kingston, Jamaica (airport); Kinross
kina monetary unit of Papua, New Guinea
Kinc Kincardinel
kind. kindergarten
kine kinema (variation of cinema)
Kines J M Keynes (pronounced as italicized)
kinesi kinesics; kinesiologist; kinesiology
King Kingston
King of the Congo Leopold II of Belgium (1835–1909)
kingd kingdom
King James King James Version of the Bible (authorized by King James I of England in 1611)
King Karls King Karl Islands in the Norwegian sector of the Arctic
King Leopolds King Leopold Ranges of northern Western Australia
Kings X Sta King's Cross Station (rail terminal)
King's King's College (Cambridge, Columbia)
Kings Canyon Kings Canyon National Park in central California
Kings Point United States Merchant Marine Academy at Kings Point, New York
Kinr Kinross-shire
kinsym kinematic synthesis
KINTEL K Laboratories (instruments and television)
Kintetsu Kinki Nippon Railway Company, Ltd
kip thousand pounds (from contraction of kilo and pound)
KIP Kennedy Institute of Politics (Harvard)
kip ft thousand foot pounds

KIPS Knowledge Information Processing System(s)
kiq (KIQ) key intelligence questions
Kir Kirghiz; Kirghizia; Kirghizian; Kiribati
Kircud Kircudbrightshire *(Kircoobrisheer)*
Kiribati Republic of Kiribati (Gilbert and Ellice islands colony in the equatorial Pacific, includes Tarawa)
Kirk Kirkham; Kirkland; Kirkudbright *(Kircoobri)*; Kirkwood
Kirov Sergei Mironovich Kostrikov; Soviet name for Viatka
KISA Korean International Steel Associates
kisc knowledge industry system concept
kismif keep it simple—make it fun
KI smog potassium-iodide smog (automobile induced)
KISO Kol Israel Symphony Orchestra
KISR Kuwait Institute for Science Research
kiss keep it simple, stupid
kiss. keep it simple, sir; keep it simple, stupid
KISS Kids In Safety Seats (automobile safety program)
KIST Korean Institute for Science and Technology
kit. key issue tracking; kitchen(ette); kitten; kitty
KIT Kentucky and Indiana Terminal (railroad); Korean International Telecommunications
KIT *Koninklijk Instituut voor de Tropen* (Dutch—Royal Institute for the Tropics)
KITCO Kwajalein Import and exporting Company
kiteoon kite + balloon
kitin' kiting (money)
kits. kittens
kitsch *kitschen* (German—thrown together)—commercial art or art objects cheapened by vulgarity; e.g., miniature reproduction of the Venus de Milo with an alarm clock set in her belly
Kitsch *Kitschmensch* (German—kitschman)—anyone creating, dealing in, or displaying artistic rubbish—junk art
KIVI *Koninklijk Instituut van Ingenieurs* (Dutch—Royal Institution of Engineers)

KIWA *Keurings Instituut voor Waterleiding Artikelen* (Dutch—Inspection Institute for Waterworks Equipment)
kj killer judo; kilojoule; kimberly joint (lumbing); knee jerk; kraut joint; krystal joint (pcp)
k-j knee-jerk(s)
kJ kilojoule
KJ Kahlil Jibran (Gibran)
KJ *King James* (version of the Bible)
KJB Korea-Japan Board
KJC Kaiser Jeep Corporation; Keystone Junior College
K John *Life and Death of King John*
K.J.St.J. Knight of Justice, Order of Saint John of Jerusalem
KJV King James Version
kk killer karate
k-k knee-kicks (knee-jerks)
kK kilokelvin
KK Karen Keltner; Karen Kerchelich
K-K Krupp-Koppers
K of K Kitchener of Khartoum
KK *Kabushiki Kaisha* (Japanese—joint stock company of shareholders with limited liability); *Kaiserlich Königlich* (German—Imperial Royal)
K.K. *Kahal Kadosh* (Hebrew—Holy Congregation)
KKASSR Kara-Kalpak Autonomous Soviet Socialist Republic
KKI *Keren Kayemeth le Israel* (Hebrew—National Fund of Israel)
kkk kikes (Jews), koons (Afro-Americans), katholics (Roman Catholics) according to the Ku Klux Klan
KKK Ku Klux Klan (secret organization antagonistic to certain racial & religious groups)
KKK *Kataas-taasan Kagalang-galan-gang Katipunan* (Filipino—Mightiest Warriors Fighting for Freedom); *Kinder, Kirche, Küche* (German—Children, Church, Kitchen)—traditional three Ks of Teutonic womanhood
KKKK Kansai Kisen Kabushiki Kaisha; Kawasaki Kisen Kabushiki Kaisha (steamship lines)
KKKK *Koenhavns Kul og Koks Kompagne* (Danish—Copenhagen Coal and Coke Com-

pany)
KKKKs Knights of the Ku Klux Klan
KKKUK Ku Klux Klan in the United Kingdom
KKL Karlander Kangaroo Line
kklmg *koom, koom, lauz mere gain* (Yiddish—come, come, let us go)
KKMKI *Kungliga Karolinska Mediko-Kirurgiska Institutet* (Sweden—Caroline Medico-Surgical Institute-Stockholm)
KKO *Korps Kommando* (Bahasa-Indonesian Malay—Commando Corps)—marine corps
Kkr Karlskrona
kkv (KKV) kinetic kill vehicle
kl key length; kiloliter
kl *klasse* or *klokken* (Dano-Norwegian class or o'clock); *klockan* (Swedish—o'clock)
Kl *Klarinette* (German—clarinet); *Klasse* (German—class); *Klein(e)* (German—little, small)
KL Key Largo; Klebs-Loeffler; Knutsen Line; Kuala Lumpur; Kwik Lok
KL *King Lear*
kla Klavier; klystron amplifier
KLA Kansas Library Association; Karachi Library Association; Kentucky Library Association; Korean Library Association
Klamaths Klamath Mountains bordering California and Oregon
Klan Ku Klux Klan (*q.v.* KKK)
Klar *Klarinette* (German—clarinet)
Klaus Nikolaus
klax klaxon
K-L bacillus Klebs-Loeffler bacillus (diphtheria)
KLC Kaingaroa Logging Company
kld *kaelder* (Dano-Norwegian—basement or cellar)
Klebs Klebisella
Klem Klemens; Klement; Klementi; Klement
klepto kleptomania(c, al)
kl Fl *kleine Flöte* (German—piccolo)
Klg Keelung
klh keyhole limpet hemocyanin (KLH)
KLIAU Korean Land Improvement Association Union
klic key letter in context
klicks kilometers
klieg klieg light (named for

German-American inventor brothers Anton and J H Kliegl)

klim (milk spelled backwards) dried milk

K Line Kawasaki Kisen Kaisha

KLM Koninklijke Luchtvaart Maatschappij (Dutch—Royal Dutch Airlines)

Klmpb Klampenborg

Kln Köln (Cologne)

klo klystron oscillator

k-lo kello (Finnish—hour, o'clock)

KLPA Knuckeys Lagoon Protected Area (Australian Northern Territory)

KLr Kuala Lumpur

kls key lock switch

k-l-s kidney-liver-spleen

KLSE Kuala Lumpur Stock Exchange

klt kiloton (nuclear equivalent, 1,000 tons of high explosives)

klto knurling tool

Kluxer member of the Ku Klux Klan (*q.v.* KKK)

km kilometer

kM kilomega (10⁹ giga)

Km Kingdom

KM Kaffrarian Museum; Kearny Mesa; Khedivial Mail (steamship line)

K-M Krauss-Maffei

K.M. Knight of Malta

K&M King and Martyr (Charles Ist's sobriquet)

km² square kilometer

km³ cubic kilometer

KMA Kalgoorlie Mining Associates; Kinematograph Manufacturers Association

KMAG United States Military Advisory Group to the Republic of Korea

KMB Kowloon Motor Bus

kmc kilomegacycle

KMD Kentucky Manpower Development

kmef keratin, myosin, epidermin, fibrin (proteins)

KMF Koussevitzky Music Foundation

km/h kilometers per hour

KMH Kleinhans Music Hall (Buffalo)

KMI Kentucky Military Institute

KMIDC Korean Marine Industry Development Corporation

KMIT four-letter name bestowed by Trotsky on the Bolshevik ministry of foreign affairs; when foreign journal-

ists demanded to know what KMIT stood for his aides confided it was Yiddish for *küss mir im tuchus* (kiss my ass)

km/l kilometers per liter

KMMA Korean Merchant Marine Academy

KMO Kobe Marine Observatory

KMP Kaiser Metal Products; Kearny Mesa Plant (Convair)

KMPA Korean Maritime and Port Administration

kmph kilometers per hour

kmps kilometers per second

Kmr Khorramshahr

KMR Kwajalein Missile Range

kms kilometers

KMS Kansas Medical Society; Keeve M Siegel

KMT Kuomintang

KMTC Korea Marine Transport Company

KMUB Karl-Marx-Universitäts Bibliothek (German—Karl-Marx University Library)—on Beethovenstrasse in Leipzig

KMUL Karl Marx Universität Leipzig (German—University of Leipzig)

kmv killed measles-virus vaccine

kmw kilomegawatt

KMW Karlstads Mekaniska Werkstad (Swedish iron foundry)

kmwhr kilomegawatt-hour

kn kilonewton; knife (philatelic stationery); knot; krone; kronen

Kn Knight

KN Koninkrijk der Nederlanden (Kingdom of the Netherlands); *Kongeriket Norge* (Kingdom of Norway)

K-N Know-Nothing (political party)

K of N King(dom) of Naples; King(dom) of Navarre; King(dom) of Norway

KNA Kenya News Agency; Korean National Airlines

KNA Kongelig Norsk Automobilklub (Norwegian—Royal Norwegian Automobile Club)

KNAC Koninklijke Nederlandse Automobiel Club (Dutch—Royal Dutch Automobile Club)

KNAN Koninklijke Nederlandse Akademie voor Naturwetenschappen (Dutch—Royal Netherlands Academy

of Sciences)

K'naw Kanawha River

KNB Kita-Nihon Broadcasting

KNC Kalamazoo Nature Center

Knd Kandla

K-NEA Kansas-National Education Association

KNGR Kruger National Game Reserve

Kng X King's Cross (rail terminal)

Knick Knickerbocker Club

knickers knickerbockers

KNIP Komite Nasional Indonesia Pusat (Indonesian Central National Committee)

knit. knitted; knitting

KNK Kita Nippon Koku (Northern Japan Airlines)

Knls Knolls

KNM Katmai National Monument; Kenya National Museum; *Kongelige Norske Marine* (Norwegian—Royal Norwegian Navy)

KNMI Koninkliji Nederlands Meteorologisch Instituut (Dutch—Royal Netherlands Meteorological Institute)

KNMR Kenai National Moose Range (Alaska)

KNO Kano, Nigeria (tracking station)

KNOC Kuwait National Oil Company

Knockmealdowns Knockmealdown Mountains of southern Ireland

KNOJ Korpus Narodne Odbrane Jugoslavje (Serbo-Croat—National Defense Corps of Yugoslavia)

Knopf Alfred A Knopf

knork knife + fork (combination utensil)

Knott's Knott's Berry Farm

Know-Nothings American political party members (1853–56 attempted to keep control of the government in hands of native-born citizens; members professed ignorance of the party's activities)

KNP Kafue National Park (Zambia); Kalahari NP (South Africa); Kalbarri NP (Western Australia); Kanha NP (India); Kejimkujik NP (Nova Scotia); Kinabalu NP (Sabah); Kinchega NP (New South Wales); Kootenay NP (British Columbia); Korean National Party; Kosciusko NP (New South Wales); Kruger NP (South Africa)

KNPC Kuwait National Petroleum Company
KNPI Kundu's Neurotic Personality Inventory
KNR Kinki Nippon Railway; Korean National Railroad
KNSM Koninklijke Nederlandsche Stoomboot Maatschappij (Dutch—Royal Netherlands Steamship Company)
kn sw knife switch
Knt Knight
KNT Knight-Knott Hotels (stock-exchange symbol)
KNT Koninklijke Nederlandsche Toeristenbond (Dutch—Royal Netherlands Touring Club)
KNTC Korean National Tourism Corporation
knu knuckle
KNUFNS Kampuchean National United Front for National Salvation
KNUST Kwame Nkrumah University of Science and Technology
KNVD Koninklijk Nederlands Verbond van Drukkerijen (Dutch—Royal Netherlands Printing Association)
KNVL Koninklijke Nederlandse Vereniging voor Luchtvaart (Dutch—Royal Netherlands Aero Club)
KNWR Kirwin National Wildlife Refuge (Kansas)
KNX Kinney Company (stock-exchange symbol)
Knxv Knoxville
ko keep off; keep out; kilohm; kit order; knockout (KO)
k-o knockout
Ko Korea; Korean
KO kickoff (football); knockout (boxing); Kodiak Airways (2-letter coding)
KO Komische Oper (German—Comic Opera)—Berlin opera company and opera house
K o A Kampgrounds of America
KOA Kentucky Opera Association
Kob Kobe (British Maritime contraction)
Køb København (Dano-Norwegian—Copenhagen)
Koblenz Coblenz
kobol keystation on-line business-oriented language
Kobuk Kobuk Valley National Park, Alaska
KOC Kollmorgen Optical Corporation; Kuwait Oil Company

kod kickoff drift
ko'd knocked out
KODAK trade name for Eastman Kodak photographic products
k-o drops knockout drops (chloral hydrate sedative)
koe knotted one end
kOe kiloOersted(s)
KOEX Korean Exhibition Center
kog kindly old gentleman
KOG Kansas, Oklahoma & Gulf (railroad)
KOH potassium hydroxide
KOHEMA Korean Heavy Machinery Industries
kohm kilohm
KOJ Kagoshima (airport)
Kok Cochrane
KOKS Dul og Koks Selskab (Danish—Coal and Coke Company)
Kol Kolonia, Ponape (Trust Territory of the Pacific)
KOLA Keep Old Los Angeles
Kolloid Z Z Polm Kolloid Zeitschrift und Zeitschrift Polymere (German—Colloidal and Polymer Periodicals)
Kol Nid Kol Nidre (Aramaic—all promises and vows be nullified and forgiven)—prayer recited and sung in synagogues on eve of Yom Kippur or played by 'cello and orchestra as arranged by Max Bruch
KOM Knight of the Order of Malta
Komei (Japanese—Komeito)—Buddhist party
komm kommunal or kommune or kommunistisk (Dano-Norwegian—communal or commune or communist)
Komp Kompanie (German—company)
Kon Konstant; Konstantin
Kon Bel Koninkrijk België (Flemish—Kingdom of Belgium)
Kon Dan Kongeriget Danmark (Danish—Kingdom of Denmark)
Kon Den Kongeriget Danmark (Danish—Kingdom of Denmark)
Kong Kristian Kong Kristian stod ved højen Mast (Danish—King Christian stood by the lofty mast)—national anthem of Denmark
Kon Ned Koninkrijk der Nederlanden (Dutch—Kingdom of

the Netherlands)
Kon Nor Kongerik Norge (Norwegian—Kingdom of Norway)
Konr Konrad
KONR Komitet Osvobozhdeniya Narodov Rossii (Russian—Committee for the Liberation of the Peoples of Russia)
kons konservativ (Dano-Norwegian—conservative)
Konst Konstantin
Kon Sver Konungariket Sverige (Swedish—Kingdom of Sweden)
konz konzentriert (German—concentrated)
KOON Knight of the Order of Orange Nassau
Kootenay Kootenay Lake or Kootenay National Park in British Columbia; Kootenay River flowing from British Columbia to Idaho and Montana
kop kopeck(s)
Kop Kopenhagen (Dutch, Flemish, German—Copenhagen)
KOP Koppers (company)
KOP Kansallis-Osake-Pankii (Finnish—National Bank)
kops keep off pounds sensibly
KOPS Keep Our Precinct Safe
kor knowledge of results
kor koreanisch (German—Korean)
Kor Korea; Korean; The Koran
Kör Körfez(i) (Turkish—bay, gulf)
KORDI Korean Ocean Research and Development Institute
Korea Democratic People's Republic of Korea (North Korea) and Republic of Korea (South Korea)—*Chosen Minchuchui Inmin Konghwa-Guk* (North Korea) and *Daehan-Minkuk* (South Korea)
Korin Ogata Korin (early 18th-century Japanese decorative artist)
KORR King's Own Royal Regiment
KORSTIC Korean Scientific and Technological Information Center
koruna monetary unit of Czechoslovakia
kos kilograms; kilos
KOSB King's Own Scottish Borderers

Kosci Mount Kosciusko (Australia's highest mountain)
kosmo kosmonavt (Russian—cosmonaut)
Koste André Kostelanetz
Kot Kotakinablalu
KOTRA Korea Trade Promotion Corporation
Koussi Serge Koussevitzky
kov key-operated valve
Kow Koweit (Spanish Kuwait)
KOW Knight of the Order of William
KOWACO Korean Water Resources Development Corporation
KOYLI King's Own Yorkshire Light Infantry
kp key personnel; kick plate; kill probability; kilopond; king post; kitchen police (KP); knotty pine
kp (KP) keypunch
Kp Kochpunkt (German—boiling point)
KP Korean People's (Republic)
K.P. Knight of St Patrick
K of P King(dom) of Poland; King(dom) of Portugal; King(dom) of Prussia; Knights of Pythias
KP Kommunistische Partei (German—Communist Party); *Komsomolskaya Pravda* (Russian—Young Communist League Truth) Moscow newspaper; *Kuvendi Popullore* (Albanian People's Assembly)
kPa kilopascal (pressure unit)
KPA Korean People's Army; Kraft Paper Association
kpc keypunch cabinet; kiloparsec
KPC Kangaroo Protection Committee; Koblenz Procurement Center; Korean Productivity Council
kp & d kick plate and drip
KPD Kommunistische Partei Deutschland (German—Communist Party of Germany)
KPDR Korean People's Democratic Republic
KPFC Kuwait Pacific Finance Company
KPFSM King's Police and Fire Service Medal
KPGA Kansas Personnel and Guidance Association; Kentucky Personnel and Guidance Association
kph kilometers per hour; knots

per hour
kpi kips per inch
kpic key phrase in context
KPJ Kommunisticka Partija Jugoslavije (Serbo-Croat—Communist Party of Yugoslavia)
kpl kilometers per liter
KPL Knoxville Public Library
kpm kathode pulse modulation
KPM King's Police Medal
KPM Koninklijke Paketvaart Maatschappij (Dutch—Royal Packet Company)—inter-island shipping line
KPMB Kunming Phosphorus Mining Bureau (China)
Kpmtr Kapellmeister (German—conductor)
KPNO Kitt Peak National Observatory
KPNWR Kern-Pixley National Wildlife Refuge (California)
kpo keypunch operator
KPO Korean Post Office
kpos keep pounds off sensibly (weight-reduction program)
KPP Keeper of the Privy Purse
kpps kilopulses per second
kpr keeper; knots per revolution
Kpr Kodak photo resist
KPR Korean Presidential Ribbon
KPRA Korean People's Revolutionary Army (North Korea)
kps kips (thousand pounds) per square foot
kpsi kips (thousand pounds) per square inch
KPSS Kommunisticheskaya Partiya Sovetskovo Soyuza (Russian—Communist Party of the Soviet Union)—CPSU
Kpt Kaptajn (Danish—captain)
Kpt Kapitein (Dutch—captain)
KPU Kenya People's Union (party)
kq line squall
K & Q king and queen
kr keel rider; kiloroentgen
k & r kidnapping and ransom (insurance)
Kr krypton
KR Khmer Rouge (Red Cambodian)—Communist insurgents in Cambodia; King's Regulations; Korean Registry (of ships); krona (Icelandic or Swedish monetary unit); krone (Danish or Norwegian monetary unit)
KR III King Richard III
krad kilorad
Krag Krag-Jörgensen rifle

Krak Krakatoa, Indonesia
Krakow Cracow
K-ration Calorie ration (lightweight emergency meal)
Kraut(s) Anglo-American slang for German(s)
K-R bb Krebs-Ringer bicarbonate buffer
KRC Knight of the Red Cross
KREEP K (potassium) REE (rare-earth elements) P (phosphate)—yellow brown glassy lunar material
Krema Krematorium (German—crematory)
kreml (Russian—fortress)—the Moscow Kremlin, a citadel and seat of Soviet government
Krete Crete
Kreuzb Kreuzberg
KRF Kentucky Research Foundation
Krh Karachi
KRI Kyle Railway Inc
KRIC Korean Reinsurance Company
Kriegies Kriegsgefangenen (German—war prisoners)
KRIM Danish association for penal reform
Kripo Kriminalpolizei (German—Criminal Investigation Department)
Krist Kristian; Kristijonas; Kristmann; Kristofer
Krist Kristallnacht (German—night of broken glass)—night when Nazi gangs broke glass shop windows and killed or captured Jewish merchants in Germany and Austria
KRN Knight-Ridder Newspapers
kroat kroatisch (German—Croatian)
KROM the Norwegian association for penal reform
krona monetary unit of Sweden
krone monetary unit of Denmark
kroner monetary unit of Norway
kronur monetary unit of Iceland
krp key resource people
KRP Keogh Retirement Plan
KRR King's Royal Rifles
krs kurus (Turkish—piastre)
Krs Kristiansand
KRs (see CRs)
KRS Kerato-Refractive Society; Kinematograph Renters Society
KRSB Kindergarten Reading

Screening Battery
krt cathode-ray tube
KRT Khartoum, Sudan (airport)
KRU Krueger Brewing (stock-exchange symbol)
Kruger Kruger National Park (South Africa's big game and wildlife reservation named for Oom Paul Kruger, the President of the Transvaal)
KRUM the Swedish association for penal reform
Krung Thep Bangkok
Krungthep Krungthep Mahanakhon Bovorn Ratanakosin Mahintharayutthaya Mahadilokpop Noparatratchanthani Burirom Udomratchanivetmahasathan Amornpiman Avatarnsathit Sakkathattiyavisnukarmprasit (Siamese—full name of the capital city of Bangkok, Thailand)
Krung Thep (Siamese—Bangkok)
Krupp Krupp von Bohlen (German armament and steel firm)
Krupskaya Nadezhda Konstantinovna Krupskaya Lenin
ks drifting snowstorm (symbol); keep (type) standing; knowledge structure(s)
k's kilobytes; kilograms; kilometers; kilos; kilowatts; New Zealand slang for kilometers per hour (pronounced *kays*)
Ks Kaposi's sarcoma; kyats (Burmese money)
K-s King-size
KS Kansas; King's Scholar; King's School; Kipling Society; *Konungariket Sverige* (Swedish—Kingdom of Sweden); Korea Shipping
K of S King(dom) of Scotland; King(dom) of Siam (Thailand); King(dom) of Spain; King(dom) of Sweden
ksa kite-supported antenna
KSA Kingdom of Saudi Arabia; Knitwear and Sportswear Association
KSA Kommission für die Sicherheit von Atomlagen (German—Commission for the Safety of Nuclear Power Plants)
KSAA Keats-Shelley Association of America
KSB Kypriakos Synthesmos Bibliothicarion (Modern Greek—Library Association of Cyprus)
KSBA Kentucky School

Boards Association
KSC Kansas State College; Kennedy Space Center; Kentucky State College; Korean Shipping Corporation; Kutztown State College
KSC Komunistická Strana Ceskoslovenska (Czechoslovak Communist Party)
KSEAFA Korea and South-East Asia Forces Association
KSEC Korea Shipbuilding and Engineering Corporation
ksf kips (thousand pounds) per square foot
KSF Kulkyne State Forest (Victoria, Australia)
KSFUS Korean Student Federation of the United States
KSG Kennedy School of Government
K.S.G. Knight of Saint Gregory the Great
K sh Kenya shilling(s)
ksi kips (1000 pounds) per square inch
KSI Keshvare Shahanshahiye Iran (Iran—Persia); Kingdom of Saudi Arabia
ksia thousand square inches absolute
K-size King-size
k/sk (truncated fixed) keel (and) swing keel (combined)
KSK ethyl iodoacetate (tear gas)
ksl kidney, spleen, liver
KSL Kinsel Drug (stock-exchange symbol)
KSLI King's Shropshire Light Infantry
KSM King's Service Medal; Korean Service Medal; Kungliga Svenska Marinen (Royal Swedish Navy)
KSM Kommunisticheskii Soyuz Molodozhi (Russian—All-Union League of Communist Youth)—Komsomol or Young Communist League—YCL
ksml kosher meal
KSN Kit Shortage Notice
KSNP Khao Salob National Park (Thailand)
KSO Kalamazoo Symphony Orchestra; Knoxville Symphony Orchestra
ksoc key symbol out of context
ksr keyboard send-receive (set)
K.S.S. Knight of Saint Sylvester
KSS Kommission zur Stahlenschutz (German—Radiation Protection Commission); *Komunisticka Strana Sloven-*

ska (Communist Party of Slovakia)
KSSR Kazak Soviet Socialist Republic; Kirghizian Soviet Socialist Republic
KSSU Kiev IG Shevchenko State University (University of Kiev)
kst keyseat
KST King-Seeley Thermos (company)
KSTC Kansas State Teachers College
K-S Test Kveim-Siltzbach Test
K.St.J. Knight of the Order of Saint John of Jerusalem
ksu key service unit
KSU Kansas State University; Kent State University; Kentucky State University
KSUAAS Kansas State University of Agriculture and Applied Science
ksv kinetic safety vehicle
KSY King Seeley (stock-exchange symbol)
kt karet (caret); key telephone; kiloton (nuclear equivalent, 1000 tons of high explosives); knot
Kt Knight
K$_t$ stress concentration factor
KT Kärntnerthor-Theater (Vienna); Kentucky & Tennessee (railway); Knight of the Order of the Thistle; Knight Templar; Missouri-Kansas-Texas (Katy Route Railroad)
K-T Kazin-Turkic
K.T. Knight of the Thistle
K/T Kretaceous/Tertiary (boundary between Cretaceous and Tertiary geologic periods)
K of T Kingdom of Tonga
KTA Kindergarten Teachers Association; Knitted Textile Association; Korean Traders Association
KTAAK Korea Trading Agents Association
ktas knots true airspeed
Ktb Kriegstagebuch (German—war diary)
KTB Kluwer Technical Books
Kt. Bach. Knight Bachelor
KTC Key Telephone System; Keystone Tankship Corporation; Kindergarten Teachers College; Kodiak Tracking Station
KTH Kungliga Tekniska Högskolan (Royal Institute of Technology, Stockholm)

k through 12 kindergarten through high school

ktl *kai ta loipa* (Greek—et cetera)

KTM *Keretapi Tanah Melayu* (Malayan Railway)

KTN Ketchikan, Alaska (Annette Island airport)

Kto *Konto* (German—account)

K-Town Knoxville, Tennessee

ktr keyboard typing reperforator

Ktr *Katorzhane* (Russian—compound)—prison compound reserved for people sentenced to hard labor

K-truss K-shaped truss

kts knots

KTS Kagoshima Television Station; Key Telephone Systems; Kwajalein Test Site

KTSA Kahn Test of Symbol Arrangement

KTTC Kingston-upon-Thames Technical College

ktu kill the umpire

KTX Keith Railway Equipment (railway code)

Ku Karmen unit(s)

Kü *Küçk* (Turkish—little, small)

KU Kalmar Union; Kansas University; Keio University; Kutztown University; Kuwait Airways (2-letter symbol)

KU *Københavns Universitet* (Danish—Copenhagen University)

kub kidney(s)-ureter(s)-bladder

ku'd knocked up (made pregnant)

KUD *Koperasi Unit Desa* (Indonesian—Village Cooperative Unit)

Ku'dam *Kurfürstendamm* (main street of Berlin)

KUED Kodak's Unitized Engineering Drawing (system)

K u H Kingston upon Hull (official name for Hull)

KUK Kollege of Universal Knowledge

KUL Kabul University Library (Kabul, Afghanistan); Karachi University Library (Pakistan); Kyoto University Library (Japan)

Kun Kunsan

K unit Kimball unit

Kunluns Kunlun Mountains of Tibet

Kur (British maritime contraction of Kure); Kurdish; Kurile Islands

Kuria Murias Kuria Muria Islands in the Arabian Sea

Kuril Cur Kurile Current (Oyashio)

Kuriles Kurile Islands in the northwest Pacific

Kurland Courland

Kuro *Kuroshio* (Japanese—Black Salt)—warm ocean current of the North Pacific Ocean

KURRI Kyoto University Research Reactor Institute

Ku's Karmen units

kutd keep up to date

KUU *Kungliga Universitet i Uppsala* (Swedish—Royal University of Uppsala)

Kuw Kuwait

Kuwait State of Kuwait (Persian Gulf gas-and-oil producer), *Dowlat al Kuwait*

Kuyb Kuybyshev

kv kilovolt

kv *kvinde* (Dano-Norwegian—woman)

kV kilovolt; kV/s

KV *Köchel-Verzeichnis* (German—Kochel Catalog)—catalog of Mozart's compositions

KV-107 Japanese-made Sea-Knight-type helicopter built by Kawasaki

kva kilovolt ampere

kVA kilovolt(s)/ampere

KVA Korean Veterans Association; *Kungliga Vetenskaps Akademien* (Swedish—Royal Swedish Academy of Sciences)

kvah kilovolt-ampere-hour

kvam kilovolt ampere meter

Kvan Elektron *Kvantovaya Elektronika* (Russian—Journal of Quantum Electronics)

kvar kilovar; kilovolt ampere reactive

kvarh kilovar hour

kvcp kilovolt constant potential

K-Vets Korean War Veterans of the United States

kvg keyed video generator

KVHS Kanawha Valley Historical Society

KvK Kill van Kull

KVL Kirchoff's Voltage Law

kvm kilovolt meter

KVNP Kidepo Valley National Park (Uganda)

kvp kilovolt peak

KVP *Katholieke Volkspartij* (Dutch—Catholic People's Party)

KVW Kansas City Kaw Valley (railroad)

kw killer weed (pcp sprinkled on leaves and smoked); kilowatt

kw Zambian kwacha(s)—monetary unit(s)

kW kilowatt(s)

KW Kellogg West; Key West; King-Wilkinson

K-W Keith-Wagener

kwac key word and context

K-W AG Kitchener-Waterloo Art Gallery

Kwaj Kwajalein

Kwan Kwantung

kwat key well allowable transfer

KWB Keith, Wagener, Barker (classification)

KWC Kentucky Wesleyan College

kwe kilowatts electrical

KWest Key West

K-W findings Keith-Wagener (ophthalmoscopic findings)

kwh kilowatt hour

kWh kilowatt hour

kwhr kilowatt hour

Kwi Kuwait

kwic key word in context

kwip keyword in permutation

kwit key word in text; key word in title

kwm kilowatt meter

KWMA Kirtland's Warbler Management Areas (Michigan)

KWNWR Key West National Wildlife Refuge (Florida)

kwoc key word out of context

kwot key word out of title

KWP Korean Workers Party

KWPL Kitchener-Waterloo Public Library

kwr kilowatts reactive

KWS Kaziranga Wildlife Sanctuary (India)

KWSM Korean War Service Medal

kwt key word in title; kilowatts thermal

Kwt Kuwait

KWT King William's Town

KWU Kansas Wesleyan University

kwuc keyword and universal decimal classification

KWVZAB *Ko-operative Wijnbouwers Vereeninging van Zuid Afrika Beperkt* (Dutch—Cooperative Wine Farmers Association of South Africa, Limited)

kwy keyway

kxu kilo-x-unit

ky cocoa; key; keyer; keying (device); kyat

Ky Kentuckian; Kentucky

KY Kentucky (zip code); Kol Yisrael (Israel Broadcasting Service); (underground slang—federal hospital in Lexington, Kentucky where drug addicts are treated)
kybd keyboard
kybo keep your bowels open (keep healthy)
KYC Klan Youth Corps; Knickerbocker Yacht Club
kyd Kilo yard
kyeri know your endorsers—require identification (advice to all who cash checks)
Kyle Kyle Railway
kymo kymograph; kymography
KYNP Khao Yai National Park (Thailand)
Kyo Kyoto
Kyocera Kyoto Ceramics
Kyot Univ Kyoto University
Kypros Cyprus
Kyr. *Kyrie eleison* (Greek—Lord, have mercy upon us)
kyrie *kyrie eleison* (Latin—Lord have mercy on us)
kytoon kite balloon
kz duststorm or sandstorm
kz *konzentrations* (Dano-Norwegian—concentration camp)
Kz Kazakh(stan)
KZ *Konzentrationslager* (German—concentration camp)
Kz's unlicensed citizen's-band radio-interference jammers
K z S *Kapitan zur See* (German—Sea Captain)—naval rating

L

l azimuthal or orbital quantum number (symbol); elbow (plumbing); land; large; late; latent heat per unit mass (symbol); lateral; latitude; law; leaf; league; left or port (L or P); length; levorotatory; liaison; light(ning); lignite; line; link; lire; liter; locus; loose; losing pitcher; low; lumen

l (L) luminosity

l/ *letra* (Spanish—letter)

l* lumen

l *lectio* (Latin—reading); *links* (German—left); units of labor (microeconomics)

L Bell Aircraft (symbol); center line (symbol); coil or inductance (symbol); elevated railroad (EL); inductance (symbol); kinetic potential (symbol); Labor; Labour; lactobacillus; lago; Lagrange function; lake; lake vessel; Lamar State College of Technology; lambert; lameness; langmuir; Latin; launching; law (the police, task force, vice squad, etc.); left (port side); lempira (Honduran currency unit); Leo; Leon; Liberal; lift (symbol); lift force; light; Lima—code for L; Linnaeus; Lions International; loch; London; longitude; loran; Lorentz unit; lottery; lough; Luckenbach Lines; Luxembourgh (auto plaque); Lykes Lines; rolling moment (symbol)

L *(L)* demand for money (macroeconomics symbol)

L *lähteä* (Finnish—departure); Lago (Spanish—lake); *läm-min* (Finnish—warm); *länsi* (Finnish—wheat); (Latin—Lucius); *laudes* (Latin—praises); *levato* (Italian—raised); *libra* (Latin—pound); *Life Magazine; links* (German—left); *llegada* (Spanish—arrival); *Loteria* (Spanish—lottery)

L-1 first language

l₁ first lumbar vertebra

l₂ second lumbar vertebra

L-2 second language

l/3 lower third

l₃ third lumbar vertebra

L-3 lousy trio

L-4 military version of the Piper Cub

l₄, l₅, etc. fourth lumbar vertebra; fifth lumbar vertebra

L7 Hollywood slang for old-fashioned person or *square* as capital-letter L and figure 7 may be combined to form a square

L-19 Cessna Bird Dog liaison aircraft

L-100 Lockheed four-engine transport aircraft for civilian use

L-188 Lockheed Electra turbo-prop transport plane

L-1011 Lockheed's jumbo jet-liner

la landing account; large aperture(s); lava; lead antimony; leave allowance; left angle; left atrium; left auricle; light alloy; light amphibian; lighter than air; lightning arrestor; long-acting; low alcohol; low altitude

la (LA) linoleic acid; longitudinal acoustic

l/a landing account; letter of advice; letter of authority; lighter than air

l & a left and above; light and accommodation

la A in fixed-do system; (Italian—the); sixth tone in diatonic scale

l.a. *lege artis* (Latin—according to the art)—as directed

l/a *lettre d'avis* (French—letter of advice)

La Lane; lanthanum; Lao; Laos; Laotian; Louisiana; Louisianian

La *Lebensalter* (German—chronological age); *Luisiana* (Spanish—Louisiana)

LA Latin America(n); Leasehold Area; Legal Advisor; Legislative Assembly; Leschetizky Association; Letter of Activation; Library Association; License Application; Licensing Act; Licensing Authority; Lieutenant-at-Arms; light-alcohol beer brewed by Anheuser-Busch; Local Authority; Los Angeles; Louisiana; Louisiana & Arkansas (railroad); Louvain Association; Lower Alabama (jocular place-name)

L-A Loire-Atlantique (formerly Loire-Inférieure)

L/A Launch Area; Lloyd's Agent

L & A Louisiana & Arkansas (railroad)

LA *Linea Aéria de Chile* (Spanish—Air Line of Chile)

LA 400 400 women of Los Angeles who raised 4 million dollars for its music center

laa light anti-aircraft

LAA League of Advertising

Agencies; Library Association of Australia; Life Assurance Advertisers; Los Angeles Airways

LAACC Light Antiaircraft Control Center

LAAD Latin America Agribusiness Development

LAADS Los Angeles Air Defense Sector

LAAF Libyan Arab Air Force

LAAG Latin American Anthropology Group

laam (LAAM) levo-alpha acetylmethadol (alternative to methadone for treatment of drug addiction)

laar liquid-air accumulator rocket

LAAS Los Angeles Air Service

lab label; labeling; labor; laboratory; load/lorry aboard barge

Lab Laboratory; Labour(ite); Labrador

LAB Labor; Labour; Labour Party; Licquor [*sic.*] Administration Board; Liquor Administration Board; Lloyd Aereo Boliviano (Bolivian airline); low-altitude bombing

LAB Lloyd Aéreo Boliviano (Spanish—Bolivian Air Lines)

LABA Laboratory Animal Breeders Association

LABAN Lakasang Bayan (Filipino—Peoples Power)

Lab Cur Labrador Current—cold Arctic current flowing southward along Atlantic coast of Canada and northern New England

LABEN Laboratori Elettronici e Nucleari (Italian—Electronic and Nuclear Laboratories—Milan)

labe(s) label(s)

labi (Latin prefix—lip)—labial

Labor US Department of Labor

Laborers' Union Laborers' International Union of North America

lab proc laboratory procedure(s)

labrador label-address routine

labrv (LABRV) large ballistic reentry vehicle

labs laboratories

Lab(s) Labrador retriever(s)

LABS Low-Altitude Bombing System

labv (LABV) large ballistic (reentry) vehicle

lac lacquer; lacrimal; lactation; large-aperture component(s); linear amplitude continuous; lunar aeronautical chart; shellac

lac (LAC) load accumulator

Lac Lacerta; Lacertilia

LAC Laboratory Animals Center; Leading Aircraftsman; League of Arab Countries; Liberty Amendment Committee; Library Association of China; Lockheed Aircraft Corporation

LAC Lineas Aéreas Chaqueñas (Spanish—Aero Chaco)

LACA Latin American Coffee Agreement

LACAC Latin American Civil Aviation Commission

LACAP Latin American Cooperative Acquisitions Project

LACATA Laundry and Cleaners Allied Trades Association

LACBW La Crosse Boiling-Water (reactor)

lacc lathe chuck

LACC Los Angeles City College

Laccadives the Laccadive Islands in the Arabian Sea

LACE liquid-air cycle engine

lacertiliol lacertiliologic(al)-(ly); lacertiliologist; lacertiliology

LACES London Airport Cargo Electronic Scheme; Los Angeles Council of Engineering Societies

Lachie Lachlan

LACIE Large Area Crop Inventory Experiment

LACIRS Latin American Communication Information Retrieval System

LACJ Los Angeles County Jail

LACL Latin American Citizens League

LACM Latin American Common Market; Los Angeles County Museum

LACMA Los Angeles Conservatory of Music and Arts; Los Angeles County Museum of Art

LACMedA Los Angeles County Medical Association

LACO Los Angeles Chamber Orchestra

LA Co Art Mus Los Angeles County Art Museum

laconiq laboratory computer on-line inquiry

Lacp Lloyd's anchors and

chains proved

LACP London Association of Correctors of the Press

lacr low-altitude coverage radar

lacri (Latin prefix—tears)—lacrimose

LACSA Líneas Aéreas Costarricenses (Spanish—Costa Rican Airlines)

lact lease automatic custody transfer

LACTC Los Angeles County Transportation Commission

lactns lactations

lacto-ovo(s) lacto-ovo vegetarian(s) (confining diet to milk, milk products, eggs, and vegetables)

lactovegan vegetarian whose diet includes milk products

lacv (LACV) light-amphibious air-cushion vehicle

LACW Leading Aircraftswoman

lad. ladder; liquid agent detector; logarithmic analog-to-digital; logistic approval data; lunar atmosphere detector

lad. (LAD) language-acquisition device

Lad Ladino

Lad (Spanish dialect spoken by many persons of Judaic origin who were forced to flee Spain during the Inquisition)

LAD Library Administration Division (American Library Association); Light Air Detachment

ladar laser detection and ranging

ladd low-altitude drogue delivery

ladder. life-assurance direct entry and retrieval

LADE Líneas Aereas del Estado (Spanish—State Airlines, Argentina)

LADECO Línea Aéreo del Cobre (Spanish—Copper Air Line)

LADIES Life After Divorce Is Eventually Sane (Hollywood ex-wife group); Los Alamos Digital-Image-Enhancement Software

Ladies' Garment Workers International Ladies' Garment Workers Union

Ladies Home J Ladies Home Journal

ladir low-cost arrays for detection of infrared

LADO Latin American Defense Organization; Latin

American Development Organization

Ladoga Lake Ladoga east of Leningrad and called Ladoshskoye Ozero by the Russians

ladp ladyship

Ladrones Marianas Islands

Lad(s) Ladino(s)—Europeanized Central American mestizo(s) of Spanish descent; their Judeo-Spanish dialect developed by refugees from the Holy Inquisition (1487 to 1834), also called Judeo-Spanish, Spagnuolo, or Spanish Yiddish

LADSIRLAC Liverpool and District Scientific, Industrial, and Research Library Advisory Council

L Adv Lord Advocate

LADWP Los Angeles Department of Water and Power

LAE Leadership Ability Evaluation

laetrile laevo-mandelonitrile-beta-glucuronic acid

laev. laevus (Latin—left)

laf laminar air flow

Laf Lafayette

LaF Louisiana French

LAF *L'Académie Française* (French—The French Academy); Living Arts Foundation

Lafayette Marie Joseph Paul Yves Roch Gilbert du Motier (Marquis de Lafayette)

LAFB Lincoln Air Force Base

LAFC Latin-American Forestry Commission

LAFCO Local Agency Formation Commission

LAFD London Association of Funeral Directors

Lafe Lafayette

LAFE *Laboratorio de Fisica Espacial* (Portuguese—Space Physics Laboratory)

La Fen *La Fenice* (Italian—The Phoenix)—Venice opera house

La Font La Fontaine

LAFS Los Angeles Funeral Society

LAFTA Latin American Free Trade Area; Latin American Free Trade Association

lafts laser and flir test set

lafv (LAFV) light-armored fighting vehicle

lag. lagan

lag. lagena (Latin—bottle, flask)

Lag Lagoon; Laguna

Lag Laguna (Spanish—lagoon or lake)

La G La Guaira

LAG Layton Art Gallery; Librarians Automation Group

LAGB Linguistics Association of Great Britain

LAGE Los Angeles Grain Exchange

LAGEOS Laser Geodetic Satellite

LAGIC Life and General Insurance Committee

lags. (LAGS) laser-activated geodetic satellite

Lags Lagunas

LAGS Los Angeles Geographic Society

Lagunas Laguna Mountains of California

Lah Lahore

LAH Licentiate Apothecaries Hall

LAHAWS Laser Homing and Warning System

LAHC Los Angeles Harbor College; Los Angeles Harbor Commission

LAHD Los Angeles Harbor Department

lahs low-altitude high speed

LAHS Local Authority Health Services

lai leaf area index

LAI Library Association of Ireland

LAI Linea Aèree Italiane (Italian—Italian Air Lines)

LAIA Latin American Integration Association

LAIC Lithuanian-American Information Center

LAICA Los Angeles Institute of Contemporary Art

LAIICS Latin American Institute for Information and Computer Sciences

LAINS Low-Altitude Inertial Navigation System

LAIRS Labor Agreement Information Retrieval System

LAIS Loan Accounting Information System (AID)

LAIT Logistics Assistance and Instruction Team

LAIV Loss Adjusters Institute of Victoria

La J La Jolla

LAJ Los Angeles Junction (railroad)

Lake Clark national park in Alaska

laks lakrids (Danish—licorice)

LAL Langley Aeronautical Laboratory (Langley Research Center)

LAL Laboratoire de l'Accelerateur Linéaire (French—Linear Accelerator Laboratory)

LA-LB Los Angeles-Long Beach (ports)

La Leche La Leche League International

lali lonely aged of low income

lalsd language for automated logic and system design

LALUCS Local Authority Land Use Classification System

lam laminate

lam (LAM) load accumulator with magnitude

Lam The Book of Lamentations; Lamarck; Lambretta

Lam Lamentations

La M La Monnaie (French—the mint) Brussels theater

LAM Lamarck; Lambert; Latin American Mission; London Academy of Music

L.A.M. *Liberalium Artium Magister* (Latin—Master of Liberal Arts)

LAMA Latin American Manufacturers Association; Lead Air Materiel Area; Library Administration and Management Association; Locomotive and Allied Manufacturers' Association

Lamb Lambert; Lamberto; Lambertus; Lambeth

LAMBC Los Angeles Motor Boat Club

lambsan lamb sandwich

lambwich lamb sandwich

LAMC Letterman Army Medical Center; Los Angeles Metropolitan College; Los Angeles Music Center

LAMCO Liberian-American-Swedish Mineral Corporation

LAMDA London Academy of Music and Dramatic Art

LAME Licensed Aircraft Maintenance Engineer

LAMIDA Lancashire and Merseyside Industrial Development Association

LAMM Los Angeles Master Morticians; Lutheran-American Melancthon Movement

lamma laser microprobe mass analyser

LA MOCA Los Angeles Museum of Contemporary Art

Lamp Lampeter

LAMP Library Additions and Maintenance Program; Low-Altitude Manned Penetration; Lunar Analysis and Mapping

Program
LAMPP Los Alamos Molten Plutonium Program (AEC)
Lamps ship's lamp trimmer
LAMPS Center for the Study of Legal Authority and Mental Patient Status; Light Airborne Multi-purpose System
LAMS Launch Acoustic Measuring System
LAMSACC Local Authorities Management Services and Computer Committee
lamsim launcher-and-missile simulator
lan listing agent's name
Lan Lancaster; Lansing
L An Los Angeles
LAN *Línea Aérea Nacional de Chile;* Local Apparent Noon; Local Area Network
LAN *Latin American Newspapers* (bibliographic reference)
lanac laminair air navigation and anti-collision
Lanarks Lanarkshire
Lan Bag Lansing Bagnall
lanby large automatic-navigation buoy
Lanc Lancaster
Lance Lancelot; Ling-Temco-Vought MGM-52A surface-to-surface tactical missile
Lance Cpl Lance Corporal
Lancs Lancashire
land. landscaping
LandCraB landing craft and bases
LANDJUT Land Forces (Schleswig-Holstein and) Jutland (NATO)
LANDNONOR Land Forces Northern Norway (NATO)
LANDNORTH Land Forces Northern Europe (NATO)
'Lando Orlando, Florida
Land of Peace Thailand
Land of Religion India
Land of Shining Mountains Montana
landprop landlord proprietor
Lands Landsmaal (Norwegian national language)
Lands Down Under Australia and New Zealand
LANDSONOR Land Forces Southern Norway (NATO)
Land Time Forgot Australia
Landw Landwirtschaft (German—agriculture)
LANDZEALAND Land Forces Zealand (NATO)
Lan Fus Lancashire Fusiliers
lang language
Lang Langbridge; Langdon; Lange; Langer; Langford;

Langhorne; Langlois; Langson; Langston; Languedoc
Langley Langley, Virginia headquarters of the CIA
LANICA Líneas Aéreas de Nicaragua (Spanish—Air Lines of Nicaragua)
LANL Los Alamos National Laboratory
Lan Reg Lancashire Regiment
LANS Land Area Navigation System
LANSA Líneas Aéreas Nacionales SA (Spanish—National Airlines Corporation)—Peru
Lant Atlantic (naval short form)
'Lanta Atlanta
LANTIRNS Low-Altitude Navigation, Targeting Infra-Red Navigation System
LANWR Laguna Atascosa National Wildlife Refuge (Texas); Lake Andes National Wildlife Refuge (South Dakota)
LANY Linseed Association of New York
Lao Laotian
LAO Legal Assistance Office(r); Licentiate of the Art of Obstetrics
LAOAR Latin American Office of Aerospace Research
LAOD Los Angeles Ordnance District (USA)
Laos Lao People's Democratic Republic (Indo-Chinese country)
LAOT Los Angeles Opera Theater
lap. laparotomy; launch analyst's panel; learning activity package; left atrial pressure
Lap Lapland; Lappish
LaP Las Palmas (British maritime abbreviation)
La P La Paz
LAP La Paz, Mexico (airport); Laboratory of Aviation Psychology (Ohio State University); Library Awareness Program; *Líneas Aéreas Paraguayas* (Spanish—Paraguayan Air Lines)
laparo laparoscope; laparoscopic (sterilization); laparotomy
LAPC Los Angeles Pacific College; Los Angeles Pierce College
LAPCO Lavan Petroleum Company
LAPD Los Angeles Police Department
LAPDis Los Angeles Procure-

ment District (US Army)
LAPES Low-Altitude Parachute-Extraction System
L A Phil Los Angeles Philharmonic
LAPI Los Angeles Philharmonic Institute
lapid. lapideum (Latin—stony)
LAPL Los Angeles Public Library
LAPO Los Angeles Philharmonic Orchestra
Lapp Lappish; member of a people of northern Scandinavia, Finland, and Russia
LAPS List Assembly Programming System
LAPT London Association for the Protection of Trade
LAPTA Local Authorities Passenger Transport Association
La Pucelle La Pucelle d'Orléans (French—The Maid of Orleans)—Joan of Arc
laput light-activated programmable unijunction transistor
laq lacquer
lar left arm reclining; liquid argon; local-acquisition radar
lar (LAR) light-artillery rocket; long-range radar
Lar Lara; Larina; Larry
LAR Lease Application Request; Library Association of Rhodesia; Life Assurance Relief; Limit Address Register
lara (LARA) light armed reconnaissance aircraft
LARA League of Americans Residing Abroad; Licensed Agency for Relief of Asia
laram line-addressable random-access memory
larat low-altitude radar altimeter
La Raza (Hispanic-American Spanish—The Race) *La Raza Unida* (Spanish—The United Race)—Mexican-American political organization
larbord (Norwegian—larboard)—originally a vessel's loading side opposite starboard, hence the portside
larc lighter, amphibious, resupply, cargo (vehicle); logic alarm radio clock
Larc Lovermore automatic research computer
LARC Langley Research Center; League Against Religious Coercion; Library Automation and Consulting; Libyan-American Reconstruction Commission; Lind-

heimer Astronomical Research Center; Local Alcoholism Reception Center(s)
larct last radio contact
larf low-altitude radar fuzing
larg *largamente* (Italian—broadly); *largeur* (French—width); *largo* (Italian—slow)
Large Print Large Print Publications
largo. larghetto (Italian—moderately slow)
LARIAT Laser-Radar Intelligence-Acquisition Technology
LARIC Later Reading In-Service Course
LARO Latin American Regional Office (FAO)
larp line automatic reperforator; local and remote printing
La R-P La Rochelle-Pallice
lar rep larceny report
lars laminar angular-rate sensor
Lars Lawrence
LARS Laboratory for Applications of Remote Sensing (Purdue); Light Artillery Rocket System; Low-Altitude Radar System
LART Los Angeles Rapid Transit
larv (LARV) low-angle reentry vehicle
larva (LARVA) low-altitude research vehicle
larvi larvicidal; larvicide
laryng laryngological; laryngologist; laryngology
laryngol laryngology
las large astronomical satellite; liberal arts and sciences; lookout aiming sight; low-alloy steel; lower airspace
las lassú (Hungarian—slow introductory passages leading to fast section, *friss*, of a csárdas or rhapsody)
LAs Latin Americans
LAS Land Agents Society; large astronomical satellite; Las Vegas, Nevada (airport); League of Arab States; Lebanese-American Society; Legal Aid Society; Library Association of Singapore; Lord Advocate of Scotland
LA & S Liberal Arts and Sciences
lasa large-aperture seismic array
LASA Latin American Shipowners Association
LASAIL Land-Sea Interaction Laboratory

LASBO Louisiana Association of School Business Officials
LASC Los Angeles State College
LASCO Latin American Unesco Science Cooperation Office
lascot large-screen color television
lascr light-activated silicon-controlled rectifier
lascs light-activated silicon-controlled switch
LASEORS London and South Eastern Operational Research Society
laser light amplification by stimulated emission of radiation; lucrative approach to support expensive research
LASER London and South Eastern Library Region
LASER Los Angeles Skeptics Evaluative Report
LASERS London and South Eastern Regional Library System
LASH Legislative Action on Smoking and Health; Lighter Aboard Ship (cargo system)
lasi landing-site indicator
LASIE Library Automated System Information Exchange
LASL Los Alamos Scientific Laboratory
LASMCO Liberian American-Swedish Minerals Company
LASMO London and Scottish Marine Oil
LASO Latin American Solidarity Organization; Los Angeles Sheriff's Office; Los Angeles Society of Ophthalmology
lasp low-altitude space platform
LASRA Leather and Shoe Research Association
lasrm (LASRM) low-altitude short-range missile
lass. lighter-than-air submarine simulator (LASS)
LASS launch-area support ship; Local Authority Social Services; Los Angeles Special Services (telephone)
LASSCO Los Angeles Steamship Company
Lassen Volcanic California national park containing Lassen Peak
'lasses molasses
lassie low-airspeed sensing and indicating equipment
lasso laser search-and-secure observer

LASSO Latin American Student Studies Organization
La State U Pr Louisiana State University Press
LASUSSR Library of the Academy of Science of the USSR (Leningrad)
lasv (LASV) low-altitude surface vehicle; low-altitude supersonic vehicle
lat lateral; latitude
lat (LAT) lowest astronomical tide
lat lateinsich (German—Latin); *latin* (Dano-Norwegian—Latin); *latitud* (Spanish—latitude)
lat. latus (Latin—wide)
Lat Latin; Latvia; Latvian
Lat Latin (classical language of Roman antiquity and the base of Romance languages)
LAT Latex Agglutination Test(ing); Linseed Association Terms; Local Apparent Time; Taxader Airport (Bogotá)
LAT Los Angeles Times
LATA Local Access and Transport Areas; Los Alamos Technical Associates
lat. admov. lateri admoveatum (Latin—apply to the side)
LATARS Laser-Augmented Target Acquisition and Recognition System
LATCC London Air Traffic Control Center
LATCRS London Air Traffic Control Radar Station
lat. dol. lateri dolenti (Latin—to the painful side)
later (Latin prefix—side)—lateral
LATH Laos and Thailand Military Assistance
lat ht latent heat
Latin America Portuguese-speaking Brazil and Spanish-speaking countries such as the Central American republics, Cuba, the Dominican Republic, Mexico, Puerto Rico, the South American republics; understood variously as: Spanish-speaking America plus Brazil; all of the Americas south of the United States; Mexico, Central and South America and islands of the Caribbean
latoff lowest astronomical tide of the foreseeable future (assuming anyone can foresee the future)
latom lowest astronomical tide

of the month

latoy lowest astronomical tide of the year

lats long-acting thyroid stimulator

LATS Los Angeles Times Syndicate

latssn late season

LATTC Los Angeles Trade-Technical College

LATUF Latin American Trade Union Federation

Latv Latvia; Latvian

LATWPNS Los Angeles Times Washington Post News Service

lau laundry

Lau Laura; Lauretta

LAUA Lloyd's Aviation Underwriters' Association

Lauch Lauchlin

laughing gas nitrous oxide (N_2O)

LAUK Library Association of the United Kingdom

Lau Lib Laurentian Library (Florence)

laun launched

Laun Launceton, Tasmania

Launce Lancelot

laund launder; laundry

laundromat automatic coin-operated laundry

Laur Laurence

Laurentians Laurentian Mountains of southern Québec where they are also called the Laurentides as is Laurentides Park

Laurie Laurence

LAUSC Linguistic Atlas of the United States and Canada

LAUSD Los Angeles United School District

lav lavatory; lymphadenopathy virus

LAV Lalomalava (Western Samoan airport); light-armored vehicle; *Línea Aeropostal Venezolana* (Spanish—Venezulean Airmail Line); lymphadenopathy-associated virus

Lava Beds Lava Beds National Monument in northern California

LAVC Los Angeles Valley College

lavm loran automatic vehicle monitoring

law. lawyer; light assault weapon; low-altitude weapon

Law Lawrence

LAW Lawyers Against Wigs; League of American Wheelmen; League of American Writers; Legal Aid Warranty;

light anti-tank weapon; Local Air Warning

Law Lat Law Latin

Lawr Lawrence; Lawrencian

Law Rept Law Report(s)

Lawrie Lawrence

LAWRS Limited Airport Weather Reporting System

LAWS Leadership and World Society

lax. laxative

LAX Los Angeles, California (International Airport)

Lax-Chi Los Angeles—Chicago

Lax-NO Los Angeles—New Orleans

Lax-NY Los Angeles—New York

Lax-San Los Angeles—San Diego

Lax-Sea Los Angeles—Seattle

Lax-Sfo Los Angeles—San Francisco

Lax-Tor Los Angeles—Toronto

Laz Lazarus

LAZ Los Angeles Zoo

lb landing barge; lavatory basin; left back; letter box; lifeboat; link belt(ing); linoleum base; local battery; low band; lumen band; pound

lb (LB) line buffer

l-b lemon-and-butter (sauce)

l & b land and building(s); left and below

lb libra (Latin—pound)

l.b. lectori benevolo (Latin–to the kind reader)

LB landing barge; large burgh; Leonard Bernstein; light bomber; Lloyd Brasileiro (Brazilian Steamship Line); Local Board; Long Beach; Longview Bridge (Columbia River, Washington); Luther Burbank

L-B Link-Belt

L,B Little, Brown Publishing Company

L.B. Baccalaureus Litterarum (Latin—Bachelor of Letters)

lba lifting-body airship

Lba Luba (formerly San Carlos)

LB & AL Lever Brothers and Associates Limited

L-band 390–1550 mc

lb ap apothecaries' pound

L-bar capital-L-shaped bar

lb av avoirdupois pound

LBB Lubbock, Texas (airport)

lbbb left bundle branch block

lb/bhp-hr pounds per brake horsepower hour

lbbsb left bundle branch sys-

tem block

Lbc Lübeck

LBC Liberian Broadcasting Corporation; London Broadcasting Company; Lutheran Baptist Convention

lb cal pound calorie

LBCC Long Beach City College

lbcd left border of cardiac dullness

LBCH London Bankers' Clearing House

lb chu pound centigrade heat unit

LBCL Liberty Baptist College

LBCM Licentiate of Bandsmen's College of Music

lb/cu ft pounds per cubic foot

lbd learning and/or behavior disordered; left border of dullness; lifeboat deck; little black dress; lower bovine distemper; lower-back disorder

LBD League of British Dramatists

L/Bdr Lance Bombardier

L-beam capital-L-shaped beam

LBEB Laboratory of Brain Evolution and Behavior

lbf lactobacillus bulgaricus factor; pound-force

LBF Louis Braille Foundation (for blind musicians)

lbf-ft pound-force foot

lbf/in.² pound-force per square inch

lb ft pound foot

lb ft² pound per square foot

lb ft³ pound per cubic foot

lbg liquefied butane gas

LBG Paris, France (Le Bourget Airport)

lbh length, breadth, height

LBHD Long Beach Harbor Department

lb/hr pounds per hour

LBHS Luther Burbank High School

LBI Library Binding Institute; Licensed Beverage Industries; Lloyds Bank International

LBI Lands Bókasafn Islands (Icelandic—National Library of Iceland)—in Reykjavik

lb in. pound inch

lb in.² pound per square inch

lb in.³ pound per cubic inch

lbir laser-beam-image reproducer

LBJ Lyndon Baines Johnson—thirty-sixth President of the United States

LBJL Lyndon Baines Johnson

Library (Austin)
LBJSHP Lyndon B Johnson State Historic Park (Texas)
LBJTMC Lyndon B Johnson Tropical Medical Center (American Samoa)
LBK landing barge, kitchen
lbl label (flow chart)
LBL Lawrence Berkeley Laboratories; Lloyds Bank Limited
lb/lb pound per pound
lbld labelled
lblg labelling
lbm lean body mass; liquid (loose) bowel movement
lb m pound mass
lb/m pounds per minute
Lbm Lifeboatman
lb-mol pound-mole (mass)
LBMS London Boroughs Management Services
Lbn Libano (Spanish—Lebanon)
lbnpd lower-body negative-pressure device
lbo (LBO) leveraged buyout
LBO Lima, Peru (Limatambo Airport)
lboe lime-base oil emulsified
lbp length between perpendiculars; low back pain; low blood pressure
LBP Lanier Business Products; Lester Bowles Pearson (Canada's eighteenth Prime Minister); London Borough Polytechnic
LBPL Long Beach Public Library
lbr labor; laser-beam recorder; lumber
Lbr Labrador; Librarian
Lbr Liberia (Spanish abbreviation)
LBR Library Bill of Rights
lbs pounds (from the Latin—*Librae*)
lb/s pounds per second
l.b.s. lectori benevolo salutem (Latin—to the kind reader, greetings)
LBS landing barge support; Lean Burn System; Libyan Broadcasting Service; Lifeboat Station; London Boroughs Association; London Botanical Society
LBSC Long Beach State College
LB & SCR London, Brighton and South Coast Railway
lbs H₂O mm pounds of water per million standard cubic feet (natural gas)
LBSM Licentiate of Birming-

ham and Midland Institute School of Music
lb/sq in. pounds per square inch
lbs sq ft pounds per square foot
lbt laser-beam transmissiometer
lb t pound(s) thrust; pound(s) troy
LBTF Long Beach Test Facility
LBTS Land-Based Test Site
Lbu Labuan
LBV landing barge, vehicle
lbw leg before wicket; low body weight; low-speed black-and-white (photography)
lc label clause; laundry chute; lead-covered; leading case(s); left center; legal currency; letter card; level crossing; light case; light change; line-carrying; liquid crystal; load carrier; load center; localized corrosion; locked-closed; locus caeruleus; low calorie; low carbon; lower case; single acetate single cotton
l-c launch control; low calorie; low carbohydrate
l/c letter of credit; lower center
l.c. loco citato (Latin—in the place cited)
Lc corrected middle latitude
LC Lackawanna College; Ladycliff College; Lafayette College; Lake Central Airlines; Lakehead College; Lakeland College; Lambuth College; Lance Corporal; Lander College; landing craft; Lane College; Laredo College; Lassen College; L'Assumption College; launch(ing) control; Law Court(s); Lawrence College; Lee College; Legal Committee; Legislative Council; Lesley College; Letter Contract; Lewis College; Library of Congress; Lieutenant Commander; Limestone College; Lincoln Center; Lincoln College; Lindenwood College; line of communication; Linfield College; Livingstone College; London Clause; London Club (of criminologists, freelance investigators, and members of the media concerned with controversial trials and unsolved crimes); Longwood College; Loras College; Louisburg College;

Louisiana College; Loyola College; Luther College; Luzon College(s); Lycoming College; Lynchburg College
L-C Liquid-Carbonic (Division of General Dynamics)
L/C Letter of Credit
L of C Library of Congress
LC A Lover's Complaint
lca low-cost automation
LCA Lake Carriers Association; Lake Central Airlines; landing craft—assault; Launcher Control Area; Learning Corporation of America; Library Club of America; Licensed Company Auditor; Lutheran Church in America
LCAC Landing-Craft Air Cushion
LCACS Landing-Craft Air Cushion Ships
LC-ADD Library of Congress—American Doctoral Dissertations
lcal lowercase alphabet length
lcao linear combination of atomic orbitals
lcat (LCAT) lecithin-cholesterol acyltransferase
L & C ATA Laundry and Cleaners Allied Trades Association
L Cav Lucy Cavendish Collegiate Society, Cambridge
lcb longitudinal position of center of buoyancy
LCB Liquor Control Board; London and Continental Bankers
LCBBC Liquor Control Board of British Columbia
LCBM Liquor Control Board of Manitoba
LCBO Liquor Control Board of Ontario
LCBS Liquor Control Board of Saskatchewan
lcc lateral center of gravity; ledger card computer; life cycle cost; light curtain closed
LCC landing craft, control (3-letter symbol); Lands Conservation Council; Lansing Community College; Launch Control Center; League of California Cities; Life Cycle Center; London County Council; Lower Columbia College
L & C C Lewis and Clark College
LCcc Library of Congress catalog card

LCCC Library of Congress Computer Catalog; Lorain County Community College; Lucas County Corrections Center (Ohio)

lcce life-cycle cost estimate

LCCI London Chamber of Commerce and Industry

LCCJ Louisiana Council on Criminal Justice

LCCMARC Library of Congress Current MARC (file)

lccs low cervical caesarian section

LCCS Launcher Captain Control System; Lucy Cavendish Collegiate Society (Cambridge)

lccv (LCCV) large crude-carrying vessel (oil tanker)

lcd liquid crystal diode; liquid crystal display; lowest common denominator

lcd (LCD) liquid-crystal display

LCD Lord Chamberlain's Department; Lord Chancellor's Department

l/c derv lowercase derivative (angora, axminster, bakelite, bunsen burner, canada balsam, castile soap, china clay, congo red, cordovan leather, delftware, etc.)

LCDHWIU Laundry, Cleaning, and Dye House Workers International Union

lcdo *licenciado* (Spanish—licensed)

Lcdo *Licenciado* (Spanish—lawyer)

LCDs Lower Court Decisions

lcdtl load-compensated diode-transistor logic

lce lance; left center entrance

LCE Licentiate in Civil Engineering

lces least-cost estimating and scheduling

lcf least common factor; liquid complex fertilizer; longitudinal position of center of flotation; low cycle fatigue; lowest common factor

LCF landing craft, flak; launch control facility

LCFA Lower California Fisheries Association

l-c f-s pr last-come first-served preemptive resumé

LCFTA London Cattle Food Trade Association

lcg liquid-cooled (under) garment; longitudinal position of center of gravity

LCG British armored landing craft; Load Classification Group

LCGB Locomotive Club of Great Britain

LCGC London Community Gospel Choir

lch launch

L Ch Licentiate in Surgery

LCH landing craft—heavy

LCHQ Local Command Headquarters (NATO)

lchr launcher

lci legally-correct interpretation; locus of control interview

LCI landing craft, infantry; Learner-Centered Instruction; Liquid Crystal Institute (Kent State University); Livestock Conservation Incorporated

LCIP London Centre for International Peacebuilding

lcircon letter of credit irrevocable and confirmed

LCIS Library of Computer and Information Sciences

LCJ Lord Chief Justice

LCJ *Louisville Courier-Journal*

LCJC Lake City Junior College

Lcks Locks (postal abbreviation)

lcl less than carload lot; less than container lot; lifting condensation level; local-(izer); loose container load; lower card lever; lower control limit

lcl (LCL) lowest charge level

LCL Liberal and Country League (Australia); Licentiate in Canon Law; Licentiate in Canonic Law; Licentiate in Common Law

L-C-L Levinthal-Coles-Lillie (bodies)

LCL *La Casa del Libro* (Spanish—House of the Book)— Puerto Rico's typographic arts museum on San Juan's Calle del Cristo

LCLA Lutheran Church Library Association

L-C-L bodies Levinthal-Coles-Lillie bodies

lcl/ci limited calendar life/controlled item

LCLs Liverpool Central Libraries

LCLS Livestock Commission Levy Scheme

lcm lead-coated metal; least common multiple; left coastal margin; limit-cycle monitor; liquid curing media; logistics composite model; lost circulation materials (oil well); lowest common multiple

lcm (LCM) large-core memory; lymphocytic choriomeningitis

LCM landing craft, mechanized; landing craft—medium; London College of Music

LCMARC Library of Congress MARC (files beginning in 1968)

lcmm life-cycle management model

lcmp launcher control and monitoring panel

LCMS Launch Control and Monitoring System; Lutheran Church Missouri Synod

lcn local civil noon

Lcn Lincoln

LCN *La Cosa Nostra* (Italian—Our Thing)—The Mafia; Load Classification Number(ing)

LCNM Lehman Caves National Monument (Nevada)

LCNN Land Commander Northern Norway (NATO)

LCNY Linguistic Circle of New York

LCNYC Lincoln Center (New York City)

LCO Landing Craft Officer; Launch Control Officer; London College of Osteopathy

lcoc launch control officer's console

L Col Lieutenant Colonel

lcos lead computing optical sight

lcp language conversion program; last complete program; local coastal plan; low-cost production

LCP landing craft, personnel; Latvian Communist Party; Liberal Country Party (Australia); Library Company of Philadelphia; Licentiate of the College of Preceptors; Livable Cities Program; London College of Printing

LCPA Lincoln Center for the Performing Arts

L Cpl Lance Corporal

LCPL landing craft, personnel, large (naval symbol)

LCPR landing craft, personnel, ramped (naval symbol)

lcps last card program start

LCPS Licentiate of the College of Physicians and Surgeons

LCP & SA Licentiate of the College of Physicians and Surgeons of America

LCP & SO Licentiate of the College of Physicians and Surgeons of Ontario

lcr limited carrier(s) risk

l/cr letter of credit

l/cr *lettre de crédit* (French— letter of credit)

L Cr Lieutenant Commander

LCR landing craft, rubber; Line Condition Report

LCRA Lower Colorado River Authority

l/crt lamb crutchings

LCRT Lincoln Center Repertory Theater

lcs launch-control simulator

lcs (LCS) large-core storage

lc's (LCs) liquid crystals

LCs Langerhans cells

LCS Laboratory of Computer Sciences (M.I.T.); landing craft, support (naval vessel); Library Computer System

lcsa letter of credit (negotiable by drafts at) sight airmail

LCSA Lewis and Clark Society of America

LCSH *Library of Congress Subject Headings*

lcss land combat support set

LCSS London Council of Social Service

LCST Licentiate of the College of Speech Therapists

LCSW Licensed Counseling Social Worker

lct last card total; less than truckload lot

LCT *Laboratoire Central de Télécommunications* (French—Central Telecommunications Laboratory); landing craft—tank; latest closing time; less than truckload lot; Local Civil Time; Loughsborough College of Technology

LCTC Langlade County Teachers College; Leicester College of Technology and Commerce; Lewis and Clark Trail Commission

lctp launcher control test panel

lcty locality

lcu launch-control unit; lower control unit

lcu (LCU) large closeup

LCU landing craft, utility

LCUSA Lutheran Council in the United States of America

lcv low calorific value

LCV landing craft, vehicle

LCVP landing craft, vehicle,

personnel

lcv's light commercial vehicles

LCWS Landing Configuration Warning System

lcx launch complex

lcxt large cosmic-X-ray telescope

LCY League of Communists of Yugoslavia

LCYC Lemon Creek Yacht Club

LC zone land conservation zone

ld ladies day; land; lead; learning disabled; lethal dose; library development; lid; lifeboat deck; light difference; line of departure; line of duty; list(ing) date; load; load draft; Lord; low density; low door; lower deck

ld (LD) laser disc; learning-disabled; lethal dose

l-d low-density

l/d length to diameter (ratio); life to drag (ratio)

l & d labor and delivery; loans and discounts; loss(es) and damage(s)

l.d. *lepide dictum* (Latin—wittily related)

Ld Leopold; Limited

LD Labor (US department); lighting director; line of departure; line of duty; London Docks; Low Dutch; lower berth (double occupancy)

L-D Leishman-Donovan (bodies)

L/D Letter of Deposit

L.D. *Litterarum Doctor* (Latin—Doctor of Letters)

LD-3 lower-deck container

ld₅₀ median lethal dose

LD-10 lethal dose for 10 percent of the animals tested

LD-50 lethal dose for 50 percent of the animals tested

lda land development aircraft; landing distance available; learning disabilities average; left dorso-anterior; line drawing amplifier; localiser direction(al) aid

lda (LDA) light defense aircraft

L^{da} *Limitada* (Portuguese or Spanish—Limited); *Licenciada* (Spanish—lawyer)—feminine form of *L^{do}—Licenciado*

LDA Ladies Darts Association; Laser Disc Association; Lead Development Association

ldac lunar-surface data-acquisition camera

L da P Lorenzo da Ponte (orig.

Emanuele Conegliano)

L da V Leonardo da Vinci

ldb light distribution box

LDBHS Louis D Brandeis High School

ldc large document copier; leased directive circuits; long-distance call; lower dead center

ldc (LDC) latitude data computer

LDC Late Developing Country; Laundry and Dry Cleaning (union); Less Developed Countries; Light Direction Center; List of Design Changes; Local Defense Center; Local Development Corporation

LD & C Louis Dreyfus & Compagnie

LDCMMA Laundry and Dry Cleaners Machinery Manufacturers Association

ldc's (LDCs) least-developed countries; less-developed countries

ldd laser development device

lddo long-distance diesel oil

LDDS Low-Density Data System

LDEF Long-Duration Exposure Facility

ldel land development encouragement loans

L Dent Sci Licentiate in Dental Science

Lderry Londonderry

L de V Lope de Vega

LDF Legal Defense Fund (NAACP); Local Defense Force(s)

ldg landing; loading; lodging

Ldg Lodge (postal abbreviation)

ldg & dly landing and delivery

Ldge Lodge

ldg gr landing gear

ldglts leading lights

ldgs lodgings

ldh lactic-acid dehydrogenase

L d'H *Légion d'Honneur*— [French—Legion of Honor (decoration)]

Ld'H *Légion d'Honneur* (French—Legion of Honor)

LDH *Ligue des Droits de l'Homme* (French—League for the Rights of Man)

ldhc locker-door hydraulic cylinder

ldi lecture-driven instruction

ldk lower deck

LDKG *Life and Death of King John*

ldl learning disabilities limited;

501

loudness discomfort level; low-density lipoprotein

ld lmt load limit

LDMA London Discount Market Association

Ld May Lord Mayor

ld mk landmark

ldmwr limited depot maintenance work requirements

Ldn London; Londoner

LDN Long-Distance Network

ldo light diesel oil; long-distance oil

L^{do} *Licenciado* (Spanish—lawyer, licentiate holding master's degree)

LDO Licensed Deck Officer; Limited-Duty Officer

L-dopa levodihydroxyphenylalanine (Parkinson's disease treatment drug)

LDOS Lord's Day Observance Society

ldp left dorso-posterior; logistics data package

Ldp Ladyship; London daily price; Lordship

LDP landed duty paid; Liberal Democratic Party (Japanese)

ldpe low-density polyethylene

ldr launder; laundry; leader; lecture-discussion-recitation; ledger; light-dependent resistor; list(ing) date received; lodger

l-d-r labor-delivery-recovery

LDR Large Deployable Reflector; London Digital Recording

l/d ratio length to diameter ratio; lift-to-drag ratio (of aircraft)

LDRC Lumber Dealers Research Council

ldri low data rate input

L-drivers learner-drivers

Ldr's London depository receipts

LDRTA Long-Distance Road Transport Association

ldry laundry

lds loads

lds (LDS) large disc store; large disk storage

Lds Leeds

LDs Learning Disabilities

LDS Latter Day Saints (Church of Jesus Christ of); Licentiate in Dental Surgery; Line Drawing System

LDSc Licentiate in Dental Science

ld/sd look down/shoot down

LDSR League of Distilled Spirits Rectifiers

LDSRA Logistics Doctrine

Systems and Readiness Agency (USA)

LDSRCPS Licentiate in Dental Surgery of the Royal College of Physicians and Surgeons

LDSRCS Licentiate in Dental Surgery of the Royal College of Surgeons

l&d store liquor and delicatessen store

ldt logic design translator

LDTO Long-Distance Telegraph Office

ldtr long-dwell-time radar

LDV Local Defense Volunteer

ldw loss damage waiver

ldx long-distance xerography

ldy laundry

Ldy Londonderry

LDY Lancashire and Derbyshire Yeomanry; Leicestershire and Derbyshire Yeomanry

le launch(ing) equipment; leading edge; left eye; library edition; limit of error; limited edition; low explosive

l/e lifetime earnings

l.e. *lupus erythematosus* (skin disease)

Le Lebanese; Lebanon

LE Labor Exchange; Labour Exchange; light equipment; Limited Edition; low explosive

lea leather

LEA Local Education Authority; Local Education(al) Agency; Locomotive Engineers Association; Loss Executives Association; Lutheran Education Association

LEAA Lace and Embroidery Association of America; Law Enforcement Assistance Administration

LEAA *Law Enforcement Assistance Act*

LEAD Law Students Exposing Advertising Deception

LEADER Lehigh Automatic Device for Efficient Retrieval

LEADS Law Enforcement Agencies Data System (Illinois)

Lea & F Lea & Febiger

LEAF Law, Equality, and Freedom (association)

leaf(s) leaflet(s)

LEAJ Law Enforcement and Administration of Justice (President's Commission on)

Leamington Royal Leamington Spa in central England

Leander British class of all-purpose frigates

Leanu Nikolaus Lenau (Nikolaus Niembsch von Strehlenau)

leap. liftoff elevation and azimuth programmer

LEAP Lambda Efficiency Analysis Program; Language for the Expression of Associative Procedures; Law, Education, and Participation; Loan and Educational Aid Program

LEAPS Laser Engineering and Applications for Prototype Systems; Law Enforcement Agencies Processing System (Massachusetts); London Electronic Agency for Pay and Statistics

Lear *The Tragedy of King Lear*

Lear 23 Lear jet transport

leas lower-echelon automatic switchboard

Leasat Leased Satellite (communications service)

Leavenworth U.S. Penitentiary at Leavenworth, Kansas

leaverats leave rations

Leb Lebanese; Lebanon

LEB London Electricity Board

Lebanese Lebanon-grown brownish-red hashish; people of Lebanon

Lebanon Republic of Lebanon (Middle East country), *al-Jumhoūriya al-Lubnaniya*

le bodies lupus erythematosus bodies (LE bodies)

LEBS London Emergency Bed Service

lec lunar equipment conveyor

L Ec Ecclesiastic Latin

LEC Lake Erie College; Law and Economics Center (University of Miami); Law Enforcement Center; Livestock Equipment Council

LECE *Ligue Européenne de Coopération Economique* (French—European League for Economic Cooperation)

le cells lupus erythematosus cells (LE cells)

lech lecher; lecherous; lechery

LECLU Law Enforcement Civil Liberties Unit

Le Corbu Le Corbusier

Le Corbusier Charles Edouard Jeanneret-Gris

lect lecture

lect. *lectio* (Latin—lesson)

Lect Lecturer

lectr lecturer

'lectric electric

led (LED) light-emitting diode

led. light-emitting dial; light-

emitting diode(s)
Led Ledbetter; Ledyard
L Ed Lawyer's Edition (US Supreme Court Reports)
LED Library Education Division (American Library Association)
LEDC League for Emotionally Disturbed Children
LEDETS Law-Enforcement Detachments (Coast Guard)
led's light-emitting diodes
lee. laser energy evaluator
Lee Leroy
LEE Low Expressed Emotion
LEEA Law Enforcement Education Agency
LEEGS Law Enforcement Explorer Girls
Lee I Leeward Islands
LEEP Law Enforcement Education Program
Leeward Islands Anguilla, Antigua, Barbuda, British Virgin Islands, Montserrat, Nevis, Redonda, Saint Christopher (St Kitts)
Leewards Leeward Islands
lef leading-edge flap; light-emitting film
LEF Life Extension Foundation; Lincoln Educational Foundation
LEF Liberté, Egalité, Fraternité (Liberty, Equality, Fraternity—slogan of the French Revolution)
LEFTA Labour (Party) Economic, Finance, and Taxation Association
leg. legal; legate(e); legation; legislation; legislative; legislature
leg. (LEG) liquefied energy gas
leg. legato (Italian—smoothly flowing)
Leg Leghorn
Leg Legierung(German—alloy)
LEG Law Enforcement Group
LEG (UN) Legal Affairs (department of United Nations)
legat FBI agent or office working in an overseas legation of the United States; legation
LEGCO Legislative Council
leg com legally committed
legcrit legal critic(ism)
legg leggiero (Italian—lightly and rapidly)
legis legislative; legislature
legit legitimate
LEGT Lycée d'Enseignement Général et Technologique

(French—High School of General Education and Technology)
legumes legumbres (Spanish-American truncation—beans, greenstuff, vegetables)
leg. wt legal weight
LEH Licentiate in Ecclesiastical History
LEHI Lehigh University
Lehman Caves Lehman Caves National Monument in eastern Nevada
lei land exclusive of improvements
LEI Leading Economic Indicator; Life-Expectancy Inventory; Life Extension Institute
Leic Leicester
leichtl leichtlöslich (German—readily soluble)
Leics Leicestershire
Leip Lepzig
lei's (LEIs) leading economic indicators
Le Is Leeward Islands
LEIS Law Enforcement Information System
Leit Leitrim
LEIU Law Enforcement Intelligence Unit
lej longitudinal expansion joint
lek monetary unit of Albania
lel lower explosive limit
LEL Labor Electoral League; Laureate in English Literature; Letitia Elizabeth Landon
LELDC Law Enforcement Legal Defense Center
lem lateral eye movements; layered-earth model; lemon-(ade); logical end of media
lem (LEM) lunar excursion module
Lem Lemuel
LeM Le Monde (The World)—Paris
LEM Lunar Excursion Module
LEMA Lifting Equipment Manufacturers' Association
lemac leading edge mean aerodynamic chord
lemo lemonade
lempira monetary unit of Honduras
LEMSIP Laboratory for Experimental Medicine and Surgery in Primates
Lemurio Lemuriologic(al)(ly); Lemuriologist(ic)(al)(ly); Lemuriology
len length
Len Leningrad, formerly Petrograd, formerly St Petersburg; Lensky; Leonard

LEN Law Enforcement News
LENA Lower Eastside Neighborhoods Association
LENA Labatorio Energia Nucleare Applicata (Italian—Applied Nuclear Energy Laboratory)
LENDS Library Extends Catalog Access and New Delivery System
Lenin Vladimir Ilich Ulyanov
Leninpor Lenin Port (Leningrad Harbor)
lenit. leniter (Latin—gently)
Len Lib Lenin Library (Moscow)
Lenny Leonard; Leonard Bernstein
Lenson Levensohn; Levenson; Levinson; Levinsky
lento lentando (Italian—increasingly slow)
Leo Leonard; Leonese; Leonidas; Leonine; Leopold; Leopoldville
LEO Leopoldville, Congo (airport)
leone monetary unit of the Seychelles
LEOPARD Law Enforcement Operations and Activities to Reduce Drugs
leopon leopard + lioness (hybrid offspring of male leopard and lioness)
LEOT Laser/Electro Optics Technology
lep lepton (collective term embracing anti-neutrino, electron, neutrino, photon, positron); lowest effective power
lep (LEP) large-electron positron collider
Lep Lepus (constellation)
LEP Labor Education Project; Library of Exact Philosophy
LEP Lycée d'Enseignement Professionnel (French—High School of Professional Education)
LEPA Law Enforcement Planning Agency
lep. dict. lepide dictum (Latin—well said)
lepid(s) lepidopterist(s)
LEPMA Lithographic Engravers and Plate Makers Association
Lepmus Lepramuseet (Norwegian—Leprosy Museum)—Bergen museum reflecting Dr Armanser Hansen's struggle against leprosy also known as Hansen's disease
Lepontines Lepontine Alps along the Italo-Swiss border

LEPORE Long-Term and Expanded Program of Oceanic Research and Exploration
LEPRA Leprosy Relief Association (British)
le prep lupus erythematosis preparation
lepro leprology; leprosarium; leprose; leprosy
lep(s) lepidopterist(s)
Leps Lepus (constellation)
lept (LEPT) long-endurance patrolling torpedo
Lepto Leptospira
ler life expectancy reduction
Ler Lerida
LeRC Lewis Research Center (NASA)
LERC Laramie Energy Research Center; Law Enforcement Resource Center
les lesbian; local excitatory state
les (LES) lower esophageal sphincter
lEs limited English speaking
Les Lescombe; Lesley; Leslie; Lester
LES Launch Escape System; Lincoln Experimental Satellite
lEsa limited English-speaking ability
LESA Licensing Executives Society of Australia; Lunar Exploration System—Apollo
lesb lesbian(ism)
lesbie lesbian
lesbo lesbian (Lesbos-type woman); lesbianism
Leschy Theodor Leschetizky (Teodor Leszetycki)—Polish composer-pianist
Les L Licensie es Lettres (French—Licentiate in Letters)
Les Lip Leslie Lipson
Leso Lesotho (formerly Basutoland)
Lesotho Kingdom of Lesotho (formerly called Basutoland, landlocked South African country)
les Ricains les Americans (French slang—The Americans)
LESS Least-cost Estimating and Scheduling Survey
Les Sc Licensie es Sciences (French—Licentiate in Science)
Lesser Antilles Leeward and Windward Islands extending from the Netherlands Antilles (Aruba, Bonaire, Curaçao) to the Virgin Islands

Lesser Sundas Lesser Sunda Islands east of Bali in Indonesia
lessie(s) lesbian(s)
Lester Leicester
let. letter; linear energy transfer
Let Lettish (Latvian)
LET Leader Effectiveness Training; Logical Equipment Table
LETAC Law Enforcement Training Advisory Council
L-et-C Loir-et-Cher
letch slang shortcut—lecher; lecheress; lecherous; lecherous feeling for; lechery
letfo letter follows
L-et-G Lot-et-Garonne
let's let us
LETS Low-Energy Telescope System
lett letter(s)
lett letteratura (Italian—literature); *letterlijk* (Dutch—literally); *lettisch* (German—Latvian)
Lett Lettish
Letter Carriers Union National Association of Letter Carriers
letterk letterkunde (Dutch—literature)
Lettie Leticia
Lett Nuovo Cimento Lettere al Nuovo Cimento (Italian—Communications Regarding New Findings)
Letts Lettish peoples (Latvians)
Letty Leticia; Letitia
Letz Letzeburgesch (Flemish dialect of Luxembourg)
leu monetary unit of Romania
leu (LEU) leucine (amino acid)
leuk (Latin prefix—white)—leukocytes
leuko (Greek *leukos*—white)—leukemia, leukocyte(s), leukorrhea—the whites
lev lever; monetary unit of Bulgaria
lev (LEV) lunar excursion vehicle
lev levert (Norwegian—delivered)
lev. levis (Latin—light)
Lev The Book of Leviticus; Leo; Leon
Lev Leviticus
levant levant or morocco leather (also called levant morocco and characterized by its prominent grain and high quality prized by book-

binders and book lovers alike)
Levant eastern Mediterranean lands such as Israel, Lebanon, and Syria
levis Levi Strauss' reinforced denim workclothes but particularly dungaree trousers with heavily-stitched-and-riveted pockets
levit. leviter (Latin—lightly)
Lew Lewis; Llewellyn
le'ward leeward
Lewisburg U.S. Penitentiary at Lewisburg, Pennsylvania
lex lexical; lexicographer; lexicography; lexicon
Lex Lexington
LEX Lexington, Kentucky (airport)
Lex-Fay Lexington-Fayette, Kentucky
lexi lexical; lexicographer; lexicographic(al)(ly); lexicography; lexicolater; lexicological(ly); lexicon(s)
lexic lexical(ic)(ally)
lexico lexicographer
lexicog lexicographer, lexicography
lexicon. lexiconist(ic)(al)(ly)
lexig(s) lexigram(s)—word symbol(s)
Lexington Lexington, Kentucky's U.S. Public Health Service Hospital for narcotic addicts
Lexington Bks Lexington Books—Division of DC Heath
lexiphan lexiphanic(al)(ly); lexiphanicism; lexiphanist
Lex Phil Lexington Philharmonic (Kentucky)
lexis (LEXIS) legal data base on-line to head
LEXIS Lexicography Information Service
l/ext lower extremity
Ley Leyden
LEY Liberal European Youth
Leyd Leyden
lez(es) lesbian(s)
lezz lesbian
lf lawn faucet; leaf; ledger folio; left field; left front; life float; light face type; line feed; linear feet, linear foot; linoleum floor; load factor; low frequency (30-300 kc)
lf (LF) line feed (data-processing character); line feed character (data processing)
l/f left front; light fittings
Lf Loaf (postal abbreviation)
LF Launch(ing) Facility; Lind-

bergh Field; Local Force (Red China)
lfa last field address; left fronto-anterior
LFA Land Force, Airmobility (NATO); Light Freight Agent; Low Flying Area
LFB Licensed Fishing Boat; London Fire Brigade
LFBC London Federation of Boys' Clubs
lfc laminar flow control; level of free convection; low-frequency current
l-fc low-frequency current
LFC Lutheran Free Church
lfd least fatal dose; low fat diet
lfd laufend (German—current, consecutive)
Lfd Laufend (German—current)
LFE Laboratory For Electronics
lffp laser fusion feasibility project
lfg live foal guaranteed
Lfg (Lfrg) Lieferung (German—installment, party delivery)
LFI Lethal Force Institute
LFICS Landing Force Integrated Communications System (USMC)
lfl lower flammable limit
LFL Lesbian Feminist Liberation (society); Lithuanian Freedom League
lfm low-power fan marker
lf/mf low-frequency medium-frequency
lfo light fuel oil; low-frequency oscillator
LFO Licentiate of the Faculty of Osteopathy
lfp left fronto-posterior
LFP Lindbergh Field Plant (Convair)
LFPC Louisiana Foundation for Private Colleges
LFPP Louisiana Family Planning Program
LFPS Licentiate of the Faculty of Physicians and Surgeons
lfq light-foot quantisizer
L fr Luxembourg franc(s)
LFR inshore fire-support ship (naval symbol)
LFRC League for Fighting Religious Coercion
lfrd low-friction reliability deviation
lfred liquid-fuel ramjet engine
LFS amphibious fire-support ship (naval symbol)
lft leaflet; left fronto-transverse; linear feet; linear foot

l/ft² lumens per square foot
LFTB Liquid Fuels Trust Board
LFTU Landing Force Training Unit
LFU Light Fighting Unit; Lunar Flying Unit
lg lagoon; landing; landing gear; languages(s); large; large grain; leg bye; length; long; long grain; low grade
l/g locked gate
Lg Landgrave; Landgraviate
LG Landing Ground; Leipzig Gewandhaus; Lloyd George; Low German
L/G Letter of Guarantee
LGA La Guardia, New York City airport; Local Government Association
LGAA Local Government Auditors Association
LGAT Local Government Appeals Tribunal
lgb laser-guided bomb
Lgb Long Beach, California
LGB Long Beach, California (airport)
L-G-B Landry-Guillain-Barré (syndrome)
LGC Laboratory of the Government Chemist; Local Government Commission
L-G C Lockheed-Georgia Company
LGCC Letchworth Garden City Corporation
LGCR Location Geological Characterization Report
lgd leaderless group discussion
LGD London Gaol Delivery
lge large
LGEA Local Government Electricity Association
LGEB Local Government Examination Board
L-Gen Lieutenant-General
L Ger Low German
Lg of H-D Landgrav(iate) of Hesse-Darmstadt
Lg of H-K Landgrav(iate) of Hesse-Kassel
LGIO Local Government Information Office
LGk Late Greek
lgm little green men (supposedly inhabiting extraterrestrial planets)
LGM Lloyd's Gold Medal
LGM Laboratorium voor Grondmechanica (Dutch—Soil Mechanics Laboratory)
LGMB Lady Godiva Marching Band
lgm's land-gobbling monsters (airports)

lgn lateral geniculate nuclei
Lgn Lagoon; Leghorn
LGO Lamont Geological Observatory (Columbia University); Local Government Office(r); Local Government Ordinances
LGOC London General Omnibus Company
lgp liquefied petroleum gas; low ground pressure
lgp (LGP) lasergraphic plotter
Lgp Legaspi Albay
LGPA Livestock and Grain Producers Association
lgr leasehold ground rent; ligroin
L Gr Late Greek
LGRA Local Government Reports of Australia
L Gr Ec Ecclesiastic Late Greek
LGRs Local Government Reports
lgs (LGS) laser geodynamics satellite
Lgs Lagos
LGS Landing Guidance System
LGSM Licentiate of the Guildhall School of Music
Lgt Light (postal abbreviation)
LGT Liggett Group (stock-exchange symbol)
LGTB Local Government Training Board
lgth length
lg tn long ton
lg tpr long taper
lg-type ed large-type edition
LGU Ladies Golf Union
lgv lymphogranuloma venereum (venereal disease)
LGW London, England (Gatwick Airport); Longines-Wittnauer (watches)
lg wh & br landing gears, wheels, and brakes
lh left hand; lighthouse; lightly hinged; lower half; lower hold
lh (LH) lactogenic hormone; lateral hypothalamus; left hand; luteinizing hormone
l/h labor hour; lamp holder; liters per hectare; low to high
lH linke Hand: (German—left hand)
LH Liberty House; lighthouse; Lufthansa (airline)
L + H Lamport & Holt (Line)
L.H. left hand
LH₂ liquid hydrogen
lha lateral hypothalamic area; lower-half assembly
LHA landing ship, helicopter,

assault; local hour angle

lhams lower hour angle of the mean sun

LHAR London-Hamburg-Antwerp-Rotterdam (range of ports)

LHAs multipurpose amphibious-warfare ships (naval symbol)

lhats lower hour angle of the true sun

lhb left halfback; lost heartbeat (attractive woman)

LHC Lease Housing Coordinator; Lord High Chancellor

LHCJEA London and Home Counties Joint Electric Authority

lhd left-hand drive; load-haul-dump(ing) machinery

L.H.D. *Litterarum Humanorum Doctor* (Latin—Doctor of Human Letters); *In Litteris Humanioribus Doctor* (Latin—Doctor in Humane Letters)

lhdc lateral homing depth charge

lh dr lefthand drive

lHe liquid helium

LHe liquid helium

L Heb Late Hebrew

L'heed Lockheed

LHG Library History Group

LHI Library of the Hoover Institution (on War, Revolution, and Peace)—Stanford, California

LHI Ligue Homeopathique Internationale (French—International Homeopathic League)

L-hinge capital-L-shaped hinge

lhm letterhead memo(randum)

LHMC London Hospital Medical College

LHNCBC Lister-Hill National Center for Biomedical Communications

LHO Local Health Office(r); Livestock Husbandry Office(r)

l hold(er) lease hold(er)

LHOOQ Elle à Chaud au Col (French slang—She of the hot crotch)–Marcel Duchamp's crude nickname for da Vinci's *Mona Lisa*

lhr lumen hour(s)

l/hr liters per hour

LHR London, England (Heathrow Airport)

L & HR Lehigh and Hudson River (railroad)

lhrf (LHRF) luteinizing hormone releasing factor

lhrh (LHRH) luteinizing hor-

mone-releasing hormone

LHRT Library History Round Table

lhs lefthand side

LHS Lafayette High School

LHSC Lock Haven State College

lhsv liquid hourly space velocity

LHT Lord High Treasurer

lh th lefthand thread

lhtl luxury class hotel

LHW League of Hispanic Women; lower high water

LHWI lower high water interval

lhwnt lowest high water neap tides

li line item; link; lira; lithograph; lithographer; lithography; longitudinal interval

li (Li) liability

Li lithium

Li *Limburg* (Dutch province)

LI Leeward Islands; Letter of Introduction; Liberia; Liberian; Lions International; Locksmithing Institute; Long Island (L.I.)

L-I Loire-Inférieure

LI Lingvo Internacia (Esperanto—International Language); *Lydveldid Island* (Icelandic—Republic of Iceland)

Li-2 Soviet Lisunov transport plane called Cab

lia liaison

LIA Laser Institute of America; Lead Industries Association; Leather Industries of America; Lebanese International Airways; *Ligue Internationale d'Arbitrage* (French—International Arbitration League); Livestock Improvement Association; Long Island Association

LIA Ligue Internationale d'Arbitrage (French—International Arbitration League)

LIAA Life Insurance Association of America

liab liability

LIALS Long Island Airport Limousine Service

LIAMA Life Insurance Agency Management Association

LIAT Leeward Islands Air Transport

lib liberal; liberalism; liberation(ist); libertarian(ism); liberty; librarian; library

lib libretto (Italian—operatic text)

lib. liber (Latin—book); *libra*

(Latin—pound)

Lib Liberal; Liberal Party; Liberation(ist); Liberty Party; Libra (constellation); Libya; Libyan

Lib Libano (Italian—Lebanon); *Libano* (Portuguese or Spanish—Lebanon)

LIB Let's Ignite Bras

Lib Auto Res Con Library Automation Research Consulting Associates

LIBBA Long Island Beach Buggy Association

Libby Elizabeth

lib cat. library catalog

libcon libertarian conservative (term invented by John Chamberlain to describe the liberal-conservative views of Pablo Casals, Milovan Djilas, John Dos Passos, Max Eastman, James T Farrell, Sidney Hook, Alberto Moravia, Allen Tate, Edmund Wilson, etc.)

LIBCON/E Library of Congress/English

Lib Cong Library of Congress

Lib Dem Liberal Democratic Party (formerly Social Democratic Party of the United Kingdom)

libe librarian; library

LIBE Ligo Internacia de Blindaj Esperantisol (International League of Blind Esperantists)

libec light behind camera

lib ed library edition

Liberace Wladziu Valentino Liberace

Liberia West African coastal country adjacent to Sierra Leone—founded by United States in 1822 and settled by freed American Negroes

Libert Libertarian

LIBGIS Library General Information Survey

Lib-Lab Liberal-Labour (Australian coalition)

Libn Librarian

Libor London interbank-offered rate

Lib Parl Library of Parliament

libr librarian; library

libr libretto (Italian—opera or oratorio text)

LIBRA Living In the Buff Recreational Associates

LIBRE Living In the Buff Residential Enterprises (nudist apartments and beaches)

Lib Res Library Research Associates

LIBRIS Library Information System (Swedish on-line retrieval system)

Lib(s) Liberal(s)

LIBS Library Information and Bibliographic System

Lib Soc Sci Library of Social Science

libst librettist (Italian—libretto author)

Libs Unl Libraries Unlimited

Lib UN Library of the United Nations (New York headquarters)

Libya People's Socialist Libyan Arab Republic (North African country populated by Arab Berbers), *Al-Jumhuria al-Arabia allibya*

lic license; linear integrated circuit; low-intensity conflict

Lic Licentiate

Lic Licenciado (Spanish—lawyer, licentiate holding master's degree)

LIC Lands Improvement Company; Liquor Industry Council; Local Import Control

LICA Ligue Internationale Contre le Racisme et l'Antisemitisme (French—International League Against Racism and Antisemitism)

LICC London Institute for Contemporary Christianity; Long Island Council of Churches

licd licensed

Lic D Licenciado Don (Spanish—Sir Lawyer)

lic dlr licensed dealer

LICeram Licentiate of the Institute of Ceramics

licm left intercostal margin

Lic Med Licentiate in Medicine

Lic Phil Licentiate in Philosophy

LICTBOSS Life-Cycle Theory of Bureaucratic Ossification

Lic Theol Licentiate in Theology

LID League for Industrial Democracy

L & ID London and India Docks

lidar laser radar device (for measuring wind direction and speed); laser-impulsed radar; light detection and ranging (laser-beam air pollution or smog measuring device)

LIDB Logistics Ingelligence Data Base

LIDC Lead Industries Development Council; Livestock

Industry Development Council

Liddie Lydia

LIDH Ligue Internationale des Droits de l'Homme (French—International League for the Rights of Man)

LIDO Logistics Inventory Disposition Order

lidoc lidocaine (xylocain)

Lie Liepaya

LIE Liberal Intellectual Establishment (Philip Wylie's description of the befuddled and often nonsensical liberals of his time)

LIEAP Low-Income Energy-Assistance Program

Liech Liechtenstein

Liechtenstein Principality of Liechtenstein (Alpine country), *Fürstentum Liechtenstein*

Lief Lieferung (German—issue)

LIEJA Long Island Equal Justice Association

LIEMA Long Island Electronics Manufacturers Association

LIESA Long Island Episcopal Schools Association

Lieut Lieutenant

Lieut Col Lieutenant Colonel

Lieut Comdr Lieutenant Commander

Lieut Gen Lieutenant General

Lieut Gov Lieutenant Governor

lif left iliac fossa

LIF Lone Indian Fellowship

LIFA Life Insurance Federation of Australia

life. laser-induced fluorescence of the environment

life see LIFFE

LIFE Ladies Involved For Education; League for International Food Education; Love Is Feeding Everyone

lifes laser-induced fluorescence and environmental sensing

Life Sta Lifeboat Station (US Coast Guard)

liff laser-induced fluorescence fluorimetry

LIFFE London International Financial Futures Exchange

LI Fire Eng Licentiate of the Institution of Fire Engineers

lifmop linearly frequency-modulated pulse

lifo last in, first out

LIFPL Ligue Internationale de Femmes pour la Paix et la

Liberté (French—International League of Women for Peace and Freedom)

LIFR Low Instrument Flight Rules

lift. logically-integrated fortran translator

LIFT London International Festival of Theatre; London International Freight Terminal

lig ligament; ligature

Lig Liguria(n); Limoges

Lige Elijah

liger offspring of lion and tigress

light. lighting; lightning

lightex searchlight illumination exercise

lign lignende (Dano-Norwegian—similar)

lignite brown coal

liguid. liguidation

Liguori Liguori Publications

Ligurians Ligurian Alps or Ligurian Apennines of northwestern Italy or the people of the region around the Gulf of Genoa

lih left inguinal hernia; light intensity high

LIHDC Low Income Housing Development Corporation

Lik Obs Bull Lick Observatory Bulletin(s)

lil light intensity low; lilliputian; little

li'l little

Lil Lilian; Lillian; Lily

LIL Lunar International Laboratory (proposed in 1961 by Dr Theodore von Karman)

lila life insurance logistics automated

lilangeni monetary unit of Swaziland

LILCO Long Island Lighting Company

Lilly Lilian; Lillian

lilo last in, last out

LILS Lead-in-Light System (airport term)

lim light intensity marker; limber; limit(er); line induction motor; line insulation monitor; line interface module; linear induction motor; linear-induction motor(s); liquid injection moulding; locator inner marker

Lim Limburg (Dutch province); Limerick

LIM Lima, Peru (Callao International Airport)

Lima letter L radio code; (pronounced *leema*)

LIMA Long Island Museum Association
LIMAC Linden Industrial Mutual Aid Council
lim dat limiting date
lime calcium oxide (CaO)
limestone calcium carbonate $(CaCO_3)$
limewater calcium-hydroxide solution—$Ca(OH)_2$; limejuice and water mixture
lim-lib(s) limousine liberal(s)
limnol limnology
limo lemonade; limousine
limon lime-and-lemon (hybrid citrus fruit)
Limón Puerto Limón, Costa Rica
limos limousines
limp limp cloth binding; limp cloth bound
Limpopo the Limpopo or Crocodile River of East Africa where in 1497 Vasco da Gama named it Rio do Espiritu Santo
LIMRA Life Insurance Marketing and Research Association
LIMRF Life Insurance Medical Research Fund
LIMS Logistic Inventory Management System
limvr linear-induction motor vehicle research
lin lineal; linear
lín *línea* (Spanish—line)
Lin Lincoln; Linda; Lindenberg(er); Lindley; Limdolfo; Limdon; Lindsay; Linley; Linnaeus; Linsley, Linton; Linus
L i N *Lokalhistorisk institutt Norge* (Norwegian Local History Institute)
LIN *Linjeflyg* (Swedish airline); Milan, Italy (Linate Airport)
Lina Angelina; Carolina; Caroline
linac linear accelerator
Linacre Linacre College, Oxford
linc laboratory instrument computer
Linc Lincoln; Lincoln College, Oxford
LINC Learning Institute of North Carolina; Logic and Information Network Coupler
Linc Coll Lincoln College—Oxford
LINCO Linearly-Organized Chemical Code (for computer system)
Lincoln Abraham Lincoln,

16th President of the United States; Nebraska's capital
Lincoln's Abraham Lincoln's Birthday (February 12)
lincompex linked compressor and expander
Lincs Lincoln automobiles; Lincolnshire
LINCS Language Information Network and Clearinghouse System
Lindbergh Charles A Lindbergh; Lindbergh Field (San Diego's international airport)
LINDE Linde Air Products
Lindy Colonel Charles A Lindbergh
Line The Line—the Equator
Lines Line Islands in the equatorial mid-Pacific Ocean where they include Caroline, Christmas, Fanning, Flint, Kingman Reef, Malden, Palmyra, Starbuck, Vostock, and Washington Islands
lines/m lines per minute
lines/mm lines per millimeter
lines/s lines per second
L-Infre Loire-Inférieure
lin ft linear feet; linear foot
ling linguist(ics)
Linguis Linguistics
linim liniment
Linlithgow West Lothian, Scotland
Linn Linné; Linnaeus
lino linoleum; linotype; linotypist
linol linoleum
lino oper linotype operator
LINOSCO Libraries of North Staffordshire in Cooperation
LINS Laser Inertial Navigation System
L Inst Phys Licentiate of the Institute of Physics
LINTAS Lever's International Advertising Service
L'Intran *L'Intransigeant*
LINWR Lake Ilo National Wildlife Refuge (North Dakota)
lin yd linear yard
LIO Lionel Corporation (stock exchange symbol); Lions International Organization; Livestock Improvement Organization
LIOB Licentiate of the Institute of Building
LIOCS Logical Input/Output Control System
lip. lease in perpetuity; life insurance policy
lip (Latin prefix—fat)—lipectomy

LIP Local Initiatives Program
Lipari Islands Italian penal colony northeast of Sicily; islands include Stromboli and Vulcano; also called Aeolian Islands
Liparis Lipari Islands
LIPC Livestock Industry Promotion Corporation
LIPI *Lembaga Ilmu Pengetahuan Indonesia* (Indonesian Academy of Sciences)
lipl (LIPL) linear information programming language
LIPM Lister Institute of Preventive Medicine
lipo lipogram(matic)
Li Po Li T'ai-po
Lippincott J B Lippincott Company
lips. logical inferences per second
lip sync lip synchronization (in sound films)
lipup backward pupil (pupil spelled backwards)
liq liquid; liquor
liq f rkt liquid fuel rocket
liqn *liquidación* (Spanish—liquidation)
liqt liquid transient
LIR Liaison Investigation Report; Library of International Relations
L & IR Legislation and Intergovernmental Relations
lira monetary unit of Italy, San Marino, Turkey, and Vatican City
lira. loft-type infrared analysis
LIRA Linen Industry Research Association; Logging Industry Research Association
lirbm liver, iron, red bone marrow
LIRES Literature Retrieval System
LIRES-MS Literature Retrieval System-Multiple Searching
LIRI Leather Industries Research Institute
lirl low-intensity runway lights
liroc last instruction readout cycle
lirod lightweight radar and optronic director
LIRR Long Island Railroad
LIRS Lutheran Immigration and Refugee Service
LIRT Library Instruction Round Table
lis laser isotope separation; lobar in situ
Lis Lisbon
LIS Liberian Information Ser-

vice; Library and Information Science; Light Industry Services; Lisbon, Portugal (airport); Livestock Incentive Scheme; Locate in Scotland; Lockheed Information System(s); Long Island Sound

lisa (LISA) low-input sustainable agriculture

LISA Linear Systems Analysis; Long Island Schizophrenia Association

LISA Library and Information Science Abstracts

Lisb Lisboa (Portuguese or Spanish—Lisbon); *Lisbona* (Italian—Lisbon)

LISC Lions International Stamp Club; London Institute for the Study of Conflict

LISD Library Information Science Division (World Information Systems Exchange)

LISM Licentiate of the Incorporated Society of Musicians

LISO Library Information Service Office(r)

lisp. list processor (computer language)

LISP List Processing (for text manipulation)

LISPA Long Island Sound Pilots Association

LISR Line Information Storage and Retrieval

LISS London Institute of Strategic Studies

list. laser and isotope separation technology

LIST Library and Information Services—Tees-side

LIST Library and Information Science Today

'listed enlisted

LISTEN Low-Income-Schools Teacher Education

'listment enlistment

lit literacy; literate

lit. liter; literal; literally; literary; literature; litter; little

l lit italiane (Italian lire)

lit litauisch (German—Lithuanian)

lit. litterae (Latin—letters)

Lit Litvak (Yiddish—Lithuanian)—person of Judaic origin from Lithuania or nearby regions

LIT Light Intratheater Transport (aircraft); Little Rock, Arkansas (airport)

LITA Library and Information Technology Association

litcrit literary critic(ism)

litcy literacy (ability to read and write)

lite light

LITE Legal Information Through Electronics

litex searchlight illumination exercise

lith lithograph; lithography; lithology

Lith Lithuania; Lithuanian

litharge lead oxide (PbO)

litho lithograph

lithol lithology

LITINT Literacy International

litr lighter

litrg literage

Lits Lithuanians; Litvaks

litt litteratur or *litteraer* (Dano-Norwegian—literature or literary)

Litt.B. Litterarum Baccalaureus (Latin—Bachelor of Letters)

Litt.D. Litterarum Doctor (Latin—Doctor of Letters)

Little Little, Brown

Litt. M. Master of Letters

litur liturgical; liturgy

liturg liturgical; liturgistic; liturgy

Litvak (Yiddish—Lithuanian Jew)

Litvak(s) Lithuanian(s)

Litz Litzendraht (German—wire)

LIU Long Island University

LIUNA Laborers International Union of North America

LIUP Long Island University Press

liv liver

liv le livre (French—book); *la livre* (French—pound)

Liv Liverpool

Liv Titus Livius (Roman historian often referred to as Livy)

LIV Light Infantry Volunteers

Liver Liverpool; Liverpudlian(s)

Liverpool Liverpool Prison (also called LP)

livex live exercise (military)

Liv Phil Liverpool Philharmonic

livr livraison (French—issue of a journal, part of a book or serial)

liv st livre sterling (French—pound sterling)

Liv St Liverpool Street (rail terminal)

lix lixiviation

liz Lizard; lizzie (as in *tin lizzie,* an old Ford Automobile)

Liza Liza Minnelli

Lizard Lizard Head, Lizard Peninsula, Lizard Point, Liz-

ard Town—at the tip of southwest Cornwall

LIZARDS Library Information Search and Retrieval Data System

lj life jacket

LJ La Jolla; Law Judge; Libby, McNeil & Libby (stock exchange symbol); Library Journal; Lord Justice; Sierra Leone Airways (2-letter coding)

LJ laufen Jahre (German—current year); *Law Journal; Library Journal*

LJC Lackawanna Junior College; Laredo Junior College; Lincoln Junior College; London Juvenile Court

LJMCA La Jolla Museum of Contemporary Art

ljp localized juvenile periodentitis

LJR Law Journal Reports

LJ/SLJ Library Journal/School Library Journal

LJT Lear jet airplane

LJTSA Library of the Jewish Theological Seminary of America (NYC)

LJU La Jolla University

lk link

Lk Lake; Luke

LK Lockheed Aircraft Corporation (stock exchange symbol)

LKAB Luossavaara-Kiirunavaara Aktiebolag (iron-ore mines in Luossa-Kiiruna range of northern Sweden)

LKB Link-Belt Company (stock exchange symbol)

lkd locked

lked linkage editor

lkg locking

lkg & bkg leakage and breakage

lkge leakage

LKGR Lake Kyle Game Reserve (Rhodesia)

lk-n lock-in

LKPA Landeskriminalpolizeiamt (German—Land Criminal Office)—Prussian organization (1940s)

LK & PRR Lahaina-Kaanapali and Pacific Railroad

LKQCPI Licentiate of the King and Queen's College of Physicians of Ireland

lkr locker

Lkr Landskrona

lks links; liver, kidney, spleen

Lks Lakes

Lksde Lakeside

lkt lookout

lk up lock up
Lkw Lastkraftwagen (German—lorry, truck)
lkwash lockwasher
ll land lines; light lock; limited liability; live load; long lead; lower left; low(er) level; lower lid; lower limit
ll (Ll) landlord
l/l library labels; line-by-line; looseleaf; lower left; lower limit
l & l leave and liberty; look and listen
'll (contraction of till and will)
ll lectiones (Latin—readings); *llegada* (Spanish—arrival)
LL Language Laboratory; Language Lessons; Law List-(ing); Lebanese pound; Lending Library; Linda Love; Little League (baseball); Loftleidir (Icelandic Airlines); Lord Lieutenant; Low Latin
LL (Ec) Ecclesiastic Late Latin
L/L Lutlang (Norwegian—limited company)
lla left lower arm; limiting lines of approach
LLA Latin Liturgy Association; Lend-Lease Administration; Louisiana Library Association; Luther League of America
llama. long-life atmospheric-motoring airship
Llanfairp Llanfairpwllgwyng-llgogershwyrndro-bwllabtysi-liogogoch (Welsh—Church of St Mary near the Raging Whirlpool and the Church of St Tysilio by the Red Cave)—the longest place-name in the world
L Lat Late Latin; Low Latin
llb long-leg brace
LLB Little League Baseball
LL.B. Legum Baccalaureus (Latin—Bachelor of Laws)
LLBA Language and Language Behavior Abstracts
llbcd left lower border of cardiac dullness
LLBO Liquor License Board of Ontario
l-l brace long-leg brace
llbs low-level bombsight
llc lower left center
LLC Libertarian Law Council; Library Learning Center; Living Learning Center (Indiana University)
ll. cc. locis citatis (Latin—in the places cited)

llcca long-life cycle-cost avionics
LLCM Licentiate of the London College of Music
LLCO Licentiate of the London College of Osteopathy
Ll & C's Lloyd's and Companies
LLCUNAE Law Library of Congress United Association of Employees
LL.D. Legum Doctor (Latin—Doctor of Laws)
LLDEF Lambda Legal Defense and Education Fund
LlD factor *Lactobacillus lactis* Dorner factor (vitamin B_{12})
lle left lower extremity
lle llegada (Spanish—arrival)
LLE Laboratory for Laser Energetics (University of Rochester)
LLEI Lincoln Library of Essential Information
L Lett Licentiate of Letters
L-L f Laki-Lorand factor
LLF Laubach Literacy Fund
llfm land line frequency modulation
lli latitude and longitude indicator; long lead items
LLI Laubach Literacy International; Lord Lieutenant of Ireland
LLIL Long Lead Item List-(ing)
LLJ Leaf Library of Judaica
LLJJ Lords Justices
lll left lower limb; left lower lobe; light load line; looseleaf ledger; low-level logic
l/ll line-by-line libretto
LLL Lawrence Livermore Laboratories; Lutheran Laymen's League
LLL Love's Labour's Lost
lllb left long-leg brace
LLLI La Leche League International
llll left lower lung lobe
lllp leased long-lines program
llltv low-light-level television
llm localized leucocyte mobilization
LL. M. *Legum Magister* (Latin—Master of Laws)
LLN League for Less Noise
LLNL Lawrence Livermore National Laboratory
LLNNR Loch Leven National Nature Reserve (Scotland)
LLNWR Long Lake National Wildlife Refuge (North Dakota)
lloc land line of communications

Lloydbras Lloyd Brasileiro
Lloyd's Lloyd's Register of Shipping
Lloyd's Bank Lloyd's Bank International Ltd
LLP Lifetime Learning Publications
L L & P of H Life, Liberty, and the Pursuit of Happiness (original draft of the *Declaration of Independence* read: "Life, Liberty, and the Pursuit of Profit")
LLPI Linen and Lace Paper Institute
llps low-level pumping station
llq left lower quadrant
llqa limiting lines of quiet approach
llr lender of last resort; line of least resistance; load-limiting resistor; log-likelihood ratio; long-length record(ing)
llr (LLR) latent lethality of radiation
LLRS Laser Lightning-Rod System
llrv (LLRV) lunar landing research vehicle
lls low-level solids
l & l's losers and lunatics
LLS Lunar Logistics System
LLSS Low-Level Sounding System
llsv (LLSV) lunar logistics system vehicle
llt long lead time
LLT London Landed Terms
llti long lead time items
lltruw low-level transuranic waste(s)
lltv low-light-level television
llu lending library unit
LLU Loma Linda University
LLUU Laymen's League—Unitarian Universalist
llv (LLV) long-life vehicle (postal vans); lunar landing vehicle
llw lower low water (LLW); low-level waste
LLWI lower low water interval
llwl light load water line
LLWM Low-Level Waste Management
LLWSAS Low-Level Wind Shear Alert System
Lly Llanelly
llyp long-leaf yellow pine
llz localizer
lm land mine; light metal(s); liquid metal(s); long meter; longitudinal muscle; lower motor; lumen(s)
l/m lines per minute

lm livello del mare (Italian—sea level)

l.m. locus monumenti (Latin—place of the monument)

Lm middle latitude

LM Legion of Merit; Life Master (bridge); Liggett Myers Tobacco (stock exchange symbol); Lincoln Memorial; Lord Marquis; Lord Mayor; Lourenço Marques; Lunar Module

L.M. Licentiate in Midwifery

L & M Linotype and Machinery

LM Lacus Mortis (lunar area)

LM-1 Fuji Heavy Industries trainer plane

lma left mento-anterior

LMA Last Manufacturers Association; League for Mutual Aid; Lingerie Manufacturers Association; Linoleum Manufacturers' Association; London-Midlands Association

LMAA Lift Manufacturers Association of Australia

LMAC Labor-Management Advisory Committee

lmad let's make a deal

LMAF Live Missile Assembly Facility

LMAGB Locomotive and Allied Manufacturers' Association of Great Britain

lmb local message box

LMBA London Master Builders' Association

LMBC Liverpool Marine Biological Committee

l & m bond labor and material bond

LMBP Lake Manyas Bird Paradise (Turkey)

lmc liquid-metal cycle; low middling clause

lmc (LMC) large magellanic cloud

LMC Lake Michigan College; Liberia Mining Company; Lloyd's Machinery Certificate; Lutheran Medical Center

LMC (LMC) Lloyd's Machinery Certificate (temporarily suspended when enclosed in parentheses)

LMCA Lorry Mounted Crane Association

LMCC Licentiate of the Medical Council of Canada

LMCT Licensed Motor Car Trader

lmd leafmould; local medical doctor

LMD Laboratory of Meteorological Dynamics

LMDC Lawyers Military Defense Committee

lme liquid-metal embrittlement

LME Late Middle English; London Metal Exchange

LMEC Liquid Metal Engineering Center (AEC)

L Med Licentiate in Medicine

L Med Ch Licentiate in Medicine and Surgery

LMEE Light Military Electronic Equipment (department of General Electric)

lmf language media format

l/mf low and medium frequency

lmfbr liquid-metal fast-breeder reactor

lmfr liquid metal fuel reactor

lm/ft² lumen per square foot

lmg liquefied methane gas

Lmg Leichtesmaschinengewehr (German—light machine gun)

LMG light machine gun

LMH Lady Margaret Hall, Oxford

LMHC Lady Margaret Hall College (Oxford)

lm hormone lipid mobilizing hormone

lm-hr lumen-hour

L Mi Leo Minor (constellation)

LMI Lawn Mower Institute; Logistics Management Institute

l/min liters per minute

LMIS Labor Market Information System

lml left mediolateral

LML Lankard Materials Laboratory; Lerner Marine Laboratory

LMLA Lizzadro Museum of Lapidary Arts

LMLI Liberty Mutual Life Insurance

lm/lm² lumen per square meter

lmlr load memory lockout register

lm/lrv lunar module/lunar roving vehicle (LM/LRV)

lmm localiser middle marker; locator at middle marker (compass)

l/mm lines per millimeter

LMM Library Microfilms and Materials; Luis Muñoz Marin, first native governor of Puerto Rico

lmmi like mamma made it

lmn lineman; lower motor neuron

LMNP Lake Manyara National Park (Tanzania)

LMNRA Lake Mead National Recreation Area (Arizona and Nevada)

lmo lens-modulated oscillator; light machine oil; limousine

LMO Local Medical Officer; Logistics Management Office (USA); London Meteorological Office

lmp last menstrual period; left mento-posterior; lunar module pilot

LMP Linea Mexicana del Pacifico; London Metropolitan Police

LMP Literary Market Place (Directory of American Book Publishers)

LMPA Library and Museum of the Performing Arts (Lincoln Center, New York City); London Master Printers' Alliance

LMPT Logistics and Material Planning Team

L Mq Lourenço Marques

LMR Lifetime Merit Register; London Midland Region—British Railways

LMRC London Medical Research Council

LMRCP Licentiate in Midwifery of the Royal College of Physicians

LMRI Living Marine Resources, Inc

lmrp lower marine riser package (oil well)

LMRSH Licentiate Member of the Royal Society for the Promotion of Health

LMRU Library Management Research Unit (Cambridge)

lms lambs; least mean square

lms (LMS) lunar mass spectrometer

lm's lunar modules (LMs)

lm/s lumen per second

LMS Lelean Memorial School; Licentiate in Medicine and Surgery; London Mathematical Society; London Medical Schools; Lotto Management Services

LMSA Labor Management Services Administration

LMSC Lockheed Missiles & Space Company

LMSD Lockheed Missile and Space Division

LMSSA Licentiate in Medicine and Surgery of the Society of Apothecaries

lmst loom state

lmt left mento-transverse; length, mass, time; limit

LMT Local Mean Time

LMTA Language Modalities Test for Aphasis; Library Media Technical Assistant; London Master Typefounders' Association
lmtd limited; logarithmic mean temperature difference
lmtg limiting
LMU Loyola Marymount University
LMUM *Ludwig-Maximilians-Universität München* (German—University of Munich)
L Mus Licentiate in Music
L Mus TCL Licentiate in Music—Trinity College of Music
LMVD Licensed Motor Vehicle Dealer
LMVUS League of Men Voters of the United States
lm/w lumen per watt
lm/W lumen(s) per watt
ln liaison; logarithm (natural, base *e*)
Ln Lane; London; Lyttelton
LN Air Liban (Lebanese Airlines); League of Nations; Napierian logarithm (symbol)
L & N Leeds & Northrup; Louisville & Nashville (railroad)
L of N League of Nations
LN *Liga Nacional* (Spanish—National League)
LN₂ liquid nitrogen
LN₂cou liquid-nitrogen clip-on unit
LN₂ trailer liquid-nitrogen trailer
lna low noise amplifier
LNA Liberian National Airways; Libyan News Agency
lnb (LNB) large navigational buoy
lnc loran navigation chart(s)
LNC League of Nations Covenant; Leith Nautical College; Libertarian National Committee
lnchr launcher
L-N CP Liberal-National Country Party (Australian)
LNDC Lesotho National Development Corporation
Lndg Landing
lndh local nationals direct hire
lndkjng *landkjenning* (Norwegian—land ho)—in sight of land
lndrs laundress
lndry laundry
lndscp landscape; landscaping
L & NE Lehigh & New England (railroad)
LNER London and North Eastern Railway
lng length (flow chart); lining; liquefied natural gas; lounge
lng (LNG) liquefied natural gas
lngc (LNGC) liquefied natural gas carrier
LNG tanker liquid-natural-gas tanker
LNHS London Natural History Society
LNI *Lega Navale Italiana* (Italian Naval League)
LNLA Lithuanian National League of America
lnlw lowest normal low water
lnmp last normal menstrual period
LNNP Lake Nakuru National Park (Kenya)
LNNR Lindisfarne National Nature Reserve (England)
LNOC Libya National Oil Company
L-note $50 bill
lnp (LNP) lunar neutron probe
LNP Lamington National Park (Queensland); Lincoln NP (South Australia); London Northern Polytechnic
lnpf lymph node permeability factor
lnr liner; low noise receiver
LNR Loteni Nature Reserve (South Africa)
Lnrk Lanark
LNS Land Navigation System; Liberation News Service
LNSW Library of New South Wales
LNT Leo Nicholas Tolstoy
Lntl lintel
LNTS *League of Nations Treaty Series*
lntwta low-noise travelling-wave tube amplifier
lnu last name unknown
LNU League of Nations Union
LNWR Lacassine National Wildlife Refuge (Louisiana); Lacreek NWR (South Dakota); London and North Western Railway; Lostwood NWR (North Dakota); Loxahatchee NWR (Florida)
lo light; light open; local; local oscillator; locked open; longitudinal optical; low; low gear; low lights; low(er) order; lubricating oil; lubrication order
lo (LO) longitudinal optic
lo' look
'lo hello
lo *loco* (Italian—place)—in music, return to original pitch

Lo low (gear)
Lo Lordag (Danish—Lord's Day)—Saturday
LO Land Office; Launch Operator; Liaison Office(r); Lick Observatory (Mount Hamilton, California); Livestock Office; London Office; Louisville Orchestra; Lowell Observatory (Flagstaff, Arizona); Lubrication Order; Polish Airlines (2-letter symbol)
L/O Letter of Offer
LO *Landsorganisationen* (trade union in Norway and Sweden)
LO₂ liquid oxygen
loa leave of absence; left occiput anterior; length overall
loa (LOA) light observation aircraft
LOA Letter of Agreement; Letter of Offer and Acceptance; Letter Officers Association; Letter Offices Association; Light Observation Aircraft; Lithuanian Organists Alliance
loadex loading exercise
loadg & dischg loading and discharging
loadicator computerized ship-loading indicator
LOAF Lesbians Over the Age of Forty
loan/A vessel(s) loaned to Army
loan/C vessel(s) loaned to Coast Guard
loan/m vessel(s) loaned to miscellaneous governmental activities (Maritime Academy)
loan/s vessel(s) loaned to states
LOANZ Life Offices Association of New Zealand
LOAP List of Applicable Publications
lob left on base; line of balance
lob. logs on board; lumber on board
LOB Launch Operations Building; Loyal Order of the Boar; Loyal Order of Boors; Loyal Order of Bores
lobal long base-line buoy
lobar long baseline radar
loboto lobotomy
lob(s) lobster(s)
loc lines of communication; locate; location; locus of control; logistics other charges
l-o-c letter of credit
LoC Library of Congress
LOC Launch(ing) Operations

Center; Launch(ing) Operations Complex; Letter of Certification; Louisiana Office of Conservation; Lyric Opera of Chicago

loca loss of coolant accident (nuclear reactor)

local. load on call; local area network

lo-cal low calorie

locals local people; local trains

locat location; locative; low-altitude clear-air turbulence

LOCATE Library of Congress Automation Techniques Exchange

LOCC Logistical Operations Control Center

loc.cit. loco citato (Latin—in the place cited)

loc. dol. loco dolenti (Latin—to the painful spot)

loci logarithmic computing instrument

loc. laud. loco laudato (Latin—cited in the approved place)

locn location

loco locomotion; locomotive

locp launcher operation control panel

locport lines of communications ports

loc. primo cit. loco primo citato (Latin—in the place first cited)

loc pr pnl local control purge panel

locpuro local purchase order

LOCS Librascope Operations Control System

loc. supra cit. loco supra citato (Latin—in the place cited above)

locum tens. locum tenens (Latin—temporary position)

locuz locuzione (Latin—phrase)

lod limitation on dividends; line of duty

lo-d low-density

Lod Lödose

LOD Launch Operations Directorate

LOD Little Oxford Dictionary

lodestone magnetic iron oxide; Fe_3O_4: magnetite

lodg loading; lodging

lodif long-distance infrared flash (camera)

lodor loaded (vessel) awaiting orders or assignment

loe level of effort

LOE Light-Off Examination (USN)

lof lecherous old fool; lowest

operating frequency

LOF Lloyd's Open Form (insurance policy); Lloyd's Open-Form (contract); London and Overseas Freighter

L-O-F Libbey-Owens-Ford

lo-fi low fidelity (low-quality sound reproduction)

Lofotens Lofoten Islands

loft. low-frequency radio telescope

lofti low-frequency trans-ionosphere (research satellite)

Loftleidir Icelandic Airlines

log. logarithm; logic; logical; logistic(s)

Log Longview

LOG Legion of Guardsmen

log.$_{10}$ logarithm to the base 10

logair logistics transport by air

logairnet logistics air network

logal logical algorithmic language

Logan Logan International Airport (named for WW-II hero General Edward Lawrence Logan who gave land to the city of Boston now served by this airport)

logands logic commands

logan(s) loganberry; loganberries

LOGC Logistics Center (USA)

LOGCMD Logistical Command

logcom logistic communications

Log Com Logistical Command; Logistics Command

LOGDESMAP Logistics Data Element Standardization and Management Program (DoD)

LOGDESMO Logistics Data Element Standardization and Management Office (DoD)

LOGDIV Logistics Division

log.$_e$ logarithm to the base e

logel logic-generating language

logest logistics estimate

logg loggerhead; loggia; logging; log glass

logie killogie

logipac logical processor and computer

loglan logical language

logland logistics transport by land

logo logogram [initial letter, number, or symbol used as an abbreviation or as part of an abbreviation as in Q & A (question and answer), 3M (Minnesota Mining and Manufacturing Company), c (cents)]; logotype (two or more type characters cast as

one piece of type)

LOGOIS Logistics Operating Information System

logol logological; logologically; logologist; logology

logophi logophilia(c)—lover(r) of words

logother logotherapeutic(al)(ly); logotherapist; logotherapy

logp logistics plans

logr logistical ration; logistics ratio

Logr Logroño

logram logical program

Log Rep Logistics Representative (USN)

logsea logistics transport by sea

logsup logistical support; logistics support

logsvc logistics service

loh (LOH) light observation helicopter

L o H Library of Hawaii (Honolulu)

loi loss on ignition

loi (LOI) letter of intent

LOI Lunar Orbit Insertion

loib lunar orbit insertion burn

loid celluloid (strip used by burglars to unlock doors)

LOIS Library Order Information System

lo-J low inertia

loktal locked octal tube

lol length of lead (actual); little old lady

LOL Lobitos Oilfields Limited; Loyal Orange Lodge

lola lollapalooza (excellent or extraordinary person or thing)

LOLA Library On-Line Acquisition

lolita language on-line investigation/transformation of abstractions; library on-line information and text access

lolli lollipop

lo-lo load on-load off

lolw laid off–lack of work

lom locater at outer marker (compass)

Lom Columbus

LOM League of Mothers; List of Modifications; Loyal Order of Moose

LOMA Life Office Management Association; Lutheran Outdoor Ministry Association

LOMAC Logical Machine Corporation

Lomb Lombard; Lombardian; Lombardy

lo mi low mileage

Lompoc Federal Correctional Institution at Lompoc, California also site of a Federal Prison Camp

lon *longitud* (Spanish—longitude)

Lon Alonso; London

L o N League of Nations

LON London, England international airports (London-Central Airport)

Lon Brg London Bridge (rail terminal)

Lond *Londen* (Dutch—London); London; Londonderry; Londoner(s); *Londra* (Italian—London); *Londres* (French, Portuguese, Spanish—London)

long longeron; longitude

'long along

Long Longfellow; Longford; Long Island; Longjumeau; Long Key; Longmeadow; Longview

longl longitudinal

Long Lane Girls Long Lane School for (delinquent) Girls at Middletown, Connecticut

Longshoremen's Union International Longshoremen's Association

'longside alongside

Long Straight The Long Straight—297-mile-long (478-kilometer-long) straight stretch of railway track laid across Australia's Nullarbor Plain—the world's longest straight stretch of railroad

longv longevity

long vac long vacation

Lon Phil London Philharmonic

LONRHO London and Rhodesian Mining and Land Company Limited

Lon Sym London Symphony

loo British term for toilet; looker; looker-after; looker-on

LOOE Loyal Order of Overtime Experts

looktr lookout tower

LOOM Loyal Order of Moose

Loop The Loop—Chicago's business section

LOOP Louisiana Offshore Oil Port

LOOS League of Older Students

lop. launch operator's panel; left occiput posterior

l-o-p line-of-position

LOP lunar orbiting photographic (vehicle)

lopar low-power acquisition radar

l'Opera (French—the opera)—Paris Opera House

L O P & G Live Oak, Perry & Gulf (railroad)

lopkgs loose or in packages

l'Op Mont l'Opera de Montreal

lopo local post

LOPS Lloyd's Ocean Platform System

lopuro local purchase order

lo-q low iq (IQ)

loq *loquitur* (Latin—he speaks)

lor level of repair; lunar orbital rendezvous

Lor Lorenzo; Lorong

LOR *L'Osservatore Romano* (Papal Roman Observer)

lorac long-range accuracy

lorad long-range active detection

loran long-range aid to navigation

LORAPHS Long-Range Passive Homing System

LORAS Low-Range Omnidirectional Airspeed System

lord. long-range and detection (radar); lordosis

Lord Laurence Sir Laurence Olivier

Lords House of Lords

LORIDS Long-Range Iranian Detection System

lorl (LORL) large orbital research laboratory

Lorraine Lorraine De Sola Chervin

lorv (LORV) low orbital reentry vehicle

Lor²⁰ Lorenzo

los length of stay; liaison operating sheet; loss of signal

l-o-s line-of-sight

Los (Mexican-American truncation—Los Angeles)

LoS *Language of Sport* (Tim Considine's splendid guide)

LOS Lagos, Nigeria (airport); Latin Old Style; Law of the Sea; Little Orchestra Society; Lockheed Ocean Systems

Losa Los Angeles

Los Alamos Los Alamos National Laboratory

losam (LOSAM) low-altitude surface-to-air missile

Los Ang Los Angeles

Los Desastres *Los Desastres de la Guerra* (Spanish—The Disasters of War)—Francisco Goya's etchings

Los Guilucos Los Guilucos School for (delinquent) Girls

at Santa Rosa, California

lösl löslich (German—soluble)

LOSS Large Object Salvage System; Line Operation Status System

los sys landing observer's signal system

lostf line-of-sight test fixture

lot. large orbiting telescope; lateral olfactory tract; left occipito-transverse; load on top

lot. *lotio* (Latin—lotion)

LOT Polish Air Lines (3-letter symbol)

LOTADS Long-Term Air Defense Study (USA)

LOTCIP Long-Term Communications Improvement Plan

lote lesser of two evils

lo-temp low temperature

Lot-et-Gar Lot-et-Garonne

Loth Lothian

Lothians East Lothian, Midlothian, West Lothian

lotis logic, timing, sequencing

LOTUS Ladies Organized to Unfetter Sexuality

lotw loaded on trailers or wagons

Lou Lewis; Loualta; Louanna; Louanne; Loudella; Louella; Louina; Louis; Louisa; Louise; Louisetta; Louisette; Louisiana; Louisville; Loula; Loulou; Loura; Lourane; Lourene; Lourette; Louvilla; Louvina

Lou French Louisiana French

louh light observation utility helicopter

Louie Louis; Louisa; Louise; St Louis, Missouri

Louis Louisville

Louisvillain(s) native(s) of Louisville

Lou Orc Louisville Orchestra

Louv Louvain

l'Ouverture Toussaint l'Ouverture—founder and first president of Haiti

lov limit of visibility

LOVE League of Victims and Emphathizers (pro-capital-punishment group)

lovisim low-visibility landing simulation

LoW Launch on Warning (during nuclear warfare)

lo wat low water

'lowed allowed

Lowell Florida Correctional Institution at Lowell

lower 48 48 continental United States

lower 49 lower 48 plus Hawaii

Low L Low Latin
low. log. lower loge
lowpro low protein (diet)
Low Tatras Low Tatra Mountains of Czechoslovakia, Hungary, and Yugoslavia
low tec(h) low technology
low-Z low-impedance
lox also the name for smoked salmon; liquid oxygen; liquid-oxygen explosive
lox-sox liquid oxygen, solid oxygen
loxygen liquid oxygen
loy loyalty
LOYA League of Young Adventurers
Loyalists Loyalist American Colonists (Tories); Loyalist Episcopalian Traditionalists; Loyalist Spanish Republicans
Loyola Saint Ignatius de Loyola (Iñigo de Oñez y Loyola)
loz liquid ozone
Loz Lozère
lp lambing percentage; landplane; last paid; latent period; launch(ing) platform; light perception; linear programming; line pair (postage stamp); liquefied petroleum; liquid propellant; liquefied propane; list(ing) price; litter patient; local procurement; long primer (type); longplay; long-playing; low pass; low point; low power; low pressure; lumbar puncture
lp (LP) long-play (record)
l-p low-pressure
l/p lactate/pyruvate ratio; launch platform; letterpress; life policy; listening post
Lp Ladyship; Lordship
LP Aeralpi (2-letter symbol); Labor Party; Labour Party; Liberal Party; Libertarian Party; Liberty Press; Library of Parliament; Licensing Plan; Limited Partnership; litter patient; Liverpool Prison; long-play (record); Lower Peninsula
L-P Lionel-Pacific
LP *lunga pausa* (Italian—long pause)
lpa low-power amplifier
LPA Labor Party Association; Labor Policy Association; Little People of America
LPAA London Poster Advertising Association
L-pam L-phenylalanine mustard (anti-cancer drug)
LPAP Local Planning and As-

sessment Process
LPB La Paz, Bolivia (airport)
lpc leaf protein concentrate; least-preferred co-worker; linear predictive coding; low-pressure chamber; low-pressure compressor
lpc (LPC) linear-power controller
LPC Livestock and Pastoral Company; Lockheed Propulsion Company; Low Price Center
LPCC Lamb Promotion Coordination Committee
LPCG Laser Planning and Coordination Group (ERDA)
LPCM London Police Court Mission
lpcp launcher preparation control panel
lpcw long-pulse continuous wave
lp cyl low-pressure cylinder
lpd least perceptible difference; liquid protein diet; local procurement direct; low performance drone
LPD amphibious transport dock ship (naval symbol); Local Procurement District; low performance drone
LPE London Press Exchange
LPEA Licentiate of the Physical Education Association
L Ped Licentiate in Pedagogy
lpf leukocytosis-promoting factor; low-power field
LPF Latvian Popular Front
lpg liquid propane gas
lpg (LPG) liquefied petroleum gas
LPGA Ladies Professional Golfers Association; Liquefied Petroleum Gas Association; Louisiana Personnel and Guidance Association
lp gas liquefied petroleum gas
lph landing personnel helicopter; lines per hour
lph (LPH) landing personnel helicopter
L Ph Licentiate of Philosophy
lpi launching position indicator; lines per inch; low-power indicator
LPI Lifetime Productivity Index; Lightning Protection Institute; Louisiana Polytechnic Institute
lpicbm (LPICBM) liquid-propellant intercontinental ballistic missile
L-pills cyanide L-pills (deadly poisonous)
LPIU Lithographers and Pho-

toengravers International Union
LPIW Lumber, Production and Industrial Workers
LPKS Lone Pine Koala Sanctuary (Queensland)
LPKTF London Printing and Kindred Trades' Federation
lpl lightproof louver; list processing language
LPL Liverpool Public Libraries; London Public Library; Louisville Public Library; Lunar and Planetary Laboratory (University of Arizona)
LP & L Louisiana Power and Light
LPL *Lembaga Penelitian Laut* (Indonesian—Institute for Marine Research)—Jakarta
L-plane US Army liaison aircraft
LP & LC Louisiana Power & Light Company
L Plms Las Palmas
lplr lock pillar
LPLs Liverpool Public Libraries
lpm lines per millimeter; lines per minute; liters per minute
LPM Licensing Project Manager
LPMES Logistics Performance Measurement and Evaluation System
LP/MOSS Linear Programming/Mathematical Optimization Subroutine System
LPN Licensed Practical Nurse; Longview, Portland and Northern (railway)
LPN *Lembaga Padi Negara* (Malay—National Rice Paddy)
LPNA Licensed Practical Nurses Association; Lithographers and Printers National Association
LPNI Langley Porter Neuropsychiatric Institute
lpo liquid phase oxidation; local purchase order
LPO Licensing Project Office(r); London Philharmonic Orchestra; London Post Office
Lpool Liverpool
LPPM *Lembaga Penilitian Pertanian Maros* (Indonesian—Department of Agriculture)
LPPTFS London and Provincial Printing Trades Friendly Society
lpr (LPR) liquid-propellant rocket
LPR Lauritzen Peninsula

Reefer (steamship line)
LPRC Library Public Relations Council
lps lightproof shade; line program selector; liters per second; low primary sequence; low-pressure sodium
lps (LPS) lipopolysaccharide
lp(s) loop(s)
lp's (LPs) long-playing records
LPS Laboratory for Planetary Studies (Cornell); Lanterman-Petris-Short Act; Lebanese Press Syndicate; Light Photo Squadron; Linear Programming System; London Philharmonic Society; Lord Privy Seal; Lyceum Performing Society
LPSA Liberal Party of South Africa
LPS Act Lanterman-Petris-Short Act (commitment procedures covering mental patients in California)
LPSO Lloyd's Policy Signing Office
LPSS amphibious transport submarine (naval symbol)
lpstt low-power schottky transistor-transistor logic
lpt limited-production test
LPT Licensed Physical Therapist
LPTA Louisiana Parent-Teacher Association
LPTB London Passenger Transport Board
lptv low-power television
lptv (LPTV) large payload test vehicle
lpu limited-production urgent
Lpud Liverpudlian (native to or inhabitant of Liverpool)
lpv launching point vertical; lightproof vent
lpw low-power window; lumens per watt
l & p wood lumber and plywood wood
LPYS Labour Party Young Socialists
Lpz Leipzig
L Pz La Paz
LPZG Lincoln Park Zoological Gardens
LPZS Lincoln Park Zoological Society
lq last quarter; linear quantifier; lowest quartile
l.q. *lege quaeso* (Latin—please read)
lqdr liquidator
LQR *Law Quarterly Review*
lqss liquid steady state
LQST Leadership Q-Sort Test

lr latency relaxation; leave rations; letter report; lire; log run; long range; long run; lower
l/r left right; lower right
l-to-r left-to-right (photo caption abbreviation)
l R *laufen Rechnung* (German—current account)
Lr lawrencium
LR Laboratory Report; Land Registry; Landing Report; Lee Rubber (stock exchange symbol); Letter Report; Liaison Report; Little Rock
LR *Law Reports; Libertarian Review; Lloyd's Register*
lra long-range aviation
LRA Labor Research Association; Landing Rights Airport; Libertarian Republican Alliance; Lithuanian Regeneration Association
lraam (LRAAM) long-range air-to-air missile
lrac long-run average cost
LRAD Licentiate of the Royal Academy of Dancing
LRAFB Little Rock Air Force Base
LRAM Licentiate of the Royal Academy of Music
LRB Labor Research Bureau; Laboratory of Radiation Biology (University of Washington); Legislative Reference Bureau; Loyalty Review Board
LRBA *Laboratoire de Recherches Balistiques et Aérodynamiques* (French—Laboratory for Ballistic and Aerodynamic Research)
LRBC Lloyd's Register Building Certificate
lrbm long-range ballistic missile
lrc longitudinal redundancy check(ing); long-range communication; lower right center
lrc (LRC) longitudinal redundancy check character (data processing)
LRC Labor Representation Committee; Ladies Recreation Club; Langley Research Center (NASA); Law Reform Commission; Learning Resource Center; Lesbian Resource Center; Lewis Research Center (NASA); Library Resource Center; Linguistics Research Center; Logistics Research Center; London Rowing Club

LRCA London Retail Credit Association
LRCE Little Rock Cotton Exchange
lrcm (LRCM) long-range cruise missile
LRCM Licentiate of the Royal College of Music
lrco limited remote communications outlet
LRCP Licentiate of the Royal College of Physicians
LRCPE Licentiate of the Royal College of Physicians of Edinburgh
LRCPI Licentiate of the Royal College of Physicians of Ireland
lrcr longitudinal redundancy check register
LRCS League of Red Cross Societies; Licentiate of the Royal College of Surgeons
LRCSE Licentiate of the Royal College of Surgeons of Edinburgh
LRCSI Licentiate of the Royal College of Surgeons of Ireland
LRCT Licentiate of the Royal Conservatory of Toronto
LRCVS Licentiate of the Royal College of Veterinary Surgeons
lrd labelled radar display; long-range data
L-rd Lord (Hebraic contraction)
LRDC Learning Research and Development Center
LRDP Library Research and Demonstration Program
lrdr last revision date routine
lre law-related education
LREA Licensed Real Estate Agent
lrecl logical record length
LRES Linear Rocket Engine System
lrew long-range early warning
lrf latex and resorcinol formaldehyde; liver residue factor
lrf (LRF) luteinizing hormone-releasing factor
LRFI League for Religious Freedom in Israel
LRFPB Louisiana Rating and Fire Prevention Bureau
LRFPS Licentiate of the Royal Faculty of Physicians and Surgeons
LRFPSG Licentiate of the Royal Faculty of Physicians and Surgeons of Glasgow
lrg large; liquefied refinery gas; long range

lrg grdn large garden
lrh (LRH) luteinizing releasing hormone
LRHL Law Reports—House of Lords
lri left-right indicator; long-range input; long-range interceptor; lower respiratory infection
LRI Library Resources Incorporated
LRIBA Licentiate of the Royal Institute of British Architects
LRIC Licentiate of the Royal Institute of Chemistry
lrim long-range input monitor
lrip language research in progress
lrir limb radiance inversion radiometer
LRIS Lloyd's Register Industrial Services
LRJC Lake Region Junior College
LRKB Law Reports—King's Bench
LRL Lawrence Radiation Laboratory; Lunar Receiving Laboratory
LRLA La Raza Legal Alliance
lrl's living-room liberals
LRLS London Regional Library System
LRLSA La Raza Law Students Association
LRLTRAN Lawrence Radiation Laboratory Translator
lrm length register mark; liquid radiation monitor
LRMC Lloyd's Refrigerating Machinery Certificate
lrmg (LRMG) lockless-rifle machine gun
lrmp long-range maritime patrol
LRMS Liquid Radwaste Management System
LRN Landslaget for Reiselivet i Norge (Norwegian—Norway Travel Association)
LRNC Long Reference Number Code
lrp launching reference point; long-range planning
LR-P La Rochelle-Pallice
LRP Law Reports–Probate
lrpa long-range patrol aircraft
lrpg long-range proving ground
LRPGR Long-Range Planning Ground Rules
LRPL Liquid Rocket Propulsion Laboratory; Little Rock Public Library
LRPS Long-Range Planning Service

LRQB Law Reports—Queen's Bench
lrr lower reduced rate
LRR Location Recommendation Report
lrra low-range radio altimeter
lrrd long-range reconnaissance detachment
lrrmf long-range resource and management forecast
lrrp lowest required radiated power
LRRS Long-Range Radar Station
LRRT Library Research Round Table
LRRTS Light-Rail Rapid-Transit System
lrs long-range search; long-run supply
lr's leave rations; light refreshments
l/r/s library rubber stamps (used-book trade abbreviation indicating book may belong or may have belonged to a public library)
lRs lactated Ringer's solution
Lrs Lancers
LRS Land Registry Stamp; London Research Station (British Gas)
LRS Lloyd's Register of Shipping
lrsam (LRSAM) long-range surface-to-air missile
lrsm long-range seismic measurement
LRSM Licentiate of the Royal Schools of Music
LRSS Long-Range Survey System
lrt laser ray tube; launch, recovery, and transport
lrt (LRT) light rail transit
lrtc long-run total cost
LRTgt last resort target
LRTL Light Railway Transport League
LRTS Library Resources and Technical Services
lru least recently used; line replacement unit
lrv (LRV) light-rail vehicles (rapid transit); lunar roving vehicle
LRWES Long-Range Weapon Experimental Station
LRWRE Long-Range Weapons Research Establishment
LRY Liberal Religious Youth
ls landing ship; left side; light vessel; lightship; limestone; liminal sensitivity; limit switch; local sunset; long shot; long site; long sleeves;

loud-speaker; low secondary; low speed; lump sum
ls (LS) legal seal
l-s lumbo-sacral
l's losers (gambling short form)
l/s liters per second
l & s launch(ing) and servicing
l.s. locus sigilli (Latin—place of the seal)
Ls Lopes; Louis
LS Lamson & Sessions; Law Society; Leading Seaman; Letter Service; Licensed Surveyor; Linnaean Society
L-S Lewis-Shepard
L & S Lands and Survey (department or office)
lsa left sacro-anterior; logistic support analysis; low specific activity
lsa (LSA) lichen sclerosus et atrophicus
LSA Labor Services Agency; Labour Services Association; Land Service Assistant; Land Settlement Association; Leukemia Society of America; Licentiate of the Society of Apothecaries; Lighthouse Society of America; Limbless Soldiers Association; Linguistic Society of America; Liquor Stores Association; Lithuanian Society of America; London Salvage Association; London School of Accountancy
L&SA Law and Society Association
LSA Library Science Abstracts
LSAA Linen Supply Association of America
LSAC Law School Admission Council; London Small Arms Company
lsar local storage address register
LSAS Law School Admission Service
LSAT Law School Admission Test; Law School Aptitude Test
lsb left sternal border; lower sideband
lsb (LSB) least significant bit
LSB Launch Service Building; London School Board; Louisiana School Board
LSBA Leading Sick-Bay Attendant; Louisiana School Boards Association
LSBR Large Seed-Blanket Reactor (AEC)
lsc linear sequential circuit; logistic support cost
l.s.c. loco supra citato (Latin—

in the foregoing place cited)
LSC Laser Systems Center;
Legal Services for Children;
Legal Services Corporation;
Lie Scale for Children; Lower School Certificate
lsca left scapulo-anterior
LSCA Library Services and Construction Act
LSCC Library of the Supreme Court of Canada
lscp left scapuloposterior
LSCRS Law School Candidate Referral Service
lscs lower segment caesarean section
lsct low-speed compound terminal
LSCT Lamar State College of Technology
lsd laser-support detonation; last significant data; last significant digit; leadless sealed device; least significant difference; least significant digit; library system(s) development; liquid scale disintegrator; logarithmic-series distribution; long, slow, distance (jogging); loss, short, and damage
ls & d liquor store and delicatessen
l s d librae, solidi, denarii (Latin—pounds, shillings, pence)
LSD landing ship, dock (naval symbol); League for Spiritual Discovery; lysergic acid diethylamide—dangerous psychedelic drug nicknamed *acid*
L.S.D. Doctor of Library Science
LSD Lyserginsaure Diathylamid (German—lysergic acid diethylamide)
LSDAS Law School Data Assembly Service
LSDI Logistics Support Departmental Instruction
lsd li leased line (telephone)
LSDS Low-Speed Digital System
lse limited signed edition; limited special edition
LSE London School of Economics; London Stock Exchange; Louisiana Sugar Exchange
LSECS Life Support and Environmental Control System
l sect longitudinal section
LSEL London School of Economics Library
LSE & PS London School of

Economics and Political Science
lse skds loose (or on) skids
LSET Logistics Supportability Evaluation Team
LSEU La Salle Extension University
lsf log super feet
LSF Literary Society Foundation; Lloyd Shaw Foundation; Lock Security Force (Panama Canal)
lsfa logistic system feasibility analysis
lsg list set generator
Lsg Lösung (German—solution)
lsgd lymphocyte specific gravity distribution
L Sgt Lance Sargeant
LSH Latter-day Saints Hospital
L-shape ell-shaped
LSHTM London School of Hygiene and Tropical Medicine
lsi large-scale integration; lateral shear interferometer
LSI Labor and Socialist International; Lake Superior & Ishpeming (railroad); landing ship—infantry; Law of the Sea Institute; Law-Science Institute (University of Texas); Lear Siegler Incorporated; Logistic Shipping Instruction(s); Lunar Science Institute
LS & I Lake Superior & Ishpeming (Railroad)
LSIA Lamp and Shade Institute of America
lsic large-scale integrated circuitry
LSIO Labor Standards Inspection Office(r)
LSIS Laser-Scan Inspection System
lsk liver, spleen, kidney
lsl left sacrolateral; low-speed logic
LSL landing ship, logistic; Life Sciences Laboratory; Linnaean Society of London; Lucy Stone League
lslb left short-leg brace
lsm linear synchronous motor; lysergic acid morpholide
lsm (LSM) lysergic acid morpholide
l.s.m. litera scripta manet (Latin—the written word remains)
LSM Laboratory for the Structure of Matter (USN); Lancastrian School of Manage-

ment; landing ship, medium; Liberation Support Movement; Logistic Support Manager
LS/mft Leopold Stokowski/ means fine tone; Lucky Strike/means fine tobacco
LSMI Lake Superior Mining Institute
LSMP Logistic Support and Mobilization Plan
LSMR rocket ship
LSMSC Lake Superior Mines Safety Council
LSNR League of Struggle for Negro Rights
LSNSW Linnaean Society of New South Wales
LSNY Linnaean Society of New York
LSO Landing Signal Officer; Leningrad Symphony Orchestra; London Symphony Orchestra
lsp left sacro-posterior; logical signal processor
LSP Launch Pad; Logistic Support Plant
LSPOJC La Salle-Peru-Oglesby Junior College
LSPR Library Society of Puerto Rico
L-square capital-L-shaped square; carpenter's square
lsr launch signal responder; lens shutter reflex
Lsr Luftschutzraum (German—air raid shelter)
LSR landing ship, rocket; landing ship, support
Lsr Ant Lesser Antilles (Leeward and Windward Islands)
lss liquid scintillation spectroscopy
LSS Life Saving Service; Life Saving Station; Life Support System; Lockheed Space Systems; Logistic Support Squadron
L.S.S. Licentiate of Sacred Scripture; Leopold-Sedar Senghor
lssc logistic support system characteristics
LS Sc Licentiate in Sacred Scriptures; Licentiate in Sanitary Science
LSSC Logistic System Support Center (USA)
LSSF Land Special Security Force (USA)
LSSG Logistics Studies Steering Group; Logistics Studies Support Group
lssm local scientific surface module

LSSR Latvian Soviet Socialist Republic; Lithuanian Soviet Socialist Republic
LSSS London School of Slavonic Studies
L S St L Louis Stephen St Laurent (Canada's sixteenth Prime Minister)
lst large space telescope; laser spot tracker; left sacro-traverse; liquid storage tank; liquid-oxygen start tank; living structures tank
Lst Launceston
LST landing ship, tank; Local Sidereal Time; Local Standard Time
LST London Sunday Times (newspapers)
lstc low-speed trim compensation
LSTM Liverpool School of Tropical Medicine
lst wk last week
lsu launcher selector unit; livestock unit
LSU landing ship, utility; Louisiana State University
LSU-IES Louisiana State University Institute of Environmental Studies
LSUNO Louisiana State University (New Orleans)
LSUP Louisiana State University Press
lsuv lunar surface ultraviolet (camera)
LSV landing ship, vehicle
LSVP landing ship, vehicle, and personnel
lsw least significant word; limit switch(ing)
LSW Licensed Shorthand Writer
lsw lt landing signal wand light
LSWR London and South Western Railway
LSWY League of Socialist Working Youth
LSZ Limited Speed Zone; Local Slow Zone
lt landed terms; language translation; latch trip; laundry tray; lid tank; light; light trap; line terminator; local time; long ton; loop test; low temperature; low tension; low torque
lt (LT) lymphotoxin
l/t loop test
lt laut (German—according to)
l.t. locum tenens (Latin—substitute)
Lt Lieutenant
LT Land Transfer; landing team; large tug; Lloyd

Triestino; local time; London Transport
lta lighter-than-air
LTA Lawn Tennis Association; Library Technical Assistant; lighter-than-air; Listening Transit Analysis; Logistic Task Authorization
LTA Life of Timon of Athens
LTAA Lawn Tennis Association of Australia
ltadl launcher tube azimuth datum line
LTAS Lighter-Than-Air Society
ltb laryngo-trachael bronchitis; line terminating battery; low-tension battery
Lt. B. Bachelor of Literature
LTB Lepers Trust Board; London Tourist Board; London Transport Board
LTBP London Tanker Broker Panel
LTBT Limited Test Ban Treaty (prohibiting nuclear testing in certain environments)
ltc long-term care
ltc (LTC) locking torque converter
LTC Land Transport(ation) Commission; Lawn Tennis Club; Le Tourneau College; Library of Trinity College; Loop Test(ing) Conference
LTCB Long-Term Credit Bank (Japan)
Lt Cdr Lieutenant Commander
LTCL Licentiate of Trinity College of Music (London)
Lt Cmdr Lieutenant Commander
Lt Col Lieutenant Colonel
ltc's long-term contracts
ltd long-term disability
lt/d long tons per day; lower tween deck
Ltd Limited
Ltda Limitada (Spanish—limited)
ltd ed limited edition
lte large table electroplotter; linear threshold element
Lte (French—Limité)—limited
LTE London Transport Executive
lted letter to the editor
LTEU Liquor Trades Employees Union
ltf (LTF) lipotrophic factor
LTF Lithographic Technical Foundation; Logistic Task Force; tropical fresh water load line (Plimsoll mark)
ltfrd lot tolerance fraction reli-

ability deviation
ltg lighting
LTG Leadership Training Graduate
ltgc lithographic
ltge lighterage
Lt Gen Lieutenant General
ltgh lightening hole
Lt Gov Lieutenant Governor
lth lath; lathing; less than honest (crooked, dishonest); luteotrophic hormone (LTH)
Lth Leith
L Th Licentiate in Theology
lthr leather
lti land training installation(s)
lti (LTI) light transmission index
Lti Laotian
LTI Ladder Towers Incorporated; London Taxis International; Louisiana Training Institute; Lowell Technological Institute
LTIB Lead Technical Information Bureau
Lt Inf Light Infantry
Lt JG Lieutenant Junior Grade
ltl listing time limit
ltl (LTL) less than truckload
Ltl Little (postal abbreviation)
ltla launcher tube longitudinal axis
ltm laser target marker; long-term memory; long-term mortgage; low thermal mass
ltm (LTM) long-term memory
LTM Licentiate of Tropical Medicine; Little Theatre Movement
ltmr laser target marker ranger
ltng lightning
ltng arr lightning arrester
lto landing takeoff
Lto lento (Italian—slowly)
LTO Land Transfer Office; Leading Torpedo Operator
ltof low-temperature optical facility
LTon long ton
ltp limit on tax preferences; low-temperature passivation
LTP Library Technology Program
ltpd lot tolerance percent defective
ltpp lipothiamide-pyrophosphate
ltr letter; lighter; liter; long-term reserve
LTR Long Term Reserve
LTR Library Technology Reports
Lt RN Lieutenant—Royal Navy
LtrO letter order

LTRP Long-Term Requirement Plan

ltrs (LTRS) letters shift (data processing)

LTRS Laser Target Recognition System

lts lights

l'ts let's

LTs *Legal Times*

LTS Landfall Technique School; London Transport System; London Typographical Society

LTSB London Trustee Savings Bank

LT & SR London, Tilbury and Southend Railway

ltt liquid toning transfer

LTT Lymphocyte Transformation Test

ltta long-tank thrust augmented

LTTC Lowry Technical Training Center

LTTE Liberation Tigers of Thamil Eelam

lttr latter

L-T Trade Agreement Liao-Takasaki Trade Agreement

ltu line terminating unit

LTU La Trobe University

ltv long tube vertical

lt/v light vessel

L-T-V Long-Temco-Vought (corporation)

ltvc launcher tube vertical centerline

ltwt lightweight

lu lock up; logic unit; logistical unit; lumen

lu. lues (Latin—contagious disease)—plague or syphilis

Lu Lorentz unit; Lugano; Lugo; lutetium

LU Langston University; Lawrence University; Laurentian University; Laval University; Lefèvre-Utile (French bakers); Lehigh University; Lethbridge University; *Ligue Universelle* (French—Universal Esperantist League); Lincoln University; Liverpool University; London University; Loyola University

lu. I lues I—primary syphilis

lu. II lues II—secondary syphilis

lu. III lues III—tertiary syphilis

lua left upper arm

LuA Launch under Attack (during nuclear warfare)

LUA Life Underwriters Association; London Underwriters Association

LUAA Life Underwriters Association of Australia

LUAC Land Use Advisory Council; Life Underwriters Association of Canada

LUANZ Life Underwriters Association of New Zealand

lub lubricant, lubricate; lubrication

lub (LUB) logical unit block

lube lubricate; lubrication

lubed lubricated (intoxicated)

lub oil lubricating oil

lubs large undisturbed bottom sampler

Luc Lucan; Lucifer; Lucretius; Lucullus

LUC Land Use Commission; Land Use Committee; Louisiana University Center

LUCB Library of the University of California at Berkeley

LUCHIP Lutheran Church and Indian People

luchtv luchtvaart (Dutch—aviation)

Luci Lucifer

Lucia St Lucia

Lucia di Lucia di Lammermoor (Donizetti opera)

lucid. language used to communicate information system design

Luck Lucknow

lucom lunar communication

luc. prim. luce primo (Latin—at daybreak)

Lucr The Rape of Lucrece

Lucretius Roman poet-philosopher Titus Lucretius

Lucrezia Bori Lucrecia Borja y Gonzalez de Riancho

lud liftup door

Lud Ludlow; Ludo; Ludolf; Ludolph; Ludovic; Ludovica; Ludovick; Ludovico; Ludovicus; Ludvig; Ludwell; Ludwig; Ludwik

luda land use data

Luddy Ludlow (*see* Lud)

lude quaalude (a depressant drug)

Ludendorff General Erich Friederich Wilhelm Ludendorff

Lud(s) Luddite(s)

Ludwig van Ludwig van Beethoven

lue left upper entrance; left upper extremity

LUER Land Use and Environmental Regulation

lues I primary syphilis

lues II secondary syphilis

lues III tertiary syphilis

luf lowest useful high frequency

LUFTHANSA *Deutsche Lufthansa* (German Airline)

lug luggage; lugger; lugging; lugsail; lugworm

lug luglio (Italian—July)

Lugd. Bat. Lugdunum Batavorum (Latin—Leiden)—Leyden

lu h lumen hour(s)

luhf lowest usable high frequency

LUI Labor Union Insurances

Luigi Cherubini María Luigi Carlo Zenobio Salvatore Cherubini

LUIP London University Institute of Psychiatry

Luis Buñuel Luis Buñuel Portolés

lujb left umbilical junction box

Luke The Gospel according to St Luke

lul left upper limb; left upper lobe

LUL London University Library

LULA Loyola University of Los Angeles

LULAC League of United Latin-American Citizens

LULAC La Liga de Ciudadanos Latinoamericanos Unidos (Spanish—The League of United Latin-American Citizens)

LULOP London Union List of Periodicals

Lulu Louise

lum lumbago; lumbar; lumber; lumen; luminosity; luminous

lum (LUM) lunar excursion module

Lum Columbus

LUMAS Lunar Mapping System

lumb lumber; lumbering

LUMC Laval University Medical Center

LUMIS Land Use Management Information System

lumpec lumpectomy, surgical removal of a tumor with a limited amount of associated tissue

Lumpen Lumpenproletariat (German—ragged bums, ragged street people)

lun lune; lunette

lun lundi (French—Monday); *lunedi* (Italian—Monday); *lunes* (Spanish—Monday)

Lunar Lunar Society (Birmingham, England)

lunar caustic silver nitrate ($AgNO_3$)

lunch luncheon
LUNCO Lloyds Underwriters Nonmarine Claims Office
lun int lunitidal interval
Lunnon London
luo *luogo* (Italian—place)—in music, return to original pitch
Lup Lupus (constellation)
LUP Liverpool University Press; Loyola University Press
lupa lupanar (Latin—brothel)
Lupe Guadalupe
Lupe Vélez Maria Guadalupe Vélez de Villalobos
luq left upper quadrant (abdomen)
Luqa Malta's main airport
LUS Land Utilization Survey
LUSB Land Utilization Survey of Britain
lus fin luster finish
Lu-shun (Chinese—Port Arthur)
lusi lunar surface inspection
lusing lusingando (Italian—coaxing)
Luso Lusotania (Portuguese—Lusitania)
lust. lustrous
lut launcher umbilical tower (LUT)
lut. luteum (Latin—yellow)
LUT Launcher Umbilical Tower; Loughborough University of Technology; Ludwig Universe Tankships
LUTA Library of the University of Texas at Austin
LUTC Life Underwriter Training Council
Lutetia (Latin—Paris)—more fully Lutetia Parisiorum
LUTFCSUSTC Librarians United to Fight Costly, Silly, Unnecessary Serial Title Changes
Luth Luther(an)
Lutia St Lucia, West Indies
Lutz Lucien
luv let us vote (popular teenage plea); lightweight utility vehicle (pickup truck)
LUV Love Uniting Volunteers
lux luxurious; luxury
Lux Luxembourg; Luxembourger; Luzon
lux aet lux aeterna: (Latin—everlasting light)
LUXAIR Luxembourg Airlines
Luxem Luxembourg
Luxembourg Grand Duchy of Luxembourg (European lowland), *Grand-Duché de Luxembourg*
Lux Fr Luxembourger franc

Luz Luzon
lv land valuation; largest vessel; launch vehicle (LV); leave; left ventricle; light and variable (wind); low viscosity; low voltage; lumbar vertebra; luncheon voucher; lyric vocalist
l-v lacto-vegetarian
l/v light vessel (lightship)
lv *livre* (French—book)
Lv Latvia; Latvian; lev (Bulgarian currency unit)
LV Las Vegas; Laser Vision; launch vehicle; Lehigh Valley (railroad); Licensed Victualler; light vessel (light ship); Linda Vista; Lindholmens Varv (Lindholmens Shipyard)
LV-3 Atlas launch vehicle (Convair)
lva landing vehicle airoll; left visual acuity
lva (LVA) landing vehicle—assault
LVA Licensed Victuallers Association; Literary Volunteers of America
lvad left ventricular assist device
L v B Ludwig van Beethoven
L v Bthvn Ludwig van Beethoven
lvcd least voltage coincidence detection
lvd louvered door
LVD laser video-disc player
lvda launch vehicle data adapter
lvdc launch vehicle digital computer
lvdt linear variable-differential transformer; linear variable-displacement transducer
lved left ventricular end diastolic
lvet left ventricular ejection time
lvf left ventricular failure; left visual field
Lvfa low-voltage fast activity
lvgo light vacuum gas oil
lvh left ventricular hypertrophy
lvh (LVH) landing vehicle hydrofoil
lvhv low volume high velocity
lvi low viscosity index
LVI Local Veterinary Inspector
lvl laminated veneer lumber; level
LVL La Verendrye Line (Hall Corporation); Linda Vista Library
LVLO Local Vehicle Licens-

ing Office
lvn light virgin naphtha
LVN Licensed Visiting Nurse; Licensed Vocational Nurse
LVNM Lava Beds National Monument (California)
LVNP Lassen Volcanic National Park (California); Luangwa Valley National Park (Zambia)
lvo (LVO) leveraged buyout
lvp low-voltage protection
lvp (LVP) left ventricular pressure
LVP Launch Vehicle Program(s)
lvp dr leverpak drum
lvr line voltage regulator; longitudinal recorder; longitudinal video recording; low-voltage release
LVRB Launch Vehicle Reliability Board
lvrj low-volume ramjet
Lvrpl Liverpool
LVRS Longitudinal Video Recording System
lvs leaves
lv's lunch(eon) vouchers
LVs launch vehicles
LVS Licentiate in Veterinary Science
LVT landing vehicle, tracked
LVTC landing vehicle, tracked command
lvupk leave and upkeep
LVUSA Legion of Valor of the USA
lvw linked vertical wall
lw light warning; lightweight; live weight; long wave; low water
l/w in lieu of weighing; lumens per watt
l & w living and well
l W *lichte Weite* (German—internal diameter)
Lw lawrencium (element 103)
LW light warning; lower berth
L-W Lee-White (method)
lwar lightweight attack and reconnaissance
L-wave long wave (usually the third major earthquake shock wave)
l'way leeway
lwb long wheelbase
lwc lightweight concrete
LWCA London Wholesale Confectioners Association
LWCF Land and Water Conservation Fund
lwcp lightweight coated paper
lwd larger word; leeward; left wing down; lewd; lowered
LWD Laser Welder/Driller;

Liquid Waste Disposal
lwest low water equinoctial spring tide
lwf lightweight fighter
LWF Lutheran World Federation
LWFB Lake Washington Floating Bridge
lwf & c low water full and change
lwg last we've got; live-weight gain
LWG Logistic Work Group (NATO)
lwgr (LWGR) light-water-cooled graphite-moderated reactor
L-w-H Lewis-with-Harris (Outer Hebrides)
lwic lightweight insulating concrete
lwir long-wave infrared
LWJ Lowell Weicker, Jr
lwl length at waterline; load waterline; low-water line (tidal marking)
LWL Limited War Laboratory (US Army)
lwld light-weight laser designator
lwm low-water mark
LWM Leonard Wood Memorial (American Leprosy Foundation)
LWMAT Locke-Wallace Marital Adjustment Test
LWMEL Leonard Wood Memorial for the Eradication of Leprosy
LWNWR Lake Woodruff National Wildlife Refuge (Florida)
lwont low water ordinary neap tide
lwop leave without pay
lwos low-water ordinary spring
lwost low-water ordinary spring tide
lwp leave with pay; load water plane
lwpf long-wave pass filter
lwr lightweight radar; lower
lwr (LWR) light water reactor
Lwr Lower (postal abbreviation)
LWR Light Water Reactor
l'wrd leeward
lwrm (LWRM) lightweight radar missile
lwrs light-warning radar set
lwru lightweight radar unit

LWS Late West Saxon; Letter Writing System
LWSI *Lloyd's Weekly Shipping Index*
lwst low-water spring tides
lwt lightweight
Lwt Lowestoft
LWT amphibious warping tug (naval symbol); London Weekend Television
lwta laser window test apparatus
LWTMA London Wool Terminal Market Association
lwtvp laser window technology validation program
LWU Leather Workers Union
LWUI Longshoremen's and Warehousemen's Union International
LWV Lackawanna & Wyoming Valley (railroad); League of Women Voters
LWVEF League of Women Voters Education Fund
LWVUS League of Women Voters of the United States
lww launch window width
lwyr lawyer
lx lux
lx. *lux* (Latin—light)
LX Lox Angeles Airways (2–letter coding)
Lx^a *Lisboa* (Portuguese—Lisbon)
Lxmbrg Luxembourg
lXr limb X-ray
LXX Septuagint (70)
lxxx love and kisses
ly langley (solar heat unit); last year; last year's model
ly (LY) lethal yellowing (coconut-palm disease)
Ly Lyceum Theatre, English Opera House; Lyman; Lyon
LY Light Yeomanry; Love Year
LYC Larchmont Yacht Club
Lyd Lydia; Lydian
Lýd Isl *Lýdhveldidh* (Icelandic—Republic of Iceland)
lye potassium hydroxide (KOH) or sodium hydroxide (NaOH)
LYK Lykes Brothers Steamship company (stock exchange symbol)
LYKU Lykes Lines (container) Unit
Lylis Lilian; Lilly
ly & lt lastex yarn and lactron

thread
lym last year's model(s); lymph; lymphatic(s)
lymphoks lymphokines
lympho(s) lymphocyte(s)
lymphs lymphocytes
lyn lynch (named for Captain William Lynch, also called Judge Lynch, who advocated hanging on the basis of mob action rather than legal procedure; this type of violence is also called lynch law)
Lyn Lynch; Lynde; Lyndon; Lyne; Lynn; Lynx (constellation)
Lynwood Lynwood (delinquent) Girls Center at Anchorage, Alaska
Lyo Lyons (British maritime contraction)
Lyo Isl *Lýoveldio Island* (Icelandic—Republic of Iceland)
lyr lyric; lyrical; lyricism; lyricist; lyrics
Lyr Lyra (constellation)
L & YR Lancashire and Yorkshire Railway
lyric. language for your remote instruction by computer
lys lysine
LYs Light Years
lysis (Greek—dissolution or loosening)—analysis, catalytic, electrolyte, hydrolysis, paralysis; (Latin suffix—dissolve or solution)—hemolysis
lysog lysogen(ic)(al)(ly); lysogenization; lysogenize(d); lysogenizing; lysogeny
Lyt Lyttelton, New Zealand
Lytt Lyttelton
Lyttelton Lyttelton Harbour (Christchurch, New Zealand's port)
Lz Lopez
LZ Landing Zone
lzm lysozyme
LZOA Labor Zionist Organization of America
lzp left zero point
LZSU Leningrad AA Zhdanov State University (University of Leningrad)
LZT Local Zone Time
L-Zug *Luxus-Zug* (German—luxury railroad train)
lzy lazy

M

m difference of meriodional parts (symbol); magnetic dipole moment (symbol); main; maintainability; male; malignant; manual; married; masculine; mass; mature; mean (arithmetical); measure; mediator (chemical); mega; meg-ohm; member; memory; mentum; meridian; mesh; metabolite; meter; mile; mill; milli-(thousandth); minim; minor; minute; minutes; modulus; molal (concentration); molar; molecular weight; molecule; monkey; month; moon; morning; morphine; mother; motile; mucoid; murmur (heart); muscle; myopia

m (M) marijuana; morphine

m/ merged into

μ micron (symbol); micro

'm (contraction—am)—as in I'm here

m mass (symbol); *Mazda* (Japanese auto with German Wankel rotary engine); *metro* (Portuguese or Spanish—meter); *mijnheer* (Dutch—mister); *murió* (Spanish—died)

m. macerare (Latin—macerate)

m/ med (Norwegian—with)—as in *varm aplepai m/is* (hot apple pie with ice cream)

M bending moment (symbol); Mach (Austrian physicist); mach number; mach speed; magnaflux; magnetic inspection; maintainability; Majesty; Malay; Malaya; Malaysia; March; mark; Martin; materiel; Matson Navigation Company; median; medium; mega-(million); megacycle;

Member; metal; Metro(politan); metropolitan; Mike—code for letter M; Min; missile; mixture; mobile; Mohammedan; Mohammedanism; molal (concentration); molecular weight (symbol); moment; Monday; *Monsieur* (French—Mister); Montour (railroad); Moore-McCormack (steamship lines); Moslem; muscle; pitching moment (symbol); thousand (symbol)

M *(M)* money supply (macroeconomics symbol)

M' *Mac* (Gaelic—son of)

M der Mörder (German—the murderer)—Fritz Lang film; *Marcus* (Latin); *Missa* (Latin—Mass); *mujeres* (Spanish—women)

m_1 mitral first sound

M1 money supply including currency and trading bank demand deposits

M-1 basic money aggregate (funds for spending); mutual inductance symbol; U.S. semi-automatic service rifle used in Vietnam

m1s matte one side

M2 M1 plus savings and small-denomination time deposits under $100,000; money supply (M1) plus demand deposits of savings & loan banks as well as stock firms

M-2/M-3 White half-track armored-personnel carrier

m2s matte two sides

m/3 middle third (long bones)

m^3 cubic meter(s)

$μ^3$ cubic micron

M3 M2 plus large-denomination time deposits, term repurchase agreements, and investments of institutions in money-market mutual funds

M-3 sterling deposits of British residents plus coins and notes

M-3A1/M-5 Stuart tank armed with a 37mm gun

$m^{3/m}$ cubic meters per minute

$m^{3/s}$ cubic meters per second

M-4 Sherman medium tank with 76mm gun

M-6 American-made armored car

m8 medium octavo

M-8 Greyhound 6-wheeled armored car carrying a 37mm gun and made in the U.S.A.

M-14 U.S. fully-automatic or semi-automatic service rifle used in Vietnam

M15 gasoline extender (15% methanol plus 85% gasoline)

M-15 British secret service charged with counterespionage and security operations at home and overseas

M-16 British Foreign Service Military Intelligence (secret intelligence service); U.S. fully-or semi-automatic light-weight small-bore service rifle used in Vietnam

M-18 Hellcat 76 mm gun mounted on a tracked chassis made in the U.S.A.

M-20 Mystère 20 aircraft; unarmed Greyhound 6-wheeled armored car produced in the U.S.A. during World War II

M-36 U.S.-made Slugger tank destroyer

M-44 U.S.-made self-propelled

155mm howitzer
M-47 U.S.-made Patton tank carrying a 90mm gun
M-48 later version of the M-47 medium tank
M-56 U.S. self-propelled 90mm antitank gun called Scorpion
M-60 Patton main battle tank carrying 105mm gun
M-61 20mm Vulcan aerial machine gun firing 6000 rounds per minute
M100 100% methanol
M-107 U.S.-made self-propelled 175mm gun
M-113 U.S.-made 13-man amphibious armored personnel carrier
M-198 155mm howitzer (USA)
M-551 U.S.-built Sheridan assault vehicle armed with a 152mm gun
ma machine account; machine accountant; maleic anhydride; manpower allotment; manufacturing assembly; map analysis; mechanical advantage; medium amphibian; menstrual age; mental age; microscopic agglutination; mill annealed; mill-anneal-(ing); milliampere; mixed ages; mixed-age (ewes); monthly account(ing)
ma (MA) maleic anhydride
m/a mechanical airframe; my account
m & a maintenance and assembly
μa micro+ampere
mA milliampere(s)
mÅ milliangstrom(s)
μA microampere(s)
Ma Malayalan; Mama; Manchuria; Manchurian; María; masurium (symbol)
Mᵃ María
Ma Mandag (Danish—Monday)
MA Magma Arizona (railroad); Magnesium Association; Mahogany Association; Maintainability Analysis; Manpower Administration; Manpower Authorization; Maritime Administration; Marshaling Area; Marshalling Area; Massachusetts; Material Authorization; May Department Stores (stock exchange symbol); Mediterranean Area; Menorah Association; Merchandising Assessment; Metric Associa-

tion; Military Academy; Military Attaché; Mountaineering Association
M-A Miller-Abbott (tube)
M.A. *Magister Artium* (Latin—Master of Arts)
M & A Missouri & Arkansas (railroad)
MA Maison d'Arrêt (French—jail, lockup, prison); *Modern Age*; *Musical America*
maa maximum authorized altitude
maa (MAA) macroaggregated albumin
Maa Madras
Maa Maandag (Dutch—Monday)
MAA Manufacturers Aircraft Association; Master Army Aviator; Master-at-Arms; Master-of-Arms; Mathematical Association of America; Medical Assistance for the Aged; Medieval Academy of America; Microfilm Association of Australia; Museum of African Art (Smithsonian); Museums Association of Australia; Mutual Aid Association; Mutual Assurance Association
MA of A Motel Association of America
ma ac machine accessory
MAAC Major Additions Adjustment Clause; Medical Assistance Advisory Council; Metropolitan Area Advisory Committee; Mutual Assistance Advisory Committee
MAAEE Ministero degli Affari Esteri (Italian—Ministry of Foreign Affairs)
MAAF Mediterranean Allied Air Force; Mediterranean Army Air Force
MAAG Military Assistance Advisory Group
MAAGB Medical Artists' Association of Great Britain
MAAH Museum of African-American History
MAAI Modeling Association of America, International
ma'am madam
ma'amselle mademoiselle
MAAN Mutual Advertising Agency Network
MAAN Much Ado About Nothing
maap maintenance and administration panel
MAAP Minority Association for Animal Protection
M.A.Arch. Master of Arts in

Architecture
maarm (MAARM) memory-aided anti-radiation missile
MAARS Multi-Access Airline Reservation System
MAAS Member of the American Academy of Arts and Sciences
MAATC Mobile Antiaircraft Training Center
mab multibase arithmetic blocks
Mab Mabel
MAB Magazine Advertising Bureau; Malfunction(ing) Analysis Board; Man and the Biosphere (UNESCO); Maracaibo Oil Exploration (stock exchange symbol); Marine Air Base; Medical Advisory Board; Metropolitan Asylum Board; Missile Assembly Building; Monetary Affairs Branch; Munitions Assignment Board
MAB Manufacture d'Armes Automatiques Bayonne (French—Bayonne Automatic Arms Factory)
M.A.B.E. Master of Agricultural Business and Economics
MABF Major Adjustment Billing Factor
mabflex marine amphibious brigade field exercise
mablex marine amphibious brigade landing exercise
MABO Marianas-Bonin (islands)
mabp mean arterial blood pressure
MABRON Marine Air Base Squadron
MABs Marine Amphibious Brigades
MABS Marine Air Base Squadron; Marine Automation Bridge System
MABSC Management and Behavioral Science Center
MABYS Metropolitan Association for Befriending Young Servants
mac macadam(ize)(d); macerate; machine-aided cognition; mackintosh; maximum allowable concentration(s); mean aerodynamic chord; motion analysis camera; multiple-access computer
mac (MAC) mycobacterium avium-intracellulare
mac macizo (Spanish—strong, solid); *maquereau* (French—mackerel)—pimp

mac. _macerare_ (Latin—macerate)

Mac Freddie Mac, Federal Home Loan Mortgage Corp.; Macao, Portuguese China; nickname of anyone whose surname begins with Mac

M.Ac. Master of Accountancy

MAC Maintenance Advisory Committee; Maintenance Analysis Center; Major Air Command; Management Aggregation Code; Marine Amphibious Corps; Maritime Advisory Committee; Material Availability Commitment; McDonnell Aircraft Corporation; Mediterranean Air Command; Message Authentication Code; Miami Aviation Corporation; Middle Atlantic Conference; Military Airlift Command; Mineralogical Association of Canada; Multidimensional Actuarial Classification; Municipal Assistance Corporation; Musical Arts Center

MACA Maritime Air Control Authority; Mental After-Care Association

MACAE Minnesota Association of Continuing Adult Education

MACAIR Macao Air Transport

Macanese of Chinese and Portuguese descent

MACAP Major Appliance Consumer Action Panel

MACAS Magnetic Capability and Safety System

Macb _Macbeth_

MACBETH Memphians Against Culture Buffs Exposing Themselves Heedlessly

Macc Maccabees

MACC Mexican-American Cultural Center; Military Aid to the Civilian Community

MACCS Manufacturing Cost-Collection System; Manufacturing and Cost-Control System; Marine Air Command and Control System

m. accur. _misce accuratissme_ (Latin—mix very accurately)

MACD Member of the Australian College of Dentistry

MACDC Military Assistance Command Director of Construction

Macdonnells short form for the Macdonnell Ranges of Australia's Northern Territory

MACE Machine-Aided Composition and Editing; Massachusetts Advisory Council on Education; Military Aircraft Capability Estimator(s); Minnesota Association for Childhood Education; Missile and Control Equipment (North American Aviation); trade name for tear gas used by policemen and postmen

M.A.C.E. Master of Air-Conditioning Education; Master of Air-Conditioning Engineering

Maced Macedonia; Macedonian

MACG Marine Air Control Group

Macgillicuddy's Macgillicuddy's Reeks (Ireland's highest mountain range)

mach machine; machinery; machinist

Mach _The Tragedy of Macbeth_

machdiprt machine discharge print(ed)

machflshd machine fleshed (skins)

Machinists Union International Association of Machinists and Aerospace Workers

MACHO Memphians Against Chest Hair at Operas

machovprt machine overprinted

machprt machine printed

machsltd machine salted (skins)

Machu Machu Picchu (ancient Incan sanctuary and stronghold in the high Andes near Cuzco, Peru)

machwsh machine washed (skins)

maci military adaptation of commercial items

MACJC Minnesota Association of Community and Junior Colleges; Missouri Association of Community and Junior Colleges

mack mackinaw; mackintosh; maststack (marine superstructure containing mast and smokestack)

Mackenzies Mackenzie Mountains of the Canadian Northwest

mack(es) mackintosh(es)

Mackinac or Mackinaw formerly Michilimackinac

Macmillan Macmillan Publishing Co

MACOM Major Army Command; Mayor's Committee of

Welcome

MACR Missing Air Crew Report

macrf macroformat(ion)

macro (Greek _makro_—great or large)—macrocyte, macromolecule, macromutation, macroscopic, macrophage

macrobio macrobiologic(al); macrobiology; macrobiotic(s)

macrobop macrobopper (underground slang—older teenager in sympathy with the modern scene)

macrocephs macrocephalics (large-headed people)

macroeco macroeconomics

macrol macrologic(al)(ly); macrologist(s); macrology

macros macroinstructions

MACRS Modified Accelerated Cost-Recovery System

MACS Marine Air Control Squadron; Military Airlift Command Service

macship merchant aircraft ship (merchant vessel fitted with a flight deck)

MACSS Medium-Altitude Communication Satellite System; Montana Association of County School Superintendents

Mactan air field near Lapu-Lapu, Cebú Island, Philippines

MACTU Mines and Countermeasures Tactical Unit (USN)

MAC/V Military Assistance Command, Vietnam

Macy's RH Macy's

mad. magnetic airborne detector; magnetic anomaly detector; maintenance, assembly, and disassembly; mathematical analysis of downtown (computer); mean absolute deviation; midpoint air dose; mind-altering drug

mad. (MAD) music and dance (festival); mutual(ly) assured destruction (via nuclear warfare)

Mad Madam(e); Madeira; Madison; Madras; Madrid

M. Ad. Master of Administration

MAD Madrid, Spain (airport); Maintainability Analysis Data; Manufacturing Assembly Drawing; Marine Air Detachment; Marine Aviation Detachment; Memphians Against Degeneracy; Michigan algorithmetic decoder;

Mine Assembly Depot; Mississippians Against Disposal (of nuclear wastes); Mongolian Asiatic Development (plan)

MAD Militarischer Abschirmdienst (German—Military Screening Service)—German counterintelligence corps

MADA Muda Agricultural Development Authority

MADAEC Military Application Division of the Atomic Energy Commission

Madag Madagascar

Madagascar Democratic Republic of Madagascar (Indian Ocean island country long a French colony)

MADAM Manchester Automatic Digital Machine

madar malfunction analysis detection and recording

MADARS Maintenance Analysis, Detection, and Reporting System

Mad Av advertising and communications enterprises (located on Madison Avenue, New York City)

MADD Manufactured Artificial Dog Dung; Mothers Against Drunk Drivers; Mothers Against Drunk Driving

maddam macromodule and digital differential analyzer; multiplexed analog-to-digital digital-to-analog multiplexed

MADDER Memphians Against Damsels Doing Ecdysiast Routines

maddida magnetic-drum digital-differential analysis

MADE Multichannel Analog-Digital Data Encoder

MADECO Manufacturas de Cobre (Spanish—Copper Manufacturers)

Madeiras Madeira Islands in the North Atlantic off Morocco

madevac medical evacuation

madex magnetic anomaly detection exercise

madge microwave aircraft digital guidance equipment

Mad I Madeira Islands

MADIAR Societé Nationale Malgache des Transports Aériens (French—Madagascar Air Transport)

MADIS Manual Aircraft Data Input System; Millivolt Analog-Digital Instrumentation System

Mad Isl Madeira Islands

Madison James Madison, author of the *Bill of Rights* and fourth President of the United States; capital of Wisconsin

madm medium atomic demolition munition

M. Admin.S. Master of Administrative Studies

MAD Policy Mutually-Assured Destruction Policy (nuclear warfare)

madr minimum adult daily requirement

Madr Madrid; Madrileño

madre magnetic-drum receiving equipment

madrec malfunction detection and recorder

mads mind-altering drugs

MADs Mothers Against Drugs

MADS Maintainability Analysis Data System; Modular Army Demonstration System

madt microalloy diffused-base transistor

mae mean absolute error; motion aftereffect

Mae Fannie Mae, Federal National Mortgage Association; Mary

Ma.E. Master of Engineering

MAE Medical Air Evacuation; Museum of Atomic Energy

M.A.E. Master of Aeronautical Engineering; Master of Art Education; Master of Arts in Education; Master of Arts in Elocution

M.A.Econ. Master of Arts in Economics; Master of Arts in Economic and Social Studies

MAECON Mid-America Electronics Convention

M.A.Ed. Master of Arts in Education

MAEE Marine Aircraft Experimental Establishment

MAEF Master Asphalt Employer's Federation

MAELU Mutual Atomic Energy Liability Underwriters

M.Aero.E. Master of Aeronautical Engineering

MAES Mexican American Engineering Society

maesto maestoso (Italian—majestically)

ma ewes mixed-age ewes

maf macrophage activation factor; major academic field; manpower authorization file; million-acre feet; minimum audible field; multiplanar angular forces

MAF MacArthur Foundation; Marine Air Facility; Middle Atlantic Fisheries; Midland, Texas (airport); Minister of Armed Forces; Ministry of Agriculture and Fisheries; Ministry of Agriculture and Forestry; Missile Assembly Facility; Mission Aviation Fellowship; Mobile Air Force; Mutual Adjustment Fund(ing); Mutual Asset Fund(ing)

MA & F Ministry of Agriculture and Fisheries

MAFA Manchester Academy of Fine Arts; Museum of American Folk Art

MAFAC Marine Fisheries Advisory Council

MAFB Mitchell Air Force Base

MAFC Major Army Field Command

MAFCA Model-A Ford Club of America

MAFDAL Miflaga Datit Le'umit (Hebrew—National Religious Party)

mafe magnesium + iron (Ma + Fe)

MAFF Minister of Agriculture, Fisheries and Food; Ministry of Agriculture, Forestry, and Fisheries

MAFFS Modular Airborne Fire Fighting System (USAF)

MAFI Medic-Alert Foundation International; Ministry of Agriculture, Forestry, and Irrigation

MAFIA Morte Alla Francia Italia Anela (Italian—Death to France Is Italy's Cry), acronym devised when the secret society was organized in the 1860s, to combat French forces of intervention

mafr merged accountability and fund reporting

MAFS Mobilization Air Force Specialty

MAFVA Miniature Armored Fighting Vehicles Association

maf/yr million-acre feet per year

mag magazine; magnesia; magnesium; magnet; magnetic; magnetism; magneto; magnetron; magnum

mag. magnus (Latin—great)

Mag Magallanes (Punta Arenas); Magallanic; Magyar; Margaret

Mag. Magnificat (Latin—it

magnifies)—song of the Virgin Mary
MAG Magnavox (stock exchange symbol); magnesium (machine shop style); Marine Aircraft Group; Marine Aviation Group; Military Advisory Group
maga magazine
Mag.Agg. Magister Aggregatus (Latin—Master of Aggregation)—Head Master
mag ampl magnetic amplifier
MAGB Microfilm Association of Great Britain; Mining Association of Great Britain
Mag Bay Magdalena Bay, Baja California
mag cap magazine capacity
mag card magnetic card
magcheck magneto check
mag ci magnetic cast iron
magcon magnetic concentration
mag cs magnetic cast steel
Magd Magdalene (pronounced *Modlin*) College (Cambridge or Oxford)
Magda Magdalen(a)
Magdalens Magdalen Islands in the Gulf of Saint Lawrence
Mag Dav Magen David (Hebrew—Shield of David; Star of David)—six-pointed star, symbol of Judaism
Magd Coll Magdalen College—Oxford
M.Ag.Ec. Master of Agricultural Economics
M.Ag.Ed. Master of Agricultural Education
MAGERT Map and Geography Round Table
magg maggio (Italian—May); *maggiore* (Italian—major)
magic. modern analytical generator of improved circuitry; modern analytical generator of improved circuits
MAGIC Madison Avenue General Ideas Committee; Men's Apparel Guild In California; Midac Automatic General Integrated Computation; Mothers Against Gangs In our Communities; Mozambique, Angola, and Guinea Information Center
magid magnetic intrusion detector
maglev magentically levitated (linear motor-propelled railroad trains); magnetic levitation
maglevs magnetically-levitated

superfast trains; magnetically levitated vehicles
magloc magnetic logic computer
mag mod magnetic modulator
magn magnetism
magn. magnus (Latin—large)
magna material and geometrically nonlinear analysis
magnalium magnesium + aluminum (alloy)
magneform magnetic forming (process)
Mag Nép Magyar Népköztárasság (Hungarian—People's Republic)
magnesia magnesium oxide (MgO)
magnet. magnetics
MAG^{ni} Magazini (Italian—warehouse)
magno manganese-nickel alloy
magnox magnesium oxide
magnum high-powered cartridge or weapon for firing magnum ammunition; $2/5$-gallon champagne bottle
mag.op. magnum opus (Latin—major work)
M.Agr. Master of Agriculture
mags magazines; magnesium wheels
MAGSAT Magnetic Field Satellite
mag tape magnetic tape
magtig allemagtig (Afrikaans—almighty)—Almighty God
mah mahogany
µAh microampere hour
MAHA Malaysian Agri-Horticultural Association
MAHE Michigan Association for Higher Education
mahog mahogany
mai machine-aided index(ing); marriage adjustment inventory; mean annual increment; minimum annual income
MAI Member Appraisal Institute; Military Assistance Initiative (Philippines); Military Assistance Institute; Multilateral Assistance Initiative; Museum of the American Indian
MAI Moskovskiy Aviatsionny Institut (Russian—Moscow Aviation Institute)
MA.I. Magister in Arte Ingeniaria (Latin—Master of Engineering)
MAIBL Midland and International Banks Limited
maid. maintenance automatic integration detector

M-Aid Marshall-Plan Aid (given European countries by the United States after World War II)
MAIG Matsushita Atomic Industrial Group
Maimon Maimonides
MAIN Medical Automation Intelligence System
Mainbocher Main Rousseau Bocher
Maine St Mus Maine State Museum
Maine Turn Maine Turnpike
Maine Yankee Maine Yankee Atomic Power Company
Mainichi Mainichi Shimbun (Japanese—Everyday Newspaper)—modern Japan's oldest periodical
maint maintenance
maintnce maintenance
maip memory access and interrupt processor
MAIS Maine Association of Independent Schools; Maintenance Information System; Minnesota Adaptive Instructional System
MAIT Maintenance Assistance and Instruction Team; Multidiscipline Accident Investigation Team
maitre d' maitre d'hotel (French—head waiter)
Maître d'Hôtel of Philosophy Baron Paul-Henri-Dietrich d'Holbach (1723—1789)
maj major; majority
Maj Major
MAJ Majuro (Marshall Islands airport); Muhammad Ali Jinnah
majac maintenance antijam console
Maj Com Major Command
maj dem(s) major demon(s)—Asmodeus (lechery), Beelzebub (gluttony), Belphegor (sloth), Leviathan (envy), Lucifer (pride), Mammon (avarice), Satan (anger)
MAJECA Malaysia-Japan Economic Association
Maj Gen Major General
Mak Makdougall; Makoto; Maksim; Maksimovich
makbsctr make best counteroffer
Mák-i-no Mackinac (island, river, or strait as pronounced locally)
MAKN Mongol Ardyn Khuv's-galt Nam (Kalkha Mongol—Mongolian People's Revolutionary Party)

maksutsub make suitable substitutions
mal (Latin prefix—abnormal, bad, disorder)—malignant; *malayisch* (German—Malayan)
Mal The Book of Malachi; Malaga; Malagueña(o); Malay; Malayan; Malaysia; Malta; Maltese
Mal Malay (the basic language of the Indo-Malayan islands and nations including Indonesia); *Maréchal* (French—Marshal)
MAL Malaysian Airways Limited; Material Allowance List
Mala Malaya; Malayan; Malaysia; Malaysian
malac malacology
malachite hydrated copper carbonate
malaprop *mal à propos* (French—out of place, unappropriate)
Malaspina Malaspina Glacier on Yukutat Bay, Alaska
Malawi Republic of Malawi (formerly Nyasaland, East African nation)
Malaya Malay Peninsula also called Malaysia
Malaysia Malay Peninsula (countries formerly comprising British Malaya plus Saba and Sarawak but minus Singapore)
Malaysian Federation Johore, Kedah, Kelantan, Malacca, Negri Sambilan, Pahang, Penang, Perak, Perlis, Sabah, Sarawak, Selangor, Trengganu; up to 1965 included Singapore
Mald Maldive Islands
Mal$ Malaysian dollar
Mal $ Malaya dollar
M.A.L.D. Master of Arts in Law and Diplomacy
MALDEF Mexican-American Legal Defense and Educational Fund
Maldives Republic of Maldives (Indian Ocean island nation)
MALEV (Hungarian Airline)
MALEV Magyar Legikolekedesi Vallat (Hungarian Airlines)
malfunc malfunction
Malg Rep Malagasy Republic
Mali Republic of Mali (landlocked West African country), *République du Mali*
MALI Air Mali
malig malignant

Malindo Malaysia-Indonesia
Mal Isl Maldive Islands
mall malleable
Mall Mallorca
Mallows (British slang shortcut—St Malo)
MALODES Modern Army Logistics Data Exchange System
malor mortar-and-artillery-locating radar
maloti monetary unit of Lesotho
malpais (Spanish—badlands)—basaltic-lava wastelands
Malpartidas (Spanish—Badly Divided Lands)—short form for Malpartida de Cáceres, Malpartida de la Serena, and Malpartida de Plasencia—all in western Spain near the Portuguese borderlands
Malpaso Alto de Mal Paso (highest point on Hierro in the Canary Islands)
Mal-Port Malay-Portuguese (East African patois)
malprac(s) malpractice(s); malpractitioner(s)
MALRA Malaysian Leprosy Relief Association
MALS Medium-intensity Approach Light(ing) System
M.A.L.S. Master of Arts in Liberal Studies; Master of Arts in Library Science; Master of Arts in Library Service
Mal Sam Sis Malotuto'atasi o Samoa i Sisifo (Samoan—Independent State of Western Samoa)
MALSCE Massachusetts Association of Land Surveyors and Civil Engineers
Mal St Malay States
malt. malted milkshake
Malt(s) Maltese sailor(s)
Malukus Maluku or Moluccas Islands of Indonesia also called the Spice Islands
Malvinas Malvinas Islands (Falklands)
mam medium automotive maintenance; milliampere; minute(s)
ma'm madam
m + am (compound) myopic astigmatism
mam mot a mot (French—word for word)
MAM Military Assistance Manual; Missile Acceptance Meeting; Montclair Art Museum

MAMA Mobile Air Materiel Area; Middletown Air Materiel Area
Mam Arab Saud al-Mamlaka al-'Arabiya as Sa'udiya (Arabic—Kingdom of Saudi Arabia)
MAMB Military Advisory Mission—Brazil
MAMBO Mediterranean Association of Marine Biology and Oceanography
MAMC Madigan Army Medical Center
MAME Michigan Association for Media in Education
MAMENIC Marina Mercante Nicaraguense (Nicaraguan Merchant Marine—Mamenic Line)
mami machine-aided manufacturing information
mamie minimum automatic machine for interpolation and extrapolation
Mamie Margaret
Mam Mag al-Mamlaka al-Maghrebia (Arabic—Kingdom of Morocco)
mammal. mammalogist; mammalogy
mammax machine-made and machine-aided index
mammog mammogram; mammograph; mammographer; mammographic(al)(ly); mammography
Mammoth any mammoth caves in Australia, California, and Kentucky; Mammoth Hot Springs in Wyoming; Mammoth Lakes in California; Mammoth Onyx Cave in Kentucky; Mammoth Spring in Arkansas; Mammoth Village in Arizona
Mammoth Cave national park in Kentucky
mamos marine automatic meteorological observing station
MAMS Missile Assembly and Maintenance Shop
Mam Urd Hash al Mamlaka al Urduniya al Hashemiyah (Arabic—Hashemite Kingdom of Jordan)
M.A.Mus. Master of Arts in Music
Mamzel Mademoiselle
man. manhold; manifest; manifold; manual; manufacture; manure
man. *manipulus* (Latin—handful)
m A n *meiner Ansicht nach* (German—in my opinion)

Man management; Manager; Isle of Man in the Irish Sea or the Man canal and river in Burma or La Mancha in Spain, or Manchester, Mangalore, Manhattan, Manila, and Manitoba

MAN Managua, Nicaragua (airport); Motorcyclists Against Noise

M-A-N Maschinefabrik-Augsburg-Nurnberg

MAN Movemento Antillese Nuevo (Papiamento—New Antillean Movement)—based in Curaçao

Man¹ Manuel (Spanish—Emanuel)

MANA Manufacturers' Agents National Association; Mexican-American National Women's Association

M.Anaes. Master of Anaesthesiology

Man Brdg Manhattan Bridge (New York City)

manc mancando (Italian—gradually softer)

Manc Machester; Mancunian—inhabitant of Manchester

Manch Manchuria

Manch Guard Manchester Guardian (newspaper)

mand mandamus; mandate; mandatory; mandible; mandibular

mand mandamus (Latin—we command) writ issued by a superior court commanding the performance of a specified official act or duty

Mand Mandarin

MANDFHAB Male and Female Homosexual Association of Great Britain

Man Dir Managing Director; Managing Directress

mando mancando (Italian—gradually softer)

mandy man day

Man Ed Managing Editor

manf manifold; manufacture; manufacturer; manufacturing

MANFED Manufacturers Federation

MANFEP Manitoba Finite Element Program

MANFORCE Manpower for a Clean Environment

mang management

manganim manganese-copper-nickel alloy

mang b manganese bronze

manglish mangled English

manhr manhour

MANI Minister of Agriculture for Northern Ireland

maniac. (MANIAC) mechanical and numerical integrator and computer

manif manifest

MANIFILE Manitoba File (of worldwide nonferrous metallic deposits)

Manil Marcus Manilius (Roman poet)

manip. manipulus (Latin—handful)

manit man minute

Manit Manitoba

Manitoulins Manitoulin Islands in Lake Huron

Manley Norman Manley International Airport serving Jamaica and named for its first native-born chief minister—an Irish-Negro lawyer

manmam manufacturing management

Man Med Dept Manual of the Medical Department (USN)

manmo man month

Manny Emanuel; Manuel

mano monograph; manometer

MANO Mexican-American Neighborhood Association

MANO Movimiento Argentina Nacional Organizado (Spanish—National Organized Movement of Argentina)—terrorist group

Mano Blanca (Spanish—White Hand)—Cuban exile groups

man. one first-degree manslaughter

manop manually operated; manual operation

Man Op Manitoba Opera Association; Manual of Operation(s)

Manor The Manor (British underworld slang for London)

manova multivariate analysis of variance

man. p. mane primo (Latin—early in the morning, first thing in the morning)

MANP Masai Amboseli National Park (Kenya); Mount Apo NP (Mindanao, Philippines); Mount Arayat NP (Luzon, Philippines)

Man Ray Emmanuel Radnitsky

Man Rtg Manual Rating (code)

mans. mansions

Mans Mansfield College, Oxford; Mansion

MANS Map Analysis System

mansat manned satellite

mansec man second

Mansf Coll Mansfield College-Oxford

man(s) rep(s) manufacturer(s) representative(s)

Man Sym Manila Symphony

mant (MANT) mantissa (calculator)

MANTECH Manufacturing Technology (USN)

MANTIS Manchester Technical Information Service

MANTRAP Machine and Network Transients Program

manuf manufacture(r); manufacturing

Manutius Aldus Manutius—Latinized version of Aldo Manuzio—inventor of italic type

manuv maneuvering

MANWEB Merseyside and North Wales Electricity Board

manwich man-sized sandwich

manwk man week

manx manx cat (almost tailless breed of cat originating on the Isle of Man); manx shearwater (small black-and-white oceanic bird of the eastern North Atlantic); month after next

manyr man year

MANZ Medical Association of New Zealand; Montreal-Australia New Zealand (Line); Motel Association of New Zealand

mao (MAO) monoamine oxidase

mao med andra ord (Swedish—in other words); *med andre ord* (Dano-Norwegian—in other words)

Mao Mao Zedong

MAO Master of the Art of Obstetrics; Musica Aeterna Orchestra

MAO Magyar Allami Operhaz (Hungarian State Opera)

MAOF Mexican-American Opportunity Foundation

maoi (MAOI) monoamine oxidase inhibitor

Maoriland New Zealand

MAOs monoamine oxidase inhibitors

maot medium-aperture optical telescope

MAOT Member of the Association of Occupational Therapists; Military Assistance Observer Team

Mao Zedong (Pinyin Chinese—Mao Tse-tung)—Chinese leader 1949–1976

map. manifold absolute pressure; manifold air pressure; mapping; maximum average price; micro-assembly program; minimum association price; minimum audible pressure; missed approach point; missed approach procedure; multiple-aim point

map. (MAP) machine automated protocol (system); mean arterial pressure

MAP *Maghreb-Arabe Presse* (Maghreh Arab Press Agency); Maintenance Analysis Program; Material Analysis Plan(ning); Media Access Project; Medical Aid Post; Medical Assistance Program; Melanesian Alliance Party; Microprocessor Application Project; Middle Atmosphere Program; Military Aid Program; Military Assistance Program; Military Association of Podiatrists; Mini-Activity Plan; Ministry of Aircraft Production; Ministry of Aircraft Production; Multiple Application Procedure; Mutual African Press (agency); Mutual Agreement Processing (for parolees); Mutual Assistance Program

M-A-P Modified American Plan (breakfast and dinner included)

MAPA Malayan Agricultural Producers Association; Malaysian Airlines Pilots Association; Mexican-American Political Association

MAPAD Map Address Directory

MAPAG Military Assistance Program Advisory Group

MAPAI *Miflaget Poaley Israel* (Hebrew—Israel Labor Party)

MAPAM *Miflaget HaPaolim HaMe'uchedet* (Hebrew—United Workers Party)

MAPC Minnesota Association of Private Colleges

mapche mobile automatic programmed checkout equipment

MAPCO Mid-America Pipeline Company

mapd maximum allowable percent defective

MAPDA Mid-American Periodical Distributors Association

MAPDFA Media-Advertising Partnership for a Drug-Free America

maped machine-aided program for the preparation of electrical power

MAP-ga Military Assistance Program—grant aid

maph manned ambient-pressure habitat

MAPHILINDO Malaysia, Philippines, Indonesia (proposed unification of these Malayan countries)

MAPI Machinery and Allied Products Institute; Millon Adolescent Personality Inventory; Mitsubishi Atomic Power Industries

mapid machine-aided program for the preparation of instruction(al) data

MAPL Manufacturing Assembly Parts List

Maple Lane Maple Lane School (for juvenile delinquents) in Centralia, Washington

MAPNY Maritime Association of the Port of New York

MAPOM MAP-owned materiel

mapp methylacetylenepropadiene

mapple macro-associative processor-programming language

M.App.Sc. Master of Applied Science

MAPR Manufacturing Aids Program Requirements

mapros maintain production schedule(s)

maps Mothers of AIDS Patients

maps. monopropellant accessory power supply

MAPS Major Assembly Performance System; Management Analysis and Planning System; Middle Atlantic Planetarium Society; Military Products and Systems (RCA); Miniature Air Pilot System; Monetary and Payments System; Multiple Address Processing System; Multiple Aiming Point System; Multivariate Analysis and Prediction of Schedules (USA)

MAPU *Movimiento de Acción Popular Unitaria* (Spanish—Movement for United Popular Action)—Chile

MAPW Medical Association

for the Prevention of War

maq monetary allowance in lieu of quarters

maq *maquereau* (French—mackerel)—a pimp (also abbreviated *mac*)

MAQ Measures for Air Quality (NBS)

MAQC Metropolitan Air Quality Council

maquils *maquiladoras* (Spanish—corn grinders)—manufacturing plants on the Mexican Border

mar. marine; maritime; married; marry; memory address register; minimal angle resolution; minimum acceptable rate (of return); multiarray radar; multifunction array radar

mar *marokkanisch* (German—Moroccan)

mar. *mardi* (French—Tuesday); *martedi* (Italian—Tuesday); *martes* (Spanish—Tuesday)

Mar Marathi; March; Marseilles; Marshall Islands

Mar *Marseillaise* (national anthem of France)

M.Ar. Master of Architecture

MAR Manistee and Repton (railroad); Maracaibo, Venezuela (airport); Maritime Central Airways; Mars Excursion Module; Material Availability Report; Material Availability Request

MARA Mexican-American Research Association

MARA *Majilis Amanah Raayat* (Malay—Handicraft Center for the Development of Malaysian People)

MARAD Maritime Administration (US Department of Commerce)

MARAIRMED Maritime Air Mediterranean

marb marbling

Marb Marblehead; Marbleheart; Marbury

marbi machine-readable bibliographic information

marble calcium carbonate ($CaCO_3$)

marc monitoring and results computer

marc *marcato* (Italian—marked)

Marc Marcus

MARC Machine-Readable Cataloging (Library of Congress magnetic-tape catalog system); Manpower Authori-

zation Request for Change; Matador Automatic Radar Command; Metropolitan Applied Research Center; Micronesia Area Research Center (Guam); Model-A Restorers Club (Model-A Ford autos)
MARCA Mid-Continent Area Reliability Coordination Agreement
MarCad Marine Cadet
Mar Cad Marine Cadet
MARCEP Maintainability and Cost-Effectiveness Program
March. Marchioness
March Marchese (Italian—Marquis)
M.Arch. Master of Architecture
MARCHA Methodists Associated Representing the Cause of Hispanic Americans
Marcha Real (Spanish—Royal March)—anthem of Spain
Marchbanks (British contraction—Marjoribanks)
March^{sa} Marchesa (Italian—Marchioness)
MARC(LC) Machine-Readable Catalog(ing) (Library of Congress)
MARCO Marine Construction and Design Company
MARCOM Maritime Command (Canadian)
MARCONFOR Maritime Contingency Force
MARCONFORLANT Maritime Contingency Forces—Atlantic
MARCOR US Marine Corps
MARCS Marine Computer System
MARC(S) Machine-Readable Catalog(ing) for Serials
MARC(UK) Machine-Readable Catalog(ing) in the British Library of the United Kingdom
Marcus Mark
mardan marine digital analyzer
MARDEC Malaysian Rubber Development Corporation
MARDI Malaysian Agricultural Research and Development
MARECS Marine Communications Satellites
mar eng marine engineer(ing)
MARFOR Marine Forces
marg margarine; margin; marginal; marginalia
Marg Margrave; Margravine
marge margarine (oleomargarine); margin

margen management report generator
Mar Gils Area Marshalls-Gilberts (island) Area
Marg^{ta} Margarita (Spanish—Margaret)
marg trans marginal translation
marhelilex marine helicopter landing exercise
MARI Middle America Research Institute
María Félix Maria de los Angeles Félix Guereña
Marianas Mariana Islands; once called Ladrones (thieves)
Maribo Paramaribo, Surinam
Marichu (Spanish-American nickname—María de Jesús)
maricult mariculture; mariculturist
mariculture marine culture (growing food in the sea)
marifarm maritime farm
marifex marine firine exercise
mariholic marijuanaholic (addict)
Marinsky Marinsky Theater in St Petersburg (Leningrad), renamed Kirov
Mariol Mariolatry; Mariology
Marion Federal Prison Camp at Marion, Illinois; Mary; Maryjane; U.S. Penitentiary at Marion, Illinois
MARIS Maritime Research Information Service
marisat maritime industry satellite
marit maritime
marita maritime airfield
Marit Admin Maritime Administration
Marit Com Maritime Commission
Maritime Alps *Alpes Maritimes* (French)—AM
Maritimes Canada's Maritime Provinces; the Maritime Alps between France and Italy; the Soviet Union's Maritime Territory along the Sea of Japan
maritrain(s) maritime train(s)—articulated sea-going barges
Marj Marja; Marjan; Marjorie; Marjory
mark monetary unit of Germany
mark. market; marketing
Mark The Gospel according to St Mark
Marka Markarian (galaxy thirteen times larger than the

Milky Way)
markkaa monetary unit of Finland
mark twain leadline sounding of two fathoms (12 feet or 3.66 meters); leadsmen announcing *mark twain* meant there was enough water to keep the average shallow-draft paddlewheel river steamer afloat; Sam(uel) L(anghorne) Clemens; *mark three* is three fathoms and *quarter twain* is two-and-a-half fathoms
Marlag Marinenlager (German—sailor's camp for prisoners of war)
Marlene Mary + Helena
marlex marine reserve landing exercise
MARLF Middle Atlantic Regional Library Federation
mar lic marriage license
MARLIS Multi-Aspect Relevance Linkage System
Marm Marmaduke
marmap marine resources monitoring, assessment, and prediction
Marmara Sea of Marmara (connecting the Black Sea with the Mediterranean via the Dardanelles Strait and separating Asiatic Turkey from European Turkey)—called Marmara Denizi by the Turks
mar merc marina mercantile (Italian—merchant marine)
mar mil marina militare (Italian—navy)
MARNR Ministerio del Ambiente y los Recursos Naturales Renovables (Spanish—Ministry of the Environment and Renewable Natural Resources)—Venezuela
Maro Marocco (Italian—Morocco)
MARO Maritime Air Radio Organization
marops maritime operations
marots (MAROTS) marine orbital technical satellite
MARPEX Management of Repair Parts Management (USA)
Marpril March and April
Marq Marquesas Islands
Marquesas Marquesas Islands of the South Pacific or the Marquesas Keys west of Key West, Florida in the Gulf of Mexico
Marquis Marquis Who's Who Books

marr marriage; minimum acceptable rate of return
Marr Marranic; Marranism; Marranoism; Marrano(s)
Marr Marruecos (Spanish—Morocco)
MARRES Manual Radar Reconnaissance Exploitation System
marr lic marriage license
Marro Marrocos (Portuguese—Morocco)
marr sett marriage settlement
Marru Marruecos (Spanish—Morocco)
mars master attitude reference system; mathematics anxiety rating scale; military affiliated radio system
Mars Marseilles
Mars' Marshals' Offices
Mars Ares (Latin—god of war); *Marselha* (Portuguese—Marseilles); *Marsella* (Spanish—Marseilles); *Marsiglia* (Italian—Marseilles)
MARs Middle American Radicals
MARS Magnetic-Electronic Automatic Reservation System; Maintenance Action Reporting System; Manned Astronautical Research Station; Master Altitude Reference System; Military Affiliate Radio System; Miniature Accurate Ranging System; Mobile Atlantic Range Station; Modern Architectural Research Society; Monitored Atherosclerosis Regression Institute
MARSAP Mutual Assistance Rescue and Salvage Plan
MARSAS Marine Search and Attack System (USMC)
marsat maritime satellite
MARSATE Maintenance and Repair of Scientific and Technical Equipment
MARSATS Maritime Satellite System
M.Ar.Sci. Master of Arts and Sciences
mar settl marriage settlement
MARSH Matching Aid to Restore States' Habitats (for waterfowl)
Marshalls Marshall Islands in the western Pacific
marsh gas methane (CH_4)
MA/RSO Mobilization Augmentee/Reserve Supplement Officer (USAF)
mart. maintenance analysis and review technique; mean

active repair time
mart martes (Spanish—Tuesday)
Mart Martinique
Mart. Martyrology
Mart Marcus Valerius Martialis (Roman poet)
MART Metropolitan Area Rapid Transit
MARTA Metropolitan Atlanta Rapid Transit Authority
Marth Martha
Martí Aeropuerto José Martí (Havana, Cuba's airport named for the founder of the Cuban Revolutionary Party who did much to organize resistance to Spanish rule but was killed in 1895, three years before his island was liberated by American and Cuban forces)
Martin St Martin or Sint Maarten in the Leeward Islands
M.Art RCA Master of Art of the Royal College of Art
mart(s) market(s)
MARTS Master Radar Tracking Station
Mart(y) Martin
marv maneuvering reentry vehicle (MaRV in Salt Talk reports, also MARV); marvel; marvelous
Marv Marvin
Marylebone St Marylebone
mas masculine; masonry; mathematics anxiety scale; metal angle slots; military assistance sales; milliampere second; moved aboard ship
mas Malaysian Airline System's trademark
Mas Massachusetts; Massachusettsan
MAs Mothers Anonymous
MAS Malaysian Airline System; Marine Acoustical Services; Maryland Academy of Sciences; Master Activation Schedule; Military Agency for Standardization; Ministry of Aviation Supply; Missile Assembly Site; Monetary Authority of Singapore; Municipal Art Society; Mutual Assured Security
M.A.S. Master of Applied Science
M & AS Music and Art School
MAS Motoscafi Anti Sommergibli (Italian—antisubmarine lotor torpedo boat); *Movimiento al Socialismo* (Span-

ish—Movement toward Socialism)—Venezuelan's leftist party
MASA Mail Advertising Service Association; Malaysian Shipowners Association; Member of the Acoustical Society of America; Michigan Association of School Administrators; Military Automotive Supply Agency; Minnesota Association of School Administrators; Mississippi Association of School Administrators; Montana Association of School Administrators
MASANYC Mail Advertising Service Association of New York City
masar microwave-accurate-surface antenna reflector
MASB Michigan Association of School Boards
MASBO Minnesota Association of School Business Officials
masc masculine
masc. masculus (Latin—male)
M.A. Sc. Master of Applied Science
MASC Massachusetts Association of School Committees
MASCA Middle Atlantic States Correctional Association
Mascarenes Mascarene Islands
mascon massive concentration
MASCOT Meteorological Auxiliary Sea Current Observation Transmitter
MASCS Marriage Adjustment Sentence Completion Survey
MASEA Midwest Association of Student Employment Administrators
maser microwave amplification by stimulated emission of radiation
mash. mashed potatoes
MASH Medical Aid for Sick Hippies; Memphians Against Social Harassment; Mobile Army Surgical Hospital; Multiple Accelerated Summary Hearing (for alien deportation); Mutual Aid Self Help
MASHAE Member of the American Society of Heating and Air Conditioning Engineers
mash(ed) mashed potatoes
MASHVE Member of the Australian Society of Heating and Ventilating Engineers

MASIS Management and Scientific Information Service

mask maskulinum (Dano-Norwegian—masculine)

MASL Military Assistance Articles and Services List

MASME Member of the American Society of Mechanical Engineers; Member of the Australian Society of Mechanical Engineers

MASO Munition-Accountable Supply Office(r)—USAF

M.A. Soc. Stud. Master of Arts in Social Studies

mas. pil. massa piluarum (Latin—pill mass)

mass. masseter; multiple-access sequential selection

Mass Massachusetts; Massachusettsan(s)

MASS Marine Air Support Squadron; Massachusetts Association of School Superintendents; Michigan Automatic Scanning System

M.A.S.S. Master of Arts in Social Science

Massachusetts General Massachusetts General Hospital

Massanutteas Massanuttea Mountains (on the Appalachian Trail in northern Virginia between Charlottesville and Culpepper Court House)

masscult mass culture (culture for the masses)

massdar modular analysis, speedup, sampling, and data reduction

Massey Massey-Harris; Massey University (Palmerston North, New Zealand)

MASSP Michigan Association of Secondary School Principals; Minnesota Association of Secondary School Principals; Missouri Association of Secondary School Principals; Montana Association of Secondary School Principals

MASSPIRG Massachusetts Public Interest Research Group

MASSR Mari Autonomous Soviet Socialist Republic; Mordovian Autonomous Soviet Socialist Republic

Mass Turn Massachusetts Turnpike

mast. masthead; missile automatic supply technique

MAST Metro Arson Strike Team (San Diego); Metropolitan Arson Strike Team; Michigan Alcoholism

Screening Test; Military Assistance to Safety and Traffic

MAST Minimum Abbreviations of Serial Titles

MASTARS Mechanical and Structural Testing and Referral Service (NBS)

Ma State New South Wales, Australia

master. matching available student time to educational resources; multiple-access shared-time executive routine

MASTIF Multiple Axes Space Test Inertia Facility

mastir microfilmed abstract system for technical information referral

MASUA Mid-America State Universities Association

mat. machine-aided translation; material; materiel; maternity; matins; maturity; microalloy transistor; molankothane (molybdenum disulfide urethane)

mat matemática (Spanish—mathematics); *matematik* (Dano-Norwegian—mathematics)

Mat Matadi; Matanzas; Matthew

MAT Manual Arts Therapy; Mechanical Aptitude Test; Metropolitan Achievement Tests; Military Air Transport; Miller Analogies Test

M.A.T. Master of Arts in Teaching

mata multiple-answering teaching aid

MATA Motorcycle and Allied Trades Association; Museums Association of Tropical Africa

Mata Soc Mattachine Society

MA&TB Missile Assembly and Test Building

MATCALS Marine Air Traffic Control and Landing System (USN)

match. medium-range antisubmarine torpedo-carrying helicopter

MATCH Manpower and Talent Clearinghouse

MATCOM Materiel Command (USA)

MATCOMEUR Materiel Command, Europe

MATCOMTELNET MATS Command Teletype Network

matcon microwave aerospace terminal control

Mate the Mate (Chief Officer)

Mat.E. Materials Engineer

Ma Tec Maintenance Technician

MATELO Maritime Air Telecommunications Organization

matern maternal; maternity

Mater Res Bull Materials Research Bulletin(s)

MATFA Meat and Allied Trades Federation of Australia

math mathematical(ly); mathematician; mathematics

Math Mathematics; Matthew; Matthews; Mathewson; Mathias; Mathieu; Mathilde; Mathurin; Mathys; Mattias

Math.D. Doctor of Mathematics

M.A. Theol. Master of Arts in Theology

mathn mathematician

maths mathematicians; mathematics; mathematics majors

math soc science mathematical social science

MATI Moskovskiy Aviatsionnyy Teknologicheskiy Institut (Russian—Moscow Aviation Technology Institute)

MATIC Multiple and Technical Information Center

Mat Lab Material Laboratory

mat.med. materia medica

matnav mathematics for navigators

matp masking template

MATP Military Assitance Training Program

matr. matrimonium (Latin—marriage)

MATRESS Money, Advancement, Training, Recreation, Security, Satisfaction (USAF recruiting acronym)

matric matriculate; matriculation

mats. maintenance analysis test set

MATS Military Air Transport Service

Mat Soc Mattachine Society

Matt Matthew; Matthewtown, Great Inagua; The Gospel according to St Matthew

Matt Matthew

mattergy matter + energy

MATTS Multiple Airborne Target Trajectory System

Mattw Matthew

Matty Matthew

matut. matutinus (Latin—in the morning)

matv master antenna television

MATVS Master Antenna Tele-

vision System
matw metal awning-type window
MATZ Military Air Traffic Zone
mau maintenance analysis unit; marine amphibious unit
Mau Mauritius
Maud Mathilda
Maude Morse automatic decoder
M.Au.E. Master of Automotive Engineering
maulex marine amphibious unit landing exercise
MAUM Movement Against Uranium Mining
Maur Mauritius
M.A. Urb. Plan. Master of Arts in Urban Planning
Maurit Mauritania (Islamic Republic of)
Mauritania Islamic Republic of Mauritania (West African nation)
MAUS Master of Arts in Urban Studies; Metric Association of the United States
mauw maximum all-up weight
mav manpower authorization voucher; maverick
mav (MAV) maleic anhydride value
maverick. manufacturers assistance in verifying identification in cataloging
mavica magnetic video camera
MAVIS Medical Audio-Visual Information Service
MAVOGA Malaysian Vocational Guidance Association
maw medium assault weapon
maw *met andere woorden* (Dutch—in other words)
Maw Mama
MAW Marine Aircraft Wing; Mission Adaptive Wing
mawb master air waybill
mawec maritime exercise weather code
MAWLOGS Models of the Army Worldwide Logistics System (USA)
MAWS Marine Air Warning Squadron
max maximal; maximum
m'ax (American contraction—my ax)
Max Maxene; Maxie; Maxim; Maxime; Maximilian; Maximiliano; Maxine; Maxwell; Maxy; Soviet Yakovlev trainer aircraft designated Yak-18
max cap. maximum capacity
maxi maximum

maxibop maxibopper (underground slang—fatter or older woman wearing miniskirts)
maxid maximize indefinite delivery (contracts)
maxill maxilla; maxillary
maximin maximum + minimum
maxis maximum-length garments (coats, skirts, etc.)
maxnet modular application executive for computer networks
max q maximum aerodynamic pressure per square foot
maxr maximum room rate desired
max trq maximum torque
May Maybelle; NATO nickname for Soviet Ilyushin transport designated Il-38
MAYA Maya Airways (British Honduras); Mexican-American Youth Association
mayday international distress call (from the French *m'aidez*—help me)
May Day May 1 (Morris Dancers in England, international worker's day in communist and socialist lands)
Mayjun May and June
May^mo Mayordomo (Spanish—butler, estate manager, steward)
mayn't may not
mayo mayonnaise
MAYO Mexican American Youth Organization
mayoralection mayoral election
maz mazda
Maz Mazatlan
mazh missile azimuth heading
mb machine blended; macrobiotic (MB); magnetic bearing; main battery; master bundles; may be; medium bomber; megabyte(s)—two million bytes; megabyte; meter band; methyl bromide; methylene blue; midbody; millibar(s); minimum bid; motor barge; motorboat
mb (MB) megabyte; memory buffer
m/b make-break; master batch
m & b matched and beaded; metes and bounds
m.b. *misce bene* (Latin—mix well)
mB male Black
m/B male Black
Mb million bytes; myoglobin
MB magnetic bearing; Manitoba; March-Bender (factor);

Marine Barracks; Marine Base; Maritime Board; Marketing Board; Mechanized Battalion; Medical Board; Meridian & Bigbee (railroad); Miami Beach; middle of bow; municipal borough; Munitions Board; Music for the Blind; Myrtle Beach; Sir Mackenzie Bowell (Canada's sixth Prime Minister)
M-B Mercedes-Benz
M.B. *Medicinae Baccalaureus* (Latin—Bachelor of Medicine)
M/B Master Barber
M & B metes and bounds
Mba Mombasa
MBA Make or Buy Analysis; Make or Buy Authorization; Marine Biological Association; Master Builders Association; Men's Basketball Association; Military Bases Agreement; Military Benefit Association; Monterey Bay Aquarium; Monument Builders of America; Mortgage Bankers of America; Mortgage Bankers Association; Mortgage Brokers Association
M.B.A. Master of Business Administration
MBAA Master Brewers Association of America; Mortgage Bankers Association of America
MBAC Member of the British Association of Chemists
M-Bahn *Magnetbahn* (German—magnetic levitation transit system)—train cars travel on a thin cushion of air providing high-speed transportation quieter than wheel-on-rail systems
MBAL Master Bookbinders' Alliance of London
m bale 1000 bales
mbar millibar
MBAR Multiple Beam Acquisition Radar
MBARI Monterey Bay Aquarium Research Institute
MBAUK Marine Biological Association of the United Kingdom
MBAWS Marine Base Warning System
mbb make before break; mortgage-backed bonds
MBB Messerschmitt-Bülkow-Blom; Museum of the Borough of Brooklyn
m bbl 1000 barrels

mbc maximum breathing capacity; multiple burst correction
MBC Malawi Broadcasting Corporation; Malaysian Building Construction; Mauritius Broadcasting Corporation; Mercantile Bank of Canada; Metropolitan Borough Council; Metropolitan Business College(s); Miname Nihon Broadcasting (Japan); Mitsubishi Bank of California
MBCA Motor Boat Club of America
MBCC Massachusetts Bay Community College; Migratory Bird Conservation Commission
MBCMC Milk Bottle Crate Manufacturers Council
mbd macro-block design; management by decision; million barrels per day; minimal brain dysfunction; minimum brain damage
MBDA Minority Business Development Agency
m bd ft 1000 board feet
mbe minority business enterprise; missile-borne equipment; my own bloody efforts
MBE Mail Boxes Etc.
M.B.E. Member of the Order of the British Empire
M. B. Ed. Master of Business Education
Mbf thousand board feet
MBF Master Builders Federation; Medical Benefits Fund(ing); Military Banking Facility; Milk Bottlers Federation
M-B factor Marsh-Bender factor
MBF et H Magna Britannia, Francia et Hibernia (Latin—Great Britain, France, and Ireland)
MBFR Mutual Balanced-Forced Reduction
MBG Midland Bank Group; Missouri Botanical Garden
mbge missileborne guidance equipment
mbh manual bomb hoist
mbH mit beschränkter Haftung (German—limited liability)
mbi may be issued
mbi (MBI) magnetic resonance infarction
Mbi Mbini (formerly Rio Muni)
MBI Molecular Biosystems Inc

MBIA Malting Barley Improvement Association; Municipal Bond Insurance Association
M. Bi. Chem. Master of Biological Chemistry
M. Bi. Eng. Master of Biological Engineering
MBII Minority Business Information Institute
M. Bi. Phy. Master of Biological Physics
M. Bi. S. Master of Biological Science
MBJ Montego Bay, Jamaica (airport)
mbk missing, believed killed
mbl missile baseline; mobile; mobile branch library; model breakdown list(ing); model breastline
m & bl meat and bonemeal
Mbl Monatsblatt (German—monthly report)
MBL Marine Biological Laboratory (Woods Hole, Massachusetts); Mobile, Alabama (airport)
MBLIC Mutual Benefit Life Insurance Company
mbm thousand feet board measure
mbm (MBM) magnetic bubble memory
MBM Mac Bride Museum
M.B.M. Master of Business Management
MBMA Master Boiler Makers' Association; Metal Building Manufacturers Association
MBMHC Malcolm Bliss Mental Health Center (St Louis)
MBMS Magnetic Bubble Memory System
MBNA Monument Builders of North America
MBNBR Mount Bruce Native Bird Reserve (North Island, New Zealand)
mbo management by objectives
Mbo Maracaibo (inhabitants called Maracaiberos or Maracuchos)
MBO Mutual Benefit Organization
MbO₂ oxymyoglobin
MBOC Minority Business Opportunity Committee(s)
MBOU Member British Ornithologists Union
mbp mean blood pressure
MBP Marine Biotelemetry Project; Minnesota Business Partnership
MBPA Metropolitan Bicycle

Polo Association; Military Blood Program Agency
mbp antigen melitensis bovin porcine antigen
mbpfo make best possible firm offer
MBPO Military Blood Program Office(r)
MB & PR MacMillan, Bloedel & Powell River
mbps megabits per second; million bits per second
MBPXL Missouri Beef Packers Express Line
mbr member; memory buffer register
MBR Minerações Brasileiras Reunidas (Brazil—Brazilian Mining Reunited)
MBRBA Motor Body Repairers and Builders Association
mbr/e memory buffer register—even
M Bret Middle Breton
MBRF Mission Bay Research Foundation
mbrl million barrels
mbr/o memory buffer register—odd
Mbro Middlesbrough
mbrt methylene-blue reduction time
mbruu may be retained until unserviceable
mbrv maneuverable ballistic reentry vehicle (MBRV)
mbs magnetron beam switching; main bang suppressor; megabits per second
mb's milk brothers (two or more males who have had sexual relations with the same female)—*see* ms's
MBS Macquarie Broadcasting Service; Mainichi Broadcasting System; Miami Beach Symphony; Motor Bus Society; Music Broadcasting Society; Mutual Broadcasting System
MBSA Modular Building Standards Association; Munitions Board Standards Agency
M. B. Sc. Master of Business Science
mbsd multi-barrel smoke discharger
mbsi missile battery status indicator
MBSI Musical Box Society International
MBSJC Metropolitan Boroughs Standing Joint Committee (of librarians)
MBSM Mexican Border Service Medal

MBSSM Maxfield-Buchholz Scale of Social Maturity
mbt main ballast tank; mean body temperature; mechanical bathythermograph; metalbase transistor; murder before treason
MBT Minimum Blood Test; Modified Boiling Test
MBT-70 Main Battle Tank
MBTA Massachusetts Bay Transportation Authority; Metropolitan Boston Transit Authority; Midwest Book Travelers Association; Migratory Bird Treaty Act
MBTI Manpower Business Training Institute; Myers-Briggs Type Indicator
MBTs main battle tanks; Municipal Bond Trusts
MBTS Meteorological Balloon Tracking System
MBtu thousand British thermal units
mbu mobile tracking unit; thousand bushels
MBUCV *Museo de Biología de la Universidad Central de Venezuela* (Spanish—Biology Museum of the Central University of Venezuela)
M Build Master of Building
M. Bus. Ed. Master of Business Education
MBV Mexican Border Veterans
MBW Metropolitan Board of Works
MBYC Manhasset Bay Yacht Club; Mission Bay Yacht Club
mbz must be zero
mc magnetic center (MC); magnetic course (MC); main color; marginal check megacycle(s); marked capacity; marker card; market(ing) capacity; married couple; material control; message composer; metal case; meter candle; metric carat; miles on course; military characteristic(s); millicurie(s); mission control; moisture content; momentary contact; monkey cells; motorcycle; motor-(ized) contact; moving coil(s); multiple contact
mc (MC) Maltese cross; marginal cost
m-c medico-chirugical (surgical); mineralo-corticoid (hormones)
m/c machine(ry); marginal credit; metalling clause; mid-

dle center
m&c manufacturers and contractors; morphine and cocaine
mc *mois courant* (French—current month)
m/c *mi cargo* (Spanish—my debt, my responsibility); *mi casa* (Spanish—my home, my house); *mi cuenta* (Spanish—my account)
μC microcoulomb
Mc Mac (Gaelic—son of)
M/c metallic currency
MC Macalester College; Machinery Certificate; Madison College; Madonna College; Magistrates Court; magnetic course; Mailet College; Maine Central (railroad); Malin College; Malone College; Manatee College; Manchester College; Manhattan College; Manhattanville College; Manpower Commission; Maria College; Marian College; Marietta College; Marine Corps; Marion College; Marist College; Maritime Carrier(s); Maritime Commission; Marlboro College; Marriage Certificate; Martin College; Mary College; Marycrest College; Maryglade College; Marygrove College; Marylhurst College; Marymount College; Maryville College; Marywood College; Master of Ceremonies; Master Commandant; Material Center; Materiel Center; Materiel Command; Maunaolu College; Medical Center; Medical Certificate; Medical College; Medical Corporation; Medical Corps; Member of Congress; Member of Council; Memorial Commission; Memphis College; Menlo College; Mesa College; Mess Committee; Michigan Central (railroad); Microfilm Corporation; Microstat Corporation; Middlebury College; Midland College; Miles College; Military Committee; Military Cross; Milligan College; Mills College; Milsaps College; Milton College; Misericordia College; Mitchell College; Mitsubishi Corporation; Monmouth College; Monticello College; Moravian College; Morehouse College; Morris Col-

lege; Morse Code; Morse College; Muhlenberg College; Multnomah College; Mundelein College; Munitions Command; Muskingum College; Muskogee College
M-C Magovern-Cromie (prosthesis)
M.C. Military Cross
M/C Machinery Certificate
MC *Maison Centrale* (French—Central Prison); *Mercado Común* (Spanish—Common Market)
M.C. *Magister Chirurgiae* (Master of Surgery)
mca minimum control airspeed; minimum crossing altitude; monetary compensatory account(ing); money compensatory account(ing); mud cleaning agent
mca (MCA) maximum credible accident
Mca Macassar
MCA Malacca Consumers Association; Malayan Chinese Association; Malaysian Chinese Association; Manufacturing Chemists Association; Maritime Central Airways; Maritime Control Area; Massachusetts Correctional Association; Material Control Area; Material Coordinating Agency; Maternity Center Association; Mechanical Contractors Association; Mechanization Control Area; Media Credit Association; Medical Correctional Association; Medical Council of Australia; Millinery Credit Association; Minnesota Correctional Authority; Minnesota Corrections Association; Missouri Corrections Association; Movers Conferences of America; Multiple Classification Analysis; Muscat Control Agency; Music Corporation of America; Music Critics Association; Musicians Club of America
MCA *Metric Conversion Act*
MCAA Mason Contractors Association of America; Mechanical Contractors Association of America; Medical Consumers Association of Australia; Military Civil Affairs Administration
MCAAA Midland Counties Amateur Athletic Association
MCAB Marine Corps Air Base
MCAD Military Contracts Ad-

ministration Department

MCADO Micronesian Community Action Development Organization

MCAF Marine Corps Air Facility; Marine Corps Air Field; Military Construction, Air Force

MCAIR McDonnell Aircraft Company

m car 1000 carats

MCAR Material Corrective Action Report(s)

mcarquals marine-carrier qualifications

mca's money compensatory amounts

MCAS Marine Corps Air Station

MCAT Medical College Admission Test; Midwest Council on Airborne Television

m. cau. misce caute (Latin—mix cautiously)

MCAUSA Military Chaplain's Association of the U.S.A.

MCAUTO McDonnell-Douglas Automation

mcb membranes cytoplasmic bodies; miniature circuit breaker

McB McBurney's (point)

MCB Marine Corps Base; Metric Conversion Board; Metric Conversion Bureau; Mobile Construction Battalion

M.C.B. Master of Clinical Biochemistry

MCBA Master Car Builders' Association

mcc maintenance of close contact; midcourse correction; modified close control; multilayer ceramic chip

mcc (MCC) main communication(s) center; multi-component circuit(s)

MCC Maintenance Control Center; Manual Combat Center; Marine Corps Commandant; Marine Corps Commander; Marylebone Cricket Club; Massachusetts Council of Churches; Materials Characterization Center (Battelle, Richland); Mesta Machine Company (stock exchange symbol); Meteorological Communications Center; Metropolitan Community Church; Metropolitan Correctional Center; Microelectronics and Computer Technology Corporation; Microfilm Card Catalog; Missile Con-

trol Center; Mission Control Center; Monroe Community College; Mortgage Credit Certificate; Munitions Carriers Conference; Music Critics Circle

MCCA Michigan Community College Association; Minor Counties Cricket Association; Mortgage Capital Company of America

MCCA Mercado Común Centro Americano (Spanish—Central American Common Market)

MCCB Multinational Configuration Control Board

MCCC Metropolitan Correctional Center (Chicago); Muskegon County Community College

MCCCA Marine Corps Combat Correspondents Association

MCC-H Mission Control Center—Houston (NASA)

MCCISWG Military Command, Control, and Information Systems Working Group

McCL McCabe Library (Swarthmore)

mccp maintenance console control panel

mccs missile critical circuit simulator

MCCs Metropolitan Correctional Centers (of the Bureau of Prisons in Chicago, New York, and San Diego); Military Committee in Chiefs-of-Staff Session(s)

MCCS Modular Communications Control Systems

mccu multiple communications control unit

MCCW Miami Citizens Crime Watch; Military Council of Catholic Women

mcd magnetic crack detector; mean corpuscular diameter; median control death; metal-covered door; mine clearance dive; mine clearance diving; minimum cash deposit; miscellaneous cash deposit

mcd minimo comune denomiatore (Italian—least common denominator)

Mc D Mc Donald; Mc Donald's

MCD Melbourne College of Divinity

M.C.D. Doctor of Comparative Medicine; Master of Civic Design

McDA McDonnell Aircraft

MCDA Manpower and Career

Development Agency; Motor Car Dealers Association

McDAC McDonnell Aircraft Corporation

MCDC Montgomery County Detention Center

McD Obs McDonald Observatory

MCDS Management Control Data System

mcd/slv minimum-cost-design/space launch vehicle (MCD/SLV)

mcdt mean corrective down time

mce mean chance expectation; military characteristics equipment

MCE Memphis Cotton Exchange; Montgomery Cotton Exchange

M.C.E. Malaysian Certificate of Education; Master of Civil Engineering

MCE Mercado Común Europeo (Spanish—European Common Market); *Mercato Comune Europeo* (Italian—European Common Market)

MCEB Military Communications Electronics Board

M.C.Eng Master of Civil Engineering

MCEO Malayan Council of Employers Organization; Malaysian Council of Employers Organization

M. Cer. E. Master of Ceramic Engineering

MCET Mississippi Center for Educational Television

MCEWG Multinational Communication-Electronics Working Group (NATO)

mcf magnetic card file; medium corpuscular fragility; thousand cubic feet

MCF Master Code File; Michigan Colleges Foundation; Missouri Colleges Fund

mcfd 1000 cubic feet of gas per day

mcfh 1000 cubic feet of gas per hour

MCFI Malaysian Chamber of Film Industries

mcfim microfilm(ing)

mcflm microfilm; microfilming

mcfm 1000 cubic feet of gas per month

MCFP Medical Center for Federal Prisoners (Springfield, Missouri); Member of the College of Family Physicians

mcfshe microfische

mcg microgram

mc & g mapping, charting, and geodesy

MCG Mandalay Coral Gardens (Queensland)

McG-H McGraw-Hill

McGill-Queens U Pr McGill-Queens University Press

McGraw McGraw-Hill

MCGS Microwave Command Guidance System

McG U McGill University

McGUL McGill University Library

mch mail chute; mean corpuscular hemoglobin (MCH)

Mch Manchester; March

M. Ch. *Magister Chirurgiae* (Latin—Master of Surgery)

MCH Maternal and Child Health

mcha merchandise

mchan multichannel

mchc mean corpuscular hemoglobin concentration

M.Ch.D. *Magister Chirugiae Dentalis* (Latin—Master of Dental Surgery)

M.Ch.E. Master of Chemical Engineering

M. Chem. E. Master of Chemical Engineering

M.Chir. *Magister Chirugiae* (Latin—Master of Surgery)

M.Ch. Orth. *Magister Chirurgiae Orthopaedicae* (Latin—Master of Orthopedic Surgery)

M.Ch.Otol. Master of Otorhinolaryngological Surgery

MCHP Maternal and Child Health Program

mc hr millicurie hour(s)

MCHR Medical Committee for Human Rights

MCHRD Mayor's Committee for Human Resources Development

M.Chrom. Master of Chromatics

M Ch S Member of the Society of Chiropodists

MCHS Maternal and Child Health Service

mcht merchant

Mchter Manchester

M. Chur. *Magister Chirugiae* (Latin—Master of Surgery)

mchy machinery

mci malleable cast iron; megacurie; mottled cast iron; multichip integration

mCi millicurie(s)

MCI Kansas City Airport (symbol); Marine Corps Institute; Maximum Credible Incident; Massachusetts Correctional Institution (Framingham); Mexican Coffee Institute; Microwave Communications, Inc; Milk Can Institute; Motor Coach Industries; Motor Coach Institute

MCIC Metals and Ceramics Information Center (DoD)

mcid multipurpose concealed intrusion detection (device)

MCIE Midland Counties Institution of Engineers

McINP McIwaine National Park (Rhodesia)

MCIS Management Control and Information System; Multi-Currency Intervention System

MCIT Member of the Chartered Institute of Transport

M.C.J. Master of Comparative Jurisprudence

MCJC Mason City Junior College

MCJCC Mayor's Criminal Justice Coordinating Council (New York City)

McKay David McKay

McKinley Mount McKinley or Mount McKinley National Park in Alaska between Anchorage and Fairbanks, containing North America's highest mountain named for President William McKinley

McKnight McKnight Publishing Co

McKS (Sir Colin) McKenzie Sanctuary (Victoria, Australia)

McKVHS McKee Vocational High School

mcl macro-creation language; midclavicular line; midcostal line; most comfortable level

Mc L Marshall McLuhan

MCL Manchester Central Library; Marine Corps League; Master Configuration List(ing); Master Control Log; Metal Control Laboratories; Metropolitan Central Library; Mid-Canada Line (radar warning fenceline); Moore-McCormack Lines; Mushroom Canners League

M.C.L. Master of Civil Law

MCLA Marine Corps League Auxiliary

McLaughlin Youth McLaughlin Youth Center (for delinquents) at Anchorage, Alaska

mclg maximum contaminant level goal

MCLI Meiklejohn Civil Liberties Institute

M.Clin.Psychol. Master of Clinical Psychology

mcll missile compartment, lower level

MCLO Medical Construction Liaison Office

M.Cl.Sc. Master of Clinical Science

mcm military characteristics motor vehicles; mine counter-measures; minimum commitment method(ology); missile-carrying missile; thousand circular mils

mcm (MCM) missile-control module

mcm *minimo comune multiple* (Italian—least common multiple, lowest common multiple)

McM McMahon; McManus; McMaster; McMillan; McMurry

MCM Manual for Courts-Martial; Marine Corps Manual; Mine Countermeasure (minesweeper); Monte Carlo Method

MCMA Machine Chain Manufacturers Association; Marine Corps Memorial Commission; Metal Cookware Manufacturers Association

MCMC Marine Corps Memorial Commission

MCMF Marie Curie Memorial Foundation

MCMI Millon Clinical Multiaxial Inventory

mcml missile compartment, middle level

mcmops mine countermeasures operations

MCMS Marin County Medical Society

MCMSP Material Command Maintenance Service Publication

MCM&T Michigan College of Mining & Technology

McM U McMaster University

McMUL McMaster University Library

McMUMC McMaster University Medical Centre (Hamilton)

MCN Management Control Number; Manual Control Number; Master Control Number(ing)

MCN *Maternal Child Nursing* (journal)

McNally McNally & Loftin

McNeil Island U.S. Penitentiary at McNeil Island, Washington

mcng meaconing

MCNLP Museo de Ciencias Naturales de La Plata (Spanish—La Plata Natural Sciences Museum)—La Plata, Argentina

MCNP Mammoth Cave National Park (Kentucky); Mount Cook NP (South Island, New Zealand)

MCNY Museum of the City of New York

MCNZ Medical Council of New Zealand

mco main civilian occupation; mills culls out; miscellaneous charges order

mco março (Portuguese—March)

Mco Morocco

MCO Materials Characterization Organization; Michigan Corrections Organization (prison wardens); Movement Control Office(r)

MCOAG Marine Corps Operations Analysis Group

MCODA Motor Cab Owner-Drivers' Association

mcol musicological; musicologist; musicology

M.Com. Master of Commerce; Minister of Commerce

MCOM Mobility Command (US Army)

M. Com. Adm. Master of Commercial Administration

M.Comm.H. Master of Community Health

M. Comp. Law Master of Comparative Law

M. Com. Sc. Master of Commercial Science

MCON Military Construction—Navy

MCOO Monte Carlo Opera Orchestra

MCOP Marine Corps Ordnance Publication

mcos marcos (Spanish—marks), German coins

MCOW Medical College of Wisconsin

mcp main control panel; male chauvinist pig; manual control panel; mode control panel; multi-component plasma; multiple chip package

mcp (MCP) master control program

mCp my Cadillac payment

MCP Maine Coastal Program; Malaysian Communist Party;

Management Control Plan; Maritime Company of Philadelphia; Maritime Company of the Philippines; Massachusetts College of Pharmacy; Master Control Program; Military Construction Program; Minerals and Chemicals Philipp; Model Cities Program

M.C.P. Master of City Planning

MCPA Member of the College of Pathologists of Australasia

MC Path Member of the College of Pathologists

mcph metacarpal-phalangeal

MCPO Master Chief Petty Officer

mcps megacycles per second

MCPS Mechanical Copyright Protection Society; Member of the College of Physicians and Surgeons

MCPT Maritime Central Planning Team

McQ-E McQuaid-Ehn (grain size)

mcr master control routine; metabolic clearance rate; micrographic computer retrieval; military compact reactor; modular circuit reliability; monitor control routine; mother-child relationship

MCR Manufacturing Change Request; Marine Corps Reserve; Mass Communications Research; Master Change Record; Master Charge Record; Mobile Control Room

M.C.R. Master of Comparative Religion

MCRA Member of the College of Radiologists of Australasia

MCRC Mass Communications Research Center (University of Wisconsin)

MCRD Marine Corps Recruit Depot

MCRE Mother-Child Relationship Evaluation

MCREL Mid-Continent Regional Educational Laboratory

mcrfsch microfische

MCRHS Mid-Continent Railway Historical Society

MCRL Master Cross-Reference List(ing)

MCRML Midcontinental Regional Medical Library (University of Nebraska)

MCROA Marine Corps Reserve Officers Association

MCRS Micrographic Computer Retrieval System; Military Command Research System

mcrt multichannel rotary transformer

mcrwv microwave

mcs meridian control signal; meter-candle second; missile checkout set; motor circuit switch; multiple column selector

mc/s megacycles per second

MCs Military Characteristics

MCS coastal minesweeper (naval symbol); Maintenance Control Section; Major Component Schedule; Management Computing Services; Management Control System; Marine Cooks and Stewards (union); Marine Corps School; Marine Corps Station; Message Control System; Military College of Science; mine counter-measures support ship (naval symbol); Missile Commit Sequence; Mobile Checkout Station; Mobile Coastal Service

M.C.S. Master of Commercial Science

MCSA Marble Collectors Society of America; Medical Computer Services Administration

MCSB Motor Carriers Service Bureau

MCSC Medical College of South Carolina; Military College of South Carolina (The Citadel); Monell Chemical Senses Center

mc/sec megacycles per second

MCSH Manhattan College of the Sacred Heart

MCSL Marine Corps Stock List; Marine Corps Supply List

MCSP Member of the Chartered Society of Physiotherapy

mc spec motorcycle specialist(s); motorcycle specification(s)

MCSs Memorial Cremation Societies

mcst magnetic card selectric typewriter

MCST Member of the College of Speech Therapists

MC S & T Manchester College of Science and Technology

MCSTB Motor Carriers Service Tariff Bureau

MCSWG Multinational Command Systems Working Group

mct maximum continuous thrust; multiple-compressed tablet

mct (MCT) modular computing typewriter

MCT Maritime Crew Trainer; Master Cycle Trader; Mechanical Comprehension Test; Minimum-Competency Test(ing)

m/cta *mi cuenta* (Spanish—my account)

MCTA Metropolitan Commuter Transportation Authority; Motor Carriers Traffic Association

MCTC Maritime Cargo Transporation Conference; Microelectronics and Computer Technology Corporation (Austin, Texas)

MCTE Michigan Council of Teachers of English

MCTI Metal Cutting Tool Institute

mctow maximum certificated takeoff weight

mctp missile control test panel

MC & TS Monotype Casters' and Typefounders' Society

MCTSSA Marine Corps Tactical System Support Activity

mcu mechanical condition unknown; median control unit; medium closeup; microprocessor control unit; monitor control unit

m & cu monitor and control unit

MCU Modern Churchmen's Union

MCUG Military Computer Users Group

mcul missile compartment, upper level

mcv mean corpuscular volume; model control volume

MCV Medical College of Virginia

mcvf multichannel voice frequency

mcw metal casement window; modulated continuous wave

m & cw maternity and child welfare

MCW Mallinckrodt Chemical Works

m cwt 1000 hundredweight

mcx maximum-cost expediting

mcy machinery

MCZ Museum of Comparative Zoology

md main deck; malicious damage; manufacturing day; map distance; mathematics disabled; maximum demand; maximum design; mean deviation; memorandum of deposit; mental(ly) defective; mentally deficient; mentally disordered; message dropping; milliard; minor defect(s); minute difference(s); mitral disease; month's date; motorized damper; movement directive; muscular dystrophy

m-d manic-depressive

m/d market day; memorandum of deposit(s); messages per day; missile driver; modulator-demodulator; month(s) after date

m & d medicine and duty

md *main droite* (French—right hand); *mano derecha* (Spanish—right hand); *mano destra* (Italian—right hand); *marchand* (French—good value, marketable); *milliard* (French—1000 million)

m d *mano destra* (Italian—right hand)

Md Maid; Maryland; Marylander; mendelevium

M$ Malaysia dollar (Singapore dollar)

MD Management Directive; Managing Director; Marine Detachment; Maryland; Material Division; Medical Department; Medical Discharge; Mess Deck; Meteorological Department; Middle Dutch; Military District; Mine Depot; Music Director; Musical Director

M.D. *Medicinae Doctor* (Latin—Doctor of Medicine)

M D *mano destra* (Italian—right hand)

mda maintenance depot assistance; minimum detectable activity; monochrome display adaptor; multiple-docking adapter

mda (MDA) methyldiamphetamine (stimulant); minimum descent altitude

Mda Mérida (inhabitants—Meridanos)

MDA Marking Device Association; Master Dyes Association; Material Department Amendments; Material Disposal Authority; Middle Depot Activity; Minor Deviation Authorization; Multiple-

Docking Adapter; Mural Decorators Association; Muscular Dystrophy Association; Mutual Defense Agency; Mutual Defense Assistance

MDAA Muscular Dystrophy Association of America; Mutual Defense Assistance Act

MDAC McDonnell Douglas Astronautics Company; Mutual Defense Assistance—China area

MDAGT Mutual Defense Assistance, Greece and Turkey

MDAIKP Mutual Defense Assistance, Iran, Republic of Korea, and the Philippines

MDAN Material Department Administrative Notices

MDANAA Mutual Defense Assistance, North Atlantic Area

MDAP Mutual Defense Assistance Program

MDAPT Machover Draw-a-Person Test

MDAs Medical Doctor Anesthesiologists

MDAS Multispectral Data Analysis System

Mda Vle Maida Vale

M-day manufacturing day; mobilization day; moratorium day

mdb multilateral development bank(ing)

Mdb Middlesbrough

M d B *Mitglied des Bundestages* (German—member of the Bundestag)

MDB *Movimiento Democrático Brasileiro* (Portuguese—Brazilian Democratic Movement)—political party

MDBVHS Mabel D Bacon Vocational High School

mdc maintenance data collection; more developed country

M d C *Maestro di Cappella* (Italian—Chapel Master); *Maître de Chapelle* (French—Chapel Master)—titles often meaning conductor or musical director

MDC Manhattan Drug Corporation; Manufacturing Development Council; McDonnell Douglas Corporation; Metropolitan District Commission; Minnesota Department of Corrections; Moncure Daniel Conway

MDCA Master Diamond Cutters Association

MDC-W McDonnell Douglas Corporation—West

mdd mechanical dough development; milligrams per square decimeter per day
MDDC Military Dependents Dental Clinic
mddpm magnetic-drum data-processing machine(ry)
Mddx Middlesex
mde matrix difference equation
M.D.E. Master of Domestic Economy
MDE *Modern Drug Encyclopedia*
m'dear my dear
M de C Maître de Chapelle (French—conductor)
M.Dent.Sci. Master of Dental Science
M.Des. Master of Design
mdf mild detonating fuze; minimum diversion fuel
mdf (MDF) main distributing frame (data processing); manual direction finder
MDF Manitoba Development Fund; Modderfontein Dynamite Factory
MDFC McDonnell Douglas Finance Corporation
mdfd modified
mdfg modifying
mdfn modification
mdfy modify
MDG Medical Director-General
mdh minimum descent height; multidirectional harassment
mdh (MDH) malate dehydrogenase
MDHB Mersey Docks and Harbour Board (Liverpool)
Md Hist Maryland Historical Society
Mdhv Marek's disease herpes virus
mdi magnetic detection indicator
MDI Material Department(al) Instruction; Material Development(al) Instruction
MDIA Material Department-(al) Instruction Amendment(s)
MDIB Material Departmental Instruction Bulletin
m. dict. more dictu (Latin—in the manner directed)
M.Did. Master of Didactics
M.Di.Eng. Master of Diesel Engineering
MDIG Multipurpose Display Indicator Group
M.Dip. Master of Diplomacy
M dis Marek's disease
mdise merchandise
M.Div. Master of Divinity

MDJC Miami-Dade Junior College; Mississippi Delta Junior College
m dk main deck
mdl master design layout; middle; model; modular design language
Mdl Middle (postal abbreviation)
M d L Mitglied des Landtages (German—member of the Landtag)
MDL Mine Defense Laboratory
MDL Master Drug List
Mdlle Mademoiselle (French—Miss)
mdm (MDM) middiastolic murmur
Mdm Madam
MDM Mass Democratic Movement (South African); Material Division Manual; Movement (for a) Democratic Military
mdma (MDMA) methylene-dioxy-methamphetamine (also known as Adam or Ecstacy)
Mdme Madame (French—Missus)
mdn median
m$n moneda (pesos) nacional [Spanish—national monetary unit(s)—Argentinian peso(s)]
MDNA Machinery Dealers National Association
mdnb metadinitrobenzene
mdngt midnight
MDNS Modified Decimal Numbering System
mdnt midnight
mdo medium-density overload-(ing); monthly debit ordinary
MDO MARC Development Office (Library of Congress)
M-dog mine dog (trained to find buried mines)
mdp minimum-distance principle; multi-disc player
m.d.p. mento-dextra posterior (Latin—right mento-posterior)
MDP Mainland Data Processing; Manufacturing Development Program
MDPR Manufacturing Development and Process Request
mdr magnetic disk recorder; master-clock generator; memory data register; minimum daily requirement; multichannel data record(er)
Mdr Madras
MDR Master Discrepancy Report; Material Deficiency

Report(ing)
mdrc maximum distance for radiological consequences
MDRC Manpower Demonstration Research Corporation
MDRSF Multi-Dimensional Random Sea Facility
mds minimum discernible signal; mission design and series
Mds Mesdames (French—Ladies)
M$S peso (*moneda nacional*—Argentine letter symbol)
MDS mail distribution schedule; mail distribution scheme; Main Dressing Station; Manufacturing Data Series; Medical-Dental Service; meteoroid detection satellite; Model Designator Series; Multipoint Distribution Systems (microwave)
M.D.S. Master of Dental Surgery
mdsa multiple disc-sampling apparatus
M.D.Sc. Master of Dental Science
mdse merchandise
MDSF Mission to Deep Sea Fishermen
mdsg merchandising
MDSI Manufacturing Data Systems Inc
MDSOs mentally disordered sex offenders
MDST Mountain Daylight Saving Time
mdt mean down time; moderate
m.d.t. mento-dextra transversa (Latin—right mento-transverse)
MDT Multidisciplinary Team; Mutual Defense Treaty
MDTA Manpower Development and Training Act
mdtm medium duty target mechanism
MDTS Modular Data Transmission System
M Du Middle Dutch
MDU Medical Defence Union; Mine Disposal Unit; Mobile Development Unit
M du N Magasin du Nord (Copenhagen's leading department store)
MDUS Medium Data Utilization Station
Mdv Marek's disease virus
M.D.V. Doctor of Veterinary Medicine
mdw measured day work
MDW Chicago, Illinois (Mid-

way Airport); Military Defense Works; Military District of Washington; Minnesota, Dakota & Western (railroad)

MDWC Mississippi Department of Wildlife Conservation

Mdws Meadows

Mdx Middlesex

mdy magnetic deflection yoke

MDY Midland Oil (stock exchange symbol)

me. male employee; marbled edges; marbled edging; mathematics education; maximum effect; maximum effort; mechanical equipment; metabolizable energy; methyl; mill edge; milligram equivalent; miter end; most excellent; multi-engine; multiple exposure; muzzle energy

me. (ME) measles encephalitis

m/e mechanical/electrical; mobility equipment

m & e mechanical and electrical; music and (sound) effects

m E meines Erachtens (German—in my opinion)

Me Maine; Mainers; Mexican(s); Mexico

Me Maître (French—Master)—advocate; attorney

ME Mail Early; Maine; Managing Editor; Marine Engineer; Mechanical Engineer(ing); Medical Examiner; Methodist Episcopal; Middle East(ern)(er); Middle English; Military Engineer; Mining Engineer; Morristown and Erie (railroad); *Mouvement Europeen* (French—European Movement)

M.E. Master of Education; Mechanical Engineer

mea measure(s); measuring; minimum enroute altitude; monoethanolamine (MEA)

MEA Maintenance Engineering Analysis; Malaysian Economics Association; Maritime Employers Association; Medical Exhibitors Association; Member of the European Assembly; Michigan Education Association; Middle East Airlines; Minnesota Education Association; monoethanolamine; Montana Education Association; Municipal Employees Association; Music Educators Asso-

ciation; Musical Educators Association

M.E.A. Master of Engineering Administration

MEA Municipalidades En Acción (Spanish—Municipalities in Action)

MEAF Middle East Air Force

meal. master equipment allowance

MEAL Master Equipment Authorization List(ing)

mean max mean maximum; mean maximum temperature

MEAR Maintenance Engineering Analysis Record

meas measure; measurement

Meas for M Measure for Measure

M-East Middle-East

M.E. Auto. Master of Automobile Engineering; Master of Automotive Engineering

meb military early bird

MEB Marine Expeditionary Brigade; Master Electronics Board; Medical Board; Melbourne, Australia (airport); Midlands Electricity Board (UK)

MEBA Marine Engineers' Beneficial Association

mec main engine cutoff; marine extension clause; measuring equipment control

méc mécanique (French—mechanical)

M. Ec. Master of Economics

MEC Maine Central (railroad); Maintenance Engineering Change(s); Marine Expeditionary Corps; Master Executive Council; Member of the Executive Council; Methodist Episcopal Church; Monetary and Economic Council

M.E.C. Master of Engineering Chemistry; Member of the Executive Council

meca maintainable electronics component assembly; malfunctioned equipment corrective action; mercury evaporation and condensation; multi-element component array

MECA Manufacturers of Emission Controls Association; Molecular Emission Cavity Analysis

mecano mechanotherapy

MECAS Middle Eastern College for Arabic Studies (Beirut, Lebanon)

mecc meccanica (Italian—mechanic)

mecca master electrical com-

mon connector assembly

MECCA Minnesota Environmental Control Citizens Association

mech mechanic; mechanical; mechanism

Mech Mechanics

ME Ch Methodist Episcopal Church

MECHA Movimiento Estudiantil Chicano de Aztlán (Mexican-Spanish—Chicano Student Movement of Aztlán)

Mechai (Thai—condom)—named for Thailand's Mr Contraception, Mechai Viravaidya

mechanochem mechanochemical; mechanochemistry

M.E. Chem. Master of Chemical Engineering

Mech Eng Mechanical Engineering

Mech Illus Mechanix Illustrated

MEC/IES Merrimac Education Center's Institute for Educational Services

mecl monolithic emitter-coupled logic (computer)

meco main engine cutoff

MECO Metropolitan Edison Company

mecom marine engine condition monitor(ing)

M.Econ. Master of Economics

MECON Metallurgical and Engineering Consultants

MEC/PA Maintenance Engineering Change/Problem Analysis

MECR Maintenance Engineering Change Request

MECS Middle East Container Service

mecu main engine control unit

MECU Municipal Employees Credit Union

mecz mechanized

med medal; medalist; medallion; median; median erythrocyte diameter; medic; medical; medication; medicinal; medicine; medieval; medievalism; medievalist; medium; minimal effective dose; minimal erythema dose; mobile engine diagnosis

Med Medicine; medieval; Mediterranean

Med Médico (Italian, Portuguese, Spanish—doctor); *Méditerranee* (French—Mediterranean); *Mediterraneo* (Ital-

ian—Mediterranean); *Mediterrâneo* (Portuguese—Mediterranean); *Mediterráneo* (Spanish—Mediterranean)
Méd médicine (French—medicine)
M.Ed. Master of Education
MED Maintenance Engineering Data; Manhattan Engineer District (cover name used by developers of the first atomic bomb); Metalworking Equipment Division (US Department of Commerce); Military Electronics Division (Motorola); Municipal Electricity Department
M.E.D. Master of Elementary Didactics
medac medical accounting
medal. micromechanized engineering data for automated logistics
Med C Medical Corps
Med CAP Medical Civil Action Program
medcat medium clear-air turbulence
MEDCO Meat Export Development Company
med col medium color
MEDCOM Mediterranean Communications System
medcrit medical critic(ism)
medda mechanized defense decision anticipation
MED-DENT Medical-Dental Division (USAF)
medevac medical evacuation
medex medical expert
medex medecin extension (French—doctor's aides, medics)
Medfly (Med fly) Mediterranean fruit fly
Med Gr Medieval Greek
medi (Latin prefix—middle)—median
Medi (British seamen's short form—Mediterranean)
Media Magnavox electronic data-image apparatus
MEDIA Manufacturers Educational Drug Information Association; Missile Era Data Integration Analysis; Move to End Deception in Advertising
mediacrit media critic(ism)
mediaese cultivated English spoken by many entertainment, radio, and television personalities
mediator. media time-orienting and reporting

medic medical corpsman; medical doctor; medical student
medicaid medicinal aid (free medicine for the needy)
Medicaid Medical Aid (federal and state health insurance for people unable to afford medical care)
MEDI-CAL Medical Aid of California
Medical Exam Medical Examination Publishing Company
medicare medical care
Medicare Medical Care (federal health insurance for aged and disabled persons)
MEDICO Medical International Corporation
medifraud medical fraud
Medigap uninsured medical costs often requiring supplemental coverage
mediog mediograph(ic)(al); mediography
Med Isr Medinat Israel (Hebrew—State of Israel)
Medit Mediterranean
Mediterranean Mediterranean Sea
Mediterranean Countries Albania, Algeria, Cyprus, Egypt, France, Greece, Israel, Italy, Lebanon, Libya, Malta, Monaco, Morocco, Spain, Syria, Tunisia, Turkey, Yugoslavia
MEDIUM Missile Era Data Integration Ultimate Method
medivac medical evacuation
medix medical students
med juris medical jurisprudence
med lab(s) medical laboratories; medical laboratory
MEDLARS Medical Literature Analysis and Retrieval System
Med Lat Medieval Latin
MEDLINE Medical On-Line (computer retrieval system)
M. Ed. L. Sc. Master of Education in Library Science
med nec medically necessary (abortion)
Med Phys Medical Physics
med ray medullary ray
med ray par medullary ray parenchyma
med ray trac medullary ray tracheids
MEDRC Medical Reserve Corps
MEDRECO Mediterranean Refining Company
MEDRESCO Medical Re-

search Council
meds medicaments; medicines
MEDS Maintenance Engineering Data Sheets
MEDSAC Medical Service Activity (USA)
Med. Sc. D. Doctor of Medical Science
Med Sch Medical School
Med Sea Mediterranean Sea
med show medicine show (carnival slang)
Med Supt Medical Superintendent
med tech medical technologist; medical technology
Med Tech Medical Technician; Medical Technologist
med trans medical transcriptionist
mee methylethyl ether
M.E.E. Master of Electrical Engineering
MEECN Minimum Essential Emergency Communications Network
M.E. Eng. Master of Electrical Engineering
meerschaum hydrated magnesium silicate
MEES Middle East Economic Survey
mef maximal expiratory flow
Mef Mefisto
MEF Marine Expeditionary Force; Meal Export Federation; Mesopotamian Expeditionary Force; Middle East Forces; Musicians Emergency Fund
Mefisto Mefistofele
mef's morality enhancing factors
mefv maximum expiratory flow volume
meg megacycle; megaton; megawatt; megohm
Meg Margaret; Megan
MEG Management Evaluation Group
mega 10^6
mega megas (Greek—great, large, powerful)—acromegaly, megacycle, megaspore, megaton
megabuck one million bucks (dollars)
megacorpses one million corpses (atomic bomb unit)
megacurie one million curies
megacycle one million cycles
megadeaths million deaths
megajoule one million joules
megal megalopolis (industrial urban area)
megameter one million meters

megamouse one million mice (statistical unit—experimental biology)
megaton one million tons
megawatt one million watts
megger megohmmeter
Meggie Margaret
Meglin Megliola
mego megaphone; megohm(s)
MEGO mine eyes glaze over
megohm one million ohms
megs megacycles
megv million volts
megw megawatt
megwh megawatt-hour
mei mathematics in education and industry
mei (MEI) marginal efficiency of investment
MEI Maintenance and Engineering Inspection; Manual of Engineering Instructions; Marine Ecological Institute; Metals Engineering Institute; Middle East Institute
MEIC Member of The Engineering Institute of Canada
MEIS Military Entomology Information Service
MEIU Management Education Information Unit
mej mejikanisch (German— Mexican); *mejuffrouw* (Dutch—Miss)
Mej Mejuffrouw (Dutch— Miss)
mek methyl ethyl ketone
mel melanesisch (German— Melanesian)
Mel Melanesia; Melanesian; Melanesian Pidjin English (Bêche de Mer); Melanie; Melba; Melbourne; Melissa; Melvil; Melville; Melvin; Melvina; Melvyn
MEL Master Equipment List-(ing); Minimum Equipment List (FAA); Music Education League
M.E.L. Master of English Literature
MELA Middle East Librarians Association
Melan Melanesia; Melanesian
Melanesia islands in the western Pacific whose natives inhabit the Bismarck Archipelago, Fiji, New Caledonia, New Hebrides, and the Solomon Islands
Melb Melbourne
Melba Nellie Armstrong (of Melbourne)
MELCO Mitsubishi Electric Corporation
Meld melt + weld

MELF Middle East Land Forces
melg most European languages
melo melodrama; melody
M. Elo. Master of Elocution
melos melodic lines
melt. pt melting point
mem member; memoirs; memorial
mem (MEM) memory map (calculator)
mem. memoria (Latin—memory)
Mem Member (of Congress, Parliament, etc.); Memorial (postal abbreviation)
MEM Mars Excursion Module; Member; memorial; Memphis, Tennessee (airport)
MEMA Marine Engine Manufacturers' Association
MEMAC Machinery and Equipment Manufacturers Association of Canada
memb membrane
'members remembers
MEMC Marathon Electric Manufacturing Corporation
memci model ewe for microclimate integration
MEML Master Equipment Management List
memo memoranda; memorandum
MEMO Medical Equipment Management Office
Memp Memphis
Mem Pkwy Memorial Parkway
Mem Roy Astron Soc Memoirs of the Royal Astronomical Society
Mem Soc Assn Memorial Society Association
Mem SP Memorial State Park
MEMSPO Michigan Elementary and Middle School Principals Organization
men. menses; menstruation; mensuration
men meno (Italian—less)
Men Mensa (constellation)
M.En. Master of English
MEN Manasco (stock-exchange symbol)
MEN Middle East News
MENA Middle East News Agency
MENC Music Educators National Conference
mend. macro end
MEND Medical Education for National Defense; Mothers Embracing Nuclear Disarmament

Mendl Lib Mendelssohn Library
Mendy Mendelssohn
M.Eng. Master of Engineering; Mining Engineer
M. Eng. P.A. Master of Engineering and Public Administration
Menn Menninger
Mennon Mennonite
meno menopausal; menopause; menorrhoea
MENP Mount Elgon National Park (Kenya)
MENS Mission Element Needs Statement
mens san. mens sana in corpore sano (Latin—a sound mind in a sound body)
menst menstrual; menstruation
mensur mensuration
ment mental; mentalis
M. Ent. Master of Entomology
mentd mentioned
mentholyptus menthol + eucalyptus
meo (MEO) manned earth observatory
Meo Bartolomeo
MEO Maintenance Engineering Order; Marine Engineer(ing) Office(r)
MEOA Malaysian Estate Owners Association
meoh methanol; methol
MEOW Moral Equivalent of War (President Carter's energy program)
mep mean effective pressure
MEP Main European Port; Management Engineering Program; Management Evaluation Program; Member of the European Parliament; Micro-Electronic Education Program; Middle East Perspective
M.E.P. Master of Engineering Physics
MEP Movemiento Electoral Popular (Papiamento—Popular Electoral Movement)— Aruba's pro-independence party; *Movimiento Electoral del Pueblo* (Spanish—People's Electoral Movement)— Venezuelan political party
M.E.P.A. Master of Engineering and Public Administration
MEPC Maritime Environment Protection Committee; Medical Examination Publishing Company (imprint of Elsevier Science Publishing Company); Member of the

Educational Publishers' Council; Metropolitan Estate and Property Corporation

MEPCOM Military Enlisting Processing Command

M.E.P.H. Master of Public Health Engineering

Mephisto Mephistopheles (The Devil)

MEPS Means-Ends Problem-Solving Test; Multilanguage Electronic Phototypesetting System

mepu (MEPU) monopropellant emergency power unit

meq/l millequivalents per liter

mer meridian; minimum energy requirement(s)

mer (MER) methanol extraction residue

m & er mechanical and electrical room

mer mercoledi (Italian—Wednesday); *mercredi* (French—Wednesday); *meros* (Greek—part)—blastomere, centromere, isomer, polymer

Mer Mercury; Merino

MER Metropolitan Elevated Railroad; Ministry of Energy Resources

mera minimum enroute altitude

MERA Michigan Educational Research Association

MERADO Mechanical Engineering Research and Development Organization

MERAG Middle-East Research and Action Group

MERB Mechanical Engineering Research Board

merc mercury

Merc Mercantile Exchange; Mercator; Mercedes; Mercedes-Benz; Mercury

MERC Music Education Research Council

MERCEDO Merseyside Economic Development Office

merch merchantable

Merch V Merchant of Venice

'mercial commercial

mercs mercenaries

Mercurys Mercury Islands off New Zealand's North Island

MERDL Medical Equipment Research and Development Laboratory (USA)

Meredith Meredith Press

meres matrix of environmental residuals for energy systems

merfin mercerized finish

Merguis Mergui Islands off Lower Burma

Meri Merionethshire

'Merica(n) [Cockney contraction—America(n)]—in the Far East, the South Seas, and other parts of the world this sometimes comes out as *'Mellica(n)*

merid meridian

Merigrove the *New Grove Dictionary of American Music*

MERIP Middle East Research and Information Project

MERIT Medical Relief International

meritoc meritocracy; meritocrat(ic)(al)(ly)

MERL Mechanical Engineering Research Laboratory

MERLIN Machine Readable Library Information

Merriam G & C Merriam

Merritt Pkwy Merritt Parkway

Merry W Merry Wives of Windsor

mer's multiple ejection racks

Mers Merseyside

mersar merchant ship search and rescue

mersex merchant shipping exchange

Mert Merton College, Oxford

MERT Maintenance Engineering Review Team; Milwaukee Electric Railway and Transit

Mert Coll Merton College—Oxford

MERU Mechanical Engineering Research Unit

Merv Mervin

merzd meridian zenith distance

MERZONE Merchant Shipping Control Zone

mes main engine start; main equipment supplier; missile engineering station; motor end support; mud estuary slick; mutual energy support

mes meseta (Spanish—plateau; tableland); *mesos* (Greek—middle)—mesentery, mesoderm(ic), Mesopotamia, Mesozoic

Mes Mesozoic; Messina

Mes Mesdames (French—ladies)

MES Michigan Engineering Society; Midwest Electronic Society

mesa (MESA) mathematics, engineering, and scientific achievement

mesa. modularized equipment storage assembly

MESA Malarial Eradication Special Account; Marine Ecosystems Analysis; Mechanics Educational Society of America; Mining Enforcement and Safety Administration

Mesabi Mesabi Range of iron ore in Minnesota

Mesa Verde Mesa Verde National Park in Colorado

mesbic (MESBIC) minority enterprise small business investment companies

mesc mescal; mescaline

M. E. Sc. Master of Engineering Science

MESC Middle Eastern Solidarity Council

MESCO Middle East Science Cooperative Office (UNESCO)

Mesd Mesdames (French—Ladies)

MESF Mobile Earth Station Facility

mesfet metallized semiconductor field-effect transistor

mesh. medical headings

MeSH Medical Subject Heading (National Library of Medicine's thesaurus)

Meslier Jean Meslier—obscure parish priest whose name was used by Voltaire to escape persecution (*see* Jean Meslier); even the names of the parishes he served—Entrepigny and But—are not to be found in most atlases and gazetteers

meso (Latin prefix—middle, moderate)—mesoderm, Mesopotamia

Mesoamer Mesoamerica(n)—Middle America(n) (Central America, Mexico, and the West Indies)

Mesop Mesopotamia (Iraq)

MESP More Effective Schools Program

MESPA Minnesota Elementary School Principals Association

Mespot Mesopotamia (Iraq)

mess. maximum effective sonar speed

Mess Messidor (French—Harvest Month)—beginning June 19th—tenth month of the French Revolutionary Calendar

Messner Julian Messner

messplex multiplex emission sensors

Messrs Messieurs (French—Gentlemen)

mest mestizo

mestranol methyl + estrogen + pregnane (synthetic oral contraceptive)
met. metal; metallic; metallize; metaphor; metaphysics; meteorology; methionine (amino acid) (MET); metronome; metropolitan
met *metropolitana* (Spanish—metropolitan)
Met Metro; Metropolitan Correction Center; Metropolitan Museum of Art; Metropolitan Opera
MET Mobile Examining Team; Multi-Environment Trainer
meta (Greek—after or beyond)—metabolism, metacarpal, metastasis, metatarsal
Meta Margarita
META Megachannel Extraterrestrial Assay; Metropolitan Educational Television Association (Canadian); Model Engineering Trade Association
metab metabolism
metadex metal abstracts index
metall metallurgy
metallog metallography
METALMA Metalúrgica Matarazzo (Brazilian company)
metaph metaphor(ical)(ly); metaphysical(ly); metaphysician; metaphysics
metaphys metaphysics
metaplan methods of abstracting text automatically programming language
metas metastasis; metastasize
metath metathesis
metb metal base
met bor metropolitan borough
metc metal curb; mouse embryo tissue culture
Met Cen Lib Metropolitan Central Library
METCO Metropolitan Council for Educational Opportunity
metd metal door
mete multiple engagement test environment
Met. E. Metallurgical Engineer
metec meteoroid technology
METEI Medical Expedition to Easter Island
meteor. meteorology
Meteorol Meteorology
meteorolo meteorology
meteosat meteorological satellite
metf metal flashing
metg metal grille
meth methadone; methamphet-

amine; methane; methedrine; methyl; methylated; methyprylon
Meth Methodist
methan methanol
methanol methyl alcohol or wood alcohol (CH_3OH)
Meth Epis Methodist Episcopal
meth freak methedrine freak (underground slang—habitual user of methedrine)
meth head methedrine head (underground slang—methedrine addict)
metho methodology; methyl alcohol
meths methylated spirits (denatured alcohol)
methu methuselah (8-bottle capacity)
meti major engineering test item; metal jalousie
M-et-L Maine-et-Loire
meth lab methamphetamine laboratory
metical monetary unit of Mozambique
Met Lith Assn Metropolitan Lithographers Association
metm metal mold
M-et-M Meurthe-et-Moselle
Met Man Metro Manila
Met Mus Metropolitan Museum
m. et n. *mane et nocte* (Latin—morning and evening); *mane et nocte* (Latin—morning and night)
meto maximum except takeoff
Met O Meteorological Office(r)
METO Middle East Treaty Organization
metob meteorological observation
metol methyl-p-aminophenol (photographic developer)
meton metonomy
metp metal partition
Met Pol Metropolitan Police
metr metal roof
Met R Metropolitan Railway
METR Manufacturing Engineering Test Report
metro metropolitan
métro *chemin de fer métropolitain* (Paris subway system)
Metro Metromedia; Metropolitan Life Insurance Company
Metro *Metropolitan* (Paris and Madrid subway systems—originally stood for Metropolitan District Railway—the London *Underground*)
METRO New York Metro-

politan Reference and Research Library Agency
metroc meteorological rocket
metrocenter metropolitan center
metrocomplex metropolitan complex
metrocore metropolitan core
metroframe metropolitan framework
metrol metrology
METROMEX Metropolitan Meteorological Experiment
metrop metropolis; metropolitan
metroplex metropolitan complex
metropol metropolis; metropolitan
METRRA Metropolitan Toronto Residents' and Rate Payers' Association
mets metal strip
METS Mayne's Exchange Transfer Systems; Mechanized Export Traffic System (USA)
metsats meteorological satellites
m. et sig. *misce et signa* (Latin—mix and write a label)
metso sodium metasilicate
mett (METT) manned evasive-target tank (USA)
Met Tec Meteorologist Technician
METU Middle East Technical University (Ankara)
Met-Vic Metropolitan-Vickers (electrical company)
MEU Marine Expeditionary Unit; Mental Evaluation Unit; Municipal and Shire Employees Union
MEU *Modern English Usage*
mev million electron volts
meV mega-electron volt(s); milli-electron volt(s); million-electron volt(s)
Mev *Mevrouw* (Dutch—Missus)
MeV megaelectronvolt; million electronvolt
mevr *mevrouw* (Dutch—Mrs)
Mevr *Mevrouw* (Dutch—Missus)
MEW Microwave Early Warning; Ministry of Economic Warfare
MEWA Motor and Equipment Wholesalers Association
MEWS Maintenance Engineering Work Sheet; Missile Early Warning Station
MEWTA Missile Electronic Warfare Technical Area

mewttos main engine working to telegraph orders
mex military exchange
Mex Mexican; Mexico
Méx México (Spanish—Mexico)
MEX Mexico City, Mexico (airport)
Mex C Mexico City
Mex Cy Mexican currency
Mex$ Mexican peso
Mexi Mexicali
MEXICANA Compañía Mexicana de Aviación (Spanish—Mexican Aviation Company)
Mexicanos Mexicanos al grito de guerra (Spanish—Mexicans, the war cry)—national anthem of Mexico
Mexico United Mexican States (Middle America's largest and most populated nation), *Estados Unidos Mexicanos*
mexit macro exit
MEXSM Mexican Service Medal
Mex Sp Mexican Spanish
Mexsur Mexican (automobile) insurance
mez mezcal(ine)
mez mezzo (Italian—half)
MEZ mitteleuropäische Zeit (German—Central European Time)
mezz mezzanine; mezzotint
Mezzogiórno (Italian—south)—southern Italy including Naples, Palermo, Salerno
mezzo(s) mezzosoprano(s); mezzotint(s)
mf machine finish; main feed; main force; maintenance factor; male-to-female (ratio); manufacture(d); manufacturing; mastic floor; medium frequency (300–3,000 kc); microfarad(s); microfiche; microfilm; mill finish; millifarad(s); mother fucker; motor field; motor freight; multiplying factor
m/f maintenance-to-flight (ratio); male or female; manifest; marked for; marked for; milk fat
m & f male and female
μf micro + farad
mf mezzo-forte (Italian—half loud, moderately loud)
m/f mi favor (Spanish—my favor); *motorfaerge* (Dano-Norwegian—motor ferry)
μF microfarad(s)
M⸍ Massif (French—mountain mass)
MF Magazines for Friendship; Marshall Field (stock exchange symbol); Medal of Freedom; Middle Fork (railroad); Millard Fillmore (13th President U.S.)
M-F Massey-Ferguson; Monday through Friday
M.F. Master of Forestry
MF medlem af Folketinget (Danish—Member of Parliament)
mfa malicious false alarm; multi-fiber arrangement
MFA Master Fencers Association; Metal Fixing Association for Building Insulation; Military Flying Area; Ministry of Foreign Affairs; Motor Factors' Association; Museum(s) of Fine Arts
M.F.A. Master of Fine Arts; Museum of Fine Arts
MFA Movimento das Forças Armadas (Portuguese—Armed Forces Movement)—military dictatorship
M Fac Hom Member of the Faculty of Homeopathy
M-factor mobility, movement, migration (automotive Americans on the move)
MFAH Museum of Fine Arts of Houston
M.F.A. Mus. Master of Fine Arts in Music
MFAR Michigan Foundation for Advanced Research
mfb message from base; metallic foreign object; moisture-free basis
mfb (M) median forebrain bundle
MFB Metropolitan Fire Brigade; MFB Mutual Insurance (Manufacturers, Firemen's and Blackstone combined)
mfbm thousand foot board measure(ment)
mfc magnetic-tape field scan(ning); medicated face conditioner; membrane fecal coliform; microfilm frame card; microfunction circuit
mfc (MFC) marginal factor cost
MFC Master Facility Census
MFCA Master Fruit Carriers Association
MFCC Marriage, Family, Child Counsellor
MFCCA Master Floorcovering Contractors Association
m/fcha meses fecha (Spanish—months dated)

mfcm multifunction card machine
MFCM Member of the Faculty of Community Medicine
MFCMA Magnuson Fishery Conservation and Management Act
mfco manual fuel cutoff
mfcs mathematical foundations of computer science
MFCS Manual Flight-Control System (NASA)
mfcu multifunction card unit
mfd manufactured; microfarad; minimum fatal dose (MFD)
mfdf medium-frequency direction finder
mfdp maintenance float-distribution point
MFDT Memory for Designs Test
mfe multiflow evaluator
MFE Mouvement Fédéraliste Européen (French—European Federalist Movement); *Movimento Federalista Europeo* (Italian—European Federalist Movement)
MFECS Mediterranean Far East Container Service
M Fed Miners' Federation
MFED Manned Flight Engineering Division (NASA)
M. F. Eng. Master of Forest Engineering
mff mighty fine fuckin'
MFF Master Freight File
mfg manufacturing; molded fiber glass
MFGT Multiple Family Group Therapy
mfh military family housing
MFH Master of Fox Hounds; Mobile Field Hospital
mfi machine feature index; melt-flow index
MFI Master Facility Inventory; Multi-port Fuel-Injected (engine); Musicians Foundation Incorporated
MFIANE Mutual Fire Insurance Association of New England
MFIBNE Mutual Fire Inspection Bureau of New England
MFIC Military Flight Information Center
MFIT Manual Fault Isolation Test
mfkp multifrequency key pulsing
m fl med flere (Dutch—and others)
Mfl Monfalcone
MFL Master Facility List; Missile Firing Laboratory;

Mobile Field Laboratory; Mutual Funds Limited

MFL *My Fair Lady* (musical adaptation of George Bernard Shaw's *Pygmalion*)

MFLA Midwest Federation of Library Associations

MFLDA Malaysian Federal Land Development Authority

M Flem Middle Flemish

mflops million floating-point operations per second

MFM Miracle Food Mart

MFM *Measure For Measure*

MFMA Master Fish Merchants Association

mf method membrane or millipore filter method

mfn (**MFN**) most-favored nation

mf(n) microfiche (negative)

MFNO Midland Federation of Newspaper Owners

MFNP Mount Field National Park (Tasmania); Murchison Falls NP (Uganda)

MFNZ Music Federation of New Zealand

mfo missile firing order

MFOA Municipal Finance Officers Association

mfopp missile firing order patch panel

M.For. Master of Forestry

MFOWW Marine Firemen, Oilers, Watertenders, and Wipers

mfp minor forest produce

mfp (**MFP**) monoflurophosphate

mf(p) microfiche (positive)

mfpa monolithic focal-plane array

MFPB Mineral Fiber Products Bureau

MFPS Mobile Field Photographic Section

mfr manufacture; manufactured; manufacturer; missile firing range (MFR)

M Fr Mali franc(s); Middle French; Moroccan franc(s)

MFR Military Force Reduction(s)

mfrd manufactured

mfrg manufacturing

mfrn manufacturer's number

MFRP Midwest Fuel Recovery Plant (AEC)

mfrr manufacturer(s)

mfs magnetic-tape field search; maximum file size; missile firing simulator

mf & s magazine flooding and sprinkling

Mf's Moslem fanatics

MFS Malleable Founders' Society; Manned Flying System; Medal Field Service; Military Flight Service; Missile Firing Station; Mountain Fuel Supply; steel-hulled fleet minesweeper (3-letter naval symbol)

M.F.S. Master of Food Science; Master of Foreign Service; Master of Foreign Study

MFSA Master Floor Sanders Association; Metal Finishing Suppliers' Association

mfsk multiple-frequency shift keying

mfso main fuel shutoff

mfsov main fuel shutoff valve

MFSS Missile Flight Safety System(s)

mfst manifest

mft major fraction thereof; mechanized flamethrower; motor freight tariff; multiprogramming with a fixed number of tasks

m. ft. *mistura fiat* (Latin—make a mixture)

MFT Microflocculation Test; Muscle Function Test; Musical Fundamentals Test(ing)

M.F.T. Master of Foreign Trade

MFTB Motor Freight Tariff Bureau

mftbf mean flight time between failure(s)

MFTD Mobile Field Training Detachment

mftl millifoot lamberts

m.ft.m. *misce fiat mistura* (Latin—mix to make a mixture)

mftv mechanical fit test vehicle

mfu military fuckup

MFURB Maryland Fire Underwriters Rating Bureau

MFUSYS Microfiche File Update System

mfv magnetic field vector; microfilm viewer; motor fleet vessel

MFV Mars Flyby Vehicle

MfVB Museum für Volkerkunde, Berlin

MFW Maritime Federation of the World

mfy manufactory

mfz *mezzo forzando* (Italian—with moderate force)

mg machine gun; marginal; milligram; motor generator; multigauge

mg (**MG**) myasthenia gravis

mg % milligrams percent

m-g machine glazed

m/g motor generator

m & g mapping and geodesy

μg microgram

mg *main gauche* (French—left hand)

m/g *mi giro* (Spanish—my check, my draft)

mG *méridien de Greenwich* (French—Greenwich meridian)

Mg magnesium; Margrave; Margraviate; megagram (metric ton)

Mg *Molekulargewicht* (German—molecular weight)

MG machine gun; Maintainability Group; major general; Marine Gunner; Middle German; Military Government; Minas Gerais; Minister General; Morris Garage (M-G); Murray Grey (cattle)

M-G Morris-Garage (British sports car)

M & G Mobile & Gulf

MG *Manchester Guardian* (newspaper); *Maschinegewehr* (German—machine gun)

Mga Malaga; Mongolia

Mga *Mongolia* (Spanish—Mongolia)

MGA Managua, Nicaragua (Las Mercedes airport); Member of the General Assembly; Military Government Association; Monongahela (railroad); Mushroom Growers Association

mgal milligal

M-gauge meter gauge (39.37-inch) railroad track

mgawd make good all works disturbed

mgb (**MGB**) missile gunboat

Mg of B Margrave of Breslau; Margraviate of Breslau

MGB motor gunboat (British naval symbol); Soviet Ministry of State Security (see *VOT*)

MGB *Ministerstvo Gosurdastvennoi Bezopasnosti* (Russian—Ministry of State Security)—Soviet secret police

mgc manual gain control

MGC Machine Gun Corps; Machinery of Government (committee); Marriage Guidance Council; Marriage Guidance Counsellor

mg/cig milligrams (of nicotine tar) per cigarette

mgcir master ground-control-ler-interception radar

mgcr maritime gas-cooled reactor

mg/cu m milligrams (dust, fume, or mist) per cubic meter of air

mgd magnetogasdynamics; million gallons per day

mg/d million gallons per day

Mgd Magdeburg

MGD Military Geographic Documentation

mge (MGE) maintenance ground equipment

M.G.E. Master of Geological Engineering

M.Geol.Eng. Master of Geological Engineering

MGES Maintenance Ground Equipment Specification

mgf macrophage growth factor

MGF Myasthenia Gravis Foundation

mgg mouse gamma globulin

mgh milligram hour(s)

MGH Massachusetts General Hospital

mgi military geographic(al) intelligence

MGI Managed Growth Initiative (urban planning); Media Group Inc; Mining and Geological Institute of India

MGIC Mortgage Guarantee Insurance Corporation

MGICA Mortgage Guaranteed Insurance Corporation of Australia

MGID Military Geographic Information and Documentation

MGk Medieval Greek

mg/l milligrams per liter

MGL Morris Geneological Library

mgm (MGM) mobile guided missile

Mg of M Margrave of Moravia; Margraviate of Moravia

MGM Metro-Goldwyn-Mayer

MGM-18A Lacrosse surface-to-surface missile

MGM-29A Sperry Sergeant surface-to-surface missile

MGM-31A Pershing surface-to-surface missile made by Martin

MGMI Mining, Geological, and Metallurgical Institute

MGMS Manchester Geological and Mining Society

mgmt management

mgn micrograin

M^{gna} *Montagna* (Italian—mountain)

M^{gne} *Montagne* (French—mountain)

Mgo Mormugao

MGP Marcus Garvey Park (formerly Mount Morris Park); Mountain Gorilla Project

mgr mined geological repository

mgr (MGR) mobile guided rocket

Mgr Manager; Monseigneur (French—Monsignor); Monsignore (Italian—Monsignor)

M Gr Middle Greek

MGR Matusadona Game Reserve (Rhodesia); MicroGraphic Recording

mgress manageress

mgs milligrams; missile guidance set (system); money-grubbing scum

mg('s) machine gun(s)

m-g-s meter-gram-second

Mg's Malay gavials, also called Malay gharials

MGS Minnesota Geological Survey

MGSA Military General Supply Agency

MGSMTC Mid-Gulf Seaports Marine Terminal Conference

mgt management

MGTB Mexican Government Tourist Bureau

MGTC Morgan Guaranty Trust Company

MGTD Mexican Government Tourist Delegation

mgtrn magnetron

MGU *Moskovskiy Gosudarstvenny Universitet* (Russian—Moscow State University)

M Gun Sgt Master Gunnery Sergeant

mgw maximum gross weight

MGW *Manchester Guardian Weekly*

m'gwd my gawd (my god)

mh magentic heading; main hatch; manhole; man-hour; marital history; materials handling; menstrual history; mental health; millihenries; millihenry; murine hepatitis

μh microhenry

mH millihenry

Mh *Monatsheft* (German—monthly magazine)

MH magnetic heading; Marshall Islands; Master Hosts; Medal of Honor; Military Hospital; Ministry of Health; Mission Hills; Most Honorable; Most Honourable

M-H Minneapolis-Honeywell (stock exchange symbol and trademark)

M & H Mason and Hamlin

MH *Mo'etzet Hapo'alot* (Hebrew—Woman Workers Council)

MH^2 *Mary Hartman, Mary Hartman* (tv show)

mha manhour accounting; maximum holding altitude

MHA auxiliary minehunter (naval symbol); Marine Historical Association; Medal for Humane Action; Member of the House of Assembly; Mental Health Administration; Mental Health Association; Mental Health Authority; Mining Houses of Australia; Multiple Handicapped Association

M.H.A. Master of Hospital Administration

MHANY Mutual Housing Association of New York

MHATA Mental Health Assistant Therapy Aide

MHb Mueller-Hinton broth

MHB Material Handling Bureau; Mental Health Branch

M-H B Mid-Hudson Bridge

mhc major histocompatibility complex

MHC coastal minehunter (naval symbol); Massachusetts Historical Commission

MHCO Mine-Hunting Control Office(r)

MHCOA Motor Hearse and Car Owners Association

mh cp mean horizontal candlepower

mhcv (MHCV) manned hypersonic cruise vehicle

mhd magnetohydrodynamics

MHD Mental Health Department; Military History Detachment

mhdg magnetohydrodynamic generator

mhdl magnetohydrodynamic laser

mhd lt masthead light

mhe materials handling equipment

MHE Mechanical Handling Engineering

M.H.E. Master of Home Economics

MHEA Mechanical Handling Engineers Association

M Heb Middle Hebrew

MHEDA Material Handling Equipment Distributors Association

M.H.E.E. Master of Home Economics Education

M. H. E. Ed. Master of Home

Economics Education
mhf medium high frequency
M-H-F Massey-Harris-Ferguson
MHFNZ Mental Health Foundation of New Zealand
mhg message-header generator
μHg microns of mercury
MHG Middle High German
mhhw mean higher high water
MHI Manufactured Housing Institute; Material Handling Institute; Metal Hydrides Incorporated; Mitsubishi Heavy Industry
mhic microwave hybrid integrated circuit
M.Hi.E. Master of Highway Engineering
M.Hi.Eng. Master of Highway Engineering
MHII Material Handling Institute Incorporated
MHJC Mary Holmes Junior College
MHK Member of the House of Keys (Isle of Man)
mhl metal halide lamps
MHL Manaus Harbour Limited; Mission Hills Library
M.H.L. Master of Hebrew Literature
MHLF Mutual Home Loan Funds
MHLG Ministry of Housing and Local Government
mhls metabolic heat-load simulator
mhlw mean higher low water
Mhm Mannheim
MHM Mill Hill Missionary
MHMA Mobile Homes Manufacturers Association
MHMC Mercy Hospital and Medical Center; Montefiore Hospital and Medical Center
mho unit of conductance or reciprocal ohm
M. Hor. Master of Horticulture
M. Ho. Sc. Master of Household Science
MHP Missouri Highway Patrol
mhpg (MHPG) 3-methoxy-4-hydroxy phenylethylene
MHQ Maritime Headquarters; Mediterranean Headquarters
mhr manhour(s); maximum heart rate; microwave hologram radar
mhr (MHU) mental health unit
MHR Member of the House of Representatives
MHRA Modern Humanities Research Association
MHRF Mental Health Research Fund

MHRI Mental Health Research Institute (University of Michigan)
MHRT Mental Health Review Tribunal
mhs medical history sheet
MHS Massachusetts Historical Society; Measurement Handicap System; Morris High School; Musical Heritage Society
MHSA Military Historical Society of Australia
MHSc Master (Mistress) of Household Science
MH strain Mill Hill (viral) strain
mht mean high tide; mild heat treatment; military hospital trainee
mht (MHT) missile-handling trailer
MHT Museum of History and Technology (Smithsonian Institution)
MHTA Mental Hygiene Therapy Aide
MHTC Manufacturers Hanover Trust Company
MHTF Manhattan Homicide Task Force (NYPD)
MHTG Marine Helicopter Training Group
MHTGR Modular High Temperature Gas Reactor
mhtl mean high tide line
Mhtn Manhattan
M. Hu. Master of Humanities
mhv mean horizontal velocity; murine hepatitis virus
mhw mean high water
MHW Mental Health Worker; Ministry of Health and Welfare
mhwli mean high water lunitidal interval
mhwlr mobile hostile-weapon-locating radar
mhwn mean high water neaps
mhws mean high water springs
M. Hy. Master of Hygiene
MH y C Miguel Hidalgo y Costilla
M. Hyg. Master of Hygiene
mhz (MHz) megahertz(es), formerly megacycle(s) per second
mi malleable iron; manual input; marginal inscription; mentally ill; metabolic index; middle initial; mildew; mile(s); mill; minor; minute(s); mitral; mitral insufficiency; mutual inductance
mi (MI) myocardial infarction
m & i modernization and im-

provement; municipal and industrial
m of i moment of inertia
mi (Italian—third tone in diatonic scale, E in fixed-do system)
Mi Mach indicated; Mach speed indicated; Miami; Minor; Mitte
Mi Michel (stamp catalog)
MI Maintenance Instruction; Mare Island; Marshall Islands; Match Institute; Mauritius Institute; Meat Inspection (US Department of Agriculture); Mellon Institute; Member of the Institute; Member of the Institution; Metal Industries; Michigan; Military Intelligence; Ministry of Information; Missouri-Illinois (railroad); Mounted Infantry
M-I Missouri-Illinois (railroad)
M & I Manpower and Immigration (Canada)
Mi-1 Soviet utility helicopter nicknamed Hare
mi² square miles(s)
mi³ cubic mile(s)
MI 5 (British) Military Intelligence Security Service (somewhat equivalent to American FBI)
MI-6 Military Intelligence 6 (British external intelligence organization)
Mi-8 Soviet transport helicopter nicknamed Hip
Mi-10 Soviet heavy-transport helicopter nicknamed Harke
Mi-12 Soviet heavy helicopter nicknamed Homer by NATO and in the mid-1970s allegedly the world's heaviest and largest aircraft of its kind
Mi-16 40-ton Soviet helicopter
mia missing in action (military personnel)
mia (MIA) missing in action
MIA Malaysian Institute of Art; Malleable Ironfounders' Association; Manila International Airport; Manitoba Institute of Agrologists; Marble Institute of America; Masonry Institute of America; Media Information Australia; Miami, Florida (airport); Mica Industry Association; Millinery Institute of America; missing in action; Montgomery Improvement Association; Murrumbidgee Irrigation Area (Australia)
M.I.A. Master of International

Affairs
MIAA Mortgage Insurers Association of Australia
MIAC Manufacturing Industries Advisory Council
MIA-CHI Miami—Chicago
MIA-LAX Miami—Los Angeles
MIA-NY Miami—New York
MIAP Member of the Institution of Analysts & Programmers
MIAPD Mid-Central Air Procurement District
MIARS Maintenance Information Automated Retrieval System (USN)
MIAS Major-Item Automated System (USA)
MIA-SAN Miami—San Diego
MIA-SFO Miami—San Francisco
MIASI Moore Institute of Art, Science, and Industry
MIA-TOR Miami—Toronto
MIB Management Improvement Board; Maritime Index Bureau; Meat Inspection Branch; Medical Information Bank; Mental Information Bureau; Metal Information Bureau; Michigan Inspection Bureau; Military Intelligence Branch; Military Intelligence Bureau; Missouri Inspection Bureau; Sir Marc Isambard Brunel
mibk methyl isobutyl ketone
mic machine index card; micrometer; microphone; microwave integrated circuit; military-industrial complex; minimum ignition current
mic (MIC) methyl isocyanate (poison gas)
Mic The Book of Micah; Microscopium (constellation)
MIC Malayan Indian Congress; Management Information Center; Marshall Islands Congress; Medical Information Center; Mitsubishi International Corporation; Monaco Information Centre; Motorcycle Industry Council; Motors Insurance Corporation; Music Industry Council
mica. macro instruction compiler assembler
mica (Spanglish—migration card)—issued by the U.S. Immigration and Naturalization Service
MICA Maternity and Infant Care Association; Medical Imaging Centers of America;

Medicare Insurance Counseling and Advocacy; Moscow Institute for Complex Automation
micbm (MICBM) mobile inter-continental ballistic missile
micc mineral-insulated copper-covered (cable); miniature integrated circuit computer
MICC Malaysian International Chambers of Commerce
MICE Member of the Institution of Civil Engineers
Mich Michael; Michelle; Michigan; Michiganite; Michoacan; Mitchell
MI Chem E Member of the Institution of Chemical Engineers
Michl Michael
mick manufacturer's item correlation key
Mick Michael
Mickey Mickey Mouse
Mickey D's McDonalds
MICLASS Metal Institute Classification
MICMA National Ice Cream Mix Association
MICMD Milwaukee Contract Management District
mic-min micro-mini (automobiles)
MICOM Missile Command(er)
micpac molecular integrated circuit package
mic. pan. mica panis (Latin—bread crumb)
MICPS Microfiche Interface Controller-Processor System
micr magnetic ink character recognition; microscope; microscopic; microscopy
micr (MICR) magnetic-ink character recognition
Micr Microscopium
MICR Magnetic Ink Character Recognition
micro 10^{-6}
micro (Greek *mikros*—small) microbe, microbiology, microcephalic, micrometer, microscope
Micro Micronesia (Trust Territory of the Pacific); Micronesian
microbiol microbiology
microbop microbopper (underground slang—very young person attuned to the modern scene)—see macrobop
Microcard Microcard Editions
microcephs microcephalics (small-headed people)

microcom microcomputer (pocket calculator)
microdoc microphotography and document (reproduction)
microeco microeconomics
microelectro microelectronic(s)
Microg Microgramma
micro-id microscopic identification disk
micro-in. micro-inch
micromation microfilm + automation
micromoms micromomentaries (split-second facial expressions)
micron millionth of a meter
Micron Micronesia; Micronesian
Micronesia U.S. Trust Territory of the Pacific including the Caroline, Mariana, and Marshall islands
micropaleo micropaleontology
micropros microprocess(ing); microprocessor
micros microscopy
micro(s) microcomputer(s)
MICROSIFT Microcomputer Software Information for Teachers
microt microtome
microwav microwave(able)
micr's magnetic ink characters
MICRS Magnetic Ink Character Recognition System
mics metal-insulated copper-sheathed (cable)
mic's military-industrial complex executives; military-industrial complex salesmen
MICS Museum of the International College of Surgeons
MICTI Ministerio de Industria, Comercio, Turismo, e Integración (Spanish—Ministry of Industry, Commerce, Tourism, and Integration)—Peru
micu (MICU) medical intensive care unit; mobile intensive care unit
micv (MICV) mechanized infantry combat vehicle(s)
mid. mentioned in dispatches; middle; midnight
mid. (MID) minimal inhibiting dose; minimum infective dose; multi-infarct dementia; multiple-infarct dementia
Mid Midshipman
MID Merida, Yucatan (airport); Midway Islands (in mid-Pacific); Military Information Division; Military Intelligence Division; Multi-

Infarct Dementia

M.I.D. Master of Industrial Design

midac management information for design and control

Midac Michigan digital automatic computer

midas modified-integration digital-analog simulator (USAF)

MIDAS Maintenance Integrated Data Access System; Materials for Industry Data and Applications Service; Media Investment Decisions Analysis Systems; Meteorological Information and Dose Acquisition System; Missile Defense Alarm System; Missile Detection Anti-Surveillance

midcult middle-class culture

Middlx Middlesex

Middx Middlesex

Middy Midshipman

Middys Midshipmen

MIDEASTFOR Middle East Air Force (USN)

MIDEC Middle East Industrial Development Corporation

MIDELEC Midlands Electricity Board

MIDF Major Item Data File

MIDFL Malayan Industrial Development Finance Limited

Mid-Glam Mid-Glamorgan

midi (MIDI) musical instrument digital interface

midis mid-length (below-the-knee) skirts

Midl Midlands; Midlothian

Midland Midland Bank Limited

Mid Lat Middle Latin

MIDLNET Midwest Region Library Network

Mid Loth Midlothian

midmo middle of the month

Midn Midshipman

MIDP Major-Item Distribution Plan

midr mandatory incident and defect report(ing)

mids middies (middieblouses, midshipmen); missile ignition and destruct simulator

Mids Midlands

mid. sag. midsagittal

Mids N D A Midsummer-Night's Dream

midssn mid-season

'mid(st) amid(st)

Midsummer Midsummer Day (Saturday nearest June 21 or 22); Midsummer Eve or Mid-

summer Night

MIDU Mineral Investigation Drilling Unit

midw midwestern

midwk midweek

mie military-industrial establishment

mie miércoles (Spanish—Wednesday)

M.I.E. Master of Industrial Engineering

MIECO Marshall Islands Import-Export Company

MIEE Member of the Institution of Electrical Engineers

MIEL Malaysian Industrial Estates Limited

mierc miércoles (Spanish—Wednesday)

Mies Ludwig Mies van der Rohe (1886–1969)

mif merthiolate-iodine-formaldehyde (fecal examination technique); modulus irregularity factor

mif (MIF) migratory inhibitory factor

MIF Market Intervention Fund; Milk Industry Foundation; Miners International Federation

MIFCT Moscow Institute of Fine Chemical Technology

MIFI Moskovskiy Inzhenerno Fizicheskiy Institut (Russian—Moscow Engineering Physics Institute)

mifil microwave filter

MIFT Manchester International Freight Terminal

mig magnesium-inert gas; metal inert gas (welding)

MIG Marine Industry Group; Mikhail Ivanovich Glinka; Soviet jet fighter aircraft named for designers Mikoyan and Gurevich

MIGA Make It Go Away

MIGB Millinery Institute of Great Britain

mightn't might not

Mig¹ Miguel (Spanish—Michael)

Mignon opera by Thomas

migra migración (Mexican-American slang—immigration or migration)—*la migra* means the U.S. Border Patrol or the Immigration and Naturalization Service

mi/h mile(s) per hour

M.I.H. Master of Industrial Health

mihn-baau (Cantonese Chinese—bread)

mihped microwave-induced

helium-plasma emission detection

MIHS Marshall Islands High School

MIIA Medical Information and Intelligence Agency

MIIDS Maine's Integrated Inservice Delivery System

MIIF Master Item Intelligence File

MI Inf Sci Member of the Institute of Information Scientists

MIIS Marshall Islands Intermediate School

mij maatschappij (Dutch—company, society)

M i J Made in Japan

MIJ Muhammad Ali Jinnah

miji meaconing, interference, jamming, intrusion

Mik Mikhail

mike micrometer; microphone

Mike letter M radio code; Michael

Mikhail Mikhaylovich

Mikimotos Mikimoto cultured pearls

mikos mindervärdighets komplex (Swedish—inferiority complex)

Mikrop Mikropunkt (German—microdot)—microfilm marvel of World War II when a page of top-secret information could be reduced to a dot no larger than the dot over a letter i and then could be enlarged when needed

mil mileage; military; militia; milieme; million; 1/1000 inch; 1/10 cent; 1/1000 Palestinian pound (currency formerly used in Israel)

m-i-l mother-in-law

Mil Milan; Milford Haven (British maritime abbreviation); Military; Milwaukee

MIL Malaya Indonesia Line; Member of the Institute of Linguists; Microsystems International Limited; Milan, Italy (Malpensa Airport); Wisconsin

MILA Merritt Island Launch Area

MilAdGru Military Advisory Group

Mil Att Military Attaché

milc military characteristics

MILC Midwest Inter-Library Center

MILCAP Military Civic Action Program; Military Civil Action Plan(ning); Military Contract Administration Pro-

cedure(s)
MILCASE Military Career Awareness Course for Educators
milcomsat military communication satellite
milcrit military critic(ism)
MILDAT Military Damage Assessment Team
mildec(s) military decision(s)
Mil Dist Military District
Mil Dist 1 Virginia from 1869 to 1870
Mil Dist 2 North Carolina from 1868 to 1870; South Carolina from 1868 to 1876
Mil Dist 3 Alabama from 1868 to 1874; Florida from 1868 to 1877; Georgia from 1870 to 1871
Mil Dist 4 Arkansas from 1868 to 1874; Mississippi from 1870 to 1876
Mil Dist 5 Louisiana from 1868 to 1877; Texas from 1870 to 1873
mile mille passuum (Latin— 1000 paces), a pace being a double step
MILES Multiple Integrated Laser Engagement System
Mil-Hndbk Military Handbook
MILIMETS Military Meteorological System
milit military; militia
Mil Jrn Milwaukee Journal
milk of magnesia magnesium hydroxide—Mg(OH)$_2$
mill. millinery; milling; million(s)
Mill Million(en) [German— million(s)]
milli 10^{-3}
Millo Escamillo
mil m/t million metric tons
milob military observer
milpac military personnel accounting activity
MILPERCEN Military Personnel Center
mil pers military personnel
MILPO Military Personnel Office
M.I.L.R. Master of Industrial and Labor Relations
milrep military representative
mils missile impact locator system
mil-s milling specification(s)
MILSATCOM Military Satellite Communications
MILSIMS Military Standard Inventory Management System
milspec military specification

Mil-Spec Military Specification(s)
milstac military staff communication
milstam military staff memorandum
MILSTAMP Military Standard Transportation and Movement Procedures
Mil-Std (MIL-STD) Military Standard
MILSTRAP Military Standard Requisitioning and Accounting Procedures
MILSTRIP Military Standard Requisitioning and Issue Procedures
Mil Sym Milwaukee Symphony
Milt Milton
Milw Milwaukee
Milw German Milwaukee German
MILW Milwaukee Route (Chicago, Milwaukee, St Paul & Pacific Railroad)
mim micro-impulse mosaic, mimeograph(ing; y)
mim (MIM) mobile interceptor missile
M i M Morality in Media
MIM Maintenance Instruction Manual (DoD); Material Inventory Master; Mount Isa Mines (Queensland)
MIM-3A Douglas Nike-Ajax surface-to-air missile
MIM-10 military designation of the Boeing Bomarc missile
MIM-14A Douglas surface-to-air missile called Nike-Hercules and armed with a heavy-explosive or nuclear warhead
MI Mar E Member of the Institute of Marine Engineers
MIMB Malaysia International Merchant Bankers
MIME Midland Institute of Mining Engineers
MI Mech E Member of the Institution of Mechanical Engineers
mimeo mimeograph(ed)
mimic. microfilm-information master-image converter
mi/min miles per minute
mimo man in-machine out; many-input many-output (computer)
MIMR May Institute of Medical Research
mims mineral-insulated metal-sheathed (cable)
MIMS Major Item Manage-

ment System
MIMS Monthly Index of Medical Specialties
mimsy miserable and flimsy
min minim; minimum; minor; minority; minute
min minore (Italian—minor); *minuto* (Portuguese or Spanish—minute)
Min Minister; Ministry; Minoan
Min Ministerio (Portuguese or Spanish—Ministry); *Ministro* (Portuguese or Spanish— Minister)
MIN Media Industry Newsletter
Min Agric Ministry of Agriculture
min b/l minimum bill of lading
M-in-C Matron-in-Chief
MINCEX Ministerio de Comercio Exterior (Spanish— Ministry of Foreign Trade)
Min Counc Mining Councillor
mind. magnetic integrator neutron duplicator
Mind Mindanao
mindac miniature inertial navigation digital automatic computer
mindd minimum due date
Min Def Ministry of Defence
MINDUR Ministerio de Desarallo Urbano (Spanish— Ministry of Urban Development)
Min. E. Mining Engineer
MINE Minnesota Information Network for Educators
Mineap Minneapolis
minec military necessity
minelco miniature electronic component
mineola orange + tangerine (hybrid citrus fruit)
mineral. mineralogy
Mineral Soc Mineralogical Society
minex minelaying, minesweeping, and minehunting exercise
MINFAR Ministerio de las Fuerzas Armadas Revolucionarias (Spanish—Ministry of the Revolutionary Armed Forces)—Cuba
minfin minicare finish(ing)
Min Fuel Ministry of Fuel and Power
mingy mean and stingy
Minho Entre Douro e Minho (Porgutuese—Between the Douro and the Minho)— province of Portugal
Min Hous Ministry of Housing

MINI Minicomputer Industry National Interchange

minibop minibopper (underground slang—older child attuned to the modern scene)— *see* macrobop

minibra(s) miniature brassiere(s)

minibus miniature autobus

minicam lightweight miniature camera; miniature camera

minicane miniature hurricane

minidoc miniature documentary (radio or tv)

minimax minimize maximum possible losses

MININT Ministerio del Interior (Spanish—Ministry of the Interior)

mininuke(s) miniature nuclear-explosive device(s)

miniskirt(s) short skirt(s)

minisym miniature symphony

minium red lead (lead oxide)

Mink Minkus (stamp catalog)

Minkies Minquier Islands (Rocks)

min/mc minimum material condition

Minn Minnesota; Minnesotan

Minne Minnesota

Minn Geol Surv Minnesota Geological Survey

Minn Hist Soc Minnesota Historical Society

Minn Orch Minnesota Orchestra

Minn Trib Minneapolis Tribune

Min° Ministro (Spanish—Minister, Ministry)

Min P Minister Plenipotentiary

MINP Mallacoota Inlet National Park (Victoria, Australia); minpac; Mine Warfare Forces, Pacific (USN)

MINPAC Mine Warfare Forces, Pacific (USN)

Min PBW Ministry of Public Building and Works

Min Plenip Minister Plenipotentiary

min prm minimum premium

Min PW Ministry of Public Works

Minquiers Minquier Islands (Rocks)—also called the Minkies

Minquiers (French—The Minkies)—semi-submerged reefs and rocks in Gulf of St Malo between Jersey and port of St Malo on the English Channel

minr minimum room rate desired

Min^r Minister

Min Res Minister Residentiary

MINREX Ministerio de Relaciones Exteriores (Spanish—Ministry of Foreign Relations)

min rnfl minimum rainfall

MINRON Mine Squadron

mins minutes

mins (MINS) minor(s) in need of supervision

M Inst BE Member of the Institute of British Engineers

M Inst Met Member of the Institute of Metals

M Inst SP Member of the Institution of Sewage Purification

MINTACTS Mobile Integrated Telemetry and Tracking System

Min Tech Ministry of Technology

mintie minimum test instrumentation equipment

min till. minimum tillage

M. Int. Med. Master of Internal Medicine

min trq minimum torque

MINWR Merritt Island National Wildlife Refuge (Florida)

min wt minimum weight

mio meteoritic impact origin; minimum identifiable odor

MIO Marine Inspection Office; Metric Information Office; Mobile Issuing Office; Movements Identification Order

Mioc Miocene

MIOUDO Museo del Instituto Oceanográfico de la Universidad de Oriente (Spanish—Museum of the Oceanographic Institute of the University of Oriente)

mip magnetic-induced polarization; malleable iron pipe; marine insurance policy; mean indicated pressure; missile impact predictor; modulated interference plan; monthly investment plan; mortgage insurance premium(s)

MIP Manufacturers of Illumination Products; Marine Interdiction Program; Material Improvement Program; Metals Investigation Proprietary; Methods Improvement Program; Military Improvement Program

mipe modular information-processing equipment

mipir missile precision instrumentation radar

MIPL Mauritius Institute Public Library (Port Louis)

mip/ma missile in place/missile away

MIPO Multiple Item Purchase Order

Miporn Miami pornography (FBI investigation's code name covering billion-dollar pornographic racket)

MIPR Member of the Institute of Public Relations; Military Interdepartmental Purchase Request

MIPRO Manufactured Imports Promotion Organization (of Japan)

mips male iron-pipe size; million instructions per second

MIPS Modular Integrated Pallet System

MIPTC Men's International Professional Tennis Council

Mipu Mikropunkt (German—microdot)—espionage technique assuring transmission of microscopic messages no bigger than a dot

Miq Maiquetía (Venezuela's principal airport)

mir memory information register; minimum implementation requirement; mirror; music information retrieval

M Ir Middle Irish

MIR Manufacturing Inspection Record; Medical Inspection Room; Missile Intelligence Report; Movement for International Reconciliation

MIR Movimiento de Izquierda Revolucionaria (Spanish—Movement of the Revolutionary Left)—active in Bolivia, Chile, Ecuador, Peru, and Venezuela

MIRA Motor Industry's Research Association

MIRA Monthly Index of Russian Accessions

mirac microfilmed reports and accounts

MIRAC Management Information Research Assistance Center

miracl mid-infrared advanced chemical laser

mirad monostatic infrared intrusion detector

MIRADOR Multinational Investment Review Agency and Department of Research

MIRADS Marshall Information Retrieval and Display System

miras mortgage interest relief at source
MIRC Member of the Idle Rich Class
mird medium internal radiation dose
MIRE Media Information Research Exchange; Member of the Institution of Radio Engineers
mirfac mathematics in recognizable form automatically compiled
Miri Miranda; Miriam(ne)(ah)
MIRINZ Meat Industry Research Institute of New Zealand
mirl medium-intensity runway lights
MIRPL Major Item Repair Parts List
MIRR Materials Inspection and Receiving Report
MIRROS Modulation-Inducing Reactive-Retrodirective Optical System
mirv multiple independent reentry vehicle
mirv (MIRV) multiple independently-targeted reentry vehicle (warhead)
mirving fitting missiles with multiple warheads
mis metal insulator semiconductor; miscarriage; missing; mistake(n)
Mis. Miserere (Latin—have mercy); Misisipi (Spanish—Mississippi)
MIS Management Information Services; Management Information System; Master Integrated Schedule; Material Inspection Service; Met Path Information System; Migrant Information Service; Military Intelligence Service; mine issuing ship (naval symbol); Mining Institute of Scotland; Minstrel Instruction Society; Modified Initial System
M.I.S. Master of International Service
misa (MISA) medium-income sustainable agriculture
MISAA Middle-Income Student Assistance Act
mis. accur. misce accuratissme (Latin—mix very intimately)
misad misadventure
misc miscarriage; miscellaneous; miscible
MISC Malaysian International Shipping Corporation
mis. caute misce caute (Latin—mix cautiously)

miscend. miscendus (Latin—to be mixed)
miscg miscarriage
miscld miscalculated
miscln miscalculation
miscon misconduct
miscy miscellaneously
mis doc miscellaneous documents
MISE Member of the Institution of Sanitary Engineers
miser. microwave space relay
MISER Management Information System for Expenditure Reporting; Methodology of Industrial System Energy Requirements; Moorfields Information System Exception Reporting
mis. et seg. misce et signa (Latin—mix and write a label)
misg missing
Misha (Russian—Mikhail)—Michael
MISHAP Missile High-Speed Assembly Program
MISI Member of the Iron and Steel Institute
MISL Major Indoor Soccer League
MISLIC Mid-Staffordshire Libraries in Cooperation
mismed mismedication
mis. mei miserere mei (Latin—have mercy on me)
misn misnumbered; mistaken
Misn Bch Mission Beach
MISO Military Intelligence Service Organization
misog misogynic(al)(ly); misogynist;misogynistic(al)(ly); misogyny
misol misologist; misology
MISP Member of the Institution of Sewage Purification
mispo mission summary printout
Misr (Arabic—Egypt)
M I Sr Muy Ilustre Señor (Spanish—Very Illustrious Sir)
MISR Macauley Institute for Soil Research; Major Item Status Report; Material Item Status Report(ing)
miss. mission; missionary
Miss. Mississippi; Mississippian
MISS Management and Information System Staff; Man In Space Soonest; Medical Information Science Section; Mississippi
missilese engineering jargon of guided-missile experts

missilex missile firing exercise
Missini Mussolini's neo-fascist followers
MISSIS Mississippi Student Information System
missy missionary
mist. mistura (Latin—mixture)
Mist Mistress
MIST Manchester Institute of Science and Technology; Medical Information Service (via) Telephone
mistr management of items subsequent to repair
MISTRAM Missile Trajectory Measurement System
mistrans mistranslation
misudstd misunderstand
misudstdg misunderstanding
misudstod misunderstood
mit master instruction tape; milled in transit; minimum individual training; mono-iodotyrosine
mit. mitte (Latin—send)
Mit Mittwoch (German—Wednesday)
M i T Made in Taiwan
M It Middle Italian
MIT Mara Institute of Technology (Kuala Lumpur); Maritime Institute of Technology; Massachusetts Institute of Technology (M.I.T. preferred as periods set it apart from all other MITs); Massachusetts Investors Trust; Materials Interaction Test(s); Military Intelligence Translator; Milwaukee Institute of Technology; Miracidial Immobilization Test
M.I.T. Massachusetts Institute of Technology
MITAGS Marine Institute of Technology and Graduate Studies
MITC Magdalen Island Transportation Company
mite. master instrumentation timing equipment
MITERS Minor Traffic Engineering and Road Safety Improvements
MITGS Marine Institute of Technology and Graduate Studies
MITI Ministry of International Trade and Industry; Ministry of International Trade and Industry (Japan); Ministry of International Trade and Investment (Japan)
mit insuf mitral insufficiency
mito minimum interval takeoff; miscellaneous tool

mito (Latin prefix—thread)— mitosis
mi tp miniature template
MITRE Massachusetts Institute of Technology Research Establishment
Mitrop Dimitri Mitropoulos
Mitropa Mitteleuropäische Schlafund Speisewagen Aktiengesellschaft (German— Middle-European Sleeping Car and Dining Car Company)
MITS Missouri-Illinois Traffic Service; Museum Institute for Teaching Science
mit. sang. mitte sanguinem (Latin—bleed)
Mitsubishi Mitsubishi Bank; Mitsubishi Corporation; Mitsubishi International Corporation
mitt mittente (Italian—sender)
Mitt Mitteilungen (German— communications)
MITT Management Implications of Team Teaching
mit. tal. mitte tales (Latin— send such)
mitt(s) mitten(s)
mitz mitzvah (Yiddish from Hebrew *miswah*—a good deed)
MIU Maharishi International University; Micronesian Insurance Underwriters
MIV Moody's Investor Service (stock exchange symbol)
MIWE Member of the Institution of Water Engineers
MIWMA Member of the Institute of Weights and Measures Administration
mix. mixture
mixt mixture
Mizrachi Merkaz Ruchani (Hebrew—Spiritual Center)—orthodox organization
mizzle mist + drizzle
mj marijuana; megajoule
mJ millijoule
MJ Mary Jane; (underground slang—marijuana); megajoule; Ministry of Justice
M.J. Master of Journalism
M & J sexologists William Masters and Virginia Johnson
MJ Military Justice Reporter
MJA Manuel José Arce; Mortimer J Adler
MJC Manatee Junior College; Masters and Johnson Center; Metropolitan Junior College; Moberly Junior College
MJCA Mississippi Junior College Association

mjd management job description
mjg management job guide
MJI Member of the Journalists Institute
MJ/kg megajoules per kilogram
MJme megajoules metabolizable energy
MJQ Modern Jazz Quartet
MJS Member of the Japan Society
MJV Mojud Hosiery (stock exchange symbol)
mk mark (British equivalent of type)
mk (MK) master clock
mK millikelvin(s)
Mk Mark; markka (Finnish monetary unit)
Mk Manualkoppler (German— manual coupler)—organ
MK Mackey Airlines; Member of Knesset; Mishima-Kisumi (steel company)
M-K Morrison-Knudsen
M/K Member of the Knesset
M.K. Multi-Kontact (electrical accessories)
Mk₃nbc Mark 3 nuclear; biological, chemical (coverall suit)
MKC Kansas City, Missouri (airport)
mkd marked
MKE Milwaukee, Wisconsin (airport)
mkg meter kilogram
MKGM Milli Kütüphane Genel Müdürlügü (Turkish—National Library General Directorate)—Ankara
MKH Mackintosh-Hemphill (stock-exchange symbol)
MKK Mitsubishi Kakoki Kaishi
mkm marksman
Mkm Mohammedan-killed meat
MKM Manawatu Knitting Mills
MKNP Malawi Kasungu National Park (Malawi); Mount Kenya National Park (Kenya)
MKO Mauna Kea Observatory; Muskogee Company (stock exchange symbol)
MKPL Modification Kit Parts List
mkr marker
mkr mikroskopisch (German— microscopic)
Mkr million Swedish kroner
MKR Mkuzi Game Reserve (South Africa)

mks meter, kilogram, solar second system of fundamental standards
mksa meter, kilogram, second, ampere system
mkt market
Mkt Market
MKT Missouri-Kansas-Texas (railroad)
mktg marketing
mktl marketable
Mkt Mgr Marketing Manager
mk tp mark template
MKU Mary Kathleen Uranium (Australian firm)
MKW Military Knight of Windsor
MKY McKee and Company (stock-exchange symbol)
ml machine language; mean level; millilambert(s); milliliter(s); mine layer; mixed lengths; mold line; molder; money list; mother language; motor launch; muzzle-loading
ml (ML) maximum load
m/l middle left; missile lift
m:l monocyte-lymphocyte (ratio)
m or l more or less
µl microliter
ml moneda legal (Spanish—legal tender)
m/l mi letra (Spanish—my letter)
mL millilambert(s)
Ml Malay; Malaya; Malayan; Malaysia; Manuel; marl
ML Manuel; Maori Land; Martin-Marietta (stock exchange symbol); Micro Log; Middle Latin; Military Liaison; minelayer; Mineral Lease; Missile Launcher; Mitchell Library; motor launch; small minesweeper (naval symbol)
ML (Ec) Ecclesiastic Middle Latin
M/L MacNeil-Lehrer, PBS newscasters Robert MacNeil and Jim Lehrer; Maersk Line
M.L. Medicinae Licentiatus (Latin—Licentiate in Medicine)
mla magnetic lens assembly; manpack loop antenna; microwave linear accelerator
MLA Maine Library Association; Maintenance Level Analysis; Manitoba Library Association; Marine Librarians Association; Maryland Library Association; Massachusetts Library Association;

Master Locksmiths Association; Medical Library Association; Member of the Landlord's Association; Member of the Legislative Assembly; Michigan Library Association; Minnesota Library Association; Mississippi Library Association; Missouri Library Association; Modern Language Association; Montana Library Association; Music Library Association

M-LA Mont-Laurier Aviation

M.L.A. Master of Landscape Architecture

MLAA Modern Language Association of America

m'lady my lady

mlaf missile-loading alignment fixture

MLAP Migrant Legal Action Program

ml ar mill arbor

M. L. Arch. Master of Landscape Architecture

mlases molasses

MLAT Modern Language Aptitude Test

mlb multilinear board

mlb (MLB) major league baseball

Mlb thousand pounds

MLB Major League Baseball; Marginal Lands Board; Maritime Labor Board; Multiple Listing Board; Multiple Listing Bureau

ML & BC Montana Logging and Ballet Company

mlbf mean life before failure

mlbm (MLBM) modern large ballistic missile

Mlbo Malabo (formerly Santa Isabel)

MLBPA Major League Baseball Players Association

mlc machine level control; machine location card; main lobe clutter; mesh level control; microelectric logic circuit; missile launch computer (MLC); mixed leucocyte culture; motor load control; multilayer circuit; multilens camera; multiplanar chain link

MLC Maori Land Court; Meat and Livestock Commission; Member of the Legislative Council; Military Liaison Committee; Mutual Life and Citizens (insurance company)

mlca machine level control address

MLCAEC Military Liaison Committee to the Atomic Energy Commission

MLCRA Model Litter Control and Recycling Act

ml cu mill cutter

mld mailed; middle landing; minimum lethal dose; minimum line of detection; molded

mld (MLD) metachromatic leukodystrophy

MLD Missile Launch(ing) Detection

mld₅₀ minimum lethal (radioactive) dose

M. L. Des. Master of Landscape Design

mldg moulding

mldr molder

mle maximum likelihood estimate; maximum loss expectancy; microprocessor language editor

mle *modèle* (French—model, pattern)

Mle Mile

M. L. Eng. Master of Landscape Engineering

MLES Multiple-Line Encryption System

MLEU *Mouvement Libéral pour l'Europe Unie* (French—Liberal Movement for a United Europe)

mlf media language and format

m/lf medium/low frequency

Mlf thousand linear feet

MLF Mobile Land Force(s); motor launch, fast (naval symbol); Multi-Lateral Force

MLF *Mouvement de Libération de la Femme* (French—Feminine Liberation Movement)

ml fx mill fixture

mlg mailing; main landing gear; most languages

MLG Middle Low German

MLG *Ministry of Labour Gazette*

mlge mileage

mlg(s) mailing(s)

mlgt marine light

Ml'H *Musée de l'Homme,* Paris

MLHA Master Ladies Hairdressers Association

mlhw mean lower high water

mli minimum line of interception

M-Li Muller-Lyer (illusion)

M. Lib. Master of Librarianship

M. Lib. Sci. Master of Library Science

MLIRB Multi-Line Insurance Rating Bureau

M. Lit. Master of Letters; Master of Literature

MLK Jr Martin Luther King, Junior

mlk mag milk of magnesia

MLL Manchester Lines Ltd; Manned Lunar Landing; Music Lovers League

Mlle Mademoiselle (French—Miss)

Mlles Mesdemoiselles (French—Misses)

mllw mean lower low water

mllws mean lower low water springs

mlm million locomotive miles

MLMA Metal Lath Manufacturers Association; Miners' Lamp Manufacturers' Association

mln million

Mln Milan

MLN *Movimiento Liberación Nacional* (Spanish—National Liberation Movement)— Guatemalan political party; *Movimiento de Liberacion Nacional* (Spanish—National Liberation Movement)—Uruguay

MLNP Malawi Lengwe National Park

mlnr milliner

MLNR Ministry of Land and Natural Resources

mlns mucocutaneous lymphnode syndrome

MLNS Ministry of Labour and National Service

MLNWR Medicine Lake National Wildlife Refuge (Montana)

MLO Midland Light Orchestra; Military Liaison Office(r)

m'lord my lord

Mloth Midlothian

M Low G Middle Low German

mlp metal lath and plaster; multiple-line printing

m.l.p. *mento-laeva posterior* (Latin—left mento-posterior)

MLP Master Logistics Plan

mlpwb multilayered printed wiring board

MLQ *Modern Language Quarterly*

mlr main line of resistance; minimum lending rate; mortar-locating radar; multiple linear regression; multiple-line reading; multiple-location risk; muzzle-loading rifle

mlr (MLR) minimum lending rate; missile launch(ing) re-

sponse; mixed lymphocyte response
m-l r muzzle-loading rifle
MLR Main Line Rail(way); Marine Life Resources (program)
MLR Modern Labor Review; Modern Law Review
MLRB Master Logistics Review Board; Mutual Loss Research Bureau
mlrc multi-level railway car
m-l rg muzzle-loading rifled gun
MLRP Marine Life Research Program
MLRS Multiple Launch(ing) Rocket System
mls machine literature search-(ing); median longitudinal section; medium life span; medium long shot; milliliters
ml's magnetically levitated railroad trains
ml's (MLs) mine layers
Mls Mills
MLS Microwave Landing System; Mixed Language System; Moon Landing Site; Multi-Language System; Multiple Listing Service
M.L.S. Master of Library Science
M & LS Manistique & Lake Superior (railroad)
MLSA Ministry of Labour Staff Association
mlsc measured logistic support cast
MLSU Moscow MV Lomonosov State University (University of Moscow)
mlt mean low tide; median lethal (radioactive) time (MLT)
mlt (MLT) master library tape; median lethal (radioactive) time
Mlt Malta
mltl mean low tide line
mltn 1000 long tons
mltu missile loop test unit
mlty military
mlu mean length of utterance
m'lud my lord
mlv membrane light valve; murine leukemia virus
mlv(M) murine leukemia virus (Moloney)
mlv(R) murine leukemia virus (Rauscher)
ml vs mill vise
mlw maximum landing weight; mean low water; medium-level waste
MLW Monrovia, Liberia (airport)

M.L.W. Master of Labour Welfare
mlwli mean low water lunitidal interval
mlwn mean low water neaps
mlws mean low water springs; minimum level water stand
mlx millilux
mly multiply
MLYC Moosehead Lake Yacht Club
mm made merchantable; megameter(s); merchant marine; metronome marking; middle marker; millimeter(s); millimicron; mismated; modified mercalli (scale); mucous membrane
mμ millimicron(s)
m'm madam
m/m millimeter(s)—small-arms ammunition term meaning the diameter of a weapon's bore expressed in millimeters
m & m make and mend
μm micrometer; micron(s)
mm med mera (Swedish—and so forth, etc.)
m.m. mutatis mutandis (Latin—with the necessary changes)
mm² square millimeter
mm³ cubic millimeter
mM millimole; millimore
m/M male Mexican
μM micromole(s)
Mm Martyres (Greek—witnesses, martyrs)
MM Master Mason; Machinist's Mate; Maintenance Manual; Majesties; Manufacturing Manual; Marilyn Monroe; Marine Midland (stock exchange symbol); Mariner Mars (NASA project); Martin Marietta; Martyres (martyrs); Maryknoll Missionary; Material Management; maximum misfit; Medal of Merit; mercantile marine; merchant marine; Methadone Maintenance; Messageries Maritimes; *Messieurs* (French—gentlemen); Metropolitan Museum; Mickey Mouse; Military Medal; Minister of Munitions; Mira Mesa; Moral Majority
M-M Marshall-Marchetti
M.M. Master of Music
M/M Mr and Mr; Mr and Mrs; Mr and Ms; Mrs and Mrs; Mrs and Ms
M&M Marxism and Market economy

M & M Merton and Morden
M for M Measure for Measure
M of M Ministry of Munitions; Museum of Man
MM *Marine Marchande* (French—Merchant Marine); *Modern Medicine*
M.M. Maelzel's Metronome
M & M Morbidity and Mortality (Center for Disease Control's weekly report)
mma major maladjustment; multiple module access
MMA Maine Maritime Academy; Malaysia Medical Association; Manitoba Medical Association; Maritime Museum of the Atlantic (Halifax); Massachusetts Maritime Academy; Material Manufacturing Authorization; Merchandise Marks Act; Meter Manufacturers' Association Metropolitan Museum of Art; Missile Manufacturers Area; Monorail Manufacturers Association; Museum of Modern Art; Music Masters Association
MM of A Minute Men of America
mmac multiple model adaptive control
MMAC Material Management Aggregate Code
m-machine marijuana machine
MMAJ Metal Mining Agency of Japan
MMAL Mitsubishi Motors Australia, Ltd
MMAS Manufacturing Management Accounting System
M. Math. Master of Mathematics
MMB Marine Midland Bank; Milk Marketing Board; Mitsui Manufacturers Bank
mm bat main missile battery
MMBC Maryland Motor Boat Club
mmbd million barrels per day
mmBtu/hr million Btu's per hour
mmc marine moisture control; maximum metal condition
MMC Malaysian Marketing Corporation; Malaysian Mining Corporation; Marine Moisture Control; Materiel Management Code; Meharry Medical College; Mitsubishi Motors Corporation; Monopolies + Mergers Commission
MMCB Midwest Motor Carriers Bureau

mmcfpd million cubic feet (of gas) per day

MMcKNP Mount McKinley National Park (Alaska)

MMCL Major Missile Component List(ing)

MMCNY Marine Museum of the City of New York

mmc's money-market certificates

MMCT maritime mobile coastal telegraphy

mmd mass median diameter; master monitor display; moving map display

MMD minelayer, fast (naval ship symbol)

m mde marine marchande (French—merchant marine)

mme maximum maintenance effort

Mme Madame (French—Missus)

MME Manned Mars Expedition; Midland Mathematics Experiment

M.M.E. Master of Mechanical Engineering; Master of Music Education

M. Mech. Eng. Master of Mechanical Engineering

M. Med. Master of Medicine

MMEG Meter Manufacturers' Export Group

Mmes Mesdames (French—ladies)

M. Met Master of Metallurgy

M. Met. E. Master of Metallurgical Engineering

mmf magnetomotive force; micromicrofarad

μμF micromicrofarad(s)

MMF fleet mine layer (naval symbol); *Maggio Musicale Fiorentino* (Italian—Florence May Festival); Milbank Memorial Fund

MMFA Montreal Museum of Fine Arts

mmfds microfarads

MMFI Moravian Music Foundation, Incorporated

MMFPI Man-Made Fiber Producers Institute

mmg medium machine gun

MMGR Masai Mara Game Reserve (Kenya)

MMGS Mount Muhavura Gorilla Sanctuary (Uganda)

M.Mgt.Eng. Master of Management Engineering

mmh/fh maintenance man-hours per flight hour

mmHg millimeter of mercury

mmi management and maintenance inspection; microphage

migration inhibition; modified mercalli intensity

Mmi Miami

MMI Malaysian Marine Industries; Manufacturers Mutual Insurance; Micro-Magnetic Industries; Moslem Mosque Incorporated (formerly American Mohammedan Society)

M. Mic. Master of Microbiology

M. Mi. Eng. Master of Mining Engineering

MMIJ Mining and Metallurgical Institute of Japan

MMIS Master of Management Information Systems; Medicaid Management Information System; Modified Mercalli Intensity Scale

MMJC Meridian Municipal Junior College

m mk material mark

mml multimaterial laminate

mm/l millimols per liter

MMLES Map-Match Location-Estimation System

MMLME Mediterranean, Mediterranean Littoral, and/or Middle East (sector of conflict)

mmm merchandising, marketing, management; military medical mobilization; millimicron(s)

mMm mobile Minuteman missile

MMM Mauritian Militant Movement; Merseyside Maritime Museum (Liverpool); Minerals, Mining, and Metallurgy; Modern Music Masters

MMM Membre de l'Ordre du Mérite Militaire (French—Member of the Order of Military Merit)

MMMA Maine Merchant Marine Academy; Metalforming Machinery Makers Association

MMMC Medical Materiel Management Center

MMMF Multinational-Mixed Manned Force(s)

mmm/fhr maintenance man minutes per flight hour

mmmrpv (MMMRPV) modular multi-mission remotely piloted vehicle

MMMS Modern Music Masters Society

MMM & SA Master Monumental Masons and Sculptors Association

MMN Museum of Man and Nature (Winnipeg)

MM & N Museum of Man and Nature (Winnipeg, Manitoba)

MMNP Mount McKinley National Park (Alaska)

mmo medium machine oil

Mmo Malmö

MMO Maine Meteorological Office; Music Minus One

MMOB Military Money Order Branch

MMOW Machinist's Mate of the Watch (USN)

mmp (MMP) maritime mobile phone

MMP Major Medical Plan-(ning); Masters, Mates and Pilots (union); Military Mounted Police

MM & P Masters, Mates and Pilots

MMPA Marine Mammal Protection Act; Midland Master Printers' Alliance

MMPC maritime mobile phone coastal

MMPDC maritime mobile phone distress and calling

MMPI Minnesota Multiphase Personality Inventory

MMPNC Medical Materiel Program for Nuclear Casualties

mmpp millimeters partial pressure

MMPP Moose Mountain Provincial Park (Saskatchewan)

mmq minimum manufacturing quality

mmr mass miniature radiography; measles, mumps, rubella (vaccine); minimum maintenance requirement; minimum management requirement

mm & r maintenance modification(s) and repair(s)

MMR Main Machinery Room (USN); Mass Media Research; Method of Mixed Ranges

MMRA Maritime Marshland Rehabilitation Administration (Canada)

MMRB Master Material Review Board

mmrbm (MMRBM) mobile medium-range ballistic missile

MMS Manpower Management System; Mass Memory System; Metabolic Monitoring System; Microfiche Management System; Minerals Management Service; Mobile Monitoring System; Modulation Measuring System; motor minesweeper; multimission ship (naval symbol); Multiplex Modulation System

M.M.S. Master of Management Studies; Master of Medical Science

MMSA Materials and Methods Standards Association; Mercantile Marine Service Association; Mining and Metallurgical Society of America

M.M.S.A. Master (Mistress) of Midwifery of the Society of Apothecaries

MMSC Mediterranean Marine Sorting Center

mmscfd million standard cubic feet per day

m&m session morbidity and mortality session

MMSR Master Material Source Record

MMSS Missile Motion Subsystem

MMSW Mine, Mill and Smelter Workers (union)

mmt manual muscle test(ing); maritime mobile telegraphy; memory test(er); missile mate test(ing); multicomponent mass transport; multiple-mirror telescope

MMT Manual Muscle Test; maritime mobile telegraphy

MMTC maritime mobile telegraphy calling

MMTDC maritime mobile telegraphy distress and calling

MMTP Methadone Maintenance Treatment Program

mmtv mouse mammary tumor virus; murine mammary tumor virus

mmu millimass unit(s)

MMU Manned Maneuvering Unit; McMaster University

M.Mus. Master of Music

MM&W McKim, Mead & Wright (American architects)

MMWD Marin Municipal Water District

mmx memory multiplexer

mmy military man years

MMY *Mental Measurements Yearbook*

mn manual; million

m(n) microfilm negative

mn *maison* (French—house)

m.n. *mutato nomine* (Latin—the name being changed)

m/n *moneda nacional* (Spanish—national currency)

Mn Main; manganese

MN Magnetic North; mega-

newtons; Merchant Navy; Minnesota

M.N. Master of Nursing

MN *Magyar Nepkoztarsasag* (Hungarian People's Republic); *Musee Nationale* (French—National Museum)

mna (MNA) multi-network area (tv)

MNA *Matematikmaskinnämnden* (Swedish—Swedish Computing Machinery Board); Multi-National Account(s)

M.N.A. Master of Nursing Administration

MNAG *Museo Nacional de Antropología* (Spanish—National Anthropology Museum), Guatemala

MNAM *Museo Nacional de Antropología* (Spanish—National Anthropology Museum), Mexico

MNAOA Merchant Navy and Airline Officers' Association

M. N. Arch. Master of Naval Architecture

MNAs Members of the National Assembly (Québec)

MNAS Member of the National Academy of Sciences; Military Navigational Aids System

Mnasi Mnasidika

MNB Macias Nguema Biyogo (formerly Fernando Po); Moscow Narodny Bank

M-N BA Multi-National Business Association

MNC Major NATO Commanders; Media News Corporation; Multinational Corporation

mncpef meaning not clear; please explain fully

MNCR *Mouvement National Contre le Racisme* (French—National Movement Against Racism)

MNCRR Metro-North Commuter Railroad

mnc's multinational corporations

MNCS Multipoint Network Control System

mnd minimum necrosing dose

Mnd Mound

MND Ministry of National Defence

M-N D *A Midsummer Night's Dream* (Shakespeare)

mndth mean depth

MNDTS Member of the Non-Destructive Testing Society

M.N.E. Master of Nuclear Engineering

MNEA Merchant Navy Establishment Administration

mnem mnemonic

Mnemo Mnemosyne (goddess of memory and mother of the nine muses)

mnemon minimum unit of information; mnemoneutic(al)-(ly); mnemonic(al)(ist); mnemonician(s); mnemonicon; mnemonic(s); mnemonist(s); mnemonization(al)(ly); mnemonize(r); mnemotechnic(al)(ly); mnemoteechny

M. N. Eng. Master of Naval Engineering

MNF Menagasha National Forest (Ethiopia); Multilateral Nuclear Force (NATO navy)

mnfe missile not fully equipped

mnfg manufacturing

MNFP Multi-National Fighter Program

mnfrs manufacturers

mng managing; meaning

Mng Mongolia(n)

mnging managing

mngmt management

Mngr manager

mngt midnight

mnh mint never hinged

MNH Museum of Natural History (Smithsonian)

MNHN *Museo Nacional de Historia Natural* (Spanish—National Natural History Museum)—Uruguay; *Musée National d'Histoire Naturelle* (French—National Natural History Museum)—Paris

MNI Malaysian National Insurance; Member of the Nautical Institute; Ministry of National Insurance

MNIMH Member of the National Institute of Medical Herbalists

mnl marine navigating light

Mnl Manila; Manuel

MNL Main North Line; Manila, Philippines (airport)

MNLF Malayan National Liberation Front; Moro National Liberation Front

MNLL Malaysian National Liberation League

MNLO Merchant Navy Liaison Officer

MNLOA Merchant Navy and Air Line Officers Association

mnls modified new least squares

MNLS Marine Navigating Light System

mnm minimum; mnemonic (*see* mnemon)
MNM Museum of New Mexico
MNNP Malawi Nyika National Park
mnos metallic nitrogen-oxide semiconductor
M-note $1000 bill
MNP Malay National Party; Marsabit National Park (Kenya); Meru National Park (equatorial Kenya); Mikumi National Park (Tanzania); Mushandike National Park (Rhodesia)
MNPL Machinist Non-Partisan Political League
mnpo main port
MNPS Minimum Navigational Performance Specification
mnpz monopolize
mnpzd monopolized
mnpzg monopolizing
mnpzn monopolization
mnr massive nuclear retaliation; mean neap rise
Mnr Manor
Mnr Mijnherr (Dutch—Mr, Sir)
MNR Mozambique National Resistance (Renamo)
MNR Movimiento Nacionalista Revolucionario (Spanish— National Revolutionary Movement)
MNRJ Museo Nacional de Rio de Janeiro (Portuguese—National Museum of Rio de Janeiro)
MNRS Mobile Neutron Radiography System
MNRU Medical Neuropsychiatric Research Unit
mns metal-nitride-semiconductor (transistor)
Mns Manaus; Mines
M.N.S. Master of Nutritional Science
M. N. Sc. Master of Nursing Science
m'ns'l mainsail
Mnstr Munster
mnt mean neap tide
MNT Minnesota and Ontario Paper (stock exchange symbol)
mntmp minimum temperature
mntn maintain; maintenance
mntnc maintenance
mntnd maintained
mntng maintaining
MNTO Moroccan National Tourist Office
mntr monitor
MNU Maniti Sugar (stock ex-

change symbol)
M.Nurs. Master of Nursing
MNV Marion Power Shovel (stock exchange symbol)
MNWEB Merseyside and North Wales Electricity Board
MNWR Malheur National Wildlife Refuge (Oregon); Mattamuskeet NWR (North Carolina); Merced NWR (California); Mingo NWR (Missouri); Minidoka NWR (Idaho); Mississiquoi NWR (Vermont); Modoc NWR (California); Montezuma NWR (New York); Moosehorn NWR (Maine)
mnx (short-order slang contraction—ham and eggs)
Mnx Manx (Manx Gaelic)
Mnzlo Manzanillo
mo mail order; manual operation; manually operated; masonry opening; mass observation; master oscillator; method of operation; moment; money order(s); monthlies; monthly; month(s); moth eaten; motor operated; mustered out
mo (MO) molecular orbital
mo' more; morning
m-o months old
m/o maintenance-to-operation (ratio)
m & o maintenance and overhaul(ing); management and organization
m.o. modus operandi (Latin— manner, method, or mode of operating, way of working)
m/o mi orden (Spanish—my order)
m/O male Oriental
Mo Missouri; Missourian; molybdenum; Monday; Morris; Moselle; Moses; Mozelle
Mo' Moses
Mo Maestro (Italian—master, title given any great artist, composer, conductor, or teacher)
MO Mail Order; Marketing Organization; Mass Observation; Medical Officer; Meteorological Office; Missouri; Mobile Station; Mohawk Airlines (2-letter coding); Money Order; Monthly Order; Morale Branch (of Secret Service); Movement Order(s); Municipal Office(r)
M-O Morris-Oxford
M & O Muscat and Oran
moa. medium observation air-

craft; minute of angle; missile optical alignment; mud on airstrip
MoA Ministry of Agriculture
M o A Memorandum of Agreement
MOA Marine Office of America; Metropolitan Oakland Area; Metropolitan Opera Association; Metropolitan Opera Auditions; Military Operations Area; Ministry of Aviation; Minnesota Orchestral Association; Municipal Officers Association; Music Operators of America
MOADS Montgomery Air Defense Sector
MOAMA Mobile Air Materiel Area
MOARS Mobilization Assignment Reserve Section
moat. missile-on-aircraft testing)
moAt mainstream of American thought
mob. make or buy; mobile; mobilization; mobilize(d)
mob. mobile vulgus (Latin— disorderly group of people)
Mob Mobile, Alabama (maritime abbreviation)
MOB Main Operating Base; Mobile, Alabama (airport); Montreux-Oberland-Bernois (railway)
Mo' Bay Mobile Bay, Alabama; Montego Bay, Jamaica
mobcom mobile communications
MOBCOM Mobile Command (Canadian)
mobeu mobile emergency unit
MOBIDACS Mobile Data Acquisition System
mobidic mobile digital computer
mobil mobility
mobilarian mobile branch librarian
mobilary mobile library
mobiles motion sculptures (plastic forms in motion)
Mobil Wl Mobil World
mobl macro-oriented business language; macro-oriented business language
mobl möbliert (German—furnished)
moblas mobile laser satellite tracking station
mob lib mobile librarian; mobile library
mob lt man overboard and breakdown light
mobot(s) mobile robot(s)

MOBS Mobile Ocean Basing System; Multiple Orbit Bombardment System

MOBTA Mobilization Table of Distribution and Allowances

mobula model-building language

moc manufacturing other charges; master operation(al) control(ling); mission operations computer; mocassin

MOC Maintenance Operation Center; Makapuu Oceanic Center (Hawaii); Mauna Olu College (Maui)

moca minimum obstruction clearance altitude

MOCA Museum of Contemporary Art

mocamp motor camp; motorists camp

MOCCC Massachusetts Organized Crime Control Council

MOCI Ministry of Commerce and Industry

Mo City Motor City (Detroit)

mocktail(s) mock cocktail(s)—free from alcohol

MoCom Mobile Command

MOCOM Mobile Command (US Army)

mocp missile out of commission for parts

mocr mission operation control room

mocs mocassins

mod magneto-optical disc; manned orbital development (MOD); mesial-occlusaldistal (dental cavities); model; moderate; modern; modernize(d); modification; modify; modular; module

m-o-d mesial-occlusal-distal (inlay)

Mod Modern

M o D Ministry of Defence (British)

MOD Mail Order Department; Medical Officer of the Day; Ministry of Defense; Ministry of Overseas Development; Miscellaneous Obligation Document

modasm modular air-to-surface missile

m-o-d-b mesial-occlusal-distal-buccal (inlay)

modcom modernity commercialized

mod cons modern conveniences

mod-cons modern-construction houses

moddem modulator-demodulator

mod/demod modulate-demodulate; modulating-demodulating (units)

ModE Modern English

MODE Mid-Ocean Dynamic Experiment

modem modulating-demodulating; modulator-demodulator

Modern Lib Modern Library

Modern Nihilist Jean Genet

modf modification; modify

ModGr Modern Greek

ModHeb Modern Hebrew

mod/iran modification, inspection, and repair as necessary

ModL Modern Latin

modo. moderato (Italian—moderately)

mod. pres. modo prescripto (Latin—in the manner prescribed)

modr moderate room rate desired

mods mesial-occlusal-distal (dental cavities); models; moderates; moderators; moderns; modification; modifiers; modulators; modules

MODS Manned Orbital Development Station (or System); Manned Orbiting Development Station (or System); Medically Oriented Data System

modto moderato (Italian—moderately)

moe measure of effectiveness

Moe Moses

MoE Ministry of Education

M o E Ministry of Energy

MOE Major Organizational Entity

MOEA Ministry of Economic Affairs

mof maximum observed frequency; member of (the police) force; metal oxide film

M o F Ministry of Finance

MOF Ministry of Food

mo' fr mother fucker

Mog Margaret

MOG Metropolitan Opera Guild

M.O.G. Master of Obstetrics and Gynaecology

mogas motor gasoline

moggy mongrel

moh material overhead; maximum operating hours

M o H Ministry of Health

MOH Medical Officer of Health; Ministry of Health; Mohawk Airlines

Moham Mohammedan

MOHATS Mobile Overland Hauling and Transport System (USAF)

MOHLG Ministry of Housing and Local Government

μohm microhm

mohms milliohms

moho Mohorovicic discontinuity

Mohole a hole to the Mohorovicic discontinuity, the boundary between the earth's crust and mantle

mohs mud, oil, hooks, slings (oil well insurance)

moi maximum obtainable irradiance; military occupational information; multiplicity of infection

MoI Ministry of the Interior

MOI Military Operations and Intelligence; Ministry of Information

MOIC Medical Officer in Command

M.O.I.G. Master of Occupational Information and Guidance

moip missile on internal power

MOIS Minnesota Occupational Information System

Moish Moishe

Moish Moishe (Yiddish—Moses)

moiv mechanically operated inlet valve

Mok Mokpo

MOK Mohawk Carpet Mills (stock exchange symbol)

mol machine-oriented language; maximum output level; molecular; molecule

mol. mollis (Latin—soft)

Mol Mollendo

M o L Minister of Labour; Ministry of Labour

MOL Manned Orbiting Laboratory; Mitsui-OSK Lines

M.O.L. Master (Mistress) of Oriental Languages

molab mobile laboratory

MOLAB Mobile Lunar Laboratory

Mol Crys Liq Crys Molecular Crystals and Liquid Crystals

MOLDS Management On-Line Data System

Moldv Moldavia; Moldavian

mole. molecular; molecule

molecom molecularized computer

Molink Moscow link (teletype cable circuit linking Moscow's Kremlin with Washington, D.C.'s White House), The Hot Line

moll metallo-organic liquid laser

mol/l molecules per liter

mollie mollienisia (tropical fish)

mollie(s) mare mule(s)—*see* hinny

Mollus Mollusca

MOLLUSA Military Order of the Loyal Legion of the U.S.A.

MOLNS Ministry of Labour and National Service

MOLOC Ministry of Labour Occupational Classification

Mol Phys *Molecular Physics*

MOLS Mirror Optical Landing System

molt. molten

Moluccas Maluku or Spice Islands of Indonesia

mol wt molecular weight

moly molybdenum

mom military ordinary mail; milk of magnesia

mom (MOM) micromation online microfilmer

m-o-m middle of month; milk of magnesia

m/ o m/ *más o menos* (Spanish—more or less)

Mom Momma

MOM *Musée Océanographique Monaco* (French—Monaco Oceanographic Museum)

MoMA Museum of Modern Art

m-o-m in a.m. if no bm by p.m. milk-of-magnesia in the morning if no bowel movement by evening

momar modern mobile army

momau mobile mine assembly unit

Moml Moslem meal

Mo-Mo MF Grant—pseudonym

MOMR Mayor's Office of Manpower Resources

moms missile operate mode simulator

moms *mervaerdiomsaetningsskat* (Danish—value-added tax); *mervardesomsattningsskat* (Swedish—value-added tax)

MOMS Mothers for Moral Stability

MOM/WOW Men Our Masters/Women Our Wonders (anti-feminist acronym reading the same upside down)

mon monetary; monsoon; monument; motor octane number

mon *maison* (French—house)

Mon Monaco; Monday; Monegasque; Mongol(ia)(n); Monitor; Monmouthshire; Monoceros (constellation); Monongahela; Monsieur (French—Mister); monument

Mon *Mónaco* (Spanish—Monaco); *Montag* (German—Monday)

MON *Ministrstwo Obrony Naradowej* (Polish—Ministry of National Defense)

Mona Madonna (Italian—Lady, Our Lady); (Manx—Isle of Man)

Monaco Principality of Monaco (tiny Mediterranean country famed for its gambling casino)

Monag Monaghan

Monas Monastic(ism); Monastery

Monashees Monashee Mountains of British Columbia

monbas monobasic

MONC Metropolitan Opera National Council

Mondale Walter E Mondale, 42nd Vice President of the United States

mon/dir monitoring direction

MONEVAL Monthly Evaluation Report (USA)

monex monsoon experiment

mong *mongolisch* (German—Mongolian)

Mong Mongol; Mongolia(n)

Mongolia Mongolian People's Republic (landlocked Asiatic nation of great antiquity), *Bügd Nayramdakh Mongol Ard Uls*

Mongoose Mongoose Gang (secret police in Grenada)

'mongst amongst

mon-H monohydrogen

Moni Monica; Monika

monic monocular

monik moniker

Monitor *Christian Science Monitor*

Mon Not Roy Astron Soc. *Monthly Notices of the Royal Astronomical Society*

mono mononucleosis; monophonic; monopoly; monopropellant; monorail(road); monotype; monotyper

mono (Latin prefix—alone, one, single)—monograph, monorail

Mono Monocerus (constellation)

monob (MONOB) mobile noise barge

monocl monoclinic

monocot(s) monocotyledon(s)

Monod Monon Railroad

monog monogram; monograph

monokini one-piece topless bikini (swimsuit)

monos monitor out of service

monot monotonous; monotony; monotype; monotypic

monpl monopoly

Mon River Monongahela River

mons (Latin prefix—mountain)—monstrosity

Mons *Monsieur* (French—Mister)

Mons Cur Monsoon Current

Monsig *Monseigneur* (French—My Lord)

monsoons seasonal storms of southern Asia

monstro(s) monstrosity; monstrosities

Mont Montana; Montanan; Monterrey; Montevideo; Montgomery; Montgomeryshire; Montpelier; Montreal

Montagne *Ce qu'on entend sur la Montagne* (French—What one hears on the mountain)—Liszt's Symphonic Poem No 1

Monte Montague; Monte Carlo; Montebianco (Mont Blanc); Montefiore; Montevideo; Montgomery

Montesquieu Baron de La Brède et de Montesquieu, Charles-Louis de Secondat (1689—1755)

Montezuma Castle Montezuma Castle National Monument in central Arizona

Montgom Montgomeryshire

Montie Montgomery

Montparno Montparnasse

Montr Montreal

montrg monitoring

Mont S *Montreal Star*

Monty Montagu; Montague; Montana; Montgomery; Montmorency

Mony monastery

MONY Music Operators of New York; Mutual Life Insurance Company of New York

MOO Money Order Office

Moody's *Moody's Investors Service*

Moody and Sankey Dwight Lyman Moody and Ira David Sankey—an evangelist preacher and his organist partner

moop mechlorethamine, vincristine, procarbazine, prednisone (Hodgkin's disease treatment)
MOOP Ministerstvo Okhranenia Obshehestvennogo Poriadka (Russian—All-Union Ministry for the Preservation of Public Order)—secret police agency
Moor Dartmoor Prison, Devon, England
Moore's Adj Moore's International Adjudications
Moore's Arb Moore's International Arbitrations
Moore's Dig Moore's Digest (of international law)
MOOSE Move Out of Saigon Expeditiously (USA)
moot. move(d) out of town; moving out of town
mop mother-of-pearl; mustering-out pay
mop. medical outpatient; mother-of-pearl; mustering-out pay
M o P Member of Parliament; Minister of Pensions; Ministry of Pensions; Minister of Power; Ministry of Power; Minister of Production; Ministry of Production
MOP Migrant Opportunity Program
MOP Ministerio de Obras Publicas (Spanish—Ministry of Public Works)
mopa master oscilator power amplifier
MOPA Museum of Photographic Arts (San Diego)
MoPac Missouri Pacific—Texas & Pacific (railroad)
mopar master oscillator-power amplifier radar
mopb manually operated plotting board
mopeds motorized pedals (bicycles containing auxiliary motors)
mopf missile onloading prism fixture
MOPH Military Order of the Purple Heart
mopr manner of performance rating; mop rack
MOPS Merchandise Order-(ing) Processing System; Missile Operations System
MOPSS Multispectral Opium Poppy Sensor System
M. Opt. Master of Optometry
mor middle of the road; morocco; mortar
mor (MOR) middle-of-the-

road (tv program)
mor morendo (Italian—dying away, gradual softening of tone and slowing of tempo)
Mor Morelia; Morelos; Morisco; Moroccan; Morocco
M o R Ministry of Reconstruction
MOR Mandatory Occurrence Report(ing); Military Operations Research
Morav Moravia; Moravian
Morb Morbihan
MORC Medical Officers Reserve Corps; Midget Ocean Racing Club (smallest racing cruisers)
Mord Mordehai
Mordhy Mordehai
mor. dict. more dicto (Latin—as directed)
Mordy Mordechai
MORE Mission for Outreach, Renewal, and Evangelism
Moreau Louis Moreau Gottschalk (1829—1869)
moreps monitor station reports
morf hermaphrodite
mor fib moral fiber
morf(ie) morphine
MORG Museo Oceanografico de Rio Grande (Portuguese—Oceanographic Museum of Rio Grande)—Brazil
morg mar morganatic marriage
MORI Market Opinion and Research International
moritzer mortar howitzer
MORL Manned (or Medium) Orbital Research Laboratory
Mor Lib Morgan Library
Morm Mormon
Morm Mormon, Book of
Mor Maj Moral Majority
morn morning
Moro. Book of Moroni
Moroc Moroccan; Morocco
Morocco Carolina Varga Dinicu; Kingdom of Morocco (North African Arab nation), *al-Mamlaka al-Maghrebia*
morph morphine; morphology
morph (Latin prefix—form, shape)—morphological
morpha hermaphrodite
morpheme smallest sound unit (linguistics)
morphophysio morphophysiologic(al)(ly); morphophysiologist; morphophysiology
Morrow William Morrow
MORS Midland Operational Research Society
mor. sol. more solito (Latin—

in the usual manner)
mort mortal; mortality; mortar; mortgage; mortician; mortuary
mor t Morse taper
moRt mainstream of Republican thought
Mort Mortemart; Mortimer; Morton
mortal. mortality
mos metal-oxide semiconductor; metal-oxide-silicon (compound); missile on stand; mit-out sound (silent film); months; mosaic
mos (MOS) military occupational specialty
Mos Moscow
Mos. Book of Mosiah
Mos Mosca (Italian—Moscow); *Moscou* (French or Portuguese—Moscow); *Moscu* (Spanish—Moscow); *Moskau* (German—Moscow); *Moskou* (Dutch—Moscow)
Moslem (Arabic—True Believer)
MOs Military Observers (UN)
MOS Magneto-Optical System; Management Operating System; Manned Orbital Station; Marine Observation Satellite; Ministry of Supply
MOS Ministwo Opieki Spotecznes (Polish—Ministry of Social Welfare)
MOSA Medical Officers of Schools Association
Mosbas Moscow Basin
Mosby C V Mosby
mosc manned orbital systems concepts
MOSC Midland-Odessa Symphony and Chorale
Mose Moisés; Mosè; Moseley; Mosen; Moses; Moshe
MOSES Manned Open Sea Experimentation Station
mosfet metal-oxide semiconductor field-effect transistor
mosic metal-oxide-semiconductor integrated circuit(s)
MOSID Ministry of Supply Inspection Department
Mosk Moscovici; Moscowitz; Moskowitz
mosm milliosmol(s)
MOSOP Missouri Sexual Offender Program
moss. maintenance-operations support set
MOSS Manned Orbital Space Station; Market Opening Sector Specific
MOSST Ministry of State for Science and Technology (Ca-

nadian)

most. metal-oxide semiconductor transistor

'most almost

MOST Michigan Opportunities and Skills Training

mostl metal-oxide semiconductor transistor logic

mot mean operating time; mechanical operability test; member of our tribe; middle of target; motor; motorized

M o T Minister of Transport; Ministry of Transport

MOT Military Ocean Terminal

MOTAT Museum of Transport and Technology

MOTC Ministry of Transit and Communications (Philippines)

M o TCP Ministry of Town and Country Planning

motel hotel for motorists

moth mother

moth-in-law mother-in-law

MOTNE Meteorological Operational Telecommunications Network, Europe (NATO)

motoboard(s) motorized skateboard(s)

motocross cross-country motorcycle race

mot op motor operated

motorcade motorized-vehicle parade

motorcross motorcycle cross (country race)

MOTOREDE Movement To Restore Decency

Mo' Town Motor Town (Detroit, Michigan)

mots minitrack optical tracking system

MOTU Mobile Technical Unit

mou memorandum of understanding

MoU Memorandum of Understanding

Mountbatten of Burma Admiral of the Fleet and last Viceroy of India known until 1917 as Prince Louis Francis Albert Victor Nicholas of Battenberg (1900–1979)

mounties mounted policemen (especially Royal Canadian Mounted Police)

MOUSE minimum orbital unmanned satellite

mov movable; movement; moving; multiple-orifice valve

mov movimento (Italian—movement)

movem movement overseas

verification of enlisted members (of the USA)

moverep movement report

Move Short Soc Movement Shorthand Society

movi movie; moving pictures

movies moving pictures

MOVIMS Motor Vehicle Information Management System

movord movement order

M o W Minister of Works; Ministry of Works

MOW Moscow, USSR (Vnukovo Airport); Movement for the Ordination of Women

mowasp mechanization of warehousing and shipment processing

MoWD Ministry of Works and Development

MOWOS Meteorological Office Weather Observing System

M o WT Minister of War Transport; Ministry of War Transport

MOWW Military Order of the World Wars

mox mixed oxides (platinum and uranium); oxidized metal explosive

moy money

Moz Mozambique

MOZ Mezhdunarodnaya Organizacia Zhurnalistov (Russian—International Organization of Journalists)

Mozam Mozambique

Mozambique People's Republic of Mozambique (formerly Portuguese East Africa)

Moz Cur Mozambique Current (Natal)

mozza mozzarella

mp mail payment; maintenance part(s); manifold pressure; medium pressure; meeting point; melting point; *mezzopiano* (Italian—half soft, moderately soft); milepost; motion picture; multipole; multipurpose

mp (MP) marginal product

m(p) microfilm positive

m-p metal-point (bullet)

m/p milk powder

m & p materials and processes

m.p. mille pasuum (Latin—thousand paces)—the Roman mile of 1000 paces

mP polar maritime air

MP Member of Parliament; Mercator's Projection; Metropolitan Police; Military Police; Mining Permit; Minister

Plenipotentiary; Minister Provincial; Miscellaneous Proposal; Missouri Pacific (railroad); Mitsubishi Plastics; Mounted Police; Northern Mariana Islands

M/P Memorandum of Partnership

M & P Maryland & Pennsylvania (railroad)

MP Maschinenpistole (German—submachine gun, tommy gun)

mp₁ marginal product of labor

mpa megapascal; multiple product (television) announcement

mpa (MPa) megapascal

mpa (MPA) maritime patrol aircraft

mpa Maryland Port Authority's italicized logotype

MPA Magazine Publishers of America; Magazine Publishers Association; Main Propulsion Assistant (USN); Maryland & Pennsylvania (railroad); Master Photographers Association; Mechanical Packing Association; Medical Procurement Agency; medroxyprogesterone; Metal Powder Association; Metropolitan Pensions Associations; Midwestern Psychological Association; Military Police Association; Mobile Press Association; Modern Poetry Association; Motion Picture Alliance; Music Publishers Association

M.P.A. Marine Physician Assistant; Master of Professional Accounting; Master of Public Administration; Master of Public Affairs

MPAA Motion Picture Association of America; Musical Performing Arts Association

MPAC Master Plan for Academic Computing

MPACS Management Planning and Control System

mpad maximum permissible annual dose

MPAGB Modern Pentathlon Association of Great Britain

mpai maximum permissible annual intake

mpam maritime polar air mass

m part movable partition

mpas millipascal second

MPAS Maryland Parent Attitude Survey

MPAUS Music Publishers Association of the United States

m payl maximum payload
mpb male pattern baldness
MPB Maintenance Parts Breakdown (spare parts); Miniature Precision Bearings; Missing Persons Bureau; Montpelier & Barre (railroad)
mpbb maximum permissible body burden (of radiation)
MPBC Memphis Power Boat Club
mp br multipunch bar
MPBS Mutual Permanent Building Society
MPBW Ministry of Public Buildings and Works
mpc marine protein concentrate; material program code; mathematics, physics, chemistry; maximum permissible concentration; military payment certificate; minimal planning chart; multipurpose carrier
mpc (MPC) marginal propensity to consume
MPC Manpower and Personnel Council; Manpower Priorities Committee; Manufacturing Plan Change; Marine Policy Center (Woods Hole, Mass.); Member of Parliament of Canada; Metropolitan Police College; Metropolitan Police Commissioner; Military Payment Certificate; Military Pioneer Corps; Military Police Corps; Military Police Force; Model Penal Code; Montana Power Company
MPCA Magnetic Powder Core Association; Marine and Ports Council of Australia; Master Pastry Cooks Association
MPCAG Military Parts Control Advisory Groups
MPCB Manufacturing Plan Control Board
mpc black medium-processing channel black
MPCC Minnesota Private College Council
MPCL Movimiento Patriótico Cuba Libre (Spanish—Free Cuba Patriotic Movement)
mpcp missile power control panel
MPCS Master Plan for Computing Services
mpcur maximum permissible concentration of unidentified radionuclides
mpd magnetoplasmadynamics;

maximum permissible dose; missile purchase description; multiple personality disorder
M. Pd. Master of Pedagogy
MPD Metropolitan Park District; Metropolitan Police Department; Military Pay Division
MPDA Motion Picture Distributors Association
MPDFA Master Photo Dealers' and Finishers' Association
mp di multipunch die
MPDPIS Master Plan for Data Processing and Information Systems
MPDS Message Processing Distribution System
MPDSA Master Painters, Decorators, and Signwriters Association
MPDT Minnesota Perception Diagnostic Test
mpe maximum permissible exposure (to radiation)
M.P.E. Master of Physical Education
MPEA Motion Picture Exhibitors Association
MPEAUS Master Printers and Engravers Association of the United States
M. Pe. Eng. Master of Petroleum Engineering
M Pen Minister of Pensions; Ministry of Pensions
MPers Middle Persian
MPES Mathematical, Physical, and Engineering Science (NSF)
mpf motion-picture film; multi-purpose food
MPF Malaysian Peasants Front; Metallurgical Plantmakers Federation; Metropolitan Police Force (London)
mpfg 1000 proof gallons
mpg miles per gallon
MPG Magazine Promotion Group; Max Planck Gesellschaft
MPGA Maine Personnel and Guidance Association; Maryland Personnel and Guidance Association; Metropolitan Public Gardens Association; Michigan Personnel and Guidance Association; Minnesota Personnel and Guidance Association; Missouri Personnel and Guidance Association
mpgn membrano proliferative glomerulonephritis

MPGR Mana Pools Game Reserve (Rhodesia)
MPGS Mobile-Protected Gun System
mph miles per hour
M.Ph. Master of Philosophy
MPH Meat Packing House; Methodist Publishing House
M.P.H. Master of Public Health
MPH Maintenance Parts Handbook
M. Phar. Master of Pharmacy
M. Pharm. Master of Pharmacy
MPHEC Maritime Provinces Higher Education Commission
M. Ph. Ed. Master of Public Health Education
M. P.H. Eng. Master of Public Health Engineering
M.Phil. Master of Philosophy
M. Pho. Master of Photography
mphps miles per hour per second
M. Ph. Sc. Master of Physical Science
M.P.H.T.M. Master of Public Health and Tropical Medicine
M. Phy. Master of Physics
M Phys A Member of the Physiotherapists Association
mpi magnetic particle inspection; maximum point of impulse; mean point of impact; multiphasic personality inventory; multiphoton ionization
mpi (MPI) marginal propensity to invest
MPI Material Process Instruction; Max Planck Institute; Medicine in the Public Interest; Mitsui Petrochemical Industries; Museum of the Plains Indians
MPI Movimiento Pro-Independencia (Spanish—Pro-Independence Movement)— Puerto Rico
M-pill menstruation pill
MPIRO Multiple Peril Insurance Rating Organization
MP & IS Material Process and Inspection Specification
mPk polar maritime air colder than underlying surface
mpl mathematical programming language; maximum payload; maximum permissible language; maximum permissible level; message processing language; mul-

tiple-position lock
MPL Maintenance Parts List; Memphis Public Library; Metropolitan Police Laboratory; Miami Public Library; Milwaukee Public Library; Minnesota Power and Light; Missouri Pacific Lines; Montreal Public Library
M.P.L. Master (Mistress) of Patent Law
MPLA Mountain Plains Library Association
MPLA Movimento Popular Liberação Angola (Portuguese—Popular Movement for the Liberation of Angola)
MPLP Marxist Progressive Labor Party
Mpls Minneapolis
mpm meters per minute; missile power monitor; molepercent metal; multipurpose meal
MPM Milwaukee Public Museum; Modest Petrovich Mussorgsky (1839–1881)
MP-M Museum Plantin-Moretus (Antwerp's museum devoted to book production and typography of Plantin and Moretus)
MPMI Magazine and Paperback Marketing Institute
mpn most probable number
MPNA Midwest Professional Needlework Association
MPNI Ministry of Pensions and National Insurance
mpo memory printout
MPO Memorandum Purchase Order; Metropolitan Police Office (Scotland Yard); Miami Philharmonic Orchestra; Military Pay Order; Military Planning Office(r); Military Post Office; Mobile Post Office; Mobile Printing Office
MPOIS Military Police Operating Information System
M.Pol. Econ Master (Mistress) of Political Economy
MPOLL Military Post Office Location List(ing)
mpp marginal physical product; most probable position
MPP Mailer's Postmark Permit; Maintainability Program Plan(ning); Member Provincial Parliament (Canada); Mothers in Prison Projects
M.P.P. Master (Mistress) of Physical Planning
M & PP Manitou & Pikes Peak (Railroad)
MPPA Music Publishers Pro-

tective Association
MPPCA Maryland Probation, Patrol and Corrections Association
mppcf millions of particles per cubic foot of air
mp pl multipunch plate
mpps million pulses per second
MPPWCOM Military Police Prisoner of War Command
mpq manpower-planning quota(s)
mpr 1000 pair; medium-power radar
MPR Maintainability Program Requirements; Military Pay Record; Mongolian People's Republic
MPR Madjelis Permusjawaratan Rakat (Indonesian—People's Deliberative Assembly); *Maritime Provinces Reports*
MPRC Military Personnel Records Center
mpress medium pressure; medium pressurization
MPRL Master Parts Reference List
M. Prof. Acc. Master of Professional Accountancy
MPRP Mongolian Peoples Revolutionary Party; Muslim Peoples Republican Party
mp & rs motive power and rolling stock
mps marbled paper sides; megacycles per second; meters per second; motor parts stock
mps (MPS) marginal propensity to save; mucopolysaccharidosis
Mp's Minneapolis pimps
MPs Members of Parliament; plural of military police or mounted police
M.Ps. Master of Psychology
MPS Mail Preference Service; Manufacturing Process Specification; Marriage Prediction Schedule; Master Project Summary; Mathematical Programming System; Microprocessor System; Military Postal Service; Milwaukee Public Museum; Minimum Property Standards; Minister of Public Security; Mont Pelerin Society; Motor Products Corporation; Multiprogramming System
M.P.S. Member of the Pharmaceutical Society
MPSA Military Petroleum Supply Agency
MPSC Military Provost Staff

Corps
mpsh mean pressure suction head
MPSM Master Problem Status Manual
M.Ps.O. Master (Mistress) of Psychology Orientation
MPSP Mathematical Problem-Solving Project; Military Personnel Security Program
MPSS Multiple Protective Structure System
M.P.S.W. Master (Mistress) of Psychiatric Social Work
M.Psych. Master (Mistress) of Psychology
M. Psy. Med. Master of Psychological Medicine
mpt male pipe thread; melting point; microprocessing programmable terminal; midpoint; multiple pure tone; multipower transmission
mpt (MPT) miles per tankful
Mpt Maryport
MPT Marquis Public Theater; Maryland Public Television; Minister of Posts and Telecommunications
mpta main propulsion test article
MPTA Machine Power Transmission Association; Municipal Passenger Transport Association
MPTP Music Preference Test of Personality
mpu microprocessor unit (MPU); monitor printing unit
MPU Medical Practitioners Union; Mental Parents Union; Missing Persons Unit (of a police department)
M. Pub. Adm. Master of Public Administration
mpv (MPV) multipurpose vehicle
M-P v Mason-Pfizer virus
mPw polar maritime air warmer than underlying surface
MPW Minneapolis-Moline (stock exchange symbol)
MPWBS Master Plan Works Breakdown Structure
mpx multiplex
mpxr multiplexor (flow chart)
mpy multiply
Mpy Maatschappij (Dutch—company)
MPZ Mid-Continent Petroleum (stock exchange symbol)
mq metol-quinol (MQ); multiple quotient (register); multiplier quotient

mq (MQ) memory quotient; metol-quinone

Mq mosque

MQ merit quotient

M&Q Mines and Quarries

MQA Manufacturing Quality Assurance; Medical Quality Assurance

MQAB Medical Quality Assurance Board

Mqe Martinique

mqf mobile quarantine facility

MQI Maiquetía (Venezuelan airport)

mqil miniature quartz incandescent lamp

mql miniature quartz lamp

MQM Master of the Queen's Music

MQO Marksmanship Qualification Order

MQS Mobile Quality Services

MQT Model Qualification Test

M Quad Charles Bertrand Lewis

MQV *Ministére de la Qualité de la Vie* (French—Ministry of the Quality of Life)

mqyco minimum quantity yards per color

mqyds minimum quantity yards per design

mr machine record(s); machine rifle; map reference; medium range; mental retardation; mentally retarded; metabolic rate; methyl red; mill run; mineral rubber; milliroentgen; mine run; motivational research (MR)

mr (MR) marginal revenue

m/r map reading; middle right

m & r maintainability and reliability; maintainability and repairs; maintenance and repair

mr *meester* (Dutch—master)—attorney-at-law; *mi remesa* (Spanish—my remittance)

mR milliroentgen

Mr Master; Mister; Mother

MR Machinery Repairman; Magnetic Resonating; Marketing Research (division, US Department of Agriculture); Mark Russell; Master of the Rolls; Medical Record; Memorandum for Record; Memorandum Report; Michigan Reformatory; Military Railroad; Military Requirement; Minister Residentiary; Ministry of Reconstruction; Miscellaneous Report; Mobi-

lization Regulation; Monon Railroad; Monthly Report; Morning Report; Multifamily Residential zone; Municipal Reform

M/R map reading; Mates Receipt(s)

M & R maintenance and repairs

M of R Minister of Reconstruction; Ministry of Reconstruction

MR *Marca Registrada* (Spanish—Registered Trademark); *Mobilización Republicana* (Spanish—Republican Mobilization)—Castro-controlled political party in the Nicaraguan underground; *Motormannes Riksforbund* (Swedish—Motorists' Association)

MR-13 *Movimiento Revolucionario de 13 de Noviembre* (Spanish—Revolutionary Movement of 13 November)—Guatemala

mra medium-powered radio range (Adcock); minimum reception altitude

mra (MRA) metro rating area (tv)

MRA Maritime Royal Artillery; Master Retailers Association; Materials Review Area; Moral Rearmament

mraam (MRAAM) medium-range air-to-air missile

mrac manifold-regulator accumulator charging

MRACP Member of the Royal Australasian College of Physicians

mrad megarad; millirad

M. Rad. Master of Radiology

M. Ra. Eng. Master of Radio Engineering

MRAF Marshal of the Royal Air Force

Mr Air Brake George Westinghouse

MRAM Multimission Redeye Air-launched Missile

MRAP Management Review and Analysis Program

MRAS Manpower Resources Accounting System (USAF)

mrasm (MRASM) medium-range air-to-surface missile

mrat medium-range applied technology

MRAUSCAN Masonic Relief Association of the United States and Canada

mrb marble base

MRB Material Review Board; Mileage Rationing Board;

Modification Review Board; Mutual Reinsurance Bureau

MRBA Mississippi River Bridge Authority

mrbm medium-range ballistic missile (MRBM)

MRBP Missouri River Basin Project

mrc magnetic rectifier control

Mrc *Mauricio* (Spanish—Mauritius)

MRC Maintenance Requirements Cards; Marine Research Committee; Market Research Council; Marlin-Rockwell Corporation; Material Redistribution Center; Material Review Crib; Materials Research Corporation; Measurement Research Center; Medical Research Center (Council); Medical Reserve Corps; Men's Republican Club; Metals Reserve Company; Methods Research Corporation; Minnesota Restitution Center; Mississippi River Commission; Model Railway Club; Modern Railroad Club; Motor Racing Club; Movement Report Center

mrca multirole combat aircraft

MRCA Market Research Corporation of America

MRCC Medical Research Council of Canada

MRCF Module Repair Calibration Facility

MRCGP Member of the Royal College of General Practitioners

MRCI Medical Registration Council of Ireland; Medical Research Council of Ireland

MRCIU Men's Residence Center (Indiana University)

MRCo Malaysian Refrigerator Company

MRCO Member of the Royal College of Organists

MRCOG Member of the Royal College of Obstetricians and Gynaecologists

MRCP Maoist Revolutionary Communist Party; Member of the Royal College of Physicians

MRC Path Member of the Royal College of Pathologists

MRCPE Member of the Royal College of Physicians of Edinburgh

MRCPI Member of the Royal College of Physicians of Ire-

land
MRC Psych Member of the Royal College of Psychiatrists
MRCPUK Member of the Royal College of Physicians of the United Kingdom
MRCS Member of the Royal College of Surgeons
MRCSE Member of the Royal College of Surgeons of Edinburgh
MRCSI Member of the Royal College of Surgeons of Ireland
MRCVS Member of the Royal College of Veterinary Surgeons
MRCWA Midland Railway Company of Western Australia
mrd metal rolling door; metal-(lic) roof(ing) deck(ing); minimum reacting dose (MRD)
MRD Main Roads Department; Medical Records Department; Medical Reference Department; Microbiological Research Department; Motorized Rifle Division
MRDC Military Research and Development Center
MR & DC Medical Research and Development Command (US Army)
mrdf machine-readable data files
MRDF maritime radio direction finding
MR&DF Malleable Research and Development Foundation
mrdhd maximum recommended daily human dose
MRDN Material Receipt Discrepancy Notice
MRDs Motorized Rifle Divisions
MRDTI Metal Roof Deck Technical Institute
mre mean radial error
mre (MRE) meal ready to eat (freeze-dried field ration)
MRE Microbiological Research Establishment (UK)
M.R.E. Master of Religious Education
M.Ref.Eng. Master of Refrigeration Engineering
MREI Marriage Role Expectation Inventory
MRELB Malaysian Rubber Exchange and Licensing Board
mrem milliroentgen equivalent man

mrep milliroentgen equivalent physical
mrf maintenance replacement factor; marble floor
MRF Mayo Research Foundation; Meteorological Rocket Facility; Music Research Foundation
MRFB Malayan Rubber Fund Board
MRFIT Multiple Risk-Factor Intervention Trial
MRFL Master Radio Frequency List
mr flight meteorological research flight
mrg magnetic radiation generator; margin; marginal; marginalia; methane-rich gas
MRG Maintainability Requirements Group; Material Review Group; Minorities Research Group; Minorities Rights Group
MRGO Mississippi River Gulf Outflow
MRGS Member of the Royal Geographical Society
MRH Member of the Royal Household
mrhm milliroentgens per hour at one meter
MRHMC Michael Reese Hospital and Medical Center
mr/hr milliroentgens per hour
MRHS Midwest Railway Historical Society
mri magnetic rubber inspection; mean rise interval; medium-range interceptor; milstrip routing identifier; monopulse resolution improvement
mri (MRI) magnetic-resonance imager
MRI Magazine Research Incorporated; Magnetic Resonance Imaging; Marine Research Institute; Marital Roles Inventory; Meat Research Institute; Medical Records Index(ing); Mental Research Institute; Meteorological Research Institute; Meuse-Rhine-Issel (cattle breed); Midwest Research Institute; Military Reform Institute; Missile Range Index; Motor Repair Insurance
MRINA Member of the Royal Institution of Naval Architects
MRINZ Meat Research Institute of New Zealand
MRIPHH Member of the Royal Institute of Public

Health and Hygiene
mrir medium resolution infrared
MRIS Maritime Research Information System; Market Research Information System; Material Readiness Index System; Medical Research Information System; Mobile Range Instrumentation System
MRIW Medical Research Institute of Worcester
mrkd marked
mrkg marking
mrkr marker
Mrkt-Deli Market-Delicatessen
Mrkts Markets
mrl medium-powered radio range (loop radiators); motor refrigerator lighter; multiple rocket launcher (MRL)
MRL Materiel Requirements List; Medical Records Librarian; Medical Records Library; Mineral Research Laboratories
MRLA Malayan Races Liberation Army (Chinese-communist guerrillas)
mrm mail readership measurement; mechanically recovered meat; miles of relative movement
MRM Maintenance Reporting and Management
MRMVA Master Retail Milk Vendors Association
Mrn Martin
MRN Material Recorder Notice; Meteorological Rocket Network
mRNA messenger RNA (ribonucleic acid)
mrng mooring; morning
MRNP Mount Rainier National Park (Washington); Mount Revelstoke National Park (British Columbia)
Mrnz Martínez
mro maintenance, repair, and operating
Mro Maestro
MRO Maintenance, Repair, and Operation(s); Materiel Release Order
MROAR Modification and Repair Order and Acceptance Record
M-roof M-shaped roof
mrov moreover
mrp machine-readable passport; manned reusable payload; manned reusable product; marginal revenue prod-

uct; maximum resolving power; maximum retail price
mrp (MRP) marginal revenue product
MRP Mobile Repair Party
M.R.P. Master in Regional Planning
MRPA Metropolitan Region Planning Authority
MRPP Maoist Reorganization Movement of the Party of the Proletariat
MRPRA Malaysian Rubber Producers Research Association
M rps Mauritius rupee(s)
Mr Q Marquardt Corporation
MRQ Marquardt Corporation (stock exchange symbol)
mrr medical research reactor
MRR Material Rejection Report; Mechanical Reliability Report(ing)
MRRAS Murder Release Risk Assessment Scale
MRRC Mechanical Reliability Research Center
MRRDB Malaysian Rubber Research and Development Board
mrs (MRS) marginal rate of substitution
Mrs Missus; Mistress
MRs Maintenance Reports
MRS Market Research Society; Marseilles, France (airport); Master Repair(ing) Schedule; Material Request Summary; Material Requirement Summary; Military Railway Service; Ministry of Recreation and Sport; Monitored Retrievable Storage; Mountain Rescue Service
MR & S Materials Research and Standards
mrsa (MRSA) medium-range surveillance aircraft
MRSA methicillin-resistant *Staphylococcus aureus* (hospital-acquired infection)
MR San Asn Member of the Royal Sanitary Association
M.R.Sc. Master (Mistress) of Rural Science
MRSH Member of the Royal Society of Health
MRSL Member of the Royal Society of Literature
MRSM Member of the Royal Society of Medicine
MRSMGB Member of the Royal Society of Musicians of Great Britain
MRSP Myakka River State Park (Florida)

mrsss manned revolving space systems simulator (MRSSS)
MRST Member of the Royal Society of Teachers
mrt mean radiant temperature; mid-range trajectory; mildew-resistant thread; military-rated thrust(ing); mission readiness tester; music 'riter typewriter
Mrt Martinique
Mrt Maart (Dutch—March)
MRT Maintainability Review Team; Mass Rapid Transit; Mass Rapid Transport; Metropolitan Readiness Test; Military Review Team; Modulus of Rupture Test(ing)
MRTA Maintenance Requirements Task Analysis
mrtm maritime
Mrtnz Martinez
mrto miscellaneous reference tool(ing)
MRTPI Member of the Royal Town Planning Institute
mrts (MRTS) marginal rate of technical substitution
Mrts Mauritius
MRTS Mass Rapid Transit System; Master Radar Tracking Station
mru minimal reproductive units; mobile radio unit (MRU)
mru (MRU) mass radiography unit; mobile radio unit
MRU Medical Rehabilitation Unit; mobile radio unit; mobile repair unit
MRUA Mobile Radio Users' Association
mrv material receipt voucher; missile re-entry vehicle (MRV); mixed respiratory vaccine; multiple re-entry vehicle (MRV)
MRV missile recovery vessel
mr V-P methyl red Voges-Proskauer
mrw morale, recreation, and welfare
mr/w multiple read/write
MRWA Midland Railway of Western Australia
mrwc multiple reading, writing, compiling; multiple read, write, compute
Mrylb Marylebone (railway terminal)
mrytm must have reply here by tomorrow morning
mrz marzo (Spanish—March)
ms machine screw; machine steel; main switch; maintenance and service; major

subject; manuscript; margin of safety; mass spectrometric; master switch; matched set; maximum stress; mean square; medium shot; medium steel; meters per second; metric system; microseismic; mild steel; millisecond; minimum stress; mint state; mitral stenosis; months after sight; multiple sclerosis; multiple starters; muscle strength
ms (MS) morphine sulfate; moving and storage; multiple sclerosis
m/s marking and stenciling; metal shank; meters per second; milestone; miniature sheet of stamps; month after sight
m & s maintenance and supply; model and series; mud and snow
μs microsecond(s)
ms. manuscript
m s mano sinistra (Italian—left hand)
m/s motorskib (Norwegian—motorship)
mS millisiemens (millimho)
Ms mature motion pictures (for adults); Mendes; mesothorium; (pronounced *Miz*)—feminine title replacing Miss and Mrs; Seattle Mariners baseball team
MS Machinery Survey; magnetic south; Mail Steamer; major subject; Manuscript Society; Master Sergeant; Material Specifications; Medical Survey; Metallurgical Society; Meteoritical Society; Michigan State University of Agriculture and Applied Science; Military Service; Military Standard; Ministry of Shipping; Ministry of Supply; Misair (Egyptian Airline); Mississippi; Motorship
M-S Material Service (division of General Dynamics); Monday through Saturday
M.S. Master of Science; Master of Surgery
M/S Mannlicher-Schoenauer; motorship
M&S Marks and Spencer; Maternity and Surgical; Medical and Surgical
M & S Maintenance and Supply; Medicine and Surgery
MS Material Standard (usually followed by a number); *Mittelsatz* (German—middle

clause) central theme in a sonata

M-S Minshu-Shakaito (Japanese—Democratic Socialist Party)

msa method of steepest ascent; minimum safe altitude; mission system avionics

m.s.a. misce secundum arten (Latin—mix skillfully)

MSA Major Systems Acquisition; Malaysia Singapore Airlines; Management Selection Australia; Marine Safety Agency; Maritime Safety Agency; Medical Statistics Agency (US Army); Metropolitan Statistical Area; Middle States Association (colleges and schools); Mine Safety Appliances (company); Mineralogical Society of America; Motor Schools Association; Mutual Security Agency; Mutual Society of Arts

M-S-A Mine Safety Appliances

MSA Marine Sanctuaries Act; Merchant Shipping Act

MSAA Mower Specialists Association of Australia

MSAAB Military Services Ammunition Allocation Board

msac most seriously affected countries

MSAC Moore School Automatic Computer

MSA/CHE Middle States Association of Colleges and Schools Commission on Higher Education

M.S.Agr.Eng. Master of Science in Agricultural Engineering

MSA Inst MM Member of the South African Institute of Mining and Metallurgy

MSAIT Member of the South African Institute of Translators

m^s a^s muchos años (Spanish—many years)

MSAS Mandel Social Adjustment Scale; Modal Suppression Augmentation System

MSAT Marine Services Association of Texas

MSAUS Masonic Service Association of the United States

msaw (MSAW) minimum safe altitude warning

msb main switchboard; marginal social benefits; most significant bit

msb (MSB) minority small business; missile storage building

MSB Mackinac Straits Bridge (Michigan); Marine Safety Board; minesweeping boat (naval symbol)

MSBA Maine School Boards Association; Minnesota School Boards Association; Missouri School Boards Association; Montana School Boards Association

M.S.B.A. Master of Science in Business Administration

MSB-COD Minority Small Business–Capital Ownership Development

MSBLS Microwave Scanning-Beam Landing System

MSBO Michigan School Business Officials; Mooring and Salvage Office(r)

M.S.Bus. Master (Mistress) of Science in Business

msc marginal social costs; millisecond; miscellaneous; moved, seconded, and carried

m.s.c. mandatum sine clausula (Latin—authority without restriction)

M. Sc. Master of Science

MSC coastal minesweeper (3-letter naval symbol); Maine Sardine Council; Manchester Ship Canal; Manned Spacecraft Center (NASA); Manpower Services Commission; Maple Syrup Council; Marine Safety Council; Marine Science Center (Lehigh University); Medical Service Corps; Medical Specialist Corps; Mediterranean Sub-Commission; Melbourne Steamship Company; Meteorological Service of Canada; Metropolitan Special Constabulary; Military Sealift Command; Missile and Space Council; Missile System Checkout; Mississippi Central (railroad); Mountain Safety Council

M & SC Missile and Space Council

MScA Make or Subcontract Authorization

MSCA McCarthy Scales of Children's Abilities; Moore School of Automatic Computers; Mount Saint Agnes College; Murray State Agricultural College

M Scand Middle Scandinavian

mscc magnetic-strip credit card

M.S.C.E. Master of Science in Civil Engineering

M.S.Ch.E. Master of Science in Chemical Engineering

Mschr Monatsschrift (German—monthly magazine)

MSCIC Maryland State Colleges Information Center

MSCKC Measurement of Self-Concept in Kindergarten Children

M. Sc. L. Master of the Science of Law

mscn misconnection

mscnd misconnected

MSCNY Marine Society of the City of New York

MSC(O) old coastal minesweeper (naval symbol)

M.S. Conv. Master of Science in Conservation

M. Sc. Ost. Master of Science in Osteopathy

M Scot Middle Scottish

mscp mean spherical candle-power

MSCP Master Shielding Computer Program

MSCRB Margaret Sanger Clinical Research Bureau

mscrbl manuscribble (hand-scribbled manuscript)

mscrg miscarriage

MSCT Member of the Society of Cardiological Technicians

MSCW Mississippi State College for Women

msd missile system development; most significant digit; multiple spark discharge

MSD Management Services Department; Marine Sanitation Devices; Merck, Sharp & Dohme

M.S.D. Master (Mistress) of Scientific Didactics; Master (Mistress) Surgeon Dentist; Medical Science Doctor

M & SD Missile and Space Division (General Electric)

MSDA Marconi Space and Defence Systems

MSDC Mass Spectrometry Data Center; Molten Salts Data Center

M.S. Dent. Master of Science in Dentistry

M.S. Derm. Master of Science in Dermatology

MSDF Maritime Self-Defense Force (Japanese Navy)

M & SDI Mayonnaise and Salad Dressing Institute

MSDO Major Systems Development Organization

MSDS Multi-Spectral-Scanner

Data System
mse manufacturing support equipment; mean square error; military stressful era(s)
MSE Malaysia Shipyard and Engineering; Mental Status Examination; Midwest Stock Exchange; Mississippi Export Railroad (stock exchange symbol); Montreal Stock Exchange
M.S.E. Master of Sanitary Engineering; Master of Science in Education; Master of Science in Engineering
m sec millisecond
µsec microsecond
M.S.Ed. Master (Mistress) of Science in Education
MSED Mobile Source Enforcement Division (EPA)
M.S.E.E. Master of Science in Electrical Engineering
M.S.E.M. Master of Science in Engineering Mechanics
M.S. Eng. Master of Science in Engineering
MSEO Marine Services Engineering Office(r)
m/seq master sequencer
MSER Manufacturing Support Equipment Request
mses marchandises (French—goods)
MSET Maintenance Supportability Evaluation Team
MSEUE Mouvement Socialiste pour les États Unis d'Europe (French—Socialist Movement for the United States of Europe)
msf minimum sector fuel; muscle shock factor
Msf thousand square feet
MSF fleet minesweeper (naval symbol); Maintenance Support Flight; Minesweeping Flotilla; mobile striking force; Motorcycle Safety Foundation; Multiple Shops Federation
M.S.F. Master of Science in Forestry
MSF Médecins Sans Frontères (French—doctors without borders)—international group of volunteer physicians
MSFC Marshall Space Flight Center
ms fm master form
M & SFM Maintenance and Supply Facility Manager
msfn manned space flight network
ms fx master fixture
msg machine stress grading;

message; monosodium glutamate
msg (MSG) monosodium glutamate
MSG Madison Square Garden; Marine Systems Group (General Dynamics)
ms ga master gauge
M.S.G.E. Master of Science in Geological Engineering
msgfm messageform
MSGp Mobile Support Group
msgr messenger
msgs messages
M Sgt Master Sergeant
msg/wtg message waiting
msh melanocyte-stimulating hormone (MSH)
Msh Islas Marshall (Spanish—Marshall Islands)
MSH Music Society for the Handicapped
M.S.H. Master of Science in Horticulture; Master of Science in Hygiene
MSHA Mine Safety and Health Administration
M.S.H.A. Master of Science in Hospital Administration
M.S.H.E. Master of Science in Home Economics
MSHFA Multiservice Health Facility Association
MSHI Mitsubishi Singapore Heavy Industries
Mshl Marshal
M. S. Hort. Master of Science in Horticulture
M. S. Hyg. Master of Science in Hygiene
msi maintenance supply item-(ization); management system indicator; medium-scale integration; military standard item; missile status indicator
MSI minesweeper, inshore (naval symbol); Motor Specialties Industries; Museum of Science and Industry
MSI Movimento Sociale Italiano (Italian Social Movement)—neo-fascist militants known as Missini
Msia Malaysia
MSIA Movement of Spiritual Inner Awareness
Msian Malaysian
MSIB Mountain States Inspection Bureau
m'sieur monsieur (French—mister, sir)
M.S.Ind.Eng. Master of Science in Industrial Engineering
MSIRI Mauritius Sugar Industry Research Institute

M.S.J. Master of Science in Journalism
msk mission support kit
MSK Mitsubishi Shoji Kaisha
MS-K Memorial Sloan-Kettering (cancer center)
MS-KCC Memorial Sloan-Kettering Cancer Center
MSKK Mitsui Sempaku Kabushiki Kaisha (Mitsui Line)
Mskr Manuskript (German—manuscript)
msl mean sea level; midsternal line; missile
msl mesela (Turkish—for example)
Msl Marseilles
MSL Marine Science Laboratories; minesweeping launch (naval symbol); Mulla Sadra Library (Shiraz, Iran); Munitions Supply Laboratories
M.S.L. Master of Science in Linguistics
MSLC Manufacturing Specification Liaison Change
ms lo master layout
mslp mean sea level pressure
MSLS Military Standard Logistics Systems
M.S.L.S. Master (Mistress) of Science in Library Science
msm maximum safety margin; modern school mathematics
MSM Manhattan School of Music; Meritorious Service Medal; Montana School of Mines; Mystic Seaport Museum (Conn)
M.S.M. Master of Science in Music
MSMA Maine School Management Association; Master Sign Makers' Association
MSMC Marie Stopes Memorial Centre
M.S.M.E. Master of Science in Mechanical Engineering
M.S.Med. Master (Mistress) of Medical Science
MSMM Missouri School of Mines and Metallurgy
msmq mild steel—merchant quality
MSMS Mutual Security Military Sales
msmt measurement
M.S. Mus. Master of Science in Music
M.S. Mus. Ed. Master of Science in Music Education
msn mission
Msn Mission
MSN Madison, Wisconsin (airport); Master Serial Number-(ing); Material Supply No-

tice(s)

M.S.N. Master of Science in Nursing

MSNB Machine Screw Nut Bureau

M.S.N. Ed. Master of Science in Nursing Education

M.S.Nucl.Eng. Master of Science in Nuclear Engineering

MSNY Mattachine Society of New York

mso (MSO) multiple-systems operator (tv)

MSO Manila Symphony Orchestra; Marine Safety Office(r); Melbourne Symphony Orchestra; Memphis Symphony Orchestra; Milwaukee Symphony Orchestra; Minneapolis Symphony Orchestra; Monetary Statistics Ordinance; Montreal Symphony Orchestra; Morale Support Office(r); ocean minesweeper (naval symbol)

M.Soc.Sci. Master (Mistress) of Social Science

M. Soc. Wk. Master of Social Work

m-sop mezzo-soprano

M.S. Ophthal Master (Mistress) of Ophthalmological Surgery

m sopr mezzo-soprano

MSORS Mechanical Solvent Oil-Spill Recovery System

M.S.Ortho Master (Mistress) of Orthopedic Surgery

msp metal splash pan

msp (MSP) missile support plane

MSP Material Support Plan-(ning); Maximum Security Prison; Medical Services Plan(ning); Memorial State Park; Minneapolis, Minnesota (airport); Mutual Security Program

M.S.P. Master (Mistress) of Science in Pharmacy

MSPB Merit Systems Protection Board

MSpC Medical Specialist Corps

MSPE Master of Science in Physical Education

M.S. Pet. Eng. Master of Science in Petroleum Engineering

M.S.P.H. Master of Science in Public Health

M.S.Pharm. Master of Science in Pharmacy

M.S.P.H.E. Master of Science in Public Health Engineering

M.S.P.H.Ed. Master of Sci-

ence in Public Health Education

ms pl master plate

mspr master spares positioning resolver

MSPRB Meteorological Satellite Program Review Board

MSPU Massachusetts State Prostitutes Union

msr main supply route; mean spring rise (tides); mechanical strain recorder; mineral-surface roof; missile site radar

ms & r merchant shipbuilding and repairs

m & sr missile and surface radar

MSR Manufacturing Specification Request; Material Stores Requisition; mean spring tide

MSRA Multiple Shoe Retailers' Association

M.S. Rad. Master of Science in Radiology

MSRB Mississippi State Rating Bureau

M.S. Rec. Master of Science in Recreation

M.S. Ret. Master of Science in Retailing

MSRG Member of the Society for Remedial Gymnasts

MSRN Manufacturing Specification Revision Notice

msrp manufacturer's suggested retail price; massive selective retaliatory power

MSRP Manufacturer's Suggested Retail Price

msrpp multidimensional scale for rating psychiatric patients

MSRs Marketing Service Representatives

MSRS Missile Strike Reporting System

Msrte Misroute

MSRTS Migrant Student Record Transfer System

msry masonry

mss magnetic storm satellite; manual safety switch; manuscripts; message switching station; missile select(ion) switch; missing sea stores; mode selection switch(ing); multispectral scanning (radar)

mss (MSS) magnetic storm satellite

ms's milk sisters (two or more females who have had sexual relations with the same male)—*see* mb's

ms's (MSs) mine sweepers

mss. manuscripts

Mss Misses; Mizzes (plural of

Miz written Ms)

MSS Manufacturers Standardization Society of the Valve and Fittings Industry; Mass Storage Systems; Master Supporting Schedule; Maximum Security System; Medical Service School; Medical Service School (USAF); Medical Superintendents Society; Metropolitan Security Service(s); Movement Shorthand Society; Multiple Sclerosis Society; Multispectral Scanner Subsystem

M.S.S. Master of Social Science

MSS Museo Storico degli Spaghetti (Italian—Historical Museum of Spaghetti)—close to the Italian Riviera in Pontedassio

MSSA Maine School Superintendents Association; Maintenance Supply Services Agency; Manchester Scales of Social Adaptation

MSSAtl Military Sealift Service Atlantic

M.S.Sc. Master of Sanitary Science; Master of Social Science

MSSC Metropolitan School Study Council

msscc multicolor spin-scan cloudcover camera

MSSCS Manned Space Station Communications System

MSSD Model Secondary School for the Deaf

M & SSD Missile & Space System Division (Douglas Aircraft)

M.S.S.E. Master of Science in Sanitary Engineering

M.S. S. Eng. Master of Science in Sanitary Engineering

MSSGB Motion Study Society of Great Britain

MSSH Massachusetts Society for Social Hygiene

MSSInd Military Sealift Service Indian

MSSMS Munition Section Strategic Missile Squadron

MSSNY Medical Society of the State of New York

MSSPac Military Sealift Service Pacific

MSSR Moldavian Soviet Socialist Republic

MSSRC Mediterranean Social Science Research Council

MSSS Maintenance Supply Services System

M.S.S.S. Master (Mistress) of

Science in Social Service

MSSST Meeting Street School Screening Test

M.S. St.Eng. Master of Science in Structural Engineering

mssu midstream specimen of urine

MSSVD Medical Society for the Study of Venereal Diseases

MSSVFI Manufacturers Standardization Society of the Valve and Fittings Industry

M.S.S.W. Master (Mistress) of Science in Social Work

mst mean solar time; mean spring tide(s); mean survival time; measurement

m(st) metal-stabilized (runway)

M'st' Mister

MST Marconi Telecommunications Systems; Maximum Service Telecasters; Military Science Training; Mountain Standard Time

M.S.T. Master of Science in Teaching

MSTA Maryland State Teachers Association; Michigan State Teachers Association; Missouri State Teachers Association

M.Stat. Master (Mistress) of Statistics

mstb 1000 stock tank barrels

mstc mastic

MSTC Maryland State Teachers College; Massachusetts State Teachers College

MSTD Member of the Society of Typographic Designers

M.S. T.Ed. Master of Science in Teacher Education

msth mesothorium

M & ST L Minneapolis & St Louis (railroad)

mstn 1000 short tons

ms tp master template

M ST P & SSM Minneapolis, St Paul & Sault Ste Marie Railroad (Soo Line)

mstr master

Mstr Master

M.S.Trans Master of Science in Transportation

M.S. in Trans.E. Master of Science in Transportation Engineering

Mstr Mech Master Mechanic

MSTRP Military, Strategic, Tactical, and Relay Program

msts (MSTS) missile static test site

MSTS Military Sea Transport

Service; Missile Static Test Site; Mobile System Test(ing) Site

msty mostly

msu main storage unit; maximum space use; maximum space utilization; mode selector unit

msu (MSU) maximum security unit

MSU Mature Students Union; Memphis State University; Michigan State University; Mississippi State University; Montana State University

MSUC Middle South Utilities Company

msud maple-syrup urine disease

MSUL Medical Schools of the University of London; Memphis State University Library; Michigan State University Library; Mississippi State University Library; Montana State University Library

MSU Lond Medical Schools of the University of London

M. Surgery Master of Surgery

M.Surv Master of Surveying

msus midstream urine specimen

msv (MSV) magnetically-supported vehicle; Martian surface vehicle; mean square velocity; miniature solenoid valve; molecular solution volume; murine sarcoma virus

MSV Medical Society of Victoria

MSVC Mount Saint Vincent College

MSVD Missile and Space-Vehicle Department (General Electric)

msv(M) murine sarcoma virus (Moloney)

Msw Massawa

M Sw Middle Swedish

MSW Master of Social Work; Medical Social Worker

M.S.W. Master of Social Welfare; Master of Social Work

mswa main storage work area (address)

MSX Seaboard Oil (stock exchange symbol)

msy (MSY) maximum sustainable yield

MSY New Orleans, Louisiana (airport)

msyd 1000 square yards

mt empty; machine translation; mail transfer; maximum torque; mean tide; mean

time; measurement ton; mechanical translation; mechanical transport; medical technology; megaton (MT); membrana tympani; metal(lic) tape; metatarsal; metric ton; miniature tube; missile test; modified term; motor terminal; motor transport; mount torque; mount(ed); mounting

mt (MT) motor tanker

m/t mail transfer; manual transmission; measurement tons

m & t maintenance and test; movements and transports

mT tropical maritime air

Mt Mount; Mountain; tympanic membrane

MT Machine Translation; Mail Transfer; Mandated Territory; Manning Table; Mark Twain (Samuel Clemens); Masoretic Text; Mechanical Translation; Medical Technologist; Meteorological Aids; Military Training; Military Transport; Mining Tenement; Ministry of Transport; Montana; Moscow Time; Motor Transport; Mountain Time; Muscat Transport

MT-6 mercaptomerin (diuretic)

mta maximum time aloft; microwave transistor amplifier

mt/a million tons per annum

m^{ta} muita (Portuguese—much)—feminine form

MTA Maine Teachers Association; Manpower Training Association; Market Technicians Association; Massachusetts Teachers Association; Master Tilers Association; Metropolitan Transit Authority; Mica Trade Association; Mississippi Teachers Association; Mississippi Test Area; Motor Trade Association; Music Teachers Association

mtac mathematical tables and other aids to computation

MTACCS Marine Tactical Command and Control System

MTAG Manufacturing Technology Advisory Group

MTAI Minnesota Teacher Attitude Inventory

MTAK Magyar Tudományos Akadémia Könyvtára (Hungarian—Library of the Hungarian Academy of Sciences)—Budapest

mtam maritime tropical air mass

MTAMR Metropolitan Toronto Association for the Mentally Retarded

MT/AMT Mail Transfer—Airmail Transfer (funding)

MTASCP Medical Technologist of the American Society of Clinical Pathologists

mtb maintenance of true bearing

MTB Major Trading Bank; Malayan Tin Bureau; Malaysian Tin Bureau; Materials Transportation Bureau; Medium Tank Battalion; Miyagi Television Broadcasting; motor torpedo boat

MTBA Machine Tool Builders' Association

MTBC Mitsubishi Trust and Banking Corporation

mtbd mean time between demand

mtbe (MTBE) methyl tertiary butyl ether (octane-booster additive)

mtbf mean time before failure; mean time between failures

mtbfa mean time between false alarms

mtbff mean time between first failure

mtbfl mean time between function loss

mtbm mean time between maintenance

MTBRON Motor Torpedo Boat Squadron

mtbsf mean time between system failure

mtc memory test computer; more to come

m & tc mission and traffic control

MTC Malayan Tobacco Company; Marcus Tullius Cicero (106–43 B.C.); Marine Technology Center (Electric Boat); Maritime Transport Committee; Massachusetts Treatment Center; Materiel Testing Command; Mechanical Transport Corps; Medical Training Center; Men's Tennis Council; Metropolitan Transportation Committee; Military Training Cadets; Missile Test Center; Monsanto Chemicals (stock exchange symbol); Montreal Trust Company; Morse Telegraph Club; Motor Transport Corps; Mystic Terminal (rail-

road)

M.T.C. Master of Textile Chemistry

MTC *Ministerio de Transporte y Comunicaciones* (Spanish—Ministry of Transportation and Communication)

MTCA Ministry of Transport and Civil Aviation

MTCB Metropolitan Taxicab Board

mtce maintenance; million tons of coal equivalent

mt & ce missile test and checkout equipment

MTCL Metropolitan Toronto Central Library

MTCP Minister of Town and Country Planning; Ministry of Town and Country Planning

mtcu magnetic tape control unit

mtd manufactured technological demonstrator; mean temperature difference; midpoint tissue dose; mounted

mtd (MTD) maximum tolerated dose

m.t.d. *mitte tales doses* (Latin—send such doses)

Mtd Marstrand

MT$ Maria Theresa dollar (Yemeni currency unit)

MTD Mobile Training Detachment

M.T.D. Master of Transport Design; Midwife Teachers' Diploma

MTDB Metropolitan Transit Development Board

MTDDA Minnesota Test for Differential Diagnosis of Aphasia

mtde maritime tactical data exchange

MTDE Maintenance Technique Development Establishment

MTDS Marine Tactical Data System

mte manufacturing test(ing) equipment; maximum temperature engine; maximum thermal energy; multiple-track error

M^{te} *Monte* (Italian, Portuguese, Spanish—mountain)

MTE Marine Technical Education

MTEA Metal Trades Employers Association

M. Tech. Master (Mistress) in Technology

M.Tel.Eng. Master (Mistress) of Telecommunication Engi-

neering

MTER Manufacturing Test Equipment Request

M^{tes} *Montes* (Italian, Portuguese, Spanish—mountains)

M.Text. Master (Mistress) of Textiles

mtf mechanical time fuze; modulation transfer function; multiple technical force

MTF Medical Treatment Facility; Metal Trades Federation; Mississippi Test Facility; Multiracial Training Facility

mtfex mountain field exercise

mtg main turbogenerator(s); meeting; methanol to gasoline; mortgage; mounting

Mtg Meeting (postal abbreviation)

mtgc mounting center

mtgd mortgaged

mtge mortgage

mtgee mortgagee

mtgor mortgagor

mth microptic theodolite; month

M. Th. Master of Theology

MTH Master of Trinity House

mthly monthly

mthm metric tons of heavy metal

mths months

mthv must have

mthw medium-temperature hot water

mti moving target identification; moving target indicator(s); moving target information

M^{ti} *Munti* (Romanian—mountain)

MTI Metal Treating Institute; Motorola Teleprograms Incorporated

MTI *Magyar Távirati Iroda* (Hungarian Press Agency)

MTIA Metal Trades Industry Association

MTIB Malaysian Timber Industry Board

mtik missile test installation kit

mtime meantime

MTIRA Machine Tool Industry Research Association

mTk tropical maritime air colder than underlying surface

mtks many thanks

mtl material; materiel; mean tide level; merged transistor logic; metal(lic); mixed thermoluminescence

mtl *monatlich* (German—monthly)

Mtl Montreal; Motel
MTL mean tide level; Modern Terminals Limited; Motor Traders Limited
MTLA Micropublishers Trade List Annual
mtlp metabolic toxemia of late pregnancy
mtl(s) material(s)
mtlz materialize
mtlzd materialized
mtm method, time, and motion; methods time measurement(s)
MTM Mary Tyler Moore
MTMA Modern Teaching Methods Association
MTMC Military Traffic Management Command (USA); Mother Teresa's Missionaries of Charity
Mt McK NP Mount McKinley National Park
MTMCTEA Military Traffic Management Command Transportation Engineering Agency
MTMS Multi-Terminal Modular System
MTMTS Military Traffic Management and Terminal Service
mtn motion; mountain; mutton
Mtn Mountain
MTN Medical Television Network; Multilateral Trade Negotiations
MTNA Music Teachers National Association
mtnt must not
mtn vu mountain view
MTNWR Mark Twain National Wildlife Refuge (Illinois)
mto modification task outline
m^{to} muito (Portuguese—much)—masculine form
MTO Mississippi Test Operations
mtoe million tons oil equivalent
Mton Moncton
mtons metric tons
mTorr millitorr(s)
mtp minimum tour price
Mt P Mount Palomar (observatory)
MTP Management Training Program; Mobilization Training Program; Modification Task Proposal; Mount Tom Price
M.T.P. Master (Mistress) of Town Planning
mtpa million tons per annum
MTPCNA Metal Tube Pack-

aging Council of North America
Mt P O Mount Palomar Observatory
mtpp missile-to-target patch panel
mtpy millions of tons per year
mtr materials testing reactor; matter; mean time to restore; meter; minimum time rate; missile-tracking radar; motor; moving target reactor; multiple track radar
mtr (MTR) marginal tax rate
Mtr Meinicke turbidity reaction; Montrose
MTr meridian transit
MTR Mass Transit Railroad; Mass Transit Railway; Mass Transportation Railroad; Materials Testing Report; Montour (railroad)
MTRB Motor Truck Rate Bureau
MTRC Mass Transit Railway Corporation
mtrcl motorcycle
mtre missile test and readiness equipment
Mt Rev Most Reverend
MTRF Mark Twain Research Foundation
mtrg metering
mtri missile test range instrumentation
mtrl material
Mt R NP Mount Rainier National Park
Mtro Maestro (Spanish—Master)
M.T.R.P. Master (Mistress) of Town and Regional Planning
mtr rdr meter reader
MTRS Magnetic Tape Recording System
mtr vlu meter(ing) value
mts mobile training set; motorship twin screw; mountains
mt's empties
Mts Mountains
Mts Montes (Spanish—mountains)
MTS Machine Tractor Station; Marine Technology Society; *Mashinno-Traktornye Stantsii* (Russian—Machine Tractor Stations); Melanesian Tourist Services; Member(s) of the Technical Staff; Metropolitan Transit System; Middlebare Technical School; Missile Test Stand; Missile Test Station; Money Transfer Service
mt/sc magnetic-tape selectric composer

MTSC Middle Tennessee State College
mtst maximum treadmill stress time
mt/st magnetic-tape selectric typewriter
MTSU Middle Tennessee State University
mtt magnetic tape terminal; mean transit time; moving target tracking
MTT Maintenance Training Team; Metropolitan Transport Trust; Mobile Training Team (USN); Municipal Tramways Trust
MTTA Machine Tools Trades' Association
MTTAGB Machine Tool Trades Association of Great Britain
mtte magnetic tape terminal equipment
mttf mean time to failure
mttff mean time to first failure
mttms magnetic tape transmissions
mttr mean time to repair
mtu metric tons of uranium; mobile tracking unit; mobile training unit
mtu (MTU) multiplexer and terminal unit
MTU Maintenance Training Unit; Michigan Technological University; Michigan Training Unit (reformatory); Missile Training Unit; Mobile Training Unit
MTU Motoren und Turbinen Union (German—Motors and Turbines United)—corporation
M. tuberc. Mycobacterium tuberculosis
MTUOP mobile training unit out for parts
mtv mammary tumor virus; motor test(ing) vehicle; multichannel tv sound
mtv (MTV) music television channel
MtV Mount Vernon
M.Tv. Master of Television
MTV Motor Test Vehicle; motor torpedoboat (British naval symbol)
MTVs Motor Torpedo Vessels
mtw main trawl winch
mTw tropical maritime air warmer than underlying surface
mt we must we
Mt W O Mount Wilson Observatory
mtx methotrexate

MTX Morrell Tank Line (railway symbol)
mtxs military traffic expediting service
Mty Monterrey (inhabitants—Regiomontanos)
MTY Monterrey, Mexico (airport)
mtz motorize
MTZS Metropolitan Toronto Zoological Society
mu machine unit; mail unit; marijuana user; monetary unit; mouse unit; multiple unit
mu (MU) marginal utility (microeconomics symbol); mark-up (calculator)
m/u makeup; mockup
MU Macquarie University; Maintenance Unit; Marquette University; Marshall University; Massey University; Mercer University; Mercy University; Mercyhurst University; Meredith University; Merrimack University; Mesa University; Messiah University; Methodist University; Miami University; Midwestern University; Milliken University; Monash University; Mothers' Union; Murdoch University; Musicians Union
MUA Machinery Users' Association; Malayan Union Association; Monotype Users' Association; Musicians Union of Australia
MUAC Metropolitan Universities Admissions Center
muap motor unit action potential(s)
muat mobile underwater acoustic unit
muc mucilage
muc. mucilago (Latin—mucilage)
MUC Magee University College; Meritorious Unit Citation; Muchea, Australia (tracking station); Munich, Germany (Riem airport)
mu car multiple-unit (railroad) car
MUCC Michigan United Conservation Clubs
Much Ado Much Ado About Nothing
MUCIA Midwest Universities Consortium for International Activities
MUCM Medical University College of Medicine
MUCO Material Utilization Control Office

MUD Municipal Utility District
'muda Bermuda
'muda grass bermuda grass
MUDPAC Melbourne University Dual-Package Analog Computer
MUDPIE Museum and University Data, Programs, and Information Exchange
muf material unaccounted for; maximum usable frequency
MUFON Mutual UFO Network
Muh Muharram (Arabic—first month of the Mohammedan year)
Mühlhausen Mülhausen in Thüringen (Mülhausen in Thuringia)
Mühlheim Mühlheim am Main (Mühlheim on the Main River) or Mühlheim an der Donau (Mühlheim on the Danube)—Germany
Mujib Mujibur Rahman
Muk Mukden
mul multiply
mul mulig(vis) (Dano-Norwegian—eventual, probable, or possible, perhaps)
MUL Makerere University Library (Kampala, Uganda)
mulat mulatto
Mulatas Mulatas Islands
mule (see hinny)
mule. modular universal laser equipment
MULES Missouri Uniform Law Enforcement System
Mülheim Mülheim an der Ruhr (Mülheim on the Ruhr River) or Mülheim am Rhein (Mülheim on the Rhine)—Germany
MULS Minnesota (University) Union List of Serials
mult multiplication
MULTEWS Multiple-Target Electronic-Warfare System
multi (Latin prefix—many or much)—multitude
multics multiplexed information and computing service
multitran multiple translation (translating one language into several target languages)
multr multimeter
mulv murine complex leukemia
mum mumble(d); mumbling; mummed; mummer(s); mummery
mum. (MUM) (robot) methodology for unmanned manufacturers

Mum Mumford
MUMC McMaster University Medical Center
MUMMS Marine Corps Unified Management System
MUMPS Multi-Programming System (Massachusetts General Hospital)
mums chrysanthemums
mun munition
Mun Müngo; Munro; Munroe; Munster
Mün München (German—Munich)
MUN Memorial University of Newfoundland; Model United Nations
Mund Edmund
muni municipal; municipality
muni bond(s) municipal bond(s)
munic municipal; municipality
munis municipal bonds
munit munitions
Muñoz Marín Luis Muñoz Marín—democratic leader and first governor of Puerto Rico
muo myocardiopathy of unknown origin
MUO Municipal University of Omaha
muon mu meson (Siamese—town)
MUP Manchester University Press; Melbourne University Press
M.U.P. Master of Urban Planning
Mur Murcia
MUR Mouvements Unis de la Résistance (French—United Movements of the Resistance)
MURA Midwestern Universities Research Association
Murasaki Baroness Murasaki Shikibu (The Tale of the Genji)
MURFAAMCE Mutual Reduction of Forces and Armaments in Central Europe
Murgie Murgatroyd
muriatic acid hydrochloric acid (HCI)
Murph Murphy
mus multiunit school(s); musculoskeletal; museum; music; musical; musician
Mus Musca (constellation); Muscat; museum; music; Muslim
MUS Magnetic Unloading System; Manned Underwater Station
musa multiple-unit steerable

antenna
MUSA Medical University of Southern Africa
Mus Anthro Mo University of Missouri Museum of Anthropology
Mus Art RI Museum of Art—Rhode Island
Mus. Bac. Bachelor of Music
Mus Bks Museum Books
musc muscle; muscular
Muscov Muscovite(s)
muscrit music critic(ism)
mus dir music(al) director
Mus. Doc. Doctor of Music
Mus.Ed.B. Bachelor of Music Education
Mus.Ed.D. Doctor of Music Education
Mus.Ed.M. Master of Music Education
museo museography; museological; museologist; museololgy
Music Academy of Music; High School of Music
MUSIC Maryland University Sectored Isochronous Cyclotron
musicol musicological; musicologist; musicology
mus id's musical identifications (of radio or tv feature programs)
MusiMus Musikkhistorisk Museum (Norwegian—Music History Museum)
muskie muskellunge
Mus.M. Master of Music
Mus Northern Ariz Museum of Northern Arizona
Musso Mussolini
Mus Sys Museum Systems
must. manned undersea station
MUST Medical Unit, Self-contained, Transportable
mustargen mustard-nitrogen (poison compound)
mustn't must not
MUSTRACS Multiple Simultaneous-Target Steerable Telemetry-Tracking System
mut mutation; mutilate(d); mutton; mutual
mutil mutilate; mutilated; mutilation
Mutiny Mutiny on the Bounty
mut. mut. mutatis mutandis (Latin—necessary changes were made)
mut. nom. mutatis nomine (Latin—name was changed)
mutt muttonhead
mutt (MUTT) military utility tactical truck
Mutton Birds Mutton Bird Is-

lands off the southwest coast of New Zealand's Stewart Island where they are also called the Titis
mutu mutual; mutualism
Mutual Mutual Association for Professional Services; Mutual Benefit Life Insurance Company of Newark, New Jersey; Mutual Broadcasting System; Mutual Life Insurance Company of New York; Mutual Nurses Registry; Mutual of Omaha; Mutual Protection Trust; Mutual Security Life Insurance; Mutual Trust Life Insurance
muu mouse uterine units
muw music wire
MUWS Manned Underwater Station
mux multiplex
mux/aro multiplex-automatic error correction
muz muziek (Dutch—music)
Muz Muzio
Muz Muzlim (Arabic—True Believer)
Muzel (Cornish-English—Mousehole)—fishing port resort in Cornwall
muzh muzzle hatch
mv main verb; mean variation; mercury vapor; millivolt; monochromatic vision; moving magnet; multivibrator; muzzle velocity
m & v meat and vegetable
μv microvolt
mv meervoud (Dutch—plural)
m v mezzo voce (Italian—middle voice)
m V megavolt(s); millivolt(s)
Mv megavolt; mendelevium
M/V motor vessel
MV Maria Vergine (Italian—Virgin Mary); *Merchant of Venice*
M.V. Medicus Veterinarius (Veterinary Physician)
M of V Merchant of Venice
MV-678 Agricultural Research Service chemical for fighting fire ants by mimicking hormones to create drones
mva mean vertical acceleration; megavolt ampere; motor vehicle accident
MVA Machinists Vise Association; Mississippi Valley Association; Missouri Valley Authority
MVAS Milwaukee Vocational and Adult School
MVB Martin Van Buren (8th President U.S.)

MVBA Mercado de Valores de Buenos Aires (Buenos Aires Stock Exchange)
mvbd multiple V-belt drive
MVBL Mississippi Valley Barge Line
mvc manual volume control; manufacturing variation control
MVC Military and Veterans Code
MVCC Mount Vernon Community College
MVD Montevideo, Uruguay (Carrasco Airport); Motor Vehicles Department
MVD Ministerstvo Vnutrenniy Delo (Russian—Ministry of Internal Affairs)—*(q.v.—VOT)*
MVDA Motor Vehicle Dismantlers' Association
MVe Murray Valley encephalitis
MVE Metropolitan Vickers Electrical
M.V.E. Master of Vocational Education
MVEMJSUNP *M*en *V*ery *E*asily *M*ake *J*ugs *S*erve *U*seful *N*octurnal *P*urposes (acrostic mnemonic for remembering the order of planets from the sun—Mercury, Venus, Earth, Mars, Jupiter, Saturn, Uranus, Neptune, Pluto); My Very Energetic Mother Just Served Us Nine Pickles—phrase used to remember names of planets in order of their distance from the sun (every 216 years the order of Neptune and Pluto changes)
M.Vet.Med. Master (Mistress) of Veterinary Medicine
M.Vet.Sci. Master (Mistress) of Veterinary Science
mvg most valuable girl
MVG Medal for Victory over Germany
MVHS Mergenthaler Vocational High School
mvi multi-vitamin infusion
M/video Montevideo
MVJC Mount Vernon Junior College
mvlf (MVLF) motor-vehicle license fee(s)
MVM Motor Vehicle Mechanic
MVMA million vehicle miles; Motor Vehicle Manufacturers Association
MVMFB Mississippi Valley Motor Freight Bureau
mvmt movement

MVNP Mesa Verde National Park
MVNWR Monte Vista National Wildlife Refuge (Colorado)
Mvo Montevideo
MVO Member of the Victorian Order
mvp maximum-value package; most valuable performance; most valuable player
MVP Manpower Validation Program
MVPBA Mississippi Valley Power Boat Association
MVPCB Motor Vehicle Pollution Control Board
mvps (MVPS) mitral valve prolapse syndrome
MVPT Motor-Free Visual Perception Test(ing)
mvri mixed vaccine—respiratory infections
mvs multiple virtual storage
Mvs Maldivas (Spanish—Maldive Islands)
MVS Mennonite Voluntary Service; Multi-Vest Securities
M.V.Sc. Master of Veterinary Science
MVSS Motor Vehicle Safety Standard
mvt moisture-vapor transmission; multiprogramming with variable number of tasks
MVT Motor Vehicle Technician
MV & THS Manhattan Vocational and Technical High School
MVTI Mohawk Valley Technical Institute
MVUS Merchant Vessels of the United States
mvv maximum vacation value; maximum voluntary ventilation
mvw most valuable wife
mw male white; megawatt; milliwatt; molecular weight
m/w manufacturing week
m/w (M/W) midwife
mw mevrouw (Dutch—Mrs)
mW megawatt(s)
m/W male White
mW meines Wissens (German—as far as I know)
Mw megawatt
MW Middle Welsh; Montgomery Ward
M-W Merriam-Webster
M of W Ministry of Works
MWA Metal Window Association; Modern Woodmen of America; Mystery Writers of

America; Mystery Writers Association
MWAA Movers and Warehousemen Association of America
MWAI Mystery Writers of America, Incorporated
M-way Motorway (superhighway)
mwb motor whale boat
MWB Metropolitan Water Board; Minister of Works and Buildings; Ministry of Works and Buildings
MWC Ministry of War Communications; Missile Warning Center; Motorola Western Center
MWCG Metropolitan Washington Council of Governments
MWCOG Metropolitan Washington Council of Government
mwd megawatt day
MWD Metropolitan Water District; Military Works Department; Ministry of Works and Development; Mutual Weapons Development
mwdm multiwavelength distance measuring (instrument)
mwd/mtu megawatt day per metric ton of uranium
MWDP Mutual Weapons Development Program
mwe megawatts of electricity; meters of water equivalent
MWF Medical Women's Federation
mwg music wire gauge
MWGCP Most Worthy Grand Chief Patriarch
MWGM Most Worshipful Grand Master; Most Worthy Grand Master
mwh milliwatt hour
m & whm missile and warhead magazines
MWHS Martha Washington High School
mwi message-waiting indicator
MWIA Medical Women's International Association
MWJC Marjorie Webster Junior College
mwk millwork
Mwl Malawi (Spanish abbreviation)
MWL Minimum Wage Law(s); Mutual Welfare League
MWLP Meadowview Wild Life Preserve
MWMCA Michigan Women for Medical Control of Abor-

tion
MWMFB Midwest Motor Freight Bureau
MWN Medical World News
MWNM Muir Woods National Monument
mwnt mean water neap tide
MWO Marshallese Women's Organization; Midwest Oil; Modification Work Order; Mount Wilson Observatory
mwp maximum working pressure; membrane waterproofing
MWP Most Worthy Patriarch
MWPA Married Women's Property Act
mwr mean width ratio
MWR Morton Wildlife Refuge (New York)
mws magnetic weapon sensor
MWS Manas Wildlife Sanctuary (India); Mudamalai Wildlife Sanctuary (India)
MWSC Midwestern Simulation Council
MWSG Marine Wing Support Group
M.W.T. Master of Wood Technology
mwth megawatts thermal
mwv maximum working voltage
MWV Mineralöl Wirtschafts Verband (German—Petroleum Industry Association)
mww manual wire wrap; municipal waste water
MWW Merry Wives of Windsor
MWZ Manischewitz (stock exchange symbol)
mx maxwell; motocross (rough-terrain motorcycle race); multiplex
mx (MX) missile experimental
Mx maxwell; Middlesex
Mx (MX) experimental intercontinental ballistic missile (designed for in-the-air, on-the-ground or under-the-sea launching)
MX Mexicana de Aviación (2-letter code)
mxa mobile exercise area
MXC Minnesota Experimental City
mxd mixed
mxd cl mixed carload
mxdth maximum depth
Mxl Mexicali (inhabitants—Cachanias)
mxm maximum
MXP Milan, Italy (Malpensa Airport)
mxpst maximum possible

storm
mxr mask index register
mXr mass X-ray
mx rnfl maximum rainfall
MXS Missile Experimental System
mxtmp maximum temperature
mxwnd maximum wind
my. million years; myopia; myopic
m/y man-year
My Malayalam; Milo; Mylan
MY Medinat Yisrael (State of Israel); motor yacht
mya millions of years ago
Mya Myasishchev
Mya-4 Soviet heavy bomber named Bison by NATO
Myanmar Burma
mybp million years before present
Myc Mycenaean
MYC Manchester Yacht Club; Middletown Yacht Club; Milwaukee Yacht Club; Minnetonka Yacht Club; Mobile Yacht Club
myco mycobacterium
mycol mycology
myel(s) myelocyte(s)
Myf's *Mayflower* descendants
myg myriagram
myl myrialiter
mylo mylohyoid

mym myriameter
myn million
myo (Greek *mys*—mouse or muscle)—myocardial infarction, myocardium, myoma; *mayo* (Spanish—May)
myob mind your own business
myobb mind your own bloody business
myodyn myodynamics
myoelectric myoelectrical(ly)
myo inf myocardial infarction
myol myology
myop myopia
mypo multiyear procurement objective
Myr Myriopeda
Myrt Myrtle
Mys Mysore
Mys Sea Mystic Seaport
myst mystagogue; mystagogy; mysteries; mysterious; mystery; mystic; mystical; mysticism; mystics
Mystery Queen Agatha Christie
myth. mythological; mythologist; mythology
Myth Mythology
mz monozygotic
ms Mangelszahlung (German—for non-payment)
Mz Méndez
MZ Mail Zone; Museum of

Zoology; RH Macy and Company (stock exchange symbol)
M & Z Mombasa and Zanzibar
MZ Moskauer Zeit (German—Moscow Time)
MZA Madrid, Zaragoza, Alicante
mzm multiple-zone monitor
MZMA Moskva Zavod Maloitrazhkaya Automobili (Russian—Moscow Small-Engine Car Factory) producing the Moskvich auto
MZn magnetic azimuth
MZNP Mountain Zebra National Park (South Africa)
mzo marzo (Spanish—March)
M-zone manufacturing zone
MZP Marwell Zoological Park
mzs mezzo-soprano
M.Z.Sc. Master of Zoological Science
Mzt Mazatlán (inhabitants—Mazatlecos)
mz twins monozygotic twins (genetically identical)
MZUSP Museo de Zoologia de Universidade de São Paulo (Portuguese—Zoology Museum of the University of São Paulo)

N

n name; nasal; national; nautical; naval; neap; negative; nerve; neuter; neutral; neutron; new; night; nominative; noon; norm; normal; noun; nuclear; number; refractive index (symbol); shear modulus of elasticity (symbol); transport number (code)
n' and
n/ and; number
'n' and (as in fish 'n' chips, rock 'n' roll)
'n' and
n index of refraction (symbol); load factor (symbol); nació (Spanish—born); revolutions per second (symbol); rotative speed (symbol)
n. haploid generation; numerus (Latin—number)
n/ nuestro (Spanish—our); número (Spanish—number)
N International Nickel (stock exchange symbol); national; nautical; naval; Navy; Negro; neon; neuroticism index; neutral; neutron(s); newton(s); night; nimbus; Nippon; nitrogen; noon; normal; Norse; north; northern; Norway (auto plaque); November—code for letter N; nuclear-propelled vessel (naval symbol); nucleus
N naira (Nigerian currency)
N (N) employment of labor (microeconomics symbol)
(N) nuclear-powered ship (naval symbol, as in CL[N]—nuclear-powered cruiser)
N avogadro constant or number (symbol); natus (Latin—born); Nebenstimme (German—secondary line or mo-

tif)—12-tone term; neer (Dutch—down); noord (Dutch — north); nord(Danish, French, Italian, Norwegian, Swedish—north); Nord (German—north); norre (Danish—north); norte (Portuguese or Spanish—north); north; number of propeller rotation (symbol); revolutions per minute (symbol); yawing moment (symbol)
N- nuclear
N1, N2, etc. North One, North Two, etc. (London postal zones)
n.₂ diploid generation
N₂ nitrogen
N₂O nitric oxide
N¹⁴ radioactive nitrogen
n/30 net (payment) in 30 days
na naturally aspirated; negative attitude; next assembly; nicotinic acid; no account; not absolutely; not applicable; not appropriated; not authorized; not available; nucleic acid (NA); numerical aperture
n/a navigation and attack; next assembly; no account; no advise; not applicable
na nestre ar (Norwegian—next year)
nA nanoampere
Na nadir; Napier; natrium (sodium); sodium (symbol)
Nᵃ Nuestra (Spanish—our)
NA Narcotics Anonymous; National Academician; National Academy; National Airlines; National Archives; National Association; Nautical Alma-

nac; Naval Academy; Naval Architect; Naval Attaché; Naval Auxiliary; Naval Aviator; Netherlands Antilles (Aruba, Bonaire, Curaçao, Saba, Sint Eustatius, Sint Maarten); Neurotics Anonymous; North America; North American; Northrup Aircraft; Nurse's Aide
NA Nautical Almanac; Nederlandse Antillen (Dutch—Netherlands Antilles)—Aruba, Bonaire, Curaçao, Saba, Sint Eustatius, Sint Maarten—the Dutch West Indies; Nomina Anatomica (Latin—Anatomical Names)—official nomenclature adopted by the International Congresses of Anatomists
Na₂CO₃ sodium carbonate (sal soda)
Na²⁴ radioactive sodium
naa neutron activation analysis; not always afloat
NAA National Academy of Arbitrators; National Aeronautic Association; National Alumni Association; National Apple Association; National Arborist Association; National Archery Association; National Association of Accountants; National Auctioneers Association; National Automobile Association; Naval Association of Australia; Naval Attache for Air; North American Aviation; North Atlantic Assembly; Nurses Auxiliaries Association
NAAA National Alliance of Athletic Associations; National Association of Ameri-

can Academicians; National Association of Arab Americans; National Auto Auction Association
NAAB National Architectural Accrediting Board
NAABC National Association of American Business Clubs
NAABI National Association of Alcoholic Beverage Importers
naabsa not always afloat but safe aground
NAAC National Agricultural Advisory Commission; National Association of Agricultural Contractors
NAACC National Association for American Composers and Conductors
NAACO North American Arms Corporation of Canada
NAACOG Nurses Association of the American College of Obstetrics and Gynecology
NAACP National Association for the Advancement of Colored People
NAADC North American Area Defense Command
NAADS New Army Automatic Data System
NAAF North African Air Force (World War II)
NAAFA National Association to Aid Fat Americans
NAAFI Navy, Army, and Air Force Institutes
NAAG National Association of Attorneys General; NATO Army Advisory Group
NAAMM National Association of Architectural Metal Manufacturers
NAAN National Advertising Agency Network
NAANACM National Association for the Advancement of Native American Composers and Musicians
NAAO National Association of Amateur Oarsmen; Navy Area Audit Office
NAAP National Association of Advertising Publishers
NAAPPA North American Association for the Protection of Predatory Animals
NAAQS National Ambient Air Quality Standards
NAARI National Aero- and Astronautical Research Institute
NAARPR National Alliance Against Racist Political Repression

NAAS National Agricultural Advisory Service; National Apprentice Assistance Scheme; Naval Area Audit Service; Naval Auxiliary Air Station
NAASC North American Aviation Science Center
NAASFEP National Association of Administrators of State and Federal Education Programs
NAA S & ID North American Aviation Space and Information Division
NAASRA National Association of Australian State Road Authorities
NAASS North American Association of Summer Sessions
NAATI National Accreditation Authority for Translators and Interpreters
NAATP National Association of Alcoholism Treatment Programs
NAATS National Association of Air Traffic Specialists
NAAUC National Association of Australian University Colleges
NAAUS National Archery Association of the United States
NAAV National Association of Atomic Veterans
NAAW National Association of Accordion Wholesalers
NAAWS North American Association of Wardens and Superintendents
NAB National Aborigine Conference; National Alliance of Businessmen; National Assistance Board; National Association of Broadcasters; National Association of Businessmen; Naval Advanced Base; Naval Air Base; Naval Amphibious Base; News Agency of Burma; Newspaper Advertising Bureau
NAB New American Bible (Roman Catholic)
NABA North American Benefit Association
NABACO National Association for Bank Audit, Control, and Operation
NABAE National Association of Black Adult Educators
NABB National Association for Better Broadcasting
NABBC National Association of Brass Band Conductors
NABC North American Bridge Championships; National As-

sociation of Boys' Clubs
NABD North American Band Directors
NABDC National Association of Blueprint and Diazotype Coaters
NABE National Association of Bilingual Education; National Association for Bilingual Education; National Association of Book Editors; National Association of Business Economists
NABEO National Association of Black Elected Officials
NABET National Association of Broadcast Employees and Technicians
NABEWD North American Board for East-West Dialogue
NABIM National Association of Band Instrument Manufacturers
NABISCO National Biscuit Company
NABJ National Association of Black Journalists
NABLT National Association of Business Law Teachers
NABMA National Association of British Market Authorities
NABMO NATO Bullpup Management Office(r)
nabor neighbor
NABP National Association of Book Publishers
NABPO NATO Bullpup Production Office(r); NATO Bullpup Production Organization
Nabrico Nashville Bridge Company
NABRT National Association for Better Radio and Television
NABS National Association of Barber Schools; National Association of Black Students; nuclear-armed bombardment satellite
NABSE National Alliance of Black School Educators
NABSP National Association of Blue Shield Plans
NABT National Association of Biology Teachers; National Association of Blind Teachers
NABTE National Association for Business Teacher Education
nabu non-adjusting ballup (unsolvable confusion)
NABUG National Association of Broadcast Unions and

Guilds
NABW National Association of Bank Women
nac nacelle; negative air cushion; nozzle area control
NAC National Achievement Clubs; National Advisory Council; National Agency Check; National Agriculture Centre; National Airways Corporation (New Zealand); National Americanism Commission (American Legion); National Amusements Council; National Archives Council; National Arts Club; National Association of Cemeteries; National Association of Chiropodists; National Association of Choirs; National Association of Coroners; National Association of Counties; National Aviation Club; National Aviation Corporation; National Can Corporation (stock exchange symbol); Native American Church; Naval Academy; Naval Air Center; Naval Aircraftman; Non-Airline Carrier; North Atlantic Council; Northeast Air Command; Norwegian American Cruises; Norwegian-American Council; Nuclear Assurance Corporation; (*see* PNAC)
NACA National Advisory Committee for National Aeronautics; National Agricultural Chemicals Association; National Air Carrier Association; National Armored Car Association; National Association of Children of Alcoholics; National Association of Cost Accountants; National Association of County Administrators; New Australian Cultural Association
NACAC National Ad Hoc Committee Against Censorship; National Association of College Admissions Counselors
NACADA National Academic Advising Association
NACAE National Advisory Council for Art Education
NACAM National Association of Corn and Agricultural Merchants
NACASBVH National Accreditation Council for Agencies Serving the Blind and Visually Handicapped
NACATTS North American

Clear-Air Turbulence-Tracking System
NACB National Association of Convention Bureaus
NACC National Association for Core Curriculum
NACCA National Association for Creative Children and Adults
NACCAM National Coordinating Committee for Aviation Meteorology
NACCC National Association of Citizens Crime Commissions
NACCD National Advisory Commission on Civil Disorders
NACCG National Association of Crankshaft and Cylinder Grinders
N-accident(s) nuclear-power accident(s)
NACCT Nebraska Association of Community College Trustees
NACD National Association for Community Development; National Association of Conservation Districts; National Association of Corporate Directors
NACDL National Association of Criminal Defense Lawyers
NACDR National Association of College Deans and Registrars
NACE National Association for Career Education; National Association of Corrosion Engineers
NACEEO National Advisory Council on Equality of Educational Opportunity
NACEL Naval Air Crew Equipment Laboratory
NACF National Agricultural Cooperatives Federation; National Art Collections Fund; Navy Air Combat Fighter (plane)
NACFI North American Council on Fishery Investigations
nach (nAch) need for achievement
Nach Nachman
NACH National Advisory Council for the Handicapped
NACHA National Automated Clearinghouse Association
NACHEPO National Advisory Commission on Higher Education for Police Officers
Nachf Nachfolger (German—successor)
nachm nachmittags (Ger-

man—afternoon, p.m.)
NACHM National Advisory Committee on Health Manpower
Nachr Nachrichten (German—bulletin)
NACHRI National Association of Children's Hospitals and Related Institutions
Nachtr Nachtrag (German—appendix, supplement)
NACIAD National Council on Integrated Area Development
NACIDA National Cottage Industries Development Authority (Philippines)
NACILA National Council of Indian Library Associations
NACIMFP National Advisory Council on International Monetary and Financial Problems
Nacional El Nacional (Venezuela periodical published in Caracas)
NACISA NATO Communications and Information Systems Agency
NACISO NATO Communications and Information Systems Organization
NaCl sodium chloride (salt)
NACL National Advisory Commission on Libraries; Navy-Arpa Chemical Laser; Nippon Aviotronics Company Limited
NACLA North American Congress on Latin America
NACLIS National Commission on Libraries and Information Science
NACM National Association of Chain Manufacturers; National Association of Charcoal Manufacturers; National Association of Colliery Managers; National Association of Credit Management
naco night-alarm cutoff
NACO National Arts Centre Orchestra (Ottawa); National Association of Counties; Noise Abatement and Control Office, San Diego
NACOA National Advisory Committee on Oceans and Atmosphere
NACOC National Arts Centre Orchestra of Canada
NACODS National Association of Colliery Overmen, Deputies, and Shotfirers
NACOM National Communications
NACOR National Advisory

Committee on Radiation

NACP National Academy of Cable Programming; National Association of Chiefs of Police

nacro night-alarm cutoff

NACRO National Association for the Care and Resettlement of Offenders

NACS National Association of College Stores; National Association of Cosmetology Schools

NACSE National Association of Civil Service Employees

NACSIM NATO Communications Security Information

NACSW National Action Committee on the Status of Women (Canadian)

NACT National Association of Careers Teachers; National Association of Craftsman Tailors; National Association of Cycle Traders; National Association of Cycle Trades

NACTA National Association of Colleges and Teachers of Agriculture

NACTST National Advisory Council on the Training and Supply of Teachers

NACUA National Association of College and University Administrators; National Association of College and University Attorneys

NACUBO National Association of College and University Business Office Associations

NACUFS National Association of College and University Food Services

NACUSS National Association of College and University Summer Sessions

NACV National Association of Concerned Veterans

NACVE National Advisory Council on Vocational Education

NACW National Advisory Committee on Women

NACWC National Association of Colored Women's Clubs

NACWPI National Association of College Wind and Percussion Instruments

nad nadir (lowest point); network addressing device; no apparent defect; no appreciable difference; no appreciable disease; not on active duty; nothing abnormal detected; nothing abnormal dis-

covered

nad (NAD⁺) nicotinamide adenine dinucleotide; (same as DPN)

Nad Nadine; Nedezhda

NAD National Academy of Design; National Alliance for Democracy; National Association of the Deaf; Naval Air Depot; Naval Air Division; Naval Ammunition Depot; North Atlantic Division

NADA National Association of Dealers in Antiques; National Association of Drug Addiction; National Automobile Dealers Association

NADABB National Alzheimer's Disease Autopsy and Brain Bank

NADAC National Anti-Drug Abuse Campaign

NADAP National Association of Drug Abuse Problems

NADAR North American Data Airborn Recorder

NADB National Aerometric Data Bank

NADC National Anti-Dumping Committee; Naval Aide-de-Camp; Naval Air Development Center; Northern Agricultural Development Corporation

NADD National Association of Diemakers and Diecutters

NADDIS Narcotics and Dangerous Drugs Intelligence File (computerized criminal file)

NADEE National Association of Divisional Executives for Education

NaDefCol Nato Defense College

NADEM National Association of Dairy Equipment Manufacturers

NaDevCen Naval Air Development Center

NADF National Alzheimer's Disease Foundation

NADFAS National Association of Design and Fine Art Societies

NADFS National Association of Drop Forgers and Stampers

NADGE NATO Air Defense Ground Environment Organization

NADGEMO Nato Air Defense Ground Environment Management Office

nadh (NADH) dihydronicotinamide adenine dinucleotide;

(same as dpnh or DPNH)

NADL National Association of Dental Laboratories; Navy Authorized Data List

NAD/NADH, nicotinamide adenine dinucleotide (coenzyme system affecting hydrogen transfer in biological oxidation-reduction reactions)

NADO Navy Accounts Disbursing Office

NADOP North American Defense Operational Plan

NADOT North Atlantic Deepwater Oil Terminal

NADOW National Association for Training the Disabled in Office Work

nadp (NADP⁺) nicotinamide adenine dinucleotide phosphate; (same as tpn or TPN)

NADPAS National Association of Discharged Prisoners' Aid Societies

nadph (NADPH) dihydronicotinamide adenine dinucleotide phosphate

NADSA National Association of Dramatic and Speech Arts

NADUS National Association of Doctors in the United States

NADWARN National Disaster Warning System

nae national administrative expenses; not always excused

nAe no American equivalent

Naₑ exchangeable body sodium

NAE National Academy of Education; National Academy of Engineering; National Association of Evangelicals; Naval Aeronautical Establishment (Canadian); Naval Aircraft Establishment

NAEA National Art Education Association; National Association of Enrolled Agents; National Association of Estate Agents

NAEB National Association of Educational Buyers; National Association of Educational Broadcasters

NAEBM National Association of Engine and Boat Manufacturers

NAEC National Aerospace Education Council; National Aviation Education Council

NAECA National Appliance Energy Conservation Act

NAEd National Academy of Education

NAED National Association of Electrical Distributors

NAEDS National Association of Engraversand Die Stampers

NAEE National Association for Environmental Education

NAEF Naval Air Engineering Facility

NAEFTA National Association of Enrolled Federal Tax Accountants

NAEIR National Association for the Exchange of Industrial Resources

NAEMT National Association of Emergency Medical Technicians

NAEN National Association of Educational Negotiators

NAEP National Assessment of Educational Progress; National Association of Educational Programs (Carnegie Foundation)

NAES National Association of Educational Secretaries; National Association of Episcopal Schools; Native American Educational Services

NAESP National Association of Elementary School Principals

NAEST National Archive of Electrical Science and Technology

NAESU Naval Aviation Engineering Service Unit

NAEW NATO Airborne Early Warning

NAEYC National Association for the Education of Young Children

naf nonappropriated funds

NAF National Abortion Foundation; National Amputation Foundation; National Arts Foundation; Naval Aircraft Factory; Naval Air Facility; Netherland-America Foundation; Northern Attack Force

NAF Norges Automobil Forbund (Norwegian—Norway Automobile Association)

NAFA National Academy of Foreign Affairs; National Aerobic Fitness Award; National Association of Fleet Administrators

NAFAC National Association for Ambulatory Care

NAFAG NATO Air Force Advisory Group; NATO Air Force Armaments Group

NAFAS National Association of Flower Arrangement Societies

NAFB National Association of Franchised Businessmen

NAFBRAT National Association for Better Radio and Television

NAFC National Association of Food Chains

NAFCA North American Family Campers Association

NAFCU National Association of Federal Credit Unions

NAFD National Air Forwarding Division (Institute of Freight Forwarders); National Association of Funeral Directors

NAFEC National Association of Free-standing Emergency Centers; National Aviation Facilities Experimental Center

NAFEO National Association for Equal Opportunity (in higher education)

naff (nAff) need for affiliation

NAFF National Association For Freedom

NAFFBIA National Association of Former FBI Agents

NAFFP National Association of Frozen Food Producers

NAFI National Association of Fire Investigators; Naval Avionics Facility

NAFINSA Nacional Financiera (Spanish—National Finance Corporation)

NAFM National Armed Forces Museum; National Association of Furniture Manufacturers

NAFMB National Association of FM Broadcasters

NAFO National Association of Fire Officers; Northwest Atlantic Fisheries Organization

NAFP New Armed Forces of the Philippines

NAFPC National Academy for Fire Prevention and Control

N Afr North Africa

NAFRC National Association of Fiscally Responsible Cities

NAFRLG National Alliance of Financially-Responsible Local Governments

NAFS National Association of Foot Specialists; National Association of Forensic Sciences

NAFSA National Association of Foreign Sudent Advisers; National Association of Foreign Student Affairs

NAFT National Alternative Fuel Test(ing)

NAFTA New Zealand-Australia Free Trade Agreement; North Atlantic Free Trade Area (Canada, United Kingdom, United States)

NAFWR National Association of Furniture Warehousemen and Removers

nag. net annual gain

Nag Nagasaki; Nagoya

NAG National Action Group; National Association of Gag Writers; National Association of Gardeners; National Association of Goldsmiths; National Association of Groundsmen; Naval Advisory Group; Naval Applications Group (USN); Negro Actors Guild; Neighborhood Action Group

NA & G Norgulf Lines (North Atlantic & Gulf)

NAGARD NATO Advisory Group for Aeronautical Research and Development

Nagas Naga Hills (mountains on the Burmese border of India); Nagasaki, Japan

NAGC National Association for Gifted Children

NAGCP National Association of Greeting Card Publishers

N-age nuclear age

NAGE National Association of Government Employees

NAGLDCs National Association of Gay and Lesbian Democratic Clubs

NAGM National Association of Glove Manufacturers; National Association of Glue Manufacturers

Nagp Nagpur

NAGPM National Association of Grained Plate Makers

NAGRA Nationalen Genossenschaft für die Lagerung Radioaktiver Abfaelle (German—National Cooperative Society for the Storage of Radioactive Wastes)

NAGS National Allotments and Gardens Society

NAGT National Association of Geology Teachers

NAGWS National Association for Girls and Women in Sport

Nah The Book of Nahum

Nah Nahum

NAH National Association of Homebuilders

NAHA National Association of Handwriting Analysts; National Association of Health Authorities

Nahal *Na'or Halutsi Lohem* (Hebrew—Fighting Pioneer Youth)—youngest section of the Israeli army

NAHB National Association of Home Builders

NAHC National Advisory Health Council; National Anti-Hunger Coalition

NAHCAC National Ad Hoc Committee Against Censorship (sometimes abbreviated NACAC)

NAHE National Association for Humanities Education

NAHFO National Association of Hospital Fire Officers

NAHSA National Association for Hearing and Speech Action; National Association of Hearing and Speech Agencies

NAHSTA National Hiking and Ski Touring Association

NAHT National Association of Head Teachers

nai no action indicated; no address instruction

NAI National Agricultural Institute

NAI New Acronyms and Initialisms

NAIA National Association of Insurance Agents; National Association of Intercollegiate Athletics

NAIB National Association of Insurance Brokers

NAIC National Association of Insurance Commissioners; National Association of Investment Clubs; National Association of Investors Corporation; Naval Aircraft Investigation Center

NAICU National Association of Independent Colleges and Universities

NAIDS North Atlantic Institute for Defense Studies (NATO)

NAIEC National Association for Industry-Education Cooperation

NAIES National Association of Interdisciplinary Ethnic Studies

NAIG Nippon Atomic Industry Group

NAII National Association of Independent Insurers

NAIL National Association of Independent Lumbermen; Naval Aircraft Inventory Log; Neurotics Anonymous International Liaison

NAILS National Automated

Immigration Lookout System

NAILSC Naval Air Integrated Logistics Support Center

naiop navigational aids inoperative for parts

NAIR National Arrangements for Incidents Involving Radioactivity

naira monetary unit of Nigeria

NAIRD National Association of Independent Record Distributors & Manufacturers

NAIRE National Association of Internal Revenue Employees

Nairns Nairnshire

NAIRS National Athletic Injury/Illness Reporting System

NAIS National Association of Independent Schools

NAISC National American Indian Safety Council

NAISS National Association of Iron and Steel Stockholders

NAIT Northern Alberta Institute of Technology

NAITTE National Association of Industrial and Technical Teacher Educators

naivnik naive person or politician

NAIW National Association of Insurance Women

NAIWA North American Indian Women's Association

NAJ National Association for Justice

NAJC National Assessment of Juvenile Correction (University of Michigan); Northern Australia Jockey Club; Northwest Alabama Junior College

NAJCA National Association of Juvenile Correctional Agencies

NAJE National Association of Jazz Educators; National Association of Jazz Education

nak negative knowledge; nothing adverse known

nak (NAK) negative acknowledge character (data processing)

nakl naklad (Polish—edition, publisher); *nakladatel* (Czech—edition, publisher)

NAL National Acoustics Laboratory; National Aerospace Laboratory; National Agriculture Library (US Department of Agriculture); National Airlines; National Association of Laity (Catholic); National Astronomical League; Nigeria America Line; Norwegian America Line

NAL New American Library

NALA National Association of Legal Assistants

NALC National Association of Letter Carriers; National Association of Litho Clubs

NALCC National Automatic Laundry and Cleaning Council

NALCO Newfoundland and Labrador Corporation

NALCON Navy Laboratory Computer Network

NALDEF Native American Legal Defense and Education Foundation

NALEAO National Association of Latino Elected and Appointed Officials

NALED National Association of Limited Edition Dealers

NALGG National Association for Lesbian and Gay Gerontology

NALGO National and Local Government Officers Association

NALLA National Long-Lines Agency

NALM National Association of Lift Makers

NALP National Association for Law Placement

NALS National Association of Legal Secretaries

NALSA North American Land Sailing Association

NALSAT National Association of Land Settlement Association Tenants

NALU National Association of Life Underwriters

nam network access machine; non-aligned movement

Nam Vietnam; Namibia (South-West Africa)

N Am North America(n)

NAM National Aero Manufacturing; National Air Museum (Smithsonian Institution); National Association of Manufacturers; Naval Aircraft Modification; Newspaper Association Managers; Non-Alignment Movement; North America(n); North American Movement

NAM Nederlandsche Aluminium Maatschappij (Dutch—Netherlands Aluminum Company)

NAMA National Automatic Merchandising Association; New Amsterdam Musical Association; North American Maritime Agencies

NAMAC National Association of Merger and Acquisition Consultants

NAMB National Association of Master Bakers

NAMBLA North American Man-Boy Love Association (of child molesters)

NAMBO National Association of Motor Bus Operators

NAMC Naval Air Materiel Center; Naval Air Materiel Command; Nihon Aeroplane Manufacturing Company

NAMCC National Association of Mutual Casualty Companies

NAMCO Naval and Mechanical Company

NAMDI National Marine Data Inventory

NAME National Association of Marine Engineers; National Association of Media Educators; National Association of Medical Examiners; National Association of Metal Name Plate Manufacturers

NAMEB National Association of Marine Engine Builders

NAMESU National Association of Music Executives in State Universities

NAMF National Association of Metal Finishers

NAMFI NATO Missile Firing Installation

NAMFREL National Citizens Movement for Free Elections

NAMH National Association for Mental Health; Norwegian-American Historical Museum

NAMIA National Association of Mutual Insurance Agents

Namib Namibia or Namib Desert of South-West Africa

NAMIC National Association of Mutual Insurance Companies

NAMilCom North Atlantic Military Committee

naml namligen (Swedish—namely)—viz.

NAMM National Association of Music Merchandisers; National Association of Music Merchants

NAMMA NATO Multi-Role Combat Aircraft Development and Production Management Agency

NAMMC Natural Asphalt Mineowners' and Manufacturers' Council

NAMMO NATO Multi-Role Combat Aircraft Development and Production Management Organization

NAMMW National Association of Musical Merchandise Wholesalers

NAMOA National Association of Miscellaneous Ornamental and Architectural Products Contractors

NAMOS National Art Museum of Sport

NAMP National Association of Magazine Publishers; National Association of Married Priests; National Association for Mature People; Naval Aviation Maintenance Program

NAMPA NATO Maritime Patrol Aircraft Agency

nampg nautical air miles per gallon

NAMPMW Vietnam Prisoners of War (organization)

namppf nautical air miles per pound of fuel

NAMS National Association of Marine Surveyors

NAMSA NATO Maintenance and Supply Agency

NAMSB National Association of Mutual Savings Banks

NAMSO NATO Maintenance and Supply Organization

NAMT National Association for Music Therapy

NAMTA National Art Materials Trade Association

NAMTC Naval Air Missile Test Center

NAMTRADET Naval Air Maintenance Detachment

NAMTRAGRU Naval Air Maintenance Training Group

n.a.n. *nisi aliter notetur* (Latin—unless it is otherwise noted)

Nan Anna; Nancy; Nanette; Nanking

NAN Nandi, Fiji Islands (airport)

nana (NANA) N-acetylneuraminic acid

NANA National Advertising News Association; North American Newspaper Alliance; Northwest Alaska Natives Association

NANAC National Aviation Noise Abatement Council

Nancy Anne Frances "Nancy" Robbins Davis Reagan, wife of President Ronald Reagan

NAND NOT AND (data-processing logic operator)

Nando Fernando

NANE National Association for Nursery Education

Nanga Nanga Parbat, India

NANM National Association of Negro Musicians

nano 10^{-9}

nanova non-orthogonal analysis of variance

NANTIS Nottingham and Nottinghamshire Technical Information Service

NANVH&SWO National Assembly of National Voluntary Health and Social Welfare Organizations

NANWEP Navy Numerical Weather Prediction; Navy Numerical Weather Problems (USN)

NAO National Accordion Organization; National Association of Outfitters; Noise Abatement Office

NAOA Navy Officers Accounts Office

NAOC Nigerian Agip Oil Company

NaOH sodium hydroxide (caustic soda)

NAOP National Association of Operative Plasterers

NAORPG North Atlantic Ocean Regional Planning Group

NAOT National Association of Organ Teachers

NAOTC National Association of Over-the-Counter Companies

NAOTS Naval Aviation Ordnance Test Station

nap. knapsack; napalm (naphthalene and coconut oil—jellied gasoline incendiary mixture); naphtha; naval aviation pilot (NAP); non-agency purchase; not at present

Nap Naples; Napoleon; Napoleonic

NAP Naples, Italy (airport); Narragansett Pier (railroad); National Aerospace Plane (hypersonic aircraft capable of flying up to 25 times the speed of sound); National Association for the Paralysed; National Association of Parliamentarians; National Association of Postmasters; National Association of Publishers; Naval Auxiliary Patrol; Naval Aviation Pilot

N.A.P. Neighborhood Awareness Program

NAP Nomina Anatomica, Pa-

ris; Nuclei Armati Proletari (Italian—Armed Proletarian Nucleus)—terrorists

NAPA National Asphalt Paving Association; National Association of Performing Artists; National Association of Purchasing Agents

NAPAC National Program for Acquisitions and Cataloging

napalm naphthene palmitate (napththalene plus coconut oil—jellied gasoline used in flame-throwers)

NAPAN National Association for the Prevention of Addiction to Narcotics

NAPARE National Association for Perinatal Addiction Research and Education

NAPATMO NATO Patriot Management Office

NAPBL National Association of Professional Baseball Leagues

napc non-adherent peritoneal cells

NAPC National Association of Precancel Collectors

NAPCA National Air Pollution Control Administration

NAPCAE National Association for Public Continuing and Adult Education

NAPCRO National Association of Police Community Relations Officers

NAPD National Association of Police Driving

NAPDEA North American Professional Driver Education Association

NAPE National Alliance of Postal Employees; National Association for Professional Educators; National Association of Port Employees; National Association of Power Engineers

NAPECW National Association for Physical Education of College Women

NAPF National Association of Pension Funds

NAPFE National Alliance of Postal and Federal Employees

naph naphtha; naphthyl

NAPH National Association of Professors of Hebrew

NAPHC National Association of Plumbing/Heating/Cooling Contractors

NAPIA National Affiliate of Printing Industries of America

NAPIM National Association of Printing Ink Manufacturers

NAPL National Association of Photo Lithographers; National Association of Printers and Lithographers

NAPLP National Association of Para-Legal Personnel

NAPLPS North American Presentation Level Protocol Syntax

NAPM National Association of Paper Merchants; National Association of Punch Manufacturers; National Association of Purchasing Management

NAPMO NATO Airborne Early Warning and Control Programme Management Organisation

NAPN National Association of Physician Nurses

NAPNAP National Association of Pediatric Nurse Associates and Practitioners

NAPNES National Association for Practical Nurse Education and Service

NAPO National Association of Performing Artists; National Association of Probation Officers; National Association of Property Owners; National Association of Purchasing Agents

Napoleon Napoleon Bonaparte

Napoléon I Napoléon Bonaparte (1769–1821)

Napoléon II *l'Aiglon* (French—the Eaglet) François-Charles-Joseph Bonaparte (1811–1832)

Napoléon III *Napoléon le Petit* (French—Little Napoleon) Louis-Napoléon Bonaparte (1808–1873)

NAPPH National Association of Private Psychiatric Hospitals

nap(py) napkin

na pr na priklad (Czech—for example)

NAPR National Association for Pastoral Renewal

NAPRA National Association of Progressive Radio Announcers

NAPRC National Association for the Prevention of Rape by Castration

NAPS National Alliance of Postal Supervisors; Nissan Air Pollution System

NAPSA National Association

of Pretrial Service Agencies

NAPSAE National Association for Public School Adult Education

Nap's bones Napier's bones (first slide rule)

NAPSS Numerical Analysis Problem Solving System

NAPT National Association of Physical Therapists; National Association for the Prevention of Tuberculosis

NAPTC Naval Air Propulsion Test Center

NAPTIC National Air Pollution Technical Information Center

NAPUS National Association of Postmasters of the United States; Nuclear Auxiliary Power Unit System

NAPV National Association of Prison Visitors

NAPVD National Association for the Prevention of Venereal Disease

NAQI National Air Quality Index

NAQP National Association of Quick Printers

nar narrow; net assimilation rate; no apparent reason

Nar Narragansett

NAR National Association of Realtors; National Association of Rocketry; Nelson Aldrich Rockefeller; North American Rockwell; North American Royalties; Northern Alberta Railway

NARA Narcotic Addict Rehabilitation Act; National Archives and Records Administration; Nippon Australian Relations Agreement

NARAA National Association of Recruitment Advertising Agencies

NARAD Navy Research and Development

NARAL National Abortion Rights Action League; National Association for the Repeal of Abortion Laws

NARAS National Academy of Recording Arts and Sciences

NARB National Advertising Review Board

narc narcotic; narcotics agent; narcotics; narcotics officer

narc (Latin prefix—numbness or stupor)—narcotic

NARC National Agricultural Research Center; National Archives and Records Service; National Association for Retarded Children; National Association of Retired Catholics

narco narcotic; narcotics hospital; narcotics officer; narcotics treatment center

NARCO United Nations Narcotics Commission

narcocard narcotic-addict registration card

narcodollars narcotic (traffic) dollars

Narconon Narcotics Anonymous

narcos narcotics; narcotics police officers

narcot narcotic, narcotize(d), narcotiz(ing)

narcotest narcotics test

narco-traf narcotics traffick(er); narcotics trafficking

narcs narcotics; narcotics agents; narcotics hospitals; narcotics officers; narcotics treatment centers

nard spikenard

NARD National Association of Regimental Drummers; National Association of Retail Druggists

NARDIC Naval Research and Development Information Center

Nar Div Narodni Divadlo (Czech—National Theater)— Prague opera house

NAREB National Association of Real Estate Boards; National Association of Real Estate Brokers

narec naval research electronic computer

NAREE National Association of Real Estate Editors

NAREIF National Association of Real Estate Investment Funds

NARELLO National Association for Real Estate License Law Officials

narf natural axial-resonant frequency

NARF National Association of Retail Furnishers; Native American Rights Fund; Naval Air Rework Facility; Nuclear Aircraft Research Facility

NARFE National Association of Retired Federal Employees

NARGA National Association of Retail Grocers of Australia

NARI National Association of Recycling Industries; National Atmospheric Research

Institute; Native American Research Institute

Nar Inv Narcotics Investigation

narist. naristillae (Latin—nasal drops)—nosedrops

nark narcotics agent or law-enforcement officer

NARK Nikolai Andreyvich Rimsky-Korsakov

Narkomvneshtorg Narodny Komissariat Vneshney Torgovli (Russian—People's Commissariate of Foreign Trade)

NARL National Aero Research Laboratory; Naval Arctic Research Laboratory

NARM National Association of Relay Manufacturers; National Association of Retail Merchants

NARMCO National Research and Manufacturing Company

N-armed nuclear-armed (aircraft, bomb, missile, submarine, etc.)

N-arm(s) nuclear armament(s); nuclear arms

N-arms control nuclear arms control

N-arms race nuclear arms race

NARO North American Regional Office

NAROCTESTSTA Naval Air Rocket Test Station

NARP National Association of Railroad Passengers; Nuclear Weapons Accident Report Procedures

NARPA National Air Rifle and Pistol Association

Nar Rep Bul Narodna Republika Bulgaria (Bulgarian People's Republic)

Narrows waterway between Brooklyn and Staten Island, New York—spanned by 4260-foot-long Verrazano Bridge, world's longest suspension bridge; strait in the Dardanelles near the Aegean; strait between American and British Virgin Islands

Narrow Seas the Channel between England and France as well as the southern end of the North Sea between England, Belgium, and the Netherlands

NARS National Archives and Records Service; National Association of Radiation Survivors; Non-Affiliated Reserve Section

NARSIS National Association

for Road Safety Instruction in Schools

NARST National Association for Research in Science Teaching

NARTB National Association of Radio and Television Broadcasters

NARTC North America Region Test Center

NARTEL North Atlantic Radio Telephone Committee

NARTM National Association of Rope and Twine Merchants

NARTS National Association of Reporter Training Schools; Naval Air Rocket Test Station

NARTU Naval Air Reserve Training Unit

NARU North Australian Research Unit

NARUC National Association of Regulatory Utility Commissioners

NARVRE National Association of Retired and Veteran Railroad Employees

NARWACL North American Regional World Anti-Communist League

nas nasal; nasalis; nasology

n-a-s no added salt

NAS Nassau, Bahamas (airport); National Academy of Sciences; National Adoption Society; National Advocates Society; National Aerospace Standard(s); National Agricultural Society; National Aircraft Standard(s); National Airspace System; National Association of Sanitarians; National Association of Scholars; National Association of Schoolmasters; National Association of Stevedores; National Association of Supervisors; National Audubon Society; Native American Studies; Naval Air Station; Nursing Auxiliary Service

N A S Noise Abatement Society

N^{a}S^a Nuestra Señora (Spanish—Our Lady)

NASA National Acoustical Suppliers Association; National Aeronautics and Space Administration; National Appliance Service Association; National Association of Schools of Art; National Association of Securities Ad-

ministrators; National Automobile Salesmen's Association; North American Sailing Association
NASAA National Aeronautics and Space Administration Act; National Assembly of State Arts Agencies
NASABCA National Aeronautics and Space Administration Board of Contract Appeals
NASA-CF National Aeronautics and Space Administration—Cocoa Beach, Florida
NASA-CO National Aeronautics and Space Administration—Cleveland, Ohio
NASA-EC National Aeronautics and Space Administration—Edwards, California
NASAEN National Association for State-Enrolled Assistant Nurses
NASA-GM National Aeronautics and Space Administration—Greenbelt, Maryland
NASA-HA National Aeronautics and Space Administration—Huntsville, Alabama
NASA-HT National Aeronautics and Space Administration—Houston, Texas
Nasakom Nationalist-Communist
NASA LST National Aeronautics and Space Administration Large Space Telescope
NASA-LV National Aeronautics and Space Administration—Langley Field, Virginia
NASA-MC National Aeronautics and Space Administration—Moffett Field, California
NASAO National Association of State Aviation Officials
NASAP Nonproliferation Alternative System Assessment Program; Nuclear Alternative Systems Assessment Program
NASAPR National Aeronautics and Space Administration Procurement Regulations
NASAR National Association of Search and Rescue
NASARR North American Searching and Ranging Radar
NASA-SC National Aeronautics and Space Administration—Santa Monica, California
NASA STAR National Aeronautics and Space Adminis-

tration Scientific and Technical Aerospace Reports
NASA/STIF National Aeronautics and Space Administration/Scientific and Technical Information Facility
NASBE National Association of State Boards of Education
NASC National Aeronautics and Space Council; National Aircraft Standards Committee; National Alliance of Senior Citizens; National Association of Student Councils; NATO Supply Center; Naval Air Systems Command; North American Supply Council; Northwest Association of Schools and Colleges
NASCAR National Association of Sports Car Racing; National Association for Stock Car Advancement and Research
NASCO National Academy of Sciences Committee on Oceanography; National Automotive Service Company; North American Students of Cooperation
NASCom Naval Air Systems Command
NASCOM National Aeronautics and Space Administration tracking network, also performing command and control functions
NASCP North American Society for Corporate Planning
NASCUS National Association of State Credit Union Supervisors
NASD National Amalgamated Stevedores and Dockers; National Association of Securities Dealers; Naval Aviation Supply Depot; Nippon Advanced Ship Design(ing)
NASDA National Association of State Development Agencies; National Space Development Agency (Japan)
NASDAQ National Association of Security Dealers Automated Quotation (system)
NASDAQS National Association of Security Dealers Automated Quotation System
NASDCD National Association of State Directors of Child Development
NASDS National Association of Scuba Diving Schools
nase neutral atom space engine (sputtering engine)
NASE National Academy of

School Executives; National Academy of Stationary Engineers; National Association of Stationary Engineers; National Association of Steel Exporters
NASEES National Association for Soviet and East European Studies
NASEN National Association for State-Enrolled Nurses
NASF National Aboriginal Sports Foundation; National Association of State Foresters
NASFAA National Association of Student Financial Aid Administrators
NAS & FCA National Automatic Sprinkler and Fire Control Association
NASFO National Asset Seizure and Forfeiture Office
NAS-GB Noise Abatement Society of Great Britain
Nash Nashville
NASH National Association of Specimen Hunters
NASHA North American Survival and Homesteading Association
NASIS National Association for State Information Systems
NASL North American Soccer League
NASM National Air and Space Museum (Smithsonian); National Association of School Magazines; National Association of Schools of Music; Naval Aviation School of Medicine
NASM Nederlandsche-Amerikaansche Stoomvaart Maatschappij (Dutch—Holland-American Line)
NASML National Air and Space Museum Library (Smithsonian Institution)
NASMV National Association for a Standard Medical Vocabulary
NASN National Air Sampling Network; National Association of School Nurses
NASNI Naval Air Station, North Island (Halsey Field, San Diego, California)
NAS-NRC National Academy of Science—National Research Council
NASOH North America Society for Oceanic History
NASP National Aero-Space Plane; National Airport Systems Plan; National Associa-

tion of School Psychologists; Negro Anglo-Saxon Protestant

NASPA National Society of Public Accountants

Nas Par Nasionale Party (Afrikaans—National Party)—South Africa's Apartheid party

NASPD National Association of Steel Pipe Distributors

Nas Pers Nasionale Pers (Afrikaans—National Press)—publisher of apartheid books and periodicals

NASPM National Association of Seed Potato Merchants

NASQAN National Stream-Quality Accounting Network

Nᵃ Srᵃ Nossa Senhora (Portuguese—Our Lady); *Nuestra Señora* (Spanish—Our Lady)

NASRC National Association of State Racing Commissioners

NASRP National Association of Special and Reserve Police

Nass Nassau

NASS National Association of School Superintendents; National Association of Summer Sessions; Naval Air Signals School

NASSAM National Association for the Self-Supporting Active Ministry

NASSC National Alliance on Shaping Safer Cities

NASSCO National Steel and Shipbuilding Company

NASSD National Association of School Security Directors

NASSL National Association of Spanish-Speaking Librarians

NASSO National Association of Socialist Students' Organizations

NASSP National Association of Secondary-School Principals

NASSR Nahichevan Autonomous Soviet Socialist Republic

NAST Nuclear Accident Support Team

NASTBD National Association of State Text Book Directors

NASTI Naval Air Station, Terminal Island

NASTL National Anti-Steel-Trap League

NASU National Adult School Union

NASULGC National Associa-

tion of State Universities and Land-Grant Colleges

NASW National Association of Science Writers; National Association of Social Workers

NASWM National Association of Scottish Woolen Manufacturers

nat nation; national; nationalist; native; natural; naturalist; naturalization; naturalize(d); nature; normal allowed time

nat natuurkunde (Dutch—natural science)

Nat Natalia; Natalie; Nathalie; Nathan; Nathanael; Nathaniel; Natasha; Nation; National; Nationalist; naturalized

Nat Hof und Nationaltheater (German—Court and National Theater, Munich); *Naturkunde* (German—natural science)

NAT National Air Transport; National Arbitration Tribunal

NATA National Association of Tax Accountants; National Association of Tax Administrators; National Association of Testing Authorities; National Association of Transportation Advertisers; National Athletic Trainers Association; National Automated Transportation Association; National Aviation Trades Association; North American Telephone Association; North Atlantic Treaty Alliance

Nat Absten National Abstentionalist

NATAPROUBU National Association of Professional Bureaucrats

Nat Arc National Archives

NATAS National Academy of Television Arts and Sciences

Nat Assn National Association

natat natation

NATB National Automobile Theft Bureau; Naval Air Training Base

Nat Bur Econ Res National Bureau of Economic Research (Columbia and Princeton)

Nat Bur Stand Circ National Bureau of Standards Circular(s)

Nat Bur Stand Misc Pub National Bureau of Standards Miscellaneous Publication(s)

Nat Bur Stand Spec Pub National Bureau of Standards Special Publication(s)

NATC National Air Traffic Controllers; National Air Transportation Conferences; National Alcohol Tax Coalition; Naval Air Training Command

NATCA National Air Traffic Controllers Association

NATCG National Association of Training Corps for Girls

natch naturally

Natch Natchez

Natchez Trace national parkway serving Alabama, Mississippi, and Tennessee

NATCO National Automatic Tool Company; National Tank Company

natcol natural color(ing)

natcom national communications

NATCOM NATO communication

Nat Con Nature Conservancy

NATCS National Air Traffic Control Service; National Air Traffic Control System

NATD National Association of Teachers of Dancing

Nat Dem National Democrats

Nate Nathan(iel)

NATE National Association for Teachers of Electronics; National Association for the Teaching of English; Native American Teacher Education

Nat Fed National Federation

NATFHE National Association of Teachers in Further and Higher Education

Nat Gal National Gallery

Nat Geog Mag National Geographic Magazine

Nath Nathan(iel)

Nath B Nathaniel Bowditch

nat hist natural history

Nathl Nathaniel

NATIDC Netherlands-Australia Trade and Industrial Development Council

NATIE National Association of Trade and Industrial Education

nation. nationality

National National Gallery in London or the National Gallery of Art in Washington, D.C.

NATIONAL National Cash Register

Nations Bus Nations Business

NATIS National Information System(s); North Atlantic Treaty Information Service

Nativ Nativity

NATKE National Association

of Theatrical and Kine Employees
natl national
N Atl North Atlantic
N Atl Cur North Atlantic Current
Nat Lib National Liberal; National Library of Canada (Ottawa)
NATLIBCAN National Library of Canada
NATLIBNZ National Library of New Zealand
NATMAP National Mapping
Nat Mon National Monument
Nat Mus Natal Museum; National Museum
nato no action—talk only
NATO National Association of Taxicab Owners; National Association of Theater Owners; National Association of Trailer Owners; National Association of Travel Organizations; North Atlantic Treaty Organization (Belgium, Canada, Denmark, Greece, Iceland, Italy, Luxembourg, Netherlands, Norway, Portugal, Spain, Turkey, United Kingdom, United States, Germany)
NATO-AGARD North Atlantic Treaty Organization—Advisory Group for Aeronautical Research and Development
Nat Obs National Observer
NATO Council Belgium, Canada, Denmark, Federal Republic of Germany, Greece, Iceland, Italy, Luxembourg, Netherlands, Norway, Portugal, Spain, Turkey, United Kingdom, United States
NATODC NATO Defense College
NATO-ELLA North Atlantic Treaty Organization—European Long Lines Agency
NATO-LRSS North Atlantic Treaty Organization—Long-Range Scientific Studies
NATOMILOCGRP North Atlantic Treaty Organization Military Oceanography Group
NATOPS Naval Air Training and Operating Procedures Standardization
Nat Ord Natural Order
NATO-RDPP North Atlantic Treaty Organization—Research and Development Production Program
NATOs National Association of Theatre Owners

NATPE National Association of Television Program Executives
nat phil natural philosophy
Nat Pk National Park
natr. *natrium* (Latin—sodium)
Nat Rev National Review
Nats National Party members; Nationalists; naturalized citizens
Nats Natsionalnyii (Russian—national)
NATS National Association of Teachers of Singing; National Association of Temporary Services; Naval Air Test Station; Naval Air Transport Service
Nat. Sc.D. Doctor of Natural Science
Nat Sci Natural Science(s)
Nat Sci Fdn National Science Foundation
Nat Sec Soc National Secular Society (founded in 1866 by Charles Bradlaugh)
NATSEMI National Semiconductor Incorporated
NATSF Naval Air Technical Service Facility
NATSJA National Association of Training Schools and Juvenile Agencies
NATSOPA National Society of Operative Printers and Assistants
NATSPG North Atlantic Systems Planning Group
Nat Sup National Superannuation
N Att Naval Attaché
N-attack nuclear attack
NATTC National Tank Truck Carriers; Naval Air Technical Training Center
NATTKE National Association of Theatrical, Television, and Kine Employees
NATTS National Association of Trade and Technical Schools; Naval Air Turbine Test Station
Nat U Nations Unies (French—United Nations)
Nat Uni National University
natur naturalist
Natural Bridges Natural Bridges National Monument in Southeastern Utah
NATUSA North African Theater of Operations
Nat West National Westminster (British bank)
naty naturally
náu náutica (Spanish—nautical)

Nau Nauruan(s); Nauru Island
NAU Naval Administrative Unit; Northern Arizona University
NAUA National Aircraft Underwriters' Association; National Auto Underwriters Association
nauga naugahide (plastic upholstery)
Naughty Island Pulau Sajahat (resort offshore Singapore)
NAUPA National Association of Unclaimed Property Administrators
Nauru Republic of Nauru (western Pacific Ocean island nation), Pleasant Island
NAUS National Association for Uniformed Services
naut nautical
NAUW National Association of University Women
n aux b new auxiliary boiler
nav naval; navigable; navigate; navigation; navigational; navigator
n/a/v net asset value
Nav Navaho; naval; Navarra; Navarre; Navassa Island between Haiti and Jamaica
NAV Net Asset Value
NAVA National Audio-Visual Association; Net Asset Value; North American Vexillological Association
navaco navigation action cutout (switchboard)
NAVAE National Association for Vietnamese-American Education
NAVAERORECOVF Naval Aerospace Recovery Facility
navaid(s) navigation aid(s)
NAVAIR Naval Air (Systems Command)
NAVAIRLANT Naval Air Forces, Atlantic
NAVAIRPAC Naval Air Forces, Pacific
NAVAIRREWORKF Naval Air Rework Facility
NAVAIRSYSCOM Naval Air Systems Command
Nav. Arch. Naval Architect
Navarraise La Navarraise (French—The Girl of Navarre)—Massenet opera
NavAus navigation in Australian waters
NAVBALTAP Naval Forces, Baltic Approaches (NATO)
NAVBASE Naval Base
navbm (NAVBM) naval ballistic missile
nav brz naval bronze

Nav Bs Naval Base
Navcad Naval Cadet
NAVCAMS Naval Communication Area Master Station
NAVCENT Allied Naval Forces, Central Europe
NAVCJ National Association of Volunteers in Criminal Justice
Navcm navigation countermeasures and deception
navcom navigation communication
Nav.Const. Naval Constructor
NAVCOSSACT Naval Command Systems Support Activity
navdac navigation data assimilation computer
NAVDAC Navigation Data Assimilation Center
Nav Dep Naval Deputy (NATO)
Nav.E. Naval Engineer
NavEams navigation in the eastern Atlantic and the Mediterranean
NavEast navigation along the east coast of Asia
NAVEDTRASUPPCEN Naval Education and Training Support Center
NAVELEX Naval Electronic (Systems Command)
NAVEOFAC Naval Explosive Ordnance Disposal Facility
Navesink Highlands of the Navesink also called Atlantic Highlands on the New Jersey coast around Sandy Hook
navex navigation exercise
NAVFE Naval Forces Far East
NAVFEC Naval Facilities
NAVFECENGCOM Naval Facilities Engineering Command
NAVFOR Naval Forces
NAVFORJAP Naval Air Forces, Japan
NAVFORKOR Naval Air Forces, Korea
NAVH National Aid to Visually Handicapped; National Association for the Visually Handicapped; National Association of Voluntary Hostels
Nav I Navassa Island (uninhabited American islet in north Caribbean close to Windward Passage between Cuba and Hispaniola, navigational light maintained by U.S. although recently Haiti claimed the islet)
NAVIC Navy Information Center

navicert naval inspection certificate (allowing neutral vessels to proceed through blockades)
navicert(s) navigation certificate(s)
Navidad Natividad (Spanish—Nativity)—Christmas
navig navigation
Navigators Navigator Islands (American Samoa)
NavInd navigation in the Indian Ocean
NAVINFO Navy Information Offices
NAVINTCOM Naval Intelligence Command
NAVINTCOMINST Naval Intelligence Command Instructions
NAVISTAR formerly International Harvester
NAVLIS Navy Logistics Information System
NAVMACS Naval Modular-Automated Communications Systems
NAVMAR Naval Forces, Marianas
NAVMAT Naval Materiel Command (USN)
NAVMED Naval Medicine
NAVMEDIS Naval Medical Information System
NavMisCen Naval Missile Center
NAVNON Naval Forces, Northern Norway (NATO)
NavNoPac navigation in the North Pacific
NavNorlant navigation in the North Atlantic
NAVNORTH Allied Naval Forces, Northern Europe
NavOceanO Naval Oceanographic Office (USN)
NAVOCFORMED Naval On-Call Force, Mediterranean (NATO)
NAVOCS Naval Officer Candidate School
NAVORD Naval Ordnance
NAVORDSYSCOM Naval Ordnance Systems Command
NAVPACEN Navy Public Affairs Center
NAVPERS Naval Personnel
NAVPERSRANDLAB Naval Personnel Research and Development Laboratory
NAVPHIBSCOL Naval Amphibious School
NAVPHIL Naval Forces—Philippines
NAVPORCO Naval Port Control Officer

NAVPRO Naval Plant Representative Office(r)
NAVPUB Naval Publications
NAVREGMEDCEN Naval Regional Medical Center
NAVROM Romanian merchant marine
NAVS National Anti-Vivisection Society; North American Vegetarian Society
navsat navigational satellite
NavSat navigation in the South Atlantic
NAVSCAP Naval Forces, Scandinavian Approaches (NATO)
NAVSCOLCOM NORVA Naval Schools Command, Norfolk, Virginia
NAVSEA Naval Sea Systems Command (USN)
NAVSEACENLANT Naval Sea Support Center—Atlantic
NAVSEACENPAC Naval Sea Support Center—Pacific
NAVSEC Naval Ship Engineering Center
NAVSHIPCOM Naval Ship Systems Command
NavShipyd Naval Shipyard
NAVSMO Navigation Satellite Management Office
NavSoPac navigation in the South Pacific
NAVSOUTH Naval Forces, Southern Europe
NAVSPASUR Naval Space Surveillance (USN)
NAVSPECWARGRU Naval Special Warfare Group
NAVSTA Naval Station
NAVSTAR Navigation System using Time and Ranging
NAVSUPGRU Naval Support Group
NAVSUPORANT Naval Support Forces, Antarctica
navtac (NAVTAC) navigation tactical (aircraft)
NAVTELCOM Naval Telecommunications Command
NAVTIS National Vessel Traffic Information System
NAVTRACEN Naval Training Center
NAVTRACOM Naval Training Command
NAVTRADEVCEN Naval Training Device Center
NAVUWSEC Naval Underwater Weapons Systems Engineering Center
navvies navigators (unskilled canal builders, unskilled laborers)
NAVWAG Naval Warfare

Analysis Group
NAVWEASERV Naval Weather Service
NAVWUIS Naval Work Unit Information Service
NAW National Association of Wholesalers; National Association for Women; North African Waters
NAWA National Association of Women Artists
NAWAC National Weather Analysis Center
NAWAPA North American Water and Power Alliance
NAWAS National Air Warning Service
NAWB National Association of Workshops for the Blind
NAWCC National Association of Watch and Clock Collectors
NAWCH National Association for the Welfare of Children in Hospitals
NAWDAC National Association for Women Deans, Administrators, and Counselors
NAWDC National Association of Women Deans and Counselors
NAWESA Naval Weapons Engineering Support Activity
NAWF National Aborigine Welfare Fund; North American Wildlife Foundation
NAWIC National Association of Women in Construction
NAWK National Association of Warehouse Keepers
NAWM National Association of Wool Manufacturers
NAWMP North American Waterfowl Management Plan
NAWND National Association of Wholesale Newspaper Distributors
NAWPA North American Water and Power Alliance
NAWS National Association of Women Students; National Aviation Weather System
NAWSA National American Woman's Suffrage Association
Naxas Naxalites (Maoist extremists active in India)
Nay Nayarit
NAYBC North American Youth Bridge Championship
NAYC National Association of Youth Clubs
NAYE National Association of Young Entrepreneurs
NAYRU North American Yacht Racing Union

naz nazionale (Italian—national)
Naz Nazaire
Naze (Old Norse—Nose)— southern tip of Norway at Lindesnes
Nazi adherent of the former National Socialist German Workers' Party *(Nationalsozialistische Partei)*
nb nanobar(n); narrow band; new boiler(s); newborn; no ball (cricket); no bias (relay)
n/b narrow beam; no balls (lacking nerve); no brands; north-bound
n.b. *nota bene* (Latin—note well)
Nb nimbus; niobium (formerly columbium)
Nb *Noordbrabent* (Dutch—North Brabant)
NB National Bank; National Battlefield; Naval Base; Navy Band; New Brunswick; Niagara Frontier Tariff Bureau; North Borneo
NB *Nauchnaya Biblioteka* (Russian—Scientific Library) —in Leningrad; *Naviera Boliviana* (Spanish—Bolivian Shipping); *Norsk Bibliotekforening* (Norwegian Library Association)
Nb⁹⁴ radioactive niobium
NBA National Band Association; National Bank of Australia; National Bankers Association; National Banking Association; National Bar Association; National Basketball Association; National Boat Association; National Bowling Association; National Boxing Association; National Button Association
NBAA National Business Aircraft Association
NBAC National Black Alcoholism Council
NBAD National Bank of Abu Dhabi
N balance nitrogen balance
NBBB National Better Business Bureau
NBBC National Brass Band Club
NBBS New British Broadcasting Station
NBBU New Brunswick Board of Underwriters
nbc non-battle casualty; nothing but chaos
n-b-c (NBC) nuclear-biological-chemical (warfare)
NBC Nagasaki Broadcasting

Company; National Ballet of Canada; National Baseball Congress; National Beagle Club; National Beef Council; National Biscuit Company; National Book Committee; National Book Council; National Bowling Council; National Braille Club; National Broadcasting Commission; National Broadcasting Company; National Broadcasting Corporation; National Bulk Carriers; National Bus Company; Navy Beach Commando; Nigerian Broadcasting Corporation
NB & C Norfolk, Baltimore and Carolina Line
NBCA National Baseball Congress of America; National Beagle Club of America
NBCC National Book Critics Circle
NBCCA National Business Council for Consumer Affairs
nbccw nuclear, biological, chemical, conventional warfare
nbcd nuclear, biological, and chemical defense
NBCDA *National Black Child Development Act*
nbcdx nuclear, biological, and chemical defense exercise
NBCSO National Broadcasting Company Symphony Orchestra
NBCU National Bureau of Casualty Underwriters
NBCUSA National Baptist Convention U.S.A.
nbd negative binomial distribution
NBD National Bank of Detroit
NBDA National Bicycle Dealers Association
NB & DA National Barrel & Drum Association
NBDC National Bomb Data Center (Washington, D.C.); National Book Development Council
NBE National Bank Examiner(s)
NBEA National Business Education Association
NBER National Bureau of Economic Research; National Bureau of Engineering Registration
NBET National Business Entrance Test(s)
NBF National Bank of Fiji; National Boating Federation

NBFA National Baseball Fan Association; National Bricklayers Foundation of Australia; National Business Forms Association

NB & FAA National Burglar and Fire Alarm Association

nbfi's non-bank(ing) financial intermediaries

NBF Life National Ben Franklin Life Insurance

nbfm narrow-band frequency modulation

nbfr neutral balance force reductions

NBFU National Board of Fire Underwriters; Newfoundland Board of Fire Underwriters

nbg no bloody good

NBG National Bank of Georgia; Naval Beach Group

NBGC National Ballet Guild of Canada

NBH National Bellas Hess

NBHA National Builders Hardware Association

NBHC New Broken Hill Consolidated

NBHS National Bureau for Handicapped Students

nbi no bone(y) injury

NBI Nathaniel Branden Institute; National Benevolent Institution

NBI Norges Byggforskningsinstitutt (Norwegian—Norwegian Building Research Institute)

NBIPP National Black Independent Political Party

NBIS Narcotics Border Interdiction System

NBIT New Bedford Institute of Technology

nbl not bloody likely

NBL National Basketball League; National Book League

NBLC Nederlands Bibliotheek en Lektuur Centrum (Dutch —Netherlands Center for Public Libraries and Literature)

NBL & P National Bureau for Lathing and Plastering

nbm no bowel movement; non-book material(s); normal bowel movement; nothing by mouth

nbm (NBM) nuclear ballistic missile

NBM New Brunswick Museum

NBME National Board of Medical Examiners

NBMG Navigation Bombing

and Missile Guidance System

NBMGS Navigation Bombing and Missile Guidance System

NBMV & NSL New Bedford, Martha's Vineyard, and Nantucket Steamship Line

nbn new bad news

nbn (NBN) national book number

NBN Nagoya Broadcasting Network

NBNZ National Bank of New Zealand

NBO Nairobi, Kenya (airport); Navy Bureau of Ordnance

n-bomb neutron bomb

N-bomb neutron bomb; nuclear bomb

nbp normal boiling point

NBP National Battlefield Park; National Business Publications; Neighborhood Beautification Program; New Brooklyn Philharmonic

NBPA National Bark Producers Association; National Basketball Players Association; National Black Police Association

NBPC National Border Patrol Council

NBPI National Board for Prices and Income

NBPRP National Board for the Promotion of Rifle Practice

nbp's nude beach pests (prurient snoopers and voyeurs)

nbq no broken quantities

nbr nitrile-based rubbers; nitrilebutadiene rubber

n br naval brass; naval bronze

n Br nördliche Breite (German—north latitude)

NBR National Bison Range (Montana); Nightly Business Report (tv)

NBR National Business Review

nbre noviembre (Spanish—November)

NBRF National Biomedical Research Foundation

NBRI National Building Research Institute

NBRMP National Board of Review of Motion Pictures

NBRPC New Brunswick Research and Productivity Council

NBRS National Beef Recording Scheme; National Beef Recording Service

NBRT National Board for Respiratory Therapy

nbs normal burro serum

NBS Nagano Broadcasting System; National Battlefield

Site; National Broadcasting Service (NZ); National Bureau of Standards; New British Standard

NBSA National Bank of South Africa; Netherlands Bank of South Africa

NBSBL National Bureau of Standards Boulder Laboratory

NBSCCST National Bureau of Standards Center for Computer Sciences and Technology

NBSRS Narodna Biblioteka Socijalisticke Republike Srbije (Serbo-Croatian—National Library of the Socialist Republic of Serbia)—Belgrade

NBS-SIS National Bureau of Standards—Standard Information Services

nb st nimbo-stratus

NBST National Board for Science and Technology

NBT National Book Trust (India)

NBTA National Baton Twirlers Association; National Business Teachers Association

NBTC New Brunswick Teachers College

NBTL Naval Boiler Test Laboratory

NBTS National Blood Transfusion Service

nbuf not buffed (leather)

n butt national buttress (thread)

nbv net book value

nbw noise bandwidth

NBW National Book Week

NBWA National Beer Wholesalers Association

Nby Newbury

NBYWCAUSA National Board of the Young Women's Christian Association of the U.S.A.

nc national coarse (thread); natural convector; nitrocellulose; no change; no charge; no connection; noise criteria; normally closed; nose cone; not cataloged; not catalogued; not complete; nuclear capability; numerical control(s)

n-c numerical control (automation)

n/c new charter; new crop; no charge; numerical control (automation)

nc non chiffre (French—un-

numbered)
nc na Christus (Dutch—after Christ)
NC Napa College; Nashville, Chatanooga & St. Louis (railroad); Nasson College; Natchez College; National Cash Register (stock exchange symbol); National Center; National Certificate; National Coarse (screw threads); National Congress; National Council; National Fire Waste Council; Nature Conservancy; New Caledonia; New College; Newark College; Newberry College; Newcomb College; Newnham College; Nicholls College; Nichols College; Norfolk College; Norman College; North Carolina; North Carolinian; Northern Counties; Northern County; Northland College; Northwestern College; Nuclear Congress; Nuffield College (Oxford); Nurse Corps
N.C. NC Wyeth
NC Norske Creditbank (Norwegian Credit Bank)
NC-17 no children under 17 admitted
nca neurocirculatory asthenia; no copies available
NCA Narcotics Control Act; National Camping Association; National Canners Association; National Capital Award; National Cashmere Association; National Cattlemens Association; National Charcoal Association; National Cheerleaders Association; National Chiropractic Association; National Civic Association; National Club Association; National Coal Association; National Coffee Association; National Command Authority; National Commission on Accrediting; National Committee for Adoption; National Composition Association; National Confectioners Association; National Constructors Association; National Contesters Association; National Costumers Association; National Council for the Arts; National Council on the Aging; National Council on Alcoholism; National Coursing Association; National Cranberry Association; National Creameries Association; National

Credit Association; National Cricket Association; Naval Communications Annex; Navy Contract Administrator; Nebraska Correctional Association; Nevada Correctional Association; Ngorongoro Conservation Area (Tanzania); North Central Airlines; North Central Association (of colleges and schools); Northern Consolidated Airlines
N C A National Cricket Association
NCAA National Children Adoption Association; National Collegiate Athletic Association
NCAAA National Center of Afro-American Artists
NCAB National Cancer Advisory Board
NCAB National Cyclopedia of American Biography
NCAC National Copyright Advisory Committee (Library of Congress)
NCACME National Center for Adult, Continuing, and Manpower Education
ncad net cash against documents
NCAE National Center for Audio Experimentation; National College of Agricultural Engineering; National Council of Adult Education; National Council of Agricultural Employers; North Carolina Association of Educators
NCAI National Clearinghouse for Alcohol Information; National Congress of American Indians; National Council on Alcoholism Inc
NCAICU North Carolina Association of Independent Colleges and Universities
NCAIR National Center for Automated Information Retrieval
NCALI National Clearinghouse for Alcohol Information (USPHS)
NCAM National Center for Advanced Materials
NCAMP National Coalition Against the Misuse of Pesticides
NCAN National Coalition of American Nuns
NCANH National Council for the Accreditation of Nursing Homes
ncap nematic curvilinear aligned phase

N-CAP Nurses Coalition for Action in Politics
NCAPC National Center for Air Pollution Control
NCAR National Center for Atmospheric Research; National Committee for Antarctic Research
NCARB National Council of Architectural Registration Boards
NCARL National Committee Against Repressive Legislation
NCARMD National Commission on Arthritis and Related Musculoskeletal Disease
N-carrier(s) nuclear-powered aircraft carrier(s)
NCAS National Collegiate Association for Secretaries
NCASF National Council of American-Soviet Friendship
NCAT National Center for Alternative Technology; National Center for Audiotape; Northampton College of Advanced Technology
NCATE National Council for the Accreditation of Teacher Education
NCAVAE National Council for Audio-Visual Aids in Education
NCAW National Council for Animal Welfare
NCAWE National Council for Administrative Women in Education
ncb narcotic-centered behavior; new crime buffer; nickel-cadmium battery; no claim bonus
Ncb Norrlands Skogsägaves Cellulosa AB
NCB National Cargo Bureau; National Coal Board; National Conservation Bureau; Nippon Credit Bank; Nippon Cultural Broadcasting
NCBA National Cattle Breeders' Association; National Clydesdale Breeders Association; Northern California Booksellers Association
NCBD National Council for Balanced Development
NCBE National Clearinghouse for Bilingual Education; National Conference of Bar Examiners
NCBFAA National Customs Brokers and Forwarders Association of America
NCBH National Coalition to Ban Handguns

NCBIAE National Council of Bureau of Indian Affairs Educators

NCBL National Conference of Black Lawyers

NCBM National Conference of Black Mayors

NCBMP National Council of Building Material Producers

NCBR National Council of Black Republicans

NCBS National Cattle Breeding Station (Australian)

NCBVA National Concrete Burial Vault Association

ncc numerical control code

NCC Namhae Chemical Corporation (Korean); Nassau Community College; National Cadet Corps; National Carloading Corporation; National Castings Council; National Certified Counselors; National Civic Council; National Climatic Center; National Coaches Council; National Computer Center; National Computer Council; National Conference on Citizenship; National Consumer Council; National Container Committee; National Cotton Council; National Council of Churches; National Council of Churches of Christ in the USA; National Cultural Center; Nature Conservation Council; Navajo Community College; Newhouse Communications Center (University of Syracuse); Newspaper Comics Council; Noise Control Committee; Non-Combatant Corps; NORAD Control Center; Northwest Community College

NCC Nederlands Cultureel Contact (Dutch—Netherlands Cultural Contact)

NCCA National Coil Coaters Association; North Carolina Correctional Association

NCCAN National Center on Child Abuse and Neglect

NCCAS National Center of Communication Arts and Sciences

NCCAT National Committee for Clear Air Turbulence

NCCB National Consumer Cooperative Bank; National Council of Catholic Bishops

NCCC Niagara County Community College

NCCCA National Coordinating Council for Constructive

Action

NCCCC Navy Command, Control, and Communications Center

NCCCD National Center for Computer Crime Data

NCCCLC Naval Command Control Communications Laboratory Center (formerly NEL—Navy Electronics Laboratory)

NCCCUS National Council of the Churches of Christ in the United States

NCCCUSA National Council of the Churches of Christ in the U.S.A.

NCCD National Council on Crime and Delinquency

NCCE National Commission for Cooperative Education; National Council for Catholic Evangelization

NCCEOA National Coordinating Council of Educational Opportunities Associations

NCCF National Committee to Combat Fascism (Black Panther front); National Commission on Consumer Finance

NCCG National Council on Compulsive Gambling

NCCH National Council to Control Handguns

NCCI National Committee for Commonwealth Immigrants; National Council on Compensation Insurance

NCCIHE North Carolina Center for Independent Higher Education

NCCIS NATO Command, Control, and Information System

NCCJ National Coalition for Children's Justice; National Conference of Christians and Jews

NCCJPA National Clearinghouse for Criminal Justice Planning and Architecture

NCCL National Council for Civil Liberties; National Council of Canadian Labor

NCCLS National Committee for Clinical Laboratory Standards; National Consumer Center for Legal Services

NCCNHR National Citizens' Coalition for Nursing Home Reform

NCCOP National Corporation for the Care of Old People

nccp nagivation control console panel

NCCP National Center for Children in Poverty (Columbia University)

NCCPA National Council of College Publications Advisers

NCCPG National Council for the Conservation of Plants and Gardens

NCCPL National Community Crime Prevention League

NCCPV National Commission on the Causes and Prevention of Violence

NCCR National Council for Civic Responsibility

NCCS National Command and Control System; National Council for Civic Responsibility; National Council for Constitutional Studies

NCCU National Conference of Canadian Universities; North Carolina Central University

NCCVD National Council for Combating Venereal Diseases

NCCW National Council of Catholic Women

NCCY National Council of Catholic Youth

ncd no can do; not considered disabling

NCD National Commission on Diabetes; Naval Construction Department; Naval Construction Depot; New Community Development

NCD New Collegiate Dictionary

NCDA National Center for Drug Analysis; National Council on Drug Abuse

NCDAD National Council for Diplomas in Art and Design

NCDAI National Clearinghouse for Drug Abuse Information

NCDC National Capital Development Commission; National Center for Disease Control; National Communicable Disease Center; National Community Development Corporation; National Council on Crime and Delinquency; National Curriculum Development Center; New Community Development Corporation

NCDL National Canine Defence League

NCDs Negotiable Certificates of Deposit

NCDS National Center for Dispute Settlement (American

Arbitration Association)
ncdu navigation control and display unit
nce normal curve equivalent
NCE Newark College of Engineering; Nice, France (Côte d'Azur airport)
NCE New Catholic Encyclopedia
NCEA National Catholic Educational Association; National Catholic Evangelization Association; National Center for Economic Alternatives; National Community Education Association; North Carolina Education Association
NCEB National Center for Educational Brokering; NATO Communications Electronic Board
NCEC National Committee for an Effective Congress; National Community Education Clearinghouse
NCECA National Council on Education for the Ceramic Arts
NCECS North Carolina Educational Computing Service
NCED National Center for the Employment of the Deaf
NCEDT National Council to Eliminate Death Taxes
NCEE National Commission on Excellence in Education; National Congress for Educational Excellence; National Council of Engineering Examiners
NCEER National Center for Earthquake Engineering Research
ncef national calling and emergency frequencies
NCEFT National Commission on Electronic Fund Transfers
NCEI National Commission on Emerging Institutions
NCEL Naval Civil Engineering Laboratory
NCEMP National Center for Energy Management and Power
NCEN National Commission on Egg Nutrition
NCEP National Cholesterol Education Program
NCER National Center for Earthquake Research; National Council on Educational Research
NCERT National Council for Educational Research and Training

nces necessary; normal curve equivalent scores
NCES National Center for Educational Statistics
NCET National Council for Educational Technology
ncf nerve cell food
NCF National Consumer Federation
NCFA National Cat Fanciers Association; National Commission of Fine Arts; National Consumer Finance Association; Navy Campus for Achievement
NCFC National Council of Farmer Cooperatives
NCFDA National Council on Federal Disaster Assistance
NCFILP National Coalition for Fair Immigration Laws and Practices
NCFIRB North Carolina Fire Insurance Rating Bureau
NCFM National Commission on Food Marketing
NCFP National Conference on Fluid Power
NCFPC National Center for Fish Protein Concentrate
NCFR National Council on Family Relations
NCFSU Naval Construction Force Support Unit
NCFT National College of Food Technology
NCG National Council for the Gifted; National Cylinder Gas (division of Chemotron)
NCGA National Council on Governmental Accounting
NCGE National Council for Geographic Education
NCGG National Council for Geodesy and Geophysics
NCGRC National Church Growth Research Center
nch number changed (telephone)
NCH National Children's Home; National Clearing House
NCHA National Campers and Hikers Association; National Capital Housing Authority; National Culling Horse Association
NCHCS National Council for Health Care Services
NCHEE National Council for Home Economics Education
NCHELP National Council of Higher Education Loan Programs
N Chem L National Chemical Laboratory

NCHEMS National Center for Higher Education Management Systems
n chg normal charge
NCHI National Council of the Housing Industry
NCHMT National Capitol Historical Museum of Transportation
NCHP Nouvelle Compagnie Havraise Peninsulaire (de Navigation) (French—Havre Peninsula Navigation Line)
n Chr *nach Christus* (German—after Christ, A.D.)
NCHS National Center for Health Statistics
NCHSR & D National Center for Health Services Research and Development (HEW)
NCHVRFE National College for Heating, Ventilating, Refrigeration, and Fan Engineering
nci naphthalene-creosote-iodiform (lice-control powder); no common interest; no-cost item
NCI National Cancer Institute; National Casing Institute; National Cello Institute; National Cheese Institute; Naval Cost Inspection; Naval Cost Inspector; Naval Court of Inquiry
NCIA National Council of Instructional Administrators; National Council for Islamic Affairs
NCIAC National Consumer Information and Advisory Center
NCIC National Cancer Institute of Canada; National Career Information Center; National Crime Information Center
NCIES National Center for the Improvement of Educational Systems
NCIJC National Council of Independent Junior Colleges
NCILT National Centre for Industrial Language Training
NCIO National Council on Indian Opportunity
NCIP Northeast Corridor Improvement Program
nci powder naphthalene creosote iodoform powder (for killing lice)
NCIS National Chemical Information System; National Council of Independent Schools
NCISC Naval Counterintelli-

gence Support Center
NCIT National Council on Inland Transport
NCJA National Criminal Justice Association
NCJAVM National Council on Jewish Audio-Visual Materials
NCJISS National Criminal Justice Information and Statistics Service
NCJMS National Center for Job Market Studies
NCJR National Coalition for Jail Reform
NCJRS National Criminal Justice Reference Service
NCJSC National Criminal Justice Statistics Center
NCJW National Council of Jewish Women
Nck Neck
N Cl New Caledonia(n)
NCL National Carriers Limited; National Central Library; National Chemical Laboratory; National Consolidated Limited; National Consumers League; National Culture League; Norwegian Caribbean Line; Norwegian Cruise Lines
NCLA National Council of Local Administrators (of vocational education and practical arts); North Carolina Library Association
NCLAN National Crop Loss Assessment Network
N-class a Soviet class of nuclear-powered attack submarines
NCLC National Caucus of Labor Committee; National Consumer Law Center; National Council of Labour Colleges; National Council of Local Administrators
NCLE National Center for Labored Enterprises
NCLIS National Commission on Library and Information Sciences
NCLR National Council of La Raza
NCLS National Clearinghouse for Legal Services
ncm non-corrosive metal; non-crew member
n.c.m. *non compos mentis* (Latin—of unsound mind)—insane
NCM National Congress for Men; Nippon Calculating Machine
NCMA National Catalog Man-

agers Association; National Council of Music Associations; North Carolina Museum of Art
NCMC National Center on Missing Children; NORAD Cheyenne Mountain Complex
NCMDA National Commission on Marijuana and Drug Abuse
NCME National Council on Measurements in Education; Network for Continuing Medical Education
NCMEA National Catholic Music Educators Association
NCMEC National Center for Missing and Exploited Children
NCMH National Committee on Maternal Health; National Committee for Mental Hygiene
NCMHE National Clearinghouse for Mental Health Education
NCMLB National Council of Mailing List Brokers
NCMP National Commission for Manpower Policy
NCMU National Commission on Marijuana Use
NCN National Council of Nurses; New Caledonian Nickel
NCNA National Council on Noise Abatement; New China News Agency (mainland China)
NCNC National Captive Nations Committee; National Council of Nigeria and the Cameroons
NCNE National Campaign for Nursery Education
NCNP National Conference for New Politics; North Cascades National Park (Washington)
NCNW National Council of Negro Women
nco no-cost option
NCO Noncommissioned Officer
NCOA National Council on the Aging; Noncommissioned Officer Academy
NCOAUSA Non-Commissioned Officers Association of the U.S.A.
NCOC National Commission on Organized Crime; National Council on Organized Crime
ncod net cash on delivery

NCOES Noncommissioned Officer Education System
NCOIC Noncommissioned Officer in Charge
NCOIL National Conference of Insurance Legislators
NCOLS Noncommissioned Officers Leadership School
N/COM Navy/Chief of Naval Operations
NCOMP National Catholic Office for Motion Pictures
NCOR National Committee on Oceanographic Research
ncos non-commissioned officers
ncp national cycling proficiency; nitrogen charge panel; normal circular pitch; number of channel programs
NCP National Capital Parks; National Country Party; National Customs Police (Philippines); *Naviera Chilena del Pacífico* (Spanish—Chilean Pacific Line); Navy Capabilities Plan; Noise Control Plan; Nutrition Center of the Philippines
NCP *Naviera Chilena del Pacífico* (Spanish—Chilean Pacific Line)
NCPA National Crime Prevention Association
NCPAC National Conservative Political Action Committee
NCPC National Capital Planning Commission; National Consumer Protection Council; National Crime Prevention Coalition; National Crime Prevention Council; Northern Canada Power Commission
NCPCA National Center for the Prosecution of Child Abuse
NCPERL National Coalition for Public Education and Religious Liberty
NCPGA North Carolina Personnel and Guidance Association
NCPI National Clay Pipe Institute; National Crime Prevention Institute; Navy Civilian Personnel Instructions
NCPL National Center for Programmed Learning
NCPPL National Committee on Prisons and Prison Labor
NCPRV National Council of Puerto Rican Volunteers
NCPS National Cat Protection Society; National Commission on the Public Service;

National Commission on Product Safety; Non-Contributory Pension Scheme

NCPSM National Center of Preventive and Stress Medicine

NCPSSM National Commission to Preserve Social Security and Medicare

NCPT National Congress of Parents and Teachers

NCPTSD National Center for Post-Traumatic Stress Disorder

NCPTWA National Clearinghouse for Periodical Title Word Abbreviations

NCPV National Commission on the Prevention of Violence

NCQR National Council for Quality and Reliability

ncr natural circulation reactor; no calibration required; no carbon required; not combat ready

n Cr novo Cruzeiro (Portuguese—new cruzeiro)—Brazilian monetary unit

NCR National Capital Region; National Cash Register; National Consumer Research; National Council of Reconciliation (in Vietnam); New Christian Right; Non-Communist Resistance

NCR National Catholic Record; National Catholic Reporter

NCRA National Correctional Recreation Association

NCRC National Condor Research Center

NCRCL National Civil Rights Clearinghouse Library

NCRD National Council for Research and Development; National Council for Resource Development

NCRE Naval Construction Research Establishment

NCRFCL National Commission on Reform of Federal Criminal Laws

NCRFP National Council for a Responsible Firearms Policy

NCRI National Red Cherry Institute

NCRL National Chemical Research Laboratory

NCRLC National Committee on Regional Library Cooperation

NCROPA National Campaign for the Repeal of the Obscene Publications Act (British)

ncrp narrow cold-rolled products; nonreinforced concrete pipe

NCRP National Committee on Radiation Protection; National Council on Radiation Protection; National Council for Research and Planning

ncr paper no-carbon-required paper

NCRPM National Committee on Radiation Protection and Measurements

NCRR National Center for Resource Recovery

NCRS National Committee for Rural Schools

NCRT National College of Rubber Technology

NCRVE National Center for Research in Vocational Education

NCRY National Commission on Resources for Youth

ncs naval control of shipping; navigation control simulator

NCS National Cartoonists Society; National Cemetery System; National Chrysanthemum Society; National Communications System, Naval Communication Station; National Computer Systems; Net Control Station; Numerical Control Society

NCSA National Carl Schurz Association; National Council of Seamen's Agencies; National Crushed Stone Association; National Customs Service Association; North Carolina School of the Arts; North Coast of South America

NCSAW National Catholic Society for Animal Welfare

NCSBA North Carolina School Boards Association

NCSBCS National Conference of States on Building Codes and Standards

NCSBEE National Council of State Boards of Engineering Examiners

NCSC National Cargo Security Council; National Center for State Courts; National Companies and Securities Commission (Australia); National Council for Senior Citizens; National Council of Senior Citizens

NCSCEE National Council of State Consultants in Elementary Education

NCSCT National Center for School and College Television

NCSDCJC National Council of State Directors of Community and Junior Colleges

NCSE National Commission on Safety Education

NCSEA National Council of State Education Associations

NCSF National College Student Foundation

NCSGC National Council of State Garden Clubs

NCSH National Clearinghouse for Smoking and Health

NCSI National Council for Stream Improvement; National Council of Savings Institutions

NCSJ National Conference on Soviet Jews

NCSL National Center for Service Learning; National Civil Service League; National Conference of Standards Laboratories; National Conference of State Legislators; Naval Code and Signal Laboratory

NCSMC National Council for the Single Mother and Her Child

NCSNE Naval Control of Shipping in the Northern European Command Area of NATO

NCSNP National Council for a Sane Nuclear Policy

NCSO Naval Control of Shipping Office(r); North Carolina Symphony Orchestra

NCSP National Conference on State Parks

NCSPA North Carolina State Ports Authority

NCSPS National Committee for the Support of Public Schools

NCSR National Center for Systems Reliability; National Council for Scientific Research

NCSRC National Centre for Social Research and Criminology (Cairo)

ncsry necessary

NCSS National Center for Social Statistics; National Council for Social Studies; National Council of Social Service

NCSSA National Community Service Sentencing Association; Nature Conservation Society of South Australia;

Naval Command Systems Support Activity
NCSSC Naval Command Systems Support Center
NCSSFL National Council of State Supervisors of Foreign Languages
NCSTAS National Council of Scientific and Technical Art Societies
NC & ST L Nashville, Chattanooga & St Louis (railroad)
NCSTRC North Carolina Science and Technology Research Center
NCSW National Conference on Social Welfare
NCSWCL National Commission on State Workmen's Compensation Laws
NCSWD National Center for Solid Waste Disposal
NCSWR National Conference on Solid Waste Research
nct natural contour theory; no charge for terms; no civil twilight
NCT National Chamber of Trade; National Childbirth Trust; National Culture Trust
n/cta *nuestra cuenta* (Spanish—our account)
NCTA National Cable Television Association; National Capital Transport Agency; National Community Television Association; National Committee for Technological Awards; National Council for Technological Awards
NCTAEP National Committee on Technology, Automation, and Economic Progress
NCTC National Collection of Type Cultures
NCTE National Council of Teachers of English
NCTEC Northern Counties Technical Examinations Council
NCTEPS National Commission on Teacher Education and Professional Standards
NCTI Nationwide Consumer Testing Institute
NCTJ National Council for the Training of Journalists
NCTM National Council of Teachers of Mathematics
NCTR National Center for Toxicological Research; National Council on Teacher Retirement
NCTS National Council of Technical Schools
ncu navigation(al) computer

unit; nitrogen control unit
NCU National Communications Union; National Cyclists' Union
NCUA National Credit Union Administration; National Credit Union Association
NCUC National Commission on Unemployment Compensation
NCUF National Computer Users Forum
NCUMC National Council for the Unmarried Mother and her child
ncup no commission until paid
NCUPUFUB National Clean-up, Paint-Up, Fix-Up Bureau
NCURA National Council of University Research Administrators
NCUSA Navy Club of the U.S.A.
NCUSIF National Credit Union Share Insurance Fund
NCUTLO National Committee on Uniform Traffic Laws and Ordinances
ncv no commercial value
NCVA National Center(s) for Volunteer Action
NCVAE National Council for Audio-Visual Aids in Education
NCVO National Council for Voluntary Organizations
NCVOTE National Center for Vocational, Occupational, and Technical Education
NCVT National Crime and Violence Test
ncw nosecone warhead
NCW National Council of Women; North City West
NCWA National Council of Women of Australia
NCWC National Catholic Welfare Conference
NCWSA National Council of Women of South Africa
NCWSB National Council of Wool Selling Brokers
NCWUS National Council of Women of the U.S.
NCY National Cylinder Gas (stock-exchange symbol)
NCYC National Council of Yacht Clubs
NCYMCA National Council of Young Men's Christian Associations
NCYRE National Council on Year-Round Education
nd national debt; natural draught; neutral density; new deck(ing); new drugs; next

day; no date; no decision; no deed; no delay; no discount-(ing); no drawing; non-deliv-ery; non-directional; not dated; not deeded; not determined; not drawn; nothing doing; nuclear detonation
n-d non-drying
n/d neutral density
nd *niederdruck* (German—low pressure); *no hay datos* (Spanish—no data)
Nd neodymium; refractive index (symbol)
ND Environment Near Death; Narcotics Division (NYPD); National Dairy Products (stock exchange symbol); National Debt; Naval District; Navy Department; New Drugs; North Dakota; Notre Dame
N.D. Doctor of Naturopathy
ND *New Drugs*; *Notre Dame* (French—Our Lady)
nda new drug application; non-destructive analysis; non-destructive assay
nda (NDA) new drug applications
N d A *Nota dell'Autore* (Italian—Author's Note)
NDA National Dairy Association; National Dairymens' Association; National Dental Association; National Diploma in Agriculture
ndaa not dated at all
NDAA National District Attorneys Association
NDAB Numerical Data Advisory Board
NDAC National Defense Advisory Committee; National Defense Advisory Commission; Nuclear Defense Affairs Committee (NATO)
ND Agr Eng National Diploma in Agricultural Engineering
N Dak North Dakota; North Dakotan
NDANZ National Dairy Association of New Zealand
n da r *nota da redação* (Portuguese—author's note)
NDAS *New Dictionary of American Slang*
NDASSP North Dakota Association of Secondary School Principals
ndb national development bond(ing); new domestic boiler; new donkey boiler; nondirectional beacon
NDB National Development

Bank; Navy Department Bulletin; Niue Development Board
NDBC National Data Buoy Center; National Duckpin Bowling Congress
NDBI National Dairymen's Benevolent Institution
NDBO NOAA Data Buoy Office
NDBS National Data Buoy System
NDC National Dairy Council; National Defense Contribution; National Defense Corps; National Democratic Club; National Development Company; National Development Corporation; National Development Council; NATO Defence College; Naval Dental Clinic; Nippon Decimal Classification; Nuclear Development Corporation
NDCA National Dry Cleaners Association
NDCC National Defense Cadet Corps; National Democratic Congressional Committee; National Drug Control Center
NDCD National Drug Code Directory
NDCP National Drug Control Policy
NDCS National Deaf Children's Society
N d D Nota della Direzione (Italian—Director's Note)
NDD National Diploma in Dairying
nddad net demand draft against documents
ndd(s) narcotic-detection dog(s)
NDDT National Diploma in Dairy Technology
nde near-death experience; non-destructive evaluation; non-linear differential equation(s)
NDEA National Defense Education Act
N-defense nuclear defense
NDEI National Defense Education Institute
n del a nota del autor (Spanish—author's note)
n del e nota del editor (Spanish—editor's note)
n del t nota del traductor (Spanish—translator's note)
N de M Nacional de México (railroad)
N de M Ferrocarriles Nacionales de México (Spanish—

National Railways of Mexico)
NDER National Defense Executive Reserve
ndf nacelle drag efficiency factor
NDF National Diploma in Forestry
ndg nedenfor (Dano-Norwegian—beneath)
NDG National Dance Guild
NDGMH National Development Group for the Mentally Handicapped
NDGS National Defense General Staff; National Duncan Glass Society
NDH Delhi, India (airport); National Diploma in Health; National Diploma in Horticulture
NDHA National District Heating Association
NDHS New Drop High School
ndi numerical designation index
NDI National Dance Institute; National Death Index
NDIB National Drug Intelligence Bureau
NDICF North Dakota Independent College Fund
NDIRS North Dakota Institute for Regional Studies
NDIS National Drug Information Service
ndl network definition language
ndl niederländisch (German—Netherlandic)
Ndl Nederland (Dutch—The Netherlands)
NDL National Development Loan; Nuclear Defense Laboratory
NDL Norddeutscher Lloyd (German—North German Lloyd)
NDLA North Dakota Library Association
NDLB National Dock Labour Board
NDMB National Defense Mediation Board
ndml never during my lifetime
NDN National Diffusion Network
ndo negotiable delivery order
NDO National Debt Office (and Office for the Payment of Government Life Annuities); Natural Disasters Organization; Northern Dance Orchestra
ndp net domestic product; normal diametric pitch

NDP National Dairy Products; National Democratic Party; National Detective Police; National Drug Policy; New Democratic Party (Canada)
NDP Nationaldemokratische Partei Deutschlands (German—German National-Democratic Party)—neo-Nazi oriented
NDPA National Decorating Products Association; National Democratic Party of Alabama
NDPBC National Duck Pin Bowling Congress
NDPD National-Demokratische Partei Deutschlands (National Democratic Party of East Germany)
NDPGA North Dakota Personnel and Guidance Association
NDPH National Diploma in Poultry Husbandry
NDPP National Drug Prevention Program
NDPR NATO Defense Planning Review
NDPs Narcotic Detention Pens (NYC)
NDPS National Data Processing Service
ndr net discount(ed) revenue
N^{dr} Neder (Dutch or Swedish—lower); Nieder (German—lower)
N d R Nota della Redazione (Italian—Editor's Note)
NDR Norddeutscher Rundfunk (German—North German Radio)
NDRC National Defense Research Committee
NDRG NATO Defense Research Group
NDRI Naval Dental Research Institute
ndro nondestructive readout
NDRSWG NATO Data Requirements and Standards Working Group
nds national development strategy
nds (NDS) nuclear detection satellite
NDs Northern Districts
NDS National Directory Service
NDSB Narcotic Drugs Supervisory Body
NDSBA North Dakota School Boards Association
NDSF North Dakota School of Forestry
NDSK Nippon Dendo Sharyo

Kyokai (Japanese—Japan Electric-Powered Vehicle Association)
NDSL National Direct Student Loan
NDSM National Defense Security Medal
NDSs Nuclear Delivery Systems
NDSSS North Dakota State School of Science
ndt nondestructive testing
ndt *nota del traductor* (Spanish—translator's note); *nota del traduttore* (Italian—translator's note); *note du traducteur* (French—translator's note)
NDT *Ferrocarril Nacional de Tehuantepec* (Spanish—National Railroad of Tehuantepec—symbol); National Diet Library (Tokyo); National Driver's Test; Newfoundland Daylight Time; Nichigeki Dancing Team; Nuclear Defense Laboratory
NDTA National Defense Transportation Association; Non-Destructive Testing Association
NDTAA Non-Destructive Testing Association of Australia
NDTC Nottingham and District Technical College
NDTI *National Disease and Therapeutic Index*
NDTS Non-Destructive Testing Standard(s)
ndu navigation display unit; nuclear data unit
NDU National Defense University; Notre Dame University
N-dump(ing) nuclear-waste dump(ing)
N-dump(s) nuclear (waste-disposal) dump(s)
ndup nonduplication; nonduplicate
Ndv Newcastle disease virus
ndw net deadweight
NDW Naval District Washington (D.C.)
NdYAG neodymium, yttrium, aluminum, garnet (laser components)
ne new edition; new engine(s); nital etch(ing); not enlarged; not entitled; not essential; not exceeding
ne (NE) norepinephrane
n/e no effects
ne *non ebarbe* (French—untrimmed)

Ne neon; Nepal; Nepalese; Netherlander; Netherlands
NE National Emergency; National Estate(s); Naval Engineer(ing); Nebraska (postal code); new edition; New England(er); News Editor; northeast; Northeast Airlines (2-letter coding); Nuclear Engineer(ing)
N.E. Nuclear Engineer
NE *Navio Escola* (Portuguese—Schoolship); *Noreste* (Spanish—northeast)
ne/4 mos new edition expected in four months
ne/6m new edition in preparation, expected in 6 months (for example)
ne/6 mos new edition expected in six months
nea net energy analysis
NEA National Education Association; National Electrification Administration; National Endowment for the Arts; Net Energy Analysis; New England Aquarium (Boston); Newspaper Enterprise Association; Northeast Airlines; Northern Electric Authority; Nuclear Energy Agency (UN)
N.E.A. Newspaper Enterprise Association
NEAC National Energy Advisory Committee; New English Art Club
NEACAP National Emergency Air Command Post
NEACH New England Automated Clearing House
NEACSS New England Association of Colleges and Secondary Schools
NEAF Near East Air Force; New Era Aboriginal Fellowship
NEAFC Northeast Atlantic Fisheries Commission
NEAG New English Art Gallery
NEAHI Near East Animal Health Institute
NEAL National Electron Accelerator Laboratory
NEAP National Assessment of Educational Progress
NEA-PAC National Education Association Political Action Committee
Neapolitans islands off Naples; natives of Naples
NEAR National Emergency Aid Radio; National Emergency Alarm Repeater
NEARA New England Antiq-

uities Research Association; New England Archeological Research Association
Near East Libya, Egypt, Sudan, Ethiopia, Jordan, Israel, Lebanon, Syria, Saudi Arabia, the United Arab Emirates, Oman, Yemen, Iraq, Iran, Turkey, Afghanistan, Pakistan
Near North Australian equivalent of the Far East
Nears the Near Islands of the outermost Aleutians in southwestern Alaska, including Agattu and Attu
NEAS National Engineering Aptitude Search
NEASC New England Association of Schools and Colleges
NEAT National (Cash Register) Electronic Autocoding Technique; National Employment and Training
NEATE New England Association of Teachers of English
'neath beneath; underneath
NEATO Northeast Asian Treaty Organization
neb nembutal
neb *nebbisch* (Yiddish—colorless, plain, retiring, socially ill at ease)
neb. *nebula* (Latin—spray)
NEB National Electricity Board; National Energy Board (Canada); National Enterprise Board (United Kingdom); North Equatorial Belt
NEB *New English Bible*
NEBAC National Ethnic Broadcasting Advisory Council
nEbC no-European-before-Columbus school of historic discovery despite Irish and Viking claims to the contrary
NEBHE New England Board of Higher Education
Nebr Nebraska; Nebraskan
NEBSS National Examinations Board for Supervisory Studies
nebuchad nebucdhadnessar (20–quart-capacity champagne bottle)
nebul. *nebula* (Latin—spray)— nebulizer
nec necessary; no error check(ing); not elsewhere classified
Nec (NEC) Navy enlisted classification
NEC National Economic Com-

mission; National Economic Council; National Egg Council; National Electrical Code; National Equestrian Centre; National Equity Corporation; National Exchange Club; National Exhibition Centre (Birmingham, England); Negro Ensemble Company; Netherlands Electrochemical Committee; New England Conservatory of Music; New England Council; Nippon Electric Company; Nippon Electric Corporation

NECA National Electrical Contractors' Association; Near East College Association; Numismatic Error Collectors of America

NECAA National Entertainment and Campus Activities Association

NECAP NASA Energy-Cost Analysis Program

NECC National Education Computer Center

NECCC New England Correctional Coordinating Council

NECCO New England Confectionary Company

NECEL New England Coalition of Educational Leaders

NECLC National Emergency Civil Liberties Committee

NECM New England Conservatory of Music

NECMD Newark Contract Management District

NECO Nuclear Engineering Company

NECOS Northern Europe Chiefs of Staff (NATO)

NECP New England College of Pharmacy

NECPA National Energy Conservation Policy Act

necr necrosis

necro (Latin prefix—corpse or dead)—necrophilia, necropholia, necrosis

necrol necrology

necropo necropolis; necropolitan(ic)

NECS National Electrical Code Standards

necy necessary

ned normal equivalent deviation

Ned Edmund; Edward; Edwin

Ned Nederland (Dutch—the Netherlands); *Nederlands* (Dano-Norwegian—the Netherlands)

NED National Endowment for Democracy; Nuclear Energy Division (GE)

NED New English Dictionary (Oxford English Dictionary)

NEDA National Economic and Development Authority; National Economic Development Association; National Electronic Distributors Association; National Electronics Development Association

Ned Ant Nederlandse Antillen (Dutch—Netherlands Antilles)–Aruba, Bonaire, Curaçao, Saba, Sint Eustatius, half of Sint Maarten

Ned Buntline Edward Zane Carroll Judson

NEDC National Economic Development Council (of Great Britain where it is nicknamed Neddy); Near East Development Council

nedela network definition language

nederl nederlandsk (Dano-Norwegian—Dutch)

Nederl Nederland (Dutch—Netherlands)

NEDICO Netherlands Engineering Consultants

NEDL New England Deposit Library

Nedlloyd Netherlands Line

NEDO National Economic Development Office; New Energy Development Organization

Ned Opera Nederlandse Operastichting (Dutch—Netherlands Opera Foundation)

NEDT National Educational Development Tests

NedThTs Nederlands Theologisch Tijdschrift (Dutch—Netherlands Theological Periodical)

NEDU Navy Experimental Diving Unit

NEEB North Eastern Electricity Board (UK)

NEEC National Export Expansion Council

need. needlework

NEED National Environmental Education Development

Needle Park open-air hangout of addicts, pushers, pimps, and prostitutes

needn't (contraction—need not)

NEEDS New England Electronic Data System

ne'er never (contraction)

Néerl Néerlandais (French—Dutch)

NEES Naval Engineering Experiment Station; New England Electric Service

neev (NEEV) natural energy electric vehicle

NEEWSSOP NATO-Europe Early-Warning-System Standard Operating Procedures

nef national extra fine (screw thread); net energy for fattening; noise exposure forecast; nuclear energy factor(s)

NEF Naval Emergency Fund; Near East Foundation; New Education Fellowship

nefa nonesterified fatty acid

NEFA Northeast Frontier Agency

NEFC Near East Forestry Commission

NEFEN Near and Far East News

NEFIRA New England Fire Insurance Rating Association

NEFMO NATO European Fighter Aircraft Development, Production, and Logistics Management Organization

NEFO National Electronic Facilities Organization

Nefos New Emerging Forces

NEFP National Educational Finance Project

NEFSA National Education Field Service Association

neg negation; negative; negligent; negotiable; negotiate; negritude

nég négation (French—negation)

Neg Negro; Negroid

neg ad negative advertisement; negative advertising

negatron negative electron

negistor negative resistor

Negley Farson James Scott Negley Farson

nego negotiate

negobl negotiable

negod negotiated

negoin negotiation

negotn negotiating

negotng negotiating

Negrasian(s) person(s) of African and Asian parents such as Afro-Chinese, Afro-Indian, Afro-Japanese, etc.

NEGRO National Economic Growth and Reconstruction Organization

négt négociant (French—merchant)—wholesaler

negtax negative (income) tax

Neh The Book of Nehemiah

Neh Nehemiah

NEH National Endowment for

the Humanities
NEHA National Environmental Health Association; National Executives Housekeepers Association
NEHC National Extension Homemakers Council
nehi knee-high
Nehm Nehemiah
nei not elsewhere included; not elsewhere indicated
n.e.i. *non est inventus* (Latin— it is not found)
NEI National Eye Institute; Netherlands East Indies; New England Institute
NEIC National Earthquake Information Center; National Energy Information Center
NEIDP National Electronic Industries Procurement
NEISS National Electronic Injury Surveillance System
NEISSS National Electronics Injury Surveillance Safety System
NEJA National Equal Justice Association
NEJM New England Journal of Medicine
nek nekton
NEK Norsk Electrotecnisk Komite (Norwegian—Norwegian Electrotechnical Committee)
nekolim neocolonialist-colonialaist-imperialist (Indonesian acronym)
nel noise-exposure level
NEL National Electronics Laboratory; National Engineering Laboratory; Navy Electronics Laboratory (USN)
NEL New English Library
NELA National Electric Light Association; New England Library Association
NELC Naval Electronics Laboratory Center (formerly NEL)
NELDIC Nippon (Electric Company) Electric Layout Design (System) for Integrated Circuits
NELH National Energy Laboratory of Hawaii
NELIA Nuclear Energy Liability Insurance Association
NELIAC Navy Electronics Laboratory International Algol Compiler
NELINT New England Library Information Network
NELL North East Lancashire Libraries
NELMA Northeastern Lumber

Manufacturers Association
Nel-Mar Nelson-Marlborough (NZ)
NELP North East London Polytechnic
NELPIA Nuclear Energy Liability Property Insurance Association
Nels Nelson
NELS National Environmental Laboratories
Nelson Horatio Nelson; Knute Nelson; Nelson Olsen Nelson; Thomas Nelson
NELSON New Editing and Layout System of Newspaper
NELTAS North East Lancashire Technical Advisory Services
NEly north-easterly
nem net energy for milk; not elsewhere mentioned
NEM New Economic Mechanism
NEMA National Eclectic Medical Association; National Electrical Manufacturers Association; National Electrical Motors Association
nemat nematology
Nemat Nemathelminthes
NEMC New England Medical Center
NEMCA NATO Electromagnetic Compatability Agency
nem. con. *nemine contradicente* (Latin—no one contradicting)
nem. dis. *nemine dissentiente* (Latin—no one dissenting)
NEMI National Elevator Manufacturing Industry
NEMLA New England Modern Language Association
NEMO Naval Edreobenthic Manned Observatory (for sedentary sea bottom research); Naval Experimental Manned Observatory
NEMPA North-Eastern Master Printers' Alliance
NEMPS National Environmental Monitoring and Prediction System
NEMRB New England Motor Rate Bureau
nems (NEMS) near-earth magnetospheric satellite
nen noise and exposure number
NEN New England Nuclear (corporation)
nencl nonenclosed; nonenclosure
ne/nd new edition in preparation—no date can be given

N-energy nuclear energy
N Eng Naval Engineer(ing); New England; North England
N-engine(s) nuclear engine(s)
N Engl J Med New England Journal of Medicine
nenmld not enameled
NENP New England National Park (New South Wales)
neo near earth orbit
neo (Latin prefix—new or young)—neonatal
NEOA National Entertainers and Operators Association
NEOB New Executive Office Building (D.C.)
neobych neobychny (Russian— incomparable)
NEOC National Emergency Operations Center
Neo-Cath Neo-Catholic(ism)
Neo-Christ Neo-Christian(ity)
neoclas neoclassical; neoclassicism
neocol neocolonial(ism)
neocolim neocolonial-colonialimperialist
Neo-Conf Neo-Confucian(ist)
neo-con(s) neo conservative(s)
Neo-Dar Neo-Darwinian; Neo-Darwinist(ic)
neo-dhc neohesperidin dihydrochalcone (sweetener)
NEODTC Naval Explosive Ordinance Disposal Technical Center
Neogaea landmass including Central and South America
Neo-Goth Neo-Gothic
Neo-Heg Neo-Hegelian
neo-imp neo-impressionism; neo-impressionistic
Neo-Kant Neo-Kantian(ism)
neol neologism(s); neologistic(al)(ly), neologize(r)(s)
Neo-Lam Neo-Lamarckian; Neo-Lamarckism; NeoLamarckist
Neo-Lat Neo-Latin(ism)
Neo-Luth Neo-Lutheran(ism)
Neo-Mel Neo-Melanesian (pidgin English of Melanesia, New Guinea, and North-East Australian islanders)
Neo-Nor Neo-Norwegian
Neo-Plas Neo-Plastic(ism)
Neo-Plat Neo-Platonic; Neo-Platonism
Neo-Pyth Neo-Pythagorean(ism)
Neo-Real Neo-Realism; Neo-Realistic
Neorican(s) New York American(s)
Neo-Ricans repatriated Puerto Ricans

Neo-Rom Neo-Romantic(ism)
Neo-Schol Neo-Scholastic(ism)
neotrop neotropical
neotwy (last-letter mnemonic—when, where, who, what, how, why)
nep new edition pending; noise equivalent power; not elsewhere provided; nude-encounter parlor (brothel)
Nep Nepal; Nepomucene; Nepomuceno; Nepomuk; Neptune
Nep Cornelius Nepos (Roman biographer)
NEP National Education Program; National Energy Plan; New Ecological Paradigm; New Economic Policy; New England Pathology; New England Power (company); Nixon Economic Policy
nepa (NEPA) nuclear energy for the propulsion of aircraft
NEPA National Electric Power Authority; National Endowment Policy Act; National Environmental Policy Act
Nepal Kingdom of Nepal (Himalayan mountain nation)
NEPAL National Egg Packers' Association, Ltd
NEPC National Employers Policy Committee
NEPCO New England Provision Company
NEPE National Emergency Planning Establishment (Canada)
neph nephew
nepho nephograph; nephological; nephologist; nephology
nephro (Latin prefix—kidney)—nephritis
NEPIA Nuclear Energy Property Insurance Association
NEPLEX New England Power Exchange
NEPMU Navy Environmental and Preventive Medicine Unit
NEPOOL New England Power Pool
NEPR Nato Electronic Parts Recommendation
Nep Rs Nepalese rupees
nep's nude-encounter parlors
NEPSC National Employee Participation Steering Committee
Nep Soc Neptune Society
NEPSS Naval Environmental Protection Support Service (USN)
Nept Neptune
N Equ Cur North Equatorial

Current
ner nervous system
NER National Educational Radio; National Elk Refuge (Wyoming); North Eastern Railway (England)
NERA National Economic Research Associates; National Emergency Relief Administration
NERAIC Northern European Region Air Information Center
NERBC New England River Basins Commission
NERC National Electronic Reliability Council; National Environmental Research Center; Natural Environment Research Council
NERDDC National Energy Research Development and Demonstration Council
ne rep. ne repetatur (Latin—do not repeat)
NERO Near East Regional Office (FAO); Nutrition Education Research Organization
NERPG Northern European Regional Planning Group (NATO)
nerv nervous; nuclear emulsion recovery vehicle (NERV)
nerva nuclear engine for rocket vehicle application
NE-Rx Northeast Regional Exchange
nes not elsewhere specified
nEs non-English speaking
Nes Nesta; Nestor
NES National Emergency Services; National Extension Service; Naval Education Service; News Election Service; Nucleus Estate and Smallholders
NESA National Environmental Study Area; Near East and South Asia; New England School of Art
NESBIC Netherlands Student's Bureau for International Cooperation
NESC National Electric Safety Code; National English Syllabus Committee; National Environmental Satellite Center
NESCO National Energy Supply Corporation
NESDA National Electronics Service Dealers Association
NESDB National Economic and Social Development Board

NESDEC New England School Development Council
NESNE New England Society of Newspaper Editors
NESO Naval Electronics Supply Office
Ness Agnes
NESS National Environmental Satellite Service
nest. node execution selection table
NEST Naval Experimental Satellite Terminal; Nuclear Emergency Search Team
nestor neutron source thermal reactor
net. network; not earlier than; nuclear electronic transitor
Net Antoinette; Nettie; Netty
NET National Educational Television; Nippon Educational Television; Noise Enforcement Team (police anti-noise team)
NETA Northwest Electronic Technical Association
netanal network analysis
NETE Navel Engineering Test Establishment (Canadian)
NETF Nuclear Engineering Test Facility
Neth Netherlands
Neth Ant Netherlands Antilles
Netherlands Kingdom of the Netherlands (North Sea nation created and enlarged by reclamation of salt marshes and lowland waters), *Koninkrijk der Nederlanden*
netic nonretentive nonshocksensitive (alloy made for high-level attenuation)
n. et m. nocte et mane (Latin—night and early morning)
netma nobody ever tells me anything
NETRANZ National Endurance and Trail Riding Association of New Zealand
NETRB New England Territory Railroad Bureau
NETRC National Educational Television and Radio Center
nets. network techniques
NETSO Northern European Transshipment Organization (NATO)
Netza Netzahualcoyotl (Aztec—Hungry Coyote)
neu neuter; neutral; neutrality
neubarb neubearbeitet (German—revised)
Neuk Neuköln
neur neuralgia; neurasthenia; neuritis; neurology
neuro neurotic

neuro *neuron* (Greek—nerve, sinew, tendon)—neurasthenia, neuroanatomy, neurosis

neurobio neurobiological; neurobiologist; neurobiology

neurol neurological; neurologist; neurology

neuropath neuropathology

neurophys neurophysiological

neuropsychiat neuropsychiatry

neuropsycho neuropsychological

neurosci neuroscientific(al) (ly); neuroscientist

neurosurg neurosurgeon; neurosurgery; neurosurgical

neurs neurosis

NEUS Northeastern United States

neut neuter; neutral; neutralize; neutralizer; neutron bomb (mini-hydrogen bomb releasing neutrons and producing the minimum radioactive blast, fallout, and heat)

neutron neutral ion

Nev Nevada; Nevadan; Neville

Nevado del Ruiz (Spanish—Snowpeak of Ruiz)—snow-capped volcano, central Colombia

Never *Never on Sunday*

Never Never Never Land Cape York Peninsula, Australia

nevrls nevertheless

new newton

new. net economic welfare; newton

New New College, Oxford

New Am Lib New American Library

Newark Newark-upon-Trent near Nottingham, England

Newberry Newberry Library (Chicago)

Newc Newcastle-upon-Tyne

New Cal New Caldonia

New Castile (see *Castilla la Nueva*)

Newcastle Newcastle Emlyn, Newcastleton, Newcastle-under-Lyme, Newcastle-upon-Tyne, Newcastle Waters, and Newcastle West

New Col New Columbia (proposed name for the 51st state, formerly Washington, DC)

new cruzado monetary unit of Brazil

New Eng New England

New England Maine, New Hampshire, Vermont, Massachusetts, Rhode Island, and Connecticut

New England Colonies Mas-sachusetts, New Hampshire, Rhode Island, Connecticut

Newf Newfoundland

Newfie(s) Newfoundlander(s)

New H New Hall College, Oxford

New Haven New York, New Haven, and Hartford Railroad

New Heb New Hebrides (Anglo-French island condominium in the South Pacific)

New Heb Con New Hebrides Condominium

New Hebrides New Hebrides Islands, *Nouvelles Hébrides*

new kip monetary unit of Laos

New Lib Newberry Library

New Lon New London, Connecticut

New London U.S. Coast Guard Academy at New London, Connecticut

New Mex New Mexico

Newn Newnham College, Oxford

NEWO National Energy Waste Office

New Orl New Orleans

new par new paragraph

New Phil Orch New Philharmonia Orchestra

NEWRADS Nuclear Explosion Warning and Radiological Data System

NEWRIT Northeast Water Resources Information Terminal

news. naval electronic warfare simulator; news agency; news agent; new standards

NEWS New England Wildflower Society

New Sarum Salisbury, capital of Wiltshire, England northwest of Southhampton

newscast(er) news broadcast(er)

newscomp newspaper composition

New Sib New Siberian Islands

New Siberians New Siberian Islands in the Arctic (Novosibirskiye Ostrova)

Newt Newton

new Taiwan dollar monetary unit of Taiwan

New Test. New Testament

NEWWA New England Water Works Association

New Year's New Year's Day (January 1)

New Yorican New York Puerto Rican

New Zealand Dominion of New Zealand (western Pacific Ocean nation); New Zealand flax and New Zealand wineberry

nex not exceeding

nexis (NEXIS) news data base on-line to head

N-explosion(s) nuclear explosion(s)

N-exports nuclear exports

next. near-end crosstalk

NEXT NATO Experimental Tactics

nez (NEZ) northern economic zone

NEZs New Economic Zones (Vietnamese forced-labor camps)

nf national fine; near face; near field; no fool; no funds; noise factor; non-ferrous; non-fiction; non-fundable; nose fuze; not fordable

n-f nonfordable

n/f neutrons per fission; no funds

n & f near and far

nf *nouveau franc* (French—new franc)—issued in 1960

n.f. *ny foljd* (Swedish—new series)

n/f *nuestro favor* (Spanish—our favor)

n.F. *neue Folge* (German—new series)

NF National Fine (threads); National Formulary; National Foundation; National Front; Newfoundland; Nieman Foundation; Norfolk, Virginia (airport); Norman French; Nutrition Foundation

N-F Norman-French

NF *Neue Folge* (German—new series); *Nuestra Familia* (Spanish—Our Family)—prison racketeers also called *La Nuestra Familia*

nfa no further action

nfa (NFA) net financial assets

NFA National Faculty Association; National Farmers Association; National Federation of Anglers; National Flute Association; National Food Administration; National Food Authority; National Foundry Association; Nature Friends of America; Naval Fuel Annex; New Farmers of America; Night Fighters Association; Northwest Fisheries Association

NFAA National Field Archery Association; Navy Fighter Attack Aircraft

NFAC National Food and Agriculture Council; Native Forests Action Council

NFAH National Foundation for the Arts and the Humanities

NFAIS National Federation of Abstracting and Indexing Services

NFAL National Foundation of Arts and Letters

N-fallout nuclear fallout (radioactive fallout)

nfb nacelle fuselage base; narrow flange beam; no feedback

NFB National Federation of the Blind; National Film Board (Canada)

NFB Nippon Fudosan Bank (Japan Real Property Bank)

NFBC National Film Board of Canada; Newfoundland Base Command

NFBF National Farm Bureau Federation

NFBPM National Federation of Builders' and Plumbers' Merchants

NFBPWC National Federation of Business and Professional Women's Clubs

NFBTE National Federation of Building Trades' Employers

NFBTO National Federation of Building Trades Operatives

nfc not favorably considered

NFC National Fitness Council; National Football Conference; National Foundry College; National Freight Corporation; Navy Finance Center; Newspaper Features Council

NFCA National Federation of Community Associations

NFCC National Foundation for Consumer Credit

NFCG National Federation of Consumer Groups

nfcs night fire-control sight

NFCSA National Finance Corporation of South Africa

NFCTA National Federation of Corn Trade Associations; National Fibre Can and Tube Association

NFCU Navy Federal Credit Union

NFCUS National Federation of Canadian University Students (now NUS)

nfd no further description

Nfd Newfoundland

NFD National Federation of Doctors; National Fisheries Development; Naval Fuel Depot

NFD National Faculty Direc-

tory

NFDA National Food Distributors Association; National Funeral Directors Association

nfdm non-fat dry milk

nfd(m) non-fat dry (milk)

NFDRS National Fire Danger Rating System

NFDS National Fire Data Center

nfe net funds employed; nose-fairing exit; not fully equipped

NFE National Front of England (racists advocating immediate deportation of all non-whites to wherever they originated)

NFEA National Federated Electrical Association; Newspaper Farm Editors of America

n fem feminine form of a noun

NFEMC National Federation of Export Management Companies

NFER National Foundation for Education Research

NFF National Farmers Federation; National Froebel Foundation; Naval Fuel Facility

NFFA National Farmers Federation of Australia; National Freight Forwarders Association

NFFC National Film Finance Corporation

NFFE National Federation of Federal Employees

NFFF National Federation of Fish Friers; National Firearms Freedom Fund

NFFPC National Foundation to Fight Political Corruption

NFFPT National Federation of Fruit and Potato Trades

NFFS National Foundation for Funeral Services; Non-Ferrous Founders' Society

NFFTR National Federation of Fishing Tackle Retailers

Nfg Nachfolger (German—successor)

NFGCA National Federation of Grandmother Clubs of America

NFHS National Federation of Housing Societies

nfi non-bank financial intermediaries

NFI National Federation of Ironmongers; National Fisheries Institute; National Flood Insurance; Nature Friends of Israel

NFIB National Federation of

Independent Business; National Foreign Intelligence Board

NFIC National Foundation for Ileitis and Colitis

NFIE National Foundation for the Improvement of Education

NFIP National Flood Insurance Program; National Foundation for Infantile Paralysis

NFIU National Federation of Independent Unions

NFJC National Foundation for Jewish Culture

NFK Norfolk Island

Nfl Newfoundland

Nfl Nachfolger (German—successor)

NFL National Film Library; National Football League; National Forensic League; National Foresters League

NFLCC National Fishing Lure Collectors Club

Nfld Newfoundland

NFLPA National Football League Players Association

NFLPN National Federation of Licensed Practical Nurses

NFLS Niagara Falls

NFLSV National Front for the Liberation of South Vietnam

NFLTA National Federation of Language Teachers Associations

nfm next full moon

NFMA National Forest Management Act

NFMC National Federation of Music Clubs; National Food Marketing Commission

NFMD National Foundation for the March of Dimes

NFME National Fund for Medical Education

NFMLTA National Federation of Modern Language Teachers Association

NFMPS National Federation of Master Printers in Scotland

NFMTA National Federation of Meat Traders' Associations

NFND National Foundation for Neuromuscular Diseases

nfnshd not finished

NFO National Farmers Organization; National Freight Organization; Naval Flight Officer

NFOIO Naval Field Operational Intelligence Office(r)

NFOO Naval Forward Observ-

ing Officer
nfou number of fourier coefficients
nfp not file protect(ed)
NFP National Federation of Parents (for drug-free youth); National Federation Party; Natural Family Planning
NFPA National Fire Protection Association; National Flaxseed Processors Association; National Flexible Packaging Association; National Fluid Power Association; National Forest Products Association; Niagara Frontier Port Authority
NFPC National Federation of Priests Councils; Niagara Falls Power Company
NFPCA National Fire Prevention and Control Administration
NFPDB NATO Force Planning Data Base
NFPEX NATO Force Planning Exercise
NFPW National Federation of Press Women; National Federation of Professional Workers
nfq night frequency
nfr no further requirement
NFRC National Forest Reservation Commission
NFRN National Federation of Retail Newsagents, Booksellers, and Stationers
NFRW National Federation of Republican Women
nfs not for sale
NFS National Fire Service; National Forest Service; National Forest System; Network File System; Nuclear Fuel Services
NFSA National Fertilizer Solutions Associations
NFSA & IS National Federation of Science Abstracting and Indexing Services
NFSE National Federation of the Self Employed
NFSG National Federation of Students of German
NFSHSA National Federation of State High School Associations
NFSID National Foundation for Sudden Infant Death
NFSM National Fraternity of Student Musicians
NFSNC National Federation of Settlements and Neighborhood Centers
NFSNO National Federation

for Specialty Nursing Organizations
NFSO Navy Fuel Supply Office
nft no fixed time; no forwarding time; nutrient film technique
NFT National Film Theatre
NFTA National Film Theatre of Australia; Niagara Frontier Transportation Authority
NFTB Nuclear Flight Test Base
NFTC National Foreign Trade Council
NFTS National Federation of Temple Sisterhoods
nfu not for us
NFU National Farmers Union; National Film Unit; National Froebel Union
n-fuel nuclear fuel
N-fuel nuclear fuel
NFUW National Farmers' Union of Wales
nfv no further visits
nfva net free ventilation area
nfw new field wildcat (oil well)
NFWA National Farm Workers Association; National Furniture Warehousemen's Association
NFWI National Federation of Women's Institutes
NFYFC National Federation of Young Farmers' Clubs
nfyg notifying
nfz no fire zone
ng narrow gauge; nasogastric; new genus; nitroglycerine; no go; no good; no gum (on back of stamps); not given; not good; not ground; nut grounds
ng (NG) natural gas
n-g nitro-glycerine
n/g *nuestro giro* (Spanish—our draft)
Ng Norwegian
NG National Gallery; National Guard; National Gypsum; New Guinea
nga (NGA) non-gonococcal urethritis
Nga Nagoya
NGA National Gallery of Art; National Glider Association; National Governors Association; National Grains Authority; National Graphical Association; National Guard Association; Needlework Guild of America; Never Go Away (travel club dedicated to seeing America first)

NGAA National Gift and Art Association; Natural Gasoline Association of America
NGAC National Guard Air Corps
Ngaragba Ngaragba Prison in Bangui (capital of the Central African Republic)
N-gauge narrow gauge (railroad track less than standard gauge, gauge: 4 feet 8-1/2 inches)
NGAUS National Guard Association of the United States
ngb negative guard board
NGB National Garden Bureau; National Guard Bureau
NGC National Gallery of Canada; National Gambling Commission; National Gypsum Company; Natural Gas Corporation
NGC *New Galactic Catalog; New General Catalog* (astronomical)
ngcil nice guys come in last
NGCM Navy Good Conduct Medal
NGCMS National Guild of Community Music Schools
NGCSA National Guild of Community Schools of the Arts
NGDA National Glass Dealers Association
NGDC National Geophysical Data Center
NGDM & M *New Grove Dictionary of Music and Musicians*
NGE New York State Electric & Gas (stock exchange symbol)
NGEC National Gypsy Education Council
n gen new genus
ngf naval gunfire
ngf (NGF) nerve growth factor
NGF National Genetics Foundation; National Golf Foundation; Naval Gun Factory; Nordic Gunners Federation
NGFLO Naval Gunfire Liaison Officer
NGFLT Naval Gunfire Liaison Team
NGI National Garden Institute; Norwegian Geotechnical Institute
NGI *Navigazione Generale Italiana* (Italian—Italian General Navigation Line)
NGJA National Gymnastics Judges Association
NGJC North Greenville Junior College

N Gk New Greek
NGK *Nihon Gakujutsu Kaigi* (Japanese—Japan Research Council)
ngl natural gas liquids
NGL North German Lloyd Line
NGLTF National Gay and Lesbian Task Force
nglzd not glazed
N Gmc North Germanic
NGMEX Northern Gulf of Mexico
NGMP New Guinea Marine Products
ngo national gas outlet (thread); nongovernmental organization
Ngo Nagoya
NGOs Nongovernmental Organizations (UN)
NGPA Natural Gas Processors Association
NGPT National Guild of Piano Teachers
ngr narrow gauze roll; non-grain rating
ngr *neugriechisch* (German—modern Greek)
Ngr *Niger* (Spanish abbreviation)
NGr New Greek
NGR Ndumu Game Reserve (Zululand); Newbold General Refractories
NGRA National Gay Rights Activists
NGRI National Geophysical Institute
NGRS Narrow Gauge Railway Society; National Greyhound Racing Society
ngs national gas straight (threading); net gas sand (oil well)
NGS National Geodetic Survey; National Geographic Society; Nuclear Generating Station
NGSA National Gallery of South Africa; Natural Gas Supply Association
NGSDC National Geophysical and Solar-Terrestrial Data Center (NOAA)
NGSIC National Geodetic Survey Information Center (NOAA)
NGSR Nizam's Guaranteed State Railway
ngt national gas taper (threading)
ngt *negociant* (French—merchant)—wholesaler
NGT National Guild of Telephonists; North German

Traders
NGTE National Gas Turbine Establishment
NGTF National Gay Task Force
ng tube nasogastric tube
ngu nongonococcal urethritis
ngultrum monetary unit of Bhutan
NGUS National Guard of the United States
NGUT National Group of Unit Trusts
ngv nongonococcal vulvovaginitis
NGV Natural Gas Vehicles
NGV *Nederlands Genootschap van Vertalers* (Dutch—Netherlands' Translators Society)
nh never hinged; no hurry (hospitalese); non-hygroscopic
Nh *Noordholland* (Dutch—North Holland)
NH Naval Home; Naval Hospital; New Hampshire; New Hampshirite; New Haven, Connecticut; New Hebrides; New York, New Haven & Hartford (railroad); Nippon Airways (2-letter code); North Holland(er); Nursing Home
N & H Nedlloyd & Hoegh (steamship lines)
NH *Norges Hjemmenfrontmuseum* (Norwegian—Norwegian Home-Front Museum) —Oslo exhibit recalling anti-German resistance from 1940 to 1945; *Nueva Hampshire* (Spanish—New Hampshire)
N-H *Noord-Holland* (Dutch—North-Holland)
NH$_3$ ammonia
NH$_4$ ammonium radical
NH$_4$CL ammonium chloride; sal ammoniac
NH$_4$OH ammonium hydroxide (ammonia)
nha never has anything; next higher assembly; next higher authority
NHA National Hay Association; National Health Association; National Hide Association; National Hockey Association; National Housing Act; National Housing Administration; National Housing Agency; National Housing Association; Neighborhood House Association; New Homemakers of America; Nigerian Housing Administration; Nursing Home

Administration
NHAGB National Horse Association of Great Britain
NHAIAC National Highway Accident and Injury Analysis Center
NHAL National Hellenic American Line
NHANES National Health and Nutrition Examination Survey
NHAS National Hearing Aid Society
NHB National Harbours Board (Canada); Northland Harbour Board (New Zealand)
NHBRC National House Builders' Registration Council
NHBU New Hampshire Board of Underwriters
NHC National Health Council; National Hurricane Center; Naval Historical Center; New Hall College
NHCA National Hairdressers and Cosmetologists Association
NHCBS New Hampshire Council for Better Schools
NHCIC National Hazardous Chemicals Information Center
N.H.D. Doctor of Natural History
NHDC Naval Historical Display Center
nh di notch die
nhe nitrogen heat exchange
NHEA National Higher Education Association; New Hampshire Education Association
N-head(s) nuclear warhead(s)
N Heb New Hebrew
NHEF National Health Education Foundation
NHESA National Higher Education Staff Association
NHF National Hairdressers' Federation; National Headache Foundation; National Health Federation; National Health Foundation; National Heart Foundation; National Heart Fund; National Hemophilia Foundation; National Horse Festival; National Humanities Faculty; Naval Historical Foundation
NHFA National Heart Foundation of Australia
NHF Bull *National Health Federation Bulletin*
NHFNZ National Heart Foundation of New Zealand
NHFPL New Haven Free Pub-

lic Library
NHG New High German
NHGA National Hang Gliding Association
NHHS New Hampshire Historical Society
nhi (NHI) no humans involved
NHI National Health Institute; National Health Insurance; National Heart Institutes
NHIC National Health Insurance Commission; National Home Improvement Council
NHK Nippon Hoso Kyokai (Japanese—Japan Broadcasting Corporation)
NHKTV Nippon Koso Kyokai (Japanese—Japanese Television Broadcasting Corporation)
NHL National Hockey League
NHLA National Hardwood Lumber Association; National Home Library Association
NHLBAC National Heart, Lung, and Blood Advisory Council (NIH)
NHLBI National Heart, Lung, and Blood Institute
NHLI National Heart and Lung Institute
NHMA National Housewares Manufacturers Association
NHMRCA National Health and Medical Research Council of Australia
NHMS New Hampshire Medical Society
nhn neither help nor hinder
NHO National Hospice Organization; Navy Hydrographic Office
NHOS National Hellenic Oceanographic Society
nhp nominal horsepower
NHP National Corporation for Housing Partnerships; National Historic(al) Park; Natural History Park (Calgary, Alberta); Natural History Press; New Haven Police; New Hebrides Protectorate; Nursing Home Placement
NHPA National Horseshoe Pitchers Association
NHPC National Historical Publications Commission
NHPGA New Hampshire Personnel and Guidance Association
NHPL New Haven Public Library
NHPLO NATO Hawk Production and Logistics Organiza-

tion
NHPMA Northern Hardwood and Pine Manufacturers Association
NHPRC National Historical Publications and Records Commission
NHQ National Headquarters
NHR National Housewives Register; National Hunt Rules; National Hurricane Research
nhra next higher repairable assembly
NHRA National Hot Rod Association
NHRC Naval Health Research Center
NHRE National Hail Research Experiment
NHRL National Hurricane Research Laboratory
NHRP National Hurricane Research Project
NHRR New Haven Railroad
NHRU National Home Reading Union
nhs net hydrocarbon sand (oil well); normal human sera
NHS National Health Service; National Historic(al) Site; National Historical Society; National Honor Society; Newport Historical Society
NHSA National Head-Start Association; National Heart Savers Association; Negro Historical Society of America
NHSAA New Hampshire School Administrators Association
NHSB National Highway Safety Bureau
NHSBA New Hampshire School Boards Association
NHSC National Health Service Corps; National Health Statistics Center; National Highway Safety Council; National Home Study Council
NHSF National Hispanic Scholarly Fund
NHSO New Haven Symphony Orchestra
NHSR National Hospital Service Reserve
NHTI New Hampshire Technical Institute
NHTPC National Housing and Town Planning Council
NHTSA National Highway Traffic Safety Administration
NH Turn New Hampshire Turnpike
NHUC National Highway Us-

ers Conference
Nhv Newhaven
NHV New Haven Clock and Watch (stock exchange symbol)
NHYC New Haven Yacht Club; Newport Harbor Yacht Club
ni new impression; night
ni (NI) inversion of the note series (12-tone); national income; net income
Ni Nica; Nicaragua; Nicaraguan; Nicaragüense; Nicas; nickel
NI National Insurance; Native Infantry; Nautical Institute; Naval Instructor; Naval Intelligence; Negotiation Institute; Netherlands Indies; Neutralization Index; News International; Nicaraguan Airways (2-letter code)—LANICA; North Island (New Zealand); North Island, San Diego, California; Northern Ireland; Northern Island (New Zealand); Numerical Index; other North Islands
NI ampere turns (symbol)
nia nearest international airport
nia (NIA) noise-impact area
NIA National Institute on Aging; National Intelligence Authority; National Irrigation Administration; Neighborhood Improvement Area; Neighborhood Improvement Association
NIAA National Industrial Advertising Association; National Institute of Animal Agriculture
NIAAA National Institute on Alcohol Abuse and Alcoholism
NIAB National Institute of Agricultural Botany
NIABC Northern Ireland Association of Boys' Clubs
NIAC Nissho-Iwai American Corporation; Nuclear Insurance Association of Canada; Nutritional Information and Analysis Center
NIAE National Institute of Agricultural Engineering; National Institute for Architectural Education
NIAG NATO Industrial Advisory Group
Niagara Fort Niagara, Niagara Falls, Niagara-on-the-Lake, Niagara River, Niagara University
NIAID National Institute of

Allergies and Infectious Diseases

NIAL National Institute of Arts and Letters

NIAMD National Institute of Arthritis and Metabolic Diseases

NIAMDD National Institute of Arthritis, Metabolism, and Digestive Diseases (formerly NIAMD)

NIASA National Insurance Actuarial and Statistical Association

NIASE National Institute for Automotive Service Excellence

nib noninterference basis

NIB National Information Bureau; Nebraska Inspection Bureau

NIBA National Insurance Buyers Association

NIBM National Institute of Business Management

nibo nibonitschjo (ni boga ni tschjorta) (Russian—neither in god nor the devil)—materialist sceptics

NIBS National Institute of Building Sciences

NIBSC National Institute for Biological Standards and Control

nic negative impedance converter; newly industrializing country; not in contact

Nic Nicaragua; Nicolayev; Nicosia

N i C Nurse in Charge

Nic Nicola (Italian—Nicholas)

NIC Natick Industrial Centre; National Incomes Commission; National Indications Center; National Industrial Council; National Information Center; National Institute of Corrections; National Institute of Creativity; National Institute of Credit; National Insurance Certificate; National Insurance Contributions; National Interfraternity Conference; National Inventors Council; National Investors Council; Navigation Information Center; Neighborhood Info(rmation) Center(s); Niagara International Centre; Nicosia, Cyprus (airport); Nineteen-hundred Indexing and Cataloging; Nippon International Containers

Nica Nicaragua(n)

nicad nickel cadmium

NiCad battery nickel-cad-

mium (rechargeable) battery

NICAP National Investigations Committee on Aerial Phenomena

Nicar Nicaragua(n)

Nicaragua Republic of Nicaragua (Spanish-speaking two-coast Central American country), *Républicia de Nicaragua*

Nicas Nicaraguans

NICB National Industrial Conference Board

Ni-Cd nickel-cadmium (rechargeable storage battery)

nice. normal input/output control executive

Nice Eunice

NICE National Institute of Ceramic Engineers

nice cuppa nice cup of tea

NICEIC National Inspection Council for Electrical Installation Contracting

NICEM National Information Center for Educational Media

NICF Nebraska Independent College Foundation; Northern Ireland Cycling Federation

Nich Nicholas

NICHA Northern Ireland Chest and Heart Association

NICHHD National Institute of Child Health and Human Development

nichrome nickel-chromium alloy

NICIA Northern Ireland Coal Importers' Association

NICJ National Institute of Consumer Justice

nick. name information correlation key

Nick Nicholas; Nichols; Nicodemus; Nikos

Nick-Pack (*see* NICPAC)

Nicky Nicholas; Nicole; Nikos

NICM Nuffield Institute of Comparative Medicine

Nico Nicobar Islands

NICO National Insurance Consumer Organization; Navy Inventory Control Office(r)

Nicobars Nicobar Islands in the Indian Ocean

Nicolass Sint Nicolaas, Aruba

NICOP Navy Industry Cooperation Plan

Nicos Nicosia, Cyprus

NICP National Inventory Control Point

NICPAC National Independent Conservative Political Action Committee

NICRA Northern Ireland Civil

Rights Association

NICRAD Navy-Industry Cooperative Research and Development

nic's newly industrializing countries

NICs National Institute of Corrections

NICS NATO Integrated Communications System

NICS COA NICS Control Operating Authority

NICSEM/NIMIS National Information Center for Special Education Material/National Instructional Material Information System

NICSO NATO Integrated Communications System Organization

NICSS Northern Ireland Council of Social Science

NICSSE National Information Center for Social Science Education

nicu (NICU) neonatal intensive-care unit

NICU Nippon International Container Unit

NICUFO National Investigations Committee on Unidentified Flying Objects

nid network in dial

NID National Institute for the Deaf; National Institute of Drycleaning; Naval Intelligence Department; Northern Ireland District

NID National Intelligence Daily; New International Dictionary (Webster's Third New International Dictionary of the English Language Unabridged)

nida numerically integrated differential analyzer

NIDA National Institute on Dramatic Art; National Institute on Drug Abuse; National Investment and Development Authority; Northern Ireland Development Agency

NIDC National Institute of Dry Cleaning; National Investment Development Corporation; Northern Ireland Development Council

NIDCD National Institute on Deafness and other Communication Disorders

NIDER Nederlands Instituut voor Documentatie en Registratuur (Dutch—Netherlands Institute of Documentation and Filing)

NIDFA National Independent

Drama Festivals Association
NIDH National Institute of Dental Health
NIDM National Institute for Disaster Mobilization
NIDR National Institute of Dental Research
nie not included elsewhere
NiE Newspaper in Education
NIE National Institute of Education; National Intelligence Estimate; Newspaper In Education
NIEA National Indian Education Association
NIECC National Industrial Energy Conservation Council
niedr niedrig (German—low)
NIEHS National Institute of Environmental Health Sciences
NIEM National Industrial Energy Management
NIEO New International Economic Order; Non-Incorporated Engineering Order
NIER National Industrial Equipment Reserve
NIEs Newly Industrializing Economies
NIESR National Institute for Economic and Social Research
NIEU Negro Industrial Economic Union
nif nickel-iron film
NIF Navy Industrial Fund
NIFA National Islamic Front of Afghanistan
NIFC National Income Forecasting Committee
nife nickel + iron (Ni + Fe)
NIFES National Industrial Fuel Efficiency Service
NIFI National Inland Fisheries Institute
nifti near-isotropic flux-turbulence instrument
nig. niger (Latin—black)
Nig Nigeria
Nig Niger (Spanish—Niger)
niga nuclear-induced ground radioactivity
NIGC National Iranian Gas Company
Niger Republic of Niger (landlocked North African nation)
Nigeria Federal Republic of Nigeria (West African country)
nightie(s) nightdress(es); nightgown(s)
nightsoap(s) nighttime (tv) soap opera(s)
NIGMS National Institute of General Medical Sciences

NIGP National Institute of Governmental Purchasing
NIGRO Northern Ireland General Register Office
nig(s) nigger(s); renege(s); revoke(s)
nigyysob now I've got you, you SOB
nih not invented here
NIH National Institute of Hardware; National Institutes of Health
NIH 204 antimalarial drug
NIHB National Indian Health Board
NIHBC Northern Ireland House Building Council
NIHE Northern Ireland Housing Executive
nihil nihil obstat quominus imprimatur (Latin—nothing hinders it from being printed)
nihil obs. nihil obstat (Latin—nothing stands in the way)—official Catholic publications must obtain this before their publication
NIHR National Institute of Handicapped Research
NIHT Northern Ireland Housing Trust
NII Netherlands Industrial Institute
NIIC National Injury Information Clearinghouse
NIIG NATO Item Identification Guide
NIIN National Item Identification Number
NIIP National Institute of Industrial Psychology
NIIS Niagara Institute for International Studies
NIJ National Institute of Justice
NIJC North Idaho Junior College
NIJFCM National Institute of Jig and Fixture Component-Manufacturers
NIJRs National Institute of Justice Reports
nik narcotic identification kit
Nik Nikolayev
Niki Nicholas
Nik Nik Nicholas Nickleby
Niko (Russian nickname—Nikolai)—Nicholas; Nick; Nicky
Nikola Nikola Tesla (1856–1943)
Niky Nicholas; Nicole; Nickerson; Nikerson
nil not in labor
NIL National Instrument Laboratories; National Investment

Library
NILA National Industrial Leather Association
NI Lab Northern Ireland Labour (party)
NILECJ National Institute of Law Enforcement and Criminal Justice
NILI Netzach Israel Lo Ishakare (Hebrew—The eternity of Israel will not die)—acronymic password of the Nili spies who aided Britain by facilitating Turkish defeat in an effort to establish a homeland for Jews in Palestine
'nilla vanilla
N Ill U Pr Northern Illinois University Press
NILOJ National Institute for Law/Order/Justice
NILP National Institute for Labor Policy; Northern Ireland Labour Party
NILQ Northern Ireland Legal Quarterly
nil sig nothing significant
NILT National Institute for Lay Training
nim newspaper(s) in microfilm; newspaper(s) in microform
NIM Neurological Impress Method; North Irish Militia
NIMA National Insulation Manufacturers Association
NIMAC National Interscholastic Music Activities Commission
nimby not in my backyard
nimby's not in my backyarders
NIMFR National Institutes of Marriage and Family Relations
NIMH National Institute of Mental Health
NIMLO National Institute of Municipal Law Officers
nimm nuclear-induced missile malfunction
n imp new impression
NIMP National Intern Matching Program
nimphe nuclear isotope mono-propellant hydrazine engine
NIMR National Institute for Medical Research; National Institute for the Mentally Retarded
NIMT National Institute for Music Theater
nimto not in my term of office
NIMU North Island Mutual Insurance (New Zealand)
NIN Narcotics Intelligence Network; National Informa-

tion Network; Neighbors In Need

NINA No Irish Need Apply

NINB National Institute of Neurology and Blindness

NINCD National Institute of Neurological and Communicative Disorders

NINCDS National Institute of Neurological and Communicative Disorders and Stroke

NINDB National Institute of Neurological Diseases and Blindness

NINDS National Institute of Neurological Diseases and Stroke

NIO National Institute of Oceanography; National Institute of Oceanology; National Intelligence Office(r); National Intelligence Organization; National Iranian Oil; Naval Institute of Oceanology; Northern Ireland Office

NIOC National Iranian Oil Company

niod network in-out dial

NIOSH National Institute for Occupational Safety Hazards; National Institute of Occupational Safety and Health

nip. nipper; nipple; not in possession

Nip Nippon (Japan); Nipponese (Japanese)

NIP National Industrial Policy; Neighborhood Improvement Program; Northern Ireland Parliament

NIP Norges Kommunistiske Parti (Norwegian—Norwegian Communist Party)

NIPA National Institute of Public Affairs

NIPCC National Industrial Pollution Control Council

NIPDOK Nippon Documentesyon Kyokai (Japanese—Japanese Documentation Society)

NIPE National Intelligence Programs Evaluation

NIPG Nederlands Instituut voor Praeventieve Gneeskunde (Dutch—Netherlands Institute for Preventive Medicine)

NIPH National Institute of Public Health

niphl noise-induced permanent hearing loss

nip nip(s) nipple nipper(s)

nipo negative input—positive output

NIPO Nederlands Instituut

voor Publick Opinie (Dutch—Netherlands Institute for Public Opinion)

NIPPORO Nihon Hoso Rodo Kumiai (Japan Broadcasting Workers Union)

NIPR National Institute for Personnel Research

ni pri nisis prius (Latin—unless before)

nips. nippers; non-impact printers

Nip(s) Nippon(ese)

NIPs Not In Profile students

NIPS National Information Processing System; National Institute of Police Science (Japanese)

NIPSSA Naval Intelligence Processing Systems Support Activity

nipts noise-induced permanent threshold shifts

N Ir Northern Ireland

NIR Northern Ireland Railways

NIRA National Industrial Recovery Act; National Industrial Recovery Administration; Newspaper Industries Research Association

NIRC National Industrial Relations Court

NIRD National Institute of Research in Dairying

N Ire Northern Ireland

NIRI National Investor Relations Institute

NIRMP National Intern and Resident Matching Program

NIRNS National Institute for Research in Nuclear Science

NIROP Naval Industrial Reserve Ordnance Plant (USN)

NIRR National Institute for Road Research

NIRRA Northern Ireland Radio Retailers' Association

NIRs Norfolk International (container) Terminals

NIRS National Institute of Radiological Science; Nuclear Information and Resource Service

ni & rt numerical index and requirement table(s)

NIRT National Iranian Radio and Television

nis not in stock

n i s not in stock

Ni s nickel steel

NIS National Information System; National Institute of Science; National Insurance Scheme; National Intelligence Service; National In-

telligence Survey; National Investment Strategy; Naval Intelligence Service; Naval Investigative Service; News and Information Service (NBC)

NISA National Impacted Schools Association; National Intelligence Security Authority (Philippines)

NISBS National Institute of Social and Behavioral Science

NISC National Independent Study Center; National Industrial Safety Committee; Naval Intelligence Support Center

NISGAZ National Intelligence Survey Gazetteer

NISIR National Institute of Scientific Industrial Research

NISM National Iron and Steel Mills

NISO National Industrial Safety Organization; Naval Investigative Service Office(r)

NISP National Information System for Psychology

NISRA Naval Investigative Service Resident Agent

NISS National Institute of Social Sciences

NISSPO NATO Identification System Special Project Office

NIST National Institute of Science and Technology

NISUCO Nigerian Sugar Company

nit. negative income tax; none in town

nit. (NIT) nautical industrial technology; negative income tax

nit unit of luminance (symbol)

NIT National Institute of Tourism; National Instructional Television; National Intelligence Test; National Invitation Tournament; Negative Income Tax; Northrop Institute of Technology; Northrup International Terminals

Nita Juanita

NITA National Industrial Television Association

NITC National Information Transfer Center; National Iranian Tanker Company

nite night

NiteDevRon Night Development Squadron

NITEP Native Indian Teacher Education Programme (Canadian)

niter potassium nitrate

NITHC Northern Ireland Transport Holding Company

NITL National Industrial Traffic League

ni tp nibbling template

NITR National Institute for Telecommunications Research

nitre potassium nitrate (KNO_3)

nitric acid HNO_3

nitro nitrocellulose; nitroglycerine

nitros nitrostarch

nitts noise-induced temporary threshold shift

NITV National Iranian Television

NIU Northern Illinois University; Northern Interparliamentary Union

Niv Nivose (French—Snowy Month)—beginning December 21st—fourth month of the French Revolutionary Calendar

NIV New International Version (Zondervan Bible)

NIVE Nederland Instituut voor Efficiency (Dutch—Netherlands Institute for Efficiency)

NIW National Industrial Workers Union

NIWAAA Northern Ireland Women's Amateur Athletic Association

NIWL National Institute for Work and Learning

NIWR National Institute for Water Research

NIWW National Institute for Working Women (prostitutes)

nix (NIX) nuclear inclusion X (clam parasite)

NIYC National Indian Youth Council

Nizh Nizhen (Bulgarian—lower); *Nizhni* (Russian—lower)

Nizim Nizmennost (Russian—lowland)

n J nächstes Jahr (German—next year)

NJ New Jersey; New Jerseyite

NJA National Jail Association; National Jewellers Association; National Jogging Association

NJAC National Joint Advisory Council

NJACU New Jersey Association of Colleges and Universities

NJAIS New Jersey Association of Independent Schools

NJASBO New Jersey Association of School Business Officials

NJASSPS New Jersey Association of Secondary School Principals and Supervisors

njb nice Jewish boy

NJC Natchez Junior College; National Joint Council; National Judicial College; National Junior College; Navarro Junior College; Newton Junior College; Norfolk Junior College

NJCAA National Junior College Athletic Association

NJCC Northeastern Junior College of Colorado

NJCCC New Jersey Casino Control Commission

NJCF New Jersey Conservation Foundation

NJCMS New Jersey Chamber Music Society

NJDA National Juvenile Detention Association

NJDL New Jewish Defense League

NJE Network Job Entry; New Jersey Experiment

NJEA New Jersey Education Association

NJF Nordiske Jordburgsforskeres Forening (Nordic Agricultural Research Workers' Association)

NJFR National Joint Fiction Reserve

njg nice Jewish girl

NJH National Jewish Hospital

NJHA National Junior Horticultural Association

NJ Hist Soc New Jersey Historical Society

NJHS National Junior Honor Society; New Jersey Historical Society

NJIT New Jersey Institute of Technology

njk not just kidding

NJLA New Jersey Library Association

NJLC National Juvenile Law Center

NJLJ New Jersey Law Journal

NJMA National Jail Managers Association

NJMP New Jersey Marine Police

NJPBA New Jersey Public Broadcasting Authority

NJPC National Joint Practices Commission

NJPGA New Jersey Personnel and Guidance Association

NJROTC Naval Junior Reserve Officers' Training Corps

NJRW New Jersey Reformatory for Women (Clinton)

NJSA New Jersey Student Association

NJSBA New Jersey School Boards Association

NJSD National Joint Service Delegate; National Joint Service Delegation

NJSO New Jersey Symphony Orchestra

NJSP New Jersey State Police

NJ Turn New Jersey Turnpike

NJWB National Jewish Welfare Board

NJZ New Jersey Zinc

nk neck; not known; not ours (publishing)

NK Nihon Kyosanto (Japanese Communist Party); *Nippon Gakushiin* (the Japanese Academy); *Nippon Kokan Steel* (Japanese Steel Exchange) *Nomenklatur Kommission* (Anatomical Nomenclature Commission); *Nordiska Kompaniet* (the Norse Company, Stockholm's leading department store); North Korea(n)

NKA National Kindergarten Association

NKBA National Kitchen and Bath Association

NKCA National Kitchen Cabinet Association

NKDR National Key Deer Refuge (Florida)

NKF National Kidney Foundation

NKG Nordiska Kommissionen for Geodesi (Nordic Commission for Geodesy)

NKGB People's Commissariat for State Security (*q.v.* VOT)

NKK Nippon Kokan Steel (Japan)

NKK Nippon Kaiji Kyokai (Japanese—Japanese Marine Classification Society)

NKL Norges Kooperative Landsforening (Norwegian—Norwegian Consumer Cooperative)

nklc nickel copper

NKM New Park Mining (stock exchange symbol)

N.K. Naomi code for chemical and biological warfare

NKOA National Knitted Outerwear Association

NKP Nickel Plate Railroad (stock exchange symbol for New York, Chicago & St Louis Railroad)—locomo-

tives on this line gleamed with nickel-plated ornaments
NKP Norges Kommunistiske Parti (Norwegian Communist Party)
NKPA National Kraut Packers Association
NKr Norwegian krone(r)
nks necks (woolen)
NKS Norge Kjemisk Selskap (Norwegian—Norwegian Chemical Society)
NKSO Narodniy Kommissariat Sotsialnogo Obespecheniya (Russian—People's Commissariat of Social Security)
NKT Nihon Kai Telecasting
Nkv Nakskov
NKVD Narodnyi Kommissariat Vnutrennikh Del (Russian—People's Commissariat for Internal Affairs, Soviet secret police, *q.v.* VOT)
NKZ Narodnyi Kommissariat Zdravokhranenia (Russian—People's Commissariat of Health)
nl new line; no liability; nonlubricant; not listed
nl (NL) new line character (data processing); not licensed (to sell liquor)
nl nemlig (Dano-Norwegian—namely); *nicht löslich* (German—not soluble); *non longue* (French—not so far)
n.l. non licet (Latin—not permitted)
n/l nuestra letra (Spanish—our letter)
Nl National
NL National Lakeshore; National League (of Professional Baseball Clubs); National Liberal; National Library; naval lighter (naval symbol); Navy (US department) Library; Navy League; Navy List(ing); Netherlands (auto plaque); New Latin; New London, Connecticut; Night Letter; North Latitude; Nuevo León
NL Nederland (Dutch—Netherlands); *Norddeutscher Lloyd* (North German Lloyd Line)
N.L. non liquet (Latin—unclear)
nla net lettable area
NLA National Leukemia Association; National Liberation Army; National Librarians Association; National Libraries Authority; National Library of Australia (Canberra);

National Lumbermen's Association; Nevada Library Association; Nigerian Library Association
NL-A Nationaal Luchtvaartlaboratorium (Dutch—National Airline Laboratory)-Amsterdam
NLAA National Legal Aid Association
NLA & DA National Legal Aid and Defender Association
NLAE National Laboratory for the Advancement of Education
NLAPW National League of American Pen Women
N Lat north latitude
NLB National Labor Board; National Library for the Blind; Northern Lighthouse Board
NLC National Lead Chemicals; National League for Cities; National Leathersellers College; National Legislative Conference; National Legislative Council; National Liberal Club; National Library of Canada; New Liberal Club; New Location Code; New Orleans & Lower Coast (railroad); Northern Land Council
NLCA Norwegian Lutheran Church of America
NLCIF National Light Castings Ironfounders' Federation
NLCMDD National Legal Center for the Medically Dependent and Disabled
NLD National Legion of Decency
NLDC Native Land Development Corporation
NLEC National Lutheran Educational Council
NLETS National Law Enforcement Telecommunications System
nlf nearest landing field
NLF National Labour Federation; National League of Families (of men missing in action); National Liberal Federation; National Liberation Front; nearest landing field
nlg nose landing gear
NLG National Lawyers Guild; National Library of Greece (Panepistemiou Street in Athens); Netherlands Guilder(s); Numismatic Literary Guild
NLGHF National Lesbian and

Gay Health Foundation
NLGI National Lubricating Grease Institute
NLHE National Laboratory for Higher Education
NLHS Neo Lao Hak Sat (Lao Patriotic Front)
NLI National Lead Inc; National Library of India (Calcutta); National Library of Ireland (Dublin); National Lifeboat Institution
NLJ National Law Journal
NLL National Lending Library; Nature Lovers League; Nedlloyd Lines
NLL cards *National Luchtenruimtevaart Laboratorium* (international card catalog devised in Amsterdam)
NLLST National Lending Library for Science and Technology (UK)
nl lt net-laying light
NLM National Liberation Movement; National Library of Medicine
NLMA National Lumber Manufacturers Association
NLMC National Labor Management Council
nln no longer needed
NLN National League for Nursing
NLNE National League of Nursing Education
NLNP Naujan Lake National Park (Philippines)
NLNZ National Library of New Zealand
NLO Naval Liaison Office(r); Neighborhood Law Office(r)
NLOGF National Lubricating Oil and Grease Federation
nlp (NLP) neuro-linguistic programmers
NLP National League of Postmasters; Neighborhood Loan Program; Neuro-Linguistic Programming
nlpc (NLPC) n-laurylpyridinium chloride (detergent compound)
NLPI National Loss Prevention Institute
NL POW/MIA F National League of Prisoner-of-War/Missing-in-Action Families
nlr noise load ratio
NLR Nationaal Lucht-en Ruimtevaartlaboratorium (Dutch—National Aero- and Astronautical Research Institute), Amsterdam; *Newfoundland Law Reports*; *Nigeria Law Reports*

NLRA National Labor Relations Act
NLRB National Labor Relations Board
nls new least squares; no-load start; non-linear system(s)
NLs New Leftists
NLS National Library of Scotland (Edinburgh); National Library Service (New Zealand and elsewhere); Non-Linear Systems
NLSB National League Service Bureau
NLSCS National League for Separation of Church and State
NLSI National Library of Science and Invention
NLSLS National Library of Scotland Lending Services
nlt new logic technology; not later than; not less than
NLT National Library of Thailand (Bangkok)
NLT Navigazione Libera Triestina (Italian Line)
NLTA National Lawn Tennis Association; National League of Teachers Associations
NLTB Native Land Trust Board
NLTU New London Training Unit (USN)
NLUCS National Land Use Classification System
NLUS Navy League of the United States
NLW National Lawyers Wives; National Library of Wales; National Library Week
NLWP National Library Week Program
Nly northerly
NLYL National League of Young Liberals
nm nanometer; nautical mile(s); neuromuscular; new moon; nitrogen mustards; nomenclature; nonmetallic; non-motile (bacteria); nuclear magneton; nutmeg
n/m no mark
nm nachmittags (German—afternoon, P.M.); *namiddag* (Dutch—a.m.); nanometer; nautical mile(s); nomenclature; nonmetallic
n.m. nocte et mane (Latin—night and morning)
n M nachsten Monats (German—next month)
Nm newtometer
NM Natal Museum (South Africa); National Monument;

National Mutual (life insurance); New Mexico; Nigeria Museum
N-M Neiman-Marcus
n/m² newton per square meter
n/m³ normal cubic meter
nma negative mental attitude
NMA National Management Association; National Market Authority; National Medical Association; National Microfilm Association; National Micrographics Association; National Mortgage Association; Navy Mutual Aid (Association); Needle Makers' Association; Northwest Mining Association
NMAA National Machine Accountants Association; Navy Mutual Aid Association
NMAB National Materials Advisory Board
nmac near mid-air collision
NMAC National Medical Audiovisual Center
NMACT Nuclear Materials Accounting Control Team
NMAF National Medical Association Foundation
n mar nivel del mar (Spanish—sea level)
n masc masculine form of a noun
N-materials nuclear materials
NMB National Maritime Board; National Mediation Board; Nippon Miniature Bearing
nmbr number
nmc no more credit
NMC National Manufacturers Code; National Mapping Council; National Maritime Council; National Marketing Council; National Meteorological Center; National Museum of Canada; National Museums of Ceylon; National Music Council; Naval Material Command; Naval Medical Center; Naval Missile Center; Northern Mining Corporation
NMCA National Music Camp Association; Navy Mother's Clubs of America
NMCB National Metric Conversion Board
NMCC National Military Command Center
NMCCIS NATO Military Command, Control, and Information System
NMCDA National Model Cities Directors Association

NMCO Naval Material Catalog Office
NMCP National Memorial Cemetery of the Pacific
NMCS National Military Command System
NMCSSC National Military Command System Support Center
NMDA National Metal Decorators Association; National Motorcycle Dealers Association
NMDC National Materials Development Center
NMDL Navy Mine Defense Laboratory
NMDZ NATO Maritime Defense Zone
nme noise-measuring equipment
NME National Medical Enterprises; National Military Establishment; National Mortgage Exchange
Nmea Noumea
NMEA National Marine Education Association; National Marine Electronics Association
N-medicine nuclear medicine
N-med tech nuclear-medicine technician
N Mem National Memorial
nmembler mnemonic assembler
NMERI National Mechanical Engineering Research Institute
N Mex New Mexico; New Mexican
N Méx Nuevo México (Spanish—New Mexico)
NMF National Marine Fisheries
NMFMA National Mutual Fund Managers Association
NMFO Navy Maintenance Field Office
NMFRL Naval Medical Field Research Laboratory
NMFS National Marine Fisheries Service
NMFSL National Marine Fisheries Service Laboratories
NMG National Management Game
NMGC National Marriage Guidance Council
nmh nautical miles per hour
NMH Northwestern Memorial Hospital
NMHA National Mental Health Association
NMHB National Materials Handling Bureau

NMHC National Material Handling Center
NMHSA National Mine Health and Safety Academy
nmi new (automobile) model introduction; no middle initial
n mi nautical miles
NMI National Mutual Insurance; New Mexico Military Institute; Northern Mariana Islands (Rota, Saipan, Tinian, Pagan, Guguan, Agrihan, Aguijan)
NMIA National Meteorological Institute of Athens
NMICA New Mexico Independent College Association
n mi/lb nautical miles per pound (of fuel)
NMIM & T New Mexico Institute of Mining and Technology
N-mishap nuclear mishap
N-missile(s) nuclear missile(s)
NMJ Northern Masonic Jurisdiction
NML National Measurement Laboratory; National Municipal League; National Museum Library; National Music League; National Mutual Life (insurance); Northwestern Mutual Life (insurance)
NMLA National Mutual Life Association; New Mexico Library Association
NMLRA National Muzzle-Loading Rifle Association
NMM National Maritime Museum
NMMA National Macaroni Manufacturers Association; National Marine Manufacturers Association
nmn no middle name
NMN⁺ nicotinamide mononucleotide
NMNA National Male Nurse Association
nmnc nonmercuric noncorrosive
NMNH National Museum of Natural History (DC)
NMO National Mapping Office; Navy Management Office
nmoc new man on campus
nmp navigational microfilm projector; normal menstrual period
NMP National Military Park
NMPA National Marine Paint Association; National Motorsports Press Association; National Music Publishers Association

NMPC National Maintenance Publications Center (USA); National Moratorium on Prison Construction
NMPGA New Mexico Personnel and Guidance Association
nmph nautical miles per hour
nmpm nautical miles per minute
nmps nautical miles per second
n. mque. *nocte maneque* (Latin—night and morning)
nmr normal mode rejection; nuclear magnetic resonance
nmr (NMR) nuclear magnetic resonance (imaging)
NMR Natal Mounted Rifles; National Military Representative
NMRA National Model Railroad Association
NMRI Naval Medical Research Institute
NMRL Naval Medical Research Laboratory
NMRP New Mexico Research Park
NMRTC New Mexico Research and Treatment Center
nms nuclear materials safeguards
NMS National Medal of Science; National Meteorological Service; Nobles of the Mystic Shrine
NMSA National Middle School Association
NMSC National Merit Scholarship Corporation; National Mountain and Safety Committee
NMSE Naval Material Support Establishment
NMSM New Mexico School of Mines
NMSO Naval Manpower Survey Office(r)
NMSQT National Merit Scholarships Qualifying Test
NMSRC National Middle School Resource Center
NMSS National Multiple Sclerosis Society; Nuclear Materials Safety and Safeguards
NMSSA NATO Maintenance Supply Service Agency
NMSST Naval Manpower Shore Survey Team
NMSU Naval Motion Study Unit; New Mexico State University
NMSWF National Manufac-

turers of Soda Water Flavors
nmt nor more than
NMT National Museum of Transport
NMTA National Metal Trades Association
NMTBA National Machine Tool Builders' Association
NMTF National Market Traders' Federation
NMTFA National Master Tile Fixers' Association
NMTLM Nuclear Materials Transportation Logistics Model
NMTS National Milk Testing Service
NMU National Maritime Union; National Miners Union
NMW National Museum of Wales
NMWA National Mineral Wool Association
NMWP National Migrant Workers Program
nn neutralization number; no name; nouns
n/n no number; not to be noted
nn *non numerato* (Italian—unnumbered)
n.n. *nemini notus* (Latin—known to no one); *nescio nomen* (Latin—I do not know the name)
NN Newport News; Northwestern National
N/N Northrop/Nortronics
NNA National Neckwear Association; National Needlework Association; National Newspaper Association; National Notary Association
nnad network non-addressing device
NNAG NATO Naval Advisory Group; NATO Naval Armaments Group
NNBIS National Narcotics Border Interdiction System
NNBPWC National Negro Business and Professional Women's Clubs
NNC National Negro Conference; National News Council; Naval Nuclear Club (rival members include France, the United Kingdom, the United States, and the USSR); Navy Nurse Corps
NNCVTE National Network for Curriculum Coordination in Vocational and Technical Education
NNCR North Norfolk Coast Reserves (England)

nnd neonatal death
NND New and Non-Official Drugs
NNE Net National Expenditure; north northeast
NNEB National Nursery Examination Board
NN & EB National Newark & Essex Bank
NNECH National Nutrition Education Clearinghouse
NNEU Naval Nuclear Evaluation Unit
NNF Northern Nurses Federation
NNFA National Nutritional Foods Association
NNG Netherlands New Guinea; Northern Natural Gas (company)
NNGA Northern Nut Growers Association
NNHT Nuffield Nursing Homes Trust
nni noise and number index (sound pollution)
NNI Norwegian Nobel Institute
NNI Nederlands Normalisatie Instituut (Dutch—Netherlands Standards Institute)
nnk (NNK) notify next of kin
NNL Nigerian National Line
NNLC National Negro Labor Council
nnm next new moon
NNMC National Naval Medical Center
nnn no national name; no native named
NNN Nihon News Network; Novy-Nicolle-McNeal (bacteriological culture)
NNNR Noss National Nature Reserve (Shetlands)
NNO noord noordoost (Dutch—north northeast)
NNOC Nigerian National Oil Company
n. nov. nomen novum (Latin—new name)
nnp net national product
nnp (NNP) net national product
NNP Nairobi National Park (Kenya); Ngezi National Park (Rhodesia); Nimule National Park (Sudan)
NNPA National Negro Press Association; National Newspaper Promotion Association; National Newspaper Publishers Association
n-n p-i-f never-never pay-in-full (installment plan)
NNPP Naval Nuclear Propul-

sion Program (USN)
NNPT Nuclear Non-Proliferation Treaty
NNR New and Nonofficial Remedies
NNRC Neutral Nations Repatriation Commission
NNRI National Nutrition Research Institute
NNRO Norske Nasjonalkomite for Rasjonell Organisasjon (Norwegian National Committee for Scientific Management)
nn's nubile nymphs
nnS (NNS) Navy navigation satellite
n N's nice Nellyisms, euphemisms
n-N's neo-Nazis
N Ns Newport News
NNS National Newspaper Syndicate
NNSC Neutral Nations Supervisory Commission
NNS & DDC Newport News Shipbuilding and Dry Dock Company
NNSL Nigerian National Shipping Line
nnsn no national stock number
NNSS Navy Navigational Satellite System
NNTO Netherlands National Tourist Office; Norwegian National Travel Office
NNTT National New Technology Telescope
NNW north northwest
NNW noord noorwest (Dutch—north northwest)
NNWR Necedah National Wildlife Refuge (Wisconsin); Noxubee National Wildlife Refuge (Mississippi)
NNWSI Nevada Nuclear Waste Storage Investigation
no. natural order; normally open; number
no. (NO) neuromyelitis optica
n-o not or
n/o no orders
no norsk (Dano-Norwegian—Norwegian); *nummer* (Dutch—number)
n⁰ *número* (Spanish—number)
No nobelium; Norskie (Norwegian-American); Norway; Norwegian; number
No. Numero (Latin—number)
NO Naval Observatory; Naval Officer; New Orleans; nitrous oxide; North Central Airlines; Nuffield Observatory (Jordrell Bank, England); Nursing Officer

NO noordoost (Dutch—northeast); *Nordosten* (German—northeast); *noroeste* (Spanish—northwest)
No 1 first; first quality; first rate; first person; most important; most important person; number one
No 2 next in line; next in rank; number two; second; second person; second quality; second rate
NO₂ nitrogen dioxide
No 10 Number 10 Downing Street (London residence of the British prime minister)
noa net operating assets; new obligational authority (NOA); not operationally assigned; not otherwise authorized
n-o-a not-or-and
NOA National Onion Association; National Opera Association; National Optical Association; National Orchestral Association
NOAA National Oceanic and Atmospheric Administration
NOAB National Outdoor Advertising Bureau
NO-AB New Orleans-Algiers Bridge
NOADS Newspapers Opposed to Advertising Death by Smoking
NOAL National Order of Arts and Letters
noala noise-operated automatic level adjustment
NOAOs National Optical Astronomy Observatories
NOASSR North Ossetian Autonomous Soviet Socialist Republic
nob. no open burning; nobility; noble; not on board
nob nabob (Urdu—viceroy)
nob. nobis (Latin—to us)
NOB National Oil Board; Naval Operating Base; Naval Order of Battle
NOB Nationaal Orkest van Belgie (Flemish—National Orchestra of Belgium)
Nobelst Nobelstiftelsen (The Nobel Foundation)
no biz no business
NOBLEE National Organization of Black Law Enforcement Executives
Noble Patria Noble Patria, tu hermosa bandera (Spanish—Noble country, your lovely flag)—Costa Rican anthem
noc not otherwise classified; notation of content(s)

NOC National Oceanographic Council; National Olympic Committee
nocc navigation operator's control console
NOCC New Orleans Crime Commission
NOCHA National Off-Campus Housing Association
NOCI National Organization of Circumcision Information
NOCIL National Organic Chemical Industries
No-Clo Z No-Clone Zone
NOCM Nuclear Ordnance Commodity Manager
no cn no connection
No Co Northern Counties
NOCO Nuclear Ordnance Catalog Office
No Code do not resuscitate; no cardiopulmonary resuscitation
no cpr no cardiopulmonary resuscitation
noct. nocte (Latin—by night, nocturnal)
NOCTI National Occupational Competency Testing Institute
noct. maneq. nocte maneque (Latin—night and morning)
nod. network out dial; new offshore discharge; night observation device
NOD National Organization on Disability; Naval Ordnance Depot; Navigation and Ocean Development
NODA Night Operatic and Dramatic Association
NODAC Naval Ordnance Data Automation Center
Nodaks North Dakotans
NODC National Oceanographic Data Center
NODCP National Office of Drug Control Policy
NODECA Norwegian Defense Communications Agency
nodex new offshore dischargement exercise
NODL National Organization for Decent Literature (Catholic)
no do a nota do autor (Portuguese—author's note)
no do e nota do editor (Portuguese—editor's note)
no do t nota do tradutor (Portuguese—translator's note)
noe not otherwise enumerated
NOE Notice of Exception Oceanographic Foundation; National Osteopathic Foundation
NOEB NATO Oil Executive

Board(s)
NOEL National Organization of Episcopalians for Life
NOESS National Operational Environmental Satellite System
NOF National Oceanographic Foundation; National Optical Font; National Osteopathic Foundation; Naval Ordnance Facility
NOFI National Oil Fuel Institute
noforn no foreign nationals; special handling—not to be released to foreign nationals
noft notification of foreign travel
nog noggin
Nogal Nogales, Sonora, Mexico
NOGC Nationaal Overleg voor Gewestelijke Cultuur (Dutch—National Council for Regional Culture)
Noguchi Hideyo Noguchi, orig. Noguchi Seisaku (1876–1928)
nohp not otherwise herein provided
noi net operating income; not otherwise identified
noibn not otherwise identified by name; not otherwise indexed by name
NOIC National Oceanographic Instrumentation Center; Naval Officer in Charge; Navy Opportunity Information Center
NOIM Nuclear Ordnance Inventory Manager
noise. not only inserted in (modern) symphonic epics; the other fellow's music; unwanted sound
NOISE National Organization to Insure Support Enforcement; National Organization to Insure Sound-controlled Environment
noisic noisy music
NOJC National Oil Jobbers Council
NOJTP National On-the-Job Training Program
nok next of kin
NOK Norsk Aero Klub
nol normal overload(ing)
NOL Naval Ordnance Laboratory; Neptune Orient Line; Norse Oriental Line
NOLA New Orleans, Louisiana
NOLAC National Organization of Liaison for Allocation

of Circuits
NOLC Naval Ordnance Laboratory, Corona
nol. con. nolo contendere (Latin—I do not wish to contend)
nolo nolo contendere (Latin—I do not wish to contend)
NOLPE National Organization on Local Problems in Education
nol-pros nol-prossed; nol prossing
nol. pros. nolle prosequi (Latin—to be unwilling to prosecute)
NOLS National Oceanographic Laboratory System
nol. vol. nolens volens (Latin—unwilling or willing); willy-nilly
NOLWO Naval Ordnance Laboratory, Whit Oak (Maryland)
nom nominal; nominate; nominated; nomination; nominative
no'm no madam
NOM National Organization for Men
NOMA National Office Management Association
NOMAD Navy Oceanographic and Meteorological Device (world's first nuclear-powered weather station)
nombos nonmine bottom objects
nom cap nominal capital
nom com nom commercial (French—business name, trade name)
nom. con. nomen conservandum (Latin—generic or specific name to be preserved by special sanction)
nom dam nominal damages
nom de fam nom de famille (French—family name, surname)
nom de g nom de guerre (French—assumed name)—stage-name
nom de jf nom de jeune fille (French—maiden name)
nom de p nom de plume (French—pen name, pseudonym)
nom de t nom de théâtre (French—stage name)
nom. dub. nomen dubium (Latin—doubtful name)
nomen nomenclature
nomin nominative
nom. nov. nomen novum (Latin—new name)

nom. nud. *nomen nudem* (Latin—naked name); name for an animal or plant lacking further description
NOMSS National Operational Meteorological Satellite System
nom std nominal standard
NOMTF Naval Ordnance Missile Test Facilities
Non Nonoc
NON National Organization of Non-Parenthood
non acpc non-acceptance
non arrl non-arrival
non-can non-cancellable
nonce-wd nonce-word
non-coll non-collegiate
Non-Com noncommissioned officer
noncom(s) nonconformist(s)
non-com(s) non-commissioned officer(s)
noncon(s) nonconformist(s)
non-contigs non-contiguous states, Alaska and Hawaii
non cul. *non culpabilis* (Latin—not culpable, not guilty)
non-cum non-cumulative
nondely non-delivery
none no one; not one
None Nonesuch
non est *non est inventus* (Latin—he was not found; it is wanting)
non flam non-flammable
non-flam non-flammable film (slow-burning acetate-base film)
N/ONI Navy/Office of Naval Intelligence
non negl non-negotiable
non obs. *non obstante* (Latin—notwithstanding)
non op non-operational
n-on-p negative on positive
non-par non-participating
non perf non-perforated
nonporno not pornographic
non pos. *non possumus* (Latin—we cannot)
non pros. *non prosequitur* (Latin—does not prosecute)
non pyt non-payment
N/ONR Navy/Office of Naval Research
non-rem non-rapid eye movements
non repetat. *non repetatur* (Latin—do not repeat)
non res non-resident
non-res nonresident
non rtnl non-returnable
NONSAP Nonlinear Structural Analysis Program
non seq. *non sequitor* (Latin—

it does not follow)
non-sked non-scheduled
nonstand nonstandard
non std nonstandard
nontax nontaxable
non-U not upper class
nonum national number
NOO Navy Oceanographic Office (formerly Hydrographic Office, USN)
NOOA New Orleans Opera Association
noodle-noodle-noodle-noodle tremolo passages played by the strings, called noodling by many musicians
no op no opinion
no op (NO OP) no operation (data processing)
Noor-Brab Noord-Brabant (Dutch province)
Noor-Hol *Noordholland* (Dutch—North Holland)—province including Amsterdam, Haarlem
nop navigating operating procedure; normal operating procedure; not on production; not open (to the) public; not otherwise provided; not our publication
NOP National Oceanographic Program; National Opinion Poll; Naval Oceanographic Program; North Oscura Peak
NOPA National Office Products Association; National Organization of Police Associations
no par. no paragraph (matter runs on)
NOPE New Orleans Port of Embarkation
NOPEC non-members of OPEC (Angola, China, Egypt, Malaysia, Mexico, Oman)
N O Phil New Orleans Philharmonic
NOPHN National Organization for Public Health Nursing
NOPL New Orleans Public Library
nopn normally open
NOPO New Orleans Philharmonic Orchestra
NOPS New Orleans Public Service
NOPWC National Old People's Welfare Council
NOQUIS Nucleonic Oil Quantity Indication System
nor not otherwise rated
nor. normal; not or
nor' norther (Middle English

contraction); north
nør *nørre* (Danish—north)
Nor Norma (constellation); Norway; Norwegian
Nor *Norge* (Norwegian—Norway); *Norr* (Swedish—north)
NoR Notice of Readiness
NOR North Central Airlines; NOT OR (data-processing logic-operator equivalent)
NORAD North American Air Defense
NORAID Northern Aid (to IRA and other groups in Northern Ireland); Norwegian Agency for International Development
Nor Ant Norwegian Antarctica (Bouvet Island, Peter I Island, Queen Maud Land)
Nor Arc Norwegian Arctic (Bear, Edge, and Hope islands in Barents Sea, Jan Mayen Island in Norwegian Sea, Svalbard or Spitsbergen in Arctic Ocean)
nor'ard northward
NORASDEFLANT North American Antisubmarine Defense Force, Atlantic
Nor Atl North Atlantic
norc national ordnance research computer
NORC National Opinion Research Center (University of Chicago); Naval Ordnance Research Computer; Nippon Ocean Racing Club
Nor-Cor Northland-Coromandel (NZ)
Nor Cur Norwegian Current
nor'd northward
NORD Naval Ordnance
NORDEK Nordic Economic Community (Denmark, Finland, Norway, Sweden)
NORDEL Nordic Electricity Union
Norden (Scandinavian—the North)—Denmark, Finland, Iceland, Norway, Sweden
Nordica Lillian Nordica, Lillian Norton (1857–1914)
Nordic Council Scandinavian union including Denmark, Finland, Iceland, Norway, and Sweden
Nordic Countries Denmark, Finland, Iceland, Norway, Sweden
NORDITA Nordic Institute for Theoretical Atomic Physics
NORDSFORSK *Nordiska Samarbetsorganisationen för Teknisk-Naturventenskaplig Forening* (Nordic Council for

Applied Research)
nor'easter northeaster (storm from the northeast)
noref no reference
Norelco North American Philips Company
Norf Norfolk
Norf S Norfolk Southern, Norfolk & Western, Southern Railway (merger)
NORGRAIN North American Grain Charter
Norics Noric Alps in southern Austria
NORK New Orleans Rhythm Kings
N'Orleans New Orleans
Norlina North Carolina
norm not operationally ready (pending) maintenance
norm. normal; normalize; normalizing; not operationally ready (because of) maintenance; nuclear operational readiness maneuvers
Norm Norman
NORM National Optimism Revival Movement
Normands Norman Islands (Channel Islands)
NORML National Organization for the Reform of Marijuana Laws; National Organization for the Reinforcement of Marijuana Laws; National Organization for the Repeal of Marijuana Laws
Noroil Norwegian Oil
NORONTAIR Northern Ontario Airways
Nor Pac Northern Pacific
Nor Pol Norsk Polarinstitutt (Norwegian—Norwegian Polar Institute)
norrd no reply received
nors not operationally ready, supplies (supply)
Norsker(s) Norwegian sailor(s)
Norskie Norwegian-American
NORTEP Northern Teacher Education Program
north. northerly; northern
North Africa Africa north of the Tropic of Cancer; Algeria, Egypt, Libya, Morocco, Tunisia
NORTHAG North European Army Group
North America islands and lands extending from Canada to Colombia (Canada, Central America, Greenland, Mexico, the United States, the West Indies)
Northants Northamptonshire
North BH North Broken Hill

North Borneo Sabah and Sarawak
North Cascades North Cascades National Park in Washington
Northcliffe Viscount Northcliffe (Alfred Charles William Harmsworth)
Northeast Middle Atlantic and New England States
Northeast Corridor megalopolis extending from Boston to Washington, including Providence, New Haven, New York, Newark, Trenton, Philadelphia, Wilmington, Baltimore
Northeast Region Middle Atlantic and New England states
Northern Bear political cartoonist's symbol for Russia or the Soviet Union
Northern Institute Northern Region Correction Institute at Fairbanks, Alaska
Northern Ireland Ulster (six northern counties of Ireland)
northern lights aurora borealis
Northerns Burlington, Great Northern, and Northern Pacific railroads
Northern States northern United States in the Federal Union during the Civil War —The North
Northern Territories Kuril Islands seized from Japan by the USSR during World War II
North Jersey Coast Atlantic City to the Atlantic Highlands
Northld Northumberland
North Pole 90 degrees North latitude; zero degrees longitude; northernmost point on the globe; discovered by American explorers Frederick A Cook and Robert E Peary in 1909
Northum Northumberland
Northumb Northumberland; Northumbrian
Northwest northwestern United States (Washington, Oregon, Idaho, Montana, Wyoming)
NORTLANT North Atlantic
Norton WW Norton & Co
Nortown WW Norton
Nortraship Norwegian Trade and Shipping Mission
Norvic. Norvicensis (Latin—of Norwich)
norw norwegisch (German—

Norwegian)
Norw Norwegian
Norway Kingdom of Norway (northernmost Scandinavian country), *Kongeriket Norge*
NORWEB North Western Electricity Board
Norwegian Arctic Norway north of the Arctic Circle; Svalbard (Spitsbergen) and surrounding islands
NORWESTLANT Northwest Atlantic (project)
nos net oil sand; night operation sight; not on shelf; not otherwise specified; numbers
NOs New Orleans (British maritime abbreviation)
NOS National Ocean Survey; NATO Office of Security; Network Operating System; New Orleans; Night Observation Sight; Night Operation System
N OS New Orleans
NOS Nederlandse Omroep Stichting (Dutch—Netherlands Broadcasting Foundation)
NOSA National Occupational Safety Association
NOSC Naval Ocean Systems Center (USN); Naval Ordnance Systems Command (USN)
NOSCAF New Orleans Sickle Cell Anemia Foundation
NOSE Neighbors Opposing Smelly Emissions
NOSG Naval Operations Support Group
nosh no show
NOSIE Nurses Observation Scale for Inpatient Evaluation
no sig no signature
nosigchng no significant change
nosmo no smoking
no smoke/drugs no smoking or drugs (allowed)
Nosodak North Dakota + South Dakota—the Dakotas
NOSOPEX Northern Sumatra Offshore Petroleum Exploration
NOSSOLANT Naval Ordnance System Support Atlantic
NOSSOPAC Naval Ordnance System Support Pacific
NOSTA National Ocean Science and Technology Agency
nosub not subject to load
not. nucleus opticus tegmenti
Not Notary
nota none of the above (candi-

dates)

notal not to, nor needed by, all addressees

NOTAM Notice to Airmen

NOTB National Ophthalmic Treatment Board

notg nothing

notif notification

no-till no-tillage

noto numbering tool

noto (Latin prefix—back)—notochord

Notogaea landmass including Australasia

notox non toxic; not to exceed

NOTP *New Orleans Times-Picayune*

notr no traffic rights

Notre Dames Notre Dame Mountains of Québec

not's non-classical organizational theories

NOTS Naval Ordinance Test Station

Not(t) Nottingham

Nottm Nottingham

Notts Nottinghamshire

notwg notwithstanding

NOU Noumea, New Caledonia (airport)

'nough enough

Nou Heb Nouvelles Hébrides (French—New Hebrides)

nouv nouvelle (French—new)

nov novels; novelist; novels

nov noviembre (Spanish—November)

nov. novum (Latin—new)

Nov November

Nov Nova (Bulgarian, Italian, Portuguese, Serbo-Croatian—new); *Novaya* (Russian—new); *Novo* (Portuguese or Russian—new); *Novy* (Czechoslovakian—new)

NOVA (Latin—new) Public Broadcasting System series on scientific discoveries; National Organization of Victim Assistance; Network of Volunteer Assistance

Nova Roma (Latin—New Rome)—Constantinople, Istanbul

Novdec November and December

nov^e noviembre (Spanish—November)

November letter N radio code

nov. n. novum nomen (Latin—new name)

Novo Novosibirsk

NOVS National Office of Vital Statistics

nov. sp. novum species (Latin—new species)

Nov T Novum Testamentum (Latin—New Testament)

Novy(s) Nova Scotian(s)

now. (**NOW**) negotiable orders of withdrawal (banking accounts)

NoW News of the World

NOW National Organization for Women; Negotiable Order of Withdrawal (interest-earning checking account); New Opportunities for Women

NOWAPA North American Water and Power Alliance

NOWC National Association of Women's Clubs

NOWs Negotiable (deposits) Order of Withdrawals

NOx nitrous oxide (smog component)

noy not out yet; (unit of noisiness)

Noy Noybr (Russian—November)

noydb none of your damn business

noz nozzle

np napalm (incendiary gasoline mixture); national pipe; neap; neap range; near point; net proceeds; neuropsychiatric; neuropsychiatry; new paragraph; new pattern; new police; nickel-plated; nitroproof; no paging; no payment; no place; no place of publication; no protest; nonparticipating; nonpropelled; normal pressure; nose plug; not paginated; noun phrase; nursing procedure

np (NP) no parking; note payable

n/p net proceeds; new pence

n.p. nedsat pris (Dano-Norwegian—reduced price)

Np neap; neap range; neap tide; neper; neptunium (symbol)

N_p neper

NP Narragansett Pier; National Park; National Pipe; National(ist) Party; Naval Prison; Neighborhood Professional; New Providence, Bahama Islands; Newport, Rhode Island; no parking; North Pacific; North Park; Northern Pacific (railroad); Northern Province; not published; Notary Public; Nurse Practioner

N-P Non-Partisan

N/P nitrogen phosphorus ratio

NP Nasionale Partij (Afrikaans—National Party)

NPA National Packaging Association; National Paperboard Association; National Parenthood Association; National Parking Association; National Parks Association; National Parks Authority (NZ); National Particleboard Association; National Personnel Associates; National Personnel Authority; National Pet Association; National Petroleum Association; National Pharmaceutical Association; National Pigeon Association; National Pilots Association; National Pipeline Authority; National Planning Association; National Police Agency (Japan); National Preserves Association; National Proctologic Association; National Production Authority; Naval Procurement Account; Navy Postal Affairs; New People's Army (Filipino insurgency); Newspaper Publishers Association; Nigerian Ports Authority; Nurse Practice Act

NPABC National Public Affairs Broadcast Center

N Pac North Pacific

NPAC National Program for Acquisitions and Cataloging (Library of Congress)

N Pac Cur North Pacific Current

NPACI National Production Advisory Council on Industry

NPACT National Public Affairs Center for Television

N-panel panel of nuclear experts

NPAP National Psychological Association for Psychoanalysis

npat net profit after tax(es)

NPB National Park Board; National Parole Board (Canada); National Productivity Board (U.S.); North Pacific Bank

NPBA National Paper Box Association; National Pig Breeders' Association

NPBC National Programming Black Consortium

NPBI National Pretzel Bakers Institute

npc near point of convergence; New Process Company's trademark

NPC National Patent Council;

National Peach Council; National Peanut Council; National Peoples Congress; National Periodicals Center; National Personnel Consultants; National Petroleum Council; National Pharmaceutical Council; National Philatelic Collection; National Police Computer; National Ports Council; National Potato Council; National Power Company; National Press Club; National Productivity Center; Nauruan Phosphate Commission; Naval Photographic Center; New Peoples Center; Nigerian Population Commission; Nippon Petro-Chemicals

NPCA National Parks and Conservation Association; National Pest Control Association

NPCC National Projects Construction Corporation; Nebraska Penal and Correctional Complex

NPCFB North Pacific Coast Freight Bureau

NPCI National Potato Chip Institute

NPCP National Press Club of the Philippines

npcr no periodic calibration required

np-ct naval personnel conversion tables

npd no payroll division; north polar distance

np or d no place or date (of publication)

N-P d Neimann-Pick's disease

NPD North Polar Distance

NPD *Nationaldemokratische Partei Deutschlands* (German—National Democratic Party of Germany)

NPDC National Patent Development Corporation

NPDEA National Professional Driver Education Association

NPDES National Pollution Discharge Elimination Scheme

NPDN Nordic Public Data Network (Denmark, Finland, Iceland, Norway, and Sweden)

NPDO Non-Profit Distributing Organization

np or dp no place or date of publication

NPE Navy Preliminary Evaluation

N-peace nuclear peace

npef new product evaluation form

NP en G *Nederlandse Postcheque en Girondienst* (Dutch—Netherlands Postal Check and Transfer Service)

NPEP National Public Expenditure Plan(ning)

NPES National Printing Equipment & Supply Association

npf newsprint pulp flat; no private facilities; not provided for

NPF National Park Foundation; National Parkinson Foundation; National Piano Foundation; National Poetry Foundation; National Provident Fund(ing); National Psoriasis Foundation; National Pugilistic Federation; Newspaper Press Fund

NPFA National Playing Fields Association

NPFC Naval Publications and Forms Center; Northwest Pacific Fisheries Center

NPFFA National Police and Fire Fighters Association

NPFI National Plant Food Institute

npfid nitrogen-phosphorus flame-ionization detector

NPFSC North Pacific Fur Seal Commission

NPFT Neurotic Personality Factor Test

NPG National Portrait Gallery; NATO Planning Group

NPGA Nebraska Personnel and Guidance Association; Nevada Personnel and Guidance Association

NPGS Naval Postgraduate School; Net Profit Generator System

nph no profit here

n ph nuclear physics

npH neutral protamine Hegedorn (isoophane insulin)

NPHIS Nested Phrase Indexing System

NPHQ National Park (Ranger) Headquarters

npi no previous information

NPI National Paralegal Institute; National Penitentiary Institute; National Population Inquiry; National Productivity Institute; Neuro-Psychiatric Institute; Nippon Pulp Industry

NPIA Norfolk Port and Industrial Authority

NPIC Naval Photographic Interpretation Center

NPICPS National Policy Institute of the Center for Political Studies

NPIPF Newspaper and Printing Industries Printing Fund

NPIS National Physics Information System

N-P-K Nitrogen-Phosphate-Potash (fertilizer)

npl new processor line; new program language; nipple; no personal liability; noise-pollution level

n pl plural form of a noun

NPL Nashville Public Library; National Physical Laboratory; Newark Public Library; Norfolk Public Library

N-plant(s) nuclear plant(s); nuclear-power plant(s)

NPLGS Night Plane Guard Station

NPLO NATO Production and Logistics Organization

nplu not people like us

npm number of points in the point-matching method

NPMAA National Piano Manufacturers Association of America

npn (NPN) nonprotein nitrogen

n-p-n negative-positive-negative

NPN negative positive negative

npna no protest for nonacceptance

np/nd not published/no date (given)

N & PNWR Ninepipe and Pablo National Wildlife Refuge (Montana)

npo nothing by mouth

n.p.o. *nil per os* (Latin—nothing by mouth)—sometimes written *ne per oris*

NPO National Philharmonic Orchestra (Manila); National Program Office (for nuclear waste terminal storage); Navy Post Office; Navy Purchasing Office(r); New Philharmonia Orchestra (London); Non-Profit Organization

NPOAA National Police Officers Association of America

NPOEV Nuclear-Powered Ocean Engineering Vehicle (miniature submarine)

N-pollution nuclear pollution

NPO–NIA Nonprofit Organizations National Insurance Alliance

NP & OSR Naval Petroleum

and Oil Shale Reserve
N-power nuclear power
N-power plant(s) nuclear-power plant(s)
npp no passed proof
NPP National Potato Panel; National Prison Project (ACLU); Naval Propellant Plant; New Progressive Party (Puerto Rico); Nuclear Power Plant
NPP (ACLU) National Prison Project (American Civil Liberties Union)
NPPA National Press Photographers Association
NPPAJ *National Probation and Parole Association Journal*
nppd nitrophenylpentadiene aldehyde (spy dust)
NPPF National Planned Parenthood Federation; National Press Photographers Foundation
NPPO Navy Publications and Printing Office
NPPR Nationalist Party of Puerto Rico
NPPS National Plants Preservation Society; Navy Publication Printing Service
N-P Pubns National Press Publications
NPQ *Naviera de Productos Químicos* (Spanish—Chemical Products Shipping Line)
npr night press rate
n/p/r noise/power/ratio
Npr Napier, NZ
NPR National Public Radio; Naval Petroleum Reserves; Navieras de Puerto Rico; Nickel Plate Road (railroad)
NPRA National Parks and Recreation Association; National Parks and Reserves Authority; National Petroleum Refiners Association; Naval Personnel Research Activity; Newspaper Personnel Relations Association
NPRAC National Public Radio Association of California
NPRC National Personnel Records Center; Newspaper Production and Research Center
nprd nuclear plant reliability data
NPRDS Nuclear Plant Reliability Data System
N Pres National Preserve
NPRL National Physical Research Laboratory
NPRN National Public Radio

News
NPRO Navy Plant Representative Office(r)
NPROA National Police Reserve Officers Association
N-project nuclear-power project
N-proliferation nuclear proliferation
N-propulsion nuclear propulsion
NPR & OSR Naval Petroleum Reserves and Oil Shale Reserves
npr's nuclear-power reactors
nps non-professional staff; normal pipe size; no prior service
nps (NPS) nuclear-powered ship(ping)
NPs Notaries Public; Nurse Practitioners
NPS Narcotics Preventive Service; National Park Service; National Portrait Society; Naval Postgraduate School; Nuclear-Powered Ship(ping)
NPSB *National Prisoner Statistics Bulletin*
NPSC National Public Service Commission
npsh net positive suction head
npt normal pressure and temperature
npt (NPT) nocturnal penile tumescence
Npt Navy pointer tracker; Newport
NPT national (taper) pipe thread; Non-Proliferation Treaty
NPTA National Paper Trade Association; National Passenger Traffic Association; National Piano Travelers Association; Nevada Parent-Teacher Association
NPTC National Postal and Travelers Censorship
NPTRL Naval Personnel Training Research Laboratory
npu not-passed urine
n.p.u. *ne plus ultra* [Latin—nothing beyond (it); the summit; the ultimate]
NPU National People's Union; National Pharmaceutical Union; National Police Union; National Postal Union
npv net present value; no par value
npv (NPV) nuclear polyhedrosis virus
NPVLA National Paint, Varnish, and Lacquer Associa-

tion
npw new-pool wildcat (oil well)
NPW *Naturpark Pfalszer Wald* (German—Falls Forest Nature Park)—in western Germany near France
NPWC Navy Public Works Center
NPWS National Parks and Wildlife Service (Australia)
NPWU National Production Workers of America
NPX National Phoenix Industries (stock-exchange symbol)
NPY National Productivity Year
n-p-z negative-positive-zero
nq notes and queries
NQ North Queensland
N & Q *Notes & Queries*
nqa net quick assets
NQA Nuclear Quality Assurance
NQAPO Nuclear Quality Assurance Program Office
NQD Notice of Quality Discrepancy
nqokd not quite our kind, dear
nqos not quite our sort
nqot not quite our type
nqr nuclear quadruple resonance
NQX North Queensland Express
nr narrow resonance; natural rubber; near; net register; no risk; noise reduction; non-reactive (relay); norm-referenced; number
n-r no(n) return; non-resident
n/r no record; non-recoverable; not reported; not required; not responsible (for)
nr *non rogne* (French—untrimmed); *nummer* (Polish—issue, number); *nummer* (Dano-Norwegian or Swedish—number)
n.r. *non repetatur* (Latin—not to be repeated)
nR *neue Reihe* (German—new series)
Nr *Nummer* (German—number)
NR National Register; National River; National Riverway; *Norsk Rikskringkasting* (Norwegian Broadcasting); North Riding
N/R Notice of Readiness
NR *National Review*
nra never refuse anything; no repair action; non-representational artist

nra nuestra (Spanish—our)
NRA National Racing Authority; National Reclamation Association; National Recovery Act; National Recovery Administration; National Recreation Association; National Recreation(al) Area; National Reform Association; National Rehabilitation Association; National Research Associates; National Resistance Army (Uganda); National Restaurant Association; National Rifle Association (of America); Naval Reserve Association
NRAA National Rifle Association of America
NRAC National Research Advisory Council; National Resources Analysis Center; National Rural Advisory Council
NRACCO Navy Regional Air Cargo Control Office(r)
nrad no risk after discharge
NRAF Navy Recruiting Aids Facility
nral no risk after landing
NRAO National Radio Astronomy Observatory
nras no risk after shipment
NRAS Navy Readiness Analysis Section; Navy Readiness Analysis System
Nra Sra Nuestra Señora (Spanish—Our Lady)
Nrb Nordby
NRB National Religious Broadcasters; National Research Bureau; National Roads Board; National Rubber Bureau
NRB Narodna Republika Blgariya (Bulgarian—Bulgarian Peoples' Republic)
Nrbi Nairobi
NRBs National Religious Broadcasters
nrc noise-reduction circuitry; not recommended for children
NRC Nacorazi Railroad Company; National Racquetball Club; National Rainbow Coalition; National Referral Center (Library of Congress); National Reporting Center; National Republican Club; National Research Corporation; National Research Council; National Resources Committee; National Resources Council; National Roofing Contractors; Na-

tional Rural Center; Naval Retraining Command; Neighborhood Recovery Center (for alcoholism); Neighborhood Reinvestment Corporation; Netherlands Red Cross; Newport Research Corporation; Nuclear Regulatory Commission; Nuclear Research Council
NRC Nieuwe Rotterdamse Courant (Dutch—New Rotterdam Courant)
NRCA National Resources Council of America; National Retail Credit Association; National Roofing Contractors Association
NRCC National Republican Campaign Committee; National Republican Congressional Committee; National Research Council of Canada
NRCD National Reprographic Center for Documentation
nrcf not reconfirmed
NRCL National Research Council Library; Natural Resources Conservation League
NRC-NAS National Research Council—National Academy of Sciences
nrcp non-reinforced concrete pipe; non-residential conditional purchase
NRCPC National Rural Crime Prevention Center (Ohio State University)
NRCR Northern Railway of Costa Rica *(Ferrocarril del Norte de Costa Rica)*
NR Crit Nuclear Rocket—Critical
nrcy not received yet
NRD National Range Division; National Register of Designers; Navy Recruiting Depot; Navy Recruiting District
NRDA National Research and Development Authority (Israel); Nevada Research and Development Area
NRDB Natural Rubber Development Board
NRDC National Research Development Corporation; National Resources Defense Agency; National Resources Development Council; National Running Data Center; Natural Resources Defense Council
NRDL Naval Radiological Defense Laboratory
nrdo naval radio; Navy radio
NRDO National Research and

Development Organization
NRDS Nuclear Rocket Development Station
N-reactor(s) nuclear reactor(s)
NREB Navy Reserve Evaluation Board
NREC National Resource Evaluation Center
NRECA National Rural Electric Cooperative Association
nrem (NREM) non-rapid eye movement
nrems (NREMS) non-rapid eye-movement sleep
nrem sleep non-rapid eye-movement (spindle) sleep
NRF National Relief Fund; Naval Reactor Facility; Naval Repair Facility
NRF Nouvelle Revue Française
NRFA National Retail Furniture Association
NRFC Navy Regional Finance Center
NRFL National Rugby Football League
nrg energy
NRG National Resurrection Group (Athenian rightist terrorists); Naval Research Group
NRGA National Rice Growers Association
NRh Northern Rhodesia
NRHA National Retail Hardware Association; National Roller Hockey Association
NRHC National Rural Housing Coalition
NRHS National Railway Historical Society
NRI National Radio Institute; Nomura Research Institute (Japan)
NRIAD National Register of Industrial Art Designers
NRIC National Registration and Identity Card
NRIMS National Research Institute for Mathematical Sciences
NRIS Natural Resource Information System
Nrk Newark
NRK Nikolai Rimsky-Korsakov
NRK Norsky Rikskringkasting (Norwegian—Royal Norwegian Broadcasting)
nrl normal rated load
NRL National Radiation Laboratory; National Reference Library (of Science and Invention); National Registry for Librarians; National Re-

search Library; Naval Research Laboratory; Nelson Research Library
NRLC National Right to Life Committee
NRLCA National Rural Letter Carriers' Association
NRLDA National Retail Lumber Dealers Association
NRLM National Research Laboratory of Meteorology
NRLSI National Reference Library of Science and Invention
nrm natural remanent magnetism; next to reading matter; non-routine maintenance; normal rabbit serum
NRM Naval Reserve Medal; Northern Roller Mills
NRMA National Reloading Manufacturers Association; National Retail Merchants Association; National Roads and Motorists Association
NRMC National Records Management Council; Naval Records Management Center; Naval Regional Medical Center
NRMCA National Ready-Mixed Concrete Association
NRMG Nederlands Reken-machine Genootschap (Dutch—Netherlands Computer Society)
nrml normal
NRMM National Register of Microform Masters
NRN National Radio Network
nro nuestro (Spanish—our, m.)
NRO Narcotic Rehabilitation Office(r); National Reconnaissance Office; National Registration Office(r); Naval Research Objectives
NROO Naval Reactors Operations Office
NROTC Naval Reserve Officers Training Corps
nrp net rating points; no replacement part; normal rated power
NRP National Religious Party (Israel); National Republican Party; National Research Poll
NRPA National Recreation and Park Association
NR & PA National Recreation and Park Association
NRPB National Research Planning Board
NRPC National Railroad Passenger Corporation
NRPRA Natural Rubber Producers' Research Association

NRR Northern Rhodesia Regiment; Nuclear Reactor Research
NRRC National Restitution Resource Center
NRRE Netherlands Radar Research Establishment
NRRL Norsk Radio Relae Liga (Norwegian—Norwegian Radio Relay League)
nrs normal rabbit serum; numbers
N rs Nepalese rupee(s)
NRS National Rose Society; National Runaway Switchboard; Naval Recruiting Service; Navy Records Society; Navy Relief Society; New Reading System; Noise-Reduction System
NRS Naturhistoriska Riksmuseet Stockholm (Swedish—Royal Natural History Museum)
NRSA National Rural Studies Association
NRSCC National Registry System for Chemical Compounds
NRSFPS National Reporting System for Family Planning Services
nrt net register(ed) tonnage (tons); normal rated thrust; norm-referenced testing
NRT Tokyo-Narita (airport)
NRTA National Retired Teachers Association
NRTC Naval Recruit Training Center; Naval Reserve Training Center; Northrup Research and Technology Center
NRTI National Rehabilitation Training Institute
nrtor no risk till on rail
nrts not reparable this station
NRTS National Reactor Testing Station
nrtwb no risk until waterborne
nru nuclear reactor—universal
Nru Nauru
NR-U Nederlandsche Radio-Unie (Dutch—Netherlands Union of Radio Broadcasters)
nrv net realisable value; non-return value
NRVC National Railway Utilization Corporation
Nrvkg Nervenkrieg (German—nerve warfare)
NRVN Navy of the Republic of Viet Nam
Nrw Norwegian
NRWC National Right to

Work Committee
NRWLDF National Right to Work Legal Defense Foundation
nrwt non-resident withholding tax
nrx nuclear reactor, experimental
Nry Newry
NRYC New Rochelle Yacht Club
NR Yorks North Riding, Yorkshire
NRZ National Railways of Zimbabwe
nrz c (NRZ C) non-return-to-zero change (data processing)
nrzi non-return-to-zero IBM
nrz m (NRZ M) non-return-to-zero mark recording (data processing)
ns nanosecond; near side; neuropsychiatric; new series; nickel steel; no sparring; noise suppressor; nonstandard; nonstop; not specified
ns (NS) neurosurgery; note series (synonymous with original or prime)
n/s neutrons per second; no service; not scheduled; not stocked; not sufficient
ns nostro (Italian—our or ours); *nouvelle serie* (French—new series)
nS neue Serie (German—new series)
Ns nimbostratus; Nunes; Nuñez
NS National Seashore; National Socialist; National Society; National Special (screw threads); National Superannuation; Naval Shipyard; Naval Station; New Style; Nippon(ese) Standard; Norfolk Southern (Norfolk & Western and Southern railroads merged); North Sea; Nova Scotia; Nuclear Ship; Nuclear Submarine; Numismatic Society
N.S. New Style; Norfolk Southern (railroad)
NS Nachschrift (German—postscript); *Nasjonal Samling* (Norwegian—National Unification)—fascist collaborationists headed by Vidkun Quisling during World War II (*see* quis); *Nederlandsche Spoorwegen* (Dutch—Netherlands Railway); *Notre Seigneur* (French—Our Lord); *Nuestro Señor* (Spanish—Our Lord)

N.S. *Nuestro Señor* (Spanish— Our Lord)

nsa (NSA) nonenyl succinic acid

NSA National Sawmilling Association; National Safety Association; National Secretaries Association; National Security Adviser; National Security Agency; National Service Acts; National Shellfisheries Association; National Sheriff's Association; National Shipping Authority; National Showmen's Association; National Silo Association; National Skating Association; National Ski Association; National Slag Association; National Slate Association; National Society of Auctioneers; National Speakers Association; National Standards Association; National Stereoscopic Association; National Students Association; Naval Stock Account; Naval Supply Account; Neurological Society of America; Norwegian Seamen's Association; Nuclear Science Association; Nursery School Association; Nurses Supply Association

NSA *Nuclear Science Abstracts*

NSAA Norwegian Singers' Association of America

NSAC National Society for Autistic Children; Nova Scotia Agricultural College

NSACG Nuclear Strike Alternate Control Group

NSACS National Society for the Abolition of Cruel Sports

NSA/CSS National Security Agency/Central Security Service

NSAD National Society of Art Directors

NSADD National Society of Alcoholism and Drug Dependence

NSAE National Society of Art Education

NSAFA National Service Armed Forces Act

nsai (NSAI) non-steroidal anti-inflammatory

nsaids non-steroidal anti-inflammatory drugs

NSAM National Security Agency Memorandum; Naval School of Aviation Medicine

NSAS National Smoke Abatement Society

NSASAB National Security Agency Scientific Advisory Board

NSB National Science Board; Nippon Short-wave Broadcasting

NSB *Norges Statsbaner* (Norwegian—Norwegian State Railway)

NSBA National School Boards Association; National Sheep Breeders Association; National Small Business Association; National Sugar Brokers Association

NSBC National Student Book Club

NSBF National Scientific Balloon Facility

NSBISS NATO Security Bureau/Industrial Security Section

NSBIU Nova Scotia Board of Insurance Underwriters

NSBMA National Small Business Men's Association

nsc non-service connected

NSC National Safety Council; National Science Council; National Security Council; National Shippers Council; National Standards Commission; National Steel Corporation; NATO Steering Committee; NATO Supply Center; NATO Supply Classification; Naval School Command; Naval Supply Center; New Sessions Cases; New Solidarity Club(s); Newark State College; Nutrition Society of Canada

NSCA National Safety Council of Australia; National Society for Clean Air; National Strength Conditioning Association; Nova Scotia College of Art

NSCAR National Society of Children of the American Revolution

NSCBS National Society for the Conservation of Bighorn Sheep

NSCC National Society for Crippled Children

NSCCA National Society for Crippled Children and Adults

nscd nonservice-connected disability

NSCD National Society of Colonial Dames

NSCDRF National Sickle Cell Disease Research Foundation

NSCIA National Supervisory Council for Intruder Alarms

NSCID National Security Council Intelligence Directive

NSCLC National Senior Citizens Law Center

NSCM National Stamp Collecting Month

NSCR National Society for Cancer Relief

NSCs Network Switching Centers

NSCSP National Site Characterization and Selection Plan

NSCT North Staffordshire College of Technology

nsd no significant defect; no significant deviation; no significant difference; noise-suppression device; non-soapy detergent; normal spontaneous delivery

NSD Naval Stores Department; Naval Supply Depot; Naval Support Data

NSDA National Soft Drink Association

NSDAP *Nationalsozialistische Deutsche Arbeiterpartei* (German National Socialist [Nazi] Workers Party)

NSDB National Science and Development Board

NSDC National School Development Council; National Serials Data Center; National Space Development Center

NSDD National Security Decision Directive

nsdf naval standard distillate fuel

NSDF National Sex and Drug Forum

NSDMs National Security Decision Memorandums

NSDO National Seed Development Organisation

NSDP National Society of Dental Prosthetists

NSDS National Shut-in Day Society

NSE Nigerian Society of Engineers

NSEA Nebraska State Education Association

nsec nanosecond

n/sec neutrons per second

NSEC National Service Entertainments Council

NSEF National Student Educational Fund

NSEI Norwegian Society for Electronic Information

NSERC Natural Science and Engineering Research Council of Canada

NSERI National Solar Energy

Research Institute

NSES National Society of Electrotypers and Stereotypers

NSESG North Sea Environmental Study Group

nsf not sufficient funds

NSF National Sanitation Foundation; National Science Foundation; National Sex Forum; Naval Stock Fund; Navy Strike Fighter

NSF Norges Standardiserings Forbund (Norwegian—Norwegian Standards Institute)

NSFA National Science Faculty Association; Naval Support Force Antarctica (USN); New Settlers Federation of Australia

NSFD National Surface Freight Division (Institute of Freight Forwarders)

NSFGB National Ski Federation of Great Britain

NSFNET National Science Foundation Network

nsftd normal spontaneous fullterm delivery

nsg neurosecretory granules

NSG National Supply Group; NATO Standardization Group; Naval Security Group

NSGA National Sporting Goods Association

NSGC Naval Security Group Command

NSGD Near-Surface Geological Disposal

nsgn noise generator

NSGT Non-Self-Governing Territories; Non-Self-Governing Territory

nsh no stock on hand; not so hot

NSHA National Steeplechase and Hunt Association

NSHC North Sea Hydrographic Commission

NSHEB North of Scotland Hydro-Electric Board

N-ship(s) nuclear-powered ship(s)

nsi next sequential instruction; nonstandard item; nonstocked item; nuclear safety inspection; numeric signal insignia

NSI National Stock Exchange; Nuclear Safety Inspection

NSI Norsk Senter for Informatikk (Norwegian—Norwegian Information Center)

NSIA National Security Industrial Association

NSIBU Nova Scotia Board of Insurance Underwriters

NSIC National Small Industries Corporation; National Solar Information Center

NSID National Society of Interior Designers

n sing singular form of a noun

NSIO Nova Scotia Information Office

NSJC Nuestro Señor Jesucristo (Spanish—Our Lord Jesus Christ)

nsk not specified by kind

NSK Nihon Shimbun Kyokai (Japanese—Japan Newspapers and Publishers Association); Nippon Seiko KK (bearings)

NSKK Nito Shosen Kabushiki Kaisha (Japanese—Japanese steamship line)

nsl non-standard label; not stock listed

NSL National Science Library; National Socialist League (American Nazi Party); National Standards Laboratory; Navy Stock List; Northrop Space Laboratory; Numidian Support League

NSLA National Society of Literature and the Arts

NSLF National Socialist Liberation Front (American-Nazi student organization)

NSLI National Service Life Insurance

NSLL National Savings and Loan League

NSLS National Science Library System

nsm new smoking material (wood-substitute tobacco); noise source meter; number of similar (negative) matches

NSM National Savings Movement; National Security Medal; National Selected Morticians; Naval School of Music; Nevada State Museum; Nova Scotia Museum (Halifax)

ns/m² newton second per square meter

NSMA National Scale Men's Association

NSMC Naval Submarine Medical Center

NSMHC National Society for Mentally Handicapped Children

NSMI National Special Media Institutes

NSMM National Society of Metal Mechanics

NSMP National Society of Master Patternmakers; National Society of Mural Painters; Navy Support and Mobilization Plan

NSMPA National Screw Machine Products Association

NSMR National Society for Medical Research

NSMS National Sheet Music Society

NSMSES Naval Ship Missile Systems Engineering Station

NSN NATO Stock Number

NSNA National Student Nurses' Association

NSNC Nova Scotia Normal College

NSO Nashville Symphony Orchestra; National Symphony Orchestra; Naval Staff Officer; Navy Subsistence Office(r); Norfolk Symphony Orchestra; Northern Sinfonia Orchestra

NSOA National School Orchestra Association

NSOC Navy Satellite Operations Center

NSOEA National Stationery and Office Equipment Association

NSOSG North Sea Oceanographic Study Group

nsp non-standard part

n sp new species

NSP National Siting Plan; National Society of Professors; National Stuttering Project; Navy Standard Part; Nebraska State Patrol; Nebraska State Police; North Solomons Province; Northern States Power

NSPA National Scholastic Press Association; National Society of Public Accountants; National Soybean Processors Association; National Split Pea Association; National Standard Part Association; Naval Shore Patrol Administration

NSPB National Society for the Prevention of Blindness

NSPC National Security Planning Commission; National Society of Painters in Casein; Northern States Power Company

NSPCA National Society for the Prevention of Cruelty to Animals

NSPCC National Society for the Prevention of Cruelty to Children

NSPD Naval Shore Patrol Detachment

NSPE National Society of Professional Engineers

nspf not specifically provided for

NSPG National Security Planning Group

NSPI National Society for Performance and Instruction; National Society for Programmed Instruction; National Swimming Pool Institute

NSPLO NATO Sidewinder Production and Logistics Organization

NSPO Navy Special Projects Office; Nuclear Systems Project Office

NSPRA National School of Public Relations Association

NSPS National Sweet Pea Society; New-Source Performance Standards

NSPSE National Society of Painters, Sculptors, and Engravers

NSPWA National Society of Patriotic Women of America

nsq neuroticism scale questionnaire

nsr natural sinus rhythm; normal sinus rhythm

NSR National Scenic River-(way); National Scientific Register; National Security Regulation(s); Norfolk Southern Railway

NSRA National Shoe Retailers Association; National Shorthand Reporters Association; National Small-bore Rifle Association; National Street Rod Association; North-South Reconstruction advisors; Nuclear Safety Research Association

NSRB National Security Resources Board

NSRC Natural Science Research Council

NSRD National Standards Reference Data

NSRDC National Standards Reference Data System

NSRDF Naval Supply Research and Development Facility

NSRDL Naval Ship Research and Development Laboratory; Naval Supply Research and Development Facility

NSRDS National Standard Reference Data System

NSRF Nova Scotia Research Foundation

nsrp non-technical support real

property

NSRP National States Rights Party

nsrpie non-technical support real property installed equipment

nsrt near-surface reference temperature

nss (NSS) normal saline solution

NSS National Sample Survey(or)(s); National Sculpture Society; National Secular Society (British); National Serigraph Society; National Slovak Society; National Speleological Society; National Stockpile Site; New Shakespeare Society; Newburgh and South Shore (railroad); Nitrogen Supply System

NSSA National Sanitary Supply Association; National School Sailing Association; National Science Supervisors Association; National Skeet Shooting Association; National Sportscasters and Sportswriters Association

NSSAR National Society of the Sons of the American Revolution

NSSC National School Safety Center; National Society for the Study of Communication

NSSCC National Space Surveillance Control Center

NSS Co Northern Steam Ship Company (New Zealand)

NSSE National Society for the Study of Education; National Study of School Evaluation

NSSEA National School Supply and Equipment Association

NSSF National Shooting Sports Foundation; Navy Submarine Support Facility

NSSFC National Severe Storm Forecast Center; National Society of Student Film Critics

NSSFNS National Scholarship Service and Fund for Negro Students

NSSGA *Nicherin Shoshu Soka-Gakkai Academy* (international peace society)

NSSL National Severe Storms Laboratory

NSSMA National Spanish-Speaking Management Association

NSSN National Standard Shipping Note

NSSOPAC Naval Ordnance

System Support Pacific

NSSP National Severe Storms Project

NSSR New School for Social Research

nsss nuclear steam system supply

NSST Northwestern Syntax Screening Test

nsst(s) no-smoking seat(s)

NSSU National Sunday School Union

NSSWC National Severe Storm Warning Center

nst nonslip thread

NST National Security Team (U.S. National Security Affairs Adviser, Secretary of Defense, Secretary of State); Newfoundland Standard Time; Nigata Sogo Television

NST New Straits Times

NSTA National School Transportation Association; National Science Teachers Association

NSTAP National Strategic Targeting and Attack Policy

NSTC Nebraska State Teachers College; Nova Scotia Technical College

nstd nested

NSTF National Science Teachers Foundation; Near-Surface Test Facility

NSTI Norwalk State Technical Institute

NSTIC Naval Scientific and Technical Information

NSTL National Strategic Target Line

NSTP Nuffield Science Teaching Project

NS Tripos Natural Science Tripos

NSTS National Sea Training Schools

N-study nuclear study

nsu nitrogen supply unit; nonspecific urethritis

NSU Neckarsulmer Fahrzeugwerke (NSU Motorenwerke)

N-sub(s) nuclear-powered submarine(s)

NSUC North Staffordshire University College

N-super nuclear-powered supercarrier (naval vessel)

nsurg neurosurgeon; neurosurgery; neurosurgical

nsv nuclear service vessel

n/sv nonautomatic self-verification

NSV National Socialist Vanguard (neo-Nazi group, Or-

egon and Idaho)
NSVP National School Volunteer Program; National Student Volunteer Program
NSW New South Wales
NSWC Naval Surface Weapons Center (USN); New South Wales Centre
NSWG Naval Special Warfare Group; New South Wales Government
NSWGR New South Wales Government Railways
NSWGTB New South Wales Government Tourist Bureau
NSWHC New South Wales Health Commission
NSWIER New South Wales Institute for Educational Research
NSWIT New South Wales Institute of Technology
NSWMSB New South Wales Maritime Services Board
NSWNA New South Wales Nurses Association
NSWP New South Wales Police; Non-Soviet Warsaw Pact
NSWPAG New South Wales Prisoners Action Group
NSWPP National Socialist White People's Party (formerly American Nazi Party)
NSWPTC New South Wales Public Transport Commission
NSWR New South Wales Reports
NSWRL New South Wales Rugby League
NSWRU New South Wales Rugby Union
NSWTA New South Wales Transport Association
NSWTF New South Wales Teachers Federation
NSY New Scotland Yard
NSYF Natural Science for Youth Foundation
nt narrower term; net terms; nit (unit of luminous intensity); no trace; nontight; normal temperature; not titled
n't not
n/t net tonnage; new terms
n & t nose and throat
nt Northern Telecom
n.t. nel testo (Italian—in the text)
Nt nitron
NT National Theater; National Trust; New Territories (Hong Kong); New Testament; Northern Territory
NT National Times (of Austra-

lia); *Ny Testamente* (Dano-Norwegian—New Testament)
N.T. Novum Testamentum (Latin—New Testament)
nta net tangible assets; nitrilotriacetic (phosphate substitute for detergents); nuclear test aircraft (NTA)
NTA Narcotics Treatment Administration; National Tattoo Association; National Tax Association; National Technical Association; National Tourist Association; National Travel Association; National Trust of Australia; National Tuberculosis Association; New Territories Administration; Northern Textile Association; Northern Trade Association
NTAA National Travelers Aid Association
NTAC Nederlandse Touring en Auto Club (Dutch—Netherlands Touring and Auto Club)
NTAMS Northern Territory Aerial Medical Service
NTAs Nielsen Television Areas
NTATB Northwestern Truck Association and Tariff Bureau
ntavl not available
ntb non-tariff barrier(s); not to be
NTB National Theatre Board; National Tobacco Board
NTB Norsk Telegrambyra (Norwegian—Norwegian News Service)
ntba name(s) to be advised
NTBL Nuffield Talking Book Library (for the blind)
NtBuStnds National Bureau of Standards
ntc negative temperature coefficient
NTC National Teacher Corps; National Theatre Conference; National Training Council; National Travel Club; Naval Training Center; Nigerian Tobacco Company
NTCA National Training Council of Australia; National Tribal Chairmen's Association; National Tuberculosis and Chest Association
NTCC Nimbus Technical Control Center
NTCCL Northern Territory Council for Civil Liberties
ntd non-tight door; noted

NT$ New Taiwan dollar
NTD National Theater of the Deaf
NTDA National Tire Distributors Association; National Trade Development Association; National Tyre Distributors Association
NTDC Naval Tactical Data System; Naval Technical Data System; Naval Training Device Center
NTDPMA National Tool, Die, and Precision Machining Association
NTDS Naval Tactical Data System; Naval Technical Data System
NTDSC Nondestructive Testing Data Support Center
nte not to exceed
nte norte (Spanish—north)
NTE National Teacher Examination
N-tec nuclear technology
NTEC Naval Training Electronics Center(s); Naval Training Equipment Center
ntep not to exceed price
ntepq not to exceed price quoted
N-terror(ism)(ist) nuclear terrorism; nuclear terrorist
N-test nuclear test(ing)
NTETA National Traction Engine and Traction Association
NTEU National Treasury Employees Union
NTF Narcotics Task Force; National Turkey Federation; Navy Technological Forecast
NTFA National Track and Field Association
NTFP National Task Force on Prostitution
ntfy notify
ntg nontoxic goiter
NTGB North Thames Gas Board
NT Gk New Testament Greek
Nth Netherlands
NTH Norges Tekniske Hogskole (Norwegian—Norwegian Technical University, Trondheim)
Nthb Northumberland
Nth BHH North Broken Hill Holdings
Nth country next country of a series acquiring nuclear power
nthn northern
NTHP National Trust for Historic Preservation
N-threat nuclear threat

nti noise-transmission impairment

NTI National Theatre Institute; Nielsen Television Index (tv rating)

NTIA National Telecommunications and Information Administration

NTIAC Nondestructive Testing Information Analysis Center

NTIATA National Tax Institute of America Tax Association

NTIC National Training and Information Center (Chicago-based housing group); Nondestructive Testing Information Center (Battelle)

NTID National Technical Institute for the Deaf

NTIS National Technical Information Service (U.S. Department of Commerce); Nippon Technical Information Service

NTISBDF National Technical Information Service Bibliographic Data File

NTISearch National Technical Information (on-line computer) Search Service

NTK Nippon Toshokan Kyokai (Japanese—Japan Library Association)

ntl no time lost

NTL National Tennis League; National Training Laboratories

NTLC National Tax Limitation Committee

NTLF National Taxpayers Legal Fund

NTLS National Truck Leasing System

ntm net ton mile; non-tariff measure(s)

ntm (NTM) national technical means

Ntm Nottingham

NTMs National Technical Means (for verifying compliance with arms-control treaties)

NTMS National Topographic Map Series; Northern Territory Medical Service

NTNP Natchez Trace National Parkway

nto not taken out; not tried on

nto neto (Spanish—net)

NTO National Tenants Organization; National Theatre Organisation (South Africa); Naval Transport Office(r)

ntp normal temperature and pressure; no title page

NTP National Transportation Policy

NTPC National Technical Processing Center; Navy Training Publications Center

ntpl nut plate

ntr noise temperature ratio; non-typing reperforator

NTR National Tape Repository; Northern Test Range

Ntra Sra Nuestra Señora (Spanish—Our Lady)

NTRB Northern Territory Reserve Board (Australia)

NTRDA National Tuberculosis and Respiratory Disease Association

NTRL Naval Training Research Laboratory

NTRS National Therapeutic Recreation Society

N Tr Z North Tropical Zone

nts not to scale

nts (NTS) navigation(al) technology satellite

Nts Nantes

NTS National Technical School(s); National Traffic System; National Trust for Scotland; Naval Telecommunications System; Naval Torpedo Station; Naval Transport Service; Naval Transportation System; *Nederlandse Televisie Stichting* (Dutch—Netherlands Television Foundation); Nevada Test Site; Nitroglycerin Transdermal System

NTS Narodnyi Trudovoy Soyuz (Russian—National Labor Union)—anti-communist Russian exiles

NTSA National Traffic Safety Agency

NT & SA National Trust and Savings Association

NTSB National Transportation Safety Board

NTSC National Television Standards Committee; National Television System Committee; North Texas State College

NTSK Nordiska Tele-Satelit Kommitten (Nordic Committee for Satellite Telecommunications)

NTSSO Nevada Test Site Safety Office

NTSWG National Training School for Women and Girls

NTT New Technology Telescope; Nippon Telegraph and Telephone

NTTC National Tank Truck Carriers

NTTS Northern Territory Teaching Service

NTT & TTI National Truck Tank and Trailer Tank Institute

ntu nuts to you

NTU National Taiwan University; National Taxpayers Union; National Teachers Union; Navy Toxicology Unit

NTUC National Trades Union Congress

ntv nerve tissue vaccine

NTV Nippon Television

NTVLRO National TV License Records Office (Bristol)

ntwistdg notwithstanding

nt wt net weight

NTWU National Textile Workers Union

NTX Navy Teletype Exchange

nty not this year

N-type Jungian intuitive type

nt yt not yet

NTZ Neutral Zone; North Temperate Zone

Ntzrm Nutzraum (German—cubic capacity)

nu name unknown; new; nose up; nuclear; number unobtainable

Nu Nusselt number

NU *Naciones Unidas* (Spanish—United Nations); National Union; *Nations Unies* (French—United Nations); Naval Unit; Niagara University; Northeastern University; Northern Union; Northwestern University; Norwich University

NU Nahdatul Ulama (Indonesian—Muslim Scholars Party)

NUAAW National Union of Agricultural and Allied Workers

NUAK Nordisk Union for Alkoholfri Traffic (Scandinavian Union for Alcohol-free Traffic)

NUAUS National Union of Australian University Students

NUAW National Union of Agricultural Workers

NUB National Union of Blastfurnacemen; National Unity Board

NUBE National Union of Bank Employees

nube(s) nubile(s)

NUBSO National Union of Boot and Shoe Operatives

nuc not under command; nuclear; nucleated; nucleus

NUC National University Consortium; National Urban Coalition; Naval Undersea Center

NUC National Union Catalog

nu-car prep new-car preparation

Nuc.E. Nuclear Engineer

NUCEA National University Continuing Education Association

nucex nuclear exercise

nuc(l) nuclear; nucleus

Nucl Data Nuclear Data

nuclex nuclear loadout exercise

Nucl Fusion Nuclear Fusion

Nucl Instrum Nuclear Instruments

Nucl Instrum Methods Nuclear Instruments and Methods

Nucl Phys Nuclear Physics

Nucl Sci Eng Nuclear Science and Engineering

NUCMC National Union Catalog of Manuscript Collections

nuco numerical code; numerical coding

NUCO National Union of Co-operative Officials

NUCOM National Union Catalog of Monographs

nuc phy nuclear physics

nucpwrd nuclear powered

Nuc Reg Com Nuclear Regulatory Commission

NUCS National Union of Christian Schools

NUCSTAT Nuclear Operational Status Report

NUCUS National Union of Conservative and Unionist Associations

NUCW National Union of Commercial Workers

nud nudism; nudist

nud nudnick (Yiddish—nuisance, pest)

NUDBTW National Union of Dyers, Bleachers, and Textile Workers

NUDE National Union of Domestic Employees

NUDET Nuclear Detonation Report

NUDETS Nuclear Detonation, Detection, and Reporting System

nudies nude films; nude magazines; nude shows

NUE Nuremberg, Germany (airport)

NUEA National University Extension Association

nuevo lek monetary unit of Albania

'nuf enough

NUF National Urban Fellows

NUFCOR Nuclear Fuels Corporation

NUFCW National Union of Funeral and Cemetery Workers

Nuff Nuffield College, Oxford

NUFLAT National Union of Footwear, Leather, and Allied Trades

nufp not used for production

'nuf said enough said

NUFTIC Nuclear Fuels Technology Information Center

NUFTO National Union of Furniture Trade Operatives

nug nuggar (cargo boat used on the Nile)

NUGMW National Union of General and Municipal Workers

NUHS New Utrecht High School

NUHW National Union of Hosiery Workers

NUI National University of Ireland (Ollscoil na h-Eireann); Norwegian Underwater Institute

NUIC National Urban Indian Council

NUIW National Union of Insurance Workers

NUJ National Union of Journalists

nuke nuclear (slang)

nuke leak nuclear radioactive leak

nukes nuclear explosives; nuclear power plants

nul no upper limit

nul (NUL) null character (data processing)

NUL National Union for Liberation; National Urban League; Northwestern University Library

null null idle

Nulla Nullarbor Plain of southern South Australia and Western Australia

nullies nullifiers

NULWAT National Union of Leather Workers and Allied Trades

num number; numbered; numbering; numeracy (*see* numcy); numeral(s); numeration(s); numerical; numerolo-

gist; numerology

num numero(s) [Portuguese or Spanish—number(s)]

Num The Fourth Book of Moses, called Numbers

Num Numbers

NUM National Union of Mineworkers; New Ulster Movement

NUMA National Underwater and Marine Agency

NUMAS National Multifactor Assessment System

numb. numbered

Number 10 Number 10 Downing Street, official residence of the British Prime Minister

numcy numeracy (ability to count)

NUMEC Nuclear Materials and Equipment Corporation

numer numeral; numerative; numerical

numer order numerical order

numis numismatics

numism numismatic(s); numismatist

NUMMI New United Motor Manufacturing Inc

num order numerical order

NUMW National Union of Mine Workers

nuna not used on next assembly

NUOS Naval Underwater Ordnance Station

NUP Negro Universities Press

NUPAC Nuclear Packaging Inc

NUPBPW National Union of Printing, Bookbinding, and Paper Workers

NUPE National Union of Public Employees

NUPGE National Union of Provincial Government Employees

NUPI Norsk Utenrikspolitisk Institutt (Norwegian—Norwegian Foreign Policy Institute)

nuplex nuclear-powered complex (of manufacturers)

NUPSA National Union of Pharmaceutical Students of Australia

NUPT National Union of Press Telegraphists

NUPW National Union of Planning Workers; National Union of Plantation Workers

NUR National Union of Railwaymen

NURA National Union of Ratepayers' Associations

NURC National Union of Re-

tail Confectioners
NURDA National Urban and Regional Development Authority
NURE National Uranium Resource Evaluation (ERDA program)
NURT National Union of Retail Tobacconists
nus nuclear upper stage
NUS National Union of Seamen; National Union of Students; National University of Singapore; Nuclear Utility Service(s)
nusar nuclear sweep and radar
NUSAS National Union of South African Students
NUSC Naval Underwater Systems Center
NUSEC National Union of Societies for Equal Citizenship
NUSL Navy Underwater Sound Laboratory
NUSMWCHDE National Union of Sheet Metal Workers, Coppersmiths, Heating and Domestic Engineers
NUSRL Navy Underwater Sound Reference Laboratory
NUSS National Union of School Students; National Union of Small Shopkeepers
nusum numerical summary
Nu T Newcastle-upon-Tyne
N-u-T Newcastle-upon-Tyne
NUT National Union of Teachers (Great Britain)
NUTAT *Nordisk Union for Alkoholfri Trafic* (Nordic Union for Alcohol-free Traffic)
NUTAW National Union of Textile and Allied Workers
nu-tec nuclear detection (radiation monitoring device)
NUTGW National Union of Tailors and Garment Workers
NUTI Northwestern University Traffic Institute
NUTIS Numerical and Textual Information System
NUTN National Union of Trained Nurses
nutr nutrition
nuts. (NUTS) nuclear-utilization theories
NUU New University of Ulster
nuv *nuvaerende* (Dano-Norwegian—present)
NUVB National Union of Vehicle Builders
NUWA National Unemployed Workers' Association
NUWAX Nuclear Weapon Accident Exercise
NUWC Naval Undersea Warfare Center
NUWT National Union of Women Teachers
NUWW National Union of Women Workers
Nuyorican(s) New York Puerto Rican(s)
nv naked vision; needle valve; new version; number of variables
nv (NV) nuclear vitrification
n-v non-vaccinated; non-veteran; non-voting
n/v nuclear vessel
n & v nausea and vomiting
nv. novicius (Latin—new, recent)
NV Nevada; Nevada Operations Office; Nord-Viscount
NV *Naamloze Vernootschap* (Dutch—corporation); *Naviera Vascongada* (Basque Navigation Company); *Norske Veritas* (Norwegian—Norwegian Register of Shipping)
nva near visual acuity
nva *nueva* (Spanish—new)
NVA North Vietnamese Army
NVAiO *Norske Videnskaps-Akademi i Oslo* (Norwegian—Norwegian Academy of Science and Letters in Oslo)
NVATA National Vocational Agricultural Teachers Association
NVB National Volunteer Brigade
NVB *Nederlandse Vereniging van Bedrijfsarchivarissen* (Dutch—Netherlands Association of Business Archivists); *Nederlandse Vereniging van Bibliothekarissen* (Dutch—Netherlands Library Association)
NVBF *Nordisk Videnskabeligt Bibliotekarieforbund* (Nordic Federation of Research Librarians)
nvc non-verbal communication
NVC National Violence Commission
nvd night-viewing device; night-vision device
NVDA National Volunteer Defense Army
nvebw non-vacuum electron beam welding
NVF National Volunteer Force
NVFC National Vulcanized Fibre Company
nvg null voltage generator
NVGA National Vocabulary Guidance Association; National Vocational Guidance Association
nvh noise, vibration, hardness (problems)
NVI Nordic Volcanological Institute
Nvk Narvik
NVL Night Vision Laboratory
nvm non-volatile matter
NVMA National Veterinary Medical Association
NVNS *Naamloze Vernootschap Nederlandsche Spoorwagen* (Dutch—Netherlands Railway Corporation)
nvo non-vessel operating
NVO Nevada Operations Office; Northern Variety Orchestra
nvocc (NVOCC) non-vessel operating common carrier
NVOCC New Version Ocean Container Control; New Version Overseas Container Control; Non-Volatile Ocean Container Control
NVOILA National Voluntary Organization for Independent Living for the Aging
NVOO Nevada Operations Office
nvp natural vegetable powder (powdered psyllium seed and dextrose laxative)
NVPA National Visual Presentation Association
NVPO Nuclear Vehicle Projects Office (NASA)
nvr no voltage release
NVRC National Victims Resource Center
NVRS National Vegetable Research Station
nvs neutron velocity selector
NVS Night Vision System
NVT National Veld Trust
NVTS National Vocational Training Service
NVV *Nederlands Verbond van Vakverenigingen* (Dutch—Netherlands Trade Union Federation)
NV/VC North Vietnamese/Vietcong
nw nanowatt; net worth; no wind; number of weeks
n/w net weight
Nw New (sometimes confused with NW—Northwest)
NW Chicago & North Western Railway; Noah Webster; Norfolk & Western (railroad); Northern Wings Ltd; North Wales; Northwest; Northwest Airlines

N & W Norfolk & Western (railroad)

NW noordwest (Dutch—northwest); *Nordwesten* (German —northwest)

NW1, NW2, etc. Northwest One, Northwest 2, etc. (London postal zones)

NWA National Wrestling Alliance; Northwest Airlines; Northwest(ern) Australia

NWAA National Wheelchair Athletic Association

nwab necks with anybody

NWAC National Womens Advisory Council

NWAF New World Archeological Foundation

NWAH & ACA National Warm Air Heating and Air Conditioning Association

N-war nuclear war(fare)

NWASCO Nation Water and Soil Conservation Organization

N-waste nuclear (radioactive) waste

nwb non-weight bearing

nWb nano Weber

NWB National Westminster Bank

NWBA National Wheelchair Basketball Association

nwc nuclear war capability

Nwc Newcastle-upon-Tyne

NWC National Wages Council; National War College; National Water Commission; National Writers Club; Naval War College; Naval Weapons Center

NWCC Northern Wyoming Community College

NWCCL Naval Weapons Center—Corona Laboratories

NWCF Naval War College Foundation; New Waste Calcining Facility

NWCS NATO-wide Communications System

NWCTU National Woman's Christian Temperance Union

NWD New World Dictionary

nwdc navigation weapon-delivery computer

NWDR Nordwestdeutscher Rundfunk (German—North-West German Broadcasting System)

N-weapon(ry) nuclear weapon(ry)

N-weapon(s) nuclear weapon(s)

NWEB Northwestern Electricity Board (UK)

NWEF National Women's

Education Fund; Naval Weapons Evaluation Facility; Naval Weapons Evaluation Force

NWES New World Exploration Society

NWF National Welfare Fund; National Wildlife Federation

NWF National War Formulary

Nwfld Newfoundland

NWFP North-West Frontier Province

nwg national wire gauge

NWG National Welfare Growth; Neighborhood Watch Group

NWGA National Wheat Growers Association; National Wool Growers Association

nwh normal working hours

NWI Netherlands West Indies

NWIDA North West Industrial Development Association

NWIP Naval Warfare Instruction Publication

NWIRP Naval Weapons Industrial Reserve Plant

NWJA National Wholesale Jewelers Association

nwl natural wavelength

NWL Naval Weapons Laboratory

NWLB National War Labor Board

NWLC National Women's Law Center

NWLEE Northwest Law Enforcement Equipment

NWLF New World Liberation Front (terrorists)

NWly northwesterly

nwm nuclear waste materials

nw/m net words per minute

NWM Nuclear Waste Management

NWMC Northwest Michigan College

NWMCC Nuclear Waste Materials Characterization Center

N/Wmn Night Watchman

NWMPA North Wales Master Printers' Alliance

NWMS Northwest Medical Service

Nw Ned Nieuw Nederland (Dutch—New Netherlands)

NWNT North Wales Naturalists' Trust

NWO Nuclear Weapons Office(r)

nwoc new woman on campus

NWOO NATO Wartime Oil Organization

n-word nonce word (word coined for the occasion)

NWORG North Western Operational Research Group

NWP Naval Weapons Plant; North West Provinces

NWPAG NATO Wartime Preliminary Analysis Group

NWPC National Women's Political Caucus

NWPF New Waste Processing Facility

NWPFC Northwest Pacific Fisheries Commission

Nwprt News Newport News

NWPSC Northwestern Public Service Company

NWQAO Naval Weapons Quality Assurance Office

NWQI National Water Quality Inventory (EPA)

NWQSS National Water Quality Surveillance System (EPA)

nwr next word request

NWR National Waste Repository; National Welfare Rights; National Wildlife Refuge; National Wildlife Reserve; Nuclear Weapon Report

NWRB National Waste Repository Basalt

NWRC National Weather Records Center; National Wildhorse Research Center; Naval War Research Center (USN)

NWREL Northwest Regional Educational Laboratory

NWRF Naval Weather Research Facility

NWRLF New World Radical Liberation Front

NWRO National Welfare Rights Organization

NWRS National Wildlife Refuge System

nws normal water surface; nosewheel steering

NWS National Weather Service; Naval Weapons Station; New World Symphony (Miami, Florida); Nimbus Weather Satellite; Norfolk & Western Southern (railways)

NWSA National Welding Supply Association; National Woman Suffrage Association

NWSC National Weather Satellite Center; Naval Weather Service Command

NWSCA National Water and Soil Conservation Authority

NWSCO National Water and Soil Conservation Organization

NWSF Nuclear Weapons Stor-

age Facility (USA)
NWSO Naval Weapons Services Office
NWSS Nuclear Weapons Support Section (USA)
NWSY Naval Weapons Station—Yorktown, Va
nwt net weight; nonwatertight
NWT Northwest Territories
nwtb new water-tube boiler(s)
NWTB Northwestern Tariff Bureau
nwtd nonwatertight door
nwtdb new water-tube donkey boiler(s)
NWTEC National Wool Textile Export Corporation
NWTS National Waste Terminal Storage; Naval Weapons Test Station
NWTSR-1 NWTS Repository No 1 (high-level waste in a dome)
NWTSR-2 NWTS Repository No 2 (spent fuel in bedded salt)
NWTS-RSP NWTS Repository Sealing Program
nwu nosewheel up
NWU National Workers Union; National Writers Union; Nebraska Wesleyan University
NWUS Northwestern United States
NWVP Nuclear Waste Vitrification Program
NWWA National Water Well Association
NWWDA National Wood Window and Door Association
nwy newly
nx nonexpendable
NX Notice to Marines
NXD Non-Executive Director
NXDO Nike-X Development Office (USA)
NXMIS Nike-X Management Information Office
nx mo next month
NXPM Nike-X Project Manager
NXPO Nike-X Project Office
nxr non-crossing rule
NXSO Nike-X Support Office
nxt next
nxt ssn next season
nx wk next week
nx yr next year
ny new year; no year; nylon
Ny Niles; Nylan
NY New York; New York Airways (2-letter code); New Yorker; North Yorkshire
NY Neuyork (German—New

York); *New Yorker* (magazine); *Nieuw York* (Dutch—New York); *Nova Iorque* or *Nova York* (Portuguese—New York); *Nueva York* (Spanish—New York)
Nya Nyasaland
NYA National Youth Administration; Neighborhood Youth Association; New York Aquarium
NYAB National Youth Advisory Board
NYAC New York Athletic Club
NYADS New York Air Defense Sector
NYAM New York Academy of Medicine
NYANA New York Association for New Americans
NYAO New York Assay Office
NYAP New York Assembly Program
Nyas Nyasaland
NYAS New York Academy of Science; New York Asian Society
NYATI New York Agricultural and Technical Institute
NYBFU New York Board of Fire Underwriters
NYBG New York Botanical Garden
NYBSBC New York Bureau of State Building Codes
NYC National Yacht Club; Neighborhood Youth Corps; Newburgh Yacht Club; New York Central (railroad); New York City; New York Coliseum
NYCA National Youth Council of Australia; New York City Affiliate (of the National Council on Alcoholism)
NYCB New York City Ballet
NYCC New York Cultural Center
NYCCC New York City Community College
NYCCCC New York City's Citizens Crime Commission
NYCCIW New York City Correctional Institution for Women
NYCDC New York City Department of Correction
NYCE New York Cash Exchange; New York Cocoa Exchange; New York College of Education; New York Commodity Exchange; New York Cotton Exchange

NYCERS New York City Employees Retirement System
NYCHA New York City Housing Authority; New York Clearing House Association
NY-CHI New York—Chicago
NYCJG Nikka Yuko Centennial Japanese Garden (Lethbridge, Alberta)
NYCMA New York City Metropolitan Area
NYCMD New York Contract Management District
NYCMEO New York City Medical Examiner's Office
NYCMSL New York County Medical Society Library
NYCNHA New York City Nursing Home Association
NYCO New York City Opera
NYCOC New York City Opera Company
NY Col New York Coliseum
NYCPB New York Consumer Protection Board
NYCPD New York City Police Department
NYCPM New York City Police Museum
NYCS New York Chamber Symphony; New York Choral Society
NYCSCE New York Coffee, Sugar, and Cocoa Exchange
NYCSE New York Coffee and Sugar Exchange
NYC & ST L New York, Chicago & St Louis (Nickel Plate Line)
NYCT New York Community Trust
NYCTA New York City Transit Authority
NYCTN New York Cotton Exchange
NYCWRU New York Cooperative Wildlife Research Unit
nyd not yet dead; not yet diagnosed; not yet dressed (poultry)
NYDCC New York Drama Critics Circle
NYDMC New York Downstate Medical Center
NYDR New York Dock Railway
NYF New York Foundation
NYFCC New York Film Critics Circle
NYFDM New York Fire Department Museum
NYFE New York Futures Exchange
NYFH New York Foundling

Hospital

NYFIRO New York Fire Insurance Rating Organization

NYFUO New York Federation of Urban Organizations

NYGASP New York Gilbert and Sullivan Players

NYGC New York Governor's Conference

NYGS New York Graphic Society

NYHA New York Heart Association (classification)

NYH-CMC New York Hospital—Cornell Medical Center

NYHD New York House of Detention

NY Hist Soc New York Historical Society

NYHS New York Herpetological Society; New York Historical Society

NYI New York Institute (of Photography)

NYIAS New York Institute of the Aerospace Sciences

NYIBS New York International Bible Society

NYIE New York Insurance Exchange

NYIH New York Institute for the Humanities

NYIT New York Institute of Technology

N Yk New York

NYK Nippon Yusen Kaisha Line

NYKU Nippon Yusen Kaisha (container) Unit

nyl nylon

NYLA New York Library Association

NY-LAX New York—Los Angeles

NY & LB New York & Long Branch (railroad)

nylfin nylon finish

NYLS National Yacht Listing Service; New York Law School

NYLTI National Youth Leadership Training Institute

nym *nymon* (Greek—name)— as in antonym, homonym, pseudonym, synonym, etc.

NYMC New York Maritime College

NYME New York Mercantile Exchange

NY Met New York City Metropolitan Correctional Center

NY-MIA New York—Miami

nympho nymphomania; nymphomaniac; nymphomaniacal

NYNEX New York/New England Exchange (telephone

company)

N Y N H & H New York, New Haven and Hartford (railroad)

NY-NO New York—New Orleans

NYNR *New York National Review*

nyo not yet out

NYO National Youth Orchestra

NYOC New York Opera Company

NYOGB National Youth Orchestra of Great Britain

NYOL New York Opera Library

N Yorks North Yorkshire

NYORT New York Opera Repertory Theatre

NYOSL New York Oceans Science Laboratory

NYOTBC New York Off-Track Betting Corporation

NYOW National Youth Orchestra of Wales

NYO & W New York, Ontario and Western (railroad)

nyp not yet published

NYP Neighborhood Youth Program; New York Philharmonic (orchestra)

NYPA New York Port Authority

NYPD New York Police Department

NYPDis New York Procurement District (U.S. Army)

NYPE New York Port of Embarkation; New York Produce Exchange

NYPFO New York Procurement Field Office (USAF)

NYPHR New York Physicians for Human Rights

NYPIRG New York Public Interest Research Group

NYPL New York Public Library

NYPLA New York Patent Law Association

NYPM New York Pro Musica

NYPO New York Philharmonic Orchestra

NYPs Neighborhood Youth Programs

NYPS New York Paleontological Society; New York Psychiatric Society; New York Publishing Society

NYPSS New York Philharmonic-Symphony Society

NYPUM National Youth Program Using Minibikes

Nyq Nyquist (data-processing time or rate)

nyr not yet returned; nuclear yield requirement

NYR National Young Republicans

NYRA National Yacht Racing Association; New York Racing Association

NYRB *New York Review of Books*

NYRF National Young Republican Federation

NYRG New York Rubber Group

NYRM New York Reformatory for Men

NYRPG New York Rights and Permission Group

NYRs National Young Republicans

NYRW New York Reformatory for Women (Westfield Farm)

NYS New York Shavians; New York State

NYSA New York Shipping Association; New York State Assembly

NYSAA New York State Aviation Association

NYSAC New York State Athletic Commission

NYSAIS New York State Association of Independent Schools

NYSAJC New York State Association of Junior Colleges

NY-SAN New York—San Diego

NYSASBO New York State Association of School Business Officials

NYSASDA New York State Atomic and Space Development Authority

NYSAVC New York State Audio-Visual Council

NYSBA New York State Bar Association

NYSBB New York State Banking Board

NYSBC New York State Barge Canal (modern extension of Erie Canal)

NYSC New York Shipbuilding Corporation

NYSCC New York State Crime Commission

NYSCCJ New York State Coalition for Criminal Justice

NY Sch Indus Rel New York State School of Industrial Relations (Cornell University)

NYSCSDA New York State Council of School District Administrators

NYSDCS New York State Department of Correctional Services

NYSE New York Stock Exchange

NYSERDA New York State Energy Research and Development Authority

NYSES New York State Employment Service

NYSF New York Shakespeare Festival

NY-SFO New York—San Francisco

NYSILL New York State Inter-Library Loan (network)

NYSL New York Society Library

NYSM New York State Museum

NYSMM New York State Maritime Museum (New York City)

NYSNA New York State Nurses' Association

NYSNACC New York State Narcotic Addiction Control Commission

NYSNC New York State Narcotics Commission

NYSNI New York State Nutrition Institute

NYSO New York String Orchestra

NYSP New York School of Printing; New York State Police

NYSPA New York State Power Authority

NYSPGA New York State Personnel and Guidance Association

NYSPI New York State Psychiatric Institute

NYSSILR New York State School of Industrial and Labor Relations

NYSSMA New York State School Music Association

NYSTA New York State Teachers Association; New York State Thruway Authority

NY Sup New York Supreme Court Reports

NYSUT New York State United Teachers

NYS & W New York, Susquehanna and Western (railroad)

NYT National Youth Theatre

NYT The New York Times

NY Thru New York Thruway

NY Times Bk R New York Times Book Review

NYTNS New York Times News Service

NY-TOR New York—Toronto

NYT/TS New York Turtle/Tortoise Society

NYTU New York Theological Union

NYU New York underworld (used in law-enforcement circles); New York University

NYUL New York University Library

NYUMC New York University Medical Center; New York Upstate Medical Center

NYUP New York University Press

NYUSM New York University School of Medicine

NYWASH Navy Yard, Washington

NYYC New York Yacht Club

NYZP New York Zoological Park

NYZS New York Zoological Society

Nz Nuñez

NZ New Zealand; New Zealand dollar; New Zealand National Airways (2-letter coding); Novaya Zemlya

N-Z Nike-Zeus

NZ Nueva Zelandia (Spanish—New Zealand)

N-Z Nouvelle-Zélande (French—New Zealand)

NZAA New Zealand Amateur Athletic Association; New Zealand Antique Arms Association; New Zealand Auto Association

NZAB New Zealand Association of Bacteriologists

NZABC New Zealand Audit Bureau of Circulation

NZABM New Zealand Anglican Board of Missions

NZAC New Zealand Accommodation Council; New Zealand Alpine Club

NZACA New Zealand Amateur Cycling Association

NZACAU New Zealand Athletics, Cycling, and Axemens Union

NZACE New Zealand Association for Community Education

NZACU New Zealand Auto Cycle Union

NZADS New Zealand Association for Disabled Skiers

NZAEC New Zealand Atomic Energy Committee

NZAEI New Zealand Agricultural Engineering Institute

NZAF New Zealand Authors Fund; New Zealand Aviation Federation

NZAHBS New Zealand Arab Horse Breeders Society

NZAHPER New Zealand Association of Health, Physical Education, and Recreation

NZALT New Zealand Association of Language Teachers

NZAPA New Zealand Airline Pilots Association

NZARA New Zealand Amateur Rowing Association

NZARE New Zealand Association for Research on Education

NZARP New Zealand Antarctic Research Programme

NZART New Zealand Amateur Radio Transmitters Association

NZAS New Zealand Aluminium Smelters; New Zealand Antarctic Society; New Zealand Arthritis Society; New Zealand Association of Scientists

NZASA New Zealand Amateur Swimming Association; New Zealand Asian Studies Association

NZASC New Zealand Administrative Staff College; New Zealand Army Service Corps; New Zealand Association of Soil Conservators

NZASF New Zealand Association of Small Farmers

NZASW New Zealand Association of Social Workers

NZATD New Zealand Association of Training and Development

NZAWA New Zealand Air Women's Association

NZb New Zealand black (mice hybrids)

NZB New Zealand Ballet

NZBA New Zealand Bankers Association; New Zealand Bowling Association

NZBC New Zealand Ballet Company; New Zealand Book Council; New Zealand Broadcasting Corporation

NZBCSO New Zealand Broadcasting Corporation Symphony Orchestra

NZBF New Zealand Basketball Federation

NZBIE New Zealand Bureau of Importers and Exporters

NZBS New Zealand Broadcasting Service

NZBTO New Zealand Book Trade Organisation

NZC New Zealand Certificate
NZCAR New Zealand Civil Aviation Regulations
NZCAS New Zealand Clean Air Society
NZCC New Zealand Chamber of Commerce; New Zealand Cricket Council
NZCD New Zealand Certificate in Draughting
NZCDC New Zealand Cooperative Dairy Company
NZCE New Zealand Certificate in Engineering
NZCEA New Zealand Combined Educational Associations
NZCER New Zealand Council for Educational Research
NZCF New Zealand Cadet Forces; New Zealand Cat Fancy association; New Zealand Cycling Federation
NZCG New Zealand Chemists Guild
NZCGF New Zealand Coast Guard Federation
NZCGP New Zealand College of General Practitioners
NZCGS New Zealand Standard Classification of all Goods and Services
NZCH New Zealand Cement Holdings
NZCLA New Zealand Childrens Literature Association
NZCLS New Zealand Certificate of Land Surveying
NZCMA New Zealand Cable Makers Association; New Zealand Concrete Masonry Association
NZCMF New Zealand Coal Merchants Federation
NZCO New Zealand Concert Orchestra
NZCOSS New Zealand Council of Social Services
NZCRA New Zealand Coal Research Association; New Zealand Concrete Research Association
NZCRS New Zealand Council for Recreation and Sports
NZCS New Zealand Certificate in Science; New Zealand Certificate in Statistics; New Zealand Computer Society
NZCSS New Zealand Council of Social Services
NZCTF New Zealand Cycle Traders Federation
NZCTOA New Zealand Container Terminal Operators Association
NZCUL New Zealand Credit

Union League
NZCWI New Zealand Country Women's Institutes
NZd New Zealand dollar
NZ$ New Zealand dollar(s)
NZD New Zealand Division; New Zealand Dollar
NZDA New Zealand Dairy Association; New Zealand Dental Association; New Zealand Department of Agriculture; New Zealand Dietetic Association
NZDB New Zealand Dairy Board
NZDC New Zealand Dental Corps
NZDCMBA New Zealand Dairy Confectionary and Mixed Biscuits Association
NZDCS New Zealand Department of Census and Statistics
NZDE New Zealand Department of Education
NZDFA New Zealand Deer Farmers Association
NZDLS New Zealand Department of Lands and Survey
NZDRI New Zealand Dairy Research Institute
NZDS New Zealand Drama School
NZDSIR New Zealand Department of Scientific and Industrial Research
NZDT New Zealand Daylight Time
NZDVA New Zealand Dunkirk Veterans Association
NZDXRA New Zealand DX Radio Association
NZE New Zealand Engineers
NZEA New Zealand Esperanto Association
N Zeal New Zealand(er)
NZEAS New Zealand East Asia Service; New Zealand Educational Administration Society
NZEB New Zealand Electricity Board
NZECF New Zealand Electrical Contractors Federation
NZED New Zealand Electricity Department
NZEF New Zealand Employees Federation; New Zealand Expeditionary Force
NZEI New Zealand Educational Institute; New Zealand Electronics Institute
NZer New Zealander
NZERF New Zealand Engine Research Foundation; New Zealand Equine Research Foundation

NZES New Zealand Ecological Society
NZESA New Zealand Education Standards Association; New Zealand European Shipping Association
nzf near zero field
NZFA New Zealand Football Association
NZFB New Zealand Foundation for the Blind
NZFCA New Zealand Farmers Cooperative Association; New Zealand Freezing Companies Association
NZFCDC New Zealand Farmers Cooperative Distributing Company
NZFCMA New Zealand Ferro-Cement Marine Association
NZFF New Zealand Farmers Fertiliser (company); New Zealand Federated Farmers; New Zealand Fruitgrowers Federation
NZFHA New Zealand Finance Houses Association
NZFKTA New Zealand Free Kindergarten Teachers Association
NZFKU New Zealand Free Kindergarten Union
NZFL New Zealand Federation of Labor
NZFMA New Zealand Ferrocement Marine Association
NZFMC New Zealand Federation of Master Cleaners
NZFMRA New Zealand Fertiliser Manufacturers Research Association
NZFP New Zealand Forest Products
NZFPA New Zealand Family Planning Association
NZFRI New Zealand Forest Research Institute
NZFS New Zealand Film Service; New Zealand Forest Service
NZFUW New Zealand Federation of University Women
NZFWA New Zealand Farm Workers Association
nzg near zero gravity
NZG New Zealand Government
NZG New Zealand Gazette
NZGA New Zealand Gliding Association; New Zealand Golf Association; New Zealand Grasslands Association
NZGenS New Zealand Genetical Society
NZGR New Zealand Govern-

ment Railways
NZGS New Zealand Geographical Society; New Zealand Geological Society; New Zealand Geological Survey
NZGTB New Zealand Government Tourist Bureau
NZGTC New Zealand Government Travel Commissioner
NZGTO New Zealand Government Tourist Office
NZH New Zealand Helicopters
NZH New Zealand Herald
NZHA New Zealand Hockey Association
NZHC New Zealand High Commission
NZHF New Zealand Heart Foundation
NZHGA New Zealand Hang-Gliding Association
NZHI New Zealand Horological Institute
NZHPT New Zealand Historic Places Trust
NZHS New Zealand Horse Society
NZI New Zealand Insulators; New Zealand Insurance
NZIA New Zealand Institute of Architects; New Zealand Irrigation Association
NZIAS New Zealand Institute of Agricultural Science
NZIC New Zealand Institute of Chemistry; New Zealand Intelligence Council
NZICFM New Zealand Institute of Credit and Financial Management
NZICM New Zealand Institute of Credit Management
NZID New Zealand Institute of Draughtsmen
NZIDA New Zealand Invention Development Authority
NZIDC New Zealand Industrial Design Council
NZIE New Zealand Institute of Engineers
NZIELEC New Zealand Institute of Electricians
NZIEPC New Zealand Indonesia Economic Promotion Council
NZIER New Zealand Institute of Economic Research
NZIET New Zealand Institute of Engineering Technicians
NZIF New Zealand Institute of Foresters
NZIFST New Zealand Institute of Food Science and Technology

NZIG New Zealand Institute of Gases
NZIH New Zealand Institute of Horticulture
NZIHVE New Zealand Institute of Heating and Ventilation Engineers
NZIIA New Zealand Institute of International Affairs
NZIIS New Zealand Institute of Industrial Safety
NZILA New Zealand Institute of Landscape Architects
NZIM New Zealand Institute of Management; New Zealand Institute of Mining
NZIME New Zealand Institute of Mechanical Engineers
NZIMP New Zealand Institute of Medical Photography
NZIP New Zealand Institute of Printing
NZIPA New Zealand Institute of Public Administration
NZIPM New Zealand Institute of Personnel Management
NZIPRA New Zealand Institute of Parks and Recreation Administration
NZIPS New Zealand Institute of Purchasing and Supply
NZIRE New Zealand Institute of Refrigeration Engineers
NZIS New Zealand Information Service; New Zealand Institute of Surveyors
NZISM New Zealand Institute of Safety Management
NZIT New Zealand Institute of Travel
NZIUW New Zealand Industrial Union of Workers
NZIW New Zealand Institute of Welding
NZJCB New Zealand Joint Communications Board
NZJPA New Zealand Japan Parliamentary Association
NZJU New Zealand Journalists Union
NZK Noord Zee Kanaal (Dutch—North Sea Canal)— linking the Atlantic with Amsterdam
NZKC New Zealand Kennel Club
NZKVA New Zealand Korean Veterans Association
NZL New Zealand Line
NZLA New Zealand Legal Association; New Zealand Library Association; New Zealand Loggers Association
NZLCC New Zealand Litter Control Council
NZLF New Zealand Literary

Fund
NZLIRA New Zealand Logging Industry Research Association
NZLL New Zealand Light Leathers
NZLP New Zealand Labour Party
NZLR New Zealand Law Reports
NZLS New Zealand Land Securities; New Zealand Law Society; New Zealand Library School; New Zealand Library Service
NZLTA New Zealand Lawn Tennis Association
NZMA New Zealand Medical Association; New Zealand Modelling Association; New Zealand Motel Association; New Zealand Motorcycle Association
NZMAF New Zealand Ministry of Agriculture and Fisheries
NZMB New Zealand Meat Board
NZMBF New Zealand Master Builders Federation
NZMC New Zealand Maori Council; New Zealand Medical Corps
NZMCA New Zealand Motor Caravan Association
NZMF New Zealand Manufacturers Federation; New Zealand Military Forces; New Zealand Motel Federation; New Zealand Music Federation
NZMFA New Zealand Master Floorcovering Association
NZMGA New Zealand Mountain Guides Association
NZMGC New Zealand Marriage Guidance Council
NZMJ New Zealand Medical Journal
NZMOT New Zealand Ministry of Transport
NZMPH New Zealand Meat Packing House
NZMRC New Zealand Medical Research Council
NZMS New Zealand Mapping Service; New Zealand Meteorological Service
NZMSC New Zealand Mountain Safety Council
NZMSS New Zealand Marine Sciences Society
NZMTCB New Zealand Motor Trade Certification Board
NZMTMA New Zealand Methods Time Measurement

Association
NZMWA New Zealand Maori Wardens Association
NZM & WB New Zealand Meat and Wool Board
NZMWU New Zealand Meat Workers Union
NZNA New Zealand Nurserymens Association; New Zealand Nurses Association
NZNAC New Zealand National Airways Corporation
NZNCC New Zealand Nature Conservation Council
NZNCOR New Zealand National Committee on Oceanic Research
NZNF New Zealand Neurological Foundation
NZNFU New Zealand National Film Unit
NZNPA New Zealand Newspaper Proprietors Association
NZNTA New Zealand National Travel Association
NZOA New Zealand Optometrical Association
NZOC New Zealand Opera Company
NZOCGA New Zealand Olympic and Commonwealth Games Association
NZOI New Zealand Oceanographic Institute
NZ£ New Zealand pound
NZP National Zoological Park; New Zealand Pacific; New Zealand Players; New Zealand Police
NZPA New Zealand Police Association; New Zealand Ports Authority; New Zealand Press Association
NZPARS New Zealand Prisoners Aid and Rehabilitation Society
NZPB New Zealand Pony Breeders; New Zealand Potato Board
NZPBA New Zealand Power Boat Association; New Zealand Publishers' Association
NZPBR New Zealand Pony Breeders Register
NZPBS New Zealand Pony Breeders Society
NZPC New Zealand Peace Council; New Zealand Planning Council; New Zealand Pony Club; New Zealand Press Council; New Zealand Print Council
NZPCA New Zealand Pony Club Association; New Zealand Portland Cement Association

NZPCI New Zealand Prestressed Concrete Institute
NZPEA New Zealand Port Employers Association
NZPGMF New Zealand Post Graduate Medical Federation
NZPM New Zealand Paper Mills
NZPMS New Zealand Plumbers Merchants Society
NZPO New Zealand Post Office
NZPOA New Zealand Purchasing Officers Association
NZPPA New Zealand Professional Photographers Association
NZPPTA New Zealand Post Primary Teachers Association
NZPS New Zealand Park Service; New Zealand Police Service
NZPSA New Zealand Political Studies Association; New Zealand Public Service Association
NZPsS New Zealand Psychological Society
NZPTA New Zealand Parent-Teachers Association
NZPTO New Zealand Public Trust Office
NZQHA New Zealand Quarter-Horse Association
NZR New Zealand Railways
NZRC New Zealand Red Cross
NZRDXL New Zealand Radio DX League
NZRFU New Zealand Rugby Football Union
NZRL New Zealand Rugby League
NZRLS New Zealand Railway and Locomotive Society
NZRMA New Zealand Ready-Mix Concrete Association
NZRMTA New Zealand Retail Motor Trade Association
NZRN New Zealand Registered Nurse
NZRNC New Zealand Radio Navigation Chart
NZRRS New Zealand Railways Road Services
NZRTA New Zealand Road Transport Association
NZS New Zealand Standards Institute
NZSA New Zealand Statistical Association
NZSB New Zealand Speech Board; New Zealand Survey Board
NZSBG New Zealand South

British Group
NZSC New Zealand Securities Commission; New Zealand Settlers Club; New Zealand Squid Company; New Zealand Staff Corps; New Zealand Standards Council
NZSCA New Zealand Sheep and Cattlemens Association; New Zealand Society of Customs Agents; New Zealand Soil Conservation Association
NZSCC New Zealand Standard Country Code
NZSCES New Zealand Society of Certified Executive Secretaries
NZSCHA New Zealand Society of Custom House Agents
NZSCI New Zealand Standard Classification of Imports
NZS Co New Zealand Shipping Company
NZSCO New Zealand Standard Classification of Occupations
NZSCS New Zealand Senior Citizens Service
NZSDA New Zealand Sign and Display Association; New Zealand Stamp Dealers Association
NZSDST New Zealand Society of Dairy Science and Technology
NZSE New Zealand Stock Exchange
NZ Sea Fron New Zealand Sea Frontier (NZSEAFRON)
nzsg non-zero-sum game
NZSI New Zealand Seismological Institute
NZSIA New Zealand Security Industry Association
NZSIC New Zealand Standard Industrial Classification
NZSID New Zealand Society of Industrial Designers
NZSL New Zealand Steel Limited
NZSL New Zealand Shipping Line
NZSLO New Zealand Scientific Liaison Office
NZSNA New Zealand Society of National Accounts
NZSO New Zealand Symphony Orchestra
NZSRA New Zealand Surf Riders Association
NZSS New Zealand Social Security; New Zealand Speleological Society; New Zealand Standard Specification(s)
NZSSS New Zealand Society

of Soil Science
NZST New Zealand Standard Time
NZSWWS New Zealand Spinning, Weaving, and Woolcrafts Society
nzt non-zero test(ing)
NZTC New Zealand Trade Commission
NZTCA New Zealand Teachers College Association
NZTCB New Zealand Trade Certification Board
NZTCI New Zealand Technical College Institute; New Zealand Technical Correspondence Institute
NZTF New Zealand Territorial Force(s); New Zealand Theatre Federation

NZUA New Zealand Underwater Association; New Zealand Underwriters Association
NZUE New Zealand Unit Express
NZV New Zealand Victoria (insurance)
NZVA New Zealand Veterinary Association
NZw New Zealand white (mice hybrids)
NZWA New Zealand Woolbuyers Association
NZWB New Zealand Wool Board
NZWEA New Zealand Workers Educational Association
NZW & PCS New Zealand Weed and Pest Control Society

NZWRAC New Zealand Womens Royal Army Corps
NZWS New Zealand Wildlife Service
NZWSA New Zealand Water Ski Association
NZWSC New Zealand Water Safety Council
NZWTA New Zealand Wool Testing Authority
NZWWC New Zealand Working Womens Council
NZWWF New Zealand Waterside Workers Federation
NZYF New Zealand Yachting Federation
NZYHA New Zealand Youth Hostels Association
NZZ Neue Züricher Zeitung (New Zurich Newspaper)

O

o observer; occasional; occidental; octavo; ohm; oil; oiliness; Olivetti; opium; orange; oriental; overcast
o' (Gaelic contraction—of, on)
'o (Gaelic contraction—also)
o (Japanese—big, great, large); omkring (Dano-Norwegian—about or around)
o. oculus (Latin—eye); oeste (Portuguese or Spanish—west); oost (Dutch—east); op (Dano-Norwegian or Dutch—up); os (Latin—bone); ouest (French—west); ovest (Italian—west)
o/ order (Spanish—order)
ö (Dano-Norwegian or Swedish—island); öster (Swedish—east)
ø øst (Dano-Norwegian—east)
O absence of perception of sound (symbol); New Orleans Mint (coin symbol); observation; Observer; ocean; Oceanic Steamship Company; October; office; officer; Ohio; Olsen Line; Omaha; Ontario; order; Oregon; ortho; Oscar—code for letter O; oxygen; Ohio
O' (Gaelic prefix meaning of)
Ø shortage (symbol)
O center of the earth (symbol); observer (symbol); oeste (Portuguese or Spanish—west); oost (Dutch—east); optimus (Latin—best possible); organo (Italian—organ); Ost (German—east); ouest (French—west); ovest (Italian—west)
Ö Österreich (German—Eastern Empire)—Austria; Östre (Swedish—East); Öy (Swed-

ish—island)
Ø Øst (Dano-Norwegian—East); Øy (Dano-Norwegian—island)
O1 organized seagoing naval reserve
O-1 Cessna Bird Dog liaison aircraft
O2 organized naval reserve aviation
O-2 Cessna liaison-utility aircraft
O_2 oxygen
O_2cap oxygen capacity
O_2sat oxygen saturation
O^3 ozone
oa occiput anterior; old age; on account; on or about; osteoarthritis; overall
o/a on account; on or about
oa och andra (Swedish—and others)
o/A oro Americano (Spanish—American gold, American money)
OA Obligation Authority; Office of Applications; Office Automation; Olympic Airways; Operations Analysis; Osborne Association; overall noise level (symbol); Overeaters Anonymous; Overtime Authorization
O/A Office of Administration (EPA)
O & A October and April
O of A Office of Administration
OA Océan Atlantique (French—Atlantic Ocean)
oaa (OAA) oxalo-acetic acid
OaA Office of Aging
OAA Office of Air Accidents; Old Age Assistance; Older Americans Act; Organisation

des Nations Unies pour l'Alimentation et l'Agriculture (French—United Nations Organization for Food and Agriculture); Organization of Athletic Administrators; Orient Airlines Association
OAA Organización de las Naciones Unidas para la Agricultura y la Alimentación (Spanish—United Nations Organization for Agriculture and Food)
OAAA Outdoor Advertising Association of America
OAAB Objective-Analytic Anxiety Battery
oaad ovarian ascorbic acid depletion
OAAI Office of Air Accidents Investigation
OAAU Organization of Afro-American Unity
OAB Old Age Benefits
OABA Outdoor Amusement Business Association
OABETA Office Appliance and Business Equipment Trades Association
oac on approved credit; optical aberrations compensation; outer approach channel
OAC Oceanic Affairs Committee; Operating Agency Code; Ordnance Ammunition Command; Oregon Agriculture College
OACA Ontario Arms Collectors Association (of Beamsville near Toronto)
OACI Organisation de l'Aviation Civile Internationale (French—International Civil Aviation Organization); Or-

ganización de Aviación Civil Internacional (Spanish—International Civil Aviation Organization)

OACJC Oklahoma Association of Community and Junior Colleges

OACLD Ontario Association for Children with Learning Disabilities

OACT Ohio Association of Classroom Teachers

oad overall depth

OAD ordered, adjudged, and decreed

OADAP Office of Alcoholism and Drug Abuse Prevention

oadc oleic acid, albumin, dextrose, catalase

OAE Orzeck Aphasia Evaluation

OAEC Organization for Asian Economic Cooperation

OAESA Ohio Association of Elementary School Administrators

oaf open-air factor; overhaul attrition factor

OAFB Orfutt Air Force Base (Nebraska)

OAFIE Office of Armed Forces Information and Education

OAG Office of the Adjutant General; Office of the Attorney General

OAG Official Airline Guide

OAGB Osteopathic Association of Great Britain

oah overall height

OAH Organization of American Historians

OAHE Ohio Association for Higher Education

OAI Office of Aeronautical Intelligence; Opera America, Incorporated; Osborne Association, Incorporated

OAIA Organisation des Agences d'Information d'Asie (French—Organization of Asian News Agencies)

OAICU Oklahoma Association of Independent Colleges and Universities

oaide operational assistance and instructive data equipment

oais opinion, attitude, and interest survey

oak. oakum

Oak Oakland, Oak Park, Oak Ridge, etc.

OAK Oakland, California (Metropolitan International Airport)

Oak Sym Oakland Symphony

oal overall length

OAL Office of Administrative Law; Ordnance Aerophysics Laboratory

OALJ Office of Administrative Law Judges

OALMA Orthopedic Appliance and Limb Manufacturers Association

o. alt. hor. omnibus alternis horis (Latin—every other hour)

OAM Office of Aviation Medicine; Order of Australia Medal

OAMA Ogden Air Material Area; Oil Appliance Manufacturers Association

oamce optical alignment, monitoring, and calibration equipment

oame orbital attitude and maneuvering electronics

OAMS Orbital Attitude and Maneuvering System

ÖAMTC Österreichischer Automobil-Motorrad und Touring Club (German—Austrian Automobile Motoring and Touring Club)

OANA Organization of Asian News Agencies

o-and-o one-and-only

oao off and on

OAO Orbiting Astronomical Observatory

oap ophthalmic artery pressure

OAP Office of Aircraft Production; Old-Age Pension

OAPC Office of the Alien Property Custodian

OAPEC Organization of Arab Petroleum Exporting Countries

OAPEP Organisation Arabe des Pays Exportateurs de Petrole (French—Arab Organization of Petroleum Exporting Nations)

OAPs Old-Age Pensioners

OAPU Old Age Pension Union

oapwl overall power watt level

O Ar Old Arabic

OAR Offender Aid and Restoration; Office of Aerospace Research; Order of Augustinian Recollects; Organized Air Reserve

OARAC Office of Aerospace Research Automatic Computer

OARP Old Age Revolving Pensions (Townsend Plan)

OARS Offender's Aid Rehabilitation Services

OART Office of Advanced Research and Technology (NASA)

oas old-age security; on active service

OAs older adults

OAS Office of Advanced Studies; Office of Appalachian Studies; Old Age Security; Ordinary Ammunition Storage; Organization of American States

OAS Organisation de l'Armée Secrete (French—Organization of the Secret Army)—counter-revolutionary group attempting to crush Algerian independence

OASBO Ohio Association of School Business Officials; Oregon Association of School Business Officials

OASD Office of the Assistant Secretary of Defense

OASD-AE Office of the Assistant Secretary of Defense, Application Engineering

OASDHI Old-Age, Survivors, Disability, and Health Insurance Social Security

OASDI Old Age, Survivors, and Disability Insurance

OASD-R & D Office Assistant Secretary of Defense, Research and Development

OASD-S & L Office Assistant Secretary of Defense, Supply and Logistics

OASD-T Office of the Assistant Secretary of Defense—Telecommunications

OASHDS Office of the Assistant Secretary for Human Development Services

OASI Old-Age and Survivor's Insurance

OASIS Office of Academic Support Instructional Services; Office for Academic Support in Service; Ohio (chapters) of the American Society for Information Science; Older Adult Service and Information System; Overseas Access Service for Information Systems

oasp organic acid-soluble phosphorus

oaspl overall sound pressure level

OASSO Operational Applications of Satellite Snowcover Observations (NASA)

oat. outside air temperature

OAT Office of Advanced Technology (USAF)

OATC Oceanic Air Traffic Control
OATS Office of Air Transportation Security; Old-Age Theatre Society (Great Britain)
oau (OAU) optical alignment unit
OAU Organization for African Unity
OAVTME Office of Adult, Vocational, Technical, and Manpower Education
oaw old abandoned well; over-all width
OAWM Office of Air and Water Measurement (NBS)
Oax Oaxaca
OAYR Outstanding Airman of the Year Ribbon
ob obligation; oboe; obsolete; obstetric; obstetrical; obstetrician; old boy; on board; operational base (OB); ordered back; out of bounds; outboard buffer; outbound; output buffer; outward bound; over bought; overboard (vent line)
ob (OB) outside broadcast (TV from a remote location)
o/b opening of books; outboard (engine)
ob (Latin prefix—against, in front of, toward)—obstruction
ob. obit (Latin—died)
o B off Broadway
o-B off-Broadway; off-Broadway theater
o B ohne Befund (German—without findings)
Ob object art (art accented with real objects, *e.g.*, a real watch chain dangling between two pockets of a man's vest in a painting); 3500-mile Siberian river entering Arctic Ocean at Gulf of Ob
Ob Obadiah; Ober (Germany—higher, upper)
OB Ocean Beach; Old Bailey; Operating Base; Operational Base; Order of Battle; Ordnance Battalion; Ordnance Board; Ormond Beach; Ox Box (corporation)
O.B. obstetrical; obstetrician; obstetrics
O'B O'Brien; O'Bryan
OB Oranjeboom (Dutch—orange tree)—Amsterdam-brewed beer
oba optical bleaching agent
OBAA Oil-Burning Apparatus

Association
Obad The Book of Obadiah
OBAN Operação Bandeirantes (Portuguese—Operation Bandeirantes)—Brazilian Intelligence Service
OBAR Ohio Bar Automated Research
OBAWS On-Board Aircraft Weighing System
obb obbligato
OBB battleship, old (3-letter naval symbol)
ÖBB Österreichische Bundesbahnen (Austrian Federal Railways)
obbl obbligato
obc old brutal con(vict); on-board checkout; outer back cover
OBC Old Boys Club; Osaka Broadcasting Corporation; Outboard Boating Club
obce on-board checkout equipment
obd omnibearing distance
ob d'am oboe d'amore
OBDC Otago Business Development Centre
ob dk observation deck
obdt obedient
obe open both ends; operating basis earthquake; other bugger's efforts
OBE Office of Business Economics; Officer of the British Empire; Order of the British Empire
O.B.E. Officer of the Order of the British Empire
OBEMLA Office of Bilingual Education and Minority Languages Affairs
Oberst Oberstimme (German—soprano, treble, descant)
Oberw Oberwerk (German—highest organ bank; upperwork)
OBES Office of Basic Engineering Sciences
OBEV Oxford Book of English Verse
obf operating basis flood
obfusc obfuscated
obg oldie but goodie (musical hits)
Ob-G Obstetrician-Gynaecologist
OBGA Office of Block Grant Assistance
obgn obligation
obᵍᵒ obrigado (Portuguese—thank you)
ob-gyn obstetrical-gynecological; obstetrician-gynecologist
obi omnibearing indicator

Obie off-Broadway; off-Broadway theater; Off-Broadway Theater Award
OBIPS Optical Band Imager and Photometer System
obit obituary
obits obituaries
obj object; objective
object. objective(ly)
objn objection
obl obligation; oblique; oblong; obloquy
ob/l ocean bill of lading
OBL Ocean Beach Library; Ohio Barge Line; Order of the Brave Librarian
oblg obligate; obligation
OBLI Oxford and Birmingham Light Infantry
oblig obligation(s); obligatory
obln obligation
obm oil-base mud (oil well)
obo oil/bulk freight/ore (multipurpose seagoing carrier)
oboe. offshore buoy-observing equipment
ob ph oblique photograph(y)
OBRA Overseas Broadcasting Representatives' Association
obre octubre (Spanish—October)
Obre Octobre (French—October)
obro outubro (Portuguese—October)
obs observation; observe; observed; observer; obsolete; obstacle; obstetrical; obstetrician; obstetrics; ocean bottom suspension (oil well); omnibearing selector
obs (OBS) organic brain syndrome
obs oboes
Obs The Observer
OBS Oita Broadcasting Service; Organization Breakdown Structure
obs alt observed altitude
obsc obscure(d)
obsd observed
observ observation; observatory
obsn observation
obsol obsolescent
ob & sol objection and solution
ob. s.p. obiit sine prole (Latin—died without issue)
OBSP Old Bailey Sessions Papers
obss ocean bottom scanning sonar
obs spot observation spot
obst obstacle; obstruction
obstet obstetrical; obstetrician;

obstetrics
obstl obstruction light(s)
obstr obstruction
obsv observation; observatory; observer
ob syn organic brain syndrome
o b syn organic brain syndrome
obt obedient
obt. obiit (Latin—he died)
OBT Overseas Branch Transfer
OBTA Oak Bark Tanners' Association
obtd obtained
obts offender-based transaction statistics
OBU One Big Union; Operative Bootmakers Union
ÖBUB Öffentliche Bibliothek der Universität Basel (German—Public Library of the Basel University)—founded in 1460
O Bul Old Bulgarian
obv obverse; obvious; ocean boarding vessel; octane blending value
obvy obviously
obw observation window
Obw Oberwerk (German—highest organ bank; upperwork)
oc ocean; odor control; on camera; on center; open charter; oral contraceptive
oc (OC) obstetrical conjugate; on camera; open cup; overdraft charge
o-c open-circuit
o'c o'clock (of the clock)
o/c open charter; open cover; organized crime; overcharge
o & c onset and course (disease)
oc (Latin prefix—against)—occlusion
o.c. opere citato (Latin—in the work cited)
Oc Ocean
OC Air California (airline code); Oakland City; Oakwood College; Oberlin College; Oblate College; Occidental College; Odessa College; Office of Censorship; Office of the Commissioner; Office Consultation; Officer Candidate; Officer in Charge; Officer Commanding; Ohio College; Okolona College; Olivet College; Olympic College; Opera Company; Optometric Corporation; Order in Council; Oriel College; Orlando College; Otero Col-

lege; Overseas Chinese; Overseas Commands
O.C. Officer Commanding
O of C Order of the Coif
OC Opéra-Comique (French—Comic Opera)—Paris
O.C. Organo Corale (Latin—choir organ)
OC-5 Organizing Committee for a Fifth Estate
oca ocarina (flutelike clay instrument nicknamed "sweet potato")
OCA Oceanic Control Area; Office of the City Attorney; Office of Computing Activities (NASA); Office of Consumer Affairs (ombudsman function of the U.S. Postal Service); Ohio College Association; Oil Company of Australia; Old Comrades Association; Ontario College of Art; Open Communications Architecture; Oregon Corrections Association
OCA Organización de las Cooperativas de America (Spanish—Organization of American Cooperatives)
OCAA Oklahoma City-Ada-Atoka (railroad); Organization of Central American Armies
OCAC Office of Chief of Air Corps
OCADS Oklahoma City Air Defense Sector
OCAFF Office Chief of Army Field Forces
ocal on-line cryptanalytic aid language
OCAL Overseas Containers of Australia Limited
OCAM Organisation Commune Africaine et Malgache [French—Organization of the African and Malagasy Community (of former French colonies)]
OCAMA Oklahoma City Air Materiel Area
O Canada O Canada! terre de nos aïeux (French—O Canada! Land of our forefathers)—national anthem sung in English and in French
OCAS Office of Civil Aviation Security; Organization of Central American States
O Cat Old Catalan
OCAT Optometric College Aptitude Test; Optometry College Admissions Test
OCAW Oil, Chemical and Atomic Workers (union)

ocb oil circuit breaker
OCB Officer Career Brief (DoD résumé)
OCBC Overseas Chinese Banking Corporation
oc b/l ocean bill of lading
occ occasionally; occupation
o & cc order and change control
Occ occulting (light)
OCC Office of the Comptroller of the Currency; Oklahoma Crime Commission; Olney Community College; Onondaga Community College; Orange Coast College
OCCA Oil and Colour Chemists Association
occas occasional(ly)
OCCC Oil Control Coordination Committee; Orange County Community College; Organized Crime-Control Commission (California)
Oc C Cm O Office of the Chief Chemical Officer
occd occupied
OCCDC Oregon Coastal Conservation and Development Commission
OCC-E Office of the Chief of Communications—Electronics (USA)
OCCF Oklahoma City Community Foundation
occip occipital; occiput
OCCIS Operational Command and Control Intelligence System (USA)
occl occlude(d); occluded front; occluding; occlusal; occlusion; occlusive(ness)
OCCL Ontario Community College Librarians
OCCM Office of Commercial Communications Management
OCCO Office of the Chief Chemical Officer
OCCP Outside Communications Cable Plant
OCCS Office of Computer and Communications Systems (U.S. National Library of Medicine)
OCCSA Ohio Correctional and Court Services Association
occ th occupational therapy
occup occupation(al)
ocd obsessive compulsive disorder; on-line communications driver; operational capability date; optical character definition; ovarian cholesterol depletion
oc/d other cargo damage

OCD Office of Child Development; Office of Civil Defense; Office of Collection and Dissemination (CIA)

OCDA Ordnance Corps Detroit Arsenal

OCDE Organización Común Africana, Malgache y Mauriciana (Spanish—African Common Organization including Madagascar and Mauritius); *Organización de Cooperación y Desarrollo Económico* (Spanish—Organization of Cooperation and Economic Development)

OCDM Office of Civil and Defense Mobilization

OCDQ Organizational Climate Description Questionnaire

OCDR Office of Collateral Development Responsibility

OCDS Overseas College of Defense Studies (UK)

O/Cdt Officer-Cadet

oce operational control equipment

OCE Office of Career Education; Office of the Chief of Engineers; Ontario College of Education

OC & E Oregon, California, and Eastern (railroad)

Ocean Ocean Transport and Trading Limited; The Ocean (Antarctic, Arctic, Atlantic, Indian, Pacific)

OCEAN Oceanographic Coordination Evaluation Analysis Network

OCEANAV Oceanographer of the U.S. Navy

oceaneer(ing) ocean engineer(ing)

Ocean Inst Oceanografiska Institute (Swedish—Oceanographic Institute)—Göteborg, Sweden

oceano oceanologic(al)(ly); oceanologist; oceanology

oceanog oceanography

OCED Office of Comprehensive Employment Development

OCEE Organisation de Coopération économique Européene (French—European Economic Cooperation Organization)

OCEL Optical Coating Evaluation Laboratory

OCEL Oxford Companion to English Literature

O Celt Old Celtic

ocf originally cultured formulation

OCF Officiating Chaplain to the Forces; Ossining Correctional Facility (Sing Sing); Owens-Corning Fiberglass

OC of F Office of the Chief of Finance

OCFR Oxford Committee for Family Relief

OCFT Office of Curriculum Frameworks and Textbooks

ocg omnicardiogram

ÖCG Österreichische Computer Gesellschaft (German—Austrian Computer Society)

och ochre

OCHAMPUS Office for the Civilian Health and Medical Program of the Uniformed Services

OCHS Old Colony Historical Society

oci organization conflict of interest

OCI Office of Computer Information (U.S. Department of Commerce); Office of the Coordinator of Information; Office of Current Intelligence (CIA); Operational Checkout Instruction

OCIB Organized Crime Intelligence Bureau

OCID Organized Crime Intelligence Division (LAPD)

OCIMF Oil Companies International Marine Forum

OCIS Organized Crime Information System (FBI)

OCJA Oklahoma Criminal Justice Association

OCJP Office of Criminal Justice Planning

ocl operator control language; optical communications link(age)

OCL Ocean Cargo Line; Overseas Container Line; Overseas Containers Limited

OCL/ACT Overseas Container Lines and Associated Container Transport

OCLAE Organización Continental Latino-Americana de Estudiantes (Spanish—Continental Organization of Latin American Students)

OCLC Ohio College Library Center; On-Line Computer Library Center

OCLI Optical Coating Laboratory, Inc

o'clock of the clock

OCLU Overseas Container Line (container) Unit

ocm oil content monitor

OCM Oxford Companion to Music

OCMA Oil Companies' Material Association

OCMH Office of the Chief of Military History

OCMMINST Office of Civilian Manpower Management Instruction (USN)

OCMS Optional Calling Measured Service (telephone)

OCN Operation Completion Notice

Ocn Bch Ocean Beach

ocnl occasional(ly)

OCNM Oregon Caves National Monument (limestone caverns near Medford, Oregon)

OCNUAD Oficina del Coordinador de las Naciones Unidas para la Ayuda en los Desastres (Spanish—Office of the Coordinator of the United Nations for Help in Disasters)

oco open-close-open

OCO Office of the Chief of Ordnance; Ontario College of Ophthalmology; San José, Costa Rica (El Coco Airport)

OCOA Organismo Coordinador de Operaciones Antisubversivas (Spanish—Coordinating Organism of Antisubversive Operations)—Uruguay's secret service

o'coat overcoat

OCOM Oficina Central de Organización y Metodos (Spanish—Central Office of Organization and Methods)

OComS Office of Community Services

OConUS outside continental limits of the United States

OCORA Office de Coopération Radiophonique (French—Office of Radiophonic Co-operation)—French overseas radio help for former colonies

O Corn Old Cornish

ocp output control pulses; overland common points

OCP Office of the Chief of Protocol (US Department of State); Office of Consumer Protection; Office of Cultural Presentations

OCP Oficina Central de Personal (Spanish—Central Personnel Office)

OCPAC Orange County Performing Arts Center, Segerstrom Hall, Costa

Mesa, California
OCPCJR Office of Crime Prevention and Criminal Justice Research
OCPD Officer-in-Charge Police District
OCPL Oklahoma City Public Library
ocr optical character reader; optical character recognition
ocr (OCR) optical character reader
OCR Office of Civil Rights; Office of Civilian Requirements; Office of Coal Research; Office of Collateral Responsibility; Office of Coordinating Responsibility; Office of the County Recorder; Organization Change Request; Organization for the Collaboration of Railways
OCRA *Organisation Clandestine de la Révolution Algerienne* (French—Secret Organization of the Algerian Revolution)
OCRD Office of the Chief of Research and Development
ocre optical character recognition equipment
OCRE Organizations Concerned about Rural Education
ocrit optical character-recognizing intelligent terminal
OCRS Organized Crime and Racketeering Section (Dept of Justice)
OCRSF Organized Crime and Racketeering Strike Force (U.S. Dept of Justice)
OCRU Office of Communication and Research Utilization
OCRWM Office of Civilian Radioactive Waste Management
ocs obstacle clearance surface; on company service; outer continental shelf
oc's obscene (telephone) callers; obscene (telephone) calls
OCS Office of Civilian Supply; Office of Commercial Services; Office of Contact Settlement; Officer Candidate School; Officers' Chief Steward; Outer Continental Shelf; Overseas Civil Servants; Overseas Courier Service
OCS' Overseas Civil Servants (members of the British Overseas Civil Service)
OCS *Organe de Controle des Stupéfiants* (French—Narcotic Drug Control Organiza-

tion)
OC of SA Office, Chief of Staff, Army
OCSE Office of Child Support Enforcement
ocsf office contents special form (insurance)
OCSIGO Office of the Chief Signal Officer
ocsn occasion
ocsnl occasional
ocsnly occasionally
OCSPC Outer Continental Shelf Policy Committee (California)
ocst overcast
oct octagon; octal; octane; octave; octavo; octet
Oct Octans (constellation); Octavius; October
OCT Office of the Chief of Transportation; Overseas Countries and Territories
octe optical component testing and evaluation
oct^e *octubre* (Spanish—October)
October October and November
OCTI *Office Central des Transports Internationaux par Chemins de Fer* (French—Central Office for International Railway Transport)
Octn Octanus (constellation)
October October Railway (Leningrad-Moscow); October Revolution (Bolshevik insurrection of October 1917)
octo(s) octoroon(s)
oct. pars *octava pars* (Latin—eighth part)
octr prot *octrooi protectie* (Dutch—patent protected)
OCTU Officer-Cadet Training Unit
octupl octuplicate
octup. *octuplus* (Latin—eightfold)
octv open-circuit television
ocu operational conversion unit
OCUA Ontario Council on University Affairs
OCUC Oxford and Cambridge Universities' Club
OCUFA Ontario Confederation of University Faculty Associations
ocul. *oculis* (Latin—to the eyes)
oculent. *oculentum* (Latin—eye ointment)
ocv open-circuit voltage
OCVs Overseas Cooperation Volunteers

oc vu ocean view
OCZ Ocean Container (terminal) Zebrugge
OCZM Office of Coastal Zone Management (NOAA)
od olive-drab; on demand; optical density; optic(al) disc; organization(al) development; original design; outside diameter; outside dimension; oven dried; overdose; overdrive
od (OD) overdrawn
o/d on demand; overdraft
o & d origin and destination
od och dylika (Swedish—and the like); *odur* (German—or)
o.d. *oculus dexter* (Latin—right eye)
Od *Odyssey*
OD Aerocondor (Aerovias Condor de Colombia); external grinding; officer of the day; Office of the Director; olive drab; Operational Directive; Ordnance Department; original design; outside dimension
O.D. Doctor of Optometry
oda occipito-dextra anterior
Oda Odessa
ODa Old Danish
ODA Office of Debt Analysis; Office of the District Administrator; Office of the District Attorney; Office of Drug Abuse; Overseas Development Administration; Overseas Development Assistance
ODALE Office of Drug Abuse Law Enforcement
ODAS Ocean Data Acquisition System
odat one day at a time
odb opiate-directed behavior; output to display buffer
odc other direct costs; outer dead center
ODC Old Dominion College; Overseas Development Corporation; Overseas Development Council
ODCSRDA Office of the Deputy Chief of Staff for Research, Development, and Acquisition (USA)
ODCTI Old Dominion College Technical Institute
odd (ODD) operator distance dialing
od'd overdosed
ODDRE Office of the Director of Defense Research and Engineering
ode one-day event
O^{de} *Oude* (Afrikaans, Dutch,

Flemish—old)
ODE Oil Drilling and Exploration
ODEC Ocean Design Engineering Corporation
ODECA *Organización de Estados Centroamericanos* (Spanish—Organization of Central American States)
ODECO Ocean Drilling and Exploration Company
od'ed overdosed
ODEE *Oxford Dictionary of English Etymology*
ODEPLAN *Oficina de Planificación Nacional* (Spanish—Office of National Planning)
ODESSA Ocean Data Environmental Sciences Services Acquisition
ODESSA *Organisation Der Ehemaligen SS Angehörigen* (German—Organization of Former Members of the SS)
ODESY On-Line Data Entry System
ODF Old Dominion Foundation; Operational Deployment Force
odfc outside diameter of female coupling
ODFI Open Die Forging Institute
O d G *Ordine del Giorno* (Italian—Order of the Day)
ODGSO Office of Domestic Gold and Silver Operations
ODH Ontario Department of Health
ODI Office of Defense Investigation (U.S. Department of Justice); Open-Door International (championing economic emancipation of women workers)
ODIL Overseas Development Institute Limited
Odin Scandinavian equivalent of Wotan, the supreme god of the Norse gods
od'ing overdosing
o-d-ing overdosing
O Div Ontario Division (RCMP)
ODJB Original Dixieland Jazz Band
o dk orlop deck
ODL Office of Defense Lending
odm ophthalmodynamometry
ODM Office of Defense Mobilization; Order of De Molay; Overseas Development Ministry
ODMA Optical Distributors and Manufacturers Associa-

tion
odmc outside diameter of male coupling
ODMC Office for Dependents Medical Care
odn own doppler nullifier
Odn Odense; Odin; Odinist (member of Nordic-supremacy sect)
ODO Outdoor Office(r)
ODOE Oregon Department of Energy
odom odometer
odont odontology
odop offset doppler
odoram. *odoramentum* (Latin—perfume)
odorat. *odoratus* (Latin—odorous, perfuming)
odorl odorless
ODOTS One-Day One-Trial System (for jurors)
odp occipito-dextra posterior; order despatched
ODP Office of Disaster Preparedness; Operational Deployment Plan(ning); Orbit Determination Program; Orderly Departure Program
odr order
ODR Office of Defense Resources
ODRC Office of Disaster Relief Coordinator (UN)
o'drive overdrive
ods oxide dispersion strengthened
o d's other denominations
ODS Ocean Data Station; Office of Defender Services; Orton Dyslexia Society
odsd overseas duty selection date
ODSE Open-Door Student Exchange
ODSI Ocean Data Systems Inc
ODSR Office of the Director of Scientific Research
odt occipito-dextra transverse; octal debugging technique; odor detection threshold; one-day trials; on-line debugging technique
ODT *Otago Daily Times*
ODTF Operational Development Test(ing) Facility
ODTS Operational Development Test Site
ODU Old Dominion University
od units optical-density units
ODWIN Opening Doors Wider in Nursing
ODWSA Office of the Directorate of Weapon Systems Analysis (USA)

oe oersted; omissions expected; open end(ed); outdoor education
oe (**OE**) organizational effectiveness
o/e on examination; otitis externa
o & e operations and engineering
oe organo espressivo (Italian—swell organ)
öe öesterreichisch (German—Austrian)
Oe oersted
OE Office of Education; Old English; Old Etonian; Oregon Electric (railroad)
OEA Oahu Education Association; Office of Economic Adjustment (USA); Office Education Association; Office of Environmental Affairs; Office Executives Association; Office of Export Administration; Ohio Education Association; Oregon Education Association; Outdoor Education Association; Overseas Education Association
OEA *Organización de los Estados Americanos* (Spanish—Organization of American States)
OEAA Oil Engineering Apprentices Association
OEB Oregon Educational Broadcasting
oec organizational entity code
OEC Office of Energy Conservation; Ohio Edison Company; Oil Exporting Countries
ÖEC Österreichischer Aero-Club (German—Austrian Aero Club)
OECC Office for Educational Credit and Credentials
OECCNU Organización para la Educación, la Ciencia, y la Cultura (Organization for Education, Science, and Culture)—UN
OECD Organization for Economic Cooperation and Development
OECE Organisation Européenne de Coopération économique (French—Organization for European Economic Cooperation)
OECF Overseas Economic Cooperation Fund
oeco outboard engine cutoff
OECON Offshore Exploration Conference
OECQ Organisation Euro-

péene pour la Contrôle de la Qualité (French—European Quality-Control Organization)

OECS Organization of East Caribbean States

oecu outboard engine cutoff

OED Oxford English Dictionary

OEDA Office of Energy Data and Analysis

OEDP Office of Employment Development Programs

oee outer enamel epithelium

OEEC Organization for European Economic Cooperation

OEEO Office of Equal Educational Opportunities

OEF Osteopathic Educational Foundation

OEG Operations Evaluation Group

oegt observable evidence of good teaching

OEGT Office of Education for the Gifted and Talented

oei organizational entity identity

OEI Offshore Ecology Investigation

OEI Oficina de Educación Ibero-americana (Spanish—Office of Ibero-American Education)

OEIPS Office of Engineering and Information Processing (NBS)

OEIU Office Employees International Union

o-e-l owner's risk of leakage

OEL Organization Equipment List

OEL/MA Ohio Educational Library/Media Association

oem oil-emulsion mud (oil well); original equipment manufacturer

oem (OEM) optical electron microscope

OEM Office of Environmental Mediation; Office of Executive Management

OEMA Office Equipment Manufacturers Association

oemcp (OEMCP) optical effects module electronic controller and processor

OEMs Original Equipment Manufacturers

oen oenanthic; oenanthyl; oenolyn; oenology; oenological; oenologist; oenomancy; oenomel (wine and honey); oenometer; oenophilist; oenophobist; oenopoetic

oeo officer's eyes only

OEO Office of Economic Opportunity; Ordnance Engineer Overseer

OEOB Old Executive Office Building (D.C.)

OEP Office of Emergency Planning; Office of Emergency Preparedness; Optional Educational Programs; Organization, Education, and Personnel

OEPP Organisation Européenne et Méditerranéenne pour la Protection des Plants (French—European and Mediterranean Organization for the Protection of Plants)

OEPS Office of Educational Programs and Services

OEQ Order of Engineers of Québec

OEQC Office of Environmental Quality Control

oer oersted (unit of magnetic force); original equipment replacement

o'er over

OER Office of Aerospace Research (USAF); Office of Energy Research; Officer Effectiveness Report; Officer Efficiency Report; Officer Engineering Reserve; Officers Emergency Reserve; Organization for European Research

oerc optimum earth-reentry corridor

OERPA Office of Exploratory Research and Problem Assessment (National Science Foundation)

OERS Organisation Européenne de Recherches Spatiales (French—European Space Research Organization)

OES Office of Economic Stabilization; Office of Emergency Services; Official Experimental Station; Order of the Eastern Star; Organization of European States

OES Organización de Estados Americanos (Spanish—Organization of American States)

oesbr oil-extended styrene-butadiene rubber

OESL Oceanographic and Environmental Service Laboratory (Raytheon)

oesoph oesophagus

OESP O Estado de São Paulo (State of Sao Paulo)—Brazil newspaper

OESS Office of Engineering Standards Services

OET Office of Education and Training; Office of Emergency Transportation; Overseas Exchange Transactions

OETA Occupied Enemy Territory Administration

OETB Offshore Energy Technology Board

OEVE Office of Earthquakes, Volcanoes, and Engineering (U.S. Geological Survey)

OEW Office of Economic Warfare

OEWG Open-Ended Working Group; Operation and Evaluation Wartime Group

OEX Office of Educational Exchange

OEZ osteuropäische Zeit (German—East European Time)

of. old face (type); optional form; outside face; oxidizing flame

o/f oxidation/fermentation; oxidizer to fuel ratio

Of Ovenstone factor

OF Oceanographic Facility; Odd Fellows; Old French; Operating Forces; Ophthalmological Foundation; Osteopathic Foundation; Oxbow Falls; Oxenstierna Foundation; Oxford Foundation

OFA Office of Financial Analysis; Old Folks Association; Orthopedic Foundation for Animals

OFAC Owens Fine Arts Center (Dallas)

O-factor oscillation factor

ofc office

OFC Overseas Food Corporation

OFCA Ontario Federation of Construction Associations

OFCC Office of Federal Contract Compliance

OFCCP Office of Federal Contract Compliance Programs

ofcl official

ofd one-function diagram; optical fire detector

OFDA Office of Foreign Disaster Assistance (U.S.)

OFDI Office of Foreign Direct Investments

OFE Office of Fuels and Energy

OFEMA Office Français d'Exportation de Matériel Aéronautique (French—French Office for the Exportation of Aeronautical Materiel)

O'Fest October Fest (Munich)

off. office(r); official

Off Officer
OFF Office for Families
OFFAR Office of Fuel and Fuel Additive Registration (EPA)
offen offensive (ammunition)
offeq office equipment
Offeq-1 Horizon-1, Israel's first satellite
offer. offertories; offertory
offg offering
offic official(ly)
Office Pubns Office Publications
off-st pkg off-street parking
OFHA Oil Field Haulers Association
ofhc oxygen-free high conductivity; oxygen-free high-carbon (copper)
OFI Office of the Federal Inspector
ofic oficial (Spanish—official)
OFIC Ohio Foundation of Independent Colleges
ofl official
Oflag Offizierlager (German—officer's prison camp)
OFlem Old Flemish
Ofly Offaly
OFM Office of Flight Missions (NASA); Office of Foreign Missions (State Dept.); Order of Franciscan Minors
OFNS Observer Foreign News Service
OFPA Order of the Founders and Patriots of America
OFPM Office of Fiscal Plans and Management
OFPP Office of Federal Procurement Policy
ofr off frequency rejection
OfR Office for Research
O Fr Old French
OFR Office of the Federal Register
OFR-ALA Office of Recruitment—American Library Association
OFris Old Frisian
O Frk Old Frankish
ofs one-function sketch
OFS Ontario Federation of Students; Orange Free State
OFSPS Office of Federal Statistical Policy and Standards
OFST Office of the Secretary of the Air Force
OFT Office of Fair Trade; Office of Fair Trading; Ohio Federation of Teachers; Optimal Foraging Theory
OFTC Overseas Finance and Trade Corporation
OFTS Office of Technical Services; Office of Transportation) Security; Officers Training School; Overseas Fixed Telecommunications System
OFY Opportunities for Youth (Canada)
og oh gee; oil gland; old girl; on ground; on guard; original gum
o-g orange-green
o/g opto-graphic; outgoing
OG Officer of the Guard; Old Gaelic; Olympic Games
O/G Opto/Graphic
ÖG Österreichische Galerie (Austrian Gallery)
OG O Globo (Rio de Janeiro's Globe)
O Gael Old Gaelic
OGAMA Ogden Air Materiel Area
Ogasawaras Ogasawara Islands (Bonins)
O-gauge 1-¼-inch track gauge (model railroads)
OGB Österreichischer Gewerkschaftsbund (German—Austrian Trade Union Federation)
OGC Office of General Counsel
OGCMD Ogden Contract Management District
Ogd Ogdensburg
OGDC Oil and Gas Development Corporation
oge (OGE) operational ground equipment
OGE Office of Government Ethics
OGES Operating Ground Equipment Specification
ogf option growth fund
ogg oggetto (Italian—object)
OGI Opera Guilds International
ÖGI Österreichische Gessel-schaft für Informatik (German—Austrian Society for Information Processing)
OGJ Oil and Gas Journal
ogl obscure glass
OGL Open General Licence
OGMC Ordnance Guided Missile Center
OGNR Oribi Gorge Nature Reserve (South Africa)
OGO Orbiting Geophysical Observatory
OGPU Obiedinennoye Gosu-darstvennoye Politicheskoye Upravlenie (Russian—United State Political Administration)—*q.v.m.*—VOT
OGR Ontario Government Railway (Ontario Northland)
OGR Official Guide of the Railways
ogse operational ground-support equipment
OGSEL Operational Ground-Support Equipment List
OGSM Office of General Sales Manager
OGSR Office of Graduate Studies and Research
o-g stain orange-green stain
ogt on-going thing; outlet gas temperature
OGTT Oral Glucose Tolerance Test(ing)
OGU Occupational Guidance Unit
ogv outlet guide vane
oh (OH) ocular herpes
oh. office hours; on hand; open hearth; out home; oval head; overhead; over-the-horizon-(communication)
o/h overhaul
o.h. omni hora (Latin—hourly)
o-H on-Hudson
OH hydroxyl radical (symbol); Ohio; Olduvai Hominid; Omega House; opera house; San Francisco and Oakland Helicopter Airlines (2-letter code)
O/H Overzuche Handelsmaat-schappij (Dutch—Overseas Trading Company)
OH-6 Hughes observation helicopter called Cayuse
OH-13 Bell Sioux helicopter
OH-23 Hiller Raven utility helicopter
OH-58 Bell Kiowa turbine-powered helicopter
oha outside helix angle
OHA Occupational Health Administration; Office of Hearings and Appeals; Oriental Herb Association
O'Hare O'Hare International Airport (Chicago)—named for navy pilot Edward H (Butch) O'Hare killed during World War II
OH-B Ocean Hill-Brownsville
OHBMS On Her (His) Britannic Majesty's Service
ohc outer hair cells; overhead cam
OHC Office of Humanities Communication; Ottumwa Heights College; Overseas Hotel Corporation
OHCS Office of Home Care Services
ohd organic hearing disease; organic heart disease; over-

head drive
OHDETS Over-Horizon Detection System
OHDS Office of Human Development Services (HEW)
OHD & W Outer Harbor Dock and Wharf
OHE Office of Health Economics
oheat overheat
ohf overhaul factor
Ohf Omsk hemorrhagic fever
OHG Old High German
OHG Offene Handelsgesellschaft (German—ordinary partnership)
Oh Gloria Oh Gloria inmarcesible (Spanish—Oh Unfading Glory)—Colombian anthem
ohi ocular hypertension indicator
OHI Oil Heat Institute
OHI Organisation Hydrographique Internationale (French—International Hydrographic Organization)
OHIA Oil Heat Institute of America
Ohio Turn Ohio Turnpike
Ohio U Pr Ohio University Press
OHIP Ontario Hospital Insurance Plan
OHJ Old House Journal
OHK Okayama Hoso KK
OHL Oberste Herresleitung (German—Supreme Headquarters)
ohm. ohmmeter
ohm-cm ohm-centimeter
OHMO Office of Hazardous Materials Operations
OHMR Office of Hazardous Materials Regulation
OHMS On Her (His) Majesty's Service; Onboard Health Monitoring System
OHN Occupational Health Nurse
OHNC Occupational Health Nursing Certificate
OHNO Occupational Health Nursing Officer
OHNS Occupational Health Nursing Sister
oho out-of-house operation
ohp overhead projection; oxygen at high pressure
oh Ped ohne Pedale (German—without pedals)
ohrf overhaul replacement factor
OHRG Official Hotel and Resort Guide
ohs open-hearth steel

ohs (OHS) hydroxy-steroids
OHS Office of Highway Safety; Ontario Humane Society; Oral Hygiene Service; Oregon Historical Society; Organization of Historical Studies; Overland Highway Society
OHSGT Office of High-Speed Ground Transportation
OHSIP Ontario Health Services Insurance Plan
OHSPAC Occupational Health-Safety-Programs Accreditation Commission
oht overheating temperature
OHTE Ohmic Heating Toroidal Experiment
ohv overhead valve; overhead vent
ohv's off-highway vehicles
oi oil-immersed; oil-immersion
o-i orgasmic impairment
o/i opsonic index
o & i organizational and intermediate
OI Office of Information; Office Instruction; Operating Instruction; Optimist International; Oriental Institute
O-I Owens-Illinois
OIA Ocean Industries Association; Office of Impact Analysis; Office of Industrial Associates; Office of International Administration; Oil Import Administration; Oil Insurance Association; Outboard Industry Associations
OIA Organización Internacional de Azucar (Spanish—International Sugar Organization)
OIAA Office of Inter-American Affairs; Office of International Aviation Affairs
OIAB Oil Import Appeals Board
OIAC Organización Internacional de la Aviación Civil (Spanish—International Civil Aviation Organization)
OIAJ Office for Improvements in the Administration of Justice
OIAS Occupational Information Access System
OIB Ohio Inspection Bureau; Oklahoma Inspection Bureau
oic oil cooler
O-i-C Officer-in-Charge
OIC Oceanographic Instrumentation Center; Office of the Insurance Commissioner; Officer in Charge; Ohio Improved Chester (white

swine); Oil Industry Commission; Opportunities Industrialization Centers; Overseas Investment Commission
OIC Organisation Interafricaine du Café (French—Inter-African Coffee Organization); *Organisation Internationale du Commerce* (French—International Trade Organization)
OICA Ontario Institute of Chartered Accountants; Oregon Independent Colleges Association
OICD Office of International Cooperation and Development
OIcel Old Icelandic
OICF Oklahoma Independent College Foundation; Oregon Independent College Foundation
OICI Organización Interamericana de Cooperación (Spanish—Inter-American Cooperation Organization)
OICJ Office of International Criminal Justice
oic's one-issue callers (call-in programs)
OICS Office of Interoceanic Canal Studies
oid (Latin suffix—resembling)—sigmoid; original issue discount
Oid mortales Oid, mortales, el grito sagrado (Spanish—Hear, mortals, the sacred cry)—Argentine anthem
OIE Office of Indian Education; Office of International Epizootics
OIE Organisation Internationale des Employeurs (French—International Organization of Employers)
OIEA Organismo Internacional de Energia Atómica (Spanish—International Atomic Energy Agency)—IAEA
OIER Office of International Economic Research
OIF Office for Intellectual Freedom (ALA)
OIG Office of the Inspector General
OIG Organisation Intergouvernementale (French—Intergovernmental Organization)
OIGS On Indian Government Service
oih (OIH) ovulation-producing hormone
OIHP Office International

d'Hygiene Publique (French
—International Office of
Public Health)—UN
OII Office of Invention and In-
novation
*OIJ Organisation Internatio-
nale des Journalistes*
(French—International Orga-
nization of Journalists)
OIL Operation Inspection Log;
Orbiting International Labo-
ratory
*OIL Organizzazione Interna-
zionale del Lavoro* (Italian—
International Labor Organiza-
tion)
oiloff oil ripoff
OILSR Office of Interstate
Land Sales Registration
OIM Oriental Institute
Museum (University of Chi-
cago)
OINA Oyster Institute of
North America
O-in-C Officer-in-Charge
OINC Officer in Charge
*OING Organisation Internatio-
nale Non-Gouvernementale*
(French—Non-Governmental
Organization)
oint ointment
OIO Oklahomans for Indian
Opportunity
oip oil in place; oxford india
paper
OIP Office for Information
Programs (NBS); Office of
International Programs; Op-
erations Improvement Pro-
gram
*OIPC Organisation Internatio-
nale de Police Criminelle*
(French—International Crim-
inal Police Organization)—
also known as Interpol; *Or-
ganisation Internationale de
Protection Civile* (French—
International Civil Defense
Organization)
OIPH Office of International
Public Health
OIr Old Irish
OIR Office of Inter-American
Radio
OIRB Oregon Insurance Rat-
ing Bureau
OIRM Office and Industrial
Records Management
*OIRSA Organism Internacio-
nal Regional de Sanidad
Agropecuaria* (Spanish—In-
ternational Regional Associa-
tion for Healthy Land and
Cattle)
*OIRT Organisation Internatio-
nale de Radiodiffusion et*

Télévision (French—Interna-
tional Radio and Television
Organization)
OIS Office Information Sys-
tem; Overseas Investors Ser-
vices
OISA Office of International
Scientific Affairs
OISE Ontario Institute for
Studies in Education
OISS Online Information
Search Service
*OISS Organisation Ibéro-amé-
ricaine de Securite Sociale*
(French—Iberian-American
Social Security Organiza-
tion)
*OISTV Organisation Interna-
tionale pour la Science et la
Technique du Vide* (French—
International Organization for
Vacuum Science and Tech-
nology)
O i T Officer in Training
(rookie police officer)
O It Old Italian
OIT Organic Integrity Test
*OIT Organisation Internatio-
nale du Travail* (French); *Or-
ganización Internacional del
Trabajo* (Spanish)—Interna-
tional Labor Organization
also known as ILO
OITF Office of International
Trade Fairs
OIUC Oriental Institute of the
University of Chicago
*OIVV Office International de
la Vigne et du Vin* (French—
International Office of Vines
and Wines)
OIW Oceanographic Institute,
Wellington (New Zealand)
OIWP Oil Industry Working
Party
OIWR Office of Indian Water
Rights
oj open-joint; open-joist(ed)
orange juice
oJ ohne Jahr (German—with-
out year)—no date
OJAJ October, January, April,
July
OJARS Office of Justice As-
sistance, Research, and Sta-
tistics
*OJC Organisation Juive de
Combat* (French—Jewish
Combat Organization)
OJD Office de Justification de
la Diffusion
OJDYD Office of Juvenile De-
linquency and Youth Devel-
opment
*OJEC Official Journal of the
European Communities*

oji on-the-job injuries
OJJ Office of Juvenile Justice
OJJDP Office of Juvenile Justice and Delinquency Prevention
oJr old Jamaica rum
ojt on-the-job training
OJT (National) On-the-Job
Training (Program)
ok all correct; okay; optical
klystron; outer keel
ok ohne kosten (German—
without cost); *ola kala*
(Greek—all is fine, all is
good)—believed to be the
original okay used by Greek
sailors of antiquity
OK all correct; okay; Oklahoma; Old Kinderhook
(birthplace and home of President Martin Van Buren),
Democratic OK Club be-
lieved to have started practice of putting "OK" on deals
and documents they ap-
proved; Old Kingdom
(Egypt); Oskar Kokoschka
(1866–1980)
O & K Orenstein & Koppel
Ø K Østasiatiske Kompagni
(Danish—East Asiatic Com-
pany)
oka otherwise known as
OKA Okinawa, Ryukyu Islands (airport)
OKC Oklahoma City, Oklahoma (airport)
OKd okayed
Okefinokee Okefinokee Na-
tional Wildlife Refuge and
the Okefinokee Swamp be-
tween northern Florida and
southern Georgia
*OKH Oberkommando des Hee-
res* (German—Army High
Command)
Okhotsk Sea of Okhotsk be-
tween Kamchatka Peninsula,
Sakhalin Island, and eastern
Siberia
Okin Okinawa(n)
*OKL Oberkommando der Luft-
waffe* (German—Air Force
High Command)
Okla Oklahoma; Oklahoman
OklaC Oklahoma City
*OKM Oberkommando der Ma-
rine* (German—Naval High
Command)
Okt Oktober (German—Oct-
ober); *Oktyabr* (Russian—
October)
*OKT Oslo Kommune Tunnel-
banekontoret* (Oslo subway
system)

Oktronics Oklahoma Electronics (corporation)

OKW Oberkommando der Wehrmacht (German—Armed Forces High Command)

ol oil level; operating license; or less

ol' old

o/l operations/logistics; outlook

ol. oleum (Latin—oil)

o.l. oculus laevus (Latin—left eye)

ö L östlich Längengrad (German—east longitude)

Ol olive; olympiad, four-year period between the Olympic Games

OL October League (communist group active in U.S.); Old Latin; Olsen Line; Oranje Line (Orange Line)

ola occipito-laeva anterior

OLA Office of Legislative Affairs; Ohio Library Association; Oklahoma Library Association; Ontario Library Association; Osteopathic Libraries Association

OLADE Organización Latinamericana de Energía (Spanish—Latin American Energy Organization)

o'land overland

OLAPEC Organization of Latin American Petroleum Exporting Countries

OLAS Office of Arid Land Studies (University of Arizona) Organization of Latin American Solidarity; Organization of Latin American Students

OLAS Organización Latinoamericana de Solidaridad (Spanish—Latin American Solidarity Organization)

Olav Tryg Olav Trygvason

olbm (OLBM) orbital launched ballistic missile

OlBr olive brown

olc on-line computer

OLC Oak Leaf Cluster; Office of Legal Counsel

olcc optimum life-cycle costing

O L Cr Ordnance Lieutenant-Commander

OLCS On-Line Computer System

OLD Office of Legislative Development

Old Bailey London's Central Criminal Court

old-fash old-fashioned

Oldfos Old Established Forces

Old Maid's Old Maid's Day (June 4)

Old Point Old Point Comfort, Virginia

old pro(s) old professional(s)

old rep old repertory; old reprobate

Olds Oldsmobile

OLDS On-Line Display System

Old Territorial Old Territorial Penitentiary (Santa Fé, New Mexico)

Old Test. Old Testament

OLE Office of Library Education (American Library Association)

OLEA Office of Law Enforcement Assistance

oleo oleomargarine; oleoresins; oleum

OLEP Office of Law Enforcement and Planning

olericult olericulture

'oleum petroleum

O-levels ordinary levels (of educational tests)

olf olfactory; on-line filing

OLF Ohio Library Foundation; Orbital Launch Facility; Organ Literature Foundation

Olg Olga

OlG olive green

OLG Old Low German

Olgas The Olgas—mountain range west of Ayers Rock in Australia's Northern Territory

OLHMIS On-Line Hospital Management Information System

Oli Oliver

OLI Ocean Living Institute

O-license operator's license

Olig Oligocene

oligo (Latin prefix—few or small)—oligarchy

Olive Olivera; Olivia

OLIVER On-Line Instrumentation Via Energetic Radioisotopes

Oliver P Oliver (Cromwell) Protector

OLL Office of Legislative Liaison

OLMAT Otis-Lennon Mental Ability Test

olmr (OLMR) organic liquid-moderated reactor

OLMR Office of Labor Management Relations

OLMS Office of Labor Management Standards

ol'n olden

OLOGS Open-Loop Oxygen-Generating System

ol ol olive oil

olos out of line of sight

OLOS Office for Library Outreach Services

olow orbiter liftoff weight

olp occipito-laeva posterior; original list price

OLP Organización para la Liberación Palestina (Spanish—Palestinian Liberation Organization)—the PLO terrorists

olpar other large phased-array radar

OLPR Office for Library Personnel Resources (ALA)

OLPS On-Line Programming System

olq officer-like qualities

olr overload relay

OLRB Ontario Labor Relations Board

ol res oleoresin

olrt on-line real time

ols ordinary least squares

ol's office ladies (divorcees and spinsters); old girls

OLS Optical Landing System

olsc on-line scientific computer

OLSD Office for Library Service to the Disadvantaged (ALA)

olt occipito-laeva transverse

ol & t owners, landlords, and tenants

Olt Old Italian

oltt on-line teller terminal

olv olivaceous; olive; on-line validation

OLV Onze Lieve Vrouw (Dutch—Our Lady)

o-l v's ovo-lacto vegetarians

Oly Olympia; Olympic

Olym Olympia

Olympic Olympic National Park, Washington

Olympics Olympic Games; Olympic Mountains, Washington

om old man; old measurement; old men; operational monitor; organic matter; our memo; outer marker

o & m (O & M) operation and maintenance

o.m. omni mane (Latin—every morning)

Om Omaha; Oman

Om. Book of Omni

OM Occupational Medicine; Old Man (colloquial); Ordnance Map

O.M. Order of Merit

O & M Organization and Methods

OM Obermanual (German—upper manual keyboard); *Ostmark* (East German mark)

O.M. *Optimus Maxum* (Latin—best and greatest)-title given Jupiter by the Romans who worshipped him

oma orderly marketing arrangement

oma (Greek—swelling or tumor)—carcinoma, glaucoma, hematoma, lipoma, sarcoma

Oma Omaha, Nebraska

OMA Ocean Mining Administration (USDI); Office of Maritime Affairs; Oklahoma Military Academy; Omaha, Nebraska (airport); Ontario Medical Association; Overall Manufacturers' Association

OMAI *Organisation Mondiale Agudas Israel* (French—Agudas Israel International Organization)

Oman Sultanate of Oman (Arab oil-producing nation on Arabia's southeast coast), *Saltanat Oman*

omarb *omarbetad* (Swedish—revised)

OMARS Outstanding Media Advertising by Restaurants

OMAT Office of Manpower, Automation, and Training

Omb Ombudsman

OMB Office of Management and Budget Ontario Municipal Board

OMBAC Old Mission Beach Athletic Club

OMBE Office of Minority Business Enterprise

om. bid. *omnibus bidendis* (Latin—every two days)

OMC Office of Munitions Control; Outboard Marine Corporation

omd off-market date

OMD Organic Mental Disorder

omdr off-market date received

'ome (Cockney contraction—home)

OME Office of Manpower Economics; Office of Minerals Exploration; Ordnance Mechanical Engineer(ing)

OMEF Office Machines and Equipment Federation

OMEGA Optimal Missile Engagement Guidance Algorithm (worldwide navigational system)

OMEL Orient Mid-East Lines

OMEP *Organisation Mondiale pour l'éducation Préscolaire* (French—World Organization for Pre-school Education)

O-Mess Officer's Mess

OMF Office of Management and Finance

omfp obtaining money by false pretenses

OMG Ophthalmology Medical Group

OMGE *Organisation Mondiale de Gastro-Entérologie* (French—World Gastro-Enterological Organization)

OMGUS Office of Military Government, United States

OMH Office of Mental Health

OMI Olympic Media Information; Operation Move-In

O.M.I. Oblate of Mary Immaculate

OMI *Organización Maritima Internacional* (Spanish—International Maritime Organization)

OMII Oxy Metal Industries International

omiom original meaning is the only meaning

omit. orinthine-decarboxylase, motility, indole, trytophandeaminase

omkr *omdring* (Norwegian—about)

oml outside mold line

OML Ontario Motor League; Orbiting Military Laboratory

OMM Office of Minerals Mobilization

OMM *Organisation Météorologique Mondiale* (French), *Organización Meteorologica Mundial* (Spanish—World Meteorological Organization)—WMO

OMMA Outboard Motor Manufacturers Association

OMMS Office of Merchant Marine Safety (USCG)

omn. bih. *omni bihora* (Latin—every two hours)

omn. hor. *omni hora* (Latin—every hour)

omni omnidirectional; omnirange; omnivisual

omn. man. *omni mane* (Latin—every morning)

omn. noct. *omni nocte* (Latin—every night)

omn. quad. hor. *omni quadrante hora* (Latin—every quarter of an hour)

omor one man, one responsibility

omp organo-metallic polymer(s)

ompa one-man pension arrangement

OMPD Office of Mineral Po-

licy Development

OMPER Office of Manpower Policy Evaluation and Research

ompf omphaloskepsis

OMPI *Organización Mundial de la Propiedad Intelectual* (Spanish—World Intellectual Property Organization)

OMPO Oahu Metropolitan Planning Organization

ompr optical mark page reader

OMPRA Office of Minerals Policy and Research Analysis

OMPSA *Organisation Mondiale pour la Protection Sociale des Aveugles* (French—World Organization for the Welfare of the Blind)

OMPU *Oficina Municipal de Planeamiento Urbano* (Spanish—Municipal Office of Urban Planning)

omr office methods research; optical mark reader; optical mark recognition

OMR Officer Master Record

OMRD Overseas Mineral Resource Development

OMRs Optical Mark Readers

oms output per man shift

OMS Office of Management Studies; Orbital Maneuvering System

OMS *Organisation Mondiale de la Santé* (French), *Organización Mundial de la Salud* (Spanish)—World Health Organization—WHO; *Otdel Mezdunarodnyk Svyazey* (Russian—International Relations Section)—network of overseas agents

OMSA Orders and Medals Society of America

OMSF Office of Manned Space Flight (NASA)

OMSIP Ontario Medical Surgical Insurance Plan

omt orthomode transducer

OMT Old Merchant Taylors

OMTS Organizational Maintenance Test Station

OMV Orbital Maneuvering Vehicle

on. octane number

o/n own name

on. *onomastikon* (Greek—lexicon)

o.n. *omni nocte* (Latin—every night)

On *Onorevole* (Italian—Honorable); *Onsdag* (Danish—Wednesday)

ON Official Number; Ogden Nash; Old Norse; Ontario

Northland (railway); Operation Notice
O.N. Orthopedic Nurse
O/N Order Number
O & N Oregon & Northeastern (railroad)
ÖN Österreichische National-bibliotek (Austrian National Library)
ona optical navigation attachment
ONA Office of National Assessment; Office of Noise Abatement; Overseas National Airways; Overseas News Agency
on a/c on account
ONAC Office of Noise Abatement and Control
ONAP Orbit Navigation Analysis Program
on approv on approval
onbep onbepaald (Dutch—indefinite)
O-N Border Oder–Neisse Border separating Germany and Poland
onc operational navigational chart(s)
ONC Office of New Careers; Oficina Nacional del Café (Spanish—National Coffee Administration—Honduras); Oregon-Nevada-California (fast freight truck line)
oncol oncologic(al)(ly); oncologist(ic)(al)(ly); oncology; oncolysis; oncolytic(al)(ly)
OND Ophthalmic Nursing Diploma
ONDC Office of National Drug Control
ONDCP Office of National Drug Control Policy
ONE Office of National Estimates (CIA)
Oneg Onegin
Onega Lake Onega northeast of Leningrad, called Ozero Onezhskoye by the Russians
Oneida Oneida Community noted for the silverware and steel traps produced while practicing complex marriage and common care of their offspring in Oneida, New York; in 1881 the commune was incorporated
ONEO Office of Navajo Economic Opportunity
ONERA Office National des Etudes et des Recherches Aérospatiales (French—space research agency)
one-spot $1 bill
ONF Offensive Nuclear Forc-

es; Old Norman-French
onfm on nearest full moon
ong ongaku (Japanese—music); ongeveer (Dutch—about, approximately, roughly)
ONG Old North German
ONG Organisation Non-Gouvernementale (French—Non-Governmental Organization)
on hol(s) on holiday(s)
ONI Office of Naval Intelligence; Office of NWTS Integration
ÖNJ Österreichische National-bibliothek Josefsplatz (German—Josefsplatz Austrian National Library)
ONM Ocmulgee National Monument; Office of Naval Material
ONMSS Office of Nuclear Material Safety and Safeguards
ONNI Office of National Narcotics Intelligence
onnm on nearest new moon
ono or near offer
o-'n'-o one and only
ONO Oesnoroeste (Spanish—west northwest); oost noord (Dutch—east northeast)
oost (Dutch—east northeast)
onomast onomastic(al)(ly); onomastics; onomatologist; onomatology
onomat onomatologic(al)(ly); onomatologist(ic)(al)(ly); onomatology; onomatopoeia
O Norw Old Norwegian
O Norm F Old Norman French
O North Old Northumbrian
O Norw Old Norwegian
o noz oil nozzle
onp operating nursing procedure
ONP Office of National Programs; Olympic National Park (Washington)
ONR Office of Naval Research; Official Naval Reporter
ONRL Office of Naval Records and Library
ONRRR Office of Naval Research Resident Representative
ON Rwy Ontario Northland Railway
ONSR Ozark National Scenic Riverways (Missouri)
On Sta On Station
ont ontology; ordinary neap tide
Ont Ontario
ONT Our New Thread

(Clark's trademark)
ONTC Ontario Northland Transportation Commission
Ont Pen Ontario Penitentiary
Ont Sci Cen Ontario Science Center
ONU Organisation Nations Unies (French—United Nations Organization); Organización de las Naciones Unidas (Spanish—United Nations Organization)—UNO; Organizzazione Nazioni Unite (Italian—United Nations Organization)
ONUC Operation des Nations Unies, Congo (French—United Nations Operation in the Congo)
ONUDI Organización de la Naciones Unidas para el Desarollo Industrial (Spanish—United Nations Organization for Industrial Development)
ONUESC Organisation des Nations Unies pour l'Education, la Science et la Culture Intellectuelle (UNESCO)
ONULP Ontario New Universities Library Project
on w onovergankelijk werkwoord (Dutch—intransitive verb)
ONW Oregon and Northwestern (railroad)
ONWI Office of Nuclear Waste Isolation
ONWM Office of Nuclear Waste Management
ONWR Okefinokee National Wildlife Refuge (Florida and Georgia); Ottawa National Wildlife Refuge (Ohio); Ouray National Wildlife Refuge (Utah)
ony onymous (opposite of anonymous)
oo (OO) office of origin
o/o oil/ore (carrier); on order
o & o owned and operated
o-to-o out-to-out
oo (Latin prefix—egg)—oocyte, oology
o(O) original
O/o Order of
OO Observation Officer; Oceanic Operators; Oceanographic Office
O/O Office of Oceanogrphy (UNESCO)
O of O Order of Owls
ooa on or about
OOA Office of Ocean Affairs
OOAA Olive Oil Association of America

OOAMA Ogden Air Materiel Area
oob opening of business; out of bed
o-o B off-off Broadway; off-off Broadway theater(s)
OoB Order of Battle
OOB Old Orchard Beach
oobe out of body experience
OoC Office of Censorship
OOCH Orient Overseas Container Holdings
OOCL Orient Overseas Container Line
OOD Officer of the Day; Officer of the Deck
oodep owners, officers, directors, and executive personnel
Oody Eunice
OO/Eng out of stock but on order from England (for example)
OoF Office of Facilitation
OOG Office of Oil and Gas; Officer of the Guard
OOH *Occupational Outlook Handbook*
OOHA Operation Oil Heat Associates
ooj obstruction of justice
ool oology; operator-oriented language
OOL Odessa Ocean Line; Orient Overseas Line
oolhmd optimized optical-link helmet-mounted display
oolr ophthalmology, otology, laryngology, rhinology
OOM Officers Open Mess
OO McIntyre Oscar Odd McIntyre (newspaper columnist: *New York By Day*)
o/o/o out of order
O o O One on One (tv program)
OOO-gauge ¾-inch track gauge (model railroads)
oop out of pocket (expenses); out of print (book)
OOP Oceanographic Observations of the Pacific
ooparts out-of-place artifacts
oops off-line operating simulator; offshore oil-pollution sleeve
OOPS Organization of Oil Producing States
OOQ Officer of the Quarters
OOR Office of Ordnance Research
oos orbit-to-orbit shuttle; orbit-to-orbit stage; out of stock
o & o's owned and operated (tv broadcast) stations (controlled by a network)
OOSC Olfactronics and Odor Sciences Center (IITRI)
oot out of tolerance; out of town
OOT Office of Operational Testing
oote out-of-town executive
ootg one of the greats
Ooty Ootacamund, Madras
OOW Officer On Watch
op oil pressure; old prices; open policy; opera; operating point; operation; operation plan(s); operational; operational priority; operetta; opium; opposite prompt (stage left); optical probe; opus; ordinary pay; other people's (possessions); out of print; outer panel; outside production; overproof; overprune; overpuff
op (OP) outpatient
o/p off peak; optional; output; overpriced
o & p ova and parasites
Op optical art (art accented with or based on optical illusions); Oregon pine
Op. *Opus* (Latin—composition, literary or musical work)
OP Observation Post; Office of Preparedness; Office of Protocol (US Department of State); Open Policy (floating cargo insurance); Oregon pine
O-P Oppenheimer-Phillips (process)
O.P. *Optimus Maximus* (Latin—supreme and best)— Jupiter's title as he was believed to be the king of the gods and the ruler of all rulers
opa optical plotting attachment; optoelectric pulse amplifier
OPA Office of Population Affairs; Office of Price Administration; Office of Public Affairs; Overall Payments Agreement
OPA Oficina Postal Ambulante (Spanish—Mobile Post Office)
opal hydrous silica (SiO_2 .nH_2O)
opal. optical platform alignment linkage
op amp operational amplifier
OPANAL Organismo para la Proscripción de las Armas Nucleares en la América Latina (Spanish—Organization for the Prohibition of Nuclear Weapons in Latin America)

op art optical art (art involving optical illusion)
OPB Occupational Pensions Board
OPBE Office of Planning, Budgeting, and Evaluation (NIE)
OPBMA Ocean Pearl Button Manufacturers Association
opc office percentage; ordinary portland cement
OPC Office of Price Control; Office of Public Communication; Ohio Power Company; Out-Patient Clinic; Overseas Press Club
OPCA Overseas Press Club of America
op. cit. opere citato (Latin—in the work cited); *opus citato* (Latin—in the work cited)
OPCNM Organ Pipe Cactus National Monument
opco operating company
op code operation code (data processing)
op com opéra-comique (French—comic opera; operetta)
opcon(s) operation control(s)
OPCS Office of Population Censuses and Surveys
opd optical path difference
o-p-d oto-palato-digital (syndrome)
OPD Office of Policy Development (White House); Officer Personnel Directorate; Out Patient Department
opdar optical direction and ranging
OPDD Operational Plan Data Document
op dent operative dentistry
OPDR Oldenburg–Portugiesische–Dampfschifs–Reiderei (steamship company)
ope open-point expanding; opium; oxidation pond effluents
OPE Office of Planning and Evaluation (FBI); Operations Project Engineer
O P & E Oregon, Pacific & Eastern (railroad)
OPEC Oil Producer's Economic Cartel; Organization of Petroleum Exporting Countries
op ed opposite the editorials (newspaper page usually reserved for readers' letters and syndicated columns)
opef overall plume-enhancement factor
OPEI Outdoor Power Equipment Institute

OPEIU Office and Professional Employees International Union
open. open circuit; opening
opens. open circuits (electrical parlance); openings
opep (OPEP) orbital plane experiment package
OPEP Organisation des Pays Exportateurs de Pétrole (French—Organization of Petroleum Exporting Countries)
OPEP/OPEC Organización de Paises Exportadores de Petróleo (Spanish—Organization of Petroleum Exporting Countries)
oper operational
O Per Old Persian
OPER Office of Policy, Evaluation, and Research
Opéra Paris Opera
Opera-Com Opéra-Comique (Paris)
operg operating
OPers Old Persian
opex operational (and) executive (personnel)
OPEX Operational, Executive (and Administrative Personnel Program of the United Nations)
opfor opposition force
opg opening
O Pg Old Portuguese
OPG Overseas Project Group
OPGA Ohio Personnel and Guidance Association; Oregon Personnel and Guidance Association
oph office phone; ophicleide; ophthalmologist; ophthalmology; ophthalmoscope; ophthalmoscopic
Oph Ophiuchus (constellation)
Oph.D. Doctor of Ophthalmology
ophidiol ophidiologic(al)(ly); ophidiologist; ophidiology
OPHS Operational Propellant Handling System
ophth ophthalmologist; opthalmology
ophthal ophthalmic; ophthalmologist; ophthalmology
Ophthalmias Ophthalmia Range of mountains in Western Australia near Jiggalong and Mundiwindi
ophthalmol ophthalmologic-(al)(ly); ophthalmologist; ophthalmology
OPI Office of Primary Interest; Office of Programs Integration (ERDA); Office of Protective Intelligence (U.S. Secret Service); Office of Public Information; Office of Public Inquiry; Offsite Production (Purchase) Inspection; Omnibus Personality Inventory; Ordnance Procedure Instrumentation; Outside Production (Purchase) Inspection
OPIC Overseas Private Investment Corporation
opim order processing and inventory monitoring
opis opisometer
OPIS Operational Priority Indicating System
opl operational
opl oplag (Danish—edition)
OPL Omaha Public Library; Orlando Public Library; Ottawa Public Library
OPLA Offshore Pollution Liability Agreement
OPLP Office of Program and Legislative Planning
opm operations per minute; operator programming method; optically-projected map; orthophoto map; other people's money
OPM Office of Personnel Management; Office of Production Management
OPMA Office Products Manufacturers Association
OPMAC Operation for Military Aid to the Community
OPMCS Otto Pre-Marital Counselling Schedules
opn open (flow chart); operation
o.p.n. ora pro nobis (Latin—pray for us)
OpNav Office of the Chief of Naval Operations
OPNAVINST Office of the Chief of Naval Operations Instruction
opnd opened (flow chart)
opng opening
OPNL Osaka Prefectural Nakanoshima Library (Japan)
opnn opinion
Op. no. opus number
opo one-person operation; one price only; other programmed operations
Opo Oporto
OPO Office of Personnel Operations (US Army)
O Pol Old Polish
OPOR Office of Public Opinion Research
O por O Ojo por Ojo (Spanish—Eye for an Eye)—

Guatemalan terrorists
O Port Old Portuguese
OPOs one-person-operated buses
opp opportunity; opposed; opposite; opposition; out of print at present
OPP Office of Pesticide Programs; Ontario Provincial Police; Otago Press and Produce
OPPE Office of Programming, Planning, and Evaluation; Operational Propulsion Plant Examination (USN)
OPPI Organization of Pharmaceutical Producers of India
opplan operating plan
oppor opportunity
oppo's opposite numbers
oppy opportunity
opq opaque
opr operate; operator; optical pattern recognition
OPr Old Provençal
OPR Office of Planning and Research; Office of Population Research (Princeton); Office of Primary Responsibility; Office of Professional Responsibility (FBI); Office of Professional Responsibility (INS)
oprad operations research and development
oprex operational exercise
O Prov Old Provençal
opr's old prices riots
OPruss Old Prussian
ops operations; opposite prompter's side (of stage)
op's other people's
OPS Office of Price Stabilization; Office of Product Standards; Oxygen Purge System
OPS Organisation Panaméricaine de la Santé (French—Pan-American Health Organization); *Organización Panamericana de la Salud* (Spanish—Pan-American Health Organization)
ops analysis operations analysis
Ops Atts Gen Opinions of the Attorneys-General of the United States
opscan optical scanning
Ops Comms Opinions of the Commissioners
OPSM Optical Prescriptions Spectacle Makers
OPSP Office of Product Standards Policy
OPSR Office of Pipeline Safety Regulations

opstat operational status

opt optic; optical; optician; optics; optimal; optimum; option; optional

OPT Office of Promotion and Tourism

OPTA Organ and Piano Teachers Association

optacon optical-to-tactile converter

Opt Acta Optica Acta (Latin—Optics Gazette)

Opt Commum Optics Communications

Opt.D. Doctor of Optometry

OPTEVFOR Operational Test and Evaluation Force

OPTEVG Operational Test and Evaluation Group

opti optimist(ic); optimize; optimum

optic. optical(ly); optician; opticociliary; opticopupillary

opticon optical tactical converter

optim optimization(al(ly), optimize(d), optimum

Opt Lett Optics Letters

Opt-Mekh Prom Optika-Mekhanicheskaya Promyshlennost (Russian—Journal of Optical Technology)

optmrst optometrist

optn optician

Opt News Optics News

optoel optoelectronics

optom optometer; optometric-(al)(ly); optometrist; optometry; optomyometer

optr optryk (Dano-Norwegian—reprint)

optrak optical tracking

Opt Spektrosk Optika i Spektroskopiya (Russian—Optics and Spectroscopy)

optul optical pulse transmitter using laser

OPU Unemployed Peoples Union

opur objective program utility routines

OPUS Older People United for Service; Open University System; Operating Utility System; Organization for Promoting the Understanding of Society

opv oral polio vaccine; oral polio virus

OPW Office of Public Works

oq oil quench; overmation quotient

OQ Officers Quarters

oqe objective quality evidence

oql on-line query language

OQMG Office of the Quarter-

master General

OQR Officer's Qualification Record

or. operationally ready; operations research; other ranks; out of range; outside radius; outside right; overseas replacement; owner's risk; oxidation-reduction

or. (OR) orienting reflex; released on one's own recognizance

o/r on request; other ranks

o & r ocean and rail; overhaul and repair

or (Latin prefix—mouth)—loral

or. oratio (Latin—speech, discourse)

Or Oregon; Orient(al)

Ór Óri (Modern Greek—mountains); *Óros* (Modern Greek—mountain)

OR Oak Ridge; Officer Records; Official Records; omnidirectional radio range (symbol); Operating Room; Operational Requirement; Operation Rescue; Operations Requirement; Operations Research; Operations Room; Ordinance Report; Oregon; Owasco River (railroad); Oyster River

O.R. Operating Room

O/R Owner's Risk

O of R Office for Research (ALA)

ÖR *Österreichischer Rundfunk* (Austrian Radio and Television)

OR Ontario Reports; Operations Research

Ora Orabel(le)

ORA Oil Refiners Association; Operations Research Analyst

oracle. optical reception of announcements of coded-line electronics

ORACLE Optimum Record Automation for Courts and Law Enforcement (Los Angeles, CA)

ORAD Office of Rural Areas Development

ORAM Office for Research in Academic Methods

orang orangutan

orang orangutan (Malay—forest person)—one of the great anthropoid apes found in Borneo and Sumatra

orange light change approaching; potential danger

Oranges New Jersey's East Orange, Orange, South Or-

ange, and West Orange, may also refer to the Orange Mountains also called the Watchungs

ORASS Offender Risk Assessment Scoring System

orat oration; orator; oratorio; oratory

ORAU Oak Ridge Associated Universities

orb owner's risk of breakage

orb. omnidirectional radio beacon

orb. (ORB) oceanographic research buoy

orb owner's risk of breakage

orbatrep order of battle report

orbic orbicular; orbicularis

Orbis Polish Travel Office

ORBIT On-line Retrieval of Bibliographic Information Timeshared

Orbiter half-plane half-satellite space shuttle

ORBIS Orbiting Radio Beacon Ionospheric Satellite

orbs off-reservation boarding school

ORBS Orbital Rendezvous Base System

orc owner's risk of chafing

Orc Orcadian (inhabitant of or pertaining to Orkney Islands)

ORC Ocean Racing Club; Officers Reserve Corps; Offshore Racing Council; Opinion Research Corporation; Overseas Research Council; Ozarks Regional Commission

ORCA Ocean Resources Conservation Association

Orcades Orkney Islands

ORCAP Oficina Regional para Centroamérica y Panamá (Spanish—Regional Office for Central America and Panama)

ORCB Order of Railway Conductors and Brakemen

ORCC Orangutan Research and Conservation Center, Tanjung Puting National Park, Borneo

orch orchestra; orchestral; orchestration

Orch Orchard

orch circ orchestra circle

Orch Consv Orchestre de la Société des Concerts du Conservatoire de Paris (French—Concert Society Orchestra of the Paris Conservatory)

Orch de l'Opera de Paris Orchestre du Theéâtre Na-

tional de l'Opera de Paris (French—National Theater Orchestra of the Paris Opera)

orches orchestration

Orch H Orchestra Hall

orchi (Latin prefix—testicles)—orchid, orchiectomy

ORCHIS Oak Ridge Computerized Hierarchical Information System

orchl orchestral

Orch Nat Orchestre National de la Radiodiffusion Française (French—National Orchestra of French Broadcasting)

Orch Suisse Rom Orchestre de la Suisse Romande (French—Swiss Canton Orchestra)

Orch Symp de Mont Orchestre Symphonique de Montreal (French—Montreal Symphony Orchestra)

ORCMD Orlando Contract Management District

orcon organic control

ORCS Organic Rankine Cycle System

ORCUP Ontario Region Canadian University Press

ord operational ready date; order(s); ordinal; ordnance

o-r-d owner's risk of damage

Ord Order; Orderly; Ordinary Seaman

ORD Chicago, Illinois (O'Hare Airport); Office of Research and Development

ORDA Oceanographic Research for Defense Application

ORD-ALA Office of Research and Development—American Library Association

Ord Bd Ordnance Board

OrdC Ordnance Corps

Ord Dept Ordnance Department

ordfin ordinary finish

ordinst ordnance instruction

Ord Man Ordnance Manual

ordn ordnance

Ordn Surv Ordnance Survey

Ordo Ordovician

ORDP Office of Rural Development Policy

ords ordinary shares

Ord Sgt Ordnance Sergeant

ordvac ordnance variable automatic computer

Ore Oregon(ian)

ORE Ocean Research Equipment; Operational Research Establishment

OR & E Office of Research and Engineering

ORE *Office de Recherches et d'Essais* (French—Office of Research and Testing)

OREAM *Organisation d'études d'Aires Métropolitaines* (French—Organization for the Studies of Metropolitan Areas)

Oreg Oregon; Oregonian

Oregon Caves Oregon Caves National Monument

Ore-Ida pots Oregon-Idaho potatoes

o/r enema oil-retention enema

ORES Office of Research and Engineering Services

ORESCO Overseas Research Council

orf orifice; overhaul replacement factor

o-r-f owner's risk of fire

ORF Norfolk, Virginia (airport); Oceanic Research Foundation

ÖRF *Österreichischer Rundfunk* (Austrian radio and TV network)

Or F S Orange Free State

org organ; organic; organization; organize; organizer

ORG Operations Research Group

organ. organic; organization

Organ Pipe Cactus Organ Pipe Cactus National Monument in Arizona

org art organic art(ist)

Orgburo Organizational Bureau of the Central Committee (of the Communist Party)

org exp organo espressivo (Italian—expressive organ part)

Org Gard Organic Gardening

orgl organizational

org-man organization man

orgn organization

ORGS Operational Research Group of Scotland

orgst organist

ori orientation inventory

Ori Orient(al)(ism); Oriente; Orion (constellation)

ORI Ocean Research Institute; Ocean Resources Institute; Office Research Institute; Operation Readiness Inspection

ORIC Oak Ridge Isochronous Cyclotron

oride override

ORIEL Oriel College, Oxford

orient. oriental; orientation

ORIENT Orient Airways

Orient(al) Asia(tic)

Orientales Orientales—la patria o la tumba (Spanish—

Eastern landsmen, our country or the tomb)—anthem of Uruguay

Orient Express (*see* Ori Exp)

Ori Exp Orient Express (formerly between Paris and Istanbul via Vienna but now called Central Kingdom Express running from London to Hong Kong via Paris, Berlin, Warsaw, Moscow, Irkutsk, Peking, Nanking, and Canton)

orif open reduction with internal fixation

orig origin; original; originally; originator

O-ring O-shaped ring

ORINS Oak Ridge Institute of Nuclear Studies

orion on-line retrieval of information over a network

oris orismological; orismologist; orismology

ORIT Operational Readiness Inspection Test

ORIT Organización Regional Interamericana de Trabajadores (Spanish—Interamerican Regional Labor Organization)

or j orange juice

Ork Orkney Islands

Orkneys Orkney Islands

orl orlon (synthetic fiber); owner's risk of leakage

ORL Orbital Research Laboratory; Ordnance Research Laboratory; Orlando, Florida (Herndon Airport)

ORL Outlook on Research Libraries

ORLA Optimum Repair Level Analysis

Orleans New Orleans

Órm Órmos (Modern Greek—bay)

ORM Ohio Reformatory for Men

ORMA Office of Refugee and Migration Affairs

ORMAK Oak Ridge Tokamak

orml oriental meal

orm('s) off-road motorcycle(s)

orn orange; ornament

orn orne (French—decorated, ornamented)

Orn Oran (British maritime contraction)

ORN Operating Room Nurse

ornith ornithology

ornithol ornithologic(al)(ly); ornithologist; ornithology

ORNL Oak Ridge National Laboratory

ORNLL Oak Ridge National

Laboratory Library

ORO Oak Ridge Operations Office; Operations Research Office (Johns Hopkins University)

or. obliq. *oratio obliqua* (Latin—indirect speech, oblique speech)

orog orographer; orographic; orographical; orography

ORP Okret Rzecypospolitej Polskiej (Polish—Ship of the Polish Republic)

ORPA Office of Regional and Political Affairs (CIA)

ORPC Office of Rail Public Counsel

orph orphan; orphanage; orphaned; orphans

orpiment arsenic sulfide

o-r pot. oxidation-reduction potential

orr operations research research (ORR)

o-r-r owner's risk rates

o-r release own-recognizance release

ORRRC Outdoor Recreation Resources Review Commission

ORRT Operational Readiness Reliability Test

ors owner's risk of shifting

ors (ORS) orbiting research satellite; orthopaedic surgery

or's onion rings; orienting responses

ors. orationes (Latin—speeches)

ORS Office of Refugee Settlement; Office of Research and Statistics; Official Rate Standard; Old Red Sandstone; Operational Research Society

ORSA Operations Research Society of America

ORSANCO Ohio River Valley Water Sanitation Commission

ORSE Operational Reactor Safeguard Examination

ORSJ Operations Research Society of Japan

ORSTOM Office de la Recherche Scientifique et Technique d'Outre Mer (French—Overseas Office of Scientific and Technical Research)

ort odor recognition threshold; operational readiness training

ORT Operating Room Technician; Operational Readiness Test; Oral Rehydration Therapy; Order of Railroad Telegraphers; Organization for Rehabilitation through Train-

ing; Overage Retirement Training (program)

ORTF Office de Radiodiffusion Télévision Française (French—French Office of Television Broadcasting)

ortho orthochromatic; orthographic; orthography; orthopedic(s)

ortho (Latin prefix—normal or straight)—orthopedic

Ortho Greek Orthodox

orthog orthography

ortho-k orthokeratological(ly); orthokeratologist; orthokeratology

orthokera orthokeratologist; orthokeratology

orthomol orthomolecular; orthomolecularologist; orthomolecularology

orthop orthopedics

orthor orthorhombic

ORTO Occupational Rehabilitation Training for Overseas

ORTPA Oven-Ready Turkey Producers' Association

ORTS Optional Residence Telephone Service

ORTU Other Ranks Training Unit

ORU Oral Roberts University

ORuss Old Russian

ORV Ocean Range Vessel (naval symbol)

Orvidius Orvidius Naso, Roman poet

orv('s) off-road vehicle(s)

orw owner's risk of wetting

ORW Ohio Reformatory for Women

ory (Latin suffix—pertaining to)—sensory

ORY Paris, France (Orly Airport)

os oil solvent; oil switch; old series; old style; on station; out of stock; output secondary; outside; outsize; overseas; oversize

os (OS) operating system (data recording)

o/s out of service; out of stock

o & s over and short

o.s. oculus sinister (Latin—left eye)

o-S on-Sea

Os osmium

OS Ocean Station; Office Surgery; Old Saxon; Old Series; Operating System; Operation Sandstone; Operation Snapper; Optical Society; Ordinary Seaman; Ordnance Specifications; Ordnance Survey; Overseas Service

O.S. Old Style

OS/2 Operating System 2

osa oil-soluble acid; order for simple alert

Osa Osaka

Osa (Russian—Bee)—a Soviet class of guided-missile patrol boats

OSA Office of the Secretary of the Army; Official Secrets Act; Omnibus Society of America; Optical Society of America; Osaka, Japan (airport); Overseas Sterling Area; Overseas Supply Agency; Oyster Shell Association

OSA Ocean Shipping Act; Official Secrets Act

osac orifice spark advance control

OSAF Office of the Secretary of the Air Force

OSAHRC Occupational Safety and Health Review Commission

OSAP Office of Substance Abuse Prevention in the Alcohol, Drug Abuse, and Mental Health Administration; Ontario Student Awards Program

OSAS Overseas Service Aid Scheme

O Sax Old Saxon

OSB Order of Saint Benedict; Otago Savings Bank; Overseas Service Bureau

O.S.B. Order of St Benedict

OSB Occupational Safety Bulletin

OSBA Ohio School Boards Association; Oregon School Boards Association

OSBM Office of Space Biology and Medicine

osc oscillator

Osc Oscan

OSC Office of Special Counsel; On-Scene Commander; Ontario Science Centre; Ontario Securities Commission; Order of St Clare; Ordnance Systems Command (formerly Bureau of Weapons); Overseas Shipping Company

O.S.C. Oblate of Saint Charles

O of SC Order of Scottish Clans

OSCA Office of Senior Citizens Affairs

OSCA Officine Specializzate Costruzione Automobili (Italian—Special Office of Automobile Construction)

OSCAA Oil-Spill Control As-

sociation of America
O Scan Old Scandinavian
oscar orbital-satellite-carrying amateur radio (OSCAR); oxygen steelmaking computer and recorder
Oscar letter O radio code
OSCAR On-Line System for Controlling Activities and Resources; Optimum System for the Control of Aircraft Retardation
Oscar(s) Motion Picture Academy Award(s)
ÖSCG *Österreichische Studiengesellschaft für Kibernetik* (German—Austrian Society for Cybernetic Studies)
OSCO Oil Service Company of Iran; Oil Shipment Corporation
oscope oscilloscope
oscp oscilloscope
OSCP Ocean Sediment Coring Program (NSF)
OSCT Office of Scholarly Communication and Technology
osd on-line systems driver; open shelter deck; optical scanning device; out-of-station designation
o s & d over, short, and damaged
OSD Office of the Secretary of Defense; Operational Support Directive; Ordnance Supply Depot; Original Sponsoring Distributor
OSDBMC Office of the Secretary of Defense, Ballistic Missile Committee
OSDNRL Ocean Science Division—Naval Research Laboratory
osdocs over-the-shore discharge of container ships
osdp on-site data processing
OSDP Operational System Development Program
OSDSA Office of the Secretary of Defense, Systems Analysis
OSDSAC Office of the Secretary of Defense, Scientific Advisory Committee
ose operational support equipment
ose (Latin suffix—full of)—adipose
OSE Ocean Shipping and Enterprises; Office of Science Education; Office of of Sex Equity (HEW); Office of Systems Engineering
OS & E Ocean Science and Engineering
OSEAP Oil Shale Environmental Advisory Panel
o'seas overseas
OSEB Orissa State Electricity Board
OSerb Old Serbian
osf operational service fee; ordinary shareholders funds
OSF Open Software Foundation; Order of St Francis
OSFI Open Steel Flooring Institute
O.S.F.S. Oblate of Saint Francis of Sales
osg outstanding
OSG Office of Sea Grant (NOAA); Office of the Secretary General (UN)
OSG *Official Steamship Guide*
OSGP Office of Sea Grant Programs
OSGS On Sudan Government Service
o.s.h. *omni singula hora* (Latin—every hour)
Osh Ossian
OSH Office on Smoking and Health
OSHA Occupational Safety and Health Act; Occupational Safety and Health Administration
o/sheep odd sheep
o/ship ownership
OSHPD Office of Statewide Health Planning and Development
OSHRC Occupational Safety and Health Review Commission
OSHS Occupational Safety and Health Scheme
osi out of stock indefinitely
OSI Office of Samoan Information; Office of Special Investigation (USAF); Office of Special Investigations (Dept. of Justice and US Army); Off-Site Instruction; Ohio Scientific Incorporated; Open Systems Interconnect; Other Service Investigation
OSIA On-Site Inspection Agency; Order of the Sons of Italy in America
osie operational support integration engineering
OSIP Operational and Safety Improvement Program
osis (Greek—condition or state of being)—arteriosclerosis, cirrhosis, halitosis, tuberculosis
OSIS Office of Science Information Service; On-Site Inspection System
Osk Oskarshamm
OSK Osaka Syosen Kaisha (Osaka Mercantile Steamship Company)
OSK *Országos Széchényi Könyvtár* (Hungarian—National Széchényi Library)—Budapest
Osl Oslo
OSl Old Slavonic
OSL Office of the Secretary of Labor; Orbiting Solar Laboratory; Oslo, Norway (airport)
OSLat Old-Style Latin
OSlav Old Slavic
osm osmosis; osmotic
Osm osmol(s)
O s M *Orchestre symphonique de Montréal* (French—Montreal Symphony Orchestra)
OSM Office Service Manual; Office of Surface Mining; One of the Swinish Multitude (Philip Freneau, poet of the American Revolution, used this three-letter device after his name, thereby deriding similar-looking British titles); *Overzees Scheepvaart Maatschappij* (Overseas Shipping Company)
OSMA Otago-Southland Manufacturers Association
OSMM Office of Safeguards and Materials Management (AEC)
osmol osmosis + mol (standard unit of osmotic pressure)
osmos own ship's motion simulator
OSMRE Office of Surface Mining Reclamation Enforcement
OSN Office of the Secretary of the Navy
OSN *Orquesta Sinfónica Nacional* (Spanish—National Symphonic Orchestra)
OSNC Orient Steam Navigation Company
OSNY Oratorio Society of New York
oso (OSO) orbiting solar observatory
OSO Offshore Supplies Office; Offshore Supply Office; Omaha Symphony Orchestra; Ordnance Supply Office(r); Oregon Symphony Orchestra
OSO *Oessudoeste* (Spanish—west southwest); Orbiting Solar Observatory; Ordnance Supply Office
OSODS Office of Strategic

Offensive and Defensive Systems (USN)
osp optimum sustainable population; outside purchased
o-sp off-street parking
o.s.p. obiit sine prole (Latin—died without issue)
OSp Old Spanish
OSP Open-Space Program; Order of St Paul
OSP Oficina Sanitaria Panamericana (Spanish—Pan-American Sanitation Office)
OSPA Overseas Pensioners' Association
OSPA Organisation de la Santé Panaméricaine (French—Pan-American Health Organization)
OSPAAL *Organización de Solidaridad de los Pueblos de Asia, Africa, y Latino-América* (Spanish—Organization of Solidarity of the Peoples of Asia, Africa, and Latin America)
OSPIC Overseas Private Investment Corporation
OSPJ Offshore Procurement, Japan
osprd(s) oblate spheroid(s)
OSQ Office of Safety Quality Assurance and Safeguards
OSQ Orchestre Symphonique de Québec (French—Quebec Symphonic Orchestra)
OSQAS Office of Safety Quality Assurance and Safeguards
osr own ship's roll
OSR Office of Scientific Research; Office of Security Review; Office of Strategic Research; Oil Shale Reserves; Operational Support Requirement(s); Oversea Returnee
OSR Orchestre de la Suisse Romande (French—Orchestra of French Switzerland)
OSRA Overseas Shipping Representatives Association
OSRB Overseas Service Resettlement Bureau
OSRD Office of Scientific Research and Development; Office of Standard Reference Data
OSRO Office of Scientific Research and Development
OSROK Office of Supply—Republic of Korea
OSRS Organization of Senegal River States (Guinea, Mali, Mauritania, Senegal)
OSRTN Office of the Special Representative for Trade Negotiations
oss order short shipped
oSS operates Saturday and Sunday
OSS Object-Sorting Scales (psychological test); Office of Space Science; Office of Strategic Services; Office of Support Services; old submarine (3-letter code); Operating Supply Specification; Operation Safe Streets; Operational Storage Site; Optical Surveillance System; Orbital Space Station; Orient Shipping Services; Overseas Shipping Services
OSSA Office of Space Sciences and Applications (NASA)
OSSBA Oklahoma State School Boards Association
OSSC Oregon School Study Council
OSSNSS Ordnance Supply Segment of the Navy Supply System (USN)
oss(OSS) orbiting space station
OSSP Oregon Small Schools Program
OSSS Orbital Space Station Studies
OSSTF Ontario Secondary School Teachers' Federation
ost objectives, strategies, tactics; oldest; on same terms; optical star tracker; ordinary spring tides
öst österreichisch (German—Austrian)
Ost Ostend
Ost Ostrów (Polish—island)
Öst Österreich (German—Austria)
Øst Øterrike (Norwegian—Austria)
OST Office of Science and Technology; Old Spanish Trail (US 90); Operational Suitability Test
OS & T Office of Science and Technology
osteo osteopath(ic)
osteo osteon (Greek—bone)—ossification, ossified, osteomyelitis, osteopath(ic)
osteoart osteoarthritic; osteoarthritis
osteol osteology
osteomy osteomyelitis
osteop osteopath(ic); osteopathy
osteoporo osteoporosus
Österreich (German—Eastern Empire)—Austria (modern remnant of the once great

Austro-Hungarian Empire); *Österreichische Bundeshymne* (German—Austrian Confederation)—national anthem
OSTF Operational System Test Facility
OSTI Office for Scientific and Technical Information
OSTIV Organisation Scientifique et Technique Internationale du Vol à Voile (French—International Scientific and Technical Organization for Soaring Flight)
O.St.J. Officer of the Order of Saint John of Jerusalem
Østland (Norwegian—Eastland)—eastern and southeastern Norway
OSTP Office of Science and Technology Policy
Ostpr Ostpreussen (German—East Prussia)
OSTS Office of State Technical Services; Official Seed Testing Station
OSU Ohio State University; Oklahoma State University; Oregon State University
OSUAS Ohio State University (College of) Administrative Science
OSUK Ophthalmological Society of the United Kingdom
OSUL Ohio State University Library; Oklahoma State University Library; Oregon State University Library
OSUP Ohio State University Press
osv och sa vida (Swedish—and so forth); *og sa videre* (Dano-Norwegian—and so forth)—etc.
Osv Osvald; Osvaldo
OSV Ocean Station Vessel
OSV Orquesta Sinfonica Venezuela (Spanish—Venezuela Symphony Orchestra); *Our Sunday Visitor*
Osv Rom Osservatore Romano (Italian—Vatican newspaper)
osw operational switching
Osw Oswald
OSw Old Swedish
OSW Office of Saline Water
OSWA Off-Shift Work Authorization
osy (OSY) optimum sustainable yield
os & y outside screw and yolk
ot observer target; oiltight; old terms; old tuberculin; on time; on track; ordinary tide(s); otitis; otology; our

telegram; overtime; ovum transfer

ot (OT) occupational therapy; otolaryngology; overtime; original transposed (in a 12-tone row)

o't (Gaelic contraction—of it)

o/t on truck; overtime

'ot hot

o-T on-Thames

O/t old term (grain market)

OT Occupational Therapist; Occupational Therapy; Ocean Transportation; Office of Territories; Office of Transportation; Old Testament; Operational Training; Oregon Trunk (railroad); Organization Table; Otis Elevator (stock exchange symbol); Overseas Tankship (Caltex Line); Overseas Trade

O of T Office of Telecommunications (OT)

OT *Organisation Todt* (German—Death Organization)—Hitler's extermination corps

OTA Occupational Therapists Association; Office of Technology Assessment; Office of Territorial Affairs; Outer Transport Area

OTA *Organisation mondiale du Tourisme et de l'Automobile* (French—World Tourism and Automobile Organization)

OTAC Ordnance Tank and Automotive Command

otadl outer target azimuth datum line

OTAF Office of Technology Assessment and Forecast

OTAG Office of the Adjutant General (USA)

Otago Otago Harbour (Dunedin, New Zealand's port); Otago Peninsula (southeast of the port)

OTAN *Organisation du Traite del l'Atlantique Nord* (French—NATO); *Organizacion del Tratado del Atlántico Norte* (Spanish—NATO)—North Atlantic Treaty Organization

OTAR Overseas Tariffs and Regulations

OTAS *Organización del Atlántico Septentrional* (Spanish—North Atlantic Treaty Organization)—NATO

OTASE *Organisation du Traite de l'Asie du Sud-Est* (French—Southeast Asia Treaty Organization)—

SEATO

OTAT Office of Technical Assistance and Training; Orthotoluidine Arsenite Test

OTATO One-Trip Air Travel Orders

otb off-track betting

OTB Overseas Trust Bank

OTBA Owners, Traders, Breeders Association

otbd outboard

otc objective, time, and cost; ocean transshipment cargo; one-stop charter; outer tube centerline; over the counter

otc (OTC) over the counter

OTC Officer in Tactical Command; Officers Training Corps; Organization ior Trade Cooperation; Ottawa Transit Commission; Overseas Telecommunications Commission

OTC *Office de Tourisme du Canada* (French—Canadian Government Office of Tourism)

otch obedience trial champion

otd organ tolerance dose

OTD Ocean Technology Division

otda other-than-defined adult

otdc optical target designation computer

OTDC Observational Test and Development Center (NWS)

OTD & SP Office of Technical Data and Standardization Policy

ote operational test and evaluation; overtaken by events

ote *oriente* (Spanish—east)

otec (OTEC) ocean thermal energy conversion

OTEC Ontario Teacher Education Colleges

OTECS Ocean Thermal Energy Conversion System

otel our telegram

OTeut Old Teutonic

otf optical transfer function

otf (OTF) off-the-film (camera metering system)

o-t-f off-the-film (light measurement)

OTF Ontario Teachers Federation

oth over the horizon

Oth *Othello, The Moor of Venice*

othb over-the-horizon backscatter

OTH-B Over-the-Horizon Backscatter

othf over-the-horizon forward scatter

oti official test insecticide

OTI Oregon Technical Institute

OTI *Organización de Televisión Iberoamericana* (Spanish—Ibero-American Television Organization)

OTIA Ordnance Technical Intelligence Agency

OTID Office of Talented Identification and Development (Johns Hopkins)

OTIG Office of the Inspector General

OTIS Occupational Training Information System; Oregon Total Information System

OTIU Overseas Technical Information Unit

otj on the job

otk old tuberculin Koch

otl out to lunch; output transformerless; over the line (softball)

OTL Operating Time Log

otlx our telex

otm other than Mexican; other track material

OTM Office of Telecommunications Management; Old Turkey Mill

otml oatmeal

otno our telegram number

oto one time only (tv); otorhinolaryngologist

oto (Latin prefix—ear)—otology

OTO Operational Testing Office

otol otology

otolaryngol otolaryngology

OTO/Neth only to order from Netherlands (for example)

otorhinol otorhinolaryngology

otp obstacle to progress; order to plan; oxygen tanking panel

OTP Office of Telecommunications Policy

otr on the rag (underground slang—on the menstrual cycle)

OTR Ovarian Tumor Registry; Owning-the-Realty; Registered Occupational Therapist

otrac oscillogram trace

OTRACO *Office de l'Exploitation des Transports Coloniaux* (Congolese railway and river transportation administration)

OTRAG Orbital Transport and Rocket AG (German rocket company)

otran ocean test range and instrumentation

otrt operating time record tag

ots (OTS) orbital technical satellite
OTS Office of Technical Services; Office of Traffic Safety; Office of Thrift Supervision; Officers Training School; Operational Test Site; Organization of Tropical Studies
otsdg outstanding
OTSG Office of the Surgeon General,
otsr optimum track ship routing
OTSS Operational Test Support System
ott one-time tape; otter; outgoing teletype
ott ottava (Italian—octave); *ottobre* (Italian—October)
Ott Ottawa
OTT Ocean Transport and Trading; Office of Traffic and Transportation
ottb owner to take back
otu operational taxonomic unit
otu (OTU) operational training unit
OTU Office of Technology Utilization (NASA)
O Turk Old Turkish
OTUS Office of the Treasurer of the United States
otv orbital transfer vehicle; outer television
otvct outer tube vertical centerline target
otw over the wing
oty over to you
ou oat unit; official use
o & u over and under
'ou thou
o.u. oculus uterque (Latin—either eye)
OU Oglethorpe University; Ohio University; Oklahoma University; Open University; Otago University; Ottawa University; Otterbein University; Owen University; Owosso University; Oxford University
OUA Order of United Americans
OUA Organisation de l'Unité Africaine (French—OAU); *Organización de Unidad Africana* (Spanish—OAU)—Organization of African Unity
OUAC Oxford University Appointments Committee; Oxford University Athletic Club
OUAFC Oxford University Association Football Club
OUAM Order of United Amer-

ican Mechanics
OUAS Oxford University Air Squadron
OUBC Oxford University Boat Club
OUCC Oxford University Cricket Club
OUDP Officer Undergraduate Degree Program (USA)
OUDS Oxford University Dramatic Society
Ouga Ougadougou, Upper Volta
OUGC Oxford University Golf Club
oughta (American slang—ought to)
oughtn't ought not
ouguiya monetary unit of Mauritania
OUHC Oxford University Hockey Club
OUHS Oxford University Historical Society
OULC Oxford University Lacrosse Club
OULCS Ontario Universities Library Cooperative System
OULTC Oxford University Lawn Tennis Club
OUM Oxford University Mission
OUN Organizatsia Ukrainiskikh Nationalistiv (Russian—Ukrainian Nationalist Organization)—anti-communist
OUP Oxford University Press
oupt output
OUR Office of University Research
OURC Oxford University Rifle Club
OURFC Oxford University Rugby Football Club
o/us over-under shotgun
o/US oro US (Spanish—American gold, American money)
OUS Oxford Union Society
OUSA Open University Students Association
OUSC Oxford University Swimming Club
'ouse douse; house; kouse; louse; mouse; rouse; souse; touse
OUSF Oxford University School of Forestry
OUSL Office of the Undersecretary of Labor
out. outlet; output
outbd outboard
outran output translator
out of sync out of synchronization
ouv ouvrage (French—work)
Ouviram Ouviram do Ypiranga

(Portuguese—Listen to the Heart of Brazil)—Brazil's national anthem
ov observed velocity; optimum value; orbiting vehicle (OV); ovary; over; overture
ov oi vay (Yiddish—alas)
ov. ovum (Latin—egg)
Ov Ovid; Oviedo
Ov Over (Dano-Norwegian or Dutch—upper); *Overijssel* (Dutch province above the Ijssel River)
Öv Över (Swedish—upper)
OV Office Visit
OV Oranje Vrystaat (Afrikaans—Orange Free State); Orbital Vehicle
ÖV Österreichische Volkspartei (German—Austrian People's Party)
OV-10 North American-Rockwell Bronco counterinsurgency aircraft
oᵛᵃ ottava (Italian—octave)
Oᵛᵃ Ostrova (Bulgarian, Czechoslovakian, Russian—island)
OVA Office of Veterans' Affairs
OVAC Overseas Visual Aids Center
ovbd overboard
ovc other valuable consideration(s); overcast
ovcst overcast
ove on vehicle equipment
ÖVE Österreichischer Verband für Elektrotechnik (German—Austrian Society for Electro-technology)
over. overture
Over Overijssel (Dutch province)
overmation
overs overshoes
ovf ovenfor (Dano-Norwegian—over)
ovfl overflow
ovflow overflow
ovh overhead; overheat
ovhd oval head; overhead
ovhdld overhandled
ovhl overhaul
ovh p overhead projector
ovht overheat
OVIS Ohio Vocational Interest Survey
ovk overkill
OVKOT On Various Kinds of Thinking (essay by James Harvey Robinson)
ovld overload
ovly overlay
ovm on-vehicle material
ovm oi vayz mir (Yiddish—woe unto me)

ovo (Latin prefix—egg)—ovo-vegetarian

ovolactos ovolactovegetarians (confining their diet to eggs, milk and milk products, as well as vegetables)

ovos ovovegetarians (confining their diet to eggs and vegetables)

ÖVP Österreichische Volkspartei (German—Austrian People's Party)

ovpd overpaid

ovprt overprinted

ovpt overprint

OVPUS Office of the Vice President of the United States

OVR Office of Vocational Rehabilitation

OVRA Opera Voluntaria per la Repressione dell' Antifascismo (Italian—Voluntary Work for the Repression of Anti-Fascism)

ovrd override

ovsl overslow

ovsp overspeed

ovstfd overstuffed

ovstk overstock(ed)

OVSVA Oranje Vrystaatse Veld Artillerie (Afrikaans—Orange Free State Field Artillery)

ovtr operational videotape recorder

ov w overgankelijk werkwoord (Dutch—transitive verb)

ow off white; old woman (slang for wife); one way; ordinary warfare (OW); out of wedlock (born of unmarried parents); outer wing; over water

o-w oil-in-water

o/w oil/water ratio (oil well)

o:w oil-water ratio

oW ohne Wert (German—without value)

öW österreichische Währung (German—Austrian currency)

OW Observation Ward; Old Welsh

O & W Oldest and Wisest (newspaper reporter's nickname for Ronald Reagan)

OW Oberwerk (German—swell organ)

OWAA Outdoor Writers Association of America

OWAEC Organization for West African Economic Cooperation

owc owner will carry

OWC Ordnance Weapons Command; Outline of World Cultures

O-WC Oil-Water Contract (oil well)

OWCP Office of Workers' Compensation Programs

owe operating weight empty

Owen Stanleys Owen Stanley Mountains of New Guinea

owf optimum working frequency

owgl obscure wire glass

OWH Office of the War on Hunger

OWHA Oliver Wendell Holmes Association

OWI Office of War Information; Office of Waste Isolation

owise otherwise

owl. outlined-white letter (on tires)

OWL Ocotillo Water League; Older Women's League; Older Women's Liberation; Order of Women Legislators; Other Woman, Limited; Overland Western Limited

owm over without marks

OWM Office of Weights and Measures

OWMA Oscar Wells Museum of Art (Birmingham, Alabama)

OWO OWI Washington Office

owp outer wing panel

OWPP Office of Welfare and Pension Plans

owpr ocean wave profile recorder

OWPS Offshore Windpower System

OWPT Overpaid Windfall Profits Tax

OWR Ouse Washes Reserve (England)

OWRR Office of Water Resources Research

OWRT Office of Water Research and Technology

ows (OWS) operational weapon satellite

OWS Ocean Weather Station

OWSS Ocean Weather Ship Service

OWU Office Workers Union; Ohio Wesleyan University

OWWS Office of World Weather Systems

ow/ym older woman/younger man

ox. our telex; oxalic; oxide; oxygen

Ox. Oxford

OX oxygen (commercial symbol)

oxa oxalic acid

oxalic acid $(COOH)_2$

Oxbridge Oxford and Cambridge universities (the ultimate in British formal education)

oxd oxidation; oxidize(d)

Oxf Oxfordshire

OXFAM Oxford Committee for Famine Relief

Oxf & Bucks Oxfordshire and Buckinghamshire (light infantry)

Oxford Oxford English Dictionary published by Oxford University Press

Oxford UP Oxford University Press

oxim oxide-isolated monolithic technology

Oxm Oxmantown

Ox M OUP Oxford Medical (division) Oxford University Press

OXOCO Offshore Exploration Oil Company

Oxon Oxfordshire

Oxon. Oxonia (Latin—Oxford); *Oxoniensis* (Latin—Oxonian)

oxr oxidizer

oxwld oxyacetylene weld

oxy oxygen

Oxy Occidental College; Occidental Petroleum Corporation; Oxy Metal Industries International

oxycephs oxycephalics (pointed skulled people)

oxym oxymel (honey-water-vinegar solution); oxymoron (figure of speech creating a seeming self-contradiction)

oy (OY) optimum yield

OY orange yellow

O/Y Osakeytiö (Finnish—limited company)

OYA Oy Yleisradio Ab (Finnish Broadcasting Company)

Oya Cur Oyashio Current (Kurile or Okhotsk or Oyasiwo)

OYD Office of Youth Development

oyo own your own (apartment, house, yacht)

oys oysters

OYS Outstanding Young Singaporeans

oystersan oyster sandwich

oysterwich oyster sandwich

oz ounce

oz onza (Spanish—ounce)

Oz Aussie(s); Australia(n); ooze; Osborn(e)

OZ Ozark Airlines (two-letter designation)

OZ *Ozean* (German—ocean);
Ozero (Russian—lake)
OZA Ozark Airlines
oz ap apothecaries' ounce(s)
ozarc ozone-atmosphere rocket
Ozarks Ozark Mountains of
Arkansas, Missouri, and Ok-
lahoma

oz avd avoirdupois ounce(s)
ozd observed zenith distance
ozf ounce-force
oz-ft ounce-foot
oz-in. ounce-inch
OZNa *Odelejenje Zastite Narodna* (Serbo-Croat—Department for National Protec-

tion)—Yugoslav
OZO *oost zuidoost* (Dutch—
east southeast)
ozone O_3
ozs ounces
oz t ounce troy
ozws otherwise
Ozy Ozzie

P

p fluid density (symbol); page; pamphlet; paragraph; park; parking; part; participle; pass(ed); past; paste; pawn; pebbles; pectoral; pence; *pengü* (Hungarian monetary unit); penny; *per* (Latin—by); percentile; perceptual (speed); percussion; perforate; perforated; perforation; perimeter; period; perishable; peseta; peso; peta (P)—10^{15} (one quadrillion); peyote; *piano* (Italian—softly); piaster; piastre; picot; pie; pilaster; pimp; pink; pint; pipe; pitch; pitcher; plasma; plaster; plate; plus; point; polar; pole; pond; poor; population; porcelain; port, or left side of an airplane or vessel when looking forward (P or L); position; positive; post; postage; posterior; postpartum; power; predicate; predict(ion); premolar; presbyopia; present; pressure; primary; primitive; principal; principle; probability (ratio); product; prompter; proprionate; proton; publication; pulse; pupil

p (P) prime; program

£ pound sterling

p. *pagina* (Italian, Latin, Portuguese, Spanish—page); *parte* (Latin—part); *pater* (Latin—father); *per* (Latin—by); *pondere* (Latin—by weight); *proximum* (Latin—near); *pugillus* (Latin—fistful)—handful

p % *por ciento* (Spanish—per hundred, percent)

P Pacific; pamphlet; Panama Line; Papa—code letter for

P; Paris; Parisian; passenger vessel (symbol); patrol; Pennzoil; permeance (symbol); Philadelphia Mint (symbol); phosphorus; Piasecki; plate; Pleyel; poise; polar; polarization; pole; Police; poor; Pope; port; Portugal (auto plaque); power; present value; President; Prince Line; principal; priority; project; propulsion; Protestant; protozoa; pulse

P. protein(s) (dietary symbol)

P (Latin—Publius); pilot (white *P* on a blue flag flown on a pilot boat); *Pilot* (German); *pilota* (Italian); *pilote* (French); *piloto* or *practico* (Spanish); *Policía* (Spanish—police)

p00 program zero-zero

P₁ first parental generation

P 1/C Private First Class

P 1/C M Private First Class Marine

P-2 Lockheed Neptune antisubmarine and reconnaisance naval aircraft

P₂ pulmonic second sound

P2 *Panzer* (German—armor, armor plated, tank)

P-2J Kawasaki version of the Lockheed Neptune antisubmarine and reconnaissance aircraft

P-3 Lockheed Orion antisubmarine and patrol aircraft

P-4 Soviet Komsomolets motor torpedo boats

P-5 Marlin twin-engine all-weather seaplane for long-range antisubmarine patrol and electronic reconnaissance

P-5M Martin Marlin flying

boat

P-6 Soviet motor torpedoboats used in many communist satellite countries

P.08 German marking denoting the so-called luger service pistol

P³³ radioactive phosphorus

P-38 U.S. pursuit aircraft

P.38 German 9mm service pistol (World War II)

P₅₅ partial pressure of O₂ wherein hemoglobin is half saturated with O₂

P-60 60-minute parking

P-149 Piaggio trainer aircraft built in Italy

P-166M Piaggio Albatross coastal patrol aircraft

P-333C Lockheed antisubmarine patrol plane

pa intensity of atmospheric pressure (symbol); paper; paper advance; paralysis agitans; participial adjective; particular average; patient; pattern analysis; pending availability; performance analysis; performance appraisal; permanent appointment; pernicious anemia; personal appearance; piaster; piastre; point of aim; position approximate; power amplifier; power approach; power of attorney; press agent; pressure altitude; private account; provisional allowance; psychoanalyst; public address (system); public assistance; publication announcement; purchasing agent

pa (Pa) pascal

pa (PA) posteroanterior

p-a psychogenic aspermia

p/a paid annually; payment authority; per annum; power of attorney

p & a percussion and auscultation; plugged and abandoned (oil well); price and availability

p in the a pain in the ass

p.a. *per abdomen* (Latin—by the abdomen); *per annum* (Latin—by the year)

p/a per adres (Dutch—care of)

pA picoampere

p A por autorización (Spanish—in care of)

Pa Panama; Panamanian; *Panameña*; *Panameño*; Papa; Para; *Pará (Belem do Pará)*; Pascal; Pennsylvania; Pennsylvanian; protactinium

PA Pan American; Parents Anonymous; Passenger Agent; Pennsylvania; Pennsylvania Railroad (stock exchange symbol); Philippine Army; Philippine Association; Physician's Assistant; Piedmont Airlines; Polled Angus (cattle); Port Agency; Post Adjutant; Prefect Apostolic; Press Agent; Press Association; Prince Albert (coal); Procurement Authorization; Production Assistant; Proprietary Association; Prosecuting Attorney; Prothonotary Apostolic; psychological age; Public Act; Public Affairs; Publishers Association; Puppeteers of America; Purchasing Agent

P-A Pacific-Atlantic Line; Pan-Atlantic Line

P/A Picatinny Arsenal

P & A Professional and Administrative

P of A Port of Anchorage

PA Priok Administration (Malay—Port Administration); *Psychological Abstracts*

P A Partij van de Arbeid (Dutch—Party of Labor)

PA₀₂ alveolar oxygen pressure

p.a.a. *parti affectae applicetur* (Latin—apply to the affected parts or region)

PAA Pacific Alaska Airways; Pan American World Airways System (3-letter designation); Paper Agents Association; Pharmaceutical Association of Australia; Phonetic Alphabet Association; Photographers Association of America; Plywood Association of Australia; Potato As-

sociation of America; Prisoners Aid Association; Purchasing Agents Association

PAAA Premium Advertising Association of America

PAAB Public Arts Advisory Board

PAAC Product Assurance Action Center; Program Analysis Adaptable Control; Public Arts Advisory Council

PAADC Principal Air Aide-de-Camp

PAAE Pennsylvania Association for Adult Education

PAAF Professional Actors Association of Florida

PAAM Prisoners Aid Association of Maryland

paanga monetary unit of Tonga

PAAO Pan-American Association of Ophthalmology

pab per acre bonus

pab (PAB) p-aminobenzoic acid

PAB Panair do Brasil (airline); Petroleum Administrative Board; Price Adjustment Board

PAB (CIA) Problems Analysis Branch of the CIA

paba para-amino benzoic acid

pabla problem analysis by logical approach

pabst primary adhesively-bonded structure

pabx private automatic branch telephone exchange

pac packaged assembly circuit; personal analog computer; person in addition to the crew; phenacetin-aspirin-caffeine (all-purpose capsule); prearrival confirmation; production acceleration capacity; project analysis and control; pursuant to authority contained (in); put and call (stock exchange jargon)

pac (PAC) premature atrial contraction

p-a-c parent-adult-child (ego states)

p A c pure Argentinian cocaine

Pac Pacific

Pac Pacifico (Italian—Pacific); *Pacífico* (Portuguese or Spanish—Pacific); *Pacifique* (French—Pacific)

PAC Pacific Air Command; Pacific Automotive Corporation; Pacific Telephone & Telegraph (stock exchange symbol); Palo Alto Clinic; Pan-Africanist Congress;

Pan-American Congress; Performing Arts Center; Pharmaceutical Advertising Club; Philbrook Art Center; Political Action Committees; Public Access Catalog(ue); Public Affairs Committee; Public Assistance Cooperative

PACA Picture Agency Council of America

PACAF Pacific Air Force

Pacaraimas Pacaraima Mountains forming the Brazil-Guyana and Brazil-Venezuela borders

PACAS Patient Care System; Psychological Abstracts Current Awareness Service

PACB Pan-American Coffee Bureau

Pac Bch Pacific Beach

Pac Bell Pacific Bell (telephone)

PACC Product Administration and Contract Control; Project Administration Contact Control

PACCS Post Attack Command and Control System

PacD Pacific Division

PACDA Personnel and Administration Combat Development Activity (USA)

pace (PACE) package-crammed executive; performance and cost evaluation; precision analog computing equipment; pre-launch automatic check-out equipment; program to advance creativity in education; programmed automatic communications equipment; projects to advance creativity in education

pace. pacemaker

PACE Pacific America Container Express; Professional and Administrative Career Examination; Professional Association of Consulting Engineers; Professional Athletes Career Enterprises; Program for Afloat College Education (USN); Public Access Cabletelevision by and for the Elders; Public Awareness Communication Exchange

PACECO Pacific Coast Engineering Company

PACED Program for Advanced Concepts in Electronic Design

pacer. planning automation and control for evaluating requirements

PACFACS Programmed Appropriation Commitments—Fixed-Asset Control System

PACFLT Pacific Fleet

PACFORNET Pacific Coast Forest Research Information Network

Pac Gas & El Pacific Gas and Electric

'pache Apache

Pacif Pacific

Pacific Pacific Ocean

Pacific Basin Countries Australia, China, Hong Kong, Indonesia, Japan, Malaysia, New Zealand, Philippines, Singapore, South Korea, Taiwan, Thailand

pack. packing

pacm pulse amplification code modulation

PACMD Philadelphia Contract Management District

PacO Pacific Ocean

PACO Polaris Accelerated Change Operation

Pa$_{CO2}$ arterial carbon dioxide pressure

Pac Ocean Terr Pacific Ocean Territories (Pitcairn Island, Ducie, Henderson, and Oeno)

PACOM Pacific Command

pacor passive correlation and ranging

PACOS Package Operating System

PACR Performance and Compatability Requirements

PACRA Pottery and Ceramic Research Association

Pac Rail Missouri Pacific, Union Pacific, Western Pacific (railroads merged)

PACRNB President's Advisory Commission on Recreation and Natural Beauty

PACs Political Action Committees

PACS Pacific Area Communications System

Pac Ship *Pacific Shipper*

pacsim performance achievement computer model for waste package

Pac Sym Pacific Symphony

pact. production analysis control technique; programmed automatic circuit tester

PACT Prisoner and Community Together (for offenders and victims); Production Analysis Control Technique; Project for the Advancement of Coding Techniques

Pac Tel Pacific Telephone (company)

Pac-Tex Pacific-Texas (pipe-line)

Pac T & T Pacific Telephone and Telegraph

PACU Pennsylvania Association of Colleges and Universities

pacv (PACV) personnel air-cushion vehicle

PACV Patrol Air-Cushioned Vehicle (naval)

PACW President's Advisory Committee on Women

PACX Private Automatic Computer Exchange

pad padding; padlock; para-aminobenzoic acid

pad. packet assembler/disassembler; padding; padlock; para-aminobenzoic acid (PAD); payable after death; pitch axis definition; provisional assembly date

pad. (PAD) peripheral arterial disease

Pad Padre; Padstow

P Ad Port Adelaide

PAD Pacific Australia Direct (steamship line); Passive Air Defence; Patient Accounts Department; People Against Displacement (caused by urban redevelopment); Performance Analysis Department (ONWI); Pontoon Assembly Depot; Port of Aerial Debarkation; Provisional Air Division; Public Administration Division; Public Affairs Department

padal pattern for analysis, decision, action and learning

PADAP Philippine-Australian Development Assistance Program

padar passive detection and ranging

PADAT Psychological Abstracts Direct Action Terminal

PADC Pennsylvania Avenue Development Corporation

PADD Pedestrians Against Dangerous Drivers; Pedestrians Against Drunken Drivers; Political Art Documentation and Distribution

Paddo Paddington, Australia

PADDS Petroleum Administration for Defense Districts

PADF Pan-American Development Foundation

PADI Professional Association of Diving Instructors

p adj participial adjective

PADL Pilotless Aircraft Development Laboratory

padloc passive detection and location of countermeasures

PADMIS Patient Administration Information Information System

PADPAO Philippine Agency Detective Protective Association

p Adr *per Adresse* (German—in care of)

padre. portable automatic data-recording equipment

Padre Island National Seashore, Texas

PADS Precision Azimuth Determination System

Pad Sta Paddington Station (rail terminal)

pae public affairs event

p. ae. *partes aequales* (Latin—equal parts)

PAE Peoria and Eastern (railroad); Port of Aerial Embarkation

PAEC Pakistan Atomic Energy Commission; Philippine Atomic Energy Commission

paect pollution abatement and environmental control technology

paed paediatric

paei perisocope azimuth error indicator

PAESP Pennsylvania Association of Elementary School Principals

paf peripheral airfield; pulmonary arteriovenous fistula; punishment and fine

paf (PAF) personal article floater (baggage insurance policy); Polaris accelerated flight

pa & f percussion, auscultation, and fermitus

paf *puissance au frein* (French—brake horsepower)

PAF Pacific Air Force(s); Pakistan Air Force; Palestine Arab Fund (for terrorists); Pet Assistance Foundation; Philippine Air Force; Ports Authority of Fiji

PAFA Pennsylvania Academy of Fine Arts

PAFB Patrick Air Force Base

PAFCO Pacific Fishing Company

PAFMECA Pan-African Freedom Movement of East and Central Africa

PAFS Primary Air Force Specialty

PAFSC Primary Air Force Specialty Code

PAFTA Pacific Area Free

Trade Association
pag periaqueductal gray matter
pag pagaré (Spanish—I will
pay); *pagina* (Italian—page)
pág página (Spanish—page)
Pag pagoda
PaG Pennsylvania-German
PAG Planning Advisory
Group; Primary Analysis
Group; Prince Albert's Guard
PAGASA Philippine Atmospheric Geophysical and Astronomical Services Administration
PAGB Proprietary Association of Great Britain
PAGEL Priced Aerospace Ground Equipment List
pageos (PAGEOS) passive geodetic satellite
Pa Ger Soc Pennsylvania German Society
pagg segg pagine seguenti (Italian—following pages)
pAgmk primary African green monkey kidney
pág(s) página(s) [Spanish—page(s)]
PAGT Port Authority Grain Terminal
pah polynuclear aromatic hydrocarbon(s)—(photochemical smog ingredient)
pah (PAH) para-aminohippuric acid
Pah Pahlavi
PAH Pan-American Highway (also called Inter-American Highway)
PAHC Pan American Highway Congress
PAHO Pan-American Health Organization
PAHOCENDES Pan-American Health Organization Center for Development Studies
pai parts application information; personal accident insurance; personal adjustment inventory; please airmail immediately; prearrival inspection
PAI Panama Airways Incorporated; Piedmont Airlines (3-letter coding)
PAIGCV Partido Africano da Independencia da Guine e Cabo Verde (Portuguese—African Party for an Independent Guinea and Cape Verde)
PAIGH Pan-American Institute of Geography and History
PAILS Projectile Airburst and Impact Location System

PAIN Pan-American Institute of Neurology
paint. painter; painting
Painters Union International Brotherhood of Painters and Allied Trades of the United States and Canada
pair. performance and integration retrofit
PAIR Psychological Audit for Interpersonal Relations
PAIRC Pacific Air Command
PAIRS Private Aircraft Inspection Reporting System
PAIS Pennsylvania Association of Independent Schools; Project Analysis Information System (AID); Public Affairs Information Service
PAIT Program for the Advancement of Industrial Technology
PAJU Pan-African Journalists Union
Pak Pakistan
PAK Pëtr Alekseevich Kropotkin
Paki(s) Pakistani(s)
Pakistan Islamic Republic of Pakistan (Moslem country between Afghanistan and India); *Pakistan* in Urdu means Land of the Pure; *Pak* (Persian—holy) plus *tan* (Urdu—land)—Holy Land
PAKISTAN Punjab, Afghan Border states, Kashmir, Sind, and *tan* from Baluchistan
pal. paleontology; peripheral availability list(ing); permissive action link; phase-alteration line (color tv system); prescribed action link; professional adjustable ladder; program assembly language; programmed application library
pal. (PAL) phase alternate line; products and area locator
p-a-l prisoner-at-large
Pal Palace; Palencia; Paleozoic; Palermo; Palestine
Pal Palacio (Spanish—palace); *Palácio* (Portuguese); *Palais* (French—palace); *Palazzo* (Italian—palace)
PAL Pacific Aeronautical Library; Pacific Aluminium; Pakistan Airlines; Pan Asia Line; Pensioners Advancement League; phase-alternating (television) line; Philippine Air Lines; Podiatry Arts Laboratory; Police Athletic League; Polynesian Airlines

Limited; Prison Atheist League; prisoner-at-large; Public Archives Library
PALA Polish-American Librarians Association
Palat Palatinate
P Alb Port Alberni
PALC Point Arguello Launch Complex
paleo paleography
paleob paleobotany
paleon paleontology
Palestine southern Syria, according to many Arabs; Turkish province containing what is now Israel plus adjacent Arab countries in the Jerusalem area often called the Holy Land
PALI Pacific and Asian Linguistics Institute (University of Hawaii)
palimony alimony awarded a former common-law partner
palin palindrome; palindromic
PALINET Pennsylvania Area Library Network
PALIS Property and Liability Information Systems
Palisades Palisades Interstate Park along the west bank of the Hudson River washing the shores of New Jersey and New York; Palisades (amusement) Park near Englewood, New Jersey; Palisades Peaks in Kings Canyon National Park, California
pall. pallet
palm. palmist(ry); precision attitude and landing monitor
Palma Palma de Mallorca (capital of the Balearic Islands and the island of Mallorca)
Palmach Plugot Machatz (Hebrew—Spearhead Units)—commando units active in the establishment of Israel when still called Palestine
Palmas Las Palmas de Gran Canaria (capital and main seaport of the Canary Islands)
Palmn Palmerston
PALMS Propulsion Alarm and Monitoring System
Pal Obs Palomar Observatory
Palomar Mt. Palomar Mountain Observatory near San Diego, California
Palos Palos de la Frontera (port of departure of Columbus in 1492)
palp palpable; palpitation
palpi palpitation

PALs Parcel Air Lifts (U.S. Post Office parcel-post service for servicemen)

PALS Permissive Action Link Systems

PALSG Personnel and Logistics Systems Group

PALTC Pacific Asian and Latino Training Center

pam pamphlet; primary amoebic meningoencephalitis; procurement aircraft and missiles; pulse amplified modulation; pulse amplitude modulation

Pam Lord Palmerston; Pamela

PAM Palestine Archeological Museum; Pasadena Art Museum; Portland Art Museum

PAMA Pan-American Medical Association; Professional Aviation Maintenance Association

pamac parts and materials accountability control

PAMC Pakistan Army Medical Corps

PAMCO Pacific Annuity Marketing Company

PAMÉTRADA Parsons Marine Experimental Turbine Research and Development Association

pamf programmable analog-matched filter

pam file pamphlet file

PAMIPAC Personnel Accounting Machine Installation Pacific Fleet

pamirasat (PAMIRASAT) passive microwave radiometer satellite

Pamirs Pamir Mountains of Soviet Central Asia

PAML Pan American Mail Line

PAMO Port Air Materiel Office

PAMPA Pacific Area Movement Piority Agency (DoD)

pamph pamphlet

pams pamphlets

PAMS Plan Analysis and Modeling System

PAMT Port Authority Marine Terminal

pan (PAN) peroxyacetyl nitrate (smog ingredient)

pan. panchromatic; panorama; panoramic; pantomime; pantry

Pan Panama; Panamanian; Panameño

PAN Pan American Navigation; Parents Against Narcotics; peroxyacetylnitrate (air-

pollutant poison)

PAN *Partido Acción Nacional* (Spanish—National Action Party)—Mexican; *Polska Akademia Nauk* (Polish Academy of Sciences)

PANA Pan-African News Agency; Pan-Asia Newspaper Alliance

PANAFTEL Pan-African Telecommunications (network)

PANAGRA Pan American-Grace Airways

PANAIR *Panair do Brasil* (Brazilian airline)

Pan-Am Pan-American World Airways

Panama Republic of Panama (Spanish-speaking Central American country bisected by the Panama Canal), *República de Panamá*

Panamax maximum size Panama Canal locks can accommodate

Panamints Panamint Mountains of eastern California along the Death Valley border of Nevada

PANANEWS Pan-Asia Newspaper Alliance (Hong Kong)

pan b panic bolt

panc pancreas

Pan Can Panama Canal

Pan Canal Panama Canal

pand *panderazo* (Spanish—blow struck with a tambourine); *panderetero* (Spanish—tambourine maker or player); *pandero* (Spanish—tambourine)

PANDA Prestel Advanced Network Design Architecture; Professional Association of Numismatic Dealers of Australasia

pandex *pan* (Greek—all) + *dex* (from index)—all-inclusive index

pandg people are no damn good

p-and-p struggle prude-and-prurient struggle

PANEES Professional Association of Naval Electronics Engineers and Scientists

P Ang Port Angeles

Pango (naval argot—Pago Pago, American Samoa)

panjan panjandrum

Pank Pankow

panol panology

panorams panoramas

PANPA Pacific Area Newspaper Production Association

Pan pan *Pan paniscus* (pygmy chimpanzee found south of the Congo River)

pans. peroxyacetylnitrates

PANS Procedures for Air Navigation Services

PANSDOC Pakistan National Scientific and Technical Documentation Center

Pan Sea Fron (PANSEAFRON) Panama Sea Frontier

PANSY Programme Analysis System

P Ant Port Antonio

panth pantheism; pantheist; pantheistic(al)(ly)

panto pantograph(ic); panto-mime; pantomimic

Pan trog *Pan troglodytes* (common chimpanzee found north of the Congo River)

pants pantaloons

PANY Power Authority of the State of New York

pao product assurance operations

PAO Public Affairs Officer

Pa$_{O2}$ arterial oxygen pressure

PAOA Pan-American Odontological Association

PAODAP President's Action Office for Drug Abuse Prevention

pap (PAP) pension administration plan

pap. papa; papacy; papal; paper; papyrus

pap *prêt à porter* (French—ready to wear)

p-a-p *poco a poco* (Italian—little by little)

Pap Papa; Papeete; Papist; Pappie; Papua; Papuan

PAP Pacific Automation Products; People's Action Party; Performance Assessment Plan; *Polska Agencja Prasowa* (Polish News Agency); Port-au-Prince, Haiti (airport); Progressive Australia Party

papa parallax aircraft parking aid

Papa letter P radio code

PAPA Parents As Partners Associated

PAPAs Parents as Partners Associates

PAPAS Pennsylvania Association of Private Academic Schools

PAPC Philological Association of the Pacific Coast

Pap diag Papanicolaou diagnosis

Papermac paperback book

published by Macmillan
PAPF Pan-American Philatelic Federation
papi precision path indicator
PAPI Pacific Automation Products Incorporated
papil papilla; papillae
Pap Inf Papal Infallibility
Pap Lib Paperback Library
Pap NG Papua New Guinea
p app puissance apparente (French—apparent power)
Pap(s) [Irish-Protestant English-Papist(s)—*see* Prod(s)]
PAPS Periodic Armaments Planning System
Pap smear Papanicolaou smear
Papsom Papaver somniferum (opium poppy)
PAPSS Procurement and Production Status System
Pap Sta Papal States
PAPTE President's Advisory Panel on Timber and the Environment
Pap Ter Papua Territory
Pap Test Papanicolaou Test (for cervical cancer)
Papua Indonesian island called Papua New Guinea in the eastern sector and West Irian on the western sector
paq position-analysis questionnaire
par (PAR) perimeter acquisition radar
par. parabolic aluminized reflector; paragraph; parallax; parallel; parenthesis; per acre rental; planed all around (timber); precision approach radar; pulse acquisition radar
par (Latin prefix—bear or give birth to)—parturition
Par Paris; Parish
Par Parigi (Italian—Paris);
Parijs (Dutch—Paris)
PAr Punta Arenas
PAR Parental Awareness and Responsibility; Paris, France (Orly airport); Program Appraisal and Review; Protect Abortion Rights
PAR Partido Acción Revolucionario (Spanish—Revolutionary Action Party)
para parachute; paragraph; parallel; perceiving and recognition automation
para (Greek—alongside, beside, beyond)—paralysis, paramedic, parameter, parasite, parathyroid
Para Paraguay(an)
Pará Belém do Pará, Brazil

PARA Program for At-Risk Addicts
para I; para II; para III; etc. unipara; bipara; tripara; etc.—having given birth to one child, to two children, to three children, etc.
parab parabola
parabat parachute battalion
Paracels Paracel Islands in the South China Sea east of Vietnam
paracent paracentesis
parad paradicholorobenzene; paradigm(atic)(al)(ly); paradisiac(al)(ly); paradisal; paradise; paradisiacal(ly); paradox(ical)(ly); paradoxicalness
paradrop parachute airdrop
par. aff. pars affecta (Latin—to the part affected)
Paraguay Republic of Paraguay (Spanish-speaking South American country), *República del Paraguay*
Paraguayos Paraguayos, república o muerte (Spanish—Paraguayans, republic or death) Paraguay national anthem
paral parallax; paralysis
param parameter(s); parametric
Parami Parsons active ring around miss indicator
paramp parametric amplifier
parapsy parapsycholic(al)(ly); parasychologist; parapsychology
parapsych parapsychologist; parapsychology
paraquat paraquat-tainted marijuana
paras parasite(s); parasitic; parasitism; paratroopers
parasail parachute sail (steerable parachute)
parasitol parasitologic(al)(ly); parasitologist; parasitology
parasym div parasympathetic division
parasyn parametric synthesis
Parbo Paramaribo
parc progressive aircraft repair cycle
PARC Princeton Applied Research Corporation; Public Archives Records Centre
PARCA Pan American Railway Congress Association
parch. parchment
PARCS Parking and Revenue Control System (for autos); Perimeter Acquisition Radar Characterization System
pard partner

PARD Personnel Actions and Records Directorate
pardac parallel digital-to-analog converter
pardop passive-ranging doppler
PARDS Precision-Annotated Retrieval Display System
paregoric compound tincture of opium
paren parenthesis
parens parentheses
parent. parental(ly)
Parents Parents Magazine
parex programmed accounts-receivable extra (service)
par for par for the course (average, typical, usual)
PARFR Program for Applied Research on Fertility Regulation (Northwestern University)
Parg Paraguay; Paraguayan
pari parietal
Pariñas Pariñas Point (westernmost point of South America)
Paris O Paris Opera
PARKA Pacific Acoustic Research (Kaneoche, Alaska)
parkade parking arcade
parl parallel
Parl Parliament
PARL Palo Alto Research Laboratory (Lockheed)
Parl Agt Parliamentary Agent
Parl Const Parliamentary Constituency
par light parabolic aluminized reflector lamp
Parlor Irish well-to-do people
Parl Sec Parliamentary Secretary
parm (PARM) precision anti-radiation missile
PARM Partido Autentico de la Revolución Mexicano (Spanish—Authentic Party of the Mexican Revolution)
PARMA Public Agency Risk Managers Association
parm(s) parameter(s)
parochiaid parochial-school aid (provided by tax monies)
paros passive ranging on submarines
parot parotid
parox paroxysm(al)
PARPRO Peacetime Aerial Reconnaissance Program
parq parquet
Parra Parramatta estuary, Sydney, Australia
pars paragraphs
PARS Passenger Airlines Reservation System; Prisoners

Aid and Rehabilitation Society; Private Aircraft Reporting System; Programmed Airlines Reservation System
parsec parallax second (3.26 lightyears or 19.2 trillion miles)
Parsee (Arabic—Iranian, Persian)—Indian Zoroastrian descended from refugees who came to India to escape Muslim persecution
parsq pararescue
parsyn parametric synthesis
part. parterre; partial; participate; participle; particle; partition; partner; partnership
part. partim (Latin—part)
PART Part Allocation Requirements Technic
part. aeq. partes aequales (Latin—equal parts)
partan parallel tangents
Partas Partagas cigars
part.dolent. partes dolentes (Latin—painful parts)
PARTEI Purchasing Agents of Radio, TV, and Electronics Industries
parth parthenogenesis
Parthia (Latin—parts of Assyria and Persia in northeastern Iran)
parti participle
partic participle; particular
partic exh particulate exhaust (soot)
Partisan Partisan Review
partit partitive
partn partition
partner. proof of analog results through numerical equivalent routines
Partrys Partry Mountains of western Ireland
part. vic. partibus vicibus (Latin—in divided doses)
paru postanesthetic recovery unit
par uni party unity (political utopia)
parv paravane
parv parvus (Latin—small)
PARVO Professional and Academic Regional Visits Organization
pas passive; photoacoustic spectrometer; power-assisted steering; public-address system
pas (PAS) para-aminosalicylic acid; periodic acid Schiff; photo-acoustic spectroscopy
pa's public appearances
paS periodic acid Schiff
Pas Pasadena; Pascagoula;

Pashto; Passage; Passaic; Passau
Pa s Pascal second
Pas. Paschae (Latin—Easter)
PAs Parents Anonymous; Police Agents
PA's purchasing agents
PAS Percussive Arts Society; Pioneer Air System; Pontifical Academy of Science; Pregnancy Advisory Service; Primary Alerting System; Probation and Aftercare Service; Professor of Air Science; Public Address System
pasa (PASA) para-aminosalicylic acid
PASA Pennsylvania Association of School and Administrators; Pipelines Authority of South Australia
pasar psychological abstracts search and retrieval
PASAR Philippine Associated Smelting and Refining
PASB Pan-American Sanitary Bureau
PASBO Pennsylvania Association of School Business Officials
PASC Pacific Area Standards Congress; Palestine Armed Struggle Command (controlled by El Fatah); Pan-American Standards Committee
PASCAL Philips Automatic Sequence Calculator
PASCAL Program Appliqué à la Selection et la Compilation Automatique de la Litterature (French—Program Applied to the Selection and the Automatic Compilation of Literature)
PASCO Pan American Sulfur Corporation
p'ase alkaline phosphatase
PASF Photographic Art and Science Foundation
PASG Programs Activities and Services Guide
PASGT Personal Armor System—Ground Troops (new helmet of the U.S. Army)
pasim pasimological; pasimologically; pasimologist; pasimology (study of gestures as means of communication)
PASL Pakistan Association of Special Libraries
PASLIB Pakistan Association of Special Libraries
PASO Pan-American Sanitary Organization; Pan-American Sports Organization

PASOK Pan-Hellenic Socialist Commune
pass. passage; passenger; passitive; passivate; passive; passport
pass. passim (Latin—far and wide, here and there, up and down)
Pass Passover
PASS Passengers Automatic Selection System; Procurement Automated Source System; Prototype Artillery Subsystem
PASSIM President's Advisory Staff on Scientific Information Management
Pass Kristyen Pass Christian, Louisiana
PASSP Pennsylvania Association of Secondary School Principals
Past Pasteurella
PASTIC Pakistan Scientific and Technological Information Center
pastram passenger traffic management
pastramasan pastrami sandwich (pickled corned-beef sandwich)
pastramwich pastrami sandwich (pickled corned-beef sandwich)
PASWEPS Passive Antisubmarine Warfare Environmental Protection System
p-a system public-address system
pat. patent(s); patrol(s); pattern; points after touchdown
pat. (PAT) paroxysmal atrial tachycardia
p-à-t pied-à-terre (French—small occasional lodging)
Pat Patricia; Patrick
Pat Patrone (German—cartridge, round of ammunition)
PaT Parents as Teachers
PAT Pacific Air Transport; Pacific Automobile Train; Philippine Aerial Taxi; Postavailability Trials; Prescription Athletic Turf; Production Assessment Test; Progressive Achievement Tests
PATA Pacific Area Travel Association
Patag Patagonia(n)
PATAS Publications Automated Task Analysis System
PATCA Panama Air Traffic Control Area; Professional and Technical Consultants Association
PATCO Port Authority Transit

Corporation; Professional Air Traffic Controllers Association
PATCRA Papua New Guinea—Australia Trade and Commercial Relations Agreement
patd patented
PATE Philippine Association of Technological Education
path. pathological; pathologist; pathology; pituitary adrenotrophic hormone (PATH)
path (Latin prefix—disease)—pathologist, pathology
PATH Port Authority Trans-Hudson (Hudson Tubes)
Pathét Pathétique
patho pathological
patho (Greek—suffering)—osteopath(ic), pathological, pathologist, pathology
pathogen pathogenic
pathol pathologic(al)(ly); pathologist; pathology
pathomorph pathomorphologic(al)(ly); pathomorphologist; pathomorphology
pathy (Latin suffix—abnormality or disease)—neuropathy, psychopathy
Patk Patrick
pat. med patent medicine
PATMO Patent and Trademark Office
patn pattern
PATO Pacific-Asian Treaty Organization
Pat Off Patent Office
PATOLIS Patent On-Line Information System
pat pend patent pending
PATRA Printing, Packaging, and Allied Trades Research Association (also appears as PPATRA)
PATRIC Pattern Recognition and Information Correlation (police computer)
PATRICIA Practical Algorithm to Receive Information Coded in Alphanumeric
patron. patronym(ic)(al)(ly)
Patronat (French equivalent of National Association of Manufacturers in United States)
pats. patents
PATs Pre-Authorized (bank deposit) Transfers
PATS Philippine Aeronautics Training School; Portable Acoustic Tracking System; Proof and Transit System
patt pattern
PATTERN Planning Assistance Through Technical Evaluation of Relevance

Numbers
Patti Adelina Patti (1843–1919)
PATWAS Pilot's Automatic Telephone Weather Answering Service
PATX Private Automatic Telex Exchange
PATY Private Annuity for Term of Years
pau pattern articulation unit; programmer's analysis unit
Pau Pablo
PAU Pan American Union; Pan American University; Police Airborne Unit
P-au-P Port-au-Prince
pav paving
p/av particular average
Pav Luciano Pavarotti; pavilion; Pavo (constellation)
PAV Personnel Allotment Voucher
PAV Poste Avion (French—airmail)
PAVAA Polish Army Veterans Association of America
pave. position and velocity extraction
PAVE Professional Audiovisual Education (study)
PAVEPAWS Precision Acquisition of Vehicle-Entry Phased-Array Warning System
PAVE-PAWS Precision Acquisition of Vehicle Entry—Phased Array Warning System (early-warning radar system against submarine-launched missiles)
PAVM Potential Acquisition Valuation Method
PAVN Peoples Army of Viet Nam
pav. noc. pavor nocturnus (Latin—nightmares, night terrors)
PAVPAWS Precision Acquisition of Vehicle-Entry Phased-Array Warning System
pavt power-adjusted variable track(ing)
paw. portable auxiliary workroom; powered all the way
Paw Papa
PAW People for the American Way; Pets and Wildlife; Poetic Allusion Watch
PAWA Pan American World Airways; Pan-American Womens Association
PAWO Pan-African Women's Organization
pawob passengers arriving

without baggage
PAWS Phased Array Warning System; Programmed Automatic Welding System; Progressive Animal Welfare Society
pax. passenger(s); private automatic exchange
Pax Paxon; Paxton
Pax Am Pax Americana (Latin—American Peace)
Pax Brit Pax Britannica (Latin—British Peace)—a long period of peaceful stability imposed throughout the British Empire and many adjacent parts of the world
Pax Por Pax Porfiriana (Latin—Porfirian Peace)—imposed on Mexico by its dictator-general-president—Don Porfirio Díaz—from 1876 to 1910 when ousted by Madero
Pax River Patuxent River Naval Air Station, Maryland
Pax Rom Pax Romana (Latin—Roman Peace)—imposed throughout the Roman Empire
pax vob. pax vobiscum (Latin—peace be with you)
Pay Paymaster; Paymistres
Pay Cmdr Paymaster Commander
paye (PAYE) pay as you earn, pay as you enter
payld payload
Paymr Paymaster; Paymistress
PAYS Patriotic American Youth Society
payt payment
Paz Ladislao Pazmany
pb painted base; paper base; passed ball; patrol bombing; permanent ballast; permanent bunker(s); petrol bomb(ing); plate block; plotting board; plugged back (oil well); polybutylene; poor bastard; ports and beaches; power brake; pull box; pulse beacon; push button
p/b paperback; pass book; poor bastard; pushbutton
pB purplish blue
Pb *plumbum* (Latin—lead)
PB Pacific Beach; Packard Bell; Palm Beach; patrol boat; patrol bomber; patrol bombing; Permian Basis; Personnel Board; Planning Board; Pocket Book; police boat; Pompano Beach; Presiding Bishop; Public Bath; Publication Bulletin
P-B Pitney-Bowes

PB *Paises* *Bajos* (Spanish—Netherlands); *Planta* *Baja* (Spanish—ground floor), elevator pushbutton designation; *Prayer Book*

P.B. *Pharmacopeia Britannica*

P & B *Porgy* *and* *Bess* (George Gershwin's opera)

pba poor bloody assistant; pressure-breathing assister; published by arrangement

PBA Patrolmen's Benevolent Association; Philadelphia Bar Association; Port of Brisbane Authority; Port of Bristol Authority; Professional Billiards Association; Professional Bookmen of America; Professional Bowlers Association; Provincetown-Boston Airline; Public Buildings Administration

PBAA Periodical and Book Association of America; Public Broadcasting Association of Australia

p'back paperback

pbai *proyectil* *balístico* *de* *alcance* *intermedio* *(PBAI)*— (Spanish—intermediate range ballistic missile)

P-band 225–390 mc

PBAS Post Block Aerial Survey(ing)

pbb push-button banking

PBBH Peter Bent Brigham Hospital (Boston)

pbb's (PBBs) polybrominated biphenyls

pbc peripheral bus computer; point of basal convergence; pregnancy and birth complication(s)

pBc pure Bolivian cocaine

PBC Palisade Boat Club; Pen and Brush Club; Philadelphia Blood Clinic; Philadelphia Book Clinic; Power Boat Club; Provincial Bank of Canada

pbcb plugboard circuit breaker

pbd particle board

PBD Public Buildings Department

pbdndb perceived barking dog noise decibels

pbe present-barrel equivalent

Pbe Perlsucht bacillen emulsion

PBEC Pacific Basin Economic Council; Public Broadcasting Environment Center

P.B.Ed. Bachelor of Philosophy in Education

PBEIST Planning Board for European Inland Surface

Transport (NATO)

pbf permalloy-bar file

PBF fast patrol boat (naval symbol)

PBF *Prins* *Bernhard* *Fonds* (Prince Bernhard Fund)

PBFG guided missile fast patrol boat (naval symbol)

PBFL Planning for Better Family Living (UN)

Pbg Pittsburgh

PBGC Pension Benefit Guaranty Corporation

pbh partial bulkhead; primary borehole

pbhp pounds per brake horsepower

pbi please book immediately; polybenzimidazole (space-age fabric); poor bloody infantry; protein-bound iodine

pbi (PBI) polybenzimidazole

pbi *proyectil* *balístico* *intercontinental* *(PBI)* (Spanish—intercontinental ballistic missile)

PBI Paper Bag Institute; Paving Block Institute; Pitney-Bowes Incorporated; Plant Breeding Institute; Plumbing Brass Institute; Projected Books Incorporated; West Palm Beach, Florida (airport)

PBiB *Paperback* *Books* *in* *Print*

pbip pulse beacon impact predictor

PBJC Palm Beach Junior College

pbk paperback

PBK Phi Beta Kappa

PBKTOA Printing, Bookbinding, and Kindred Trades Overseers' Association

pbl planetary boundary layer; probable

PBL Pacific Beach Library; Public Broadcast Laboratory

pb list phonetically balanced (word) list

P Blr Port Blair

pbm performance-based management

pbm (PBM) permanent bench mark

PBM Mariner twin-engine Navy bomber built by Martin; Paramaribo, Surinam (airport); Principal Beach Master; Production Bill of Material

PBMA Peanut Butter Manufacturers Association

PBMR Provisional Basic Military Requirements

PBN Provisional Buy Notice

PBN *Producto* *Bruto* *Nacional* (Spanish—National Bulk Products)

pbo packed by owner; polite brushoff

P-boat Patrol Boat

P. Bor. *Pharmacopoeia Borussica* (Latin—Prussian Pharmacopoeia)

PBOS Planning Board for Ocean Shipping (USA)

pbp pushbutton panel

pbpGinfwmy please be patient; God is not finished with me yet

pb/ps power brakes/power steering

PBPS Program Budgeting and Planning System

pbr payment by results

pbr (PBR) power breeder reactor; precision bombing range

PBR Project Budget Report(ing)

pbs paginated by sections; polarizing beamsplitter; production base support; program breakdown structure

pb's paperback books; petrol bombs; poor bastards

p-bs phosphate-buffered saline (solution)

PBS Pacific Biological Station (Canada); Panama Bureau of Shipping; Permanent Building Societies; Pharmaceutical Benefits Scheme; Philippine Broadcasting Service; Prevent Blindness Society; Public Broadcasting Service; Public Broadcast(ing) Station; Public Buildings Service

PBSA Partially Blinded Soldiers Association; Permanent Building Societies Association

PB & SC Power Boat and Ski Club

PBSCMA Peanut Butter Sandwich and Cookie Manufacturers Association

PBSE Philadelphia-Baltimore Stock Exchange

pbsp prognostically bad signs during pregnancy

pbt performance-based teaching; profit(s) before tax(ation)

pbt (PBT) polybutylene terephthalate

PBT President of the Board of Trade

PBTB Paper Bag Trade Board; Paper Box Trade Board

pbte performance-based teacher education

Pburg Pittsburgh

pbv predicted blood volume; pulmonary blood volume

pbw parts by weight; posterior bite wing

pbw (PBW) particle-beam weapon

PBWSE Philadelphia-Baltimore-Washington Stock Exchange

pbx private branch exchange

pbx's (PBXs) personal business exchanges (computerized telephones)

PBY Consolidated-Vultee PBY flying boat; vacation island near Long Beach, California—Santa Catalina

pbz phosphor bronze

pbz (PBZ) pyribenzamine (anti-histamine)

pc paper copy; parent cells; parsec; pay clerk; paycheck; percent; percentage; percentile; personal correction; petty cash; pica(s); piece(s); pitch circle; point of curve; port of call; postcard; prices current; printed circuit; privileged character; pull chain; pulsating current; purchasing and contracting purified concentrate

pc (PC) personal computer; pitch class; programme counter

p-c phophlogistic-corticoid; printed circuit

p/c percent; percentage; processor controller; programmer-comparator; pulse counter

p & c put and call

pc *point de congélation* (French—freezing point)

p.c. *per centum* (Latin—by the hundred; percent); *post cibum* (Latin—after a meal, after meals)

Pc Phillips curve (macroeconomics)

PC Pace College; Pacific Airlines; Pacific Coast (railroad); Pacific College; Paine College; Palmer College; Palomar College; Panama Canal; Panola College; Paris College; Park College; Parsons College; Pasadena College; Peace Corps; Pembroke College; Pepperdine College; personnel carrier; Pfeiffer College; Pharmacy Corps; Philadelphia College; Philip-

pine Constabulary; Phoenix College; Piedmont College; Pikeville College; Pilotage Chart(s); Pineland College; Pittsburgh Corning; Plane Commander; Police College; Police Commissioner; Pomona College; Population Council; Porterville College; Presbyterian College; Principal Celebrant (at the Mass); Principia College; Privy Council; Privy Councillor(s); Procurement Command; Producers Council; Professional Corporation; Providence College; submarine chaser patrol vessel (naval symbol)

P-C Penn-Central (railroad)

P.C. Penal Code; Plaid Cymru (party)

P&C Parents and Citizens Association

P & C Pickpocket and Confidence (police department squad)

PC *Parti Communiste* (French—Communist Party); *Partido Colorado* (Spanish—Colorado Party)—the reds; *Partido Comunista* (Spanish—Communist Party); *Partido Conservador* (Spanish—Conservative Party); *Patres Conscripti* (Latin—Senators); *Penal Code*; *Plaid Cymru* (Welsh—Party of Wales); *Poder Chicano* (Spanish—Chicano Power); *Première Classe* (French—First Class); *Privy Council* (British Law Reports); *Publishers' Circular*

pca permanent change of assignment; physical configuration audit(ing); Porsche Club of America (uses lowercase initials); principal component analysis

pca (PCA) p-chloraphenylalanine

Pca Pensacola

PCA Parachute Club of America; Permanent Court of Arbitration (The Hague); Pest Control Association; Pennsylvania Council on the Arts; Photogrammetric Consultants Association; Plaster Contractors Association; Police Complaints Authority; Pollution Control Agency; Pony Club Association; Portland Cement Association; Positive Control Area; Primary Coverage Area; Printers' Costing

Association; Production Code Administration; Production Credit Association; Proprietary Cremation Association

PCA *Partido Comunista Argentina* (Spanish—Argentine Communist Party)

PCAC Professional Classes Aid Council

pcam punchcard accounting machine; punchcard accounting method

PCAO Presidential Complaints and Action Office(r); President's Commission on Americans Outdoors

PCAPA Pacific Coast Association of Port Authorities

PCAPK President's Commission on the Assassination of President Kennedy

PCARS Point Credit Accounting and Reporting System

PCAs Progressive Citizens of America

PCAST Presidential Commission on Aviation Security and Terrorism

PCAT Philippine College of Arts and Trades

pcb petty cash book; printed circuit board

pcb (PCB) polychlorinated biphenyls

PCB Pest Control Bureau; Program Control Board

PCB *Partido Comunista Boliviano* (Spanish—Bolivian Communist Party); *Partido Comunista Brasileiro* (Portuguese—Brazilian Communist Party)

pcbb primary commercial blanket bond(ing)

PCB-BKB *Parti Communiste Belge* (French—Belgian Communist Party); *Kommunistischem Partij* (Flemish—Communist Party)

PCBL Pacific Commercial Bank Limited

pcb's (PCBs) polychlorinated biphenyls (industrial pollutants)

p-c b's printed-circuit boards

pcc phosphate carrier compound; pitch of cone to cone; portland concrete cement; program-controlled computer

pCc pure Colombian cocaine

p¢c *plus ça change, plus c'est la même chose* (French—the more it changes, the more it stays the same)

PCC Pacific Coast Conference; Pacific Conference of

Churches; Palmer Community College; Panama Canal Commission; Panama Canal Company; Pennsylvania Crime Commission; Philippine Cotton Corporation; Poison Control Center; Polynesian Cultural Center; Port of Corpus Christi; Portland Community College; Postal Customer Council; Presidents' Conference Committee; Price Control Council; Program Control Center

PCC Partido Comunista Cubano (Spanish—Cuban Communist Party)

PCCC Pakistan Central Cotton Committee

PCCD PACE Center for Career Development

PCCEMRSP Permanent Commission for the Conservation and Exploitation of the Maritime Resources of the South Pacific

PCCI President's Committee on Consumer Interests

PCCNY Penal Code of the City of New York

PCCR Publishing Center for Cultural Resources

PCCT Percept and Concept Cognition Test

pccu (PCCU) progressive coronary care unit

PCCU President's Commission on Campus Unrest

pcd pitch circle diameter; pounds per capita per day

PCD Planned Community Development; Principal Criteria Document

PCDA Post Card Distributors Association

PCDG Prestressed Concrete Development Group

pc di pierce die

PC-DOS Personal Computer Disk Operating Sytem

P Cdr Paymaster Commander

PCDS Program Control Display System (NATO)

pce pyrometric cone equivlent

pce (PCE) pseudocholinesterase

PCE patrol craft escort (3-letter coding); Personal Consumption Expenditure

PCE Partido Comunista Española (Spanish—Spanish Communist Party)

PCEA Pacific Coast Electrical Association; President's Council of Economic Advisors

PCEH The President's Committee on Employment of the Handicapped

PCEM Parliamentary Council of the European Movement

PCEQ President's Council on Environmental Quality

PCER rescue escort (naval symbol)

pcf pistol center fire (greater than a 22); pounds per cubic foot; power per cubic foot

Pcf Pacifico (Italian—Pacific); *Pacífico* (Portuguese or Spanish—Pacific); *Pacifique* (French—Pacific)

PCF Personnel Control Facility; Program Checkout Facility

PCF Parti Communiste Français (French—French Communist Party)

PCFAP The President's Committee on the Foreign Aid Program

PCFLIS President's Commission on Foreign Language and International Studies

pcg phonocardiogram

PCG guided-missile coastal-escort vessel (naval symbol)

PCGG Presidential Commission on Good Government (Philippines)

PCGN Permanent Committee on Geographical Names

pch paroxysmal cold hemoglobinuria

P Ch Parish Church

PCH hydrofoil submarine chaser (3-letter coding)

pchbd patchboard

p Ch c pure Chilean cocaine

pci pattern correspondence index; pellet-cladding interaction; peripheral command indicator; perpetual cost index; photochromic micro-image; picocurie; potential criminal informant; programmed-controlled interruption

PCI Packer Collegiate Institute; Pilot Club International; Planning Card Index; Prestressed Concrete Institute; Program of Correctional Institutions (Puerto Rico)

PCI Partito Comunista Italiano (Italian—Italian Communist Party)

PCIB Pacific Cargo Inspection Bureau

PCIC Polaris Control and Information Center

PCIFC Permanent Commission of the International Fisheries Convention

PCII Potato Chip Institute International

PCIJ Permanent Court of International Justice

PCIM Presidential Commission on Income Maintenance

PCjr IBM's small personal computer

pck polycystic kidney

Pck conditional probability of kill (armament)

pckt printed circuit

pcl parcel; precancel; printed-circuit lamp

PCL Pacific Coast Line; Peoples College of Law; Police Crime Laboratory

PCLA Project Coordination and Liaison Administration

PCLEAJ President's Commission on Law Enforcement and the Administration of Justice

p-c lens perspective-correction lens

pclk pay clerk

PCLO Police Community Liaison Office(r)

PCLTT Permanent Committee on Land Transportation and Telecommunications (ASEAN)

pcm phase-change material(s); plug-compatible manufacturer(s); protein-calorie malnutrition; pulse-code modulation; pulse-count modulation; punchcard machine(s)

PCM Peabody Conservatory of Music; President's Certificate of Merit

PCM Partido Comunista Mexicano (Spanish—Mexican Communist Party)

PCMA Post Card Manufacturers Association; Professional Convention Management Association

pcmb (PCMB) parachloromercuric benzoic (acid)

pcmi photographic micro-image(s)—microdot photos

PCMIA Plasterers and Cement Masons International Association (U.S. and Canada)

PCMO Principal Colonial Medical Officer

PCMP Progressive Car Manufacturing Program

pcm/pl pulse-code modulated/polarized light

pcmr patient computer medical record; photochromic micro-reproduction

PCMR President's Committee

on Mental Retardation
PCMSER President's Commission on Marine Science, Engineering, and Resources
pcmx (PCMX) parachlorometaxylenol (antiseptic)
pcn parent-country national(s); printed control number; processing control number
PCN Part Control Number; Pharmaceutical Case Network; Primary Care Network; Procurement Control Number
PCN Partido de Conciliación Nacional (Spanish—National Conciliation Party)
PCNB Permanent Control Narcotics Board
PCNG President's Commission on National Goals
PCNR Part Control Number Request
PCN's Planning Change Notices
PCNV Provisional Committee on Nomenclature of Viruses
PCNY Proofreaders Club of New York
pco post checkout operation(s)
pc/o por ciento (Spanish—percent)
PCO Pacific Chamber Opera; Printing Control Office(r); Procuring Contracting Office(r); Public Carriage Office(r)
P/CO Purser/Catering Officer
P$_{CO2}$ carbon dioxide pressure (or tension)
PCOB Permanent Central Opium Board (UN)
PCOOS Pacific Coast Oto-Ophthalmological Society
PCOP President's Commission on Obscenity and Pornography
PCOS Primary Communication Operating System
pcp passenger control point; phosphor-coated paper; pimp-controlled prostitution; polychloroprene (rubber); production change point
pcp (PCP) pentachlorophenol; phencyclidine (sometimes called Pure California Poison)
PCP Peking Central Philharmonic; Postgraduate Center of Psychotherapy; Program Change Proposal; Progressive Conservative Party
PCP Partido Comunista Pan-ameño (Spanish—Panamanian Communist Party);

Partido Comunista Para-guayo (Spanish—Paraguayan Communist Party); *Partido Comunista Peruviano* (Spanish—Peruvian Communist Party); *Partido Communista Portugues* (Portuguese Communist Party)
PCPA Panama Canal Pilots Association; parachlorophenylalanine; Philadelphia College of the Performing Arts
PC(PBC)R Pedestrian Crossings (Push-Button Control) Regulations
PCPD Portland Commission of Public Docks
PCPF President's Council on Physical Fitness
PCPI Parent Cooperative Preschools International
PCPJ People's Coalition for Peace and Justice
PCPM Program Control Procedures Manual
PCPP President's Commission on Pension Policy
PCPS Philadelphia College of Pharmacy and Science
PC & PS Professional Credentials and Personnel Service (nursing)
pcpt perception
pcpv prestressed concrete pressure vessel
pcq production-control quantometer
PCQ Personal Control Questionnaire
pcr photoconductive relay
pcr (PCR) program control register
P Cr Paymaster Commander
PCR Program Change Request; Publication Contract Requirement
PCR Partido Comunista Revolucionario (Spanish—Revolutionary Communist Party)—Chile; *Partidul Comunist Roman* (Roman Communist Party)
PCRB Pollution Control Revenue Bond
PCRC Paraffined Carton Research Council; Primary Communications Research Center (University of Leicester)
pcrca pickled, cold rolled, and closely annealed
PC R & D C Pomona Colleges Research and Development Center
PCRI Papanicolaou Cancer

Research Institute
PCRs Pedestrian Crossings Regulations; Planning and Compensation Reports
PCRS Poor Clergy Relief Society
PCRU Pacific Coast Rugby Union
pcrv prestressed concrete reactor vessel
pcs permanent change of station; phonocardioscan; picas; pieces; planning control sheet; program counter storage; program counter store
pc's protective clothes
pc's (PCs) personal computers
PCs Police Constables; Progressive Conservatives
PCS 136-foot submarine chaser (3-letter coding); Parent's Confidential Statement; Permanent Committee on Shipping (ASEAN); Petrochemical Corporation of Singapore; Pharmaceutical Card System; Polytechnic Certificate in Shipping; Program Control System; Punch(ed) Card System; Punjab Cooperative Society
PCSA Polish Cultural Society of America
PCSCA Permanent Committee on Socio-Cultural Affairs (ASEAN)
PCSE Pacific Coast Stock Exchange; President's Council on Scientists and Engineers
PCSFA Potato Chip/Snack Food Association
pc sh pierce shell
PCSIR Pakistan Council of Scientific and Industrial Research
PCSP Permanent Commission for the South Pacific; Polar Continental Shelf Project; Princeton Cooperative School Program
PCSS Platform Check Subsystem
PCST Permanent Commission on Science and Technology (ASEAN)
PCSW President's Commission on the Status of Women
pct percent
pct (PCT) portable camera transmitter
pct procent (Dano-Norwegian—percent)
Pct Precinct
PCT Patent Corporation Treaty; Portsmouth College of Technology; Potash Core

Test(ing)
PCT *Partido Conservador Tradicional* (Spanish—Traditional Conservative Party)— Nicaragua; *Programa de Cooperación Tecnica* (Spanish—Technical Cooperation Program)
PCTB Pacific Coast Tariff Bureau
pctfe polychlorotrifluoroethylene
pc tp pierce template
PCTs Panama Canal Treaties
PCTS President's Committee for Traffic Safety
pcu photocopy unit; power-control unit; pressurization-control unit
pcu (PCU) palliative care unit (for terminal patients); portable checkout unit; protective custody unit
pcur pulsating current
PCUS Propeller Club of the United States
PCUS *Partido Comunista de la Unión Sovietica* (Spanish—Communist Party of the Soviet Union)
PCUSA Presbyterian Church in the U.S.A.
PCU-USA Portuguese Continental Union of the U.S.A.
pcv packed-cell volume; passenger-controlled vehicle; physical control volume; pollution-control valve; positive crank-case ventilation
PCV Peace Corps Volunteer(s); Pestalozzi Children's Village; President's Commission on Violence
PCV *Partido Comunista Venezolana* (Spanish—Venezuelan Communist Party)
PCVC Public Citizen Visitor's Center
PC virus Port Chalmers (New Zealand) type of influenza virus
PCVs Peace Corps Volunteers
pcv valve positive crankcase ventilation valve
PCWPC Permanent Council of the World Petroleum Congress
pcx periscope convex
PCY coastal yacht (3-letter naval symbol); Pittsburgh, Chartiers & Youghiogheny (railroad)
PCYC Port Credit Yacht Club
PCZ Panama Canal Zone
PCZST Panama Canal Zone Standard Time

pd interpupillary distance; paid; paralysing dose; passed; period; permanent dunnage; physical distribution; pitch diameter; point detonating; poop deck; port dues; position doubtful; post date; post dated; postage due; potential difference; pound; pour depressant; preliminary design; preventive detention; prism diopter; procurement directive; property damage; public domain; pulse duration; purchase description
p-d prism diopter
p/d post dated
p& d pickup and delivery
pd *prima donna* (Italian—first lady)
p.d. *per diem* (Latin—by the day)
Pd palladium; Parade; Parkinson's disease; Pick's disease
PD Parish District; Parliamentary Debates; Pharmacopoeia Dublin; Phelps-Dodge; Physics Department; Police Department; Port of Debarkation; Port Director; Port Dues; position doubtful (navigation chart marking); Preliminary Design; Presidential Directive; Production Department; Program Director; Public Defender
P-D Parke-Davis
P&D Probate and Divorce; Promotion and Development (program)
P of D Port of Duluth
PD *Partido Democrático* (Spanish—Democratic Party); *Cleveland Plain Dealer*; *Probate Division* (British Law Reports)
P-D *St Louis Post-Dispatch*
P.D. *Pharmacopoeia Dublinensis* (Latin—Dublin Pharmacopoeia)
pda patient distress alarm; personal death awareness; predicted drift angle; public display of affection
pda (PDA) probability distribution analyzer
pda *pour dire adieu* (French—to say goodbye)
Pda *Prima donna assoluta* (Italian—absolute first lady) —a principal female singer in an opera or concert organization
PDA Photographic Dealers' Association; Plywood Dis-

tributors Association; Port Development Authority; Pregnancy Discrimination Act
PDAD Probate, Divorce, and Admiralty Division
P Dal Port Dalhousie
P Dar Port Darwin
PDARS Pulsed Doppler Acoustic Radar System
PDAS Police Department American Samoa
P-day day when rate of production of an item or military consumption equals rate required by armed forces
pdb paradichlorobenzene
Pd.B. *Pedagogiae Baccalaureus* (Latin—Bachelor of Pedagogy)
pdc preliminary diagnostic clinic; private diagnostic clinic
p&d c premium and dispersion credit(s)
Pdc probability of detection and conversion
PDC Pacific Development Corporation; Penang Development Corporation; Periodical Distributors of Canada; Petroleum Development Corporation; Pregnancy Distress Center; Prevention of Deterioration Center (National Academy of Sciences); Primary Distribution Course; Project Development Corporation; Proposal Development Center
PDC *Partido Democrático Cristiano* (Spanish—Christian Democratic Party)
PDCL Provisioning Data Check List
pdd pre-dental discomfort
Pd.D. *Pedagogiae Doctor* (Latin—Doctor of Pedagogy)
PDD Petty Delinquency Detention; Public Documents Department (GPO)
pdda power-driven decontaminating apparatus
PDDS Parasitic Disease Drug Service
pde paroxysmal dyspnea on exertion; partial differential equations
Pde Parade
PDE Post-test Disassembly Examination; Projectile Development Establishment
P de C *Pas de Calais* (French—Strait of Calais)— Dover Strait
P-de-D Puy-de-Dôme

PDEIS Preliminary Draft Environmental Impact Statement

P del E Penitenciario del Estado (Spanish—State Penitentiary)

P de M Principauté de Monaco (French—Principality of Monaco)

pdes pulse-doppler elevation scan

P des L Parc des Laurentides (French—Laurentian Mountains Park)—Québec

pdf point detonating fuse; probability distribution function

PDF Panamanian Defense Force; Parkinsons' Disease Foundation

PDFA Partnership for a Drug-free America

PDFLP Popular Democratic Front for the Liberation of Palestine

PDG Paymaster Director-General

pdga (PDGA) pteroyldiglutamic acid

pdgf platelet-derived growth factor

PDGW Principal Director of Guided Weapons

pdh past dental history

pdi point-diffraction interferometer; powered-descent initiation; pre-delivery inspection

PDI Printing Developments Incorporated

pdic periodic

PDID Public Disorder Intelligence Department (LAPD); Public Disorder Intelligence Division

PDIN Pusat Dokumentasi Ilmiah Nasional (Bahasa Indonesian—National Scientific and Technical Documentation Center)

PDIS Pusat Dokumentasi Ilmu-Ilmu Sosial (Bahasa Indonesian—Social Sciences Documentation Center)

p dk poop deck

pdl poundal; poverty datum line

PDL Patent Depository Library

PDLP Patent Depository Library Program

pdm pulse-delta modulation; pulse-duration modulation

Pd.M. Master of Pedagogy

PDMS Point Defense Missile System

pdn production

pdnes pulse-doppler non-el-

evation scan(ning)

pdo pasado (Spanish—past)

Pdo Partido (Spanish—Party)—political party

PDO Petroleum Development Oman; Property Disposal Office(r); Publication Distribution Office(r)

p/doz per dozen

pdp plasma display panel; power distribution panel; project definition phase

Pdp Paradip

PDP Popular Democratic Party (Puerto Rico); Prescription Drug Program; Program Definition Phase; Program Development Plans

PDPA People's Democratic Party of Afghanistan

PDPS Parts Data Processing System

pd pt production pattern

pdq (PDQ) programmed data quantizizer

p d q pretty damn (or darn) quick

PDQB P.D.Q. Bach

pdr pounder; powder; precision depth recorder (PDR)

pdr polder (Dutch—dike-protected lowland reclaimed from the sea or other body of water

PDR People's Democratic Republic; Philippine Defense Ribbon

PD & R Policy Development and Research

PDR Physicians' Desk Reference

PDRK People's Democratic Republic of Korea (North Korea)

PDRL Permanent Disability Retirement List

pdrm payload distribution and retrieval mechanism

PDRP Power Distribution Reactor Program

PDRY People's Democratic Republic of Yemen

pds point detonating self-destroying; programming documentation standards

pd's public defenders

PDs Police Departments; Program Directors

PDS Pacific Data Systems; Passive Defense System; Passive Detection System; Personnel Data System; Philadelphia Divinity School; Priority Distribution System; Project Data Sheet; Proposed Delivery Schedule

PDSA People's Dispensary for Sick Animals

PDSC Performers and Teachers Diploma—Sydney Conservatorium

PDSOC Police Department Superior Officers' Council

pdsq point detonating super-quick fuze

PDSR Principal Director of Scientific Research

PDST Pacific Daylight Saving Time

pdt power distribution trailer; practice delivery torpedo

pdt (PDT) potentially dangerous taxpayer

PDT Pacific Daylight Saving Time

PDT-1 Picatinny Arsenal Detonation Trap 1

PDTC Plymouth and Devonport Technical College

Pdte Presidente (Spanish—President)

PDTLO Pierre Dominique Toussaint l'Ouverture

PDTS Police Detective Training School; Program Development Tracking System

pdu power distribution unit

pdv pure dried vacuum (salt)

pdv (PDV) pyrotechnic development vehicle

PDV Petroleos de Venezuela (Spanish—Venezuelan Petroleum)—state oil company

PDVSA *Petroleos de Venezuela Sociedad Anónima* (Spanish—Petroleum of Venezuela Corporation)

pd work public domain work (of art, history, literature, publication, etc.)

PDX Portland, Oregon (airport)

PDZ Parachute Dropping Zone

pe period entry; personnel equipment; private eye (private investigator); probable error; professional equipment; program element; printer's error

pe (PE) physical education; physical examination

p-e precipitation-environment (index)

p/e porcelain enamel; price earning

p & e planning and estimating

pe par exemple (French—for example); *per esempio* (Italian—for example); *por ejemplo* (Spanish—for example)

Pe Pecltet number; Pernambuco

Pᵉ Padre (Spanish—father)
PE Pacific Electric (railroad); patrol vessel (naval symbol); Petroleum Engineer(ing); Philadelphia Electric; Pistol Expert; Plant Engineer(ing); Port Elizabeth; Port of Embarkation; Port Everglades, Florida; Post Exchange; Prince Edward Island; probable error; Production Engineer(ing); Professional Engineer; Protestant Episcopal
P-E Perkin-Elmer
P & E Peoria & Eastern (railroad)
P of E Port of Entry
P.E. Pharmacopoeia Edinburgensis (Latin—Edinburgh Pharmacopoeia)
pea. (PEA) primary expense account
PEA People Express Airlines; Physical Education Association; Plastics Engineers Association; Policewomen's Endowment Association; Potash Export Association; Public Education Association; Publication Effectiveness Audit
PEAB Professional Engineer's Appointments Bureau
PEACE People Emerging Against Corrupt Establishments; Project Evaluation and Assistance in Civil Engineering (USAF)
PEACESAT Pan-Pacific Education and Communications Experiments using Satellites
PEAL Professional Engineers Association Limited
Pea Mus Peabody Museum
PEAP Personal Egress Air Pack
PEAQ Personal Experience and Attitude Questionnaire
Pearl Pearl Harbor—Oahu, Hawaii
PEARL (Committee for) Public Education and Religious Liberty
Pearls Pearl Islands (Las Perlas)
PEAS Production Engineering Advisory Service
PEAT Programmer Exercised Autopilot Test(ing)
PEAT Programme Élargi d'Assistance Technique (French—Enlarged Technical Assistance Program)—UN
peb phototype environmental buoy
Pe. B. Pediatriae Baccalaureus (Latin—Bachelor of Pediat-

rics)
PEB Physical Evaluation Board; Propulsion Examining Board (USN); Public Examination Board
pebb public employees blanket bond(ing)
pebd pay entry base date
pec photoelectric cell; position error correction; program element code
p E c pure Ecuadoran cocaine
PEC Plain English Campaign; Presidential Ethics Commission; Production Equipment Code; Protestant Episcopal Church; Psychology Examining Commission; Psychology Examining Committee
pecan. pulse envelope correlation air navigator
PECE President's Emergency Committee for Employment
PECI Projects and Equipment Corporation of India
'pecker woodpecker
PECM Preliminary Engineering Change Memorandum (USAF)
PECO Philadelphia Electric Company
PECP Preliminary Engineering Change Proposal
PECS Plant Engineering Check Sheet
pecto pectoral
pecul peculated; peculating; peculation; peculator
Peculiar Institution Mount Holyoke College founded by Mary Lyon and described by her as a peculiar institution as it was for women
PECUSA Protestant Episcopal Church of the U.S.A.
ped pedagogue; pedagogy; pedal; pedestal; pedestrian; personnel equipment data
ped (Latin prefix—children)—pediatrics
Ped pedal (music); Pediatrics
P Ed Physical Education
pedag pedagogue; pedaguese (patois of pedants)
pedageese pedagogue jargon
Ped.B. Bachelor of Pedagogy
Ped.D. Doctor of Pedagogy
pediat pediatric(al)(ly); pediatrician; pediatrics
PE Dir Physical Education Director
Ped.M. Master of Pedagogy
pedo pedologic(al)(ly); pedologist(ic)(al)(ly)
pedobap pedobaptism; pedobaptist

pedog pedograph(ic); pedography
pedogen pedogenesis
pedol pedologic(al)(ly); pedologist(ic)(al)(ly); pedology
pedom pedometer; pedometric(al)(ly)
pedont pedodontic(al)(ly); pedodontist(ry)
pedop pedophile; pedophilia(c)
Pedralvez Pedro Alvarez
Pedrarias Pedro Arias
Pedro San Pedro, California
peds pediatrics
pedstl pedestal(s)
PED XING pedestrian crossing
pee. photoelectric emission; pressure environment equipment; urine
PEE Proof and Experimental Establishment (British Ministry of Defence)
P & EE Proving and Experimental Establishment
Peeb Peebles
Peebl Peebleshire
peed off pissed off
peep. positive and expiratory pressure
peep. (PEEP) pilot's electronic eye-level presentation
Peer Peer Gynt
PEER Planned Environment and Education Research Institute
pees South Vietnamese piasters
pef peak expiatory flow; personal effects floater (policy)
PEF Palestine Exploration Fund; Personality Evaluation Form; Plastics Education Foundation; Presidential Election Fund; Psychiatric Evaluation Form
peg (PEG) pneumoencephalogram
peg. polyethylene glycol
Peg Pegasus (constellation); Peggy
Peg Pegunungan (Malay—mountain range)
PEG Petrochemical Energy Group; Pittsburgh Elderly Gay
PEGE Program for Evaluation of Ground Environment
Pegs Pegasus (constellation)
pei pointless electronic ignition; precipitation-efficiency index
PEI Porcelain Enamel Institute; Preliminary Engineering Inspection; Prince Edward Island
PEINP Prince Edward Island

National Park
PEIP Presidential Executive
Interchange Program
PEIS Preliminary Environmental Impact Statement
pej premolded expansion joint
p ej por ejemplo (Spanish—for
example)
PEJO Plant Engineering Job
Order(s)
pejor pejorative(ly)
pek pig embryo kidney
Pek Peking; Pekinese
peke pekinese dog
pel pelagic; pellet; pelvis; picture element
P El Port Elizabeth
PEL Petroleum Exploration
License; Physics and Engineering Laboratory
P EL Port Elizabeth
Pelagies Pelagian Islands in
the Mediterranean between
Sicily and Tunisia
P Eliz Port Elizabeth
Pellews Pellew Islands in Australia's Gulf of Carpentaria
Pellys Pelly Mountains of the
Yukon
PELNI Pelajaran Nasional Indonesia (National Shipping
Company of Indonesia)
pem photoelectromagnet(ic);
program element monitor
Pem Pembrokeshire
PEM Production Engineering
Measures; Project Engineering Memo
Pemb Pembroke College, Oxford; Pembrokeshire
Pemb Coll Pembroke College—Cambridge
Pemex Petróleos Mexicanos
(Spanish—Mexican Petroleum)
PEMR Petroleum Engineering
Monthly Report
PEMS Portable Environmental
Measuring System
PE Mus Port Elizabeth Museum
Pem Yeo Pembroke Yeomanry
pen. penal; penetrate; penology; peninsula; penitentiary;
penmanship
pen (Latin prefix—lack or
need)—penicillin
Pen Penang; Penarth; Peninsula; Penitentiary
Pen Península (Portuguese or
Spanish—peninsula); *Péninsule* (French—peninsula);
Penisola (Italian—peninsula)
PEN Poets, Playwrights, Editors, Essayists, and Novelists
(international organization

often referred to as the P.E.N.
Club)
PEN Presse Etudiante Nationale (French—Student National Press)—Québec's student news cooperative
pen. aids penetration aids
Pen Ala Peninsula de Alaska
(Spanish—Alaska)
*Pen BC Peninsula de Baja
California* (Spanish—Lower
California Peninsula)
pencil. pictorial encoding language
PEN Club (*see* PEN)
*Pen de Yuc Peninsula de
Yucatan* (Spanish—Yucatan
Peninsula)
Pene Penelope
*P/E News Petroleum/Energy
News*
P Eng Professional Engineer(ing)
Pen Fla Peninsula de la Florida (Spanish—Florida Peninsula)
PENGEM Penetrating the
Gray Electronic Market (FBI
undercover operation)
Penguin Norwegian surface-to-surface missile; Penguin
Books
peni penicillin
Pen Ib Peninsula Ibérica
(Spanish—Iberian Peninsula)
—Portugal and Spain
penic penicillin
penic. penicillum (Latin—
brush)
penic. cam. penicillum camelinum (Latin—camel's-hair
brush)
Penit Penitentiary
Pen Lab Peninsula de Labrador (Spanish—Labrador Peninsula)
Penn Pennsylvania; Pennsylvanian
Penna Pennsylvania
Pennamite(s) Pennsylvanian(s)
Penn Central Pennsylvania
New York Central Transportation Company (merger of
Pennsylvania, New York
Central, New Haven, and
Lehigh Valley railroads)
PennDOT Pennsylvania Department of Transportation
Penney JC Penney Company
Penn German Pennsylvania
German
Pennines Pennine Alps between Italy and Switzerland;
Pennine Hills ranging from
southern Scotland to central
England—the Pennine Chain

Penn State Pennsylvania State
University
Pennsy Pennsylvania; Pennsylvania Railroad
Penn Turn Pennsylvania
Turnpike
penol penological; penologist;
penology
Peñon de Veléz Peñon de
Veléz de la Gomera (rocky
islet belonging to Spain in
the western Mediterranean)
penrad penetration radar
pens pensioneret or *pensionist*
(Dano-Norwegian—retired or
pensioner)
pensad pension administration
PENSADS Pension Administration System
Pension-expanding President
Benjamin Harrison
Pensy Pensacola, Florida
pent. penetrate; penetration;
pentode
Pent Pentagon; Pentecost
Pent Pentateuch
PENT Project for the Education of Native Teachers
Pentlands Pentland Hills
southwest of Edinburgh or
the Pentland Skerries comprising the southernmost
Orkneys
pento (sodium) pentothal
penval penetration evaluation
PeO President ex-Oficio
PEO Philanthropic Educational
Organization; Plant Engineering Order; Protect Each
Other (secret women's organization)
PEOC Publishing Employees
Organizing Committee
pep pepper; pep pill; peptide
pep. pepper; peppermint;
peppy; personal effects protection
pep. (PEP) phosphoenolypyruvate; polyestradiol phosphate;
Public Employment Program
P e P Partija e Punes (Albanian—Workers Party)
PEP Parent Effectiveness Program; P.E.P. Deraniyagala;
Pepsi-Cola (stock-exchange
symbol); Performance Evaluation Process; Personalized
Engineering Program; Personalized Exercise System;
Petroleum Electric Power;
Political and Economic Planning; Positron-Electron Project; Preventive Enforcement
Patrol; Professional Enhancement Program; Proficiency in
English Program; Program

Evaluation Procedure; Public Employment Program
PEPA Petroleum Electric Power Association
Pep-Bis Pepto-Bismol
Pepco Potomac Electric Power Comapny
pepg piezo-electric power generator
PEPG Port Emergency Planning Group (NATO)
PEPIC Public Education Project on the Intelligence Community
PEPLAN Polaris Executive Plan (UK)
pep materials propellants, explosives, pyrotechnics
PEPP Professional Engineers in Private Practice
Peppe Giuseppe (Italian—Joe)
pepr precision encoder and pattern recognizer
peps pep pills; peptides
peps. pepsin
PEPs Public Employment Programs
Pepsi Pepsi Cola
PEPSU Patiala and East Punjab States Union
PEQC President's Environmental Quality Council
per period; periodic; periodicity; person; personal; personate
per perito (Italian—expert)
Per Perseus (constellation); Persia; Persian
Per Perciles, Prince of Tyre; Pereval (Russian—mountain pass); *Perevoz* (Russian—crossing, ferry); Persian
PER Perth, Australia (airport)
PE&R Policy, Evaluation, and Research
PERA Production Engineering Research Association
per agrim perito agrimensore (Italian—surveyor)
per an. per annum (Latin—by the year); *per anum* (Latin—by the anus)
per art perito artistico (Italian—art expert)
p/e ratio price-earning ratio
PERB Personnel Evaluation Research Bureau; Planning and Environmental Review Board; Professional Engineers Registration Board; Public Employment Relations Board
perc perchloroethylene; percolate; percussion
PERC Peace on Earth Research Center

per call perito calligrafo (Italian—handwriting expert)
per cent. per centum (Latin—by the hundred)—percent
Percept Psychophys Perception and Psychophysics
perco percobarg (barbiturate synthetic morphine derivative); percodan (synthetic morphine derivative)
per con. per contra (Italian—on the other side)
PERCOS Performance Coding System
perd perdenosi (Italian—dying away)
PERDDIMS Personal Development and Distribution Management System
perden. perdendosi (Italian—dying away)
perdi per diem
Peregil Pedro Gil
perestroi perestroika (Russian—economic restructuring)
perf perfect; perfection; perforate; perforation; perform; performance; performer; perfume(d)
PERF Planetary Entry Radiation Facility (NASA); Police Executive Research Forum; Police Executive Resource Form
perfect calc perfect calculation(s)
perfs perforations; performances; performers; perfumers
perg pergamino (Spanish—parchment)
Pergamon Pergamon Press
perh perhaps
peri perigee; perimeter
peri (Greek—around)—pericardial, pericarp, perimeter, periosteum, peripheral, peritoneum
PERI Platemakers Educational and Research Institute
periap periapical
Peric Periclean
Perico Pedro
peridot yellow-green tourmaline
perig perigee
perih perihelion
PERINTREP Periodic Intelligence Report
period menstrual period; period of rotation; period of revolution
period. periodical
periodontol periodontology
peris periscope

perjy perjury
perk percolate
perk. payroll earnings record keeping
PERK Prospective Evaluation of Radial Kerotomy
perk(s) perquisite(s)
perl pupils equal and reactive to light
perla pupils equal—react to light and accommodation
perm permanent
Perm Permian
permaflowers permanent (plastic) flowers
permafrost permanent frost
permafruit permanent (plastic) fruit
Perm Ct of Arb Permanent Court of Arbitration
permed permanently waved
PERMIS Public Employees Retirement Management Information System
PERMREP Permanent Representation to the North Atlantic Council (NATO)
perms permanents; permanent waves
per nav per navale (Italian—ship expert)
PERO President's Emergency Relief Organization
per. op. emet. peracta operatione emetici (Latin—when the emetic action is over)
peroxide hydrogen peroxide (H_2O_2)
perp perpendicular; perpetrator
Perp Perpignan
per pro. per procurationem (Latin—by proxy)
PERPS Police Executive Forum (DC)
perq(s) perquisite(s)
per rec per rectum (Latin—through the rectum)
perrla pupils equal, round, react to light and accommodation
pers person; personal; personality; personnel; persons
Pers Perseus (constellation); Persia(n)(s)
Pers Aulus Persius Flaccus (Roman satiric poet)
PERS Public Employees' Retirement System
Per.Sac.Lit. Peritus in Sacred Liturgy
persian white fantanyl (more powerful than morphine, also known as china white)
Persimfans Pervi Simfonichesky Anseabl (Russian—conductorless symphonic en-

semble)
PERSIS Personnel Information System
pers n personal noun
'personation impersonation
personi personification; personified; personifier; personifying
persp perspective
pers pron personal pronoun
Persymfans Pervyi Symfonitchesky Ansamble (Russian—First Symphonic Ensemble)—conductorless orchestra organized in 1922 in Moscow
pert. pertaining
pert. pertussis (Latin—whooping cough)
PERT Program Evaluation and Review Technique/Critical Path Technique
PERTCO Program Evaluation and Review Technics (plus) Cost Analysis
per tecn comm perito tecnico-commerciale (Italian—estimator)
pertest percolation test(ing)
Perths Perthshire
PERTVS Perimeter Television System
Peru Republic of Peru (Andean nation containing monumental structures left by the Incas), *República del Perú*
Peru Cur Peruvian Current
Peruv Peruvian
perv perversion; pervert; perverted
perv show pervert show
pes photoelectric scanner
pe's printer's errors
pe & s parts engineering and standardization
P es per esempio (Italian—for example)—e.g.
PEs Professional Engineers
PES Philosophy of Education Society
PESA Petroleum Equipment Suppliers Association
PESC Public Expenditure Survey Commission
pescado pez pasado (Spanish—past fish)—dead fish or fish out of water
Pescadores (Portuguese or Spanish—Fishermen)—islands off Taiwan and part of the Republic of China
peseta monetary unit of Spain
Pesh Peshawar
peso monetary unit of Bolivia, Chile, Colombia, Cuba, the Dominican Republic,

Guinea-Bissau, Mexico, and the Philippines
PEST Pressure for Economic and Social Toryism
pet. paper equilibrium tester; personal electronic translator; petroleum; petrological; petrologist; petrology; point of equal time
pet. (PET) positive-emission tomography; positron emission tomography (scan)
Pet Peter; Peterhead; Peterhouse College, Oxford; Peterkin; Petronius
Pet Peters' United States Supreme Court Reports
PET Parent Effectiveness Training; Pet Milk Company (stock-exchange symbol); Pierre Elliott Trudeau (former Canadian Prime Minister); Positron Emission Computed Tomography; Production Environmental Test(ing,s); Production Evaluation Test(ing,s); Prostitution Enforcement/Team (police versus pimps and prostitutes)
PETANS Petroleum Training Association—North Sea
Pete Peter; St Petersburg
Petersburg Saint Petersburg (later called Petrograd and Leningrad); nickname for the Federal Reformatory at Petersburg, Virginia
peth petroleum ether
petn petition
petr petrifaction; petrified
Petr Petronius Arbiter (Roman satirist)
PETR Preliminary Flight Test Report
petri petroleum
Petriburg. Petriburgensis (Latin—Peterborough)
Petrified Forest Petrified Forest National Park in Arizona's Painted Desert
Pëtr Makadonski (Russian—Peter the Great)—Pyotr Alekseyevich Romanov (1672–1725)
petro petrochemical; petroleum; petrology
Petro Petrograd (Russian—City of Peter)—Leningrad in the Russian Revolution
PETROBAS Petróleo Brasileiro (Portuguese—Brazilian Petroleum Corporation)
petro-chem petroleum-chemical
petrodollars petroleum-controlled dollars

Petrofina Compagnie Financiere Belges des Pétroles (Belgian Financed Petroleum Company)
petrog petrography
Petrograd (Russian—Peter's City)—Saint Petersburg renamed Petrograd in 1914 and Leningrad in 1924
petrol. petroleum; petrological; petrologist; petrology
Petronas Petroliam Nasional (Malay—National Petroleum)
Petropolis (Greek—City of Peter)—St Petersburg, Petrograd, Leningrad
petros petrochemicals
PETROVEN Petróleos de Venezuela (Spanish—Venezuelan Petroleum Corporation)
pets. prior to expiration of term of service
PETS Posting and Enquiry Terminal System
pet/spect (PET/SPECT) positron-emission tomography/ single photon-emission-computed tomography
pett (PETT) positron emission transaxial tomography
peua pelvic examination under anesthesia
pev propeller-excited vibration
PEVE Prensa Venezolana (Venezuelan press service)
pewter lead-tin alloy containing some antimony
p ex par exemple (French—for example
p ext por extensão (Portuguese—by extension)
pf page footing; perfect; performance factor; pfennig; picofarad; pneumatic float; power factor; preferred; preflight; profile; profiled; proximity fuse; public funding; public funds; pulse frequency; pyrolysis fluorescence
pf (PF) page footing; page formatter; personal finance; punch-off character
p/f portfolio
pf pro forma (Latin—for the sake of the form), an advance declaration for a financial statement or overseas invoice
pf. piano e forte (Italian—soft and then loud)
p f piu forte (Italian—more loudly)
pF picofarad(s)
Pf Pfennig (German—penny)
PF frigate—patrol escort ves-

sel (naval symbol); Packaging Facility; Physician's Forum; Pioneer & Fayette (railroad); Police Foundation; Procurator Fiscal

P/F Peace and Freedom (political party)

pfa psychologic-flight avoidance; pulverized fuel ash

PFA *Policía Federal Argentina* (Spanish—Argentine Federal Police); Press Foundation of Asia; Private Fliers Association

P factor hypothetical pain-producing substance produced in ischemic muscle; preservation factor

PFAPC Psychological Factor Affecting Physical Condition

PFAS President of the Faculty of Architects and Surveyors

pfb prefabricate(d); preformed beam(s); pseudofillicutitis barbae

PFBMF Polaris Fleet Ballistic Missile Force

PFBrg pneumatic float bridge

pfc passed flying college; passed (with) flying colors; plaque-forming cell(s); privately financed consumption

Pfc Private first class

PFC Pusan Fisheries College

PFCCT Pennsylvania Federation of Community College Trustees

pfce performance

pfc's perfluorocarbons

PFCS Primary Flight Control System

pfd personal flotation device; preferred, present for duty; primary flash distillate

Pfd Pfund (German—pound)

PFDF Petroleum Fuel Development Facility

pf di progressive die

pfd s preferred spelling

PFEFES Pacific and Far East Federation of Engineering Societies

PFEL Pacific Far East Line

pff pie-fed farmer; plaque-forming factor

PFF Police Field Force

pffb pie-fed farm boy

PFFBI Pacific Fire Fighters Burn Institute (Sacramento)

PFFF Plutonium Fuel Fabrication Facility

pffg pie-fed farm girl

PFF Inc Police-FBI Fencing Incognito (traffickers in stolen goods)

pf fx profiling fixture

PFGM guided missile patrol escort vessel (naval symbol)

PFGX Pacific Fruit Growers Express

pfi physical fitness index (PFI)

PFI Pacific Forest Industries; Pet Food Institute; Photo Finishing Institute; Picture and Frame Institute; Pie Filling Institute; Pipe Fabrication Institute; Police Foundation Institute

PFIAB President's Foreign Intelligence Advisory Board

PFJM *Policía Federal Judicial Mexicana* (Spanish—Mexican Federal Judicial Police)

pfk (PFK) phosphofructokinase

pfl pressed-for-life (dress materials)

PFL Pacific Freight Lines

PFLG Parents and Friends of Lesbians and Gays

PFLO Popular Front for the Liberation of Oman

PFLP Popular Front for the Liberation of Palestine (Marxist)

pfm power factor meter; pulse frequency modulation

PFMA Plumbing Fixture Manufacturers Association; Pressed Felt Manufacturers' Association

pfn prefinish(ed)

PFN *Partido Frente Nacional* (Spanish—National Front Party)—Costa Rica

PFNM Petrified Forest National Monument

PFNP Petrified Forest National Park

pfo patent foramen ovale

PFOBA Paso Fino Owners and Breeders Association

PFOC Prairie Fire Organizing Committee (communist)

PFP Progressive Federal Party (South African)

PFP *Progresief Federaal Partij* (Afrikaans—Progressive Federal Party)

pfr peak flow rate; peak flow reading; programmable film reader; prototype fast reactor (PFR)

PFRB Pacific Fire Rating Bureau

PFRS Programmed Film Reader System

PFRT Performance Flight-Rating Test; Preliminary Flight-Rating Test

pfs porous friction surface(d)

pfsa pour faire ses adieux

(French—to say goodbye)

PFS Professional Software

PFSO Postal Finance and Supply Office(r)

pfst pianofortist (pianist)

P-F Study Picture-Frustration Study (Rozensweig)

pft portable flame thrower

PFT Pet-Facilitated Therapy

pft acct pianoforte accompaniment

PFTAW People For The American Way

PFTC Pestalozzi Froebel Teachers College

pfte pianoforte (piano)

pfu pock-forming units; preparation for use

P Fu Port Fuad

pfv physiological full value

pfv pour faire visite (French—to make a call)

PFV *Pestalozzi-Froebel Verband* (Pestalozzi-Froebel Association)

pfx prefix

PFX Pacific Fruit Express

pg page; paregoric; paris granite; pay group; paying guest; permanent grade; pistol grip; postgraduate; pregnant program guidance; proving ground; public gaol; pure gin

pg (PG) parental guidance; prostaglandin

pg pago (Portuguese—paid)

p.g. persona grata (Latin—an acceptable person)

Pg Paraguay; Paraguayan; Portugal; Portuguese

PG gunboat patrol vessel (naval symbol); Pan American-Grace Airways; Pennsylvania-German; Post Graduate; Proctor & Gamble; Project Group (NATO); Provincial Government

P.G. Preacher General

P & G Proctor & Gamble

P of G Port of Galveston

PG *Prisonnier de Guerre* (French—prisoner of war)

P.G. *Pharmacopoeia Germanica* (Latin—German pharmacopoeia)

PG-13 parental guidance suggested (movie rating)

pga pressure garment assembly

pga (PGA) pteroylglutamic acid (folic acid)

PGA Pharmacy Guild of Australia; Professional Golfers Association

PG-AC Professional Group—Automatic Control

PGAH Pineapple Growers Association of Hawaii

p-gal(s) proof gallon(s)

PGA-NOC Permanent General Assembly—National Olympic Committees

PGB patrol gunboat (naval symbol)

pgbd pegboard(s)

PG-BTS Professional Group—Broadcast Transmission System

pgc per gyro compass

PGC Peoples Gas Company; Punxsutawney Groundhog Club

PGCE Post-Graduate Certificate of Education

PGCOA Pennsylvania Grade Crude Oil Association

PG-CS Professional Group-Communication System

PG-CT Professional Group—Circuit Theory

pgd paged; paradigm

PGD Past Grand Deacon

PGDF Pilot Guide Dog Foundation

pgdo pagado (Spanish—paid)

pge phenyl glycidyl ether

PGE Pacific Great Eastern (railroad); Portland Grain Exchange

PG-E Professional Group—Education

PG & E Pacifc Gas and Electric

PG-EC Professional Group—Electronic Computers

PG-ED Professional Group—Electronic Devices

PG-EM Professional Group—Engineering Management

PGER Pacific Great Eastern Railway

pgh (PGH) pituitary growth hormone

Pgh Pittsburgh, Pennsylvania

PGH patrol gunboat—hydrofoil (naval); Philadelphia General Hospital; Philippine General Hospital

PG-HFE Professional Group—Human Factors in Electronics

PG-I Professional Group—Instrumentation

PG-IE Professional Group—Industrial Electronics

PGIM Professional Group on Instrumentation and Measurement (NBS)

Pgio Poggio (Italian—hill, hillock, hilltop)

P-girls pub girls (waitresses in British barrooms)

PGIS Project Grant Informa-

tion System

PGIT Professional Group on Information Theory (IEEE)

PGJD Past Grand Junior Deacon

pgk phosphoglycerate kinase

pgl (pronounced *pee-gul*) puppy beagle

PGL Provincial Grand Lodge

P GL Port Glasgow

P Glg Port Glasgow

pglin page and line (flow chart)

pgm porous glass matrix (method of immobilizing nuclear waste); program

pgm (PGM) phosphoglucomutase

PGM motor gunboat (3-letter naval symbol); Past Grand Master

PGMA Private Grocers' Merchandising Association

PGmc Proto-Germanic

PG-ME Professional Group—Medical Electronics

PG-MITT Professional Group—Microwave Theory and Technics

pgm's precision-guided munitions

pgn pigeon

pgn (PGN) proliferative glomerulonephritis

PGNP Pagsanjan Gorge National Park (Philippines)

PGNS Primary Guidance and Navigation System

pgo pyrolysis gas oil

PGOC Philadelphia Grand Opera Company

P of GP Pearl of Great Price

PGPR Provincial Guild of Printers' Readers

pgr population growth rate; psychogalvanic reaction; psychogalvanic response

pgr (PGR) precision graph record(er)

pg rating parental-guidance rating (of a motion picture or television program)

PGRO Pea Growing Research Association

pgrv (PGRV) precision-guided reentry vehicle

pgs predicted ground speed

pg's (PGs) prostaglandins

PGS Pennsylvania-German Society; Pidaung Game Sanctuary (Burma); Power Generation System; Primary Guidance System

PGSC Panel on Geological Site Criteria

PGSD Past Grand Senior Dea-

con

PGSW Past Grand Senior Warden

pgt per gross ton

PGT Pacific Gas Transmission (company); Program Global Table

PGTB Pierre Gustave Toutant Beauregard

Pgu Pagalu (formerly Annobon)

PGU Pontifical Gregorian University

pgut (PGUT) phosphogalactose uridyl transferase

PGWA Pottery and Glass Wholesalers' Association

ph page heading; pharmacopoeia; phase; phone; phosphor; phot; photon; physically handicapped; power house; precipitation hardening; previous hardening

ph (PH) past history

p/h per hour

p & h postage and handling

pH hydrogen-ion concentration

Ph Pahari; phenyl; Philosophy

PH Parachute Handler; Paradise Hills; Pearl Harbor; Philharmonic Hall; Plane Handler; Power House; Public Health; Purple Heart

P-H Prentice-Hall

pha (PHA) phytohemagglutinin

PHA Public Housing Administration

PHADS Phoenix Air Defense Sector

Phaedr Phaedrus (Roman fabulist-poet)

phag (Latin prefix—to eat)—phagocytes

phage(s) bacteriophage(s)

phal phalange; phalanx

'phant elephant

phar pharmacy

P Har Port Harcourt

Phar. B. Bachelor of Pharmacy

Phar. C Pharmaceutical Chemist

Phar. D. Doctor of Pharmacy

pharm pharmaceutical; pharmacist; pharmacology; pharmacopoeia(s); pharmacy

Phar. M. *Pharmaciae Magister* (Master of Pharmacy)

Pharmaceutical Pharmaceutical Press

pharmacol pharmacology

pharm chem pharmaceutical chemistry

Pharm.D. *Pharmaciae Doctor* (Latin—Doctor of Pharmacy)

PHAs polyclitic aromatic hydrocarbons; Public Housing Agencies

'phasia aphasia

Ph.B. *Philosophiae Baccalaureus* (Latin—Bachelor of Philosophy)

Ph. B.J. Bachelor of Philosophy in Journalism

ph brz phosphor bronze

Ph. B. Sp. Bachelor of Philosophy in Speech

Ph. C. Pharmaceutical Chemist

PHC Patrick Henry College

PHCC Plumbing, Heating, Cooling Contracters

PHCIB Plumbing-Heating-Cooling Information Bureau

ph const phase constant

phd panty-hose distributor; piled higher and deeper

Ph. D. *Philosophiae Doctor* (Latin—Doctor of Philosophy)

PHD Port Huron and Detroit (railroad)

P.H.D. Public Health Doctor

PHDA Presently Hasn't Done Anything

PhD cameras Push-here-Dummy cameras designed so anyone can take a picture

Ph. D. Ed. Doctor of Philosophy in Education

Phe Phoenix (constellation)

PHE phenylalanine (amino acid)

P.H.E. Public Health Engineer

PHEAA Pennsylvania Higher Education Assistance Agency

P-head pinhead small-minded person, user of amphetamine)

phency phencyclidine (angel dust)

pheno phenobarbital; user of phenobarbital

pheno/d phenomenological death

phenolp phenolphthlein

phenom phenomena; phenomenal; phenomenon

phf's (PHFs) paired helical filaments

PHF Potomac Horse Fever

Ph. G. Graduate in Pharmacy

Ph. G. *Pharmacopoeia Germanica* (Latin—German Pharmacopoeia)

PHG Postman Higher Grade

phgt package height

PHHS Patrick Henry High School

phi philosophy

Phi Philips

Ph I *Pharmacopoeia Internationalis*

PHI Public Health Inspector

phial. *phiala* (Latin—bottle)

PHIBLANT Amphibious Forces—Atlantic (USN)

PHIBPAC Amphibious Forces—Pacific (USN)

phil philosophy

phil (Latin suffix—having an affinity for)—neutrophiliac

Phil Philadelphia; Philadelphian; Philbert; Philharmonia; Philharmonic; Philip; Philippa; Philippine; Philippines; Phillip; Phillipa; The Epistle of Paul to the Philippians

Phil *Philippians*

Phila Philadelphia; Philadelphian

Philada Philadelphia

Phila Free Lib Philadelphia Free Library

PHILAG Public Health Institute—London Action Group

philat philately

PHILDis Philadelphia Procurement District (US Army)

Philem The Epistle of Paul to Philemon

Philem *Philemon*

PHILEX Philadelphia Exchange

Phil Hung Philharmonica Hungarica

Philippines Republic of the Philippines, *Republika ñg Pilipinas* (Pilipino) or *República de Filipinas* (Spanish)

Philips' Philips' Gloeilampenfabrieken (Dutch—Philips' Electric Lamp Factory)

Philips Res Rept *Philips Research Reports*

PHILIRAN Philips Petroleum Iran

Phil Is Philippine Islands

Phillies Philadelphians

Phil Lip Philosophical Library

PHILLIPS Phillips Petroleum Company

Philly Philadelphia

Phil Mag *Philosophical Magazine*

philocrit philosopher critic; philosophical criticism

philol philology

Phil Orch Philadelphia Orchestra

philos philosophy

Philos Philosophy

philos educ philosophy of education

Philos Lib Philosophical Library

Philos Mag *Philosophical Magazine*

philosoph philosopher; philosophical; philosophy

Philos Pub Philosophical Publishing Co

Philos Res Philosophical Research Society

Philos Trans R Soc London *Philosophical Transactions of the Royal Society of London*

PHILSA Philippine Standards Association

Phil Soc Philharmonic Society

PHILSOM Periodical Holdings in the Library of the School of Medicine

Phil Sp Philippine Spanish

PHILSUCOM Philippine Sugar Commission

PHILSUGIN Philippine Sugar Institute

Phil Trans *Philosophical Transactions* (Royal Society of London)

phiz physiognomy

phk cells postmortem human kidney cells

Phl (Port of) Philadelphia

Ph. L. Licentiate in Philosphy

PHL Philadelphia, Pennsylvania (airport)

PHLAGS Philipps Petroleum Load-and-Go System

phlebo (Latin prefix—vein)—phlebitis

phleg phlegm(atic)(al)(ly); phlegmaticalness; phlegmaticness; phlegmatize(d); phlegmier; phlegmiest

phl h phillips head

PHLS Public Health Laboratory Service

phm phase meter

phm (PHM) patrol hydrofoil missile

Ph. M. *Philosophiae Magister* (Latin—Master of Philosophy)

PHM patrol-combat missile (hydrofoil craft)

Phm. B. Bachelor of Pharmacy

PHMC Pennsylvania Historical and Museum Commission

Phm. G. Graduate in Pharmacy

PHMS Patrol Hydrofoil Missile Ship(s)

PHN Public Health Nurse; Public Health Nursing

PHO Public Hazards Office

phobe (Latin suffix—abnormal fear or dread)—felinophobic, hydrophobia

phocis photogrammetric circulatory surveys

phocl photo-initiated chemical laser
PhOD Philadelphia Ordnance Depot
Phoen Phoenix
Phoen Phoenicians
Phoenix Arizona's capital; Phoenix Islands in the equatorial mid-Pacific Ocean where they are claimed by the UK and the U.S.A.—included are Birnie, Canton, Enderbury, Gardner, Hull, McKean, Phoenix, and Sydney islands
PHOENIX Plasma Heating Obtained by Energetic Neutral Injection Experiment
phofl photoflash
phon phonetics; phonology
phone telephone
phone book telephone book
phoneme smallest sound unit (linguistics)
phonet phonetic(s)
Phonet Phonetics
phono phonograph
phonorecord(s) phonograph record(s)
phonos phonoscopy (voiceprint analysis and identification)
phonovision telephone television
Phons Alphonse
Phor Phoronida
phos phosphate; phosphorescent
phot. photograph; photographer; photographic; photography; photon; photostat; photostatic
phot photographie (French—photography)—plus all derivatives such as *photocopie* (photostat), *photographe* (photographer), *photogravure, phototype*, etc.
Phot Photographie (German—photography)—plus all derivatives
photac photographic typesetting and composing (AT & T)
photex photographic exercise
photint photographic intelligence
photo photograph; photographer; photography
photocomp photocomposed; photocomposition
photog photograph; photographer; photographic; photography
photogeog photogeography
photogeol photographic geol-

ogy
photograv photogravure
photog(s) photographer(s)
photom photometry
photo op photo opportunity
photosyn photosynthesis
phot r photographic reconnaissance
p'house steak porterhouse steak
php pounds per horsepower; propeller horsepower
ph&p peace, heath, and prosperity
PHP Psychologists Helping Psychologists (self-help group); Public Health Plan
phr phrase; pounds per hour; preheater
Ph R Photographic Reconnaissance
PHRA Poverty and Human Resources Abstracts
phraseo phraseogram; phraseograph; phraseological(ly); phraseologist; phraseology
phren phrenic; phrenology
PHRI Public Health Research Institute
ph & ru pubic hair and revealing underwear
Ph S Philosophical Society of England
PHS Pennsylvania Historical Society; Printing House Square; Prison Health Services; Pubic Hair Society; Public Health Service
PHSO Postal History Society of Ontario
phsp phase splitter
pht phototube; pitch, hit, and throw
Ph T putting husband through (college or university)
PHt Port Harcourt
PHT Passive Hemagglutination Test(ing)
PHTF Pearl Harbor Training Facility
PHTS Psychiatric Home Treatment Service
Phu Port Hueneme
P Hur Port Huron
phv phase velocity
phw pressurized heavy water
PHWA Protestant Health and Welfare Assembly
phwr (PHWR) pressurized heavy-water-moderated reactor
PHX Phoenix, Arizona (airport)
phy physical; physics
phyce photocopy-control electronics unit

phylo phylogeny
phys physic; physical; physician; physics
phy s physiological saline
Phys Chem Solids Physics and Chemistry of Solids
phys dis physical disability
phys ed physical education
Phys Ed Physical Education
physexam physical examination
Phys Fluids Physics of Fluids
PHYSH Physicians and Surgeons Hospital
physiat physiatric(s); physiatrical; physiatrist
physiog physiognomy
physiogr physiography
physiol physiology
Physiol Physiology
Phys Konden Mater Physik der Kondensierten Materie (German—Physics of Condensed Materials)
physl physiological
Phys Lett Physics Letters
phys med physical medicine
physocean physical oceanography
physog physiognomy
Phys Rev Physical Review
Phys Rev Lett Physical Review Letters
Phys S Physical Society
phys sci physical science; physical sciences
Phys Status Solidi Physica Status Solidi (Latin—Solid-State Physics)
Phys Teach Physics Teacher
phys ther physical therapy
Phys Today Physics Today
Phys Z Physikalische Zeitschrift (German—Physics Journal)
Phys Z Sowjetunion Physikalische Zeitschrift der Sowjetunion (German—Physics Journal of the Soviet Union)
phytopath phytopathologic(al)(ly); phytopathologist; phytopathology
pi personal income; photo interpreter; photo interpretation; pigeon trainer; pig iron; pilotless interceptor; pimp; point initiating; point insulating; point of interception; point interception; poison ivy; position indicating; position indicator; present illness; private investigator; production interval; programmed instruction; protamine insulin; protocol international (international protocol); public in-

vestigation
pi (PI) point of inversion
p & i principal and interest; protection and indemnity
pi Greek-letter symbol π indicating ratio of circumference of a circle to its diameter; the ratio itself; expressed as a number, *pi* is approximately 3.14159
Pi piaster
P$_i$ inorganic orthophosphate
PI Packaging Institute; Paducah and Illinois (railroad); Party Islam; Pasteur Institute; Paul Isnard (Mana River settlement, French Guiana); Perlite Institute; Philippine Islands; physical instruction; Piedmont Airlines; Plastics Institute; Popcorn Institute; Pratt Institute; Principal Investigator; Productivity Index; Public Information
PI Printer's Ink
P-I Seattle Post-Intelligencer
P.I. Pharmacopoeia Internationalis
pia peripheral interface adaptor
pia (PIA) primary insurance amount
PIA Pakistan International Airlines; Photographic Importers' Association; Plastics Institute of America; Printing Industries of America; Prison Industry Authority; Professional Insurance Agents
PIAA Pacific Index of Abbreviations and Acronyms
PIAI Printing Industry of America, Incorporated
PIANC Permanent International Association of Navigation Congresses
piang piangendo (Italian—mournful, plaintive)
pianiss pianissimo (Italian—very softly)
pianocorder piano recorder and reproduction system
PIARC Permanent International Association of Road Congresses
pias piaster
PIASA Polish Institute of Arts and Sciences in America
piat projector infantry antitank (weapon)
PIAT Peabody Individual Achievement Test(ing)
pib power ionosphere beacon
pib producto interno bruto (Spanish—gross internal product)

PIB Petroleum Information Bureau; Polytechnic Institute of Brooklyn; Prices and Incomes Board; Prison Industry Board
PIBA Primary Industry Bank of Australia
PIBAC Permanent International Bureau of Analytical Chemistry of Human and Animal Food
pibal pilot balloon
pic (French—peak); piccolo; picture; polymer-impregnated concrete; positive-impedance converter; production inventory control; pulse-indicating cartridge; pulse-induced collapse
pic (PIC) program-interrupt control(ler)
Pic Pictor (constellation)
PIC Physics International Company; Piedmont Interfaith Council; Poison Information Center (Cleveland Academy of Medicine); Poisons Information Centre (Australia); Private Industry Council
PICA Palestine Israel Colonization Association; Police Insignia Collector's Association; Printing Industry Computer Associates; Printing Industry Craftsmen of Australia
picar picaresque
PICC Peoples Insurance Company of China; Philippine International Convention Center
Piccy Piccadilly
PICGC Permanent International Committee on Genetic Congresses
PICIC Pakistan Industrial Credit and Investment Corporation
pick. part information correlation key
Pick Pickens Railroad
PICL President's Intelligence Checklist
PICM Permanent International Committee of Mothers
Picnic City Mobile, Alabama
pico 10^{-12}
PICO Person In Column One (census-taker euphemism for head of household)
PICOE Programmed Initiations, Commitments, Obligations, and Expenditures
PICOP Philippine Industries Corporation of the Philippines

pics pictures; publishers information cards
PICS Pacific Islands Central School; Personnel Information Communication System; Pharmaceutical Information Control System
pict pictorial; picture
Pictorial Satirist Supreme William Hogarth
Pictured Rocks Michigan's national lakeshore
PICUTP Permanent and International Committee of Underground Town Planning
pid pelvic inflammatory disease; prolapsed intervertebral disk
p-i-d poverty-ignorance-disease syndrome of society
PID Police Intelligence Detail; Procurement Information Digest
PID Partido Institucional Democrático (Spanish—Institutional Democratic Party)
pida payload installation and deployment aid
PIDA Pet Industry Distributors Association
PIDC Pakistan Industrial Development Corporation
PIDE Policia Internacional e de Defesa do Estado (Portuguese—International Police and Defense of the State)—security police
Pid Eng Pidgin English (hybrid dialect)
PIDO Primitive Indian Development Organization
pidp pilot information display panel
PIDS Parameter Inventory Display System
pie. pulmonary infiltration (with) eosinophilia
pie. (PIE) plug-in electronics
PIE Pacific Intercultural Exchange; Pacific Intermountain Express (fast freight); Partners In Education; St Petersburg, Florida (airport)
PIEA Petroleum Industry Electrical Association
PIEC Public Interest Economics Center
Piedmont Piedmont Plateau or Piedmont Triad (Greensboro, High Point, and Winston-Salem, North Carolina) or placename found in Alabama, California, South Carolina, or West Virginia
PIERS Port Import/Export Recording Service

pif (PIF) prolactin inhibiting factor
PIF Paper Industry Federation; Pilot Information File
pig. pigment; pigmentation
PIG Pride, Integrity, Guts (acronym adopted by the Chicago police)
pigmi positron-indicating general measuring instrument
pigmt pigment(ation)
PIGS Poles, Italians, Greeks, Slavs
pigu pendulous integrating gyroscope unit
pik payment in kind
pil payment in lieu; percentage increase in loss
pil (PIL) procedure implementation language
pil. pilula (Latin—pill)
Pil Pitt interpretive language
PIL Pacific International Lines; Pest Infestation Laboratory
pilc paper-insulated lead covered
PILCOP Public Interest Law Center of Philadelphia
pill the pill (birth-control pill)
pills. particulate instrumentation by laser light scattering
pilnav piloting navigation
PILO Public Information Liaison Officer
pilot. printing industry language for operations of typesetting
PILOT Piloted Low-speed Test; Programmed Inquiry, Learning, or Teaching
pilot-on-board flag signal flag consisting of a white and a red vertical band; letter H or Hotel in the international code
pilot-wanted flag yellow-and-blue vertically striped signal flag flown to indicate a pilot is wanted; letter G or Golf in the international code
pilp parametric integer linear program
pils pilsner
Pil Sta Pilot Station
pim penalties in minutes; pulse-interval modulation
PIM Pacific Islands Monthly
PIMA Paper Industry Management Association
PIMI Preinactivation Material Inspection
pimola pimento olive (pimento-stuffed olive)
pimpmobile pimp's vehicle
PIMPS Program for Interac-

tive Multiple Process Simulation
pin. page and item number; piece identification number; plan identification number; position indicator
pin. (PIN) personal identification number (for computer protection)
pin. pinguis (Latin—fat, grease)
PIN Police Information Network
p/in.^{2} parts per square inch
p/in.^{3} parts per cubic inch
PINA Pacific Islands News Association; Permaculture Institute of North America
PINAC Permanent International Association of Navigation Congresses
Pind Pindar
pines. pineapples
pino positive input—negative output
pins person(s) in need of supervision
pins. person in need of supervision
PINS Padre Island National Seashore (Texas); Palletized Inertial Navigation System
PINWR Pungo National Wildlife Refuge (North Carolina)
pinx. pinxit (Latin—he painted it)
PINY Polytechnic Institute of New York
Pinyin (Chinese—phonetic sound)—official spelling system adopted in 1979 for Chinese words written in Roman letters
PINZ Plastics Institute of New Zealand
pio precision-interpret operation
PIO Photographic Interpretation Office(r); Public Information Office(r)
PIOA Pacific Index of Abbreviations and Acronyms in Common Use in the Pacific Basin Area
PIOB President's Intelligence Oversight Board
PIOCS Physical Input-Output Control System
pi-on pi-meson; pioneer
PIOSA Pan Indian Ocean Science Association
pip. peripheral interchange package; precise installation position; predicted intercept(ion) point; project initiation period; proximal interphalan-

geal; public and institutional property
pip. (PIP) picture in picture (tv)
Pip. Philip
PiP Proceedings in Print
PIP Peripheral Interchange Program; Permatite Instant Plastic; Personal Identification Program; Personnel Identification Project; Product Improvement Plan; Product Improvement Program; Product Information Package; Psychotic Inpatient Profile
PIP Policia de Investigación del Peru (Spanish—Peruvian Investigation Police)
PIPA Pacific Industrial Property Association; Pacific Islands Producers Association
PIPEF Pacific Islands Polynesian Education Foundation
piper. pulsed intense plasma for exploratory research (PIPER)
pipe(s). pipe bomb(s)
pipi pipizintzintli
pipit. peripheral-interface and programme—interrupt translator
Pipo Filippo
PIPR Polytechnic Institute of Puerto Rico
pips. pulsed integrating pendulums
piq property in question
PIQ Performance IQ
Pir Piraeus
PIR Philippine Independence Ribbon; Phillip Island Reserve (Victoria, Australia); Preliminary Information Report
PIRA Paper Industries Research Association; Printing Industry Research Association; Provisional Irish Republican Army
pirb position-indicating radio beacon
pirf perimeter-insulated raised floor
PIRF Petroleum Industry Research Foundation
PIRG Public Interest Research Group
PIRGs Public Interest Groups
pirid passive infrared intrusion detector
PIRL PRISM Information Retrieval Language
pi rm pilot reamer
PIRS Personal Information Retrieval System; Poseidon Information Retrieval System

Pis Pisces
PIS Postal Inspection Service; Public Insurance Service
P Isb Port Isabel
PISC Philippine International Shipping Corporation; Phoenix International Science Center
PISCES Production Information Stocks and Cost Enquiry System
Pish Parish
PISO Philippines Investment Systems Organization
piss pissoir (French—urinal); *pissotière* (French—public urinal)
pissoirs pissotières (French—public urinals for men)
pistaz piss-tinted topaz
pisw process-interrupt status word(ing)
pit. pitot static; progressive inspection tag
pit. (PIT) principal, interest, and taxes
Pit Pitanga; Pitcairn; Pitkin; Pitman; Piton; Pittsboro; Pittsburg; Pittsburgh; Pittsfield; Pittsford; Pittston; Pittsylvania
PIT Pasadena Institute of Technology; Petr Ilich Tchaikovsky; Pittsburgh, Pennsylvania (airport)
PITA Petroleum Industry Training Association; Provincial Intermediate Teachers Association (Canadian)
PITAC Pakistan Industrial Technical Assistance Center
PITAS Petroleum Industry Training Association—Scotland
PITB Pacific Inland Tariff Bureau; Pacific Island Teachers Board
PITC Pacific International Trust Company
pitchblende uraninite ore (chief source of radium and uranium)
PITDC Pacific Islands Tourism Development Council
piti principal, interest, taxes, insurance
PITL Pacific Islands Transport Line
pit. log pitot-static log
PITO Portuguese Information and Tourist Office
Pitons Piton Mountains (St Lucia)
pitr plasma iron turnover rate
pits. payload integration test set

PITS Pacific Islands Training School
PITT Polaris Integrated Test Team
Pitts Pittsburgh, Pennsylvania
pitu piping or tubing
PIU Public Inspection Unit (vice squad)
PIUS Process-Inherent Ultimately Safe (nuclear reactor)
piv peak inverse voltage; post indicator valve
PIV Positive Infinity Variable
pivs particle-induced visual sensations
PIW Petroleum Intelligence Weekly
pix photographs; pictures
pixel picture element
pix/sec pictures per second
PIYA Pacific International Yachting Association
pizz. pizzicato (Italian—plucked)
pj prune juice
PJ Police Judge; Presiding Judge; Probate Judge
P of J Port of Jacksonville
PJ Police Judiciare (French—criminal investigators, detective division)
PJA Pipe Jacking Association
P Jac Port Jackson
PJB Patrick J Buchanan
PJBD Permanent Joint Board on Defense (Canada-US)
PJC Paducah Junior College; Paris Junior College; Polydox Jewish Federation
pjex parachute jumping exercise
pjm postjunctional membrane
pj's physical jerks
Pjs Pasajes
pk pack; park(ing); peak; peck; psychokinesis
pk (PK) packed tight, kept right; probability of kill; sugar-coated chewing gum (symbol)
pK negative logarithm of the dissociation constant (symbol)
Pk Park; Peak; pink
Pk Pedalkoppel (German—pedal coupler); *Pauken* (German—kettledrums)
PK Principal Keeper; probability of kill (symbol)
PK Panama Kanaal (Dutch—Panama Canal); *Posta Kutusu* (Turkish—post office box)
P Ka Port Kembla
P-K antibodies Prausnitz-Küstner antibodies

pkb photoelectric keyboard
PKbanken Post-och Kreditbanken (Swedish—Post and Credit Bank)
pkd packed (flow chart); partially knocked down
PKD Parker Drilling Company (stock-exchange symbol)
pkdom pack(ed) for domestic use
pkg package; packing
Pkg Port Kelang (also written Port Klang and formerly Port Swettenham)
pkge package
PKI Partai Komunis Indonesia (Communist Party of Indonesia)
Pkl Port Kelang (Port Klang formerly Port Swettenham)
PKL Possum Kingdom Lake
pkm perigee kick motor
PKN Polski Kometet Normalizacyny (Polish—Polish Standards Committee)
pknghse packinghouse
PKNP Pu Kradeung National Park (Thailand)
pkp pre-knock pulse
pKp purple K powder (purple potassium-bicarbonate powder)
PKP Partido Komunista Pilipinas (Pilipino—Communist Party of the Philippines)
pkr packer
PKR Parker Pen (stock exchange symbol)
Pk Rdg Park Ridge
P-K reaction Prausnitz-Küstner reaction
pkrg parking
pks packs; pecks
PKS Photo-Kit System (criminal identification)
pksea pack(ed) for overseas use
PKSRP Possum Kingdom State Recreation Park (Texas)
pkt packet
P-K test Prausnitz-Küstner test
PKTF Printing and Kindred Trades Federation (UK)
pkts packets
pku phenylketonuria
pkv killed poliomyelitis vaccine
Pkw Personenkraftwagen (German—automobile, passenger vehicle)
Pkwy Parkway
pky pecky
Pky Parkway
pl parting line; party line; per-

ception of light; phase line; pipeline; place; plastic; plate; plural

pl (PL) party line; phone line; private telephone line; product liability

p/l partial loss; payload; pipeline; plain language

p & l profit and loss

pl. plenarius (Latin—complete, fully attended)

£L pound Lebanese

Pl Place

Pl Place (French—place, plaza); *plantage* (Dutch—plantation); *plass* (Scandinavian—place, plaza); *Platz* (German—place, plaza); *plaza* (Spanish—place, plaza); *plein* (Dutch—place, plaza); Titus Maccius Plautus (Roman writer of comedies)

PL perception of light (symbol); Place; Pluto; Point Loma; Poland (auto plaque); Port Line; Public Law; Public Library

P.L. Poet Laureate

PL Paradise Lost; Partido Liberal (Spanish—Liberal Party); *Pharmacopoeia Londinensis* (Pharmacopoeia of London)

PL 1 Programming Language 1

PL/1 Programming Language/version 1

pla plasma resin activity; probation and rehabilitation of airmen

Pla Plaza; Pula (Pola)

P^{la} Playa (Spanish—beach, strand)

PLA Palestine Liberation Army; Pedestrian's League of America; People's Liberation Army (Chinese communist); Philadelphia Library Association; Philatelic Literature Association; Port of London Authority; Port of Los Angeles; Private Libraries Association; Public Library Association; Pulverized Limestone Association

P of LA Port of Los Angeles

place. programming language for automatic checkout equipment

Place Pig Place Pigalle in Paris

PLADs Price Level Adjusted Deposits

PLADS Parachute Low-Altitude Delivery System

plam plastic laminate

plam (PLAM) price-level adjusted mortgage

plame (PLAME) propulsive lift aerodynamic maneuvering entry

plan. planet; planetarium

Plan Planina (Bulgarian or Serbo-Croatian—mountain, mountain range)

PLAN Paterson Looks Ahead Now; Prevent Los Angelization Now; Program for Learning in Accordance with Needs

Plan A North Atlantic Treaty Reginal Planning Group

plane(s) airplane(s)

PLANES Programmed Language-based Enquiry System

planet. planetary

Planets The Planets, Gustav Holst's tone poem for large orchestra

Planet Space Sci Planetary and Space Science

planex planning exercise

PLANNET Planning Network

PLANS Programming Language for Allocation and Network Scheduling (NASA)

Plan Soc Planetry Society

plantflex plantar flexion

plantk plantkunde (Dutch—botany)

PLAP Port of London Authority Police

PLAR Partido Liberal Autentica Radical (Spanish—Authentic Radical Liberal Party)—Paraguay

plarbage plane-floor garbage

PLARS Position-Locating-and-Reporting System

plas plaster

plasm (Greek—something formed or molded)—chromoplast, dermoplasty, plasma, plasmasol, protoplast

Plasma Phys Plamsa Physics

plastique (French—plastic)—plastic bomb(s)

plasty (Latin suffix—reconstruction of)—rhinoplasty

plat. plateau; platinum; platoon

platf platform

PLATO Port Lincoln Advancement Trust Organization; Programmed Logic for Automatic Teaching Operations

platy *Platypoecilus* (genus of tropical fishes); platysma

platy (Latin prefix—flat or side)—platypus

Platy Platyhelminthes

Plaut Plautus

PLAV Polish Legion of

American Veterans

plb plumber; plumbing; publisher's library binding(s); pull button

plb (PLB) publisher's library binding

PLB Poor Law Board

plbd plugboard

plc power-line carrier; prelaunch computer

PLC Pacific Lighting Corporation; Point Loma College; Probe Launch Complex; Preferred Line of Credit; Products List Circular; Public Limited Company

P of L C Port of Lake Charles

P.L.C. Poeta Laureatus Caesareus (Latin—Imperial Poet Laureate)

PLCA Pipe Line Contractor's Association

plcs propellant-loading control system

plcu propellant-level control unit

plcy policy

pld payload

Pld Portland, Oregon

PLD Paul Lawrence Dunbar

PLDG Portuguese Language Development Group

PLDTC Philippine Long Distance Telephone Company

pldx polydox; polydoxy

ple preliminary logistics evaluation; primary loss expectancy; prudent limit of endurance; puerile light entertainment

P & LE Pittsburgh & Lake Erie (railroad)

plea. prototype language for economic analysis

PLEA Poverty Lawyers for Effective Advocacy

plebe plebeian

plebs plebeians

pled pleaded

plegia (Latin suffix—paralysis or stroke)—paraplegic

PLEI Public Law Education Institute

Pleis Pleistocene

plem pipeline end manifold

Plen Plenary; Plenipotentiary

plenipo plenipotentiary

Plenum Plenum Publishing Corp

pleon pleonastical(ly)

plex plant experiment(ation)

plf polyforming

PLF Pacific Legal Foundation; Palestine Liberation Front

pl x fe plastic to female

plff plaintiff

plftr please furnish transportation requests
plfur please furnish
plg piling
Plg Porto Alegre
PLG Poor Law Guardian
PLGC Pension Loan Guarantee Corporation
plgl plateglass
p-lgv psittacosis-lymphogranuloma venereum
plh (**PLH**) palaemontes-lightening hormone
PLHS Public Library of the High Seas (American Merchant Marine Library Association)
pli preload indicating
PLI Pacific Law Institute; Plant Location International; Photo Library Inc
PLI Partido Liberal Independiente (Spanish—Independent Liberal Party); *Partito Liberale Italiano* (Italian—Italian Liberal Party); *Photo-Lab-Index*
p'lice police
PLIDCO Pipe Line Development Company
Plim 1 Plimsoll line
P Lin Port Lincoln
Plin C Gaius Plinius Secundus major (Roman naturalist often referred to as Pliny the Elder)
Plin L Plinius Caecilius Secundus minor (Roman writer often referred to as Pliny the Younger)
Plioc Pliocene
plis propellant-level indicating system
plk plank
PLK Phi Lambda Kappa; Poincare-Lighthill-Kuo (mathematical method)
p lkr peacoat locker
pll phase-locked loop
PL/I Programming Language 1
PLL Prince Line Limited
PLLS Portable Landing Light System
pllt pallet
plltn pollution
plm pulse-length modulation
Plm Palembang
P l M Pépé le Moko
P-L-M Paris-Lyon-Méditerranée (famous French railway)
plmb plumber; plumbing
pl mo plastic mould
Plms Palms (postal abbreviation)
pln posterior lymph node

pl-n place-name
Pln Plain (postal abbreviation)
PLN (aviation flight) Plan
PLN Partido Liberación Nacional (Spanish—National Liberation Party); *Partido Liberal Nacionalista* (Spanish—National Liberal Party)
plng planning
PLNP Port Lincoln National Park (South Australia)
Plns Plains
plo phase-locked oscillator
PLO Palestine Liberation Organization; Passenger Liaison Office(r); Peoples Liberation Organization; Plans Office(r); Presidential Libraries Office (Library of Congress)
PLO Pairti Lucht Oibre (Irish—Labour Party); *Polskie Linie Oceaniezne* (Polish—Polish Ocean Lines)
plom prescribed loan optimization model
Plosk Ploskogorye (Russian—plateau)
plot. plotting
plp plastic-lined pipe
plp (**PLP**) pyridoxal phosphate
PLP Parliamentary Labour Party; Partners for Liveable Places; Progressive Labor Party
pl & pd personal loss and personal damage
PLPG Publishers' Library Promotion Group
plpgrndg pulp grinding(s)
pl x pl plastic to plastic
PLPP Pennsylvania League for Planned Parenthood
PLP-PVV Parti pour la Liberté et le Progrès (French—Party of Liberty and Progress); *Partij voom Vrijheid en Vooruitgang* (Flemish—Party of Freedom and Progress)—Belgium
PLQ Public Library Quarterly
plr pillar; primary loss retention
Plr Pillar (postal abbreviation)
PLR Philippine Liberation Ribbon; Public Lending Right
P L & R Postal Laws & Regulations
PLR Partido Liberal Radical (Spanish—Radical Liberal Party)
PLRA Photo Litho Reproducers' Association
PLRE Partido Liberal Radical Ecuatoriano (Spanish—Ecuadorean Liberal-Radical

Party)
PLRS Position-Location Reporting System
plry poultry
pls plates; please
PLS Purnell Library Service
plsd promotion list service date
plsfc part load specific fuel consumption
Pl Sgt Platoon Sergeant
plshd polished
plshr polisher
PLSS Portable Life-Support System
plstc plastic
plstr plasterer
plt personal leave time; pilot; primed lymphocyte typing; psittacosis-lymphogranuloma trachoma
pltc political
pltf plaintiff
pltry poultry
PLTS Point Loma Test Site (Convair)
plu people like us; plural; plurality; price look-up
P Lu Port Luis
PLU Patrice Lumumba University (Moscow)
PLUG Public Law Utilities Group
plumb. plumber; plumbing
plumb. plumbum (Latin—lead)
plumcot plum plus apricot (hybrd)
plumr plumber
PLUNA Primeras Líneas Uruguayas de Navegación Aérea (Spanish—First Uruguayan Aerial Navigation Lines)
Plunket Plunket Society (Royal New Zealand Society for the Health of Women and Children)
pluperf pluperfect
plur plural
PLUS Physically-Limited United Students; Project Literacy United States; Professional Learning Unit System
plute(s) plutocrat(s)
pluto (**PLUTO**) pipeline under the ocean
Pluv Pluviôse (French—Rainy Month)—beginning January 20th—fifth month of the French Revolutionary Calendar
plwd plywood
plx plexus; propellant-loading transfer
Ply Plymouth
PLYMCHAN Plymouth Subarea Channel (NATO)

Plz Plaza
pm paramilitary, post mortem; premium; premolar; presystolic murmur; preventive maintenance (PM); program manager; project manager; publicity man; pulse modulation; pumice
pm (PM) primary memory
p-m permanent magnet; phase modulation
p.m. *post meridiem* (Latin—after noon, night)
p/m pounds per minute
p&m probate and matrimonial
pm poids molículaire (French—molecular weight)
Pm promethium
PM Pacific Mail; Past Master; Pattern Maker; Pay Master; Peabody Museum; Pére Marquette (railroad); *Petróleos Mexicanos*; Physical Medicine; Police Magistrate; Pontifex Maximum; Postmaster; Prime Minister; Provost Marshal; publicity man
P.M. *post meridiem* (Latin—after noon); Prime Minister
P/M Pacific Molasses; Physical Medicine
PM *Pistol Makarov* (Russian—Makarov pistol); *Policía Metropolitana* (Spanish—Metropolitan Police)
P.M. *Piae Memoriae* (Latin—of pious memory); *Pontifex Maximus* (Latin—Supreme Pope)
pma positive mental attitude
pma (PMA) paramethoxyamphetamine
PMA Pacific Maritime Association; Parts Manufacturing Associates; Peat Moss Association; Pencil Makers Association; Pharmaceutical Manufacturers Association; Philadelphia Museum of Art; Philippine Mahogany Association; Phonograph Manufacturers Association; Photo Marketing Association; Police Management Association; Politico-Military Affairs; Precision Measurements Association; Primary Mental Abilities (test); Production and Marketing Administration
PMA *Programa Mundial de Alimentos* (Spanish—World Food Program)
PMAA Petroleum Marketers Association of America
PMAC Provisional Military

Administrative Council; Purchasing Management Association of Canada
PMAD Public Morals Administrative Division (New York City Police Department)
PMAE Peabody Museum of Archeology and Ethnology
PMAF Pharmaceutical Manufacturers' Association Foundation
PMAs Power-Marketing Administrations
PMAS Purdue Master Attitude Scales
PMATA Paint Manufacturers' and Allied Trades Association
pmb post-menopausal bleeding
PMB Potato Marketing Board
PMBC Pacific Motor Boat Club; Portland Motor Boat Club (Oregon)
pmbo participative management by objectives
pmbx private manual branch exchange
pmc precision mirror calorimeter; preventive maintenance contract(or)
pMc pure Mexican cocaine
PMC Pacific Medical Center; Pennsylvania Military Academy; Princeton Microfilm Corporation; Project Management Committee
pmcs process monitoring and control systems
pmd post-mortem dumps; projected map display
Pmd Portmadoc
PMD/BMI Project Management Division/Battelle Memorial Institute
PMDC Pakistan Minerals Development Corporation
PMDD Personnel Management Development Directorate
pmds projected map display set
PMDS Property Management and Disposal Service
pme performance-measuring equipment; planning, management, evaluation; protective multiple earthing
P Me Portland, Maine
PMEA Powder Metallurgy Equipment Association
PMEL Pacific Marine Environmental Laboratory; Precision Measuring Equipment Laboratory
pmest personality, matter, energy, space, time (Raganathan's fundamental catego-

ries)
pmet painted metal
pmf probable maximum flood(ing); progressive massive fibrosis
PMF Presidential Medal of Freedom
PmG Paymaster General; Postmaster General
PMG Provost Marshal General
PMG *Pall Mall Gazette*
pmh past medical history; probable maximum hurricane
PMHP Primary Mental Health Project
pmi photographic micro-image; point of maximum impulse; private mortgage insurance
PMI Palma de Mallorca, Balearic Islands, Spain (airport); Pre-Marital Inventory
PMI *Partai Muslimin Indonesia* (Indonesian Muslim Party)
PMIA Presidential Management Improvement Award
PMIC President's Management Improvement Council
PMIG Political-Military Interdepartmental Group
PMIS Personnel Management Information System; Planning Management Information System; Product Management Information System
PMJC Pine Manor Junior College
pmk pitch mark; postmark(ed)
pml probable maximum loss
PML Pacific Micronesian Line; Pierpont Morgan Library
Pmla Parmelia
PMLA Publications of the Modern Language Association of America
PMLO Principal Military Landing Officer
pmm pulse mode multiplex
pmma (PMMA) polymethylmethacrylate
PMMI Packaging Machinery Manufacturing Institute
pmmu paged memory-management unit
pmn polymorphonuclear neutrophil
pmn producto material neto (Spanish—net material product)
PMNA Pacific Mountain Network Association; Parkers Marsh Natural Area (Virginia)
PMNH Peabody Museum of

Natural History
pmnl polymorphonuclear leukocyte
pmnr periadenitis mucosa necrotica recurrens
pmo printed matter only
pmo *pianissimo* (Italian—very softly)
PMO Palomar Mountain Observatory; Polaris Material Office; Principal Medical Officer; Provost Marshal's Office
PM & OA Printers' Managers and Overseers Association
PMOLANT Polaris Material Office, Atlantic
PMOPAC Polaris Material Office, Pacific
P Mor Port Moresby
PMOSC Primary Military Occupational Code
pmp per-member payment; precious metal plating; previous menstrual period; probable maximum precipitation
PMP Preliminary Management Plan; Procurement Methods and Practices (manual)
pmr pressure-modulated radiometer
pm & r physical medicine and rehabilitation
Pmr Paymaster
PMR Pacific Missile Range
PMRAFNS Princess Mary's Royal Air Force Nursing Service
PMRC Parents' Music Resource Center
PMRL Pulp Manufacturer's Research League
PMRM Periodic Maintenance Requirements Manual
PMRS Physical Medicine and Rehabilitation Service
PMRY Presidio of Monterey
pms poor miserable soul; postmenopausal syndrome; pregnant mare's serum
pms (PMS) phenazine methosulphate; pollution-monitoring satellite; pre-menstrual syndrome
pm's push monies
p-m-s processors-memories-switches
PMS Pantone Matching System; Peabody Museum of Salem; Performance Management System; Permanent Manual System; Planned Missile System; Preventive Maintenance System; Project Management System; Project Manager, Ships; Public Man-

agement Sources; Public Message Service
PMSA Pacific Merchant Shipping Association
pmsg pregnant mare's serum gonadotrophin
PMSP Plant Modelling System Program
pm specialists paramilitary specialists
PMSSMS Planned Maintenance System for Surface Missile Ships
PMST Professor of Military Science and Tactics
pmt payment; photomultiplier tubes; positive matte technique; premenstrual tension; programs, materials, techniques
PMT Perceptual Maze Test; photo mechanical transfer
PMTB Pacific Motor Tariff Bureau
PMTS Predetermined Motion Time System
pmu performance monitor(ing) unit; physical mockup; productive man work unit
PMU Pattern Makers Union
PMUSAOAS Permanent Mission of the United States of America to the Organization of American States
PMVB Pocono Mountain Vacation Bureau
pmvi periodic motor vehicle inspection
pmvp *precio maximo de venta al publico* (Spanish—maximum price charged the public)
p mvr prime mover
pmv's parcel mail vans (British railways)
pmx private manual exchange (telephone)
pmyob please mind your own business
pn partition; part number; percussion note; percussive note; please note; position; promissory note; psychiatry-neurology; psychoneurotic
pn (PN) punch-on (computer character)
p-n positive-negative
p/n part number; promissory note
p & n psychiatry and neurology
Pn North Pole; North Celestial Pole; perigean range
PN Pacific Northern (airline); Pan-American World Airways (stock exchange sym-

bol); part number; plasticity number; point of no return; Practical Nurse
P/N Part Number
P & N Piedmont and Northern (railroad)
PN *Partido Nacional* (Spanish—National Party); *Partido Nacionalista* (Spanish—Nationalist Party)
pna (PNA) pentosenucleic acid
Pna Panama
PNA Pacific Northern Airlines; Pakistan National Alliance; Philippines News Agency; Project Network Analysis
PNAC President's National Advisory Committee
PNAI Provincial Newspapers Association of Ireland
pnavq positive-negative ambivalent quotient
pnb *producto nacional bruto* (Spanish—gross national product)
PNB Philippine National Bank
PNB *Produto National Bruto* (Portuguese—Gross National Product)
PNBA Pacific Northwest Booksellers Association
PNBB *Parc National de la Boucle du Baoule* (French—Baoule River Bend National Park)—in the highlands of Mali
PNBC Pacific Northwest Bibliographic Center (American and Canadian libraries)
PNBP *Parc National de la Boucle de la Pendjari* (French—Penjari River Bend National Park)—in northwestern Dahomey
pnbt paranitroblue tetrazoleum
pnc penicillin; plate number coil; premature nodal contraction
P 'n C Picnic 'n Chicken
PNC Palestine National Council; People's National Congress; Prohibition National Committee
PNC *Parque Nacional Canaima* (Spanish—Canaima National Park)—encloses Venezuela's Angel Falls—world's tallest waterfall; *Prairie Home Companion* (National Public Radio program)
PNCC President's National Crime Commission
pnch punch (flow chart)
Pncla Pensacola
pnd paroxysmal noctural dyspnoea; postnasal drip

Pnd Pandjang

pndb perceived noise decibels

pndg pending

P-N-D-L-R park-neutral-drive-low-reverse (positions on automatic transmission gauge)

Pndo Pinedo

pne practical nurse's education

pne (PNE) peaceful nuclear explosion

PNe Pointe Noire

PNE Pacific National Exchange (Vancouver); Pacific National Exhibition (Vancouver)

PNEA Parque Nacional El Avila (Spanish—El Avila National Park)—between Caracas and the Caribbean, encloses the Humboldt National Monument

P Ned Pharmacopee Nederlandsche (Dutch—Netherlands' Pharmacopeia)

PNERL Pacific Northwest Environmental Research Laboratory

Pnes Pines (postal abbreviation)

PNET Peaceful Nuclear Explosion Treaty

pneu pneumatic(s)

PNEU Parents' National Education Union

pneumato (Latin prefix—breathing)—pneumonia

pneumoccon pneumocconiosis (lung fibrosis due to dust-particle inhalation)

pneumog pneumograph; pneumographer; pneumographic-(al)(ly); pneumography

pneumonoultra pneumonoultra-microscopicsilicovolcanoconiosis (miner's lung disease)

pnf proprioceptive neuromuscular facilitation

pnfd present not for duty

p.n.g. persona non grata (Latin—an unacceptable person)

Png Penang

PNG Papua New Guinea; Professional Numismatists Guild

PNG Papua Nueva Guinea (Spanish—Papua New Guinea); *Parque Nacional Guatopo* (Spanish—Guatopo National Park)—near Caracas, Venezuela

PNGL Papua New Guinea Line

pnh (PNH) paroxysmal nocturnal hemoglobinuria

PNH Phnom-Penh, Cambodia

(airport)

PNHA Physicians National Housestaff Association

PNHP Parque Nacional Henri Pittier (Spanish—Henri Pittier National Park)—near Maracay, Venezuela

pni positive noninterfering (alarm); psychoneuroimmunology; pulsed neutron interrogation

PNI Pharmaceutical News Index

PNI Parque Nacional Iguazu (Spanish—Iguazu National Park)—surrounding the Iguazu Falls shared by Argentina, Brazil, and Paraguay

P Nic Port Nicholson

PNITC Pacific Northwest International Trade Council

pnl panel

PNL Pacific Naval Laboratories; Pacific Northwest Laboratories; Philippine National Line

PNLA Pacific Northwest Library Association; Pacific Northwest Loggers Association

PNM Pinnacles National Monument (California)

pno piano

pno pergamino (Spanish—parchment)

Pno Pantano (Spanish—bog, marsh)

P 'n' O P and O (Peninsular and Occidental Steamship Company, Peninsular and Oriental Line)—P & O

PNO Port of New Orleans; Principal Nursing Officer

PNO Parque Nacional Ordesa (Spanish—Ordesa National Park)—near Spain's French frontier

PNOC Philippine National Oil Company; Proposed Notice of Change

pnp positive negative positive

p 'n' p pimping and pandering

PNP Pediatric Nurse Practitioner; People's National Party; Platt National Park (Oklahoma)

PNP Partido Nuevo Progresista (Spanish—New Progressive Party)—Puerto Rico

pnpn positive-negative positive-negative

pnpr positive-negative pressure respiration

pnr point of no return; prior notice required

Pnr Pioneer

PNR Passenger Name Record (airlines); Philippine National Railways; Pittsburgh Naval Reactor; Pulletop Nature Reserve (New South Wales)

PNRP Philadelphia Pulmonary Neoplasm Research Project

pns parasympathetic nervous system; peripheral nervous system

PNS Pacific Navigation Systems; Pakistan Naval Ship; Philadelphia Naval Shipyard; Philippine News Service; Professor of Naval Science

PNSN Parque Nacional Sierra Nevada (Spanish—Sierra Nevada National Park)—encloses Venezuela's Mount Bolívar

PNSTDC Pakistan National Scientific and Technical Documentation Center

PNSY Portsmouth Naval Shipyard

pnt paint(ed)

Pnt Pentagon

PNT Parque Nacional Tijuca (Portuguese—Tijuca National Park)—near Rio de Janeiro, Brazil

Pnt Anx Pentagon Annex

PNTBT Partial Nuclear Test Ban Treaty

pntd painted

Pnte Pointe (French—point)

PNTO Principal Naval Transport Officer

pntr painter

PNU Pneumatic Scale Corporation (stock-exchange symbol)

pnutbutsan peanut-butter sandwich

p-nut butter peanut butter

pnutbutwich peanut-butter sandwich

p-nut(s) peanut(s)

PNVS Pilot's Night-Vision System

PNW Parc National du W (W-shaped park on the borders of Dahomey, Niger, and Upper Volta)

PNWD/BMI Pacific Northwest Division/Battelle Memorial Institute

PNWL Pacific Northwest Laboratory (AEC)

PNWR Piedmont National Wildlife Refuge (Georgia); Presquile National Wildlife Refuge (Virginia); Pungo National Wildlife Refuge (North Carolina)

pnx pneumothorax

pnxt. *pinxit* (Latin—he or she painted it)
PNYA Port of New York Authority
PNYCTC Pennsylvania New York Central Transportation Company (merger of Pennsylvania and New York Central railroads)
Pnz Penzance
po piss off; poetry; polarity; power oscillator; power-operated; previous orders
po' poor
p-o postoperative
p/o part of
p & o paints and oil; pickled and oiled
p.o. *per os* (Latin—by mouth)
Po polonium; Portugal; Portuguese
P⁰ Pedro
PO Parole Officer; Passport Office; Patent Office; Personnel Office(r); Petty Officer; Philadelphia Orchestra; Police Officer; Port Office(r); Post Office; Probation Officer; Project Office; Province of Ontario; purchase order
P-O Pyrénées-Orientales
P/O Parole Officer; Pilot Officer; Probation Officer
P & O Peninsular & Occidental Steamship Company; Peninsular & Oriental Line
PO *Portland Oregonian*
PO 1/C Petty Office First Class
pO₂ oxygen pressure
PO--2 Soviet minesweeping launch; Soviet trainer aircraft nicknamed Mule by NATO
PO 2/C Petty Office Second Class
PO 3/C Petty Officer Third Class
poa place of acceptance; primary optical area; primary optic atrophy
P o A Power of Attorney
POA Police Officers Association; Portland Opera Association; Prison Officers Association
POAC Peace Officers Association of California; Post Office Advisory Council
POADS Portland Air Defense Sector
POAG Peace Officers Association of Georgia
POAU Protestants and Other Americans United for Separation of Church and State
pob persons on board; pilot on board; point of beginning; prevention of blindness
pob *población* (Spanish—population)
PoB Port of Baltimore
POB post office box
Pobeda Pobeda Peak (highest mountain between China and the USSR in the Tien Shan range where it attains 24,406 feet)
po'-boy poor-boy (sandwich)
pobra pony + zebra (hybrid)
PO BX Post Office Box
poc point of contact; principal operating component; privately owned conveyance
poc (POC) process operator console
POC Pittsburgh Opera Company; port of call; Prison Officer's Club; Public Oil Company
Pocahontas (Algonquin—Tomboy)—nickname of Matoka the daughter of Chief Powhatan; her married name was Rebecca Rolfe
Poca(loo) Pocatello, Idaho
po'ch porch
pocill. *pocillum* (Latin—small cup)
pock pocket
Pocket Bks Pocket Books
Pocket State Luxembourg (pocketed between Belgium, France, and Germany)
Poconos Pocono Mountains of eastern Pennsylvania
poc's ports of call
POCS Patent Office Classification System
pocul. *poculum* (Latin—cup)
pod. payable on (or upon) death; point-of-origin device; port of debarkation; port of departure; probability of detection
pod. (POD) process-oriented design
p.o.d. paid on delivery
pod (Greek—foot)—anthropod, cephalopod, gastropod, podiatrist, podiatry, pseudopod
POD Port of Debarkation; Post Office Department; Professional and Organizational Development (higher education network)
POD *Pocket Oxford Dictionary*
PODAPS Portable Data Processing System
Pod D Doctor of Podiatry
podex photographic exercise
podia podiatrist(ic)(al)(ly); podiatry

poe (POE) polyoxyethylene
POE Pacific Orient Express; port of embarkation; port of entry
poe buoy plank-on-edge buoy
poecrit poetry critic(ism)
p o'ed put out
POED Post Office Engineering Department
poet. poetical(ly); poetry
Poet Poetry
POETS Phooey On Everything—Tomorrow's Saturday
POEU Post Office Engineering Union
pof please omit flowers
pof (POF) pyruvate oxidation factor
POF Philharmonic Orchestra of Florida
POFI Pacific Oceanographic Fisheries Investigation
POG Pacific Oceanographic Group (British Columbia)
POGO Pennzoil Offshore Gas Operators; Polar Orbiting Geophysical Observatory
poh pull out of hole (oil well)
pOH alkalinity factor
Poh Pohang
POHMA Project for the Oral History of Music in America
poi poison; poisonous (spelled out and symbolized with skull and crossbones on labels)
POI Personal Orientation Inventory; Program of Instruction
Point Point of Air, Alcock, Arena, Arguello, Ayre, Baker, Barber, Cairndoon, Chevalier, Chicot, Conception, Fortin, George, Harbor, Hueneme, Judith, Lay, Leamington, Lobos, Loma, Lookout, Pedro, Pleasant, Reyes, Sal, San Luis, San Pedro, Sur; The Point—West Point, U.S. Military Academy at West Point, New York
POINTER Particle Orientation Interferometer
Point Reyes national seashore in California
pois poison
POIT Power-of-Influence Test
pol petroleum-oil-and-lubricants (POL); polar; polarize(d); police; political; politician; problem-oriented language
Pol Poland; Polish
Pol *Polen* (Norwegian—Poland); Polish (Slavic language); *Polonia* (Italian, Lat-

in, Portuguese, Spanish—Poland)
POL Pacific Oceanography Laboratories; Patent Office Library; petroleum-oil-and-lubricants; Polish Ocean Lines
p-ola payola (kickback, bribe)
POLA Prostitutes of Los Angeles (protective association)
Pol Ad Political Adviser
polad(s) political adviser(a)
Poland Polish People's Republic (North-European country between Germany and Russia), *Polska Rzeczpospolita Ludowa*
polang polarization angle
polar. polarity; polarization; polarize(d)
Polar BEAR Polar Beacon Experiments and Auroral Research (satellite)
POLARS Pathology On-Line Logging and Reporting System
Pol Col Police College
pol com political committee
Pol Com Police Commissaire (Interpol); Police Commissioner
polcrit political critic(ism)
poldamr petroleum, oil, and lubrication installations damage report
pol econ political economy
polem polemic; polemicist; polemical(ly); polemicize
POLEX Polar Experiment (weather)
polf parents of large families
Pol Fed Police Federation (London)
POLFER *Polizia Ferroviaria* (Italian—Railroad Police)
Pol Found Police Foundation (Washington, D.C.)
poli politician
pol ind pollen index
polio poliomyelitis
POLIS Parliamentary On-Line Information System
poli sci political science
polish polish sausage (*kielbasa*)
polit political; politician; politics
Politburo *Politicheskoe Byuro* (Russian—Political Bureau of the Central Committee)
polka. petroleum, oil, and lubricants out-of-kilter algorithm
poll. pollution
pollie(s) politician(s)
POLLS Parliamentary On-

Line Library Study
pol in the pen politician in the penitentiary
poln *polnisch* (German—Polish)
Polon *Polonais* (French—Polish)
Pol Rze Lud *Polska Rzeczpospolita Ludowa* (Polish People's Republic)
pols political prisoners; politicians
pol(s) political prisoner(s); politician(s); politician(s); poll parrot(s)
POLs Problem-Oriented Languages (computer)
pol sci political science; political scientist(s)
POLSTRADA *Polizia Stradale* (Italian—Highway Police)
polwar political warfare
poly polyethylene; polymer; polytechnic; polytechnical; polyvinyl
po'ly poorly
poly (Greek—many)—polydactyly, polygenic, polymer, polymorphism, polypeptide
Poly Polynesia; Polynesian; Polytechnic (institute or school)
Polyb Polybius
poly bot polyethylene bottle
polyg polygraph(er); polygraphic(al)(ly); polygraphy (lie detection)
polymorph polymorphous
poly sci political science
polysex polysexual(ity)
polytech polytechnic(al)
polywater polymerized water
pom polycyclic organic matter; pomeranian; pomological; pomology; pom-pom; preparation for overseas movement
pom (POM) polyoxymethylene
pom pomeridiano (Italian—afternoon, p.m.)
PoM Port of Miami
POM Port Moresby, New Guinea (airport)
pomato potato-tomato hybrid vegetable
pomcus (POMCUS) prepositioned material configured in unit sets
POME Prisoners of Mother England—Pommies; early convict immigrants (Australian slang)
POMFLANT Polaris Missile Facility, Atlantic
pomol pomologic(al)(ly); pomologist(ic)(al)(ly); pomol-

ogy
Pomp Pompey
POMPAC Polaris Missile Facility, Pacific
Pompadour Jeanne-Antoinette Poisson—Marquise de Pompadour (1721–1764)
pom-pom antiaircraft gun
POMR Problem-Oriented Medical Record
POMS Panel on Operational Meteorological Satellites
pomsee preparation, operation, maintenance, shipboard electronics equipment
POMSIP Post Office Management Service Improvement Program
pon pontoon
'pon upon
Pon Ponce
PON Program Opportunity Notice; Program Opportunity Notification
pona paraffin, olefin, naphthene, aromatic (test for petroleum octane rating)
PonBrg pontoon bridge
pond. *pondere* (Latin—by weight)
Pondo Pondoland
p-on-n positive on negative
pons profile of nonverbal sensitivity (body language)
Pont Pontevedra
pont b pontoon bridge
Ponti Pontiac
Pontines Pontine Islands off Anzio, Italy or the Pontine Marshes of Italy
Pont. Max. *Pontifex Maximus* (Latin—Supreme Pontiff—the Pope)
PONY Prostitutes of New York (protective association)
p & oo pianistic and orchestral orgasm (as in the finale of Rachmaninoff's Concerto No 3 in D minor for piano and orchestra)
Poo Poole
POO Post Office Order
pood poodle dog; (Russian—36-lb. weight)
POOD Provisioning Order Obligation Document
poof. peripheral on-line-oriented function
Pool The Pool (the Thames just below London Bridge around Billingsate Market)
poop. nincompoop
Poor's Poor's *Register of Corporations, Directors, and Executives*
POOS Priority Order Output

System

poosslq person of opposite sex sharing living quarters

POoW Petty Officer on Watch

pop carbonated beverage; pop-pet; popular; population

pop. carbonated beverage; perpendicular ocean platform (POP); persistent occipito-posterior; plasma osmotic pressure; plaster of paris; popliteal; poppet; popular; population

pop. (POP) public offering price

p-op post-operative

p-o-p plaster of paris; printing-out-paper

Pop Poppa

POP Palletizing Optimization Potential; Panoramic Office Planning; Portuguese Overseas Province (Macao, China); Post Office Plan(ning)

Popa Popayan, Colombia

POPA Property Owners Protection Association

pop. advertising point-of-purchase advertising

pop art popular art (advertising displays, comic strips, posters)

popb proposed operating plan and budget

POPE Product Oriented Procedures Evaluation

popex population explosion

popf prepared-on-premises flavor

popi post office position indicator (navigation system developed by British post office)

poplit popliteal

pop music popular music

Popo Popocatepetl (Aztec— Smoking Mountain)

pop psych popular psychiatry

popr pilot overhaul provisioning review

pops popular concerts; popular tunes

POPS People Opposed to Pornography in Schools

Pop Sci Popular Science

POPSER Polaris Operational Performance Surveillance Engineering Report

poq periodic order quantity

POQ Public Opinion Quarterly

por porosity; porous; public opinion research

p-o-r pay-on-receipt; payable-on-receipt

Por Porifera; Portland; Portugal; Portuguese

Por Porogi (Russian—rapids, waterfall)

POR Partido Obrero Revolucionario (Spanish—Revolutionary Workers' Party); Policy, Organisation, and Rules (of the Girl Guides and Scouts)

PORA Police Officers Research Association

PORAC Peace Officers Research Association of California

porc porcelain

PORC Peralta Oaks Research Center

Porcupines Porcupine Islands east of Bar Harbor, Maine

'pore Singapore

PORIS Post Office Radio Interference Station

porksan pork sandwich

porkwich pork sandwich

porm plus or minus

porn pornographic; pornography (see porno)

pornette(s) pornographic cassette(s)

pornfilm pornographic motion picture film

porno pornofilm; pornographer; pornographic; pornographically; pornographic bookshop; pornography

pornobio pornographic biography

pornofilm pornographic motion picture

porno mag pornographic magazine

pornos pornographic books, moving pictures, photographs, recordings, etc.

pornovel pornographic novel

pornovelist pornographic novelist

porn pub(s) pornographic publication(s); pornographic publisher(s)

Porn Squad Pornographic (Publication) Squad

porny pornographic

pornzines pornographic magazines

porp (PORP) printed on recycled paper

porp(s) porpoise(s)

PORS Post Office Research Station

port. portable; portrait; portraiture

port. (PORT) photo-optical recorder tracker

port portugiesisch (German— Portuguese)

Port Portland; Portugal; Portu-

guese

Port Portuguese

Port Ade Port Adelaide, South Australia

Port Alb Port Alberni on Vancouver Island, British Columbia

portalet portable toilet

Port Alex Port Alexander, Alaska

Portañol Portuguese-Spanish

Port Ant Port Antonio, Jamaica

Port Art Port Arthur (Manchuria, Ontario, Tasmania, or Texas)

portashed portable shed(ding)

Port Chi Port Chicago; Portuguese China (Macao)

Port Dal Port Dalhousie, Ontario

porteños (Spanish—port people)—in Argentina means the people of Buenos Aires and in Chile those of Valparaiso

Port Ind Portuguese India

Port Jack Port Jackson (seaport of Sydney, New South Wales, Australia)

Port Jeff Long Island, New York; Port Jefferson

Port Liz Port Elizabeth, New Jersey; Port Elizabeth, South Africa

Port Nick Port Nicholson (Wellington, New Zealand's harbor)

Port Phil Port Phillip, Melbourne, Victoria, Australia

Port Rich Port Richmond, Staten Island, New York

port side left side of an airplane, ship, or other craft when looking forward, symbolized by a fixed red light

Portsmouth U.S. Naval Disciplinary Command at Portsmouth, New Hampshire—the U.S. Naval Prison

Port Sud Port Sudan (Sudanese harbor on the Red Sea)

Port Swett Port Swettenham, Malaysia

Port Talb Port Talbot, Wales

Port Tew Port Tewfik (Egypt's Port Taufiq at the southern end of the Suez Canal)

Port Tim Portuguese Timor

Portug Portugais (French— Portuguese)

Portugal Republic of Portugal (Iberian country once ruling a vast colonial empire),

República Portuguesa

Port Wash Port Washington, Long Island, New York

Port Wel Port Weller, Ontario

pos point of sale; position; positive; possibility; possible; product of sums

PoS Point of Sale; Port of Service; Port of Spain

POs Police Officers; Postal Orders

POS Patent Office Society; Port-of-Spain, Trinidad (airport); Primary Operating System; Problem-Oriented System

posa payment outstanding suspense accounts

POSB Post Office Savings Bank

POSC Problem-Oriented System of Charting

POSD Post Office Savings Department

posdcorb planning–organization –staffing–directing–coordinating–reporting–budgeting (mnemonic device for remembering the functions of management)

posdsplt positive displacement

posh permuted on subject headings; port side out, starboard side home (British slang)

posistor positive resistor

posit position; positive; positron

positron positive electron

POSIX Portable Operating Systems for Computer Environments

posm patient-operated selected mechanisms

posn position

POSNY People of the State of New York

pos pron possessive pronoun

poss possession; possessive

P o S S Point-of-Sale System

POSS Passive Optical Satellite Surveillance (System)

P-O-S S Point-of-Sale System; Point-of-Service System

posses possessive

posslq person of the opposite sex (in) same living quarters

posslq's persons of the opposite sex sharing living quarters

'possum(s) opossum(s)

post. postage; postal; posterior; post mortem

post (Latin prefix—after or behind)—postwar; *posterior* (Spanish abbreviation)

POST Frederick Post Drafting Equipment; Peace Officers Standards and Training; Police Officer Student Training; Processes of Science Test

post-Aug post-Augustan

post aur. *post aurem* (Latin—behind the ear)

post.d posterior diameter

poster. posterior

pos terminal point-of-sale terminal

postgangl postganglionic

Postgrad Med Inst Postgraduate Medical Institute

postgrad(s) postgraduate(s)

posth posthumous

postl postlude

post-mort post mortem (autopsy)

post-op post-operative

post part. *post partum* (Latin—afterbirth)

post-sync post-synchronization of a sound track made after a motion-picture film has been shot

POSWG Poseidon Software Working Group

pot. point of tangency; portable outdoor toilet; potash; potassa (potassium hydroxide); potassium; potential; potentiometer; marijuana

pot. potaguaya (Mexican Indian—marijuana); *potio* (Latin—dose, draft, potion)

'potamus(es) hippopotamus(es)

potash potassium carbonate (K_2CO_3)

potash alum potassium aluminum sulfate

potass potassium

POTASWG Poseidon Test Analysis Software Working Group

potats potatoes

POTC PERT *(q.v.)* Orientation and Training Program

P o TD Port of The Dalles

POTIB Poseidon Technical Information Bulletin

potosslq persons of the opposite sex sharing living quarters (*sometimes appears as* posslq)

potr potrero (Spanish—cattle ranch, pasture)

pots lobster pots; plain old telephone service

pots. potentiometers

pott pottery

Potteries The Potteries (Stoke-on-Trent)

PotUS Lyndon Johnson's acro-

nym meaning President of the United States

POTUS President of the United States (address name used by Churchill when communicating with Roosevelt, later used by President Johnson—PotUS)

pot w potable water

pou piss on you

poul poultry

POUM *Partido Obrero de Unificación Marxista* (Spanish—Workers Party of Marxist Unification)

POUNC Post Office Users' National Council

pound monetary unit of Cyprus, Egypt, Ireland, Lebanon, Malta, Sudan, Syria, the United Kingdom, British colonies and dominions

POUR President's Organization for Unemployment Relief

pov privately owned vehicle

p-o-v point-of-view

P_{ov} *Poluostrov* (Russian—peninsula)

POV Pend Oreille Valley (railroad)

pov's privately owned vehicles

pow power; prisoner of war (POW)

P o W Prince of Wales; Prisoner(s) of Watergate

POW Country Potash, Oil, and Wheat Country around Saskatoon, Saskatchewan

powd powder; powdered; powered

power. programmed operational warshot evaluation and review

POWER Professionals Organized for Women's Equal Rights

po'white poor white person

pows (POWS) prisoners of war

POWS Pyrotechnic Outside Warning System

pox police (journalists' abbreviation)

poy pre-oriented yarn

Poz Poznan

pozn poznamka (Czech-footnote)

pp baby-talk for urinate(d); pages; painful pissing; panel point; parcel post; part paid; partial pay; partially paid; passive participle; past participle; pellagra preventive (factor); per person; perceptual performance; permanent party; petticoat peeping; physical profile; physical properties; pickpocket; postage paid; postpaid; present position; pressure-proof; private property; privately printed; professional paper; purchased part(s); push-pull; urination; urine

pp (PP) planning permission; planning permit

p-p peak-to-peak; pee-pee (urine); push-pull; pussy-power (feminine wiles)

p/p peepee (urinate, urine)

p&p payments and progress

p & p parsimonious and penurious (miserly and stingy)

p-to-p peak-to-peak; point-to-point

pp pianissimo (Italian—very softly)

p.p. piena pelle (Italian—full leather); *post partum* (Latin—afterbirth)

Pp. Papa (Latin—father or Pope)

PP Pacific Petroleum; Parcel Post; Parish Priest; Past President; Planned Parenthood; Power Plant; Proletarian Party (Communist)

P-P pellagra-preventive factor

PP The Passionate Pilgrim; *Patres* (Latin—Fathers); *Polizei Pistole* (German—police pistol)

P.P. Pater Patriae (Latin—Father of his Country)

PP¹ inorganic pyrophosphate

Ppa palpitation, percussion, auscultation; photo-peak analysis; program, project, activity

ppa (PPA) phenylpropanolamine

pp & a palpitation, percussion, and auscultation

p. pa. per procura (Latin—by proxy)

p.p.a. phiala prius agitate (Latin—bottle having first been shaken)—shake well before using

PPA Pakistan Press Association; Paper Pail Association; Paper Plate Association; Parcel Post Association; People for Prison Alternatives; Periodical Publishers Association; Personnel Pool of America; Popcorn Processors Association; Poultry Publishers Association; President's Professional Association; Produce Packaging Associa-

tion; Professional Photographers of America; Proletarian Party of America; Public Personnel Association; Purple Plum Association

PPAB Program and Policy Advisory Board (UN)

PPAC Pesticide Policy Advisory Board (EPA)

PPATRA Printing, Packaging, and Allied Trades Research Association (also appears as PATRA)

ppb parts per billion

ppb (PPB) polybrominated biphenyl (cattle poison)

pp&b paper, printing, and binding; planning, programming, and budgeting

Ppb Pappband (German—boards, hard cover)

PPBAS Planning-Programming-Budgeting-Accounting System

PPBC Portland Problem Behavior Checklist

PPBES Planning-Programming-Budgeting-Evaluation System

PPBMIS Planning, Programming, and Budgeting Management Information System

PPBS Planning-Programming-Budgeting System

ppc picture postcard; plain-paper copier; progressive patient care

p p c pour prendre congé (French—to take leave)

pPc pure Peruvian cocaine

PPC Penang Port Commission(er)(s); Personal Productivity Center; Pet Population Control; Policy Planning Council (U.S. Department of State); Positive Peer Culture; Purchase Price Control

ppca plasma prothrombin conversion accelerator

PPCAA Parole and Probation Compact Administrators Association

PPCD Plant Pest Control Division

ppcf plasma prothrombin conversion factor

PPCLI Princess Patricia's Canadian Light Infantry

PPCS Personnel Protection and Communication Services (British anti-terrorist organization); Primary Producers' Cooperative Society

ppd prepaid; purified protein derivative (tuberculin)

PPD Party for Peace and De-

mocracy; Paranoid Personality Disorder; Petroleum Production Division; Portland Public Docks; Propulsion and Power Division

PPD Partido Popular Democrático (Spanish—Popular Democratic Party)

PPDA Produce Packaging Development Association

PPDC Polymer Products Development Center

ppdi pilot's projected-display indicator

ppdo per person, double occupancy

p p_{do} próximo pasado (Spanish—last month)

PPDP Preprogram Definition Phase

PPDS Publishers' Parcels Delivery Service

PPDSE Plate Printers, Die Stampers, and Engravers (union)

ppe philosophy, politics, and economics

PP & E Program Planning and Evaluation

PPES Pilot Performance Evaluation System

ppf personal property floater (policy)

PPF Panamanian Public Force (police); Plumbers and Pipefitters (union)

PPFA Planned Parenthood Federation of America

p-p factor pellagra-preventive factor

ppg planning and programming guidance

PPG Pago Pago, Samoan Islands (airport); Pittsburgh Plate Glass

ppga post-pill galactorrhea-amenorrhea

PPGA Pennsylvania Personnel and Guidance Association

pph pamphlet; post-partum hemorrhage; pounds per hour; pulses per hour

P Php Port Phillip

pphpm parts per hundred parts of mix; pints per hundred parts of mix

pphr parts per hundred parts of rubber

ppi pages per inch; parcel post insured; plan position indicator; policy proof of interest

PPI Plastic Pipe Institute; Producer Price Index; Project Public Information; Protective Packaging Inc; Pulp and Paper International

PPIC Plumbing and Piping Industry Council
ppif photo-processing interpretation facility
p-pille praeventivpille (Dano-Norwegian—preventive pill)—contraceptive
pp/in. pages per inch
P Ping Pulau Pinang (Malay—Penang Ferry)
PPIQ Personality and Personal Illness Questionnaire(s)
pPk purplish pink
ppl pipeline
PPL Philadelphia Public Library; Phoenix Public Library; Pittsburgh Public Library; Planned Parenthood League; Police Protective League; Portland Public Library; Private Pilot's License; Providence Public Library; Provisioning Parts List
PP&L Pennsylvania Power and Light (company)
PP & L Pacific Power and Light
P-plane pilotless airplane (explosive carrying and reaction propelled)
PPLC Patients Protection Law Commission
pple past participle
p-p letters poison-pen letters
pplo pleuropneumonia-like organism(s)
ppm parts per million; pounds per minute; pulse position modulation
ppm (PPM) peak program meter
PPM Peter, Paul, and Mary (singing group)
PPM Partido Proletario de México (Spanish—Proletarian Party of Mexico)—Chinese-trained guerrilla active in Mexico and from California to Texas in the Chicano community; *Persutuan Perpustakaan Malaysia* (Malay—Library Association of the Federation of Malaya)
ppma post-polio muscular atrophy
PPMS Plastic Pipe manufacturers' Society
ppn proportion(al)
PPNA Pupil-Perceived-Needs Assessment
ppng (PPNG) penicillinase-producing Neisseria gonorrhoeae; penicillin-resistant gonorrhea-producing enzyme that inactivates most penicillins

PPNP Point Pelee National Park (Ontario)
PPNW Physicians for the Prevention of Nuclear War
ppo polyphenylene oxide; prior permission only
PPO Preferred-Provided Organization
p-p-ola political plugola (media plugging or touting of a candidate or an ideological issue)—propaganda device in disrepute
ppom particulate polycyclic organic matter
ppo's (PPOs) preferred provider organizations
ppp petty political pismire
p & pp pull and push plate
ppp piu pianissimo (Italian—very very softly)
PPP Peoples Party of Pakistan; Peoples Progressive Party (Guyana); Petroleum Production Pioneers; Pickford Projective Pictures; Population Policy Panel (Hugh Moore Fund); Private Patients Plan
pppp piu piu piu pianissimo (Italian—very, very, very softly)
pp & p's perverts, pimps, and prostitutes
ppq (PPQ) polyphenylquinoxaline
ppr present participle; printed paper rate; prior permission required
PPr Port Pirie
PPR Permanent Pay Record; Permanent Personal Registration; Procurement Problem Report
PPRA Past President of the Royal Academy
pprbd paperboard
PPRICA Pulp and Paper Research Institute of Canada
pps pictures per second; pounds per second; private parliamentary secretary; pulses per second
pp's payless paydays
p-ps post-polio syndrome
PPs Prairie Provinces (Alberta, Manitoba, Saskatchewan)
PPS Pacific Passenger Services; Paper Publications Society; Pennsylvania Prison Society; Petroleum Press Service; Program Policy Staff (UN)
PPS Partido Popular Salvadoreño (Spanish—Salvadoran Popular Party); *Partido Popular Socialista* (Span-

ish—Popular Socialist Party); *Persatuan Perpustakaan Singapura* (Malay—Library Association of Singapore)
P.P.S. post postscriptum (Latin—additional postscript)
PPSA Pan-Pacific Surgical Association
PPSAWA Pan Pacific and Southeast Asia Women's Association
PPSB Periodical Publishers' service Bureau
PPSEAWA Pan-Pacific and South-East Asia Women's Association
ppsn present position
ppso per person, single occupancy
PP Society (*see* PPTPP)
ppt precipitate
PPT Papeete, Society Islands (airport); Pre-Production Test(ing)
PPT Pericles, Prince of Tyre
pptd precipitated
pptn precipitation
PPTPP Promulgators of Public Toilets in Public Parks (also known as the PP Society)
ppty property
ppu platform position unit
PPU Peace Pledge Union; Primary Producers Union
P & PU Peoria and Pekin Union (railroad)
ppv pay-per-view (cable tv channel); people-powered vehicle(s)
PPVT Peabody Picture Vocabulary Test
PPWC Pines to Palms Wildlife Committee; Pulp, Paper, and Woodcutters of Canada
PPWP Planned Parenthood-World Population
pq peculiar; permeability quotient; personality quotient (PQ); previous question; punishment quarters
p-q phenol-hydroquinone (photographic developer)
p & q peace and quiet (solitary confinement)
PQ personality quotient; Province of Quebec; South Pacific Airlines of New Zealand (2-letter code)
PQ Parti Quebecois (French—Québec Party)
pqa procurement quality assurance
PQAP Procurement Quality Assurance Program
PQC Production Quality Control

PQD Plant Quarantine Division
PQD *Partido Quisqueyano Demócrata* (Spanish—Democratic Quisqueyan Party)—Dominican Republic
pqe post-qualification education
pqi professional qualification index
PQIH Plant Quarantine Inspection House
PQLI Physical Quality of Life Index
PQR Personnel Qualification Roster; Program Quality Review
pqrs productivity increases, quality control, robotization, and savings (Japanese formula for economic success)
PQS Percentage Quota System; Personnel Qualification Standard(s)
pr pair; parcel receipt; payroll; percentile rank; peripheral resistance; public relations
pr (PR) proctosigmoidoscopy
p/r per rectum
p & r parallax and refraction
pr *protestants* (Dutch—Protestants)
p.r. *per rectum* (Latin—by the rectum); *punctum remotum* (Latin—remote point)—far point of vision
pR purplish red
Pr Panama-red marijuana; Parana; Prairie; prandtl number; praseodymium; presbyopia; Press; Prince; Proctoscopy; propyl
Pr *Praca* (Portuguese—plaza, square); *Presbyter* (Latin—elder or priest)
PR Parachute Rigger; Park Ranger; Performance Rating; Performance Report; Photoreconnaissance; Pinar del Rio; Plant Report; Problem Report; Progress Report; Psychiatric Record; Public Relations; Puerto Rican(s); Puerto Rico; river gunboat (2-letter naval symbol)
P-R Pennsylvania-Reading (Seashore Lines)
P/R payroll
P & R Parks and Recreation
PR *Paradise Regained* by John Milton (1671); *Partido Republicano* (Spanish—Republican Party); *Partisan Review; Peking Review; Pipe Rolls; Polish Register* (of shipping); *Polskie Radio*

(Polish Radio); *Puerto Rico* (Porto Rico)
P.R. *Populus Romanus* (Latin—Roman People)
pra payroll audit(or); plasma renin activity; probation and rehabilitation of airmen; progressive retinal atrophy
pra (PRA) print alphanumerically
Pra Pará (British maritime abbreviation)
Pra *Prachtausgabe* (German—de luxe edition)
PRA Pay Readjustment Act; Paymaster Rear Admiral; Personnel Research Activity; Popular Rotocraft Association; Postal Reorganization Act; Psoriasis Research Association; Psychological Research Association; Public Roads Administration; Puerto Rico Association
P.R.A. President of the Royal Academy
prac practice; practitioner
pracl page-replacement algorithm and control logic
pract practical; practice; practitioner
Prado El Prado (Madrid museum)
Praeger Frederick A Praeger
praen praenomen
prag pragmatic; pragmatism
pragma processing routines aided by graphics for manipulation of arrays
PRAI Pre-Reading Assessment Inventory
PRAICO Puerto Rican American Insurance Company
Prair *Prairial* (French—Meadowy Month)—beginning May 20th—ninth month of the French Revolutionary Calendar
prais passive-ranging interferometer sensor
pral *principal* (Spanish—principal)
pram perambulator
pram. productivity, reliability, availability, and maintainability
Pram Poseidon random-access memory
Pr of An Principality of Ansbach
prand. *prandium* (Latin—dinner)
PRANG Puerto Rico Air National Guard
PRAT Prattsburgh (railroad)
p. rat. aet. *pro ratione aetatis*

(Latin—in proportion to age)
PRATRA Philippines Relief and Trade Rebilitation Administration
PRAY Paul Revere Associated Yeoman
prb principal borehole
PRB People's Republic of Benin; Personnel Review Board; Population Reference Bureau; Pre-Raphaelite Brotherhood
PRB *Partido de la Revolución Boliviana* (Spanish—Bolivian Revolutionary Party)
prc packed red cells; procedure
prc (PRC) polysulphide rubber compound
PRC Pain Rehabilitation Center; Palestine Red Crescent; Pay-Raise Commission; Pension Research Council; People's Republic of China; Picatinny Research Center (Picatinny Arsenal); Planning Research Corporation; Postal Rate Commission; Public Relations Club
P.R.C. *Post Roman Conditam* (Latin—after the founding of Rome)—753 Before the Christian Era
PRCA Professional Rodeo Cowboys Association; Puerto Rico Communications Authority
PRCB Program Requirement Control Board (NASA)
prcd priced
Pr Ch Parish Church
prchst parachutist
prcht parachute
PRCP President of the Royal College of Physicians
prcs process; processing
PRCS President of the Royal College of Surgeons
prcst precast
prcu power regulation and control unit
prd partial reaction of degeneration; pro-rata distribution
prd (PRD) printer dump(ing)
PRD Pesticides Regulation Division (USDA); Planned Residential Development (permit); Program Requirement Document
PRD *Partido Revolucionario Democrático* (Spanish—Revolutionary Democratic Party); *Partido Revolucionario Dominicano* (Spanish—Dominican Revolutionary Party)
PRDA Program Research and

Development Announcement
PRDC Personnel Research and Development Center (USN); Power Reactor Development Corporation
PRDL Personnel Research and Development Laboratory (USN)
PRDS Processed Radar Display System
prdx paradox
pre prefix (computer character); progressive resistance exercise
pre (Latin prefix—before)— prenatal, presuppose
PRE Psychophysiological Reeducation
prealateen program for children below teen age who are affected by an alcoholic family (*see* alateen)
preamp(s) preamplifier(s)
preb prebend
PREBS Pennsylvania Real Estate Brokers and Salesmen's (licensing examinations)
prec precedence; preceding; precision
Prec Precentor
precip precipitate; precipitation
PRECIS Preserved Context Index System
precomdet pre-commissioning detail
pred predicate; prednisolone
PREDA Puerto Rico Economic Development Administration
pre-design preliminary design
predic predicate; predicative; prediction
pre-em preeminence; preeminent; preempt; preemptible; preemption; preemptive; preemptory
preemies premature babies
preemy premature baby
pref preface; prefatory; prefecture; preference; prefix
Pref Prefect
prefab prefabricated
Pref-Ap Prefect-Apostolic
prefaz prefazione (Italian— foreword)
prefd preferred
preframo prepare fleet rehabilitation and modernization overhaul (USN)
preg pregnancy; pregnant
pregang preganglionic
prehis prehistoric
prej prejudice
prel prelude
prelim preliminary

prelim diag preliminary diagnosis
prelims preliminaries; preliminary pages (frontmatter)
prem premature; premium
pre-med premedical
premie premature baby
premies premature babies
Prensa La Prensa (Buenos Aires' Press)
'prentice apprentice
Prenzl Bg Prenzlauer Berg
pr enzyme prosthetic-group removing enzyme
pre-op preoperation; preoperational
prep preparation; preparatory; prepare; preposition
PREP Personal Radio-Equipped Police; Predischarge Education(al) Program; Preparation Rehabilitation Education Program; Pupil Record of Educational Progress
prepd prepared
prep'ed prepared
prepn preparation
prepr precracovane (Czech— rewritten)
pre-pub pre-publication
pres present
Pres President
PRES Puerto Rico Employment Service
presby presbyopia; presbyopic
Presby Presbyterian
presc prescription
Presc Prescott
Presd_{te} Presidente (Spanish— President)
preserv preservation
presilection presidential election
press. pressure
PRESS Pacific Range Electromagnetic Signature Studies
Presse Die Presse (Neue Freie Presse)—Vienna's Press
presstitute poison-pen prostitute of the press (columnist skilled in writing defamatory articles)
prestmo. prestissimo (Italian— very quickly)
PRESTO Program Reporting and Evaluation System for Total Operations
presv preservation; preserve
pret preterit
Pret Pretoria
pre-Teut pre-Teutonic
PRETTYBLUEBATCH Philadelphia Regular Exchange Tea Total Young Belles Lettres Universal Experimental

Bibliographical Association To Civilize Humanity (initialism contrived by Edgar Allan Poe to satirize all such pseudo-intellectual devices)
pretz pretzel
prev previous
prevan precompiler for vector analysis
preven preventive
prevoc prevocational
prex(y) president (usually college or university)
prez president
prf proof; pulse recurrence frequency; pulse repetition frequency
prf (PRF) priority-reserved flight (air cargo); prolactin-releasing factor
prf. praefatio (Latin—introduction, preface)
PRF Personality Research Form; Petroleum Research Fund; Plywood Research Foundation; Porpoise Rescue Foundation; Public Relations Foundation; Puerto Rican Forum
PRF Publications Reference File (GPO)
prfe polar-reflection faraday effect
prfg proofing
prfnl professional
prfr proofreader
PRFT Portable Rod-and-Frame Test
PRG Prague, Czechoslovakia (airport); Provisional Revolutionary Government (of South Vietnam)
PRHS Port Richmond High School
pri photographic reconnaissance and interpretation; primary; primer; primitive; priority; priority repair induction; private; pulse recurrence interval
PRI Paleontological Research Institute; Plastics and Rubber Institute
PRI Partido Revolucionario Institucional (Spanish—Institutional Revolutionary Party); *Partito Repubblicano Italiano* (Italian—Italian Republican Party)
PRIA Proceedings of the Royal Irish Academy
Pribilovs Pribilov Islands in the Bering Sea off Alaska
Price Stern Price, Stern, Sloan
P Rich Port Richmond
PRIDCO Puerto Rico Indus-

trial Development Company
PRIDE Parents Resource Institute for Drug Education; Personal Responsibility in Defect Elimination; Professional Recruiting (with) Integrity, Determination, and Enthusiasm; Protection of Reefs and Islands from Degradation and Exploitation
Prieta Agua Prieta (Spanish—Dark Water)—Mexican border town
prim. primary
prim (Latin prefix—first)—primitive, primordial
primaries primary colors—blue, red, yellow
prime. precision recovery including maneuvering entry
PRIME Philadelphia Regional Introduction for Minorities to Engineering; Program Independence, Modularity, Economy; Program Research in Integrated Multi-ethnic Education; Programmed Instruction for Management Education
PRIMES Pennsylvania Retrieval of Information in Mathematics Education System; Productivity Integrated Measurement System (USA)
primip primipara, woman bearing or who has borne her first child
primo primero or supremo (French, Italian, Portuguese, or Spanish—first, first place, top quality, supreme)
primogen primogeniture, exclusive inheritance belonging to the eldest son or the eldest daughter if there is no son
primo temp primo tempo (Italian—first tempo)—return to original tempo
prin principal
Prin Principal; Principality
PRIN Partido Revolucianario de Izqueirda (Spanish—Revolutionary Party of the Left)—Bolivian
PRINAIR Puerto Rico International Airlines
PRINCE Parts, Reliability, and Information Center (NASA)
Prin d'And Principat d'Andorra (Catalan—Principality of Andorra)
Prin Monaco Principauté d'Monaco (French—Principality of Monaco)
prin pts principal parts

print. printed; printing
print.(PRINT) preedited interpreter (computer language)
PRINUL Puerto Rico International Undersea Laboratory
Prinz Prinzregententheater (German—Prince Regent Theater) Munich
PRINZ Public Relations Institute of New Zealand
prio priority
PRIO Peace Research Institute, Oslo (Norway)
prions proteinaceous infectious particles (believed by some to cause Alzheimer's disease)
prior. priority
PRIP Park Restoration Improvement Program; Puerto Rican Independence Party
prir parts reliability improvement route; parts reliability improvement routing
PRI & RB Puerto Rico Inspection and Rating Bureau
pris prison(er)
Prisca Priscilla
prise program for integrated shipboard electronics
PRISE Pennsylvania's Regional Instruction System for Education (intercollegiate network)
pris g prisonnier de guerre (French—prisoner of war)
prism. prismatic
PRISM Personnel Record Information System; Program Reliability Information System for Management
Prisoner The Prisoner; Prisoner No 1; The Prisoner of Shark Island; The Prisoner of Zenda; The Prisoners; La Prisonnière
Pris(sy) Priscilla
pritac primary tactical radio circuit
Pritch Pritchard
prithee I pray thee
priv privacy; private; privateer(ing); privation; privative; privet; privilege(d); privily; privy
priv pr privately printed
priv pub privately published
prix de fque prix de fabrique (French—manufacturer's price)
PRIZM Potential Rating Index for ZIP Markets
PRJC Puerto Rico Junior College
PRK People's Republic of Kampuchea (Cambodia)
pr kassa per kassa (Norwe-

gian—for cash)
prkng parking
prl periodical; pick-resistant lock
Pr of L Prince of Liechtenstein; Principality of Liechtenstein
PRL Personnel Research Laboratory; *Polska Rzeczpospolita Ludowa* (Polish Republic); Prairie Research Laboratory (Canada); Precision Reduction Laboratory; Price Reduction League; Project Records List
Prl Cmm Parole Commission
prld pick-resistant locking device
prls prepaid rental-listing service
prm parameter; portable radiation monitor; prime
Prm Promenade
PRMA Puerto Rican Maritime Authority
p-r man public-relations man
prmld premolded
prm's presidential review memorandums
prn print numerically
p.r.n. pro re nata (Latin—as needed, for an emergency)
PRN Physicians Radio Network; Private Registered Nurse(s)
PRNC Potomac River Naval Command
PRNL Pictured Rocks National Lakeshore (Michigan)
PRNS Point Reyes National Seashore
prntr printer
PRNWR Parker River National Wildlife Refuge (Massachusetts)
pro procedure; proceed; procure; procurement; professionally; prophylactic
pro (PRO) print octal; proline (amino acid)
pro (Latin prefix—before or in favor of)—prosection, protribal
Pro Provost
PRO Peer Review Organization; Personnel Relations Office(r); Plant Representative's Office; Professional Review Organization; Public Record Office; Public Relations Office(r)
PROA Public Record Office Archives
pro-am professional-amateur
prob probability; probable;

probably; problem; problematic; problematical
Prob Probate
probcost probabilistic budgeting and costing; probable cost
PROBES Processes and Resources of the Bering Sea Shelf
probie(s) probationer(s)
Prob Off Probation Officer
probs problems
proc procedure; proceeding(s); procure; procurement
proc (Latin prefix—anus)—proctologist
Proc Procedure; Proceedings; Proctor
Proc Cambridge Philos Soc Proceedings of the Cambridge Philosophical Society
Procd procedure
pro-celeb professional celebrity
Proc-Gam Proctor-Gamble
Proc IEEE Proceedings of the IEEE
Proc IRE Proceedings of the IRE
proclib procedure library
Proc Nat Acad Sci U.S.A. Proceedings of the National Academy of Sciences of the United States of America
proco programmed combustion (auto engine)
Procoll Proletarian Collective of Soviet Musicians
procomm program communication
Procop Procopius
Proc Phys Soc, London Proceedings of the Physical Society, London
procrast(s) procrastinator(s)
Proc Roy Soc Proceedings of the Royal Society
Proc R Soc London Proceedings of the Royal Society of London
procsim processor simulation language
procstep procedure step
procto proctocolitis; proctocolonoscopy; proctologist; proctology; proctosigmoidoscopy; proctosigmoidectomy; proctoplegia
PROCTOR Priority Routine, Computer Transfers, and Register Operations
prod product; production
prodac programmed digital automatic control
PRODAC Production Advisers Consortium

PRODFINA Protection et Defense de la Nature (French—Protection and Defence of Nature)
Prod(s) Irish-Catholic English—Protestant—[*see* Pap(s)]; Protestant(s)
prof profession; professional; professor
prof (PROF) pupil registering and operational filling
Prof Professor
PROF Peace Research Organization Fund
profac propulsive fluid accumulator
Prof D Profesor Don (Spanish—Sir Professor)
Prof D^{na} Profesora Doña (Spanish—Madam Professor)
Prof Eng Professional Engineer
Proff Professori (Italian—Professors)
Profintern Red international of Trade Unions
profit. program for financed insurance technic; programmed reviewing, ordering, and forecasting
Prof Lib Pr Professional Library Press
profs professionals; professors
PROFS Professional Office System
prog progenitor; progeny; prognose; prognosis; prognostic; prognostication; prognosticator; program; programmer; progress
Prog Gro Progressive Grocer
proglang(s) progressive language(s)—usually euphemistically slanted
progr program(mer); programme
Prog(s) Progressive(s)
Prog Theor Phys Progress of Theoretical Physics
prohib prohibit(ion)
proi project return on investment
proj project; projectile; projection; projector
PROJACS Project Analysis and Control System
Prol prologue
prolan processed language
prole(s) proletarian(s)
proletcult proletarian culture
pro-lifer person against abortion
PROLLAP Professional Library Literature Acquisition Program
prolog programming in logic

prolong. prolongatus (Latin—prolonged)
ProLt procurement lead time
prom programmable read-only memory; promenade (concert or dance); prominent; promontory; promote; promoter; promotion; promotional; prompter
prom promedio (Spanish—average)
Prom The Prom—Wilson's Promontory—national park at the southernmost tip of Australia
promex productivity measurement experiment
PROMIS Problem-Oriented Medical Information System; Prosecution Management Information System (U.S. Attorney's Office—Washington, DC)
proml promulgate
promo promotional
promo(s) promotional announcement(s)
PROMPT Project Management and Production Team
PROMS Projectile Measurement System (USA)
PROMSTRA Production Methods and Stress Research Association
Pro Mus Orc Pro Musica Orchestra
Promy Promontory
pron pronoun; pronounced; pronunciation; pronunciator(y)
PRON Procurement Request and Order Number (USA)
prond pronounced
prong(s) pronghorn(s)—pronghorn antelope(s)
pronom pronominal
pro note promissory note
PRONTO Program for Numeric Tool Operation
PRONTOS Programmable Network Telecommunications Operating System
pronun pronunciate; pronunciation
pronunc pronunciation
PROOF Parole Resource Office and Orientation Facility (Jersey City, New Jersey)
prop propaganda; propeller; property; proportion(al); proposed; proprietary
prop. proper(ly)
Prop Sextus Propertius (Roman poet)
PROP Panel Review of Products; Planetary Rocket Ocean

Platform; Portland Regional Opportunities Program; Preservation of the Rights of Prisoners

propaed propaedutic(al)(ly); propaedutics

Propaedia outline of knowledge in *The New Encyclopaedia Britannica*

prop art propaganda art

propay proficiency pay

pro.per. in *propria persona* (Latin—acting as one's own attorney)

proph prophetic; prophylactic; prophylaxis

Prophète Le Prophète (French—The Prophet)— Meyerbeer opera

propjet propeller turned by jet engine (same as turboprop)

propl proportional

propn proportion(al)

props (theatrical) properties

prop wash propeller wash

pro rat.aet. pro ratione aetatis (Latin—according to age)

pro rect. pro recto (Latin—by rectum)

PRORM Pay and Records Office—Royal Marines

pros professionals; prosody; prostitute(s)

Pros Atty Prosecuting Attorney

prosc proscenium

PROSE Personal Record of School Experiences

prosig procedure signal

prosine procedure sign

prosp prospecting

pross(ie) prostitute

prost prostate; prothetics; prostitution

prosth prosthesis

prostie(s) prostitute(s)

prot protective; protectorate; protein; protestant; protozoa; protractor

prot (PROT) protein anion

Prot Protectorate; Protestant; Protozoa

protag protagonist

Prot-Ap Protonotary-Apostolic

Protec Protectorate

pro tem. pro tempore (Latin—for the time being)

PROTEUS Propulsion Research and Open-Water Testing of Experimental Underwater Systems

prothrom prothrombin

pro time prothrombin time

Protoch Protochorda

Protocols Protocols of the Learned Elders of Zion

(fraudulent document created and distributed in 1905 by the czarist secret police to incite pogroms against Russia's Jews; since used by antisemitic bigots in defense of their cause)

protozool protozoologic(al)(ly); protozoologist; protozoology

protr protractor

pro us.ext. pro uso externo (Latin—for external use)

prov proverb(ial)(ly); provide; provision; provisional; proviso

prov provincia (Spanish— province)

Prov Provençal; Provence; Proverbs, The (book of the Bible); Providence; Province

Prov Provençal (Romance language); *Proverbs; Provinz* (German—province)

Prov Eng Provincial English

prover procurement-value-economy-reliability

Prov GM Provincial Grand Master

Providence Plantation Rhode Island, full name—Rhode Island and Providence Plantation

Providence Plantations Rhode Island and Providence Plantations

provin provincial

Provincias Vascas Provincias Vascongadas (Spanish— Basque Provinces)—Álava, Guipúzcoa, and Vizcaya

provn provision

Provo city in Utah; Providenciales island and town in the Turks and Caicos Islands; Provisional (member of the IRA)

provos provokers (Dutch— street people engaged in militant tactics to provoke the police)

Provos Provisionals (Provisional Sinn Fein party members of Northern Ireland)

PROVOST Priority Research and Development Objectives for Vietnam Operations Support

proword procedure word

prox proximal; proximity

prox. proximo (Latin—next, adv.)

proxi protection by reflection optics of xerographic images

prox. luc. proxima luce (Latin—the day before)

prp peak radiated power; pickup (zone) release point; present participle; pseudo random pulse; pulse recurrence period; pulse repetition period

prp (PRP) platelet-rich plasma; polyribophosphate

Prp Principality

PRp Puerto Rican pimp

PRP People's Revolutionary Party (Tanzania); Production Requirements Plan; Production Reserve Policy; Public Relations Personnel

PRPA Puerto Rico Ports Authority

PRPC Public Relations Policy Committee (NATO)

PRPG Political Resident Persian Gulf (British)

PRPGA Puerto Rico Personnel and Guidance Association

prpln propulsion

prpp (PRPP) 5-phosphoribosyl 1-pyrophosphate

pr. pr. praeter propter (Latin— about, nearly)

PRp('s) Puerto Rican pimp(s)

PRPUC Philippine Republic Presidential Unit Citation

prr pulse repetition rate

PRR Pennsylvania Railroad

PRRI Puerto Rico Rum Institute

p&rr's patriotic and religious racketeers

PRRWO Puerto Rican Revolutionary Workers Organization (communist)

prs pairs; printers

Prs Preston

PRs Pakistani rupees; Problem Reports; Puerto Ricans

PRS Park and Ride Scheme; Pattern-Recognition System; Pennsylvania-Reading Seashore (railroad); Performing Rights Society; Precision Ranging System; Property Recovery Squad (of a police department); Protective Research Section (U.S. Secret Service); Protestant Reformation Society; Public Radio Stations; Public Rehabilitation Scheme; Pupil Rating Scale

PRSA Public Relations Society of America

prsd pressed

prsd met pressed metal

prsfdr pressfeeder

prsmn pressman

PRSO Puerto Rico Symphony Orchestra

PRSP Puerto Rican Socialist Party (communist)
PRSS Pennsylvania-Reading Seashore Lines
PRSSA Public Relations Student Society of America
PRST Puerto Rican Standard Time
Pr strain Prague (viral) strain
prsvn preservation
PRSY People's Republic of Southern Yemen
prt parachute radio transmitter; personnel research test; publication requirement table(s); pulse repetition time
prt (PRT) personal rapid transit; printer (flow chart); program reference table
p & rt physical and recreational training
Prt Port
PrT   Prinzregentheater (Munich)
PRT Personnel Research Test; Philadelphia Rapid Transit; Prison Reform Trust; Production Re-evaluation Testing
PRT Partido Revolucionario de los Trabajadores (Spanish—Revolutionary Party of the Workers)—Mexican socialists; *Prinzregententheater* (German—Prince Regent Theater)—Munich
prtd printed
prtg printing
prtlsp printer line spacing
prtot prototype real-time optical tracker
prtov printer overflow
PRTS Personal Rapid Transit System
prty priority
pru peripheral resistance unit; prude; prudence; prudent
Pru Prudence; Prudential Life Insurance Company
PRU Polish-Russian Union
Prue Prudence
pru pru(s) prurient prude(s)
Prus Prussia; Prussian
prv peak reverse voltage; pressure-reducing valve; pressure-reduction valve
prv   pour   rendre   visite (French—to return a call)
Prv Pravda (Russian—truth)— daily newspaper published in Moscow by Central Committee of the Communist Party
prw percent rated wattage
PRWAD Professional Rehabilitation Workers with the Adult Deaf
prx pressure regulator exhaust

PRY Pittsburgh Railways Corporation (stock exchange symbol)
PRZ People's Republic of Zanzibar
ps parlor snake; parts shipped; parts shipper; passenger service; passing scuttle; patient's serum; penal servitude; picosecond; pieces; piper syndrome; plastic surgery; point of switch; point of symmetry; power steering; proof shot; pseudo; pseudonym(s); pull switch; pulmonary stenosis
p-s pressure-sensitive
p's pennies
p/s paddle steamer; point of shipment; port or starboard
p & s paracentesis and suction; piss and shit; port and starboard
p.s. post scriptum (Latin— postscript)
Ps Psalms, The (book of the Bible); South Pole; South Celestial Pole; static pressure
Ps Posaunen (German—trombones); *Psalms*
PS Pacific Southwest Airlines; Paleontological Society; Palm Society; Palm Springs (California); Paymaster Sergeant; Pennsylvania State University; Pharmaceutical Society; Philippine Scouts; Photo(graphic) Service; picket ship(s); Pistol Sharpshooter; Pittsburg & Shawmut (railroad); Planetary Society; Plastic Surgery; Privy Seal; Public Safety; Public School; Puget Sound
P-S Pullman-Standard
P.S. paddle steamer; public school
P & S Physicians and Surgeons; Pittsburg & Shawmut (railroad)
P of S Port of Spain
PS Parti Socialiste (French— Socialist Party); *Pferdestärke* (German—horsepower); *Pubblica Sicurézza* (Italian— public security) police
P.S. post scriptum (Latin— written after)
PS 166 Public School 166 (for example)
ps3 plate strip of 3
ps5 plate strip of 5
psa parametric sound amplifier; passed staff college; pressure-sensitive adhesive;

psychoanalytic(al)
psa (PSA) public service announcement (radio or television)
PsA Pisces Austrinus (constellation)
PSA Pacific Science Association; Pacific Southwest Airlines; Packers and Stockyards Administration; Phobia Society of America; Photographic Society of America; Play Schools Association; Poetry Society of America; Port of Singapore Authority; Poultry Science Association; Pretrial Service Agency; Program Study Authorization; Property Services Agency; Public Service Administration; Public Service Announcement; Public Service Association
P & SA Program and Systems Analysis
PSA Proceedings of the Society of Antiquaries
psaa post-stimulatory auditory adaptation
PSAB Public Schools Appointments Bureau
p sac pericardial cavity
PSAC President's Science Advisory Committee; Public Service Alliance of Canada
PSACPOO President's Scientific Advisory Committee Panel On Oceanography
psad prediction-simulation-adaptation-decision (data processing)
PSAI Play Schools Association, Inc
PSAL Public School Athletic League
Psalt. Psalterium (Latin— Book of Psalms)
PSAMPP Philadelphia Society for Alleviating the Miseries of Public Prisons (founded by Benjamin Franklin, William Rush, and others)
ps an psychoanalysis; psychoanalyst; psychoanalytic(al)-(ly); psychoanalyze
p's and q's pints and quarts (in British pubs)
PSAODAP Presidential Special Action Office for Drug Abuse Prevention
PSAR Preliminary Safety Analysis Report
PSAT Palm Springs Aerial Tramway; Preliminary Scholastic Aptitude Test(ing)
psb please send a boat; public service band (radio)

PSB Paradox Salt Basin; Psychological Strategy Board; Public Service Board

P & SB Portland & South Bend (railroad)

PSBA Pennsylvania School Boards Association; Public Schools Bursars' Association

PSBLS Permanent Space-Based Logistics System

PSBO Public Savings Bond Office

PSBR Public-Sector Borrowing Requirement

psc passed staff college; per standard compass; port service charge; prestressed concrete

ps & c program scheduling and control

Psc Pisces (constellation)

P-S c Porter-Silber chromogen

PSC Pacific Sea Council; Pakistan Shipping Corporation; Peralta Shipping Corporation; Pittsburgh Steel Company; Point Shipping Company; Police Staff College; Porcelain-on-Steel Council; Potomac State College; Procurement Strategy Corporation; Product Safety Commission(er); Professional Services Corporation; Program Structure Code; Public Service Careers; Public Service Commission

PSC *Partido Social Cristiano* (Spanish—Social Christian Party)—Catholic actionists

pscb padded sample collection bag

PSCC Public Service Commission of Canada

PSCD Patrol Service Central Depot

PSCFB Pacific Southcoast Freight Bureau

pscg power supply and control gear

PSCNI Public Service Company of Northern Illinois

PSCO Personnel Survey Control Office(r)

PSCP Public Service Careers Program

PSCPT Preschool Self-Concept Picture Test(ing)

PSCS Pacific Scatter Communications System

Psc's calculator Pascal's calculator (first adding machine)

pscu power-supply control unit

psd power spectral density; pre-shipment document; prevention of significant deterio-

ration; promotion service date

P Sd Port Said

PSD Pittsburgh Steamship Division (United States Steel); Port of San Diego; Prevention of Significant Deterioration (of air quality); Public Safety Division (Texas)

PSD *Partido Social Democrático* (Portuguese and Spanish—Social Democratic Party)

ps detn particle size determination

PSDI *Partito Socialista Democratico Italiano* (Italian—Italian Social Democratic Party)

ps distn particle size distribution

psdo pseudo; pseudonym

psdp phrase structure and dependency parser

PSDS Primary Solar Duct(ing) System

psdu power-switching distribution unit

PSDUPD Port of San Diego Unified Port District

pse please; point of subjective equality

pse (PSE) psychological stress evaluator (voice-analysis lie detector)

PSE Public Service Employment

PSEA Pennsylvania State Education Association; Physical Security Equipment Agency

psec picosecond

PSE & G Public Service Electric and Gas Company

PSE & GC Public Service Electric and Gas

psen pupils with special educational needs

Pseo *Paseo* (Spanish—boulevard)

pser production support and equipment replacement

pset permanent service on earth tides

pseud pseudandry (women using male names as pseudonyms); pseudepigraphy (attributing false names to artists, authors, or composers); pseudograph (falsely attributing a work to an artist, author, or composer); pseudo-jyn (men using female names as pseudonyms); pseudonym (false name, nom de plume, pen name); pseudonyma (pseudonymous works)

pseudo (Latin prefix—false)—pseudonym

psf payload-structure-fuel (ratio); point-spread function; pounds per square foot

PSF Phelps-Stokes Fund; Presidio of San Francisco

P & SF Panhandle and Santa Fe (railroad)

P of SF Port of San Francisco

PSFC Pacific Salmon Fisheries Commission

PSFL Puget Sound Freight Lines

PSFS Philadelphia Savings Fund Society

psg production system generator; psychogalvanometer; psychogalvanometric(al)(ly)

PSGBI Pathological Society of Great Britain and Ireland

psgi permanent service on geomagnetic indices

psgr passenger

psgr lng passenger lounge

psgr(s) passenger(s)

P-Shaw George Bernard Shaw (also GBS)

PSHFA Public Servants Housing and Finance Association

pshr pusher

psi posterior saggital index; pounds per square inch; public school(s) investigation

PSI Pacific Semiconductors Incorporated; Personalized System of Instruction; Personnel Security Investigation; Pharmaceutical Society of Ireland; Physician's Services Incorporated; Pollutants Standards Index; Population Services Incorporated; Population Services International; Private Sector Initiatives; Public Services International

PSI *Partito Socialista Italiano* (Italian Socialist Party); *Pollution Standards Index*

psia pounds per square inch absolute

PSIC Pacific Scientific Information Center (Bernice Pauahi Bishop Museum, Honolulu)

psid pounds per square inch differential

PSIDC Punjab State Industrial Development Corporation

psig pounds per square inch gage

psil preferred-frequency speech interference level

PSIP Poultry Stock Improvement Plan; Private Sector Initiative Program

PSIUP *Partito Socialista Italiano di Unita Proletaria* (Italian—Italian Socialist Party of Proletarian Unity)

psk phase shift keying

p sl pipe sleeve

PSL Pacific Star Line; Peruvian State Line; Philharmonic Society of London; Pretoria State Library

PSL *Patterson Strategy Letter*

p-slips old-fashioned postcard-size (3- X 5-inch) slips of paper used for filing

psl sol potassium, sodium chloride, sodium lactate solution

ps lt port side light

PSLT Picture Story Language Test

psm passed school of music

psm (PSM) presystolic murmur

PSM People for Self Management; Product Sales Manager

psma progressive spinal muscular atrophy

PSMA Power Saw Manufacturers Association; Pressure-Sensitive Manufacturers Association

PSMFC Pacific States Marine Fisheries Commission

psmr parts specification management for reliability

psmsl permanent service for mean sea level

psn position; pulse-shaping network(ing)

PSn Port Sudan

PSN *Partido Socialista de Nicaragua* (Spanish—Socialist Party of Nicaragua)

PSNA Phytochemical Society of North America

PSNC Pacific Steam Navigation Company

PSNH Public Service New Hampshire

PSNS Puget Sound Naval Shipyard

Pˢᵒ Passo (Italian—pass)

PSO Pad Safety Officer; Pasadena Symphony Orchestra; Phoenix Symphony Orchestra; Pilot Systems Operator; Pittsburgh Symphony Orchestra; Portland Symphony Orchestra; Prague Symphony Orchestra

PSOE *Partido Socialista Obrero Español* (Spanish Socialist Workers' Party)

p sol partially soluble; partly soluble

pson person

psp paralytic seafood poisoning; phenolsulfonaphthalein (test); pierced-steel plank; positive screened print

psp (PSP) progressive supernuclear palsy

PSP Pacific Security Pact; Pocahontas State Park (Virginia); Price-Subsidy Program; Price-Support Program; Programs Support Plan; Public School Pronunciation (British)

PSP *Pacifistisch Socialistische Partij* (Dutch—Pacifist-Socialist Party); *Policia de Segurança Pública* (Portuguese—police force)

PSPA Professional Sports Photographers Association

PSPCD Puget Sound Pollution-Control District

PSP & L Puget Sound Power and Light (company)

PSPMW Pulp, Sulphite and Paper Mill Workers

PSPP Proposed System Package Plan

ps & ps pimps and prostitutes

PSPS Paddle Steamer Preservation Society; Primary Solar Piping System

PSQC Philippine Society for Quality Control

psql process-screening quality level

p's & q's minding your p's & q's originated when printers instructed apprentices about similarity of lowercase p's and q's when handsetting type; also used in saloons to keep count of the number of pints and quarts of beer consumed

psr pain-sensitivity range; passenger space ratio; plow-steel rope

PSR Pacific School of Religion; Physicians for Social Responsibility

PSRC Public Service Research Council

PSRF Profit Sharing Research Foundation

PSRI Public Systems Research Institute (UCLA)

PSRM Pacific Southwest Railway Museum

PSRMA Pacific Southwest Railway Museum Association

psro passenger standing route order

PSRO Professional Services (Standards) Review Organization; Professional Standards Review Organization

pss packet-switching service; physiological saline solution; progressive systemic sclerosis

Pss Princess

PSS Pad Safety Supervisor; Parents for Student Safety; Personal Security System; Personal Signalling System; Pre-School Screening (program); Printing and Stationery Service; Professional Services Section; Public Service System

P.S.S. Professor of Sacred Scripture

P.S.S. *postcripta* (Latin—postscripts)

pssbb public school system blanket bond(ing)

PSSC Personal Social Services Council; Physical Science Study Committee (NSF); Pious Society of Saint Charles; Public Service Satellite Consortium

PS & SC Public Service and Safety Committee

P.S.S.C. Pious Society of Saint Charles

PSSNY Philharmonic Symphony Society of New York

psso passed slip stitch over (knitting)

PSSS Philosophic Society for the Study of Sport

PSST Prairie State Standard Time; Public Sector Standardization Team

pst polished surface technique

pst (PST) prefrontal sonic treatment

PST Pacific Standard Time

PSTA Public Safety and Training Association

PSTB Picture Story Test Blank

PSTBC Puget Sound Tug and Boat Company

PSTC Pressure Sensitive Tape Council

PSTD Prison Service Training Depot (Pretoria)

£ sterling pound sterling

p stg c per steering compass

psth peristimulus time histogram

PSTIAC Pavements and Soil Trafficability Information Analysis Center (USA)

pstl postal

PSTMA Paper Stationery and Tablet Manufacturers Association

PSTO Principal Sea Transport

Officer
P-strip P-shaped strip
P-stuff pcp (PCP)
pstz pasteurize
pstzd pasteurized
pstzg pasteurizing
psu package size unspecified; power supply unit; primary sampling unit
PSU Pennsylvania State University; Portland State University; Public Security Unit (Ugandan secret police)
PSU Parti Socialiste Unifé (French—Unified Socialist Party); *Partito Socialista Unitario* (Italian—Unitary Socialist Party)
p-substance protein substance
PSUC Pennsylvania State University Center(s)
PSUC Partido Socialista Unificado de Cataluña (Spanish—Unified Socialist Party of Catalonia)
P Sud Port Sudan
PSU-MRL Pennsylvania State University—Materials Research Laboratory
PSUP Pennsylvania State University Press
p surg plastic surgeon; plastic surgery
psv polished-stone value; public service vehicle
PSV Petit St Vincent (Grenadines in the West Indies); Project Salt Vault
PSW Psychiatric Social Worker
PSWB Plateau State Water Board
pswbd power switchboard
PSWC Pacific Tsunami Warning Center (Honolulu)
P Swet Port Swettenham (now Port Kelang or Port Klang)
PSWFA Prestige Saltwater Fly Anglers
PSWO Picture and Sound World Organization
psy psychological
Psy Paisley
psych psychiatry; psychology; psychopathology
psych/d psychological death
psychedeli psychedelicatessen (store selling the paraphernalia of drug addicts)
psychiat psychiatric; psychiatry
psycho dangerous lunatic; a psychiatric hospital or ward; a psychoneurotic personality; a psychotic individual (pseudo-scientific slang)

psycho (Latin prefix—mental)—psychologist
psychoan psychoanalytic; psychoanalysis; psychoanalyst
psychobab psychobabble(r)—psychological patter(er)
psychobio psychobiological; psychobiologist; psychobiology
psychobiog psybiographer; psychobiographic(al)(ly); psychobiography
psychochron psychochronic-(al)(ly); psychochronicle(r); psychochronologer; psychochronologic(al)(ly); psychochronology
psychochronicle psychiatric chronicle; psychological chronicle
psychodelics hallucinogenic drugs
psychodels psychodelics (hallucinogens)
psychogeog psychogeographer; psychogeographic(al); psychogeography
psychohist psychohistorian; psychohistorical; psychohistory
psychol psychological; psychologist; psychology
Psychol Psychology
psychomet psychometric
psychopathol psychopathological; psychopathologist; psychopathology
psychophys psychophysical; psychophysics; psychophysicist
psychophysiol psychophysiology (and derivatives)
psychoprison psychiatric hospital prison (USSR)
psychosurg psychosurgeon; psychosurgery; psychosurgical(ly)
psychot psychotic
psychother psychotherapist; psychotherapeutic(al,s); psychotherapy
psycho ward psychopathic ward
Psych Qtly Psychoanalytic Quarterly
psych test. psychological testing
psydoc psychiatrist doctor
psyk psykologi or *psykologist* (Dano-Norwegian—psychology or psychologist)
psyop psychological operation
psypath psychopath(ic)
psysom psychosomatic
psywar psychological warfare
psz (PSZ) partly stabilized zir-

conia
pt part; part time; personal trade; physical therapy; physical training; pint(s); plenty tough; plenty trouble; pneumatic tube; point; point of tangency; point of turn; point of turning; primary target; private terms; prothrombin time
p & t personnel and training; posts and timbers
pt partie (French—part)
pt. perstetur (Latin—let it be continued)
p.t. protempore (Latin—temporarily)
£T pound Turkish
Pt part; platinum; Point; Port; Porto; Puerto
P^t Petit (French—little, small); *Pont* (French—bridge)
PT motor torpedo boat (naval symbol); Pacific Time; Peninsula Terminal (railroad); Philadelphia Transportation; Physical Therapist; physical therapy; physical training; Postal Telegraph; primary trainer; Provincetown-Boston Airline (2-letter coding)
P & T Pope & Talbot (steamship line)
P & T The Phoenix and the Turtle
PT-76 Soviet Amphibious tank
pta plasma thromboplastin antecedent; posttraumatic amnesia; primary target area; prior to admission; proposed technical approach; peseta (Spanish monetary unit, diminutive of peso)
pta peseta (Spanish—monetary unit)
Pta Pretoria
Pta Punta (Spanish—Point)
P^ta Ponta (Portuguese—point); *Puerta* (Spanish—gate, gateway, mountain pass); *Punta* (Spanish—point)
Pt A Port Arthur, Ontario
PTA Paper and Twine Association; Parent-Teacher Association; Pet Traders Association; Philippine Travel Authority; Pope and Talbot; Postal Transportation Association; Prevention of Terrorism Act; Protestant Teachers Association
PTA Prevention of Terrorism Act (British)—provides for seven days detention of suspects
ptacv (PTACV) prototype air-

cushioned vehicle
P Tal Port Talbot
Pt Alb Port Alberni
Pt Ant Port Antonio
PTAR Prime Time Access Rule
Pt Art Port Arthur
ptas pesetas
pta's part-time alcoholics
PTAs Passenger Transport Authorities
PTAS Productivity and Technical Assistance Secretariat
P Tau Port Taufiq (formerly Port Tewfik)
ptb patellar-tendon bearing
PTB *Partido Trabalhista Brasileiro* (Portuguese—Brazilian Workers Party); *Physikalisch-Technische Bundesanstalt* (German—Physical Technical Institute)
ptbl portable
PTBM PT Barnum Museum (Bridgeport, Connecticut)
PT-boat patrol torpedo boat
ptbr punched-tape block reader
PTBT Partial Test Ban Treaty
ptc personnel transfer capsule; positive temperature coefficient
ptc (PTC) phenylthiocarbamide; plasma thromboplastin component (clotting factor IX)
PTC Pacific Theological College; Pacific Tin Consolidated; Paisley Technical College; patrol vessel (naval symbol); Peoria Terminal (railroad); Philadelphia Transportation Company; Pine Tree Camp; Pipe and Tobacco Council; Power Transmission Council; Press Trust of Ceylon; Private Truck Council
PTCA Private Truck Council of America
ptcldy partly cloudy
ptd painted
PTDA Power Transmission Distributors Association; Professional Tournament Directors Association
ptdl programmable tapped-delay line
PTDP Preliminary Technical Development Plan
PTDR Post-Test Disassembly Report
PTDS Photo Target Detection System
pte parathyroid extract; *poriente* (Spanish—west)

pte (PTE) pulmonary thromboembolism
p^**te** *parte* (Spanish—part)
Pte *Pointe* (French—Point); *Presidente* (Portuguese or Spanish—President)
PTE Passenger Transport Executive
pt ed patient education
pt ex part exchange
ptf plasma thromboplastin factor
PTF fast patrol boat (naval symbol); Propulsion Test Facilities
ptfe polytetrafluorethylene
ptfp prime-time family programming
ptg printing
Ptg Portugal; Portuguese
PTG Piano Technician's Guild; Polaris Task Group
ptgt primary target
pth parathormone
Pth Perth
PTH hydrofoil motor torpedo boat (naval symbol); parathyroid hormone
pti persistent tolerant infection; physical training instructor (PTI)
PTI Philips Telecommunicatie Industrie; Pictorial Test of Intelligence; Press Trust of India; Protect the Innocent (anti-crime lobby)
PTIDG Presentation of Technical Information Discussion Group
PTIS Piano Teachers Information Service
PTJ *(Cuerpo) Técnico de Policía Judicial* (Spanish—Technical Corps of the Judicial Police)—Venezuela
Pt K Port Klang (also written Kelang and formerly Port Swettenham)
ptl partial total loss; pass the loot; pintle; primary target line
Pt L Point Loma
P t L Praise the Lord
PTL People That Live; Photographic Technology Laboratory; Praise the Lord; People That Love (the Lord)
PTLA *Publishers' Trade List Annual*
PTLL Pittsburgh Toy Lending Library
ptm proof test model; pulse-time modulation
Ptm Pietermaai
Ptm (PTM) Polaris tactical missile

ptma phosphotungstomolybdic acid
PTMTCS Power-Tape-to-Magnetic-Tape Conversion System
ptn partition
PTNA Professional Travel Nurses Association
Ptnr Partner
pto please turn over; power take-off
Pto Porto; Puerto; Punto
P^**to** *Ponto* (Italian—sea)—poetic term; *Porto* (Italian, Portuguese, Spanish—port); *Puerto* (Spanish—port); *Punto* (Italian—point)
PTO Patent and Trademark Office; Public Trustee Office(r); Purdue Teacher Opinionaire
Pto Blvr Puerto Bolívar
Pto Cab Puerto Cabello
Pto Cast Puerto Castilla, Honduras
ptol peacetime operating level
Ptol Ptolemaic; Ptolemy
Ptolemy Alexandrian astronomer Claudius Ptolemaeus
Pto Rico Puerto Rico
P-town Provincetown
P Town Port Townsend
ptp paper-tape printer; part-time pimp
p-t-p point-to-point
PTP Pointe à Pitre, Guadeloupe (airport); Productive Thinking Program
ptpg participating
pt/pt point-to-point
ptr printer; pupil-teacher ratio
ptr (PTR) photoelectric tape reader
PTR pool test reactor
ptrf peacetime rate factor(s)
ptry pantry; poetry; pottery
pts *pesetas* (Spanish—plural of peseta); pints; *puntos* (Spanish—degrees; periods)
Pts Portsmouth
PTS Philatelic Traders Society; Postal Transportation Service; Potential Tax Savings; Princeton Theological Seminary; Public Television Station(s)
PT & S Pacific Towboat and Salvage (tugs)
ptsd (PTSD) post-traumatic stress disorder
PTSD Post-Traumatic Stress Disorder
pts/hr parts per hour; pieces per hour
Ptsmth Portsmouth
Pt Sp Port of Spain

PTSS Princeton Time-Sharing System
PTSTV Prime Time School Television
ptt push to talk
ptt (PTT) partial thromboplastin time
PTT Posta, Telgraf ve Telefon (Turkish—Post, Telegraph, and Telephone); *Postes, Télégraphes, Téléphone* (French—national postal, telegraph, and telephone system)
PTTA Philippine Tourist and Travel Association; Postal Telegraph and Telephone Authority
ptti precise time and time interval
PTTI Postal, Telegraph, and Telephone International
pt-tm part-time
pttnmkr patternmaker
ptu propylthiouracil
PTU Plumbers' Trade Union; Plumbing Trade Union; Psychiatric Treatment Unit
PTUC Philippine Trade Unions Council
ptv passenger transfer vehicle; public television
ptv (PTV) propulsion test vehicle
ptv's personal transportation vehicles (three-wheeled vehicles for city driving)
PTVs Public Television Stations
ptw per thousand words
Pt W Port Weller
PTWC Pacific Tsunami Warning Center
Pty Party; Proprietary
P-type Jungian perceptive type
pu passed urine; peptic ulcer; pickup; plant unit; pregnancy urine; propellant utilization; propulsion unit; pump(ing) unit; pump unit
p-u phew (what a stench)
p.u. *plus ultra* (Latin—beyond the pinnacle, beyond the ultimate)
Pu plutonium
PU Pacific University; Phillips University; Princeton University; Prisoner's Union; Purdue University
P U Peter Ustinov (1921—)
PUA Punta de la Unidad Africana (Spanish—Point of African Unity)—formerly Fernanda Point
PUAS Postal Union of the Americas and Spain

pub public; publican; publication; public house; publicity; publish; published; publisher; publishing
púb públicas (Spanish—publications)
Pub Publican; Public House; Publisher's Announcement
PUB Public Utilities Board
pub. aff public affairs
pub aide publication aide
pubbl pubblicità (Italian—advertising, publicity)
Pub Cit public citizen
Pub Doc Public Document
pub ed publication editor
pubinfo public information
publ publication; publicity; publisher; publishing
Publ Astron Soc Pac Publications of the Astronomical Society of the Pacific
pub(s) public house(s) (British short form)
Pub Sect Lab Rel Public Sector Labor Relations Conference Board
pubtronics publication electronics that convert electronically-produced manuscripts into type
Pub W Publishers Weekly
Pub Wks Public Works
puc papers under consideration; pickup car
PUC Peoples University of China; Presidential Unit Citation; Public Utilities Code; Public Utilities Commission; Public Utilities and Corporations
PUC Post Urbem Conditam (Latin—after the foundation of the city)—city usually means Rome
PUCC Port Users Consultative Committee
pucf polyurethane-coated fabric; polyurethane-coated fibers
pud puddle; pudding
pud (PUD) planned unit development
pu & d pickup and delivery
PUD Planned Unit Development
Pue Puebla
Puebla Puebla de Zaragoza (in central Mexico)
Puerto Rico islands of Puerto Rico, Culebra, Vieques
PUF Presses Universitaires de France (University Presses of France)
pufa polyunsaturated fatty acid
PUFF People United to Fight

Frustrations
PUFFT Purdue University Fast Fortran Translator
pug. print under glaze; puggy; pugilism; pugilist
PUHS Phoenix Union High School
PUK Pechiney Ugine Kuhlmann
puka pukalolo (Hawaiian Polynesian—crazy tobacco)—marijuana also known locally as Kauai electric, Kona gold, Maui wowie, Puna butter
pul pulley
PUL Princeton University Library (New Jersey); Punjab University Library (Lahore, Pakistan)
pula monetary unit of Botswana
pulchris pulchritudinous
'Pulco Acapulco
pulg pulgadas (Spanish—inches)
pulheems physical capacity, upper and lower limbs, hearing, eyesight, emotional capacity, mental stability
pul ins pulmonary insufficiency
pulm pulmonary
pulm. *pulmentum* (Latin—gruel)
pulm a pulmonary artery
pulm emb pulmonary embolism
pulmo pulmoaortic(al)(ly); pulmology; pulmometer; pulmometric(al)(ly); pulmometry; pulmonary; pulmonectomy; pulmonic(al)(ly); pulmonitis; pulmonologist(ic)(al)(ly); pulmotor
pulmotor pulmonary motor
pulsar pulse star (pulsed radiowave-emitting star); pulsing astronomical signal (received from outer space)
pul sten pulmonary stenosis
pulv pulverize(r)
pulv. pulvis (Latin—powder)
pulv. gros. pulvis grossus (Latin—coarse powder)
pulv. subtil. pulvis subtilis (Latin—smooth powder)
pulv. tenu pulvis tenuis (Latin—very fine)
pum pop-up mechanism
PUM Postal Union Mail
puma catamount, cougar, mountain lion, panther
puma. (PUMA) programmable universal mechanical assembly (robot)

PUMA Prostitutes Union of Massachusetts
Pumfret Pontefract
pump. pumping
PUMP Protesting Unfair Marketing Practices
pums permanently unfit for military service
pun. puncheon
puN plasma-urea Nitrogen
PUN *Partido Union Nacional* (Spanish—National Union Party)
punc punctuation
pundonor *punta de honor* (Spanish—point of honor)
Punj Punjabi
puo pneumonia; pyrexia of unknown origin
pup puppy
pup. (PUP) peripheral unit processor
Pup Puppis (constellation)
PUP People's United Party (Belize); Polytechnic University of the Philippines; Princeton University Press
Pupp Puppis (constellation)
puppie(s) pregnant urban professional(s)
pups puppies
pup(s) pregnant urban professional(s)
p'up(s) pickup(s)
pur purchase; purchaser; purchasing; purifier; purification; purify; purple; purplish; pursuant; pursuit
Pur Purim
purch purchasing
Purdue Purdue University Press
pure mat. pure machine-aided translation
pure mt pure machine translation
pureq purchase requisition
PUREX Plutonium Uranium Extraction Plant
purg. *purgativus* (Latin—purgative)
purp purple
purv powered underwater research vehicle
pus. permanently unfit for service
Pus Pusan
PUs Public Utilities
PUS Parliamentary Under-Secretary; Permanent Under-Secretary
PUS *Pharmacopoeia of the United States*
Push Pushtu
PUSH People United to Save Humanity; People United to

Serve Humanity
puss pussy; pussycat
puta(s) *prostituta(s)* [Portuguese or Spanish—prostitute(s)]
Putnam GP Putnam; Putnam
putty linseed oil and powdered chalk mixture
puva psoralen (drug) + ultraviolet-A (light)
PUVAS Plutonium Value Analysis System
puvep propellant-utilization vehicle-borne electronic package
Puy-de-D Puy-de-Dôme
pv par value; paravane; pave(d); paving; plasma value; position value; prime vertical; public voucher
p/v peak-to-valley; per vagina; pressure vacuum; pressure valve; profit volume (ratio)
p & v pressure and velocity
pv *por vida* (Spanish—for life); *prossimo venturo* (Italian—next month)
p v *petite vitesse* (French—slow train); *piccola velocity* (Italian—slow train)
p.v. *per vaginam* (Latin—by vagina)
Pv Peru; Peruvian
PV Eastern Provincial Airways (2-letter coding); patrol vessel; Post Village; post-Virgil; Priest Vicar; Puerto Vallarta
P. V. *Procès verbaux* (French—official report); *Processi verbali* (Italian—official report)
PV-2 Lockheed maritime reconnaissance bomber
pva polyvinyl acetate
PVA Paralyzed Veterans of America; Prison Visitor's Association
p.vag *per vaginam* (Latin—by the vagina)
pval polyvinyl alcohol
P-value probability value
pvb potentiometer voltmeter bridge
PVB Prison Visitors' Board
pvc polyvinyl chloride (thermoplastic)
pvc (PVC) premature ventricular contractions
PVC Philippine Volconology Commission; Precision Valve Corporation
PVCC Piedmont Virginia Community College
pvccf polyvinyl-chloride-coated fabric; polyvinyl-chlo-

ride-coated fibers
pvd peripheral vascular disease; pulmonary vascular disease
PVD Providence, Rhode Island (airport)
PvdA *Partij van de Arbeid* (Dutch—Labor Party)
pvdc polyvinyl dichloride
pvem pulse-vector emittance meter
pvf polyvinyl fluoride
pvH propane-vacuum hydrogen
pvi point of vertical instersection
PVI Personal Values Inventory
pvis pneumatic vertical-indicating scale
pvm polyvinyl methyl
PVM Process Evaluation Module
PVMNM Perry's Victory Memorial National Monument
Pvmnt Pavement
pVns post-Vietnam syndrome
pvnt prevent; preventive
PVO Principal Veterinary Officer
pvp photovoltaic power; polyvinylpyrrolidone (plasma extender)
pvp *precio máximo de venta al publico* (Spanish—maximum price charged the public)
PVP President's Veterans Program
PVPMPC Perpetual Vice President and Member of the Pickwick Club
pvpp polyvinyl-polypyrrolidone
pvq personal-value questionnaire
pvr portable volume-controlled respirator; precision voltage reference
PVR Police Volunteer Reserves; Premature Voluntary Requirement
PVRC Pressure Vessel Research Committee (NBS)
pvs persistent vegetative state
PVS Pecos Valley Southern (railroad); Periventricular (fiber) System; Personal Value System
pvt page view terminal; pressure volume temperature; private
pvt *par voie télégraphique* (French—by telegraph)
Pvt Private
Pvt 1/C Private First Class
PVU Prairie View University
pvw pure virgin wool

pw packed weight; passing window; pivoted window; postwar; prisoner of war; private wire; projected window; psychological warfare; public works; pulse width
p/w parallel with
p & w pension and welfare (retirement benefits)
PW Pacific Western Airlines; Palau; Philadelphia & Western (railroad); Pittsburgh & West Virginia (railroad); prisoner of war; Public Works
P-W Prader-Willi (syndrome)
P & W Pratt and Whitney Aircraft Division, United Aircraft Corporation
PW Petroleum Week; Publishers' Weekly
PWA Pacific Western Airlines; Prison Wardens Association; Professional Writers of America; Psychic Workers Association; Public Works Administration
PWA Papierwerke Waldhof-Ashaffenburg (German— Waldhof-Ashaffenburg Paper Works)
pwafrr present worth of all future revenue requirements
PW AIDS person(s) with AIDS
P Wash Port Washington
p-wave pressure wave
p waves primary (earthquake) waves
pwc physical working capacity
pwc (PWC) pulse-width coded; pulse-width coding
PWC Parents Who Care; Prisoner of War Convention; Public Works Canada; Public Works Center (USN)
pwd plywood; powered
pwd (PWD) pulse-width discriminating; pulse-width discriminator
PWD Psychological Warfare Division; Public Works Department
pwdrd powdered
PWDS Protected Wireline Distribution System
pwe (PWE) pulse-width encoder; pulse-width encoding
PWE Political Warfare Executive; Prisoner of War Enclosure
P Wel Port Weller
pwf pregnancy without fear (pillow-simulated pregnancy); present-worth factor
PWFP Prince William Forest

Park (Virginia)
PWG Permanent Working Group (NATO); Province Working Group
PWHS Public Works Historical Society
PWI Physiological Workload Index
P & W I Poets and Writers Incorporated
PWIF Plantation Workers' International Federation
PWJC Piney Woods Junior College
pwl power watt(age) level
PWLB Public Works Loan Board
pwm pokeweed mitogen (PWM); pulse width modulation
pwm (PWM) pulse-width modulating; pulse-width modulator
PWM Partnership for World Mission
PWMS Public Works Management System (USN)
pwmsp people with multiple social problems; person(s) with multiple social problems
PWNDA Provincial Wholesale Newspaper Distributors' Association
PWNP Parra Wirra National Park (South Australia)
PWO Principal Weapons Officer; Public Welfare Office(r); Public Works Office(r)
pwp picowatt power
PWP Parents Without Partners
pwr power; pressurized water reactor (PWR)
PWR Police War Reserve
PWRS Pacific War Research Society
pwr sup power supply
pws paddlewheel steamer
pw's prisoners of war
PWs Professional Warriors
PWS Periyar Wildlife Sanctuary (India); Private Wire System
pwt pennyweight; propulsion wind tunnel
PWT Picture World Test
pwtn power train
pwtr pewter
pwv pulse-width valve
P & WV Pittsburgh & West Virginia (railroad)
Pwy Poway
px past history; physical examination; please exchange; pneumothorax; press; prognosis
PX Aspen Airways (2-letter

code); Post Exchange
PXCMD Phoenix Contract Management District
pxe (PXE) pseudoxanthoma elasticum
px in time of arrival
pxl (PXL) patrol experimental land-based aircraft
pxlst passenger list(ing)
px me report my arrival and departure
pxo próximo (Spanish—next)
px out takeoff time
PX-S Japanese reconnaissance flying boat
pxt. pinxit (Latin—he painted it)
py pitch and yaw
p/y pitch or yaw
PY commissioned and armed yacht (2-letter naval symbol); program year; Surinam Airways (2-letter symbol); yacht (naval symbol)
Pya Pyatnitsa (Russian—Friday)
PYA plan, year, age (insurance)
pyc proteose-yeast castione
PYC Philadelphia Yacht Club; Portland Yacht Club (Maine); Poughkeepsie Yacht Club
PYE Protect Your Environment
Pyg Pygmalion (George Bernard Shaw play)
pyg broth proteose-yeast-glucose broth
Pyi Soc Tha Mya Nai Pyidaungso Socialist Thammada Myanma Naingngandaw (Burmese—Socialist Republic of the Union of Burma)
pyo pick your own (flowers, fruit, vegetables)
pyo (Latin prefix—pus)—pyorrhea
pyof pick your own fruit
pyph polyphase
pyr pyridine
p-y-r pitch-yaw-roll
pyramid pyramid investment scheme
Pyrenees Pyrenees Mountains between France and Spain
pyrite fool's gold; iron disulfide; iron pyrites
pyrites copper, iron, tin pyrite; also known as fool's gold
pyrmd pyramid(ed)
pyro pyromaniac; pyrotechnic(s); pyroxylin
pyroglu pyroglutamic acid
pyrolag pyrolagnia(c)
pyrom pyrometer; pyrometry

Pyr-Or Pyreneés-Orientales

pyrot pyrotechnics

pyt pretty young thing

pyx. pyxis (Latin—box, vessel)

Pyx Pyxis (constellation)

pz pancreozymin

PZ Paolei Zion(ist); Pickup Zone; Police Zone

PZ-61 Swiss medium tank armed with a 105mm gun

pza pyrazinamide

pza pieza (Spanish—piece)

Pza Plaza (Italian or Spanish—Plaza)

P^{za} Piazza (Italian—Square)

pzc point of zero charge

PZC *Partido Zapatista Comunista* (Spanish—Zapatist Communist Party) group active along the Mexican Border

pz-cck pancreozymin-cholecystokinin

pzi protamine zinc insulin

pzm pressure-zone microphone

PZM *Polska Zegluga Morska* (Polish—Polish Merchant Marine)

P^{zo} Pizzo (Italian—peak, summit)

PZPR *Polska Zjednoczona Partia Robotnicza* (Polish—Polish United Workers Party)

PZS President of the Zoological Society

pzt photographic zenith tube

P^{zza} Piazza (Italian—Square)

Q

q coefficient of association (statistical symbol); cue; dynamic pressure (symbol); electric charge (symbol); quality factor; quart; quarter; quarterly; quartile; quarto; queer; quench; quenching; queries; query; question(s); quick; quintal; quire; semi-interquartile range (symbol); stagnation pressure (symbol)

q quaque (Latin—each, every)

Q bankruptcy or receivership (stock exchange symbol); electric quadruple moment of atomic nucleus (symbol); Fairchild (symbol); Polaris correction (symbol); prison at San Quentin, California; quadrillion; Quaker Line; quality factor; quantity; quarantine; Quartermaster; quartile variation (symbol); Québec—code for letter Q; Queen; Queensland; question(s); *quetzal* (Guatemalan monetary unit); quotient; radio inductive reactance to resistance (symbol); semi-interquartile range (symbol); target or drone (symbol); thermoelectric power (symbol)

Q *(Q)* quantity (microeconomics)

Q (Latin—Quintus); pseudonym for Sir Arthur Quiller-Couch; *Quai* (French—embankment or quay); *quetzal* (Guatemalan monetary unit); torque (symbol)

Q. Quintus (Latin—fifth time)

Q1 quintal (Spanish—hundredweight)

Q₁, Q₂, Q₃, Q₄ first quartile, second quartile, third quar-

tile, fourth quartile

q²h quaque secunda hora (Latin—every two hours)

q³h quaque tertia hora (Latin—every three hours)

q⁴h quaque quarta hora (Latin—every four hours)

qa quality assurance; quick-acting; quick assembly; quiescent aerial

q & a question and answer

QA Qualification Approval; Qualified Acceptance; Quality Assurance; Quarters Allowance

Q-A Quint-A

Q & A question and answer

QAA Quality Assurance Assistant

QAB Quality Assurance Board; Quality Assurance Bulletin; Queen Anne's Bounty (for indigent clergymen)

qac quaternary ammonium compound

QAC Quality Assurance Check(ing); Quality Assurance Code; Quality Assurance Coding; Queensland Arts Council

QACA Queensland Amateur Cyclists Association

QACAD Quality-Assurance Corrective-Action Document

QACC Queensland Automobile Chamber of Commerce

qad quick-attach-detach

QAD Quality Assurance Data; Quality Assurance Department; Quality Assurance Directive; Quality Assurance Division

QADC Queen's Aide-de-Camp

QADI Quality Assurance De-

partment Instruction

qadk quick attach-detach kit

QADS Quality Assurance Data Summary; Quality Assurance Data System

QAE Quality Assurance Engineer(ing)

qaf quality-assurance firing

QAFCO Quatar Fertilizer Company

QAFL Queensland Australian Football League

qafo quality-assurance field operation(s)

QAG Quaker Action Group

QAGA Queensland Amateur Gymnastic Association

qagc quiet automatic gain control

Qahira El Qahira (Egyptian Arabic—Cairo)

QAI Quality Assurance Instruction; Queen's Award to Industry

QAICG Quality Assurance Interface Coordination Group

QAIMNS Queen Alexandra's Imperial Military Nursing Service

QAIP Quality Assurance Inspection Procedure

qak quick-attach kit

qal quartz aircraft lamp; quaternary alluvium

qal quintal (French—hundredweight)

QAL Quality Assurance Laboratory; Quarterly Accession List; Quebec Airways Limited; Queensland Alumina Limited

QALAS Qualified Associate of the Land Agents' Society

QALD Quality-Assurance Liaison Division (DNA)

qall quartz aircraft landing lamp
QALTR Quality Assurance Laboratory Test Request
qam quadrature amplitude modulation; queued access method
QAM Quality Assurance Manager; Quality Assurance Manual; Quality Assurance Monitor
QAM Quality Assurance Manual
QAMIS Quality Assurance Monitoring Information System
QAMS Quad-Phase Amplitude Modulation System
QANTAS Queensland And Northern Territories Aerial Services
qao quality assurance operation
QAO Quality Assurance Office (USN)
QAOC Quality Assurance Overview Contractor
QAOP Quality Assurance Operating Procedure
qap quinine, atebrin, plasmoquine (malaria treatment)
QAP Quality Assurance Planning; Quality Assurance Procedure(s); Quality Assurance Program
QA & P Quanah, Acme & Pacific (railroad)
QAPL Queensland Airlines Proprietary Limited
QAPP Quality Assurance Program Plan(ning)
QAPS Queensland Association of Personnel Services
qar quick-access recording
QAR Quality Assurance Representative
QAR Quality Assurance Report
QARAFNS Queen Alexandra's Royal Air Force Nursing Service
QARANC Queen Alexandra's Royal Army Nursing Service
QARNNS Queen Alexandra's Royal Naval Nursing Service
qas quick-acting scuttle
QAs Queen Alexandra's
QAS Quality Answering System; Quality Assurance Service; Quality Assurance System; Question Answering System
QASA Queensland Amateur Swimming Association
QASAR Quality Assurance Systems Analysis Review
QASP Quality Assurance

Standard Practice
QAST Quality Assurance Service Test(s)
Qat Qatar
QAT Qualification Approval Test; Quantitative Assessment and Training Center
Qatar State of Qatar (oil-producing Persian Gulf country)
QATB Queensland Ambulance Transport Brigade
QATP Quality Assurance Technical Publication(s); Quality Assurance Test Procedure(s)
Qattara Qattara Depression in northern Egypt's Libyan Desert
qavc quiet automatic volume control
QAVT Qualification Acceptance Vibration Test
QAWA Queensland Amateur Wrestling Association
qax quacks
qb qualified bidders; quarterback; quick break
QB Queensboro Bridge (New York City); Quiet Birdmen (glider enthusiasts)
Q.B. Queen's Bench
QBA Quebecair; Queensland Bowling Association
QBAA Quality Brands Associates of America
QBAC Quality Bakers of America Cooperative
Q-band 36,000–46,000 mc
Q-bar second output of a flip-flop
QBB Queensland Butter Board
Qbc Quebec
QBD Queen's Bench Division; Queensland Book Depot
qbi quite bloody impossible
QBI Queen's Bureau of Investigation
QBL Qualified Bidder's List
Q-boats mystery ships used in antisubmarine warfare by the British in World War I
qbop quality basic-oxygen process
QBRs Queen's Bench Reports
qb's quarterbacks
QBSM que besa su mano (Spanish—who kisses your hand)—used in closing personal letters
QBSP que besa sus pies (Spanish—who kisses your feet)—used in closing personal letters
qc qualification course; quality control; quantitative command; quantum counter;

quartz crystal; quick connect; quit claim
q/c quick change
qc qualcosa (Italian—something)
Qc impact pressure (symbol)
QC Quadrantal Correction(s); Quality Control; Quartermaster Corps; Québec Central (railroad); Quebec City; Queens College; Queen's College; Quezon City; Quincy College; Quinnipiac College; Quit Claim
Q.C. Queen's Counsel
QCA Queen Charlotte Airlines; Queensland Coal Associates; Queensland Cricket Association; Queensland Croquet Association
Q-cab quiet (tractor) cab
Q-card qualification card
qcb (QCB) queue control block (data processing)
QCB Quality Control Bulletin
QCBC Queen's Commendation for Brave Conduct
qcbm quick-connects bulkhead mounting
qcc qualification correlation certification; quick-connect coupling(s)
QCC Queensborough Community College; Queensland Conservation Council; Quinsigamond Community College
QCCA Queensland Cleaning Contractors Association
QCCARS Quality Control Collection Analysis and Reporting System
qcd quality-control data; quantum chromodynamics; quitclaim deed
QCD Quality Control Directive; quit claim deed
QCDI Quality Control Departmental Instruction
QCDR Quality Control Deficiency Report
QCE Quality Control Engineering
QCEU Queensland Colliery Employees Union
qcf quartz-crystal filter
QCF Quality Control Form
qcfo quartz-crystal frequency oscillator
QCGC Queensland Cane-Growers Council
qch quick-connect handle
QCH Queen Charlotte's Hospital
qci quality-control information
QCI Quality Conformance In-

spection; Queensland Confederation of Industry; Quota Club International

Q Cic Quintus Tullius Cicero (the brother of the Roman orator Marcus Tullius Cicero)

QCIM Quarterly Cumulative Index Medicus

QC Isl Queen Charlotte Islands

Q City Quezon City, Philippines

qck quick-connect kit

qcl quality-control level

Q-class Soviet Québec-type submarines

Q-clearance Department of Energy's highest security classification; highest security clearance from the FBI

QCM Quality Control Manager; Queensland Coal Mining

QCM Quality Control Manual

QCMA Queensland Cooperative Milling Association

QCMP Queens' Council Member of Parliament

QCNIC Quad-Cities Nuclear Information Center

qco quartz-crystal oscillator

Q Co Queens County

QCO Quality Completion Order; Quality Control Officer

QCOP Quality Control Operating Procedure

Q country Qatar

QCP Quality Control Procedure; Queens College Press

QCPE Quantum Chemistry Program Exchange

Qc/Ps impact/static pressure ratio (symbol)

QCPSA Quaker Center for Prisoner Support Activities

qcr quick-change response

qcr (QCR) quality control/reliability

QCR Quality Control Representative

QC/R Quality Control/Reliability

QC & R Quality Control and Reliability

QCRC Québec Central Railway Company

QC Rep Quality-Control Representative

QC Rept Quality-Control Report

qcrt quick-change real time

QC Ry Québec Central Railway

QCS Quality Control Standard; Quality Control System;

Quality Cost System

QCSO Quality Control Stop Order

QCSR Quaker Committee on Social Rehabilitation

QCSSO Queensland Council of State School Organizations

QC Stand Quality-Control Standard

qct quiescent carrier telephony; questionable corrective task

QCT Quality Control Technology

QC & T Quality Control and Test

QCTR Quality Control Test Report

qcu quartz crystal unit; quick-change unit

qcus quartz crystal unit set

qcvc quick-connect valve coupler

qcw quadrant continuous wave

QCWA Quarter-Century Wireless Association; Queensland Country Women's Association

Qcy Quincy

QCYC Queen City Yacht Club; Queensland Cruising Yacht Club

qd quarterdeck; quartile deviation; questioned document; quick delivery; quick detachable (weapon)

q-d quick-disconnect

q & d quick and dirty

q.d. quater in die (Latin—four times a day)

QD *Sadios Transportes Aéreos*

qda quantity discount agreement

qdc quick dependable communication(s); quick-disconnect cap; quick-disconnect connector; quick-disconnect coupling

qdcc quick-disconnect circular connection

qdc's quick, dependable, communications

qdd qualified for deep diving; quantized decision detection

QDG Queen's Dragoon Guards

QD/GD Quincy Division/General Dynamics

qdh quick-disconnect handle

qdk quick-disconnect kit

qdn quick-disconnect nipple

qdo quadripartite development objective

q^{do} *quando* (Portuguese or Spanish—when)

Qd'O Quai d'Orsay

QDO Queensland Dairymens Organisation

qdp quick-disconnect pivot

QDR Quality Deficiency Report

QDRI Qualitative Development Requirements Information (program)

qdrnt quadrant

qds quick-disconnect series; quick-disconnect swivel

qd's questioned documents

QDS Quality Data System; Quantitative Decision System

qdta quantitative differential thermal analysis

qdv quick disconnect valve

qe quadrant elevation; quick estimate

qe (QE) quantum electronics

q.e. quod est (Latin—which is)

QE Quality Engineer(ing); Quality Evaluation; Quebec

QE2 Queen Elizabeth 2 (passenger vessel)

QEA Qantas Empire Airways

qeav quick—exhaust air valve

qec quick engine change

QEC Queen Elizabeth College

QECC Queen Elizabeth Chemical Center

qecu quick engine-change unit

qed quantitative evaluative device; quantum electrodynamics; quick-reaction dome

q.e.d. quod erat demonstrandum (Latin—that which was to be proved)

QED Quality, Efficiency, Dependability (reliability program)

qee quadruple expansion engine

qeev quantum electrodynamics electron volts

q.e.f. quod erat faciendum (Latin—that which was to be done)

QEF Queensland Employers Federation

QEFD Queen Elizabeth's Foundation for the Disabled

QEH Queen Elizabeth Hall

q.e.i. quod erat inveniendum (Latin—that which was to be discovered)

qel quiet extended life

QEL Quality Evaluation Laboratory

qem quadrant electrometer

QEM Quality Education for Minorities; Qualified Export Manager

QENP Queen Elizabeth National Park (Uganda)

qeo quality engineering opera-

tions
QEONS Queen Elizabeth's Overseas Nursing Service
QEOP Quartermaster Emergency Operation Plan
QEP Quality Evaluation Program; Quality Examination Program; Queen Elizabeth Park; Queen Elizabeth Planetarium; Queensland Environmental Program
qer qualitative equipment requirements
QER Quarterly Economic Review
qescp quality engineering significant control points
QESP Queen Emma Summer Palace
QEST Quality Evaluation System Test(s)
QESTS Query, Update Entry, Search, Time Sharing
QET Queen Elizabeth Theatre (Vancouver)
qev quick exhaust valve
QEW Queen Elizabeth Way (highway linking Buffalo with Toronto)
qf quality factor; quench frequency; quick freeze; quick frozen
QF quick-firing
qfa quality per final article
Q-factor quality rating
q-fastener(s) quick-fastener(s)
qfc quantitative flight characteristics
qfcc quantitative flight characteristics criteria
qfe quartz fiber electrometer
Q-fellows quartermaster fellows; queer fellows
Q fever query fever (of uncertain cause); Balkan grippe or nine-mile fever (viral disease with pneumonial symptoms caused by rickettsia)
qff quadruple flip-flop
QFGA Queensland Farmers and Graziers Association
QFI Qualified Flight Instructor
qfirc quick-fix interference-reduction capability
qfl quasi-fermi level
qfm quantized frequency modulation
qfo quartz frequency oscillator
qfp quartz fiber product
QFP Quick-Fix Program
QFR Quarterly Force Revision (USN)
Q-fract quick fraction (membrane potentials)
QFRI Queensland Fisheries Research Institute

QFS Queensland Fire Service; Queensland Fisheries Service
QFSM Queen's Fire Services Medal
qft quantized field theory
qg quadrature grid
QG Quartermaster General
QG Quartier Général (French— Headquarters; *Quartier Generale* (Italian—Headquarters)
qgb searchlight sonar (symbol)
QGGA Queensland Grain Growers Association
qgm quarter-girth measure
QGM Queen's Gallantry Medal
QGPO Qatar General Petroleum Organization
QGTB Queensland Government Tourist Bureau
qgv quantized gate video
qh quartz helix
q-h quartz-halogen (lights)
q.h. quaque hora (Latin—every hour)
QH Queen's Hall
QHC Queen's Honorary Chaplain; Queensland Housing Commission
QHDS Queen's Honorary Dental Surgeon
QHM Queen's Harbour Master
QHNS Queen's Honorary Nursing Sister
QHO Queen's Hall Orchestra
QHP Queen's Honorary Physician
QHS Queen's Honorary Surgeon; Queensland Historical Society
QHV Queen's Honorary Veterinarian
qi quality improvement; quality indices
QI Queensland Insurance; Quota International
QI Quality Index; Quarterly Index
QIA Queensland Institute of Architects
qiam queued indexed access memory
qic quality inspection criteria; quartz-iodine crystal
QIC Quality Information Center
q.i.d. quater in die (Latin— four times a day)
QIDN Queen's Institute of District Nursing
qie quantitative immuno-electrophoresis
QIE Qualified International Executive

QIER Queensland Institute for Educational Research
QIH Quality International Hotels
qil quartz incandescent lamp; quartz iodine lamp
QIMR Queensland Institute of Medical Research
qip quartz insulation part
QIP Quality Inspection Point
Q.I.P. Quiescat in Pace (Latin—Rest in Peace)
QIPA Queensland Institute of Public Affairs
QIPS Qualitative Incentive Procurement Service
QIR Quechan Indian Reservation (originally Fort Yuma)
qisam queued-indexed sequential-access method
qit qualification information and test (system)
QIT Queensland Institute of Technology
QITS Quality Information and Test System
QJC Quincy Junior College
QJSA Quarterly Journal of Studies in Alcohol
qjump queue(d) jump(ing)
qk quick
Qk Fl quick flashing (light)
qkly quickly
qkm Quadratkilometer (German—square kilometers)
ql quarrel; query language; quick look; quintal
ql (QL) quantum leap
ql quilate (Portuguese—carat)
q.l. quantum libet (Latin—as much as you like)
QL Queen's Lancers; Queensland (airline code)
Q/L Quarantine Launch
Q'land Queensland
QLAP Quick Look Analysis Program
QLCS Quick Look and Checkout System
Qld Queensland
qlfy qualify
qlfyg qualifying
qlfyn qualification
QLGA Queensland Local Government Association
qli quality of life index
qlii quasi-laser-intensity interferometer
qlit quick-look intermediate tape
qll quartz landing lamp
qlm quasi-laser machine
QLOC Queensland Light Opera Company
QLPC Queensland Library Promotion Council

QLR Queen's Lancashire Regiment
QLR Québec Law Reports
QLS Queensland Law Society; Queensland Littoral Society; Quick Law Systems; Quick Loading System
qlsm quasi-laser sequential machine
qlt quantitative leak test
QLTA Queensland Lawn Tennis Association
qlty quality
qm (QM) quantum mechanics; query message
qm Quadratmeter (German—square meter); *quintal métrico* (Spanish—metric quintal, 220 pounds)
q.m. quaque mane (Latin—every morning); *quo modo* (Latin—in what manner)
QM Decca navigation system; Quartermaster; Queen's Messenger; Queens Museum
qma qualified military available; quality material approach
QMA Quartermasters Association; Qatar Monetary Agency
QMAAC Queen Mary's Army Auxiliary Corps
QMAC Quadripartite Material and Agreements Committee
qmao qualified for mobilization ashore only
Q-max quarantine maximum
qmb quick make-and-break
QMBA Queensland Master Builders Association
QMC Quartermaster Corps; Queen Mary's College (London)
QMC & SO Quartermaster Cataloging and Standardization Office
QMDEP Quartermaster Depot
qmdk quick mechanical disconnect kit
QMDO Qualitative Materiel Development Objective
QMDPC Quartermaster Data Processing Center
qme queueing matrix evaluation
QME Quantock Marine Enterprises
QMEPCC Quartermaster Equipment and Parts Commodity Center
QMFCI Quartermaster Food and Container Institute
QMFCIAF Quartermaster Food and Container Institute for the Armed Forces
QMG Quartermaster General

QMGF Quartermaster-General to the Forces
QMGMC Quartermaster General—Marine Corps
QMH Queen Mary Hospital
QMI Qualification Maintainability Inspection
QMIA Queensland Motor Industry Association
QMIMSO Quartermaster Industrial Mobilization Services Offices
qmo qualitative material objective
QMORC Quartermaster Officers Reserve Corps
QMP Quezon Memorial Park (Philippines)
QMPA Quartermaster Purchasing Agency; Queensland Master Painters Association
QMPCUSA Quartermaster Petroleum Center US Army
qmqb quick-make quick-break (connection)
qmr qualitative materiel requirement
Qmr Quartermaster
QMRC Quartermaster Reserve Corps
QMR & E Quartermaster Research and Engineering
QMRL Quartermaster Radiation Laboratory
QMs Quarterly Meetings (Quakers); quartermasters
QMS Quartermaster School (US Army)
Qm Sgt Quartermaster Sergeant
QMSO Quartermaster Supply Office(r)
qmsw quartz metal sealed window
QMT Queens-Midtown Tunnel
QMTOE Quartermaster Table of Organization and Equipment
qmw quartz metal window
qn question; quotation
q.n. quaque nocte (Latin—every night); *quid nunc* (Latin—what now?)—person eternally interested in getting the latest news
Qn Queen
qna quality per next assembly
QNI Queen's Nursing Institute
QNP Quezon National Park (Philippines)
qns quantity not sufficient
Qns Queens
QNS Queen's Nursing Sister
Qns Coll Queen's College
Qnsd Queensland
Qnsk Quensk (language of the

Quains)
QNS & L Québec North Shore and Labrador Railway
Qnsld Queensland
Qns Pk Queens Park
qnt quantisizer; quintet
qnty quantity
QNWR Quivira National Wildlife Refuge (Kansas)
qo quick opening; quick outlet
QO Quaker Oats; Qualified in Ordnance; Quartermaster Operation; Queen's Own (regiment)
Q & O Quebéc and Ontario (transportation company)
qo₂ oxygen quotient
QO₂ oxygen consumption (or quota)
QOA Quasi-Official Agencies
QOCH Queen's Own Cameron Highlanders
qod quick-opening device
QOD Québec Order of Dentists
QOF Quaker Oats Foundation
QOH Queen's Own Hussars
QOIC Quarantine Officer in Charge
QOMY Queen's Own Mercian Yeomanry
qon quarter ocean net
qopri qualitative operational requirement(s)
qor qualitative operational requirement
Qor Qoran (Koran)
QOR Queen's Own Royal (regiment)
QORC Queen's Own Rifles of Canada
QOS Quick On System
qot quote
qotn quotation
qp queen post; quick process(ing)
q.p. quantum placet (Latin—at discretion)
q-P quanti-Pirquet (reaction)
QP Qualification Proposal; Queen's Printer
qpa qualitative point average; quantity per article; quantity per assembly
QPA Queensland Police Academy; Queensland Polynesian Association
QPB Quality Paperback (book club)
QPC Qatar Petroleum Company
qpei quality per end item
qpf quantitative precipitation forecast
QPF Québec Police Force
QPFC Queen's Park Football

Club
QPFL Queensland Profession-
al Fishermens League
qpi quadratic performance in-
dex
QPIS Quality Performance In-
struction Sheet
QPL Qualified Parts List;
Qualified Product(s) List-
(ing); Queens Public Library
qplt quiet propulsion lift tech-
nology
QPM Queen's Polar Medal;
Queen's Police Medal
QPP Québec Provincial Police;
Quetico Provincial Park (On-
tario)
QPR Quality Progress Report;
Quantity Progress Report;
Quarterly Progress Report;
Queen's Park Rangers
QPRI Qualitative Personnel
Requirements Information
qps quantitative physical sci-
ence
QPS Quick Program Search
Q P & S Quaker Peace and
Service
qpsk quad-phase shift key
qq quartos; questionable questionnaires; questions
qq quelques (French—some);
quintales (Spanish—quintals)
qq. quaque (Latin—each);
quoque (Latin—every)
QQ Celestial Equator; Qara
Qash in Sinkiang province of
China; Qara Qum, also in
Sinkiang province of China,
but sometimes spelled Kara
Kum; Que Que, Rhodesia
q.q.d. quantum quatra die
(Latin—every fourth day)
qqf quelquefois (French—
sometimes)
q.q.h. quantum quatra hora
(Latin—every four hours)
qq. hor. quaque hora (Latin—
every hour)
qqma quality qualified military
availability
qqpr quantitative and qualitative personnel requirements
q.q.v. quae vide (Latin—which
see)
q/qy question/query
qr qualifications record; quar-
ter(ly); quick reaction; quick
receipt; quire
qr (QR) quick response
qr. quadrans (Latin—farthing)
q.r. quantum rectus (Latin—
quantity is correct)
QR Queensland Railways;
Quintana Roo; Quotation Request

Q & R Quality and Reliability
QR Quarterly Review
qra quality reliability assurance; quick reaction alert
QRA Queensland Rifle Association
qrbm quasi-random band model
qrc quick reaction capability
QRC Queensland Rubber
Company
qrcg quasi-random code generator
QRCUP Québec Region Canadian University Press (now
CUPBEQ)
QRDC Quartermaster Research and Development
Command
QRDEA Quartermaster Research and Development
Evaluation Agency
*QRDS Quarterly Review of
Drilling Statistics*
qrg quick response graphic
qrga quadrupole residual gas
analyzer
qri qualitative requirements information
qric quick reaction installation
capability
QRICC Quick Reaction In-
ventory Control Center
QRIH Queen's Royal Irish
Hussars
QRL Quadripartite Research
List; Queensland Research
League
QRMF Quick-Reacting Mo-
bile Force
Qrmr Quartermaster
qro quick reaction operation
Qro Queretaro
QRO Quality Review Organi-
zation; Quick Reaction Op-
eration; Quick Reaction Or-
ganization
Q Roo Quintana Roo
Q-room cue room (billiard
room)
QRPA Quartermaster Radia-
tion Planning Agency
QRPS Quick Reaction Pro-
curement System
QRR Queen's Royal Rifles
QRRR extreme emergency
amateur radio call signal
QRRs Qualitative Research
Requirements (for nuclear
weapons effects information)
qrs quarters
QR's Quality Reports
Q.R.S. trademark of Q.R.S
Music Rolls
qrt quarter
QRT Quick Reaction Team

qrtg quartering
qrtly quarterly
qrtmstr quartermaster
qrv quick-release valve
QRV Qualified Real-estate
Valuer
QRX Queensland Railfast Express
qry quality and reliability year
QRZ Quaddel Reaktion Zeit
(German—lump reaction
time, rash reaction time,
wheal reaction time)
qs quarter section; quarter sessions
qs (QS) quadraphonic stereo;
quiet sleep
q.s. quantum satis (Latin—as
much as is sufficient); *quantum sufficit* (Latin—as much
as suffices)
Qs Conquistadores; Conquistadors; quartzes; questions
QS Quarantine Station; Quarter Section; Quarter Sessions;
Quartermaster Sergeant;
Quarternote Society; Queen's
Scarf; Queen's Scholar;
Queensland Society; Queueing System
QS Quecksilbersäule (German—mercury column)
QSA Queensland Shopkeepers
Association
QSAL Quadripartite Standardization Agreements List
qsam queued sequential access
method
qsbg quasi-stellar blue galaxies
qsbo quasi-stellar blue objects
QSC Quebec Securities Commission
QSD Quality Surveillance Division (USN); Quincy Shipbuilding Division—General
Dynamics
qse qualified scientists and engineers
qsf quasi-static field; quasi-sta-
tionary front
QSF Queensland Soccer Federation
qsg quasi-stellar galaxy
Q-ship disguised man-of-war
used to decoy enemy vessels
qsi quality salary increase
qsic quality standard inspection
criteria
Q-size Queen-size
QSJM Queen's Silver Jubilee
Medal
qs & l quarters, subsistence,
and laundry
QSL Queensland State Library
Q & SL Qualifications and
Standards Laboratory

qsm quadruple-screw motor-ship; quarter-square multipliers

qsm (QSM) Queen's Service Medal

QSMO Quaker State Motor Oils

qso quasibiennial stratospheric oscillation; quasistellar object

QSO Québec Symphony Orchestra; Queen's Service Order; Queensland Symphony Orchestra

QSOP Quadripartite Standing Operating Procedure(s)

qsp quality search procedure

QSPP Québec Society for the Protection of Plants

qsr quick-strike reconnaissance

QSR Quarterly Status Report; Quarterly Summary Report

QSR Quartier de Securité Renforcée (French—Maximum Security Prison)

qsra quiet short-haul research aircraft

QSRIG Quantity Surveyors Research and Information Group

qsrs quasi-stellar radio sources

qss quasi-stellar source

QSS quadruple-screw ship; Quota Sample Survey

qssa quasi-stationary-state approximation

QSSCT Queensland Society of Sugar Cane Technologists

qssp quasi-solid-state panel

QSSR Quarterly Stock Status Report

QST Québec Standard Test

QSTAG Quadripartite Standardization Agreement

Q-star quiet observation aircraft

qstn question

qstnr questionnaire

qstol quiet-and-short takeoff and landing

qsts quadruple-screw turbine steamship

q. suff. quantum sufficit (Latin—as much as needed, as much as will suffice)

Q-switch quantum switch

qsy quiet sun year

qt quality test(ing); quantity; quarry tile; quart; quarter; quick test; quiet (see q.t.)

qt (QT) quality test(ing); queuing theory

q.t. quiet (as "on the q.t.")

q & t quenched and tempered

QT Qualification Test(ing); Quick's Test (pregnancy or prothrombin)

qta quadrant transformer assembly

q^{ta} quanta (Portuguese or Spanish—how much)

QTAC Queensland Tertiary Admissions Centre

qtam queued telecommunication access method

qtaux quintaux (French—quintals)

qtb quarry-tile base

QTB Queensland Timber Board; Queensland Trotting Board

QTC Québec Teaching Congress; Queensland Turf Club

qtd quartered

QTDGs Quaker Theological Discussion Groups

qte quote

qted quick text editor; quoted

Q-Test(ing) Quality Test(ing)

qtf quarry-tile floor

QTF Québec Teachers' Federation

qtg quoting

QTIB Québec Tourist Information Bureau

QTLC Queensland Trades and Labor Council

qtly quarterly

QTM Quechon Tribal Museum (Yuma, Arizona)

qtn quotation

qto quarto

q^{to} quanto (Portuguese or Spanish—how much)—masculine form

qtol quiet takeoff and landing

Q'town Queenstown

qtp quantum theory of paramagnetism

QTP Qualification Test Procedure

qtr quarry-tile roof; quarter; quarterly

QTR Quality Technical Report; Quality Technical Requirement; Quarterly Technical Report

qtrs quarters

qts quarts; quick turn stock

QTTC Queensland Tourist and Travel Corporation

qtte quartette

QTTP Q-Tags Test of Personality

qtt(s) quartette(s)

QTU Queensland Teachers Union

qty quantity

qtydesreq quantity desired or requested

qtz quartz

qtze quartzose

qtzic quartzitic

qtzt quartzite

qu quart; quarter; quarterly; query; question

qu. quasi (Latin—as it were, like)

Qu Queen

QU Queen's College (Cambridge, Oxford); Queen's University

qua quadrate; quadratus

quaal quaalude

quaalude trade name of methaqualone (hypnotic and sedative drug)

quack quacksalver (person pretending to be a doctor)

quacks quacksalvers (sixteenth-century doctors who used quicksilver or mercury in treating syphilis)

Quacks CWACs [City-Wide Anti-Crime (units of the New York City Police Department)]

quack-u-p's quack acupuncturists

quackupunc quackupuncture; quackupuncturist

quad quaalude; quadrangle; quadrangular; quadrant; quadraphonic; quadrat; quadruplet(s); quadruplex; quadruplicate(s); quadruplication

quad (Latin prefix—fourfold)—as in quadrille or quadriplegic

Qu-AD Quality-Assurance Department; Quality-Assurance Division

quad .50's quadruple .50-caliber machine guns

quad c quadripod cane

Quad Cities adjacent and across-the-river cities of Rock Island, East Moline, and Moline, Illinois, plus Davenport, Iowa

quadplex quadriplex

quadradar four-way radar (surveillance)

Quadrangle Quadrangle/The New York Times Book Company

quadrap quadraphonic(al)(ly)

quadrip quadriplegia

quadrivium the four liberal arts—arithmetic, astronomy, geography, and music

quadro quadroon

quadrup quadruped(s); quadruple

quadrupl. quadruplicato (Latin—four times as much)

quads quadraphonic records; quadruplets

QUADS Quality-Assurance

Data System
quag quagmire
Quaker Quaker Oats; Quaker Press
Quaker MMs Quaker Monthly Meetings
quake(s) earthquake(s)
'quake(s) earthquake(s)
qual qualification; qualify; quality
qual anal. qualitative analysis
quals qualifying examinations; qualifying tests
qual(s) qualification(s)
quam quadrature-amplitude modulation
Quandary Quandary Peak in central Colorado
quango (QUANGO) quasi-autonomous non-governmental organization
quant quantity; quantum
quant anal. quantitative analysis
Quantico Quantico, Virginia's FBI Academy and U.S. Marine Base
quantras question analysis transformation and search (data processing technique)
quant. suff. quantum sufficit (Latin—sufficient quantity)
quaops quarantine operations
QUAP Questionnaire Analysis Program
QUAPs Quality Assurance Publications
Quaq Quaquero (Spanish— Quaker)
quar quarantine; quarter
quar. pars quarta pars (Latin—one-fourth part)
quarpel quartermaster water-repellent (cloth or clothing)
quarr quarries; quarry; quarrying
quart quarter gallon; quarterly
quart. quartet; quartette; quartile
Quart Quarterly
QUART Quality Assurance and Reliability Team
Quart Ital Quartetto Italiano (Italian quartet)
- **quartz** crystalline silica (SiO₂)
quartzite granular quartz rock
quasar quasi-stellar radio (object)
quaser quantum—amplification—by—stimulated-emission—of—radiation (acronym covering irasers, lasers, and masers varying only in operational frequency)
Quash Quashey; Quashley
quat quaternary; quaternary

era
quat. quattuor (Latin—four)
Quat Quaternary
Quathlamba Quathlamba Mountains of Lesotho and South Africa where it is called Drakensberg
Quattrocento (Italian—four hundred) 15th century artistic and cultural development in Italy
Quayle Dan Quayle, 44th Vice President of the United States
QUB Queen's University of Belfast
QUD Queen's University of Dublin
Que Québec (inhabitants— Québecois); Quechua; Quechuan
Que Quênia (Portuguese— Kenya)
QUE Quebecair
Québec letter Q radio code
Queen Charlottes Queen Charlotte Islands off British Columbia
Queen Elizabeths Queen Elizabeth Islands in the Canadian Arctic
Queensboro' Queensborough
Queensl Queensland
Quen Quentin
Quent San Quentin (California State Prison)
Quer Querétaro
ques question
quest. quality electrical system test; questioned
QUEST Quality Electrical Systems Test; Queens Educational and Social Team
questal quiet, experimental, short-takeoff-and-landing (program of NASA)
questar quantitative utility evaluation suggesting targets for the allocations of resources
quester quick and efficient system to enhance retrieval
questn questionnaire
quetzal monetary unit of Guatemala
qufyd qualified
QUGA Queensland United Graziers Association
QUI Queen's University of Ireland; Quincy (railroad)
QUIC Question and Information Connection (telephone reference department of the St Louis Public Library)
Quich Quichua
quicha quantitative inhalation challenge apparatus

QUICK Queens University Interpretative Code
QUICKTRAN Quick Fortran (programming language)
quico quality improvement through cost optimization
QUIDS Quick Interactive Documentation System
quiktran quick fortran (programming language)
Quilmas San Quilmas (Mexican-Americanism—San Antonio, Texas)
quim química (Portuguese or Spanish—chemistry)
Quimigal Química de Portugal
quin quintet; quintette; quintuplet; quintuplicate; quintuplication
Quin Quincy; Quinten; Quintilianus; Quintilius; Quintillian; Quintin; Quintino; Quintius; Quintus
quinq quinque (Latin—five)
quins quintuplets
quint quintuplicate
quint. quintus (Latin—fifth)
Quint. Quintilian—Roman critic and rhetorician Marcus Fabius Quintilianus
Quintilian Marcus Fabius Quintilianus
quint(s) quintet(s); quintuplet(s); quintuplicate(s)
quintupl quintuplicate
quip. query interactive processor; questionnaire interpreter program
quis quisling (term for traitor derived from Vidkun Quisling who during World War II headed Norway's puppet government set up by the German invaders)
Quitmans Quitman Mountains of west Texas
quix quixote; quixotic(al)(ly); quixotism; quixotry
Quix Quixote
QUJ true course to station
QUL Queen's University Library
qume cue me
Q-unit one quintillion (1 x 10¹⁸)—equal to 38.46 billion tons of coal or 172.4 billion tons of oil or 968.9 trillion cubic feet of natural gas
QUNO Quaker United Nations Office
quo' quoth
quod. quodlibet (Latin—as you please)
Quoddy Passamaquoddy Bay between Maine and New Brunswick; Passamaquoddy

Indians
Quoins Gunners Quoin and Quoin Channel north of Mauritius in the Indian Ocean; other Quoins in Australia, Burma, and South Africa
quok(s) quokka(s)
Quon Pt Quonset Point, Rhode Island
quonset quonset hut (originally built during World War II at Quonset, Rhode Island)
quor quorum
quor. *quorum* (Latin—of which)
quot quotation
quot. *quotidie* (Latin—daily)
quotes quotation marks
quote(s) quotations(s)
quote-unquote quotation marks (slang—phrase or word in quotation marks)
quotid. *quotidie* (Latin—every day)
qup quantity per unit pack

Qur Quran (Malay—Koran)
QUSA "Q" Airways
qv quality verification
q-v q-value
q.v. *quantum vis* (Latin—as much as is desired); *quod vide* (Latin—which see)
QVM Queen Victoria Museum (Launceston, Tasmania)
QVM ¡*Que Viva Mexico*! (Spanish—Long Live Mexico!)
QVR Queen Victoria's Rifles
QVS Quality Verification Surveillance
qvt quality verification test
qw quarter wave
qwa quarter-wave antenna
qwd quarterly world day
q-wedge quartz wedge
qwerty standard typewriter keyboard
QWG Quadripartite Working Group
QWGCD Quadripartite Working Group for Combat Development (American, Australian, British, and Canadian armies)
qwl quality of working life; quick weight loss
QWMP Quadruped Walking Machine Program (US Army)
qwot quarter-wave optical thickness
qwp quarter-wave plate
qx *quintaux* (French—hundredweights)
qy quantum yield; query
Qy Quay
QYC Quincy Yacht Club
QYO Queensland Youth Orchestra
qz quartz
Qz quartz
QZ Zambia Airways (2-letter coding)
QZS Québec Zoological Society

R

r angle of reflection (symbol); position vector (symbol); racemic; racket(eer); radius; rain; range; rare; rate of interest; received; recipe; reconnaissance; recto; red; redetermination; refraction; registered; relative; relative humidity; report; reprint; research; reserve; resistance; restricted; retard; retarded; right or starboard side; ring; ringer; riser; robotics; rod; rook; rough; rubbed; rule; rules; runs; rupee (Indian monetary unit); solubilizing agent (symbol); symbol for reluctance, resistance, or resistor; thermal resistance

r (R) restricted (movie rating—children under 17 must be accompanied by parent or guardian); retrograde

r angular yaw velocity (symbol); front of the sheet (recto); *remotum* (Latin—far, remote)

R acoustic resistance (symbol); annual rent; electrical resistance; gas constant; ohmic resistance; product moment coefficient of statistical correlation; Rabbi; radioactive range; radiolocation; Rankine; rare; ratio; Réaumur; received solid; reconnaissance; Regina (Queen); registered; Reiz; report(s); Representative; reprint; Republic Republican; research; reserve; resistance; respiration; restricted; Rex (King); rial (Iranian monetary unit); Richfield Oil; right; ring; river; Road; Robin Line; rocket; Rocketdyne Division of North American Avia-

tion; Roentgen; Roger—radio slang meaning all right or okay; Roma; Roman; Rome; Romeo—code for letter R; Rotary International; Royal; ruble (Russian monetary unit); rupee (Indian monetary unit); Rwanda; Rydberg; US Rubber Company

R *(R)* economic rent (microeconomics)

R. rand (South African monetary unit)

R' radius of circle in minutes of arc

R'' radius of circle in seconds of arc

-R Rinne's hearing test negative

+R Rinne's hearing test positive

R rechts (German—right); *Reka* (Bulgarian, Czechoslovakian, Russian, Serbo-Croatian—river); resultant force (symbol); *rett* (Danish—right); *Ría* (Portuguese, Spanish—river mouth); *Rio* (Portuguese—river); *Río* (Spanish—river); *Rivière* (French—river); rogue branded on British convicts transported overseas in the early 1800s); *Romanus* or *Rufus* (on Latin inscriptions); *rua* (Portuguese—street); *rubeus* (Latin—red); *Rud* (Persian—river); *rue* (French—street); *Rzeka* (Polish—river); on the flag of Rwanda it stands for Rwanda, a Republic born of Revolution and confirmed by Referendum; The Book of Ruth

R₁ primary roots

R₂ secondary roots

R-4 Recovery and Reuse of Refuse Resources (USN)

ra radio; radioactive; radioactivity; reduced area; right angle; right angulation; right ascension; right atrium; right auricle; robbery committed while armed (RA); rubber-activated; ruling action

ra (RA) retrograde amnesia; rheumatoid arthritis

r/a radioactive; return to author

r & a right and above

Ra radium; Range

RA Argentina (auto plaque); Coast Radar Station (symbol); high-powered radio range (Adcock symbol); Rabbinical Assembly; Rdeca Armada (Yugoslav—Red Army); Rear Admiral; Reduction of Area; Regular Army; Rehabilitation Act; Remington Arms; Rental Agreement; Republic Aviation; *República Argentina*; Resident Adviser; Resident Auditor; Right Arch; right ascension; Rotogravure Association; Royal Academician; Royal Academy; Royal Arcanum; Royal Artillery; Royal Nepal (2-letter airline code)

RA(A) Rear Admiral (Aircraft Carriers)

RA (D) Rear Admiral (Destroyers)

R.A. right ascension

R/A Redstone Arsenal

R & A Research and Analysis; Royal and Ancient Golf Club at St Andrews, Fife, Scotland

RA República Argentina (Spanish—Argentine Republic)

r.a.a. reductio ad absurdum (Latin—reduction to an absur-

dity)—in mathematics sometimes appears as raa or RAA
RAA Rabbinical Alliance of America; Rockport (Mass.) Art Association; Royal Academic Association; Royal Academy of Arts; Royal Australian Artillery
RAAA Red Angus Association of America; Relocation Assistance Association of America
RAAC Regional Affirmative Action Clearinghouse; Royal Australian Armoured Corps
RAADC Royal Australian Army Dental Corps
RAAEC Royal Australian Army Educational Corps
RAAF Royal Afghan Air Force; Royal Australian Air Force
RAAFMS Royal Australian Air Force Medical Service
RAAFNS Royal Australian Air Force Nursing Service
RAAFPO Royal Australian Air Force Post Office
RAAM Race Across America (bicycle)
RAAMC Royal Australian Army Medical Corps
RAAMS Remote Anti-Armor Mine System
RAANC Royal Australian Army Nursing Corps
RAANS Royal Australian Army Nursing Service
RAAOC Royal Australian Army Ordnance Corps
raap residue arithmetic-association processor
RAAPS Resource Allocation and Planning System
RAAS Royal Amateur Art Society
RAASC Royal Australian Army Service Corps
rab rabbet(ing)
Rab François Rabelais (c.1483–1553); Rabat, Morocco; Rabaul, New Britain; Rabbi; Rabbinic Hebrew
RAB Radio Advertising Bureau
RAB *Republik Arab Bersatu* (Malay—United Arab Republic)
rabar Raytheon advanced battery acquisition radar
rabb rabbinate; rabbinic; rabbinical
rabbi rapid-access blood-blank information
Rabbit Ears short form for Rabbit Ears Mountain or Rabbit Ears Pass in northwestern Colorado

RABDF Royal Association of British Dairy Farmers
RABFM Research Association of British Flour Millers
RABI Royal Agricultural Benevolent Institution
RABPCVM Research Association of British Paint Colour and Varnish Manufacturers
rac racemic; radiometric area correlator; relative address coding; rhomboidal air controller
racr accommadage(s) [French —repair(s)]
RAC Railway Association of Canada; Rear Admiral Commanding; Recombinant (DNA) Advisory Committee; Reliability Action Center; Reliability Analysis Center; Rent-Adjustment Commission; Republic Aviation Corporation; Research Advisory Council; Research Analysis Corporation; Royal Air Cambodge; Royal Arch Chapter; Royal Armoured Corps; Royal Automobile Club; Rubber Allocation Committee; Rubber Association of Canada
RACA Recovered Alcoholic Clergy Association; Royal Automobile Club of Australia
RACAN Rubber Association of Canada (also RAC)
RACB Royal Automobile Club of Belgium
racc radiation and contamination control
racc raccomandata (Italian—registered letter)
RACCA Refrigeration and Air Conditioning Contractors Association
race. random-access computer equipment; rapid automatic checkout equipment
RACE Railways of Australia Container Express; Research on Automatic Computation Electronics
RACE *Real Automóvil Club de España* (Spanish—Royal Automobile Club of Spain)
racep random access and correlation for extended performance
races. (RACES) radio amateur civil emergency service
racfire tactical fire-direction (system)
RACGP Royal Australian College of General Practitioners
Rachl Rocketdyne advance

chemical laser
RACI Royal Australian Chemical Institute
RACIC Remote Area Conflict Information Center
racon radar beacon
RACP Royal Australasian College of Physicians
RACS Remote Access Computing System; Royal Australasian College of Surgeons
RACT Royal Australian Corps of Transport
RACUK Royal Aero Club of the United Kingdom
RACV Royal Automobile Club of Victoria
rad radar; radian; radiation; radiation-absorbed dose; radiator; radical; radicalism; radio; radioactive; radius; radix; rapid-access disc; released from active duty; return to active duty; roentgen-administered dosage; roentgen-administered dose
rad (RAD) rapid access disc
rad. radix (Latin—root)
Rad Radnor; Radnorshire
RAD Royal Academy of Dancing; Royal Albert Docks; Rural Area Development
RA(D) Rear Admiral (Destroyers)
rada radioactive; random-access discrete address
RADA Royal Academy of Dramatic Arts
radac rapid digital automatic computing
radal radio detection and location (system)
radan radar doppler automatic navigator
radant radome antenna
radar radio detection and ranging
RADAR Royal Association for Disability and Rehabilitation
RADARS Receivable Accounts Data-entry and Retrieval System
RADAS Random Access Discrete Address System (battlefield communications system)
radat radar data transmission and ranging; radiosonde observation data
radata radar automatic data transmission assembly
RADATS Radar Data-Transmission System
RADC Rome Air Development Center; Royal Army Dental Corps
RADCC Rear Area Damage

Control Center
rad-ch radical-changing
RADCM radar countermeasures and deception
RADCOLS Rome Air Development Center on-Line Simulator
radcon radar data converter
RADD Royal Association in Aid of the Deaf and Dumb
raddef radiological defense
raddol raddolcendo (Italian—growing calmer)
radem (RADEM) random access data modulation
rad encl radiator enclosure
radep radar departure
radex radiation exclusion plot (actual or predicted fallout)
radfac radiating facility
radf(s) rapid-access data file(s)
radhaz radiation hazard(s)
radi radiological inspection
radiac radioactivity-detection-indication-and-computation
radial-ply radial-ply tire
Radiat Eff Radiation Effects
radic radical; radicle; radicotomy; radiculalgia; radicular; radiculectomy; radiculitis; radiculomeningomyelitis; radiculomyelopathy; radiculoneuritis; radiculopathy
RADIC Research and Development Information Center
radic-lib radical liberationist
radic-lib(s) radical-liberal(s); radical-liberationist(s)
radint radar intelligence
Radio 1 British disc jockey commentary and teenage pop music station
Radio 2 British family phone-in and pop programs station
Radio 3 British classical music programs station
Radio 4 British station featuring educational and informational programs on cooking, farming, and the theater
Radio City Radio City Book Store, Radio City Music Hall
radiog radiography
radiol radiology
Radio Sci Radio Science
radir random access document indexing and retrieval
radist radar distance indicator
RA Dks Royal Albert Docks
radl radiological
rad lab radiation laboratory
radlfo radiological fallout
Rad Lib Radio Liberty
radlib(s) radical liberal(s)
radlic radio link
RADLO Radiological Defense

Officer
radlop radiological operations
radlsafe radiological safety
Rad Lux Radio Luxembourg
radlwar radiological warfare
R Adm Rear Admiral
RADMAPS Radiological Monitoring Assessment Prediction System
radmon radiological monitor(ing)
radn radiation
radnote ratio note
RADOC Regional Air Defense Operations Center
radome radar dome
radon daughter deadly microscopic radioactive uranium particles
radop radar operator
rad op radio operator
radose radiation dosimeter satellite
radot real-time automatic digital-optical tracker
RadPropCast radio propagation forecast
RADR Royal Association for Disability and Rehabilitation
rad rec radiator recess
RADRON Radar Squadron (USAF)
radru rapid-access data-retrieval unit
rad/s radians per second
Rad(s) Radical(s)
RADS Ryukyu Air Defense System
radsab radiator sabotage
radscat radiometer-scatterometer sensor
radsick radiation sickness
RadSo Radiological Survey Officer
radss radar alphanumeric-display subsystem
radsta radio station
radtel radar telescope
radtt radio teletypewriter
radu radar analysis and detection unit
radvs radar altimeter and doppler velocity sensor
radwar radiological warfare
rae (RAE) radio astronomy explorer
Rae Rachel; Raquelle
RAE Royal Aircraft Establishment; Royal Australian Engineers
RAE Real Academia Española (Royal Spanish Academy)
R Ae C Royal Aero Club
RAEC Royal Army Educational Corps
RAEL Real Academia Española

de la Lengua (Royal Spanish Academy of Language)
RAEME Royal Australian Electrical and Mechanical Engineers
RAeS Royal Aeronautical Society
raet range-azimuth-elevation-time
Raf Rafael; Rafe; Rafelsz; Raffaele; Raffaello
RAF Red Army Fraction (Baader-Meinhof terrorists); Regular Air Force; Royal Aircraft Factory; Royal Air Force
RAF Rote Armee Fraktion (German—Red Army Faction) terrorist group
RAFA Royal Air Force Association; Royal Australian Field Artillery
rafar radar-automated facsimile reproduction; radio-automated facsimile and reproduction
rafax radar facsimile transmission
RAFB Randolph Air Force Base
RAFBF Royal Air Force Benevolent Fund
RAFC Royal Air Force College
Rafe (Ralph)
RAFES Royal Air Force Educational Service
raff raffiné (French—exquisite, polished, refined)
Raffles Raffles Hotel; Raffles Institution (Singapore Institution and Library); Raffles Place; Sir Thomas Stamford Raffles (founder of Singapore)
RAFGSA Royal Air Force Gliding and Soaring Association
Raf¹ Rafael
RAFMS Royal Air Force Medical Services
RAFO Reserve of Air Force Officers
rafos long-range navigation system (sofar reversed)
RAFR Royal Air Force Regiment
RAFRO Royal Air Force Reserve of Officers
RAFS Royal Air Force Station
RAFSAA Royal Air Force Small Arms Association
RAFSC Royal Air Force Staff College
RAFSE Royal Air Force School of Education
raft. recom algebraic formula translation; recom algebraic formula translator

RAFT Regional Accounting and Finance Test
RAFTC Royal Air Force Technical College; Royal Air Force Transport Command
RAFVR Royal Air Force Volunteer Reserve
rag. ragtime; ring airfoil grenade; runaway arresting gear
rag ragioniere (Italian—accountant)
RAG Red Army Group (see *B-M B*); River Assault Group; Royal and Ancient Game (of golf)
RAGA Royal Australian Garrison Artillery
RAGB Refractories Association of Great Britain
RAGC Royal and Ancient Golf Club (St Andrews, Scotland)
RAGE Radio Amplification of Gamma Emissions
rah hurrah
RAH Royal Albert Hall
RAHS Royal Australian Historical Society
rai radioactive interference; random access and inquiry
RAI Reading Association of Ireland (actually the International Reading Association); Royal Australian Infantry
RAI Radiotelevisione Italiana (Italian—Italian Radio-Television)—broadcasting system; *Réseau Aérien Interinsulaire* (Tahiti); Royal Albert Institution; Royal Anthropological Institute
RAIA Royal Australian Institute of Architects
RAIAD Reverse Acronyms, Initialisms, and Abbreviations Dictionary
RAIC Royal Architectural Institute of Canada
raidex raiders exercise
rail. railroad; railway
RAIL Religion In American Life
rails. runway alignment indicator lights
railwayac railway + maniac (railway fan)
Railway Employees Union Brotherhood of Railway, Airline, and Steamship Clerks, Freight Handlers, Express, and Station Employees
Rainbow Bridge Rainbow Bridge National Monument (world's largest natural bridge located in southern Utah on the Colorado River close to the Arizona border)

RAIOMA Resource Assessment Investigation of the Mariana Archipelago
rair remote access/immediate response
rair (RAIR) ram-augmented interstellar rocket
RAIRS Recordak Automated Information Retrieval System
RAI-TV Radio Audizioni Italiane—TV (Italian—Italian Radio Audition—TV)
raiu radioactive iodine uptake
Raj Rajasthan
Raj Rajah (Arabic—seventh month of the Mohammedan year); *Rajah* (Hindi—king, prince, ruler); Rajasthani (culture, language, or people); the period of British rule in India
RAJ Royal Association of Justices
ra k raised keel
RAK Rikets Allmanna Kartverk (Swedish—Geographical Survey Office)
ral resorcyclic acid lactone
Ral Raleigh
RAL Resort Airlines; Royal Air Laos
Ralegh Sir Walter Raleigh (who spelled his name *Ralegh*)
Raliks Ralik Chain of Islands in the west-central Pacific, including Bikini, Eniwetok, Jaluit, Kwajalein, Rongerik
RALIP Resource and Land Information Program; Resources and Land Investigations Program
RALLA Regional Allied Long-Lines Agency
rallo. rallentando (Italian—slower by degrees)
ralph reduction and acquisition of lunar pulse heights
RALPH Royal Association for the Longevity and Preservation of the Honeymooners
ralu register and arithmetic logic unit
ralv rat leukemia virus
ram. radio attenuation measurement; random access memory; rapid area maintenance; right ascension of the meridian
ram. (RAM) research and applications module; reverse-annuity mortgage; rolling airframe missile
Ram Raman effect in spectrum analysis; Ramona; Ramsgate
RAM Reliability, Availability, Maintainability (program); Reverse Amortization Mortgage; Revolutionary Action Move-

ment; Rodrigo A Muñoz; Royal Academy of Music; Royal Air Maroc; Royal Arch Masons; Royal Australian Mint
RAMA Rome Air Materiel Area
ramac random access memory accounting
Ramapos Ramapo Mountains of New Jersey and New York
Rama's Bridge also called Adam's Bridge; 18-mile chain of shoals between Coromandel Coast of India and Mannar Island off Ceylon; Hindus relate Rama built causeway across these shoals so his Indian army could invade Ceylon and rescue his wife Sita from the demon king Ravana; Moslems insist building this bridge was Adam's first task after his expulsion from paradise
ramb(s) rambler(s)
ramc rob all my comrades
RAMC Royal Army Medical College; Royal Army Medical Corps
ramd reliability, availability, maintainability, durability
RAMIS Rapid-Access Management Information System; Rapid-Automatic Malfunction-Isolation System
ramit rate-aided manually implemented tracking
ramont radiological monitoring
ramp. rate-acceleration measuring pendulum
RAMP Radar Mapping of Panama; Radiation Airborne Measurement Program; Resource Allocation and Management Program; Reverse Annuity Mortgage Program
RAMPAC Realty and Mortgage Investors of the Pacific
rampallion ramp + rapscallion
RAMPC Raritan Arsenal Maintenance Publication Center
RAMPI Raw Material Price Index
ramps. resources allocation and multiproject scheduling
RAMPS Resources Allocation and Multiproject Scheduling
rams. right ascension of mean sun
Rams Ramsgate
RAMS right ascension mean sun
RAMSA Radio Aeronáutica Mexicana S.A.
RAMSS Royal Alfred Mer-

chant Seamen's Society
ramt rudder-angle master transmitter
ran. reconnaissance-attack navigator; request for authority to negotiate
Ran Rangoon
RAN Royal Australian Navy
Ranally Rand McNally
RANAS Royal Australian Naval Air Squadron
ranc radar attenuation, noise, and clutter
RANC Royal Australian Naval College
Rance Ransom(e)
rancom random communication satellite
rand monetary unit of South Africa
Rand Rand McNally; Witwatersrand (Johannesburg)
randam random-access nondestructive advanced memory
RAND Corporation Research and Development Corporation
randid rapid alphanumeric digital indicating device
Random Random House
RANF Royal Australian Nursing Federation
'rang(s) boomerang(s)
Ranier Ranier Bancorporation (National Bank of Commerce of Seattle)
RANN Research Applied to National Needs
RANR Royal Australian Naval Reserve
RANRL Royal Australian Navy Research Laboratory
ran's revenue anticipation notes
RANSA Royal Australian Naval Sailing Association
RANSA Rutas Aéreas Nacionales (Spanish—National Airlines)
RANT Reentry Antenna Testing
RANVR Royal Australian Naval Volunteer Reserve
RANZCP Royal Australian and New Zealand College of Psychiatrists
rao radio astronomical observatory
RAO Regional Administrative Office(r); Regional Airways Office(r); Rudolf A Oetker (steamship line)
RaOb radiosonde observation
RAOC Royal Army Ordnance Corps
raomp report of accrued obli-

gations—military pay
raot rocker-arm oiling time
RAOU Royal Australasian Ornithologists' Union
rap. talking frankly about any topic; rapid; rapport; reactive atmosphere processing; rear area protection; relative accident probability; rupees, annas, pies (Indian currency)
rap. (RAP) random access programming; random access projector
rap rapido (Spanish—rapid)—fast train
Rap H Rap Brown; Rapids
RAP Radical Alternatives to Prison; Radiological Assistance Plan (AEC); Regimental Aid Post; Release Aid Plan; Royal Army Post
RAPC Royal Army Pay Corps
RAPCAP Radar Picket Combat Air Patrol
rapcoe random access programming and checkout equipment
rapcon radar approach control
RAPCs Regional Action Planning Commissions
rapec rocket-assisted personnel ejection catapult
rape rep rape report
Raph Raphael
Raphael Raffaello Sanzio
RAPI Royal Australian Planning Institute
rapid. random-access personnel information device; relative address programming implementation device; retrieval through automated publication and information digest-(ing)
RAPID Register for the Ascertainment and Prevention of Inherited Diseases; Rocketdyne Automatic Processing of Integrated Data
RAPIDS Random-Access Personnel Information System
RAPM Russian Association of Proletarian Musicians
rapp rapport; rapporteur; rapprochement
RAPP Radical Alternatives to Prison Plan; Radiologists, Anesthesiologists, Pathologists, and Psychiatrists
rappelling rapidly lowering
rappi random-access plan-position indicator
RAPPORT Rapid-Alert Programmed-Power-Management of Radar Targets
rapr radar processor
RAPRA Rubber and Plastics

Research Association
rap's rocket-assisted projectiles
RAPS Radar Automatic Plotting System; Risk Appraisal of Programs System
rap. & sup. rapport and support
raptap random access parallel tape
raptus. rapid thorium-uranium-sodium (reactor)
rar radio acoustic ranging; rapid-access recording; right arm reclining
RAR Reliability Action Report; Rhodesian African Rifles; Royal Australian Regiment(s)
rarad radar advisory
RARDE Royal Armament Research and Development Establishment
rare. ram air rocket engine
RARE Rare Animal Relief Effort; Rehabilitation of Addicts by Relatives and Employers
rarep radar report
RARG Regulatory Analysis Review Group
RARO Regular Army Reserve of Officers
ras radome antenna structure; radula sinus; rapid audit summary; rectified air speed; requirements allocation sheet; rheumatoid arthritis serum
ras (RAS) reticular activating system
ras. rasurae (Latin—shavings)
RAs Resident Agencies; Resident Agents
RAS Report Audit Summary; Royal Aeronautical Society; Royal Agricultural Society; Royal Asiatic Society; Royal Astronomical Society; Rubber Association of Singapore
RASA Railway and Airline Supervisors Association
RASAR Resource Allocation System for Agricultural Research
RASB Royal Asiatic Society of Bengal
RASC Royal Army Service Corps; Royal Astronomical Society of Canada
RASC/DC Rear Area Security and Damage Control
RASD Reference and Adult Services Division (American Library Association)
rase rapid automatic-sweep equipment
RASE Royal Agricultural Society of England
raser range and sensitivity ex-

tending resonator
rash. rain shower(s)
Rash Rashomon
RASK Royal Agricultural Society of Kenya
rasn rain and snow
RASNZ Royal Agricultural Society of New Zealand
RASP Reliability and Aging Surveillance Program (USAF)
RASPB Royal and Ancient Society of Polar Bears (Hammerfest, Norway's town-hall club)
RASS Rock Analysis Storage System; Royal Alfred Seafarers' Society
rastac random access storage and control
rastad random access storage and display
Rastafians Rastafarians
RASTAS Radiating Site Target Acquisition System
rat rotational automatic tester
rat. ram air turbine; ratchet; rate; rating; ration(s); rocket-assisted torpedo (RAT)
rat. (RAT) repeat-action tablet
RAT Remote Associates Test
ratac radar analog target acquisition computer
ratan radio television aid to navigation
RATAS Research and Technical Advisory Services (Lloyd's Register of Shipping)
ratc radar-aided tracking computer
RATCC Radar Air Traffic Control Center
RATCF Regional Air Traffic Control Facility
ratcon radar terminal control
rate. remote automatic telemetry equipment
ratel radiotelephone
ratelo radio telephone operator
ratepayer(s) [Canadian English—taxpayer(s)]
rat/epr ram air temperature/engine pressure ratio
RATER Raytheon Acoustic Test and Evaluation Range
ratfor rational fortran
ratg radiotelegraph
rato rocket-assisted takeoff
Ratons Raton Mountains of Colorado and New Mexico
RATP Régie Autonome des Transports Parisiens (Le métro—Paris subway system)
RATR Reliability Abstracts and Technical Reviews
rats. repeat-action tablets

Rats Rat Islands (Amchitka, Kiska, Rat, etc.)
RATS Ram Air Turbine Systems
ratscat radar target scatter site
RATSEC Robert A Taft Sanitary Engineering Center
ratt radioteletypewriter
RAU Rand Afrikaans University; River Assault Unit (USN)
RAU Repubblica Araba Unita (Italian—United Arab Republic)—Egypt
RAUS Retired Association for the Uniformed Services
'raus mit i'm *heraus mit ihm* (German—out with him)
R Aux AF Royal Auxiliary Air Force
Rav Roux-associated virus
RAVA Rochester Audiovisual Association
RAVC Royal Army Veterinary Corps
rave. radar acquisition vocal-tracking equipment
rave. (RAVE) research aircraft for visual environment (USA)
RAVE Register And Vote Easily
RAVEC Regional Adult and Vocational Education Council
raven. ranging and velocity navigation
RAVES Rapid Aerospace Vehicle Evaluation System
ravir radar video recorder; radar video recording
RAW Reconnaissance Attack Wing (USN)
RAWA Renaissance Artists and Writers Association
Rawal Rawalpindi
rawarc radar and warning coordination
RAWI Radio American West Indies (Virgin Islands)
rawin radar wind sounding
raws radar altimeter warning set
RAWs Replenishment Agricultural Workers
rawx returned account of weather (aviation)
rax random access (computing system)
'ray hurray
Ray Rachel; Raymond
RAYCI Raytheon Controlled Inventory
razel range, azimuth, elevation
razon range and azimuth only
razz razzberry (slang for raspberry)
rb read backward; read buffer; relative bearing; return to bias;

rigid boat; road bend; rubber-base(d)
r/b reentry body
r & b rhythm and blues; right and below; room and board
Rb rubidium
RB Rancho Bernardo; reconnaissance bomber; Regiment Botha; Renegotiation Board; Republica Boliviana (Bolivian Republic); Republic of Burma; Rifle Brigade; *Ritzaus Bureau* (Danish news agency); Robert Burchfield, editor of the *Oxford English Dictionary* supplement; *Royaume de Belgique* (Kingdom of Belgium)
R.B. Robert Browning
R$_B$ Rockwell hardness (B-scale)
Rb-08 Saab surface-to-surface missile
RBA Rabat, Morocco (airport); Reserve Bank of Australia; River Boards Association; Road Bitumen Association; Roads Beautifying Association; Roadside Business Association; Royal Brunei Airlines
RBAF Royal Belgian Air Force
rbb room, board, and beverages
RBB Richard Bedford Bennett (Canada's fourteenth Prime Minister)
RBB Reference Books Bulletin
rbbb right bundle branch block
rbbsb right-bundle-branch system block
rbc red blood cell; red blood cell (count); red blood corpuscle
RBC Rhodesian Broadcasting Corporation; Richard Bland College; Roller Bearing Company; Royal Bank of Canada
RBCA Russian Book Chamber Abroad
rbcd right border of cardiac dullness
RBCM Royal British Columbia Museum (Victoria)
rbd rapid beam deflector; right border of dullness (heart response to percussion)
RBD Rittenhouse Book Distributors
rbde radar bright-display equipment
rbe relative biological effectiveness
RBEC Roller Bearing Engineering Committee
rbelet relative biological effectiveness linear energy transfer

R Bern Rancho Bernardo
rbf renal blood flow
RBF Rockefeller Brothers Fund
RBFC Rural Banking and Finance Corporation
RBG Royal Botanic Gardens (Kew Gardens)
RBGS Radio Beacon Guidance System
RBH Rutherford Birchard Hayes (19th President U.S.)
rbi reply by indorsement; request better information; runs batted in
rbí recibí (Spanish—I received)
RBI Reserve Bank of India; Rochester Business Institute
rb imp rubber-base impression
RBK Royal Borough of Kensington
rbl ruble
RBL Royal British Legion
RBLC Royal British Legion Club
R Bn radio beacon
RBN Registry of Business Names
RBNA Royal British Nurses' Association
RBNM Rainbow Bridge National Monument (Utah)
RBNSW Rural Bank of New South Wales
RBNZ Reserve Bank of New Zealand
rbo right back outside
RBO Russian Brotherhood Organization
rboc rapid-bloom off-board chaff
RBOT Rotating Bomb Oxidation Test
rbox rail box car (rolling-stock pool)
rbp ration breakdown point
RBP Registered Business Processor
RBP *Raffinerie Belge de Petroles* (French—Belgian Petroleum Refinery)
RBPP Rotor-Burst Protection Program (NASA)
rbr risk-to-benefit ratio; rubber
rBr reddish brown
RBR Renegotiation Board Regulation
RBR *Reference Book Review*
RBRF Reproductive Biological Research Foundation
rbs radar bomb score; radar bomb scoring; request blocks
Rbs Rutherford back-scatter(ing)
RBS Ranganthittoo Bird Sanctuary (India); Research for Better Schools; Royal Bank

of Scotland; Royal Botanical Society
RBSA Royal Birmingham Society of Artists
rbsn (RBSN) reaction-bonded silicon nitride
rbt rabbet; rabbit; resistance bulb thermometer; roundabout
RBT Rational Behavior Therapy; Rose Bengal Test(ing)
rbtwt radial-beam travelling-wave tube
RBU Rabindra Bharati University
rbv return-beam videcon
rc radio code; radio coding; rate of change; ready calendar; red cell; red corpuscle; regional controller; reinforced concrete; resin coat(ed); resin coating; resistance capacitance; resistor-capacitor; respiratory center; reverse course; right center; rigid center; rock-crushed; rubber-cushioned
r/c reconsign(ed); recredit(ed)
r & c rail and canal
r/c *rés-do-chão* (Portuguese—ground floor)
R_e Rockwell hardness (C-scale)
RC Radcliffe College; Radio City; Radio Code; Reception Center; Reconstruction Commission; Red China; Red Cross; Regina College; Regional Commissioner; Regis College; Reinhardt College; Renison College; *República de Chile*; *República de Colombia*; *República de Cuba*; Ricker College; Ricks College; Rider College; *Río Colorado*; Ripon College; Rivier College; Roanoke College; Rockefeller Center; Rockford College; Rockhurst College; Rockmount College; Rollins College; Roman Catholic; Rosary College; Rosemount College; Rosenwal College; Rust College
R, C Cauchy constant
R of C Republic of China
RC *República Centroafricana* (Spanish—Central African Republic)
R.C. *Rendiconti* (Italian—proceedings or reports)
rca replacement cost accounting
R^ca *Rocca* (Italian—rock; tower)
RCA Rabbinical Council of America; Radio Club of America; Radio Corporation

of America; Radio Council of America; Reformed Church in America; Rocket Cruising Association; Rodeo Cowboys Association; Roofing Contractors Association; Royal Canadian Academician; Royal Canadian Academy; Royal Canadian Artillery; Royal College of Art; Rug Corporation of America; Rural Crafts Association
RCA *République Centrafricaine* (French—Central African Republic)
RCAA Royal Cambrian Academy of Art; Royal Canadian Academy of Arts
RCAC Radio Corporation of America Communications
RCACS Readiness Command and Control System
RCAF Royal Canadian Air Force
RCAM Royal Canadian Artillery Museum
R Cam A Royal Cambrian Academy of Art
RCAMC Royal Canadian Army Medical Corps
R Can Rio Canario
RCAR Religious Coalition for Abortion Rights
RCA Rev RCA Review
RCAS Royal Central Asian Society; Rutgers Center of Alcohol Studies
RCA Satcom RCA Domestic Communications Satellite
RCASC Royal Canadian Army Service Corps
rcat remote-controlled aerial target
RCAT Royal College of Arts and Technology
RCA Vic RCA Victor
RCB Ready-Crew Building; Regiment Christiaan Beyers; Retail(ers) Credit Bureau
RCBB Royal Commission on Bilingualism and Biculturalism (Canada)
rcbf (RCBF) regional cerebral blood flow
rcc read(er) channel continue-(d); reader common contact; remote communications complex; rough combustion cut-off
r & cc riot and civil commotion
RCC Radio-Chemical Center; Radiological Control Center; Rag Chewers Club; Rape Crisis Center; Reply Coupon Collector(s); Rescue Control

Center; Rescue Coordination Center; Rockland Community College; Roman Catholic Church; Royal Crown Cola
R & CC Ross and Cromarty Constabulary
RCCA Rickenbacker Car Club of America
RCCC Regular Common Carrier Conference; Republican County Central Committee
RCCE Regional Congress of Construction Employers
rC Ch Roman Catholic Church
RCCL Royal Caribbean Cruise Line
RCCLS Resource Center for Consumers of Legal Services
RCCP Royal Commission on Criminal Procedure
rccs revenue consequences of capital schemes; riots, civil commotions, and strikes
rcd received; relative cardiac dullness
rcd (RCD) record(ing)
RCD Regional Cooperation for Development (Pakistan, Iran, Turkey)
RCDA Retail Coin Dealers Association
RCDC Royal Canadian Dental Corps
RCDEP Rural Civil Defense Education Program
RCDI Reliability Control Departmental Instruction
RCDMS Reliability Central Data Management System
RCDs Royal Canadian Dragoons
RCDS Royal College of Defence Studies (UK)
rce rapid circuit etch(ing); remote-controlled equipment; right center entrance
RCE Reliability Control Engineering
RCEEA Radio Communications and Electronic Engineers Association
RCEME Royal Canadian Electrical and Mechanical Engineers
RCEP Royal Commission on Environmental Pollution
RCET Royal College of Engineering Technology; Rugby College of Engineering Technology
rcf recall finder; recall finding; relative centrifugal force
RCF Remote Call Forwarding (telephonic); Residential Care Facility
RCFA Reliability Control Fail-

ure Analysis; Royal Canadian Field Artillery
RCFCA Royal Canadian Flying Clubs Association
rcfm radiocommunication failure message
RCG Reception Guidance Center
RCGA Royal Canadian Golf Association
RCGP Royal College of General Practitioners
RCGS Royal Canadian Geographical Society
Rch Rochester
RCH Railway Clearing House; Resource Center for the Handicapped
RCHM Royal Commission on Historical Monuments (England)
rci radar coverage indicator; read channel initial(ize)
RCI Radio Canada International; Range Communications Instructions; Reichold Chemicals Incorporated; Research Council of Israel; Resident Cost Inspection; Resident Cost Inspector; Royal Canadian Institute
RCIA Retail Clerks International Association; Retail Credit Institute of America
RCIC Rumor Control and Information Center
R-C IP Roosevelt-Campobello International Park near Eastport, Maine in southern New Brunswick
rcirc recirculate
RCIs Recontres Culturelles International (French—International Cultural Meetings)
RCIU Retail Clerks International Union
rcj reaction-control jet
RCJ Royal Courts of Justice
RCJCLDS Reorganized Church of Jesus Christ of Latter Day Saints
RCK Research Centrum Kalkzandsteen Industrie (Dutch—Research Center for the Calcium Silicate Industry)
rcl runway center line
RCL ramped cargo lighter (naval designation); Royal Canadian Legion
R-class Soviet submarines named Romeo by NATO
rclm reclaim; reclamation
rcm radar countermeasure(s); radio-controlled mine; radio countermeasure(s); right costal margin

RCM Reliability Control Manual; Royal College of Midwives; Royal College of Music
RCMF Royal Commonwealth Military Forces
RCMP Royal Canadian Mounted Police
RCMPM Royal Canadian Mounted Police Museum, Regina, Saskatchewan
rcn reticulum cell neoplasms
RCN Reactor Centrum Nederland; Record Control Number; Republic of China Navy; Royal Canadian Navy; Royal College of Nursing
RCN Radio Cadena Nacional (Spanish—National Radio Chain)—Mexican broadcasting system
RCNC Royal Corps of Naval Constructors
RCNM Russell Cave National Monument
RCNR Royal Canadian Naval Reserve
RCNT Registered Clinical Nurse Teacher
RCNVR Royal Canadian Naval Volunteer Reserve
rco rendezvous compatible orbit
rco (RCO) remote-control oscillator; representative calculating operation
RCO Radio Control Office; Royal College of Organists
RCOA Radio Club of America; Record Club of America; Royal Concertgebouw Orchestra of Amsterdam
RCOC Royal Canadian Ordnance Corps
RCOG Royal College of Obstetricians and Gynecologists
R-complex reptilian complex (evolutionarily most recent part of the forebrain)
rcp recording control panel; reinforced concrete pipe; remote communications processor; reserved circuits program
RCP Regional Community Physician; Revolutionary Communist Party; Royal College of Pathologists; Royal College of Physicians; Royal College of Psychiatrists
RCPA Royal College of Pathologists of Australia; Royal College of Physicians of Australia
RCPI Royal College of Physicians—Ireland
RCPL Realtors Co-op Photo

Listing
RCPS Royal College of Physicians and Surgeons
rcpt receipt
rcr reader control relay; reverse contactor
RCR República de Costa Rica
RCRA Resource Conservation and Recovery Act
RCRBSJ Research Council on Riveted and Bolted Structural Joints
rcrd record
rcs radar cross-section; reloadable control storage
RCs Roman Catholics
RCS Reaction Control System; Rearward Communications System; Reentry Control System; Reliability Control Standard; Report Control Symbol; Residential Conservation Service; Royal College of Science; Royal College of Surgeons; Royal Commonwealth Society (formerly Royal Empire Society)
RCSB Royal Commonwealth Society for the Blind
RCSD Regional Council for Social Development
RCSE Royal College of Surgeons—Edinburgh
RCSI Royal College of Surgeons—Ireland
RCSS Random Communication Satellite System
RCST Royal College of Science and Technology
rct reversible counter
Rct Recruit
RCT Regimental Combat Team(s); Registered Clinical Teacher; Rorschach Content Test; Royal Corps of Transport
rctl rectal; resistor capacitor transistor logic
RCTT Regional Center for Technology Transfer (UN)
rcu remote control unit; research coordination unit
RCU Road Construction Unit
RCUEP Research Center for Urban and Environmental Planning (Princeton U)
rcv receive
rcv (RCV) radar control van; remote-controlled vehicle
rcvr receiver
RCVS Royal College of Veterinary Surgeons
RCWP Rural Clean Water Program
RCYB Revolutionary Communist Youth Brigade (Trotsky-

ite)
RCYC Royal Canadian Yacht Club; Royal Corinthian Yacht Club; Royal Cork Yacht Club
R Cy N Royal Ceylon Navy
RCYP Revolutionary Communist Youth Brigades
RCZ Radiation Control Zone; Rear Combat Zone
rd reaction of degeneration; readiness date; renal disease; required date; research and development (R & D); respiratory distress; restricted data; retinal detachment; roof drain; round; rutherford
rd (RD) red devil (seconal tablet)
r & d reamed and drifted; research and development
Rd Road
RD Air Lift International; Radio Denmark; República Dominicana; Restricted Data; Royal Dragoons; Royal Dutch Petroleum (stock exchange symbol); Rural Dean; Rural Delivery
R.D. Royal (Naval Reserve) Decoration
R/D Research/Development
R & D research and development
R of D Report of Debate
rda recommended daily allowance; recommended dietary allowance; right dorso-anterior
rd a (Rd A) reading age
RDA Railway Development Association; Reliability Design Analysis; Respiratory Diseases Association; Royal Docks Association
R & D A Research and Development Association
RDA Reader's Digest Almanac; República Democrática Alemana (Spanish—German Democratic Republic)—East Germany
RDAF Royal Danish Air Force
Rdam Rotterdam
RDAR Reliability Design Analysis Report
rdb research and development bond
rdb (RDB) radar decoy balloon
RDB Ramped Dump Barge; Research and Development Board; Royal Danish Ballet
rdbl readable
RDBMS Relational Database Management System
rd bot rubber diaphragm (stoppered) bottle

rdc rail diesel car; repository design condition; running down clause
RDC Rand Development Corporation; Research Diagnostic Criteria; Rural District Council
RDCA Rural District Councils' Association
rd/chk read/check
RDCO Reliability Data Control Office
rdd required delivery date
rd & d (RD & D) research, development, and demonstration
RD$ República Dominicana peso (Dominican currency)
rde receptor-destroying enzyme
r d & e research, development, and engineering (usually R D & E)
RDE Research and Development Establishment
R de C Radiodiffusion du Cameroun (French—Radio Network of Cameroon)
R de F Republica de Filipinas
R de J République de Djibouti (formerly French Somaliland or the Territory of Afars and Issas); Rio de Janeiro
R de O Rio de Oro (Spanish Sahara)
R de P República de Panamá; República del Paraguay; República Portuguesa
R de T Ralph de Toledano
rdf radio direction finder
RDF Rapid Deployment Force; Royal Dublin Fusiliers
Rdg Reading; Ridge
RDG Reading Railroad
R d'H République d'Haiti
rd hd round head
rdi recommended daily intake
RDI Royal Designer for Industry
RDL Radiocarbon Dating Laboratory (Florida State University); Ritter Dental Laboratories
RDLI Royal Durban Light Infantry
rdline read a line
RdlR Regiment de la Rey
rdm root drum
RDM Rand Daily Mail (Johannesburg)
Rdm3c Radarman, third class
rdmu range-drift measuring unit
rdn resource decision network
RDN Royal Danish Navy
rdo research and development objectives
RDO Radiological Defense Office(r)

rdo('s) regular day(s) off; research and development objective(s)
rdp radar detector processor; right dorso-posterior
RDP Regional Development Program(s); Repository Development Plan(ning)
RDPC Research Data Publication Center
rdpe radar data-processing equipment
RDP Lao República Democrática Popular Lao (Spanish— Lao Popular Democratic Republic)
RDPP Repository Development Program Plan(ning)
rd/q reading quotient
rdr radar
rdr (RDR) receiver data register
r dr rive droite (French—right bank)
RDR Reliability Diagnostic Report; Research and Development Report
rdr rel radar relay
rdrsmtr radar transmitter
rds respiratory distress syndrome
Rds Rixdllar; Roads; Roadstead
RDs Revolutionary Development teams; Royal Dockyards
RDS Research Defence Society; Royal Dublin Society; Rural Development Service; Rural Development Society
RD/S Royal Dutch/Shell
RD & S Research, Development, and Studies (USMC)
RD/SG Royal Dutch/Shell Group
rdt reserve duty training
rdt (RDT) remote data transmitter
RDT Regiment Danie Theron; Reliability Demonstration Test
R.D.T. Registered Dental Technician
RDT Repubblica Democratica Tedesca (Italian—German Democratic Republic)—East Germany
rdt & e (RDT & E) research, development, test, and evaluation
RDTF Rapid Deployment Task Force (US Marines)
rdu research and development utilization
RDU Royal Development Unit
RDUP Research and Development Utilization Project
R du Z République du Zaïre

(French—Republic of Zaire)
rdvu rendezvous
RDW Regiment De Wet
Rdwy Roadway
rdx cyclonite (research department explosive)
RDX Research and Development Exchange
rdy ready
RDY Royal Dock Yard
RDZ Radiation Danger Zone
RDZ République Démocratique du Zaïre (French—Democratic Republic of Zaire)—formerly the Belgian Congo
rdz(s) (RDZ or RDZs) radiation danger zone(s)
re radium emanation; real estate; reinforce(d); reinforcing; research and engineering (R & E); reticulo-endothelium; right eye
re (RE) revised edition
r/e rate of exchange
re B in diatonic scale, *D* in fixed-do system; (Italian— second tone) (Latin prefix— again or back)—reflect, repair, restate
Re real part (symbol); Reno; Reynold's Number; rhenium; rupee (Ceylon, India, Pakistan currency)
R_e récipe (Spanish—recipe; prescription)
RE Radio Eireann (Radio Ireland); Reformed Episcopal (church); Reliability Engineering; Religious Education; *República de Ecuador*; Rifle Expert; Right Excellent; Royal Engineers; Royal Exchange
rea right ear advantage
REA Railway Express Agency; Request for Engineering Authorization; Rubber Export Association; Rubber Export Association; Rural Education Association; Rural Electrification Administration (US Department of Agriculture)
reac reactor
REAC Reeves electronic analog computer; Reliability Engineering Action Center
REACH Rape Emergency Aid and Counseling for Her; Retired Executives Action Clearinghouse
reack receipt acknowledged
react reactance; reaction; reactor; register-enforced automated-control technique
REACT Radio Emergency Associated Citizens Team;

Register-Enforced Automated Control Technique; Resource Allocation and Control Techniques
READ Real-Time Electronic Access and Display
Read Dig Reader's Digest
readi rocket-engine-analyzer-and-decision-instrumentation
readm readmission
READS Reno Air Defense Sector
Reagan Ronald Reagan, 40th President of the United States
Reaganomics economic policy of the administration of President Reagan
REAL Rape Emergency Assistance League; *Real-Aerovias do Brasil* (Portuguese—Brazil Air Lines); Residential Experience in Adult Living
realcom real-time communication(s)
real est real estate
realgar arsenic sulfide
Realm of the Chinese Alligator lower Yangtze River valley
ream. rapid excavation and mining
REAMS Ramond Electronically Applied Maintenance Standards
REAP Revenue Enforcement and Protection Program; Rural Environmental Assistance Program
reapt reappoint; reappointment
REAR Reliability Engineering Analysis Report
Rear Adm Rear Admiral
reasm reassemble
REAT Radiological Emergency Assistance Team
Réau(m) Réaumur
reb rebel; rebellion
Reb Reba; Rebecca; Rebekah
REB Regional Examining Body
REB Revised English Bible
Reba Rebecca
rebar reinforcing (steel) bar
Rebilds Denmark's Rebild Hills including the Rebild National Park
reb(s) rebel(s)
rec receipt; receive; recessed; record; recreation
rec. recens (Latin—fresh)
Rec Recife
REC Recife, Brazil (airport); Rural Electrification Corporation
R & EC Research and Engineering Council
reca repetitive-element column

analysis
recap recapitulate; recapitulation
RECAP Reliability Evaluation Continuous Analysis Program
RECC Rhine Evacuation and Control Command (NATO)
rec chg record change(r)
recco reconnaissance
recd received
recep reception
recg radioelectrocardiograph
R & ECGAI Research and Engineering Council of the Graphic Arts Industry
rech *recherche* (French—research)
rec hall recreational hall
reci recitation
recid recidivism; recidivist(ic); recidivous
recids recidivists
recip reciprocating
recipe. recomp computer interpretive program expeditor
recip & lp turb reciprocating steam engine and low-pressure turbine
recirc recirculate; recirculation
recit. *recitativo* (Italian—recitative)
reclam reclamation
recm recommend
recmark record mark(ing)
RECMF Radio and Electronic Component Manufacturers Federation
recncln reconciliation
recog recognition; recognize
recol retrieval command language
recom recommendation; recommend(ed)
recomp recomplement(ary); repairs completed; retrieval composition
recompen recognized company pension
recon reconcentration; reconciliation; recondite; recondition; reconduction; reconnaissance; reconnoiter; reconsign; reconsigned; reconsignment; reconstruct; reconstructed; reconstruction; reconversion; reconvert; reconverted; reconvey; reconveyance; reconveyed
RECON Regional Communication Outreach Network; Retrospective Conversion of Bibliographic Records (Library of Congress)
recond recondition
R Econ S Royal Economic Society

RECONS Reliability and Configurational Accountability System
reconst reconstruct
recov recover; recovery
recp receptacle; reciprocal; reciprocating
RECP Rural Environmental Conservation Program
recpt receptionist
recr receiver
rec room receiving room; reception room; record room; recreation room
recryst recrystallize
Rec S Record of Survey
RECSAM Regional Center for Education in Science and Mathematics
Rec Sec Recording Secretary
RECSTA Receiving Station
recsys recreational systems analysis
rect (Latin prefix—straight)—rectified; rectifier; rectify; rectitude
rect. *rectificatus* (Latin—rectified)
Rect Rector(y)
recto obverse; right-hand page
rectr recommends transfer
recur. recurrence; recurrent; recurring
rec vehicle(s) recreation vehicle(s)—campers, dune buggies, snowmobiles, trailers, vans, etc.
red. reduce; reduction
red *redaktör* (Swedish—editor); *redigé* (French—compiled; edited)
Red *Rederi* (Scandinavian—shipowners)
RED Real Estate Department
REDAR R E Darling (Company)
red burgee red signal flag flown when explosives or flammable fuel is being loaded aboard a vessel; letter B or Bravo in the international code
redcape readiness capability
redcat readiness requirement
redcon readiness condition
Redcraft Red aircraft (communist-controlled aircraft)
redec redecorate
redig *redigerat* (Swedish—edited)
redig. in pulv. *redigatur in pulverem* (Latin—reduce to powder)
REDIS Redistricting System
redisc rediscount
redist redistilled

REDLARS Reading Literature Analysis and Retrieval Service
red light danger signal; port side; stop signal; warning signal
red ochre reddle (hematite red)
redox reduction oxidation
red. in pulv. *reductus in pulverem* (Latin—reduced to a powder)
red ru red kangaroo
redsg redesign; redesigned; redesigning
redsh reddish
redup(l) reduplicate; reduplication
redux reduction
Redwood Redwood City, Redwood Empire, Redwood National Park
ree rare-earth elements
REE Regional Economic Expansion (Canada)
REECO Reynolds Electrical and Engineering Compay
Reed *Reederei* (German—shipowners)
reef The Reef—Australia's Great Barrier Reef off the coast of Queensland
Reefer(s) inhabitant(s) of the Great Barrier Reef
reeg radioelectroencephalograph
REEGT Registered Electroencephalographic Technicians
Reen Irene
reenl reennlist
reep range estimating and evaluation procedure
ref refer; referee; reference; reformatory; refraction; refresher
ref (REF) renal erythropoietic factor
ref *refondue* (French—reorganized)
Ref reference
Ref *Referate* (German—abstract, compedium)
REF Railway Engineers Forum; Reject Errors in Football; Romanian Engineers Forum
refash refashion(ed)
Ref Ch Reformed Church
refcom refuse conversion to methane
REFCORP Resolution Funding Corporation
refd refund
refd conc reinforced concrete
ref dent referring dentist
refd met reinforced metal
ref doct referring doctor
refd ply reinforced plywood

ref eso reflux esophagitis
reffo refugee from Europe
refg refrigerating; refrigeration
refl reflection; reflective; reflector; reflex; reflexive
ref l reference line
reflecs retrieval from literature on electronics and computer science
Ref Libr Reference Librarian
refl pron reflexive pronoun
Reform Reformatory
Reforma National Association of Spanish-Speaking Librarians in the United States
reforst reforestation
refphocon reference to telephone conversation
ref phys referring physician
ref press reference pressure
refr refraction; refractory; refrigerate; refrigerator
refrg refrigerate; refrigeration; refrigerator
refrig refrigeration; refrigerator
Refrig Eng Refrigerating Engineering
refs references
ref temp reference temperature
reftra refresher training
refurb refurbish(ed)
refy refinery
Ref Zhu Referativnyi Zhurnal (Russian—Abstract Journal)
reg region; regular; regulate; regulation
reg (REG) register (flow chart)
Reg Registered
Reg Regina (Latin—queen)
RegAF Regular Air Force
regal. range and elevation guidance for approach and landing; remote generalized application language
Reg Arch Registered Architect
Reg Bez Regierungsbezirk (German—administrative district)
reg bot regular bottle (3/4-liter of wine)
regd registered
regen regenerate; regeneration
Regg Reggimento (Italian—Regiment)
Reg Gen Registrar General
Reggie Regina(ld)
Reg(gie)(y) Reginald
Reggio Reggio di Calabria; Reggio nel'Emilia
regis register; registered; registration; registry
reg'lar regular
Regnery Henry Regnery
Reg P Regent's Park College, Oxford
Reg Prof Regius Professor

Regr Registrar
regs regions; regulars; regulations
regt regiment
Reg TM Registered Trade Mark
regu regulable; regular; regularize; regularly; regulate; regulation; regulator
regurg regurgitant; regurgitate; regurgitation
REGY Regional Employment Growth (program for) Youth
reh rehearsal
rehab rehabilitate
Rehab Department of Rehabilitation
Rehab Dept Rehabililation Department
rehob rehoboam (6-bottle capacity)
REI Régie Aérienne Interinsulaire
R & EI Religion and Ethics Institute
REIC Radiation Effects Information Center; Rare Earth Information Center (Atomic Energy Commission, Ames Laboratory, Iowa State University)
Reichenhall Bad Reichenhall
reig rare-earth iron garnets
reils runway end identification lights
reimb reimburse; reimbursement
reincorp reincorporate(d)
reinf reinforce(d); reinforcing
reinfmt reinforcement
reins. radio-equipped inertial navigation system
REINS Radio-Equipped Inertial Navigation System
reit reiteration
REIT Real Estate Investment Trust
REIWA Real Estate Institute of Western Australia
rej reject; rejected; rejection
rejase re-using junk as something else
rejn rejoin
REK Reykjavik, Iceland (airport)
rekenk rekenkunde (Dutch—arithmetic)
rel rate of energy loss; relation; relative; relay; release; relief; relieve; religion; religionist
rel relie; reliure (French—bound, binding)
REL Radio Engineering Laboratories; Robert Edward Lee (1807–1870)
RELACS Radar Emission Lo-

cation Attack Control System
rel adv relative adverb
RELC Reformation Evangelical Lutheran Church; Regional Educational Laboratory for the Carolinas
RELCV Regional Educational Laboratory for the Carolinas and Virginia
RELHS Robert E Lee High School
rel hum relative humidity
reliab reliability
relig religion; religious
reliq. reliquus (Latin—remainder)
reloc relocate; relocated; relocation
rel pron relative pronoun
Rel R Reliability Report
RELS Rapidly Extensible Language System
rem rapid eye movements; remain(ing); remission; remit; remittance; removable; remove; removed; roentgen equivalent, man
Rem Remington; roentgen equivalent, man
REM Registered Equipment Management
REMA Refrigeration Equipment Manufacturers Association
remab radiation equivalent manikin absorption
remad remote magnetic anomaly detection
Remarkables Remarkable Range of mountains in New Zealand's South Island
remc resin-encapsulated mica capacitor
REMC Radio and Electronics Measurements Committee; Regional Educational Media Center
remcal radiation equivalent manikin absorption
remd rapid eye movement (sleep) deprivation
REME Royal Electrical and Mechanical Engineers
REML Radiation Effects Mobile Laboratory
REMP Research Group for European Migration Problems
rem(s) rémora(s)
rems (REMS) rapid-eye-movement sleep
REMS Registered Equipment Management System
REMSA Railway Engineering Maintenance Suppliers Association
rem sleep rapid-eye-movement

(paradoxical) sleep
remstar remote electronic microfilm storage transmission and retrieval
REMT Radiological Emergency Medical Teams
Rem-UMC Remington-Union Metallic Cartridge (company)
remus routine for executive multi-unit simulation
ren. renovetur (Latin—renew)
Ren Renaissance
RENAMO Resistência Nacional Moçambicana (Portuguese—Mozambique National Resistance)
rene rocket-engine nozzle ejector
Renf Renfrew
RENFE Red Nacional de los Ferrocarriles Españoles (Spanish—National Network of Spanish Railroads)
RENS Reconnaissance Electronic Warfare and Naval Intelligence System
ren. sem. renovetum semel (Latin—renew only once)
rent. reentry nose tip
renv renovate; renovation
reo rare-earth oxide; regenerated electrical output
Reo early American automobile named after initials of its maker, Ransom E Olds of Oldsmobile fame
REO Ransom Eli Olds, automobile inventor and manufacturer (1864–1950); Regional Education Officer
reoc report when established on course (aviation)
reopt reorder point
REORG reorganization; reorganize; reorganized
reorgn reorganization
REOs Real-Estate-Owned banking departments
REOS Reflective Electron Optical System
reo viruses respiratory-enteric-orphan viruses
rep repair; repeat; repertory; represent; representative; reputation
rep. reparation; report; representative;
r-ep rational-emotive psychotherapy
rep reparto (Italian—department)
rep. repetatur (Latin—let it be repeated)
Rep Representative; Republic; Republican; Republican Party; roentgen equivalent, physical

REP Radical Education Project; Recovery and Evacuation program; Republic Corporation (stock exchange symbol); Research Expenditure Proposal; Reserve Enlisted Program; River Engineering Program
Rep Árabe Yem República Árabe del Yemen (Spanish—Arabic Republic of Yemen)—formerly British Crown colony of Aden
Rep Arg República Argentina (Spanish—Argentine Republic)
Rep de Bol República de Bolivia (Spanish—Republic of Bolivia)
Rep Bot Republic of Botswana
Rep Cabo Verde República de Cabo Verde (Portuguese—Cape Verde Republic)
Rép Cent République Centrafricaine (French—Central African Republic)
Rep Chile República de Chile (Spanish—Republic of Chile)
Rep Col República de Colombia (Spanish—Republic of Colombia)
Rép Côte d'Ivoire République de la Côte d'Ivoire (French—Republic of the Ivory Coast)
Rep CR República de Costa Rica (Spanish—Republic of Costa Rica)
Rep Day Republik Dayti (Haitian Creole—Republic of Haiti)
Rep de Cuba República de Cuba (Spanish—Republic of Cuba)
Rep Dem Mal Repoblika Demokratika Malagasy (Malagasy—Democratic Republic of Madagascar)
Rep Dem Pop Yem República Democratica Popular del Yemen (Spanish—Popular Democratic Republic of Yemen)
Rep Dem Sao Tome Prin República Democrática de Sao Tome e Principe (Portuguese—Democratic Republic of Sao Tome and Principe)
Rép d'Haiti République d'Haiti (French—Republic of Haiti)
Rep Dom República Dominicana (Spanish—Dominican Republic)
Rep Ecu República del Ecuador (Spanish—Republic of Ecuador)
Rep El S República de El Salvador (Spanish—Republic of El Salvador)

El Salvador)
Rep Fed Bra República Federativa do Brasil (Portuguese—Federative Republic of Brazil)
Rép Fran République Francaise (French Republic)
Rép Gab République Gabonaise (French—Gabonese Republic)
Rep Ghana Republic of Ghana
Rep Gua República de Guatemala (Spanish—Republic of Guatemala)
Rép Gui République de Guinée (French—Republic of Guinea)
Rep Gui-Bis República de Guiné-Bissau (Portuguese—Republic of Guinea-Bisau)
Rep Gui Ecu República de Guinea Ecuatorial (Spanish—Republic of Ecuatorial Guinea)
Rep Hond República de Honduras (Spanish—Republic of Honduras)
Rep Ind Republik Indonesia (Malay—Republic of Indonesia)
Rep Isl Maur République Islamique de Mauritanie (French—Islamic Republic of Mauritania)
Rep Ital Repubblica Italiana (Italian Republic)
Rep Kiri Republic of Kiribati
Rep Lib Republic of Liberia
Rep Mal Republic of Malawi
Rép Mali République du Mali (French—Republic of Mali)
Rep Malta Repubblika ta' Malta (Maltese—Republic of Malta)
Rep Nauru Republic of Nauru
Rep Nic República de Nicaragua (Spanish—Republic of Nicaragua)
Rép Nig République du Niger (French—Republic of Niger)
Rep Ori Uru República Oriental del Uruguay (Spanish—Oriental Republic of Uruguay)
Rep Öst Republik Österreich (German—Austrian Republic)
Rep de Pan República de Panamá (Spanish—Republic of Panama)
Rep Para República del Paraguay (Spanish—Republic of Paraguay)
Rep Peru República del Peru (Spanish—Republic of Peru)
Rep Phil Republic of the Philippines
Rep Pop de Ang República Popular de Angola (Portuguese—People's Republic of

Angola)
Rép Pop Ben *République Po-
pulaire du Benin* (French—
People's Republic of Benin)
Rép Pop Con *République Po-
pulaire du Congo* (French—
People's Republic of the
Congo)
Rep Pop Moç *República Popu-
lar de Moçambique* (Portu-
guese—People's Republic of
Mozambique)
Rep Pop Soc e Shq *Republika
Popullore Socialiste e Shqipë-
risë* (Albanian Popular Social-
ist Republic)
Rep Port *República Portugue-
sa* (Portuguese—Republic of
Portugal)
Rep Rwa *Republika y'u Rwan-
da* (Swahili—Republic of
Rwanda)
Rép Seneg *République du
Sénégal* (French—Republic of
Senegal)
Rep Sey Republic of the Sey-
chelles
Rep Sierra Leone Republic of
Sierra Leone
Rep Singa Republic of Singa-
pore
Rep Soc Rom *Republica So-
cialista România* (Romanian
—Socialist Republic of Ro-
mania)
Rep Suid-Afrik *Republiek van
Suid-Afrika* (Afrikaans—Re-
public of South Africa)
Rep Sur *Republiek Suriname*
(Dutch—Suriname Republic)
Rep Suri Suriname Republic
Rép Tch *République du Tchad*
(French—Republic of Chad)
Rep The Gam Republic of The
Gambia
Rép Togo *République Togo-
laise* (French—Republic of
Togo)
Rep T & T Republic of Trini-
dad and Tobago
Repub Republic; Republican
Rep Ugan Republic of Uganda
Rep V Repair Locker 5 (Engi-
neering)—USN
Rep Ven *República de Venezu-
ela* (Spanish—Republic of
Venezuela)
Rep y'Ub *Republika y'Uburun-
di* (Rundi—Republic of Bu-
rundi)
Rép Zaire *République du Zaire*
(French—Republic of Zaire)
Rep Zambia Republic of Zam-
bia
REPA Research and Engineers
Professional Employees Asso-

ciation
REPC Racial Ethnic Parent
Councils; Regional Economic
Planning Council
repcon rain repellant and surface conditioner
REPE Radio Engineering Europe
reperf reperforator
repl replace(d); replacement;
replacing
repltr report (by) letter
repm repairman; repairmen
REPM Representatives of Electronic Products Manufacturers
repo repossess; repossessed; repossession
repo men repossession men
repop repetitive operation(s)
repo(s) repurchase agree-
ment(s)
reppac repetitively-pulsed plasma accelerator
Rep Prog Phys *Reports on
Progress in Physics*
repr repairman; representative;
reprint; reprinted; reprinting
repro reproduce; reproducing;
reproduction
reprosex reproductive sex
repro typ reproduction typist;
reproduction typing
reps repetitive electromagnetic
pulse simulator; representatives
Rep(s) Republican(s)
REPS Rail(way) Express Parcel Service
rep. sem. *repetatur semel* (Latin—let it be repeated once)
rept report; reprint; reptile;
reptilia(n)
rept (Rept) report
rept. *repetatur* (Latin—let it be
repeated)
Rept Reptilia
repub republication; repub-
lish(ed)
REPUBLIC Republic Aviation
Corporation
Republocrat Republican Democrat
Repubs Republicans
req request; require
reqafa request advise as to further action
reqd required
reqdi request disposition instructions
reqfolinfo request following information
reqid request if desired
reqmad request mailing address
reqmt requirement
reqn requisition

reqrec request(ed) recommendation
reqs requires
reqssd request supply status
(and expected delivery) date
reqsupstafol request supply status of following
reqt requirement
reqtat requested that
requint request interim (reply)
rer (RER) radar effects reactor
RER Railway Equipment Register
RERC Real Estate Research
Corporation
REREI Redwood Empire Research and Education Institute
rereq reference requisition
RERF Radiation Effects Research Foundation
rerl residual equivalent return
loss
RERO Royal Engineers Reserve of Officers
res rescue; research; researcher;
reservation; reserve; reservoir;
residence; resilient; resistant;
respiratory; reticuloendothelial
system (RES)
res (RES) restore (computer
character)
Res Reservation; Reservoir
RES *República de El Salvador*;
Royal Economic Society; Royal Entomological Society
RESA Regional Educational
Service Agencies; Regional
Educational Service Areas;
Research Society of America
ResAF Reserve of the Air
Force
Res Aud Resident Auditor
resc rescue
RESC Regional Educational
Service Centers
RESCAM Regional Center for
Education in Science and
Mathematics
rescan reflecting satellite com-
munication antenna
RESCO Refuse Energy Sys-
tems Company
rescu rocket-ejection seat cata-
pult upward
RESCU Radio Emergency
Search Communications Unit
rescue. remote emergency sal-
vage and cleanup equipment
Res & Educ Research and Education Association
reser reentry system evaluation
radar
resgnd resigned
resid residual; residual oil
resig resignation
RESIG Research and Engineer-

ing System Integration Group
resil resilient
resist. resistance; resistor
resistojet resistance-connective jet engine
resojet resonant pulse jet
resp respective; respelling; respiration(s); respirator; respire; responder; responsibility; responsible; responsive
RESPA Real Estate Settlement Procedures Act
Res Phys Resident Physician
respir respiration; respiratory
respirol respirologic(al)(ly); respirologist; respirology
respirom respirometer; respirometric(al)(ly); respirometrist; respirometry
RESPO Responsible Property Officer
RESPONSA Retrieval of Special Portions from Nuclear Science Abstracts
respub responsible Republican(ism)
Resrt Resort
RESS Radar Echo-Simulation Study; Radar Echo-Simulation System
Res Sec Resident Secretary
RESSI Real Estate Securities and Syndication Institute
rest restrict; restricted; restriction
rest. (REST) regressive electric shock therapy
REST Radar Electronic-Scan Technique; Reentry Environment and Systems Technology; Reentry System Test Program; Routine Execution Selection Table
resta reconnaissance, surveillance, and target acquisition
restr restaurant
ResTraCen Reserve Training Center
RESTTA Restitution Education, Specialized Training, and Technical Assistance program, funded by the Office of Juvenile Justice and Delinquency Prevention, U.S. Department of Justice
resub resublimed
resup resupply
resvr reservoir
RE system reticuloendothelial system
ret rational emotive therapy; retainer; retaining; retire; retirement
ret (RET) rational-emotional therapy; return (flow chart)
r-et rational-emotive psycho-

therapy
Ret Reticulum (constellation); retired
RET R. Emmett Tyrrell, Jr
RET *Rotterdamse Elektrische Tram* (Dutch—Rotterdam Electric Tramway)—electric surface car and subway system
reta retrieval of enriched textual abstracts
RETA Refrigerating Engineers and Technicians Association
Retail Clerks Union Retail Clerks International Association
retain. remote technical assistance and information network
retard. retardation; retarded
retc railroad equipment trust certificate
RETC Regional Employment Training Center; Regional Employment and Training Consortium
retd retired
rete (Latin prefix—network)—retinal
R. et I. *Regina et Imperatrix* (Latin—Queen and Empress)—title of Victoria—Queen of England and Empress of India—The Queen
retic reticulate(d); reticulation; reticule
retic count reticulocyte count
retics reticulocytes
retl retail
RETL Rocket Engine Test Laboratory
RETMA Radio-Electronics-Television Manufacturers Association
retng retraining
retnr retainer
retort. (RETORT) reason and equity in tort
ret p retired pay
RETP Reliability Evaluation Test Procedure
retpd retention period
retr retractable
RETRA Radio, Electrical, and Television Retailers Association
Ret Res Retirement Research
retro retroactive; retrofit; retrograde; retrorocket
retro (Latin prefix—backward or behind)—retroactive, retrograde
retros retrogrades; retrorockets
RETS Renaissance English Text Society
Reun Reunion Island

Reuter's Reuter's international news agency
rev reverse; reversed; review; revise; revised; revision; revolute; revolution
rev (REV) reentry vehicle
rev revisado (Spanish—revised)
Rev Reverend; Review, Revised, Revue; The Revelation of St John the Divine
Rev Revelation
reva recommended vehicle adjustment
rev a/c revenue account
Revd Reverend
Rev d'Opt *Revue d'Optique* (French—Optics Review)
rev ed revised edition
revel. reverberation elimination
Revell Fleming H Revell
revid reviderad (Swedish—revised)
Revilla Gigedos Revilla Gigedo Islands off Mexico's west coast, not to be confused with Revilla Gigedo Island off Alaska
rev/min revolutions per minute
Rev Mod Phys *Reviews of Modern Physics*
revocon remote volume control
revolving-door revolving-door criminal-justice system persisting in returning dangerous defendants to their communities again and again
revr reviewer
Rev Rul Revenue Ruling
revs revolutions
rev(s) revolution(s)
rev/s revolutions per second
REVS Rotor-Entry Vehicle System
Rev Sci Instrum *Review of Scientific Instruments*
rev/sec revolutions per second
Rev Stat Revised Statutes
rev of sym review of symptoms
Rev Ver Revised Version of the Bible
rew reward; rewind(ing)
rewdac retrieval by title words, descriptors, and classifications
rewk reword
rewrc report when established well to right of course
REWSON Reconnaissance Electronic Warfare Special Operations and Naval Intelligence Processing System(s)
rex real-time executive routine; reduced exoatmospheric cross-section
Rex Reginald
REX Rexall Drug and Chemical (stock exchange symbol)

rexs (REXS) radio-exploration satellite

Rex trem *Rex tremendae* (Latin—King of Tremendous Majesty)

Reykjvk Reykjavik

Reynall Reynal & Co

rf radiofrequency; range finder; rapid fire; rat fink; reception fair; reflight; relative flow; replacement factor; representative fraction; rheumatic fever; rheumatoid factor; right field; right full-back; rim fire; rubber-free

r-f radiofrequency

r/f right front

r_f rate of flow

rf *rinforzando* (Italian—reinforcing)

Rf Reef; rutherfordium, also known as unnilquadium

RF *République Française*; Reserve Force; Rockefeller Foundation; Rocky Flats; Rodeo Foundation; Royal Fusiliers

R-F Reitland-Franklin (unit)

rfa radiofrequency attenuator; radiofrequency authorization(s); request further airways; right fronto-anterior

RFA *Repúblique Fédérale Allemande* (French—Federal Republic of Germany); Royal Field Artillery; Royal Fleet Auxiliary

RFA *República Federal de Alemania* (Spanish—Federal Republic of Germany)

RFAC Royal Federation of Aero Clubs; Royal Fine Arts Commission

R factor resistance factor

rfad release for active duty

rfa's return(ed) for alterations (tailoring)

rfb request for bid

RFB Recording for the Blind

RFB *República Federativa do Brasil* (Portuguese—Federal Republic of Brazil)

rf black reinforcing furnace black

rfc radiofrequency choke

RFC Rare Fruit Council; Reconstruction Finance Corporation; River Forecast Center; Royal Flying Corps; Rugby Football Club

RFCL Referral Form Checklist

rfcs radio-frequency carrier shift

RFCWA Regional Fisheries Commission for Western Africa

rfd raised foredeck; reentry flight demonstration; refund; reinforced; reporting for duty

RFD Radio Frequency Devices; Rural Free Delivery

rfd con reinforced concrete

rfd met reinforced metal

rfd ply reinforced plywood

rfdr rangefinder

RFDS Royal Flying Doctor Service

RFE Radio Free Europe

RFED Research Facilities and Equipment Division (NASA)

rff remote-fiber fluorimetry

R f F *Rat für Formgebung* (German—Fashion Council)

RFF *Rede Ferroviária Federal* (Portuguese—Federal Railway System)—Brazil

RFFS River and Flood Forecasting Service

RFFSA *Rede Ferrocarril Federal Sedada Anonima* (Portuguese—Federal Railway Route Company)—Brazil

rfg roofing

RFH Royal Festival Hall

rfi radiofrequency interference; ready for issue

rfing royal fucking

rf/ir radiofrequency/infrared

R Fix running fix

rfl refuel(ing); right frontolateral

RFL Refrigerated Freight Lines; Rugby Football League

Rflmn Rifleman

RFLPs Restriction-Fragment-Length Polymorphisms

rfls rheumatoid factor-like substance

rfm radio frequency management

r-f m ripple-flow mill (grain)

RFMA Reliability Figure of Merit Analysis

RFMF Royal Fiji Military Forces

Rfn Rifleman

RFN Registered Fever Nurse

rfna red-fuming nitric acid

rfnip reduced-flow nominal-inlet pressure

rfnop reduced-flow nominal-output pressure

RFNZJ Royal Federation of New Zealand Justices

rfo request for factory order

RFO Regional Fisheries Office(r)

rfp right frontoposterior

RFP Request for Proposal

RF & P Richmond, Fredericksburg and Potomac (railroad)

RFPs Requests for Proposals

RFPS(G) Royal Faculty of Physicians and Surgeons of Glasgow

RFQ Request for Quotation

rfr refraction; reject failure rate; required freight rate

R fr Ruanda franc(s)

RFR Royal Fleet Reserve

rfrd referred

rfs radio-frequency surveillance; ready for sea; regardless of future size

Rfs Reefs

RFS Registry of Friendly Societies; Royal Forestry Service

rf scale representative fraction scale

rfs/ecm radio-frequency surveillance/electronic countermeasures

RFSU *Riksføbundet før Sexuall Upplysning* (Norwegian—National League for Sexual Education); Rugby Football Schools' Union

rft right frontotransverse

RFT Rod and Frame Test

RFT *Repubblica Federale Tedesca* (Italian—German Federal Republic)

rfts radiofrequency test set

rfu ready for use

RFU Rugby Football Union

R-F unit Reitland-Franklin unit

rfw rapid-filling wave

RFW Radio Free Women

rfwe ring-finished with engines

Rfy Refinery

RFYC Royal Forth Yacht Club

rfz restrictive fire zone

rfz *rinforzando* (Italian—with extra emphasis)

rg real girl (not a birl); re-gummed; repetitive group(ing)

r g *rive gauche* (French—left bank)

Rg *Ruckgang* (German—retrogression)—in sonatas

RG *República de Guatemala*; Reserve Grade

RG *Reader's Guide to Periodical Literature; Rive Gauche* (French—Left Bank)

rga rate gyro assembly

Rga Riga

RGA Republican Governors Association; Royal Garrison Artillery; Rubber Growers' Association

RGAHS Royal Guernsey Agricultural and Horticultural Society

R-gauge Russian gauge (5-foot) railroad track

rgb red-orange, green, blue-vio-

let (television's triad of primary colors)
RGC Reception and Guidance Center
rgd reigned
R Gd Rio Grande
RGDATA Retail Grocery, Dairy, and Allied Trades Association
RG do S Rio Grande do Sul
rge relative gas expansion
Rge Range; Ridge
RGE Rat der Gemeinden Europas (German—Council of European Municipalities); *República de Guinea Ecuatorial* (Spanish—Republic of Equatorial Guinea)
RGEB Rockefeller General Education Board
R Gen Registrar General
RGEPS Rucker-Gable Educational Programming Scale
rgf range-gated filter
RGF Red Guerrilla Family (black terrorists)
RGG Royal Grenadier Guards
RGH Royal Gloucestershire Hussars
RGI Robert G Ingersoll; Royal Glasgow Institute of Fine Arts
RGJ Royal Green Jackets
rgl regulate; regulation; regulatory
rgm residential growth management
RGM Revenue Generation and Management
rgn region
Rgn (Port of) Rangoon
RGN Rangoon, Burma (airport); Registered General Nurse
RGNR Rugged Glen Nature Reserve (South Africa)
RGO Royal Greenwich Observatory
RGP Riegel Paper Company (stock-exchange symbol)
RGPL Readers' Guide to Periodical Literature
RGPM Regional Geological Project Manager
rgr reference geological regime
rgs radar ground stabilization
RGS Rio Grande do Sul; Royal Geographical Society
RGSA Royal Geographical Society of Australasia
Rgt Regiment
RGTC Robert Gordon's Technical College
RGTF Royal General Theatrical Fund
Rgtl Regimental
rg tp rough template

RGV Rio Grande Valley Gas Company (stock exchange symbol)
rgz recommended ground zero
RGZ Rio Grande Zoo (Albuquerque)
rh rheumatic; rheumatism; rheumatoid; righthand (RH); roundhead
r/h relative humidity; roentgens per hour
rh. rhonchi (Latin—rales)
Rh Rhesus factor (symbol); rhodium
Rh+ Rhesus positive
Rh– Rhesus negative
Rh Rhein (German—Rhine)
RH Air Rhodesia; Random House; República de Honduras; Round House; Royal Highlanders; Royal Highness; Ryan Herco
RH Rechte Hand (German—right hand); *Research Highlights*
RH¹⁰⁶ radioactive rhodium
RHA Regional Health Authority; Road Haulage Association; Royal Hibernian Academy; Royal Humane Association; Rural Housing Alliance
RHAF Royal Hellenic Air Force
R Hamps Royal Hampshire (regiment)
rhap rhapsody
RHAWS Radar Homing and Warning System
RHB Regional Hospital Board; Robin Hood's Bay
rhbdr rhombohedral
rhc respirations have ceased; rubber hydrocarbon
RHC Road Haulage Cases; Rosary Hill College
RHC Radio Habana Cuba (Spanish—Havana, Cuba Radio)
RHCSA Regional Hospitals Consultants' and Specialists' Association
rhd radioactive health data; relative hepatic dullness; rheumatic heart disease
RHD Robin Hood Dell (Philadelphia)
RHD Random House Dictionary
RHDEL Random House Dictionary of the English Language
RHDO Robin Hood Dell Orchestra
rhe reversible hydrogen electrode
RHE Reliability Human Engineering

RHEL Rutherford High-Energy Laboratory
rheo rheostat
rheol rheological; rheology
rhet rhetoric; rhetorical; rhetorician
rheu rheumatic; rheumatism; rheumatoid
rheu fev rheumatic fever
rheu ht dis rheumatic heart disease
rheum rheumatic; rheumatism
rhf right heart failure
RHF Royal Highland Fusiliers
Rh factor Rhesus group of red cell agglutinogens
RHG Royal Horse Guards
RHGPS Rhodesian Hunters and Game Preservation Society
RHHI Royal Hospital and Home for Incurables
rhi range height indicator
RHIB Rain and Hail Insurance Board; Rain and Hail Insurance Bureau
rhin (Latin prefix—nose)—rhinitis
rhino range height indicator not operating
rhinol rhinologic(al)(ly); rhinologist; rhinology
rhino(s) rhinoceros(es)
rhip rank has its privileges
rhir rank has its responsibilities
R Hist S Royal Historical Society
RHIT Rose-Hulman Institute of Technology
RHK Radio Hong Kong
RHKAAF Royal Hong Kong Auxiliary Air Force
RHKP Royal Hong Kong Police
RHKPF Royal Hong Kong Police Force
RHKR Royal Hong Kong Regiment
RHKTV Royal Hong Kong Television
RHKYC Royal Hong Kong Yacht Club
rhl rectangular hysteresis loop
RHL Radiological Health Laboratory; Rape Help Line
rhm roentgen per hour per meter
RHMG Rogers House Museum Gallery
RHMS Royal Hibernian Military School
RHN Royal Hellenic Navy
Rho Rhoda
RHO Regional Hospital Office(r); Rickwell Hanford Operations; Rural Health Of-

fice(r)
RHOB Rayburn House Office Building
Rhod Rhodesia
Rhoda Rhodacella; Rhodacelle
Rhodes Cecil John Rhodes (1853–1902)
rhodies rhododendrons
rhodo(s) rhododendron(s)
Rhodos (Greek—Rhodes)—island in the Aegean
RHOFLIGHT Rhodesian Air Services
rhom rhombic; rhomboid; rhombus
rhp rated horsepower
RHQ Regimental Headquarters
rhr roughness height reading
r/hr roentgens per hour
RHR Royal Highland Regiment (Black Watch)
rhs righthand side; round-headed screw
RHS Radio Ham Shack (amateur radio operator's station); Royal Historical Society; Royal Horticultural Society
RHSI Royal Horticultural Society of Ireland
RHSNZ Royal Humane Society of New Zealand
RHSV Royal Historical Society of Victoria
Rhumba (stock exchange form for Royal McBee Company whose symbol is RMB)
RHV République de Haute-Volta (French—Republic of Upper Volta)
rh & w radar homing and warning
RHYP Runaway and Homeless Program
ri random interval; reflective insulation; refractive index; reliability index; require identification; respiratory illness; retroactive inhibition; rubber-insulated; rubber insulation
ri (RI) retrograde inversion
RI Recruit Instruction; Refractories Institute; Religious Instruction; Republic of India; *Republik Indonesia*; Rhode Island (R.I.); Rhode Islanders; Rice Institute; Rock Island (Chicago, Rock Island & Pacific Railroad); Rotary International; Royal Institute
R & I Rural and Industries (bank)
RI Readers International; Registro Italiano (Italian—Italian Register)—of shipping; *Repubblica Italiana* (Italian—Italian Republic); *Républic-*

ains Independants (French—Independent Republicans);
Ring Index
ria radioimmunoassay
RIA Railroad Insurance Association; Recording Industry Association; Research Institute of America; Robot Institute of America; Rock Island Arsenal; Royal Irish Academy
RIAA Record Industry Association of America; Recording Industry Association of America
RIAC Research Information Analysis Corporation
RIAEC Rhode Island Atomic Energy Commission
RIAF Royal Indian Air Force; Royal Iranian Air Force; Royal Iraqui Air Force
riah monetary unit of Iran
riah (RIAH) radioimmunoassay of hair (for drug detection)
RIAI Royal Institute of Architects of Ireland
rial monetary unit of Yemen
rial (RIAL) revised individual allowance list
RIAL Religion In American Life; Rock Island Arsenal Laboratory
rial omani monetary unit of Oman
RIAM Royal Irish Academy of Music
RIANZ Record Industry Association of New Zealand
RIAS Research Initiation and Support (National Science Foundation); *Rundfunk im amerikanischen Sektor* (German—Radio in the American Sector), Berlin
RIASBO Rhode Island Association of School Business Officials
RIASC Rhode Island Association of School Committees
RIASLP Rattlesnake Island Air Service Local Post
RIASSP Rhode Island Association of Secondary School Principals
rib. range in a box; ribbon
RIB Railway Information Bureau; Referee in Bankruptcy; Roanoke Iron & Bridge; Rural Industries Bureau
RIB Rijksinkoopbureau (Dutch—Government Purchasing Office)
Rib^a Ribeira (Portuguese—brook; creek; riverside; river valley, stream); Ribera (Span-

ish—bank, beach, riverside, shore)
RIBA Royal Institute of British Architects
RIBNY Republic International Bank of New York
RIBS Restructured Infantry Battalion System
ric radar intercept calculator; ritual infant circumcision; routine infant circumcision
ric ricevuta (Italian—receipt)
Ric Ricardo; Richard; Richmond
RIC Republic Industrial Corporation; Republic of the Ivory Coast; Richmond, Virginia (airport); Royal Institute of Chemistry; Royal Irish Constabulary
RICA Research Institute on Communist Affairs (Columbia University)
RICASIP Research Information Center and Advisory Service on Information Processing
RICE Rhode Island College of Education
Rich Richard; Richards; Richardson; Richford; Richmal; Richmond
Rich II King Richard II
Rich III King Richard III
Richd Richard; Richmond
Rich-Pete Turn Richmond-Petersburg Turnpike (Virginia)
Rick Richard
ricksha(w) *jinrikisha* (Japanese—man-drawn two-wheeled carriage)
Ricky Richard
ricm right intercostal margin
RICM Registre International des Citoyens du Monde (French—International Registry of World Citizens)
RICMD Richmond Contract Management District
RICMO Radar Input Countermeasures Officer
'Rico Enrico; Puerto Rico; Ricardo
RICO Racketeer-Influenced Corrupt Organization; Racketeer-Influenced and Corrupt Organizations
RICS Royal Institute of Chartered Surveyors
RICU Russian Institute, Columbia University
RID Regimented Inmate Discipline (program for educating felons); Registry of Interpreters for the Deaf; Remove Intoxicated Drivers; Riddle Aviation

RIDA Rural and Industrial Development Authority
ridac range interference directing and control
RIDE Research Institute for Diagnostic Engineering
Riders Riders of the Purple Sage
Riding Mountain Riding Mountain National Park in southwestern Manitoba
Ridley Henry Nicholson Ridley (1855–1956), established rubber industry in Malaya, developed Botanic Gardens in Singapore, and for whom a Pacific Ocean sea turtle is named
ridp radar-iff (if friend or foe) data processor
rie range of incentive effectiveness; resources in education
RIE Royal Institute of Engineers
RIEC Royal Indian Engineering College
RIEI Republic Industrial Education Institute (Republic Steel)
riel monetary unit of Cambodia
RIEM Research Institute for Environmental Medicine
rif reading is fundamental; reduction in force; right iliac fossa
rif (RIF) resistance-inducing factor
rif rifatto (Italian—restored; repaired)
RIF Reading Is Fundamental; Royal Irish Fusiliers
RIFA Royal Institute of Foreign Affairs
Rif Brig Rifle Brigade
rifc rat intrinsic factor concentrate
Riff mountainous region of northern Morocco opposite Straits of Gibraltar
riffed reduced in force (dismissed or fired)
rifi radio interference field intensity
rifl random item file locater
rifma roentgen-isotope-fluorescent method of analysis
rift. (RIFT) reactor-in-flight test
RIFT Rhode Island Federation of Teachers
Rig Riga
RIG Restricted Interagency Group
RIGB Royal Institution of Great Britain

RIGHT Rhodesian Independence Gung-Ho Troops
right on right on the nose (exactly)
rih repetition-induced hypnosis; right inguinal hernia
RIH Royal Institute of Horticulture
RIHS Rhode Island Historical Society
rihsa radioactive iodinated human serum albumin
RIIA Royal Institute of International Affairs
RIIC Research Institute on International Change
RIISOM Research Institute for Iron, Steel, and Other Metals
ril record input length
RIL Royal Interocean Lines
RILSS Rapid Integrated Logistic Support System
rim. radar input mapper; receiving, inspection, and maintenance; rubber insulation material
RIM Relevant Instructional Material; Resident Industrial Manager
RIMAC Research Institute for Mediciane and Chemistry
RIMB Roche Institute of Molecular Biology
RIMR Rockefeller Institute for Medical Research
RIMV Registrar and Inspector of Motor Vehicles
Rin Rintintin
RIN Royal Institute of Navigation
RIN Registro Italiano Navale (Italian—Italian Naval Register)—bureau of shipping
rina reinitiation
RINA Royal Institution of Naval Architects
RINA Registro Italiano Navale e Aeronautico (Italian—Italian Air and Shipping Registry)
RIND Research Institute of National Defense
rinf rinforzando (Italian—with additional emphasis)
Ring Ring Lardner
Ring Ringstrasse (German—Ring Street)—tree-lined boulevard encircling inner Vienna
ringgit monetary unit of Malaysia
ringkasan (Malay—abbreviation)—also called *kependekan* or *singkatan*
RINM Resident Inspector of Naval Material
rin (RIN) report identification

number
RINS Research Institute for the Natural Sciences
RINSMAT Resident Inspector of Naval Stores and Materiel
rint rap in the nuts (kick in the scrotum)
Rio Rio de Janeiro, Brazil
RIO Reporting In and Out; Rhodesian Information Office; Rio de Janeiro (Galeao Airport)
Rioj La Rioja
riometer relative ionospheric opacity meter
RIOP Royal Institute of Oil Painters
RIOPR Rhode Island Open-Pool Reactor
Ríos originally Rodríguez
riot. real-time input-output transducer (translator); retrieval of information by on-line terminal (data processing)
rip. radar identification point; radioisotope precipitation
rip ripieno (Italian—filling up)
RIP Reduction in Implementation Panel; Reduction in Personnel (layoffs); Reliability Improvement Program; Reserve Intelligence Program; Riker's Island Penitentiary; Rockefeller Institute Press
R.I.P. requiesca[n]t in pace [Latin—may one (they) rest in peace]
RIPA Royal Institute of Public Administration
Rip Blong Van Ripablik Blong Vanuatu (Bislama—Republic of Vanuatu)—formerly New Hebrides
RIPGA Rhode Island Personnel and Guidance Association
RIPH Royal Institute of Public Health
RIPHH Royal Institute of Public Health and Hygiene
R I Phil Rhode Island Philharmonic
RIPO Rhode Island Philharmonic Orchestra
ripple. radioactive isotope-powered pulsed-light equipment (RIPPLE)
RIPPR Reliability Improvement Program Progress Report
RI & Prov Plant Rhode Island and Providence Plantation (Rhode Island's official name)
ripr viet riproduzione vietata (Italian—reproduction forbidden)
RIPS Radar-Impact Prediction

System; Range-Instrumentation Planning Study; Range-Instrumentation Planning System

rip viet riproduzione vietata (Italian—reproduction forbidden)

RIPWC Royal Institute of Painters in Water Colours

RIQS Remote Information Query System

rir reduction in requirement

rir (RIR) receiver input register

RIR Riverside International Raceway

R Ir AM Royal Irish Academy of Music

rirb radio-iodinated rose bengal

ririg reduced-excitation inertial reference-integrating gyroscope

ris (RIS) racially isolated school(s)

RIS Radio Information Service; Range Instrumentation Ship; Redwood Inspection Service; Regulatory Information System; Research Information Service; Royal Imperial Society; Royal Infantry Society

risa radioactive iodinated serum albumin

RISB Rotter Incomplete-Sentence Blank

RISC Reduced Instruction Set Computer; Rockwell International Science Center

RISCO Rhodesian Iron and Steel Company

RISCOM Rhodesian Iron and Steel Commission

RISD Rhode Island School of Design

rise. reliability improvement selected equipment; reusable inflatable salvage equipment

RISE Research Information Services for Education

RISM Research Institute for the Study of Man (USA)

RISOS Research in Secured Operations Systems; Research in Secured Operating Systems

risp rispettivamente (Italian—respectively)

RISP Ross Ice Shelf Project

RISS Range Instrumentation and Support System

RISSA Rhode Island School Superintendents Association

RISW Royal Institution of South Wales

rit ritard; ritardando; ritornello; ritual; ritualism; ritualistic; ritualization; ritualize

rit (RIT) retrograde inversion transposed (12-tone)

rit ritardando (Italian—holding back, retarding)

RIT Radio Information Test; Radio Network for Inter-American Telecommunication; Rochester Institute of Technology; Rorschach Ink-blot Test; Royal Institute of Technology

RIT Red Interamericana de Telecomunicaciones (Spanish—Inter-American Telecommunication Network)

Rita Margaret; Margarita

RITA Rand Intelligent Terminal Agent; Rural Industrial Technical Assistance

ritard ritardando (Italian—holding back, retarding)

RITC Rehabilitation Investment Tax Credit

Ritchie Ward Ritchie Press

RITE Rapid Information Technique for Evaluation

riten ritenuto (Italian—retaining the tempo)

RITES Rail India Technical and Economics Services

RITR Rework Inspection Team Report

RITS Rapid Information Transmission System; Reconnaissance Intelligence Technical Squadron

RITU (Profintern) Red International of Trade Unions

ritz ritzier; ritziest; ritziness; ritzy

riv radio influence voltage; river; rivet(ed)

riv riveduto (Italian—revised)

Riv River; Riviera; Rivington; Rivke

Riv Rivke (Yiddish—Rebecca)

Rivadavia Comodoro Rivadavia, Argentina

Riverside Riverside County Jail (California)

rivu river view

RIW Reliability Improvement Warranty

riyal monetary unit of Qatar and Saudi Arabia

RIZ Radio Industry Zagreb

rj (RJ) ramjet

RJ Rio de Janeiro; Royal Jordanian (airlines)

R & J Romeo and Juliet

RJA Reform Jewish Appeal; Retail Jewelers of America; Royal Jersey Artillery

RJAF Royal Jordanian Air Force

RJAS Royal Jersey Agricul-

tural Society

RJC Rochester Junior College; Rosenwald Junior College; Roswell Junior College

rje remote job entry

RJIS Regional Justice Information System

Rjk Reykjavik

RJM Royal Jersey Militia

rjp realistic job preview

RJR RJ Reynolds

RJR Nab R J Reynolds Nabisco

rk radial keratotomy; rock; run of kiln

r/k (R/K) radial keratotomy

rk rooms-katholiek (Dutch—Roman Catholic)

r-k rooms-katholiek (Dutch—Roman Catholic)

Rk Rock

RK Air Afrique (2-letter coding); Radio Kabul

RK Rdeci Kriz (Yugoslavian—Red Cross)

Rka Rijeka

rkg radiocardiogram

RKN Republic of Korea Navy

RKO Radio-Keith-Orpheum (theater circuit)

rkp record key position

rkt rocket

Rkt Sta Rocket Station

RKU Ruprecht-Karl-Universität (Heidelberg)

RKV Rose Knot (tracking station vessel)

rkva reactive volt-ampere

rky rocky; roentgen kymography

rl coarse rales; radiation length; rail(ing); reduction level; rocket launcher

r/l radio location; random length(s)

r & l rail and lake

r-to-l right-to-left (photo caption abbreviation)

Rl Raphael

RL high-powered radio range loop radiator(s); Radiation Laboratory; Radio Liberty; Reading List; Record Librarian; Record Library; Regent's Line; Republic of Liberia; Research Laboratory; Richfield Oil (stock exchange symbol); River Lines (railroad); Roland Line; Rupert Line; Rutland Line

RL Rape of Lucrece; *Rijksuniversiteit Limburg* (Dutch—State University of Limburg)

rl$_1$ few line rales

rl$_2$ moderate number of rales

rl$_3$ many coarse rales

rla restricted landing area; right lower arm

RLA Religious Liberty Association

RLAA Red Light Abatement Act

rladd radar low-angle drogue delivery

RLAF Royal Laotian Air Force

RLB Sir Robert Laird Borden (Canada's ninth Prime Minister)

rlbcd right lower border of cardiac dullness

rlbm (RLBM) rearward-launched ballistic missile(s)

RLC Radio Liberty Committee

RLCA Rural Letter Carriers' Association

RLCS Radio-Launch Control System

rld radar laydown delivery; rolled

rld (RLD) relocation list dictionary

RLD Raymond L Ditmars

RLDPAS Royal London Discharged Prisoners' Aid Society

rld's retail liquor dealers

rle relative luminous efficiency; right lower extremity

Rle Ramble

rl est real estate

rletfl report leaving each thousand-foot level

rlf relief; retrolental fibroplasia

RLF Royal Literary Fund

rlg railing

rlg *rilegato* (Italian—bound)

RLG Research Library Group; Royal Laos Government

rlgn realign; religion

rlgn dfld religion defiled

RLHTE Research Laboratory of Heat Transfer in Electronics (MIT)

RLI Realtors Land Institute; Rhodes-Livingstone Institute

RLIN Research Libraries Information Network

rll right lower limb; right lower lobe (lung)

rllb right long-leg brace

RLM Regional Library of Medicine (PAHO)

rlmd rat-liver mitochondria

RLNWR Rice Lake National Wildlife Refuge (Minnesota); Ruby Lake National Wildlife Refuge (Nevada)

RLO Regional Liaison Office(r); Returned-Letter Office

rlp rail loading point

RLPAS Royal London Prisoners' Aid Society

RLPO Royal Liverpool Philharmonic Orchestra

rlq right lower quadrant (abdomen)

rlr right lateral rectus (eye muscle)

rls reels (flow chart)

Rls rial (Iranian currency unit)

RLS Robert Louis Stevenson; Royal Lancastrian Society

rlse release

RLSS Royal Life Saving Society

rltr realtor

RLTS Radio-Linked Telemetry System

rltv relative

rlty realty

rlv relieve

Rlv Rauscher leukemia virus

rly relay

Rly Railway

rm range mark(s); raw material; ream; receiving memorandum; respiratory movement; ring micrometer; room; rubber marker(s)

rm (RM) record mark (flow chart)

r/m revolutions per minute

r & m redistribution and marketing; reliability and maintainability; reports and memoranda

Rm Romania (Rumania); Romanian (Rumanian)

RM Radioman; Raybestos-Manhattan; Registered Magistrate; Registered Mail; Registered Midwife; Reichsmark (German currency); Research Memorandum; Ringling Museum; Royal Mail; Royal Malta; Royal Marine; Royal Marines

R/M Raybestos/Manhattan

R & M Robbins & Myers

rma right mento-anterior

RMA Radio Manufacturers Association; Regional Manpower Administration; Retread Manufacturers Association; Rice Millers Association; Ringling Museum of Art; Robert Morris Associates (Bank Loan Officers and Credit Men's Association); Royal Marine Artillery; Royal Military Academy; Royal Musical Association; Rubber Manufacturers Association

RMADB Reactor Maintenance and Disassembly Building

RMAF Royal Malaysian Air Force; Royal Moroccan Air Force

RMAG Rocky Mountain Association of Geologists

RMAI Radio Manufacturers' Association of India

rm ar reaming arbor

RMAS Rochester Museum of Arts and Sciences

r mast radio mast

RMB Royal McBee

RMBAA Rocky Mountain Business Aircraft Association

RMBN Rocky Mountain Broadcasting Network

rmc rod memory computer

RMC Radio Monte Carlo; Revolutionary Military Council; Reynolds Metal Company; Rochester Manufacturing Company; Royal Military College

RMCC Royal Military College of Canada

RMCM Royal Manchester College of Music

RMCPA Rocky Mountain College Placement Association

RMCS Royal Military College of Science

rmct rat mass cell technique

RMCU Royal Mail Container Unit

rmd ready money down; retromanubrial dullness

RMD Reaction Motors Division (Thiokol Chemical Corporation); Research Management Division (D of E)

RMEA Rubber Manufacturing Employers' Association

R-meter radiation meter

R Met S Royal Meteorological Society

RMFVR Royal Marine Forces Volunteer Reserves

rmi radio magnetic indicator; reliability maturity index(ing)

RMI Rack Manufacturers Institute; Reaction Motors Incorporated; Reactive Metals Incorporated; Republic of the Marshall Islands; Roll Manufacturers Institute

rmicbm (RMICBM) road-mobile intercontinental ballistic missile

r/min revolutions per minute

RMIS Resource Management Information System

RMIT Royal Melbourne Institute of Management

RMJC Robert Morris Junior College

rmks remarks

rml right mediolateral; right middle lobe

RML Rand Mines Limited;

Royal Mail Lines; Royal Malta Library (Valetta)
RMLF Robert M La Folette
RMLI Royal Marine Light Infantry
RMM & EA Rolling Mill Machinery and Equipment Association
RMMNH Regar Memorial Museum of Natural History (Anniston, Alabama)
RMN Registered Maternity Nurse; Registered Mental Nurse; Richard Milhous Nixon (37th President of the United States and first to resign the presidential office); Royal Malaysian Navy
RMNP Rhodes Matopos National Park (Rhodesia); Riding Mountain National Park (Manitoba); Rocky Mountain National Park (Colorado)
RMNS Royal Merchant Navy School
RMO Regimental Medical Officer; Regional Medical Officer; Resident Medical Officer; Royal Marine Office
RMOGA Rocky Mountain Oil and Gas Association
R'mond Richmond
rmp right mento-posterior
RMP Radio Motor Patrol; Reentry Measurement Program; Regional Medical Program; Research Management Plan; Research and Microfilm Publications; Royal Marine Police; Royal Mounted Police
RMPA Royal Medico-Psychological Association
rmpc rubber-mold plaster casting
RMQ Records Management Quarterly
RMR Royal Marines Reserve
RMRA Royal Marines Rifle Association
Rmrs Ramirez
RMRS Rocky Mountain Radiological Society
rms root mean square
RMS Radiation Monitoring System; Railway Mail Service; Records Management System; Remote Manipulator System; Resources Management System; Royal Mail Ship; Royal Microscopical Society
RMSA Rural Music Schools Association
RMSC Royal Marines Sailing Club
RM Sch Mus Royal Marines

School of Music
rmsd root-mean-square deviation
rmse root mean square error
RMsf Rocky Mountain spotted fever
RMSM Royal Marines School of Music; Royal Military School of Music
RMSP Royal Mail Steam Packet (company)
rmt right mento-transverse
rmte remote
rmu remote maneuvering unit
rmv respiratory minute volume
RMWC Randolph-Macon Woman's College
Rm-W/MB Rijksmuseum Meermanno-Westreenianum/Museum van het Boek (Dutch—Merrmanno-Westreenianum Royal Museum and the Museum of the Book)—The Hague
rn reception nil; removal note; research note; round-nose (bullet); running noose; running nose
r of n range of neap (tides)
Rn radon; Rangoon
RN radionavigation; Registered Nurse; *República de Nicaragua*; Reynold's number; Royal Navy
RN Registered Nurse (periodical)
rna (RNA) ribonucleic acid
RNA Registered Nurse Anesthetist; Research Natural Area; Romantic Novelists' Association; Royal Naval Association
R/NAA Rocketdyne/North American Aviation
RNAC Royal Nepal Airline Corporation
RNADC Royal Netherlands Air Defense Command
RNAF Royal Naval Air Force
RNAFF Royal Netherlands Aircraft Factories Fokker
RNAO Registered Nurses Association of Ontario
RNAS Royal Naval Air Station
rnase ribonuclease
RNAV Royal Naval Artillery Volunteers
RNAW Royal Naval Aircraft Workshop
RNAY Royal Naval Aircraft Yard
rnb received—not billed
RNB Royal Naval Barracks
RNBT Royal Naval Benevolent Trust
RNC Republican National Com-

mittee; Royal Naval College (Greenwich)
Rnch Ranch
Rnchs Ranches
RNCM Royal Northern College of Music
RN & CR Ryde, Newport, and Cowes Railway
RNCS Royal Netherlands Chemical Society
RNCSRL Ralph Nader Center for the Study of Responsive Law
rnd round
RND Royal Naval Division
RND Rijksnijverheidstdienst (Dutch—Government Industrial Advisory Service)
rnd(s) round(s)
RNE Radio Nacional de España (Spanish—National Radio Broadcasting System)
RNEC Royal Naval Engineering College
RNES Radiodifusora Nacional de El Salvador (Spanish—National Radio Network of El Salvador)
rnf receiver noise figure
Rnf Renfrew
RNF Royal Northumberland Fusiliers
rnfp radar not functioning properly
RNFU Rhodesia National Farmers' Union
rng range
R ng P Republika ng Pilipinas (Pilipino—Republic of the Philippines)
rngt renegotiate
RNIB Royal National Institute for the Blind
RNID Royal National Institute for the Deaf
rnit radio noise interference test
RNL Raffles National Library (Singapore); Royal Netherlands Line
RNLAF Royal Netherlands Air Force
RNLI Royal National Lifeboat Institution
RNLO Royal Naval Liaison Office(r)
rnm radionuclide migration
rnm (RNM) radionavigation mobile
RNMC Royal Naval Medical Corps
RNMD Registered Nurse for Mental Defectives
RNMDSF Royal National Mission to Deep-Sea Fishermen
RNMI Realtors National Marketing Institute

RNMS Registered Nurse for the Mentally Subnormal; Royal Naval Medical School

RNMWS Royal Naval Mine-watching Service

RNN Royal Nigerian Navy

RNNP Royal Natal National Park (South Africa)

RNO Resident Naval Officer

RNoAF Royal Norwegian Air Force

RNOC Royal Naval Officers Club

R No N Royal Norwegian Navy

RNP Redwood National Park (California); Rondane National Park (Norway); Ruaha National Park (Tanzania); Ruhana National Park (Ceylon)

R.N.P. Registered Nurse Practitioner

RNP Radio Nacional de Peru (Spanish—National Radio of Peru)

RNPFN Royal National Pension Fund for Nurses

RNPL Royal Naval Physiological Laboratory

RNPS Royal Naval Patrol Service; Royal Navy Polaris School

rnr runner

r-'n'-r rock-and-roll

RNR Royal Naval Reserves

RNRA Royal Naval Rifle Association

RNRRA Royal Naval Reserve Rifle Association

RNRS Royal National Rose Society

rns radar netting station

RNS Religious News Service; Royal Naval School; Royal Numismatic Society

RNSA Royal Naval Sailing Association

RNSC Royal Naval Staff College; Royal Netherlands Steamship Company

RNSR Royal Naval Special Reserve

RNSS Royal Naval Scientific Service

RNSYS Royal Noval Scotia Yacht Squadron

rnt residual nitrogen time; roentgenologist; roentgenology

RNT Registered Nurse Tutor

RNT Revised New Testament (Roman Catholic)

RNTE Royal Naval Training Establishment

rnth raised non-tight hatch

RNTU Royal Naval Training Unit

rnu radar netting unit; radio noise voltage

rnvc reference number variation code

RNVR Royal Naval Volunteer Reserve

RNW Radio Navigational Warning

RNWMP Royal Northwest Mounted Police

RNWR Ravalli National Wildlife Refuge (Montana)

rnwy runway

RNYC Royal Northern Yacht Club; Royal Norwegian Yacht Club

RNZ Radio New Zealand

RNZAC Royal New Zealand Aero Club; Royal New Zealand Armoured Corps

RNZAEC Royal New Zealand Army Education Corps

RNZAF Royal New Zealand Air Force

RNZAMC Royal New Zealand Army Medical Corps

RNZAOC Royal New Zealand Army Ordnance Corps

RNZAS Royal New Zealand Astronomical Society

RNZASC Royal New Zealand Army Service Corps

RNZCD Royal New Zealand Chaplains Department

RNZC Sigs Royal New Zealand Corps of Signallers

RNZDC Royal New Zealand Dental Corps

RNZE Royal New Zealand Engineers

RNZEME Royal New Zealand Electrical and Mechanical Engineers

RNZIH Royal New Zealand Institute of Horticulture

RNZ Inf Royal New Zealand Infantry Corps

RNZIR Royal New Zealand Infantry Regiment

RNZN Royal New Zealand Navy

RNZNC Royal New Zealand Nursing Corps

RNZNR Royal New Zealand Naval Reserve

RNZNVR Royal New Zealand Naval Volunteer Reserve

RNZPC Royal New Zealand Provost Corps

RNZSHWC Royal New Zealand Society for the Health of Women and Children (Plunket Society)

RNZYS Royal New Zealand Yacht Squadron

ro rancho; receive only; recto (frontside of page); reddish orange; right opening; right orifice; road oil; rough opening; runover

ro (RO) readout (flow chart)

r/o roll out (final turn of an interceptor); routing order; rule out

r & o rail and ocean

ro. recto (Latin—front of the page, right-hand page)

r° *recto* (Portuguese—face of page; right-hand page; this side)

RO Radar Observer; Radar Operator; Radio Observer; Radio Operator; Recorder's Office; Recruiting Officer; *Republik Osterreich* (Republic of Austria); Reserve Order

R-O Reporting Officer; Ritter-Oleson (technique)

RO Resedentie Orkester (Dutch —Resident Orchestra)—The Hague

R-O Residentie-Orkest (Dutch —Residency Orchestra)—at The Hague

roa received on account; return on assets; right occiput anterior

roa (ROA) rights of accumulation

RoA Record of Acquisition

ROA Reserve Officers Association; Retired Officers Association; Royal Order of Altruists

ROA Russkaya Osvoboditelnaya Armiya (Russian—Russian Liberation Army)

ROAD Reorganization Objective Army Division; Re-Organize Army Division

roads. roadstead

Roads ports of Hampton Roads (Portsmouth, Newport News, Norfolk, Sewells Point)

ROA/I Received On Account/ Insurance

roam. return of assets managed (banking)

ROAMA Rome Air Materiel Area

ROA/P Received On Account/ Private

roar. right of admission reserved

ROAR Royal Optimizing Assembly Routine

ROARE Reeducation of Attitudes and Repressed Emotions

roast-beefsan roast-beef sandwich

roast-beefwich roast-beef sandwich

ROAUS Reserve Officers Association of the United States
rob. remaining on board (aircraft or ship cargo)
Rob Robert; Robinson College, Oxford
ROB Regional Office Building
Robby Robert(a)
robc readiness objective code
robe. wardrobe
robeps radar operating below prescribed standards
Roberts Roberts International Airport serving Monrovia, Liberia
robin. (ROBIN) rocket-balloon instrument
robo rocket orbital bomber
robrep robbery report
Robt Robert
roc rate of climb; receiver operating characteristic (curve); required operational capabilities; return on capital; rotatable optical cube; run on crap (fuel of the future)
RoC Register of Copyrights
RoC (ROC) Republic of China (offshore China); Republic of the Congo (formerly the French Congo)
R o C Republic of Congo
ROC Regional Occupation Center; Rochester, New York (airport); Royal Observer Corps
R o Cam Republic of Cameroon
ROCAPPI Research on Computer Applications for the Printing and Publishing Industries
roc coc rock cocaine
roce return on capital employed
Roch Rochester
R o Ch Republic of Chad
Rochambeau Cayenne, French Guiana's airport; Count Jean Baptiste Donatien de Vimeur de Rochambeau
Roch Phil Rochester Philharmonic
rocid reorganization of combat infantry divisions
rock rock'n'roll music; a form of cocaine
rock-a-billy rock-'n'-roll + hillbilly (music)
Rockaways Long Island, New York's south shore beaches—Far Rockaway, Rockaway Beach, Rockaway Park, Rockaway Point—plus other Rockaways in California, New Jersey, and Oregon
Rockefeller Nelson A Rock-

efeller, 41st Vice President of the United States
rockex rocket exercise
rockfest rock music festival
Rockies Rocky Mountains—major mountain system of western North America extending from Alaska and Canada to central New Mexico
rockoon(s) balloon-supported rocket(s)
ROCMD Rochester Contract Management District
Rocosas Rocallosas (Spanish—Rockies)—Rocky Mountains
rocp radar (or radio) out of commission for parts
rod. required operational data; required operational date; rusting, oxidation, discoloration
Rod Roderick; Rodney; Rodrigo; Rodrigues; Rodriguez
RoD Record of Decision
ROD Rosskoye Osvoboditelnoye Dvizheniye (Russian—Russian Liberation Movement)
Rodale Rodale Books
rodar rotor-blade radar
Roddy Roderick; Rodney
rodeocade rodeo parade
Rodg Roger
rodiac rotary dual input for analog computation
roe. (ROE) reflector orbital equipment
RoE Rules of Engagement
ROE Royal Observatory—Edinburgh
roentgen roentgenology
ROEP Refugee Orientation and Employment Program
rof reporting organizational-file
ROF Royal Ordnance Factory
ROFA Radio of Free Asia
rofor route forecast
roft radar off target
rog rise-off-ground
R o G Republic of Guinea
roger your message received and understood
r o/h regular overhaul(ing)
ROH Royal Opera House (Covent Garden)
roi return on investment
ROI Range Operating Instructions
Rois Rodrigues
Roiz Rodriguez
roj range on jamming
Rok a South Korean
RoK Republic of Kiribati in the western Pacific between Micronesia and the Solomon Islands

ROK Republic of Korea
ROKA Republic of Korea Army
ROKAF Republic of Korea Air Force
ROKAMS Republic of Korea Army Map Service
ROKN Republic of Korea Navy
ROKPUC Republic of Korea Presidential Unit Citation
roksonde rocket sounding
rol record output length; right occipitolateral
rolet reference our letter
Rolf Rudolf; Rudolph
rol k rolling keel
Rolls Rolls-Royce
ROLS Recoverable Orbital Launch System
rom radar operator mechanic; range of motion; range of movement; roman (type); rough order of magnitude
rom (ROM) read-on memory; read-only memory; roll-over mortgage
Rom The Letter of Paul to the Romans; Roman; Romance language; Romania(n)
Rom Book of Romans (New Testament); (German—Rome) —capital of Italy; Romanian (Romance language)
RoM Republic of the Marshall Islands
ROM Rome, Italy (Fiumicino airport); Royal Ontario Museum
R O M Republic of Malagasy
roman remotely operated mobile manipulator (acronym); roman candle (firework display); roman number (I, II, III, IV, V, etc.); roman type
Roman Empire in A.D. 117 comprised Hispania (Spain and Portugal); Gallia (France and the Lowlands); Germania (much of Germany and Austria); Illyricum (the Balkans); Italia (Italy); Turkey, the Middle East, Egypt, North Africa; and Britannia (some of England, Wales, and southern Scotland)
Romania Socialist Republic of Romania (Balkan state), *Republica Socialista Romania*; also spelled Rhumania or Rumania
Rom Ant Roman Antiquities
ROMBI Results of Marine Biological Investigations
Rom Cath Roman Catholic
romemo refer to our memoran-

dum
Romeo *Romeo and Juliet* (Shakespearean tragedy inspiring a dramatic symphony by Berlioz, a five-act opera by Gounod, a ballet by Prokofiev, an overture-fantasia by Tchaikovsky)
Rom Hist Roman History
Rom & Jul *Romeo and Juliet* by William Shakespeare
romom receiving-only monitor
ROMT Range-of-Motion Test
romv return on market value
ron remain overnight; research octane number
Ron Ronald
Ronald Ronald Press
rond rondeau; rondeaux; rondel; rondels
RONDA Royal Oriental Nut Date Association
roo kangaroo
Roo Roosevelt
ROO Range Operations Office(r)
rooi return on original investment
roor released on own recognizance
roo(s) kangaroo(s)
'roo(s) kangaroo(s)
root. relaxation oscillator optically tuned
rop right occiput posterior; run of press
RoP Republic of Palau, located between Papua, New Guinea and the Philippines
ROP Regional Occupational Program
ropeval readiness-operational evaluation
ropp receive-only page printer
Roques Los Roques Islands
ror rocket-on-rotor (device for assisting helicopter takeoffs)
ror (ROR) release on recognizance
Ror Rorschach (inkblot test)
RORA Reserve Officer Recording Activity
RORC Royal Ocean Racing Club
rord return on receipt of document
roreq reference our requisition
ro/ro roll on/roll off
RoRoRo *Rowolt Rotations Romäne* (German—Rowalt's Rotary Romances)
ros reduced operational status
ros (ROS) run of schedule (radio or television)
Ros Rosamund; Roscommon; Rosemary; Rostock

R o S Republic of Senegal
ROS Range Operating Station; Range Operation Station; Royal Order of Scotland
rosa recording optical-spectrum analyzer
RÖSAT *Röntgensatellite* (German—Röntgen satellite)
Rosc Roscommon
ROSCOE Remote Operating System Conversational Operating Environment
ROSCOP Report on Observations/Samples Collected by Oceanographic Programs
rose. residuum-oil supercritical extraction; rising observational sounding equipment; rose cut(ting); rose engine; rose fever; rose gum; rose hips; rose lathe; rose leaf; rose leaves; rose mill; rose oil; rose quartz; rose reamer; rose window; rose wine; rose worm; rose wort; roseate; rosebud(s); rose-cake; rose-colored; rosemary; rosette; rosewood
Rosh Hash *Rosh Hashanah* (Hebrew—New Year)
Rosh Hod *Rosh Hodesh* (Hebrew—beginning of the new month beginning at the new moon)
rosie (ROSIE) reconnaissance by orbiting ship-identification equipment
Rosie Rosa; Rosamund; Rose; Rosemarie; Rosemary
rosla raising of school-leaving age
ROSPA Royal Society for the Prevention of Accidents
Ross Ross and Cromarty
Rostov Rostov-on-Don
rot. remedial occupational therapy; right occipito-transverse; rotary; rotate; rotation; rotor
rot. (ROT) rate of return(ing)
Rot Rotterdam
Rotary Flight Pioneer Juan de la Cierva
ROTC Reserve Officers Training Corps
rotcc receiver-off-hook-tone connecting circuit
roti recording optical tracking instrument
rotis rotisserie
rotmh raised oil-tight manhole
rotn rotation
roto rotary press; rotogravure
Rot Phil Rotterdam Philharmonic
rotr (ROTR) receive-only typing reperforator (data processing)

ROTS Reusable Orbital Transport System
rotsal rotate and scale
Rou Rouen
ROU *República Oriental del Uruguay*
roul roulette
Roum Roumanian
'round around
rout routine
rov remotely-operated vehicle
Rov Rover(s)
Rover(s) Coloradan(s)
rov's (ROVs) remotely-operated vehicles
row. reverse-osmosis water; risk of war
RoW (ROW) Right of Way
R-O-W disease Rendu-Osler-Weber disease
Rox Roxburgh; Roxburghshire; Roxbury
Roxy Roxana; SL Rothafel
Roy Royal
Roy Alb Hall Royal Albert Hall
Royal Society The Royal Society of London for Improving Natural Knowledge (incorporated 1662)
Roy Bel *Royaume de Belgique* (French—Kingdom of Belgium)
Roy Com Soc Royal Commonwealth Society (formerly Royal Empire Society; formerly Royal Colonial Institute)
Roy Fest Hall Royal Festival Hall
ROY G BIV (acronymic mnemonic for recalling spectral colors—red, orange, yellow, green, blue, indigo, violet)—*see* vibgyor
Roy Liv Phil Orch Royal Liverpool Philharmonic Orchestra
Roy Opera Royal Opera House Orchestra (Covent Garden)
Roy Phil Royal Philharmonic Orchestra
Roy Soc Royal Society
Rozh Rozhdestvensky (the admiral or the conductor)
rp plate resistance (symbol); raid plotter; rally point; rearscreen projection; received pronunciation (RP); reception poor; redundancy payment; relay paid; release point; reply paid; reporting post; reprint; response pattern; retained personnel; return of post; return premium; rhodium plating; rhodium-plated;

rocket projectile (RP); rocket propellant; role playing; rust preventive
rp (RP) retinitis pigmentosa
r-p reprint; reprinting
rP reddish purple
Rp *Rappen* (Swiss—centime); *rupiah* (Indonesian currency unit)
RP remote pickup (broadcast); *República de Panamá; República del Paraguay; República del Peru; República Portuguesa* (Portugal); rocket projectile; Rules of Procedure
R-P Rhône-Poulenc
R/P Registered Plumber; Reporting Person; Royal Provincial (Tory American troops)
RP *Radiotelevisāo Portugesa* (Portuguese—Radio-Television); *Révérend Père* (French —Reverend Father)
RP-1 rocket-propellant type-1 fuel (kerosene)
rpa radar performance analyzer; random phase approximation
RPA Rationalist Press Association; Record of Personal Achievement; Regional Planning Association; Registered Plumbers' Association; Rubber Proofers' Association
R & PA Rifle and Pistol Association
RPAA Radiata Pine Association of Australia
rpar rebuttable presumption against registration (dangerous substance examination)
RPB Regional Preparedness Board; Research to Prevent Blindness (fund)
rpc radar planning chart; remote position control; reply postcard; request (the) pleasure (of your) company; reversed phase column
RPC Reliability Policy Committee; Republican Party Conference; Royal Pay Corps; Royal Pioneer Corps
RPC *République Populaire du Congo* (French—Popular Republic of the Congo)—formerly the French Congo
RPCC Reactor Physics Constants Center
RPCFT Reiter Protein Complement Fixation Test
RP China *República Popular China* (Spanish—People's Republic of China)
RPCV Returned Peace Corps Volunteer

rpd radar planning device
RPD Regional Port Director; Regius Professor of Divinity; Rocket Propulsion Department; Rocket Propulsion Division
R.P.D. *Rerum Politicarum Doctor* (Latin—Doctor of Political Science)
RPD Cor *República Popular Democratica de Corea* (Spanish—People's Democratic Republic of Korea)—North Korea
RPDL Radioisotope Process Development Laboratory
Rpds Rapids
rpe range probable error; related payroll expense
RPE Radio Propagation Engineering; Rocket Propulsion Establishment
RPEA Regional Planning and Evaluation Agency
rpedl repetitively pulsed electric-discharge laser
rpf radiometer performance factor; relaxed pelvic floor; renal plasma flow
RPF *Rassemblement du Peuple Français* (French—Rally of the French People)—de Gaulle's party; Gaullists
RPFMA Rubber and Plastics Footwear Manufacturers' Association
rpfod reported for duty
RP-FS Rozensweig Picture-Frustration Study
rpg radiation protection guide; report program generator; rifle-propelled grenade; rocket-propelled grenade; rounds per gun
RPG Regional Planning Group; Report Program Generator
rph revolutions per hour
rph (RPH) remotely piloted helicopter
RPh Registered Pharmacist
RPH Royal Perth Hospital
rpha reversed passive hemmagglutination
R Phil S Royal Philharmonic Society
RPHST Research Participation for High School Teachers
rpi radar precipitation integrator; random procedure information; rated position identifier; real progress index(ing)
RPI Railway Progress Institute; Rensselaer Polytechnic Institute; Retail Price Index; Rose Polytechnic Institute; Royal Pakistan Institute; Ryerson

Polytechnical Institute
RPIA Rocket Propellant Information Agency
RPIC Rock Properties Information Center (Purdue)
rpie (RPIE) real property installed equipment
rp index respiratory rate index; respiratory pulse index
RPK Regiment President Kruger
rpl running program language
RPL Radiation Physics Laboratory (NBS); Regina Public Library; Repair Parts List; Richmond Public Library; Roanoke Public Library; Rochester Public Library; Rocket Propulsion Laboratory; Rockhampton Public Library; Rockport Public Library
rplca replica
rpm radiation polarization measurement; reliability performance measure(ment); remote performance monitoring; repairman; revenue passenger miles; revolutions per minute; rotations per minute
RPM Raven's Progressive Matrices (test); Regional Plant Manager; Regulatory Project Manager; Rustenburg Platinum Mines
R & PM Research and Program Management (NASA)
rpmb (RPMB) remotely piloted miniature blimp
RPMF Radiation Pattern Measurement Facility
rpmi revolutions-per-minute indicator
RPMI Roswell Park Memorial Institute
RPN Registered Psychiatric Nurse
rpo revolutions per orbit
RPO Railway Post Office; Repository Program Office(r); Rochester Philharmonic Orchestra; Rotterdam Philharmonic Orchestra; Royal Philharmonic Orchestra
RPO *Rotterdams Philharmonisch Orkest* (Dutch—Rotterdam Philharmonic Orchestra)
rpoa recognized private operating agencies
rpoc report proceeding on course
rpp radar power programmer; reply paid postcard; request present position; return paid postal
RPP Radio Propagation Physics

rppe research, program, planning, evaluation

rppi repeater plan-position indicator

RPPI Rubber and Plastics Processing Industry

RPPMP Repair Parts Program Management Plan

RPQ Request for Price Quotation

rpr read printer

rpr (RPR) rapid plasma reagin

RPR Rassemblement pour la République (French—Assembling for the Republic); Republica Populara Romana (Romania)

RPRAGB Rubber and Plastics Research Association of Great Britain

rprt report

RPRT Rapid Plasma Reagin Test

rps revolutions per second; rotational position scanning; rotational position sensing

rp's rice planters; rubber planters

RPs repurchase agreements at commercial banks

RPS Radiological Protection Service; Railway Progress Society; Rapid Processing System; Registered Publication Section; Reliability Problem Summary; Republika Popullore Socialiste e Shqipërisë (Albania); Roadway Package System; Royal Philatelic Society; Royal Philharmonic Society; Royal Photographic Society

rp shortly reprinting shortly

RPSM Resources Planning and Scheduling Method

RPSs Reliability Problem Summary Cards; Republic of the Philippines Ships

rpt repeat

Rpt Report

RPT Registered Physical Therapist

rpt's (RPTs) rapid-phase transformations

RPU Radio Propagation Unit (USA)

rpv remotely-piloted vehicle; remote pilotless vehicle

rpw ranked positional weight

RPYC Royal Perth Yacht Club

rq remoulded-regraded quality (tires)

rq (RQ) respiratory quotient

RQ romance quotient

R/Q Request for Quotation

R & QA Reliability and Qual-

ity Assurance

RQAS Royal Queensland Art Society

RQBA Royal Queensland Bowls Association

rqd rock-quality designation; rock-quality determination

r qd raised quarterdeck(ing)

rqdcz request clearance to depart control zone

rqecz request clearance to enter control zone

rqiac requires immediate action

rql reference quality level

RQMS Regimental Quartermaster Sergeant

rqmt requirement

rqr require; requirement

rqs ready qualified for standby

RQS Rate Quoting System

rqtao request time and altitude over

rqto request travel order

RQYS Royal Queensland Yacht Squadron

rr radiation response; radio range; radio ranging; railroad; rapid rectilinear; rear; rearward; respiratory rate; rifle range; rural route; rush release; rush and run

r/r right rear

r & r rape and robbery; rate and rhythm (pulse); rest and recreation; rest and recuperation; rest and rotation (of military personnel); rock and roll; rock and rye (whiskey); rush and run

r & r (R & R) rape and ruin; rest and recovery; rest and recreation

RR Railroad; Raritan River (railroad); Recommendation Report; Recovery Room; Recruit Roll; Reliability Requirements; Remington Rand; Renegotiation Regulations; Research Report; Rifle Range; Right Reverend; Rolls-Royce; Ronald Reagan—fortieth President of the United States whose full initials are RWR (Ronald Wilson Reagan); Rural Route

R-R Rolls-Royce

rra (RRA) radio relay aircraft; ready reserve advances

rRA specific acoustic resistance

RRA Radiation Research Associates

R/RA Repair/Rework Analysis

RRAF Royal Rhodesian Air Force

R-rated moving picture restricted to adults

RRB Race Relations Board;

Railroad Retirement Board; RR Bowker

RRBC RR Bowker Company

RRBS Rapid-Response Bibliographic Service

rrc radar return code; reference repository conditions; reports of rating cases

rr & c records, reports, and control

RRC Race Relations Commission; Race Relations Conciliator; Recruit Reception Center; Regional Resource Center; Requirements Review Committee; Rocket Research Corporation; Royal Red Cross; Rubber Reserve Committee; Rubber Reserve Company; Rubber Reserve Corporation; Russian Research Center (Harvard)

R.R.C. Lady of the Royal Red Cross

rrcc reduced-rate contribution clause

RRCC Redwood Region Conservation Council

rr cells radiation reaction cells

rrd receive, record, display

rr & d reparations, removal, and demolition

RRD Reliability Requirements Directive

rrda rendezvous retrieval, docking, and assembly (of orbital station or space vehicle)

RRDS Rough Rock Demonstration School

rr & e round, regular, and equal (eye pupils)

RRE Railroad Enthusiasts; Royal Radar Establishment

RREA Rural/Regional Education Association

rr/eo race relations/equal opportunity

R Rep Records Repository (USAF)

RRF Reading Reform Foundation; Refrigeration Research Foundation

rrhage (Latin suffix—excessive flow)—hemorrhage

rrhea (Greek-derived suffix—to flow)—diarrhea, gonorrhea

rri range rate indicator

RRI Radio Republik Indonesia; Rocket Research Institute; Rubber Research Institute

RRIC Rubber Research Institute of Ceylon

rrid reverse radial immunodiffusion

RRIM Rubber Research Institute of Malaya; Rubber Re-

search Institute of Malaysia
RR-IM Research and Reports-Intelligence Memo
RRIS Remote Radar Integration Station
rrl reference repository location
RRL Regimental Reserve Line; Registered Record Librarian; Reserve Retired List; Road Reserve Laboratory
R.R.L. Registered Record Librarian (hospital)
RRLNWR Red Rock Lakes National Wildlife Refuge (Montana)
RRLs Registered Record Librarians
rrm('s) renegotiable-rate mortgage(s)
rrm(s) [RRM(s)] renegotiable-rate mortgage(s)
rrna (RRNA) ribosomal ribonucleic acid
rRNA ribosomal RNA (ribonucleic acid)
rrp reader and reader-printer; recommended retail price
RRP Reduced Repayment Program; Riot Reinsurance Program; Rotterdam-Rhine Pipeline
RRPC Reserve Reinforcement Processing Center (USA)
RRPS Ready Reinforcement Personnel Section (USAF)
rrr rebel, resist, riot (New Left student-activist program in abbreviated form); risk-reward ratio
r & rr range and range rate
RRRA Regional Rail Reorganization Act
rrr's rapid runway repairs
RRS Radiation Research Society; Reaction Research Society; Resource and Referral Service; Retired Reserve Section; River and Rainfall Station (NWS); Royal Research Ship
RRSP Registered Retirement Savings Plan (Canadian)
rrt rendezvous radar transponder
RRTA Railroad Retirement Tax
RRU Radio Research Unit (USA); Road Research Unit
rrv rate of rise of voltage
RRW Royal Regiment of Wales
rs radio station; reading of standard; ready service; rear spar; receiver station; receiving ship; receiving station; reception station; record separator; regulating station; reinforcing

stimulus; response stimulus; right side; road space; rubble stone
rs (RS) report separator character (data processing)
r/s range safety; rejection slip; revolutions per second
r & s rapport and support; re-enlistment and separation; research and study
Rs restricted motion pictures (adults only); rupees
RS Radio Station; Receiving Ship; Receiving Station; Reception Station; Reconnaissance Squadron; Reconnaissance Strike; Recording Secretary; Recruiting Station; Regular Station; Regulating Station; Regulation Station; Republic Steel; Research Summary; Revised Statutes; Ringer's Solution; Rio Grande do Sul; Roberval & Saguenay (railroad); Royal Scots; Royal Society
RS Rengo Sekigun (Japanese—United Red Army)—urban guerrilla group active in the Middle East
RS-70 reconnaissance-strike bomber (formerly B-70)
rsa radar signature analysis; remote station alarm; right sacro-anterior
'r SA around South America
RSA Railway Supervisors Association; Railway Supply Association; Redstone Arsenal; Regional Science Association; Rehabilitation Services Administration; Renaissance Society of America; Rental Service Association; *Republiek van Suid-Afrika* (Republic of South Africa); Returned Services Association; Road Safety Act; Royal Scottish Academy; Royal Society of Arts; Royal Society of Australia
RSA (AFL-CIO) Railway and Airline Supervisors Association
RSAA Remote-Sensing Association of Australia
rsac radar significance analysis code
RSAC Reactor Safety Advisory Committee (Canada)
RSAF Royal Saudi Air Force; Royal Small Arms Factory; Royal Swedish Air Force
RSA/HEW Rehabilitation Services Administration—HEW
RSAI Royal Society of Antiquaries of Ireland

rsalt running, signal, and anchor lights
RSAM Royal Scottish Academy of Music
RSAS Royal Sanitary Association of Scotland; Royal Surgical Aid Society
RSASA Royal South Australian Society of Arts
rsb range safety beacon
RSB Regimental Stretcher Bearer; Revolutionary Student Brigade
RSBA Rail Steel Bar Association; Royal Society of British Artists
rsbe ring-standby engines
RSBS Radar Safety Beacon System
rsbt rhythmic sensory bombardment therapy (RSBT)
rsc range-safety command; range-safety control; rational self-counseling; rigid-steel conduit
RSC Range Safety Command; Records Service Center; respiratory syncytial virus; Richard Strauss Conservatory (Munich); Royal Shakespeare Company; Royal Society of Canada
rsca right scapuloanterior
rscd request to start contract definition
RSCDS Royal Scottish Country Dance Society
rsch research
RSCM Royal School of Church Music
RSCN Registered Sick Children's Nurse
rscp right scapuloposterior
RSCS Rate Stabilization and Control System
RSCT Rhode Sentence Completion Test
rsd robustness semantic differential; rolling steel door
rs & d receipt, storage, and delivery
RSD Riverside Drive; Royal Society of Dublin
RSD-ALA Reference Services Division—American Library Association
RSDLP Russian Social-Democratic Labor Party
rsdp remote-site data processor
RSDS Range Safety Destruct System
RSE Royal Society of Edinburgh
rsea reference sensing-element amplifier
RSEC Regional Science Expe-

rience Center
rseu remote scanner-encoder unit
RSF Religious Society of Friends; Royal Scots Fusiliers; Russell Sage Foundation; Russian Socialist Forces; Russian Soviet Forces
RSFPP Retired Serviceman's Family Protection Plan
RSFS Royal Scottish Forestry Society
RSFSR *Rossiskaya Sovietskaya Federatvnaya Sotsialisticheskaya Respublika* (Russian—Russian Soviet Federal Socialist Republic)
rsg reassign; receiver of stolen goods; receiving stolen goods; regional seat of government
RSG Royal Scots Greys
RSGB Radio Society of Great Britain
RSGS Royal Scottish Geographical Society
rsh radar status history
Rsh Rosyth
RSH Recreational Services for the Handicapped; Royal Society for the Promotion of Health
RSHA *Reichssicherheitshauptampt* (German Secret Police headed by Heinrich Himmler); *Reichssicherheitshauptamt* (German—Reich Central Security Department)—combined Gestapo, Kripo, and SD secret police
RSHWC Royal Society for the Health of Women and Children (New Zealand's Plunkett Society)
rsi radarscope interpretation; radial-shear interferometer; reflected signal indication; replacement stream input
rs & i rules, standards, and instructions
RSI Research Studies Institute; Royal Sanitary Institute
rsia reference site initial assessment
RSIC Radiation Shielding Information Center; Radiation Standards Information Center; Redstone Scientific Information Center
RSID Recruiting Station Identification
R Sigs Royal Signals
RSIS Reference, Special, and Information Section (Library Association)
rsivp rapid sequence intravenous pyelogram

rsj rolled-steel joist
rsl right sacrolateral
RSL Radio Standards Laboratory; Red Star Line; Revolutionary Socialist League; Royal Society of Literature; Royal Society of London
rsla range safety launch approval
rslb right short-leg brace
rslt result
rsm (RSM) reconnaissance strategic missile
RSM Regimental Sergeant Major; Royal Scottish Museum; Royal Society of Medicine; Royal Society of Musicians
RSM *Repubblica di San Marino* (Italian—Republic of San Marino)
RSMA Railway Systems and Management Association; *Republica di San Marino* (San Marino—world's smallest republic); Royal School of Mines; Royal Society of Medicine
rsn reason
RSN Radiation Surveillance Network (USPHS)
RSNA Radiological Society of North America
RSNC Royal Society for Nature Conservation
RSNP Rancho Seco Nuclear Plant; Registered Student Nurse Program
RSNZ Royal Society of New Zealand
rso railway sorting office; railway suboffice; research ship of opportunity
RSO Radiation Safety Office(r); Range Safety Officer; Rehabilitation Service Office(r); Research Ships of Opportunity; Richmond Symphony Orchestra
RSO *Radio-Symphonie-Orchester* (German—Radio Symphony Orchestra)—Berlin
rso's regional sharing organizations
RSOs Resident Surgical Officers
rsp rain stops play; rear-screen projection; right sacro-posterior
RSP Repository Sealing Program
RSPA Research and Special Programs Administration; Royal Society for the Prevention of Accidents
RSPB Royal Society for the Protection of Birds

RSPCA Royal Society for the Prevention of Cruelty to Animals
RSPCC Royal Society for the Prevention of Cruelty to Children
RSPE Royal Society of Painter-Etchers and Engravers
RSPH Royal Society for the Pro motion of Health
rspl radar significant power line
rspp radio simulation patch panel
RSPP Royal Society of Portrait Painters
RSPR Royal Society for the Protection of Rats (mythical society created by Hans Werner Henze for his opera *The English Cat*)
RSPWC Royal Society of Painters in Water Colours
rsq rescue
r-sq r-squared
rsr regular sinus rhythm; required supply rate
RSR Range Safety Report; Request for Scientific Research; Research Study Requests
r-s ratio response-stimulus ratio
R-SR B Richmond-San Rafael Bridge
RSRC Remote Sensing Research Center (UCB)
RSRE Radar and Signals Research Establishment
RSROAA Roller Skating Rink Operators Association of America
RSRS Radio and Space Research Station
rsrv (RSRV) rotor systems research vehicle
rss ready service spares; remote safing switch; root-sum square; rotary stepping switch
R s-s Russian spring-summer (encephalitis)
RSS Range Safety System; Reactant Service System; Regional Support System; Rehabilitation Support Schedule; Remote Sensing Society; Remote Sensing System; Resource Security System; Royal Security Service; Royal Statistical Society; Rural Sociological Society
RSSA Royal Society of South Africa; Royal Society of South Australia
RSSAILA Returned Sailors, Soldiers, and Airmen's Imperial League of Australia
RSSC Rand School of Social

Sciences

R s-s e Russian spring-summer encephalitis

RSSF Retrievable Surface Storage Facility

Rssl Raytheon Scientific simulation language

RSSL Recruitment Subsidy for School Leavers

RSSPCC Royal Scottish Society for the Prevention of Cruelty to Children

RSSRT Russell Sage Social Relations Test

RSSS Regiae Societatis Socius Sodalis (Latin—Fellow of the Royal Society)

RSST Recruiter-Salesman Selection Test

rst radius of safety trace; reinforcing steel; right sacro-transverse

r-s-t readability–signal strength –tone (amateur radio signal)

Rst Rest (postal abbreviation)

RST Royal Society of Teachers

RST Republica Socialista Romania (Romanian Socialist Republic)

R Sta radio station

RSTMH Royal Society of Tropical Medicine and Hygiene

rstr restricted

rstrt restart

RSTS Registry of Scientific and Technical Services

rsu road safety unit

RSU Radical Student Union; Regional Service Unit; Road Safety Unit

rsv respiratory syncytial virus

rsv (RSV) research safety vehicle

Rsv Rous sarcoma virus

RSV Revised Standard Version (Bible)

rs virus respiratory synctial virus

rsvp rapid serial visual presentation; research-selected vote profile; restartable solid variable pulse

RSVP Response System with Variable Prescription; Retired Senior Volunteer Persons; Retired Senior Volunteer Program

R.S.V.P. répondez s'il vous plaît (French—please reply)

rsvr reservoir

rswc (RSWC) right side up with care

R Sw N Royal Swedish Navy

RSWS Royal Scottish Water-

Colour Society

rt radio telephone; radio telephony; rate; reaction time; real time; receive-transmit; reduction table(s); related term; remote terminal; right; rocket target; room temperature; round table; round trip; runup & taxi

rt (RT) recreational therapy; respiratory therapy; transposed retrograde (of a 12-tone row)

r/t radar trigger; radiotelephone

r/t (R/T) radiotelephone

RT Radio Technician; Ranger Tab; Reading Test; Recreational Therapy; Registered Technician; Registered X-ray Technician; *République Togolaise* (French—Togo Republic); River Terminal (railroad); Rubber Technician

R/T Record of Trial

RT radio en televisie (Dutch—radio and television); *République Togolaise* (French—Togo Republic)—Togo

rta reliability test(ing) assembly; road traffic accident; rumor told about

RTA Rail Travel Authorization; Railway Tie Association; Reciprocal Trade Agreements; Refrigeration Trade Association; Road Traffic Act; Royal Thai Army; Rubber Trade Association

RTA Radiodiffusion et Télévision Algérienne (French—Algerian Radio and Television Network)

RTAC Regional Technical Aids Center.

rt ad router adapter

RTAF Royal Thai Air Force

RTAM Resident Terminal Access Method

rtb return to base

RTB Rural Telephone Bank-(ing)

RTB Radiodiffusion-Télévision Belge (French—Belgian Radio-Television Network)

RTB/BRT Radifussion-Télévision Belge/Belgische Radio den Televisie (French and Dutch—Belgian Radio and Television Network)

RTBL Richard Thomas and Baldwins Limited

rtc ratchet; reader tape contact(ing)

RTC Rail Travel Card; Real Time Command; Replacement Training Center; Re-

serve Training Corps; Resolution Trust Corporation; Revenue and Taxation Code; Rochester Telephone Corporation; Royal Trust Company

RTCA Radio Technical Commission for Aeronautics

rtcc real-time computer complex

RTCEG Rubber and Thermoplastic Cables Export Group

rtcp radio transmission control panel

rtcu real-time control unit

rt cu router cutter

rtd remote temperature detector; resistance temperature detector(s); returned; righted

RTD Rapid Transit District; Research and Technology Division

RTD/CCS Resources and Technical Services Division/Cataloging and Classification Section (American Library Association)

rtdd real-time data distribution

RTDHS Real-Time Data Handling System

rtd ht retired hurt

rt dr returnable-trip drum

RTDS Real-Time Data System

rte route

r-t-e ready-to-eat (foods)

Rte Route

RTE Research Training and Evaluation

RTE Radio Telefis Eireann (Irish Radio Television)

RTEB Radio Trades Examination Board

RTECS Registry of Toxic Effects of Chemical Substances

R te G Rijksuniversiteit te Groningen (Dutch—State University at Groningen) Netherlands

rtel radiotelemetry; radio telephone; radiotelephony

R te L Rijksuniversiteit te Leiden (Dutch—State University of Leiden)

rtem radar tracking error measurement

RTES Radio and Television Executives Society

RTESO Radio Telefis Eireann Symphony Orchestra (Irish Radio Television Symphony Orchestra)

R test reductase test

R te U Rijksuniversiteit te Utrecht (Dutch—State University of Utrecht)

rtf radiotelephone; resistance-transfer factor; rubber-tile

floor(ing); rubber-tile foundation
RTF Radiodiffusion-Télévision Française (French—French Television Broadcasting)
rt fm router form
RTFR Reliability Trouble and Failure Report
rtfv radar target folder viewer
rtg radioactive thermal generator; rare tube gas; reusable training grenade
RTG Royal Thai Government
RTG Radiodiffusion Télévision Gabonaise (French—Gabonese Television Broadcasting)
rtgd room temperature gamma detector
rt gu router guide
rtgv real time generation of video
Rt Hon Right Honourable
RTHPL Radio Times Hulton Picture Library
rti respiratory tract infection; rise time indicator; rotor temperature indicator
RTI Reliability Trend Indicator; Research Triangle Institute; Roanoke Technical Institute
RTI Radiodiffusion Télévision Ivoirienne (French—Ivorian Television Broadcasting)—Ivory Coast
rtip radar target identification point
RTIR Reliability Trend Indicator Report
RTITB Road Transport Industry Training Board
RTK Ras Tafari Makonnea (Haile Selassie)
R Tks Royal Tank Regiment; Royal Tanks
rtl reinforced tile lintel; resistor transistor logic
rtl (RTL) register-transfer language; resistor-transistor logic
RTL Right to Life (party)
RTLA Road Transport Lighting Act
RTLO Regional Training Liaison Office(r)
rtls return to launch site
rtm running time meter
RTM Rotterdam, Netherlands (airport)
RTM Radiodiffusion Télévision Marocaine (French—Moroccan Television Broadcasting)
RTMA Radio and Television Manufacturers Association
RTMS Radar Target Measuring System
rtmso real-time multiprogramming support operation

rtn retain; return
rtn (RTN) routine (flow chart)
RTN registered trade name; Royal Thai Navy
RTNA Radio and Television News Association
RTNDA Radio-Television News Directors Association
Rtnst Rottnest
rto radio-telephone operator
RTO Railway Transport Office
r-to-d forms right-to-die forms
rtol restricted takeoff and landing
rtor right turn on red (traffic light)
rtp records turnover package; reinforced thermoplastic
R Tp radio telephone
RTP Rehabilitation Through Photography; Request for Technical Proposal (DoD)
RTP Radiotelevisão Portuguesa (Portuguese Radio Television)
RTPI Royal Town Planning Institute
rtpr (RTPR) reference theta-pitch reactor
rtqc real-time quality control
rtr returning to ramp
R Tr radio tower
RTR Reliability Test Requirement(s); Royal Tank Regiment
RTR Radiodifuziunea Televisiunea Romana (Romanian Radio-Television Network)
RTRA Radio and Television Retailers' Association; Road Traffic Regulation Act
rtrc radio telemetry and remote control
RTRC Regional Technical Report Centers
Rt.Rev. Right Reverend
r/t room radio/telegraph room (radio shack)
rtrsw rotary switch
rts radar target simulation; radar tracking station
rt's rubber tappers
RTS Repair Technical Service (tractor stations—USSR); Repair Tracking Service; Royal Television Society; Rubber Traders Society
RTSA Retail Trading Standards Association
RTSD Resources and Technical Services Division (American Library Association)
RTSRS Real-Time Simulation Research System
rtt radiation tracking transducer
RTT Radiodiffusion Télévision Tunisienne (French—Tunisian

Broadcasting)
RTTC Road-Time Trials Council
RTTDS Real-Time Telemetry Data System
rt tp router template
RTTPS Real-Time Telemetry-Processing System
rttv research target and test vehicle
r-ttv real-time television
rtty (RTTY) radio-teletypewriter communication(s)
rtu remote terminal unit; returned to unit
RTU Rahway Treatment Unit; Railroad Telegraphers Union; Reinforcement Training Unit; Reserve Training Unit
rtv reentry test vehicle (RTV); room-temperature vulcanizing
rtv (RTV) radio television
RTVE Radio Televisión Española (Spanish—Spanish Radio Television)
RTVHK Radio-Television Hong Kong
RTVS Royal Television Society
rtw ready to wear
rtx rapid-transit experimental (bus); report time crossing
rty rarity; realty
RTYC Royal Thames Yacht Club
rtz return to zero
RTZ Rio Tinto Zinc
ru are you?; radium unit; rat unit; roentgen unit; rusted
ru (RU) railroad underwriter; railway underwriter
Ru Rumania (Romania); Rumanian (Romanian); Russia; Russian; ruthenium
RU Radford University; Readers Union; Revolutionary Union (communists active in Puerto Rico and the United States); Rice University; Rhodes University; Roosevelt University; Rugby Union; Rumanian Union; Rutgers University
RU Regno Unito (Italian—United Kingdom); *Reino Unido* (Spanish—United Kingdom)
R-U 486 contraceptive preparation—Mifepristone—made in France by Reussel-Uclaf
rua right upper arm
RUA Royal Ulster Academy
RUAS Royal Ulster Agricultural Society
rub. rubber
rub rubato (Italian—with vary-

ing tempo); *ruber* (Latin—red)
Rub Rubbestadneset
RUB Radio Ulan Bator
rubbers rubber bullets
rubd rubberized
Rube Reuben
rubel rubella (german measles)
Rubg Rummelsburg
rubisco ribulose-1,5-bisphosphate carboxylase-oxygenase (plant protein)
ruble monetary unit of the USSR
RUBN Russian, Ukrainian, and Belorussian Newspapers
RuBPCase ribulose-1,5 bisphosphate
rub. **rm** rubber room (padded cell)
ruby spinel red spinel gemstone
RUC Royal Ulster Constabulary
RUC République Unie du Caméroun (French—United Republic of Cameroon)
RUCA Rijksuniversitair Centrum Antwerpen (Flemish—Antwerp State University Center)
Ruch Ruchel
Ruch Ruchel (Yiddish—Rachel)
Rucos Russian Communists
RUCR Royal Ulster Constabulary Reserve
rud rudder
Rud Rudd; Rúdiger; Rudolf; Rudolph; Rudulph; Rudyard
Rud(dy) Rudyard
rudis reference your dispatch
Rud Kip Rudyard Kipling
Rudy Rudolf; Rudolph
rue. right upper entrance; right upper extremity
RUE Regional Urban Environment
RUFAS Remote Underwater Fisheries Assessment System
Rufe Rufus
rufiyaa monetary unit of the Maldives
rug red under gold
rugger rugby football
RUI Royal University of Ireland
RUKBA Royal United Kingdom Beneficent Association
rul right upper limb; right upper lobe (lung); rule
RUL Rutgers University Library
rulet reference your letter
rum **(RUM)** remote underwater manipulator
rum rumänisch (German—Ro-

manian)
Rum Rumania (Romania); Rumanian (Romanian)
RUM Ranger Uranium Mines; Royal University of Malta
rumem reference your memo
rumnog rum-flavored eggnog
run. rewind(ing) and unload(ing)
RUN Revolutionary United Nations
runcible revised unified new computer with its basic language extended
R und J Romeo und Julia (German—Romeo and Juliet)
R unit millimeter of mercury divided by milliliters per second; unit of resistance in the cardiovascular system
RUP Rice University Press; Rockefeller University Press; Rutgers University Press
rupee monetary unit of India, Mauritius, Nepal, Pakistan, Seychelles, Sri Lanka
rupho reference your telephone (call)
rupiah monetary unit of Indonesia
rupp road used as public path
rupt rupture(d)
ruq right upper quadrant (abdomen)
rur reliably unreliable
RUR Rossum's Universal Robots (acronym-titled play by Karel Capek)
Rural Educ Rural Education Association
rureq reference your requisition
rur's rural and urban reformers
rurti recurrent upper-respiratory-tract infection
Rus Russ; Russia; Russian
Rus Russian (Slavic language)
Rusdic Russian dictionary
rush. remote use of shared hardware
Rush Rushdi; Rushmore; Rushton; Rushworth
RuSHA Rasse und Siedlungshauptamt (German—Race and Resettlement Department)
RUSI Royal United Service Institution
RUSM Royal United Service Museum
russ russet; russian (leather)
russ russisch (German—Russian); *russisk* (Dano-Norwegian—Russian)
Russ Russia(n)

Russ Russland (Norwegian—Russia)
Russell Cave Russell Cave National Monument in northeastern Alabama
rúst rústico, a la (Spanish—paperback, paperbound)
rust of iron iron oxide
rut. are you there
Rut Rutland Railroad; Rutlandshire
rutile titanium dioxide
RUU Ryksuniversiteit Utrecht (Dutch—Utrecht State University)
RUWS Remote Underwater Work System
rv rear view; recoil velocity; recreation vehicle; reentry vehicle; relief valve; residual volume; retroversion; right ventricle
rv **(RV)** recreational vehicle; reentry vehicle
r/v reentry vehicle
Rv Rendezvous
RV Rahway Valley (railroad); Reading and Vocabulary Test; *República de Venezuela*; Revised Version; Rifle Volunteer(s)
R/V rendezvous; research vessel
RV Radikale Venstre (Danish—Radical Left)—Radical Liberal Party; *Revised Version*
rva reactive volt-ampere (meter); right visual acuity
RvA Rouva (Finnish—Madam)
RVA Regular Veterans' Association
R & VA Rating and Valuation Association
rvb radar video buffer; red venous blood
rvbr riveting bar
rvc random vibration control; relative velocity computer
RVC Rifle Volunteer Corps; Royal Veterinary College
RVCI Royal Veterinary College of Ireland
rvd radar video digitizer; residual vapor detector; right vertebral density
RVDA Recreational Vehicle Dealers of America
rvdo right ventricular diastolic overload
rvdp radar video data processor
rve radar video extractor
rvedp right ventricular end diastolic pressure
rvedv right ventricular end diastolic volume

rvf rate variance formula; right visual field
RVFN Report of Visit of Foreign Nationals
rv fx riveting fixture
rvh right ventricular hypertrophy
RVH Royal Victoria Hospital (Belfast)
RV(H)R Road Vehicles (Headlamps) Regulations
RVI Recreational Vehicle Institute
RVIA Recreation Vehicle Industry Association; Recreational Vehicle Institute of America; Royal Victoria Institute of Architects
Rvik Reykjavik
RVL Royal Viking Line
RVLP Rift Valley Lakes Park (Ethiopia)
RVLR Road Vehicles Lighting Regulations
rvm reactive voltmeter
Rvn Ravenna
RVN Republic of Vietnam
RVNAF Republic of Vietnam Air Force; Republic of Vietnam Armed Forces
RVNF Republic of Vietnam Forces
rvo relaxed vaginal outlet; runway visibility observer
R v O Rijksinstituut voor Oorlogsdocumentatie (Dutch—Netherlands State Institute for War Documentation)
RVO Regional Veterinary Officer; Royal Victorian Order
rvp radar video preprocessor
Rvp Reid vapor pressure
rvpa rivet pattern
RVPA Rape Victims Privacy Act
rvr runway visual range
R v R Rembrandt van Rijn
R & VR Rating and Valuation Reports
rvrse reverse
rvs reported visual sensation
rv's recreation vehicles
Rvs Riverside
RVS Relative Value Scale; Relative Value Study
rvsc reverse self check
RVSN Raketny Voiska Strategicheskovo Naznacheniya (Russian—Strategic Rocket Forces)
R.V.S.V.P. répondez vite, s'il vous plaît (French—please reply at once)
rvsz riveting squeezer
rvtd riveted
Rvtn Riverton
rvtol rolling vertical takeoff and landing

rvu relief valve unit
rvx reentry vehicle—experimental
RVYC Royal Vancouver Yacht Club; Royal Victoria Yacht Club
rw radiological warfare; railwater (transport); random widths; raw water; recreation and welfare; recruiting warrant; rotary wing; runway
r/w read/write; right-of-way
r & w rail and water
Rw Rwanda
RW radiological war; radiological warfare; Recruiting Warrant; redwood; Richard Wagner; Right Worshipful; Right Worthy; Royal Welsh
rwa (RWA) rotary-wing aircraft
Rwa Rwanda
RWA Railway Wheel Association; Regional Water Authority
RWAFF Royal West African Frontier Force
Rwanda Republic of Rwanda (landlocked East African country)
R War R Royal Warwick Regiment
RWAS Royal Welsh Agricultural Society
rwb rear wheel brake
RWB Rand Water Board; Royal Winnipeg Ballet
rwbh records will be hand-carried
rwc rainwater conductor; read, write, compute; read, write, continue; receive with code
RWC Roberts Wesleyan College
rwc's round-wire cables
RWCS Royal Water Colour Society
rwd rearward; rear wheel drive; rewind(ing); right wing down; right word(ing)
RWDGM Right Worshipful Deputy Grand Master
RWDSU Retail, Wholesale, and Department Store Union
RWEMA Ralph Waldo Emerson Memorial Association
RWF Royal Wholesalers' Federation; Royal Welch Fusiliers
rwg rigid waveguide
RWG Radio Writers' Guild; Reliability Working Group; Roebling Wire Gage
rwgl rough wire glass
RWGM Right Worshipful Grand Master

RWGR Right Worthy Grand Representative
RWGT Right Worthy Grand Templar; Right Worthy Grand Treasurer
RWGW Right Worthy Grand Warden
rwh radar warning and homing
rwi read, write, initial; real world interval; remote weight indicator
rwi (RWI) radar warning installation
R Wilts Yeo Royal Wiltshire Yeomanry
RWJC Roger Williams Junior College
RWJF Robert Wood Johnson Foundation
RWJGW Right Worthy Junior Grand Warden
rwk rework
RWK Royal West Kent (regiment)
rwl relative water level
r/w/l random widths and lengths
rwlr relative water-level recorder
rwm rectangular wave modulation; resistance welding machine; roll wrapping machine
RWMA Resistance Welding Manufacturers' Association
r/w memory read/write memory
rwms radioactive-waste management site
rwp radio wave propagation
RWQCB Regional Water Quality Control Board
rwr radar-warning receiver
r-w-r rail-water-rail
RWR rail-water-rail; Ronald Wilson Reagan—40th President of the United States
rwrc remain well to right of course
rws range while search; reaction wheel scanner; reaction wheel system; release with services
rws (RWS) release with services
RWS Regional Weather Service; Royal Water Colour Society
RWSGW Right Worshipful Senior Grand Warden
r/w storage read/write storage
rwt read-write-tape
R-W Test Rideal-Walker Test
rwth raised watertight hatch
rwv read-write-versify
rwy railway; runway
RWY Royal Wiltshire Yeomanry

rx reverse; rix dollar; tens of rupees
r/x receiver
Rx recipe; prescription
rxb roxburgh (binding)
rxp radix point
rxs radar cross-section
ry railway; relay; rydberg
Ry railway; rydberg(s); Ryukyu (islands)
RY Royal Air Lao (coding); Royal Yeomanry
RYA Railroad Yardmasters of America; Royal Yachting Association
Ry Age Railway Age
ryal relay alarm
Ryan Ryan Aeronautical Company (coding)
Rybinsk Andropov
RYC Richmond Yacht Club; Rochester Yacht Club; Royal Yacht Club
Ry I Ryukyu Islands
rym refer to your message
RYM Revolutionary Youth

Movement
RYM-I Revolutionary Youth Movement (Weathermen)
RYM-II Revolutionary Youth Movement (Marxist-Leninist)
Ryojun Japanese equivalent of Port Arthur
ryrqd reply requested
Rys Railways
RYS Royal Yacht Squadron
ryt reference your telegram; reference your telex
Ryu Ryukyu; Ryukyuan
Ryukyus Ryukyu Islands between Japan and Taiwan
R y'u R Republika y'u Rwanda (Kinyarwanda—Rwanda)
rz return to zero
Rz Rodriguez
RZ Pacific Seaboard Airlines (2-letter symbol) doing business as Bay Area Helicopter Airlines and Los Angeles Helicopter Airlines; *République du Zaire* (formerly Belgian Congo)

R–Z Royal Australian New Zealand
R of Z Republic of Zambia
RZ Referativnyi Zhurnal (Russian—Reference Journal)—printed in Russian with abstracts also in their original language
RZA Religious Zionists of America
rzl return to zero level
rzm return to zero mark
RZMA Rolled Zinc Manufacturers Association
RZn relative azimuth
RZS Royal Zoological Society
RZ S Royal Zoological Society of Scotland
RZSI Royal Zoological Society of Ireland
RZSNSW Royal Zoological Society of New South Wales
RZSS Royal Zoological Society of Scotland
RZSSA Royal Zoological Society of South Australia

S

s displacement (symbol); sacral; saline; sand; saves; schilling (Austrian currency); scuttle; sea-air temperature difference correction (symbol); second; secret; section; sections; sedimentation (coefficient); self timer; sen (Japanese currency unit); sensation; sensitive; separate; separation; share(s); shilling (British monetary unit); ship; sign; silicate; silver; simultaneous transmission of range signals and voice (symbol); slope; slow; small; smooth; snow; soft; sol (Peruvian monetary unit); solo; soluble; son; sou (French monetary unit); space; spar; specific; specific factor; speed; spherical; spherical lens; start; steel; stere; stimulus; stock; string; subject; substrate; succeeded; sucre (Ecuadorian monetary unit); sum; summary; summer; supravergence; surface; surgeon; symbol; symbol surface; syphilis (sometimes indicated in reports by a Greek sigma)

's (contraction—does, has, is)

s signa (Latin—write); signetur (Latin—label; let it be written); sinister (Latin—left)

s. sinister (Latin—left)

S antisubmarine (symbol); sailing vessel (symbol); San; San Francisco mint (coin symbol); Santa; Santo; satisfactory; Saturday; Saturn; Saxon; Schilling (Austrian currency); school; Schweitzer; Schweizer Aircraft; Scotland; Seaman; seaplane; search and rescue; Sears, Roebuck (stock exchange symbol); Seatrain Lines; secondary and source electrode (symbol); secondary winding (symbol); secret; Section; See; sen (Japanese currency); Senate; Senate Bill; Senator; Shinto; Shintoism; Shintoist; ship; siemens (mho); Sierra—code for letter S; Sigma; sign; Signor (Italian—mister); Sikorsky; silver; Silver Lines; Sinclair; Sister; Socialist; sol (Peruvian monetary unit); solo; solubility; son; soprano; south; southern; spar buoy; specific factor; specification(s); Sperry; Staff; Statesman's Party; Statute; steamer; steamship; Steinway; stokes; stop; subject; sucre (Ecuadorian monetary unit); summer; sun; Sunday; sune; sunur; Surgery; Sweden (auto plaque); Sylvania; total entropy (symbol); wing plan area (symbol)

S (S) supply (microeconomics)
S/ sol (Peru); sucre (Ecuador)
:/S/ sign (music)

S general area (symbol); Sábado (Spanish—Saturday); Sacrum (Latin); San or Santo (Italian, Spanish—saint, m); Santa (Italian, Portuguese, Spanish—saint, f); São (Portuguese—saint, m); semis (Latin—half); sinister (Latin—left); sisälle (Finnish—in); söder (Swedish—south); sor (Norwegian—south); south; strada (Italian—street); subir (Span-

ish—to go up, mount); sud (French or Italian—south); Süd (German—south); sul (Portuguese—south); sur (Spanish—south); syd (Danish—south)

s₁ first heart sound
S-1 military personnel; personnel officer
S1c Seaman, first class
s 1 s 1 e smooth 1 side 1 edge; surfaced on one side and one edge (lumber)
S1, S2, S3, etc. first sacral nerve, second sacral nerve, third sacral nerve, etc.
s₂ second heart sound
S-2 Grumman Tracker antisubmarine search-and-attack aircraft; intelligence officer; military intelligence
S2F Tracker twin-engine antisubmarine aircraft flown from carriers
s2s surfaced two sides
S-21 Khmer Rouge name for Tuol Sieng near Phnom Penh, Cambodia where only 7 of the 20,000 prisoners escaped the mass executions
S-3 military operations and training; military operations and training officer
S³ Systems, Science, and Software
S-4 military logistics; military logistics officer
s 4 s smooth 4 sides; surfaced on four sides (lumber)
S-35 Saab double-delta-wing supersonic fighter or fighter bomber built in Sweden and named Draken (Dragon)
S³⁵ radioactive sulfur
S-51 Sikorsky four-seat heli-

copter

S-60 Soviet antiaircraft system consisting of one 57mm cannon mounted on a towed carriage

S-61 Sikorsky civilian or military helicopter

sa sack(s); sail area; semi-annual(ly); semiautomatic; sex appeal; shaft alley; sino-atrial; small arms; software applications; soluble in alkaline; special activities; spectrum analyzer; stone arch; subject to approval; subsistence allowance; sun-affected; superabnormal; supra-abdominal; sustained action

s-a single-action (handgun); sinoatrial

s/a safe arrival; storage area; subject to approval

s & a safety and arming (mechanism)

sa siehe auch (German—see also)

s.a. secundum artem (Latin—according to the art); *sine anno* (Latin—undated)

sª Señora (Spanish—Madam)

Sa samarium; Sara; Sarah; Sarita; Saturday; Serra; Sierra

Sa Summa (German—total)

Sª Serra (Portuguese or Spanish—mountain range); *Sierra* (Spanish—mountain range)

SA Safeway Stores (stock exchange symbol); Salvation Army; Saudi Arabia; Saudia Arabian; Savage Arms; Savannah & Atlanta (railroad); Seaman Apprentice; search amphibian; second attack (lacrosse); Secretary of the Army; sex appeal; Shipping Authority; Society of Actuaries; Society of Authors; South Africa; South African; South African Airways (2-letter coding); South America; South American; South Australia; South Australian; Southern Association; (Spanish—National Air Routes Corporation); Special Agent; Special Artificer; Springfield Armory; State's Attorney; Studio Address (public address system); Sugar Association; Supplemental Agreement; Supplementary Agreement; Surgery Assist

S-A Stokes-Adams (disease)

S/A Special Agent; State Agent

S of A Society of Actuaries

SA Société Anonyme (French—

limited company); *Sudáfrica* (Spanish—South Africa)

S.A. Sociedad Anónima (Spanish—corporation); *Sturmabteilung* (German—Storm-troopers); *Sucursales Asociados* (Spanish—associated branches)

S/A Societa Anonima (Italian—limited company)

SA-2 Soviet surface-to-air missile called Guideline by NATO

SA-3 Soviet air-defense missile system nicknamed Goa by NATO

SA-4 Soviet missile system nicknamed Ganef by NATO

SA-5 Soviet surface-to-air missile called Griffon by NATO

SA-6 Soviet air-defense missile system nicknamed Gainful by NATO

SA-7 Soviet shoulder-fired surface-to-air missile called Grail by NATO

SA-8 Soviet missile system nicknamed Guideline by NATO

SA-9 Soviet air-defense missile system nicknamed Gaskin by NATO

SA-315 Aerospatiale helicopter made in France and called Lama

SA-341 Aerospatiale observation helicopter built in Brazil by Embraer

saa small arms ammunition

SAA Saudi Arabian Airlines; Shakespeare Association of America; Signal Appliance Association; Singapore Aftercare Association (for ex-convicts); Society for Academic Achievement; Society for American Archeology; Society of American Archivists; Society for Applied Anthropology; Society for Asian Art; South African Airways; Southern Ash Association; Speech Association of America; State Archeological Area; Surety Association of America; Swedish-American Association; Systems Application Architecture

SAA Single-Article Announcement (American Chemical Society)

SAAA Salvation Army Association of America

SAAARNG Senior Army Advisor, Army National Guard

SAAAS South African Association for the Advancement of Science

SAAASE South African Association for the Administration and Settlement of Estates

SAAB Svenska Aeroplan Aktiebolaget (Swedish—Swedish Airplane Company)

saac simulator for air-to-air combat

SAAC Sciences and Arts Camps; Seismic Array Analysis Center (IBM); Special Assistant for Arms Control (DoD)

SAAD Sacramento Army Depot; Small-Arms Ammunition Depot; Society for the Advancement of Anesthesia in Dentistry

SAAEB South African Atomic Energy Board

SAAF Saudi Arabian Air Force; South African Air Force

SAAL Syrian Arab Airlines

SAALIC Swindon Area Association of Libraries for Industry and Commerce

saam simulation, analysis, and modelling

SAAMA San Antonio Air Materiel Area

SAAMI Sporting Arms and Ammunition Manufacturers Institute

SAAN South African Associated Newspapers

SAANYS School Administrators Association of New York State

SAAP Saturn-Apollo Applications Program; South Atlantic Anomaly Probe (NASA)

sa ar saw arbor

SAARC South Asia Association of Regional Cooperation—Bangladesh, Bhutan, Maldives, Nepal, India, Pakistan, Sri Lanka

SAAS Science Achievement Awards for Students; Society of African and Afro-American Students; Southern Association of Agricultural Scientists; Standard Army Ammunition System

SAAT Society of Architects and Allied Technicians

SAAU South African Agricultural Union

SAAVS Submarine Acceleration and Velocity System

SAAWK Suid Afrikaanse Akademie vir Wetenskap en

Kuns (Afrikaans—South African Academy for Science and Art)
sab sabbath; sabbatical; soprano, alto, baritone (SAB)
s-a b steel-arch bridge
sáb *sábado* (Portuguese or Spanish—Saturday); *sabato* (Italian—Saturday)
Sab Sabah; Sabbatarian; Sabbatarianism; Sabbath; Sabelian; Sabine; Sabra(s)
Sab Sabkhat (Arabic—salt flats)—also appears as *Sebkhat*
S-A b South-American blastomycosis
SAB Sabena; School of American Ballet; Scientific Advisory Board; Society of American Bacteriologists
SAB Sveriges Allmänna Biblioteksforening (Swedish Library Association)
s-aba science—a basic approach
Saba Sheba
SABA Scottish Amateur Boxing Association; South African Black Alliance
sabbat sabbatical
SABC Scottish Association of Boys Clubs; South African Broadcasting Corporation
SABCO Society for the Area of Biological and Chemical Overlap
SABCOA Screw and Bolt Corporation of America
SABE Society for Automation in Business Education
SABENA Société Anonyme Belge d'Exploitation de la Navigation Aérienne (Belgian World Airlines)
saber (SABER) semiautomatic business environment research
sabh simultaneous automatic-broadcast homer
SABHATA Sand and Ballast Haulers and Allied Trades Alliance
SABIC Saudi Basic Industries Cooperation
sabin square-foot unit of absorption
sabir semi-automatic bibliographic information retrieval
Sable Cape Sable; Sable Island
SABMIS Seaborne Anti-Ballistic Missile Intercept System (USN)
SABMS Safeguard Anti-Ballistic Missile System
sabo sabotage

sabot sabotage; saboteur
SABR Society for American Baseball Research
SABRA South African Bureau of Racial Affairs
sabre self-aligning boost and reentry
SABS South African Bureau of Standards
SABTS Shared-Aperture Breadboard Test System
Sabu Sabu Dastagir
SABW Society of American Business Writers
sac sacral; sacrament; sacramental; sacred; surface air consumption
Sac Sacramento, California
SAC Sacramento, California (airport); San Angelo College; San Antonio College; Science Applications Corporation; Sexual Assault Center; Society of Analytical Chemistry; Southwest Automotive Company; Special Agent in Charge (FBI); Statistical Analysis Center; Strategic Air Command; Suburban Authorization Committee; Swedish-American Co-operative
SAC Service Action Civique (French—Civil Action Service); *Sociedad de Albizu Campos* (Puerto Rican terrorists); *Sveriges Arbetares Centralorganisation* (Swedish—Swedish Workers Central Organization)
saca store and clear accumulator
SACA Steam Automobile Club of America; Supreme Allied Commander Atlantic
SACA Servicio Aereo Colombiano (Spanish—Colombian Airline Service)
sacad stress analysis and computer-aided design
SACANGO Southern Africa Committee on Air Navigation and Ground Operation
SACARTS Semi-Automated Cartographic System
Sacate Sacatepéquez, Guatemala
SACB Subversive Activities Control Board
S Acc Societá in Accomandita (Italian—limited partnership)
SACC South African Council of Churches; Supplemental Air Carrier Conference; Supporting Arms Coordination Center

saccm slow-access charge-coupled memory
SACCR Southeastern Association of Community College Researchers
SACCS Strategic Air Command Control System
sace systems acceptance check-out equipment
SACEM Société des Auteurs, Compositeurs et Éditeurs de la Musique (French—Society of Authors, Composers, and Editors of Music)
SACEUR Supreme Allied Command, Europe
sach solid ankle cushion heel (prosthetic foot)
SACH Small Animal Care Hospital
sach foot solid-ankle-and-cushion-heel foot
saci secondary address code indicator
SACI South Atlantic Cooperative Investigations
Sackpig *See* SACPG
SA & CL South Atlantic & Caribbean Line
SACLant Supreme Allied Commander, Atlantic
SACLANTCEN SACLANT Anti-Submarine Warfare Research Centre (NATO)
sacm simulated aerial combat maneuver
SACM South African College of Music; South African Corps of Marines; South Arabian Common Market
SACMA Société Anonyme de Construction de Moteurs Aéronautiques (French—Aeronautical Engine Construction Corporation)
SACMP South African Corps of Military Policy
SACNAS Society for the Advancement of Chicano and Native American Scientists
saco select address and contract operate
SACO Sino-American Cooperative Organization
SACO Sveriges Akademikers Centralorganisation (Swedish—Swedish Central Professional Organization)
SACP South African Communist Party
SACPG Senior Arms Control Policy Group (pronounced "Sackpig")
Sacr Sacramento
Sacramentos Sacramento Mountains of New Mexico

and Texas

SACRO Scottish Association for the Care and Resettlement of Offenders

SACs Solar Appliance Centers

SACS South African College System; South African Corps of Signals; Southern Association of Colleges and Schools

Sac-San Sacramento—San Diego

SACSEA Supreme Allied Command South-East Asia

Sac-Sfo Sacramento—San Francisco

SACSIR South African Council for Scientific and Industrial Research

Sacto Sacramento

SACTU South African Congress of Trade Unions

SACU Service for Admission to College and University

SACUBO Southern Association of College and University Business Officers

SACVT Society of Air Cushion Vehicle Technicians

sad. safety analysis document; safety, arming, destruct; safety and arming device; situation attention display; somatosensory affectional deprivation

SAd (SAD) St Augustine decline (grass virus)

SAD Seasonal Adjustment Deficiency; Seasonal Affective Disorder; simple, average, or difficult; Social Affairs Department (Communist China's espionage agency)

S & AD Science and Applications Directorate (NASA)

SAD South African Digest

sadap simplified automatic data plotter

SADC Sector Air Defense Commander; Singapore Air Defense Command

SADCC South African Development Coordination Conference

SADD Students Against Drunk Drivers, Students Against Drunk Driving

sade sensitive acoustic-detection equipment

SADE Sociedad Argentina de Escritores (Spanish—Argentine Writers' Society)

SA de CV Sociedad Anónima de Capital Variable (Spanish—Variable Capital Society)

SADF South African Defense Forces

sadic solid-state analog-to-digital computer

sadie scanning analog-to-digital input equipment; semi-automatic decentralized intercept environment

sadm special atomic demolition munition

sado-maso sado-masochism; sado-masochist

sado-sex sado-sexual(ity)

SADS Schedule for Affective Disorders and Schizophrenia; Swiss Air Defense System

sadsac sampled data simulator and computer

sadsact self-aligned descriptors from self and cited titles (automatic index)

sad sam (SAD SAM) sentence appraiser and diagrammer—semantic analyzer machine

SADTC Shape Air Defense Technology Center

s-a-d test sugar-acetone-diacetic acid test

sae San Diego Aircraft Engineering (corporate symbol); self-addressed envelope; standard average European

sae (SAE) stamped, addressed envelope

SAE Sigma Alpha Epsilon (fraternity); Society for the Advancement of Education; Society of American Etchers; Society of Automotive Engineers; Solar Atmospher(ic) Explorer

S.A.E. Société Anonyme Egyptienne (French—Egyptian limited company)

SAEA Southeastern Adult Education Association

saeb self-adjusting electric brake

SAEB Spacecraft Assembly and Encapsulation Building (NASA); Special Army Evaluation Board

saec. saeculum (Latin—century)

SAEC South African Engineer Corps; Sumitomo Atomic Energy Commission (Japan)

SAEH Society for Automation in English and the Humanities

SAEI Sumitomo Atomic Energy Industries (Japan)

SAEL South African Emergency League

SAemc South African endomyocardiopathy

SAEMR Small Arms Expert Marksmanship Ribbon

SAEST Society for the Advancement of Electrochemical Science and Technology

SAET Spiral Aftereffect Test

saew ship's advanced electronic warfare

saf safety

SAF Secretary of the Air Force; See America First; Singapore Air Force; Social Affairs Federation; Society of American Florists; Society of American Foresters; Strategic Air Force

SAF Svenska Arbetsgivareforeningen (Swedish—Swedish Employers' Confederation)

safa solar-array failure analysis; soluble-antigen fluorescent antibody

SAFA School Assistance in Federally Affected Areas; Society for Automation in the Fine Arts

SAFAA South African Fine Arts Association

SAFB Scott Air Force Base; Shaw Air Force Base

saf black super-abrasion furnace black

SAFC South African Flying Corps

SAFCA Safeguard Communications Agency

SAFCB Secretary of the Air Force Correction Board

SAFCMD Safeguard Command (USA)

SAFCO Standing Advisory Committee on Fisheries in the Caribbean Organization

safe. satellite alert force employment; stamped, addressed foolscap envelope; system, area, function, equipment

SAFE Braathens South American & Far East Air Transport; Security Against Fatal Encounter; Security Assured For Each; Solvent Abuse Foundation for Education; South African Friends of England; Survival and Flight Equipment Association; System for Automated Flight Efficiency

S.A.F.E. Society of Aeronautic Flight Engineers

SAFEORD Safety of Explosive Ordnance Databank (USN)

SAFE TRIP Students Against Faulty Tires Ripping in Pieces

SAFI Senior Air Force Instructor

SAFMARINE South African Marine (corporation)

SAFO Senior Air Force Officer (present)

SAFOH Society of American Florists and Ornamental Horticulturists

SAF£ South African pound

S Afr South Africa(n)

SAFR Senior Air Force Representative

SAFRAS Self-Adaptive Flexible-Format Retrieval And Storage System

S-Afr Du South-African Dutch (Afrikaans)

SAFS Secondary Air Force Specialty; selective automatic feed stripe (knitting machine)

SAFSL Secretary of Air Force Space Liaison

SAFSO Safeguard System Office(r)

SAFSR Society for the Advancement of Food Service Research

SAFTI Singapore Armed Forces Training Institute

SAFU Scottish Amateur Fencing Union

SAFUS Secretary of the Air Force, United States

sa fx saw fixture

Sag Sagittarius

SAG Scientific Advisory Group; Screen Actors Guild; Society of Arthritic Gardeners; Surface Action Group (USN); Systems Analysis Group

SAGA Sand and Gravel Association; Scout and Guide Activity; Society of American Graphic Artists

SAGB Spiritualist Association of Great Britan

sag. d saggital diameter

sage. semi-automatic ground environment (for defense against air attack); solar-assisted gas energy (for heating)

SAGE Senior Action in a Gay Environment; Senior Actualization and Growth Exploration; Skylab Advisory Group for Experiments (NASA); South African General Electric; Stratospheric Aerosol and Gas Experiment

SAGE/BUIC Semi-Automatic Ground Environment and Back-Up Interceptor Control (systems)

SAGGA Scout and Guide Graduate Association

SAGP Society for Ancient Greek Philosophy

SAGS Semiactive Gravity Gradient System (NASA)

SAGSET Society for Academic Gaming and Simulation in Education and Training

sagt systematic approach to group technology

SAG & U San Antonio, Gulf & Uvalde (railroad)

sah subarachnoid hemorrhage

SAH Society of American Historians; Society of Automotive Historians; Supreme Allied Headquarters

SAHAND Societ Against *Have A Nice Day*

Sah Esp Sahara Español (Spanish Sahara)

sahf semiautomatic height finder

SAHR Society for Army Historical Research

SAHSA Servicio Aéreo de Honduras SA (Spanish—Air Service of Honduras Inc)

sahyb simulation of analog and hybrid computers

sai self-appraisal instrument; self-appraisal inventory; sell (sold) as is

sai (SAI) standard advertising invoice

Sai Saigon

SAI Schizophrenics Anonymous International; Science Applications Inc; Science Applications International; Self-Analysis Inventory; Social Adequacy Index; South African Irish (regiment); Stern Activities Index

SAI Societá Anonima Italiana (Italian—Italian Incorporated Company); *Son Altesse Impériale* (French—Her or His Imperial Highness); *Su Alteza Imperial* (Spanish—Your Imperial Highness)

SAIA South Australian Institute of Architects

SAIC Science Applications International Corporation; Special Agent in Charge (Secret Service)

said. speech auto-instructional device

Said Port Said, Egypt

SAIDET Single-Axis Inertial-Drift Erection Test

SAIF South African Industrial Federation; South African Institute of Foundrymen

sail. structural analysis input language

SAIL Sea-Air Interaction Laboratory

Sails ship's sailmaker

SAILS Software-Adaptable Integrated-Logic System

SAIM South African Institute of Management

SAIMC South African Institute for Measurement and Control

SAIMENA South African Institute of Marine Engineers and Naval Architects

SAIMR South African Institute for Medical Research

SAIMS Selected Acquisition Information and Management System

SAINT Systems Analysis of an Integrated Network of Tasks (USAF)

Saint-Barth Saint-Barthélemy (French—Saint Bartholomew)—a Caribbean island; a massacre of 3,000 Protestants instigated by Catherine de Médicis on the August 23, 1572

Saint-Ex Antoine de Saint-Exupéry

Saint Joe Saint Joseph, Missouri

Saint John St John, New Brunswick

Saint Johns St Johns, Antigua

Saint John's St John's, Newfoundland; St John's University, New York

Saint Kitts and Nevis Saint Christopher and Nevis (West Indies islands)

Saint-Laurent (French—Saint Lawrence)—Canadian-American river

Saint Lawrence St Lawrence River

Saint Lawrence Islands Saint Lawrence Islands National Park on the Canadian islands and nearby shore of the Saint Lawrence River

Saint Loo Saint Louis, Missouri

Saint Lucy St Lucia, West Indies

Saint P St Pancras (London railway station); St Paul, Minnesota

Saint Patrick's Saint Patrick's Day (March 17)

Saint Pete St Petersburg, Florida

Saint-Pierre (French—Saint

Peter)—island; Rome's basilica

Saint Stephen's Saint Stephen's Day (December 26)

Saint Vince St Vincent (West Indies)

SAIRR South African Institute of Race Relations

SAIS School of Advanced International Studies (Johns Hopkins University)

SAIT Southern Alberta Institute of Technology

SAIT Service D'Analyse de l'Information Technologique (French—Technological Information Analysis Service)

SAJ Shipbuilders Association of Japan; Society for the Advancement of Judaism

SAJ Suomen Ammattijärjestö (Finnish—Finnish Federation of Trade Unions)

SAJAC South African Jewish Association of Canada

SAJC Southern Association of Junior Colleges

sa ji saw jig

SAK Serge Alexandrovich Koussevitsky

SAK Suomen Ammattilittojen Keskulitto (Finnish—Finnish Trade Union Confederation)

Sakura-3A Japanese CS-3A communications satellite

sal salary; salt; salicylate; saloon

sal (SAL) surface and airlift

sal salida (Spanish—departure)

s.a.l. *secundum artis leges* (Latin—according to the rules of art)

Sal Salamanca; Salaverry; Salem; Sallie; Sally; Salomon

Sal Islas Salomón (Spanish—Solomon Islands); *salida* (Spanish—departure; exit); *Salmonella*

SAL San Salvador, El Salvador (airport); Seaboard Airline Railroad; Society of Antiquaries of London; South African Library (Cape Town); Symbolic Assembly Language

SAL Svenska-Amerika Linien (Swedish-American Line)

SALA Scientific Assistant Land Agent; South African Library Association; Southwest Alliance for Latin American(s)

SALALM Seminars on the Acquisition of Latin American Library Materials

salam salamanzar (12-bottle capacity)

sal ammoniac ammonium chloride (NH_4Cl)

SALB South African Library for the Blind (Grahamstown)

sale. simple algebraic language for engineers

sal gal saloon girl

salic salicional (French—soft string-toned organ stop)

salicyl salicylate

SALINET Satellite Library Information Network

Salisbury Harare, Zimbabwe; plain in southern England on which Stonehenge is located

SALJ South African Law Journal

Sall Gaius Sallustius Crispus (Roman historian often referred to as Sallust)

Sallust Roman historian Gaius Sallustius

Sally Army Salvation Army

salm single-anchor leg mooring

Salm Salamon

Salm Salmonella

SALM Society of Airline Meteorologists

salmiak sal ammoniac (ammonium chloride)

salmonsan salmon sandwich

salmonwich salmon sandwich

Salomons Salomon Islands in the Chagos Archipelago in the Indian Ocean

SALP South African Labour Party

salpingect salpingectomic (sterilization); salpingectomy (removal of the fallopian tubes)

salr saturated adiabatic lapse rate

SALR South African Law Reports

SALRC Society for the Assistance of Ladies in Reduced Circumstances

salt sodium chloride (NaCl)

salt. suggestive-accelerative learning and teaching

SALT Society for Applied Learning Technology; Strategic Arms Limitation Talks

Saltees Saltee Islands in St George's Channel off Wexford, Ireland

Salt 'Uman Saltanat 'Uman (Arabic—Sultanate of Oman)

salts of lemon oxalic acid

Saludemos Saludemos la patria (Spanish—We salute the country)—anthem of El Salvador

salut salutation; sea-air-land-and-underwater targets (SALUT)

salv salvage

Salv Salvador

Salvador Brazilian port of Bahia or São Salvador de Todos os Santos; Central American republic of El Salvador

Salv Army Salvation Army

Salve Salve Oh Patria (Spanish—Hail, oh country)—anthem of Ecuador

Salve a tí Salve a tí Nicaragua (Spanish—Hail Nicaragua)—national anthem of Nicaragua

Salz Salzburg, Austria

sam scanning auger microscope; scuba ascent method; self-advising materials; serial access memory; served available market; small (secondary) annular mirror; space-available mail (SAM); student accountability model; surface-to-air missile (SAM); synchronous amplitude modulation

sam (SAM) shared-appreciation mortgage; standard academic monograph

sam (SaM) (SAM) sales and marketing

sam samedi (French—Saturday)

Sam Samoa; Samoan; Samson; Samoyed; Samuel; Samuelito

S-a-m S-adenosyl-methionine

Sam Samstag (German—Saturday); *Samuel*

SAM School Administrators of Montana; School of Aerospace Medicine; Society for the Advancement of Management; Society of American Magicians; Special Air Mission; Student Accountability Model

SAM Societa Aerea Mediterranea (Italian—Mediterranean Airline)

SAMA Sacramento Air Materiel Area; Saudi Arabian Monetary Agency; Scientific Apparatus Makers Association; Student American Medical Association

SAMANTHA System for the Automated Management of Text from a Hierarchical Arrangement

Samarians Samarian Mountains

SAMB School of Aviation Medicine—Brooks AFB

SAMBA Special Agents Mu-

tual Benefit Association (FBI); Systems Approach to Managing Bureau of Ships Acquisitions (USN)
SAMC South African Marine Corporation; South African Medical Corps
SAM/CAR South America/Caribbean
SAM-D surface-to-air missile for field air defense
SAME Society of American Military Engineers
SAMECS Structural Analysis Method for Evaluation of Complex Structures
Sam'el Samuel
S Am(er) South America(n)
samex surface-to-air missile exercise
SAMF Seaborne Army Maintenance Facilities; Seaborne Army Materiel Facilities
samfu self-adjusting military fuckup
SAMH Scottish Association for Mental Health
sami socially-acceptable monitoring instrument
SAMI Scanner-Augmented Market Intelligence; System Acquisition Management Inspection
samizdat samizdatel'stvo (Russian—self-published and self-distributed)—literature suppressed by the Soviet government
Sam J Dr Samuel Johnson
SAMJ South African Medical Journal
Saml Samiel; Samuel
SAML Standard Army Management Language
SAMLA South Atlantic Modern Language Association
samm semi-automatic measuring machine
Samml Sammlung (German—collection)
SAMNS South African Military Nursing Service
Samoa national park on American Samoa, features a paleotropical rain forest
Samoas Samoa Islands
samos (SAMOS) satellite and missile observation system
SAMPAM System for Automation of Materiel Plans for Army Materiel
SAMPE Society of Aerospace Material and Process Engineers
SAM & PE Society for the Advancement of Material

and Process Engineering
sample. simulation and modeling of profiles in lithography and etching
SAMR Special Assistant for Materiel Readiness (USA)
samrt shared-aperture medium-range tracker
sams stratospheric and mesospheric sounder
Sams Howard W Sams and Company
SAMS Sample Method Survey; Satellite Automation System; Satellite Auto-Monitor System; South American Missionary Society; Standard Army Maintenance System
SAMSA Silica and Moulding Sands Association
SAM-SAC Special Aircraft Modification for Strategic Air Command
SAM/SAT South America/South Atlantic
SAMSO Space and Missile System Organization (USAF)
SAMSON Strategic Automatic-Message-Switching Operational Network
SAM/SPAC South America/South Pacific
sam(s) [SAM(s)] shared-appreciation mortgage(s)
SAMTC South Atlantic Marine Terminal Conference
SAMTEC Space and Missile Test Center
SAMU Service Aide Médicale Urgente (French—Urgent Medicaid Service)
SA Mus South African Museum (Cape Town)
san sandwich; sanitary; styreneacrylonitrile copolymer
San Santos (British maritime abbreviation)
SAN San Diego, California (Lindbergh Field); South African Navy
SAN Space Age News
SANA Scientists Against Nuclear Arms; Scottish Association of Nurse Administration; State (Department), Army, Navy, Air (Force)
San Andreas San Andreas Fault of western California
sanat sanatoria; sanatorium
San Augustins San Augustin Mountains of southern New Mexico
SANB South African National Bibliography
Sanc. Sanctus (Latin—holy)
SANCAD Scottish Association

for National Certificates and Diplomas
SANCAR South African National Council for Antarctic Research
San Carlo Teatro di San Carlo—Naples' opera house
San Carlo Teatro San Carlo (Italian—San Carlo Theater)—Naples opera-house theater
Sanche St Charles
San-Chi San Diego—Chicago
San Clem San Clemente
SANCOB South African Foundation for the Conservation of Birds
SANCOG San Diego Council of Governments
SANCOR South African National Committee for Oceanographic Research
SANCOT South African National Commission on Tunnelling
sand silicon dioxide—SiO_2
Sand Sandford's New York Reports
San. D. Doctor of Sanitation
SAND Sampling Aerospace Nuclear Debris
SANDA Supplies and Accounts
SANDAG San Diego Association of Governments
Sand Eng Sandalwood English (Polynesian Pidgin English)
SANDER San Diego Energy Recovery
SANDERP San Diego Energy Recovery Project (garbage and trash converted to energy)
San Di San Diego
SANDIA Sandia National Laboratories
San Domingo Santo Domingo (Dominican Republic)
sand(s) sandwich(es)—invented by the Earl of Sandwich, who disliked leaving the gaming table to eat, and had thin slices of cheese or meat brought to him between two pieces of bread; his culinary invention was devised around 1776 when he was First Lord of the Admiralty
Sands the Sands, the Godwin Sands off England's Channel coast of Kent
SANDT School of Applied Non-Destructive Testing
sane. severe acoustic noise environment
SANE National Committee for

a Sane Nuclear Policy; Security Against Nuclear Extinction; South African National Antarctic Expedition

SANER San Diego Energy Recovery

Sa Nev Sierra Nevada(s)

San Fran San Francisco

SANFREE San Diegans for Fiscally Responsible Elected Employees

San Gabriels San Gabriel Mountains of southern California

Sangre de Cristos Sangre de Cristo Mountains extending from Colorado to New Mexico

sanguin (Latin prefix—blood)—sanguine

San Insp Sanitation Inspection; Sanitary Inspector(ate)

sanit sanitar; sanitation; sanitize

San Jac San Jacinto

San Jo (Mexican-American—San Jose, California)

San Juans San Juan Islands (Washington); San Juan Mountains (Colorado and New Mexico)

sanka sans kaffeine (coffee without caffeine)

San-Lax San Diego—Los Angeles

San Le San Leandro, California

San Lucas Cabo San Lucas, Baja California

s-a-n man stop-at-nothing man (dangerous criminal)

San Marco D atmospheric research satellite carrying experiments from Italy, Germany, and the United States

San Marino Most Serene Republic of San Marino (tiny country surrounded by Italy, on the slopes of Mount Titano near the Adriatic), *La Serenissima Repubblica di San Marino*

San Martin José de San Martin—patriot-soldier who fought to liberate Argentina, Chile, and Peru from the Spanish rule

San Met San Diego Metropolitan Correctional Center

S Ann St Anne's College, Oxford

San-NO San Diego—New Orleans

San-NY San Diego—New York

s-a node sino-atrial node

Sanpaolo Instituto Bancario San Paolo di Torino (Italian—San Paolo Banking Institute of Turin)

SANPAT San Diego Plans for Air Transportation

sanr subject to approval—no risks

sans sans serif

Sans Sanskrit

SANS South African Naval Service

San-Sac San Diego—Sacramento

Sansan San Diego to San Francisco (city complex)

San-Sfo San Diego—San Francisco

Sansk Sanskrit

Sant Santander; Santiago

S Ant St Antony's College, Oxford

SANTA South African National Tuberculosis Association; Souvenir and Novelty Trade Association

Santa Barbaras Santa Barbara Islands off Santa Barbara, California

Santa Fe Atchison, Topeka & Santa Fe (Railway)

Santa Monicas Santa Monica Mountains of southern California

Santa Ritas Santa Rita Mountains of southeastern Arizona

SANTAS Send A Note To A Serviceman

Santa See (Spanish—Holy See)—Vatican City, Rome

Santayana George Santayana, Jorge Augustín Nicolás Ruiz de Santayana (1863–1952)

Santiago (Portuguese or Spanish—Saint James)—Santiago do Boqueirão, Brazil; Santiago de Calatrava and Santiago de Compostela, Spain; Santiago de Chile and Santiago de Cuba, Santiago de los Cabelleros, Dominican Republic; Santiago Ixcuintla, Mexico; Santiago Sacatepéquez, Guatemala; Santiago-Zamora, Ecuador

Santiagos Santiago Mountains in the Big Bend National Park in Texas

San-Tor San Diego—Toronto

SANU Sudanese African National Union

San-Van San Diego—Vancouver

SANWR Santa Ana National Wildlife Refuge (Texas)

San Ysidros San Ysidro

Mountains of southern California

SANZ Standards Association of New Zealand

SAO São Paulo, Brazil (airport); Secret Army Organization; Smithsonian Astrophysical Observatory

Sa$_{O2}$ arterial oxygen saturation

SAODAP Special Action Office for Drug Abuse Prevention

SAORC Supreme Assembly of the Order of the Rainbow for Girls

SAOS Scottish Agricultural Organization Society

São Tomé and Principe Democratic Republic of São Tomé and Principe (West African coastal islands)

sap. saphead; scruple, apothecaries; simplified astro pattern; soon as possible; strong anthropic principle

SA£ South African pound

SAP Safety Assessment Plan; San Pedro Sula, Honduras (airport); Scottish Academic Press; Share Assembly Program; Society for Applied Spectroscopy; South African Police; Symbolic Assembly Program; Systems Assurance Program

s-apa sciences—a project approach

SAPA South African Press Association; South African Publishers' Association

SAPARLI Saudi Arabian Parsons Limited

SAPAT South African Picture Analysis Test

SAPE Society for Automation in Professional Education

SAPF South African Police Force

sapfu surpassing all previous foul ups

sapi semi-armor-piercing incendiary

SAPL San Antonio Public Library; Society for Animal Protective Legislation; South African Public Library

SAPM Scottish Association of Paint Manufacturers; Society for the Aid of Psychological Minorities

sap. no. saponification number

sapon saponification; saponify

saponite soapstone (hydrous magnesium aluminum silicate)

sapp sapphic; sapphist(ic)(al)-

(ly)
SAPRI South African Plain Research Institute
SAPS South African Price Schedule
sar search and rescue; semiautomatic rifle; short-term acquisition and retrieval; submarine advanced reactor
Sar Saracen; Saracenic; Sardinia; Sardinian
SAR Safety Analysis Report; Society of Authors' Representatives; Solar Aircraft (company); Sons of the American Revolution; South African Railways; South African Republic; South Australian Railways
S-AR *Sud-Africane République* (French—South African Republic)
Sara Sarah; Saratoga
SARA Superfund Amendments and Reauthorization Act
sarac steerable array for radar and communications
SARAH Search and Rescue and Homing (radio lifesaving beacon)
Sarasate Pablo de Sarasate (Pablo Martín Melitón Sarasate y Navacuez)
Saraw Sarawak
SARB South African Reserve Bank
SARBE Search and Rescue Beacon Equipment
SARBICA Southeast Asian Regional Branch of the International Council on Archives
sarc (Latin prefix—flesh)—sarcoma
SARC Sexual Assault Referral Centre (Australia)
SARCCUS South African Regional Committee for the Conservation and Utilisation of the Soil
sarcol sarcological; sarcologist; sarcology
SARD Special Airlift Requirement Directive
SARDA State and Regional Disaster Airlift
sardonyx chalcedony consisting of alternate layers of onyx and sard
sardsan sardine sandwich
sardwich sardine sandwich
sare self-addressed return envelope
sarge sergeant
Sargent Porter Sargent, Inc
SAR & H South African Railways and Harbours

SARHA South African Railways, Harbours, and Airways
SARHWU South African Railway and Harbor Workers Union
sarie selective automatic-radar-identification equipment
SARL *Sociedade Anónima de Responsabilidadem Limitada* (Portuguese—Limited Liability Corporation)
SARLANT Search-and-Rescue, Atlantic
Sarmiento Domingo Faustino Sarmiento—Argentinian educator and early president
SARMS Self-Adapting Account Receivable Management System
SARPAC Search-and-Rescue, Pacific
sarps standards and recommended practices
sarra short-arc reduction of radar altimetry
SARs Selected Acquisition Reports
SARS Ship Attitude Record System
sarsat search-and-rescue astronomical satellite system
SARSATS Search-and-Rescue Satellite and Tracking System
SART St Alban's Repertory Theater; Strategic Arms Reduction Talks
sartac search radar device
sartel search and rescue telephone
SARTS Switched-Access Remote Test System
SARU Systems Analysis Research Unit
sas so and so
sas (SAS) small astronomy satellite; supersonic attack seaplane; surface-air-surface (second-class international mail service)
Sas Sasebo
SAs Special Agents (FBI)
SAS Scandinavian Airlines System; Science Attitude Scale; Seattle Audubon Society; Sherwood Anderson Society; Sklar Asphasia Scale; Special Air Service; Special Armed Service(s); Statistical Analysis System; Studio Address System; Systems Assessment Survey
SAS *Societa in Accomandita Semplice* (Italian—Limited Partnership Company)
SASA South African Sugar

Association
SASBO Southeastern Association of School Business Officials
SASC Senate Armed Services Committee; Small Arms School Corps (UK); South African Staff Corps
SASCOM Special Ammunition Support Command (USA)
SASD School Administrators of South Dakota
sase self-addressed stamped envelope
SASI Society of Air Safety Investigators
SASIDS Stochastic Adaptive Sequential Information Dissemination System
SASIS Semi-Automatic Speaker-Identification System
SASJ South African Society of Journalists
Sask Saskatchewan
SASL South American Saint Line
SASLO South African Scientific Liaison Office
SASM Smithsonian Air and Space Museum
SASMIRA Silk and Artificial Silk Mills Research Association
SASO San Antonio Symphony Orchestra; Saudi Arabia Standards Organization; South African Students Organization; South Australia Symphony Orchestra
sasol South African (coal-based synthetic) oil
SASOL South African Coal, Oil, and Gas Corporation
SASP State Agencies for Surplus Property
SASR Special Air Service Regiment
Sass *Sassenach* (Gaelic—English, Saxon)
SASS San Antonio Symphony Society; Shanghai Academy of Social Sciences; Society for the Advancement of Scandinavian Study; Swarthmore Afro-American Students Society
SASSO Senior Air Staff Officer
SASSY Supported Activity Supply System
sast single asphalt-surface treatment
SAST Society for the Advancement of Space Travel
sat. sampler address translator;

satellite; satisfactory; saturate; saturation; service acceptance trials; system alignment tool; systems approach to training

sat. (SAT) satellite; systematic assertive therapy

Sat Satan; Satanic; Saturday; Saturn

S At South Atlantic

SAT San Antonio, Texas (airport); Scholastic Aptitude Test; Scholastic Assessment Team; School of Applied Tactics; Security Air Transport; Sound-Apperception Test; Southern Air Transport; Specific Aptitude Test; Spiral Aftereffect Test; Stanford Achievement Test; Support Analysis Test

SATA *Sociedade Acoriana de Transportes Aéreos* (Portuguese—Azores Air Transport Line)

SATAF Site Activation Task Force

satan satellite automatic tracking antenna; sensor for airborne terrain analysis

satanas semi-automatic analog setting

Satanic names Abaddon, Amon, Apollyn, Asmodeus, Azazel, Balaam, Beelzebub, Behomoth, Coyote, Dagon, Diabolus, Dracula, Fenriz, Gorgo, Hecate, Ishtar, Kali, Lilith, Loki, Mammon, Marduk, Mephisto (Mephistopheles), Moloch, Nija, Pluto, Prosepine, Samiel, Shiva, Tezcalipica, Typhon, Yaotzin

satar (SATAR) satellite for aerospace research

satb (SATB) soprano, alto, tenor, bass

SATC South African Tourist Corporation

satco signal automatic air traffic control

SATCO Senior Air Traffic Control Officer

satcom satellite communication

SATCOM Satellite Communications Agency (US Army)

satd saturated

satel satellite

SATENA *Servicio Aeronavegación a Territorios Nacionales* (Spanish—National Air Service)—Bogotá, Colombia

SatEvePost *Saturday Evening Post*

satex semi-automatic telegraphic exchange

sat. fix. (SAT FIX) satellite (aircraft or ship position) fix

satfy satisfactory

SATGA *Société Aérienne des Transports Guyane Antilles* (French—Guinea Air Transport Society)

satgci satellite ground-controlled interception

SAT-HI Stanford Achievement Test for the Hearing Impaired

SATIF Scientific and Technical Information Facility (NASA)

SATIN Sage Air Traffic Integration

SATIRE Semi-Automatic Technical Information Retrieval

SAtk strike attack

S Atl South Atlantic (Falkland Islands and Dependencies) - British

S Atl Cur South Atlantic Current

SAT-M Scholastic Aptitude Test—Mathematical

satn saturation

satnav satellite navigation; satellite navigator

SATO South American Travel Organization; Southern Africa Treaty Organization

SATOUR South African Tourist Corporation

Sat Pax Pax Lao *Sathalanalat Paxathipatai Paxaxōn Lao* (Lao People's Democratic Republic)

satpic satellite picture

SATRA Shoe and Allied Trade's Research Association

Sat Rev *Saturday Review*

sats (SATS) short airfield for tactical support

SATs Scholastic Aptitude Tests

SATS Satellite Antenna Test System (NASA); South African Transport Services

satsim saturation countermeasures simulator

sat sol saturated solution

sattr satisfactory to transfer

SATU Singapore Air Transport Union; South African Typographical Union

SAT-V Scholastic Aptitude Test—Verbal

SATW Society of American Travel Writers

saty satyagraha; satyriasis; satyr(ic)(al)(ly); satyrid

sau (SAU) standard advertising unit

Sau Saudi Arabia

Sau Arab Saudi Arabia(n)

SAUCERS Saucer and Unexplained Celestial Events Research Society

Saudi Saudi Arabian(s)

Saudia Saudi Arabia

Saudi Arabia Kingdom of Saudi Arabia (largest Middle Eastern country), *al-Mamlaka al-'Arabiya as-Saudiya*

Saudis Saudi Arabians

'sault assault

'sault & assault and battery

Saunders W.B. Saunders Co

SAUS *Statistical Abstract of the United States*

S Austral South Australia(n)

sav savings; stock at valuation

sa/v surface area/volume

Sav Savannah

SAV Savannah, Georgia (airport)

SAVAK *Sazemane Etelaat va Aminate Kechvar* (Persian—Iranian Security and Intelligence Organization)

SAVC Society for the Anthropology of Visual Communication

SAVE Service Activities of Volunteer Engineers; Society of American Value Engineers; Stop Addiction through Voluntary Effort; Student Action Voters for Ecology; Systematic Alien Verification for Entitlements

savi science activities for the visually impaired

SAVICOM Society for the Anthropology of Visual Communications

savor single-actuated voice recorder

SAVS Scottish Anti-Vivisection Society

Savus Savu Islands of Indonesia

saw. sample assignment word; space at will; squad automatic weapon

SAW Society of Architects in Wales; Society of Australian Writers; Special Agricultural Workers; Special Air Warfare

SAWA Screen Advertising World Association; Soil and Water Management Association

SAWARA Southern Arizona Water Resources Association

SAWAS South African Wom-

en's Auxiliary Services
Sawatches Sawatch Mountains of central Colorado
SAWC Special Air Warfare Center
sawd surface acoustic(al) wave device
SAWE Society of Aeronautical Weight Engineers
SAWF Special Air Warfare Force
SAWG Special Advisory Working Group; Special Air Warfare Group
SAWI Society Against World Imperialism (Beirut-based Arabic terrorists)
sawo surface acoustic-wave oscillator
s-a-w q seeking-asking-and-written questionnaire
saw(s) (SAWs) special agriculture worker(s)
SAWS Satellite Attack Warning System; Small Arms Weapons Study; Squad Automatic Weapon System
Sawtooths Sawtooth Mountains of south-central Idaho
SAWTRI South African Wool Textile Research Institute
Sawy Sawyer's United States Circuit Court Reports
sax saxophone; strong anion exchange
Sax Saxon
Sax Duc Saxon Duchies; Saxon Dukes
saxist saxophonist
SAY Salisbury, Rhodesia (airport)
SAYCO South African Youth Conference
saye save as you earn
SA y P San Andrés y Providencia (Spanish—San Andres and Providence)—Caribbean island possessions of Colombia
SAZF South African Zionist Federation
sb simultaneous broadcast-(ing); single-bayonet (lamp base); single-breasted (coat or jacket); small business; smooth bore; solid body; southbound; special bibliography; stolen bases; stove bolt; stretcher bearer; subbituminous; submarine (fog) bell; switchboard
s/b should be; surface based
sb styrbord (Swedish—starboard; right side of a vessel looking forward, from Viking steering oar on right side of

their long boats)
Sb stibium (Latin—antimony)
SB Air Caledonia International; Savannah Beach; Savings Bank; scouting-bombing (aircraft); Seaboard World Airlines (2-letter coding); Secondary Battery; Section Base; Selection Board; Senate Bill; Service Bulletin; Short Bill (payable on demand); shipbuilding; Signal Battalion; Signal Boatswain; Small Borough; South Bronx; South Buffalo (railroad); Soviet Bloc; Soviet Branch; Special Branch; Standard Brands (stock exchange symbol); Stanford-Binet (intelligence test); Submarine Base
S-B Stanford-Binet (intelligence test)
S & B sterilization and bath
SB San Bartolomeo (Italian—Saint Bartholomew)—Naples opera house; *Schweizerischer Bankverein* (German—Swiss Bank); *Sitzungbericht* (German—report of a proceeding)
S.B. Scientiae Baccalaureus (Latin—Bachelor of Science)
Sba Surabaya
SBA School Bookshop Association; School of Business Administration; Sick Bay Attendant; Small Business Administration; Small Businesses Association
SBAC Society of British Aerospace Companies
sbae stabilized bombing approach equipment
S-bahn Stadt-Schnellbahn (German—State Rapid Transit)—Berlin's electric railway system
SBAMA San Bernardino Air Materiel Area
S-band 1550—5200 megahertz radio-frequency band
SBAs Sick Bay Attendants
SBAW Santa Barbara Academy of the West
Sbb. Sabbatum (Latin—Sunday)
SBB Schweizerische Bundesbahnen (German—Swiss Federal Railways)
SBBNF Ship and Boat Builders' National Federation
sbc silicon blue cell; small business computer
SBC Senate Budget Committee; Service Bureau Corporation; Small Business Council;

Southern Baptist Convention; Southern Building Code; Sumitomo Bank of California; Supplementary Benefits Commission; Surinam Bauxite Company; Sweet Briar College; Swiss Bank Corporation
SBCC Santa Barbara City College
SBCCI Southern Building Code Conference International
SBCCOE State Board for Community College and Occupational Education
SBCPO Sick-Bay Chief Petty Officer
SBCR State Board of Charities and Reform (Wyoming); Stock Balance Consumption Report
sbd standard bibliographic description
sbdt surface-barrier diffused transistor
sbe soft-boiled egg(s); standby engine(s); subacute bacterial endocarditis
s-b-e standby engine(s)
SBE State Board of Equalization
SBEA Southern Business Education Association
S-bend S-shaped bend
sbf surface burst fuze
SBFA Small Business Foundation of America
sbfc standby for further clearance
sbg selenite brilliant green
Sbg Solvesborg
SBGI Society of British Gas Industries
sbh supermassive black hole
SBH Scottish Board of Health; State Board of Health
sbi space-based interceptor
SBI Security Bureau Incorporated; Southern Burn Institute (Baton Rouge); State Bank of India
sbic's small business investment companies
SBII Serikat Buruh Islam Indonesia (Central Islamic Labor Union of Indonesia)
SBIR Small Business Innovation Research
sbis (SBIS) satellite-based interceptor systems
SBIW Sybil Brand Institute for Women (Los Angeles correctional facility)
Sbl Setubal
SBL Stephen B(utler) Leacock

SBLI Savings Bank Life Insurance

sblo strong black liquor oxidation

sbm submission; submit

SBM school-based model

SBM Société Anonymes des Bains de Mer et du Cercle des Etrangers à Monaco (company managing gambling casino of Monte Carlo)

SBMA Santa Barbara Museum of Art

SBME Society of Business Magazine Editors; State Board of Medical Examiners

SBMF Santa Barbara Mariculture Foundation

SBMI School Bus Manufacturers Institute

SBMNH Santa Barbara Museum of Natural History

sbn standard book number(ing)

Sbⁿ Sebastián (Spanish—Sebastian)

SBN South Bend, Indiana (airport); Standard Book Number

SBNA Standard Book Numbering Agency

SBNB Subic Bay Naval Base

S Bno San Bernardino

SBNO Senior British Naval Officer

SBNS Society of British Neurological Surgeons

sbo secure base of operations; specific behavioral objectives

Sbo Sasebo

s'board starboard

sbom soy bean oil meal

sbp slotted-blade propeller; sugar-beet pulp; systolic blood pressure

SBP Society of Biological Psychiatry

SBPIM Society of British Printing Ink Manufacturers

sbr space-based radar; styrene-butadiene rubber

s Br *südliche Breite* (German—south latitude)

SBR Society of Biological Rhythm

SBRC Santa Barbara Research Center

sbre septiembre (Spanish—September)

SBRI Simon Baruch Research Institute

sbs simulated borehole specimen; surveyed before shipment

sbs (SBS) small business satellite

sb's sonic booms; space brothers

s-b-s side-by-side (double-barrel shotgun)

SBS Satellite Business System(s); Singapore Bus Service; Special Boat Squadron; Swiss Broadcasting Society

SBS Société de Banque Suisse (French—Swiss Bank)

SBSA Standard Bank of South Africa

Sbsc Schottky-barrier solar cell

SBSUSA Sport Balloon Society of the United States

sbt screening breath tester (for drunken drivers); segregated ballast; surface-barrier transistor

SBT Screening Breath Test

sbtg sabotage

sbti soy bean trypsin inhibitor

sbtow standby tow(ing) ship

sbt's segregated ballast tanks

sbv sea-bed vehicle

SBV Space Biospheres Ventures

SBW Seaboard & Western (Airlines); single-engine scout bomber (3-letter naval symbol)

SBWR Seal Beach Wildlife Refuge (near Long Beach, California); South Bay Wildlife Refuge (south end of San Francisco Bay)

sbx S-band transponder

SBX Student Book Exchange

sby standby

sc sad case; same case; scratched; separate cover; shaped charge; single circuit; single contact; single crochet; sized and calendered; slow cool; small caps (small capital letters); smooth contour; statistical control; supercycle; superimposed current

sc (SC) service charge; site contractor; spinal cord; systolic click

s/c short circuit (electrical); single-column (bookkeeping); suspicious circumstances

s & c search and clear; shipper and carrier; sized and calendered

sc. scilicet (Latin—mainly)

s/c su cuenta (Spanish—your account)

Sc scandium; Scotch (whiskey); Scot(s); Scottish; stratocumulus

Sc La Scala (Teatro alla Scala)—Milan's opera house; *Scoglio* (Italian—reef; rocky reef)

SC Sacra Congregatio (Sacred Congregation); Sacramento City; Salem College; San Carlos; Sandia Corporation; Sanitary Corps; Santa Claus; Scripps College; Seamen's Center; Security Council (United Nations); Selwyn College; Service Club; Service Command; Shasta College; Shaw College; Shell Transport; Shelton College; Shenandoah College; Shepherd College; Sheridan College; Shimer College; Ship's Cook; Shorter College; Siena College; Sierra Club; Sierra College; Signal Corps; Simmons College; Simpson College; Sinclair College; Sister(s) of Charity; Skidmore College; Smith College; Somerville College; South Carolina; South Carolinian; Southern California; Southern Californian; Southern Conference; Southwestern College; Special Constable; Spellman College; Springfield College; Staff College; Staff Corps; Stephens College; Sterling College; Stockton College; Stonehill College; Stratford College; Strike Command; submarine chaser; Sullins College; Summary Court; Sumter & Choctaw (railroad); Suomi College; Supply Corps; Support Command; Supreme Court; Surgical Corporation; Swarthmore College; Systems Command

S-C Serbian-Croatian (people); Serbo-Croat (language); Stromberg-Carlson

S/C Star & Crescent (excursion steamer, ferry, towing, water-taxi service)

S&C search and clear;

SC Scott Catalog; Statistics Canada

sca sequencer control assembly; small-caliber ammunition; subchannel adapter

sca (SCA) supersonic cruising aircraft

SCA Schipperke Club of America; School and College Ability (test); Science Clubs of America; Screen Composers Association; Senior Citizens of America; Shipbuilders Council of America; Shipbuilders Council of

America; Society of Consumer Affairs; Soybean Council of America; Speech Communication Association; Standard Consolidated Area; State Commemorative Area; Stock Company Association; Student Conservation Association; Sub-Contract Authorization; Suez Canal Authority; Survey of College Achievement; Svenska Cellulose AB; Switzerland Cheese Association; Synagogue Council of America
SCAA State Communities Aid Association
SCAAP Special Commonwealth African Assistance Plan
SCAC School and College Advisory Center; Sunrise Cultural and Art Center (Charleston, West Virginia)
SCACOP Southern California Area Construction Opportunity Program
scad schedule, capability, availability, dependability
SCAD State Commission Against Discrimination (New York)
scadar scatter detection and ranging
SCADS Sioux City Air Defense Sector
SCADTA *Sociedad de Colombo-Alemana de Transportes Aereos* (Spanish—Colombia-German Air Transport Society)
SCAF Supreme Commander of Allied Forces
SCAG Sandoz Clinical Assessment—Geriatric; Southern California Association of Governments; Supplier Corrective Action Group
SCAGL *Société Cinématographique des Auteurs et Gens de Lettres* (French—Cinematic Society of Authors and Writers)
sc al steel-cored aluminum
SCALA Society of Chief Architects of Local Authorities
scaler statistical calculation and analysis of engine removal (USN)
scama (SCAMA) switching, conferencing, and monitoring arrangement
scams scanning microwave spectrometer
scan. self-correcting automatic navigation; suspected child

abuse and neglect; switched-circuit automatic network
Scan Scandinavia; Scandinavian
SCAN Scheduling and Control by Automated Network; Selected Current Aerospace Notices (NASA-computerized dissemination of information); Self-Correcting Automatic Navigator; Service Center for Aging Information; Southern California Answering Network; Switched-Circuit Automatic Network
SCANCAP System for Comparative Analysis of Community Action Programs
Scand Scandinavia; Scandinavian
ScanDoc Scandinavian Documentation Center
scanit scan-only intelligent terminal
scan. mag. *scandalum magnatum* (Latin—defamation of high-placed persons)
SCANNET Scandinavian (computer) Network
SCANPED System for Comparative Analysis of Programs of Educational Development
SCANs Southern California Answering Networks (cooperative library information-retrieval system)
SCANS Scheduling and Control Automation by Network Systems; Stockmarket Computer Answering Service
scantie submersible-craft acoustic-navigation and track-indication equipment
SCAO Senior Civil Affairs Office(r); Standing Conference on Atlantic Organizations
scap scapula; scapular; scapuloid
SCAP Supreme Commander, Allied Powers
Scapa Scapa Flow naval anchorage in the Orkney Islands off Scotland's north coast between Hoy, Orkney, and South Ronaldsay
SCAPA Society for Checking the Abuses of Public Advertising
'scape escape(ment); landscape; seascape; skyscape
scaphocephs scaphocephalics (narrow-skulled people)
s caps small capital letters
SCAQMD South Coast Air

Quality Management District (California)
scar. subcaliber aircraft rocket; submarine celestial altitude recorder
SCAR Scandinavian Council for Applied Research; Scientific Committee for Antarctic Research; Supersonic Cruise Airplane Research (NASA)
scarab. (SCARAB) submersible craft assisting repair and burial (of underwater telephone cables)
Scarboro' Scarborough
scard signal conditioning and recording device
scare. sensor-control anti-anti-radiation-missile radar evaluation
SCARF Special Committee on the Adequacy of Range Facilities
scarp escarpment
S-car(s) Swedish car(s)
sca's subsidiary communications authorizations
SCAS Senior Citizen Audiological Service
scat. share compiler assembler and translator
scat. (SCAT) speed-control attitude range; supersonic commercial air transport
scat. scatula (Latin—box)
SCAT School and College Ability Test; Science College Ability Test(ing); Service Command Air Transportation (USN)
scata survival sited casualty treatment assemblage
SCATANA Security Control of Air Traffic and Air Navigational Aids
SCATE Stromberg-Carlson automatic test equipment
scatha spacecraft charging at high(er) altitude(s)
scato (Greek—excrement)—scatologic(al); scatology
scat. orig. *scatula originalis* (Latin—original box or package)
scats (SCATS) sequentially-controlled automatic transmitter start (data processing)
scat's supersonic commercial air transports
SCATs Southern California Acrobatic Teams
SCATS Simulation, Checkout, and Training System
SCAULWA Standing Conference of African University Libraries—Western Area

(Ghana)
scav scavenge
Scaw Fells Scaw Fell (or Scafell) Mountains of the Cumbrians in England's Lake District
scb state-capacity building; strictly confined to bed (q.v. fob)
sc b screw base (lamp)
Sc.B. Scientia Baccalaureus (Latin—Bachelor of Science)
SCB Sawyer College of Business; Sierra Club Books; Southern California Bookbuilders
SCB Sociedad Bolivariana de Venezuela (Spanish—Bolivarian Society of Venezuela)
SCBA Southern California Booksellers Association
SCBC Somerset Cattle Breeding Centre
SCBCA Small Claims Board of Contract Appeals
scbf spinal-cord blood flow
SCBQ Science Classroom Behavior Q-sort
scbu (SCBU) special-care baby unit
SCBW Society of Children's Book Writers
scc single-channel controller; specific clauses and conditions; stress corrosion cracking
Sc C Scottish Command
SCC Sea Cadet Corps; Security Coordination Committee; Select Cases in Chancery; Ship Control Center; Shoreline Community College; Sitka Community College; Society of Cosmetic Chemists; Spokane Community College; Standard Commodity Classification; Standing Consultative Commission (U.S.–USSR group created to monitor disputes); Stromberg-Carlson Corporation; Student Coordinating Council; Surveillance Coordination Center
S&CC Suicide and Crisis Counseling
SCCA Society of Company and Commercial Accountants; South Carolina Correctional Association; Southeastern Cottonseed Crushers Association; Sports Car Club of America
SCCAPE Scottish Council for Commercial, Administrative, and Professional Education

SCCC Singapore Chinese Chamber of Commerce; Suffolk County Community College; Sullivan County Community College
SCCCI Singapore Chinese Chamber of Commerce and Industry
SCCF Security Clearance Case Files
SCCG Southern California Culinary Guild
SCCOP State Consulting Company for Oil Projects
SCCPG Satellite Communications Contingency Planning Group
SCCPT Subcommittee on Computer Program Terminology (Association for Computing Machinery)
sccrt sub-zero cooled, cold-rolled, and tempered
scd screen door; screwed; service computation date; standard change dispenser
scd (SCD) security coding device
Sc.D. Scientiae Doctor (Latin—Doctor of Science)
SCD Specification Control Drawing
SCD Standard College Dictionary
scda scapula-dextra anterior
SCDA Scottish Community Drama Association
SCDC Senior Citizen's Dental Clinic; South Carolina Department of Correction
scde's schools, colleges, and departments of education
SCDL Scientific Crime Detection Laboratory
scdp scapula-dextra posterior
SCDS Shipboard Chaff-Decoy System
sce situationally caused error; standard calomel electrode
SCE Schedule Compliance Evaluation; Society for Clinical Ecology; Southern California Edison
S.C.E. Scottish Certificate of Education
SCEA South Carolina Education Association
SCEI Safe Car Educational Institute; Special Libraries Committee on Environmental Information
SCEL Signal Corps Engineering Laboratories
scen scenario(s); scenarist(s); scenographic(al)(ly)
SCEPC Senior Civil Emer-

gency Planning Committee (NATO)
SCES State Cooperative Extension Service
SCET Scottish Council for Educational Technology
SCF Save the Children Federation; Sectional Center Facility (USAF); Station Code File; Stephen Collins Foster
sc f & a screw forward and aft
SCFA Southern California Fishermen's Association
scfd standard cubic feet per day
scfh standard cubic feet per hour
SCFIC South Carolina Foundation of Independent Colleges
scfm standard cubic feet per minute
scfs standard cubic feet per second
scg scoring
SCG Screen Cartoonists Guild; Social Credit Group; Society of the Classic Guitar; Special Consultative Group
Sc Gael Scottish Gaelic
SCGB Ski Club of Great Britain
SCGC Southern California Gas Company; Southern Counties Gas Company
SCGR Sale Common Game Refuge (Victoria, Australia)
SCGRL Signal Corps General Research Laboratory
SCGSA Signal Corps Ground Signal Agency
SCGSS Signal Corps Ground Signal Service
sch school
sch (SCH) schedule
Sch Schiedam; School (postal abbreviation)
Sch Schauspielhaus (German—playhouse, theater)
SCHAVMED School of Aviation Medicine (USN)
Schbg Schönberg
schd scheduled; scheduling
sched schedule
scheepv scheepvaart (Dutch—navigation, shipping)
scheik scheikunde (Dutch—chemistry)
schem schematic
Schen Schenectady
scherz scherzando (Italian—jesting, in a sportive manner)
schilling monetary unit of Austria
Schipol Amsterdam airport
Schirmer EC Schirmer (Bos-

ton); G Schirmer (New York)
Sc Hist Scottish History
schizo schizoid; schizophasia; schizophrenia; schizophrenic
schizzy schizoid; schizophrenia; schizophrenic
SCHLA School of Latin America
Schlags *Schlagobers* (Austrian German—whipped cream)
schlem *schlemiel* (Yiddish—person afflicted with bad luck)
Sch Lib Sci School of Library Science
schm schematic
Sch M School Master
Schmarg Schmargendorf
Sch Mist School Mistress
schmoo space cargo handler and manipulator for orbital operations
schol schola cantorum; scholar(ly); scholarship; scholastic(ally); scholasticate; scholasticism; scholiast(ic); scholium
SCHOLAR Schering-Oriented Literature Analysis and Retrieval System
Schotl *Schotland* (Dutch—Scotland)
schott *schottisch* (German—Scottish)
schr schooner
Schr *Schriften* (German—publication; script, text, writing)
SCHS Senior Citizen Hospital Service
Schupo *Schutzpolizei* (German—defense police used as a paramilitary force by Hitler)
Schwann *Schwann-1 Record & Tape Guide*
Schweitzerpsalm (German—Swiss Psalm)—national anthem of Switzerland also sung in French, Italian, and Romansh
schweiz *schweizerisch* (German—Swiss)
Schwyz Schwyzer(tütsch)
Schwyzd *Schwyzerdütsch* (Swiss-German language)
sci science; scientific; scientist
sci (SCI) secret confidential informant; sensitive compartmented information
sci *scientifique* (French—scientific)
SCI School of Counter-Insurgency; Science Citation Index; Seamen's Church Institute; Service Civil International; Shipping Container

Institute; Shipping Corporation of India; Simulation Councils Incorporated; Society of Chemical Industries; Society of the Chemical Industry; Sponge and Chamois Institute; State Commission of Investigation; Supervisory Cost Inspector
SCI *Science Citation Index; Servicio Central de Inteligencia* (Spanish—Central Intelligence Service)
SCIA Signal Corps Intelligence Agency
Sci Am *Scientific American*
SCI/ARC Southern California Institute of Architecture
scicrit scientific critic(ism)
scics semiconductor integrated circuits
scid severe combined immune deficiency
Sci D Doctor of Science
Sci D Com Doctor of Science in Commerce
Scidgie Sicilian-Italian (dialect)
Sci D Met Doctor of Science in Metallurgy
Science Academy of Science; High School of Science
scient scientific; scientist
sci-fi science-fiction
sci-fic science-fiction
SCII Strong-Campbell Interest Inventory
scil. *scilicet* (Latin—namely)
SCIL Support Center International Logistics (USA)
Scillies Scilly Islands, Isles of Scilly, the Sorlings
scim standard cubic inches per minute
Sci M Science Master
SCIM Selected Categories in Microfiche
Sci Mist Science Mistress
scimp. self-contained-imaging microprofiler
scinti scintillate; scintillation
SCIO Staff Counterintelligence Officer
scioneer scientist + engineer
SCIOP Social Competence Inventory for Older Persons
SCIP School Campus Interaction Programme
SCIPA *Servicio Cooperativo Interamericano de Producción de Alimentos* (Spanish—Interamerican Cooperative Service for the Production of Food)
sci-phi science-philosophy
scipp sacrococcygeal-to-infe-

rior pubic point
SCIPP Santa Cruz Institute for Particle Physics
SCI & RB South Carolina Inspection and Rating Bureau
Sci Res Assoc Science Research Associates
SCIRP Select Commission on Immigration and Refugee Policy
SCI(s) Success Motivation Institutes
SCIS Science Curriculum Improvement Study
SCISP *Servicio Cooperativo Interamericano de Salud Pública* (Spanish—Interamerican Cooperative Public Health Service)
Sci-Tec Science-Technology Division (American Libraries Association)
SCITEC Association of the Scientific, Engineering, and Technological Community of Canada
SCI-TECH-SLA Science-Technology Division of the Special Libraries Association
sc&j signal collection and jamming
SCJ School of Criminal Justice
scl scleroderma; space charge limited
Scl Sculptor (constellation)
SCL Santiago, Chile (airport); Scottish Central Library; Seaboard Coast Line; Society of County Librarians; Southeastern Composers' League; Springfield City Library
scla scapula-laeva anterior
SCLC Southern Christian Leadership Conference
SCLED South Carolina Law Enforcement Division
SCLERA Santa Catalina Laboratory for Experimental Relativity by Astrometry
SCLH Standing Committee for Local History; Standing Conference for Local History
SCLI Seaboard Coast Line Industries; Somerset and Cornwall Light Infantry
sclp scapulo-laeva posterior
SCLS Serra Cooperative Library System
scm samarium cobalt magnet; small-core memory; soluble cytotoxic mediator; steam-cure mortar
scm (SCM) specification change memo(randum); strategic cruising missile

Sc.M. *Scientiae Magister* (Latin—Master of Science)

SCM Section Communication Manager; Smith-Corona-Marchant; Society of Connoisseurs in Murder; Southampton City Museum; Special Court-Martial; Summary Court-Martial

S.C.M. State Certified Midwife

SCM Su Católica Majestad (Spanish—Your Catholic Majesty)

SCMA Southern California Marine Association; Southern Cypress Manufacturers Association

SCMAI Staff Committee on Meditation, Arbitration, and Inquiry (ALA)

SCMC Senior Citizen's Medical Clinic

SCMES Society of Consulting Marine Engineers and Ship Surveyors

SCMP South China Morning Post

scn scan (flow chart); suprachiasmatic nuclei

Scn Scunthorpe

SCN System Control Number

SCNAWAF Special Category Navy with Air Force

SCNM Sunset Crater National Monument (Arizona)

SCNO Senior Canadian Naval Officer

SCNR Scientific Committee of National Representatives (NATO)

scns self-contained navigation system

scn/sin sensitive command network/sensitive information network

SCNUL Standing Conference of National and University Libraries (UK)

SCNVYO Standing Conference of National Voluntary Youth Organisations (UK)

SCNWR Squaw Creek National Wildlife Refuge (Missouri)

sco subcarrier oscillator; sustainer cutoff

Sco Scorpius (constellation)

ScO Scientific Officer

SCO Sales Contracting Office(r); Statistical Control Office(r)

SCOC Senior Citizen Otolaryngological Clinic; Support Command Operations Center

scoda scan coherent doppler attachment

SCODS Standing Committee on Ocean Data Stations

SCOFF Society for the Conquest of Flight Fear

SCOGS Select Committee on Generally-Regarded-As-Safe Substances

SCOLCAP Scottish Libraries Cooperative Automation Project

SCOLE Standing Committee on Library Education

S Coll Staff College

SCOLLUL Standing Conference of Librarians of Libraries of the University of London

SCOLMA Standing Conference on Library Materials on Africa

SCOM Scientific Committee (NATO)

scon self-contained

scond semiconductor

SCONMEDLIB Standing Conference of Mediterranean Libraries

'Sconsin Wisconsin

SCONUL Standing Conference of National and University Libraries

scoop. scientific computation of optimum procurement

scop (SCOP) single copy order plan

scope microscope; oscilloscope; periscope; telescope; telescopic gunsight

SCOPE Scholarly Communication—Online Publishing and Education; School-to-College Opportunity for Post high-school Education; Scientific Committee on Problems of the Environment; Selected Contents of Periodicals for Educators; Simple Checkout-Oriented Program Language; Special Committee on Problems of the Environment (ICSU); Student Council on Pollution and Environment

SCOPES Squad Combat Operations Exercise Simulation (USA)

scor skin-conductance orienting response

Scor Scorpio

SCOR Scientific Committee on Oceanographic Research; Standing Conference on Refugees

score. signal communications by orbiting relay equipment;

spectral combinations by reconnaissance exploitation

SCORE Service Corps of Retired Executives; Special Covert Operations for Resale; System Capability over Requirement Evaluation

SCORES Scenario-Oriented Recurring-Evaluation System (USA)

scorpio subject-content-oriented retrieval for processing information on-line

SCOS Scottish Certificate in Office Studies; Senior Citizen Optometrical Service

scot steel car of tomorrow

Scot Scotch; Scotland; Scots; Scotsman; Scotswoman; Scottie(s); Scottish; Scotty

SCOTAPLL Standing Conference of Theological and Philosophical Libraries in London

SCOTBEC Scottish Business Education Council

ScotGael Scots Gaelic

Scotiabank Bank of Nova Scotia

ScotNats Scottish Nationalists

Scots wha ha'e Scots wha ha'e wi' Wallace bled (Scottish—Scots who have with Wallace bled)—national anthem

Scott Scott, Foresman; Scott Publications; William R Scott

Scotts Bluff Scotts Bluff National Monument in western Nebraska on the Oregon Trail

SCOTUS Supreme Court of the United States

Scot virus Scottish type of influenza virus sometimes called Scotland virus

'scouse lobscouse (sailor's stew)

scp secondary control point; single-cell protein; spherical candlepower; supervisor's control panel

SCP Sea Containers Pacific; Senior Companion Program; Site Characterization Plan; Social Credit Party; Survey Control Point

SCP (AFL-CIO) Sleeping Car Porters

SCPA Scottish Chick Producers' Association; South Carolina Ports Authority

SCPAs State Criminal-Justice Planning Agencies

scpc single channel per carrier

SCPCU Society of Chartered

Property and Casualty Underwriters

SCPD Staff Civilian Personnel Division (USA)

SCPE State Committee on Public Education

SCPEA Southern California Professional Engineering Association

SCPGA South Carolina Personnel and Guidance Association

SCPI Structural Clay Products Institute

SCPL Social Credit Political League (New Zealand Party)

SCPN Society of Certified Professional Numismatists

SCPO Senior Chief Petty Officer

SCPR Scottish Council of Physical Recreation

SCPS Senior Citizen Podiatric Service; Society of Civil and Public Servants (British)

SCPt security control point

SCPU Sea Containers Pacific Unit

scpv (SCPV) silkworm cytoplasmic polyhedrosis virus

SCQ Coastal Sentry (tracking station vessel—naval symbol)

scr screw; scruple; silicon-controlled rectifier

s-c r short-circuit radio

SCR Signal Corps Radio; Site Characterization Report; Standardized Casualty Rate

SCRA Southern California Restaurant Association; Stanford Center for Radar Astronomy

scram self-contained radiation monitor; supersonic combustion ramjet (engine)

SCRAM Special Criteria for Retrograde Army Materiel; Synanon Committee for Responsible American Media

scrap. simple-complex reaction-time apparatus

SCRAP Society for Completely Removing All Parking (Meters); Students Challenging Regulatory Agency Proceedings

SCRATA Steel Castings Research and Trade Association

scr bh screen bulkhead

SCR brick Structural Clay Research brick

SCRC Southern California Renewal Communities; Southern California Research Council

SCRCC Soil Conservation and Rivers Control Council

SCRDT Stanford Center for Research and Development in Teaching

SCRE Scottish Council for Research in Education

SCREAM Society for the Control and Registration of Estate Agents and Mortgage Brokers

SCREAMS Society to Create Rapprochement among Electrical, Aeronautical, and Mechanical Engineers

screenex screening exercise

SCRF Scripps Clinic and Research Foundation; Small Craft Repair Facility (USN)

Scribner Charles Scribner's Sons

SCRID Southern California Registry of Interpreters for the Deaf

scrim scrimmage

scrip scriptural; scripture

script manuscript; prescription

Script Scriptural; Scripture

SCRIPT Stanford Computerized-Researcher Information-Profile Technique

SCRIS Southern California Regional Information Study (Bureau of the Census)

SCRL Signal Corps Radar Laboratory

SCRLC South Central Research Library Council

scrn screen; screening; screens

scr's silicon-controlled rectifiers

Scrt Sanskrit

SCRTD Southern California Rapid Transit District

Scrtrt the Secretariat (UN)

Scrubs Wormwood Scrubs

scrum scrummage

scs satellite control system; secret cover sheet; space command station; stabilization control system

scs (SCS) sea-control ship

sc & s strapped, corded, and sealed

SCS Scientific Control System(s); Screening and Costing Staff (NATO); Secondary Control Ship (USN); Serving Christian Scientists; Society of Civil Servants; Society of Clinical Surgery; Society for Computer Simulation; Soil Conservation Service; Southern California Skeptics; Student Counseling Service

SCSA Soil Conservation Soci-

ety of America; Southern California Symphony Association

SCSBM Society for Computer Science in Biology and Medicine

SCSC South Carolina State College

sc-se smooth curve-smooth earth

SCSE Society of Casualty Safety Engineers

SCSEA Southern California Solar Energy Association

SCSEP Senior Community Service Employment Program

SCSF Surface Cask Storage Facility

Sc.Soc.D. Doctor of Social Science

SCSP Site Characterization and Selection Plan; State Center Service Program; System Calibration Support Plan (USAF)

SCSPA South Carolina State Ports Authority

SCSS Scottish Council of Social Service

SCSU Southern Connecticut State University

sct structural clay tile; sub-zero cooled and tempered

sct (SCT) subroutine call table; surface charge transistor

Sct Scutum (constellation)

Sc & T Science and Technology

S Ct *Supreme Court Reporter*

SCT Society of Commercial Teachers

s/cta *su cuenta* (Spanish—your account)

SCTA Steel Carriers Tariff Association

SCTE Society of Cable Television Engineers

sctl short-circuited transmission line

Sctl Schottky coupled-transistor logic

sctr sector (flow chart)

SCTR Standing Conference on Telecommunications Research

sctrd scattered

sct's sugar-coated tablets

SCTS Sycamore Canyon Test Site (Convair)

Sctsmn *The Scotsman* (Edinburgh)

sctt submarine-command team trainer

SCTTF Small-Core Triaxial Test Facility

scty security
SCU Santa Clara University; Selector Checkout Unit; Special Care Unit; Sharecroppers' Union
SCUA Suez Canal Users' Association
scuba self-contained underwater breathing apparatus
scubasub scuba-diver's submarine; scuba-diver's submersible
S-cubed serial-signalling scheme; serial-signalling system
SCUK South Coast of the United Kingdom
sculp sculptor; sculpture
sculp. sculpsit (Latin—he carved or engraved it)
SCUM Society (for) Cutting Up Men
scup scupper
SCUP Society for College and University Planning
S-curve S-shaped curve
SCUS Supreme Court of the United States
'scutcheon escutcheon
scv single concave
s-c-v single-capsulated-virulent (bacteria)
s & cv stop and check valve
SCV Sons of Confederate Veterans
SCV Santa Città Vaticana (Italian—Holy Vatican City)
S.C.V. Stato della Città del Vaticano (Italian—Vatican City State)
SCVANYO Standing Conference of Voluntary Youth Organizations
scvtr scan-converting video tape recorder
SCW State College of Washington
SCWC Special Commission on Weather Modification
SCWPH Students Concerned With Public Health
scwr (SCWR) supercritical water reactor
SCWS Scottish Co-operative Wholesale Society
scx single convex
SCXU Sea Containers Atlantic Unit
SCYC South Coast Yacht Club
S Cz Salina Cruz
sd second defense (lacrosse); self-destroying; semidiameter; septal defect; serum defect; shell-destroying; shit disturber (troublemaker); sight draft; silver dollar;

single deck; skin dose; sound; special duty; spontaneous delivery; stage door; standard deviation; storm detection; storm drain(age); streptodornase; sudden death; system demonstration; systolic to diastolic; systolic discharge
s-d slow-drying
s/d sea-damaged; systolic-to-diastolic
s & d search and destroy; song and dance
sd siehe dies (German—see this)
s.d. sine die (Latin—without date)
sD samme Dato (Danish—same date)
Sd Sound
S$ Singapore dollar
S^d Sound
SD Salt Domes; San Diegan; San Diego; Secretary of Defense; Senior Deacon; Sight Draft; snare drum; Specification for Design; Spectacle Dispenser (oculist); Standard Oil Company of California (stock exchange symbol); State Department; Superintendent of Documents; Supply Depot
SD Sicherheitsdienst (German—Intelligence Service); Social(ist) Democrat(ic) (party); *Stofarts Directoratet* (Norwegian—Directorate of Shipping); *Stronnictwo Demokratyczne* (Polish—Democratic Party)
sda sacro-dextra anterior; source data acquisition; source data automation; specific dynamic action; succinic dehydrogenase activity
SDA Scottish Development Agency; Scottish Dinghy Association; Scottish Diploma in Agriculture; Seventh Day Adventist; Ship Destination Authority; Soap and Detergent Association; Social Democratic Alliance; Source Data Automation; Students for Democratic Action
SDAA San Diego Apartment Association
SDAC San Diego Art Center
SDACCLRC San Diego Area Community Colleges Library Resources Cooperative
Sdad Sociedad (Spanish—Society)
SDAE San Diego Adult Edu-

cators
SD & AE RR San Diego & Arizona Eastern Railroad
SDAF San Diego Architectural Foundation
SDAG San Diego Association of Governments
S Dak South Dakota; South Dakotan
SDAM San Diego Aerospace Museum
sdaml send by airmail
SDAP Systems Development Analysis Program; System Development and Performance
SDASBO South Dakota Association for School Business Officials
sdAt (SDAT) senile dementia of the Alzheimer's type
SDAT Senile Dementia of the Alzheimer Type
S-day submarine-deployment day (NATO)
SDB Salesian of Don Bosco; Society for Developmental Biology
SDB Sluzba Drzavne Bezbednosti (Serbo-Croat—State Security Service)—Yugoslav
sdbl sight draft bill of lading
sd bl sandblast
SDBL Sight Draft with Bill of Lading
SDBRI San Diego Biomedical Research Institute
sdby standby
sdc shipment detail card; single drift connection; submersible decompression chamber
sdc (SDC) signal data converter
SDC Southern Defense Command; Space Development Corporation; Special Devices Center; State Defense Council; State Department of Corrections (Alabama, Colorado, Virginia); Strategic Defense Command; Support Design Change; Systems Development Corporation
SDCA Society of Dyers and Colourists of Australia
SDCB State Dissemination Capacity Building
SDCC San Diego City College
SD/CC Security Designation/Custody Classification
SDCCD San Diego Community College District
SDCCs San Diego Community Colleges
SDCE Society of Die Casting Engineers

SDCF San Diego Community Foundation
SDCINTF San Diego County Integrated Narcotic Task Force
SDCJ San Diego County Jail
SDCL System Distress Check List
SD Class. Superintendent of Documents Classification
SDCMD San Diego Contract Management District
SDCMS San Diego County Medical Society
SD Co San Diego County
SDCS San Diego City Schools
SDCSO San Diego County Symphony Orchestra
s-d curve strength-duration curve
SDCWA San Diego County Water Authority
sdd store-door delivery
SDD Scottish Diploma in Dairying; System Definition Directive; System Design Description
sddl saddle(d); sorted data-definition language
sde self-disinfecting elastomer; simple designational expression
SDE Society of Data Educators; State Department of Education
SDEA South Dakota Education Association
's'death god's death
S de B Simone de Beauvoir
S de C *Société des Cuisiniers* (French—Society of Cooks)
SDEC San Diego Ecology Center; San Diego Engineering Council; San Diego Evening College
SDECE *Service de la Documentation Extérieure et du Contre-Espionage* (French equivalent of American CIA)
SDEE *Société de la Diffusion d'Equipements Electroniques* (French—Electronic Broadcasting Society)
SDEI San Diego Eye Institute
S de M Salvador de Madariaga
SDEO Salt Domes Exploration Office
sdf single-degree-of-freedom (gyroscope)
sdf *sans domicile fixe* (French—without address; without a fixed living place)
SDF Louisville, Kentucky (airport); Self-Defense Forces (Japan)
SDFD San Diego Fire Depart-

ment
SDFMC San Diego Foundation for Medical Care
SDFS San Diego Federal Savings
sdg siding
Sdg Siding
SDG Sacred Dance Guild; Self-Development Group
S.D.G. *Solo Deo Gloria* (Latin—Glory to God Alone)
SDG & E San Diego Gas & Electric
SDGP State Dissemination Grant Program
sdh (SDH) sorbitol dehydrogenase
SDH Scottish Diploma in Horticulture
SDHA San Diego Hospital Association
SDHC San Diego Housing Commission
sdhe spacecraft data-handling equipment
SDHRC San Diego Human Relations Commission
sdi selective dissemination of information
SDI Saudi Arabian Airlines; Secret Diplomatic Initiative; Selective Dissemination of Information; Senior Drill Instructor (USMC); Standard Data Interface; Strategic Defense Initiative (Star Wars)
SDIBM San Diego Institute for Burn Medicine
SDIC San Diego Improvement Association
S Diego San Diego
sdiline selective dissemination of information on-line
SDIO Strategic Defense Initiative Organization
SD & IV San Diego & Imperial Valley Railroad
SDJC San Diego Junior Colleges
sdk shelter deck
Sdk (SDK) San Diego (container symbol)
sdl saddle
sdl (SDL) state-dependent learning
SDL Special Duties List(ing); Systems Dimensions Limited
SDLA South Dakota Library Association
sdlc synchronous data-link communication(s)
SDLP Social Democratic and Labour Party
sdm selective discrimination on microfiche
SDM *Su Divina Majestad*

(Spanish—Your Divine Majesty)
SDMA San Diego Museum of Art; Surgical Dressing Manufacturers' Association
SDMC San Diego Mesa College
SDMICC State Defense Military Information Control Committee
SDMJ September, December, March, June
sdml seaward defense motor launch
SDMM San Diego Museum of Man
SDMNH San Diego Museum of Natural History
SDMS San Diego Memorial Society
sdn sedan
SDN System Designation Number
SDN *Société des Nations* (French—League of Nations)
Sdn Bhd *Sendirian Berhad* (Malay—Private Limited)—limited corporation
SDNHM San Diego Natural History Museum
SDNS Scottish Daily Newspaper Society
SDO San Diego Opera; Santo Domingo (Dominican Republic); Squadron Duty Office(r); System Design Objectives
S Doc Senate Document
SDOC Space Defense Operations Center (Cheyenne, Wyoming)
sdof single degree of freedom
SDOG San Diego Opera Guild
Sdom Sodom
SDOP San Diego Organizing Project
sdp sacro-dextra posterior; social, domestic, and pleasure
Sd £ Sudanese pound (currency unit)
SDP Social(ist) Democratic Party; Subseabed Disposal Program
SDP *Sozialdemokratische Partei Deutschlands* (German—German Social-Democratic Party)
SDPCC San Diego Poison Control Center
SDPD San Diego Police Department; Schizoid Personality Disorder
SDPGA South Dakota Personnel and Guidance Association
SDPL San Diego Public Li-

brary
S Dpo Station Depot
SDPO Site Defense Project Office(r)
SDPOA San Diego Police Officers Association
SDPT Structured Doll Play Test
SDQ Santo Domingo, Dominican Republic (airport)
sdr scientific data recorder; self-decoding readout; simple detection response; size-to-diameter ratio; sodium deuterium reactor; sonar data recorder; splash-detection radar; strip domain resonance; successive discrimination reversal
SDR Special Despatch Rider; Special Dispatch Rider; Special Drawing Right; Special Drilling Rights; Strategic Defense Response
SdRng sound ranging
SDRs Special Drawing Rights; Special Drilling Rights
sds self-directed search; speech discrimination score; sudden death syndrome
s-d s single-day surgery
SDS San Diego Symphony; Scientific Data Systems; Sons and Daughters of the Soddies; Spatial Data System(s); Special District Services; Students for a Democratic Society
SDSC San Diego State College; San Diego Steamship Company; San Diego Supercomputer Company
sd sms clsd side seams closed
SDSMT South Dakota School of Mines and Technology
SDSNH San Diego Society of Natural History
SDSO San Diego Symphony Orchestra
SDSRU Soil Data Storage and Retrieval Unit
SDSS Self-Deploying Space Station
SDSU San Diego State University
sdt sacro-dextra transversa; scientific distribution technique; sea depth transducer; serial data transmission; serial data transmission; source distribution technique; surveillance data transmission
SDT San Diego Transit; Society of Dairy Technology
SDTC San Diego Transit Corporation

SDTD San Diego Transit District
sdtdl saturating drift transistor diode logic
sdti selective dissemination of technical information
SDTI San Diego Technical Institute
SDTS Satellite Data Transmission System
SDTTS San Diego Turtle and Tortoise Society
SDTU Sign and Display Trades Union
sdu shelter decontamination unit; signal display unit; spectrum display unit; sub-carrier display unit
SDU Rio de Janeiro, Brazil (Santos Dumont Airport); Social Democratic Union
SDU San Diego Union
SDUK Society for the Diffusion of Useful Knowledge; Spoiled Duck (according to Edgar Allan Poe in his essay on *How to Write a Blackwood Article*)
SDUPD San Diego Unified Port District
SDUSD San Diego Unified School District
sdv slowed-down video; swimmer delivery vehicle
sdw swept delta wing
SDWA Safe Drinking Water Act
SDWAP San Diego Wild Animal Park
SDX Stromberg DatagraphiX; Sunray Mid-Continent Oil Company
SDX Sigma Delta Chi, society of professional journalists
SDYC San Diego Yacht Club
SDYS San Diego Youth Symphony
SDZ San Diego Zoo
SDZS San Diego Zoological Society
se second entrance; semiannual; service entrance; single end; single-ended; single engine; single entry; special equipment; spherical equivalent; standard error; straight edge
se (sem) standard error of the mean
s/e standardization/evaluation
s & e services and equipment
sE standard English
Se selenium
SE Sanford & Eastern (railroad); Sanitary Engineering; Select Edition; Sero-

tonin; Servel (stock exchange symbol); Site Exploration; Southeast; Sports Edition; Stock Exchange; Student Engineer
S-E Starr-Edwards (prosthesis)
SE Son Eminence (French— His Eminence); *Sureste* (Spanish—southeast)
SE1, SE2, etc. Southeast One, Southeast Two, etc. (London postal zones)
s-e 22 silencer-equipped 22-caliber revolver
sea. sheep erythrocyte agglutination; spontaneous electrical activity
Sea (Port of) Seattle; Sea of Arabia, Galilee, Islands, Japan, Marmora, Okhotsk, Rybinsk, the Plain, Straw; The Sea (Andaman, Baltic, Bering, Black, Caribbean, Japan, Mediterranean, North, Okhotsk, South China)
SEA Safety Equipment Association; Science and Education Administration; Sea Containers Inc; Sea Education Association; Seattle, Washington (Seattle-Tacoma Airport); Senior Executives Association; Ships Editorial Association; Society for Education through Art; Society of Evangelical Agnostics; Southeast Airlines; Southeast Asia; Southern Economic Association; Special Equipment Authorization; State Education Agencies; State Education Agency; Students for Ecological Action; Subterranean Exploration Agency
SEA Sociedad Española de Automoviles (Spanish—Automobile Society of Spain)
SEAAC South-East Asia Air Command
seac standards electronic automatic computer
SEAC Save Europe's Asiatic Colonies; Southeast Asia Command
seacel silver-chloride/magnesium cell (battery)
SEACOM South East Asia Commonwealth Cable
seacon seafloor construction
SEADAC Seakeeping Data Analysis Center
SEADAG Southeast Asia Development Advisory Group
seadex seaward defense exercises

SEADS Seattle Air Defense Sector

Sea H Seaforth Highlanders

seal. sea-air-land

SEAL South-East Area Libraries

sealab sea laboratory (underwater research vessel)

SEALF South-East Asia Land Forces

SEALs Sea, Air, Land commandos; Sea, Air, and Land Teams (US Navy frogmen engaged in covert infiltration and surveillance)

SEALS Sea-Air-Land Forces (counterinsurgents)

SEAM Seattle Environmental Arts Museum

SEAM Servicios de Equipos Agricolas Mecanizados (Spanish—Mechanized Agricultural Equipment Service)

SEAMEC Southeast Asian Ministers of Education Council

SEAMEO South East Asian Ministers of Education Organisation

seamount sea mountain

SEAP Scientific Event Alert Program; South-East Asia Peninsula

searam semi-active radar missile

SEARCC South-East Asia Regional Computer Conference

SEARCH System for Electronic Analysis and Retrieval of Criminal Histories; Systematized Excerpts, Abstracts, and Reviews of Chemical Headlines

searchex sea/air search exercise

SEARS Sears, Roebuck; Socioeconomic Assessment for Repository Siting

SEAS Senior Emergency Alert System; Strategic Environmental Assessment System

seasat sea satellite

seascarp undersea escarpment

S-E Asia Southeast Asia (Burma, Cambodia, Hong Kong, Indonesia, Laos, Malaysia, Philippines, Singapore, Thailand, Vietnam)

Sea Sym Seattle Symphony

SEAT Sociedad Español de Automoviles de Turismo— (Spanish Society of Touring Automobiles)—manufacturer's name

Seatac Seattle-Tacoma (area)

seatainer(s) seagoing container(s)—theftproof steel containers for overseas cargo

Seatl Seattle

SEATO Southeast Asia Treaty Organization

SEAU Sea Containers Incorporated Unit

seb static error band

seb (SEB) surface-effect boat

Seb Sebastian(o)

Seb Sebjet or *Sebkhat* or *Sebkra* (Arabic—salt flats)— also appears as *Sabkhat*

SEB Society for Experimental Biology; South Equatorial Belt; Southern Electricity Board

SEB Skandinaviska Enskilda Banken (Swedish—Scandinavian Loan Bank)

S & EBC Ship and Engine Building Company

SEBM Society of Experimental Biology and Medicine

SEBT South-Eastern Brick and Tile (federation)

sec secant; second; secondary; secret; section; security

sec. secundum (Latin—according to)

Sec Secretary; section

SEC Section Emergency Coordinator; Securities and Exchange Commission; State Electricity Commission; State Energy Commission; Strategic Economic Council; Supreme Economic Council (USSR)

S.E.C. Springfield Equipment Company

SecA Secretary of the Army

SECA Southern Educational Communications Association

SecAgi Secretary of Agriculture

Sec Air Secretary of the Air Force

SECAIR Secretary of the Air Force

secam séquential couleur à mémoire (French—sequential color memory)—Franco-Soviet television color transmission standard, sometimes translated as the system contrary to the American method (SECAM)

SECAM Séquential à Mémoire (French—sequence and memory color television system)

secar secondary radar

sec. art. secundum artem (Latin—according to the art)

SecCom Secretary of Commerce

secd second

SECDA Southeastern Community Development Association

SecDef Secretary of Defense

SECDEF Secretary of Defense

SecEdu Secretary of Education

SecEne Secretary of Energy

secesh secessionist

SECFLT Second Fleet (Atlantic)

Sec-Gen Secretary-General

sech hyperbolic secant

SecHHS Secretary of Health and Human Services

SecHou Secretary of Housing and Urban Development

secinsp security inspection

SecInt Secretary of the Interior

SecLab Secretary of Labor

sec. leg. secundum legem (Latin—according to law)

Sec Leg Secretary of the Legation

SECMA Stock Exchange Computer Managers Association

sec. nat. secundum naturam (Latin—according to nature)

SecNav Secretary of the Navy

SECNAV Secretary of the Navy

seco second-stage engine cutoff; sustainer engine cutoff

seco (SECO) self-regulating error-correct coder-decoder

secondaries secondary colors—green, orange, violet

Second International Second International Workingmen's Association (of socialists convening in Paris in 1889)

secor (SECOR) sequential collation of range

secr secret

SE & CR Southeastern and Chatham Railway

sec. reg. secundum regulam (Latin—according to regulations, according to rule)

secret[a] secretaria (Spanish—secretariat)

secs secants; seconds; sections

sec's soft elastic capsules

Secs sections

Sec Soc Foun Second Society Foundation

SecSta Secretary of State

sect section; sector

sect (Latin suffix—cut)—dissect

sectemp temporary secretary

SecTra Secretary of Transportation

SecTre Secretary of the Trea-

sury
Secty Secretary
SECUS Sex Education Council of the United States
SecWar Secretary of War
SecWel Secretary of Welfare
Sec'y Secretary
sed sedative; sediment; sedimentation; severely emotionally disturbed; skin erythema dose
sed. sedes (Latin—a chair; a stool)
SeD Socioeconomic Democracy
SED Scientific Equipment Division (Westinghouse); Scottish Education Department; Special Enforcement Detail (law enforcement team)
SED Sozialistische Einheitspartei Deutschlands (Germany's Socialist Unity Party)—Soviet-oriented East German Party
sedar submerged electrode detection and ranging
SEDEIS Société d'Etudes et de Documentation Economiques, Industrielles et Sociales (French—Society of Economic, Industrial and Social Studies and Documentation)—Paris
sedi sediment(ation)
SEDIS Surface-Emitter-Detection Identification System
sedi time sedimentation time
sed rate sedimentation rate
sed('s) seeing-eye dog(s)
sedtn sedimentation
see. secondary electron emission; stop-everything environmentalists; survival, evasion, and escape; systems efficiency expert(ise)
SEE Society of Environmental Engineers; Society of Explosives Engineers
SEE Société des Eléctriciens, des Electroniciens, et des Radioélectriciens (French—Society of Electricians, Electronicians, and Radio Electricians)—electric, electronic, and radio technicians
SEEA Société Européenne d'Energie Atomique (French—European Atomic Energy Society)
SEEB Southeastern Electricity Board (UK)
Seec Saburo exhaust-emission control
SEECA State Environmental Education Coordinators Association

SEECB Solar Energy and Energy Conservation Bank
SEECC Standards for Educators of Exceptional Children in Canada
seecom sensible, economical, electrical commuter (electric automobile)
SEECTS Subaru Exhaust Emission-Control Thermal System
seed. summer of experience, exploration, and discovery
SEED Scientists and Engineers in Economic Development (National Science Foundation); Skills Escalation and Employment Development; Special Elementary Education (for the underdeveloped)
SEEJ Slavic and East European Journal
SEEK Search for Elevation and Educational Knowledge (NY State dropout program); Sooner Exchange for Educational Knowledge; Systems Evaluation and Exchange of Knowledge
seeo sauf erreur et omission (French—excepting errors and omissions)
s.e.e.o. salvis erroribus et omissis (Latin—excepting errors and omissions)
seep seagoing jeep
seer. submarine explosive echo ranging
SEER System for Electronic Evaluation and Retrieval
seex systems evaluation experiment
sef small end first
SEF Shipbuilding Employers' Federation; Southern Education Foundation; Space Education Foundation
SEFA Scottish Educational Film Association
SEFT Society for Education in Film and Television
seg segment; segmentation; segmented; segments; segregate; segrated; segregation; segregationist; special-effects generator
seg (SEG) sonoencephalogram
seg segno (Italian—sign); *segue* (Italian—comes after; follows)
Seg Segovia
SEG Screen Extras Guild; Society of Economic Geologists; Society of Exploration Geophysicists; Systems Engi-

neering Group
SEGB South Eastern Gas Board
SEGBA Servicios Eléctricos del Gran Buenos Aires (Spanish—Electrical Services of Greater Buenos Aires)
seggy secobarbital sedative, also nicknamed seccy)
segm segmented
Segr Segretario (Italian—Secretary)
Segr^to Segretariato (Italian—Secretariat)
segs segmented neutrophils; segments
SEH St Elizabeth's Hospital
SEH Société Européenne d'Hématologie (French—European Society of Haematology)
seha specific emotional hazards of adulthood
sehc specific emotional hazards of childhood
SEHMF South of England Hat Manufacturers' Federation
SEI Scientific Engineering Institute; Self-Employment Income
SEIA Security Equipment Industry Association; Solar Energy Industries Association; Solar Energy Institute of America
SEIC Solar Energy Information Center; System Effectiveness Information Center
SEIF Secretaria de Estado da Informação e Turismo (Portuguese—Secretariat of Information and Tourism)
SEIFSA Steel and Engineering Industries' Federation of South Africa
Seiji Seiji Ozawa
seis seismograph; seismography; seismology; submarine emergency identification signal (SEIS); submarine-escape immersion suit
SEISA South Eastern Intercollegiate Sailing Association
Seiscor Seismograph Service Corporation
seismo seismograph(er); seismographic(al)(ly); seismologist; seismology
seismol seismology
SEIT Search for Extra Terrestrial Intelligence
SEIU Service Employees International Union
sel select(ed); selectee; selectivity; selector; socioeconomic level; sound-exposure

level (SEL)
sel (SEL) socio-economic level
Sel Selby
SEL Seoul, Korea (airport); Signal Engineering Laboratories; Southeastern Education Laboratory; Stanford Electronics Laboratories; Systems Engineering Laboratories
SELA Southeastern Library Association
SELA *Sistema Económica Latino Americana* (Spanish— Latin American Economic System)
SELC South Eastern Louisiana College
selcall selective calling
sel-cl self-closing
SELDAMS Selective Data Management System
seleac standard elementary abstract computer
selen selenography; selenology
self-prop self-propelled
Selk Selkirk
Selkirks Selkirk Mountains of British Columbia
SELMA SEL Maduro
SELNEC South-East Lancashire North-East Cheshire
S/ELPS Spanish/English Language Performance Screening
sels selsyn
selsyn self-synchronous
Selvagens Selvagen Islands between the Canaries and Madeira
Selw Selwyn College, Oxford; Selwyn College—Cambridge
Sely southeasterly
SEly south-easterly
sem scanning electron microscope; semi; semicolon; seminal; slow eye movements; standard error of mean; systolic ejection murmur
sem (SEM) scanning electron microscope; systolic ejection murmur
sem. semen (Latin—seed); *semper* (Latin—always, ever)
Sem Semarang; Seminary; Semitic
SEM Society for Ethno-Musicology
SEMA Spray Equipment Manufacturers' Association; Storage Equipment Manufacturers Association
seman semantic(s)
semcor semantic correlation
SEMDA Surveying Equipment Manufacturers and Dealers

Association
SEMFA Scottish Electrical Manufacturers' and Factors' Association
semi semicolon
semi- semi-detached house (town house)
semi (Latin prefix—half)— semilunar
semicol semicolon
semidr. semidrachma (Latin— half drachma)
semidur semiduration
semih. semihora (Latin—half hour)
Seminex Seminary in Exile
semiot semiotic(al)(ly); semiotician(s); semiotics (study of signs and symbols)
semipro semiprofessional(ly)
semis semifinished; semitrailers
SEMKO *Svenska Elektriska Materielkontrollanstalten* (Swedish—Swedish Institute for Testing and Approval of Electrical Equipment)
semp self-erecting marine platform
semp sempre (Italian—always)
Semper Fi Semper Fidelis (Latin—Always Faithful)— U.S. Marine Corps motto
Semper Paratus (Latin—Always Ready)—U.S. Coast Guard motto
sems screw and washer assemblies
SEMT Société d'Etudes des Machines Thermiques (French—Society for the Study of Thermal Machines)
SEMTA Southeastern Michigan Transportation Authority
sem ves seminal vesicle
sen sense (flow chart)
sen seno (Italian—sine); *senza* (Italian—without)
Sen Senate; Senator
Sen Marcus (or Lucius) Seneca (Roman rhetorician) or his second son Lucius Annaeus Seneca (Roman author); *Senatore* (Italian—senator)
SEN State-Enrolled Nurse
S en C Sociedad en Comandita (Spanish—limited partnership)—silent partnership; *Socieété en Commandite* (French—limited partnership)
Sen Clk Senior Clerk
Sen Doc Senate Document
Seneg Senegal; Senegalese
Senegal Republic of Senegal (West African nation), *République du Sénégal*

Senegambia Senegal + Gambia
senel single-event noise-exposure level
S Eng O Senior Engineering Office(r)
SENI Society for the Encouragement of National Industry
senior dent senior-citizen dental care
senior(s) senior citizen(s)
Sen M Senior Master
Sen Mist Senior Mistress
S en NC Société et Nom Collectif (French—joint stock company)
senr senior
Sen Rept Senate Report
Senr Tech Weld I Senior Technician of the Welding Institute
sens sensitivities (test)
sensistor semiconductor resistor
sent. sentence
Sent Sentyabr (Russian—September)
SENTAC Society for Ear, Nose, and Throat Advances in Children
sentimiento trágico sentimiento trágico de la vida (Spanish—tragic perception of life)— conflict between faith and reason
Sen Wt O Senior Warrant Officer
seo (SEO) satellite for earth observation
seo salvo errori e omissioni (Italian—excepting errors and omissions)
Seo Seoul
SEO Senior Experimental Officer; Snake Ender's Organization
SEODSE Special Explosive Ordnance Disposal Supplies and Equipment (USA)
SEOG Supplemental Educational Opportunity Grant
seoo sauf erreurs ou omissions (French—excepting errors and omissions)
SEOOs State Economic Opportunity Offices
seos (SEOS) synchronous earth observation satellite
seou salve error u omisión (Spanish—except for error or omission)
sep separate; separation
sep (SEP) solar electric power; somatosensory-evoked potential
Sep September

SEP Selective Employment Payments (UK); Self-Employment Plan (retirement plan); Simplified Employee Pension; Society of Engineering Psychologists; Society of Experimental Psychologists; Source Evaluation Panel; Student Expense Program

SEP Saturday Evening Post; *Secretaría de Educación Publica* (Spanish—Secretary of Public Education)

SEPA Southeastern Power Association; State Elementary Principals Association

separ. separatum (Latin—separately)

SEPB Southern Europe Ports and Beaches

SEPD Scottish Economic Planning Department

SEPE Seattle Port of Embarkation

SEPEL Southeastern Plant Environment Laboratories

Seph Sephardim (Hebrew—Jews from Portugal and Spain)

Sephard Sephardim (Hebrew—Jews from Portugal and Spain who were forced to emigrate during the Inquisition)

SEPO Space Electric Power Office (AEC)

SEPP Simplified Employee Pension Plan

SEPP Société d'Étude de la Prévision et de la Planification (French—Society for the Study of and Planning for the Future)

SEPR Société pour l'Etude de la Propulsion par Réaction (French—Society for the Study of Jet Propulsion)

SepRos separation processing

Seps (SEPS) Smithsonian earth physics satellite

SEPs Simplified Employee's Pensions

SEPSA Society of Educational Programmers and Systems Analysts

sept. septem (Latin—seven)

Sept September

SEPTA Southeastern Pennsylvania Transportation Authority

sept^e septiembre (Spanish—September)

septel separate telegram

septen septentrionale (northern)

septi septicos (Greek—infected or rotten)—antiseptic, aseptic, septic, septicemia

Septober September and October

seq sequence

seq. sequens (Latin—the following); *sequente* (Latin—what follows); *sequitur* (Latin—it follows)

seq. luce sequenti luce (Latin—the following day)

Seq NP Sequoia National Park

S Equ Cur South Equatorial Current

Sequoia Sequoia National Park in east-central California

ser serial; series

ser (SER) serine (amino acid)

ser série (French—series)

Ser series; Serpens (constellation)

Ser Serranía (Spanish—mountains)

SER Safety Exploration Report; Service, Employment, Redevelopment; Society for Educational Reconstruction; Soil Erosion Service; Student Eligibility Report

SER Sociaal Economische Raad (Dutch—Social Economic Council); *Sociedad Española Radiodifusión* (Spanish—Spanish Broadcasting Society)

Sera Seraphim

SERA Services, Education, Rehabilitation for Addiction

Serb Serbia; Serbian

Serb-Croat Serbo-Croatian (slavic language spoken in Yugoslavia)

SERC Science and Engineering Research Council

SERCH State Education Research Clearinghouse

SERE Survival, Evasion, Resistance, and Escape (U.S. Naval Training Base)

SEREB Société pour l'Etude et la Réalisation d'Engins Balistiques (French—Society for the Study and Development of Ballistic Missiles)

serendip serendipitous(ly); serendipity

Serengeti Serengeti Plains of Tanzania

Serg Sergente (Italian—Sergeant)

Sergey Sergeyevich

Serg Magg Sergente Maggiore (Italian—Sergeant Major)

Serg(t) Sergeant

SERI Solar Energy Research

Institute; Solar Energy Research Institute (ERDA)

serj space electric ramjet

SERL Services Electronics Research Laboratory

SERLANT Service Forces, Atlantic (USN)

serline serials on-line

serm sermon

SERM Society of Early Recorded Music

serol serology

serp simulated ejector-ready panel

SERPAC Service Forces, Pacific (USN)

SERPLANT Service Forces, Atlantic (USN)

serr serrate

Ser Rep San Mar Serenissima Repubblica di San Marino (Italian—Most Serene Republic of San Marino)

SE-RRT Southern Europe-Railroad Transport (NATO)

ser sect serial sections

sert space electronic rocket test

SE-RT Southern Europe-Road Transport (NATO)

SERTOMA Service To Mankind

serv service

serv. serva (Latin—keep; preserve)

Serv Servia(n)

serv chge service charge

serv clg service ceiling

SERVE Serve and Enrich Retirement by Volunteer Experience

servo device using a servomechanism; servoamplifier, servocontrol, servodyne, servomotor, servosystem

serv^o servicio (Spanish—service)

Servomation Service America Corporation

serv^or servidor (Spanish—servant)

servos servomechanisms

ses secondary engine start; single-ended scotch (boilers); socio-economic status; socioeconomic strata; solar environment stimulator; surface-effect ship

ses (SES) surface-effect ship

SES Seafarers' Education Service; Self-Esteem Score(s); Senior Executive Service; Society of Engineering Science; Solar Energy Society; Standards Engineers Society; State Employment Service; Steam Engine Systems; Suit-

ability Evaluation Scale
SES *Service des Études Scientifiques* (French—Scientific Studies Service)
SESA Social and Economic Statistics Administration; Society for Experimental Stress Analysis; Solar Energy Society of America
SESAC Society of European Stage Authors and Composers
sesame. service, sort, and merge
SESAME Search for Excellence in Science and Mathematics Education
sesco secure submarine communications
SESL Space Environment Simulation Laboratory
SESO Senior Equipment Staff Office(r); Ship Environmental Support Office(r)
sesoc surface-effects ship for ocean commerce
SESPO Space Environmental Support Project Office(r)
sesquih sesquihora (Latin—an hour and a half)
sesquilin sesquilingual (ability to use one-and-a-half languages)
sess session
SESS Society of Ethnic and Special Studies; Space Environmental Support System; Summer Employment for Science Students
sest short effective-service time
set. settlement
set septiembre (Spanish—September); *setembro* (Portuguese—September)
SET Scientists, Engineers, Technicians; Security Escort Team; Selective Employment Tax(ation); Senior Electronic Technician; Senior Evaluation Treatment; Simplified Engineering Technique; Synchro Error Tester
S.E.T. Selective Employment Tax
seta set arithmetic (value)
SETAF Southern European Task Force
setb set binary (value)
setc set character (value)
SETCO Summit and Elizabeth Trust Company
se/td system engineering/technical direction (SE/TD)
set^e septiembre (Spanish—September)

SETEP Science and Engineering Technician Education Program
SETI Search for Extraterrestrial Intelligence
SETIL Société de l'Equipement de Tahiti et des Iles (French—Equipment Company of Tahiti and the Islands)
S-et-L Saône-et-Loire
S-et-M Seine-et-Marne
S-et-O Seine-et-Oise
SETP Society of Experimental Test Pilots
SETS Solar Energy Thermionic Conversion System
sett settling
sett settembre (Italian—September)
seu smallest executable unit
SEU Southeastern University
SEUA South Eastern Underwriters Association
seuo salvo error u omisíon (Spanish—errors and omissions excepted)
SEUS Southeastern United States
sev seven; sevenfold; seventeen(th); seventy; sever; several; severally; severance; severe; severity; surface-effect vehicle
sev sever (Russian—north)
Sev Sevilla; Seville
Sev Sever or *Severnaya* (Russian—north, northern)
SEV Soviet Ekonomischeskoy Vzaimopomoschchi (Russian—Soviet Council for Mutual Economic Aid)—the COMECON
Seven Continents Africa, Antarctica, Asia, Australia, Europe, North America, South America
Seventh Seventh Seal
Severnaya Zemlya (Russian—North Land)—*Zemlya Imperatora Nikolaya* II (Russian—Emperor Nicholas IInd Land)
SEVFLT Seventh Fleet, Pacific (USN)
sevocom secure voice communications
sew. safety equipment workers; sewage; sewer; sewerage
sewido surface electromagnetic-wave-integrated optics
SEWT Simulator for Electronic Warfare Training
sex. sextet; sexual
Sex Sextans (constellation)
Sexag Sexagesima

sexational sexually sensational
sexcite excite sexually
sexcitement sexual excitement
sexclusive sexually exclusive
sex ed sex(ual) education
sexercises sexual exercises
sexgregation sexual segregation
sexhibit sex exhibit
sexhibitors sex exhibitors
SExO Senior Experimental Officer
sexones sex odors
sexorgies sexual orgies
sexpensive sexually expensive
sexperience sexual experience
sexpert sex expert; sexual expert; sexpertise
sexpionage sexual exploitation in espionage
sexplanatory sexually explanatory
sexplicit sexually explicit
sexploitation sex(ual) exploitation
sexploiter sex exploiter
sexploit(s) sexual exploit(s)
sexplosion sexual explosion
SEXPOL Sexual Equality and Politics (German communist movement originated by Wilhelm Reich)
s. expr. sine expressione (Latin—without expressing; without pressing)
sex psycho sexual psychopath(ic)
sexquisite sexually exquisite
sexsation sexual sensation
sexslanguage sexual slang
sext sextant
Sext Sextans (constellation)
Seychelles Seychelle Islands
sez (SEZ) southern economic zone
SEZ Special Economic Zone (China)
sf sacrifice fly; safety factor; salt free; science fiction; semifinished; single feeder; single-feed; sinking fund; sound and flash; special facilities; spent fuel; spinal fluid; spotface; square foot; standard form; stress formula; sulphation factor; sunkface
s/f shift forward; store and forward
s & f stock and fixtures
sf sans frais (French—without expense); *sforzando* (Italian—accented strongly; forced; reinforced)
s.f. sub finem (Latin—near the end)

Sf Svedberg flotation (units)
SF San Franciscan; San Francisco; Santa Fe (Atchison, Topeka & Santa Fe Railway); Santa Fe, New Mexico; Scouting Force; Security Force(s); Shipfitter; Soumi Finland; Special Facilities; Special Forces; Standard Frequency; State Facilitator; Swedenborg Foundation; Swiss Federation (auto plate); Syrian Forces
SF Slovenska Filharmonica (Serbo-Croat—Slovene Philharmonic)—Ljubljana, Yugoslavia); *Socialistisk Folkeparti* (Dano-Norwegian—Socialist People's Party); *Système français* (French—French system)—screw threads
S/F Sinn Fein (Irish Gaelic—Ourselves Alone)
SF-5 Spanish version of the F-5 Northrup Freedom Fighter
sfa simulated flight automatic; slow flying aircraft; spatial frequency analyzer
sfa (SFA) serum folate; suppressive factor of allergy
s & fa shipping and forwarding agent
SFA Saks Fifth Avenue; Scandinavian Fraternity of America; Scientific Film Association; Scottish Football Association; Show Folks of America; Slide Fastener Association, *Société Française d'Astronautique* (French—French Astronautical Society); Solid Fuels Administration; Soroptimist Federation of the Americas; Southeastern Fisheries Association; Speech Foundation of America; Symphony Foundation of America
SFAAW Stove, Furnace, and Allied Appliance Workers (International Union of North America)
SFAC Société des Forges et Ateliers du Creusot (French—Schneider-Creusot Forges and Factories)
SFAD Society of Federal Artists and Designers
SFAI San Francisco Art Institute; Steel Furnace Association of India
SFAO San Francisco Assay Office
sfar sound fixing and ranging
SFAR System Failure Analysis Report
SFB San Francisco Ballet
SFB Sender Freies Berlin (German—Free Berlin Broadcasting Station); Spencer Fullerton Baird
SFBARTD San Francisco Bay Area Rapid Transit District
sf bh surface broach
SFBMS Small Farm Business Management Scheme
SFBNS San Francisco Bay Naval Shipyard
sfc S-bank frequency converter; sight fire control; specific fuel consumption; supercritical fluid chromatography; switching filter connector; synchronized framing camera
sfc (SFC) spinal fluid count
Sfc Sergeant First Class
SFC Saint Francis College; Sioux Falls College; Small Faith Communities; Space Flight Center
SFC San Francisco Chronicle
SFCA Southwest Flight Crew Association
SFCC San Francisco City College
SFCI State Farms Corporation of India
SFCJ San Francisco City Jail
SFCM San Francisco Conservatory of Music
SFCMD San Francisco Contract Management District
SFCP Shore Fire Control Party
SFCS Survivable Flight Control System
SFCTA San Francisco Classroom Teachers Association
sfcw search for critical weakness
SFCW San Francisco College for Women
Sfd San Fernando
sfd/algol system function description/algol (language)
SFDS Spent-Fuel Disposal System
sfe safety function earthquake; stacking fault energy; surface-energy
SFE Society of Fire Engineers
SFE Société Française des Electriciens (French—French Society of Electricians)
SFEA Survival and Flight Equipment Association
SFEL Standard Facility Equipment List
SFEN Société Française d'Energie Nucléaire (French—French Nuclear

Energy Society)
sff se faz favor (Portuguese—please)
SFF Solar Forecast Facility
sfff salt-free fat-free (diet)
SFG Studien und Förderungsgesellschaft (German—Studies and Advancement Society)
sfga single floating-gate amplifier
sfgd safeguard
SFGGB San Francisco Golden Gate Bridge
SFGH San Francisco General Hospital
SFHP Spent-Fuel Handling and Packaging; Spent-Fuel Handling Project
SFHR San Francisco Historic Records
SFHS Stephen Foster High School
sfi sequential fuel injection
SFI Sport Fishing Institute
SFI Société Financière Internationale (French—International Finance Corporation)
SFIAE San Francisco Institute of Automotive Ecology
SFIB Southern Freight Inspection Bureau
SFIO Section Française de l'Internationale Ouvriere (French—French section of the Worker's International)—former name of the French Socialist Party
SFIS Small Firms Information Service
SFIT Standard Family Interaction Test
sfl sequenced flashing lights (airport runways)
s fl Surinam florin
SFL Scottish Football League; Sexual Freedom League; Society of Federal Linguists
sfm surface feed per minute; surface feet per minute
SFMA San Francisco Museum of Art; School Furniture Manufacturers' Association; Southern Furniture Manufacturers Association
SFMC San Francisco Medical Center (University of California)
SFMR San Francisco Municipal Railway (operates the cable cars)
SFMS Shipwrecked Fishermen and Mariners (Royal Benevolent Society)
SF & NV San Francisco & Napa Valley (railroad)

sfo simulated flame out; submarine fog oscillator

S Fo (Port of) San Francisco

SFO San Francisco, California (airport); San Francisco Opera; San Francisco Operations (office); San Francisco-Oakland Airlines; Santa Fe Opera; Senior Flag Officer; Service Fuel Oil; Space Flight Operations

SF-OBB San Francisco-Oakland Bay Bridge (Transbay Bridge)

SFOD San Francisco Ordnance District; Special Forces Operational Detachment

SFOF Space Flight Operations Facility

SFOLDS Ship-Form On-Line Design System

SFP Sherbrooke Forest Park (Victoria, Australia)

SFP *Société Française de Photogrammétrie* (French—French Society of Photogrammetry)

SFPD San Francisco Police Department

SFPDis San Francisco Procurement District (US Army)

sf pe surface plate

SFPE San Francisco Port of Embarkation; Society of Fire Protection Engineers

SFPF Spent-Fuel Packaging Facility

SFPL San Francisco Public Library

sfpm surface feet per minute

SFPO Spent-Fuel Project Office (Savannah River Operations Office)

SFPR Society of Friends of Puerto Rico

sfprf semifireproof

SFPs Sinn Fein Provisionals (Provos)

sfqa (**SFQA**) structurally fixed question-answering system

sfr sinking fund rate (of return); star-formation rate

sfr (**SFR**) submarine fleet reactor

SFR Safety of Flight Requirement

SFRA Science Fiction Research Association

S Fran San Francisco

SFRJ *Socijalisticka Federativna Republika Jugoslavija* (Socialist Federated Republic of Yugoslavia)

sfrr sinking fund rate of return

sfr(s) *schweizerfranc(s)* [Dano-Norwegian—Swiss franc(s)]

SFRS Sea Fisheries Research Station (Haifa)

sfs strictly for suckers; surfaced four sides

SFs Special Forces (Green Berets); State Facilitators

SFS San Francisco Symphony; Senior Foreign Service; Society of Fleet Supervisors

SFSA Scottish Field Studies Association; Steel Founders' Society of America

SFSAFBI Society of Former Special Agents of the Federal Bureau of Investigation

SFSC San Francisco State College

SF & SC Standard Fruit & Steamship Company

SFSE San Francisco Stock Exchange

SFSO San Francisco Symphony Orchestra

SFSP Spent-Fuel Storage Program

SF/SP Santa Fe/Southern Pacific (merged railroads)

SFSS Satellite Field Services Stations (National Oceanographic and Atmospheric Administration)

SFSSP Society of the Friendly Sons of St Patrick

sft soft; specified financial transactions; stop for tea; superfast train

SFT Society of Forensic Toxicologists; Spent-Fuel Testing (ing)

SFTA Scientific Film Television Award; Society of Film and Television Arts

SFTAA Short-Form Test of Academic Aptitude

SFTB Southern Freight Tariff Bureau

SFTI San Fernando Technical Institute (Trinidad)

SFTP Science For The People

SFTW Stamps For The Wounded

sftwd softwood

sftwr software

SFU Signals Flying Unit; Simon Fraser University

S_f units Svedberg flotation units

sfv sight feed valve

SFv Semliki Forest virus

SFVAH San Francisco Veterans Administration Hospital

SFVSC San Fernando Valley State College

SFWA Science Fiction Writers of America

sfwd slow forward

SFWR Stewardesses for Women's Rights

sfx sound effects

sfxd semifixed

sfxr superflash X-ray

sfy standard facility year(s)

SFYC San Francisco Yacht Club

sfz *sforzando* (Italian—emphasized chord or note)

sfz p *sforzato piano* (Italian—emphasis followed by a soft note or chord)

sg screen grid; single groove; singular; smoke generator; soluble gelatin; specific gravity; steam generator; steel girder; structural glass; swamp glider

s-g sub-generic; sub-genus

sg *selon grandeur* (French—according to size); on menus, sg or SG indicates an item is priced according to the size of the serving

s.g. *salutis gratia* (Latin—for safety's sake)

s/G *sur Garonne* (French—on the Garonne)

Sg spring range of tide; Surgeon

SG *Aerotransporte Litoral Argentino* (Spanish—Argentine Coastal Air Transport); Scots Guards; Solicitor General; South Georgia (railroad); Standing Group; sub-group; Sudan Government; Sunset Gun; Surgeon General

S-G Sachs-Georgi (test); Saint-Gobain; Space-General (Corporation)

SG *Stanley Gibbons* (stamp catalog)

SGA Saskatchewan Government Airways; Society of the Graphic Arts; Southern Gas Association; Special Grant Application; Standards of Grade Authorization; Student Government Association

SGAE *Sociedad General de Autores de España* (Spanish—General Society of Authors of Spain)

S-gauge standard gauge (4-foot 8 1/2-inch) railroad track

SGB *Société Générale de Banque* (Belgian Bank); *Société Générale de Belgique* (French—General Society of Belgium)

SGBIP *Subject Guide to Books in Print*

sgc screen grid current; simulated generation control;

spartan guidance computer (SGC); spherical gear coupling; stabilizer gyro circuit

Sg C Surgeon Captain

SGC Saint Gregory College; South Georgia College

S-G C Space-General Corporation

SGCA Secrétariat Général à l'Aviation Civil (French—Secretariat General of Civil Aviation)

Sg Cr Surgeon Commander

sgd signed

SGD Senior Grand Deacon

sgdg sans garantie du gouvernement (French—patent issued without government guarantee)

sg di swaging die

Sge Sagitta

S Ge South Georgia

sgemp system-generated electromagnetic pulse

SGF Scottish Grocers' Federation

SGF Sveriges Gummitekniska Forening (Swedish—Swedish Rubber Industry Association)

sgg sustainer gas generator

sghwr steam-generating heavy-water reactor

SGI Spring Garden Institute

SGINDEX System Generation Cross-Reference Index (NASA)

SGIO State Government Insurance Office

sgl signal; single

S Glam South Glamorgan

Sg L Cr Surgeon Lieutenant Commander

SGLI Servicemen's Group Life Insurance

SGLS Space-Ground Link Subsystem

SGM Sea Gallantry Medal; Society of General Microbiology

sg md swaging mandrel

SGMEX Southern Gulf of Mexico

SGMT Société Générale des Transports Maritimes (French—General Society of Maritime Transport)

sgn scan gate number; signum function

Sgn (Port of) Saigon

SGN Saigon, Vietnam (airport); Surgeon General of the Navy

Sgno Stagno (Italian—pond; pool)

sgnr signature

sgo surgery, gynecology, and obstetrics

SGO Surgeon General's Office

sgot serum glutamic oxaloacetic transaminase

sgp starch graft polymers

SGP Shell Gasification Process; Society of General Physiologists

SGP Secretario General del Partido (Spanish—Secretary General of the Party); *Staatkundig Gereformeerde Partij* (Dutch—Political Reformed Party)

SGPA Scottish General Publishers Association

sgpt serum glutamic pyruvic transaminase

sgr steam gas recirculation (oil-from-shale removal process)

Sgr Sagittarius (constellation)

SGR Sumbu Game Reserve (Zambia)

Sg RA Surgeon Rear Admiral

SGRS Stockton Geriatric Rating Scale

SGS Society of General Surgeons; Sunderbans Game Sanctuary (Bangladesh)

SGSB Stanford Graduate School of Business

SGSR Society for General Systems Research

sgt special gas taper (threading)

Sgt Sergeant

SGT Society of Glass Technology

Sgt 1/C Sergeant First Class

S-G Test Sachs-Georgi Test

SGTIA Standing Group Technical Intelligence Agency (NATO)

Sgt Maj Sergeant Major

SGU Scottish Gliding Union; Scottish Golf Union; Singapore Golfers Union

SGU Sveriges Geologiska Undersokning (Swedish—Swedish Geological Survey)

SGUs Special Guerrilla Units

Sg VA Surgeon Vice Admiral

SGVHS Samuel Gompers Vocational High School

SGW Senior Grand Warden

SGX Seeger Refrigerator Express (stock exchange symbol)

sh sacrifice hits (bunts); scleroscope hardness; serum hepatitis; shelf; shelving; ship's heading; shop; shopping; short; sick in hospital; social history; somatotrophic hor-

mone speech handicapped; surgical hernia

sh (SH) sexual harassment

s/h shorthand

Sh shells; shilling (British East Africa)

Sh Sh'aib (Arabic—ravine; road); *Shatt* (Arabic—river; riverbank); *Shima* (Japanese—island); *Suid Holland* (Dutch—South Holland)

SH Schenley Industries (stock exchange symbol); Soldier's Home; Station Hospital; Symphony Hall

S-H Schleswig-Holstein; Scripps-Howard

S & H Sperry & Hutchinson; Sundays and Holidays

SH Sa Hautesse (French—Her or His Highness)

sha (SHA) sidereal hour angle

Sha Shanghai

SHA Safety and Health Administration; Society for Humane Abortion; Southern Historical Association; State Historic Area

SHAA Society of Hearing Aid Audiologists

shab soft and hard acids and bases

sh abs shock absorber

SHAC Seale-Hayne Agricultural College; Shelter Housing Aid Center

shaco shorthand coding

SHAD Sharpe Army Depot

shade. (SHADE) shielded hot-air-drum evaporator

SHAEF Supreme Headquarters, Allied Expeditionary Forces

shag. simplified high-accuracy guidance

shags shaggy carpets or rugs

Shah Shahanshah (Persian—King of Kings)

Shak(e) Shakespeare

Shakes Shakespeare

shale. standoff high-altitude long endurance

Sham Shamrock

shamateur(s) sham amateur(s)

shamburger hamburger containing more additives and adulterants than meat

SHAME Save, Help Animals Man Exploits; Society to Humiliate, Aggravate, Mortify, and Embarass Smokers

shandy shandygaff (beer-and-ginger-ale mixture)

Shang Shanghai

shan't shall not (colloquial)

Shanty Irish poor people

SHAPE Scanning Hartmann Aperture Plate Experiment; Supreme Headquarters, Allied Powers, Europe
SHARE Scottish Health Authorities Revenue Equalisation; Self-Help and Resources Exchange
SHARP Senior-High Assessment of Reading Performance; Ships Analysis and Retrieval Project
SHARPS Ship/Helicopter Acoustic Range-Prediction System (USN)
SHAS Shared Hospital Accounting System
Shav Shavuot
SHAWCO Students Health and Welfare Centers Organization
SHB Svenska Handelsbanken (Swedish—Swedish Bank of Commerce)
shbd serum X-hydroxy-butyrate dehydrogenase
shbg sex-hormone-binding globulin
shc spontaneous human combustion
SHC Sacred Heart College; Seton Hall College; Siena Heights College; Spring Hill College; Streets and Highways Code; Surveillance Helicopter Company
SHCC Statewide Health Coordinating Council
SHCJ Society of the Holy Child of Jesus
shco sulfonated hydrogenated castor oil
sh con shore connection
SHCS School of Health Care Sciences (USAF)
shd should
Sh.D. Doctor of Showbizology (Frank Sinatra)
SHD Scottish Home Department; State Hydroelectric Department
she. signal handling equipment; standard hydrogen electrode
she. (SHE) sodium heat engine
SHE Shelter for Help and Emergency
Sheba Saba
she'd she had; she would
Shedd Shedd Aquarium (Chicago)
Sheed Sheed & Ward
SHEEO State Higher Education Executive Officers
Sheet Metal Workers Union Sheet Metal Workers Interna-

tional Association
Sheff Sheffield; Sheffield Scientific School (Yale)
shekel monetary unit of Israel
shelf. super-hardened extremely low frequency
she'll she will
SHELL Royal Dutch Shell Oil Company; Shell Oil Company
shellrep shelling report
SHELREP Shelling Report
Shenandoahs Shenandoah Mountains of Virginia and West Virginia
Shen NP Shenandoah National Park
Shep Shep(p)ard; Shepton
Sher Sherbrooke
Shere Shirley
Sherm Sherman
sherm(s) sherman(s)—pcp-soaked marijuana cigarette(s)
she's she has; she is
SHES School Health Education Study
Shet Shetland
Shetland Shetland Island, Shetland Islands called the Zetlands
Shetlands Shetland Islands off northern Scotland
Shets Shetland Islands, Scotland
Shex Sundays and holidays excepted
shf super high-frequency—300 –30,000 mc
Shf Sheffield
SHF Soil and Health Foundation
SHFF Scottish House Furnishers' Federation
shftg shafting
S-H-G diet Sauerbruch-Herrmannsdorfer-Gerson (tubercular) diet
SHH Sociedad Honoraria Hispánica (Spanish—Honorary Hispanic Society)
SHHV Society for Health and Human Values
Shi Shanghai
SHI Strategic Homeporting Initiative (USN)
Shickshocks Shickshock Mountains of the Gaspé Peninsula of New Brunswick
SHIELD Sylvania High-Intelligence Electronic Defense
Shig Shigella
shil (SHIL) shillelagh (surface-to-surface missile of the U.S. Army)
shilling monetary unit of Kenya, Somalia, Tanzania,

Uganda
Shim Shimonoseki
Shin Bet Israel's domestic security agency
Shin Bet Sherut Habitachon (Hebrew—Security Department)—Israel
S & h inc Sundays and holidays included
shinerium shoe-shine stand
ship. shipment; shipping
SHIP Self-Help Improvement Program
shipcon shipping control; shipping convoy
ShipDTO ship on depot transfer order
shipmt shipment
SHJ Society for Humanistic Judaism
SHJC Sacred Heart Junior College
shk shank
Shl Shields; shoal
Sh L Shipwright Lieutenant
SHL Society for Humane Legislation
shld shoulder
shl dk shelter deck
SHLM Society of Hospital Laundry Managers
shlp shiplap
Shls Shoals (postal abbreviation)
shm simple harmonic motion
Shm Shimizu; Shoreham
SHM Service Hydrographique de la Marine (French—Naval Hydrographic Service)
SHMO Senior Hospital Medical Officer; Social Health Maintenance Organization
shmt shock mount
SHNC Scottish Higher National Certificate
SHND Scottish Higher National Diploma
SHNHP Signal Hill National Historical Park, St John's, Newfoundland
SHNHS Sagamore Hill National Historic Site
SHNNR Studland Heath National Nature Reserve (England)
shnoz shnozzle; shnozzola; nose
ShNP Shenandoah National Park
sho shore(d); shoring
SHO Senior House Officer; Student Health Organization
SHOC Self-Help Opportunity Center
SHOCK Students Hot on Conserving Kilowatts

shocks. shock absorbers
S Holmes, Esq Sherlock Holmes
shootin shooting
shop-op shopping opportunity
SHORADS Short-Range All-Weather Air-Defense System
shoran short-range navigation
SHORS School/Home Observational Referral System
shorted short circuited
shortg shortage
short(s) short circuit(s)
SHOT Society for the History of Technology
shouldn't should not
show biz show business
show exhibs show exhibitions
shp shaft horsepower
Shp Sharpness
SHP Sandy Hook Pilots; Society of Hospital Pharmacists; State Historic Park
SHPBG Small Horticultural Production Business Grant
SHPC Scenic Hudson Preservation Conference
SHPDA State Health Planning and Development Agency
shpng shipping
shpng/hndlg shipping & handling (charges)
SHPO State Historic Preservation Office
shps seahead pressure simulator
shpt shipment
SHP Test Strongin-Hinsie-Peck (salivary secretion) Test
SHQ Station Headquarters
shr share(s)
Shr Shore (postal abbreviation)
shram (SHRAM) short-range air-to-surface missile
shrap shrapnel
shrd shredded
shrimpsan shrimp sandwich
shrimpwich shrimp sandwich
SHRMA South Hampton Roads Metropolitan Area (Norfolk, Portsmouth, Chesapeake, and Virginia Beach)
Shrops Shropshire
Shrs Shores
shrtg shortage
shs ship's heading servo
SHS Sacred Heart Seminary; Scottish History Society; Senior High School; *Srba, Hrvata, i Slovenaca* (Serbo-Croatian—Serbs, Croats, and Slovenes)—Yugoslavia; State Historic(al) Site; Stuyvesant High School
SHSA Steamship Historical Society of America

SHSL Sherlock Holmes Society of London
SHSLB Street and Highway Safety Lighting Bureau
SHSN Sod House Society of Nebraska
SHSP Sam Houston State Park (Louisiana)
SHSS Sanford Hypnotic Susceptibility Scale
SHSSI Steamship Historical Society of Staten Island
SHSW State Historical Society of Wisconsin
shswc sample-and-hold square-wave converter
sht sheet(ing)
SHT Society for the History of Technology
shtg shortage
shth sheathing
sht irn sheet iron
sht mtl sheet metal
sh tn short ton
SHU Security Housing Unit; Seton Hall University
Shula Shulamite; Shulamith
SHUR System of Hospital Uniform Reporting
shv solenoid hydraulic valve
s.h.v. *sub hoc voce* (Latin—under this work)
shvg shaving(s)
shw safety, health, and welfare
SHW Sherwin-Williams (stock exchange symbol)
shwrs showers
S & H x Sundays and Holidays excepted
SHYC Sachem's Head Yacht Club
si salinity indicator; shift in; short interest; slight imperfection; spark ignition; straight-in (aircraft landing approach); subicteric; subindex; subinguinal; surface interval
si (SI) shift-in character (data processing)
s-i semiconductor-integrated (circuits)
s/i signal/intermodulation; subject issue
s & i stocked and issued
Si Silas; silicon (symbol); Simon; Simone
Sⁱ Sidi (Arabic—My Lord)— title of honor also written *Saiyidi*
SI Sandwich Islands; Saturday Inspection; Secret Information; Secret Intelligence; Serra International; Sertoma International; Service Instruction; Shipping In-

struction(s); Smithsonian Institution; Society of Illustrators; Solomon Islands; South Island (New Zealand); Spokane International (railroad); Staff Inspector; Staten Island; Stevens Institute; Sulfur Institute; Summer Institute; Survey Instruction(s); *Système International des Unités* (French—International System of Units)
S-I Spokane International (railroad)
SI *Scheepvaart Inspectie* (Dutch—Shipping Inspection); *Sports Illustrated*; *Système International des Unités* (French—International System of Units)
sia subminiature integrated antenna
sia (SIA) storage instantaneous audimeter
SIA Sanitary Institute of America; Scaffolding Industry Association; School of International Affairs (Columbia University); Securities Industries Association; Self-Insurers Institute; Singapore Airlines; Ski Industries of America; Society of Insurance Accountants; Soroptimist International Association; Special Interest Area; Spinal Injuries Association; Sprinkler Irrigation Association; Standard Instrument Approach; Strategic Industries Association; Structural Insulation Association
SIA *Schweizerischer Ingenieur und Architekten Verein* (German—Swiss Institute of Engineers and Architects); *Société Internationale d'Acupuncture* (French—International Acupuncture Society)
SIAC Securities Industry Automation Corporation
SIAD Society of Industrial Artists and Designers
SIAE *Società Italiano degli Autori ed Editori* (Italian— Italian Society of Authors and Editors)
sial silicon + aluminum (Si + Al)
siam signal information and monitoring
SIAM Society for Industrial and Applied Mathematics
SIAO Smithsonian Institution Astrophysical Observatory
SIAP *Sociedad Interamericana*

de Planificación (Spanish—Interamerican Planning Society)

sib satellite ionospheric beacon(s); sibilant; sibling; sibship

Sib Siberia; Siberian; Sibyl; Sybil

SIB Securities and Investment Board; Shipbuilding Industry Board; Society of Insurance Brokers; Soviet Information Bureau; Special Investigation Branch (Police)

SIB *Sveriges Investeringsbank* (Swedish—Swedish Investment Bank)

SIBC *Société Internationale de Biologie Clinique* (French—International Society of Clinical Biology)

SIBIS Smithsonian Institution Bibliographic Information Service

Sib Or *Sibylline Oracles*

Sibr Siberia

sibs siblings

SIBS Salk Institute for Biological Studies

sic semiconductor integrated circuits; specific inductance capacity

sic. siccus (Latin—dry)

Sic Sicilian; Siciliana; Siciliano; Sicily

SIC Scientific Information Center; Secret Intelligence Command; Security Intelligence Corps; Sisters in Crime (mystery writers); *Société Intercontinental des Containers* (French—Intercontinental Container Society); *Société International de Cardiologie* (French—International Cardiology Society); *Société Internationale de Chirurgie* (French—International Surgery Society); Standard Industrial Classification; Survey Information Center

SIC *Société Internationale de Cardiologie* (French—International Cardiology Society); *Société Internationale de Chirurgie* (French—International Surgery Society); *Société Internationale de Criminologie* (French—International Criminology Society)

SICA Society of Industrial and Cost Accountants

sicbm (SICBM) super-intercontinental ballistic missile

SICC Staten Island Community College

Sic Chan Sicilian Channel between Sicily and Tunisia

sicklemia sickle-cell anemia

SICOT *Société Internationale de Chirurgie Orthopédique et de Traumatologie* (French—International Society of Orthopedic Surgery and Traumatology)

SICR Specific Intelligence Collection Requirement

sicsva sequential-impaction cascade-sieve volumetric air (sampler)

sic transit sic transit gloria mundi (Latin—so passes away the glory of the world)

sicu (SICU) surgical intensive care unit

Sic Vesp Sicilian Vespers (massacre of the French in 1282)

sid sidereal; standard instrument departure; sudden infant death; sudden ionospheric disturbance

s & id surveillance and identification

Sid Sidney; Sidney Sussex College, Oxford; Sydney

S.i.D. *Spiritus in Deo* (Latin—His Spirit is with God)—he's dead

SID Security and Intelligence Department; Society for Information Display; Society for International Development; Society for Investigative Dermatology; Standard Instrument Departure; Sudden Ionospheric Disturbance Division

SIDA Swedish International Development Agency

sidar selective information dissemination and retrieval

sidase significant data selection

SIDEC Stanford International Development Education Center

Siding Spring Siding Sprin Observatory, Australia

SIDINSA *Siderurgia Integrada SA* (Spanish—Integrated Iron-and-Steel Industry Corporation)

SIDOR *Siderúrgica del Oriente* (Spanish—Oriente Iron and Steel Industry)—Venezuela

Sidro San Ysidro, California

sids sudden infant-death syndrome

SIDs Sports Information Directors

SIDS Ships Integrated Defense System; Shrike Improved Display System; Space Identification Device System; Space Investigations Documentation System

SIDS *Société Internationale de Défense Sociale* (French—International Society of Social Defense)

sie single instruction execute

SIE Science Information Exchange (Smithsonian); Scientific Information Exchange; Society of Industrial Engineers; Southwestern Industrial Electronics

SIEC Scottish Industrial Estates Corporation

SIECUS Sex Information and Educational Council of the United States

SIEE Student of the Institution of Electrical Engineers

Siem Siemensstadt

Sierra letter S radio code

Sierra Leone Republic of Sierra Leone (West African nation established by the British as a native home for freed slaves)

Sierra Madre high mountains of western Mexico

Sierra Nevadas Sierra Nevada Mountains in California, Nevada, Spain, and Venezuela

Sierras Sierra Nevada Mountains; Sierra Mountains

SIES Soils and Irrigation Extension Service

SIETAR Society for Intercultural Education, Training, and Research

Sieur de Cadillac (French—Mr Cadillac)—Antoine de la Mothe Cadillac (1658—1730)

SIEX *Superintendencia de Inverciones Extranjeras* (Spanish—Superintendence of Foreign Investments)

SI Exy Staten Island Expressway

sif selective identification feature

SIF Society for Individual Freedom

SiF₄ silicon tetrafluoride

SIFA *Seguridad e Inteligencia de las Fuerzas Armadas* (Spanish—Security and Intelligence of the Armed Forces)—Venezuela

SIFE Society of Industrial Furnace Engineers

SIFF Suomen Illmailuliitto Finlands Flygforbund (Finnish—Finnish Aeronautical Association)

sif/iff selective identification feature/identification friend or foe

sifl standard industry fare level

SIFO Statens Institut för Opinionsundersökning (Swedish—State Institute for Opinion Research)

SIFs Stock Index Futures

SIFS Special Instructors Flying School

sift. share interval fortran translator; simplified input for toss

sig signal; signaling; signature

sig. signetur (Latin—mark with directions)

Sig Siegfried; Sieglinde; Sigdrifa; Sigmund; Sigmunt; Sigsbee; Sigurd; Sigyn

Sig Signor (Italian—Mister; Sir); *Signore* (Italian—Gentlemen, Our Lord, Sir); *Signori* (Italian—Gentlemen, Lords)

SIG Secret Intelligence Group; Senior Interagency Group (Space); Snowy Irrigation Scheme (Snowy Mountains Authority—Australia); Special Interest Group

SIG Schweizerische Industrie Gesellschaft (German—Swiss Industry Society)

siga sigatoka (banana leaf spot disease)

Sig^a Signora (Italian—Missus)—Mrs

SIGACT Special Interest Group on Automata and Computability Theory

SIGARCH Special Interest Group on Architecture of Computer Systems

SIGART Special Interest Group on Artificial Intelligence

SIG/BDP Special Interest Group on Business Data Processing

SIGBIO Special Interest Group on Biomedical Computing

SIG/BIOM Special Interest Group on Biomedical Information Processing

SigC Signal Corps

SIGCAPH Special Interest Group on Computers and the Physically Handicapped

SIGCAS Special Interest Group on Computers and So-

ciety

SIGCOMM Special Interest Group on Data Communication

SIGCOSIM Special Interest Group on Computer Systems Installation Management

SIGCPR Special Interest Group on Computer Personnel Research

SIGCSE Special Interest Group on Computer Science Education

SIGCUE Special Interest Group on Computer Uses in Education

SIGDA Special Interest Group on Design Automation

Sig Div Signal Division

sigex signal exercise

Sigg Signori (Italian—Messrs)

SIGGRAPH Special Interest Group on Computer Graphics

SIGI System of Interactive Guidance and Education

sigill. sigillum (Latin—seal)

sigint signals intelligence

sigint (SIGINT) signals intelligence (facsimile communications)

SIGIR Special Interest Group on Information Retrieval

Sig L Signal Lieutenant

SIGLASH Special Interest Group on Language Analysis and Studies in the Humanities

SIGLE System for Information on Grey Literature in Europe

siglun signal-light gun

Sigm Sigmund

SIGMA Science in General Management

SIGMAP Special Interest Group on Mathematical Programming

SIGMETRIC Special Interest Group on Metrication

SIGMICRO Special Interest Group on Microprogramming

SIGMINI Special Interest Group on Minicomputers

Sigmn Signalman

SIGMOD Special Interest Group on Management of Data

sigmoido sigmoidoscopy

sign. signature

Sig^{na} Signorina (Italian—Miss)

signif signifiable; signifiably; significance; significancy; significant(ly); signification; significative(ly); signifier; signify

sig. nom. pro. signa nomine proprio (Latin—label with

the proper name)

SIGNUM Special Interest Group on Numerical Mathematics

Sig O Signal Officer

SIGOPS Special Interest Group on Operating Systems

SIGPLAN Special Interest Group on Programming Languages

SIG/REAL Special Interest Group on Real-Time Processing

SIGs Special Interest Groups

SIGS Sandia Interactive Graphics System

SIGSAM Special Interest Group on Symbolic and Algebraic Manipulation

Sig Sam Lib Sigmund Samuel Library (Toronto)

SIGs-ASIS Special Interest Groups of the American Society for Information Science—AH: Arts and Humanities; ALP: Automated Language Processing; BSS: Behavioral and Social Sciences; BC: Biological and Chemical; CB: Costs, Budgeting, Economics; CR: Classification Research; ED: Education for Information Science; FS: Foundations of Information Science; IAC: Information Analysis Centers; IP: Information Publishing; ISE: Information Services to Education; LAN: Library Information and Networks; LAW: Law and Information Technology; MGT: Management Information Activities; MR: Medical Records; NDB: Numerical Data Bases; NPM: Non-Print Media; PPI: Public-Private Interface; RT: Reprographic Technology; SDI: Selective Dissemination of Information; TIS: Technology, Information, Society; UOI: User On-line Interaction

Sig Saus Sig Sauer (pistols)

SIGSDI Special Interest Group on Selective Dissemination of Information

SIGSIM Special Interest Group on Simulation

SIGSOC Special Interest Group on Social and Behavioral Science Computing

SIGSPAC Special Interest Group on Urban Data Systems, Planning, Architecture, and Civil Engineering

Sig Sta signal station
SIG/TIME Special Interest Group on Time Sharing
SIGUCC Special Interest Group on University Computing Centers
Sig Und Sigrid Undset
SIG/UPACE Special Interest Group on Urban Planning, Architecture, and Civil Engineering
SIH Samuel Ichiye Hayakawa
SIH Société Internationale d'Hématologie (French—International Hematology Society)
SIHS Society for Italian Historical Studies
SII School Interest Inventory; Security-Insecurity Inventory; Self-Interview Inventory; Standards Institution of Israel; Staten Island Institute; Structural Impediments Initiative (Japan and the United States)
SIIA Stevenson Institute of International Affairs
SIIAS Staten Island Institute of Arts and Sciences
SIIP Systems Integration Implementation Plan
SIIRS Smithsonian Institution Information Retrieval Service
SIIS Shanghai Institute of International Studies
SIJD Subcommittee to Investigate Juvenile Delinquency (U.S. Senate)
Sik Sikkim
siks single income, kids
sil silver; speech interference level
s-i-l sister-in-law
Sil Silesia; Silesian; Silurian
SIL Society for Individual Liberty; Society for International Law; Summer Institute of Linguistics; System Implementation Language
SIL Société International de la Lèpre (French—International Leprosy Society)
Silas Silvanus
silcads silver-cadmium batteries
Sile Cecilia
SILI Standard Item Location Index
SILIA South Island Livestock Improvement Association
silic silicate; siliceous
silica silicon dioxide (SiO_2)
silicos silicosis (sickness caused by stone-dust inhalation)

silkool silk + wool (Japanese synthetic textile combining qualities of silk and wool)
silos side-looking sonar
sils silver solder
sil(s) speech interference level(s)
SILs Smithsonian Institution Libraries
silv silver; silvery
silvercel silver-zinc cell (battery)
silvicult silviculture
sim self-inflicted mutilation; similar; simile; simple; simulate; simulated approach
Sim Simm(s); Simmy; Simon(d); Sims; Syme(s); Symme; Syms; etc.
SIM Society for Industrial Microbiology
SIM Servicio Inteligencia Militar (Spanish); *Servizio Informazioni Militari* (Italian—Military Intelligence Service); *Société Internationale de Musicologie* (French—International Musicological Society)
SIMA Scientific Instrument Manufacturers' Association; Steel Industry Management Association; Suburban Insurance Managers' Association
SIMAGB Scientific Instrument Manufacturers Association of Great Britain
SIMAJ Scientific Instrument Manufacturers Association of Japan
SIMBAD Set of Identifications, Measurements, and Bibliography for Astronomical Data
SIMC Société Internationale pour la Musique Contemporaine (French—International Society for Contemporary Music)
SIMCA Société Industrielle de Mécanique et Carosserie Automobile (French—Industrial Society of Automobile Manufacturers)
simch single mach change
simcon simplified control; simulated control
simd single-instruction multiple-data stream
SIME Security Intelligence Middle East (British)
sim excu simulated execution
SIMG Societas Internationalis Medicinae Generalis (Latin—International General Medicine Society)

SIMHA Société Internationale de Mycologie Humaine et Animale (French—International Society for Human and Animal Mycology)
SIMILE Simulator of Immediate Memory in Learning Experiments
'simmon(s) persimmon(s)
SIMNET Simulation Network
Simons Simonstown; Simonstown naval base near the Cape of Good Hope in South Africa
simp simpleton
simp. simplex (Latin—simple)
simpac simulated package
SIMPL Scientific, Industrial, and Medical Photographic Laboratories
Simplon Simplon Pass in the Swiss Alps
Simpson Simpson Desert of in the southeast sector of Australia's Northern Territory
sims secondary ion mass spectroscopy
SIMS Surface-to-Air Intercept Missile System
simstrat simulation strategy
sim sui simulated suicide
simul simulation
simula simulation language
simulcast simultaneous broadcast (am & fm)
simulcast(ing) simultaneous broadcast(ing) of the same program on radio and television
Simyens Simyen Mountains of Ethiopia
sin. sewer in; sine; single
sin. sinfonia (Italian—symphony); *sinister* (Latin—left)
sin' sino (Italian—as far as; until)
Sin Sinaloa (inhabitants—Sinaloens); Singapore; Sinhalese
SIN Singapore (airport); Society for International Numismatics; Stop Inflation Now
SIN Scientific Information Notes (National Science Foundation); *Société Industrielle et Navale* (French—Industrial and Naval Society); Spanish International Network (tv)
SINB Southern Interstate Nuclear Board
S-in-C Surgeon-in-Chief
SINCGARS Single-Channel Ground-Air Radio System
Sind Sindhi
S Ind Cur South Indian Cur-

rent
sin Dich sinfonische Dichtung (German—symphonic poem)
sinf sinfonia (Italian—symphony)
S Infre Seine-Inférieure (Lower Seine River)
sing. singer; single; singing; singular
sing. singulorum (Latin—of each)
Sing Singapore
Singa Singapore
singan singularity analyzer
Singapore Republic of Singapore (island nation at the southernmost tip of the Malay Peninsula)
Singer Isaac Bashevis Zinger
Sing U Singapore University
sinh hyperbolic sine
Sinh Sinhalese
Sinjent St John
Sink Sinkiang
sins. ship-inertial-navigation systems
SINS Ship's Inertial Navigation System; Situational Inertial Navigation System (USN)
s int senza interruzione (Italian—without interruption)
Sint Eust Sint Eustatius
Sint Maart Sint Maarten
SINTO Sheffield Interchange Organisation
si n. val. si non valet (Latin—if of no value)
sio satellite in orbit; staged in orbit
si/o star input/output
SIO Scripps Institution of Oceanography; Ship's Information Office(r); Special Intelligence Office(r)
sioh supervision, inspection, and overhead
SIOP Single Integrated Operations Plan
si op. sit si opus sit (Latin—if necessary)
SIOR Society of Industrial and Office Realtors
Sioux Falls Pen South Dakota Penitentiary at Sioux Falls
sip. standard inspection procedure; step in place
SIP Share of Impact Panel (measuring alcohol consumption); Smithsonian Institution Press; *Sociedad Interamericana de la Prensa* (Spanish—Inter-American Press Association—IAPA); Society of Integral Psychoanalysis; Standard Inspection Proce-

dure; State Improvement Plan(ning); Street Improvement Program
SIP Sociedad Interamericana de la Prensa (Spanish—Interamerican Press Association); *Société Interaméricaine de Psychologie* (French—Interamerican Society of Psychology)
SIP/AG Sri Lanka, India, Pakistan/Arabian Gulf (freighter route)
SIPC Securities Investor Protection Corporation
SIPE System Internal Performance Evaluation
SIPG Société Internationale de Pathologie Géographique (French—International Society of Geographical Pathology)
SIPI Southwestern Indian Polytechnic Institute
sipl scientific information processing language
Sipo security police (Nazi)
Sipo Sicherheitspolizei (German—State Security Police)
SIPRC Society of Independent Public Relations Consultants
SIPRE Snow, Ice, and Permafrost Research Establishment
SIPRI Stockholm International Peace Research Institute
SIPROS Simultaneous Processing Operation System
SIPS State Implementation Plan System
siq superior internal quality
sir. selective information retrieval
sir. (SIR) submarine intermediate reactor; supplemental inflatable restraint (automobile air bag)
Sir Siria (Italian, Latin, Spanish—Syria); *Síria* (Portuguese—Syria)
SIR Society for Individual Responsibility; Society for Individual Rights; Society of Industrial Realtors; Staten Island Rapid Transit (railroad code)
SIR Società Italiana Resine (Italian—Italian Resin Association)
SIRA Scientific Instrument Research Association
Sir Ambrose Sir John Ambrose Fleming (1849–1945)
SIRC Spares Integrated Reporting and Control System
SIRCS Shipboard Intermedi-

ate-Range Combat System
Sir David Sir David Lean
SIRE Small Investors Real Estate (plan); Society for the Investigation of Recurring Events
Sir Edmund Sir Edmund Hillary
Sir Ernest Sir Ernest Henry Shackleton
Sir Francesco Sir Francesco Paolo Tosti
Sir George Sir George Grove (1820–1900)
Sir Henry Sir William Henry Hadow (1859–1937)
Sir Isaac Sir Isaac Newton (1642–1727)
Sirin Vladimir Nabokov (1899–1977)
Sir Jehudi Sir Jehudi Menuhin
Sir John Sir John Pritchard
Sir Neville Sir Neville Marriner (1924–)
Sir Noël Sir Noël Coward (1899–1973)
SIRP Salon International de la Recherche Photographique (French—International Photographic Research Show)
Sir Peter Sir Peter Pears (1910–1986)
SIRR Spokane International Railroad
Sir Rex Sir Rex Harrison (1908–1990)
SIRS School Information and Research Service; Security Information Retrieval System; Ship-Installed Radiac System; Sorption Information-Retrieval System; Student Information Record System
Sir Stephen Sir Stephen Spender
Sir Steven Sir Steven Runciman
SIRT Staten Island Rapid Transit
SIRTF Spacelab Infrared Telescope Facility
Sir William Sir William Schwenck Gilbert
Sir Yehudi Sir Yehudi Menuhin
sis sister; shock insulation support; sterile injectable suspension
sis (Latin suffix—action or process)—dialysis
Sis Cecilia; sister
SIs Sandwich Islands; Service Instructions; Shipping Instructions; Solomon Islands;

Survey Instructions
SIS School of Information Studies; Secret Intelligence Service(s); Shut-In Society; Signal Intelligence Service; Special Industrial Services (UN); Special Isotope Separation; Standard Indexing System (DoD); Standards Information Service; Stockholm Information Service; Strategic Intelligence School; Strategic Intelligence Summary; Student Information System; Submarine-Integrated Sonar (system)
S & IS Space and Information System(s)
SISAL Società Italiana Sistemi a Lotto (Italian—Italian Lotteries)
SISGAP Scottish Industrial Safety Group Advisory Council
sisi short-increment sensitivity index
Sisister (British contraction—Cirencester)
sis-in-law sister-in-law
sisp sudden increase of solar particles
siss single-item single-source
SISS Semiconductor-Insulation Semiconductor System; Submarine Improved Sonar System; System Integration Support Service
SISS Société Internationale de la Science du Sol (French—International Society of Soil Science)
SISTER Special Institution for Scientific and Technological Education and Research
Sistine pertaining to Pope Sixtus
SISUSA Scotch-Irish Society of the United States of America
sit. silicon intensifier target; situation; spontaneous ignition temperature; statement of inventory transaction; stopping in transit
SIT Senate Intelligence Committee; Singapore Improvement Trust; Slosson Intelligence Test; Society of Industrial Technology; Stevens Institute of Technology; Street Interface Transmission; Sugar Industry Technicians
SITA Students International Travel Association
SITA Société Internationale de Télécommunications Aero-

nautiques (French—International Society of Aeronautic Telecommunications)
SITC Standard International Trade Classification
SITCEN Situation Center (NATO)
sitcom situation comedy (tv)
site. shipboard information, training, entertainment
SITE Satellite Instructional Television Experiment; Society of Incentive Travel Executives
SITES Smithsonian Institution Traveling Exhibition Service
SITF Shuttle Infrared Telescope Facility
sitol sitological; sitologist; sitology
sitp scheduled into production
SITP Shipyard Installation Test Procedure
sitpro simplification of international trade procedures
sitr silent treatment
SITRA South India Textile Research Association
sitrag situation tragedy
sitrep situation report
SITS Securities Instruction Transmission System; *Société Internationale de Transfusion Sanguine* (French—International Organization for Blood Transfusion)
SITSUM Situation Summary (NATO Intelligence)
sitt sitting room
SITU Society for the Investigation of the Unexplained
sitv (SITV) system-integration test vehicle
SIU Seafarers International Union; Southern Illinois University; Special Investigating Unit (NY Police Bureau of Narcotics)
SIU Société Internationale d'Urologie (French—International Urological Society)
SIUE Southern Illinois University at Edwardsville
SIUL Southern Illinois University Library
SIUM Southern Illinois University Museum
SI unit Système International unit (French—International System of Units)
SIUP Southern Illinois University Press
siv survey of interpersonal values
siv (SIV) simian immune virus

si vir. perm. si vires permittant (Latin—if the strength will permit)
siw (SIW) self-inflicted wounds
Six-Day Six-Day War (between Israel, Egypt and Syria)—June 5 to 10, 1967
SIXFLT Sixth Fleet (USN)
six-pac six-pack
SIXPAC System for Inertial Experiment Pointing to Attitude Control
SIYC Shelter Island Yacht Club; Staten Island Yacht Club
SIZ Security Identification Zone
SIZS Staten Island Zoological Society
sj slip joint; subject(s)
s.j. sub judice (Latin—under judicial consideration)
SJ San José; San Juan; Society of Jesus (S.J.—Jesuits); *Statens Järnvägar* (Swedish State Railways)
S-J Stevens-Johnson (syndrome)
SJ Solicitors' Journal
SJAA St John Ambulance Association
SJAC Society of Japanese Aircraft Constructors
SJB Sluzba Javne Bezbednosti (Serbo-Croat—State Security Service)—Yugoslav
SJC San Jose, California (airport); San Juan Carriers; Snead Junior College; Spartanburg Junior College
SJCC San Jose City College
S.J.D. Scientiae Juridicae Doctor (Latin—Doctor of Juridical Science)
sje swivelling jet engine
sJf single Jewish female
Sjf Sandefjord
Sjf Sjofartsverket (Swedish—Shipping Inspection Bureau)
SJI Steel Joist Institute
SJIs San Juan Islands
SJJC Sheldon Jackson Junior College
SJLA Studies in Judaism in Late Antiquity
sJm single Jewish male
S Jn San Juan
SJO San José, Costa Rica (La Sabana Airport)
SJPC South Jersey Port Commission
SJPL San Jose Public Library
S-J-R Shinawora-Jones-Reinhart (units)
SJSC San Jose State College

SJSO San Jose Symphony Orchestra
SJU San Juan, Puerto Rico (airport); St John's University
sk sick; skein; sketch; skip (knitting instruction); skip (punched card)
sk (SK) streptokinase
SK Saskatchewan
Sk Skizze (German—sketch)
SK end of transmission (telegraphic symbol); South Korea(n)
S-K Sloan-Kettering
SK Stuttgarter Kammerorchester (German—Stuttgart Chamber Orchestra); *Suomen Kansallisoopera* (Finnish—Finnish National Opera)
SK-37 Saab Thunderbolt or Vigen multimission combat aircraft also known as AJ-37, JA-37, and S-37
SK-60 Saab attack-type jet aircraft design based on the A-60
s-ka spolka (Polish—association, company)
SKA Switchblade Knife Act
skachet skinning knife, hammer, hatchet, and hunting knife all-purpose utility tool
Skag Cape Skagen or The Skaw; Skagway, Alaska
Skager Skagerrak (North Sea between Denmark and Norway)
skamp station keeping and mobile platform
Skaw Cape Skagen or The Skaw—northernmost Denmark
skb skindbind (Dano-Norwegian—leatherbound)
SKBF Svensk Kärnbränsleförsörjning (Swedish Nuclear Fuel Supply Company)
skc sky clear
SKC Scottish Kennel Club
SKCC Sloan-Kettering Cancer Center
SkCsr Státní knikhovna Ceské socialistické republiky (Czechoslovakian—State Library of the Czech Socialist Republic)—Prague
skd skilled
skdn shakedown
sked schedule
skedcon schedule conference
skel skeletal; skeleton
S Ken South Kensington
skep skeptic(al)(ly); skepticism
SKF Svenska Kullagerfabriken (Swedish ball-bearing fac-

tory)
SK & F Smith Kline & French
SKI Sloan-Kettering Institute
skil science keyboard input language
skill. satellite kill; skin, kidneys, intestines, liver, lungs
skinmag magazine featuring nudes of both sexes
Skins Skinheads (neo-Nazi organization)
SKIP Skimmer Investigation Platform
skiv skiver
SKJ Savez Komunista Jugoslavije (Yugoslavian Communist League)—political party
SKKCA Supreme Knight of the Knights of Columbus of America
skl skylight(ing); spleen, kidney, liver
Skm Stockholm
SKM Süleymaniye Kütüphahesi (Turkish—Suleiman Mosque Library)—Istanbul
skmr (SKMR) hydroskimmer
S^knoll Seaknoll
skort short skirt
skp station-keeping position
skp (SKP) skip
SKP Suomen Kommunistinen Puolue (Finnish—Communist Party)
skpo slip one, knit one, pass slipped-stitch over (knitting)
SKQ Sexual Knowledge Questionnaire
skr standardized kill rate; station-keeping radar
skr sanskrit (German—Sanskrit)
Skr Sanskrit; Saturn kilometric radiation; Skipper; Skire (Thursday)
Skr Skrifter (Swedish—publication)
SKr Swedish krona (kronor)
SKR South Korea Republic
sks sacks
SKS Soren Kierkegaard Society; station-keeping ship
SKS Savvezna Komisija za Standardizacija (Serbo-Croatian—Federal Commission for Standardization)
Skt Sanskrit
Skt Sankt (German—saint)
SKU(s) stock-keeping unit(s)
SKY Skyways Limited (aviation symbol)
skyjack skyjacked; skyjacker; skyjacking (aircraft hijacking)
Skynet 4B British military communications satellite

skys'l skysail
Sky & Tel Sky & Telescope
sl liability; safety lighting; sales letter; sand-loaded; sea level; searchlight; shipowner's; slightly; slip (knitting instruction); sound locator; stock length; support line
sl (SL) sprinkler leakage; standard label; straight line
s-l short-long (signals); sound-locator sublease
s/l self-loading
s & l savings and loan; signed & limited; supply and logistics
s.l. secundum legem (Latin—according to law); *sensu lato* (Latin—in the broad sense); *sine loco* (Latin—no place of publication)
s/l sobreloja (Portuguese—mezzanine floor); *su letra* (Spanish—your letter)
s/L sur Loire (French—on the Loire)
Sl Slovak; Slovakian; small diurnal range
SL San Luis Obispo; Sandia National Laboratories; Savings and Loan; Sea-Land (America's seagoing motor carrier); Sierra Leone; Solicitor-at-Law; Squadron Leader; Sub-Lieutenant; Support Line; Sydney & Louisburg (railroad)
S-L Sea-Land (Line); short-long
S&L Savings and Loan;
S & L Supply and Logistics
SL Schweizerische Landesbibliothek (German—Swiss State Library)—in Bern; *Sierra Leone* (Spanish—Sierre Leone)
sla sacro-laeva anterior; single-line approach
SLA Sandia Laboratories—Albuquerque; School Library Association; Scottish Library Association; Showmen's League of America; Sleep-Learning Association; Southeastern Library Association; South Lebanon Army; Southwestern Library Association; Special Libraries Association; Standard Life Association; State Liquor Authority; Supply Loading Airfield; Supply Loading Airport; Symbionese Liberation Army
SLAA Surf Lifesaving Association of Australia
SLAB Students for Labelling

Alcoholic Beverages
SLAC Stanford Linear Acceleration Center
SLAD Society of London Art Dealers
SLADE Society of Lithographic Artists, Designers, Engravers, and Process Workers
slado system library activity dynamic optimiser
SLAET Society of Licensed Aircraft Engineers and Technologists
slam. (SLAM) scanning-laser acoustic microscope; supersonic low-altitude (nuclear-powered) missile
s.l.a.m. sine loco, anno, nomine (Latin—without place, year, or name)
SLAM Society's League Against Molestation (of children)
SLANG Systems Language
slanguage slang language (according to Carl Sandberg language which takes off its coat, spits on its hands—and goes to work); slum language
S Lan R South Lancashire Regiment
SLANT Student League Against Narcotic Traffic
SLAP Student Loan Abuse Prevention
slar side-looking airborne radar
slarg *slargando* (Italian—broadening of the tempo)
S Lat south latitude
slate. small lightweight altitude-transmission equipment
SLATE Structured Learning and Teaching Environment; Systems for Learning by Applications of Technology to Education
SLATS Safe, Loft, and Truck Squad (of a police department)
slav *slavonisch* (German—Slavic)
Slav Slavic; Slavonic
slax slacks
slb short-leg brace
slbm (SLBM) submarine-launched ballistic missile
slc searchlight control; shift left and count (instructions); straight-line capacity
sl & c shipper's load and count
SLC Salt Lake City, Utah (airport); Scout Launch Complex; Space Launch Complex; Stanford Linear Col-

lider (pronounced *slick*)
SLCL Sierra Leone Council of Labour
slcm (SLCM) sea-launched cruise missile
SLCMD St Louis Contract Management District
SLCPL Salt Lake City Public Library
SLCR *Scottish Land Court Reports*
SLCS Sea Level Canal Study
sld sailed; solid; specific learning disability
sld (SLD) serum lactate dehydrogenase
Sld Sunderland
SLD Special Low Dispersion
sldf solidification
sl di slot die
S Ldr Squadron Leader
sld's specific learning disabilities
SLDVS Scanning Laser Doppler Vortex System
sle systemic lupus erythematosus
sle (SLE) systemic lupus erythematosus
S le Sierra Leone leone(s)—monetary unit(s)
SLe St Louis encephalitis
SLE Society of Logistics Engineers
SLEAT Society of Laundry Engineers and Allied Trades
SLED State Law Enforcement Division (South Carolina)
Sleepers Sleeper Islands in Hudson Bay just north of the Belchers
Sleeping Bear Sleeping Bear Dunes national lakeshore in Michigan
slent *slentando* (Italian—slackening of the tempo)
SLEP Service Life Extension Program (USN)
s.l. et a. sine loco et anno (Latin—without place and year)
S level scholarship level
slew. static load error washout
slf straight-line frequency; symmetric filter
SLF Scottish Landowners' Federation; Silcock and Lever Feeds
SLF *Skandinaviska Lacktknickers Förbund* (Federation of Scandinavian Paint and Varnish Technicians)
SLFCS Survivable Low Frequency Communications System
S-L Fl short-long flashing

(light)
SLFP Sri Lanka Freedom Party
slg state or local government
SLGB Society of Local Government Barristers
SLGLW St Lawrence and Great Lakes Waterway
slh severe legislative hypocrisy
SLHC St Luke's Hospital Center
sli suppressed-length indication
Sli Sligo
SLI Slick Airways
slic selective listing in combination
SLIC Sober Live-In Center (for drug-troubled teenagers); Supreme Life Insurance Company
SLICE Southwestern Library Interstate Cooperative Endeavor; Surrey Library Interactive Circulation Experiment
slickums sea-launched cruise missiles (SLCMs)
slid. scanning light-intensity device
SLID Student League for Industrial Democracy
Slide Slide Mountain (highest in the Catskills)
slim. (SLIM) submarine-launched inertial missile
SLiM Selected Library in Microfiche
SLIM South London Industrial Mission; Spanish Language Immersion (teaching program)
slip. symmetric(al) list processor
SLIP Skills Level Improvement Plan
slithy lithe and slimy (Lewis Carroll's portmanteau word from *Through the Looking Glass*)
SLJ *School Library Journal*
SLKP Supreme Lodge of the Knights of Pythias
SLL Socialist Labour League
SLLA Scottish Ladies Lacrosse Association; Sri Lanka Library Association
slld specific language and learning disability
slm single-level masking
slm (SLM) ship-launched missile
slm *sul livello del mare* (Italian—at sea level)
SLMC Scottish Ladies' Mountaineering Club
SLMR Speed Limit on Mo-

torways Regulations
slms selective level measuring
set
SLMSU Scientific Library of
Moscow State University
SLMTA St Louis Municipal
Theatre Association
sln standard library number
slnd sans lieu ne date
(French—without place or
date of publication)
s.l.n.d. sine loco nec data
(Latin—without indication of
date)
SLNM Statue of Liberty National Monument
SLNSW State Library of New
South Wales (Sydney)
SLNWR Sand Lake National
Wildlife Refuge (South Dakota); San Luis NWR (California); Swan Lake NWR
(Missouri)
Slo Saltillo (inhabitants—
Saltilleños or Saltilleros);
Slovak; Slovakia; Slovene(s)
SLO San Luis Obispo; Senior
Liaison Officer
SLOA Steam Locomotive Operators Association
slob. satellite low-orbit bombardment
Slob Sloboda (Russian—big
village, suburb)
SLOBB Stop Littering Our
Bays and Beaches
sloc sea lanes of communication
SLOE Special List of Equipment
slomar space logistics, maintenance, and rescue
s'long so long (from the Arabic *salaam* or the Hebrew
shalom, both meaning *peace
be with you*)
s'loon saloon
SLOS Stabilized Long-Range
Observation System
s/loss salvage loss
Slot The Slot—San Francisco's
downtown Mission Street off
Market Street
slotperson head copy reader
on a newspaper
Slov Slovene; Slovenian
Slov Phil Slovenian Philharmonic
SLOWPOKE Safe Low-Power Critical Experiment (AEC)
slp sacro-laeva posterior; super
long play
s.l.p. sine legitima prole (Latin—without legitimate issue)
SLP San Luís Potosí; Scottish
Labour Party; Socialist Labor

Party
Slphr Sulphur
SLPL St Louis Public Library
slr self-loading rifle; side-looking radar; single-lens reflex
(camera)
slr (SLR) storage limits register
s-l r sea-level resident(s)
SLR State Liaison Representative
S & LR Sydney and Louisburg
Railway
SLR Scottish Land Reports
SLRB State Labor Relations
Board
SLRC San Luis Rey College
sl rd searchlight radar
SL Rev Scottish Law Review
SLRP Society for Long-Range
Planning
SLRP St Lawrence River Pilot
SLRs Scottish Land Reports
sls sequential light switch;
slide set
S&Ls Savings and Loan banks
SLS School of Library Science; School of Library Service; School of Library Studies; Sea-Land Service;
Stephenson Locomotive Society; St Lawrence Seaway;
St Louis Symphony
sl sa slotting saw
SLSA Saint Lawrence Seaway
Authority; Surf Life Saving
Association
SLSC Surf Life Saving Club;
Swedish Lloyd Steamship
Company
SLSDC Saint Lawrence Seaway Development Corporation
sls2e surface(d) long side and
two edges
SLSENY School Librarians of
Southeastern New York
S L S F St Louis-San Francisco (railroad)
SLSFC Severe Local Storm
Forecast Center
slsmgr salesmanager
slsmn salesman; salesmen
sl st slip stitch
SLST Sierra Leone Selection
Trust
s-l stil spring-loaded stiletto
sl st(s) slip stitch(es)—knitting
SLSU Sea Land Service (container) Unit
SLS-UBC School of Library
Science—University of British Columbia
slt sacro-laeva transversa;
searchlight
sl&t shipper's load and tally

SLT Solid-Logic Technology;
Stress Limit Test(ing)
SLT Scots Law Times
SLTA Scottish Licensed Trade
Association
SLTAN Società Lloyd Triestino per Azioni di Navigazione (Lloyd Triestino)
SLTC Society of Leather
Trades Chemists
slto sea-level takeoff
sl tr silent treatment
Slu slough
SLU Saint Lawrence University; Saint Louis University,
Southeastern Louisiana University; Southern Labor
Union
slug. superconducting low-inductance undulatory galvanometer
slumlord slum landlord
slumpfla slumpflation (high
inflation coupled with high
unemployment)
slumpflation slump + inflation
(economic decline coincident
with rising inflation)
slurb slum suburb; suburban
slum
slurp. self-levelling unit to remove pollution
SLUs Special Liaison Units
SLUSSR State Library of the
USSR (Lenin Library, Moscow)
slutt surface-launched underwater transponder target
slv satellite launching vehicle;
space launch vehicle; standard launch vehicle (SLV)
SLV-3 Atlas standard launch
vehicle (Convair)
sly slowly
sly. safety, liquidity, yield;
slowly
Sly southerly
slyp short-leaf yellow pine
SLZG St Louis Zoological
Gardens
sm service member; service
module; servomechanism;
sheet metal; small; statute
mile; strategic missile (SM);
streptomycin; sustained
medication; systolic murmur;
syzygy mathematical
sm (SM) secondary memory
s-m sadist-masochist; sadomasochism
s/m sensory-to-motor (ratio)
s&m surface and matched
stock and machinery; sadism
and masochism; sausages and
mashed potatoes
s/M sur mer (French—by the

sea); *sur Marne* (on the Marne River, France); *sur Maroni* (on the Maroni River, French Guiana); *sur Meurthe* (on the Meurthe River, France); *sur Moselle* (on the Moselle River, France)

Sm samarium

Sm *Seemeile* (German—nautical mile)

SM mine-laying submarine; Salvage Mechanic; San Marino; Santa Monica; Scientific Memorandum; Senior Magistrate; Sergeant-Major; Service Module; Shipment Memorandum; Signalman; Society of Mary; Society of Medalists; Soldier's Medal; Special Memorandum; Spiritual Mobilization; Staff Memorandum; State Militia; State Monument; States Marine (steamship lines); Structures Memorandum; submarine; Summary Memorandum; *Suomi Merivorma* (Finnish Seapower); Supply Manual; Svenska Metallverken (Swedish Metal Works)

S-M Seine-Maritime (formerly Seine-Inférieure)

S.M. *Scientiae Magister* (Latin—Master of Science)

S.M. *Sanctae Memoriae* (Latin—of sacred memory); *Su Majestad* (Spanish—Her/His Majesty)

SM-4 Polish three-place helicopter

SM-65 Atlas intercontinental ballistic missile (Convair)

SM-68 Titan intercontinental ballistic missile (Martin)

SM-75 Thor intermediate-range ballistic missile (Douglas)

SM-78 Jupiter intermediate-range ballistic missile (Chrysler)

SM-80 Minuteman intercontinental ballistic missile (Boeing)

sma small-motion accelerometer; subject matter area

SMA Safe Manufacturers Association; San Miguel Arizona (railroad); Santa María, Azores (airport); Scale Manufacturers Association; Screen Manufacturers Association; Senior Military Attaché; Service Merchandisers of America; Sheffield

Metallurgical Association; Socialist Medical Association; Society of Makeup Artists; Society of Motor Auctions; Solder Makers Association; Squadron Maintenance Area; Steatite Manufacturers Association; Steel Manufacturers Association; Stoker Manufacturers Association

SMAA Submarine Movement Advisory Authority

SMAB Solid Motor Assembly Building

SMAC Scientific Machine Automation Corporation

SM & ACCNA Sheet Metal and Air Conditioning Contractors National Association

s mach sounding machine

SMACNA Sheet Metal and Air Conditioning Contractors National Association

SMAE Society of Model Aeronautical Engineers

SMAJ Sugar Manufacturers' Association of Jamaica

smalgol small computer algorithmic language

SMAMA Sacramento Air Materiel Area

smap surprised middle-aged person

S Mar San Marino

smarea (SMAREA) squadron maintenance area

smart. special methods for attacking the right targets

SMART Scheduled Maintenance At Regular Times; School Management Appraisal and Rating Technique; Silent Majority Against Revolutionary Tactics; Software for Market Analysis and Restriction on Trade; Stop Marketing Alcohol on Radio and Television; Supersonic Military Air Research Track; Supersonic Missile and Rocket Track; System for the Mechanical Analysis and Retrieval of Text

smartie simple-minded artificial intelligence

SMASH Students Mobilizing on Auto Safety Hazards

smashex search for simulated submarine casualty exercise

s-m-a showing suggested-for-mature-adult showing (motion picture producers code)

smat see me about this

smatv (SMATV) satellite mas-

ter antenna television

smaw shoulder-launched multipurpose automatic weapon

smaze smoke + haze (*see* smog)

SMB Straits of Mackinac Bridge

SMB *Sa Majesté Britanique* (French—Her/His Britannic Majesty)

SMBA Scottish Marine Biological Association

smbl semimobile

SMBO Small and Minority Business Office

SMBW Society of Mineral and Battery Works

smc sheet-molding compound; small magellanic cloud; sperm (spore) mother cell; standard mean chord

Smc Samic (Lapp)

SMC Saugus Marine Corporation; Scientific Manpower Commission; State Medical Society; Strategic Military Council

S & MC Supply and Maintenance Command (US Army)

smca suckling-mouse cataract agent

sm caps small capital letters

SMCC Saint Mary's College of California; Santa Monica City College

SMCCL Society of Municipal and County Chief Librarians

SMCL Southeastern Massachusetts Cooperating Libraries

smcln semicolon

SMCRC Southern Motor Carriers Rate Conference

smd submanubrial dullness; surface-mounted device

smd (SMD) senile macular degeneration

SMD Submarine Mine Depot

SMDA Sewing Machine Dealers' Association

SMDC Saint Mary's Dominican College

SME School of Military Engineering; Society of Manufacturing Engineers; Society of Mining Engineers; Special Minister of the Eucharist; Standard Medical Examination

S.M.E. *Sancta Mater Ecclesia* (Latin—Holy Mother Church)

SMEAR Span/Mission Evaluation Action Request

SMEC Snowy Mountains Engineering Corporation; Strategic Missile Evaluation Committee

SMEG Spring Makers' Export Group

smel single and multiengine license

smelerience smell(ing) experience

smelt. smelter; smelting

sm-er (SM-ER) surface missile—extended range

s/mer sur mer (French—by the sea)

SMERC San Mateo Educational Resource Center

SMERSH Smert Shpionam (Russian—Death to Spies)— Soviet organization for murdering political enemies

smes superconducting magnetic energy storage

S Met O Senior Meteorological Officer

SMF Sacramento, California (airport); Shaker Museum Foundation; Snell Memorial Foundation; South Moluccan Force; System Management Facility

SMfVL Stuttgart Museum för Volker and Landerkunde (German—Stuttgart Museum for National and Regional Studies)

smg speed made good; submachine gun

Smg Samarang

SMG Stato Maggior Generale (Italian—General Staff)

SMH Sydney Morning Herald

SMHEA Snowy Mountains Hydro-Electric Authority

smi standard measuring instrument

s mi statute mile(s)

SmI Solidaritet med Israel (Dano-Norwegian—Solidarity with Israel)

SMI Scale Manufacturers Institute; School Management Institute; Secondary Metal Institute; Shippers Management International; Spring Manufacturers Institute; Success Motivation Institute; Super Market Institute

SMI Sa Majesté Imperiale (French—Her/His Imperial Majesty)

SMIA Sheet Metal Industries Association

SMIAC Soil Mechanics Information Analysis Center (Corps of Engineers)

SMIC Study of Man's Impact on Climate

smicbm (SMICBM) semi-mobile intercontinental ballistic missile

smice smoke + ice (ice-crystal-laden fog)

SMIG Sergeant-Major Instructor of Gunnery

SMILE Something Meaningful In Local Effort (predelinquency file kept in Orange County, California); Space Migration, Intelligence, and Life Extension (achieved by settling on other planets)

S-mine shrapnel-filled mine

SMIS Society for Management Information Systems

smist smoke + mist

smit spin-motor interruption technique

SMIT Sherman Mental Impairment Test

SMITES State-Municipal Income-Tax Evaluation System

Smith Coll Smith College

Smith Coll Lib Smith College Library

Smith Inst Smithsonian Institution

Smithsonian Smithsonian Institution (United States National Museum)

Smitty Smith

SMJ Southern Masonic Jurisdiction

SMJAB State Medical Journal Advertising Bureau

SMJC Saint Mary's Junior College

smk smoke

Smk Shimonoseki

smk gen smoke generator

smkls smokeless

smkstk(s) smokestack(s)

sml simulate; simulation; simulator; small; symbolic machine language

sml sammenlign (Danish—compare)

Sml Samuel

SML Science Museum Library; States Marine Lines

SMLA Samoa Muamua Le Atua (Samoan—In Samoa God Is First)

SMLC Save Mono Lake Committee

SMLE short-model Lee Enfield (British service rifle used in both world wars)

sml grdn small garden

smlm simple-minded learning machine

SMLM Soviet Military Liaison Mission

smls seamless

SMLS Saint Mary of the Lake Seminary; Seaborne Mobile Logistic System

smm standard method of measurement

smm (SMM) solar maximum mission

SMM Science Museum of Minnesota; Solar Maximum Mission

S.M.M. Sancta Mater Maria (Latin—Holy Mother Mary)

SMMA Small Motor Manufacturers Association

s:m::m:b: soybean is to milk as margarine is to butter

SMMB Scottish Milk Marketing Board

smmc system maintenance monitor console

smmp screw machine metal part

smmr (SMMR) surface missile—medium range

SMMT Society of Motor Manufacturers and Traders

SMN Société Maritime Nationale (French—National Maritime Society)

SMNA Safe Manufacturers National Association

SMNH Saskatchewan Museum of Natural History

SMNO Singapore Malays National Organization

SMNP Simien Mountains National Park (Ethiopia)

SMNRA Shadow Mountain National Recreation Area (Colorado)

Smnry Seminary

SMNWR Saint Marks National Wildlife Refuge (Florida)

SMO Senior Medical Officer

SMO Servicio Militar Obligatorio (Spanish—Compulsory Military Service)

SMOA Ships Material Office—Atlantic

smog smoke + fog (*see* smaze); smoky air (with or without fog)

smogway smog-polluted automobile freeway

SMOH Society of Medical Officers of Health

smoker smoking car

smoketaz smoke-tinted topaz

Smokies Smoky Mountains between North Carolina and Tennessee

smokin' smoking

smokin' pot smoking marijuana

SMOM Sovereign Military

Order of Malta (claiming to be the world's smallest country, founded in 1048 before the first crusade and located at 68 Via Conditti in downtown Rome where in 1981 it reported a population of 80)

smon subacute myelo-optic neuropathy

SMOOSA Save Maine's Only Official State Animal (the moose)

smoothies smooth ones

SMOP Ships Material Office—Pacific

SMOPS School of Maritime Operations

smor standard mean ocean water

smörgas smörgåasbord (Swedish appetizers or delicatessen-style meal)

smorz smorzando (Italian—dying away)

smotherlove smothering mother love

smow standard mean ocean water

smp scanning measuring projector; social marginal productivity; sound motion picture(s)

smp (SMP) special multi-peril (insurance) policy

s.m.p. sine mascula prole (Latin—without male issue)

SMP Science Manpower Project; School Mathematics Program; Society of Mural Painters; St Martin's Press

SMP Soviet Military Power (published by the Pentagon); *Suomen Masseudun Puloue* (Finnish Rural Party)

SMPC Saint Mary of the Plains College

SMPR Supply and Maintenance Plan and Report

smps switched-mode power supply

SMPS Society of Master Printers of Scotland

SMPSD Systematic Management Plan for School Discipline

SMPTE Society of Motion Picture and Television Engineers

smpx smallpox

smr somnolent metabolic rate; standard mortality rate; submucous resection

smr (SMR) standard Malaysian rubber

SMR Student Master Record; South Manchurian Railway

SMŘ Sa Majesté Royale (French—Her/His Royal Majesty)

SMRA Spring Manufacturers' Research Association

SMRC South Manchurian Railway Company

smrd spin-motor rotation-detector

SMRE Safety in Mines Research Establishment

SMRI Sugar Milling Research Institute

SMRL Submarine Medical Research Laboratory

SMRMIS Supply, Maintenance, and Readiness Management Information System

sms silico-manganese steel; subject matter specialist; synchronous meteorological satellite (SMS)

sm's (SMs) submarines

SMs subway musicians

SMS Sacramento Medical Society; Sequence Milestone System; Software Monitoring System

SMS Seine Majistäts Schiffe (German—His Majesty's Ship)

smsa standard metropolitan statistical area

SMSA Standard Metropolitan Statistical Area (any city and surrounding suburbs with a population of 50,000 or more)

SMSB Strategic Missile Support Base

SMSG School Mathematics Study Group

SMSgt Senior Master Sergeant

SMSO Senior Maintenance Staff Officer

SMSP Spring Mill State Park (Indiana)

SMSSS Sheet Metal Screw Statistical Society

smstrs seamstress

smt ship's mean time

Smt Summit

S^mt Seamount

SMT Scottish Motor Traction; Shipboard Marriage Test; Stabilized March Technique; System Maintenance Test

SMTA Scottish Motor Trade Association

SMTF Scottish Milk Trade Federation

smti selective moving target indicator

SMTO Senior Mechanical Transport Officer

SMTRB Ship and Marine

Technology Requirements Board

SMTS Scottish Machinery Testing Station

SMTWTFS Sunday, Monday, Tuesday, Wednesday, Thursday, Friday, Saturday

SMU Southern Methodist University

SMUD Sacramento Municipal Utility Department; Sacramento Municipal Utility District

Smu Gul Smuggler's Gulch (Monument Road, San Diego, California—the last road in the southwestern United States)

SMUN Soviet Mission to the United Nations

SMUP Southern Methodist University Press

SMUSE Socialist Movement for the United States of Europe

smust smoke + dust

smutcom smut communication

smw standard metal window

SMW Society of Magazine Writers; Society of Military Widows

SMWIA Sheet Metal Workers International Association

SMWIU Steel and Metal Workers Industrial Union

smx serial microxerography; submultiplexer unit

smx (SMX) sulphamethoxazole

sn sanitation; sanitary; service number; solid neutral; stock number; supernovae

sn (Sn) snow

s/n serial number; service number; signal-to-noise ratio

s-n sin número (Spanish—unnumbered, without number)

s.n. secundum naturam (Latin—according to nature); *sine nomine* (Latin—without name)

Sn Sn; Santa, Santo; *stannum* (Latin—tin)

S_n labor supply (macroeconomics)

S^n San (Spanish—saint)

SN Sacramento Northern (railroad); Scientific Note; Scope Note; Secretary of the Navy; Serial Number; Service Number; Standard Oil (stock exchange symbol)

S/N Serial Number; Service Number; Shipping Note; stress versus number of cycles (to failure); successes

versus total number of trials
S of N Sons of Norway
SN Sûreté Nationale (French—National Security)—law-enforcement agency
S-N stress versus number of cycles
sna systems network architecture
SNA Society of Naval Architects; Suburban Newspapers of America; System of National Accounts (UN); Systems Network Architecture
SNAC Syndicat National des Auteurs et Compositeurs (French—National Union of Authors and Composers)
SNACS Share News on Automatic Coding Systems; Society for the North American Cultural Survey
snafu situation normal, all fouled up; situation normal—all fucked up
SNAI Standard Nomenclature of Athletic Injuries
SNAM Società Nazionale Metanodotti (Italian—National Natural Gas Company)
SNAME Society of Naval Architects and Marine Engineers
snap. simplified numerical automatic processor; simplified numerical automatic programmer; subroutine(s) for natural actuarial processing
SNAP Shelter Neighborhood Action Project; Society of National Association Publishers; Society of National Publications; Space Nuclear Auxiliary Power; Student Naval Aviation Pilot; Suffolk Network on Adolescent Pregnancy; Systems for Nuclear Auxiliary Power
Snapp Serviços de Navegação da Amazonia e de Administração do Porto do Pará (Portuguese—Amazon Navigation and Administration of the Port of Pará)
snapper(s) snapping turtle(s)
snappies snappy stories
snap(s) snapshot(s)
snark snake and shark (Lewis Carroll)
snc severe noise environment; standard navigation computer
SNC Sistema Nervioso Central (Spanish—Central Nervous System); *Société Navale Caennaise (Lamy et Cie)* (French—Caen Naval Soci-

ety
SNCASCO Société Nationale de Constructions Aéronautique de l'Ouest (French—National Society of Western Aeronautical Construction)
SNCC Student Nonviolent Coordinating Committee (also called SNIC)
SNCFB Société Nationale des Chemins de Fer Belges (Belgian State Railways)
SNCFF Société Nationale des Chemins de Fer Français (French—State Railways)
snd sound
SND Society of Newspaper Design
SNDA Sunday Newspaper Distributing Association
SNDO Standard Nomenclature of Diseases and Operations
sndp sin nota de precio (Spanish—without indication of price)
sndv (SNDV) strategic nuclear delivery vehicle
SNE Society for Nutrition Education
SNEA Student National Education Association
sneaks. sneakers (tennis shoes)
SNEC Sub-Group on Nuclear Export Coordination
SNECMA Société Nationale d'Etude et de Construction de Moteurs d' Aviation (French—National Air Motors Research and Construction Company)
SNEMSA Southern New England Marine Sciences Association
SNEP Saudi Naval Expansion Program
snf solids-non-fat
SNF Serbian National Federation; Short-range Nuclear Forces; Skilled Nursing Facility
SNFA Standing Naval Force, Atlantic
SNFCC Shippers National Freight Claim Council
SNFU Scottish National Farmers' Union
sng synthetic natural gas
sng sans notre garantie (French—without our guarantee)
Sng Singapore
sngl single (flow chart)
SNHM Stanford Natural History Museum

sni sequence-number indicator
SNI San Nicolas Island; Selective Notation of Information; Selective Notification of Information; Sports Network Incorporated
SNI Secretariado Nacional da Informação (Portuguese—State Tourist Bureau); *Syndicat National des Instituteurs* (French—National Union of Teachers)
SNIC Student Non-Violent Coordinating Committee (SNCC)
SNICER State of Nebraska Information Center for Educational Resources
SNIE Special National Intelligence Estimate
SNIEs Special National Intelligence Estimates
sniffex sniffer exercise
snirt snort of laughter
SNL Sandia National Laboratories; Singapore National Library; Standard Nomenclature List
SNL Science News Letter
snlr services no longer required
SNLS Society for New Language Study
snlv (SNLV) strategic nuclear-launch vehicle
snm signal-to-noise merit; special nuclear material(s)
snm sobre el nivel del mar (Spanish—above sea level)
SNM Saguaro National Monument (Arizona); Senior Naval Member; Sitka National Monument (Alaska); Society of Nuclear Medicine
SNMT Society of Nuclear Medical Technologists
SNN Shannon, Eire (airport)
sno snow
s no serial number
SNO Scottish National Orchestra; Senior Naval Officer; Senior Nursing Officer; Singapore National Orchestra
snob *sine nobilitate* (Latin—without nobility)—anyone trying to outdo the manners and style of the nobility
SNOB Senior Naval Officer on Board
snobol string-oriented symbolic language
snoe smart noise equipment
snok secondary next of kin
Snooks surname contracted from Seven Oaks
SNOOP Students Naturally

Opposed to Outrageous Prying
snoopervise snoop and supervise
SNOP Standard Nomenclature of Pathology
snorkex snorkel exercise
SNORT Supersonic Naval Ordnance Research Track
Snowys Snowy Mountains of New South Wales
Snowy Scheme Snowy Mountains Scheme (Australian hydroelectric and irrigation system)
snp soluble nucleoprotein
SNP Salorp National Park (Thailand); Scottish Nationalist Party; Sebakwe NP (Rhodesia); Sequoia NP (California); Serengeti NP (Tanzania); Shenandoah NP (Virginia); Sivpuri NP (India); Sitka NP (Alaska); Snowdonia NP (Wales); Swiss NP (Switzerland)
SNPA Scottish Newspaper Proprietors' Association; Southern Newspaper Publishers Association
SNPO Space Nuclear Propulsion Office
snr signal-to-noise ratio
Snr Senhor (Portuguese—Mister)
Sñr Señor (Spanish—Mister)
SNR Society for Nautical Research
Snra Senhora (Portuguese—Missus)
Sñra Señora (Spanish—Missus)
SNRA Sanford National Recreation Area (Texas)
Snro Senhoro (Portuguese—Mister)
Snrta Senhorita (Portuguese—Miss)
Sñrta Señorita (Spanish—Miss)
Sñrto Señorito (Spanish—Master)
sns sympathetic nervous system
SNS Senior Nursing Sister
SNSC Scottish National Ski Council
S'n Simons Saint Simons Island off the coast of Brunswick, Georgia
SNSN Standard Navy Stock Number
SNSO Superintending Naval Stores Officer
SNSP Syrian National Social Party

SNSPS Scottish National Sweet Pea Society
snt sealant
snt so nota (Japanese—and so forth)—etc.
Snt Santander
SNT Society for Nondestructive Testing
snto spinning tool
SNTO Spanish National Tourist Office; Swedish National Tourist Office; Swiss National Tourist Office
SNTPC Scottish National Town Planning Council
SNUPPS Standardized Nuclear Unit Power Plant System
SNVBA Scottish National Vehicle Builders Association
SNVDO Standard Nomenclature of Veterinary Diseases and Operations
SNW Symphony of the New World
SNWMA Stillwater National Wildlife Management Area (Nevada)
SNWR Sabine National Wildlife Refuge (Louisiana); Sacramento NWR (California); Santee NWR (South Carolina); Savannah NWR (South Carolina); Seedskadee NWR (Wyoming); Seney NWR (Michigan); Sherburne NWR (Minnesota); Shiawasse NWR (Michigan); Slade NWR (North Dakota)
snwt steel non-watertight
Sn Ysdr San Ysidro
so shift out; soiled
so (SO) shift-out character (data processing)
so. seller's option; senior officer; sex offender; shipping order; ship's option; shop order; show off; south(ern); special order; staff officer; standing order; strikeout; suboffice; supply office(r)
s-o shutoff
s/o shipping order; solvent-to-oil (ratio); son of
so. siehe oben (German—see above)
s/o su orden (Spanish—your order)
So. Somali(a)
So Sondag (Danish—Sunday)
SO Scottish Office; Scottish Opera; Scouting-Observation (naval aircraft); Seattle Orchestra; Secretary's Office; Senior Officer; Shipment Order; Shipping Order; Shop Order; somalo (Somalian

currency unit); Southern Airways (letter coding); Southern Company (stock exchange symbol); Special Order(s); Staff Officer; Standard Oil; Standing Order(s); Stationery Office; Supply Office(r); Steward Observatory
SO (I) Staff Officer (Intelligence)
SO (O) Staff Officer (Operations)
S/O Station Officer
SO Staatsoper (German—State Opera); *sudoeste* (Spanish—southwest); *Südösten* (German—southeast); *Suroeste* (Spanish—southwest)
SO₂ sulfur dioxide
SO₄ sulfate
soa speed of advance; speed of approach; state of the art
SOA Seattle Opera Association; Shoe Corporation of America (stock exchange symbol); Staff Officer, Administration
soaa state-of-the-art advancement
SOAA Solus Outdoor Advertising Association
SOAD Staff Officer—Air Defence
So Afr South Africa(n)
soap. symbolic optimum assembly programming
SOAP Society of Airway Pioneers; Student Opportunity and Access Program
SOAPD Southern Air Procurement District
soaps. suction, oxygen, apparatus, pharmaceuticals, saline (anesthetist's mnemonic for checking equipment)
soap(s) soap opera(s)
SOAR Save Our American Resources; Society of Authors' Representatives
SOAs Separate Operating Agencies
SOAS School of Oriental and African Studies (University of London)
SOASIS Southern Ohio (chapter of) ASIS
sob. see order blank; shipped on board; shortness of breath; souls on board (passengers and crew aboard an aircraft); still on board; suboccipito-bregmatic
s-o-b son of a bitch
SOB Senate Office Building; State Office Building; Soci-

ety of Bookmen; son of a bitch

sobe sober; sobriety

SOBHD Scottish Official Board of Highland Dancing

soblin self-organizing binary-logic network

sob's silly old buggers; sons of bitches; souls on board

SOBs Sons of Bosses

SOBS Society for Office-Based Surgery

soc social; society; sociology; socket; state of consciousness (SoC)

Soc Socialist; Society; Socrates

Soc Sociedad (Spanish—society); *Sociedade* (Portuguese—society); *Società* (Italian—society); *Société* (French—society)

S o C Society of Cyprus

SOC Save Our Children; Scottish Ornithologists Club; Servicemen's Opportunity College; Southwestern Oregon College; Space Operations Center (Colorado Springs, Colorado); Special Operations Command (counterinsurgency forces); Stamp Out Crime

So Ca South Carolina's old abbreviation

SOCAD School of Crafts and Design

SOCAL Standard Oil of California (Chevron)

Soc An Société Anonyme (French—corporation)

SOCAP Society of Consumer Affairs Professionals (in business)

SOCARE Seniors Only Comprehensive Assessment and Retirement Evaluation

Soc. Chr. Societas Christi (Latin—Christian Society)

soc/d social death; sociological death

Soc-Dem Social-Democrat(ic) (Party)

SOCEM Save Our City from Environmental Mess; Society of Objectors to Compulsory Egg Marketing

Soc Fed Rep Jugo Socijalisticka Federativna Republika Jugoslavija (Serbo-Croatian—Socialist Federal Republic of Yugoslavia)

SOCGPA Seed, Oil Cake, and General Produce Association

Soc I Society Islands

Societies Society Islands of Polynesia in the South Pa-

cific

Socinus Laelius Socinus (Lelio Sozzini, Italian theologian and anti-Trinitarian whose nephew Faustus Socinus developed Socinianism, the forerunner of Unitarian-Universalism)

sociobio sociobiologic(al)(ly); sociobiologist; sociobiology

socioecol socioecologic(al)(ly); socioecologist; socioecology

sociol sociological; sociologist; sociology

Soc Isl Society Islands

socks. soccer teams

SOCMA Synthetic Organic Chemical Manufacturers Association

SOCMEU Special Operations Capable Marine Expeditionary Unit

Soc Mining Eng Society of Mining Engineers

Soc NC sociedad en nombre colectivo (Spanish—general partnership under a collective name)

So Co Southern Counties

SOCO Standard Oil Company of California

socom solar communication

SOCONY Standard Oil Corporation of New York

soc psych social psychology

SOCRATES System for Organizing Content to Review and Teach Educational Subjects

Socred Social Credit (party of Canada)

socrit social critic(ism)

socs survey of clerical skills

soc sci social science; social scientist

Soc Sec Social Security

sod. sodium; sodomite; sodomy

Sod West Virginia town named after its first postmaster—Samuel Odell Dunlap—SOD

S o D Society of Dilettanti

SOD Some of the Digits; Special Operations Division (CIA)

soda (SODA) source-oriented data acquisition

SODAC Society of Dyers and Colourists

sodar sound-detecting and ranging

soda water water charged with carbon dioxide (CO_2)

SODOMEI Nihon Rodo Kumiai Sodomei (Japanese—Japanese Trade Union Fed-

eration)

SODRE Servicio Oficial de Difusión Radio Eléctrica (Spanish—Uruguayan radio and tv network)

SoE Secretary of Energy

SOE Special Operations Executive (World War II British intelligence operation for rescuing scientists from Hitler)

SOED Shorter Oxford English Dictionary

SOEEA Saskatchewan Outdoor and Environmental Education Association

SOE/F SOE in France

SOEKOR Southern Oil Exploration (South Africa)

soep (SOEP) solar-oriented experiment package

sof sound on film

sof (SOF) succinic oxidase factor

Sof Sofia

SoF Soldier of Fortune

S o F Society of Friends

SOFA Socially Oriented For Action; Strongly Oriented For Action; Student Overseas Flights for Americans

SOFAA Society of Fine Art Auctioneers

sofar sound fixing and ranging

SOFCS Self-Organizing Flight-Control System

SOFINA Société Financiére de Transports et d'Entreprises Industrielles (French—Belgian investment syndicate)

sofnet solar observing and forecasting network

SOFRATOME Société Francaise d'Études et de Réalisation Nucléaires (French—French Society for Nuclear Study and Realization)

soft. signature of fragmented tanks

SOFT Status of Forces Treaty; Swedish Orienteering Federation

softech software technology

softlenses soft contact lenses

soft porn soft-core pornography

sog speed over (the) ground

sog sogenannt (German—so called)

SOG Seat of Government (Washington, D.C.); Special Operations Group

SOGAT Society of Graphical and Allied Trades

SOGC Society of Gynecologists and Obstetricians of Canada

SO & GC Signal Oil and Gas Company

sogg soggettivo (Italian—subjective); *soggetto* (Italian—subject)

soh (SOH) start of heading character (data processing)

soha soft hard

SOHIO Standard Oil of Ohio

SoHo South of Houston Street (New York City artist's colony in lower Manhattan)

SOHO Save Our Heritage Organization

SOHYO Nihon Rodo Kumiai Sohygikai (Japanese—Japanese General Council of Trade Unions)

soi space object identification

SOI Signal Operation Instruction(s); Southern Indiana (railroad); Specific Operating Instruction(s); Staff Officer—Intelligence

soit soitenly (New Yorkese—certainly)

SoJ Sea of Japan

sok sokak (Turkish—lane; street)

sol solar; soldier; solenoid; soluble; solubility; solution; solvent(s)

sol (SOL) shipowners liability; simulation-oriented language

s-o-l short of luck

sol. solutio (Latin—solution)

Sol Solicitor; Solomon; Solomon Islands

SoL Secretary of Labor; Solicitor of Labor

SOL Slightly Older Lesbians; Systems Optimization Laboratory

SOL Svenska Orient Line (Swedish—Swedish Orient Line)

SOLACE Sales Order and Ledger Accounting (using) Computerline Environment

SOLAR Semantically Oriented Lexical Archive; Shop Operations Load Analysis Report(ing)

SoLaS Safety of Life at Sea (international conference)

solb start of line block

sold. solder; soldering

Sol Gen Solicitor General

sol hgt solid height

sol htg solar heating

solidif solidification

Solid-State Commun Solid-State Communications

Solid-State Electron Solid-State Electronics

Solid-State Phys Solid-State

Physics

SOLINET Southeastern Library Network

solion solution of ions

Sol Isl Solomon Islands

SOLIT Society of Library and Information Technicians

Sol J Solicitors' Journal

SOLL Selma Ottiliana Louisa Lagerlöf

Sol(ly) Solomon

soln solution

solo. status of logistics offensive

SOLO System for Ordinary Life Operations

SOLog standardization of certain aspects of operations and logistics

sologs standardization of operations and logistics

solomon simultaneous-operation-linked ordinal modular network

Solomons Solomon Islanders; Solomon Islands (nation in the western Pacific)

Solovetskis Solovetski Islands (penal colonies in the Archangelsk Region of the USSR—part of the Gulag Archipelago)

Solovki Solovetski Islands

Sol Phys Solar Physics

solr solicitor

solrad solar radiation

solut solution

solv solvent

solv. solve (Latin—dissolve)

Solv Solveig

soly solubility

som serous otitis media; somatology; start of message

som (SOM) standoff missile

Som Somali(a); Somaliland(er); Somerset; Somerville College, Oxford

SoM School of Musketry

SOM Society of Occupational Medicine; Standing Group on Oil Markets

SOMA Sharing of Ministries Abroad; Society of Mental Awareness

Somal Somali(a)(n)—Somalia formerly British and Italian Somaliland

Somalia Somali Democratic Republic (East African nation)

somat somatic

somat or some (Greek—body)—centrosome; chromosome, somatic

SOME Senior Ordnance Mechanical Engineer

Somerset Somersetshire

SOMEX Sociedad Mexicana de Credito Industrial (Spanish—Mexican Industrial Credit Society)

som-h start of message—high precedence

som-l start of message—low precedence

Som LI Somerset Light Infantry

somm (SOMM) standoff modular missile

somnam somnambulant, somnambulating, somnambulation, somnambulism, somambulistic(al(ly), somnambulist(s)

SOMOS Society of Military Orthopedic Surgeons

Somos libres Spanish—We are free)—Peruvian national anthem

SOMPA System of Multicultural Pluralistic Assessment

SOMS Standing-Order Microfiche Service

Som sh Somali shilling

son. sonata

Son Sonora

Son Sonntag (German—Sunday)

S o N Spear of the Nation (armed force of the African National Congress)

SON Snijders-Oomen Nonverbal (intelligence scale)

sonac sonacelle (sonar nacelle)

SONAP Sociedade Nacional de Petroleos (Portuguese—National Petroleum Company)

sonar sound navigation and ranging

Sonbrit Simfonischen orkestur na bulgarskoto radio i televiziya (Bulgarian Radio and Television Symphony Orchestra)

SONDE Society of Non-Destructive Examination

SONGS San Onofre Nuclear Generating Station

Song Sol The Song of Solomon

Sonia Sophia

sonmc sonar countermeasures and deception

Sonn Sonnets of Shakespeare

Sonny George Bernard Shaw (1856–1950)

sono sonobuoy; sonography

sonoan sonic noise analyzer

SONPP San Onofre Nuclear Power Plant

son(s) sonata(s)
Sons Sonnets
SOO Staff Officer Operations
SO(O) Staff Officer (Operations)
SOOP Submarine Oceanographic Observation Program
soot. solar optical observing telescope
sop. sleeping-out pass; soprano; sum of products; surgical outpatient
s-o-p standard operating procedure
SOP Senior Officer Present; Standard Operating Procedure; Study Organization Plan
SOPA Senior Officer Present Afloat
SOPAC Southern Pacific; South Pacific
sop glock soprano glockenspiel
Soph Sophocles
SOPHE Society of Public Health Educators
soph(s) sophomore(s)
SOPL Save Our Public Libraries
SOPLASCO Southern Plastics Company
sop met soprano metallophone
Soppnata Sociedade Portuguese de Navios Tanques (Portuguese Tankers)
sop xyl soprano xylophone
sor sale or return; sequential occupancy rate; sorority; specific operating requirement(s)
s-o-r stimulus-organism-response
Sor Soerabaya; Sorong
Sor Señor (Spanish—Mister)
Sor Sênior (Portuguese—Mister)
SOR Sandia Optical Range; Special Order Request; Specific Operational Requirement
SORB Subsistence Operations Review Board
sord submerged object recovery device
SORD Southeastern Order Retrieval and Distribution Center
SORDID Summary of Reported Defects, Incidents, and Delays
Sores Señores (Spanish—gentlemen)
SORG Southern Operations Research Group
SORI Southern Research Institute
Sor Juana Sor Juana Inés de

la Cruz (1651–1695)
Sorlings Sorling Islands (Isles of Scilly)
SORO Special Operations Research Office
Sorolla Spanish painter Joaquin Sorolla y Bastida (1863—1923)
SORT Ship's Operational Readiness Test(ing); Slosson Oral Reading Test; Staff Organizations Round Table; Structured-Objective Rorschach Test
sorti satellite orbital track and intercept
sos same old stew; same old stuff; same only softer (musical direction); slag on a shingle (military description of creamed chicken or beef on a slice of toast)
s.o.s. si opus sit (Latin—if necessary)
sos sostenuto (Italian—sustained)
SoS Secretary of State; Source(s) of Supply
S-o-S Southend-on-Sea
S o S Society of Separationists (of church from state)
SoS Song of Solomon
SOs Sheriff's Offices
SOS Safety Observation Station; Save Our School(s); Save Our Shore; Senior Opportunities and Services; Share Our Spectacle(s); Ships Ordnance Summary; Squadron Officer School(ing); Stamp Out Smog; Student-Oriented Studies (National Science Foundation); Supervisor of Shipbuilding; Supplementary Ophthalmic Service(s)
SOS international distress signal—three dots, three dashes, three dots; popularly translated as meaning Save Our Souls
sosc safety observation station display console
SOSC Smithsonian Oceanographic Sorting Center; Source of Supply Code
So sh somali shilling(s)
SOSIAC Singapore-Soviet Shipping Agency
SOSS Shipboard Oceanographic Survey System
sost sostenuto (Italian—sustained)
Sost Sostavitel (Russian—compiler)
SOSTAC Scottish Industrial

Safety Training Advisory Council
SOSUS Sound and Surveillance System
sot. shower over tub; sound on tape
SoT Secretary of Transport(ation); Secretary of the Treasury; Sons of Temperance
SOT Solar Optical Telescope; Special Operations Team
sota state of the art
SOTA Statewide Organization of Third-world Artists
SOTAA State-of-the-Art Association
SOTAS Stand-Off Target-Acquisition System
sotd stabilized optical tracking device
SOTDAT Source Test Data System (EPA)
sotim sonic observation of the trajectory and impact of missiles
SOTO Society Of Transporter Owners
Soton Southampton
SOTP Ship(yard) Overhaul Test Program; System Overhaul Test(ing) Program
SoTT School of Technical Training
sotus (SOTUS) sequentially-operated teletypewriter universal selector (data processing)
Sou Southampton
SOU Southern Airways
Sou Afr South Africa(n)
Sou Amer South America(n)
Sou Aus South Australia(n)
Sound The Sound (Arctic straits in the Canadian sector such as Lancaster Sound, Smith Sound, Viscount Melville Sound; Long Island Sound; Block Island, Rhode Island, Nantucket, and Vineyard Sounds; North Carolina's Albemarle, Bogue, Currituck, and Pamlico Sounds; Sundet—also called Öresund between Denmark and Sweden)
soundamp sound amplification; sound amplifier
Sound River old name for New York City's East River—an extension of Long Island Sound linking the Sound with New York Bay, the Harlem River, and the Hudson
SOUP Students Opposed to

Unfair Practices

Sou Pac Southern Pacific

SOUR Stamp Out Urban Renewal

source the source (nickname for a baton with a rechargeable flashlight on one end and shock terminals on the other, used for jolting criminals with a 10,000-volt charge)

Source of the Sun Japan (called Nihon by the Japanese as it means Source of the Sun and is emblazoned on their flag)

Souse America booze-cruise destination of many alcohol addicts

soussa steady, oscillatory, and unsteady, subsonic, and supersonic aerodynamics

s/out sleep out (porch)

South southern American states from Virginia to Texas

South Africa Republic of South Africa (*Republiek van Suid-Afrika* in Afrikaans), formerly Union of South Africa

South African inhabitant of South Africa; Afrikaans language or people; South African Republic now the Transvaal

South African Commonwealth South Africa and its territories

South African Dominion South Africa and its territories

South-African Dutch Afrikaans

South African Ports (large, medium, and small from west to south to east) Walvis Bay, Luderitz, Cape Town, Simontown, Mosselbaai, Port Elizabeth, East London, Port St Johns, Durban

South Africa's Principal Port Cape Town

South Africa's Spine Drakensburg Mountains

South America islands and lands extending from Cape Horn to Colombia (Argentina, Bolivia, Brazil, Chile, Colombia, Ecuador, French Guiana, Guyana, Paraguay, Surinam, Uruguay, Venezuela)

South American Welfare State Uruguay

South America's Largest Country Brazil

South Arabia Southern Yemen

South Atlantic ocean between South America and Africa

South Atlantic States Delaware, Florida, Georgia, Maryland, North Carolina, South Carolina, Virginia, and West Virginia

South Britain England and Wales

South Carolina Port Charleston

South Carolina's Capital City Columbia

South Central States Arkansas, Louisiana, Oklahoma, Texas

South China Sea between Indochina, Indonesia, and Philippines

SouthCom Southern Command (U.S. Air Force, Army, and Navy bases straddling the Panama Canal for its protection)

Southeast southeastern United States (North Carolina to Florida, Atlantic Coast to Mississippi River)

South Eastern Region South Eastern Region Correctional Institute at Juneau, Alaska

Southeast Sun Belt Alabama, Arkansas, Florida, Georgia, Louisiana, Mississippi, North Carolina, South Carolina, Tennessee, and Virginia

South End Boston, Massachusetts slum

souther storm from the south

Southern Southern Railway

Southern Alplands Albania, France, Italy, Yugoslavia

Southern Alps mountain range on South Island of New Zealand

Southern California California south of the Tehachapis

Southern Colonies Virginia, Maryland, North Carolina, South Carolina, Georgia

Southern Cone Argentina, Chile, Paraguay, Uruguay

Southern Cross outstanding constellation of the Southern Hemisphere where it is emblazoned on the flags of Australia, Brazil, New Zealand, Papua New Guinea, the Solomon Islands, and Western Samoa as well as the state of Victoria in southern Australia

Southerns Southern Alps of New Zealand's South Island

South Ken South Kensington Imperial Institute (London's museum of science and industry)

South Orkneys South Orkney Islands in British Antarctica

South Sandwiches South Sandwich Islands

South Shetlands South Shetland Islands off British Antarctica

sou'wester southwester (waterproof oilskin hat and/or coat); southwestern wind

sov shutoff valve; special orientation visit

Sov Soviet; Sovietic; Soviets

SOVA Savant of Virginia

Soviet Arctic the USSR north of the Arctic Circle from Europe to Siberia

Soviet Baltics Estonia, Latvia, Lithuania

Sovinformburo Soviet Information Bureau

s-o vlv shutoff valve

Sov Medron Soviet Mediterranean Squadron

Sov muz *Sovetskaya muzyka* (Russian—Soviet music)

Sov strike attack by the Soviet Union

sow. sent on (their) way

SoW Statement of Work

SOW Sunflower Ordnance Works

SOWC Senior Officers War Course (UK)

SOWETO Southwestern Townships (South Africa)

SOWSD Statement of Work, Specifications, and Design

sox socks; solid oxygen; stockings

SoX School of Xerography (Xerox)

SOXAL Singapore Oxygen Air Liquids

Soyuz nerushi *Soyuz nerushi mi respublik svobod nikh* (Russian—Unbreakable Union of freeborn Republics) USSR national anthem

SOZ Soviet Occupied Zone

sp self-propelled; selling price; shear plate; single-phase; single-pole; single-purpose; small paper; smokeless powder; solid-propellant; space; spare; spare part; special; special paper; special performance (airliners); special propellant(s); special-purpose; specie; species; specific; speed; standard play; starting point; starting price;

static pressure; stop payment; summary plotter; summary programmed

sp (SP) space character (data processing); spelling; super-performance

s-p sequential-phase

s/p soft-point (bullet with lead core exposed to increase expansion)

s & p systems and procedures

sp sans prix (French—without price); *spanisch* (German—Spanish)

sp. species (Latin—species)

s.p. sine prole (Latin—without issue)

s p senza pedale (Italian—without pedal)

Sp Space (trailer-court address); Spain; Spanish; Spring(s)

Sp Spalten (German—column; division); Spanish; *Spanje* (Dutch—Spain); Spitz (German—point)—pointed high-velocity bullet; *Spitze* (German—point)

SP San Pedro, California; Sâo Paulo, Brazil; Scientific Paper; Section Control; Security Publication; Shore Party; Shore Patrol; Shore Police; Socialist Party; Society of Protozoologists; Sociolinguistics Program; Southern Pacific (railroad); Special Publication; Standard Practice(s); State Park; State Police; Strategic Plan(ning); subliminal perception; Submarine Patrol; subprofessional (civil service rating)

S-P Studebaker-Packard

S & P Standard & Poor's Corporation

S of P Society of Philaticians

SP Senterpartiet (Norwegian—Centrist party); *Socialdemokratiet Parti* (Danish—Social Democratic Party); *Sozialistische Partei* (German—Socialist Party)

S.P. Sanctissimus Pater (Latin—Most Holy Father); *Summus Pontifex* (Latin—Supreme Pontiff, the Pope)

Sp/1 Specialist, 1st class

Sp3c Specialist, third class

spa (SPA) stimulation-produced analgesia

spa. subject to particular average; sudden phase anomaly

S p A Società per Azioni (Italian—joint stock company)

SPA *Société Protectrice des*

Animaux (French—Society for the Protection of Animals); Salt Producers Association; School of Performing Arts; *Società per Azioni* (Italian—joint stock company); Society of Participating Artists; Society for Personnel Administration; Society of Philatelic Americans; Songwriters Protective Association; Software Publisher's Association; South Pacific Area; Southern Pine Association; Southwestern Power Administration; Standard Practice Amendment(s); State Principals Association; Systems and Procedures Association

SPAA Systems and Procedures Association of America

SPAAMFAA Society for the Preservation and Appreciation of Antique Motor Fire Apparatus in America

SPAB Society for the Protection of Ancient Buildings

spac spatial computer

SPAC Saratoga Performing Arts Center

S Pac South Pacific

S Pac Cur South Pacific Current

Spacenet III communications satellite network

SPACES Scheduling Package and Computer

spad (SPAD) space patrol air defense

SPAD Seafarers Political Activity Donation; Space Patrol Air Defense; Support Planning and Design

SPADETS Space Detection and Tracking System

Spag Spagnuolo [*see* Lad(s)]

SPAG Society for the Preservation of American Grandchildren

SPAI Screen Printing Association International

Spain Spanish State (Iberian nation once the center of an almost global colonial empire), *Estado Español*

spal stabilized platform airborne laser

spam spiced pork and meat (canned meat introduced during World War II)

SPAM Society for the Publication of American Music

spams spiced hams

SPAMS Ship Position and Altitude Measurement System

span. space navigation

Span Spanish

Span Spania (Norwegian—Spain)

SPAN Solar Particle Alert Network; South Pacific Action Network; System for Procurement and Analysis

SPANA Society for the Protection of Animals in North Africa

SPANC Society for St Peter the Apostle for Native Clergy

Spandan Spanish dance (castanets clipping, feet stomping, fingers snapping, guitars thumbing, hands clapping) originated in Andalucia

spandar space-and-range radar

Spanglish Spanish + English (Latin American mixture of the two tongues)

Spanish Africa Ceuta, Peñón de Vélez de la Gomera, Peñón de Alhucemas, Villa Sanjurjo, Mellila, Islas Chafarinas, Isla Alborán; formerly Spanish Guinea, Spanish Morocco, and Spanish Sahara

Spanish Creole localized Spanish spoken in many former Spanish colonies; person of Spanish ancestry

Span Neth Spanish Netherlands

span(s) spaniel(s)

SPANS Sealift Procurement and National Security

Spansule span + capsule (prepared so different drugs encapsulated are released at various times)

Spantran Spanish translation (programming language)

spar. (SPAR) space processing applications rocket; store port allocations register; submersible pipe-alignment rig; super-precision-approach radar

SPAR Seagoing Platform for Acoustics Research; Selection Program for ADMIRAL Runs (*see* ADMIRAL); Society of Photographer and Artists Representatives

sparc steam power automation and results computer

SPARC Space Program Analysis and Review Council

Sparks ship's radio operator

sparm (SPARM) sparrow anti-radiation missile

sparr steerable paraboloid altazimuth radio reflector (Jordrell Bank Radio-Telescope,

Cheshire, England)
SPARS Women's Coast Guard Reserve (from the Coast Guard motto, *Semper Paratus*—Always Ready)
SPARTAN Special Proficiency at Rugged Training and National Building (Green Beret training program); System for Personnel Automated Reports, Transactions, and Notices (NASA)
SPAS Societatis Philosophicae Americanae Socius (Latin—Fellow of the American Philosophical Society)
SPASM Society for the Prevention of Asinine Student Movements
spasur space surveillance
spat. self-protective antitank (weapon); silicon precision alloy transistor
spat. (SPAT) self-propelled anti-tank gun
SPAT Submarine Processing Action Team
SPATC South Pacific Air Transport Council
spats spatterdashes
spau signal processing arithmetic unit
S Pau São Paulo
Spauld Turn Spaulding Turnpike
spb special boiling point
SPB Space Science Board; Special Branch Policeman (British English—detective); State Personnel Board
spbd springboard
SPBF Scientific Peace Builders Foundation
spc salicylamide-phenacetin-caffeine; special fuel consumption; suspended plaster ceiling
SPC Service Processing Center (Immigration and Naturalization Detention Center); Society of Photographers in Communications; Society for the Prevention of Crime; Software Productivity Consortium; Solar Power Corporation (Exxon); South Pacific Commission; Space Projects Center; Standard Products Committee; State Planning Council; Subcontract Plans Committee
SPCA Society for the Prevention of Cruelty to Animals
spcat special category
SPCB Structural Pest Control Board

SPCC Ships Parts Control Center; Society for the Prevention of Cruelty to Children; Standardization, Policy, and Coordination Committee (NATO)
sp cd spinal cord
SPCH Society for the Prevention of Cruelty to Homosexuals
SPCK Society for Promoting Christian Knowledge
spcl special
SPCM Special Court-Martial
SPCMO Special Court-Martial Order
SPCO St Paul Chamber Orchestra; St Paul Civic Opera
spcr spacer
SPCs Suicide Prevention Centers; Suicide Prevention Clinics
Sp Cttee 24 Special Committee of 24 (United Nations' 24-member Special Committee concerning Granting Independence to Colonial Countries and Peoples)
SPCW Society for the Prevention of Cruelty to Women
spd separation program designator; ship pays dues; silicon photo diode; silver plated; speed; subject to permission to deal; surface potential difference
Spd Spandau
SPD Sales Promotion Department; South Polar Distance; *Sozialdemokratische Partei Deutschlands* (German—Social Democratic Party of Germany); System Program Director
spda single-premium deferred annuity
SPDC Spare Parts Distributing Center
sp del special delivery
spdl spindle
spdltr speedletter
sp dt single pole, double throw
spdtdb single-pole double-throw double-break (switch)
spdtncdb single-pole double-throw normally closed double-break (switch)
spdtno single-pole double-throw normally open (switch)
spdtnodb single-pole double-throw normally open double-break (switch)
spdtsw single-pole double-throw switch
SPDV Site Preliminary Design Validation (program)

spe special purpose equipment
spe (SPE) sucrose polyester
Spe San Pedro
SPE Society of Petroleum Engineers; Society for Photographic Education; Society of Plastics Engineers; Society for Pure English
SPEA Southeastern Poultry and Egg Association
SPEAK Society for Preserving and Encouraging Arts and Knowledge
SPEARS Satellite Photo-Electronic Analog Rectification System
SPEBSQSA Society for the Preservation and Encouragement of Barber Shop Quartet Singing in America
spec special(ly); specialty; specie; species; specific(ally); specification; specimen; spectacle; speculation; speech-predictive encoded communication(s)
's'pec' suspect
Spec Speculative Society (of debaters)
SPEC Society for Pollution and Environmental Control; South Pacific Bureau for Economic Cooperation; Systems and Procedures Exchange Center
spec appt special appointment
specat special category
spec emp specially employed
special. specialization; specialized
special ops special operations (assassinations and sabotage)
specif specific; specifically
specl specialist; specialize
specs specifications; spectacles
SPECS School Planning, Evaluation, and Communication Service
spect (SPECT) single photon-emission-computed tomography
SPECTRE Single-Pulse CO_2 Transient Experiment; Special Executive for Counterintelligence, Terrorism, Revenge, and Extortion (fictional organization created by Ian Fleming for his James Bond books)
spectrog spectrography
SPECTROL Scheduling, Planning, Evaluation, and Cost Control (USAF)
spectrophotom spectrophotometry
spectros spectroscopy

SPECWAR Special Warfare
SPEDE System for Processing Educational Data Electronically
S Pedro San Pedro
Speech Comm Assn Speech Communication Association
speed methamphetamine
speed. (SPEED) simplified profile enlargement from engineering drawing(s)
SPEED Signal Processing in Evacuated Electronic Devices; Systematic Plotting and Evaluation of Enumerated Data
speedalyzer automatic radar-controlled automotive-vehicle speed analyzer (for detecting speeders on byways and highways)
speedo speedometer
spef single-program-element fund(ing)
SPELL Society for the Preservation of the English Language and Literature
Spelman Spelman College
spelpat spelling pattern(s)
Spel Soc Am Speleological Society of America
Spen Spencer; Spencerian
Spence Spencer
Sperrins Sperrin Mountains of Northern Ireland
Sperry Sperry Rand Corporation
SPERT simplified program evaluation and review task (technique)
S Pete St Petersburg
Spett Spettabile (Italian—Dear Sir)
Spett ditta Spettabile ditta (Italian—Messrs)
Spezia La Spezia naval station near Genoa in northern Italy
spf sun-protection factor
s-p-f spruce-pine-fir
SPF Science Policy Foundation; Society for the Propagation of the Faith; South Pacific Forum
spf/db superplastic forming/diffusion bonding
sp fl spinal fluid
spg specific gravity; sponge; spring; sprung
spg (SPG) sex-hormone-binding globulin
Spg Spring
SPG Society for the Propagation of the Gospel; Special Patrol Group
SPGA Scottish Professional Golfers' Association

SPGB Socialist Party of Great Britain
Spgfld Springfield
spgg solid-propellant gas generator
sp gr specific gravity
Spgs Springs
SPGS Spare Guidance System
sph sphenoidal
SPH Special Psychiatric Hospital
sphd special pay for hostile duty
sp hdlg special handling
SPHE Society of Packaging and Handling Engineers
SP & HE Society of Packaging & Handling Engineers
sphen sphenodon (tuatara lizard); sphenoid; sphenoidal
spher spherical; spheroid
sp—hl sun present—horizon lost
SPHS Seward Park High School; Swedish Pioneer Historical Society
sp ht specific heat
sphyg sphygmic(al)(ly); sphygmogram(atic)(al)(ly); sphygmograph(ic)(ly); sphymography; sphygmoid; sphygmology; sphygmomanometer; sphygmomanometric(al)(ly); sphygmometer; sphygmometric(al)-(ly); sphygmophone; sphygmoscope; sphygmus
spi scientific performance index; ships plan index; solid propellant information; specific polarization index
spi (SPI) serum precipitable iodine
SPI Sisters of Perpetual Indulgence; Smoking Policy Institute; Society of Photographic Illustrators; Society of the Plastics Industry; Society of Professional Investigators; Southern Police Institute; Spanish Paprika Institute; Strategic Planning Institute; Superintendent of Public Instruction
SPI Secrétariats Professionnels Internationaux (French—International Professional Secretariats); *Service Pédagogique Interafricain* (French—Inter-African Teaching Service)
spia single-premium insurance annuity
SPIA South Pacific Island Airways
SPIB Society of Power Indus-

try Biologists
spic ship position-interpolation computer
SPIC Society of the Plastics Industry of Canada; Society for the Promotion of Identity on Campus
spicbm (SPICBM) solid-propellant intercontinental ballistic missile
SPICE Scientific Personal Interactive Computing Environment; Spacelab Payload Integration and Coordination in Europe
spid submerged portable inflatable dwelling
spidac specimen input to digital automatic computer
SPIDR Society of Professionals in Dispute Resolution
spids sensor personnel intrusion devices
spie self-programmed individualized education
SPIE Society of Photographic Instrumentation Engineers
SPIL Society for the Promotion and Improvement of Libraries
SPIN Searchable Physics Information Notes; Searchable Physics Information Notices; Submarine Program Information Notebook
sp. indet. species indeterminata (Latin—species indeterminate)
spindex selective permutation index(ing)
SPIndex Subject Profile Index (ABC-Clio's innovative new indexing system)
spinel magnesium aluminum oxide
sp. inquir. species inquirendae (Latin—species of doubtful status)
spins. special inquiries (FBI)
SPINSTRES Spencer Information Storage and Retrieval System
spintcomm special intelligence communication(s)
spip special position identification pulse
s'pipe standpipe
spir spiral
spir. spirituoso (Italian—spirited)—animated; *spiritus* (Latin—spirits)
Spirals Spiral Tunnels of the Canadian Pacific in Yoho National Park
spire. space inertial reference equipment

SPIRES Standard Personnel Information Retrieval System
SPIRGs Student Public Interest Groups
Spirid Spiridione
spirit. sales processing interactive real-time inventory technic
spirit spiritoso (Italian—spirited)
Spirit Spiritualism
spirt solar-powered isolated radio transceiver
spis service packaging instruction sheet
spis spissus (Latin—dried)
spit. selective printing of items from tape
Spit Spithead Channel joining The Solent and Southampton Water between the Isle of Wight and Portsmouth
spital (Early English contraction—hospital)
Spits Spitalsfields, England; Spitsbergen Islands in the Norwegian Arctic
spiu ship position-interpolation unit
spiw special-purpose infantry weapon
SPJ Society of Professional Journalists (Sigma Delta Chi)
SPJC Saint Petersburg Junior College
spk speckled
Spk Spokane
Sᵖᵏ Seapeak
SPK Staatsbibliothek Prevssicher Kulturbesitz (German—Prussian Culture Treasure State Library)—Berlin
spklr(s) sprinkler(s)
spkr speaker
spl simplex; sound pressure level; special; spelling
s.p.l. sine prole legitima (Latin—without legitimate offspring)
Spl Sevastopol
SPL Sacramento Public Library; Saskatoon Public Library; Scan Pacific Line; Seattle Public Library; Space Programming Language; Spokane Public Library; Springfield Public Library; Syracuse Public Library
SPLA Sudan People's Liberation Party
splad (SPLAD) self-propelled light air-defense gun
SPLAN School Organization Budget-Planning System
SPLASH Special Program to List Amplitudes of Surges

for Hurricanes
SPLC Southern Poverty Law Center; Standard Point Location Code
splcf sustained-peak low-cycle fatigue
splf simplification
SPLIT Sundstrand Processing Languages Internally Translated
SPLMPR State Public Library of the Mongolian People's Republic (Ulan-Bator)
splsm single-position letter-sorting machine
spm self-propelled mount(ing); sequential processing machine; set program mask; single-point mooring; source program maintenance; strokes per minute
s.p.m. sine prole mascula (Latin—without male issue)
SPM Saint-Pierre et Miquelon
SPM Scuola Professionale Marittima (Italian—Professional Maritime School)
SPMA Sewage Plant Manufacturers' Association
Sp Mor Spanish Morocco
SPMRL Sulfite Pulp Manufacturers' Research League
SPMS System Program Management Surveys
SPMU Society of Professional Musicians in Ulster
sp/mva status post motor vehicle accident
spn sponsor; spoon; stop-press news
sp. n. species nova (Latin—new species)
Spn Spain; Spaniard; Spanish
SPN Saipan, Trust Territory of the Pacific (airport); Satellite Program Network; Separation Program Number; Student Practical Nurse
SPNB Security Pacific National Bank
SPNB&S Solitary, Poor, Nasty, Brutish & Short (legal counsel of *The American Spectator*)
SPNC Society for the Promotion of Nature Conservation
SPNEA Society for the Preservation of New England Antiquities
SPNI Society for the Protection of Nature in Israel
SPNM Society for the Promotion of New Music
sp. nov. species novum (Latin—new species)
SPNR Society for the Promo-

tion of Nature Reserves
Spn Riv Spoon River in central Illinois
SPNS Standard Product Numbering System
SPNWR Salt Plains National Wildlife Refuge (Oklahoma)
spo sausages, potatoes, and onions
S Po São Paulo
SPO Sea Post Office; Senior Press Officer; Site Program Office(r); Special Project(s) Office; Staff Planning Office(r); System Program Office(r)
SPO Socialistische Partei Osterreichs (German—Austrian Socialist Party)
spoc single-point orbit calculator
SPOE Society of Post Office Engineers
SPOIE Society of Photo-Optical Instrumentation Engineers
Spoke Spokane, Washington
spoke(s). spokesperson(s)
Spol Fest USA Spoleto Festival USA (Charleston, SC)
spont spontaneous
SPOOK Supervisory Program Over Other Kinds
spool. simultaneous peripheral operation on-line
Spoon River Spoon River Anthology by Edgar Lee Masters
spoorw spoorwegen (Dutch—railway car)
S por A Sociedad por Acciones (Spanish—limited liability company)
Sporades Sporades Islands
Spore Singapore
spork spoon + fork (combination utensil)
spork(s) spoon-shaped fork(s)
sport. sporting; sportsman; sportsmanship; sportswoman
sportscast(er) sports broadcast(er)
s'pose suppose
spot. spotlight
spots spotlights
spp species; surplus personal property
spp. species (Latin—two or more species) singular is *sp.*
SPP Southern Pacific Properties; Suicide Prevention Program; System Package Program
SPPA Society for the Preservation of Poultry Antiquities
SPPL St Paul Public Library; St Petersburg Public Library

sppo scheduled program printout

SPPPQCT Society for the Protection, Preservation, and Propagation of the Queensland Cane Toad

spps stable plasma protein solution (SPPS)

Sp Pt Sparrows Point

spqr small profits and quick returns

S.P.Q.R. Senatus Populusque Romanus (Latin—the Senate and People of Rome)

spr solid-propellant rocket (SPR); spring

spr (SPR) strategic petroleum reserve

Spr Spring; Springfield; Spruce

SPR Simplified Practice Recommendation(s); Society for Pediatric Research; Society for Psychical Research; solid-propellant rocket; Special Project Report; Strategic Petroleum Reserve; Supplementary Progress Report

sprat. small portable radar torch

spr Bog springender Bogen (German—bouncing bow)

SPRC Society for the Prevention and Relief of Cancer

SPRD Science Policy Research Division (Library of Congress)

sprdng spreading

SPRDO Service Parts Repairable Disposition Order

spre siempre (Spanish—always)

SPRE Society of Park and Recreation Educators

spread. spring evaluation analysis and design

SPREd Society of Picture Researchers and Editors

SPRI Scott Polar Research Institute

Spring Bank Spring Bank Holiday (last Monday in May in Great Britain)

Springs The Springs (Palm Springs, California)

sprint (SPRINT) solid-propellant rocket-intercept missile

SPRITE Sequential Polling and Review of Interacting Teams of Experts

sprklg sparkling; sprinkling

SPRL Société de Personnes à Responsibilité Limitée (French—limited company)

spr's small parcels and rolls

Sprs Springs

SPRs Strategic Petroleum Reserves

SPRS Sate Police Radio System (South Dakota)

sps ship program schedule; student-paced statistics; super proton synchrotron (for smashing atoms)

sps (SPS) service propulsion system

s.p.s. sine prole supersite (Latin—without surviving issue)

SpS Special Services

SPs Shore Patrol vans; single-premiums (life insurance)

SPS Scottish Painters' Society; Service Propulsion System; Society of Pelvic Surgeons; Society of Plastic Surgeons; Society of Saint Patrick; Southwestern Public Service; Special Public School; Spokane, Portland & Seattle (railroad); Standard Pressed Steel; Steam Power Systems; String Process System; Submerged Production System; Symbolic Programming System; System of Procedure Specifications

SP & S Spokane, Portland & Seattle (railroad)

SPSA Senate Press Secretaries Association

SPSC Scottish Prison Service College

SPSE Society of Photographic Scientists and Engineers

SPSHS Stanford Profile Scales of Hypnotic Susceptibility

SPSI Senate Permanent Subcommittee on Investigations

SPSL Society for the Protection of Science and Learning

SPSO Senior Principal Scientific Officer

SPSS Statistical Package for the Social Sciences

spst single-pole single-throw (switch)

SPST Symonds Picture-Story Test

spstnc single-pole single-throw normally closed (switch)

spstno single-pole single-throw normally open (switch)

spstsw single-pole single-throw switch

spt seaport; soldered piezoelectric transducer; strength-probability-time; support

spt. spiritus (Latin—alcohol; spirits)

Spt Split (Yugoslavia)

sptc specified period of time contract

sptg sporting

sptl (SPTL) superconducting power transmission line

SptL support line

SPTL Society of Public Teachers of Law

sptr spectrum

sptt single-pole triple-throw (switch)

spu swimmer propulsion unit

SPUC Society for the Protection of Unborn Children

spud speech perception under distraction/distortion

spud. solar power unit demonstrator

SPUD St Paul Union Depot

SPUK Special Projects—United Kingdom

SPUR Space Power Unit Reactor; Special People United to Ride

SPURT Short Public Responsibility Theory

spurv self-propelled underwater research vehicle

sputnik iskustvennyi sputnik zemli (Russian—artificial fellow-traveler around the earth)—Soviet satellite launched October 4, 1957

SPV Society for the Prevention of Vice

SPVA Self-Propelled Vehicles Association

SPVD Society for the Prevention of Venereal Disease

Sp Vly Spring Valley

SPW Sillonian Plant Watchers; Society for the Protection of Whitey; Society of Protestant Wardens

SpWAfr Spanish West Africa

spwl single-premium whole life (insurance)

SPWLA Society of Professional Well Log Analysts

spx simplex(ed); stepped piston crossover

spx circuit simplex circuit (data processing)

Spz Spezia

sq squadron; square; stereo-quadraphonic; superquick

sq. sequens, sequentia (Latin—what follows; result; sequel)

Sq Square

SQ stereo-quadraphonic (discs and recordings)

SQ Secondo Quantità (Italian—according to the quantity consumed)—menu abbreviation

sq3r survey, question, read, review, recite (psychological

sequence)
sqa stereo-quadraphonic amplifier
sqc self-quenching control; statistical quality control
sq cell ca squamous cell carcinoma
sq cm square centimeter(s)
SQCP Statistical Quality Control Procedure
sqd squad
sqdc special quick-disconnect coupling
Sqdn Ldr Squadron Leader
sq ft square foot (feet)
sq hd square head
sq in. square inch (inches)
sq km square kilometer
SQL Standard Query Language
sq m square meter; square mile
SQMS Staff Quartermaster Sergeant
sqn squadron
Sqn Ldr Squadron Leader
SqNP Sequoia National Park
Sq O Squadron Office(r)
SQP San Quentin Prison (California)
sqr square; square root; supplier quality rating
SQR Site Qualification Report
sq rd square rod
sq rt square root
sq's stereo-quadraphonic recordings; stereo-quadraphonic records
SQS Stochastic Queuing System; Supplier Quality Services
Sqs SM Squadron Sergeant-Major
sqt square rooter
SQT Ship Qualification Test (USN)
squa squamoid; squamous
squak squall and squeal
square symbol of four corners of the earth; four points of the compass; male symbol; quadrature; slang term for s o m e -
one with unsophisticated tastes
squares'l square-sail
s quark strange quark
squarson squire + parson
squidsan squid-cutlet sandwich
squidwich squid-cutlet sandwich
SQUIRE System for Quick Ultra-fiche-based Information Retrieval
'squitoes mosquitoes
sq yd square yard

sr scaling ratio; scientific research; sedimentation rate; selective ringing, semi regular; sensitization response; separate rations; sex ratio; shipment request; short range; sigma reaction; single-reduction (geared turbine); sinus rhythm; slow release; sound ranging; spares requirement; split ring; *srovnej* (Czech—compare); standard range (aviation landing); steradian; stimulus response
sr (SR) saturable reactor; surveillance radar
s/r (S/R) safety representative
s/R sur Rhone (French—on the Rhone)
Sr Saudi Arabia; Saudi Arabian; Senior; strontium
Sr Señor (Spanish-mister, sir);
Sredniy (Russian—mid; middle)
Sr Sønder (Danish—southern);
Söndre (Swedish—southern)
SR saturable reactor; Savannah River (Operations Office); Scientific Report; Scottish Rifles; Seaman Recruit; seaplane reconnaissance (naval aircraft); Section Report; Senate Resolution; Senior Registrar; Service Record; Service Report; Shipping Receipt; Simulation Report; Signed Road; Society of Radiologists; Society of Rheology; Sons of the Revolution; Sound Report; Southern Railway; Special Regulation(s); Special Report; Specification Requirement(s); Staff Report; Standardization Report; Star Route (rural postal delivery); *S t a t s j a n s t e m a n n e n s Riksforbund* (Swedish—National Association of Salaried Government Employees, Sweden); Status Report; Study Requirement; Summary Report; Supporting Research; surveillance radar; *Sveriges Radio* (Swedish radio broadcast network); Swissair
S-R Saunders-Roe; stimulus-response; Schopper-Riegler (paper-pulp scale)
SR Saudi Arabian riyal (currency unit)
SR-71 Lockheed Blackbird jet reconnaissance aircraft
Sr^{85} radioactive strontium
sra sulforicinuleic acid
sra (Sra) sierra

Sra Señora (Spanish—Missus; Mistress)
SRA Science Research Associates; Screw Research Association; Society of Residential Appraisers; Special Refractories Association; Spelling Reform Association; Station Representatives Association
SRAA Senior Army Advisor
SRAB Sveriges Radio AB (Swedish Broadcasting Corporation)
srac short-run average-cost curve
srac (SRAC) short-run average cost
SRAC Social Research Applications Corporation
Sra D^{na} Señora Doña (Spanish—Lady Madam)
s'raight straight
sram static random access memory
sram (SRAM) short-range attack missile
sran short-range aids to navigation
Sranangtong Sranangtongo Surinam dialect spoken by former slaves
Sras Señoras (Spanish—ladies)
SRAs Senior Resident Agents
srats (SRATS) solar radiation and thermospheric structure (satellite)
sr auth senior author
srb selective reenlistment bonus
srb (SRB) short-range booster; solid-rocket booster
srbc sheep red-blood cell
SRBC Susquehanna River Basin Compact
Srb-Crt Serbo-Croat (Yugoslavian)
srbm (SRMB) short-range ballistic missile
srbp synthetic resin-bonded paper
src sample return container; solvent-refined coal
SRC Science Research Council; Signal Reserve Corps; Southern Regional Council; Southwest Research Corporation; Space Research Corporation; Standard Requirements Code; Strict Regime Camp (for Soviet prisoners); Sul Ross State College; Swiss Red Cross
SRC Santa Romana Chiesa (Italian—Holy Roman Church)

srcc strikes, riots, and civil commotions
SRCD Society for Research in Child Development
s-r cells sensitization-response cells
srch search (computer)
srcr sonar control room
SRCs Strict-Regime Camps (Soviet imprisonment centers)
SRCS Special Reverse Charge Service
srd single radial diffusion
Sr D Señor Don (Spanish—Sir Mister)
SRD Secret Restricted Data; Society for the Right to Die; State Registered Dietician; Systems Requirements Definition
SRD Standard Rate and Data
SRDA Scottish Retail Drapers Association
SRDC Standard Reference Data Center
SRDE Signals Research and Development Establishment
Sr Dr Señor Doctor (Spanish—Mister Doctor)
SRDS Standard Reference Data Service
SRDT Single Radial Diffusion Test
sre single-round effectiveness; single-round effectivity
Sre Sreda (Russian—Wednesday)
SRE Society of Reproduction Engineers
S.R.E. Sancta Romana Ecclesia (Latin—Holy Roman Church)
SR EB Southern Regional Education Board
Sr Ed Senior Editor
SRED Scientific Research and Experiments Department
SREEC Southern Regional Environmental Education Council
SREL Savannah River Ecology Laboratory
srem sleep with rapid eye movements
Sres Señores (Spanish—Messrs)
srev slow reverse
srf self-resonant frequency; semi-reinforced furnace; solar radiation flux; stable radio frequency; submarine range finder; supported ring frame; system recovery factor
SRF Self-Realization Foundation; Ship Repair Facility

(USN)
srf black semireinforcing furnace black
srg sound ranging
SRG System Review Group
SRGM Solomon R Guggenheim Museum
srh single radial hemolysis
SRHE Society for Research into Higher Education
SRHL Southwestern Radiological Health Laboratory
Sr HS Senior High School
sri servo repeater indicator; silicone rubber insulation; spectrum resolver integrator; surface roughness indicator
Sri Sri Lanka (Ceylon); Srinagar, India
SRI Scientific Research Institute; Southern Research Institute; Southwestern Research Institute; Space Research Institute; Stanford Research Institute
SRI Sacro Romano Impero (Italian—Holy Roman Empire)
Sria Secretaria (Spanish—secretariat)
srif somatotropin release-inhibiting factor
Sri Lan Sri Lanka (Singhalese—Resplendent Land)—Ceylon
Sri Lanka Republic of Sri Lanka (Asian island off India's southern tip)
Sri Lan Pra Sam Jan Sri Lanka Prajathanthrika Samajavadi Janarajaya (Sinhala—Democratic Socialist Republic of Sri Lanka)—formerly Ceylon
SRILTA Stanford Research Institute Lead Time Analysis
srim selected research in microfiche
Sri Nep Sar Sri Nepala Sarkar (Nepali—Kingdom of Nepal)
SRINF Shorter Range Intermediate-Range Nuclear Force
Srio Secretario (Spanish—Secretary)
SRIS Safety Research Information Service; School Research Information Service
srj self-restraint joint; static round jet
SRJC Santa Rosa Junior College
srl (SRL) systems reference library
Srl Sorel
SRL Savannah River Laboratory; Save-the-Redwoods

League; Science Reference Library (Chancery Lane, London); Scientific Research Laboratory; Study Reference List
SRL Saturday Review of Literature; sociedad de responsabilidad limitada (Spanish—limited liability company)
Srls Saudi Arabian riyal(s)
srm speed of relative movement; spontaneous rupture of membrane; survey radiation monitor
srm (SRM) short-range missile
SRM Society for Range Management; Standard Reference Material
SR & M Safety, Reliability, and Maintenance
SRM Su Real Majestad (Spanish—Your Royal Majesty)
SRME Society for Research in Music Education
Sr M Sgt Senior Master Sergeant
SRMU Space Research Management Unit
SRN State Registered Nurse; Student Registered Nurse
SRN-6 British Hovercraft hovercraft designation
srna (SRNA) soluble ribonucleic acid
sRNA soluble or transfer RNA (same as tRNA)
SRNA Shipbuilders and Repairers National Association
SR NC Severn River Naval Command
SRNP Stirling Range National Park (Western Australia)
sro sex-ratio organism
sro (SRO) single-room occupancy; standing room only
SRO Savannah River Operations Office(r); standing room only; Superintendent of Range Operations
srob short-range omnidirectional beacon
s rod stove rod
sro's single-room-occupancy hotels
SROTC Senior Reserve Officers Training Corps
srp supply refuelling point
SRP Saturday Review Press; Savannah River Plant; Scientific Research Proposal; Stratospheric Research Program
SRPA Southwestern Regional Planning Association
s-r psychology stimulus-response psychology

srr survival, recovery, and reconstitution
srr (SRR) skin resistance response
SRR Site Recommendation Report; Supplementary Reserve Regulations
SRRA Scottish Radio Retailers' Association
SRRC Sperry Rand Research Center
srrcs surface raid reporting control ship
Srrnto Sorrento
SRRS Social Readjustment Rating Scale
SRRT Social Responsibilities Round Table
srs slow reacting substance
srs (SRS) short-run supply
SRs Socialist Revolutionaries (moderates in czarist Russia)
SRS Scoliosis Research Society; Seat Reservation System; Sight Restoration Society; Social and Rehabilitation Service; Special Revenue Sharing; Sperry Rail Service; Sperry Rand Service; Statistical Reporting Service; Structural Research Series; Structural Research Service; Supplemental Restraint System
S.R.S. *Societatis Regiae Sodalis* (Latin—Fellow of the Royal Society)
SRSA Scientific Research Society of America
SRSC Sul Ross State College
SRSM *Serenissima Repubblica di San Marino* (Italian—Most Serene Republic of San Marino)—official name of San Marino
SRSNY Sons of the Revolution in the State of New York
S-R strain Schmidt-Ruppin (viral) strain
srt speech reception threshold
SRT Short Range Transport (aircraft); Social Relations Test(ing); Speech Reception Test(ing); Strategic Rocket Troops; Stroke Rehabilitation Technician; System Reliability Test(ing)
SRT *Standard Radio och Telephon* (Swedish—Standard Radio and Telephone)
Srta *Señorita* (Spanish—Miss)
SRTC Salford Royal Technical College
SRTN Solar Radio Telescope Network
Srto *Señorito* (Spanish—master; young gentleman)

SRTOS Special Real-Time Operating System
SRTS Science Research Temperament Scale
sru servo(mechanism) repeat unit; shop-replaceable unit
SRU Scottish Rugby Union
SRUBLUK Society for the Reinvigoration of Unremunerative Branch Lines in the United Kingdom
srv (SRV) submarine research vehicle
SRV Socialist Republic of Vietnam
srvlv servovalve
SRW Sherwin-Williams Company of Canada (stock exchange symbol); State Reformatory for Women
SRY Sherwood Rangers Yeomanry
SR y C Santiago Ramón y Cajal
srypu (SRYPU) sour puss
ss saline soak; sample size; semi-steel; setscrew; shortstop; simplified spelling; single signal; single source; single strength; single-seated; software systems; sole source; sparingly soluble; spin-stabilized; stainless steel; sterile solution; straight shank; superspeed; supersport (automobile model); sword stick; sworn statement
ss (SS) suspended sentence
ss (s/s)(SS)(S/S) steamship
s-s surface-to-surface missiles
s-s. solid-state
s/s same size; souvenir sheet; steamship; suspended sentence
s & s signs and symptoms; slings and springs
s of s source of sex (also appears as sos)
s to s ship-to-shore; station-to-station
ss *senza sordini* (Italian—without mutes); *siglos* (Spanish—centuries)
ss. *scilicet* (Latin—namely); *semis* (Latin—one-half); *supra scriptum* (Latin—written above)—usually printed to left of signature line in sworn statements)
s.s. *sensu stricto* (Latin—in the strict sense)
sS *siehe Seite* (German—see page)
s/S *sur Seine* (French—on the Seine)
Ss students; subjects

SS diesel-powered attack submarine (naval symbol); Saints; Science Service; Secret Service; Secretary for Scotland; Secretary of State; Selective Service; Sharpshooter; Ship Service; Ship's Stores; Silver Star; Social Security; Special Service; Special Staff; Specification(s) for Structure; Standard Score; steamship; Straits Settlements; Submarine Studies; Sunday School; supersonic; Support Services; Support System; Surveillance Station; sworn statement
S-S Camille Saint-Saëns (1835–1921); Sans-Serif
S & S Simon & Schuster; Steen & Strom
S of S Society of Separationists
SS *Saints; Schutzstaffel* (German—Nazi blackshirt elite corps); *Santa Sede* (Spanish—Holy See); *Seitensatz* (German—second theme in a sonata or symphony); *Statens Skipstilsyn* (Danish—State Shipping Inspection)
SS. *Sanctissimus* (Latin—most holy)
SS-4 Soviet medium-range ballistic missile called Sandal by NATO
SS-5 Soviet intermediate-range ballistic missile called Skean by NATO
SS-6 Soviet intercontinental ballistic missile nicknamed Sapwood by NATO
SS-7 Soviet intercontinental ballistic missile called Saddler by NATO
SS-8 Soviet two-stage intercontinental ballistic missile named Sasin by NATO
SS-9 Soviet intercontinental ballistic missile called Scarp by NATO and capable of releasing warheads below early-warning radar range
SS-10 Soviet three-stage intercontinental ballistic missile named Scrag by NATO
SS-11 Nord antitank missile built in France where its air-launched version is called AS-11; Soviet liquid-fuel intercontinental ballistic missile; U.S. antitank missile called AGM-22A
SS-12 Nord antitank missile with greater range than the SS-11
SS-13 Soviet three-stage inter-

continental ballistic missile code-named Savage by NATO

SS-14 Soviet two-stage inter-continental ballistic missile code-named Scapegoat by NATO

SS-18 Soviet intercontinental ballistic missile

SS-20 intermediate-range nuclear missile developed by the USSR

SS-21 tactical nuclear missile developed by the USSR

ssa smoke-suppressant additive; solid-state amplifier

ssa (SSA) skin-sensitizing antibodies

SSA Scottish Schoolmasters' Association; Secretary of State for Air; Seismological Society of America; Semiotic Society of America; Smallest Space Analysis; Soaring Society of America; Social Security Administration; Society of Scottish Artists; Society for the Study of Addiction; Southern Surgical Association; Subscriber Savings Account

S-S A Self-Sufficiency Association

SSA Secretaría de Salubridad y Asistencia (Spanish—Secretariat of Health and Assistance)

ssaa screened-shift-and-add

SSAC Soldier's, Sailor's, and Airmen's Club

SSAFA Soldiers', Sailors', and Airmen's Families Association

SS agar Shigella and *Salmonella* agar

SSAGO Student Scout and Guide Organisation

SSAP Statement of Standard Accounting Practice(s)

ss ar spotface arbor

SSAR Society for the Study of Amphibians and Reptiles

SSARR Streamflow Synthesis and Research Regulation; Streamflow Synthesis and Reservoir Regulation

SSAS Special Signal Analysis System; Static Stability Augmentation System

SSASA Social Services Association of South Africa

SSAT Secondary School Admission Test(s)

SSATB Secondary School Admission Test Board

ssb single side band; single-

strength B (quality glass); subseabed

S Sb San Sebastian

SSB fleet ballistic missile submarine (3-letter naval symbol); Security Screening Board; Selective Service Board; Society for the Study of Blood; Source Selection Board; Space Science Board; Subseabed (project)

S-S B Sino-Soviet Bloc

SSBN nuclear-powered fleet ballistic missile submarine (4-letter naval symbol)

SSBS S-2 French intermediate-range ballistic missile launched from an underground silo

ssc safe-shielded cask; sealed storage cask; shape-selective cracking

ssc (SSC) station-selection code (data processing)

s & sc sized and supercalendered

SSC Sacramento State College; Sarawak Shipping Company; Sculptors' Society of Canada; Ships Systems Command (formerly Bureau of Ships); Sidney Sussex College (Cambridge); Straits Steamship Company; Superconducting Supercollider (atom smasher); Supply Systems Command (formerly Bureau of Supplies and Accounts); Surveillance Support Center

S.S.C. Societas Sanctae Crucis (Latin—Society of the Holy Cross)

SSCA Southern Speech Communication Association; Southern States Correctional Association

sscc spin-scan cloud camera

SSCC Space Surveillance Control Center

S.Sc.D. Doctor of Social Science

SSCDS Small Ship Combat Data System

SSCI Steel Service Center Institute

SSCI Social Sciences Citation Index

SSCNS Ship's Self-Contained Navigation System

SSCQT Selective Service College Qualification Test

ss cr stainless-steel crown

sscrn silkscreen

sscrng silkscreening

sscs strain-sensitive cable sen-

sor

SSCS Sea Shepherd Conservation Society; Shipboard Satellite Communications System

ssd source skin distance

ssd (SSD) sentence-structure determination

SSD Schizophrenic Spectrum Disorders; Science Services Department; Scientific Services Department; Social Services Department; Space Systems Division (USAF); System for System Development

SSD Staatssicherheitsdienst (German—State Security Service)—former East German political police

SS.D. Sanctissimus Dominus (Latin—Most Holy Lord)—the Pope

S.S.D. Sacrae Scripturae Doctor (Latin—Doctor of Sacred Scripture)

SSDA Self-Service Development Association

SSDC Social Science Documentation Center (UNESCO)

SSDHPER Society of State Directors of Health, Physical Education, and Recreation

SSDL Society for the Study of Dictionaries and Lexicography

SSDO Social Security District Office

ssdr subsystem development requirement

SSDS Ship Structural Design System

sse safe-shutdown earthquake; signal security element; surface support equipment; switching single element

SSE Scale of Socio-Egocentrism; south southeast; Support System Evaluation

S.S.E. Society of Saint Edmund

SSEB South of Scotland Electricity Board

ssec selective-sequence electronic calculator

SSEC Secondary School Examination Council; Social Science Education Consortium; Solar System Exploration Committee

SSEES School of Slavonic and East European Studies

ssef solid-state electro-optic(al) filter

SSEL Space Science and Engineering Laboratory

SSEN Senior State Enrolled Nurse

ss enema soap-suds enema

SSET Steady-State Emission Test(ing)

ssf saybolt seconds furol; single-seated fighter; standard saybolt furol (viscosity)

SSF Service Storage Facility; Seven-Step Foundation; Ship's Service Force; Social Science Foundation (University of Denver); Society of Saint Francis; Special Service Force

SSFA Scottish Schools' Football Association; Scottish Steel Founders' Association

SSFC Severe Storms Forecast Center (Kansas City, Missouri)

SSFF Solid Smokeless Fuels Federation

ss fx spotface fixture

ssg second-stage graphitization

SSG guided missile submarine (3-letter naval symbol)

SSGN nuclear-powered guided-missile submarine (4-letter naval symbol)

SSgt Staff Sergeant

ssgw (SSGW) surface-to-surface guided weapon

SSH Sailor's Snug Harbor

S Sh A Soyedinennye Shtaty Ameriki (Russian—United States of America)

SSHA Scottish Special Housing Association

SSHRC Social Sciences and Humanities Research Council (Canadian)

ssi sites of scientific importance; small-scale integration

Ssi Surekasi (Turkish—company)

SSI Saint Simon's Island; Social Security Income; Society of Scribes and Illuminators; Space Studies Institute; Supplemental Security Income

SSI Service Social International (French—International Social Service); Social Sciences Index

SSIB Seaway Skyway International Bridge

ssic small-scale integrated circuit

SSIC Southern States Industrial Council; Standard Subject Identification Code

SSIDC Small-Scale Industries Development Corporation (Indian)

SSIE Smithsonian Science Information Exchange

SSIG State Student Incentive Grant(s)

SSIH Société Suisse pour l'Industrie Horlogère (French—Swiss Society of the Horological Industry)

SSI/ITL SSI Container Corp/ITEL

ssip system setup indicator panel

SSIS Squibb Science Information System

SSISI Statistical and Social Inquiry Society of Ireland

SSI/SSP Social Security Income/State Supplemental Program

ssit (SSIT) semi-submarine icebreaking tanker

SSJ Sisters of St Joseph

ssk set storage key; soil stack; solid-state keyboard

ssl spent sulfite liquor

SSL Saguenay Shipping Limited; Sapphire Steamship Lines; Seven Stars Line; Space Science Laboratory (Convair); Space Sciences Laboratory (GE)

S.S.L. Sacrae Scripturae Licentiatus (Latin—Licentiate of Sacred Scripture)

s sleep synchronized sleep

S-sleep slow-wave sleep

SS loran sky-wave synchronized loran

SSLS Solid-State Laser System

ss lt starboard side light

sslv (SSLV) standard space-launched vehicle

ssm set system mask; solid-state material(s); spread spectrum modulation

ssm (SSM) surface-to-surface missile

SSM Saturday(s), Sunday(s), Monday(s); Singer Sewing Machine; System Support Management; System Support Manager

ssma solid-state microwave amplifier

SSMA School Science and Mathematics Association; Stainless Steel Manufacturers' Association

ssme space-shuttle main engine

ssmm space station mathematical model

SS MM Sus Majestades (Spanish—Their Majesties; Your Majesties)

SSMS Submarine Safety Monitoring System

ssmt supersonic magnetic (railroad) train

SSN Space Surveillance Network; Social Security Number; Standard Serial Number; Station Serial Number

SS(N) nuclear-powered submarine (3-letter naval symbol)

SSNC Scindia Steam Navigation Company

ssnd solid-state neutral dosimeter

ssnf source spot noise figure

SS^{no} escribano (Spanish—court clerk, notary, scribe)

SSNP Syrian Social Nationalist Party

SSNS Standard Study Numbering System

SSO Sacramento Symphony Orchestra; Savannah Symphony Orchestra; Seattle Symphony Orchestra; Shanghai Symphonic Orchestra; Shreveport Symphony Orchestra; Source Selection Official; Spokane Symphony Orchestra; Springfield Symphony Orchestra; Sydney Symphony Orchestra; Syracuse Symphony Orchestra; System Staff Office(r)

SSO Seguro Social Obligatorio (Spanish—Obligatory Social Security); sudsudoeste (Spanish—south southwest)

SSOA Subsurface Ocean Area

SSOFS Smiling Sons of the Friendly Shillelaghs

S of Sol Song of Solomon

s sord senza sordini (Italian—without mutes)

SSORM Standard Ship's Organization and Regulations Manual (USN)

ssorts ship's systems operational requirements

ssos (SSOS) severe-storm-observing satellite

SSOs Student Services Organization members

ssp seismic section profiler; ship's stores profit; single-shot probability; standby-status panel; steam service pressure; subspecies; sustained superior performance

ssp (SSP) statutory sick pay

S-S p Sanarelli-Schwartzman phenomenon

SSP scouting seaplane (3-letter naval symbol); Seashore State Park (Virginia); Site Selection Report; Society for Scholarly Publishing; Society of St Paul; Source Selection

Panel; Species Survival Plan; S.S. Pierce; Sunshine State Parkway

S.S.P. Society of Saint Paul

sspc solid-state power controller

SSPC Steel Structures Painting Council

SSPCA Scottish Society for the Prevention of Cruelty to Animals

sspe subacute sclerosing panencephalitis

sspe (SSPE) subacute sclerosing panencephalitis

SSPFC Stainless Steel Plumbing Fixture Council

SSPHS Society for Spanish and Portuguese Historical Studies

SSPN Satellite System for Precise Navigation

SSPP Society for the Study of Process Philosophies

S-spring S-shaped spring

SSPU Self-Service Postal Unit

SSPV Scottish Society for the Prevention of Vivisection

ssq simple sinusoidal quantity

SSQ Station Sick Quarters

SSQT Selective Service Qualification Test

ssr secondary surveillance radar

SSR Site Safety Report; Soviet Socialist Republic(s)

SSR *Sovétskaya Sotsialiísticheskaya Respúblika* (Russian—Soviet Socialist Republic)

SSRA Scottish Squash Rackets Association

SSRB Soil Survey Research Board

SSRC Social Science Research Council

SSRCAS Secondary-Surveillance-Radar Collision-Avoidance System

SSRCC Social Science Research Council of Canada

SSRI Social Science Research Institute

SSRL Systems Simulation Research Laboratory

SSRP Stanford Synchrotron Radiation Project

SSRs Safe Secure Railcars

SSRS Society for Social Responsibility in Science; Submarine-Sand Recovery System

sss single-screw ship; specific soluble substance; sterile saline soak

s/ss sector/subsector

sss *(SSS) su seguro servidor* (Spanish—your sure servant; yours truly)

s.s.s. *stratum super stratum* (Latin—layer upon layer)

SSS Secretary of State for Scotland; Selective Service System; Special Social Services; System Safety Society

S-S-S *Schweiz-Suisse-Svizzera* (Switzerland in the three languages of the country)

S.S.S. *Societas Sanctissimi Sacramenti* (Latin—Congregation of the Most Blessed Sacrament)

SS-20s Soviet supersonic medium-range nuclear missiles

SSSA Simplified Spelling Society of America; Soil Science Society of America

S-S SA Singapore-Soviet Shipping Agency

SSSB System Source Selection Board

sssc soft-sized super-calendered (paper)

sss&c sin, syph(ilis), sulfa, and cystoscopes

SSSC Space Science Steering Committee (NASA)

sssd second-stage separation device; solid-state solenoid driver

sssi sites of special scientific importance

SSSI Science Supervisory Style Inventory

SSSJ Student Struggle for Soviet Jewry

SSSL Solid State Sciences Laboratory (USAF)

sssm site space surveillance monitor

sssm (SSSM) standard surface-to-surface missile

SSSM South Street Seaport Museum (New York City)

SSSP Space Shuttle Synthesis Program

SSSR Society for the Scientific Study of Religion; *Soyuz Sovetskikh Sotsialisticheskikh Respublik* (Russian—Union of Soviet Socialist Republics)

SSSR *Soyuz Sovietskikh Sotsialisticheskikh Respublik* (Russian—Union of Soviet Socialist Republics)

SSSRU School Safety and Security Resource Unit

SSSS Society for the Scientific Study of the Sea; Society for the Scientific Study of Sex

ssst symbol for the sound of an aerosol spray

SSSU Seaspeed Sea Services Unit

SSSWP Seismology Society of the South-West Pacific

sst safe-secure trailer(s); solid-state triangulation (automatic focusing system); stainless steel; supersonic transport (airplane)

SST Samoan Standard Time; Society of Silver Collectors; Source Selection Team; Space Systems Center (Douglas); Submarine Supply Center; supersonic transport (airplane); target and training submarine (naval symbol)

SSTA Scottish Secondary Teachers' Association; Secondary School Theatre Association; Special Services Transportation Agency

SSTAR Society for Sex Therapy and Research

SSTC Specialized System Test Contractor

SSTEP System Support Test Evaluation Program

ssto single-stage to orbit

SSTO Superintending Sea Transport Office(r)

SSTP Student Science Training Program

sst's safe (and) secure trailers used to haul nuclear weapons on American highways

sstu seamless steel tubing

ssu saybolt seconds universal; self-serving unit

ssus spring-solid upper stage

ssv (SSV) semi-submersible support vessel; ship-to-surfacevessel; submarine support vessel

s.s.v. *sub signa veneni* (Latin—under a poison label)

SSV ship-to-surface vessel

SSvd Selective Service

SSV/GC & N Space Shuttle Vehicle/Guidance, Control and Navigation

ssvs slow-scan video simulator

ssw safety switch

SSW south southwest; S.S. White

SSWA Scottish Society of Women Artists

SSWS Seismic Sea Wave Warning System

SSX South Coast Corporation (stock exchange symbol)

SSXTF Solar Soft X-ray Telescope Facility

ssz specified strike zone

ssz (SSZ) pocket submarine; specified strike zone

SSZ Society of Systematic Zoology

st sedimentation time; service test; short ton; single tire; single-throw; slight trace; sounding tube; space telescope; special text; special translation; stained; statement(s); steel; steel truss; stock transfer; stone; strata; surface tension; survival time; syncopated time

s & t science and technology; sink and laundry tray; supply and transport

st. stet (Latin—let it stand)—usually referring to what has been mistakenly crossed out

St Saint; Sainte; Stanton number; State; status; Street; strontium

Sᵗ Sint (Afrikaans, Dutch, Flemish—saint); *Staryy* (Russian—old)

ST Seaman Torpedoman; Service Test(ing); Shipping Ticket; Sons of Temperance; Speech Therapist; speech therapy; Standardized Test; Summer Time; *Suomen Tsavalta* (Finnish—Finland); Syrian Territory

S.T. sidereal time

S & T Supply and Transport

S of T Sons of Temperance

sta static; station; stationary; stationery; stator; submarine tender availability

Sta Santa (Italian, Portuguese, Spanish—Saint)—feminine; *Señorita* (Spanish—Miss)

STA Scottish Typographical Association; Society of Typographic Arts; Southern Textile Association; Supersonic Tunnel Association

STAA Survey Test of Algebraic Aptitude

STAAS Surveillance and Target Acquisition Aircraft System

stab. stabilizer

STAB Svenska Tandsticks Aktiebolaget [Swedish—Match (stick) Company]

Sta'b'd starboard

stabiles static abstract sculptures

stac sensor transmitter automatic choke

stac staccato (Italian—separately and with great distinction)

St AC Saint Anne's College; Saint Anthony's College

STAC Science Teacher's

Adaptable Curriculum; Science and Technology Advisory Committee (NASA)

STACO Society of Telecommunications Administrative and Controlling Officers

STACS Satellite Telemetry and Computer System

sta eng stationary engineer

stafex staff exercise(s)

STAFF Stellar Acquisition Flight Feasibility (guidance system)

Staffs Staffordshire

staflo stable-flow (free-boundary electrophoresis apparatus)

stag. stagger; staggered

STAG Special Task Air Group; Standards Technical Advisory Group; Strategy and Tactics Analysis Group

stagfla stagflation (high inflation coupled with high unemployment)

stagflation stagnant economy marked by rising unemployment and spiralling inflation; stagnation and inflation

stagmag magazine featuring nude women

STAI State-Trait Anxiety Inventory

STAIFA St Anselm's International Friendship Association

Stairs Storage and Information Retrieval Systems

Sta L Santa Lucía (Spanish—St Lucia)

Stalag Stammlager (German—base camp, for military prisoners)

stam sequential thermal anhysteric magnetization; stammer(er); stammering

sta mi statute miles

stamp. small tactical aerial-mobility platform

STAMP Systems Tape Addition and Maintenance Program

Stampa La Stampa (Turin's Press—one of Italy's leading newspapers)

STAMPS Structural Thermal and Meteorite Protection System

stan stanchion; standard; standing

Stan Standard; Stanford; Stanley; Stanleyville; Stanton

STANAG Standardization Agreement (NATO)

stanal statistical analysis

STANAVFORCHAN Standing Naval Force Channel (NATO)

STANAVFORLANT Standing Naval Force Atlantic (NATO)

St And St Andrews

standard. standardization

Standard and Poor's Standard and Poor's Corporation Records

Stand Engl Standard English

STANDINAIR Standing Instructions for Air Attachés

stanine score standard-nine score (USAF standard psychological score)

Stan Psychiat Nomen Standard Psychiatric Nomenclature

STANVAC Standard Vacuum (oil company)

STAO Science Teachers Association of Ontario

STAPFUS Stable Axis Platform Follow-Up System

staph staphylococcus

staq security-traders automatic quotations

star symbol of perfection

star. (STAR) special tactics against robbery (police program)

STAR Selective Training and Retention (program); Serial Titles Automated Record (National Agricultural Library); Ship-Tended Acoustic Relay; Space Thermionic Auxiliary Reactor; Special Tactics Against Robberies; Special Tactics and Response; submersible test and research (Electric Boat)

STAR Scientific and Technical Aerospace Reports

STARFIRE System to Accumulate or Retrieve Financial Information Random Extract

STARLAB Space Technology Applications and Research Laboratory (NASA)

starquake star + earthquake

stars. specialized training and reassignment students; stationary automotive road stimulator (Toyota)

STARs Scientific and Technical Aerospace Reports

STARS Satellite Telemetry Automatic Reduction System; Small Tethered Aerostat Relocatable System; Student Tuition and Repayment System

Stars and Stripes Stars and Stripes Forever (American march); American military newspaper

START Space Technology and Reentry Test(s); Space Transport and Reentry Test(s); Spacecraft Technology and Advance Reentry Test; Strategic Arms Reduction Talks (U.S.A.-USSR); Strategic Arms Reduction Treaty (U.S.A.—USSR)
STARTS Safety Technology Applied to Rapid Transit Systems
Star War(s) Strategic Defense Initiative(s), SDI
stas staff-to-arm signal
Stash Stanislas; Stanislaus
STASH Student Association for the Study of Hallucinogens
Stasia Anastasia
stasis or stat or stato (Greek—stand)—colonic stasis, electrostatic, hydrostatic, metastasis, thermostat
stat electrostat; electrostatic; microstat; photostat; static; stationary; statistic(al); statuary; statue; statute
stat. statim (Latin—immediately, right now)
Stat Publius Papinius Statius (Roman poet)
Stat Can *Statistics Canada*
state. simplified tactical approach and terminal equipment (STATE)
Statesman's Statesman's Year Book
Statesville Statesville Correctional Center (Joliet, Illinois)
Stat Hall Stationers' Hall
STATIC Student Taskforce Against Telecommunication Concealment
STATLIB Statistical Computing Library (Bell System)
stat mux statistical multiplexor
Stat Off Her (His) Majesty's Stationery Office
stats statistics
Stats statutes
Statsbib Statsbiblioteket (Dano-Norwegian—State Library)
STATUS Subscriber Traffic and Telephone Utilization System
St AU University of St Andrew
STAUK Seed Trade Association of the United Kingdom
Stav Stavanger, Norway
St A YC St Augustine Yacht Club
Sta Ysbl Santa Ysabel
s-t b steel-truss bridge

St B Státni Bezpečnost (Czech—State Security)—secret police
STB Surinam Tourist Bureau
STB Sandatahang Tanod ng Bayan (Filipino—People's Home Defense Guard)
S.T.B. *Sacrae Theologiae Bacalaureus* (Latin—Bachelor of Sacred Theology)
stba selective top-to-bottom algorithm
St Bart's Day Saint Bartholomew's Day—date in 1587 when French Huguenots were attacked by Catholics as they left their churches
stbd starboard
St Ben St Benet's Hall, Oxford
st brz statuary bronze
stbt steamboat
stc said to contain; security time control; sensitivity time control; short time constant; sound transmission class; stepchild
STC Satellite Television Corporation; Satellite Test Center; Satellite Tracking Committee; Scandinavian Travel Commission; Short Title Catalog; Society for Technical Communication; Southwestern Technical College; Standard Telephone and Cables; Standard Transmission Code; Sunderland Technical College
S.T.C. Samuel Taylor Coleridge
STC Short Title Catalogue
STCA Stereo Tape Club of America
St Cat St Catherine's College, Oxford
STCC Springfield Technical Community College
STCCM Sistema de Transporte Colectivo Ciudad de México (Spanish—Mexico City Collective Transportation System)
Stckhlm Stockholm
st cl storage closet
St C & N Saint Christopher and Nevis (Leeward Islands)
STCS Society of Technical Civil Servants
std salinity, temperature, depth; sexually-transmitted disease; short-term disability; skin test dose; standard; standard test dose; state-of-the-technology design; subscriber trunk dialing
St D Stage Director

STD Society for Theological Discussion; Subscriber Trunk Dialing
S.T.D Sacrae Theologiae Doctor (Latin—Doctor of Sacred Theology)
std by stand by
St DC St David's College
STDC Society of Typographic Designers of Canada
Stde Stunde (German—hour)
stder social introversion, thinking introversion, depression, cycloid tendencies, rhathymia (personality traits)
st diap stopped diapason (organ)
stdn standardization
Std Oil Cal Standard Oil of California
std p stand pipe
stdr steam turbine double reduction
st dr single-trip drum
std's sexually transmitted diseases
STDSD Solar-Terrestrial Data Services Division (NOAA)
Stdy Saturday
Ste Suite
Ste Sainte (French—saint, *f.*)
Sté Société (French—Society)
St E St Etienne
STE Society of Telecommunications Engineers; Society of Tractor Engineers; Support of Theological Education
steakwich steak sandwich
steamers steamed clams
STECC Scottish Technical Education Consultative Council
steelie steel ball-bearing playing marble
steeving stevedoring (loading or unloading a ship's cargo)
Stef (Joseph) Lincoln Steffens; Stefan(i)(e); Vilhjalmur Stefansson (William Stevenson)
STEFER Società della Tranvia e Ferrovia Elettrica di Roma (Italian—Rome transportation system)
STEG Supersonic Transport Evaluation Group
St E H St Elizabeth's Hospital
St EHC Saint Edmund's Hall College (Oxford)
ste/ice simplified test equipment/internal combustion engines
Steiermark Styria, Austria
Stein Steinway
Steinbeck Grosssteinbeck (original name of author John Steinbeck's family)

STEL *Studenta* *Tutmonda* *Esperantista* *Liga* (Esperanto—Worldwide Esperanto Students League)
Stell Estella; Estelle
Stella Estella; Estelle
stellar. star tracker for economical-long-life attitude reference
STELO *Studenta* *Tutmonda* *Esperantista* *Ligo* (Esperanto—World League of Esperanto Students)
stem. storable tubular extendible member
STEM stay time excursion module
sten stencil
Sten (Swedish—cliff); *Stenón* (Greek—pass, strait)
Sten gun Sheppard and Turpin Bren gun (submachine gun)
steno stenographer; stenography; stenotype; stenotypy
steno (Latin prefix—narrow)— stenosis
stent *stentando* (Italian—delaying)
Step Stephen
STEP Safety Test Engineering Program; Scientific and Technical Exploitation Program; Secondary Teachers Education Program; Sequential Tests of Educational Progress; Short-Term Elective Program; Solutions to Employment Problems; Systematic Training for Effective Parenting; Systems to Encourage Potential
Steph Stephen
STEPS Solar Thermionic Electric Power System; Specialized Training and Employment Placement Service
ster stereoscope; stereotype; sterilization; sterilize; sterilizer; sterling
stereo stereophonic; stereoprojection; stereoprojector; stereoscope; stereoscopic
STERILE System of Terminology for Retrieval of Information through Language Engineering
stet let stand what has been crossed out; stetted; stetting
stet (Latin—let it stand)— proofreader's mark
STETF Solar Total Energy Test Facility (ERDA)
Stetson Stetson hat (broadbrim high-crown hat made by John B Stetson of Philadelphia, Pa)

stev stevedore; stevedoring
Steve Stephan; Stephen; Steven
Stew Stewart
stewbum man sexually attracted to flight attendant(s)
stew(s) steward(esses), flight attendant(s)
STEWS Shipboard Tactical Electronic Warfare System
stewzoo hotel catering to flight attendants resting between flights
St Ex Stock Exchange
stf soluble thymic factor; staff
STF Salt Test Facility; Sycamore Test Facility
STF *Svenska* *Turisforeningen* (Swedish Swedish—Tourist Information)
s-t fibers slow-twitch fibers
st fm stretcher form
stg seating; stage; staging; steering; sterling; storage
STG Study Group
STG *Schiffbautechnische* *Gesellschaft* (German—Shipbuilding Technical Association)
stg ar staging area
stge storage; strings
stgg staging
Stgo Santiago
Stgo de C Santiago de Chile (Compostela, Cuba)
stgr stringer
STgt secondary target
STGWU Scottish Transport and General Workers' Union
sth straight to hell
sth (STH) somatotrophic hormone
Sth Stockholm
St Hel St Helena; St Helens; St Helier
St Hil St Hilda's College, Oxford
Sthlm Stockholm
St Hug St Hugh's College, Oxford
sti service and taxes included; sure to inquire; sure to investigate; surface transfer impedance
sti (STI) scientific and technical information
s & ti scientific and technical information
St I St Ives
STI Service Tools Institute; Space Technology Institute; Steel Tank Institute
STIA Scientific, Technological, and International Affairs Directorate (National Science Foundation)

STIAD Scientific, Technological, and International Affairs Directorate (NSF)
stic serum trypsin inhibitory capacity
STIC Scientific and Technical Intelligence Center
STICAP Stiff Circuit Analysis Program
stiction static friction
STID Scientific and Technical Information Division (NASA)
STIF Scientific and Technical Information Facility (NASA)
stiff. stiffener; stiffened corpse
Stikines Stikine Mountains of British Columbia
stillat. *stillatim* (Latin—by drops, in small amounts)
stilli stillicide; stillicidium; stilliform
stillson stillson wrench
stim stimulant
stimn stimulation
STIMS Scientific and Technical Modular System
stinfo scientific and technical information
STING Stellar Inertial Guidance (System)
STINGS Stellar Inertial Guidance System (USAF)
stink. stinkage; stinkerino (offensive odor)
stinkerette(s) odoriforous cigarette(s)
stinkerino(s) smelly cigar-(ette)s; tobacco product(s)
stip stipend(iary); stipulation
STIP Science Teaching Improvement Program; Skills Training Improvement Program
STIPIS Scientific, Technical, Intelligence, and Program Information Service (HEW)
Stir Stirling
Stirner Max Stirner whose original name was Kaspar Schmidt
STIS Scientific and Technological Information Services; Specialized Textile Information Service
STISS Scientific and Technical Information Services and Systems
St J St John (New Brunswick)
STJ Special Trial Judge
STJC South Texas Junior College; Southwest Texas Junior College
St Joe St Joe Minerals Corporation (energy and metals plus natural gas)

St Joh St John's College (Oxford)

StJU St John's University

stjw stretcher jaws

stk sticky; stock

Stk Stockton

STK Standard Test Key

St Kitts West Indian islands of Anguilla, Nevis, and St Christopher

St K-N St Kitts officially St Christopher; St Kitts-Nevis (West Indian island nation gained independence in 1983)

St K-N-A St Kitts-Nevis-Anguilla (Caribbean island federation)

stl steel; studio transmitter link

Stl Schottky transistor logic

St L St Louis

STL Seatrain Lines; Space Technology Laboratories (Thompson-Ramo-Wooldridge); Speech Transmission Laboratory; Standard Telecommunication Laboratories; St Louis, Missouri (airport); studio transmitter link (FM); Swedish Transatlantic Line

StLe St Louis encephalitis

StLGR Saint Lucia Game Reserve (South Africa)

STLL Submarine Tender Load List

St Lo St Louis

STLO Scientific and Technical Liaison Office(r)

STLOs Scientific/Technical Liaison Offices

STLOUISPDis St Louis Procurement District (US Army)

St L P-D St Louis Post-Dispatch

stlr semi-trailer

St L SW St Louis Southwestern (railroad)

STLT studio transmitter link-TV

St LU St Lucia; St Louis University

STLU Seatrain Line (container) Unit

St L YC St Louis Yacht Club

St L ZG St Louis Zoological Garden

stm (STM) scanning tunneling microscope; scientific, technical, and medical; shielded tunable magnatron; short-term memory; special test missile; surface-to-target missile; synthetic timing mode

St M St Malo

STM Science Teaching Museum (Franklin Institute);

System Training Mission

S.T.M. *Sacrae Theologiae Magister* (Latin—Master of Sacred Theology)

St Martin's St Martin's Press

stmev storm evasion

stmftr steamfitter

stmn stimulation

Stmn The Statesman (Calcutta)

stmnt statement (flow chart)

STMP Scientific, Technical, and Medical Publishers

stmrs steamers

STMSA Scottish Timber Merchants' and Sawmillers' Association

stmt statement

stn stain

Stn Station

St N St Nazaire

stnd stained

stnry stationary

stnwr stoneware

sto standard temperature and pressure; standing order; stoker; stop; stoppage

Sto *Santo* (Spanish—saint); *Señorito* (Spanish—master; young gentleman)

Stº Santo (Portuguese or Spanish—Saint)

STO Sea Transport Office(r); Stockholm, Sweden (Arlanda Airport)

STO Service Travail Obligatoire (French—Obligatory Labor Service)—Vichy-instituted law giving the Germans a massive labor force during World War II

Stock Stockholm

Stokowski Leopold Antoni Stanislaw Boleslawowicz (1882–1977)

stol short takeoff and landing

stolport short-takeoff-and-landing airport

stol/ved short takeoff and landing/vertical climb and descent

stom stomach

stoma (Greek—mouth or opening)—cyclostome, protostome, stomatic

stomat stomatology

STOMP Short-Term Offshore-Measurement Program

stomy (Latin suffix—surgical opening)—tracheostomy

S'ton Southampton

STon short ton

stoners gangs engaged in throwing stones; people who have taken an overdose of alcohol or other drugs

Stonys Stony Mountains (early

American name for the Rockies)

S'toon Saskatoon

stop. slight touch on pedal; spin tires on pavement

STOP Single Title Order Plan; Strategic Orbit Point; Study of Protection (against nuclear warfare)

STOPP Society of Teachers Opposed to Physical Punishment

stops. stabilized-terrain optical-position sensor

STOPS Self-contained Tanker Offloading System

stor storage; stored

STOR Scripps Tuna Oceanographic Research

storet storage and retrieval

Storm Storm Over Asia

STORM Stormscale Operation and Research Meteorology

Stormont Stormon Castle—official Belfast residence of Northern Ireland's prime minister; Northern Ireland's capital district near Belfast, contains the home and office of the governor general, the House of Commons, and the Senate

stovl short takeoff with vertical landing

stow. stowage

stp service time prediction; solar-terrestrial physics; solar-terrestrial probe; step; stop

stp (STP) seawater treatment plant; solar thermal power; standard temperature and pressure

St P St Paul

St & P São Tome and Principe

STP nickname of dangerous psychedelic drug—methylmethoxyamphetamine; Scientifically Treated Petroleum (gasoline additive); sodium tripolyphosphate (water softener); Space Test Program (USAF); State Testing Program(s); stop the police

STP Santo Tomé y Principe (Spanish—St Thomas and Prince)—islands off Africa, formerly São Tomé

S.T.P. Sacrae Theologae Professor (Latin—Professor of Sacred Theology)

st part steel partition

St Pat Saint Patrick; Saint Patrick's Day (March 17)

STPB Singapore Tourist Promotion Bureau

stpd standard temperature and

pressure—dry (0°C, 760mm Hg)

St Pet St Peter's College, Oxford

St Pete St Petersburg

St-P-et-M *Saint-Pierre et Miquelon* (French—Saint Pierre and Miquelon)—French islands off Newfoundland

STPL Space Tracking Pty Ltd

St P & M St Pierre and Miquelon Islands

stpr short taper; stumper

s tpr short taper

stps specific thalamic projection system

StP Sta St Pancras Station (rail terminal)

str steamer; straight; strainer; strait; strength; structural; structure; submarine test reactor (STR)

str (STR) synchronous transmitter receiver (data processing)

str *strana(y)* [Czech—page(s)]

Str Strait; Stranraer; Street

Str *Strasse* (German—street); *Streichinstrumente* (German—stringed instrument); *Streptococcus*

STR Science and Technical Research; section, township, range; Society for Theatre Research; Southern Test Range; Stuttgart, Germany (airport); submarine test reactor

STRA State Teacher's Retirement System

strabad strategic base air defense

STRAC Strategic Army Corps

STRACNET Strategic Rail Corridor Network

STRACS Surface Traffic Control System

strad stradivarius (violin made by Antonio Stradivari or his sons Francesco and Omobono)

strad (STRAD) signal transmitting—receiving and distributing

stradav storm radar data processor

STRADS Switching, Transmitting, Receiving, and Distribution System

STRAF Strategic Army Forces

strag straggler; strategic; strategist; strategy

StragL straggler line

Strait Strait of Bab el Mandab, Bali, Bass, Belle Isle, Bering, Bosporus, Canso, Dardanel-

les, Denmark, Dover, Florida, Formosa, Georgia, Gibraltar, Hainan, Juan de Fuca, Korea, Lombok, Luzon, Magellan, Makassar, Malacca, Messina, Molucca, Otranto, Palk, Sunda, Tiran, Torres

Straits Straits Settlements (Malaysia and Singapore); Straits of Tiran (at entrance to the Gulf of Aqaba or Eilat)

Strangeways Strangeways Prison in Manchester, England

STRAP Stretch Assembly Program

Stras Strasbourg

STRATAD Strategic Aerospace Division (USAF)

STRATCOM Strategic Communications Command (USA); Stratospheric Composition (program)

Strath Strathclyde

stratig stratigraphy

strato stratosphere

straw strawberry

STRAYS Society To Rescue Animals You've Surrendered

strbd *stuurboord* (Dutch—starboard)

STRC Science and Technology Research Center; Scientific, Technical, and Research Center; Scientific, Technical, and Research Commission

STREAK Surfaces Technology Research in Energetics, Atomistics, and Kinetics

Stream the Stream (Gulf Stream)

Street The Street—London's Fleet Street (center of periodical publishing); New York's Wall Street (financial center)

Streetcar *Streetcar Named Desire*

strep streptococcus

STREP Ship's Test and Readiness Evaluation Procedure

stress. (STRESS) structural engineering system solver

STRESS Stop the Robberies, Enjoy Safe Streets (program of the Detroit Police Department)

stret *stretto* [Italian—squeezed together; more rapid (as musical notes), strait]

STRI Smithsonian Tropical Research Institute

STRICOM Strike Command (US Army)

strikeops strike operations

strikex strike exercise

STRIKFLANTREPEUR Striking Fleet Atlantic Representative in Europe (NATO)

STRIKFORSOUTH Striking and Forces Support, Southern Europe (USN)

string. string-processing systems, technics, languages

string *stringendo* (Italian—accelerate)

strings stringed instruments: balalaikas, banjos, bass viols, 'cellos, dulcimers, guitars, harps, mandolins, samisens, vinas, violins, violas, zithers

strip. standard taped routines for image processing

strip. (STRIP) string processing language

Strip The Strip—main street of Las Vegas, Nevada

STRIPS Separate Trading of Registered Interest and Principal of Securities

strl straight line

S-t-R L Save-the-Redwoods League

str lgths straight lengths

Strm Stream

STRN Standard Technical Report Number

strobe satellite tracking of balloons and emergencies

strobed stroboscopically illuminated; stroboscopically measured

strobes. shared-time repair of big electronic systems

strobo stroboscope

strobotron stroboscope + electron (tube)

str off fixt store (or) office fixtures

STRS State Teachers Retirement System

struc structure

struct structural

STRUT Safe TRU Transit

's' truth god's truth

Strv-74 Swedish light tank armed with 75mm gun

Strv-S Bofors-built Swedish medium tank with 105mm gun

strwbrd strawboard

sts scour the shower; ship-to-shore; special treatment steel; surfaced two sides

st's sanitary towels

Sts Streets

STS Science Talent Search; Scottish Text Society; Serological Test for Syphilis; Space Tranport(ation) System; Standard Test for

Syphilis; Stockpile-to Target Sequence

STS-26 26th space-shuttle mission

STS-27 27th space-shuttle mission

STSA State Technical Services Act

STSC Southwest Texas State College

STScI Space Telescope Science Institute

STSD Society of Teachers of Speech and drama

stsg split-thickness skin graft-(ing)

STSI Space Telescope Science Institute

STSO Senior Technical Staff Officer

st st stocking stitch (knitting)

stt scrub the tub

St T (Port of) St Thomas

STT Medical Stenographer (USN); St Thomas, Virgin Islands (airport); Sensitization Test

S-T T Skin-Temperature Test-(ing)

STTA Scottish Table Tennis Association

STTC Sheppard Technical Training Center

sttch(es) stitch(es)

ST T NHS St Thomas National Historic Site

sttr stator

STTT Space Telescope Task Team (NASA)

stu service trials unit; skin test unit; student; submersible test unit

Stu Stewart; Stuart

STU Seatrain (container) Unit

STU Styrelsen foer Teknisk Utveckling (Swedish—Board for Technical Development)

STUC Scottish Trades Union Congress

stucco calcium sulfate

stud. student

Stud Studebaker; Studies

Stud Studii (Russian—studies)

stude(s) student(s)

stud(s). student(s)

stuff. system to uncover facts fast

Stuka Sturzkampfflugzeug (German—dive bomber)

stump. submersible, transportable, utility marine pump-(ing)

stuns'l studdingsail

stupidental(ly) stupidly accidental(ly)

Sturt Sturt Desert in the north-

west sector of New South Wales, Australia

stuvs standard unit variance scale

stv subscription television

stv (STV) submersible transport vehicle; subscription television

St V Stavanger; St Valentine; St Vincent

STV Scottish Television; Separation Test Vehicle

STV Solidaridad de Trabajadores Vascos (Spanish—Solidarity of Basque Workers)

St Val Saint Valentine; St Valentine's Day

St Val's Day Saint Valentine's Day (February 14)

stvd r stevedore

St V & G St Vincent and the Grenadines

s tv i subliminal television intoxication

STVPS Salinity, Temperature, Sound Velocity, and Pressure-Sensing System

st w storm water

STW Society of Technical Writers

ST WAPNIACLE old abbreviation mnemonic for U.S. departments in order of their creation before new ones were added and some were consolidated: State, Treasury, War, Attorney General (Justice), Post Office, Navy, Interior, Agriculture, Commerce, Labor, Education

Stwd Steward

STWE Society of Technical Writers and Editors

STWP Society of Technical Writers and Publishers

stwy stairway

stx start of test (data processing); static test stand

STX St Croix, Virgin Islands (airport)

Sty Stymie

STYCAR Screening Tests for Young Children and Retardates

S-type Jungian sensate type

Styria Steiermark, Austria

STZ South Temperate Zone; South Tropical Zone; Sterling Drugs (stock exchange symbol)

su sensation unit(s); service unit(s); setup; strontium unit(s); sulfur unit(s)

su. sumat (Latin—let him take)

s u siehe unten (German—see

below)

Su Sudan; Sudanese; Sunday

SU Saybolt Universal; Scripture Union; Seattle University; Shaw University; Skinner Union; Southeastern University; Southwestern University; Soviet Union; Standord University; Stanford University; Stetson University; Student Union; Suffolk University; Sydney University; Syracuse University

SU Stati Uniti (Italian—United States)

SU-7 Soviet ground-attack fighter aircraft designated Fitter by NATO

SU-9 Soviet all-weather jet fighter aircraft called Fishpot by NATO

SU-11 Soviet delta-wing fighter aircraft called Flagon-A by NATO

SU-76 Soviet 76mm assault gun used in World War II and thereafter in Korea and Vietnam

SU-85 Soviet 85mm assault gun

SU-100 Soviet 100mm assault gun

SU-122 Soviet 122mm assault-gun howitzer also designated JSU-122

SU-152 Soviet 152mm assault-gun howitzer (JSU-152)

sua shipped unassembled

sua (SUA) serum uric acid

S-u-A Stratford-upon-Avon

SUA Shan United Army (Burma revolutionary force); Silver Users Association; State Universities Association

SUA Stati Uniti d'America (Italian—United States of America)

SUAB Svenska Utvecklinasaktiebolaget (Swedish—Swedish Development Corporation)

SUADPS Shipboard Uniform Automatic Data Processing System (USN)

sub subcontract(or); submarine; submerse; subordinate; substitute; suburb; subway

sub (SUB) substitute character (data processing)

sub (Latin prefix—below, beneath, under)—subterranean, subway

Sub Subic Bay; Subway

SUB Supplemental Unemployment Benefit (fund)

SUB Subbota (Russian—Satur-

day)
subac subacute
SUBACLANT Submarine Allied Command, Atlantic (NATO)
SUBAN Scottish Union of Bakers and Allied Workers
subassy subassembly
Sub Base Submarine Base
sub-bell submarine fog bell
sub chap subchapter
subcontr subcontract(or)
subcrep subcrepitant
subcut subcutaneous(ly)
subd subdivide; subdivision
subdeb subdebutante
SUBDIV Submarine Division (naval)
SUBDIZ Submarine Defense Identification Zone
sub-ed sub-editor
subex submarine exercise; submerged exercise
sub. fin. coct. sub finem coctionis (Latin—at the end of boiling)
subfusc subfuscous (dark and dingy)
subgen. subgenus (Latin)
subic (SUBIC) submarine integrated control program
Subic Bay US Naval Base, Luzon Island, the Philippines
subing substituting
subj subject; subjunctive
subject. subjective(ly)
subl sublimes
SUBLANT Submarine Forces, Atlantic (USN)
subling sublingual
sublse sublease
Sub Lt Sub-Lieutenant
subm submission; submit
submand submandibular
SUBMED Submarines Mediterranean (NATO)
SUBMEDNOREAST Submarines—Northeast Mediterranean (NATO)
submgd submerged
submtl submittal
subn substitution
SUBNOTE Submarine Notice (USN)
subor subordinate
sub-osc submarine oscillator
subot submarine bottom
SUBPA Submarine Patrol Area (USN)
SUBPAC Submarine Forces, Pacific (USN)
sub para sub paragraph
subplane submersible seaplane
sub-pro subprofessional
subprog subprogram(ming)
sub pub(s) subsidy pub-

lisher(s) [vanity publisher(s)]
SUBPZ Submarine Patrol Zone (USN)
subq subsequent
subroc (SUBROC) submarine rocket
subrog subrogation
Subron Submarine Squadron
subrqmt subrequirement
subs submarines; subscription(s); subsistence; substantial violations; substitutes
subsafe submarine safety (program)
subsan submarine sandwich (also called sub)
sub sec subsection
subseq subsequent(ly)
subset subscriber set
subsis subsistence
subsp. subspecies (Latin)
SUBSS Submarine Schoolship (USN)
subst substantive
substa substation
substance P polypeptide found in the brain
substand. substandard
substd substandard
substr substructure
subsunk submarine sunk
subsys subsystem
SUBTACGRU Submarine Tactical Group(ing)
subtopia suburban utopia
subtr subtraction
SUBTRAFAC Submarine Training Facility
sub u substitute unit
suburb suburban; suburbanite; suburbia; suburbian
SUBWESTLANT Submarine Force—Western Atlantic (NATO)
suc succeed; success; successor
suc. succus (Latin—juice)
SUC Society of University Cartographers; Sussex University College
Succ Successori (Italian—Successors); *Succursale* (Italian—Branch)
Sucr Sucursal (Spanish—subsidiary, branch)
sucre monetary unit of Ecuador
Sucre Antonio José de Sucre—South American liberator fighting with Bolívar for freedom of Venezuela, Colombia, Ecuador, Peru, and Bolivia from Spanish rule; Mariscal Sucre (Quito, Ecuador's airport named for Marshal Sucre)
suct suction

SUCU Society for Universal Cosmic Uncertainty
sud sudden unexpected death; sudden unexplained death
Sud Sudan; Sudanese
SUD Aerovias Sud Americanas (3-letter airline coding)
Sudaf Sudáfrica (Spanish—South Africa)
SUDAM Superintêndencia do Desenvolvimento da Amazonia (Portuguese—Superintendency for the Development of Amazonia)
Sudan Democratic Republic of Sudan (Africa's biggest country), *Jumhuryat es-Sudan Al Democratia*—formerly the Anglo-Egyptian Sudan known as Nubia in Roman times
SUDAN Sudan Airways
SUDENE Superintêndencia do Desenvolvimento do Nordeste (Portuguese—Superintendency for the Development of North-East Brazil)
SUDS Silhouetting Underwater Detecting System; Submarine Detecting System
Sud Tas Sudmen Tasavalta (Finnish—Republic of Finland)
Sue Susan; Susannah; Suzanne
suec suéco (Spanish—Swedish); *sueco* (Portuguese—Swedish)
Suec Suecia (Spanish—Sweden); *Suécia* (Portuguese—Sweden)
SUEL Sperry Utah Engineering Laboratory
Suet Gaius Suetonius Tranquillus (Roman biographer)
suf sufficient; suffix
Suff Suffolk
suffoc suffocating
sug suggest(ion)
SUG Southern California Gas Company (stock exchange symbol)
SUGAR Services, (to diabetics through) Understanding, Grants, Assistance, Recreation
SuH Sundays and Holidays
SUI State University of Iowa
suicidol suicidologist(ic); suicidology
suid sudden unexplained infant death (crib death)
sui rep suicide report
SUIT Scottish and Universal Investment Trust
suiv suivant (French—following)

Suk Sukkot
Suky Susan; Suzanne
sul simplified user logistics; small university libraries
Sul Suleiman (Arabic—Solomon)
SUL Stanford University Libraries
Sula Sulawesi (Celebes)
sulcl set up in less than carloads
sulf sulfate; sulfur
sulfa sulfanilamide
sulfd sulfide(s)
sulfuric acid H_2SO_4
Sulli Sullivan
Sult Sultan(a)
Sulu Jolo
Sulus Sulu Islands between Indonesia and the Philippines
sum (SUM) surface-to-underwater missile
sum. summary; surface-to-underwater missile (SUM)
sum. sume (Latin—take)
Sum Sumatra; Sumatran; Sumer; Sumeria; Sumerian
SUM Servicio Universitario Mundial (Spanish—World University Service)
SUMCMO Summary Court-Martial Order
Sumi Sumitomo Bank
Sumitomo Sumitomo Shoji America; Sumitomo Shoji
summ summarization; summarize; summarizing
Summer Bank Summer Bank Holiday (last Monday in August in Great Britain)
SUMOC Superintendencia da Moeda e do Crédito (Portuguese—Superintendency of Money and Credit)
sumr summer
sums. summons
SUMS Sperry Univac Material System
sum. tal. sumat talem (Latin—take one like this)
sun. symbolic unit number (SUN)
Sun Sunday
Sun The Baltimore Sun
SUN Solar Usage Now; Symbols, Units, and Nomenclature Commission
SUNA Sudan News Agency
Sund Sunda Islands; Sundanese
Sundarbans Sundarban creeks, half-reclaimed islands, marshes, rivers, and swamps in the Ganges delta country between Bangladesh and India

Sundas Sunda Islands of Indonesia
SUNFED Special United Nations Fund for Economic Development
Sungaria Dzungaria or Zungaria region between Mongolia and Russia
SUNOCO Sun Oil Company
SUNS Sonic Underwater Navigation System
Sunset Crater Sunset Crater National Monument in north-central Arizona
SUNY State University of New York
SUNYAB State University of New York at Buffalo
Sun Yat-Sen's Dr Sun Yat-Sen's Birthday (November 12)
sup superb; superfine; superior; superlative; supersede(s); supine; supplement(ary); supplies; supply; support; supposition; supreme
sup supérieure (French—higher; superior, upper)
sup. supra (Latin—above)
SUP Sailors Union of the Pacific; Socialist Unity Party; Southern University Press; Stanford University Press; Sussex University Press; Syracuse University Press
SUPCE Syracuse University Publications in Continuing Education
supchg supercharger
Sup Ct Superior Court; Supreme Court
supdel superdelicious
Sup Dpo Supply Depot
supe (slang) superintendent; supernumerary
super superficial; superfine; superheterodyne; superimposition; superintendent; superior; supermarket; supernumerary; supersede; supersession
super (Latin prefix—above, beyond, upper)—superficial, superior, superpower; *supermercado* (Spanish—supermarket)
SUPER Skills Upgrading Program for Educational Reinforcement
superaero superaerodynamics
superan superannuated
superconduct superconductive; superconductivity; superconductor(s)
superf superficie (Italian—area; surface, surface area)

superhet superheterodyne
superjet(s) supersonic jet airplane(s)
superl superlative
super(s) supercargo(s); supercharger(s); superheater(s); superheterodyne(s); superhighway(s); superhuman(s); superintendent(s); superior(s); superior court(s); superior planet(s); superlative(s); superliner(s); supermarket(s); superorganism(s); superpatriot(s); superpower(s); superscript(s); supersonic(s); superstition(s); superstructure(s); supervisor(s)
superstr superstructure
Super^{te} Superintendente (Spanish—superintendent)
superv supervisor
supgon super gonorrhea (resistant to all antibiotics)
suphtr superheater
SUPIR Supplementary Photographic Interpretation Report
sup. lint. super linteum (Latin—on lint)
Sup O Supply Office(r)
SUPOPS Supply Operations (DoD)
supp supplement; suppuration
supp. suppositorium (Latin—suppository)
Sup P Supply Point
suppl supplement (French—supplement)
Suppl supplement
suppos suppository
supps supplementary procedures; supplements
SupPt supply point
suppy supplementary
supr superior; supreme
supra (Latin prefix—above or over)—suprarenal
supra cit. supra citato (Latin—cited above)
supsd supersede(d)
Sup Ship Supervisor of Shipbuilding
supt superintend; superintendent
Supt Docs Superintendent of Documents
supv supervise; supervisor
supvr supervisor
supvry supervisory
sur surface; surfacing
Sur surgery; Surinam (Netherlands Guiana)
Sur Surabaya (Indonesian—Soerabaya)
Suralco Surinam Aluminum Company
surano surface radar and navi-

gation operation
sur art surrealistic art
surbage subway-floor garbage
surcal surveillance calibration (satellite)
Sur Cdr Surgeon Commander
SURE Symbolic Utilities Revenue Environment
sureq submit requisition
surf. spent unreprocessed fuel
Sur f Surinam florin (guilder)
surf. a surface area
SURFF Spent Unreprocessed Fuel Facility
SURFPA Surface Patrol Area
SURFPZ Surface Patrol Zone
surg surgeon; surgery; surgical
Surg Cdr Surgeon Commander
surge. sorting, updating, report generating
Sur Gen Surgeon General
Surg Gen Surgeon General
surgiserv surgical service(s)
Surg Lt Cdr Surgeon Lieutenant Commander
Surg Maj Surgeon Major
Suri Surinam (formerly Dutch Guiana)
suric surface ship integrated control
Surinam formerly Dutch or Netherlands Guiana
surpic surface picture
surr surrender
Surr Surrogate
surrept surreptitious(ly)
SURS Surface Export Cargo System
SURSAN Superintendência de Urbanismo e Saneamento (Portuguese—Superintendency of Urbanism and Sanitation)
SURTASS Surveillance-Towed-Array Sonar System
surv survey; surveying; surveyor
SURV Standard Underwater Research Vessel
Surv Gen Surveyor General
survll surveillance
sus supressor sensitive; suspect(ed); suspected person; suspend(ed)
Sus Saybolt universal second; Susanna, The (Apocryphal) History of Sussex
SUS Scottish Union of Students; Society of University Surgeons
SUSA Scouting USA (formerly the Boy Scouts of America—BSA)
susfu situation unchanged—still fouled up

susie surface and underwater ship-intercept equipment
susp suspect(ed); suspend; suspend(ed)
susp b suspension bridge
sus. per coll. suspensio per collum (Latin—hanging by the neck)
suspn suspension
suspnd suspending
susp(s) suspect(s) [person(s) suspected]
Susque Susquehanna River flowing from western New York through Pennsylvania and Maryland before entering Chesapeake Bay
Süss Franz Xaver Süssmayr
SUSS Society of Utah School Superintendents
sust sustainer
SUSTA Southern United States Trade Association
Susx Sussex
SUT Society for Underwater Technology
Suth Sutherland
s'uth'ard southward
SuU Staats und Universitätsbibliothek (German—State and University Library)—Hamburg
suud sudden unexpected unexplained death
SUV Saybolt Universal Viscosity; Suva, Fiji Islands (Nandi Airport)
SUVCW Sons of Union Veterans of the Civil War
SUX Sioux City, Iowa (airport)
Suz Suez
sv (RCA patent); safety valve; sailing vessel (SV); security violator; selectavision (SV); sheet vinyl; simian virus; single vibrations; sinus venosus; stroke volume; survey; surveyor
s/v surrender value; survivability/vulnerability
sv sotto voce (Italian—in an undertone, in a whisper); *svacek* (Czech—volume); *svensk* (Dano-Norwegian—Swedish)
s.v. spiritus vini (Latin—alcohol); *sub verbo* or *sub voce* (Latin—under the word; under the voice)
Sv Svaty (Czechoslovakian—holy); *Sveti* (Serbo-Croatian—holy)
SV sailing vessel; Selective Volunteer; Sons of Veterans
S & V Sinclair and Valentine

SV Standard Version
sv 40 simian virus 40
Sva Suva
SVA Schweizerische Vereinigung für Atomenergie (German—Swiss Association for Atomic Energy)
Sval Svalbard (Spitsbergen)
Svalbard (Norwegian—Spitsbergen)—Arctic islands
Svb Svendborg
SVB Stephen Vincent Benét
svc service; superior vena cava
svc (SVC) service (flow chart); supervisor call(ing)
SVC Skagit Valley College; Society of Vacuum Coaters
svcbl serviceable
SVCP Special Virus Cancer Program
svcs superior vena cava syndrome
svd spontaneous vaginal delivery; spontaneous vertex delivery; swine vesicular disease
SVD Schweizerische Vereinigung für Dokumentation (German—Swiss Documentation Association)
sve secure voice equipment
Sve Sveits (Norwegian—Switzerland)
SVE Society for Visual Education
Sven Akad Svenska Akademien (Swedish Academy)
Sver Sverdlovsk; Sverige (Swedish Academy); *Sverige* (Norwegian—Sweden)
svg saving
SVG San Vicente y las Granadinas (Spanish—St Vincent and the Grenadines)
s.v. gal. spiritus vini gallici (Latin—brandy)
svi stroke volume index
s.v.i. spiritus vini industrialis (Latin—industrial alcohol)
svib strong vocational interest blank
SVIOC South Varanger Iron Ore Company
SVL Scripps Visibility Laboratory
s.v.m. spiritus vini methylatus (Latin—methyl alcoholic)
SVN Student Vocational Nurse
Svn Dag Svenska Dagbladet (Swedish Daily Blade)
SVnese South Vietnamese
SVNV Societa Veneziana di Navigazione a Vapore (Venetian Steamship Company)
SVO Moscow, USSR (Sheremetyevo Airport); Special

Vehicle Operation
SVP Society of Vertebrate Paleontology
S V P s'il vous plaît (French—if you please)
SVPs Senior Vice Presidents
svr super video recorder
s.v.r. spiritus vini rectificatus (Latin—rectified spirit of wine)
SVR Suomen Valtion Rautatiet (Finnish—Finnish State Railways)
SVRA Sportscar Vintage Racing Association
sv's security violators
SVS Society for Vascular Surgery; Society for Visiting Scientists; Still-camera Video System
SVS Sveriges Standardiseringkommission (Swedish—Swedish Standards Commission)
s.v.t. spiritus vini tenuis (Latin—proof alcohol; proof spirit)
SVT Self-Valuation Test
SVTL Services Valve Testing Laboratory
svtol (SVTOL) short/vertical takeoff and landing
svtp sound, velocity, temperature, pressure
svtt surface-vessel torpedo tube
s.v.v. sit venia verbo (Latin—forgive the expression)
svy survey
sw salt water; sea water; sent wrong; shipper's weights; short wave; shotgun wedding; single weight; slow wave; special weapon; spotweld; spotwelding; station wagon; steelworker; stock width; swell organ; switch; switchband wound
s-w shortwave
s/w salt water; sea water; seaworthy; standard weight
s & w salaries and wages; surveillance and warning
Sw Sweden; Swedish
SW Sadler's Wells (London theater); Secretary of War; Security Watch; Senior Warden; Shelter Warden; Ship's Warrant; South Wales; southwest; Southwest Airways (2-letter coding); Stone & Webster (stock exchange symbol)
S-W Sherwin-Williams
S & W Seaboard & Western (airlines); Smith & Wesson; Stone & Webster

SW1, SW2, etc. Southwest One, Southwest Two, etc. (London postal zones)
swa single-wire armored; superwide angle
Swa Swahili
SWA Seaboard World Airlines; South-West Africa; Southwest Airways
SWAA Southwestern Aeronautical Association
swabk sealed with a big kiss
swac special warhead arming control
Swac Standards western automatic compiler (NBS)
SWAC South-West Africa Company
SWACS Space Warning and Control System
SWAFAC Southwest Atlantic Fisheries Advisory Commission
swag(s) scientific wild-assed guess(es)
SWAI South-West African Infantry
swak sealed with a kiss
SWALCAP South-West Academic Libraries Cooperative Automation Project
swalk sealed with a loving kiss
swami. software-aided multifont input
Swans Swan Islands off Honduras
SWANU South-West Africa National Union
SWANUF South-West Africa National United Front
swap. selective wide-area paging
SWAPO South-West Africa People's Organization
swash sea wash (scouring surf running up a beach after a wave breaks)
SWAT Special Weapons and Tactics (team of law-enforcement officers trained to combat guerrillas and terrorists)
swath small waterplane-area twin hull
swath (SWATH) small-waterplane-area twin-hull (naval craft)
SWATH Small Waterplane-Area Twin Hull (craft designed for stability in rough seas)
swatson so what's on?
s waves secondary (earthquake) waves
S-waves shear waves
Swaz Swaziland
Swaziland Kingdom of Swazi-

land (landlocked South African country)
swb short wheelbase; single with bath; swing bridge
SWB South Wales Borderers
swbd switchboard
swbld switchblade (knife or stiletto)
swbm still-water bending moments
S & W bracelets Smith and Wesson handcuffs
SWBRC Southwest Border Regional Commission
swc specific water content
SWC Simon Wiesenthal Center; Soil and Water Conservation (US Department of Agriculture); Special Weapons Command; Supreme War Council
SWCEL Southwestern Cooperative Educational Laboratory
Swch Switch
SWCHS Simon Wiesenthal Center for Holocaust Studies (Yeshiva University)
SWCLR Southwest Council of La Raza
swd sawed; sewed; short-wave diathermy
SWD South Wales Docks
SWDA Scottish Wholesale Druggists' Association; Solid-Waste Disposal Act (EPA)
Swe Swede(n); Swedes; Swedish
SWE Society of Wine Educators; Society of Women Engineers
sweatl student work experience and training
sweat(s) sweatshirt(s)
SWEB South Wales Electricity Board; South West Electricity Board
SWEC Stone & Webster Engineering Corporation
Swed Swede; Sweden; Swedish
Swed Swedish (Germanic language)
Sweden Kingdom of Sweden, *Konungariket Sverige*
SWEDL Southwest Educational Development Laboratory
Sweetwaters Sweetwater Mountains of California and Nevada
SWETM Society of West End Theatre Managers
swf single white female
SWF Stockholders for World Freedom

SWFB Southwestern Freight Bureau
Sw Fr Swiss franc
sw fx spotweld fixture
SWG Society of Women Geographers; Standard Wire Gauge
Sw-Ger Swiss-German (derived from Alemannic)
swi stroke work index
Swi Swietochlowice
SWI Spring Washer Institute
SWIE South Wales Institute of Engineers
swife sexual wife
swift. selected words in full title
SWIFT Society for Worldwide Interbank Financial Telecommunication
swift. lass. signal word index of field and title—literature abstract specialized search
swift. sir. signal word index of field and title—scientific information retrieval
SWINE Students Wildly Indignant (about) Nearly Everything (cartoonist Al Capp's contribution to contemporary acronyms)
Swinglish Swedish-English
SWIO SAClant War Intelligence Organization
SWIR Special Weapons Inspection Report
SWIRL South Western Industrial Research Limited
SWIRS Solid Waste Information Retrieval System
SWISSAIR Swiss Air Transport
Swiss Confed Swiss Confederation
switch switchblade knife
Switz Switzerland
Switzerland Swiss Confederation of Cantons (Alpine nation of great productivity and high-quality workmanship)— Schweiz (German or Romansch), Suisse (French), Svizzera (Italian)
swives sexual wives
swJf single white Jewish female
swJm single white Jewish male
Sw kr Swedish krona (monetary unit)
swl short wave listener
SWL safe working load (for cargo booms and derricks; SWL 5T 15 deg means the safe working load is 5 tons at 15 degrees off the horizon-

tal); Swedish American Line
SWLA Southwestern Library Association
SWLI Southwestern Louisiana Institute
swlolak's sealed with lots of love and kisses
SWly south-westerly
swm single white male; standards, weights, and measures
SWM Southwest Museum
SWMA Steel Wool Manufacturer's Association
swmbo she who must be obeyed
SWMF South Wales Miners' Federation
SWMFB Southwestern Motor Freight Board
Swn Swinoujscie
SWN Synoptic Weather Network
Swnbne Swanbourne
SWO Solid Waste Office (Environmental Protection Agency)
SWOA Scottish Woodland Owners' Association
swoc subject word out of context
swog special weapons overflight guide
SWOPSI Stanford Workshops on Political and Social Issues
SWORCC Southwestern Ohio Regional Computer Center
's' word god's word
SWORDS Shallow-Water Oceanographic Research Data System
SWORL Southwestern Ohio Regional Libraries
swot strengths, weaknesses, opportunities, threats
's' wounds god's wounds
swp safe working pressure; sewer(age) planned; sweep; sweeper; sweeping
SWP Saskatoon Wheat Pool; Sherwin-Williams Paints; Socialist Workers Party; South Wales Ports; Southwest Pacific; Special Weapons Project
SWPA Southwest Pacific Area; Southwestern Power Administration; Surplus War Property Administration
swpf short wave-pass filter
swr serum wasserman reaction; sewer(age); standing-wave ratio; steel-wire rope; switch rails
swrf sine wave response filter
S-W RI Sterling-Winthrop Research Institute

swrj split wing ramjet
SWRL Southwest Regional Laboratory
sws seam-welding system; service-wide supply; slow-wave sleep; solar-wind spectrometer; still water surface
Sws Swansea
SWS Sariska Wildlife Sanctuary (India); Space Weapons System; Special Weapons System
SWSC Schlumberger Well Surveying Corporation
S & W S C Space and Warfare Systems Command (USN)
s-w sleep slow-wave sleep
swt short-wave transmission; short-wave transmitter; single weight; spiral(ly)-wrap(ped) tubing; steel watertight; switch(ing)
SWT School of Welding Technology; Scottish Wildlife Trust
SWTB Surface Wellbore Test Bank
SWTC Scottish Woolen Technical College
swtchmn switchman
SWTEA Scottish Woolen Trade Employers' Association
swtg switching
SWTMA Scottish Woolen Trade Mark Association
SWTS Seabury Western Theological Seminary
SWUS Southwestern United States
swv swivel
SWWJ Society of Women Writers and Journalists
swy slipway; stopway
swymmd see what you made me do
sx section; simplex
Sx (medical) signs and symptoms
SX Essex; sex; Southern Pacific (stock exchange symbol)
sxa stored index to address
SXC Saint Xavier College
sxl short-arc xenon lamp
SXM St Maarten, Netherlands Antilles (airport)
sxn section
SXO Senior Experimental Officer
sxr soft X-ray region
sxrm straight reamer
sxs stellar X-ray spectra
SXS Sigma Xi Society
sxt sextant; stable X-ray transmitter

sy shipyard; square yard; sticky; supply; sustainer yaw
Sy Shipyard; Syria; Syrian
SY San Ysidro; South Yorkshire; steam yacht (naval symbol); (U.S. State Department) Security Office
Syb Sybil
SYB *Statesman's Year-Book*
SYC Sandusky Yacht Club; Savannah Yacht Club; Seattle Yacht Club; Springfield Yacht Club; Stamford Yacht Club
SYCATE Symptom-Cause Test
sycom synchronous communication(s)
sy crs sundry creditors
syd see your doctor; sum of the year's digits
Syd Sydney
Syd sydlig (Danish—southerly)
SYD Scotland Yard; Sydney, Australia (airport)
S Yem South Yemen
SYEP Summer Youth Employment Program
syf syphilis
syfa system for application
SyG Secretary General
syh see you home
SYHA Scottish Youth Hostels Association
Sy'kat Syarikat
syl syllogism
syla-iawc see you later, alligator—in a while, crocodile
syll syllabication (syllabification)
syllo syllogism; syllogistic(al)-(ly); syllogist; syllogize(d); syllogizing
SYLP Support Your Local Police
Sylv Sylva; Sylvain; Sylvan(der); Sylvanus; Sylvester; Sylvius
sym symbol; symbolic; symbolism; symmetric; symmetrical; symmetry; symphonic; symphony
sym (Latin prefix—together)— symphony; *symphonie* (French—symphony)
sym. *symbolus* (Latin—token; sign)
symb symbol; symbolic; symbolism
symbal symbolic algebra
Sym Fan Symphonie Fantastique
symp symposia; symposium
sympac symbolic program for automatic control
sympath sympathetic; sympa-

thy
Symph Mont Orchestre symphonique de Montréal (French—Montreal Symphony Orchestra)
symphon symphonia (Greek or Latin—symphony)
symps symptoms
sympt symptom(s)
SYMRAP Symbolic Reliability Analysis Program
SYMRO System Management Research Operation
SYMS Symmetrical System
Sym & Signs Symbols & Signs
SYMWARR System for Estimating Wartime Attrition and Replacement Requirements
syn synagogue; synesthesia; synonym; synonymous; synonymy; syntax; synthetic
syn (SYN) synchronous idle character (data processing)
syn (Greek—together or with)—synapsis, syndrome
Syn Synagogue
Synanon anti-drug addiction group
sync synchronize; synchronous
synchro synchronize; synchronous
synchros synchronous devices
synco syncopate(d); syncopation; syncopative; syncopator
syncom synchronous communication (satellite)
syncon synergistic convergence
syncop syncopate(d); syncope
syncrude(s) synthetic crude oil(s)
synd syndicalism; syndicate
syndet(s) synthetic detergent(s)
syndro syndrome
syne syntactic elements
synec synecdoche
Synfuel U.S. Synthetic Fuel Corporation
synfuel(s) synthetic fuel(s)
syn gas synthetic gas
SYNMAS Synchronous Missile Alarm System
syn oil synthetic oil
synon synonymous; synonym
synonym. synonymous
synop synopsis; synoptic
synroc synthetic rock
syns synopsis
synscp synchroscope
synt syntax
syntan synthetic tanning
synth synthesis; synthetic
synth-pop synthesized popular music
syntol syntagmatic organiza-

tion of language
syntrain synthetic training (aviation)
syntran syntax translation
S Yorks South Yorkshire
SYP Society of Young Publishers
syph syphilis; syphilitic
syphil syphilology
Sy PO Supply Petty Officer
SYPR Southern Yemen People's Republic
syr syrup
syr. *syrupus* (Latin—syrup)
Syr Syracusan; Syracuse; Syria; Syriac; Syrian
SYR Syracuse, New York (airport)
Syrac Syracusan; Syracuse
syrg syringe
Syria Syrian Arab Republic (Middle Eastern nation), *al-Jamhouriya al Arabia as-Souriya*
syrm save-your-rear memorandum
sys system; systematic; systematization; systematize; systemic; systems
SYS Sun Yat-sen
sysabend system abnormal end(ing)
syscp system card punch(ing)
sysda system direct access
sysgen systems generation
sysin system input
syslib system library
syslined system linkage editor
syslmod system load module
sysop systems operator
sysout system output
SYSP Sixth-Year Specialist Program (library science)
Sys PO Systems Program Office(r)
syssq system sequential
syst system; systematic; systemic; systems
System ABC System of Automation of Bibliography through Computerization
systol systolic
systran systems analysis translator
sysut system utility (data sets)
syt sweet young thing
syz syzgetic; syzygial; syzygium; syzygy (alignment of the Earth, Moon, and Sun resulting in unusually high tides and weather disturbances)
sz schizophrenia; schizophrenic; seizure; size; stratum zonal
s Z seinerzeit (German—at that

time)
Sz Swiss; Switzerland
sza solar zenith angle
SZA Student Zionist Association
Szb Salzburg
SZG Salzburg, Austria (airport); Soviet Zone (in) Ger-

many
Szle *Szemle* (Hungarian—journal, review)
Szn Szczecin (formerly Stettin—Stn)
SZO Student Zionist Organization
SZOG Soviet Zone of Occupa-

tion in Germany
szr (SZR) sodium-cooled zirconium-hydride moderated reactor
szvr silicon zener voltage regulator

T

t airfoil temperature thickness (symbol); hour angle (symbol); meridian angle (symbol); table; tabulated (loran); tackle; tardy; tare; teaspoon; teeth; telephone; temperature; temporary; tenor; tense; tensor; tentative; tentative target; thunder; thunderstorm; tide; tide rips; time; title; tons; tonnage; torn; toward; town; trace of precipitation; transferred; transit; transitive; translation; tread; tropical; troy; true; tug; tugline

t *(t)* units of land (microeconomics)

t (T) tea (marijuana)

t' the; to

't it

t *tome* (French—volume); *tomo* (Spanish—volume)

t. *ter* (Latin—three times; thrice)

't *het* (Dutch—the)

T Northrup Aircraft (symbol); Pacific Transport Lines; propeller thrust (symbol); tablespoon; tactical; Tango—code for letter T; tanker; Taoism; Taoist; T-bar; tee; teletype; temperature; temple; temporary magnitude; tension of eyeball; Tesla; Testla; Texaco; Texas; Texas Company; Thursday; thymine; torpedo; town(ship); trainer; training; Transamerica (airline); transducers; transport number; triangle; triple bond; true; truss; Tuesday; turboprop; Turk; Turkey; Turkish

T (Latin—Titus); tea (slang—marijuana or Texas tea); *Teil* (German—division, part);

thrust (symbol); *Time* (magazine); transformer (symbol); *tulo* (Finnish—arrival)

T-1 absolute temperature or transformer symbol; Canadian income-tax return

t½ radioactive half life

T - 1, T - 2, T - 3, etc. decreasing stages of interocular tension

T + 1, T + 2, T + 3, etc. increasing stages of interocular tension

T₁, T₂, T₃, etc. first thoracic vertebra, second thoracic vertebra, third thoracic vertebra, etc.

T2 stabilized

T-2 North American-Rockwell Buckeye trainer aircraft; tricothecenes used in biochemical warfare

T2g Technician (second grade)

T₃ triiodothyronine

t-4 therefore

T4 heat treated

T-4 Canadian statement of employment income recorded for tax purposes

T₄ thyroxine

T6 heat treated and aged

T-6 North American-Rockwell Harvard or Texan trainer aircraft

T7 heat treated and stabilized

T-7 Beechcraft navigational-training aircraft

T-10 Soviet heavy tank armed with a 122mm gun

T-11 Beechcraft bomber-training aircraft

T-28 North American Trojan trainer aircraft

T-29 Convair military transport also called Samaritan

T-33 Lockheed Shooting Star trainer aircraft

T-34 Beechcraft Mentor trainer aircraft; Soviet medium tank armed with an 85mm gun

T-37 Cessna Dragonfly twin-engine jet trainer

T-39 North American Sabreliner transport aircraft

T-41 Cessna 172 Mescalero trainer-utility aircraft

T-42 Beech Cochise transport aircraft

T-43 Boeing navigational trainer and transport aircraft; Soviet fleet minesweeper

T51 specially aged

T-54 Soviet medium tank

T-55 Soviet medium tank armed with a 100mm gun

T-59 mainland-China-made medium tank modeled after Soviet T-54 tank

T-62 Soviet medium tank with a 115mm gun

T-64 Soviet medium tank with a 120mm gun

T-104 Tupolev 104 aircraft

T-144 Tupolev 144 (Soviet supersonic transport)

T-301 Soviet coastal minesweeper

T-1824 Evans blue

ta target area; temperature, axillary; test accessory; third attack (lacrosse); time and attendance; toxin-antitoxin; transactional analysis; transverse acoustic; travel allowance; true altitude; tuberculin, alkaline

ta (TA) teaching assistant; terephthalic acid; transactional analysis

t-a toxin-antitoxin

t/a trading as

t & a taken and accepted; time and attendance; tonsillectomy and adenoidectomy; tonsils and adenoids

t of a terms of agreement

ta transit authority (New York City Transit Authority)

t.a. testantibus actis (Latin—as the records show)

Ta tantalum; Tasmania; Tasmanian

TA Table of Allowances; tactical air (missile); Tax Amortization; Teaching Assistant; Technical Assistance; Tel Aviv, Israel; Territorial Army; Trade Agreement(s); Transactional Analysis; Trans-Air; Trans-America Corporation (stock exchange symbol); Transit Authority; Truth in Advertising; Turkish Army

T-A Tacna-Arica (on the border of Peru and Chile)

T/A Teaching Assistant; Temporary Assistant

T of A Timon of Athens

taa turbine-alternator assembly

TAA Technical Assistance Administration; Temporary Assistance Authority; Trade Adjustment Assistance; Trade Agreements Act; Trans-Australia Airlines; Transit Advertising Association; Transportation Association of America

TAACOM Theater Army Area Command

TAAF Terres Australes et Antarctiques Françaises (French —French Austral and Antarctic Territories)—Adélie Land in Antarctica plus the islands of Amsterdam and St Paul, the Crozets, and the Kerguelans in the south Indian Ocean

TAAG Transportes Aéreos de Angola (Portuguese—Air Transports of Angola)

taalk taalkunde (Dutch—linguistics)

TAALODS The Army's Automated Logistic Data System

TAALS The American Association of Language Specialists

TAAP Total Action Against Poverty

TAARS The Army Ammunition Reporting Service

taas three-axis attitude sensor

TAAS Telfair Academy of

Arts and Sciences (Savannah)

TAASA Tool and Alloy Steels Association

TAASP The Association for the Anthropological Study of Play

tab. table; tablet; tabulate(d); tabulation; tabulator; technical assistance broker(age); therapeutic abortion

tab. tabella (Latin—small board; tablet)

Tab Tabascan; Tabasco

Tab Tabelle (German—table; index)

TAB Technical Assistance Board (UN); Tobago (airport); Totalisator Agency Board; Totalizator Board

TAB Technical Abstract Bulletin

TABA The American Book Award(s)

TABA Transportes Aéreos Buenos Aires (Spanish—Buenos Aires Air Transport)

tabasco tabasco sauce

Tabby Tabitha

tabc typhoid-paratyphoid A, B, and C vaccine (TABC)

tabel tabella (Latin—tablet)

TABL Tropical Atlantic Biological Laboratory

tabl(s) tablet(s)

tab run tabulator run

tab(s) tablet(s)

Tabs Cantabrigians or Cantabs —Cambridge University undergraduates

TABS Transatlantic Book Service

tabsim tabulating simulator

TABSO Transport Aerien Civil Bulgare (Bulgarian Civil Air Transport)

tabsol tabular systems-oriented language

tabt tab vaccine plus tetanus toxoid (TABT)

tabtd combined tab vaccine plus tetanus and diphtheria toxoid

TAB vaccine typhoid plus paratyphoid A and B vaccine (triple vaccine)

tabwx tactical air base weather

tac tactic; tactical; tactician; tactics; total automatic color (tv); try and collect

Tac Tacitus; Tacoma

TAC Tactical Air Command; Talent-Assistance Cooperative; Technical Advisory Committee; Technical Assistance Center; Terrain Analy-

sis Center; Thai Airways Company; Trade Agreements Committee

TACA Texas and Central American Airlines

TACAMO Take Charge And Move Out (USN)

tacan tactical air navigation

Tac Brdg Tacoma Bridge

TACC Tacna-Arica Copper Consortium; Tactical Air Command Center; Tactical Air Control Center; Technology Assessment Consumerism Center

taccar time-averaged clutter-coherent airborne radar

tacco tactical coordinator

TACCP Tactical Command Post (USA)

TACCTA Tactical Air Commander's Terrain Analysis

tacden tactical data-entry device

TACELIS Transportable Emitter Location and Identification System

tacelron tactical electronic warfare

TACEST Tactical Test(ing)

TACG Tactical Air Control Group

tach tachometer

Tacho Anastasio

tachy tachygraphy (shorthand)

tachy (Latin prefix—rapid or swift)—tachycardia

tachycard tachycardia

tacit. tacitus (Latin—unmentioned)

tacjam tactical jammer; tactical jamming

TACL Tactical Air Command Letter

taclan tactical landing system

TACLET Tactical Law Enforcement Team (Coast Guard)

tacmar tactical malfunction-array radar

tacnav tactical navigation

tacnuc tactical nuclear (weapon)—also written taknuk

TACO Tactical Coordinator

tacoda target coordinate date

tacol thinned-aperture computed lens

TACOM Tank-Automotive Command (USA)

TACOMEWS Tactical Communications Electronic Warfare Systems

Taconics Taconic Mountains ranging from New York to Vermont but called the Berkshires in Connecticut and

Massachusetts
TACOS Tactical Airborne Countermeasures or Strike (USAF); Tactical Air Command Simulation
TACP Tactical Air Control Party
tacpol tactical procedure-oriented language
TACR Tactical Air Command Regulation
TACRON Tactical Air Control Squadron
TACs Technical Assistance Committees (UN)
TACS Tactical Air Control System
t-a-c salad turkey-avocado-cheese salad
tacsatcom tactical satellite communications
TACSS tactical schoolship (USN)
tact. technological aids to creative thought
TACT Texas Association of College Teachers; Truth About Civil Turmoil
TACTIC Technical Advisory Committee to Influence Congress (Federation of American Scientists)
TACTICS Technical Assistance Consortium to Improve College Services
tacv tracked air-cushion vehicle
tad tadpole; telemetry analog-to-digital (information converter); terminal area distribution (processing); traffic analysis and display; transaction application driver; throwaway detector; time available for delivery
tad (TAD) temporary additional duty
Tad Thaddeus; Theodore
TAD Thrust-Augmented Delta
TAD The Anglican Digest
TADA Teletypewriter Automatic-Dispatch System
TADARF Toronto Alcoholism and Drug Addiction Research Foundation (Canadian)
TADARS Tropo Automated Data Analysis Recorder System
TADC Tactical Air Direction Center; Texas Association of Developing Colleges; Training and Distribution Center
tadic telemetry analog-to-digital information computer
tad(s) tadpole(s)
TADS Teletypewriter Auto-

matic Dispatch System
TADSYS Turbine Automated Design System
Tadz Tadzhik; Tadzhikistan; Tadzhikistanian
Tadzhik SSR Tadzhik Soviet Socialist Republic (Tadzhikistan)
TAE National Greek Airlines; Trans-Antarctic Expedition
TAEA Texas Art Educators Association
TAEC Turkish Atomic Energy Commission
TAEDS Texas Association for Educational Data Systems
TAEG Training Analysis and Evaluation Group (USN)
TAEHS Thomas A Edison High School
Tae Kin Taehan Min'guk (Korean—Republic of Korea) - South Korea
ta'en taken
TAERF Texas Atomic Energy Research Foundation
taf terminal aerodrome forecast
taf (TAF) toxoid-antitoxin floccules
Taf Tessar auto focus
Taf Bildtafel (German—list of illustrations)
TAf Tuberculin Albumose frei (German—albumose-free tuberculin)
TAF Tactical Air Force
TAFA Territorial and Auxiliary Forces Association
tafcsd total active federal commissioned service date
Taf-d Tessar auto focus dating
tafg two-axis free gyro
TAFI Technical Association of the Fur Industry
tafmsd total active federal military service date
tafor terminal aerodrome forecast
TAFSEA Technical Applications for Southeast Asia
TAFSONOR Tactical Air Force, Southern Norway (NATO)
tafubar things are fouled up beyond all recognition
ta fx tapping fixture
tag. the acronym generator (RCA device)
Tag Tagalog (the language of the Philippines)
TAG The Adjutant General; The Alzheimer Group; The Association for the Gifted; Test Analysis Guide; Timken Art Gallery
T A & G Tennessee, Alabama

& Georgia (railroad)
TAG Transports Aeriens Guyanais (French—Guiana Air Transport)
TAGA Technical Association of the Graphic Arts
Tagal Tagalog
tagawi try and get away with it
TAGCEN The Adjutant General's Center (USA)
TAGG Taxpayers Against Government Giveaways
tagl täglich (German—daily; per day)
TAGP Transportes Aéreos do Guine Portuguesa (Portuguese—Air Transport of Portuguese Guinea)
TAGS Time-Automated Grid System
tagw takeoff gross weight
tah temperature, altitude, humidity; total abdominal hysterectomy
Tahiti formerly Otaheite
Tah Pac Tahitian Pacific (area around Tahiti)
TAHq Theater Army Headquarters
TAHRI Tobacco and Health Research Institute
tai taiga (coniferous evergreen forests of subarctic America, Asia, and Europe)
Tai Taipei; Taiwan (Formosa)
Tai Tailandia (Spanish—Thailand)—Siam
TAI Thai Airways International; *Transports Aériens Intercontinentaux* French—Intercontinental Air Transport; Travel Agents International
TA & IC Texas Arts and Industries College
TAICH Technical Assistance Information Clearinghouse
taid (TAID) thrust-augmented improved delta
TAIDET Triple-Axis Inertial-Drift Erection Test
TAIDHS Tactical Air Intelligence Handling System
tail tailpiece
'taint it aint
Taipas Taipa Islands off Macao in the South China Sea
TA-ISSA Travelers Aid—International Social Service of America
TAJAG The Assistant Judge Advocate General (USA)
Taju Tajumulco
take 5 take a rest
take 10 take a rest

tako terms and conditions of employment

tal traffic and accident loss

tal (TAL) tetra-alkyl lead

tal. talis (Latin—such)

Tal Talcahuano

Tal Talmud (Hebrew canon and civil lawbook)

TAL Transair Limited

tala monetary unit of Western Samoa

TALA The American Lyceum Association (currently the International Platform Association)

Talamancas Talamanca Mountains of Costa Rica

talar tactical landing-approach radar

talbe talk and listen beacon

talc hydrous magnesium silicate (agalmatolite); take a look see

TALC Tank-Automotive Logistics Command (USA); Texas Association for the Advancement of Local Culture

Talco Talcahuano

talff total allowable level of foreign fishing

TALIC Tyneside Association of Libraries for Industry and Commerce

talisman. transfer accounting and lodgment for investors and stock management for jobbers (London Stock Exchange)

talkies talking motion pictures

Talla Tallahassee

Talladegas Talladega Mountains of Alabama

Tallahassee Institution Tallahassee Correctional Institution in Florida

TALMA Truck and Ladder Manufacturers Association

'Talo Italo

TALOA Transocean Airlines

'talpa(s) catalpa(s)

tal. qual. talis qualis (Latin—as they come, average quality)

TALUS Transportation and Land Use Study

tam tambourine; tam-o'-shanter; tam-tam; total available market

tam (TAM) tactical air missile

t-a m toxoid-antitoxin mixture

Tam Tamar; Tamara; Tamil; Tampa; Tampan; Tampico (inhabitants—Tampiqueños); Tamualipas (inhabitants—Tamualipecos)

TAM Tel Aviv Museum; Television Audience Measurement; *Transporte Aéreo Militar* (Spanish—Paraguayan Military Air Transport)

TA & M Texas A & M University

TAMA Third Avenue Merchants' Association; Training-Aids Management Agency (USA)

tamb tambor (Spanish—drum or drummer)

tambo tambourine

TAMC Tripler Army Medical Center

tamco training aid for morbidic console operations

TAME Television Accessory Manufacturers Institute

tami tip air mass injection

Tamiami Tampa-Miami area

Tamiami Trail trans-Florida highway between Tampa and Miami

TAMIS Technical Meetings Information Service

Tammies Tamburitzans

Tamp Tampa, Florida; Tampico; Tampico, Mexico

Tamps Tamaulipas

TAMRC Tank-Automotive Materiel Readiness Command (USA)

TAMS Token and Medal Society

Tam Shrew Taming of the Shrew

TAMTU Tanzania Agricultural Machinery Testing Unit

TAMU Texas A & M University

tan. tangent; tangential; tannery; tanning; total ammonia nitrogen; twilight all night

Tan Tanganyika; Tangier

TAN Transportes Aéreos Nacionales

Tanan Tananarive

tan. bkt tangency bracket

tandel tandem + parallel

TANESCO Tanzania Electric Supply Company

Tang Tanganyika; Tangier

Tangas Tanga Islands in the southwest Pacific near New Ireland

tangelo tangerine + pomelo (tangerine-grapefruit hybrid citrus fruit)

tanglo(s) tangelo(s)

Tango letter T radio code

tanh hyperbolic tangent

Tania Tatiana

Tanimbars Tanimbar Islands of Indonesia

Tanjug Telegrafska Agencija Nova Jugoslavija (Serbo-Croat—New Yugoslav Telegraph Agency)

Tano Cayetano

tan's tax anticipation notes (TANs)

TANS Terminal Area Navigation System; Territorial Army Nursing Service

tanstaafl there aint no such thing as a free lunch

TANU Tanganyika African National Union

TANY Typographers Association of New York

Tanz Tanzania (Tanganyika + Zanzibar)

Tanzam Tanzania-Zambia (railway)

Tanzania United Republic of Tanzania (East African country combining Tanganyika and Zanzibar; includes the island of Pemba north of Zanzibar island

tao tactical air observation; thromboangiitis obliterans

TAO Tactical Air Office(r); Technical Assistance Operations; Test Analysis Outline; The Athenaeum of Ohio

TAO Taxi Aéreo Opita (Spanish—Opita Air Taxi)—Bogotá, Colombia

TAOC Tactical Air Operations Center

TAOCC Tactical Air Operations Control Center

TAOI Tactical Area of Interest

tap. telephone tap(ping); transient analysis program

TAP Table of Authorized Personnel; Tax Action Planning; Technical Advisory Panel; Telephone-A-Partner; Test Analysis Program; Timesharing Assembly Program; Total Action Against Poverty; Trans-Alaska Pipeline; Trend Analysis Program; Tuition Assistance Program

TAP Transportes Aéreos Portugueses (Portuguese—Portuguese Air Transport)—airline; *Tunis Afrique Presse* (French—Tunis Africa Press)

tapa. three-dimensional antenna-pattern analyzer

tapac tape automatic positioning and control

tape. tape automatic-preparation equipment

TAPE Target Profile Examination (USAF); Transactional Analysis of Personality and

Environment; Trust for Agricultural Political Education
taphon taphonomist(ic)(al)(ly); taphonomy
TAPLAN Tax Action Planning
TAPLine Trans-Alaska Pipe Line
TAPLINE Trans-Arabian Pipeline
Taplinger Taplinger Publishing Co
TAPPI Technical Association of the Pulp and Paper Industry
taps tapaderos (Spanish—leather hoods covering stirrups to protect the feet while riding through thorny cactus or mezquite); the last bugle call, the *taptoo*, meaning *lights out* or sounding the last honors at a military funeral
TAPS Teacher Audio Placement System; Trajectory Accuracy Prediction System (USAF); Trans-Alaska Pipeline System
TAPSC Trans-Atlantic Passenger Steamship Conference
tapvc total anomalous pulmonary venous connection
TAQ *The African Queen*
tar. **(TAR)** tariff(s); tarpaulin(s); terminal area radar; terrain-avoidance radar
TAR Technical Action Request (USA); Trans-Australian Railways
TARA Technical Assistant—Royal Artillery; Territorial Army Rifle Association
taran test and replace as necessary
TARC Tactical Air Reconnaisance Center
TARDC Tank-Automotive Research and Development Command (USA)
tare. transistor analysis recording equipment
tarex target exploitation
tarfu things are really fouled up
targ target
TARGET Team to Advance Research for Gas Energy Transformation
tarmac tar plus macadam (tarred road or runway)
Tar-Man Taranaki-Manawatu (NZ)
tarn. tarnish; tarnishes; tarnishing
TARO Territorial Army Reserve Office(r)(s)
TAROM *Transporturile*

Aeriene Romine (Romanian Air Transport)
TARP Test and Repair Processor; Transitional Aid Research Project
tarp(s) tarpaulin(s)
Tarr Tarragona
Tarryalls Tarryall Mountains of central Colorado
tars. **(TARS)** three-axes reference system
TARS Technical Assistance Recruitment Service
tart. tartaric
TART Test Analysis Reduction Technique (USN)
tart. a tartaric acid
tas true airspeed
tas **(TAS)** torpedo anti-submarine
Tas Tasmania
TAs teaching assistants
TAS The Asia Society; Texas Academy of Science; Traveler's Aid Society; *Turk Anonim Sirketi* (Turkish Joint Stock Company)
TAS *The American Spectator*
tasa test area support assembly
TASA Texas Association of School Administrators
TASAMS The Army Supply and Maintenance System
tasc terminal area sequence and control; treatment alternatives to street crimes
TASC The Alumni Service Cooperative; The Analytic Sciences Corporation; Telecommunications Alarm Surveillance and Control; Test Anxiety Scale for Children; Treatment Alternatives to Street Crime
tascon television automatic sequence control
TASD Terminal (Railway) Alabama State Docks
TASDC Tank-Automotive Systems Development Center (USA)
tase tactical support equipment
taser taser gun (electronically activated stunning device used by law-enforcement officers)
TASES Tactical Airborne Signal Exploitation System
TASF Teachers Association of San Francisco
Tash Tashkent
TASHAL *Tseva Hagana Le-Israel* (Hebrew—Defense Army of Israel)
tasi time-assignment speech interpolation

TASK Test of Academic Skills (Stanford)
TASKFLOT task flotilla; Task Flotilla (NATO) (USN)
TASKFORNON Task Force—Northern Norway (NATO)
tasm **(TASM)** tactical air-to-surface missile
Tasm Tasman; Tasmania; Tasmanian
Tasmans Tasman Mountains of New Zealand's South Island
TASO Television Allocations Study Organization; Training Aids Service Office (USA)
TASP The Army Studies Program; The Army's Study Program
taspac total analysis system for production accounting and control
tasr terminal area surveillance radar
tass technical assembly
TASS *Telegrafnoie Agenstvo Sovietskavo Soyuza* (Russian—Soviet News Agency)
TASSO Tactical Special Security Office(r)
TASSq Tactical Air Support Squadron (USAF)
TASSR Tartar Autonomous Soviet Socialist Republic; Tuva Autonomous Soviet Socialist Republic
TAST Tactical Assault Supply Transport
tat. **(TAT)** tetanus antitoxin; tyrosine amino transferase
t & at tank and antitank
Tat Tatar (Turkestan)
TAT tetanus antitoxin; Thematic Apperception Test; Thrust-Augmented Thor; Touraine Air Transport; Trans-Atlantic Telephone; *Transportes Aéreos de Timor* (Timor Air Transport)
TATA Tobacco Accessories Trade Association (formerly PTA—Paraphernalia Trade Association)
Tat Aut Sov Soc Rep Tatar Autonomous Soviet Socialist Republic
TATC Tactical Air Traffic Control; Trans-Atlantic Telephone Cable
tatce terminal air-traffic-control element
TATCO Tactical Automatic Telephone Central Office
'tater(s) potato(es)
TATL Trust for Appalachian

Trail Lands
TATPAC Trans-Atlantic Trans-Pacific (telecommunications network linking London, Montreal, New York, Tokyo, Hong Kong, and Sydney)
Tatras Tatra Mountains of Czechoslovakia
TATSA Transportation Aircraft Test and Support Activity
Tatts Tattersalls
TATU Tanganyika African Traders Union
Tau Taurus
TAU Tel Aviv University
Taughannock Taughannock Falls State Park on Cayuga Lake in central New York
TAUN Technical Assistance of the United Nations
taurom tauromachia
taurom tauromaquia (Spanish—art of bullfighting)
TAUSA Tea Association of the U.S.A.
taut. tautology
tav tavern
tav (TAV) transatmospheric vehicle (McDonnell-Douglas aircraft designed to travel at up to 20 times the speed of sound)
T-a-v Tout-à-vous (French—Yours truly)
Tave Octave; Octavius
Tavia Octavia
TAVINA Trans-Colombiana de Aviación (Spanish—Trans-Columbian Aviation)
Tavita Octavita
T Aviv Tel Aviv
T & AVR Territorial and Army Volunteer Reserve
tav(s) tavern(s)
TAVSS Toward, Away, Versus Selection System
Tavy Octavius
taw thrust-augmented wing; twice a week
T A & W Toledo, Angola & Western (railroad)
TAW Times Atlas of the World
TAWACS Tactical Airborne Warning and Control System
TAWC Tactical Air Warfare Center
TAWG Target Acquisition Working Group
tax. taxation; taxes; taxonomic; taxonomy
Taxco Taxco de Alarcón
taxi taxicab; taxiing
taxid taxidermy
taxir taxonomic information

retrieval
taxis or taxo (Greek—to arrange in an orderly manner)—geotaxia, phototaxis, taxonomy
taxon taxonomic(al)(ly); taxonomist(ic)(al)(ly); taxonomy
Tay Tayside
Taz Tazmania(n)
TAZ Tactical Alert Zone
taz(es) topaz(es)
tb temporary buoy; terminal board; thymol blue; tile base; time bank; total bouts; tractor biplane; trial balance; true bearing; tubercle bacillus; tuberculosis; turbine; turretbase; turretbased
t/b title block
t & b top and bottom; turned and bored
Tb terbium
TB Tank Battalion; temporary buoy; Torpedo Boat; Treasury Bill; Troop Basis; Twin Branch (railroad); Tyburn (reports)
TB Technical Bulletin
tba to be announced; to be approved; to be assigned; to be audited; terminal board assembly; tires-batteries-accessories
TBA Tables of Basic Allowance; Television Bureau of Advertising; Torrey Botanical Association; Triborough Bridge Authority
tbab to be approved by
tbab (TBAB) tryptose blood agar base
tban to be announced
T-bar T-shaped bar
tbawrba travel by aircraft, military and/or naval water carrier, commercial rail and/or bus is authorized (USA)
tbb to be billed
TBB tenor, baritone, bass
TBB Television Blue Book
tbc to be crated; to be culled
tbc (TBC) time base corrector
TBC The British Council; Trinidad Broadcasting Company
TBC Co Tropical Belt Coal Company (invented by Joseph Conrad for use in his novel *Victory*)
tbcf to be called for
tbd to be determined; to be discontinued; thousand barrels daily
TBD torpedo-boat destroyer
TBDS Test Base Dispatch Service

tbe to be edited; to be encoded; to be executed; to be expanded; to be expended; to be expired; to be expunged; time base error
tbe (TBE) tuberculin bacillen emulsion
TBE Toronto Board of Education
T-beam T-shaped beam
tb ex tube expander
tbf to be furnished
TBF single-engine torpedo bomber (naval symbol); Teachers Benevolent Fund
tbfx tube fixture
tbg to be garnished; to be gathered; testosterone-binding globulin; thyroxine-binding globulin
t & bg top and bottom grille
Tbg Tönsberg
tbh to be had (sexually available); to be held
tbi to be invented; to be inventoried; tooth-brushing instruction; traditionally black institutions
TbI Tax-based Income
TBI Tennessee Bureau of Investigation; Texas Board of Insurance; The Business Institute; The Tobacco Institute
T-bill(s) Treasury bill(s)
T-bird Thunderbird
tbj to be joined
tbk to be killed
t-bk talking-book
tbl to be labelled; table; tablet; through back of loops (knitting); through bill of lading
tb lc term birth, living child
tbm tactical ballistic missiles; temporary bench marks; to be manufactured; to be monitored; tuberculous meningitis
tbm (TBM) terabit memory; tired businessman
TBM Ten Broeck Mansion (Albany)
TBMA Timber Building Manufacturers' Association
tb md tube mandrel
TBMD Terminal Ballistic Missile Defense (USA)
TBMS Turtle Bay Music School
tbmt transmitter buffer empty
tbn to be named; to be nominated; to be ordered
TBN Trinity Broadcasting Network
tbo to be ordered; time between overhaul(s)
TBO Test Base Office
tboip tentative basis of issue

plan(ning)

T-bolt bolt with T-shaped square head

tbone trombone

T-bone T-bone steak; T-shaped bone; trombone

T-bowl toilet bowl

tbp to be promoted; to be purchased; true boiling point

tbp (TBP) telephone-bill payment

tbpa thyroxine-binding prealbumin

tbq to be queried

tbr to be rented; to be restored; to-be-remembered (word)

TBR Test of Behavioral Rigidity; Treasury Bill Rate; Treasury Bond Receipt

TBRI *Technical Book Review Index*

tbs to be sold; tablespoon; talk-between-ships (radiotelephone)

tb's tuberculosis patients; tuberculosis victims

tb & s top, bottom, and sides

TBs Torpedo Boats (World War I)

TBS Tokyo Broadcasting System; Turner Broadcasting System

tb sa tube saw

tbsd thermal-blooming slow dither

TBSI The Baker Street Irregulars

tbsn tablespoon

tbsp tablespoon

tbt to be tested; target-bearing transmitter; tolbutamide test-(ing); total bottom time; tracheobronchial toilet; tributyltin

tbt (TBT) torpedo-bearing transmitter

TBT Terminal Ballistic Track

TB & TA Triborough Bridge & Tunnel Authority

tbto (TBTO) tributyl tin oxide

TBTS Tracker Breadboard Test System

tbu to be used

tbv to be vacated; to be vented; tubercle bacillus vaccine

TB & VD C Tuberculosis and Venereal Diseases Clinic

tbw to be weighed; to be withheld; total body washout; total body water

tbx to be x'd (out)

tby to be young

tbz to be zoned; to be zonked

tc temperature classification; temperature controlled; terra cotta; tetracycline; thermo-

couple; thermocoupled; thermocoupling; thrust chamber; tierce(s); time check; time closing; top chord; transportation cask; trash compactor; trial color; trip coil; true course (TC); type certification

tc (TC) total cost

t/c tabulating card; temperature coefficient; thermocouple; transformer rectifier; trim coil; type certificate

t & c threads and couplings; turn and cough

tc tre corde (Italian—three strings)

Tc technetium; tropic tides

TC Air Canada (formerly TCA); The Citadel; Tabor College; Taft College; Talladega College; Tank Corps; Tariff Commission; Tarkio College; Tax Court; Tea Council; Teachers College; Technical Circular; Technical Communication; Tennessee Central (railroad); Texarkana College; Texas College; Thiel College; Tift College; Time Charter; Training Center; Training Circular; Transaction Code; Transportation Corps; Transylvania College; Trial Counsel; Trinity College; Tri-State College; troop carrier; Trucial Coast (Arabian sheikdoms); True Course; Trusteeship Council; Turret Captain; Tusculum College

T & C Turks and Caicos Islands

T of C Tournament of Champions

TC *Technical Communications; Tragedy of Coriolanus; Tre Corde* (Italian—three strings)

T & C *Troilus and Cressida*

TC 1 Traffic Conference 1—North and South America, Greenland, Bermuda, West Indies, Hawaiian Islands

TC 2 Traffic Conference 2—Europe, adjacent islands, Ascension Island, Africa, and Asia west of and including Iran

TC 3 Traffic Conference 3—Asia, adjacent islands, East Indies, Australia, New Zealand, Pacific Islands except Hawaiian

tca telemetering control assembly; terminal control area

(TCA); to come again; track crossing angle; trichloro-acetate

tca (TCA) tri-cyclic anti-depressant

TCA Tanners Council of America; Technical Cooperation Administration; Tele-Communications Association; Television Consumer Audit; Television Corporation of America; Temporary Change Authorization; Tennessee Correctional Association; Terminal Control Area; Texas Corrections Association; Textile Converters Association; Theater Commander's Approval; Thoroughbred Club of America; Tile Council of America; Tissue Culture Association; Trailer Coach Association; Trans-Canada Airlines; tricyclic antidepressant

TCAA Technical Communication Association of Australia

tcam telecommunications access method

TCAs Terminal Control Areas (establishing airfield-safety flight paths)

TCAS The College of Advanced Science; Traffic-alert and Collision-Avoidance System

tcb take care of business

tcb (TCB) task-control block

TCB Thames Conservancy Board

TCBC Ty Cobb Baseball Commission

TCBI Television Center for Business and Industry

tcbs (TCBS) thiosulfate-citratebile salt sucrose

tcc tatical control computer; television control center; test conductor console; topical cocaine compound

tcc (TCC) transitional cell carcinoma

Tcc Tagliabue closed cup

TCC Telecommunications Coordinating Committee; Transcontinental Corps; Transport and Communications Commission; Transport Control Center; Transportation Control Committee; Troop Carrier Command

T-C C Tri-Continental Corporation

TCCA Textile Color Card Association

TCCB Test and County

Cricket Board

TCCP Thirteen College Curriculum Program

TCCS Texaco Controlled-Combustion System; Tide Communication-Control Ship

tcd task completion date; ternary coded decimal; tungsten carbide depositing

TCD Trinity College, Dublin

TCDA Texas Civil Defense Agency

tcdb turning, coughing, and deep breathing

tcdd (**TCDD**) tetrachloro-dibenzo-p-dioxin

tcdf (**TCDF**) tetrachlorodibenzofurans

tcd's time certificates of deposit (TCDs)

tce ton-coal equivalent; total composite error

tce (**TCE**) trichloroethylene

Tce Terrace

TCE Tax(ation) Counseling for the Elderly

T-cell thymus-derived cell

tcet transcerebral electrotherapy

tcf trillion cubic feet (natural gas)

TCF 20th-Century Fox; Transparent Computing Facility; Twentieth Century Fund

TCF *Touring Club de France* (French—Touring Club of France)

TCFB Transcontinental Freight Bureau

tcfy trillions of cubic feet per day

TCG Theatre Communications Group

T C & G B Tucson, Cornelia & Gila Bend (railroad)

tcgf (**TCGF**) T-cell growth factor

tch travel counselor's handbook

Tch (**TCH**) Tacoma (container symbol)

TCH Trans-Canada Highway

tchg teaching

TcHHW tropic higher high water

TcHHWI tropic higher high water interval

TcHLW tropic higher low water

tchr teacher

Tchrs Coll Pr Teachers College Press

TCI Takeda Chemical Industries; Technical Correspondence Institute; The Combustion Institute; The Container-

ization Institute; Theoretical Chemistry Institute

T & CI Turks and Caicos Islands

TCI *Touring Club Italiano* (Italian—Italian Touring Club)

tcj terminal coaxial junction

TCJC Texas Criminal Justice Council

tcl transfer chemical laser; transistor-coupled logic

Tcl Tymshare conversational language

TCL Tokyo Commercial University; Transatlantic Carriers Limited; Trinity College Library; Turkish Cargo Lines

TcLHW tropic lower high water

TcLLW tropic lower low water

TcLLWI tropic lower low water interval

tcm terminal-to-computer multiplexer

TCM Texas Citrus Mutual; Trinity College of Music

TCMA Telephone Cable Makers' Association

TCMB Tomato and Cucumber Marketing Board

TCMP Taxpayer Compliance Measurement Program (IRS)

TCN Transportation Control Number

TCNA Turks and Caicos National Airline

TCNCO Test Control Noncommissioned Officer

TCNM Timpanagos Cave National Monument (Utah)

tco thrust cutoff

TCO Termination Contracting Office(r); Test Control Office(r); Trinity College—Oxford

TCO *Tjänstemännens Centralorganisation* (Swedish—Salaried Employees' Central Organization)

tcoc transverse cylindrical orthomorphic chart

TCOC Tri-Cities Opera Company (Binghamton)

TCOM Tethered Communications

T-conn T-shaped connection

TCOS Toronto Classroom Observation Schedule

tcp timing and control panel; traffic control panel; traffic control post; training control(ler) panel

tcp (**TCP**) trichlorophenyliodomethylsalicylates; tricres-

ylphosphate

TCP Task Change Proposal; Task Control Proposal; Technical Cooperation Program (between Australia, Canada, the United Kingdom, and the United States); Temporary Change Proposal; Traffic Control Post; Transitional Community Placement; Transmission Control Protocol

TCPA Town and Country Planning Association

tcpc tab card punch control

TCPC Tennessee Council of Private Colleges

TCP/IP Transmission Control/Internet Protocol

TCPL Trans-Canada Pipe Lines

tcr temperature coefficient of resistance; total controlled return

TCR Tennessee Central Railway

TCRB *Touring Club Royal de Belgique* (French—Royal Belgian Touring Club)—automobile club

TCRMG Tripartite Commission for the Restitution of Monetary Gold (American-British-French commission, headquartered in Brussels)

tcs temporary change of station; terne-coated stainless steel; tierces

TCs Tax Cases; transit cops

TCS Target Cost System; The Costeau Society; Torpedo Control System; Twin-City Secularists; Typesetting Consultation Service

T & CS Transportation and Communication Service

TCS *Touring Club Suisse* (French—Swiss Touring Club)

tcsa (**TCSA**) tetrachlorosalicylanilide

tcsev (**TCSEV**) twin-cushion surface-effect vehicle

TCSO Tri-City Symphony Orchestra

tct total-controlled tabulation

TCTA Texas Classroom Teachers Association

tctl tactical

TCTO Time Compliance Technical Order(s)

TCTS Trans-Canada Telephone System

tcu tape-control unit; teletypewriter control unit; test(ing) computer unit; threshold con-

trol unit; training combustion unit; typewriter control unit
TCU Texas Christian University; Tokyo Commercial University
TCUS Tax Court of the United States
T-cushion T-shaped cushion
tcv temperature-control valve
TCV Terminal-Configured Vehicle (NASA)
TCVA Terminal Configured Vehicles and Avionics (NASA program)
tcvr transceiver
tcw time code work
TCWG Telecommunications Working Group
TCWH Teamsters, Chauffeurs, Warehousemen and Helpers (union)
TCWIB Trans-Continental Weighing and Inspection Bureau
TCWP Texas Committee for Wildlife Protection
tcxo temperature-compensated crystal oscillator
td tank destroyer; technical data; test data; tetanus-diphtheria; third defense (lacrosse); tile drain; time delay; time of departure; time disintegration; tod (28 pounds of wool); tool design; tool disposition; touchdown (football); transmitter distributor; trust deed; turbine drive; 'tween deck
td (TD) tardive dyskinesia; technical director; tracking dog
t/d table of distribution; telemetry data; time deposit; transmission and distribution
t & d taps and dies
t.d. *ter die* (Latin—thrice daily)
Td townsend
T$ Taiwan dollar(s)
TD Table of Distribution; Tactical Division; tank destroyer; Teachers Diploma; Technical Director; Territorial Decoration; Testing and Development (USCG); Topographic Draftsman; Town District: Castletown, Douglas, Peel, and Ramsey in the Isle of Man; Training Detachment; Treasury Decision; Treasury Department; Treasury Division; Trinidad and Tobago; Typographic Draftsman
TD *Teachta Dala* (Gaelic— Member of the House of Commons)

tda tax-deferred annuity; tunnel-diode amplifier
t & da tracking and data acquisition
TDA Timber Development Association; Toa Domestic Airlines; Train Dispatchers Association
tdana time-domain automatic-network analysis; time-domain automatic-network analyzer
T-day day for time schedule testing; truce day
T-Day Transition Day
tdb total disability benefit (TDB)
TDB Toronto-Dominion Bank; Toxicology Data Bank; Trade Development Bank; Trade and Development Board
tdc top dead center; total distributed control; total distribution costs; transverse directional control
tdc (TDC) through-deck cruiser; torpedo-tracking computer
TDC Telemetry Data Center; Texas Department of Corrections; The Discovery Channel (cable tv)
td cu tinned copper
tdd telecommunication device for the deaf
TDD Diploma in Tubercular Diseases
tddl time-division data link(age)
tddlpo time division data link printout
TDDS Teacher Development in Desegregating Schools
TDE Technology Development and Engineering
T del F Tierra del Fuego
T de M *Teléfonos de México* (Spanish—Telephone System of Mexico)
T de S *Teatro della Scala* (La Scala)
tdf two-degree-of-freedom (gyroscope)
TDF 1 first French direct-broadcast satellite
TDF *Télédiffusion de France* (French—French Television Broadcasting)
TDFS Terminal Digit Fitting System
tdg twist drill gauge
tdg (TDG) test data generator
TDG Test Documentation Group; Transport Development Group
tdh total dynamic head

Tdh Trondheim
tdi toluene di-isocyanate
TDI Target Data Inventory; Tool and Die Institute; Transportation Displays Incorporated
tdic target data input computer
TDIS Travel Document and Issuance System (for processing passports)
tdiu target data input unit
TDK *Turk Dil Kurumu* (Turkish Language Association)
t dk(s) 'tween deck(s)
tdl total damn loss; translation definition language
TDL Topographic Developments Laboratory
tdlr terminal-descent-landing radar
tdm tandem; teacher-developed materials; time division multiplexing
tdma time division multiple access
tdmg telegraph(ic) and data message generator
tdm/pcm time-division multiplex (using) pulse-code modulation
tdn totally digestible nutrients
tdo tornado
TDO Technical Development Objective
tdol (TDOL) tetradecanol
TDOP Truck Design Optimization Program
TDOT Thorndike Dimensions of Temperament
tdp target director post; technical data package; technical development plans; thermal death point
TDP Technical Development Plan; Trade and Development Program
td passes touchdown passes
tdpfo temporary duty pending further orders
tdpj truck discharge point jet
tdr time-delay; time domain reflectometry
tdr (TDR) transmit data register
tdr *tous droits réservés* (French —all rights reserved)
TDR Technical Deficiency Report; Technical Documentary Report; tender (naval symbol)
TDRL Temporary Disability Retired List
t/d rly time-delay relay
tdrs (TDRS) tracking and data relay satellite
TDRSS Tracking and Data Re-

lay Satellite System

tds telemetering decommutation system; total dissolved solids

tds (TSS) temperature, depth, salinity

t.d.s. *ter die sumendum* (Latin —to be taken three times daily)

TDS Tanami Desert Sanctuary (Northern Territory, Australia); Telemetering Decommunication System; Tennessee Department of Safety; Transaction-Driven System

TDS *Toronto Daily Star*

tdsa telegraphic data signal analyzer

TDSCC Tidbinbilla Deep Space Communication Complex

TDSTS Tidbinbilla Deep Space Tracking Station

tdt thermal death time

TDT Transport Department Tasmania

tdtcu target designation transmitter and control unit

tdtl tunnel diode transistor logic

TDTS Technical Data Transfer System

tdu target detection unit

TDU Teamsters for a Democratic Union

TDUP Technical Data Usage Program

TdV *Teatro dal Verme* (Milan)

tdw tons deadweight (tare of a ship)

tdwy treadway

tdy temporary duty; toady

TDZ Touch-Down Zone

te table of equipment; tank element; task element; technical exchange; tenants; tenants by the entirety; thermal efficiency; tight end; tinted edge; trailing edge; transverse electric; transverse wave (symbol); trial and error; turbine electric; turboelectric; twin engine

t/e time expired

t & e testing and evaluation; thorough and efficient; training and evaluation; travel and entertainment; trial and error

Te tellurium

TE Table of Equipment; Task Element; Technical Exchange; *Telefis Eireann* (Television Ireland); Topographical Engineer

T & E Toledo & Eastern (railroad)

tea. triethanolamine

TEA Tennessee Education Association; Tucson Education Association

TEAA Tax Equity for American Abroad

teac turbine engine analysis check(ing)

teach. teacher; teaching

TEAL Tasman Empire Airways, Limited

TEAM Technique for Evaluation and Analysis of Maintainability; Terminology, Evaluation, and Acquisition Me-thod; The European–Atlantic Movement; Trend Evaluation and Monitoring

Teamsters Teamsters Union (International Brotherhood of Teamsters, Chauffeurs, Warehousemen, and Helpers of America)

TEAS Texas Energy Advisory Council; Threat Evaluation and Action Selection (program)

tease tracking errors and simulation evaluation (radar)

teatr *teatrale* (Italian—theatrical)

Teatro Colón (Spanish—Columbus Theater)—South American opera house

teb tape error block

TEB Tax Exemption Board; Textile Economics Bureau

tec technic; technical; technician; technics; technological; technology; total environmental control; total estimated cost

'tec detective

Tec Tecate

TEC Technical Education Council; Technician Education Council

TECAUS Temporary Emergency Court of Appeals of the United States

TECE Trans-Europe Container Express (train)

tech technic; technical; technician; technics; technique(s); technological; technology

Tech. CEI Technician of the Council of Engineering Institutions

tech ed technical editing; technical editor

Tech Eng Technical English (application of good English to any technical writing task)

techie(s) technician(s); technologist(s)

tech memo technical memo-

randum

techn technician

technocrit technological criticism; technology critic

technol technological; technologist; technology

techno-pop technolosized popular music

tech rep technical representative

tech rept technical report

Tech Weld Inst Technician of the Welding Institute

tech writer technical writer

TEC-NACS Teachers Educational Council—National Association of Cosmetology

TECOM Test and Evaluation Command (US Army)

tecquinol hydroquinone

tecr technical reason

'tecs detectives

TECS Treasury Enforcement Communications System; Treasury Enforcement Computer File

Tec Sgt Technical Sergeant

tecspert technical expert

ted transferred electron device

ted (TED) turtle excluder device

ted *tedesco* (Italian—German)

TEDS Tactical Electronic Decoy System

TEE Telecommunications Engineering Establishment; Theological Education by Extension; Trans Europe Express

TEEM Trans-Europe Express Merchandise (train)

'teens thirteen through nineteen

teenybop teenybopper (slang—young child attuned to the modern scene)—*see* macrobop

TEEP Teacher Education Examination Program

teeto teetotaler

TEFL teaching English as a foreign language

teflon tetrafluoroethylene (polymerized synthetic plastic resin)

TEFRA Tax Equity and Fiscal Responsibility

teg top edge gilt

Teg Tegel

te ga taper gauge

TEGMA Terminal Elevator Grain Merchants Association

teg(s) thermoelectric generator(s)

Teh Teheran

Tehachipis Tehachipi Moun-

tains traversing south-central California
TEI Texaco Experiment Incorporated
TEJA Tutmonda Esperantista Jurnalista Asocio (International Association of Esperantist Journalists)
TEJO Tutmonda Esperantista Junulara Organizo (International Organization of Esperantist Youth)
tekn teknisk (Dano-Norwegian—technical)
tel telegraph; telegraphic; telegraphy; telephone; telephonic; telephony; teletype; teletypewriter; television; tetraethyl lead
tel (TEL) transporter-erector launcher
Tel Telefunken; Telescopium (constellation); Telugu
Tel Teluk (Indonesian or Malay—bay, bight, riverbend)
TEL Tests for Everyday Living
TELAM Telenoticiosa Americana (Argentine press service)
telaut telautograph; telautography
TELBRAS Telecommunicaões Brasileiras (Portuguese—Brazilian Telecommunications)
Tel Can Television Canada
telco telephone company
telcos telephone companies
TELDEC Telefunken + Decca (video disc)
tele television
tele (Latin prefix—far)—telegraph
Tele Telescopium (constellation)
telec thermo-electronic laser energy converter
telecast(er) television broadcast(er)
telecom telecommunication
Telecom 1C French commercial and communications satellite
telecon telephone communication
teleconcert televised concert; television concert
telecopy telephonic copying process (developed by Xerox)
telecourse television-constructed course
teledis teletypewriter distribution
teledrama televised drama;

television drama
telef *telefone* (Portuguese—telephone)
telef *telefon* (Norwegian—telephone)
telefac television facsimile
telefilm television film
teleg *telegramas* (Portuguese—telegrams); telegrapher; telegraphy
telegen telegenic
telegr *telegrafie* (Dutch—telegraphy)
Tel Eir Telefis Eireann (Gaelic —Irish Television)
telemark telemarketing
telemorality television morality
teleol teleology
teleopera televised opera; television opera
teleosts teleostomist fishes (bony fishes)
telep telephathic(ally); telepathy
telepak telemetering package
teleph telephony
teleplay televised play; television play
teleran televised radar aerial navigation
telesex telephone(d) sex (fantasy conversations contracted for and carried out by telephone)
telesurance television insurance
tele tape television tape
telethon television marathon
teletrial television trial
telev (TV) television
telev televisão (Portuguese—television); *televisión* (Spanish—television)
televangelist television evangelist
telex (tex) teletype exchange
Tel-Law Telephone-Law (free over-the-telephone answers to many legal questions provided by many county bar associations in the U.S.)
tellie(s) television (sets)
telly television
Telly Telegonus; Telemachus; Telemus; Telephus; Telesphorus
Tel-Med Telephone-Medical (free over-the-telephone answers to many medical questions provided by many hospitals and county medical societies in the U.S.)
tel no telephone number
TELOPS Telemetry On-Line Processing System

TELS Tokyo English Language Society
telsat telecommunications satellite
tel sec telephone secretary
telsim teletypewriter simulator
tel sur telephone survey
telw telwoord (Dutch—word count)
tem technical error message; temporal; temporary; transverse electromagnetic
tem. tempus (Latin—time); *tempo* (Italian-time)
tem 1º tempo primo (Italian—tempo at the start of a musical composition)
Tem temple
TEM Territorial Efficiency Medal
TEMA Telecommunications Engineering and Manufacturing Association
temadd temporary additional duty
temar thermoelectric marine application
TEMIS Targets Engineering Management Information System (USN)
temp temper; temperature; tempered; tempering; template; temporary; temporize
temp. tempo (Italian—time)— musical time; *tempore* (Latin —in the time of)
Temp Tempest, The
temp. dext *tempori dextro* (Latin—to the right temple)
temping substituting
tempistors temperature compensating resistors
Temple central London's lawcourt area
tempo. total evaluation of management and production output
TEMPO Technical Military Planning Operation
tempos temporary buildings, houses, offices, officials, workers, et cetera
temp prim tempo primo (Italian—tempo or time in the musical sense as at the start)
temps tempests; temperatures; temporary workers; transportable electromagnetic pulse simulator
temp(s) temporary worker(s)
temp sec temporary secretary
temp. sin. tempori sinistro (Latin—to the left temple)
tempy temporary
ten. tenant; tender; tenderize(d); tenement; tenor

ten. (TEN) toxic epidermal necrolysis; trans-European night (flight)

ten. tenuto (Italian—to hold, a chord or tone)

Ten Ten Commandments; *Tenente* (Italian or Portuguese); *Teniente* (Spanish)—Lieutenant

T(en) Col Tenente Colonnello (Italian); *Tenente Coronel* (Portuguese); *Teniente Coronel* (Spanish)—Lieutenant Colonel

ten. com tenant(s) in common

tency tenancy

tend. tendon

ten. ent tenant(s) by the entireties

TENES Teaching English to Non-English Speaking

Teng Teng Hsiao-ping

Ten Gen Tenente General (Portuguese); *Tenente Generale* (Italian); *Teniente General* (Spanish)—Lieutenant General

Tenn Tennessee; Tennessean

tenna(s) antenna(s)

Tenneco Tennessee Gas Companies

Tenn–Tom Tennessee-Tombigbee Waterway

Tennyson Alfred, Lord Tennyson (1809–1892)

TENOC ten years of oceanography (1961–1970)

tenot tenotomy

TENRAC Texas Energy and Natural Resources Advisory Council

tens tensile; tension

tens (TENS) transcutaneous electrical nerve stimulation

tens (Latin prefix—stretch)—tensor

ten-spot $10 bill

tens str tensile strength

tent. tentative

Ten^te Teniente (Spanish—Lieutenant)

Ten Vasc Tenente di Vascello (Italian—Lieutenant of the Vessel)—Navy Lieutenant

Teol Teología (Portuguese, Spanish—Theology)

TEOO Territorial Economic Opportunity Office(r)

Teor Teoretyczna (Russian—Theoretical)

TEOSS Tactical Emitter Operational Support System (USAF)

tep transparent electrophotographic process(ing); transparent electrophotography

TEP Teacher Education Program; Tucson Electric Power

tepi training equipment planning information

TEPIAC Thermophysical and Electronic Properties Information Analysis Center

TEPIGENS Television Picture Generation System (computer-controlled)

TEPS Teacher Education and Professional Standards

ter terminal; terminate; termination; terrace; terrazzo; territory; teritary

ter. tere (Latin—rub)

Ter Terrace; Territory; Teruel

Ter Terence (Publius Terentius Afer)—Roman writer of comedies

tera 10^{12}

TERA The Electrical Research Association

terat teratology

TERC Technical Education Research Centers

terco telephonic rationalization by computer

tercom terrain contour matching

t & e rec time and events recorder

TERL Transit Expressway Revenue Line (mass transportation)

term terminal; terminate; terminology

te rm taper reamer

Term Terminal

Terminal Terminal Island (Bureau of Prisons correctional facility between Long Beach and San Pedro, California)

TERMS Terminal Management System

tern. terminal and enroute navigation

TERPACIS Trust Territory of the Pacific Islands

TERPES Tactical Electronic Reconnaissance Processing and Evaluation System

terps elixir of terpin hydrate and codeine—cough mixture and codeine combination

terps (TERPS) terminal instrument approach

TERPS Terminal Inquiry/Response Programming System

terr terrace; territory; terrorist

Terr Terrace

TERRA Terricide Escape by Rethinking, Research, Action; The Earth Regeneration and Reforestation Association

Terra Nova Terra Nova National Park in Newfoundland

terrs terrorists

terry terrycloth(ing)

Terr^y Territory

tersab terrorist sabotage; terrorist saboteur

tersabs terrorist saboteurs

Tersch Terschelling

ter. sim. tere simul (Latin—rub together)

TERSSE Total Earth Resources System for the Shuttle Era (NASA)

Tert. Tertiary

Tertullian Quintus Septimus Florens Tertullianus

tes tesorero (Spanish—treasurer)

TES Telemetering Evaluation Station

TES Times Educational Supplement

TESA Television and Electronic Service Association

tesac temperature-salinity-currents

tesl (TESL) teaching English as a second language

tesla technical standards for library automation

TESM Trinity Episcopal School for the Ministry

TESO Texel's Eigen Stoomboot Onderneming (Dutch—Texel's Own Steamship Society)

TESOL Teachers of English to Speakers of Other Languages

tess tessili (Italian—textiles)

tessit tessitura (Italian—texture, tissue, weaving)—range of a musical work

test. test-oriented engineering symbol(ic) translator

TEST Thesaurus of Engineering and Scientific Terms

TESTCOMDNA Test Command Defense Nuclear Agency

test^{mto} testamento (Spanish—testament)

test° testigo (Spanish—witness)

testran test translator (data processing)

TESYS Terminal Editing System

tet test equipment tool; tetanus; tetrachloride

TET Teacher of Electrotherapy; Teacher Evaluation Testing

TETAM Tactical Effectiveness Testing of Antitank Guided Missiles (USA)

T-et-G Tarn-et-Garonne
tetmtu (TETMTU) tetramethyl thiourea
TETOC Technical Education and Training for Overseas Countries
tetr tetragonal
tetra (Latin prefix—four or four-fold)—tetrachord
tetrac tetraiodothyroacetic acid
tetrah tetrahedral
tetroon tetrahedral balloon
tet tox tetanus toxin
teu twenty-foot equivalent unit(s) (container measurement)
TEU Test of Economic Understanding
Teut Teuton; Teutonic
tev tevatron (atom smasher speeding electrons to an energy level of 1000-billion electron volts)
tev (TeV) trillion electron volts
teV tetra-electron volt(s)
tew (TEW) tactical early warning; tactical electronic warfare (aircraft)
tewa threat evaluation and weapons assignment
TEWDS Tactical Electronic Warfare Defense System
tews tactical electronic warfare suite
TEWS Tactical Electronic Warfare System
TEWT Tactical Exercise Without Troops
tex telex (teletype exchange); textile(s)
t ex till exempel (Swedish—for example)
Tex Texan; Texas
TEX Corpus Christi, Texas (tracking station)
TEXACO The Texas Company
Tex A&M Texas Agricultural and Mechanical University
Tex A&M Pr Texas Agricultural and Mechanical University Press
TEXAS Trained Experienced Area Specialist
Texas RRC Texas Railroad Commission
Tex Chr U Texas Christian University
Tex Chr U Pr Texas Christian University Press
Texcoco Texcoco de Mora
Texhoma Texas + Oklahoma
Texican Texas-Mexican or anyone from the Texas side of the Mexican Border
Texico Texas + New Mexico

Tex Instr Texas Instruments (Corporation)
Tex-Mex Texan-Mexican; Texas-Mexico
Texola Texas + Oklahoma
Texoma Lake Texoma between Texas and Oklahoma
texp time exposure
text. textile
text ed text edition
Textel Trinidad and Tobago External Telecommunications Company
Textile Mus Textile Museum
textir text indexing and retrieval
text. rec. textus receptus (Latin—received text)
Tex W Pr Texas Western Press
tf tabulating form; tactile fremitus; temporary fix; thin film; tile floor; till forbidden (run ad until stopped by advertising client); transfer function; tuberculin filtrate
t/f true/false
TF Tallulah Falls (railroad); Task Force; Tax Foundation; Test Flight; Tolstoy Foundation; torpedo-fighter (airplane); trainer-fighter (airplane); training film; tropical freshwater (vessel loadline marking); Twentieth Century-Fox Films (stock exchange symbol)
TF Travail Forcé (French—penal servitude)
TF-1 Télévision 1 Français (French tv network)
tfa total fatty acids; transfer function analyzer
TFA Task Force on Alcoholism; Textile Fabrics Association; Tie Fabrics Association; Tobacco-Free America; Trout Farmers Association
TFAA Track and Field Athletes of America
TFAI Territoire Français des Afars et des Issas (French—French Territory of Afars and Issas)—formerly French Somaliland
TFB Thatcher Ferry Bridge (over Panama Canal)
tfc traffic
TFCF Twenty-First Century Foundation
TFCNN Task Force Commander—Northern Norway (NATO)
TFCRI Tropical Fish Culture Research Institute
tfcsd total federal commis-

sioned service date
tfd target-to-film distance
tfe tetrafluoroethylene (halon or teflon plastic)
TFF Tropical Fish Farm
TFFW The Foundation for Wellness
tfg typefounding
TFI Table Fashion Institute; Tax Foundation Incorporated; Textile Foundation Incorporated; Traditional Family Ideology scale
tfio thin film integrated optics
tfis theft from an interstate shipment
TFL Trans Freight Line
TFLA Texas Foreign Language Association
TFLC Tulane Factors of Liberalism-Conservatism
tfm transmit frame memory
TFNS Territorial Force Nursing Service
tf/p tubular fluid divided by plasma concentration (concentration of a substance in renal tubular fluid divided by its concentration in plasma)
TFP Trees for People
TFP Tradicion, Familia, y Propiedad (Spanish—Tradition, Family, and Property)—rightwing movement
tfr terrain-following radar; transfer
TFr Tunisian franc
TFR Territorial Force Reserve
TFR/CAR Trouble and Failure Report/Corrective Action Report
tfs time and frequency standard
TFS Transport Ferry Service
TFSK Turkish Federated State of Kibris
TFSR Tools for Self-Reliance
tft thin-film technology; thin-film transistor
TFT Transfer Factor Test(ing)
TFTA Textile Finishing Trades Association
tfu telecommunications flying unit
TFX variable geometry supersonic fighter-bomber
tg tail gear; telegram; telegraph; tollgate; tongue and groove; transformational grammar; transformational generative; type genus
tg (TG) transformational generative; transformational grammar
t/g tracking and guidance
t & g tongue and groove
tg tangente (Italian—tangent)

Tg Tanjung (Malayan—cape)
TG Task Group; Tate Gallery; Texas Gulf Sulphur (stock exchange symbol); Thai Airways International (airline code); Theatre Guild; Torpedo Group; Traffic Guidance; Translators' Guild
T & G Traveres & Gulf (Florida railroad); Tremont & Gulf (Louisiana railroad)
tga thermal gravimetric analysis; thermogravimetric analysis
Tga Tonga (Spanish abbreviation)
TGA Toilet Goods Association; Turpentine Growers of America
t'gallant topgallant (sail)
t'gal'n't topgallant (sail)
t'gansail topgallant sail
tgarq telegraphic approval requested
tgb tongued, grooved, and beaded
TGC Travel Group Charter(s)
tgca transportable ground-control approach
tgd (TGD) thiodiglycol
tge transmissible gastroenteritis
TGF Transonic Gasdynamics Facility (USAF)
TGG temporary geographic grid
TGH Toronto General Hospital
tgif thank goodness it fits; toes go in first
tGiF thank God it's Friday (TGIF)
tgl toggle
TG loran traffic guidance loran
TGM Thomas G Masaryk; Torpedo Gunner's Mate
TGMLI Tussock Grasslands and Mountain Lands Institute
tgn tangent
Tgo Tsingtao
TGO Timber Growers' Organization
TGP Terminal Guidance Program
TGPLC Transcontinental Gas Pipe Line Corporation
TGR Tiger International; Total Gross Receipts
T-Group Training Group
tgs thermal growing season
TGS Taxiing Guidance System; Translator Generator Service; Turkish General Staff
tgt target; teams-games-tournaments; turbine gas tempera-

ture
TGT Tennessee Gas Transmission
TGU Tegucigalpa, Honduras (airport)
tgurq telegraphic authority requested
TGV Train de Grande Vitesse (French—Train of Great Speed)—high-speed railroad train; *Two Gentlemen of Verona*
TGWU Transport and General Workers' Union
th tee handle
th' the
t & h transportation and handling
Th Thai (Siamese); Thailand (Siam); Thomas; thorium; Thursday
Th Theil (German—part)
TH Town Hall; Toynbee Hall; Transport House; Trinity House; true heading
T-H Taft-Hartley
T & H Thames and Hudson
T H Technische Hochschule (German—technical college)
tha thiodical hydrocarbon analyzer
tha (THA) tetrahydroaminoacridine
Th A Theological Association
THA Transvaal Horse Artillery
Thad Thaddeus
THAI Thai Airways International
Thailand Kingdom of Thailand (*Muang-Thai* or *Prathes Thai*) formerly Siam
Thaler (German abbreviation—Joachimsthaler)—Joachim's dollar—Bohemian coin struck in 16th century at Czech town of Jachymov (Joachimsthal) —its name has become *dollar*
THANACAP Funeral Service Consumer Action Program (*Thana* is the Greek word for death)
thanat thanatology
thanatol thanatologic(al)(ly); thanatologist(ic)(al)(ly); thanatology
than ever than ever before
Thang-Pho (Vietnamese—city)—Saigon, Ho Chi Minh City
Thanksgiving Thanksgiving Day (fourth Thursday in November in the United States)
Thatcherism policy and work of Britain's former prime minister, Margaret Thatcher
that's that is

that's 30 (journalistic jargon—that's all)—the end of the article, report, or story
Th.B. *Theologiae Baccalaureus* (Latin—Bachelor of Theology)
TH & B Toronto, Hamilton and Buffalo (railroad)
TH & BA Toll, Highways and Bridge Authority
thc tetrahydrocannabinol (active ingredient in psychedelic drugs such as hashish, indian hemp, and marijuana)
THC Toledo House of Correction; Toronto Harbour Commission; Toronto Harbour Commissioners; Tourist Hotel Corporation (NZ); Trinity Hall College (Cambridge)
thccre tetrahydrocannabinol cross-reacting cannabinoids
thd thread; threaded; threads; total harmonic distortion
Th.D. *Theologiae Doctor* (Latin—Doctor of Theology)
THD Technisch Hogeschool te Delft (Dutch—Technological University of Delft)
th di thread die
the. (THE) tetrahydrocortisone
The. Theodora; Theodore
THE Technical Help to Exporters
thea theater
Thea Theadora; Theodeline; Theodosia; Theresa
T-head Texas-tea head (slang —marijuana user)
theat theater; theatrical
theatcrit theatrical criticism
THEC Tennessee Higher Education Commission
the E the Equator
THEN Those Hags Encourage Neuterism
theo theoretical; theoretician
Theo Theobald; Theobold; Theocritus; Theodoor; Theodor; Theodora; Theodore; Theodorus; Theodosia; Theodosius; Theodoric; Theodric; Theodule; Theophil; Theophile; Theophilus; Theophraste; Theophrastus
THEO They Help Each Other
Theoc Theocritus
theod theodolite
Theodore Roosevelt Park national park in North Dakota
theol theologian; theological; theologist; theology
Theol Theology
Theoph Theophrastus
theophilanthro theophilanthropic(al)(ly); theophil-

anthropist; theophilanthropy (Thomas Paine's deistic religion combining belief in a god with service to mankind)

theor theorem; theoretical; theory

Theor Theorique (French—theoretical)

theos theosophical; theosophist; theosophy

Theo Soc Theosophical Society

ther therapy

therap therapeutic; therapeutics; therapy

there's there is

therm thermometer; thermostat(ic)

therm (Latin prefix—heat)—thermometer

Therm Thermidor (French—Hot Month)—beginning July 19th—eleventh month of the French Revolutionary Calendar also called the *Fervidor*

thermistor thermal resistor

thermo thermostat

thermoc thermocouple

thermochem thermochemical; thermochemistry

thermodyn thermodynamics

thermonuc thermonuclear

THES Times Higher Education Supplement

THESIS Thematic Elementary Science Individualized Studies

thesp(s) thespian(s)

Thess Thessalonians

thetcrit theater critic; theatrical criticism

The Troubles Ireland's efforts to separate from the United Kingdom

they'd they had; they would

they'll they will

they're they are

they've they have

thf (THF) tetrahydrocortisol

t$_h$f Trust Houses Forte (British motel chain)

THF Berlin, Germany (Tempelhof Airport)

THG Technische Hochschule Graz (German—Technical University of Graz)

th ga thread gauge

THHS Townsend Harris High School

THhwm Trinity House high-water mark

thi temperature-humidity index

THI Texas Heart Institute

Thief The Thief of Bagdad (Douglas Fairbanks' motion picture)

Thim Thimbu, Bhutan

things. three-dimensional input of graphical solids

THIWRP The Hoover Institution on War, Revolution, and Peace

thixo thixotropic

Th:J Thomas Jefferson (initials written by him as shown)

thk thick(ness)

THK Turk Hava Kurumu (Turkish Air Association)

Th. L. Theological Licentiate

THlwm Trinity House low-water mark

thm (THM) trihalomethane

Th.M. Theologiae Magister (Latin—Master of Theology)

thms trihalomethanes

Thn Trollhättan

tho' though

'tho' although

Tho Thomas; Thorshavn

Thomas St Thomas, American Virgin Islands

Thomas' Thomas' Register of American Manufacturers

THOMIS Total Hospital Operating and Medical Information System

thor thorax; thoracic

Thor medium-range ballistic missile

THOR Tandy High-Intensity Optical Recording

thorac (Latin prefix—chest)—thoracic

Thoreau Foun Thoreau Foundation

thoro thorough

thoro' thorough

Thoro thoroughfare

Thos Thomas

Thos Jeff Thomas Jefferson

thou. thousand

Thousands Thousand Islands in the St Lawrence River between New York and Ontario

thp thrust horsepower; track history printout

THPD Tragedy of Hamlet, Prince of Denmark

THq theater headquarters

thr target heart rate; their; threonine (amino acid) (THR); through; thrust

THR Teheran, Iran (airport)

Three Kings Three Kings Islands bird sanctuary in the South Pacific off New Zealand's North Island

Three King's Three King's Day (January 6—Epiphany)

thrmst thermostat

thro' through

THRO Throw the Hypocritical Rascals Out

thro' b/l through bill of lading

Throgs Throgs Neck (site of New York State Maritime College)

Throg Street Throgmorton Street, London

thrombo thrombosis

thrombo (Latin prefix—clot or lump)—thrombosed

thrombol thrombolytic therapy

throt throttle

thru through

Thru Thruway

THS Technical High School; Titanic Historical Society; Tiwi Hot Springs (Philippines); Tottenville High School

tht (THT) tetrahydrothiopen

THT Teacher of Hydrotherapy

th ta thread tap

thtr theater

THTRA Thorium High-Temperature Reactor Association

Thu Thursday

THU The Hebrew University (Jerusalem)

Thuc Thucydides

THUMS Texaco, Humble, Union, Mobil, Shell (oil-drilling complex dominating Long Beach, California)

Thur Thuringia(n); Thursday

Thurs Thursday

Thursday Thursday Island pearl-shell fishery in Torres Strait near Cape York, Australia

Thus (nickname—Calcutta Steam Tug); Thursday

thv thoracic vertebra

Thv Thorvald(sen)

THW Technische Hochschule Wien (German—Technical University of Vienna)

THwm Trinity House water mark

Thwy Thruway

THY Turk Hava Yollari (Turkish airline)

thz (tHz) tetraherz

ti target identification; temperature indication; temperature indicator; termination instruction; tricuspid insufficiency

t/i target identification; target indicator

ti Texas Instruments (trademark); *tudni illik* (Hungarian—that is)

Ti titanium

Ti Tirsdag (Danish—Tuesday);

(Latin—Tiberius)
TI Technical Inspection; Technical Institute; Technical Intelligence; Terminal Island; Termination Instruction; Terrorist International; Texas Instruments; Textile Institute; Thread Institute; Title Insurance (and Trust Company); Toastmasters International; Tobacco Institute; Tonga Islands; Training Instruction; Treasure Island; Tungsten Institute; Tuskegee Institute
TI Théâtre-Italien (French—Italian Theater)—Paris
T of I Times of India
TI-67 Israeli designation for captured built-in-the-USSR tanks (T-54 and T-55 models armed with 100mm guns)
tia transient ischemic attack
tia (TIA) trading investment area; transient ischemic attack
TIA Tax Institute of America; Tobacco Institute of Australia; Trans International Airlines; Travel Industry of America; Tricot Institute of America; Trouser Institute of America; Typographers International Association
TIA Tutukuvul Isukul Association (Melanesian—United Farmers Association)—Papua New Guinea coconut planters united
TIAA Teachers Insurance and Annuity Association of America
TIAC Thrift Institutions Advisory Council
TIAP Total Ischemia Awareness Program
TIAS Treaties and Other International Acts Series (U.S. Department of State)
tib tibia(l); trimmed in bunkers
tib tibetisch (German—Tibetan)
Tib Isabel; Tibet; Tibetan
Tib Albius Tibullus (Roman poet)
TIB Technical Information Bulletin; Tennessee Inspection Bureau; Thousand Islands Bridge; Tourist Information Bureau
tibc total iron-binding capacity
Tibet. Tibetian
tic total information card
tic. target intercept computer
TIC Teacher Information Center; Technical Information Center; Technical Institute

Council; Technical Intelligence Center; Texas Industrial Commission; Tyne Improvement Commission
TICA Technical Information Center Administration; The Independent Cat Association; The International Cat Association
TICACE Technical Intelligence Center Allied Command Europe (NATO)
TICC Technical Intelligence Coordination Center
TICCI Technical Information Center for the Chemical Industry
ticcit time-shared interactive computer-controlled information television
TICF Tennessee Independent Colleges Fund; Transient Installation Confinement Facility
tick. tickler
tick(er) ticker tape
tictac time compression tactical communications
TICUS Tidal Current Survey System
tid task initiation date
t.i.d. tres in die (Latin—thrice a day)
tideda time-dependent data analysis
tidskr tidskrift (Swedish—periodical)
TIDU Technical Information and Documents Unit
tidy. teletypewriter integrated display
tie. technical integration and evaluation
tie. (TiE) (TIE) telephone interconnect equipment
TIE Technology Information Exchange; The Institute of Technology; Total Interlibrary Exchange (California Library Network); Traveler's Information Exchange; Truck Insurance Exchange
Tiempo El Tiempo (Time—Bogotá newspaper)
Tien Tientsin
Tien Shan high mountain ranges north of Pamirs and Himalayas between Siberia and Turkestan
tier. tierce
tier tierce (French—third)
Tierg Tiergarten
TIES Transmission and Information Exchange System
tif telephone influence factor; telephone interference factor;

tumor inducing factor
Tif Tiflis
TIF Turtle Island Foundation
TIFA Tamburitzan Institute of Folk Art
Tiff Tiffany
Tiff Tiffany's Reports
TIFI Technology Insight Foundation Incorporated
tifr total investment for return
TIFR Tata Institute of Fundamental Research
tifs technology for instructional feedback
tig time in grade; tungsten-inert gas
TIG The Inspector General
TIGER Topologically-Integrated Geographic-Encoding-and-Reference System (Bureau of the Census)
TIGERS Telephone Information Gathering for Evaluation and Review System
tigon offspring of tiger and lioness
TIGRs Treasury Investment Growth Receipts
tigt turbine inlet-gas temperature
TIH Their Imperial Highnesses
TII Texas Instruments Incorporated; Toastmasters International Incorporated
TIIAL The International Institute of Applied Linguistics
TIIRS Title-I Information and Reporting System
TIJ Tijuana, Mexico (airport)
'til until
TIL Taylor Institution Library (Oxford); Tube Investments Limited
tili translunar injection
TILS Technical Information and Library Service
tim technical information on microfiche; technical information on microfilm; time is money
tim (TIM) transient intermodulation
Tim Timor; Timothy
Tim Timon of Athens; Timothy
Tima Fatima
TIMA Thermal Insulation Manufacturers Association
timation time navigation
timations time navigation artificial satellite
timb timbales (French—kettledrums)
TIMC The Industrial Management Center
TIME Telecommunication Information Management Ex-

ecutive

time imm time immemorial (time beyond memory; time out of mind)

Time-Life Time-Life Books

Times The New York Times; The Times (leading British newspaper, published in London); local designation for all other newspapers containing *Times* in their title

TIMES The Institute of Mining and Engineering Surveyors

Times Roman Times Roman type (sometimes abbreviated T-R)

timet titanium metal(s)

timms thermionic integrated micromodules

Timmy Timothy

timp timpani (Italian—kettledrums)

Timpanogos Timpanogos Cave National Monument in northcentral Utah or Mount Timpanogos in the same area

timps timpani (kettledrums)

TIMS The Institute of Management Sciences; Thermal Infrared Multispectral Scanner

Tim-Tim (Portuguese—Timor, Timur)—former colony in the Lesser Sunda islands of Indonesia

tin pipe-tobacco tin filled with marijuana

TIN Taxpayer Identification Number; tinnitic-induced noise; Transaction Identification Number

tinc tincture

tin can(s) submarine(s)

tinct tincture

tinct. tinctura (Latin—tincture)

TINFO Tieteellisen Informoinnin Neuvosto (Finnish—Council for Scientific and Research Libraries)

'tini Martini (cocktail)

tin in tinnitus instrument

tins thermal-imaging navigation set

TINs Temporary Instruction Notices

tint international practical temperature

Tintagel Tintagel Head, Cornwall (legendary birthplace of King Arthur)

tiny terrs tiny terrorists (children used by terrorists to run errands or spot their enemies)

tio take it off; time interval optimization; time in office

(TIO)

TIO Target Indication Office(r); Television Information Office(r); Test Integration Office(r); Troop Information Office(r)

tip tax information plan; theory in practice; to insure promptness (a gratuity given to insure promptness); translation-inhibiting protein (TIP)

tip tipografia; tipografico (Italian—printing firm; typographic); truly important person (TIP)

Tip Thomas P (Tip) O'Neill, former Speaker of the U.S. House of Representatives

TIP Target Interactive Project; The Institute of Physics; Tax-based Income Policy; Technological Institute of the Philippines; Terrorist Information Project; Trans-Israel Pipeline; Transportation Improvement Program; Trauma Intervention Program; Tripoli, Libya (airport); Troop Information Program(s); truly important person(age)

TIP Türkiye Işçi Partisi (Turkish Labor Party)

TIPAC Texas Instruments Programming and Control

tip.bkt tipping bracket

Tipp County Tipperary, Ireland

TIPRO Texas Independent Producers and Royalty Owners

tips. to insure prompt service (gratuities); topical information packages; truly important persons (TIPS)

TIPs Tax-based Income Policies

TIPS Technical Information Processing System; Telemetry Impact Prediction System; Total Integrated Pneumatic System; truly important persons

tiptap target input panel (and) target assign panel

tiptop tape input—tape output

TIP & TPS The Institute of Physics and The Physical Society

tir total indicator reading

TIR Transport International des Marchandises par la Route (French—International Transport of Merchandise by Road)—twenty-six nation custom agreement permitting trucks marked TIR to avoid

customs until reaching their final destination; *Transport Internationale Routiers* (French—International Truck Routes)

TIRB Transportation Insurance Rating Bureau

TIRC T Tauri Infrared Companion (pronounced *turk*); Tobacco Industry Research Committee

tire burner hot pursuit of one vehicle by another

T-iron T-shaped iron or steel section

Tiros American meteorological satellite designed to observe cloud coverage and infrared heat radiation of the earth; television and infrared observation satellite

TIRR Texas Institute of Rehabilitation and Research

tirs thermal infrared scanner

tis tissue(s); total integrated scattering

'tis it is

TIs Thousand Islanders; Thursday Islanders; Tonga Islanders; Turks Islanders

TIS Technical Information Service; Total Information System; Transactional Information Systems; Transdermal Infusion System

TISC Technology Information Sources Center

TISI Thai Industrial Standards Institute

ti-slash tire slash(ing)

TISPM Territorie des Iles St Pierre et Miquelon (French territory offshore Canada)

TISS Title-I Support System

tit *título* (Spanish—title)

tit. title; titular; titulary; transitive inference training

tit títre (French—title)

Tit Titus, The Epistle of Paul to

Tit Titus

TIT Tokyo Institute of Technology; Tustin Institute of Technology

Tit A Titus Andronicus

titanox titanium dioxide

Titis Titi Islands also called the Mutton Birds, off the southwest coast of New Zealand's Stewart Island

tit⁰ titulo (Spanish—title)

TITUS Textile Information Treatment Users Service

tiu trigger inverter unit

TIU Telecommunications International Union; Tokyo Im-

perial University
tiv total indicator variation
Tiv Tivoli
tix ticket(s)
tixi turret-integrated xenon illuminator
TIYC Thousand Island Yacht Club
tj tomato juice; triceps jerk; turbojet (TJ)
tj (TJ) talk jockey
tj to jest (Polish—that is)
Tj Tijuana, Baja California, Mexico
TJ Thomas Jefferson—third President of the United States
TJAG The Judge Advocate General
tjc trajectory
TjC trajectory chart
TJC The Jockey Club; Trenton Junior College; Tyler Junior College
TJC Tragedy of Julius Caesar
TjD trajectory diagram
TJHS Thomas Jefferson High School
Tji Tjirebon (Cheribon)
TJM The Jewish Museum; Thomas Jefferson Memorial
tjp (TJP) turbojet propulsion
TJPOI Twisted Jute Packing and Oakum Institute
tjs tight-jean syndrome (impotence and sterility)
TJS Tactical Jamming System
TJSUSA Thomas Jefferson Society of the United States of America
tjt tactical jamming transmitter
TJTA Taylor-Johnson Temperament Analysis
tk track; truck; trunk
tk (TK) transkelotase
tk to kum (printer's expression meaning material is *to come*)
Tk Turkmenian; Turkmenistan
Tk Teluk (Malay—bay; bight; riverbend)
tkbd tackboard(ing)
tkd tokodynamometer
tkg tanking; tokodynagraph(y)
Tki Takoradi
TKK Teikoku Kaiji Kyokai (Imperial Japanese Marine Corporation, ship classifiers)
TKL Tragedy of King Lear
tko technical knockout
TKP Turkiye Komünist Partisi (Turkish Communist Party)
tkr tanker; terrestrial kilometric radiation
T KR II Tragedy of King Richard II
tks thanks
TKTF Tanker Task Force

tkt(s) ticket(s)
tl terminal limen; test link; thrust line; time length; time limit; total load; transmission level; transmission line; truckload; truck loading
t-l trade last (slang, a compliment)
t/l total loss
t.l. tukus lecker (Yiddish—ass licker)—flatterer; sycophant
Tl thallium
TL Technical Letter; Technical Library; Texas League; The Leprosarium (U.S. Public Health Service, Carville, Louisiana); Torpedo Lieutenant; Townland (UK); Turk lirasi (Turkish pound)
T/L Telegraphist/Lieutenant; Torpedo Lieutenant
TL Teatro Lirico (Italian—Lyric Theater)—Milan; *Théâtre-Lyrique* (French—Lyric Theater)—Paris
T-L Time-Life (books, magazines, recordings)
tla translumbar aortogram
TLA The Library Association (of the United Kingdom); Texas Library Association; Theatre Library Association; Trial Lawyers Association; Trinidad Lake Asphalt
Tlax Tlaxcala (inhabitants—Tlaxcaltecas)
TLB temporary lighted buoy
tlbl tape label
TLBs Time-Life Books
tlc talcum; tender loving care; thin-layer chromatography; total lung capacity
TLC Television Licensing Center; Thin-Layer Chemistry; Total Life Care; Total Life Center; Trades and Labour Club
TLCPA Toledo-Lucas County Port Authority
TLCs Tire and Lube Centers
TLC for Seniors Total Life Care for Senior Citizens
tld thermoluminescent detector; thermoluminescent dosimeter; tooled
tl dating thermoluminescent dating
tle theoretical line of escape; thin-layer electrochemistry
tlf telefon (Norwegian—telephone)
TLFB Texas-Louisiana Freight Bureau
tlg tail landing gear; telegraph
TLG Theatrical Ladies' Guild; Tiger Leasing Group

TLH Tallahassee, Florida (airport)
tlli tank liquid-level indicator
tlm telemeter; telemetry
TLMA Tag and Label Manufacturers Association
Tln Tallinn
tlo total loss only
TLO Technical Liaison Officer
tlp tension-leg platform; term-limit pricing; threshold learning process; time-lapse photography
tlp (TLP) tension-leg petroleum (oil rig)
TLP Theraputic Learning Program
TLP Telefones de Lisboa e Porto (Portuguese—Lisbon and Oporto Telephone Company)
tlr trailer; twin-lens reflex (camera)
TLR Tool Liaison Request
tls testing the limits for sex; typed letter signed
TLS Technical Library Service; Technical Library System; Terminal Landing System; The Law Society; Trinity Lighthouse Service
TLS Times Literary Supplement
tlt transportable link terminal
TLTB Trunk Line Tariff Bureau
tltr translator
tlu table look up
tlv threshold limit value(s)
tlv (TLV) tracked levitated vehicle
TLV Tel Aviv, Israel (airport)
tlvsn television
tly tally
tlz titanium, lead, zinc; transfer on less than zero
tm standard mean temperature; tactical missile (TM); team; temperature meter; time modulation; tractor monoplane (TM); trademark; transport mechanism; transverse magnetic; true mean; twisting moment
t/m test and maintenance
t & m time and material(s)
tm tonelada métrica (Spanish—metric ton, 2,200 pounds)
Tm thulium
TM tactical missile; Technical Manual; Technical Memoranda; Technical Memorandum; Technical Minutes; Technical Monograph; Telemetering; Test Manual;

Texas Mexican (railroad); The Maccabees; Toledo Museum; tractor monoplane; trademark; Training Manual; Training Mission(s); Trainmaster; Transcendental Meditation; Tropical Medicine

T/M (t/m) trailmobile (automobile trailer)

TM Technical Manual; Turk Mali (Turkish—Made in Turkey); *Théâtre de la Monnaie* (French—Theater of the Currency)—Brussels; *Tragedy of Macbeth*

tma total material assets; total military assets

Tma Tema

TMA Texas Maritime Academy; Theatrical Mutual Association; Tile Manufacturers Association; Tobacco Merchants Association; Toiletery Merchandisers Association; Toy Manufacturers Association; Trans-Mediterranean Airways (Lebanese); Twine Manufacturers Association

TMAMA Textile Machinery and Accessory Manufacturers' Association

T-man Treasury Department special agent of the IRS

tmar trial marriage

TMAS Taylor Manifest Anxiety Scale

TMB Travelling Medical Board

TMBC Toronto Motor Boat Club

tmbr timber

tmc total market coverage

TMC Tata Memorial Center; Technical Measurement Corporation; Texas Medical Center (Houston); Trans Mar de Cortés (Mexican airline); Transportation Materiel Command

TMCA Tabulating Card Manufacturers Association; Titanium Metals Corporation of America

tmcd tetramethylcyclobutanediol

tmcp trimethylenecyclopropane

TME Teacher of Medical Electricity

T-men Treasury Department law-enforcement officers

TM-Eng Technical Manual—Engineering

t'ment tournament

tmf the mushroom factor

TMF The Menninger Foundation

tmh thermomechanical hydraulic; tons per manhour

tmi technical market index (TMI)

Tmi Tsurumi

TMI Telemeter Magnetics Incorporated; Three-Mile Island, Pennsylvania—site of nuclear reactor meltdown March 28, 1979; Tool Manufacturing Instruction; Trucking Management Incorporated; Tube Methods Incorporated; Turkish Military Institute; Turkish Military Intelligence

TMI Technical Manual Index (USN); *Technical Market Index*

TMIC Toxic Materials Information Center

TMIF Three-Mile Island (nuclear-power) Facility

TMIS Technical Meetings Information Service

tmj temporomandibular joint

tmj (TMJ) temporomandibular joint (syndrome)

TMJ Trade Marks Journal

tmkpr timekeeper

tml (TML) three-mile limit

TML Transport Managers License; Transmanche Link (Eurotunnel construction)

TMM Transportación Maritima Mexicana (Spanish—Mexican Maritime Transportation)

TMMC Theater Materiel Management Center

TMMG Teacher of Massage and Medical Gymnastics

tmn transmission (flow chart)

Tmn Tamano

TMNP Tamborine Mountain National Parks (Queensland)

TMNT Teenage Mutant Ninja Turtle

tmo (TMO) telegraph money order

TMO Table Mountain Observatory; telegraph money order; Traffic Management Officer

TMORN Texaco Metropolitan Opera Radio Network

tmos the man on the street

tmp temperature; temporary; thermomechanical pulp(ing); trimethoprim; trimethyl phosphate (male contraceptive)

tmp (TMP) total mind power

Tmp Tampico

tmpry temporary

tmp's transcedental meditation

practitioners

TMPS Trans-Mississippi Philatelic Society

tmr timer; total materiel requirement; trainable mentally retarded (semi-autistic children)

TMRB Tropical Medicine Research Board

tmrbm (TMRBM) transportable midrange ballistic missile

tms terms; type, model, and series

tms tai muuta semmoista (Finnish—and so on)

TMs Temporomandibular Disorders

TMS Tactical Missile Squadron; Technical Museum, Stockholm; Tramway Museum Society; Transmatic Money Service

TMS Tribunal Maritime Special (French—Special Maritime Court)—disciplinary prison court in French Guiana

TMSA Technical Marketing Society of America

tmsd total military service date

tmt turbine-motored train

T^{mt} Tablemount

TMT transonic model tunnel

TMTB The Malayan Tin Bureau

tmtc through-mode tape converter

TMU Tokyo Metropolitan University

TMUS Toy Manufacturers of the United States

tmv true mean value

tmv (TMV) tobacco-mosaic virus

TMV Transportadora Maritima Venezolana (Spanish—Venezuelan Line)

tmw thermal megawatts; tomorrow

TMW Textile Machine Works

TMWC Trial of the Major War Criminals

tn tariff number; telephone number; thermonuclear; train; true north

Tn thoron (chemical symbol); Ton

TN Task Number; Technical Note; Tennessee

T & N Turner and Newhall

TN Twelfth Night

TNA The National Archives; Thermal Neutron Analysis (system for detecting bombs in baggage)

tnad thermal-neutron activation device

TNAS Tuberculosis Nursing Advisory Service

TNB *Tsentral'naya Nauchnaya Biblioteka* (Russian—Central Scientific Library)—Kiev

tnc total numerical control

TNC Thai Navigation Company; The Nature Conservancy

tnct tincture

tnd tender

TNDC Thai National Documentation Center

tndr tendered

TNEC Temporary National Economic Committee

t*nes* tonnes (French—tons)

tnf transfer on no overflow

tnf (TNF) tumor necrosis factor (natural substance killing cancer cells)

TNF Theater Nuclear Forces (NATO); Toiyabe National Forest

tng training

Tng *Tandjung* (Malay—Cape)

TNG Tangier, Morocco (airport); The National Grange; The Newspaper Guild

TNG *The New Grove Dictionary of Music and Musicians* (20-volume 1981 edition)

tnge tonnage

TNI *Tentara Nasional Indonesia* (Indonesian National Army)

TNIAU *Tentara Nasional Indonesia Angkatan Udara* (Bahasa Indonesian—Indonesian Armed Forces—National Air Force)

Tn IOB Technician of the Institute of Building

tnm tumor, node, metastasis

tnm (TNM) tactical nuclear missile

TNM Texas-New Mexican; Texas-New Mexico; Tokyo National Museum; Tumacacori National Monument

TNM *Telégrafos Nacionales de México* (Spanish—National Telegraph of Mexico)

TNNP Taman Negara National Park (Malaysia); Terra Nova National Park (Newfoundland)

Tno Taranto

T No (TNO) Track Number

T & NO Texas and New Orleans (railroad)

t no c threads no couplings

tnp (TNP) trinitrophenol

TNP Tarangire National Park (Tanzania); Taroba National Park (India); Tonariro National Park (North Island, New Zealand); Tsavo National Park (Kenya)

TNP *Théâtre National Populaire* (French—Popular National Theater)

tnpg trinitrophloroglucinol

TNPG The Nuclear Power Group

Tnpk Turnpike

TNPO Terminal Navy Post Office

tnr trainer

tnr *toneladas de registro neto* (Spanish—net registered tonnage)

TNR Tananarive, Malagasy (airport); Tucki Nature Reserve (New South Wales)

TNR *The New Republic*

TNRIS Transportation Noise Research Information Service

Tnry Tannery

tns transcutaneous nerve stimulator

Tns Townsville; Tunis

TNS Tennessee Nuclear Specialties; Transit Navigation System

tnt (TNT) trinitrotoluene

t-n t trans-national terrorism; trans-national terrorist

t'n't tequila and tonic (mixed drink)

TNT Tactical Narcotics Team (NYC)

tntc too numerous to count

TNTC Thames Nautical Training College

t-n t's trans-national terrorists

tntv tentative

tnw (TNW) tactical nuclear warfare

tn wep(s) thermonuclear weapon(s)

TNWR Tamarac National Wildlife Refuge (Minnesota); Tewaukon NWR (North Dakota); Tishomingo NWR (Oklahoma)

tnx thanks

tnz transfer on non zero

to. telephone order (TO); time off time opening; tool order (TO); transverse optic; turn off; turn over

t/o (TO) takeoff

t & o taken and offered; technical and office (workers)

t.o. *tinctura opii* (Latin—tincture of opium)

t° tomo (Spanish—volume)

To Togo; Toronto

To *Torsdag* (Danish—Thursday)

TO Table of Organization; takeoff; Technical Observer; Technical Order(s); Theater of Operations; The Order (neo-Nazi organization); Third Order (of a religious congregation); Toledo, Ohio; Tool Order; Trans Arabian Air Transport (cargo); Transportation Office(r); Travel Order

T/O Table of Organization

TO *Technical Order*

toa total obligational authority

TOA Theater Owners of America; The Orchestral Association; Toledo Opera Association

toac tool accessory

tob tobacco

Tob Tobago; The (Apocryphal) Book of Tobit

T o B Tour of Britain (bicycle)

tobac tobacco; tobacconist

Tobaccos Tobacco Root Mountains of southwest Montana

TOBE Test of Basic Education

Tobio Gorria anagrammatic pseudonym of Arrigo Boito (1842–1918)

TOBWE Tactical Observing Weather Element (USAF)

toc table of contents; top-blown oxygen converter; top of concrete; total organic carbon

Toc Tagliabue open cup

TOC Tactical Operations Center; Technical Order Compliance; Television Operating Center

TOCCWE Tactical Operations Control Center Weather Element (USAF)

Toch Tocharian

TOCHR Terminal Operators Conference of Hampton Roads

Toco Tocopilla, Chile

TOCS Terminal Operating Control System

tod technical objective document(s); time of day; time of delivery

Tod Todhunter

TOD Technical Objective Document; The Open Door

to'ds toads; towards

toe. term of enlistment; ton-oil equivalent; total operating expense

TOE Table of Equipment

T O & E Texas, Oklahoma & Eastern (railroad)

TOEFL Test of English as a Foreign Language

TOES The Other Economic Summit; Tradeoff Evaluation System

TOET Test of Elementary Training

tof time of flight

tofc trailer on flatcar (or piggyback)

tog. together; toggle; to order grog

TOGA Tests of General Ability; Tropical Ocean/Global Atmospheric Program

to'gal'nt topgallant (mast or sail)

Togo Admiral Togo Heihachiro (victor of the Battle of Tsushima where his forces annihilated the Russian fleet in 1905); Republic of Togo (West African coastal country) *République Togolaise*

Togoland (German—Togo)— West African colony under German domination from 1884 to 1916

togr together

togw takeoff gross weight

tog/wi together with

tohp takeoff horsepower

toj track on jamming

Tojo Premier Tojo Hideki (Japanese general and premier during World War II)

Tok Tokyo

Tokelaus Tokelau Islands of the Pacific also called the Union Islands including Atafu, Fakaofu, and Nukunono

toke(s) token(s)

Tok Uni Tokyo University

tol tolerance; toluene

Tol Toledo; Toledan

T o L Tower of London

TOL Toledo, Ohio (airport); Trans-Ocean Leasing (corporation)

tol'able tolerable

TOLCCS Trends in On-Line Computer Control Systems

Tolly Tolliver

Tol Orc Toledo Orchestra

to lt towing light

TOLU Transocean Leasing (container) Unit

t-o-m the old man (the boss; the captain, the chief, the father)

tom *tomo* (Spanish—volume)

Tom $2 bill; Thomas

TOM *Territoire d'Outre-Mer* (Overseas Territory)

tom(at) tomato

tomats tomatoes

tomb. technical organizational memory bank

Tombigbee Tombigbee River of Alabama and Mississippi

tomcat (TOMCAT) theater-of-operations missile continuous-wave anti-tank (weapon)

tome (Greek—a cutting or a slice)—anatomy, dichotomy, lobotomy, microtome

Tommie Thomas

toms male turkeys; males of various animals; tired old movies

TOMS Total Ozone Mapping Spectrometer; Total Ozone Mapping System

TOMV *Tragedy of Othello, the Moor of Venice*

tomy (Latin suffix—cut)—appendectomy

ton *toneel* (Dutch—scene, set, stage); *toneladas* (Spanish—tonnage, tons); *tyurma osobogo naznacheniya* (Russian —special-purpose prison)

Ton Tonga or Friendly Islands

TON *Top of the News*

TONACS Technical Order Notification and Completion System

Tonga Kingdom of Tonga (South Pacific island nation)

Tongariro Tongariro National Park in New Zealand's North Island or an active volcano in the same area

Tongas Tonga Islands in the South Pacific

Tongass Tongass National Forest in southern Alaska

tonguesan tongue sandwich

tonguewich tongue sandwich

Ton Isl Tonga Islands

tonk honky tonk

Tonka Minnetonka, Minnesota

tonn tonnage

'toon(s) cartoon(s)

top. temporarily out of print; topographer; topographic(al)-(ly); topographica (three-dimensional) art; topography; torque oil pressure

t-o-p temporal-occipital-parietal (lobes of the brain)

Top Topeka; Topology

ToP Taxonomy of Programs

TOP Targeted Outreach Program; Third Order of Penance

top 10 top 10 best sellers (books or recordings of classical, jazz, or popular music)

topa tooling pattern

topaz hydrous aluminum fluorosilicate

TOPAZ Technic for the Optimum Placement of Activities in Zones

TOPCOPS The Ottawa Police Computerized On-line Processing System (Canada)

Top End northernmost Australia

TOPICS Tables of Periodical Indices Concerning Schools; Test of Performance in Computational Skills

to po topographic; topography

Topo Topolobampo, Sonora, Mexico

TopoCom Topographic Command (USA)

topog topography

topol topology

topon toponym

topony toponym(ic)(al); toponymist; toponymy

topo(s) toponym(s)

TOPP Terminal-Operated Production Program

tops. (TOPS) take off pounds sensibly

TOPS Task-Oriented Processing System; Teen-age Opportunity Programs in Summer; Tested Overhead Projection Series; Total Operations Processing System; Training Opportunities Scheme

Top Sec Top Secret

tops'l topsail

TOPSTAR The Officer Personnel System—The Army Reserve (USA)

Toquemas Toquemas Mountains of central Nevada

tor time of receipt; torque; torquing; torquing up

tor (TOR) teletype on radio; transmitter output register

Tor Toronto

TOR Third Order Regular

Toray Tokyo Rayon Company (tradename)

TORCH Toronto Orthopaedic Recreational Center's Headquarters

Torch of Liberty Statue of Liberty

Tor-Chi Toronto—Chicago

Tor Dep Torpedo Depot

Tor Dom Toronto Dominion (bank)

Tor Int Air Toronto International Airport

Tor-Lax Toronto—Los Angeles

Tormentine Cape Tormentine —easternmost point in New Brunswick, Canada

Tor-Mia Toronto—Miami

torn. tornado
Torngats Torngat Mountains of Labrador
Tor-NY Toronto—New York
torp torpedo; torpedoman
Torport Toronto (container) Port
torr 1mm of mercury
Torr toor
Tor-San Toronto—San Diego
Tor-Sea Toronto—Seattle
Tor-Sfo Toronto—San Francisco
Tor Sym Toronto Symphony Orchestra
Tortugas Tortuga Islands (Dry Tortugas and Wet Tortugas)
tos term of service; truck operational surcharge; temporarily out of stock
TOS Tape Operating System; The Orton Society; Tiros Operational Satellite
Tosa Tsunetaka
TOSBAC Toshiba Scientific and Business Automatic Computer
tosc toscano (Italian—Tuscan)
TOSCA Toxic Substances Control Act
TOSCO The Oil Shale Corporation
tose tooling samples
Toshiba Tokyo Shibaura Electric
toss. takeoff safety speed
TOSS Tiros Operation Satellite System
tot time on (over) target; total; totalize; totalizer
t o t *tukus om tisch* (Yiddish—put your cards on the table)
t-o-t tip-of-the-tongue
tot tuchis afn tish (Yiddish—buttocks on the table)—put up or shut up, put your cards on the table
TOT Tourist Occupancy Tax; Tourist Organization of Thailand; Transient Occupancy Tax
totalism totalitarianism
TOTCO Technical Oil Tool Corporation
tote. totalizator
TOTE Task-Oriented Teacher Education; Totem Ocean Trailer Express (to Alaska)
TOTES Test-Operate-Test-Exit System
t'other the other
TOTO Tongue of the Ocean (deep-water channel in Great Bahama Bank); Totable Tornado Observatory (National Severe Storms Laboratory,

Norman, Oklahoma)
totp tooling template
Tou Toulon
TOU The Open University; Tractor Oils Universal
tour. tourism, tourist
tourn tournament
TOUS Test on Understanding Science
tov ten opzichte van (Dutch—with regard to)
TOVALOP Tanker Owner's Voluntary Agreement concerning Liability for Oil Pollution
tow. tug of war
tow. (TOW) tube-launched optically-tracked wire-guided (anti-tank missile)
TOW Top Antitank Weapon
Tower The Tower of London (formerly a prison and now a museum by the Thames in London); Tower Publications
TOWER Testing, Orientation, and Work Evaluation in Rehabilitation
Towers Charters Towers
townet towing net
tox toxemia; toxic; toxicant; toxicologist; toxicology
tox (Latin prefix—poison)—toxicology
toxback toxicology information backup
toxicol toxicology
toxline toxicology hot line (public information program); toxicology on-line
TOXLINE Toxicology On-Line (computer retrieval system)
toz tidelands overlay zone
tp target practice; teaching practice; technical paper; telephone; teleprinter; title page; toilet paper; total points; total protein; transport pilot; treaty port; turning point
tp (TP) tape (computer flow chart); teleprocessing; total product; transaction processing
t/p test panel
t & p theft and pilferage
tp tempo primo (Italian—speed as at the outset)
Tp Township; Troop
TP Technical Pamphlet; Technical Paper; Technical Problem; Technical Publication; Technographic Publication; Texas & Pacific (railroad); Thomas Paine; Thompson Products; Torrey Pines (Insti-

tute); True Position
T & P Texas and Pacific (railroad)
T.P. Tempore Pachale (Latin —Easter time); *Tribunicia Potestas* (Latin—Tribune of the People)
tpa third-party adjuster; travel by privately owned conveyance authorized
tpa (TPA) tissue plasminogen activator
TPA Tampa, Florida (Tampa International Airport); Tampa Port Authority; Telephone Pioneers of America; Trans-Pacific Airlines (Aloha Airline); Travelers' Protective Association
TPAC Thomas Performing Arts Center (Akron)
TPAO Türkiye Petrolleri Anonim Ortakligi (Turkish Petroleum Corporation)
tpb tryptone phosphate broth
TPB Transportation Programs Bureau
TPBA Transit Patrolmen's Benevolent Association
TPBC Toledo Power Boat Club
tpc treated-paper copier
TPC The Peace Corps (US Department of State)
TPC/JCA Texas Public Community/Junior College Association
TPCNA Titanium, Palladium, Copper, Nickel, Au (gold) telephone circuit-board plating
TPCP Test Plan Change Procedure
tpd tons per day
tp'd toilet papered (some teenager's idea of house-and-garden decoration)
TPDC Tanjong Pagar Dock Company (Singapore)
TPE Taipei, Formosa (airport)
TPEQ Task of Public Education Questionnaire
TPF Tactical Police Force; Thomas Paine Foundation
TPFH Tasmanian Pulp and Forest Holdings
TPGA Texas Personnel and Guidance Association
tpgh tons per gang hour
tph tons per hour
TPH Theosophical Publishing House
TPH Television Production Handbook
Tpha Treponema pallidum hemagglutination

tphasap telephone as soon as possible

tphayc telephone at your convenience

TPH & PCA Toy Pistol, Holster, and Paper Cap Association

TPHS Thomas Paine High School

tpi teeth (threads, tons, or turns) per inch; treponema pallidum immobilization (test)

t-p i title-page, index

Tpi Taipei; Treponema pallidum immobilization

TPI Tax-and-Price Index; Tennessee Polytechnic Institute; Torrey Pines Institute; Truss Plate Institute

Tpi test *Treponema pallidum* immobilization (for the detection of syphilis)

Tpk Turnpike

Tpke Turnpike

TPL Tallahasee Public Library; Tampa Public Library; Toledo Public Library; Toronto Public Libraries; Tucson Public Library; Tulsa Public Library

TPLA Turkish People's Liberation Army

tplab tape label

TPLF Tigré People's Liberation Front; Togrean People's Democratic Front (Ethiopia); Turkish People's Liberation Front

TPLs Trust for Public Lands

tpm tape preventive maintenance; tons per minute

tpmark tapemark(ing)

tpn trigger price mechanism

tpn (TPN) triphosphopyridine nucleotide; (same as nadp or NADP+)

TPN Total Parental Nutrition

TPN *Tatrzanskiego Parku Narodowego* (Polish—High Tatra National Park)—in the Tatra Mountains of Poland

tp & nd theft, pilferage, and non-delivery

tpnh (TPNH) reduced triphosphopyridine nucleotide

TPNHA Thomas Paine National Historical Association (New Rochelle, NY)

TPNHS Thomas Paine National Historical Society

tpnl test panel

tpo terminal-performance objective; transmitter (signal) power output

tpo *tiempo* (Spanish—time)

TPO Travelling Post Office; Tulsa Philharmonic Orchestra

tpob true point of beginning

TPOR Teacher Practices Observation Record

tpp (TPP) thiamine pyrophosphate

TPP Tax Preparers Program; Technical Program Plan-(ning); Total Package Procurement

TPPC Total Package Procurement Concept; Trans-Pacific Passenger Conference

TP-PL Technical Publications Planning (USN)

TP-PU Technical Publications —Public Utilities (USN)

tpqi teacher-pupil question inventory

tpr tape programmed raw; telescopic photographic recorder; temperature profile recorder; temporary price reduction; thermal plastic rubber; thermoplastic recording

tpr (TPR) temperature, pulse, respiration

Tpr Trooper

TPRC Thermophysical Properties Research Center

tpri teacher-pupil relationship inventory

TPRI Tropical Pesticides Research Institute

T & P Ry Texas and Pacific Railway

tps technical problem summary; technopolymer structures; terminals per station; text processing service; throttle position sensor; tree-pruning system (computer language)

tp's taxpayers

TPS Technical Preservation Services; Technical Publishing Society; Telephone Preference Service; Text Processing Service; The Physical Society

tpt tetraphenyl tetrazolium; total protein tuberculin; transport; trumpet

TPT Tactual Performance Test(ing); Toy Preference Test; Transonic Pressure Tunnel (NASA)

tptg turned plate turned grid

tptn toilet partition

tpto tripropyl tin oxide

tptr trumpeter

tpu tape preparation unit; thermoplastic urethane

TPU Travel Planning Unit

TPUS Transportation and Public Utilities Service

tpw title page wanting

TP & W Toledo, Peoria & Western (railroad)

t.q. *tale quale* (Latin—as is)

TQCA Textile Quality Control Association

tqcm thermoelectric quartz-crystal microbalance

TQE Technical Quality Evaluation

t quark top quark

tr temperature, rectal; test run; time remaining (at depth); tons registered; toothed ring; trace; trace; tracer (bullet); tracking radar; translation; transmit-receive; transmitter-receiver; transpose; tuberculin R

tr (TR) total revenue

t-r transmit-receive

t/r transmit(ter)/receive(r)

tr *tinctura* (Latin—tincture); *trillo* (Italian—rolled or shaken, as in drumming or when shaking a tambourine); *traduit* (French—translated); *trykkeri* (Dano-Norwegian—printing office); *tryckt* (Swedish—printed); *trykt* (Dano-Norwegian—printed)

Tr Transcript; Trench; Trieste; Trough

TR Tasmanian Railway; Technical Regulation; Technical Report; Test Report; Texas Gulf Production Company (stock-exchange symbol); Theodore (Teddy) Roosevelt (26th President U.S.); therapeutic radiology; torpedo reconnaissance (naval aircraft); Training Regulation(s); Transportation Request; Travel Request; Trieste; Trip Report; Triumph (British auto or motorcycle); Turkey (auto plaque)

T-R Times-Roman

tra transformer-reactor assembly

Tr A Triangulum Australe (constellation)

TRA Tactical Response Association; Technical Report Authorization; Teledyne Ryan Aeronautical; Textile Refinishers Association; Theodore Roosevelt Association; Thoroughbred Racing Associations; Tire and Rim Association; Trade Relations Association; Travel Research Association

traac transit-research and altitude-control (satellite)

trac text-reckoning and compiling (computer language); tracer; tracing; tractor

TRACALS Traffic Control and Landing System

tracap transient circuit-analysis program

tracdr tractor-drawn

trace. tape-controlled recording and automatic checkout equipment; task reporting and current evaluation; time-shared routines for analysis, classification, and evaluation; total-risk assessing-cost estimate(s)

TRACE Task Reporting and Current Evaluation; Trane Air Conditioning Economics

trach trachea; tracheal; tracheate; tracheation; tracheoscopy; tracheostomy; tracheotomy

TRACIS Traffic Records and Criminal Justice Information System (Iowa)

trackex tracking exercise

tracon terminal radar control

TRACON Terminal Radar Approach Control

TRACS Telemetry Receiver Acoustic Command System; Telescoping Rotor Aircraft System; Total Royalty Accounting and Copyright Systems

tract (Latin prefix—drag or draw)—traction

tractorcade tractor vehicle parade

trad tradition(al)

trad traducido (Spanish—Translated); *traduzione* (Italian—translation)

TRADA Timber Research and Development Association

tradex target resolution and discrimination experiment

tradic transistor digital computer

traf traffic

Trafalgar Cape Trafalgar in southwestern Spain at the western entrance to the Strait of Gibraltar

TRAFFIC Trade Records Analysis of Fauna and Flora in Commerce (endangered species)

trafphobia traffic phobia (fear of driving in traffic)

trag tragedy

T-rail T-shaped rail

train. trainee; trainer; training

TRAIN Telerail Automated In-

formation Network; To Restore American Independence Now

TRAIS Transportation Research Activity Information Service (Department of Transportation)

trai vai (Vietnamese—lychee) —tasty fruit

tram. tracking radar automatic monitoring; tramcar; trammel; tramway

TRAM Test Reliability and Maintenance Program (USN); Treatment Rating Assessment Matrix; Treatment Response Assessment Method

tramp. temperature regulation and monitor panel

tramps. temperature regulator and missile power supply

tran transient

tran (TRAN) transmit (data processing)

tran tranvia (Spanish—tramway)—streetcar or streetcar line

trandir translation director (computer language)

tranny transistor radio

trans transactions; transfer; transit; translation; translator; transport; transportation; transpose; transposition

trans (Latin prefix—across or over)—transalpine, transatlantic

Trans Transactions; Transmission; Transvaal

transac transaction(s)

Trans Am Cryst Soc Transactions of the American Crystallographic Society

Trans Am Geophys Union Transactions of the American Geophysical Union

Trans Am Inst Min Metall Pet Eng Transactions of the American Institute of Mining, Metallurgical, and Petroleum Engineers

Trans Am Nucl Soc Transactions of the American Nuclear Society

Trans Am Soc Mech Eng Transactions of the American Society of Mechanical Engineers

Trans Am Soc Met Transactions of the American Society for Metals

Transan Transandean Railway

Transandine Transandean Railway connecting Argentina and Chile

transatl transatlantic

Transbai Transbaikal Railway

Trans Br Ceram Soc Transactions of the British Ceramic Society

transc transcription

Trans-Carib Trans-Caribbean Airways

Trans-Caspian Trans-Caspian Railroad linking the Caspian Sea region with the southern Urals of the USSR

Transcau Transcaucasian Railway

transceiver transmitter-receiver

TRANSCOM Transportation Coordinated Management

transcrit transportation critic-(ism)

trans d transverse diameter

TRANSDEC Transducer Electronic Center

transec transmission security

transf transfer; transference; transformer

Trans Faraday Soc Transactions of the Faraday Society

transfax facsimile transmission

transie(s) transvestite(s)

TRANSIS Transportation Safety Information System

Transisthmian Transisthmian Highway (flanking the Panama Canal and the Panama Railroad)

transistor transfer resistor

transit. transitive

Transj Transjordan; Transjordanian

Transk Transkei

Trans-Ky Exp Trans-Kyusho Expressway

transl translation; translator

translit transliteration

translu translucent

translun translunar; translunarian; translunarite

transm transmission

Transmark Transportation Systems and Market Research (British rails)

Trans Metall Soc AIME Transactions of the Metallurgical Society of the American Institute of Mechanical Engineers

transmog transmogrification; transmogrify(ing)

Transnistria Trans-Dniestria

Transocean California-Hawaii-Orient Airline; Transoceanic

transp transparent

transpac transpacific

transpl transplant(ation); transplanted

transport. transportation

Transron Transport Squadron

trans sect transverse section

transsexual(s) transvestite homosexual(s)

Trans-Sib Trans-Siberian Railroad linking European Russia with its North Pacific coast

Trans Soc Rheol Transactions of the Society of Rheology

TRANSUB Translation and Publishing Corporation (China)

transv transverse

Transv Transvaal

transvest transvestic(al)(ly); transvestism; transvestite

transv sect transverse section

Transylvanians Transylvanian Alps of Romania

transyt traffic network study tool

trany transparency

trap. trapdoor; trap drums; trapeze; trapezoid(al); trapezium

TRAP Tracker Analysis Program

traps. trap drums; trap drummer(s)

tratel tracking through telemetry; trailer motel

tratt trattenuto (Italian—detained or held back)

trau traumatic

TRAUS Thoroughbred Racing Association of the U.S.

trav. travel

Trav Travancore; Travis

Trav Travessa (Portuguese—Lane)

TRAWL Tape Read-and-Write Library

trb tribunal; tribune; trombone

trb toneladas de registro bruto (Spanish—gross registered tonnage)

TRB *New Republic's* pseudonymic initials standing for columnist Richard Strout; Transportation Research Board

trc total response to crisis

Tr & C Troilus and Cressida

TRC Tape Relay Center; Technical Review Committee; Technology Reports Center; Telegram Retransmission Centre; Trans-Caribbean Airways; Transportation Research Command

TRCA Toronto Region Coordinating Agency (Hamilton to Oshawa)

trccc tracking radar central

control console

Tr Co Trust Company

tr coil tripping coil

Tr Coll Training College

TRCS Trade Relations Council of the United States

TRCUD Technical Review Committee on Underground Disposal of Radioactive Wastes

trcver transceiver

Trd Trinidad

TRD Test Requirements Document

TRD Teatro Regio Ducal (Italian—Royal Ducal Theater) - Milan

TRDA Timber Research and Development Association

TRDCOM Transportation Research and Development command

trdto tracking radar data takeoff

T^{re} Torre (Italian or Portuguese —tower)

TRE Telecommunications Research Establishment

treas treasure; treasurer; treasury

Treas Treasurer

trec tracking radar electronic components

TRECOM Transportation Research and Engineering command

tree trustee

trees. (TREES) transient radiation effects on electronic systems

TREF Trauma Research and Education Foundation

trem tremolando (Italian—trembling)

trem card transport or truck emergency card

Tren Trenton

trend. tropical environment data

treph trephining (trepanning)

Trep. pal. Treponoma pallida—the spirochete of syphilis

Tres Hermanas Tres Hermanas Mountains of southwestern New Mexico

TREVI Terrorisme, Radicalisme, et Violence International (French—Terrorism, Radicalism, and International Violence)—EEC police network

trf transfer; tuned radio frequency

trf (TRF) thyrotropin-releasing factor

TRF Task Request Form; Teacher Rating Form; Transportation Research Foundation; Tuna Research Foundation; Turf Research Foundation

trg training; triangle

tr&g transmit, receive, and guard

trgt target

trh (TRH) thyrotrophin-releasing hormone

Tr H Trinity Hall, Oxford

TRH Their Royal Highnesses

TRHS Theodore Roosevelt High School

tri total response index (TRI); triangle; triangulation; tricolor; tricycle; triode

tri (Latin prefix—three)—triangle

Tri Triangulum (constellation); Trieste

Tri Tohtori (Finnish—doctor)

TRI Technical Report Instruction; Television Reporters International; Textile Research Institute; The Rockefeller Institute; Tin Research Institute; Tire Retreading Institute; total response index

triad air, sea, and land defense

TRIAL Technique for Retrieving Information from Abstracts of Literature

trian triangle; triangulation

Trias Triassic

trib tribade; tribadism; tribal; tribalism; tribalist; tribasic; tribology; tribunal; tribune; tributary

Trib Tribune

Tri B Triborough Bridge

TRIB Tire Retread(ing) Information Bureau

tribas tribasic

TRIBE Teaching and Research in Bicultural Education

trib^l Tribunal (Spanish—tribunal; court of justice)

tric trachoma inclusion conjunctivitis; trichloroethylene

tricaphos tricalcium phosphate

trice. transistorized real-time incremental computer expandable; trichomoniasis (protozoan vaginal infection)

trich (Latin prefix—hair)—trichosis

Trich Tiruchchirappalli or Trichinopoly (famous for its Indian cigars)

Trichi Trichinopoly, Hindustan; cigar from that area

trick (slang—trichomoniasis)

tricl triclinic

trico trichomoniasis
Tri Com Trilateral Commission (Council of Foreign Relations)
TRICON Tri-Service Container (program)
trid. *triduum* (Latin—three days)
Trident Trident Region (Berkeley, Charleston, and Dorchester counties comprising the Charleston, South Carolina area)
tridundant triple redundant
TRIEA Tea Research Institute of East Africa
trig trigamist; trigamy; trigger-(man); trigonal; trigonometric; trigonometry
Trig trigonometry
triga trigger reactor
trihem trihemeral; trihemirer
tri ins tricuspid insufficiency
trik trichloroethylene
trike tricycle
trilat trilateral; Trilateral Commission (Council on Foreign Relations); trilateralist(ic)-(al)(ly)
trillion *American*—a million million—10¹²; *British*—a million million million—10¹⁸
tril(s) trillion(s)
trim. trimetric
trim. (TRIM) test rules for inventory management
trim. *trimestre* (Latin—quarter; three months)
TRIM Targets, Receivers, Impacts, and Methods; Technical Requirements Identification Matrices; Tax Reform Immediately
TRIMIS Tri-Service Medical Information System
TRIMMS Total Refinement and Integration of Maintenance Management Systems (USA)
TRIMS Texas Research Institute of Mental Sciences
trimtu (TRIMTU) trimethyl thiourea
Trin Trinidad(ian); Trinitarian(ism); Trinity; Trinity College
Trinco Trincomalee
Trin Col Trinity College
Trin H Trinity Hall
Trinity Trinity Christian College; Trinity Church; Trinity College; Trinity House (Pilot Service); Trinity Parish; Trinity Parish School; Trinity School; Trinity University; Trinitytide

triol triolism; triolist
triols triolists (also called troilists)
trip. triple; triplicate; triplication; tripos
trip. (TRIP) technical reports indexing project
TRIP The Road Improvement Program
triphib triphibian; triphibious (land, sea, air)
tripl triplication; triplicate
Triple-A (*see* AAACE)
triple-A S AAAS (American Association for the Advancement of Science)
tris tris (hydroxymethyl) aminomethane
Tris Tristán; Tristõ; Tristram
trishaw tricycle rickshaw
trisk triskelion
TRISNET Transportation Research Information Services Network
TRISTAN Tri-Ring Intersecting Storage Accelerators in Nippon
Tristan da Cunha Tristan da Cunha Islands (Gough, Inaccessible, Nightingale, Tristan da Cunha)
tri sten tricuspid stenosis
trisyll trisyllable
trit. *tritura* (Latin—triturate)
TRI-TAC Tri-Services Tactical Communications Program (DoD)
tritic tritical (trite); triticale (*Triticum* + *Secale* hybrid between wheat and rye); triticeous; triticeum; tritish; triticum; tritium
triv trivia(l)
TRJ *Tragedy of Romeo and Juliet*
trk track; truck; trunk
Trk Turk; Turkey; Turkic; Turkish
trkdr truck-drawn
trkg tracking
trkhd truckhead
TRK T Track Time
trl trailer
Trl Trail
TRLB temporarily replaced by lighted buoy
trlfsw tactical-range landing-force support weapon
trlr trailer
Trlr Trailer (postal abbreviation)
trm task response module (engineer's desk area); thermoremanent magnetism
Trm Trincomalee
trml terminal

trmn trainman
trmr trimmer
TRMS Technical Requirements Management System
trmt treatment
trn transfer
Trn Troon
TRN Technical Research Note
tRNA transfer RNA (same as sRNA)
trnbkl turnbuckle
trng training
TRNMP Theodore Roosevelt National Memorial Park
trnsp transport; transportation
TRO Technical Reviewing Office; Temporary Restraining Order
TROA The Retired Officers Association
troch troche
troch *trochiscus* (Latin—cough drop, lozenge, troche)
Troch Trochelminthes
troil troilism; troilist
Troj Trojan
trol tapeless rotorless on-line cryptographic equipment
trom tromba; trombone
T Rom Times Roman
trombst trombonist
tromp *trompette* (French—trumpet)
troms trombones
T-room (American slang—tiolet)
trop tropic; tropical; tropics
trop *tropos* (Greek—to turn or to turn toward)—entropy, geotropism, phototropism, tropic(al), tropism
troparium tropical aquarium
Trop Can Tropic of Cancer—23½°N Lat
Trop Cap Tropic of Capricorn—23½°S Lat
tropec tropical experiment
trophe (Greek—nutrition)—atrophy, autotrophe, heterotrophe, trophic level
trophy (Latin suffix—relating to nutrition)—hypertrophy
tropic (Latin prefix—pertaining to a turn)—tropical, tropicolitan; (Latin suffix—turning toward)—gonadotropic
TROPICS Tour Operators Integrated Computer System
trop med tropical medicine
troposcatter beyond-the-horizon communication
TROSCOM Troop Support Command (USA)
Trots Trotskyite(s)
trp troop

trp (TRP) tryptophan
Trp Tripoli
tr pl treatment plan
trr teaching and research reactor; train repetition rate
TRRA Terminal Railroad Association (of St Louis)
TRRB Test Readiness Review Board (NASA)
TRRG Tax Reform Research Group
trs target range servo(mechanism); transfer; transparency; transpose; tropical revolving storm; trustees
trs (TRS) tetrahedral research satellite
TRs Tax(ation) Reports; Technical Reports; Temporary Reserves
TRS Ticket Reservation System; Transair Limited
TRSA Terminal Radar Service Area
trsb time reference scanning beam
trsd total rated service date
tr sh trim shell
TrSMS triple-screw motor ship
trsp transport
TRSP Turtle River State Park (North Dakota)
trsr taxi and runway surveillance radar
TrSS triple-screw steamer
trssgm tactical range surface-to-surface guided missile
trsv (TRSV) tobacco-ringspot virus
trt total response to trauma; treatment; turret
TRTA Trader's Road Transport Association
TRTC Tropical Radio Telegraph Company
trtch tape-recording technic
tru (TRU) transuranic (contaminated) waste
Tru Trucial; Trucial Sheikdoms; Truman; Truman Capote (1924–1984); Truro
Tru Truman's Railway Reports
TRU The Rockefeller University
TRUB temporarily replaced by unlighted buoy
Tru Cst 1 Trucial Coast Number 1
Tru Cst 2 Trucial Coast Number 2
trud time remaining until dive (of satellite into Earth's atmosphere)
TRUE Teachers Resources for Urban Education

Truemid Movement for True Industrial Democracy
tru-fi tru fidelity (sound reproduction)
Truman Harry Truman Field (U.S. Virgin Islands airport near Charlotte Amalie on St Thomas)
trump. trumpet
TRUMP Target Radiation Measurement Program
trumps trumpets
trun trunnion
trunc truncate; truncated; truncation
trunch truncheon
tr unit turbidity reducing unit
Truron (Church Latin—Truro)
tru(s) trustee(s)
trust. trusteeship
truthsayer(s) honest person(s)
trv torpedo recovery vessel
trveh tracked vehicle
trw trawler
trwov transit without visa
TRW SL TRW Space Log
trxrx transmitter-receiver
try. truly
try. (TRY) tryptophan
TRY Teens for Retarded Youth (juvenile correctional program)
Tryg Trygve Lie
tryp (TRYP) tryptophan
tryp(s) trypanosome(s)
ts taper shank; temperature switch; tensile strength; terminal sensation; test solution; tilt and shift; time shack; time sharing; too short; tool steel; tough situation; traffic signal; transit storage; transmitter station; triple strength; tubular sound; type specification(s); typescript
ts (TS) thesis
t's twins
t/s test stand; third stage; transship(ed)(ment)
t/s (T/S) thyroid serum
t & s toilet and shower
TS Tasmania (airline code); Tasmanian Steamers; Tentative Specification; Terminal Service; Test Summary; Theosophical Society; Thoreau Society; Tidewater Southern (railroad); top secret; Topical Search; Training Ship; Transmittal Sheet; Type Specification
T S Taming of the Shrew; tasto solo (Italian—play without accompaniment)
tsa tax-sheltered annuity; total survey area; two-step antenna

tsa (TSA) total survey area (radio and tv)
TSA Teacher on Special Assignment; Tourist Savings Association; Track Supply Association; Transportation Service, Army; Transportation Standardization Agency; Transuranic Storage Area
tsac title, subtitle, and caption
TSAC Target Signature Analysis Center
tsar time scanned array radar
TSB Trustee Savings Bank(s)
TSBA Trustee Savings Banks Association
TSBD Texas School Book Depository
TSBI Texas Social Behavior Inventory Form
TSBR Thomas Stamford Bingley Raffles
tsc (TSC) transmitter start code (data processing)
TSC Texas Southmost College; Transamerican Steamship Corporation; Transportation System Center
TSCA Tactical Satellite Communications System; Top Secret Control Agency; Toxic Substance Control Act
TSCC Telemetry Standards Coordination Committee
tscf top secret cover folder
TSCO Thomas Scherman's Concert Opera; Top Secret Control Officer
TSCS Tennessee Self-Concept Scale
t-s curve temperature-salinity curve
tsd tactical simulator display; target skin distance; treatment, storage, and disposal
Tsd Tausend (German—thousand)
TSd Tay-Sachs disease (TSD)
TSD Tay-Sachs Disease; Technical Services Division (CIA); towed submersible drydock (naval symbol)
TSD-CIA Technical Services Division—Central Intelligence Agency
tsdd temperature-salinity-density-depth
tsds two-speed destroyer sweeper
tse (TSE) test support equipment
TSE Texas South-Eastern (railroad); T(homas); S(tearns) Eliot; Tokyo Stock Exchange; Toronto Stock Exchange

TSE *Tribunal Supremo de Elecciones* (Spanish—Supreme Election Tribunal)

T-sect cross-section; transverse section

TSES Thumb-Signature Endorsement System

tsf tower shield facility

tsf *telegrafia sem fios* (Portuguese), *telegrafo senza fili* (Italian), *télégraphie sans fil* (French)—radio or wireless telegraphy

TSF Tertiary of the Society of St Francis

tsfr transfer

TSG Television and Screen Writers' Guild; Tri-Service Group

TSGEE Tri-Service Group on Communications and Electronic Equipment

TSgt Technical Sergeant

tsh (TSH) thyroid stimulating hormone

tsh *telegrafía sin hilos* (Spanish—wireless telegraphy)—radio

T sh Tanzanian shilling(s)

TSH Their Serene Highnesses

TSHA Texas State Historical Association

T-shirt T-shaped shirt; T-shaped undershirt

t-shower thundershower

tsi The Socialist International; test structure input; tons per square inch

TSI Test of Social Insight; Test of Social Intelligence; Theological School Inventory; Transport(ation) Safety Institute

T&SI Technical and Scientific Information (UN)

tsi agar triple sugar (glucose, lactose, sucrose) iron agar

tsiaj this scherzo is a joke (abbreviation devised and used by composer Charles Ives)

TSID Technical Service Intelligence Detachments

Tsj *Tsjeko-Slovakia* (Norwegian—Czechoslovakia)

TSJC Trinidad State Junior College

TSKK *Tsentralnya Kontrolnaya Komissiya* (Russian—Central Control Commission)

TSL Terrestrial Sciences Laboratory; Texas Short Line (railroad)

TSLNP Tung Slang Luang National Park (Thailand)

TSM *Treasure of the Sierra Madre*

tsms twin-screw motor ship

tsmt transmit

TSMTS Tri-State Motor Tariff Service

Tsn Tientsin

TSN Tape Serial Number

TSNHS Touro Synagogue National Historic Site

tso time-sharing option

Tso Tsingtao

TSO Taiwan Symphony Orchestra; Tasmania Symphony Orchestra; Teheran Symphony Orchestra; Toronto Symphony Orchestra; Tucson Symphony Orchestra

TSOR Tentative Specific Operational Requirements

TSOS Time-Sharing Operating System

tsp teaspoon; total suspended particles; tracking station position

TSP thyroid-stimulating (hormone of) prepituitary; trisodium phosphate (Na_3PO_4)

tspa tally and special precinct analysis

tspn teaspoon

T-square T-shaped ruler for making right angles

tsr temperature-sensitive resistor

TSR Sir Thomas Stamford Raffles (founder of Singapore as well as the London Zoo); Trans-Siberian Railway

TSRB Top Salaries Review Body

T & SRC Tubular and Split Rivet Council

tss tangential-signal sensitivity; target-selector switch(ing); time-sharing system(s)

tss (TSS) toxic shock syndrome

t/ss turbine steamship

TSS Time-Sharing System(s); Traffic Safety Service; Trident Submarine System; turbine steamship; twin-screw ship

tssa (TSSA) tumor specific surface antigen

tssm total ship simulation model

tsspar time-sharing system-performance activities record(s)

TSSR Tadzhikistan Soviet Socialist Republic; Turkmenistan Soviet Socialist Republic

tst test (computer flow chart)

tsta tumor specific transplantable antigen (TSTA)

TSTA Texas State Teachers Association

t-storm thunderstorm

TSTP Test of Selected Topics in Physics

tstr tester

tstrms thunderstorms

t's t's & t's tortoises, terrapins, and turtles [tortoises are terrestrial chelonians with domed shells and elephantine feet; terrapins are semiaquatic chelonians with depressed shells, rudder-like tails, and webbed feet; turtles are marine chelonians with streamlined shells and paddle-like flippers; the term turtle(s) is often applied to all the chelonians]

Ts & Ts Trinidadians and Tobagonians

tsu tape search unit; this side up

tsu (TSU) triple sugar urea (agar)

TSU Texas Southern University; Tulsa-Sapulpa Union (railway)

tsu's thermosetting urethanes

TSUS *Tariff Schedule of the United States*

Tsushima Tsushima Current flowing northeasterly between Japan and Korea; Tsushima Strait where in 1905 Admiral Togo's Japanese fleet defeated Admiral Rozhdesvenski's Russian fleet

tsvp *tournez s'il vous plaît* (French—please turn over)

TSW tropical summer winter (load line mark)

TSWE Test of Standard Written English

tsx time-sharing executive

TSX Telecommunications Satellite Experiment

tt tablet triturate; technical term(inology); technical test(ing); teetotaler; telegraphic transfer; teletype; teletypewriter; tetanus toxoid; torpedo tube(s); transit time; tree top(s); tuberculin tested

tt (TT) telephone transfer; train time

t-t tube-in-tube

t/t time to turn

t&t time and temperature

tt. *tantum* (Latin—fixed allowance, so much)

t.t. *totus tuus* (Latin—all yours)

TT tam-tam (Chinese gong); target-towing (naval aircraft);

technical test(ing); Tidningarnas Telegrambyra (Swedish News Agency); Toledo Terminal (railroad); Trailer Train; Trans-Texas (Airways); Troop Test

T/T twin turbine (steamship)

T & T Trinidad and Tobago

tta test target array

TTA Taiwan Telecommunication Administration; Trans-Texas Airways; Travel Time Authorization; Travel Trade Association

TTA *Tragedy of Titus Andronicus*

ttab Trademark Trial and Appeal Board (US Patent Office)

ttac tracking, telemetry, and command; tracking, telemetry, and control

TTAF Technical Training Air Force

TTBT Threshold Test Ban Treaty

ttc temperature test chamber; tetrazolium chloride; tight tape contact; tin telluride crystal; tow target cable; transient temperature control; tube temperature control

TTC Tariff Trade Code; Technical Training Command; Teletypewriter Center; Texas Technological College; Tobacco Tax Council; Tokyo Tanker Company; Toronto Transit Commission; Transportation Technology Center; Tuition Tax Credits

ttce tooth-to-tooth composite error

ttci transient temperature-control instrument

TTCS Truck Transportable Communications Station

ttd transponder transmitter detector

ttdr tracking telemetry data receiver

tte temporary test equipment; trailer test equipment

Tte Teniente (Spanish—Lieutenant)

TTE Tropical Testing Establishment

Tte Cnel *teniente coronel* (Spanish—Lieutenant Colonel)

TTEX Trailer Train Express

ttf time to failure; tone telegraph filter; transistor text fixture

ttf (TTF) tetrathiafulvalene

TTF Timber Trade Federation;

Townsend Thoresen Ferry

ttfn ta-ta for now

ttg time to go

TT-gauge Tiny Tim Gauge—¼ inch track gauge (model railroads)

ttgd time-to-go engine dial

tth thyrotropic hormone

tti time-temperature indicator; trait treatment interaction

TTI The Technological Institute; Transition Technology, Inc

T-time takeoff time

TTIO Turkish Tourism and Information Office

TTJC Tyne Trade Joint Committee

ttk two-tone keying

ttl to take leave; transistor-transistor logic

ttl (TTL) through-the-lens (camera-flash monitor)

TTL Tokaido Trunk Line (Japanese railroad running trains at 125 miles per hour)

ttm two-tone modulation

TTMA Trinidad-and-Tobago Medical Association; Truck-Trailer Manufacturers Association

tto this transaction only

Tto Toronto

TTO Tanzania Tourist Office

ttp time-temperature parameter; total taxable pay

TTPI Trust Territory of the Pacific Islands

ttr type token ratio

ttr (TTR) target-tracking radar; thermal test reactor

TTRI Telecommunication Technical Training and Research Institute

T & T RR Tijuana and Tecate Railroad

tts teletypesetter (TTS); teletypesetting; temporary threshold shift

tts (TTS) teletypesetting

TTS Terminal Transparent System; Transdermal Therapeutic System

TTS *The Truth Seeker*

TTSU Taxi-Truck Surveillance Unit (NYPD)

ttt telemetry time transposition; time to target; time to think; time to turn

t t&t tortoise, terrapin, and turtle (*see* t's t's & t's)

TTT Transamerica Trailer Transport; Tyne Tees Television

TT & T Texas Transport and Terminal

t't'ta triple-note trumpet flourish

TTTB Trinidad and Tobago Tourist Board

TTTC Technical Teachers Training College

T & T TS Trinidad and Tobago Television Service

ttu timing terminal unit

TTU Texas Technological University

TTUT Through-Transmission Ultrasonic Test(ing)

TTV Taiwan Television (offshore China)

ttvm thermal transfer voltmeter

ttw total temperature and weight

TTW Tennessee-Tombigway Waterway

ttwl twin-tandem wheel loading

ttx tritated tetrodotoxin

tty teletypewriter

T-type Jungian thinking type

tu tape unit; thermal unit; toxic unit; trade union (TU); traffic unit; transfer unit; transmission unit; turbidity unit

Tu thulium; Tudor; Tuesday; Turkey; Turkish

TU Taylor University; Temple University; Tiffin University; Trade Union; transmission unit; Trinity University; Tufts University; Tulane University; Tunis Air; Typographical Union

T.U. tuberculin unit(s)

TU *Technische Universität* (German—technical university); *temps universel* (French—universal time)

Tu-4 Soviet Tupolev bomber inspired by the Boeing B-29 Superfortress aircraft

Tu-16 Soviet Tupolev bomber code-named Badger by NATO

Tu-20 Soviet Tupolev heavy bomber named Bear by NATO

Tu-22 Soviet Tupolev bomber named Blinder by NATO

Tu-28 Soviet Tupolev long-range interceptor aircraft named Fiddler by NATO

Tu-104 Soviet Tupolev medium-range transport aircraft called Camel by NATO

Tu-114 Soviet Tupolev long-range transport plane named Cleat by NATO

Tu-124 Soviet Tupolev jet-transport aircraft named Cookpot by NATO

Tu-144 Tupolev supersonic transport
Tu-154 Tupolev 154 supersonic aircraft
TUAC Trade Union Advisory Committee
Tuamotus Tuamotu Islands of Polynesia in the South Pacific, once called the Dangerous Islands
tu ar turning arbor
tub. tubing
TUB temporary unlighted buoy
TUBA Tubists Universal Brotherhood Association
Tu bandera Tu bandera es un lampo del cielo (Spanish— Your flag is a lamp of the sky) —Honduran anthem
tube subway; television; tunnel
Tube The Tube (London's Underground subway system)
TUBE Terminating Unfair Broadcasting Excesses
tuberc tuberculosis
tublr tubular
Tubuais Tubuai Islands of Polynesia in the South Pacific, also called the Australs
tuc transportation, utilities, communications
Tuc Tucana (constellation); Tucson
TUC Trades-Union Congress (British)
tu ca turning cam
TUCC Temple University Community College; Triangle Universities Computation Center
TUCGC Trades Union Congress General Council
TUCSA Trade Union Council of South Africa
Tucsons Tucson Mountains of southeastern Arizona
tudor two-door
Tue Tuesday
TUEL Trade Union Educational League
Tues Tuesday
TUF Tamil United Front; Tokyo University of Fisheries; Trade Union Federation (British)
TUFEC Thailand-Unesco Fundamental Education Center
tuff tape update of formatted files
tu fx turning fixture
tug. tape update and generator
TUG Transac Users Group
tugrik monetary unit of Mongolia
tug(s) tugboat(s)
TUH Taiwan University Hos-

pital
TUI Trade Union International
TUIAFW Trade Unions International of Agricultural and Forestry Workers
tuifu the ultimate in foulups
Tul Tulsa
TUL Tokyo University Library; Tulane University of Louisiana; Tulsa, Oklahoma (airport)
Tularosas Tularosa Mountains of western New Mexico
TULF Tamil United Liberation Front
Tul Phil Tulsa Philharmonic
TULRA Trade Union and Labor Relations Act
tum tummy (stomach); tumor
TUM Panama City, Panama (Tocumen Airport); Trades Union Movement
Tumacacori Tumacacori National Monument south of Tucson, Arizona
Tumuc-Humacs Tumuc-Humac Mountains between Brazil and the Guianas
tun tuning
Tun Tunis; Tunisia; Tunisian; Tunnel
Tun Túnez (Spanish—Tunisia)
tunasan tuna sandwich
tunawich tuna sandwich
tung tungsten
Tunic Tunicata
Tunisia Republic of Tunisia (North African Arab country), Al-Djoumhouria Attunusia—called Carthage in Roman times
Tunl Tunnel
tuos trained under other schemes
TUP Temple University Press; Trinity University Press; Tulane University Press
Tupper Tupper Creek in eastern British Columbia or Tupper Lake in northern New York
Tupun Tupungato
tur transurethral resection (TUR); turbine; turret
Tur Turin; Turkish
turb transurethral resection of the bladder (TURB); turbine
TURB Trainer Update Review Board
turbid. turbidity
turboalt turboalternator
turbo-elec steam turbine connected to electric motor
turbogen turbogenerator
turbojet turbine-driven jet (airplane engine)

turboprop turbine-driven jet engine (moving the) propeller
turbosuch trubosupercharter
turbotrain turbine-driven railroad train
turbpmp turbopump
turbu turbulence; turbulent
Turch Turchia (Italian—Turkey)
turk turkey
türk türkisch (German—Turkish)
Turk. Turkey; Turkish
Turk Turkish language
Turkana Lake Turkana (formerly East Rudolf)
Turk Cum Turkiye Cumhuriyeti (Turkish—Republic of Turkey)
Turkey Republic of Turkey (formerly the center of the Ottoman Empire extending from Morocco to Persia), *Türkiye Cumhuriyeti*
Türk-Is Tükiye Isçi Sendikalari Konfederasyonu (Turkish Confederation of Trade Unions)
Turkish instruments bass drum, cymbals, kettledrums, and triangle
Turkmen Turkmenia; Turkmenian
Turkmen SSR Turkmen Soviet Socialist Republic (Turkmenistan)
turks turkeys
Turks Turkish people; Turks Islands east of the Bahamas and northeast of the Windward Passage
Turks and Caicos Turks and Caicos Islands northeast of the Windward Passage between Cuba and Haiti
Turk-Sib Turkestan-Siberian (railroad)
Turk-Tat Turko-Tataric
Turku formerly Abo
turn. turning
Turn Turnpike
TURN Toward Utility Rate Normalization
Turner Turn Turner Turnpike
turp transurethral resection of the prostate (TURP); turpentine
turps elixir of terpin hydrate; turpentine
TURPS Terrestrial Unattended Reactor Power System
turq turquoise
Turq Turquía (Spanish—Turkey)
Turtles Turtle Islands in the Sulu Sea south of the Philip-

pines; Turtle Islands off Africa's Sierra Leone; Turtle Mountains between northern North Dakota and southern Manitoba

TUs Tenant's Unions

TUS Tuscon, Arizona (airport)

TUSAFG The United States Air Force Group (American Mission for Aid to Turkey)

TUSC Technology Use Studies Center

Tuscans Tuscan people; Tuscan Islands

Tushars Tushar Mountains of central Utah

TUSLOG The United States Logistic Group

TUSM Tufts University School of Medicine

tuss. *tussis* (Latin—cough)

tut tutor; tutorial

Tut Tutankhamen

TUT The University of Tokyo

Tut Books Charles E Tuttle's books

TUTF Technology Use Task Force

TUTI Temple University Technical Institute

Tutu Tutuila, American Samoa

TUUL Trade Union Unity League

Tuv Tuvalo (Ellice Islands); Tuvalu Islands

TUV *Technischer Uberwachungs Verein* (German—Testing Organization)—Berlin

tuwr turning wrench

tux tuxedo (dinner jacket)

Tuzigoot Tuzigoot National Monument in central Arizona

tv transvestite

tv (**TV**) television; terminal velocity; test vehicle; tetrazolium violet; total volume; transverse; trichomonas vaginalis; true view; tuberculin volution

t/v thrust-to-weight

t & v terrorism and vandalism

TV television; test vehicle; Tidewater Oil (stock exchange symbol); transport vehicle

TV *Totenkopfverbände* (German—Death's Head formations)—concentration-camp guard units

tva thrust vector alignment

tva *taxe sur la valeur ajoutée* (French—tax value added)

TVA Temporary Variation Authorization; Tennessee Valley Authority

tvac time-varying adaptive correlation

tv a.m. morning television

TVAs Temporary Variation Authorizations

TVB Television (Advertising) Bureau

TVBS Television Broadcast Satellite

tvc temperature valve control; thermal voltage converter; throttle valve control; thrust vector control; time-varying coefficient; timed vital capacity; torsional vibration characteristics

tvc (**TVC**) total variable cost

TVC Technical Valve Committee

TVCC Treasure Valley Community College

tvcrit television critic(ism)

tvd toxic vapor damper; toxic vapor detector; tuned viscoelastic damper

tvdc test volts—direct current

TVDC Tidewater Virginia Development Council

tv'dict(s) television addict(s)

tvdp thrust-vector display (unit)

tvdy television deflection yoke

tve test vehicle engine; thermal vacuum environment

TVE Televisión Española (Spanish TV network)

tvel track velocity

TVERS Television Evaluation and Renewal Standards

tvft television flyback transformer

tvg television video generator; threshold voltage generator; triggered vacuum gap

TVG *T V Guide*

tvhh (**TVHH**) television households

TV household television-equipped home

tvi television interference

TVIC Television Interference Committee

tvid televised identification; television identification; television identity

tvig television and inertial guidance

tvist television information-storage tube

tvk terminal volume kill

T v K Theodore von Karman

tvl tenth value layer; travel

Tvl Transvall

tvm tachometer voltmeter; track via missile; trailer van mount; transistorized voltme-

ter

TVM *Television Martí*

TVN *Television News*

TVNZ Television New Zealand

tvop television observation post

tvor terminal visual omnirange; very high frequency terminal omnirange station

tvp television poor (viewers who have never learned how to read or who have lost the faculty; textured vegetable protein; time-varying parameter

tvp (**TVP**) textured vegetable protein

TVPA Thames Valley Police Authority

tv p.m. evening television

tvq top visual quality

tvr temperature variation of resistance; textured vegetable protein

TVRB Tactical Vehicle Review Board (USA)

TVRI Television Rating Inventory

TV-RI *TV-Republik Indonesia* (Bahasa Indonesia—Republic Indonesia Television)

tv rm television room

Tvrn Tavern

tvr's television recordings

tvs tactical vocoder system; telemetry video spectrum; television viewing system

tv's television dinners; transvestites

TVSAT Television Satellite

tvsd time-varying spectral display

tvsg television signal generator

tvsm time-varying sequential measuring (apparatus)

tvso television space observatory

TVSTI Thames Valley State Technical Institute

tvsu television sight unit

tvt television typewriter

tvu total volume urine

tw tail warning; tail water; tail wheel; tail wind; tankwagon; taxiway; tempered water; terrawatt; tile wainscot; torpedo water; total weight; traveling wave; twice a week; twin(s)

tw (**TW**) typewriter (computer flow chart)

tw *tussenwerpsel* (Dutch—interjection)

Tw Twaddell

TW Trans World Airlines (2-

letter coding)
T&W Tyne and Wear
twa time-weighted average; trailing-wire antenna
TWA Textile Waste Association; Thames Water Authority; Tooling Work Authorization; Toy Wholesalers Association; Trans World Airlines
TWAD Twadell
'twas it was
twb twin with bath
twbp transcribed weather broadcast program
TWC Tail Waggers' Club; Texas Wesleyan College
TWC Trials of War Criminals
TWCIS Transuranic-contaminated Waste Container Information System
twcrt travelling-wave cathode ray tube
TWCS Test of Work Competency and Stability
twd tail wags dog
twds tradewinds
twe tap-water enema
TWE Textile Waste Exchange
TWEA Trading With the Enemy Act
'tween between
Twel N Twelfth Night
Twelve Tribes Twelve Tribes of Israel—Reuben, Simeon, Judah, Zebulun, Issachar, Dan, Gad, Asher, Nephtali, Benjamin, Ephraim, and Manasseh
twens twenties (store catering to people in their twenties)
'twere it were
twerl tropical wind, energy conversion, and reference level
TWF Twin Falls airport
TW & FS The Wine and Food Society
twh typically wavy hair
twhl tailwheel
twi training within industry
TWI The West Indies
'twill it will
twimc (TWIMC) to whom it may concern
'twixt betwixt
twi zn twilight zone
twk typewriter keyboard
twl top water level
twm traveling-wave maser
Twn Taiwan—Republic of China consisting of offshore islands; Town
twn hse town house
two. this week only
two-0 $20 bill
two-fer two for the price of

one
Two Gent Two Gentlemen of Verona
T-word tax
two-spot $2 bill
twot travel without troops
'twould it would
Twp Township
TWP True Whig Party (Liberia)
TWPD Tactical and Weapons Policy Division
twr tower
Twr Tower
TWR Trans-World Radio
tws timed wire service; track while scan
tw/s twin-screw (ship)
TWSO Transuranic Waste Systems Office(r)
twsr track-while-scan radar
twsrs track-while-scan radar simulator
twt torpedo water tube; traveling-wave tube; travel with troops
t/wt tare weight
TWT Toy World Test(ing); Transonic Wind Tunnel
twta travelling-wave-tube amplifier
TWU Tata Workers Union; Transport Workers Union
TWUA Textile Workers Union of America; Transport Workers Union of America
T WW Thick Weather Watch (Coast Guard)
twx time-wire transmission
twx (TWX) teletypewriter exchange (message)
TWX teletypewriter exchange (message)
TWX (TWXS) Teletypewriter Exchange Service
twy taxiway; twenty
twych travel with your children
twyl taxiway link(age)
twzo trade-wind-zone oceanography (term of derision by experts or about armchair oceanographers)
tx tax(ation); telex; time; torque transmitter; traction
tx (TX) transmitter
Tx treatment
TX Texas
txclk (TxCLK) transmit data clock
txe telephone exchange electronic
txh transfer on index high
txi transfer on index incremented
txl transfer on index low

Txl Texel
txn taxation
txt text; textbook; textile; textual(ly); textualism; textualist; textuary; texture(d); texturize; texturizing
ty territory; thank you; truly; type
ty tysk (Dano-Norwegian—German)
Ty Territory; Tybalt; Tyler; Tyndall; Tyonek; Tyrone; Tyrus Raymond Cobb
tyc tycoon
TYC Thames Yacht Club; Thomas Y Crowell; Toledo Yacht Club
TYCOM Type Commander (USN)
tydac typical digital automatic computer
tyg (TYG) trypticase yeast glucose
tylenol acetaminophen (trade name for an analgesic)
tymp tympanic(ity); tympany
tymp memb tympanic membrane
tyng topping
tyo two-year-old (horse)
TYO Tokyo, Japan (airport)
typ typical; typing; typist; typographer; typography; typewriter
TYP Ten-Year Plan; Twenty-Year Plan; etc.
type. typewriter; typewriting
typer typewriter
typewriters Chicago-gangster (Scarface) Al Capone's nickname for submachine guns
typh typhoon
typo typographical (error)
TYPOE Ten-Year Plan for Ocean Exploration
typog typographer; typographical; typography
typol typological(ly); typologist; typology
typout typewriter output
typr typewritten
typw typewriter
tyr (TYR) tyrosine (amino acid)
Tyr Tyrol; Tyrolean; Tyrolese; Tyrone
Tyr Tyrkia (Norwegian—Turkey)
Tyrol Tyrol(ean); Tyrolese
tys tensile yield strength
TYS Knoxville, Tennessee (airport)
tysd total years service date
Tysk Tyskland (Norwegian—Germany)
Tyskl Tyskland (Danish—Ger-

many)
tytipt tape training in port (USN)
tyurzak tyuremnoye zakyucheniye (Russian—prison confinement)
tyvm thank you very much
TYZ Toronto, Ontario airport
tz terrazzo; tidal zone; time zero

Tz tuberculin zymoplastiche (symbol)
TZ Tactical Zone; Transair Limited, Canada (2-letter code)
tzd true zenith distance
tze transfer on zero
tzg thermofit zap gun
TZIK Tzentralny Ispolnitelny Kommitet (Russian—Central Executive Committee)
tzj tubular zippered jacket
TZm true azimuth
TZM titanium-zirconium-molybdenum (alloy)
tzp time zero pulse
tzt te zijner tijd (Dutch—in due time)
tzv tetrazolium violet

U

u density of radiant energy (symbol); ugly threatening weather (symbol); umpire; unified atomic mass (symbol); unit(s); unknown; unoccupied; unsymmetrical; unwatched; upper; velocity (symbol); you

u & lc upper and lowercase

u und (German—and); viscosity (symbol)

U Chance Vought Aircraft (symbol); kilourane (1000 uranium units—symbol); overall co-efficient of heat transfer (symbol); potential energy (symbol); total internal energy (symbol); U Thant; U-boat; unclassified; Underground (London's subway system); Uniform—code for letter U; University; up; uranium; Utah; Utahans; utility; you

U Uad (Arabic—wadi)—gulley, ravine, riverbed; ud (Danish—out); uit (Dutch—out); ulos (Finnish—out); Université (French—University); unter (German—down); up; upp (Swedish—up); ute (Swedish—arrival); violaceus (Latin—violet-color)

U-1A American version of De Haviland Otter utility aircraft

u 1 b unit 1 bedroom

u 1 r unit 1 rental

u-2 you too

u 2 b unit 2 bedrooms

u 2 r unit 2 rentals

U-2 high-altitude high-performance photo-reconnaissance airplane

u/3 upper third

U-3 Cessna 6-passenger aircraft

U_3O_8 uranium oxide

U-4 Aero Commander transport aircraft

U 4 T union (coupling) 4 tons

U-6 De Havilland Beaver transport aircraft

U-8 Beech Seminole transport aircraft

U-17 Cessna Skywagon aircraft

U-17A Cessna 6-passenger Skywagon

U-22 Beech Bonanza trainer aircraft

U234 trace component of natural uranium

U235 0.7 percent of natural uranium (atomic energy source)

U238 99.3 percent of natural uranium (atomic energy source)

ua unauthorized absence; unauthorized absentee; underage; uniform allowance; upper arm; urine aliquot; user area

ua (UA) urinalysis

u/a unit of account

ua unidad(es) astronómica(s) (Spanish—astronomical unit(s)

u a uden ar (Dano-Norwegian—without date); und andere(s) (German—among other things, and others, inter alia); und ähnliche(s) (German—and the like)

u.a. usque ad (Latin—as far as; up to)

uA und andere (German—and others)

U/a underwriting account

UA Ulster Association; Underwater Association; United Aircraft; United Air Lines (2-letter coding); United Artists; University of the Americas; University of Auckland

U-A Universal-American

U of A University of Aberdeen; University of Adelaide; University of Akron; University of Alabama; University of Alaska; University of Alberta; University of the Americas; University of Arizona; University of Arkansas

UA Universidad de las Americas (Spanish—University of the Americas)

UAA United Arab Airlines; University Aviation Association

UAAGM University of Alberta Art Gallery and Museum

UAASUS Ukrainian Academy of Arts and Sciences in the United States

UAB Underwriters Adjustment Board; Unemployment Assistance Board; United Asian Bank; Universities Appointments Board; University of Aston in Birmingham

UABS Union of American Biological Societies

uac underwriters adjusting company

UAC Ulster Automobile Club; United Africa Company; United Aircraft Corporation; Urban Affairs Council; Utility Aircraft Council

UACC Upper Area Control Center

UACL United Aircraft of Canada, Limited

uacte universal automatic control and test equipment

UADPS Uniform Automatic Data Processing System

UADW Universal Alliance of Diamond Workers

UAE United Arab Emirates (Trucial Sheikdoms of Trucial States)

UAEMS University Association for Emergency Medical Services

UAESP Utah Association of Elementary School Principals

uaf unit authorization file

uafs/t universal aircraft flight simulator/trainer

UAFT United Agency for Fair Treatment

UAG Universidad Autónoma de Guadalajara (Spanish—University of Guadalajara)

UAHC Union of American Hebrew Congregations

uai universal azimuth indicator

UAI Urban America Incorporated (Action Council for Better Cities)

UAI União Astronomica Internacional (Portuguese—International Astronomical Union); *Union Académique Internationale* (French—International Academic Union); *Union des Associations Internationales* (French—Union of International Associations); *Union Astrónomica Internacional* (Spanish—International Astronomical Union); *Union Astronomica Internazionale* (Italian—International Astronomical Union)

uaide uses of automatic information display equipment

UAISEGR University of Alaska Institute of Social, Economic, and Government Research

UAJAPPFI United Association of Journeymen and Apprentices of the Plumbing and Pipe Fitting Industry (U.S. and Canada)

UAK University of Alaska

ual upper acceptance limit

UAL United Air Lines; University of Aberdeen Library; University of Akron Library; University of Alabama; University of Alabama Library; University of Alaska Library; University of Alberta Library; University of the Americas Library; University

of Arizona Library; University of Arkansas Library; University of Auckland Library

UALL University of Arizona Lunar Laboratory

U of Alla University of Allahabad

uam (UAM) underwater-to-air missile

uam und andres mehr (German—and so forth)

UAM *Union Africaine et Malgache* (French—African and Malagasy Union); United American Mechanics

UAMC United Arab Maritime Company

UAMPT Union Africaine et Malagactie des Postes et Telecommunications (French—Union of African and Malagasy Postal Service and Telecommunication)

uan uric-acid nitrogen

UANA Unión Amateur de Natación de las Americas (Spanish—Amateur Swimming Alliance of the Americas)

UANC United African National Council

uao unexplained aerial object

UAOD United Ancient Order of Druids

UAOS Ulster Agricultural Organisation Society

uap unexplained atmospheric phenomenon

Uap Micronesian name for Yap

UAP Union of American Physicians; Union of Associated Professors; United Australia Party

UAPD Union of American Physicians and Dentists

U of A Pr University of Alabama Press; University of Alaska Press; University of Arizona Press

uar underwater acoustic resistance; underwater angle receptacle, upper air route; upper atmosphere research

UAR Uniform Airman Record; United Arab Republic; University of Arkansas

UARAEE United Arab Republic Atomic Energy Establishment

UARL United Aircraft Research Laboratories

UARRSI Universal Aerial Refuelling Receptacle Slipway Installation

uart universal asynchronous receiver-transmitter

UARTO United Arab Republic Tourist Office

uas unmanned aerial surveillance; upper air space

UAS Unit Approval System; University Air Squadron

UASC United Arab Shipping Company

UASCS United States Army Signal Center and School

UASIF Union des Associations Scientifiques et Industrielles Françaises (French—Union of French Scientific and Industrial Associations)

UASM University of Arkansas School of Medicine

UASS Unmanned Aerial Surveillance System

UASSP Utah Association of Secondary School Principals

UASSR Udmurt Autonomous Soviet Socialist Republic

uat ultraviolet acquisition technique

UAT Union Aéromaritime de Transport

UATI Union des Associations Techniques Internationales (French—Union of International Technical Organizations)

UATO United Airlines Tour Order

UATP Universal Air Travel Plan

UAU Universities Athletic Union

auv urban assault vehicle

UAW United Automobile Workers

uAwg um Antwort wird gebieten (German—reply requested)

uax (UAX) unit automatic exchange

UAZ University of Arizona

UAZEES University of Arizona Engineering Experiment Station

ub up(ward) bound; urine bilirubin

Ub Universiteitsbibliotheek (Dutch—University Library)—Amsterdam

UB Union Bank; Union of Burma; United Bank (of Arizona); United Biscuit; *Universität Basel* (Basel University); *Universität Berne* (Berne University)

U of B University of Baltimore; University of Bath; University of Birmingham;

University of Bombay; University of Bradford; University of Bridgeport; University of Bristol; University of Buffalo
UB The University Bookman
uba undenatured bacterial antigen
UBA Union of Burmah Airways; United Business Associates
UBA Universidad de Buenos Aires (Spanish—University of Buenos Aires)
UBAF Union de Banques Arabes et Francçaises (French—Union of Arab and French Banks)
U-bahn Untergrundbahn (German—underground road)—subway system
UBAV United Buddhist Association of Vietnam
UBB Union Bank of Bavaria
UBBA United Boys' Brigades of America
ubc universal buffer controller
UBC Uniform Building Code; United Baltic Corporation; Universal Bibliographic Control; University of British Columbia
U of BC University of British Columbia
UBC Uniform Building Code (legal); *Universidad de Baja California* (Spanish—University of Baja California)
UBC & J United Brotherhood of Carpenters and Joiners
UBCL University of British Columbia Library
UBCP Union Bag-Camp Paper; University of British Columbia Press
ubd utility binary dump
UBD Universal Business Directories
ubdi underwater battery director indicator
UBEA United Business Education Association
U-beam U-shaped beam
UBEM Union Belge d'Enterprises Maritimes (French—Belgium Union of Maritime Enterprises)
ubers übersetzt (German—translated)
ubf universal boss fitting
UBF Union of British Fascists
ubfc underwater battery fire control
ubi ultraviolet blood irradiation; universal battlefield identification

UBI United Business Investments
UBI Unione Bocciofila Italiana (Italian—Italian Bocce-Ball (Bowling) Association); *Unione Bibliografica Italiana* (Italian—Italian Bibliographical Society)
Ubib Wien Universitätsbibliothek Wien (German—Vienna University Library)
ubip ubiquitous immunopoietic polypeptide
ubitron undulating beam interaction electron tube
UBL Union Barge Line; United Benefit Life
UBLS University of Botswana, Lesotho, and Swaziland
ubm ultrasonic bonding machine; unit bill of material
UBM United Biscuit Manufacturing (company)
U-boat Unterseeboot (German—submarine)
U-bolt capital-U-shaped bolt
U-bomb uranium-cased atomic or hydrogen bomb
U Books University Books
UBP United Bermuda Party; United Business Publications
UBR University Boat Race
UBS United Bank of Switzerland; United Bible Societies; United Business Service
UBSA United Business Schools Association (formerly American Association of Commercial Colleges)
UBSO Uinta Basin Seismological Observatory
ubt universal book tester
ubu you be you
ubv ultraviolet
UBVS Ultraviolet-Blue Visual System
uc underclass; undercover (agent); universal coarse (screw thread); upper case (capital letters); upperclass
u/c upper center
UC Ulster College (Northern Ireland); Umpqua College; Union Carbide; Union College; University of California; University of Canterbury; University of Ceylon; University of Cincinnati; University College; University of Colorado; University of Connecticut; Upland College; Upper Canada; Upsala College; Urban Council; Urgent Care; Ursinus College; Ursuline College; Utica College

U of C University of Calcutta; University of Calgary; University of California; University of Cambridge; University of Chattanooga; University of Chicago; University of Cincinnati; University of Colorado; University of Connecticut; University of Corpus Christi
UC una corda (Italian—one string)—soft pedal
uca upper control area
UCA United Chemists' Association; United Conservatives of America; United Consumers of America; University of California; Utah Correctional Association
UCAB Universidad Católica Andrés Bello (Spanish—Andrés Bello Catholic University)
UCAE Universities Council for Adult Education
UCAF You See America First
UCAN Utilities Consumer Action Network
UCAR Union of Central African Republics; University Corporation for Atmospheric Research
UCAS Uniform Cost Accounting Standards; Union of Central African States
UCATT Union of Construction, Allied Trades, and Technicians
ucb unless caused by
UCB Unemployment Compensation Board; United California Bank; University of California at Berkeley; University College at Buckingham
UCBHM United Church Board for Homeland Ministries
UCBILR University of California at Berkeley—Institute of Library Research
UCBR University of California Board of Regents
ucc unadjusted contractual changes; universal copyright convention
UCC Uniform Commercial Code; Union Carbide and Carbon; Union Carbide Corporation; *Union de la Critique Cinématographique* (French—Society of Cinema Criticism); United Cancer Council; United Church of Christ; United Community Campaign; United Electric Coal Companies (stock exchange symbol); Universal

Copyright Convention; University College (Cork)

U-CC Upper Canada College

UCCA United Citizens Concerned with America; Universities Central Committee for Admissions; Universities Central Council on Admissions

UCCC Ulster County Community College; Uniform Consumer Credit Code

UCCD United Christian Council for Democracy

UCCELLO Paolo di Dono

UCCCM University of Cincinnati College Conservatory of Music

UCC-ND Union Carbide Corporation—Nuclear Division

UCCS Universal Camera Control System

ucd usual childhood diseases

UCD University of California at Davis; University College, Dublin

UCDA University and College Designers Association

U c de L Université Catholique de Louvain

UC de L Université Catholique de Louvain (French—Catholic University of Louvain)

ucdp uncorrect data processor

uce unforseen circumstances excepted

UCEA University College of East Africa (Makerere College); University Council for Educational Administration

UCEMT University Consortium in Education Media and Technology

U of Cey University of Ceylon

UCF United Cat Federation; United Community Funds; University of Central Florida

UCFE Unemployment Compensation for Federal Employees

UCFGB University Catholic Federation of Great Britain

UCFH University College of Fort Hare

UCG University College, Galway; University College of Ghana

UCGSM University of California Graduate School of Management

UCH University College Hospital

U-channel U-shaped channel

UCHCIS Urban Comprehensive Health Care Information System

uchd usual childhood diseases

U Chi University of Chicago

U Chi Lib University of Chicago Library

UCHS University City High School

uci unit construction index

UCI Union Cycliste Internationale (French—Cyclists International Union)

UCIDT University Consortium for Instructional Development and Technology

UCIIR University of California Institute of Industrial Relations

UCIIS University of California Institute of International Studies

UCIrv University of California at Irvine

UCIW Union of Commercial and Industrial Workers

UCIWP United Cannery and Industrial Workers of the Pacific

ucj unsatisfied claim and judgment

ucl upper control limit; upper cylinder lubricant; urea clearance test

UCL Union Castle Line; Union Central Life; Union Oil Company of California (symbol); Universal Color Language; University of California Library; University College, London

UCLA University of California at Los Angeles; University of California Caucasians Lost Among Asians

U-class upperclass

uc & lc upperclass and lowerclass

UCM University Christian Movement

UCM Universidad Complutense de Madrid (Spanish—Alcalá de Henares University of Madrid)

u-c man undercover narcotics agent

UCMC University of Colorado Medical Center

UCMEA Ufficio Centrale di Meteorologia e di Ecologia Agraria (Italian—Central Office of Meteorology and Agrarian Ecology)

UCMJ Uniform Code of Military Justice

UCMP University of California at Berkeley Museum of Paleontology

UCMS Unit Capability Measurement System

U-C M S Union-Castle Mail Steamship

UCN University College of Nigeria

UCNL University of California Nuclear Laboratory

UCNW University College of North Wales

uco universal code; universal coding

UCO University of Colorado

U Conn University of Connecticut

UCOR Uranium Enrichment Corporation

UCP Unified Command Plan; United Cerebral Palsy; United Country Party; Universal Citizen Plan; University of California Press

UCPA United Cerebral Palsy Associations

UCPP Urban Crime Prevention Program

U of C Pr University of California Press; University of Chicago Press

ucr unconditioned response

UCR Uniform Crime Reports; University of California at Riverside; Utah Coal Route (railroad)

UCRA University Centers for Rational Alternatives

UCRC Underground Construction Research Council

UCRG Uniform Contractor Reporting Guidelines

UCRI Union Carbide Research Institute

UCRL University of California Radiation Laboratory

UCRN Unique Consignment Reference Number

UCR & N University College of Rhodesia and Nyasaland

UCRS Uniform Contractor Reporting System; Uniform Crime Reporting Section (FBI); University, College, and Research Section (Library Association)—also appears as UCR

ucs unconditioned stimulus; unconscious; unit-count system; universal card scanner; universal character set

uc's uterine contractions

UCs Urban Coalitionists

UCS Union of Concerned Scientists; United Community Service(s); Universal Child Survival; Universal Classification System; Universal-Cyclops Steel; University Com-

puter Systems (computerized real estate listings); Upper Clyde Shipbuilders
UCSB University of California at Santa Barbara
UCSC University of California at Santa Cruz; University City Science Center
UCSD University of California at San Diego
UCSF University of California at San Francisco
UCSL University College of Sierra Leone
U of C SL University of California School of Law
UCSW University College of South Wales
uct unit compatability test(ing)
UCT United Commercial Travelers; University of Cape Town; University of Connecticut
UCTA United Commercial Travellers' Association
UC & U Union College and University
uc & uc underclass and upperclass
UCUC University College of the University of Cincinnati
ucv uncontrolled variable
UCV Universidad Central de Venezuela (Spanish—Central University of Venezuela)
UCVs United Confederate Veterans
UCW Union of Communication Workers; University College of Wales
UCWC University College of the Western Cape
UCWI University College of the West Indies
UCWP University College of the Western Province
ucwr upon completion will return
UCWRE Underwater Countermeasures and Weapons Research Establishment
UCX Unemployment Compensation for Ex-Servicemen
UCY United Caribbean Youth
UCZ University College of Zululand
ud unfair dismissal; upper berth (double occupancy); upper deck; urethral discharge; uroporphyrinogen decarboxylase (UD)
ud (UD) utility dog
u/d under deck
u.d. *ut dictum* (Latin—as directed)
Ud Udjung (Malay—point);

usted (Spanish—you)
UD Underground (London's subway); Undesirable Discharge; United Dairies; University of Denver; University of Detroit; Urban District
U of D University of Dallas; University of Dayton; University of Delaware; University of Delhi; University of Denver; University of Detroit; University of Dublin; University of Dubuque; University of Dundee; University of Durham
UD Unlisted Drugs
UDA Ulster Defence Association (Protestant counterpart of the IRA); Urban Development Authority
udaa unlawfully driving away auto
U da C Uriel da Costa (Uriel Acosta)
UDAG Urban Development Action Grants
UDAL Union de Universidades de América Latina (Spanish—Union of Latin American Universities)
udam universal digital of avionics module
udarg udarbeidet (Danish—prepared)
UDB Uprava Drzavne Bezbednosti (Serbo-Croat—Administration for State Security)—Yugoslavian Secret Service
udc universal decimal classification (UDC); upper dead center; usual diseases of childhood
U d C Universidad de Carabobo (Spanish—Carabobo University)—Venezuela
UDC United Daughters of the Confederacy; United Dye & Chemical; universal decimal classification; Urban District Council
UDCA Urban District Councils' Association
UDD Ulster Diploma in Dairying
'Uddersfield (Cockney contraction—Huddersfield)
udd's undisposed diapers; undumped diapers
UDE Underwater Development Establishment; *Union Douanière Equatoriale* (French—Equatorial Customs Union); University of Delaware
U de A Universidad de Alcala (Spanish—Alcala University); *Universidad de*

Antioquia (Spanish—Antioch University)
UDEAO Union Douanière des Etats de l'Afrique de l'Ouest (French—Customs Union of West African States)—former French colonies
U de B Universidad de Barcelona (Spanish—University of Barcelona); *Université de Bâle* (French—University of Basel)
U de BA Universidad de Buenos Aires (Spanish—University of Buenos Aires
udec unitized digital electronic calculation
U de C Universidad de Cartagena; Universidad de Cauca; Universidad de Chile; Universidad de Córdoba; Universidad de Cuzco; Universidad de Coímbra
U de CR Universidad de Costa Rica
U de F Université de Fribourg
U de G Universidad de Granada; Universidad de Guadalajara; Universidad de Guanajuato; Université de Genève; Université de Grenoble
U de H Universidad de la Habana (Havana)
U de L Universidad de Lérida; Universidad de Lima; Universidade de Lisboa (Lisbon); *Université de Lausanne*
UDEL Union des Editeurs de Littérature (French—Literature Editors Union)
U de LA Universidad de Los Andes
U de M Université de Montreal
U de Monc Université de Moncton
U de O Universidad de Oviedo
U de Pan Universidad de Panamá
U de Q Universidad de Quito (Universidad Central)
U de S Universidad de Salamanca; Universidad de San Andrés (La Paz); *Universidad de San Augustín* (Arequipa); *Universidad de San Javier* (Panama); *Universidad de San Marcos* (Lima); *Universidad de Santiago; Universidad de Santo Tomaás* (Bogotá or Santo Domingo)
U de SC de G Universidad de San Carlos de Guatemala

U de SD Universidad de Santo Domingo
U de SM Universidad de San Marcos (Lima, Peru)
U de SP Universidade de São Paulo
U de ST Universidad de Santo Tomás (Manila)
U de T Universidad de Toledó; Universidad de Trujillo (Peru)
U de V Universidad de Valencia; Universidad de Valladolid
U de Z Universidad de Zaragoza
udf und die folgende (German—and the following)
UDF Ulster Defence Force; Union Defence Force; United Democratic Front (South Africa)
UDF Union pour la Démocratic Françoise (French—Union for French Democracy)
udg udgave (Danish—edition)
u dgl (m) und dergleichen (mehr) (German—and the like)
U of D GSIS University of Denver Graduate School of International Studies
Ud'H Université d'Haiti (French—University of Haiti)
UDI Unilateral Declaration of Independence
UDI Unione Donne Italiane (Italian—Italian Women's Alliance)
U di A Università di Arezzo
UDIA United Dairy Industry Association
U di B Università di Bologna
U d F Università di Firenze (University of Florence)
U di G Università di Genova (Genoa)
U di N Università di Napoli (Naples)
U di P Università di Padova (Padua); *Università di Perugia; Università di Piacenza; Università di Pisa*
U di R Università di Roma
U di S Università di Siena
U di T Università di Torino
U di V Università di Venezia (Venice); *Università de Vicenza*
u dk upper deck
udk udkom (Dano-Norwegian—Published)
udl up-data link
udm upright drilling machine
udM unter dem Meeresspiegel (German—below sea level)

UDM United Merchants and Manufacturers (stock exchange symbol); Universal Drafting Machine (corporation)
Udm Aut Sov Soc Rep Udmurt Autonomous Soviet Socialist Republic
udmh (UDMH) unsymmetrical dimethyl hydrazine
udn ulcerated dermal necrosis
UDN Underwater Doppler Navigation
UDN União Democrática Brasileira (Portuguese—Brazilian Democratic Union)
udo unwilling drop-out
U d O Universidade de Oriente (Spanish—Oriente University)—Venezuela
U do B Universidade do Brasil (Portuguese—University of Brazil)—Brasilia
udom udometer; udometric; udometrical
U do P Universidade do Pôrto (University of Oporto)
UDP United Democratic Party
udpg (UPDG) uridine diphosphoglucose
UDP-gal uridine diphosphate galactose
UDP-glu uridine diphosphate glucose
UDPH Ulster Diploma in Poultry Husbandry
UDPS Utah Department of Public Safety
udr universal data report(er); universal digital readout; usage data report; utility data reduction
UDR Ulster Defence Regiment
UDR Union des Democrates pour la cinquième Republique (French—Union of Democrats for the Fifth Republic)
udrc utility data retrieval control
UDRI University of Dayton Research Institute; University of Denver Research Institute
UDRI-A University of Dayton Research Institute—Albuquerque
udro utility data retrieval output
Uds ustedes (Spanish—you, pl.)
UDS Ultraviolet Detection System; Underwater Demolition School; United Drapery Stores
UdSSR Union der Sozialistischen Sowjetrepubliken (Ger-

man—Union of Soviet Socialist Republics)—USSR
udt underdeck tonnage
UDT Underwater Demolition Team; Union for a Democratic Timor; United Dominions Trust
UDTC University of Dublin Trinity College
U of D TC University of Dublin Trinity College
UDU Underwater Demolition Unit
udw ultra-deep water
UDW United Domestic Workers
UD-W University of Durban-Westville
Udy Oodie; Uddevalla
UDY United Dye and Chemical Corporation (stock exchange symbol)
ue unexpired; unit equipment; unit exception; unit extremity; upper entrance
u E unseres Erachtens (German—in our opinion)
UE United Electrical Workers; University Extension
U of E University of the East (Manila); University of Edinburgh; University of Essex; University of Exeter
uea unattended equipment area
UEA Universal Esperanto Association; University of East Africa; University of East Anglia; University Entrance Examination; Utah Education Association
U of EA University of East Anglia
ueac unit equipment aircraft
ueb ultrasonic epoxy bonder
UEB Union Économique Benelux (French—Benelux Economic Union)
UEC United Engineering Center (NYC)
UEC Union Européene de la Carrosserie (French—European Union of Coachbuilders)
UECC United Electric Coal Companies
UECM Union Electric Company of Missouri
UECU Union for Experimenting Colleges and Universities
uee unit essential equipment
UEE Unione Economica Europea (Italian—European Economic Union)
uef universal extra fine (screw thread)
UEF Union Européenne des

Féderalistes (French—European Union of Federalists); *Union Européenne Féminine* (French—European Union of Women)
UEFA Union of European Football Associations
UEI Union of Educational Institutions
UEIC United East India Company
uel upper explosive limit
UEL Unilever Export Limited; United Empire Loyalists
u enr uranium enrichment
UEO Union de l'Europe Occidentale (French—Western European Union); *Universala Esperanto-Asocio* (Universal Esperanto Association)—Rotterdam
uep underwater electrical potential; uniform external pressure
UEP Union Electric Power Company; *Union Européenne des Payements* (French—European Payments Union—EPU)
UEPA Utility Electric Power Association
UEPMD Union Européenne des Practiciences en Médécine Dentaire (French—European Union of Practitioners of Dentistry)
UER University Entrance Requirements; Unsatisfactory Equipment Report
UER Unione Europea di Radiodiffusione (Italian), *Union Européenne de Radiodiffusion* (French)—European Broadcasting Union
UERD Underwater Explosives Research Division (USN)
UERMWA United Electrical, Radio, and Machine Workers of America
UES Underground Experiment Subcommittee (AECL); United Engineering Societies
uesk unit essential spares kit
uet unattended earth terminal
UET United Engineering Trustees
ueta (UETA) universal engineer tractor—armored
uetrt (UETRT) universal engineer tractor—rubber-tired
UEW United Electrical Workers
uex unexposed
u/ext upper extremity
Uey U-turn (traffic)
uf urea-formadehyde; under-

ground feeder; used for
u/f urea-formaldehyde resin
UF Uniformed Force (police); United Fruit
U-F Ugro-Finnic
U of F University of Florida
UF$_6$ uranium hexafluoride
ufa until further advised
ufa (UFA) unesterified free fatty acid
UFA Uniformed Firefighters Association; University Film Association
UFA Universum-Film-Aktiengesellschaft (German—Universe Film Company)
ufac unlawful flight to avoid custody
UFACCC United Faculty Associations of California Community Colleges
ufaed unit forecast authorization equipment data
ufap unlawful flight to avoid prosecution
ufat unlawful flight to avoid testimony
UFAW Universities Federation for Animal Welfare
ufc uniform freight classification
UFC Uni-Flex Container(s); United Fruit Company
UFCc United Free Churches
UFCE Union Fédéraliste des Communautés Ethniques Européennes (French—Federal Union of European Nationalities)
UFCS Underwater Fire-Control System
UFCT United Federation of College Teachers
UFCU Uni-Flex Container Unit
ufe (UFE) unducted fan engine
UFEL United Farmers Educational League
UFERP Union Fraternelle Entre les Races et les Peuples (French—Fraternal Union Between Races and Peoples)
uff ufficiale (Italian—officer; official); *ufficio* (Italian—bureau, office); *und folgende* (German—and the following)
UFF Ulster Freedom Fighters; University Film Foundation
uffi urea-formaldehyde foam insulation
UFH University of Fort Hare
UFI University Foundation International
UFI Union des Foires Internationales (French—Union of

International Fairs)
UFIPTE Union Franco-Ibérique pour la Production et le Transport de l'Électricité (French—Franco-Iberian Union for the Production and Transmission of Electricity)
UFIRS Uniform Fire-Incident Reporting System
ufl upper flammable limit
UFL United Farmers League; University of Florida
UfM University for Man
UFMCC Universal Fellowship of Metropolitan Community Churches
ufn until further notice
ufo unfiltered oil; unforeseen obstacle; unidentified flying object
UFOA Uniformed Fire Officers Association
UFOD Union Française des Organismes de Documentation (French—French Union of Documentary Organizations)
ufol ufologic(al)(ly); ufologist-(ic)(al)(ly); ufology
UFON Unidentified Flying Object Network
UFORA Unidentified Flying Objects Research Association
ufo's unidentified flying objects
uf p unemployed full pay
UFP United Federal Party
UFPA University Film Producers Association
UFPC United Federation of Postal Clerks
UFPO Underground Facilities Protective Organization
U-frame U-shaped frame
UFS University Film Society
UFT United Federation of Teachers
UFTAA Universal Federation of Travel Agents Associations
UFTM Ulster Folk and Transport Museum
UFU Ulster Farmers' Union
UFW United Farm Workers; United Furniture Workers
UFWU United Farm Workers Union
ug undergraduate; underground; urogenital
Ug Uganda; Ugandan; Ugric; Ugus
Ug Udjung (Malay—point)
UG Underground Railroad—secret system to aid slaves seeking freedom; United Gas
U of G University of Georgia;

University of Glasgow; University of Guam; University of Guelph; University of Guyana

UG Universität Graz

UG3RD Upgraded Third-Generation System (for air-traffic control)

uga unity gain amplifier

UGA University of Georgia

Ugan Uganda

Uganda Republic of Uganda (East African country)

ugb unity gain bandwidth

ugc ultrasonic grating constant; unity grain crossover

UGC United Gas Corporation; University Grants Committee

UG & CW United Glass and Ceramic Workers

UGDP University Group Diabetes Program

UGE Unified Global Enterprises

UGEQ Union Generale des Estudiants du Québec (French—General Union of Students of Québec)

ugf unidentified growth factor

UGGI Union Géodésique et Géophysique Internationale (French—International Geodesic and Geophysical Union)

ugi upper gastrointestinal

UGI Unione Geografica Internazionale (Italian), *Unión Geografica Internacional* (Spanish), *Union Géographique Internationale* (French)—International Geographical Union

UGLE United Grand Lodge of England

UGLIAC United Gas Laboratory Internally-Programmed Automatic Computer

U of G Lib University of Georgia Libraries

UGM Union of Graduates in Music

UGMA Unified Gift to Minors Act

ugmit you got me into this

UGMS Utah Geological and Mineral Survey

UGPL United Gas Pipe Line

U of G Pr University of Georgia Press

ugr ultrasonic grain refinement; universal graphic recorder

UGR Umfolozi Game Reserve (South Africa)

UGRR Underground Railroad (Quaker-organized means of

aiding fugitive slaves escaping from southern slave states to Canada and northern free states)

ugs uniaxial gyrostabilizer; urogenital system

Ugs Ugus

UGS United Girls' School

ugt urgent; urogenital tract

UGT Unión General de Trabajadores (Spanish—General Union of Workers)—Socialist trade union

ugtl ugentlig (Dano-Norwegian—weekly)

UGU University of Guam

UGW United Garment Workers

uh upper half

uh (UH) utility helicopter

U of H University of Hartford; University of Hawaii; University of Houston; University of Hull

UH University Heights

UH Universidad de la Habana; Universität Hamburg

UH-1 Bell 204B Iroquois military helicopter

UH-19 Sikorsky transport helicopter called H-19 or Chickasaw

UH-23 Hiller Raven utility helicopter H-23

uha upper-half assembly

UHA Union House of Assembly

UHAA United Horological Association of America

UHAB Urban Housing Assistance Board

uhc under honorable conditions

UHCBCN United Hebrew Congregations of the British Commonwealth of Nations

UHCC Upper House of the Convocation of Canterbury

uhcs ultra-high-capacity storage

UHCY Upper House of the Convocation of York

uhel ultra-high-efficiency lamp

uhf ultra-high frequency—300–3000 mc

UHF United Health Foundation; United Holyland Fund (for Arab terrorists); United Hospital Fund

uhfdf ultra-high-frequency direction finder

uhff ultra-high-frequency filter

uhfg ultra-high- frequency generator

uhfj ultra-high-frequency jam-

mer

uhfo ultra-high-frequency oscillator

uhfr ultra-high-frequency receiver

UHI University of Hawaii

UHK University of Hong Kong

U of HK University of Hard Knocks

uhl user header label

uhmw ultra-high molecular weight

UHOIA University of Houston Office of International Affairs

uhp ultra-high purity

UHP University of Hawaii Press

uhr ultra-high resistance; ultra-high resolution

uhrn ultra-high radio navigation

uhs ultra-high speed

UHS International Union of the History of Science; Union High School; University for Humanistic Studies

uht ultra-high temperature; ultrasonic hardness tester; universal hand tool

uht milk ultra-high-temperature milk (capable of keeping without refrigeration)

uhtv unmanned hypersonic test vehicle

UHU Unhappy Hookers United

uhv ultra-high vacuum

uhvc ultra-high vacuum chamber

UHVS Ultra-High Vacuum System

ui ultrasonic industries; unit indicator; you (and) I

u/i unit of issue

ui unidades internacionales (Spanish—international units)

u.i. ut infra (Latin—as below)

UI Ube Industries; Unemployment Insurance; Universität Innsbruck; Urban Institute

U of I University of Idaho; University of Illinois; University of Iowa; University of Israel; University of Istanbul

UIA Ultrasonic Industry Association; Union of International Associations; United Israel Appeal; University of Iowa; Urban Intervention Associates

UIA Union Internationale des Architects (French—International Alliance of Architects);

Union Internationale des Avocats (French—International Alliance of Attorneys)

UIAA *Union Internationale des Associations d'Alpinisme* (French—International Union of Alpinism Associations)

UIAB Unemployment Insurance Appeals Board

UIACM *Union Internationale de Automobile-Clubs Médicaux* (French—International Union of Medical Auto Clubs)

UIAS Union of Independent African States

UIATF United Indians of All Tribes Foundation

U i B Universitet i Bergen

UIB Unemployment Insurance Benefits; United International Bank

UIB Union Internationale des Maîtres Boulanger (French—International Union of Master Bakers)

uibc unsaturated iron-binding capacity

uic ultraviolet image converter

UIC Unemployment Insurance Code; Union International Company; University of Illinois in Chicago; Utah Innovation Center

UIC Unio Internationalis Contra Cancrum (International Union Against Cancer)

UICA Union of Independent Colleges of Art

UICC Union Internationale Contre le Cancer (French—International Union Against Cancer); *Unione Internazionale Contro il Cancro* (Italian—International Union for the Control of Cancer)

UICF Union Internationale des Chemins de Fer (French—International Union of Railways)

UICIO Unit Identification Code Information Office(r)

UICN Union Internationale pour la Conservation de la Nature (French—International Union for the Conservation of Nature)

UI Comm Unemployment Insurance Commission

UICPA Union Internationale de Chimie Pure et Appliquéee (French—International Union of Pure and Applied Chemistry)

UICPS Uniform Inventory Control Points System

UICR Union Internationale des Chauffeurs Routiers (French—International Coach and Lorry Drivers' Association)

UICT Union Internationale Contre la Tuberculose (French—International Union Against Tuberculosis)

UID University of Idaho

UIE UNESCO Institute for Education

UIEIS Union Internationale pour l'Etude des Insectes Sociaux (French—International Union for the Study of Social Insects)

UIEO Union of International Engineering Organizations

UIES Union Internationale pour l'Education Sanitaire (French—International Union for Health Education)

uif ultraviolet interference filter; unfavorable information file; universal intermolecular force

UIF Unemployment Insurance Fund

UIFI Union Internationale des Fabricants d'Imperméables (French—International Union of Rainwear Manufacturers)

Uig Uighur; Uigur

U i G Universitet i Göteborg

UIHL Union Internationale de l'Humanisme Laïque (French—International Union for Ethical Humanism)

UIHPS Union Internationale d'Histoire et de Philosophie des Sciences (French—International Union of the History and Philosophy of Science)

UIIG Union Internationale de l'Industrie du Gaz (French—International Gas Industry Union)

UIII Urban Information Interpreters Incorporated

U i L Universitet i Lund

UIL University of Idaho Library; University of Illinois; University of Illinois Library; University of Indiana Library; University of Iowa Library

UIL Unione Italian del Lavoro (Italian—Italian Labor Union)—republican and social-democrat

U of Ill Lib Sci University of Illinois Graduate School of Library Science

U of Ill Pr University of Illinois Press

UIM Union Industrielle & Maritime (Société Française de l'Armement) (French—Industrial and Maritime Union, French Ordnance Company); *Union Internationale Motonautique* (French—International Motorboating Union)

UIMC Union Internationale des Services Médicaux des Chemins de Fer (French—International Union of Railway Medical Services)

UIMNH University of Illinois Museum of Natural History

UIN United States and International Securities (stock exchange symbol); University of Indiana

UINF Union Internationale de la Navigation Fluviale (French—International Union for River Navigation)

Uintas short form for the Uinta Mountains of northeastern Utah and southwestern Wyoming

U i O Universitet i Oslo; Universitetsbiblioteket i Oslo (Norwegian—University Library in Oslo)

UIO Union Internationale des Orientalistes (French—International Union of Orientalists)

UIOOT Union Internationale des Organismes Officiels de Tourisme (French—International Union of Official Travel Organizations)

U of Iowa Pr University of Iowa Press

UIP United Irish Party; University of Illinois Press

UIP Union Internationale de Patinage (French—International Skating Union); *Union Internationale de Physique* (French—International Union of Physics)

UIPC Utah Industrial Promotion Commission

UIPC Union Internationale de la Presse Catholique (French—Catholic Press International Union)

UIPD Ulrich's International Periodicals Directory

UIPE Union Internationale de Protection de l'Enfance (French—International Union for the Protection of Children)

UIPM Union Internationale de la Presse Médicale (French—International Union

of the Medical Press)
UIPVT Union Internationale contre le Péril Vénérien et les Tréponématoses (French—International Union against the Peril of Venereal Diseases and Syphilis)
uir upper information region
UIR University Industrial Research
uis (UIS) urban industrial society
U i S Universitet i Stockholm
UIS Unemployment Insurance Service; Unit Identification System
UIS Union International de Secours (French—International Relief Union)
UISAE Union Internationale des Sciences Anthropologiques et Ethnologiques (French—International Union of Anthropological and Ethnological Sciences)
UISB Union Internationale des Sciences Biologiques (French—International Union of the Biological Sciences)
uisc unreported interstate shipment of cigarettes
UISE Union Internationale de Secours aux Enfants (French—International Child Welfare Union)
UISN Union Internationale des Sciences de le Nutrition (French—International Union of Nutritional Sciences)
UISP Union Internationale des Syndicats de Police (French—International Union of Police Trade Union)
uit unit impulse train
uit uitgaaf (Dutch—publication)
UIT Unión Internacional de Telecomunicaciones (Spanish), *Union Internationale des Télécommunications* (French), *Unione Internazionale Telecomunicazione* (Italian)—International Telecommunications Union—ITU
UITAM Union Internationale de Mécanique et Appliquée (French—International Union of Theoretical and Applied Mechanics)
uitg uitgegeven (Dutch—published)
UITS Unione Italiana Tiro e Segno (Italian—Italian Rifle Association)
U i U Universitet i Uppsala
UIU Quito, Ecuador (airport)

UIUNA Upholsterers' International Union of North America
UIUPGWA United International Union of Plant Guard Workers of America
UJ Union Jack (United Kingdom flag incorporating crosses of St Andrew for Scotland, St George for England, and St Patrick for Northern Ireland); University of Judaism
U of J University of Judaism
UJ Universidad Javeriana (Bogotá and Sucre)
UJA United Jewish Appeal
UJC Union Jack Club
U.J.D. Utriusque Juris Doctor (Latin—Doctor of Civil and Canon Law)
ujf unsatisfied judgment fund(ing)
U-joint(s) U-shaped joint(s)
ujr unijunction rectifier
UJSCs Union Jack Services Clubs
ujt unijunction transistor
uk unknown
uk (UK) urokinase
UK United Kingdom; *Universita Karlova* (Karl University—University of Prague)
U of K University of Kansas; University of Keele (formerly University College of North Staffordshire); University of Kent; University of Kentucky
UK Universiti Kebangsaan (Malay—National University)
UKA Ulster King of Arms; United Kingdom Alliance; United Klans of America
UK(A) United Kingdom All-comers (athletics)
UKAC United Kingdom Automation Council
UKADR United Kingdom Air Defense Region (NATO)
UKAEA United Kingdom Atomic Energy Authority
UKAPE United Kingdom Association of Professional Engineers
ukb universal keyboard
UKBC United Kingdom Bomber Command
UKBG United Kingdom Bartenders' Guild
UKC United Kennel Club; University of Kent at Canterbury
U of KC University of Kansas City; University of King's

College
UKCA United Kingdom Citizens Association
UKCATR United Kingdom Civil Aviation Telecommunications Representative
UKCBDA United Kingdom Carbon Block Distributors' Association
UKCSBS United Kingdom Civil Service Benefit Society
UKCTA United Kingdom Commercial Travellers' Association
UKDA United Kingdom Dairy Association
uke ukelele
UK fo United Kingdom for orders
UKGBNI United Kingdom of Great Britain and Northern Ireland
UKGPA United Kingdom Glycerine Producers' Association
UKHH United Kingdom-Havre-Hamburg (range of ports)
UKHS United Kingdom Hovercraft Society
UKIAS United Kingdom Immigrants Advisory Service
UKIRT United Kingdom Infrared Telescope
UKISC United Kingdom Industrial Space Committee
UKITO United Kingdom Information Technology Organisation
UKJGA United Kingdom Jute Goods Association
UKKKK United Kingdom Ku Klux Klan
UKL University of Kansas Library; University of Khartoum Library
UKLF United Kingdom Land Force
UKLFS United Kingdom Low-Flying System
UKM University of Kansas Museums
UKMC University of Kansas Medical Center
UK(N) United Kingdom National (athletics)
UKOP United Kingdom Oil Pipelines
U K£ United Kingdom pound
UKPA United Kingdom Pilots' Association
Ukr Ukraine; Ukrainian
Ukr Acad Pr Ukrainian Academic Press
Ukraine Ukrainian Soviet Socialist Republic
Ukrainian SSR Ukrainian Soviet Socialist Republic (Uk-

raine)
UKRAS United Kingdom Railway Advisory Service
UKS University of Kansas
UKSATA United Kingdom South Africa Trade Association
UKSM United Kingdom Scientific Mission; University of Kansas School of Medicine
UKSMA United Kingdom Sugar Merchants' Association
UKSMT United Kingdom Sea Mist Test(ing)
UKSTC United Kingdom Strike Command
Ukulele (UK) stock exchange slang for Union Carbide
ukv underground keybox vault
UKW Ultra-Kurzwellen (German—ultra-short wave)
UKY University of Kentucky
ul up link; upper left; upper leg; upper level; upper lid
ul (UL) user language
u/l upper left; upper limit
u & l upper and lower
UL Urban League; Underwriters Laboratories; Universal League; University Libraries; University Library
U of L University of Lancaster; University of Laval; University of Leeds; University of Leicester; University of Lethbridge; University of Liverpool; University of London; University of Louisville
UL Union List
ula uncommitted logic array
ULA Ulster Launderers' Association; United Labor Agency; University of Louisiana
ULA Uniform Laws Annotated; Universidad Los Andes (Spanish—Andes University)—Venezuela
ULAA Ukrainian Library Association of America
Ulaan Ulaanbaatar (Khalkha Mongol—Red Hero)—capital city of the People's Republic of Mongolia; formerly called Ulan Bator or Urga
ULAD Unilever Limited Accounts Department
u-land udviklingsland (Dano-Norwegian—development land)
ULAP University-wide Library Automation Program (University of California)
ULAST Unión Latino Américana de Sociedades de

Tisiology (Spanish—Latin-American Union of Societies of Phthisiology)
ulb universal logic bloc
ULB Université Libre de Bruxelles (French—Free University of Brussels)
ulc unsafe lane change (vehicular code); upper left center
u & lc upper and lower case
ULC Ulster Loyalist Council; Underwriters' Laboratories of Canada; Urban Library Council
ULCA United Lutheran Church of America
ULCC Ultra Large Cargo Carrier (bulk freighter or tanker of 400,000 or more tons)—superfreighter or supertanker
ULCI Union of Lancashire and Cheshire Institutes
ULD Unit Load Device
uldb ultralight-displacement boat
uldest ultimate destination
ule ultra-low expansion
ulf ultra-low frequency; unfair labor practice
uli ultra-low interstitial
ULI Urban Land Institute
ULI Union pour la Langue Internationale Ido (French—Union for the International Language Ido)
ULICS University of London Institute of Computer Science
ULII Union pour la Langue Internationale Ido (French—Union for the International Language Ido)
ull ullage
'Ull (Cockney contraction—Hull)
ULL Unitarian Laymen's League; University of Liverpool Library; University of London Library; University of Lund Library
ullv (ULLV) unmanned lunar logistics vehicles
ulm ultrasonic light modulator; universal logic module
ULM University Library of Manchester (includes John Rylands Library)
ULMS Underwater Long-range Missile System
ULO United Licensed Officers (union); Unmanned Launch Operations
ULP Université Louis Pasteur; University of London Press
ULPA Uniform Limited Part-

nership Act
ulpr ultra low-pressure rocket
ULPZ Upper Limits for the Prescriptive Zone
Ulrich Ulrich's Books
uls unsecured loan stock
Uls Ulsan; Ulster (ancient Irish province comprising 6 of the 9 original counties)
ULS Universities Libraries Section (Association of College and Research Libraries)
ULS Union List of Serials
ulsi ultra-large-scale integration
ult ultimate; ultimo
ult. ultimo (Latin—at last)
ULT United Lodge of Theosophists
Ult Bod Ultra Bodoni
ulto ultimo
ult° *último* (Spanish—last)
ult. praes. ultimum praescriptus (Latin—last prescribed)
ultra (Latin prefix—beyond or in excess)—ultramontane, ultrasonic
ultracom ultraviolet communications system
ultra hi-fi ultra-high fidelity
ultralight ultralight flying machine; ultralight luggage; ultralight wearing apparel
ultrason ultrasonic(s)
ultra-x universal language for typographic reproduction applications
ULTS Universal Lifeline Telephone Service
ult ts ultimate tensile strength
U of Luck University of Lucknow
ULUCLA University Library of the University of California at Los Angeles
ULUM University Library, University of Michigan (Ann Arbor)
ulv ultra-low volume
um umpire; unmarried
u/m unit of measure
üM über dem Meeresspiegel (German—above sea level)
UM Universal Match; Universal Mill; University of Malaysia (University of Malaya—Raffles Institute); University of Manitoba; University of Melbourne; University Museum(s)
U of M University of Maine; University of Malaysia; University of Manchester; University of Manitoba; University of Maryland; University of Massachusetts; University

of Miami; University of Michigan; University of Minnesota; University of Mississippi; University of Missouri; University of Montreal

UM Universiti Malaya (University of Malaya)—Raffles Institute

U Ma Ursa Major (Big Bear)

UMA Ultrasonic Manufacturers Association; *Union de Mujeres Americanas* (Spanish—United Women of the Americas); Union Maghreb Arabe; University of Massachusetts

UM-A University of Mid-America

U-magnet U-shaped magnet

U of Mand University of Mandalay

UMAS United Mexican-American Students

umass unlimited machine access from scattered sites

UMass University of Massachusetts

U of Mass Pr University of Massachusetts Press

umb umber; umbilical; umbilicus

Umb Umbrian

UMB Union Mondiale de Billard (French—World Billiards Union)

UMBIR University of Michigan Bureau of Industrial Relations

umbl umbilical

UMBR Umbria(n)

UMC Uniform Mechanical Code; United Metallic Cartridge (company); United Methodist Church; Universal Match Corporation; Upstate Medical Center; Urban Mobility Corporation

UMCA Urabá, Medellín and Central Airways

umd unitized microwave device

UMD Unit Manning Document; University of Maryland

UMDA United Micronesian Development Association

U of Md Lib Serv University of Maryland School of Library and Information Services

U of Mdrs University of Madras

UME University of Maine

umf ultramicrofiche

UMF Umbrella-Makers' Federation

UMFC United Methodist Free Churches

umgearb umgearbeitete (German—revised)

UMHK Union Miniére du Haut-Katanga (French—United Mines of Upper Katanga)

UMHP Union Mondiale des sociétiés d'Histoire Pharmaceutique (French—World Union of Pharmaceutical History Societies)

U Mi Ursa Minor (Little Bear)

UMI University of Michigan; University of Microfilms Incorporated; University Microfilms International; Utah Management Institute

U of Miami Pr University of Miami Press

U of Mich Bus Res University of Michigan Graduate School of Business Research

U of Mich Inst Labor University of Michigan Institute of Labor and Industrial Relations

U of Mich Pr University of Michigan Press

U of Mich Soc Res University of Michigan Institute for Social Research

U/min Umdrehungen in der Minute (German—revolutions per minute)

U of Minn Bell Mus University of Minnesota Bell Museum of Pathology

U of Minn Pr University of Minnesota Press

UMIST University of Manchester Institute of Science and Technology

UML University of Michigan Library; University of Minnesota Library; University of Missouri Library

umler universal machine language

UMLS University Microfilm Library Service

UM & M United Merchants and Manufacturers

UMMS University of Maine (or Manchester, Manitoba, Maryland, Massachusetts, Michigan, Minnesota, Mississippi, Missouri, Montana) Medical School

UMMZ University of Michigan Museum of Zoology

umn upper motor neuron

UMN University of Minnesota

UMNO United Malay National Organization

UMO University of Maine at

Orono; University of Missouri

umoc ugly man on campus

U of Monc University of Moncton

U of Mo Pr University of Missouri Press

ump umpire

UMP Upper Mantle Project; Upper Merion and Plymouth (railroad); University of Massachusetts Press

'Umphrey (Cockney contraction—Humphrey)

UMPO Upper Manhattan Planning Office

umr under main roof

U MR Umvoti Mounted Rifles

UMREL Upper Midwest Regional Educational Laboratory

UMRRC Universities Mobile Radio Research Corporation (Bath, Birmingham, Bristol)

UMRWFR Upper Mississippi River Wildlife and Fish Refuge (Minnesota)

ums unmanned machinery space

UMS Undersea Medical Society; Unfederated Malay States; Universal Military Service; University of Mississippi

UMSM University of Michigan School of Music

UMSU University of Malaya Student's Union

UMT Universal Military Training; University of Montana

UMT Union Marocaine du Travail (French—Moroccan Labor Union)

UMTA Urban Mass Transportation Administration

umtd using mails to defraud

UMTRAP Uranium Mill Tailings Remedial Action Program

UMTS Universal Military Training and Service

UMW United Mine Workers

UMWA United Mine Workers of America

U of Mys University of Mysore

un (UN) unsatisfactory

Un Union (postal abbreviation)

UN Union Twist Drill (trademark); United Nations; University of the North; unsatisfactory

U of N University of Natal; University of Nebraska; University of Nevada; University of Newcastle; University of Nottingham

UN União Nacional (Portuguese—National Union); *Unificación Nacional* (Spanish—National Unification)—Costa Rica

UNA United Nations Association; United Native Americans; United Natives Association

UNAA United Nations Association of Australia

UNAAF Unified Action Armed Forces

unab unabridged

unabr unabridged

UNAC United Nations Appeal for Children

UNACC United Nations Administrative Committee on Coordination

unaccomp unaccompanied

UNACIL United Africa Commercial and Industrial Limited

UNACOMS Universal Army Communications System

UNAIS United Nations Association International Service

unalot unallotted

UNAM Universidad Nacional Autónoma de Mexico (Spanish—National University of Mexico)

unamace universal automatic map compilation equipment

unan unanimous

UNAPO United National Association of Post Office (Craftsmen)

UNARCO United Nations Narcotics Commission

unasgd unassigned

unatt unattached

UNAUS United Nations Association of the United States

UNAUSA United Nations Association of the United States of America

unauthd unauthorized

unb unbound; universal navigation beacon

UNB United Nations Bookshop; University National Bank; University of Nebraska

U of NB University of New Brunswick

UN Bank International Bank for Reconstruction and Development

unbd unbound

Unbib van Amsterdam Universiteitsbibliotheek van Amsterdam (Dutch—Amsterdam

University Library)

unblkng unblanking

UNBRO United Nations Border Relief Operation

unb's unbelievers

UNBSA United Nations Bureau of Social Affairs

unc uncertain; unconscious; undercurrent; unified coarse (thread)

unc (UNC) unconditional (computer flow chart)

Unc Uncle

UNC United Nations Command; United Nuclear Corporation; University of North Carolina; University of Northern Colorado

U of NC University of North Carolina

UNC Union Nationale Camerounaise (French—Cameroon National Union)—party; *Universidad Nacional de Colombia* (Spanish—National University of Colombia)

UNCA United Natioins Correspondents Association

UNCAST United Nations Conference on the Applications of Science and Technology

UNCC United Nations Cartographic Commission

UNCCP United Nations Commission on Crime Prevention; United Nations Conciliation Commission for Palestine

UNCF United Nations Children's Fund (formerly UNICEF); United Negro College Fund

unch unchanged

U of NC Inst Gov University of North Carolina Institute of Government

UNCIO United Nations Conference on International Organization

UNCIP United Nations Commission on India and Pakistan

uncir uncirculated

UNCIRSS University of North Carolina Institute for Research in Social Science

UNCITRAL United Nations Commission on International Trade Law

UNCIWC United Nations Commission for the Investigation of War Crimes

UNCL University of North Carolina Library

unclas unclassified

U.N.C.L.E. United Network

Command for Law Enforcement (fictional organization created for television)

UNCLOS United Nations Conference on the Law of the Sea

UNCMAC United Nations Command Military Armistice Commission

unco uncouth

UNCO United Nations Civilian Operations Mission (to the Congo)

UNCOK United Nations Commission on Korea

uncol universal computer-oriented language

uncomp uncompensated

uncond unconditioned

UNCOPUOS United Nations Committee on the Peaceful Uses of Outer Space

uncor uncorrected

uncov uncover; uncovered; uncovers

U of NC Pr University of North Carolina Press

un cs unconditioned stimulus

UNCs United Neighborhood Centers

unct. unctus (Latin—smeared)

UNCTAD United Nations Conference on Trade and Development

UNCURK United Nations Commission for the Unification and Rehabilitation of Korea

und under

UND University of National Defense; University of North Dakota

U of ND University of North Dakota; University of Notre Dame

UNDAT United Nations Development Advisory Team

UN Day United Nations Day (October 24)

UNDC United Nations Disarmament Commission

UNDCC United Nations Development Cooperation Cycle

undeco underground economy (composed of persons who report less than they earn, or who file no income tax returns)

unded underdeduction

undercover narc undercover narcotics agent

undergrad undergraduate

Under Sec Nay Nav Under Secretary of the Navy

Undex United Nations Index

UNDI *United Nations Document Index*
undies underthings (underwear)
undoc(s) undocumented alien(s)—illegal alien(s)
UNDct United Nations Document
UNDOF United Nations Disengagement Observer Force
UNDP United Nations Development Program
und pkng underground parking
U of ND Pr University of Notre Dame Press
undrgrnd underground
UNDRO United Nations Disaster Relief Office
undrwrld underworld
undsgd undersigned
UNDSM University of North Dakota School of Medicine
undtkr undertaker
undw underwater
undwrtr underwriter
UNE University of New England (New South Wales)
UNEAS Union of European Accountancy Students
U of Neb Pr University of Nebraska Press
UNEC United Nations Education Conference
UNECA United Nations Economic Commission for Asia
UNECLA United Nations Economic Commission for Latin America
UNECOLAIT *Union Européenne du Commerce Laitier* (French—European Milk Trade Union)
UNEDA United Nations Economic Development Association
unef unified national extra fine (screw thread)
UNEF United Nations Emergency Forces
UNEF *Union Nationale des Étudiants Français* (French—National Union of French Students)
UNEO United Nations Emergency Operation
UNEP United Nations Environment(al) Program
UNESCO United Nations Educational, Scientific, and Cultural Organization
UNESEM *Union Européenne des Sources d'Eaux Minérales du Marché Commun* (French—European Union of Natural Mineral Water

Sources of the Common Market)
UNETAS United Nations Emergency Technical Aid Service
U of Nev Pr University of Nevada Press
unex unexecuted
unexpl unexplained; unexploded; unexplored
unexpur unexpurgated
UNEXSO Underwater Explorers Society
unf unfinished; unfuzed; unified fin thread
UNF United National Front
U of NF University of North Florida
UNFAO United Nations Food and Agricultural Organization
unfav unfavorable
UNFB United Nations Film Board
UNFC United Nations Food Conference
unfd unfurnished
UNFDAC United Nations Fund for Drug Abuse Control
UNFICYP United Nations (Peace-Keeping) Force in Cyprus
unfin unfinished
UNFPA United Nations Fund for Population Activities
UN Fund International Monetary Fund
unfurn unfurnished
ung unguent
ung *ungarische* (German—Hungarian)
ung. *unguentum* (Latin—ointment)
Ung Ungava; Ungavan
Ung *Ungarn* (Norwegian—Hungary)
UNGA United Nations General Assembly
Ungar Frederick Ungar Publishing Company
UNH University of New Hampshire; University of New Haven
U of NH University of New Hampshire
UNHCR United Nations High Commissioner for Refugees
UNHQ United Nations Headquarters (Geneva, New York, Vienna)
uni (Latin prefix—one)—unilateral
Uni University
UNI United News of India; United Nuclear Industry
UNI *Unione Naturista Italiana* (Italian—Italian Naturist As-

sociation); *Ente Nazionale Italiano di Unificazione* (Italian Unification Council)
UNIA Universal Negro Improvement Association (Garveyites)
UNIC United Nations Information center
UNICCAP Universal Cable Circuit Analysis Program
UNICE *Union des Industries de la Communauté Européenne* (French—Industrial Union of the European Community)
UNICEF United Nations International Children's Emergency Fund
unicike unicycle
UNICIS Unit Concept Indexing System; University of Calgary Information Systems
unicom underwater integration communication; universal communication
UNICOM aeronautical advisory station operating on 122.8 mc
UNIDIR United Nations Institute for Disarmament Research
UNIDO United Nations Industrial Development Organization
unif uniform; uniformity
unif coef uniformity coefficient
Unif Gift Min Act Uniform Gifts to Minors Act
UNIFIL United Nations Interim Force in Lebanon
Unif L Ann Uniform Laws Annotated
Uniform letter U radio code
unihedd universal head-down display
UN I–I MOG UN Iran–Iraq Military Observer Group
unilat unilateral
Unilatcorps Unilateral Corps
UNIMA *Union Internationale de grands Magasins* (French—International Union of Department Stores)
UNIMERC Universal Numeric Coding System
UNIMS Univac Information Management System
UNINCO *Union Internationale des Corps Consulaires* (French—International Consular Corps Union)
unincorp unincorporated
UNIO United Nations Information Organization
Union Coll Pr Union College

Press
Unions Union Islands of the Pacific also called Tokelaus
Union of Soviet Socialist Republics (world's largest nation occupying much of Asia and Europe), *Soyuz Sovyetskikh Sotsialisticheskikh Respublik*
UNIP United Independence Party
uniparse universal parser
UNIPEDE Union Internationale des Producteurs et Distributeurs d'Energie Electrique (French—International Union of Producers and Distributors of Electric Energy)
unipol universal procedure-oriented language
uni(s) unisexual(s)
unis unisoni (Italian—unison)
unis 8va (Italian—in unison with the octave)
UNIS United Nations International School; Univac Industrial System
UNISCAN United Kingdom and Scandinavia
UNISIST Universal System for Information in Science and Technology
UNISOMI Universal Symphony Orchestra and Music Institute
UNISTAR User Network for Information Storage Transfer
UNISYM Unified Symbolic Standard Terminology for Mini Computer Instructions
Unit Unitarian
UNIT Union Nationale des Ingénieurs Techniciens (French—National Union of Engineers and Technicians)
UNITA Unión Nacional para la Independencia Total de Angola (Portuguese—National Union for the Total Independence of Angola)
UNITAR United Nations Institute for Training and Research
United Auto Workers International Union, United Automobile, Aerospace, and Agricultural Implement Workers of America
United Kingdom United Kingdom of Great Britain and Northern Ireland (England, Scotland, Wales, the Isle of Man, colonies and dependencies such as Belize; Bermuda; British Antarctica; the British Indian Ocean Terri-

tory; the British West Indies; the Channel Islands; Gibraltar; Hong Kong; the Gilberts, New Hebrides; Pitcairn; Ascension, the Falklands, St Helena, Tristan da Cunha)
United Mine Workers United Mine Workers of America
United Provinces United Provinces of the Netherlands (Friesland, Gelderland, Groningen, Holland, Oberyssel, Utrecht, Zeeland)—the Seven Provinces
United Rubber Workers United Rubber, Cork, Linoleum, and Plastic Workers of America
United States United States of America (North American nation); United States of Brazil, the United States of Colombia, the United States of Indonesia, the United States of Mexico, the United States of North America (the U.S.A.), the United States of Venezuela
UNITS United Nations Information for Teachers
univ universal
Univ Universal; Universalist; University; University College, Oxford
univac universal automatic computer
univar universal valve action recorder
Univ-Buchdr Universitats-Buchdrukerei (German—university press)
Univ C University College (Oxford)
Univ. D. Doctor of the University (degree)
universal languages mathematics and music
Univ Mus of UP University Museum of the University of Pennsylvania
UNIX Universal Inner-Active Executive (IBM)
unjc united national J-series coarse (thread)
unjef united national J-series extra fine (thread)
unjf united national J-series fine (thread)
unjs united national J-series special (thread)
unk unknown
Unk Uncle
unkn unknown
UNKRA United Nations Korean Reconstruction Agency

UNL University of Nairobi Library
UNLA Unione Nazionale per la Lotta contro l'Analfabetismo (Italian—National Association for the Fight Against Illiteracy)
UNLC United Nations Liaison Committee
unld unload (flow chart)
unldh underloading
unlib unliberated
unliq unliquidated
unlk unlock
UNLL United Nations League of Lawyers
UNLOS United Nations Law of the Sea (conference)
UNLOSC United Nations Law of the Sea Conference
unltd unlimited
UNLV University of Nevada at Las Vegas
unlwfl unlawful(ly)
unm unmarried
UNM Ukrainian National Museum (Chicago); University of New Mexico
U of NM University of New Mexico
UNMC United Nations Mediterranean Commission; University of Nebraska Medical Center
UNMEM United Nations Middle East Mission
U of NM Gen Lib University of New Mexico General Library
UNMOGIP United Nations Military Observer Group in India and Pakistan
U of NM Pr University of New Mexico Press
UNMSC United Nations Military Staff Committee
UNMSM de L Universidad Nacional de San Marcos de Lima (Spanish—University of Lima)
unmtd unmounted
unnilhexium unnamed element 106
unnilseptium unnamed element 107
UNO United Nations Organization; United Neighborhood Organization; University of Nebraska at Omaha; University of New Orleans
UNO Union Nacional Odría (Spanish—Odria National Union)—political party; *Unión Nacional Oposición* (Spanish—National Opposition Union)—Nicaraguan political

party
UNOC United Nations Operations in the Congo
UNOCAL Union Oil Company of California
unodir unless otherwise directed
unof unofficial
UNOID United Nations Organization for Industrial Development
unoindc unless otherwise indicated
UNOLS University-National Oceanographic Laboratory System
U or non-U upperclass or not upperclass
unop unopposed
unoreq unless otherwise requested
unp unpaged; unpaid
UNP United National Party; University of Nebraska Press; Urewara National Park (North Island, New Zealand)
UNPA United Nations Postal Administration
UNPC United Nations Palestine Commission
unpd unpaid
UNPF United Nations Population Fund
UNPHU Universidad Nacional Pedro Henriquez Urena (Spanish—Pedro Henriquez Urena National University)—Dominican Republic
unpkd unpacked (flow chart)
unpleas unpleasant
UNPOC United Nations Peace Observation Commission
UNPP United Nations Partition Plan
unpub unpublished
unqte unquote
unqual unqualified
UNR & EC United Nuclear Research and Engineering Center
UNREF United Nations Refugee Emergency Fund
unrel unreliable
unrep unreported; unrepresented
UNRISD United Nations Research Institute for Social Development
UNRPR United Nations Relief for Palestine Refugees
UNRRA United Nations Relief and Rehabilitation Administration
UNRWA United Nations Relief and Works Agency
uns unified special (thread);

unsymmetrical
UNS Unified Numbering System
UNSA United Nations Specialized Agencies; University of Nottingham School of Agriculture
unsat unsatisfactory; unsaturated
unsatfy unsatisfactory
unsatis unsatisfactory
UNSC United Nations Security Council
UNSCC United Nations Standards Coordinating Committee
UNSCCUR United Nations Scientific Conference on the Conservation and Utilization of Resources
UNSCEAR United Nations Scientific Committee on the Effects of Atomic Radiation
UNSCOB United Nations Special Commission on the Balkans
UNSCOP United Nations Special Commission on Palestine
unscv unserviceable
UNSDRI United Nations Social Defense Research Institute
UNSF United Nations Special Fund for Economic Development
UNSG United Nations Secretary General
unsgd unsigned
unskd unskilled
UNSM United Nations Service Medal; University of Nebraska State Museum
UNSO United Nations Sahel Office (*see* Sahel); United Sabah Organization
UNSR United Nations Space Registry
unst unstable
un stim unconditioned stimulus
unsus-look(ing) unsuspicious-look(ing)
unsvc unserviceable
UNSvM United Nations Service Medal
UNSW University of New South Wales
UNSY United Nations Statistical Yearbook
unsym unsymmetrical
Unt Unter (German—lower; under)
UNTA United Nations Technical Assistance
UNTAA United Nations Technical Assistance Administra-

tion
UNTAB United Nations Technical Assistance Board
UNTAG United Nations Transition Assistance Group
UNTAM United Nations Technical Assistance Mission
UNTC United Nations Trusteeship Council
unthd unthreaded
UNTS United Nations Treaty Series
UNTSO United Nations Truce Supervision Organization
UNTT United Nations Trust Territory
UNTTA United Nations Trust Territory Administration
UNU United Nations University
UNUP United Nations University Press
UNUSA United Nations Association of the United States of America
UNV University of Nevada
UNWCC United Nations War Crimes Commission
unwmk unwatermarked
UNYNJSHPBA United New York and New Jersey Sandy Hook Pilots Benevolent Associations
u/o used on
u & o use and occupancy
uo und öfters (German—and often)
UO Ulster Orchestra (Belfast); University of Otago (at Dunedin, New Zealand); University of Ottawa
U of O University of Ohio; University of Oklahoma; University of Omaha; University of Oregon; University of Ottawa; University of Oxford
uoa use of other automobiles
UOB United Overseas Bank
U of O B University of Oregon Books
uoc ultimate operational capability
UOC Uniform Offense Classification
UOCO Union Oil Company
uod ultimate oxygen demand
UOFS University of the Orange Free State
UOH University of Ohio
uohc under other than honorable conditions
UOJCA Union of Orthodox Jewish Congregations of America
UOK University of Oklahoma

U of Okla Pr University of Oklahoma Press
uol underwater object locator
uoo undelivered orders outstanding
UOP Universal Oil Products
UOPH Unaccompanied Officer Personnel Housing
UOPWA United Office and Professional Workers of America
UOR Uniform Officer Record; University of Oregon; Unusual Occurrences Report
UORI University of Oklahoma Research Institute
uos Underwater Ordnance Station (USN)
uo's undelivered orders
uot uncontrolled overtime
UOT United Ocean Transport (Daido Line)
UOTS United Order of True Sisters
uov unit of variance
up. underproof; underproofed; underproofing; unpaged; upper
u/p urine-plasma concentration
u & p uttering and publishing
Up Upper
UP Ulster Parliament; Ulster Party; Union Pacific (railroad); Union Postale (Postal Union); United Party; United Presbyterian; United Press; United Province; University of Paris; University of Pennsylvania; University of Pittsburgh; University Press; Uttar Pradesh; Uptown Planners
U of P University of the Pacific; University of Pennsylvania; University of Pittsburgh; University of Portland; University of Pretoria; University of Puget Sound
UP Unidad Popular (Spanish—Popular Unity)—political party; *Unión Panamericana* (Spanish—Pan-American Union); *Unión Patriótica* (Spanish—Patriotic Union)—old Colombian Communist party; *Union Postale* (French—Postal Union)—international mail organization
UPA United Productions of America; University of Pennsylvania; University Photographers Association
UPA Union Postale Arabe (French—Arab Postal Union); *Unions Professionnelles Agricoles*

(French—Professional Agricultural Unions)
UPAA University Photographers Association of America
UPAC Union of Pan-Asian Communities
UPADI Unión Panamericana de Asociaciones de Ingenieros (Spanish—Pan-American Union of Engineers Associations)
UPAE Union Postal de las Américas y España (Spanish—Postal Union of the Americas and Spain); *Union Postale des Amériques et de l'Espagne* (French—Postal Union of the Americas and Spain)
UPAO University Professors for Academic Order
UPASI United Planters Association of South India
upc universal product code
UPC Unesco Publications Center; Uniform Plumbing Code; United Power Company; United Presbyterian Church; Universal Postal Convention; Universal Product Code
upd unpaid
UPD Unified Port District; Urban Planning Directorate
UPD Union Periodistas Democratica (Spanish—Union of Democratic Journalists)—Mexico
UPDW United Piece Dye Works
U of PE University of Port Elizabeth
UPE Union Parlementaire Européenne (French—European Parliamentary Union)
UPEP Undergraduate Preparation of Educational Personnel
UPGA Utah Personnel and Guidance Association
uphd uphold
uphol upholsterer; upholstery
UPI United Press International (merger of United Press and International News Service)
UPICA University of Pennsylvania Institute of Contemporary Art
UPIGO Union Professionnelle Internationale des Gynécologistes et Obstétriciens (French—International Professional Union of Gynecologists and Obstetricians)
UPIN United Press International Newsfeatures
upk's unpopped kernels
upl unauthorized practice of

law
UPL United Philippine Line; University of Pensylvania Library; University of the Philippines Library (Quezon City); University of Pittsburgh Library; University of Portland Library
up. log. upper loge
upm uninterruptible power module; units per mile
UPNE Unversity Press of New England
UPNG University of Papua and New Guinea
upo undistorted power output; unidentified paleontological object
UPO United Partisans' Organization; Unit Personnel Office(r)
UPOA Ulster Police Officers Association; Ulster Public Officers Association
UPOV Union for the Protection of New Varieties of Plants
UPOW Union of Post Office Workers
upp upplaga (Swedish—edition)
UPP University of Pennsylvania Press; University of Pittsburgh Press
UPPC Union Pacific Petroleum Corporation
Upper Volta Republic of Upper Volta (landlocked West African country), *République de Haute-Volta*; name changed in 1984 to Bourkina Fasso
UPPPP Underprivileged Peoples' Public Pool
upr (most); unsaturated polyester resin; upper
Upr Upper
U Pr University Press (Washington, DC)
UPR Union Pacific Railroad; University of Puerto Rico
UPREAL Unit Property Record and Equipment Authorization List
U Presses Fla. University Presses of Florida
U Pr Hawaii University Press of Hawaii
U Pr Kan University Press of Kansas
U Pr Ky University Press of Kentucky
U Pr Miss University Press of Mississippi
U Pr NE University Press of New England

U Pr Va University Press of Virginia
U Pr Wash University Press of Washington
ups uinterrupted power supply; United Parcel Service
UPS Underground Press Syndicate; Underground Publication Society; Underwater Production System(s); United Parcel Service; United Publishers' Services; Universal Press Syndicate; University of Puget Sound
UPSA Ukrainian Political Science Association
UPSEB Upper Pradesh State Electricity Board
UPSG universal polar stereographic grid
UPSM University of Pennsylvania School of Medicine
UPSS United Postal Stationery Society
UPSTC Upper Pradesh State Textile Corporation
UPSTEP Undergraduate Pre-Service Teacher Education Program
Up Swn Upper Swan
upt (UTP) uridine triphosphate
up tor upper torso
up tr up train
UPU United Prisoners Union; Universal Postal Union
UPU Unión Postal Universal (Spanish—Universal Postal Union)
Up V Upper Volta
UPV Ulster Protestant Volunteers (paramilitary counterpart of the IRA)
UPW Union of Postal Service Workers
UPWA United Public Workers of America
UPWIU United Paper Workers International Union
uq upper quartile
UQ University of Queensland
U of Q University of Québec; University of Queensland
UQP University of Queensland Press
u quark up quark
ur unconditioned response; up right (stage direction); upper right; urinal; urinary; urine; utility rectifier
ur (UR) unemployment rate
u/r upper right
ur ouron (Greek—urine)— urea, uremia, ureter, urethra, urine, urology
Ur Urania; uranium; Uranus; Urdu; Uruguay; Uruguayan

UR Uganda Railway; Uniform Regulations; Unsatisfactory Report; Urban Renewal; Utilization Review
U of R University of Reading; University of Redlands; University of Richmond; University of Rochester
UR Universidad de la República (Spanish—University of Uruguay)
URA United Republicans of America; Universities Research Association; Urban Redevelopment Authority; Urban Renewal Administration
urad your radio (message)
u-rail U-shaped rail
Urals Ural Mountains dividing Asia from Europe in the USSR
URAMEX Mexican-government's uranium company
uran (Latin—tail)—anuran, urochordate
Uran Uranus
ur anal. urine analysis
U of Rang University of Rangoon
uranog uranographer; uranographic; uranography
urb urban; urbanism; urbanist; urbanistic; urbanite; urbanization; urbanize; urbicultural; urbiculture
Urb Urbanización (Spanish— Urbanization)
Urban Inst Urban Institute
Urbank Urban Bank (National Development Bank)
urbanol urbanologic(al); urbanologist; urbanology
urb guer(s) urban guerilla(s)
urbm (URBM) ultimate-range ballistic missile
urbol urbanologist; urbanology
urb ter urban terrorism; urban terrorist(s)
urc upper right center
URC United Republic of Cameroon; Universal Resources Corporation; Urban Renewal Commission; Urban Research Center (NYU)
urclk universal receiver clock
URCLPWA United Rubber, Cork, Linoleum, and Plastic Workers of America
urd upper disease (head cold)
Urd Urdu (literary language of Pakistan)
Ur$ Uruguayan peso
URD Unión Republicana Democrática (Spanish—Democratic Republican Union)—

political party in Venezuela
urdis your dispatch
ure unintentional radiation exploitation
URE Undergraduate Record Examination
URESA Uniform Reciprocal Enforcement of the Support Act (for the collection and enforcement of child support)
uret urethra(l)
urf (URF) uterine-relaxing factor
URF Union des Services Routiers des Chemins de Fer Européens (French—Union of European Railways Route Services)
urg urgent
Urga Ulaanbaatar, Mongolia
uri upper respiratory illness (head cold)
Uri not an abbreviation but a Swiss canton
URI Union Research Institute (Hong Kong); University of Rhode Island
U of RI University of Rhode Island
uria (Latin prefix—urine)— urinal, urinalysis; (Latin suffix—urine)—polyuria
URIMA University Risk and Insurance Managers Association
urinalysis urine analysis
URISA Urban and Regional Information System Association
url (URL) user requirements language
URL Underground Research Laboratory (Canadian); Unilever Research Laboratory; University of Rhodesia Library (Salisbury)
urltr your letter
urmgm your mailgram
urmsg your message
urn. ultra-high radio navigation; urnal, urnary, urned, urnement(al)(ly), urnful
Urn Urnest (Ernest)
uro urological; urology
uro (Latin prefix—urine)—urinal, urinary
URO United Restitution Organization
urodyn urodynamic(s)
urogen urogenital
urol urologic(al)(ly); urologist; urology
U-room U-boat room (petty officer's quarters)
uro or uran (Greek-derived prefix or suffix from *oura*

meaning tail)—anuran or urochordate
URP Unit Reporting Program; United Revolutionary Party
Urq Urquhart
urr (URR) ultra-rapid reader (computer program)
URR Union for the Resurrection of Russia
URRVS Urban Rapid-Rail-Vehicle Systems
urs unit reference sheet
URs Unsatisfactory Reports; University Rationalists
UR's Unsatisfactory Reports
URS Universal Reporting System; Universal Reference System
urser your serial (number or reference)
URSI *Union Radio Scientifique Internationale* (French—International Scientific Radio Union)
urspr *ursprünglich* (German—originally)
URSS *União das Repúblicas Socialistas Soviéticas* (Portuguese—Union of Socialist Soviet Republics)—the USSR; *Unión de Repúblicas Socialistas Soviéticas* (Spanish—Union of Soviet Socialist Republics); *Union des Républiques Socialistes Soviétiques* (French—Union of Socialist Soviet Republics)—the USSR
urt upper respiratory tract; utility radio transmitter
URT United Republic of Tanzania (Tanganyika and Zanzibar)
urtel your telegram
urti upper respiratory tract infection (common cold; influenza)
URTI *Université Radiophonique-Télévisuelle Internationale* (French—International Radio-Television University)
URTU United Road Transport Union
Uru Uruguay; Uruguayan
Uruguay Oriental Republic of Uruguay, *República Oriental del Uruguay*
urv underseas research vehicle
URWA United Rubber Workers of America
us. ultrasound; under seal; undersize; uniform sales
us. (US) unconditioned stimulus
u-s upper-stage
u/s unserviceable

u.s. ubi supra (Latin—where mentioned above); *ut supra* (Latin—as above)
US Uncle Sam; United States; University of Stellenbosch
U.S. United States
U of S University of Salford; University of Saskatchewan; University of Scranton; University of Sheffield; University of Sherbrooke; University of the South (Sewanee, Tennessee); University of Southampton; University of Stirling; University of Strathclyde; University of Sudbury; University of Surrey; University of Sussex; University of Swansea; University of Sydney
U.S. Ufficio Stampa (Italian—Press Agency)
USA Underwriters Service Association; Union of South Africa; United States of America (more correctly U.S.A.); United States Army; United States Attorney; United Steelworkers of America; United Swaziland Association; United Switzerland Association; University of South Africa
US of A United Steelworkers of America; United Synagogue of America
U.S.A. United States of America
U.S. of A. United States of America; United Secularists of America
U of SA University of South Africa
USA *Unser Shtickel Arbeit* (Yiddish—Our Bit of Work)—rifle grenade produced in Palestine
U.S.A. (title of trilogy by John Dos Passos—*42nd Parallel, 1919, The Big Money*—describing first three decades of American life in the twentieth century)
USAA United Services Automobile Association
USAAA US Army Audit Agency
USAABMDA United States Army Advance Ballistic Missile Defense Agency
USAAC United States Army Air Corps (now USAF)
USAACDA United States Army Aviation Combat Development Agency
USAAD US Army Airmobile

Division
USAADC United States Army Air Defense Center
USAADEA US Army Air Defense Engineering Agency
USAAF United States Army Air Forces
USAAFINO United States Army Aviation Flight Information and Navigation Aids Office
USAAFO US Army Avionics Field Office
USAAMR & DL United States Army Air Mobility Research and Development Laboratory
USAAPSA United States Army Ammunition Procurement and Supply Agency
USAASD United States Army Aeronautical Service Detachment
USAASO United States Army Aeronautical Services Office
USAAVNC United States Army Aviation Center
USAAVNS United States Army Aviation School
USAAVSCOM United States Army Aviation Systems Command
USAB United States Activities Board
USABAAR United States Army Board for Aviation Accident Research
USABRL US Army Ballistic Research Laboratories
USAC United States Aircraft Carriers (air cargo line); United States Auto Club; US Air Conditioning Corporation
USACAA United States Army Concepts Analysis Agency
USA CAC United States Army Continental Army Command
USACC United States Army Communications Command; U.S.-Arab Chamber of Commerce
USACDA United States Arms Control and Disarmament Agency; United States Army Catalog Data Agency
USACDC US Army Combat Developments Command
USACDCCA United States Army Combat Development Command Combined Arms Agency
USACDCEC United States Army Combat Development Command Experimentation Command
USACDCFAA United States

Army Combat Developments Command Field Artillery Agency

USACDCNG United States Army Combat Developments Command Nuclear Group

USACDCOA United States Army Combat Developments Command Ordnance Agency

USACDCQA United States Army Combat Developments Command Quartermaster Agency

USACDCSWCAG United States Army Combat Developments Command Special Warfare and Civil Affairs Group

USACE US Army Corps of Engineers

USACENDCDSA United States Army Corps of Engineers National Civil Defense Computer Support Agency

USACIC United States Army Criminal Investigation Command

USACIDC United States Army Criminal Investigative Command

USACMA United States Army Club Management Agency

USACMR United States Army Court of Military Review

USACPEB United States Army Central Physical Evaluation Board

USACRR United States Army Crime Records Repository

USACSA US Army Combat Surveillance Agency

USACSLA United States Army Communications Security Logistics Agency

USACSSEA United States Army Computer Systems Support and Evaluation Agency

USAD US Army Dispensary

USADIP United States Army Deserter Information Point

USADSC US Army Data Services and Administrative Systems Command

USAE United States Army Engineer(s); United States Army, Europe

USAEC United States Army Engineer Command; United States Atomic Energy Commission; US Army Electronics Command

USAECA United States Army Engineer Construction Agency

USAECBDE United States Army Engineer Center Brigade

USAECLRA United States

Army Electronics Command Logistics Research Agency

USAED United States Army Engineer Division

USAEDC United States Army Engineer Division—Caribbean

USAEDH United States Army Engineer Division—Huntsville, Alabama

USAEDLMV United States Army Engineer Division—Lower Mississippi Valley

USAEDM United States Army Engineer Division—Mediterranean

USAEDMR United States Army Engineer Division—Missouri River

USAEDNA United States Army Engineer Division—North Atlantic

USAEDNC United States Army Engineer Division—North Central

USAEDNE United States Army Engineer Division—New England

USAEDNP United States Army Engineer Division—North Pacific

USAEDOR United States Army Engineer Division—Ohio River

USAEDPO United States Army Engineer Division—Pacific Ocean

USAEDSA United States Army Engineer Division—South Atlantic

USAEDSP United States Army Engineer Division—South Pacific

USAEDSW United States Army Engineer Division—Southwest

USAEEA United States Army Enlistment Eligibility Activity

USAEL US Army Electronic Laboratories

USAEMA US Army Electronics Materiel Support Agency

USAEMCA United States Army Engineer Mathematical Computation Agency

USAEMSA United States Army Electronics Materiel Support Agency

USAENGCOM United States Army Engineer Command

USAENPG United States Army Engineer Power Group

USAEPG US Army Electronic Proving Ground

USAERA United States Army

Electronic Command Research Agency

USAERC United States Army Enlisted Records Center

USAERDAA United States Army Electronics Research and Development Activity (Fort Huachuca, Arizona)

USAERDL US Army Electronics Research and Development Laboratory

USAERG United States Army Engineer Reactor Group

USAES United States Association of Evening Students

USAETDC U.S. Army Engineer Topographic Data Center (D.C.)

USAEUR United States Army Europe

USAEVD United States Alliance for the Eradication of Venereal Disease

U S Af Union of South Africa

USAF United States Air Force

USAFA US Air Force Academy

USAFABD United States Army Field Artillery Board

USAFAC United States Army Finance and Accounting Center

USAFACS US Air Force Aircrew School

USAFAGOS US Air Force Air Ground Operations School

USAFAPS US Air Force Air Police School

USAFAS United States Army Field Artillery School

USAFB United States Army Field Bank

USAFBMS US Air Force Basic Military School

USAFBS US Air Force Bandsman School

USAFC United States Army Forces Command

USAFD United States Air Force Dictionary

USAFE US Air Forces in Europe

USAFECI United States Air Force Extension Course Institute

USAFESA United States Army Facilities Engineering Support Agency

USAFEURPCR United States Air Force European Postal and Courier Region

USAFFGS US Air Force Flexible Gunnery School

USAFFSR US Air Force Flight Safety Research

USAFHRC US Air Force His-

torical Research Center
USAFI United States Armed Forces Institute
USAFIGED United States Armed Forces Institute Tests of General Educational Development
USAFIT US Air Force Institute of Technology
US AFLANT US Air Force, Atlantic
USAFMPCR United States Air Force Mideast Postal and Courier Region
USAFNS US Air Force Navigation School
USAFO United States Army Field Office
USAFOCS US Air Force Officer Candidate School
USAFOF United States Army Flight Operations Facility
USAFPACPCR United States Air Force Pacific Postal and Courier Region
USAFPS US Air Force Pilot School
USAFSAAS United States Air Force School of Applied Aerospace Sciences
USAFSAB US Air Force Scientific Advisory Board
USAFSACS United States Air Force School of Applied Cryptologic Sciences
USAFSAM US Air Force School of Aerospace Medicine
USAFSAWC US Air Force Special Air Warfare Center
USAFSC US Air Force Systems Command; United States Army Food Service Center
USAFSE US Air Force Supervisory Examination
USAFSG United States Air Field Support Group
USAFSO US Air Forces, Southern Command
USAFSOC United States Air Force Special Operations Center
USAFSOF United States Air Force Special Operations Force
USAFSOS United States Air Force Special Operations School
USAFSS US Air Force Security Service
USAFSTC United States Army Foreign Science and Technology Center
USAFSTDS US Army-Air Force Standards

USAFSTRIKE US Air Force Strike Command
USAFTS US Air Force Technical School
USAGETA United States Army General Equipment Test Activity
USAGMPC United States Army General Materiel and Parts Center
USAH United States Army Hospital
USAHAC United States Army Headquarters Area Command
USAHC United States Army Health Clinic
USAHSC United States Army Health Services Command
USAHSDSA United States Army Health Services Data Systems Agency
USAIA United States Army Institute of Administration
USAIC US Army Infantry Center; US Army Intelligence Corps
USAICA US Army Interagency Communications Agency
USAICS United States Army Intelligence Center and School
USAID United States Aid for International Development
USAIG United States Aircraft Insurance Group
USAIIA United States Army Imagery Interpretation Agency
USAIIG United States Army Imagery Interpretation Group
USAILG United States Army International Logistics Group
USAIMS United States Army Institute for Military Systems
USAINTA United States Army Intelligence Agency
USAINTS US Army Intelligence School
USAIPSG US Army Industrial and Personnel Security Group
USAir formerly Allegheny Airlines
USAirA United States Air Attaché
USAIRE United States of America Aerospace Industries Representatives in Europe
USAir MilComUN United States Air Force Representative, UN Military Staff Committee
USAISC United States Army Intelligence and Security

Command
USAJ United States Army, Japan
USAJPG United States Army Jefferson Proving Ground
USAK United School Administrators of Kansas
USALC United States Army Logistics Center
USALEA United States Army Logistics Evaluation Agency
USALSA United States Army Legal Services Agency
USAMAA United States Army Memorial Affairs Agency
USAMBRDL United States Army Medical Bioengineering Research and Development Laboratory
USAMCFG United States Army Medical Center—Fort Gordon
USAMC-ITC United States Army Materiel Command—Interim Training Center
USAMDRC United States Army Materiel Development and Readiness Command
USAMDW United States Army Military District of Washington
USAMEDCOM United States Army Medical Command
USAMEOS United States Army Medical Equipment and Optical School
USAMFSS United States Army Medical Field Service School
USAMIDA United States Army Major Item Data Agency
USAMIIA United States Army Medical Intelligence and Information Agency
USAML United States Army Medical Laboratory
USAMMA United States Army Medical Materiel Agency
USAN United States Adopted Name
USA NC United States Army Nurse Corps
USAO United States Assay Office
USAPA United States Army Procurement Agency
USAPACDA United States Army Personnel and Administration Combat Development Activity
USAPDC United States Army Petroleum Distribution Command
USAPEB United States Army Physical Evaluation Board
USAPEQUA United States

Army Production Equipment Agency

USAPHC United States Army Primary Helicopter Center

USAPIA United States Army Personnel Information Activity

USAPO United States Antarctic Projects Office

USAPRO United States Army Personnel Research Office

USAR United States Army Reserve

USARA United States Army Reserve Affairs

USARADCEN United States Army Air Defense Center

USARADCOM United States Army Air Defense Center; United States Army Air Defense Command

USARAE United States Army Reserve Affairs—Europe

USARAL United States Army, Alaska

USARB United States Army Retraining Brigade

USARC United States Army Recruiting Command

USARCS United States Army Claims Service

USAREC United States Army Recruiting Command

USAREUR United States Army, Europe

USARIBSS United States Army Research Institute for the Behavioral and Social Sciences

USARIEM United States Army Research Institute of Environmental Medicine

USARJ United States Army, Japan

USARP United States Antarctic Research Program

USARPA United States Army Radio Propagation Agency

USARPAC United States Army, Pacific

USARPACINTS United States Army Pacific Intelligence School

USARSA United States Amateur Roller Skating Association

USARSC United States of America Roller Skating Confederation

USARSO United States Army, Southern Command

usart universal synchronous-asynchronous receiver-transmitter

USARV United States Army, Vietnam

USAS United States Army South; United States of America Standard

US ASA United States Army School of the Americas; United States Army Security Agency

USASACDA United States Army Security Agency Combat Development Activity

USASADEA United States Army Signal Air Defense Engineering Agency

USASAE United States Army Security Agency—Europe

USASAFO United States Army Signal Avionics Field Office

USASATCOMA United States Army Satellite Communications Agency

USASC United States Army, Southern Command—Caribbean; United States Army Support Center

USASCAF United States Army Service Center for Army Forces

USASCC United States Army Strategic Communications Command

USASCII USA Standard Code for Information Interchange (data processing)

USASCSA United States Army Signal Communications Security Agency

USASG United States Army Standardization Group

USASI United States of America Standards Institute

USA Sig C United States Army Signal Corps

USASMC United States Army Supply and Maintenance Command

USASMSA United States Army Signal Corps Material Support Agency

USASRDL United States Army Signal Research and Development Laboratory

USASSA United States Army Signal Supply Agency

USASSG United States Army Special Security Group

USAT United States Army Transport

USATA United States Army Transportation Aviation

USATC United States Army Traffic Command

USATDC United States Army Training and Doctrine Command

USATEA United States Army Transportation Engineering Agency

USATEC United States Army Test and Evaluation Command

USATECOM United States Army Test and Evaluation Command

USATIA United States Army Transportation Intelligence Agency

USATISU United States Army Troop Information Support Unit

USATL United States Army Technical Library

USATMACE United States Army Traffic Management Agency—Central Europe

USATopoCom United States Army Topographic Command

USATRATCOM United States Army Strategic Communications Command

USATSC United States Army Terrestrial Sciences Center

USATTC United States Army Tropic Test Center

USATTU United States Army Transportation Terminal Unit

USAU United States Aviation Underwriters

usaw (USAW) underwater security advance warning

USAWC United States Army War College; United States Army Weapons Command

USAWES United States Army Waterways Experiment Station

USAWF United States Amateur Wrestling Society

usb unified S-band; upper sideband

USB United States Borax (company)

USBA United States Billiard Association; United States Boomerang Association; United States Brewers Association

USBC United States Bureau of the Census; United States Bureau of Customs

USB & C United States Borax and Chemical (company)

USBCSC United Society of Believers in Christ's Second Coming (Shakers)

USBE Universal Serials and Book Exchange (formerly United States Book Exchange)

USBG United States Botanic Garden

USBGN United States Board on Geographic Names

USBH United States Bureau Highways

USBIS United States Border Inspection Station

USBLS United States Bureau of Labor Statistics

USBM United States Bureau of Mines

USBP United States Board of Paroles; United States Border Patrol; United States Bureau of Prisons

USBPA United States Bicycle Polo Association

USBPR United States Bureau of Public Roads

USBS United States Border Station; United States Bureau of Standards

USBTA United States Board of Tax Appeals

USBuStand United States Bureau of Standards

usc under separate cover; upper stage center

USC Underwater Systems Center (Groton, Conn.); United Services Club; United Shipping Company; United States Code; United States Congress; United States of Colombia; United Steamship Company; University of South Carolina; University of Southern California

USC United States Catalog; United States Code (legal)

USCA Ulster Special Constabulary Association; United States Copper Association; United States Courts of Appeals

USCA United States Code Annotated

USCAC US Continental Army Command

USCANS Unified S-band Communication and Navigation System

USCB United States Customs Bonded

USCC United States Catholic Conference; United States Chamber of Commerce; United States Circuit Court; United States Commercial Company; United States Customs Court

USCCA United States Circuit Court of Appeals

USCCAN United States Code–Congressional and Administrative News

USCCPA United States Court of Customs and Patent Appeals

USCCR United States Commission on Civil Rights

USCE US Coast Guard Reserve; US Commissioner of Education

USCF United States Chess Federation; United States Churchill Foundation

USCG United States Coast Guard

USCGA US Coast Guard Academy

USCGAD United States Coast Guard Air Detachment

USCGAS United States Coast Guard Air Station

USCG Aux United States Coast Guard Auxiliary

USCGC United States Coast Guard Cutter

USCGI United States Coast Guard Institute

USCGMSC United States Coast Guard Marine Safety Council

USC & GS United States Coast and Geodetic Survey

USCHS United States Capitol Historical Society; United States Catholic Historical Society

USCI United Satellite Communication Inc

USCIIC United States Civilian Internee Information Center (USA)

USCINCEUR United States Commander-in-Chief, Europe

USCINSO United States Commander-in-Chief, Southern Command

USCM United States Conference of Mayors

USCMA United States Coal Mines Administration; United States Court of Military Appeals

USCMI United States Commission of Mathematical Instruction

usco underwriters salvage company

USCO Union Steel Corporation (South Africa)

US Comm UNICEF United States Committee for UNICEF

USCONARC US Continental Army Command

US Const Constitution of the United States

USCP University of South Carolina Press; University of Southern California Press; U.S. Capitol Police (DC)

USCP United States Coast Pilot

USCPSC US Consumer Product Safety Commission

USCR United States Committee for Refugees

USCRC United States Civil Rights Commission

USCRS United States Cotton Research Station

USCS United States Civil Service; United States Claims Service; United States Conciliation Service; United States Customs Service; Universal Ship Cancellation Society

USCSC United States Civil Service Commission

USCSup United States Code Supplement

USCT United States Colored Troops (1862–1865)

USCUN United States Committee for the United Nations

USCUNICEF United States Committee for UNICEF

USCWF United States Council for World Freedom

USCWHO United States Committee for the World Health Organization

US Cy United States currency

usd ultimate strength design; uninhibited sexual desire

US $ American dollar(s); United States dollar

USD Unified School District; University of San Diego; University of South Dakota

USD United States Dispensatory

USDA United States Department of Agriculture; United States Disarmament Agency

USDA/CRIS US Department of Agriculture/Current Research Information System

USDB United States Disciplinary Barracks

USDC United States Department of Commerce; United States District of Columbia; United States District Court

USDCFO US Defense Communication Field Office

USDEA United States Drug Enforcement Administration

USDF United States Dressage Federation

USDHEW United States Department of Health, Education, and Welfare (HEW)

USDHUD United States Department of Housing and Ur-

ban Development
USDI United States Department of the Interior
USDJ United States District Judge
USDL United States Department of Labor
USDLGI United States Defense Liaison Group—Indonesia
USDOCO United States Document Officer
USDoD United States Department of Defense
USDP University of San Diego Press; University of South Dakota Press
USDR United States Divorce Reform
USDSA United States Deaf Skiers Association
USDSEA United States Dependent School European Area
USDT United States Department of Transportation
USE United States Envelope (corporation); Univac Scientific Exchange; U.S. English
usea undersea
u/Sec Under Secretary
USELMCENTO United States Element Central Treaty Organization
USEP United States Escapee Program
USERC United States Environment and Resources Council
USES United States Employment Service
USEUCOM United States European Command
usf und so fort (German—et cetera)—and so forth
USF United States Forces
U of SF University of South Florida
USFA United States Fire Administration; United States Food Administration (World War I); United States Forces in Austria (World War II)
USFAA United States Fronton Athletic Association
USFC United States Foil Company
USFET United States Forces—European Theater
USFF United States Flag Foundation
USF & G United States Fidelity—Guaranty (insurance underwriters)
USFGC United States Feed Grains Council

USFIS United States Foundation for International Scouting
USFJ United States Forces, Japan
USFL United States Football League
USForAz United States Forces in the Azores
USFPL United States Forest Products Laboratory
USfs United States frequency standard
USFS United States Foreign Service; United States Forest Service
USFSA United States Figure Skating Association
USFWS United States Fish and Wildlife Service
USG Ulysses Simpson Grant (18th President U.S.); United States Government; United States Gypsum (company)
U.S.G. United States Government (railroad)
USGA United States Golf Association
US gal United States gallon
USGC United States Gold Commission
USGLI United States Government Life Insurance
USGM United States Government Manual
USGOM United States Government Organization Manual
USGPO United States Government Printing Office
USGRDR United States Government Research and Development Report(s)
USGRR United States Government Research Reports
USGRS United States Graves Registration Service
USGS United States Geological Survey
USGSMMS United States Geological Survey and Minerals Management Service
ush usher
Ush Ugandan shilling(s)
USHA United States Handball Association; Utah System of Higher Education
USHCC U.S. Hispanic Chamber of Commerce
USHDA United States Highland Dancing Association
U of Sherb University of Sherbrooke
USHGA United States Hang Gliding Association
USHHFA United States Hous-

ing and Home Finance Agency
USHL United States Hygienic Laboratory
USHMC U.S. Holocaust Memorial Council
USHR United States Highway Research
USHS United States Hospital Ship
USI United States of Indonesia; United States Industries
USIA United States Information Agency
USian United Statesian
USIAS Union Syndicale des Industries Aéeronautiques et Spatiales (French—Aeronautic and Space Industry Union)
USIB United States Intelligence Board
USIBR United States Institute of Behavioral Research
usic undersea instrument chamber
USIC United States Industrial Chemicals; United States Industrial Council; United States Instrument Corporation
USICA United States International Communication Agency
USIF United States Investment Fund
USIH United States Indian Health Service
USILA United States Intercollegiate Lacrosse Association
USI & NS United States Immigration and Naturalization Service
USIOSLCC United States Inter-Oceanic Sea-Level Canal Commission
USIP University of Stockholm Institute of Physics
USIS United States Information Service
USISL United States Information Service Library
USISS United States Institute of Space Studies
USITA United States Independent Telephone Association
USITC United States International Trade Commission
USITT United States Institute for Theater Technology
USIU United States International University
USJ United States Jaycees
USJC United States Job Corps
USJCC United States Junior

Chamber of Commerce
USJF United States Judo Federation
USJPRS United States Joint Publications Research Service
USL Union Steamships Limited; United States Legation; United States Lines; University of Singapore Library; University of Sydney Library
U-slag upperclass slang
USLant United States Atlantic Subarea
USLANTCOM United States Atlantic Command
USLO United States Liaison Office(r); University Students for Law and Order
USLP U.S. Labor Party
USLSA United States League of Savings Associations; United States Livestock Sanitary Association
USLSI United States League of Savings Institutions; U.S. League of Savings Institutions
USLTA United States Lawn Tennis Association
USLU United States Lines (container) Unit
usm (USM) underwater-to-surface missile
USM United States Shoe Machinery; United States Mail (U.S.M.); United States Marines; United States Marshall; United States Mint; University of Southern Mississippi; Unlisted Securities Market
USMA United States Maritime Administration; United States Metric Association; United States Military Academy (West Point); Utah State Medical Association
USMACTHAI United States Military Assistance Command, Thailand
USMACV United States Military Assistance Command, Vietnam
USMB United States Metric Board
USMBPHA United States-Mexico Border Public Health Association (of American and Mexican Public health officials)
USMC United States Marine Corps; United States Maritime Commission; United States Microfilm Corporation (company)
USMCR United States Marine

Corps Reserves
USMD United States Medical Doctor
USMeMilComUN United States Military Members, UN Military Staff Committee
USMH United States Marine Hospital
USMICC United States Military Information Control Committee
USMilComUN United States Delegation, UN Military Staff Committee
USMilLias United States Military Liaison Office
USMILTAG United States Military Technical Advisory Group
USML U.S. Marxist-Leninists (left-wing youth party)
USMM United States Merchant Marine
USMMA United States Merchant Marine Academy
USMMCC United States Merchant Marine Cadet Corps
USMO United States Marshal's Office
USMS United States Maritime Service; United States Marshalls Service
USMSMI United States Military Supply Mission to India
USMSPB United States Merit Systems Protection Board
USMUN United States Mission to the United Nations
usn ultrasonic nebulizer
Usn Ulsan
USN United States Navy
USNA United States Naval Academy; United States Naval Archives
USNAM United States Naval Academy Museum
USNARS United States National Archives and Records Service
USNAS United States Naval Amphibious School
USNB United States National Bank; United States Naval Base
USNC United States National Committee; United States Navigation Company (North German Lloyd—Hamburg-American Line); United States Nuclear Corporation
USNCB United States National Central Bureau (Interpol); United States Naval Construction battalion (Seabees)
USNCC United States Naval Correction Center

USNCCC United States National Council of Churches of Christ
USND United States Navy Department
USNDCS United States National Drug Control Strategy
USNDRC United States Navy Drug Rehabilitation Center
USNEC United States National Earthquake Center, Golden, Colorado
USNEL United States Naval Electronics Laboratory
U.S. News U.S. News and World Report
USNFEC United States National Fruit Export Council
USNG United States National Guard
USNH United States Naval Harbor; United States Naval Hospital; United States North of Hatteras
USNHO United States Navy Hydrographic Office
USNI United States Naval Institute
USNIAAA United States National Institute on Alcohol Abuse and Alcoholism
USNII United States National Indian Institute
USNIS United States Naval Investigative Service
USNJA United States National Jogging Association
USNL United States Navy League
USNLM United States National Library of Medicine
usnm United States National Museum (Smithsonian Institution)
USNMR United States National Military Representative
USNO United States Naval Observatory
USNOO United States Naval Oceanographic Office
USNPC United States Naval Photographic Center
USNPS United States Naval Postgraduate School
USNR United States Naval Reserve
USNRC United States Nuclear Regulatory Commission
USNRDL United States Naval Radiological Defense Laboratory; United States Navy Research and Development Laboratory
USNS United States Naval Ship (Military Sea Transport Service); United States

Nuclear Ship

USNSA United States National Student Association; United States Naval Sailing Association

USNSMC United States Naval Submarine Medical Center

USNTAF United States Navy Training Aids Facility

USNTS United States Naval Torpedo Station

USNUSL United States Navy Underwater Sound Laboratory

USNWD United States Naval War College

USNWR Union Slough National Wildlife Refuge (Iowa); Upper Souris NWR (North Dakota)

USN & WR U.S. *News & World Report*

uso unmanned seismological observatory

USO United Service Organizations; Utah Symphony Orchestra

U-soc upperclass society

USOC United States Olympic Committee

USOE United States Office of Education

USOEO United States Office of Economic Opportunity

USofAF Under Secretary of the Air Force

USOICP United States Oil Import Control Program

USOID United States Oversea Internal Defense (USA)

USOM United States Operations Mission

usp unique selling proposition

USP U.S. Penitentiary (Atlanta, Georgia; Leavenworth, Kansas; Lewisburg, Pennsylvania; Marion, Illinois; McNeil Island, Washington; Terre Haute, Indiana); United States Plywood (company); University of the South Pacific (Fiji)

USP United States Pharmacopeia

USPA United States Philatelic Agency; United States Polo Association

USPACAF United States Pacific Air Forces

US Pat United States Patent

USPB United States Parole Board

USPC Ulster Society for the Preservation of the Countryside; United States Parole Commission; United States

Peace Corps

USPCA United States Police Canine Association

USPCS US Philatelic Classics Society

USPDO United States Property and Disbursing Office(r)

U-speech upperclass speech

USP & F United States Pipe and Foundry (company)

USPFO United States Property and Fiscal Officer

USPG United Society for the Propagation of the Gospel

US Phar United States Pharmacopeia

USPHS United States Public Health Service

USPHSC United States Public Health Service Clinic

USPHSH United States Public Health Service Hospital

USPIS United States Postal Inspection Service

USPLS United States Public Land Surveys

USPO (**U.S.P.O.**) United States Post Office

USPP United States Probation and Parole

USPPS US Possessions Philatelic Society

USPQ United States Patents Quarterly

U.S. Pros United States Prostitutes collective

USPs United States Penitentiaries

USPS United States Postal Service; United States Power Squadron

USPTO U.S. Patent and Trademark Office

USPUN United States People for the United Nations

USPWIC United States Prisoner of War Information Center

usque. usquebaugh (Old Scottish—whisky)

usr underwater search and recovery; unheated serum reagin

USR United States Reserves; United States Rubber

USR United States Supreme Court Reports

USRA United States Racquetball Association; United States Railway Association; United States Revolver Association; Universities Space Research Association

USRB United States Renegotiation Board

USRD Underwater Sound Ref-

erence Division (USN)

USRDA United States Recommended Daily Allowance

USREDCOM United States Readiness Command

USRepMilComUN United States Representative, UN Military Staff Committee

usr/grp user groups for UNIX operating systems

USRL Underwater Sound Reference Laboratory

USRS United States Rocket Society

USRS United States Revised Statutes

usrt universal synchronous receiver/transmitter

USS Under-Secretary of State; Underway Saturdays and Sundays (destroyer crews' definition of USS); Union Switch and Signal; United Scholarship Service; United States Senate; United States Ship (U.S.S.); United States Shoe (company); United States Standard; United States Steel (company)

US & S Union Switch and Signal

U of SS University of the Seven Seas (Chapman College)

USS Union Syndicale Suisse (French—Swiss Federation of Trade Unions); *Union Syndicale Suisse* (French—Swiss Trade Union Syndicate)

USSA United States Salvage Association; United States Ski Association; United States Student Association

USSAF United States Strategic Air Force

USSB United States Savings Bond(s); United States Shipping Board (World War I)

USSBD United States Savings Bonds Division

USSC United States Secret Service; United States Sentencing Commission; United States Space Command, Paterson Air Force Base, Colorado Springs, Colorado; United States Strike Command; United States Supreme Court

USSCC United States Senate Computer Center

USS Co Ulster Steam Ship Company; Union Steam Ship Company (New Zealand)

USSCS United States Soil Conservation Service

USSDP Uniformed Services

Savings Deposit Program
USSEI United States Society of Esperanto Instructors
USSF United States Science Foundation; United States Soccer Federation; United States Steel Foundation; United States Special Forces (Green Berets)
USSFA United States Soccer Football Association
USSFC United States Synthetic Fuel Corporation
USSG United States Standard Gauge
USSIC United States Sex Information Council
USS & LL United States Savings & Loan League
US Soc Fed United States Soccer Federation
USSOUTHCOM United States Southern Command
USSPA United States Student Press Association
USSR Union of Soviet Socialist Republics comprising Russia, Ukraine, Kazakhstan, Byelorussia, Uzbekistan, Georgia, Azerbaijan, Lithuania, Moldavia, Latvia, Kirghizia, Tadzhikistan, Armenia, Turkmenistan, and Estonia
USSRA United States Squash Rackets Association
USSS United States Secret Service; United States Steamship
USSSA United States Social Security Administration
USSSM United States Sinai Support Mission
USSST United States Salt Spray Test(ing)
USSTA United States Sail Training Association
USSTRICOM United States Strike Command
USSTS United States Student Travel Service
ust. ustus (Latin—burnt)
UST undersea technology; United States Treaties; University of Santo Tomás (Manila)
UST UnderSea Technology; The Magazine of Oceanography, Marine Sciences, and Underwater Defense
U of St A University of St Andrews
USTA United States Tennis Association; United States Trademark Association; United States Trotting Association

USTC United States Tariff Commission; United States Tax Court; United States Testing Company
USTCRDWWA United Slate, Tile, and Composition Roofers, Damp, and Waterproof Workers Association
USTD United States Transportation Department
USTDC United States Taiwan Defense Command
USTEMC United States Territorial Expansion Memorial Commission
USTES United States Training and Employment Service
USTF United States Tuna Foundation
USTFF United States Track and Field Federation
USTIS Ubiquitous Scientific and Technical Information System
USTMA United States Trade Mark Association
USTOA United States Tour Operators Association
ustol ultra short takeoff and landing
USTR United States Trade Representative
USTS United States Travel Service
USTTA United States Table Tennis Association; United States Travel and Tourism Administration
usu usual; usually
USU Uniformed Services University; Utah State University
usuf usufruct(uary)
USUHS Uniformed Services University of the Health Sciences
USUN United States Mission to the United Nations
usurp. usurpandus (Latin—to be used)
USV United States Volunteers
USVA United States Veterans Administration; United States Volleyball Association
USVB United States Veterans Bureau (former name of the Veterans Administration)
USVH United States Veterans Hospital
USVI United States Virgin Islands (St Croix, St John, St Thomas)
USVIDT United States Virgin Islands Division of Tourism
USVMS Urine Sample Volume Measurement System
usw ultra short wave; underwa-

ter submarine warfare
usw und so weiter (German—and so forth)
USW United Show Workers
USWA United Steel Workers of America
USWAC United States Women's Army Corps
USWACC United States Women's Army Corps Center
USWACS United States Women's Army Corps School
USWB United States Weather Bureau
USWD Undersurface Warfare Division
USWGA United States Wholesale Grocers' Association
USWI United States West Indies (Virgin Islands—St Thomas, St John, St Croix, and smaller islands)
USWLS United States Wild Life Service
USWP Ultra-Short-Wave Propagation Panel
USWV United Spanish War Veterans
USX United States Steel and Marathon Oil, Texas Oil and Gas, US Diversified Group
USY United Synagogue Youth
USYRU United States Yacht Racing Union
usysf United States Youth Symphony Federation
ut universal trainer; urinary tract; user test; utilitarian; utility
u/t untrained
UT Union Terminal (railroad); United Territories; United Territory; United Utilities (stock exchange symbol); Universal Time (Greenwich Mean Time); Universal Tubes; Utah; Utilities Man
U.T. U Thant
U of T University of Tampa; University of Tasmania; University of Tennessee; University of Texas; University of Toledo; University of Toronto; University of Tulsa
U of T (Austin) University of Texas in Austin
U of T (El Paso) University of Texas in El Paso (also UTEP)
U-T Union-Tribune (newspapers)
uta upper terminal area
UTA Ulster Transport Authority; *Union des Transports Aériens* (French—Air Trans-

port Union); United Tribes of Alaska; United Typothetae of America; University of Texas—Austin; Urban Transportation Administration
utacv (UTACV) urban-tracked air-cushion vehicle
UTAD Utah Army Depot
Utagawa Utagawa Toyokuni
Utah St Hist Soc Utah State Historical Society
Utah St U Pr Utah State University Press
Utamaro Kitagawa Utamaro
utarb utarb eidet (Norwegian—prepared)
UT/AT Underway Trial/Acceptance Trial (USN)
UTB United Tariff Bureau; United Technocratic Board; Universal Technological Bureau
utc unit type code; unit type coding
utc (UTC) universal time coordinated
UTC United Tank Car; United Technology Center (United Aircraft); United Transformer Corporation; Universe Tankships Corporation (National Bulk Carriers); University Training Corps
UT-C University of Tennessee—Chattanooga
utclk universal transmitter clock
utd united
UTD University of Texas—Dallas
UTDA Ulster Tourist Development Association
UTDC Urban Transportation Development Corporation
ut dict. ut dictum (Latin—as ordered)
utdne. mor. sol. utendus more solito (Latin—use in the usual way)
Utd Tech United Technology
ute Australian utility truck
UTE underwater tracking equipment
uten utensil(s)
utend. utendus (Latin—to be used)
U of Tenn Pr University of Tennessee Press
UTEP University of Texas—El Paso
U of Tex Pr University of Texas Press
UTF Underground Test Facility
utg utgave (Norwegian—edition)

uti urinary tract infection
uti (UTI) urinary tract infection
UTI Union Title Insurance; Unit Trust of India
UTIAS University of Toronto Institute for Aerospace Studies
util utility; utilization
utilit utilitarian(ism); utilities
ut inf. ut infra (Latin—as below)
utl universal transpor(er) loader; user trailer label
UTL University of Tampa Library; University of Tennessee Library; University of Texas Library; University of Tokyo Library; University of Toronto Library; University of Tulsa Library
UTLAS University of Toronto Library Automation System
utm universal testing machine; universal test(ing) module; universal transverse mercator
UTN University of Tennessee
UTO United Thank Offering; United Town Organisation
U of Tok University of Tokyo
utop utopia (from the Greek *utopia*—no place)—ideal place
U of Tor Pr University of Toronto Press
UTP Unified Test Plan; University of Texas Press; University of Toronto Press
UTQG Uniform Tire Quality Grading
utr (UTR) university training reactor
Utr Utrecht (Dutch province)
UTR United Tire and Rubber
UTRC United Technologies Research Center
uts ultimate tensile strength; unit training standard
UTS Underwater Telephone System; Unified Transfer System (Russian-to-English translation); Uniform Thread Standard; Union Theological Seminary; Universal Time-Sharing System; University of Toronto Schools
UTSSM University of Texas-Southwestern School of Medicine
ut sup. ut supra (Latin—as above)
UTTAS Utility Tactical Transport Aircraft System
uttc universal tape-to-tape converter
UTTR Utah Test and Training

Range
UTU United Transportation Union
U-tube U-shaped tube
U-turn U-shaped turn
utv (UTV) underwater television
UTV Universal Test Vehicle
utw under the wing
UTWA United Textile Workers of America
UTX 4-engine jet utility transport; University of Texas
uty utilities
uu (UU) urine urobilinogen
u U unter Umständen (German—circumstances permitting)
UU Ulster Unionist; Union University; Universal Underclass; University of Utah
U-U Unitarian-Universalist
U & U Underwood and Underwood
U of U University of Uppsala; University of Utah
UU Uppsala Universitetsbiblioteket (Swedish—Uppsala University Library); *ustedes* (Spanish—you, pl.)
UUA Unitarian-Universalist Association; Univac Users Association
UUCM University of Utah College of Medicine
uue use until exhausted
uuf micromicrofarad
UUI United Utilities Incorporated
UUIP Uppsala University Institute of Physics
uum (UUM) underwater-to-underwater missile
UUP Ulster Unionist Party
uu's universal undevelopers
UUSC Unitarian Universalist Service Committee
uut unit under test
U of Utah Pr University of Utah Press
UUUC United Ulster Unionist Coalition
uuv unter üblichen vorbehalt (German—errors and omissions excepted)
UUWF Unitarian Universalist Women's Federation
uv ultraviolet; umbilical vein; under voltage; urinary volume
u-v ultraviolet
UV Ulster Vanguard; Unadilla Valley (railroad); Upper Volta
U of V University of Vermont; University of Victoria; Uni-

versity of Virginia
UV Una Via (Spanish—One Way)
UVA University of Virginia
U van A Universiteit van Amsterdam
U van A Universiteit van Amsterdam (Dutch—Amsterdam University, University in Amsterdam)
uvas ultraviolet astronomical satellite (UVAS)
uvaser ultraviolet amplification by stimulated emission of radiation
UVC United Veterans Council
u-v camera ultraviolet evidence camera
UVCM University of Vermont College of Medicine
UVCT University of Vermont College of Technology
uvd undervoltage device
UVDC Urban Vehicle Design Competition
UVE Unión Velocipédica Española (Spanish—Spanish Bicycle Union)
UVF Ulster Volunteer Force
UVH University of Virginia Hospital
UVI Unione Velocipedistica Italiana (Italian—Italian Cycling Association)
uviol ultraviolet
uvl ultraviolet light
UVL University of Virginia Library
UVM University of Vermont
U-vocab upperclass vocabulary
uvr ultraviolet radiation
uvs ultraviolet spectrometer; universal versaplot software
UVSA Unie van Suid Afrika (Union of South Africa)
uvsc ultraviolet solar constant
UVSM University of Virginia School of Medicine
UVT University of Vermont
uw unconventional warfare; underwater; underwing; underwriter; unwound
u/w underwater; underway; underwear; underwriter; used with
U/w Underwriter

UW University of Waikato; Uppity Women
U of W University of Wales; University of Warwick; University of Washington; University of Waterloo; University of Wichita; University of Windsor; University of Winnipeg; University of Wisconsin; University of Witwatersrand; University of Wollongong; University of Wyoming
UW Universität Wien (German—University of Vienna)—see *Ubib Wien*
UWA United Way of America; United World Atheists; University of Washington; University of Western Australia
U of Wash Pr University of Washington Press
UWC University of the Western Cape
UWCE Underwater Weapons and Countermeasures Establishment
U-wear underwear
U-weld U-shaped weld
UWF United World Federalists
UWFL University of Washington Fisheries Laboratory
UWGB University of Wisconsin at Green Bay
UWH University of Washington Hospital
UWI University of the West Indies (Jamaica); University of Wisconsin
UWIL University of the West Indies Library (Kingston, Jamaica)
U of Wis Pr University of Wisconsin Press
UWIST University of Wales Institute of Science and Technology
UWIUB undrinkable wine in unusable bottles
UWL University of Wales Library; University of Washington Library; University of Wichita Library; University of Wisconsin Library; University of Witwatersrand Library; University of Wyo-

ming Library
UWM United World Mission; University of Wisconsin at Milwaukee
UWMI University of Wisconsin Management Institute
UWO University of Western Ontario
uwoa unclassified without attachments
UWP University of Wales Press; University of Washington Press; Up With People
UWSM University of Washington School of Medicine
uwtr underwater
UWTU Underwater Training Unit
UWUA Utility Workers Union of America
UWV University of West Virginia
UWW University Without Walls (Antioch College)
UWY University of Wyoming
ux. uxor (Latin—wife)
uxb (UXB) unexploded bomb
uxgb unexploded gas bomb
uxib unexploded incendiary bomb
'Uxley (Cockney contraction—Huxley)
uxor uxoricide
UY Universal Youth
U of Y University of York
UYA University Year for Action
UYL United Yugoslav Lines
Uz Uzbek; Uzbekistan; Uzbekistani
Uz Uhrzuender (German—clockwork fuze)
UZ University of Zululand
UZ Universität Zürich
Uzbek SSR Uzbek Soviet Socialist Republic (Uzbekistan)
Uzi Uziel Gal
UZM Universitet Zoologiske Museum (Copenhagen)
UZRA United Zionist Revisionists of America
U zu B Universität zu Berlin
U zu G Universität zu Göttingen
Uzz Uzziah

V

v vacuum; vacuum tube; vagabond; vagrant; Valium; value; valve; van; vapor; variable; variation; vector; vein; velocity; vent; ventilator; ventral; verb; verbal; verse; version; vertex; vertical; very; vice; vincinal; violent (motion picture); violet; violin; virus; viscosity; vise; visibility; vision; visual acuity; voice; volt; voltage; voltmeter; volume; volunteer; vowel; vox

v *van* (Dutch—of); *verso* (Latin—back of page or sheet; left-hand page); *versus* (Latin—against); vibrational quantum number; *voltare* (Italian—turn, turn the page); *von* (German—of; from; used in titles)

v/ *vostra* (Italian—your)

V coefficient of vibration (symbol); five-dollar bill; Lockheed (symbol); potential (symbol); relative wind velocity (symbol); stalling velocity (symbol); Standard Fruit & Steamship Company (Vaccaro Line); vanadium; Venerable; Ventzke; Venus; Verdet constant; Vicar; Vice; Victor—code for letter V; Victory; Village; Village District; volt; Volta; volume (symbol)

V airspeed, forward velocity (symbol); speed (symbol); vacuum tube (symbol); *varm* (Dano-Norwegian or Swedish—hot); *väst* (Swedish—west); *Venstre* (Danish or Norwegian—Left)—Liberal Party; *vertrek* (Dutch—departure); *vest* (Dano-Norwegian—west); *Via* (Italian—highway road, way); *Villa* (Spanish—village); *violaceus* (*Latin*—violet color); *viridis* (Latin—green); *vrouw* (Dutch —woman)

v-1 vernier engine 1

V1 Voyager 1 satellite with close Titan flyby

V₁ decision speed (go-no-go) for aircraft to continue take-off run or abort flight; valve-current voltage

V¹ *violino primo* (Italian—first violin)

v-1 p vernier engine 1 pitch

V-1, V-2 rockets launched by the Germans in World War II

v-1 y vernier engine 1 yaw

V2 Voyager 2 with Uranus option

V₂ aircraft takeoff speed or position where nose is lifted so plane becomes airborne

V² *violino secondo* (Italian—second violin)

V-4 four-cylinder engine with two cylinders in each side of V-shaped engine block

V-6 six-cylinder engine with three cylinders in each side of V-shaped engine block

V-8 eight-cylinder engine with four cylinders in each side of V-shaped engine block

V-10 Viscount 10 jet airplane

v 26 d M *von 26 dieses Monats* (German—of the 26th instant; of the 26th of this month)

va variable; variance; verb active; verbal adjective; viola; voltampere(s)

v-a volt-ampere(s)

v/a verbal auxiliary; voucher attached

v/a (VSA) vulnerable area

v.a. *vixit—annas* (Latin—he lived—years)

Va Virginia; Virginian

Va *Vila* (Portuguese—Villa; Village); *Villa* (Italian or Spanish—Villa, Village)

Vᵃ *Vila* (Portuguese—small town, villa); *Viuda* (Spanish—widow)

VA Veterans Administration (United States); Veterans' Affairs (Canada); Virginia; Voice of America; voltaic alternative (symbol); Volunteers of America

V-A Vickers-Armstrong Limited

V.A. Order of Victoria and Albert; Vicar Apostolic

V & A Victoria and Albert (Museum)

V of A Volunteers of America

V & A *Venus and Adonis*

VAA Vaccination Assistance Act; Vietnamese-American Association

VAACR Vietnamese Association for Asian Cultural Relations

vab voice answer back

VAb Van Allen belt (zone of high-intensity radiation surrounding the earth at altitudes of about 500 miles)

VAB Vandenberg Air Force Base; Vertical Assembly Building (world's largest all-steel structure of its type; used for assembling missiles and space exploration vehicles on Merritt Island at Cape Kennedy, Florida)

VAbd Van Allen belt dosimeter
Va Bk Virginia Book Company
VABM vertical angle bench mark
vac vacant; vacate; vacation; vacuum; volts alternating current (*volts AC* preferable)
VAC Victor Analog Computer; Video Amplifier Chain; Voluntary Action Center; Volunteer Advisor Corps
VACAB Veterans Administration Contract Appeals Board
vacc vaccination; vaccine; value-added common carrier
Vaccaro Standard Fruit & Steamship Company
vacci vaccinate; vaccination; vaccine
vac-dist vacuum-distilled
Vaclav (Czech—Wenceslas)
vac pmp vacuum pump
VACRP Victorian Association for the Care and Resettlement of Prisoners
vacs vacuum cleaners
v/act. verb active
vad variable abbreviated dialing; velocity azimuth display; voltmeter analog-to-digital converter
VAd Veterans Administration
VAD Voluntary Aid Detachment
vada versatile automatic data exchange
V Adm Vice Admiral
vad. mec. *vade mecum* (Latin—go with me)—companion volume; handbook; manual; ready reference for readers and reference librarians
vae vinyl-acetate ethylene
VAEA Virginia Adult Education Association
VAF Vendor Approval Form; Vincent Astor Foundation
VAFB Vandenberg Air Force Base
vag vagabond; vagina; vaginal; vaginitis; vagrant; vagrancy
vag charge vagrancy charge
vag hist vaginal hysterectomy
vagonzak *vagon zaklyuchennykh* (Russian—railroad prisoner car)
vags vagabonds; vagrants
VAH Veterans Administration Hospital
VAHS Victorian Aboriginal Health Service
vai video-assisted instruction; vorticity area index
va & i verb active and intransitive

VAI Video Artists International
VAIS Virginia Association of Independent Schools
vakt visual-auditory-kinesthetic and tactual (imagery applied to teaching reading)
val valance; valence; valenciennes (lace); valentine; valise; valley; valuation; value; valued; valve; valvular
val (VAL) valine (amino acid)
Val Valencia; Valentina; Valentine; Valentino; Valerie; Valerius
VAL Vehicle Authorization List; Veterans Administration Library
VALA Viewers and Listeners Association
VALB Veterans of the Abraham Lincoln brigade
valc visual approach and landing chart
Vald Valdivia
Val Fl Gaius Valerius Flaccus (Roman epic poet)
valid. validate; validation
Valka Valentin
Vall Valladolid
VALNET Veterans Administration Library Network
Valpo Valparaiso
valsas variable-length word symbolic assembly system
valt vtol approach-and-landing technic
VALUE Visible Achievement Liberates Unemployment (Air Force program for disadvantaged youth)
val vu valley view
vam volt ammeter
Vam Vogel's approximation method
VAMCO Village and Marketing Corporation
vamp vampire; vampirism; volume, area, mass properties
VAMP Voluntary Association of Master Pumpers (mid-19th-century English firefighters)
vamps vampires; seductive enemy agents, also called swallows
vam's vision-aid magnifiers
van (VAN) value-added network
van. caravan; value-added network; vanguard; vanilla; vanillin
Van Vanessa
Van (VAN) Vancouver, British Columbia

VAN Value-Added Network
VAN *Vereniging van Archivarissen in Nederland* (Dutch—Association of Archivists in the Netherlands)
Vanc Vancouver
Vancoram Vanadium Corporation of America
Vandy Vanessa
Vanechka (Russian nickname —Ivan)
Vang Vickers-Armstrong Vanguard (aircraft)
Vang Esp *Vanguardia Española* (Barcelona's Spanish Vanguard)
Vanguard Vanguard Press
Vanier Centre Vanier Centre for Women (criminals) at Brampton, Ontario
Vanka (Russian diminutive— Ivan)
Van-Lax Vancouver—Los Angeles
Van-Mia Vancouver—Miami
Van Op Vancouver Opera
van. pub. vanity publisher; vanity publishing
VANS Value-Added Network Service(s)
Van-San Vancouver—San Diego
Van-Sea Vancouver—Seattle
Van-Sfo Vancouver—San Francisco
Van Sun Vancouver Sun
Van Sym Vancouver Symphony Orchestra
Van-Tor Vancouver—Toronto
Vanu Vanuatu (island republic in the southwestern Pacific)
VAP Victims Assistance Program; Victims Assistance Project
vapi visual approach path indicator
vapor. vaporization
vap prf vaporproof
var variable; variant; variation; variety; variometer; visual-aural range; volt-ampere reactive
var (VAR) value-added reseller; vertical air rocket
var *variazione* (Italian—variation)
Var Varna
VAR Volunteer Air Reserve
varactor variable capacitor
varad varying radiation
var con variable condenser
var dial. various dialects
var ed & trans various editions and translations
VARES Vega-Aircraft Radar-Enhanced System

vari VariType(r)
VARIG Empresa de Viação Aérea Rio Grandense (airline in southern Brazil)
varistors variable resistors
varizistor variable resistor
var. lect. *varia lectio* (Latin—variant reading)
varn varnish
VARP Veterans Administration Procurement Regulations
varr variable-range reflector
Varr Marcus Terentius Varro (Roman writer on agriculture and natural history)
vars varieties
Vars Varsavia (Italian or Latin—Warsaw); *Varsovia* (Spanish—Warsaw); *Varsóvia* (Portuguese—Warsaw)
VARS Vertical and Azimuth Reference System
vas vasectomy
vas (Latin prefix—vessel)—vasoconstriction
Vas Vasteras
VAs Voluntary Aids
VAS Virginia Academy of Science; Vocational Advisory Service
VAS Vedette Anti-Sommergibile (Italian—Anti-Submarine Sentry)—naval craft; *Vereniging van Accountancy Studenten* (Dutch—Society of Accountancy Students)
VASA Virginia Association of School Administrators
vas bund vascular bundle
vasc vascular
VASC Verbal Auditory Screen for Children
VASCA (electronic) Valve and Semi-Conductor (manufacturers') Association
vascar visual average-speed computer recorder
VASCO Vanadium-Alloys Steel Company
VASEC vasectomy
vasi visual approach slope indicator
vasim voltage and synchro-interface module
VASP Viação São Paulo (São Paulo airline)
VASSP Virginia Association of Secondary School Principals
VASSS Van Allen Symplified Scoring System
vast. vibration and static analysis
Västtyskland (Swedish—Germany)
vas vit. vas vitrium (Latin—glass vessel)
vat. value-added taxes (VAT); ventricular activation time; vinyl asbestos tile; vinyl asbestos tiling
Vat Vatican
VAT Value-Added Tax; Vertical Assembly Tower; Veterinary Aptitude Test; Visual Apperception Test
vate versatile automatic test equipment
VATI Vermont Agricultural and Technical Institute
Vatic Vatican
Vatican Bank Institute for Religious Works, Vatican City, which failed in 1982 when Archbishop Paul C Marcinkus and two other officials made off with its funds
Vat Lib Vatican Library (Rome)
VATLS Visual Airborne Target Location System
vatpayer value-added taxpayer
VATS Vertical-lift Airfield for Tactical Support; Video-Augmented Tracking System
Vat Sta Vatican State
VATTR Value-Added-Tax Tribunal(s)
vaud vaudeville
v aux verb auxiliary
vav variable air volume
vavbd vavband (Swedish—clothing)
v/a v/e value-analyst value-engineer
vavp variable-angle variable-pitch
vax virtual address extension
vb valence band; verb; verbal; vertical bomb (VB); vibration
v/b vehicle-borne
VB Navy bomber (2-letter naval symbol); Vero Beach; very bad; Virginia Beach; Vulgate Bible
vba verbal adjective
VBA Veterans Benevolent Association
V-band 46,000–56,000 mc
vbc ventrobasal complex
VBC Vancouver British Columbia
VBCO Vector Biology and Control Office (California)
vbcr vanished black community remnant
VBEC Venezuelan Basic Economy Corporation
V-belt V-shaped belt (cross-section of belt is V-shaped)
VBFNPVGFPMTF Véndemai-

re, Brumaire, Frimaire, Nivôse, Pluviôse, Ventôse, Germinal, Floréal, Prairial, Messidor, Thermidor, Fructidor (as abbreviated on the French Revolutionary Calendar—*see Vend, Brum, Frim, Niv, Pluv, Vent, Germ, Flor, Prair, Mess, Therm, Fruc*)
VBI Venetian Blind Institute
vbl verbal
V-block V-shaped block
VBMA Vacuum Bag Manufacturers Association
V-bomb German long-range missile-type bomb used during World War II; designated as V-1 and V-2
vbos veronal-buffered oxalated saline
V-bottom V-shaped bottom
VBP Vietnam Boat People
vbr (VBR) ventricular-brain ratio
V B R Virginia Blue Ridge (highway)
VBRA Vehicle Builders' and Repairers' Association
VBS Vedanthangal Bird Sanctuary (India); Vocabulary Building System
vc valuation clause; venereal case; violoncello; visual communication
vc (VC) variable cost; vital capacity
vc vuelta de correo (Spanish—by return mail)
v/c vuelta de correo (Spanish—return mail)
vC voor Christus (Dutch—Before Christ)
Vc Vietcong
VC acuity of color vision (symbol); Vassar College; Vatican City; Vehicle Code; Vennard College; Ventura College; Vermont College; Veterinary Corps; Vice Consul; Victoria College; Victoria Cross; Viterbo College; Volusia College
VC Vehicle Code
VC-10 British BAC long-range transport aircraft
VC-137 USAF designation of the Boeing 707
vca voltage-controlled amplifier
VCA Virginia Correctional Association; Volunteer Civic Association
VCAR Vendor Corrective Action Request
VCAS Vice-Chief of Air Staff

VCB Victim Compensation Board

vcc vasoconstrictor center; video compact cassette; vocational career concept

Vcc supply voltage

VCC Value Control Coordinator; Vancouver Community College; Variable-Cycle Control; Visual Communications Congress

vc card index (or reader) visual coincidence index (or reader)

vcce variable contrast/constant exposure

VCCL Victoria Council for Civil Liberties

vccs voltage-controlled current source

vcd variable-capacitance diode

v-c d voluntary-closing device

vce (VCE) variable-cycle engine

Vce Venice

VCE Venice, Italy (airport)

vcf vaginal contraceptive film; voltage-controlled filter

vcg vectorcardiogram; vertical line through center of gravity

VCG Vice-Consul General

vch vehicle; vinyl cyclohexane (VCH)

VCH Victoria County History

vchp variable-conductance heat pipe

v Chr vor Christis (German—before Christ)

vci visual communication instructor; volatile corrosion inhibitor

VCI Variety Clubs International; Vision Conservation Institute

VCIC Vermont Crime Information Center

VCIGS Vice-Chief of the Imperial General Staff

VCIP Veterans Cost-of-Instruction Program

VCK Verenigo Cargodoorskantoor

vcl vertical center line; visual comfort light(ing)

VCL Vancouver Public Library

vcllo violincello

VCLU Virginia Civil Liberties Union

vcm vacuum; vinyl chloride monomer

VCN Vendor Contact Notice

VCNS Vice-Chief of Naval Staff

vcnty vicinity

vco voltage-controlled oscilla-

tor

vcod vertical-carrier onboard delivery

vcoi veterans cost of instruction

v coul volt coulomb

vcp vitrified clay pipe

VCP Vendor Change Proposal; São Paulo, Brazil (Viracopas Airport)

vcr variable compression ratio

vcr (VCR) videocasette recorder

Vcr Vancouver

VCR Victor Comptometer (stock exchange symbol)

vcr's video cassette recorder owners

vcr('s) video cassette recorder(s)

vcs vasoconstrictor substances; voices

vc's viejos cristianos (Spanish—old Christians)—Spaniards who believe they are without Jewish or Moorish blood

VCs Viet Congs; Vigilance Committeemen; Vigilant Committeemen; Vigilante Committeemen

VCS Vernier Control System; Veterans Canteen Service; Vice Chief of Staff; Video Cassette System

V & C S Virginia & Carolina Southern (railroad)

vcsr voltage-controlled shift register

vct vinyl-composition tile

Vct Victoria

vctv vocative

vcty vicinity

VCU Virginia Commonwealth University

vcxo voltage-controlled crystal oscillator

V Cz Vera Cruz

vd vapor density; various dates; venereal disease (VD); videodisc; void

v/d vandyke reproduction

Vd vanadium

Vd usted (Spanish—you)

VD Village District, Isle of Man

V.D. Volunteer Officer's Decoration

vda venereal disease awareness; video distribution amplifier; visual discriminatory acuity

Vda Venda (Spanish abbreviation) Bantu homeland, South Africa; **Viuda** (Spanish—widow)

VDA Vermont Department of Agriculture

VDA Verband der Automobilindustrie (German—Automobile Industry Association); **Volksbund für das Deutschtum im Ausland** (German—League for Germans Abroad)

V-day day of victory

vdB velocity decibel

VDB Venereal Disease Branch (US Public Health Service); **Verband Deutscher Biologen** (German—Association of German Biologists)

VDBC Vertol Division, Boeing Company (helicopter design and manufacturing)

vdc volts direct current (*volts DC* preferable)

vdc (VDC) vinylidene chloride

VDC Venereal Disease Clinic; Virginia Department of Corrections

vdcm (VDCM) vinylidene chloride monomer

VDE Verband Deutscher Elektrotechniker (German—Association of German Electrical Engineers)

v def verb defective

VDEH Verein Deutscher Eisenhüttenleute (German—German Foundry Society)

VDEL Venereal Disease Experimental Laboratory

vdem vasodepressor material

v dep verb deponent

V De S Vittorio De Sica

VdF Vigili del Fuoco (Italian—Fire Brigade)

vdfg variable diode function

vdg vertical display generator

vd-g venereal disease—gonorrhea

vdh very-deep hold

vdh (VDH) valvular disease of the heart

vdi vegetation draught index; vehicle deformation index; venereal disease inhibition; video display input

VDI Verein Deutscher Ingenieure (German—Association of German Engineers)

V-dies V-shaped dies

vdif very difficult

V di R Virtuosi di Roma

VdK Verband der Kriegsbeschadigten (German—League of War Invalids)

Vdkhr Vodokhranilishche (Russian—reservoir)

vdl ventilation deadlight

VDL Van Dieman's Land

(Tasmania)
vdm vector-drawn map
vdm (VDM) vasodepressor material
Vdm Veendam
VDMA Verein Deutscher Maschinenbau Anstalten (German—Mechanical Engineering Association)
vdm('s) video disc machine(s)
VDN Varudeklarationsnamnden (Swedish—Institute for Informative Labelling); *Vin Doux Naturel* (French—fortified wine, natural sweet wine)
vdo video
vdp vehicle deadlined for parts; vertical data processing; videodisc player
vdr variable-diameter rotor
VDRL Venereal Disease Research Laboratories
VDRS Verdun Depression Rating Scale
VDRT Venereal Disease Reference Test
vds variable depth sonar
vd-s venereal disease—syphilis
Vds ustedes (Spanish—you, plural)
VDSCRC Very Dirty and Small Coal Railway Company (created by Dickens for service in *The Uncommercial Traveller*)
VDSI Verein Deutscher Sicherheits Ingenieure (German—Association of Safety Engineers)
vdt variable density (wind) tunnel; video data terminal
vdt (VDT) video display terminal
VDT Visual Display Terminal; Visual Distortion Test
vdt's video display terminals
vdu visual display unit
ve vaginal examination; varicose eczema; vernier engine; very excellent
've have
ve veuve (French—widow)
Ve Venezuela; Venezuelan
VE Value Engineer(ing); *Vasileion tis Ellados* (Greek—Kingdom of Hellas—Greece)
V-E Verzhbolovo-Eydtkuhnen (Russo-German railway frontier for passengers and freight changing from wide gauge to standard European gauge rolling stock and tracks)
ve/a value engineering/analysis (program)

VEA Valve Engineering Association; Vermont Education Association; Veterans Education Administration; Virginia Education Association; Vocational Education Act
VEA-H Vocational Education Act—Handicapped
vealsan veal sandwich
vealwich veal sandwich
VEAP Veterans' Educational Assistance Program
veb variable elevation beam
VEB Volks Eigener Betriebe (German—Peoples-Owned Companies)
vec vector
veco vernier engine cutoff
vecp visually evoked cortical potential
VECP Value Engineering Change Proposal
VECR Vendor's Engineering Change Request
VECS Vocation Education Curriculum Specialists
vecto vectograph; vectographic; vectographical
ved vedova (Italian—widow)
Ved Vedic
VED Vickers Electric Division
VEDA Victorian Eastern Development Association
V-E Day Victory in Europe Day—May 8, 1945
VEDC Vitreous Enamel Development Council
vedr vedrorende (Danish—concerning)
VEDS Vocational Education Data System
Vee Venezuelan equine encephalomyelitis
vee dee venereal disease; visiting dignitary
VEENAF (South) Vietnamese Air Force
VEEP Voluntary Ethnic Enrollment Program
veg vegetable; vegetarian; vegetarianism; vegetation
vegan (extreme) vegetarian; vegetarian(ism)
vegans vegetarians
vegan(s) strict vegetarian(s)
Vegas Las Vegas
veggies vegetables
Veg Soc Vegetarian Society
vegtan vegetable tanning
veh vehicle; vehicular
VEH Vocational Education for the Handicapped
vehic. vehiculum (Latin—vehicle)
vehic manslgtr vehicular manslaughter

veh pt(s) vehicle part(s)
VEIN Vocational Education Information Network
VEIS Vocational Education Information System
vel vellum; velocity; velvet
Vel Vela (constellation)
Vel Velikiy (Russian—large)
Velázquez Diego Rodriguez de Silva Velázquez (1599–1660)
Vell Gaius Velleius Paterculus (Roman historian)
velo velodrome
veloc velocity
vem vasoexciter material
ven veneer; veneering; venerable; venereal; venery; venetian; venetian blind(s); venison; venom; venomous; ventral; ventricle
ven vendredi (French—Friday); *venerdi* (Italian—Friday); *venesianisch* (German—Venetian)
Ven Venetian; Venice; Venus
vend vending; vending machine; vendor(s)
Vend Vendémaire (French—Vintage Month)—beginning September 22nd—first month of the French Revolutionary Calendar
vend. mach vending machine
Venez Venezuela; Venezuelan
Venezuela Republic of Venezuela (oil-producing Spanish-speaking South American nation), *República de Venezuela*
Venezuelan Islands Aves, Blanquilla, Cubagua, Hermanos, Margarita, Monjes, Orchila, Testigos, Tortuga
V-engine V-shaped engine
VENISS Visual Education National Information Service for Schools
vent. ventilate; ventilating; ventilation; ventilator; venting; ventral; ventricle; venture
Vent Ventôse (French—Windy Month)—beginning February 19th—sixth month of the French Revolutionary Calendar
vent.fib. ventricular fibrillation
ventric ventricular
vents. ventilators
vent. tachy ventricular tachycardia
Venus (Latin—Aphrodite)—goddess of beauty and love
vep visual-evoked potential
VEP Veterans Education Project; Voter Education

Project
VEPCO Virginia Electric and Power Company
VEPM Value Engineering Program Manager
ver verification; verify; verse(s); versine; vertex (Ver)
Ver Vera Cruz
Ver Verband; Verein (German—association)
Vera Veratchke; Veronica
VERA Vision Electronic Recording Apparatus (videotape)
verand verandert (German—revised)
verb verbesserte (Dutch or German—improved)
verb. et lit. verbatim et literatim (Latin—exact copy; word for word)
verb. sap. verbum satis sapienti (Latin—a word to the wise is sufficient)
Verdi Giuseppe Verdi; Victor Emmanuel Re d'Italia
verdigris copper acetate
Verds Cape Verde Islands
verdt verdict
Verf Verfasser (German—author)
Verg Publius Vergilius Maro (Roman poet often referred to as Virgil)
Vergl Vergleische (German—compare)
Verh Verhandlungen (German—proceedings)
VERIC Vocational Educational Research Information Center
verisim verisimilar; verisimilitude; verisimilitudinous
Veritas Det norske Veritas (The Norwegian Bureau of Shipping)
Verkh Verkhniy (Russian—upper)
verk v verkorting van (Dutch—abbreviation, abridgement, shortening)
Verl Verlag (German—publisher)
Verlagshdlg Verlagshandlung (German—book-publishing house)
verlort very-long-range tracking (radar)
verm vermiculite
verm (Latin prefix—worm)—vermiform; *vermehrte* (German—enlarged)
Verm Vermont
vermilion mercury sulfide
vern vernacular

Vern Vernay; Verne; Verney; Vernon
Vern Vernon's Law Reports
vernac vernacular(ism); vernacularly
vers versed sine; verses; versification; versine (versed sine)
versine versed sine
verso reverso (left-hand page; reverse side of a page)—opposite of recto
Ver St Vereinigte Staaten (German—United States)
vert vertebra; vertebrate; vertical; vertigo
verticam vertical camera
ves vertical electric soundings; vessel
ves. vesica (Latin—bladder)
VES Veterans Employment Service; Voluntary Euthanasia Society
VESC Vehicle Equipment Safety Commission
vesca(s) vessel(s) and cargo
vesda (VESDA) very early smoke detection apparatus
VESIAC Vela Seismic Information Analysis Center
vesic. vesicula (Latin—blister)
VESO Value Engineering Services Office
vesp. vesper (Latin—evening)
vesper. vehicles, equipment, and spares provision—economics and repairs
VESPER Voluntary Enterprises and Services and Part-time Employment for the Retired
vest vestibule
VEST Volunteer Engineers, Scientists, and Technicians (organization)
Vesters Vester Islands
ves. ur. vesica urinaria (Latin—urinary bladder)
vet veteran; veterinarian; veterinary
v. et. vide etiam (Latin—also see)
Vet Veterinary Medicine
VET Verbal Test
Vet Admin Veterans' Administration
Veterans Veterans Day (November 11)—commemorating armistice to end World War I on the 11th hour of the 11th day of the 11th month of 1918, originally called Armistice Day
Vet M. B. Bachelor of Veterinary Medicine
vet med veterinary medicine
VETMIS Vehicle Technical

Management Information System (USA)
vet reg veterans' regulations
vet rep veteran's representative
vets veterans; veterinaries
vet sci veterinary science
Vets Info Veterans Information Service
Vet Surg Veterinary Surgeon
vett vetted; vetting
'vette corvette
vetted (English contraction—veterinary inspected)—inspected and investigated
vev voice-excited vocoder
VEV Vietnam Era Veterans
V Ex^a Vossa Excelência (Portuguese—Your Excellency)
vexdex vexation index
vexil vexillogical; vexillologist; vexillology
vf vertical file; very fair; very fine; video frequency; visual field; voice frequency, vulcanized fiber
Vf Verfasser (German—author)
VF fixed-wing fighter airplane (2-letter naval symbol); Valley Forge
V.F. Vicar Forane
VF Vigili del Fuoco (Italian—Fire Brigade)
V f A Voice for America (Alistair Cooke)
VFA Video Free America; Voluntary Foreign Aid
V-FA Vietnamese-France Association
V-factor verbal (comprehension) factor; violence factor
v-f band voice-frequency band
vfc video frequency carrier; video frequency channel; visual field control; voice frequency carrier
VFC Victorian Film Commission
VFD Volunteer Fire Department
vfdr viewfinder
vfet vertical field-effect transistor
vff black very-fine furnace black (rubber filler)
VFHS Valley Forge Historical Society
vfi visual field information
VFI Vocational Foundation Incorporated
VFI Vertical File Index
VFIC Virginia Foundation for Independent Colleges
vfl variable focal length
vfl (VFL) variable field length

VFMJC Valley Forge Military Junior College
vfn very-flowery no
VFNP Victoria Falls National Park (Rhodesia)
vfo variable-frequency oscillator
VFOAR Vandenberg Field Office of Aerospace Research (USAF)
vfp variable-factor programming
VFP Volunteers for Peace
vfr vehicle fuel refinery
VfR *Verein für Raumschiffahrt* (German—Space Travel Society)
VFR Visual Flight Rules
vfr's visiting friends and relatives
VFSTC Valley Forge Space Technology Center (General Electric)
vftg voice frequency telegraph
vfu vertical format unit
VFU Vancouver Free University
VFW *Vereinigte Flugtechnische Werke*; Veterans of Foreign Wars
vfy verify
vg variable geometry; velocity gravity; very good (VG)
v.g. *verbi gratia* (Latin—for example)
vg *verbigracia* (Spanish—for example); *virgen* (Spanish—virgin)
Vg. *Virgo* (Latin—virgin)
VG Vocational Guidance
V.G. Vicar General
VG *Vaisseau de Guerre* (French—warship)
vga variable gain amplifier; videographics array
VGA Victor Gruen Associates
VGAA Vegetable Growers Association of America
VGB Vandenberg Air Force Base
vgc viscosity gravity constant
vge visual gross error
VGH Vancouver General Hospital
VGIK *Vsesoyuznyi Gosudarstvenyi Institut Kinematografi* (Russian—All-Union State Institute of Cinematography)
V-girl vice girl
Vgk Vegesack
vgl *vergelijken* (Dutch—compare); *vergleiche* (German—compare)
VGLI Veterans Group Life Insurance
Vgm Vizagapatam

vgo vacuum gas oil
Vgo Vigo (British maritime abbreviation)
VGP Van Gelder Papier; Volunteer Grandparent Program
vgpi visual glide-path indicator(s); visual ground-position indicator
Vgr Voyager (robot spacecraft)
V gr *verbigracia* (Spanish—for example)
V-groove V-shaped groove
VGSA Viola da Gamba Society of America
vgu *vorgelesen-genehmigt-unterschrieben* (German—read, confirmed, signed)
vgw *voegwoord* (Dutch—conjunction)
vh very high
v/h vulnerability/hardness
v/h *vorheen* (Dutch—formerly)
v H *vom Hundert* (German—percent; per hundred)
VH Veterans Hospital
VHA Vermont Headmasters Association; Voluntary Hospitals of America
vhb very heavy bombardment
vhc very highly commended
vhcl vehicle
vhclr vehicular
vhd video high density
vhf very high frequency (30,000 kc–300 mc)
VHF very high frequency (British educational tv)
vhf/df very high frequency direction finding
vhf/fm very high frequency/frequency modulated
vhf/uhf very high and ultra high frequency
VHIS Vaal-Hartz Irrigation Scheme
VHMCP Voluntary Home Mortgage Credit Program
vhmwpe very-high-molecular-weight polyethylene
Vhn Vickers hardness number
vho very high output
vhocm very-heavy oil-cut mud
vhp very high performance
vhs very high speed; video helical scan; video home system(s)
VHS Vocational High School
vhsbw very-high-speed black-and-white (photography)
VHSIC Very High Speed Integrated Circuit
vhtr very-high-temperature reactor
V-hut inverted V-shaped hut (sometimes called A-hut)
vi variable interval; verb in-

transitive; viscosity index; visual editor; volume index
v/i verb intransitive
v.i. *vide infra* (Latin—see below)
Vi Viola; Violet; Virginia; Vivian
VI Vancouver Island; Vermiculite Institute; Victoria (airline code); Virgin Islander(s); Virgin Islands (V.I.)
VI *Veiligheids Institut* (Dutch—Safety Institute)
via virus inactivating agent
Via Viaduct
VIA Vancouver, British Columbia's Vancouver International Airport; VIA Rail Canada; Vision Institute of America; Visually Impaired Association; Vocational Interests and Aptitudes
viad viaduct
vi antigen virulence antigen
VIAR Volcani Institute of Agricultural Research (Israel)
Via Rail Canadian National + Canadian Pacific
VIARCO Venezuelan International Airway Reservations Computerized
VIAs Vocational Information Agencies
VIAS Voice Interference Analysis System
VIASA Venezolana Internacional de Aviación SA
vib vibrate; vibration; vibratory
VIB Vertical Integration Building
vibes vibraphones; vibrations
vibgyor (mnemonic for remembering the spectral colors—violet, indigo, blue, green, yellow, orange, red)—*see* ROY G BIV
vibra vibraphone
vibs vocabulary-information-block-design similarities
vib/s vibrations per second
VIBS Virgin Islands Broadcasting System
vic convict; value-incentive clause; vicinal; vicinity; victim; victor; victorious; victory (V)
vic *vices* (Latin—times)
Vic RCA Victor; Vicar; Victor; Victoria; Victorine
VIC Virginia Intermont College; Virgin Islands Corporation
VICA Vocational Industrial Clubs of America
Vic Adm Vice Admiral

vicci voice-initiated cockpit control and integration
Vic Hist Victoria History of the Counties of England
vicoed visual communication education
vicom visual communication management
VICORP Virgin Islands Corporation
Vic Pk Victoria Park
vic(s) convict(s)
Vic Sta Victoria Station (rail terminal)
Vict Victor(ia)
Vic^ta Victoria (Spanish)
Vic^te Vincente (Spanish—Vincent)
victimol victimological(ly); victimologist; victimology
Victoria Alexandrina Victoria, Queen of the United Kingdom of Great Britain and Ireland, and Empress of India (1819–1901); La Victoria (Santo Domingo City prison of the Dominican Republic)
vid video
vid. vide (Latin—see); *Viuda* (Spanish—widow)
VID Volunteers for International Development
vidac visual information display and control
vidat visual data acquisition
VIDC Virgin Islands Department of Commerce
VIDD Virgin Islands Development Department
videocomp videocomposition (highspeed phototypesetting controlled by programmed digital-control unit)
videot(s) video (television) idiot(s)
vidiac visual information display and control
vidisc video disc
Viditel Videotelevision viewdata system (Dutch)
vie viernes (Spanish—Friday)
VIE Vienna, Austria (airport)
VIEDS Virgin Islands Educational Dissemination System
Vien Vienna
vier viernes (Spanish—Friday)
Viet Vietnam
Viet Vietnamese (oriental language)
Viet Cong Vietnam Congsan (Vietnamese—Vietnamese Communists)
Vietminh Vietnam Doc Lap Dong Ming (League for the Independence of Vietnam)
Vietnam Socialist Republic of

Vietnam (Indo-Chinese country), *Cong Hoa Xa Chu Nghia Viet Nam*
Vietnam congsam Vietnamese communist (see *congsam*)
Vietsyn Vietnam syndrome
Vietvet(s) Vietnam veteran(s)
Vieux Henri Vieuxtemps; *Vieux Carré* (French—Old Quarter) —French Quarter of New Orleans
VIEW Vital Information for Education and Work (education-on-microfilm program)
vig video image generator; vigilante; vigorish
vig (VIG) vaccine-immune globulin
VIG Video Integrating Group; Virgin Islands Government
Vig Com Vigilance Committee (men); Vigilant(e) Committee (men)
VIGIC Virgin Islands Government Information Center
vigilant. (VIGILANT) visually guided infantry light antitank (missile)
vign vignette
VIGOPRI Virgin Islands Government Office of Public Relations and Information
vigs vigilantes
vii viscosity index improver
VIJ Vera Institute of Justice
Vik Vickers; Vikelas; Vikenti; Vikentievich; Viki; Vikie; Viking; Viktor; Viktoria; Vikramaditya; Viktorovich
Vikes Vikings
Viking Pr Viking Press
vil vertical injection logic; village
Vil Las Villas (Santa Clara)
vill village
Villa Pancho Villa, orig. Doroteo Arango (1878–1923)
VIM Venture in Missions; Vertical Improved Mail; Virgin Islands Museum; Visible Impact Management
VIMI Virgin Islands Medical Institute
v imp verb impersonal
v imper verb imperative
VIMS Vertical Improved Mail Service; Virginia Institute of Marine Science
vim/var vacuum-induction melt/vacuum-arc remelt
vin vehicle identification number; vinegar; vinyl
vin. vinum (Latin—wine)
Vin Vincent
VIN Vehicle Identification Number

Viña Viña del Mar, Chile
VINB Virgin Islands National Bank
Vince St Vincent, West Indies; Vincent
Vincent Vincent Van Gogh
vind vindicate; vindication
VINES Virtual Networking Software
VINHS Virgin Islands National Historic Site
vini viniculture
VINITI Vsesoyuznyi Institut Nauchnoi Tekhnicheskoi Informatsii (Russian—All Union Institute of Scientific and Technical Information)
VINP Virgin Islands National Park (West Indies)
Vinson Vinson Massif (Antarctica's highest mountain)
VIO Veterinary Investigation Office(r)
viol violino (Italian—violin)
vip value improving product(s); variable information processing; variable input phototypesetting (VIP); vasoactive intestinal peptide; very important passenger; very important people; very important person; visual identification point; visual inspection protection
vip Virgil I Partch
VIP Value Improvement Project(s); Variable Information Processing; Very Important Person; Very Important Program; *Vías Internacionales de Panamá* (Panamanian airline); Virgin Islands Police; Visitor Information Phone; Vocabulary Improvement Program; Volunteer in Parks; Volunteers in Probation
VIPAC Virgin Islands Public Affairs Council
VIPI Volunteers in Probation, Incorporated
vipp variable-information processing package
vipre visual precision
VIPRE Very-Intense Pulsed-Radiation Experiment
vips voice interruption priority system
VIP-VIP Value in Performance through Very Important People (motivational program)
viq verbal iq
vir vertical interval reference (automatic television color system)
vir. viridis (Latin—green)

Vir Virgil; Virgo
VIR Vendor Information Request
V.I.R. Victoria Imperatrix Regina (Latin—Victoria Empress and Queen)
vira vehicular infrared alarm
VIRB Virginia Insurance Rating Bureau
Virg Virgil; Virgin; Virginia
Virgil Roman poet Publius Virgilius Maro
Virginias Virginia and West Virginia
Virgins Park Virgin Islands National Park, Saint John Island
virol virology
virr verb irregular
v/irr verb irregular
vis viscera; visible; visibility; visual
Vis Visayan; Vista (postal abbreviation)
VIs Virgin Islands
VIS Veterinary Investigation Service; Visual Instrumentation Subsystem
VISAR Visual Inspection System for the Analysis of Reports
visc viscosity
Visc Viscount(ess)
viscer (Latin prefix—organ)—visceral
Viscount Bolingbroke Henry St John (1678–1751)
VISIT Visit to Innovative Schools for Interested Teachers
vismins visual minorities (Africs, Asiatics, racially mixed Hispanics)
vispa virtual storage productivity aid(s)
vissr visible infrared spin-scan radiometer
vista. viewing instantly security transactions automatically
VISTA Volunteers in Service to America
vit vital; vitamin; vitreous
vit (Latin prefix—life)—vitamins
vit A carotene vitamin
VITA Volunteers for International Technical Assistance; Volunteers In Tax Assistance
vit A₁ nutritive vitamin found in egg yolk, milk, and milk products such as butter
vit A₂ freshwater fish-liver-oil vitamin
VITAL Variably-Initialized Translator for Algorithmic

Languages
vitamin(s) vital amine(s)
VITAP Voluntary Income Tax Assistance Program
vit B nutritive vitamin essential to digestive and nervous systems; found in breads, egg yolk, lean meats, fruits, nuts, green vegetables
vit B₁ thiamine
vit B₂ riboflavin
vit B₃ nicotinamide
vit B₆ pyridoxine
vit B₁₂ cobalmine-cyancobalmine
vit B₁₂b hydroxycobalmine
vit Bc folic-acid
vit B cx vitamin B complex (water-soluble vitamins B_1, B_2, etc.)
vit C ascorbic acid
vit cap. vital capacity
vit D antirachitic
vit D₁ calciferol and lumisterol
vit D₂ calciferol
vit D₃ cholecalciferol (natural vitamin D)
vit E antisterility vitamin; tocopherol
vitel. vitellus (Latin—egg yolk)
vit G riboflavin
vit H biotin
viti viticulture
vit K coagulant
vit K₁ blood-clotting vitamin
vit M folic-acid vitamin
vit. ov. sol. vitello ovi solutus (Latin—dissolved in egg yolk)
vit P permeability vitamin (bioflavonoid found in paprika)
vit PP pellagra-preventive vitamin (nicotinamide nicotinic acid)
vitr vitreous
Vitr Vitruvius Pollio (Roman writer on architecture)
vit rec vital records
vitriol concentrated sulfuric acid (oil of vitriol); copper sulfate (blue vitriol); ferrous sulfate (green vitriol); zinc sulfate (white vitriol)
vits vertical-interval test signals
vits & mins vitamins and minerals
vit stat vital statistics
vit U cabagin (anti-ulcer vitamin)
VIUS Virgin Islands of the United States
viv vivace
viv vivienda (Spanish—apartment house; dwelling)

Viv Vivian; Vivien; Vivienne; Vivyan; Vivyanne
VIV Virgin Islands View
VIVA Virgin Islands Visitors Association; Voices in Vital America (organization)
VIVB Virgin Islands Visitors Bureau
Vivette Genevieve
vivi vivisection
vix. vixit (Latin—he/she lived)
viz. videlicet (Latin—namely)
Viz Vizcaya (Biscay); Vizcayan (Biscayan)
Viz Bay Bahía Sebastían Vizcaíno, Mexico
Vizc Vizcaya
vj jet velocity
v J vorigen Jahres (German—last year)
V-J agar Vogel-Johnson agar
VJC Vallejo Junior College
V-J Day Victory in Japan Day—August 15, 1945
V-joint angular V-shaped masonry joint
vj's video jockeys
Vjschr Vierteljahrschrift (German—quarterly)
vk vertical keel; volume kill
V of K Voice of Kenya (radio-television network)
VKC Von Karman Center
VKI Von Karman Institute
VKIFD Von Karman Institute for Fluid Dynamics
VKM Van Kampen Merritt (U.S. Government Fund)
VKO Moscow, USSR (Vnukovo Airport)
vkr video kinescope recording(s)
VKR Vodennaya Kontr Rozvedka (Russian—Counter-Infiltration Organization)
vl vision, left
v/l vapor-to-liquid
Vl Ville
V/l vapor-liquid ratio
Vl Violino (Italian—violin); *Vlaanderen* (Dutch—Flanders)
VL Vaasa Line; Vaasan Laiva; Venezuelan Line; Viking Line; Volcano Line; Vulgar Latin
vla very low altitude; very-large array (radio telescopes)
vla viola (Italian—viola)
Vla Venezuela; Vlaardingen
VLA Very Large Array (Radio Astronomy Observatory); Veterans' Land Administration (Canada); Volunteer Lawyers for the Arts
VLAA Volunteer Lawyers and

Accountants for the Arts
Vlad Vladimir; Vladivostok
Vlad Vladivostok (Russian—
Rule the East)
vladd visual low-angle drogue
delivery
Vlad(i) Vladimir
vlb very long baseline
v-l b vertical-lift bridge
vlba very long baseline array
VLBC very large bulk carrier
vlbi very-long baseline interferometry
VLCC very large cargo carrier
(bulk freighter or tanker)
vlchv (VLCHV) very-low-cost
harassment vehicle
vlcs voltage-logic-current switching
vld visual laydown delivery
vldl (VLDL) very-low-density
lipoproteins
vldz Valdez
Vle Vale
V^le Viale (Italian—Avenue;
Boulevard)
vlf vertical linear foot; very
low frequency (to 30 kc)
vlf (VLF) vectored lift fighter
Vlg Village
vlh very lightly hinged
VLI Port Vila, Vanuatu (airport)
vllo violoncello (Italian—cello)
vln very low nitrogen; violin
Vln Valenciennes
vlnt van links naar rechts
(Dutch—from left to right)
vlo vertical lockout
vlp video long play(er) (video-
disc)
vlr very long range
vlrc very long range commuter
vls vertical liquid spring
VLS Vertical Launching System
vlsi very-large-scale integration
vlt very large telescope; violet
vltg voltage
vlv valve; valvular
vl/vs voltage logic/voltage
switching
Vly Valley
vm voltmeter
vm (VM) virtual machine
v/m various marks; volts per
meter
vm voormiddag (Dutch—P.M.);
vormittags (German—forenoon; A.M.)
v M vorigen Monats (German—last month)
VM Value Management; Viet
Minh; Vulcan Materials
V & M Virgin and Martyr
VM Völkerkundemuseum der

Universität Zürich (German
—Ethnographic Museum of
Zurich University)
V.M. Votre Majesté (French—
Your Majesty); *Vuestra Ma-
jestad* (Spanish—Your Maj-
esty); *Vuestra Merced* (Span-
ish—Your Worship)
vma vanillymandelic acid
VMA Valve Manufacturers
Association
VMAG Vanderpoel Memorial
Art Gallery
V-Mann Vertrauensmann
(German—Trusted Man) in-
telligence agent
vmap video map equipment
V max maximum flight velocity
vmc visual meteorological con-
ditions
VMC Viet Montagnard Cong
VMCCA Veteran Motor Car
Club of America
vmd vertical magnetic dipole
*V.M.D. Veterinariae Medi-
cinae Doctor* (Latin—Doctor
of Veterinary Medicine)
VMDP Veterinary Medical
Data Program
vmh (VMH) ventromedial nucleus of the hypothalamus
VMH Victoria Medal of Honour
vmi visual motor integration;
visual motor interaction
VMI Video Music Inc; Virginia Military Institute
v/mil volts per mil
V min minimum flight velocity
VMLI Veterans Mortgage Life
Insurance
vmm virtual machine monitor
v & mm vandalism and malicious mischief
VM Molotov Vyacheslav M
Skryabin
vmos V-groove metal-oxide
semiconductor
vmos (VMOS) vertical metal-
oxide semiconductor
VMOS Virtual Memory Operating System
vmp value of the marginal
product
vm & p varnish makers and
painters
vms vertical-motion simulator
VMS Veterinary Medical Society; Vigital Memory System;
Virtual Memory System;
Voluntary Medical Services
vmt vehicle miles travelled;
very many thanks; video matrix terminal
vn vulnerability number

v/n verb neuter
vn vellón (Spanish—copper-
silver alloy)
VN Václav Neumann; Viet-
nam; Vietnamese; Vladimir
Nabokov (1899–1977); Vo-
cational Nurse
vna (VNA) ventral noradrener-
gic bundle
Vna Vienna
VNA Air Vietnam; Viet-
nam News Agency; Visiting
Nurses Association
VNAF Vietnamese Air Force
vnav volumetric area naviga-
tion (three-dimensional)
VNB Valley National Bank
V-N B Verrazano-Narrows
Bridge
Vnc (VNC) Vancouver, Wash-
ington
VN$ Vietnamese dollar
VN de B Vasco Nuñez de
Balboa (first European to dis-
cover the Pacific Ocean)
V-neck V-shaped neck (line)
Vnese Vietnamese
vnf very near field
Vng Vereeniging
vni variable name initialization
Vni Violini (Italian—violins)
Vnla Venezolana (Spanish—
female Venezuelan)
Vnlo Venezolano (Spanish—
male Venezuelan)
VNM Victoria National Mu-
seum (Ottawa)
VNMC Vietnam Marine Corps
VNN Vietnam Navy
VNNBS Vietnamese National
Broadcasting Service
Vno Violino (Italian—violin)
VNO Vital National Objective
V-note $5 bill
VNP Vietnamese piastre; Vo-
yageurs National Park (Min-
nesota)
vnr variable navigation ratio
VNR Van Nostrand Reinhold
VNRC Vegetarian Nutritional
Research Center
VNs Vietnamese
*VNS Vereenigde Nederlands
Scheepvaartmaatschappij*
(Dutch—United Netherlands
Navigation Company)
vnw voornaamwoord (Dutch—
pronoun)
VNWR Valentine National
Wildlife Refuge (Nebraska)
vo voluntary opening
vo. verso (Latin—back of the
page, lefthand page); *violino*
(Italian—violin)
v/o vossa ordem (Portuguese—
your order)

vᵒ verso (Portuguese—lefthand page, other side, over, reverse)

VO Valuation Officer(r); verbal order(s); very old; Veterinary Office(r); Victorian Order; voice over

VO Volksoper (German—People's Opera)—Vienna

voa vetoed on arrival (at the President's desk)

V o A Voice of America

VOA Vancouver Opera Association; Vasa Order of America; Virginia Opera Association; Voice of America; Volunteers of America

VOA Vereeniging Ontwikkeling Arbeidstechniek (Dutch—Work Study Association)

vo-ag vocational agriculture (educators' jargon)

vob vacuum optical bench

vobanc voice band compression

VᵒBᵒ vista bueno (Spanish—approved, okay)

Vᵒ Bᵒ visto bueno (Spanish—okay)

voc vocal; vocalist; vocation; vocational; vocative; volatile organic compound

VOC Vereenigde Oostindische Compagnie (Dutch—United East India Company)—often called the Very Old Company

VOCA Visiting Orchestras Consultative Association (London)

vocab vocabulary

VOCAL Vessel Ordnance Allowance List

vocat vocation(al); vocative

voc ed vocational education

Voc Foun Vocational Foundation

vocg verbal orders—commanding general

voco verbal order—commanding officer

vocoder voice coder

VOCOSS Voluntary Organisations Cooperating in Overseas Social Service

vocs verbal orders—chief of staff

voctl vocational

vod vision of right eye (d standing for *dexter*—Latin for right)

v-o d voice-operated device; voluntary-opening device

vodacom voice data communication(s)

vodactor voice data compactor

vodaro vertical ozone distribution (from) absorption and radiation of ozone

vodat voice-operated device for automatic transmission

voder voice-operated demonstrator

VÖEST Vereinigte Österreichische Eisen and Stahlwerke (United Austrian Iron and Steel Works)

vof variable-operating frequency

vog volcano smog

Vog Vogue

VoG Voice of Germany

VOG Vanguard Operations Group

vogad voice-operated gain-adjusting device (data processing)

VOHI Vancouver Oral Health Index

VOICE Voice of Informed Community Expression

VOICES Voice-Operated Identification and Computer Entry System

VOIS Visual Observation Instrumentation Subsystem

voit voiture (French—railroad coach, truck, wagon, etc.)

vol volume; volunteer

vol % volume percent

vol. volatilis (Latin—volatile)

Vol Volans (constellation); Volcán; Volcano; volume

Vol Volcan (French—volcano); *Volcán* (Spanish—volcano); *Vulcano* (Italian—volcano)

VOLAG Voluntary Agency

VOLAR Volunteer Army

vol ash volcanic ash

volat volatile; volatizes

volc volcanic; volcano; volcanology

Vol Isl Volcano Islands (south of Japan and Bonin Islands)

Volks Volkswagen

volkst volkstaal (Dutch—slang; vernacular)

vollst vollstandige (German—complete)

Voln Volans (constellation)

vols volumes

VOLS Voluntary Overseas Libraries Service

Volta Voltaic Republic (Republic of the Upper Volta)

volts AC volts alternating current

volts DC volts direct current

volum volumetric

volvar volume variety

volvend. volvendus (Latin—to be rolled)

Volvo (Latin—I roll)—Swedish automobile

voly voluntary

vom volt milliammeter; volt-ohm microammeter; volt-ohm milliammeter; vomer; vomerine; vomit; vomitory; vomitus

vom. vomitus (Latin—vomit)

VOM Vereniging voor Oppervlaktetechnieken Metalen (Dutch—Metal Finishing Association)

VOMI Volksdeutsche Mittelstelle (German—Racial Assistance Office for Germans Abroad)

vom neg vomito negro (Spanish—black vomit)—last stage of yellow fever

VON Victorian Order of Nurses (public health)

vona vehicle of the new age (computer-controlled rapid-transit shuttle)

von K Herbert von Karajan (1908–1989); Theodor von Kármán (1881–1963)

V.O.N.O. Vendor of Oysters in New Orleans (Walt Whitman's invention)

vop valued as in original policy

VOP very oldest procurable

Vo-Po Volks Polizei (East German Police)

VOQ Visiting Officer's Quarters

vor very high frequency omni-directional range (VOR); visual omnirange

vordme very-high-frequency-omnirange distance-measuring equipment

vorm vormals (German—formerly); *vormittags* (German—forenoon, A.M.)

Vor Mus Voortrekker Museum (Pietermaritzburg)

VORP Victim Offender Reconciliation Program

Vors Vorsitzender (German—chairman)

vort vortex; vortices

vortac visual omnirange and tacan

vos vision of left eye (s standing for *sinister*—Latin for left)

vo('s) verbal order(s)

vos vostok (Russian—east, as in Vladivostok)

v.o.s. vitello ovi solutus (Latin—dissolved in egg yolk)

Vos Voskresene (Russian—Sunday)
VOS Victims of Superstition; visual observation airplane (naval symbol)
Vost Vostochnyy (Russian—eastern)
vot voice on set time; voluntary overtime
vot. votivus (Latin—promissory or votive)
VOT Foreign Operational Center of Soviet Intelligence forces (formerly called MGB, MVD, NKGB, NKVD, OGPU, GPU, VECHEKA, and CHEKA—founded in December 1917)
votc volume table of contents
VOTE Voters Organized to Think Environment
votem voice-operated typewriter employing morse
vou voucher
vow vowel(s)
VOW Voice of Women
VOWS Vilas-Oneida Wilderness Society
vox voice-operated transmission
vox pop. vox populi (Latin—voice of the people)
voy voyage
Voyageurs Voyageurs National Park on the Canadian border of Minnesota
Vozv Vozvyshennost' (Russian—uplands)
vp vanishing point; variable pitch; verb phrase; vertically polarized; vistaphone
v/p verb passive; verb phrase
v & p vagotomy and pyloroplasty
V$_p$ valve-position voltage
VP British United Air Ferries fixed-wing fighter airplane; Valencia Park; Ville de Paris; Vice-President
VP (NSC) Verification Panel (National Security Council)
V-P Voges-Proskauer (reaction)
VP Vigilancia de la Pesca (Spanish—Fishery Patrol)
VPA Vancouver Public Aquarium (British Columbia); Videotape Production Association; Virginia Port Authority
v pag various paging
VP & B Veterinary Pharmaceuticals and Biologicals
vpc volume-packed cells
VPCP Volunteer Probation Counseling Program

vpd vapor-phase degrease; variation per day; vehicles per day
vpe vapor-phase epitaxy
vpg very pregnant guppy (NASA); voltage pressure gradient
VPGA Vermont Personnel and Guidance Association; Virginia Personnel and Guidance Association
vph variation per hour; vehicles per hour; vertical photography
vpi vapor-phase inhibitor
VPI Veterinary Pet Insurance; Virginia Polytechnic Institute; Vocational Preference Inventory
VPIRG Vermont Public Interest Research Group
vpl visible panty line
VPL Van Pelt Library (University of Pennsylvania)
vpm vehicles per mile; versatile packaging machine; vertical panel mount; vibrations per minute; volts per meter; volts per mile
VPM Vendor Part Modification
Vpn Vickers pyramid number
V P/N vendor('s) part number
vpo vapor-phase oxidation
Vpo Valparaiso
VPO Vienna Philharmonic Orchestra
vpp viral porcine pneumonia
vpr vacuum pipette rig
VPR Vanguarda Popular Revolucionaria (Portuguese—Popular Revolutionary Vanguard) —Brazilian terrorist organization
V Pres Vice President
v-prez vice-president
vps vibrations per second; volume pressure setting
VPS Visual Programme Systems
VPSA Vertebrate Paleontological Society of America
V-P test Voges-Proskauer test
vq virtual quantum; visual quotient
vqa vendor quality assurance
vqc vendor quality certification
vqd vendor quality defect
VQMG Vice Quartermaster General
vqzd vendor quality zero defects
vr variable ratio; variable response; ventilated rib; vintage racing; vision, right; voltage regulator; vulcanized

rubber
vr (VR) voluntary return (voluntary deportation of illegal aliens)
v/r verb reflexive
vr vedi retro (Italian—please turn over)
VR fixed-wing transport airplane; Victoria Railways (Australia)
V-R Veeder-Root
V.R. Victoria Regina
VR Valtionrautatiet (Finnish—State Railways)
V.R. Victoria Regina (Latin—Queen Victoria)
vra vuestra (Spanish—your, *f.*)
VRA Vocational Rehabilitation Administration; Voluntary Restraint Agreement
vras vuestras (Spanish—your, pl.)
vrb voice rotating beacon
vrbl variable
vrbl mnmncs verbal mnemonics (abbreviations and acronyms)
vrc vertical redundancy check(ing); visible record computer
VRC Vehicle Research Corporation
VRCAMS Vehicle/Road Compatibility Analysis and Modification System
v-r'd voluntarily returned (deported)
VRD (Royal Naval) Volunteer Reserve Decoration
vre voltage-regulator exciter
vr&e vocational rehabilitation and education
v refl verb reflexive
VR et I Victoria Regina et Imperatrix (Latin—Victoria, Queen and Empress)
V Rev Very Reverend
VRF Vehicular Research Foundation
vrg veering
Vrg Varig (Brazilian Airlines)
vri virus respiratory infection
vri (VRI) visual rule instrument landing
Vri Vrijdag (Dutch—Friday)
VRI Vehicle Research Institute; Victorian Railways Institute
V-ring V-shaped ring
VRIS Vietnam Refugee and Information Services
vrm variable-rate mortgage(s)
v rms volt(s) root mean square
Vroni Veronica
vros vuestros (Spanish—your, pl)

vrp valuable-record protector; very reliable product
VRP Volta River Project
vrps voltage-regulated power supply
vrr visual radio range
VRR Veterans Reemployment Rights
vrs velocity response shape
VRS Van Riebeeck Society; Vanguard Recording Society; Video Response System
V & RS Vocational and Rehabilitation Service
vrt visual recognition threshold
vru voltage readout unit
vr vnw vragend voornaamwoord (Dutch—interrogative pronoun)
vrx virtual resource executive
Vry Viceroy
vs variable speed; vein shot (intravenous injection); venesection; ventricles; versus; volumetric solution
vs (VS) vital signs; voluntary simplicity
v.s. very soluble
vs. ve soire (Turkish—and so forth); *versus* (Latin—against)
v.s. vide supra (Latin—see above)
VS scouting airplane (2-letter symbol); Vancouver Symphony; Victoria Symphony
V.S. Veterinary Surgeon
V & S Valley & Siletz (railroad)
VS Vereinigte Staaten (German—United States); *Verenigde Staten van Amerika* (Dutch—United States of America); *Vostra Signoría* (Italian—Your Honor)
V S volti subito [Italian—turn (music page) swiftly]
VSA Victorian Society of America; Volunteer Services to Animals
vsam virtual storage access method
VSAP Vehicle Structure Analysis Program
vsb vestigial sideband
vs. b. venesectio brachii (Latin—bleeding in the arm)
VSBA Vermont School Boards Association; Virginia School Boards Association
vsby visibility
vsc virtual speech control
v.s.c. *vidi siccam cultam* (Latin—I have seen a dried cultivated specimen)—botanic term

VSC Virginia State College; Vocations for Social Change
VSCC Vintage Sports Car Club
vscf variable-speed constant-frequency
VSCU Vatican Secretariat for Christian Unity
vsd ventricular septal defect
VSD Vancouver School of Design; Vendor's Shipping Document(s); Veteran Services Division; Veterans Affairs Department
VSE Vancouver Stock Exchange
vsep very superior extra pale
vsff volte, se faz favor (Portuguese—please turn over)
VS f U Vatican Secretariat for Unbelievers
VSGLS Vehicle Space Ground Link Subsystem
V-shape V-shaped
vshps vernier solo hydraulic power supply
vsi variable-speed indicator; very seriously ill; very slight imperfection; very slight inclusion
V-sign victory sign (raised index and middle fingers)
v signs vital signs (blood pressure, pulse, temperature, respiration)
vs jw vise jaws
vsl variable safety level
VSL Venture Scout Leader
vsm vibrating-sample magnetometer
vsmf visual search microfilm file
VSMF Vendor Spec Microfilm File
VSMS Vermont State Medical Society; Vineland Social Maturity Scale
vsn vision
V S/N vendor('s) serial number
VSNAP Vermont State Nuclear Advisory Panel
vso very special old; very superior old
VSO Vancouver Symphony Orchestra; Victoria Symphony Orchestra; Victor Symphony Orchestra; Vienna State Orchestra; Vienna Symphony Orchestra
VSOE Venice-Simplon Orient Express
vsop very superior old pale (cognac)
vsp vertical seismic profile
Vsp. Vespertina (Latin—Vespers)

VSP VS Pritchett
VSPA Virginia State Port Authority
vspc virtual storage personal computing
vsq very special quality (VSQ)
vsr variable speed reversible; very short range; visual security range
vss versions
vss (VSS) vstol support ship
v.s.s. vidi siccam spontaneam (Latin—I have seen a dried wild specimen)—botanic term
VSS Vancouver Symphony Society; Vermont State Symphony; Voluntary Social Services
VSSSN Verification Status Social Security Number
vst violinest
V St A Vereinigte Staaten von Amerika (German—United States of America)
vstol vertical and/or short take-off and landing
vsula vaccination scar upper left arm
vsv vesicular stomatitis virus
vsw vitrified stoneware
vswr voltage standing wave ratio
VSX heavier-than-air antisubmarine warfare carrier-based aircraft (naval symbol)
vt vacuum technology; vacuum tube; variable time; velocity; verb transitive; vinyl tile; vinyl tiling; voice tube
vt (VT) vertical tabulation character (data processing)
v-t vacuum technology; variable time (fuse); velocity-time (diagram)
v/t verb transitive
v & t volume and tension (of the pulse)
vt vaart (Dutch—canal); *viz tez* (Czech—see also)
v T vom Tausend (German—per thousand)
Vt Vermont; Vermonter; Vietnam(ese)
VT fixed-wing trainer-type airplane; *Reseau Aérien Interinsulaire* (Tahiti airline); Vermont; Virgil Thomson (1896–1989)
V.T. *Vetus Testamentum* (Latin—Old Testament)
vta ventral tegmental area
v^{ta} vuelta (Spanish—turn)
VTA Virginia Teachers Association
VTA Voenno-Transportnayavi-

atsiya (Russian—Air Transport Aviation)

v/tab vertical tabulation

VTAE Vocational Technical and Adult Education (System)

VTB *Vereniging voor het Theologisch Bibliothecariaat* (Dutch —Association of Theological Librarians)

vtc voting trust certificate

VTC Vermont Technical College

vte vertical-tube evaporator (for producing freshwater from the sea); vicarious trial and error

Vte Vicomte

V-TECS Vocational-Technical Education Consortium of States

Vtesse Vicomtesse

V-test Voluter test

vtf vertical test fixture

vt fuse variable-time fuse

vtg voting

VTG Vehicle Technology Group

vti volume thickness index

VTI Valparaiso Technical Institute

vtl variable threshold logic; vertical turret lathe

VTLs *Vehicular Traffic Laws*

VTM Victorian Tourist Ministry (Australia)

VTN Video Tape Network; Voorheis, Trindle, and Nelson

vto vertical takeoff; viable terrestrial organism

v^{to} *vuelto* [Spanish—change (money)]

Vto Vtornik (Russian—Tuesday)

vtoc volume table of contents (data processing)

vtohl vertical takeoff and horizontal landing

vtol vertical takeoff and landing

vtolport vertical-takeoff-and-landing airport

vtovl vertical takeoff vertical landing

vtp voluntary termination of pregnancy (abortion)

vtp viajes todo pagado (Spanish—all trips paid)

vtpr vertical temperature profile radiometer

vtr video tape recorder; video tape recording

vtr. vitreum (Latin—glass)

VTR Vermont Railway

VTRS Video Tape Recorder System

vtr sot. videotape recorder sound on recorder tape

VTS Viewfinder Tracking System; Virginia Theological Seminary

VTSRS Verdun Target Symptom Rating Scale

VTTA Veteran's Time Trial Association

VTU Volunteer Training Unit

vtvm vacuum-tube voltmeter

vu varicose ulcer; view; voice unit; volumetric unit; volume unit

vu von untem (German—from the bottom)

VU Air Ivoire; fixed-wing utility airplane; Valparaiso University; Vanderbilt University; Vice Unit (police); Victoria University; Villanova University; Vincennes University

VU Vigile Urbano (Italian—Traffic Policeman)

VUA Valorous Unit Award

VUA Vrije Universiteit, Amsterdam (Dutch—Free University —Amsterdam)

vue d'opt vue d'optique (French —optical view)—multidimensional art

VUH Vanderbilt University Hospital

vu indicator volume-unit indicator (data processing)

Vul Vulgate; Vulpecula (constellation)

vulc vulcanize(d,r)

vulcan vulcanization; vulcanize; vulcanizer; vulcanizing

vulg vulgar; vulgar fraction; vulgarian; vulgarism; vulgarist; vulgarization

Vulg Vulgar Era (Christian Era); Vulgar Latin; Vulgate

vulp vulpine

Vulp Vulpecula (constellation)

v-u meter volume-unit meter

VUNC Voice of United Nations Command

v u p (VUP) very unimportant person

VU-PD Vice Unit-Police Department

VU Pr Vanderbilt University Press

VUSM Vanderbilt University School of Medicine

vuv vacuum ultraviolet

VUW Victoria University of Wellington, New Zealand

vv vagina and vulva; verbs; verses; vice versa

v/v volume for volume

v&v verification and validation

v & v vintage and veteran (automobiles)

v.v. vice versa (Latin—conversely); *violini* (Italian—violins)

Vv. Virgines (Latin—Virgins)

VV Villa Viscaya (Dade County Art Museum, Miami, Florida); Voice of Vietnam (Hanoi)

VV ustedes (Spanish—you, pl.); *Viva Verdi!; Viva Vivaldi!* (Italian—Long live Verdi; Long live Vivaldi)

VVA Vietnam Veterans Association; Vietnam Veterans of America

VVAW Vietnam Veterans Against the War

v.v.c. vidi vivam cultam (Latin—I have seen a living cultivated specimen)—botanic term

VVCP Victims of Violent Crimes Program

VVD Volkspartij voor Vrijheid en Democratie (Dutch—People's Party for Freedom and Democracy)—Liberal Party

vvds video verter decision storage

Vve Veuve (French—widow)

vv hr vibration velocity per hour

Vvl Varavel

vv. ll. variae lectiones (Latin—variant readings)

VVLP Vietnam Veterans Leadership Program

VVMF Vietnam Veterans Memorial Fund

VVN Verein der Verfolgten des Naziregimes (German—League of Victims of Naziism)

VVO very, very old

vvr variable-voltage rectifier

vvrm vortex valve rocket motor

vvs very, very superior

vv's varicose veins

v.v.s. vidi vivam spontaneam (Latin—I have seen a living wild specimen)—botanic term

V v V Volkspartij voor Vrijheid en Democratie (Dutch—Peoples' Party for Freedom and Democracy)

VVS Veteran's Vigil Society (Vietnam-era veterans)

V-VS Voenno-Vozdushniye Sily (Russian—Air Forces of the USSR)

vvsf very very slightly flawed

(gems)

vvsi very very slight imperfection; very very slight inclusion

vvsop very very superior old pale (cognac)

vvt variable valve timing

VVT Visual-Verbal Test

VV UU Vigili Urbani (Italian—Traffic Police)

v.v.v. veni, vidi, vici (Latin—I came, I saw, I conquered)

VVV Vasili Vasilievich Vereschagin

vw vessel wall

vw voegwoord (Dutch—conjunction)

Vw View (postal abbreviation)

VW Very Worshipful; Volks-wagen (People's Car)

vWd von Willebrand's disease

VWD Vereinigte Wirtschafte Dienst (German—German News Agency)

vWf von Willebrand factor

vwg vibrating wire gage

vwl variable word length

VWOA Volkswagen of America

vwp variable width pulse

VWP Victim/Witness Project

VWPI Vacuum Wood Preservers Institute

vws ventilated wet suit; vibrating-wire stressmeter(s)

VWWI Veterans of World War I

vx vertex

VX Experimental Squadron (symbol)

vxo variable crystal oscillator

Vxtmps Vieuxtemps

vy various years; very

VY Air Cameroun; Victualling Yard

vyd vydani (Czech—edition)

Vygr Voyager (robot spacecraft)

Vy Rev Very Reverend

vyt vytah (Czech—abstract)

vz virtual zero

v-z varicella-zoster

vzd vendor zero defect(s)

VZP Venezuelan Petroleum Company (stock exchange symbol)

vzt visual zenith tube

W

w loading (symbol); transverse acoustical displacement (symbol); wall; war; warm; waste; water; water vapor constant; watt; weather; week; weekly; weight; wet; white; wide; widow; widowed; width; wife; win; wind; wine; winning pitcher; with; won; wood; word; work; work (symbol); worn; wrong

w % weight percent

w + weakly positive

w − weakly negative

w *vatios* (Spanish—watts)

W Canadian Car & Foundry (naval designator symbol); College of Wooster; gross weight (symbol); irradiance (symbol); tungsten (Wolfram); very wide (symbol); Wales; Ward; Ward Line; warning; Washington; water; Waterman Steamship Line; watt(s); weather reconnaissance; Wednesday; Welsh; west; western; Westinghouse; Weyerhaeuser; Whiskey—code for letter W; Willys-Overland; Woolworth; Wu

W (W) wage rate (microeconomics)

W *Wadi* (Arabic—gulley, ravine, riverbed); *Wald* (German—forest, wood); *Wan* (Chinese or Japanese—bay; bight); *warm* (Afrikaans, Dutch, German—hot); west; *west* (Afrikaans, Dutch, German—west); Wilhelmsen (steamship line); women

W1, W2, etc. West One, West Two, etc. (London postal zones)

wa warm air; wire armored; with average; work energy; writing ability

w/a welded assembly

Wa *Waffenamt* (German—Ordnance Department)—Third Reich marking followed by a code number and stamped on all military equipment

WA Wabash Railroad (stock exchange symbol); Washington; Watchmen's Association; Welfare Administration; West Africa; West African; Western Airlines; Western Approaches (to British Isles); Western Australia; Wheeler Airlines; Wire Association; Workshop Assembly

W of A Western of Alabama (railroad)

W A *World Almanac and Book of Facts*

waa wartime aircraft activity; welded aluminum alloy

waa (w/a/a) with added adhesive

WAA War Assets Administration; Warden's Association of America; Western Amateur Astronomers; Women's Auxiliary Association

WAA *World Aluminum Abstracts*

WAAA Women's Amateur Athletic Association

WAAC West African Airways Corporation

WAACs Women's Auxiliary Army Corps

WAADS Washington Air Defense Sector

WAAE World Association for Adult Education

WAAF Women's Auxiliary Air Force

WAAFB Walker Air Force Base

waaj water-augmented air jet

WAAP World Association for Animal Production

waapm wide-area anti-personnel mine

WAAS Women's Auxiliary Army Service; World Academy of Art and Science

WAAVP World Association for the Advancement of Veterinary Parisitology

wab water-activated battery; when authorized by

WAB Wabash (railroad); Wage Adjustment Board; Wage Appeals Board; Western Actuarial Bureau; Westinghouse Air Brake; White American Bastion (neo-Nazi group); Wine Advisory Board; Women's Abolition Bureau (for the abolishment of adultery, alcoholism, and discrimination)

WABCO Westinghouse Air Brake Company

wablics waterborne logistical craft (junks, sampans, wallawallas)—Hong Kong harbor craft

wac wage analysis and control; waste acceptance criteria; weapon assignment console; write address counter

WAC Western Athletic Conference; Women's Army Corps (USA); Worked All Continents; World Aeronautical Chart; World Affairs Council

WACA World Airline Clubs Association

WACB Women's Army Classification Battery

WACC Washington Association of Community Colleges
WACCC Worldwide Air Cargo Commodity Classification
wack. wait before sending positive acknowledgement
WACL World Anti-Communist League
WACM Western Association of Circuit Manufacturers
waco written advice of contracting officer
WACO World Air Cargo Organization
WACRI West African Cocoa Research Institute
WACSM Women's Army Corps Service Medal
WACSSO Western Australian Council of State School Organisations
WACVA Women's Army Corps Veterans Association
Wad Wadham College, Oxford
WAD World Association of Detectives; Wright Aeronautical Division (Curtiss-Wright Corporation)
WADC Western Air Defense Command; Wright Air Development Center
wadd with added (costs, freight, etc.)
WADD Westinghouse Air Arm Division; Wright Air Development Division (USAF)
Wade-Giles Sir Thomas Wade (1818–1895) and Herbert Giles (1845–1935) (English lexicographers)
wadex word and author index
WADF Western Air Defense Force
Wadh Wadham College, Oxford
WADS Wide Area Data Service; Wide Area Dialing Service
Wadsworth Wadsworth Atheneum (Hartford)
wae when actually employed
WAED Westinghouse Aerospace Electrical Division
WAES Workshop on Alternative Energy Strategies
waf with all faults
WAF Women in the Air Force
WAFB Warren Air Force Base
WAFC West African Fisheries Commission
WAFF West African Frontier Force
waffle. wide-angle fixed-field locating equipment
W Afr West Africa(n)
waf(s) waffle(s)

WAG Walters Art Gallery; Winnipeg Art Gallery; Writers' Action Group
W A & G Wellsville, Addison & Galeton (railroad)
WAGBI Wildfowlers' Association of Great Britain and Ireland
WAGGGS World Association of Girl Guides and Girl Scouts
Wag hrn Wagner horn
wagr windscale advanced gas-cooled reactor
WAGR Western Australian Government Railways
WAGRO Warsaw Ghetto Resistance Organization
wags. weighted agreement scores
wai walk-round inspection; water installed
WAI Work in America Institute
WAIF World Adoption International Fund
WAIS Wechsler Adult Intelligence Scale
WAIT Wechsler Adult Intelligence Test(ing); Western Australian Institute of Technology
WAITR West African Institute for Trypanosomiasis Research
waj water-augmented jet
WAJ World Association of Judges
wak water analyzer kit; wearable artificial kidney; with all knowledge
Wakefield Wakefield Prison south of Leeds in Yorkshire, England
wal walnut; wide-angle lens
Wal Wallace; Wallach; Wallachian; Wallsend-on-Tyne
WAL Western Airlines; Westinghouse Astronuclear Laboratory; Westland Aircraft Limited
W-AL Westinghouse-Astronuclear Laboratory
WALA West African Library Association
WALDO Wichita Automatic Linear Data Output (Boeing)
Wales section of Great Britain; The Wales—The Bank of New South Wales
Wal I Wallops Island
WALIC Wiltshire Association of Libraries of Industry and Commerce
walk-in robes walk-in wardrobe closets
Wall Walloon
Wall *Wallace* (U.S. Supreme Court Reports)

Wallenstein Albrecht Wenzel Eusebius von Wallenstein (Bohemian general)
Wallis and Futuna Wallis and Futuna Islands in the southwest Pacific near Samoa
Wall-Wall prison in Walla Walla, Washington
Wally Wallace; Walter
Walnut Canyon Walnut Canyon National Monument in north-central Arizona
walopt weapons allocation optimizer
Walpurgis Walpurgis Night (April 30 in Finland and Sweden)
WALST Western Alaska Standard Time
Wal Sta Wallops Station
Walt Walter; Walton
wam walk-around money; wife and mother; words a minute; wrap-around mortgage
wAm white American male
WAM We Aint Metric; Western Australian Museum (Perth); Wolfgang Amadeus Mozart; Women Against Men; Worcester Art Museum; Working Association of Mothers
WAMI Washington, Alaska, Montana, Idaho
WAML Watertown Arsenal Medical Laboratory
wamoscope wave-modulated oscilloscope
WAMP Wire Antenna Modelling Program
wampum. wage and manpower process utilizing machines
WAMRU West African Maize Research Unit
WAMY World Assembly of Muslim Youth
WAN West Africa Navigation (steamship line)
WANA We Are Not Alone
WANAP Washington National Airport
Wand Wanderers
WAND Women's Action for Nuclear Disarmament
WANDPETLS Wandsworth Public Educational and Technical Library Services
Wandsworth Wandsworth Prison, London, England
Wankie Wankie National Park in Rhodesia
WANL Westinghouse Astronuclear Laboratories
wanna (American slang) want to
wannabe (American slang)

want to be
WANR Wadi Amud Nature Reserve (Israel)
WANS Women's Australian Nursing Society
WANYNJ Warehousemen's Association of New York and New Jersey
wao wet-air oxidation
WAO Weapons Assignment Office(r)
WAOB World Agricultural Outlook Board
WAOS Welsh Agricultural Organization Society; Wide-Angle Optical System
wap water planned; weak anthropic principle; wide-angle panorama
WA£ West African pound
WAP Women Against Pornography; Work Assignment Plan; Work Assignment Procedure; Writing Associates Program
WAPA Western Area Power Administration; White American Political Association
WAPC Women's Auxiliary Police Corps
WAPD Westinghouse Atomic Power Division
WAPET Western Australia Petroleum Pty Ltd
WAPO White American Political Organization
WAPOR World Association for Public Opinion Research
WAPPRI World Association of Pulp and Papermaking Research Institutes
WAPs Work Assignment Plans
WAP's Work Assignment Plans
WAPS World Association of Pathology Societies
WAPSD Westinghouse Electric Corporation Advanced Power Systems Division
WAPT Wild Animal Propagation Trust
WAPV gunboat (USCG symbol)
war. warrant; with all risks
War War Department; Warsaw; Warwickshire
WAR White Aryan Resistance (neo-Nazi group); William A Rusher; Women Against Rape
WARC Western Air Rescue Center; World Alliance of Reformed Churches
warcat workload and resources correlation analysis technique(s)
WARDA West African Rice

Development Association
WARES Workload and Resources Evaluation System
warex (WAREX) we have a warrant and will extradite
warf warfare
Warf Warfarin (rodenticide)
WARF Wisconsin Alumni Research Foundation
WARFI Western Alumni Research Foundation Institute; Wisconsin Alumni Research Foundation Institute
wargasm war + orgasm; sudden outbreak of war
warhd warhead
WARI Waite Agricultural Research Institute
Warks Warwickshire
warla wide-aperture radio location array
WARLOCE Wartime Lines of Communication—Europe
warn. warning
warr warranty
WARRS West African Rice Research Station
WARS Worldwide Ammunition Reporting System
was. wide-angle sensor; wideband antenna system
WAS Worked All States
WASA Washington Association of School Administrators; Wyoming Association of School Administrators
WASAL Wisconsin Academy of Sciences, Arts, and Letters
WASAMA Women's Auxiliary to the Student American Medical Association
Warsaw Pact Former mutual defense alliance among Bulgaria, Czechoslovakia, East Germany, Hungary, Poland, Romania, and the USSR
was wheel alignment stopper
WASB Wisconsin Association of School Boards
WASBO Washington Association of School Business Officials; Winconsin Association of School Business Officials
WASC Western Association of Schools and Colleges
wascala wide-angle scanning-array lens antenna
WASCO War Safety Council
WASDA Wisconsin Association of School District Administrators
Wash Washington; Washingtonian
WASH White Anglo-Saxon Hebrew
Wash Corr Cen Washington

Correctional Center
Wash DC Washington, D.C.
Washington's George Washington's Birthday (February 22)
Washmic Washington, (DC) military-industrial complex
WASHO Western Association of State Highway Officials
Wash Post The *Washington Post*
Wash St Hist Soc Washington State Historical Society
Wash St U Pr Washington State University Press
Wash U Med Lib Washington University School of Medicine Library (St Louis)
WASI Wage and Salaries Index
wasn't was not
WASO West Australian Symphony Orchestra
wasp. weightless analysis sounding probe; window atmosphere sounding projectile
WASP War Air Service Program; Water and Steam Program; White Anglo-Saxon Protestant; Williams Aerial System Platform; Women Against Soaring Prices; Women's Air Force Service Pilots; Workshop Analysis and Scheduling Program; Wyoming Atomic Simulation Project
WASP(S) White Anglo-Saxon Protestant(s)
Wass Wasserman
WASS Washington Association for Scientific Security
Wassermann August von Wassermann—German bacteriologist who devised test to determine diagnosis of syphillis
WASSP Washington Association of Secondary School Principals
WAST Western Australian Standard Time
WASU West African Student's Union
wat weight, altitude, temperature
Wat Waterford
WAT Word Association Test; World Airport Technology
WATA World Association of Travel Agencies
watashi watakushi (Japanese— I, me, myself)
Watchungs Watchung Mountains of northern New Jersey
WATDA Western Australia Tourist Development Authority

water H_2O

WATER Working Alliance To Equalize Rates

Water Gap Delaware Water Gap between New Jersey and Pennsylvania

water res water resistant

watertec water technologist; water technology

Waterton Waterton-Glacier International Peace Park on the Alberta-Montana border or Waterton Lakes National Park in the same area

watg wave-activated turbine generator

WATPL Wartime Traffic Priority List

wats wide-area telephone service

WATS Wide Area Telecommunications Service

Wat Sta Waterloo Station (rail terminal)

watt's wide-area telephone transmission lines

W Aust Western Australia

W Aust Cur West Australian Current

WAVA World Association of Veterinary Anatomists

WAVAW Women Against Violence Against Women

WAVES Women Accepted for Volunteer Emergency Service (USN)

WAVFH World Association of Veterinary Food Hygienists

WAW Warsaw, Poland (airport)

WAwa West Africa wins again

WAWF World Association of World Federalists

wax. weapon assignment and target extermination

'way away

WAY World Assembly of Youth

WAYC Welsh Association of Youth Clubs

Wayne St U Pr Wayne State University Press

'ways always

wb warehouse book(ing); water ballast(ing); waybill; weber; wheelbase; whole blood; widebeam; wingback; winner's bitch

w/b westbound; will be

Wb weber

WB Wage Board; Warner Brothers; Weather Bureau; Women's Bureau; World Bank for Reconstruction and Development (UN)

W-B Wilkes-Barre, Pennsylvania

wba wideband amplifier

WBA Washington Booksellers Association; Wisconsin Booksellers Association; World Boxing Association

WBAA Wholesale Booksellers Association of Australia

WBAFC Weather Bureau Area Forecast Center

WBAMC William Beaumont Army Medical Center

WBAN Weather Bureau, Air Force-Navy

wbar wing bar (lighting or lights)

wbat wideband adapter transformer

WBAWS Weather, Briefing, Advisory, and Warning Service

wbc white blood cell; white blood cell (count); white blood corpuscle

WBC Welsh Books Council; World Boxing Commission; World Boxing Council

wbco waveguide below cutoff

wbct wideband circuit transformer

wbd wideband data

WBD *Webster's Biographical Dictionary*

wbdl wideband data link

WBEA Western Business Education Association

WBED Women's Business Enterprise Division

WBF World Bridge Federation

wbfp wood-burning fireplace

wbgt wet-bulb globe temperature; wet-bulb globe thermometer

WBH Welsh Board of Health

wbi will be issued

WBI Wooden Box Institute

WBINA Wreck and Bone Islands Natural Area (Virginia)

WBIT Wechsler-Bellevue Intelligence Test

wbl wideband laser; wood blocking

Wbl Whitstable

Wbl *Wochenblatt* (German—weekly publication)

WBL Western Biological Laboratories; World Basketball League

wblc waterborne logistics craft

wblo weak black liquor oxidation

Wb/m² webers per square meter

WBMA Wirebound Box Manufacturers Association

WBMC William Beaumont Medical Center (El Paso)

wbn well-behaved net

w/bndr(s) with binder(s)

wbnl wideband noise limiting

WBNM Wright Brothers National Monument

WBNP Wood Buffalo National Park (northwest Territories, Canada)

WBNR Wadi Bezet Nature Reserve (Israel)

wbns water boiler neutron source

wbnv wideband noise voltage

wbo wideband oscilloscope; wide band overlap; wide bridge oscillator

w/bo(s) with blowout(s)

wbp weather and boilproof

WBP Wartime Basic Plan; Water Bank Program

WBPA Western Book Publishers Association

wbr water boiler reactor; whole body radiation; wideband receiver

W Branch Wireless Branch (British intelligence)

wbrbn will be reported by notam (Notice to Airmen)

wbrs wrought brass

wbs without benefit of salvage; work breakdown structure

WBSEB West Bengal State Electricity Board

WBSF Water Basin Storage Facility

WBSI Western Behavioral Sciences Institute

WB Sig Sta Weather Bureau Signal Station

wbt wet-bulb temperature; wet-bulb thermometer; wideband transformer; wideband transmitter

WBT World Board of Trade

WBTA Webb-Pomerene Trade Association

WBTS Watchtower Bible and Tract Society

W B T & S Waco, Beaumont, Trinity & Sabine (railroad)

wbtv weather briefing television

wbv wideband voltage

wbvco wideband voltage-controlled oscillator

W By Walvis Bay

wc wadcutter; wage change; water closet (English euphemism for *lavatory*); weapon carrier; wheelchair; will call; without charge; wood casing; working capital; working circle; workmen's compensation

w/c wave change; with corrections (correct proof before printing)

WC Wabash College; Wagner College; Waldorf College; Walker College; Walsh College; Wartburg College; Washington College; Waynesburg College; Weatherford College; Webber College; Weber College; Webster College; Wellesley College; Wells College; Wesley College; West African Airlines; West Coast Airlines; Westmar College; Westminster College; Westmont College; Wheaton College; Wheeling College; Wheelock College; Whitman College; Whittier College; Whitworth College; Wiley College; Wilkes College; Williams College; Wilmington College; Wilson College; Windham College; Winthrop College; Wofford College; Woodbury College; Woodstock College; World Court; Wycliffe College

W/C Weapons Controller; Wing Commander

WC1, WC2, etc. West Central One, West Central Two, etc. (London postal zones)

wca wideband cassegrain antenna; worst case analysis

WCA Wackenhut Corrections Corporation; Washington Correctional Association; Washingtonian Center for Addiction; Western Correctional Association; Wisconsin Correctional Association; Women's Correctional Association; World Calendar Association

WCAA West Coast Athletic Association

w cab wall cabinet

WCAC West Coast Athletic Conference; Women's Crusade Against Crime (St Louis)

WCAFS Wideband Cassegrain Antenna Feed System

WCAP Westinghouse Commercial Atomic Power

WCAT Welsh College of Advanced Technology

WCB Workmen's Compensation Board

WCBA West Coast Bookmen's Association; Western College Bookstore Association

WCBHS William Cullen Bryant High School

wcc water-cooled copper; wilson cloud chamber

WCC War Crimes Commission; Wayne County Community College; Westchester Community College; Western Cartridge Company; Westminster Choir College; White Citizens' Council; World Council of Churches

wcca worst-case circuit analysis

WCCE West Coast Commodity Exchange

WCCI World Council for Curriculum and Instruction

WCCU World Council of Credit Unions

WCD Workers Compensation Department

wcdb wing control during boost

wcdo war consumable distribution objective

wce weapon control equipment

WCED World Commission on Environment and Development

WCEMA West Coast Electronic Manufacturers' Association

WCEU World's Christian Endeavor Union

wcf white cathode follower

WCF Waste Calcining Facility; Winchester Center Fire (rifle shell designation)

WCFPR Washington Center of Foreign Policy Research

WCFST Weigl Color-Form Sorting Test

WCFTB West Coast Freight Tariff Bureau

WCG Women's Cooking Guild

wci white cast iron; wind chill index

WCI Wildlife Conservation International

WCIA Watch and Clock Importers Association

WCIR Workers Compensation and Insurance Report(ing)

WC & IR Workmen's Compensation and Insurance Report(ing)

WCJE World Council on Jewish Education

WCK West Virginia Coal and Coke (stock exchange symbol)

wcl watercooler

WCL West Coast Line; World Confederation of Labor

W-class Soviet class of submarines named Whiskey by NATO

wcld watercooled

WCLIB West Coast Lumber Inspection Bureau

wcm welded cordwood module; wired-core matrix; wired-core memory; word combine and multiplexer

WCM Worshipful Company of Musicians

WCMA Wiping Cloth Manufacturers' Association; Wisconsin Cheese Makers' Association

WCML Women's Caucus for Modern Languages

WCMR Western Contract Management Region

WCNA West Coast of North America

WCNM Walnut Canyon National Monument

WCNP Wind Cave National Park (South Dakota)

WCNYH Waterfront Commission of New York Harbor

WCO Weapons Control Officer(r)

WCOTP World Confederation of Organizations of the Teaching Profession

wcp welder control panel; white combination potentiometer

WCP Weapon Control Plan; World Council of Peace; Work Control Panel; Work Control Plan

WCPA Western College Placement Association; World Constitution and Parliament Association

WCPS Women's Caucus for Political Science; World Confederation of Productivity Sciences

WCPT World Confederation for Physical Therapy

wcr water-cooled reactor; water-cooled rod; water cooler; wire contact relay; word-control register

WCR Western Communication Region (USAF); Women's Council of Realtors

WCRA Weather Control Research Association; Western College Reading Association; Women's Cycle Racing Association

WCRP World Council of Religion for Peace

wcs wing center section

WCS Weapons Control Station; Weapons Control System; Wisconsin Correctional Service

WCSA West Coast of South America

wcsb weapon control switchboard

wcsc weapon control system console

WCSC World Correctional Service Center

WCSI World Center for Scien-

tific Information
WCSRC Wild Canid Survival and Research Center
WC & S's S & EBC William Cramp & Son's Ship and Engine Building Company
WCT World Championship Tennis
WCTB Western Carriers Tariff Bureau
WCTL Western Center Telecommunications Laboratory
WCTU Wild Cats and Tigers United; Women's Christian Temperance Union
WCU West Coast University; Western Carolina University
WCUK West Coast of United Kingdom
wcv water check valve
WCW William Carlos Williams
WCWB World Council for the Welfare of the Blind
wd water damage; weed; well deck(ing); white dwarf (star); whole depth; wind; window; winner's dog; withdrawn; wood; word; would; wound
w/d warranted; weight-displacement ratio; wind direction
Wd weeds
WD War Department; Water Department; Waterworks Department; Western Division
wda wheeldrive assembly; withdrawal of availability; writing down allowance (tax)
WDA Warranty Disclosure Act; Welsh Development Agency
WDALMP Warehouse Distributors Association of Leisure and Mobile Products
WDC War Damage Commission; Women's Detention Center; World Data Center
WDC-A World Data Center-A (Washington, D.C.)
WDC-B World Data Center-B (Moscow, USSR)
wdd Western Development Division (USAF Air Research and Development Command)
wdf wood door and frame
wdg winding; wording
wdi water district
WDIF Women's Democratic International Federation
wdk wives don't know
WDL Western Defense Laboratories (Philco subsidiary of Ford Motor Company)
WDM Western Development Museum (Saskatoon); World Development Movement
wdmf wall-defective microbial

forms
WDNR Wadi Dishon Nature Reserve (Israel)
wdo willing dropout
wdp wood door panel
WDP Women in Data Processing
WDPC Western Data Processing Center
wdr white drum
Wdr Wardmaster
Wdr L Wardmaster Lieutenant
wds wood-dye stain; word discrimination score; words; wounds
wd sc wood screw
wdsprd widespread
wdt width
wdtahtm (**wahm, for short**) why does this always happen to me?
WDTC Western Defense Tactical Command
wdu window de-icing unit
wdv written-down value (tax)
W$W *Wall Street Week* (educational tv program)
wdwn well developed, well nourished
wdwrk woodwork
wdy wordy
we. watch error; weekend; white edge; white edging
w/e weekend
w & e windage and elevation
We Wednesday; Welsh
WE War Establishment; Western Electric; World Education
W E *Wärmeeinheit* (German—thermal unit)
wea weapon(s); weather
WEA Washington Education Association; West End Avenue; Wisconsin Education Association; Workers Educational Association; Wyoming Education Association
WEAAC Western European Airports Association Conference
WEAL Women's Equity Action League
WeAPD Western Air Procurement District
WEARCONS Weather Observation and Forecasting Control System
weat weathertight
Weaver Weaverscope (rifle telescope)
Web *Webster's Third New International Dictionary of the English Language Unabridged*
WEBA Women Exploited by Abortion

WEBDEC WEB Du Bois Club(s)
webelos we'll be loyal scouts
Webelos We'll be loyal scouts.
webrock weather buoy rocket
WEBS Weapons Effectiveness Buoy System
Webster's *Webster's Dictionary* (published in many editions by G & C Merriam of Springfield, Massachusetts); *Webster's Third New International Dictionary of the English Language Unabridged*, published by Merriam Webster
wec wide energy conversion
WEC Westinghouse Electric Corporation; Women of the Episcopal Church; World Energy Conference
WECAF Western Central Atlantic Fishery
WECAN Walking Enforcement Campaign Against Narcotics
WeCen Weather Center (USAF)
WECEP Work Experience Career Exploration Program
WECO Western Electric Company
WECOM Weapons Command (USA)
wecpnl weighted-equivalent continuous-perceived noise level
WECS Wind Energy Conversion System
we'd we had; we would
Wed Wednesday
Wed *Weduwe* (Dutch—widow)
WED Walter Elias Disney
WEDA Wholesale Engineering Distributors' Association
wedar water-damage reduction; weather-damage reduction
Wedd Wedding (Berlin borough)
Wednes Wednesday
Wedy Wednesday
Wee Western equine encephalitis
WEEA *Women's Education Equity Act*
WEEAP Women's Educational Equity Act Program
WEECN Women's Educational Equity Communications Network
WEEP Work Education Evaluation Project
WEETA Women's Educational Equity Technical Assistance
wef with effect from
WEF War Emergency Formula; World Education Fellowship
WE & FA Welsh Engineers' and Founders' Association
wefax weather facsimile

WEFC West European Fisheries Conference
weft wings, engine, fuselage, tail
weg war emergency grant
weg(s) wild-eyed guess(es)
WEH William Ernest Henley
WEHS Wadleigh Evening High School
WEI World Economic Institute; World Education Incorporated
weia wife's earned income allowance (tax)
weir wife's earned income relief (tax)
Weiss Weissensee
WEIU Women's Educational and Industrial Union
Wel Welsh
WEL Weapons Effects Laboratory (USA)
Wel Adm Welfare Administration
Wel Can Welland Canal
weld welding
Wel Dept Welfare Department
we'll we shall; we will
Well Wellington
Welly Wellington
WELS Wisconsin Evangelical Lutheran Synod
Welt Die Welt (Hamburg's World)
Welts Weltschmerz (German—world pain)—universal misery
WEMA Western Electronic Manufacturers Association
WEMSB Western European Military Supply Board (NATO)
WEMTA Wisconsin Emergency Technician's Association
Wen Wendel; Wendell; Wendy
WEN Western Educational Network; Wien-Alaska Airlines
Wenatchees Wenatchee Mountains of central Washington
Wend Wendell's Reports
WENOA Weekly Notice to Airmen (CAA)
WEOG Western European and Other Groups
wep water-extended polyester
WEP Wisconsin Electric Power Company
WEPA Welded Electronic Packaging Association
WEPCO Weather-Proof Company
wepex weapons exercise
WERA World Energy Research Authority
WERC World Environment and Resources Council
we're we are
weren't were not
WERM World's Encyclopaedia

of Recorded Music
Werner Egk Werner Mayer
WERPG Western European Regional Planning Group (NATO)
Wes Wesley; Weston
WES War Equipment Scale; Water Electrolysis System; Waterways Experiment Station (Corps of Engineers); Weather Editing Section (FAA); Women's Engineering Society; World Economic Survey
WESCOM Weapon System Cost Model
WESCON Western Electronics Show and Convention
wesentl wesentlich (German—essential, main)
WESO Weapons Engineering Service Office
Wes Pac Western Pacific
WESRAC Western Research Application Center
Wes Sam Western Samoa (formerly British Samoa)
West states west of the Mississippi; western bloc countries of Europe and North America; Western States (Mountain and Pacific Divisions); Wild West
WEST Western Educational Society for Telecommunications; Western Energy Supply and Transmission (Association); Women's Enlistment Screening Test
WESTAF Western Transport, Air Force
WESTCOMMRGN Western Communications Region
wester storm from the west
Western Isles Hebrides off Scotland's west coast
WESTIS Westinghouse Teleprocessing Interface System
WestLant Western Atlantic Area
Westlaw computerized legal research service offered by West Publishing Co
West LB Westdeutsche Landesbank (West German Land Bank)
Westm Westminister; Westmorland
Westminster British Parliament; the Palace of Westminster and Westminster Abbey
Westmld Westmorland
Westo West Countryman
West Pac Western Pacific (ocean or railroad)
WESTPAC Western Pacific
Westrain Western Australian

Trains
Westralia Western Australia
Westralia(n) Western Australia(n)
West's West's Annotated Education Code
West Sam Western Samoa
West Symp Orch Westphalia Symphony Orchestra
Wes Univ Wesleyan University
WET Water Engineering Trading; Weapon(s) Effectiveness Test(ing)
WETA Washington Educational Television Association
wetensch wetenschap (Dutch—knowledge, science)
WeTip We Turn in Pushers (of narcotics)
Wet Mary Western Maryland Railway (stock exchange slang)
wets. Tory moderates
WETS Weekend Training Site(s)
WETUC Workers' Educational Trade Union Committee
WEU Western European Union
we've we have
WEWP West European Working Party (Book Development Council)
Wex Wexford
WEX Westinghouse Electric Company (stock exchange nickname)
Wexf Wexford
Wey Weymouth
wez (WEZ) western economic zone
WEZ westeuropäische Zeit (German—West European Time); Greenwich Mean Time
wf white female; wide flange; winner's female; write forward; wrong font
w/f white female
w/f (W/F) withdrawing and failing; withdrawn/failed
w & f water and feed; wow and flutter
WF Wake Forest; Wake Forest College; Wells Fargo & Company
W-F Weil-Felix (reaction)
W.F. White Father
W & F Wallis and Futuna Islands
WFA War Food Administration (World War II); White Fish Authority; World Federalists Association; World Friendship Association
w factor will factor
WFALW Weltbund Freiheitlicher Arbeitnehmerverbände

auf Liberaler Wirtschafs-grundlage (German—World Union of Liberal Trade Union Organizations)

WFAOSB World Food and Agricultural Outlook and Situation Board

WFAW World Federation of Agricultural Workers

WFB Wells Fargo Bank; World Federation of Buddhists

WFBI Wood Fiber Blanket Institute

WFBMA Woven Fabric Belting Manufacturers Association

wfc wide field camera; wolf first class (woman chaser)

WFC Wake Forest College; Water Facts Consortium; World Food Council; World Forestry Contresses

WFCA Western Fire Chiefs Association

wfd wool forward (knitting)

WFD World Federation of the Deaf

WFDY World Federation of Democratic Youth (communist)

wfe with food element

WFEA World Federation of Educational Associations

WFEB Worcester Foundation for Experimental Biology

WFEO World Federation of Engineering Organizations

WFEX Western Fruit Express

WFF World Friendship Federation

WFFL World Federation of Free Latvians

wfg waveform generator

WFGA Women's Farm and Garden Association

WFHE Washington Friends of Higher Education

WFI Wheat Flour Institute

WFIC Wisconsin Foundation of Independent Colleges

WFJCC World Federation of Jewish Community Centers

wfl worshipful

WFL Women's Freedom League; World Football League

W Flem West Flemish

WFLRY World Federation of Liberal and Radical Youth

WFM Walter F Mondale; Western Federation of Miners

WFMH World Federation for Mental Health

WFMW World Federation of Methodist Women

wfn well-formed net

WFN World Federation of Neurology

wfna white-fuming nitric acid

WFNS World Federation of Neurosurgical Societies

wfo wide-field optics

WFO Washington Field Office (FBI)

WFOA Western Fishboat Owners of America

wfof wide-field optical filter

WFOT World Federation of Occupational Therapists

wfp warm frontal passage

WFP World Food Program (UN)

WFP Winnipeg Free Press

WFPA World Federation for the Protection of Animals

WFPMM World Federation of Proprietary Medicine Manufacturers

WFPN Women in the Future Priesthood Now

WFPT World Federation for Physical Therapy

W Fris West Frisian

WFS World Future Society

WFSA World Federation of Societies of Anaesthesiologists

WFSF World Future Studies Federation

WFSPL Wright Field Special Projects Laboratory

WFSW World Federation of Scientific Workers

wft wandering finger trouble

WFT Washington Federation of Teachers

wfttngs with fittings

WFTU World Federation of Trade Unions

WFUNA World Federation of United Nations Associations

WFW Woltföderation der Wis-senschaftler (German—World Federation of Scientific Workers)

WFWFTHI World Federation of Workers in Food, Tobacco, and Hotel Industries

WFY World Federalist Youth

wg water gauge; weighing; weight guaranteed; wing; wire gauge

Wg Wolfgang

WG Welsh Guards; West German; Western Gear (company); WG Grace (cricketer and physician); Working Group; Writers Guild

WG Westminster Gazette

wga wheat-germ agglutinin

WGA Waterfront Guard Association; Western Governors Association; Writers' Guild of America

w-gal(s) wine gallon(s)

WGAO World Guide to Abbreviations of Organizations

W-gauge wide-gauge railroad track (exceeding the standard gauge of 4 feet 8½ inches)

WGB Weltgewerkschaftsbund (German—World Federation of Trade Unions)

wgbc waveguide operating below cutoff

WGC West Georgia College; World Gas Conference

Wg-Comdr Wing-Commander

WGCTA Watson-Glaser Critical Thinking Appraisal

WGD Webster's Geographical Dictionary

WGDS Warm Gas Distribution System

W Ger West Germany

WGER Working Group on Extraterrestrial Resources

wgf waveguide filter; wound glass filter

WGGB Writers' Guild of Great Britain

WGH William Gamaliel Harding (29th President U.S.)

WGI Work Glove Institute; World Geophysical Interval

WGIPP Waterton-Glacier International Peace Park (Alberta, Canada, and Montana, U.S.A.)

wgj wormgear jack

WGJB World's Greatest Jazz Band

w gl wireglass

WGL Weapons Guidance Laboratory

W Glam West Glamorgan

WGM Worthy Grand Master

WGMA Wet Ground Mica Association

WGmc West Germanic

WGMEX Western Gulf of Mexico

WGP Western Gas Processors

WGPMS Warehousing Gross Performance Measurement System

WGPORA Western Gas Processors and Oil Refiners Association

wgr wide gauze roll

WGR War Guidance Requirements

Wg & Rgn Comdr Wing and Regional Commander

W Grnld Cur West Greenland Current

wgs waveguide glide slope; web guide system

WGs Welsh Guards

WGS Western Gerontological

Society; World Geodetic System
wgsj wormgear screw jack
WGSPR Working Group for Space Physics Research (NATO)
wgt weight
WGTA Wisconsin General Testing Apparatus
WGTW Won't Go To Wembleys (anti-imperialist group)
WGU Welsh Golfing Union
WGVN Willard Gibbs Van Name
wgw waveguide window
WGWC Working Group for Weather Communications (NATO)
WGWP Working Group for Weather Plans (NATO)
wh water heater; watt hour; white; withholding
w/h withholding
Wh Whig Party
WH White House
wha wounded by hostile action
wha' what
WHA Welsh Hockey Association; Western History Association; World Health Assembly; World Hockey Association
wham winning the hearts and minds (of the listeners)
W'hampton Wolverhampton
whap when or where applicable
WHASA White House Army Signal Agency
whate'er whatever
what's what has; what is
whatso'er whatsoever
WHC White House Conference (on libraries and information services)
WHCA White House Communications Agency; White House Correspondents Association
WHCF White House Conference on Families
WHCLIS White House Conference on Library and Information Services
WHCOA White House Conference on Aging
WHCT West Ham College of Technology
whd warehead
WHD Women's House of Detention (NYC)
whdm watt-hour demand meter
whe water hammer eliminator
Wheat *Wheaton's* (US Supreme Court Reports)
wheats wheatcakes
wheatstone wheatstone bridge

(electrical measuring device named for its inventor—Sir Charles Wheatstone—an English physicist)
whecon wheel control
whene'er whenever
where'er wherever
wheresoe'er wheresoever
whf wharf
WHFAM William Hayes Fogg Art Museum
whfg wharfage
whfr wharfinger
whf(s) white homosexual female(s)
WHH William Henry Harrison (9th President U.S.)
WHHA White House Historical Association
Whi Whitehall
WHI Western Highway Institute
Whigs Whigamores (originally a group of West Scottish revolutionaries against church and king)
WHIM Western Humor and Irony Membership; World Humor and Irony
WHIMSY *Western Humor and Irony Membership Yearbook*
whis whistle (fog)
Whiskey letter W radio code; Soviet class of diesel submarines; stock exchange (slang); Western Kentucky (coal company)
Whit Whitaker; Whitbread; Whitcomb; Whitman
Whitaker's *Whitaker's Almanac*
White Carpathians White Carpathian Mountains of Czechoslovakia
Whiteman Marjorie Millace Whiteman's 15-volume *Digest of International Law*
White Mts White Mountains
White Pines White Pine Mountains of eastern Nevada
white precipitate ammoniated mercury
White Sands White Sands National Monument in southeastern New Mexico
white snow cocaine nickname
Whitman Albert Whitman (Chicago); Whitman Publishing Company (Racine)
WHL Western Hockey League
wh lt white light
WHMA Women's Home Missionary Association
WHML Wellcome Historical Medical Library
whm(s) white homosexual male(s)

whmstr weighmaster
WHMV & NSA Woods Hole, Martha's Vineyard and Nantucket Steamship Authority
Whn Whitehaven
WHO White House Office; World Health Organization (UN)
WHOA Wild Horse Organized Assistance
WHOA? Who Hammers Out Acronyms?
who'd who had
WHODAP White House Office of Drug Abuse Prevention
WHOI Woods Hole Oceanographic Institution
WHOIRP World Health Organization International Reference Preparation
whol wholesale(r)
who'll who shall; who will
whoretel whore hotel
who's who is
who've who have
whp water horsepower; whirlpool
W & H & PC Wage and Hour and Public Contracts
wh pl whole plate (silver)
whr watt hour
WHRA Welwyn Hall Research Association; Western Historical Research Associates; Western Housing Research Association
WHRC World Health Research Center
whrlp whirlpool
whs warehouse
WHS Walton High School; Washington Headquarters Services; White Sands, New Mexico (tracking station)
whse warehouse
whsl wholesale
whsmn warehouseman
whsng warehousing
whs rec warehouse receipt
Wht White
WHT William Howard Taft (27th President of the U.S.)
WHTHS William Howard Taft High School
whtm(s) white heterosexual male(s)
WHTSO Welsh Health Technical Services Organization
whvs wharves
why. what have you?
why'd why did
whyinel why in hell
wi wrought iron
wi' (Gaelic contraction—with)
w & i weighing and inspection
WI Wake Island; Washington

Institute for Values in Public Policy; West India; West Indian; West Indies; Windward Islands; Wine Institute; Wire Institute; Wisconsin; Women's Institute

W&I Welfare and Institutions (Code)

wia (WIA) wounded in action

WIA Western Interpreters Association

WIAB Wistar Institute of Anatomy and Biology

Wib Wibbert; Wilbert

WIB War Industries Board

WIB Werkgroep Instrument Beoordeling (Dutch—Working Group on Instrument Behavior); *Wissenschaftliche Internationale Bibliographie* (German—International Scientific Bibliography)

WIBC Women's International Bowling Congress

wic women, infants, children

wic (WIC) war insurance corporation

WIC Welfare and Institutions Code; West India Committee; Women in Construction

wich sandwich

WICHE Western Interstate Commission for Higher Education

Wichitas Wichita Mountains of Oklahoma and Texas

WICI Women in Communications, Incorporated

Wick Wicklow

Wicklows Wicklow Mountains in eastern Ireland

WICP Women, Infants, and Children Program

WICS Women's Institute for Continuing Study

wid widow; widower

WID Waste Isolation Division; West India Docks

WIDF Women's International Democratic Federation

Widm Widmung (German—dedication)

WIF West India Fruit and Steamship Company; West Indies Federation

WIFL World Indoor Football League

wig periwig

Wig Wigtown(shire)

wige wing-on-ground effect

wigo what is going on?

Wigorn. Wigorniensis (Latin—of Worcester)

wih went in hole

WIHM Wellcome Institute of the History of Medicine

WIHS Washington Irving High School

wiifm what's in it for me

Wil Wilber; Wilbert; Wilbur; Wilburn; Wiley; Wilford; Wilfred; Wylie

WIL West India Lines

Wil Blvd Wilshire Boulevard

wilco will comply

WILD What I Like to Do (psychological test)

wilde wildebeest (Afrikaans—gnu)

Wilder's d novelist-playwright Thornton Wilder's dictum declaring *Who can count the prayers that have ascended to gods who do not exist? Mankind has himself created sources of help where there is no help and sources of consolation where there is no consolation.*

Wiley John Wiley & Sons

Wilhelmina Wilhelmina Helena Pauline Maria (1880–1962), Queen of the Netherlands 1890–1948

Wilhelmus Wilhelmus van Nassouwe (Dutch—William of Nassau)—national anthem of the Netherlands

Will Willard; William; Willis

Will & Mar King William and Queen Mary

Will Rogers Turn Will Rogers Turnpike

Wilm Wilmersdorf; Wilmington

WILPF Women's International League for Peace and Freedom

WILS Wisconsin Interlibrary Loan Service

Wilson HW Wilson

Wilts Wiltshire

Wilts R Wiltshire Regiment

WIM Waste Isolation Manager(s)

W I & M Washington, Idaho & Montana (railroad)

WIMA Western Industrial Medical Association; Writing Instrument Manufacturers Association

Wimb Wimborne

w i m c whom it may concern

win. window(s)

Win Winchester Arms; Winterthur

WIN Whip Inflation Now; Work Incentive Program

WINA Webb Institute of Naval Architecture

win'ard windward

WINBAN Windward Islands Banana Growers Association

WINBAN(GA) Windward Islands Banana Growers Association

Winch Winchester

wind. windlass

W Ind West Indian; West Indies

Wind Cave Wind Cave National Park in southwestern South Dakota

Wind I Windward Islands

'winds woodwind instruments: bagpipes, bassoons, clarinets, English horns, fifes, flutes, harmonicas, jew's harps, kazoos, oboes, piccolos, recorders, saxophones, whistles

Windwards Windward Islands

WINE Webb Institute of Naval Engineering

Wing Cdr Wing Commander

winkle(s) periwinkle(s)

Winn Winnipeg; Winnipegger

wino alcoholic addicted to wine

win'rd windward (pronounced *win-urd* by sailors)

WINS Western Integrated Navigation System

wint winter; wintry

Winterthur Winterthur Museum

Wint Gard Winter Garden

wintr winter

Wint T The Winter's Tale

WIO Wyoming Infrared Observatory

wip work in process; work in progress

WIP Wage Insurance Program; Wall Improvement Project; West Indian Process (for sorting ripe coffee berries); Work Incentive Program; World Internationalist Party; World International Partisan

WIPAP Waste Isolation Performance Assessment Program

WIPO World Intellectual Property Organization

WIPP Waste Isolation Pilot Plant; Wool Incentive Payment Program

WIPSEP Waste Isolation Program and System Evaluation Project

WIPTC Women's International Professional Tennis Council

WIR Weekly Intelligence Report

WIRA Wool Industry Research Association

WIRDS Weather Information Reporting and Display System

WIRE Western Installation Requirements Evaluation (DoD); Wisconsin Information Re-

sources for Education
WIRO Wyoming Infrared Observatory
WIRs West Indian Reports
Wis Wisconsin; Wisconsinite
WIS Waste Isolation System; Weizmann Institute of Science; West Indies Shipping
WISA West Indian Sugar Association; West Indies Students Association
WISAP Waste Isolation Safety Assessment Program
Wisc Wisconsin
WISC Wechsler Intelligence Scale for Children
WISCo West Indies Sugar Company
Wisconsin Dells Dells of the Wisconsin
WISC-R Wechsler Intelligence Scale for Children—Revised
Wisd of Sol Wisdom of Solomon (apocryphal book of the Bible)
WISE Weapon Installation System Engineering; World Information Systems Exchange; Worldwide Information System for Engineering
wisk wiskunde (Dutch—mathematics)
wisp. wide-range-imaging spectrometer
WISP Waste Isolation Systems Panel (NAS); Wisconsin Inventory of Science Processes; Women in Scholarly Publishing
WISPr Women in Scholarly Publishing Newsletter
Wiss Wissenschaft (German—science)
wit. witness
WIT West India Tankers; World International Tennis
WITC Women's International Tennis Council
WITCH Women's International Terrorist Conspiracy (from) Hell
WITCO What Is This Thing Called Opera? (Seattle opera association)
withdrl withdrawal
with(out) hype with(out) hyperbole [with(out) exaggeration]
witht without
witned witnessed
witneth witnesseth
wits witkars (Dutch—white cars)—drive-it-yourself two-seater electric vehicles facilitating clean inner-city transportation
WITS Weather Information Te-

lemetry System; Westinghouse Interactive Time-Sharing System; West Integrated Test Stand
Wits U Witwatersrand University
wittos women in the transition of separation
WIU Western International University
WIVAB Womens' Inter-Varsity Athletic Board
wiz wizard
WIZO Women's International Zionist Organization
WJA World Jazz Association
WJB William Jennings Bryan (1860–1925)
wjc wife's judicial separation
WJC Westbrook Junior College; World Jewish Congress
W & JC Washington and Jefferson College
WJCB World Jersey Cattle Bureau
WJCC Western Joint Computer Conference
wJf white Jewish female
WJFITB Wool, Jute, and Flax Industry Training Board
wJm white Jewish male
wjs wife's judicial separation
wk walk; warehouse keeper; weak; week; well-known; work; wreck
Wk Walk; wreck
WK Western Alaska Airlines
W-K-B Wentzel-Kramers-Brillouin
wkbk workbook
wkd worked
W-K disease Wilson-Kimmelstiel disease
wkds weekdays
wkg working
WKKC Who Killed Kennedy Committee
wkly weekly
wkn weaken
WKNR Wadi Kziv Nature Reserve (Israel)
wkr workers; wrecker
wks weeks; white-knuckle seminarian; works; workshop(s)
Wks Works; wreckage (navigational abbreviation)
WKSC Western Kentucky State College
wkshp workshop
Wk/Site Work Site
wkt wicket
wk vb weak verb
WKY Western Kentucky (coal company); Wall Street slang for this company is *Whiskey*
W Ky Pkwy Western Kentucky

Parkway
wl wall lavatory; water level; waterline; waterplane coefficient; wavelength; working level
w L westlichst Längengrad (German—west longitude)
WL Sir Wilfred Laurier (Canada's eighth Prime Minister); Waiting List; West Lothian; Women's Liberation; Women's Lobby
W-L Westfal-Larsen Line
W & L Washington and Lee University
WL Wagon Lits (French—sleeping cars)
WLA Washington Library Association; Welsh Library Association; Western Literature Association; Wisconsin Library Association
wlb wallboard
WLB War Labor Board; Women's Liberation Party
WLB Werkgroep Instrument Beoordeling (Dutch—Working Group on Instrument Behavior); *Wilson Library Bulletin; Wissenschaftliche Internationale Bibliographie* (German—International Scientific Bibliography)
WLC World Liberty Corporation (Niarchos)
WLCJ Women's League for Conservative Judaism
WL & Co Westfal-Larsen & Company (steamship line)
wl coef waterline coefficient
wld west longitude date; would
wld ch world championship
WLDF Women's Legal Defense Fund
wldmt weldment
wldr welder
WLF Washington Legal Foundation; Women's Liberation Front; World Law Fund
WLFA Wildlife Legislative Fund of America
WLFNWR William L Finley National Wildlife Refuge (Oregon)
wl fwd wool forward
WLG Wellington, New Zealand (airport)
WLGS Women's Local Government Society
WLHB Women's League of Health and Beauty
wli workload index
WLI Women's Law Institute; Wyoming Law Institute
WLJBP William Langer Jewel-Bearing Plant

Wlk Walk
W-L LL Washington-Lincoln Laurels for Leaders
wlm working level month
WLM Women's Liberation Movement
WLMI Wildlife Management Institute
WLMK William Lyon Mackenzie King (former Canadian Prime Minister)
WLMO Worldwide Logistics Management Office (USA)
Wlmsbrg Brdg Williamsburgh Bridge
Wln Wellington
WLN Washington Library Network
W Long west longitude
W'loo Waterloo
W Loth West Lothian
WLP Wallops Island, Virginia (tracking station)
WLPB War Labor Policies Board
WLPS Wild Life Protection Society
WLPSA Wild Life Preservation Society of Australia
wlr wrong-length record(ing)
Wlr Walter
WLR Weekly Law Reports
WLRI World Life Research Institute
WLRs Weekly Law Reports
Wls Wells
WLS Wild Life Sanctuary
WLSC West Liberty State College
WLSP World List of Scientific Periodicals
WLSR Wild Life Society of Rhodesia; World League for Sexual Reform
WLTBU Watermen, Lightermen, Tugmen, and Bargemen's Union
WLU Wilfrid Laurier University; World Liberal Union
W & LU Washington and Lee University
WLUS World Land Use Survey
WLW Women Library Workers
WLWH Workshop Library on World Humour
Wly westerly
wlz waltz
wm wattmeter; wavemeter; white male; white metal; winner's female; wire mesh; wordmark (flow chart)
w/m weight or measure; white male
w & m weight and/or measurement

Wm William
WM Western Maryland (railroad); White Motors; William McKinley (25th President of the U.S.); Women Marines; Worshipful Master
W & M College of William and Mary; War and Marine; Washburn & Moen (wire gauge)
WM World Monitor (Christian Science Monitor magazine and television program)
WMA Wildlife Management Area; Women Marines Association; World Medical Association
WMAA Whitney Museum of American Art
WMAC Waste Management Advisory Council
WMARC World Maritime Administrative Radio Conference
WMATA Washington Metropolitan Area Transit Authority
WMATC Washington Metropolitan Area Transit Commission
WMB War Mobilization Board
WMBL Wrightsville Marine Biomedical Laboratory
WMC Ways and Means Committee; Western Maryland College; World Meteorological Center (WMO); World Methodist Council
WMCCA Washington Metropolitan Coalition for Clean Air
WMCE Western Montana College of Education
WMCIU Working Men's Club and Institute Union
WMcK William McKinley (25th President of the U.S.)
WMCL William Mitchell College of Law
WMCP Women's Medical College of Pennsylvania
wmd wind measuring device
Wmd Willemstad
WMD Water Management District; Weights and Measures Division
WMECO Western Massachusetts Electric Company
Wmg Cal Wilmington, California
Wmg, Del Wilmington, Delaware
Wmg NC Wilmington, North Carolina
WMI Webbing Manufacturers Institute; Wildlife Management Institute
W Mid West Midlands

wmk watermark
w/m°k watt per meter degree kelvin (thermal conductivity unit)
WMM World Movement of Mothers
WMMA Woodworking Machinery Manufacturers' Association
Wmn Wilmington, North Carolina
WMNF White Mountain National Forest
WMO World Meteorological Organization
WMOAS Women's Migration and Overseas Appointments Society
W of Mormon Words of Mormon
wmp with much pleasure
WMR Wasatch Mountain Railway
WMS Waste Management System; Webster Memory Scale; Women in Medical Service; Women's Medical Specialist; Work Measurement System; World Magnetic Survey
W & MS Wisconsin & Michigan Steamship (company)
WMS Willem Mengelberg Stichting (Dutch—Willem Mengelberg Foundation)
WMSC Women's Medical Specialist Corps
WMSDI Western Mood & Sleep Disorders Institute
W & M SS Co Wisconsin & Michigan Steamship Company
wmt weighing more than
WMT Wilson Marine Transit
WMTB Western Motor Tariff Bureau
WMTC Women's Mechanized Transport Corps
Wmth Westmeath
WMU Western Michigan University
WMUSE World Markets for US Exports
W M W & NW Weatherford, Mineral Wells & Northwestern (railroad)
WMWR Wichita Mountains Wildlife Refuge (Oklahoma)
w/n well-nourished
WN Worlds of Nature (Amarillo botanical and zoological gardens)
WN Weekly Notes
WNA Washington, DC, National Airport; winter North Atlantic (loadline marking for ships crossing the North At-

lantic in winter)
WNAP Washington National Airport
wnb will not be
WNBA Women's National Book Association
WnBanc Western Bancorporation
WNCCC Women's National Cancer Control Campaign
wndml windmill
WNDO Weather Network Duty Officer
wndp with no down payment
WNE Welsh National Eisteddfod
wng warning
WNGA Wholesale Nursery Growers of America
wnl within normal limits
WNLF Women's National Liberal Federation
wnm white noise making
WNM Washington National Monument
WNMC Weather Network Management Center (USAF)
WNNP Walpole-Nornalup National Park (Western Australia)
WNO Welsh National Opera
WNP Wankie National Park (Zimbabwe); Warrumbungle NP (New South Wales); Welsh National Party; Westland NP (South Island, New Zealand); Wilpattu NP (Ceylon); Wyperfeld NP (Victoria, Australia)
WNRE Whiteshell Nuclear Research Establishment
WNS Washington National Symphony (District of Columbia); Women's News Service
WNSB White Nile Scheme Board (Sudanese cotton production)
WNW west northwest
WNW *west noordwest* (Dutch—west northwest)
WNWDA Welsh National Water Development Authority
WNWR Wapanocca National Wildlife Refuge (Arkansas); Washita National Wildlife Refuge (Oklahoma); Wheeler National Wildlife Refuge (Alabama); Willapa National Wildlife Refuge (Washington)
WNY West New York, NJ
WNYNRC Western New York Nuclear Research Center
WNYNSC Western New York Nuclear Service Center
wo wait order; water-in-oil

(emulsion); *wie oben* (German—as previously mentioned); wine of origin; without; work order; write out; written order
wo' war; wore
w/o without
WO War Office; Warrant Officer; Welsh Office
WO *World Oil*
WOA Wharf Owners' Association
WOAR Women Organized Against Rape
wob washed overboard
wobndr(s) without binder(s)
wobo(s) without blowout(s)
Wobs Wobblies
woc without compensation
WOCCI War Office Central Card Index
wocg weather outline contour generator
WOCIT We Oppose Computers In Tournaments
WOCL War Office Casualty List
W & O D Washington & Old Dominion (railroad)
WODA World Dredging Association
WODECO Western Offshore Drilling and Exploration Company
woe. without equipment
Woe *Woensdag* (Dutch—Wednesday)
WOFIWU World Federation of Industrial Workers Unions
wofttngs without fittings
wog golliwog; polliwog; water or gas (valve); wily oriental gentleman [a confidence man from the Far East]; with other goods
'wog golliwog; polliwog
WOG Wily Oriental Gentleman (applied to Farouk I of Egypt and similar monarchs of the area)
WOGA Western Oil and Gas Association
wogs workers on government service
wogs (British slang—wily oriental gentlemen; wily oriental peoples)
WOGS War Office General Stores
WOGSC World Organization of General Systems and Cybernetics
woh work on hand
WOHC Warrant Officer, Hospital Corps
WOIS Wisconsin Occupational

Information System
WOJG Warrant Officer, Junior Grade
WOK Warren O Kessler
wol wharf owners' liability
WOL War Office Letter
WOLA Washington Office on Latin America
Wolf Wolfgang; Wolfmar; Wolfrad; Wolfram; Wolfred
Wolfs Wolfson College, Oxford
wom wireless operator mechanic
WOM Woomera, Australia (tracking station)
WOMAN World Organization of Mothers of All Nations
Women's Lib Women's Liberation Movement
womi women on words and images
womlib women's liberation
won monetary unit of Korea
won. wool on needle (knitting)
W-o-N Walton-on-Naze
WONARD Women's Organization of the National Association of Retail Druggists
wong weight on nose gear
won't will not
WOO Western Operations Office (NASA); World Oceanographic Organization
Wood Woodbine; Woodbridge; Woodburn; Woodbury; Woodfield; Woodfin; Woodhill; Woodley; Woodrow; Woodruff; Woodson; Woodville; Woodward; Woodworth
Wood Buffalo Wood Buffalo National Park in northern Alberta
woof(s) woofer(s)
Wool *Woolworth's* (Circuit Court Reports)
Woolwich Royal Arsenal at Woolwich on the south bank of the Thames near London
woool words out of ordinary language
Wooster(sheer) (British contraction—Worcestershire)
wop. with other property; without (immigration) papers; without personnel
wopar(s) without partition(s)
wope without personnel or equipment
WOPN Women Officers Professional Network
wopo without purchase order
WOQT Warrant Officer Qualification Test
wor without our responsibility
Wor Worshipful
worbat wartime order of battle

Worc Worcester College
WORC Washington Operations Research Council
Worc Coll Worcester College—Oxford
Worc Reg Worcester Regiment
Worcs Worcestershire
word proc word processor
Words Wordsworth
WORDS Western Operational Research Discussion Society
words/min words per minute
words/sec words per second
WORK Wanted Older Residents (with) Knowhow
Work. Comp Workmen's Compensation
workfare working for welfare (alternative to high-cost-assistance welfare)
workfare welfare programs aimed at returning people to the workforce
workh workhouse
workingclass English modern cockney
Workmen's Workmen's Circle; Workmen's Compensation
World World Almanac
World Bank International Bank for Reconstruction and Development (IBRD)
worm write once, read many times
Wormald Wormald International Security
WORSAMS Worldwide Organization Structure for Army Medical Support
worse word selection
WOS Washington Opera Society; Western Orchestral Society; Wilson Ornithological Society
wosac worldwide synchronization of atomic clocks
WOSB War Operations Selection Board
WOSD Weapon Operational Systems Development
WOSL Women's Overseas Service League
wot wide-open throttle
WOTAG Women's Taxation Action Group
W-o-t-N Walton-on-the-Naze
wott wolves on the track (prowling males)
wouldn't would not
W & O V Washington & Ouachita Valley (railroad)
wow waiting on weather; worst of worst (prisoners)
w-o-w worst-on-worst (worst on top of the worst possible disaster, etc.)

W o W War on Want
WOW Wider Opportunities for Women; Women's Opportunity Week; Woodmen of the World
w/o wn without winch
wp waste package; waste pipe; water repellency; waterproof; water repellent; way point; weather permitting; white phosphorus; wild pitch; will proceed; working paper; working party; working point; working pressure
wp (WP) word processing; word processor
w-p waterproofed
w/p without prejudice
w/p (W/P) withdrawing and passing; withdrawn/passed
Wp Worship(ful)
WP War Plan(s); Warsaw Pact; Western Pacific (railroad); West Point; West Virginia Pulp and Paper (stock exchange symbol); Worthington Pump; Worthy Patriarch
WP Wiener Philharmoniker (German—Vienna Philharmonic Orchestra); Winkler Prins Encyclopedieen (Dutch—Winkler Prins Encyclopedia)
wpa with particular average
WPA Western Pine Association; Western Psychological Association; William Penn Association; Women's Prison Association; Works Progress Administration; World Parliament Association; World Psychiatric Association
WPAFB Wright-Patterson Air Force Base
WPA & H Women's Prison Association and Home
wpar(s) with partition(s)
WPAS Work Package Authorization System
wpb wastepaper basket
WPB War Plan Basic; War Production Board (World War II)
WPBA Western Power Boat Association; Women's Professional Billiard Association
WPBIC Walker Problem Behavior Identification Checklist
WPBL Women's Professional Basketball League
WPBS Welsh Plant Breeding Station
wpc water pollution control; watts per candle; wood plastic combination; world planning chart
WPC Washington Press Club; William Penn College; Wom-

en's Press Club; World Peace Council
WPCA Water Pollution Control Act
WPCC Wage and Price Control Council; Western Pharmaceutical and Chemical Corporation
WPCF Water Pollution Control Federation
WPD Work Package Department (ONWI)
wpe white porcelain enamel
WPEC World Plan Executive Council
w/p equipment word-processing equipment
WPF World Peace Foundation
WPFC Western Pacific Fisheries Commission; World Press Freedom Committee
Wpfl Worshipful
wpg waterproofing
WPg West Point graduate
WPG gunboat (USCG symbol)
WPGA Wisconsin Personnel and Guidance Association; Women's Professional Golfers' Association; Wyoming Personnel and Guidance Association
WPGR Willem Pretorius Game Reserve (South Africa)
WPGT Women's Professional Golf Tour
WPHC Western Pacific High Commissioner
WPHI Western Pennsylvania Horological Institute
WPHOA Women Public Health Officers Association
wpi wholesale price index
WPI Wall Paper Institute; Western Psychiatric Institute (Pittsburgh); Worcester Polytechnic Institute; World Press Institute; Waxed Paper Institute
WPI World Port Index
W pk Ward's (mechanical tissue) pack
WPK Workers Party of (North) Korea
wpl warning point level
WPL Weapons Propulsion Laboratory; Wichita Public Library; Winnipeg Public Library; Worcester Public Library
WPLC Wisconsin Power and Light Company
WPLO Water Port Liaison Office(r)
wpm words per minute
WPMSF World Professional Marathon Swimming Federa-

tion
wpn weapon
WPN West Penn Traction (stock exchange symbol)
WPN *World Press News*
wpns weapons
wpo world public opinion
wpo (WPO) without purchase order
WPO Water Programs Office (Environmental Protection Agency); Wiener Philharmonic Orchester (Vienna Philharmonic Orchestra); World Ploughing Organization
WPOD Water Port of Debarkation
WPOE Water Port of Embarkation
wpp waterproof paper packing
WPP West Penn Power Company; Witness Protection Program; Work Package Plan(ning)
WPPC West Penn Power Company
WPPDA Welfare and Pension Plans Disclosure Act
WPPO Work Package Program Office(r)
WPPP Work Package Program Plan(ning)
WPPSI Wechsler Preschool and Primary Scale of Intelligence
WPPSS Washington Public Power Supply System
wp & r work-planning-and-review (discussions)
WPRA Wallpower and Paint Retailers' Association; Waste Paper Recovery Association; Women's Professional Rodeo Association
WPRL Water Pollution Research Laboratory
wpr's wartime personnel requirements
WPRS Wittenborn Psychiatric Rating Scale
wps with prior service; words per second
WPs Warsaw Pact members; Warsaw Pact nations
WPS Waveform Processing System; Wildlife Preservation Society; Wildlife Preserve Society; World Peanut Syndicate; World Porpoise Society
WPSA World Professional Squash Association; World's Poultry Science Association
WPSL Western Primary Standard Laboratory
WPSP White People's Socialist Party
WPT Windfall Profits Tax

WPTB Wartime Prices and Trade Board
WPTF World Peace Tax Fund
wpu with power unit; write punch
wpwod will proceed without delay
W-P-W syndrome Wolff-Parkinson-White syndrome
WPY World Population Year (1974)
WP & Y White Pass & Yukon (railroad)
WP & YR White Pass & Yukon Route
WPZ Woodland Park Zoo (Seattle)
wq water quench
WQA Water Quality Association
WQCB Water Quality Control Board
WQF Wider Quaker Fellowship
wr war risk; water repellent; write (flow chart); write out
w/r water and rail; water resistant
w & r water and rail; welfare and recreation
Wr Walter
WR Ward Room; Warehouse Receipt; War Reserve; Wassermann Reaction; Western (railway) Region; West Riding
WR *Weekly Reporter*
W.R. *Wilhelmus Rex* (Latin—King Wilhelm, King William)
WRA War Relocation Authority; Water Research Association; Western Railway of Alabama; Winchester Repeating Arms (company)
WRA *Water Resources Abstracts*
WRAAC Women's Royal Australian Army Corps
WRAAF Women's Royal Australian Air Force
WRAC Women's Royal Army Corps
wraceld wounds received in action combat with enemy or in line of duty
WRAF Women's Royal Air Force
WRAIN Walter Reed Army Institute of Nursing
WRAIR Walter Reed Army Institute of Research
WRAMA Warner-Robins Air Material Area
WRAMC Walter Reed Army Medical Center
Wrangell Wrangell Island (*Vrangelya Ostrov*) in the Russian

Arctic, once an American possession
Wrangell–St Elias largest-area national park in the U.S.
WRANS Women's Royal Australian Naval Service
WRAP Weapons Readiness Analysis Program; Weighted Record Analysis Program
WRAT Wide-Range Achievement Test
WRB War Refugee Board; Water Resources Board
WRBC Weather Relay Broadcast System
wrc water-retention coefficient
WRC Water Research Center; Water Resources Commission; Weather Relay Center; Welding Research Council
WRCB Water Resources Control Board
wrcr wife's restitution of conjugal rights
WRCUP Western Region Canadian University Press
WRDC Western Rural Development Center; Westinghouse Research and Development Center
WRE Weapons Research Establishment (Woomera, Australia)
WREE Women for Racial and Economic Equality
WREEC Western Regional Environmental Education Council
w ref with reference
w reg with regard (to)
WREN Women's Royal Naval Service
wresat weapons research establishment satellite
W-response whole response
WRF World Rehabilitation Fund; World Research Foundation
wrfg wharfage
WRGH Walter Reed General Hospital
WRH Walter Reed Hospital
WRHS Western Reserve Historical Society
wri war risk insurance
WRI War Resisters' International; Weatherstrip Research Institute; Wellcome Research Institute; Will Rogers Institute; Wire Reinforcement Institute; Wire Rope Institute; Women's Rural Institute; World Reserves Institute; World Resources Institute
WRI *World Research INK* (monthly publication)

WRIR Walter Reed Institute of Research
WRIT Waste-Rock Interactions Technology (program)
wrk work (flow chart)
Wrk Workington
WR Knottman (abbreviated signature—we are not man and wife)
wrkshp workshop
wrl wing reference line
WRL Wantage Research Laboratory; War Readiness Material; War Resisters League; Westinghouse Research Laboratories; Willow Run Laboratories (University of Michigan)
WRLC World Role of Law Center (Duke University)
wrm war readiness materiel
WRM Wasatch Railway Museum
wrmn wireman
wrms weighted root mean square
WRMT Woodcock Reading Mastery Test
wrn wool round needle (knitting)
WRNGA William Rockhill Nelson Gallery of Art (Kansas City)
WRNR Women's Royal Naval Reserve
WRNS Women's Royal Naval Service
wrnt warrant
WRNWR White River National Wildlife Refuge (Arkansas)
wro war risk only
WRO Washington Regional Office; Weed Research Organization
wros with right of survivorship
WRP Workers' Revolutionary Party (British communists)
WRPA Water Resources Planning Act
WRPC Weather Records Processing Center(s)
WRRA Women's Road Records Association (cycling)
WRRC Willow Run Research Center
WRRI Water Resources Research Institute
WRRR Walter Reed Research Reactor
WRRS Wire Relay Radio System
wrs war reserve stock(s)
WRS Warning and Report(ing) System; Worldwide Reference Sources
WRSA Western Regional Science Association

WRSIC Water Resources Scientific Information Center
wrsk war-readiness spares kit
WRSP *World Register of Scientific Periodicals*
wrt wrought
wrtd warranted
WRTF Waste Retrieval and Treatment Facility
wrtr writer
wru who are you?
WRU Western Reserve University
wrv water relief valve
WRVS Women's Royal Voluntary Service
WRVT Wide-Range Vocabulary Test
wr(w) war reserve (weapon)
WRX Western Refrigerator Express (railroad code)
WRY World Refugee Year
W Ry A Western Railway of Alabama
WR Yorks West Riding, Yorkshire
ws water supply; weather station; wobbling slowly; working space; working storage
w/s weapon system; weather ship
w & s whiskey and soda
WS Wallops Station (NASA); Ware Shoals; Warner & Swasey; weapon system(s); Western Samoa; West Saxon(y); West Sussex; Wilderness Society; Wildlife Society; windspeed; Writer to the Signet (Scottish lawyer)
WS *Wiener Stadtbibliothek* (German—Vienna State Library)
W S *Washington Star*
wsa weapons system analysis
WSA Weed Society of America; Worker-Student Alliance
WSA War Shipping Administration; Western Soccer Alliance
WSA *Wasser und Schiffahrtsampt* (German—Water and Ship Canal Authority)
WSAC West of Scotland Agricultural College
WSAD Weapon System Analysis Division (USN)
WSAG Washington Special Action Group (personnel in Situation Room in White House basement)
W Sam Western Samoa
WSAO Weapon System Analysis Office
WSAP Weighted Sensitivity

Analysis Program (EPA)
WSAVA World Small Animal Veterinary Association
wsb water-soluble base; wheatsoy blend; will send boat
WSB Wharton School of Business (University of Pennsylvania); World Scout Bureau
WSBA Wyoming School Boards Association
WSBI Western Sciences Behavioral Institute
wsc weapon system contractor
WSC Western Simulation Council; Western Society of Criminology; Winona State College; Winston Spencer Churchill; Wisconsin State College; World Series Cricket; Writing Services Center
WSCC Western State College of Colorado
WSCF World Student Christian Federation
Wschr *Wochenschrift* (German—weekly magazine)
WSCS Woman's Society for Christian Service
wsd working stress design
wsdb world studies data bank
WSDC Women's Self-Defense Council
WSDL Weapons System Development Laboratory
WSEC Washington State Electronics Council
WSECL Weapon System Equipment Component List
wsed weapon system electrical diagram(s)
WSED Weapon Systems Evaluation Division
WSEG Weapons Systems Evaluation Group
WSEL Weapons System Engineering Laboratory
WSEP Waste Solidification Engineering Prototype Plant (AEC); Weapon System Evaluation Program
WSET Writers and Scholars Educational Trust
wsev (WSEV) winged surface-effect vehicle
WSF Washington State Ferries; Western Sea Frontier; Women's Strike for Peace; World Sephardic Federation
WSFI Water Softener and Filter Institute
WSFR Worcestershire and Sherwood Foresters Regiment
wsg worthiest soldier in the group
WSG Wesleyan Service Guild; Wire Service Guild

WSGE Western Society of Gear Engineers

wshr/dryr washer/dryer

WSHS Wisconsin State Historical Society

WSI Writers and Scholars International

WSIN Western State Information Network

WSJ Wall Street Journal

WSL Warren Spring Laboratory; Washington State Library

WSLA World Savings and Loan Association

WSLCB Washington State Liquor Control Board

WSLF Western Somali Liberation Front (communist)

WSLO Weapon System Logistics Office(r)

Wsm Wesermünde

WSM Weapon System Manager; Western Society of Malacologists; W Somerset Maugham

WSM Weapon System Manual

WSMAC Weapon System Maintenance Action Center

WSMC Western Space and Missile Center, Vandenberg Air Force Base, California; Western States Movers Conference

WSMO Weapon System Materiel Office(r)

WSMR White Sands Missile Range

WSMSA Washington Standard Metropolitan Statistical Area

WSN Washington, DC

WSNA Washington State Nurses Association

WSNM White Sands National Monument

WSO Warrant Stores Office(r); Weapon System Office(r); Western Support Office (NASA); Wichita Symphony Orchestra; Winnipeg Symphony Orchestra; World Simulation Organization

WSO Wiener Symphonisches Orchester (German—Vienna Symphony Orchestra)

WSOC Wider Share Ownership Council

wsp water supply point; working steam pressure

WSP Witness Security Program; Women Strike for Peace; Work Study Program; Work Systems Package (naval salvage device); Work Systems Program; World Series of Philately; Wyoming State Parks

WSPACS Weapon System Program and Control System

WSPB Western Society of Business Publications

WS Pen Washington State Penitentiary

WSPG White Sands Proving Ground

WSPL Winston-Salem Public Library

WSPO Weapon System Project Officer

WSPOP Weapon System Phase-Out Procedure

WSPU Women's Social and Political Union

wsr (WSR) weapon system reliability

w/sr watt(s) per steradian

Wsr Wesermünde

W & S R Warren & Saline River (railroad)

WS & RB Washington Surveying and Rating Bureau

WSRI World Safety Research Institute

w/srm² watt(s) per steradian square meter

WSS Warfare Systems School; Winston-Salem Southbound (railroad); World Ship Society

WSSA Weapon System Support Activities; World Secret Service Association

WSSC Weapon System Support Center

WSSCA White Sands Signal Corps Agency

WSSO Winston-Salem Symphony Orchestra

WSSS Weapon System Storage Site

WSSSP Western States Small Schools Projects

WSS & YP White Sulphur Springs & Yellowstone Park (railroad)

WST Whitworth Standard Thread

WSTA Washington State Trustees Association; White Slave Traffic Act

WSTC Winston-Salem Teachers College

WSTF White Sands Test Facility (NASA)

WSTI Waterbury State Technical Institute; Welded Steel Tube Institute

WSTNRA Whiskeytown-Shasta-Trinity National Recreation Area (California)

WSU Washington State University; Wayne State University; Western State University; Wichita State University

w sup water supply

W Sus West Sussex

WSUSM Wayne State University School of Medicine

WSV Wiener Stadtwerke Verkehrsbetriebe (Vienna transportation system)

wsw white sidewall (tires)

WSW west southwest

WSWA Wine and Spirits Wholesalers of America

WSWL Warheads and Special Weapons Laboratory

WSWMA Western States Weights and Measures Association

WSWS Wexford Slobs Wildfowl Sanctuary (Ireland)

wt watch time; waterproof(ed); waterproofing; watertight; weight; withholding tax (WT)

wt % weight percent

w/t wireless telegraph(y)

w/t (W/T) walkie/talkie

w & t wear and tear

WT war time; wealth tax; winterization test; withholding tax

W & T Wrightsville & Tennille (railroad)

WT The Winter's Tale

WTA Washington Technological Associates; Women's Tennis Association; World Trade Association; World Transport Agency

WTAA World Trade Alliance Association

WTAU Women's Total Abstinence Union

wtawtar where there's a will there's a relation

w/tax withholding tax

Wtb Whitby

Wtb Wörterbuch (German—dictionary)

WTBA Washington Toll Bridge Authority; Water-Tube Boilermakers' Association

WTB & TS Watchtower Bible and Tract Society (Jehovah's Witnesses)

WTC Women's Timber Corps; World Tanker Corporation (Niarchos); World Trade Center; World Trade Commission

wtchmn watchman

wtd watertight door; weighted

WTD World Trade Directory

WTDAOT What to Do About Old Town

wte wartime extension

wte (WTE) world time equivalent

WTE World Tapes for Education

Wt Eng Warrant Engineer

wtf waterfront; will to fire
Wtf Waterford
WTFDA Worldwide TV-FM-DX Association
WTFP Wolf Trap Farm Park (Vienna, Virginia)
WTG Welt-Tierärztegesellschaft (German—World Veterinary Association)
wthr weather
WTIC World Trade Information Center
WTIS World Trade Information Service
WTJ Westminster Theological Journal
WTL Wyle Test Laboratories
wtm write tape mark
WTMA West Texas Museum Association
wtmh watertight manhole
WTNID Webster's Third New International Dictionary
WTNR Wadi Tabor Nature Reserve (Israel)
WTO Warsaw Treaty Organization; World Tourism Organization
Wt Ofcr Warrant Officer
w/t office wireless/telegraph office (aboard ships in the 1920s became the radio room)—the radio shack
WTP Weapons Testing Program; Western Tropical Pacific
wtqad watertight quick-acting door
wtr waiter; winter; writer
Wtr Water
WTR Western Test Range (formerly Pacific Missile Range)
WTRC Wool Textile Research Council
wtrz winterize
wtrzn winterization
wts word terminal synchronous
WTS Watchtower Society; William Tecumseh Sherman (1820–1891); Women's Transport Service
WTSC West Texas State College
wtspt waterspout
WTT World Tennis Team
WTTA Wholesale Tobacco Trade Association
WTU Washington Theological Union
WTUC World Trade Union Conference
WTVN Worldwide Television News
wtw wet tissue weight
wu work unit
WU Washington University;

Weather Underground Organization; Wesleyan University; Western Union; Wilberforce University; Wittenberg University
W/U Western Union
WUA Western Underwriters Association
wuaa wartime unit aircraft activity
WUAA Wartime Unit Aircraft Activity
wuc work unit code
WUCM Work Unit Code Manual
WUCOS Western European Union Chiefs of Staff
WUCT World Union of Catholic Teachers
WUCWO World Union of Catholic Women's Organizations
WUD Water Utility Department
WUDO Western European Defense Organization
WUF World Underwater Federation; World Union of Free Thinkers
WUI Western Union International
WUIS Work Unit Information System
WUJS World Union of Jewish Students
Wuli Xuebao (Acta Phys Sin) Acta Physica Sinica (Chinese Journal of Physics)
WULTUO World Union of Liberal Trade Union Organizations
WUM Women's Universal Movement
WUMP(S) White Urban Middleclass Protestant(s)
WUNS World Union of National Socialists
WUO Weather Underground Organization
WUOSY World Union of Organizations for Safeguarding Youth
Wupatki Wupatki National Monument in northern Arizona
WUPJ World Union for Progressive Judaism
WUPO World Union of Pythagorean Organizations
WUS Western United States; World University Service
WUSL Women's United Service League
WUSM Washington University School of Medicine
wut warmup time
WUT Washburn University of

Topeka
wuts work-unit time standard
WUX Western Union (teleprinter) Exchange
wv wall vent; whispered voice; wind velocity; with view (room with view)
w/v weight in volume
WV West Virginia; West Virginia Pulp and Paper Company
W Va West Virginia; West Virginian
WVA World Veterinary Association; Wyoming Vocational Association
WVAESP West Virginia Association of Elementary School Principals
WVAS Wake-Vortex Avoidance System
W Va Turn West Virginia Turnpike
WVa U Lib West Virginia University Library
WVAWRD West Virginia Water Resources Division
WVC Wenatchee Valley College
wvd waived
WVD Werelverbond van Diamantbewerkers (Dutch—World Alliance of Diamond Workers)
wvdc working voltage—direct current
WVEA West Virginia Educational Association
wveh wheel(ed) vehicle
wvem water-vapor electrolysis module
WVF World Veterans' Federation
WVFIC West Virginia Foundation for Independent Colleges
WVIT West Virginia Institute of Technology
WVL Warfare Vision Laboratory (USA)
WVLA West Virginia Library Association
WVMA Women's Veterinary Medical Association
Wvn Wivenhoe
W V N West Virginia Northern (railroad)
WVPA World Veterinary Poultry Association
WVRB West Virginia Rating Bureau
WVRO World Vision Relief Organization
WVS Women's Voluntary Service
WVSBA West Virginia School Boards Association

WVSC West Virginia State College; Wisconsin Vocational Studies Center
WVSP West Virginia State Police
WVSSPC West Virginia Secondary School Principal's Commission
wvt water vapor transfer; water vapor transmission
WVT Watervliet Arsenal
wvtr water vapor transmission rate
w/vu with view
WVU West Virginia University
WVWC West Virginia Wesleyan College
ww warehouse warrant; water white; waterworks; wirewound; wrong word
w/w wall-to-wall (carpet, floor covering, linoleum, tile); weight for weight
ww werkwoord (Dutch—verb)
Ww Witwe (German—widow)
WW Walworth (trademark); Woodmen of the World; Woodrow Wilson (28th President of the U.S.); world war; world wide
W-W Winchester-Western
W & W Waynesburg & Western (railroad); Winchester & Western (railroad)
WW Who's Who
WW I World War I (1914–1918)
WWIVM World War I Victory Medal
WW II World War II (1939–1945)
WWIIHSLB World War II Honorable Service Lapel Button
WWIIVM World War II Victory Medal
wwa with the will annexed
WWA Western Writers of America
WWABNCP Worldwide Airborne Command Post (USAF)
wwap worldwide asset position
WWB Walt Whitman Bridge
WWBA Walt Whitman Birthplace Association; Western Wooden Box Association
WWBCN World-Wide Business Centers Network
wwc wall-to-wall carpeting
WWC Walla Walla College; Warren Wilson College; William Woods College; World Weather Centers (Melbourne, Moscow, Washington, D.C.)
WWCP Walking Wounded Collecting Post

WWCTU World's Women's Christian Temperance Union
wwd weather working days; windward
WWD Women's Wear Daily
WWDC World War Debt Commission
W Wdr Warrant Wardmaster
wwdShex weather working days Sundays and holidays excluded
Wwe Weduwe (Dutch—widow); *Witwe* (German—widow)
wwf welded wire fabric
WWF Welder Wildlife Foundation; Woodrow Wilson Foundation; World Wildlife Fund
WWG World Wildlife Guide
WWHS Wilbur Wright High School; Woodrow Wilson High School
wwi whirlwind computer
WWI Weight Watchers International; World Watch Institute
WWIB Western Weighing and Inspection Bureau
WWICS Woodrow Wilson International Center for Scholars
wwio worldwide inventory objective
WWJC Western Wyoming Junior College
WWM WW Morrow
WWMB Woodrow Wilson Memorial Bridge
WWMC Woodrow Wilson Memorial Commission
WWMCCS Worldwide Military Command and Control System
WWMHA World-Wide Miniature Horse Association
WWMMP Western Wood Moulding and Millwork Producers
w/wn with winch
W Wnd Drft West Wind Drift (Antartic)
WWNFF Woodrow Wilson National Fellowship Foundation
WWNSSS World-Wide Network of Standard Seismograph Stations
WWNT West Wales Naturalists Trust
w/wo with or without
WWO Wing Warrant Officer; World Weather Organization
wwp water wall peripheral; working water pressure; write without program
WWP Washington Water Power company; Workers World

Party (leftwing)
WWPA Western Wood Products Association; World Wide Philatelic Agency
WWR Washington Week in Review (educational television)
ww's walla wallas (Hong Kong harbour launches)
WWSA Walt Whitman Society of America
WWSC Western Washington State College
WWSN World-wide Seismology Net (NBS)
WWSPIA Woodrow Wilson School of Public and International Affairs (Princeton University)
wwss water wall side skegs
WWSSN World-Wide Standardized Seismograph Network
WWSU World Water Ski Union
wwt whitewall tires
WWTP Waste Water Treating Process
WWTVN World-Wide Television News
WWTVS World-Wide Television Service
WWU West Washington University
W W V call letters of United States Bureau of Standards worldwide radio time signal; Walla Walla Valley (railroad)
WWVH World Wide Time (US Bureau of Standards, Hawaii)
WWW World Weather Watch; World Without Walls; World Without Wars
WWW Who Was Who
W. W. Wash Walla Walla, Washington
WWWF Worldwide Wrestling Federation
WWWV Women World War Veterans
WWWVA Wild, Wonderful West Virginia
WWWW Women Who Want to be Women
WWWW Worldwide What & Where—geographic glossary and traveller's guide
wwwwwh who, what, when, where, why, how—reporters' mnemonic for encompassing elements of a news story
WWY Warwickshire and Worcestershire Yoemanry
wx watts second; waxy; weather report
Wx weather; Wilcox (formation)

wxb wax bite
WXD meteorological radar station
wxg warning
wxp wax pattern
wy wey (14 pounds of wool)
Wy Way; Wyatt; Wycliffe
Wy *Wy-dit-Joli-Village* (French —Wy called Pretty Village)— near Paris
WY West Yorkshire; Wyoming
Wya Whyalla
WYACL World Youth Anti-Communist League
wyaio will you accept (the po-

sition) if offered
Wyantskill Wyantskill Center for (delinquent) Girls at Wyantskill, New York
Wyc Wycliffe; Wycliffe College
WYC Washington Yacht Club; Winthrop Yacht Club
WYCF World Youth Crusade for Freedom
Wycl Wycliffe
wye Y (as in wye circuit)
WYF World Youth Forum
wyo what's your opinion
Wyo Wyoming; Wyomingite

Wyo Sem Wyoming Seminary (college preparatory school)— Kingston, Pennsylvania
W Yorks West Yorkshire
WYR West Yorkshire regiment
wysiwyg what you see is what you get
WZ *Welt Zeit* (German—world time)
WZO World Zionist Organization
WZOA Womem's Zionist Organization of America
WZW *west zuidwest* (Dutch— west southwest)

X

x an abscissa (symbol); an unknown quantity (symbol); any point on a great circle; by (in measurements); cross; cross reactance (symbol); exchange; execute(d); extra; frost; gang territorial mark(er) or place where one gang will fight another; mole ratio; nowind distance; parallactic angle; specific acoustic reactance; hoarfrost (meteorology), kiss, mechanical defect, motion picture not suitable for viewing by minors, the spot where a crime was committed (*x* marks the spot), the position of objects on a chart or map, the signature of the illiterate; by (as in 3 x 5 file card)

x (X) $10 bill; Christ; Christian; Christianity; cross; execution(er); experiment; experimental (symbol); explosive (symbol); extra; extract(ed); Kienbock unit (symbol); magnification power; movie rating—no one under 17 admitted; reactance (symbol); research aircraft (symbol); single strength; times (multiplied by); univalent negative (symbol); unknown quantity; U.S. Steel Corporation (stock exchange symbol); X ray; Xavier; X ray —code for letter X

X longitudinal axis

X-2 counterintelligence

X-15 rocket-propelled research aircraft

X 17 mortality table

X.25 International Protocol for Packet-Switched Networks

xa chiasma; extended architecture; transmission adapter

XA Crucible Steel (stock exchange symbol); experimental (USAF symbol)

xaam experimental air-to-air missile

xact exact(ly); X (in any computer) automatic code translation

XAE merchant ammunition ship (naval symbol)

xafh X-band antenna feed horn

XAK merchant cargo ship (naval symbol)

XAKc merchant coastal cargo ship, small (naval symbol)

xal xenon arc lamp

Xalapa Jalapa

Xalisco Jalisco

Xalostoc San Cosme Xalostoc, Tlaxcala, Mexico

Xaltocan San Martin Xaltocan, Tlaxcala, Mexico

XAM merchant ship converted to minesweeper (naval symbol)

x-a mix. xylene-alcohol mixture (insect larva killer)

x-a mixture xylene-alcohol mixture

xan xanthic; xanthine; yellow

Xan Xanthe; Xanthian; Xanthippe; Xanthus

Xana Xanadu

xanth xanthoma(tosis)

Xantip Xantippe (archetype of the scolding termagent shrew as she was the peevish wife of Socrates)

XAP merchant transport (naval symbol)

XAPc merchant coastal transport, small (naval symbol)

x arm cross arm

XAS X-band Antenna System

xasm experimental air-to-surface missile (XASM)

xat X-ray analysis trial

Xav Xaver; Xavier; Xaviera

XAV auxiliary seaplane tender (naval symbol)

X-axis horizontal axis on a chart, graph, or map

xb crossbar; exploding bridgewire

XB experimental bomber

xbag excess baggage

Xbal Cristobal

X-band 5,200–10,900 mc

xbar crossbar

X bear grizzly bear

Xber December

xbr experimental breeder reactor

X-bracing cross bracing

X^{bre} *décembre* (French—December)

xbt expendable bathythermograph

xbts exhibits

xc cross country; ex coupon; X-chromosome

X-c X-chromosome

X$_c$ capacitive reactance

XC experimental cargo aircraft (naval symbol); Xavierian College

Xca Xcalac, Quintana Roo, México

xcar from the railroad car

XCG experimental cargo glider (naval symbol)

xch exchange

xchgr exchanger

X-chromosome female-producing gene found in male sperm

xcit excitation

X-City site of UN Headquarters along New York's East River between 42nd and 49th streets

xcl excess current liabilities
XCL armed merchant cruiser (naval symbol)
xclu exclusive; exclusivity
x-con ex-convict
xconn cross connection
xcp without coupon
XCR Extraterrestrial Research Center
X-craft midget submarines
xcs cross-country skiing
xct X-band communications transponder
xcu excuse; extra-care unit; extreme closeup
xc & uc exclusive of covering and uncovering
xcvr transceiver
x cy cross country
xd ex dividend; expiration date
x'd executed
X'd crossed out
XD Executive Development
X-day launching day
xdcr transducer
xder transducer
xdf X-band flow detection
X & DFLOT Experimental and Development Flotilla
xdh xanthine dehydrogenase
xdis ex distribution, without distribution
xdiv without dividend
X division branch of society consisting of swindlers and thieves
x'd out crossed out, deleted
xdp X-ray density probe; X-ray diffraction powder
xdpc X-ray diffraction powder camera
xdps X-band diode phase shifter
xdr expanded dynamic range; transducer
Xdr Crusader
x drs ex drawings
XDS Xerox Data Systems; X-ray Diffraction System
xdt xenon discharge tube
xe ex entitlement
Xe experimental engine; xenon
xecf x-e cold-flow engine
xeg X-ray emission gage
XEG Xerox Education Group
Xen Xenia; Xenik; Xenocratic; Xenophon (c.434–c.355); Xenos
xeno xenodiagnosis; xenodiagnostic; xenogenic; xenograft; xenolith; xenolithic; xenophile; xenophilia; xenophobe; xenophobia
Xeno Xenocrates; Xenophanes; Xenophon
xenobio xenobiologic(al)(ly); xenobiologist; xenobiology

xenodiag xenodiagnosis
xenop xenophile—person attracted to what is foreign
Xenop Xenophon
xenop(s) xenophobe(s); xenophobia(s); xenophobic(s)
XEP Xerox Educational Publications
xer Xerox reproduction
Xer Xerxes
xerocops xerocopies (books reproduced by xerography)
xerodups xerographic duplicates
xerog xerograph(ic)(al)(ly); xerography
xeromamo xeromammograph (also called xerox mammograph—xerographic process used in diagnosis of breast cancer)
xerorads xerographic radiographs
xes X-ray emission spectra
xf ex offer; extra fine
XF experimental fighter (naval symbol)
xfa crossed-field acceleration; X-ray fluoresence absorption
xfc X-band frequency converter
xfd crossfeed; X-ray flow detection
xfer transfer
xfh X-band feed horn
Xfher Christopher
xflt expanded flight-line tester
xfm X-band ferrite modulator
xfmr transformer
xformer transformer
xfqh xenon-filled quartz helix
xfrmr transformer
xft xenon flash tube
xg crossing
xgam experimental guided air missile (XGAM)
XGP Xerox Graphic Printer
xh extra hard; extra heavy; extra high
Xh Xhosa
XH experimental helicopter (naval symbol)
x heavy extra heavy
X-height height of central portion of lowercase letters exclusive of ascenders and descenders
xhf extra high frequency
x-high of a height equal to a lowercase x of the same face and size
xhil xenon high-intensity light
xhm X-ray hazard meter
xhmo extended huckel molecular orbit
xhr extra-high reliability
X-hr X-hour (when shipping

evacuation is ordered from major ports by NATO)
Xhs Xhosa
xhst exhaust
xhv extremely high vacuum
x hvy extra heavy
xi ex interest; xi particle
xia X-band interferometer antenna
Xianggang (Chinese—Victoria) —on Hong Kong island
Xian(s) Christian(s)
xic transmission interface converter
Xico Xicoténcatl
xil xilography; xilogravure (woodcuts)
xim X-ray intensity meter
xin without interest
Xin Xingu
Xina Christina
XING crossing (highway or railroad)
xio execute input-output
xiph xiphoid; xiphoidal
Xipho Xiphosura
Xiq-Xiq Xique-Xique, Brazil
xirs xenon infrared searchlight
xis xenon infrared searchlight
xist xistoma; xistomiasis
xistor transistor
xk X-band klystron
xl crystal; crystalline; extra large; extra long
Xl inductive reactance
xla X-band limiter anntenuator
xlam cross-laminate(d)
xlc xenon lamp collimator
xld experimental laser device
xldt xenon laser discharge tube
xlf ex lady friend
xli extra-low interstitial
xlnt excellent
XLO Ex-Cell-O (precision products, trade name)
xlps xenon lamp power supply
xlr experimental liquid rocket
xls extreme long shot; xenon light source
XLSS Xenon Light-Source System
xlt cross-linked polyethylene; excellent; xenon laser tube
xltn translation
xl & ul exclusive of loading and unloading
xlwb extra-long wheelbase
xm crossmatch; examine
xm (XM) experimental missile
xmm X-ray multiple mirror
Xm Christmas
XM experimental missile
XM-1 main battle tank (USA)
XM-706 Cadillac-Gage amphibious armed car and military personnel carrier called

the Commando
XM-723 cavalry of infantry fighting vehicle (USA)
Xma$ Christmas (commercialized)
Xmas Christmas
X-matching—cross matching
xmfr transformer
xmit transmit
xmitter transmitter
x mod experimental module
xmp (XMP) experimental extraordinary multiprocessor
xms X-band microwave source
XMS Experimental Development Specification; Xavier Mission Sisters
xmsn transmission
xmt exempt; transmit; X-band microwave transmitter
xmtg transmitting
xmtl transmittal
xmtr transmitter
xmt-rec transmit-receive
xmtr-rec transmitter-receiver
xn ex new
Xn Christian
XN experimental (USN)
Xndu Xanadu
xnor gate exculsive-nor gate
X-note $10 bill
XNS Xerox Network Services
xnt excellent
Xnty Christianity
xo crystal oscillator
XO Executive Officer; Experimental Office(r); Turner's syndrome wherein one of the sex-determining pair of XX chromosomes is missing
X-O cross-out test
xob xenon optical beacon
Xochi Xochimilco
x-off transmitter off
xoloiz xoloizcuintli (pronounced *sholloizquintly*)—Mexican hairless dog both hotblooded and flealess as well as faithful
x-on transmitter on
xon/xoff transmitter on/transmitter off
xoophorec xoophorectomic (sterilization); xoophorectomy (removal of the ovaries)
xor exclusive or (data processing)
xos extra outside clothing; extra outsize (clothing)
X-out cross out; delete; strike out
xover cross over
X-over cross over
xp express paid; xerodema pigmentosum
Xp fire-resistive protected cabi-

net, safe, or vault
XP (Greek—chi rho)—first two letters of the Greek word for Christ
xpa X-band parametric amplifier; X-band passive array; X-band planar array; X-band power amplifier
xpaa X-band planar-array antenna
XPARS External Research Publication and Retrieval System
XPC inshore patrol cutter (naval symbol)
xpd cross-polarization discrimination; cross-pollination discrimination; expedite(d)
xper without privileges
Xper Christopher
xpert expert
XPG converted merchant ship (naval symbol)
xpl explain; explanation; explosion; explosive
xplo explosion
xplos explosive
xplt exploit
XPM Xerox Planning Model
xpn expansion
Xpo Cristo (Spanish—Christ)
xpond transponder
xpp exprès payé lettre (French —express-paid letter)
xppa X-band pseudo-passive array; X-band pulsed-power amplifier
xpr ex privileges; without privileges
X-press Express
xprs express
xprts expertise; experts
xprtz expertize
xps X-band phase shifter; X-ray photoelectric spectroscopy; X-ray photoelectron spectroscopy
xps (XPS) X-ray photomission spectroscopy
xpt except; X-band pulse transmitter
xpt exprès payé télégraphe (French—express-paid telegraph)
Xpto Cristóbal (Spanish—Christopher)
X-punch punch in X row (11th row) of an 80-column punchcard
xq cross-question
XQ Experimental Target Drone
xqh xenon quartz helix
xr ex rights; Xerox radiography
Xr Christopher; examiner
XR External Relations (UNESCO)
X-rated movie moving picture

not recommended for minors
X-rated shops sex-oriented establishments such as massage parlors and adult bookstores
xray execution recorder analyzer
Xray letter X radio code
X-ray letter X radio code; photograph or photography made by X-rays; radiograph; radiography; roentgenograph; roentgenography; roentgen ray
xrb X-band radar beacon
xrcd X-ray crystal density
xrd X-ray diffraction
X rds crossroads
X-rea X-ray events analyzer
xref cross-reference
xrep auxiliary report
xrf X-ray fluorescence
xrfs X-ray fluorescence spectrometer
xrii X-ray image intensifier
xrl extended-range lance (missile)
xrm X-ray microanalyzer
xro xeroradiography
X-roads crossroads
xrp X-ray and photofluorography
xrpm X-ray projection microscope
xrpt X-ray and photofluorography technician
XRPT X-Ray and Photoflurorography Technician (USN)
xrspec X-ray spectograph
xrt ex-rights; without rights; X-ray technician
Xrx Xerox (corporation or copying process)
xs cross-section; excess; extra strength; extra strong
Xs atmospherics
xsa X-band satellite antenna
xsal xenon short arc lamp
XSB Xavier Society for the Blind
X-scale scale of a line parallel to the horizon
x sec extra sec *(très sec)*—dry champagne
xsect cross-section
xsf X-ray scattering facility
xsistor transistor
XSL Experimental Space Laboratory
xsm experimental strategic missile; experimental surface missile
xsoa excess speed of advance authorized
X-sonad experimental sonic azimuth detector
X-spot $10 bill
xspv experimental solid-propel-

lant vehicle
xsr X-band scatterometer radar
XSS Experimental Space Station
xsta X-band satellite-tracking antenna
xstd X-band stripline tunnel diode
xstda X-band stripline tunnel diode amplifier
xstr transistor
x str extra strong
xstrat cross-stratified
xt crosstalk; X-ray tube
xt (XT) extra technology (computer)
Xt Christ
xta chiasmata; X-band tracking antenna
xtal crystal
XTE X-ray Timing Explorer (satellite)
Xth tenth
Xtian Christian
xtk cross track
xtlo crystal oscillator
xtnd extend
xtnd wknd extended weekend
xtntn extension
xto X-band triode oscillator
xtr extra (computer flow chart)
XTR Xtra Inc
xtra extra
xtran experimental language; experimental translation
xtrm extreme
XTRU Xtra Inc (container) Unit
xtry extraordinary
xtwa X-band traveling-wave amplifier
xtwm X-band traveling-wave masser
Xty Christianity

xu X-ray unit; x-unit
Xu fire-resistive unprotected cabinet, safe, or vault
XU Xavier University
XUL Xavier University of Louisiana
Xulla formerly the Sula Islands of Indonesia
xuloc *xuloctzcuintle* (Aztec—hairless dog)—breed imported to Mexico from China in the late 16th century; popular among ranchers for its loud bark and its two-degree warmer body useful in the feet of sleeping bags
XUM Xerox University Microfilm
xut crosscut
xuv extreme untraviolet
xva X-ray videocon analysis
xvers transverse
XVP Executive Vice President
xvtr transverter
xw experimental warhead; ex warrants; without warrants
X-wave extraordinary wave
Xway (XWAY) Expressway
X-way expressway
X-weld X-shaped weld
X/Windows X-Window System
XWS Experimental Weapon System
xx without securities or warrants
XX doublecross; double strength; female (see X chromosome); twenty (Roman numerals)
X-X Xai-Xai (Mozambique seaport)
XX Dos Equis (Spanish—Two X)—Mexican beer
XXer doublecrosser

xxh double extra hard; double extra heavy
xxl cancel
XX-note (double-X note) $20 bill
xxos extra-large outside (clothing)
xxs extra-extra strength
xxx international urgency signal
XXX triple strength; triple-X; triple X syndrome; thirty (Roman numerals)
XXXX quadruple strength
XXXXX quintuple strength
XXY Klinefelter's syndrome wherein the sex-determining chromosomes are XXY instead of the normal XY
xy xylography
XY male
xya x-y axis
xyat x-y axis table
xyl ex young lady; xylene; xylography; xylophone
xylo xylophone
xyloc xylocain (lidocaine)
xylog xylography
xyp x-y plotter
xyr x-y recorder; xyridaceae; xyridaceous; xyridales; xyris
x yr dev ten-year device (US Army service badge)
xyt x-y table
xyv x-y vector
xyz examine your zipper (your fly is open)
XYZ XYZ Affair leading to undeclared naval war between France and the United States from 1798 to 1800
X zone adrenal cortex inner zone (of some young mammals)

Y

y altitude (symbol); an ordinate (symbol); an unknown quantity (symbol); depth or height (symbol); yacht; yard; year; yellow; yen (Japanese monetary unit); you; young(est)

y income (microeconomics); (Spanish—and)

Y admittance (symbol); Convair (symbol); service test (symbol); yacht; Yankee—code for letter Y; Yard (The Yard—Scotland Yard); yen (Japanese money unit); YMCA; YMHA; YWCA; YWHA

Y admittance (symbol); lateral axis (symbol); *ylös* (Finnish—up)

Y1C Yeoman First Class

Y2C Yeoman Second Class

Y3C Yeoman Third Class

Y-18 Ilyushin 18 aircraft

Y-40 Yak 40 aircraft

Y-5 molars lower molars characteristic of apes, hominoids, and humans

Y62 Ilyushin Il-Y62 jet airplane

ya yaw axis; young adult

YA Yasser Arafat; Young Adults; Youth Aliyah; Youth Authority

Y/A York-Antwerp Rules

YAA Yachtsmen's Association of America

YAAP Young Americans Against Pollution

YABA Yacht Architects and Brokers Association

YAC Young Adult Council; Young Alumni Club; Youth Advisory Council

YACA Youth and Adult Correctional Agency

YACC Young Adults Conser-

vation Corps

YACH Yugoslav-American Cooperative Home

yactoff yaw-actuator offset

YAD Youth Aid Division; Youth Authority Department

Yad Fiz Yaderna Fizika (Russian—Journal of Nuclear Physics)

yadh yeast alcohol dehydrogenase (YADH)

YAEC Yankee Atomic Electric Company

YAF Young Americans for Freedom

YAF-PAC Young Americans for Freedom—Political Action Committee

yag yttrium aluminum garnet

yag (YAG) yttrium, aluminum, garnet (surgical laser)

YAG district auxiliary miscellaneous (3-letter naval symbol)

yagl yttrium-aluminum garnet laser

yag laser yttrium-aluminum-garnet laser

YAI Young Adult Institute

YAIC Young American Indian Council

Yak Yakolev; Yakov; Yakovlevich

Yak Yakarta (Spanish—Djakarta)

YAK Yakovlev aircraft (named for its designer)

Yak-11 Soviet Yakovlev two-place trainer aircraft named Moose by NATO

Yak-12 Soviet Yakovlev two-place trainer aircraft

Yak-18 Soviet Yakovlev two-place aircraft used as a trainer and named Max by NATO

Yak-25 Soviet Yakovlev all-weather interceptor fighter aircraft named Flashlight by NATO

Yak-26 Soviet Yakovlev tactical reconnaissance aircraft named Mangrove by NATO

Yak-28 Soviet Yakovlev tactical bomber aircraft named Brewer by NATO

Yak-28P Soviet Yakovlev all-weather interceptor aircraft named Firebar by NATO

Yakutia Yakut Autonomous Soviet Socialist Republic (eastern Siberia)

yal yttrium-aluminum laser

YAL Young Australia League

YALDS Young Australia Language Development Scheme

Yale LJ Yale Law Journal

Yallo Ballys short form for the Yallo Bally Mountains of northern California

Yam Yamaha

YAM Yates American Machine (company)

YAN Yancey (railroad); Young American Nazis

YANCON Yankee Conference (intercollegiate sports)

Yangpat Yangtze Patrol

Yank Yankee; Yankel

Yank Yankel (Yiddish—Jacob)

YANK Youth of America Needs to Know

Yankee letter Y radio code; Soviet class of nuclear-powered submarines—Yankee or Y-class—similar to U.S. Polaris-type subs

Yanks British nickname for Americans

yap. yaw and pitch

yaps yaw and pitch sensor

Yar Yarmouth
YAR Yemen Arab Republic (Sana—capital); York-Antwerp Rules (insurance)
YARA Young Americans for Responsible Action
yard prison yard
'yard shipyard
Yard Scotland Yard; Yardley
YARD Yarrow-Admiralty Research Department
yarden yard + garden
YARDS Yard Activity Reporting and Decision System
YARN Young Adult Resource Notebook
yas yaw-attitude sensor
YA's Young Adults (young people)
YAS Yorkshire Agricultural Society
YASD Young Adult Services Division (ALA)
YASSR Yakut Autonomous Soviet Socialist Republic
Yat Yatyiopia (Amharic—Ethiopia)
yavis young, attractive, verbal, intelligent, and successful
YAWF Youth Against War and Facism
Y-axis vertical axis on a chart, graph or map
yb yardbird (confined to a military camp)
Yb ytterbium
YB yearbook; Youngstown Sheet & Tube (stock market symbol)
YBA Young Buddhist Association; Youth Basketball Association
YBC Yerba Buena Center
Ybk Yearbook
ybr yellowish brown
YBR sludge-removal barge (naval symbol)
YBRA Yellowstone-Bighorn Research Association
Y-branch Y-shaped pipe fitting
YB(RS) Year Books (Rolls Series)
YBs Young Boys Inc—drug ring using young boys to promote street sales
YBS Yale Bibliographic System (computer cataloging)
yc yaw channel; yaw coupling; yellow chrome
Y-c Y-chromosome
YC open lighter (naval symbol); Yacht Club; Yankton College; Yeomanry Cavalry; York College; Youth Club; Yuba College
YCA Yachting Club of Amer-

ica; Young Citizens' Army; Young Concert Artists; Youth Camping Association; Youth Correction Act
YCC Youth Conservation Corps; Youth Correctional Center; Yuba Community College
YCCA Youth Council on Civic Affairs
YCCC Yui Chui Chan Club
YCCIP Youth Community Conservation and Improvement Projects
YCD feuling barge (naval symbol); Youth Correction Division (U.S. Dept Justice)
YCF car float (naval symbol); Yankee Critical Facility; Young Calvinist Federation
YCGJ Young Christians for Global Justice
Y-chromosome male-producing gene found in male sperm
YCI Young Communist International; Youth Correctional Institution
YCI Yacht Club Italia (Italian Yacht Club)
YCia Ybarra Compañía (steamship line)
YCK open cargo lighter (naval symbol)
YCL Yarmouth Cruise Lines; York City Library; Young Communist League
YCLA Young Circle League of America
Y-class Soviet class of nuclear-powered submarines, also called Yankee, similar to U.S. Polaris-type subs
YCM Young Christian Movement
YCNM Yucca House National Monument
ycp yaw-coupling parameter
YCP Youth Challenge Program
YCS Young Catholic Students; Young Christian Students; Youth and Community Services
YCSM Young Christian Student Movement
yct yacht
YCTF Younger Chemists Task Force
YCU aircraft transportation lighter (naval symbol)
YC & UO Young Conservative and Unionist Organisation
YCV aircraft transportation lighter (naval symbol)
ycw you can't win
YCW Young Christian Workers

ycz yellow caution zone (airport runway lighting)
yd yard
y/d yaw damper
YD floating derrick (naval symbol); Young Democrat; Yugoslav dinar
Y & D Yards and Docks (USN)
yd² square yard(s)
yd³ cubic yard(s)
yda yesterday
YDA Dawson City, Yukon Territory (airport)
ydaa yellow dinitrophenyl aspartic acid
yday yesterday
ydb yield-diffusion bonding
ydc yaw-damping computer
YDCA Youth Democratic Clubs of America
YDF floating drydock (naval symbol)
ydg yardage; yarding
YDG degaussing vessel (naval symbol)
ydi yard drain inlet
YDI Youth Development Incorporated
YDL Young Development Laboratories
ydmn yardman
ydmstr yardmaster
yds yards
Yds Yards
YDs Young Democrats
YDS Yale Divinity School
YDSD Yards and Docks Supply Depot (USN)
YDSO Yards and Docks Supply Office
YDT diving tender (naval symbol)
Y-duct Y-shaped duct
ye yellow edges; yellow edging; yellow enzyme; yellow-edged
yᵉ (Early English—thou)—also written ye
YE aircraft homing system
yea. yaw-error amplifier
YEA Yale Engineering Association
yearb yearbook
YEB Yorkshire Electricity Board
YEDPA Youth Employment and Demonstration Projects Act
YEFC Youth Education for Citizenship (American Bar Association)
yeg yeast extract—glucose
YEG Edmonton, Alberta (International Airport)
yegg yeggman (burglar specializing in opening safes and vaults)
yel yellow

Yell Yellowstone National Park
Yellowjackets Yellowjacket Mountains of eastern Idaho
Yellow Peril danger to Western civilization held to arise from the influx of Oriental laborers willing to work for low wages
Yellowstone Yellowstone County in Montana, Yellowstone Lake in Wyoming, Yellowstone National Park (Idaho, Montana, and Wyoming), Yellowstone River (Montana, North Dakota, and Wyoming)
Yel NP Yellowstone National Park
yelsh yellowish
yem yeast extract—malt
Yem Yemen; Yemenite
Yemen Republic of Yemen *(Jumhurijah al-Yemen)*
Yem RA *República Árabe del Yemen* (Spanish—Arab Republic of Yemen)
Yem RDP *República Democrática Popular del Yemen* (Spanish—Popular Democratic Republic of Yemen)
yen monetary unit of Japan
Yeo Yeoman
YEO Youth Employment Office(r)
Yeoman F Yeoman Female (naval rating)
yeomn yeomanry
yep your educational plans
yepd yeast extract—peptone, dextrose
YES Youth Educational Services; Youth Education Systems; Youth Employment Service; Youth to End Smoking; Youth Engaged in Service
yesty yesterday
YETP Youth Employment Training Program
YEWTIC Yorkshire, East and West Ridings, Technical Information Centre
YEX-ZA World's Largest Blimp
yf wife (orthographic contraction proposed by Benjamin Franklin)
yf (YF) yellow fever
YF covered lighters (naval symbol)
YF-16 air-superiority single-engine lightweight-fighter aircraft (USAF)
YFB ferryboat or launch (naval symbol)
YfC Youth for Christ
YFC car float (naval symbol); Young Farmers' Club

YFCU Young Farmers' Clubs of Ulster
YFD yard floating drydock (naval symbol)
YFFP Yarrawonga Flora and Fauna Park (Australian Northern Territory)
YFN covered lighter, nonself-propelled (naval symbol)
YFNB large covered lighter (naval symbol)
YFND drydock companion craft (naval symbol)
YFNX special-purpose lighter (naval symbol)
YFP floating power barge (naval symbol); Youth For Progress
YFR self-propelled refrigerated covered lighter (naval symbol)
YFRN refrigerated covered lighter, nonself-propelled (naval symbol)
YFRT covered lighter, range tender (naval symbol)
YFT torpedo transportation lighter (naval symbol)
yfu yard freight unit
YfU Youth for Understanding (teenage exchange program)
YFU harbor utility craft (naval symbol)
y fwd yarn forward (knitting)
yG yellowish green
Yg *Young's Literal Translation of the Holy Bible*
YG garbage lighter (naval symbol); yellow green
YGC Youth Guidance Center
YGH Yankee Go Home
ygl yttrium-garnet laser
ygmd yaw-gimbal command
YGN garbage lighter, nonself-propelled (naval symbol)
YGR Yankari Game Reserve (Nigeria)
YGS Young Guard Society
Y-gun Y-shaped gun used aboard ships for firing depth charges
YH Youth Hostel
YHA Youth Hostels Association
Yhama Yokohama
YHANI Youth Hostel Association of Northern Ireland
YHB houseboat (naval symbol)
YHLC salvage lift craft, heavy (naval ship symbol)
YHt Young-Helmholtz theory
YHT heating scow (naval symbol)
YHVH *see* JHVH
Yi Yiddish
YIC Yardney International Cor-

poration
Yid Yiddish; Yiddish-speaking person
Yid Yiddish (German dialect spoken by many persons of Judaic origin and augmented by the languages of the countries where they have emigrated)
Yidgin-English Yiddish + English
Yie Young interference experiment
YIEP Youth Incentive Entitlement Project
yig yttrium iron garnet (ferrite)
yigib your improved group insurance benefits
YIIJS Young Israel Institute for Jewish Studies
YIJR Yivo Institute for Jewish Research
YIJS Young Israel Institute for Jewish Studies
YIKOR *Yidishe kultur-organizatsye* (Polish-Yiddish Culture Organization)
yil yellow indicator lamp
Yinglish Yiddish-English
YIP Detroit, Michigan (Willow Run Airport); Youth International Party
YIR Yearly Infrastructure Report
Yis *Yisroel* (Yiddish—Israel)
YI & S Yawata Iron and Steel
Yivo Inst Yivo Institute for Jewish Research
yj radar homing beacon (map symbol)
YJC York Junior College
Y-joint Y-shaped joint
yk radar beacon (map symbol)
Yk Yakut; York
YK Yankee Airlines
Yka Yokohama
YKAA Young Keyboard Artists Association
YKF *Yiddisher Kulture Farband* (Yiddish Culture Club)
YKKK *Yamashita Kisen Kabushiki Kaisha* (Japanese—steamship line)
Ykn Yukon
Yko Yokosuka
Yks Yorkshire
Ykt Yakut
yl yellow; yield limit; young lady
Y & L York and Lancaster
YLA open landing lighter (naval symbol)
YLC Young Life Campaign
YLI Young Ladies Institute; Yorkshire Light Infantry
YLJ *Yale Law Journal*

YLL Yerkes Language Laboratory
YLLC salvage lift crane, light (naval ship symbol)
YLM Yale Literary Magazine
YLO Young Lords Organization
YLP Young Lords Party
Y & LR York and Lancaster Regiment
yl's young ladies
Ylstn Yellowstone
ym yacht measurement; yawing moment; yellow metal; your measurement; your message
YM dredge (naval symbol); Yehudi Menuhin
YMA Yarn Merchants Association
ymb yeast malt broth
YMBA Yacht and Motor Boat Association
YMCA Young Men's Christian Association
YM Cath A Young Men's Catholic Association
YMCU Young Men's Christian Union
ymd your message date
Yme Young's modulus of elasticity
YMF Young Musicians Foundation
YMFS Young Men's Friendly Society
YMHA Young Men's Hebrew Association
YMHAL Young Men's Hebrew Association Library
YMI Young Men's Institute
YML Yang Ming Line
YMLC salvage lift craft, medium (naval ship symbol)
YMLU Yamashita Line (container) Unit
YMP motor mine planter (naval symbol); Young Management Printers
YMPA Young Master Printers' Alliance
yms yield measurement system
YMs Yearly Meetings (Quakers)
YMS motor minesweepers (naval symbol)
YMT motor tug (naval symbol)
Ymu Ymuiden
YMV Montreal, Quebec airport; Yazoo and Mississippi Valley (railroad)
YM & YWHA Young Men's and Young Women's Hebrew Association
yn yen
y-n yes-no
Yn Yeoman; Yeowoman

YN net tender (naval symbol); Youngstown & Northern RR
Y network wye network
yng young
YNG gate vessel (naval symbol)
YNHA Yosemite Natural History Association
Y-NHH Yale-New Haven Hospital
ynhl why in hell
YNP Yellowstone National Park (Idaho, Montana, Wyoming); Yoho NP (British Columbia); Yosemite NP (California); Youth National Party
YNSO Yomiuri Nippon Symphony Orchestra
YNT net tender, tug (naval symbol)
Ynv Ynvar (Russian—January)
YNWR Yazoo National Wildlife Refuge (Mississippi)
yo yarn over (knitting); year old
yo' yore; you; your
y/o years old
YO fuel-oil barge (naval symbol); Yerkes Observatory
YOAN Youth Of All Nations
yob year of birth
YOB Youth Opportunities Board
YOC Youth Opportunity Campaign; Youth Opportunity Center(s); Youth Opportunity Corps; Youth Ornithologists' Club
YOC-RSPB Young Ornithologists' Club—Royal Society for the Protection of Birds
yod year of death
YOG gasoline barge, self propelled (naval symbol)
YOGN gasoline barge, nonself-propelled (naval symbol)
Yok Yokohama
Yoko Yokohama; Yoko Ono
Yokuska (navalese—Yokosuka, Japan)
yom year of marriage
Yom Yomiuri (Japanese—News Crier)—Tokyo newspaper
YOM yellow oxide of mercury
Yom Kip Yom Kippur (Hebrew—Day of Atonement)
yon yonder
YON fuel-oil barge, nonself-propelled (naval symbol)
yood (slang pronunciation—iud)—intrauterine device
YOP Youth Opportunity Program
YOPB Youthful Offender Parole Board
York New York; New York State

York Yorkshire Post
Yorkie Yorkshire terrier
Yorks Yorkshire
Yos Yosu
YOS oil storage barge (naval symbol)
Yosemite Yosemite National Park in California; natural attractions such as the Yosemite Falls or the Yosemite Valley
Yoshino-kumano Yoshino-kumano National Park in southern Honshu, Japan
Yos NP Yosemite National Park
yot (YOT) youthful offender treatment
YOU Youth Opportunities Unlimited; Youth Organizations United; Youthful Offender Unit
you'd you had; you would
you'll you shall; you will
Youngs Youngstown
you're you are
youthploit youth exploitation
you've you have
YOW Ottawa, Ontario (airport)
yp yellow pine; yield limit; yield point (psi)
YP patrol craft (naval symbol); yellow peril; young people; young person(s)
ypa yaw-precession amplifier
YPA Young Pioneers of America
YPCS Young Peoples Computer Society
ypd yaw-phase detector
YPD floating pile driver (naval symbol)
YPEC Young Printing Executives Club
YPF Yacimientos Petroliferos Fiscales (Spanish—Government Oil Deposits)—Argentina
YPFB Yacimientos Petroliferos Fiscales Bolivianos (Spanish—Bolivian Government Oil Deposits)
YPFP Yunnan Phosphate Fertilizer Plant
YPG Yuma Proving Ground
yPk yellowish pink
YPK pontoon stowage barge (naval symbol)
YPM Yale Peabody Museum
YPO Young Presidents' Organization; Youth Programs Office (Bureau of Indian Affairs)
Yps Ypsilanti
YPSCE Young People's Society of Christian Endeavor
YPSL Young People's Socialist League

Y-punch punch in Y row (12th row) of an 80-column punchcard

YQX Gander, Newfoundland (airport)

yr year; younger; your

y-r yaw roll

YR district patrol vessel (naval symbol); floating workshop (naval symbol); Young Republican(s)

YRA Yacht Racing Association

yrb year built

Yr B Year Book

YRB submarine repair and berthing barge (naval symbol)

yrbk yearbook

YRBM submarine repair—berthing and messing barge (naval symbol)

YRC submarine rescue chamber (naval symbol)

YRD submarine repair and berthing vessel (naval symbol)

YRDH floating drydock hull workshop (naval symbol)

YRDM floating drydock machinery workshop (naval symbol)

YRL covered repair lighter (naval symbol)

yrly yearly

YRNF Young Republican National Federation

Yr obt servt Your obedient servant

YRR radiological repair barge (naval symbol)

yrs years; yours

Yrs Yours

YRs Young Republicans

YRS Yugoslav Relief Society

YRST salvage craft tender (naval ship symbol)

yrs ty yours truly

yrt yearly renewable term (insurance)

YRU Yacht Racing Union

ys yellow spot (on retina); yield strength

Ys Yugoslavia; Yugoslavian

YS Yard Superintendent; Young Socialists

Y-S Yamashita-Shinnihon

Y & S Youngstown & Southern (railroad)

YS-11 Japanese medium-range transport plane

YSA Young Socialist Alliance; Youth Service America; Youth Services Administration (District of Columbia)

ysb yield-stress bonding

YSB Yacht Safety Bureau; Youth Service Bureau

YSC Yugoslav Seamen's Club; Youth Studies Center (juvenile correctional facility in Philadelphia)

YSD seaplane wrecking derrick (naval symbol); Youngstown Steel Door (company); Youth Services Division

ysdb yield-stress diffusion bonding

yse yaw-steering error

ysh yellowish

YSI Yellow Springs Instrument (company)

Ysl Ysrael

YSL Young Socialist League; Yves Saint Laurent

Y-S Line Yamashita-Shinnihon Line (steamships)

YSM Yale School of Music; Yangtze Service Medal

yso young stellar object

YSO Youngstown Symphony Orchestra

ysp years service for severance pay purposes

YSP pontoon salvage vessel (naval symbol)

ysr you're so right

YSR sludge-removal barge (naval symbol)

YSS Young Scots Society

YSSAS Yale Summer School of Alcohol Studies

yst youngest

YST Yukon Standard Time

YS & T Youngstown Sheet & Tube

YSTO Yugoslav State Tourist Office

YSU Youngstown State University

yt yoke top

Yt yttrium

Y' *Ytre* (Dano-Norwegian or Swedish—outer)

YT harbor tug (naval symbol); Yukon Territory

Y & T Tale & Towne

YTA Yiddish Theatrical Alliance

ytb yarn to back

YTB large-harbor tug (naval symbol)

YTCA Yorkshire Terrier Club of America

ytd year to date

YTEP Youth Training and Employment Project

ytf yarn to front

YTL small-harbor tug (naval symbol)

YTM medium-harbor tug (naval symbol)

YTP *Yeni Türkiye Partisi* (New Turkish Party)—socialist oriented

YTPM Yuma Territorial Prison Museum

YTS Youth Training Scheme; Youth Training School; Yuma Test Center; Yuma Test Station

YTT torpedo-testing barge (naval symbol)

Y-tube Y-shaped tube

YTV Yokohama Television

Yu Yugoslav; Yugoslavian

YU Yale University; Yeshiva University; York University; Youngstown University; Yugoslavia (auto plaque)

YUAG Yale University Art Gallery

yuan monetary unit of China

Yuc Yucatan (natives nicknamed boxitos—Maya term meaning darks)

Yuca Yucatan, Mexico

YUCA Young Upscale Cuban-American

Yud Yudel

Yud *Yudel* (Yiddish—Judah)

yuffie(s) young urban failure(s)

Yug Yugoslavia(n); Yugoslavic

Yugo Yugoslav; Yugoslavia; Yugoslavian

Yugoslavia Socialist Federal Republic of Yugoslavia (central-European nation), *Socijalisticka Federativna Republika Jugoslavija*

Yuk Yukon

YUK Youth Uncovering Krud (antipollution society)

Yukon Canadair version of the Britannia designated CC-106; Yukon River; Yukon Territory

YUL Montreal Quebec (airport); Yale University Library

YULRC Yale University Lung Research Center

yumpie(s) young upwardly-mobile professional(s)

YUN *Yearbook of the United Nations*

YUO Yale University Observatory

yup you're uncommonly perceptive

YUP Yale University Press

yuppie flu debilitating chronic fatigue syndrome

yuppie(s) young urban professional(s)

yup(s) young urban professional(s)

Yur Yuri; Yurievich

Yus Yussel

Yus *Yussel* (Yiddish—Joseph)

YUSM Yale University School

of Medicine
Yuzh *Yuzhnaya* (Russian—southern)
Yv Yvette; Yvonne
Yv *Yvert et Tellier* (stamp catalog)
YV Young's Version
YVA Young Volunteers in Action
yvc yellow-varnish cambric
YVC Yakima Valley College
YVF Young Volunteer Force
YVHS Yorkville Vocational High School
YVJC Yakima Valley Junior College
YVP Youth Voter Participation
YVR Vancouver, British Columbia (airport)
YVRL Yakima Valley Regional Library

YVT Yakima Valley Transportation (railroad)
y v v y *viaje vuelta* (Spanish—and return trip)
YW water barge (naval symbol); Yreka Western RR
YWAA Youth Welfare Association of Australia
YWAM Youth With A Mission
YWCA Young Women's Christian Association
YWCAUSA Young Women's Christian Association of the U.S.A.
YWCTU Young Women's Christian Temperance Union
ywd you would
YWF Young World Federalists
YWFD Young World Food and Development (UN)
YWG Winnipeg, Manitoba (air-

port)
YWHA Young Women's Hebrew Association
YWHS Young Women's Help Society
YWLL Young Workers Liberation League
YWN nonself-propelled barge (naval symbol)
YWPG Young World Promotion Group
YWS Young Wales Society
YWU Yiddish Writers Union
YX Yannis Xenakis
y-y yaw axis
y & y yin and yang
YY pseudonymous initials of Robert Lynd noted for his *New Statesman* essays
YYC Calgary, Alberta (airport)
YYZ Toronto, Ontario (airport)

Z

z cracked (moving and storage symbol); ounce (truncation of oz—ounce); complex variable (symbol); z-bar; zed (British usage); zee (American usage); zero; zinc; zone

z (Z) zloty (Polish currency unit)

z *zu* (German—closed, shut)

Z atomic number (symbol); azimuth (symbol); azimuth angle (symbol); gram equivalent weight (symbol); impedance (symbol); lighter-than-air aircraft (symbol); obsolete (symbol); radius of circle of least confusion (symbol); zenith; zenith distance; zero meridian time; Zionism; Zionist; Zoroaster; Zoroastrian; Zoroastrianism; Zulu—code for letter Z

Z normal axis (symbol); *Zeit* (German—time); *Zeitschrift* (German—periodical publication); *zuid* (Dutch—south)

Z^1, Z^2, Z^3 first degree of contraction, second degree of contraction, third degree of contraction

Z39 Library Work, Documentation, and Related Publishing Practices (American National Standards Institute Standards Committee)

za B-flat (Tartini's scale); zero absolute; zero and add

za *zirka* (German—about; approximately)

Za *Zéro absolu* (French—absolute zero)

ZA *Zuid Afrika* (Afrikaans or Dutch—South Africa)

Zaa Zeeman-effect atomic absorption (spectrometry)

zaap zero anti-aircraft potential

zab zabaglione; zinc-air battery

zab *zabaglione* (Italian—egg-yolk-and-wine dessert)

Zab Greater Zab or Lesser Zab river in Iraq; Zaboj

Zab *Zabriskie's Reports*

Zac Zacatecas

ZAC Zale Award Committee; zinc ammonium chloride

Zacatecas purple Zacatecas-purple marijuana from central Mexico

Z-account Zurich account (bank deposits in Zurich, Switzerland identified only by number, not by the depositor's name)

Zach Zachary; Zachariah; Zacharias; Zachary; Zachris

Zack Zachariah; Zacharias; Zachary

'zactly exactly

Zad Zadar; Zadock

ZADCA Zinc Alloy Die Casters' Association

ZADCC Zone Air Defense Control Center

ZAED *Zentralstelle für Atomkernenergie Dokumentation* (German—Atomic Energy Documentation Center)

zaf zero-alignment fixture

Zafarinas Zafarinas Islands (also spelled Chafarinas)—off the Mediterranean coast of Morocco

Z-Afrika *Zuid-Afrika* (Dutch—South Africa)

zag *zaguán* (Spanish—passageway from street door to central patio of homes in Mexico and American Southwest)

Zag Zagreb

ZAG Zagreb, Yugoslavia (airport)

Zahal *Zva Hagana Leyisrael* (Hebrew—Israel Defense Forces)

Zahlentaf *Zahlentafeln* (German—table of illustrations)

zai zero address instruction

zai *zaibatsu* (Japanese—money clique)—plutocratic oligarchy of wealthy families

Zai Zaire

zaire monetary unit of Zaire

Zaire Republic of Zaire (central African nation formerly the Belgian Congo), *République de Zaire*

zak *zaklyuchenny* (Russian—prisoner)—pronounced *zek*

zal *zaliv* (Russian—bay)

Zal *Zalmen* (Yiddish—Solomon)

ZALIS Zinc and Lead International Service

zam Z-axis modulation; zinc, aluminum, magnesium

Zam Zambia; Zamboanga; Zamiel; Zamora

ZAMAL *Zva Maganah Le Israel* (Hebrew—Israel Defense Force)

Zamb Zambia

Zambia Republic of Zambia (landlocked southern African nation formerly Northern Rhodesia)

Zambo Zamboanga

Zamp Zampa

ZAMPA Zanzibar and Madagascar Peoples Airway

zams zero-age main sequence

zam(s) examination(s)

Zan Zanzibar

Z Anal Chem *Zeitschrift für Analytische Chemie* (German—Analytical Chemistry Periodical)

ZANC Zambia National Congress

ZANLA Zimbabwe African National Liberation Army

ZANU Zimbabwe African National Union

Zanzi Zanzibar

zap zero and add packed; zero antiaircraft potential

zap zapad (Russian—west)

Zap Zapotec; Zapotecan

zapb zinc-air primary battery

zapp zygo automatic pattern processor

ZAPU Zimbabwe Africa People's Union

zar zeus acquisition radar

Zar Zaragoza

Zara Sara(h); Zarathustra (Zoroaster)

Zara (Italian—Zadar)—Yugoslavian port city

ZARPS Zuid-Afrikaansche Republiek Polisie (Afrikaans—South African Republic Police)

zas zero-access storage

ZASM Zuid Afrikaansche Spoorweg Maatschappij (Afrikaans—South African Railway serving the Transvaal at the turn of the century)

Z Astrophys Zeitschrift für Astrophysik (German—Astrophysics Periodical)

zasts zastrugas

zat zinc atomspheric tracer

ZAT Zaterdag (Dutch—Saturday)

Z-A test Zondek-Ascheim test (for pregnancy)

ZAW Zuid-Afrikaansche Weehuis (Afrikaans—South African Orphan Asylum)

Zazen Zen meditation

zb zero beat

z B zum Beispiel (German—for instance)

ZB Zen Buddhist

ZBA Zero Bracket Amount

Z-bar Z-shaped bar

zbb zero-base budget(ing)

ZBBS Zero-Based Budgeting System

zbe zinc battery electrode

zbl zero-based linearity

Zbl Zentralblatt (German—central publication)

zbr zero-base review; zero-beat reception; zero-bend radius

ZBS Zambia Broadcasting Services

zbSd zero-bias Schottky diode

zc zone capacity

z of c zones of communication

ZC Zale Corporation; Zinc Cor-

poration; Zionist Congress; Zoning Commission; Zonta Club; Zouave Corps; Zuñian Club

ZCA Zirconium Corporation of America

Z-car police car (British slang)

zcb zinc-coated bolt

ZCBC Zambian Consumer Buying Corporation

zcc zirconia-coated crucible

zcd zero crossing detector

zcic zirconia-coated iridium crucible

ZCL Zona di Commercio Libero (Italian—Free Trade Zone)

Z-class Soviet class of submarines named Zulu by NATO

Z-clip Z-shaped clip

zcm zero cerebral muscle

ZCMI Zion's Cooperative Mercantile Institution

zcn zinc-coated nut

ZCNP Zion Canyon National Park

Z country Zimbabwe

zcr zero-temperature coefficient resistor

zcs zinc-coated screw

ZCS Zim Container Service

ZCSU Zim Container Service Unit

zcw zinc-coated washer

ZCX Zone Center Exchange

zd zener diode; zenith distance; zero defects; zipper duffle (garment bag); zonal depot; zone description

Zd zenith description; zenith distance

ZD zenith description; zero defects (quality-control goal); zond description

ZDA Zero Defects Association; Zinc Development Association

zdc zinc die casting

ZDC Zero Defects Council

Z de T Zulano de Tal (Spanish—so and so)

zdg zinc-doped germanium

Zdm Zaandam

ZDP Zero Defects Program; Zero Defects Proposal

zdpa zero defects program audit

zdpg zero defects program guideline

zdpo zero defects program objective

zdpr zero defects program responsibility

zdr zeus discrimination radar

ZDR Zentraldeutsche Rundfunk (German—Central German

Radio)

ZDS Zinc Detection System

ZDSI Zung Depression Status Inventory

zdt zero-ductility transition

ze zero effusion; zone effect

zE zum Exempel (German—for example)

Ze José

ZE Zenith Radio (stock-exchange symbol)

Z-E Zollinger-Ellison (syndrome)

zea zero-energy assembly

Zeb Zebedee; Zebulon

zebra. zero-energy breeder reactor assembly

zebrass zebra + ass—hybrid of zebra and jenny ass or zebress and jackass

zebroid zebra + horse (hybrid)

Zebrule zebra + horse—hybrid of male zebra and domestic mare

zeb(s) zebra(s)

zec zero-energy coefficient

zecc zinc electrochemical cell

Zech. Zechariah (book of the Bible)

Zech Zechariah

zed z; zero

Zed New Zealand; Zedekiah

Zedong Mao Zedong

Zee Zeeland (Dutch province); Zellerbach

Zee Zeeland (Dutch—Sea Land)

zeep zero energy experimental pile

zeg zero economic growth

zei zero environmental impact

Zeichn Zeichnung(en) [German—drawing(s)]

zel (ZEL) zero-length launcher

Zel Zelia; Zelide

Z Elektrochem Zeitschrift für Elektrochemie (German—Electrochemistry Periodical)

zell zero-length launching

Zem Zemlya (Russian—earth; land)

Zemlya Imperatora (Russian—Emperor Land)—Severnaya Zemlya

Zempo Zempoaltepetl (11,142-foot peak near Oaxaca in southern Mexico)

zen nickname for lsd (LSD); zenith (highest point)

Zen Zen Buddhism; Zen Buddhist; Zengo; Zenith; Zenobe; Zenobia; Zenobio; Zenón; Zenophon; Zentippe; Zenus

ZEN EIEN Zenkoku Eiga Engeki Rodo Kumiai (Japanese—National Movie and

Theater Workers Union)
Zenga *Zengakuren* (Japanese leftwing students)
zenith zero-energy nitrogen-heated thermal reactor
ZENKO *Zen Nihon Kinsoku Kozan Rodo Kumiai Rengokai* (Japanese—All-Japan Federation of Metal Miners Union)
ZENRO *Zen Nihon Rodo Kumiai Kaigi* (Japanese—All-Japan Trade Union Congress)
ZENTEI *Zen Teishin Rodo Kumiai* (Japanese—Postal Workers Union)
Zentr *Zentralblatt* (German—journal)
zeony zebra + pony (hybrid)
Zep Giuseppe
Zeph. Zephaniah (book of the Bible)
Zeph *Zephaniah*
zephyr warm westerly breeze
ZEPHYR Zero-Energy Plutonium-Fueled Fast Reactor
zepp zeppelin
zep(s) zeppelin(s)
zer zero-energy reflection
ZERA Zero-Energy Critical Assemblies Reactor(s)
zerc zero-energy reflection coefficient
zero-g zero gravity (weightlessness)
zert zero-reaction tool
ZES Zenith Enterprise Service (telephonic); Zero Energy System
zet zetetic(s)
zeta. zero energy thermonuclear assembly
Zetland Zetland Island or the Zetland Islands; the Shetlands
zetr zero-energy thermal reactor
zeug zeugma; zeugmatic; zeugmatically
ZEUS Zero-Energy Uranium System
zf zero frequency
z/f zone of fire
Z-F Zermelo-Fraenkel (set theory)
ZF *Zagrebacka Filharmonija* (Croatian—Zagreb Philharmonic)
zfb signals fading badly
zfc zirconia fuel cell
ZFGBI Zionist Federation of Great Britain and Ireland
ZFMA Zip Fastener Manufacturers' Association
Z ∫ N *Zeitschrift für Namenforschung* (German—Journal for the Study of Place-names)
ZFO *Zone Francaise d'Occu-*

pation (French—French Occupation Zone)
zfp zonal flow pattern; zyglo-fluorescent penetrant
zfpt zyglo-flurescent penetrant testing
zfs zero field splitting
Zf's Zionist fanatics
ZFV *Zentrale für Fremdenverkehr* (German—Central Tourist Association)
zg zap gun
z/g zoster-immune globulin
Zg Zug
ZG Zoological Gardens
Z-gauge super-miniature model railway scale
Z-gas Zyklon-B gas (deadly)
zge zero-gravity effect; zero-gravity environment; zero-gravity expulsion
zget zero-gravity expulsion technique
ZGF Zero Gravity Facility
zgg zero gravity generator
zgh zero-gravity harmonic
ZGM Zeitner Geological Museum
zgs zero-gravity simulator
zgs (ZGS) zero gradient synchrotron
zgt (ZGT) zero-gravity trainer (NASA)
Z-gun anti-aircraft rocket gun
zh zinc heads (freight); zonal harmonic; zone heater
zH *zu Händen* (German—care of, deliver to)
Zh *Zuidholland* (Dutch—South Holland)
ZH lighter-than-air search and rescue aircraft (naval symbol)
ZH *Zone d'Habitation* (French—residential area)
Zh Eksp Teor Fiz *Zhurnal Eksperimental'noi i Teoreticheskoi Fiziki* (Russian—Journal of Experimental and Theoretical Physics)
Zh Fiz Khim *Zhurnal Fizicheskoi Khimii* (Russian—Journal of Physical Chemistry)
Zhg *Zhongguo* (Chinese—China)
Zho Ren Gon Guo *Zhonghua Renmin Gonghe Guo* (Mandarin Chinese—People's Republic of China)
Zh Prik Spektrosk *Zhurnal Prikladnoi Spektroskopii* (Russian—Journal of Applied Spectroscopy)
zhr zenith hourly rate; zirconium hydride reactor
Z hr zero hour

ZHRC Zinsmaster Hol-Ry Company
zhs zero hoop stress
ZHS Zion Historical Society
Zh Tekh Fiz *Zhurnal Tekhnicheskoi Fiziki* (Russian—Journal of Technical Physics)
zi zero input; zonal index
Zi Zollner illusion
ZI Zim Israel (steamship line); Zinc Institute; Zone of the Inferior; Zone of the Interior; Zonta International
Z of I Zone of the Interior
ZI *Zone Industrielle* (French—industrial zone); *Zone Interdite* (French—prohibited zone)
ZIA Zone of the Interior Armies
zic zirconia-iridium crucible
ZID Zionist Immigration Depot
Zier Ziervogel process
zig zero immune globulin
zig (ZIG) zero immune globulin; zoster-immune globulin
Zig Ziegfield; Zigfield; Zigfrid; Zigfrids
ziggur(s) ziggurat(s)
zig(s) *zigaboo(s)* [British West Indian—Black(s)]
zig-zag zig-zag cigarette paper; zig-zag rule(r); zig-zag sewing machine attachment for making zig-zag stitches
ZiJ *Zeitschrift für Instrumentenbau* (German—Journal for Instrument Builders)
zil zillion (a number beyond belief)
ZIL (Russian—*Zavod Imeni Likhatov*)—Likhatov Auto Factory producing a Packard-like luxury car formerly named for Stalin—the ZIS (*Zavod Imeni Stalin*)
Zilw Zilwaukee
zim zero-interest mortgage; zimbalon; zonal interdiction missile
Zim Zimmerman(n)
Zim *Zimbel* (German—cymbal); *Zi Mischari* (Hebrew—merchant fleet)
ZIM *Zeitschrift der Internationalen Musikgesellschaft* (German—Journal of the International Music Society)
ZIMA Zimbabwe Medical Aid
Zimb Zimbabwe
Zimb *Zimbalon* (German—cimbalom or dulcimer)
Zimbabwe formerly Zimbabwe-Rhodesia; Rhodesia; Southern Rhodesia
Zimco Zambia Industrial and

Mining Company

Zim-Rho Zimbabwe-Rhodesia (formerly Rhodesia or Southern Rhodesia)

Zim Tim Zimbabwe Times

zin zinfandel (grapes or wine)

ZINC Zim Israel Navigation Company (Zim Israel Line)

zinco zincograph

ZINCO Zim Israel Navigation Company

zincog zincography

zinc white zinc oxide (ZnO)

zineb zinc ethylenebis (fungicide)

zine(s) magazine(s)

Zingi Zingari (Italian—Gypsies)

Zinj Zinjanthropus

Zion Zion National Park, Utah

zip zero (slang); zinc impurity photodetector; zipper (slide fastener or similar device)

ZIP Zone Improvement Plan (US Post Office Zip Code)

ZIPA Zimbabwe People's Army

ZIPRA Zimbabwe People's Revolutionary Army

zir zero internal resistance

ZIR Zug Island Road (Delray Connecting Railroad)

ziram zinc dimethyldithiocarbamate (fungicide)

zircaloy zirconium alloy

ZIRCOA Zirconium Corporation of America

zircon zirconium silicate $(ZrSiO_4)$

Zirk Hagen Zirkus Hagenbeck (German—Hagenbeck Circus)

zirox zirconium oxide (ZrO_2)

ZISS Zebulon Israel Seafaring Society

zith zither

zix zinc isopropyl xanthate

zj zipper(ed) jacket

zj zonder jaartel (Dutch—without date of publication)

ZKD Zagreb Kajkavian Dialect (Serbo-Croatian)

zkrat zkratka(y) [Czech—abbreviation(s)]

Z Kristallog Kristallgeom Krystallphys Kristallchemie Zeitschrift für Kristallographie, Kristallgeometrie, Kristallphysik, Kristallchemie (German—Periodical for Crystallography, Crystallographic Geometry, Crystallographic Physics, Crystallographic Chemistry)

ZKSK Zentrale Kommission für Staatliche Kontrolle (German—Central Commission for State Control)—commu-

nist

zl freezing drizzle (meteorological symbol); zero lift

Zl zloty (Polish ruble)

ZL freezing drizzle (symbol)

ZLA Zambia Library Association

zlc zero lift cord

zld zero level drift; zero lift drag; zodiacal light device

Zld Zeeland (Dutch—Sea Land)—old province made of land captured from the sea

zlg zero line gap

zll zero length launch

zloty monetary unit of Poland

Zlsm Zeiss light-section microscope

zm zoom; zoomar (variable focus lens)

ZM Zubin Mehta

Z-M Zuckerman-Moloff (sewage treatment)

ZM Zeevaart Maatschappij (Dutch—navigation company); *Zona Militare* (Italian—Military Zone)—restricted area

Z-man U.S. Army reserve

zmar zeus malfunction array radar

Z-marker zone marker

Zmbbw Zimbabwe (Rhodesia; Southern Rhodesia)

ZMC Zion Mule Corps

Zmd Zung measurement of depression

Z Metallk Zeitschrift für Metallkunde (German—Metallurgy Periodical)

zmkr zone marker

ZMMD Zurich, Mainz, Munich, Darmstadt (algol processor joint effort of universities in those cities)

ZMRI Zinc Metals Research Institute

zmrz zmrzlina (Slovak—ice cream)

ZMT Zip (Zone Improvement Plan) Mail Translator (post office sorting device)

ZMW Zeitschrift für Musikwissenschaft (German—Journal for Musical Knowledge)

zn zenith; zone (computer flow chart)

zn zelfstandig naamwoord (Dutch—substantive noun)—any group of words or a pronoun serving as a noun

Zn true azimuth (symbol); zinc

ZN Zuid-Nederlands (Dutch—South Netherlands)—Belgium

Znak (Polish—Sign)—Roman

Catholic pro-government party

Z Naturforsch Zeitschrift für Naturforschung (German—Natural History Periodical)

zng zoning

ZnO zinc oxide

ZNP Zanzibar Nationalist Party; Zimbabwe National Park; Zion National Park (Utah)

Zn_{pgc} azimuth per gyro compass

ZNPM Zion National Park Museum

ZNPP Zanzibar and Pemba People's Party

ZNPS Zion Nuclear Power Station

znr zinc resistor; zirconium nitride

ZNS Zodiac News Service

ZNZ Zanatska Nabarnoproajna Zadruga (Yugoslavian—Procurement Sales Cooperative)

zo zero output; zobo (yak + zebu hybrid)

Zo Zoa; Zoe(belle); Zoela; Zoeta; Zofia; Zohora; Zohra; Zoila; Zona; Zonula; Zora(bel); Zorah; Zoraida; Zorana; Zorayda; Zore; Zorica; Zoril; Zorislava; Zorna; Zoruna; Zosa; Zosia; Zosimia; Zowart

ZO Zionist Organization

ZO Zone Occupée (French—Occupied Zone); *zuidoost* (Dutch—southeast)

zoa zero-ohms adjustment

ZOA Zionist Organization of America

ZOB Zentral Omnibus Bahnhof (German—Central Bus Depot); *Zydowska Organizacja* (Polish—Jewish Fighting Organization)—anti-Nazi ghetto forces in World War II

zoba bull + yak—hybrid offspring of common bull and yak cow

zobo cow + yak—hybrid of yak bull and common cow

zoc zócalo (Mexican Spanish—public square)

Zoc Zocalo (Mexico City's great plaza)

zod. zodiacus (Latin—circle of animals)—the zodiac

zoe zero energy; zinc-oxide eugenol

Zoe Zoebelle; Zoela; Zoeta; Zofia

zof zone of fire

Zog Ahmed Zogu

ZOG Zionist Occupation Government

Zoh Zohar (The Book of Splendor)

Zolá Émile Zolá—French novelist (1840–1902)

zon zoning

Zon Zondag (Dutch—Sunday)

Zondervan Zondervan Publishing House

Zone Panama Canal Zone

zoo zoological (garden); zoology

zoo (Latin prefix—animal)—zoological, zoologist, zoology

zoochem zoochemistry

zoogeog zoogeography

zool zoologic; zoological; zoologist; zoology

zool zoologi(sk) (Dano-Norwegian—zoology or zoologist)

Zool Zoology

zoomorph zoomorphic initial letter

zoopal zoopaleontology

zoopar zooparasitology

zoopath zoopathology

zooph zoophytology

zoopharm zoopharmacology

zop zero-order predictor; zinc-oxide pigment

zopi zero-order polynomial interpolator

zopp zero-order polynomial predictor

zor zinc-oxide resistor; zone of reconnaissance

Zor Zoram(ites); Zoroastrian

Zora Zorabel(la); Zorah; Zoraida; Zorana; Zorayda

Zorba Zorba the Greek

zos zoster; zosteriform; zosteriformal

ZOS Zapata Corporation (stock exchange symbol)

zot (slang—zero)

zounds (euphemistic contraction—god's wounds)

zox zirconium oxide

zoz zie ommezijde (Dutch—plesse turn the page)

ZP lighter-than-air patrol and escort aircraft (naval symbol); Zellerbach Paper

Z & P Zanzibar and Pemba

ZP Zagrebian Philharmony (Yugoslavian—Zagreb Philharmonic Orchestra)

zpa zeus program analysis; zone of polarizing activity

ZPA Zeus Program Analysis; Zoological Parks and Aquariums

zpar zeus-phased array (radar)

zpb zinc primary battery; zone portion of the byte (leftmost four bits of an eight-bit byte)

ZPC Zellerbach Paper Company

ZPDA Zinc Pigment Development Association

zpe zero-point energy

ZPEN Zeus Project Engineer Network

zpg zero population growth

ZPG Zero Population Growth

ZPH Zondervan Publishing House

Z Phys Zeitschrift für Physik (German—Physics Periodical)

Z Phys Chem Zeitschrift für Physikalische Chemie (German—Physical Chemistry Periodical)

zp & j zonder plaats en jaar (Dutch—without place of publication or date)

ZPKK Zentrale Parteikontrollkommission (German—Central Control Commission of the Party)—communist

zpl zonder plaats (Dutch—without place of publication)

Z Plz Zellerbach Plaza

zpo zinc peroxide

ZPO Zeus Project Office

Zpp Zeiss projection planetarium

zppr zero-power plutonium reactor

zpr zero-power reactor

zprf zero-power reactor facility

ZPRSN Zurich Provisional Relative Sunspot Number

zpt zero-power test(ing); zoxazolamine paralysis time

ZPT Zero Power Test

ZPU-4 Soviet antiaircraft weapon combining fire power of four 14.5mm heavy machine-guns

zr freezing rain (meteorological symbol); zone refined

Zr zirconium

ZR freezing rain (symbol); Zenith Radio

Z-R Zimbabwe-Rhodesia (Zimbabwe; formerly Rhodesia or Southern Rhodesia)

Z/R Zone of Responsibility

Zr⁹⁵ radioactive zirconium

zrc zirconium carbide

ZRC Zenith Radio Corporation

ZRCL Zlac Rowing Club Limited

ZRH Zurich, Switzerland (airport)

zrn zirconium nitride

zrp zero radial play

zrt zero-reaction tool

ZRU Zone de Rénovation Urbaine (French—Urban Redevelopment Zone)

zrv zero relative velocity

zs zero shift; zero and subtract;

zero surpress; zero suppression (of non-significant zeros in computer-printed numerals)

z S zur See (German—of the navy)

Zs Zeitschrift (German—periodical)

ZS Zoological Society

zsa zero-set amplifier

zsat zinc-sulfide atmospheric tracer

zsb zinc storage battery

zsc zinc sub-carrier chromaticity; zinc silicate coat(ing)

ZSC Zeeland Shipping Company; Zoological Society of Cincinnati

Z-scale height determination scale

zsd zebra-stripe display; zinc sulfide detector

ZSDS Zinc Sulfide Detection System

ZSE Zagreb Soloists Ensemble (Solisti di Zagreb)

zsf zero skip frequency

zsg zero-speed generator

zsi zero-size image

ZSI Zoological Society of Ireland

Zsig Zsigmond

ZSL Zoological Society of London

ZSL Zjednoczone Stronnictwo Ludowe (Polish—United Peasant Party)

ZSM Zoar State Memorial

ZSN Zoological Station of Naples

ZSP Zoological Society of Philadelphia

zspg zero-speed pulse generator

ZSS Zero-Sum Society; Zinc Sulfide System

ZSSD Zoological Society of San Diego

Zssg(n) Zusammensetzung(en) [German—compound word(s)]

zst zero strength time (measurement); zinc-sulfide tracer

ZST Zone Standard Time

ZSU-23 Soviet self-propelled antiaircraft gun including quadruple 23mm cannon

ZSU-23-4 Soviet antiaircraft system mounted on a tank and carrying four 23mm cannons

ZSU-57 Soviet self-propelled antiaircraft gun including twin 57mm cannon

zt zero tolerance (drugs); zipper tube; zipper tubing

z T zum Teil (German—partly)

Zt Zeit (German—time)

ZT lighter-than-air training aircraft (naval symbol); Zachary Taylor (12th President U.S.); zero time; zone time

ZT *Zone Torride* (French—torrid zone)

ZTA Zulu Territorial Authority

Z-table mortality table

Z-test Zulliger test

Ztg *Zeitung* (German—newspaper)

Z-time zebra time or zulu time (jargon for Greenwich Mean Time)

ztlp zero-transmission level point

ZTO Zone Transportation Office(r); *Zürich Tonhalle Orchester* (Zurich Concert Hall Orchestra)

ztp zero temperature plasma

Ztr *Zentner* (German—hundred-weight)

Ztschr *Zeitschrift* (German—periodical)

Z-TWIST Z-shaped open-band twist

Zu Zublena; Zudegi; Zula; Zuleika; Zulena; Zulima; Zulu; Zuma

ZU lighter-than-air utility aircraft (naval symbol)

ZU-23 Soviet antiaircraft system having a maximum fire power of 2000 rounds per minute

Zuck *Zuckung* (German—contraction)—sometimes abbreviated *Z*

'zuco bazuco

ZUF Zapata Urban Front (Mexican terrorist group)

Zulo Félix Zuloaga (Mexican president and soldier); Ignacio Zuloaga (Spanish painter)

Zulu code word for Greenwich mean time (Zulu time); letter Z radio code; Soviet Z-class attack submarines

ZUM Zimbabwe Unity Movement (ZANU and ZAPU united); Zone Usage Measurement

Z und Z *Zar und Zimmermann* (German—Czar and Carpenter)—opera Albert Lortzing

ZUP *Zone à urbaniser en priorité* (French—Priority Urbanization Zone)—slum cleanup or demolition zone

Zur Zurab; Zürich; Zuriel; Zurr

Zur Col Mus Zurich Collegium

Musicum (Latin—Zurich Music College)

Zurl Zuriel

zus *zusammen* (German—together)

Zus *Zusammenfassung* (German—summary)

Zuschr *Zuschrift(en)* [German—communication(s)]

Zut Zutphen

zuverl *zuverlassig* (German—authentic)

zuzzur *zuzzurullóne(a)* (Italian—overgrown child, just a big kid)

zv zika virus

zv *zu verfugung* (German—at disposal)

Zv *Zolverein* (German—customs union)

Z-value degrees Fahrenheit required to reduce thermal death time by 90% or one log value

ZVEI *Zentralverband der Elektrotechnischen Industrie* (German—Central union of the Electrotechnical Industry)

zvr zener voltage regulator

zvrd zener voltage regulator diode

zw zero wear

zw *zwart* (Dutch—black); *zwischen* (German—between; within)

Zw *Zwischensatz* (German—insertion or interpolation or parenthesis)

ZW *zuidwest* (Dutch—southwest)

zwc zone wind computer

zwitt zwitterion (diplole ion)

zwl zero wave length

ZWO *Zuiver Wetenschappelijk Onderzoek* (Dutch—Netherlands Organization for the Advancement of Pure Research)

Zwol Zwolle

zwp zone wind plotter

zwv zero wave velocity

Zy Zylota; Zyma

ZYA Zionist Youth Association

zyg zygote

Zyg Zygmunt

zygo zygomatic; zygomaticus

zygo (Latin prefix—join or union)

zym zymbalon; zymurgy

zymb *zymbalum* (Hungarian—dulcimer)—also known as the

cimbalom, cymbalon, czimbalom, zimbalon

Zymb *Zymbel* (German—cymbal)

zymo zymogen(esis); zymogenic(al)(ly); zymogenous; zymologic(al)(ly); zymologist; zymology; zymolysis; zymometer; zymoplastic; zymosis; zymosthenic(al)(ly); zymotic(al)(ly); zymotic disease

zymol zymology

ZYP Zefkrome Yarn Program

zyr zyrian(s)

Zyr Zyrian (Finno—Ugric language spoken by Zyrians in Komi SSR)

zyth zythum (ancient beer beverage)

zythep zythepsary (obsolete term for brewery)

zythia zythiaceae

zyz zyzzyva

zyzo zyzogeton; zyzomys

zz increasing degrees of contraction (symbol); zigzag

z-z longitudinal axis/roll axis

zz. *zingiber* (Latin—ginger)

z *Z zur Zeit* (German—at present, for the time being)

ZZ Ariana Afghan Airlines; longitudinal or roll axis (symbol); zed-zed; zz-approach

Z&Z Zulch and Zulch

ZZ *Zentralbibliothek Zürich* (German—Zurich Central Library)—combines the canton state, and university libraries

zza zamack zinc alloy

ZZB Zanzibar (tracking station)

zzc zero-zero condition

zzd zig-zag diagram

z-z fold zig-zag fold (concertina fold)

ZZO *zuidzuidoost* (Dutch—south southeast)

zzr zig-zag rectifier

Z-z's Zionist zealots

z *Zt zur Zeit* (German—at present, for the time being)

zzv zero-zero visibility

ZZV Zanesville, Ohio (airport)

ZZW *zuidzuidwest* (Dutch—south southwest)

ZZZ Zayda, Zorayda, Zorahayda—The Three Beautiful Princesses in Washington Irving's *Alhambra*

ZZZ-ZZZ-ZZZ sawing or snoring (cartoonist symbol)

Airlines of the World

Many of the following entries are in past editions but many more are new or revised.
An open space after a two-letter entry means an airline so coded has been discontinued or the code is available for new airlines.

AA American Airlines
AB Falcon Airlines
AC Air Canada
AD Exec Express
AE Air Europe
AEI Air Express International
AF Air France
AG Air Bridge Carriers (cargo)
AH Air Algerie
AI Air India
Air Canada AC (Canadian international airline)
Air France AF
Air India AI (international Indian airline service)
Air NZ NZ, TE
AIRPAC Air Pacific
Air UK United Kingdom airlines (Air Anglia and BIA)
AJ All Island Air
AK Altair Airlines
AL Allegheny Airlines (now USAir)
Alaska Alaska Airlines
Alitalia AZ (international Italian airline)
AM Aeroméxico
American AA
AN Ansett Airlines of Australia
AO Avisco
AP Aspen Airways
AQ Air Anglia; Aloha Airlines
AR Aerolineas Argentinas
AS Alaska Airlines
AT Royal Air Maroc
AU Austral Lineas Aéreas
AV Avianca
Avensa VE
Avianca Aerovias Nacionales de Colombia (Spanish—National Airlines of Colombia)
AW Aeroquetzal
AX Connectair (passenger)
AY Finnair
AZ Alitalia
BA British Airways
BAX Burlington Air Express
BB Air Great Lakes
BC Brymon Airways
BCAL British Caledonian Airways
BD British Midland Airways
BE Enterprise Airlines
BF Iowa Airlines and Horizon Airways
BG Bangladesh Biman

BH Augusta Airways
BI Royal Brunei Airlines
BJ Bakhtar Afghan Airlines
BK Chalk's International Airline
BL Air BVI
BM Aero Transporti Italiani
BN Braniff International Airways
BO Bouraq Indonesia Airlines
BOAC British Overseas Airways Corporation
BP Air Botswana
BQ Business Jets
BR British Caledonian Airways
Braniff BN
British European British European Airways
BS Auxaire-Bretagne
BT Air Martinque (Satair)
BU Braathens SAFE Airtransport
BV Northwest Skyways
BW BWIA International
BX
BY Britannia Airways; Burlington
BZ Capital Airlines (passenger); Skyfreighters (cargo)
CA CAAC (Civil Aviation Administration of China)
CB Commuter Airlines
CC Crown Aviation
CD Trans-Provincial Airlines
CE Air Virginia
CF Faucett
CG Clubair
CH Express Airways
CI China Airlines
CJ Colgan Airways
CK Connair
CL Capitol International Airways
CM COPA (Compañia Panameña de Aviacíon)
CN Tropic Air
CO Continental Airlines (Air Micronesia)
Continental CO
CP CP Air
CP Air Canadian Pacific Airlines
CQ Aero-Chaco; Aerolinea Federal Argentina
CR
CS Air Toronto
CT Command Airways

CU Cubana Airlines
CV Associated Airlines (passenger)
CW St Andrews Airways
CX Cathay Pacific Airways
CY Cyprus Airways
CZ Cascade Airways
DA Dan-Air Services
DB Brittany Air International
DC Trans Catalina Airlines
DD Command Airways
DE Downeast Airlines
Delta DL
DF Air Nebraska
DG Darien Airlines
DH Discovery Airlines; Tonga Air Service
DI Delta Air (Germany)
DJ Air Djibouti
DK Decatur
DL Delta Air Lines
DM Meersk Air
DN Skystream Airlines
DO Dominicana de Aviacion
DP Cochise Airlines
DQ Coastal Air Transport
DR Advance Airlines
DS Air Senegal
DT TAAG-Angola Airlines
DU Roland Air
DV Nantucket Airlines
DW DLT Deutsche Regional
DX Danair
DY Alyemda Democratic Yemen
DZ Douglas Airways
EA Eastern Airlines
Eastern EA
EB Eagle Airlines
EC Air Ecosse
ED Sunbird
EE Eagle Commuter Airlines
EF Far Eastern Air Transport
EG Japan Asia Airways
EH Roederer Aviation
EI Air Lingus (Irish)
EJ New England Airlines
EK Masling Commuter Services
EL Nihon Kinkyori Airways
El Al LY
EM Empire Airlines; Hammond's Air Service
EN Air Caravane
EO Aeroamérica; Air Nordic Sweden
EP Tropic Air Services

EQ TAME
ER
ES Air Atlantique
ET Ethiopian Airlines
EU Empresa Ecuatoriana de Aviación
EV Atlantic Southeast
EW East-West Airlines
EX Eagle Aviation; Emirates Airlines
EY Europe Aero Service
EZ Sun-Air of Scandinavia
FA Finnaviation
FB
FC Chaparral Airlines
FD Wiscair
FE Florida Airlines and Air South
FEDEX Federal Express
FF Air Link
FG Ariana Afghan Airlines
FH Mall Airways
FI Flugfelag-Icelandair
Finnair Finnish Airlines
FJ Air Pacific
FK Flamenco Airlines; Geelong Air Travel
FL Frontier Airlines
FM
FN Air Carolina
FO Southern Nevada
FP Simmons
FQ Air Aruba; Compagnie Aerienne du Languedoc
FR Susquehanna
FS Key Airlines
FT Flying Tiger Line
FU Air Littoral
FV Frisia Luftverkehr
FW Isles of Scilly Skybus
FX Express Air; Mountain West Airlines
FY Metroflight Airlines and Great Plains Airline
FZ Air Chico
GA Garuda Indonesian Airways
GB Air Inter Gabon
GC Lina-Congo
GD Air North
GE Guernsey Airlines
GF Gulf Air
GG Gem State Airlines
GH Ghana Airways
GI Air Guinee
GJ Ansett Airlines of South Australia
GK Laker Airways
GL Gronlandsfly
GM Air America
GN Air Gabon
GO Gambia Air Shuttle
GP Hadag Air Seebaederflug
GQ Big Sky Airlines
GR Aurigny Air Services
GS BAS Airlines
GT Gibraltar Airways

GU Aviateca
GV Talair
GW Golden West Airlines
GX Great Lakes Airlines
GY Guyana Airways
GZ Air Rarotonga
HA Hawaiian Air Lines
HB Air Melanesiae
HC Haiti Air International
HD New York Helicopter
HE Green Bay Aviation
HF First Air
HG Harbor Airlines
HH Somali Airlines
HI Papillon Airways
HJ
HK South Pacific Island Airways
HL
HM Air Mahe
HN NLM-Dutch Airlines
HO Airways International; Charterair
HP Air Hawaii; America West Airlines
HQ Business Express; Heussler Air Service; New York Helicopter
HR Air Bremen
HS Marshall's Air
HT Air Tchad
HU Trinidad and Tobago Air Services
Hughes Hughes Air West
HV Air Central
HW Havasu Airlines; North-Wright Air Ltd.
HX Hamburg Airlines
HY Metro Airlines
HZ Henebery Aviation
IA Iraqi Airways
IB Iberia Air Lines of Spain
Iberia IB
IC Indian Airlines
ID Apollo Airways
IE Solomon Islands Airways
IF Interflug
IG Alisarda
IH Channel Flying; Itavia
II Imperial Airlines; London City Airways
IJ Touraine Air Transport
IK Eureka Aero Industries
IL Island Air
IM Jamaire
Imperial II
IN East Hampton Air
IO Air Paris
IP Airlines of Tasmania
IQ Caribbean Airways
IR Iran Air; Iran National Airlines
Irish EI
IS Eagle Air
IT Air Inter
IU Midstate Airlines

IV British Island Airways
IW International Air Bahama
IX Flandre Air
IY Yemen Airways
IZ Arkia-Israel Inland Airlines
JA Bankair
JAL JL
Japan JL
JB Pioneer Airways
JC Rocky Mountain Airways
JD Toa Domestic Airlines
JE Yosemite Airlines
JF LAB Flying Service
JG Swedair
JH Nordeste-Lineas Aéreas Regionais
JI Gull Air; Jet Express
JJ Coddair Air East
JK Sun World
JL Japan Air Lines
JM Air Jamaica
JN Japan Air Commuter
JO Nordeste Airlines
JP Indo-Pacific International
JQ Trans-Jamaican Airlines
JR Air California; Delta Air
JS Korean Airways
JT Iowa Airways
JU Yugoslav Airlines
JV Bearskin Lake; Jersey European Airways
JW Polar Avia; Royal American
JX Bougair
JY Jersey European
JZ Alamo Commuter Airlines
KA Coastal Plains Commuter
KB Burnthills
KC Cook Islands International
KD Kendell Airlines
KE Korean Air Lines
KF Catskill Airways
KG Catalina Airlines
KH Cook Islandair
KI Air Atlantique
KJ Air Guyane
KL *KLM (Koninklijke Luchtvaart Maatschappij)*—Royal Dutch Airlines
KLM KL
KM Air Malta
KN Air Kentucky; Temsco Airlines
KO Kodiak Western Alaska Airlines
KP Safair
KQ Kenya Airways
KR Kar-Air (Finland)
KS Peninsula Airways
KT Turtle Airways
KU Kuwait Airways
KV Transkei Airways
KW Carnival Air Lines; Dorado Wings
KX Cayman Airways
KY Sun West

KZ Oriens & King
LA LAN Chile
LB Lloyd Aereo Boliviano
LC Loganair
LD LADE (Lineas Aéreas del Estado)
LE Magnum Airlines
LF Linjeflyg
LG Luxair (Luxembourg Airlines)
LH Lufthansa German Airlines
LI LIAT (Leeward Islands Air Transport)
LJ Sierra Leone Airways
LK Letaba Airways
LL Bell-Air
LM *ALM (Antillianaanse Luchtvaart Maatschappij)*—Dutch-Antillean Airline Company
LN Libyan Arab Airlines
LO LOT (Polish Airlines)
LP Air Alpes
LQ Inland Empire Airlines
LR LACSA *(Lineas Aéreas Costarricenses)*—Costa Rican Airlines
LS Marco Island Airways
LT Great Sierra
LU Theron Airways
Lufthansa LH
LV El Al Israel Airlines; LAV *(Linea Aeropostal Venezolana)*—Venezuelan Aeropost Lines
LW Air Nevada
LX Crossair
LY El Al Israel Airlines
LZ Balkan (Bulgarian Airlines)
MA *MALEV (Magyar Legikolekedesi Vallat)*—Hungarian Air Lines
MB Countrywide; Western Airlines
MC Rapidair
MD Air Madagascar
ME Middle East Airlines/Air Liban
MF Red Carpet Flying Service
MG MGM Grand Air; Pompano Airways
MH Malaysian Airline System
MI Mackey International Airlines
MJ Lineas Aereas Privadas Argentinas
MK Air Mauritius
ML Aviation Services
MM Sociedad Aeronautica Medellin
MN COMAIR (Commercial Airways)
MO Calm Air International
MP Atlantis Airlines
MQ Magnum Airlines; Simmons Airlines
MR Air Mauritanie
MS Egyptair

MT Mac Knight Airlines
MU China Eastern Airlines; Misrair
MV MacRobertson-Miller Airline Service
MW Maya Airways
MX Mexicana de Aviación
MY Air Mali
MZ Merpati Nusantara Airlines
NA Executive Air Charter; National Airlines
National NA
NB New Haven Airways
NC Newair
ND Nordair
NE Air New England
NF EJA/Newport
NG Green Hills Aviation
NH All Nippon
NI LANICA *(Lineas Aéreas de Nicaragua)*—Nicaraguan Airlines
NJ Namakwaland Lugdiens
NK NORCANAIR
NL Air Liberia
NM Mt Cook Airlines
NN Air Trails
NO Air North
Northwest NW
NP Desert Pacific
NQ Cumberland Airlines
NR NORONTAIR
NS Nuernberger
NT Lake State Airways
NU Southwest Airlines
NV Northwest Territorial Airways
NW Northwest Orient Airlines
NX New Zealand Air Charter
NY New York Airways
NZ Air New Zealand (domestic)
OA Olympic Airways
OB Opal Air
OC Air California
OD Emerald Airlines
OE Samoan
OF Noosa Air
OG Air Guadeloupe
OH Comair
OI TAVINA *(Trans-Colombiana de Aviación)*
OJ Air Texana
OK Czechoslovak Airlines
OL ÖLT *(Östfriesische Lufttransport)*—German—East Frisian Air Transport
OM Air Mongol (MIAT)
ON Air Nauru
OO Sunaire Lines
OP Air Panamá Internacional
OQ Royale Airlines
OR Air Comores
OS Austrian Airlines
OT Evergreen Helicopters of Alaska
OU City Express; Otonabee

Airways
OW Trans Mountain Airlines
OX Air Atlantic Airlines
OY New Jersey Airways
OZ Ozark Air Lines
PA Pan American World Airways
Pan Am PA
PB Air Burundi
PC Fiji Air
PD Pem Air
PE People Express
PF Trans Pennsylvania Airlines
PG Florida Commuter
PH Polynesian Airlines
Philippine PR
PI Piedmont Aviation
PJ Air St Pierre
PK Pakistan International
PL Aero Peru
PM Pilgrim Airlines
PN Coastal Airways; Princeton Aviation
PO Aeropelican Intercity Commuter Air Services
PP Phillips Airlines
PQ Pacific Coast Airlines; PRINAIR (Puerto Rican International Airlines)
PR Philippine Airlines
PS Central States Airlines; PSA (Pacific Southwest Airlines)
PT Provincetown-Boston Airline
PU PLUNA *(Primeras Lineas Uruguayos de Navegación Aérea)*—Spanish—First Uruguayan Aerial Navigation Lines
PV Eastern Provincial Airways
PW Pacific Western Airlines
PX Air Niugini (Air New Guinea)
PY Surinam Airways
PZ LAP *(Lineas Aéreas Paraguayas)*
QA Air Caribe
Qantas QF
QB Quebecair
QC Air Zaire
QD Trans-Brasil
QE Air Tahiti
QF Qantas Airways
QG Sky West Aviation
QH West African Airways
QI Cimber Air
QK Air Nova; Mexico Air Service
QL Lesotho Airways
QM Air Malawi
QN Bush Pilots Airways
QO Bar Harbor Airlines
QP Sunbird
QQ Michigan Airways
QR Air Satellite
QS Cal Sierra
QT Vaengir (Wings Air Ice-

land)
QU Uganda Airlines
QV Lao Aviation
QW Air Turks and Caicos
QX Century Airlines; Horizon Air
QY Aero Virgin Islands
QZ Zambia Airways
RA Royal Nepal Airlines
RB Syrian Arab Airlines
RC Atlantic Airways—Faroe Islands; Republic
RD Aviona
RE Aer Arann Teo
RF Rossair; Travelair Goteborg
RG VARIG *(Viação Aérea Rio Grandense)*—Portuguese—Rio Grande Airlines
RH Air Zimbabwe; Regal Bahamas International Airlines
RI Eastern Airlines
RJ Royal Jordanian Airlines (ALIA)
RK Air Afrique
RL Aerolineas Nicaraguenses; Crown International Airlines
RM Wings West
RN Royal Air International
RO TAROM (Romanian Air Transport)
Route of the Red Baron LH
RP Precision Airlines
RQ Maldives International Airlines
RR Royal Air Force
RS Aeropesca
RT Norving
RU Britt Airways
RV Reeve Aleutian Airways
RW Republic
RX British Independent Airways; Capitol Air Service
RY Air Rwanda; Perkiomen Airways
RZ Arabia (Arab International)
SA South African Airways
SABENA SN
SAS SK
SB Air Caledonie International
SC Cruzeiro do Sul
SD Sudan Airways
SE Southeast Skyways; Wings of Alaska
SF Scruse Air
SG Atlantis
SH SAHSA (Servicio Aéreo de Honduras SA)
SI Air Sierra
SJ Stewart Island
SK SAS (Scandinavian Airlines)
SL Rio-Sul
SN SABENA (Belgian Airlines)
SO Austrian Air
SP *SATA (Sociedade Açoriana de Transportes Aéreos)*—Portuguese—Azores Air Trans-

port Line
SQ Singapore Airlines
SR Swissair
SS South Coast Airlines
ST Belize Airways
SU Aeroflot (Soviet Union Airlines)
SV Saudi Arabian Airlines
SW Namib Air
Swissair SR
SX Christman Air System
SY Air Alsace
SZ China Southwest Airlines; ProAir Services
TA Taca International
TAP TP
TB Tejas Airlines; Trump Shuttle
TC Air Tanzania
TD Tansavio
TE Air New Zealand (international)
TF Veeneal
TG Thai Airways (international)
TH Thai Airways (domestic)
TI Texas International Airlines
TJ Oceanair
TK Turk Hava Yollari
TL Trans Mediterranean (cargo)
TM *DETA (Direçãao de Exploração dos Transportes Aéreos)* —Portuguese—Directorate of Exploration of Aerial Transport— (Mozambique Airline)
TN Trans-Australia Airlines
TO Alkan Air Ltd
TP *TAP (Transportes Aéreos Portugueses)*—Portuguese Air Transport; Tap Air Portugal
TQ Las Vegas Airlines
TR Royal Air; Transbrasil
TS Samoa Air
TT Royal West
TU Tunis Air
TV Haiti Trans Air; Transamerica
TW Trans World Airlines
TWA TW
TX Transportes Aéreos Nacionales
TY Air Caledonie
TZ American Trans Air; (SANSA) Services Aereos Nacionales
UA United Airlines
UB Burma Airways
UC *LADECO (Linea del Cobre)* —Spanish—Copper Line
UD Georgian Bay; Lloyd Your Trans-Australian Airline
UE Air La; United Air
UF Sydaero
UG Norfolk Island Airlines
UH Austin Airways
UI *Flugfelag Nordurlands*—

Northlands Air
UJ Air Sedona
UK British Island Airways (Air UK)
UL Air Lanka
UM Air Zimbabwe
UN East Coast Airlines
United UA
UO Direct Air
UP Bahamas Air
UQ Suburban Airlines
UR British International Helicopters; Empire Airlines
US Usair Express
USAir airline absorbing PSA; formerly Allegheny Airlines
UT UTA *(Union de Transports Aeriens)*—Union Transport Airline
UU Reunion Air
UV Universal Airways
UW Perimeter Airlines
UX Air Illinois
UY Cameroon Airlines
UZ Air Resorts Airlines; Nefertiti
VA *VIASA (Venezolana Internacional de Aviación)*—Spanish—Venezuelan International Aviation)
VB Birmingham European Airways; Westair Commuter Airlines
VC TAC *(Transportes Aéreos del Cesar)*
VD
VE *AVENSA (Aerovias Venezolanas)*—Spanish—Venezuelan Airlines
VF British Air Ferries; Golden West
VG City Flug
VH Air Burkina; Air Volta
VI Vieques Airlink
VJ Trans-Colorado
VK Air Tungaru
VL Mid-South Commuter Airlines
VM Ocean Airways
VN Hang Khong Vietnam
VO Tyrolean Airways
VP *VASP (Viação São Paulo)* — Portuguese—São Paulo Airline
VQ Oxley Airlines
VR Transportes Aéreos de Cabo Verde
VS Virgin Atlantic Airways
VT Air Polynesie
VU Air Ivoire
VV Semo Aviation
VW Ama-Flyg
VX Aces
VY Coral Air
VZ Aquatic Airways
WA Western Airlines
WB *SAN (Servicios Aéreos Na-*

cionales)—Spanish—National Air Services
WC Wien Air Alaska
WD Ward Air
WE Votec
Western WA
WF Wideroes Flyveselskap
WG *ALAG (Alpine Luft Transport AG)*—German—Alpine Air Transport Company
WH China Northwest Airlines; Southeastern Commuter Airlines
WI Swift-Aire Lines
WJ Labrador Airways; Torontair
WK Westkuestenflug
WL Bursa Hava Yollari
WM Windward Island Airways International
WN Southwest Airlines
WO World Airways
WP Aloha Islandair; Princeville Airways
WQ Wings Airways
WR Wheeler Flying Service
WS Northern Wings (Québec-air)
WT Nigeria Airways
WU Netherlines; Rhine Air
WV Midwest Aviation
WW Scottish European Airways; Trans-West
WX Ansett Airlines of New South Wales
WY Indiana Airways
WZ Berlin European; Trans Western Airlines of Utah
XC Caribbean Air Transport
XE South Central

XF Cobden Airways
XG Air North
XJ Mesaba Aviation
XK *AEROTAL (Aerolineas Territoriales de Colombia)*—Spanish—Territorial Airlines of Colombia
XO Rio Airways
XP Avior
XQ Caribbean International
XT Executive Transportation
XU Trans Mo Airlines
XV Mississippi Valley Airways
XW Walker's Cay Air Terminal
XX Valdez Airlines
XY Munz Northern
XZ Air Tasmania
YB Hyannis Aviation
YC Alaska Aeronautical Industries
YD Ama Air Express
YE Grand Canyon Airlines; Pearson Aircraft
YH Trans New York
YI Intercity
YJ Commodore
YK Cyprus Turkish Airways
YL Long Island Airlines; Montauk Caribbean Airways and Ocean Reef Airways
YM Mountain Home Air Service
YN Nor-East Commuter Airlines
YO Heli-Air-Monaco
YP Pagas Airlines
YQ Lakeland
YR Scenic Airlines
YS San Juan Airlines
YT Sky West

YU Aerolineas Dominicanas
YV Mesa Aviation
YW Stateswest Airlines
YX Midwest Express Airlines; Société Aeronautique Jurassiènne
YZ Linhas Aéreas da Guine-Bissau
ZA Alpine Aviation
ZB Air Vectors
ZC Royal Swazi National Airways
ZD Ross Aviation
ZE Air Caribe International; Pacific National
ZF Berlin U.S.A.
ZG Silver State
ZH Royal Hawaiian Airways
ZI Lucas Air Transport
ZK Great Lakes Aviation; Shavano Air
ZL Hazelton Air Services
ZM Trans-Central
ZN Tennessee Airways
ZO Trans-California
ZP Virgin Air
ZQ Ansett New Zealand
ZR Star Airways
ZS Hispaniola Airways
ZT *SATENA (Servicio Aeronavegación a Territorios Nacionales)*—Spanish—Aeronavigation Service to National Territories
ZU Zia Airlines
ZV Air Midwest
ZW Air Wisconsin
ZX Air West Airlines
ZY Air Pennsylvania

Air-Pollution Index

psi (PSI) pollution standards index

0 to 50 good air
above 50 moderately good air
above 100 hazardous and unhealthful air
200 to 299 very hazardous and unhealthful air

300 and over hazardous enough to cause premature death to elderly and sick persons
Stage I Alert unhealthful air pollution (200 to 270 psi)
Stage II Alert unhealthful and

hazardous air pollution (275 to 390 psi)
Stage III Alert unhealthful and extremely hazardous air pollution (400 to 500 psi), often deadly

Airports of the World

AAA Ararangúa, Brazil
AAB Arrabury, Queensland
AAE Annaba, Algeria

AAG Alto Araguaia, Brazil
AAI Arraias, Brazil
AAL Aalborg, Denmark

AAO Anaco, Venezuela
AAQ Aqiq, Saudi Arabia
AAR Aarhus-Randers, Den-

mark
AAU Arua, Uganda
AAX Araxá, Brazil
AAY Al Ghaydah, Aden
ABA Ababa, Ethiopia
ABD Abadan, Iran
ABE Allentown-Bethlehem-Easton, Pennsylvania
ABG Abingdon, Queensland
ABI Abilene, Texas
ABJ Abidjan, Ivory Coast
ABQ Albuquerque, New Mexico
ABR Aberdeen, South Dakota
ABW Abau, Papua
ABX Albury, New South Wales
ABY Albany, Georgia
ABZ Aberdeen, Scotland
ACA Acapulco, Mexico
ACC Accra, Ghana
ACE Arrecife, Canary Islands
ACI Alderney, United Kingdom
ACK Nantucket, Massachusetts
ACN Mbala, Zambia
ACT Waco, Texas
ACV Arcata-Eureka, California
ACY Atlantic City, New Jersey
ADA Adana, Turkey
ADD Addis Ababa, Ethiopia
ADE Aden, Southern Yemen
ADH Ada, Oklahoma
ADL Adelaide, South Australia
ADM Ardmore, Oklahoma
ADN Aydin, Turkey
ADP Anuradhapura, Ceylon
ADZ San Andrés, San Andrés Island (Caribbean)
AEH Abecher, Chad
AEO Aioun el Atro, Mauritania
AER Adler/Sochi, USSR
AES Aalesund, Norway
AEY Akureyri, Iceland
AFA San Rafael, Argentina
AFI Amalfi, Colombia
AFY Afyon, Turkey
AGA Agadir, Morocco
AGH Angelhome-Hälsingborg, Sweden
AGN Angoon, Alaska
AGP Málaga, Spain
AGQ Agrinion, Greece
AGR Agra, India
AGS Augusta, Georgia
AGX Araguacema, Brazil
AGZ Agri, Turkey
AHB Abha, Saudi Arabia
AHN Athens, Georgia
AHO Alghero-Sassari, Italy
AHU Al Hoceima, Morocco
AIA Alliance, Nebraska
AIM Salima, Malawi
AIT Aitutaki Island, Cook Islands
AJA Ajaccio, Corsica

AJF Jouf, Saudi Arabia
AJJ Akjoujt, Mauritania
AJN Anjouan, Comoro Islands
AJU Aracaju, Brazil
AJY Agades, Niger
AKF Kufra, Libya
AKH Akhisar, Turkey
AKL Auckland, New Zealand
AKN King Salmon-Naknek, Alaska
AKY Akyab, Burma
ALA Alma-Ata, USSR
ALB Albany-Schenectady, New York
ALC Alicante, Spain
ALF Alta, Norway
ALG Algiers, Algeria
ALH Albany, Western Australia
ALJ Alexander Bay, South Africa
ALL Albenga, Italy
ALM Alamogordo-Holloman, New Mexico
ALO Waterloo, Iowa
ALP Aleppo, Syria
ALQ Alegrete, Brazil
ALR Alexandra, New Zealand
ALS Alamosa, Colorado
ALW Walla Walla, Washington
ALY Alexandria, Egypt (UAR airport)
AMA Amarillo, Texas
AMB Ambilobe, Malagasy
AMC Am-Timan, Chad
AMD Ahmedabad, India
AMF Amparafaravola, Malagasy
AMH Arba Mintch, Ethiopia
AMI Ampenan, Indonesia
AMJ Almenara, Brazil
AML Puerto Armuellas, Panama
AMM Amman, Jordan
AMO Mao, Chad
AMP Ampanihy, Malagasy
AMQ Ambon, Indonesia
AMS Amsterdam, Netherlands
ANB Anniston, Alabama
ANC Anchorage, Alaska
AND Anderson, South Carolina
ANF Antofagasta, Chile
ANH Nhill, Victoria, Australia
ANK Ankara, Turkey
ANM Antalaha, Malagasy
ANN Annette Island, Alaska
ANO Antonio Enes, Mozambique
ANR Antwerp, Belgium
ANU Antigua, West Indies
ANX Andenes, Norway
AOD Abou Deia, Chad
AOI Ancona, Italy
AOL Paso de los Libres, Argentina
AON Arona, New Guinea

AOO Altoona-Martinsburg, Pennsylvania (airport serving both communities)
AOR Alor Star, Malaysia
AOT Abbottabad, Pakistan
APA Amapá, Brazil
APL Nampula, Mozambique
APS Anápolis, Brazil
APV Apple Valley, California
APW Apia, Western Samoa
APY Alto Parnaíba, Brazil
AQI Qaisumah, Saudi Arabia
AQJ Aqaba, Jordan
AQP Arequipa, Peru
AQR Alenquer, Brazil
AQU Aquidauana, Brazil
ARD Andradina, Brazil
ARH Atharan Hazari, Pakistan
ARI Arica, Chile
ARM Armidale, New South Wales
ARN Stockholm, Sweden (Arlanda airport)
ARS Aragarças, Brazil
ART Watertown, New York
ARU Araçatuba, Brazil
ARY Arusha, Tanzania
ASA Assab, Ethiopia
ASD Andros Town, Bahamas
ASE Aspen, Colorado
ASG Asiago, Italy
ASM Asmara, Ethiopia
ASO Asosa, Ethiopia
ASP Alice Springs, Northern Territory, Australia
ASU Asunción, Paraguay
ASW Aswan, Egypt
ATB Atbara, Sudan
ATH Athens, Greece
ATL Atlanta, Georgia
ATM Altamira, Brazil
ATQ Amritsar, India
ATR Star, Mauritania
ATV Ati, Chad
ATY Watertown, South Dakota
ATZ Assiut, Egypt
AUA Aruba, Netherlands Antilles
AUC Arauca, Colombia
AUD Augustus Downs, Queensland
AUG Augusta, Maine
AUH Abu Dhabi, Trucial Oman
AUM Austin, Minnesota
AUS Austin, Texas
AUW Wausau, Wisconsin
AVL Asheville-Henderson, North Carolina
AVP Scranton-Wilkes-Barre, Pennsylvania
AVX Avalon Bay, Santa Catalina Island, California
AWK Wake, Wake Islands, Pacific Ocean
AWZ Ahwaz, Iran
AXA Anguilla, Leeward Islands

AXC Aramac, Australia
AXD Alexandroupolis, Greece
AXK Attak, South Arabia
AXM Armenia, Colombia
AXU Axum, Ethiopia
AXV Aramac, Queensland
AYA Ayapel, Colombia
AYC Ayacucho, Colombia
AYQ Ayers Rock, Northern Territory, Australia
AYR Ayr, Queensland
AYS Waycross, Georgia
AYT Antalya, Turkey
AYU Aiyura, New Guinea
AZD Yazd, Iran
AZO Kalamazoo, Michigan
AZR Adrar, Algeria
BAH Bahrain Island, Persian Gulf
BAK Baku, USSR
BAL Baltimore, Maryland
BAQ Barranquilla, Colombia
BAT Barth, Germany
BAU Bauru, Brazil
BAV Balovale, Zambia
BAZ Barcelos, Brazil
BBI Bhubaneswar, India
BBN Bario, Malaysia
BBO Berbera, Somalia
BBQ Barbuda, Leeward Islands
BBT Berbérati, Central African Republic
BBV Bereby, Ivory Coast
BBY Bambari, Central African Republic
BCA Baracoa, Cuba
BCD Bacolod, Philippines
BCN Barcelona, Spain
BCO Baco, Ethiopia
BCR Boco do Acre, Brazil
BCY Bulchi, Ethiopia
BDA Hamilton, Bermuda
BDB Bundaberg, Queensland
BDC Barra do Cordo, Brazil
BDH Bandar Lengeh, Iran
BDI Barbados, West Indies
BDJ Bandjarmasin, Indonesia
BDL Hartford, Connecticut-Springfield, Massachusetts-Windsor Locks, Connecticut
BDM Bandirma, Turkey
BDN Badana, Saudi Arabia
BDO Bandung, Indonesia
BDQ Baroda, India
BDR Bridgeport, Connecticut
BDS Brindisi, Italy
BDU Bardufoss, Norway
BEB Benbecula, United Kingdom
BEG Belgrade, Yugoslavia
BEI Beica, Ethiopia
BEJ Benjamin Constant, Brazil
BEL Belém do Pará, Brazil
BEN Benghazi, Libya
BEP Bhola, Bangladesh

BER Berlin (Tempelhof airport)
BES Brest, France
BEU Bedourie, Queensland
BEW Beira, Mozambique
BEY Beirut, Lebanon
BFD Bradford, Pennsylvania
BFF Scottsbluff, Nebraska
BFI Seattle, Washington (Boeing Field)
BFL Bakersfield, California
BFN Bloemfontein, South Africa
BFO Buffalo Range, Rhodesia
BFS Belfast, Northern Ireland
BFU Beaufort West, South Africa
BGA Bucaramanga, Colombia
BGB Booue, Gabon
BGD Borger, Texas
BGF Bangui, Central African Republic
BGG Bogra, Bangladesh
BGH Boghe, Mauritania
BGI Bridgetown, Barbados
BGM Binghamton-Endicott-Johnson City, New York
BGO Bergen, Norway
BGR Bangor, Maine
BGU Bangassou, Central African Republic
BGW Baghdad, Iraq
BGX Bagé, Brazil
BHA Bahía, Ecuador
BHB Bar Harbor, Maine
BHD Kabwe, Zambia
BHH Bisha, Saudi Arabia
BHI Bahía Blanca, Argentina
BHJ Bhuj, India
BHM Birmingham, Alabama
BHN Beihan, South Arabia
BHO Bhopal, India
BHQ Broken Hill, New South Wales
BHS Bathurst, New South Wales
BHT Brighton Downs, Queensland
BHU Bhavnagar, India
BHX Birmingham, England
BHZ Belo Horizonte, Brazil
BIA Bastia, Corsica
BIF The Bight, Cat Island, Bahamas
BIK Biak, Indonesia
BIL Billings, Montana
BIM Bimini, Bahamas
BIO Bilbao, Spain
BIQ Biarritz, France
BIR Biratnagar, Nepal
BIS Bismarck, North Dakota
BIV Bria, Central African Republic
BIX Franz Josef Glacier, New Zealand
BJC Bartica, Guyana
BJD Birjand, Iran

BJI Bemidji, Minnesota
BJM Bujumbura, Burundi
BJO Brejo, Brazil
BJP Begumganj, Pakistan
BJR Bahar Dar, Ethiopia
BJZ Badajoz, Spain
BKE Baker, Oregon
BKG Boke, Guinea
BKI Kota Kinabalu, Malaysia
BKK Bangkok, Thailand
BKM Moscow, USSR (Bykovo Airport)
BKN Birni, Nkoni, Niger
BKO Bamako, Mali
BKQ Blackall, Queensland
BKR Bokoro, Chad
BKS Bengkulu, Indonesia
BKU Betioky, Malagasy
BKW Beckley, West Virginia
BKX Brookings, South Dakota
BKY Bukavu, Zaire
BKZ Bukoba, Tanzania
BLA Barcelona, Venezuela
BLB Balboa, Panama Canal
BLD Boulder City, Nevada
BLE Borlange, Sweden
BLF Bluefield-Princeton, West Virginia
BLG Belaga, Malaysia
BLH Blythe, California
BLI Bellingham, Washington
BLJ Batna, Algeria
BLL Billund, Denmark
BLO Blonduos, Iceland
BLQ Bologna, Italy
BLR Bangalore, India
BLX Belluno, Italy
BLY Bled, Yugoslavia
BLZ Blantyre, Malawi
BMA Stockholm, Sweden
BMB Bumba, Congo
BMD Belo, Malagasy
BME Broome, Western Australia
BMI Bloomington, Illinois
BML Berlin-Milan, New Hampshire
BMM Bitam, Gabon
BMN Batman, Turkey
BMO Bhamo, Burma
BMP Brampton Island, Queensland
BMV Banmethuot, South Vietnam
BMZ Belmonte, Brazil
BNA Nashville, Tennessee
BNB Boende, Zaire
BND Bandar Abbas, Iran
BNE Brisbane, Queensland
BNI Benin City, Nigeria
BNJ Bonn, Germany
BNM Barrancas, Venezuela
BNN Brönnöysund, Norway
BNP Bannu, Pakistan
BNS Barinas, Venezuela
BNX Banaras, India

BNZ Banz, New Guinea
BOB Bora-Bora, Society Islands
BOD Bordeaux, France
BOG Bogotá, Colombia
BOH Bournemouth, England
BOI Boise, Idaho
BOJ Burgas, Bulgaria
BOM Bombay, India
BON Bonaire, Netherlands Antilles
BOO Bodo, Norway
BOP Bouar, Central African Republic
BOS Boston, Massachusetts
BOY Bobo Dioulass, Volta
BPN Balikpapan, Indonesia
BPT Beaumont-Port Arthur, Texas
BPY Besalampy, Malagasy
BQL Boulia, Queensland
BQQ Barra, Brazil
BQR Butare, Rwanda
BRA Barreiras, Brazil
BRC San Carlos de Bariloche, Argentina
BRE Bremen
BRF Bradford, England
BRI Bari, Italy
BRL Burlington, Iowa
BRM Barquisimeto, Venezuela
BRN Bern, Switzerland
BRO Brownsville, Texas
BRQ Brno, Czechoslovakia
BRR Barra, United Kingdom
BRS Bristol, England
BRU Brussels, Belgium
BRW Barrow, Alaska
BRZ Bruzual, Venezuela
BSB Brasilia, Brazil
BSE Sematan, Sarawak, Malaysia
BSG Bata, Spanish Guinea
BSH Brighton, England
BSK Biskra, Algeria
BSL Basel, Switzerland (also spelled Basle)
BSN Bossangoa, Central African Republic
BSR Basra, Iraq
BSS Balsas, Brazil
BSU Basankusu, Zaire
BTC Batticaloa, Ceylon
BTD Brunett Downs, Northern Territory, Australia
BTH Bathurst, Gambia
BTJ Banda Atjeh, Indonesia
BTL Battle Creek, Michigan
BTM Butte, Montana
BTN Brunei Town, Brunei
BTO St Barthelemy, Leeward Islands, West Indies
BTR Baton Rouge, Louisiana
BTS Bratislava, Czechoslovakia
BTU Bintulu, Sarawak, Malaysia

BTV Burlington, Vermont
BTX Betoota, Queensland
BTZ Bursa, Turkey
BUA Buka, Solomon Islands
BUC Burketown, Queensland
BUD Budapest, Hungary
BUE Buenos Aires, Argentina
BUF Buffalo, New York
BUG Benguela, Angola
BUH Bucharest, Romania
BUJ Buno Bedelle, Ethiopia
BUK Bolu, Turkey
BUL Bulolo, New Guinea
BUN Buenaventura, Colombia
BUO Burao, Somalia
BUQ Bulawayo, Zimbabwe
BUR Burbank, California
BUX Bunia, Zaire
BUZ Bushire, Iran
BVA Beauvais, France
BVB Boa Vista, Brazil
BVH Vilhena, Brazil
BVI Birdsville, Queensland
BVO Bartlesville, Oklahoma
BVS Bela Vista, Brazil
BWG Bowling Green, Kentucky
BWP Bahawalpur, Pakistan
BXE Bakel, Senegal
BXO Bissãu, Portuguese Guinea
BYI Burley-Rupert, Idaho
BYK Bouake, Ivory Coast
BYV Baiyer River, New Guinea
BZE Belize, British Honduras
BZI Balikesir, Turkey
BZL Barisal, Bangladesh
BZN Bozeman, Montana
BZO Bolzano, Italy
BZV Brazzaville, Congo
BZY Brasileia, Brazil
CAB Cabinda, Angola
CAC Cascavel, Brazil
CAE Columbia, South Carolina
CAF Carauari, Brazil
CAG Cagliari, Italy
CAI Cairo, Egypt
CAJ Canaima, Venezuela
CAK Akron-Canton, Ohio
CAL Campbeltown, Scotland
CAN Canton, China
CAS Casablanca, Morocco
CAT Cat Island, Bahamas
CAX Carlisle, England
CAY Cayenne, French Guiana
CBB Cochabamba, Bolivia
CBE Cumberland, Maryland
CBG Cambridge, England
CBH Colomb Bechar, Algeria
CBL Ciudad Bolívar, Venezuela
CBN Cabarien, Cuba
CBQ Calabar, Nigeria
CBR Canberra, Australian Capital Territory
CBY Canobie, Queensland

CBZ Cucui, Brazil
CCH Chile Chico, Chile
CCK Cocos Island, Keeling Islands, Australia
CCM Criciuma, Brazil
CCP Concepción, Chile
CCQ Cachoeira do Sul, Brazil
CCS Caracas, Venezuela (Maiquetia Airport)
CCU Calcutta, India
CCX Caceres, Brazil
CDC Cedar City, Utah
CDD Castle Donington-East Midlands, England
CDF Cortina d'Ampezzo, Italy
CDJ Conceição, Brazil
CDP Chandpur, Bangladesh
CDQ Croydon, Queensland
CDR Chadron, Nebraska
CDV Cordova, Alaska
CDX Codo, Brazil
CDZ Codajaz, Brazil
CEB Cebu, Philippines
CEC Crescent City, California
CEG Chester, England
CEN Ciudad Obregon, Mexico
CEP Concepción, Bolivia
CEQ Cannes, France
CER Cherbourg, France
CEZ Cortez, Colorado
CFD Bryan, Texas
CFE Clermont-Ferrand, France
CFG Cienfuegos, Cuba
CFR Caen, France
CFU Corfu, Greece
CGB Cuiaba, Brazil
CGC Cape Gloucester, New Britain, New Guinea
CGH São Paulo, Brazil (Congonhas Airport)
CGI Cape Girardeau, Missouri
CGN Cologne, Germany
CGO Chengchow, China
CGP Chittagong, Bangladesh
CGR Campo Grande, Brazil
CHA Chattanooga, Tennessee
Changi Singapore airport
CHC Christchurch, New Zealand
CHI Chicago, Illinois (airports)— C-CGX (Meigs); O-ORD (O'Hare); M-MDW (Midway)
CHL Chalna, Bangladesh
CHO Charlottesville, Virginia
CHQ Chania, Greece
CHS Charleston, South Carolina
CHT Chita, USSR
CHU Chabua, India
CHW Charleston, West Virginia
CIC Chico, California
CID Cedar Rapids-Iowa City, Iowa
CII Chitipa, Malawi
CIR Cairo, Illinois

CIS Canton Island, Phoenix Islands
CIX Chiclayo, Peru
CIY Comiso, Italy
CJB Coimbatore, India
CJC Calama, Chile
CJL Chitral, Pakistan
CJS Ciudad Juarez, Mexico
CJU Cheju, South Korea
CJZ Cajazeiras, Brazil
CKB Clarksburg, West Virginia
CKG Chungking, China
CKV Clarksville, Tennessee-Hopkinsville, Kentucky
CKY Conakry, Guinea
CKZ Canakkale, Turkey
CLA Comilla, Bangladesh
CLE Cleveland, Ohio (airports —L-BKL (Lakefront); C-CLE (Hopkins); Y-CGF (Cuyahoga)
CLF Clear, Alaska
CLH Coolah, New South Wales
CLN Carolina, Brazil
CLO Cali, Colombia
CLT Charlotte, North Carolina
CLY Calvi, Corsica
CLZ Cristalandia, Brazil
CMA Cunnamulla, Queensland
CMB Colombo, Ceylon
CME Ciudad del Carmen, Mexico
CMG Corumba, Brazil
CMH Columbus, Ohio
CMI Champaign, Illinois
CML Camooweal, Queensland
CMQ Clermont, Queensland
CMV Cayo Mambi, Cuba
CMW Camagüey, Cuba
CMX Hancock-Houghton, Michigan
CNA Cananea, Mexico
CND Constanta, Romania
CNJ Cloncurry, Queensland
CNM Carlsbad, New Mexico
CNN Carinhanha, Brazil
CNQ Corrientes, Argentina
CNR Chañaral, Chile
CNS Cairns, Queensland
CNV Canavieras, Brazil
COC Concordia, Argentina
COE Coeur d'Alene, Idaho
COG Condoto, Colombia
COH Cooch Behar, India
COK Cochin, India
CON Concord, New Hampshire
COO Cotonou, Dahomey
COR Córdoba, Argentina
COS Colorado Springs, Colorado
COU Columbia, Missouri
COW Corowa, New South Wales
CPA Cape Palmas, Liberia
CPE Campeche, Mexico

CPG Carmen de Patagones, Argentina
CPH Copenhagen, Denmark
CPL Chaparral, Colombia
CPO Copiapó, Chile
CPQ Campinas, Brazil
CPR Casper, Wyoming
CPT Cape Town, South Africa
CPV Campina Grande, Brazil
CQF Calais, France
CQQ Crato, Brazil
CQS Vitória da Conquista, Brazil
CQZ Cherokee Sound, Bahamas
CRD Comodoro Rivadavia, Argentina
CRF Carnot, Central African Republic
CRI Crooked Island, Bahamas
CRP Corpus Christi, Texas
CRQ Carasinho, Brazil
CRV Caravelas, Brazil
CRW Charleston, West Virginia
CSG Columbus, Georgia
CSJ Cape St Jacques, South Vietnam
CSO Montevideo, Uruguay (Carrasco Airport)
CSX Changsa, China
CTA Catania, Sicily
CTB Cut Bank, Montana
CTC Catamarca, Argentina
CTG Cartagena, Colombia
CTH Crateus, Brazil
CTL Charleville, Queensland
CTM Chetumal, Mexico
CTN Cooktown, Queensland
CTQ Santa Vitoria, Brazil
CTU Chengtu, China
CTW Crotone, Italy
CUC Cúcuta, Colombia
CUE Cuenca, Ecuador
CUJ Canutama, Brazil
CUK São Paulo, Brazil (Combika Airport)
CUL Culiacán, Mexico
CUM Cumaná, Venezuela
CUP Carupano, Venezuela
CUQ Coen, Queensland
CUR Curaçao, Netherlands Antilles
CUU Chihuahua, Mexico
CUV Casigua, Venezuela
CUZ Cuzco, Peru
CVG Cincinnati, Ohio
CVM Ciudad Victoria, Mexico
CVN Clovis, New Mexico
CVO Albany-Corvallis, Oregon
CVQ Carnarvon, Western Australia
CVT Coventry, England
CWB Curitiba, Brazil
CWE Cromwell, New Zealand
CWL Cardiff, Wales

CWP Campbellpore, Pakistan
CXA Caicara de Orinoco, Venezuela
CXB Cox's Bazar, Bangladesh
CXJ Caxias do Sul, Brazil
CXS Caxias, Brazil
CXT Charters Towers, Queensland
CXY Cat Cay, Bahamas
CYB Cayman Brac, West Indies
CYS Cheyenne, Wyoming
CZA Coari, Brazil
CZB Cruz Alta, Brazil
CZE Coro, Venezuela
CZL Constantine, Algeria
CZM Cozumel, Mexico
CZS Cruzeiro do Sul, Brazil
CZU Corozal, Colombia
CZY Cluny, Queensland
DAB Daytona Beach, Florida
DAC Dacca, Bangladesh
DAD Danang, South Vietnam
DAH Dathina, Yemen
DAL Dallas-Fort Worth—D-DAL (Love Field); F-FTW (Meacham); G-GSW (Greater Southwest International)
DAM Damascus, Syria
DAN Danville, Virginia
DAR Dar-es-Salaam, Tanzania
DAU Daru, Papua
DAY Dayton, Ohio
DBL Dabolim, India
DBM Debra Markos, Ethiopia
DBO Dubbo, New South Wales
DBQ Dubuque, Iowa
DBT Debra Tabor, Ethiopia
DBV Dubrovnik, Yugoslavia
DBY Dalby, Queensland
DCA Washington, D.C. (National Airport)
DDC Dodge City, Kansas
DDL Dodollo, Ethiopia
DDU Dadu, Pakistan
DDZ Dedza, Malawi
DEC Decatur, Illinois
DEM Dembidolo, Ethiopia
DEN Denver, Colorado
DEQ Daydream Island, Queensland
DET Detroit, Michigan
DEZ Deir Ezzor, Syria
DGO Durango, Mexico
DGU Dedougu, Upper Volta
DHA Dhahran, Saudi Arabia
DHD Durham Downs, Queensland
DHL Dhala, Aden
DHN Dothan, Alabama
DIA Dulles International Airport (Washington, D.C.)
DIC Dili, Zaire
DIE Diego Suarez, Malagasy
DIJ Dijon, France
DIL Dilly, Portuguese Timor

DIP Diapaga, Upper Volta
DIR Dire Dawa, Ethiopia
DIS Dolisie, Congo
DIY Diyarbakir, Turkey
DJB Djambi, Indonesia
DJE Djerba, Tunisia
DJG Djanet, Algeria
DJJ Djajapura, West Irian, Indonesia
DJM Djambala, Congo
DJO Daloa, Ivory Cost
DJR Dajarra, Queensland
DJV Dabajuro, Venezuela
DKR Dakar, Senegal
DLA Douala, Cameroon
DLB d'Albertis, Australia
DLH Duluth, Minnesota-Superior, Wisconsin
DLI Dalat, South Vietnam
DLS The Dalles, Oregon
DMD Doomadgee Mission, Queensland
DMT Diamantina, Brazil
DND Dundee, Scotland
DNF Deschutes National Forest, Oregon
DNG Danghila, Ethiopia
DNI Wad Medani, Sudan
DNJ Dinajpur, Bangladesh
DNO Dianopolis, Brazil
DNQ Deniliquin, New South Wales
DNR Dinard, France
DNT Natitingou, Dahomey
DNV Danville, Illinois
DOA Doany, Malagasy
DOD Dodoma, Tanzania
DOG Dongola, Sudan
DOH Doha, Qatar
DOK Donetzk, USSR
DOL Deauville, France
DOM Dominica, Leeward Islands
DOR Dori, Upper Volta
DOU Dourados, Brazil
DOV Dover, Delaware
DOY Deboyne, Louisiade Archipelago, Papua
DPE Dieppe, France
DPO Devonport, Tasmania
DPS Denpasar, Bali, Indonesia
DPU Dumpu, New Guinea
DRB Derby, Western Australia
DRM Drama, Greece
DRN Dirranbandi, Queensland
DRO Durango, Colorado
DRR Durrie, Queensland
DRS Dresden, Germany
DRT Del Rio, Texas
DRW Darwin, Northern Territory, Australia
DSE Dessie, Ethiopia
DSK Dera Ismail Khan, Pakistan
DSL Daru, Sierra Leone
DSM Des Moines, Iowa

DTM Dortmund, Germany
DTT Detroit, Michigan (airports)—D-DET (City Airport); M-DTW (Metropolitan); R-YIP (Willow Run); W-WQG (Windsor, Ontario)
DUB Dublin, Ireland
DUD Dunedin, New Zealand
DUG Bisbee-Douglas, Arizona
DUI Duisburg, Germany
DUJ Dubois, Pennsylvania
DUR Durban, South Africa
DUS Düsseldorf, Germany
DUT Dutch Harbor, Alaska
DVO Davao, Philippines
DVP Davenport Downs, Queensland
DWB Soalala, Malagasy
DWP Dalbandin, Pakistan
DXB Dubai, Trucial Oman
DXT Dhoxaton, Greece
DXY Derby, England
DYM Diamantina Lakes, Queensland
DYU Dushanbe, USSR
DYW Daly Waters, Northern Territory, Australia
DZA Dzaoudzi, Comoro Islands
EAH El Arish, Egypt
EAM Nejran, Saudi Arabia
EAS San Sebastian, Spain
EAT Wenatchee, Washington
EAU Eau Claire, Wisconsin
EBA Elba, Italy
EBB Entebbe-Kampala, Uganda
EBD El Obeid, Sudan
EBG El Bagre, Colombia
EBU St Etienne, France
EBW Ebolowa, Cameroon
ECG Elizabeth City, North Carolina
EDI Edinburgh, Scotland
EDL Eldoret, Kenya
EEN Brattleboro, Vermont-Keene, New Hampshire (airport serving two places in two states)
EFK Newport, Vermont
EGL Neghelli, Ethiopia
EGN El Geneina, Sudan
EGS Egilsstadir, Iceland
EIN Eindhoven, Netherlands
EIS Beef Island, British Virgin Islands
EIT Eilat, Israel
EJA Barrancabermeja, Colombia
EJH Wedjh, Saudi Arabia
EJO Nejo, Ethiopia
EKN Elkins, West Virginia
EKO Elko, Nevada
EKT Eskilstuna, Sweden
ELB El Banco, Colombia
ELD El Dorado, Arkansas

ELE El Adem, Libya
ELF El Fasher, Sudan
ELG El Golea, Algeria
ELH Eleuthera Island, Bahamas
ELM Corning-Elmira, New York
ELP El Paso, Texas
ELQ Gassim, Saudi Arabia
ELS East London, South Africa
ELU El Oued, Algeria
ELY Ely, Nevada
EMN Nema, Mauritania
ENA Kenai, Alaska
ENK Enniskillen, Northern Ireland
ENS Enschede, Netherlands
ENU Enugu, Nigeria
EOR El Dorado, Venezuela
EOZ Elorza, Venezuela
EPR Esperance, Australia
EQS Esquel, Argentina
ERA Erigavo, Somalia
ERC Erzincan, Turkey
ERF Erfurt, Germany
ERG Eromanga, New Hebrides
ERI Erie, Pennsylvania
ERM Erechim, Brazil
ERN Eirunepe, Brazil
ERZ Erzurum, Turkey
ESA Esa Ala, New Guinea
ESB Ankara, Turkey
ESC Escanaba, Michigan
ESF Alexandria, Louisiana
ESH Shoreham, England
ESM Esmeraldas, Ecuador
ESN Easton, Maryland
ETE Metema, Ethiopia
ETH Eilat, Israel (Elath)
EUG Eugene, Oregon
EUN El Aaiún, Morocco
EUX Sint Eustatius, Netherlands Antilles
EVN Erevan, USSR
EVV Evansville, Indiana
EWB Fall River-New Bedford, Massachusetts
EWI Enarotali, West Irian, Indonesia
EWN New Bern, North Carolina
EWR Newark, New Jersey
EXT Exeter, England
EYW Key West, Florida
EZS Elazig, Turkey
FAG Fagurholsmyri, Iceland
FAI Fairbanks, Alaska
FAN Farsund, Norway
FAR Fargo, North Dakota
FAT Fresno, California
FAY Fayetteville-Ft Bragg, North Carolina
FBA Fonte Boa, Brazil
FBM Lubumbashi, Zaire
FBU Oslo, Norway (Fornebu Airport)

FCA Kalispell, Montana
FDA Fundación, Colombia
FDB Forte Princip, Brazil
FDF Ft de France, Martinique
FDP Faridpur, Bangladesh
FDU Bandundu, Zaire
FEI Feijo, Brazil
FEZ Fez, Morocco
FFT Frankfort, Kentucky
FFU Futaleufú, Chile
FGD Ft Derik, Mauritania
FGL Fox Glacier, New Zealand
FHU Ft Huachuca, Arizona
FIG Fria, Guinea
FIH Kinshasa, Zaire
FIN Finschhafen, New Guinea
FIT Fitchburg, Massachusetts
FJM Chipata, Zambia
FJO Ft Johnston, Malawi
FKI Kisangani, Zaire
FKL Franklin-Oil City, Pennsylvania
FKQ Fak Fak, West Irian, Indonesia
FLA Florencia, Colombia
FLB Floriano, Brazil
FLG Flagstaff, Arizona
FLL Ft Lauderdale, Florida
FLN Florianopolis, Brazil
FLO Florence, South Carolina
FLR Florence, Italy (Firenze Airport)
FLS Flinders Island, Tasmania
FMA Formosa, Argentina
FMI Kalemie, Zaire
FMN Farmington, New Mexico
FMY Ft Myers, Florida
FNA Freetown, Sierra Leone
FNC Funchal, Madeira Islands
FNE Finnsnes, Norway
FNG Fada Ngourma, Upper Volta
FNI Nimes, France
FNJ Feng Yang-Pyongyang, North Korea
FNT Flint, Michigan
FOG Foggia, Italy
FOM Foumban, Cameroon
FOO Noemfoor, New Guinea
FOR Fortaleza, Brazil
FOU Fougamou, Gabon
FPC Ft Polignac, Algeria
FPO Freeport, Bahamas
FRA Frankfurt, Germany
FRJ Frejus, France
FRL Forli, Italy
FRV Franceville, Gabon
FRW Francistown, Botswana
FSD Sioux Falls, South Dakota
FSM Ft Smith, Arkansas
FSS Ft Sandeman, Pakistan
FTF Ft Flatters, Algeria
FTL Ft Lamy, Chad
FTR Ft Archambault, Chad
FTU Ft Dauphin, Malagasy
FTV Ft Victoria, Zimbabwe

FTX Ft Rousset, Congo
FUE Fuerteventura, Puerto del Rosario, Canary Islands
FUK Fukuoka, Japan
FUN Funafuti Atoll, Ellice Islands
FWA Ft Wayne, Indiana
FXO Nova Freixo, Mozambique
FYA Faya Largeau, Chad
FYV Fayetteville, Arkansas
FZB Mansa, Zambia
GAD Gadsden, Alabama
GAO Guantanamo, Cuba
GAQ Gao, Mali
GAR Garaina, New Guinea
GAS Gach Saran, Iran
GAU Gauhati, India
GBD Great Bend, Kansas
GBE Gaberones, Botswana
GBG Galesburg, Illinois
GBK Gbangbatok, Sierra Leone
GBU Khasm el Girba, Sudan
GCI Guernsey, Channel Islands, United Kingdom
GCK Garden City, Kansas
GCM Grand Cayman Island, Cayman Islands
GCN Grand Canyon, Arizona
GDH Sargodha, Pakistan
GDL Guadalajara, Mexico
GDN Gdansk-Gdynia, Poland
GDO Guasdualito, Venezuela
GDQ Gondar, Ethiopia
GED Georgetown, Delaware
GEG Spokane, Washington
GEL Santo Angelo, Brazil
GET Geraldton, Western Australia
GFF Griffith, New South Wales
GFK Grand Forks, North Dakota
GFL Glens Falls, New York
GFN Grafton, New South Wales
GFR Granville, France
GFX Ghuraf, South Arabia
GFY Grootfontein, South-West Africa
GGD Gregory Downs, Queensland
GGG Gladewater-Kilgore-Longview, Texas
GGQ Gagnoa, Ivory Coast
GGS Gobernador Gregores, Argentina
GGT George Town, Great Exuma Island, Bahamas
GHA Ghardaia, Algeria
GHB Governor's Harbour, Bahama Islands
GHO Grahamstown, South Africa
GHU Gualeguaychu, Argentina
GIB Gibraltar
GII Siguiri, Guinea

GIL Gilgit, Pakistan
GIM Miele Mimbale, Gabon
GIR Giradot, Colombia
GIS Gisborne, New Zealand
GIZ Gizan, Saudi Arabia
GJB Marie-Galante Island, Guadeloupe
GJM Guajará Mirim, Brazil
GJT Grand Junction, Colorado
GJX Guaira, Brazil
GKA Goroka, New Guinea
GKO Kongo Boumba, Gabon
GLA Glasgow, Scotland
GLB Gilbues, Brazil
GLG Glengyle, Queensland
GLH Greenville, Mississippi
GLO Cheltenham-Gloucester, England
GLS Galveston, Texas
GMA Gemena, Zaire
GMB Gambela, Ethiopia
GMM Gamboma, Congo
GNB Grenoble, France
GND Grenada, West Indies
GNJ Genjem, West Irian, Indonesia
GNL Greenwood, Mississippi
GNN Ghinnir, Ethiopia
GNV Gainesville, Florida
GNZ Ghanzi, Botswana
GOA Genoa, Italy
GOB Goba, Ethiopia
GOE Gore, New Zealand
GOM Goma, Zaire
GOO Goondiwindi, Queensland
GOP Gorakhpur, India
GOR Gore, Ethiopia
GOT Gothenburg, Sweden
GOU Garoua, Cameroon
GOY Gal Oya, Ceylon
GPO General Pico, Argentina
GPP Guarapuava, Brazil
GPT Biloxi-Gulfport, Mississippi
GPZ Grand Rapids, Minnesota
GRA Gamarra, Colombia
GRB Green Bay, Wisconsin
GRG Georgetown, Guyana
GRI Grand Island, Nebraska
GRJ George, South Africa
GRN Grand Forks, British Columbia
GRO Gerona, Spain
GRQ Groningen, Netherlands
GRR Grand Rapids, Michigan
GRS Grosseto, Italy
GRT Gujrat, Pakistan
GRU Grajau, Brazil
GRX Granada, Spain
GRZ Graz, Austria
GSA Gusau, Nigeria
GSO Greensboro-High Point, North Carolina
GSP Greenville-Spartanburg, South Carolina

GST Gustavua, Alaska
GSU Gedaref, Sudan
GSW Greater Southwest (international airport serving Ft Worth and Dallas, Texas)
GTF Great Falls, Montana
GTR Great Barrier Island, New Zealand
GTW Gottwaldov-Holesov, Czechoslovakia
GUA Guatemala City, Guatemala
GUC Gunnison, Colorado
GUD Goundam, Mali
GUG N'Guigmi, Niger
GUI Guiria, Venezuela
GUM Guam, Marianas
GUP Gallup, New Mexico
GUU Gulu, Uganda
GUY Guymon, Oklahoma
GUZ Guiratinga, Brazil
GVA Geneva, Switzerland
GVD Gravdal, Norway
GVL Gainesville, Georgia
GVR Governador Valadares, Brazil
GWD Gwadar, Pakistan
GWE Gwelo, Rhodesia
GWL Gwalior, India
GXQ Coyhaique, Chile
GXX Yagoua, Cameroon
GYE Guayaquil, Ecuador
GYM Guaymas, Mexico
GYN Goiania, Brazil
GZT Gaziantep, Turkey
HAD Halmstad, Sweden
HAE Hatia, Bangladesh
HAG The Hague, Netherlands
HAJ Hanover, Germany
HAM Hamburg, Germany
HAN Hanoi, North Vietnam
HAR Harrisburg-New Cumberland, Pennsylvania
HAS Hail, Saudi Arabia
HAV Havana, Cuba
HBA Hobart, Tasmania
HBG Hattiesburg, Mississippi
HBI Harbour Island, Bahamas
HCA Big Spring, Texas
HDA Honda, Colombia
HDD Hyderabad, Pakistan
HDM Hamadan, Iran
HEA Herat, Afghanistan
HEL Helsinki, Finland
HER Heraklion, Greece
HEZ Natchez, Mississippi
HFA Haifa, Israel
HFN Hofn, Iceland
HFT Hammerfest, Norway
HFW Haverfordwest, Wales
HGA Hargeisa, Somalia
HGD Hughenden, Queensland
HGH Hangchow, China
HGO Korhogo, Ivory Coast
HGR Hagerstown, Maryland
HGU Mt Hagen, New Guinea

HIB Chisholm-Hibbing, Minnesota
HIJ Hiroshima, Japan
HIR Honiara, Solomon Islands
HIU Higuerote, Venezuela
HJR Khajuraho, India
HKG Hong Kong, British Crown Colony
HKK Greymouth-Hokitka, New Zealand
HKN Hoskins, New Britain
HKP Kaanapali, Maui
HKY Hickory, North Carolina
HLF Hultsfred, Sweden
HLG Wheeling, West Virginia
HLN Helena, Montana
HLS St Helens, Tasmania
HLT Hamilton, Victoria, Australia
HLZ Hamilton, New Zealand
HME Hassi Messaoud, Algeria
HMO Hermosillo, Mexico
HNL Honolulu, Oahu, Hawaii
HNM Hana, Maui, Hawaii
HNO Hercegnovi, Yugoslavia (also spelled Herzegovina)
HNS Haines, Alaska
HOD Hodeida, Yemen
HOF Hafuf, Saudi Arabia
HOM Homer, Alaska
HON Huron, South Dakota
HOO Quang Duc, South Vietnam
HOS Hosana, Ethiopia
HOT Hot Springs, Arkansas
HOU Houston, Texas
HPN White Plains, New York
HPO Hippo Valley, Zimbabwe
HQM Aberdeen-Hoquiam, Washington
HRA Haura, South Arabia (Yemen)
HRB Harbin, China
HRD Harstad, Norway
HRK Kharkov, USSR
HRL Harlingen, Texas
HSD Harnosand, Sweden
HSP Hot Springs, Virginia
HSR Hot Springs, South Dakota
HSV Decatur-Huntsville, Alabama
HTS Ashland, Kentucky—Huntington, West Virginia
HUE Humera, Ethiopia
HUF Terre Haute, Indiana
HUH Huahaine, Society Islands
HUI Hue, South Vietnam
HUL Houlton, Maine
HUN Hualien, Taiwan
HUR Hurn, England
HUT Hutchinson, Kansas
HUY Hull, England
HVA Analalava, Malagasy
HVK Holmavik, Iceland
HVN New Haven, Connecticut

HWN Haldwani, India
HXX Hay, Australia
HYA Hyannis, Massachusetts
HYD Hyderabad, India
HYG Hydaburg, Alaska
HYT Humaita, Brazil
HZK Husavik, Iceland
HZL Hazleton, Pennsylvania
IAG Niagara Falls, New York
IAH Houston, Texas—I-IAH (Intercontinental); H-HOU (Hobby)
IAM In Amenas, Algeria
IBA Ibadan, Nigeria
IBE Ibagué, Colombia
IBU Itambacuri, Brazil
IBZ Ibiza, Balearic Islands, Spain
ICA Icabarú, Venezuela
ICR Nicaro, Cuba
ICT Wichita, Kansas
IDA Idaho Falls, Idaho
IDP Independence, Kansas
IDR Indore, India
IEV Kiev, USSR
IFF Iffley, Queensland
IFJ Isafjordur, Iceland
IFL Innisfail, Queensland
IFN Ishfahan, Iran
IGA Great Inagua Island, Bahamas
IGH Ingham, Queensland
IGM Kingman, Arizona
IGR Iguassu Falls, Argentina
IGU Iguassu Falls, Brazil
IGZ Iguatu, Brazil
IHU Ihu, Papua, New Guinea
IJU Ijuí, Brazil
IKL Ikela, Zaire
IKT Irkutsk, USSR
ILF Milford Haven, Wales
ILG Wilmington, Delaware
ILM Wilmington, North Carolina
ILO Iloilo, Philippines
ILP Île des Pins, New Caledonia
ILY Islay, Inner Hebrides, Scotland
IMF Imphal, India
IMP Imperatriz, Brazil
IMT Iron Mountain, Michigan
INA Içana, Brazil
IND Indianapolis, Indiana
ING Lago Argentino, Argentina
INH Inhambane, Mozambique
INL International Falls, Minnesota
INM Innamincka, South Australia
INN Innsbruck, Austria
INO Inongo, Zaire
INT Winston-Salem, North Carolina
INV Inverness, Scotland

INW Winslow, Arizona
INX Inanwatan, West Irian, Indonesia
INZ In Salah, Algeria
IOA Ioannina, Greece
IOM Isle of Man, United Kingdom
ION Impfondo, Congo
IOS Ilheus, Brazil
IOW Iowa City, Iowa
IPC Easter Island, Pacific Ocean
IPG Phoolbagh, India
IPH Ipoh, Malaysia
IPI Ipiales, Colombia
IPL El Centro-Imperial, California
IPO Ipora, Brazil
IPT Williamsport, Pennsylvania
IPW Ipswich, England
IQQ Iquique, Chile
IQT Iquitos, Peru
IRD Ishurdi, Bangladesh
IRG Iron Range, Queensland
IRI Iringa, Tanzania
IRJ La Rioja, Argentina
IRO Birao, Central African Republic
IRP Isiro, Zaire
ISA Mt Isa, Queensland
ISB Nisab, South Arabia
ISC Isles of Scilly, England
ISH Ischia, Italy
ISI Isisford, Queensland
ISK Iskenderon, Turkey
ISN Williston, North Dakota
ISO Kinston, North Carolina
ISP Islip, New York
IST Istanbul, Turkey
ISW Wisconsin Rapids, Wisconsin
ITA Itacoatiara, Brazil
ITH Ithaca, New York
ITI Itapetinga, Brazil
ITJ Itajai, Brazil
ITN Itabuna, Brazil
ITO Hilo, Hawaii
ITQ Itaqui, Brazil
ITT Wittenoom Gorge, Western Australia
IVA Ambanja, Malagasy
IVC Invercargill, New Zealand
IVL Ivalo, Finland
IWA Iwakuni, Japan
IWD Ironwood, Michigan
IXA Agartala, India
IXB Bagdogra, India
IXC Chandigarh, India
IXD Allahabad, India
IXE Mangalore, India
IXG Belgaum, India
IXH Kailashahar, India
IXI Lilabari, India
IXJ Jammu, India
IXK Keshod, India
IXL Leh, India

IXM Madurai, India
IXN Khowai, India
IXP Pathankot, India
IXQ Kamalpur, India
IXR Ranchi, India
IXS Silchar, India
IXT Pasighat, India
IXU Aurangabad, India
IXV Along, India
IXW Jamshedpur, India
IXY Kandla, India
IXZ Port Blair, Andaman Islands
IZM Izmir, Turkey
IZT Ixtepec, Mexico
JAC Jackson, Wyoming
JAE Jacksonville, Illinois
JAF Jaffna, Ceylon
JAI Jaipur, India
JAN Jackson-Vicksburg, Mississippi
JAQ Jacquinot Bay, New Britain
JAX Jacksonville, Florida
JCB Joacaba, Brazil
JCK Julia Creek, Queensland
JCS Jaicos, Brazil
JDH Jodhpur, India
JDO Juàzeiro do Norte, Brazil
JED Jeddah, Saudi Arabia
JEF Jefferson City, Missouri
JER Jersey, Channel Islands
JFA Jaffa, Israel
JGA Jamnagar, India
JHW Jamestown, New York
JIB Djibouti, French Somaliland
JIM Jimma, Ethiopia
JIN Jinja, Uganda
JIP Jipijapa, Ecuador
JIW Jiwani, Pakistan
JJU Julienhaab, Greenland (also spelled Julianehab)
JKG Jönköping, Sweden
JKH Chios Island, Greece
JKR Janakpur, Nepal
JKT Djakarta, Java, Indonesia
JLN Joplin, Missouri
JLO Jesolo, Italy
JLP Juan-les-Pins, France
JLR Jabalpur, India
JMK Mikanos Island, Greece
JMS Jamestown, North Dakota
JNA Januaria, Brazil
JNB Johannesburg, South Africa
JNP Jasper National Park (Alberta)
JNU Juneau, Alaska
JNX Jackson, Michigan
JOE Joensuu, Finland
JOG Jogjakarta, Java, Indonesia
JOI Joinvile, Brazil
JOM Njombe, Tanzania
JON Johnston Island, Pacific

Ocean
JOS Jos, Nigeria
JPA João Pessôa, Brazil
JPO Pomona, California
JRH Jorhat, India
JRS Jerusalem, Israel
JSI Skiathos, Greece
JSR Jessore, Bangladesh
JST Johnstown, Pennsylvania
JSU Sukkertoppen, Greenland
JTI Jatai, Brazil
JTR Thira, Greece
JUB Juba, Sudan
JUJ Jujuy, Argentina
JUN Jundah, Queensland
JVA Ankavandra, Malagasy
JXN Jackson, Michigan
JYV Jyvaskyla, Finland
KAA Kasama, Zambia
KAB Kariba, Zimbabwe
KAC Kamishli, Syria
KAD Kaduna, Nigeria
KAE Kake, Alaska
KAJ Kajaani, Finland
KAM Kamaran Island, South Arabia (Yemen)
KAN Kano, Nigeria
KAR Kars, Turkey
KAT Kaitaia, New Zealand
KAU Kauhava, Finland
KBA Bení Abbès, Algeria
KBK Kirkjubaejar, Iceland
KBL Kabul, Afghanistan
KBO Kabalo, Zaire
KBP Koala Bear Park (Adelaide)
KBR Kota Bharu, Malaysia
KBS Bo, Sierra Leone
KBU Kotabaru, West Irian, Indonesia
KCH Kuching, Malaysia
KCK Kansas City, Kansas
KDA Kolda, Senegal
KDH Kandahar, Afghanistan
KDI Kendari, Sulawesi, Indonesia
KDJ Ndjole, Gabon
KDL Koronadal, Mindanao, Philippines
KDN Ndende, Gabon
KDR Kandrian, New Britain
KDU Skardu, Pakistan
KED Kaedi, Mauritania
KEF Keflavik, Iceland
KEM Kemi, Finland
KEN Kenema, Sierra Leone
KEP Nepalgang, Nepal
KEQ Kebar, Indonesia
KER Kerman, Iran
KFA Kiffa, Mauritania
KGG Kedougou, Senegal
KGI Kalgoorlie, Western Australia
KGJ Karonga, Malawi
KGL Kigali, Rwanda
KGO Kasongo, Zaire

KGS Kos, Greece
KGU Keningau, Malaysia
KHE Kherson, USSR
KHH Kaohsiung, Taiwan
KHI Karachi, Pakistan
KHK Khark Island, Iran
KHL Khulna, Bangladesh
KHN Nanchang, China
KHS Kushtia, Bangladesh
KHV Khabarovsk, USSR
KIA Kaiapit, New Guinea
KID Kristianstad, Sweden
KIE Kieta, Bougainville, Solomon Islands
KIK Kirkuk, Iraq
KIM Kimberley, South Africa
KIN Kingston, Jamaica
KIS Kisumu, Kenya
KIU Kainantu, New Guinea
KIV Kishinev, USSR
KIW Kitwe, Zambia
KIY Kilwa, Tanzania
KKD Kokoda, New Guinea
KKN Kirkenes, Norway
KKO Kaikohe, New Zealand
KKW Kikwit, Zaire
KLA Kampala, Uganda
KLB Kalabo, Zambia
KLC Kaolack, Senegal
KLE Kaele, Cameroon
KLH Long Akha, Malaysia
KLR Kalmar, Sweden
KLU Klagenfurt, Austria
KLV Karlovy Vary, Czechoslovakia
KLX Kalamata, Greece
KLY Kalima, Zaire
KMA Kerema, Papua
KME Karl-Marx Stadt, Germany
KMG Kunming, China
KMK Makabana, Congo
KML Kamileroi, Queensland
KMN Kamina, Zaire
KMP Keetmanshoop, South-West Africa
KMS Kumasi, Ghana
KMU Kismayu, Somalia
KND Kindu, Zaire
KNG Kaimana, West Irian
KNN Kankan, Guinea
KNS King Island, Tasmania
KNT Sanandaj, Iran
KNU Kanpur, India
KNX Kununurra, Western Australia
KOA Kona, Hawaii
KOB Koutaba, Cameroon
KOE Kupang, Timor, Indonesia
KOI Kirkwall, Orkney Islands
KOJ Kagoshima, Japan
KOK Kokkola, Finland
KON Kontum, South Vietnam
KOO Kongolo, Zaire
KOX Kokonao, West Irian, Indonesia

KPU Khapalu, Pakistan
KRA Kerang, Victoria, Australia
KRB Karumba, Queensland
KRI Kikori, Papua
KRK Cracow, Poland
KRN Kiruna, Sweden
KRP Karup, Denmark
KRS Kristiansand, Norway
KRT Khartoum, Sudan
KSC Kosice, Czechoslovakia
KSD Karlstad, Sweden
KSE Kasese, Uganda
KSH Kermanshah, Iran
KSI Kissidougou, Guinea
KSL Kassala, Sudan
KSN Sam Neua, Laos
KSO Kastoria, Greece
KST Kosti, Sudan
KSU Kristiansund, Norway
KTI Kratie, Cambodia
KTL Kitale, Kenya
KTM Katmandu, Nepal
KTN Ketchikan, Alaska
KTR Katherine, Northern Territory, Australia
KTU Kutaisi, USSR
KTW Katowice, Poland
KUA Kuantan, Malaysia
KUD Kudat, Malaysia
KUL Kuala Lumpur, Malaysia
KUO Kuopio, Finland
KUS Kulusuk Island, Greenland
KUT Kutahya, Turkey
KUU Kulu, India
KVA Kavalla, Greece
KVG Kavieng, New Ireland
KWA Kwajalein, Marshall Islands
KWE Kweiyang, China
KWI Kuwait, Kuwait
KWU Kawau Island, New Zealand
KWZ Kolwezi, Zaire
KXU Kastamonu, Turkey
KYA Konya, Turkey
KYS Kayes, Mali
KYZ Kayseri, Turkey
KZI Kozani, Greece
KZR Khuzdar, Pakistan
LAD Luanda, Angola
LAF Lafayette, Indiana
LAG La Guaira, Venezuela
LAL Lakeland, Florida
LAN Lansing, Michigan
LAP La Paz, Mexico
LAQ Al Bayda, Libya
LAR Laramie, Wyoming
LAS Las Vegas, Nevada
LAW Lawton, Oklahoma
LAX Los Angeles, California—L-LAX (International Airport); B-BUR (Burbank); O-ONT (Ontario)

LAZ Bom Jesus da Lapa, Brazil
LBA Leeds, England
LBB Lubbock, Texas
LBF North Platte, Nebraska
LBL Liberal, Kansas
LBQ Lambarene, Gabon
LBR Lábrea, Brazil
LBS·Lambasa, Fiji Islands
LBU Labuan, Malaysia
LBV Libreville, Gabon
LBY La Baule, France
LCE La Ceiba, Honduras
LCG La Coruña, Spain
LCH Lake Charles, Louisiana
LCI Laconia, New Hampshire
LCM La Cumbre, Argentina
LDB Londrina, Brazil
LDE Lourdes, France
LDI Lindi, Tanzania
LDR Lodar, South Arabia
LDU Lahad Datu, Malaysia
LDZ Lodz, Poland
LEA Learmonth, Western Australia
LEB Hanover-Lebanon, New Hampshire-White River Junction, Vermont
LED Leningrad, USSR
LEG Aleg, Mauritania
LEH Le Havre, France
LEI Almería, Spain
LEJ Leipzig, Germany
LEK Labe, Guinea
LEN León, Mexico
LET Leticia, Colombia
LEW Auburn-Lewiston, Maine
LEX Lexington, Kentucky
LFK Lufkin, Texas
LFR La Fria, Venezuela
LFT Lafayette-New Iberia, Louisiana
LFW Lome, Togo
LGB Long Beach, California
LGG Liege, Belgium
LGH Leigh Creek, South Australia
LGI Deadman's Cay, Long Island, Bahamas
LGL La Gloria, Colombia
LGU Logan, Utah
LGY Lagunillas, Venezuela
LHE Lahore, Pakistan
LHV Lock Haven, Pennsylvania
LHW Lanchow, China
LHX La Junta, Colorado
LIA Lima, Ohio
LIE Libenge, Zaire
LIK Likasi, Zaire
LIL Lille, France
LIM Lima, Peru
LIO Limón, Costa Rica
LIQ Lisala, Zaire
LIS Lisbon, Portugal
LIT Little Rock, Arkansas

LJA Lodja, Zaire
LJU Ljubljana, Yugoslavia
LJZ Lajes, Brazil
LKL Lakselv, Norway
LKM Nekempt, Ethiopia
LKO Lucknow, India
LKW Larkana, Pakistan
LLA Lulea, Sweden
LLB Luluabourg, Zaire
LLI Lalibella, Ethiopia
LLJ Lalmonirhat, Bangladesh
LLM Long Lama, Malaysia
LLW Lilongwe, Malawi
LMM Los Mochis, Mexico
LMN Limbang, Malaysia
LMQ Marsa Brega, Libya
LMT Klamath Falls, Oregon
LNH Lengeh, Iran
LNK Lincoln, Nebraska
LNL Land O'Lakes, Wisconsin
LNS Lancaster, Pennsylvania
LNY Lanai, Hawaii
LNZ Linz, Austria
LOA Lorraine, Queensland
LOB Lobito, Angola
LOH Loja, Ecuador
LOI Laredo, Texas
LON London, England
LOO Laghouat, Algeria
LOP Loanda, Brazil
LOQ Lobatsi, Botswana
LOS Lagos, Nigeria
LOV Monclova, Mexico
LOZ London, Kentucky
LPA Las Palmas, Canary Islands
LPB La Paz, Bolivia
LPI Linköping, Sweden
LPL Liverpool, England
LPP Lappeenranta, Finland
LPQ Luang Prabang, Laos
LQM Puerto Leguizamo, Colombia
LRA Larisa, Greece
LRE Longreach, Queensland
LRH La Rochelle, France
LRI Lorica, Colombia
LRT Lorient, France
LSC La Serena, Chile
LSE La Crosse, Wisconsin
LSI Lerwick, Scotland
LSM Long Semado, Malaysia
LSP Las Piedras, Venezuela
LST Launceston, Tasmania
LTC Lai, Chad
LTD Ghadames, Libya
LTL Lastourville, Gabon
LTN Luton, England
LTO Loreto, Mexico
LTQ Le Touquet, France
LUL Laurel, Mississippi
LUM Lourenço Marques, Mozambique
LUN Lusaka, Zambia
LUO Luso, Angola
LUQ San Luis, Argentina

LUT Miri, Malaysia
LUX Luxembourg, Luxembourg
LUY Lushoto, Tanzania
LVB Livramente, Brazil
LVI Livingstone, Zambia
LWH Lawn Hill, Queensland, Australia
LWM Lawrence, Massachusetts
LWO Lwów, USSR
LWS Lewiston, Idaho
LWT Lewistown, Montana
LWY Lawas, Sarawak, Malaysia
LXG Luong Namtha, Laos
LXR Luxor, Egypt
LXS Lemnos, Greece
LXU Lukulu, Zambia
LYH Lynchburg, Virginia
LYM Lympne, England
LYP Lyallpur, Pakistan
LYS Lyon, France
LYX Lydd, England
MAA Madras, India
MAB Maraba, Brazil
MAD Madrid, Spain
MAF Midland-Odessa, Texas
MAG Madang, New Guinea
MAH Mahon, Minorca
MAI Marianna, Florida
MAJ Majuro, Marshall Islands
MAK Malakal, Sudan
MAL Malone, New York
MAM Matamoros, Mexico
MAN Manchester, England
MAO Manaus, Brazil
MAQ Sena Maduereira, Brazil
MAR Maracaibo, Venezuela
MAS Manus Island, Bismarck Archipelago
MAT Matadi, Zaire
MAU Mastung, Pakistan
MAX Matam, Senegal
MBA Mombasa, Kenya
MBH Maryborough, Queensland
MBI Mbeya, Tanzania
MBJ Montego Bay, Jamaica
MBM Mambone, Mozambique
MBN Mombo, Tanzania
MBR Mbout, Mauritania
MBS Bay City-Midland-Saginaw, Michigan
MBZ Maués, Brazil
MCA Macenta, Guinea
MCE Merced, California
MCJ Maicao, Colombia
MCM Monte Carlo, Monaco
MCN Macon, Georgia
MCO Orlando, Florida
MCP Macapá, Brazil
MCS Monte Caseros, Argentina
MCT Muscat, Oman
MCW Mason City, Iowa
MCZ Maceió, Brazil

MDC Menado, Sulawesi, Indonesia
MDD Puerto Maldonado, Peru
MDE Medellín, Colombia
MDI Makurdi, Nigeria
MDK Mbandaka, Zaire
MDL Mandalay, Burma
MDQ Mar del Plata, Argentina
MDV Medouneu, Gabon
MDX Mercedes, Argentina
MDZ Mendoza, Argentina
MEB Melbourne, Victoria, Australia
MEC Manta, Ecuador
MED Medina, Saudi Arabia
MEF Melfi, Chad
MEI Meridian, Mississippi
MEK Meknes, Morocco
MEM Memphis, Tennessee
MEP Mendi, Ethiopia
MES Medan, Sumatra, Indonesia
MEU Marromeu, Mozambique
MEX Mexico City, Mexico
MEZ Merces, Brazil
MFA Mafia Island, Tanzania
MFD Mansfield, Ohio
MFE McAllen-Mission, Texas
MFF Moanda, Gabon
MFN Milford Sound, New Zealand
MFQ Maradi, Niger
MFR Medford, Oregon
MFU Mfuwe, Zambia
MGA Managua, Nicaragua
MGB Mt Gambier, Australia
MGM Montgomery, Alabama
MGN Magangué, Colombia
MGO Mato Grosso, Brazil
MGQ Mogadishu, Somalia
MGW Morgantown, West Virginia
MHD Meshed, Iran
MHE Mitchell, South Dakota
MHH Marsh Harbour, Great Abaco Island, Bahamas
MHK Manhattan, Kansas
MHO Mohanbari, India
MHQ Mariehamn, Finland
MHT Manchester, New Hampshire
MIA Miami, Florida
MID Mérida, Mexico
MIE Muncie, Indiana
MIL Milan, Italy
MIM Merimbula, New South Wales
MIQ Maiquetia, Venezuela (airport serving Caracas and La Guaira)
MIR Monastir, Tunisia
MIU Maiduguri, Nigeria
MIX Mores Island, Bahamas
MJA Manja, Malagasy
MJC Man, Ivory Coast
MJD Mohenjodaro, Pakistan

MJG Mayajigua, Cuba
MJH Majma, Saudi Arabia
MJI Maji, Ethiopia
MJL Mouila, Gabon
MJM Mbuji-Mayi, Zaire
MJN Majunga, Malagasy
MJP Mastuj, Pakistan
MJT Mytilene, Greece
MJV Murcia, Spain
MJX Masjed Soleyman, Iran
MJZ Mahfid, South Arabia
MKC Kansas City, Missouri
MKE Milwaukee, Wisconsin—
M-MKE (Mitchell); T-MWC
(Timmerman)
MKG Muskegon, Michigan
MKJ Makoua, Congo
MKK Hoolehua-Kaunakakai,
Molokai, Hawaii
MKL Jackson, Tennessee
MKM Mukah, Sarawak, Ma-
laysia
MKQ Merauke, West Irian, In-
donesia
MKR Meekatharra, Western
Australia
MKT Mankato, Minnesota
MKU Makokou, Gabon
MKW Manokwari, West Irian,
Indonesia
MKX Mukalla, South Arabia
MKY Mackay, Queensland
MKZ Malacca, Malaysia
MLA Valetta, Malta
MLB Melbourne, Florida
MLC McAlester, Oklahoma
MLH Mulhouse, France
MLI Davenport, Iowa-Moline,
Illinois
MLN Melilla, Morocco
MLU Monroe, Louisiana
MLW Monrovia, Liberia
MLX Malatya, Turkey
MMA Malmö, Sweden
MMC Ciudad Mante, Mexico
MME Middlesborough, En-
gland
MMF Mamfe, Cameroon
MMP Mompos, Colombia
MMX Miracema do Norte, Bra-
zil
MNB Moanda, Zaire
MNC Nacala, Mozambique
MNE Mentone, France
MNI Montserrat, Leeward Is-
lands, West Indies
MNJ Mananjary, Malagasy
MNK Mankoya, Zambia
MNL Manila, Philippines
MNM Menominee, Michigan
MNO Manono, Zaire
MNR Mongu, Zambia
MNX Minia, Egypt
MOB Mobile, Alabama
MOC Montes Claros, Brazil
MOD Modesto, California

MOF Maumere, Flores, Indo-
nesia
MOJ Muong Sing, Laos
MON Mt Cook, New Zealand
MOQ Morondava, Malagasy
MOW Moscow, USSR
MOY Monterrey, Colombia
MOZ Moorea, Society Islands
MPD Mpanda, Tanzania
MPK McKinley Park, Alaska
MPL Montpellier, France
MPV Barre-Montpelier, Ver-
mont
MQG Milgarra, Queensland
MQL Mildura, Victoria
MQQ Moundou, Chad
MQT Marquette, Michigan
MQU Mariquita, Colombia
MQX Makale, Ethiopia
MRD Mérida, Venezuela
MRE Manicore, Brazil
MRG Mesters Vig, Greenland
MRH Beaufort-Morehead City,
North Carolina
MRO Masterson, New Zealand
MRS Marseilles, France
MRU Mauritius, Indian Ocean
MRV Mineralnye Vody, USSR
MRX Mineiros, Brazil
MRY Carmel-by-the-Sea-Mon-
terey, California
MSD Mossoro, Brazil
MSE Manston, England
MSI Moshi, Tanzania
MSJ Misawa, Japan
MSK Mastic Point, Andros Is-
land, Bahamas
MSL Florence-Muscle Shoals-
Sheffield, Alabama
MSN Madison, Wisconsin
MSO Missoula, Montana
MSP Minneapolis-St Paul,
Minnesota—D-JDT (Minne-
apolis Heliport), I-MSP (Inter-
national)
MSQ Minsk, USSR
MSR Makassar, Sulawesi, Indo-
nesia
MSS Massena, New York
MST Maastricht, Netherlands
MSU Maseru, Lesotho
MSW Massawa, Ethiopia
MSX Mascota, Mexico
MSY New Orleans, Louisiana
MSZ Mossamedes, Angola
MTB Monte Libano, Colombia
MTE Monte Alegre, Brazil
MTF Mizan Teferi, Ethiopia
MTJ Montrose, Colorado
MTO Mattoon, Illinois
MTQ Mitchell, Queensland
MTR Montería, Colombia
MTS Manzini, Swaziland
MTT Minatitlan, Mexico
MTW Manitowoc, Wisconsin
MTY Monterrey, Mexico

MUA Munda, Solomon Islands
MUB Maun, Botswana
MUC Munich, Germany
MUD Murchison Falls, Uganda
MUE Kamuela, Hawaii
MUH Mersa Matruh, Egypt
MUR Marudi, Sarawak, Malay-
sia
MUW Mutarara, Mozambique
MUX Multan, Pakistan
MUZ Musoma, Tanzania
MVD Montevideo, Uruguay
MVN Mt Vernon, Illinois
MVO Mongo, Chad
MVR Maroua, Cameroon
MVU Mulege, Mexico
MVY Martha's Vineyard-Vine-
yard Haven, Massachusetts
MWA Marion, Illinois
MWE Merowe, Sudan
MWL Mineral Wells, Texas
MWP Mangla, Pakistan
MWZ Mwanza, Tanzania
MXD Marion Downs, Queens-
land
MXK Metekel, Ethiopia
MXL Mexicali, Mexico
MXM Morombe, Malagasy
MXR Moussoro, Chad
MXT Maintirano, Malagasy
MYB Mayoumba, Gabon
MYC Massenya, Chad
MYD Malindi, Kenya
MYG Mayaguana Island, Baha-
mas
MYH Rosh-Pina, Israel
MYP Montgomery, Pakistan
MYV Marysville, California
MYW Mtwara, Tanzania
MYZ Mayoko, Gabon
MZB Mocímboa da Praia,
Mozambique
MZC Mitzic, Gabon
MZG Makung, Formosa
MZI Mopti, Mali
MZL Manizales, Colombia
MZM Metz, France
MZN Minj, New Guinea
MZO Manzanillo, Cuba
MZQ Mozambique, Mozam-
bique
MZR Mazar-i-Sharif, Afghani-
stan
MZT Mazatlan, Mexico
MZU Muzaffarpur, India
MZX Massio, Ethiopia
MZY Mzimba, Malawi
MZZ Marion, Indiana
NAG Nagpur, India
NAN Nandi, Fiji Islands
NAP Naples, Italy
NAS Nassau, New Providence
Island, Bahamas
NAT Natal, Brazil
NAV Natividade, Brazil
NBO Nairobi, Kenya

NCE Nice, France
NCG Nueva Casas Grandes, Mexico
NCH Nachingwea, Tanzania
NCL Newcastle, England
NCM New Moon, Queensland
NCT Nicoya, Costa Rica
NDD Novo Redondo, Angola
NDE Notodden, Norway
NDH Delhi, India
NDR Nador, Morocco
NEV Nevis, Leeward Islands, West Indies
NGE Ngaoundere, Cameroon
NGO Nagoya, Japan
NHA Nha-Trang, South Vietnam
NHB Kodiak, Alaska
NIC Nicosia, Cyprus
NIM Niamey, Niger
NIO Nioki, Zaire
NIX Nioro, Mali
NKC Nouakchott, Mauritania
NKG Nanking, China
NKL Nkolo, Zaire
NLA Ndola, Zambia
NLD Nuevo Laredo, Mexico
NLK Norfolk Island, Pacific Ocean
NMR Nappamerrie, Queensland
NNG Nanning, China
NNI New Nickerie, Surinam
NNU Nanuque, Brazil
NOG Nogales, Arizona-Nogales, Mexico
NOS Nossi-Be, Malagasy
NOU Noumea, New Caledonia
NOV Nova Lisboa, Angola
NPA Napan, West Irian, Indonesia
NPE Hastings-Napier, New Zealand
NPL New Plymouth, New Zealand
NPT Newport, Rhode Island
NQN Neuquen, Argentina
NQY Newquay, England
NRA Narrandera, New South Wales
NRK Norrköping, Sweden
NRM Nara, Mali
NSM Norseman, Western Australia
NSN Nelson, New Zealand
NSO Scone, New South Wales
NTE Nantes, France
NTL Newcastle, New South Wales
NTN Normanton, Queensland
NUD En Nahud, Sudan
NUE Nuremberg, Germany
NVA Neiva, Colombia
NVE Nova Esperança, Brazil
NVK Narvik, Norway
NWA Moheli, Comoro Islands

NYC New York City; New York, New York—E-EWR (Newark); J-JFK (Kennedy); L-LGA (La Guardia)
NZE Nzerekore, Guinea
OAG Orange, New South Wales
OAK Oakland, California
OAM Oamaru, New Zealand
OAX Oaxaca, Mexico
OBI Obidos, Brazil
OCF Ocala, Florida
OCJ Ocho Rios, Jamaica
ODA Ouadda, Central African Republic
ODB Cordoba, Spain
ODD Oodnadatta, South Australia
ODE Odense, Denmark
ODJ Ouanda Djalle, Central African Republic
ODL Cordillo Downs, South Australia
ODS Odessa, USSR
OER Ornskoldsvik, Sweden
OFK Norfolk, Nebraska
OGD Ogden, Utah
OGG Kahului, Maui, Hawaii
OGR Bongor, Chad
OGS Ogdensburg, New York
OGX Ouargla, Algeria
OHD Ohrid, Yugoslavia
OJO Outjo, South-West Africa
OJW Otjiwarongo, South-West Africa
OKA Okinawa, Ryukyu Islands, Japan
OKC Oklahoma City, Oklahoma
OKK Kokomo-Logansport, Indiana
OKN Okondja, Gabon
OKY Oakey, Queensland
OLB Olbia, Italy
OLE Olean, New York
OLF Wolf Point, Montana
OLG Nordmaling, Sweden
OLM Olympia, Washington
OLN Colonia Sarmiento, Argentina
OLO Olomouc, Czechoslovakia
OLU Columbus, Nebraska
OMA Omaha, Nebraska
OME Nome, Alaska
OMO Mostar, Yugoslavia
OMS Omsk, USSR
ONA Winona, Minnesota
ONB Monkey Bay, Malawi
ONG Mornington Island, Queensland
ONM Condamine, Queensland
ONO Ontario, Oregon
ONR Monkira, Queensland
ONS Onslow, Western Australia

ONT Ontario, California
ONU Kongoussi, Upper Volta
OOL Coolangatta, Queensland
OOM Cooma, New South Wales
OOR Mooraberrie, Queensland
OPA Kopasker, Iceland
OPO Oporto, Portugal
OPU Balimo, Papua
ORA Oran, Argentina
ORE Greenfield, Massachusetts
ORF Norfolk, Virginia
ORH Worcester, Massachusetts
ORK Cork, Ireland
ORL Orlando, Florida
ORM Northampton, England
ORN Oran, Algeria
ORO Porto Seguro, Brazil
ORP Ormara, Pakistan
ORU Oruro, Bolivia
ORW Orange Walk, British Honduras
ORX Oriximina, Brazil
OSA Osaka, Japan
OSD Ostersund, Sweden
OSH Oshkosh, Wisconsin
OSL Oslo, Norway
OSM Mosul, Iraq
OSR Ostrava, Czechoslovakia
OST Ostend, Belgium
OSY Namsos, Norway
OSZ Koszalin, Poland
OTA Mota, Ethiopia
OTC Bol, Chad
OTG Worthington, Minnesota
OTH North Bend, Oregon
OTL Boutilimit, Mauritania
OTM Ottumwa, Iowa
OTV Otavi, South-West Africa
OTZ Kotzebue, Alaska
OUA Ouagadougou, Upper Volta
OUD Oujda, Morocco
OUE Ouesso, Congo
OUG Ouahigouya, Upper Volta
OUH Oudtshoorn, South Africa
OUI Ban Houei Sai, Laos
OUL Oulu, Finland
OUR Batouri, Cameroon
OUT Bousso, Chad
OVA Bekily, Malagasy
OVD Oviedo, Spain
OWB Owensboro, Kentucky
OXC Waterbury, Connecticut
OXF Oxford, England
OXO Orientos, Queensland
OXR Oxnard-Ventura, California
OXY Morney Plains, Queensland
OYE Oyem, Gabon
OYK Oiapoque, Brazil
OZC Ozamiz City, Mindanao, Philippines
PAA Pa-An, Burma
PAB Pedro Afonso, Brazil

PAF Paraburdoo, Western Australia
PAG Panjim, India
PAP Port-au-Prince, Haiti
PAR Paris, France
PAU Pauk, Burma
PAV Paulo Afonso, Brazil
PBD Porbandar, India
PBE Puerto Berrio, Colombia
PBI West Palm Beach, Florida
PBL Puerto Cabello, Venezuela
PBM Paramaribo, Surinam
PBN Porto Amboin, Angola
PBR Puerto Barrios, Guatemala
PBS Plettenberg Bay, South Africa
PBY Pillars Bay, Alaska
PCC Puerto Rico, Colombia
PCE Palm Island, Queensland
PCH Pari-Cachoeira, Brazil
PCR Puerto Carreño, Colombia
PCZ Panama Canal Zone
PDG Padang, Sumatra, Indonesia
PDO Prado, Brazil
PDP Punta del Este, Uruguay
PDS Piedras Negras, Mexico
PDT Pendleton, Oregon
PDX Portland, Oregon
PDZ Pedernales, Venezuela
PEC Pelican, Alaska
PEI Pereira, Colombia
PEK Peking, China
PEN Penang, Malaysia
PER Perth, Western Australia
PET Pelotas, Brazil
PEW Peshawar, Pakistan
PFB Passo Fundo, Brazil
PFJ Patreks Fjordur, Iceland
PFN Panama City, Florida
PFR Port Francqui, Zaire
PGA Page, Arizona
PGF Perpignan, France
PGH Pantnagar, India
PGK Pangkalpinang, Bangka, Indonesia
PGM Palenque, Mexico
PGT Porangatu, Brazil
PHB Parnaiba, Brazil
PHC Port Harcourt, Nigeria
PHE Port Hedland, Western Australia
PHF Hampton-Newport News-Williamsburg, Virginia
PHH Phan Thiet, South Vietnam
PHL Camden, New Jersey-Philadelphia, Pennsylvania— P-PHL (International); P-PNE (Northeast)
PHN Port Huron, Michigan
PHS Phitsanuloke, Thailand
PHV Pahlavi, Iran
PHX Phoenix, Arizona
PIA Peoria, Illinois

PIC Picos, Brazil
PIE Clearwater-St Petersburg, Florida
PIH Pocatello, Idaho
PIK Prestwick, Scotland
PIN Parintins, Brazil
PIR Pierre, South Dakota
PIT Pittsburgh, Pennsylvania
PIU Piura, Peru
PJG Panjgur, Pakistan
PKB Marietta, Ohio-Parkersburg, West Virginia
PKC Phuket, Thailand
PKK Pakokku, Burma
PKU Pakanbaru, Sumatra, Indonesia
PKY Pak Lay, Laos
PKZ Pakse, Laos
PLB Plattsburgh, New York
PLC Planeta Rica, Colombia
PLF Pala, Chad
PLH Plymouth, England
PLM Palembang, Sumatra, Indonesia
PLN Cheboygan-Pellston, Michigan
PLO Port Lincoln, South Australia
PLS Providenciales, Turks and Caicos Islands, West Indies
PLW Palu, Sulawesi, Indonesia
PLZ Port Elizabeth, South Africa
PMA Pemba Island, Tanzania
PMC Puerto Montt, Chile
PME Portsmouth, England
PMG Ponta Pora, Brazil
PMH Portsmouth, Ohio
PMI Palma de Mallorca, Spain
PMJ Porto Murtinho, Brazil
PMO Palermo, Sicily
PMQ Perito Moreno, Argentina
PMR Palmerston North, New Zealand
PMV Porlamar, Venezuela
PMZ Palmar, Costa Rica
PNA Panna, India
PNB Porto Nacional, Brazil
PNG Popondetta, New Guinea
PNH Phnom Penh, Cambodia
PNI Ponape, Caroline Islands
PNJ Paterson, New Jersey
PNK Pontianak, Borneo, Indonesia
PNL Pantelleria, Italy
PNQ Poona, India
PNR Pointe Noire, Congo
PNS Pensacola, Florida
PNZ Petrolina, Brazil
POA Porto Alegre, Brazil
POD Podor, Senegal
POG Port Gentil, Gabon
POI Potosí, Bolivia
POL Porto Amelia, Mozambique
POM Port Moresby, Papua,

New Guinea
POR Pori, Finland
POS Port-of-Spain, Trinidad
POT Port Antonio, Jamaica
POU Poughkeepsie, New York
POV Presov, Czechoslovakia
POX Port Alexander, Alaska
POY Lovell-Powell, Wyoming
POZ Poznan, Poland
PPB Presidente Prudente, Brazil
PPF Parsons, Kansas
PPG Pago Pago, American Samoa
PPI Port Pirie, South Australia
PPN Popayan, Colombia
PPP Proserpine, Queensland
PPR Pirapora, Brazil
PPT Papeete, Tahiti, Society Islands
PPZ Puerto Paez, Venezuela
PQC Phuquoc, South Vietnam
PQI Presque Isle, Maine
PRB Paso Robles, California
PRC Prescott, Arizona
PRG Prague, Czechoslovakia
PRJ Capri, Italy
PRM Puerto Lopez, Colombia
PRQ Presidente Roque Sáenz Peña, Argentina
PRS Puerto Lempira, Honduras
PRU Paranagua, Brazil
PRX Paris, Texas
PSA Pisa, Italy
PSB Bellefonte-Clearfield-Philipsburg, Pennsylvania
PSC Pasco, Washington
PSD Port Said, Egypt
PSE Ponce, Puerto Rico
PSF Pittsfield, Massachusetts
PSG Petersburg, Alaska
PSI Pasni, Pakistan
PSL Perth, Scotland
PSM Portsmouth, New Hampshire
PSO Pasto, Colombia
PSP Indio-Palm Springs, California
PSR Pescara, Italy
PSS Posadas, Argentina
PST Preston, Cuba
PSY Port Stanley, Falkland Islands
PSZ Puerto Suárez, Bolivia
PTE Nouadhibou, Mauritania
PTJ Portland, Victoria, Australia
PTL Pietermaritzburg, South Africa
PTM Palmarito, Venezuela
PTP Pointe-à-Pitre, Guadeloupe
PTR Port Macquarie, New South Wales
PTY Panama City, Panama
PUB Pueblo, Colorado
PUD Puerto Deseado, Argen-

tina
PUF Pau, France
PUK Paducah, Kentucky
PUN Punia, Zaire
PUP Po, Upper Volta
PUQ Punta Arenas, Chile
PUR Puerto Rico, Bolivia
PUS Pusan, South Korea
PUT Putao, Burma
PUU Puerto Asís, Colombia
PUW Pullman, Washington
PUY Pula, Yugoslavia
PUZ Puerto Cabezas, Nicaragua
PVD Providence, Rhode Island
PVH Porto Velho, Brazil
PVI Paranavai, Brazil
PVK Preveza, Greece
PVO Portoviejo, Ecuador
PVR Puerto Vallarta, Mexico
PWM Portland, Maine
PWR Port Walter, Alaska
PXA Paraná, Brazil
PXK Parakou, Dahomey
PXO Porto Santo, Madeira
PXU Pleiku, South Vietnam
PXX Porto Afonso, Brazil
PYH Puerto Ayacucho, Venezuela
PYR Pyrgos, Greece
PZA Paz de Ariporo, Colombia
PZE Penzance, England
PZO Puerto Ordaz, Venezuela
PZU Port Sudan, Sudan
PZY Piestany, Czechoslovakia
QIL Qillainau, Afghanistan
QMM Marina di Massa, Italy
QSM South Molle Islands, Queensland
QUI Quirindi, New South Wales
QUN Qutdligssat, Greenland
QUP Quincemil, Peru
RAB Rabaul, New Britain, New Guinea
RAH Rafha, Saudi Arabia
RAJ Rajkot, India
RAK Marrakech, Morocco
RAL Riverside, California
RAP Rapid City, South Dakota
RAR Rarotonga, Cook Islands, Polynesia
RAU Rangpur, Bangladesh
RBA Rabat, Morocco
RBF Raba Raba, New Guinea
RBG Roseburg, Oregon
RBL Red Bluff, California
RBO Roboré, Bolivia
RBQ Rurrenabaque, Bolivia
RBR Rio Branco, Brazil
RBU Roebourne, Western Australia
RCH Ríohacja, Colombia
RCM Richmond, Queensland
RCS Rochester, England

RCU Río Cuarto, Argentina
RDD Redding, California
RDG Reading, Pennsylvania
RDM Bend-Redmond, Oregon
RDS Rio Grande do Sul, Brazil
RDT Richard-Toll, Senegal
RDU Raleigh-Durham, North Carolina
REC Recife, Brazil
REG Reggio Calabria, Italy
REH Rehoboth Beach, Delaware
REK Reykjavik, Iceland
REL Trelew, Argentina
REP Siem Reap, Cambodia
RER Potrerillos, Chile
RES Resistencia, Argentina
REX Reynosa, Mexico
RFD Rockford, Illinois
RFH Río Mayo, Argentina
RFP Raiatea, Leeward Islands, Society Islands
RFW Robinhood, Queensland
RFX Roxborough, Queensland
RGA Río Grande, Argentina
RGI Rangiroa, Tuamotu Islands, Polynesia
RGL Río Gallegos, Argentina
RGN Rangoon, Burma
RGO Rosella Plains, Queensland
RGR Rio Grande, Brazil
RGT Rengat, Sumatra, Indonesia
RHD Río Hondo, Argentina
RHE Rheims, France
RHI Rhinelander, Wisconsin
RHL Roy Hill, Western Australia
RHO Rhodes, Dodecanese Islands, Greece
RIA Santa Maria, Brazil
RIB Riberalta, Bolivia
RIC Richmond, Virginia
RIO Rio de Janeiro, Brazil
RIW Riverton, Wyoming
RIX Riga, Latvia, USSR
RIY Riyan Mukalla, South Arabia (Yemen)
RJH Rajshahi, Bangladesh
RJK Rijeka, Yugoslavia
RKD Rockland, Maine
RKS Rock Springs, Wyoming
RKT Ras-al-Khaima, Trucial Oman
RMA Roma, Queensland
RMG Rome, Georgia
RMI Rimini, Italy
RMK Renmark, South Australia
RMT Rocky Mount, North Carolina
RNB Ronneby, Sweden
RNN Ronne, Denmark
RNO Reno, Nevada
RNS Rennes, France

RNU Ranau, Malaysia
ROA Roanoke, Virginia
ROB Robertsfield, Liberia
ROC Rochester, New York
ROF Rose Hall, Guyana
ROK Rockhampton, Queensland
ROM Rome, Italy
RON Rondón, Colombia
ROO Rondonopolis, Brazil
ROP Rota, Marianas Islands
ROS Rosario, Argentina
ROT Rotorua, New Zealand
ROV Rostov, USSR
ROW Roswell, New Mexico
RPR Raipur, India
RRK Rourkela, India
RRS Roros, Norway
RSA Santa Rosa, Argentina
RSD Rock Sound, Eleuthera Island, Bahamas
RSK Ransiki, West Irian, Indonesia
RSO Remanso, Brazil
RSS Roseires, Sudan
RST Rochester, Minnesota
RSU Río Sucio, Colombia
RTB Roatan Island, Honduras
RTF Rutherford, New Jersey
RTM Rotterdam, Netherlands
RTP Rutland Plains, Queensland
RTS Rottnest Island, Western Australia
RUH Riyadh, Saudi Arabia
RUN St Denis, Réunion Island, Indian Ocean
RUP Rupsi, India
RUT Rutland, Vermont
RUY Ruinas de Copan, Honduras
RVA Farafangana, Malagasy
RVC River Cess, Liberia
RVK Rorvik, Norway
RVN Rovaniemi, Finland
RVY Rivera, Uruguay
RWL Rawlins, Wyoming
RWP Rawalpindi, Pakistan
RXA Raudha, South Arabia
RXS Roxas City, Philippines
RYK Rahimyar Kahn, Pakistan
RYO Río Turbio, Argentina
RZA Santa Cruz, Argentina
RZB Roseberth, Queensland
RZE Rzeszow, Poland
RZR Ramsar, Iran
RZY Rezayeh, Iran
SAC Sacramento, California
SAD Safford, Arizona
SAF Santa Fe, New Mexico
SAH Sana, Yemen
SAJ Sirajgang, Bangladesh
SAK Saudarkrokur, Iceland
SAL San Salvador, El Salvador
SAM Saba, Netherlands Antilles

SAN San Diego, California—G-SEE (Gillespie); M-MYF (Montgomery); S-SAN (Lindbergh)
SAO São Paulo, Brazil
SAP San Pedro Sula, Honduras
SAQ San Andros, Andros Island, Bahamas
SAT San Antonio, Texas
SAV Savannah, Georgia
SAY Salisbury, Zimbabwe
SBA Santa Barbara, California
SBB Santa Barbara-Barinas, Venezuela
SBJ Simla, India
SBK St Brieuc, France
SBL Santa Ana, Bolivia
SBN South Bend, Indiana
SBP San Luis Obispo, California
SBQ São Borja, Brazil
SBR Santa Barbara, Monagas, Venezuela
SBT San Bernardino, California
SBW Sibu, Sarawak, Malaysia
SBY Ocean City-Salisbury, Maryland
SCA Santa Catalina, Colombia
SCC Stamford, Connecticut
SCD Sulaco, Honduras
SCK Stockton, California
SCL Santiago, Chile
SCN Saarbrücken, Germany
SCQ Santiago de Compostela, Spain
SCU Santiago, Cuba
SDD Sá da Bandeira, Angola
SDE Santiago del Estero, Argentina
SDF Louisville, Kentucky
SDG São Domingos, Brazil
SDK Sandakan, Malaysia
SDL Sundsvall, Sweden
SDQ Santo Domingo City, Dominican Republic (formerly Ciudad Trujillo)
SDR Santander, Spain
SDS Serondela, Botswana
SDT Sandy Point, Great Abaco Island, Bahamas
SDW Sandwip, Bangladesh
SDY Sidney, Montana
SEA Seattle, Washington—BBFI (Boeing); S-SEA (Seattle-Tacoma)
SEB Sebha, Libya
SEL Seoul, South Korea
SEN Southend, Scotland
SES Selma, Alabama
SET San Esteban, Honduras
SEU Seronera, Tanzania
SFA Sfax, Tunisia
SFD San Fernando, Venezuela
SFJ Sondre Stromfjord, Greenland
SFK Safia, Papua

SFN Santa Fé, Argentina
SFO San Francisco, California—S-SFO (International); O-OAK (Oakland)
SFP Sherbrooke Forest Park (Victoria, Australia)
SFS Sharm es-Sheikh, Israel
SFT Skelleftea, Sweden
SFU Surfdale, Waiheke Island, New Zealand
SFX San Félix, Venezuela
SFZ Pawtucket-Woonsocket, Rhode Island
SGD Sonderborg, Denmark
SGF Springfield, Missouri
SGG Simanggang, Malaysia
SGH Springfield, Ohio
SGN Saigon, South Vietnam
SGO St George, Queensland
SGU St George, Utah
SGX Songea, Tanzania
SGY Skagway, Alaska
SGZ Singora, Thailand
SHA Shanghai, China
SHB Shark Bay, Western Australia
SHD Staunton, Virginia
SHE Shenyang, China
SHJ Sharjah, Oman
SHR Sheridan, Wyoming
SHU Shute Harbour, Queensland
SHV Shreveport, Louisiana
SIA Sian, China
SIC San Antonio do Ica, Brazil
SID Ilha do Sal, Cape Verde Islands
SII Sidi Ifni, Morocco
SIL Sao Hill, Tanzania
SIN Singapore, Singapore
SIO Smithtown, Tasmania
SIP Simferopol, USSR
SIQ Singkep Island, Indonesia
SIT Sitka, Alaska
SIW Samarai, Papua
SJA San Juan de Arama, Colombia
SJB San Joaquin, Bolivia
SJC San Jose, California
SJE San José del Guaviaro, Colombia
SJF St John, Virgin Islands
SJH St Johns, Antigua, Leeward Islands, West Indies
SJI San José, Philippines
SJJ Sarajevo, Yugoslavia
SJO San José, Costa Rica
SJP San Juan, Peru
SJQ Sesheke, Zambia
SJR San Juan de Uraba, Colombia
SJS San José, Bolivia
SJT San Angelo, Texas
SJU San Juan, Puerto Rico—ISIG (Isla Grande); S-SJU (International)

SJV San Javier, Bolivia
SKB St Kitts, Leeward Islands, West Indies
SKD Samarkand, USSR
SKE Skien, Norway
SKG Salonika, Greece
SKI Skilda, Algeria
SKO Sokoto, Nigeria
SKP Skoplje, Yugoslavia
SKR Skogar, Iceland
SKS Skrydstrup, Denmark
SKZ Sukkur, Pakistan
SLA Salta, Argentina
SLC Salt Lake City, Utah
SLD Sliac, Czechoslovakia
SLE Salem, Oregon
SLF Sulayel, Saudi Arabia
SLK Lake Placid-Saranac Lake, New York
SLN Salina, Kansas
SLO San Luis Obispo; Santa Ana, Colombia
SLP San Luis Potosí, Mexico
SLU St Lucia, Windward Islands, West Indies
SLV São Paulo de Olivença, Brazil
SLW Saltillo, Mexico
SLZ São Luiz, Maranhao, Brazil
SMA Santa Maria, Azores
SMB Cerro Sombrero, Chile
SMF Sacramento, California
SMI Samos, Greece
SMJ Santa Margherita, Italy
SML Stella Maris, Long Island, Bahamas
SMM Semporna, Malaysia
SMO Santa Monica, California
SMR Santa Marta, Colombia
SMS St Marie, Malagasy
SMX Santa Maria, California
SMY Simenti, Senegal
SNA Laguna Beach-Santa Ana, California
SND Seno, Laos
SNF San Felipe, Venezuela
SNG San Ignacio de Velasco, Bolivia
SNI Sinoe, Liberia
SNK Snyder, Texas
SNL Sand Creek, Guyana
SNM San Ignacio de Moxos, Bolivia
SNN Shannon, Ireland
SNS Salinas, California
SNU Santa Clara, Cuba
SNV Santa Elena, Venezuela
SNW Sandoway, Burma
SNY Sidney, Nebraska
SOA Soc Trang, South Vietnam
SOB Sobral, Brazil
SOF Sofia, Bulgaria
SOM San Tome, Venezuela
SON Espiritu Santo, New Hebrides

SOP Southern Pines, North Carolina
SOQ Sorong, West Irian, Indonesia
SOU Southampton, England
SPC Santa Cruz de la Palma, Canary Islands
SPE Sepulot, Malaysia
SPF Spearfish, South Dakota
SPI Springfield, Illinois
SPK Sapporo, Japan
SPL San Pedro de Jagua, Colombia
SPN Saipan, Marianas Islands
SPP Serpa, Portugal
SPS Wichita Falls, Texas
SPU Split, Yugoslavia
SPX San Pedro, Colombia
SPY San Pedro, Ivory Coast
SRA Santa Rosa, Brazil
SRE Sucre, Bolivia
SRG Semarang, Java, Indonesia
SRL Santa Rosalía, Mexico
SRQ Bradenton-Sarasota, Florida
SRR Sorreisa, Norway
SRS San Marcos, Colombia
SRZ Santa Cruz, Bolivia
SSA Salvador, Brazil
SSG Santa Isabel, Spanish Guinea
SSI Brunswick, Georgia
SSJ Sannessjoen, Norway
SSM Sault Sainte Marie, Michigan
SSN Auburn, New York
SSX Samsun, Turkey
SSY São Salvador, Angola
SSZ Santos, Brazil
STA Stauning, Denmark
STB Santa Barbara, Zulia, Venezuela
STD Santo Domingo, Venezuela
STE Stevens Point, Wisconsin
STF Setif, Algeria
STG Santiago, Brazil
STI Santiago, Dominican Republic
STJ St Joseph, Missouri
STL St Louis, Missouri
STM Santarem, Brazil
STO Stockholm, Sweden
STR Stuttgart, Germany
STS Santa Rosa, California
STT St Thomas, American Virgin Islands
STV Staverton, England
STX St Croix, American Virgin Islands
STZ Santa Terezinha, Brazil
SUA Stuart, Florida
SUB Surabaja, Java, Indonesia
SUG Surigao, Philippines
SUI Sukhumi, USSR

SUL Sui, Pakistan
SUM San Juan de César, Colombia
SUN Sun Valley, Idaho
SUR Starcke, Queensland
SUS Surkhet, Nepal
SUV Suva, Fiji Islands
SUX Sioux City, Iowa
SVB Sambava, Malagasy
SVC Silver City, New Mexico
SVD St Vincent, Windward Islands, West Indies
SVG Stavanger, Norway
SVJ Svolvaer, Norway
SVN Saravena, Colombia
SVQ Seville, Spain
SVU Savusavu, Fiji Islands
SVZ San Antonio, Venezuela
SWH Swan Hill, Victoria, Australia
SWL Spanish Wells, Bahamas
SWP Swakopmund, South-West Africa
SWQ Sumbawa, Indonesia
SWS Swansea, Wales
SWW Sweetwater, Texas
SXB Strasbourg, France
SXC Santa Catalina Island, California
SXE Sale, Victoria, Australia
SXG Senanga, Zambia
SXM Sint Maarten, Netherlands Antilles
SXR Srinagar, India
SXU Soddu, Ethiopia
SYA Shemya, Alaska
SYC Sanday, Scotland
SYD Sydney, New South Wales
SYI Shelbyville, Tennessee
SYR Syracuse, New York
SYY Stornoway, Outer Hebrides, Scotland
SYZ Shiraz, Iran
SZA Santo Antonio do Zaire, Angola
SZB Santa Barbara, Honduras
SZG Salzburg, Austria
SZI Soroti, Uganda
SZU Segou, Mali
SZZ Szczecin, Poland
TAB Tobago, Trinidad and Tobago, West Indies
TAC Tacloban, Philippines
TAI Taiz, Yemen
TAJ Taracua, Brazil
TAK Takamatsu, Japan
TAM Tampico, Mexico
TAP Tapachula, Mexico
TAR Taranto, Italy
TAS Tashkent, USSR
TAT Tatry/Poprad, Czechoslovakia
TAU Tauramena, Colombia
TBB Tuy Hoa, South Vietnam
TBL Tableland, Western Australia

TBN Ft Leonard Wood, Missouri
TBO Tabora, Tanzania
TBP Tumbes, Peru
TBS Tbilisi, USSR
TBT Tabatinga, Brazil
TBU Tongatabu, Nukualofa, Tonga Islands
TBZ Tabriz, Iran
TCA Tennant Creek, Northern Territory, Australia
TCB Treasure Cay, Bahamas
TCH Tchibanga, Gabon
TCI Santa Cruz de Tenerife, Canary Islands
TCO Tumaco, Colombia
TCQ Tacna, Peru
TDA Trinidad, Colombia
TDD Trinidad, Bolivia
TDM Palmyra, Syria
TEA Tela, Honduras
TEB Teterboro, New Jersey
TEE Tyee, Alaska
TEG Tenkodogo, Upper Volta
TES Tessenei, Ethiopia
TET Tete, Mozambique
TEU Te Anau, New Zealand
TEY Thingeyri, Iceland
TEZ Tezpur, India
TFF Tefe, Brazil
TFR Tarbes, France
TFY Tarfaya, Morocco
TGD Titograd, Yugoslavia
TGG Trengganu, Malaysia
TGN Tarragona, Spain
TGR Touggourt, Algeria
TGS Tuxtla Gutierrez, Mexico
TGT Tanga, Tanzania
TGU Tegucigalpa, Honduras
TGX Taguatinga, Brazil
TGY Punta Gorda, British Honduras
THA Tullahoma, Tennessee
THE Teresina, Brazil
THG Thangool, Queensland
THJ Theodore, Queensland
THK Thakhek, Laos
THO Thorshofn, Iceland
THR Teheran, Iran
THY Thylungra, Queensland
THZ Tahoua, Niger
TIA Tirana, Albania
TIE Tippi, Ethiopia
TIF Taif, Saudi Arabia
TIJ Tijuana, Mexico
TIN Tindouf, Algeria
TIP Tripoli, Libya
TIR Tiree Island, Scotland
TIS Thursday Island, Queensland
TIU Timaru, New Zealand
TIV Tivat, Yugoslavia
TIW Tacoma, Washington
TIX Titusville, Florida
TJA Tarija, Bolivia
TJI Trujillo, Honduras

TJQ Tandjungpandan, Billiton, Indonesia
TKC Tiko, Cameroon
TKD Takoradi, Ghana
TKG Telukbetung, Sumatra, Indonesia
TKI Turks Islands, West Indies
TKK Truk, Caroline Islands
TKL Tak, Thailand
TKR Thakurgaon, Bangladesh
TKU Turku, Finland
TKY Turkey Creek, Western Australia
TLB Tortola, British Virgin Islands
TLE Tulear, Malagasy
TLG Tres Lagoas, Brazil
TLH Tallahassee, Florida
TLL Tallinn, Estonia, USSR
TLN Hyeres-Toulon, France
TLP Talpa, New Mexico
TLR Talgarno, Western Australia
TLS Toulouse, France
TLU Tolu, Colombia
TLV Tel Aviv, Israel
TLW Talasea, New Britain, New Guinea
TMD Timbedra, Mauritania
TME Tame, Colombia
TMH Tanahmerah, West Irian, Indonesia
TML Tamale, Ghana
TMM Tamatave, Malagasy
TMP Tampere, Finland
TMQ Tambao, Upper Volta
TMR Tamanrasset, Algeria
TMT Temora, New South Wales
TMW Tamworth, New South Wales
TMX Timimoun, Algeria
TMZ Termez, USSR
TNA Tsinan, China
TND Trinidad, Cuba
TNG Tangier, Morocco
TNJ Tanjungpinang, Bintan Island, Indonesia
TNN Taiwan, Formosa
TNQ Tongo, Sierra Leone
TNR Tananarive, Malagasy
TNS Tønsberg, Norway
TOA Tromsø, Norway
TOB Tobruk, Libya
TOL Toledo, Ohio
TOM Tombouctou, Mali (better known as Timbuctoo)
TOP Topeka, Kansas—A-TPB (Allen); P-TOP (Billard)
TOQ Tocopilla, Chile
TOU Touraine, South Vietnam
TOW Tororo, Uganda
TOX Tocantina, Goiás, Brazil
TPA Tampa, Florida
TPE Taipei, Formosa
TPG Taiping, China

TPH Tonopah, Nevada
TPL Temple, Texas
TPS Trapani, Italy
TPU Taputuquara, Brazil
TPY Tocantinopolis, Brazil
TQS Tres Esquinas, Colombia
TQV St Moritz, Switzerland
TRB Turbo, Colombia
TRC Torreon, Mexico
TRD Trondheim, Norway
TRG Tauranga, New Zealand
TRI Tri-City Airport—Bristol, Tennessee; Bristol, Virginia; Johnson City-Kingsport, Tennessee
TRK Tarakan, Indonesia
TRN Turin, Italy
TRO Taree, New South Wales
TRQ Tarauaca, Brazil
TRR Trincomalee, Ceylon
TRS Trieste, Italy
TRT Tiaret, Algeria
TRU Trujillo, Peru
TRV Trivandrum, India
TRW Tarawa, Gilbert Islands
TRY Treviso, Italy
TRZ Trichinopoly, India
TSB Tsumeb, South-West Africa
TSH Tshikapa, Zaire
TSN Tientsin, China
TSS Tebessa, Algeria
TSV Townsville, Queensland
TTC Taltal, Chile
TTD Palm Island, Windward Islands, West Indies
TTG Tartagal, Argentina
TTM Tablón de Tamara, Colombia
TTN Trenton, New Jersey
TTS Tsaratanana, Malagasy
TTU Tetuan, Spanish Morocco
TUB Tubarao, Brazil
TUC Tucuman, Argentina
TUD Tambacounda, Senegal
TUF Tours, France
TUI Turaif, Saudi Arabia
TUL Tulsa, Oklahoma
TUN Tunis, Tunisia
TUO Taupo, New Zealand
TUP Tupelo, Mississippi
TUQ Tougan, Upper Volta
TUS Tucson, Arizona
TUU Tabuk, Saudi Arabia
TUV Tucupita, Venezuela
TUX Tuxpan, Mexico
TVA Morafenobe, Malagasy
TVC Traverse City, Michigan
TVF Thief River Falls, Minnesota
TVU Taveuni, Fiji Islands
TWB Toowoomba, Queensland
TWF Twin Falls, Idaho
TWU Tawau, Malaysia
TXA Texeira, Portugal
TXG Taichung, Formosa

TXK Texarkana, Arkansas
TXM Teminabuan, West Irian, Indonesia
TXR Tanbar, Queensland
TXU Tabou, Ivory Coast
TYA Yalova, Turkey
TYB Tibooburra, New South Wales
TYL Talara, Peru
TYN Taiyaun, China
TYO Tokyo, Japan
TYR Tyler, Texas
TYS Knoxville, Tennessee
TZG Waha Leaf, British Honduras
TZX Trabzon, Turkey
UAK Narsarssuak, Greenland
UAQ San Juan, Argentina
UBA Uberaba, Brazil
UBG Limón, Honduras
UBI Buin, Solomon Islands
UBJ Ube, Japan
UBK Port Augusta, South Australia
UBP Ubol, Thailand
UBS Columbus, Mississippi
UCA Rome-Utica, New York
UCN Buchanan, Liberia
UDD Cuddapan, Queensland
UDI Uberlandia, Brazil
UDR Udaipur, India
UEL Quelimane, Mozambique
UET Quetta, Pakistan
UGA Ugashik, Alaska
UIB Quibdó, Colombia
UIH Qui Nhon, South Vietnam
UIN Quincy, Illinois
UIO Quito, Ecuador
UIP Quimper, France
UKI Ukiah, California
UKR Mukeiras, South Arabia
ULA San Julian, Argentina
ULN Ulan Bator, Mongolia
ULP Quilpie, Queensland
ULQ Tulua, Colombia
UME Umeaa, Sweden
UMK Umanak, Greenland
UMU Umuarama, Brazil
UMW Mumbwa, Zambia
UNC Unguia, Colombia
UND Kunduz, Afghanistan
UNE Unst, Shetland Islands, Scotland
UNI União da Vitoria, Brazil
UNK Unalakleet, Alaska
UON Muong Sai, Laos
UOX University, Mississippi
UPA Upala, Costa Rica
UPN Uruapan, Mexico
UPP Upolu Point, Hawaii
UPV Upernavik, Greenland
URB Urubupunga, Brazil
URF Urfa, Turkey
URG Uruguaiana, Brazil
URI Uribia, Colombia
URY Gurayat, Saudi Arabia

USH Ushuaia, Argentina
UTA Umtali, Zimbabwe
UTB Muttaburra, Queensland
UTI Uttaradit, Thailand
UTL Utila Island, Honduras
UTN Upington, South Africa
UTO Utopia, Alaska
UTW Queenstown, South Africa
UUP Uaupes, Brazil
UVL New Valley, Egypt
UYL Nyala, Sudan
UZU Curuzu Cuatia, Argentina
VAA Vaasa, Finland
VAE Ciudad de Valles, Mexico
VAG Vagar, Faeroe Islands
VAN Van, Turkey
VAR Varna, Bulgaria
VAS Sivas, Turkey
VAT Vatomandry, Malagasy
VAV Vaxjo, Sweden
VBN Vrnjacka Banja, Yugoslavia
VBY Visby, Sweden
VCE Venice, Italy
VCT Victoria, Texas
VCW Victoria West, South Africa
VDM Viedma, Argentina
VDO Vadso, Norway
VDP Valle de la Pascua, Venezuela
VDR Villa Dolores, Argentina
VDZ Valdez, Alaska
VEL Vernal, Utah
VER Veracruz, Mexico
VEY Vestmannaeyjar, Iceland
VFA Victoria Falls, Zimbabwe
VGA Vijayawada, India
VGO Vigo, Spain
VGR Virgin Gorda, British Virgin Islands
VHO Vila Coutinho, Mozambique
VHY Vichy, France
VIC Vicenza, Italy
VIE Vienna, Austria
VIL Villa Cisneros, Spanish Sahara
VIQ Violetvale, Queensland
VIS Visalia, California
VIV Vivigany, New Guinea
VIX Victória, Brazil
VJB Vila de João Belo, Mozambique
VKS Vicksburg, Mississippi
VLC Valencia, Spain
VLD Valdosta, Georgia
VLI Port Vila, New Hebrides
VLK Viqueque, Timor
VLN Valencia, Venezuela
VLR Vallenar, Chile
VLV Valera, Venezuela
VMU Baimuru, Papua
VNO Vilnius, Lithuania, USSR
VNR Vanrook, Queensland

VNX Vilanculos, Mozambique
VOG Volgograd, USSR
VOH Vohemar, Malagasy
VOL Volos, Greece
VPA Silver Plains, Queensland
VPS Eglin Air Force Base, Florida
VPY Vila Pery, Mozambique
VPZ Valparaiso, Indiana
VRA Varadero, Cuba
VRB Vero Beach, Florida
VRN Verona, Italy
VRZ Voronezh, USSR
VSA Villahermosa, Mexico
VSO Phuoc Long, Vietnam
VTE Vientiane, Laos
VTL Vittel, France
VTZ Vishakhapatnam, India
VUP Valledupar, Colombia
VVB Mahanoro, Malagasy
VVC Villavicencio, Colombia
VVO Vladivostok, USSR
VXC Vila Cabral, Mozambique
WAB Wabag, New Guinea
WAC Waca, Ethiopia
WAD Andriamena, Malagasy
WAE Aoulef, Algeria
WAG Wanganui, New Zealand
WAI Antsohihy, Malagasy
WAK Ankazoabo, Malagasy
WAM Ambatondrazaka, Malagasy
WAN Waverney, Queensland
WAP Alto Palena, Chile
WAQ Antsalova, Malagasy
WAS Washington, D.C.—N-DCA (National); D-IAD (Dulles); B-Bal (Friendship)
WAW Warsaw, Poland
WBA Washington Bay, Alaska
WBD Befandriana, Malagasy
WBE Bealanana, Malagasy
WBG Wichabai, Guyana
WBM Wapenamanda, New Guinea
WBO Beroroha, Malagasy
WCA Castro, Chile
WCH Chaiten, Chile
WCJ Caleta Josefina, Chile
WCO Coolullah, Australia
WDA Wadi Ain, South Arabia
WDG Enid, Oklahoma
WDH Windhoek, South-West Africa
WEI Weipa, Queensland
WEL Welkom, South Africa
WEN Papa Westray, Orkney Islands, Scotland
WFI Fianarantsoa, Malagasy
WGA Wagga-Wagga, New South Wales
WGM Wilmington, California
WGP Waingapu, Sumba, Indonesia
WHA Wadi Halfa, Sudan
WHK Whakatane, New Zea-

land
WIC Wick, Scotland
WIN Winton, Queensland
WIS Central, Wisconsin
WJF Lancaster-Palmdale, California (airport serving both places in the Mojave)
WKB Warracknabeal, Victoria, Australia
WKI Wankie, Zimbabwe
WKM Wankie Game Reserve, Zimbabwe
WKN Wakunai, Solomon Islands
WKP Wrotham Park, Queensland
WLG Wellington, New Zealand
WLS Wallis Island, Polynesia
WMA Mandritsara, Malagasy
WMB Warrnambool, New South Wales
WMD Mandabe, Malagasy
WML Malaimbandy, Malagasy
WMN Maroantsetra, Malagasy
WMP Mampikony, Malagasy
WMR Mananara, Malagasy
WMV Madirovalo, Malagasy
WMX Wamena, West Irian, Indonesia
WNR Windorah, Queensland
WNS Nawabshah, Pakistan
WNY Burnie-Wynward, Tasmania
WON Wondoola, Queensland
WOQ Wooroona, Queensland
WPA Puerto Aysen, Chile
WPB Port Berge, Malagasy
WPR Porvenir, Chile
WPU Puerto Williams, Chile
WRE Whangarei, New Zealand
WRG Wrangell, Alaska
WRL Worland, Wyoming
WRO Wroclaw, Poland
WSM Wiseman, Alaska
WSP Waspam, Nicaragua
WSR Wasior, West Irian, Indonesia
WSZ Westport, New Zealand
WTA Tambohorano, Malagasy
WTD West End, Grand Bahama Island, Bahamas
WTS Tsiroanomandidy, Malagasy
WUG Wau, New Guinea
WUH Wuhan, China
WUU Wau, Sudan
WVA Alexandria, Virginia
WVB Walvis Bay, South-West Africa
WVK Manakara, Malagasy
WVL Waterville, Maine
WVV Volovan, Malagasy
WWD Cape May, New Jersey
WWK Wewak, New Guinea
WYA Whyalla, South Austra-

lia
WYC Yes Bay, Alaska
WYD Wyandotte, Queensland
WYE Yengema, Sierra Leone
WYN Wyndham, Western Australia
XAP Xapecó, Brazil
XAY Xapuri, Brazil
XIE Xieng Khouang, Laos
XIQ Xique-Xique, Brazil
XLS St Louis, Senegal
XMC Malacoota, New South Wales, Australia
XMM Mamaia, Romania
XNG Quang Ngai, Vietnam
XRY Jerez de la Frontera, Spain
XSC South Caicos, Caicos Islands, West Indies
XTG Thargomindah, Queensland
XTN Qatn, South Arabia
XTO Taroom, Queensland
XTR Tara, Queensland
YAC Yacuiba, Bolivia
YAK Yakutat, Alaska
YAM Sault Sainte Marie, Ontario
YAO Yaounde, Cameroon
YAP Yap, Caroline Islands
YBC Baie, Comeau, Québec
YBE Uranium City, Saskatchewan
YBG Saguenay, Québec
YBR Brandon, Manitoba
YCG Castlegar, British Columbia
YCL Charlo, New Brunswick
YDA Dawson City, Yukon
YDF Deer Lake, Newfoundland
YEG Edmonton, Alberta
YEV Inuvik, Northwest Territories, Canada
YFB Frobisher, Northwest Territories, Canada
YFC Fredericton, New Brunswick
YFE Forestville, Québec
YGR Magdalen Island, Québec
YGT Thunder Bay (formerly Ft William), Ontario
YHZ Halifax, Nova Scotia
YJT Stephenville, Newfoundland
YKA Kamloops, British Columbia
YKL Schefferville, Québec
YKM Yakima, Washington
YKN Yankton, South Dakota
YLE Yule Island, New Guinea
YLW Kelowna, British Columbia
YMA Mayo, Yukon Territory, Canada
YME Matane, Québec
YML Murray Bay, Québec
YMV Manicouagan, Québec

YND Yandina, Solomon Islands
YNG Sharon, Pennsylvania-Warren-Youngstown, Ohio
YNK Gagnon, Québec
YOL Yola, Nigeria
YOR Yoro, Honduras
YOW Ottawa, Ontario
YPA Prince Albert, Saskatchewan
YPR Prince Rupert, British Columbia
YQB Québec City, Québec
YQG Windsor, Ontario
YQH Watson Lake, Yukon
YQI Yarmouth, Nova Scotia
YQJ Porquis Junction, Ontario
YQL Lethbridge, Alberta
YQM Moncton, New Brunswick
YQR Regina, Saskatchewan
YQT Thunder Bay, Ontario— serving communities formerly known as Ft William and Port Arthur
YQU Grande Prairie, Alberta
YQV Yorkton, Saskatchewan
YQX Gander, Newfoundland
YQY Sydney, Nova Scotia
YQZ Quesnel, British Columbia
YRF Ross Bay, Newfoundland
YRI Riviere du Loup, Québec
YRQ Trois Rivières, Québec
YRX Rimouski, Québec
YSB Sudbury, Ontario
YSJ Saint John, New Brunswick
YSM Ft Smith, Northwest Territories, Canada
YSU Summerside, Prince Edward Island
YTS Timmins, Ontario
YUL Montreal, Québec
YUM Yuma, Arizona
YUY Noranda-Rouyn, Québec
YVA Moroni, Comoro Islands
YVO Burlamaque-Val d'Or, Québec
YVR Vancouver, British Columbia
YWG Winnipeg, Manitoba
YWH Whalehead, Québec
YWK Wabush, Newfoundland
YWL Williams Lake, British Columbia
YWY Wrigley, Northwest Territories
YXC Cranbrook, British Columbia
YXD Edmonton, Alberta—l-YEG (International); Y-YXD (Industrial)
YXE Saskatoon, Saskatchewan
YXH Medicine Hat, Alberta
YXJ Ft St John, British Columbia

YXR Earlton, Ontario
YXS Prince George, British Columbia
YXT Terrace, British Columbia
YXU London, Ontario
YYB North Bay, Ontario
YYC Calgary, Alberta
YYD Smithers, British Columbia
YYE Ft Nelson, British Columbia
YYF Penticton, British Columbia
YYG Charlottetown, Prince Edward Island
YYJ Victoria, British Columbia
YYN Swift Current, Saskatchewan
YYR Goose Bay, Labrador
YYT St Johns, Newfoundland
YYU Kapuskasing, Ontario
YYY Mont Joli, Québec
YYZ Hamilton-Toronto, Ontario
YZF Yellow Knife, Northwest Territories
YZP Sandspit, British Columbia
YZV Sept-Îles (Seven Islands), Québec
ZAD Zadar, Yugoslavia
ZAG Zagreb, Yugoslavia
ZAH Zahedan, Iran
ZAL Valdivia, Chile
ZAM Zamboanga, Mindanao, Philippines
ZAN Zanderij, Surinam
ZAP Zaporozhe, USSR
ZAR Zaria, Nigeria
ZAZ Zaragoza, Spain
ZBO Bowen, Queensland
ZBY Sayaboury, Laos
ZCO Temuco, Chile
ZDK Zonguldak, Turkey
ZED Pakatoa, New Zealand
ZGL South Galway, Queensland
ZGM Ngoma, Zambia
ZHM Shamshernagar, Bangladesh
ZIC Victoria, Chile
ZIG Ziguinchor, Senegal
ZIH Zihuatanejo, Mexico
ZKB Kasaba Bay, Zambia
ZKL Steenkool, West Irian, Indonesia
ZKM Sette Cama, Gabon
ZLG La Guera, Morocco
ZLO Manzanillo, Mexico
ZMD São Madureira, Brazil
ZND Zinder, Niger
ZNG New Glasgow, Nova Scotia
ZNZ Zanzibar, Tanzania
ZOM Zomba, Malawi
ZON Queenstown, New Zea-

land
ZOS Osorno, Chile
ZQN Queenstown, New Zealand
ZRH Zurich, Switzerland
ZRI Serui, West Irian, Indonesia

ZSA San Salvador Island, Bahamas
ZSS Sassandra, Ivory Coast
ZTK Stokmarknes, Norway
ZUD Ancud, Chile
ZUL Silfi, Saudi Arabia
ZVA Miandrivazo, Malagasy

ZVG Springvale, Queensland
ZVK Savannakhet, Laos
ZWA Andapa, Malagasy
ZYL Sylhet, Bangladesh
ZZU Mzuzu, Malawi
ZZV Zanesville, Ohio

American States and Capitals

Ala. (Alabama) Montgomery
Alas. (Alaska) Juneau
Ariz. (Arizona) Phoenix
Ark. (Arkansas) Little Rock
Calif. (California) Sacramento
Colo. (Colorado) Denver
Conn. (Connecticut) Hartford
Del. (Delaware) Dover
Fla. (Florida) Tallahassee
Ga. (Georgia) Atlanta
Ha. (Hawaii) Honolulu
Ia. (Iowa) Des Moines
Id. (Idaho) Boise
Ill. (Illinois) Springfield
Ind. (Indiana) Indianapolis
Kans. (Kansas) Topeka
Ky. (Kentucky) Frankfort
La. (Louisiana) Baton Rouge
Mass. (Massachusetts) Boston
Md. (Maryland) Annapolis
Me. (Maine) Augusta

Mich. (Michigan) Lansing
Minn. (Minnesota) St Paul
Miss. (Mississippi) Jackson
Mo. (Missouri) Jefferson City
Mont. (Montana) Helena
N.C. (North Carolina) Raleigh
N.D. (North Dakota) Bismarck
Nebr. (Nebraska) Lincoln
Nev. (Nevada) Carson City
N.H. (New Hampshire) Concord
N.J. (New Jersey) Trenton
N.M. (New Mexico) Santa Fe
N.Y. (New York) Albany
Oh. (Ohio) Columbus
Okla. (Oklahoma) Oklahoma City
Oreg. (Oregon) Salem
Pa. (Pennsylvania) Harrisburg
R.I. (Rhode Island) Providence
S.C. (South Carolina) Colum-

bia
S.D. (South Dakota) Pierre
Tenn. (Tennessee) Nashville
Tex. (Texas) Austin
Ut. (Utah) Salt Lake City
Va. (Virginia) Richmond
Vt. (Vermont) Montpelier
Wash. (Washington) Olympia
Wis. (Wisconsin) Madison
WVa (West Virginia) Charleston
Wyo. (Wyoming) Cheyenne
foregoing fifty states plus two possible states if their citizens approve and Congress concurs

D.C. (District of Columbia) Washington
P.R. (Puerto Rico) San Juan

Astronomical Constellations, Stars and Symbols

And Andromeda (Princess Enchained), also called Mirach
Ant Antlia (Air Pump)
Aps Apus (Bird of Paradise)
Aql Aquila (Eagle); contains Altair
Aqr Aquarius (Water Carrier)
Ara (Altar)
Ari Aries (Ram); contains Hamal
Aur Auriga (Charioteer); contains Capella
Boö Boötes (Herdsman); contains Arcturus
Cae Caelum (Chisel)
Cam Camelopardalis (Giraffe)
Cap Capricornus (Horned Goat)
Car Carina (Keel of Argo); contains Canopus

Cas Cassiopeia (Queen Enthroned); contains supernova 1572
Cen Centaurus (Centaur); contains Alpha Centauri, Proxima Centauri
Cep Cepheus (Monarch)
Cet Cetus (Whale); contains Mira
Cha Chameleon
Cir Circinus (Compasses)
CMa Canis Major (Great Dog); contains Sirius
CMi Canis Minor (Little Dog); contains Procyon
Cnc Cancer (Crab); contains Praesepe
Col Columba (Dove)
Com Coma Berenices (Berenice's Hair)
CrA Corona Australis (South-

ern Crown)
CrB Corona Borealis (Northern Crown), also called Gemma
Crt Crater (Cup)
Cru Crux (Southern Cross); Black Magellanic Cloud nearby
Crv Corvus (Crow)
CVn Canes Venatici (Hunting Dogs); contains Cor Caroli
Cyg Cygnus (Swan); contains Deneb, Northern Cross
Del Delphinus (Dolphin)
Dor Dorado, also called Xiphies (Swordfish); Large Magellanic Cloud
Dra Draco (Dragon)
Equ Equuleus (Colt)
Eri Eridanus (Great River); contains Achernar
For Fornax (Furnace)

Gem Gemini (The Twins); contains Castor, Pollux
Gru Grus (Crane)
Her Hercules; contains Ras Algethi
Hor Horologium (Clock)
Hya Hydra (Marine Monster); contains Alphard
Hyd Hydrus (Water Snake)
Ind Indus (Indian)
Kif Aus Kiffa Australis (Southern Breadbasket); contains Zuben el Genubi
Kif Bor Kiffa Borealis (Northern Breadbasket); contains Zubeneschamali
Lac Lacerta (Lizard)
Leo (Lion) contains Regulus, Denebola
Lep Lepus (Hare)
Lib Libra (Balance or Scales)
LMi Leo Minor (Little Lion)
Lup Lupus (Wolf)
Lyn Lynx
Lyr Lyra (Lyre); contains Vega
Men Mensa (Table), also called Mons Mensae (Table Mountain)

Mic Microscopium (Microscope)
Mon Monoceros (Unicorn)
Mus Musca (Fly)
Nor Norma (Rule)
Oct Octans (Octant)
Oph Ophiuchus (Serpent Bearer); contains supernova 1604
Ori Orion (Hunter); contains Betelgeuse, Rigel
Pav Pavo (Peacock)
Peg Pegasus (Winged Horse)
Per Perseus (Rescuer or Champion); contains Algol
Phe Phoenix
Pic Pictor (Painter's Easel)
PsA Piscis Austrinus (Southern Fish); contains Formalhaut
Psc Pisces (Fishes)
Pup Puppis (Stern), in Argo
Pyx Pyxis (Mariner's Compass Chest or Binnacle), in Argo
Ret Reticulum (Net)
Scl Sculptor (Sculptor's Workshop)
Sco Scorpius (Scorpion); con-

tains Antares
Sct Scutum (Shield)
Ser Serpens (Serpent)
Sex Sextans (Sextant)
Sge Sagitta (Arrow)
Sgr Sagittarius (Archer), Center of Galaxy
Tau Taurus (Bull); contains Hyades—Aldebaran; Pleiades
Tel Telescopium (Telescope)
TrA Triangulum Australe (Southern Triangle)
Tri Triangulum (Triangle)
Tuc Tucana (Toucan); Small Magellanic Cloud
UMa Ursa Major (Great Bear—Big Dipper); contains Dubhe, Mizar
UMi Ursa Minor (Little Bear —Little Dipper); contains Polaris (Pole Star)
Vel Vela (Sail), in Argo
Vir Virgo (Virgin)
Vol Volans (Flying Fish)
Vul Vulpecula (Little Fox); also called Vulpecula cum Ansere (Little Fox with Goose)

Astronomical Symbols

⊖☾ : center
☄ : comet
● : crescent moon (first quarter)
◐ : crescent moon (last quarter)
⊕ : Earth (symbol shows globe bisected by meridian lines into four quarters)
○ : full moon
◖ : gibbous moon (first quarter)
◗ : gibbous moon (last quarter)
◑ : half moon (first quarter)
◐ : half moon (last quarter)
♃ : Jupiter (symbol said to represent a hieroglyph of the eagle, Jove's bird, or to be the initial letter of Zeus with a line drawn through it to indicate its abbreviation)
☾☽ : lower limb
♂ : Mars (symbol represents

shield and spear of the god of war, Mars; it is also the male or masculine symbol)
☿ : Mercury (symbol represents head and winged cap of Mercury, god of commerce and communication, surmounting his caduceus)
♆ : Neptune (symbolized by the trident of Neptune, god of the sea)
● : new moon
☽ : moon (symbol depicts crescent moon in last quarter)
♇ : Pluto (symbol is monogram made up of P and L; initials of astronomer Percival Lowell, who predicted its discovery)
♄ : Saturn (symbol thought to represent an ancient scythe or sickle, as Saturn was the god of seed sowing and hence also of

time)
☆ : star
☆-P : star-planet altitude correction
☉ : sun (symbolized by a shield with its boss; some believe this boss represents a central sunspot)
☉☽ : upper limb
♅ : Uranus (symbolized by combined devices indicating the sun plus the spear of Mars, as Uranus was the personification of heaven in the Greek mythology, dominated by the light of the sun and the power of Mars)
♀ : Venus (designated by the female symbol, thought to be the stylized representation of the hand mirror of this goddess of love)

Bafflegab Divulged

Bafflegab consists of the ambiguous and euphemistic fig leaves of language and literature. *Bafflegab* begins where abbreviations, acronyms, and other short forms leave off.

abandoned woman prostitute
abbot nembutal
above critical in danger of exploding; out of control
absolute tripe official gibberish
abstinence syndrome withdrawal from alcohol or other addictive substances
acapulco gold gold-tinted, high-priced Mexican marijuana
acceptable deception(s) half truth(s)
accommodation house(s) whorehouse(s)
account executive(s) pimp(s) offering prostitutes to executives on liberal expense accounts
accounting aberrations profit-making mistakes made at the expense of the customer
ace marijuana cigarette
achieved status social position gained by ability, accomplishments, and personal effort
aching stiffness of the middle leg an erection
acid freak habitual LSD user who exhibits bizarre behavior
acid head habitual LSD user
acid house a building where LSD is sold and used
acid indigestion heartburn, acidosis, acute indigestion, cardialgia, colic, dyspepsia, gripes, pyrosis, stomach condition, tormina, water qualm
acid rock synthesized music associated with drug use
acoustical problem deafness; loss of hearing
active defense offense
activism agitation
activist agitator
activities incompatible with diplomatic status spying
activity booster pep pill
acute heroin-morphine intoxication death due to an overdose of the drugs
acute irregularity constipation; irregular menstrual flow
adjustment center(s) prison;

solitary confinement cell(s)
administrative action Soviet euphemism for imprisonment without trial
administrative domain collectivist system of managing and taxing distilleries, factories, and farms
administrative error bureaucratic bungle
administrative segregation solitary confinement
adult bookstore emporium of sexually explicit publications
adult briefs disposable diapers for incontinent adults
adult entertainment erotic dancing
adult fiction sexually explicit novels
adult institution(s) prison(s)
adult movies motion-picture theaters featuring sexually explicit films; sexually explicit films
adult relaxation center bathroom
adult retreat bedroom
adult undergarments diapers for adults
advanced in years old
advance to a strategic position retreat
advantaged gainfully employed; rich; successful
aerialist trapeze artist; trapeze performer; trapezist
aerial mishap(s) aviation accident(s)—airplane crashes, explosions
aerial visitation nuclear attack
affaire affaire d'amour (French—love affair)—usually illicit
affaire de coeur (French—affair of the heart)—love or lust adventure
affaire de voyage (French—love trip)—shipboard romance
affinals affinal relatives (in-laws)—*see* consanguineals
affirmative action programs designed to admit minority people to jobs and schools
affordable merchandiser's term for anything attractive enough to make potential customers become buyers although they may not be able to pay in full
Afro-American American black; American Negro, African-American

after he (she) left after he (she) died
aftershock post-release prisoner program(s)
A-funk acid (LSD) funk, drug depression
agates nembutal
ageism prejudice against the elderly
agelaugh person who never laughs
Agent Orange deadly defoliant used in tropical warfare believed to cause cancer and birth defects in humans
agricultural laborer(s) farmhand(s)
A-head acid head; LSD addict; amphetamine addict
airsickness nausea followed by vomiting when travelling in an aircraft
aisle manager floorwalker
Alaska sable skunk fur
Albany beef Hudson River sturgeon
alcoholic beverage alcoholic drink, liquor
Alice B Toklas brownie cookie with baked-in marijuana
a little sensitive neurotic; very touchy
all about the birds and the bees sex education
all-devouring element fire
all-out strategic exchange atomic bombs in nuclear warfare
all this and more, right after this important message please stay tuned
alpha alcoholic person who drinks in an effort to drown problems
alter caponize; castrate; desex; denut; geld; remove or ligate procreative organs; sterilize
alternative four-letter term for copulation fuck; screw
alternative to birth abortion
Alzheimer's disease characterized by gradual loss of memory and reasoning resulting in confusion, brain degeneration, and death; a form of presenile dementia
amalgamation interracial mixing; miscegenation
amatory exercises calisthenic in-bed exercises frequently culminating in lovemaking

971

Bafflegab Divulged

ambisexuals persons without preference as to the gender of their sexual partners
amenity center public toilet
Americaid welfare
American broadtail domestic lamb's wool
American tweezers burglar's tools
amply proportioned fat
amply rewarded well paid
amusement park entertainer organ grinder; mouse swallower and vomiter; snake charmer and swallower; sword swallower
anal pers anal retentive personality (marked by excessive orderliness and extreme meticulousness)
an association sexual contact
androgen effeminate homosexual male; fairy; girlish male
Angel of the Bottomless Pit the Devil
angel dust phencyclidine (PCP)
animal controller(s) dogcatcher(s)—*see* dog warden(s)
animal shelter(s) dog pound(s)
anointing the sick Roman Catholic euphemism for unction in extremis (service rendered at the point of death); last rites
anonyma nameless women (nineteenth-century euphemism for prostitutes)
answering the call of nature defecating and/or urinating
answer the final summons die; drop dead
anthropogenically-derived acidic substances acid rain
anticipatory retaliation first strike in a conflict
anti-gods marijuana cigarettes
anti-Semite anti-Jewish person
antisocial offender(s) convict(s); criminal(s)
antisocial restructuring candidates convicts; felons
anti-Zionist often anti-Semite
antsy nervous feeling, jitters
arbitrageur person who profits by ordering the simultaneous purchase and sale of securities and gaining from their price discrepancies
arbitrary deprivation of life extermination; murder
arcanum mysterious knowledge known only to initiate members of a society
archivist(s) library clerk(s)
ardent spirits alcohol; alco-

holic drinks
area of operations battlefield
Argtec Argentine technique of disposing of political prisoners by casting them, shackled, from aircraft flying over the La Plata estuary
Armageddon the final conflict, predicting the last and completely destructive battle
armed emergency small-scale war
Armenian mink dyed cat fur
arms-reduction specialists arms-control agents
Arsetralia waterfront slang for Australia
articulating on paper writing
artificial dentures false teeth
artillery injection paraphernalia
artistic success box-office failure
asbestosis degenerative lung disease caused by inhalation of asbestos particles
assignation point(s) place(s) where pimps and streetwalkers accost potential customers
atherosclerosis hardening of the arteries, may lead to angina pectoris
athletic supporter(s) jockstrap(s)
at that future juncture when
at that point in time then
at this juncture now
at this moment of history now
at this point in time now
at this present juncture now
audience augmentation device for producing a capacity audience; attained by generous distribution of complimentary tickets, and cut-rate prices
audiovisually qualified able to run a motion-picture projector
auditorially handicapped deaf
auditory problem deafness; loss of hearing
au naturel naked
Australian buck processed rabbit fur
Australian steak mutton
Australian treat kangaroo meat
authoritarian anti-communist, totalitarian regime denying human rights to its people and persecuting all dissidents
autocompress automobile metal compression device for reducing junk vehicles into a square mass of metal ready for the smelter

auto insurance problem revocation of driver's license and subsequent loss of auto insurance
automanipulation masturbation
automotive internist auto mechanic
avian propagation facility an incubator
avoid polarizing people don't rock the boat
awardee(s) person(s) imprisoned in a ship's brig also known as the ccu or correctional custody unit
away from one's desk absent; gone to the toilet; not accepting calls
awful acronym AIDS
axed fired
Aztec two-step loose bowels acquired in Mexico, the land of the Aztecs
bachelor girl old maid; spinster; single woman
back-alley butcher abortionist
backhouse privy; toilet
backward country poor country; third-world country
bad actors cesium, plutonium, and strontium wastes from high-level fusion in nuclear power plants
bad scene unhappy experience
bafflegab bewildering language composed of words created to confuse listeners and readers
bag drugs packaged in balloons or paper bags
bagno (Italian—bath)—prison or whorehouse
balderdash bullshit
bale 100 to 500 pounds of marijuana compressed
balloons drugs such as heroin packaged in toy balloons
Baltic leopard spotted-cat fur dyed to look like leopard
Baltimore beefsteak broiled liver
banheiro (Portuguese—bath)—men's toilet
baños de caballeros (Spanish—gentlemen's baths)—men's toilet(s)
baños de damas (Spanish—ladies' baths)—women's toilet(s)
bargain one down haggle, negotiate a price
bar hustler(s) male prostitute(s) plying trade in cocktail bars
bathroom toilet (in the United States, although in the British Isles and on the Continent a bathroom is fitted primarily for bathing although it may

contain a toilet)

bath(room) tissue toilet paper

battering domestic violence

battle fatigue shell shock; a variety of psychotic and neurotic disorders associated with the stress of combat

Bauhaus German school of modern industrial design linking art and technology, headed by architect Walter Gropius

bc's birth-control pills

bears law-enforcement officers

beaut book beautiful book

beautician beauty parlor owner; hairdresser; make-up artist

beauty culturist beauty parlor operator; make-up artist

beauty parlor hairdresser's shop

beaverette processed cat or rabbit fur

bebop music of the 1950s

bedded sexually connected

beddies attractive companions suitable for bedding

bedroom activities sex

bedroom behavior sex

bedswever unfaithful spouse

been inside served time in prison

behind the iron door behind bars; jailed

bennies benzedrine pills

Bess o'Bedlam female lunatic

bevy of pulchris group of attractive women

beyond the black stump Australia's far outback; beyond the boondocks

bidet (French—hygienic apparatus)—crotch cleaner installed in many homes, hotels, and steamships

big O orgasm

billiard academy pool hall

billiard lounge pool hall

binary weapons two chemicals kept separate until mixed for firing

biological soldiering use of biological (bacterial) agents against an enemy

biological urge lust

birth attendant midwife

birthing options choice of place and method of childbirth (home v. hospital, midwife v. physician-attended, natural v. caesarean)

black-and-whites skunks

Blackbeard's curse botulism, diarrhea, or dysentery contracted in the Caribbean

black beauties biphetamine

black mayonnaise poisonous sludge contaminating bays, beaches, harbors, and oceans

black propaganda lies repeated until they seem believable

black tar Mexican heroin, also called Mexican mud or tootsie roll

Blau's theory the larger the organization, the greater the proportion of total resources devoted to management, hence the proliferation of the bureaucracy

blemishes acne; blackheads; pimples

blister agents chemicals that burn the skin and cause blisters

blood agents chemical substances that enter the body via the respiratory system and attack the blood cells

blood box ambulance

blood disease syphilis

blue amytal (barbiturate)

blue acid pale-blue liquid LSD-25

blue angels amytal (barbiturate)

bluebirds capsules of sodium amytal

blue-collar labor factory workers

blue devils amobarbital capsules

bodily appetite lust

body moisture sweat

body odor stinking sweat

Bombay duck dried and salted lizardfish

bone box ambulance

book of pseudonyms hotel register

boomer member of the baby-boom generation

boot camp minimum-security jail

borborygmus bowel rumblings

bordello whorehouse

borne by the stork born

born into a better world died

born out of wedlock illegitimate

borrow without intent to return steal

bosco South American narcotic compound

bosom breast; female breasts

bottle baby alcoholic addict

bottom ass; backside

boutique (French—small specialty shop)—expensive specialty shop

bovine stool bullshit or cowshit

boys men; old men; young men

boys on the border South African euphemism for soldiers

defending the borders

boy's room men's toilet

breach of the peace to agitate, to arouse, to assemble unlawfully, to awaken, to hinder, to incite to riot, to molest, to obstruct traffic, to trespass

break the news inform, tell

break wind release flatus via the anus; vulgarly referred to as fart

breathe one's last breathe one's last breath—die

brief encounter one-night stand

bright lighters city people

Broadmoor patients criminally insane convicts

broken-down woman old whore

bromidrosis stinking sweat

bronzing sunbathing

bubo Low Latin for a swelling in the armpit or crotch, usually associated with gonorrhea; clap

buff use of solvents to remove graffiti

building engineer janitor

bulldung bullshit

bullflop bullshit

bummer bad experience

bunny a rabbit-costumed whore

bureaucrap bureaucratic bafflegab

bush telegraph (Australian—word of mouth)

business moratorium lockout

business slowdown depression

butch lesbian who plays male role

buy the farm die

by-product of the arts of peace war

caballeros (Spanish—gentlemen)—men's toilet

cabinet (French—closet; small room; water closet)—toilet

cabinet d'aisance (French—cabinet of comfort)—public toilet

cabrito young goat meat

caca (baby-talk euphemism—excrement)

cacá (Brazilian-Portuguese—excrement)

caducity senility

call boy(s) high-priced male prostitute(s); gigolo(s)

call girl(s) high-priced female prostitute(s)

call of nature need to go to the toilet

Cambodian Incursion Cambodian Invasion

cameo bit part in a dramatic production

campo santo (Spanish—sacred

field)—cemetery

can going to the can; going to the toilet

candlestine affair mispronunciation of clandestine affair

cangrejo (Spanish—crab)—a homosexual sodomite

canine seculsion habitat doghouse

canine stool dogshit

canned cow condensed milk

Cape Cod turkey codfish

capitalism economic system characterized by corporate or private ownership of goods, investments, and services with distribution controlled by competition in a free market unfettered by state control

carbon monoxdied killed by carbon-monoxide poisoning

cardiac heart attack; heart failure

cardiac arrest cessation of heart beat

cardiovascular accident stroke

career deceleration loss of job opportunities or jobs

career girl professional woman

carminitive laxative

carnal acquaintance copulation

carnal connection copulation

carnal desire sexual passion; lust

carnal enjoyment copulation with sexual pleasure

carnal intercourse copulation

carnal parts genitals

carnal passion lust

carnal stump penis

carnal trap vagina

carnifex hangman

carnificate to hang a person

carsickness nausea followed by vomiting when riding in an automobile or other vehicle

cash in your chips die

casket coffin

casketing putting a dead person in a coffin

cast up one's cookies vomit

cast up your accounts vomit

casual companion prostitute

casual sex promiscuity

cat a prostitute

catastrophic reaction psychiatric euphemism for angry frustration marked by crying, pulling, pushing, yelling, and even trying to commit suicide or kill others; often includes cataleptic convulsions, loss of memory and vocabulary, coronary disorders, and death

categorical imperative absolute and binding moral law

cathouse house of prostitution

cavalheiros (Portuguese—gentlemen)—men's toilet

cavalry police reinforcements

cease to purchase boycott

cemetery worker gravedigger

cerebral insufficiency brain deficiency

cerebrovascular accident stroke

cf cystic fibrosis

cf's confessions of fornication (colonial abbreviation used by the Puritans before the American Revolution)

C-girl call girl; hundred-dollar girl

chalet de nécessité (French—public toilet)

chamber pisspot; thundermug

channels trails of interoffice memos

Chapter 7 liquidation

Chapter 11 bankruptcy

character deficiency fault

charlady charwoman

cheapies cheap goods; cheap merchandise; cheap stocks

check out die

chemical abuse alcohol or drug addiction

chemical agent deployment throwing tear gas

chemical, electrical, or physical duress chemical, electrical, or physical torture

chicken manure chickenshit

chicken ranch house of prostitution

chic sale country privy; outhouse; shithouse

children's tutor governess

children with latent ability students with underdeveloped intellects

children with untapped potential students with underdeveloped intellects

chinchillette processed rabbit fur

chippy prostitute

chirtonsor(s) barber(s)

chop shop place where stolen autos are dismantled and sold by the part

chronic alcoholic sot

chronic irregularity constipation

chubby fat

churchyard cemetery

chutzpah (Yiddish—brazen effrontery, impudence, nerve)

circular protector condom

circumorbital haematoma black eye

civilian irregular mercenary

civilian irregular defense soldier mercenary

Civil War the War Between the States, the War of the Secession

clandestine affair secret love affair

clandestine connection sex in the shadows

clandestine exhumation body snatching; grave robbing

clap gonorrhea

clean bombs bombs that destroy structures rather than people; nonradioactive bombs; neutron bombs

cleanup policy burial insurance

client(s) prison inmate(s); users of professional services

client(s) of the correctional system convict(s)

climacterium change of life in men and women; menopause or cessation of menses in women

climb the golden stairs die

cloakroom toilet (British)

closet water closet (toilet)

cocktail lounge saloon, bar

cocotte (French—little chick; paper hen; saucepan)—prostitute in western Europe and the United States

Code Blue medical jargon frequently used to mean cardiac arrest

cognitive behavioral modification using task analysis to define mental problems

cognitive services any professional service involving counseling, talking, or thinking

cohabitees unmarried lovers who live together

coke spoon device designed for inhaling drugs

cold war deterrent threat

collaboration assisting and co-operating with the enemy

collation light meal

collection correspondent bill collector

colonic infirmity cancer of the colon

colonic irregularity constipation

colonic stasis constipation

colored folks elderly African-Americans

colored man black male; African-American male

coloureds persons of mixed race in South Africa

combat emplacement evacuator shovel

combat fatigue (*see* battle fatigue)

comfort station toilet

command of nature feeling it's time to defecate and/or urinate

commercial statement advertisement; sales pitch

commercial travellers travelling salespersons

commission agent bookmaker

commit no nuisance do not defecate or urinate

commits a nuisance defecates in a public place

commode toilet

communication arts reading, speaking, and writing

communication problem deaf and dumb

communications engineer tv repairman

community treatment center(s) prison(s)

companionate marriage free love; trial marriage

companion in misfortune fellow prisoner

Company The Company (Washington euphemism for the Central Intelligence Agency—CIA)

complete elimination mass murder

complete evacuation bowel movement

complete fabrication a lie

complete liquidation mass murder

compromising situation caught with your pants down

concernitis more concern than necessary; worry

conchie conscientious objector

concubine prostitute; part-time prostitute; unmarried sexual partner of a married man

concupiscence lust; sensual longing; sexual desire

condominium (Spanish euphemism—jail or prison)

conflict with the law breaking the law, committing a misdemeanor or felony

confrontation facing the enemy

congal cuarto con gal (Mexican-Spanish—room with girl) —house of prostitution

conjugal relations sex within marriage

conjugal visits sexual visits to prisoners by spouses

conjunctivitis sty on the eye

conned ambushed; ripped off; taken in; fooled

connubial bliss happy marriage; sex within marriage

connubial rites marriage ceremony

connubial substitute prostitute; concubine

consanguineals consanguineal relatives (related biologically)—*see* affinals

constant companion mistress; sexual pal

constant interruption chronic constipation

constructive engagement interchange that has positive results

Contragate 1980s political scandal involving the secret sale of arms to Iran in order to fund the Contra uprising in Nicaragua after Congress decided the U.S. should no longer support the Contras

contributor taxpayer

controlled substances addictive drugs; dangerous drugs; narcotics

convalescent home nursing home

convenience toilet

coordinated national intelligence spying program aimed at radicals

coordinator person whose desk is flanked by those of two expediters

copacetic fine, okay, satisfactory

cop a plea plead guilty to a lesser crime; plea bargain

cop-out convenient excuse to avoid unwanted responsibility

coronary heart attack; heart failure

corpocracy corporate bureaucracy, middle management

corpulent fat

correctional custody facility brig, jail, lockup, penitentiary, prison, reformatory, etc.

correctional custody unit ship's brig

correctional facilities brigs, jails, prisons

correctional institution penitentiary; prison

correctional officer(s) prison guard(s); prison warden(s)

corrective labor camps Soviet forced-labor prison camps

correct within an order of magnitude incorrect; wrong

cosmetician beauty parlor operator

cosmopolite(s) Soviet term for Jew(s)

costive constipated; constipating foods

couch potatoes sedentary people who watch a lot of television

counterproductive hindrance

courier service technician messenger

courtesan paramour; prostitute associating with members of the court and other notables

Cousins the Cousins (British secret service term for American intelligence organizations such as the CIA and the FBI)

cover up conceal the truth

cover-up words euphemisms

cover your bottom cover your ass; cover your backside

cowchips cattle dung

cow chips dried cow dung

cowflop cow excreta; cow droppings

cow flops cow dung

crab ugly girl

crack cheap, but deadly, cocaine, also called rock

crack peddling selling cocaine

crank. underworld nickname for methamphetamine, a mind-altering drug

crapola cover-up phrases and words characterized by their ambiguity, insincerity, and mendacity; nonsense

crapper toilet (believed by some to be named for an English turn-of-the-century plumber named Crapper and inventor of the water-flush system)

crash and burn fail utterly

creative conflict civil rights demonstration; nonviolent confrontation

creative financing where the seller loans some of the costs to the buyer in order to facilitate a sale; questionable accounting practices

creative unresponsiveness sullen indifference

credibility gap widespread disbelief and distrust due to impossible promises and lies uttered by bureaucrats and politicians

crib(s) whorehouse(s)

criminal conversation British euphemism for fornication

criminally attacked raped

criticalese language and style of professional critics

croak mixture of cocaine and crack; to die

crocd crocodiled; cast to the crocodiles as is still the custom in some areas

croton bug cockroach

crowd engineer(s) police dog(s)

cruising in the corn drinking and driving

culminating experience death by nuclear weapons

culturally deprived poor

culturally deprived environment ghetto; poor section; slum

cunctator procrastinator; putteroffer; thief of time

cupid's itch venereal disease

curse the curse (menstruation)

Curse of Balboa botulism, diarrhea, or dysentery contracted in Panamá

Curse of Cabral loose bowels contracted in Brazil

Curse of Cairo diarrhea plus summertime heat

Curse of Columbus botulism, diarrhea, or dysentery contracted in the American tropics

Curse of Cortez Mexican-acquired diarrhea or dysentery

Curse of Cromwell Cromwell's Irish campaign and the massacres in Clonmel, Drogheda, and Wexford

Curse of Pizarro loose bowels contracted in Perú

cuspidor spittoon

custodial care janitorial services; providing services to hospital and nursing home patients

custodial engineer janitor

custodian janitor

custom-fitted clothes ready-to-wear garments with cuffs are cut to size and sleeves shortened to suit the customer

cynical ailment cynicism, doubt, disbelief

dailies chambermaids or charwomen paid on a daily basis (British)

damas (Italian, Portuguese, Spanish—ladies)—women's toilet

Damen (German—ladies)—women's toilet

damer (Danish or Swedish—ladies)—women's toilet

dames (Afrikaans, Dutch, French—ladies)—women's toilet

damska (Polish—ladies)—women's toilet

dancing on air hanging by the neck until dead

dang(ed) damn(ed)

dark gentleman man from India

dark meat legs

darn it damn it; goddamn it

data processing electronic paperwork

dating street slang term meaning prostituting

daughter of Bilitis female homosexual, lesbian

daughter of joy prostitute

d___d damned (used in polite prose)

deaccessioning selling; divesting

death list persons to be assassinated

death machinery Hitler's extermination practices

debris disposal technician dustman; garbage man; trash collector

debt of nature death

decapitation beheading

decapitator beheader; guillotineur; guillotine

deceased dead

decimate destroy; kill off

decollation beheading

deek detect; look at; observe

defecate drop dung; move one's bowels; shit

defenestration jumping-out-the-window suicide

defense officer(s) soldier(s)

defensive aggression hitting the enemy before you are hit

defensive aircraft bombers

deferred maintenance putting off to tomorrow what should be repaired today

deferred schedule slowdown

deinstitutionalization discharge of felons from prisons and patients from mental institutions into surrounding communities

delicate condition pregnant

dementing illness Alzheimer's disease; intellectual deterioration caused by the slow death of the brain

demise death

den of iniquity house of prostitution

den of vice house of prostitution

dental discomfort toothache

dentures artificial teeth; false teeth; dental plates

deodorant of language euphemism

departed died

department of defense war department

depart this mortal coil die

depopulate(d) slaughter(ed)

depopulating slaughtering

depopulation death

Deputy Director for Opera-

tions chief of the clandestine service of the Central Intelligence Agency

dermasurgeon restorative embalming artist in a morgue

derrière (French—backside; behind)—ass

derun deruncinate (to prune)

designer drugs steroids (synthetic narcotics)

desktop publishing process of producing inexpensive professional-looking publications on a personal computer

detained in federal custody jailed in a federal penitentiary

détente easing of discord and warlike threats

détente (French—slackening of tensions)

detention center prison

deterrence policy of scaring the enemy by maintaining a larger armed force in the hope it will deter the enemy from war

developing nations have-not countries; poor countries; underdeveloped nations

deviate(s) homosexual(s)

deviation from the truth lying; untruth

devil's breath poison gas

devious woman furtive female

dezinformatsia (Russian—disinformation)—branch of the KGB

diagnostic misadventure of high magnitude a patient's death

dicey chancy; unclear

died of lead poisoning died from bullet wounds; shot

died of target practice executed by a firing squad

diet pills weight-reducing preparations

dilution reduction in earnings

diminished capacity slightly insane; impaired mental functioning

ding-dong a ding-a-ling, nitwit, kook

dining-room attendant busboy

dinks double-income no-kids couples

dipsomaniac alcoholic

directory assistance telephone information service

direct reduction killing off excess

dirty dishes evidence planted to incriminate another or others

dirty old man lecher, woman chaser

dirty tricks sinister activities

disadvantaged poor; unemployed
disappeared into the dust died
disbeliever an agnostic, atheist, doubter, freethinker, unbeliever
discontinuance of student populations school closures
disengage retreat
disinflation falling inflation, reduced rate of inflation
disinform mislead
disinterested free from bias
dislike the cut of one's jib dislike one's looks
disorderly house house of prostitution
disposable underclothes diapers
disposable undergarments diapers
disposal area dump
disposal center junkyard
disreputable woman prostitute
dissemble hide under a false appearance
dissolution death; end
distanced themselves ran for cover, left the scene; remained uninvolved
ditchweed Mexican marijuana
diversion embezzlement
diverted stolen from
D-notice death notice; death report; obituary
documented immigrant legal alien
doesn't know his elbow from his nalgo British way of declaring he doesn't know his ass from his elbow
dog behavior modification dog training program for guard or police work
doggydo dog dung
dog nuisance dog dung
dog tutor dog trainer
dog warden dogcatcher
doing one's business defecating; moving one's bowels
domestic functionary servant
domestic violence child beating; wife beating
dones (Catalan—ladies)—women's toilet
don't polarize people don't rock the boat
do one's business move one's bowels
do the dutch act commit suicide
doublespeak the language of contradiction, coverup distortion, misdirection, and omission
doubter an agnostic, atheist,

freethinker, skeptic, unbeliever
downers nickname for sedative drugs also called sleeping pills or tranquilizers
down the drain down the toilet (beyond recovery); too late
down the latrine down the toilet (beyond recovery); too late
down the tube down the toilet (beyond recovery); too late
down the waterslide down the toilet
downward revision decrease in prices or wages
dreamless sleep death
drinking problem alcohol addiction
dropout person living outside conventional society; person who does not complete an educational program; spot in a magnetic tape indicating loss of information
druggie drug addict
drugola bribes given law-enforcement officers by narcotics dealers in exchange for protection from discovery and prosecution
drug problem narcotic addiction
dub double
dumb insolence farting while marching; unspoken impertinence
duplication of the double number 4
dustbin garbage can
dustbinman garbage collector (British)
dust of desuetude neglect seldom
dutch gold imitation gold leaf
dutch metal gold-leaf-like ornamental coating
dutch treat each person pays own portion of the bill
dweeb small, harmless person
dyke lesbian
dysgenic tending to promote reproduction by less well-adapted individuals; biologically deficient
dyspepsia acid indigestion; heartburn
ease oneself defecate and/or urinate
easy money stolen goods
ebonies ebony phonetics (black English)
ecdysiast striptease artist
ecological receptacle garbage can; trash barrel; trash container
economic action picket line;

sitdown; slowdown; strike
economical with the truth untruthful; misleading
economically deprived poor
economic decline depression; recession
economic lull depression
economic regression loss
economic return profit
economic slowdown depression
educationist(s) bureaucratic educator(s) believed to be good at communicating with students
eduflation rising cost of education
eeble eyeball; look at
effect a separation fire (from a job); dismiss
effecting linkages coordination
effluents contaminating gaseous discharges, industrial wastes, and liquid sewage pollution
effluvium stench; stink
egress exit
Egyptian calisthenics getting into bed and staring up at the ceiling or sky until the forces of sleep prevail
electrolethe electric chair
electronic surveillance wire tapping
electronic technician electrician
eleemosynary organization charity fund-collecting business
elegancies words believed by some to be more genteel than the words they replace
elevated intoxicated (by drugs and liquor); high
elevated to a lower level demoted; lowered
eliminate kill; murder
ellas [Spanish—they (feminine)] —women's toilet
ellos [Spanish—they (masculine)]—men's toilet
elopement risk psychiatric euphemism for patients escaping or trying to escape from the hospitals, institutions, or wards to which they are confined
embarazada (Spanish—embarrassed)—pregnant
embarrassing noises belchings, burpings, and fartings
emerging nations backward countries; poor nations
emolument salary; wage
employment center commercial district
emporium department store
Endlösung die Endlösung (Ger-

man—final solution)—death; Hitlerian euphemism masking the annihilation of the Jews

end of the day when all is said and done

end-of-the-streeter aged prostitute

energetic disassembly explosion

engine opium pipe

engineering landfill(s) garbage dump(s)

English glass the Spanish euphemism for dog dung

enlargement of the lower back big bottom; fat ass

enlisted dining facility mess deck aboard American naval vessels

ensanguined undergarment bloody shirt

entertainment consumers casino gamblers

entombment burial

entomology specialist(s) bug sprayer(s)

entrapment luring anyone into the commission of a crime

entrepreneur imaginative and independent person usually under 40 who starts a business selling a new product or service

entrepreneurial ladies of the night prostitutes

enuresis bedwetting

environmental control specialist(s) janitor(s)

environmentally handicapped people ghetto people; poor people; street people; vagrants

environmental services janitorial work

epistemology capabilities and limits of the mind to acquire and understand knowledge

equal-opportunity ailments venereal diseases

equine fertilizer horse manure

equine paradox the existence of more horses asses than horses

Equipes d'Action (French—action teams) to prevent the traffic of women and children

equitable compensation living wage

erase assassinate; kill; murder

ergonomics designing equipment for maximum user comfort

erminette dyed rabbit fur

eros centers brothels

erotic portrayal pornography, when the depth of the dirt exceeds the width of the humor

erring brethren Confederate soldiers and states as defined by their Northern neighbors

erring sister prostitute

error in overamplification an error in the cover-up of the cover-up arranged by Nixon's aides after Watergate

eruct(ation) belch(ing)

escalated interpersonal altercation murder

escort prostitute in the jargon of the convention, entertainment, and travel world

escort vessel(s) destroyer(s); frigate(s)

eternal rest death

ethical lackers those with no scruples

ethnic evacuation forcible removal of a people from their native homeland

ethnic minority in the United States usually means African-Americans, Hispanics, and native Americans

ethnophaulisms derogatory and offensive slurs aimed against various ethnic groups

Eufemia Greek goddess of fair speech or good report—Euphemia

eugenics improvement of hereditary qualities by control of mating—see dysgenic

eugeria happy old age

euphemia softening of phrases or words thought to be coarse or offensive

euphemian euphemistic

euphemism affected niceness; goody-goodyism; overrefinement; purism; substitution of an auspicious phrase or word for inauspicious ones

euphuisms high-flown terms in the style of the English author John Lyly

euthanasia medically induced death for persons suffering from incurable and painful diseases

euthanize kill

eutrophication freshwater polluted by chemical plant or sewage runoff

Eve's curse menstruation

evidentiary material evidence

excellent compensation good pay

exceptional child retarded child

excrement dung

excreta dung

excusado (Spanish—reserved; set apart)—toilet

executive action removal of an

executive by assassination

executive explanation explaining the unexplainable by bending the truth

exfiltration retreat

exotic entertainer striptease dancer

expanding the circle of love mate swapping

expansion of the lower back big bottom; fat ass

expecting pregnant

expectorate spit

experienced automobiles second-hand autos

experienced tires recapped tires; retreads

expire die

expulsion of intestinal gas fart

exsanguination death due to loss of blood

exterior landscaper youngster mowing lawns

exterminating engineer bug and rat killer; termite remover; hired killer

extinguishment death

extrajudicial executions murders

extramarital sex fornication with a married person

extrapolation educated guess; guesstimate

extreme penalty death; execution

extreme power ordered killings of anti-government people

eye service working when the boss is looking

fabrication lie

fabricator liar

facial blemishes pimples

facial dew sweat

fag bash(ing) physical and verbal attacks on homosexuals

faggot habit homosexual practice

failed to reply accurately, completely, and fully lied

fairy a male homosexual

fall asleep die

far out crazy; wild

fashion stylist(s) dress designer(s)

featherbedding employing excess employees

fecu feculate(d); feculation

federalese federal bafflegab

feisty abusive; outspoken

feline effluvia the penetrating smell of cat excrement

feline stool catshit

fell in battle killed in action

female above position wherein woman is above man

female below position wherein

woman is below man

feminine bosom breasts; dugs (if a domestic animal such as a dog or cat); tits

feminine protection menstrual cloth, napkin, or tampon

fenestration arrangement of doors, openings, and windows in a building

ferfak (Hungarian—men)—men's toilet

ferly wonderfully great

fertilizer manure

fiber dietitianese for bowel-movement bulker, formerly called roughage, and present in bran, celery, whole wheat, etc.

fig leaves expressions contrived to conceal anything deemed to be indecorous, offensive, or of questionable taste

fig leaves of language and literature euphemisms

fille de joie (French—daughter of joy)—prostitute

filteration radiation

final injection execution by lethal drug injection

finalization conclusion

finalize end

final solution Hitler's euphemism masking his plan to murder all the Jews imprisoned in concentration camps

financially embarrassed broke; without funds

financially undernourished broke; without funds

finite period of future time later

fir (Gaelic—men)—men's toilet

first strike first attack

first use initial firing of nuclear weapons

fiscal policy government spending and tax collection

flamboyant flouncer conspicuous homosexual

flatulent given to farting

flatulents fart-filled pompous persons; fart-productive foods such as beans and cabbage

flight attendant airline stewardess or steward

flight host(ess) airline steward-(ess)

floral tribute wreath

food-order expediter short-order cook

food preparation center kitchen

food science specialist(s) short-order cook(s)

forensic laboratory morgue

foreplay genital pleasuring

fornicatrix prostitute

four-letter words so-called unmentionable Early English terms such as cock, crap, crud, cunt, dung, fart, fuck, lust, piss, puke, scum, shit, snot, spit, suck

freedom fighter terrorist; rebel; revolutionary

free enterprise capitalism; corporate control; multinational corporations imposing cartels and other controls

free society where it safe to be unpopular; where freedom of expression is protected

freethinker agnostic; atheist; disbeliever; *esprit fort* [French—strong spirit]; humanist; iconoclast; latitudinarian; *libero pensatore* [Italian—liberal thinker]; *librepensador* [Spanish—liberal thinker]; nonbeliever; rationalist; secularist; skeptic; truthseeker; unbeliever

freeway tax-supported toll-free expressway; tax-supported toll-free highway

French disease syphilis

French welcome smallpox or syphilis

friendly discussion argument

friendly house British euphemism for brothel

front-and-rear-end wipers babies' diapers

front-parlor girl the most attractive woman in a house of prostitution

fruits of treason convict being strung up, cut down when nearly strangulated, then disemboweled, beheaded, and the torso cut into four quarters by an ax

fuckin' morphadite fucking hermaphrodite (damn homosexual)

fuller figure fatter

full-figured fat; obese

funart funerary art

functions under the auspices of takes orders from; is sponsored by

funding creativity using taxes to support the arts

funeral decorations flowers

funeral director undertaker

fusilated executed by a firing squad

future points in time the future; when

future unpleasantness anticipated war

fuzz police

gabinetto (Italian—cabinet)—

toilet

Gallic disease syphilis

Gang of Five see Group of Five

Gang of Six Canada plus the Gang of Five

garbologist garbage collector

gastric distress belching and farting

gathered gathered unto death, died

gathered to his fathers died

gay bashing physical and verbal assaults on homosexuals

gay bit prostitute

gay bowel syndrome intestinal disruptions affecting homosexuals

gay boy (girl) (Australian euphemism) homosexual

gay deceiver cunt chaser, lecher, old goat, rapist, ravisher, whorehound, whoremaster, whoremonger, or woman chaser

gay house brothel (Victorian)

gay(s) homosexual(s)

G_d God

gelding castrated male horse

general paresis of the insane central nervous system syphilis

generously proportioned fat

genteelism euphemism

gentleman cow's excreta bullshit

gent's gentlemen's toilet

genuine simulated imitation

geriatric(s) old person(s)

German disease syphilis

gestation control abortion

get off your butt get off your buttocks (originally); get off your ass (currently)—stop loafing and get to work

get rid of kill

getting along in years aging

getting your girl in trouble impregnating your unmarried girlfriend

ghetto blaster portable radio/tape player

giblets edible entrails of fowl

GI bride foreign-born wife of an American soldier

gifted children intelligent and studious children

ginkitis ailment of the aged; general breakdown of mental and physical faculties; geriatric syndrome; old-gink's disease

girls women (in the United States such terms are considered uncomplimentary)

girl's room women's toilet

given the pink slip fired

given special treatment exterminated (in the vocabulary of Nazis as well as Soviet commissars)

give up the ghost die

give us the benefit of your present thinking what do you think?

giving informational numbers writing parking tickets

giving the whole picture describing entire situation fully

glasnost (Russian—openness)

glow Victorian euphemism for a woman's sweat

goatish dirty-old-man or unwashed-male smell

God-Damnation of the Gods Wagner's *Götterdämmerung*

god's acre cemetery

godsquad paramedic ambulance and crew; evangelists

going out with staying in bed with; dating

going to see a man about a dog going to the toilet

go lay a cable go to the toilet to defecate

gold gold-tinted Acapulco marijuana

golden gauntlet early retirement

golly God

gone dead

gone on vacation suspended from active duty

gone to a better world died

gone to brush their teeth gone to the toilet

gone to Davy Jones' locker drowned

gone to heaven died

gone to rest died

go(ne) to the bank to make a deposit go(ne) to the toilet

go(ne) to the bathroom go(ne) to the toilet

gone to the great beyond died

gongorisms overly ornate expressions in the style of the Spanish poet Luis de Gongora y Argote

goody-goodyism affected niceness

goo-goo eyes amorous glances

gorilla sexually-aggressive person

gosh God

gosling person callow and foolish as a young goose

go to one's reward die

got to be destroyed must be killed

go under die

governmentalese federal bafflegab

government scheme govern-

ment-sponsored program

grant-in-aid diplomatic euphemism for a handout; giveaway

grayfish shark meat

gray propaganda mixture of half truths and truths whose source is hidden

Greek love pederasty

Greek Row fraternity and sorority houses

greengoods (man) paper money; counterfeiter or passer of counterfeit money

green goose beginning prostitute

greenhouse greenhouse effect (carbon dioxide release causing atmospheric changes)

grief therapist funeral director skilled in comforting the family and friends of the dead

grisette (French—salesgirl; working girl)—prostitute

Group of Five Britain, France, Japan, the US, and Germany (concerned with stabilization of the dollar and mutually-profitable trade agreements)

Group of Seven Britain, Canada, France, Italy, Japan, the US, Germany

group sex orgy

gu genitourinary

guest house boarding house

guest of the governor inmate in a state penitentiary

gunk aerosols, glues, and solvents inhaled by would-be addicts

gustatorial distinctions bitter, salty, sour, or sweet

had an accident defecated or urinated in bed or clothes

hairologist barber; hair stylist; hairdresser

hair stylist barber

halitosis bad breath

hamburger ground meat

handicapped crippled or retarded people

hardbody targeted missile hidden from infrared detection by huge plume of gas and heat

hardcore to the max sexually explicit to the maximum

hard drugs addictive substances causing psychological and/or physical harm

hard liquor beverage high in alcoholic content such as akavit, arrack, brandy, cognac, gin, pisco, rum, tequila, vodka, or whisky

hares edible rabbits

harlot prostitutes

harmonica mouth organ

harmonize it musician's phrase meaning fake it

harvesting marine mammals killing dolphins, dugongs, manatees, polar bears, porpoises, sea lions, seals, walruses, and whales

harvest worker fruit-and-vegetable picker

hash hashish

have to sharpen my skates have to go to the toilet

haystack haycock

head the head (the toilet)—originally the head of a ship where human wastes dropped over the vessel's bow from an open or partially enclosed toilet bench

health alteration assassination

hearing problem deafness

heated argument fist fight

heaven-forsaken idiot goddamned fool

heavy metal driving and loud music produced electronically

Hebrew ancient Semitic language; Jewish person

heli-hiking helicopter transport to remote areas where one wants to hike

hemped hanged

hempen collar hangman's noose

hempen cravat hangman's noose

hemp stretcher hangman

hepcat person knowing all the answers

her women's toilet

here (Afrikaans—gentlemen) —men's toilet

heren (Dutch—gentlemen)— men's toilet

Her (His) Majesty's carriage prison van

heroic measures life-sustaining medical procedures

herrar (Swedish—gentlemen) —men's toilet

Herren (German—gentlemen) —men's toilet

herrer (Danish—gentlemen)— men's toilet

hers women's toilet

high intoxicated on alcohol or other drug

high achiever(s) good student(s)

high coefficient of slip slippery

highly scenario-dependent success of a plan depends on whatever develops

high negatives aspects that detract from a candidate's popu-

larity

high positives aspects that enhance a candidate's popularity

high-risk activities life-threatening practices

high-risk behavior life-threatening practice

hirondelle de nuit (French—night swallow)—prostitute

hirsute adornments hairy appendages such as beards, mustaches, long tresses, sideburns (also called louseladders); unisexual hairdos

his men's toilet

his crown was shorn he was beheaded

Hispanic time slower than standard time; also known as mañana time or Mexican time

histrionic art acting

holiday headache alcohol-induced ailment

Holy Toledo Holy Jesus or Holy Moses

hombres (Spanish—men)—men's toilet

homes (Catalan—gentlemen)—men's toilet

homeys persons from the same hometown or region

hommes (French—gentlemen)—men's toilet

homogenized adulterated or watered

honorarium fee; payment for services

hooker prostitute

hookshop brothel

horizontalist prostitute

hospital environmental services janitorial work

hot blooded lustful

house apes other people's unhousebroken children

house guest boy friend; girl friend; live-in lover

household technician cook or other domestic servant

house of all nations brothel offering women of every color and nationality

house of assignation brothel wherein rooms were assigned to prostitutes and their clients

house of confusion brothel

house of correction criminal reformatory; reform school

house of ill fame brothel

house of ill repute brothel

house of joy brothel

house trailers mobile homes

housitosis household halitosis (unpleasant odors)

Hudson seal muskrat fur

humane shelter dog pound; place for stray cats, dogs, and other creatures

human-factor engineering designing equipment for maximum user comfort; ergonomics

humanism doctrine based on doing good without the necessity of believing in God or receiving rewards in the hereafter

humanities cultural arts, languages, literature, and philosophy

human resources people

hustler male prostitute

hygienic apparatus *see* bidet

hygienic paper toilet paper

hype hyperbole (gross exaggeration)

hypersexualist oversexed person

iatrogenic disorders ailments inadvertently induced by a physician

ice crystal methamphetamine (smokable form of speed)

ice-cream fruit Peruvian cherimoya

I'd appreciate it very much do not expect me to pay you or to reciprocate

I have a little problem with what you did I dislike what you did

I hear you I understand what you are saying

illegitimate bastard; child(ren) born of illegitimate parents

illicit love adultery; prostitution; rape

illicit lover prostitute; anyone not married to the person he or she is having sex with

I'll let you go now shut up and get off the telephone

imbibed too much drank too much

immoderate voluptuary sex fiend

immolation a fiery death symbolizing extreme sacrifice (*see* self-immolation)

immurement burial

I'm not crazy about it I don't like it

impaired person person suffering from Alzheimer's, Parkinson's, Pick's, or pre-senile dementia involving loss of memory and even everyday words; drunk(s)

impecunious broke; penniless; without funds

impregnated knocked up;

made pregnant; pregnant

impure extraneous matter scum

inactive colon atonic constipation

in an interesting condition pregnant

incarcerate imprison; jail

inclusionary building low-cost housing included in building-development projects funded in part or wholly by federal, state, or local funds

incontinence inability to control the bladder

increase your equity increasing your indebtedness in the hope of profit

incrustation build-up of crud (filth, grease, refuse of any kind)

incursion hostile entrance into a territory; invasion; raid

indisposed recovering from an alcoholic hangover; suffering menstrual-cycle pains; unavailable

indisposed at the moment on the toilet

individual confinement solitary imprisonment

individualized learning center student's classroom desk

in durance vile imprisoned

industrial action sit-in; slowdown; strike

industrial consultant(s) lobbyist(s)

industrial disharmony strike

industrial espionage stealing company formulas or secrets pertaining to manufacture or sale

industrialist(s) successful speculator(s)

inebriated drunk

inflation depression in the purchasing power of your currency; cost increase

informant a snitch, informer

information processing duplicating and typing pool

information scientist librarian

information specialist librarian

infrastructure fundamental framework of an organization or system; permanent facilities such as airports, bridges, gas and electric lines, highways and streets, railways, sewers, tv cables, water mains, etc.

inheritance tax death duty

inmate(s) prisoner(s)

inner city economically deprived residential areas formerly called barrios, ghettos,

or slums; also called the innercore city
inoperate broken down
inoperative statement a lie
in pocket to hold drugs
in point of fact in fact
in reasonable supply available
in short supply scarce
insinuendo insinuated
instantaneous respiratory arrest death
institutional superintendent(s) prison warden(s)
intact virginal
intellectually underprivileged uninformed
intelligence acquisition spying
intensive care unit locked unit for juvenile delinquents
inter bury
interactive discrimination system transmitting neutral particle beams to distinguish decoys from warheads emitting gamma rays
interdic interdiction (destruction of an enemy's line of supply)
interment burial
intermodal interface when you get off the plane, ship, or train a bus awaits
internal exile Soviet term meaning exile in Siberia or some other remote area
international disposal man hired killer working for a government espionage agency
internationalization of norms conscience
interpretive dancer(s) striptease dancer(s)
interrogation equipment devices and drugs used to obtain confessions
intestinal distress cramps and/or gas and/or loose bowels
intestinal flatus see borborygmus
intestinal fluidity loose and watery bowel movements; botulism, diarrhea, or dysentery
intestinal fortitude guts; stomach
in the can in the toilet
in the education field teacher
in the family way pregnant
in the time period when
intimacy adultery; sexual intercourse
intimate relations sexual intercourse
intimate wear underwear
intoxicated drunk; under the influence of drugs

intoxication drunkenness
in trouble about to be arrested; pregnant
intrusion detector burglar alarm
intrusion device burglar's tool
inverted homosexual
investment consultants stockbrokers
investor-owned hospitals hospitals run for profit
invisible handicap deaf and/or dumb
invisible risk deafness-producing noise generated by aircraft, traffic, loud music and other sounds
involuntary audience captive audience
Iranamok see Contragate
Iranscam see Contragate
ironmongery department Her (His) Majesty's prison
irregularity irregular menstrual period
irritable colon spastic constipation
isolation booth(s) prison cell(s)
it girl woman with sex appeal
it has long been known take my word for it
it is believed I think
it is generally believed two other guys agree with me
it is thought I think
I wobbly wobbly Chinese immigrants' name for the IWW (Industrial Workers of the World) in 1905
jackfruit durian
jagger tattoo artist
jakes toilet (British)
janets middle-aged widows (see jennifer)
jappy derived from Jewish-American Prince(s) or Princess(es)—young generation living in luxury provided by doting parents
jaundiced eyes hateful eyes; prejudiced eyes
jennifer young woman who marries an older man
jezebel abandoned and shameless woman; a prostitute
jive bitches female troublemakers
job action sit-down or strike
job problem demoted or fired
john prostitute's male client; toilet
joint brothel; prison
journey's end death
Judas priest Jesus Christ!
juice joints beer bars, cocktail lounges, saloons, etc.

junglish jungle English, bafflegab
junior executive(s) clerk(s)
junior wolf teenage philanderer
junkie drug addict
kaleidoscopic career checquered career
Katzenjammer (German—alcoholic hangover headache; tuning-up sounds produced by a symphony orchestra)
kept in custody jailed
keyboard type; input on a computer
key money bribery (to get an apartment)
kick the bucket commit suicide; die
kick over the traces cast off all restraint
kick upstairs promote to a higher but less sought-after position
kick up your heels have a good time
kit injection paraphernalia
kleptomaniac shoplifter
Klosett das Klosett (German—the closet)—the water closet or toilet
knocked up pregnant
kvinner (Norwegian-women)—women's toilet
labor unrest picket lines; sabotage; sit-downs; slowdowns; wide-scale discontent resulting from layoffs, poor wages, and poor working conditions
lack of suppression of paradoxical sleep unsuppressed drowsiness due to bad side effects of some pills
ladies ladies in waiting; old women; older women; most women
ladies' ladies' toilet
ladies' limbs women's legs
ladies' lunch lament the man shortage (all the good men are either gay or married)
ladies' man womanizer; gigolo
ladies of the night prostitutes
ladies' room women's toilet
lady cow's excreta cowshit
lady dog bitch
lady of the evening prostitute
laid back easy-going; relaxed
laid off fired
laid to rest buried
la langue du Coca-Cola (French—the Coca-Cola language)—English
land developer land despoiler; builder
landfill garbage dump
landscape architect gardener

landscaping specialist gardener
lane bowling alley
language arts reading, speaking, and writing
lapin rabbit fur (French *lapin*— rabbit)
lapsus calami (Latin—slip of the pen)—error exposed and reduced to writing
lapsus linguae (Latin—slip of the tongue)—saying what you meant to say but didn't want to say; Freudian slip
lapsus memoriae (Latin—lapse of memory)—in acute instances this is called Alzheimer's disease or pre-senile dementia
large maritime targets oil tankers
large naval targets aircraft carriers
lascar East Indian sailor
last curtain call death
last debt death
last mile walk from the prison cell to the place of execution
last obsequies funeral
last roundup death
last sleep death
last taboo incest
last waltz condemned prisoner's march to the place of execution
late lamented remains corpse
late unpleasantness American Revolution; Civil War or War of the Secession
latrine outdoor toilet
laugh lines facial wrinkles between the nose and mouth
launched into eternity died
lav lavatory (toilet)—common British euphemism
lavabo (Spanish—washroom) —toilet
Lavabos de Homens (Portuguese—gentlemen's lavatory) —men's toilet
Lavabos de Senhoras (Portuguese—ladies' lavatory)— women's toilet
lavatorium toilet
lawmen detectives, investigators, law-enforcement officers, police, secret service, sheriffs, or other security forces
lazy colon constipation
learning facilitator teacher
lecher satyr; sex fiend, wolf
left the scene died
legal problems drunken-driving arrests
legislative work lobbying
less affluent people people lacking financial resources

lessie(s) lesbian(s)
less than able unable
less than accurate inaccurate
less than adequate inadequate
less than appetizing repulsive; unappetizing
less than artistic inartistic; ugly
less than attractive unattractive
less than bearable unbearable
less than beautiful plain or ugly
less than believable improbable; incredible; unbelievable
less than broadminded bigoted; narrowminded
less than candid crafty; disingenuous; misleading; scheming; untruthful
less than charismatic uncharismatic; without charm or the look of leadership
less than charming disagreeable
less than convincing unconvincing
less than decisive undecisive
less than delicious burned; half baked; nauseating; sour; unwholesome
less than diligent lazy, sloppy, careless
less than effective ineffective
less than elegant inelegant; shoddy
less than enthusiastic unenthusiastic
less than exciting dull as dishwater; unexciting
less than fastidious dirty; sloppy
less than gentlemanly ungentlemanly, impolite
less than honest crooked; dishonest
less than honorable (discharge) dishonorable (discharge)
less than hospitable inhospitable; standoffish; unfriendly
less than ideal awful; terrible; unbearable
less than informative cryptic; obscure; uninformative
less than intelligent dumb; stupid; unintelligent
less than interesting dull; uninteresting
less than knowledgeable uninformed; unknowledgeable
less than ladylike unladylike
less than lawful illegal
less than legal illegal
less than lovable hateful; unlovable
less than navigable impassible
less than palatable inedible; repulsive

less than perfect short of ideal
less than quenched unquenched, thirsty
less than receptive unreceptive
less than reliable unreliable
less than sensitive cool, brusque
less than sheltered unsheltered, open
less than strong weak, ineffective
less than sublime shabby
less than a success a failure
less than supportive uncooperative
less than sympathetic hostile; uncomprehending; unsympathetic
less than thrilled depressed; disappointed; uninterested
less than total enthusiasm lukewarm, cool, indifferent, halfhearted
less than totally cooperative reluctant
less than truthful false; lying; untruthful
lethal assistants deadly weapons
liaison illicit sexual connection
libation alcoholic drink
liberal shibboleths liberal delusions
liberal spender spendthrift
liberty cabbage sauerkraut
lieu lieux d'aisance (French— comfort room)—toilet (or in British slang, *loo*)
lifeblood of the industrialized world oil
lift attendant elevator operator
light-fingered gentry pickpockets; thieves
likvidatsiya (Russian—liquidation)—execution
linguistic fig leaves euphemisms
liquidate kill
liquidation assassination
liquidation of undesirable elements murder
liquid refreshment alcoholic drink, beverage
literary agent author's representative
literary fig leaves euphemisms
little boy's room men's toilet
little girl's room women's toilet (*see* little boy's room)
live-in live-in friend, unmarried lover
live-in companion a concubine, gigolo, kept man or woman, mistress, paramour
live-in lover boyfriend; girlfriend

living wage equitable compensation
loaded under the influence of a drug
loan expert pawnbroker
locus of evaluation classroom
lonely couch companionless bed; half-empty bed; solitary bed
loo toilet (British)
lorette prostitute
lose the number of one's mess die
lost at sea presumed drowned
Lost Generation people who came of age just after World War I and lacked cultural, political, or social stability
lost her husband her husband died; was widowed
lost his wife his wife died
lounge toilet
love child child born of illegitimate parents
loved one corpse
love handles flesh over hips
low achiever poor student
low economic background poor
lower ability group slow learners
lower back ass; backside; behind; buttocks
lower chest abdomen; belly; guts
lower the boom fire
low frustration tolerance easily irritated
low-income poor
low-income neighborhood barrio; ghetto; slum
lowing herd cattle
low priced cheap
lubritorium automotive service station
lues cholera, the plague, syphilis
luetic infection of the central nervous system central nervous system syphilis
lung affliction tuberculosis
L-word liberal
machine population owners of tape and video disc players
machinery drug-injection paraphernalia
madam the operator or owner of a brothel
madamas (Italian—ladies)—women's toilet
maiden lady old maid
make one's exit die
make-out artist(s) fornicator(s)
make water pee; piss; urinate
making water leaking; urinating

maladjusted youths backward or troublesome youngsters
male above position wherein man is above woman partner
male below position wherein man is below woman partner
male organ penis
mañana time see Hispanic time
mandate League of Nations euphemism covering the awarding of territory formerly German or Turkish to the Japanese or the Russians
manufactured houses mobile homes
Marble City a small town in Oklahoma; the cemetery
marketing analysis sales promotion
marketing engineer sales person
marketing manager sales manager
marketing representative salesperson
marquee tent
mass vigilance widespread police-informant system
measurable end products results
meat-eater's euphemisms beef—cow or steer meat; cabrito—young goat meat; lamb or mutton—sheep meat; pork—pig meat; veal—calf meat
media access center the library
media coordinator someone who operates a cassette player-recorder, motion-picture camera or projector, tv equipment, etc.
medical center hospital
medical examiner coroner
meditation solitary confinement
megacity metropolis with a million or more people
mellowspeak soft-spoken euphemisms
member of the lower socioeconomic bracket poor person
member(s) of a minority group non-white(s)
memorial park cemetery
mendacious tendency given to lying
mendacity lie
menn (Norwegian—men)—men's toilet
men's men's toilet
men's lounge men's toilet
men's restroom men's toilet
mental institution insane asylum; psychiatric hospital
mentally disturbed crazy; psychotic
meowsic language of cats

meretricious traffic prostitution
méski (Polish—gentlemen)—men's toilet
messieurs (French—gentlemen)—men's toilet
metallic age attained by older people who discover they have silver in their hair, gold in their teeth, copper pennies in their purse, iron rust in their gut, and lead in their bottom
met in an unfriendly manner collision of aircraft, automobiles, boats or other vehicles
met one's maker died
metropolitan rehabilitation center downtown penitentiary or federal prison
Mexicali Revenge loose bowels in border cities such as Mexicali, Baja California
Mexicancellation Mexican divorce
miced all crytee Christ Almighty
micturate urinate
midlifers 35- to-55-year-old people also called the sandwich generation with responsibilities for their children and elderly parents
mild irregularity constipation; irregular menstrual flow
minimal prescriptiveness with fewest number of restraining rules blocking monetary grants to interest groups or states
minor misstatement lie
misadventure shootout resulting in death
misalliance mismatch
miscarriage spontaneous abortion
misconduct adultery
Miss Emma morphine
misspeaking inadvertently mislead; blunder
mistress kept woman
mixed bag assortment
mixer(s) bartender(s)
mná (Gaelic—women)—women's toilet
mobile home house trailer
moderately priced cheap
modified limited hangout temporary office or headquarters
molest make indecent sexual advances
money laundering concealing source of funds
monthly pains menstrual pains, cramps
monument memorial tombstone

moonchild(ren) person(s) born under the zodiacal sign of Cancer

moon rock mixture of crack and heroin

mopping up military operation involving capturing or killing of enemy personnel

moral renovation the honest way; truthfully

moral turpitude depravity; immorality

morbus gallicus (Latin—French disease)—syphilis

mortical surgeon undertaker

mortician burier; funeral director; undertaker

most effective painkiller death

mother homosexual dope pusher; madam of a brothel; motherfucker

motion discomfort motion sickness; nausea; seasickness

motion discomfort container vomit bag; vomit bucket

motivating through fear coercion

motor vehicle operator chauffeur; driver

mountain herring whitefish

mountain oysters bull, hog, or sheep testicles used as food and famed for their imagined aphrodisiac powers

ms multiple sclerosis

mujeres (Spanish—women)—women's toilet

multiprisoner transportation unit black lady; black maria; paddy wagon; police patrol van

multiversities large universities

mumblespeak mumbled bafflegab

Murphy's Law anything that can go wrong will go wrong, nothing is lost until you look for it

muzhay (Bulgarian—men)—men's toilet

muzi (Czechoslovakian—men)—men's toilet

my back teeth are almost floating I must urinate

myrmidons warlike people

my statement is inoperable I lied

mythomaniac liar

nappies babies' napkins; diapers

narco narcotics officer

narco-military military system protecting narcotics trafficking

nasal discharge snot

nasal exudate snot

nasal mucus snot

nasal problem loss of smell

natal day birthday

national assistance financial aid and surplus food for the poor and the unemployed

national razor the guillotine of France

national test bed high-tech computer and video operation simulating space-battle scenarios

natural euthanasia death by starvation

naturally fed breast-fed

nature break time to go to the toilet

naughty house brothel; house of prostitution; whorehouse

nautch girl dancing girl of India; prostitute

navigational errors bombing friendly territory

Neapolitan pox syphilis (*see* French disease)

neck oil whiskey

necktie hangman's noose

necktie party lynching; lynch mob

necropsy autopsy, post mortem

negative patient outcome death

negative savers people who spend more than they earn

negotiate the price bargain

neighborhood revitalization slum clearance

neither here nor there irrelevant, nowhere, unimportant

neoplasm tumor

nerve agents lethal chemical agents attacking the body's nervous system

nervous need nervous need to go to the toilet

nervous stomach bad breath augmented by belching, burping, chronic constipation, and farting

nether garments underwear

netherworld underworld; world of the dead

neuter castrate; denut; remove or tie off procreative organs

neutralize assassinate; kill

new ro new romantic music

news specialist(s) reporter(s)

Nicaraguan eggs iguana lizard eggs dried, pickled, or fried

Nicaraguan nut nipper German shepherd dog trained for police work in quelling riots and hard-to-handle mobs

nice-nellyisms euphemisms

night house brothel

night-soil excreta; feces; human

dung

No 1 urine

No 2 excrement

No Code do not resuscitate

nocturnal emissions wet dreams

noisic ear-splitting and monotonous modern music

nök (Hungarian—women)—women's toilet

no longer with us dead

non-ambulatory bedridden; unable to walk

nonch no chance; objectionable

non compos mentis (Latin—not of sound mind)—mentally incompetent

nondiscernible microbionoculator concealed poison-dart gun

nonmarital sex sex before marriage

nonterrorists unarmed and uninvolved civilian victims of terrorism

nonwhite African-American; Hispanic mestizo or mulatto; oriental

nose candy cocaine

nose paint alcohol (whose prolonged use breaks the arterioles of the skin and leaves a reddened nose)

nose powder a powdered drug such as cocaine, any of several hallucinogens, or heroin when snorted

not acceptable forbidden; not tolerated; unlawful

not all that impressive unimpressive

not an exact statement, to put it mildly a lie, to put it bluntly

not completely accurate inaccurate, false

not completely truthful untruthful, disingenuous

not entirely accurate inaccurate

not entirely safe risky

not entirely satisfied dissatisfied

not entirely truthful untruthful

Notice of Reduction in Force pink slip inserted in the pay envelope to announce layoffs

not my favorite person someone I dislike intensely

not overly bright stupid

not precisely accurate inaccurate; untrue, false

not precisely truthful untruthful, lying

not quite kosher dishonest (in the ethical sense); unclean (in the culinary or physical sense)

not so young elderly; old
not too accurate inaccurate
not too attractive ugly
not too profitable unprofitable
not too tall from average height to short
not too truthful untruthful
nude-encounter parlor brothel; massage parlor
nudie film featuring nudity
nuptial couch marriage bed
nuptials marriage ceremonies
nut factory insane asylum; mental hospital; psychiatric hospital or ward
nuthouse insane asylum or psychiatric hospital
nutpicker(s) psychiatrist(s)
nutria coypu fur (from a giant South American rodent)
nuttery insane asylum
nymphomaniac lustful female
oblique love adultery; clandestine action; secret love
obsequies funeral ceremonies
obstetric paramedic midwife
obstitute part-time prostitute; substitute for a prostitute; whore
occasional interruption acute constipation
OD overdose
odoriferous smelling; stinking
official gibberish bureaucratic bafflegab
offshore production overseas investments in foreign countries
old goat a dirty old man; a satyr; a whorehound; a woman chaser; lecher
old hare rabbit meat
old sparky electric chair
olfactoristic smelly
one-armed bandits slot machines
one's spirit departed one died
on the nod sleepy sensation following use of drugs
opportunity school school for the handicapped or retarded
optical problem blindness; failing eyesight
ordure shit (often used to include urine, corruption, contamination, and snot)
organ male organ; penis
organic receptors sense organs
orgone box orgasm box
orthographical errancy misspelling
orthotics orthopedic appliances used to support and brace weak or ineffective joints or muscles
osculation kiss(ing)

osmidrosis foul-smelling sweat
osteoporosis condition characterized by porosity and fragility of bones
other prisoner(s) prison guard(s)
Our Crowd New York City's German-Jewish elite
our wayward sisters Confederate States
outdoorsmen vagrants; homeless
outhouse backyard toilet; country privy
out in left field disoriented; out of touch with reality
outlaw party illegal nightclub
out of the ballpark out of touch; unaware of reality
out of town in prison
outplaced fired (from a job)
outright fabrication lie
out to lunch daffy
overachiever embarrassingly brilliant student; successful person
overindulgence gluttony
overkill possessing more than enough weapons needed to destroy enemy targets
over the blue wall confined to a hospital for the criminally insane
over the river dead
over the side buried at sea
overwrought drunken; agitated
oxometric analysis imaginary method for measuring the depth of the bullshit
oxymoron(s) combination of contradictory or incongruous words; seeming contradiction(s) e.g., peacekeeper missile(s)
p pee (urine)
pacification elimination of disturbances or disturbers of the peace; peace at the point of a bayonet
package store liquor store
paid the debt of nature died
pain-compliance device nunchaku
pain in the butt pain in the ass
Panamá chicken chicken-flavored iguana lizard tail
panel house whorehouse (whose bedrooms had sliding panels close to the bedstead in easy reach of an accomplice who could snatch a man's wallet or watch while he was preoccupied)
Pap smear Papanicolaou smear (cancer-screening procedure use to analyze cells removed

from the cervix)
paramnesia *déjà vu*
paramour (French—for love; illicit lover; a man's mistress)—prostitute
paranormal phenomena supernatural events; not scientifically explainable
paraphernalia papers, pipes, scales, and spoons used by drug addicts
park under construction dump
parrhesia freedom of speech; outspokenness; the opposite of euphemia
parsimonious frugal; niggardly; pennypinching; stingy
pass away die
passed over died
pass gas fart
pass into the unknown die
passive euthanasia letting the patient die
passive resistance response deliberate procrastination
pass on die
pass stool defecate
pass water empty the bladder; urinate
patriarchal structure social organization marked by male supremacy, dependence of women and children, and reckoning of descent and inheritance through the male line
pavement princess prostitute
paying guest boarder; lodger
peaceful person lazy or lethargic person
Peacemaker MX missile's euphemistic name
peace pill pcp (phencyclidine)
peculator thief
peculiar entertainment opera
peculiar institution slavery in the United States
peculiar members lexicographer Noah Webster's euphemism for testicles
peculiar service espionage
pecuniary distress poverty
pedophiliac adult molester attacking children sexually
pee-pee baby-talk for urine
peeping tom voyeur
pejorative demeaning or derogatory
pentagonese Pentagon bafflegab
people expressways sidewalks
people mover light-rail rapid-transit system
people's republic political prefix in many totalitarian nations run by communists, not

by the people
people with conspicuous access to wealth in their own right rich people
per capita income total income of a nation divided by the number of its inhabitants
perform one's ablutions shampoo, shave, and shower; wash oneself
period menstrual period
period of reconsideration cooling-off period
permanent color hair dye
perpetrator criminal; offender; suspect
personal emergency must go to the toilet immediately
personal protection dog attack dog
personal relationship sexual relationship
perspiration sweat
perspire sweat
pertussis whooping cough
Peruvian perfume cocaine
pet snatcher dogcatcher
p funk synthetic heroin
phallicism sex worship
phantom proliferators African, Asian, and South American would-be atomic powers
phased out brought to a halt; cancelled; closed; concluded; discontinued; ended; finished; retired; stopped; terminated
phased withdrawal scheduled retreat
Philippine Pacification Philippine Subjugation, according to Filipino historians
phone con(s) telephone conversation(s)
pigeon painting splashed droppings of pigeons
pimpmobile automobile converted for use by pimps and prostitutes
pink slip a dismissal notice
p in p piss in your pants
pixillated alcoholically befuddled
planned parenthood family planning
planned withdrawal defeat; retreat
plant food chickenshit; manure
plaque bacteria-laden mucus-hardening film endangering one's teeth
plaque invaders dental hygienists
plastic money credit cards
platform pushers persons who push others from the platform to the tracks of a railway or

subway when the train is coming into the station
pleasingly plump fat
plight one's troth pledge marriage; promise to marry
plongeur (French—diver or plunger)—dishwasher
ploy clever maneuver
plump fat
plunger hypodermic syringe
plural marriage polygamy
plural relations South African euphemism for *apartheid*
pneumatic bliss breast-cushioned fun-filled fornication
podium a platform designed to elevate a conductor or a speaker above the ensemble or the audience; a lectern
point in time moment; time
police state law and order imposed by law-enforcement agencies who overlook civil liberties
political unrest assassination; bombing; martial law; mass demonstrations; riots; sabotage; terrorist attacks; widescale subversion often leading to revolution
polychromatic alphasite colored chalk
polygraph test(ing) lie-detection test(ing)
poopoo cat or dog droppings; excrement; shit
poorly compensated poorly paid
poor man's nuclear weapon chemical weaponry; fuel—air bomb
popular price cheap
population-control equipment riot-control equipment such as high-pressure water hoses, rubber bullets, and tear gas
population limitation birth control
pop-up fast launching of a missile carrying a nuclear weapon to generate an X-ray laser able to shoot from space
portable john portable toilet
posey vest straitjacket
position situation job; job task
posterior backside; behind
post-traumatic stress syndrome condition characterized by a variety of neurotic and psychotic symptoms associated with the stress caused by any traumatic experience
potty chamber pot
poule de luxe (French—highly paid prostitute)—elegantly kept woman

poultry fertilizer chickenshit
powder her nose euphemistic phrase meaning a woman is going to the toilet
powder room toilet
power outage power failure
precursor bursts nuclear explosions in space designed to foil defense systems by creating a background of nuclear emissions, magnetic pulses, and heat capable of fooling sensors
predacide chemical agent for killing predatory animals
predigested mass bit of shit; hunk of crap; mass of manure
preemptive strike bombing the enemy before being attacked
pregnancy termination abortion
prelighten bleach the hair
premarital sex intercourse before marriage
prematurely retired fired
preorgasmic females frigid females
pre-owned previously owned; secondhand; used
preparation room undertaker's embalming room
press the flesh shake hands
prevaricate lie
prevarication lying
prevaricator(s) liar(s)
preventive detention jail without bail to assure defendants appear for trial and commit no crimes in the meantime
preventive war sneak attack
previously owned auto secondhand automobile; used car
prices you can afford prices salesmen tell you you can afford
pricey expensive
primary degenerative dementia Alzheimer's disease or presenile dementia characterized by progressive loss of memory
printouts computer-published results
prioritize arrange in chronological order; choose; select
prison officer gaoler (U.K.); jailer (U.S.)
private enterprise business
privates private parts (sexual and urinary organs)
privy outdoor toilet
problem child(ren) juvenile delinquent(s)
problem drinker alcoholic
problem pregnancy clinic abortion center

problem skin marred by acne, eczema, pimples, psoriasis, syphilitic scars, or their combination

pro-choice in favor of a woman's right to choose when to bear children

proglang(s) progressive language(s)

proliferation the spread of nuclear weapons

promenade princess prostitute

property control officer watchman

prophylactic device for preventing venereal infection; condom; a contraceptive device

proprietary institutions institutions, such as clinics and hospitals, run for profit

prosecutorial agent attorney general; city attorney; county attorney; district attorney; federal attorney; magistrate; public prosecutor; state prosecutor

protect cover up

protective bombing bombing the enemy first

protective coating engineer house painter

protective reaction bombing raids

protective reaction strike attack

protective residence fallout shelter

provisional period probation

prurience lust

pseudologist liar

psychiatric center insane asylum

psychic reader fortune teller

pub date publication date

public convenience outdoor or public toilet (British)

public house a bar where drinks and food are sold

public property property controlled by politicians

public relations adviser press agent

public relations assistant office receptionist

public utilities contraception and family-planning clinics (Dutch)

puffer auction booster

pulling the plug euthanasia; mercy killing; turning off life-support systems

pupil station desk

purloin steal

purple passage flowery speech or writing

purveyors of meat butchers

puta *prostituta* (Portuguese or Spanish—prostitute)

put one down bring about death as in the case of a pet; put to sleep

put to sleep kill

putt out of misery kill

quacksalver charlatan; fake physician; quack

quadruplication of the quadruplicate number 16

quality-control engineer inspector

quandong Australian fruit whose soft exterior covers a hard interior; prostitute

quarks elementary and unseen particles making up electrons, neutrons, and protons

quarter 25 dollars; 25-dollar narcotics-filled balloon

quean (Middle English—prostitute)

queen male homosexual

queen-size bigger than large but smaller than king-size

queer homosexual

quiet room solitary-confinement cell

quietus death

quite a few many, several

rabelaisian writer robust humorist noted for naturalism

racial hygiene eugenics

racial unrest race riots

raffs raffish bohemians

'rak arrack also called rack or raki

rap buddies easy-to-talk-to people

rap parlor place of prostitution

ratepayer taxpayer

rate payer tax payer

rather elderly old

rational defense planned warfare

ravisher one who captures and carries off another; one who is filled with strong emotions; one who uses force to carry off and rape a woman

realistic fees low fees

realtor real estate agent

rear end ass; buttocks

rear yard environment back yard

receipts strengthening raising taxes

recession depression

recommend the matter be studied action to move controversial issues to the back burner

reconciliation room church confessional

reconditioned secondhand

reconnaissance in force search out and destroy the enemy

rectification of the front retreat

redeployment retreat; withdrawal of armed forces

red-light district brothel section

redneck ultraconservative person

Red Team SDI scientists charged with developing and analyzing possible Soviet countermeasures

reduced circumstances poor

reduced to ashes cremated

reduction in the work force firing; layoff

re-education camps forced-labor work camps

reemployable annuitant specialist called back from retirement

reflation ascending inflation

registered warrant government-issued i-o-u issued when money is not available to pay for goods or services

regurgitated threw up; vomited

regurgitation upchucking; vomiting

rehabilitation center jail; lockup; penitentiary; prison

rehabilitation laboratory modern prison

relationship sexual affair

relocation specialist realtor

remains cadaver; corpse; dead person

remand institution jail; penitentiary; prison

remediation remedial reading

remittance person living on monies sent from home; ticket-of-leave man

remove exterminate

rent boys British teenage male prostitutes

repos (REPOS) repurchase agreements (Eurodollar deposits, foreign bank deposits, repurchase agreements)

repose beneath the sod die

repossessed taken back by creditor

residuals leftover funds, monies, or profits

resource center library

resources control bombing of dams, defoliation of forests, poisoning of sources of drinking water during the course of a war

rest home convalescent or retirement home

resting in marbletown buried in the cemetery
restroom toilet
resurrectionist grave robber
retire for the night go to bed
retirement allowance pension
retirement home institution catering to retired people
retiring room toilet
retreat toilet
retrete (Spanish—retreat)—toilet
returnee defector
revenue enhancement higher tax; tax increase
reverse engineering industrial espionage achieved by taking apart the products of competitors to find out how they work and how they may be imitated
revised upward increased
revolverize to shoot someone
rhumba steam sweat
riffed reduced in force (dismissed from duty; fired)
right decision decision favored by the public despite massive opposition by bureaucrats and lobbyists
right in the ballpark feasible
right-wing survivalists Ku Klux Klan and neo-Nazi groups
rip off cheat; overprice; rob; steal
roach cockroach; marijuana butt
Robin Hood rationing using revenues from the middle and upper classes to provide social programs for those in need
rock crack, cocaine
rodent operators rat catchers
rolling in the rye making love in the grass
rolling through the rye drinking and driving
roll in the hay copulation
roommates unmarried lovers
rooster(s) barnyard cock(s)
rotund fat
R-U 486 abortion pill made in France by Roussel-Uclaf
rubbed out assassinated
rubber mirror computerized thin-glass mirror on honeycomb panel controlled by mechanical arms and microchips enabling them to compensate for the distorting effects of the earth's atmosphere on laser beams
ruddy-faced drunk
run around the corner go to the toilet
sacrament of penance the rite of reconciliation (Roman

Catholic)
sailor's curse constipation
Saint James Version King James Version of the *Bible*
Salisbury steak World War-I euphemism for hamburger
saliva spit
sanctuary city any community hospitable to refugees
sandwich generation adults responsible for the support and care of both their children and their parents
sanitary engineer garbage collector
sanitation engineer garbage collector
sanitation person garbage and trash collector
sanitation smell odor of ordure; stench of ordure and urine
satyr satyrus (Latin—lecher)—person afflicted with satyriasis manifested by abnormal and uncontrollable sexual desire
sauce sanctuaries beer bars; cocktail lounges; saloons
sauna massage brothel (Danish)
scam non-violent confidence game involving pilferage, robbing, stealing
screw fornicate; fuck; copulate
screw and bolt love 'em and leave 'em
search and clear search and destroy
seasickness dizziness followed by nausea and vomiting
seclusion solitary confinement
secret disease syphilis
Section 8 discharged from the armed forces because of insanity or intoxication; government-subsidized decent, safe, and sanitary housing
secular humanism agnostic or atheistic humanism
secular humanist person believing in doing good outside the bonds of religious teachings; freethinkers, unbeliever
secularist agnostic; atheist; disbeliever; doubter; humanist; iconoclast; nonbeliever; rationalist; skeptic; truthseeker; unbeliever
security zone occupied territory
see a man about a dog go to the toilet
seed ox bull
selected out dismissed from duty; fired
selective use of violence to neutralize an intended victim assassination plan

self-enhancement masturbation
self-help masturbation
self-immolation sacrifice attained by setting oneself on fire or plunging into a fire
self-pleasuring masturbation
selling spring prostituting (Japanese)
senile exhibiting a loss of mental faculties
senior(s) senior citizen(s)—older person(s)
señoras (Spanish—ladies)—women's toilet
señores (Spanish—gentlemen)—men's toilet
sensual gratification sexual intercourse
sensuous desire lust
senyores (Catalan—ladies)—women's toilet.
senyors (Catalan—gentlemen)—men's toilet
separate from school expel
separate from the payroll fire
serious headache gunshot wound in the head
servicio (Spanish—service)—toilet
servicios de caballeros (Spanish—gentlemen's services)—men's toilet
servicios de damas (Spanish—ladies' services)—women's toilet
sexed to death fucked out
sextasy sexual ecstasy
sexual expressionist pornographer; purveyor of smut
sexual intercourse coitus; copulation; fucking; genital contact
sexually active lustful
sexually molested raped
sexual minorities perverts
sexual variety promiscuity
shading the truth lying
sharing group sex or orgy
sharpen one's skates go to the toilet
sharps needles, scalpels, pointed objects; glass pipettes, slide pipettes
she dog bitch
sheer prevarication outright lie
shell shock battle fatigue (in World War I); variety of psychotic and neurotic disorders associated with the stress of combat
shero female hero
sherutim legvarim (Hebrew—men's services)—men's toilet
sherutim lenashim (Hebrew—women's services)—women's toilet
she's a plain girl but she

means well she's ugly but not malicious
shibo shit
shinjuku (Japanese—brothel)
shizos Soviet solitary-confinement cells
shochu (Japanese—distilled saki) —beverage with high alcoholic content
shoe rebuilder shoe repairman
short shorts hot pants
short term 180 days or less; an IOU due in 6 months or less
shredding party destroying documents
shuffle off this mortal coil die
signore (Italian—ladies)
signori (Italian—gentlemen)— men's toilets
sin against chastity birth control
single unmarried
sink of iniquity brothel
sin tax fee on alcohol and tobacco products
sips beer, liquor, wine
skeptic agnostic; atheist; nonbeliever; unbeliever
skidlid motorcycle helmet
skillful inveracity artful lie
skimming secretly diverting unreported profits
skin doctor dermatologist, syphilologist
skinheads neo-Nazi groups that promote interracial violence
skin problems acne; blackheads; pimples; syphilitic sores
skin(s) *see* skinheads
skin trade peep shows, sexually explicit literature, and X-rated videos
slatternly, slovenly, sluttish, or untidy individual slob
sleep engineer(s) mattress maker(s)
sleepingly warm recently dead
sleeping partner of life death
sleep with copulate
slender skinny
slip out of life die
slippery slope difficult-to-defend argument
slumbermobile hearse
slumber robe funeral shroud
slumber room area in a funeral establishment where the embalmed may be viewed
sly language of evasion euphemisms
smack hard drugs such as heroin
small untruth little lie
smart rocks small kinetic-energy projectiles designed to

be hurled at missiles or warheads
smilers and defilers, reekers and leakers Ambrose Bierce's euphemistic description of dogs
smoking gun observable evidence
smust smog plus dust
snuff spoon tobacco snuff spoon—cocaine sniffing spoon, also coke spoon
s.o.b. son of a bitch
social ailments venereal diseases
social disease venereal disease
social diseases AIDS (Acquired Immune Deficiency Syndrome), chancroid, gonorrhea, granuloma inguinale, lymphogranuloma venereum, and syphilis
social indiscretion belching or farting in a public place within earshot of other people
socially disadvantaged poor
socially underprivileged poor
social maladjustment crime
social promotion advancing poor pupils so they will not suffer the embarrassment of sitting in classrooms with brighter and younger students; built-in time bomb of education
social prophylaxis Soviet euphemism for execution or imprisonment of dissidents or suspects
social safety net Medicare, Social Security, and unemployment compensation
social unrest demonstrations, hunger marches, hunger strikes, race riots, terrorism
sociobiology genetics
soft money currency of third-party barter; money available for only a short time; unrestricted political contributions
soiled linen dirty clothes
solecisms grammatical errors
somewhat advanced in years old(er)
somewhat less than credible incredible; unbelievable
somewhat less than honest dishonest
somewhat less than reliable unreliable
somewhat less than truthful untruthful
somewhat unlettered rather ignorant
sonderbehandlung (German— special treatment)—Nazi eu-

phemism for killing by gas
son of Onan masturbator
so perceived seen
souvenir hunters vandals
souvenir hunting vandalism
space mines orbiting explosives designed to threaten satellite defenses
Spanish disease syphilis, according to the Portuguese
speaks a low level of colloquialisms talks slang
special action murder
special agent(s) FBI or insurance agent(s)
specialized vehicle army tank
special psychiatric hospitals Soviet term for prisons where dissenters are kept
special transaction gambling bargain
sperm barrier condom
Spetsnaz Soviet undercover specialists in assassination and behind-the-lines counterinsurgency operations
spontaneous abortion miscarriage
sporting house brothel
spouse abuse domestic violence
Starspeak Star Wars language
Star Wars Strategic Defense Initiative
state chemist executioner charged with poisoning convicts in a gas chamber
state electrician executioner charged with electrocuting convicts in an electric chair
steatopygous fat-assed
steel-collar labor robots
steer manure bullshit, cow flops
stepped out for a few minutes gone to the toilet
still warm died recently
stomach condition acid indigestion; constipation; dyspepsia; heartburn
stomach distress acute constipation, diarrhea, nausea, or vomiting
stones testicles
stonewall act or argue defensively; conceal; cover-up; hide the truth
stool dung; fecal discharge
straightforward and aboveboard blunt, honest
straitened circumstances poor; of limited resources
Strategic Defense Initiative Star Wars
strategic deterrents weapons of war
strategic hamlet refugee camp

strategic misrepresentation lying; speaking with a forked tongue
strategic redeployment retreat
strategic withdrawal retreat
street hustler(s) male prostitute(s)
street of fallen women red-light area
street orderlies garbage and trash collectors
street people the homeless
street sweepers large magazine semi-automatic shotguns
streetwalker prostitute
stretches the truth exaggerates; lies
stretch hemp hang by the neck until dead
stretch the truth lie
struggle for preeminence in the Sicilian community Mafia-type gang warfare
submarine below street-level prison cell
subsidy publisher vanity press
substitute worker scab; strike-breaker
subversive delinquent political prisoner
subway pushers *see* platform pushers
succulent viands appetizing food
succumbed to hypertensive cardiovascular disease died of a heart attack
suddenly retired fired without warning
sugar shit; diabetes
suicycles all-terrain three-wheel vehicles
summer complaint diarrhea; loose bowels
sunshine units nuclear radiation units
supreme measure capital punishment; execution
supreme sacrifice giving one's life; losing one's virginity
surreptitious entry break-in
sweetbreads animal pancreas or thymus prepared as a gastronomic delicacy
swinger promiscuous person
swish effeminate male homosexual
symphonic plight also known as pp's or payless paydays
synergistic(al) advance(s) when the whole is said to be worth more than the sum of its parts
take a leak urinate
take a powder leave hurriedly; leave suddenly
take leave of abstinence get

drunk
takeoff inject narcotics
take one's last sleep die
take out of the picture die
take the electric cure be electrocuted
take to the tall tules evade arrest; hide out
talent promoter high-class pimp
tavern beer bar; cocktail lounge; saloon
tax reform often means tax increase
tax withholding earning sharing
tb tuberculosis
technical correction downward skidding of the stock market
technology transfer assistance giving advice
techspeak technical talk
teena teenage girl
teeno teenage boy
telegenic tv attractive
temporary work cessation layoff
termcert terminal certificate (death certificate)
terminal causing death or occurring at the end of life
terminal cell prison cell designed to kill the inmate(s)
terminal communication death certificate; death notice; order of execution; death warrant
terminal illness fatal illness
terminate fire; remove from job or office
terminate pregnancy have an abortion
termination of pregnancy abortion
termination with extreme prejudice assassination
terminological inexactitude lie
terpsichorean trauma dancer's pain
territorial extermination mass murder of Jews and others in German-occupied countries during World War II
Texas turkey armadillo meat
that point in time then
that time frame then
theater of war battlefield
The Company the CIA
Theory X managerial assumption that employees dislike work and must be coerced, controlled, or threatened to motivate them
Theory Y managerial assumption that employees do not dislike work and under proper conditions will accept and

seek out responsibilities to fulfill their esteem, self-actualization needs, and social needs
Theory Z an organization with a record of long-term employment, shared decision making, slow promotions, and non-specialized but varied job assignments
the pill oral contraceptive for women
therapeutic accident drug-induced death due to an overdose
therapeutic correctional community prison
therapeutic misadventure operation resulting in the patient's death
therapeutic termination abortion
thermalicide killing with nuclear weapons
these points in time now
thinking the unthinkable contemplating nuclear warfare
thinning out killing wildlife
third age old age (the first age is infancy through adolescence, the second age is post-adolescence through middle age)
Third Programme BBC radio or television schedule of classical music and intellectual talk shows
this is a serious issue a sin
this job offers great experience the pay is lousy and working conditions are the worst
this point in time now
thoroughly soiled defiled with dung, urine, and vomit
those points in time then
threat dog dog trained to bark and snarl at intruders
three-letter words Victorian arbiters of taste felt some three-letter words were as bad as four-letter words; these included leg (if a woman's), pee, pus, nut (if a testicle), sex, sin, and tit
three-wayer(s) person(s) engaging in sex anally, naturally, or orally
thrice threefold; tree times
thundermug chamberpot; pottie
time to retire time to go to bed
tissue toilet paper
tocador (Spanish—boudoir; dressing room; dressing table; vanity)—often means toilet
toilet powder scented talcum

powder
toilet tissue toilet paper
toilet water perfume-scented cologne
tonsorial artist barber
tonsorial expert barber
tonsorialist barber
tonsorial parlor barbershop
tootsie roll Mexican-made heroin, also called black tar or Mexican mud
tossed up the cookies vomited
total and complete immobilization assassination
totalitarian state tyranny
touched insane
tp toilet paper
trade of kings war, as defined by John Dryden
trainsickness nausea followed by vomiting; motion sickness
transient street person; vagrant
transient population vagrants
traumatic amputation of the lower extremities loss of the legs in an accident
tree surgeon tree trimmer
trendy chic
trial marriage free love
tribal scarification circumcision
trick prostitute's customer
triplication of the triple number 9
tropical eggs iguana lizard eggs dried, pickled, or fried
trots the trots (botulism, diarrhea, dysentery)
truthseeker agnostic; atheist; freethinker; nonbeliever; secularist; skeptic; unbeliever
truth stretcher liar
tubal ligation severing the female's Fallopian tubes (*see* vasectomy)
tube steak frankfurter; hotdog
turistas (Portuguese or Spanish—tourists)—traveller's diarrhea caused by toxic-producing bacteria found in contaminated food or water; if prolonged, a protozoan invasion may occur and produce amebic dysentery
turn on sell narcotics
turn one's face to the wall die
twenty toes ten toes up and ten toes down (British euphemism for a bedroom diversion of some antiquity)
two pc two-physician certification (of insanity)
ubornaya (Russian—adornment place)—toilet
udd's undisposed diapers; undumped diapers

umbles deer entrails made into umble or humble pie (*to eat humble pie* comes from the custom of assigning inferior parts of a game carcass to the servants)
umbrage to take offense
unaccompanied officer personnel housing bachelor officers quarters
unannounced social explosions belches, burps, and farts
unbeliever agnostic, atheist, disbeliever, doubter, freethinker, skeptic
uncharismatic without ability to command or lead
uncivil rude
unclothed naked, bare
underachiever unsuccessful person
underarm damp sweat
underdeveloped countries poor countries
underdeveloped nation(s) poor nation(s)
underfunded broke; in debt; lacking money
undergoing emotional retraining under psychiatric treatment
under par feels or looks awful; looks like death warmed over
underprivileged poor; unemployed
under the auspices of sponsored by
under the influence drunk, high, inebriated
under the sod dead and buried
under the weather drunk; sick
undesirable books banned or censored books
undies underwear
undocumented immigrant illegal alien
undocumenteds undocumented aliens
unemancipated minor teenager
unfortunate activity invasion of a country; takeover of a country; warlike preparations
unfortunate female prostitute
unimpressive height short
unit (Texan English) one hundred million dollars
unlawful deprivation of life killing
unlearned ignorant
unmarried wives concubines; kept women; paramours
unmentionable disease syphilis
unmentionables underwear
unobscene song(s) outspoken

songs
unpedigreed alley cat or mutt; one-hundred-percent pure cat or dog
unpeople depopulate
unpredictably nonplussed I've nothing to say
unprosperous poor
unrenewed discontinued
unsavory scandal morally offensive disgrace
unshreddables electronic documents
unslim fat; heavy; overweight
unstructured conversations friendly chats; informal talks
untoward clinical event bad side effect
untruism falsehood; lie
untruth lie
unwell menstruating
unwhisperables underwear
unworthy servant slave
upchuck vomit
upper amphetamine
upper frontal superstructures a woman's breasts
upset stomach nausea, acute constipation, or loose bowels
up the river Sing Sing prison, up the Hudson River from New York City
up to your elbows in alligators up to your ass in alligators
upward adjustment rise, increase
upward revision increase
urinal of the planets rain-filled Ireland
urinary stress incontinence leakage of urine during coughing, laughing, sneezing, or sexual arousal
urning(s) male homosexual(s)
utilitarian theory the right act is most likely to bring the greatest amount of pleasure to the most people
utter an untruth lie
uttering inexactitudes telling lies
utter peer rejection failure to get along with family, friends, neighbors, fellow workers
vacationing near Chappaqua Sing Sing prison
validated learning package approved textbook
vasectomy severing the male's vas deferens connecting the testes and seminal vesicles so no sperm can reach the ejaculatory duct (*see* tubal ligation)
vashzimmer (Yiddish—washroom)—toilet
vd venereal disease (AIDS,

chancroid, gonorrhea, granuloma inguinale, lymphogranuloma venereum, nongonococcal urethritis, or syphilis)

velvet prison censorship

venereal desire lust

venerist prostitute

venery pursuit of sexual pleasure; sexual intercourse

Venus's curse venereal disease

verification the process of confirming compliance with a treaty limiting nuclear arms

vertical transportation corps elevator operators

victims of target practice persons executed by a firing squad

vidrio inglés (Spanish—English glass)—dog dung

viewing the remains visiting a funeral parlor to see the deceased

violated raped

visceral reaction gut feeling

visiting spouse pill Chinese contraceptive

visually handicapped blind

visually impaired blind

vntr variable number tandem repeat

voice informant

volume reduction unit city or town dump

volumetrics number of square feet of a house

voluntary compliance paying taxes

voluntary return voluntary deportation (of illegal aliens)

voluntary termination of pregnancy abortion

voluptuarial excess sexual excess

waistcloth loincloth

walking papers dismissal notice

wallet biopsy examination of a patient's insurance coverage and financial status before admission to a hospital

wanaume (Swahili—men)—men's toilet

wanawake (Swahili—women)—women's toilet

war baby child born of illegitimate parents during wartime

War Between the States Civil War, also called the War of the Secession

warm glow hot sweat

wash one's hands go to the toilet

washroom facilities toilets

waste kill; rape

water methamphetamine

wáter (Spanish—water closet)—toilet

water closet(s) toilet(s)

Watergate English terminology invented by President Nixon's aides to cover up lies and to spar for time when undergoing congressional examination; *at this point in time* meant *now, in point of time* equivalent to *then* or *when,* and *my statement is inoperable* really meant *don't believe a word I say*

water qualm acid indigestion; heartburn; sudden faintness sometimes accompanied by nausea

wc (WC) water closet (toilet)

wc babati (Romanian—male water closet)—men's toilet

wc femei (Romanian—female water closet)—women's toilet

weed of wisdom hashish; marijuana; other cannabis products

weights dumbbells

well-filled out fat

well-nigh impossible almost impossible

well placed at the present level of management will never be promoted

welsh rabbit dish made of melted cheese, beer, and other ingredients poured over toast or served as a dip; incorrectly called welsh rarebit

went away died

went to sleep died

went west died; was killed

wet affairs (Soviet euphemism) assassinations

we will no longer detain you you're dismissed; you're fired; you may leave

when I'm gone when I'm dead

where the King goes alone bathroom

white-collar labor office workers

white goods bleached cotton and linen fabrics; colorless high alcohol-content drinks (aguardiente, akavit, brandy, cognac, gin, rum, schnapps, tequila, vodka); large household appliances (refrigerators, stoves, washing machines, etc.)

white meat breast

white propaganda identifiable and truthful dissemination of facts

white slave prostitute

wilderness health strip mining

wimp ineffectual weakling; in-

sipid and spineless person

winding cloth burial shroud

winning the West killing the Indians

wipe your bottom wipe your ass; wipe your backside

with balls between parentheses Shakespeare's description of a bow-legged man

with child pregnant

with facilities cabin (aboard ship) or hotel room with toilet, tub and/or shower, and washstand

without benefit of clergy living together but unmarried

without female friends male homosexual

without male friends female homosexual

wizzeewig with a wig

wolf to betray or deceive

wolf ticket bad report

womanizer cunt chaser; philanderer; skirt chaser; whoremaster; whoremonger; wolf, lecher

woman of the street prostitute

women's women's toilet

wooden interdental stimulator and particle remover toothpick

word from our sponsor radio or television advertisement

word processor text-editing computer technology program allowing electronic collecting, creation, and writing of articles, books, letters, manuscripts, and periodicals

work cessation strike; walkout; work stoppage

work cessation on the premises sit-down strike

working girl(s) prostitute(s)

write out a check go to the toilet

writing instruments expensive pens including fountain pens and pencils

Xanadu dream girl prostitute (term recalling the stanzas in *Kubla Khan* by Samuel Taylor Coleridge)

xawfully thanks awfully

X-chaser mathematician in search of the value of X; aspiring naval officer

xd deleted

xerotic wrinkled

Xeroxpoxed copier paper marred by black blotches

yellow peril Oriental conquest of Western civilization by economic penetration, immigration, and war

yellow rain biochemical poison dropped by enemy aircraft; deadly weapon consisting of chemicals and myotoxins

yellows nembutal

yellow suppuration pus

yesterlingo language of the recent past

you know where to go go to hell

you know where to put it shove it up your ass

young adult teenager

your eyes are changing your sight is failing

your services are no longer required you're fired

youthful offender juvenile delinquent

yuppies young upperly mobile professionals

yvette French prostitute

zap advertising jargon for alerting a customer by subliminal messages; kill quickly (as with a burst of machine gun or tommy gun bullets)

Zarzuela girl Hispanic playgirl; operetta performer

zealous Zelia a hardworking pavement-pounding prostitute

zeny (Czechoslovakian—women)—women's toilet

zero-defect system foolproof

system

zhenny (Bulgarian—women)—women's toilet

zilch nothing; zero

zilches misspellings and typographical errors

zipcuffed caught in your pants by a faulty zipper

zonk stun, stupefy

zonked drugged; exhausted

zoophilia erotica (Latin—erotic love of animals)—sexual intercourse with animals (usually men with cattle and women with dogs)

zymurgic outburst alcoholic belch; beer belch

Bell Code from Bridge or Pilothouse to Engine Room

The bell codes are used on ferries, launches, tugs, and other powered vessels.

1 bell ahead
2 bells stop

3 bells astern
4 bells full speed

Birthstones—Ancient and Modern

Relative Values. Diamonds, emeralds, rubies, and sapphires are termed precious stones; all the rest are semiprecious. Precious gems are minerals enhanced by the lapidary's art. The pearl, although not a stone, is classed with the gems and, depending on its beauty and size, may be as valuable as any of the precious stones.

	Ancient	*Modern*
January	garnet	garnet
February	amethyst	amethyst
March	jasper	aquamarine or bloodstone
April	sapphire	diamond
May	agate	emerald
June	emerald	alexandrite, moonstone, or pearl
July	onyx	ruby
August	carnelian	peridot or sardonyx
September	chrysolite	sapphire
October	aquamarine	opal or tourmaline
November	topaz	topaz
December	ruby	turquoise or zircon

British Counties Abbreviated

England, Northern Ireland, Scotland, and Wales

Adeen Aberdeenshire
Ang Angus
Angle Anglesey
Ant Antrim
Arg Argyll

Arm Armagh
Ayrs Ayrshire
Banffs Banffshire
Beds Bedfordshire
Ber Berwickshire

Berks Berkshire
Brecon Brecknock
Bucks Buckinghamshire
Bute County Bute
Caern Caernarvonshire

Caith Cathiness
Cambs Cambridgeshire and Isle of Ely
Card Cardiganshire
Carm Carmarthenshire
Ches Cheshire
Clack Clacmannanshire
Corn Cornwall
Cumb Cumberland
Denb Denbighshire
Derbys Derbyshire
Dev Devon
Dor Dorset
Down County Down
Dumf Dumfriesshire
Dunb Dunbarton
Dur County Durham
E Lothian East Lothian
E R Yorks East Riding, Yorkshire
Ess Essex
E Suffolk East Suffolk
E Sussex East Sussex
Ferm Fermanagh
Fife Fifeshire
Flints Flintshire
Glam Glamorgan
Glos Gloucestershire
Great Lon Greater London

Hants Hampshire
Herefs Herefordshire
Herts Hertfordshire
Hunts Huntingdon and Peterborough
Iness Inverness-shire
Kent County Kent
Kinc Kincardineshire
Kinross Kinross-shire
Kircud Kircudbrightshire
Lanarks Lanarkshire
Lancs Lancashire
Leics Leicestershire
Lincs Lincolnshire (Holland, Kesteven, Lindsey)
Lond County Londonderry
Merion Merioneth
Mloth Midlothian
Mon Monmouthshire
Mont Montgomeryshire
Moray County Moray
Nairns Nairnshire
Norf Norfolk
Northants Northamptonshire
Northld Northumberland
Notts Nottinghamshire
N R Yorks North Riding, Yorkshire
Ork Orkney Islands

Oxon Oxfordshire
Peebl Peebleshire
Pemb Pembrokeshire
Perths Perthshire
Rad Radnor
Renf Renfrewshire
Ross Ross and Cromarty
Rox Roxburghshire
Rut Rutland
Selk Selkirkshire
Shet Shetland Islands
Shrops Shropshire
Som Somerset
Staffs Staffordshire
Stir Stirlingshire
Sur Surrey
Suth Sutherland
Tyr Tyrone
Warks Warwickshire
Westmld Westmorland
Wig Wigtownshire
Wilts Wiltshire
W Lothian West Lothian
Worcs Worcestershire
W R Yorks West Riding, Yorkshire
W Suffolk West Suffolk
W Sussex West Sussex

Canadian Provinces

Alb Alberta (inhabitants called Albertans)
BC British Columbia (British Columbians)
Man Manitoba (Manitobans)
NB New Brunswick (New Brunswickers)

Nfld Newfoundland (Newfies, Newfoundlanders, or Labradorans)
NS Nova Scotia (Nova Scotians)
NWT Northwest Territories (Territorials)
Ont Ontario (Ontarians)

PEI Prince Edward Island (Prince Edward Islanders)
Qué Québec (Québecois)
Sask Saskatchewan (Saskatchewanians)
Yuk Yukon Territory (Yukoners)

Capitals of Nations, Provinces, Places, and States

Afghanistan Kabul
Aguascalientes Aguascalientes
Alabama Montgomery
Alaska Juneau
Albania Tirana
Alberta Edmonton
Alderney Alderney
Algeria Algiers
American Samoa Pago Pago
American Virgin Islands Charlotte Amalie
Andorra Andorra la Vella
Angola Luanda
Anguilla The Valley
Antigua and Barbuda St John's
Argentina Buenos Aires

Arizona Phoenix
Arkansas Little Rock
Armenia Yerevan
Aruba Orangestad
Australia Canberra
Australia's Northern Territory Darwin
Austria Vienna
Azerbaijan Baku
Azores Angra do Heroísmo, Horta, and Ponta Delgada
Bahamas Nassau
Bahrain Manama
Baja California Mexicali
Baja California Sur La Paz (capital of the Southern Territory of Baja California

—Territorio Sur—abbreviated BC Sur)
Balearic Islands Palma de Mallorca
Bangladesh Dhaka
Barbados Bridgetown
Belgium Brussels
Belize Belmopan
Benin Porto-Novo
Bermuda Hamilton
Bhutan Thimphu
Black Forest Freiburg, Germany
Bolivia La Paz (de facto) and Sucre (legal)
Bophuthatswana Mmabatho
Botswana Gaborone

Brazil Brasilia
British Columbia Victoria
British Virgin Islands Road Town
Brunei Bandar Seri Begawan
Bulgaria Sofia
Burkina Faso Ouagadougou
Burundi Bujumbura
Byelorussia Minsk
California Sacramento
Cambodia Phnom Penh
Cameroon Yaoundé
Campeche Campeche
Canada Ottawa
Canada's Northwest Territories Yellowknife
Canary Islands Las Palmas
Cape Verde Islands Praia
Cayman Islands Georgetown
Central African Republic Bangui
Chad N'djamena
Chiapas Tuxtla Guitiérrez
Chihuahua Chihuahua City
Chile Santiago
China Beijing (Peking)
Christmas Island Flying Fish Cove Settlement
Ciskei Zwelitsha
Coahuila Saltillo
Cocos Islands West Island
Colima Colima
Colombia Bogotá
Colorado Denver
Comoros Moroni
Congo Brazzaville
Connecticut Hartford
Cook Islands Rarotonga
Corsica Ajaccio
Costa Rica San José
Côte d'Ivoire Abidjan
Croatia Zagreb
Cuba Havana
Cyprus Nicosia
Czechoslovakia Prague
Delaware Dover
Denmark Copenhagen
Distrito Federal México City
Djibouti Djibouti
Dominica Roseau
Dominican Republic Santo Domingo
Durango Durango
Ecuador Quito
Egypt Cairo
El Salvador San Salvador
England London
Equatorial Guinea Malabo
Estonia Tallinn
Ethiopia Addis Ababa
Faeroe Islands Thorshavn
Falkland Islands Stanley
Fiji Suva
Finland Helsinki
Florida Tallahassee
France Paris

French Guiana Cayenne
French Polynesia Papeete
Gabon Libreville
Gambia Banjul
Georgia Atlanta
Georgia (Soviet) Tbilisi
Germany Bonn
Ghana Accra
Gibraltar Gibraltar
Greece Athens
Greenland Nuuk
Grenada St George's
Guadeloupe Basse-Terre
Guam Agaña
Guanajuato Guanajuato
Guatemala Guatemala City
Guernsey St Peter Port
Guerrero Chilpancingo
Guinea Conakry
Guinea-Bissau Bissau
Guyana Georgetown
Haiti Port-au-Prince
Hawaii Honolulu
Hidalgo Pachuca
Highlands Inverness, Scotland
Honduras Tegucigalpa
Hong Kong Victoria
Hungary Budapest
Iceland Reykjavik
Idaho Boise
Illinois Springfield
India New Delhi
Indiana Indianapolis
Indonesia Jakarta
Iowa Des Moines
Iran Teheran
Iraq Baghdad
Ireland Dublin
Isle of Man Douglas
Israel Jerusalem
Italy Rome
Ivory Coast Abidjan
Jalisco Guadalajara
Jamaica Kingston
Japan Tokyo
Jersey St Helier
Jordan Amman
Kansas Topeka
Kazakhstan Alma-Ata
Kentucky Frankfort
Kenya Nairobi
Kirghizia Frunze
Kiribati Tarawa
Korea (North) Pyongyang
Korea (South) Seoul
Kuwait Kuwait
Laos Vientiane
Latvia Riga
Lebanon Beirut
Lesotho Maseru
Liberia Monrovia
Libya Tripoli
Liechtenstein Vaduz
Lithuania Vilnius
Louisiana Baton Rouge
Lower Saxony Hannover, Ger-

many
Luxembourg Luxembourg
Macau Macau
Madagascar Antananarivo
Madeira Funchal
Maine Augusta
Malawi Lilongwe
Malaysia Kuala Lumpur
Maldives Malé
Mali Bamako
Malta Valletta
Manitoba Winnipeg
Martinique Fort-de-France
Maryland Annapolis
Massachusetts Boston
Mauritania Nouakchott
Mauritius Port Louis
Mayotte Dzaoudzi
Mexico Mexico City
Michigan Lansing
Michoacán Morelia
Minnesota Saint Paul
Mississippi Jackson
Missouri Jefferson City
Moldavia Kishinev
Monaco Monaco-Ville
Mongolia Ulaanbaatar
Montana Helena
Montserrat Plymouth
Morelos Cuernavaca
Morocco Rabat
Mozambique Maputo
Myanmar (Burma) Yangon
Namibia Windhoek
Nauru Yaren
Nayarit Tepic
Nebraska Lincoln
Nepal Katmandu
Netherlands Amsterdam
Netherlands Antilles Willemstad
Nevada Carson City
New Brunswick Fredericton
New Caledonia Nouméa
Newfoundland St John's
New Hampshire Concord
New Jersey Trenton
New Mexico Santa Fé
New South Wales Sydney
New York Albany
New Zealand Wellington
Nicaragua Managua
Niger Niamey
Nigeria Lagos
Niue Alofi
Norfolk Island Kingston
North Carolina Raleigh
North Dakota Bismarck
Northern Ireland Belfast
Norway Oslo
Nova Scotia Halifax
Nuevo León Monterrey
Oaxaca Oaxaca
Ohio Columbus
Oklahoma Oklahoma City
Old California Monterey

Oman Muscat
Ontario Toronto
Oregon Salem
Orkneys Kirkwall on Pomona Island
Pakistan Islamabad
Panama Panama City
Panama Canal Balboa Heights
Papua New Guinea Port Moresby
Paraguay Asunción
Pennsylvania Harrisburg
Peru Lima
Philippines Manila
Pitcairn Adamstown
Poland Warsaw
Portugal Lisbon
Portuguese China Macao
Prince Edward Island Charlottetown
Puebla Puebla
Puerto Rico San Juan
Qatar Doha
Québec Québec City
Queensland Brisbane
Querétaro Querétaro
Quintana Roo Chetumal
Réunion Saint-Denis
Rhode Island Providence
Romania Bucharest
Rwanda Kigali
Saint Helena Jamestown
Saint Kitts Basseterre
Saint Lucia Castries
Saint Pierre and Miquelon St Pierre
Saint Vincent Kingstown
Samoa Apia
San Luis Potosí San Luis Potosí
San Marino San Marino
São Tomé and Principe São Tomé
Sark La Collinette
Saskatchewan Regina
Saudi Arabia Riyadh
Scotland Edinburgh

Senegal Dakar
Seychelles Victoria
Sierra Leone Freetown
Sinaloa Culiacán
Singapore Singapore
Slovenia Ljubljana
Solomon Islands Honiara
Somalia Mogadishu
Sonora Hermosillo
South Africa Bloemfontein (judicial), Cape Town (legislative), Pretoria (administrative)
South Australia Adelaide
South Carolina Columbia
South Dakota Pierre
Spain Madrid
Sri Lanka Colombo
State of México Toluca
St Christopher (Kitts) and Nevis Basseterre
St Kitts and Nevis Basse-Terre
St Vincent and Grenadines Kingstown
Sudan Khartoum
Suriname Paramaribo
Swaziland Mbabane
Sweden Stockholm
Switzerland Bern
Syria Damascus
Tabasco Villa Hermosa
Tadzhikistan Dushanbe
Taiwan Taipei
Tamaulipas Ciudad Victoria
Tanzania Dar es Salaam
Tasmania Hobart
Tennessee Nashville
Texas Austin
Thailand Bangkok
Tlaxcala Tlaxcala
Togo Lomé
Tonga Nuku'alofa
Transkei Umtala
Trinidad and Tobago Port-of-Spain
Trust Territory of the Pacific Saipan
Tunisia Tunis

Turkey Ankara
Turkmenistan Ashkhabad
Turks and Caicos Islands Grand Turk
Tuvalu Funafuti
Uganda Kampala
Ukraine Kiev
United Arab Emirates Abu Dhabi
United Kingdom London
Uruguay Montevideo
U.S.A. Washington, DC
USSR Moscow
Utah Salt Lake City
Uzbekistan Tashkent
Vanuatu Vila
Vatican City State Vatican City
Venda Thohoyandou
Venezuela Caracas
Veracruz Jalapa
Vermont Montpelier
Victoria Melbourne
Vietnam Hanoi
Virginia Richmond
Virgin Islands Charlotte Amalie
Wales Cardiff
Wallis and Futuna Matautu
Washington Olympia
Western Australia Perth
Western Kentucky Paducah
Western Sahara El Aaiún
Western Samoa Apia
Westphalia Münster, Germany
West Virginia Charleston
Wisconsin Madison
Wyoming Cheyenne
Yemen Sana
Yucatán Mérida
Yugoslavia Belgrade
Yukon Whitehorse
Zacatecas Zacatecas
Zaire Kinshasa
Zambia Lusaka
Zimbabwe Harare

C-B (Citizen's-Band) Radio Frequency Shortwave Call Signs

10-1 receiving you poorly
10-2 receiving you well
10-3 stop transmitting, channel in use
10-4 OK, message received
10-5 relay message
10-6 busy, can't talk now, stand by
10-7 out of service; going off air
10-8 in service, subject to call,

working well
10-9 repeat message
10-10 transmission completed, standing by
10-11 talking too rapidly
10-12 visitors are present
10-13 advise weather, road conditions
10-14 time by the clock
10-16 make pickup at
10-17 urgent business

10-18 anything for us?
10-19 nothing for you, return to base
10-20 my location is:
10-21 contact me by phone
10-22 make personal contact with
10-23 standby
10-24 assignment completed
10-25 contact another station by radio

10-26 disregard last transmission
10-27 I am moving to channel
10-28 identify your station
10-29 time is up for contact
10-30 violates regulations
10-31 no longer in violation of regulations
10-32 I will advise re signal readability
10-33 EMERGENCY TRAFFIC AT THIS STATION
10-34 TROUBLE AT THIS STATION, HELP NEEDED
10-35 matter of urgency but cannot discuss it by radio
10-36 transmission or event is scheduled for
10-37 send tow truck
10-38 ambulance needed at
10-39 your message was delivered
10-41 please tune to channel
10-42 traffic accident at
10-43 traffic congestion at

10-44 I have a message for
10-45 stations on this channel please identify
10-46 assist motorist
10-50 break channel
10-55 intoxicated driver
10-60 what is next message number?
10-62 unable to copy, use phone
10-63 network directed to
10-64 network is clear
10-65 awaiting your next message
10-66 cancel message
10-67 all units comply
10-68 repeat message
10-69 message received
10-70 fire at
10-71 proceed with transmission in sequence
10-73 speed trap at
10-74 negative
10-75 you are causing interference

10-77 negative contact
10-81 reserve hotel room for
10-82 reserve room for
10-84 my telephone number is
10-85 my address is
10-88 advise phone number of
10-89 radio repairman needed at
10-90 I have tv interference
10-91 talk closer to mike
10-92 your transmitter is out of adjustment
10-93 check my frequency on this channel
10-94 please give me a long count
10-95 transmit dead carrier for 5 seconds
10-97 check test signal
10-99 mission completed, all units secure
10-100 rest-room stop
10-200 police needed at

Chemical Element Symbols, Atomic Numbers, and Discovery Data

Symbol	Element	Atomic Number	Discovered
Ac	actinium	89	1899 by Debierne
Ag	silver *(argentum)*	47	Before the Christian Era
Al	aluminum	13	1825 by Oersted
Am	americium	95	1944 by Seborg and others
Ar or A	argon	18	1894 by Raleigh and Ramsay
As	arsenic	33	13th century by Magnus
As	astatine	85	1940 by Corson and others
Au	gold *(aurum)*	79	Before the Christian Era
B	boron	5	1808 by Davy
Ba	barium	56	1808 by Davy
Be	beryllium	4	1798 by Vauquelin
Bi	bismuth	83	15th century by Valentine
Bk	berkelium	97	1949 by Thompson, Ghiorso, and Seborg
Br	bromine	35	1826 by Balard
C	carbon	6	Before the Christian Era
Ca	calcium	20	1808 by Davy
Cd	cadmium	48	1817 by Stromeyer
Ce	cerium	58	1803 by Klaproth
Cf	californium	98	1950 by Thompson and others
Cl	chlorine	17	1774 by Scheele

Cm	curium	96	1944 by Seborg and others
Co	cobalt	27	1735 by Brandt
Cr	chromium	24	1797 by Vauquelin
Cs	cesium	55	1861 by Bunsen and Kirchoff
Cu	copper *(cuprum)*	29	Before the Christian Era
Dy	dysprosium	66	1886 by Boisbaudran
Er	erbium	68	1843 by Mosander
Es	einsteinium	99	1952 by Ghiorso and others
Eu	europium	63	1901 by Demarcay
F	fluorine	9	1771 by Scheele
Fe	iron *(ferrum)*	26	Before the Christian Era
Fm	fermium	100	1953 by Ghiorso and others
Fr	francium	87	1939 by Perey
Ga	gallium	31	1875 by Boisbaudran
Gd	gadolinium	64	1886 by Marignac
Ge	germanium	32	1886 by Winkler
H	hydrogen	1	1766 by Cavendish
Ha	hahnium	105	1970 by Ghiorso and others
He	helium	2	1895 by Ramsay
Hf	hafnium	72	1923 by Coster and Hevesy
Hg	mercury *(hydrargyrum)*	80	Before the Christian Era
Ho	holmium	67	1879 by Cleve
I	iodine	53	1811 by Courtois
In	indium	49	1863 by Reich and Richter
Ir	iridium	77	1804 by Tennant
K	potassium *(kalium)*	19	1807 by Davy
Kr	krypton	36	1898 by Ramsay and Travers
La	lanthanum	57	1839 by Mosander
Li	lithium	3	1817 by Arfvedson
Lu	lutetium	71	1907 by Welsbach and Urbain
Lw	lawrencium	103	1961 by Ghiorso and others
Md	mendelevium	101	1955 by Ghiorso and others
Mg	magnesium	12	1830 by Bussy and Liebig
Mn	manganese	25	1774 by Gahn
Mo	molybdenum	42	1782 by Hjelm
N	nitrogen	7	1772 by Rutherford
Na	sodium	11	1807 by Davy
Nb	niobium (formerly columbium)	41	1801 by Hatchett
Nd	neodymium	60	1885 by Welsbach
Ne	neon	10	1898 by Ramsay and Travers
Ni	nickel	28	1751 by Cronstedt
No	nobelium	102	1958 by Ghiorso and others
Np	neptunium	93	1940 by Abelson and McMillan
O	oxygen	8	1774 by Priestley and Scheele
Os	osmium	76	1804 by Tennant
P	phosphorus	15	1669 by Brandt
Pa	protactinium	91	1917 by Hahn and Meitner

Pb	lead *(plumbum)*	82	Before the Christian Era
Pd	palladium	46	1803 by Wollaston
Pm	promethium	61	1945 by Glendenin and Marinsky
Po	polonium	84	1898 by P and M Curie
Pr	praseodymium	59	1885 by Welsbach
Pt	platinum	78	1735 by Ulloa
Pu	plutonium	94	1940 by Seborg and others
Ra	radium	88	1898 by P and M Curie
Rb	rubidium	37	1861 by Bunsen and Kirchoff
Re	rhenium	75	1925 by Noddack and Tacke
Rf	rutherfordium	104	1969 by Ghiorso and others
Rh	rhodium	45	1803 by Wollaston
Rn	radon	86	1900 by Dorn
Ru	ruthenium	44	1845 by Claus
S	sulfur	16	Before the Christian Era
Sb	antimony *(stibium)*	51	1450 by Valentine
Sc	scandium	21	1879 by Nilson
Se	selenium	34	1817 by Berzelius
Si	silicon	14	1823 by Berzelius
Sm	samarium	62	1879 by Boisbaudran
Sn	tin *(stannum)*	50	Before the Christian Era
Sr	strontium	38	1790 by Crawford
Ta	tantalum	73	1802 by Eckeberg
Tb	terbium	65	1843 by Mosander
Tc	technetium	43	1937 by Perrier and Segre
Te	tellurium	52	1782 by von Reichenstein
Th	thorium	90	1828 by Berzelius
Ti	titanium	22	1789 by Gregor
Tl	thallium	81	1861 by Crookes
Tm	thulium	69	1879 by Cleve
U	uranium	92	1789 by Klaproth
V	vanadium	23	1830 by Sefström
W	tungsten (wolfram)	74	1783 by d'Elhuyar brothers
Xe	xenon	54	1898 by Ramsay and Travers
Y	yttrium	39	1794 by Gadolin
Yb	ytterbium	70	1878 by Marignac
Zn	zinc	30	Before the Christian Era
Zr	zirconium	40	1789 by Klaproth

Civil and Military Time Systems Compared

Civil	Military		Civil	Military
12.01 A.M. =	0001		12.01 P.M. =	1201
12.02 A.M. =	0002		12.02 P.M. =	1202
12.03 A.M. =	0003		12.03 P.M. =	1203

12.04 A.M.	=	0004		12.04 P.M.	=	1204
12.05 A.M.	=	0005		12.05 P.M.	=	1205
12.15 A.M.	=	0015		12.15 P.M.	=	1215
12.30 A.M.	=	0030		12.30 P.M.	=	1230
12.45 A.M.	=	0045		12.45 P.M.	=	1245
1.00 A.M.	=	0100		1.00 P.M.	=	1300
1.15 A.M.	=	0115		1.15 P.M.	=	1315
1.30 A.M.	=	0130		1.30 P.M.	=	1330
1.45 A.M.	=	0145		1.45 P.M.	=	1345
2.00 A.M.	=	0200		2.00 P.M.	=	1400
3.00 A.M.	=	0300		3.00 P.M.	=	1500
4.00 A.M.	=	0400		4.00 P.M.	=	1600
5.00 A.M.	=	0500		5.00 P.M.	=	1700
6.00 A.M.	=	0600		6.00 P.M.	=	1800
7.00 A.M.	=	0700		7.00 P.M.	=	1900
8.00 A.M.	=	0800		8.00 P.M.	=	2000
9.00 A.M.	=	0900		9.00 P.M.	=	2100
10.00 A.M.	=	1000		10.00 P.M.	=	2200
11.00 A.M.	=	1100		11.00 P.M.	=	2300
12.00 noon	=	1200		12.00 midnight	=	2400

Climatic Region Symbols

Typical Climatological Regional Divisions Worldwide

Climatic Symbols	Climatic Regions	
Af Am	*Tropical Rainforest*	Tropical rainforests of the Amazon and Middle America from southern Mexico to Colombia and the West Indies; Congo and the Guinea Coast of Africa; jungles of Ceylon, India, Indonesia, Madagascar, Malaya, the Philippines, Southeast Asia
Aw	*Tropical Dry and Wet*	Grassy savannas of Middle America; llanos of eastern Colombia and southern Venezuela; campos of south-central Brazil; damp lowland savannas of Africa and its dry uplands; plains of northern Australia, Burma, India, Pakistan, Southeast Asia
Bsk Bwk	*Midlatitude Dry*	Great plains and prairies of Canada and the United States; arid plains of Patagonia; pampas of Argentina, Bolivia, Paraguay, and Uruguay; Gobi and Takla Makan desert dunes of Asia; Kirghizian steppe of Turkestan; Ukrainian steppe
Bwh	*Tropical Dry*	Afghan, Arabian, Atacaman, Australian, Kalihari, Sahara, Somali, Sonoran, and other subtropical and tropical desertlands of the world
Caf	*Humid Subtropical*	Southeastern United States; northern Argentina; southern Brazil, Paraguay, Uruguay; southeast Africa; southeastern China; southern Japan; eastern Australia
Cfb	*West Coast Marine*	Pacific Northwest of Canada and the United States; southern Chile; west coast of Norway and south coast of Sweden; British Isles and northwestern Europe including northern

		Spain; south coast of South Africa; southeast coast of Australia; New Zealand
Csa	*Mediterranean Subtropical*	Southern California; central Chile; Mediterranean region including Portugal and most of Spain, southern France, Italy, Yugoslavia, Albania, Greece, Turkey, parts of Morocco and Algeria, much of Israel; Cape of Good Hope area around Cape Town, South Africa
Daf	*Humid Continental*	Southern Canada and the northeastern United States plus much of the Midwest; much of the Soviet Union and the eastern section of China
Dcf	*Continental Subarctic*	Alaska and northern Canada; Siberia and the northern USSR from the Arctic Ocean to the North Pacific Ocean
E	*Tundra*	Arctic coasts of Alaska, Canada, Greenland, northernmost Europe and Asia from northern Norway to easternmost Siberia
Ef	*Polar Icecap*	Interior of Greenland; Antarctica's northernmost tip
H	*Highland*	High valleys and mountainous areas of the world where climatic conditions are so variable they almost defy classification

Climatic Symbols Explained

A	Hot and moist equatorial or tropical climate
B	Dry climate with evaporation greater than precipitation
C	Moist and warm with well-defined summer and winter seasons
D	Cold and snowy subarctic with northern boundary the northern limit of forest growth—the taiga
E	Ice climates of the icecaps where ice and snow are perpetual or of the tundra where the growing season above the permafrost is very short
H	Highland climates in mountainous regions where weather conditions are extremely variable and difficult to classify
a	Long and hot summers
b	Short and wet winters
c	Cool or short and moderate summers
d	Very cold and dry winters
f	Moist the year around
h	Hot and moist most of the year
k	Cold and dry most of the year
m	Monsoon conditions
s	Dry summers and wet winters
w	Wet summers and dry winters

Corrections Facilities and Organizations

"Am I my brother's keeper?"
—**Genesis 4:9**

This section covers correctional and penal institutions of every kind, ranging from custodial schools for delinquent juveniles to halfway houses; included are police-station lockups, county jails, and penitentiaries as well as penal colonies and rehabilitation centers. Parole and probation services are included, as are the slang names given by many inmates or former convicts. Many entries are toponyms—place-names used to describe the institution.

All of the more than 150 nations of the world are mentioned and many have several entries. All entries, except for numbered ones, are in alphabetical order.

A-to-z ready reference to correctional facilities, halfway houses, jails, penitentiaries, prisons, and reformatories around the world; sequel to the *Crime Dictionary* compiled by Ralph De Sola.

AACFO American Association of Correctional Facility Officers (publishes *The Correction Officer* newsletter)

AACP American Association of Correctional Psychologists (publishes *Journal of Criminal Justice and Behavior*)

AACTP American Association of Correctional Training Personnel

AAEOCJ American Association of Ex-Offenders in Criminal Justice

AAMHPC American Association of Mental Health Professionals in Corrections

AARC Association for the Advancement of Released Convicts

AAWS American Association of Wardens and Superintendents (formerly Wardens Association of America founded in 1870)—publishes *The Grapevine*

Abbotsford town southeast of Vancouver and location of the Matsqui Institution

Aber Aberdeen, Scotland and its prison

Aberdeen Aberdeen jail in Hong Kong, Scotland, South Dakota, and the state of Washington

Abidjan (*see* Ivory Coast's prisons)

absent without leave or permission awol (AWOL)

Abu Dhabi (*see* United Arab Emirates)

AC Administration of Correction (Puerto Rico)

ACA American Correctional Association (publishes books and the periodical *Corrections Today*)

Acapulco Mexican prison in a popular seaside resort

ACCA American Correctional Chaplains Association

Accra (*see* Ghana's prisons)

ACFSA American Correctional Food Service Association

ACHSA American Correctional Health Services Association

aci (ACI) adult correctional institution

ACJ Arlington County Jail (Virginia)

Acklington HM Prison at Acklington, Northumberland, England

ACRIM Association for Correctional Research and Information Management

ACTO Advisory Council on the Treatment of Offenders

Acuña jail in Ciudad Acuña, México

Adana (*see* Turkey)

Addis Ababa (*see* Ethiopia's prisons)

Adee Adelaide, South Australia or its jail

Adelaide Adelaide Gaol in South Australia's capital city

Aden (*see* Yemen)

Adirondack Camp Adirondack (minimum-security prison in the Adirondack Mountains near Lake Placid, New York)

adjustment center segregated center of any prison; used for the protection of inmates who refuse to be intimidated by prison gangs but cannot defend themselves

administrative segregation solitary confinement

Adobe (often pronounced *'Dobe*) Adobe Mountain School for juvenile delinquents in Phoenix, Arizona

Adrian Michigan city site of the Adrian Training School

Adrian Training School medium-security coeducational penal facility in Adrian, Michigan

Adult Diagnostic and Treatment Center Avenal, New Jersey

Adult Training Centre Milton, Ontario

Aeolian Islands Isole Eolie or the Lipari Islands used as Italian convict colonies in the Mediterranean

aerial surveillance (*see* ASTREA)

AFOSP Air Force Office of Security Police

Afyon Afyonkarahisar (Turkish —Black Castle of Opium) —prison in western central Turkey, where opium is grown

Agaña capital of Guam and location of the Adult Correctional Facility, the Community Correction Center, Cottage Homes, Juvenile Hall, the Juvenile Justice School, and the Agña Lockup as well as the U.S. Navy Brig

Agaña Lockup Guam's jail

Agassiz British Columbia town east of Vancouver containing the Kent Institution as well as the Mountain Institution

Agra Agra Central Prison in

Uttar Pradesh, India

Agua Prieta Agua Prieta, Sonora and its jail across the border from Douglas, Arizona

Aguascaliente prison in Aguascaliente, Mexico

AI Adult Institutions (New Hampshire); Amnesty International (London-based international organization concerned with the release of political prisoners)

Aiea Halawa High Security Facility at Aiea on Oahu Island, Hawaii

air dancing hanging (execution also known as air jigging, air polka dancing, air rhumba dancing, etc.)

aislamiento penal (Spanish—penal isolation)—solitary confinement

AJA American Jail Association

AJCA Association of Juvenile Compact Administrators

AJIS Automated Jail Information System

Akron Akron, Ohio or its Summit County jail

Alabama jails scattered throughout the state's 67 counties from Prattville, Autauga County, to Double Springs in Winston County; each county has jail a plus those in Birmingham, Huntsville, Mobile, and Montgomery

Alabama State Training School at East Lake for female delinquents

Alameda Alameda County Prison and Rehabilitation Center near San Francisco, California

Alaska (*see* Anchorage, Eagle River, Fairbanks, Juneau, Ketchikan, Nome, Palmer)

Alaska lockups the state's 29 divisions each has a lockup plus the jails of Anchorage and Fairbanks

Albania prisons are located in Tirana and elsewhere

Albany names for prisons in the Albanys of California, Georgia, Indiana, Kentucky, Missouri, New York, Oregon, and Texas as well as in England and Australia

Albany, Georgia (*see* USMC)

Albany overseas prisons in England, the original Albany, and southwest Australia

Alberta Alberta Institution for Girls (Canadian prison facility)

Albertslund Herstedvester Detention Centre at Albertslund, Denmark

Albion Albion State Institution and Western Correctional Facility at Albion, New York

Albuquerque site of the New Mexico Youth Diagnostic Center and the Re-Integration Center

Albuturkey slang for Albuquerque, New Mexico and its jail

Alcalá de Henares women's prison northeast of Madrid in central Spain

Alcatraz former maximum-security prison of the United States on Alcatraz Island in San Francisco Bay; today National Park Service guides conduct sight-seeing tours through its old cell blocks

Alcatraz replaced U.S. Penitentiary at Marion, Illinois

Aldershot British military-training post and prison holding soldiers discharged from service with ignominy

Alderson Federal Correctional Institution at Alderson, West Virginia (for women serving lengthy sentences)

Aleppo (*see* Syria)

Alex slang for Alexandria, and its jail or prison

Alexander Alexander Youth Service Center in Alexander, Arkansas with a capacity for 96 coed juvenile delinquents

Alexandria place-name of a jail, lockup, or prison in Australia, Egypt, Louisiana, or Virginia

Alexis Ravelin maximum-security section of the Peter-Paul fortress-prison of St Petersburg during czarist times

Algeria prisons dating from the years of French-colonial domination may be seen in Algiers, Oran, Constantine, Annaba, and even in smaller places such as Arzew

Algiers capital city of Algeria, its prison

Alhambra Alhambra Reception and Treatment Center for incoming adult felons, in Phoenix, Arizona

Alice The Alice—Alice Springs, Northern Territory, Australia—and its jail

Allentown Lehigh County, Pennsylvania, courthouse and jail

Allenwood Federal Prison Camp at Allenwood, Pennsylvania

alley corridor or hallway between cell rows

Almacen Tia Moreno Quito, Ecuador's penitentiary

a l'ombre (French—in the shadow)—in jail or in prison

Alston Wilkes Society organization aiding families of inmates in South Carolina

Alyce D McPherson School for coeducational juvenile delinquents in Ocala, Florida

Amache Japanese-American relocation center near Granada, Colorado, where they were interned after Pearl Harbor

Amarillo north Texas city, Potter County courthouse and jail

Amenia Amenia Center for Girls at Amenia, New York

American Association of Correctional Facility Officers (*see* AACFO)

American Association of Correctional Psychologists (*see* AACP)

American Association of Wardens and Superintendents (*see* AAWS)

American Correctional Association originally the National Prison Association and later the American Prison Association (*see* ACA)

American Journal of Correction formerly *Prison World*

American People for American Prisoners (*see* APAP)

American prisons (*see* individual entries by Bureau of Prisons name; city, county, or state name; or nickname)

America's Devil's Island post–Civil-War nickname of the military prison at Fort Jefferson on the Dry Tortugas about 65 miles (105 kilometers) west of Key West in the Gulf of Mexico

Am Jour Corr American Journal of Correction

Amman (*see* Jordan's prisons)

Amnesty International (*see* AI)

Amsterdam (*see* Netherlands)

'Aña Agaña, Guam's capital or its jail

Anamosa site of The Men's Reformatory in Iowa, east of Cedar Rapids

Anchorage Anchorage Correctional Center in Anchorage, Alaska, together with the An-

nex holding maximum-security felons

Andersonville Confederate prisoner-of-war camp in Georgia where nearly 14,000 Union prisoners lost their lives due to overcrowded conditions and lack of good food

Andrade Andrade, Baja California and its jail

Angleton Texas location of Retrieve Unit opened in 1919

Angola Southwest African country with old prison in Luanda and jails in smaller places; site of the Louisiana State Penitentiary

Angolite bimonthly publication by prisoners in Angola, Louisiana state prison

Ankara capital of Turkey and nickname of its several jails and prisons

ankles ankle shackles

Annaba (*see* Algeria)

Annadale New Jersey site of the Youth Correctional Institution opened in 1929

Anna's Hope Anna's Hope Detention Center on St Croix in the American Virgin Islands

Anniston Calhoun County seat and jail in Alabama

Antakya city or prison in southernmost Turkey

Antananarivo (*see* Madagascar's prisons)

Anteroom of Auschwitz nickname of Dutch transit camp established by the Nazis in World War II (*see* Lager Westerbork)

anti-penetration glazing built to resist blowtorches, gunfire, and sledgehammers (*see* detention glazing)

antiquity of Newgate site of London gaol as early as 1190

antisocial offender(s) convict(s); criminal(s)

Anto Antofagasta, Chile jail

APA Adult Parole Authority; American Prison Association; Association of Paroling Authorities

Apalachee Apalachee Correctional Institution in Sneads, Florida

APAP American People for American Prisoners (in overseas jails and prisons)

APFO Association on Programs for Female Offenders

Apia (*see* Samoa)

APPA American Probation and Parole Association

Appleton Thorn HM Prison at Appleton Thorn in Lancashire, England

Ararat Ararat Prison in Victoria, Australia

Arcadia south-central Florida site of the De Soto Correctional Institution

Archambault Institution maximum-security facility at Ste Anne des Plaines, Québec

Argentina is replete with prisons and jails—Buenos Aires, Cordoba, Rosario, La Plata, etc.

Arizona (see Adobe, Alhambra, Catalina, Florence, Fort Grant, Phoenix, Tucson)

Arizona Girls School correctional facility at Phoenix for juvenile delinquents from 8 to 21 years of age

Arizona jails the state's 14 county lockups are augmented by jails in Phoenix and Tucson; the old Territorial Prison outside Yuma is a tourist attraction

Arkansas (see Alexander, Cummins, Pine Bluff, Tucker)

Arkansas lockups the state's 75 counties are served by jails for the county seat; city facilities serve Little Rock

Armagh HM Prison at Armagh in Northern Ireland and the penal facility for female offenders

Armley jail in Leeds, Yorkshire, England

Arohata Arohata Women's Borstal Institution, Wellington, New Zealand

Arrowhead Arrowhead Juvenile Detention Center in Duluth, Minnesota

Arthur Kill Arthur Kill Correctional Facility in Staten Island, New York—once the Drug Rehabilitation Center

Aryan Brotherhood California prison gang including members of the American Nazi Party involved in narcotics activity and racial confrontations

Arzew (see Algeria)

ASCA Association of State Correctional Administrators (publishes *Correctional Memo* quarterly)

Asheville Ashville, North Carolina, the seat of Buncombe County and the site of its jail

Ashford remand prison center in the London area

Ashland Federal Youth Center at Ashland, Kentucky

Ashwell HM Prison at Ashwell, Leicestershire, England

Asilo Toribio Durán Barcelona, Spain's reformatory

Asinara Italian penal colony, prison, and prison farm on Asinara Island off the northwest coast of Sardinia

ASJJA Association of State Juvenile Justice Administrators

Askham Grange HM Prison at Askham Grange, Yorkshire, England, for female offenders

Asmara (see Ethiopia's prisons)

Associated Marine Institutes Florida's federation of correctional programs for young offenders

Association of State Correctional Administrators (see ASCA)

ASTREA Aerial Support to Regional Enforcement Agency (helicopter surveillance)

Asunción Paraguay's capital containing jails and a prison filled with persons charged with crimes against the regime

Atascadero institution for the criminally insane and mentally disordered sex offenders at Atascadero, California

Atchison Kansas Youth Center

Atheist Penologist Jeremy Bentham (1748–1832)—English philosophical radical remembered for his Panopticon —a prison designed so every cell and interior area would have natural light and air; he argued the function of punishment was not revenge but the prevention of crime; opposed the death penalty

Atlanta Atlanta Youth Development Center in Atlanta, Georgia holding female juvenile delinquents up to the age of 17; U.S. Penitentiary at Atlanta, Georgia—maximum-security; also location of the Staff Training Center of the Bureau of Prisons in Atlanta

Atlantic Avenue Brooklyn House of Detention for Men at 275 Atlantic Avenue in Brooklyn, New York

Atlantic City Atlantic City, New Jersey jail and police-station lockups

Atmore Atmore State Prison Farm northeast of Bay Minette, Alabama; Fountain Cor-

rectional Center at Atmore, Alabama northeast of Mobile

ATPE Association of Teachers in Penal Establishments

Attica Correctional Facility at Attica, New York

Auburn Auburn Correctional Facility (maximum-security institution formerly named Auburn Prison in Auburn, New York)

Auburn cell-block plan keep the most hardened convicts in solitary confinement in separate cells, keep less hardened criminals in solitude until they give evidence of repentance, and keep the so-called least guilty in separate cells at night but working in silence in workshops during the day; popular during much of the 19th century

Auburn system characterized by enforced silence at all times for all inmates (also called the silent system)

Auckland Auckland Prison in Auckland, North Island, New Zealand

Augie Augusta, Georgia jail

Auk Auckland, New Zealand or its jail

Aurora Staff Training Center of the Bureau of Prisons in Aurora, Illinois

Auschwitz-Birkenau (German —Oswiecim-Brzezinka)— Nazi concentration camp

Austin seat of Travis County jail

Australian prisons in the Australian Capital Territory containing Canberra, New South Wales—Sydney, Northern Territory—Darwin, Queensland—Brisbane, South Australia—Adelaide, Tasmania —Hobart, Victoria—Melbourne, Western Australia —Fremantle

Austrian prisons Vienna, Graz, Linz, Salzburg, and Innsbruck

Austro-Hungarian Empire (see Austria, Hungary, and Yugoslavia)

Avenel New Jersey site of the Adult Diagnostic and Treatment Center

Avon Park Avon Park Correctional Institution in south-central Florida

awa absent without authority

Awaiting Trial Facility in Cranston, Rhode Island (formerly the Providence County

Jail)

away away from here, away from home, away from work (imprisoned)

awol (AWOL) absent without leave or permission

AWS (*see* Alston Wilkes Society)

axe and block decapitating equipment broad-bladed axes and hardwood blocks with slightly hollowed-out neck rests

Aylesbury HM Prison at Aylesbury, Buckinghamshire, England

BA any of Buenos Aires, Argentina's jails and prisons

Babi Yar concentration camp outside Kiev where more than seventy thousand Jews and several hundred thousand Russian troops were killed by German forces during World War II

back-gate exit dying in jail or prison and being carried out the back gate

Back Home convicts' nickname for the Tombs Prison in downtown New York City

back time unserved portion of a prison sentence any parole violator must serve once apprehended

baddie(s) bad guy(s)—incorrigible criminal(s)

bad rap(s) long prison sentence(s)

Baffin Correctional Centre in Frobisher Bay, Northwest Territories

Baghdad Iraq's oldest prison

bagne (French—convict prison; convict ship; penal servitude)

Bahamas Nassau—on New Providence Island contains an old gaol built in British colonial times

Bahrain its capital in Manama houses the old prison (*see* Manama)

Baird Andrew C Baird Detention Center (Wayne County Jail, Detroit, Michigan)

Baird House residence of the Quaker Committee on Social Rehabilitation at 135 Christopher Street in New York City

Baker Correctional Institution offers inmates counseling, recreation, schooling, and work in Olustee, Florida

Bakersfield Kern County courthouse and jail in central southern California

Bakirkoy hospital for the criminally insane in Istanbul, Turkey

Balearic Islands (*see* Spain)

Ball site of the Louisiana Training Institute for female delinquents

Balti Baltimore, Maryland penitentiary and other penal facilities

Baltimore site of the Maryland Penitentiary, the Maryland Training School for Boys, the Reception Center, and local lockups managed by the police and the sheriff

Bamako (*see* Mali)

Banana City Brisbane, Queensland, Australia jail

banasto (Spanish—basket)—cell or prison

bandbox county workhouse

Bandung (*see* Indonesian prisons)

Bangkok (*see* Thailand)

Bang Kwang Bangkok's maximum-security prison and its oldest

Bangladesh its principal prison is in its capital city of Dacca

Bangui (*see* Central African Republic)

banishment exile in another country or to some far place belonging to the land of the person banished

banishment upheld some penologists argue banishment is better for prisoners and society than imprisonment even if this means exile to distant deserts or remote islands

Banning Banning Rehabilitation Center in Banning, California

Barbados its capital—Bridgetown—has an old prison dating back to the years of British control

barbecue stool electric chair

Barcelona (*see* Spain)

Barlinnie Glasgow, Scotland's prison and its Young Offenders Institution

Barna Barcelona, Spain, jail

Barquisimeto (*see* Venezuela)

barracoon(s) temporary prison(s)

Barranquilla Colombian prison on the Magdalena

Bartholomew Fair nickname of the solitary-confinement section of London's Fleet Prison in Elizabethan times

Bartons Mills medium-security prison near Perth, Australia

Basel (*see* Switzerland)

Basil Basil Health Systems

Bastille (French—small fortress)—La Bastille, the infamous royalist prison of Paris, destroyed by French revolutionaries on July 14, 1789—synonym for prison holding political prisoners

Bastille by the Bay inmates' nickname for San Quentin Prison in San Francisco Bay

Bastrop Federal Youth Center at Bastrop, Texas (cares for inmates under 21 years of age)

Batavia (*see* Indonesian prisons)

Bath Ontario site of the Millhaven (maximum-security) Institution

Baton Rouge East Baton Rouge Parish jail, also known for the Louisiana Juvenile Reception and Diagnostic Center for delinquent and neglected juveniles

Bat Rou Baton Rouge, Louisiana jail or Juvenile Reception and Diagnostic Center

Baumes Law New York State statute requiring life imprisonment for anyone convicted four times of felonies

Baumettes France's cobblestone-walled prison in Marseilles

Bay Botany Bay penal settlement in New South Wales

Bayamón Puerto Rican city holding the Metropolitan Regional Institution

Bay ship British prison ship destined for Botany Bay in New South Wales, Australia

Bay State Correctional Center at Norfolk, Massachusetts for long-term minimum-security male felons

BC Baja California, México or British Columbia, Canada jails and prisons.

BCC Bureau of Charities and Corrections (South Dakota)

BCI Bureau of Correctional Institutions (Iowa)

BC Pen British Columbia Penitentiary in New Westminster

BCS Bureau of Criminal Statistics

Beaconsfield Marian Hall for delinquent English-Catholic juvenile offenders at Beaconsfield, Québec

bear den police station and its lockup

Beaune-la-Rolande site of a

French concentration camp southeast of Pithiviers

Beccaria (*see* Father of Criminology)

Bedford HM Prison at Bedford, Bedfordshire, England

Bedford Hills Bedford Hills Correctional Facility at Bedford Hills, New York

Bedlam nickname of Saint Mary of Bethlehem, the celebrated lunatic asylum of old London, where many inmates were criminally insane

Beersheba Israel's largest prison located between Jerusalem and the Negev desert

behavioral control unit solitary-confinement cell or dungeon

beheading form of capital punishment; decapitation

behind the iron door behind bars; jailed

behind the iron house in jail

Beira (*see* Mozambique)

Beirut (*see* Lebanon)

Belfast Welfare Unit 14 euphemism for Her Majesty's Prison Camp outside Belfast, Northern Ireland

Belgian Congo (*see* Zaire)

Belgrave (*see* Yugoslavia)

Belize HM Gaol facilities are in Belize as well as Belmopan

Bellefonte site of the Pennsylvania State Correctional Institution at Rockview

Belle Glade site of the Glades Correctional Institution between Lake Worth and Lake Okeechobee, Florida

Belle Isle Confederate prisoner-of-war camp in the James River near Richmond, Virginia

Bellevue Bellevue Hospital Prison Ward at First Avenue and 30th Street in New York City

Belmopan (*see* Belize)

Belsen Nazi concentration camp near Hannover, Germany

Belzec German extermination camp in this Polish village on the railway line running through Lublin province

Benghazi Libyan port city and site of a prison built almost a century ago

Benin prisons Cotonou and Porto-Novo on the north coast of the Gulf of Guinea

Bentham Jeremy Bentham (1748–1832)—English penologist-philosopher who wrote

about the need for prison reform and devised the panopticon-type prison where all cells could be observed from a central site

Berdoo San Bernardino, California jail

Bergen Nazi concentration camp of Bergen-Belsen in Lower Saxony, Germany

Bergen-Belsen Nazi concentration camp in Lower Saxony, Germany

Berhala Island prisoner-of-war camp run by the Japanese in North Borneo during World War II

Berlin Confinement Facility of the U.S. Army in Germany

Bermuda (*see* Casemates)

Bern (*see* Switzerland)

Bernalillo Bernalillo County Detention Home in Albuquerque, New Mexico

Bess Bessemer, Alabama or its jail southwest of Birmingham

Beth Bethlehem, Pennsylvania or its city jail

Beto Unit (*see* Tennessee Colony)

Betty's Place St Elizabeth's Hospital in Washington, DC

Bexar Bexar County Jail in San Antonio, Texas

BHD Bronx House of Detention in New York City

Bhutan its capital, Thimphu, houses its principal prison

Bialoleka Polish prison camp southeast of Bialystok

Bialystok Nazi concentration camp northeast of Treblinka in Poland

Big A nickname for the Federal Penitentiary in Atlanta, Georgia

Big D Dallas, Texas or its jail

big day visiting day in a prison

big gate(s) prison(s)

Big H Big House (any penitentiary or prison)

Big House up the River Sing Sing Prison at Ossining, New York

big pasture penitentiary

Big Spring Federal Prison Camp at Big Spring, Texas

Bilbao (*see* Spain)

bilbo(s) leg shackle(s) consisting of an iron bar fitted with adjustable fetters

Bilibid maximum-security Philippine prison at Muntinlupa in Rizal Province

Billeshave youth home for delinquent boys, on the western

edge of Denmark's island of Funen

Billings Billings, Montana, county jail

Biloxi seat of Harrison County and its jail

bing solitary confinement

Binghampton south-central New York jail

birdcage prison cell

Birkenau (German—Birch Grove)—concentration camp next to Auschwitz

Birmingham prison in Birmingham, Alabama or HM Prison in Birmingham, England plus prisons in smaller Birminghams in Iowa, Michigan, and Saskatchewan

Biscuit Factory nickname of old Reading Gaol in Berkshire, England, where it adjoined Huntley & Palmer's biscuit factory

Bismarck site of the North Dakota Penitentiary and the North Dakota State Farm for felons and first offenders

Bissau (*see* Guinea-Bissau)

bit time served in prison

BJC Bureau of Juvenile Correction (Delaware)

BKA *Bundeskriminalamt* (German—Federal Criminal Ministry)—contains computerized files of criminal histories maintained at its center in Wiesbaden

black-and-white stripes old-fashioned convict uniforms

black book prison register of its inmates

Blackburn Correctional Complex at Lexington, Kentucky, provides vocational training release programs

Black Flower of Society Nathaniel Hawthorne's nickname for any jail, penitentiary, prison, or other place of imprisonment

Black Guerrilla Family gang involved in drug trafficking within California prisons

Black Hole of Calcutta (*see* Indian prisons)

black lock solitary confinement

black maria prison van

black peter Australian slang —solitary-confinement prison cell

Blackwell's Island early name of Roosevelt Island (formerly Welfare Island) in New York City's East River under the Queensboro Bridge and long the site of correctional institutions

Bland Correctional Center at Bland, Virginia

Bledsoe Bledsoe County Regional Correctional Facility in Pikeville, Tennessee

Blonde Beast of Belsen wardress Irma Grese

Bluefields (*see* Nicaragua)

blue lights in front of all of London's police stations and lockups except at Bow Street, near the Royal Opera

Blue Ridge Pre-Release Work-Release Center in Greenville, South Carolina

Blundeston HM Prison at Blundeston, Suffolk, England

Blythe Blythe Branch of the Riverside County Jail in California

board(ed) blindfold(ed)

B o C Bureau of Correction (Pennsylvania); Bureau of Corrections (Virgin Islands)

body shake search down to the skin and into the body

Bogotá federal prison in Colombia's capital

boiling boiling in oil; boiling people alive, boiling was replaced by hanging

Boise site of the Idaho Security Medical Facility and the Idaho State Correctional Institution county seat, courthouse, and jail of Ada County, Idaho

Bolivar Bolivar County Jail in Cleveland, Mississippi

Bolívar Carcel Nacional de Ciudad Bolívar (Spanish— National Prison of Ciudad Bolívar, Venezuela)

Bolivia La Paz and Sucre each has an old prison and there are prisons in Cochabamba and Santa Cruz

Bolshoi Dom (Russian—Big House)—prison

bolt cutter heavy-duty hardware tool used to cut bolts, chainlink fencing, handcuffs, steel bars, etc.

Bom Bombay, India, prison

Bombay (*see* Indian prisons)

Bon Air Bon Air Learning Center southwest of Richmond, Virginia

boneyard graveyard

Boniato Cuban prison in Oriente Province

Bonneville Bonneville Community Corrections Center in Salt Lake City, Utah; a work-release facility

boob (Australian slang—jail; prison)

booking formal logging of inmates when they are received in jail or prison and are fingerprinted and photographed

Boonville Missouri location of the Training School for Boys for juvenile delinquents from 12 to 17 years of age

B o P Bureau of Prisons (United States Department of Justice)

Boquillas Boquillas del Carmen, Coahuila and its small jail

Bordentown New Jersey community holding the Youth Correctional Center, originally a prison farm and later a reformatory

Boron Federal Prison Camp at Boron, California in the Mojave Desert

borstal British name for a juvenile-delinquent reformatory

Borstal Borstal Prison in Kent, England where the first juvenile-delinquent reformatory was established in 1902 for boys from 16 to 21

borstals British correctional and detention centers such as Bullwood Hall in Essex, Deerbolt, Dover, East Sutton Park in Kent, Everthorpe, Humberside, Feltham and Finnamore Wood, Gaynes Hall, Glen Parva, Guys Marsh, Hatfield, Hewell Grange, Hindley, Hollesley Bay Colony, Huntercombe, Lowdham Grange, Onley, Portland, Rochester, Stoke Heath, Usk, Wellingborough, Wetherby

Bosnia (*see* Yugoslavia)

Boston Pre-Release Center in Dorchester, southwest of Boston, Massachusetts

bot (BOT) balance of time to be served by anyone violating parole and returned to prison

Botany Bay penal settlement of Sydney, New South Wales, Australia, where some 700 British convicts were landed in 1788 after an eight-month voyage from England

Botswana South African country, has prisons in its capital, Gaborone, and in Franciston

Bouaké (*see* Ivory Coast's prisons)

Bourgoin maximum-security prison in Bourgoin, France

Bournemouth seaside site of four new prisons on England's coast southwest of London

Bowden Institution medium-security penal facility in Innisfail, Alberta

boxcar(s) prison cell(s)

boxed up (New Zealand slang—imprisoned; jailed; locked up)

box(es) prison cell(s)

Boydton Virginia location of the Mecklenburg Correctional Center

Boys Ranch Group Home for Boys at Agaña, Guam (juvenile correctional facility)

Boys Totem Town in Saint Paul, Minnesota where delinquent boys are held in minimum security while being given training

BP Board of Parole; Bureau of Prisons

B of P Bureau of Prisons (U.S. Department of Justice)

BPT Board of Prison Terms

B of R Bureau of Rehabilitation (Washington, DC)

bracelets handcuffs

Bradford federal correctional institution in Pennsylvania

Brampton Vanier Centre for Women at Brampton, Ontario

Brandenburg concentration-camp subcamp west of Berlin

Brandon Correctional Institution in Brandon, Manitoba west of Winnipeg

brank leather or rubber head harness fitted with a gag and used to prevent a prisoner from shouting or talking

Brasilia Brazil's new prison whose facilities are better than those in Belém, Belo Horizonte, Manaus, Recife, Rio de Janeiro, Salvador, Santos, or São Paulo

Braunschweig Nazi concentration camp

Brazil has as many prisons as it has cities from Belem to Rio Grande; prisons go by such generic names as *cadeia, cárcere, penitenciária estadual,* or *prisão*

Brazoria Texas site of the maximum-security Central Unit

Brazzaville the main prison of the People's Republic of the Congo

breadand bread-and-water

(prison fare for those in solitary confinement)

Bread Street London prison on Bread Street in Elizabethan times

bread and water traditional diet fed to difficult-to-handle prisoners as a form of punishment

bread and whiskey last meal fed to prisoners held by the Republic of China before they are shot

Brevard Brevard Correctional Institution in Sharpes, Florida, opened in 1975 to help first offenders up to age 25 complete their education, learn a vocation, and overcome alcohol and drug abuse

briar hacksaw

bridewell British synonym for a house of correction such as the infamous Bridewell, its site may be found on London's New Bridge Street at what was St Bride's Well near the Thames

Bridge City site of the Louisiana Training School for delinquents under 17 years of age

Bridge House detention home for juvenile delinquents in Wilmington, Delaware

Bridgeport Connecticut Community Correctional Center in Bridgeport; the Fairfield County Jail is also here

Bridge of Sighs originally the enclosed passageway connecting the Doge's Palace with the prison dungeons of Venice (*Ponte dei Sospiri*); many such bridges connect courtrooms with prisons; in London it means Waterloo Bridge; in New York it is a high-level passage linking the Criminal Courts with the Tombs Prison

Bridgetown nickname of HM Prison in Barbados

Bridgewater site of the Massachusetts Correctional Institution as well as the Southeastern Correctional Center of Massachusetts

brig ship's prison

Brighton Brighton Prison in Brighton, England

brig rat(s) naval prisoner(s); shipboard prisoner(s)

Brigs (USN) (*see* USN Brigs)

Brisbane Brisbane Prison Complex (Queensland, Australia)

Brissie Brisbane, Queensland, Australia or its jail

Bristol HM Prison at Bristol, Somerset, England as well as prisons in Bristols in Connecticut, Colorado, Florida, Georgia, Indiana, New Brunswick, New Hampshire, Pennsylvania, Québec, Rhode Island, South Dakota, Vermont, Tennessee and Virginia

British detention centers Adlington, Blantyre House, Buckley Hall, Campsfield House, Eastwood Park, Erlestoke House, Foston Hall, Haslar, Kirklevington, Medomsley, New Hall, North Sea Camp, Send, Werrington House, Whatton

British Guiana (*see* Guyana)

British Honduras (*see* Belize)

British remand centers Ashford, Brockhill, Latchmere House, Low Newton, Pucklechurch, Risley, Thorp Arch

Brix Brixham, England, jail

Brixton one of London's largest prisons where the greatest jail break in British history took place in 1973

Brno prison within the old castle of Czechoslovakia's second largest city

Bromberg Nazi concentration camp in East Prussia

Brooklyn Brooklyn Detention Center of the Immigration and Naturalization Service in New York City; Connecticut Community Correctional Center

Brookolino (Italian-American slang—Brooklyn, New York)

Brookwood Brookwood Center for juvenile offenders at Claverack, New York

Brookwood Center for Girls at Claverack, New York

Broome Broome Regional Prison (Western Australia)

Broward Broward Correctional Center at Pembroke Pines, Florida; emphasis is on educational and vocational programs

Brownwood Brownwood, Texas and the Brownwood State School for male and female juvenile delinquents

Brushes Wormwood Scrubs Prison, a suburban penal facility London

Brushy Mountain Brushy Mountain Penitentiary at Petros, Tennessee

BSSR Bureau of Social Science

Research

bt's (prison) building tenders (porters, turnkeys, wing floor tenders, etc.)

bubbling executing by injecting air bubbles in the veins

Buchanan (*see* Liberia)

Bucharest (*see* Romania)

Buchenwald Nazi concentration camp near Weimar, Germany

Buckeye Youth Center in Columbus, Ohio was started in 1914 to diagnose juvenile delinquents

Buda Budapest, Hungary, prison

Budapest capital of Hungary and nickname of its old prison built during the Austro-Hungarian Empire after the defeat of the Turks in 1697

Buenaventura Colombian prison in the Pacific seaport city

Buena Vista Buena Vista Correctional Facility at Buena Vista, Colorado

Buenos Aires (*see* Argentina and Villa Devoto)

Buffalo Erie County Jail, Buffalo, New York

Buford site of the Georgia Training and Development Center

bughouse insane asylum or prison for the criminally insane

bug trap bed or cot in a jail or prison

Bujumbura (*see* Burundi)

Bukitduri women's prison near Djakarta, Indonesia

Bulawayo (*see* Zimbabwe)

Bulgaria capital city—Sofia—has a large prison and smaller ones are in Plovdid and Varna

bullpen(s) place(s) of temporary confinement while awaiting arraignment, trial, or imprisonment; large common cell(s)

Bullwood Hall a borstal in Essex, England

Bulu women's prison near Semarang on the island of Java in Indonesia

Buna forced-labor camp near Auschwitz erected by the Nazis to aid in the production of artificial rubber

Bunbury Bunbury Rehabilitation Centre (Bunbury, Western Australia)

Bundeskriminalamt (German—Central Criminal Council)—German Interpol headquarters in Wiesbaden

Bureau of Prisons U.S. Department of Justice, Bureau of Prisons, administering U.S. penitentiaries, federal correctional institutions, a medical center for federal prisoners, federal prison camps, community treatment centers, metropolitan correctional centers, a federal detention center, and four staff training centers
buried serving a long sentence
Burlington Burlington County Jail at Mount Holly, New Jersey; largest city in Vermont and seat of Chittenden Correctional Facility
Burnaby Lower Mainland Regional Correctional Centre in Burnaby, British Columbia
burn(ed) electrocute(d)
Burrus Burruss Correctional Complex at Forsyth, Georgia
Buru Indonesian penal colony
Burundi Bujumbura prison was built during the years of German control
Bushnell Bushnell, Florida; site of the Sumter Correctional Institution
bush parole escape from confinement; escape from jail or prison
Butner Federal Correctional Institution at Butner, North Carolina; prisoners are hard-to-rehabilitate repeat offenders compelled to work at a prison job and to attend group discussions about all phases of prison life and outside life styles
butt last period of a prisoner's sentence
Butte Butte, Montana, county jail
Butterworth site of a Japanese prisoner-of-war camp
Butyrskaya Moscow's block-long four-story prison hidden behind an eight-story department store on Novoslobodskaya Street
Bu-Tyur Butyrskaya Tyurma (Russian—Butyrki Prison)— major prison in Moscow
BVR Bureau of Vocational Rehabilitation
b & w bread and water (diet often imposed on prisoners in solitary confinement)
C.3.3. Ocar Wilde's identification number while imprisoned and when he wrote his poem, *The Ballad of Reading Gaol*, and his prose apologia for being

in jail, *De Profundis*; C.3.3. stood for gallery C, 3rd landing, 3rd cell
caballo (Spanish—horse)—in Mexican-American slang a person who carries drugs into jails and prisons
Cabbage Patch Victoria, Australia or its jail or prison
cabo general (Spanish—chief corporal)—head inmate or trusty in a prison
CAC Commission on Accreditation for Corrections
CACA Central After-Care Association (British society handling prisoners on parole)
cachot (French—underground prison cell)
cage jail; lockup; imprison
CAGE Convicts Association for a Good Environment
cage and key men jailers; prison guards
Cairo capital of Egypt and its prison facilities
cala calabozo (Spanish—cell, dungeon, jail)
calaboose prison
calabozo (Spanish—prison)— also called *cárcel, celda, mazmorra, presidio, prisión*
Calc Calcutta, India or its prison
Calcutta (*see* Indian prisons)
Caledonia and Odum Complex North Carolina penal facility in Tillery
Caliente Nevada Girls Training Center (correctional facility) in Caliente
California (*see* Chino, Corona, Folsom, Frontera, Jamestown, San Luis Obispo, San Quentin, Soledad, Susanville, Tehachapi, Tracy, and Vacaville entries)
California Institute for Women California's only state prison for women called Frontera
California jails 58 counties are served by local lockups, city jails serve Fresno, Los Angeles, San Diego, San Francisco, and other big cities
California Medical Facility opened in 1955 at Vacaville
California Men's Colony state prison in San Luis Obispo
California State Prisons San Quentin and Folsom
Calle Marina (*see* Marine Street)
Camaguey (*see* Cuba)
Camarillo Ventura Reception Center and Clinic at

Camarillo, California; near the Ventura School for Girls, also a correctional facility
Cambridge seat of Middlesex County, Massachusetts, across the Charles River from Boston; both cities have a county jail and many police-station lockups
Cameroon West African nation with complete-with-prison cities such as Douala and Yaounde
camisole straitjacket
camp confinement or correctional facility; prison camp
Campbellford Ontario town containing the Warkworth Institution
Campbell Work Release Center in Columbia, South Carolina for felons and misdemeanants in minimum custody
Camp Boiro Conakry's jail in Guinea
Camp de Drancy (French— Drancy Camp)—concentration and transit camp maintained by French collaborationists and the Gestapo in World War II
Camp de Gurs (*see* Gurs)
Camp de la Transportation official name of the penal colony headquartered at Saint Laurent du Maroni in French Guiana
Camp Douglas in the 1860s a prisoner-of-war camp near Lake Michigan
Campeche prison in Campeche, Mexico
Camp Harmony Japanese-American assembly center at Puyallup, Washington
Camp Hill British reformatory on the Isle of Wight; Pennsylvania state correctional institution
Camp Iyar prison camp in Israel
Camp Lejeune (*see* USMC)
Camp O'Donnell American military encampment on Luzon in the Philippines
Campo Numero Uno (1) (Spanish—Camp No 1)— maximum-security military prison near Mexico City
Camp Pendleton (*see* USMC)
Camp San Jose California detention camp near Mount Palomar
Camp Smedley D Butler (*see* USMC)

Camp Topaz Camp Topaz—the Jewel of the Desert (Japanese-American nickname for the relocation center where they were held at Topaz, Utah)

campus prison grounds

Camp West Fork correctional center for juveniles, near Warner Springs, California

Camp Westway detention center for juveniles, near Warner Springs, California

Canada every province provides penal facilities ranging from penitentiaries and prisons to halfway houses and local lockups

canine shamus(es) dog detective(s)—used for their keen sense of smell

cannery slang for prison

Canon City Centennial Correctional Facility in Canon City, Colorado; the Colorado Territorial Correctional Facility is also here along with the Colorado Women's Correctional Facility, the Fremont Correctional Facility for medium-security felons; the Reception and Diagnostic Center for maximum-security convicts; the Shadow Mountain Correctional Facility

Can Pen Ser Canadian Penitentiary Service

Canterbury HM Prison at Canterbury, Kent, England

Canto Grande Peruvian high-security prison in Chorillos, south of Lima

Canton Canton, Ohio or its Stark County Jail

CANY Correctional Association of New York (City)

CAP Comité d'Action des Prisonniers (French—Prisoner's Action Committee)

Cape Town (see South Africa)

Cape Verde island republic with its prison in Praia, off the West African coast

Cap-Haitien (see Haitian prisons)

capital punishment the death penalty

Capron site of the Capron Correctional Unit as well as the Deerfield, St Brides, and Southampton Correctional Centers, Virginia

Cap-Rouge the Maison Notre-Dame de la Garde facility for juvenile delinquents at Cap-Rouge, Québec

captain warden of a road prison with its road-gang guards and inmates

capun capital punishment

Caracas Venezuela location of several jails and lockups

cárcel (Spanish—prison)—also called *calabozo, celda, mazmorra, presidio, prisión*

Cárcel de Mujeres (Spanish—Women's Prison)—also name of the Instituto Nacional de Orientacion Femenina situated in los Teques Ejido Miranda, Venezuela

Cárcel de Valparaiso Valparaiso, Chile's jail

carceleras (Spanish—prisoner songs)—a flamenco song form developed by prisoners incarcerated in Ronda

Cárcel Modélo (Spanish—Model Prison)—many Hispanic places throughout Latin America and the Iberian Peninsula have a so-called model prison

Cárcel Nacional de Ciudad Bolívar (Spanish—National Prison of Ciudad Bolívar) in eastern Venezuela

Cárcel Nacional de Maracaibo (Spanish—National Prison of Maracaibo)—in western Venezuela

Cárcel Nacional de Trujillo (Spanish—National Prison of Trujillo)—northeast of Merida, Venezuela

Carceri d'Invenzione (Italian—Imaginary Prisons)—series of etchings created by Giambattista Piranesi in the mid-1700s to show the oppressive frustration of confinement as well as the instruments of torture

Cardiff HM Prison at Cardiff, Wales

Carraca La Carraca (Cadiz, Spain's infamous prison)

Carranza Venustiano Carranza Penitentiary in Tepic, the capital of Nayarit, Mexico

Carson City Nevada State Prison together with the Northern Nevada Correctional Center, the Nevada Women's Correctional Center, the Northern Nevada Honor Camp, all in Carson City

Cartagena Colombian and Spanish prisons in seaport cities of the Caribbean and the Mediterranean

Casablanca Buenaventura, Colombia prison; (see Morocco)

casa de correción (Spanish—house of correction)—reformatory

Casa de Reeducacion y Trabajo Artesanal (Spanish—House of Reeducation and Artisan Work)—Venezuelan penal facility in Maracaibo as well as Caracas where it is nicknamed La Planta

casa di correzione (Italian—house of correction)—reformatory

Casemates Bermuda's prison island, formerly a fortress

Caserta women's prison north of Naples, Italy

Cassidy Lake Technical School at Chelsea, Michigan for felons over 21 and under 30 in need of academic and vocational training

Castieau's hotel nickname given the old jail in Melbourne, Victoria, Australia honoring the jail's governor—JB Castieau

Castle The Castle—U.S. Disciplinary Barracks at Fort Leavenworth, Kansas

Castle Huntly Scottish borstal east of Dundee

Castle Thunder Castle Thunder Prison in Richmond, Virginia where political prisoners were held during the Civil War

Castries (see Saint Lucia)

Castro's Prison nickname applied to communist-controlled Cuba

Catalina Catalina Mountain School near Tucson, Arizona where it offers a program for male juvenile delinquents

Catete former royal palace used as Rio de Janeiro's immigration prison

Caves The Caves (HM Prison at Rockhampton in Queensland, Australia)

Cayahoga Hills Boys School juvenile-delinquent penal facility in Warrensville Heights, Ohio

cayenne (French slang—prison ship)

Cayenne French Guiana prison on the Rue Francois-Arago in Cayenne, the capital city

cc condemned cell

CCA California Correctional Association; Colorado Correctional Association

CCC Central Correctional Center in Macon, Georgia

CCD & C Commission on Crime, Delinquency, and Corrections (Nevada)

CCHS Computerized Criminal Histories System (FBI)

CCI Coastal Correctional Institution at Garden City, Georgia; Connecticut Correctional Institution at Niantic on Long Island Sound

CCOA California Correctional Officers Association; County Court Officers Association

CCTF California Correctional Training Facility

CD Corrections Department (New Mexico); Corrections Division (Hawaii, Oregon)

CEA Correctional Education Association

Cedar City north of Jefferson City, Missouri and site of the Renz Correctional Center

cela (Portuguese—prison cell)

celda (Spanish—prison cell)

Celery City Clink Kalamazoo, Michigan's jail

cella (Italian—prison cell)

cell smell usually a mixture of excreta, sweat, stale tobacco, unaired bedding, and vomit

cement tomb prison cell

Center City Detention Center Eastern State Penitentiary in Philadelphia, Pennsylvania

Central downtown San Diego's Central Detention Facility; Hong Kong's Central Police Station and lockup; North Carolina's Central Prison in Raleigh

Central African Republic Bangui, the capital, has an old prison built by the French

Central Community Center halfway house for released prisoners in Los Angeles, California

Central Community Corrections Center in Salt Lake City, Utah

Central Correctional Center (*see* Macon)

Central Correctional Institution in Columbia, South Carolina

Central Detention Facility District of Columbia's maximum-security prison

Centralia Centralia Correctional Center east of East Saint Louis; Washington community and address of the Maple Lane School (*see entry*)

Central Missouri Correctional

Center in Jefferson City

Central Ohio Regional Forensic Unit in Columbus established to provide psychiatric treatment for mentally ill offenders

Central Ohio Training Institution in Columbus for male juvenile delinquents in maximum custody

Central Oklahoma Juvenile Treatment Center in Tecumsah, Oklahoma, housing delinquent females

Central Prison in Raleigh, North Carolina, a reception center for male felons

Central Unit Texas maximum-security facility in Sugar Land

Centro de Reeducacion Agropecuario (Spanish—Center of Cattle and Land Reeducation)—penal farm facility in El Dorado, Venezuela

Centro Penitenciario de Occidente's Ejido Trujillo (Spanish—Central Penitentiary of the West in the Ejido Trujillo of Venezuela)

Centro Penitenciario de Oriente (Spanish—Central Penitentiary of the East)—Venezuela

Centro Penitenciario de Valencia, Ejido Carabobo (Spanish—Central Penitentiary of Valencia in the Ejido Carabobo)—Venezuela

Ceuta (*see* Spain)

ceza evi (Turkish—house of punishment)—prison

CFA Correctional Facilities Association

CGIC Comisaria General de Investigacion Criminal (Spanish—Commisariat General of Criminal Investigation)—Spain's Interpol office

Chad formerly part of the French Sudan when its prison was built in N'Djamena

chain gang prisoners chained together during periods of outdoor work or transportation

chair electric chair

chamber gas chamber

Champerico Pacific coast port of Guatemala and its jail

Changi Up Changi Road (Singapore's maximum-security prison)

Channings Wood HM Prison at Channings Wood, Devonshire, England

Charleston South Carolina's principal city and its jail; West

Virginia's capital city and Kanawha County seat with its jail

Charlestown Charlestown, Massachusetts jail

Charlotte Charlotte, North Carolina, seat of Mecklenburg County with its jail

Charlottetown capital of Prince Edward Island, Canada and location of the Sleepy Hollow Correctional Centre

chaser(s) prison guard(s)

Chateau d'If island prison off the port of Marseilles in the Mediterranean

Chatham Island easternmost of the Galápagos long used as a penal settlement

Chattahoochee in northwest Florida and site of the River Junction Correctional Institution

Chattanooga Tennessee city in Hamilton County with its jail

CHC Chicago House of Correction

cheats gallows (Elizabethan English)

checas (Spanish slang—communist prisons)—term derived from *Cheka*—the communist secret police active in Spain during the Spanish Civil War

check out commit suicide

Cheesebox nickname for the Statesville, Illinois penitentiary

Chelmo Nazi concentration camp near Lublin, Poland

Chelmsford English prison in Chelmsford, northeast of London

Chelsea Michigan community and site of the Cassidy Lake Technical School

Cherry Hill nickname of the Eastern Penitentiary designed to insure the solitary confinement of each prisoner

Chesapeake Virginia location of the Chesapeake Correctional Unit

Cheshire Connecticut Correctional Institution at Cheshire with age limits from 16 to 21

Chetumal prison in Ciudad Chetumal, capital of the Mexican territory of Quintana Roo

Cheyenne seat of Laramie County with its jail

Chicago toponym for six divisions of Department of Corrections or Metropolitan Correctional Center

Chihuahua prison in Chihuahua, Mexico

Chile principal prisons are the Cárcel de Valparaiso and the Penitenciario de Santiago

Chillicothe Training School for (delinquent) Girls at Chillicothe, Missouri; Chillicothe Correctional Institute at Chillicothe, Ohio (for males with mental ailments)

Chillicothe Correctional Center Missouri penal facility for female felons

Chilpancingo prison in Guerrero, Mexico

Chi Ma Wan Chi Ma Wan Prison, Lantau Island, New Territories, Hong Kong

Chi Met Chicago Metropolitan Correctional Center

China, People's Republic of has a number of forced-labor camps and many prisons filled with political prisoners

China, Republic of maintains prisons in Taipei as well as in Kaohsiung, Taichung, and Tainan

Chinde (*see* Mozambique)

Chino California Institution for Men at Chino

Chistopol prison 500 miles east of Moscow

Chittenden Correctional Facility in South Burlington, Vermont

choke hold used to restrain the unruly by wrapping a forearm around the neck (the bar hold) or by putting pressure on the carotid artery in the neck (the carotid hold); both types of choke hold cut off the air supply and halt the flow of oxygen to the brain

chokey punishment or solitary-confinement cell

choky (English slang—jail)— term believed to be derived from the Hindustani *chauki* also meaning jail

chow hall prison mess hall

chow line prisoners lined up while waiting to be served food

Christchurch Christchurch Prison in Christchurch on South Island, New Zealand

Christianstadt Nazi subcamp in Germany

chronophobia fear of time

Chula Vista Chula Vista Staging Center of the INS at San Ysidro, California on a hill overlooking the Mexican Bor-

der; holds undocumented aliens awaiting deportation or admission to the U.S.

Chuna Soviet forced-labor camp in Siberia

ci (CI) cooperative individual (informant)

CIA Central Intelligence Agency; Correctional Industries Association

Cincinnati Cincinnati, Ohio or the Hamilton County jail

Cincy Cincinnati, Ohio or its jail

City College British euphemism for Newgate Gaol—an old London lockup; New Yorker slang for The Tombs prison

city watchhouse police station; police station lockup

Ciudad Acuña Ciudad Acuña, Coahuila and its jail

Ciudad Juárez (*see* Juárez)

Ciudad Miguel Aleman Ciudad Miguel Aleman, Tamaulipas and its jail

Ciudad Trujillo capital of the Dominican Republic during the dictatorship of Trujillo and nickname of its many prison facilities

Ciudad Victoria prison in Ciudad Victoria, capital of the Mexican state of Tamaulipas

Civic Center San Jose, California's Civic Center Jail

CIW California Institution for Women at Frontera

clandestine prison(s) improvised jail(s) frequently found in Africa, Latin America, the Middle East, and Southeast Asia

Clarkson Kings County Hospital Prison Ward at 435 Clarkson Avenue in Brooklyn, New York

classification center correctional unit where inmates are held while awaiting commitment to a prison or rehabilitation program

Claverack Brookwood Center for Girls at Claverack, New York

Claymont Delaware's site for the Women's Correctional Institution and the Woods Haven-Kruse School for Girls

Clemens Unit Texas maximum-security penal facility in Brazoria

Clementina nickname of the San Michele reformatory for boys from 14 to 18 on Rome's Piazza di Porta

Portese

Clermont-Ferrand French prison at 1 rue de la Prison in Clermont-Ferrand where a Franco-fascist government had its capital during World War II

Cleve Cleveland, Ohio or its jail

Cleveland Cleveland, Ohio, seated in Cuyahoga County with its jail and police-station lockups

client(s) person(s) on probation

client(s) of the correctional system convict(s)

Clink London prison formerly dominating the south bank of the Thames near London Bridge; generic nickname for all prisons

Clink The Clink—Tasmanian hotel trading on the island's convict past

Clinton New Jersey location of the Correctional Institution; Clinton Correctional Facility at Dannemora, New York

close custody (*see* maximum security)

CMCC Chicago Metropolitan Correctional Center

CMS Correctional Medical Systems

CNPB Canadian National Parole Board

CO Correctional Officer

COA Correctional Officers Association

Coastal Correctional Institution (*see* CCI)

Coeur d'Alene Kootenai County's jail in northern Idaho

Coffield Unit Texas maximum-security prison in Tennessee Colony

Coiba Coiba Island Panamanian prison near the western shoreline of Panama Bay

cold Auschwitzes of the North dissident Soviet poet Yuri Galanskov's phrase describing the Arctic death camps of the USSR

Coldingley HM Prison at Coldingley, Surrey, England

Colima prison in Colima, Mexico

college reformatory

Collins Bay Institution medium-security penal facility in Kingston, Ontario

Colombia almost every city has a so-called Cárcel Modelo and there are other prisons

such as Gorgona on an island in the Pacific and La Pieota in Bogotá
Colombo (*see* Sri Lanka)
Colonia Dignidad Chilean political prison in Parral, south of Santiago
Colorado (*see* Canon City; Buena Vista)
Colorado jails 63 counties are served by jails; Denver and Colorado Springs have local detention facilities
Colorado Springs El Paso County courthouse and jail in central Colorado
Colorguard trade name for a fabric-coated pre-galvanized steel fencing system, which the Colorguard Corporation of Raritan, New Jersey claims cannot be penetrated by gun muzzles, knives, or rocks
Columbia Mississippi site of Columbia Campus controlled by the Department of Youth Services; South Carolina's site of Campbell Work Release Center, the Central Correctional Institution, the Kirkland Correctional Institution, the Manning Correctional Institution, the Maximum Security Center, and the Walden Correctional Institution
Columbia Campus (*see* Columbia)
Columbus Ohio's capital containing the Buckeye Youth Center, the Central Ohio Regional Forensic Unit, the Central Ohio Training Institution, the Columbus Correction Facility, the Training Center for Youth, the Women's Correctional Admission Center
Columbus Correctional Facility formerly the Ohio Penitentiary
Columbus Fire occurred in the Ohio State Penitentiary in Columbus in April 1930
Combinado Combinado del Este (large prison outside Havana, Cuba)
Combinado del Este (Spanish —Eastern Combination)— Cuban prison in Havana Province
Community Correctional Center (*see* Bridgeport, Brooklyn, Hartford, Litchfield, New Haven, Uncasville)
community facility adult, juvenile, or nonconfinement facility where residents are al-

lowed to depart, unaccompanied by any official, to hold or seek employment or to go to school for treatment programs
Community Treatment Centers halfway houses for male and female offenders treated by the U.S. Bureau of Prisons in Atlanta, Georgia; Chicago, Illinois; Dallas, Texas; Detroit, Michigan; Houston, Texas; Kansas City, Missouri; Long Beach, California; New York, New York; Oakland, California; Phoenix, Arizona
Comoros the Federal and Islamic Republic of the Comoros, in the Indian Ocean, has its capital and prison in the port city of Moroni
compash compassionate probation officer, prison chaplain, social worker, or prison visitor
Compiègne French concentration camp controlled by the Gestapo during the German occupation in World War II
Complex The Complex (Brisbane Prison Complex in Queensland, Australia augmented by prisons in Rockhampton, Townsville, and Woodford)
con convict
Conakry (*see* Guinea prisons)
concerning the criminal element Eugene Victor Debs, convicted for opposing World War I, had this to say— *"While there is a lower class, I am in it; while there is a criminal element, I am of it; while there is a soul in prison, I am not free."*
Conciergerie (French—porter's lodge)—the great prison of Paris on the Ile de la Cité
Concord capital of New Hampshire and location of the Concord Community Corrections Center, the New Hampshire State Prison, and the New Hampshire State Prison Community Corrections Center
Concord Community Corrections Center Concord, New Hampshire
concrete womb(s) prison(s)
condao (Mexican-American Spanish—county jail)—corruption of the Spanish word for county (*condado*)
Condemned Rock Macquarie Island, Tasmania's nickname for Grummet Island when it

was a penitentiary
conditional release parole
condominio (Spanish—condominium)—euphemism for jail or prison
confinee(s) prisoner(s)
confinement imprisonment
Congo Brazzaville, its capital, has an old prison built by the French
conjugal visit plan whereby a prisoner may enjoy a marital relationship with a spouse
conk a screw club a guard
Connecticut (*see* Bridgeport, Brooklyn, Cheshire, Enfield, Hartford, Litchfield, New Haven, Niantic, Somers, Uncasville)
Connecticut Correctional Institution (*see* Cheshire, Enfield, Niantic, Somers)
Connecticut lockups the state's eight counties provide detention facilities for Hartford, Bridgeport, New Haven, New London and Litchfield, Stamford, and Norwalk
Connor Correctional Center in Hominy, Oklahoma
con(s) convict(s)
Constantine (*see* Algeria)
Constanza (*see* Romania)
convict goods things produced by convict labor (vehicle license plates, mail sacks, school benches, etc.)
convict labor work performed by prisoners as part of their program of rehabilitation (public works such as ecology conservation, farming, road building, vehicle registration, etc.)
convict(s) convicted felon(s) serving a prison term
convict ships (plying between British and ports in New South Wales and Van Diemen's Land in the late 18th and early 19th centuries) *Aboukir, Active, Admiral Barrington, Admiral Gambier, Adrian, Albemarle, Albion, Alexander, Alibi, Almorah, America, Ann, Ann and Amelia, Anson, Arab, Asia, Asiatic, Atlantic, Atlas, Atwick, Augusta Jessie, Aurora, Bardaster, Baring, Barossa, Batavia, Bellona, Bengal Merchant, Blenheim, Britannia, Bussorah Merchant, Cadet, Calcutta, Canada, Castle Forbes, Chapman, Charlotte, Circassian, Commodore Hayes, Coromandel, Countess of Harcourt, Cressy,*

Dromedary, Earl Cornwallis, Earl Grey, Earl St Vincent, Eden, Edward, Egyptian, Eliza, Elizabeth, Elizabeth and Henry, Elphinstone, Emma Eugenia, Emily, Emperor Alexander, Eolus, Equestrian, Experiment, Fanny, Fortune, Frances Charlotte, Friendship, Ganges, General Hewart, General Stewart, Gilbert Henderson, Gilmore, Glatton, Grenada, Guildford, Harmony, Hector, Henry, Henry Porcher, Hillsborough, Hindostan, Hyderabad, Indefatigable, Indian, Indispensable, Isabella, Jane, John, John Barry, John Brewer, John Renwick, Kinnear, Lady Harewood, Lady Juliana, Lady of the Lake, Lady Rowena, Lloyds, Lord Auckland, Lord Lynidoch, Lord William Bentinck, Margaret, Maria, Maria Soames, Marion, Marmion, Marquis of Hastings, Marquis of Huntley, Mary Anne, Medina, Mexborough, Minerva, Minorca, Moffat, Morley, Navarino, Neptune, New Grove, Norfolk, North Briton, Ocean, Orator, Oriental Queen, Pestonjee Bomanjee, Pitt, Portland, Prince Regent, Rajah, Ratcliffe, Recovery, Rodney, Royal Admiral, Royal Charlotte, Runnymede, St Vincent, Scarborough, Second Fleet, Sidmouth, Sir Godfrey Webster, Sir Robert Peel, Sir Robert Seppings, Sir William Bensley, Somerset, Southworth, Speke, Stakesby, Surprize, Surrey, Susan, Sydney Cove, Tasmania, Tenasserim, Tortoise, Tottenham, Triton, Waterloo, Waverley, Westmorland, William Jardine, William Miles, Woodford, etc., including vessels with names such as *Nile, Perseus, Persia*

con wise convict who knows what's going on in the prison including illicit operations

Cook County covering the entire area of Chicago, Illinois has a Department of Corrections with six divisions; the Cook County Jail; the old House of Correction; the Women's Division; the Men's Correctional Center; an intake and reception facility; the Training Academy

cooler(s) brig(s), guardhouse(s), jail(s), lockup(s)

coop(s) prison(s)

Coos Coos County Correctional Facility, Coos Bay, Oregon

cop a broom leave in a hurry

cop a drill leave at a normal walking pace so as not to attract the attention of guards

cop a heel escape from law-enforcement officers or from correctional facilities

cop a moke escape from a jail, prison, or other facility

cop a sneak escape from confinement

Copenhagen Denmark's modern prison noted for its design insuring inmates the maximum of light and air

cop house(s) police station(s) or lockup(s)

Copper John Auburn Prison's nickname

copper shop police station and lockup

Cordoba (*see* Argentina)

Corinto Nicaragua's seaport city with a jail

Cork (*see* Irish prisons)

Cornton Vale Scottish prison

Corona California Rehabilitation Center at Corona

corralón (Mexican-Spanish —corral)—detention camp where illegal entrants await deportation

corre correccional (Mexican-Americanism—correctional institution, penitentiary, reformatory)

correctional agency federal, state, or local criminal-justice agency charged with the investigation, intake screening, supervision, custody, confinement, or treatment of adjudicated or alleged offenders

Correctional Center for Women in Raleigh, North Carolina

Correctional Clinic U.S. Bureau of Prisons Correctional Clinic in New York City

correctional custody facility brig, jail, penitentiary, prison, reformatory

correctional custody unit U.S. naval euphemism for a ship's brig or lockup

Correctional Development Centre maximum-security facility in Laval, Québec

Correctional Education Association publishes quarterly *Journal of Correctional Education*

correctional institution long-term confinement facility

correctional officer(s) prison guard(s); prison warden(s)

Correctional Service Federation American wing of the International Prisoners Aid Association

correctional training school reformatory

Correction Camp Program for male and female felons housed at Grass Lake, Michigan

Correction Division Release Center in Salem, Oregon

Correction and Rehabilitation Squadron U.S. Air Force at Lowry Air Force Base, Colorado

corrections caseload number of clients registered with a correctional agency or agent during a specified time limit

corrective labor camp communist euphemism for forced-labor camp

Corrective Services all of the gaols and prisons in Australia's New South Wales are managed by the Department of Corrective Services in Sydney

CO(s) Correction Officer(s); Correctional Officer(s)

Costa Rica Central American country with its principal prison in the capital city of San José; smaller ones are in the ports—Limón on the Caribbean and Puntarenas on the Pacific

Coti Martinez clandestine Argentine prison's pseudonym

count prison population inventory

Counterpoint bimonthly publication of the National Juvenile Detention Association

country club minimum security prison

county cooler(s) county-supported mental hospital(s) frequently confining the criminally insane

county hotel(s) county jail(s)

Cowansville Cowansville Institution holding young offenders in Cowansville, Québec

Coxsackie Coxsackie Correctional Facility (medium-security prison in Coxsackie, New York)

CPA Connecticut Prison Association

CPPCA California Probation, Parole, and Correctional Association

CPS Canadian Penitentiary Service

CPSM Colonial Prison Service Medal (British decoration)

Cracow (*see* Poland)

Cracow/Plaszow concentration camp holding many forced laborers during World War II

Cradle of the Penitentiary Philadelphia, Pennsylvania's Walnut Street Jail built in the late 1700s and providing congregate as well as individual cells and workhouses

cranky hatch(es) cell(s) reserved for mentally deranged prisoner(s)

Cranston Rhode Island city containing eleven penal facilities

crash escape from jail or prison

crazy alley(s) cell block(s) reserved for the insane

crazy hospital(s) psychiatric ward(s)

CRD Civil Rights Division, U.S. Department of Justice

creative conflict demonstration or riot

creative sentence sentence created to fit the crime (for example, making graffiti artists clean the walls they have defaced)

crime playwright George Bernard Shaw defines crime in his play *Man and Superman*, "Crime is only the retail department of what, in wholesale, we call penal law."

crimeless society Thomas Paine, author of *The Age of Reason* and *The Rights of Man*, wrote: *"When it shall be said of any country in the world, 'My poor are happy; neither ignorance nor distress is to be found among them; my jails are empty of prisoners, my streets of beggars; the aged are not in want; the taxes are not oppressive; the rational world is my friend, because I am a friend of its happiness'—when these things can be said, then may that country boast of its constitution and its government."*

criminoso inveterado (Portuguese—inveterate criminal) —jailbird

CRMT Community Resources Management Team (parole and probation)

Croatia (*see* Yugoslavia)

Crockett Crockett State School for Girls at Crockett, Texas

cross-bar hotel(s) jail(s) or prison(s)

Cross City Cross City Correctional Institution offers academic and vocational training

Crown Point Indiana jail south of Gary

CRS Correction and Rehabilitation Squadron (U.S. Air Force Base at Lowry Air Force Base, Colorado)

cruel and unusual punishment being boiled in oil, being set afire, ducking, mutilating, pilloring, and whipping were common practices in a number of countries; as recently as the mid-1920s whipping was a punishment applied publicly in Dover, Delaware

Crumlin Crumlin Road Prison in Belfast, Northern Ireland

Crumlin Road 14 address and nickname of one of Her Majesty's prison camps outside Belfast, Northern Ireland

CSCA Central States Corrections Association

CSD Correctional Services Department; Corrective Services Department

CSF Correctional Service Federation

CSM Correctional Service of Minnesota (Minneapolis)

Cuba a *cárcel modelo* can be found in Havana, Santiago de Cuba, Camaguey, Matanzas; there is a large penal settlement on the Isle of Pines

Cuernavaca prison in Cuernavaca, capital of the Mexican state of Morelos

Cueva Panamanian island prison

cuff(s) handcuff(s)

Culiacan prison in Culiacan, capital city of the Mexican state of Sinaloa

Cummins Cummins Prison Farm (facility of the Arkansas State Penitentiary holding girls and women convicted of illegal drug use and prostitution); Cummins Unit in Grady, Arkansas

Custer South Dakota location of the Youth Forestry Camp

custodial officer(s) prison guard(s)

Cux Cuxhaven, Germany or its jail

Cuyahoga Cuyahoga County Juvenile Detention Home in Cleveland, Ohio

CYC Colorado Youth Center at Denver

Cyprus island republic whose capital, Nicosia, has an old Turkish prison and a nearby camp site where Jews were detained to keep them out of Palestine from 1945 to 1948

Czechoslovakia Prague, the capital, and Brno, Bratislava, and Ostrava all have old prisons

CZ Pen Canal Zone Penitentiary at Gamboa

DAC Department of Adult Corrections (Alaska)

Dacca principal prison of Bangladesh in the capital city of Dacca

Dachau (Old German—marsh) —site of a large Nazi concentration camp near Munich

daddy tank prison cell reserved for lesbians to keep them from being attacked by other prisoners

Dade Correctional Center at Homestead, Florida offers offenders counseling, education, and vocational training

Dago slang for San Diego, California or its jail, lockups, or prison (*see* Diego)

Dallas Texas county jail; State Correctional Institution at Dallas, Pennsylvania

Dalmatia (*see* Yugoslavia)

Damascus (*see* Syria)

Damon Israeli medium-security prison for adult repeat offenders and Arab juvenile delinquents from 14 to 20 years old

Da Nang (*see* Vietnam)

Danbury Federal Correctional Institution at Danbury, Connecticut

dance death by hanging

dance hall cell or hallway leading to an execution chamber where the condemned seems to dance when the current is applied to the electric chair

dance of death hanging

dance on air death by hanging

dangerous aliens Canadian government designation of Japanese-Canadians in the post-Pearl-Harbor period

Dannemora Clinton Correctional Facility at Dannemora, New York

Danville Youth Development Center southwest of Lexington, Kentucky with special units for male delinquents

darbies handcuffs

Dar-es-Salaam (*see* Tanzania)

Darrington Unit Texas penal facility holding maximum-security felons within its confines in Rosharon

Dartmoor Her (His) Majesty's

prison near Princetown and the Dartmoor Forest of southwest England

Davao port city of Mindanao in the Philippines and nickname of its old prison

Day Dayton, Ohio or its jails and lockups

Daytona Beach Florida site of the Tomoka Correctional Institution

Dayton Forensic Hospital in Dayton, Ohio opened in 1980 to give psychiatric help to mentally ill felons

Daytop Daytop Lodge in Staten Island, New York where male narcotic violators are offered treatment

db dirt bag (an undocumented Hispanic alien)

DB Disciplinary Barracks

dbd death by drugs (execution by lethal injection)

dc death cell

DC District of Columbia jail in Washington, DC

D of C Department of Corrections; District of Columbia

DCI Donovan Correctional Institute, jail on Otay Mesa south of San Diego

DCJ Dade County Jail in downtown Miami, Florida

DCS Department of Correctional Services (Nebraska, New York)

DCWDC District of Columbia Women's Detention Center

DEA Drug Enforcement Administration

Dead Men's Cove San Diego Police Department headquarters and lockup

Dear John letter written communication to a prisoner from a lover or wife informing him their engagement or marriage is over

death by injection a massive overdose of sodium thiopental to stop breathing while pavulon, a muscle relaxer, and potassium chloride are added to stop heartbeat

death penalty execution

death row cell block reserved for prisoners awaiting execution

decap decapitation

Deerbolt HM Borstal at Deerbolt in England

Deerfield Correctional Center in Capron, Virginia

Deer Lodge site of the Montana State Prison

defective delinquent(s) criminally insane person(s)

DeHoCo Detroit House of Correction

Delaware (*see* Claymont, Dover, Georgetown, Smyrna, Wilmington)

Delaware Correctional Center in Smyrna north of Dover

Delaware jails three counties provide lockups for communities around Wilmington, Dover, and Georgetown

Delhi (*see* Indian prisons)

Delle Stinche Florentine prison dating from the early 1300s and famous for its advanced methods of inmate handling and segregation

Del Norte prison site in northwestern California

Denmark (*see* Herstedvester and Ringe); Copenhagen and Arhus have central prisons

Denver largest city in Colorado and nickname of its penal institutions

De Quincy Louisiana Correctional and Industrial School at De Quincy in Calcasieu Parish

Descanso Descanso Detention Facility near Alpine, California

Des Moines Iowa's county seat, courthouse, and jail of Polk County; well known for its role in correctional institution planning

Des Moines Plan innovative alternative to building new prisons; less risky convicts are steered to programs allowing them to help themselves by working or attending school under probationary supervision

De Soto De Soto Correctional Institution in south-central Florida

detention legal confinement of a person subject to criminal or juvenile court proceedings until commitment to a correctional facility or release

detention center jail housing prisoners awaiting trail

detention facility generic term for county farm, detention center, honor farm, jail, juvenile hall, road camp, work camp, etc.

detention glazing chemically strengthened and plastically bonded glass used in place of bars or walls in modern penal facilities

detention screening stainless-steel screening allowing air, light, and sound to enter but preventing the entrance of contraband such as drugs and weapons

detention windows designed to admit air and/or light but to prevent the escape of prisoners (*see* detention glazing)

Detoxification Center (*see* Inebriate Reception Center)

Detroit Wayne County Jail in downtown Detroit, Michigan

Deuel Vocational Institute *see* Tracy

devil's front porch prison

Devil's Island penal colony in French Guiana and its group of offshore islands in use as late as 1950 for convicts and political prisoners serving lifetime or long-duration sentences

DHC Detroit House of Correction

DHS Department of Health Services

DIA Department of Institutions and Agencies (governing New Jersey's prison system)

Diagnostic and Evaluation Center in Lincoln, Nebraska holds male felons in maximum security while under diagnosis and evaluation

Diagnostic Unit in Huntsville, Texas where all incoming inmates are diagnosed and sent to appropriate prisons

Diego (Mexican-American truncation—San Diego)—California border city or its jail, lockups, or prison (*see* Dago)

Diego Met San Diego Metropolitan Correctional Center (Hispanic appellation)

die in the hot seat be electrocuted

die of lead poisoning killed by lead bullets

die of throat trouble hanged

Directory of Juvenile Detention Homes published by the National Juvenile Detention Association

Directory of Prisoners Aid Agencies published by the International Prisoners Aid Association

Directory of Residential Treatment Centers published by the International Halfway House Association

dirt prisoner's nickname for

sugar

dirty towel prison barbershop or beauty parlor

Dismas House halfway house in St Louis, Missouri

District of Columbia site of the Central Detention Facility, and some ill-reputed prisons used during and after the Civil War but replaced by modern facilities in the District and in Lorton, Virginia

District of Columbia lockup the nation's capital lockup facilities in addition to those of federal agencies

dite (Mexican-Americanism— detention hall)—place where illegal aliens must wait while being investigated

Division No 1 Chicago's Cook County Jail

Division No 2 Chicago's House of Correction

Divisions No 1 through 6 (*see* Cook County)

Dix Dorothea Lynde Dix (1802–1887)—American reformer and pioneer in securing better treatment for the insane in asylums, poor houses, and prisons throughout New England; wrote books for children and served as superintendent of women nurses during the Civil War

Dixon Correctional Institute medium-security penal facility in northern Louisiana at Jackson

Diyarbakir prison in southeastern Turkey

DJ Department of Justice (*see* U.S. Department of Justice); Don Jail in County Donegal, Northern Ireland; Don Jail in the Don Mills section of Toronto, Ontario

Djibouti formerly known as French Somaliland and later the French territory of Afars and Issas; its old prison attracts strange characters

DLPS Department of Law and Public Safety (New Jersey)

do a bit serve a sentence in jail

do a dime serve a 10-year prison sentence

do a nickel serve a five-year prison term

do a pound serve a five-year prison term

do a quarter serve a 25-year prison sentence

Dobbs School North Carolina penal facility for male and fe-

male juvenile delinquents up to 18 years of age, in Kinston

do bird do time in prison; serve a prison sentence

D o C Department of Correction (Arkansas, Connecticut, Delaware, Indiana, Massachusetts, North Carolina, Tennessee); Department of Corrections (Arizona, California, District of Columbia, Florida, Guam, Idaho, Illinois, Kansas, Kentucky, Louisiana, Maine, Michigan, Minnesota, Mississippi, Missouri, New Jersey, Rhode Island, South Carolina, Texas, Vermont, Washington, West Virginia); Division of Corrections (Utah, Wisconsin)

DOCS Department of Correctional Services (New York)

Doftana Bucharest's prison

dog detective(s) [*see* canine shamus(es)]

Doha (*see* Qatar)

D o I Department of Institutions (Montana); Director of Institutions (North Dakota); Division of Institutions (Oklahoma)

Dominica large Windward island; its capital, Roseau, contains a small gaol built by the British

Dominican Republic prisons notorious during the dictatorship of Trujillo and before intervention by the U.S. Marines (*see* Ciudad Trujillo, Santo Domingo)

Don Don Jail in County Donegal, Northern Ireland; Don Jail in Don Mills section of Toronto, Ontario

Donovan Richard J Donovan Correctional Facility, Otay Mesa, California

D o P Department of Prisons (Nevada)

DOR Department of Offender Rehabilitation (Georgia)

Dora nickname for the Nazi concentration camp west of Leipzig, Germany at Nordhausen

Dorandordhausen Nazi concentration camp west of Leipzig, Germany

Dorchester Boston Pre-Release Center in Dorchester, Massachusetts; Dorchester Penitentiary in New Brunswick, Canada; HM Prison at Dorchester, England

Dorpat (*see* Tartu)

dossier (French—information

file)—usually about a case history or individual(s) involved in some criminal or political action

do the book serve a life sentence

do time serve a prison sentence

Douala (*see* Cameroon)

double-ceiling placing two prisoners in the same cell

Dover Delaware seat of New Castle County south of Wilmington; contains the Kent Correctional Institution and police-station lockups as well as the county jail; in Kent, England, is HM Borstal in Dover on the Channel

Down Down Home (Afro-American slang—federal penitentiary in Atlanta, Georgia)

Down Home (Afro-American slang—Manhattan House of Detention long known as the Tombs)—in downtown New York City

Downstate Downstate Correctional Facility at Fishkill, New York

Down South (Afro-American nickname—U.S. Federal Penitentiary in Atlanta, Georgia)

Dozier Arthur G Dozier School at Marianna, Florida for delinquents from 12 to 17 years of age

DPA Discharged Prisoners Association (Great Britain)

DPAS Discharged Prisoners' Aid Society (Great Britain)

DPs Detention Pens in downtown New York City, also known as the Manhattan Detention Pens

DPS Department of Public Safety (American Samoa)

DPSCS Department of Public Safety and Correctional Services (Maryland)

Drake Hall HM Prison at Drake Hall, Staffordshire, England

Drancy (*see Camp de Drancy*)

Draper Draper Correctional Center at Elmore, Alabama; Utah State Prison at Draper south of Salt Lake City

DRC Department of Rehabilitation and Correction (Ohio)

dropped in the bucket jailed

drowning capital punishment usually reserved for witches and other women accused of crime; drowning was replaced by beheading and hanging as less equipment was required

Drumheller Drumheller Institution in Drumheller, Alberta

drunk tank jail cell reserved for persons arrested while under the influence of alcohol or other drugs

Dry Tortugas U.S. Military Prison on one of the Dry Tortugas in the Gulf of Mexico, west of Key West

Dubai (see United Arab Emirates)

Dublin (see Irish prisons)

Dubrolag complex of some fifteen prison camps close to Potmu in the Moldavian Republic of the USSR

Ducato Milanese Milan, Italy's old prison

Duffy of San Quentin motion picture inspired by three semi-autobiographical books by Warden Clinton T Duffy of San Quentin Prison in California where he abolished airless and dungeon-like cells, fired guards for cruelty, introduced a cafeteria, a newspaper written and printed by the prisoners, and a night school; he insisted the death penalty never deterred murder and never will

Dumfries Dumfries Young Offenders Institution near the west coast of Scotland

dummy prisoner's nickname for bread

dump truck depressed, slow moving, or torpid prisoner

dungeon underground cell or prison

Durango prison in Durango, Mexico; jail in Durango, Colorado

Durban (see South Africa)

Durchgangslager (German—transit camp)—established in German-occupied countries for the transport of Jews and others to concentration camps

Durham HM Prison at Durham, England; prison in Durham, North Carolina and in smaller Durhams

Dutchman Correctional Center in Enoree, South Carolina, holding felons in maximum and minimum security

Dwight Dwight Correctional Center west of Kankakee, Illinois details female felons in maximum, medium, and minimum security

DYA Department of the Youth Authority (California)

DYS Department of Youth Services (Alabama); Division of Youth Services (Arkansas, Florida)

Eagle River Alaska Women's Facility next door to the Eagle River Correctional Center east of Anchorage

Eagle Springs North Carolina site of the Samarkand Manor youth service facility

East Block No 7 address of Interpol India on Rama Krishna Puram in New Delhi

East County Regional Center El Cajon, California, jail

Eastern Hong Kong's Eastern Police Station (and lockup); Eastern Correctional Facility at Napanoch, New York

Eastern New York Correctional Facility at Napanoch for male juvenile delinquents

Eastern State Penitentiary in North Philadelphia, Pennsylvania for more than a century but now replaced by the State Correctional Institution in Graterford

Eastham Unit maximum-security penal facility in Lovelady, Texas

East Lake Alabama State Training School for female juvenile delinquents

East London (see South Africa)

East Moline East Moline Correctional Center in upper Illinois opened in 1980 as a minimum security penal facility

East Palatka East Palatka Road Prison southeast of St Augustine; opened in 1961 with programs in alcohol and drug abuse

East Sutton Park borstal for delinquent girls in Kent, England

Ebensee Austrian concentration-camp subcamp during World War II

e by i execution by injection (of air or poison)

Echo Glen Children's Center in Snoqualmie, Washington (see entry)

ECJ Erie County Jail (Buffalo, New York)

Ecole Notre-Dame de Laval at Laval-des-Rapides in the Province of Québec (this and the following Ecole-type entries are for juvenile reformatories)

Ecole Ste-Agnes Montreal, Québec

Ecole Ste Domitille Montreal, Québec

Ecole Ste Helene correctional school in Montreal, Québec

ECS Episcopal Community Services (job-entry program for ex-offenders)

Ecuador its Andean capital, Quito, and its port city, Guayaquil, have so-called model prisons; former penal colony at Villamil on Isabela Island in the Galápagos has been replaced by the Penitenciario Litoral, a maximum-security prison near Guayaquil

Eddyville site of the Kentucky State Penitentiary

Edinburgh prison in Scotland's capital city and its Young Offenders Institution

Edison medicine nickname for electroconvulsive shock treatment

Edmonton Institution maximum-security facility in Edmonton, Alberta

education of correctional personnel and jailers (see Staff Training Centers)

EFEC Efforts From Ex-Convicts (Washington, DC's parole program)

Eglin Federal Prison Camp at Eglin Air Force Base in Florida

Egypt the Arab Republic of Egypt has penal facilities in Cairo, as well as Alexandria, Giza, and other cities

Elazik Turkish prison in the central eastern part of the country

El Cajon El Cajon Detention Facility, El Cajon, California

El Centro El Centro Detention Center of the Immigration and Naturalization Service in El Centro, California

El Chipote (Spanish—the box) —Managua's, Nicaragua's underground jail beneath a hill in the center of the capital

Eldora Training School medium-security juvenile-delinquent facility southwest of Waterloo, Iowa

electrocution method of execution in Alabama, Arizona, Arkansas, Colorado, Florida, Georgia, Illinois, Indiana, Kentucky, Louisiana, Massachusetts, Nebraska, New York, Pennsylvania, South Carolina, South Dakota, Tennessee, Vermont, Virginia

El Frontón (Spanish—coastal cliff)—barren island prison off Callao, Peru

Elko site of the Nevada Youth Training Center

Ellis Unit maximum-security unit complete with a death row in Huntsville, Texas

el Met (Spanish—the Metropolitan Correctional Center)

Elmira Elmira Correctional Facility in New York State; Elmira Reception Center for male prisoners

El Paso El Paso Detention Center of the Immigration and Naturalization Service in El Paso, Texas

El Pavon (Spanish—the peacock)—Guatemala's largest prison, near Guatemala City

El Reno Federal Correctional Institution at El Reno, Oklahoma

El Retén de Catía (Spanish—The Remand of Catía)—Venezuelan house of detention in the Catía suburb of Caracas

El Salvador has a prison in its capital city, San Salvador, and smaller ones in its seaports along the Pacific

El Sexto (Spanish—The Sixth Book of Canonical Decrees)—Lima, Peru prison noted for the Easter plays staged by its inmates

Endsville place convicts dream about when they imagine life outside prison

Enfield Connecticut State Prison at Enfield includes a prison farm

England (see United Kingdom and individual entries)

Englewood Federal Correctional Institution at Englewood, Colorado

English nicknames for penal colonies Andaman Islands, Botany Bay, Devil's Island, French Guiana, Lipari Islands, Moreton Bay, New Caledonia, Norfolk Island, Port Blair, Port Philip, Siberia, Solovetski Islands, Sydney Cove, Tasmania, Van Diemen's Land

Enoree South Carolina site of the Dutchman Correctional Institution

Ensenada regional jail in Ensenada, California

Ensisheim French penitentiary near Mulhouse

Equatorial Guinea Fernando Po island and Rio Muni on the nearby West African coast both have old prisons built by the Portuguese

ERC Elmira Reception Center (for male prisoners) at Elmira, New York

escape unlawful departure of a lawfully confined person

Esmeralda barkentine-rigged Chilean naval-training vessel used for a short time in 1973 as a prison holding communist activists

ESP Eastern State Penitentiary (Philadelphia)

Essen concentration-camp subcamp operated by the Nazis during World War II

Essex Essex County Jail in Newark, New Jersey

Etcher of Prisons Giambattista Piranesi (see Carceri d'Invenzione)

Ethan Allen School in Wales, Wisconsin serves as a reception center and training school

Ethiopia's prisons Addis Ababa, capital of Ethiopia, as well as Asmara and Massawa have prisons

Eugene Eugene, Oregon or its Lane County Jail

euphemisms of penology prison or penitentiary became reformatory, which became correctional center, and now is called a rehabilitation facility

Eureka Humboldt County courthouse and jail in California

Evanston Wyoming site of the Johnson Hall Forensic Unit and the Wyoming Womens Center

even-handed justice democratic doctrine advocating equal justice for criminals irrespective of their former position in society, their race, their religion, or their wealth

Evin Evin Prison in Teheran, Iran

Évreux maximum-security prison in the French city of évreux

EXCEL Ex-offender Coordinated Employment Lifeline (Indiana's parole project)

ex-con(s) ex-convict(s); former convict(s)

execution box container holding a body belt, a hangman's rope, a hood, and restraining straps for fastening the limbs

of the condemned

Execution Dock formerly on the muddy foreshore of a bend in the Thames at East Wapping below the Tower Bridge and the Tower of London; at first all convicted pirates were pegged down at low water so the incoming tide would drown them slowly but in later times they were hanged from a tall gallows and left to the mercy of seagulls as they decomposed within the chains suspending them above the river's reach

execution by injection injection of air or poison; less complicated and less costly than more conventional methods; abbreviated as e by i

Execution of Maximilian Edouard Manet's painting depicting the Emperor Maximilian and his Mexican generals standing between a wall and a uniformed firing squad

execution methods by states electrocution in Alabama, Arizona, Arkansas, Colorado, Florida, Georgia, Illinois, Indiana, Kentucky, Louisiana, Massachusetts, Nebraska, New York, Pennsylvania, South Carolina, South Dakota, Tennessee, Vermont, Virgina; hanging in Delaware, Montana, New Hampshire, Utah, Washington; lethal gas in California, Maryland, Mississippi, Missouri, Nevada, North Carolina, Oregon, Rhode Island, Wyoming; lethal intravenous injection in Idaho, New Mexico, Oklahoma, Texas; in Utah the prisoner chooses the method of execution or the sentencing judge decides whether execution is by firing squad or hanging

execution pennant small black flag flown over British prisons when an execution is taking place

execution shed gallows area within a prison

exercise in a cooler climate translation of a Soviet euphemism for forced labor in Siberia

Exeter HM Prison at Exeter in Devonshire, England; or jails in smaller Exeters in California, Illinois, Maine, Missouri, Nebraska, New Hampshire, Ontario, Pennsylvania, and

Rhode Island

exile (*see* banishment)

EXIT Ex-offenders in Transit

Ex-offender Coordinated Employment Lifeline (EXCEL) Indiana's parole project

ex-offender(s) former offender(s) no longer under the jurisdiction of any criminal-justice agency

Ex-offenders in Transit (EXIT) Maine's parole project

expunge purge or seal arrest, criminal, or juvenile-delinquent records

expungement legal ablution of a criminal's record made in an effort to assist in rehabilitation and remove any prejudice from the mind of a potential employer

extreme penalty death by execution

Eye Opener inmate publication of the Oklahoma State Prison

Fabrica de Hombres Nuevos (Spanish—Factory of New Men)—Mexico City prison constructed to permit prisoner's wives to stay overnight

Fadiffolu Fadiffolu Atoll (Maldivian island used for the banishment of lawbreakers)

Fagatoa Lockup Pago Pago jail on the island of Tutuila in American Samoa

fag factory homosexual-filled prison

Fairbanks Fairbanks Correctional Center in Alaska (a maximum-security facility for male and female felons)

Fairfield Solano County courthouse and jail in California

Fairfield School for Boys (*see* Southeastern Ohio Training Center)

Fairton federal correctional institution near Bridgeton, New Jersey

Falklands Falkland Islands capital, Stanley, houses HM Gaol

Fängelse (Swedish—Prison)— title of Ingmar Bergman's thought-provoking motion picture whose English name is *Prison*

Fannie Bay Fannie Bay Labour Prison, Fannie Bay, Australian Northern Territory

Fargo Fargo, North Dakota or its Cass County jail

farm confinement or correctional facility in a rural area;

prison farm

Farmingdale Turrell Residential Group Center at Farmingdale, New Jersey

Father of Criminology Cesare Beccaria (1738–1794), author of *Delitte e della Pene* (Italian—Crimes and Punishment)—advocated an end to capital punishment and widespread prison reform

Father of Parole Captain Alexander Maconochie, the governor of the Norfolk Island penal colony from 1840 to 1844

Father of Penitentiary Science Jean Jacques Vilain founded the Maison de Force built in Ghent in 1773

Fayetteville Fayetteville (locally called *Fitzville*), North Carolina, seat of Cumberland County with its jail

FBI Federal Bureau of Investigation

FBP Federal Bureau of Prisons (U.S. Department of Justice bureau)

FCCD Florida Council on Crime and Delinquency

FCF Federal Correctional Facility

FCI Federal Correctional Institute of the U.S. Bureau of Prisons (*see* Staff Training Centers)

FCI(s) Federal Correctional Institution(s)

FCIS Foreign Counterintelligence System (FBI)

FDH Federal Detention Headquarters in Florence, Arizona

featherbed padded cell

Featherstone HM Prison at Featherstone, north of Wolverhampton, Staffordshire, England

Federal Center for Correctional Research U.S. Bureau of Prisons facility in Butner, North Carolina

Federal Correctional Institutions Alderson, West Virginia; Ashland, Kentucky; Bastrop, Texas; Butner, North Carolina; Danbury, Connecticut; El Reno, Oklahoma; Englewood, Colorado; Fort Worth, Texas; La Tuna, Texas; Lexington, Kentucky; Memphis, Tennessee; Miami, Florida; Milan, Michigan; Morgantown, West Virginia; Otisville, New York; Oxford, Wisconsin; Petersburg, Vir-

ginia; Pleasanton, California; Ray Brook, New York; Sandstone, Michigan; Tallahassee, Florida; Talladega, Alabama; Terminal Island, California; Texarkana, Texas

Federal Detention Center in Florence, Arizona's desert country

Federal Law Enforcement Training Center (*see* Fleetsie)

Federal Penitentiaries (U.S.) Atlanta, Georgia; Leavenworth, Kansas; Lewisburg, Pennsylvania; Marion, Illinois; Terre Haute, Indiana

Federal Prison Camps Allenwood, Pennsylvania; Big Spring, Texas; Boron, California; Eglin Air Force Base, Florida; Maxwell Air Force Base, Montgomery, Alabama; Safford, Arizona; Seagoville, Texas

Federal Prison Industries (*see* UNICOR)

Federal Probation Officers Association founded in 1955 to build and maintain enlightened public interest in parole, probation, and related services

Federal Training Centre in Laval, Québec for juvenile offenders

Fed Ref Federal Reformatory

Fellowship of First Fleeters Australian society whose members must prove they were descended from the first shipment of convicts landed in Botany Bay in 1788

felon anyone who has committed a felony

felonry prison-colony population

felon swell upper-class convict

felony any crime punishable by imprisonment for more than a year or by death

felony tank jail cell reserved for felons

Feltham HM Borstal at Feltham in Middlesex, England

female penal institutions of the U.S. Bureau of Prisons Alderson, West Virginia; Pleasanton, California

Ferguson Unit Texas penal facility holding first offenders in maximum security at Midway

Fernando de Noronha Brazilian penal settlement existing since the 18th century, on an island of the same name in the South Atlantic

Fernando Po prisons in the

small island nation of Equatorial Guinea in its seaport cities, Malabo and Bata

fettered restrained by ankle or leg fetters or both

fetters steel cuffs placed on the ankles or legs of prisoners to keep them from escaping

Fez (*see* Morocco)

fiebre carcelaria (Spanish—prison fever)—fear of imprisonment

Fiesta de la Merced (Spanish—Mercy Fiesta)—celebrated on September 24, also known as Prisoners' Saint's Day

Fijian prisons except for drunks in lockups, anti-social criminals are segregated and forced to live in the most remote parts of this island nation (*see* Suva)

Filicudi Italian isle of exile in the Lipari Islands

finishing school euphemism for a women's prison, especially one for young women

Finnamore Wood HM Borstal at Finnamore Wood in Buckinghamshire, England

Finnish prisons Finland's capital city, Helsinki, and cities such as Tampere and Turku have prisons, but the emphasis is on rehabilitation rather than incarceration

Firlands minimum-security facility at Firlands, Washington

First American Penitentiary Walnut Street Jail in Philadelphia, built in 1790

First Execution by Electrocution August 6, 1890 in New York State's Auburn Prison

First Halfway House Isaac T Hooper Home opened in New York City in 1845 by the Society of Friends—Quakers

First Nazi Concentration Camp established in Dachau, near Munich, Germany, on March 23, 1933, less than three months after President von Hindenburg appointed Hitler as Reich Chancellor (prime minister); by April 26 the Gestapo was formed, and by May 10 all books by Jews, and all books opposing Nazism, were burned; the playing of music by composers of Judaic origin was forbidden

First Prison Newspaper *The Summary* published by the inmates of the New York State Reformatory at Elmira on No-

vember 22, 1883—Thanksgiving Day

fish(es) newly arrived inmate(s); pimp(s)

Fishkill Fishkill Correctional Facility near Beacon-on-Hudson, New York

Fitzville (*see* Fayetteville)

five spot five-year prison term (or a five-dollar bill)

FKL *Frauen Konzentration Lager* (German—Women's Concentration Camp)— Hitler-era prison

Flagstaff Coconino County jail in Arizona

flat bit prison sentence for a definite period of time (*see* split bit)

Fleet Fleet Prison—London's jail dating from 14th century was finally demolished in 1846, after years of service as a debtor's prison

Fleetsie nickname of the Federal Law Enforcement Training Center in Brunswick, Georgia

FLETC Federal Law Enforcement Training Center (*see* Fleetsie)

Fleury Mérgois Europe's largest prison and France's largest high-security facility

flex-cuf ties flexible plastic ankle cuffs or handcuffs for trussing prisoners

floating hells British prison ships bound for Australia and Tasmania during the late 18th and early 19th centuries; French prison ships bound for Algeria, French Guiana, and New Caledonia, and to the convict colony in French Guiana as recently as 1950

Florence Arizona State Prison; Florence Detention Headquarters holding persons awaiting trial or serving short sentences; South Carolina site of the Palmer Work Release Center

Florida penal facilities are numerous and widespread from Apalachee and Avon Park to Vero Beach and Zephyrhills as well as in Arcadia, Belle Glade, Broward, Bushnell, Chattachoochee, Clermont, Cross City, Daytona Beach, East Palatka, Homestead, Immakalee, Lake Butler, Lantana, Lowell, Niceville, Olustee, Pembroke Pines, Polk City, Raiford, Riverview,

Sneads, Sharpes, Starke, and Trenton

Florida Correctional Institution provides custody for female felons and a measure of rehabilitation through education and vocational training at Lowell

Florida lockups 67 counties are served from Gainesville in Alachua County to Chipley in Washington County; Miami, Jacksonville, and Tampa have their own facilities

Florida School Florida School at Okeechobee holding delinquents from 12 to 17 years of age

Florida State Prison at Starke, offers inmates academic and vocational training as well as on-the-job training

Flossenbürg German prison and town near Nürnberg

Floyd Floyd County Jail in Rome, Georgia

fly a kite smuggle a letter out of prison

fly the coop excape from jail or prison

Folsom California State Prison at Folsom

forçat (French—convict)—prisoner

forced-labor prison camps Germany listed more than 1600 forced-labor concentration camps erected during Hitler's regime; they are still popular in the USSR although long abolished in Germany

Ford HM Prison at Ford in Sussex, England

Fordland Missouri site of the Ozark Correctional Center

Forest Hill Forest Hill Prison in Georgetown, District of Columbia, during the Civil War and shortly after

Fort Apache nickname of the police station and lockup in New York City's south Bronx, also nicknamed the Little House on the Prairie

Fort Benning Fort Benning Confinement Facility at Fort Benning, Georgia

Fort Campbell Fort Campbell Confinement Facility at Fort Campbell, Kentucky

Fort Carson Fort Carson Confinement Facility at Fort Carson, Colorado

Fort Christian Fort Christian Detention Center on St Thomas island in Charlotte

Amalie **Fort-de-France** Martinique's prison in the capital and port city of this French West Indian island

Fort Dimanche Haiti's infamous prison

Fort Gordon Fort Gordon Confinement Facility at Fort Gordon, Georgia

Fort Grant Fort Grant Training Center in Arizona, offers educational and vocation training to felons

Fort Hood Fort Hood Confinement Facility at Fort Hood, Texas

Fort Jefferson former military prison on an island in the Dry Tortugas near Key West, Florida

Fort Knox Fort Knox Area Confinement Facility at Fort Knox, Kentucky

Fort Lauderdale Broward County, Florida, jail

Fort Leavenworth U.S. Disciplinary Barracks at Fort Leavenworth, Kansas

Fort Lewis Fort Lewis Confinement Facility at Fort Lewis, Washington

Fort Liquordale Fort Lauderdale, Florida's jail

Fort Madison Iowa State Penitentiary at Fort Madison

Fort Margherita old Sarawak prison near Kuching on the island of Borneo or Kalimantan served as a Japanese detention camp for Australian and British prisoners during World War II

Fort Meade Fort Meade Confinement Facility, Fort Meade, Maryland

Fort Montluc old prison in Lyon, France used during World War II as Gestapo headquarters by occupying Germans

Fort Ord Fort Ord Confinement Facility, Fort Ord, California

Fort Pillow Fort Pillow State Prison Farm on Cold Creek, Tennessee

Fort Polk Fort Polk Confinement Facility at Fort Polk, Louisiana

Fort Richardson Fort Richardson Confinement Facility at Fort Richardson, Alaska

Fort Riley U.S. Army Correctional Training Facility at Fort Riley, Kansas

Fort Saskatchewan Correctional Institution northeast of Edmonton also runs forest camps

Fort Savage nickname of New York City's East Harlem police station and lockup

Fort Sill Fort Sill Confinement Facility at Fort Sill, Oklahoma

Fort Wayne Indiana seat, courthouse, and jail of Allen County

Fort Worth Federal Correctional Institution at Fort Worth, Texas west of Dallas

Fossoli di Carpi transit camp north of Modena, Italy, built by Mussolini's Black Shirts as a depot for Jews and others enroute to concentration camps

Fountain GK Fountain Correctional Center at Holman Station, Alabama

four-time loser(s) criminal(s) convicted four times of felonies and hence imprisoned for life under the New York State statute called the Baumes Law

Fox Hill Her Majesty's Prison at Fox Hill on New Providence Island in the Bahamas

Fox Lake Wisconsin Correctional Institution (medium-security prison)

FPC Federal Prison Camp (Allenwood, Pennsylvania; Eglin Air Force Base, Florida; Lompoc, California; Marion, Illinois; Montgomery, Alabama; Safford, Arizona)

FPI Federal Prison Industries

FPOA Federal Probation Officers Association

FR Federal Reformatory (El Reno, Oklahoma and Petersburg, Virginia)

Framingham Massachusetts Correctional Institution (for women) in Framingham

free for all fight wherein all present participate

Freetown (see Sierra Leone)

free world outside prison walls

Fremantle Fremantle Gaol (Western Australian penal institution built by convicts)

French blade the guillotine

French Congo (see Congo)

French leave act of slipping away quietly and secretly

French prisons metropolitan France has many prisons and overseas prison colonies once were notorious in French Guiana (popularly named Devil's Island) and in New Caledonia; (see Bourgoin, Cayenne, Chateau d'If, Clermont-Ferrand, Conciergerie, Devil's Island, Évreux, Fort-de-France, Grand Hotel, Lisieux, *maison*, Mende, Natzweiler, *prison*, Salpetriere, Santé, Tarbes, Tulle, Vincennes)

Frentes Abiertos (Spanish— Open Fronts)—minimum-security prisons in Cuba

fresh fish(es) new prisoner(s)

fresh and sweet just out of jail

Fresnes southern suburb of Paris and location of Fresnes penitentiary

Fresno Fresno County courthouse and jail, in central California

Friarton Friarton Young Offenders Institution at Perth, Scotland

fried badly burned; electrocuted; intoxicated

Friends of Assata and Sundiata underground prison-support movement in New York City

Frisco San Francisco, California or its jails and lockups

frisk search a body for concealed drugs, contraband, weapons, etc.

Frobisher Bay on the southeast coast of Baffin Land in the Canadian Arctic's Northwest Territories contains the Baffin Correctional Centre

frog's march nickname for a method of conveying hard-to-handle prisoners (four officers each grab an arm or leg and carry the prisoner along face downward)

Frontera California Institution for Women at Frontera

Frontón Peruvian maximum-security offshore prison on an island close to Callao

Frostbite Fairbanks, Alaska or its jail and lockup

FRW Federal Reformatory for Women (Alderson, West Virginia)

fry burn badly; electrocute

Fuchu Fuchu Prison (Japan's unheated maximum-security facility in Greater Tokyo)

Fukuoka (see Japan's prisons)

Funafuti (see Tuvalu)

Fungus Corners Bremerton, Washington and its jail as well as police-station lockup

funny farm insane asylum;

psychiatric ward

Futility Hill prisoner's graveyard at San Quentin, California

FYC Federal Youth Center (Ashland, Kentucky and Englewood, Colorado)

Gabonese prisons Gabon has prisons in the capital and coastal city of Libreville as well as in Port-Gentil

Gadsden Etowah County seat and jail northea... of Birmingham, Alabama

Gainesville Texas site of Brownwood State School and the Statewide Reception Center

Galle (see Sri Lanka)

gallery 13 prisoner's grave-(yard)

gallows metal or wooden framework used for the execution of criminals by hanging

gallows bird(s) criminal(s)

Galvy Galveston, Texas or its jail or lockup

Gambian prison Gambia's capital, Banjul, has a jail built by the British

Gamboa rehabilitation center flanking the Panama Canal's dredging division close to the Gaillard Cut

Gamle Swedish prison on a Baltic inlet near Vastervik

gaol (British spelling—jail)— term introduced into Britain during the Norman Conquest and equivalent of the French *geôle*

gaolage a gaoler's fee (bribe)

gaoler (British—jailer)

Garbage Dump nickname of the Great Meadow Correctional Facility at Comstock, New York and of California's San Quentin Prison

Gardner Massachusetts location of the North Central Correctional Institution

garlic and glue convict slang for beef stew

garnish(es) bribe(s) given prison guards by inmates

garrote capital-punishment device of Spanish origin wherein the prisoner is strangled with an ever-tightening iron collar

Gartree Gartree Prison in Leicestershire, England

gas chamber specially built room where prisoners are executed by poison gas

Gasre Tehran, Iran's great

prison also called Ghasre or Qsar

gassing execution in a lethal gas chamber

Gates of Hell old nickname for the entrance to Macquarie Harbour on the Indian Ocean coast of Tasmania when it was a penal settlement in Van Diemen's Land

Gatesville Texas site of the Gatesville Unit holding females in maximum security

Gatun Gatun Prison for Women and Juveniles at Gatun in the Panama Canal Zone

Gavle Swedish experimental prison where inmates receive personal visits from members of their families

gcg gas-chamber green (nickname given a bilious green prevalent on the walls of many penal institutions)

GD Gaol Delivery (see jail delivery)

Geisenkirchen Nazi subcamp

Geneva Girls Training School, Geneva, Nebraska; Illinois State Training School for Girls at Geneva; (see Switzerland)

Georgetown Sussex Correctional Institution in Delaware; also the capital of Guyana and site of a prison built by the British

George Town (see Malayan prisons)

Georgia Diagnostic and Classification penal facility in Jackson, Georgia

Georgia Industrial Institute (see GII)

Georgia jails the state's 159 counties are served by prison facilities from Baxley in Appling County to Sylvester in Worth County; Atlanta, Columbus, Macon, and Savannah have local lockups

Georgia State Prison (see Reidsville)

Georgia Training and Development Center (see Buford)

geriatric institution nickname for an old prison

German prisons all of the concentration camps, extermination camps, and forced-labor camps listed elsewhere were erected and used during Hitler's regime; prisons still in use such as Berlin's Moabit or Spandau date back to the time of the kaisers

get the wind escape; take off

get the works be given a death sentence

Ghana's prisons during the 113 years of British rule jails were erected in Accra and Kumasi

ghost trains(s) late-night railroad train(s) used to transport prisoners from one place to another

Gib Gibraltar and its convicts formerly sentenced to hard labor on The Rock (an HM Gaol)

gibbet gallows for hanging prisoners from a projecting arm

Gibraltar British dependency off the south coast of Spain contains an HM Gaol

Gig Harbor Purdy Treatment Center for Women at Gig Harbor, Washington

Giglio Italian isle of exile in the Tuscan Islands; a law, enacted in post-World-War-II Italy, permits judges to exile suspected criminals to remote places such as Giglio

GII Georgia Industrial Institute at Alto

Gila Bend Indian reservation in Arizona used after Pearl Harbor as a relocation center for Japanese-Americans

Girls' Cottage School Chambly, Québec

Girls Rehab Girls Rehabilitation Facility (GRF) in San Diego, California

girl's school nickname for a reformatory for young female offenders

Girls' Town correctional facility for misdemeanants at Tecumseh, Oklahoma

Gitmo Guantanamo, Cuba (U.S. Naval Base jail or the jail in the nearby town of Guantanamo)

give a permanent wave electrocute

Giza (see Egypt)

GK Gaol Keeper

Glades Glades Correctional Institution at Belle Glade, Florida; offers English classes to Hispanic inmates as well as educational and vocational training

Glasgow prison in Scotland's seaport city, also called Barlinnie; includes the Barlinnie Young Offenders Institution

Glass House nickname for the

Los Angeles County Jail in California

glazing (*see* detention glazing)

Glenochil Glenochil Detention Centre in Scotland

glop unappetizing prison food

Gloucester HM Prison at Gloucester in Gloucestershire, England; city jail in Gloucester, Massachusetts

Golden Jefferson County courthouse and jail in Pueblo, Colorado

Golden Grove St Croix penal facility on the island of St Croix in the American Virgin Islands (*see* Anna's Hope *and* Fort Christian)

Golden Prison of Paris the Louvre—the great art museum formerly a fortress whose underground vaults held hunting dogs and political prisoners

Goldsboro North Carolina site of the Wayne Correctional Center

golpe final (Portuguese or Spanish—final blow)—death blow, execution

Goochland site of the Virginia Correctional Center for Women

Goodman Correctional Institution in Columbia, South Carolina cares for geriatric and handicapped inmates

good time time taken off a prisoner's sentence in return for good behavior

go over the hill escape

go over the wall escape

Goree Goree Island off Dakar (westernmost tip of Africa) served as the shipping point for slaves headed to the New World from the 1500s to the mid-1800s; today tourists visit its dungeons and are shown the Doorway of No Return in the House of Slaves

Goree Unit maximum-custody unit in Huntsville, Texas

Gorgona (*see* Colombia)

Gorki transit prison camp in the Russian city of Gorki, formerly Nizhni Novgorod

Goshen secure center for juvenile delinquents in Goshen, New York

go stir bug go crazy while imprisoned

Göteburg (*see* Sweden)

government men Australian euphemism for former convicts

Governor British penal equivalent of Warden

Grafton site of the West Virginia Industrial School for Boys for delinquents from 10 to 18 years of age in minimum custody

Graham Graham Correctional Center in Hillsboro, Illinois; medium-security facility

Grand Forks Grand Forks, North Dakota county jail

Grand Hotel nickname of French Polynesia's prison in Tahiti

Grand Island Grand Island, Nebraska county jail

Grand Mount Custodial School for Girls at Grand Mount, Washington

Granite site of the Oklahoma State Reformatory for maximum-security inmates

Grass Lake Michigan site of the Correction Camp Program

Graterford Pennsylvania Corrections Department prison southeast of Pottstown

graybar hotel jail, lockup, prison of any kind

Great Falls Great Falls, Montana county jail

Great Jailer of the Caribbean Fidel Castro

Great Meadow Great Meadow Correctional Facility—a maximum-security prison at Comstock, New York

Grecian prisons such structures in the Hellenic Republic date back to classical times in Athens, Piraeus, Patras, or smaller places

Green Bay Wisconsin location of the Green Bay Correctional Institution, holding first-offender juvenile delinquents and serving as a reception center for young adult males

Greencastle location of the Indiana State Farm

Green Haven Green Haven Correctional Facility—maximum-security prison near Stormville, New York

green lights in front of all New York City police stations and their lockups (*see* blue lights)

Greenock prison for female offenders, in Greenock, Scotland

Greensboro Greensboro, North Carolina jail in Guilford County

Greensburg Pennsylvania location of the Regional Correctional Facility

green triangle criminal identification badge worn in concentration camps controlled by the Nazis in World War II

Greenville South Carolina location of the Blue Ridge Pre-Release Work-Release Center

Grenada gaol in St George's where it was called HM Gaol during the years of British colonial administration

Grendon HM Prison Grendon at Grendon Underwood; the first psychiatric prison in the United Kingdom to have a full-time psychiatrist as its medical superintendent

GRF (*see* Girls Rehab)

Grimes County Texas facility in Navasota

Gross Rosen Nazi concentration camp known for the number of forced-labor slaves it furnished German industry

ground animal meat prison nickname for hamburgers, hot dogs, sausages

group home nonconfining residential facility for adjudicated adults or juveniles (*see* halfway house)

Gruenheid East Berlin prison

Guadalajara Mexican prison in Jalisco's capital city of Guadalajara

Guadalcanal (*see* Solomon Islands)

Guadalupe Guadalupe, Chihuahua and its jail

Guadeloupe an overseas department of France with lockups in Basse Terre and Guadeloupe

Guamanian prison Ordot is the penitentiary on the island of Guam

Guanajay Cuban prison southwest of Havana

Guanajuato prison in Guanajuato, Mexico

Guantanamo U.S. Naval Station at Guantanamo and its brig

guardhouse military jail or lockup

Guatemala City capital and site of a large prison

Guatemalan prisons Guatemala City has its *cárcel modelo* as well as lockups maintained by the military forces

Guaya Guayaquil, Ecuador or its jail

Guaymas prison in Guaymas, Mexican port on the Gulf of

California

guest of the city prisoner confined to a city jail

guest of the governor prisoner in a state penitentiary

guest of the nation prisoner in a federal penitentiary

guest of the realm prisoner in any HM gaol

guest of the state prisoner in a state penal institution

guillotine (*see* louisette)

guillotineur (French—guillotiner)—executioner using the guillotine

Guinea-Bissau former Portuguese colony with an old, unimproved prison in the capital and port city of Bissau

Guinea prisons Conakry, Labe, N'Zerekore, and Kankan have prisons built by the French

Gulag Archipelago Soviet author Alexander Solzhenitsyn's name for the forced labor camps and prisons throughout the USSR

Gulag of China northwest province of Qinghai where many forced labor camps are located

gurney hospital bed on wheels used to convey a death-sentenced criminal from death row to the chamber where lethal gas or injection is used to carry out the sentence

Gurs French internment camp for Spanish Nationalists fleeing Franco's troops and later for Jews interned by the Petain-Laval police

Gusen forced-labor camp run by the Nazis in Austria

Guy Guyana (nickname for any jail in this country once named British Guiana)

Guyama Guyama Regional Detention Center on the southeast coast of Puerto Rico

Guyana has a prison in its capital city of Georgetown as well as lockups in the interior

Guyane française (French Guiana)—extends along the tropical Atlantic coast of northern South America between 1 and 6 degrees north of the Equator; it qualifies as a fever-infested hellhole and until the early 1950s was notorious for its extensive penal colony nicknamed Devil's Island

Guy's Marsh HM Borstal at Guy's Marsh in Dorset, England

gyves handcuffs or fetters

Habana Havana has prisons dating back to the years of Spanish domination from 1492 to 1898; Morro Castle at the entrance to Habana Harbor contains an old Spanish prison still in use

hack(s) jailer(s); prison guard(s)

Haddam site of a Connecticut jail built in 1786 now occupied by the Connecticut Justice Academy

Hagerstown site of the Maryland Correctional Institution and the Maryland Correctional Training Center

Hague the Hague (*see* Netherlands)

Haiphong (*see* Vietnam)

Haitian prisons Haiti has prisons built by the French in Port-au-Prince and Cap-Haitien between 1677 and 1804

Halawa Halawa High Security Facility at Aiea on the island of Oahu, Hawaii (*see* Honolulu Jail)

half a stretch six month's imprisonment

halfway house nonconfining residential facility for adjudicated adults or juveniles; facility providing an alternative to confinement for persons not suitable for probation or needing a period of readjustment to the community after confinement

Hall Hall of Justice

Hallowell the Stevens School for female juvenile delinquents at Hallowell, Louisiana

Hamilton Hamilton County Jail in Cincinnati, Ohio

Hampton Road Fremantle, West Australian prison on Hampton Road

handcuffed and fettered held or restrained by handcuffs and fetters

handcuffs adjustable metal bracelets connected by a chain and used to restrain

hanging method of execution in Delaware, Montana, New Hampshire, Utah, Washington

hangman executioner

hangman's day customarily Friday

Hanoi (*see* Vietnam)

Hanoi Hilton nickname of Hanoi's Hoa Lo prison

Hanover Learning Center at

Hanover, Virginia

Harbison Harbison Correctional Institution for Women at Irmo, South Carolina

hard labor sentence involving imprisonment plus useful labor such as road building or maintenance

Hardwick site of the Middle Georgia Correctional Complex, also the site of the Youthful Offender Unit

Harlem Valley Harlem Valley Secure Center for juvenile and youthful offenders in Poughkeepsie, New York

Harris Harris County Juvenile Detention Center (Houston, Texas)

Harrisburg Dauphin County seat of Pennsylvania's capital city; its jail is one of many through the state

Hartford Connecticut Community Correctional Center in Hartford, also the site of the county jail

Hatfield HM Borstal at Hatfield in Yorkshire, England

Hattiesburg Hattiesburg, Mississippi or its jail

Havana (*see* Habana *and* Cuba)

Haverigg HM Prison at Haverigg on the coast of England opposite the Isle of Man

Hawaii's lockups four counties comprise this state and each has it own lockup; one in Honolulu serves Honolulu County on Oahu; another in Hilo serves the county and island of Hawaii; a third in Wailuku serves Maui; a fourth in Lihue is for the island of Kauai

Hawaii Youth Correctional Facility in Honolulu on Oahu Island

Hawalli (*see* Kuwait)

Hayes Hayes Prison Farm, Black Hills, Tasmania

Haynesville Louisiana site of the Wade Correctional Center

Heart of Midlothian nickname Sir Walter Scott gave the Tolbooth Prison in Edinburgh, Scotland the scene of his novel, *The Heart of Midlothian*

Heart Mountain Wyoming location of a Japanese-American relocation center set up after Pearl Harbor

Helena Montana site of the Mountain View School for female delinquents from 10 to 21 years of age

Helena State School for Boys in Helena, Oklahoma, holds delinquent boys from 15 to 17 years of age

helicoptered dropped into the ocean from a helicopter while manacled (reports from Argentina indicate political prisoners have been executed in this manner)

helicopter surveillance (*see* ASTREA)

Hellhole of the Pacific New Zealand's North Island port of Russell when it was called Kororareka

Hell of Macquarie Harbour Station nickname of an old penal colony on the coast of Tasmania

Hell's Gates Macquarie Harbour—Tasmania's first penal settlement

Helsinki (*see* Finnish prisons)

hempen four-in-hand hangman's noose

hemp stretcher euphemism for hangman

Hendry Hendry Correctional Institution at Immokalee, Florida

Hennepin refers to any of three Minnesota penal facilities: the Hennepin County Home School in Minnetonka, the Hennepin County Juvenile Center in Roseville, or the Hennepin County Workhouse in Wayzata

Hennigsdorf Nazi subcamp close to Berlin and larger concentration camps

hen pen reformatory for females

herder(s) prison guard(s)

Her (His) Majesty's Penitentiary in St John's initiated in 1859 in Newfoundland's capital

Hermes Trismegitus (Latin—Thrice-great Hermes)—Egyptian god Thoth and the code of laws he left concerning crime and punishment

Hermosillo Sonora state prison in Hermosillo, Mexico

Herstedvester Danish psychiatric prison for its advanced methods resulting in reduced recidivism

Herzegovina (*see* Yugoslavia)

Hewell Grange HM Borstal at Hewell Grange in Worcestershire, England

hierros (Spanish—irons)—handcuffs

Highland Rim School for Girls in Tullahoma, Tennessee for females from 12 to 18 years of age

Highpoint HM Prison at Highpoint in West Suffolk, England

High Security Center in Cranston, Rhode Island

Hillcrest School of Oregon in Salem holds juvenile-court commitments

Hillsboro (*see* Graham)

Hillsborough Hillsborough Correctional Institution at Riverview, Florida

Hilo site of Hawaii Community Correctional Center; also, the Kulani Correctional Facility, formerly the Kulani Honor Camp

Hilton Head old Federal prison for Confederate prisoners of war in the 1860s on the South Carolina coast

Hindley HM Borstal at Hindley in Lancashire, England

hit the fence escape from prison

hit the pit jailed

hit the sidewalk released from jail

HK Hong Kong or one of its several jails and prisons

HLPR Howard League for Penal Reform (London, England)

HMBI Her (His) Majesty's Borstal Institution

HMG Her (His) Majesty's Gaol—the royal jail

HM Gaol Her (His) Majesty's Gaol (in the British Commonwealth)

HMP Her (His) Majesty's Penitentiary; Her (His) Majesty's Prison

HM Prison Her (His) Majesty's Prison

Hoa Lo downtown prison in Hanoi, Vietnam, nicknamed the Hanoi Hilton

hobbling walking with leg irons attached

Hobo Hoboken, New Jersey or its jail

Ho Chi Minh City formerly Saigon (*see* Vietnam)

hoist to hang a person; to rob

hole(s) solitary-confinement cell(s)

Holland site of the Michigan Dunes Correctional Facility

Hollesley Bay Colony HM

Borstal at Hollesley Bay Colony in Suffolk, England

Holloway HM Prison at Holloway in Derbyshire, England near Sheffield

Holman Holman Prison at Holman Station, Alabama

Holmesburg prison in Holmesburg, Pennsylvania

Holzminden German maximum-security prison northwest of Göttingen

Homantin Homantin Girls Home (formerly Hong Kong's Matauwei Girls Home)

Homestead (*see* Dade Correctional Institution)

Hominy medium-security prison officially called the Conner Correctional Center in Tulsa, Oklahoma

Homs (*see* Syria)

Honaira (*see* Solomon Islands)

Honduran prisons Tegucigalpa, the capital city, and the port of San Pedro Sula, have *cárcel modelo* prisons

Hong Kong British crown colony off China's south coast; centers for the treatment of narcotic addicts are available as well as lockups and prisons

Honolulu site of the Conditional Release Branch, the Laumaka Conditional Release Center, the Oahu Community Correctional Center

Honolulu Jail nucleus of the Halawa High Security Facility at Aiea on Hawaii's Oahu Island

hook 'em hook them (fasten with handcuffs)

Hooper Home Isaac T Hooper Home (first halfway house in the United States, opened by the Society of Friends—Quakers—in New York City in 1845)

hoosegow jail (term may derive from the Spanish—*juzgado*—court of justice)

hoosier(s) prison visitor(s)

Hope Halls halfway houses sustained by the Volunteers of America from 1896 through the 1920s

Horsemonger Horsemonger Lane Gaol (infamous London lockup and the scene of many hangings such as the one Dickens described in a letter to *The Times*)

Hotlana Atlanta, Georgia, its jail, or federal penitentiary

hot seat electric chair

Hot Springs Garland County jail near Little Rock, Arkansas

hot squat electric chair

House 33 Soviet secret-police prison in Rostov-on-Don

House of C House of Correction

House of D House of Detention

house(s) of darkness prison(s)

house(s) of detention jail(s); lockup(s)

Houston Texas and its jails and prisons

Howard John Howard—18th-century English prison reformer who advocated vocational training and work as ways to make men honest; Rhode Island town containing the Admission and Orientation Unit, the Maximum Custody Facility, the Medium-Minimum Facility, the Rhode Island Training School for Girls, and the Rhode Island School for Boys

Hudson Hudson Correctional Facility at Hudson, New York; Hudson County Penitentiary in New Jersey; New York School for Girls at Hudson, New York

Hudson Street alimony jail on the lower west side of Manhattan where inmates were imprisoned for failure to pay alimony

Hue (*see* Vietnam)

hulk(s) prison ship(s)—usually old craft unfit for the high seas but adequate as jails

Hull HM Prison at Kingston-upon-Hull; other prisons in Canada's Hull opposite Ottawa and in Hull, Massachusetts

Humanitarian Penologist Eugene V Debs (1855–1926) (*see* Presidential Candidate and Prison Convict)

Hungarian prisons were built during the Austro-Hungarian Empire

Hunt Corrections Center St Gabriel, Louisiana's penal facility serving as an adult reception and diagnostic center together with maximum- and medium-security sections

Huntercombe HM Borstal at Huntercombe in Oxfordshire, England

Huntingdon Pennsylvania location of the State Correctional Institution

Huntsville Madison County jail north of Birmingham, Alabama; penal institutions such as the Diagnostic Unit, Ellis Unit, Goree Unit, Huntsville Unit, Wynne Unit in Huntsville, Texas

Huntsville Unit correctional facility in Huntsville, Texas

Huron Valley Huron Valley Men's Facility or Huron Valley Women's Facility in Ypsilanti, Michigan

Huron Valley Men's Facility Ypsilanti's maximum-security prison

Huron Valley Women's Facility in Ypsilanti, Michigan

Hutchinson location of the Kansas State Industrial Reformatory

hut(s) prison cell(s)

Huttonsville West Virginia site of the Huttonsville Correctional Center holding male felons

Hyderabad (*see* Pakistan)

IAPL International Association of Penal Law

ICA Illinois Correctional Association; Indiana Correctional Association; Iowa Corrections Association

ICCC International Concentration Camp Committee (Vienna, Austria)

icebox prison coroner's laboratory and office

ice(d) jail(ed)

Iceland the capital, Reykjavik, has one small prison

ICFPW International Confederation of Former Prisoners of War (Paris, France)

ICSPPR International Centre of Sociological, Penal, and Penitentiary Research (Messina, Italy)

Idaho jails facilities for incarceration are found in the 44 counties

Idaho State Idaho State Correctional Institution in Boise

if you can't do the sentence, don't do the job elderly convict's advice to anyone planning a criminal career

IHHA International Halfway House Association

Ikoyi Ikoyi Prison in Lagos, Nigeria

Iksha Soviet corrective labor colony for juvenile offenders close to Moscow's notorious prisons

Île du Diable (French—Devil's Island)—penal colony used to keep political prisoners completely segregated

Îles du Salut (French—Security Islands)—off mainland French Guiana; group consists of Île du Diable (Devil's Island), Île Royale, and Île Saint Joseph

Île Saint Louis (French—Saint Louis Island)—leper colony for convicts in French Guiana

Ilha de Flôres (Portuguese—Island of Flowers)—Brazilian prison in Rio de Janeiro

Illinois Industrial School for Boys (*see* Sheridan)

Illinois lockups the state's 102 counties offer detention facilities, the biggest are in Chicago's Cook County

Immokalee Florida location of the Hendry Correctional Institution

immurement confinement within walls

impound imprison

Imrali Turkish prison farm on Imrali Island in the Sea of Marmara

Imros Turkish prison on Imros Island in the Aegean Sea

incorrigible person who will not be corrected, reformed, rehabilitated, or made to conform to social standards

Indiana Boys School for delinquents from 12 to 21 years of age; in Plainfield

Indiana Girls School in Indianapolis for delinquents from 12 to 20 years of age

Indiana jails the state's 92 counties have facilities for holding felons; cities such as Indianapolis have their own lockups

Indianapolis Marion County's jail as well as the Indiana Girls School

Indiana State Farm medium-security facility in Greencastle

Indiana State Prison in Michigan City holds maximum- and medium-security felons

Indiana State Reformatory a maximum- and minimum-security penal facility in Pendleton

Indiana Women's Prison in Indianapolis

Indiana Youth Center a medium-security facility in Plainfield

Indian prisons the Republic of India maintains prisons in all principal cities; the Black Hole of Calcutta within Calcutta's Fort William was where, during the Indian Mutiny in 1756, some 146 Europeans were imprisoned in such cramped quarters that by the next morning only 23 remained alive

Indian River Indian River Correctional Institution at Vero Beach, Florida (*see* Vero Beach)

Indian River School for juvenile-delinquent boys in Massilon, Ohio

indic indicateur (French—informant)

Indio Indio Branch of the Riverside County Jail in California,

individual confinement solitary imprisonment

Indonesian prison island Java

Indonesian prisons one prison exists on each of the major islands along with jails on the small islands; in Jakarta, Surabaja, Bandung, Semarang, and Medan are penal institutions ranging from jails to prisons

industrial prison workshop-oriented penitentiary where inmates produce such items as highway signs, mail bags, furniture, and vehicle license plates

Industrial School for Women in Vega Alta, Puerto Rico

Indy Indianapolis, Indiana's jails

Inebriate Reception Center in San Diego, California where nonviolent abusers of alcohol and other drugs accept coffee and counseling in lieu of being jailed

informant person giving information to law-enforcement officers who may use it in the investigation of criminals

injection (*see* death by injection)

inmate convict; person in a confinement facility; prisoner

Innisfail Canadian site of the Bowden Institution

INS Immigration and Naturalization Service

In-Service Training (IST) within California state prisons

inside the tall walls inside prison

Institute The Institute (nick-

name for any penal facility having Institute in its title)

institutional capacity officially determined number of inmates a correctional facility is designed to house

institutional superintendent warden

Institution for Youthful Offenders in Santurce, Puerto Rico

Instituto Nacional de Orientacion Femenina (Spanish —National Institute of Feminine Orientation)—women's prison in los Teques of the Miranda Ejido of Venezuela

Insular Penitentiary in Rio Piedras near San Juan, Puerto Rico

Intake Service Center Cranston, Rhode Island facility for holding pre-trial detainees

intensive care unit locked unit for juvenile offenders

intermediate-term adult penal institutions of the U.S. Bureau of Prisons Danbury, Connecticut; Fort Worth, Texas; La Tuna, Texas; Lexington, Kentucky; Milan, Michigan; Sandstone, Minnesota; Terminal Island, California; Texarkana, Texas

internados judiciales (Spanish —judicial boarding houses)— penal facilities for holding convicts; in Venezuela there are 18 such institutions

internal exile Soviet euphemism for imprisonment in some remote part of the USSR

International Halfway House Association publisher of the *Directory of Residential Treatment Centers*

International Penal and Penitentiary Commission founded in 1872 and in 1950 became a part of the United Nations; originally known as the International Penitentiary Commission

International Prisoners Aid Association publishes the *Directory of Prisoners Aid Agencies*

in the grinder in jail; in prison

in the nick (Cockney English —in jail; in prison)

intimidation fear of punishment

Invercargill Invercargill Borstal Institution near Invercargill on South Island, New Zealand

Inverness prison in Inverness, Scotland

Ionia site of the Michigan Reformatory, Michigan Training Unit, and the Riverside Correctional Facility

Iowa jails 99 counties provide detention facilities; cities such as Des Moines have their own lockups

Iowa Security and Medical Facility at Oakdale

Iowa State Penitentiary at Fort Madison

Iowa Training School for Girls at Mitchellville

IPAA International Prisoners Aid Association (Louisville, Kentucky)

IPPC International Penal and Penitentiary Commission *(see entry)*

IPPF International Penal and Penitentiary Foundation (Neuchatel, Switzerland)

Iranian prisons exist in Tehran, Ishfahan, Mashad, Tabriz, and other metropolitan places

IRC (*see* Inebriate Reception Center)

Irish prisons Dublin's old prison is Kilmainham; Cork prison is on Rathmore Road; Limerick has one on Rutland Street

Irma Wisconsin site of the Lincoln Hills School

Irmo Harbison Correctional Institution for Women at Irmo, South Carolina

iron house jail; lockup; penitentiary; prison

ironmongery department Her (His) Majesty's prison

ISIS Investigative Support Information System (FBI)

Isla de la Juventud (Spanish —Isle of Youth)—Isla de Pinos (Isle of Pines)

Isla de Pinos (Spanish—Isle of Pines)—site of prisons housing political prisoners

Islamabad (*see* Pakistan)

Island Blackwell's (Welfare, now Roosevelt) Island in New York City; Parkhurst Prison on the Isle of Wight; Rikers Island in New York

Island of Hell Norfolk Island in the South Pacific Ocean, once the most dreaded of all Australian penal stations

Isle of Flowers (*see Ilha de Flôres*)

Isle of Pines (*see* Cuba, Isla de Pinos)

Isle of Wight Parkhurst Prison on the Isle of Wight in the

English Channel

isolation tank solitary-confinement cell

isolator Soviet penal colony specializing in solitary confinement; many in distant parts of the Siberian Arctic and in the White Sea east of Kem on the Solovetski Islands

Isole Eolie (Italian—Aeolian Islands)—the Lipari Islands off the north coast of Sicily used as penal colonies

ISP Idaho State Penitentiary; Institute of Social Psychiatry

IST In-Service Training (within California state prisons)

Istanbul formerly Constantinople (*see* Turkey)

Italian prisons Rome's Regina Coeli (Queen of Heaven), Genoa (Prigione d'Genova), Milan (Ducato Milanese), Naples (Caserta), and Venice (Pozzi)

Ivanovo internal prison of the Soviet secret police in Ivanovo

Ivory Coast's prisons during French occupation prisons were built that are still in use in the capital, Abidjan, and in Bouaké as well as in smaller cities

Iwakuni U.S. Marine Corps Correctional Facility of Iwakuni, Japan

Izmir Turkish city jails and prison (formerly ruled by the Greeks who called it Smyrna)

jacket prisoner's case history or dossier

Jackson site of Georgia Diagnostic and Classification Facility; Jackson, Mississippi jail; Michigan site of the world's largest walled prison enclosing 57 acres and 23 hectares, as well as the Reception and Guidance Center and the State Prison of Southern Michigan; North Carolina location of the Odom penal facility

Jacksonville Duval County's jail in Jacksonville, Florida

Jaffna (*see* Sri Lanka)

jail penal facility usually run by local law-enforcement officers; often used as pre-trial detention centers

JAIL Justice Against Identification Laws

jailage a jailer's fee, bribes

jail bait person(s) whose illegal activities lead to incarceration

jailbird ex-convict, prisoner, or recidivist

jail delivery clearing a jail of its inmates by bringing them to trial and then releasing them or remanding to prison

jail distemper jail fever (typhus); sickness brought on by incarceration or fear of incarceration

jail fever typhus fever occurring in jails and other crowded places

Jail Forum quarterly publication of the National Jail Association

jailhouse building used as jail

jailhouse lawyer convict who is well-informed about the law

jailhouse punk any prisoner who becomes a homosexual while imprisoned

jail limits area or district surrounding a jail where debtors may be at large under a bond of security

jail plant narcotics concealed on a person condemned to imprisonment or visiting a penal institution

Jakarta (*see* Indonesian prisons)

Jalapa prison in Jalapa, the capital of the Mexican state of Veracruz

Jamaica penal facilities include Richmond Farm Prison, St Catherine's District Prison, and the Tamarind Farm Prison

Jamejala Soviet psychiatric hospital

Jamesburg New Jersey location of the Training School for Boys and Girls

Jamestown Sierra Conservation Center at Jamestown, California

jamocha java + mocha (prison slang for coffee)

Janie Porter Barrett School for Girls at Hanover, Virginia

Japanese-American relocation centers camps set up after the attack upon Pearl Harbor at Manzanar and Tule Lake in California, Gila and Poston in Arizona, Topaz in Utah, Minidoka in Idaho, Heart Mountain in Wyoming, Amache in Colorado, Jerome and Rohwer in Arkansas

Japanita Santa Anita Assembly Center's nickname when the racetrack stables were used for the incarceration of Japanese-Americans

Japan's prisons penal facilities are provided by all 47 of Japan's prefectures and are to be found in its cities such as Tokyo, Osaka, Yokohama, Nagoya, Kyoto, Kobe, Sapporo, Kitakyushu, Fukuoka, and Kawasaki

jaula (Spanish—cage)—slang for a jail or lockup

Jax Jacksonville, Florida lockups

Jay State Prison at Jay, Florida

J-bird jailbird

jd juvenile delinquent

JDC Juvenile Detention Center

Jean site of the Southern Nevada Correctional Center

Jeff City Jefferson City, Missouri jail

Jefferson Jefferson County Jail in Birmingham, Alabama; Jefferson Parish Jail near New Orleans, Louisiana

Jefferson City site of the Central Missouri Correctional Center, the Missouri Intermediate Reformatory, and the Missouri State Penitentiary for men

Jefftown Journal prisoner's periodical published at Jefferson City, Missouri's prison

Jen Penjara Singapore's Remand Prison nicknamed for the road where it is located

Jerome Arkansas location of a Japanese-American relocation center during World War II

Jersey City New Jersey city lockups

Jess Dunn Correctional Center in Taft, Oklahoma

Jessup site of the Maryland Correctional Institution for Women, Maryland Correctional Pre-Release System, Maryland House of Correction

Jester Unit Texas pre-release facility in Richmond

Jesup federal correctional institution in Jesup, Georgia

Jewell Manor Jewell Manor Girls Center at Louisville, Kentucky

JHA (*see* John Howard Association)

JHAH John Howard Association of Hawaii

JHDF Juvenile Hall Detention Facility

jhj's jailhouse juvenile delinquents

Jidda (see Saudi Arabia)

JIS Jail Inspection Service

Jiuren Soviet prison near Inner Mongolia

JJC Juvenile Justice Center (Los Angeles, California)

JLS Jail Library Service (California State Library)

Joburg Johannesburg, South Africa jail

Joe Harp Correctional Center in Lexington, Oklahoma

Joelton location of the Tennessee Youth Center

Joey man who takes a prisoner's place at home while the convict is imprisoned

Johannesburg (see South Africa)

John Saint John in the American Virgin Islands, Saint John in New Brunswick, Canada and their lockups

John Howard Association founded in 1901 to honor an 18th-century English prison reformer; meets in Chicago and its members offer professional consultation services

John Howard Association of Hawaii Honolulu's prisoner-service agency

John's Saint John's, Newfoundland, jail

Johnson Hall Forensic Unit Evanston, Wyoming, for prisoners who are dangerously mentally ill or in need of extensive psychiatric treatment

Johnson's Island Confederate prisoners of war were held on Johnson's Island in Lake Erie

Joliet site of the Joliet Correctional Center and the Statesville Correctional Center

Joliet Correctional Center maximum-security prison in Joliet, Illinois

Jonestown tropical jungle commune in Guyana where the Reverend Jim Jones of San Francisco in 1978 enforced the suicide of more than 900 people

Joplin Joplin, Missouri, county jail

Jordan's prison Jordan's capital, Amman, has one of the world's worst prisons

Jorge Navarro Nicaraguan prison in Tipitapa

Journal of Correctional Education quarterly publication of the CEA (Correctional Educa-

tion Association)

Joyceville Institution in Kingston, Ontario

JPS Juvenile Probation Services

jri jail-release information

Juárez Ciudad Juárez, Chihuahua, jail

Juariles (Mexican-American —Ciudad Juárez)—or its jail

Jubilee Jubilee Lodge for Girls at Brimfield, Illinois

judas peephole in a prison-cell door constructed so the inmate(s) can be observed without knowing it

judicial execution execution in response to a court order

judicial hanging hanging performed in response to a court order

judicial murder capital punishment

judicium capitale (Latin—capital justice)—justice through execution

jug(ged) jail(ed)

Jugoslavia (see Yugoslavia)

jug(s) jail(s); prison(s)

jug tank(s) prison cell(s) for drug addicts

Julia Julia Tutwiler Prison for Women at Wetumpka, Alabama

Juneau Juneau Correctional Center in Alaska

Jungfernhof extermination facility near Riga where many Austrian Jews were murdered

junk tank prison cell for drug addicts

jus publicum (Latin—penal law; public law)

Justice U.S. Department of Justice

juve delinquent(s) juvenile delinquent(s)

Juvenile Hall holding facility for juvenile delinquents

juvenile-justice agency government agency concerned with the adjudication, care, confinement, investigation, and supervision of juvenile delinquents

Juvenile Justice Center Los Angeles, California agency designed to cope with juvenile delinquency

juvenile penal institutions of the U.S. Bureau of Prisons Ashland, Kentucky; Englewood, Colorado; Morgantown, West Virginia

juvenile record(s) official record(s) containing information concerning juvenile court

proceedings and all applicable correctional and detention processes ordered

juvie(s) juvenile delinquent(s); juvenile hall(s); juvenile law-enforcement officer(s)

K-9 Corps Canine Corps (police dogs)

Kabul capital city of Afghanistan and nickname for its prison

Kaiserwald Nazi concentration camp in Latvia during most of World War II

Kakogawa careless-driver prison serving Japan's Osaka-Kobe-Kyoto area; prisoners receive driving lessons daily; at dawn and dusk they retire to the prison's Park of Interrogation where they apologize to the memory of their victims and swear never to repeat their mistakes

Kalgoorlie Kalgoorlie Regional Prison (Kalgoorlie, Western Australia)

Kaluga Soviet prison 90 miles southwest of Moscow

Kampala (see Uganda)

Kandy (see Sri Lanka)

kangaroo court prisoners' court imposing contributions, fines, and work tasks on convicts brought before it; nickname for any small court harsh on addicts, alcoholics, and vagrants

kanga(s) kangaroo(s)—Australian prison warden(s)

Kansas City Kansas City, Kansas or Kansas City, Missouri jails

Kansas Correctional Institution in Lansing

Kansas Correctional-Vocational Training Center in Topeka

Kansas jails the state's 105 counties provide detention facilities; Wichita and Topeka also have local lockups

Kansas State Industrial Reformatory in Hutchinson

Kansas State Penitentiary in Lansing

Kansas State Reception and Diagnostic Center in Topeka

Kaohsiung (see China, Republic of)

kapidiye (Turkish—hardened criminals)—the most feared in prisons where they bribe the guards and rule the other inmates

Karachi (see Pakistan)

Karaganda and Kolyma two of the world's largest forced-labor penal camps; both in the USSR

Karnet Karnet Rehabilitation Centre (Western Australia)

Kars prison near the eastern border of Turkey

Kathmandu (*see*Nepal)

katorga (Russian—hard penal servitude)—forced labor

Kauai Kauai Community Correctional Center at Lihue on Kauai, Hawaii

Kaufering forced-labor subcamp in the south of Germany

Kawasaki (*see* Japan's prisons)

Kay-Cee Kay-Cee Honor Center in Kansas City, Missouri

KC Jackson County Jail in Kansas City, Missouri

KCCD Kentucky Council on Crime and Delinquency

KCJ Kings County Jail in Seattle, Washington

Kearney Nebraska location of the Youth Development Center

keester plant hollow suppository made of metal, plastic, rubber, or wood and used to conceal contraband in the rectum

keester stash anything hidden in the rectum

Kendall juvenile-delinquent rehabilitation center at Kendall, Florida

Kent Kent Correctional Institution in Dover, Delaware

Kent Institution maximum-security penal facility in Agassiz, British Columbia

Kentucky Correctional Institution for Women at Peewee Valley, site of the Assessment and Orientation Unit

Kentucky lockups 120 counties have jails; cities such as Louisville and Lexington have augmented facilities as well

Kentucky Manpower Development state agency charged with the task of developing a prisoner rehabilitation program

Kentucky State Penitentiary maximum-security prison at Eddyville

Kentucky State Reformatory medium-security facility at La Grange

Kenya's prisons date back to the time of British rule

Ketchikan Ketchikan Correctional Center, Alaska

Kharkov world's largest prison in the capital city of the Kharkov region of the Ukraine

Khartoum (*see* Sudan)

Kholmogori Arctic death camp established by Lenin near Archangel in 1921 for the exploitation and suppression of political prisoners in the Soviet Union

Kilby Kilby Corrections Facility at Montgomery, Alabama

Kilmainham Dublin jail

Kincheloe Michigan site of the Kinross Correctional Facility

kindergarten of vice epithet applied to many jails

Kings Kings County Jail in Seattle, Washington

King's Bench one of Southwark's seven prisons, now all gone

Kingston Canadian site of the Collins Bay Institution as well as the Joyceville Institution; (*see* Jamaica)

Kingston Pen Kingston Penitentiary (and mental hospital) in Portsmouth, Ontario

Kingstown (*see* Saint Vincent and the Grenadines)

Kinross Correctional Facility near Kincheloe, Michigan

Kinston North Carolina site of the Dobbs School for juvenile delinquents

Kiribati Tarawa, the capital of this colony what was once HM Prison

Kirikiri Lagos, Nigeria's prison

Kirkham HM Prison at Kirkham in Lancashire, England

Kirkland Correctional Institution in Columbia, South Carolina

Kitakyushu (*see* Japan's prisons)

KL Konzentrationslager (German—concentration camp)—prisoners contracted this to *KZ*

klondike solitary prison cell

KMCI Kettle Moraine Correctional Institution (near Duluth, Minnesota)

KMD Kentucky Manpower Development (*q.v.*)

knowledge factory prison school

Knoxville Knox County, Tennessee's jail

Kobe (*see* Japan's prisons)

kogus (Turkish—cell block) —see *turist kogus*

Kolyma USSR penal camp north of the Arctic Circle

Korydallos prison in Athens, Greece

Kot Lakhpat jail in Lahore, Pakistan

Kragshovhede Danish prison without high fences or walls; this type of facility is said to result in reduced recidivism; located near Skagen or Skaw

Krems site of a World-War-II concentration camp

Kresty Leningrad's central prison

KRIM the Danish association for penal reform

Kriminalstrafkunde die *Kriminalstrafkunde* (German—penology)

KROM the Norwegian association for penal reform

Krome Avenue address and nickname of the immigration and naturalization detention camp on Krome Avenue in Miami, Florida

KRUM the Swedish association for penal reform

Kryukovo USSR prison colony

Ktr Katorzhane (Russian— compound)—prison compound for people sentenced to hard labor

Kuala Lumpur (*see* Malaysian prisons)

Kuibyshev Soviet transit prison

Kulani Kulani Correctional Facility in Hilo, Hawaii

Kumasi (*see* Ghana's prisons)

Kumla Kumla Prison, institution for long-term prisoners in south-central Sweden

Kunie French penal colony in the Southwest Pacific

Kuwait has prisons in Hawalli and Kuwait City

Kyoto (*see* Japan's prisons)

LA Los Angeles penal institutions

La Cabaña (Spanish—The Cabin)—old Spanish prison at the entrance to Havana harbor still used for the execution and imprisonment of political prisoners

La Ceiba prison in a seaport city of Honduras on the Caribbean

LACJ Los Angeles County Jail (California)

La Ferté Macé French detention camp

La Force top-security prison of Paris in the late 1700s and early 1800s

lag ticket-of-leave man; transported convict of the type Britain shipped to Australia and Tasmania

Lager Westerbork transit camp established in the Drenthe province of the Netherlands by German occupation forces during World War II (see *Durchgangslager*)

lagging serving a three-year sentence in a British prison; transportation by sea of convicts from overcrowded jails in the British Isles to those in Australia and Tasmania

Lagoinha jail in Belo Horizonte, Brazil

Lagos (*see* Nigeria)

La Grange site of the Kentucky State Reformatory as well as the Luther Luckett Correctional Complex

lag(s) transported convict(s)

La Guaira Venezuelan jail

Lahore (*see* Pakistan)

Lake Butler Florida site of the Reception and Medical Center

Lake Correctional Institute penal facility at Lowell, Florida

Lakehills Lakehills Community Corrections Center in Salt Lake City, Utah

La Mesa Penitenciaria (Spanish—La Mesa Penitentiary) —Baja California's major prison, in the eastern section of Tijuana

laminated glass principal component of detention glazing (*see* detention glazing)

La Modelo La Carcel Modelo (Spanish—The Model Prison) —principal prison of Caracas, Venezuela

lamster(s) escaped convict(s)

Lancaster HM Prison at Lancaster in Lancashire, England or the jail in Lancaster, Pennsylvania, or jails in smaller Lancasters in Canada and the United States; site of the Southeastern Ohio Training Center

Lancaster Correctional Center at Trenton, Florida

Lancaster Pre-Release penal facility in Lancaster, Massachusetts

Land of Death and Chains Maxim Gorki's nickname for Siberia

'Lando Orlando, Florida jail

Land of Political Exiles

Yakutia, Siberia

Landsberg fortress prison on the Lech River in Upper Bavaria

Langholmen Swedish prison

Langi Langi Kal Kal Youth Training Centre at Trawalla, Victoria, Australia

Lansing site of the Kansas Correctional Institution and the Kansas State Penitentiary; State Industrial Farm for Women at Lansing, Michigan

'Lanta Atlanta, Georgia, jails, lockups, and federal penitentiary

Lantan Lantan Island (contains Ma Po Ping Prison serving Hong Kong)

Lantana Florida site of the Lantana Correctional Institution

La Nuestra Familia (Spanish-American jail jargon—Our Family)—prison racketeers who deal with homosexual prostitution, loan sharking, murder contracting, and narcotics; a Mexican-American Mafia

Laos has an old prison in Vientiane built during the years of French-colonial rule

La Paz Bolivia's capital high in the Andes where prisoners get cool mountain air but very little else; Mexico's La Paz, near the tip of Baja California

La Pica (Spanish—the stonecutters hammer; the pike or the long lance of the picador)—nickname of Venezuela's central penitentiary of Oriente

La Pieota prison in Bogotá, Colombia

La Planta house of reeducation and artisan work in the El Paraiso section of Caracas, Venezuela

La Plata (*see* Argentina)

La Porte Noire (French—The Black Gate)—main entrance to the penal colony and prison in Saint Laurent du Maroni, nicknamed The Gate of Hell

La Roquette Parisian prison for women

La Route Zéro (French—The Zero Route)—Road to Nowhere leading out of the Saint Laurent du Maroni headquarters built by convicts in French Guiana over 50 years

LAS Legal Aid Society

Las Colinas Las Colinas Girls' Facility in Santee, California east of San Diego

Las Palomas Las Palomas, Chihuahua jail, Mexico

last mile euphemism for the walk from the prison cell to the place of execution

last waltz condemned prisoner's march to place of execution

Las Vegas (*see* Vegas)

Las Ventas (Spanish—The Stalls)—Madrid's municipal prison

Latchmere House a remand center in Surrey, England

latrinogram latrine rumor

La Tuna Federal Correctional Institution at La Tuna, Texas

Laurel maximum-security juvenile facility near Laurel, Maryland

Laval Ville de Laval, Quebec; site of the Correctional Development Centre and the Federal Training Centre as well as the Leclerc Institution

Laval-des-Rapides Ecole Notre-Dame de Laval at Laval-des-Rapides, Québec

Lawtey Florida site of the Lawtey Correctional Institution for older inmates with medical problems

Lawton Lawton, Oklahoma, Comanche County Jail

lazaretto hospital for contagious diseases; place of quarantine; tweendecks storeroom sometimes used as a hospital or even as a lockup if a vessel is without a brig

LCCC Lucas County Corrections Center (Ohio)

LCCJ Louisiana Council on Criminal Justice

Leavenworth U.S. Penitentiary at Leavenworth, Kansas

Lebanon Lebanon Correctional Institution in Lebanon, Ohio; Middle-East republic, has prisons in its capital, Beirut, and in its port city of Tripoli

Leclerc Institution in Laval, Québec

Lecumberri-Hilton nickname of Mexico City's Lecumberri prison

Leeds HM Prison at Leeds in Yorkshire, England

Leesburg location of the Lee Correctional Institution, Georgia; New Jersey

LEF *Liberté, Égalité, Fraternité* (French—Liberty, Equality, Fraternity)—slogan of the French Revolution, inscribed in bronze over the gates of prisons built by the French

Lefortovo prison in Moscow described in *The Gulag Archipelago* by Aleksander I Solzhenitsyn

Leicester HM Prison at Leicester in Leicestershire, England

León (*see* Nicaragua)

Leopoldville (*see* Kinshasa)

Lesotho its capital, Meseru, has a lockup

lethal gas method of execution in California, Maryland, Mississippi, Missouri, Nevada, North Carolina, Oregon, Rhode Island, Wyoming

lethal intravenous injection method of execution in Idaho, New Mexico, Oklahoma, Texas

Leuven Leuven Prison (also called Louvain) in Belgium

level four death-dealing injection used by executioners

Lewes small jail just inside the Delaware River breakwater

Lewisburg U.S. Penitentiary at Lewisburg, Pennsylvania

Lewisburg Plan the plan of the Lewisburg Penitentiary, also called the telephone-pole design, providing for maximum- and medium-security cells, inside and outside, respectively

Lex any Lexington or its correctional facilities (*see* Lexington)

Lexington Federal Correctional Institution at Lexington, Kentucky; location of the Blackburn Correctional Complex as well as the special units of the Danville Youth Development Center; Oklahoma site of the Joe Harp Correctional Center and the Lexington Assessment and Reception Center; U.S. Public Health Service Hospital at Lexington, Kentucky (drug detoxification facility)

ley de fuga (Spanish—law of flight)—privilege of law-enforcement officers to kill anyone attempting to escape

Leyhill HM Prison at Leyhill in Gloucestershire, England

LGC Laminated Glass Corporation (*see* detention glazing)

LGD London Gaol Delivery

Libby Libby Prison—converted tobacco warehouse in Richmond, Virginia, used to house Union agents during the Civil War

libéré (French—liberated convict)—a prisoner free from confinement but not free to leave the country of confinement

Liberia site of Monrovia and Buchanan prisons

Liberty Center Ohio site of the Maumee Youth Camp

Liberty Street city jail on Liberty Street in Louisville, Kentucky

Libreville (*see* Gabonese prisons)

Libya Tripoli (the capital) and Benghazi contain prisons built during its Italian occupation from 1912 until the end of World War II

Liechtenstein contains an old prison in the capital, Vaduz

Lieutenant of the Tower Lieutenant of the Tower of London (its warden)

lifeboat commutation of a death sentence or a prison term; judicial order for a retrial

lifer prisoner sentenced to life imprisonment

lifer's lament *gruntin' don' git yuh nuttin'* (complaining is useless)

Lihue port city on Kauai Island, Hawaii, and location of the Kauai Community Correctional Center

likvidatsiya (Russian—liquidation)—Soviet euphemism for execution

Lilongwe (*see* Malawi)

Lima Lima, Ohio, Allen County Jail or Lima, Peru jails and prisons

Lima State Hospital in Lima, Ohio offering felons psychiatric services

Limón Costa Rican prison in Puerto Limón

Lincoln HM Prison at Lincoln in Lincolnshire, England; jails in fifty-one American cities named Lincoln; Nebraska site of four penal facilities—the Diagnostic and Evaluation Center, the Lincoln Correctional Center, the Nebraska State Penitentiary, and the Post Care Program

Lincoln Correctional Center for youthful offenders in Lincoln, Nebraska

Lincoln Hills School reception center and training school in

Irma, Wisconsin

Lino Lakes site of the Minnesota Correctional Facility

Lipari Islands Italian islands off the north coast of Sicily and called Aeolian Islands (*Isole Eolie*); for centuries, a place of exile for hardened criminals and political prisoners; islands include Alicudi, Basiluzzo, Filicudi, Lipari, Salina, Stromboli, and Vulcano

Liparis (*see* Lipari Islands)

Lisbon (*see* Portugal)

Lisieux maximum-security prison in Lisieux, France

Litchfield Community Correctional Center in Litchfield, Connecticut

Little Rock Pulaski County jail in Arkansas

Liverpool HM Prison at Liverpool in Lancashire, England; jails in other places called Liverpool as in New South Wales, New York, and Nova Scotia

Ljubljana (*see* Yugoslavia)

local lockup usually the police station or county sheriff's prison

lockup a small jail or prison

Lock-Up Tree name of a huge hollow baobab tree in Derby, Australia; its 52-foot (16-meter) girth made it a natural lockup for prisoners

locus penitentiae (Latin—place of repentance)—penitentiary

Lodz (*see* Poland)

Logan Correctional Center in Lincoln, Illinios

Lomé (*see* Togo)

Lompoc U.S. minimum-security penitentiary and prison camp at Lompoc, California

London Ohio town containing the London Correctional Institution west of Columbus; (*see* United Kingdom)

Long Bay Sydney, New South Wales, Australian prison system, including the Central Industrial Prison, Her Majesty's Training Center, and the Parramatta Gaol

Long Beach Long Beach, California jails, lockups, or Community Treatment Center

long bid(s) long prison term(s)

Long Lane Long Lane School for (delinquent) Girls at Middletown, Connecticut

Long Lartin HM Prison at Long Lartin in Worcestershire, England

Longos (Mexican-American— Long Beach, California)— nickname of the city, its lockup, and jail, and the detention pens of the Immigration and Naturalization Service

Longriggend Scotland's Longriggend Remand Institution

long stretch(es) long prison sentence(s)

long-term adult penal institutions of the U.S. Bureau of Prisons Atlanta, Georgia; Leavenworth, Kansas; Lewisburg, Pennsylvania; Lompoc, California; Marion, Illinois; Terre Haute, Indiana

loopbelt restraining device used in handling and transporting unruly prisoners

loquera (Spanish slang—prison)—also called *banasto* or *chirona*

Lorton Virginia site of the Central Facility, the Maximum Security Facility, Youth Center I and II

Los Angeles metropolitan detention center in Los Angeles, California

Los Guilucos Los Guilucos School for (delinquent) Girls at Santa Rosa, California

Los Lunas site of the Central New Mexico Correctional Facility and the Los Lunas Correctional Center

Los Lunas Correctional Center in Los Lunas, New Mexico, opened in 1940 to teach inmates farming and livestock operations

Loudonville Ohio site of the Mohican Youth Camp

Loughan House penal facility in Blacklion, Ireland

Lough Kesh location of Maze Prison near Belfast, Northern Ireland

Louie Saint Louis, Missouri, jails, lockups, and other penal facilities

louisette beheading device perfected by Dr Antoine Louis of Paris at the suggestion of his colleague Dr Joseph-Ignace Guillotin; the resulting device is called the guillotine

Louisiana Correctional and Industrial School minimum-security facility for first-time young offenders held in De Quincy

Louisiana Correctional Insti- **tute for Women** at St Gabriel for female felons 17 years of age and up

Louisiana lockups 64 parishes have holding facilities; New Orleans and Baton Rouge have extra facilities for felons

Louisiana State Penitentiary maximum-security facility at Angola

Louisiana Training Facility juvenile-delinquent penal facilities at Ball (for females), Baton Rouge, Bridge City, and Monroe

Louisville Kentucky's principal city and county seat of Jefferson County and its jail

Lourenço Marques (*see* Mozambique)

Louvain Belgium's central prison; here long-term prisoners work in open cells or in special workships; called Leuven by Flemish Belgians

Lovelady Texas location of the Eastham Unit

Lowdham Grange HM Borstal at Lowdham Grange in Nottinghamshire, England

Lowell site of the Florida Correctional Institution and the Marion Correctional Institution

Low Newton remand center in Durham, England

LP Liverpool Prison

LTI Louisiana Training Institute (branches at Baton Rouge, Monroe, and Pineville)

Luanda prison in the People's Republic of Angola

lubang buaya (Indonesian— crocodile hole)—Djakarta water hole infested with crocodiles and used as a place to dispose of people at odds with the current administration

Lubianka Moscow headquarters of the Ministry of the Interior—the Soviet secret police, named for the founder of the Cheka; located one block from the Kremlin

Lublin concentration camp in Poland

Lucasville Southern Ohio Correctional Facility at Lucasville

Lud Ludgate Prison in London long ago

Lurigancho (*see* San Pedro)

Lusaka (*see* Zambia)

Luther Luckett Correctional Complex in La Grange, Kentucky

Luxembourg capital city contains an old prison

Luzira Kampala, Uganda's prison

lynching executing someone by mob rule rather than by the rule of law; freeing a suspect in police custody (slang definition describing encounters between gangs or mobs and police)

Lynwood Lynwood (delinquent) Girls Center at Anchorage, Kentucky

M-2 Match-Two—program matching volunteers from a community with prison inmates; sponsors write to inmates and visit them regularly with the aim of establishing meaningful relationships and providing convicts with references and job support after they are paroled

Maastricht Dutch maximum-security prison

MAB Metropolitan Asylums Board (British group responsible for administration of all sorts of asylums)

MacDougall Youth Correction Center in Ridgeville, South Carolina

MacLaren School in Woodburn, Oregon holds juvenile-court commitments in medium custody

Macon Georgia location of the Central Correctional Center

Macquarie Harbour inlet on the west coast of Tasmania that in the early 1800s contained a penal colony on Settlement Island; the entrance to the inlet was nicknamed the Gates of Hell

Madagascar's prisons old prisons are in the capital, Antananarivo, and other places such as Toamasina and Majunga

Madison Dane County jail, Wisconsin

Madras (*see* Indian prisons)

Madrid (*see* Spain)

Magadan port on the Sea of Okhotsk where Soviet ships delivered prisoners enroute to forced labor camps in the Gulag; one ship arrived with all of its crew in tact but all of its prisoners frozen to death

Magdalena penal institute of the Argentine armed forces about 80 kilometers from

Buenos Aires

Magdeburg concentration-camp subcamp southwest of Berlin

Magilligan prison near Londonderry in Northern Ireland

Magilligan Camp HM prison camp outside Belfast, Northern Ireland

Ma Hang Ma Hang Prison, Hong Kong

Maidstone HM Prison at Maidstone in Kent, England

Maine Correctional Center penal facility in South Windham

Maine jails 16 counties maintain jails; cities such as Portland and Bangor have local lockups as well as county jails

Maine State Prison at Thomaston holds incorrigibles in maximum security

Maine Youth Center in South Portland holds delinquents from 11 to 18 years of age

maison (French—house)—also a jail or lockup *(maison d'arrêt)* or a borstal or reform school *(maison de correction)*

maison de arrêt (French—prison)

maison de correction (French—house of correction)—penitentiary, prison

Maison de Correction St Bernard Brussels, Belgium's house of correction

maison de force (French—workhouse)—prison where only those who work are fed

Maison Gomin women's correctional facility at St Cyrille, Québec

Maison Notre-Dame de la Garde jail for Catholic delinquents ranging from 14 to 18 years of age at Cap-Rouge, Québec

Maison Tanguay women's penal establishment in Montreal, Québec

Majdenek Nazi concentration camp near Lublin, Poland

Majunga (*see* Madagascar's prisons)

Makindye Uganda's military prison

making little ones out of big ones [*see* rock crusher(s)]

Malabar Malabar Complex of Prisons in Australia's New South Wales

Malaga (*see* Spain)

Malang women's prison in eastern Java, Indonesia

Malawi the capital, Lilongwe, and other cities, contain prisons built by the British between 1881 and 1966

Malaysia the capital, Kuala Lumpur, and cities such as George Town have prisons built by the British between 1889 and 1904

Malchow North German concentration camp

Maldives the capital, Male, contains a prison built during British dominion (1887–1965)

Male (*see* Maldives)

Mali has a prison in Bamako

Malines Belgian transit camp near the French border and the North Sea built by German forces during World War II as a way station for concentration camp prisoners; also known as Mechlin

Malmö (*see* Sweden)

Malta the capital city and seaport, Valetta, containing a prison built during British rule (1814–1964)

manacle handcuff

Managua capital city of Nicaragua replete with a prison

Manama capital city of Bahrain and its most-humid prison

Manbarco Man Barrier Corporation of Seymour, Connecticut engaged in manufacturing electronic detection systems and physical barriers made of coils of barbed wire and knife-edged wire used to keep prisoners within bounds

Manchester HM Prison at Manchester in Lancashire, England; New Hampshire location of the Manchester Community Corrections Center, the New Hampshire Youth Development Center; place of detention in other cities named Manchester in Connecticut, Georgia, Iowa, Kentucky, Massachusetts, New York, Ohio, Tennessee, and Vermont

Manchester Community Corrections Center in New Hampshire

Mandan site of the North Dakota Industrial School opened in 1903

Manhattan Manhattan Island, center of New York City and New York County; has several large correctional facilities and many police-station lockups

Manhattan House of Detention Tombs Prison in lower Manhattan; nicknamed Down Home

manhunt hunt organized to catch a criminal, an escapee, a fugitive from justice, or even a person who is lost

Manila (*see* Philippines)

Mannheim Confinement Facility of the U.S. Seventh Army in Mannheim, Germany

Manning Correctional Institution in Columbia, South Carolina

Manor English slang for London or its prisons

Mansfield site of the Ohio State Reformatory

Mansions The Mansions in Brisbane, Queensland—Australia's name for its Prison Department

Manzanar American relocation center for Japanese-Americans detained in California's Owens Valley

Manzanillo prison in Manzanillo, Mexico

Manzini (*see* Swaziland)

MAOF Mexican-American Opportunities Foundation (supporting a program to rehabilitate juvenile Chicano recidivists)

Maplehurst Correctional Centre in Milton, Ontario

Maple Lane School in Centralia, Washington south of Olympia

Ma Po Ping Ma Po Ping Addiction Centre on Lantau Island, New Territories, Hong Kong

Maracaibo Venezuela prison

Marble Hill Bollinger County jail at Marble Hill, Missouri

Marian Hall jail for English-speaking juvenile-delinquent females at Beaconsfield, Québec

Marianna federal prison in western Florida

Marigot (*see* Saint Martin)

Marina Street San Juan, Puerto Rico's district jail

Marion U.S. Penitentiary at Marion, Illinois; Marion County Detention Home for juvenile delinquents in Indianapolis, Indiana; Marion Correctional Institution, Marion, Ohio

Marion Correctional Institu-

tion in Lowell, Florida, once the Florida Correctional Institution

Marion Correctional Treatment Center in Marion, Virginia

Marquette Michigan site of the State House of Correction and Branch Prison

Marrakech (*see* Morocco)

Marshall site of a Confederate prison camp in northeast Texas

Marshalsea one of Southwark's seven prisons, now all gone, but formerly dominating much of this London district on the south bank of the River Thames

Martinez Contra Costa County courthouse and jail northeast of San Francisco, California; location of one of many clandestine prisons in Argentina

Martinière combination cargo and prison ship whose tween decks and lower holds were fitted with removable cages for transporting convicts from France to French Guiana; the vessel, built in 1911, was last seen off Devil's Island early in 1950

Maryland Correctional Institutions started in 1931 as a penal farm in Hagerstown

Maryland Correctional Institution for Women in Jessup

Maryland Correctional Pre-Release System in Jessup with supporting facilities in Baltimore, Church Hill, Hughesville, Quantico, and Sykesville

Maryland Correctional Training Center in Hagerstown

Maryland House of Correction in Jessup

Maryland jails the state's 23 counties provide lockups; Baltimore has its own lockup facilities

Maryland Penitentiary maximum-security prison in Baltimore, next to the Reception Center

Marysville Ohio Reformatory for Women in Marysville

MASCA Middle Atlantic States Correctional Association

Massachusetts Correctional Institutions Bridgewater holds alcoholics, criminally insane, and sexually dangerous felons; Cedar Junction is a maximum-security facility;

Framingham contains females; others at Norfolk, South Carver, Orange, and Concord

Massachusetts lockups 14 counties provide imprisonment facilities; Boston has penal facilities of its own

Massachusetts State Prison (*see* South Walpole)

Massawa (*see* Ethiopia's prisons)

Massilon Ohio town holding the Indian River School for male juvenile delinquents

Matamoros the Tamaulipan city maintains one of the world's dirtiest jails

Matanzas (*see* Cuba)

Matrah (*see* Oman)

Matsqui Institution British Columbia's minimum-security facility for drug addicts at Abbotsford

Matteawan Matteawan State Hospital (for the criminally insane) near Beacon-on-Hudson, New York

Maui Maui Community Correctional Center at Wailuki on Maui Island, Hawaii

Maumee Youth Camp male juvenile-delinquent penal facility in Liberty Center, Ohio

Mauritania has an old prison built during French-colonial domination

Mauritius Port Louis has an old prison built by the British that has held criminals as well as political prisoners

Mauthausen Austrian Nazi concentration camp near Linz

Mavrino Mavrino Institute on the outskirts of the Soviet capital includes the Mavrino Special Prison

maximum custody keeping prisoners in penal institutions built with tool-proof bars and cells surrounded by high walls; maximum-security prisons are manned by many guards and are run on a plan calling for rigid discipline

maximum security applied to inmates considered dangerous to correctional officers, to others, and to themselves; prisoners awaiting the death penalty are also kept under maximum security

Maximum Security Center in Columbia, South Carolina

Maximum Security Facility in Smyrna, Delaware; also in

Cranston, Rhode Island

Maxwell Federal Prison Camp at Maxwell Air Force Base near Montgomery, Alabama

Mazatlan prison in Mazatlan, México

Maze Maze (top-security) Prison outside Belfast in Northern Ireland

Mbabane (*see* Swaziland)

MCA Massachusetts Correctional Association (Boston-based); Medical Correctional Association; Minnesota Corrections Authority; Missouri Corrections Association

McAlester site of the Oklahoma State Penitentiary

MCC Metropolitan Correctional Center (in Chicago, New York, and San Diego)

McCain North Carolina Prison Sanitorium at McCain in Hoke County

MCDC Montgomery County Detention Center (Maryland)

MCI Massachusetts Correctional Institution (Framingham)

McLaughlin Youth Center Anchorage, Alaska's diagnostic-program reception center for delinquents

McNeil Island McNeil Island Correctional Center in the state of Washington (formerly a U.S. maximum-security prison)

McNeil Island Correctional Center formerly a federal prison turned over to the state of Washington in 1981

MCO Michigan Corrections Organization (of wardens)

MDC Minnesota Department of Corrections

meat wagon(s) prison van(s)

Mecca (*see* Saudi Arabia)

Mecklenburg correctional center at Boydton, Virginia

Medical Center for Federal Prisoners at Springfield, Missouri

Medina (*see* Saudi Arabia)

meditation solitary confinement

medium custody institutions designed to give freedom of movement and greater scope for positive self-direction

Medium Security Facility in Cranston, Rhode Island

Medium Security Unit at Mount Pleasant, Iowa, usually holding first-term felons within a half-year of release

Melilla (*see* Spain)

meltout escape technique used in some modern prisons where certain types of doors and windows can be melted

Memphis Federal Correctional Institution at Memphis, Tennessee; and the Women's Correctional Center in Memphis

Menard Menard Correctional Center in Menard, Illinois; the Menard Psychiatric Center

Menard Time prisoner's publication issued by the Menard branch of the Illinois State Penitentiary

Mende maximum-security prison in southern France

menschenhandel (German— trade in people)—prisoner-exchange program

Men's Reformatory in Anamosa, Iowa

Merced Merced County jail in California

Mercer Pennsylvania Corrections Department prison in Mercer County

Merida prison in Merida, capital of the Mexican state of Yucatan; or the Merida in Venezuela

meritorious good time promise of parole-induced good behavior on the part of convicts wanting to get out of prison

Meseru (*see* Lesotho)

Met Metropolitan Correctional Center in San Diego, California

metanoia change of heart and mind required for criminal rehabilitation

Metro Metro Correctional Institution in Atlanta, Georgia holds emotionally disturbed inmates in close security while providing psychiatric services

Metropolitan Correctional Centers Chicago, Illinois; New York, New York; San Diego, California

Metropolitan Regional Institution in Bayamón, Puerto Rico

Mexicali capital of Baja California, and its jail

Mexican Mafia controls much of the narcotic trafficking in California prisons such as Chino, Folsom, and San Quentin; also active in Mexican prisons

México nation with a long record of horrible prisons;

Mexico City contains some of its most formidable penal facilities

Miami city lockups, Dade County Jail, or the Federal Correctional Institution, the Miami Detention Center

Mich Michigan, Michoacan, or their jails and prisons

Michigan City Indiana State Prison at Michigan City

Michigan Dunes Correctional Facility in Holland, Michigan

Michigan jails 83 counties provide detention facilities; Detroit and other cities have additional lockups

Michigan Reformatory in Ionia

Michigan Training Unit at Ionia

Midlands Reception and Evaluation Center in Columbia, South Carolina

Mid-Orange Mid-Orange Correctional Facility at Warwick, New York

Midway Texas site of the Ferguson Unit for first offenders

Milan Federal Correctional Center at Milan, Michigan

Miles City Montana site of the Pine Hills School for juvenile delinquents

Milford Delaware's penal facility for juvenile delinquents; also known as Stevenson House

milieu therapy treatment given to aid convicts returning to society via halfway houses, pre-release guidance centers, and tranquilizing drugs

military execution execution by a military firing squad

milk van nickname for police or sheriff's van for transporting prisoners

Millhaven Institution maximum-security facility in Bath, Ontario

Milton Ontario site of the Adult Training Centre and the Maplehurst Correctional Centre

Milwaukee Milwaukee County Jail in Milwaukee, Wisconsin

Mimico Mimico Correction Center (for males) in Toronto, Ontario

M-in-C Matron-in-Chief

Mineola Nassau County Jail in Mineola, New York

Minidoka relocation center for Japanese-Americans interned after Pearl Harbor in southern

Idaho

minimum custody honor dormitories, prison camps, and prison farms offering inmates as much freedom from restraint as possible while preventing their escape

Minimum Security Facility in Cranston, Rhode Island

Minnesota Correctional Facilities at Lino Lakes, Oak Park Heights, Red Wing, Saint Cloud, Sauk Center, Shakopee, and Stillwater *(see individual entries)*

Minnesota lockups 87 counties have jails; Minneapolis, Saint Paul, and Duluth have extra facilities for felons

Minnie Minneapolis, Minnesota or its jails, lockups, or prisons

Mirikiri Lagos, Nigeria's prison

misdemeanant person convicted or guilty of misdemeanors

Missie Mississippi or its jails, lockups, and prisons

Mission Institution medium-security facility in Mission, British Columbia

Mississippi jails the state's 82 counties have prison facilities; Jackson has additional detention facilities

Mississippi State Penitentiary in Parchman; inmates are offered a supervised earned-release, work-release program of rehabilitation

Missiyahu Israeli minimum-security prison offering inmates a work-release program during their pre-release period

Missouri Eastern Correctional Center in Pacific

Missouri Intermediate Reformatory for male felons from 17 to 25 who are held in Jefferson City

Missouri jails 114 counties contain prison facilities; St Louis and other big cities have additional jails

Missouri State Penitentiary for Men in Jefferson City

Missouri Training Center for Men in Moberly

Mitchellville Iowa Training School for (delinquent) Girls at Mitchellville

Moabit Berlin's great prison in the Tiergarten section

Mob Mobile, Alabama or its penal facilities

Moberly site of the Missouri Training Center for Men

Mobile Mobile County jail in Alabama

Mobtown Baltimore, Maryland or its jails, lockups, and other penal facilities

Modelo one of Bogotá, Colombia's prisons

Modesto Stanislaus County jail east of San Francisco, California

Moengo (see Suriname)

Mogadishu (see Somalia)

Mohican Youth Camp in Loundonville, Ohio

Monaco has only a lockup

Mongolia the capital, Ulan Bator, was once known as Urga and contains a prison of ancient construction

Monowitz forced-labor subcamp close to Auschwitz, Poland

Monroe site of the Louisiana Training Institute at Monroe; site of the Washington State Penitentiary, the Washington State Reformatory, and the Washington State Special Offender Center

Monrovia Monrovia Central Prison (Liberia's largest penal facility)

Monsieur de Paris (French— Mr. Paris)—the guillotine operator

Montana jails the state's 57 counties offer detention facilities

Montana State Prison at Deer Lodge

Montenegro (see Yugoslavia)

Monterrey prison in Monterrey, capital of the Mexican state of Nuevo León

Montevideo Uruguay's capital containing jails and a prison

Montey Allenwood Federal Prison Camp at Allenwood, Pennsylvania

Montgomery Federal Prison Camp at Montgomery, Alabama or its county jail or the Kilby Corrections Facility (see Kilby)

Montjuic Montjuic Castle in Spain has been used as a court to try anarchist assassins and to hold them in its deep dungeons

Montluc French greystone prison in Lyons

Montreal Maison Tanguay correctional facility for women prisoners in Montreal, Québec; location of a number of penal institutions including those in adjacent Ville de Laval

Montreal House of Detention at 800 Boulevard Gouin

Montreal Prevention Centre on Parthenia Street

Montrose School for Girls correctional facility at Reiserstown, Maryland

Monty Montgomery, Alabama or its jail and lockups

Moon Crescent Singapore's minimum-security prison

Moondyne Joe Australian bushranger who was the first man to cross the Swan River Bridge in Fremantle near Perth while escaping from jail

Moor The Moor—Dartmoor Prison near Princetown in Devonshire, England

Moor Court HM Prison at Moor Court in Staffordshire, England

Morelia prison in Morelia, capital of the Mexican state of Michoacan

Morgan Morgan County Regional Correctional Facility in Wartburg, Tennessee

Morgantown Federal Correctional Institution at Morgantown, West Virginia

Moriah Moriah Shock Correctional Facility, New York

Morocco France and Spain built prisons in cities such as Casablanca, Fez, Marrakech, Rabat, and Tangier

Moroni (see Comoros)

Morrison Mount View School for (delinquent) Girls at Morrison, Colorado

Morro Castle (see Habana)

Morton Hall HM Borstal at Morton Hall in Lincolnshire, England

Moscow (see USSR)

Mother of Prison Reform Dorothea Dix (1802–1887), American reformer active in Massachusetts and other states

Mothers in Prison Projects organization affiliated with Women in Jails and Prisons

Motown Motor Town (Detroit, Michigan) or its penal facilities

Moundsville location of the West Virginia Penitentiary

Mountain Institution in Agassiz, British Columbia, designed for holding aged inmates

Mountain View Mountain View School for female juvenile delinquents in Helena, Montana

Mount Eden Mount Eden Prison, Auckland, New Zealand

Mountjoy Mountjoy Gaol (Dublin's jail)

Mount McGregor Mount McGregor Correctional Facility near Warwick, New York

Mount Pleasant Iowa location of the Medium Security Unit

Mount View Girls School at Morrison, Colorado

Mozambique all its coastal cities (Beira, Chinde, and its capital, Lourenço Marques) contain prisons of Portuguese origin

MPP Mothers in Prison Projects

MPPCA Maryland Probation, Parole and Corrections Association

MR Michigan Reformatory in Ionia

MRC Minnesota Restitution Center

MTU Michigan Training Unit (reform school)

mule smuggler carrying contraband such as heroin or weapons into jails and prisons

Mulegé Mexican minimum-security prison in Baja California where convicts are free to roam about during the day and many find work in local enterprises, so very few ever attempt escape

multiprisoner transportation unit paddy wagon; police patrol van; prison van

Muncy Pennsylvania site of the State Correctional Institution for female felons

Muntinlupa Philippine prison southeast of Manila (also spelled Muntinglupa)

Muscat (see Oman)

musical execution execution performed to the roll of a field drum or tenor drums

Muskegon Michigan site of the Muskegon Correctional Facility

Muskegon Correctional Facility Michigan penal facility

Mutual Welfare League convict self-government system introduced by Warden Thomas Mott Osborne at Sing

Sing and later introduced at the U.S. Naval Prison at Portsmouth, New Hampshire

MWL (*see* Mutual Welfare League)

NAAWS North American Association of Wardens and Superintendents

NAB National Alliance of Businessmen (giving ex-convicts a chance by giving them jobs)

NACRO National Association for the Care and Resettlement of Offenders

NADPAS National Association of Discharged Prisoners' Aid Societies

Nafha Israel's top-security prison

Nagoya (*see* Japan's prisons)

Nail City Wheeling, West Virginia noted for its nail factory; nickname for its detention facilities

Nairobi Kenya's capital and nickname of its prison once called HM Prison Nairobi

NAJCA National Association of Juvenile Correctional Agencies

NAJJ National Assessment of Juvenile Justice (University of Michigan)

Nakhodka port near Vladivostok and a Gulag transit center for prisoners bound for eastern Siberia

'Nam Vietnam or any of its many jails and prisons

Namibia Walvis Bay and Windhoek, the capital, have HM Gaol prisons

Napanoch site of the Eastern New York Correctional Facility

NAPO National Association of Probation Officers

NAPV National Association of Prison Visitors

Nashville Tennessee site of the Lois Deberry Correctional Institute, the Nashville Regional Correctional Facility, the Spencer Youth Center, the Tennessee Prison for Women, the Tennessee State Prison

Nassau capital of the Bahamas on New Providence Island and toponym for its jail

National Association of Training Schools and Juvenile Agencies (NATSJA) created by the merger of the National Association of Training Schools and the National

Conference of Juvenile Agencies

National Clearinghouse for Criminal Justice Planning and Architecture (NCCJPA) maintains a 10,000-volume library at the University of Illinois at Champaign

National Correctional Recreation Association (NCRA) sponsors prison postal-weight-lifting contests for inmates in Canada and the United States

National Jail Association presents annual award for the outstanding jailer and jail matron; publishes *Jail Forum* quarterly

National Jail Managers Association maintains historical archives plus an information clearinghouse and library in Eugene, Oregon

National Juvenile Detention Association publishes *Counterpoint* bimonthly and the *Directory of Juvenile Detention Homes*

National Prison Project American Civil Liberties Union's program to fix prison sentences and improve the lot of prisoners

national razor nickname for the guillotine

National Sheriff official publication of the National Sheriffs' Association

National Society of Penal Information provided some of the first aides of the Bureau of Prisons and later became part of the Osborne Association

NATSJA (*see* National Association of Training Schools and Juvenile Agencies)

Natzweiler forced-labor concentration camp in eastern France

Nauru the capital, Yaren, has a small lockup

Navasota Texas location of the Grimes County penal facility

NCA Nebraska Correctional Association; Nevada Correctional Association

NCCA North Carolina Correctional Association

NCCD National Council on Crime and Delinquency

NCCJPA National Clearinghouse for Criminal Justice Planning and Architecture

NCJRS National Criminal Justice Reference Service

NCPPL National Committee on Prisons and Prison Labor

NCRA National Correctional Recreation Association

NCW Nebraska Center for Women (in York, Nebraska)

N'Djamena (*see* Chad)

NDPS Narcotic Detention Pens in downtown New York City

Nebraska Center for Women in York

Nebraska lockups 93 counties contain jails; Omaha has additional detention facilities

Nebraska State Penitentiary in Lincoln

NECCC New England Correctional Coordinating Council

necktie hangman's noose

necktie hanger gallows

Nepal Himalayan nation with an ancient prison in Kathmandu

Netherlands has both ancient and modern prisons in cities such as Amsterdam, Rotterdam, the Hague, and Utrecht

Netherlands Antilles all islands have small lockups and a low crime rate

Neuengamme concentration camp near Hamburg

Nevada Girls Training Center in Caliente

Nevada jails the state's 16 counties have lockups; Las Vegas, Reno, and Carson City have extra detention capability

Nevada Women's Correctional Center in Carson City

Neve Tirza Israeli maximum-security prison for women

New Albany minimum-security jail in New Albany, Indiana

New Braintree medium-security prison in New Braintree, Massachusetts

Newc Newcastle-upon-Tyne, England, detention facilities

New Caledonia French penal colony in the South Pacific from 1864 to 1894, when it was moved to French Guiana

New Era prisoner's newspaper published at Leavenworth, Kansas

Newgate London prison razed by rioters more than a century ago, now the site of the Old Bailey law courts, officially called the Central Criminal Courts

Newgate's Angel Elizabeth Gurney Fry—lay visitor for London's

Newgit nickname for London's

old Newgate Prison

New Hampshire jails 10 counties maintain lockups; its cities also have extra detention facilities

New Hampshire State Prison in Concord

New Hampshire Youth Development Center in Manchester; delinquents in all stages of security between 11 and 18 years of age

New Haven Community Correctional Center in New Haven, Connecticut, also the site of the county jail

New Hebrides (*see* Vanatu)

New Jersey jails 21 counties have lockups; Newark and Jersey City have additional jails

New Jersey State Prison— Leesburg in use since 1913 for male felons

New Jersey State Prison— Rahway a prison for male felons

New Jersey State prisons in Leesburg, Rahway, and Trenton

New Jersey State Prison— Trenton built in 1798 but replaced in 1836

New Life New Life House (for prisoner rehabilitation at Tam Lung Chung, New Territories, Hong Kong)

New Lon(don) New London, Connecticut, detention facilities

New Mexico Boys School in Springer

New Mexico lockups 32 counties have jails

New Mexico Youth Diagnostic Center in Albuquerque

New Orleans parish seat has and jail plus police-station lockups

New Queens newer section of the Riker's Island Penitentiary in New York City

Newton Iowa site of the Riverview Release Center

New Westminster Canadian maximum-security facility at New Westminster, British Columbia

New York any of many penal institutions in either New York City or New York State

New York lockups the state's 62 counties contain jails; all cities have extra detention facilities

New York School for Girls at

Hudson, New York

New York State Commission of Correction inspects and monitors the many correctional facilities, community residential facilities, secure detention centers, detention institutions, sentence institutions, and county jails, municipal lockups, and state prisons

New York State Correctional Facilities Albion, Arthur Kill, Attica, Auburn, Bedford Hills, Clinton, Coxsackie, Eastern, Elmira, Fishkill, Great Meadows, Green Haven, Ossining, Taconic, Wallkill, Woodbourne, Downstate, Hudson, Mid-Orange, Mount McGregor, Otisville, Queensboro

New Zealand HM Gaols in cities such as Christchurch, Auckland, and Wellington

Nha Trang (*see* Vietnam)

Niamey (*see* Niger)

Niantic Connecticut Correctional Institution farm and prison in Niantic; the J. Bernard Gates Correctional Unit

NIC National Institute of Corrections (U.S. Department of Justice in Boulder, Colorado)

Nicaragua capital, Managua, and cities such as Bluefields and León have *cárceles modelos*

Niceville former name of the Niceville Road Prison before it was moved to Crestview, Florida where it is now known as the Okaloosa Correctional Center

Nickerie (*see* Suriname)

Nicosia capital of Cyprus as well as the popular name for its old prison

Nigeria capital, Lagos, contains an old prison built by the British

nippers chain-grip-actuated handcuffs

NJA National Jail Association

NJDA National Juvenile Detention Association

NJMA National Jail Managers Association

NJRW New Jersey Reformatory for Women at Clinton

NO New Orleans, Louisiana's jail and police-station lockups

Nogal(es) Nogales, Sonora jail in México

Nome Nome State Jail in Alaska

Noranside Noranside Borstal Institution in Scotland

Norf Norfolk (in Canada, England, the South Pacific, or the United States) or any of its detention facilities

Norfolk Massachusetts site of the Bay State Correctional Center and the Massachusetts Correctional Institution; State Prison Colony in Norfolk, Massachusetts (first community prison for male felons); Virginia seaport city or its county jail or naval brig

Northallerton HM Prison at Northallerton in North Ridging, Yorkshire, England

North Carolina Department of Correction maintains statewide coverage through its Division of Prisons and its Youth Services Division

North Carolina jails 100 counties contain lockups; cities have extra facilities

North Central Correctional Institution at Gardner, Massachusetts

North Dakota Industrial School in Mandan seeks to reform delinquents from 12 to 18 years of age

North Dakota lockups the state has 53 counties providing jails

North Dakota Penitentiary in Bismarck

North Dakota State Farm near Bismarck

Northern Ireland (*see* individual entries)

Northern Nevada Correctional Center prison farm near Carson City

Northern Nevada Honor Camp near Carson City

Northern Region Correction Institute Alaskan facility at Fairbanks

Northern Rhodesia (*see* Zambia)

Northeye HM Prison at Northeye in Sussex, England

Northside Correctional Center in Spartanburg, South Carolina

Norway modern penal facilities are in or near such port cities as Oslo, the capital, Bergen, and Trondheim

Norwich HM Prison at Norwich, England; jails in Norwich, Connecticut and smaller towns named Norwich

Not-So-Nice-Ville (*see* Niceville)

Nottingham HM Prison at

Nottingham in Nottinghamshire, England

Nou island prison of Nouvelle Calédonie in the South Pacific

Nouakchott capital of Mauretania and site of its old prison

Nouvelle Calédonie (French —New Caledonia)—penal colony from 1864 to 1894 when its prisoners were shipped to French Guiana

Nova Scotia School for Girls at Truro, Nova Scotia

NP Naval Prison

NPB National Parole Board (Canada)

NPCC Nebraska Penal and Correctional Complex

NPP (*see* National Prison Project)

NPPAJ National Probation and Parole Association Journal

NPSB *National Prisoner Statistics Bulletin*

NRTI National Rehabilitation Training Institute

NSA National Sheriffs Association

NSPI National Society of Penal Information

Nuestra Familia (Spanish —Our Family)—Mexican-American prison-based underground organization engaged in narcotic trafficking and jail breaks

Nuevo Guerrero Nuevo Guerrero, Tamaulipas, jail

Nuevo Laredo Nuevo Laredo, Nuevo León, jail

Nuku'alofa (*see* Tonga)

nullum crimen sine lege (Latin—no crime without law)— crime must be defined by law

number-one diet bread and water

Nuremberg forced-labor subcamp in the south of Germany, close to the site of the Nuremberg Trials of war criminals

Nusa Kambangan Indonesian island prison off the city of Tjilatjap (Cilacap) where a crocodile-infested marsh between island and mainland discourages escape

nut factory section of a prison where criminally insane convicts are held

nuthouse psychiatric ward

nutpicker psychiatrist

nvd night-viewing device(s)

n-v device night-viewing device

NYCCIW New York City Correctional Institution for Women

NYCDC New York City Department of Correction

NYHD New York House of Detention

Nykobing Danish prison

NYMCC New York Metropolitan Correctional Center

NY Met New York Metropolitan Correctional Center (in downtown Manhattan)

NYRM New York Reformatory for Men

NYRW New York Reformatory for Women (Westfield Farm)

NYSDCS New York State Department of Correctional Services

Oahu capital island of Hawaii containing Honolulu, Pearl Harbor, and the Oahu Community Correctional Center

Oakalla prison in Burnaby, a suburb of Vancouver, British Columbia

Oakdale Federal Detention Center, Oakdale, Louisiana; site of the Iowa Security and Medical Facility

Oakhill Correctional Center in Oregon, Wisconsin

Oakie City Oklahoma City, Oklahoma's correctional facilities

Oakland Oakland, California's jails or its Community Treatment Center

Oakley Campus (*see* Raymond)

Oak Park Heights site of the Minnesota Correctional Facility

OAR (*see* Offender Aid and Restoration)

Oaxaca prison in Oaxaca, Mexico

obc old brutal con(vict)

Obispo California Men's Colony at San Luis Obispo

Oblatos jail in Guadalajara, Mexico

OCA Oregon Corrections Association

Ocala Florida location of the Alyce D. McPherson School for juvenile delinquents

Occoquan Minimum Security Facility of the District of Columbia located in Virginia

OCCSA Ohio Correctional and Court Services Association

OCF Ossining Correctional Facility (Sing Sing) at Ossining, New York

OCIS Organized Crime Information System (FBI)

OCJA Oklahoma Criminal Justice Association

Odenville state prison northeast of Birmingham, Alabama

Odessa Soviet city on the Black Sea with typical prison facilities; its namesake in west Texas contains the Ector County jail

Odom Jackson, North Carolina penal facility

Odum Georgia site of the Wayne Correctional Institution

Offender Aid and Restoration conducts CIP (Citizens Involvement Project) to educate and train civic leaders and sheriffs in the use of volunteers in jails

Ogden Utah site of the Ogden Community Corrections Center and the Parkview Community Corrections Center

Ohio jails 88 counties have prison facilities; cities, such as Cleveland, Cincinnati, and Columbus, have additional jails

Ohio Reformatory for Women in Marysville

Ohio State Ohio State Penitentiary in Columbus, also called River House

Ohio State Reformatory in Mansfield, Ohio

Ohrdruf Thuringian town in central Germany and site of a concentration camp as well as the underground headquarters of the German army during World War II

Oil City nickname of Bartlesville or Tulsa in Oklahoma jails; name of a town in western Pennsylvania and its jail

OIPC Organisation Internationale de Police Criminelle (French—International Criminal Police Organization)— Interpol

Ojinaga Ojinaga, Chihuahua, jail

Okaloosa (*see* Niceville)

Okeechobee location of the Florida School for juvenile delinquents

Okie Oklahoma(n) or any of its penal facilities

Okinawa Okinawa Prison in the Ryukyu Islands of Japan; (*see* USMC)

Oklahoma lockups 77 counties have prisons; extra facilities are found in Oklahoma City and Tulsa

Oklahoma State Penitentiary in McAlester

Oklahoma State Reformatory in Granite

Old Capitol Old Capitol Prison in Washington, DC where political prisoners were held during the Civil War

Old Dorp Schenectady, New York or its detention facilities

old hand Australian nickname for former convict

Old Horse Bridewell Prison

old lag person serving a three-year sentence in a British prison

Old Melbourne Gaol and Penal Museum Melbourne, Victoria makes the most out of a bad beginning

Old Newgate Prison penological museum on Newgate Road in East Granby, Connecticut

Old Queens older section of the Riker's Island Penitentiary in New York City

old smoky electric chair

Old Sparkey Florida's natural-oak electric chair

Old Territorial Old Territorial Penitentiary in Santa Fe, New Mexico

Olustee Florida site of the Baker Correctional Institution

Oma Omaha, Nebraska's jail

Oman Muscat and Matrah have small prisons

Omdurman (*see* Sudan)

Omsk czarist prison near Kazakhstan; the punishments inflicted on prisoners here are described in Dostoevski's *Notes from the House of the Dead*

on the bricks out of jail and on the streets

on the ground out of jail

on ice imprisoned

on the lam escaping, evading, or hiding from the police or other law-enforcement agents

Onley HM Borstal at Onley in Warwickshire, England

Only Tennessee site of the Turney Center for Youthful Offenders from 18 to 25 years of age

Ontario Youth Training School at Ontario, California

on the shelf in solitary confinement

Ont Pen Ontario Penitentiary (Canadian)

ooze out sneak out

open prison penal facility built without bars on the windows,

locks on the doors, or walls surrounding the prison

Oran (*see* Algeria)

Oranienburg concentration camp near Berlin

Orchid Island Rehabilitation Center on Orchid Island off southeastern tip of Taiwan (Formosa)

Ordinary of Newgate Chaplain of Newgate Jail

ordinary transportation on-foot transportation of prisoners

Ordot Guam's penitentiary at Ordot

Oregon Wisconsin site of the Oakhill Correctional Institution and the Wisconsin Correctional Camp System

Oregon jails the state's 36 counties have prisons; Portland and Eugene have augmented facilities

Oregon State Correctional Institution in Salem

Oregon State Penitentiary in 1853 opened in Portland but transferred to Salem in 1866

Oregon Women's Correctional Center in Salem; began in 1965 as a section of the Oregon State Penitentiary

Oriente Mexico City prisoner-holding facility

Oroville Butte County jail in Oroville, California

ORW Ohio Reformatory for Women

Osaka (*see* Japan's prisons)

Oslo (*see* Norway)

Ossining Ossining Correctional Facility (formerly known as Sing Sing) at Ossining, New York

Osteraker Swedish prison in Osteraker near Stockholm

Ostrava (*see* Czechoslovakia)

Ostrov Vrangelya (Russian—Wrangel Island)—Soviet prison camp northwest of Alaska and north of northeastern Siberia in the Chukchi Sea

Oswiecim Polish name for Auschwitz, site of a World War II concentration camp operated by the Nazis

Otay Otay Mesa Prison in southernmost California

Otay Mesa prison Richard J Donovan Correctional Facility south of San Diego, California

other prisoner(s) correctional officer(s); prison guard(s), warden(s), etc.

Otisville Otisville Correctional Facility at Otisville, New York

oubliette (French—secret dungeon)—cell with a trapdoor in its roof so convicts can be lowered into its hole

Outlaw bimonthly publication of the Prisoner's Union

outside outside of prison

out of town in prison

over the blue wall confined to a hospital for the criminally insane

Oxford Federal Correctional Institution at Oxford, Wisconsin, a Staff Training Center of the Bureau of Prisons; HM Prison at Oxford in Oxfordshire, England; some twenty other Oxfords in America have jails

Ozark Ozark Correctional Center in Fordland, Missouri

PA Pardon Attorney (U.S. Department of Justice)

PAA Prisoners Aid Association

PAAM Prisoners Aid Association of Maryland (Baltimore)

Pachuca prison in Pachuca, Mexico

Pachuco Pachucolandia (Mexican-American—El Paso, Texas)—nickname of the jail, lockups, and detention pens of the Immigration and Naturalization Service

Pacific Missouri community containing the Missouri Eastern Correctional Center

paddy wagon police van for transporting prisoners

Pakistan prison facilities in Islamabad, its capital, were built during British rule; the same is true in cities such as Karachi, Lahore, and Hyderabad

p-a-l prisoner-at-large

palacio blanc (Spanish—white palace)—nickname of Mexico City's most modern prison; it has cement-floored steel-lined cells plus baths, a hospital, and a library

Palacio Negro (Spanish—Black Palace)—nickname for Mexico's Lecumberri prison

Palais de Justice (French—Palace of Justice)—Parisian court and prison

Palmer Palmer Correctional Center northeast of Anchorage, Alaska

Palmer Work-Release Center in Florence, South Carolina

Panamá Panamá City, Colón, David, and San Miguelito all have so-called model prisons

Panamanian prisons (*see* Cárcel Modelo, Cueva, and Gamboa)

Pango Pago Pago, American Samoa or its jail

Pankrác Prague's prison

panopticon prison where all cells are visible from a central point

panopticon pattern based on Bentham's panopticon inspection house where giant circular prison houses were manned by armed guards who could supervise the surrounding cells and their inmates; the Stateville penal establishment in Illinois is built on this pattern

Papenburg Nazi concentration camp west of Bremen, Germany

Papua New Guinea a jail is in Port Moresby, the capital seaport city

Paraguay the capital city, Asunción, has a big prison

Paramáribo (*see* Suriname)

Paranam (*see* Suriname)

Parchman site of the Mississippi State Penitentiary

Pardelup Pardelup Prison Farm in Western Australia

pardon executive-applied exemption from punishment for a crime or for a pending criminal conviction

Parkhurst top-security prison near Newport on the Isle of Wight off England's south coast

Park Row Metropolitan Correctional Center in New York City

Parkside New York State Correctional Facility in Manhattan

Parkview Parkview Community Corrections Center in Ogden, Utah

parole conditional release of an offender from a confinement facility before the expiration of his or her sentence; released offender usually put under supervision of a parole agency or officer

parole agency correctional agency supervising convicts on parole

parole authority correctional agency or officer having authority to release prisoners

committed to confinement facilities or to discharge them from parole or to revoke parole

parolee person conditionally released from a correctional institution before the expiration of his or her sentence and placed under the supervision of a parole agency or officer

parole violation parolee's failure to conform to the conditions of parole; such violation usually results in return to prison and loss of parole

Parramatta Parramatta Gaol, Sydney, New South Wales, Australia

Parris Island (*see* USMC)

Pasca Pascagoula, Mississippi, jail

Paterson northern New Jersey mill town (often called Silk City) or generic name of its jail and its lockups

Patras (*see* Grecian prisons)

Patton California State Hospital (for the criminally insane) at Patton

Patuxent Patuxent Institution for the Criminally Insane (Patuxent, Maryland); Patuxent Institution for Defective Delinquents (Jessup, Maryland)

Patuxent Institution Maryland penal facility and pre-release center in Patuxent

Pavón El Pavón (Spanish—the peacock)—largest prison in Guatemala

Pawiak Warsaw, Poland's prison

P'burg Pittsburgh, Pennsylvania's jail

PCI Program of Correctional Institutions (Puerto Rico)

pcu (PCU) protective custody unit

PDA Polizeiliches Durchgangslager Amersfoort (German—Police Concentration Camp—Amersfoort, Netherlands)—staging area for the transport of Dutch prisoners to Nazi concentration camps

P del E Penitenciaria del Estado (Spanish—State Penitentiary)

peace and quiet maximum-security cell

Pedro San Pedro, California's jail

Peewee Valley Kentucky town associated with the Kentucky Correctional Institution for Women

Pelican Bay state prison near Crescent City, California

Pembroke Pines (*see* Broward)

pen. penitentiary

penal isolation solitary confinement

penalist(s) penologist(s)

penal servitude imprisonment combined with hard labor

penalty punishment for a particular offense

Pence Springs West Virginia site of the West Virginia State Prison for Women

Pendleton town that contains the Indiana State Reformatory

Penetang Penetanguishene Provincial Establishment for the Criminally Insane on Georgian Bay, Lake Huron, northnorthwest of Toronto

peni penitenciaría (Spanish—penitentiary)

peniatrist(s) prison doctor(s); prison psychiatrist(s)

peniatry branch of medical science dealing with penal establishments and their prisoners

penitenciaría (Spanish—penitentiary)—prison

penitenciária estadual (Portuguese—state penitentiary)—each of the 22 states of Brazil maintains such a penal institution

Penitenciaría General de Venezuela Venezuela's general penitentiary situated in San Juan de los Morros Ejido Guárico

Penitenciario de Santiago Santiago, Chile's penitentiary

Penitenciario Litoral Guayaquil, Ecuador's maximum-security prison

penitentiaries federal or state maximum-security institutions

penitentiary house of correction or rehabilitation center where offenders are confined for detention, discipline, reformation, rehabilitation, or punishment; in the United States a penitentiary is a maximum-security penal facility designed to hold prisoners serving long sentences

Penitentiary of New Mexico in Santa Fé

Penninghame Penninghame Prison in southwest Scotland

Pennsylvania jails the state's 67 counties have lockups; cities such as Philadelphia and Pittsburgh have extra facilities

Pennsylvania State Correc-

tional Institutions (*see* Bellefonte, Camp Hill, Dallas, Graterford, Huntingdon, Muncy, Pittsburgh)

Pennsylvania State Correctional Institutions and Correctional Diagnostic and Classification Centers: (*see* Camp Hill, Graterford, and Pittsburgh *entries*)

Pennsylvania System (*see* solitary system)

penol penological; penologist; penolgy

penologist social scientist concerned with penal institutions and the deterrent effect of punishment decreed by law

penology scientific study of penal institutions, the deterrent effect of punishments decreed by law, the consequences of crime, the means of changing lawbreakers into law-abiding citizens, and repairing the damage done to victims of crime

pensioner(s) of the crown Australian euphemism for any former convict(s)

Pensy Pensacola, Florida's nickname or that of its jail

Pent The Pent (nickname for Pentonville Prison in the outskirts of London's Islington Parish)

Pentonville HM Prison at Pentonville, a district of London

Pentridge Melbourne, Victoria's prison

Peoria county seat, courthouse, and jail of Peoria County, Illinois

Perm Soviet labor camp region 700 miles (1127 kilometers) east of Moscow

Persian prisons (*see* Iranian prisons)

Perth Scottish prison close to Perth on the River Tay

Peru its penal facilities are typical of many others in Hispanic countries (*see* Cárcel Modelo, Lurigancho, Sepa, Sexto)

Peruvian prisons (*see* Cárcel Modelo, Lurigancho, Sepa, Sexto)

Pete St Petersburg, Florida's nickname or that of its jail

Peterhead prison near Aberdeen, Scotland

Peter-Paul Peter-Paul Fortress in St Petersburg; now a Soviet museum

Petersburg Federal Correctional Institution at Petersburg, Virginia

Petros Tennessee location of Brushy Mountain Penitentiary

Pewee Kentucky Correctional Institution for Women at Pewee Valley

Pforzheim German city and prison near the Black Forest

Philadelphia Prisons Detention Center at 8201 State Road, Holmesburg Prison on Torresdale Avenue in the 8200 block; House of Correction at 8001 State Road

Philippines has some penal institutions dating back to Spain's colonialization and more modern ones reflecting the American period

Phillipsburg capital complete with lockup in the West Indian Leeward Island of Sint Maarten (the French have their own part of the island, Saint Martin, with a capital also complete with lockup and called Marigot)

Philly Philadelphia, Pennsylvania or the name given to any of its many penal institutions

Phnom Penh (*see* Cambodia)

Phoenix Arizona site of Adobe Mountain School for juvenile delinquents, the Alhambra Reception and Treatment Center for incoming male felons, the Arizona Center for Women, the Arizona Training Facility, the Arizona Correctional Training Facility

Phoenix Correctional Facility at Plymouth, Michigan

PHP Preventive Health Programs

PHS Prison Health Services (providing a complete system of health services to prisons)

picking oakum picking apart pieces of tarred rope for use in caulking wooden ships (a century ago this was still the task given many prisoners confined to jails along the coast of Britain as well as the United States)

Piedmont Work Release Center in Spartanburg, South Carolina

Piedras Negras Piedras Negras, Coahuila and its jail

Pikeville Tennessee site of the Bledsoe County Regional Correctional Facility and the Taft Youth Center

Pinal Pinal County Jail at Florence, Arizona

Pinchgut prisoner's nickname for Fort Denison Prison in Sydney Harbour, New South Wales

Pine Bluff Pine Bluff, Arkansas with its Diagnostic Unit, the Pine Bluff Youth Service Center, and the Women's Unit

Pine Hills Pine Hills School for delinquents from 10 to 21 years of age in Miles City, Montana

Pine Street Baltimore, Maryland jail on Pine Street

Piraeus (*see* Grecian prisons)

Pithiviers concentration and transit camp in France (*see* Camp de Drancy)

Pitts nickname for Pittsburgh, Pennsylvania or its penal facilities

Pittsburgh Correctional Institution in Pittsburgh, Pennsylvania, built in 1826 as the Western Penitentiary; enlarged in 1982

PK Principal Keeper

Plainfield site of the Indiana Boys School, the Indiana Youth Center, and the Reception and Diagnostic Center

Plankinton site of the South Dakota Training School for delinquents up to their seventeenth year

Plaszow Nazi concentration camp northwest of Cracow, Poland

Pleasanton Federal Correctional Facility at Pleasanton, California

Pleasantville federal minimum-security facility at Maxwell Air Force Base in Alabama

Plummer Center Delaware's work-release center in Wilmington

Plymouth Michigan site of the Phoenix Correctional Facility

PMS Prison Management Systems (health-care plan offered in the United States); Prison Medical Services (under the Home Office in the United Kingdom)

POA Prison Officers Association

POC Prison Officers Club

Pocaloo Pocatello, Idaho's nickname or that of its jail

poetic punishment matching the punishment to fit the crime

Point Lookout Union prison and stockade close to the Potomac where many Confederate prisoners died

Point Salines prison camp at Point Salines, Grenada

pokey jail

political(s) political prisoner(s)

politico political prisoner; politician

Polk Polk Correctional Institution at Polk City, Florida

Pollington HM Borstal at Pollington in Humberside, Yorkshire, England

Polmont Scottish borstal

POME Prisoner of Mother England (also called a Pommy when in early colonial days convicts were shipped to Australia)

Ponar site of a Nazi concentration camp in what is now Lithuania

Ponce city on Puerto Rico's south coast has an old jail as well as police-station lockups

Ponte dei Sospiri (Italian—Bridge of Sighs)—heavily barred, stone-covered, two-storied bridge arching a Venetian canal, the Rio di Palazzo, and connecting the Doge's Palace with the state prisons and dungeons

Pontiac Correctional Center northeast of Normal, Illinois

poogie jail

Poolsmoor prison near Cape Town, South Africa

Poona Indian prison in Poona

poorhouses of the twentieth century Ronald Goldfarb's apt description of jails

population movement entries and exits of adjudicated persons, or persons subject to judicial proceedings, into or out of correctional facilities

Pork Dump epithet applied to the Clinton Prison near Utica, New York

porridge British slang for jail

Portage La Prairie Correctional Center for Women at Portage La Prairie, Manitoba

Port Arthur Australia's principal penal colony from 1834 to 1853 in Tasmania, then known as Van Diemen's Land; today a museum and visitor's center replace the convict's quarters

Port Augusta Port Augusta Gaol (South Australia)

Port-au-Prince capital of Haiti

whose prisons are the most noisome in the West Indies

Port Blair headquarters of a penal settlement dating from the Sepoy Rebellion of 1857 but discontinued in 1945

Porte d'Enfer (French—Gate of Hell)—convict's nickname for the prison gate at Saint Laurent du Maroni in French Guiana

Port Elizabeth (*see* South Africa)

Port Isabel Port Isabel Detention Center of the Immigration and Naturalization Service in Port Isabel, Texas

Portland HM Borstal at Portland in Dorset, England; other detention places in towns named Portland from Australia to the West Indies, including Portland Maine and Portland, Oregon

Port Laoise prison southwest of Dublin, Ireland

Port Lincoln Port Lincoln Prison (south Australia)

Port Louis (*see* Mauritius)

Port Macquarie Australian convict colony in New South Wales in the early 1800s

Port Moresby (*see* Papua New Guinea)

Porto also called Oporto (*see* Portugal)

Port-of-Spain (*see* Trinidad and Tobago)

Portsmouth site of the Royal Navy's prison; New Hampshire port and site of the U.S. Naval Prison

Port Sudan (*see* Sudan)

Portugal Lisbon, the capital, and Porto (Oporto) have some very old prisons

Post Care Program for adult felons in Lincoln, Omaha, and Norfolk, Nebraska

Poston former Indian reservation in Arizona; used as a detention camp for Japanese-Americans

Potma reputed to be the Soviet Union's largest forced-labor penal colony

Poughkeepsie seat of Duchess County and its jail

pow (POW) prisoner of war

Powell Ohio location of the Riverview School for Boys and the Scioto Village, formerly the Girls Industrial School

Poznan (*see* Poland)

Pozsony Czech name for

Bratislava and site of an old prison built long before World War I

Pozzi old Italian prison in Venice

PPCAA Parole and Probation Compact Administrators Association

PPS Pennsylvania Prison Society (Philadelphia-based)

pq punishment quarters (isolated section of many penitentiaries and reformatories)

p & q peace and quiet (solitary confinement)

Prague (*see* Czechoslovakia)

Praia (*see* Cape Verde)

Presidential Candidate and Prison Convict Eugene V. Debs, Socialist candidate imprisoned for making an anti-war speech in 1918

presidio (Spanish—military prison)—may also mean citadel, penitentiary, or prison; many were built in the American Southwest during Spanish and Mexican rule

Presidio Prisoner's publication issued bimonthly at the Iowa State Penitentiary in Fort Madison

presidio modelo (Spanish—model penitentiary or prison)

Pressburg German name for Pozsony, site of an old prison in Czechoslovakia

Pretoria (*see* South Africa)

Pretoria Central maximum-security Pretoria Central Prison in Pretoria, South Africa

Pre-Trial Annex Wilmington, Delaware's facility for those held in detention status

Preungeshim security prison in Frankfurt, Germany

prigione (Italian—prison)—also called *carcere*

Prigione d'Genova (*see* Italian prisons)

Prince Albert Canadian city containing the Saskatchewan Penitentiary

Prince George British Columbia penitentiary

Princetown Her (His) Majesty's Prison in Princetown in Devon, England

Prince William Prince William —Manassas Regional Adult Detention Center in Virginia

prisão (Portuguese—prison) —also called *cárcere*

prison confinement facility with custodial authority over adults sentenced to confinement for more than one year

Prison at the Bottom of the World Ushuaia, Argentina

Prison at the Top of the World Solovetski Island in the Soviet Union

prison bird recidivist; prisoner who has been to prison before

prison break escape from prison accompanied by force and violence

prison bug person spending most of their time in prison

prison camp minimum-security camp designed to shelter convicts assigned to farm or forestry projects, road repair work, or other federal or state projects

Prison Camps the U.S. Bureau of Prisons maintains camps at Allenwood, Pennsylvania; Eglin Air Force Base, Florida; Montgomery, Alabama; and Safford, Arizona

prison coffin plain wooden box fitted with rope handles and perforated with holes facilitating disintegration once buried

prisoner person in custody at a confinement facility or in personal custody while being transported to or between confinement facilities

prisoner at large naval prisoner confined to the barracks or the ship

prisoner-of-war camps in the Confederacy Andersonville, Georgia; others in Georgia included Camp Davidson at Savannah, Camp Oglethorpe at Macon, and one at Millen; camps in South Carolina were at Charleston, Columbia, and Florence; in North Carolina at Salisbury; in Virginia at Danville and at Libby Prison in Richmond

prisoners of the Crown old Australian euphemism meaning convicts

prisoners' rules unwritten code of conduct, contains such admonitions as *be smart, don't be a coward, don't cheat your partner or your gang, don't be disloyal, don't snitch unless you're ready to die*

Prisoner's Union publishes *Outlaw* bimonthly; seeks an end to economic exploitation of prisoners and redress for convict's grievances

prison fever typhus (usually

due to overcrowding)

prison house prison

prison hulk prison ship

prison labor work carried out by convicts such as producing furniture, road building and repairing; stamping out automobile license plates, etc.

prisonment imprisonment

prison officer British euphemism for gaoler

prison pallor bloodless-yellow paleness of many prisoners deprived of fresh air, sunshine, and vitamins

prison psychosis mental disturbance actuated by imprisonment and manifested by delusions, paranoid trends, and pseudo-hallucinations

Prisons U.S. Bureau of Prisons

prison sentence commitment to the jurisdiction of a confinement facility

prison simple mentally deranged by imprisonment

prison smell usually compounded of excreta, grease, stale tobacco, sweat, unaired bedding, and vomit

prison van black maria or paddy wagon used to transport prisoners

prison within a prison solitary confinement cell

Prison World original name of the *American Journal of Correction*

prob probation(ary); probation officer

probation conditional suspension of imprisonment of a convicted offender who must stay in the community under the supervision of a probation officer

probation agency correctional agency supervising adults and juveniles placed on probation and investigating adults and juveniles to prepare predisposition and presentencing reports to assist courts in determining sentences

probation officer employee of a probation agency or probation department

probation sentence court requirement that a person fulfill certain conditions of behavior and accept the supervision of a probation agency or department

probation violation probationer's nonconformance to the conditions of probation

PROOF Parole Resource Office and Orientation Facility (Jersey City, New Jersey)

PROP Preservation of the Rights of Prisoners

Providence Rhode Island's capital city and county jail; other correctional facilities are scattered throughout this state

Providence County Jail old name for what is now known as the Awaiting Trial Facility in Cranston, Rhode Island

Provo seat of Utah County jail

PSAMPP Philadelphia Society for Alleviating the Miseries of Public Prisons (founded by Benjamin Franklin, William Rush, and others)

PSTD Prison Service Training Depot (Pretoria, South Africa)

psychiatric prison Her Majesty's Prison Grendon Underwood in the South Midlands of England

psychoprison psychiatric hospital prison (USSR's place for dissidents)

PU Prisoner's Union

Pudu prison in Kuala Lumpur, Malaysia

Puebla prison in Puebla, México

Puerto Barrios prison in Guatemala's Caribbean seaport

Puerto Cabello Venezuelan seaport where there is a jail

Puerto Cabezas Nicaraguan prison and seaport on the Mosquito Coast

Puerto Rican District jails in Aguadilla, Arecibo, Humacao, and in Ponce

Puerto Rican Prison camps six of them were active in the mid-1980s

Puerto Rico lockups 76 municipios (municipalities) make up Puerto Rico; some have jail facilities dating back to Spain's occupation of the island

Puerto Vasco pseudonym for a clandestine prison in Argentina

'Pulco Acapulco, México, jail

Pul-i-charki prison in Kabul, Afghanistan

Punta Arenas Chilean prison in the Straits of Magellan

Puntarenas Pacific coast port and prison of Costa Rica

Punxey Punxsutawney, Pennsylvania's jail and lockup

Purdy Treatment Center for Women in Gig Harbor,

Washington
put away imprison; remove from society
put in the hole place in a solitary-confinement cell
Puyallup assembly center for Japanese-Americans brought in from other parts of Washington
PVA Prison Visitor's Association
PWA Prison Wardens Association
pw('s) prisoner(s) of war
Pyongyang capital of North Korea containing a prison built between 1910 and 1945
Q San Quentin Prison in California
Qatar the capital, Doha, is on the Persian Gulf and its old prison is primitive
QC Quezon City, capital of the Philippines, or its detention facilities
QCPSA (*see* Quaker Center for Prisoner Support Activities)
QCSR Quaker Committee on Social Rehabilitation (in New York City)
Qsar Teheran, Iran's great prison
quad prison; prison quadrangle; prison yard
quail roost women's dormitory in a house of detention
Quaker Center for Prisoner Support Activities (QCPSA) conducts nonviolent training workshops for prisoners
Quantico (*see* USMC)
quarry cure forcing addicts to work in stone quarries far from sources of alcohol and narcotics
quarter stretch three-month's sentence
Queen of Heaven (*see* Italian prisons)
Queensboro Queensboro Correctional Facility in Long Island City, New York
queen's bus prison van
queer bird jailbird
queer-ken prison
Quent San Quentin (California State Prison)
Querétero prison in Querétero, Mexico
Questore Italy's security service and the National Central Bureau of Interpol in Rome
Quezon City also called Quezon (*see* Philippines)
Quilmas San Quilmas (Mexican-American—San Antonio,

Texas)—nickname of its jail, lockups, and the detention pens of the Immigration and Naturalization Service
quod prison
R rogue (brand burned on the left shoulder of convicts transported to various British colonies from 1619 until 1868)
Rabat (*see* Morocco)
rabbit feet escaped prisoners
rabbit fever desire to break parole or leave an honor camp before completion of sentence
rabbit foot escaped prisoner
rack maximum-security cell
Radical Alternatives to Prison Plan (RAPP) British program
Rahway site of New Jersey State Prison
Raiford Florida site of the Union Correctional Institution, once called the Florida State Prison or the Raiford State Prison
railroad sending a person to jail or prison without benefit of trial or proof of guilt
Raleigh North Carolina site of the Central Prison, Correctional Center for Women, and the Triangle Correctional Center
Ramla Israeli maximum-security prison
Ranby Camp HM Prison Camp at Ranby in Nottinghamshire, England
ranch synonym for a correctional facility such as a prison camp or farm in a rural area
Rancho del Campo juvenile correctional facility for older boys in San Diego County near Campo, California
Rancho del Rayo minimum-security correctional facility for younger boys in San Diego County, California
RAP Release Aid Program
RAPP Radical Alternatives to Prison Plan
rasoir nationale (French—national razor)—the guillotine
rasphuys (Flemish—rasp house) —Ghent's workhouse-type prison—the French call this *maison de force* (workhouse)
Rathmore Road nickname derived from the address of Cork Prison, Ireland
rat row prison cells for informants
Ravensbrück Nazi concentration camp north of Berlin

Rawlins location of the Wyoming State Penitentiary
Ray Brook Federal Correctional Institution at Ray Brook, New York
Raymond Mississippi community containing the Oakley Campus run by the Department of Youth Services
Raymond Street Raymond Street Jail in Brooklyn, New York
razor ribbon razor-edged stainless-steel security fencing
RCG Reception Guidance Center
Reading HM Prison at Reading in Berkshire, England; American jails in other Readings in Kansas, Massachusetts, Michigan, Minnesota, Ohio, Pennsylvania, and Vermont
Reception and Diagnostic Center maximum-security facility of Indiana's Department of Correction in Plainfield
Reception and Diagnostic Center for Children Bon Air, Virginia's facility for delinquents from 8 to 18 years of age
Reception and Guidance Center—Jackson in Jackson, Michigan to help juvenile delinquents
Reception and Guidance Center—Riverside in Ionia, Michigan, holds male felons under 21 years of age
Reception and Medical Center in Lake Butler, Florida complete with a hospital and surgery
Reception Center Maryland penal facility in Baltimore next to the Maryland Penitentiary
reception centers World War II euphemism for camps set up to hold some 110,000 Americans of Japanese descent until they could be relocated away from the West Coast
recidivist habitual prisoner; person spending much of their life in prison
recid(s) recidivist(s)
reclusão (Portuguese—reclusion)—solitary confinement
reclusion (French—solitary confinement)
Réclusion de Saint-Joseph solitary-confinement prison on Saint-Joseph Isle off French Guiana, close to Devil's Is-

land

réclusionnaire (French—prisoner in solitary confinement)

Reclusorio Norte (Spanish—Northern Place of Retirement)—official name of Mexico City's penitentiary

reconcentrados (Spanish—concentration camps)—established by the Spanish in Cuba in 1896 but abolished by 1898 after many protests made in England, Spain and the United States

record purge complete removal of arrest, criminal, or juvenile records

Redding Shasta County courthouse and jail in California

Redención (Spanish—Redemption)—Spain's official prison publication printed in prison workshops and subscribed to by prisoners throughout the country and in Africa

Red Wing site of the Minnesota Correctional Facility at Red Wing

Redwood City San Mateo County jail in California

Reeducation of Attitudes and Repressed Emotions treatment program for sex offenders

ref reformatory

reflection cell maximum-security cell

reformatory house of correction charged with making convicts alter their life style and return to society as law-abiding citizens

Regina Coeli (Latin—Queen of Heaven)—Rome, Italy's prison

Regina Provincial Correctional Centre in Saskatchewan

Regional Correctional Facilities (*see* Greensburg, Mercer)

rehab rehabilitate; rehabilitation

rehabilitation changing the offender's character, intent, and motivation toward law-abiding conduct

rehabilitation laboratory euphemism for any modern prison

Reidsville site of the Georgia State Prison

Re-Integration Center in Albuquerque, New Mexico

Reiserstown Montrose School for Girls in Reiserstown, Maryland

Release Aid Program (*see* RAP)

rélegué (French—isolated; relegated)—convict condemned to banishment in a penal colony

relocation center camp where Japanese-Americans were confined shortly after the attack on Pearl Harbor

remand to send a prisoner back to court for a further hearing; a remand prison contains people awaiting return to court

Remand Remand Prison at Jin Penjara 3, Singapore

remand center British term for a borstal or juvenile jail where convicts undergo a period of rehabilitation before being paroled or released

remand home synonym for a remand center

remand institution jail, penitentiary, or prison

Rembert South Carolina site of the Wateree River Correctional Institution

Reno El Reno, Oklahoma's federal detention reformatory; Reno, Nevada or its jail

rent-a-con plan hiring ex-convicts so they get a fresh start in society

Renz Correctional Center in Cedar City, Missouri

reprieve executive order suspending execution of a sentence

resilient cell padded cell preventing prisoners from injuring or killing themselves

Rest and Reverie nickname of Terminal Island, California's prison in Los Angeles Harbor

restitution center small house where convicted criminals must spend every night after going out every day to work off their debts to the victims of their crimes

restraints ankle cuffs, belly chains, belt restraints, handcuffs, leg irons, straitjackets, etc.

Retrieve Unit in Angleton, Texas

Réunion French island in the Indian Ocean used for the isolation of political prisoners

Reykjavik (*see* Iceland)

Reynosa Reynosa, Tamaulipas and its jail

RGC Reception Guidance Center

Rhode Island jails five counties have lockups; Providence

has extra facilities

Rhode Island penal facilities all 11 are in Cranston

Rhode Island State Prison in Cranston, now known as the Maximum Security Facility

Rhode Island Training School for Girls at Howard, Rhode Island

Rhodesia (*see* Zimbabwe)

Richmond site of the Virginia State Penitentiary or the local jail and police-station lockups; Richmond Penitentiary in St Croix, American Virgin Islands West Indies; Texas town holding the Jester prerelease unit

Richmond Farm Jamaican prison close to Annotto Bay on the north coast

Richmond Hill prison on the island of Grenada

Richmond Village Tasmanian town noted for its convict-built bridge and old gaol

Ridgeville South Carolina site of the MacDougall Youth Correction Center

Riga main concentration camp in Riga, Latvia during World War II

Riker's complex of prisons on Riker's Island in New York City's East River; complex includes the Adolescent Remand Shelter, the Riker's Island Penitentiary, and the Riker's Island Women's Detention Center

Ringe Danish state prison at Ringe, often nicknamed the Sex Prison, as inmates of both sexes are allowed sexual freedom; on the island of Funen south of Odense

Ring-Ring Copenhagen, Denmark's penitentiary

Rio Rio de Janeiro or its jail

Rio Consumnes Correctional Facility in Elk Grove, California

Río de Oro former Spanish prison colony in northwest Africa

Riom location in France of an old and infamous prison

Río Muni Spanish Guinea on the West African mainland and the nearby island of Fernando used as a convict settlement

Rio Piedras location of Puerto Rico's State Penitentiary

RIP Riker's Island Penitentiary

Risdon Hobart, Tasmania's

prison and prison hospital

River Avenue Bronx House of Detention in New York City

River House Ohio State Penitentiary on the Scioto River

River Junction Correctional Institution in Chattahoochee, Florida

Riverside Riverside County courthouse and jail east of Los Angeles, California

Riverside Correctional Facility in Ionia, Michigan

Riverton site of the Wyoming Honor Farm

Riverview Canadian Interprovincial Home for Women (misdemeanants) in Riverside, New Brunswick; Florida site of the Hillsborough Correctional Institution

Riverview Release Center in Newton, Iowa

Riverview School for Boys in Powell, Ohio

Riyadh (*see* Saudi Arabia)

RLDPAS Royal London Discharged Prisoner's Aid Society

RLPAS Royal London Prisoners' Aid Society

Roaston, Toaston, and Duston nicknames given by the Japanese-Americans interned at prison camps in the roasting, toasting, and dusty desert Arizona

Robben Island South African penitentiary for political prisoners

Rochester detention facilities in Rochester, Minnesota and Rochester, New York; HM Borstal at Rochester on the Medway River estuary in England's Kent

Rock the Rock (nickname for the 12-acre rock occupied by Alcatraz when it served as a prison in San Francisco Bay); nickname for the Riker's Island penal facilities in New York's East River

rock crusher prisoner assigned to hard manual labor

Rockland Rockland State Hospital for the Criminally Insane in New York's Rockland County

Rock Mountain Richard J Donovan Correctional Facility at Rock Mountain, California

Rock Spring Georgia site of the Walker Correctional Institution

Rockville Training Center Indiana correction facility in Rockville

Rockwell City Women's Reformatory at Rockwell City, Iowa

Rocky Butte Portland, Oregon's jail

Roebuck Roebuck Campus Birmingham, Alabama (academic-oriented rehabilitation program for juvenile delinquents)

Rohwer Arkansas relocation center for interned Japanese-Americans

Romania Bucharest and Constanza have prisons built when the Balkan nation was a kingdom

Romanian concentration camps during World War II the most notorious were Akmecetka, Bogdnovka, and Dumanovka

roomie cellmate

Roosevelt Roosevelt Island in New York City's East River, formerly called Welfare Island and Blackwell's Island; site of lunatic asylum, prison, and workhouse

Roosevelt Roads Puerto Rico's location of the U.S. Naval Station and its brig

rope hangman's rope; marijuana; vein

Rosario (*see* Argentina)

Roseau (*see* Dominica)

Rosharon Texas site of the Darrington Unit

Rostov-on-Don Soviet state police prison, also called House 33

Rota small Spanish port in Cadiz Bay and location of a brig maintained by the U.S. Navy

rotan (Malay—rattan)—lashing stick in Malaysia and Singapore where corporal punishment is still used in jails and street riots

Rotterdam (*see* Netherlands)

Rottnest Rottnest Island, former penal colony in the Indian Ocean

Round House Fremantle, Western Australia's oldest structure built in 1830 as a jail

Rove Solomon Islands' prison in Honiara

RTU Rahway Treatment Unit for sex offenders imprisoned in New Jersey's State Prison at Rahway

rubber room padded cell for self-destructive or violent prisoners

Rutland Correctional Facility in Rutland, Vermont

Rutland Street nickname of Limerick's prison on Rutland Street in Ireland

Rwanda the capital, Kigali, contains an old prison built by the Belgians

S-21 Khmer Rouge name for Tuol Sleng near Phnom Penh, Cambodia where nearly 20,000 prisoners were executed by the Khmer Rouge

SA Salvation Army

SAA Singapore Aftercare Association (hostel for ex-convicts)

Sabanete nickname of Maracaibo, Venezuela's national prison (*Cárcel Nacional de Maracaibo*)

Sachsenhausen main Nazi concentration camp near Berlin

Sacramento Sacramento County courthouse and jail in California

SACRO Scottish Association for the Care and Resettlement of Offenders

Sacto Sacramento, California or its jail

safe house military jail; rehabilitation center for prostitutes

safekeeper felon preserved from escaping by being put in custody

safety cell padded cell for self-destructive or violent prisoners

safety vest straitjacket

Safford Federal Prison Camp at Safford, Arizona

Sagmalcilar Istanbul, Turkey's great prison

Sagmalcilar Hilton inmates' nickname for the principal prison of Istanbul

Said Port Said, Egypt, and its jail as well as police lockups

Saigon (*see* Vietnam)

Saint Anthony home of the Youth Services Center of Idaho

Saint Barts Saint Barthelemy in the French West Indies or its jail

Saint Bridget's Well original name of London's Bridewell prison, once a royal palace of King Edward VI

Saint Catherine's Jamaican district prison near Kingston

Saint Cloud Minnesota Correctional Facility at Saint Cloud;

26 rue Armengaud, Saint Cloud, Paris (headquarters of the general secretariat of Interpol—the International Criminal Police organization)

Saint Cyrille Maison Gomin correctional facility for women at Saint Cyrille, Québec

Saint Gabriel Louisiana Correctional Institute for Women at Saint Gabriel; the Hunt Correctional Center is also in Saint Gabriel

Saint George's (*see* Grenada gaol)

Saint Jean prison camp on the Maroni River of French Guiana, upstream from the penal colony headquarters at Saint Laurent

Saint Joe Saint Joseph, Missouri, its jail or lockups

Saint John jail and lockups at Saint John, New Brunswick

Saint Joseph and Saint Paul adjacent maximum-security prisons in Lyons, France

Saint-Laurent-du-Maroni French Guiana port on the Maroni River; headquarters of the penal settlement known as Devil's Island

Saint Lou Saint Louis, Missouri or its jail

Saint Louis du Maroni prison camp on the Maroni River in French Guiana upstream from Saint Laurent

Saint Lucia Castries, the capital, has an HM Gaol built during British dominion, 1814 to 1979

Saint Lucy nickname for Saint Lucia or its jail or police-station lockup

Saint Marguerite old prison close to the coast of Cannes

Saint Martin seaport capital, Marigot, has its own lockup

Saint Mary's Saint Mary's Honor Center in Saint Louis, Missouri

Saint P Saint Paul, Minnesota or its jail

Saint Paul (*see* Saint Joseph and Saint Paul)

Saint Pete Saint Petersburg, Florida or its jail

Saint Petersburg (*see* Tampa)

Saint Pierre former French penal colony on Saint Pierre Island in the Gulf of Saint Lawrence

Saint Vincent and the Grenadines the capital, Kingstown,

has an old HM Gaol

Salaspils Nazi concentration camp near Riga, Latvia

Salem capital of Oregon contains the Correction Division Release Center, the Hillcrest School of Oregon, the Oregon State Correctional Institution, the Oregon State Penitentiary, and the Oregon Women's Correctional Center; West Virginia Home for Girls at Salem

Salinas Monterey County jail south of San Jose, California

Salisbury Salisbury (North Carolina) National Cemetery containing graves of Union soldiers who died in the Confederate prison here during the Civil War; (*see* Zimbabwe)

sally port first gate to a prison

Salpetriere Paris hospital for the criminally insane

Saltillo prison in Saltillo, Coahuila, México

Salt Lake City Utah location of the Community Corrections Centers—Bonneville, Central, Lakehills, Women's; and the State Diagnostic Unit

Salt Lake Women's Community Corrections Center in Salt Lake City, Utah

Samarkand Manor North Carolina youth service facility at Eagle Springs

Samoa Apia, its capital, has a lockup

Sanaa (*see* Yemen)

San Anto (Mexican-American —San Antonio, Texas)—jail nickname

San Antone nickname for San Antonio, Texas or its jail and its police-station lockups

San Berdoo San Bernardino, California or its jail

San Bernardino San Bernardino County courthouse and jail east of Los Angeles, California (also called San Berdoo)

San Bruno San Francisco County Jail at San Bruno, California

sand prisoner's nickname for sugar

San Diego San Diego County courthouse and jail in San Diego, California (also called Dago or Diego)

Sands Sands Prison outside Beirut, Lebanon

Sandstone Federal Correctional Institution at Sandstone, Min-

nesota

San Francisco San Francisco County courthouse and jail in San Francisco, California, also known as Frisco

San Francisco (Italian—Saint Francis)—name of the prison in Parma, Italy

San Jack San Jacinto, Texas or its jail

San Jo San José, California and its jail

San Jose Santa Clara County courthouse and jail southeast of San Francisco, California

San José capital of Costa Rica and site of its main prison

San Juan capital of Puerto Rico, includes Bayamón, Carolina, Cataño, Guaynabo, Rio Piedras, and Trujillo Alto, in addition to penal facilities dating back to Spanish rule; the general penitentiary of Venezuela situated in San Juan de los Morros Ejido Guarico

San Juan Detention Center in San Juan, Puerto Rico

San Juan de Ulúa old Spanish fortress on an islet about a mile (1.6 kilometers) off the shark-infested port of Veracruz, Mexico; since colonial times the fortress has served as a dungeon for political prisoners

San Luis Obispo California Men's Colony at San Luis Obispo includes a forestry camp; also the San Luis Obispo courthouse and jail

San Luís Potosí prison in San Luís Potosí, Mexico

San Luís RC San Luís Río Colorado, Sonora and its jail

San Marino has a very small lockup

San Pedro Lima, Peru's largest penitentiary

San Pedro Sula Caribbean port of Honduras and its prison

San Quentin California State Prison at San Quentin on a small peninsula in San Francisco Bay

San Quentin Daily award-winning newspaper published by inmates at San Quentin in California

San Quilmas San Antonio, Texas or its jail or lockups

San Rafael Marin County courthouse and jail in California

Santa Ana the central penitentiary of Occidente's Ejido Trujillo in Venezuela; Orange County courthouse and jail southeast of Los Angeles, California

Santa Anita California racetrack that served as an internment camp for Japanese-Americans

Santa Barbara Peruvian prison for women in Lima's port of Callao; Santa Barbara County courthouse and jail in southern California

Santa Clarita maximum-security jail near Valencia, California

Santa Cruz Santa Cruz County courthouse and jail north of Monterey, California

Sante Fé site of the Penitentiary of New Mexico

Santa María de la Cabeza monastery in the Sierra Morena northeast of Cordoba, Spain, defended by Franco's forces during the Spanish Civil War, later was used as a prison to for Republican leaders

Santa Marta Colombian prison in the seaport of Santa Marta; federal prison outside Mexico City

Santa Marta Acatitla prison southeast of Mexico City

Santa Rita Santa Rita Rehabilitation Center in California's Alameda County

Santa Rosa Sonoma County seat and jail north of San Francisco, California

Santé Parisian prison at 42 rue de la Santé

Santiago de Cuba (*see* Cuba)

Santo Domingo capital of the Dominican Republic noted for the poor quality of its detention camps, jails, and prisons

Santurce Puerto Rican site of the Institution for Youthful Offenders

São Tome and Principe prison is a veritable antique

Sapporo (*see* Japan's prisons)

Sarah Anthony San Diego, California's school for delinquent boys and girls aged 8 to 18

Sarajevo (*see* Yugoslavia)

Sasabe Sasabe, Sonora, Mexico and its jail

Saskatchewan Saskatchewan Penitentiary in Prince Albert

sat in the hot seat died by electrocution

Saudi Arabia the capital, Riyadh, and Jidda have prisons dating back the 18th, 19th, and early 20th centuries

Saughter prison in Edinburgh, Scotland

Sauk Centre location of the Minnesota Correctional Facility

Saulsbury Tennessee site of a Civil War prisoner-of-war camp

Sav Savannah, Georgia or its jail

SBCR State Board of Charities and Reform (Wyoming)

SBIW Sybil Brand Institute for Women (correctional facility in Los Angeles, California)

scam escape from jail or prison

Scanray Scanray Corporation's X-ray system for seeing what is inside parcels brought into jails or prisons

scarce commodity prison space

SCCA South Carolina Corrections Association

Schenectady upper New York State site of county jail

Schlüsselburg Leningrad prison

school of crime epithet applied to many prisons

sci (SCI) secret confidential informant

Scioto Village Ohio penal facility for male and female delinquents; opened as the Girls Industrial School in 1869

Scotland (*see* United Kingdom and individual entries)

Scottish Association for the Care and Resettlement of Offenders SACRO

Scottish prisons Aberdeen, Castle Huntly Borstal Institution, Cornton Vale, Dumfries Young Offenders Institution, Edinburgh, Edinburgh Young Offenders Institution, Glasgow, Glasgow Young Offenders Institution, Glenochil Detention Centre, Greenock, Inverness, Longriggend Remand Institution, Low Moss, Noranside Borstal Institution, Penninghame, Perth (including the Friarton Young Offenders Institution), Peterhead, Polmont Borstal Institution

Scottish Prison Service College SPSC

scragsman British slang for a hangman

Scranton Pennsylvania city and Lackawanna County seat and jail

screening (*see* detention screening)

screw(s) prison guard(s)

Scrubs The Scrubs (Wormwood Scrubs Prison in West London's stadium area)

SDC State Department of Corrections (Alabama, Colorado, Virginia)

SDCJ San Diego County Jail in California

SDMCC San Diego Metropolitan Correctional Center (California)

SD Met San Diego Metropolitan Correctional Center in downtown San Diego, California

Seagoville Federal Correctional Institution at Seagoville, Texas

Sea-Tac Seattle-Tacoma, Washington or its jail

Seavy's Island U.S. Naval Prison near Portsmouth, New Hampshire

seclusion solitary confinement

security housing section of a prison where hardened and hard-to-handle prisoners are segregated

Security Islands Îles du Salut off the coast of French Guiana; three rocky islets—Île du Diable (Devil's Island), Île Royale, and Île Saint Joseph—surrounded by shark-infested waters; until the 1950s political prisoners were isolated here

segregation isolation of criminals from other members of society; racial segregation; solitary confinement

seis y uno (Spanish—six plus one)—*see* Six plus One

Send detention center in Surrey, England

Senegal the capital, Dakar, has an old prison built by the French

Seoul capital of South Korea contains a prison built during the Japanese occupation of Korea

Sepa maximum-security prison in the remote jungles of Peru

Serbia (*see* Yugoslavia)

serve a sentence spend time in jail or prison

serve time spend time in jail or prison

SETAF Southern European Task Force (confinement facility in Italy)

Settlement Island former penal

colony within Macquarie Harbour on the coast of Tasmania

Sevastopol Sevastopol central prison in Crimea

Seven-Step Foundation halfway house for released convicts

Seville (*see* Spain)

Sex Prison nickname for the Danish state prison at Ringe where the experiments in penology are underway in an effort to treat felons as humans

Sexto Lima, Peru jail

Seychelles the capital, Victoria, has an old HM gaol facility

SFCJ San Francisco County Jail

SFDC Santa Fe Detention Center in New Mexico

Shaker Road Albany County Penitentiary on Shaker Road in the Colonie section of Albany, New York

Shakopee location of the Minnesota Correctional Facility for female felons

shamus Irish-Gaelic nickname for a detective or other law-enforcement officer—*see* canine shamus(es)

Sharjah (*see* United Arab Emirates)

Sharpes (*see* Brevard)

Shata medium-security Israeli prison with a work-release program managed by kibbutz volunteers

Sha Tsui Sha Tsui Detention Centre on Lantau Island, New Teritories, Hong Kong

Sheff Sheffield, England or its jail

shelter confinement facility for juveniles held pending adjudication

Shelton Washington State Corrections Center or the Women's Correctional Facility at Shelton

Shelton Abbey old Irish prison near Arklow on the east coast of Ireland

Shepton Mallet HM Prison at Shepton Mallet in Somerset, England

Sheridan federal correctional institution in Oregon; Sheridan Correctional Center in Sheridan, Illinois, functions as the Illinois Industrial School for Boys (17 years old and up); the Wyoming Girls School at Sheridan

sheriff chief officer of county

law enforcement and the county jail

Shin Bet Israel's domestic security agency

shit on a raft naval and prison slang for creamed beef or creamed chicken on toast

shit on a shingle military and prison slang for creamed beef or creamed chicken on toast

shiv knife made in prison; switchblade knife

Shore Patrol Tank nickname for a lockup maintained by the Navy in many American seaports

short stretch short prison sentence

short-term adult penal institutions of the U.S. Bureau of Prisons Allenwood, Pennsylvania; Elgin Air Force Base, Florida; El Paso, Texas; Florence, Arizona; Montgomery, Alabama; Safford, Arizona

Shrewsbury HM Prison at Shrewsbury in Shropshire, England

shrouding cover feature of hard-to-pry-open locks, padlocks, and shackle locks

Siam (*see* Thailand)

Siberia Russian area given over largely to the exile and imprisonment of political prisoners

Siberia de las Américas (Spanish—Siberia of the Americas)—political prisoner's nickname for prisons on the Isle of Pines *(Isla de Pinos)*, renamed *Isla de Juventud* (Isle of Youth)—in the Caribbean off Cuba

Siberian salt mines nickname for the forced-labor camps and prisons throughout Siberia and other places in the USSR

Sierra Leone the capital, Freetown, has a prison built by the British

silence bell evening bell rung to advise prison inmates they must cease all talking and noisemaking

silent system imprisonment characterized by enforced silence and by night confinement in small solitary cells; inmates allowed to congregate with other prisoners during meals or when at work; also called the Auburn System

Silk City Paterson, New Jersey, or its jail

Simons Simonstown, South Africa, or its jail

Simsbury early American prison built within an abandoned copper mine near Hartford, Connecticut; in use from 1773 to 1827

Sin Angeles (Spanish—Without Angels)—Los Angeles, California or its jail, lockups, penitentiaries, and prisons

Singapore sentence death by hanging is the Singapore sentence for trafficking in drugs

singbird(s) informant(s)

Sing Sing New York State Penitentiary at Ossining (where the birds warble twice—sing sing), now known as the Ossining Correctional Facility

Sin Lam Sin Lam Psychiatric Centre for the criminally insane at Sin Lam in the New Territories of Hong Kong

Sint Maarten (*see* Phillipsburg)

Sioux Falls South Dakota penitentiary site

Sirkeci Istanbul, Turkey's police station and lockup

Six plus One Spanish prison sentence of six years plus one day for anyone found in possession of or using narcotics or drugs

sizzle die in the electric chair

sizzle seat electric chair

Skag Skagway, Alaska, jail

Skopje (*see* Yugoslavia)

Skowhegan Women's Correctional Center at Skowhegan, Maine

slammed jailed

slam(mer) jail or prison where steel doors slam shut on the inmates

slanguage slang language; slum language

Sleepy Hollow Trenton Prison in New Jersey

Sleepy Hollow Correctional Centre in Charlottetown, Prince Edward Island

Slovenia (*see* Yugoslavia)

slum slumgullion stew (served in many jails)

Smithfield Pennsylvania Corrections Department near Carbondale

smogged executed in a gas chamber

Smyrna the Delaware Correctional Center

Sneads Florida site of the Apalachee Correctional Institution

Snoqualmie Washington town

containing the Echo Glen Children's Center for delinquents from 8 to 14 years of age

Sobibor extermination camp near Lublin in Poland during World War II

Society for Alleviating the Miseries of Public Prisons Quaker group that planned the Eastern State Penitentiary in North Philadelphia, Pennsylvania; prisoners were kept in individual cells, each with an adjacent exercise yard; prisoners were supposed to become reformed by reading the *Bible* and reflecting on their crimes although this solitary confinement drove many of them crazy

Sofia capital city of Bulgaria and nickname of its prison

sol solitary confinement

Soledad California Training Facility at Soledad

solitary system imprisonment designed to segregate criminals from each other so as to prolong reflection and assure self-reform; many so subjected went insane; also known as the Pennsylvania System

Solomon Islands Honiara, the capital, has an HM Gaol-type prison on Guadalcanal Island

Solovetskis Solovetski Islands, penal isolators in the Archangelsk Region of the USSR

Somalia capital and its prison, both called Mogadishu

Somers Connecticut Correctional Institution in Somers

Somerville Tennessee site of the Wilder Youth Development Center

Sonderbehandlung (German —special treatment)—term meaning killing at Nazi concentration camps

Sonderkommando Umsiedlungslager (German—Special Unit Resettlement Camp)—euphemism for a quick-extermination concentration camp such as Sobibor

Sonkom Sonderkommando (German—Special Commando)—Hitler organization working at concentration camp crematoria and gas chambers where some inmates were forced to assist their armed guards in killing

weaker prisoners and removing their remains

Sonoita Sonoita, Sonora, Mexico

Soo Sault Sainte Marie (Michigan or Ontario) and their jails

Soria Madrid, Spain's great prison

Sou Southhampton, England or its jail

South Africa has prisons in all its principal cities, Cape Town, Durban, Johannesburg, Pretoria, and in smaller port cities such as East London and Port Elizabeth

Southampton Correctional Center in Capron, Virginia is used for youthful first offenders

South Bay Regional Center euphemistic name for San Diego, California's jail in Chula Vista

South Burlington Vermont site of the Chittenden Correctional Facility

South Carolina jails 46 counties ranging have lockups; Charleston, Columbia, and Greenville have extra facilities

South Carolina School for Girls at Columbia

South Dakota lockups 67 counties contain jails

South Dakota Penitentiary in Sioux Falls

South Dakota Training School serving juvenile offenders held in Plankinton

Southeastern Correctional Center in Bridgewater, Massachusetts

Southeastern Ohio Training Center in Lancaster, formerly the Fairfield School for Boys

South Eastern Region Correction Institute Juneau, Alaska

Southern Michigan Southern Michigan Prison at Jackson

Southern Nevada Correctional Center in Jean

Southern Ohio Correctional Facility in Lucasville

Southern Rhodesia (*see* Zimbabwe)

Southern Steel San Antonio, Texas firm engaged exclusively in the manufacture of detention equipment

South Lansing the South Lansing School for Girls at South Lansing, New York

South Walpole location of the Massachusetts Correctional

Institution, known as Cedar Junction

Southwark London borough on the Thames infamous for its Clink Prison for heretics; the prison is gone but the expression—*in the clink*—remains

South Windham site of the Maine Correctional Center

Soviet Kolyma deadliest penal colony in the Soviet Union

Soviet Union (*see* USSR)

Spain jails, penitentiaries, and prisons are in its capital, Madrid, and cities such as Barcelona, Bilbao, Malaga, Seville, Toledo, Valencia, and Zaragoza

Spandau Berlin's great prison in the Tiergarten section of the city

Spanish Guinea used as a convict settlement during Spain's rule

Spanish windlass straitjacket (became tighter and tighter when sprayed with water or soaked with the prisoner's sweat)

Spartanburg South Carolina site of the Northside Correctional Center and the Piedmont Work Release Center

Spassk Soviet prison camp in Kazakhstan

SPC Service Processing Center(s)—formerly called Immigration and Naturalization Detention Center(s)

Spectator prisoner's publication of the Michigan State Prison at Jackson

Spinhaus (German—workhouse)—old term for a house of correction

spinhuiz (Dutch—workhouse) —old term for a house of correction

split bit prison sentence providing for both a maximum and a minimum sentence (*see* flat bit)

Spoke Spokane, Washington or its jail or lockups

sponging house jail where debtors were kept for a day to allow them to settle their debts before being imprisoned or transported overseas

spring release from jail or prison

Springer the New Mexico Boys School, in Springer

Springfield Sangamon County courthouse and jail in Illinois; Medical Center for Federal

Corrections Facilities 1054

Prisoners at Springfield, Missouri

Spring Hill HM Prison at Spring Hill in Londonderry, Northern Ireland

Springhill Institution in Springhill, Nova Scotia

sprung released from jail on bail

SPSC Scottish Prison Service College

SQP San Quentin Prison (California)

squat to be electrocuted

squirrel cage hospital for the criminally insane

src (SRC) strict-regime (prison) camp (*see* Strict-Regime Camp)

Sri Lanka HM Gaol-type facilities are in Colombo, the capital, and in Jaffna, Kandy, and Galle

SRW State Reformatory for Women (Dwight, Illinois)

SSCA Southern States Correctional Association

Stafford HM Prison at Stafford in Staffordshire, England

Staff Training Centers the Bureau of Prisons offers training to correctional personnel and jailers from all parts of the United States in its centers in Atlanta, Georgia; Aurora, Illinois; Dallas, Texas; and Oxford, Wisconsin—the food service training center of the FCI (Federal Correctional Institute)

Stalag (German—prisoner-of-war camp)

St Albans Correctional Facility in St Albans, Vermont

Stammheim Stuttgart, Germany's maximum-security prison

Standford Hill HM Prison at Standford Hill

Stangebro the local prison at Linkoping, Sweden

Stanley Hong Kong's maximum-security prison

Stanton Thomas F Stanton Correctional Center at Elmore, Alabama (*see* Staunton)

Stapleton Staten Island Detention Pens in the Stapleton section of Staten Island, New York

Starke site of the Florida State Prison

star prisoner inmate believed susceptible to rehabilitation and even special treatment

START Special Treatment and Rehabilitation Training (program for criminals)

state chemist executioner who poisons convicts in a gas chamber

State Correctional Institution and Correctional Diagnostic and Classification Centers in Pennsylvania (*see* Camp Hill, Graterford, and Pittsburgh)

State Correctional Pre-Release Center in Tipton, Missouri

State Diagnostic Unit in Salt Lake City, Utah

state electrician executioner who operates an electric chair

State Farm Virginia site of the Deep Meadow, James River, and Powhatan Correctional Centers

State Farm Spur Illinois penal institution near Vandalia

State House of Correction and Branch Prison in Marquette, Michigan

stateliest building in all Venice the prison of the Old Republic, described in Edgar Allan Poe tale, *The Assignation*

State Penitentiary in Puerto Rico this name applies to the state penitentiary in Rio Piedras

State Prison of Southern Michigan at Jackson

State Road Philadelphia address of the Detention Center opened as Moyamensing Prison in 1835 but replaced in 1963 and also the House of Correction

Stateville Stateville Correctional Center at Joliet, Illinois

Staunton (pronounced *Stanton*) Correctional Center in Staunton, Virginia

St Brides Correctional Center in Capron, Virginia

Ste Anne des Plaines Québec town contains the Archambault institution, a maximum-security penal facility

Steilacoom Washington location of the McNeil Island Correctional Center

Stevenson House Milford, Delaware's facility for children from 8 to 18 years of age

Stevens School at Hallowell, Louisiana

stew builder jail or prison cook

Stillwater site of the Minnesota Correctional Facility

stir jail; prison

stir bug prisoner whose insanity seems linked to long confinement or the thought of long confinement

stir crazy confinement crazy; maddened by imprisonment

Stirville Ossining, New York (site of Sing Sing)

St John's capital of Newfoundland and location of HM Penitentiary

St Johnsbury Correctional Facility in St Johnsbury, Vermont

St Joseph prison in Lyons, France

St-Martin-de-Ré French island known for its old jail close to the Bay of Biscay

Stockton San Joaquin County courthouse and jail east of San Francisco, California

Stoke Heath HM Borstal in Stoke Heath, Shropshire, England

stone dump prison

Stony Mountain Institution in Winnipeg, Manitoba

Strafanstalt *die Strafanstalt* (German—the prison)—also called *das Gefängnis*

Strafrechtler *der Strafrechtler* (German—penologist)

straitjacket restraining device designed to control violent or unruly persons such as the mentally insane or hard-to-handle prisoners

Strangeways Manchester, England's prison

strapped strapped to the electric chair; penniless

Stratford Stratford, Ontario's jail

Street Haven Toronto, Ontario's center for the rehabilitation of prostitutes and wayward girls

stretch prison sentence

Stretch prisoner's periodical published at Lansing, Michigan

stretch hemp execute by hanging

Strict-Regime Camp one of 36 Soviet prison camps in the Urals

Stringtown Correctional Center in Stringtown, Oklahoma

stripes vertically striped prison clothes worn as recently as the first half of the twentieth century

Stromboli one of the Lipari Islands off of Sicily used to exile convicts and political prisoners

Stutthof main concentration camp in the north of Poland on the Bay of Gdansk

St Vincent de Paul St Vincent de Paul Penitentiary across the Riviere des Prairies from Montreal, Québec in Laval

Styal HM Prison at Styal in Cheshire, England

Subic Bay U.S. Naval Base in Subic Bay, and location of the U.S. Navy Brig in the Philippines

Sudan Khartoum and cities such as Omdurman and Port Sudan on the Red Sea, have HM Gaol-type penal facilities

Sudbury HM Prison at Sudbury

Sugamo Tokyo prison where Japanese war criminals were executed at the end of World War II

Sugar Land Texas site of the maximum-security Central Unit

Sukhanovka czarist monastery converted into a prison near Gorki

Sumpter Sumter Correctional Institution in Bushnell, Florida

supreme penalty death

Surabaja (*see* Indonesian prisons)

Suriname Paramáribo and Nickerie, Paranam, and Moengo have penal facilities

Susanville California Conservation Center at Susanville

Sussex Correctional Institution in Georgetown, Delaware

Suva capital city of Fiji contains a jail built during British colonization

swag prison jargon for contraband such as drugs, tools, fruit, pornography, or weapons

Swan Lake the Swan River Youth Forest Camp southeast of Kalispell, Montana

Swan River Swan River Youth Forest Camp near Swan Lake, Montana

Swansea HM Prison at Swansea in Glamorganshire, Wales

Swaziland Mbabane and Manzini have HM Gaol-type penal facilities

S & W bracelets Smith and Wesson handcuffs

Sweden some of the most advanced and enlightened penal facilities are found near Stockholm and Göteborg and Malmö

Swedish prisons 37 prisons, some, like Kumla, are within old fortresses while others, like Gavle, are modern and provide for connubial visits

Swift Trail Swift Trail Federal Prison Camp in the Pinaleno Mountains near Safford, Arizona

Swinfen Hall HM Prison at Swinfen Hall in Staffordshire, England

swing die by hanging

Switzerland modern penal facilities are in or near the capital, Zürich, and in other places such as Basel, Geneva, and Bern

swivels swivel nonlocking handcuffs

Sybil Sybil Brand Institute (Los Angeles county jail for women)

Syd Sydney (New South Wales, Australia) or Sydney (Nova Scotia, Canada) or their jails and lockups

Syracuse jails in Syracuse, Indiana, Kansas, Missouri, Nebraska, New York, and Ohio as well as on the island of Sicily

Syria Damacus, Aleppo and Homs have prisons built by the Turks

T-4 *Tiergarten 4* (German— Zoological Park 4)—Berlin address of Nazi medical killing facility and its director Josef Mengele

Tac Tacoma, Washington or its jail or its lockups

Taconic Taconic Correctional Facility at Bedford Hills, New York

Taft Oklahoma location of the Jess Dunn Correctional Center

Taft Youth Center in Pikeville, Tennessee

Tafuna American Samoan community containing the Territorial Correctional Facility run by the Department of Public Safety

Tai Taipei, Taiwan, jail or its prison

Tai Lam Tai Lam Addiction Department Centre or Tai Lam Centre for Women, both at Lung Chung in the New Territories of Hong Kong

Taipei (*see* China—Republic of)

take the electric cure suffer electrocution

take the pipe commit suicide

take the rap go to prison for someone else

take to the tules hide out in the bamboo-like tall grass

Talco Talcohuano, Chile, jail

Talladega Federal Correctional Institution at Talladega, Alabama

Tallahassee Federal Correctional Institution at Tallahassee, Florida

Tamarind Jamaican farm prison

Tamp Tampa, Florida; and Tampico, Mexico jails

Tampa Florida city has police-station lockups as well as the Hillsborough County jail

Tampere (*see* Finnish prisons)

Tampico prison in Tampico, Mexico

Tanforan racetrack near San Francisco used as a relocation center for Japanese-Americans during World War II

Tangerang boy's prison in western Java, Indonesia

Tangier (*see* Morocco)

Tanjong Pagar Singapore lockup and police station on Tanjong Pagar Road

tank prison cell

Tanzania Dar-es-Salaam has an HM Prison built early in the 20th century

tap code code used by prisoners who are not allowed to talk to one another but manage to communicate by tapping on cell bars or plumbing pipes

Tarawa (*see* Kiribati)

Tarbes the maximum-security Prison de Tarbes in France

Tartu Estonia city where Nazis killed more than 12,000 in a concentration camp originally known as Dorpat

Tasmania large island off Australia, once called Van Diemen's Land, served as a penal colony from 1803 until 1853

Taycheedah Correctional Institution receiving center for adult females in Taycheedah, Wisconsin

TCA Tennessee Correctional Association; Texas Corrections Association

TDC Texas Department of Corrections

Tecate Tecate, Baja California, jail

Tecumsah Oklahoma site of the Central Oklahoma Juvenile

Treatment Center

teddy boy(s) male juvenile delinquent(s) in the British Isles

teddy girl(s) female juvenile delinquent(s) in the British Isles

Teguci Tegucigalpa prison in the capital city of Honduras

Tehachapi California Correctional Institution at Tehachapi (state prison in the Tehachapi Mountains close to the Mojave Desert)

Teheran (*see* Iranian prisons)

telephone-pole design type of prison design, first introduced at Lewisburg, Pennsylvania, resembles a telephone pole with its crossarms; cellblocks and workshops are at right angles to a central corridor; this design provides flexibility in layout and coordination of elements for control and supervision of the inmates; the long connecting corridor (the telephone pole) extends from the administrative building past dining rooms and shops, and is bisected by cellblocks (*see* Lewisburg Plan)

Tel Mond medium-security Israeli prison for juvenile offenders located in the Plain of Sharon

Tennessee Colony Texas location of the maximum-custody Beto Unit; the Coffield Unit is also here

Tennessee lockups 95 counties have jails; Memphis, Nashville, Knoxville, and Chattanooga have extra facilities

Tennessee State Prison in Nashville

Tennessee Youth Center for delinquents between 15 and 18 years of age in Joelton

Tepic prison in Tepic, capital of Nayarit, Mexico

Terminal Island Federal Correctional Institution on Terminal Island opposite San Pedro, California

Terre Haute U.S. Penitentiary at Terre Haute, Indiana

Territorial Correctional Facility at Tafuna on Tutuila Island of American Samoa

Territorial(s) Territorial Prison(s)

Texarkana Federal Correctional Institution at Texarkana, Texas

Texas Department of Correction Units scattered about this state from Angleton to Sugar Land, from Brazoria and Brownwood to Rosharon and Tennessee Colony; Huntsville has the most penal facilities

Texas jails 254 counties contain lockups; cities such as Houston, Dallas, and San Antonio have local lockups

Thailand Bangkok; prison seems almost as old as the land it occupies

thana (Anglo-Indian—police station)—lockup; (Hindustani —jail)

THC Toledo House of Correction (Ohio)

theater of terror any public execution or punishment

The Castle U.S. Disciplinary Barracks at Fort Leavenworth, Kansas

The Gambia (*see* Gambian prisons)

The Pas Correctional Institution for Women at The Pas, Manitoba

therapeutic correctional community prison

Theresienstadt Nazi concentration camp about 40 miles (60 kilometers) from Prague; the Czechs called it Terezin

The Rock Otay Mesa Prison south of San Diego, also known as Rock Mountain

The Verne HM Prison at The Verne in Dorset, England

The Walls Huntsville Unit in Texas State Prison

Thieves' Palace nickname for the Surrey Prison

Thimphu Bhutan prison in the Himalayan mountains

Third Street New Jersey State Prison in Trenton on Third Street

Thomas Saint Thomas, American Virgin Islands, jail built a century ago by the Danes

Thomaston site of the Maine State Prison

Thorn Nazi concentration camp in Poland where it was known as Torun

Thorp Arch HM Prison at Thorp Arch, Yorkshire, England

Three Cs Federal Prison System logotype standing for Care, Custody, and Correction

three deuces jammed three concurrent two-year sentences

three deuces running wild three consecutive two-year sentences

three-time loser person returning to prison for the third time

throwaway juvenile delinquent or young adult criminal living in the same city as his or her parents but out of their care or control

thumbs thumbcuffs for controlling and holding unruly prisoners being transported from place to place

TI Terminal Island; Federal Correctional Institution at Terminal Island, California in Los Angeles harbor

ticket-of-leave permit allowing a convict to leave prison before the expiration of the sentence and to work under certain restrictions; parole certificate

ticket-of-leave man parolee

Tidewater Correctional Unit in Chesapeake, Virginia

Tihar central jail in New Delhi, India

Tijuana Tijuana, Baja California Norte, jail

Tijuana Hilton nickname of San Diego's multistory Metropolitan Correctional Center

Tillberga Swedish prison in Tillberga, west of Stockholm

Tillery North Carolina site of the Caledonia and Odom Complex

time off time off for good behavior (while imprisoned)

time out time out of sight (in a solitary-confinement cell)

time served total time spent in confinement before and after sentencing

Tinseltown Hollywood, California, police-station lockup as well as the nearby Los Angeles County Jail

tin throne metallic toilet; prison-cell toilet; slop bucket

tintureiro (Portuguese—dry cleaner)—Brazilian term for a prison van

Tipitapa maximum-security prison outside Managua, Nicaragua

Tipton Missouri site of the State Correctional Pre-Release Center

Tiptonville Tennessee site of the Lake County Regional Correctional Facility

Tirana prison in the capital of Albania

Tj Tijuana, Baja California, México, jail

Tjipinang maximum-security prison near Djakarta on the Indonesian island of Java

Tjirebon major prison in western Java in Indonesia

Tlaxcala prison in Tlaxcala, Mexico

TM *Therapia Magna* (Latin—Great Therapy)—euphemism invented by Nazi doctors in concentration camps where they participated in mass killings

TO Toledo, Ohio, jail or its lockups

Toamasina (*see* Madagascar's prisons)

Tobago (*see* Trinidad)

Toco Tocopilla, Chile, jail

Tocuyito nickname of the central penitentiary of Valencia in the Ejido Carabobo of Venezuela

Togo Lomé is its capital and its prison is primitive

toil factory prison workshop

Tokyo (*see* Japan's prisons)

tolbooth (Scottish—prison)

Toledo (*see* Spain, TO)

Toluca prison in Toluca, Mexico

tombas (Spanish—tombs)— solitary confinement cells

Tombs old New York City Prison in downtown Manhattan adjacent to the Criminal Court Building on the Lower East Side

Tombs Prison Manhattan's House of Detention

Tomoka Tomoka Correctional Institution (*see* Daytona Beach)

'Tona Daytona Beach, Florida, jail

Tonga an HM Gaol was built in the capital of Nuko'alofa

Tong Fuk Tong Fuk Detention Centre, Lantau Island, New Territories, Hong Kong

Toodyay an old gaol in Western Australia, now historical museum

Topaz relocation center for Japanese-Americans interned in central Utah

Topeka location of the Kansas State Reception and Diagnostic Center as well as the Kansas Correctional-Vocational Training Center and the Youth Center in Topeka

Topenish Japanese-American relocation center near

Yakima, Washington

Topo Topolobambo, Mexico, jail

toponym place-name prisoners use when telling where they were imprisoned; for example, Atlanta usually means the federal penitentiary in Atlanta, Georgia just as Trenton may be the New Jersey state prison there or even the Training School for (delinquent) Girls at Trenton

Toronto the Don Jail and the Mimico Correctional Centre are well-known penal facilities

torture chamber jail or prison where drugs are not available at any price

total segregation solitary confinement

Toulon French seaport known as a depot for convicts awaiting passage to the penal colonies of Algeria, French Guiana, and New Caledonia

Touquet Le Touquet, France, jail

Tower (*see* Tower of London)

Tower of London ancient fortress on the River Thames used as a royal residence, then a jail for political prisoners who often entered by the Traitors Gate before being beheaded; today it is an arsenal museum housing the crown jewels as well as ancient armor and many weapons; also called the Bloody Tower

Townsville HM Prison at Townsville (Queensland, Australia)

Tracy medium-security prison at Tracy, California, officially known as the Deuel Vocational Institute

Training Center for Youth in Columbus, Ohio for males from 12 to 17 years of age

Training School for Boys in Boonville, Missouri (*see* Boonville)

Training School for Boys and Girls at Jamesburg, New Jersey

Training School for Girls Trenton, New Jersey's correctional facility for delinquents from 8 to 17 years of age

training school(s) juvenile delinquent institution(s)

Traitors' Gate Thames River waterside gateway to the Tower of London, where pris-

oners were rowed in before being executed or serving long terms

tramp college nickname for county jail

Trani Italian prison near Bari on the Adriatic

Transfrisker electronically actuated hand-held no-touch personal-weapons-search device made by Federal Laboratories of Saltsburg, Pennsylvania

Transnistria Romanian-Nazi administrative region between the Bug and the Dniester rivers in the Soviet Ukraine where during World War II thousands of exiled Romanian Jews were relocated

transportation movement of prisoners to overseas penal colonies

Trautenau forced-labor subcamp in western Czechoslovakia during World War II

Treasure Island first brig for women sailors at the U.S. Navy base at Treasure Island in San Francisco Bay

Treblinka Nazi concentration camp in Poland

trembler prisoner afraid of other prisoners

Trenggalek one of Indonesia's newer prisons in eastern Java

Trenton Florida site of the Lancaster Correctional Center; site of the New Jersey State Prison and the Training School for Girls, as well as the Jones Farm at West Trenton, the St Francis Hospital Unit, and the Vroom Readjustment Unit in the Trenton Psychiatric Hospital

Tres Marías (Spanish—Three Marys)—Mexican penal settlement in the Pacific Ocean off Nayarit; prisoners are kept on María Madre Island

Trinidad and Tobago in Port-of-Spain, Trinidad, an HM Prison recalls the years of British rule from 1802 to 1976

Tripoli (*see* Lebanon and Libya)

Trondheim (*see* Norway)

Tropez St Tropez on the French Riviera or its jail

Trostyanets site of a Nazi concentration camp near Minsk

Troy New York State city north of Albany; seat of Rensselaer County and its jail

Trubetskoi bastion of the Peter and Paul Fortress, used as a prison in Leningrad

Trucial Sheikdoms (*see* United Arab Emirates)

Trujillo Cárcel Nacional de Trujillo (Spanish—National Prison of Trujillo, Venezuela)

Truk prison camp erected by the Japanese during World War II in the Caroline Islands

Truro the Nova Scotia School for Girls at Truro

trusty trustworthy convict who is allowed special privileges

TRY Teens for Retarded Youth (juvenile correctional program)

T-town Tijuana, Baja California, Mexico

tuchthuiz (Dutch—house of correction; workhouse)

Tucker Tucker Unit in Tucker, Arkansas

Tucson site of Arizona Correctional Training Facility where academic and vocational training is offered; Catalina Mountain School with a treatment program for juvenile males; Pima County seat and jail

Tule Lake waterless relocation and segregation center of northern California where many Japanese-Americans were interned after Pearl Harbor

Tullahoma Tennessee site of the Highland Rim School for Girls

Tulle maximum-security prison on Rue Souhan in Tulle, France

Tulsa Tulsa, Oklahoma, County Jail

Tulungagung Indonesian prison in eastern Java

Tunis (*see* Tunisia)

Tunisia cities, including the capital of Tunis, have French-colonial jails and prisons

Tuol Sleng *see* S-21

turist kogus (Turkish—tourist cell block)—prison section for foreigners

Turkey penal facilities are mainly from the time of the Ottoman Empire and may be seen in Ankara, Istanbul, Izmir, and other cities

Turku (*see* Finnish prisons)

Turney Center for Youthful Offenders at Only, Tennessee

turnkey anyone entrusted with the keys to a prison

Turrell Turrell Residential Group Center near Farming-

dale, New Jersey

Tuscaloosa Tuscaloosa County jail southwest of Birmingham, Alabama

Tuvalu an HM Gaol-type penal facility is in the capital, Funafuti

Tuxtla Gutiérrez prison in Tuxtla Gutiérrez, capital city of Chiapas, Mexico

Twin Maples Farm British Columbia's facility for treating women inmates with alcohol problems

twister(s) key(s)

Two Dzerzhinsky Moscow address of the KGB and the Lubyanka prison

two-time loser person going to prison for the second time

Tyburn long a favorite hanging place fitted with gallows and gibbets, between Edgware Road and the wall of Hyde Park, three miles from Newgate Jail in the City of London

UCA Utah Correctional Association

Udine Italian prison northeast of Venice

ufac unlawful flight to avoid confinement

Uganda the capital city, Kampala, has an old HM Prison as well as local lockups

ugly customer dangerously quarrelsome person

UK United Kingdom (Great Britain, Northern Ireland, and overseas colonies and territories)—or their penal facilities

Ulan Bator (*see* Mongolia)

ultimate penalty death, or life imprisonment without parole

UN United Nations and its organizations concerned with crime and punishment

unauthorized depature escape

Uncasville Connecticut Community Correctional Center at Uncasville, northeast of New London

underground kite secret message circulated throughout a prison or from one prisoner to another

underground tunnel secret systems for introducing contraband into a prison

under the gun under observation or surveillance

unhook unfasten the handcuffs or fetters

UNICOR trade name of the

Federal Prison Industries corporation maintaining 89 industrial operations in 39 penal institutions

Union Correctional Institution at Raiford, Florida

Unit 731 Japanese biological warfare complex at the Harbin Military Hospital in Manchuria during World War II

United Arab Emirates the capital, Abu Dhabi, and some older cities, such as Dubai and Sharjah have old HM Gaol-type prison facilities

United Kingdom some of the best-known gaols, penitentiaries, and prisons are in the British Isles and there are many remnants in former colonies and protectorates

United Prisoners Union revolutionary underground organization of hard-core convicts in prisons such as San Quentin or Soledad

universal staircase nickname for the treadmill once operated by felons

Unterlüss Nazi subcamp close to Bergen-Belsen near Hannover, Germany

Up Changi Road Singapore's Prison Headquarters at Kilometer 17 outside the city, nicknamed for the road on which it is located

up the river Sing Sing prison, up the Hudson River from New York City in the town of Ossining

UPU (*see* United Prisoners Union)

Urga (*see* Mongolia)

USA United States Army

U.S.A. the United States of America contains every type of penal facility from hospital prisons for the criminally insane to barless maximum-security centers for hardened criminals; *(see individual entries)*

USAF United States Air Force

USARB U.S. Army Retraining Brigade

U.S. Army confinement facilities Fort Benning, Georgia; Fort Campbell, Kentucky; Fort Carson, Colorado; Fort Gordon, Georgia; Fort Hood, Texas; Fort Knox, Kentucky; Fort Lewis, Washington; Fort Meade, Maryland; Fort Ord, California; Fort Polk, Louisiana; Fort Richardson, Alaska; Fort Riley, Kansas; Fort Sill,

Oklahoma

USBP United States Board of Parole; United States Border Patrol; United States Bureau of Prisons

U.S. Bureau of Prisons (*see* Bureau of Prisons)

U.S. Bureau of Prisons federal penitentiaries Atlanta, Georgia; Leavenworth, Kansas; Lewisburg, Pennsylvania; Lompoc, California; Marion, Illinois; Terre Haute, Indiana

U.S. Bureau of Prisons institutions for juvenile and youth offenders Ashland, Kentucky; Englewood, Colorado; Morgantown, West Virginia

U.S. Community Treatment Centers halfway houses (*see* Community Treatment Centers)

USDB U.S. Disciplinary Barracks at Fort Leavenworth, Kansas, which holds Air Force, Army, and Marine Corps prisoners whose sentences include six months or more of confinement and/or a punitive discharge

U.S. Department of Justice administers the Board of Immigration Appeals, Bureau of Prisons, Civil Division, Civil Rights Division, Criminal Division, Drug Enforcement Administration, Federal Bureau of Investigation, Immigration and Naturalization Service, Justice Management Division, Land and Natural Resources Division, Office of Legislative Affairs, Office of Public Affairs, Pardon Attorney, Tax Division, U.S. Marshals Service, U.S. Parole Division

USDJ U.S. (United States) Department of Justice *(see entry)*

U.S. Eighth Army Confinement Facility in Korea

USEP United States Escapee Program

Usk HM Borstal at Usk in England

U.S. Marshals Service maintains custody of federal prisoners, from the time of their arrest to their commitment or release, and transports federal prisoners pursuant to lawful writs and direction from the U.S. Bureau of Prisons

USMC United States Marine Corps (correctional facilities in Albany, Georgia; Camp

Smedley D. Butler in Okinawa; Camp Lejeune, North Carolina; Camp Pendelton, California; Parris Island, South Carolina; Quantico, Virginia)

USMS U.S. (United States) Marshals Service *(see entry)*

USN United States Navy

USN Brigs correctional centers and detention facilities brigs in naval parlance—are operated in or close to port cities, with the exception of a naval air station in Millington, Tennessee; seaport brigs include: Agaña, Guam; Charleston, South Carolina; Corpus Christi, Texas; Great Lakes, Illinois; Guantanamo, Cuba; Jacksonville, Florida; Long Beach, California; New London, Connecticut; Newport, Rhode Island; Norfolk, Virginia; Pearl Harbor, Hawaii; Pensacola, Florida; Philadelphia, Pennsylvania; Roosevelt Roads, Puerto Rico; Rota, Spain; San Diego, California; San Francisco, California; Seattle, Washington; Subic Bay, Philippines; Yokosuka, Japan

USNCC U.S. Naval Correction Center

USP United States Penitentiary

USPB United States Parole Board

USPC U.S. Parole Commission (formerly USPB)

USPD U.S. Parole Division of the Department of Justice

U.S. Penitentiaries Atlanta, Georgia; Leavenworth, Kansas; Lewisburg, Pennsylvania; Lompoc, California; Marion, Illinois; Terre Haute, Indiana

USPP U.S. Probation and Parole

USPs United States penitentiaries

USSR Moscow, and the capitals of its republics contain some of the most dreaded prisons in existence

Utah jails the state's 29 counties have lockups; Salt Lake City has additional facilities

Utah State Prison at Draper

Utica upstate New York prison

Utrecht Netherlands prison clinic providing psychological treatment and observation

VAC (*see* Voluntary Action Center)

Vacaresti Romanian prison; formerly a monastery on the

outskirts of Bucharest

vacationing near Chappaqua doing time in Sing Sing

Vaca Valley Star prison newspaper published by inmates in Vacaville, California

Vacaville California Medical Facility at Vacaville, psychiatric prison opened in 1955 after moving from Terminal Island

VACRP Victorian Association for the Care and Resettlement of Prisoners

VACs Voluntary Action Centers

Vaduz (*see* Liechtenstein)

Valdosta Georgia site of the Lowndes Correctional Institution

Valencia (*see* Spain and Venezuela)

Valetta (*see* Malta)

Valhalla Westchester County Penitentiary at Valhalla, New York

Valpo Valparaiso, Chile, jail

Val Verde County Clink nickname for the Val Verde County Jail in Del Rio, Texas

Vancoo Vancouver, British Columbia (or Vancouver, Washington) or their jails

Vandalia Correctional Center in Vandalia, Illinois

Van Diemen's Land former name of Tasmania and generic name for the great Australian penal settlement

Vanier Vanier Centre for Women at Brampton, Ontario

Vanuatu the old HM Gaol or Maison reflects it colonial past

VCA Virginia Correctional Association

Vega Alta Industrial School for Women at Vega Alta, Puerto Rico

Vegas Las Vegas, Nevada, jail

Vehicle City Flint, Michigan's jail

Venezuela Caracas and cities such as Barquisimeto, La Guaira, Maracaibo, and Valencia have Hispanic-type penal facilities

Ventura Ventura County jail between Los Angeles and Santa Barbara, California

Ventura Reception Center and Clinic at Camarillo, California

Ventura School for Girls at Camarillo, California

Vergennes the Weeks School

for delinquent and unmanageable convicts at Vergennes, Vermont

Vermillionville Lafayette, Louisiana's former name and that of its jail

Vermont Correctional Facilities in Burlington, Rutland, St Albans, St Johnsbury, Windsor, and Woodstock

Vermont lockups 14 counties have jails

Vernichtungslager (German—extermination camp)—concentration camp of the type built and operated by German Nazis during World War II

Vero Beach Florida site of the Indian River Correctional Institution for first-felony offenders under 20 years of age

Vesterfangsel (Danish—Western Jail)—Copenhagen prison

vettura cellulare (Italian—celled vehicle)—prison van

vic(s) convict(s)

victims of the metal age older prisoners complain they have silver in their hair, gold in their teeth, rust in their guts, steel around their cells, and lead in their asses

Victoria British Columbia location of the William Head Institution; La Victoria (Santo Domingo city prison, Dominican Republic); Victoria Reception Centre (Old Bailey Road, Hong Kong); *(see* Seychelles)

Victor Verster Cape Town, South Africa prison

Vienna Correctional Center in Vienna, Illinois

Vientiane *(see* Laos)

Vietnam southeast Asian country formerly held by French from 1858 to 1954 and by the Japanese during World War II; penal facilities built by the French and the Japanese are still used in Ho Chi Minh City (formerly Saigon) as well as in Hanoi, Haiphong, Da Nang, Hue, Nha Trang, and Vinh

Vieux Carre (French—Old Square)—French Quarter of New Orleans, Louisiana or its lockup

Vila *(see* Vanuatu)

Vila dos Remédios (Portuguese—Town of Reparation)—Brazilian settlement for ex-convicts on the isolated island of Fernando de Noronha in the Atlantic Ocean

Villa Devoto Argentinian prison in the Devoto section of Buenos Aires

Villahermosa prison in Villahermosa, capital of the Mexican state of Tabasco

Villamil Ecuador's former penal colony on the south coast of Isabela Island in the Galapagos Islands, where convicts serving life sentences were segregated with their families

Vince Saint Vincent Island in the Windward Islands of the Lesser Antilles or its jail

Vincennes French prison notorious for its many famous inmates dating back to the 14th century, when it was a castle and dungeon

Vinh *(see* Vietnam)

ViP (VIP) *(see* Volunteers in Probation)

Virginia State Department of Corrections consists of the Division of Adult Correctional Services and the Division of Youth and Community Services; these, in turn, are divided into regions and within each there are correctional centers and units as well as halfway houses, learning centers, and work-release units scattered throughout the Tidewater State

Virginia Correctional Center for Women in Goochland

Virginia jails the state's 96 counties, and its many independent cities, have lockups

Virginia State Penitentiary in Richmond

Virgin Islands jail lockup on the waterfront of Charlotte Amalie on the island of St Thomas; jail occupies an old fort built in Danish times when the islands were settled

Visalia Tulare County jail in central California

Vista town near San Diego, California or the county detention facility there

viuva alegre (Portuguese—merry widow)—prison van

viuva-alegre (Brazilian slang—prison van)

Vladimir Soviet medium-security prison halfway between Gorki and Moscow

Vladivostok transit prison port on the Pacific coast of the USSR

voiture cellulaire (French—celled vehicle)—prison van

Voluntary Action Center device for using the skills of persons convicted for minor crimes instead of putting them in prisons

Volunteers in Probation publishes *VIP Examiner* quarterly and books about probation

Voyvodina *(see* Yugoslavia)

VPCP Volunteer Probation Counseling Program

Vridsloeselille Danish state prison noted for its programs for reducing recidivism)

Vulcano southernmost island in the Liparis off the coast of Sicily where it has served since Roman antiquity as a place of penal exile

Wacol HM Prison Wacol (Queensland, Australia)—sometimes called the Bane of Brisbane

Wade Correctional Center in Haynesville, Louisiana

Wailuku port city on Maui Island, Hawaii, where the Maui Community Correctional Center is located

Wakefield jails in Wakefield, Massachusetts, Michigan, or Rhode Island

Walden Correctional Institution in Columbia, South Carolina

Wales Wisconsin site of the Ethan Allen School; *(see* United Kingdom and individual entries)

walk to be acquitted; to walk out of prison

Walker Correctional Institution in Rock Spring, Georgia

wall firing wall (place of execution by a firing squad)

Walla Walla Washington State Penitentiary at Walla Walla

Wallkill Wallkill Correctional Facility at Wallkill, New York

Wallows Walla Walla, Washington, jail

Wall-Wall nickname for the Washington State Penitentiary at Walla Walla

Walnut Street Philadelphia's oldest jail built in 1790; inmates were subjected to solitary confinement to prevent association with other prisoners and to promote reflection and self-reform

Walton Liverpool, England's Walton Prison

Walvis Bay *(see* Namibia)

Wanchai Hong Kong's Wanchai Police Station and lockup

Wandsworth prison in Wandsworth southwest of London

warden prison administrator

Ware Correctional Institution (see Waycross)

Warkworth Institution in Campbellford, Ontario

Warrensville Heights Ohio site of the Cuyahoga Hills Boys School

Warsaw (see Poland)

Wartburg Tennessee site of the Morgan County Regional Correctional Facility

Washington Corrections Center in Shelton, Washington, which serves as a reception and diagnostic center and offers a reformatory-type educational program

Washington lockups 39 counties have jails; Seattle and Tacoma have extra facilities

Washington State Funnypark Washington State Prison near Walla Walla

Washington State Penitentiary in Monroe provides all degrees of security for male felons

Washington State Penitentiary in Walla Walla holding felons 16 years of age and up in all levels of custody

Washington State Reformatory in Monroe

Washington State Special Offender Center in Monroe

watchhouse police station; police station lockup

Wateree River Correctional Institution in Rembert, South Carolina

Watergate Hilton prisoners' nickname for the District of Columbia jail in Washington, DC

Watkins Pre-Release Center in Columbia, South Carolina

Waukegan Lake County, Illinois, jail

Waupon Wisconsin site of the Dodge Correctional Institution and the Waupon Correctional Institution

Waycross Georgia location of the Ware Correctional Institution

Waymart Pennsylvania Corrections Department prison northeast of Carbondale

Wayne Wayne County Jail in Detroit, Michigan

Wayne Correctional Center in Goldsboro, North Carolina for convicts needing psychiatric care

Wayne Correctional Institution (see Odum)

WCA Wackenhut Corrections Corporator (of private jails); Washington Correctional Association; Western Correctional Association; Wisconsin Correctional Association; Women's Correctional Association

WCS Wisconsin Correctional Service (in Milwaukee)

WCSC World Correctional Service Center (q.v.)

WDC Women's Detention Center

Weeks School at Vergennes, Vermont

Weisswasser forced-labor subcamp operated by the Nazis during their occupation of Czechoslovakia

Welfare Security Program guarantees safety of prisoners who provide useful information to prison authorities

Wellingborough HM Borstal at Wellingborough in Northamptonshire, England

Wellington Wellington Prison, Wellington, New Zealand

West Concord site of the Massachusetts Correctional Institution

Westerbork Dutch site of a Nazi concentration camp during World War II

Western Hong Kong's Western Street Police Station and lockup

Western Europe's Largest Prison Fleury Mergois on the outskirts of Paris

Western State Penitentiary in Pittsburgh, Pennsylvania, replaced by the State Correctional Institution and Correctional Diagnostic and Classification Center

Westfield Westfield Correctional Center in Westfield, Indiana

Westfield Farm New York Reformatory for Women

West Palm Beach Palm Beach County jail

West Sam Western Samoa jail

West Street Federal Detention Center on New York City's waterfront in past years

Westville Westville Correc-

tional Center in Indiana

West Virginia Industrial School for Boys in Grafton

West Virginia Industrial School for Girls in Salem

West Virginia jails 55 counties have lockups; there is also a city jail in Charleston

West Virginia Penitentiary in Moundsville

West Virginia State Prison for Women in Pence Springs

Wetherby HM Borstal at Wetherby in Yorkshire, England

Wethersfield Connecticut State Prison at Wethersfield

wets. wetbacks; undocumented aliens who may get their backs wet crossing the Rio Grande to enter the United States and avoid immigration officials

Wha Wha Wha Wha Prison near Gwelo, Zimbabwe

WHD Women's House of Detention (New York City)

Wheeling Ohio County's seat in West Virginia; its jail is referred to as Wheeling

Whitehorse location of the Whitehorse Correctional Centre in Canada

White Street Manhattan House of Detention for Men at 125 White Street in New York City

Wichita Sedgwick County jail in Kansas

Wilder Youth Development Center in Somerville, Tennessee

Wilingili penal settlement on an Indian Ocean atoll of the Maldive Islands

Wilkes-Barre northeastern Pennsylvania city and seat of Luzerne County where inmates refer to the jail as Wilkes-Barre

William Head Institution in Victoria, British Columbia

Wilmas (Mexican-Americanism —Wilmington, California)— nickname for the penitentiary on Terminal Island in Los Angeles Harbor

Wilmington Delaware site of Bridge House for boys and girls in detention status, the county jail, the Ferris School for male delinquents, the Pre-Trial Annex with maximum- and medium-security cells, the Plummer (work-release) Center

Windhoek (*see* Namibia)

Windsor Correctional Facility in Windsor, Vermont

Winnipeg Manitoban site of the Stony Mountain Institution

Winson Green prison near Birmingham, England

Winston-Salem Winston-Salem, North Carolina, the seat of Forsyth County with its jail

wired addicted; electrocuted

Wisconsin Home for Women near Fond du Lac

Wisconsin jails 72 counties provide prisons; Milwaukee has local as well as county jails

Wisconsin School for Girls in the south-central part of the state

Witbank South African prison 65 miles (105 kilometers) east of Pretoria in the Transvaal

witness protection U.S. Marshals protect witnesses whose lives and those of their families are jeopardized by their testimony

Witzwil Swiss prison farm without walls in the town of Witzwil

Women's Correctional Admission Center in Columbus, Ohio

Women's Correctional Center in Columbia, South Carolina

Women's Correctional Institution Claymont, Delaware's penal facility

Women's Division current name for the Rhode Island Training School for Girls in Cranston

Women's Prison Association New York City-based service agency for prisoners; publishes *A Study in Neglect* about the plight of women in prison

Women's Reformatory officially The Women's Reformatory, in Rockwell City, Iowa

Women's Ward Oklahoma State Penitentiary at McAlester

Woodbourne Woodbourne Correctional Facility at Woodbourne, New York

Woodburn Oregon site of the MacLaren School for juvenile-court commitments

Woods Haven-Kruse School for Girls at Claymont, Delaware

Woodstock Correctional Facility in Woodstock, Vermont

Wood Street Counter one of the most notorious of London's prisons in Elizabethan times although the Clink, Newgate, and Fleet did not lag far behind in inhumane practices

Workers' Paradise derisive nickname applied to the communist-controlled USSR

workhouse originally a prison where inmates had to work if they wanted to eat; in Great Britain, they picked oakum used for caulking ships or they sewed mailbags; in the United States, a slang name for any prison

Work Release Program rehabilitation plan whereby convicts work outside of prison during the last part of their term and receive the same pay as other workers

Work Training Facility Louisiana has two—one in New Orleans and the other in Pineville

Worland site of the Wyoming Industrial Institute

World Correctional Service Center Chicago-based information clearinghouse

World's Freest and Smallest Jail San Marino's hilltop lockup, near Rimini, Italy uses a converted monastery to hold its prisoners who, if sober, are allowed to work in town providing they keep out of bars, restaurants, and other public places and avoid meeting with one another

World's Largest Prison Kharkhov in the Soviet Ukraine where more than 40,000 prisoners have been incarcerated at one time, according to the *Guiness Book of World Records*

World's Largest Walled Prison Southern Michigan Prison spanning 54 acres (22 hectares)

World's Most Gigantic Prison (*see* Most Gigantic Prison in the World)

Wormwood Scrubs English prison for young male offenders in the West London Stadium area

WPA (*see* Women's Prison Association)

WPA & H Women's Prison Association and Home

Wroclaw (*see* Poland)

WRP (*see* Work Release Program)

Wyantskill Wyantskill Center for Girls at Wyantskill, New York

Wyndham Wyndham Regional Prison (Western Australia)

Wynne Unit in Huntsville, Texas holds felons in maximum custody

Wyoming Girls School in Sheridan

Wyoming Honor Farm in Riverton, maintained by its inmates in minimum security

Wyoming Industrial Institute in Worland for 10- to 21-year-old delinquents

Wyoming lockups the state has 23 counties and jails; Cheyenne has city and county detention facilities

Wyoming School for Girls in the north-central part of the state holds delinquents from 18 to 21 years of age

Wyoming State Penitentiary in Rawlins

Wyoming Womens Center in Evanston

x'd executed

Xining capital of China's Qinghai province notorious for its forced labor camps

Xochi Xochimilco, Mexico, or its lockup

X-ray apparatus used to detect contraband, letter bombs, parcel bombs, tools, or weapons

YACA (*see* Youth and Correctional Agency)

Yakima Yakima County Jail in Yakima, Washington

Yaounde (*see* Cameroon)

yard prison yard

Yard Scotland Yard, London

yardbird convict; ex-convict; jailbird; prisoner

yard bull prison guard; railroad detective; railroad-yard policeman

Yardville New Jersey town containing the Youth Reception and Correction Center

Yaren (*see* Nauru)

YCA (*see* Youth and Correctional Agency)

YCC Youth Correctional Center

YCI Youth Correctional Institution (Bordentown, New Jersey)

YDI (*see* Youth Development Incorporated)

Yellowknife Correctional Centre in the capital of the Canadian Northwest Territories

Yemen has penal facilities built when the Turks ruled all Arabia before the end of the World War I

YGC Youth Guidance Center (San Francisco)

Yoko Yokohama, Japan, jail, lockup, and prison

Yokohama (*see* Japan's prisons)

Yokosuka Japan's naval base and lockup in Tokyo Bay south of Yokohama; U.S. Navy Fleet Activities Brig in Yokosuka; Yokosuka Prison on Japan's Honshu Island

Yokuska (*see* Yokosuka)

York community west of Lincoln, Nebraska and site of the Nebraska Center for Women

YOU Youthful Offender Unit

young-adult penal institutions of the U.S. Bureau of Prisons El Reno, Oklahoma; Lompoc, California; Milan, Michigan; Oxford, Wisconsin; Petersburg, Virginia; Seagoville, Texas; Tallahassee, Florida

young horse prisoner's name for roast beef

Youngs Youngstown, Ohio, jail

young stir boy's reformatory

Youth and Correctional Agency combines California's Board of Prison Terms, California Youth Authority, Correctional Industries Commission, Department of Corrections, Institutional Review Board, Narcotic Addict Evaluation Authority, and Youthful Offender Control Board

youth and juvenile penal institutions of the U.S. Bureau of Prisons Ashland, Kentucky; Englewood, Colorado; Morgantown, West Virginia

Youth Center at Atchison for delinquent boys from 13 to 15½ years of age

Youth Center at Topeka for juvenile delinquents in Kansas

Youth Correctional Institutions in New Jersey—one at Annadale, the other at Bordentown

Youth Development Center in Kearney, Nebraska

Youth Development Centers throughout Pennsylvania at locations such as Bensalem Heights, Loysville, New Castle, Waynesburg

Youth Development Inc juvenile education and rehabilitation program

Youth Forestry Camp in Custer, South Dakota for offenders from 15 to 21 years of age; Pennsylvania places such as Hookstown, James Creek, and Whitehaven hold delinquents from 15 to 18 years of age

Youthful Offender Unit holds offenders under age 25 in close security at Hardwick, Georgia

Youth Reception and Correction Center in Yardville, New Jersey

Youth Services Center at Saint Anthony, Idaho

Youth Studies Center Philadelphia, Pennsylvania's juvenile correctional facility

Yps Ypsilanti, Michigan, jail

Ypsilanti Michigan city containing the Huron Valley Men's Facility and the Huron Valley Women's Facility

YSA Youth Services Administration (District of Columbia)

YTPM Yuma Territorial Prison Museum

YTS Youth Training School

Yugoslavia most of the prisons recall the Austro-Hungarian Empire

Yuma Arizona site of the Territorial Prison built in 1867,

now a museum

Yuma Territorial Prison State Historic Park open daily from 8:30 to 5:30 on the banks of the Colorado River in Yuma, Arizona

Zacatecas prison in Zacatecas, Mexico

Zag Zagreb, Yugoslavia, jail

Zagreb (*see* Yugoslavia)

Zaire penal facilities date back to the days of colonial rule under Belgium

zak zaklyuchenny (Russian—prisoner)—

Zambia HM Gaol-type structures contain its convicts

Zambo Zamboanga, Mindanao, Philippines, jail

Zanzi Zanzibar, Tanzania, jail

Zaragoza Zaragoza, Chihuahua, jail (*see* Spain)

Zenith City of the Unsalted Seas Duluth, Minnesota, jail

zenkamono (Japanese—jailbirds; tramps)—the most despised elements of Japanese society; usually segregated into run-down sections of cities

Zephyrhills Zephyrhills Correctional Institution at Zephyrhills, Florida

Zero Route (see *La Route Zero*)

Zimb Zimbabwe (formerly Rhodesia) or its jails or prisons built by the British and South Africans

Zimbabwe capital city, Harare (formerly Salisbury), and Bulawayo have penal facilities superior to many on the African continent

zoo police station and its lockup

Zuchthaus das Zuchthaus (German—penitentiary; prison)

Zürich (*see* Switzerland)

Zwodau forced-labor subcamp operated by the Germans during their occupation of Czechoslovakia in World War II

Diacritical and Punctuation Marks

´ acute accent (as in Bogotá)

’ apostrophe; single quotation mark

@ commercial-letter *A* standing for *at* or *for*

[ ] brackets

ˇ breve

, cedilla (as in Curaçao)

^ circumflex (as in *rôle*)
: colon
) close parenthesis
, comma
¨ diaeresis (as in München)
. . . or . . . .
  ellipsis; leaders
! exclamation point
` grave accent (as in *funèbre*)
- hyphen
? interrogation or question mark
¯ macron (dictionary pronunciation symbol indicating long vowel, as in dāme)
(open parenthesis
( ) parentheses
. period
" " quotation marks; quotes
' ' quotation marks, single
; semicolon
~ tilde (as in São Paulo)
− vinculum (mathematics: placed above letters)

Domestic Pets

Canis domestica (Canis familiaris) (Latin—domestic dog) —see hounds, terriers, toys
cat(s) see *Felis catus*
dog breeds hounds (*see* hounds); non-sporting—bulldogs, chows, dalmatians, poodles; sporting—pointers, retrievers, setters, spaniels; terriers; toys—chihuahuas, miniature Mexican hairless dogs, Pekinese, pomeranians, pugs; working—collies, doberman pinschers, German shepherds, full-size Mexican hairless dogs, Saint Bernards
dog(s) *Canis domestica, Canis familiaris* (*see* hounds, terriers, toys)

Equus caballus (Equus domestica) (Latin—domestic horse)
Felis catus (Felis domestica) (Latin—domestic cat)—Abyssinian, Afghan, Angora, black, Burmese, gray, Himalayan, Maltese, Manx, marmalade, Persian, Russian blue, tabby, Siamese, tortoiseshell, white
horse(s) see *Equus caballus*
hounds Afghans, American or British foxhounds, bassets, beagles, bloodhounds, borzois, coonhounds, dachshunds, harriers, Irish wolfhounds, Norwegian elkhounds, otterhounds, Rhodesian ridgebacks, Scottish deerhounds, salukis, whippets—

plus an extinct breed, the spitdog
terriers airedales, Australians, Bedlingtons, borders, bull terriers, Cairnes, dandie dinmonts, fox terriers, Irish terriers, Kerry-blues, lakelands, Manchesters, Norwich terriers, Scottish, sealyhams, skyes, Staffordshires, Welsh terriers, West Highlands
toys affenpinschers; Brussels griffons; chihuahuas; dandie dinmonts; Italian greyhounds; miniature dachshunds, maltese, Mexican hairless dogs, pinschers, spaniels, terriers; Pekinese; pomeranians; poodles; pugs; shihzus

Dysphemistic Place-Names

Accident, Maryland
Atomic City, Idaho
Bad Axe, Michigan
Blue Ball, Pennsylvania
Braggadocio, Missouri
Cactus, Texas
Clam Gulch, Alaska

Cranks, Kentucky
Death Valley, California
Decoy, Kentucky
Dime Box, Texas
D'Lo, Mississippi
Dogpatch, Arkansas
Due West, South Carolina

Dusty, New Mexico
Embarrass, Minnesota
Eros, Louisiana
False Pass, Alaska
Fate, Texas
Fixer, Kentucky
French Lick, Indiana

Frogmore, Louisiana or South
  Carolina
Frozen Creek, Kentucky
Gap, Pennsylvania
Gas, Kansas
Graves, Georgia
Hanging Rock, Ohio
Hell, Michigan
Hell Gate, New York
Hell's Half Acre, Wyoming
Hemp, Georgia
Ho-Ho-Kus, New Jersey
Hungry Horse, Montana
Hygiene, Colorado
Intercourse, Pennsylvania (near
  Blue Ball)
Justiceburg, Texas
Kettle, Kentucky
Loco, Oklahoma

Mousie, Kentucky
Mud Lick, Kentucky
Nutsville, Virginia
Oblong, Illinois
Ordinary, Virginia
Panther Burn, Mississippi
Peculiar, Missouri
Pickleville, Utah
Pie Town, New Mexico
Plaster City, California
Porcupine, South Dakota
Quicksand, Kentucky
Rains, South Carolina
Shady, New York
Slick, Oklahoma
Smoketown, Pennsylvania
Speculator, New York
Stab, Kentucky
Stumptown, West Virginia

Tall Timbers, Maryland
Temperance, Michigan
Tiplersville, Mississippi
Tobaccoville, North Carolina
Tombstone, Arizona
Trussville, Alabama
Truth or Consequences, New
  Mexico
Turtletown, Tennessee
Uncertain, Texas
Vesuvius, Virginia
Volcano, California or Hawaii
Whiskeytown, California
X-ray, Texas
Yucca, Arizona
Zigzag, Oregon (former name
  of Rhodadendron)
Zilwaukee, Michigan

Earthquake Data (Richter Scale)

The Richter Scale, devised in 1935 by Dr Charles Francis Richter, seismologist of the California Institute of Technology, is a standardized scale for defining the destructive energy of earthquakes whose force is measured by seismographs. The magnitude of such earthquakes is the logarithm of the largest deflection measured and registered during an earthquake when a seismograph is 100 kilometers (62 miles) from the center of maximum shock, the epicenter of the earthquake, whose exact location is pinpointed by several scattered seismographs.

Numbers of the Richter Scale advance logarithmically and not arithmetically, so earthquakes measuring 8, for example, are ten times greater than those measuring 7, and this relationship is constant throughout the scale.

Earthquakes occurring before 1935, or before the invention of the seismograph in 1841, are approximated in terms of the Richter Scale.

Earthquake Damage and Intensity Devastation Effects Encountered Historically

0 No detectable or measurable earthquake effect although about 100,000 quakes a year can be felt and at least 1000 cause some damage

1 Very slight earthquake effects felt by sensitive persons who may experience dizziness or nausea; other creatures may appear disturbed; gentle swaying may affect bodies of water as well as buildings and trees

2 Slight earthquake effects sensed by sensitive persons as well as other creatures who display uneasiness; hanging lamps and pictures swing slightly; buildings and trees sway slightly

3 Very moderate earthquake effects sensed by a few persons as well as by the most nervous and the most sensitive; dishes on shelves may rattle as may many windows; canned goods stored on shelves may rattle and may fall off; parked vehicles may rock and this is true of shrubs and trees

4   Moderate earthquake sensed by many and sufficient to awaken light sleepers; house frames creak and houses sway slightly; shrubs and trees tremble; parked vehicles may rock and sway

5   Near medium-strength earthquake felt by everyone and frightening most persons who tend to leave buildings and run out of doors to avoid cracking ceilings and crumbling walls; in older buildings plaster falls, ceilings crack, and windows break; pictures may fall off their hangings; dishes and glasses tumble off shelves; heavy desks and tables move and many may topple; old and weak chimneys may crack off at the roofline; ornamental cornices fall from buildings; church bells toll by themselves

6   Full-strength earthquake causing general fright approaching panic; stone walls crack; steep slopes and riverbanks crack; chimneys and towers may crack apart and fall; trees shake violently and often fall as do limbs; the Los Angeles Earthquake of 1971 measured 6.6, caused considerable damage, and took the lives of some 60 persons

7   More devastating and more severe type of earthquake such as occurred in Nicaragua and Guatemala where thousands were killed in 1972 and 1976, respectively; or in the Chile Quake of 1906, preceding the San Francisco Earthquake and Fire by only two days, and causing the loss of 1550 lives in Valparaiso and 452 in San Francisco; both seismic disturbances were calculated in later years as representing 7.8 on the Richter Scale

8   Still more devastating and more severe earthquake causing general panic and marked by widespread land and water disturbances; many dams and dikes break, discharging vast volumes of flooding water; underground cables and pipelines crack and tear apart; railway rails bend and twist; brick, glass, and masonry facades peel off buildings and endanger people as they fall to the ground; loss of life quite severe as in the Peruvian Quake of 1970, accounting for the loss of some 50,000 persons, or the Alaska Quake of 1964, reported as 8.4 on the scale, and marked by heavy damage in downtown Anchorage where 131 lost their lives; earthquakes of even greater magnitude occurred in Lisbon, Portugal in 1755 when 60,000 were lost and lakes in far off Norway were disturbed violently; the Shensi Province Quake, occurring in China in 1566, cost some 830,000 lives, calculated to have been 8.9 on the Richter Scale as was Japan's great quake of 1923, destroying all of Yokohama and half of Tokyo, as well as 143,000 people; the sea bottom in Sagami Bay sank 387 meters or 1300 feet; earthquakes of this magnitude afflicted New Madrid, Missouri in 1811, Charleston, South Carolina in 1886, and are predicted as long overdue along the San An dreas Fault Zone of California extending from below the Mexican Border to San Francisco and northward; overall damage might well equal or exceed the Shinsai or Great Quake felt around Tokyo in 1923; Chinese earthquake of July 26, 1976 registered 8.2 with a 7.9 aftershock the following day; shocks affected an area in and around Peking and Tientsin and some 15 million people.

Krakatoa volcanic Indonesian island between Java and Sumatra, scene in August 1883 of one of the world's worst earthquakes

Recent Earthquakes Philippines (1991); Armenia (1989); Chile (1985); Colombia–Ecuador (1979); India (1988); Iran (1990); Japan (1931, 1946, 1948, 1978); Mexico (1985); San Francisco (1906, 1990)

9 Most devastating and most intense earthquakes, as yet unrecorded on any scale, top of the Richter Scale, extending from 0 to 9, and may never occur due to the good effects of minor earthquakes and tremors, providing stress-relief cracking of and easing the great tectonic energy tension beneath us

Eponyms, Nicknames, and Geographical Names

Aardvark US F-111 fighter-bomber

A Bachelor of Arts Phyllis Bentley's pseudonym

abandoned woman nickname for a prostitute

Abbé Sieyès Emmanuel Joseph Sieyès

abbott nembutal sleeping tablet (nicknamed for its producer, Abbott Laboratories)

Abby Abigail

Abdul the Damned Sultan Abdul–Hamid II of Turkey

Abe nickname for a $5-bill bearing Lincoln's portrait

Abigail Van Buren Pauline Esther (Popo) Phillips, better known as Dear Abby

Abolitionist Quaker Elias Hicks

Abominable Snowman big, hairy manlike creature believed by natives to inhabit the higher Himalayas (also known as yeti)

Ab-o'-th'-Yate Benjamin Brierley

Abu Zabi (Arabic—Abu Dhabi) capital of the United Arab Emirates (UAE)

Abyssinia Ethiopia

accommodation houses British nickname for brothels

Ace of Spies Sidney Reilly

Acerbic American Critic *American Mercury's* HL Mencken in the 1920s and 1930s; *American Spectator's* R Emmett Tyrell, Jr in the 1970s and 1980s

Achmed Abdullah Alexander Nicholaievitch Romanov

Aco Michel Accault (La Salle's lieutenant)

Acronym Islands Indonesia where politicians delight in creating acronyms for every occasion

Acropolis of America New York City's Morningside Heights—site of Columbia University

Acton Bell pseudonym of Anne Brontë

Ada Adelaida; Adelaide

Adam Hall Elleston Trevor's pseudonym

Adam Smith George JW Goodman's pseudonym; 18th century Scottish economist

Addie Ada; Adela; Adelaide; Adelina; Adeline

Addison's disease adrenal cortical deficiency

Admirable Doctor English author-philosopher Francis Bacon—*Doctor mirabilis*

Admiral of the Atlantic Kaiser Wilhelm II

The Admiral Doctor Roger Bacon

Admiral of the Ocean Sea Christopher Columbus

Admiral of the Pacific Czar Nicholas II

Admiralties Admiralty Islands

Adobe State New Mexico

Adonis of Fifty nickname of George IV

Adrian Hadrian

Adrian Girls (delinquent) Girls Training School at Adrian, Michigan

Adrianople former name of Eridne

Advance Agent of Emancipation Lucretia Mott

Aegean Ethicist Aristotle

Aerospace Valley California's Antelope Valley

Aesthetic Post-Impressionist Vasili Kandinski

Affable Archangel Raphael

afghan afghan blanket; afghan hound (noted for its long silklike coat and narrow head)

Africa in Miniature Cameroon

African-American Abolitionist-Author-Editor-Orator Frederick Douglass

African-American Astronomer-Inventor-Mathematician Benjamin Banneker

African-American Botanist-Chemurgist-Educator George Washington Carver

African-American Conductors Dean Dixon, Henry Lewis, James de Preist

African-American Contralto Marian Anderson

African-American Educator Booker T Washington (founder and first president of the Tuskegee Institute)

African-American Explorer Matt(hew) A Henson who pushed Peary to the North Pole as well as helping him survey the Nicaraguan canal route

African-American Historian Carter G Woodson

African-American Messiah Booker T Washington

African-American Moses Harriet Tubman

African-American Movie Pio-

neer Ralph Cooper
African-American Novelist Richard Wright
African Queen Mrs Ian Smith of Salisbury, Rhodesia
Africa's big five Cape buffalo, elephant, leopard, lion, rhinoceros
Afro-America's First Great Poet Paul Laurence Dunbar
Agate Capital Prineville, Oregon
Age of Extinction the 20th century when the human population growth infringed on wildlife habitat
Age of Reptiles early mesozoic era
Age of Uncertainty John Kenneth Galbraith's name for the late 20th century
Age of Voltaire the Enlightenment
Aggie Agatha; Agnes
Agnes Lee Mrs Otto Freer
Agnon Shmuel Yosef Agnon pseudonym of Samuel Josef Czaczkes
Agnostic Penologist Colonel Robert Green Ingersoll (1833–1899) who lectured and wrote about Crimes Against Criminals as well as Cruelty in the Elmira Reformatory
Agrarian Champion Emiliano Zapata of Mexico
Agricola Alexander Ackerman, composer; George Bauer, German minerologist; Johannes Sneider or Schnitter, Protestant reformer; Martin Sohr, composer; Roelof Huysman, Dutch musician-painter-scholar
Agricultural Wizard of Tuskegee George Washington Carver
Aguecheek Charles B Fairbanks
Air Capital of America Wichita, Kansas
Air Capital of the World Montreal, Quebéc—headquarters of the International Civil Aviation Organization and the International Air Transport Association
Air-Conditioned City Duluth, Minnesota
Air Mike Air Micronesia
Airtourer single-engine trainer plane built by Aero Engine Services of New Zealand
Ajax Annie Besant's pseudonym when working with

Charles Bradlaugh for free thought and population control in Great Britain
Akhiar formerly Sevastopol
Akmechet formerly Simferopol
Alabama Port Mobile
Alabama's Only Port Mobile
alacranes (Spanish—scorpions) persons from the Mexican state of Durango
Alain pseudonym of Emile Chartier
Alain-Fournier Henri-Alban Fournier's pseudonym
Alameda Bernardo O'Higgins Las Delicias
Alamo City San Antonio, Texas
Alan Alda Alphonso d'Abruzzo
Alan Hovhaness Alan Hovhaness Chakmakjian
Alan King Irving Kniberg
Alaric Cottin Voltaire's nickname for Frederick the Great
Alaskan Ports (south to north) Ketchikan, Wrangell, Petersburg, Sitka, Juneau, Cordova, Seward, Anchorage, Kodiak, Dutch Harbor, Adak Naval Station, Nome
Alaska's Scenic Capital Juneau
Alaska turkey Alaska salmon
Albany beef Hudson River sturgeon
Albaturkey Albuquerque, New Mexico
Albemarle Island, Galápagos Isabela
Albert the Good Prince Albert Francis Charles Augustus Emmanuel of Saxe-Coburg-Gotha, Prince Consort of Queen Victoria
Alberto Moravia Alberto Pincherle
Alberto Savinio Andrea de Chirico
Albert's disease inflammation of the bursae over the Achilles tendon
Albertville Kalima, Zaire
Albion Britain's ancient name
Albuquerque Girls New Mexico Girls Welfare Home at Albuquerque
Alburturkey Albuquerque, NM
Al Capp Alfred Gerald Caplin
Alcibiades Alfred Lord Tennyson
Alcofribas Nasier François Rabelais
Alec Waugh Alexander Raban Waugh
Aleksei Maksimovich Peshkov

Maxim Gorki
Alexander Girls Arkansas Training School for (delinquent) Girls at Alexander
Alexander of the North Charles XII of Sweden
Alexander Serafimovich Alexander Serafimovich Popov
Alexandretta Iskenderun
Alexandria Egyptian port city of El Iskandariya
Alexandrian Century the 4th century before the Christian era when Alexander of Macedonia conquered Egypt, Persia, and India as well as encouraging Greek philosophers and poets—the 300s
Alexes ten-dollar bills bearing the portrait of America's first Secretary of the Treasury, Alexander Hamilton
Alfalfa Bill Governor William Henry Murray of Oklahoma
Alfonso XIII León Fernando María Isídro Pascual António
Alfred, Lord Tennyson Alfred Tennyson (1st Baron Tennyson—poet laureate of England from 1850 until 1892)
Algerian onyx stalagmitic calcite
Algerian Ports Annaba (Bone), Skikda, Bejaia, Alger (Algiers), Mostaganem, Arzew, Oran, Mers el Kebir
Algonquin Circle F(ranklin) P(ierce) A(dams), Robert Benchley, Heywood Broun, Irvin S Cobb, Edna Ferber, George S Kaufman, Ring Lardner, Harpo Marx, Dorothy Parker, Harold Ross, Robert E Sherwood, Alexander Woolcott, and others who met informally around the bar of the Algonquin Hotel or in the offices of the *New Yorker*
Alhambra (Arabic—Red House) —ancient Moorish castle in Granada whence the Moors ruled most of Spain from 711 to 1492
The Alice Alice Springs, Northern Territory, Australia
Alice Faye Alice Leppert
Alice Markova Alice Marks
Alien-and-Sedition President John Adams
Alize Breguet carrier-based three-place antisubmarine-warfare aircraft
Al Jolson Asa Yoelson
Alkyd Winsor & Newton's trade name for alkyd-base wa-

tercolors

All-American Boy Jack Armstrong

All-American Mirror Upton Sinclair

All Bunny Albany, New York

Allenwood Camp Federal Prison Camp at Allenwood, Pennsylvania

Alligator Alley trans-Florida highway between Fort Lauderdale on the Atlantic coast and Naples on the Gulf of Mexico

Alligator State Alabama, Florida, Louisiana, Mississippi, and Texas

All-the-Talents Administration Prime Minister William Wyndham Grenville

Alma Gluck Reba Fierson

Almirante del Mar Oceano (Spanish—Admiral of the Ocean Sea), title given Christopher Columbus by Queen Isabella la Católica who also made him viceroy and governor of all the lands he discovered

Almost-Impeached President Andrew Johnson, Richard M. Nixon

Aloha State Hawaii

Aloysha (Russian nickname—Aleksei)—Alex; Alexander

Alpine Principality Liechtenstein

Alpine Republic Switzerland

Altamont, Catawba Thomas Wolfe's fictitious name for Asheville, North Carolina

Alte Fritz (German—Old Fritz) —Frederick the Great of Prussia

Alter Steffl (German—Old Stevie)—St Stephen's Cathedral in Vienna

Alto Peru (Spanish—High Peru) —Bolivia

Amazonia Brazil's Amazon River basin

Amazon of the Keyboard Teresa Carreño

Ambassador of the Air Charles A Lindbergh

Ambassador of Good Will Will Rogers

A.M. Bernard Louisa M Alcott

Ameer Baraka Lee Roy Jones

America South of the Equator American Samoa

American American aloe (century plant); American beauty (crimson rose); American buffalo (bison); American cheddar (also called American

cheese or store cheese); American cheese (cheddar); American cloth (oil-cloth); American cotton (upland cotton); American-English (American-style English); American fingering (piano); American fries (hashed brown potatoes); American leopard (jaguar); American lobster (Canadian or New England large-clawed species); American Morse (code); American plan (fixed hotel rate including board and food), (food plus room and bath); American rig (oil rig); American sable (pine marten); American school (of artists, economists, etc.); American twist (tennis) other American categories or items

American Agnostics Clarence Darrow, Felix Frankfurter, Robert G Ingersoll, Ben B Lindsey, Henry L Mencken

American Apostle of Nonviolent Disobedience Martin Luther King, Jr

American Arctic Alaska north of the Arctic Circle; the North Pole reached by Americans Robert E Peary and Matthew Henson in 1909

American Atheist-Capitalist James Hervey Johnson

American Atheist Epigrammatists Ambrose Bierce, Mark Twain, and HL Mencken

American Atheists Luther Burbank, Thomas Edison, Emanuel Haldeman-Julius, Ayn Rand, Margaret Sanger, Gordon Stein

American Ballad Composer Stephen Collins Foster

American Balladeer Paul Robeson

American Balzac William Faulkner

American Beauty Rose official flower of Washington, D.C.; nickname sometimes given young women—American Beauty Roses

American Caesar General Douglas MacArthur

American Cato Samuel Adams

American Century the 20th century marked by invention and industrial activity, discovery of the North Pole, landing of men on the moon, victory in two world wars, devotion to the democratic

ideal—the 1900s

American Chronicler John Dos Passos

American Comedians Abbott and Costello, Fred Allen, Amos and Andy, Lucille Ball, Jack Benny and Rochester (Eddie Anderson), Edgar Bergen, Milton Berle, Josh Billings, Victor Borge, Mel Brooks, (George) Burns and (Gracie) Allen, Sid Caesar, Cantinflas (Mario Moreno), Eddie Cantor, Diahann Carroll, Johnny Carson, Charlie Chaplin, Sammy Davis, Jr, Phyllis Diller, Jimmy Durante, WC Fields, Redd Foxx, Great Gildersleeve, Jackie Gleason, George B Hicks, Bob Hope, Danny Kaye, Buster Keaton, (Stan) Laurel and (Oliver) Hardy, Jay Leno, David Letterman, Sam Levinson, Harold Lloyd, Sam Lucas, Jackie (Moms) Mabley, the Marx Brothers (*see* Marx Brothers), Florence Mills, Petroleum V Nasby, Bill Nye, Will Rogers, Mark Russell, Bobby Short, Lily Tomlin, Peter Ustinov, Bert Williams

American Commoner William Jennings Bryan

American Composer-Pianist Louis Moreau Gottschalk

American Composers Samuel Barber, Leonard Bernstein, William Billings, Aaron Copland, Norman Dello Joio, Stephen Foster, George Gershwin, Louis Moreau Gottschalk, Morton Gould, Howard Hanson, Victor Herbert, Alan Hovhaness, Charles Ives, Edward MacDowell, Gian Carlo-Menotti, Walter Piston, Wallingford Riegger, William Schuman, John Philip Sousa, Virgil Thomson

American Conservationists John Muir, William T Hornaday, Williard G Van Name

American-Cowboy Comedian-Humorist Commentator-Philosopher Will Rogers

American Critic H(enry) L(ouis) Mencken

American Crusader for Religious Liberty Roger Williams

American Demosthenes Robert Ingersoll

American disease narcotic addiction

American Documentary Film

Pioneer Robert Flaherty

American (Bald) Eagle symbol of the United States

American-English Poet Thomas Stearns Eliot

American Etchers Joseph Pennell and James Abbott McNeill Whistler

American Expatriate Painters Benjamin West and James Abbott McNeill Whistler

American Filibuster William Walker (1824–1860) who invaded Baja California, Sonora, and Nicaragua attempting to become the head of Central America

American Film Pioneer David Wark Griffith

American Founder of Women's Suffrage Elizabeth Cady Stanton (founder and first president of the National Woman Suffrage Association)

American Frontier Romanticist James Fenimore Cooper

American Gateway to Alaska and the Orient Seattle

American Heartland Illinois, Indiana, Michigan, Ohio, Wisconsin

American Hellenist Frank Mortyn

American Historical Painter Emmanuel Leutzé

American Humanist Philosopher John Dewey

American Humorists George Ade, Steve Allen, Woody Allen, Steven L Anreder, Russell Baker, Robert Benchley, Ambrose Bierce, Erma Bombeck, Art Buchwald, Al Capp, Johnny Carson, Irwin B Corey, ee cummings, Finley Peter Dunne, TS Eliot, William Faulkner, Benjamin Franklin, Lewis Grizzard, Joel Chandler Harris, Bret Harte, O Henry, Oliver Wendell Holmes, Art Hoppe, Washington Irving, Vachel Lindsay, Don Marquis, Groucho Marx, HL Mencken, Gerald Nachman, Ogden Nash, George Jean Nathan, SJ Perelman, James Whitcomb Riley, Will Rogers, Leo Rosten, Damon Runyon, Morrie Ryskind, Mort Sahl, R Emmett Tyrell, Jr, Mark Twain, Artemus Ward, Diane White, Robert Yoakum

American Illustrators Anton Otto Fischer, Howard Pyle, Norman Rockwell, NC Wyeth

American Impressionist Childe Hassam

American Industrial Painter Charles Sheeler

American Infidel Colonel Robert G Ingersoll, agnostic attorney and public speaker, also known as the American Demosthenes; Luther Burbank

Americanist Americanist Press

American Karl Marx Daniel De Leon—born in Curacao, educated in Germany and the Netherlands, founded the Socialist Labor Party (SLP) and the International Workers of the World (IWW) in New York City (where he taught at Columbia University); made some of the first English translations of Karl Marx

American Landscape Painters Albert Bierstad, George Caleb Bingham, James Britton, Frederic Church, Thomas Cole, Asher Brown Durand, Edward Hopper, Henry Inman, George Inness, J Francis Murphy, Grant Wood, and Alexander Helwig Wyant

American Libertarian Philosopher, Natural Scientist, Printer, and Publisher Benjamin Franklin

American Libertarians Thomas Jefferson, Thomas Paine, Robert Ingersoll, Clarence Darrow

American Lighthouse Painter Edward Hopper

American Lithographers (Nathaniel) James (Merritt) Ives

American Medical Historian William Henry Welch

American Melodist Stephen Foster

American Modern Jackson Pollock

American Montaigne Ralph Waldo Emerson

American National Composer John Philip Sousa

American Neurologist Extraordinary Silas Weir Mitchell

American Nine best-known classical composers arranged chronologically: MacDowell, Ives, Piston, Hanson, Gershwin, Copland, Barber, Hovhaness, Bernstein

American Operetta Composers Irving Berlin, George M Cohan, Victor Herbert, Jerome Kern, Frederick Loewe,

Cole Porter, Richard Rodgers, Vincent Youmans

American Orator Extraordinary Robert G Ingersoll and Franklin D Roosevelt

American Paradise U.S. Virgin Islands

American Patriot Author Thomas Paine

American Pestalozzi Amos Bronson Alcott

American Photographer Laureate Ansel Adams

American Photographers of Distinction Berenice Abbott, Ansel Adams, Mathew B Brady, Julia Margaret Cameron, Robert Capa, Imogen Cunningham, Walker Evans, Arnold Genthe, JK Hillers, William Henry Jackson, Gertrude Käsebier, Dorothea Lange, J Ghislain Lootens, Edward Muybridge, Timothy O'Sullivan, Roy Pinney, Edward Steichen, Alfred Stieglitz, Paul Strand, Edward Weston, Clarence H White, Margaret Bourke White, James Van Der Zee, Willard Van Dyke

American Portrait Painters James Britton, John Singleton Copley, Henry Inman, Eastman Johnson, John Singer Sargent, Eugene Edward Speicher, the Peale family, Gilbert Stuart, Thomas Sully, and James Abbott McNeill Whistler

American Ports *see entries under states and territories such as* Alabama Port, American Samoan Port, California Ports, etc.

American Practical Navigator Nathaniel Bowditch

American Pragmatist Trinity John Dewey, William James, Charles Sanders Pierce

American Primitive Painters Edward Hicks, Grandma Moses

American Propagandist Novelist Upton Sinclair

American Prose-Poetry Novelist Thomas Wolfe

American Railroad Barons Jay Gould; Edward H Harriman; James J Hill; Collis P Huntington; William H Vanderbilt

American Rebel Upton Sinclair

American Samoan Port Pago Pago

American Sappho Sarah Wentworth Apthorp Morton of Braintree and Quincy, Mass
American Scott James Fenimore Cooper
American Sculptors Daniel Chester French, Borglum, Brancusi, Epstein, Lachaise, Manship, Moore, St Gaudens, Ward, and Zorach
American Skeptic Philosopher George Santayana
American Socrates Benjamin Franklin
American Spokesman for Socialism Eugene V Debs, Daniel De Leon, and Norman Thomas
American Temple of Music Carnegie Hall
Americans United Americans United for Separation of Church and State (AUSCS)
American Virgin Islands St Thomas, St John, St Croix, and other Virgin Islands
American Virgins U.S. Virgin Islands
American West Indies American Virgin Islands, Commonwealth of Puerto Rico, islands of Culebra, Vieques, and Mona, Navassa, Swan Islands, Corn Islands, and certain coral reefs in Caribbean
American Womanist Alice Walker
American Women Reformers Jane Addams, Susan B(rownell) Anthony, Elizabeth Cady Stanton, Ida M(inerva) Tarbell, Lillian D Wald, and Frances Elizabeth Willard
American Woodsman John James Audubon
American Wordsworth William Cullen Bryant
The Americas North, Central, and South America; the Western Hemisphere
America's Burns John Greenleaf Whittier
America's Cicero Richard Henry Lee
America's Dairyland Wisconsin
America's Devil's Island military prison at Fort Jefferson in the Dry Tortugas west of Key West, Florida
America's Finest Television Hour *60 Minutes*
America's First College Harvard
America's First Colonizers Roger Williams of Rhode Island and William Penn of Pennsylvania
America's First Financier Robert Morris
America's First Great Writer Washington Irving, James Fenimore Cooper, or Edgar Allan Poe
America's First Poet Philip Frenau
America's First Resort Newport, Rhode Island
America's First Suffragist Abigail Smith Adams
America's First Woman Newspaper Publisher Elizabeth Timothy of the *South Carolina Gazette* published in Charleston
America's Foremost Freethinker Robert Ingersoll
America's Forgotten Photographer Timothy O'Sullivan
America's Godfather Thomas Paine
America's Great Winter Garden Imperial Valley, California
America's Ice Box Alaska
America's Inside Fun City New York (concert halls, theaters, opera houses, museums)
America's Largest State Alaska
America's Last Frontier Alaska
America's Last Great Wilderness Alaska
America's Leading Operetta Composer Victor Herbert
America's Leading Proletarian Writer John Dos Passos
America's Most Famous Naval Hero Admiral David G Farragut
America's Most Useful Citizen Jane Addams—author of *Twenty Years at Hull-House*
America's Newest Big City Miami, Florida
America's No 1 Composer-Musician Aaron Copland
America's Nonsense Poet Ogden Nash
America's Official Poets Laureate Robert Penn Warren (1986–1987), Richard Wilbur (1987–1988), Howard Nemerov (1988–1990), Mark Strand (1990–)
America's Oldest City St Augustine, Florida
America South of the Equator American Samoa
America's Outside Fun City San Diego (bays, mountains, year-round outside sports)
America's Poet Laureate Henry Wadsworth Longfellow
America's Practical Navigator Nathaniel Bowditch—compiler of *The American Practical Navigator*
America's Premier Air Woman Amelia Earhart Putnam—first aviatrix to fly across the Atlantic
America's Premier City New York
America's Principal Port New York
America's Proudest Musical Possession Carnegie Hall
America's Safest City Lakewood, Ohio (suburb of Cleveland)
America's Second City Chicago
America's Star-Spangled Satirist Mark Russell
America's Sweetheart Mary Pickford
America's Tropical Islands Hawaiian Islands and the Virgin Islands
America's Wintergarden southern California's Imperial Valley
Amiable Atheist Paul Heinrich Dietrich von Holbach, known to friends as Baron d'Holbach
Ami des Hommes (French—Friend of Mankind)—Marquis de Mirabeau
Ami du Peuple (French—Friend of the People)—Jean Paul Marat; the revolutionary journal he edited
Ammonia King Edward Mallinckrodt
Amos and Andy Freeman F Gosden and Charles J Correll
Anarchist Geographers Prince Peter Kropotkin, Elisée Reclus
Anarchist Protagonist Michael Bakunin
Anatole France Jacques Anatole François Thibault
Anatolia Asia Minor
Anatomist of Humanity Jean Baptiste Poquelin Molière
Ancient Capital of England Winchester
Ancient Universities of England Cambridge and Oxford
Andean America (north to south) Venezuela, Colombia, Ecuador, Peru, Bolivia, Argentina, Chile
Andean Common Market Bo-

livia, Colombia, Ecuador, Peru, Venezuela (Andean Pact Nations)

Andean Group Bolivia, Colombia, Ecuador, Peru, Venezuela

Andean Lands Argentina, Bolivia, Chile, Colombia, Ecuador, Peru, Venezuela

Andrea del Sarto Andrea Domenico d'Agnolo di Francesco

Andrei Sinyavsky Abram Tertz

Andre Maurois Emile Salomon Wilhelm Herzog

Andrew Furuseth Anders Andreassen

Andrew Garve Paul Winterton

Andrew York Christopher Nicole's

Andropov Stalinist name for Rybinsk

Angel of the Battlefield Clara Barton

Angelenos natives of Los Angeles, California

Angelic Doctor Thomas Aquinas

Angolan Ports Ambriz, Luanda, Porto Amboim, Novo Redondo, Lobito, Benguela, Mocamedes, Porto Alexandre

angora a breed of long-hair cat originally from Angora (Ankara), Turkey; long hair goats or rabbits originally from Angora; long and fluffy strands of wool

Angora Turkey's old name for Ankara; long-haired breeds of cats, goats, and rabbits

Angora Goat Capital Rocksprings, Texas

Angry Eagle of Aviation General William (Billy) Mitchell

Animist Nation Dahomey

Ankara Ancyra or Angora

Anna Akhmatova Anna Andreyevna Gorenko

Annaba Bone, Algeria

Annabella Suzanne Georgette Carpentier

Anna O Bertha Pappenheim— feminist crusader against white slavery and first person to be psychoanalyzed

Annapolis of the Air Pensacola, Florida

Anna Seghers Netty Radvanyi

Anne Campbell Mrs George W Stark, contemporary American poet

Anne Morrow Mrs Charles Lindbergh

Ann Harding Dorothy Gatley

Annie Oakley Phoebe Anne Oakley Mozee

Annie's Town Anniston, Alabama

Ann Landers Esther Pauline (Eppie) Lederer

Ann Miller Lucille Ann Collier

Ann Sothern Hariette Lake

Antananarivo Tananarive, the capital of the Malagasy Republic (Madagascar)

Antarctica's Claimants Argentina, Chile, Norway, United Kingdom, U.S.A., U.S.S.R.

Antarctica's Only Known Active Volcano Mount Erebus on Ross Island in the New Zealand sector

Antelope State Nebraska

Anthony Abbot Charles Fulton Oursler

Anthony Armstrong George Anthony Armstrong Willis

Anthony Berkeley Anthony Berkeley Cox

Anthony Boucher William Anthony Parker White

Anthony Hope Sir Anthony Hope Hawkins

Anthony Quinn Anthony Rudolph Oaxaca

Anthracite City Scranton, Pennsylvania

Antid Oto Leon Trotsky

Anti-Slavery President John Quincy Adams

Anti-Trinitarian Author Faustus Socinus, originally Fausto Sozzini

Anti-Trinitarian Martyr Michael Servetus, originally Miguel Serveto

Antsirane Diégo Suarez

Apache State Arizona

Apis code name of Dragutin Dimitrijevic, chief of Serbian army intelligence, who engineered the assassination of Austro-Hungarian Archduke Franz Ferdinand providing the pretext for World War I in 1914

Apollinaire Guillaume Apollinaire, Wilhelm Apollinaris Kostrowitski

Apollyon The Devil

Apostle of Absolute Beauty Jean Sibelius

Apostle of the Anglo-Saxons Saint Augustine (first Archbishop of Canterbury)

Apostle of Caledonia Irish-born Saint Columbia

Apostle of California Padre Junipero Serra

Apostle of Christian Realism Reinhold Niebuhr

Apostle of Common Sense Voltaire

Apostle of Culture Matthew Arnold

Apostle of Dissent William Penn

Apostle of Enlightenment Thomas Paine

Apostle of Free Trade Richard Cobden

Apostle of the French Saint Denys

Apostle of the Gentiles Saint Paul (formerly Saul of Tarsus)

Apostle of Humanity Thomas Paine

Apostle of the Hungarians Saint Anastasius

Apostle of the Indians Bartolomé de Las Casas

Apostle of the Indies Saint Francis Xavier

Apostle of Ireland Saint Patrick

Apostle of Liberty Benjamin Franklin

Apostle of Mexican Rebellion Francisco I Madero

Apostle of New Zealand Reverend Samuel Marsden

Apostle of Reason Peter Abelard

Apostle Rebel Dorothy Day

Apostle of the Rights of Man Thomas Paine

Apostle of Sanity Epicurus

Apostle of Science Roger Bacon

Apostle of the Scottish Reformation John Knox

Apostles of Freedom Sam Adams and Thomas Paine

Apostle of the Slavs St Cyril

Apostle of the Sword Mohammed

Apostle of Temperance Theobold Mathew

Appeasement Premier Neville Chamberlain

Apple Capital of the World Wenatchee, Washington

Apple Island Tasmania

Apple Islanders Tasmanians

Apple Isle Tasmania

Aqueduct City Rochester, New York

Arab Africa northern Africa from Egypt to Morocco and from Mauritania to the Sudan

Arabia Deserta Desert Arabia in the northern sector of the Arabian Peninsula

Arabia Felix Fertile Arabia also known as Aden, the Hadhramaut, or Yemenite section

arabian arabian baboon; arabian camel (one-humped dromedary); arabian coffee; arabian horse

Arabian Arabian Desert; Arabian Peninsula of Arabian Sea; inhabitant of Saudi Arabia also called Saudi (plural or singular)

Arabian Gulf Persian Gulf

Arabia Petraea Rocky Arabia in the northwestern section of the Arabian Peninsula

arabic arabic numbers (1, 2, 3, 4, 5, etc., as distinct from roman numbers—I, II, III, IV, V, etc.)

Araucania region of central Chile south of Bío-Bío River

araucanos (Latin American nickname—Chileans or *chilenos*)—sobriquet recalls the liberty-loving Araucanian Indians who were never conquered by the Spaniards

Arch City St Louis, Missouri, dominated by the Jefferson National Expansion Memorial arch commemorating the Louisiana Purchase

Archeological Capital of Africa Cairo, Eqypt

Archeological Capital of North America Chichen Itza, Mexico

Archeological Capital of South America Cuzco, Peru

Archipiélago de Colón (Spanish—Columbus Archipelago) Ecuador's official name for the Galápagos Islands

Architect of the Atomic Bomb J Robert Oppenheimer

Architect in Chief of St Peters Raphael (Raffaello Santi)

Architect of the Constitution James Madison

Architect of Mexican Independence Padre Miguel Hidalgo

Architect-Naturalist-Philosopher-Statesman President Thomas Jefferson

Architect of the New Deal Franklin Delano Roosevelt

Architect of Non-Alignment Josip Broz Tito

Architect President Thomas Jefferson

Arctic Arctic Current flowing south from Baffin Bay and Greenland to cool the coasts of Labrador, Newfoundland, and most of New England; Arctic Ocean washing the north coast of Asia, Europe,

and North America including Alaska and Canada

Arctic big three muskox, polar bear, walrus

Arctic Canada Northwest Territories and the Yukon

Arctic Lands Alaska, Canada, Greenland, Iceland, Norway, Sweden, Finland, the USSR

Arctic Territories Canadian Northwest Territories

Arctogaea landmass including Africa, Asia, Europe, and North America

Ardent City Liège, Belgium

Ardent Conservationist Rachel Carson

Argentina's Principal Port Buenos Aires

Argentinian First Argentina's best known classical composer—Alberto Ginastera

Argentinian Ports Santa Fé, Rosario, Zarate, Campana, Buenos Aires, La Plata, Mar del Plata, Puerto Belgrano, Ingeniero White, Puerto Madryn, Ushuaia

Argyrol King Dr Albert C Barnes

Aristocles Plato's original name

Aristocrat of Sports billiards

The Ark HMS *Ark Royal*

Arkansawyer a native of Arkansas, also called Arkie

Arkie Arkansas (or resident; migratory worker from Arkansas)

Arkopolis Little Rock, Arkansas

Arletty Arlette-Léonie Bathiat (French actress)

Armageddon Meggido, Palestine where the British defeated the Turks in 1918

Arminius Jacobus Arminius (originally Jacob Harmensen)

Arms-control President George Bush

Arnhem Land northern end of Australia's Northern Territory

Arnold Bennett Enoch Arnold Bennett

Arrowhead Herman Melville's home in Pittsfield, Massachusetts

Arroyo del Ajo (Spanish—Garlic Gulch)—John Steinbeck's name for his home near Los Gatos, California

Arsenal of the Nation Connecticut

Art Capital New York City

Art Center of Rhode Island Wickford

Art Center of the Southwest

Taos, New Mexico

Artemus Ward Charles Farrar Browne (19th-century American humorist)

Artesian State South Dakota

Artichoke Capital Castroville, California; Watsonville, California

Artie Shaw Arthur Arshawsky

Artist of the French Revolution Jacques Louis David

Arturo de Cordova Arturo Garcia

Aruba's Ports Sint Nicolaas (Lago Refinery), Paarden Baai (Oranjestad), Druif

ARW 493 Stig Sverker Foghammar

Ashcan Artist John Sloan

Ashcan School *see* Eight

Ashenden W Somerset Maugham

Ashland Henry Clay's home in Lexington, Kentucky

Asian Subcontinent Bangladesh, Bhutan, India, Nepal, Sikkim, and Sri Lanka; Indian Peninsula

Asian Subcontinent Bangladesh, Bhutan, India, Nepal, Sikkim, and Sri Lanka; Indian Peninsula

Asia's big five elephant, leopard, rhinoceros, tiger, water buffalo

Asmus Rasmus Rasmus Meyer Museum, Bergen, Norway; Miller

Asparagus Capital Isleton, California

Asphaltic Lake the Dead Sea

Assemblyman from the Bowery Al Smith (Alfred E. Smith)

Associated States Caribbean island states (Antigua-St Kitts-Nevis, Dominica, Grenada, St Lucia, St Vincent)

Assyrian Century the 7th century before the Christian era when Assyria ruled the Middle East and conquered Egypt—the 600s

astrakhan astrakhan cloth or astrakhan wool of the type originally clipped from sheep native to Astrakhan on the Caspian in the delta of the Volga

Astrodome City Houston, Texas

Astronaut President John Fitzgerald Kennedy

Astronomical Prophet Galileo Galilei

Asylum for Talent Jacques

Copeau's *Theatre du Vieux Colombier*

Ataturk (Turkish—Chief Turk) General-President Mustafa Kemal—first president of Turkey

The Atheist Percy Bysshe Shelley

Atheist Penologist Jeremy Bentham

Atheist Poetess Ellen Prouse Mardan

Atheist Poets Heinrich Heine, Percy Bysshe Shelley, Walt Whitman

Atheist's Bible Thomas Paine's *The Age of Reason*

Athenian Century the 5th century before the Christian era when the Athenians destroyed the Persian fleet and completed the Parthenon—the 400s

Athens of Alabama Tuscaloosa

Athens of Arkansas Fayetteville

Athens of America Boston

Athens of Hispanic America Bogotá

Athens of the North Scotland

Athens of the Northwest Faribault, Minnesota

Athens of the South Nashville, Tennessee

Athens of Texas Waco

Atlantic Bitch Atlantic Beach

Atlantic Canada Labrador and Newfoundland, New Brunswick, Nova Scotia, Prince Edward Island, Québec

Atlantic Community NATO nations

Atlantic Highlands Highlands of the Navesink around Sandy Hook, New Jersey

Atlantic Narrows relatively restricted area uniting North and South Atlantic between bulge of Africa and bulge of Brazil, Freetown and Natal

Atlantic Ocean Edens West Indies before European settlement

Atlantic Provinces New Brunswick, Newfoundland, Nova Scotia, Prince Edward Island

Atlantic Scandinavia Denmark, Iceland, Norway

Atlantique (French—Atlantic Ocean)—also name of the Breguet maritime-patrol aircraft BR-1150

Atoll Soviet Sidewinder-type missile

Atoll Nation Nauru

Atomic Age Capital Los

Alamos, New Mexico

Atomic Capital of America Richland, Washington

Atomic Cities Los Alamos, New Mexico; Oak Ridge, Tennessee; Richland, Washington

Atomic City Los Alamos, New Mexico and Oak Ridge, Tennessee

Atomic Energy City Oak Ridge, Tennessee

Atterdag (Danish—Another Day)—King Valdemar IV

Attic Muse the Athenian historian Xenophon

Attorney for the Damned Clarence Darrow

Au₂H₂O Arizona Senator Barry Goldwater

Audrey Hepburn Edda van Heemstra

Augustan Age Latin literature's golden era when Horace, Livy, Ovid, and Virgil flourished during the reign of the Emperor Augustus (27 B.C. to A.D. 14)

Augustina de Aragon Augustina Domenech Zaragoza

Auld Ane (Scottish Gaelic—Old One)—the devil

Auld Brig o' Don Dundee, Scotland's Brig o' Balgownie (brig = bridge)

Auld Clootie (Scottish Gaelic—Old Cloven)—cloven-footed devil

Auld Reekie (Scottish Gaelic—Old Smelly)—smogbound Edinburgh

Auld Sod (Scottish Gaelic—Old Land)—Scotland

Aunt Jane Tia Juana (river separating Tijuana, Mexico and San Ysidro, California)

Aunty Vicky Queen Victoria

Aurelian Century the 100s—reign of Roman emperor-philosopher Marcus Aurelius—the 2nd century

Aussieland Australia

Austerlitz Slavkov, Czechoslovakia

Australia Felix (Latin—Happy Australia)—fertile central Victoria in southeastern Australia

Australian Alps mountains of New South Wales and Victoria

Australian Atheist Phillip Adams

Australian Commonwealth Australia and its territories

Australian Desert 1,300,000-

square-mile area (530,000 hectares) in central and western Australia

Australian Dominion Australia and its territories

Australian Ports Port Kennedy, Cairns Harbour, Townsville, Port of Bowen, Port of Mackay, Rockhampton, Gladstone, Maryborough, Brisbane, Clarence River, Port Waratah, Newcastle, Sydney, Port Kembla, Melbourne, Williamstown, Geelong, Portland, Port Adelaide, Port Vincent, Port Pirie, Port Augusta, Whyalla, Port Lincoln, Albany, Busselton, Bunbury, Freemantle, Perth, Geraldton, Carnarvon, Broome, Wyndham, Darwin, Gove, Hobart (in Tasmania along with Burnie, Devonport, Beauty Point, Launceton)

Australian States New South Wales, Queensland, South Australia, Tasmania, Victoria, Western Australia

Australian Territories Australian Antarctic; Australian Capital Territory (Canberra), Northern Territory, Papua New Guinea (Admiralty Islands, Heard and McDonald Islands, New Britain, New Guinea, New Ireland, the Solomons)

Australia's Little England Tasmania

Austrian Emperor of Mexico Ferdinand Maximillian Von Hapsburg, also known as Max of Baden

Austrian Quintet classical composers Haydn, Mozart, Schubert, Bruckner, Mahler

Austrian Waltz Kings Joseph Lanner, Johann Strauss, and his son Johann Strauss Jr

Author of the Declaration of Independence Thomas Jefferson

Author of the First Amendment James Madison

Author of the first draft of the Declaration of Independence Thomas Paine

Author of the Virginia Statute for Religious Freedom Thomas Jefferson

Authority on Mythology Joseph Campbell

Autocrat of All the Russias Czar Nicholas II

Autocrat of Austria Prince Clemens Wenzel Lothar von

Metternich

Autocrat of the Breakfast Table Dr Oliver Wendell Holmes

Autocrat of Commerce Venice

Automobile Capital of the World Detroit, Michigan

Automobile City Detroit, Michigan

Automobile State Michigan

Automobile Wizard Henry Ford

Auto State Michigan

Aux Cayes former name of Les Cayes, Haiti

Ava Gardner Lucy Johnson

Avalon Somerset region of southwestern England believed to be Avalon of Arthurian legend; resort port of Catalina Island off Los Angeles, California

Avenue of the Americas New York City's Sixth Avenue

Aviation Historian Octave Chanute

Avicenna Arabian astronomer-mathematician-physician Abu ibn Sina (980—1037)

Aviocar Spanish transport aircraft designated C-212

Avocado County San Diego County, California

Avon Avon Books; Avonmouth (Port of Bristol); Avon Water (flowing from Ayrshire to Lanark); Avonwick (Devonshire)

Awakener of Bulgaria George Venelin

Axis Sally Mildred E Gillars, American traitor convicted of treason for broadcasting Nazi propaganda during World War II

axminster eponymic name for good grade carpets and rugs originally made in the English town of Axminster in Devonshire; modern axminsters often copy well-known oriental designs

Ayrshire Poet Robert Burns born in Alloway, Ayrshire, Scotland

Azalea Trail City Lafayette, Louisiana

Azania African nationalist name for the Republic of South Africa

Aztecan and Incan Century the 1000s—great stone structures Mexico and Peru are mute witnesses to these indigenous American cultures—the 11th century

Aztec type microcephalic idiocy

Azure Coast Côte d'Azur on the French Riviera

Azure Sea Lake Rudolf, Kenya

Baal Shem-Tov (Hebrew—Kind Master of the Holy Name)—Israel Ben Eliezer

Babar Jean de Brunhoff's storybook elephant; Zahir ud-Din Muhammad (founder of India's Mogul dynasty)

Babe Ruth George Herman Ruth, the Sultan of Swat

Babeuf Francois Noël

Babushka (Russian—grandmother)—Ekaterina Breshkovskaya, revolutionary leader

Baby Doc Haitian dictator Jean–Claude Duvalier

Baby Langdon Harry Langdon

Baby State Arizona in 1912 when admitted to statehood

Bachelor Painter Sir Joshua Reynolds

Bachelor President James Buchanan—fifteenth President of the United States

Back Bay Boston's old residential section built on mud flats reclaimed from Boston Bay

Backbone of Asia the Himalayas

Backbone of the Confederacy the Mississippi River

Backbone of England Pennine Hills; Pennine Ridge

Backbone of Europe the Alps

Backbone of North America the Rockies

Backbone of South America the Andes

Backfire Soviet strategic supersonic bomber equivalent to the North American-Rockwell B-1

Back-of-Beyond Australia's sparsely inhabited interior

Back of the Beyond Australia's outback region

Back Home Again State Indiana

Bactrian Sage Zoroaster (founder of the Magian religion and native of Bactria)

Bad Boy of Music George Antheil

Bad Boys of 57th Street the New York Philharmonic in its post-Toscanini period

Badger NATO nickname for Soviet Tupolev medium bomber (Tu-16)

Badger(s) Wisconsinite(s)

Badger State Wisconsin

Badlands arid and eroded areas of Nebraska and South Dakota

Baffin Basin deeper parts of Arctic Ocean between Baffin Island and Greenland

Bagdad by the Bay San Francisco

Bagdad-on-Hudson New York

Bagdad on the Subway one of O Henry's nicknames for New York City. He also called it the City of Razzle Dazzle

Bag Town San Diego, California, where sailors tote seabags

Bahamian Ports Freeport (Grand Bahama), Bimini (Bimini Islands), Nassau (New Providence), Matthew Town (Great Inagua)

Bahraini Ports Al Manamah Harbour, Mina Sulman, Sitra

Baía de Guanabara (Portuguese—Guanabara Bay)—Rio de Janeiro's inner harbor

Baird Leonard Mrs Harry S Clair Zogbaum

Baked Bean State Massachusetts

Bakst Leon Bakst (originally Rozenberg)

Balanchine Georgi Balanchivadze

Baleful Prophet Cassandra

Baltic Republics Estonia, Latvia, Lithuania

Baltic Scandinavia Finland and Sweden (Denmark sometimes included although much of its coast is on the Atlantic)

Baltic States Estonia, Latvia, Lithuania

Baltimore beefsteak broiled liver's military nickname in the U.S.

Baltimore Oracle HL Mencken

Balzac Honoré de Balsa

Bambino George Herman (Babe) Ruth

Bamboo Curtain old name for the barrier between anti-communist and communist countries of Southeast Asia

Banaba Ocean Island near the Gilberts in the equatorial mid-Pacific Ocean

Banana Benders Queensland Australians

banana boat cargo vessel built to carry bananas

Banana City Brisbane

Bananaland Queensland, Australia

Bananalanders people of Queensland, Australia

Banda Oriental (Spanish —Eastern Ribbon)—Uruguay

Band City Elkhart, Indiana

Bandit Queen of the Old West Belle Starr

Bane of the Bureaucrats Parkinson's Law

banewort *Atropa belladonna's* (also called beautiful lady, deadly nightshade, or death's herb)

bangkok bangkok hat (straw hat of a type first woven in Bangkok); bangkok straw (Siamese straw used in making baskets and hats)

Bangladesh Ports Chalna Anchorage, Chittagong, Cox's Bazar

Banjo Andrew Barton Paterson of Australia

The Bank The Bank of England

Banner State Texas

Baotou (Pinyin Chinese—Paotou)

Bapu (Gujerati—father)—Mahatma Gandhi's title

Barão de Rio Branco (Baron Rio Branco) José María de Silva Paranhos

Barbadian Ports Speightstown, Bridgetown

Barbara Bel Geddes Barbara Geddes Lewis

Barbara Stanwyck Ruby Stevens

Barbara Ward Lady Jackson (wife of Sir Robert Jackson)

Barbarossa nickname of redbearded Frederick I of the Holy Roman Empire; Barbarossa I (Koruk) and Barbarossa II (Khaireddin) were Greek-born Algerian pirates

Barbary Coast North African coast; San Francisco's waterfront district a century ago

Barbary States Algeria, Libya (Tripolitania), Morocco, Tunisia

Barbellion WNP Barbellion (pseudonym of Frederick Cummings)

Barber Poet Provencal poet Jacques Jasmin—a barber by profession, also called the Last of the Troubadors

Barca the Carthaginian Maharbal

Bard of Avon William Shakespeare

Bard of Ayrshire Robert Burns

Bard of Olney William Cowper

Bard of Prose Boccaccio

Bard of Sheffield James Montgomery

Bard of the Stumblebum Nelson Algren

Bard of Twickenham Alexander Pope

Barefoot King of Cocos John Clunies-Ross (owner of the Cocos or Keeling Islands in the South Indian Ocean)

Baritone-Conductor Dietrich Fischer-Dieskau

Barney Barnato Barnett Barnato (Barnett Isaacs)

Baron Burnham Edward Levy-Lawson

Baron Corvo Frederick William Rolfe

Baron Cuvier Georges Léopold Chrétien Frédéric Dagobert

Baron de Reuter Israel Beer Josaphat (founder of Reuter's news agency)

Baroness Orczy Mrs Montagu Bartstow—author of *The Scarlet Pimpernel*

Baronet Peel Robert Peel (former Prime Minister of Great Britain)

Baron Grenville William Wyndham Grenville (former Prime Minister of Great Britain)

Baron Lugard Frederick John Dealtry Lugard

Baron Munchausen Rudolf Erich Raspe

Baron Passfield Sidney Webb

Baron Stiegel ironmaster Henry William Stiegel

Baron Tweedsmuir John Buchan

Barrington Island, Galápagos Santa Fé

Barrio Chino (Spanish—Chinese Quarter)—Barcelona's brothel area close to the waterfront

Barry Cornwall Bryan Waller Procter

Barry Fitzgerald William Shields

Barrymore family containing some of America's best beloved actors (Lionel, Ethel, John—children of Maurice and Georgiana Barrymore) —actual surname was Blythe

Barry Perowne Philip Atkey

Barry Sullivan Patrick Barry

Barton Cannon Barton Danzilio

Basil Rathbone Lawrence Northrup

Basket of Eggs hills of Downs in Northern Ireland

Basque Provinces álava, Guipúzcoa, and Viscaya, Spain

Bastille by the Bay San Quentin prison in San Francisco Bay

Bastion of the Caribbean Puerto Rico

Basutoland former name of Lesotho

Batavia Dutch name for Djakarta; former name of Djakarta, Indonesia; Roman name for the Netherlands

Batavian Republic name for the Netherlands during the French Revolutionary wars (1795 to 1806); the Netherlands during the rule of Louis Bonaparte (1806–1810)

Bath City Mt Clemens, Michigan

Bat Masterson William Barclay Masterson

Battenberg Mountbatten

Battery old seawall containing gun emplacements in Charleston, South Carolina, Manhattan Island in New York

Battle-Born State Nevada—admitted as territory in 1848 following the Mexican War

Battlefield City Gettysburg, Pennsylvania

Battlefield of the Nations Leipzig in Germany in 1813; Plain of Esdraelen in Israel between Megiddo and Nazareth; Waterloo in Belgium

Battleground of Freedom Kansas during ten years of debate over slavery

Battling Bob Robert M La Follette, Sr

Batumskaya formerly Batum

Bay Bay of Bengal, Biscay (nicknamed Biscuits by sailors), Boston, Charleston, Fundy, Holland, Islands, Laig, Mission, Naples, New York, Panama, Pigs, River, San Diego, San Francisco, State, Street, Whales; The Bay— Algoa Bay or Port Elizabeth Bay in South Africa

The Bay Hudson's Bay Company; (*see* Bay)

Bay of Biscuits (naval nickname—Bay of Biscay)

Bay Cities cities surrounding San Francisco Bay

Bay City San Francisco

Bay of Gold San Francisco Bay

Bayou City Houston, Texas

Bayou State Louisiana, Mississippi

Bay of Pigs Cuba's south coastal Bahía de Cochinos where an unsuccessful attempt was made to liberate Cubans from Castro's rule

Bay State Massachusetts

Bay Stater(s) Massachusettan(s)

Bay Street financial center of the Bahamas in Nassau on New Providence Island; financial center of Canada in Toronto, Ontario

Bean Town Boston, Massachusetts

Bear State Arkansas, Kentucky

The Beatles George Harrison, John Winston Lennon, James Paul McCartney, Ringo Starr (Richard Starkey)

Beautiful City by the Sea Portland, Maine

Beautiful Kingdom translation of China's name for the United States of America

Beaver State Oregon

Beck(y) Rebecca

Beedle General Walter Bedell Smith

Beef Barons Armour, Cudahy, Morris, Swift

Beefeaters Her (His) Majesty's Honourable Corps of Gentlemen at Arms

Beef State Nebraska

Bee Gee British Guiana (now called Guyana)

Beehive of Industry Providence, Rhode Island

Beehive State Utah

Beelzebub The Devil

Beer City Milwaukee

Beethoven Town Bonn, Germany—birthplace of Ludwig van Beethoven

Beetle Juice (American navalese—Betelgeuse)—variable giant red star of the first magnitude in the constellation of Orion

Bee Wee nickname for a British West Indian

Beggars of the Sea Dutch pirates and privateers

Bela Lugosi Bela Paul Blasko

Bel Anglais (French—Handsome Englishman)—nickname of John Churchill— Duke of Marlborough

Belgeek(s) Belgian(s)—so-called by American, British, and Canadian armed forces during World War II

belgian Belgian griffon (small black or reddish-black terrier); Belgian horse (draft horse); Belgian hare(rabbit); Belgian sheepdog

Belgian Congo colonial possession of Belgium from 1884 to 1963; Zaire

Belgian East Africa former name of Ruanda

Belgian Ports Antwerpen (Anvers), Bruxelles (Brussel), Gente (Gand), Brugge (Bruges), Zeebrugge, Blankenberge, Oostende (Ostend), Nieuwpoort (Nieuport)

Belgium Film Pioneer Jacques Feyder

Belial The Devil

Believing Unbeliever Supreme Court Justice Felix Frankfurter

Belize formerly British Honduras

Belle of Amherst Emily Dickinson

Belle City of the Bluegrass Region Lexington, Kentucky

Belle City of the Lakes Racine, Wisconsin

Belle Riviere (see *La Belle Riviere*)

Belle Starr Myra Bell Shirley

Bell Rock Inchcape Rock off the Forfarshire coast of Scotland, the subject of Robert Southey's *Ballad of Inchcape Rock*

Bell Town East Hampton, Connecticut

Beloved City of El Greco Toledo, Spain

Beloved Friend Nadejda von Meck (Tchaikovsky's patroness and lifetime friend he never met)

Beloved Infidel Colonel Robert Ingersoll

Below Sea-Level Cities Brawley and El Centro, California

Ben Block a British sailor

Ben Cur Benguela Current

Benemérito de las Américas (Spanish—Meritorious Man of the Americas)—the Mexican Indian Benito Juárez

Ben-Gurion (Hebrew—Son of a Lion)—name adopted by David Green

Benin Ports Cotonou, Kpeme

Benjies hundred-dollar bills bearing the portrait of America's Benjamin Franklin

Benny Goodman Benjamin David Goodman

bent-nail syndrome Peyronie's disease wherein the penis is

bent out of shape

Berks the Beautiful Pennsylvania's Berks County

Berkshires Berkshire Hills of western Connecticut and Massachusetts

Berlin Bach Karl Philipp Emanuel Bach—also nicknamed Hamburg Bach

Bermuda Triangle area between Bermuda, Cape Hatteras, and Key West in the western North Atlantic—also called the Devil's Triangle or the Limbo of the Lost because of the many planes and ships lost in the area; area is also said to be between Bermuda, Florida, and Puerto Rico

Bernese Oberland Bernese Alps

Bernhardt Sarah Bernhardt—stage name of Rosine Bernard

Berry City Woodburn, Oregon

Berserkeley Berkeley, California

Bertie Bertrand Russell—colossus of twentieth-century philosophy

Bert Lahr Irving Lahreim

bessel bessel equation, bessel function, or bessel method (named for the German 19th-century astronomer Friedrich Wilhelm Bessel)

bessemer bessemer converter or bessemer steel (named for its English inventor, Sir Henry Bessemer)

Bessie Love Juanita Horton

Betel Nut Island Penang, Malaysia

Betsytown Elizabeth, New Jersey

Betty Grable Elizabeth Grasle

Beverly Sills Belle Silverman

Bharat Republic of India

Biandrata Giorgio Blandrata

Bible Belt rural areas of the southern United States

Bi-City Port Gdansk-Gdynia, Poland

Bien Aime (French—Well Beloved)—Louis XV

Big A Atlanta, Georgia in the Southeast; Amarillo, Texas in the Southwest; underworld nickname of the Federal Penitentiary in Atlanta, Georgia

Big Apple New York City

Big Ben aircraft carrier USS *Franklin*; battleship USS *Franklin*; huge bell attached to clock in Parliament tower, Westminster district of London, named after Sir Benjamin Hall, commissioner of works in 1859 when bell was hung

Big Bend big bend of the Rio Grande—bounding southern section of the Big Bend National Park on the Texas border of Mexico

Big Bend State Tennessee

Big Bertha World War I howitzer capable of hurling a one-ton projectile nine miles, named for a member of the Krupp family

Big Bill Haywood William Dudley Haywood (founder of IWW)

Big Board New York City's Stock Exchange

Big Brother big government's watchful eye as described in George Orwell's *1984*

Big Burg New York City

big C cancer; cocaine

Big Charlie Charles de Gaulle

big D hallucinogen such as diethyltryptamine, dimethyltryptamine, dipropylphyptamine

Big-D Dallas, Texas

Big Dan Daniel Joseph Tobin

Big Dipper ladle-shaped constellation of Ursa Major

Big Ditch Panama Canal

Big Drink Atlantic or Pacific Ocean

Big-D of the West Denver, Colorado

Big-E aircraft carrier USS *Enterprise*

Big Finger Australia's Cape York Peninsula

Big Five (British banks) Barclays, Lloyds, Midland, National Provincial, Westminster

Big Four at the Johns Hopkins Medical School—William Howard Welch, William Osler, Howard Atwood Kelly, William Stewart Halsted; California railroad builders Charles Crocker, Mark Hopkins, Collis P Huntington, and Leland Stanford; Cleveland, Cincinnati, Chicago, and St Louis; Great Britain, France, Italy, and the United States at the end of World War I, their representatives at the Peace Conference—Lloyd George, Georges Clemenceau, Vittorio Orlando, and Woodrow Wilson

Big Four Automakers General Motors, Ford, Chrysler,

American Motors

Big Four Opera Houses Covent Garden in London, La Scala in Milan, Staatsoper in Vienna, The Met in New York

Biggest Little City Reno, Nevada

Big Heart of Texas Austin

Big Inch 24-inch pipeline carrying petroleum products from east Texas to the New York-Philadelphia area

Big Island Hawaii (largest of the Hawaiians); Vancouver Island, British Columbia

big J big John (slang—policeman or other law-enforcement officer)

Big-J battleship USS *New Jersey*

Big Jim brothel-keeper gangster James Colosimo (1877—1920); Postmaster General James Aloysius Farley

Big-M battleship USS *Missouri;* Memphis, Tennessee

Big Mac Mac Donald hamburger; New York's Municipal Assistance Corporation (MAC)

Big Mamie battleship USS *Massachusetts*

Big Minny Minnesota

Big Miss Mississippi River

Big Momma HMS *Ark Royal*

Big Muddy Missouri River

Big N Vladimir Nabokov

Big Nail translation of the Eskimo nickname for the North Pole

big O opium

Big-O attack aircraft carrier USS *IOriskany*

Big Orange Los Angeles

Big P tenor Luciano Pavarotti

Big Red racehorse Man-o'-War

Big Seven Britain, Canada, France, Germany, Italy, Japan, United States

Big Six New York City's Typographical Union Number Six

Big Sky Country Montana

Big Smoke old nickname of London, England as well as Sydney, New South Wales

Big T Tucson, Arizona

Big Three The Big Three (music publishers Robbins, Feist, Miller)—Robbins Music Corp; World-War-I peacemakers Georges Clemenceau, Lloyd George, and Woodrow Wilson; World-War-II peacemakers Winston Churchill,

Franklin Roosevelt, and Joseph Stalin

Big Three Automakers General Motors, Ford, Chrysler

Big Town Chicago

Big Two Soviet Russia, United States of America

Big Windy Chicago, Illinois

Bikini State Florida

Bilad al-Sudan (Arabic—Land of the Blacks)—Guinea, Sudan

Bill Arp Charles Henry Smith

Billie Holiday Eleanora Fagan

Bill Mauldin William H Mauldin (contemporary American cartoonist)

Bill Nye Edgar Wilson Nye

Billtown Williamstown, Kansas

Billy the Kid William Bonney alias William Wright

Billy Mitchell General William Mitchell

Billy Rose William Samuel Rosenberg

Billy Sanders Joel Chandler Harris

Bill of Rights President James Madison

Bimshire Barbados

Bing Crosby Harry Crosby

Biographer of Thomas Paine Moncure D Conway

Biologist of the Mind Sigmund Freud

Bip Marcel Marceau

Bird Dog Cessna L-19 liaison aircraft

Bird of Happiness Japan's red-crowned crane

Bird of Paradise brilliantly-plumed male perching birds in New Guinea area

Bird-of-Paradise Island Little Tobago (where birds of paradise may be seen in their wild state)

Birdofredum Sawin James Russell Lowell

Birds of Paradox flightless birds such as cassowaries, flightless cormorants, emus, kiwis, ostriches, penguins, rails, rheas

Birds of Prey flesh-eating predators such as condors, eagles, falcons, hawks, kites, ospreys, owls, vultures

Bird Woman Sacajawea

Birkenau German name for Brezinka next to Auschwitz, called Oswiecim by the Poles

Birken'ead drill maritime tradition dating from 1852 with the sinking of HMS *Birkenhead* when women and children were given first place in the lifeboats

Birthplace of American Cajuns New Brunswick

Birthplace of the American Industrial Revolution Pennsylvania's Lehigh Valley

Birthplace of American Jazz New Orleans

Birthplace of American Liberty Faneuil Hall, Boston

Birthplace of American Presidents Massachusetts (4) New York (4); Ohio (7); Virginia (8)

Birthplace of Aphrodite or Venus Cyprus

Birthplace of Aviation Dayton, Ohio where the Wright Brothers built their flying machine

Birthplace of Bach, Beethoven, and Brahms Germany (Eisenach, Bonn, and Hamburg)

Birthplace of Baseball Cooperstown, New York

Birthplace of Berlioz Cô-Saint-André, Isère, France

Birthplace of Bolívar Caracas, Venezuela

Birthplace of Brahms and Mendelssohn Hamburg, Germany

Birthplace of the British Industrial Revolution Severn Valley in England and Wales

Birthplace of Burns Alloway, Scotland

Birthplace of California San Diego

Birthplace of Camões (Camoëns) Lisbon, Portugal

Birthplace of Canada Charlottetown, Prince Edward Island

Birthplace of Cervantes Alcalá de Heneres, Spain

Birthplace of Colombia Tunja, Boyacá

Birthplace of Dante Florence, Italy

Birthplace of Democracy ancient Greece

Birthplace of the Gods Greece

Birthplace of Handel Halle, Germany

Birthplace of Hans Christian Andersen Odense, Denmark

Birthplace of Hindemith Hanau, Germany

Birthplace of Kant Kaliningrad (formerly Königsberg, East Prussia)

Birthplace of Liszt Raiding, Hungary

Birthplace of Melodramatic

Opera Italy

Birthplace of Mozart Salzburg, Austria

Birthplace of Paganini Genoa

Birthplace of Purcell London, England

Birthplace of Richard Strauss Munich, Germany

Birthplace of Saint-Saëns Paris, France

Birthplace of Schubert Vienna, Austria

Birthplace of Schumann Zwickau, Germany

Birthplace of Shakespeare Stratford-upon-Avon, England

Birthplace of Sibelius Tavastehus, Finland

Birthplace of Skiing Morgedal in the Telemark region of Norway

Birthplace of the Skyscraper Chicago

Birthplace of Spinoza Amsterdam, Netherlands

Birthplace of the Tuna Fishing Industry San Diego

Birthplace of the United States of America Philadelphia, Pennsylvania where the *Declaration of Independence* was signed July 4, 1776

Birthplace of Vaudeville Sainte-Mère-église (near Norman coast of France)

Birthplace of Villa-Lobos Rio de Janeiro, Brazil

Birthplace of Vivaldi Venice, Italy

Birthplace of Wagner Leipzig, Germany

Birthplace of Wilde, Shaw, Joyce, and Behan Dublin, Ireland

Bishop of Broadway theatrical producer David Belasco whose clerical dress earned him this title

Bishop of Rome the Pope

Bison Soviet Mya-4 four-engine jet heavy bomber (NATO)

Bison City Buffalo, New York

Bitter Bierce Ambrose Bierce

Bitter Sea Dead Sea between Israel and Jordan

Bituminous City Connellsville, Pennsylvania

Bix Leon Bismarck Beiderbecke

Bjørn **Bjørn** Bjørnstjerne Bjørnson

B Kovner Jacob Adler

Black Africa equatorial Africa from Ethiopia, Somalia, and Kenya to Gabon, the Congo,

and Zaire

Black Bart poetic stagecoach robber Charles E Bolton

Blackbeard Edward Teach —privateer-pirate also known as Edward Thatch

Black Belt black-soil growing area extending from South Carolina and Georgia to Alabama and Mississippi

Blackberry Capital McCloud, California

Blackbird Lockheed SR-71 jet reconnaissance aircraft

Black Carib Garifuna spoken in Belize

Black Castle of Opium Afyonkarahisar in western Turkey

Black Charley Sir Charles Napier

Black Country coal-mining iron-founding sections of southern Staffordshire and West Midlands of England; Midlands of England around smoke-blackened Birmingham

Black Dan Daniel Webster

Black Death bubonic plague

black diamond black or gray industrial diamond also called framesite bort; anthracite or hard coal

Black Diamond City Wilkes-Barre, Pennsylvania

black diamonds coal

black disease anthrax of sheep; braxy

Black Douglas Sir James de Douglas

Black Eagle Hubert F Julian

Black Emperor of Haiti Jean-Jacques Dessalines

black fever kala-azar (leishmaniasis)

black flag symbol of death or emblem of piracy

Black Flower of Society Nathaniel Hawthorne's nickname for any jail, penitentiary, or prison

Black Forest Schwarzwald (in mountainous south-central Germany)

Black Friday September 24, 1869 (financial panic occurred when speculators tried to corner the gold market in the U.S.)

black gold petroleum

black gold of the Amazon and Malaya rubber

black gold of the Caspian caviar

Black Governor General of Haiti Toussaint l'Ouverture

Black Hand secret terrorist society linked with the Camorra and the Mafia

Black Heart of Montana Butte

Black Irish brunette descendants of Irish women and Spanish Armada sailors

blackjack card game, bubonic plague, the black flag of pirates, blackjack chewing gum, hand-held leather-covered flexible club; zinc blende or zinc sulfide

Black Jack General John J Pershing, USA who advocated the enlistment and promotion of black troops and led the 10th Cavalry. During World War I he was commander in chief of the American Expeditionary Force fighting in France

black lead cerrusite (lead carbonate)

Black Monday October 19, 1987 when the New York stockmarket declined more than on Tuesday, October 29, 1929 triggering the Great Depression

Black Monk Grigori Efimovich Rasputin

Black Movie Pioneer Ralph Cooper

Black Muslims religious-oriented group composed of black nationalists and some militant extremists

Black Nationalist Marcus M(oziah) Garvey

Black Panthers militant black party active in the United States and overseas

Black Pope head of the Jesuit Order—the Jesuit General

Black Prince Edward—Prince of Wales—son of Edward III who always wore black armor

Black Republic Haiti and many emerging African nations

Black Rock Columbia Broadcasting System (CBS) situated in the black granite building at 51 West 52nd Street in New York City

Black Russian Writer Aleksandr Sergeyvich Pushkin (1799–1837) whose maternal grandfather was African

Black Saturday Commander's Internal Management Review

Black Sea Countries Bulgaria, Romania, Turkey, USSR

Black Sheep of Canadian Liquors 100-proof Yukon Jack

Black Shirts Mussolini's followers

black spots black settlements in white-owned South African land

Black Stream Japan Current

black tar deadly Mexican-made heroin, also called Mexican mud or tootsie roll

Black Tuesday October 29, 1929

Black Watch Royal Highland Regiment whose tartans display dark colors

blackwater fever malaria

Blackwater State Nebraska

Blackwell's Island former name of Welfare Island in New York City's East River

black widow poisonous spider *Lactrodectus mactans*

Blarney Stone Port Cork or Corcaigh in Ireland (Eire) close to Blarney Castle containing the Blarney Stone

Blessed Isles *see* Fortunate Isles

Bligh's Islands Fiji Islands

Blind Bards Homer and Milton

Blind Poet John Milton

Blind Publisher Joseph Pulitzer

Blind Tom Thomas Bethune

Bloater(s) inhabitant(s) of Yarmouth on the North Sea coast of England where herrings are salted and smoked

Blockhousers oldest Afro-American regiment whose gallant assault on a well-defended blockhouse won them this nickname during the Spanish-American War

Blonde Beauty of the Lakes Milwaukee, Wisconsin

Blonde Bombshell Jean Harlow

Blondin Charles Emile Gravele —the tightrope walker who crossed Niagara Falls in the mid-nineteenth century

blood and guts red ensign of the British Merchant Marine

Blood and Guts General George S Patton, USA

Bloodhound British surface-to-air missile

Bloody Ground Kentucky

Bloody Mary Mary I of England (Mary Tudor); tomato juice and vodka

Bloomsbury writers whose center of activities was London's Bloomsbury Square in the early 1900s; Clive Bell, EM Forster, Roger Fry, John Maynard Keynes, Lytton Strachey, V Sackville-West, Leonard and Virginia Woolf;

synonym for snobbish aestheticism

Blue-backed Speller *The American Spelling Book* by Noah Webster

Bluebeard any wife killer such as the Chevalier Raoul whose seventh wife discovered the bodies of his six previous wives

bluebook volume containing names of socially-prominent people

Blue Grass Capital Lexington, Kentucky

Blue Grass Country Kentucky

Blue Grass State Kentucky

Blue Grotto marine cavern on shore of Capri island in Bay of Naples

Blue Hen Chickens Delawareans

Bluehen(s) Delawarean(s)

Blue Hen State Delaware

Blue Law State Connecticut

Blue Light District Cartagena, Colombia's red-light district (*Distrito Luz Azul*)

Bluenose fisherman or sailor from Canada's Maritime Provinces

Bluenose Province Nova Scotia

Bluenose(s) native(s) of Canada's Maritime Provinces, especially Nova Scotia; puritan(s)

blue ointment mercurial ointment

blue peter blue signal flag with a white rectangle in its center; flown when a ship is ready to sail; letter P or Papa in the international code

Blue Steel Hawker-Siddeley air-to-surface missile

Bluff City Hannibal, Missouri; Memphis, Tennessee; and Natchez, Mississippi

Bluff King Hal Henry VIII

Boadbil Abu Abdallah—last Moorish king of Granada

boat advice boat, agent's boat, airboat, albacore boat, assault boat, bait boat, banana boat, bass boat, bum boat, bunder boat, canal boat, cargo boat, catboat, cattle boat, clam boat, cockboat, cockleboat, cockleshell boat, crab boat, crash boat, custom's boat, diesel boat, dispatch boat, eagle boat, eight-oar boat, electric boat, excursion boat, faltboat, ferryboat, fireboat, fishing boat, flyboat, flying boat, foldboat, four-oar boat, freight boat, garbage boat, gas(oline) boat, gliding boat, grain boat, guardboat, gunboat, hag boat, harbor-patrol boat, herring boat, houseboat, iceboat, immigration boat, jigboat, johnboat, jollyboat, junk boat, kelpboat, liberty boat, lifeboat, line boat, lobster boat, longboat, love boat, mail boat, market boat, menhaden boat, mosquito boat, motorboat, nefboat, oil boat, ore boat, outboard motorboat, oyster boat, packet boat, paddleboat, paddlewheel boat, passenger boat, patrol boat, patrol torpedo boat, picket boat, pilot boat, pogey boat, police boat, powerboat, public-health boat, Q-boat, railroad-car ferryboat, river boat, river gunboat, rocketboat, rowboat, sailboat, sailing boat, sardine boat, seine boat, showboat, shrimp boat, small boat, snagboat, speedboat, sponge boat, swan boat, tow boat, toy boat, trawl boat, troll boat, tugboat, tunaboat, U-boat, viking boat, waterboat, welder's boat, whaleboat, work boat, xebec boat, yawlboat, zenana boat (seafarers reserve the term *boat* for a lifeboat, a small craft, or a submarine)

Bobby Jones Robert Tyre Jones

Bob Dylan Robert Zimmerman; Robert Zimmermann

Bob Hope Leslie Townes Hope

Boer (Afrikaans or Dutch—farmer)—South African of Dutch descent whose language is Afrikaans

Bogie Maxwell Bodenheim; Humphrey Bogart

Bogland Ireland

Boglander Irishman

bogsaat bunch of guys sitting around a table

Bogside Catholic workingclass district of Derry (Londonderry)

Bohemia Czechoslovakia; habitat of gypsies and other unconventional people

Bohemian Reformer John Hus

Bojangles Bill (Bojangles) Luther Robinson

Bolingbroke Henry IV of England

Bolivarian Block Colombia, Ecuador, Peru, Bolivia

Bolivarian Republics Bolivia,

Colombia, Ecuador, Peru, Venezuela

Bolshevik Feminist Aleksandra Kollontai

Bomba (Italian—Bass Drum)—Ferdinand II—King of the Two Sicilies

bombay bombay duck (Asiatic lizardfish, dried and salted lizardfish served with curry, also called bummalo)

Bonaire's Port Kralendijk

Bonanza Beech U-22 trainer aircraft

Bonanza Land Fort Smith, Arkansas

Bonanza State Montana

Bond Street London's street of fashionable shops

Bone Algerian port city now called Annaba

boneblack animal charcoal

Boney Napoleon Bonaparte

Bon Homme Richard (French—Good Man Richard)—Benjamin Franklin

Bonnie Prince Charles Charles Edward Stuart—the Young Pretender

Bonny Johnny John Adams—second President of the United States

boobtube television

Boogie Down New York City's borough of the Bronx

boomerang bomber advanced technology bomber; stealth bomber

Boomer State Oklahoma

Boonie Daniel W Russell

Booze Bourse Brooklyn, New York

Borax King Francis Marion Smith of Death Valley, California

Border Country the U.S.-Mexican border extending from Brownsville, Texas to San Diego, California

Border Eagle State Mississippi

Border Minstrel Sir Walter Scott

Border States Delaware, Maryland, Virginia, Kentucky, and Missouri; before the Civil War they divided the North from the South

Boris Karloff William Henry Pratt

Boris Pilnyak Boris Andreyevich Vogau

Boris Savinkov Vladimir Ropshin

Borneo Kalimantan

Borodin Mikhail Markovich Grusenberg

Borscht Belt Catskill Mountain area, New York

Bosphorus strait connecting the Black Sea with the Sea of Marmora leading to the Mediterranean

Boss of Bosses Lucky Luciano (Salvatore Lucania)

Boss Kett Charles F(ranklin) Kettering

Boss Tweed William Marcy Tweed

Boston Brahmin Historian William Hickling Prescott

Boston Strong Boy John L Sullivan

Botticelli Sandro di Botticelli—palette name of Alessandro Filipepi

Bourbon Street New Orleans nightlife center

Bowery north-south thoroughfare on Lower East Side of Manhattan, New York City; notorious for its shabby hotels and saloons

Bowie State Arkansas

Box 500 British secret service headquartered on Curzon Street, London

boxer's ear hematoma auris

boxitos people from Yucatan, Mexico where in Mayan the word means dark people

Boyhood Home of Mark Twain Hannibal, Missouri

Boy Orator of the Platte William Jennings Bryan

Boy's Town Omaha, Nebraska; redlight sections of many Mexican border towns

Boz Charles Dickens

Bozzy James Boswell—biographer and friend of Dr Samuel Johnson

bra burner militant feminist

bracelets slang for handcuffs

Bragman's Bluff British pirate's name for Puerto Cabezas, Nicaragua sometimes called El Bluff

Brahmsburg Hamburg, Germany—birthplace of Johannes Brahms

Brains of the Confederacy Judah Philip Benjamin (attorney general, secretary of war, and secretary of state of the Confederate States of America)

Bram Stoker Abraham Stoker

Brandy Nan Queen Anne

Brann the Iconoclast William Cowper Brann, editor and publisher of *The Iconoclast*

Brass Butte, Montana

Brassai Gyula Halàsz

Brass City Waterbury, Connecticut

Brattle Island, Galápagos Tortuga

Bratwurst Capital Sheboygan, Wisconsin

Brazilian Aviation Pioneer Alberto Santos-Dumont

Brazilian Comedian Chico Anisio

Brazilian Composer-Conductor Heitor Villa-Lobos or Carlos Gomes

Brazilian emerald green variety of tourmaline

Brazilian Film Pioneer Alberto Cavalcanti

Brazilian National Composer Heitor Villa-Lobos

Brazilian Pianist Guiomar Novais

Brazilian Ports Manaus, Belém do Pará, São Luis, Enseada de Mucuripe, Natal, Recife, Maceio, Salvador de Bahia, Ilheus, Vitoria, Rio de Janeiro, Niteroi, Angra dos Reis, Santos, Paranagua, São Francisco, Itajai, Florianopolis, Laguna, Rio Grande, Porto Alegre

Brazilian ruby topaz altered by heating so it turns purple-red to salmon-pink and hence passes for a ruby

Brazilian sapphire blue tourmaline

Braziller George Braziller

Brazil water coffee

Bread-and-Butter State Minnesota

Bread Basket Fargo, North Dakota

Breadbasket of Canada Saskatchewan

Breadbasket of Russia the Ukraine

Breadbasket of Sweden southernmost province of Skåne

Breadbasket of the World central North America (Canada and the U.S.)

Breakfast Food City Battle Creek, Michigan

Bret Harte Francis Brett Harte

Brett Halliday Davis Dresser

Brewer Soviet Yakovlev Yak-28 tactical bomber aircraft (NATO)

Brewery Capital Milwaukee

Brick Lane London, England's East End ghetto

Bride of the Adriatic Venice

Bride of the Sea Venice

Bridewell London's old house of correction

Bridge Bum Alan Sontag

Bridge House Wilmington, Delaware's detention home for juvenile delinquents

Bridge of Sighs 16th-century Venetian bridge connecting a prison with a palace where prisoners were tried; any similar structure connecting a courthouse with a prison

Brigette Bardot Camille Jarval

Bright's disease kidney disease named for its diagnostician Dr Richard Bright of London; nephritis

Brilliant Belgian virtuoso-violinist Arthur Grumiaux

Brilliant Madman Charles XII of Sweden

Brill's disease epidemic typhus disease recurring years after the original infection and named for its American diagnostician Dr Nathan E Brill

brimstone sulfur

Bringer of Freedom's Light John Huss

Britain's First Woman Prime Minister Margaret Thatcher

Britain's Most Exclusive Club the House of Commons

Britain of the South New Zealand

Britain's Playground Blackpool (seaside resort on Irish Sea)

Britain's Premier Passenger Port Southampton

britannia britannia metal (alloy of antimony, copper, and zinc used as antifriction material and for dinnerware)

Britannia Bristol-built military transport aircraft; Britannia metal; Britannia prima (England); Britannia secunda (Wales)—symbol of Great Britain including Scotland with England and Wales; British Empire; Commonwealth of Nations; Great Britain and Northern Ireland—the United Kingdom; Roman name for Great Britain

British America British possessions in or adjacent to the Americas

British Anatomist Extraordinary Henry Gray

British Century the 19th century

British Comedian Peter Ustinov

British Guiana Guyana

British Hardware Centre Birmingham

British Honduras Belize

British Isle Ports Whitstable, Port Victoria, Chatham, Tilbury Docks, Gravesend, Woolwich, Greenwich, London, Wivenhoe, Harwich, Parkeston, Ipswich, Felixstowe, Lowestoft, Great Yarmouth, King's Lynn, Boston, Grimsby, Immingham, Kingston-upon-Hull, Goole, Whitby, Middlesbrough, Hartlepool, Seaham, Sunderland, North Shields, Newcastle, Gateshead, Blyth, Leith, Granton, Rosyth Dock Yard, Boness, Grangemouth, Alloa, Burntisland, Kirkcaldy, Methil, Perth, Abroath, Montrose, Aberdeen, Peterhead, Fraserburgh, Hopeman, Inverness, Cromarty, Invergordon, Port Mahomack, Helmsdale, Wick, Thurso, Scrabster, Stornoway, Oban, Campbeltown, Greenock, Finnart, Rothesay Dock, Glasgow, Ardrossan, Irvine, Troon, Cairnryan, Douglas, Bangor, Belfast, Larne Lough, Londonderry, Sligo, Westport, Galway, Kilrush, Limerick, Foynes, Cobh, Cork Harbour, Rosslare, Dublin, Silloth, Mayrport, Whitehaven, Barrow-in-Furness, Fleetwood, Preston, Liverpool, Manchester, Port Dinorwic, Holyhead, Caernarvon, Milford Haven, Llanelly, Swansea, Port Talbot, Barry, Cardiff, Newport, Sharpness, Gloucester, Avonmouth, Bristol, Portishead, Bideford, St Ives, Penzance, Falmouth, Fowey, Plymouth, Dartmouth, Portland, Weymouth, Poole, Cowes, Yarmouth, Southampton, Gosport, Portsmouth, Folkestone, Dover

British Isles Great Britain and Ireland

British Laugh Master Benny Hill

British Librettist Italian nickname for Shakespeare, many of his dramatic plots inspired Italian composers

British lion symbol of the British Commonwealth as well as of Great Britain

British North Borneo Sabah

British Riviera England's south coast from Land's End to Margate

British Rock Gibraltar
British West Indies island possessions or former possessions of Great Britain in or near the Caribbean: Bahamas; British Leeward, Virgin, and Windward islands; Jamaica, Tobago, Trinidad
Britons English, Scottish, and Welsh people
Brixton Brixton Prison southeast of Plymouth, England; London, England slum; one of London's largest prisons
Broad-bottomed Administration of Prime Minister Henry Pelham during the reign of George II
Broken Hill Kabwe, Zambia
Bronco North American-Rockwell OV-10 counterinsurgency aircraft
Bronze Age era of mankind when implements and weapons were forged from bronze; period marked by wars and widespread violence
Bronzino Agnolo di Cosimo
Brook Farm utopian community founded by leading American transcendentalists in 1841 but disbanded by 1847 near West Roxbury, now Boston, Massachusetts; nickname of similar ventures
Brook Farmers Albert Brisbane, Orestes Brownson, Charles A Dana, John S Dwight, Ralph Waldo Emerson, Margaret Fuller, Horace Greeley, Nathaniel Hawthorn, Isaac Hecker, George Ripley, Henry David Thoreau
Brother John John Bull—long the personification of the British Empire
Brother Jonathan British nickname for the United States and its citizens
The Brothers Rockefeller brothers—John D III, Nelson, Laurance, David
Brothers Goncourt Edmund and Jules de Goncourt—literary collaborators
Brothers Grimm Jacob and Wilhelm Grimm
Brown Bomber Joe Lewis
brown coal lignite
brown lung byssinosis or cotton-dust disease
Brown Shirts Hitler's followers
Brownstone State Connecticut
Brow of the Universe Scandinavia

BR Town Baton Rouge, Louisiana
Bruce Graeme Graham Montague Jeffries
Brummagen Birmingham (colloquial)
Bruno Jakob Bronowski
Bruno Walter Bruno Walter Schlesinger
brussels brussels carpet (woven with a raised pattern by a method first used in Brussels); brussels griffon (toy dog bred in Brussels); brussels lace (high-quality floral-pattern lace); brussels sprouts (small cabbage-like vegetable)
Brussels system universal decimal classification
B Traven Berick Traven Torscan Croves
bubbly champagne
Buccaneer Hawker-Sidddeley jet aircraft for military applications
Bückeburg Bach Johann Christoph Friedrich Bach
Buckeye North American-Rockwell trainer aircraft designated T-2
Buckeye(s) Ohioan(s)
Buckeye State Ohio
Buckie R(ichard) Buckminster Fuller
Buddy Rogers Charles Rogers
Buerger's disease chronic inflammation of the blood vessels in a limb or limbs
Buffalo De Haviland military transport aircraft
Buffalo Bill Colonel William F Cody
Buffalonians people of Buffalo, New York
Buffalo Plains State Colorado
Bughouse Square Pershing Square in Los Angeles, Union Square in New York, Washington Square in Chicago
Bugs bootlegger-burglar George Moran (1893–1957)
Bugs Baer Arthur Baer
Buhl's disease fatty degeneration associated with hemoglobinuria
Built on Oil, Soil, and Toil Ponca City, Oklahoma
Bukavu Costermansville
Bulgarian Ports Michurin, Akhtopol, Bukhta Tsiganski, Burgas, Varna, Evksinograd, Kavarna
Bulgarian Quadrilateral fortress towns of Rustchuk, Schumla, Silistria, and Varna
Bulge The Bulge of Brazil con-

sisting of South America's easternmost coast between João Pessoa and Recife
Bulldog Scottish Aviation single-engine two-place trainer
Bullion State Missouri
bull market stock market short form indicating an upward trend in securities
Bull Moose Theodore Roosevelt—twenty-sixth President of the United States
Bullpup Maxson air-to-surface missile
Bullring of Basra white-slave market close to Iraq's Persian Gulf coast
Bull Run Manassas
Bullwood Hall borstal in Essex, England
Bung Karno (Malay—Brother Karno)—Indonesian dictator Sukarno
bunsen bunsen burner (named for the German chemist Robert Wilhelm Bunsen)
Buntline Ned Buntline—nom de plume of Edward ZC Judson
Buranello Baldassare Galuppi
The Burg New York City
Burkina Faso (More—Ancestral Home of the Dignified)—Upper Volta on the southern edge of the Sahara
Burlington Route Chicago, Burlington and Quincy (railroad)
Burl Ives Icle Ivanhoe Ives
Burma former name of Myanmar
Burma's Principal Port Rangoon
burmese burmese cat (orange-eye cat originating in Burma); burmese lacquer (grayish varnish); burmese ruby (peony)
Burmese Ports Sittwe, Kyaukpyu, Bassein, Rangoon, Moulmein
Burt L Standish Gilbert Patten—creator of the American boy hero—Frank Merriwell
business-man's special 45-minute psychosis produced by dimethltryptamine
Busta Sir Alexander Bustamante
Bustees Calcutta slum
Buster Keaton Joseph Francis Keaton
Butch Fiorello H La Guardia
Butcher of the Balkans Andrija Artukovic
Butcher of Berlin Adolf Hitler
Butcher of Budapest Soviet

Prime Minister Nikita Khrushchev so named because of his brutal suppression of the Hungarian freedom fighters

Butcher of the Caribbean Rafael Trujillo

Butcher of Lyon Klaus Barbie

Butcher of Prague Reinhard Heydrich—Hitler's Reichsprotektor of Bohemia

Butter Capital Owatonna, Minnesota

Buzzard State Georgia

Byron pen name of George Gordon who used the family title of Lord Byron

Byron Janis Byron Yanks

Bytown original name of Ottawa, Canada

Byzantium (Latin—Istanbul)—Constantinople

CABAL members of the secret cabinet of Charles II of England; by coincidence their initials spelled cabal—Clifford of Chudleigh, Ashley (Lord Shaftesbury), Buckingham (George Villiers), Arlington (Henry Bennet), Lauderdale (John Maitland)

Cabbage Patch Victoria, Australia

Cabbage Town Toronto, Ontario slum

Cabo Tormentoso (Portuguese—Cape of Storms)—Cape of Good Hope

Cacos (Spanish—pickpockets, poltroons)—nickname of a Guatemalan political party

Cactus code name for French Mach 1.2 surface-to-air missile

Cactus Blossom Capital Phoenix, Arizona

Cactus Jack Vice President John Nance Garner also called the Sage of Uvalde

Cactus State New Mexico

Caesar Augusta (Latin—Zaragoza)—called Saragossa by the British

Caffarelli Gaetano Majorano

Cagliostro Giuseppe Balsamo

Cajetan Tommaso de Vio—Italian cardinal who failed to persuade Martin Luther to remain within the Catholic Church and carry on reforms from within

Cajun a descendant of French-speaking immigrants from Acadia; their dialect

Cajun Country Bayou Teche, Louisiana, also called Teche

Country

Calamity Jane Martha Jane Burke also known as Canary Jane whose activities resulted in the death of eleven of her twelve husbands

Calhoun Gulch Charleston, South Carolina district

California Channels Channel Islands south of Santa Barbara (Anacapa, San Miguel, Santa Cruz, Santa Rosa)

California Ports San Diego, Long Beach, Los Angeles, San Pedro, Monterey, San Francisco, Alameda, Oakland, Port Richmond, Mare Island, Port Chicago, Stockton, Sacremento, Eureka

California prayerbook deck of cards

California Riviera oceanside resorts ranging from San Diego to Santa Barbara

California's Cornerstone San Diego

Californicators California fornicators

Caligula Gaius Caesar

Calle de la Ballesta (Spanish—Street of the Crossbow)—Madrid's nightclub neighborhood

Calle Florida (Spanish—Florida Street)—celebrated shopping center of Buenos Aires

Calpe (Phoenician—Rock of Gibraltar)—one of the Pillars of Hercules flanking Straits of Gibraltar

Calve Emma Calvé—operahouse name of the soprano Emma de Roquer

Calvin John Calvin (originally Jean Chauvin)

Calypso Capital Port-of-Spain, Trinidad

Cambodia formerly Preah Reach Ana Chak Kampuchea or simply Kampuchea, the Khmer Republic, or Democratic Kampuchea

Cambodian-Incursion President Richard Nixon

Cambodian Ports Kampong Saom, Kampot, Phumi Phsar Ream

Cambridge Group Ralph Waldo Emerson, Oliver Wendell Holmes, Henry Wadsworth Longfellow, James Russell Lowell, John Greenleaf Whittier

Camel NATO name for Soviet Tu-104 transport aircraft

Camel-driver of Mecca the Prophet Mohammed

Camellia Capital Sacramento, California

Camellia City Greenville, Alabama

Cameroon Ports Tiko, Douala, Victoria, Kribi

Camille Erlanger Fréderic Regnal

Camille Pissarro Jacob Pizarro

camisole(s) straitjacket(s)—institutional euphemism

Campagna di Roma (Italian—Roman Campagna)—undulating lowlands around Rome

Campanella Tommaso Campanello (originally Giovanni Somenico)

Campo Alegre (Papiamento or Spanish—Happy Country)—Curaçao's controlled brothel above the hills of Willemstad

campo santo (Italian, Portuguese, Spanish—sacred ground)—name of many burial places throughout the Latin world

Canada second largest nation; Canada anemone, balsam, barberry, birch, blueberry, bluegrass, buffalo berry, field pea, fleabane, ginger, goose, hare, hemp, jay, lily, lymegrass, lynx, mayflower, mint, moonseed, pea, pitch, plum, potato, rockrose, root, tea, turpentine, violet, warbler, wormwood, yew

Canada's Breadbasket Saskatchewan

Canada's Doorstep Nova Scotia

Canada's Heartland The Province of Manitoba

Canada's Ocean Playground Nova Scotia

Canada's Principal Ports Halifax and Montreal on the Atlantic, Vancouver on the Pacific

Canada's Storied Province Québec

Canada's Wonder City Toronto

Canadian Canadian bacon (boned pork loin strips); Canadian cheddar (usually smoother and spicier than American cheddar cheese); Canadian football (rouge); Canadian-French; Canadian humorist (Stephen Leacock); Canadianism; Canadianize; Canadian whiskey (rye); prefix given such items as Canadian bacon, football, French (inhabitant or language), goldenrod, goose, hemlock, holly, lynx, whiskey

Canadian Arctic area north of the Arctic Circle including islands, the North Magnetic Pole, and the Northwest Territories

Canadian Atheist Marshall J Gauvin

Canadian black Canadian-grown marijuana

Canadian Comedians Lou Jacobi; Rich Little

Canadian Commonwealth Canada and its territories

Canadian Composer-Conductor-Educator Sir Ernest Campbell Macmillan

Canadian Dominion Canada and its territories

Canadian Freethinker Marshall J Gauvin

Canadian Gateway to the Pacific British Columbia

Canadian Humorist Stephen B(utler) Leacock

Canadian Kaleidoscope The Province of Ontario

Canadian Ports (east coast and Great Lakes) Churchill, Cartwright, Saint Anthony, Roddickton, Springdale, Baie Verte, Fortune Harbour, Botwood, Catalina, Clarenville, Harbour Grace, Wabana, St John's, Argentia, Burin, St Pierre, Grand Bank, St George's, Corner Brook, Humbermouth, Sept-Iles, Baie-Comeau, Rimouski, Tadoussac, Port Alfred, Chicoutimi, Riviere du Loup, Québec, Trois Rivieres, Sorel, Varennes, Montreal, Ottawa, Lower Lakes Terminal, Prescott, Belleville, Trenton, Cobourg, Port Hope, Oshawa, Port Whitby, Toronto, Hamilton, Port Weller, Welland, Port Colborne, Port Maitland, Rondeau Harbor, Amhertsburg, Windsor, Sarnia, Port Edward, Goderich, Owen Sound, Collingwood, Midland, Parry Sound, Little Current, Sault Ste Marie, Thunder Bay, Gaspé, Chandler, Paspébiac, Dalhousie, Bathurst, Caraquet, Chatham, Newcastle, Souris, Georgetown, Charlottetown, Summerside, Pictou, North Sydney, Sydney, Halifax, Lunenburg, Liverpool Shel-

burne Yarmouth, Digby, Parrsboro, Moncton, St John, Letang Harbor; (west coast) New Westminster, Vancouver, Horseshoe Bay, Powell River, Comox, Nanaimo, Victoria, Esquimalt, Port Alice, Ocean Falls, Prince Rupert

canadian potato artichoke

Canadian Twin Cities Fort William and Port Arthur

Canal Concessionaire Vicomte Ferdinand Marie de Lesseps —original promoter of the Suez and the Panama Canal

Canaletto Antonio Canale; Giovanni Antonio Canal

Canal of Fire Suez Canal

canalimony $25-million-dollar alimony United States paid Colombia in 1922 for separating its province of Panama in 1903 so it could proceed unhindered in constructing the Panama Canal

Canal Zone Capital Balboa Heights

Canberra British twin-jet light bomber built by BAC

Candia Erakleion

Canecutters sugar-cane-cutting Queensland, Australians

Cane Sugar States Florida, Hawaii, and Louisiana

Canned Salmon Capital Ketchikan, Alaska

Cannery City Seattle, Washington

Cannery Row Monterey, California's fishery foreshore described by John Steinbeck in his novel

Cannon City Kannapolis, North Carolina where Cannon towels are made

Cannon King Alfred Krupp —German armament manufacturer (1812–1887)

Cannon Kings of Germany the Krupp family

Canoe City Old Town, Maine

Cantabrian Surge Bay of Biscay

Cantinflas Mario Moreno

Canuck French-Canadian; two-place jet interceptor built in Canada by Avro and designated CF-100

Canuckland Canada

Cao Tú Phan Van Khoai

The Cape Cape Ann, Cape Cod, Cape of Good Hope, Cape Hatteras, Cape Horn, Cape Province (Union of South Africa); Cape Town

Cape of the Californias Cabo San Lucas (at the lower tip of Baja California)

Cape Cod turkey codfish

Cape Dutch Afrikaans

Cape Horner(s) deep-sea sailing vessel(s) rounding Cape Horn in southernmost South America

Cape Horn fever imaginary malady of sailors who complain they are ill when the sea is rough or there is too much work

Cape Kennedy Cape Canaveral, Florida

Cape Stiff Cape Horn

Cape of Storms Cape of Good Hope

Cape Verde Island Ports Mindelo, Santa Maria, Preguiça, Praia

Capirucha Mexican-Americanism for Mexico City

Capital of Black America Harlem section of New York City

Capital of the Cotswolds Cirencester

Capital of the Highlands Inverness, Scotland

Capital of Hope Brasilia

Capital of the Incan Empire Cuzco, Peru

Capital of the Pirate Coast Ras Al-Khaimak (northernmost sheikdom of Trucial Oman at the Strait of Ormuz)

Capital of Polynesia Auckland, New Zealand

Capital Province Ontario

Capital of the Rhineland Cologne (Köln), Germany

Capital of the World New York City—seat of the United Nations

Capstone of Negro Education Howard University

Captain Kidd William Kidd—privateer-pirate

Captain Sir Richard Captain Sir Richard Francis Burton

Captive of History Northern Ireland

Capucine Germaine Lefebvre

Caran d'Ache Emmanuel Poiré

Carat City diamond-mining Kimberly, South Africa

Caravaggio, Michelangelo da Michelangelo Merisio

Caravaggio, Polidoro da Polidoro Caldara

carbolic acid phenol

Carbonaro Napoleon III believed to belong to the Carboneria political society

Cargo Port of the Pacific Vancouver, British Columbia

Caribou De Havilland twin-engine stol transport designated DHC-4 in Canada and C-7A in the United States where it is also called CV-2A

Carinthia Kärnten, Austria

Carioca(s) native(s) of Rio de Janeiro

Carla De Sola Carla De Sola Eaton (liturgical dancer)

Carl Brandes Edvard Cohen

Carleton Kendrake Erle Stanley Gardner

Carl Milles Vilhelm Carl Emil Anderson

Carlo Collodi Carlo Lorenzini

Carlo Maria Carlo Maria Giulini

Carlos Arruza Carlos Ruiz Camino

Carmella Ponselle Carmella Ponzillo

Carmen Miranda Maria de Cormo Cunha

Carmen Silva Elisabeth Queen of Romania

Carnegie's Savior violinist Isaac Stern who saved Carnegie Hall from oblivion

Carol Carnac Edith Caroline Rivett

Carolina Game Cock Thomas Sumter

Carolina racehorse razorback hog

Carol Lombard Jane Peters

Carolus Magnus Charlemagne

Caronia La Coruña (northwestern Spain in Roman times)

Carpet City Amsterdam, New York

Carrion's disease Peruvian-sandfly anemia

Carry Nation Carry Amelia Moore Nation

Carter Curtain Tortilla Curtain

The Carthaginian Lion General Hannibal

Cartoonist Humorist Al Capp

Cary Grant Archibald Leach

Casanova Giacomo Girolamo

Casa Pacifica former President Nixon's Spanish-colonial seaside home at San Clemente, California

Casa Rosada (Spanish—Pink House)—Argentine president's office building in Buenos Aires

Casbah Algiers, Algeria's hillside redlight and underworld district

Casey Jones John Luther Jones

Casey Stengel Charles Dillon Stengel

Casino City Monte Carlo, Monaco

castile castile soap (mild cleaning agent originally made in Castile, Spain from olive oil and sodium hydroxide)

Cast-Iron Commodore Matthew Calbraith Perry

Catalan Cellist-Composer-Conductor Pablo (Pau) Casals

Catch-Me-Who-Can Richard Trevithick's railway engine tested in 1808

cat gold mica; yellowish mica

Cathay China

Cathedral of Learning University of Pittsburgh's 52-story building

Cathedral of Music Carnegie Hall, New York City

cat's eye chrysoberyl

catsie cat's-eye playing marble; polished agate resembling a cat's eye

Cattle Capital Willcox, Arizona

Caudillo de la Independencia de Uruguay (Spanish—Chief of the Independence of Uruguay)—José Gervasio Artigas

cauliflower ear hematoma auris

Cavalier State Virginia

caviar of drugs cocaine

Cayenne French Guiana

Cayuse Hughes OH-6 observation helicopter

C Day Lewis Nicholas Blake

Cedar Crest executive mansion of the governor of Kansas and eponym for executive government throughout the state

Cee Cee Claudia Cardinale

Celery Capital Kalamazoo, Michigan; Sanford, Florida; San Ysidro, California

Celery City Kalamazoo, Michigan

Celestial City John Bunyan's name for Heaven in *Pilgrim's Progress;* old traveller's name for Peking, China

Celestial Empire Chinese Empire

Celine Louis-Ferdinand Destouches

Celtic Fringe peoples of Cornwall, Ireland, Scotland, and Wales on the fringe of England

Celts Bretons, Cornish, Gaels, Irish Gaelics, Manx, Scots Gaelics

Cement City Allentown, Pennsylvania

Censor of the Age Thomas Carlyle

Centennial State Colorado

Center of Austria Salzburg

Center of the Copper Circle Tucson, Arizona

Center of the Nation Topeka, Kansas

Center of Scenic America Utah

Center of the Sunshine State Pierre, South Dakota

Central African Empire formerly the Central African Republic created from the Ubangi Shari territory of French Equatorial Africa

Central America land between Colombia and Mexico—Belize, Costa Rica, El Salvador, Guatemala, Honduras, Nicaragua, and Panamá

Central American States Belize, Guatemala, Honduras, El Salvador, Nicaragua, Costa Rica, Panamá

Central Bureau Amsterdam's old section for illegal & illicit activities

Central Powers Austria-Hungary; Bulgaria; Germany; and Turkey (in World War I)

Central Prairie Province Saskatchewan

Central Provinces Ontario and Québec

Central State Kansas

Centurion British tank carrying a crew of 4 and guns up to 105mm

Century of Confusion the 9th century when the empire of Charlemagne disintegrated—the 800s

Century of the Exodus the 13th century before the Christian era when Moses lead the Israelites out of Egypt and across the Red Sea—the 1200s

Ceramic City East Liverpool, Ohio

Cereal City Battle Creek, Michigan, and Cedar Rapids, Iowa

Cesspool of Crime London or Paris around the turn of the century; New York today

Cesspool of Latin America Cayenne, French Guiana

Cesspool of Pirates John F Kennedy International Airport in New York where cargo thefts are the highest in the nation

Ceylon Sri Lanka

Chagas-Cruz disease South American sleeping sickness

Chairman Mao Mao Tse-Tung

Champagne of Drugs cocaine

Champion of Copernican Cosmology Italian philosopher Giordano Bruno

Champion of Darwin Thomas Henry Huxley

Champion of Education Horace Mann

Champion of Freethought President John F Kennedy who believed in an America where the separation of church and state is absolute and who renounced the appointment of an American ambassador to the Vatican and tax support of parochial schools

Champion of Liberty English reformer Charles Bradlaugh

Champion of the Old South President John Tyler

Champions of Individualism John Stuart Mill and Herbert Spencer

Champion of States Rights John C Calhoun—U.S. Senator from South Carolina

Champion of the Underdog Clarence Darrow

Chance Personified Fortuna (Roman); Tyche (Greek)

The Channel Beagle, English, St George's

Channel City Santa Barbara, California

Channel fever the sense of excitement evident aboard ships approaching their port

Channel Islands Jersey, Guernsey, Alderney, Brechau, Great Sark, Little Sark, Herm, Jethou, Lihou

Chapino(s) Guatemalan(s)

Chappiequack Chappaqua, New York

Charcot-Marie-Tooth disease muscular atrophy

Charger Convair multipurpose short takeoff-and-landing airplane

Charioteer British World-War-II medium tank armed with an 83.4mm gun

Charles Atlas Angelo Siciliano

Charles the Bald Charles I of France

Charles B Child C Vernon Frost

Charles Blondin Jean François Gravelet (French acrobat who walked tightrope above Nia-

gara River near Niagara Falls in 1855, 1859, 1860)

Charles Bronson Charles Buchinski

Charles Dalmorès Henry Alphonse Boin

Charles de Secondat Baron de la Brède et Montesquieu

Charles the Fat Charles II of France

Charles Island, Galápagos Floreana or Santa Maria

Charles J Kenney Erle Stanley Gardner

Charles the Simple Charles III of France

Charley slang for Vietcong

Charley Car St Charles Avenue trolleycar; one of America's oldest and the last in New Orleans

Charley South Charleston, South Carolina

Charley West Charleston, West Virginia

Charlie Charles; letter C radio code; NATO name for Soviet C-class submarines built to launch missiles underwater

Charlie Chaplin Charles Spencer Chaplin

Charlot (Spanish—Charlie)— Charlie Chaplin

Charm Spot of the Deep South Mobile, Alabama

Charter Oak City Hartford, Connecticut, where the original charter was hidden in an oak tree to insure the liberty of the first settlers

Charter Oak State Connecticut

Chatham Island, Galápagos San Cristóbal

Checkpoint Charlie former international frontier between East and West Berlin

Cheesebox convict's nickname for the Illinois penitentiary at Statesville

Chelmer native of Chelm (ancient Jewish town in Poland known in folklore as the Town of Fools)

Chelmo Polish name for concentration camp called Kulmhof by the Germans

Chelsea Gang John Dos Passos, Suzanne La Follette, Sinclair Lewis, Mary McCarthy, Ben Stolberg, who were among the first to expose the totalitarian nature of the USSR; lived in the Chelsea Hotel, New York City

Chemical Capital Wilmington, Delaware

Chemnitz Karl-Marx-Stadt

Chequers British prime minister's country home

Chero(s) Salvadoran(s)—person or people of El Salvador, Central America

Chester Conklin Jules Cowles

Cheyenne Mountain air-defense complex near Pike's Peak, Colorado

Chicago Group poets and writers born in the Chicago area around 1900—Sherwood Anderson, Willa Cather, Floyd Dell, John Dos Passos, Theodore Dreiser, Finley Peter Dunne, James T Farrell, Francis Hackett, Harry Hansen, Ernest Hemingway, Vachel Lindsay, Archibald MacLeish, Edgar Lee Masters, Harriett Monroe, Frank Norris, Burton Rascoe, Carl Sandburg, Kay Boyle, TS Eliot, Scott Fitzgerald, Sinclair Lewis, Carl and Mark Van Doren, Frank Norris, Sherwood Anderson, Willa Cather and Carl Sandburg

Chicago piano submachine gun

Chi-chi naval nickname for Christchurch, South Island, New Zealand

Chickasaw Sikorsky transport helicopter designated H-19 or UH-19

Chico Marx Leonard Marx

The Chief Herbert Hoover; train on Chicago-Los Angeles run of Santa Fe

Chieftain British main battle tank armed with a 120mm gun

Child of the Mississippi Louisiana

Children of Joseph Israelites

Children of Pharoah Egyptians

Chilean Ports San Juan Bautista, Arica, Iquique, Tocopilla, Antofagasta, Taltal, Valparaiso, San Antonio, Talcahuano, Coronel, Lota, Valdivia, Puerto Montt, Puerto Quellon, Punta Arenas

Chile's Principal Port Valparaiso

Chimneyville Jackson, Mississippi

china clay kaolin (hydrous aluminum silicate)

China's Main Street *Yangtze Kiang* (Chinese—Yangtze

River)
China's Sorrow soil-eroding Yellow River
Chinatown Chinese quarter of any city outside China
china white synthetic heroin (often deadly)
China white fentanyl (more powerful than morphine, also nicknamed Persian white)
chinese chinese banana (dwarf banana), chinese cabbage *(petsai)*, chinese checkers, chinese gelatin (agar or isinglass), chinese glue (alcohol + shellac), chinese greens (vegetables), chinese ink (india ink), chinese puzzle, chinese red, chinese watermelon (wax gourd), chinese white (barium sulfate), chinese wood oil (tung oil)
chinese anesthesia acupuncture
Chinese cows soybeans
Chinese Gordon British general Charles George Gordon who suppressed the Taiping rebels; later named Gordon Pasha for similar services in the Sudan
Chinese Mainland Ports Macao, Huang-Pu, Kuang-Chou, Hong Kong (British Crown Colony), Shant-T Ou, Hsia-Men, Lo-Hsing-Ta, Mao-Ti, Ning-Po, Shanghai, Chen Chiang, Nan-Ching, Wu-Hu, Chiu-Chiang, Hang-kou, Chang-Sha, Ching-Tao, Wei-Hai, Yen-Tai, Ta-Ku, Tieng-Ching, Chin-Huang-Tao, Hu-Lu-Tao, Ying-K-Ou, Lu-Shun, Luta (Dairen)
Chinese Offshore Ports on Formosa or Taiwan Chilung (Keelung), Kaohsiung, Su-Ao, Hua-Lien, Tso-Ying, An-Ping, Tan-Shui
chinese white zinc oxide (ZnO)
Chinese white fentanyl
Chino Men California (correctional) Institution for Men at Chino
Chinook Boeing-Vertol twin-rotor helicopter designated CH-47
Chinook State Washington
Chipmunk Hawker-Siddeley trainer aircraft
Chitlin Capital of the World Salley, South Carolina
chloride of lime bleaching powder
Chocolate City Hershey, Pennsylvania
Chocolate Coast Ghana

Choctaw Sikorsky troop-transport helicopter designated H-34
Cholera Capital Calcutta
Cholly Knickerbocker Igor Cassini
Chomolungma (Tibetan— Mount Everest)
Choo-Choo Town Chattanooga, Tennessee
Chosen Apostle self-proclaimed radio preacher Herbert W Armstrong
Chotzie Samuel Chotzinoff
Christiania Oslo's medieval name
Christianna Brand Mary Christianna Milne Lewis
Christian Right American Coalition for Traditional Values (includes the Moral Majority)
Christmas Cove South Bristol, Maine
Christopher Columbus Cristóbal Colón (Spanish); Cristoforo Colombo (Italian)
Chromium Continent Africa
Chuco Mexican-Americanism for El Paso, Texas
Chuey (Spanish-American nickname—Jesus)
Cicero Marcus Tullius
Cigar Capital Key West, Florida; Tampa, Florida
Cigar City Tampa, Florida
Cigarette Josiah Flynt Willard
Cincinnati oysters pigs' feet
cinnamon stone hessonite
Cinquecento (Italian—five hundred)—Italy's 16th century artistic and cultural development
Cipango Japan, as it was called by early European explorers
Circus King John Ringling
Cisco Kid Duncan Reynaldo (Renault Renaldo Duncan)
Cissie Patterson Eleanor Medill Patterson
Cissy Loftus Mary Cecilia M'Carthy
The Cit The Citadel Military College of South Carolina
Citadel of Budapest Burgberg
Citians people of Minneapolis and St Paul also called Twin Citians
Cities of the Plain Admah, Gomorrah, Sodom, and Zeboim on the Jordan River Plain of ancient Israel near the Dead Sea
Citizen Capet Louis XVI (beheaded during French Revolution)

Citizen Composer Dmitri Shostakovich
Citizen of Geneva Jean-Jacques Rousseau
Citizen King Louis Philippe of France
Citizen Louis Capet Louis XVI
Citizen of the World Oliver Goldsmith, Thomas Paine
Citlatepetl Orizaba
citric acid $C_8H_6O_7$
Citrus Metropolis Los Angeles
The City financial, governmental, historical, and commercial core of London; including newspaper publishing district, Bank of England, Lloyd's, many famous restaurants
City of 1000 Lakes Oklahoma City, Oklahoma
City of Abraham Hebron, Israel
City of Alexander the Great Alexandria, Egypt
City of Aluminum Arvida, Québec
City of the Angel San Angelo, Texas
City of Angels Bangkok, Los Angeles
City of the Apprentice Boys Londonderry, Northern Ireland
City of the Arctic Tromso, Norway
City of the Arts Minneapolis
City of Athena Athens
City of Baked Beans Boston, Massachusetts
City by the Bay San Francisco
City of Beaches Montevideo, Uruguay
City of Beautiful Spires Copenhagen
City of Bells Strasbourg, France
City of Berwald Stockholm, Sweden
City Beside the Broad Missouri Bismarck, North Dakota
City Between Bridges Stockholm
City of Bicycles Copenhagen
City of Big Shoulders Carl Sandburg's sobriquet for Chicago
City of Birches Umeå, Sweden
City of Birm Sym Orch City of Birmingham (England) Symphony Orchestra
City of Black Diamonds Scranton, Pennsylvania
City of Blazing Lights Shanghai

City of the Blues Memphis, Tennessee (home of WC Handy)

City of Brotherly Love Philadelphia (derived from the Greek) *philos* (love) and *adelphos* (brother)

City of the Caliphs Cairo

City of the Camellias Pensacola, Florida

City of Canals and Bridges Amsterdam, Copenhagen, Leningrad, Stockholm, and Venice

City of the Carmel Haifa, Israel, on the slopes of Mount Carmel

City of Castles Copenhagen

City of Cats Venice (felines outnumber people)

City of Certainties Des Moines, Iowa

City of Champions Pittsburgh

City of Cheese Dutch cities of Alkmaar and Gouda

City of Cheese, Chairs, Children, and Churches Sheboygan, Wisconsin

City of Churches Brooklyn, New York

City College British euphemistic name for Newgate Gaol —the old London lockup; New Yorker nickname for The Tombs prison, in Manhattan

City of Coral coral reefs off the Virgin Islands

City of Corsairs St Malo, France

City of Cypresses Rome

City of David Jerusalem

City of Death Beirut, Lebanon; Kipling's name for Lahore, Pakistan

City of Destiny Tacoma, Washington

City of the Doges Venice

City of Dreadful Night Kipling's nickname for Calcutta

City of Dreaming Spires Oxford, England

City of the Dunes Dunkerque, France

City Ed City Editor

City of Elms New Haven

City of Eternal Spring Caracas

City of Fair Breezes Buenos Aires, Argentina

City of Five Seasons Cedar Rapids, Iowa

City of Flamboyants and Jacarandas Salisbury, Rhodesia

City of Flour Buffalo, Minneapolis

City of Fountains Aix-en-Provence, France; Bratislava, Czechoslovakia

City of Four Lakes Madison, Wisconsin

City of Fun and Frolic Atlantic City, New Jersey

City of Gardens Lahore, Pakistan; Victoria, British Columbia

City of Gardens and Beaches Adelaide, Australia

City of the Gods Teotihuacán —religious capital of Mexico in the fifth century before the Christian Era

City of Gold Dawson, Yukon Territory

City by the Golden Gate San Francisco

City of the Golden Horn Istanbul

City of Good Neighbors Arlington Heights, Illinois

City of Green Spires Copenhagen

City of Grieg Bergen, Norway

City Grown Too Big For Its Bridges San Francisco

City of Hans Christian Andersen Copenhagen, Denmark

City of Heat Thermopolis, Wyoming

City of Historical Charm Savannah

City of a Hundred Hills San Francisco

City of a Hundred Spires Prague

City of a Hundred Towers Pavia, Italy

City of Illicit Love Paphos on Cyprus

City of the Immortals Amarapura, Burma

City of Jade Oaxaca, México

City of Jazz and Mardi Gras New Orleans, Louisiana

City of Kielland and Bjelland Stavanger, Norway

City of Kings and Commoners Copán, Honduras

City of Lakes Dartmouth, Nova Scotia

City of Light Paris, France; Perth, Western Australia

City of Lillies Florence, Italy

City by the Lion's Gate Vancouver, British Columbia

City of Lost Angels Los Angeles, California

City of Louis Paris

City of Magnificent Distances Washington, D.C.

City of Manifold Advantages Augusta, Maine

City of Mankind Jerusalem

City of Masts Port of London

City of Millionaires Colorado Springs

City of Minarets Miknès, Morocco

City of Money Zurich, Switzerland

City of Monuments Baltimore, Maryland; Florence, Italy

City of Mosques Istanbul, Turkey

City in Motion San Diego, California

City of Mozart Salzburg, Austria

City of Music and Song Vienna

City of Nielsen Copenhagen, Denmark

City of Nine Dragons Kowloon, Hong Kong

City of Notions Boston, Massachusetts

City of Oaks Raleigh, North Carolina

City of One Hundred Hills San Francisco, California

City on the Neva Leningrad

City on the Water Amsterdam, Copenhagen, Stockholm, and Venice

City of Palaces Rome, Italy, and Vatican City

City of Palms Acajutla, El Salvador; Fort Myers, Florida; and Maracaibo, Venezuela

City of the Pampas Buenos Aires

City of Peace Brunei

City of Penn Philadelphia, Pennsylvania founded by William Penn

City of Personality Cincinnati, Ohio

City of Peter the Great Saint Petersburg (later called Petrograd and Leningrad)

City of the Plains Christchurch, New Zealand

City of Poets Jérémie, Haiti, birthplace of the father of Alexandre Dumas *(Dumas pére)* and grandfather of Alexandre Dumas *(Dumas fils)*

City of Power Peking, People's Republic of China

City of Presidents Quincy, Massachusetts

City of the Prophet Medina, Saudi Arabia, where Mohammed was protected after fleeing from Mecca

City of Quays and Grieg Bergen, Norway

City of Razzle Dazzle one of O Henry's nicknames for New York City he also called Bagdad on the Subway

City of Receptions Washington, D.C.

City by the Rivers Kansas City where the Kansas River flows into the Missouri; Minneapolis—St Paul on the Minnesota and Mississippi; Philadelphia on the Delaware and Schuylkill; Pittsburgh on the Allegheny and Monongahela; Portland, Oregon on the Columbia and the Willamette; St Louis where the Mississippi meets the Missouri

City of Rocks Nashville, Tennessee

City of Roses Portland, Oregon

City of Ruins and Roses Visby, Sweden

City of Rumors Washington, D.C.

City of Rum and Sugar Georgetown, Guyana

City of Saints Montreal

City of Salt Salzburg, Austria; Syracuse, New York

City by the Sea Newport, Rhode Island

City of the Sea Venice

City of Seven Hills Rome, Italy, built on Aventine, Caelian, Capitoline, Esquiline, Palatine, Quirinal, and Viminal

City of Seventy Isles Venice

City of Shoes Brockton, Massachusetts

City of Sibelius Helsinki, Finland

City of Silver Taxco, México

City of the Silver Gate San Diego

City of Sinbad Basra, Iraq

City of Sinding Oslo, Norway

City in the Sky Macchu Picchu, Peru

City of Skyscrapers New York

City of the Slain Arlington National Cemetery in Arlington, Virginia

City of Smells Old Delhi, India

City of Smokestacks Everett, Washington

City of Soles Lynn, Massachusetts

City of Sorrow Buchenwald (concentration camp near Weimar, Germany)

City of Southern Charm Savannah, Georgia

City of Spies Beirut, Berlin, Copenhagen, Hong Kong, London, New York, Paris, Singapore, Stockholm, Vienna, Zurich, Washington, D.C.

City of Spires Copenhagen

City State Singapore and the Vatican City State

City of Steel Pittsburgh, Pennsylvania

City of St Mark Venice

City of St Michael Dumfries, Scotland

City of St Mungo Glasgow, Scotland

City of the Straits Detroit, Michigan, on the Straits of Belle Isle

City of Suds Milwaukee

City of the Sun Baalbec, Heliopolis, and Rhodes; Campanella's utopian republic

City of Sunshine Colorado Springs, Colorado; Los Angeles, California; Tucson, Arizona

City of Surprises Amsterdam

City of Symphonies London

City of Tamales San Antonio, Texas

City of Temples Benares, India; Katmandu, Nepal

City of Ten Million Roosters Port-au-Prince, Hatti

City That Boeing Built Seattle

City That Care Forgot New Orleans

City That Knows How San Francisco

City That Swims on the Water Stockholm

City of the Thousand and One Nights Baghdad, Iraq

City of Three Capitols Little Rock, Arkansas

City of the Three Kings Cologne, Germany, where it is reputed the Magi or Three Kings are buried; Lima, Peru

City of Totems Ketchikan, Alaska

City of Trees Christchurch, New Zealand; Saratoga Springs, New York

City of the Tribes Galway, Ireland

City under Vesuvius Naples

City of the Violet Crown Athens

City of Washington Washington, D.C.

City of Walls Beijing China

City of Witches Salem, Massachusetts

City without Clocks Las Vegas, Nevada

Ciudad Blanca (Spanish— White City)—Guayaquil, Ecuador's

cemetery; (Spanish—White City)—Merida, Yucatan

Ciudad de los Reyes (Spanish —City of the Kings)—Lima, Peru

Ciudad Imperial y Coronado (Spanish—Imperial and Crowned City)—Toledo, Spain

Ciudad Loca (Spanish—Crazy City)—Barranquilla, Colombia during carnival

Civil Rights Leader Martin Luther King, Jr

Civil Service President Ulysses Simpson Grant

Civil Service Reform President Chester Alan Arthur

Civil War Photographer Matthew Brady

Clamcatcher(s) New Jerseyite(s)

Clamgrabber(s) Washingtonian(s)

Clam States New Jersey and Washington

Clam Town Norwalk, Connecticut

clap gonorrhea

clap joint brothel

Clara Covell North Clara E Ellis

Claribel Charlotte Alington-Barnard

Clarin Leopoldo Alas y Urena

Clarita Clara Elena

Clark U.S. air base, Luzon Island, Philippines

Clark Gable William Gable

Classic City Kyoto, Honshu Island, Japan

Classifier and Compiler Extraordinaire Dr Peter Mark Roget

Claude Lorraine Claude Gellée of Lorraine

Claudette Colbert Lily Cauchoin

Claudio Lars Carmen Brannon de Samayoa

Cleat Soviet Tupolev Tu-124 long-range transport (NATO)

Clifford Ashdown pseudonym shared by R Austin Freeman and John James Pitcairn

Clifton Webb Webb Parmalee Hollenbeck

Clinton's Big Ditch Erie Canal advocated by Governor De Witt Clinton of New York

Clio Joseph Addison's pseudonym; in Greek mythology the muse of history or of lyre playing

Clod Soviet Antonov 6-seat piston-powered transport plane

(NATO)

Cloud Piercer New Zealand's Mount Cook (12,349 feet or 3,764 meters)

Clown Prince of Music Danny Kaye

Clowns of the Canine World dachshunds

Clubland Pall Mall clubhouse section of London

Coach Soviet Ilyushin transport plane I1-12 (NATO)

Coal City Pottsville, Pennsylvania

Coaley Samuel Coleridge-Taylor

Coal Metropolis Cardiff, Wales

Coal State Pennsylvania

Coatzacoalcos formerly Puerto Mexico

Cobbler Poet Hans Sachs of Nuremberg, also known as Prince of the Meistersingers

Cobh Gaelic name for Queenstown

Cobra Bolkow wire-guided anti-tank missile made in Germany

Cocaine Capital Bogotá, Colombia and Jackson Heights, Queens, New York

Cocaine Capital of Colombia Medellín

Cochise Beech T-42 transport aircraft

Cockade City Petersburg, Virginia

Cockney Poet John Keats

Cockpit of the American Revolution New Jersey

Cockpit of Europe Belgium

Cockpit of the Middle East Syria

Coco Chanel Gabrielle Bonheur Chanel

Codder(s) Cape Cod resident(s)

Code Hammurabi Babylonian civil and legal code established in the 20th century before the Christian era

Code Napoléon French civil code established in 1804

Codfishland Newfoundland

Coke City Uniontown, Pennsylvania

Col Lucius Iunius Moderatus Columella (Roman writer on agriculture)

Colette Sidonie Gabrielle Claudine de Jouvenal

Collar City Troy, New York

College of New Jersey Princeton University's original name

College of Rhode Island Brown University's original name

Collodi Carlo Lorenzini

cologne cologne brown (vandyke brown); cologne spirits (highly-concentrated ethyl alcohol); cologneware (mottled brown and gray stoneware); cologne water (eau de cologne, toilet water); cologne yellow (chrome-yellow and lead-sulfate pigment)

Colombian Colombian-grown marijuana; Colombian mountain-grown coffee

Colombian Composer-Conductor Guillermo Uribe Holguín

Colombian connection network of brokers, farmers, politicians, and smugglers connected with the export of Colombian-grown cocaleaf products and marijuana to Canada and the United States

Colombian First Colombia's best-known classical composer—Guillermo Uribe Holguin

Colombian Ports Santa Marta, Barranquilla, Cartagena, Covenas, Buenaventura, and Ríohacha

Colombia's Principal Port Barranquilla

Colón formerly Aspinwall

Colonels natives of Kentucky

Colonne Edouard Judá Colonne, French conductor

Colossus of the African Continent the Sudan

Colossus of the Eurasian Continent the USSR

Colossus of Independence John Adams

Colossus of the Indian and Pacific Oceans Australia

Colossus of the North United States of America

Colossus of the North American Continent Canada

Colossus of the South American Continent Brazil

Colquhoun Calhoun

Colt Colt revolver (invented by Samuel Colt of Hartford, Connecticut); Soviet Antonov 14-passenger biplane designated AN-2 (NATO)

Columbia America; South Carolina's capital; the United States

Columbia first space shuttle launched from Cape Canaveral, Florida in April 1981

Columbia City Vancouver,

Washington

Columbia the Gem of the Ocean United States of America

Columbus Columbus Day (October 12); Cristóbal Colón (Spanish); Cristoforo Colombo (Italian); name of places in some twelve states in the United States; Ohio's capital

Columbus of the Cosmos Russian cosmonaut Yuri Alexeyevich Gagarin

Columbus of the Subconscious Sigmund Freud

Comedian Pianist Victor Borge

Comedian's Comedian Bert Williams

Comenius John Amos Komensky

Comet British medium tank built during World War II; De Haviland four-engine jet transport aircraft

Comic Symbol of the Century Charlie Chaplin

Commando C-46 Curtiss-Wright 36-passenger transport built during World War II; Cadillac-Gage amphibious armed car and military personnel carrier (XM-706); Dodge-built military personnel carrier built during World War II

Commercial Capital of Canada Toronto

Commoner The Commoner—William Jennings Bryan

Communism Peak Garmo or Stalin Peak (highest in the USSR; name subject to change with the politicians)

Communist Capitalist Friedrich Engels

Communist East countries dominated by China

Communist West occidental countries dominated by the Soviet Union

Community of True Inspiration Amana, Iowa

Comoro Island Ports Moroni, Patsy, Mutsamudu, Fomboni

Comoros Republic of the Comoros (island nation in the Indian Ocean northwest of Madagascar), *Etat Comorien*

The Company the CIA

Compton Mackenzie Edward Montagu Compton

Conca D'oro (Italian—Shell of Gold)—nickname of the hills encircling Palermo

Concha Maria de la Concepción

Conch(s) inhabitant(s) of Key West, Florida and nearby keys

Conch Town Key West, Florida

Concordance Cruden Alexander Cruden—compiler of the *Complete Concordance of the Holy Scriptures* published in 1737

Concorde Anglo-French supersonic airplane attaining cruising speeds of 1300 miles per hour

Concord Group Bronson Alcott, Ralph Waldo Emerson, Margaret Fuller, Nathaniel Hawthorn, Henry David Thoreau

Concordski nickname for the Soviet supersonic Tu-144 airplane (first civilian aircraft to break the sound barrier)

Concrete Jungle modern metropolitan center

Condemned Rock Tasmanian nickname for Grummet Island, Macquarie Harbour where there is a penitentiary

Condor North American-Rockwell air-to-surface missile (AGM-53A)

Condorcet Marie Jean Antoine Nicolas de Caritat, Marquis de Condorcet

Coney Island butter mustard

Confederate Raider Rear Admiral Raphael Semmes, CSN

Confederation Province Prince Edward Island

Confetti Canyon lower Broadway, from Bowling Green to City Hall Park, New York City where during parades confetti, and waste paper are thrown from upper-floor windows

Confucius Kung Fu-tse

Congo Ports Loango, Pointe Noire, Malongo Oil Terminal

Connecticut Ports New London, New Haven, Bridgeport

Connection City Amsterdam, Marseilles, Miami, Singapore, and major airports where drugs are smuggled

Connie Mack Cornelius McGillicuddy

Conquering Lion of Judah and King of Kings Emperor Haile Selassie of Ethiopia

Conqueror of Disease Louis Pasteur

Conqueror of Mount Everest Sir Edmund Hillary

Conqueror of Suez Ferdinand

de Lesseps

Conquerors of Yellow Fever Walter Reed and his colleagues Aristides Agramonte, James Carroll, and Jesse Lazear

Conrad Veidt Konrad Weidt

Conscience of America Norman Thomas

Conscience of American Music Gunther Schuller

Conscience of the American Theater Brooks Atkinson

Conscience of Europe Voltaire

Conscience of the Left George Orwell

Conservatory of Europe Czechoslovakia

Constable Country East Berghott in England's Sussex where James Constable's award-winning landscapes were painted

Constable of France Charles De Gaulle

Constantia Judith Sargent Murray

Constantinople Istanbul

Constant Reader Dorothy Parker

Constellation Lockheed 63-passenger transport

Constitution State Connecticut

Constructionist President James Buchanan

The Consulate France under the First Consul-Napoleon Bonaparte—1799–1804

Contadora Group Colombia, Mexico, Panama, Venezuela

Contemporary Cassandra Dorothy Thompson

Continent The Continent (usually Europe; also Africa, Antarctica, Asia, Australia, North America, South America)

Continental Divide Rocky Mountain ridge separating rivers flowing eastward to Atlantic Ocean and the Gulf of Mexico from those flowing westward to the Pacific

Continental Maritime World the Great Lakes (Erie, Huron, Michigan, Ontario, Superior)

Continental Nation Australia

Continent of Hope South America

Convair 600 Convair-Liner powered by Rolls-Royce turbo-prop engines

Cook Islands Danger, Manahiki, Penrhyn or Tongareva, Rakahanga and nearby islets in the South Pacific

Cookpot Soviet Tupolev Tu-

124 jet-transport aircraft (NATO)

Coon Dog Capital Vienna Illinois

Coot Soviet Ilyushin transport designated Moskva or Il-18 (NATO)

Copa de Oro (Spanish—Cup of Gold)—pirate's nickname for Panama

Copeia periodical of the American Society of Ichthyologists and Herpetologists named for the naturalist Edward Drinker Cope

copenhagen copenhagen blue (gray blue); copenhagen snuff; copenhagen surprise (naval attack without warning of a fleet at anchor in the manner of Nelson's sortie in 1801)

Copernicus Latinized name of Polish astronomer Nikolaus Kopernicki

Copper City Butte, Montana

Copper John Auburn Prison near Syracuse, New York

Coppernose Henry the VIII whose portrait exhibited a copper-colored nose on the coins minted during his reign

copper pyrites chalcopyrite (copper iron sulfide)

Copper State Arizona

Coral Atoll Country Nauru

Coral Coast Fiji's hotel complex between Sigatoka and Yanuca

Cora Montgomery Jane McManus

cordovan cordovan leather (originally made in Córdoba, Spain)

Corn Belt Illinois, Indiana, Iowa, and Nebraska

Corn City Toledo, Ohio

Corncob Capital Washington, Missouri

Corncracker(s) Kentuckian(s)

Corncracker State Kentucky

Corn Geneticist Barbara McClintock

Cornhusker(s) Nebraskan(s)

Cornhusker State Nebraska

Cornish Riviera English Riviera extending from Falmouth to the Isles of Scilly

Corn-Law Rhymer Ebenezer Elliott

Corno di Bassetto (Italian—basset horn)—pen name used by George Bernard Shaw when he was a music critic

Cornopolis Chicago

Corn State Illinois, Iowa

Cornubian Shore Cornwall, England

Corporal John John Churchill who became the first Duke of Marlborough; known to the Spaniards as Mambrú

Corregio Antonio Allegri

Corridor of Six Continents sobriquet given the Panama Canal, Suez Canal, and projected interoceanic sea-level canals across Mexico and Nicaragua

Corridor State New Jersey

Corrie Denison Eric Partridge

Corruption-Hating President Grover Cleveland

Corsair Chance-Vought single-engine fighter popular during World War II (F4U)

Corsican Napoleon Bonaparte who was born on Corsica

Corsican Ogre Napoleon

Corvette antisubmarine-warfare convoy escort ship

Cory Corazon Aquino

cosa nostra (Italian—our thing) —nickname for international criminal syndicate network

Cosimo palette name of Piero di Lorenzo who took the given name of his teacher Cosimo Roselli

Cosmological Popularizer Carl Sagan

Cosmopolis of the Heartland Kansas City

Cosmopolitan Canadian-built medium-range transport designed by General Dynamics and designated CC-109

Costa de la Muerte (Spanish— death coast) Spain's north-westerly stormbound coast

Costa Geriatrica (Latin—elderly coast) southeast coast of England frequented by many older people

Costa Rican Ports Limón; Puntarenas and Golfito

Costermansville former name of Bukavu

Cotton Belt cotton-growing areas of the southern United States; also known as the Cotton Kingdom

Cotton Bowl Dallas, Texas

cotton-dust disease brown lung or byssinosis

Cottonopolis Manchester, England

Cotton State Alabama

Cottonwood City Leavenworth, Kansas

Cougar Grumman carrier-based transonic fighter aircraft (F9F-

6)

Count Basie William Basie

Country of the Blacks Sudan (Arabic—black)

Country of a Thousand Hills Rwanda

Courland Kurland

Court of St James British royal court

Cousin Jack a Cornishman; a Cornish miner

Cousin Jenny Cornish girl or woman

Covent Garden The Royal Opera House in London adjacent to Covent Garden marketplace

Cowboy Artist Charles M Russell

Cowboy Capital Dodge City, Kansas

Cowboy Philosopher Will Rogers

cowboys of the sea porpoises

Cowboy State Wyoming

cow(s) cow juice; cowboy(s); cowchip(s); cowhand(s); cowpaper(s); cowpoke(s); cowpuncher(s); cowthief(s) or cowthieves; cowtown(s)

Cowtown Fort Worth, Texas; Kansas City, Kansas and Missouri; Omaha, Nebraska

Coyote Cowboy Pecos Bill

Coyote(s) South Dakotan(s)

Coyote State South Dakota

Crab Capital Crisfield, Maryland

Crabtown Annapolis, Maryland

crack low-cost smokable cocaine

Crackers rural Floridians and Georgians

Cracker State Georgia

Cradle of American Culture and Music Carnegie Hall

Cradle of American Independence Independence Hall, Philadelphia

Cradle of the American Revolution Faneuil Hall, Boston

Cradle of Aviation San Diego

Cradle of California San Diego

Cradle of Canadian Confederation Prince Edward Island

Cradle of Civilization Armenia, China, Egypt, Greece, India, Iran, Iraq, Israel, Italy, Jordan, Lebanon, Mexico, Peru, Syria, and Turkey

Cradle of Classical Civilization Greece

Cradle of the Confederacy capitol building—Montgom-

ery, Alabama

Cradle of Democracy ancient Greece

Cradle of Electrical Engineering Berlin, Germany where the first electric railroad and first large-scale power station were built

Cradle of the French Revolution Marseille and Paris

Cradle of Human Civilization Iraq

Cradle of Islam Saudi Arabia

Cradle of Japanese Art Nara, Honshu Island, Japan

Cradle of Japanese Civilization Kyoto

Cradle of Liberty Carpenters' Hall, Philadelphia; Faneuil Hall, Boston; House of Burgesses, Williamsburg, Virginia; Holland during formation of the Dutch Republic; Switzerland in William Tell's time

Cradle of Nuclear Research Los Alamos, New Mexico

Cradle of Oriental Civilization China

Cradle of Polynesia Samoa Islands

Cradle of Psychoanalysis Berlin; Vienna

Cradle of the Renaissance Florence, Italy

Cradle of the Revolution Boston, Massachusetts; Paris, France; Petrograd, Russia

Cradle of the Russian Revolution Petrograd

Cradle of Secession Charleston, South Carolina

Cradle of Texas Liberty The Alamo in San Antonio

Cradle of the Union Albany, New York where in 1754 Benjamin Franklin presented his Plan of Union to the Albany Congress

Cradle of Violent Crime the United States

Crate NATO nickname for Soviet Ilyushin transport Il-14

Crawfish Town New Orleans, Louisiana

Crawthumper(s) Marylander(s)

Crazy Alley San Quentin Prison's insane asylum

Creative Genius of American Architecture Frank Lloyd Wright

Creator of Chords Franz Liszt

Creator of the Female Language for Sexuality Anaïs Nin

Creator of French Existentialism Jean-Paul Sartre

Creator God *Viracocha* (Quechua—supreme god)—deity venerated in Incan and pre-Incan times

Creator of Modern Democracy Thomas Paine

Creator of Musical Laughter Rossini

Crébillon Prosper Jolyot

Creole Country southern counties of Alabama and Mississippi as well as coastal parishes of Louisiana where many people are of French or Spanish origin

Creole State Louisiana

Crescent City Appleton, Wisconsin; New Orleans, Louisiana

Crestwood Heights Toronto, Ontario's Forest Hill Village

Cretinsbury-on-Cesspool Tasmanian nickname for Launceston

Crime Syndicate Chief Meyer Lansky

The Crocodile W Averell Harriman

Cromwell's Curse Ireland (also called the Curse of Cromwell)

Cronian Sea Arctic Ocean

Cross of Geneva emblem of the Red Cross (red cross on a white field) used to show the neutrality of ambulances, hospitals, and hospital ships during wartime

Crossroads of Africa, Asia, and Europe Egypt

Crossroads of Africa and Europe Spain

Crossroads of Asia Afghanistan

Crossroads of Europe Belgium

Crossroads of the Pacific Oahu—the Aloha Islands, Hawaii

Crossroads of the Seven Seas Singapore

Crossroads of the South Pacific Fiji

Crossroads of the World Panama Canal; Straits of Gibraltar; Suez Canal

Crotale Thompson surface-to-air guided missile made in France

Croves Hal Croves (pseudonym of B Traven—nom de plume of Berick Traven Torsvan Croves)

Crow Eaters South Australians

Crown City Coronado, California

Crown City of the Valley Pasadena, California

Crown Jewel of the Adriatic Venice

Crown Prince of Cellists Lynn Harrell

Crown Prince of Keynesism John Kenneth Galbraith

Crown Prince of Psychoanalysis Carl Gustav Jung

Crystal City Corning, New York

Crystal Hills New Hampshire's White Mountains

Cub NATO nickname for the Soviet Antonov 100-passenger cargo plane

cuban cuban heel (broad-based heel used on women's shoes)

cubanite copper iron sulfide

Cuban Ports Bahia Honda, Cabañas, Mariel, La Habana (Havana), Matanzas, Cardenas, La Isabela, Caibarien, Nuevitas, Puerto Padre; Santiago de Cuba, Manzanillo, Cienfuegos

Cuba's Principal Port Havana

Cultural Capital of Canada Montreal

Cultural Capitals of the World Berlin, London, Moscow, New York, Paris, Vienna

Cultured Pearl Archipelago Japan

Cultured Pearl of the Orient Hong Kong

Cumberland River City Nashville, Tennessee

Curaçao's Ports Willemstad, Bullen Baai, Caracas Baai, New Port

Curator of Culture Lord Kenneth M(ackenzie) Clark; Peter Ustinov

Curmudgeon Philosopher Ambrose Bierce

Currer Bell pseudonym of Charlotte Brontë

Curse of Balboa Panamanian-style dysentery marked by acute nausea and retching

Curse of Cortez Mexican-acquired diarrhea or dysentery also called Montezuma's Revenge

Curse of Cromwell Ireland (also called Cromwell's Curse)

Curt Jurgens Curd Jurgens

Curzio Malaparte pseudonym—Curzio Suckert

Cuspidor of Europe France's nickname given it by Alexander Herzen

Crown City Coronado, California

Cuvier Georges Léopold Chrétien Frédéric Dagobert

Cyclone Coast Australia's northwest coast

Cyclone State Kansas

Cyclops of the Kremlin Josef Stalin

Cyclorama City Atlanta, Georgia with its cyclorama painting of the Battle of Atlanta

Cynic and Skeptic Par Excellence George Bernard Shaw

Cyprian Ports (counterclockwise north to south coast) Kyrenia, Xeros, Paphos, Limasol, Larnaca, Famagusta

Cypriot Apostle Barnabas—Cyprus-born companion of Paul and Mark, according to the New Testament

Czechoslovakian Capital Prague

Czechoslovakian National Composer Anton Dvořák

Czechoslovakian Operetta Composer Rudolf Friml

Czech Reformers Jan Hus, Alexander Dubcek, Vaclav Havel

Da Costa's syndrome soldier's heart

Dago Diego Garcia in the Indian Ocean (navalese for San Diego, California)

Dago Garcia navalese slang—Diego Garcia (Indian Ocean naval base)

Dahlia Dalila

Daisy Ashford Margaret Mary Ashford

Dakota Douglas DC-3 21-passenger transport also called Skytrain

Dalmatia western Yugoslavia

dalmatian dalmatian dog (black-spotted white coach dog); dalmatian cherry (marasca); dalmatian insect powder (pyrethrum dust)

Dame Cicely Dr Cicely Saunders

Dame Clara Dame Clara Butt

Dame Joan Dame Joan Sutherland

Dame Margot Fonteyn Margot Hookham

Dame Myra Dame Myra Hess

Dame Ngaio Dame Ngaio Marsh

Damnable Place Hong Kong, according to GBS

Dam on the Amstel Amsterdam

damp Spain Atlantic coasts of Spain, especially along Bay of

Biscay where rains are heaviest

Dan Beard Daniel Carter Beard

Dandy King Joachim Murat—King of Naples

Daniel Nikolai Arzhak

Daniel Stern pseudonym of Liszt's paramour, the Countess Marie d'Agoult

danish danish pastry (light pastry often filled with stewed fruit, crushed nuts, cup custard, and raisins)

Danish Agnostic Søren Kierkegaard

Danish Capital of the United States Racine, Wisconsin

Danish Caribees colonial name for what are now the U.S. Virgin Islands

Danish King of the Waltz Hans Christian Lumbye

Danish Ports Ronne *(on Bornholm),* Køge, Københavnhavne (Copenhagen), Helsingor, Frederiksvaerk, Frederikssund, Roskilde, Holbaek, Nykøbing, Kalundborg, Korsor, Skaelskor, Naestved, Vordingvorg, Stubbekøbing, Stege, Masnedsund, Nykøbing, Falster, Sakskøbing, Naskov, Rudkøbing, Marstal, Svendborg, Nyborg, Kerteminde, Odense, Middelfart, Assens, Faborg, Grasten, Sonderborg, Augustenborg, Abenra, Haderslev, Kolding, Frederkicia, Horsens, Arhus, Grena, Randers, Alborg, Frederikshavn, Skagen, Esbjerg

Danish Waltz King Hans Christian Lumbye

Danish West Indies former name of the American Virgin Islands

d'Annunzio (Gabriel) Gaetano Rapagnetta

Danny Kaye David Daniel Kominski

Danny O'Neill fictionalized name of James T Farrell *(Studs Lonigan)*

Danny Thomas Amos Jacobs

Dante Dante (Durante) Alighieri

Danube Delta Land Romania

Danube Empire Austro-Hungarian Empire

Danubian Monarchy Austro-Hungarian Empire

Dark and Bloody Ground Kentucky

Dark Continent Africa

Darkest Africa Lake Victoria's shores (19th century)

darwin glass queenstownite (silica glass)

Darwin Island, Galápagos Culpepper

Darwin's Archipelago Galápagos Islands

Darwin's Bulldog nickname of Professor Thomas Henry Huxley, president of the Royal Society, whose defense of Darwin's *Origin of the Species* pulverized the arguments of the Bishop of Oxford, Samuel (Soapy Sam) Wilberforce, who had set out to demolish Darwin's theory of evolution

Dash 8 Boeing twin-engine turboprop airplane

Date Capital Indio, California

Daughter of the Baltic Helsinki

Daughter of the Dream Emma Goldman

David Frome Zenith Jones Brown

David St John E Howard Hunt

David Wayne Wayne McKeekan

Davy Jones' Locker traditional resting place of all who are buried at sea or who are drowned in the depths of the ocean

Dawn on the Mesabi Aurora, Minnesota

Day of Infamy December 7, 1941 (when Japanese aircraft carriers attacked Pearl Harbor, Hawaii while diplomatic negotiations were in progress in Washington, D.C.)

Dazai Osamu Tsushima Shuji

Deadeye Dick Nat Love, a black cowboy of the last century who was noted for his superior marksmanship

deadly nightshade belladonna

Deadman's Cove geographic placename and nickname of the San Diego Police Department headquarters

dead President slang for American paper money

Deadwood Dick Richard W Clarke—English-born South Dakota frontier pioneer

Deaf Composer Ludwig van Beethoven

Deaf Smith Erastus (Deaf) Smith, Texan patriot-soldier

Dean of American Astronomers Henry Norris Russell

Dean of American Choral Conductors Robert Shaw

Dean of American Psychiatry

Dr Karl Menninger

Dean of Classical Guitarists Andrés Segovia

Dean of Classical Music Comedians Victor Borge

Dean of the English School Sir Edward Elgar

Dean of the French School Camille Saint-Saëns

Dean of Italian Music Gioacchini Rossini

Dean Martin Dino Crocetti

Dean of Modern American Composers Aaron Copland

Dean of Russian Music Nikolai Rimsky-Korsakov

Dear Abby Abigail Van Buren

Death Devil Charles Manson

Death Ride Charge of the Light Brigade at Balaclava in Crimea

death's head nickname of the deadly mushroom *Amanita muscaria*

Death Valley Scottie Walter Scott

Deborah Kerr Deborah Kerr-Trimmer

Debt-reduction President Calvin Coolidge

Decade of Disillusionment the 1970s

Dee Cee Washington, D.C.

Deep North Queensland, Australia

Deep South South Carolina, Georgia, Florida, Alabama, Mississippi, Louisiana, and Texas; the conservative south coast of England

Defaced City once elegant New York City where so many buildings, buses, and subways are defaced by graffiti

Defender of the Damned Clarence Darrow

Defender of Darwin Thomas Henry Huxley

Defender of Democratic Socialism John Dewey in the United States or Harold Laski in Great Britain

Defender of Freethought Thomas Jefferson, Thomas Paine, Robert Ingersoll, and Clarence Darrow

Defender of Religious Freedom Supreme Court Justice Hugo Lafayette Black

Dejerine's disease infants' interstitial neuritis

Delaware Port Wilmington

delft delft blue (characteristic of a popular china developed in the Dutch city of Delft); delft china; delftware

Delhi belly traveller's diarrhea picked up in India

Delta Dagger Convair F-102 single-engine turbojet interceptor aircraft

Delta Dart Convair F-106 supersonic-interceptor aircraft

Demetia South Wales

Democratic Kampuchea formerly Kampuchea, the Khmer Republic, or Cambodia

Demon of Deception Beelzebub

Demon of Disease Black Death (bubonic plague); hunger plague; murine plague (carried by rats); pneumonic plague; septicemic plague; sylvatic plague (carried by many species of rodents)

Demon of Misfortune and Ruin the Sphinx

Denali old Russian name for Mt McKinley also called Bol'shaya

Den Haag (Dutch—The Hague)—capital of the Netherlands and seat of the International Court of Justice

Denmark's Principal Port Copenhagen

Dental Capital of Europe Vaduz, Liechtenstein where artificial teeth are made

Dentist-Novelist Zane Grey

Den Vita Staden (Finnish—The White City)—Helsinki

Department of Construction Georgia nickname for its Department of Correction

Department-of-Labor President William Howard Taft

Depression-born Cartoonist Al Capp—creator of Li'l Abner

Der Alte *Der Alte Fritz* (German—Old Fritz)—Frederick the Great

Der alte Steffl Old Saint Stephen's Cathedral in Vienna; begun in the 12th century

Der Blaue Reiter (German—The Blue Rider)—abstract expressionism movement centered in Munich

Derbyville Louisville (home of the Kentucky Derby)

Dercum's disease subcutaneous connective-tissue dystrophy

Der Führer (German—The Leader)—Adolf Hitler

Der Meister (German—The Master)—Johann Wolfgang von Goethe

Derrick City Oil City, Pennsylvania

Descendants of Eagles the founders of Algeria, according to tradition

Desert Salt Lake City, Utah

Desert Arabia Arabian Desert in the northern Arabian Peninsula

Desert Fox Field Marshal Erwin Rommel

Desert of Ice Antarctica

Desert and Prairie Painter Georgia O'Keefe

desert roses barytes or gypsum concretions whose shapes resemble roses

Desi Arnaz Desiderio Alberto Arnaz y de Acha

Devil of Cultured Vice Mephistopheles

Devil's Chaplain Robert Taylor (1784–1844), English cleric imprisoned for blasphemy when he exposed the universality of all religious beliefs

Devil's Half Acre Augusta, Maine's slum

Devil's Island nickname for French Guiana penal colony in use up to 1950 and the Isle off its coast where Alfred Dreyfus was imprisoned from 1894 to 1899

Devil's Rock Garden California's Death Valley also called Devil's Bathtub; Devil's Parade Ground; Devil's Pulpit; Devil's Speedway and Golf Course

devil's testicle mandrake's nickname (also called mandragora or satan's apple)

Devil's Triangle (*see* Bermuda Triangle)

devil's trumpet nickname for jimson weed also called devil's apple or devil's weed

De Witt Clinton's Ditch the Erie Canal

diamondback moth; rattlesnake; terrapin

Diamond Continent Africa

Diamond Head 760-foot-high extinct crater forming cape and marking entrance to Honolulu on Oahu, Hawaii

Diamond Jim James Buchanan (Diamond Jim) Brady

Diamond Lil Mae West

Diamond State Delaware

Diamond Street New York City's 47th Street between 5th and 6th avenues

Diazpotism despotism of Por-

firio Diaz during his forty years as president of Mexico

Dice City Las Vegas, Nevada

Dick Donavan Joyce Emmerson Preston Muddock

Dicky Sam(s) inhabitant(s) of Liverpool

Dictator of Nicaragua William Walker

Dictionary Johnson Dr Sam(uel) Johnson

Diedrich Knickerbocker Washington Irving

Dief the Chief John George Diefenbaker

diesel diesel engine, diesel fuel, diesel locomotive, diesel oil (all named for the German automotive engineer, Rudolf Diesel)

Digger Land Australia

Dino Dean (Crocetti) Martin

Diplomat John Franklin Carter

Dirceu Tomaz Antonio Gonzaga

Dirk Bogarde Dirk van den Bogaerd

Discoverer of Bacteria, Blood, and Parasite Cells Anton van Leeuwenhoek

Discoverer of Capillary Blood Vessels Marcello Malpighi

Discoverer of Malaria Parasite Charles Laveran

Discoverer of Yellow Fever Parasite Hideyo Noguchi

Discoverers of How Enzymes Change Starch into Sugar Carl Ferdinand Cori and Gerty Theresa Cori

Disease of the Century AIDS; Alzheimer's disease

dismal science Carlyle's nickname for economics

Dismal Swamp City Norfolk, Virginia

Dissident Publisher Henry Regnery

Distinguished Journalism Fellow Ralph de Toledano (1989)

divi divide; dividend

Divine Poet John Donne

divine Sarah Oscar Wilde's nickname for Sarah Bernhardt, who began life as Rosine Bernard

Division No 1 Chicago's Cook County Jail

Division No 2 Chicago's House of Correction

Divorce Capital of America Reno, Nevada

Dixie southern United States; the South

Dixiecrat Southern Democrat

Dizzy Benjamin Disraeli—British Prime Minister

Dizzy Dean Jay Hanner (Dizzy) Dean

Dizzy Gillespie John Birks (Dizzy) Gillespie

Djajapura Kotabaru, formerly called Hollandia by the Dutch when they controlled western New Guinea

Djakarta Indonesian city once called Batavia

Django Jean (Django) Reinhardt

Djibouti Ports Obock, Djibouti

Doctor Angelicus (Latin—Angelic Doctor)—Thomas Aquinas also known as the *Princeps Scholasticorum* (Prince of Scholastics)

Doctor Charlie Dr Charles Horace Mayo—co-founder of the Mayo Clinic (*see* Doctor Will)

Doctor Donne John Donne

Doctor Evangelicus (Latin—Evangelical Doctor)—religious reformer John Wickliffe

Doctor Holmes John Haynes Holmes (contemporary American Universalist minister)

Doctor of the Industrial Revolution Dr Erasmus Darwin

Doctor Irrefragabilis Alexander of Hales

Doctor Jameson Sir Leander Starr Jameson

Doctor Johnson Doctor Samuel Johnson—critic, conversationalist, lexicographer

Doctor Livingston David Livingstone

Doctor Mirabilis (Latin—Admirable Doctor)—English savant Roger Bacon

Doctor of Revolution Erasmus Darwin

Doctor Rizal José Rizal (intellectual leader of Philippine insurrection against Spanish rule)

Doctor Seuss author-cartoonist Theodore S Geisel

Doctor Singularis (Latin—Singular Doctor)—William Occam

Doctor Subtilis (Latin—Subtle Doctor)—Duns Scotus

Doctor Universalis Albertus Magnus

Doctor Watson Dr John B Watson, M.D. of London; companion of Sherlock Holmes of 221-B Baker Street in the literary creations of Sir Arthur Conan-Doyle

Doctor Will Dr William James Mayo—co-founder with his brother Charles of the Mayo Foundation for Medical Education and Research at Rochester, Minnesota

Documentary Photographer Alfred Stieglitz

Dodo Islands Mascarene Islands of Mauritius, Réunion, and Rodgrigues formerly inhabited by dodo birds

Dogwood City Atlanta, Georgia

Dollar Mark Mark Hanna

Dolley (Dolly) Mrs Dorothea (Dolley) Payne Madison (wife of President James Madison); Dorothea; Dorothy

Dolores del Rio Lolita Dolores Asunsolo de Martinez

Dom Getulio President Getulio Dornelles Vargas of Brazil

Dominican Ports Pepillo Salcedo, Montecristi, Puerto Plata, Sosua, Santa Barbara de Samana, Sanchez, La Romana, San Pedro de Macoris, Andres, Santo Domingo (formerly Trujillo), Rio Jaina, Bahia de las Calderas, Azua, Barahona

Dominie Hawker-Siddeley HS-125 jet transport

The Don Don Juan (as in Mozart's opera *Don Giovanni*)

Doña Fela Felisa Rincón (female mayor of San Juan, Puerto Rico for twenty-two years)

Doña Marina Malinche (Indian interpreter-mistress of Hernán Cortés—Spanish conqueror of Mexico)

Donatello Donato di Betto Bardi

Don Emilio General Emilio Aguinaldo (fighter for Philippine independence)

Donets River City Kharkov in the Ukraine

Donetzk formerly Stalino, but originally Yuzovka

Don Francisco Francisco I Madero—Mexican president

Don Giovanni Don Juan

Don Juan Don Juan Tenorio of Seville (Mozart's *Don Giovanni*)

Don Marquis Donald Robert Perry Marquis

Don Muang Bangkok, Thailand's airport

Don Pepe José Figueres Ferrer —democratic leader of Costa

Rica

Don Porfirio Don Porfirio Diaz—Mexican dictator-president

Don Quixote pseudonym Alonso Quixano gave himself in *The Adventures of Don Quixote—Man of La Mancha*, by Cervantes

Don Romulo Romulo Betancourt—democratic leader and recent president of Venezuela

Don't Give Up The Ship nickname of Captain James Lawrence, USN

Don Venus Don Venustiano Carranza—Mexican general-president

Don Venustiano Don Venustiano Carranza—former president of Mexico

Don Vitone Vito Genovesa (1897—1969)

Doornik Flemish place-name equivalent for Tournai

Doorstep to Canada Nova Scotia

Dope Capital of Canada Vancouver

Doris Day Doris Kappelhoff

Dornford Yates Cecil William Mercer's pseudonym

Dorothy Dix Elizabeth M Gilmer

Dorothy Gish Dorothy de Guiche

Dorothy Lamour Dorothy Kaumeyer

Dorothy Malone Dorothy Maloney

Dorothy Parker Dorothy Rothchild

Dosso Dossi palette name of Giovanni de Lutero

Double-Vay Sir Henry Wilson's nickname among the French general staff of World War I

Douglas Fairbanks Douglas Ulman

Dove Hawker-Siddeley twin-engine light transport carrying up to 11 passengers

Down East Atlantic coast area extending from New York to Nova Scotia, particularly coastal New England; Maine

Down Easter person from east coast of New England or Nova Scotia

Down South nickname shared by the federal penitentiary at Atlanta, Georgia and the southern United States

Down's syndrome mongolism resulting from extra chromosome-21 material

Down Under Australia and New Zealand

Down Yonder coastal North Carolina

Doyen of European Diplomacy Prince Klemens Wenzel Nepomuk Lothar von Metternich

Doyen of Film Animators Walt Disney

Doyen of Professional Translators Ralph Manheim

Dragonfly Cessna T-37 jettrainer aircraft

Dragon Nation Bhutan

Dragon's Mouth Port-of-Spain, Trinidad's harbor entrance

Draken (Swedish—Dragon)— Saab double-delta-wing supersonic fighter or fighterbomber designated J-35 or S-35

Drapier Jonathan Swift

Dramatic Symphony Roméo et Juliette by Hector Berlioz

Dream Capital of the Western World Hollywood, California

Dream King Ludwig II of Bavaria

Dress-Rehearsal Revolution Russian Revolution of 1905

Drisheen City Cork, Ireland

Dr Jinnah Mohammed 'Ali Jinnah—president of All-India Moslem League and first governor-general of Pakistan

Dr Karl Dr Karl Augustus Menninger

Droch Robert Bridges

Droll Breughel Pieter Breughel the Elder

Dr Salazar Antonio de Oliveira Salazar—dictator and prime minister of Portugal from 1932 to 1969

Dr Seuss Theodor Seuss Geisel

Dr X Alan E Nourse

Druid City oak-tree-filled Tuscaloosa, Alabama

dry Spain Mediterranean coast of Spain

Dual Cities Minneapolis and Saint Paul, Minnesota

Dual Monarchy Austro-Hungarian Monarchy (1867–1918)

Dual Protectorate Andorra under the protection of France and Spain

Dubini's disease rapid and rhythmic muscular contraction

Duca Minimo Gabriele D'Anunzio

Ducansby Head northernmost headland in the British isles

Duce (Italian—Leader)—Dicta-

tor Benito Mussolini

Duchenne de Boulogne Guillaume-Benjamin-Amand Duchenne—father of modern neurology

Duchess of Dupont Circle Alice Roosevelt Longworth

Duchess of Windsor Bessie Wallis Warfield

Ducky Joe Medwick

Dude Ranch Capital Wickenburg, Arizona

Duke George Deukmejian (former California governor); Michael S Dukakis (former Massachusetts governor and Presidential candidate)

The Duke John Wayne

Duke of the Abruzzi Italian alpinist and arctic explorer Prince Luigi Amadeo Giuseppe Maria Ferdinando Francesco

Duke of Alba Fernando Alvarez de Toledo

Duke of Buckingham George Villiers

Duke City Albuquerque, New Mexico (named for El Duque de Alburquerque)

Duke of Devonshire William Cavendish (former Prime Minister of Great Britain)

Duke Ellington Edward Kennedy Ellington

Duke of Grafton Augustus Henry Fitzroy

Duke of Newcastle Thomas Pelham-Holles

Duke of Portland William Henry Cavendish Bentinck

Duke of Shrewsbury Charles Talbot

Duke of Vicenza Marquis Louis de Caulaincourt

Duke of Wellington Arthur Wellesley

Duke of Windsor (formerly King Edward VIII, formerly Prince of Wales when his father, George V, was king of England)

Duluthians people of Duluth

Dumas fils Alexandre Dumas (1824–1895) playright-creator of Camille (Verdi's *La Traviata*), son of the novelists

Dumas père Alexandre Dumas (1802–1870), author of *The Count of Monte Cristo, The Three Musketeers*

Du Maurier George Louis Palmella Busson

Dumb Girl of Portici Esprit Auber opera *La Muette de Portici*

Dunedin nickname for Edinburgh, Scotland and placename for a South Island, New Zealand port as well as a Florida resort near St Petersburg

Dungeness Crab Capital Newport, Oregon

Dun Laoghaire (Gaelic—Dunleary)—Kingstown on Dublin Bay

Dunleary Dun Laoghaire or Kingstown

Dunnet Head northernmost point on Scotland's mainland

Dunsany Edward John Moreton Drax Plunkett, Lord Dunsany

Dupontonia Wilmington, Delaware

Dupont Town Wilmington, Delaware (home of EI du Pont de Nemours & Co)

Durable Dictator General José de la Cruz Porfirio Díaz (Mexico, 1876–1911)

Durazzo English and Italian place-name for the Albanian port of Durrës

Durban formerly Port Natal, South Africa

dutch dutch belted (black dairy cattle with a broad body-encircling white belt of hair as originally bred in the Netherlands); dutch door (horizontally divided so either the top or bottom section may be closed or opened); dutch courage (inspired by alcohol); Dutch Harbor (U.S. naval base on Alaska's Unalaska Island); dutch lunch (cold cuts); dutch treat (where all pay their own way)

dutch act suicide

Dutch Caribees colonial name for the Netherlands Antilles

Dutch City Holland, Michigan

Dutch Cradle of U.S. Presidents the Netherlands—ancestral home of both Presidents named Roosevelt and President Van Buren

Dutch Delight Delft

Dutch East Indies Netherlands East Indies now known as Indonesia

Dutch Guiana Netherlands Guiana or Surinam

Dutch Islands Dutch West Indies—Aruba, Bonaire, Curaçao, Saba, Sint Eustatius (Statia), Sint Maarten

Dutch Masterpiece in the Caribbean Curaçao

Dutch Microscopists Zacharias

Janssen, Anton van Leeuwenhoek, and Jan Swammerdam
Dutch New Guinea West Irian, Indonesia
Dutch Ports Delfzigl, Harlingen, Den Helder, Ijmuiden, Zaandam, Amsterdam, Scheveningen, Hoek van Holland, Europoort, Maasluis, Vandelingenplaat, Vlaardingen, Schiedam, Rotterdam, Dordrecht, Middelharnis, Willemstad, Middelburg, Vlissingen, Terneuzen, Hansweert, Haven Catzand
Dutch Queens Wilhelmina, Juliana, Beatrix
Dutch Reformed Dutch Reformed Church of North America where it has been called the Reformed Church since 1867
Dutch Republic the Netherlands, sometimes called Holland
Dutch-speaking Places Flemish sections of Belgium; the Netherlands; Aruba, Bonaire, Curaçao, Saba, Sint Eustatius, Sint Maarten—the Netherlands Antilles; Netherlands New Guinea now part of Indonesia Surinam and the community around Holland, Michigan
Dutch Ultramodernist Pieter Cornelis Mondriaan
Dutch West Indies the Netherlands Antilles
Dutch William William III of Orange—Dutch-born British king
Dwellers of the Field the Poles
Eagle (see Columbia)
Eagle Forgotten Governor Peter Altgeld of Illinois
Eagle of the North Swedish statesman Count Axel Oxenstierna
Eagle Pass formerly El Paso del Aguila
Eagle and Serpent *Aguila y Serpiente* (Mexican coat of arms contains pictorialization of Aztec legend stating their people could not settle until they found an island on a lake and on that island a cactus surmounted by an eagle grasping a serpent—the lacustrine island representing Mexico City, capital of Mexico —and the two creatures the struggle between celestial and earthly elements)
Eagle State Mississippi

Earl of Aberdeen George Hamilton Gordon (former Prime Minister of Great Britain)
Earl Baldwin of Bewdley Stanley Baldwin
Earl Balfour Arthur James Balfour
Earl of Beaconsfield Benjamin Disraeli (19th-century British prime minister who declared: *A conservative government is an organized hypocrisy.*)
Earl of Bute John Stuart
Earl of Carlisle Charles Howard
Earl of Carnarvon George ESM Herbert, the Egyptologist
Earl of Chatham William Pitt
Earl of Chesterfield Philip Dormer Stanhope
Earl of Derby Edward Stanley
Earl of Godolphin Sidney Godolphin
Earl Grey Charles Grey
Earl of Guilford Frederick North
Earl of Halifax Charles Montagu
Earl of Liverpool Robert Banks Jenkinson
Earl of Lytton Edward Robert Bulwer Lytton, diplomat and poet whose pseudonym was Owen Meredith
Earl of Orford Robert Walpole
Earl of Oxford Robert Harley
Earl of Oxford and Asquith Herbert Henry Asquith
Earl of Ripon Frederick John Robinson
Earl of Rosebery Archibald Philip Primrose
Earl Russell Prime Minister John Russell, first earl; John Stanley Russell, second earl; Bertrand (Arthur William) Russell, third earl
Earl of Shaftesbury Anthony Ashley Cooper
Earl of Shelburne William Petty
Earl of Stanhope James Stanhope
Earl of Sunderland Charles Spencer
Earl of Wilmington Spencer Compton
Earth's Last Frontier Antarctica
East End congested and depressed eastern section of London
easter storm from the east
Easter Easter Island (see *Isla*

de Pascua); Easter lily; Easter Monday; Easter Sunday; Easter vacation
Easter Island English placename equivalent of Isla de Pascua whose Chilean settlers are called Pascuenses
Eastern Desert Arabian Desert
Eastern Empire Byzantine Empire
Eastern Europe Czechoslovakia, Hungary, Poland, the Soviet Union
Eastern Hemisphere half of the world containing Africa, Asia, Australia, Europe, and associated islands
Eastern Malaysia Sabah and Sarawak
Eastern Samoa American Samoa
Eastern Sea East China Sea
Eastern Shore shore of Delaware, Maryland, and Virginia comprising the Del-Mar-Va Peninsula
Eastern States states east of the Mississippi
East Germany Soviet-dominated eastern Germany, German Democratic Republic (1945—1990)
East Indies Malay Archipelago, formerly Dutch (or Netherlands) East Indies, now known as Indonesia
Eastinghouse international nickname for Soviet nuclear power plants built in Finland
East London formerly Port Rex, South Africa
East Lothian Haddington
East Malaysia Bandar Seri, Brunei, Sabah, and Sarawak now known as Kalimantan
East North Central States Indiana, Illinois, Michigan, Ohio, and Wisconsin
East Prussia old name for western Poland along the Baltic
East Rudolf Lake Turkana
East Siberian East Siberian Sea in the Arctic off East Siberia
East South Central States Alabama, Kentucky, Mississippi, and Tennessee
Eastview Canadian city now called Vanier
East Village modern euphemism for New York City's Lower East Side
Ebreo (Italian—Jew)—nickname of Salomone Rossi the composer-violinist of Mantua in the late 1500s and early

1600s

'Ebrides (Cockney contraction—Hebrides)

Eccentric Naturalist Constantine Rafinesque

Eckma Jan de Hartog's pseudonym

ECR Lorac Edith Caroline Rivett's pseudonym

Ecuadorean Ports Puerto de San Lorenzo, Esmeraldas, Bahia de Caraquez, Bahia de Manta, Puerto de Cayo, La Libertad, Salinas, Guayaquil, Puna, Puerto Bolívar, Bahia Baquerizo Moreno (on Chatham or San Cristóbal)

Ecuadorian Archipelago Galápagos Islands

Ecuador's Principal Port Guayaquil

Eddie Albert Eddie Albert Heimberger

Eddie Cantor Izzie Itskowitz

Eden of the Orient Thailand

Edgar Box Gore Vidal

Edgar Thorn(e) Edward MacDowell

Edin(a) Edinburgh's poetical name

Edinburghshire Midlothian

Edinglassie Edinburgh + Glasgow (early name of the Moreton Bay Settlement now called Brisbane)

Editor of Genius Max(well) E Perkins

Ed Lacy Len Zinberg

Ed McBain Salvatore A Lombino

Edmond Adam French author-editor Juliette Lamber

Edmund Crispin Robert Bruce Montgomery

Edna St Vincent Millay Mrs Eugen Jan Boissevain

Edo old name for Tokyo, also written Yedo

Edogawa Rampo Hirai Taro (Japan's Edgar Allan Poe)

Edoo (EDU) Lady Elgar's nickname for her husband Sir Edward Elgar (EDU is the title of the fourteenth section or finale of his *Enigma Variations on an Original Theme* scored for full orchestra)

Edouard Colonne Judas Colonne

Edu Sir Edward Elgar

Eduardo E Howard Hunt

Education President George Herbert Walker Bush

Educator-Freethinker Horace Mann

Edwardian Radical Hilaire

Belloc

Edward I Prime Stevenson Xavier Mayne's pseudonym

Edward Longshanks Edward I of England

Edward O Wilson Frank J Baird, Jr

Edward the Peacemaker Edward VII, eldest son of Queen Victoria

Edward the Rake Edward VII

Edwin Markham Charles Edward Anson Markham

Ed Wynn Isaiah Edwin Leopold

egabrag garbage spelled backwards, sometimes used as a euphemism for garbage boat or garbage scow

Egg Basket of California Petaluma

Egyptian Badlands Assiut area about 175 miles (282 kilometers) south of Cairo and notorious for narcotic raids, religious clashes, and village vendettas

Egyptian Ports (on the Mediterranean) El Iskandariya (Alexandria); Bur Said (Port Said); (on the Red Sea) El Suweis (Suez)

Egypt's Principal Port Alexandria

Eight The Eight (Ashcan School of American Art comprising Arthur B Davies, William Glackens, Robert Henri, Ernest Lawson, George Luks, Maurice Prendergast, Everett Shinn, and John Sloan)

Eighteenth State Louisiana

Eighth State South Carolina

Eighth Wonder of the World compound interest, according to Baron de Rothschild; the Panama Canal

Either/Or Søren Kierkegaard's nickname

Ekaterinburg czarist name for Sverdlovsk

El Alto (Spanish—The Tall One)—La Paz, Bolivia's airport serving the world's highest capital city

El Bosco Spanish nickname for Hieronymus Bosch

El Caballo (Spanish—The Horse)—nickname of Cuba's Communist dictator Fidel Castro Ruz

El Cabrón (Spanish—The Goat)—nickname of dissolute Dominican dictator Generalissimo Rafael Leonidas Trujillo Molino

El Caudillo (Spanish—The Chief)—sobriquet of General Francisco Franco-Bahamonde

El Cid *El Cid Campeador* (Spanish—The Lord Champion)—Rodrigo Díaz de Bivar

El Coco (Spanish—Coconut Palm)—San José, Costa Rica's airport

El Coronelazo (Spanish—the big colonel) David Siquieros

Elder Pitt William Pitt the Earl of Chatham also called the Great Commoner

Elder Statesman President Dwight David Eisenhower

Eldest Daughter of the Church France

El Discutido (Spanish—the discussed)—Diego Rivera

El Dorado the mythical land of gold sought by the Spaniards and others in the jungles and mountains of the Americas

El Dorado State California

Electric Motor City Detroit

el esmog (Spanish—the smog) Mexico City

El Español (Spanish—The Spaniard)—Giuseppe Maria Crespi—Italian painter's nickname

El Españoleto Spanish painter José Ribera

Eleventh State New York

El Fatah disease virulent antisemitism

El Fénix de España The Phoenix of Spain—Lope de Vega

El Fondo (Spanish—The Fund) —International Monetary Fund—IMF

El Gran Libertador (Spanish —The Great Liberator)— Simón Bolívar—liberated Venezuela, Colombia, Ecuador, Peru, and Bolivia from Spanish rule

El Greco (Spanish—The Greek) —Kryiakos Theotokopoulos (Domingo Theotocopuli)

El Hombre (Spanish—The Man)—nickname of Dr Arnulfo Arias de Madrid of Panama

Elia Charles Lamb

Eli Edwards Claude McKay

Elijah Muhammad Robert Poole

El Ilustre Americano (Spanish—The Illustrious American)—self-title of Venezuelan dictator Antonio Guzman Blanco

El Inca (Spanish—The Inca)— Peruvian historian Garcilaso de la Vega

Elisabethville former name of Lubumbashi, Zaire

Elizabeth Arden Florence N Graham

Ellen Glasgow Ellen Anderson Gholson

Ellery Queen Frederic Dannay and Manfred B Lee

El Libertador (Spanish—The Liberator)—Simón Bolívar

El Licenciado Tomé de Burguillos (Spanish—Attorney Tomé de Burguillos)—a pseudonym of Lope de Vega

Ellis Bell pseudonym of Emily Brontë

El Maestro (Spanish—the master)—José Clemente Orozco

El Manco de Lepanto (Spanish—the One-handed Man of Lepanto)—Cervantes whose left hand was maimed at the Battle of Lepanto

El Mar del Sur (Spanish—The South Sea)—Balboa's name for the Pacific Ocean

Elm City New Haven, Connecticut

Elmo Lincoln Otto Elmo Linkenhelt

El Mudo (Spanish—the mute)—Juan Fernández Naverette, Renaissance painter

Elmwood James Russell Lowell's home in Cambridge, Massachusetts

El Niño (Spanish—the boy)—warming of surface water off Ecuador and Peru before extending westward over the tropical eastern Pacific Ocean

El Norte (Spanish—the north)—used by Latin Americans to mean the United States

Eloquent President Abraham Lincoln

El Pisshole trucker's nickname for El Paso, Texas

El Pisso trucker's nickname for El Paso, Texas

El Precursor (Spanish—the Precursor)—Francisco Miranda —fighter for Venezuelan freedom; Antonio Nariño—fighter for Colombian freedom

El Pueblo Nuestra Señora la Reina de los Angeles de Porciúncula (Spanish—The Village of Our Lady Queen of the Angels of Porciúncula)—early Spanish name Los Angeles

Elroy American country-boy

name derived from the French for king—*Le Roi* or the Spanish equivalent—*El Rey*—or their combination

Elsa Lanchester Elizabeth Sullivan

El Salvador's Ports La Unión, Puerto El Triunfo, La Libertad, Acajutla

El Silencio (Spanish—The Silence)—downtown Caracas where bus routes start and automotive traffic is at its noisiest

El Smoggo trucker's nickname for any smog-smitten city in the Southwest from El Paso to Los Angeles

El Stinko *see* El Smoggo

El Supremo (Spanish—The Supreme)—Juan Vicente Gómez, Venezuelan dictator from 1908 to 1935; also known as *El Gran Supremo*

Elysian Fields mythological place thought by the Portuguese to be the Madeira Islands or by the Spaniards to be the Canary Islands—also known as Isles of the Blest

Emancipator of the Serfs Czar Alexander II of Russia

Emancipator of the Slaves William Wilberforce

Emanuel Swedenborg Emanuel Svedberg

Embodiment of Chilean Culture Pablo Neruda

emerald beryllium chromium aluminum silicate (gemstone variety of beryl)

Emerald Capital Bogotá, Colombia

Emerald of the Caribbean Guadeloupe Island, French West Indies

Emerald City Seattle, Washington

Emerald Country Colombia

Emerald Empire Idaho's panhandle

Emerald Isle Ireland

Emerald Necklace 18,000 acres of parks surrounding Cleveland

emerald nickel zaratite (basic hydrated nickel carbonate)

Emerald of the Spanish Main Colombia

emery aluminum oxide (Al_2O_3)

Emil Jannings Theodore Friedrich Emil Janez

Emil Ludwig Emil Cohn

Eminent Humanist Julian Huxley

Emma Calve Rosa Calvet

Emma Lathen pseudonym of Mary J Latsis and Martha Hennissart

Emmet Street Brendan Behan's pen name

Emmy award for outstanding television performances in the United States; statuette named after tv entertainer Faye Emerson

Emmy Destinn Ema Kittl

Emperor of the Cossacks Emelyan Ivanovich Pugachev (1773–1774)

Emperor of Europe Napoleon Bonaparte's self-imposed title

Emperor of Japan Akihito (1989–), Hirohito (1936–1989)

Emperor of Manchukuo Henry Pu-Yi (1934–1945)

Emperor of Mexico native revolutionary Augustín de Iturbide (1822–1823)

Emperor Philosopher Marcus Aurelius

Emperor of the West Charles the Great

emperors imperial potentates; largest of the Alaskan geese, Antarctic penguins, Central American boas, European moths, Japanese fishes

Empire Builder of British South Africa Cecil Rhodes

Empire City New York; Wellington, New Zealand

Empire State New York

Empire State of the South Georgia

Emporium of the West Indies Charlotte Amalie, St Thomas, Virgin Islands; Willemstad, Curaçao

Empress of the Blues Bessie Smith

Empress Carlota Marie Charlotte Amélie Augustine Victoire Clémentine Léopoldine—empress of Mexico under Maximilian

Empress Eugenie Eugénie Marie de Montijo de Guzman—empress of the French under Napoleon III

Empress of Hollywood Bette Davis

Empress of India Queen Victoria

Empress of Vice Mary Jeffries (who in the 1800s controlled London's most elegant brothels)

Enchanted Isles the Galápagos or Tortoise Islands originally called *Las Islas Encantadas*

by early Spanish explorers

Energy City Houston, Texas

Engineer of the Animal World the busy beaver

Engineer of Fantasy Walt Disney

Engineer-Humanitarian-Statesman Herbert Hoover

Engineer President Herbert Hoover

Engineers' Town Coulee City, Washington

England's Wooden Walls the Royal Navy during the Napoleonic wars

Englebert Humperdinck Jerry Dorsey (who took the name of Wagner's protégé)

english english horn (*cor anglais* or bass oboe); english laurel (cherry laurel); english muffin; english saddle; english setter

English Agnostics Thomas Henry Huxley, Charles Darwin, Herbert Spencer

English Alexander nickname of King Henry V

English Atheists Charles Bradlaugh, Bertrand Russell

English Bach Johann Christian Bach (who lived in London from 1759 to 1852)

English Caribees colonial name for the British West Indies

English Channel La Manche

English Cradle of U.S. Presidents England—ancestral home of both Presidents Adams, Carter, Cleveland, Coolidge, Fillmore, Ford, Garfield, Grant, Harding, Harrison, Johnson, Lincoln, Madison, Pierce, Taft, Taylor, Tyler, and Washington

English Nonsense Poet Edward Lear

English Operetta Composer Sir Arthur S Sullivan

English Opium Eater Thomas De Quincey

English penicillin a nice cuppa' tea

English Polynesia jocular nickname for the Seychelles Islands in the Indian Ocean

Eniwetok Kili Atoll

Enlightened Philanthropist Andrew Carnegie

Enlightenment (*see* The Enlightenment)

The Enlightenment Europe's 18th century when encyclopedias appeared in France and England; when Voltaire and

Lavoisier were matched across the Channel by Paine and Priestley

Enlightenment Centuries the 17th and 18th centuries (1600s and 1700s) when in America and Europe human reason prevailed in educational, political, and religious doctrine

Entac Nord wire-guided anti-tank missile made in France

Entendard Dassault single-engine jet attack aircraft made in France

Epic Poet of Ancient Greece Homer

epsom epsom salt(s)—named for the English racetrack town of Epsom Downs

epsom salt magnesium sulfate ($MgSO_4 \cdot 7H_2O$)

Equality State Wyoming

Equatorial Guinea's Main Port Bata

Era of Good Feeling administration of James Monroe—fifth President of the United States

Era-of-Good-Feeling President James Monroe

Era of Mediocrity post-World-War-II era

Erasmus Desiderius Erasmus (originally Geert Geerts or Gerard Gerardzoon)

Erckmann-Chatrian pseudonym of literary collaborators Emile Erickmann and Alexandre Chatrian

Ercoli Palmiro Togliatti

Eric Evergood King Eric I of Denmark

Erich Maria Remarque Erich Maria Kramer

Erich von Stroheim Erich Maria Stroheim von Nordenwall; Hans Maria Nordenwall

Eric the Lamb King Eric III of Denmark

Eric the Memorable King Eric II of Denmark

Eric the Red Eric Thorvaldsson

Eric Rohmer Maurice Scherer

Eric von Stroheim Eric Oswald Stroheim

Erie Canal New York State Barge Canal

Erin Ireland

Ernest Bramah Ernest Bramah Smith's pseudonym

Ernie Ernald; Ernest

Ernie Pyle Ernest Taylor Pyle

Ernst von Dohnányi Ernö Dohnányi

Eros Center sex supermarket nickname in Hamburg and Bonn, Germany

Erudite Americans Will and Ariel Durant

Escape King Harry Houdini

Escondildo dysphemistic nickname for California town of Escondido

Eskimo Village Kotzebue, Alaska

Esperanto pseudonym of Dr LL Zemenhoff—inventor of Esperanto—an artificially-contrived universal language

ESP Pioneer Dr JB Rhine who coined the term extrasensory perception, ESP

Estado Libre Asociado de Puerto Rico (Spanish—Free Associated State of Puerto Rico)—Commonwealth of Puerto Rico

Etcher of Disaster Francisco de Goya y Lucientes

Etcher of Prisons Giambattista Piranesi

Eternal City Rome

Ethel Barrymore Ethel Blythe

Ethel Leginska Ethel Liggins

Ethel Merman Ethel Zimmerman

Ethelred the Unready Ethelred II of England

Ethical Culturist Felix Adler

Ethiopian Ports Massawa, Port Smyth, Assab

Eugenie Marlitt Eugenie John's pseudonym

Eulenberg's disease congenital muscular spasms

Eumenides the Furies

European Community Belgium, Britain, Denmark, France, Greece, Ireland, Italy, Luxembourg, Germany, Netherlands, Portugal, Spain

European Twelve (arranged by population) Germany, Italy, Britain, France, Spain, Netherlands, Portugal, Greece, Belgium, Denmark, Ireland, Luxembourg

Eusebius see Florestan and Eusebius

Evangel of Liberty Thomas Paine

Eve Arden Eunice Quedons

Everglades State Florida

Evergreen City Sheboygan, Wisconsin

Evergreen State Washington

Evil Empire USSR

Evil Florist Charles Pierre Baudelaire—famous for his *Les Fleurs du Mal* (Flowers of Evil)

Excelsior State New York

Executive City Washington, D.C.

Exemplar of Feminine Fascination Cleopatra

Existenial Dane Søren Kierkegaard

Existentialist-Leftist Jean-Paul Sartre

Exocet Aerospatiale surface-to-surface missile for use aboard warships

Expounders of Utilitarianism Jeremy Bentham, James Mill, and John Stuart Mill

Exxon (formerly ESSO)—Standard Oil

"Eye" I Street in Washington, D.C.

Eye of the Baltic Gotland

Eye of England London

Eye of Greece Athens

Eye into Europe St Petersburg more recently known as Petersburg, Petrograd, or Leningrad

Eye of Italy Rome

Eyetie(s) British slang for Italian(s)

Fabien Sevitzky Fabien Koussevitzky

Fabulous Philadelphians the Philadelphia Orchestra

Fair City Perth, Scotland

Fair Deal administration of Harry S Truman—thirty-third President of the United States

Fairy of Dreams Queen Mab

Fairytale Land Denmark—home of Hans Christian Andersen

Faithful City Worcester, Massachusetts

Falcon Dassault twin-engine executive transport made in France and called Mystere 20

falcons of the sea clipper ships

Falls City Louisville, Kentucky

Falls of the Rhine Rheinfall or Schaffhausen

false topaz citrine (quartz with ferric iron)

Famous Libertarian Alexandr Pushkin

Famous Mulatto Pianists Philippa Schuyler, André Watts

Famous Mulattos Alexander Hamilton, Alexandr Pushkin

Famous Quadroons Alexander Hamilton, Alexandr Pushkin

Famous Rock-'n'-Roll Mulatto Vanity

Fanguito (Spanish—Little Muddy)—San Juan, Puerto

Rico slum

Fanny Brice Fanny Borach

Farinelli Carlo Broschi

Farmer President William Henry Harrison and George Washington

Far North northernmost Alaska, Canada, Greenland, Norway, Sweden, Finland, and the USSR

Farther India Indochina; Indochinese Peninsula

Far West the Rocky Mountain States

Fashoda former name of Kodok, Sudan

Fate American nickname for Lafayette

Father of Abolition Samuel Hopkins

Father of the A-bomb J Robert Oppenheimer

Father Abraham Abraham Lincoln

Father of Air Conditioning WH Carrier

Father of Algebra Diophantus of Alexandria

Father of All Foods alfalfa

Father of All Yankees Benjamin Franklin

Father of America Sam(uel) Adams

Father of American Anarchy Josiah Warren of Cincinati, Ohio who in 1827 advocated government activities be transferred to private citizens

Father of American Anthropology Lewis Henry Morgan

Father of the American Ballet George Balanchine

Father of American Baptists John Clarke

Father of American Botany John Bartram

Father of American Boxing William Muldoon

Father of American Church Music Lowell Mason

Father of American Conchology Thomas Say

Father of American Conservatism John Jay

Father of American Dance Ted Shawn

Father of American Egyptology James Henry Breasted

Father of American Football Walter Camp

Father of American Freethought Thomas Paine

Father of American Geography Jedidiah Morse

Father of American Geology William Maclure

Father of American History George Bancroft, William Bradford

Father of American Horticulture Peter Henderson

Father of American Independence John Adams

Father of American Lexicography Noah Webster

Father of American Literature Washington Irving

Father of American Medical Association Dr Nathan Smith Davis

Father of American Medical Botany Jacob Bigelow

Father of American Mineralogy Parker Cleaveland

Father of American Naval Architecture William A Webb

Father of American Navigation Nathaniel Bowditch

Father of the American Nuclear Navy Admiral Hyman G Rickover

Father of American Nutrition W(ilbur) O(lin) Atwater

Father of American Oceanography Matthew Fontaine Maury

Father of American Orchestral Music Johann Christian Gottlieb Graupner

Father of American Ornithology John James Audubon, Alexander Wilson

Father of American Paleontology O(thniel) C(harles) Marsh

Father of American Photography S(amuel) F(inley) B(reese) Morse

Father of American Photo-Journalism Matthew Brady

Father of American Poetry Philip Freneau

Father of American Poets William Cullen Bryant

Father of American Pragmatism Charles Sanders Pierce

Father of American Prison Reform George O Osborne

Father of American Psychiatry Dr Benjamin Rush

Father of American Psychobiology Adolf Meyer

Father of American Psychology William James

Father of American Railroads Peter Cooper

Father of the American Revolution Sam(uel) Adams

Father of American Rocketry Robert H Goddard

Father of American Surgery Dr William Halsted or Dr Philip Syng Physick

Father of American Technology Eli Whitney

Father of the American Turf Leonard Jerome

Father of American Universalism Hosea Ballou, John Murray

Father of American Zoology Thomas Say

Father of Anatomical Dissection Andreas Vesalius

Father of Andean Archeology Max Uhle

Father of Anesthetic Dentistry William T G Morton

Father of Angling Izaak Walton

Father of Annapolis George Bancroft

Father of Antarctic Whaling Captain Carl A Larsen

Father of Antiseptic Obstetrics Ignaz Semmelweis

Father of Antiseptic Surgery Sir Joseph Lister

Father of Argentina's School System Domingo Faustino Sarmiento

Father of Argentine Education Faustino Sarmiento

Father of Argentine Independence General José de San Martín

Father of the Atomic Age Enrico Fermi

Father of the Atomic Submarine Admiral Hyman Rickover, USN

Father of the Automobile Gottlieb Daimler

Father of Bacteriology Robert Koch

Father of Baseball Henry Chadwick, Alexander Cartwright, Abner Doubleday

Father of Basic Flying John Joseph Montgomery

Father of Basketball James Naismith

Father of Belgian Opera Andre Ernest Modeste Gretry

Father of Believers Mohammed

Father of the Bill of Rights James Madison

Father of Black History Chester G Woodson

Father of Blood Banks and Blood Plasma Charles R Rich

Father of the Blues W(illiam) C(hristopher) Handy

Father of Botany Aristotle

Father of Brazilian Opera Antonio Carlos Gomes

Father of British Archeology

Augustus Henry Lane-Fox Pitt-Rivers

Father of British Boxing Jack Broughton

Father of British Musicology Sir Charles Grove

Father of the British Navy Alfred the Great

Father of British Printing William Caxton

Father of British Unitarianism John Biddle

Father of Buffalo Joseph Ellicot

Father of Canadian Confederation Sir John A Macdonald and George Brown of Ontario, Sir George S Etienne Cartier and Sir Alexander Galt of Quebec, Sir Charles Tupper of Nova Scotia, Sir Samuel Leonard Tilley of New Brunswick

Father of Chemistry Robert Boyle

Father of Chemurgy George Washington Carver

Father of Chicago John Kinzie

Father of Child Psychology Leo Kanner

Father of the Chinese Revolution Dr Sun Yat-sen

Father Christmas Santa Claus; Snow King

Father of Church History Eusebius

Father of Civilian Atomic Power Admiral Hyman Rickover

Father of the Civil Rights Movement Martin Luther King, Jr

Father of Civil Service Reform George Hunt Pendleton

Father of Classical Music Franz Joseph Haydn

Father of Comedy Aristophanes

Father of Confederation John A Macdonald—Canada's first prime minister

Father of the Constitution James Madison—fourth President of the United States

Father of the Continental Congress Benjamin Franklin

Father of the Copyright William Hogarth

Father of the Cotton Gin Eli Whitney

Father of Courtesy Richard Beauchamp—Earl of Warwick

Father of the Cowboys Charles Goodnight

Father of Cuban Independence José Martí

Father of Czechoslovakian Music Bedrich Smetana

Father of Damien Joseph Damien de Veuster

Father of Danish Opera Friedrich Kuhlau

Father of Dano-Norwegian Literature Ludvig Holberg

Father of the Declaration of Independence title shared by Thomas Paine, Thomas Jefferson, John Adams and Benjamin Franklin

Father of the Detective Story Edgar Allan Poe

Father of Dinosaur Art Charles R Knight

Father Divine Morgan J Divine born George Baker

Father of Dominican Independence Juan Pablo Duarte

Father of the Dominican Republic José Nuñez de Cáceres

Father of Dutch Poetry Jakob van Maerlant

Father of the Dutch Reformed Church in America John Henry Livingston

Father of Ecclesiastical History Eusebius Pamphili

Father of Egyptian Archeology Sir Flinders Petrie

Father of Electroencephalography Hans Berger

Father of Embryology Carl Ernst von Baer

Father of the Encyclopedia Diderot

Father of English Cathedral Music Thomas Tallis

Father of English Empiricism John Locke (1632–1704)

Father of English Lexicography Dr Samuel Johnson

Father of the English Novel Henry Fielding

Father of the English Poetry Geoffrey Chaucer

Father of English Printing William Caxton

Father of English Song Caedmon

Father of English Unitarianism John Biddle

Father of Epic Poetry Homer

Father of the Erie Canal De Witt Clinton

Father of Ethical Culture Felix Adler

Father of Ethology Konrad Lorenz

Father of Euphuism John Lyly

Father of Experimental Physiology Galen

Father of the Faithful Abraham

Father of Farm Workers Cesar Chavez
Father of Fascism Benito Mussolini; Gabriele d'Annunzio
Father of the Federal Reserve System George Carter Glass
Father of the Film Industry D(avid) W(ark) Griffith
Father of Fingerprinting Alphonse Bertillon
Father of the Flivver Henry Ford
Father of the Free School System Governor James Edward English of Connecticut
Father of Free Trade Adam Smith
Father of French-Canadian Poetry Octave Cremazie
Father of French Lyric Poetry Pierre de Ronsard
Father of French Opera Jean-Baptiste Lully
Father of the French School of Neurology Jean-Martin Charcot
Father of French Surgery Ambroise Paré
Father of French Tragedy Pierre Corneille
Father of Frozen Foods Clarence Birdseye
Father of Genetics Gregor Mendel
Father of Geography Strabo the Greek Stoic who wrote seventeen books about Asia, Egypt, Libya, and Europe
Father of Geometry Pythagorus
Father of German Literature Gotthold Ephraim Lessing
Father of German Opera Christoph Willibald von Gluck
Father of the German Reformation Martin Luther
Father of German Unification Prince Otto von Bismarck
Father of Gods and Men Odin or Wotan, according to the Norse; Jove or Jupiter, according to the Romans; Zeus, according to the Greeks; etc.
Father of the Gramophone Emile Berliner
Father of Greater Philadelphia John Christian Bullitt
Father of Greek Didactic Poetry Hesiod whose poem *Theogony* describes the beginning of the world, its gods, and the five Ages of Mankind
Father of Greek Music Terpander of Lesbos

Father of Greek Sculpture Phidias
Father of Greek Tragedy Aeschylus
Father of Greenbacks Salmon Portland Chase
Father of the H-bomb Edward Teller
Father of High-Speed Photography Harold E Edgerton
Father of His Country Cicero and several Roman caesars; George Washington—first President of the United States
Father-of-his-Country President George Washington
Father of History Herodotus
Father of Homeopathy in America Dr Constantine Hering
Father of the Household Heater Benjamin Franklin
Father of Humanism in America John Dietrich
Father of the Hydrogen Bomb Edward Teller
Father of Hypnotism Franz Friedrich Anton Mesmer
Father of Independent India Mohandas Karamchand Gandhi
Father of Individual Psychology Alfred Adler
Father of Inductive Philosophy Francis Bacon
Father of Israel Chaim Weizmann
Father of Italian Landscape Painting Andrea del Verrocchio (Andrea di Michele Cione)
Father of Italian Opera Claudio Monteverdi
Father of Japanese Caricature Toba Sojo
Father of Japanese Shipbuilding Thomas Glover known to the Japanese as Kuraba
Father of Jazz Scott Joplin
Father of Jests Joseph Miller
Father of the Juvenile Court Judge Benjamin Barr Lindsey
Father of the Kindergarten Friedrich Froebel
Father of the Kiwifruit James MacLoughlin
Father of Latin Song Ennius (239–169 BCE)
Father of the Legal Code David Dudley Field
Father of Lies Satan
Father of the Locomotive Richard Trevethick
Father of Logarithms John Napier
Father of Marine Geology

Francis Shepard
Father of Massachusetts Governor John Winthrop
Father of Medicine Hippocrates
Father of Mexican Independence Miguel Hidalgo y Costilla
Father of Mexican Muralism Dr Atl (born Gerardo Murillo)
Father of Microscopy Anton van Leewenhoek
Father of Military Strategy Hannibal
Father of Mineralogy Agricola
Father of Modern Architecture Frank Lloyd Wright
Father of Modern Art Masaccio (Tommaso Guidi)
Father of Modern Astronomy Copernicus (Nikolaus Kopernicki)
Father of Modern Baseball Alexander Joy Cartwright
Father of Modern Brazil Getulio Vargas
Father of Modern Conservative Thought Edmund Burke
Father of Modern Criminology Alphonse Bertillon and Cesare Lombroso
Father of Modern Democratic Philosophy John Locke
Father of Modern Drama Henrik Ibsen
Father of Modern Economic Forecasting Trygve Haavelmo
Father of Modern Economics Adam Smith
Father of Modern English Poetry Walt Whitman
Father of Modern Existentialism Sören Kierkegaard
Father of Modern Exploration Fridtjof Nansen
Father of Modern Fingerprinting Sir Edward Richard Henry
Father of Modern French Poetry Charles Baudelaire
Father of Modern Genetics Gregor Mendel
Father of Modern Geology Sir Charles Lyell
Father of Modern German Poetry Heinrich Heine
Father of Modern Guitar Music Francisco Tárrega
Father of Modern Gynecology Dr J Marion Sims
Father of Modern Italian Poetry Gabriele D'Annunzio
Father of Modern Magnetism John Van Vleck
Father of Modern Medicine

Canadian-born Sir William Osler

Father of Modern Music Franz Liszt; Mozart

Father of Modern Navies Captain Alfred T Mahan author of *The Influence of Sea Power upon History* published in 1890

Father of Modern Neurology Guillaume-Benjamin-Amand Duchenne

Father of the Modern Novel Lion Feuchtwanger

Father of the Modern Orchestra Hector Berlioz

Father of Modern Painters Giovanni Cimabue (Cenni di Pepo)

Father of Modern Pedagogy Heinrich Pestalozzi

Father of Modern Philosophy René Descartes

Father of Modern Photography William Henry Fox Talbot

Father of Modern Physics Albert Einstein

Father of Modern Physiology William Harvey

Father of Modern Prose Fiction Daniel Defoe

Father of Modern Russian Poetry Nikolai Alekseevich Nekrasov

Father of Modern Spanish Poetry Rubén Darío

Father of Modern Surgery of the Brain Paul Broca

Father of the Modern Symphony César Franck

Father of Modern Taxonomy Carl von Linné also known Carolus Linnaeus

Father of the Modern Zoo Carl Hagenbeck

Father of Moral Philosophy Thomas Aquinas

Father of the Mormons Joseph Smith

Father of Muckrakers Upton Sinclair, Lincoln Steffens, and Joseph Flynt Williard

Father of Music Festivals Edinburgh

Father of Natural History Aristotle

Father of Naval Aviation Commander Robert C Whitton

Father of Negro History Carter G(odwin) Woodson

Father of the Neighborhood Settlement House Jacob August Riis

Father of Neurosurgery

American-born Canadian Doctor Wilder Penfield

Father of New England John Endicott—its first governor

Father of New England Transcendentalism Ralph Waldo Emerson

Father of the New Left Herbert Marcuse

Father of Niagara Power William Birch Rankine

Father of the Nuclear Submarine Admiral Hyman G Rickover, USN

Father of Oceanography Matthew Fontaine Maury

Father of Ontario Hydro Sir Adam Beck

Father of Organic Architecture Frank Lloyd Wright

Father of Organized Labor Samuel Gompers

Father of Osteopathy Dr Andrew T Still

Father of Paleontology Georges Cuvier

Father of Parole Captain Alexander Maconochie the governor of the Norfolk Island penal colony from 1840 to 1844

Father of the Patent Office John Ruggles

Father of Pathology Rudolf Virchow

Father of Penitentiary Science Jean Jacques Vilain

Father of Pennsylvania William Penn—its founder

Father of Philippine Independence Emilio Aguinaldo

Father of Philosophy Thales

Father of the Phonograph Thomas A Edison

Father of Photojournalism Alfred Eisenstaedt

Father of Physiography William Morris Davis

Father of Pianoforte Playing Muzio Clementi

Father of Pittsburgh George Washington who proposed the location and the name

Father of Poetry Orpheus

Father of Polar Exploration Fridtjof Nansen

Father of the Potteries Josiah Wedgwood

Father of Psychoanalysis Sigmund Freud

Father of Public Relations Edward L Bernays, Ivy L Lee

Father of Radio Lee De Forest

Father of Radio Broadcasting Harry P(hillips) Davis

Father of Railways George Stephenson

Father of Reform John Cartwright

Father of the Reformed Church John Henry Livingston

Father of the Religion of Reason (Unitarianism) James Martineau in England, Ralph Waldo Emerson and Theodore Parker in the United States

Father of Religious Humanism John Dietrich

Father of the Republic of China Sun Yat Sen

Father of Ridicule Rabelais

Father of the Robot Joseph Engelberger

Father of Roman Philosophy Cicero

Father of the Royal Navy King Alfred

Father of Rural Free Delivery Marion Butler of North Carolina

Father of Russian Art and Music Criticism V(ladimir) V(asilievich) Stasov

Father of the Russian Intelligentsia Vassirion Belinsky

Father of Russian Literature Alexander Pushkin

Father of Russian Marxism Georgi Valentinovich Plekhanov

Father of Russian Music Michael Glinka

Father of the Russian Navy Peter the Great

Father of Russian Opera Michael Glinka

Father of the Science of Eugenics Sir Francis Galton

Father of Science Fiction Jules Verne *(see entry)*

Father of Scientific Management Frederick W Taylor

Fathers of the Enlightenment Diderot, Rousseau, Voltaire

Father of the Sewing Machine Elias Howe

Fathers of Italian Unification Camillo Cavour, Giuseppe Garibaldi, Giuseppe Mazzini

Fathers of Kodachrome Leopold Godowsky and Leopold Mannes

Father of the Skyscraper Cass Gilbert

Father of South African Poetry Thomas Pringle

Father of the Soviet Hydrogen Bomb Nobel-Peace-Prizewinner Andrei D Sakharov

Father of Spanish Drama Lope de Vega

Father of Spanish Satire Miguel Cervantes de Saavedra

Father of the Spoletto Festivals Gian Carlo Menotti

Father of Streamlining Raymond Loewy

Father of States' Rights John Caldwell Calhoun

Father of State Universities Manasseh Cutler

Father of Steam Navigation Robert Fulton

Father of the Steam Navy Commodore Matthew C Perry —also known as Old Bruin

Father of the String Quartet and the Symphony Franz Joseph Haydn

Father of the Submarine John Philip Holland

Father of Supersonic Flight Theodor von Karman

Father of Swedish Music Johan Helmich Roman

Father of Swedish Opera Ivar Hallström

Father of Swiss Reformation Huldreich Zwingli

Father of the Symphony Franz Josef Haydn

Father of the Tablet Triturate Dr Robert Mason Fuller

Father of the Tariff Secretary of the Treasury Alexander Hamilton

Father of Taxonomy Linnaeus

Father of the Telegraph S(amuel) F(inley) B(reese) Morse

Father of the Telephone Alexander Graham Bell although he only improved the work of a German named Philipp Reis who was the inventor

Father of Television John Logie Baird

Father of Texas Stephen F Austin, Sam(uel) Houston

Father of Theoretical Chemistry Antoine Laurent Lavoisier

Father of The Pill Dr Gregory Pincus of Shrewsbury, Massachusetts—formulator of the contraceptive pill

Father Time time personified and symbolized by a bearded elder wielding a scythe

Father of Traffic Safety William Phelps Eno

Father of Tragedy Aeschylus

Father of Tropical Medicine Sir Patrick Manson

Father of the Typewriter Christopher Latham Sholes

Father of Unitarian Universalism Ralph Waldo Emerson

Father of the United States Lighthouse Service President John Quincy Adams

Father of the United States Military Academy Brigadier General Sylvanus Thayer, USA

Father of the United States National Museum John Quincy Adams

Father of the United States Naval Academy Secretary of the Navy George Bancroft

Father of the United States Naval War College Rear Admiral Stephen Bleecker Luce

Father of the United States Navy President John Adams and Commodore John Barry

Father of Universalism in the United States Hosea Ballou

Father of the University of Virginia Thomas Jefferson— third President of the United States

Father of Uruguay José Gervasio Artigas

Father of Uruguay's School System José Pedro Varela

Father of the U.S. Post Office Benjamin Franklin—author, inventor, patriot, printer, philosopher, scientist, statesman

Father of Vaccination Edward Jenner

Father of Vasectomy Sir Astley Paston Cooper

Father of Vaudeville Tony Pastor (Antonio Pastor)

Father of Venezuelan Democracy Rómulo Betancourt

Father of Verismo Giovanni Verga

Father of the Viennese Operetta Franz von Suppé

Father of Virginia Captain John Smith

Father of the Waltz Johann Strauss Sr

Father of the Waters great rivers such as the Amazon, Amur, Congo, Euphrates, Huang, Irrawaddy, Lena, Mackenzie, Mekong, Mississippi, Niger, Nile, Ob, Volga, Yangtze, Yenisei

Father of the Western Story Zane Grey

Father of West Point Sylvanus Thayer

Father of Yellowstone National Park Nathaniel Langford

Father of Zionism Theodor Herzl

Father of Zoology Aristotle

Fats Thomas (Fats) Waller

Fats Waller Thomas Waller

Fatty and Skinny Oliver Newell Hardy and Stan Laurel

Faulkner's County Yoknapatawpha (an invention of novelist William Faulkner)

Faultless Painter Andrea del Sarto

Faustus Socinus Fausto Sozzini

Fauvist Painter Raoul Dufy

Favelas Rio de Janeiro slums

Favorite Island of Columbus Jamaica

The Fed The Federal Reserve Board

Federal Capital Territory now called Australian Capital Territory (around and in Canberra)

Federal City Washington, D.C.

federalese the jargon of bureaucrats on the federal payroll

Federal Hill Providence, Rhode Island slum

Federal Republic of Germany Bundesrepublik Deutschland

Federated States of Micronesia islands of Kosrae, Phompei, Truk, and Yap

Federation of Malaysia Malaysia (formerly Brunei, Federation of Malaya, Sabah, Sarawak, and Singapore)

Federation of South Arabia Ittihad al Janub al 'Arabi— formerly Aden Colony and Aden Protectorate

Feebie (American slang—member of the Federal Bureau of Investigation)

Female Seminary Mount Holyoke College

Feminist Reformer Lucy Stone

Feminist Revolutionist Mary Wollstonecraft also known as Mary Godwin

Ferihegy Budapest, Hungary's airport

Fernán Caballero Cecilia Francisca Josefa de Arrom

Fernandel Fernand Contandin

Fernando de Magallanes (*Spanish*—Ferdinand Magellan)—originally Fernão de Magalhães

Ferret Daimler armored scout car made in Great Britain

Ferryville old name for Menzei-Bourguiba in Tunisia

Fertile Arabia Arabia Felix in the southern Arabian Peninsula, particularly in the

Yemenite lands once called Aden or the Hadhramaut

Fertile Crescent Australia's coastal plain extending from southern New South Wales to southern Queensland; agricultural region extending from the Lerant to Iraq

Fiddler Soviet long-range interceptor aircraft (Tu-28) designed by Tupolev (NATO)

Fiddler's Green traditional haven of drowned sailors supposedly filled with women, grog, fine food, and tobacco; some opine it is a suburb of Davy Jones' Locker while others insist it is a synonym for Fiddler's Grotto

Fiddler's Grotto music-filled tropical marine cavern inhabited by young women; roast turkeys fly about slowly; fine beer, whiskey, and wine cascade down its marble walls; its location, according to Captain Ed Hassel, is exactly two miles this side of Hell

Fifteenth State Kentucky

Fifteen-Year War (*see* Pacific War)

Fifth Continent Africa

Fifth Estate The Underworld of Organized Crime

Fifth State Connecticut

Fiftieth State Hawaii

Figaro Mariano José de Larra's pseudonym

Fighting Bob Rear Admiral Robley Evans; English prizefighter Robert P Fitzsimmons; Senator Robert M La Follette, Sr

Fighting Lady USS *Lexington*

Fighting Quaker General Nathanael Greene

Fijian Ports Suva and Levuka

Filatov's disease scarlatina-like exanthematous affection

Filbert Center Hillsboro, Oregon

Filipino Libertarian Emilio Aguinaldo

Film Capital Hollywood, California

Filthydelphia Philadelphia

Finality John Lord John Russell

final solution Hitlerian truncation covering the extermination of the Jews

Financial Center of the Rockies Denver, Colorado

Financial Genius of the Underworld Meyer Lansky

Financier of the Revolution Robert Morris

Finland's Principal Port Helsinki

Finnish Ports Tornio, Roytta, Kemi, Koivoluoto, Oulu, Raahe, Kokkola, Ykspihlaja, Jakobstad, Nykarleby, Vaasa, Vasklot, Kasko, Kristinestad, Pori, Mantyluoto, Reposaari, Kaunissaari, Rauma, Uusikaupunki, Turku, Storby, Mariehamn, Hango, Lappvik, Ekenas, Helsinki, Kotka, Hamina

Fiona Macleod William Sharp's pseudonym

Firebar NATO name for Soviet Yakovlev all-weather fighter interceptor aircraft Yak-28P

Firebrand of the Navy Lieutenant Stephen Decatur, USN

Firebrand of the World Tamerlane (Timur Lenk or Timur the Lame)

Fireclay Capital Mexico, Missouri

firecracker factory missile manufacturing plant

Firestreak Hawker-Siddeley air-to-air missile

First All-American Poet Walt Whitman

First American Advertiser William Penn

First American Penitentiary Walnut Street Jail in Philadelphia built in 1790

First American Poet Philip Frenau

First American Republic Iceland

First American Woman Novelist Charlotte Lennox

First American Woman Physician Elizabeth Blackwell

First Astrophysicist Johannes Kepler

First Black American Conductor Dean Dixon

First Black Member of the U.S. Supreme Court Thurgood Marshall

First California Mission San Diego de Alcalá

First Church First Church of Christ, Scientist

First Citizen of Ghana Dr WEB Du Bois

First City of the First State Wilmington, Delaware

First City of the South Savannah, Georgia

First Concentration Camp in Germany Dachau (erected in March 1933 near Munich)

First Estate The Clergy

First Execution by Electrocution August 6, 1890 in New York State's Auburn Prison

First Family family of the President of the United States

First Fleeters Australians tracing their lineage to their ancestors' arrival in 1788

First Foreign Enclave Macao, Portuguese China

first-generation money cash; currency

First Gentleman of the Land President Chester A Arthur

First Gospel Gospel according to Saint Matthew

First Great Cheerful Giver George Peabody

First Great Operatic Composer of the New World Carlos Gomes

First Great Symphonist Franz Joseph Haydn

First Halfway House Isaac T Hooper Home opened in New York City in 1845 by the Society of Friends (Quakers)

First International First International Workingmen's Association (convening in Paris in 1864)

First Internationally Prominent Albanian Poet Naim Erashëri

First Internationally Prominent American Humorist Mark Twain

First Internationally Prominent American Novelist James Fenimore Cooper

First Internationally Prominent Argentinian Author Domingo Faustino Sarmiento

First Internationally Prominent Australian Novelist Nevil Shute

First Internationally Prominent Austrian Dramatist Hugo von Hofmannsthal

First Internationally Prominent Barbadan Novelist George Lamming

First Internationally Prominent Belgian Author Maurice Maeterlinck

First Internationally Prominent Bolivian Author Ricardo Jaimes Freyre

First Internationally Prominent Brazilian Novelist Joaquim María Machado de Assis

First Internationally Prominent Bulgarian Novelist Dimiter Dimov

First Internationally Promi-

nent Canadian Author Robert W. Service

First Internationally Prominent Canadian Humorist Stephen Leacock

First Internationally Prominent Canadian Novelist Chandler Haliburton

First Internationally Prominent Chilean Poet Gabriela Mistral

First Internationally Prominent Chinese Novelist Ts'ao Msüeh-Ch'in

First Internationally Prominent Colombian Author Germán Arciniegas

First Internationally Prominent Costa Rican Writer Ricardo Fernández Guardia

First Internationally Prominent Cuban Author José Martí

First Internationally Prominent Curaçaon Political Polemicist Daniel De León

First Internationally Prominent Czech Novelist Franz Kafka

First Internationally Prominent Danish Essayist George Morris Cohen Brandes

First Internationally Prominent Danish Novelist Hans Christian Andersen

First Internationally Prominent Dutch Author Jan de Hartog

First Internationally Prominent Ecuadorean Author Juan Montalvo

First Internationally Prominent English Dramatist William Shakespeare

First Internationally Prominent English Novelist Charles Dickens

First Internationally Prominent English Poet Geoffrey Chaucer

First Internationally Prominent Finnish Novelist Mika Waltari

First Internationally Prominent French Dramatist Molière

First Internationally Prominent French Novelist Honoré de Balzac

First Internationally Prominent German Dramatist Johann Wolfgang von Goethe

First Internationally Prominent German Poet Friedrich von Schiller

First Internationally Promi-

nent Greek Novelist Nikos Kazantzakis

First Internationally Prominent Guatemalan Essayist Enrique Gómez Carillo

First Internationally Prominent Guatemalan Novelist Miguel Angel Asturias

First Internationally Prominent Guyanese Poet Jan Carew

First Internationally Prominent Haitian Novelist Oswald Durand

First Internationally Prominent Honduran Novelist Argentina Díaz Lozano

First Internationally Prominent Hungarian Dramatist Ferenc Molnar

First Internationally Prominent Icelandic Author Jon Sigurdsson

First Internationally Prominent Indian Poet Rabindranath Tagore

First Internationally Prominent Indian Political Polemicist Mahatma Gandhi

First Internationally Prominent Irish Dramatist Oscar Wilde

First Internationally Prominent Irish Novelist James Joyce

First Internationally Prominent Irish Poet William Butler Yeats

First Internationally Prominent Italian Author Dante Alighieri

First Internationally Prominent Jamaican Poet Claude McKay

First Internationally Prominent Japanese Author Yasunari Kawabata

First Internationally Prominent Mexican Author José Vasconcelos

First Internationally Prominent New Zealand Author Katherine Mansfield

First Internationally Prominent Nicaraguan Author Rubén Darío

First Internationally Prominent Norwegian Dramatist Henrik Ibsen

First Internationally Prominent Norwegian Novelist Knut Hamsun

First Internationally Prominent Panamanian Author Justo Arosemena

First Internationally Promi-

nent Paraguayan Author Juan Silvano Godoy

First Internationally Prominent Peruvian Author Manuel González Prada

First Internationally Prominent Polish Novelist Henryk Sienkiewicz

First Internationally Prominent Portuguese Poet Luís de Camões

First Internationally Prominent Puerto Rican Author Eugenio María de Hostos

First Internationally Prominent Romanian Novelist Mihail Sandoveanu

First Internationally Prominent Russian Dramatist Aleksander Sergeevich Pushkin

First Internationally Prominent Russian Novelist Fëdor Dostoevski

First Internationally Prominent Salvadoran Poet Juan José Cañas

First Internationally Prominent Scottish Poet Robert Burns

First Internationally Prominent South African Novelist Olive Schreiner

First Internationally Prominent Spanish Author Miguel de Cervantes Saavedra

First Internationally Prominent Spanish Dramatic Poet Lope de Vega

First Internationally Prominent Spanish Dramatist Pedro Calderon de la Barca

First Internationally Prominent Spanish Essayist José Ortega y Gasset

First Internationally Prominent Swedish Dramatist August Strindberg

First Internationally Prominent Swedish Novelist Pär Lagerkvist

First Internationally Prominent Swiss Novelist Johanna Spyri

First Internationally Prominent Trinidadian Novelist Vidiadhur Surayprasad Naipual

First Internationally Prominent Uruguayan Author José Enrique Rodó

First Internationally Prominent Uruguayan Poet Juana de Ibarbourou

First Internationally Prominent Venezuelan Author

Rómulo Gallegos **First Internationally Prominent Venezuelan Poet** Irma De Sola de Lovera
First Internationally Prominent Welsh Author Daniel Owens
First Internationally Prominent Yugoslav Author Milovan Djilas
First Lady wife of any American President
First Lady of the Air Amelia M(ary) Earhart
First Lady of America Pocahontas
First Lady of the American Revolution Mercy Otis Warren also known as Philomela
First Lady of Computers Lady Augusta Ada Byron Lovelace (the poet Lord Byron's daughter)
First Lady of Crime Agatha Christie
First Lady of Liberty Abigail Adams
First Lady of the Library President Millard Fillmore's wife Abigail—founder of the first library in the White House
First Lady of the Skies Amelia Earhart
First Lady of Song Ella Fitzgerald
First Lady of the World Anna Eleanor Roosevelt—wife of President Franklin D Roosevelt
First Lawyer of the Land U.S. Attorney General
First Man to Walk on the Moon Neil A Armstrong
First Mayor of Chicago William Butler Ogden
First Person to Announce His Discovery of Photography William Henry Fox Talbot on January 31, 1839
First Perspective Painter Paolo Uccello (Paolo di Dono)— known for his studies in foreshortening and linear perspective
First Picaresque Novel *Lazarillo de Tormes* (author unknown)
First Poet Laureate Ben Jonson
First Poet Laureate of the United States Robert Penn Warren (1986–1987)
First President of the United States George Washington
First Prison Newspaper *The Summary* published by the inmates of the New York State Reformatory at Elmira on November 22, 1883—Thanksgiving Day
First Quaker George Fox, founder of the Society of Friends
First Romantic Artist Giambattista Piranesi
First State Delaware
First Street in Europe Disraeli's nickname for London's Strand
First University Plato's Academy
First White House of the Confederacy temporary home of Jefferson Davis in Montgomery, Alabama in 1861
First Woman Physician Dr Elizabeth Blackwell
First Woman Reporter Anne Royale of Virginia (publisher of *Paul Pry*), Nellie Bly of Pennsylvania (reporter for the *New York World*)
First World highly industrialized countries such as Japan, the United States, and most Western European nations
First Zen First Zen Institute of America
Fish-Canning Capital of the World Stavanger, Norway
Fishpot NATO name for Soviet SU-9 all-weather jet fighter aircraft
Fitter NATO name for Soviet SU-7 jet ground-attack aircraft
Fiume Italian name for Rijeka, Yugoslavia formerly belonging to Italy
Five Arctic Countries Canada, Greenland, Norway, the United States, the USSR
Five Nations Cayugas, Oneidas, Onondagas, Mohawks, and Senecas (American Indian tribes on the English side in the French and Indian Wars)
Fjord Land Norway
Flag of Alfonso yellow-and-red emblem of Spain dating from fifteenth century when it was carried by Alfonso el Magnánimo
Flag Bay Bahía de Banderas, Mexico
Flagellum Dei (Latin—Scourge of God)—Attila the king of the Huns
Flagon-A NATO code name for Soviet SU-11 delta-wing fighter aircraft
Flashlight NATO name for Soviet Yakovlev Yak-25 two-place interceptor fighter aircraft
Flemish Colorist Peter Paul Rubens
Flemish Primitive Painter Gheeraert David
Flickertail(s) North Dakotan(s)
Flickertail State North Dakota
Flivver King Henry Ford
Flood City Johnstown, Pennsylvania
Floral Watercolorist William Demuth
Floreana Island, Galápagos Santa María
Florence Austral Florence Wilson
Florestan Robert Schumann
Florestan and Eusebius pseudonyms invented by Robert Schumann to represent the two sides of his character –Florestan was impetuous and passionate whereas Eusebius was the dreamer
Florida Ports Jacksonville, Port Everglades, Miami, Key West, Tampa, St Petersburg, Port St Joe, Panama City, Pensacola
Flour City Buffalo or Rochester, New York
Flower Capital Encinitas, California
Flower City Rochester, New York
Flower Garden of England The Sorlings or Isles of Scilly off Land's End
Flower King Carl von Linné (Linnaeus)
Flower of the Levant Zante in the Ionian Islands
Flower of Quakerism abolitionist Lucretia Mott
Flower Seed Capital of the West Santa María, California
Flower Town Brampton, Ontario
Flowertown in the Pines Summerville, South Carolina
Flower of the Transvaal Pretoria
Flowery Kingdom China
Flying Boxcar C-119 Fairchild-Hiller transport
Flying Dutchman mythical character immortalized in Richard Wagner's opera *Der Fliegende Holländer*; nickname of the baseball batting champion of the early 1900s —Honus Wagner
Flying Finn Paavo Nurmi

Foam City Milwaukee, Wisconsin

Fog City San Francisco

Fogfoundland fog-bound Newfoundland

Foggy Bottom the U.S. State Department

Football Capital of the South Birmingham or New Orleans

Foothill City Calgary, Alberta

Forbidden City Lhasa, Tibet

Forbidden Island privately-owned Niihau, westernmost island in Hawaii

Forbidden Kingdom Bhutan

Fordham Flash Frank Frisch

Ford Madox Ford Ford Madox Hueffer

Fordtown Detroit, Michigan

Foremost Left-Wing Liberal Mathematician-Philosopher Atheist of 20th Century Bertrand Russell

Forensic Psychiatrist Richard von Krafft-Ebing

Forerunner of the Reformation John Huss

Forerunner of Spanish-American Independence Francisco Miranda

Forest Cantons Swiss cantons of Lucerne, Schwyz, Unterwalden, and Uri

Forest City Cleveland, Ohio or London, Ontario

Forest City of the South Savannah, Georgia

Forest of Forests forested belt stretching from northern Norway to eastern Siberia

The Forgotten Man President Franklin D. Roosevelt's description of the American voter

Forgotten Philosopher Giordano Bruno

Former Naval Person code name of Prime Minister Churchill formerly First Lord of the Admiralty

Formidable Maestra conductor Antonia Brico

Formosa (Portuguese—beautiful, delightful, perfect)—name given Taiwan by Portuguese explorers

Fortaleza formerly Caerá

Fort Dimanche Haiti's infamous prison close to Pétionville

Fort Hill John C Calhoun's country seat in the Pendleton district of South Carolina near Anderson

Fortieth State South Dakota

Fort Liquordale Fort Lauderdale, Florida (when college students make their Easter vacation visit)

Fortunate Island Monhegan, Maine

Fortunate Islands Canary Islands

Fortunate Isles Madeira Islands—also called Isles of the Blest in classical mythology

Forty-eighth State Arizona

Forty-fifth State Utah

Forty-first State Montana

Forty-fourth State Wyoming

Forty Immortals collective nickname of the forty members of the French Academy

Forty-ninth State Alaska

Forty-second State Washington

Forty-seventh State New Mexico

Forty-sixth State Oklahoma

Forty-third State Idaho

Foster Mother of the Sciences Medicine

Founder of the AAAA Dr Charles L Andrews, founder American Association for the Advancement of Atheism

Founder of Agnosticism Thomas Henry Huxley

Founder of Agricultural Chemistry Justus von Liebig

Founder of the American Federation of Labor Sam(uel) Gompers

Founder of American Military Intelligence General Ralph H Van Deman

Founder of the American Navy John Paul Jones

Founder of Analytical Psychology Carl Gustav Jung

Founder of the Aniline Dye Industry Sir William Perkin

Founder of Antiseptic Surgery Joseph Lister

Founder of Art History and Criticism Giorgio Vasari

Founder of Bacteriology Ferdinand Cohn

Founder of Behaviorism John Watson

Founder of the Birth Control Movement Margaret Sanger

Founder of the Boy Scouts and Girl Guides Sir Robert Baden-Powell

Founder of Brazil Pedro Alvares Cabral

Founder of British Imperial India Robert Clive

Founder of Buddhism Prince Siddhartha (Gautama Buddah)

Founder of Buenos Aires Pedro de Mendoza

Founder of Cellular Pathology Rudolf Virchow

Founder of Chemistry Robert Boyle

Founder of Chicago Jean de Sable

Founder of Christian Science Mary Baker Eddy

Founder of Cleveland, Ohio Moses Cleaveland

Founder of the Columbia University School of Journalism Joseph Pulitzer

Founder of Comparative Anatomy Baron Georges Cuvier

Founder of Confucianism King Futzu (Confucius)

Founder of Conservative Surgery Sir William Fergusson

Founder of Continental Rationalism René Descartes

Founder of Cubism George Braque, Pablo Picasso

Founder of Cybernetics Norbert Wiener

Founder of Czech School of Music Bedrich Smetana

Founder of Dada Tristan Tzara

Founder of Detroit Sieur de Cadillac

Founder of Electromagnetism Michael Faraday

Founder of Electrophysiology Emil du Bois Reymond

Founder of English Empiricism Sir Francis Bacon

Founder of Episcopalianism Henry VIII

Founder of Experimental Hygiene Max von Pettenkofer

Founder of the Faculty of Physicians and Surgeons of Glasgow Peter Lowe

Founder of Fauvism Henri Emile Benoit Matisse

Founder of First Free Medical Clinic in the U.S. Benjamin Rush

Founder of the Franco-Belgian School of Violin Playing Charles Auguste de Bériot

Founder of French Colonial Empire Louis Faidherbe

Founder of French Grand Opera Daniel François Esprit Auber

Founder of French Opera Jean-Baptiste Lully

Founder of French Socialism Compte Claude Henri de Rouvroy de Saint-Simon

Founder of the Friends George Fox

Founder of Functionalism Louis Sullivan

Founder of Georgia James Oglethorpe

Founder of Gestalt Therapy Fritz Perls

Founder of Hadassah Henrietta Szold

Founder of Histology Marcello Malpighi

Founder of Homeopathy Christian Friedrich Samuel Hahnemann

Founder of Humanistic Psychology Abraham Maslow

Founder of Hungary Arpad

Founder of Iconographic and Physiologic Anatomy Leonardo da Vinci

Founder of Impressionism Claude Monet

Founder of Islam Mohammed

Founder of Jainism Mahavira also known as Vardhamana

Founder of Japanese Color-Print Making Iwasa Matabei

Founder of Judaism Moses

Founder of the Kelmscott Press William Morris

Founder of the Lutheran Church Martin Luther

Founder of Manila Miguel López de Legazpi

Founder of Medical Statistics Pierre-Charles Alexander Louis

Founder of the Methodist Church John Wesley

Founder of Modern Astronomy Nicolaus Copernicus (Nikolaus Kopernicki)

Founder of Modern Chemistry Antoine Lauret Lavoisier

Founder of Modern Existentialism Sören Kierkegaard

Founder of Modern German Sculpture Johann Gottfried Schadow

Founder of Modern Military Medicine Sir John Pringle

Founder of Modern Philosophy René Descartes

Founder of Modern Russian Literature Aleksandr Sergeyvich Pushkin

Founder of Modern Sculpture Donatello (Donato di Niccolo di Betto Bardi)

Founder of Mormanism Joseph Smith

Founder of Naturalist Movement in Literature Émile Zolá

Founder of Oklahoma Jean Pierre Chouteau

Founder of Optics Giovanni Battista della Porta

Founder of Osteopathy Andrew Taylor Still

Founder of Pakistan Dr Mohammed Ali Jinnah

Founder of Pediatric Cardiology Helen Brooke Taussig

Founder of Pennsylvania William Penn

Founder of Phenomenology Edmund Husserl

Founder of Philosophic Radicalism James Mill

Founder of Polaroid Photography Edwin Land

Founder of Positivism Auguste Compte; Immanuel Kant

Founder of Postimpressionism Paul Cézanne

Founder of Pragmatism C(harles) S(anders) Peirce

Founder President of Zambia Kenneth Kaunda

Founder of Professional Nursing Florence Nightingale

Founder of Providence, Rhode Island Roger Williams

Founder of Psychoanalysis Sigmund Freud

Founder of Psychology Wilhelm Wundt

Founder of Québec Samuel de Champlain

Founder of the Religious Society of Friends George Fox

Founder of Rhode Island Roger Williams

Founder of Rome Romulus

Founder of the Royal Navy Samuel Pepys

Founder of Russian Literature Alexander Pushkin

Founder of Salt Lake City Brigham Young

Founder of the Science of Eugenics Sir Francis Galton

Founder of Scientific History Thucydides

Founder of Scottish Presbyterianism John Knox

Founder of Secularism George Holyoake

Founder of Singapore Sir Thomas Stamford Raffles

Founder of Social Psychology Gustave Le Bon

Founder of Sociology Auguste Compte

Founder of State Socialism Louis Blanc

Founder of Swarthmore College Martha Ellicott Tyson

Founder of Taoism Lao-tse

Founder of Transcendental

Idealism Immanuel Kant

Founder of Transcendentalism Ralph Waldo Emerson

Founder of Troy Tros

Founder of Unitarianism John Biddle

Founder of the University of Pennsylvania Benjamin Franklin

Founder of the University of Virginia Thomas Jefferson

Founder of Uruguay José Gervasio Artigas

Founder of the U.S. Navy Captain John Paul Jones

Founder of the Venetian School of Painting Giovanni Bellini

Founder of Vermont Ira Allen

Founder of Victimology Hans von Hentig or Benjamin Mendelsohn

Founder of Zoroastrianism Zoroaster also known as Zarathustra

Founders of Christianity disciples of Jesus Christ

Founders of Cubism Georges Braque and Pablo Picasso

Founders of Flemish Painting the van Eyck brothers—Hubrecht and Jan

Founders of French Romantic Painting Delacroix, Géricault, and Gros

Founders of the Hudson River School (of painting) Thomas Cole and Asher Brown Durand

Founders of Neo-Impressionism Georges Seurat and Paul Signac

Founders of Scientific Socialism Karl Marx and Friedrich Engels

Founding Father City Austin, Texas named in honor of Stephen F Austin

Founding Father of Israel David Ben-Gurion

Founding Fathers of Economics Adam Smith and David Ricardo

Founding Father of the Smithsonian Joseph Henry

Founding Philosopher of Modern Capitalism Adam Smith

Fountain of Youth St Augustine, Florida

Four-C City El Paso, Texas (famous for its cattle, climate, copper, and cotton)

Four Corners boundary-line junction of Arizona, Colorado, New Mexico, and Utah

four-dimensional science geology involving the application of biology, chemistry, mathematics, and physics

Four Forest Cantons Lucerne, Schwyz, Unterwalden, and Uri—all in Switzerland

Four Lakes City Madison, Wisconsin

Four Seasons Crossroad of New England Manchester, New Hampshire

Fourteenth State Vermont

Fourth Bureau Red Army bureau in charge of overseas intelligence-gathering activities of the Soviet Union

Fourth Estate The Media—press, radio, television

Fourth Gospel Gospel according to Saint John

Fourth International Trotsky-oriented organization rejecting the Second and Third Internationals in the direction of the class struggle

Fourth State Georgia

Fourth World poorest of the Third World nations

Four Tigers Hong Kong, Singapore, South Korea, Taiwan

Four Winds Boreas (north), Eurus (east), Notus (south), Zephyrus (west)

Foxardo (naval argot—Fajardo, Puerto Rico)

Fox Populi Charles James Fox

Fra Angelico Giovanni da Fiesole

Fra Bartolommeo Baccio della Porta

Fra Elbertus Elbert Hubbard

Fragment of Eden Seychelles Islands in the Indian Ocean

Fragrant Harbor Hong Kong

Framer of the Bill of Rights James Madison

Framer of the *Declaration of Independence* Thomas Jefferson who rewrote Thomas Paine's draft with the aid of John Adams and Benjamin Franklin

Framer of the French Revolution Denis Diderot

Frances Alda Frances Davis

France's Largest Port Marseille

Franche-Compté Burgundy

Francis Beeding John Leslie Palmer's pseudonym

Franciscan Wine Capital Würzburg, Germany

Franco-Hispanic Co-Principality Andorra (between France and Spain)

Francoise Sagan pseudonym—Françoise Quoirez

François Villon François de Montcorbier

Francophone Africa Afars and Issas, Algeria, Burundi, Cameroon, Central African Republic, Chad, Congo, Dahomey, Gabon, Guinea, Ivory Coast, Madagascar, Mali, Mauritania, Mauritius, Niger, Reunion, Rwanda, Senegal, Seychelles, Togo, Tunisia, Upper Volta, Zaire

Francophone America French Guiana; Guadeloupe; Haiti; coastal parishes of Louisiana; Martinique; northern New York, Vermont, New Hampshire, Maine, and New Brunswick; Québec; St Pierre and Miquelon

Francophone Europe Andorra; French-speaking Belgium, France, Luxembourg, Monaco; French-speaking cantons of Switzerland

Francophone Pacific French Polynesia, New Caldonia, New Hebrides, Wallis and Fatuna Islands

Francophone Province Québec

Frank Leslie business name of Henry Carter

Frank Richards Charles Hamilton's pen name

Franz Josef Land Arctic islands called Zemlya Frantsa Iosifa by the Russians

Fred Alan John Sullivan

Fred Astaire Frederick Austerlitz

Frederick Douglass Frederick Augustus Washington Bailey

Frederick the Great Frederick II of Prussia

Frederic March Frederich McIntyre Bickel

Fred Niblo Frederico Nobile

Freedom Defender U.S. Supreme Court Justice William O Douglas

Freedom Fighter former name of the Northrup Tiger II or F-5 tactical fighter plane

Free and Hanseatic City Hamburg

Free State Maryland

Freestone State Connecticut

Freethinker Activist-Author-Editor-Publisher Anne Nicol Gaylor

Freethinker Activist Publisher Herbert Schapiro

Freethinker American Presidents George Washington, John Adams, Thomas Jefferson, James Madison, John Quincy Adams, Abraham Lincoln, Andrew Johnson, Ulysses Simpson Grant, James Abram Garfield, Theodore Roosevelt, William Howard Taft

Freethinker American Statesmen Benjamin Franklin and Henry Clay

Freethinker Anthropologists Franz Boas, Earnest Albert Hooten, Ales Hrdlicka, and Margaret Mead

Freethinker Astronomers Simon Newcomb, Carl Sagan, Tycho Brahe, Nicolaus Copernicus, and Galileo Galilei

Freethinker Author-Composer-Editor Carl Shapiro

Freethinker Author-Editor-Publisher Jane Kathryn Conrad

Freethinker Author-Editor-Translator Max Forrester Eastman

Freethinker Bacteriologist Hans Zinsser

Freethinker Biochemist Isaac Asimov

Freethinker Biographer Joseph Lewis

Freethinker Botanists Luther Burbank and George Washington Carver

Freethinker of Canada Marshall Jerome Gauvin

Freethinker Composers Charles Ives, Maurice Ravel, and Dmitri Shostakovich

Freethinker Conservationists John Burroughs, William Temple Hornaday, John Muir, Henry David Thoreau

Freethinker Dramatist-Critic George Bernard Shaw

Freethinker Economist and Novelist Harriet Martineau

Freethinker Editor, Printer, and Writer Elbert G(reen) Hubbard

Freethinker Editor, Satirist, and Scholar H(enry) L(ouis) Mencken

Freethinker Educators Mortimer Adler, Nicholas Murray Butler, John Dewey

Freethinker Educator-Historian-Sociologist Harry Elmer Barnes

Freethinker Electrical Engineers Charles Proteus Steinmetz and Nikola Tesla

Freethinker Encyclopedists in

France Jean le Rond d'Alembert, Denis Diderot, Charles Montesquieu, Francois Quesnay, Jean Jacques Rousseau, Anne Robert Jacques Turgot, and Voltaire

Freethinker Essayist-Physician-Poet-Novelist Oliver Wendell Holmes

Freethinker Essayist-Poet Philosopher Ralph Waldo Emerson

Freethinker Evolutionists Julian Huxley and his grandfather Thomas Henry Huxley

Freethinker Explorers William Beebe, Jacques Costeau, Alexander von Humboldt

Freethinker Fabians Beatrice and Sidney Webb

Freethinker Family Planner Margaret Sanger—founder of the American Birth Control League

Freethinker Feminist Molly Yard—president, National Organization for Women

Freethinker Geologist Sir Charles Lyell

Freethinker Herpetologist-Mammalogist Raymond L(ee) Ditmars

Freethinker Historians David Hume, Will and Ariel Durant, Harry Allen Overstreet, John Eleazer Remsburg, James Harvey Robinson, Hendrik Willem van Loon

Freethinker Horticulturalist Luther Burbank

Freethinker-Humorist-Philosopher-Television Teacher Steve Allen

Freethinker Ichthyologists Eugene Willis Gudger and David Starr Jordan

Freethinker Inventor Thomas Edison

Freethinker Inventor-Printer-Philosopher-Scientist Benjamin Franklin

Freethinker Lawyers Clarence Darrow, Josiah Quincy, and Samuel Untermeyer

Freethinker Libertarian Patriots Ethan Allen, Patrick Henry, and John Paul Jones

Freethinker Libertarian-Patriot-Pamphleteer Thomas Paine

Freethinker Literary Critics Georg Morris Cohen Brandes and Edmund Wilson

Freethinker Mathematician-Philosopher Bertrand Russell

Freethinker Naturalist-Phi-

losopher Henry David Thoreau

Freethinker Novelists Ambrose Bierce, Aldous Huxley, George Orwell, Edgar Allan Poe, Upton Sinclair, Mark Twain, H(enry) G(eorge) Wells, Thornton Wilder, Émile Zolá

Freethinker Novelist-Journalist James T(homas) Farrell

Freethinker Orator Robert Green Ingersoll

Freethinker Orator-Reformer Frances (Fanny) Wright

Freethinker Paleontologist William King Gregory

Freethinker Pathologists Simon Flexner and Hideyo Noguchi

Freethinker Philanthropists Andrew Carnegie, Peter Cooper, Stephen Girard, and James Lick

Freethinker Philosophers Socrates, Epicurus, Lucretius, John Stuart Mill, Thomas Hobbes, John Locke, Benjamin Franklin, Thomas Paine, Ralph Waldo Emerson, Karl Marx, Robert Green Ingersoll, Friedrich Nietzsche, William James, George Santayana, John Dewey, Bertrand Russell, Will and Ariel Durant, Mortimer Jerome Adler

Freethinker Philosopher-Poets Ralph Waldo Emerson and Friedrich Wilhelm Nietzsche

Freethinker Physicians Jonas E(dward) Salk and Benjamin M(cLane) Spock

Freethinker Physicist Albert Einstein

Freethinker Physicist-Chemist Marie Sklodowska Curie

Freethinker Physiologist Anton Julius Carlson

Freethinker Poets George Gordon Byron, Leigh Hunt, Rudyard Kipling, Edgar Lee Masters, Edgar Allan Poe, Percy Bysshe Shelley, Walt Whitman

Freethinker Political Reformer Charles Bradlaugh

Freethinker Psychologist Havelock Ellis *(Studies in the Psychology of Sex)*

Freethinker Publisher and Writer Emanuel Haldeman-Julius

Freethinker Quaker Minister Lucretia Mott

Freethinker Rationalist Phi-

losopher Joseph McCabe

Freethinker Reprint Publisher Dan Pezze

Freethinker Researcher Walter Reed

Freethinker Rhymer J Ashleigh Burke

Freethinker Sanitarian William Crawford Gorgas

Freethinker Satirist Voltaire

Freethinker Senator Henry Clay

Freethinker Suffragettes Susan Brownell Anthony and Elizabeth Cady Stanton

Freethinker Zoologists Charles Darwin, Ernst Heinrich Haeckel, Willard Gibbs Van Name, Alfred Russel Wallace

Freethought Author and Publisher E(manuel) Haldeman-Julius whose *Little Blue Books* sold for 5¢ a copy

Freethought Giants Charles Bradlaugh, Georg Morris Brandes, Clarence Darrow, Robert Green Ingersoll, Emanuel Haldeman-Julius, Joseph Lewis, Henry Louis Mencken, Thomas Paine, Bertrand Russell, George Bernard Shaw, Voltaire

Freeway City Los Angeles

French French bread; French bull (small breed of dog); French chalk (tailor's talc); French cuff; French curve (drafting instrument); French door; French dressing; French endive (blanched chicory); French fries; French harp (harmonica); French heel (high curved heel); French horn; French ice cream (made with cream and eggs); French kiss; French pancake (thin and sweet); French pastry; French polish (alcohol + shellac); French pox (syphilis); French roll (women's coiffure); French roof (mansard-style); French seam (completely covered seam); French system (spinning system); French tamarisk (salt cedar); French telephone; French toast

French agnostics Auguste Compte, Blaise Pascal

French Aero Pioneers Etienne and Joseph de Montgolfier

French Antilles French West Indies

French Canada French-speaking Canada, mainly the Province of Québec

French-Canadian Conductor

Sir Wilfred Pelletier
French-Canadian Freethinker Marshall Jerome Gauvin (1881—1978)
French Caribees colonial name for the French West Indies
French Century the 18th century—the 1700s
French Community metropolitan France together with its overseas departments, territories, and former territories *(Communauté française)*
French disease syphilis also known as the Italian disease or the Spanish disease as well as *morbus gallicus* (Latin—Gallic disease)—the French disease
French Equatorial Africa former colonies Benin or Dahomey, Cameroon, the Central African Empire, the French Congo, Gabon, and Guinea
French India former French possessions in India (Chandernagore, Pondicherry, etc.)
French Indo-China former name of area comprising Annam, Cambodia, Chochin China, Laos, Tonkin, and Vietnam
French Morocco eastern Morocco closest to Algeria and the Sahara
French Polynesia French island possessions in the South Seas
French Ports Dunkerque, Calais, Boulogne-sur-Mer; Le Treport, Dieppe, Fecamp, Le Havre, Rouen, Cherbourg, Granville, St Malo, Brest, Douarnenez, Aupierne, Port Louis, Lorient, Le Palais, Le Croisic, Saint Nazaire, Donges, Paimboeuf, Basse-Indre, Les Asables Dolonne, La Pallice, La Rochelle, Rochefort, Tonnay-Charente, Le Verdon, Mortagne, Trompeloup, Paulillac, Blaye, Ambes, Le Marquis, Bordeaux, Arcachon, Boucau, Bayonne, Biarritz, Port Vendres, Port La Nouvelle, Sete, Port St Louis du Rho, Port de Bouc, Berre Letang, Marseille, La Ciotat, Toulon, Cannes, Nice, Villefranche, Bastia and Ajaccio (on Corsica), Menton
French Quarter Vieux Carré in New Orleans
French Revolutionary Calen-

dar (see *Vend, Brum, Frim, Niv, Pluv, Vent, Germ, Flor, Prair, Mess, Therm, Fruc,* entries)
French Riviera resort areas along the Mediterranean from Marseilles to Menton, including Cannes, Monaco, and Nice
French Sahara former colonial areas such as Algeria, French Morocco, Mauritania, and Niger
French Shore Newfoundland's northern and western coasts
French Somaliland *Côte Française des Somalis* (French Coast of the Somalis)—now known as Djibouti
French-speaking Places (*see entries under* Francophone)
French Sudan former name of Mali
French Switzerland French-speaking areas of Switzerland
French Togoland former name of Togo
French Union France plus its overseas colonies and departments as well as all its former possessions
French West Africa former colonies of France, Algeria, Chad, French Morocco, Mali, Mauritania, Niger, Senegal, and Upper Volta
French West Indies Desirade, Guadeloupe, Les Saintes, Marie Galante, Martinique, Petite Terre, Saint Bartholomew (Barthelemy), Saint Martin (French half)
Friar Antonio Agapida pseudonym of Washington Irving
Frida and Diego Frida Kahlo and Diego Rivera
Fridjof Nansen Land formerly Franz Josef Land (Arctic island group in Queen Victoria Sea north of Barents Sea sector of Arctic Ocean)
Friend of the American Revolution Caron de Beaumarchais
Friend of Helpless Children Herbert Clark Hoover—thirty-first President of the United States
Friendliest Town in the West Geraldton, Western Australia
Friendly Island Molokai, Hawaii; St Maarten, Netherlands Antilles
Friendly Islands Tonga Islands in the South Pacific
Friendly Kingdom Tonga Is-

lands
Frog(s) Anglo-American slang for French (people)
Frontier Fighter Davy Crockett
Frontier Hero Daniel Boone
Frontier States Alaska and Hawaii
Frostbite nickname of Fairbanks, Alaska
Fruit Bowl of the Nation Yakima, Washington
Frunze modern name of Pishpek in Kirgizia
fuel of the future solar power
Führer (German—Leader)—Hitler's title
Fulton's Folly inventor Robert Fulton's steamship *Clermont*
Fum the Fourth nickname of George IV
Fun Capital of Scandinavia Copenhagen, Denmark
Fun City New York
Fundador de la República (Spanish—Founder of the Republic)—José Nuñez de Cáceres—founder and first president of the Dominican Republic (Spanish Haiti)
Fundador de Nueva Granada (Spanish—Founder of New Granada)—Francisco de Paula Santander—founder of Colombia (Nueva Granada)
funeral order maritime tradition of older persons standing back to give younger people first chance when boarding lifeboats or using other lifesaving equipment; (*see* Birken'ead drill)
Fungus Corners (naval argot—Bremerton, Washington)—a rainy port
Fun-Loving Philosopher Mark Russell
Fun and Sun Cities Acuña, México across the Rio Grande from Del Rio, Texas
Fuss and Feathers General Winfield Scott, USA
The Fuzz [slang—detective(s); law-enforcement officer(s); police]
Gabon Ports Cocobeach, Libreville, Port Gentil, Sette Cama, Gamba Oil Terminal
Gabriel Israeli surface-to-surface missile
Gabriela Mistral Lucila Godoy de Alcayaga
Gabriel d'Annunzio Gaetano Rapagnetta
Gabriel Padecopeo Lope de Vega

Gaby Gabrielle Dupont
Gail Hamilton Mary Abigail Dodge
Gaillard Cut formerly called Culebra Cut (in the Panama Canal)
Gainful Soviet surface-to-air missile also designated SA-6 (NATO)
Galeb Yugoslav two-place single-engine jet aircraft also known as the Seagull
Galen (sometimes Galin) Vasily Konstantinovich Blücher
Gambia's Port Georgetown
Gambling Capital of the Far East Macao, Portuguese China
Gambling Capital of the Far West Las Vegas, Nevada
Ganef Soviet SA-4-type mobile surface-to-air missile (NATO)
Gangland Chicago
Gannet Westland three-place early-warning aircraft developed in Britain
Garbage Dump California's San Quentin Prison; Green Meadow Correctional Facility at Comstock, NY
Garden of the Andes Mendoza, Argentina
Garden of the Antilles St Croix, Virgin Islands
Garden of Argentina Tucuman
Garden of Canada Ontario
Garden of the Caribbean Puerto Rico
Garden City of Georgia Augusta
Garden City of India Mysore
Garden City of New Zealand Christchurch
Garden of Denmark Fyn or Funen—home of Hans Christian Andersen
Garden of the East Burma, Malaysia, and Sri Lanka
Garden of England Kent and Worcester
Gardener Touched With Genius Luther Burbank
Garden of France Amboise and Touraine
Garden of God ancient eponym of Lebanon just north of the Holy Land
Garden of the Gods park near Colorado Springs, Colorado
Garden of the Gulf Prince Edward Island in the Gulf of St Lawrence
Garden of Ireland Carlow
Garden Island Kauai, Hawaii
Garden Isle of Micronesia Ponape

Garden of Italy Sicily
Garden of Love Shalimar waterside garden on Kashmir's Dal Lake
Garden of Maine Aroostook County
Garden of the Morning Breeze Naseem Bagh on Kashmir's Dal Lake
Garden of Paradise in the Sea Madeira
Garden Province Canada's Prince Edward Island
Garden of the Southwest Southern California
Garden of Spain fields of Andalucía and Valencia
Garden State New Jersey
Garden of the Sun Indonesia
Garden of Sweden Blekinge
Garden of Switzerland Thurgau
Garden of Wales southern Glamorganshire
Garden of the West California, Kansas
Garden of the World Mississippi River Valley
Gargantua François I of France (or) Henri d'Albret—King of Navarre
Garlic Capital Gilroy, California
Garry Moore Thomas Garrison Morfit
Gary Cooper Frank J Cooper
Gas House of the Nation Washington, D.C.
Gaskin Soviet SA-9 air-defense missile system contained in an amphibious armored vehicle (NATO)
Gasopolis Los Angeles (on smog-filled days); other places with similar air pollution
Gasparilla (Spanish—Little Gaspar)—nickname of José Gaspar—pirate active along west coast of Florida around 1750
Gastown waterfront area of Vancouver, BC
Gate City Keokuk, Iowa; Laredo, Texas; St Louis, Missouri
Gate City of the South Atlanta, Georgia
Gates of the Arctic national park in north-central Alaska
Gates of Hell old nickname for the entrance to Macquarie Harbour on the Indian Ocean coast of Tasmania when it was a penal settlement
Gate of Tears Bab-el-Mandeb

Strait linking Gulf of Aden and Indian Ocean with the Red Sea; many sailors call it Gate of Hell because of its desert-heated winds
Gateway to Alaska Seattle, Washington
Gateway to the Alps Zurich
Gateway to America Ellis Island immigration station, New York; New York City
Gateway Arch City St Louis, Missouri
Gateway to the Arctic Fairbanks, Alaska
Gateway to the Bahamas Bimini Island off Miami
Gateway to the Big Bend National Park Marfa, Texas
Gateway to the Caribbean Tampa, Florida
Gateway City nickname of Pittsburgh, Pennsylvania, after the Revolutionary War
Gateway to the Dakotas Sioux Falls, South Dakota
Gateway of the Day Fiji Islands near the International Date Line
Gateway to the East Port Said, Egypt
Gateway to Eastern India Calcutta
Gateway to the Golden Isles Brunswick, Georgia
Gateway to the Great Seaway Green Bay, Wisconsin
Gateway to the Gulag Magadan on the Sea of Okhotsk
Gateway to India Bombay
Gateway to Israel Haifa
Gateway to Japan Yokohama
Gateway to Lapland Rovaniemi, Finland
Gateway to Latin America Miami, Florida
Gateway to Moroland Zamboanga, Mindinao
Gateway to Mount Rainier Tacoma, Washington
Gateway to the Negev Beersheba, Israel
Gateway to the North North Bay, Ontario
Gateway to Northern Europe Göteborg (Gothenburg), Sweden
Gateway to the NY-NJ Market Bayonne, New Jersey
Gateway to Parris Island Beaufort, South Carolina
Gateway to the Rhine Valley Bonn, Germany
Gateway to the Smokies Knoxville, Tennessee
Gateway to South America

Columbia (with ports on the Atlantic and the Pacific) Gateway to Southwest Japan Kobe

Gateway States California, Louisiana, New Jersey, New York

Gateway to the West Pittsburgh, Pennsylvania, and St Louis, Missouri

Gateway to Western India Bombay

Gateway to the Yukon Skagway, Alaska

GATF Graphic Arts Technical Foundation

Gator Bowl Jacksonville, Florida

'Gator State Alligator State (Florida)

Gaucho Land Uruguay

Gautama Buddha Prince Siddhartha

Gavin Ogilvy James M Barrie

Gay City San Francisco

Gay Gateway to Europe Copenhagen (København)

Gay Paree Paris, France

Gay White Way New York City's Broadway in the 42nd Street and Times Square area

Gazelle Embracer of Brazil's version of the Aerospatiale SA-341 observation helicopter

Gecko Soviet SA-8 missile system (NATO)

Gem Beside the Amstel Amsterdam

Gem of the Mountains Idaho

Gem of the South Pacific New Zealand

Gem State Idaho

General Booth Salvation Army founder William Booth

General Bor Tadeusz Komorowski—cavalry leader of 63-day uprising of Polish underground against Germans occupying Warsaw in 1944

General Douglas pseudonym of Soviet corps commander Yakov Smuskevich while leading the Spanish Republican air force in 1936–37

General John nickname of the first Duke of Marlborough —John Churchill

General Kleber pseudonym of Soviet general Grigory Shtern while serving as chief advisor to the Spanish Republican army in 1936–37)

General's Lady Martha Washington—wife of General George Washington

General Secretary of the USSR Mikhail Gorbachev (1985–1991)

General Tom Thumb Charles S Stratton

General Tubman Harriet Ross Tubman of Underground Railroad fame

Geneva Cross (see Cross of Geneva)

Genghiz Khan (Mongolian— Prefect Warrior)—his empire stretched from the China Sea to the Dnieper and his subjects called him Ruler of the World

Genie Douglas air-to-air rocket fitted with a nuclear warhead and designated AIR-2A

Genius of the Renaissance Leonardo da Vinci

Gentleman Boss Chester Alan Arthur

Gentleman Explorer Giovanni da Verrazano

Gentleman Jim prizefighter James John Corbett

Gentleman Johnny General John Burgoyne, also a noted British playwright

Gentle Peasant-Prince—the loving Cotter-King Robert Burns, according to Robert G Ingersoll in his poem The Birthplace of Burns

Gentle Rebel of Psychoanalysis Karen Horney

Geoffrey Crayon Washington Irving

Geoffrey Homes Daniel Mainwaring's pseudonym

Geographical Center of North America Rugby, North Dakota

Geordieland Newcastle-on-Tyne area of northeastern England

Geordies people from the coalmining and industrial area of Newcastle-on-Tyne

Geordie(s) Newcastle (persons)

Georg Brandes Morris Cohen

George Arliss Augustus George Andrews

George Bellairs Harold Blundell's pseudonym

George Bizet Alexandre César Léopold Bizet

George Brent George Nolan

George Burns Nathan Birnbaum

George Eliot Mary Ann Evans Cross

George Gershwin Jacobi Gershvin

George Gissing J Storer Glous-

ton

George London George Burnstein

George Orwell Eric Blair

George Raft Georg Ranft

George Sand Amandine Aurore Lucie Dupin (Baroness Dudevant)

Georges Duhamel Denis Thevenin

George Spelvin John Chapman

Georges Simenon (pen name— Georges Sim)

George Szell György Széll

Georgia Ports Savannah, Brunswick, Fernandina Beach

Georgia's Oldest City Savannah

Georgi Vladimov Georgi Volosevich

Gerard de Nerval (pseudonym—Gerard Labrunie)

German German camomile (tea); German ivy (South African ivy); German knot (figure-8 knot); German lapis (imitation lapis lazuli); German measles (rubella); German shepherd (police dog); German silver (copper-nickelzinc alloy resembling silver)

German Aeronautical Engineering Genius Willi Messerschmitt

German Africa former colonies (Cameroons, German East Africa, German Southwest Africa, Togoland)

German Agnostic Immanuel Kant

German Cradle of U.S. Presidents Germany—ancestral home of Presidents Eisenhower and Hoover

German East Africa colonial possession of Germany from 1885 to 1916; included most of Tanganyika

German Hanseatic Seaport Cities Bremen, Danzig, Hamburg, Lübeck

German Ocean North Sea

Germanophone Countries Austria, Germany, Liechtenstein, Luxembourg, German-speaking cantons of Switzerland, and places in the United States where German dialects such as Pennsylvania German (Pennsylvania Dutch) are spoken

German Ports Warnemunde, Rostock, Wismar

Germans the Germans [Former President Nixon's Chief of Staff—HR (Bob) Haldeman and Domestic Adviser John Erlichman]

german silver 50% copper, 30% nickel, 20% zinc

German South-West Africa colonial possession of Germany from 1884 to 1915 Namibia

German-speaking Places (*see* Germanophone Countries)

Germany Federal Republic of Germany *Bundesrepublik Deutschland*

Germany's Largest Port Hamburg

Gertie Lawrence Gertrude Lawrence (Gertrud Alexandra Dagmar Lawrence Klasen)

Gettysburg Battlefield Painter Henri Emmanuel Félix Philippoteaux

Ghana Ports Takoradi and Tema

Ghirlandaio Domenico di Tomaso Bigordi

Giacomo Meyerbeer Jakob Liebmann Beer

Giant of Danish Literature Hans Christian Andersen

Giant of Freethought Charles Bradlaugh

Giant of Mexican Music Carlos Chavez

Giant in the Nursery Jean Piaget

Gilbert Roland Luis Antonio Dámaso de Alonso

Gilded Age opulent post-Civil War period in the United States

Gilois French-built scissors bridge mounted on a tank and useful in spanning canals, ditches, and small streams

Ginger Rogers Virginia McMath

Gippsland Victoria—Australia's best-endowed province

Glacier Bay Alaskan national park

Glass Capital of Massachusetts Boston

Glass Capital of New York Corning

Glass Capital of Ohio Toledo

Glass Capital of Pennsylvania Pittsburgh

Glass Center Toledo, Ohio

Glass House the glassed-in Los Angeles County Jail in California

Glass Menagerie on the East River United Nations headquarters

Glenard's disease prolapse of one or more internal organs

Glen Ford Gwyllyn Ford

Glengariff (Irish Gaelic—rough valley)—Bantry Bay

Glimmerglass James Fenimore Cooper's nickname for Lake Otsego at Cooperstown, New York

Glitter Gulch Reno, Nevada

Globemaster Douglas transport designated C-124 and built for cargo carrying or flying 200 troops

Gloomy Gus Gustav Mahler

Glorious Fifty the United States of America

Glorious Fourth July 4 (Independence Day in the U.S.A.)

Glorious Gloria Gloria Swanson

Glorious Patriarch of Esterhazy Franz Joseph Haydn

Glubb Pasha John Bagot Glubb

Gnomes of Zürich Swiss bankers

Goa Soviet SA-3 air-defense missile system

God of Animals, Crops, Fertility, Prophecy, and Rural Life Faunus (Roman); Pan (Greek)

God of Battle, Inspiration, and Death Odin (Norse)

God of Blacksmithing and Forges Hephaistos (Greek); Vulcan (Roman)

God of Blame and Ridicule Momus (Roman)

God of Bloodshed and War Greek god Ares; Roman god Mars

God of Boundaries Terminus (Roman)

God of the Christians and Jews Jehovah

God of Corn and Grain Robigus (Roman)

God of Creation and Destruction Siva (Hindu)

God of Cunning Dexterity Hermes (Greek); Mercury (Roman)

God of the Dead and the Underworld Dis (Roman); Hades (Greek); Mantus (Etruscan); Pluto (Roman)

God of Death Mors (Roman); Thanatos (Greek)

Goddess of Agriculture Ceres or Vacuna (Roman); Demeter (Greek)

Goddess of Animals, Crops, Fertility, Prophecy, and Rural Life Bona Dea or Bona Mater or Fauna (Roman)

Goddess of Arts, Crafts, and Sciences Athena (Greek); Minerva (Roman)

Goddess of Athens Athena

Goddess of Avenging Justice Nemesis (Greek)

Goddess of Beauty and Love Aphrodite (Greek); Venus (Roman)

Goddess of Birth the Roman goddess Carmenta also known as Carmentis

Goddess of the Breeze Aura (Greek)

Goddess of Bridesmaids Juno Pronuba (Roman)

Goddess of Burials, Corpses, and Funerals Libitina (Roman)

Goddess of Cattle and Pastures Pales (Roman)

Goddess of Chance Fortuna (Roman)

Goddess of Chaos, Sickness, and Death Kali (Hindu)

Goddess of Childbirth and Prophecy Carmenta, Juno Lucina, and Postverta

Goddess of the Crops Auxesia and Demeter

Goddess of the Dead Mania (Roman)

Goddess of Death Hel (Norse folklore)

Goddess of Destiny or Fate Necessitas (Roman)

Goddess of Discord and Strife Discordia (Roman) or Eris (Greek)

Goddess of Domestic Animals Bubona (Roman)

Goddess of Earth Gaea or Rhea (Greek); Cybele, Tellus, or Terra (Roman)

Goddess of Fair Speech and Good Report Eufemia (Greek)

Goddess of Faith, Honesty, and Oaths Fides (Roman)

Goddess of Fame Fama (Roman); Pheme (Greek)

Goddess of Family Harmony Verplaca (Roman) also spelled Virplaca

Goddess of Famine Fames

Goddess of the Fertile Earth Opalia or Ops (Roman)

Goddess of Fertility Frigga (Queen of Asgard and wife of Odin in Nordic mythology)

Goddess of Fertility, Love, Lust, and War Ishtar (Assyrian and Babylonian)

Goddess of Fertility and Procreation Aphrodite (Greek); Isis (Egyptian); Mylitta

(Assyrian); Venus (Roman)

Goddess of Fertility and Purity Bona Dea (Roman)

Goddess of Fire Hestia (Greek); Vesta (Roman)

Goddess of Flowers, Gardens, and Love Flora (Roman)

Goddess of Freedom Libertas (Roman)

Goddess of Fruit Trees Pomona (Roman)

Goddess of Funerals Naenia (Roman)

Goddess of the Future Antevorta (Roman)

Goddess of Gardens and Fruit Trees Pomona (Roman)

Goddess of Good Faith Fides

Goddess of Groves, Orchards, and Woods Feronia (Roman)

Goddess of Harmony Concordia (Roman)

Goddess of Healing Iaso (Greek)

Goddess of Health Hygeia (Roman); Hygieia (Greek)

Goddess of the Hearth Hestia (Greek); Vesta (Roman)

Goddess of Heaven Hera (Greek); Juno (Roman)

Goddess of the Home Hera (Greek); Juno (Roman)

Goddess of Home Security Cardea or Carna who guarded the door hinges and locks

Goddess of Horses Epona (Gallic); Hippona (Roman)

Goddess of Hunting and the Moon Artemis (Greek); Diana (Roman)

Goddess of Imposters and Thieves Laverna (Roman)

Goddess of Law and Order Eunomia or Themis (Greek); Justitia (Roman)

Goddess of Leisure and Repose Vacuna (Roman)

Goddess of Lighting Fulgora (Roman)

Goddess of Love and Lust Aphrodite (Greek); Venus (Roman)

Goddess of Magic, Sorcery, and the Underworld Hecate or Hekate (Greek—working afar); Trivia (Latin—of the three ways) and hence the Romans placed her wherever three roads met

Goddess of Married Women Juno Matronalis (Roman)

Goddess of Memory Mnemosyne (Greek)—mother of the muses; her name gives rise to mnemonic—an aid to memory

Goddess of Menstruation

Mena (Roman)

Goddess of Midwives Deverra (Roman); Eileitia (Greek)

Goddess of Mirth Thalia

Goddess of the Moon Luna (Roman); Selene (Greek)

Goddess Mother of the World Mount Everest

Goddess of Nature Cybele (Roman) or Kubele (Greek) —sometimes called Mistress of the Animals

Goddess of Newborn Babes Levana (Roman)

Goddess of Night Nux (Greek) sometimes spelled Nyx

Goddess of Nursing Mothers Rumina (Roman)

Goddess of Passion Stimula (Roman)

Goddess of the Past Postvorta (Roman)

Goddess of Peace known to the Romans as Concordia, Irene, or Pax, and to the Greeks as Eirene

Goddess of Profit Laverna (Roman)

Goddess of Public Welfare Salus (Roman)

Goddess of the Rainbow Iris (Roman)

Goddess of Robbers Furina (Roman)

Goddess of Rome Roma

Goddess of the Sea and Seaports Matuta (Roman)—originally goddess of the dawn

Goddess of Sensual Pleasure Voluptas (Roman)

Goddess of Shepherds Pales (Roman)

Goddess of Silence Muta (Roman)

Goddess of Storms and Winds Tempestes (Roman)

Goddess of Suckling Infants Rumina (Roman)

Goddess of Treachery Fraus (Roman)

Goddess of Truth Alethia (Greek); Veritas (Roman)

Goddess of the Underworld Persephone (Greek); Proserpina (Roman)

Goddess of Vice Kakia (Greek)

Goddess of Victory Nike (Greek)

Goddess of Virgins Juno Virginalis (Roman)

Goddess of War Bellona (Roman); Enyo (Greek)

Goddess of Wisdom Athena (Greek); Minerva (Roman)

Goddess of Youth Hebe (Greek); Juventus (Roman)

God of Dreams Morpheus —Greek god of dreams and sleep

God of Drinking Comus (Roman)

God of Earth Tellumo (Roman)

God of Eloquence and Oratory Hermes (Greek); Mercury (Roman)

Godfather of American Invention Thomas Jefferson

Godfather of American Liberty Thomas Paine

Godfather of Navigation Nathaniel Bowditch

Godfather of Organized Crime in the U.S. Meyer Lansky

God of Fertility Frey (Norse); Priapos (Greek); Priapus (Roman)

God of Fields, Pastures, Shepherds, and Woods Faunus (Roman); Pan (Greek)

God of Fire Agni

God of Fire and Forges Hephaestus (Greek); Vulcan (Roman)

God of Forests, Herds, Plants, and Trees Silvanus (Roman)

God of Gods and Ruler of Heaven and Earth Zeus (Greek); Jove or Jupiter (Roman)

God of Gold Mammon the Materialist

God of Good Harvests and Successful Undertakings *Eventus Bonus* (Latin—Good Results)—a Roman god

God of the Greeks Panhellenius or Zeus

God of Healing and Medicine Asclepius (Greek); Aesculapius (Roman)

God of Heaven Uranus (Greek); Coleus (Roman)

God of Heaven, Lightning, Rain, Storm, and Thunder Indra (Hindus)

God of Inanimate Dreams Phantastus (Greek)

God of the Infernal Regions Dis (Greek); Pluto (Roman); Yama (Hindu); etc.

God-Intoxicated Man Benedictus de Spinoza

God of Landmarks Terminus (Roman)

Godless Victorian Sir Leslie Stephen

God of Light Mithra (Aryan, Indian, Persian)

God of Love Cupid (Roman); Eros (Greek); Krishna

(Hindu)
God of Marriage Hymen, according to Greek mythology, also leader of the nuptial chorus and personification of the wedding feast
God of the Mohammedans Allah
God of Music Johann Sebastian Bach, according to Pablo Casals
God of Music, Poetry, and the Sun Apollo (Roman) or Apollon (Greek)
God of the Nile and Vegetation Osiris (Egyptian)
God of Purification Februus (Roman)
God of Revelry and Wine Dionysus (Greek); Bacchus (Roman)
God of the Romans Jupiter —supreme god
God of the Sea Neptune (Roman); Poseidon (Greek)
God of Ships and the Sea Nyörd (Norse)
God of Skill Hermes (Greek); Mercury (Roman)—the winged cap-and-shoes messenger of Jove or Jupiter (Zeus) in one hand he bore a rod entwined by two serpents (the caduceus)—symbol of the medical profession
God of Sleep Hypnos (Greek); Somnus (Roman)
God of Soil Fertilization Saturn or Stercutus (Roman)
God of Springs Fons (Roman)
God of the Sun Adonis (Syrian); Apollo (Roman); Apollon (Greek); Baal (Chaldean); Helios Hyperion (Greek in Homer's time); Horus (Upper Egypt); Mithras (Persian); Moloch (Canaanite); Osiris (Egyptian); Ra or Re (Egypt's Old Kingdom); Sol Invictus (Latin—Sun Invincible)—Romans shortened this to Sol; Surya (Hindu)
God of Thunder Thor (Norse)
God of Time Cronus (Greek); Saturn (Roman)
God of Trade and Travelers Hermes (Greek); Mercury (Roman)
God of the Underworld Dis (Roman); Hades (Greek); Mantus (Etruscan); Pluto (Roman)
God of Vineyards and Wine Bacchus (Roman); Dionysus (Greek)
God of War Ares (Greek);

Mars (Roman)
God of Wine Bacchus (Roman); Dionysus (Greek)
Golda Meir Goldie Mabovitch, later Golda Meyerson
Gold Coast Africa's Ghana; Australia's resort area extending from Coolangatta to Southport; Florida's resort coast extending from Key West to Palm Beach
Golden Age era of mankind, according to mythology, when people lived in contentment and peace with arts and crafts superior to those of the Silver, Bronze, Iron, and Stone ages; mankind's age of innocence where there was springtime all the time and happiness, right, and truth prevailed; there were no bodily ailments and nobody had to work—(see *Siglo de Oro*)
Golden Age of Greece 5th and 4th centuries before the Christian era when Aristotle, Euripides, Plato, and Sophocles lived
Golden Age of Opera late 1800s and early 1900s
Golden Age of Rome the reign of Augustus from 27 B.C.E. to 14 A.D.
golden beryl heliodor
Golden Century Nineteenth Century
Golden City Johannesburg
Golden City of a Hundred Spires Prague
Golden Continent Africa
Golden Crescent opium-production area along borders of Afghanistan, Iran, and Pakistan
Golden Gate entrance to San Francisco Bay
Golden Gate City San Francisco
Golden Gate to South America Cartagena, Colombia
Golden Horn Istanbul's harbor formed by the curved arm of the Bosporus
Golden Horseshoe Hamilton-Toronto-Oshawa industrial complex along Lake Ontario
Golden Hyphen Winston-Salem, North Carolina
Golden Isles Jekyll, Saint Simons, and Sea Island off Brunswick, Georgia
Golden Key to the Fjords Stavanger, Norway
Golden Mile London, New York, Tokyo

Golden Peninsula Malay Peninsula
Golden Prison of Paris the Louvre
Golden Province Canada's Ontario
Golden Rock of the Caribbean Sint Eustatius (Statia), Netherlands Antilles
Golden State California
Golden Triangle point of downtown Pittsburgh where the Allegheny and Monongahela form the Ohio River; industrialized northern Europe where the three points are Birmingham, Paris, and the Ruhr; opium-productive fields where Burma, Laos, and Thailand meet near southern Yunnan, China
Golden Triangle of Texan Ports Beaumont, Orange, Port Arthur
Golden Trombone of Abolition Frederick A Douglass
Golden-voiced Crooner Jack (Bing) Crosby
Golden-voiced Tenors Enrico Caruso and Luciano Pavarotti
Goldhunter(s) Californian(s)
Gold Rush Town Nome, Alaska
Gold-standard President William McKinley
Goldy Oliver Goldsmith's nickname bestowed him by Dr Samuel Johnson
Golftown Pinehurst, North Carolina
Goober(s) peanut grower(s), natives of Alabama, Georgia, and North Carolina
Goober State Georgia
Good Gray Poet Walt Whitman
Good Queen Bess Queen Elizabeth I of England (1558–1603)
Good Richard Benjamin Franklin
Good Samaritan City of the Mississippi Memphis
Goodwill Ambassador of Mexico Henryk Szering (Polish-born violinist)
Gopher(s) Minnesotan(s)
Gopher State Minnesota
Gorby Soviet Premier Mikhail S. Gorbachev
Gordon Holmes pseudonym shared by Louis Tracy and MP Shiel
Gordon Pasha Charles George Gordon
The Gorgeous Miliza Korjus

(Mrs Walter Schector)

Gorki Soviet name for Nizhni Novgorod

Gorki (Russian—Bitter One)— pen name of Aleksei Maxsimovich Peskov

Gothab old name for Nuuk, Greenland

Gotham New York City

Gothamite(s) native(s) of New York City; nickname derived from *The Three Wise Men of Gotham* by Washington Irving

Grace Greenwood Sara Jane Clarke Lippincott's pseudonym

Gracie Fields Grace Stansfield

Graduate of Oxford John Ruskin's pseudonym

Graffiti Capital New York City

Graffitic City New York City

Grain Coast Sierra Leone and Liberia

Granary of Canada Saskatchewan

Granary of Portugal province of Alemtejo

Granary of Russia Ukraine

Granary of Spain lower valley of the Guadalquivir

Granary of Sweden Skåne

Gran Chaco lowlands of Bolivia, Paraguay, and Argentina

Gran Colombia (Spanish— Great Colombia)—post-colonial consolidation of Colombia, Ecuador, and Venezuela

Grand Admiral of the Ocean Sea Christopher Columbus

Grand Canyon State Arizona

Grand Cham of Literature Dr Samuel Johnson

Grand Commanders Cayman Islanders

Grand Dame of the Piano Alicia de Larrocha

Grand Divide Continental Divide

Grand Duke Duke of Wellington

Grandfathers of the Welfare State Winston Churchill, Lloyd George

Grand Hotel nickname of French Polynesia's prison in Tahiti

Grand Inquisitor Tomás de Torquemada appointed first Inquisitor General by Ferdinand and Isabella, made Grand Inquisitor by Pope Innocent VIII

Grandma Moses Anne Mary Moses

Grand Master of Humorous Verse Ogden Nash

Grand Master of the String Quartet Franz Joseph Haydn

Grandmother of Boston preacher-reformer-teacher Elizabeth Palmer Peabody

Grandmother of the Russian Revolution Katherine Breshkovska

Grandmother of Shopping Streets Calle Florída in Buenos Aires

Grandmother of the Women's Movement Margaret Mead

Grand Old Lady of Fifty-seventh Street Carnegie Hall

Grand Old Lady of Opera Ernestine Schumann-Heink

Grand Old Man William Ewart Gladstone—four times Prime Minister of Great Britain

Grand Old Man of American Labor Samuel Gompers

Grand Old Man of Canadian Letters Robertson Davies

Grand Old Party Republican Party of the United States —the GOP

Grandsire of American Painting Benjamin West

Grand Zohra General Charles de Gaulle

Granger States Illinois, Iowa, Minnesota, and Wisconsin

Granite boy(s) New Hampshirite(s)

Granite Center Barre, Vermont

Granite City Aberdeen, Scotland

Granite Island Corsica

Granite State New Hampshire

Gran Libertador (Spanish— Great Liberator)—Simón Bolívar

grape sugar glucose ($C_6H_{12}O_6$)

grass marijuana

Grasshopper State Kansas

Graveyard of the Pacific Point Arguello northwest of Santa Barbara, California

Graveyards of the Atlantic Cape Hatteras, North Carolina and Sable Island off Nova Scotia

Great Agnostic Colonel Robert G Ingersoll

Great American Authors and Poets William Cullen Bryant, James Fenimore Cooper, T S Eliot, Ralph Waldo Emerson, Nathaniel Hawthorne, Ernest Hemingway, Oliver Wendell Holmes, Washington Irving, Sidney Lanier, Henry Wads-

worth Longfellow, James Russell Lowell, Herman Melville, Marianne Moore, Thomas Paine, Edgar Allan Poe, Carl Sandburg, Mark Twain, Walt Whitman, John Greenleaf Whittier, William Carlos Williams

Great American Ballad Composer Stephen Collins Foster

Great American Colonizers William Penn of Pennsylvania and Roger Williams of Rhode Island

Great American Desert great basin of western United States; Death Valley and Imperial Valley, California; Mojave Desert in California and Nevada; Sonoran Desert

Great American Freethinker Statesmen John Adams, John Quincy Adams, Benjamin Franklin, Thomas Jefferson, Abraham Lincoln, James Madison, Theodore Roosevelt, William Howard Taft, George Washington

Great American Inventors Alexander Graham Bell, Thomas A Edison, Robert Fulton, Elias Howe, Samuel F B Morse, George Westinghouse, Eli Whitney, Wilbur Wright

Great American Jurists Rufus Choate; Oliver Wendell Holmes, Jr; James Kent; John Marshall; Joseph Story

Great American Pastimes baseball, basketball, and football

Great American Sailors George Dewey, David Glasgow Farragut, John Paul Jones

Great American Sanitarian William Crawford Gorgas who freed Cuba and Panama from yellow fever

Great American Soldiers Ulysses Simpson Grant, Andrew Jackson, Thomas Jonathan (Stonewall) Jackson, Robert Edward Lee, Douglas MacArthur, William Tecumseh Sherman, George Washington

Great Apes chimpanzees, gorillas, orangutans

Great Assassin Abdul-Hamid II (notorious for his participation in the Armenian atrocities)

Great Basin Nevada national park

Great Britain England, Scotland, and Wales—GB

Great Canal waterway between Australia and the Great Barrier Reef

Great Cham of Literature Dr Samuel Johnson

Great Charter Magna Charta

Great Commoner Henry Clay, William Ewart Gladstone, William Pitt (the Elder Pitt also known as the Earl of Chatham), and Thomas Paine

The Great Commoner William Jennings Bryan

Great Compromiser Henry Clay

Great Debunker H(enry) L(ouis) Mencken—editor of *The American Mercury*

Great Destroyer syphilis

Great Disappointment October 22, 1844, when thousands of Christians expected the second coming of Jesus Christ

Great Dissenter Supreme Court Justice Oliver Wendell Holmes, Jr

Great Divide continental divide formed by Rocky Mountains; waters on western slopes flow to the Pacific, on eastern slopes flow to Gulf of Mexico

Great Duke Duke of Wellington

Great Emancipator Abraham Lincoln

Great Engineer Herbert Hoover

The Great Engineer Herbert Hoover

Greater Antilles Cuba, Hispaniola (Dominican Republic and Haiti), Jamaica, Puerto Rico

Greater Sunda Islands Borneo, Celebes, Java, Sumatra, and Indonesian islands

Greatest American Jurist John Marshall—Chief Justice of the Supreme Court from 1801 to 1835

Greatest Artist of the South Seas Paul Gauguin

Greatest Composer Haydn's name for Mozart

Greatest Heavyweight Boxer Jack Johnson

Greatest Show on Earth Barnum and Bailey–Ringling Brothers Circus

Greatest Snow on Earth Idaho

Great Fatherland War Soviet name for World War II

Great Jailer of the Caribbean Fidel Castro

Great Lakes Ontario, Erie, Hu-

ron, Michigan, Superior

Great Lakes Canada Ontario (on the northern shores of Superior, Huron, Erie, and Ontario)

Great Lakes Province Ontario

Great Lakes States New York, Pennsylvania, Ohio, Michigan, Indiana, Illinois, Wisconsin, Minnesota

Great Lake State Michigan

Great Land The Great Land—Alaska

Great Liberator Simón Bolívar

The Great Lover Rudolph Valentino

Great Moralist Dr Samuel Johnson

Great Outsider B Traven (expatriate American author)

Great Pacificator Henry Clay

Great Patriotic Struggle official Soviet name for Russia's participation in World War II

Great Plains plains and prairies of Canada and the United States east of the Rockies

Great Poet of Democracy Walt Whitman

Great River Road 3700-mile-long Mississippi-Missouri-Red Rock river system

Great Sea Biblical name for the Mediterranean

Great Separationist Paul Blanshard

Great Smoke London before air-pollution control was enforced

Great Society administration of Lyndon Baines Johnson—thirty-sixth President of the United States

Great Stink Thames River before cleanup

Great Stone Face Daniel Webster; Old Man of the Mountain also known as Profile Mountain in New Hampshire's White Mountains

Great Street State Street that Great Street in Chicago

Great Thirst Land South Africa

Great Wet Ditch British nickname for the English Channel

Great White Father (American Indian term—the President of the United States)

Great White Fleet white-hulled flotilla of United States Navy displayed in principal ports of the world during circumnavigation ordered by President

Theodore Roosevelt; white-painted ships of the United Fruit Company—also called *La Gran Flota Blanca*

Great White Strip main street of Las Vegas, Nevada

Great White Way New York City's theatrical section of Broadway

Great White Wizard Dr Albert Schweitzer

Greece's Principal Port Piraeus

Greek Century 5th century before the Christian era—the 400s BCE

Greek Isles Cyclades, Dodecanese, Ionian, Sporades

Greek Muses (*see* muses)

Greek Ports Argostolion, Patrai, Kalámai, Póros, Piraieus, Vólos, Thessaloniki, Mililini, Iraklion, Ródhos

Green Goddess Gro Harlem Brundtland (Norway's first female and youngest prime minister)

Green Hell Paraguayan Chaco

Green Irish Roman Catholics

Green Isle Ireland

Green Mountain boy(s) Vermonter(s)

Green Mountain City Montpelier, Vermont

Green Mountain State Vermont

green vitriol copperous, ferrous sulfate ($FeSO_4 \cdot 7H_2O$)

Greenwood Brooklyn's historic cemetery and eponym standing for similar burial places

Grenada's Port St George

Greta Garbo Greta Gustafson

Greyhound M-8 6-wheeled armored car carrying a 37mm gun and made in the U.S.A.

Grey Owl George S Belaney

Greytown San Juan del Norte

Griffon Soviet SA-5-type surface-to-air missile

Gröfaz *Grösster Feldherr aller Zeiten* (German—greatest general of all time)—Hitler's acronymic nickname

Groperland Western Australia

Grotius Hugo de Groot

Groucho Marx Julius Henry Marx

Groundhog State Mississippi

Guadalupe Victoria Manuel Félix Fernández

Guamanian Port Apra

Guanaco Central American nickname for a farmer or other rustic and Andean name for a member of the camel family resembling a llama

Guanahani San Salvador island, Bahamas

Guanahani (Lucayan—San Salvador or Watling Island)— first land discovered by Columbus in the New World

Guangzhou (Pinyin Chinese— southern city) Canton

Guardian Angel of Israel Michael

Guardian of the Gulf Oman

Guarnerius Giuseppi Antonio Guarneri

Guatemalan Ports Livingston, Puerto Barrios, Santo Tomas de Castillo; San José, Iztapa, Champerico

Guideline Soviet SA-2 missile system

The Guild The Newspaper Guild

Guinea-Bissau Ports Casheu, Bissau, Bolama

Guinea Port Conakry

Guitar Town Nashville, Tennessee

Gulag Archipelago Solzhenitsyn's title for the thousands of prisons the USSR

gulag gas natural gas pipeline built by Soviet forced labor and extending from Siberia to Germany

Gulf Gulf of Adalia, Aden, Alaska, Alexandretta, Aqaba (Eilat), Arabian (Persian), Boothia, Bothnia, Cadiz, California, Cambay, Campeche, Canada, Carpentaria, Cattaro (Kotor), Chihli, Chiriqui, Cutch, Darien, Eilat (Aqaba), Finland, Fonseca, Gabés, Genoa, Guayaquil, Guinea, Honduras, Izmir, Kotor (Cattaro), Kutch, Lepanto (Corinth), Lions, Maine, Manaar, Maracaibo, Martaban, Mexico, Nicoya, Oman, Panama, Persian (Arabian), Paria, Quarnero (Kvarner), Santa Catalina, Siam, Sidra, Smyra, St Lawrence, Suez, Taranto, Tehuantepec, Tonkin, Venice; Gulf Stream

The Gulf (*see* Gulf)

Gulf City Mobile, Alabama

Gulf of Mexico's Principal Port New Orleans

Gull's disease myxedema resulting from atrophy of the thyroid gland

Gumbo Coast coastal parishes of Louisiana

Gunflint(s) Rhode Islander(s)

Gussie Augusta; Augustina;

Augustine

Guyana's Ports Bartica, Georgetown, McKenzie, New Amsterdam

Guy d'Hardelot Mrs WI Rhodes (Helen Guy)

Guys Marsh borstal in Dorset, England

Gyp Gypsy; Gyp the Blood; Marie Antoinette de Riquetti de Mirabeau, Countess de Martel de Janville's pseudonym

Gyppy (British slang—Egyptian)

Gypsum City Fort Dodge, Iowa

Gypsy Rose Lee Rose Hovick

Haakon the Good King Haakon I of Norway

Haakon the Old King Haakon IV of Norway

Hab(bie) Albert; Alberta; Halbert

Habeas Corpus Howe William Frederick Howe, also nicknamed Criminal Bar Howe

Habitants (French—Inhabitants)—Canadian farmers and fishermen of French descent

Hacha Falaya (Choctaw— Long River)—Atchafalaya River

Hafun formerly Dante when Somalia was Italian Somaliland

Haggisland Scotland

Haiti's Principal Port Port-au-Prince

Hal Croves pseudonym of B Traven whose full name was Berick Traven Torsvan; his full name was concealed by his publisher as he was a fugitive from justice; best known for *The Treasure of the Sierra Madre* and *Ghost Ship*

Halévy Jacques Fromental Elie Lévy

Halfway House on the Pacific Highway Hawaii, so named by Mark Twain

Hal Meredith Harry Blyth's pseudonym

Halstern's disease endemic syphilis

hamburg hamburg brandy (beet or potato alcohol flavored to imitate grape brandy); hamburger; hamburg steak

Hamburg Bach Carl Philipp Emanuel Bach also called the Berlin Bach

Hamlet's Town Helsingør,

Denmark (called Elsinore by the English)

Hammerfestinger native of Hammerfest, Norway

Hammering Hank Henry Aaron

Hammerman John Henry

Hammer of Scotland Edward I

Hammond Innes pseudonym of Ralph Hammond-Innes

Hampton Roads Ports Newport News, Norfolk, Portsmouth

Handcuff King Harry Houdini

Hand of Fatima five-fingered heraldic symbol topping the emblem of Algeria

Hanging Judge Judge Roy Bean of Langtry, Texas—Law West of the Pecos

Hangman's Day Friday

Hangtown El Dorado, California

Hank Henry

Han Kook Republic of Korea

Hanot's disease cirrhosis of the liver accompanied by jaundice

Hansa Ports Hanseatic League ports—Bremen and Hamburg on the North Sea, Danzig and Lübeck on the Baltic, Visby on Gotland Island in the Baltic

Hansen's disease leprosy

Hans Fallada pseudonym of Rudolf Ditzen

Hap Arnold General Henry Harley Arnold, USA and USAF

Happy Chandler High Commissioner of Baseball Albert Benjamin Chandler

Happy Home of the Bulldozer Los Angeles

Happy Land Burma

Happy Valley The Vale of Kashmir in the Himalayas

Happy Warrior Franklin D Roosevelt's nickname for New York State's Governor Alfred E Smith; Hubert Horatio Humphrey

Hapsburg Empire Austro-Hungarian Empire

Harbin Russian name for Pinkiang, Manchuria

Harbor City Erie, Pennsylvania

Harbor of the Sun San Diego, California

Hard Heart of Hickland Cleveland, Ohio, according to authors Jack Lait and Lee Mortimer

Hard Rock nickname of the American Broadcasting Com-

pany (ABC)
Hardware City New Britain, Connecticut
Harke Soviet Mi-10 heavy-transport helicopter
Harmony former place-name of Ambridge, Pennsylvania
Harold Bluetooth King Harold of Denmark
Harold Harefoot Harold I of Denmark and England
Harp/Hormone Soviet Ka-20 or Ka-25k helicopter for military or commercial use (Harp is military and Hormone is commercial version)
Harpo Marx Adolph Arthur Marx
Harpoon harpoon-type aircraft command and launch subsystem missle; Lockheed maritime reconnaissance bomber
Harrier Hawker-Siddeley fixed-wing fighter aircraft; McDonnell-Douglas AV-8B jump-jet bomber
Harry Golden Herschel Goldhirsch
Harry Houdini Ehrich Weiss
Hartford Wits Joel Barlow, Timothy Dwight, Jonathan Trumbull
Harvard's Heroic Historian John Lothrop Motley
Hashish Hasan-ibn-al-Sabbah (11th-century Persian founder of the Assassins)
Hattie Harriet
Haunt of Yachtsmen British Virgin Islands
Haven for Arthritics Jacumba, California
Havercake(s) native(s) of Lancashire
Hawaiian Island Ports Hilo, Hawaii; Kahului, Maui; Honolulu, Oahu; Port Allen, Kauai; Nawiliwili Bay, Kauai
Hawaiian Pineapple King James Drummond Dole
Hawaiians Hawaiian Islanders; Hawaiian Islands
Hawkeye airborne early-warning and fighter-control aircraft—the E-2
Hawkeye(s) Iowan(s)
Hawkeye State Iowa
Head of the Adriatic Trieste
Head of the Commonwealth
Her (His) Most Excellent Majesty the Queen (King) of the United Kingdom of Great Britain and Northern Ireland and of Her (His) other Realms and Territories Queen (King)
Health City Battle Creek,

Michigan
Hearst's Castle (see *La Casa Grande*)
Heart of America Kansas City
Heart of California Sacramento
Heart of Canada Ontario
Heart of Central Alaska Fairbanks
Heart of Darkness Zaire (formerly called the Congo)
Heart of Dixie Alabama
Heart of England Warwickshire
Heart of Historic Virginia Charlottesville
Heart-of-It-All State Ohio
Heart of Kentucky Frankfort
Heartland of America the Midwest
Heartland of Catholicism Italy, the Vatican
Heartland City Kansas City
Heartland of Monarchy Grand Duchy of Luxembourg
Heart of Midlothian Tolbooth Prison in Edinburgh
Heart of Polynesia Western Samoa
Heart of Portugal Mondego Valley
Heart of the Roman Empire Italy
Heart of the South Atlanta, Georgia
Heart of South America Bolivia
Heart of Sweden Dalarna Province formerly called Dalecarlia
Hedda Hopper Elda Furry
Hedy Lamarr Hedwig Kiesler
Heel of Italy Salentine Peninsula
Heide Adelaide
Heine-Medin disease muscular atrophy sometimes followed by permanent deformity
Helena Modjeska Helena Modrejewska
Helen Hayes Helen Hayes Brown
Helen Twelvetrees Helen Jurgens
Hell Breughel Pieter Breughel the Younger who painted hellish scenes
Hellcat Grumman F6F single-seat fighter aircraft; U.S.-made 76mm gun mounted in a fully traversing turret on a tracked chasis (M-18)
Hellenic Republic Greece
Hellfire Cities Hiroshima, Nagasaki
Hell in the Hills Pittsburgh,

Pennsylvania
Hellhole of the Pacific Kororareka, New Zealand
Hell of Java Trinil (where Dr Eugene Dubois discovered *Pithecanthropus erectus*)
Hell of Macquarie Harbour Station old penal colony on the Indian Ocean coast of Tasmania
Hell and Maria Charles G Dawes
Hell's Forty Acres San Carlos, Arizona
Hell's Gates Macquarie Harbour, Tasmania's first convict settlement
Hell's Kitchen New York City's lower west side including San Juan Hill
Hell's Parlor Zanzibar
Hell's Passage Saint Helen's Passage, Oxford
Hell on Wheels Cheyenne, Wyoming
Hennie Henrietta
Henry Bolingbroke Henry IV of England
Henry Cecil Henry Cecil Leon
Henry Green Henry Vincent Yorke
Henry the K Henry Kissinger
Henry the Navigator Dom Henrique o Navegador (Prince of Portugal and patron of explorers and voyagers)
Henry Wade Henry Lancelot Aubrey-Fletcher
Herald Handley-Page turboprop transport plane
Herblock Herbert Lawrence Block
Hercules Lockheed KC-130 tanker aircraft
Herkimer diamond gem-quality quartz from New York State's Herkimer County
Hermanus Vanderdonk Washington Irving
The Hermitage Andrew Jackson's home near Nashville, Tennessee; palace museum of art in Leningrad (formerly a czarist palace)
Hermit Kingdom Korea
Hermit of Slabsides John Burroughs
Hernandarias Hernando Arias de Saavedra
Hero of a Hundred Battles the Duke of Wellington
Hero of a Hundred Fights Admiral Horatio Nelson
Hero of Antiquity Heracles or Herakles (Greek); Hercules (Roman)

Hero of Appomattox General Ulysses Simpson Grant, USA

Hero of the Cities Alfred E(manuel) Smith—usually called Al Smith

Hero of Civilization Thomas Paine

Heroes of Guanajuato Mexican priests Aldama, Allende, and Hidalgo who led the War of Independence against the Spanish

Hero of Fort Sumter Confederate General Pierre Gustave Toutant Beauregard (known to his soldiers as Old Alphabet or Old Bore)

Hero of the Frontier George Rogers Clark

Heroic City Cartagena, Colombia *(Ciudad Heroica)*

Hero of Lake Erie Commodore Oliver Hazard Perry, USN

Hero of Manila Bay Commodore George Dewey, USN

Hero of Mobile Bay Admiral David Glasgow Farragut

Hero of Modern Italy Giuseppe Garibaldi

Heron Hawker-Siddeley 17-passenger transport plane

Hero of Nacozari Jesús Garcia

Hero of New England Captain Miles Standish

Hero of New Orleans General Andrew Jackson

Hero of the Nile Lord Horatio Nelson

Hero of the Plain People Andrew Jackson

Hero of San Juan Hill Lt Col Theodore Roosevelt, USV

Hero of the Spanish-American War Admiral George Dewey, USN

Hero of Tampico General Antonio López de Santa Anna

Hero and Traitor Benedict Arnold

Hero of Upper Canada Sir Isaac Brock

Herring Chokers Newfoundlanders

Herring Pond Atlantic Ocean

Hesperides Canary and Madeira Islands

hessian hessian boots (kneehigh and tasseled); hessian fly (insect feeding on grass and wheat stems)

Hidden Empire Ethiopia

Hieronymus Bosch palette name of Hieronymus van Aeken

High Lonesome southwestern Colorado

High Plains States Arkansas, Kansas, Missouri, New Mexico, Oklahoma, Texas

High Priestess of Transcendentalism Margaret Fuller

High-Tide Province Canada's New Brunswick

Hikari (Japanese—Sunbeam)—train linking Tokyo with other coastal cities of Honshu

Hilaire Belloc Joseph Hilary Pierre Belloc

Hildegarde Neff Hildegard Knef

Hillbilly Country mountainous parts of the Carolinas, Georgia, Tennessee, Kentucky, and West Virginia

Hill City Portland, Maine

Hill of Spring Tel Aviv

Hindu Monarchy Nepal

hinny offspring of a jennet or female donkey sired by a stallion or male horse

Hi-no-maru (Japanese—Sun Flag)—emblem of Japan

Hippocrates of Pennsylvania Benjamin Rush

Hirschsprung's disease congenital colonic dilatation

Hispania (Latin—Iberian Peninsula)—land now divided between Portugal and Spain; poetic name for Spain

Hispanic America Portuguese-and-Spanish-speaking countries of Latin America

Hispanic American City Los Angeles, California; San Antonio, Texas; San Diego, California

Hispanic Places Andorra, Argentina, Azores, Balearic Islands, Bolivia, Brazil, Canary Islands, Cape Verde Islands, Ceuta and Melilla, Chile, Colombia, Costa Rica, Cuba, Dominican Republic, Ecuador, El Salvador, Equatorial Guinea, Guam, Guatemala, Honduras, Macao, Madeira, Mexico, Morocco, Nicaragua, Panama, Paraguay, Peru, Philippines, Portugal, Puerto Rico, Spain, Spanish Sahara, Uruguay, Venezuela

Hispanics people of Portuguese or Spanish descent or a study of their culture and language

Histologist of the Brain Santiago Ramón y Cajal

Historian of the American Forest Francis Parkman

Historian of Liberty Lord Acton

Historian With A Camera Mathew B Brady

Historic Center of North Carolina New Bern

Hi-Tech Capital of the World Massachusetts

Hoagy Hoagland Howard Carmichael

Hob(bie) Albert

Ho Chi Minh (Vietnamese—He Who Enlightens)—Nguyen Ai Quac whose patronym was Nguyen Van Coong but is also known as Nguyen That Thanh

Ho Chi Minh City formerly Saigon, Vietnam

hocus morphine's nickname

Hodara's disease hair splitting

Hodge nickname for the typical English farmer or for Roger

Hodgkin's disease progressive enlargement of the lymph nodes

Hog Soviet Ka-18 utility-transport helicopter

Hogarth's Act Act of Parliament passed in 1735 granting copyright protection

Hog Butcher for the World Chicago

Hog Country Arkansas

Hog and Hominy State Tennessee

Hog Lane Hoxton

Hogopolis Chicago

Holland the Netherlands; actually an old Dutch province comprising part of the Netherlands; Michigan summer resort and manufacturing center settled by the Dutch

Holland in the Caribbean Netherlands Antilles (Aruba, Bonaire, Curaçao, Saba, Sint Eustatius, Sint Maarten)

Hollie Holiday; Holladay; Hollingsworth; Hollis; Hollister; Hollway

Holstein Capital Northfield, Minnesota

Holy Alliance Austria, Prussia, Russia (in 1815)

Holy Cities Mecca and Medina in Saudi Arabia

Holy City nickname given Charleston, South Carolina by its inhabitants

Holy City of India Banaras

Holy Devil Rasputin

Holy Horatio Horatio Alger, Jr

Holy Land Israel

Holy Land of Three Religions Israel

Holy Rabble Rouser Ayatullah

Ruhollah Khomeini
Holy Three of Criminology
Enrico Ferri, Raffaele Garofalo, Cesare Lombroso
Home of Abraham Lincoln Springfield, Illinois
Home of the Alamo San Antonio, Texas
Home of Baseball Cooperstown, New York
Home of the Bean and the Cod Boston, Massachusetts
Home of the Blizzard Adelie Land, Antarctica
Home of the Blues Memphis, Tennessee
Home of the Casbah Algiers
Home of Casey Jones Jackson, Tennessee
Home of the Comstock Lode Virginia City, Nevada
Home of Contented Cows Carnation, Washington
Home of the Cotton Carnival Memphis, Tennessee
Home of Diamond Walnuts Stockton, California
Home of the Dinosaurs Glen Rose, Texas where petrified footprints of 30-foot-long dinosaurs are displayed
Home of Franklin Delano Roosevelt Hyde Park, New York
Home of George Washington Mount Vernon, Virginia
Home of the Giants Jotunheimen Mountains in Norway
Home of Goethe Heidelberg, Germany
Home of Holbein Augsburg, Germany
Home of Jesse James St Joseph, Missouri
Home of the Kentucky Derby Louisville
Homeland of the Bengalis Bangladesh
Homeland of Yogurt Bulgaria
Home of Old Miss Oxford, Mississippi—the home of The University of Mississippi
Homer Soviet heavy helicopter designated Mi-12
Homer Wilbur pseudonym of James Russell Lowell
Home of the Snow Himalaya Mountains
Home of Storms Gulf of Alaska
Home of Theodore Roosevelt Oyster Bay, Long Island, New York
Home of Thomas Jefferson Monticello, Virginia

Hometown Hamilton, Ohio as defined by author Peter Davis
Home of the Waltz Vienna
Honduran Ports Puerto Cortés, Tela, Puerto Este, La Ceiba, Trujillo, Puerto Castilla, Roatán, Amapala, San Lorenzo
Honduras Republic of Honduras (Spanish-speaking Central American nation) *República de Honduras*
Honest Abe Abraham Lincoln
Honest Harold Secretary of the Interior Harold Le Claire Ickes, also called the Old Curmudgeon
Honest John solid-sustainer motor surface-to-surface ballistic missile produced by Douglas Aircraft
Honey Capital Uvalde, Texas
Honey Fitz John F. (Honey Fitz) Fitzgerald
Honey Lulu Honolulu, Hawaii
Honeymoon City Niagara Falls, New York
Honey State Western Australia
Honorary Citizen of the United States Winston Churchill (first and only foreigner to bear this title)
Hooker Boulevard El Cajon Boulevard
Hoosier Capital Indianapolis, Indiana
Hoosier City Indianapolis, Indiana
Hoosier Poet James Whitcomb Riley
Hoosier(s) name believed to be a frontier-era contraction of *Who's there?*—native(s) of Indiana
Hoosier State Indiana
Hormone Soviet armed helicopter in naval service (KA-25)
Horn of Africa Djibouti, Ethiopia, Somalia, Sudan, particularly northeasternmost Somalia terminating in Cape Guardafui which the Arabs call the Ras Asir
Hornet F-A-18 McDonnell Douglas fighter-attack aircraft
Horse Latitudes belts of calms about 30 or 35 degrees north or south of Equator; horses were cast overboard in these places when sailing vessels were becalmed and drinking water became scarce
Horseshoe Curve Altoona, Pennsylvania close to the celebrated Horseshoe Curve built by the Pennsylvania Railroad

to cross the Alleghenies and traverse the valley of the Juniata River
Horse Thief Hollow Oak Lawn, Michigan
Hosea Biglow pseudonym of James Russell Lowell
Hostess to the Nation Dolly Madison—wife of President James Madison
Hotbed of Secession Charleston, South Carolina
Hotel letter H radio code; H-class Soviet missile-launching nuclear-powered submarines
Hotlanta Atlanta, Georgia
Hot Potato Luke Hamlin
Hot Springs Arkansas national park
Hot Springs Country southwest Arkansas
Hotspur Sir Henry Percy
Hottest Town in Arizona Quartzite
Hottest Town in Texas Presidio
Hot Water City Hot Springs, Arkansas
Hot Water State Arkansas
Houdini Harry Houdini (real name Ehrich Weiss)—America's foremost escapologist-magician
Hound Dog North American-Rockwell air-to-surface missile
house apes other people's unhousebroken children
The House Christ College, Oxford
household coal bituminous coal; soft coal
House Ruth Built New York City's Yankee Stadium in the Bronx where Babe Ruth hit many home runs
Houston haze smog of the petrochemical variety sometimes called Los Angeles haze or metropolitan mist mixed with automotive exhausts
Howie Howard; Howarth; Howe; Howell; Howland
Hsinhua New China News Agency
Huáscar's revenge loose bowels contracted in Peru where the emperor, Huáscar, was betrayed by his brother, Atahualpa, and executed by the Spaniards
Hub The Hub—Boston, Massachusetts also called Hub of American Culture, Hub of New England, and Hub of the Universe

hubby husband

Hub of the Caribbean Jamaica

Hub of the Castilian Wheel Madrid (capital of Spain, equidistant from all her boundaries)

Hub of Christianity Jerusalem

Hub City of Texas Alice

Hub of Empire London

Hubey Hubert

Hub of the Golden Mile Kalgoorlie in Western Australia where gold and nickel are found

Hub of Hinduism Benares

Hubie Hubert

Hub of Islam Mecca

Hub of Islamic Culture Cairo

Hub of Judaism Jerusalem

Hub of New England Boston

Hub of New York City Columbus Circle

Hub of the Pacific Guam

Hub of the South Pacific Fiji

Hub of the Universe nickname given by Oliver Wendell Holmes to the statehouse in Boston and later by others to the entire city

Hudson Tubes Hudson & Manhattan Railroad

Huey Cobra AH-1 gunship aircraft

Hughie Hugh

Hugo Wast Gustavo Martínez Zuviria

Hugues Capet Hugh Capet

Humanist Historian Harry Elmer Barnes

Humanitarian Scientist Louis Pasteur

Humist Philosopher David Hume

Hummer High-Mobility Multi-Purpose Wheeled Vehicle successor to the Jeep

Humorist-Pianists Steve Allen, Victor Borge, and Mark Russell

Humorists of Europe the Danes

humpday Wednesday (the middle-of-the-week day)

hungarian hungarian goulash (stew); hungarian paprika (red paprika—permeability vitamin or vitamin P)

Hungarian Ocean Lake Balaton—largest lake in central Europe

Hunkyland Hungary

Hunter Hawker jet fighter-bomber

Huntington's chorea hereditary disease marked by choreic movements and men-

tal deterioration

Huskie Kaman H-43 utility helicopter

Husky Territory the Yukon

Hustleton on the Canal Houston, Texas

Hyrcanian Caspian

I-boats Japanese transport submarines used in World War II to carry small scouting airplanes

ice bear polar bear

Iceberg Alley North Atlantic Ocean between Greenland and Labrador

Ice Capital of America Hawaii

Ice King of the Arctic the polar bear

Icelandic Ports Reykjavik, Seydhisfjordhur, Heimaey, Eyrarbakki

Iceland spar calcite (calcium carbonate)

Ice Mine City Coudersport, Pennsylvania

Iconoclast English freethought author-lecturer-publisher Charles Bradlaugh; Georg (Morris Cohen) Brandes

Iconoclast Poet Percy Bysshe Shelley

Iconoclasts dramatist-authors Becque, d'Annunzio, Duse, Gorki, Hauptmann, Herpieu, Huneker, Ibsen, Maeterlinck, Shaw, Strindberg, Sudermann

Idaho Lion Senator William E Borah

idiot pills barbiturates

IF Stone Isidor Feinstein

Ignatius Loyola Iñigo López de Recalde

Ignatz von Aschendorf pseudonym used by Joseph Conrad and Ford Madox Ford when they wrote *The Nature of a Crime*

Ignazio Silone pseudonym of Secondo Tranquilli

Ike Dwight David Eisenhower (nickname)—thirty-fourth President of the United States; Isaac

Ikey (*see* Ikie)

Ikie Isaac; Isaak; Isack; Izaak; Isaque

Il Cieco (Italian—The Blind One)—Italy's blind poet—Luigi Groto who wrote in the mid-sixteenth century

Il Duca di Spoleto (Italian—The Duke of Spoleto)—composer-impresario Gian Carlo Menotti's nickname as he directs the Spoleto Festival

Il Duce (Italian—dictator)—Benito Mussolini

Il Furioso (Italian—the Furious One)—nickname of Tintoretto

Ilich Russian patronymic often used as the popular name for Lenin—the party name of Vladimir Ilich Ulyanov

Illegal-Alien Capital of the United States Los Angeles

Illinois Ports Chicago, Wilmette, Great Lakes, Waukegan

Illinois River City Peoria

Illusion Factory Hollywood

Illustrator of Early Twentieth-Century America Norman Rockwell

Illustrator of the Russian Underground Ilya Efimovich Repin

Illustrious Infidel Colonel Robert G Ingersoll

Il Moro (Italian—The Moor)—Ludovico Sforza's nickame

Ilona Massey Ilona Hajmassy

Il Perugino (Italian—The Perugian)—Pietro Santi Bartoli

Image Maker Thomas Nast

Imamu Amiri Baraka Le Roi Jones

Immortal Beloved Nadejda von Meck (benefactress of Tchaikovsky)

Immortal Dreamer John Bunyan

Immortal Four Italian poets Dante Alighieri, Ludovico Ariosto, Francesco Petrarca (Petrarch), Bernardo Tasso

Immortal Infidel Col Robert G Ingersoll

Immortals the forty members of the French Academy

Immortal Sarah Sarah Bernhardt (originally Rosine Bernard)

Immortal Tinker John Bunyan

Immortal Trio John Caldwell Calhoun, Henry Clay, Daniel Webster

Impala South African version of Aermacchi MB-326 counterinsurgency aircraft

Imperial City Rome

Imperial Impersonation of Force and Murder Napoleon

Imperialist Composer Sir Edward Elgar

Imperialist Poet Rudyard Kipling

Imperialist Poet-Writer Rudyard Kipling

Imperial President Franklin D Roosevelt

imprisoned authors famous authors who spent time in prison include Bunyan *(Pilgrims's Progress)*, Cervantes *(Don Quixote)*, Dostoevski *(Crime and Punishment)*, Raleigh *(History of the World)*, O Henry *(Cabbages and Kings)*, Wilde *(Ballad of Reading Gaol* and *De Profundis)*

Improper Bostonian Dr Oliver Wendell Holmes

Inca Capital Cuzco, Peru

Incan and Aztecan Century the 1000s—the 11th century

Inca's curse diarrhea picked up in the Land of the Incas—Bolivia and Peru as well as Ecuador and Chile

Inchcape Rock (*see* Bell Rock)

Inchon formerly Chemulpo or Jinsen

Incomparable Infidel Voltaire

Incorruptible The Incorruptible—sobriquet given Robespierre by his followers

Indefatigable Polemicist Alexander Solzhenitsyn

india india chintz or india cotton (heavy figured fabric used by upholsterers); india ink (glue + lampblack) also called chinese ink

Indiana Dunes national lakeshore of Indiana

Indiana Ports Michigan City, Gary, Buffington, Indiana Harbor

Indian Film Pioneer Satyajit Ray

Indian Girl Guide Sacajawea who guided Lewis and Clark

Indian Ocean Eden Seychelles Islands

Indian OPEC CERT (Council of Energy Resource Tribes) holding coal, geothermal, and oil-productive lands in many parts of the United States

Indian Ports Mamdvi, Kandla, Okha, Porbandar, Bhaunagar, Bombay, Mangalor, Cochin, Alleppy, Quilon, Kolachel, Tuticorin, Negapatam, Madras, Kakinada, Vishakhapatnam, Paradip, Calcutta

Indian President Benito Juárez of Mexico

Indian Princess Pocahontas

Indian's Friend Roger Williams

Indian Territory old name of Oklahoma

India's Principal Ports Calcutta on the Bay of Bengal,

Bombay on the Arabian Sea

Indonesian Ports Tandjungpriok (Djakarta), Surabaya, Makassar

Indonesia's Largest Port Djakarta (Jakarta)—also called Tandjungpriok

Industrial Capital of Connecticut Bridgeport

Inflexible President Andrew Johnson

Inland Empire Illinois

Inland Sea Pacific Ocean inlet between Honshu and Kyusho islands, Japan

Inner City Peking's Forbidden or Tartar City containing the Palace Museum

Inner Libya the Sudan Desert extending from southern Egypt and the Sudan to Africa's west coast

Inner Mongolia northern China bordering on Mongolia

Innovators Russian composers Balakirev, Borodin, Cui, Mussorgsky, Rimsky-Korsakov

Inside Passage inland passage between southern Alaska and northern Washington; also called Inner Passage

Inspired Innovator Edgar Allan Poe

Instant Asia Singapore with its Chinese, Indian, Malay, Pakistani, and Singhalese mixtures and tongues making this a global crossroads

Instant Orient Singapore—crossroads of Asia

insultant(s) nickname for overpaid consultant(s)

Insurance Capital Hartford, Connecticut and Omaha, Nebraska

Insurance City nickname shared by Atlanta, Georgia and Hartford, Connecticut

Intellectual Emperor of Europe Voltaire

Intellectual Historian Sir Isaiah Berlin

Intellectual Seed Pod of the Nation Emerson's nickname for Concord, Massachusetts where he lived with such neighbors as the Alcotts, Hawthorne, and Thoreau

Interior Plains Canada's great plains

International Capital New York City—headquarters of the United Nations

International City Montreal, Quebec

International Functionalists

Walter Gropius and Mies van der Rohe

International Prizegiver Alfred Nobel

Interpreter of the Sea Winslow Homer

Interpreter of the Wild West Frederic Remington

Intruder Grumman electronic-intelligence-gathering aircraft (EA-6B)

Inventor of Bifocals Benjamin Franklin

Inventor of Calculus Baron Gottfried Wilhelm von Leibniz and Sir Isaac Newton

Inventor of the Detective Story Edgar Allan Poe

Inventor of Science Fiction Mary Wollstonecraft Shelley, author of *Frankenstein*

Inventor of the Stethoscope René-Théophile-Hyacinthe Laennec

Inventor of the Telephone Philipp Reis although Alexander Graham Bell is usually given credit

The Invincible Spanish Armada defeated by English vessels commanded by Sir Francis Drake

invisible disease dyslexia

Iodine State South Carolina

Iola Ida B Wells

Iolo Morgannwg bardic name of Edward Williams

Ipiranga Ypiranga

Iran-Contra 1980s political scandal involving the sale of weapons of Iran to fund Nicaraguan insurrection

Iranian Ports Khorramshahr, Abadan, Bandar-e-Mahshahr, Bandar-e-Shapur, Kharg Island Terminal, Bandar Abbas

Iran's Principal Port Abadan

Iraq Ports Al Faw and Al Basrah

Iraq's Principal Port Basra

Ireland the Great Newfoundland's name given it by Irish explorers who found it in Viking times

Ireland's Principal Port Dublin

Iris Tennessee state flower and sobriquet

Irish Irish boat (cutter-rigged fishing vessel), Irish ford (paved ford), Irish coffee (spiked with Irish whiskey and topped with whipped cream), Irish moss (edible seaweed also called carrageen), Irish pennant (unwhipped rope end flying in the breeze), Irish potato (white), Irish setter, Irish sweater (fisherman's knit), Irish terrier (red-hair terrier originally bred in Ireland), Irish tweed, Irish whiskey (originally distilled in Ireland from barley), Irish wolfhound (large breed of dog noted for its courage and originally bred in Ireland)

Irish-American Bandmaster Patrick Sarsfield Gilmore

Irish Channel New Orleans waterfront

Irish confetti bricks; thrown bricks used in street fighting

Irish Cradle of U.S. Presidents Ireland—ancestral home of Presidents Jackson, Kennedy, Nixon, and Reagan as well as Arthur, Buchanan, McKinley, Polk, Truman, and Wilson from Northern Island or Ulster

Irish Dramatist-Poet William Butler Yeats

Irish Free State Republic of Ireland

Irish Navigator Saint Brendan (formerly spelled Brandon)

Irish Ports Bangor, Belfast, Larne Lough, Londonderry, Sligo, Westport, Galway, Kilrush, Limerick, Foynes, Cobh, Cork Harbour, Rosslare, Dublin

Irish turkey corned beef

Irish whist copulation

Irish wine whiskey

Iron Age era of mankind when implements and weapons were forged from iron; period of vast degeneracy, corruption, and toil following the Stone and Bronze ages

Iron Alley corridor littered with meteorites from Portland, Oregon to Mexico City in a lane about 250 miles or 400 kilometers wide

Iron Butterfly Imelda Romauldos Marcos

Iron Chancellor Prince Otto Eduard Leopold von Bismarck-Schönhausen—first chancellor of German Empire

Iron Charles Charlemagne (Carolus Magnus)

Iron City Bessemer, Alabama and Pittsburgh, Pennsylvania

Iron Curtain Countries Albania, Bulgaria, Czechoslovakia, Estonia, East Germany, Hungary, Latvia, Lithuania, Poland, Rumania, Soviet Union, Yugoslavia

Iron Duke Arthur Wellesley the Duke of Wellington

Iron Gate narrow rapids in the Danube below Orsova in Romania

Iron Horse Lou Gehrig; a steam locomotive

Iron Lady Margaret Thatcher —Britain's first woman prime minister

Ironquill Eugene Fitch Ware

Iron Range the Mesabi Range

Ironsides Oliver Cromwell

Iron Triangle Cologne (Köln), Siegen, Solingen (all noted for their steel products)

Iroquois Bell turbo-power helicopter designated UH-1

Irving Berlin Irving Baline

Irving Stone Irving Tannenbaum

Isabela Spanish name for Albemarle Island in the Galápagos

Isak Dinesen Baroness Karen Blixen-Finecke

Isherman Israeli version of Super Sherman tank

Iskander Alexander Herzen (Aleksandr Ivanovich Yakoviev)

Iskra (Russian—Spark)—Polish single-engine jet aircraft designated TS-11

Isku Finnish guided-missile patrol boat

Isla de Juventud (Spanish—Isle of Youth)—Cuba's Isle of Pines also called *Isla de Pinos*

Isla de Pascua (Spanish—Easter Island)—Chilean island in the eastern South Pacific noted for its huge stone monuments; Polynesians call it Rapa Nui

Isla Más Afuera (Spanish—further out island) Alejandro Selkirk Island

Isla Más-a-Tierra (Spanish—island closest to land)—Robinson Crusoe Island

Islamic Century the 600s—Mohammed dies in 632; Islam begins expanding throughout the Middle East and Africa—the 7th century

Island-and-Mainland Province Newfoundland

Island at the End of the World Madagascar as described by the Malagasy

Island of Bearded Figs Barbados

Island of Betelnut Palms Penang

Island of Birds Kusadasi

Island City Manhattan, Montreal, Singapore, and Stockholm

Island of Cloves Zanzibar

Island Continent Australia

Island of Copper Cyprus

Island of Death Kahoolawe, Hawaii (used for target practice by Air Force and Navy)

Island of Dragons Komodo (home of the dragon lizards)

Island of Dreams Capri

Island of Enchantment Puerto Rico

Island of Flowers Taboga, Panama

Island Fortress Malta

Island of the Gods Bali

Island of Hell Norfolk Island in the South Pacific, the most dreaded of all Australian prison stations

Island of Hope Statue of Liberty, Liberty Island, New York

Island of Knights Hospitaliers Malta

Island of Light New Caledonia

Island of Many Faces Singapore

Island Ministate Nauru in the Central Pacific

Island of Monks and Pirates Lantau or Tai Yue Shan—largest island off Hong Kong

Island of the Moon Madagascar

Island Nation nickname shared by Australia, the Bahamas, Bahrain, Barbados, the Cape Verde Islands, the Republic of China (Taiwan), the Comoro Islands, Cuba, Cyprus, the Dominican Republic, Fiji, Grenada, Haiti, Iceland, Indonesia, Ireland, Jamaica, Japan, Madagascar, the Maldives, Malta, Mauritius, Nauru, New Zealand, Papua New Guinea, the Philippines, São Tomé and Principe, the Seychelles, Singapore, Sri Lanka (Ceylon), Tonga, Trinidad and Tobago, the United Kingdom, Western Samoa

Island of Olives Cyprus

Island of Roses Rhodes in the Dodecanese

Island of Ruins and Roses Gotland, Sweden

Island of Sages and Saints Ireland

The Islands pet name given to favorite insular groups such as the Aleutians, the Bahamas, the Balearics, the Canaries, the Hawaiians, the West India islands and even to Coney, Long, Manhattan, and Staten when referring to the New York City area

Islands of Eternal Spring the Balearics (Ibiza, Formentera, Mallorca, Menorca)

Islands of the Maoris New Zealand

Islands of Perpetual June Turks and Caicos Islands between the Bahamas and Hispaniola

Island State Tasmania, Australia

Island in the Sun Key West, Florida

Island of Venus Tahiti

Isle of Cloves Zanzibar

Isle of the Dead Böcklin painting

Isle of Destiny Ireland

Isle of Fragrant Waters Hong Kong

Isle of Pines Cuba's *Isla de Pinos,* France's *Ile des Pins,* Ibiza

Isle of Roses Rhodes

Isle of Saints Iona in the Inner Hebrides off Scotland's west coast; Ireland

Isle of Sappho Lesbos in the Aegean

Isles of the Blest the Canary Islands (*see* Fortunate Isles)

Isles of Devils Bermuda

Isle of Sleep Tasmania

Isle of Springs Jamaica

Isle of Tears Ellis Island, New York

ism of the modern world racism (according to anthropologist Ruth Benedict)

Isolator of Dysentery Kiyoshi Shiga

Isolator of Gangrene Shibasaburo Kitazato

Isolde Yseult

Israeli Ports Akko (Acre), Hefa (Haifa), Netanya, Tel-Aviv-Yafo (Jaffa), Ashdod (Azotus), Ashquelon (Ascalon), Elat, Sharm el Sheik

Israel's Largest Port Tel-Aviv-Yafo (Jaffa or Joppa)

Israfel Edgar Allan Poe

Isthmian Nation Panama

Isthmian Waterway Panama Canal linking the Caribbean-Atlantic and the Pacific; Suez Canal linking the Mediterranean-Atlantic and the Red Sea leading to the Indian Ocean

Isthmus of Panama formerly Isthmus of Darien

Italian Italian greyhound (toy dog); Italian hand (script originating in Italy in medieval times or craftiness); Italian dressing (salad dressing)

Italian Architect-Engraver-Painter Giambattista Piranesi

Italian Architect-Painter-Poet-Sculptor Michelangelo Buonarroti

Italian Architect-Painter-Sculptor Giovanni Lorenzo Bernini

Italian boot boot-shaped Italian peninsula

Italian Century the 14th century—the 1300s

Italian Classicist Sculptor Antonio Canova

Italian East Africa Eritrea, Ethiopia, and Italian Somaliland from 1936 to 1941

Italian Engraver Giambattista Piranesi

Italian Family of Sculptors the della Robbias

Italian Film Magician Frederico Fellini

Italian Goldsmith and Sculptor Benvenuto Cellini

Italian Illustrator-Painter Sandro di Botticelli

Italian Lakes Como, Garda, Isea, Lecco, Lugano, Maggiore, Orta

Italian National Composer Ottorino Respighi

Italian Naturalist Painter Michelangelo da Caravaggio

Italian North Africa Libya from 1912 to the end of World War II

Italian Ports (*on Sardinia*—La Maddalena, Olbia, Cagliari, Alghero, Porto Torres), Savona, Genova (Genoa), La Spezia, Livorno (Leghorn), Portoferraio *(on Elba),* Civitavecchia, Gaeta, Forio, Ischia, Bagnoli, Napoli (Naples), Torre Annunziata, Castellamare di Stabia, Reggio di Calabria, (*on Sicily*—Messina, Palermo, Trapani, Marsala, Licata, Siracusa, Augusta, Catania), Crotone, Taranto, Gallipoli, Brindisi, Monopoli, Bari, Molfetta, Barletta, Manfredonia, Ancona, Ravenna, Chioggia, Porto di Lido (Venice), Monfalcone, Trieste

Italian Pre-Renaissance Painter Giotto (Giotto di Bondone)

Italian Riviera resort area between La Spezia and Ventimiglia

Italian Somaliland Indian Ocean coast of what is now southern Somalia

Italy's Principal Port Genoa

It Girl Clara Bow

Itiopia Ethiopia (Abyssinia)

Ivan (nickname for the typical Russian)

Ivan Ivanovich the typical Russian

Ivan-Kremlin disease endemic antisemitism

Ivan Lermolieff Giovanni Morelli (19th-century Italian art expert, patriot, and senator)

Ivan the Terrible Czar Ivan IV Vasilievich—ruler of Russia

Ivory Coast Ports Grand-Lahou, Jacqueville, Port-Bouet, Abidjan, Grand-Bassam

Ivy League college athletic conference consisting of Brown, Columbia, Cornell, Dartmouth, Harvard, Pennsylvania, Princeton, and Yale; students and graduates of the abovementioned schools as well as their "characteristic" style of dress, which was considered "quiet and neat"

ixey morphine

Izzie Isador; Isadora; Isadore; Isidro; Isodoro; Ysidro

Jabal Tariq (Arabic—Mountain of Tarik)—Moorish name for Gibraltar in honor of their chief Jabal Tariq who settled the Rock in the year 711; (Arabic—Tariq's Mountain)—the Rock of Gibraltar

Jacaranda Capital jacaranda-tree-lined avenues and streets comprising South Africa's capital city—Pretoria

Jack Benny Benjamin Kubelsky

Jack the Dripper Jackson Pollock

Jack Frost frosty weather personified

Jack Higgins Harry Patterson's pseudonym

Jackie Jack Roosevelt Robinson; Jacqueline Kennedy Onassis

Jack London John Griffith Chaney; John Griffith London

Jack Lord John Joseph Ryan

Jack Palance Walter Paluniuk

Jacksonopolis Jackson, Michigan

Jack Soo Jack Suzuki

Jack Teagarden Weldon Leo

Teagarden
Jacky Jaqueline
Jacopo Jacopo Tatti
Jacques Halevy Jacques Francois Fromental Elias Levi
Jacques Offenbach Jakob Eberst
Jacques Tati Jacques Tatischeff
Jadotville former name for Likasi, Zaire
Jagananth Juggernaut or Puri on the Bay of Bengal
Jake Jacob; Jacobus
Jaksch's disease infantile anemia
Jamaica Jamaica ginger; Jamaica rum (heavy pungent rum originally distilled in Jamaica); Jamaica shorts (mid-thigh short pants)
Jamaica ganga Jamaica-grown marijuana
Jamaican Ports Lucea, Montego Bay, Falmouth, Rio Bueno, Dry Harbour, St Ann's Bay, Ocho Rios, Oracabessa Bay, Port Maria, Annotto Bay, Buff Bay, Port Antonio, Manchioneal, Port Morant, Morant Bay, Port Royal, Kingston, Long's Wharf, Little Pedro Point, Black River, Bluefields; Savanna la Mar
Jambalaya Capital Gonzales, Louisiana
James Gardner James Baumgardner
James Hadley Chase René Raymond's pseudonym
James Herriot author-veterinarian James Alfred Wight's pseudonym
James Island, Galápagos Bartolomé, San Salvador, Santiago
James O'Brien James Bronterre
Jane Doe the average American female
Janet Frame Janet Paterson Frame Clutha's pseudonym
Janet Gaynor Laura Gainor
Jane Welsh Mrs Thomas Carlyle
Jane Wyman Sarah Jane Fulks
Janey Canuck Judge Emily Murphy
Janie Jane; Jean
Jan Peerce Jacob Pincus Perelmuth
Jan Struther Joyce Anstruther
Jan Valtin Richard J Krebs
Japan Japan lacquer or varnish; Japan wax also called Japan

tallow or sumac wax
Japanese Japanese gelatin (agar also called Japanese isinglass); Japanese paper (high rag content quality); Japanese silk (high quality raw silk)
Japanese Drama Painter Torii Kyonobu
Japanese Lacquer Artist Korin (Ogata Korin)
Japanese Landscape Artist Supreme Sesshu
Japanese Naturalist Artist Korin
Japanese Ports Moruran, Hakodate, Otaru *(on Hokkaido);* Tokyo, Yokosuka; Shimuzu, Nagoya, Yokkaichi, Senboku, Osaka, Kobe, Fukuyama, Kure, Shimminato, Shimonoseki, Maizuru, Niigata *(on Honshu),* Kita Kyushu, Kagoshima, Nagasaki, Sasebo, Karatsu, Fukuoka *(on Kyushu)*
Japanese Riviera Enoshima Island recreation area
Japan's Back Door Sasebo
Japan's Front Gate Yokohama
Japan's Largest Port Yokohama (including Kawasaki, Tokyo, and Yokosuka)
Jascha (Russian nickname—Jacob)—Jake
Jastreb Yugoslav jet trainer aircraft called Hawk
Javelin Gloster delta-wing jet fighter aircraft
javelle water sodium hypochlorite solution (NaOCl)
Jawbone Flats Clarkston, Washington
Jayhawker(s) Kansan(s)
Jazz Ambassador Louis (Satchmo) Armstrong
Jazz Capital of the Americas New Orleans
Jazz Capital of Europe Copenhagen
The Jazz Singer Al Jolson
Jean Baptiste French-Canadian's sobriquet
Jean Baptiste Lully Giovanni Battista Lulli
Jean Crapaud (nickname for the typical Frenchman)
Jean Gabin Jean-Alexis Moncorgé
Jean Hagen Jean Verhagen
Jean Harlow Harlean Carpenter
Jean-Jacques Jean-Jacques Rousseau
Jean l'Oiseleur (French—Jean the bird tamer)—pseudonym of Jean Cocteau

Jean Meslier Voltaire's pseudonym concealing his authorship of an heretical tract whose title page reads—*Superstition In All Ages* by Jean Meslier, a Roman Catholic Priest, who after a pastoral service of thirty years at Entrepigny and But, in Champagne, France, wholly abjured religious dogmas, and left as his last will and testament the following pages entitled Common Sense *(Le Bon Sens);* Voltaire was an assumed name for François-Marie Arouet
Jean Moreas pseudonym—Jannis Papadiamantopolous
Jeannie Jane; Jean
Jean Paul Johann Paul Friedrich Richter's pseudonym
Jean-Pierre Aumont Jean-Pierre Salomons
Jean Stapleton Jeanne Murray
Jeb Stuard Major General J(ames) E(well) B(rown) Stuart, CSA
Jefferson's Country Charlottesville, Virginia
Jefferson Territory Colorado
Jelly Roll Ferdinand Joseph Morton
Jenghis Khan Genghis Khan—Mongol conqueror and grandfather of Kublai Khan
Jennie Jane; Jean; Jennifer; Lady Randolph Churchill
Jennifer Jones Phyllis Isley
Jenny Jane; Jean; Jennifer
Jenny Lind Johanna Maria Lind—the Swedish Nightingale
Jeremiah Stukeley Percy Bysshe Shelley pseudonym
Jerez Jerez de la Frontera
Jerome Hines Jerome Heinz
Jerrie(s) British slang for German(s)
Jerry Gerald(ine); Governor Edmund G Brown, Jr of California, President Gerald R Ford; Jeremiah; Jeremy; Jerome
Jerry Lewis Joseph Levitch
jersey jersey justice (reputedly efficient and speedy); jersey lightning (applejack)
Jersey Lily Lily Langtry—English actress born on the island of Jersey where her original name was Emily Charlotte Le Breton
Jersey Shore coastal New Jersey; former name of Waynesburg, Pennsylvania

Jerusalem of the West Amsterdam

Jervis Street Toronto, Ontario's red-light district

Jesselton former name of Kota Kinabalu, Sabah

Jessi James Cleveland (Jesse) Owens; Jess; Jessica

Jessup Maryland House of Corrections at Jessup

Jessup Girls Maryland Correctional Institution for Women at Jessup

Jet Provost British jet trainer aircraft designated BAC-145

Jet Ranger Bell turbine-powered helicopter also called Sea Ranger

Jet Star C-140 Lockheed light transport plane

Jet Town Seattle

Jewel of the Adriatic Venice

Jewel of Africa Lake Kivu

Jewel of the East Bali

Jewel of the Eastern Sea Sri Lanka

jewelers' putty stannous oxide

Jewel of German Cities Heidelberg

Jewel Island Ceylon

Jewel of the Kalahari Lake Okavango, Botswana

Jewell Manor Jewell Manor (delinquent) Girls Center at Louisville, Kentucky

Jewels of the Caribbean U.S. Virgin Islands

Jewish Alps Catskill Mountains, New York

Jewish ampicillin cream of chicken soup

Jewish champagne celery tonic (carbonated celery-flavored water)

Jewish Confederate Judah Philip Benjamin

Jewish Dristan horseradish (celebrated for its ability to bring about nasal decongestion)

Jewish penicillin chicken soup

Jimmy James; James Earl Carter—thirty-ninth President of the United States

Jim Thorpe formerly Mauch Chunk, Pennsylvania

Jimtown Jamestown, North Dakota

Jinx Falkenburg Eugenia Falkenburg

J.J. Connington Alfred Walter Stewart's pseudonym

Joan Crawford Lucille le Sueur

João Pessoa formerly Parahiba, Brazil

Joaquin Miller Cincinnatus Heine Miller's pen name

Jochanan John

Jock John

Joe Bananas Mafia chief Joseph Bonanno

Joe Doakes nickname for the average American man

Joe Doe name used for the average American male

Joe Louis Joseph Louis Barrow

Joe Who? Canadian Secretary of State Joe Clark

Joe Zilch the average American formerly called Joe Blow or Joe Doakes

Johann Gutenberg Johann Ganzfleisch

John I John the First (John Adams—second President of the United States)

John II John the Second (John Quincy Adams—sixth President of the United States)

John XXIII Angelo Giuseppe Roncalli

John Barleycorn personification of beer or malt liquor

John Barrymore John Blythe

John Bull Great Britain

John Bull's Other Island Ireland before its independence was declared in 1919

John Cabot Giovanni Caboto

John Calvin Jean Chauvin

John Company British East India Company's nickname

John Danger Hough Baillie (colorful journalist who rose from reporter to head of the United Press)

John Doe fictitious name used when real name is withheld or unknown

John and Emery Bonett pseudonym shared by the husband-and-wife team—John Hubert Arthur Coulson and Felicity Winifred Carter

John Ford Sean O'Fienne

John Garfield Julius Garfinkle

John Gilbert John Pringle's stage name

John Hancock signature (nickname memorializing his prominent autograph on *Declaration of Independence*)

John Houseman Jacques Haussmann

John le Carré David John Moore Cornwell

Johnny John; John M Grant, Jr; John von Neumann (Hungarian-born American mathematician)

Johnny Appleseed John (John-

ny) Chapman

Johnny Crapaud nickname for a Frenchman or a New Orleans creole

Johnny Reb(s) Johnny Rebel(s) —Confederate soldier(s)

John Oxenham William Arthur Dunkerly

John Paul Charles Henry Webb

John Paul I Albino Luciani

John Paul II Karol Wojtyla

John Rhode Cecil John Charles Street's pseudonym

John's St John's

Johns of Geneva Jean Calvin and Jean-Jacques Rousseau

John Sinjohn John Galsworthy

John Wayne Marion Michael Morrison

Jonathan Fogarty Titulescu James T(homas) Farrell

Jonathan Oldstyle (pseudonym —Washington Irving)

Jordan's Port Al Aqabah

Jordie Jordan(a)

Jordy Jordan

José Ferrer José Vicente Ferrer y Cintron

José Greco Constanzo Greco

Joseph a Guarneri violin (short form of Giuseppe Guarneri)

Joseph Bentonelli Joseph Horace Benton

Joseph Conrad Teodor Josef Konrad Korzeniowski

Joseph Hansen James Colton's pseudonym

Josephine Josephine Baker

Josephine Bell Doris Bell Collier Ball's pseudonym

Josephus Flavius Josephus, pen name of Arius Calpurnius Piso

Josephus Flavius Josephus—apostate Jew and recorder of the Roman conquests

Joseph von Sternberg Josef Stern

Josh Joshua; (pseudonym—Samuel L Clemens)

Josh Billings stage name of humorist Henry Wheeler Shaw

Josiah Flynt Josiah Flynt Willard's pseudonym

Josie Josephina; Josephine

Juana la Loca (Spanish—Juana the Mad)—queen of Castile whose lisp became the royal style known as Castilian

Juan Bimba the typical Venezuelan

Juan Gris José Victoriano González

Juanita Juana (Jane, Joan)

Juan Pablo (Spanish—John

Paul)—the Pope
Jubilee Girls Jubilee Lodge for (delinquent) Girls at Brimfield, Illinois
Judas Priest Jesus Christ (rendered as a palatable exclamation or oath)
Judy Judith
Judy Garland motion-picture-reel name of Frances Gumm
Judy Holliday Judith Tuvim
Juggernaut Jagananth or port of Puri on the Bay of Bengal
Jules Romains Louis-Henri-Jean Farigoule
Julia Marlowe Sarah Frances Frost's stage name
Julie Andrews Julia Wells
Julio Diniz Joaquim Guilherme Coelho
Jumbo Barnum's famous 6½-ton 11-foot-high trained elephant exhibited in the 'eighties
Jumbo Bill America's 27th President—300-pound William Howard Taft
Jump Jet nickname of U.S. Marine Corps AV-8B fighter-bomber capable of vertical takeoff and landing
June Allyson Ella Geisman
Jungle Novels collective name given to B Traven's books (*see* Hal Croves)
junk heroin's nickname, also called smack
Jupiter of Wall Street JP Morgan
Jusepe José de Ribera
Jussi Björling Johan Jonaton Björling
Justicia justice personified (second goddess wife of Jupiter or Themis who held the same post under the Greek god Zeus); she stands blindfolded, holding a balance in one hand and a palm frond in the other
The Just Society (nickname—Prime Minister Pierre Trudeau's administration of Canada)
Jute Port Dundee, Scotland
Jutland mainland of Denmark and Schleswig-Holstein
Juvenal Decimus Junius Juvenalis
Kabul River City Kabul, Afghanistan
Kabwe formerly Broken Hill, Zambia
Kaffir King Barney Barnato
Kahlbaum's disease dementia with muscular tension

Kahler's disease bone-marrow destruction
Kaiser Bill Wilhelm II—Emperor of Germany
Kalatdlit-Nunat (Greenlandic Eskimo—Land of the People) —Greenland's name adopted in 1979
Kalima formerly Albertville
Kalimantan Indonesian segment of Borneo
Kalinin Soviet name for Tver
Kaliningrad formerly Königsberg
Kamenev Lev Borisovich Rosenfeld
Kampuchean Ports (*see* Cambodian Ports)
Kanakalanders nickname for Queenslanders who hired many South Sea Kanakas to work on their plantations
kanaka(s) South Sea islander(s)
Kanawha River City Charleston, West Virginia
Kangaroo Soviet air-to-surface missile carried by heavy bombers
Kangarooland Australia
Kanner syndrome early infantile autism
kansas cathedrals silos
Karl Johan Jean Baptiste Jules Bernadotte
Karl Malden Karl Malden Sekulovich
Karl-Marx-Stadt formerly Chemnitz
Karl Radek Karl Sobelsohn
Kashin a class of Soviet destroyer-leader ships
Kate Catherine; Katherine; Katherine Hepburn; Katrina
Katherine Mansfield pseudonym—Kathleen Beauchamp Murry
Keeling Islands old name for Cocos in the Indian Ocean
Kelly Country Australia's northern Victoria named after the nineteenth-century outlaw Ned Kelly
Kelp Capital San Diego, California
Kelp Capital of the World offshore San Diego, California
Kelper(s) Falkland Islander(s)
Kelt Soviet air-to-surface missile carried by Tu-16 bombers
Kemal Ataturk Mustafa Kemal Pasha
Kenitra formerly Port Lyautey, French Morocco
Kennel Soviet air-to-surface ship-destroying missile carried

by Tu-16 bombers
Kenya Ports Mombasa, Takaungu, Malindi, Lamu
Kester Christopher
Ketchikan State Jail and Detention Home in Ketchikan, Alaska
Keyboard Titan Vladimir Horowitz
Key City Port Townsend, Washington; Vicksburg, Mississippi
Key to England Dover on the bay beneath the chalk cliffs of Kent flanking the English Channel
Key of the Gulf Cuba
Key of the Indian Ocean Mauritius
Key of the Mediterranean Gibraltar
Key to the New World 16th-century Havana
The Keys Florida Keys extending from Key West to Miami
Key State New South Wales
Key to Stockholm Aland Islands between Finland and Sweden
Keystone Province Manitoba
Keystoner(s) Pennsylvanian(s)
Keystone State Pennsylvania
Khazarian Way Don Volga Portage
Khazar Sea Caspian Sea
Khmer Cambodia
Khmer Republic Cambodia
Kick-'em-Jenny Diamond Island's nickname (West Indian island near Grenada)
Kiel Canal formerly Kaiser Wilhelm Canal
kieselguhr silica (SiO_2)
Kildin Soviet class of fleet destroyers
killer disease dysentery
Kilometer-high City Boone, North Carolina
Kim Novak Marilyn Novak
King of Acids sulfuric acid
King of Austrian Opera Wolfgang Amadeus Mozart
King of Bath Richard (Beau) Nash
King of Beasts the lion
King of Birds the eagle
King Bomba Ferdinand II
King of Cartoons Tex Avery
King of Cats the lion in Africa; tiger in Asia; puma in North America; jaguar in Central and South America
King of Coke Henry Clay Frick
King of Conductors Herbert von Karajan
King of the Conifers the se-

quoia

King Cotton personification of the cotton crop of the southern United States

King of Courts the forensic orator Quintus Hortensius of Rome

King of the Cowboys Roy Rogers

King Crab Capital Kodiak, Alaska

Kingdom of Death Hel (name of the Queen of Death in Nordic mythology)

Kingdom of the Hellenes Greece

Kingdom of Perpetual Night Hell

Kingdom of Sardinia the Italian Peidmont and the island of Sardinia

Kingdoms of the North Denmark, Norway, Sweden

Kingdom of the Swedes, Goths, and Wendes Sweden

Kingdom of the Two Sicilies Naples and Sicily

King of European Music Festivals Salzburg

King of Filibusters William Walker

Kingfish Senator Huey P Long of Louisiana

King of the Fjords Sogne Fjord, Norway

King of French Opera Hector Berlioz

King of Fruits the mango

King of German Opera Richard Wagner

King of the Gods Jupiter (Roman mythology)

King of the High Cs Luciano Pavarotti

King of the Huns and Scourge of God Attila

King of Instruments the human voice; the organ; the violin or other string instruments

King of Italian Opera Giuseppe Verdi

King of Jazz Louis (Satchmo) Armstrong and Paul Whiteman

King of the Jews Jesus, according to the *New Testament*

King of the Jungle the tiger

King of Kings Jehovah—God of the Christians and Jews; title of various presumptive rulers of African and Oriental lands

King of Laughter Bert Williams (Egbert Austin Williams)

King Leso Kingdom of Lesotho

King of Metals gold

King of the Missions Mission San Luis Rey, east of Oceanside, California

King of Naples Marshal Joachim Murat

King of Oceanic Scavengers the albatross

King of the Octaves Claudio José Domingo Brindis de Sala

King Oliver Joseph (King) Oliver—Doctor Jazz

King of the One-Liners Henny Youngman

King of the Opera tenor-conductor Placido Domingo

King of Ornithological Painters John James Audubon

King Penguin Roger Tory Peterson

King of the Pianists Claudio Arrau

King of the Ragtime Writers Scott Joplin

King of Rivers North America's Colorado and South America's Amazon

King of Roads John Loudon Macadam

King of Rock 'n' Roll Elvis Presley

King of Russian Opera Piotr Ilyich Tchaikovsky

King of the Seas Neptune (Roman) or Poseidon (Greek)

king's English correct English

king's evil scrofula (lymphgland tuberculosis)

King of Snakes king snake

King of Steel Andrew Carnegie

King of Swat George Herman Ruth

King Swazi Kingdom of Swaziland

King of Swing Benny Goodman; Elvis Presley

King of Tasmanian Rivers the Gordon

King of Terrors personification of death

King of Tools the lathe

King of Torts Melvin Belli

King of Trains and Train of Kings Orient Express

King Tut King Tutankhamen of Egypt

King of the Underworld Osiris (Egyptian mythology)

King of the Vagabonds François Villon whose real name was François de Montcorbier

King of Vaudeville Jimmy (Schnozzola) Durante

King of Verismo Giacomo Puccini (celebrated composer of realistic operas)

King of Waters the Amazon

King of the West Saxons Alfred the Great

King Who Lost George III, who lost the American colonies

King Who Lost America Great Britain's George III

King of Wines champagne

Kinmen Chinese name for Quemoy Island

Kinshasa formerly Leopoldville, Belgian Congo

Kiowa Bell helicopter whose civil version is called the Jet Ranger

Kipling's Khyber mountain pass between Afghanistan and Pakistan

Kipper NATO nickname for air-to-surface missile carried by Tu-16 aircraft

Kirk Douglas Issur Danielovich Demsky

Kirsty Kristina; Kristine

Kisangani formerly Stanleyville, Belgian Congo

Kit Catherine; Christopher; Kitty

Kit Carson Christopher Carson nicknamed Monarch of the Prairies as well as Nestor of the Rocky Mountains

Kit Carson City Carson City, Nevada named for the frontiersman

Kitchen Soviet air-to-surface missile carried by Tu-22 bombers

Kitchener formerly Berlin, Ontario but changed in World War I to honor Lord Kitchener

Kitchener of Khartum General Horatio Herbert Kitchener

Kitsi Kathryn

Kittie Katherine; Kitty Belairs

Kittsian(s) inhabitant(s) of St Kitts

Kitty Catherine

kiwi(s) New Zealander(s)

Klondike Country the Yukon

Klong Toey Bangkok, Thailand's waterfront area

Knickerbocker Group William Cullen Bryant, James Fenimore Cooper, Washington Irving (Diedrich Knickerbocker)

Knickerbocker(s) New Yorker(s)

Knight of La Mancha Don Quixote

Knight of the Rueful Countenance Don Quixote

Knight of the Swan Lohengrin

Know-Nothing President Millard Fillmore

Knut Hamsun Knut Pedersen

Koba Stalin's party name prior to the Bolshevik takeover of Russia

Kobarid Yugoslavian name for Caporetto

Kodak City Rochester, New York

Kodamá (Japanese—Echo)—nickname of the high-speed express train linking Kyoto with the port of Osaka

Kodok Sudanese name for Fashoda

Ko-i-noor (Persian—Mountain of Light)—Nadir Shah's name for the celebrated mountain-shaped diamond

Komar (Russian—Mosquito)—guided-missile patrol craft used by the navies of Algeria, China, Cuba, Egypt, Indonesia, Syria, etc., as well as the USSR

Komsomolets (Russian—Young Communists)—a class of Soviet torpedo boats (P-4)

kona (Hawaiian-Polynesian—lee side)—side of an island out of or protected from prevailing winds

Kona Coast gold marijuana grown on the kona or lee side of any Hawaiian or Polynesian island

Kongens By (Danish—King's City)—Copenhagen

Königsberg former name for Kaliningrad

Korea Gate nickname of the exposé involving more than a million dollars in bribes given some American politicians in return for their voting money, military aid, and supplies for South Korea

Korean Ports North Korean—Chinnampo, Wonson, Konan-Kimchaek, Chongjin, Najin Up, Unggi; South Korean—Inchon, Kunsan, Mokpo, Yosu, Masan, Pusan

Korea Strait Tsushima Strait (scene of decisive Japanese naval victory over Russian fleet during Russo-Japanese War of 1905)

Korovograd formerly Zinovievsk or Elisavetgrad

Kotlin Soviet guided-missile destroyers

Kraguj Yugoslav counterinsurgency aircraft

Kremlin seat of Soviet Govern-

ment in Moscow

Kremlin Killer Josef Stalin

Kresta Soviet guided-missile destroyer-leader warships

Kringleville Racine, Wisconsin

Krishaber's disease dizzy-and-sleepy neurosis accompanied by fainting

Kriss Kringle Santa Claus

Kristallnacht (German—Night of Broken Glass)—nights of November 8th, 9th, and 10th in 1938 when Nazi mobs broke the windows and smashed the doors of German-Jewish stores and temples before looting them, burning them, and sending their inhabitants to concentration camps

Kristiania Oslo's previous name

Kronos (Greek—Saturn)—god of time

Kronstadt Soviet subchasers

Krung Kao Ayutthaya, Thailand

Krupny a Soviet class of destroyers

Kubyshka (Russian nickname for Cuba)—a kubyshka is a jar wherein Russian peasants bury their money—Kubachka (Little Cuba) has a similar sound to Soviet taxpayers

Kulmhof Chelmo, a Polish concentration camp

Kurir Yugoslav liaison-utility aircraft

Kuwait Ports Mina abd Allah, Ash Shuaiba, Mina al Ahmadi, Abu Hulafah, Al Kuwayt

Kwok's disease Chinese restaurant syndrome produced by monosodium glutamate and resulting in dizziness, headaches, and nausea

Kynda a Soviet class of heavily-armed destroyers

La Argentina Antonia Mercé y Luque

La Belle Époque (French—The Beautiful Epoch)—1900 to 1914 (turn of the century to the start of the first world war)

La Belle Province (French—the beautiful Province)—Québec

La Belle Rivière (French—the Beautiful River)—frontier nickname of the Ohio in the days of Audubon and Boone

La Bonne Louise Louise Michel remembered for good works among the poor of

Paris

Labor Boss Samuel Gompers, John L Lewis, and George Meany

Labrador Canadian version of Boeing-Vertol CH-113 helicopter; Labrador Current; Labrador duck; Labrador jay; Labrador Peninsula; Labrador pine; Labrador retriever; Labrador Sea; Labrador spar; Labrador stone (another name for Labrador spar); Labrador tea

lace-curtain middle class

La Columna (Spanish—The Column)—Venezuela's highest mountain also called Pico Bolívar

Lacrosse USA field artillery MGM-18A surface-to-surface missile

La Cumbre (Spanish—The Summit)—nickname of the Uspallata mountain pass and tunnel in the high Andes linking Argentina and Chile

La Divina Maria Meneghini Callas

The Lady cocaine; Statue of Liberty, New York

Lady of 57th Street New York City's Carnegie Hall

Lady Bird Claudia Alta Taylor Johnson—wife of President Lyndon Johnson

Lady in the Chair Cassiopeia constellation

Lady of Faubourg Saint-Honoré Salle Pleyel, Parisian concert hall

Lady Hamilton Emma Lyon

Lady of the Harbor Statue of Liberty, New York

Lady of the Lakes Michigan

Lady of the Lamp Nurse Florence Nightingale

Lady of Laughter Erma Bombeck

Lady Snow cocaine

Lady South Charleston, South Carolina

Lady with a Lamp Santa Filomena (immortalized by Longfellow)

Lady With Lamp Statue of Liberty officially named Liberty Enlightening the World

l'Affaire (French—The Affair)—the Dreyfus Case

Lafitte Country Baratraria Bay (an old pirate settlement south of New Orleans)

La Gioconda (Italian—The Cheerful Woman)—another name for Leonardo da Vinci's portrait—the Mona Lisa

l'Aiglon (French—the Eagle)—Napoleon II, also known as the Duke of Reichstadt

Laird of Auchinleck James Boswell

Laird of Skibo Castle Andrew Carnegie

Laird of Woodchuck Lodge John Burroughs

Lake City Madison, Wisconsin

Lake of the Four Forest Cantons Lake Lucerne (Switzerland)

Lake Poets Samuel Taylor Coleridge, Robert Southey, William Wordsworth

Lake State Michigan bordering on Superior, Michigan, Huron, and Erie

Lalia Eulalia

La Lollo Gina Lollobrigida

Lama Aerospatiale observation helicopter built in France and designated SA-315

Lamb of God Jesus Christ

Lamia P L Tyraud de Vosjoli (French underground fighter and chief of intelligence)

Lamp of Heaven the Moon

Lana Turner Julia Jean Turner

Land of 10,000 Lakes Minnesota

Land of Acadie (*see* Land of Evangeline)

Land of Albert Schweitzer Gabon

Land of Alligators Florida

Land of a Million Elephants Laos

Land of Art and Mozart Austria

Land of the Aztecs Mexico

Land Between the Rivers Mesopotamia, better known as Iraq

Land Beyond the Mountains Tennessee

Land Beyond Sorrow Uganda

Land of the Bible Israel

Land of Birds Australia

Land of the Blacks Guinea and the Sudan

Land of the Bogs and the Little People Ireland

Land of Bondage Egypt in the time of Moses

Land of the Boomerang Australia

Land of the Bulgars Bulgaria

Land of Caimans swamplands of tropical America

Land of Cakes Scotland

Land of the Cedars Lebanon

LandCent Allied Land Forces, Central Europe

Land of Cheese, Trees, and Ocean Breeze Tillamook, Oregon

Land of the Cherryblossoms Japan

Land of Chopin and Copernicus Poland

Land of Clear Light the American Southwest and the Mexican Northwest (*La Tierra de Luz Clara*)

Land of the Conquistadors Extremadura, Spain where Cortez and Pizarro were born

Land of the Cornstalk Australia

Land of the Croats Croatian Yugoslavia

Land of Crocodiles Africa and Southeast Asia

Land of the Czars and the Commissars Russia—the USSR

Land of Death and Chains Siberia

Land of Desolation Antarctica and Greenland

Land Down Under Australia and New Zealand

Land of Dragons Hong Kong

Land of Dvořák and Smetana Czechoslovakia

Land of the Eagle Albania

Land of Emeralds Colombia

Land of Enchantment New Mexico

Land of Eternal Spring Guatemala

Land of Evangeline Maine east of the Kennebec River, New Brunswick, Nova Scotia, and Louisiana's coastal parishes

Land of Farmers and Fishermen Denmark

Land of the Firebird ancient Russia

Land of Five Peoples Surinam, formerly Dutch Guiana, containing black, brown, red, white, and yellow people from Africa, Indonesia, South America, Europe, and the Orient

Land of the Fjords Norway

Land of Flaming Waters Malawi

Land of Flowers Florida

Land of the Free United States of America

Land of Freedom Liberia

Land of the Gaucho Uruguay

Land of Gavials India and Malaysia

Land of Genghis Khan Mongolia

Land of Gitche Gumee Lake Superior

Land God Gave Cain Arctic Canada

Land of Gold California

Land of the Golden Lion Iran

Land of Grass Roots South Dakota

Land of Greek, Roman, and Modern Ruins Lebanon

Land of the Happy Medium Costa Rica

Land of the Heather Scotland

Land of Heroes Finland

Land of Hiawatha Michigan's Upper Peninsula

Land of Hope and Glory Great Britain

Land of Hospitality Somalia

Land of Hospitality and Charm Thailand

Land of the Hummingbird Trinidad

Land of Ice and Fire Iceland

Land of the Incas Peru

Land of the Individual and Other Endangered Species Alaska

Land of the Inland Sea Chad

Land of the Inland Seas Great Lakes country of Canada and the U.S.

Land of Instant Women Thailand

Land of Iron and Diamonds Sierra Leone

Land of Isolation Antarctica

Land of the Kangaroo Australia

Land of the Khmers Cambodia

Land of the Kiwi New Zealand

Land of Lakes Wisconsin

Land of Lakes and Fens Finland

Land of Lakes and Forests Sweden

Land of Lakes and Volcanos El Salvador

Land of the Lamas Tibet

Land of Lamentations Holy Land, Israel and Palestine

Land of Latte Stones Guam

Land of Leeks Wales

Land of Legend Canada's Yukon Territory

Land of the Lemurs Madagascar and Comoro Islands

Land of Leopold Belgium

Land of the Leprechauns Ireland

Land of Letzeburgesch Luxembourg (where the language is Letzeburgesch)

Land of Lincoln Illinois

Land of Liszt and Bartok Hungary

Land of the Llamas Peru

Landlocked South African Country Lesotho

Landlocked South American Nations Bolivia and Paraguay

Land of the Long White Cloud New Zealand

Land of the Lotus Blossom Sri Lanka (Ceylon), where the lotus blossom symbolizes Buddha

Land of the Magyars Hungary

Land of the Manchus Manchuria

Land of Many Composers Russia (birthplace of Arensky, Borodin, Bortniansky, Cui, Glazunov, Gliere, Glinka, Khachaturian, Liadov, Liapunov, Medtner, Mussorgsky, Prokofiev, Rachmaninoff, Rimsky-Korsakov, Scriabin, Shostakovich, Stravinsky, Tchaikovsky)

Land of Many Tribes Tanzania (formerly Tanganyika)

Land of the Maoris New Zealand

Land of the Marsupials Australia

Land of the Mayas Honduras

Land of Mecca Saudi Arabia

Land of Men Marquesas Islands in the South Pacific

Land of the Midnight Sun northern Alaska, Canada's Northwest Territories, Greenland, Iceland, Norway, Sweden, Finland, and Siberia

Land of Milk and Honey Israel's Jordan River Valley

Land of a Million Elephants Laos

Land of the Moors Algeria and Morocco

Land of the Mormons Utah

Land of the Morning Calm Korea

Land of Moses Israel

Land of Mountains the Austrian Tyrol, Norway, Sweden, Switzerland, Tibet

Land of My Fathers Wales

Land of Nod where Cain was exiled after killing Abel; the realm of sleep

Land o' Cakes Land of Oatmeal Cakes—nickname Robert Burns gave to Scotland

Land o' Lakes Wisconsin

Land of Opportunity Arkansas

Land of Pagodas Burma

Land of the Pentagram Morocco

Land of the People Greenland

Land of the Pharoahs Egypt

Land of the Philistines Palestine

Land of the Plastic Lotus California

Land of Plenty South Dakota

Land of the Poinciana Jamaica

Land of Political Exiles Yakutia (northeastern Siberia in the USSR)

Land of Precious Things El Salvador

Land of the Prince Wales

Land of the Prophets Israel

Land of the Quetzal Guatemala

Land of the Red People Oklahoma

Land of the Rising Sun Japan

Land of the Rolling Prairie Iowa

Land of the Rose England

Land of the Sagas Iceland (where the art of storytelling dates from the 12th century)

Land of Saints and Scholars Ireland

landsat land satellite

Land of the Sea The Netherlands

Land of Sea and Mountain Norway

Land's End Cornish cape in southwest England—westernmost England

Land of the Serbs, Croats, and Slovenes Yugoslavia (including Bosnia and Herzegovina, Croatia, Dalmatia, Macedonia, Montenegro, Serbia, and Slovenia)

Land of the Shamrock Ireland

Land of Silence Lapland

Land of Six Peoples Guyana, (formerly British Guiana) containing Africans, Amerindians, Chinese, East Indians, Spaniards, and other Europeans

Land of Skillful Farmers Lithuania

Land of the Sky North Carolina

Landslide Lyndon Senator Lyndon B Johnson

Land of Small Islands Micronesia

Land of Smiles Thailand

Land of Song Italy

Land South of the Clouds Yunnan

Land of the Southern Cross Brazil

Land of the South Slavs Yugoslavia

Land of Spring coastal southern California from San Diego to Santa Barbara

Lands Reclaimed from the Sea Netherlands

Lands of Sunlit Nights Scandinavian countries during summertime (Denmark, Finland, Iceland, Norway, Sweden)

Land of Steady Habits Connecticut

Land of Sunburned Faces Ethiopia

Land of Sunshine New Mexico, South Africa, and southern California

Land of Symphonists Austria (birthplace or home of Haydn, Mozart, Bruckner, and Mahler)

Land of the Templars Malta

Land That Time Forgot Australia

Land of the Thistle Scotland

Land of the Thousand Lakes Finland

Land of Togetherness Kenya

Land of Tomorrow Brazil

Land of the Trade Winds U.S. Virgin Islands

Land of the Vikings Norway particularly Vestfold province on the western shore of Oslo Fjord where Viking remains are plentiful

Land of Waterfalls Norway

Land of Waters Guyana

Land of the Wattle Australia

Land of the West Morocco (from Arabic *maghrib* meaning west)

Land Where The Sun Never Sets the Soviet Union

Land of the Whispering Bushes California

Land of the White Ant Australia's Northern Territory

Land of the White Eagle Poland

Land of the White Elephant Thailand

Land of the White Mountain Kenya

Land of the Winds Iran or Persia

Lane's disease chronic constipation

Langtry formerly Vinegaroon, Texas; renamed in 1882 by Judge Roy Bean to honor actress Lillie Langtry whose name also adorned his combination courthouse and saloon—*The Jersey Lily*

Lansen (Swedish—Lance)—jet

interceptor aircraft designated J-32

La Pasionaria Dolores Ibarruri famed for her impassioned speeches during the Spanish civil war

La Perla (Spanish—The Pearl) —San Juan

Lapland northernmost Finland, Norway, Sweden, and the USSR

La Popessa Mother Pascalina (nurse and confidante of Pope Pius XII for forty-one years)

Larruping Lou Henry Louis (Lou) Gehrig

Larry Laura; Laurence; Lawrence

La Salle Street Chicago's financial district

La Scala Milan's opera house

La Scala West nickname of Chicago's Lyric Opera

Las Crutches trucker's nickname for Las Cruces, New Mexico

Last Capital of the Confederacy Danville, Virginia

Last Chance Gulch gold miner's name for Helena, Montana

Last Cocked Hat James Monroe—fifth President of the United States and last to wear the cocked hat of the American Revolution

Last Continent Antarctica (last continent to be discovered)

Last Corner of Arabia Oman

Last Frontier Alaska's old nickname and current nickname of Canada's Northwest Territories; Antarctica

Last of the Incas Atahualpa

Last, Loneliest, Loveliest (city) Auckland, New Zealand, according to Kipling

Last Lovely City San Francisco

Last Outpost on the Mississippi Pilot Town (near Venice, Louisiana)

Last of the Prophets before Mohammed Jesus, according to the Moslems

Last Remaining Polynesian Kingdom Tonga—christened the Friendly Isles by Captain Cook

Last of the Romans Rienzi

Last Romantic poet and political writer Max Eastman; Richard Strauss; Sergei Rachmaninov; W Somerset Maugham

Last Son of the 18th Century

Gioacchino Rossini

Last Stronghold of the Moors Granada, Spain

La Stupenda Dame Joan Sutherland

last trump the sound of the last trumpet believers expect to hear on Judgment Day

Last Viceroy Earl Mountbatten of Burma (India's last viceroy)

La Superba (Italian—The Superb)—Genoa's proud appellation dating back to the time of Columbus

Las Vegas East Atlantic City, New Jersey

Las Villas formerly Santa Clara province in Cuba

Las Wages nickname of Las Vegas, Nevada

Latter-Day Saint Joseph Smith —author of *The Book of Mormon*

Latter-Day Saints the Mormons

Laughing Philosopher Voltaire

Launcelot Langstaff pseudonym shared by Washington Irving, William Irving, and James K Paulding when they published the *Salmagundi* essays

Laura World War II code name for Majuro, still in use by Americans and Marshallese islanders

Laura Z Hobson Laura K Zametkin

Lauren Bacall Betty Perske

Laurence Templeton Sir Walter Scott's pseudonym used in the publication of *Ivanhoe*

Lawgiver of Ancient Greece Solon of Athens

Lawrence of Arabia Thomas Edward Lawrence

Lawrence L Lynch Emma Murdock Van Deventer's pseudonym

Law West of the Pecos Judge Roy Bean of Langtry, Texas, also known as the Hanging Judge

Leadbelly Huddie Ledbetter

Leader of the East USSR (1945–1989)

Leader of French Enlightenment Denis Diderot

Leader of the Renaissance World Florence (Firenze)

Leader of the West U.S.A. (1945—1989)

Lead State nickname shared by Colorado and Missouri

League-of-Nations President

Woodrow Wilson

Leao do Mar (Portuguese— Lion of the Sea)—the stormy Cape of Good Hope

Lebanese Ports Tarabulus, Beirut, Sayda, Sur

Leber's disease congenital atrophy of the optic nerve

Le Boulevard des Princes (French nickname—Hortense Schneider—actress and intimate of European royalty)

L'Ebreo (Italian—The Hebrew)—nickname of Salomone Rossi—Renaissance composer and rabbi

Le Carré John Le Carré (pseudonym of David John Moore Cornwall)

Lecumberri-Hilton nickname of Mexico City's great prison

Le Douanier (French—The Custom House Officer)— nickname of Henri Rousseau the primitive painter

Leedsloiner(s) native(s) of Leeds

Lefty Robert Grove

Leghorn English equivalent of Livorno on Italy's west coast

Le Grand Siècle (French—The Great Century)—the 1600s

Lemnos Limnos island in the Aegean

Lemonade Lucy Mrs Lucy Ware Webb Hayes—wife of President Rutherford B Hayes —who served only non-intoxicating fruit drinks while at the White House

Lemur Paradise Madagascar

Lena River City Yukutsk, Siberia

Leninakan formerly Aleksandropol

Lens-Grinder Philosopher Benedictus de Spinoza

Leonard Holton Leonard Patrick O'Connor Wibberley's pseudonym

Leonard Q Ross (pseudonym —Leo Rosten)

Leon Bakst Leon Nikolaevich Rosenberg

Leopard West German Krauss-Maffei medium tank armed with a 105mm gun

Leopold's galloping ghost Congo-attained dysentery (African equivalent of the curse of Cortez, Montezuma's revenge, the plight of Pizarro, etc.)

Leopoldville Kinshasa, Zaire

Lepke Louis Buchalter (1897— 1944)

Le Roi Soleil (French—The Sun King)—Louis XIV
Le Sage (French—The Wise)—Charles V
Lesbian Poet Sappho, the poetess of Lesbos
Lesbos Lésvos or Mytilene in the Aegean
Lesbos Long Island Fire Island, Long Island, NY
Les Cayes modern name for Aux Cayes, Haiti also called Cayes
Leslie Charteris Leslie Charles Bowyer Yin
Leslie Ford Zenith Jones Brown's pseudonym
Leslie Howard Leslie Stainer
Lethbridge originally Coalbanks, Alberta
Leviathan of Literature Dr Samuel Johnson
Le Vigan Robert Coquillaud
Lewis Carroll Charles Lutwidge Dodgson
Lewis Grassic Gibbon pseudonym of J(ames) L(eslie) Mitchell
Liar of Biblical Antiquity Ananias, struck dead for lying, according to *The Acts* in the New Testament
The Liberator Daniel O'Connell
Liberator of Argentina General José de San Martín
Liberator Czar Alexander II (1855–1881)—abolished serfdom in Russia
Liberator of Genoa Andrea Doria
Liberator of God and Man Baruch de Spinoza
Liberator of Italy Giuseppe Garibaldi
Liberian Ports Robertsport, Monrovia, Buchanan, Harper
Libertador de Chile (Spanish—Liberator of Chile)—Bernardo O'Higgins
Libertybellsville Philadelphia
Liberty Bowl Memphis, Tennessee
Liberty Enlightening the World Statue of Liberty in New York Bay
Liberty Island formerly Bedloe's Island in Upper New York Bay where it supports the Statue of Liberty
Library Builder Andrew Carnegie
Library of Last Resort the Library of Congress in Washington, DC, where anyone can consult any book in any language

guage
Librettist-Composer Arrigo Boito
Libyan Ports Bardiyah, Tobruk, Darnah, Marsa al Hilal, Marsa Susah, Benghazi, Az Zuwaytinah, Marsa al Burayqah, As Sidr, Surt, Misratah, Tarabulus, Marsa Sabratah, Zuwarah
licorice stick clarinet's nickname
Liftmaster Douglas DC-6 92-passenger transport
Light of the Ages Moses ben Maimon of Cordoba also known as Maimonades
Light of Asia Gautama Buddha
Lighthorse Harry Major General Henry (Lighthorse Harry) Lee, USA—father of Robert E. Lee
Lighthouse of the Pacific *El Faro del Pacifico*—Izalco Volcano—whenever active, its fire can be seen from planes and ships several hundred miles away from El Salvador
Lightning British BAC all-weather supersonic jet
Ligurian Republic Republic of Genoa
Likasi formerly Jadotville in the Belgian Congo
Lilac New Hampshire state flower and nickname sometimes given New Hampshire girls
Lilac City Spokane, Tacoma, Washington
Lila Lee Augusta Appel
Lillian Gish Lillian De Guiche
Lillian Nordica Lilly Norton
Lillian Russell Helen Louise Leonard
Lillibet Elizabeth
Lillie Emily; Lillian; Lillie Langtry—the Jersey Lily—christened Emily Charlotte Le Breton
Lilli Palmer Lillie Marie Peiser
Lily of France symbolic *fleur de lis* or lily flower
Lily-Lilo Rosalie Texier
Limbo of the Lost the Bermuda Triangle, also called the Devil's Triangle
Limejuicer British sailor
Limeyland England
Limey(s) Limejuicer(s)—British sailor(s) or ship(s); nickname derived from their use of limejuice to ward off scurvy
Lincoln's Shrine Springfield, Illinois
Lincoln's State Illinois

Linnaeus Carl von Linné
Linoleum Capital of Scotland Kirkaldy
Lion of the Caribbean Sir William Alexander Bustamante
Lion City Singapore
Lionel Barrymore Lionel Blythe
Lion Flag Ceylon's emblem featuring a golden lion with an upraised scimitar in his right paw comes from the ancient name for this island
Lion Hill Norway's parliament, officially called Storting
Lion of Judah former Emperor Haile Selassie of Ethiopia
Lion of the North King Gustavus Adolphus of Sweden
Lion's Gate harbor entrance of Vancouver, British Columbia
Lion Tamer Ian Smith
Lippy Leo Ernest Durocher
Lisbeth Elisabeth; Eliza; Elizabeta; Elizabeth
Listener to the Winds Edward MacDowell
Liszt Expert Alfred Brendel
Literary Queen of Expatriate Americans Gertrude Stein
Little Alfie Alfred Austin
Little America Antarctic camp at the edge of the Ross Ice Shelf and the Bay of Whales where Admiral Byrd headquartered; London's Grosvenor Square where John Adams lived at No 9 when he was America's first ambassador to Great Britain; now site of the U.S. Embassy
Little Belt Lillebaelt (strait separating island of Fyn from Danish mainland between Baltic Sea and the Kattegat)
Little Boy A-bomb dropped on Japanese targets before the end of World War II
Little Britain Armorica or Brittany in northern France
Little Corporal five-foot-high Napoleon Bonaparte *(Le Petit Caporal)*
Little Denmark Solvang, California
Little Egypt delta country of southern Illinois around Cairo and the confluence of the Ohio and Mississippi
Little England of the Caribbean Barbados
Little Flower Fiorello H La Guardia
Little Giant Knute Nelson—populist governor of Minnesota; oratorically gifted Sena-

tor Stephen Douglas of Illinois

Little Havana Cuban-refugee-populated sections of Miami, Florida

Little Holland Garibaldi, Oregon

Little Ida Idaho

Little Inch 20-inch pipeline paralleling the Big Inch

Little India of the Pacific Fiji Islands

Little Italy Italian section of any American or Canadian city

Little Joe Apollo spacecraft booster designed and produced by General Dynamics, Convair

Little John surface-to-surface rocket produced by Emerson Electric

Little Lad of Landau American political cartoonist Thomas (Th) Nast born in Landau, Germany

Little Lady in Pants Dr Mary Walker

Little Lunnon Colorado Springs, Colorado

Little Luther Hans Kung

Little Mac General George B Mc Clellan

Little Magician Martin Van Buren—President of the United States

Little-Magician President Martin Van Buren

Little Mermaid Edvard Eriksen's bronze statue of a maiden seated atop a rock and looking out to sea from Copenhagen's harbor—immortalized in Hans Christian Andersen's fairy tale

Little Neddies Economic Development Committees

Little Netherlands Curaçao

Little New York Miami Beach, Florida's South Beach

Little Old Lady of Pennsylvania Avenue Federal Trade Commission

Little Paradise Queen Victoria's nickname for the Isle of Wight

Little Phil General Philip Henry Sheridan

Little Red Book quotations of Mao Tse-tung

Little Rhody Rhode Island

Little's disease congenital spastic paralysis

Little Sure Shot Annie Oakley (Mrs Frank Butler)

Little Tiger mainland China's

hydrofoil patrol boat called Hu Chwan

Little Tokyo Japanese section of any American or Canadian city

Little Van President Martin Van Buren

Little Van Dyke Gonzales Cocx—Flemish portrait painter who imitated the style of Van Dyke but painted family groups on small canvases

Little Van's Lady Hannah Van Buren—wife of President Martin Van Buren

Little Venice Lake Maracaibo, Venezuela

Little White House Franklin D Roosevelt's farm home near Warm Springs, Georgia

Little Yellow Book quotations of Deng Xiaoping

Live Free or Die State New Hampshire

Live Oak State Florida

Living Declaration of Independence Thomas Paine

Livy Olivia; Roman historian Titus Livius

Liz(a) Eliza(beth)

Lizard State Alabama

Lizbeth Elizabeth

Lizzie British sailors' nickname for the superliner *Queen Elizabeth*

Lizzy Elizabeth

Ljuba Welitsch Ljuba Velichkova

Lock City Stamford, Connecticut

Lock Town Stamford, Connecticut

Log Cabin-and-Hard-Cider President William Henry Harrison

Lola Dolores

Lola Montez stage name of Marie Dolores Eliza Rosanna Gilbert also known as the Comtesse de Lansfeld, Mrs Heald, Mrs Hull, and Mrs James

Lon Chaney Alonso Chaney

london london broil (thinly-sliced flank steak broiled before serving); london brown (carbuncle gemstone)

London London Bridge(s), London broil, London Company Londonderry, Londoner, Londonese, Londonesque, London forces, Londonish, London ivy, Londonization, Londonize, London particular (fog), London plane, London purple, London smoke, Lon-

dony

London Bach Johann Christian Bach

London of the Scanians King Canute the Great's name for Lund, Sweden; Scandic capital he founded to match his London of the English

London by the Sea Brighton

London-super-Mare Brighton

London Town London, England

Lone Eagle Charles A Lindbergh

Lone Lion of Idaho Senator William E Borah

Lonely Iconoclast George S Schuyler

Lone Star State Texas

Longhair Lair New York's Lincoln Center of the Performing Arts

Longhorn(s) Texan(s) named after the longhorn cattle characteristic of Texas

Longshanks Edward I of England

Longshore Philosopher Eric Hoffer

Long Tom Thomas Jefferson—third President of the United States

Lon'on Town British nickname for London

The Loop downtown commercial, financial, hotel, shopping, and theater district of Chicago

Lord Acton 1st Baron John Emerich Edward Dalberg-Acton

Lord Baltimore George Calvert

Lord Beaconsfield Benjamin Disraeli

Lord Beaverbrook William Maxwell Aitken

Lord Berners Gerald Hugh Tyrwhitt-Wilson

Lord Brougham Henry Peter

Lord Byron George Gordon Byron

Lord Chesterfield Philip Stanhope

Lord De La Warr Thomas West—Lord Delaware

Lord Desart William Ulick O'Connor Cuffe

Lord Dufferin Frederick Temple Hamilton Blackwood (Lord Rector of St Andrews University)

Lord Dunsany Edward John Moreton Drax Plunkett

Lord of the East Vladivostok

Lord Haw-Haw nickname of

William Joyce **Lord of Hell** Lucifer
Lord High Executioner of the Mafia Albert Anastasia (1903–1957)
Lord of Hokkaido Japanese red fox
Lord Kelvin William Thomson
Lord Kenneth Kenneth Clark— Lord Clark of Saltwood
Lord Keynes John Maynard Keynes
Lord Kinross John Patrick Douglas Balfour
Lord Kitchener Horatio Herbert (also known as the Earl of Khartoum)
Lord Macaulay Thomas Babington
Lord North Frederick North
Lord Palmerston Henry John Temple nicknamed Pam
Lord Passfield Sidney Webb
Lord Peter Death Brendon Wimsey Ian Carmichael
Lord Protector Oliver Cromwell—Lord Protector of England
Lord of Reason Bertrand A Russell
Lord of the Rings J(ohn) R(onald) R(euel) Tolkien
Lord Russell Bertrand A Russell
Lord Salisbury Robert Arthur Talbot Gascoyne-Cecil
Lord of San Simeon William Randolph Hearst
Lord Tweedsmuir John Buchan
Lord of War Wotan
Lorenzo da Ponte Mozart's librettist whose real name was Emanuele Corregliano
Lorenzo the Magnificent Lorenzo de Medici
Loretta Young Gretchen Young
Loris Hugo von Hofmannsthal
Lorrie Laura; Lorraine
Los Angeles' Sister City Eilat —Israel's leading oil port
Los Pinos (Spanish—The Pines) —official home of Mexico's presidents
Lost City of the Incas Machu Picchu, Peru (near Cuzco, Peru)
Lost Colony Roanoke Island, North Carolina (site of Sir Walter Raleigh's first settlement)
Lost Wages Las Vegas, Nevada (gambling resort)
Lot's Wife Japanese volcanic

islet in the North Pacific between Iwo Jima and Yokohama—resembles a pillar of salt; mountain of St Helena Island in the South Atlantic
Lottie Charlotte
Lou Costello Louis Cristillo
Lou Gehrig's disease amyotrophic lateral sclerosis
Louis Calhern Carl Vogt
Louis Capet King Louis XVI
Louise Homer Louise Dilworth Beatty
Louis-Ferdinand Céline Henri-Louis Destouches
Louis Graveure Wilfred Douthitt
Louisiana Ports Baton Rouge, Lake Charles, New Orleans
Louis Jourdan Louis Gendre
Louis le Debonnaire Louis I of France
Louis Napoleon Napoleon III— Emperor of France
Love Goddess Rita Hayworth
Lovely Felix Felix Mendelssohn-Bartholdy
love machine bedroom-on-wheels type of recreation vehicle such as a camper, trailer, or van
Lover of Liberty Thomas Paine
Low Countries Belgium, Luxembourg, and the Netherlands
Lower 48ers Alaskan name for people in the 48 contiguous states
Lower Amazon Amazon River traversing northern Brazil from Manaus to the Amazon River Delta on the Atlantic near Belém do Pará
Lower Austria southern Austria bordering on Switzerland, Italy, Yugoslavia, and Hungary
Lower Bavaria eastern Bavaria
Lower Burma coastal Burma west of Thailand
Lower California the Baja California
Lower Canada French-speaking Québec and the lower St Lawrence region during the 19th century
Lower East Side New York City's most congested section south of Washington Square
Lower Egypt Egypt's delta area north of Cairo and including Alexandria and Port Said
Lower Franconia northwestern Bavaria
Lower Galilee Israel between

the Mediterranean and the Sea of Galilee
Lower Lakes southernmost Great Lakes—Erie and Ontario
Lower Michigan peninsular Michigan south of Mackinac Strait
Lower Mississippi Mississippi River from Saint Louis to New Orleans and the Gulf of Mexico
Lower Nile Nile River flowing from Khartoum in the Sudan to Cairo in Egypt and the Nile River Delta emptying into the Mediterranean Sea
Lower Peninsula southern Michigan between Lake Michigan, Lake Huron, and Lake Erie
Lower Rhine Rhine River between Bonn, Germany and the North Sea coast of the Netherlands
Lower Saxony Neidersachsen including most of Brunswick, Hannover, Oldenburg, and Schaumburg-Lippe
Lower Silesia southern Silesia
Lowland Duchy Luxembourg
Low Newton remand center in Durham, England
Loyalist Province New Brunswick
Lozovsky Solomon Abramovich Dridzo
LT Meade Elizabeth Thomasina Meade Smith
Lubumbashi formerly Elisabethville, Belgian Congo
Lucan Roman poet Marcus Annaeus Lucanus
Lucas Luke
Lucil Gaius Lucilius (Romansatiric writer)
Lucille Ball Diane Belmont
Lucky Lucky Luciano (Salvatore Lucania)—once America's foremost gangster
Lucky Black Swan Western Australia's city of Perth where black swans swim about in Perth Water
Lucky Capital Canberra, Australia
Lucky Country Australia
Lucy Lucia; Lucilla; Lucille; St Lucia, West Indies
Lucy Stone maiden name of Mrs Henry Brown Blackwell
Lukas Foss Lukas Fuchs
Lumber Capital Tacoma, Washington
Lumber City Bangor, Maine
Lumber State Maine

Lum 'n Abner Chester Lauck and Norris Goff

Lunatic of Libya Col Muammar Qaddafi

Lungansk old name of Voroshilovgrad

Lunik Soviet cosmic rocket landed on Moon September 14, 1959

Lusians Portuguese

Lusitania (Latin—Portugal)— Roman name often used as the poetic equivalent of Portugal

L Wica pseudonym of Vilhelm the Prince of Sweden and Duke of Södermanland

Lynn Brock Alister McAllister

Lynn Doyle Leslie Alexander Montgomery

Lyric Land Wales

Lyric Poet and Literary Critic Heinrich Heine

Maastricht Corridor southernmost projection of the Netherlands between Belgium and Germany

Ma Bell Bell System telephone companies linked by AT & T

Macabre Film Robert Wiene's *Cabinet of Dr Caligari*

macadam macadam road (named for its Scottish inventor—J L McAdam); macadam stone (stone used in building macadam pavements or roads)

MacArthur's Birthplace Little Rock, Arkansas—birthplace of General Douglas MacArthur

MacGregor mechanically operated hatch cover

Machine Gun George R Kelly (1897–1954) and Jack McGurn (1904–1936)

MacRobertson Land Australian Antarctica

Madagascar's Principal Port Tamatave

Madame Bertha Sarah Bernhardt

Madame Blavatsky Helena Petrovna Hahn-Hahn

Madame Deficit Marie Antoinette's nickname

Madame de Stael Baronne Anne Louise Germaine

Madame Récamier Jeanne Françoise Julie Adélaïde Bernard

Mad Anthony Major General Anthony Wayne

Mademoiselle le Professeur Nadia Boulanger

Madge Margaret; Margarita

Mad Genius of Sex and Psy-chiatry Wilhem Reich—inventor of the orgone box

Madhouse on the Potomac the Capitol, the Pentagon, the State Department, the White House, Washington, D.C.

Mad Ludwig Ludwig II of Bavaria (Wagner's patron)

Mad Meg nickname of the Mayer van der Bergh Museum in Antwerp, Belgium

Mad Monk Gregori Rasputin

Mad Priest John Ball

madras madras cloth or madras cotton; madras kerchief

The Maestro Arturo Toscanini

Maestro of Abolition Brazilian composer-conductor Carlos Gomes who fought for the abolition of slavery

Maestro Crescendo Rossini

Maestro No No conductor-perfectionist Arturo Toscanini's

Mae West American actress of stage and screen; lifejacket named after her shape

Magallanes former name of Punta Arenas, Chile

Magazine of History *American Heritage*

Magazinist Edgar Allan Poe's self-invented title

Magda Lupescu Elena Wolff

Magellan of Modern Music Harry Partch

Maggie Margaret; Margaret Thatcher, Prime Minister of the United Kingdom of Great Britain and Northern Ireland (1979–1990); stock market nickname for Magnavox

Magic Island Haiti on Hispaniola

Magister Fouga jet trainer built in France

Magnificent 13 New York City's red-bereted teams of vigilantes determined to enforce law and order in the subways

Magnolia City Houston, Texas

Magnolia Lady Former First Lady Rosalynn Carter

Magnolia(s) Mississippian(s)

Magnolia State Mississippi

Magyar(s) Hungarian(s)

Mahatma (Hindi—Great Souled) —sobriquet of India's greatest leader, Mohandas Karamchand Ghandi

Maid of Orleans Joan of Arc —*La Pucelle d'Orleans*

Maid of Zaragoza Augustina de Aragón (Augustina Domenech Zaragoza—fighter for freedom during Spain's inva-sion by Napoleonic armies)

Mail Soviet BE-12 Beriev amphibian reconnaissance aircraft

Maimonides Moses ben Maimon

Main Drag Main Street or the main street of any American city or town

Main Drag of Many Tears 125th Street in New York City's Harlem

Maine Ports Calais to Kittery including Bangor, Searsport, Boothbay Harbor, Bath, and Portland

mainland China People's Republic of China

Mainlanders Hawaiian name for people of the 49 mainland states

Maisie Maria; Marie; Mary; Maryjane

Maison Gomin correctional facility for women at St Cyrille, Québec

Maison Tanguay Montreal's facility for women prisoners

Majocchi's disease ringlike empurplement of the lower limbs

Major Prophets of the Old Testament Isaiah, Jeremiah, Ezekiel, Daniel

Make-You-Cry Composer Puccini

Make-You-Laugh Composer Rossini

Malacañan (Filipino—Home of the Ruler)—Filipino presidential palace

Malacañan Palace official residence of the President of the Philippines

Malafon Latecoere surface-to-surface or surface-to-underwater naval missile made in France

Malagasy Republic Madagascar

Malayan Island Nation Singapore

Malaysian Ports Penang-Butter-worth; Lumut, Kelang (Port Swettenham), Port Dickson, Melaka

Malbrook (Louis XIV's mispronunciation of Marlborough—John Churchill—Duke of Marlborough)

Malcolm X Malcolm Little

Maldive Islands Port Malé

Malgache Republic Madagascar (territory includes Amsterdam, Crozet, Kerguelen, and Saint Paul islands)

Malinche (see Doña Marina)—often a Mexican synonym for Quisling or traitor—with Malinchismo meaning treachery

Maltese Ports Valetta and Marsaxlokk

Maluku (Indonesian—Moluccas)—Spice Islands

Mambru (Spanish mispronunciation of Marlborough—John Churchill—Duke of Marlborough)

Man Against Everything Henry Louis Mencken (1880–1956), editor of the *American Mercury* (1924–33)

Man for All Seasons Sir Thomas More

Mañanaland Latin America

Manassa Mauler Jack Dempsey born in Manassa, Colorado

Man of a Thousand Faces Lon Chaney

Mana-Zucca Augusta Zuckerman

Man of Blood and Iron Prince Otto von Bismarck

Man of Destiny Napoleon; William Walker, one-time dictator president of Nicaragua who planned for Central American unification and a Caribbean federation including Central America and Cuba

mandrake nickname for *Mandragora officinarum* also called devil's testicle or satan's apple

Mandrake Soviet Yakovlev strategic-reconnaissance jet aircraft

Mandy Amanda; Manda

Man from Independence President Harry S Truman

Man from Maine James G Blaine

Man from Missouri President Harry S Truman

Man of God Grigori Rasputin (so named by Czar Nicholas II—last Romanov emperor of Russia)

Mangrove Soviet Yakovlev Yak-26 tactical reconnaissance aircraft

Mangrove Coast Florida's southernmost coast between the Everglades and the Keys

Manhattan Manhattan clam chowder (minced clams plus herbs and tomatoes); Manhattan cocktail (vermouth and whiskey mix topped with a maraschino cherry); Manhat-

tan skyscraper

Manifest Destiny President James Knox Polk

Manila Manila hemp (abacá fiber used in making fabrics, rope and paper); Manila paper (buff-color paper prized for its heavy-duty applications)

Man of Monach Country Fermanagh County in Ulster, Northern Ireland

Manning Coles Cyril Henry Coles

Manny Emanuel; Manuel

Man(ny) Han(ny) Manufacturers Hanover (Bank)

Manolete Manuel Rodriguez

Man on Horseback General Georges Boulanger

Man's Oldest Disease alcoholism

Mantovani d'Annunzio Paolo

Mantua English place-name for Mantova

Manuel Emmanuel

Man Who Invented Panama Philippe Bunau-Varilla

Man Who Made the Greatest Dictionary James AH Murray who devoted his life to the creation of the *Oxford English Dictionary (OED)*

Man Who Made Ragtime Scott Joplin

Man With the Hoe The Man With the Hoe—Edwin Markham

Man of Words lexicographer Eric Partridge

Manxmen Isle-of-Man persons

Many Islands *see* Polynesia

Man You Loved To Hate Erich von Stroheim

Maple City Ogdensburg, New York

Maple Leaf Canada's flag consisting of three vertical stripes —red, white, red—with a red maple leaf on the white center stripe

Marble Capital Proctor, Vermont

Marble City Rutland, Vermont; Sylacauga, Alabama

Marble Halls of Oregon Oregon Caves National Monument

Marc Chagall Marc Segal

Marcella Sembrich Praxede Marcelline Kochanska

Marches border region of England and Wales

March King John Philip Sousa

Marco Page pseudonym of Harry Kurnitz

Marder German armored-personnel carrier fitted with a 20mm cannon

Mardi Gras Metropolis New Orleans, Louisiana

Marge Margaret; Margery

Margie Margaret

Margo Maria Margarita Guadalupe Teresa Estela Bolado Castilla y O'Donnell

Margot Margaret

Maria Callas Maria Calogeropoulos

Maria Ivogün Ilse von Günther

Maria Jeritza Mimi Jedlitzka

María Montez Maria África Vidal de Santo Solas

Marie Brema Minny Fehrman

Marie Corelli Eva Mary Mackay

Marie Curie Manya Sklodowska (1867–1934)

Marie Dressler Leila Koerber

Marie's disease chronic enlargement of the face, feet, and hands

Marilyn Monroe Norma Jean Baker

Marina Street San Juan, Puerto Rico's district jail

Mariner Venus-Mars fly-by space vehicle

Mariner Mystic Herman Melville

Mario Giovanni Matteo

Mario Lanza Alfred Arnold Cacozza

Marion Davies Marion Cecelia Douras

Mariscal de Ayacucho (Spanish—Marshal of Ayacucho)—Antonio José de Sucre—companion of Bolívar and first president of Bolivia

Maritime Capital of Canada Vancouver

Maritime Provinces New Brunswick, Nova Scotia, and Prince Edward Island

Maritzburg Pietermaritzburg

Mark Aldanov Mark Aleksandrovich Landau

Markland (Norse—Forest Land)—probably Labrador

Mark Rothko Marcus Rothkowitz

Marks & Sparks Marks & Spencer (British department store)

Mark Twain Samuel Langhorne Clemens

Mark Twain Town Hannibal, Missouri

Marlene Dietrich Magdalene von Losch

Marlin Martin P-5M reconnaissance flying boat

Marmalade Capital Dundee, Scotland

Marquis de Vaugenargues Luc de Clapiers

Marquise de Pompadour Jeanne Poisson

Marquis of Halifax George Savile (1633–1695)

Marquis of Queensbury boxing rules formulated by John Graham Chambers supervised by the 8th Marquis of Queensbury—Sir John Shollto Douglas

Marquis of Rockingham Charles Watson-Wentworth (former Prime Minister of Great Britain)

Marquis of Salisbury Robert AT Gascoyne-Cecil

Marse Robert (southern American—Master Robert)—General Robert E Lee

Marshall Plan President Harry S Truman

Martel Hawker-Siddeley in Britain and Matra in France built this AS-37 missile

Martha Albrand Heidi Huberta Freybe

Martial Roman epigrammatist Marcus Valerius Martialis

Martov Yuli Osipovich Tsederbaum

Martyr Abolitionist Elijah Parish Lovejoy

Martyr of Mexican Independence Miguel Hidalgo

Martyr for Science Giordano Bruno

Marunouchi Tokoyo's financial center

Marvel of Marble the Taj Mahal

Marx Brothers Chico (Leonard), Harpo (Arthur), Groucho (Julius), plus Gummo and Zeppo Marx

Mary Astor Lucille Langhanke

Maryland Port Baltimore

Mary Pickford Gladys Mary Smith's stage name

Mary Queen of Scots Mary Stuart

Mary Roe anonym used when the true name of the woman is unknown; nickname of the average American girl

Marysville Girls Ohio Reformatory for Women at Marysville

Masaccio Tommaso Guidi

Masaniello contracted name of Tommaso Aniello

Masha (Russian nickname—

Mary)

Mason and Dixon Line boundary between Maryland and Pennsylvania used to describe former demarcation between southern slave and northern free states

Massachusetts Ports Gloucester, Salem, Boston, Quincy, New Bedford

Massa Linkum Abraham Lincoln

Master of Color Contrasts Bartolomé Esteban Murillo

Master Composer Robert Schumann

Master of Disguises Alec Guinness

Master of Guerrilla Warfare Toussaint l'Ouverture

Master Illusionist Lawrence of Arabia, Thomas Edward Lawrence

Master of Light and Shade Rembrandt van Rijn and Leonardo da Vinci

Master Mariner of the Imagination Joseph Conrad

Master of Melancholy Andres Segovia

Master of Mexico Don Porfirio Diaz (Porfiriato lasted from 1876 to 1911)

Mastermind of Revolution VI Lenin

Master Musician/Comedian Victor Borge

Master of Neurological Anatomy Santiago Ramón y Cajal

Masterpiece of American Engineering the Panama Canal

Master Pilot Jacques Cartier *(Le Maître-Pilote)*

Master of Political Satire Mark Russell

Master of Psychic Polyphony Richard Strauss

Master of the Psychological Novel Franz Kafka

Master of Raphael Il Perugino (Pietro Vannucci)

Master Scientist-Aero Engineer Theodore von Karmen

Master of Suspense Alfred Hitchcock

Master of Swing Count Basie [originally William (Bill) Basie]

Master of Tragic Sorrow Pyotr Ilich Tchaikovsky

Master of the Yosemite Ansel Adams

Mastodon of Literature Emmanuel Swedenborg so nicknamed by Ralph Waldo Emerson

Masurca French-built surface-to-air naval missile

Mata Hari Gertrud Margarete Zelle

Mateo Boz Miguel Angel Correa's pseudonym

Mato Tepee Devils Tower National Monument in northeastern Wyoming

Matriarch of Anthropology Margaret Mead

Matsqui British Columbia's minimum-security facility for narcotic addicts at Abbotsford

Matty the Great Christy Mathewson

Matty Van Martin Van Buren

Mauch Chunk Pennsylvania place now called Jim Thorpe

Maude Adams Maude Kiskadden

Maureen Forrester Katherine Stewart

Maureen O'Hara Maureen Fitzsimmons

Maurice Barrymore Herbert Blythe

Mauritanian Port Nouakchott

Mauritius Port Port Louis

Max of Baden Maximilian Alexander Friedrich Wilhelm von Hapsburg (1867–1929)

Max Brand Frederick Faust

Maxim Gorki Aleksei Maximovich Peshkov

Maxim Litvinov Maxim Maximovich Wallach

Max Nordau Max Simon Südfeld

Max Pax Max, Prince von Baden (Maximilian Alexander Friedrich Wilhelm)—Germany's last imperial chancellor

Max Reinhardt Max Goldmann

Max Stirner Johann Kaspar Schmidt

Mayfair London's residential district

Mayor of Silicon Valley Robert N Noyce, co-inventor with Jack Kilby of the microchip

McAlester Ward Women's Ward in the Oklahoma State Penitentiary at McAlester

McCain Sanatorium North Carolina Prison Sanatorium at McCain

Meanie nickname for a mean person

Meat-Packing Capital of the World Chicago

Mecca of Spain Santiago de Compostela

Media Prophet Marshall McLuhan

Medical Essayist Oliver Wendell Holmes

Medina-Sidonia Alonso Pérez de Guzmán (Duke of Medina-Sidonia and admiral in command of the ill-fated Spanish Armada defeated by Sir Francis Drake)

Medinat Yisrael (Hebrew—State of Israel)

Meister Der Meister (German—the Master)—Johann Wolfgang von Goethe

Meister Raro Robert Schumann

Melanchthon Philipp Schwarzert

Mel Ferrer Melchor Gaston Ferry y Cintron

Melina Mercouri Maria Amalia Mercouri

Melisande Melusina

The Melting Pot New York City

Melvil Dewey Melville Louis Kossuth Dewey

Melvin Douglas Melvyn Hesselberg

Member of the Unemployed Scottish socialist leader Keir Hardie—founder of the Independent Labour Party (ILP)

Memé Remedios

Mencius Meng-tse

Menckonaclast Henry L Mencken

Men of the East Sherpas of northern India and Nepal

Meniere's disease sudden dizziness, ear ringing, and vomiting due to disturbance of the labyrinth

Menorca's Principal Port Port Mahón

Mentor Beechcraft T-34 trainer aircraft

Mentor to Parisian Intellectuals Théophile Gautier

Mercator Gerardus Mercator—real name of this 16th-century Flemish geographer is Gerhard Kremer

Merchant of Death international arms contractor Sir Basil Zaharoff

Merchant of Menace Vincent Price

Merchants of Death epithetic nickname applied to alcohol and tobacco vendors, armament makers, drug pushers, munitions makers, narcotics traffickers, and others whose business may result in the death of their customers

Merchants' Haven Copen-

hagen, Denmark

meretricious traffic prostitution; white slavery

Meritorious Man of the Americas Benito Juárez

Merle Oberon Estelle Merle O'Brien

Merry Monarch Charles II of Great Britain also nicknamed Patron of Bawdy Houses

Mescalero Cessna T-41 trainer-utility aircraft

Mesopotamia (Greek—Between Rivers)—land between the Euphrates and the Tigris; formerly Assyria, Babylonia, and Sumeria but presently Iraq

Messenger of the Gods Hermes (Greek); Mercury (Roman)

Messenger of Mercy Swiss banker Jean Henri Dunant—founder of the Red Cross

The Met Metropolitan Opera House—New York City

Metallic Age when you have silver strands in your hair, copper pennies in your purse, iron rust in your guts, and lead weights in your bottom

Metaphysician of Modern Atheism Baruch Spinoza

Metastasio Pietro Antonio Domenico Bonaventura Trapassi

Meteor Gloster twin-engine jet-fighter aircraft

Meth Capital of the World San Diego County, California

Metroland portion of London served by the Metro subway system

Metropolis of America New York City

Metropolis of the Magic Valley Brownsville, Texas on the Rio Grande

Metropolis of the Missouri Valley Kansas City

Metropolis of the South Mark Twain's nickname for New Orleans

Metropolis of the State of Oregon Portland

Metropolis of the United States New York

Metropolis of the World London

Metropolitan Museum of Art or Opera House in New York City

Metropolitan City of the Anglican Communion Canterbury

Metternich Prince Klemens Wenzel Nepomuk Lothar von

Metternich-Winneburg—Austrian statesman convening Congress of Vienna at end of Napoleonic wars

Mexican Mexican apple (white sapote); Mexican ground cherry *(tomatillo)*; Mexican hairless (dog used to herd cattle and keep ranchers company in bed as its hairlessness makes it flealess and its body temperature is higher than ours); Mexican jumping beans (inhabited by insect larvae whose movements make the beans jump about); Mexican onyx

Mexican Agrarian Reformer Emiliano Zapata

Mexican Film Pioneers José Bolanos and Luís Buñuel

Mexican Idealist Politician and Revolutionary Francisco I (dalecio) Madero

Mexican mud deadly Mexican-made heroin, also called black tar tootsie roll

Mexican Muralists José Clemente Orozco, Diego Rivera, Davíd Alfaro Siquieros, and Rufino Tamayo

Mexican National Composer Carlos Chávez

Mexican Ports Tampico, Veracruz, Coatzacoalcos, Frontera, Progreso; Salina Cruz, Acapulco, Manzanillo, San Blas, Mazatlan, Guaymas, Santa Rosalia, Ensenada

Mexican-Spanish Mexican-style Spanish enriched with more than 50,000 Mexicanisms

Mexico's Principal Port Veracruz

Meyerbeer Giacomo Meyerbeer (adopted name of Jakob Liebmann Beer)

Meyer Lansky Maier Suchowljansky

Miami Beach East Tel Aviv, Israel's nickname

Michael Angelo Titmarsh Thackeray's pseudonym

Michael Arlen Dikran Kuyumjian's pseudonym

Michael Caine Maurice Joseph Micklewhite

Michael Curtiz Michael Kertesz; Mihály Kertész

Michael Fairless Margaret Fairless Barber

Michael Field Katherine Harris Bradley and her niece Edith Emma Cooper used this pseudonym for their joint poetic efforts

Michael Innes John Innes Mackintosh Stewart

Michael Servetus Miguel Serveto

Michael Tilson Thomas Mike Thomashefsky

Michelangelo Michael Angelo Buonarroti

Michel Auclair Michal Vujovic

Michèle Morgan Simone Roussel

Michigan Ports Wyandotte, Rouge River, Detroit, Bay City, Saginaw, Alpena, Saulte Ste Marie, Frankfort, Manistee, Ludington, Muskegon, Grand Haven, Holland, St Joseph, Marquette

Mickey Mouse Walt Disney Productions (Wall Street nickname)

Mickey Rooney Joe Yule, Jr

Mickey Spillane Frank Spillane

Micky Micaela; Michael; Michelle

Mickyland Ireland

Microcosm of Canadian Life London, Ontario

Micronesia small islands of the western Pacific also known as the Trust Territory including the Federated States of Micronesia, the Marshall Islands, and Palau in the Carolines as well as many very small islands covering an area as wide as the United States

Middle America Central America, Mexico, and the West Indies

Middle Atlantic States Delaware, Maryland, New Jersey, New York, Pennsylvania, West Virginia

Middle Border Hamlin Garlin's nickname for the American Middle West

Middle Colonies New York, New Jersey, Pennsylvania, Delaware

Middle East area extending from Afghanistan to Egypt and including India, Iran, Iraq, Saudi Arabia, Syria, Lebanon, Israel, Jordan, Kuwait, and the United Arab Emirates

Middle Kingdom China

Middle Passage route of the slavers across the middle of the Atlantic between West Africa and the West Indies

Middle States New York, New Jersey, Pennsylvania, Delaware, and Maryland

Middletown Muncie, Indiana

Middletown title of Robert and Helen Lynd's study of cultural conflicts in Muncie, Indiana

Middle West United States from the Great Lakes to the northern border of the Gulf States and from the western slopes of the Appalachians to the eastern slopes of the Rockies

Midland Capital Birmingham, England

Midlands central England including counties of Bedford, Buckingham, Derby, Leicester, Northhampton, Nottingham, Rutland, Warwick

Mightiest of Rivers the Amazon

Mighty Champion of Freedom Frederick Douglass

Mighty Mo battleship USS *Missouri*

Mighty Mstislav Mstislav Rostropovich

Mihaly Munkacsy Michael Lieb

Mike Nichols Michael Peschkowsky

Mike Wallace Myron Wallace

Mildred Masters Mildred Kapilow

Mile-High or More-Than Mile-High North American Cities Butte, Montana; Cheyenne, Wyoming; Colorado Springs and Denver, Colorado; Flagstaff, Arizona; Gallup, New Mexico; Mexico City; Santa Fe, New Mexico

Mile-Square City Hoboken, New Jersey

Miliano Emiliano Zapata

Military Space Capital of the Free World United States Strike Command

Milk City Carnation, Washington

Millie Mildred; Millicent

Million-Acre Farm Prince Edward Island

Milton Berle Milton Berlinger

Milward Kennedy Milward Rodon Kennedy Burge

Mim Mimi; Miriam; Miryam (niah)

Mima Jemima

Mimico Boys Mimico Correction Centre (for males) in Toronto, Ontario

Mina Wilhelmina

Mineral Storehouse of the Nation Canada's Hudson Bay area

Ming-Ming Immingham, England

Mining Baron William A Clark

Mining State Nevada

Mini State Rhode Island

Minnesota Port Duluth

Minnie Minerva; Minneapolis; Minnesota

Minor Prophets of the Old Testament Hosea, Joel, Amos, Obadiah, Jonah, Micah, Nahum, Habakkuk, Zephaniah, Haggai, Zacharia, Malachi

Minstrel Composer James Bland

Minuteman solid-fuel intercontinental ballistic missile produced by Boeing

Minuteman III America's most advanced icbm in 1979

Minx of the Movies Betty Compson

Miracle of Fifth Avenue Guggenheim Museum

Miracle of Nature Queen Christina of Sweden

Mirage all-weather delta-wing supersonic-jet ground-support interceptor built by Dassault in France (Mirage III)

Mirage IV atomic-bomber version of the foregoing Mirage III

Mischa (Russian nickname—Michael)—Mike

Mischa Auer Mischa Ounskowsky

Missie Miss; Mississippi; Missus; Mrs

Missionary to the Lepers Father Damien

Mission City San Antonio, Texas

Mississippi Ports Pascagoula, Biloxi, Gulfport

Mississippi River Painter George Caleb Bingham

Miss New Orleans Dorothy Lamour's title in 1931

Miss Tarbarrel Ida M Tarbell

Mista Klemps conductor Otto Klemperer

Mistress of Mystery Agatha Christie

Mistress of the North the Baltic Sea, Russia, and Sweden

Mitch Mitchell; Richard Mitchell

Mitchell B-25 Bomber named in honor of General William Mitchell who in 1925 was courtmartialed for criticizing the mismanagement of the aviation service in the U.S.

Army and Navy **Mitchellville Girls** Iowa School for (delinquent) Girls at Mitchellville
Mitya (Russian diminutive—Dmitri)
Mitzi Margaret
Mizzou Missouri
mnemonic hormone vasopressin (reported to improve the memory if sniffed or sprayed into each nostril every day for three days)
Mockingbird Arkansas citizens
Model of the Common Man Benjamin Franklin
Model Republic Orange Free State
Model-T automobile once the world's most popular vehicle despite its handcranking starter and its nickname—Tin Lizzie
Modern Antigone Maria Thérèse—daughter of Louis XVI
Modernizer of Navigation Lieutenant Matthew Fontaine Maury, USN
Modern Liberal Social Philosopher José Ortega y Gasset
Modern Mother of Presidents Ohio—birthplace of Presidents Grant, Hayes, Garfield, Benjamin Harrison, McKinley, Taft, Harding; (*see* Mother of Presidents)
Mohair Capital Del Rio, Texas
Mohammed Ali Cassius Clay
Mojave Sikorsky heavy helicopter designated H-37
Molière Jean-Baptiste Poquelin
Molière of Music André Grétry
Moll Mary (slang); Molly
Molly Maria; Marie; Mary
Molly Pitcher Mrs John Hays also known as Captain Molly because she took her husband's place as cannoneer when he fell mortally wounded at the Battle of Monmouth —June 28, 1778
Molotov Vyacheslav Mikhailovich Skriabin
Mona Ramona
Monaco's Only Port Monaco
Monarch of the Mountains the elk
Monarchs of Marshes and Swamps alligators, crocodiles
Monarchy of Mount Everest Nepal
Mondrian Pieter Cornelis Mondriaan
Mongol Conqueror Genghis Khan (1162–1227)

Monk Matthew Gregory (Monk) Lewis
Monkey Trial Scopes Trial (in 1925 John Scopes, a Tennessee science teacher was tried for having taught evolution; he was prosecuted by William Jennings Bryan and defended by Clarence Darrow)
Monkey Ward Montgomery-Ward
Monk Lewis pseudonym of Matthew Gregory Lewis
Monroe Doctrine President James Monroe
Monsieur de Paris (French— Mr Paris)—guillotine operator
Monte Carlo of the East Macao, Portuguese China
Monterey Jack David Jacks and the only native cheese of California whose production he controlled
Montezuma's Revenge diarrhea or dysentery nicknamed for the last Aztec ruler of Mexico; both ailments are also nicknamed the Curse of Cortez for the Spaniard who conquered Mexico
Montgomery Camp Federal Prison Camp at Montgomery, Alabama
Monumental Intellectual John Locke
Monument City Baltimore, Maryland
Monument to Slavery Berlin Wall (1961–1989)
Moondog Louis Thomas Hardin
Moon Goddess Luna (Roman) whose Latin name means moon; Selene (Greek)
Moor Othello; The Moor (Dartmoor Prison)
Moor Court prison for female offenders in Staffordshire, England
Moose NATO name for Yak-11 Soviet aircraft
Moralist of Psychoanalysis Erich Fromm
Morand's disease paresis affecting the feet
Moravian Capital Brno
Moreton Bay Colony Queensland
Mormon City Salt Lake City
Mormon's Mecca Salt Lake City, Utah
Mormon Prophet Joseph Smith
Mormon State Utah
Moroccan Ports Tanger (Tangier), Kenitra, Casablanca,

Safi, Agadir
Morrie Maurice; Morris
Morrison Girls Mount View (delinquent) Girls School at Morrison, Colorado
Morris Rosenfeld Moshe Jacob Alter
Morton's disease metatarsal neuralgia
Moses of Her People Harriet Tubman (c.1820–1913)
Moskva (Russian—Moscow)—a Soviet class of antisubmarine —warfare cruiser and helicopter-carrier warship
Moslem India Bangladesh and Pakistan
Moslem Sultanate Oman—formerly Muscat and Oman
Mosquito Coast Caribbean coast of Honduras and Nicaragua
Mosquito State New Jersey
Mother of American Kindergartens Susan Blow
Mother of the American Legion Ernestine Schumann-Heink
Mother of the American Red Cross Clara Barton
Mother Ann Shaker leader Ann Lee
Mother of the Arts architecture
Mother of Balboa Park Kate Sessions, San Diego horticulturist
Mother of Believers Ayesha— Mohammed's favorite wife
Mother Bickerdyke Mary Ann Bickerdyke
Mother of Birth Control Margaret Sanger
Mother Bloor Ella Reeve Bloor
Mother Cabrini Frances Xavier Cabrini
Mother Carey's chickens stormy petrels
Mother Carey's geese fulmars or great white petrels
Mother of Child Education Doctor Maria Montessori
Mother of Cities Bombay, according to Kipling
Mother of Civilized Cities London
Mother of the Civil Rights Movement Coretta Scott King
Mother Earth the Greek Goddess Gaea or Ge who, according to mythology, arose out of chaos and in turn produced the sea, the sky, and the mountains; the Romans called her Tellus or Terra and some-

times called her Vesta Prisca
Mother of Exiles Statue of Liberty overlooking New York's former immigration stations at Battery Park and Ellis Island
Mother of Feminine Psychology Karen Horney
Mother of Ghosts the Roman goddess of Death—Mania
Mother Goose legendary authoress of children's rhymes and stories
Mother of Gospel Music Willie Mae Ford Smith
Mother of Her Country Queen María Theresa of Austria
Mother of Israel Golda Meir
Mother of the Japanese Novel Baroness Murasaki Shikibu *(The Tale of the Genji)*
Mother Jones Mary Harris Jones
Mother Lake Leonora Marie Kearney Barry
Mother of Libraries Alexandria, Egypt
Mother Maid The Virgin Mary
Mother of Mathematics Hypatia of Alexandria
Mother of Modern Dance Isadora Duncan
Mother of Mountains Nepal's Mount Everest
Mother of Muckrakers Ida M Tarbell (*see* Father of Muckrakers)
Mother of Parliaments British Parliament
Mother of Presidents Virginian—birthplace of Presidents Washington, Jefferson, Madison, Monroe, William Henry Harrison, Tyler, Taylor, Wilson
Mother of Prison Reform Dorothea Lynde Dix
Mother of the Red Cross Clara Barton
Mother of Rivers Tibetan Highlands
Mother of Rivers and Waves Tethys, wife of the god Oceanus, and mother of the rivers plus three thousand Oceanids—the waves
Mother of Russian Cities Kiev
Mother of the Russians Moscow
Mothers of Believers the wives of Mohammed
Mother of the Snows Mount Everest
Mother State Virginia
Mother of Storms Antarctica, the Baffin Sea, the Bay of Biscay, the Caribbean, the

Gulf of Alaska, the Gulf of Mexico, the South China Sea, and the Tasman Sea
mother tongue music (according to many great musicians, philosophers, poets, and scholars)
Mother of Trusts Standard Oil
Motion Picture Capital of the World Hollywood, California
Motion Picture Palace Potentate Roxy (SL Rothafel)
Motor City Detroit
Motor Town Detroit
Mound City St Louis, Missouri
Mountain City Chattanooga, Tennesse
Mountain Devils Tasmanians
Mountain Division States Arizona, Colorado, Idaho, Montana, Nevada, New Mexico, Utah, and Wyoming
Mountain of Fire Etna, Vesuvius, or any other active volcano
Mountain of the Lion Sierra Leone
Mountain State West Virginia
Mountain States Arizona, Colorado, Idaho, Montana, Nevada, New Mexico, Utah, and Wyoming
Mountain of Tarik The Rock of Gibraltar named for the Moorish chief Jabal Tariq
Mountain View Girls Mountain View School (for juvenile-delinquent females) at Helena, Montana
Mountbatten of Burma AF Admiral of the Fleet the Earl Mountbatten of Burma, KG (better known as Lord Mountbatten)—India's last viceroy
Mount Rainier Mount Tacoma towering over Tacoma, Washington next to Seattle
Movie Capital Hollywood, California
Movieland Hollywood, California
Mozambique Ports Moçcambique, Beira, Lourenço Marques
Mozart Town Salzburg, Austria—birthplace of Wolfgang Amadeus Mozart
Mr Automobile Gottlieb Daimler
Mr Color Television Peter Goldmark
Mr Common Sense Thomas Paine
Mr Conservative Barry M Goldwater
Mr Diesel Rudolf Diesel

Mr Dogpatch Al Capp (also known as the Mark Twain of cartoonists or the sardonic cartoonist)
Mr Dooley (pseudonym—Finley Peter Dunne)
Mr Electric Light Thomas Alva Edison
Mr Gyrocompass Elmer Ambrose Sperry
Mr Helicopter Igor Sikorsky
Mr Klemps Otto Klemperer
Mr Laser Charles Townes
Mr Linotype Ottmar Mergenthaler
Mr Long-Play Records Peter Goldmark
Mr Pilot Will Adams (nicknamed *Anjim Sama* by the Japanese because of his knowledge of navigation and shipbuilding)
Mr Radio Lee De Forest
Mr Republican Robert A Taft (1889–1953)
Mr Sam U.S. Senator Sam Ervin; former Speaker of the House Sam Rayburn (1882–1961)
Mrs Fletcher Maria Jane Jewsbury's pseudonym
Mrs Grundy nickname for the imaginary self-appointed arbiter of morality and taste
Mrs Jack Isabella Stewart Gardner
Mrs Patrick Campbell stage name of Beatrice Stella Tanner
Mr UN Carlos P Romulo
Mr Wireless Guglielmo Marconi
Mr X-Ray Wilhelm Roentgen
Muckraker of France Émile Zolá whose anticlerical, antimilitary, and antimonarchial writings forced him to flee to England
Muckrakers American crusader journalists David Graham Phillips, Charles Edward Russell, Lincoln Steffens, Upton Sinclair, Ida M Tarbell
Mudcat(s) Mississippian(s)
Mudcat State Mississippi
Mud Island San Diego's South Bay Wildlife Preserve
Muggsy Francis (Muggsy) Spanier
Muhammad (Arabic—The Praised)—Mahomet
Muhammad Ali Cassius Clay
Mule Soviet Polikarpov trainer aircraft
Mule Capital of the World Columbia, Tennessee

Mum City, U.S.A. Bristol, Connecticut

Muncy Institution State Correctional Institution at Muncy, Pennsylvania

Munich Expressionist Wassily Kandinsky

Municipal Muckraker Lincoln Steffens—author of *The Shame of the Cities*

Murihiku (Maori—End of the Tail)—New Zealand's southernmost city on South Island—Invercargill

murphy murphy bed (concealed-in-the-wall bed invented by William L Murphy); nickname for an Irish or white potato as well as for a confidence swindle

Murrumbidgee River City Canberra

Muscat and Oman former name of the Sultanate of Oman

Muse of Astronomy and Celestial Music Urania

Muse of Comedy and Pastoral Poetry Thalia

Muse of Dancing and Choral Singing Terpsichore

Muse of Epic and Heroic Poetry Calliope, who according to Horace, could play any musical instrument

Muse of Erotic Poetry Erato

Muse of History Clio

Muse of Lyric Poetry and Music Euterpe

Muse of Oratory, Rhetoric, and Sacred Song Polyhymnia

Muse of Tragedy Melpemone

Museum of Architecture Leningrad (also Petrograd or Saint Petersburg)

Museum Cities Padua, Venice, Verona, and Vicenza

Museum Metropolis London, New York, and Paris

Mushroomopolis Kansas City

Music Director of Europe Herbert von Karajan

Musky Armando Moscaritolo

Muslim Century the 8th century—the 700s when the Turkish crescent occupied the entire Hispanic Peninsula, the Levant, the Middle East, and even westernmost India

Mustang North American fighter aircraft F-51

muttnik second Soviet satellite launched in 1957; its astronaut was a mongrel dog

Myrna Loy Myrna Williams

Mystere IV Dassault jet fighter built in France

Mystere 20 Dassault twin-engine executive transport called the Falcon

Mysterious Billionaire Howard Hughes

Mytilene Aegean island of Lesbos

Nabby Abigail

Nadar Gaspard-Felix Tournachon (1820–1910)

Nail City Wheeling, West Virginia

Namby-Pamby 18th-century English dramatist-poet Ambrose Philips

Namibia modern name for South-West Africa

Nana Anna; Anne(tte); Mariana; Nanette; grandma

Nancy Agnes; Ann; Anna; Annabelle; Anne

Nannerl Maria Anna

Nanty Anthony

Nanuchka a Soviet class of guided-missile gunboats

Nanzig (German-Nancy)— French industrial center renamed by Hitler during World War II

Napoleon Bonaparte Napoleon I—Emperor of the French

Napoleon of Mexico General Antonio López de Santa Anna (1794–1876)

Napoleon of Peace Louis Philippe

Narco nickname of the U.S. Public Health Service Hospital in Lexington, Kentucky where narcotics addicts are treated

Narragansett Bay State Rhode Island

Narrow-Gauge Capital of the World Durango, Colorado

Narrow Land Between the Seas Panama

Nashville Girls Tennessee Prison for Women at Nashville

Natacha Rambova Winifred Hudnut

Natalie Wood Natasha Gurdin

National Anthem City Baltimore, Maryland

National Pastime baseball in America; cricket in Britain

National Poet of Norway Bjørnstjerne Bjørnson

National Tity navalese for National City, California

Nation of Big Cities China with at least fourteen cities each with a million people

Nation of Cities the United States with more than 150 cities containing 100,000

Nation of Gentlemen Scotland so named by King George IV

Nation's Capital District of Columbia

Nation's Front Yard The Mall in Washington, D.C.

Nation of Shopkeepers England, according to Samuel Adams as well as Napoleon

Nation's Hottest Town Quartzsite, Arizona where July temperatures average 108°F (42°C)

Nature's Wonderland Iceland

Naughty MacNaughton; McNaughton

Naughty Nineties the 1890s

Nauru Islands Ports Nauru Atoll, Saipan, Tinian, Rota

Naval Person Churchill's cover name used when addressing Roosevelt—POTUS—President of the United States

Navarino Italian name for the port of Pylos

Navel of the Nation Butte County, South Dakota (geographic center of the United States including Alaska and Hawaii); Smith County, Kansas (geographic center of the forty-eight conterminous states)

Navel of the World Easter Island according to its ancient Polynesian inhabitants

The Navigator Prince Henrique of Portugal (1394 to 1460)

Navigators' Islands Samoa Islands in the South Pacific

Navigator's Nightmare the Bermuda Islands

navy navy bean (small white bean); navy blue (dark blue); navy blue (tobacco)

Neapolitan Painter and Poet Salvator Rosa

Nearly-Impeached President Andrew Johnson, Richard M Nixon

nebbie (underground slang—nembutal)

nebbies nembutal capsules

Neddy Edgar; Edmund; Edward; Edwin; Edwina; National Economic Development Council's nickname

Nederlanders Dutch men and women

Nel Eleanor(a); Ellen; Helen(a); Nelly

Nell Eleanor(e)

Nellie Nellie McClung (pronounced *Mc Clue*)—Canadian

novelist and women's rights champion in the early 1900s

Nellie Melba Helen Porter Mitchell

Nello Emmanuel

Nelly Eleanor(a); Ellen; Helen

Nelly Bly Elizabeth Cochrane Seaman

Nelson Algren Nelson Algren Abraham

nembies nembutal (sodium pentobarbital sedative hypnotics)

nemish nembutal

nemmies nembutal capsules

Nemo Guillaume; Guillermo

Neopagan Eclectic Miguel de Unamuno

Neptune Lockheed P-2 antisubmarine and reconnaissance aircraft

Nequam Alexander Necham

Nerve Center of Alaska Anchorage

Nessa Agnes

Nessie Agnes

Nessie the Loch Ness monster's nickname

Nesta Agnes

Nestor of American Botany William Darlington

Nestor of American Pediatrics Abraham Jacobi

Nestor of Congregationalism Leonard Bacon

Nestor of Europe Leopold I of Belgium (1790–1865)

Nestor of the Rockies Kit Carson

Netherlands Antilles Aruba, Bonaire, Curaçao, Saba, Sint Eustatius, and half of Sint Maarten

Netherlands East Indies former name of Indonesia

Netherlands Guiana Dutch Guiana or Surinam

Netherlands Indies old name of Indonesia

Netherlands New Guinea former name of West Irian now part of Indonesia

Netherlands Ports (*see* Dutch Ports)

Netherlands Principal Port Rotterdam

Netherlands Timor formerly the western half of Timor now an island of Indonesia

Nettie Henrietta

Netty Henrietta

Net(ty) Antonia

Never-Never Never-Never Land (arid Australia)

Nevil Shute Nevil Shute Norway

New Albion Sir Francis Drake's name for what is now British Columbia, plus the states of Washington, Oregon, and California

New Alcatraz nickname of the maximum-security U.S. Penitentiary at Marion, Ill

New Amsterdam former name of New York City (Nieuw Amsterdam)

New Caledonia British Columbia

New Colossus Statue of Liberty's sobriquet derived from the poem by Emma Lazarus—*The New Colossus*—proclaiming: "Give me your tired, your poor, your huddled masses yearning to breathe free, the wretched refuse of your teeming shore. Send these, the homeless, tempesttossed to me, I lift my lamp beside the golden door!"

New Deal administration of Franklin Delano Roosevelt—thirty-second President of the United States

New Deal President Franklin Delano Roosevelt

New Edinburgh on the Antipodes Dunedin, New Zealand

New England New England boiled dinner (corned beef or ham with vegetables); New England clam chowder (clams, potatoes, milk, and stock); New England pine (white pine)

New England Mystic Emily Dickinson

New France old name for French Canada

New Frontier administration of John F Kennedy—thirty-fifth President of the United States

New Granada Colombia's original Spanish name—*Nueva Granada*

New Hampshire Port Portsmouth

New Holland old name for Australia discovered by Dutch navigators

New Jersey eagle big mosquito

New Jersey Ports Weehawken, Hoboken, Jersey City, Newark, Bayonne, Elizabethport, Port Socony, Grasselli, Cartaret, Chrome, Port Reading, Perth Amboy, South Amboy, Leonardo, Camden, Gloucester

New Munster old name for

New Zealand's South Island

New Netherlands old name for New York and parts of Connecticut and New Jersey

New Prometheus Immanuel Kant's nickname for Benjamin Franklin who drew lightning from the skies

new scarlet letter herpes virus (type 1 characterized by lip sores and type 2 by genital lesions)

New Sweden Sweden's shortlived colony in and around what is now Wilmington, Delaware but once called *Nya Sverige* (New Sweden)

Newton of Electricity André Marie Ampère

New Ulster old name for New Zealand's North Island

New World North and South America

New York New York aster, New York Bay, New York City, New York Curb Exchange, New York cut, New Yorker, New Yorkese, New York fern, New York Mills, New York Barge Canal, New York Philharmonic Orchestra, New York sailing barge, New York State, New York Stock Exchange, New York weevil, New York weasel, New Yorky

New York cut porterhouse steak (with the bone and fillet removed)

New York Persona New York City Mayor Edward Irving Koch

New York Ports Ogdensburg, Oswego, Rochester Harbor, Tonawanda, Buffalo, Albany, Kingston, Yonkers, Manhattan, Brooklyn, Gulfport, Port Richmond, Mariners Harbor, Stapleton, Tomkinsville

New York's Finest New York City policemen

New York State Barge Canal Erie Canal

New Zealand Commonwealth New Zealand and its territories

New Zealand Dominion New Zealand and its territories

New Zealand Ports on North Island: Auckland, Gisborne, Napier, Wellington, Wanganui, New Plymouth, Dargaville; on South Island: Port Nelson, Port Lyttelton (Christchurch), Timaru, Oamaru, Port Chalmers, Bluff Harbour,

Greymouth, Westport; plus smaller ports such as Dunedin and Invercargill

New Zealand's Garden City Christchurch

New Zealand's Principal Port Auckland

Ngaio March Edith Ngaio Marsh

Niagara Frontier Buffalo-Niagara Falls area

Niagara Fruit Belt Canadian fruit-growing region on the Niagara Peninsula between lakes Erie and Ontario

Nicaraguan Ports Cabo Gracias a Dios, Puerto Cabezas, Puerto Isabel, Bluefields, San Juan del Norte (Greytown); San Juan del Sur, Puerto Masachapa, Puerto Somoza, Corinto

Nicholas Blake C(ecil) Day Lewis' pseudonym

Nick Carter J Russell Coryell

nickel note $5 bill

Nickel-plated Paradise nickel-rich New Caledonia *(Nouvelle Calédonie)*

Nickel Plate Road New York, Chicago and St Louis Railroad Company

Nicky and Alicky Czar Nicholas II and Czarina Alexandra Feodorovna of Russia—the last of the Romanov Czars

Nicolas Copernicus Nikolay Kopernik

Nicolas-Favre disease lymphogranuloma venerea involving inguinal lymph glands and characterized by an exuding lesion

Nicolas Lenau Nikolaus Niembsch von Strehlenau

Nicolino Nicolò Grimaldi

Nidaros Trondheim's former name

Nigerian Ports Lagos, Bonny, Port Harcourt, Douala

Nightclub Aristocrat Artist Henri Marie Raymond de Toulouse-Lautrec

Night of Glass *Kristallnacht*, Nov. 9, 1938, when Nazis rampaged across Austria and Germany breaking the windows of homes, stores, and temples belonging to Jews who were rounded up and deported to concentration camps

Nightingale C-9 McDonnell-Douglas jetliner used for medical evacuation named in honor of Florence Nightingale

Nightmare of Europe Napo-

leon in the 1800s followed by Hitler in the 1900s

Night Mayor James J (Jimmy) Walker

Nikaria English place-name for Ikaría island in the Aegean

Nike-Ajax Douglas surface-to-air missile (MIM-3A)

Nike-Hercules Douglas surface-to-air missile armed with a high-explosive or nuclear warhead (MIM-14A)

Nike-Zeus one of a series of American-made anti-missile missiles

Nik-Nik affectionate nickname for the Royal Shakespeare Company's production of *The Life and Adventures of Nicholas Nickleby* by Charles Dickens

Nikolaus Lenau (pseudonym— Nikolaus Franz Niembsch von Strehlenau)

Niky Nicholas; Nicole; Nickerson; Nikerson

Nile River Cities Cairo, Egypt and Khartoum, Sudan

nimbies nembutal tablets

Nimrod Hawker-Siddeley four-engine jet transport

Nina Ann; Anna; Anne; Annette

Nine Keepers of the *Constitution* nine justices of the U.S. Supreme Court

nine old men nine justices of the United States Supreme Court

Nineteenth State Indiana

Ninon de Lenclos courtesan Anne Lenclos

Ninth State New Hampshire (*see* First State)

nissen nissen hut (designed by British military engineer P.N. Nissen for arctic use)

Niteroi formerly Nictheroy

nitrate of soda sodium nitrate (NaNO₃)

Nizhni Novgorod old name for Gorki

Noddy Nicodemus

Noll Oliver; Olivera; Oliver Cromwell—Lord Protector of England

Nolly Oliver; Olivera

Nome State Nome State Jail at Nome, Alaska

Nora Eleanora

Noratlas Norad 45-passenger transport aircraft made in France

Norfolk Norfolk coat or Norfolk jacket (made with fore-and-aft box pleats, big pock-

ets, and a belt, first produced in England's Norfolk county)

Normalcy nickname for Warren G Harding who advocated "a return to normalcy"

NORMY Norman Douglas

Norse God of Thunder Thor, whose Roman counterpart is Jove or Jupiter

North America's Largest Country Canada

North Baltic Nation Estonia

North Britain Scotland

North Carolina Ports Wilmington, Wrightsville

North Cuba Miami, Florida

Northern Ireland's Principal Port Belfast

Northernmost American Town Point Barrow, Alaska

Northern Rhodesia Zambia's former name

Northern Way Norway

Northland Riviera Sweden's summer beach on the Gulf of Bothnia and the Polar Route

North River Hudson River (Battery to 59th Street on New York waterfront)

North Sea Canal Amsterdam Ship Canal

North Slope Alaska north of the Brooks Range

North Star Minnesota

North Star City St Paul, Minnesota

North Star State Minnesota

North Western Line Chicago and North Western Railway

Norumbega historian John Fiske's name for what is now New York City (see *Norvegia)*

Norumbegaland New York to Nova Scotia including New England

Norvegia (Italian—Norway); (Latin—Norway)—appears on early maps of the east coast of North America over an area extending from the Bay of Fundy to Florida and known for its Norse viking explorations and settlements in pre-Columbian times; sometimes spelled Norvega or Norbegia as well as Norumbega

Norway's Most Popular Sculptor Adolf Gustav Vigeland

Norway's Principal Port Oslo

Norwegian Norwegian elkhound (dog originally bred for hunting elk and other game); Norwegian saltpeter (calcium nitrate)

Norwegian Expressionist Edvard Munch

Norwegian Ports Kirkenes, Vadso, Vardo, Honningsvaag, Hammerfest, Tromso, Harstad, Svolvaer, Narvik, Bodo, Mo, Mosjoen, Trondheim, Thamshamn, Kristiansund, Harosund, Molde, Ulsteinik, Alesund, Vaksdal, Bergen, Odda, Haugesund, Stavanger, Egersund, Flekkefjord, Kristiansand, Grimstad, Arendal, Tvedestrand, Langesund, Brevik, Porsgrun, Larvik, Sandefjord, Tonsberg, Horten, Drammen, Oslo, Moss, Sarpsborg, Frederikstad, Halden

nose candy cocaine for sniffing

Nostradamus Michel de Nostradame also called Michel de Notredame

Novalis Friedrich Leopold von Hardenberg

Nova Scotia Girls Nova Scotia School for (delinquent) Girls at Truro

Nuclear Falcon Hughes air-to-air missile also called Super Falcon

Nuclear-Power Admiral Hyman George Rickover, USN

Nuestra Familia (Spanish—Our Family)—Hispanic prison gang

Nuestra Señora de los Dolores de las Vegas (Spanish—Our Lady of the Sorrows of the Lowlands)—former and somewhat prophetic name of Las Vegas, Nevada

Nueva España (Spanish—New Spain)—Spanish colonial name for Mexico

Nueva Granada (Spanish—New Granada)—Spanish colonial province comprising Colombia, Ecuador, Panama, and Venezuela

Nuk (Greenlandic Eskimo—Point)—formerly called Godthaab (Good Hope) by the Danes and still the capital on the pointed peninsula on the southwest coast of Greenland

Number-One Host of the Jersey Coast Atlantic City, New Jersey

Nuoli Finnish depth-charge and mine-laying patrol boat armed with 20mm and 40mm guns

Nutmegs Connecticuters

Nutmeg State Connecticut

Nuuk correct name for Greenland's capital—Gothab

Nyasaland old name for Malawi

Nye Aneurin

NZedder(s) [En-zed-der(s)]— New Zealander(s)

Oak City Raleigh, North Carolina

Oakhill Virginia home of James Monroe

Oakie migratory farm worker or sharecropper from Oklahoma

Oasis City Roswell, New Mexico

Obediah Skinflint (pseudonym—Joel Chandler Harris)

Oberon British class of diesel submarines

Objectivist Poet William Carlos Williams

Ob River City Novosibirsk, Siberia

The Ocean The Atlantic, Antarctic, Arctic, Indian, or Pacific Ocean

Oceania islands of central and southern Pacific

Ocean Personified Oceanus (Roman); Okeanos (Greek)

Ocean State Rhode Island

Oder-Neisse Line rivers forming boundaries between Germany and Poland

Offenbach Jacques Offenbach (adopted name of Jakob Levy Eberst)

Offshore Capital of the World Aberdeen, Scotland—home port of many offshore oil exploration rigs

offshore China nationalist Republic of China headquartered on Taiwan, also called Formosa

O Henry William Sydney Porter

Ohio Ports Conneaut, Ashtabula, Fairport, Cleveland, Lorain, Huron, Sandusky, Toledo

Ohio's Beautiful Capital Columbus

Ohio Valley Ohio, West Virginia, Kentucky, Indiana, and Illinois

Oil Baron John D Rockefeller

oil of ben fine lubricant extracted from seeds of Arabian tree called *Moringa oleifera*

oil of cade juniper oil

oil cake cottonseed, linseed, or soybean mass used for cattle feed after oil is extracted

Oil Capital of Canada Edmonton, Alberta

Oil Capital of the Rockies Casper, Wyoming

Oil Capital of the United States Tulsa, Oklahoma

Oil City Bartlesville or Tulsa, Oklahoma

Oil Dorado northwestern Pennsylvania in the Oil City—Titusville area

oilies oilskin coats; oilskin garments

Oil Islands Chagos Archipelago in the Indian Ocean just north of Diego Garcia

oil of mirbane nitrobenzene

oil of palm bribe(s); palm grease

Oil Province Alberta

Oil State Pennsylvania

oil of vitriol concentrated sulfuric acid (H_2SO_4)

oil of wintergreen methyl salicylate

Okie Oklahoma; Oklahoman

Okie City Oklahoma City, Oklahoma

Olav Hunger King Olav I of Denmark

Olav the Stout Olav Haroldsson

Olav Tryggvesson King Olav I of Norway, Sweden, and Denmark

The Old King Grom of Denmark (860—935)

Old Abe Abraham Lincoln

Old Ace of Spades Lieutenant General Robert E Lee, CSA

Old Andy Andrew Jackson—seventh President of the United States

Old Bay State Massachusetts

Old Beeswax Captain Raphael Semmes, CSN

Old Billie Brigadier General William Tecumseh Sherman, USA

Old Blighty nickname for London before air-pollution control

Old Blood and Guts General George S Patton, USA

Old Blue Eyes Frank Sinatra

Old Bory General Pierre Gustave Toutant de Beauregard, CSA

Old Brown of Osawatomie abolitionist John Brown

Old Buck Admiral Franklin Buchanan; President James Buchanan

Old Buena Vista General Zachary Taylor who attacked Mexicans at Buena Vista in February 1847; later was twelfth President of the

United States
Old Bullion Thomas Hart Benton
Old Cape Stiff Cape Horn
Old Catawba fictitious name Thomas Wolfe assigned North Carolina
Old Chapultepec General Winfield Scott whose victory at Chapultepec ended Mexican War in September 1847
Old Chief Henry Clay
Old Coat Hanger Melbourne-originated nickname for the Sydney Harbour Bridge
Old Colony Massachusetts—founded in 1620
Old Corndrinking Mellifluous William Faulkner, according to Ernest Hemingway
The Old Country wherever anyone or their family originated—especially if in Europe
Old Country Lawyer Senator Sam Ervin of North Carolina
Old Creepy bankrobber-burglar Alvin Karpis
Old Curmudgeon Harold Le Claire Ickes
Old Denmark General Christian Febiger, USA
Old Dirigo Maine whose state motto is *Dirigo* (Latin—I direct)
Old Dominion Virginia—oldest English colony in America—founded in 1607
Old Dorp nickname of Schenectady, New York
Old East East Asiatic Company
Old Faithful geyser in Yellowstone National Park; spouts about every 67 minutes
Old French Town New Orleans
Old Fuss and Feathers General Winfield Scott, USA
Old Gib Gibraltar
Old Glory the American Flag
Old Greasy West Virginian nickname for the Kanawha River or K'naw
Old Guard conservatives; Napoleon's imperial guard who made the last charge at Waterloo; the establishment
Old Harry (the devil)—Satan
Old Hickory General Andrew Jackson—seventh President of the United States
Old Iron Pants Vyacheslav Molotov (1890–1986)
Old Ironsides USS *Constitution*
Old Jeb Major General J(ames) E(well) B(rown) Stuart, CSA

Old Jefferson Joseph Jefferson
Old Joe slang nickname for syphilis
Old Kinderhook Martin Van Buren—eighth President of the United States
Old Lady the boss; mother; wife
Old Lady of Eagle Bridge Grandma (Anna Mary Richardson) Moses of Eagle Bridge, NY
Old Lady of the Thames London
Old Lady of Threadneedle Street Bank of England
Old Lady White any powdered narcotic
Old Legal Lion Clarence Darrow
Old Line State Maryland
Old Man the boss; the captain; father; the skipper
Old Man Eloquent Isocrates in the opinion of Milton; John Quincy Adams in the opinion of the Congress he served after being sixth President of the U.S.
Old Man of Ferney Voltaire who lived in Ferney, France
Old Man of the Mountain New Hampshire's Profile Mountain—the Great Stone Face
Old Man of the Rhine Konrad Adenauer
Old Man River the Mississippi
Old Manse Nathaniel Hawthorne's house in Concord, Massachusetts
Old Nick (the devil)—Niccolo Machiavelli's diabolic sobriquet; Satan
Old Noll Old Oliver Cromwell
Old North State North Carolina
Old Ossawatomie John Brown
Old Pam Lord Palmerston (Henry John Temple)
Old Party W(illiam) Somerset Maugham
Old Peg Leg Petrus Stuyvesant—director-general of New Amsterdam and the New Netherlands
Old Pretender James Francis Edward Stuart (son of King James II)
Old Providence island in the Colombian West Indies off the Caribbean coast of Nicaragua
Old Pueblo Tucson, Arizona
Old Put General Israel Putnam
Old Rosey General William

Starke Rosecrans
Old Rough-and-Ready General Zachary Taylor—twelfth President of the United States
Old Salamander Admiral David Glasgow Farragut
Old Sarum Salisbury, England
Old Scratch Satan
Old Sol the sun (*see* Sun God)
Old South southern United States before 1865
The Old South Alabama, Florida, Georgia, Louisiana, Mississippi, North Carolina, South Carolina, Virginia
Old Spanish Trail Saint Augustine, Florida to San Diego, California; Gulf Coast and Mexican Border route to California
Old Sparky Florida's electric chair
Old Swamp Fox Brigadier General Francis Marion, USA
Old Tecumseh General William Tecumseh Sherman, USA
Old Three Stars General US Grant, USA
Old Tippecanoe General William Henry Harrison—ninth President of the United States
Old Ugly .45-caliber pistol
Old Vic repertory theater in London
Old Viking Norwegian-American labor leader Andrew Furuseth
Old West American West before it was settled during the 19th century
Old World Africa, Asia, and Europe
Old Zach Zachary Taylor—12th President of the United States
Ole Olaf(sen); Olav(sen)
Oleander City Galveston, Texas
Oleander City by the Sea Galveston, Texas
Ole Bull Ole Bornemann Bull
Ole Miss Old Mississippi (The University of Mississippi)
Olive Fremstad Olivia Rundquist
Oliver Hardy Oliver Norvell Hardy
Oliver Optic pseudonym of William Taylor Adams
Ollie Olive(r)
Ol' Man River Mississippi River
Ol' Miss Old Mississippi (nickname of river, state, or university)

Ol' Mo Old Missouri (the great river)

Omani Ports Masquat (Muscat) and Matrah

Omar Sharif Omar Cherif; Omar Michel Shaloub

Only Town in the U.S. with an Apostrophe in Its Name Coeur d'Alene, Idaho

onyx marble alabaster

Oom Paul (Afrikaans—Uncle Paul)—sobriquet of Stephanus Johannes Paulus Kruger—leader of Boer rebellion and president of Transvaal

Opener of Japan Commodore Matthew Calbraith Perry, USN

Operation Keelhaul Allied policy of returning escaping anti-communists to their homelands

Opium Eater Thomas De Quincey

Opium Kingdom any country where the opium poppy is cultivated for use in making heroin (Bolivia, Burma, Colombia, Ecuador, Laos, Mexico, etc.)

Opium Land poppy fields of the Golden Triangle or northeastern Burma, northern Laos, and northern Thailand

Opium's Golden Triangle opium-growing fields between borders of Cambodia, Laos, and Vietnam

Oppenheim's disease congenital lack of muscular development of the ankles and feet

Oppie Oppenheim(er); J(ulius) Robert Oppenheimer

Oppy Oppenheimer(er)

Orange Bowl Miami, Florida

orange flag potential danger signal

Orange Irish Protestants; orig., followers of William of Orange

Orangemen Protestants of Northern Ireland; orig., followers of William of Orange

Orange States California, Florida, and Texas

Orator of the American Revolution Patrick Henry

Orchard City Burlington, Iowa also called Porkopolis of Iowa

Orchard of Ireland County Armagh

Orchid Capital of Hawaii Hilo

The Orchid Island Hawaii

Orchid Set in the Sea Sulawesi (Celebes)

Oregon Girls Wisconsin

School for (delinquent) Girls at Oregon

Oregon Ports Empire, Coos Bay, Astoria, Longview, Portland, Vancouver

Organist-Medical Missionary Dr Albert Schweitzer

oriental amethyst purple corundum

oriental anesthesia acupuncture

oriental emerald green corundum

Oriental Republic Eastern Republic of Uruguay *(República Oriental del Uruguay)*

oriental topaz yellow corundum

Orient Express famed Paris-to-Istanbul train; hypersonic jet designed to fly at 25 times the speed of sound to carry passengers from Washington to Tokyo in two hours

Orient's Cleanest City Singapore

Original Glamour Girl Theda Bara (Theodosia Goodman) also called Queen of the Vampires in the early days of American motion pictures

Orinoco River City Ciudad Bolívar, Venezuela

Oriole Maryland's state bird and symbolic nickname of Marylanders—Orioles

Orion Lockheed P-3 antisubmarine and patrol aircraft

Orizaba Citialtepetl (Mexico's highest volcano)

Orlando di Lasso Roland de Lassus

Orwell's Year 1984

Oscar Wilde Oscar Fingal O'Flahertie Wills (also used the anonym: C.3.3.)

Oskar Werner Josef Bschliessmayer

Oslo modern name for Christiania or Kristiania

Osloenser native of Oslo

Oslo Fjord formerly Kristiania Fjord

Ossie Oswaldtwistle, England

Ossining Facility Ossining Correctional Facility at Ossining, New York, long nicknamed Sing Sing

Ossining-on-Hudson formerly Hunter's Landing or Sing Sing

Ossip Gabrilovich Salomonovich Gabrilovich

Ostend Manifesto President Franklin Pierce

Other Side of the Herring

Pond British nickname for America

Otter De Haviland utility aircraft (DHC-3 in Canada, U-1A in U.S.)

Ottoman Empire the old Turkish Empire extending at its height from Iran to Morocco, including the Balkans, parts of Hungary and southern Russia as well as much of Spain; Turkish Empire

Ouida pseudonym of Marie Louise de la Ramée

Ouragan (French—Hurricane) —Dassault single-engine jet fighter plane

Our Gracie Gracie Fields (created Dame Commander of the Order of the British Empire after years of entertaining many millions of Britons and others around the world)

Our Lady of the Snows Kipling's nickname for Canada

Outer Banks North Carolina's sand-dune islands separated from the mainland by Albemarle, Croatan, Pamlico, and Bogue sounds

Outer China Mongolia, Sinkiang, Tibet

Outer City metropolitan area surrounding Peking's Inner City

Outer Mongolia The Mongolian People's Republic formerly called Mongolia

Outer Ring English counties adjacent to London

Outpost of the British Empire nickname given to any remote British settlement

Outpost of the West the Philippines

Outstanding German Composer of the Early Twentieth Century Richard Strauss

Outstanding Soviet Historian and Novelist Aleksandr Isayevich Solzhenitsyn

Overthrust Belt Rocky Mountain gas-and-oil lands

Ovid Roman poet Publius Ovidus Naso

Owen Meredith Edward Robert Bulwer-Lytton's pseudonym

owlsville London's post-midnight nickname or any other place after midnight

Oxford oxford bag(s), oxford blue, oxford cloth, oxford corner(s), oxford dash, oxford double dash, oxford down,

oxford frame, oxford grey, Oxford Group (er), Oxford Group Movement, oxford india paper, oxford ochre, Oxfordshire(s), oxford shoe(s), Oxford Theory, oxford tie, oxford unit(s), oxford weed
Oyashio (Japanese—Father Current)—cold Okhotsk Current
Oyster Center Apalachicola, Florida
Oyster(s) Marylander(s)
Oyster State Maryland
Ozzie Aussy; Australian; Osborn(e); Oscar; Oswald(o)
Pablo Neruda Neftali Ricardo Reyes
Pablo Picasso Pablo Diego José Francisco de Paula Juan Nepomuceno Crispin Crispiano de la Santísima Trinidad Ruiz y Picasso
Paca Francesca
Pachuco (Mexican-Americanism—El Paso, Texas)—also called Pachucolandia
Pacific Bitch navalese for Pacific Beach, San Diego, California
Pacific Canada British Columbia and the Yukon Territory
Pacific Coast Province British Columbia
Pacific Coast States California, Oregon, Washington
Pacific Commonwealths Australia and New Zealand and their territories
Pacific Crest Trailways for hikers, historians, and naturalists—includes John Muir Trail—extends from Canada to Mexico through Washington, Oregon, and California
Pacific Division States Alaska, California, Hawaii, Oregon, and Washington
Pacific Dominions Australia and New Zealand and their territories
Pacific Northwest Alaska to California, including the Yukon, British Columbia, Washington, and Oregon
Pacific Ocean Eden Galápagos Islands
Pacific Paradise Hawaii
Pacific Province British Columbia
Pacific States Alaska, Washington, Oregon, California, Hawaii
Pacific War Japan's involvement in World War II beginning with the Manchurian Incident in 1931 when Japan in-

vaded China
Paco Pancho (Francisco)
Paddie(s) Irish person(s)
Paddy an Irishman; Patrick
Paddyland Ireland
Padre de Independencia (Spanish—Father of Independence)—José Martí—Cuban patriot, poet, and soldier
Paganini of the Double Bass Giovanni Battesini; Serge Koussevitzky
Paganini of the Guitar Fernando Sors, Andres Segovia
Paget's disease bone distortion or cancer of the nipples of women
Painted Desert petrified formations and colorful rock deposits on desert floor of northeastern Arizona
Painter of Japanese Prostitutes Kitagawa Utamaro
Painter of Light Joaquin Sorolla de Bastida
Painter of Prostitutes Henri Marie Raymond de Toulouse-Lautrec
Pakistan's Principal Port Karachi
Palatinate southwest German districts once ruled by counts palatinate of the Holy Roman Empire and referred to as Oberpfalz or Rheinpfalz
Palau Pelew (Pacific islands in Caroline area)
Paleontologist Priest Pierre Teilhard de Chardin
Palestinian Salt Sea the Dead Sea
Palestrina Giovanni Pierluigi da Palestrina
Palgrave Francis Meyer Cohen
Palma Vecchio palette name of Jacopo Negreti
Palm Coast Florida's east coast from Daytona to Jacksonville
Palmerston Henry John Temple, Viscount of Palmerston
Palmetto City Charleston, South Carolina
Palmetto(s) South Carolinian(s)
Palmetto State South Carolina
palm oil bribe(s)
Pamphleteer for American Independence Thomas Paine
Panama Panama hat (made from finely plaited young palmlike leaves)
Panama Canal Ports Balboa (Pacific terminus) and Cristóbal (Caribbean terminus)
Panama Canal President Theodore Roosevelt
Panama red high-grade mari-

juana grown in Panama and the Canal Zone
Panama's Principal Ports Colón, Cristóbal, Panamá City, Balboa, and Puerto Armuelles
Panama turkey iguana tail meat
Pancho Francisco; native nickname for Valparaiso, Chile
Pancho Villa Doroteo Arango
Pandemonium South Pacific nickname for New Hebrides islands British-French Condominium
Panhandle State West Virginia
Pantaleone patron saint of Venice; nickname for an Italian taxpayer or for a Venetian
Panther Grumman single-engine single-seat naval fighting aircraft (F9F-2)
Papa Bach Johann Sebastian Bach
Papa Doc Haiti's former dictator François Duvalier
Papa Haydn Franz Joseph Haydn
Pappas Papadmitropoulos
Pappies Papists
Papua New Guinea's Principal Port Port Moresby
Paquita Francisca (Frances)
Paracelsus Theophrastus Bombastus von Hohenheim
Paradise of New England Salem, Mass
Paradise of the Pacific Hawaii
Paradox of South America Paraguay—an affluent military dictatorship
Paraguay River City Asunción
Paraguay's Principal Port Asunción
Paraiba old name of Joao Pessoa, Brazil
Paramilitary Paradise Paraguay
Parched Heart of Australia Alice Springs, Northern Territory—The Alice
Parchman Mississippi State Penitentiary at Parchman
Paris Expressionist Henri Matisse
paris green copper acetoarsenite (poison)
Parisian Italian Luigi Carlo Zenobio Salvadore Maria Cherubini (1760–1842)
Paris of North America Montréal
Parkbench Philosopher Bernard Baruch
Park City Bridgeport, Con-

necticut

Parkinson's disease nervous tremors accompanied by muscular weakness and rigidness; also called palsy, paralysis agitans, or the shakes

Park Maker Frederick Law Olmsted

Parlour Panther *New York Review of Books*

Parmigianino Francisco Massuoli

Parrot's disease syphilitic infantile paralysis (named for a French physician—Jules Marie Parrot)

Parry's disease exophthalmic goiter

Pas de Calais (French—Calais Strait)—also called Dover Strait

Paso del Calais (Spanish—Calais Strait)—Dover Strait in the English Channel

Passionate Pilgrim John Bunyan

Passionate Skeptic Bertrand Russell

Pastoral God Pan

Pathetic Peter Pyotr Ilyich Tchaikovsky whose compositions are soaked in sorrow

Pathfinder Major General John C Frémont, USA

Pathfinder of the Seas Matthew Fontaine Maury

Path of Gold Market Street, San Francisco

Pathmaker of the West John C Frémont

Patience and Fortitude Mayor La Guardia's nickname for the couchant lions flanking the steps of the New York Public Library

Patland Ireland

Patriarca de la Independencia (Portuguese—Patriarch of Independence)—Brazil's José Bonifacio de Andrada e Silva

Patriarch of American Labor George Meany

Patriarch of Ferney Voltaire who lived from 1758 to 1778 in Ferney, France now called Ferney-Voltaire in his honor

Patriarch of the Gold Industry Nicholas Deak

Patriarch of the Modern Consumer Movement Ralph Nader

Patriarch of New England John Cotton

Patriarch of Philosophy Bertrand Russell

Patriarch of Puerto Rico Luís

Muñoz Marín

Patriarch of the West the Pope

Patricia Wentworth Dora Amy Elles Dillon Turnbull's pseudonym

Patriot Financier Robert Morris

Patriot of the Piano Polish patriot-pianist-premier Ignace Jan Paderewski

Patriot Printer of 1776 William Bradford

Patron of Bawdy House England's King Charles II, the Merry Monarch

Patron of Explorers Henry the Navigator (Dom Henrique o Navegador)—Prince of Portugal

Patron Saint of Abandoned Children Jerome Emiliani

Patron Saint of Accountants Matthew

Patron Saint of Actors Genesius

Patron Saint of Air Travelers Joseph of Cupertino

Patron Saint of All Who Work with a Hammer Cloud

Patron Saint of Alpinists Bernard of Menthon

Patron Saint of Altar Boys John Berchmans

Patron Saint of American Orchards John (Johnny Appleseed) Chapman

Patron Saint of the Americas Rosa de Lima

Patron Saint of Anesthetists Rene Goupil

Patron Saint of Angina Sufferers Swithbert

Patron Saint of Architects Thomas, Apostle

Patron Saint of Armenia Gregory the Illuminator

Patron Saint of Artillerymen Barbara

Patron Saint of Astronomers Dominic

Patron Saint of Athletes Sebastian

Patron Saint of Austria Severino

Patron Saint of Authors Francis de Sales

Patron Saint of Aviators, Porters, Seafarers, and Travellers Christopher

Patron Saint of Bakers Elizabeth of Hungary

Patron Saint of Bankers Matthew

Patron Saint of Barbers Cosmas, Damian

Patron Saint of Barren

Women Felicity

Patron Saint of Bavaria Kilian

Patron Saint of Beggers Giles

Patron Saint of Belgium Joseph

Patron Saint of Black Missions in North America Benedict the Moor

Patron Saint of Blacksmiths Dunstan

Patron Saint of the Blind Raphael

Patron Saint of the Blind and the Near Blind Clara

Patron Saint of Bloodbanks Januarius

Patron Saint of Boatmen Julian the Hospitaler

Patron Saint of Bookbinders Peter Celestine

Patron Saint of Book Collectors Jerome (credited with compiling the Latin Bible)

Patron Saint of Bookkeepers Matthew

Patron Saint of Booksellers John of God

Patron Saint of Borneo Francis Xavier

Patron Saint of Boy Scouts George

Patron Saint of Brazil Peter of Alcantara

Patron Saint of Brewers and Pawnbrokers Nicholas of Myra—the prototype of Santa Claus

Patron Saint of Bricklayers Stephen

Patron Saint of Brides Nicholas of Myra

Patron Saint of Bridgebuilders Benedict

Patron Saint of Brushmakers Anthony

Patron Saint of Builders Barbara and Vincent Ferrer

Patron Saint of Bullfighters Virgen de la Macarena (also Patron Saint of Seville)

Patron Saint of Butchers Hadrian

Patron Saint of Cab Drivers Fiacre

Patron Saint of Cabinetmakers Anne

Patron Saint of Canada Anne

Patron Saint of Candlemakers Ambrose

Patron Saint of Canonists Raymond of Peñafort

Patron Saint of Carpenters Joseph

Patron Saint of Catechists Charles Borromeo

Patron Saint of Catholic Ac-

tion Francis of Assisi
Patron Saint of the Catholic Press Francis de Sales
Patron Saint of Charitable Societies Vincent de Paul
Patron Saint of the Chase Hubert
Patron Saint of Childbirth Gerard Majella
Patron Saint of Children Nicholas of Myra, also known as Santa Claus—derived from the Dutch, Sant Nikolaas
Patron Saint of Choirboys Dominic Savio
Patron Saint of the Church Joseph
Patron Saint of Clerics Gabriel
Patron Saint of Clothiers Blaise
Patron Saint of Comedians Vitus
Patron Saint of Confessors John Nepomucene
Patron Saint of Convulsive Children Scholastica
Patron Saint of Cooks Lawrence and Martha
Patron Saint of Coppersmiths Maura
Patron Saint of Cripples Vitus
Patron Saint of the Dance Vitus, according to Washington Irving
Patron Saint of Dancers Vitus
Patron Saint of Dentists Apollonia
Patron Saint of Desperate Situations Jude
Patron Saint of Diseases of the Breast Agatha
Patron Saint of Domestic Animals Anthony
Patron Saint of Domestics Zita
Patron Saint of Domestic Workers Zita
Patron Saint of the Dominican Republic Dominic
Patron Saint of Druggists James the Less
Patron Saint of Dyers Lydia and Maurice
Patron Saint of Dying Barbara
Patron Saint of Dysentery Sufferers Matrona
Patron Saint of Earthquakes Emygdius
Patron Saint of Ecologists Francis of Assisi
Patron Saint of Editors John Bosco
Patron Saint of Emigrants Francis Xavier Cabrini
Patron Saint of Engineers Francis III
Patron Saint of England Ed-

ward the Confessor or St George
Patron Saint of Epilepsy Vitus
Patron Saint of Eucharistic Congresses Paschal Baylon
Patron Saint of Expectant Mothers Gerard Majella
Patron Saint of Eye Trouble Lucy
Patron Saint of Falsely Accused Raymond Nonnatus
Patron Saint of Farmers Isadore
Patron Saint of Fathers of Families Joseph
Patron Saint of Finland Henry of Uppsala
Patron Saint of Firemen Florian
Patron Saint of Fishermen Andrew
Patron Saint of Flagpole Sitters Simon of Stylites
Patron Saint of Florists Therese of Lisieux
Patron Saint of Foresters John Gualbert
Patron Saint of Founders Barbara
Patron Saint of Freethinkers Chapman Cohen of London
Patron Saint of French Attorneys Ives
Patron Saint of Funeral Directors Joseph of Arimathea
Patron Saint of Gardeners Dorotea
Patron Saint of Gardens Dorotea
Patron Saint of Girls Agnes
Patron Saint of Glasgow Mungo
Patron Saint of Glassworkers Luke
Patron Saint of Goldsmiths Dunstan
Patron Saint of Gravediggers Anthony
Patron Saint of Greetings Valentine
Patron Saint of Grocers Michael
Patron Saint of Hairdressers Martin de Porres
Patron Saint of the Hard of Hearing Ovidius
Patron Saint of Hatters Severus of Ravenna
Patron Saint of Haymakers Gervase
Patron Saint of Headache Sufferers Teresa of Avila
Patron Saint of Heart Patients John of God
Patron Saint of Hernias and Ruptures Drogo, according

to believers who also call him Druon
Patron Saint of Hispanic America Our Lady of Guadalupe
Patron Saint of Hospital Administrators Francis Xavier Cabrini
Patron Saint of Hospital Dieticians Martha
Patron Saint of Hospital Pharmacists Gemma Galgani
Patron Saint of Hospital Public Relations Paul
Patron Saint of Hospitals Camillus de Lellis
Patron Saint of Hotelkeepers Julian the Hospitaler
Patron Saint of Houseworkers Anne
Patron Saint of Hunters Hubert
Patron Saint of Impossible and Desperate Cases Rita of Cascia
Patron Saint of Interracial Justice Martin de Porres
Patron Saint of Invalids Roch
Patron Saint of Italy Francis of Assisi, Catherine of Siena
Patron Saint of Jewelers Eligius
Patron Saint of Journalists Francis de Sales
Patron Saint of Jurists John Capistrano
Patron Saint of Laborers Isadore
Patron Saint of Lawyers and Thieves Nicholas of Myra
Patron Saint of Librarians Jerome
Patron Saint of Lighthousekeepers Dunstan
Patron Saint of Lithuania Casimir
Patron Saint of Locksmiths Dunstan
Patron Saint of Lost Articles Anthony of Padua
Patron Saint of Lovers Valentine
Patron Saint of Maidens Catherine of Alexandria
Patron Saint of Malta Paul
Patron Saint of Mariners Michael
Patron Saint of Married Women Monica
Patron Saint of Medical Record Librarians Raymond of Penyafort
Patron Saint of Medical Technicians Albert the Great
Patron Saint of the Mentally Ill Dymphna

Patron Saint of Merchants Nicholas of Myra
Patron Saint of Messengers Gabriel
Patron Saint of Metalworkers Eligus
Patron Saint of Mexico Virgin of Guadalupe
Patron Saint of Midwives Raymond Nonnatus
Patron Saint of Millers Victor
Patron Saint of Missions Francis Xavier
Patron Saint of Monaco Devota
Patron Saint of Mothers Monica
Patron Saint of Motorists Christopher
Patron Saint of Mountaineers Bernard
Patron Saint of the Movies John Bosco
Patron Saint of Music Cecilia
Patron Saint of Nailmakers Cloud
Patron Saint of Navigators Brendan of Ireland
Patron Saint of the Netherlands Willibrord
Patron Saint of Norway Olav
Patron Saint of Notaries Mark
Patron Saint of Nurses Agatha
Patron Saint of Orators John Chrysostom
Patron Saint of Orphans Jerome Emiliani
Patron Saint of Painters Luke
Patron Saint of Paris Genevieve
Patron Saint of Parish Priests Jean-Marie-Baptiste Vianney
Patron Saint of Pawnbrokers Nicholas of Myra
Patron Saint of Persons Afflicted with Coughs and Colds Judas
Patron Saint of Persons Afflicted with Hydrophobia Hubert
Patron Saint of Persons Afflicted with Insect Stings and Snake Bites Sebastian
Patron Saint of Persons Condemned to Death Dismas
Patron Saint of Peru Rosa de Lima
Patron Saint of Pharmacists James the Greater
Patron Saint of Physicians Luke
Patron Saint of Pilots Joseph of Cupertino
Patron Saint of the Plague Roch
Patron Saint of Plasterers

Bartholomew
Patron Saint of Poets Cecilia
Patron Saint of Poland Cunegunda
Patron Saint of Policemen Michael
Patron Saint of the Poor Anthony of Padua
Patron Saint of Porters Christopher
Patron Saint of Postal Workers Gabriel
Patron Saint of Preachers John Chrysostom
Patron Saint of Pregnant Women Margareta
Patron Saint of Printers Genesius
Patron Saint of Prisons Joseph Cafasso
Patron Saint of Prisoners Barbara
Patron Saint of Public Relations Bernardino of Siena
Patron Saint of Radiologists Michael
Patron Saint of Radio Workers Gabriel
Patron Saint of Retreats Ignatius Loyola
Patron Saint of Rheumatics Gervasius
Patron Saint of Rheumatism Affliction James the Greater
Patron Saint of Russia Andrew whose cross adorns the flag of Imperial Russia
Patron Saint of Saddlers Crispin
Patron Saints of Advertisers Barnardine of Siena, John Berchmans
Patron Saints of Ailments of the Eyes Clare and Lucy
Patron Saint of Sailors Elmo
Patron Saints of Artists Catherine of Bologna, Luke
Patron Saints of Bohemia Wenceslaus, Ludmilla
Patron Saints of Canada George and Joseph
Patron Saints of Cancer Victims Michael; Peregrine Laziosi
Patron Saints of Chile Santiago (James), Virgen del Carmen
Patron Saint of Scholastic Institutions Thomas Aquinas
Patron Saint of Scotland Andrew
Patron Saint of Sculptors Claude
Patron Saints of Denmark Ansgar and Canute
Patron Saints of Doctors

Cosmas and Damian
Patron Saint of Secretaries Genesius
Patron Saint of Seminarians Charles Borromeo
Patron Saints of France Denis, Our Lady of the Assumption, Joan of Arc, Therese
Patron Saints of Germany Boniface and Michael
Patron Saints of Greece Andrew and Nicholas
Patron Saint of Shepherds Drogo
Patron Saint of Shoemakers Crispin
Patron Saints of Housewives Anne; Martha
Patron Saints of Hungary Blessed Virgin—Great Lady of Hungary and Stephen
Patron Saint of the Sick Camillus de Lellis
Patron Saint of Silversmiths Dunstan
Patron Saint of Singers Cecilia
Patron Saints of Ireland Patrick, Brigid, Columba
Patron Saint of Skaters Lidwina
Patron Saint of Skiers Bernard
Patron Saint of Skin Diseases Marculf
Patron Saints of Lawyers Genesius, Ivo, Thomas More
Patron Saints of Moravia Cyril and Methodius
Patron Saints of Morticians Dismas and Joseph of Arimathea
Patron Saint of Social Justice Joseph
Patron Saint of Social Workers Louise de Marillac
Patron Saint of the Sore Throat Blaise
Patron Saint of Spain Santiago de Compostela (St James the Greater), Teresa of Ávila
Patron Saint of Speleologists Benedict
Patron Saints of Philosophers Justin, Catherine
Patron Saints of Poland Casimir, Florian, Stanislaus
Patron Saints of Portugal Anthony Padua, Francis Padua, George, Vincent
Patron Saints of Scholars Brigida; Gregory
Patron Saints of Schoolteachers Gregory the Great and John Baptist de la Salle
Patron Saints of Scotland Andrew and Columba
Patron Saints of Stonemasons

Barbara; Stephen

Patron Saints of Surgeons Cosmas, Damian, Luke

Patron Saints of Sweden Eric, Bridget

Patron Saint of Stenographers Cassian of Tangiers

Patron Saints of Theologians Alphonsus Liguori and Augustine

Patron Saint of Stonecutters Clement

Patron Saint of Students Thomas Aquinas

Patron Saints of Weavers Anastasia; Paul the Hermit

Patron Saint of Tailors Homobonus

Patron Saint of Tanners Crispin

Patron Saint of Tax Collectors Matthew

Patron Saint of Teachers Gregory the Great

Patron Saint of Telecommunications Workers Gabriel

Patron Saint of Television Clare of Assis

Patron Saint of Theologians Alphonsus Liguori

Patron Saint of Toothaches Apollonia of Egypt

Patron Saint of Travellers Christopher

Patron Saint of Troubled Parents Monica

Patron Saint of the United States Our Lady of the Immaculate Conception, according to Roman Catholics

Patron Saint of Universities Contardo Ferrini

Patron Saint of Uruguay Our Lady of Lujan

Patron Saint of Vocations Alphonsus

Patron Saint of Wales David

Patron Saint of War Prisoners Leonard

Patron Saint of Watchmen Peter of Alcantara

Patron Saint of the West Indies Gertrude

Patron Saint of Widows Paula

Patron Saint of Winegrowers Vincent

Patron Saint of Wine Merchants Amand

Patron Saint of Writers Lucy

Patron Saint of Yachtsmen Adjutor

Patron Saint of Youth Aloysius Gonzaga

Patron of Science Ptolemy (Claudius Ptolemaeus)

Patroon Stephen Van Rens-

selaer's nickname

Patton U.S.-made M-47 or M-48 medium tanks armed with 90mm guns

Patty Martha; Patience; Patricia

Paul VI Giovanni Batista Montini

Paula Paulcela; Paulette; Paulina; Pauline; Paulita

Paul Bunyan's Capital Brainerd, Minnesota

Paul Creston Joseph Guttoveggio

Paul Éluard Eugène Grindel (1895–1952)

Paulette Goddard Marion Levy

Paul Klenovsky Sir Henry J Wood's pseudonym used when he presented his orchestral arrangement of Bach's Toccata and Fugue in D minor; pupil of Alexander Glazunov

Paul Lukas Pal Lukacs

Paul Muni Muni Weisenfreund

Paul Vesey Samuel W Allen's pseudonym

Pavel Ivanovich Jones John Paul Jones (when he served as rear admiral commanding Russia's Black Sea fleet for Catherine the Great)

Peacefield Quincy, Massachusetts home of John Adams and his son John Quincy Adams

Peace Garden State North Dakota

Peacemaker William Penn

peace pill pcp, phencyclidine

Peach Bowl Atlanta, Georgia

Peach Capital of British Columbia Penticton

Peach State Georgia

Peacracker(s) native(s) of Lowestoft

Peanut Capital of Alabama Dothan

Peanut City Suffolk, Virginia

Peanut King Amadeo Obici who organized the Planters Peanut Company in 1906

Pear City Medford, Oregon

Pearl of the Adriatic Dubrovnik, Yugoslavia

Pearl of the Antilles Cuba

Pearl of the Atlantic Madeira

Pearl of the Baltic Bornholm Island, Denmark

Pearl of the Chilean Pacific Viña del Mar

Pearl Gulf Arabian Gulf, Persian Gulf

Pearl of Ireland Saint Brigit

Pearl Island of the Caribbean

Margarita, Venezuela

Pearl King Mikimoto Kokichi (Japanese who discovered the secret of creating cultured pearls)

Pearl of the Lagoons Abidjan, Ivory Coast

Pearl of the Orient Sri Lanka (Ceylon)

Pearls of the Pacific 7,100 Philippine islands; Honolulu, Pago Pago, Papeete

Pearl of Persia Isfahan

Pearl and Petroleum Sheikdom El Qatar on the Persian Gulf

Pearl S Buck Mrs Richard J Walsh

Pearl of the Sharon Netanya, Isreal

Pearls of the South Seas Tahiti, Tonga, Samoa, and other South Sea islands

Peasant Bard Robert Burns

Peasant Breughel Pieter Breughel the Elder

Peasant With A Pen Eric Linklater's nickname

Peck's Bad Boy pseudonym of George W Peck

Pecos Bill General William Shafter, USA

Pecos Wilderness eastern New Mexico and West Texas (northern New Mexico east of Santa Fe to the Rio Grande above Del Rio, Texas)

Peggy Margaret

Peg Leg Petrus Stuyvesant, Dutch governor of Nieuw Amsterdam, who lost his right leg in the siege of Sint Maarten

Pelican Louisiana's state bird and symbolic nickname often given Louisianians—Pelicans

Pelican State Louisiana

Pen of the American Revolution Thomas Paine

Penang (Malay—Betel Nut) —formerly called George Town when British Malaya

Pence Springs West Virginia State Prison for Women at Pence Springs

Peninsular Malaysia States of the Federation of Malaysia (Federated Malay States also known as Malaya)

Peninsular State Florida

Penmen of the Revolution John Dickinson of Dover, Delaware, Thomas Jefferson, and Tom Paine

Pennie Penina

Pennsylvania Farmer John

Dickinson's pseudonym
Pennsylvania Ports Erie, Philadelphia, Chester, Marcus Hook
Penny Penelope
Penobscot River City Bangor, Maine
Peony Indiana state flower; Indiana girl's nickname
Peony Center Faribault, Minnesota
People of the Lion Singhalese of Ceylon
The People's Attorney Louis Dembitz Brandeis
People's Lawyer Associate Justice Louis Dembitz Brandeis of the Supreme Court of the United States
People's Poet Paul Lawrence Dunbar
Peory Peoria, Illinois
Pepe José (Joseph)
Pepita Josefa; Josefina
Pepper Coast Liberia
Perce Persival; Percy
Percy Percival
Père-Lachaise Paris' best known cemetery and generic eponym for other burial places
Perfector of Opalescent Glass Louis Comfort Tiffany
Peripatetic Philosopher Aristotle
Peripatetic Pope John Paul II
perks nickname for percodan
Pernambuco old name of Recife, Brazil
Pero (Russian—pen)—Trotsky nickname because of his skill in writing revolutionary tracts
Perry Como Pierino Como
Perse Percival; Percy
Pershing Martin surface-to-surface missile (MGM-31A)
Persia ancient name for Iran
Persian Persian blinds (exterior blinds); Persian carpet (hand-woven oriental rug); Persian cat (long-haired breed); Persian lamb (young lamb of karakul sheep); Persian melon (greenish muskmelon)
Persian Gulf States Bahrain, Qatar, and the United Arab Emirates
Personification of Death Thanatos (Greek)—whose brother was Hypnos or sleep
Personification of the Destroying Principle Siva
Personification of Justice (*see* Justice Personified)
Personification of the Preserving Principle Vishnu

Personification of Sleep Hypnos (Greek)—whose brother was Thanatos or death
Personification of the Soul Psyche in the Greek mythology
Persuasive Evolutionist Thomas Henry Huxley
Perugino Piero Vannucci
Peru's Principal Port Callao
Peruvian Peruvian balsam used by chocolate makers, doctors, and perfumers); Peruvian bark (*cinchona*)
Peruvian Ports Iquitos on the Amazon's headwaters; Talara, Callao, Matarani, Mollendo, Ilo, Pisco, Chimbote, and Salaverry
Pesach Hebrew Passover
Pessimistic Composer Gustav Mahler (nicknamed Gloomy Gus by some music lovers)
Pessimistic Painter Hieronymus Bosch
Pessimistic Philosopher Arthur Schopenhauer
Pesthole of the Pacific Panama City, Panama; Buenaventura, Colombia; Guayaquil, Ecuador
Peter nickname of Leningrad, formerly Petrograd and originally St Petersburg
Peter Arno Curtis Arnoux Peters
Peterhouse St Peter's College (Cambridge)
Peter Lorre Laszlo Loewenstein
Peter Martyr Pietro Martin d'Anghierra's pseudonym
Peter McGill (American slang —Pedro Miguel)—Panama Canal Locks near Balboa
Peter Mennin Peter Mennini
Peter Mikhailov pseudonym of Peter the Great, which he used while travelling and working in Dutch shipyards
Peter Pan of Politics Winston Churchill
Peter and Paul St Peter and St Paul island fortress-prison on the Neva facing Saint Petersburg
Peter Pindar Dr John Wolcot
Peter Porcupine William Cobbett's pseudonym
Peter Warlock Philip Arnold Heseltine
Petit Caporal (French—Little Corporal)—Napoleon I
Petrarch Francesco Petracco
Petr Makadonski (Russian —Peter the Great)—Peter

Alekseyvich
Petroleum Emirate Kuwait
Petroleum V Nasby David Ross Locke's pseudonym
Petya (Russian nickname—Pyotr)—Peter
Pewee Kentucky Correctional Institution at Pewee Valley
Peyronies's disease (*see* bent-nail syndrome)
Peyton Place Gilmanton, New Hampshire
Phantom F-4 fighter airplane
Philadelphia Lawyer Andrew Hamilton, Philadelphia attorney who in 1734 and 1735 successfully defended New York printer Peter Zenger whose newspaper criticized British colonial policy in America—Zenger had been unsuccessful in getting a New York lawyer to take his case; an attorney who will defend a case others are afraid to touch
Philadelphia Painter Thomas Eakins
Philander von der Linde Johann Burkhard Mencke
Philidor François Andre Danican
Philippine Ports Aparri, Port Legazpi, Cavite, Manila, Poro, Masbate, Tacloban, Cebu, Iloilo, Davao, Zamboanga, Ozamiz, Isabela, Jolo
Philippines Principal Port Manila
Philipp Melanchthon Philipp Schwarzerd
Philistine Temptress Dalila (Delilah)
Philomela Mercy Otis Warren, called the First Lady of the American Revolution
Philosopher of the Absolute Georg Wilhelm Friedrich Hegel
Philosopher of China Confucius
Philosopher of Democracy and Humanism Sidney Hook
Philosopher of Freedom John Locke
Philosopher Freethinker Elbert Hubbard
Philosopher Kung Kung Futzu (Confucius)
Philosopher Laureate John W. Gardner
Philosopher of Loyola Marymount Ronda De Sola Chervin
Philosopher of Malmesbury

Thomas Hobbes

Philosopher Physician Averroes

Philosopher of Sans Souci Voltaire's nickname for Frederick the Great

Philosopher of Sex Havelock Ellis

Philosopher of the Superman Friedrich Wilhelm Nietzsche

Philosopher of the Third Reich Alfred Rosenberg

Phiz Hablot K Browne—illustrator of the *Pickwick Papers* of Dickens—Boz

Phoenix of Spain Lope de Vega

Photographer-Editor Alfred Stieglitz

Photographer of the Himalayas Samuel Bourne (1834–1912)

Photographic Pioneer William Henry Fox Talbot

Photographic Purists Ansel Adams and Edward Weston

Photo Reporter Margaret Bourke-White

Phrasemaker of Versailles Woodrow Wilson—28th President of the United States

Phronie Sophronia

Physician to the Body Politic Émile Zolá

Physician Extraordinary Sir William Osler

Physician's Physician Jacob Mendez Da Costa

Pianist-Composer-Conductor-Singer Teresa Carreño

Pianist-Conductor Vladimir Ashkenazy; Ossip Gabrilowitsch; Daniel Barenboim; Rudolf Ganz; José Iturbi; Ethel Leginska

Pianist's Pianist Richard Buhlig

Pickle Works nickname of building occupied by Central Intelligence Agency in Langley, Virginia

Pickpocket Heroine Defoe's *Moll Flanders*

Pick's disease brain disorder characterized by loss of memory and speech

Pico Bolívar (Spanish—Bolivar's Peak)—Venezuela's highest mountain also called La Columna

Picture Island Enoshima, Yokahama

Picture-Postcard-Landscape Land Switzerland

Picture Province Canada's New Brunswick

Picture Province of Canada New Brunswick

Piedmont Plateau Appalachian Mountain region extending from Alabama to New York, including Georgia, the Carolinas, Virginia, West Virginia, western Maryland, and Pennsylvania

Pier Angeli Anna Maria Pierangeli

Piero della Francesca Piero di Benedetto de Franceschi

Pierre Loti (pseudonym—Louis-Marie Julien Viaud)

Pierre Louÿs (pseudonym—Pierre Louis)

Pierre Nord André Léon Brouillard's pseudonym

Pierre-Paul Prud'hon Pierre Prudon

Pietermaritzburg South African city also called Maritzburg

Pieter Timmerman (Dutch—Peter Carpenter)—pseudonym used by Peter the Great of Russia while working as a shipwright in Dutch shipyards

Pig Alley Place Pigalle

Pig Islander New Zealander (Australian slang)

pig's ear (Cockney English—beer)

Pillars of Hercules promontories flanking the Straits of Gibraltar—Abyla in Africa facing Gibraltar in Europe

Pilsner Country Czechoslovakia

Pineapple Island Lanai, Hawaii

Pineapple Paradise Hawaiian Islands

Pine Tree State Maine

Piney Point Harry Lundeberg School of Seamanship at Piney Point, Maryland

Pink City of Rajputana Jaipur

Pinky conductor-violinist Pinchas Zukerman's nickname

Pinturicchio Barnardino Betti

Piombo palette name of Sebastiano Luciani

Pioneer deep-space probes designed for interplanetary investigation

Pioneer Aeronautical Experimentalist Otto Lilienthal

Pioneer Airship Designer Ferdinand von Zeppelin (1838–1917)

Pioneer American Composer William Billings

Pioneer in American Science

Benjamin Franklin

Pioneer of Antisepsis Ignaz Philipp Semmelweis

Pioneer Bacteriologist Louis Pasteur; Robert Koch

Pioneer of Biblical Criticism Thomas Paine (1737–1809)

Pioneer of British Aviation, Designer, Industrialist Geoffrey De Havilland (1882–1965)

Pioneer of Child Psychoanalysis Dr Anna Freud

Pioneer Dutch Aircraft Designer Anthony Fokker (1890–1939)

Pioneer German Aircraft Designer Hugo Junkers (1859–1935)

Pioneer Heart Surgeon Daniel Hale Williams

Pioneer Inventor of Aerial and Space Photography Systems Brigadier General George Goddard—USAF

Pioneer of Law Hammurabi (1792–1750 B.C.)

Pioneer Liturgical Dancers Carla De Sola, Ruth St Denis, and Ted Shawn

Pioneer of Modern Geography Alexander von Humboldt (1769–1859)

Pioneer of Modern Orchestration Hector Berlioz (1803–1869)

Pioneer of Oceanography Sir John Murray

Pioneers Pioneer Mountains of Idaho and Montana

Pioneer of Technological Change Eli Whitney

Pioneer of Two Worlds Thomas Paine

Pioneer of University Surgery William Halsted

Pioneer of Visceral Surgery Theodor Billroth

Piotr (Russian—Peter)—nickname for Petersburg, St Petersburg, Petrograd

Piper Laurie Rosetta Jacobs

Pippa Philipa; Philippa

Pirandello Stefano Landi

Pirate City Tampa, Florida

Pirate Coast Trucial Coast of Arabia including Abu Dhabi, Ajam, Dubai, Furairah, Ras el Khaimah, Sharjah, and UMM al Quwain

Pirate of the Gulf Jean Lafitte

Pirate Port Port Royal, Jamaica 17th century

Pisanus Fraxi Herbert Spencer Ashbee

Pitcher Plant Province New-

foundland
Pitch Lake Trinidad's asphalt lake
Pitch Lake Island Trinidad
Piter (Russian nickname for Petrograd or St Petersburg)
Pitigrilli Dino Segre
Pius XII Eugenio Pacelli
Pizza (PIE) Pacific Intermountain Express (stock exchange nickname)
Place of Many Waters Walla Walla, Washington
Place of Plenty Indian name for what is now Toronto, Ontario
Place of the Seven Wells Beersheba
Place of the Winds Nouakchott in Mauritania on the coast of West Africa
Plague of the Twentieth Century AIDS
Plain Joe Canada's Prime Minister Joe Clark
Plains States Iowa, Kansas, Minnesota, Missouri, Nebraska, North Dakota, South Dakota
Plank Island Aberdeen, Washington
Planner of the New York Public Library John Shaw Billings
Plantation State Rhode Island whose official title is the State of Rhode Island and Providence Plantations
Plant Science Experimenter Luther Burbank
Plant Wizard Luther Burbank
plaster of paris calcium sulfate $(CaSO_4)_2 \cdot H_2O$
Plateau Continent Africa
Plateglasses ultra-modern style in universities
Plate River Ports Buenos Aires, Argentina and Montevideo, Uruguay
Platine States Argentina and Uruguay
Plato (Greek—broad-shouldered)—the famous philosopher's real name was Aristocles
Plato's School the Grove of Academe near Athens, later referred to by the Romans as the Academia
Plattensee (German—flat sea, level lake)—Hungary's Lake Balaton—largest lake in central Europe
Pleasant former name of Nauru
Pleasure City of the South Seas Sydney, Australia
Plein-Air Painter Manet

Plejad Swedish class of fast patrol boats
Plight of Pizarro dysentery named for the Spanish conqueror of Peru
Plough-Share City York, Pennsylvania
Plow City Moline, Illinois
Plucky Pierre Salinger
Plum Sir Pelham Warner; Sir PG Wodehouse
Plumb-line Port to Panama Charleston, South Carolina (due north of the Panama Canal)
Plumed Knight Robert G Ingersoll's name for James G Blaine when nominating him for President
Plus Brave des Braves (French —Bravest of the Brave)— Napoleon's nickname for Marshal Ney
Plymouth Rock landing place of the Pilgrims in 1620 in what is now Plymouth, Massachusetts
Poet of Affection Marianne Moore
Poet of the American Revolution Philip Freneau
Poet of the Body—Poet of the Soul Walt Whitman's self-imposed nickname
Poet of Childhood Eugene Field
Poet of the Common People James Whitcomb Riley (1849–1916)
Poet of Democracy Walt Whitman
Poet of Despair James Thomson
Poet of the Excursion William Wordsworth
Poet of Friendship Robert Burns
Poet from Jersey William Carlos Williams of Rutherford, New Jersey
Poet of Harlem Langston Hughes (1902–1967)
Poet of Imperialism Rudyard Kipling
Poet of Individuality Walt Whitman
Poet King Ossian of Ireland
Poet Laureate of England Sir John Betjeman
Poet Laureate of New England John Greenleaf Whittier
Poet Laureate of Venezuela Irma De Sola Ricardo
Poet of Liberty Johann Christoph Friedrich von Schiller

Poet Naturalist Henry David Thoreau
Poet of Nature Jean Sibelius
Poet of Passion Ella Wheeler Wilcox
Poet of the Piano Frédéric Chopin
Poet of Poets Shelley
Poet Sire of Italy Dante Alighieri
Poets Laureate of the United States Robert Penn Warren (1986–87); Richard Wilbur (1987–88); Howard Nemerov (1988–90); Mark Strand (1990–)
Poet of the Subconscious Giovanni Pascoli
Point Coma Point Loma
Poison Ivy Upton Sinclair's nickname for publicist Ivy Lee
Poitiers formerly Poictiers
Pokanoket American Indian name for what was Mount Hope and is now Bristol, RI
Poke Poughkeepsie
Pola Appolina; Policarpa Salabarrieta
Pola Negri Appolina Chapulez
Polar Bear Land the Arctic
Polaris brightest star in the constellation of Ursa Minor; usually called the Pole Star or the Seaman's Star (Stella Maris) as within a degree or two it points to true north
Polaris-Poseidon Lockheed submarine-launched missiles
Pole Star (*See* Polaris)
Polish City Hamtramck, Michigan
Polish Ports Nowy Port, Stettin, Gdynia, Ustka, Swinoujscie
Polish Siberia Nowogrodek southwest of Minsk
Polish Story Teller Isaac Bashevis Singer
Polish Town Panna Maria, Texas—America's oldest Polish settlement
Polly Mary; Pauline; Pollyanna
Polo Capital of the South Aiken, South Carolina
Polygon of Drought northeast Brazil along the Rio San Francisco and east of a line between Bahía and Fortaleza
Polynesia (Greek—Many Islands)—the South Pacific
Polynesian Kingdom Tonga (The Friendly Isles)
Polynesia's Sacred Isle Raiatea (in the South Pacific west of Tahiti)

Pommy British emigrant to Australia or New Zealand

Pompey Cneius Pompeius; nickname of Portsmouth, England

Pom(s) Prisoner(s) of Mother England [Australian nickname for person(s) newly arrived from Great Britain]

Pony Express Terminus Sacramento, California

poor-man's meat *Glycine max* —soybeans

Poor Richard Richard Saunders (pseudonym used by Benjamin Franklin in writing *Poor Richard's Almanack*)

Popcorn Capital of the World Shaller, Iowa

Pope of Geneva Calvin's nickname

Pope John XXIII Angelo Giuseppe Roncalli

Pope John Paul I Albino Luciani

Pope John Paul II Karol Wojtyla

Pope Paul VI Giovanni Battista Montini

Pope Pius XI Achille Ratti

Pope Pius XII Eugenio Pacelli

poppers amyl nitrate (also called amys, pearls, or snappers)

Pops Arthur Fiedler

Popular Porter Cole Porter

Pori Finnish name for what the Swedes call Björneborg; Polaris operational readiness instrumentation

Pork Dump nickname of Clinton Prison near Utica, New York

Porkopolis Cincinnati, Ohio

Pork Packer Philip D Armour

Porn Capital of America San Francisco

Portage La Prairie Girls Correctional Centre for Women at Portage La Prairie, Manitoba

Porter of Heaven Janus the Two-Faced

Port Everglades Fort Lauderdale, Florida's port

Portia pen name of Abigail Smith Adams—wife of President John Adams and America's First Suffragist

Port Kelang formerly Port Swettenham and also called Port Klang

Port Klang formerly Port Swettenham, Malaya

Port Lyautey former name of Kenitra, Morocco

Porto di Lido Italian—Port of

the Lido)—the port of Venice

Port o' Missing Men San Francisco

Porto Rico original name of Puerto Rico

Port of the Pilgrims Provincetown, Massachusetts

Portrait Painter of Presidents Gilbert Stuart

portsides portsiders (left-handed persons)

Port of the Southwest Galveston, Texas

Ports of Philadelphia Trenton, Camden, Gloucester City, Philadelphia, Chester, Marcus Hook, Wilmington

Port of St John of Acre Akko

Portugal's Principal Port Lisboa (Lisbon)

Portuguese America Brazil

Portuguese China Macao (near Hong Kong)

Portuguese East Africa Mozambique

Portuguese Guinea former name of Guinea Bissau on Africa's west coast

Portuguese India former name of the territories of Damão, Diu, Goa, Panjim, etc.

Portuguese Mars Affonso d'Alboquerque also called Affonso o Grande (Alphonse the Great)—Portuguese empire builder and viceroy of Portuguese India

Portuguese Overseas Province Macao

Portuguese Paradise Sintra near Lisbon

Portuguese Ports (large, medium, and small from north to south) Viana do Castelo, Porto de Leixoes, Porto (Oporto), Lisboa (Lisbon), Setubal, (*in the Azores*—Horta *and* Ponta Delgada), Funchal (Madeira)

Portuguese Republic República Portuguesa

Portuguese-Spanish Century the 16th century—the 1500s

Portuguese-speaking Places Angola, Azores Islands, Brazil, Cape Verde Islands, Guinea-Bissa, Macao, Madeira Islands, Mozambique, Portugal, São Tomé and Principe Islands, Goa and Timor

Portuguese Timor former Portuguese outpost of empire on Timor Island in Indonesia

Portuguese West Africa Angola, Portuguese Guinea,

St Thomas and Prince islands during colonial era

Port Veneris nickname for Port Vendre on the Franco-Spanish frontier

Port Wine Port Oporto, Portugal

Postage-Stamp Principalities Andorra, Liechtenstein, Luxembourg, and Monaco are so named by philatelists although Luxembourg is a grand duchy and is not ruled by a prince

Potain's disease pleural and pulmonary edema

Potash City Saskatoon, Saskatchewan

Potentate of the Pit Lucifer

Poti a Soviet class of submarine chasers

pot machine marijuana-manufacturing device

Potomac River City Washington, DC

Pot Smuggler's Paradise Florida whose waterways provide the best background for smuggling

Pott's disease vertebral inflammation

Poverty Bay Gisborne, New Zealand

Powder Keg of Europe the Balkans

Practical Political Philosopher Niccolò Machiavelli

Pragmatist Philosopher William James

Prairie Canada Alberta, Saskatchewan, and Manitoba

Prairie City Bloomington, Illinois

Prairie Provinces Alberta, Manitoba, Saskatchewan

Prairies great plains between Appalachian and Rocky mountains of North America

Prairie State Illinois

Prairie States North and South Dakota, Nebraska, Kansas, Minnesota, Iowa, and Illinois

Prayer-shawl Flag Israeli banner derived from talith or prayer shawl with horizontal blue stripes enclosing Shield or Star of David

Preacher of the Despairing Girolamo Savonarola

Preah Reach Ana Chak Kampuchea Cambodia

Precious Province Kueichow

Precursor of Dutch Painting Lucas van Leyden

Precursor of Expressionism Edvard Munch

Precursor of Japanese Art Kose no Kanaoka

Precursor of the Mexican Revolution Ricardo Flores Magon

Precursor of Pharmacology Paracelsus

Precursor of Pictorial Realism Mathias Grünewald (Mathis der Mahler)

Precursor of Sociology Charles de Secondat Baron de la Brède et de Montesquieu

Precursor of Spanish-American Emancipation Francisco Miranda

Precursor of Surrealism Hieronymus Bosch (Hieronymus van Aken)

Precursor of Venezuela Francisco de Miranda)

Pre-emption President John Tyler

Preiser's disease porosity of the wristbone

Premier Deng Deng Xiaoping (Teng Hsiao-ping)

Premier Passenger Port of Great Britain Southampton

Premier Primitive Henri Rousseau

Premier of Russia Alexander Kerenski

Pre-Raphaelite Founders Holman Hunt, Sir John Everett Millias, Daniel Gabriel Rossetti

Presbyterian Jerusalem Edinburgh

The President the President of the United States

Presidents' conference conference of presidents of major Jewish organizations (in America)

President ships American President Line vessels named after such statesmen as *President Lincoln, President Roosevelt, President Taft*

Preston K Swinehart (nickname—movie actor Alan Dinehart)

Pretender Charles Stuart

Pretty Boy Charles A Floyd (1901–1934)

Pretzel City Lancaster and Reading, Pennsylvania

Pride of the Yankees Lou Gehrig

Priest of Nature Sir Isaac Newton

Prima Donna Assoluta Dame Joan Sutherland

Primate of Italy the Pope

Prime Meridian Place Greenwich, England

Prime Minister Deng Deng Xiaoping (Teng Hsiao-ping)

Prime Minister of Hell Satan

Prime Minister Lee Prime Minister Lee Yuan Yew of Singapore

Prime Minister of Mirth Peter Sellers

Prime Minister of National Crime Frank Costello (1893–1973), born Francesco Seriglia

Prime Minister of the Underworld Frank Costello

Prince of American Letters Washington Irving

Prince of the Apostles the Pope, according to the Roman Catholics

Prince of Artists Albrecht Dürer

Prince of Comic Opera Daniel François Esprit Auber

Prince Consort Albert of Saxe-Coburg Gotha (Queen Victoria's husband)

Prince of Cranks Ignatius Donnelly

Prince of Darkness Satan

Prince of Destruction Tamerlane (Timur the Lame)

Prince of Gossips Samuel Pepys

Prince of Humbugs P(hineas) T(aylor) Barnum

Prince of Humorists Mark Twain (Samuel Langhorne Clemens)

Prince of Israel Michael

Prince of Journalists Horace Greeley

Prince of Losers Dr Frederick A Cook who claimed he reached the North Pole nearly a year before Commander Robert E Peary, who was credited with the discovery

Princely Province Prince Edward Island

Prince of the Meistersingers Hans Sachs of Nuremberg also known as the Cobbler Poet

Prince of Men Robert Louis Stevenson's nickname for Henry James

Prince of Metals silver

Prince of Music Palestrina

Prince of Orange William I of the Netherlands and his male successors

Prince of Orators Demosthenes

Prince of the Oyster Pirates Jack London

Prince of Painters Raphael

Prince of Philosophers Plato

Prince of Physicians Avicenna (Abu ibn Sina)

Prince of the Pianoforte Louis Moreau Gottschalk

Prince of Pistoleers James Butler (Wild Bill) Hickok

Prince of Poets Alexander Pushkin, according to Russian literary critics; Edmund Spenser

Prince of Prose Writers John Bunyan

Prince of the Renaissance Michaelangelo

Prince of Scoffers Voltaire

Prince of Showmen PT Barnum

Prince of Siddhartha Gautama Buddha

Prince of Skeptics Voltaire

Prince of Spanish Poetry Garcilaso de la Vega

Princess of Fruits Linnaeus' sobriquet for the pineapple

Prince of Story Tellers Giovanni Boccaccio

Prince of Trees Linnaeus' nickname for the palm

Prince of Violin Virtuosos Itzhak Perlman

Prince of Wales Island Penang's previous name

Principality of the Grimaldi Monaco

Principal Port of the United Kingdom Liverpool

Principe de la Paz (Spanish —Prince of the Peace)—Manuel Godoy y Alvarez de Faria

Printmaker to the Mexican People José Guadalupe Posada

Printmakers to the American People (Nathaniel) Currier & (James Merritt) Ives—America's most famous lithographers

Prison at the Bottom of the World Ushuaia, Argentina on Beagle Channel close to Cape Horn

Prison at the Top of the World Solovetski Island isolators in the White Sea and east of Kem in the Soviet Union

Prisoner of the Chillon François de Bonnivard

Prison of Gold The Louvre

Prison of Nations Austro-Hungarian Empire

Prison Poet Oscar Wilde (1856–1900), Irish authorplayright-poet-wit whose conviction for sodomy brought a three-year prison term vividly described in his philosophical poem *The Ballad of Reading Gaol*

Prodigy of Learning Dr Samuel Hahnemann

Professor of Blue Sky Mortimer Adler

Professor Bruno Pantoffel Jorge Mester

Professor of Earthquakes Sir William Hamilton

Professor Julius Caesar Hannibal (pseudonym—WH Levinson)

Professor Seagull Joe Gould

Profit Center of the Southwest Phoenix, Arizona

Prolific Lexicographer Eric Partridge

Prolific Professor Isaac Asimov

Prolific Rationalist English ex-priest Joseph McCabe

Prolific Typographer Frederic William Goudy

Promised Land Israel, promised to the Israelites by Moses and to the Israelis by Balfour

Promoter of Agrarian Reform Emiliano Zapata

Prophet of Allah Mohammed

Prophet of the American Way Thomas Jefferson

Prophet of Christianity John the Baptist

Prophet of Democracy William Penn

Prophet of Doom Girolamo Savonarola

Prophet of Israel Moses

Prophet of Modernity Émile Zolá

Prophet of Mythology Teiresias

Prophet Outcast Leon Trotsky

Prophet-Preacher-Hero of New England Unitarianism William Ellery Channing

Prophets of Israel Moses, Samuel, Nathan, Elijah, Elisha

Prophet of the Strenuous Life Jack London

Prose Poet of Violence Jean Genet

Prosperous Paradise of the Pacific Hawaii

Prostitution Capital of the South old nickname for New Orleans

Protector from Fever Febris (Roman goddess)

Protector from Poison Gases Mephitis—Roman goddess

venerated in volcanic lands

Protector of the Indians Rodrigo de Bastidas—Spanish navigator who founded Santa Marta; Las Casas and Eliot share the title—Protector of the Indians

Protector of Peru General José Francisco de San Martin (1778–1850)

Protector of the Protestants Marguerite de Navarre

Protector of Seafarers the Greek goddess Brizo

Protestant Hero Frederick the Great of Prussia

Provisional President of Africa Marcus Garvey

Provision State Connecticut in Revolutionary times

Provost Hunting reconnaissance-trainer aircraft built in Britain

Prune Picker(s) Californian(s)

prussic acid hydrocyanic acid

psikhushka (Russian slang—psychoprison)—for punishing and segregating dissidents

Psychedelphia San Francisco's Haight-Ashbury district inhabited by many drug addicts

Psychoanalysis Capitals Berlin, New York, and Vienna

Psychologist of the Soul Søren Kierkegaard (1813–1855)

Ptarmigan Alaska state bird; symbolic nickname given some Alaskans

Public Enemy Number One gangster Al Capone's nickname

Public Library Builder Andrew Carnegie

Publius allonymic name used by Alexander Hamilton, John Jay, and James Madison in writing *The Federalist*

Pueblo de los Angeles (Spanish—Village of the Angels)— early Spanish name for Los Angeles

Puerto Colombia formerly Savanilla

Puerto Limón Limón, Costa Rica

Puerto Rico Ports Ensenada de Honda, San Juan, Ponce, Guanica, Mayaguez

Pugetopolis industrialized urban areas surrounding Puget Sound

Puggy Booth Joseph Mallord William Turner's nickname

pukeweed *Lobelia inflata's* nickname

Puma Franco-British Aeros-

patiale-Westland transport helicopter

Punkie Town Punxsutawney, Pennsylvania

Punks' Paradise nickname given any gambling center and sometimes to the State of Nevada and the casino cities of Reno and Las Vegas

Punta Arenas (Spanish—Sand Point)—the one in Chile called Magallanes from 1927 to 1937; the one in Costa Rica is written Puntarenas

Punxey Punxsutawney

Puppet Emperor Henry P'u-yi (1906–1967) last emperor of China; made Emperor of Manchukuo by the Japanese

Puri port of Jagananth or Juggernaut on the Bay of Bengal

Puritan City Boston, Massachusetts

Puritan State Massachusetts

Purple Islands the Madeiras

Purple Land WH Hudson's sobriquet for Uruguay

purpurite iron magnesium phosphate

Pushkin modern name for Tsarskoe Selo south of Leningrad

Putrid Sea Sivash Sea (mineralized marshes along Crimea's north coast)

Pyrenean Principality Andorra

Pyrenees Principality Andorra

Qatar Ports Ad Dawhah and Musayid

Quai d'Orsay section of Paris occupied by French Foreign Ministry

Quail Californians are sometimes nicknamed Quail; California's state bird—the Golden Valley Quail; McDonnell-Douglas decoy missile

Quail Haven Cedar Vale, Kansas

Quaintest City in the U.S. Santa Fé, New Mexico (founded by the Spaniards around 1609)

Quake City San Francisco, California

Quaker Abolitionist epithet shared by Elias Hicks, Lucretia Mott, John Greenleaf Whittier, and John Woolman

Quaker Abolitionists Lucretia Mott, John Greenleaf Whittier, and John Woolman

Quaker City Philadelphia

Quaker City of the West Richmond, Indiana

Quaker Dolley Mrs Dorothea

(Dolley) Madison—wife of President James Madison

Quaker Founder George Fox —founder of the Society of Friends who were nicknamed Quakers by an English judge who persecuted them

Quaker Founder of Pennsylvania William Penn

Quaker Frontiersman Daniel Boone

Quaker Liberal Elias Hicks

Quaker Penologist Elias Hicks (1748–1830), who preached against cruelty to the imprisoned and the insane

Quaker Poet Bernard Barton in England and John Greenleaf Whittier in New England

Quaker Preacher Elias Hicks —founder of the Hicksite Friends championing the abolition of slavery and opposing creeds approved by the elders

Quaker Reformer Elizabeth Fry—noted for her campaigns to better the life of inmates in insane asylums and prisons; also worked for the betterment of education

Quakers members of the Society of Friends

Quaker State Pennsylvania

Quakertown Philadelphia

Quality City Rochester, New York

quarantine flag yellow flag flown when a vessel requests pratique; letter Q or Québec in the international signal code

quas methaqualone's nickname (also called quacks or quads)

Québec Q-class Soviet submarines

Queen of the Adriatic Venice

Queen Alice Alice Lee (Roosevelt) Longworth

Queen of the Amazons Hippolyta

Queen of the Angels the Virgin Mary

Queen of the Antilles Cuba

Queen of the Arabian Sea Cochin, India

Queen of Back Bay Isabella Stewart Gardner

Queen of Bases calcium oxide and related compounds known commercially as lime

Queen Bee Director of the Women's Air Force Auxiliary

Queen of Belgian Beaches Ostend

Queen Bess Queen Elizabeth

Queen of the Brazos Waco, Texas

Queen of the Caribbean SS *Norway* (world's largest cruise ship, formerly SS *France*)

Queen of the Caribbees Nevis

Queen of Cats the lioness in Africa; the tigress in Asia

Queen Cities of the Austro-Hungarian Empire Budapest, Prague, Vienna

Queen City Lahore (in the Punjab of Pakistan)

Queen City of Alabama Gadsden

Queen City of Canada Toronto

Queen City of the Carolinas Charlotte, North Carolina

Queen City of the Hanseatic League Lübeck

Queen City of the Hudson Yonkers, New York

Queen City of India Bombay

Queen City of the Lakes Buffalo, New York and Toronto, Ontario

Queen City of Lake Superior Marquette, Michigan

Queen City of the Lehigh Valley Allentown, Pennsylvania

Queen City of the Merrimack Valley Manchester, New Hampshire

Queen City of the Mississippi St Louis, Missouri

Queen City of the Mountains Knoxville, Tennessee

Queen City of New Zealand Auckland

Queen City of the North Edinburgh

Queen City of the Ohio Cincinnati, Ohio

Queen City of the Ozarks Springfield, Missouri

Queen City of the Pacific San Francisco, California and Seattle, Washington

Queen City of the Plains Denver

Queen City of the Rio Grande Del Rio, Texas

Queen City of the Sea Charleston, South Carolina

Queen City of the Sound Seattle, Washington on Puget Sound

Queen City of the South Atlanta, Georgia and Sydney, New South Wales

Queen City of the Southern Tier Elmira, New York

Queen City of the Trails Independence, Missouri

Queen City of Vermont Burlington

Queen City of the West Cincinnati (in the early 1800s)

Queen of the Comstock Lode Virginia City, Nevada

Queen of the Cowtowns Fort Dodge, Iowa

Queen of Crime Agatha Christie

Queen of Crossword Puzzledom Margaret Farrar

Queen of the Danube Budapest

Queen Emma Curaçao's floating bridge across Willemstad's harbor

Queen of Flowers the rose

Queen of Folk Joan Baez

Queen of the French Riviera Nice

Queen of the Goldfields Melbourne, Victoria, Australia

Queen of Gulf Ports New Orleans

Queen of Heaven Ashtoreth (Semitic); Astarte (Phoenician); Hera (Greek); Inanna (Sumerian); Ishtar (Assyrian and Babylonian); Isis (Egyptian); Juno (Roman); Virgin Mary (Christian)

Queenie Regina

Queen of the Inland Sea Chicago

Queen of Islands Cuba in the days of Columbus

Queen on the James River Richmond, Virginia

Queen of the Kingdom of Death Hel (daughter of Loki in Nordic mythology whose name is used for the Kingdom of Death)

Queen of Kings Cleopatra

Queen of Lake Malaren Stockholm

Queen of Lake Michigan Chicago

Queen of the Lakes Chicago

Queen of Long-Distance Roads the Appian Way extending from Brindisi to Rome and begun in 312 B.C.

Queen of Love and Lust Aphrodite or Venus

Queen Maud Land Norwegian Antarctica

Queen of Metals platinum

Queen of the Missions Mission San José in San Antonio, Texas and Mission Santa Barbara in Santa Barbara, California

Queen of the Mississippi St Louis

Queen of the Mountains Helena, Montana

Queen of Mystery Writers Ngaio Marsh

Queen Nef Queen Nefertiti of Egypt

Queen of the North Edinburgh

Queen of the Ohio Cincinnati

Queen of Opera Dame Joan Sutherland

Queen of the Pacific San Francisco

Queen of the Plains Regina, Saskatchewan

Queen of the Prairies Canada's Province of Saskatchewan

Queen of Queens Brutus' nickname for Cleopatra

Queen Sarah Sarah, the Duchess of Marlborough

Queensberry (*see* Marquis of Queensberry)

Queen's Birthday Queen Juliana's Birthday (April 30) —Netherlands; Queen Victoria's Birthday (May 20)— Great Britain and Commonwealth countries

Queen's College Rutgers University in colonial times

Queen's Corsair Sir Francis Drake

Queen of the Sea Islands Beaufort, South Carolina

Queen of the Seas Glasgow (reputed for its Clyde-built ships); Venice

queen's English correct English

Queen's House Buckingham Palace

Queen of Skyscrapers Empire State Building

Queen of the South New Orleans

Queen of the Spas Saratoga Springs, New York

Queen State Maryland (named for Henrietta Maria, wife of Charles I of England)

Queenstown former name of Cobh on Ireland's south coast

Queen of Summer Resorts Newport, Rhode Island

Queen of Trains Orient Express

Queen of TV Comedy Lucille Ball

Queen of the Vampires Theda Bara (Theodosia Goodman)

Queen of Watering Places Brighton, England

Queen of the West Longfellow's nickname for Cincinnati

Queermacks Cuyamaca Mountains in California's San Diego County

quicklime calcium oxide—CaO

quicksilver mercury (Hg)

Quicksilver Bob Robert Fulton's nickname

Quiet Americans soft-voiced well-mannered Canadians

Quiet Epidemic medical nickname for Alzheimer's disease

Quiet River Russia's quiet-flowing Don

Quincke's disease edema of the skin; giant hives

Quinquad's disease inflammation of the scalp resulting in bald patches

Quintessence of Africa Serengeti National Park, Tanzania

Quintuplets Herbert Morrison so nicknamed because he did the work of five

Quirinale (Italian—House of the God of War)—Ministry of War in Rome

Quisquellano(s) Santo Domingan(s)

Quisqueya Hispaniola

Rab(bie) Robert

Rabble-Rouser of the Revolution Sam(uel) Adams

Rachilde Marguerite Vallette

Radclyffe Hall Marguerite Radclyffe Hall

Radek communist-party pseudonym of Karl Sobelsohn

Radio Capital Camden, New Jersey

Radiumbad Brambach Brambach, Saxony

Raedwulf (Early English— Ralph)—alleged to be the imp of mischief in a printing house

Rafe Ralph

Raffaello Raphael

Ragnarok end of the world in Norse mythology; equivalent to Twilight of the Gods (Götterdämmerung)

Rahway New Jersey State Prison at Rahway

Railroad City nickname of cities such as Atlanta, Boston, Buffalo, Chicago, Cincinnati, Cleveland, Detroit, Edmonton, Houston, Indianapolis, Kansas City, Los Angeles, Milwaukee, Minneapolis, Montreal, New Orleans, New York, Omaha, Philadelphia, St Louis, San Antonio, San Francisco, Seattle, Toronto, Washington, D.C.,

Winnipeg; (*see* Railway City)

Railsplitter Abraham Lincoln

Railway King George Hudson

Rajah Rogers Hornsby

Ralph Connor Charles W Gordon

Ralph Iron Olive Schreiner's pseudonym

Ralph Marlowe Ralph Manheim

Ralph Rashleigh James Tucker's pseudonym

Ramón Navarro Ramón Samanlegos

Randolph Scott Randolph Crance

Randy Randolph

Ran Fiennes Sir Ranulph Twistleton-Wykeham-Fiennes

Ranger American program for investigation of the Moon and region between the Moon and the Earth; Texas state policeman

Rapier British BAC surface-to-air missile launched for low-altitude defense

Raquel Torres Paula Ostermán

Raquel Welch Raquel Tejada

Raquetball Capital San Diego, California

Ras Desiderius Erasmus

Ras Asir (Arabic—Cape Guardafui)—northeastern-most Africa on the coast of Somalia and the Gulf of Aden

Rasmus Erasmus

Rasputin (Russian—Dissolute) —nickname of the Siberian monk Gregory Efimovitch

Rassmen Jamaicans

Rastus Erastus; Theophrastus

Ratipole nickname of Napoleon III

Raven Hiller utility helicopter designated H-23 and OH-23

Ray Milland Reginald Truscott-Jones

Raynaud's disease circulatory disorder of the extremities

Raynaud's phenomenon white-finger disease brought on by long-term use of vibrating hand tools

Razor The Razor—General Hideki Tojo's nickname

Razorback(s) Arkansan(s)

Razor Clam Capital Cordova, Alaska

Realistic Recorder of Spanish Life Goya (Francisco José de Goya y Lucientes)

Realm of the American Alligator Okefenokee Swamp, Florida and Georgia

Realm of Exotic Flavors Thai-

land

Rebecca West Cecily Isabel Fairfield

Rebel City Charleston, South Carolina

Rebel of Salem Roger Williams

Rebel Unitarian Theodore Parker

Rebel of Walden Henry David Thoreau

Recafellow Andrew Carnegie's nickname for John D Rockefeller Sr

Recife (Portuguese—Reef)— Pernambuco's new name

Reclus' disease cystic growths in the breasts

Reconstruction President Rutherford Birchard Hayes

Red Sinclair Lewis

Red Baron Baron Manfred von Richthofen

Redbricks red-brick universities

Red Chamber Canadian Senate

Red China People's Republic of China

Red Crescent equivalent of the Red Cross in the Moslem world (symbolized by a red crescent on a white field)

Red Cross red cross on a white field; used on ambulances, hospitals, and hospital ships to denote their neutrality; also called the Cross of Geneva or the Geneva Cross, as its function in war is accepted by the Geneva Convention

Red Cross and Crescent Soviet equivalent of the Red Cross (symbolized by a red cross and a red crescent on a white field)

Red Dean of Canterbury The Very Reverend Doctor Hewlett Johnson—Dean of Canterbury Cathedral from 1931 to 1963

Redd Foxx John Elroy Sanford

Red Duster British flag; Red Ensign flown from British merchant vessels

Redemptorist Founder Alfonso Maria de Liguori

Red Ensign British flag

Redeye General Dynamics portable surface-to-air missile carried and fired by one man

red flag danger; stop sign

Red Gap Lone Pine, California as described in *Ruggles of Red Gap* by Harry Leon Wilson

red lead lead oxide—Pb_3O_4 (minium)

Red Lewis (Harold) Sinclair Lewis

redlight district whorehouse neighborhood; zone of prostitution

Red Lion and Sun Iran's equivalent of the Red Cross (symbolized by a red lion beneath a red sun on a white field)

Red Planet Mars

Red Priest red-headed Antonio Vivaldi

Red Rosa Rosa Luxemburg —co-founder with Karl Liebnecht of the Spartacus League later to become the Communist Party of Germany

Red Skelton Richard Bernard Skelton

Red Square in the heart of Moscow between the GUM department store, the Kremlin, and Lenin's tomb; called Krasnaya Ploschad by the Russians

red star symbol of the Soviet Union

Redtop Hawker-Siddeley air-to-air missile

reefer(s) marijuana cigarette(s); refrigerated compartment(s) in a ship; refrigerator(s)

Reginald Bliss Herbert George Wells (1866–1946)

Regiomontanus Johann Müller (1436–1476)

Region of Four Streams Szechwan Province, China

Reichmann's disease continuous and excessive gastric secretion

Reidsville Georgia State Prison Facility at Reidsville

Reistertown Girls Montrose School for (delinquent) Girls at Reistertown, Maryland

Rejectionist Front Arab countries such as Algeria, Libya, and Syria most opposed to U.S. efforts to gain Israel's acceptance by its neighbors

Religious Freedom Colony Rhode Island

Reluctant Imperialist John C Calhoun who half-heartedly supported the Mexican War

Rembrandt Rembrandt Harmenszoon van Rijn—RvR

Rembrandt of the Roman Ruins Giambattista Piranesi

Rene Irene

René Adorée Jeanne de la Fonte

Renée Clair René Chomette

Renegade Irishman James Joyce

República Oriental (Spanish— Oriental Republic)—Uruguay

Republic of the Sacred Heart Ecuador

Research Center of the Classical World Library of Alexandria, Egypt

Resplendent Land Ceylon or *Sri Lanka* (Singhalese—Resplendent Land)

The Restoration France from 1814 to 1848

Ret Marut B Traven's pen name in Bavaria after World War I

Return-to-Normalcy President Warren Gamaliel Harding

Réunion Indian Ocean island formerly called Bourbon

Reval or Revel old place-names for Tallinn, Estonia

Rex Harrison Reginald Carey

Rex Ingram Reginald Hitchcock

Rhapsodic Richard Richard Strauss

Rhineland Capital Cologne (Köln)

Rhode Island Ports Providence, Newport

Rhode Island Reds Rhode Islanders

Rhonda formerly Ystradyfodwg, Wales

Ricardo Cortez Jacob Krantz

Rice Bowl southwest Louisiana

Rice Bowl of Malaysia Kedah

Rice Center Crowley and Lake Charles in coastal Louisiana

Rice State South Carolina

Richard Arlen Van Mattimore

Richard Avalon Sir John Woodroffe

Richard Burton Richard Jenkins

Richard Coeur de Lion Richard I of England

Richard the First Richard Wagner

Richard Hull Richard Henry Sampson's pseudonym

Richard Llewellyn Richard David Vivian Llewellyn Lloyd

Richard Saunders Benjamin Franklin (*see* Poor Richard)

Richard the Second Richard Strauss

Richard Tauber Ernst Seiffert

Rich Coast Costa Rica

Richelieu Armand Jean du Plessis

Rickie Admiral Hyman George

Rickover, USN

Rideau Hall Ottawa residence of the Governor General of Canada

Rienzi Niccolo Gabrini

Rifle City Springfield, Massachusetts

Riga Latvia's capital a class of Soviet submarines

Riga's disease ulceration of the tongue

Rigg's disease inflammation of the gums with pus deposits in the tooth sockets; also called alveolar pyorrhea

Ring of Fire volcanic zone extending from Alaska to Chile and from Siberia to New Zealand via Indonesia

Ring Lardner Ringgold Wilmer Lardner

Ringo Starr Richard Starkey

Rio Rio de Janeiro, Brazil

Rio Branco José Mariá de Silva Paranhos—Baron of Rio Branco—Brazil's great statesman

Rio de la Plata Province Paraguay

Rip Rip Van Winkle; Robert; Rupert

rising sun symbol of Japan and the Japanese

Rita Hayworth Margarita Carmen Cansino

Rita Moreno Rosita Dolores Alverio

Ritter's disease skin scaling, sometimes fatal when it attacks infants

Rivalta's disease lumpy jaw

River of African Legend Congo; Nile; Orange; Zaire

River of Alaskan Legend the Yukon

River of American Legend Mississippi

River of American Midwestern Legend Missouri

River of American Northwestern Legend Columbia

River of Asian Legend Euphrates; Ganges; Tigris; Yangtze

River of the Bears Alaska's McNeil River

River of the Black Dragon Amur River on the Sino-Soviet frontier

River of Brazilian Legend Amazon

River of British Legend Severn

River of Californian Legend Sacramento

River of Canadian Legend the

St Lawrence

River of Colombian Legend Magdalena

River of Destiny Rio Grande

River of English Legend Thames

River of European Legend Danube, Rhine

River of Florida Legend Suwanee

River of French Legend Seine

River of Georgia and South Carolina Legend Savannah

River of Grass Florida's Everglades

River of Hades or Hell the Styx, according to mythology it encircles the underworld nine times and the dead are ferried over its waters by Charon

River House Ohio State Penitentiary on the Scioto River near Columbus

River of Irish Legend Shannon

River of Italian Legend Arno; Tiber

River of Kings Chao Phraya flowing through Krungthep formerly called Bangkok

River of Maine Legend Kennebec; Penobscot

River of New England Legend Connecticut; Merrimack

River of New Jersey and Pennsylvania Legend: Delaware

River of New York Legend Hudson

River of the North the Yukon

River of the Old Northwest Legend Ohio

River of Oregonian Legend Willamette

River Plate Republics Argentina, Paraguay, Uruguay (all on rivers flowing into Rio de la Plata estuary)

River of Portuguese Legend Tagus

River of Russia Legend Volga

River of Scottish Legend Tay

River of Spanish Legend Ebro

River of Venezuelan Legend Orinoco

Riverview Interprovincial Home for (misdemeanant) Women at Riverview, New Brunswick

River of Welsh Legend Towy

Riviera Mediterannean coasts of Italy, France, and Spain

Riviera of South America Uruguay

road hustler(s) card-and-dice hustler(s)

Roadrunner New Mexico state bird and nickname applied to many New Mexicans

Road of the Sun *l'Autostrade del Sole* (Italian superhighway linking Milan, Rome, and Naples)

Roaring Forties storm-tossed seas between 40 and 50 degrees south latitude

Robber Barons (*see* American Railroad Barons, Banker Barons, Mining Baron, Oil Baron, Pork Packer, Steel Baron)

Robber's Nest Berlin, according to an old German song

Robert Alda Alphonso d'Abruzzo

Robert Capa Andrei Friedmann (1913–1954)

Robert Forsythe Kyle Crichton

Robert Henri Robert Henry Cozad

Robert Merrill Merrill Miller (1919–)

Robert Rostand Robert Hopkins

Robert Taylor Spangler Arlington Brugh

Robert Weede Robert Wiedefeld

Robinson's Island Niihau, Hawaii

Rob Roy (Gaelic—Red Rob)— Robert Macgregor the Scottish freebooter

Rochedos São Paulo (Portuguese—Saint Paul's Rocks)— in the Atlantic just north of the Equator and far off Brazil

rochelle salts sodium potassium tartrate

Rocher du Diamant (French—Diamond Rock)—off Fort-de-France, Martinique; commissioned in 1800 as HMS *Diamond Rock* because here British sailors withstood a French bombardment lasting more than eighteen months

Rochers du Calvados (French—Calvados Reef)—at the mouth of the Orne in the English Channel

Rochester actor Eddie Anderson

Rock Knute Kenneth Rockne; Mount Desert Island's nickname; Rockaway; Rock of Gibraltar; The Rock Alcatraz (Federal Prison once occupying a 12-acre rock in San Francisco Bay; name now applies to Rikers Island—New York City's Correctional Facility in the East River or to San Quentin on the shores of San Francisco Bay)

The Rock Alcatraz (former prison, now museum); The Rock of Gibraltar (British crown colony on a rocky peninsula extending south from the Spanish mainland); Saba Island, Netherlands Antilles

Rock of Chickamauga General George Henry Thomas

Rock City Nashville, Tennessee

Rocket City Huntsville, Alabama

Rocket Pioneer Robert H Goddard

Rock Hudson Roy Fitzgerald

Rockie Nelson A Rockefeller

Rock Lizards Gibraltarians

Rock Music Metropolis Minneapolis

Rock of Notre Dame Knute K(enneth) Rockne

Rock-ribbed State Massachusetts; Maine

rock salt halite (sodium chloride)

Rock of Uluru Ayers Rock near Mount Olga, Australia

Rockwell Girls Women's Reformatory at Rockwell City, Iowa

Rocky Roccoforte; Rochester; Rockefeller

Rocky Arabia Arabia Petraea in the northwestern section of the Arabian Peninsula

Rocky Butte Portland, Oregon's jail

Rocky Mountain Colorado national park

Rocky Mountain States Alaska, Idaho, Montana, Wyoming, Colorado, Utah, New Mexico, and Arizona

Roger Williams City Providence, Rhode Island

Rogues' Island Rhode Island, so called by Puritans of the Massachusetts Bay Colony

Roi Citoyen (French—Citizen King)—Louis Philippe

Roi Soleil (French—Sun King) —Louis XIV

Roland Franco-German Nord-Bolkow surface-to-air missile whose name honors a medieval hero of song and story in the time of Charlemagne

Rolfe Boldrewood Thomas A Browne's pseudonym

Rolls-Royce of recreational drugs cocaine

Roloff Van Ripper Washing-ton Irving

Roman Century the 1st century before the Christian era

Romani Gypsies

Romanian Ports Mangalia, Constanta, Sulina, Isaccea, Braila, Galati, Tiglina

Romania's Principal Port Constanta

Romano Giulio Pippi de Granuzzi's palette name— Giulio Romano

Romeo letter R radio code; Soviet R-class submarines

Ronald Coleman Boris Cole Blake

Ronnie Ronald; Ronda; Veronica

Ronny Ronald

Roof Garden of Texas Alpine

Rooftop of Africa Kilimanjaro in Tanzania

Rooftop of Antarctica Vinson Massif

Rooftop of Argentina Aconcagua on the border of Chile

Rooftop of Asia Everest in China and Nepal

Rooftop of Australia Kosciusko in New South Wales

Rooftop of Austria Grossglockner

Rooftop of Bolivia Ancohuma

Rooftop of Canada Mt Logan in the Yukon

Rooftop of Chile Ojos del Salado on the border of Argentina

Rooftop of Ecuador Chimborazo

Rooftop of Europe Mont Blanc in France

Rooftop of India Mt Godwin Austen, Jammu and Kashmir

Rooftop of Italy Monte Rosa on the border of Switzerland

Rooftop of Japan Fuji

Rooftop of México Citlaltépetl also called Orizaba

Rooftop of New Zealand Mt Cook on South Island

Rooftop of North America Mt McKinley in Alaska

Rooftop of Peru Huascarán

Rooftop of South America Aconcagua in Argentina

Rooftop of Spain Mulhacén in Granada

Rooftop of Switzerland Matterhorn

Rooftop of Turkey Ararat in Armenia

Rooftop of the USSR Communism Peak, formerly Stalin, formerly Garmo in Soviet Central Asia

Roof of the World Pamir Plateau of central Asia

Roosevelt I Theodore Roosevelt—26th President of the United States

Roosevelt II Franklin D Roosevelt—32nd President of the United States

Roosevelt Island current name for New York City's East River island formerly called Welfare and originally Blackwell's

Rosa Bonheur Rosalie Mazeltov

Rosa and Carmela Ponselle Rosa and Carmela Ponzillo

roscoe slang—handgun, rifle, shotgun)

Rose Bowl Pasadena, California

Rose Capital of the World Tyler, Texas

Rose City Madison, New Jersey; Pasadena, California; Portland, Oregon

Rose-Red City Petra in Jordan across the Wadi al 'Arabah from the Negev of Israel

Rose and Shamrock England and Ireland

Rossbach's disease gastric juice secreted excessively

Ross Macdonald Kenneth Millar's pseudonym

Rosy Rosalind; Rosen; Rosenbaum; Rosenberg; Rosenfeld; Rosenthal

Rothermere Viscount Rothermere (Harold Sidney Harmsworth)

Rough Rider Theodore Roosevelt—26th President of the United States

Rough Riders First United States Volunteer Cavalry organized by colonels Theodore Roosevelt and Leonard Wood for action in the Spanish-American War

Roundheads Cromwell's followers in the Puritan Party noted for the close-cropped hair of its members

Roundup City Pendleton, Oregon

Rowan Oak William Faulkner's home near Oxford, Mississippi in Lafayette County he fictionalized as Yoknapatawpha

Royal Bob President Garfield's name for Colonel Robert G Ingersoll

Royal Brute of Great Britain King George III of Hanover

(in the opinion of Thomas Paine and many other Americans and Britons)

Royal Epicurean Emperor Hadrian

Royal Gorge Grand Canyon of the Arkansas in Colorado

Royal Martyr Charles I of England

Roy Rogers Leonard Slye

Roz Rodriguez; Rosalind(a); Rozhdestvensky

Ruanda formerly Belgian East Africa

Rub al-Khali (Arabic—Empty Quarter)—area between Oman and Yemen

Rubber Capital of the U.S. Akron, Ohio

Rubber City Akron, Ohio

rubber room padded cell

Rube Goldberg Reuben Lucius Goldberg (cartoonist creator of fantastic inventions for accomplishing simple tasks— hence any overcomplicated mechanism is termed a Rube Goldberg)

Ruben Dario Félix Rubén García Sarmiento

Rube Waddell George Edward Waddell

ruby red corundum

ruby copper cuprite (cuprous oxide)

Rudolf Valentino Rodolpho d'Antongnolla

Rudy Vallee Hubert Prior Vallee

Ruissalo Finnish class of motor gunboats and minesweepers sometimes used as patrol launches

Ruler of the East Vladivostok

Rural Garden of Eden South Carolina

Russia Russia leather (dark-red leather used for book binding)

Russian Russian dressing (salad dressing spiced with chili, chopped pickles); Russian roulette (each player pull the trigger of a revolver held against the player's head); Russian wolfhound (large breed of hound originally bred in Russia and called *borzoi*)

Russian America Alaska before its purchase from Russia in 1868 for $7.2-million

Russian-American Capital Sitka, Alaska

Russian Bear symbol of Russia or the USSR

Russian Physiologist Extraor- dinary Ivan Petrovich Pavlov

Russian Ports (*see* Soviet Ports)

Russki(s) Russian(s)

Rust Belt Detroit's automobile manufacturing area

Rust's disease tuberculosis of the upper cervical vertebrae

Rust Territory Trust Territory of the Pacific

Ryukyu Retto (Japanese— Ryukyu Islands)—also known as Loochoo or Nansei Islands

Saar River City Saarbrücken, Germany

Sabah formerly British North Borneo

Sabine River City Orange, Texas

Sabra main battle tank built by Israel Army Ordnance and armed with a 105mm gun; native-born Israeli; Sabrina

Sabre Australian-built Canadian version of the F-86 jet fighter designated CF-86 or CA-27 and originally built by North American as a single-engine jet-fighter

Sabre 32 Australian-built F-86 jet fighter

Sabreliner North American T-39 transport aircraft

Sacha Alexander

Sacha Guitry (pseudonym— Alexandre Pierre Georges)

Sacred Untouchables TS Eliot, Marcel Proust, Rainer Maria Rilke, William Butler Yeats

Saddler Soviet SS-7 liquid-fuel intercontinental ballistic missile

Sadie Sara; Sarah; Sarita

Saeta Hispano HA-200 twin-engine jet trainer built in Egypt and in Spain; also called E-14

Safford Federal Prison Camp at Safford, Arizona

Safir Saab-built training and utility aircraft also known as Saab 91-D

Saga City Stavanger, Norway

Saga Island Iceland

Sagamore Hill Theodore Roosevelt's home on Long Island at Oyster Bay, New York

Sagarmatha (Nepalese—mother of the snows)—Mount Everest

Sage of America Benjamin Franklin

Sage of Anacostia Frederick Douglass

Sage of Ashland Henry Clay

Sage of Auburn Secretary of State William H Seward

Sage of Baltimore H(enry) L(ouis) Mencken

Sagebrush Princess Sarah Winnemucca

Sagebrush State Nevada

Sage of Chappaqua Horace Greeley

Sage of Chelsea Thomas Carlyle

Sage of Concord Ralph Waldo Emerson—American philosopher-poet

Sage of East Aurora Elbert Hubbard

Sage of Ebury Street George Moore

Sage of Emporia William Allen White

Sage of Ferney Voltaire

Sage of Gramercy Park Samuel H Tilden—benefactor of the New York Public Library and governor of New York

Sage-hen(s) Nevadan(s)

Sage of Jena Ernst Haeckel

Sage of Kinderhook Martin Van Buren—eighth President of the United States

Sage of Monticello Thomas Jefferson—editor of the *Declaration of Independence*, founder of the University of Virginia, third President of the United States

Sage of Montpelier James Madison—Father of the *Constitution*, fourth President of the United States

Sage of Mount Vernon George Washington—first President of the United States

Sage of Nininger Ignatius Donnelly

Sage of Philadelphia Benjamin Franklin

Sage of Popayán Francisco José de Caldas

Sage of Princeton Grover Cleveland

Sage of Roanoke John Randolph

Sage of Samos Pythagorus

Sage of Sullivan Street Edgar Varese

Sage of Walden Pond Henry David Thoreau

Sage of Wheatland James Buchanan—15th President of the United States

Sage of Yasnaya Polyana Count Leo Nikolayevich Tolstoy (1828–1910)

Sage of Yoknapatawpha William Faulkner

Sagger Soviet antitank missile

Sahara North Africa's great desert; the Breguet 765 troop transport aircraft

Sahel Sahel Countries (Cape Verde Islands, Chad, Gambia, Mali, Mauritania, Niger, Senegal, Upper Volta)

Sailor City San Diego, California

Sailor Historian Samuel Eliot Morison

Sailor King William IV of England

Sailor on Horseback Jack London

Sailor's Poet Charles Dibden

Sailor's Friend Samuel Plimsoll

Sailor Town Norfolk, Virginia

Saint Anthony's fire gangrenous skin conditions such as ergotism, erysipelas, and hospital gangrene

Saint Augie Saint Augustine, Florida

Saint Barts Saint Barthélemy or Saint Bartholomew in the French West Indies

Saint Bart's Saint Bartholomew's Episcopal Church

Saint Croix Santa Cruz (inhabitants of this Virgin Island called Cruzans)

Saint Didacus San Diego de Alcalá de Henares

Saint Gotthard's disease intestinal hookworms

Saint of the Gutters Nobelprizewinner Mother Teresa of Calcutta

Saint-John Perse Alexis Léger

Saint Kitts Saint Christopher (Leeward Islands, British West Indies)

Saint Martin's evil dipsomania

Saint Paddy Saint Patrick

Saint Petersburg Petrograd, now Leningrad

Saint-Simon Claude-Henri de Rouvroy, Compte de Saint-Simon

Saint Vitus' dance chorea; involuntary muscular twitching

Saki Hector Hugh Munro

Salad Bowl of California Salinas

Saladin Alvis armored car built in Britain and armed with a 76mm gun

Salish Soviet surface-to-surface missile

Sallie Sarah

Sally Sara(h); South Atlantic (baseball) League (nickname)

Sally Ann Salvation Army

Sally Rand Helen Gould Beck

Salmon City Astoria, Oregon

Salonika Thessalonika, Greece

Salop(ian) Shrewsbury; Shropshire

Salt City Syracuse, New York

salt horse pickled meat served to sailors and soldiers

Salt Lake State Utah

Salton Sea formerly Salton Sink

saltpeter potassium nitrate

salt of tartar potassium carbonate

Salvatoriello Salvator Rosa

Saly Salvation Army

Salzburg Philospher Balduin V Schwarz

Samaritan Convair 48-passenger military transport adapted from 240/440 series airliners

Samian Sage Pythagoras, Greek mathematician-philosopher

Samiel The Devil

Sammy American soldier (British slang); Samuel

Samoan Ports Apia (Samoa or Western Samoa), Pago Pago (American Samoa)

Sam Slick Thomas Chandler Haliburton's nickname

Samuel Edwards Noel Bertram Gerson

Samuel Falkland Heijermans Herman

Sam(uel) Goldwyn Samuel Goldfish

San Anto San Antonio, Texas

San Antone San Antonio, Texas

San Berdoo San Bernardino, California

San Carlos de Bariloche Bariloche, Argentina

Sanctimonious City Toronto, Ontario's nickname fifty years ago

Sandal Soviet SS-4 mediumrange ballistic missile

Sandcutters Arizonans

Sanders Alexander

Sand Gropers Western Australians

Sandhill State Arizona

Sandhurst Royal Military Academy at Sandhurst on the Blackwater River in southeast Berkshire, England

Sandra Alessandra

Sandra (Russian—Aleksandra or Alessandra or Alexandra)

Sandro Alessandro

Sandstone Federal Correctional Institution at Sandstone, Minnesota

Sandwich Islands Hawaii; discovered by Capt. James Cook in 1778 and named for his sponsor John Montagu, the fourth Earl of Sandwich

Sandy San Diego, California; Sandra; Sandro; Saundra; a Scotsman

Sandy Kitty Kansas City

sangre y pus (Spanish—blood and pus)—separatist nickname for the vivid-red and golden-yellow flag of Spain

San Juan del Sur Greytown, Nicaragua

San Pedro Sucio San Pedro Sula, Honduras

San Quentin California State Prison at San Quentin

sansei (Japanese—third generation)—grandchild of Japanese immigrants to the United States; (see *issei, kibei, nisei*)

Sansovino Andrea Contucci

Santa Barbara Islands California's Channel Islands including Santa Catalina

Santa Claus originally Sint Nicolaas in Holland

Santa ships Grace Line vessels—all names begin with Santa

Santé nickname of the Parisian prison at 42 rue de la Santé

Santo Domingo City Ciudad Trujillo

San Ysidro, California formerly Tia Juana; name changed to avoid confusion with Tijuana, Mexico and Tia Juana River

São Tomé and Principe's Port São Tomé

Sapper pseudonym of Lt Col Cyril McNeile—creator of Bulldog Drummond

sapphire blue corundum gemstone

Sapwood Soviet SS-6 intercontinental ballistic missile

Saracen British Alvis armored personnel carrier

Sarah Bernhardt Rosine Bernard's stage name

Saratoga Saratoga chip (potato chip); Saratoga trunk (old-fashioned round-top trunk); Saratoga water (often laxative); Saratoga vichy (imbibed in mixed drinks or as a health potion); formerly Schuylerville, New York

Sardine Capital of Norway Stavanger

Sardine Capital of the United States Eastport, Maine

Sardonic Cartoonist Al Capp
Sarong Girl Dorothy Lamour
Sasha (Russian—Alexander or Aleksandr)
Sasin Soviet SS-8 intercontinental ballistic missile
Satanic City Devils Lake, North Dakota
Satchel Leroy Paige
Satchmo   Satchel-Mouth—Louis Armstrong
Satirist of the Mexican Revolution José Clemente Orozco
Satirist-Skeptic Writer Anatole France
Saudi Arabian Ports Jiddah or Juddah, the Red Sea approach to Mecca; Ad Dammam, Ras at Tannurah, Ras at Mishab, and Ras at Khafji on the Persian Gulf
Sauk Centre Minnesota town described as Gopher Prairie by Sinclair Lewis in *Main Street*
Savage Soviet SS-13 three-stage intercontinental ballistic missile
Savannahians natives of Savannah, Georgia
Savior of Babies Nathan Straus
Savior of Church Music Giovanni Palestrina (c.1525–1594)
Savior of England Oliver Cromwell
Savior of the Nations sobriquet earned by the Duke of Wellington at Waterloo
Savior of the Sierras John Muir
Savior of Southern Agriculture George Washington Carver
Savoyards performers in the Savoy Operas of WS Gilbert and Arthur Sullivan
Sawbuck Sears-Roebuck; a ten-dollar bill
Sawdust City Oshkosh, Wisconsin
Sawney a Scotsman
Saxe Holm Helen Hunt Jackson
Saxon Shore English coastline including Norfolk, Suffolk, Essex, Kent, Sussex, and Hampshire
Sax Rohmer Arthur Sarsfield Wade's pseudonym
Saybolt viscosity number
Say Hey Kid Willie Mays
Scandia southern Scandinavian peninsula—southern Norway and Sweden
Scandinavia Denmark, Iceland,

Norway, and Sweden (the Faeroe Islands, Finland, and Greenland are sometimes included)
Scandinavian Fun Capital Copenhagen, Denmark
Scandinavian Shield northernmost Finland, Norway, and Sweden
Scapegoat Soviet medium-range two-stage intercontinental ballistic missile SS-14
Scarface Mafia mobster Al Capone
Scarface Al Alphonse Capone
Scarmouche Tiberio Firoella
Scarp Soviet intercontinental ballistic missile capable of releasing warheads below early-warning radar range and designated SS-9
Scenic Center of the South Chattanooga, Tennessee
Sceptered Isle England; Great Britain
Schlickstadt (German—Mud Town)—German naval nickname for Wilhelmshaven
Schnozzola Jimmy Durante
schoolboy nickname for codeine
School of Europe fifteenth-century Italian states such as Florence, Mantua, Milan, and Venice
Schoolmaster in Politics Woodrow Wilson—twenty-eighth President of the United States
Schoolmaster of the Republic Noah Webster
Scillonian(s) inhabitant(s) of the Isles of Scilly
Scoop Senator Henry Martin (Scoop) Jackson
Scorpion British Alvis tracked reconnaissance vehicle; NATO armored tank running on five roadwheels and mounting an octagonal turret gun; U.S. self-propelled 90mm antitank gun designated M-56
Scotch Scotch blackface (sheep); Scotch broth (barley, mutton, and vegetable soup); Scotch mist (drizzle, fog, and mist); Scotch whisky (distilled from barley); Scotch woodcock (toast garnished with anchovy paste and scrambled eggs)
Scotch Bard Robert Burns
Scotland's Extremitude Dunnet Head the northernmost point of mainland Scotland
Scotland's Principal Port

Glasgow
Scotland Yard London police headquarters
Scotland Yard of Wildlife Crime Forensics Laboratory of the US Fish and Wildlife Service
Scots Ports Leith, Granton, Rosyth Dock Yard, Boness, Grangemouth, Alloa, Burntisland, Kirkaldy, Methil, Dundee, Perth, Arbroath, Montrose, Aberdeen, Peterhead, Fraserburgh, Hopeman, Inverness, Cromarty, Invergordon, Portmahomack, Helmsdale, Wick, Thurso, Scrabster, Stornoway, Oban, Campbeltown, Greenock, Finnart, Rothesay Dock, Glasgow, Ardrossan, Irvine, Troon, Cairnryan
Scott Fredericks Carl Shapiro
Scottish Agnostic David Hume (1711–1776)
Scottish Cradle of U.S. Presidents Scotland, ancestral home of Presidents Hayes and Monroe
Scottish Skeptic David Hume (1711–1776)
Scouce(s) Liverpool (persons)
Scourge of God Attila
Scourge of Princes Pietro Aretino
Scout Westland army helicopter built in Britain
Scrag Soviet SS-10 intercontinental ballistic missile
Scrap Iron baseball catcher Clint Courtney
Scrapple City Allentown, Pennsylvania
scream machine amusement park roller coaster
Scud Soviet mobile tactical surface-to-surface missile
Sculptor of the Colossal Frédéric Auguste Bartholdi (*Liberty Enlightening the World*)
Sculptor of Great American and French Scientists and Statesmen Jean Antoine Houdon
Sculptor-King Sculptor-King Pygmalion of Cyprus
The Sea the Baltic, Bering, Black, Caribbean, Japan, Mediterranean, North, Philippine, South China
Seabees Construction Battalion (USN)
Sea-born City Venice
Seacat Short and Harland short-range surface-to-air missile

used by naval vessels

Sea of Cortés Gulf of California also called the Vermillion Sea (*El Mar Bermejo*)

Sea of Cortez Gulf of California (*Mar de Cortés*)

Sea of Darkness Atlantic Ocean between Cape Verde Islands and west coast of Africa; area often afflicted by dusty Harmattan blowing from the Sahara seaward

Sea Devil Count Felix von Luckner

Sea Dogs originally the nickname of British pirates and privateers but more recently applied to British seamen

Seafarer William Clark Russell's pseudonym

Seafood Center Biloxi, Mississippi

Sea-girt Isle Great Britain

Sea-girt Province Nova Scotia

Sea-green Incorruptible Carlyle's nickname for Robespierre

Seagull Utah's state bird and symbolic nickname sometimes given its citizens—Seagulls; Yugoslav two-place single-engine jet aircraft called Galeb

Seahawk Armstrong-Whitworth carrier-based fighter-bomber aircraft

Sea Islands of Georgia Jekyll, Saint Simmons, Sea Island (all near Brunswick)

Sea Islands of South Carolina Edisto, Folly, Hilton Head, Hunting, Ladies, Murphy, Parris, Port Royal, Saint Helena, Wadamalaw

Sea Killer British short-range surface-to-surface missile

Sea King Sikorsky transport helicopter

Sea Knight Boeing-Vertol helicopter designated CH-46

Sea of Lot Dead Sea

Seamen's Bible Nathaniel Bowditch's *New American Practical Navigator*

Sea of the Plains Dead Sea along the Jordan River Plain of Israel

Seaport City of West Glamorgan Swansea

Seaport on the Prairie Chicago

Sea-Power Philosopher Admiral Alfred Thayer Mahan, USN

Sea Ranger Bell turbine-powered helicopter also known as

Jet Ranger

Sea of Reeds the Red Sea

Seashell Capital Sanibel Island, Florida

Seaside State New Jersey

Sea Stallion Sikorsky heavy-assault helicopter designated CH-53

Sea of Stars sparking headwaters of the Huango or Yellow River of China rising in Tibet

sea story teller (*see* Story Teller of the Sea)

Sea of Straw Tagus River estuary

Sea Venom DeHavilland carrier-based fighter-jet aircraft

Sea Vixen DeHavilland carrier-based jet-fighter aircraft

sea water 96.4% water plus 2.8% sodium chloride (common salt) and smaller quantities of magnesium chloride, magnesium sulfate, calcium sulfate, and potassium chloride; in inland seas such as the Dead Sea and the Salton Sea these percentages vary

Sebastian Melmoth name assumed by Oscar Wilde after he was released from Reading Gaol and lived in Paris until his death three years later

seccy seconal (secobarbital sedative, also nicknamed seggy)

Secession City Charleston, South Carolina

Second City Chicago

Second Estate The Nobility

second-generation money checks; cheques

Second Reich German Republic (1919–1933)

Second Republic France under the presidency of Louis Napoleon from 1848 to 1852

Second State Pennsylvania (*see* First State)

Second World highly industrialized nations of the West such as Belgium, France, Germany, Italy, the Netherlands, the United Kingdom

Section 8 military discharge due to insanity or intoxication; government-subsidized housing; mental case (military code)

Securité France's security service in Paris, also serves the National Central Bureau of Interpol

Seekers truth-seeking Quakers

Seeley Regester Metta Victoria

Fuller Victor

Seldom Ever Caught Running nickname of the Southeastern and Chatham Railway—SE & CR

Selebes (Dutch—Celebes)—Sulawesi

Selma Shulamith

Selma Lageriofland Sweden's province of Värmland where the Nobel prize-winning authoress was born

Seminole Beech U-8 light transport aircraft

Senator Sam former U.S. Senator Sam Ervin, Jr, of North Carolina

Senegal's Ports St Louis, Dakar, Rufisque; Karabane

Senior Service the British Navy

Sentimental Rebel Clarence Darrow

Sentinel of the Bolshevik Counter-Revolution Cheka

Separationist State Rhode Island

Sepia City New York City's Harlem

Sergeant Sperry MGM-29A surface-to-surface missile

serpentine hydrous magnesium silicate

Serpentine Suicide Harriet Shelley—first wife of the poet—drowned herself in the Serpentine of London's Hyde Park

Servant of the Nation Secretary of the Treasury Albert Gallatin who financed the Louisiana Purchase and found funds for the War of 1812

Servetus Michael Servetus born of Spanish parents and baptized Miguel Serveto—burned by order of Calvin in 1553

Seryozha (Russian nickname—Sergei)—Serge

Set Svenholm Karl Viktor Svanholm

Seven (telephone) Sisters Ameritech, Bell Atlantic, Bell South, NYNEX, Pacific Telesis, Southwestern Bell, US West

Seven Deadly Sins Anger, Covetousness, Envy, Gluttony, Lust, Pride, Sloth

Seven-Hill Cities Lisbon, Prague, Rome, and Valparaiso

Seven Hills of Rome Aventine, Caelian, Capitoline, Esquiline, Palatine, Quirinal or Colline, Viminal

Seven Provinces (*see* United Provinces)

Seven Sages of Greece Bias, Chilon, Cleobulus, Periander, Pittacus, Solon, Thales

Seven Seas Antarctic, Arctic, Indian, North Atlantic, South Atlantic, North Pacific, South Pacific oceans; term also applied to the Andaman, Baltic, Bering, Caribbean, Mediterranean, South China, and Yellow seas

Seven Sisters Barnard, Bryn Mawr, Mount Holyoke, Radcliffe, Smith, Vassar, and Wellesley—all colleges for women when first organized; BP (British Petroleum), Exxon (Esso—Standard Oil), Gulf, Mobil, Shell, SOCAL (Standard Oil of California —Chevron), Texaco—world's leading oil companies

Seventeenth State Ohio

Seventh Continent Antarctica

Seventh State Maryland (*see* First State)

Seven Wonders of the Ancient World Pyramids of Egypt, Lighthouse of Pharos of Alexandria, Hanging Gardens and Walls of Babylon, Temple of Artemis or Diana at Ephesus, Statue of Zeus by Phidias at Olympia, Mausoleum at Halicarnassus, Colossus of Rhodes

Seven Wonders of the Modern World Fort Peck Dam across the Missouri in Montana; Pecos, Texas oilwell; Royal Gorge Bridge, Colorado; Simplon Tunnel between Italy and Switzerland; TV Tower at Blanchard, North Dakota; Verrazano-Narrows Bridge over New York Harbor; World Trade Center in downtown New York—the biggest dam, the deepest well, the highest bridge, the longest tunnel, the tallest structure, the longest single-span bridge, the tallest buildings

Sewanee University of the South in Sewanee, Tennessee

Seward's Folly nickname given Alaska in 1867 when Secretary of State William H Seward purchased the area from Russia for $7,200,000; also called Seward's Polar Bear Garden

Sex Francisco nickname of San Francisco

Sex Isle Mykonos in the Greek Islands close to Piraeus, the port of Athens

Sex Queen of Stage and Screen Mae West

Sextan a Soviet class of trawlers

Sexyola Sixaola, Costa Rica

Sexy Rexy Rex Harrison

Seybrew Seychelles Islands brew(ery)

Seychelles Port Victoria

Shackleton Hawker-Siddeley maritime reconnaissance aircraft

Shaddock Soviet surface-to-surface missile

Shafir Israeli air-to-air missile resembling the U.S. Sidewinder

Shahada Flag Saudi Arabian green standard bearing white lettering—the Moslem shahada: "There is no god but God—and Mohammed is his prophet."

Shakopee Minnesota Correctional Institution for Women at Shakopee

Shalimar Garden of Love on Dal Lake in Kashmir

Shalom Aleichem Solomon Rabinowitz' pseudonym; based on the Hebrew greeting—*shalom alekhem*—peace be with you

Shami Shamrock; Shulamith

Shamo (Chinese—Sandy Waste)—the Gobi Desert

Shanghai mainland-China-built torpedo boat; principal port of the People's Republic of China

Shank End Cape Peninsula below Cape Town, South Africa

Shark Island Garden Key, Dry Tortugas (Fort Jefferson National Monument reached by boat from Key West)

Shaston Shaftesbury, England

Sheep Islands Faeroe Islands

Sheila Cecilia

Shelly Winters Shirley Schrift

Sheltie Shetland sheepdog

Shelty Shetland pony

Shepherd of the Ocean Sir Walter Raleigh

Sheridan M-551 assault vehicle armed with a 152mm gun; Sheridan House

Sheridan Girls Wyoming (delinquent) Girls School at Sheridan

Sherlock police computer recording and releasing essential information about many criminal activities; nickname of any good detective; investigator created by novelist Sir Arthur Conan Doyle—Sherlock Holmes; Sherlockian(s)

Sherman M-4 tank armed with high-velocity 76mm gun

Shershen Soviet class of motor torpedo boats (PT boats)

Shetland Shetland pony (small but stocky long-hair pony); Shetland sheepdog (miniature collie); Shetland wool

Shevvie(s) native(s) of Sheffield, Yorkshire

Shillelagh anti-tank surface-to-surface guided missile produced by Aeronutronic

Shining Star of the Caribbean Puerto Rico

Ship of the Desert the camel

Shirley MacLaine Shirley Beatty

Shitport Norfolk, Virginia

Shoe City Auburn, Maine; Hanover, Pennsylvania; Lynn, Massachusetts

Shooting Star Lockheed T-33 jet-fighter trainer aircraft

shorlans armored cars

Show-Me State Missouri

Showplace of the Orient Singapore

Shrike Texas Instrument air-to-surface antiradar missile

Siam former name of Thailand

Siamese Siamese cat (fawn or pale-gray breed of short-hair cat); Siamese fighting fish (*Betta*); siamese twins (congenitally connected twins)

Siberian Express nickname for a devastatingly cold polar wind afflicting Canada and much of the United States

Siberian salt mines nickname for forced-labor camps and prisons in the USSR

Siberian Seven Soviet family given refuge in U.S. embassy in Moscow from 1978–83

Sick Man of Africa nineteenth-century Ethiopia

Sick Man of the Americas fever-infested nineteenth-century Panama

Sick Man of Asia nineteenth-century China

Sick Man of Europe Turkey in the last years of the Ottoman Empire and the reign of the sultans during most of the nineteenth century and up to 1922 when the sultanate was abolished

Sick Man of South America nineteenth-century Ecuador

Siddhartha Gautama Buddha

Sidewinder air-to-air missile produced by Motorola, Philco, and Raytheon

Sierra Leone's Ports Freetown, Pepel, Bonthe

Siete Leguas (Spanish—Seven Leagues)—famous warhorse of Pancho Villa, the Mexican bandit general

Signe Hasso Signe Larsson

Silence Dogwood Benjamin Franklin's pseudonym used by him at age 15 when he wrote articles for the *New England Courant*

Silent The Silent (William I—Prince of Orange)

Silent Cal taciturn President Calvin Coolidge

silent killer high-blood pressure

silent service submarine service

Silicon Beach San Diego, California

Silicon Glen Scotland's central lowland towns specializing in microelectronics

Silicon Prairie Dallas, Texas high-technology area

Silicon Valley high-technology-oriented area around San Jose, California in Santa Clara County; nickname for Santa Clara County, California

Silk City Paterson, New Jersey, and Soochow, China

Silk Country China

Silly Billy nickname of William IV

Silver Age era when adornments and implements were made of silver; period between the Golden Age and the Bronze Age

Silver City Broken Hills, New South Wales, Australia; Taxco, México

Silver City by the Sea Aberdeen, Scotland

Silver Gate entrance to San Diego Bay, California

Silverines Coloradans

Silver Republic Argentina

Silversmith Patriot Paul Revere (also bellfounder and dentist)

Silver State official nickname of Nevada but also applied to silver-rich Colorado

Silver State of Malaysia Perak

Silver Streak the English Channel

Silver-Tongued Orator William Jennings Bryant

Simmond's disease premature senility caused by atrophy of the pituitary

Simone Pauline Benda's pseudonym

Simone Signoret Simone Kaminker

Sin Angeles (Spanish—without angels)—nickname given Los Angeles

Sin Capital Singapore's nickname

Sin City Las Vegas, Nevada

Sin City of the West Geneva, Switzerland

Singapore's Ports Serangoon, Singapore, Pulau Bukum, Pulau Sebarok

Singer of Singers Enrico Caruso

Singing Cowboy Gene Autry

Singing Nun Sister Luc-Gabrielle (Jeanine Deckers)

Singing Satellite Red China's first satellite, launched in 1970, broadcast rhymed song about Communist Party chairman Mao Tsetung

Singing Tree casuarina (sometimes called sighing tree or sobbing tree)

Sing Sing nickname of the New York State Penitentiary at Ossining formerly named Sing Sing

Sink of New England Rhode Island, uncomplimentary nickname given by Massachusetts Puritans appalled by the separation of church and state guaranteed in Rhode Island's royal charter

Sin Sin Singapore

Sin Strip San Francisco's North Beach area

sin taxes taxes on alcohol, gasoline, and tobacco

Sinyavsky Abram Tertz

Sioux Bell Model-47 helicopter built in Britain, Italy, Japan, and the United States

Sioux State North Dakota

Sir Adrian Sir Adrian Boult (conductor)

Sir Alec Sir Alec Guinness (actor)

Sir Alexander Sir Alexander Korda (motion-picture producer)

Sir Alfred Sir Alfred Hitchcock (author and film director)

Sir Arnold Sir Arnold Bax (composer)

Sir Arthur Sir Arthur Bliss (composer-conductor); Sir Arthur Conan Doyle (writer and physician); Sir Arthur S(eymour) Sullivan (com-

poser-conductor-organist)

Sir Aurel Sir Aurel Stein (archeological explorer)

Sir Benjamin Sir Benjamin Britten

Sir Bernard Sir Bernard Haitink (Dutch conductor)

Sir Charles Sir Charles Chaplin (better known as Charlie Chaplin); Sir Charles Groves (conductor); Sir Charles Hallé (conductor-pianist and founder of Manchester's Hallé Orchestra); Sir Charles Mackerras (conductor); Sir Charles Villiers Stanford (composer-conductor-organist); Sir Charles Wheatstone (physicist)

Sir Charles Morell James Ridley's pseudonym

Sir Clifford Sir Clifford Curzon (pianist)

Sir Colin Sir Colin Davis

Sir Dan Supreme Sir Dan Godfrey

Sir David Admiral Sir David Beatty—First Earl of the North Sea

Sir Edward composer-conductors Sir Edward Elgar and Sir Edward German (originally Edward German Jones)

Sirens three nymphs named Leucosia, Ligeia, and Parthenope; their seductive singing lured sailors to their death on rockbound coasts, when they failed to lure Odysseus (Ulysses) they flung themselves into the waves and perished

Sir Ernest Sir Ernest Campbell Macmillan (composer-conductor-educator-organist)

Sir Francis Sir Francis Bacon (philosopher politician); Sir Francis Drake (admiral-explorer-navigator); Sir Francis Palgrave (historian)

Sir Frank Sir Frank Athelstane Swettenham, lexicographer of Malaya and its one-time colonial administrator

Sir Freddie Sir Freddie Laker (airline organizer and operator)

Sir Frederic Sir Frederic Cowen (composer-conductor)

Sir Georg Sir Georg Solti (conductor-pianist knighted for his services at Covent Garden where he conducted from 1961 to 1971)

Sir George Sir (Isador) George Henschel (baritone-composer-conductor, founder of the Scottish Symphony Orchestra)

Sir Granville Sir Granville Bantock (composer-conductor-teacher)

Sir Guatteral (Hobson-Jobson —Sir Walter Raleigh)—as known to many Spaniards in colonial times

Sir Hamilton Sir Hamilton Harty (conductor)

Sir Henry Sir Henry Bessemer (engineer-inventor-metallurgist remembered for bessemer steel); Sir Henry Wood (conductor)

Sir Hubert Sir Hubert Parry (composer and musicologist)

Sir Isaac Sir Isaac Newton (mathematician and natural philosopher)

Sir John Sir John Barbirolli (conductor-violincellist); Sir John Betjeman (poet laureate); Sir John Gielgud (actor)

Sir John A Sir John Alexander Macdonald (Canada's first and third Prime Minister)

Sir John Mandeville Jehan de Bourgogne

Sir John Retcliffe pseudonym of author Hermann Goedsche

Sir Kenneth Sir Kenneth McKenzie Clark

Sir Landon Sir Landon Ronald (composer-conductor-pianist)

Sir Laurence Sir Laurence Olivier (actor-director-producer)

Sir Lennox Sir Lennox Berkeley (composer)

Sir Malcolm Sir Malcolm Sargent (ballet-choral-orchestral conductor who was chief conductor of the BBC and the Promenade Concerts)

Sir Max Sir Max Beloff

Sir Michael Sir Michael Costa (composer-conductor); Sir Michael Tippett (composer-conductor-educator)

Sir Pelham P(elham) G(renville) Wodehouse

Sir Ralph Sir Ralph Richardson (actor)

Sir Thomas Sir Thomas Beecham (composer-conductor-founder of the London Philharmonic Orchestra and the Royal Philharmonic Orchestra)

Sir Victor Sir V S (Victor Sawdon) Pritchett (journalist and short-story writer)

Sir Wilfred Sir Wilfred Pelletier (conductor)

Sir William Sir William Herschel (astronomer-mathematician-musician); Sir William Walton (composer)

Sir Winston Sir Winston Leonard Spencer Churchill (former Prime Minister of Great Britain)

Sissy Cecilia; sister

Sister Cities San Diego and Yokohama

Sitting Bull Tatanka Iyotanka also known as Sitting Buffalo Bill

Six Counties Northern Ireland or Ulster's counties of Antrim, Armagh, Derry, Down, Fermanagh, and Tyrone

Six Nations Five Nations plus the Tuscaroras (*see* Five Nations)

Six-Shooter Junction old name of Harlingen, Texas

Sixteenth State Tennessee

Sixth Continent Australia

Sixth State Massachusetts (*see* First State)

Skate City Northbrook, Illinois

Skean Soviet SS-5 intermediate-range ballistic missile

Skeptic-Philosopher President Thomas Jefferson

Ski Capital Aspen, Colorado

Ski Country Colorado

Skidrow on the Sound Seattle, Washington

Skipper the Captain; the Commander

Skory Soviet class of minelaying destroyers

Skowhegan Girls Women's Correctional Center at Skowhegan, Maine

Skowse Liverpool seaman

skunk nickname of a potent variety of marijuana

Skunk's Misery former name of Scranton, Pennsylvania

Skunk Works nickname of Air Force Plant 42 at Palmdale, California

Skybright Axe Paul Bunyan

Sky City Pueblo Acoma near Alburquerque, New Mexico

Sky Crane Sikorsky crane helicopter designated Ch-54 or S-64

Skymaster Douglas DC-4 44-passenger transport also called Dakota

Skyscraper Capital New York City

Skyscraper Port of the Orient Hong Kong

Skyservant Dornier utility aircraft also designated DO-27 and DO-28; both built in Germany

Skytrain Douglas DC-3 21-passenger transport also called Dakota

Skyvan turboprop transport built by Short in Great Britain

Skywagon Cessna 185E utility aircraft

Slabsides rustic cabin built by John Burroughs near Esopus, New York

slaked lime calcium hydroxide $(Ca[OH]_2)$

Slava Mstislav; Mstislav Rostropovich's nickname

Slave Capital of the South Charleston, South Carolina before the Civil War

Slave Coast West African coastal area of Togo, Dahomey, and Nigeria

Slave States former slave-holding states (Virginia, North and South Carolina, Georgia, Florida, Alabama, Mississippi, Louisiana, Texas, Arkansas, Tennessee, Delaware, Maryland, Kentucky, Missouri

Sledge and Hoe official symbol of Zaire

Sleep Personified Hypnos (Greek—sleep) whose brother was Thanatos or death

Sleepy Hollow New Jersey's Trenton Prison

Sleepy Joe Attorney General Philander C Knox

Slim Jannie Jan Christian Smuts

Slinging Sammy Sam(uel) (Adrian) Baugh of baseball and football fame

slooow seller(s) slow-selling book(s)

Slovakian Capital Bratislava called Pozsony by the Czechs and Pressburg by the Germans

Slugger U.S.-made tank destroyer designated M-36

Slumbering Giant of Capitol Hill The Library of Congress

Slumberjay Schlumberger

Slut of the North Empress Elizabeth of Russia so nicknamed by Frederick the Great of Prussia who called her *la Catin du Nord*

Sly Fox of Kinderhook Martin Van Buren

smack nickname for heroin

Smack Henderson Fletcher Henderson

Small Islands *see* Micronesia

Smelly Place Hong Kong (also called Fragrant Harbor)

Smelter City Anaconda, Montana

Smiling Jim James A Farley

Smithy Ian Smith

Smog Capital of California Los Angeles

Smokeless City Reykjavik, Iceland—heated by natural hot springs

Smokeless Coal Capital Beckley, West Virginia

Smoke that Thunders Victoria Falls (Zambia)

smokies smoked haddocks

Smoking Moses Shishaldin Volcano on South Umiak Island off southwestern Alaska

Smoky City Pittsburgh, Pennsylvania before its Renaissance Plan

Smörgåsbordland Sweden (famous for its cold-table fare)

Snapper Soviet antitank missile

Snow City Aspen, Colorado

Snow-covered Continent Antarctica

Snow King Gustavus Adolphus of Sweden

Snow Queen Christina—Queen of Sweden

snubbies snub-nosed handguns

Soap Box Derby Center Akron, Ohio

soapstone saponite (hydrous magnesium aluminum silicate)

Soapy G Mennen Williams

Soapy Sam Samuel Wilberforce, Bishop of Winchester, who debated Darwin's theory of evolution with Professor Thomas Henry Huxley, president of the Royal Society

Socialist Pope Daniel De Leon

Society Capital Newport, Rhode Island

Society of Friends the Quakers

soda ash sodium carbonate (Na₂CO₃)

Sodaks South Dakotans

Sodoma (Italian—Sodomite)— Giovanni Antonio Bazzi (1477–1549)

Sofia Loren Sofia Scicolone

Sofu-gan Japanese equivalent of Lot's Wife—volcanic islet resembling a pillar of salt in the North Pacific between Iwo Jima and Yokohama

Soho London's avant-garde district in the West End just north of Trafalgar Square

Sojourner Truth Isabella Baumfree

Solar Energy Capital Los An-

geles

Solar Energy State Arizona

Sol de Mayo (Spanish—Sun of May)—symbol of independence appearing on the flags and seals of Argentina and Uruguay

solder 50% lead, 50% tin (common solder)

soldier's heart Da Costa's syndrome

Soledad Correctional Training Facility of the State of California at Soledad

Solid City St Louis, Missouri

Solid South Southern United States usually voting as a solid bloc: Alabama, Florida, Georgia, Louisiana, Mississippi, South Carolina

Solina South Carolina

Solomon British pianist

Solomon Islands Port Honiara on Guadalcanal Island

Solomon seal six-pointed star consisting of two interlocking triangles; sometimes called the shield of David

Solon of French Prose Jean Louis Guez de Balzac

Somalian Port Berbera

Somers' Islands Bermuda

Somnolent City of the Sahara Timbuktu

Song of Songs The Song of Solomon

Son of Nature Henry David Thoreau

Son of the Ocean Yangtse River

Son of the Star Bar Kochba— military leader of the Jews who revolted against the Romans in the year 132 A.D.

Son of Valladolid José Zorilla

Sonya Sophia

Sonya (Russian nickname— Sophia)

Soo Sault Ste Marie (canal and locks)

Soo Bridge Sault Ste Marie International Bridge

Soo Canals Sault Ste Marie Canals

Soo Line Minneapolis, St Paul & Sault Ste Marie (railroad)

Sooner State Oklahoma

Sopac Southern Pacific Railroad (stock exchange nickname)

Sophia Loren Sofia Scicolone

Sophie Tucker Sophia Abuza

Sorbonne University of Paris

Sorghum Capital of the World Hawesville, Kentucky

Soul of American Law Cla-

rence Darrow

Soul City Harlem district of New York City

Southern Hemisphere the world south of the equator

Southern Ireland Republic of Ireland

southern lights aurora australis

Southern Ocean Antarctic sections of the Atlantic, Indian, and Pacific oceans

Southern Part of Heaven Chapel Hill, North Carolina

Southern Poet Sidney Lanier (*The Marshes of Glynn, The Song of the Chattahoochee, Sunrise*)

Southern Rhodesia Zimbabwe

Southern States Virginia, North and South Carolina, Georgia, Florida, Alabama, Mississippi, Tennessee, Arkansas, Louisiana, and Texas

South Jersey Coast Atlantic City to Cape May

South Pacific between Australia and South America containing South Sea Islands; South Pacific Ocean

South Pole 90 degrees South latitude; zero degrees longitude; southernmost point on the earth; discovered by Norwegian explorer Roald Amundsen in 1911

South River old name for the Delaware River

South Seas South Pacific Ocean

South Seymour Island, Galápagos Baltra

Southwest southern California and Nevada, Arizona, New Mexico, and western Texas

South-West Africa formerly German South-West Africa; Namibia

Southwest Sun Belt Arizona, California, Hawaii, Nevada, New Mexico, Ohlahoma, and Texas

South Yugoslavia formerly the kingdom of Montenegro

Soviet Central Asia Kazakh, Kirghiz, Tadzhik, Turkmen, and Uzbek Soviet Socialist Republics

Soviet Film Pioneer Sergei Eisenstein

Soviet Ports Vladivostok, Nakhodka, Sovetskaya Gavan, De-Kastrt, Nikolayevsk, Komsomolsk, Khabarovsk, Korsakov, Kholmsk, Aleksandrovsk Sakhskiy, Moskal Vo,

Magayevo, Petropavlovsk-Kamchats, Ust-Kamchatsk, Provideniya, Tiksi, Dudinka, Igarka, Mezen, Ekonomiya, Solombala, Arkhangelsk, Severodvinsk, Belomorsk, Pabocheostrovsk, Gavan Blagopoluchiya, Keret, Kovda, Guba Knyazhaya, Kandalaksha, Bolshaya Piryu, Gremikha, Vayenga, Murmansk, Kola, Vyborg, Vysotsk, Kivitokeye, Klyuchevoye, Kurkela, Leningrad, Kronshtadt, Narva Joesuv, Tallinn, Parnu, Riga, Ventspils, Liepaja, Klaipeda, Baltiysk, Kalingrad, Ilichevsk, Odessa, Nikolayev, Kherson, Bukhta Severnaya, Feodosiya, Kerch, Berdyansk, Zhdanov, Rostov, Novorossiysk, Tuapse, Poti, Batumiyskava Bukhta

Soviet Union formerly the Imperial Russian Empire

Soybean King Dwayne Orville Andreas

Soyuz Soyuz-class 32,000-ton nuclear-powered Soviet cruiser

Space Capital of the United States Colorado Springs

Space City Houston, Texas (NASA headquarters)

Space Coast Florida's Cape Canaveral area

Space Scientist Premier Krafft A Ehricke (1917–1984)

Spain's Forgotten Forest Extremadura

Spain's Largest Port Barcelona

Spaniola(s) British slang for Spaniard(s)

Spanish Spanish bayonet (*Yucca*); Spanish cedar (fragrant neotropical wood); Spanish dagger (*Yucca gloriosa*); Spanish fly (cantharides); Spanish grippe (influenza); Spanish heel (woman's high heel); Spanish influenza (highly infectious respiratory viral disease); Spanish lime (genip); Spanish mackeral (jack mackeral); Spanish moss (epiphytic plant growing in long festoons on the branches of live oak trees in the southern United States); Spanish omelet (made with green peppers, tomatoes, and seasoning); Spanish rice (made with cayenne pepper, chopped onions, and tomatoes); Spanish topaz (citrine);

Spanish trefoil (alfalfa)

Spanish Africa cities of Ceuta and Melilla; term formerly included Spanish Guinea, Spanish Morocco, and the Spanish Sahara

Spanish America Spanish-speaking countries of Latin America

Spanish Artist and Sculptor Pablo Picasso

Spanish Etcher-Lithographer-Painter Francisco José de Goya y Lucientes

Spanish Film Pioneer Luís Buñuel

Spanish Founder of the Secular School System Francisco Ferrer

Spanish Guinea former West African colony on the Gulf of Guinea; included Fernando Po, Rio Muni, and offshore islets

Spanish Honduras Republic of Honduras in Central America

Spanish Imperessionist Joaquín Sorolla y Bastida

Spanish Lithographer Francisco José de Goya y Lucientes

Spanish Main Spanish-speaking Central America and northern South America bordering the Caribbean from Mexico to Venezuela, including Guatemala, Honduras, Nicaragua, Costa Rica, Panama, and Colombia

Spanish Monastic Painter Francisco de Zurbarán

Spanish Morocco formerly all of coastal and northwestern Morocco but now only Alhucemas, Ceuta, the Chafarinas islands, Melilla, and Peñon de Vélez

Spanish National Composer Manuel de Falla

Spanish Naturalist Painter Diego Rodriguez de Silva y Velázquez

Spanish Netherlands all the Lowland Countries (Belgium, Luxembourg, and the Netherlands) when they were under Spanish rule

Spanish Ports Pasajes, San Sebastian, Zumaya, Santurce, Portugalete—Bilbao, Las Arenas—Bilbao, El Desierto—Bilbao, Castro Urdiales, Santander, Gijón, Musel, Aviles, San Esteban, El Ferrol del Caudillo, La Coruña, Villegarcia, Pontevedra, Marín,

Vigo, Santa Cruz de Tenerife and La Luz Gran Canaria (on the Canary Islands), Huelva, Bonanza, Coria del Río, Sevilla, Rota, Cádiz, Algeciras, Málaga, Motril, Adra, Almería, Cartagena, Alicante, Valencia, Castellon de la Plana, Tarragona, Barcelona, Palamós, (*and on the Balearic Islands*—Ibiza, Palma, Mahon)

Spanish Presidios Ceuta and Melilla on the Alboran coast

Spanish Riviera Spain's Mediterranean resorts

Spanish Sahara former colony on Africa's northwest coast where it included Río de Oro and Saguia el Hamra until 1976 when it was ceded by Spain and divided between Mauritania and Morocco

Spanish-speaking Places Andorra, Argentina, Balearic Islands, Bolivia, Canary Islands, Ceuta and Melilla, Chile, Colombia, Costa Rica, Cuba, Dominican Republic, Ecuador, El Salvador, Equatorial Guinea, Guam, Guatemala, Honduras, Mexico, Morocco, Nicaragua, Panama, Paraguay, Peru, Philippines, Puerto Rico, Spain, Spanish Sahara, United States, Uruguay, Venezuela, etc.

Spanish State New Mexico

Spanish Town Jamaican resort; Tampa, Florida

Spanish West Africa Sidi Ifni

Sparrow McDonnell-Douglas air-to-air missile

Speakers' Corner northeast corner of London's Hyde Park

Specie-Payment President James Abram Garfield

speed nickname for methamphetamine drugs

Spencer Spiridione

Sphinx of Concord Ralph Waldo Emerson

Spica Swedish-built patrol boat carrying a 57mm gun and six torpedo tubes

Spice Island Grenada (noted for nutmeg, cloves, and mace)

Spice Islands Indonesia's Moluccas; West Indian islands of Grenada and the Windwards

Spider of Florence Machiavelli

Spike Jones Lindley Armstrong

Spinach Capital of the World Crystal City, Texas (replete with a statue of Popeye)

Spindle City Lowell, Massa-

chusetts
Spindrift Ernest Toone
Spine of South Africa Drakensberg Mountains
Spirit of Freedom State Massachusetts
spirits of hartshorn ammonia water (NH_4OH)
spirits of salts hydrochloric acid
Spiritiual Father of the French Revolution Rousseau
Splendid Sprinter Ted Williams
Spoils-System President Andrew Jackson
Spokesman for the Negro Booker T Washington
Spokesman for the Oppressed George Meany
Sponge City Tarpon Springs, Florida
Spoon River Edgar Lee Master's poetic appelation for Lewistown, Illinois
Spoon River Poet Edgar Lee Masters
Sport of Kings (horseracing—a sport only kings can afford)
Sportsman's Paradise Louisiana
Sports Town, U.S.A. San Diego, California
Spree River City Berlin
Spruce Goose nickname of Howard Hughe's Hercules H-4 319-foot-wingspan flying boat powered by 8 wingmounted engines
Spud Island Prince Edward Island
Square Deal nickname for economic and political philosophy of Theodore Roosevelt
Square Mile London's financial center
Square Mile of Vice London's Soho and East End
Squaresville area inhabited mainly by citizens who frown on all types of criminal activity and cooperate with the police
Squatter State Kansas
Squire of Hyde Park Franklin D Roosevelt
Squire of Monticello Thomas Jefferson
Squire of Warm Springs Franklin D Roosevelt
Sri Lankan Ports Colombo, Galle, Trincomalee
Sri Lanka's Principal Port Colombo
S S Van Dine Willard Huntington Wright (1888–1939)

Stagecoach Town Fort Worth, Texas
Stagirite Aristotle the Stagirite —so named as he was born in Stagira, Macedonia)
Stagville nickname for West Hollywood, California's Santa Monica Boulevard
Stalin (Russian—steel)—Iosif Vissarionovich Dzhugashvili
Stalingrad former name of Tsaritsyn now called Volgograd
Standard Arm General Dynamics anti-radar missile
Standard Oil King John D(avison) Rockefeller, Sr
Stanislavski Konstantin Sergeevich Alekseev
Stan Laurel Arthur Stanley Jefferson
Stanley Sir Henry Morton Stanley whose original name was John Rowlands
Stanleyville former name for Kisangani, Zaire
Stan the Man Stan Musial
Stanton Forbes pseudonym of De Loris Stanton Forbes
Star City Lafayette, Indiana
Star City of the South Roanoke, Virginia
Star and Crescent Moslem symbol appearing on arms and flags of Algeria, Libya, Malaysia, Mauritania, Pakistan, Singapore, Tunisia, Turkey
Star of David Judaic symbol consisting of two superimposed equilateral triangles forming a six-pointed star; device also called the Seal of Solomon or the Shield of David
Star of the East Vladivostok
Starfighter Lockheed single-engine jet fighter aircraft built in Belgium, Canada, Germany, Italy, and the Netherlands
Star of the Indian Ocean Mauritius
Starlifter C-141 Lockheed cargo and troop transport
Star of the North King Gustavus Adolphus of Sweden; Minnesota
Stars and Bars flag of the Confederate States of America
Star Spangled Banner anthem of the United States of America and nickname of its flag
Stars and Stripes flag of the United States of America

State of Excitement Western Australia
States the United States of America
State of the Thousand Islands Maldive Islands
Statia Sint Eustatius (Netherlands Antilles)
St Dymphna's disease insanity
Steak Center Kansas City, Missouri
stealth bomber B-2 advanced technology bomber
Steel Baron Andrew Carnegie
Steel Center of the South Birmingham, Alabama
Steel City Bethlehem, Pennsylvania; Pittsburgh, Pennsylvania
Steelmaker Joe Magarac
Steel-Master **Philanthropist** Andrew Carnegie
Steel State Pennsylvania
Steiny Charles Proteus Steinmetz
Stella Maris (Latin—Seaman's Star)—*see* Polaris
Stendhal (pseudonym—Marie-Henri Beyle)
step sister of religion superstition
sterling silver 92% silver, 8% copper
Stewart Granger James Stewart
stick baton's nickname
Stinkstein (German—stinkstone)—coal-black limestone or marble giving off a fetid odor when rubbed because of its bituminous or carbonaceous inclusions; also called anthraconite
Stirville Sing Sing prison (Ossining, New York)
St-Jean-des-Puces (French—St John of the Fleas)—nickname of the Franco-Spanish border town of St-Jean-de-Luz (St John of Light or St John of the Marshes)
St-John Perse Alexis Saint-Leger's pen name
St John's evil epilepsy; old nickname for epilepsy
St Martin's evil dipsomania
St Mathurin's disease epilepsy
Stoky Leopold Stokowski (conductor-impresario-transcriber—originally named Antoni Stanislaw Boleslawowics —later adopted name of Leo Stokes but since early 1900s appeared as Leopold Stokowski

Stone Age prehistoric era when implements and weapons were made from stone—age divided into paleolithic, mesolithic, and neolithic periods

Stone of Heaven jade

Stonewall Jackson General Thomas Jonathan Jackson of the Confederate Army

Storm Norwegian high-speed motor gunboat

Stormalong Arthur Bulltop

Storm King American meteorologist James Pollard Espy

Story Teller of the Sea sobriquet shared by Conrad, Cooper, de Hartog, Forester, Innes, London, McFee, Marryat, Masefield, Melville, Nordhoff and Hall, Verne

St Paddy Saint Patrick

St Paddy's Day Saint Patrick's Day (March 17)

Strabolgi Joseph Montague Kenworthy

Strangler wrestler Ed (Strangler) Lewis originally named Robert H. Friedrich

Stratofreighter Boeing 707/720 transports adapted for military service and designated C-135

Stratotanker Boeing jet tanker designated KC-135

Strawberry Capital Hammond, Louisiana

Stream of Pleasure Thames River above London

Street Haven Toronto, Ontario's center for the rehabilitation of prostitutes

Street of Ink Fleet Street, London with its many newspaper offices

Street of Sorrows New York City's Wall Street; old-fashioned nickname for any thorofare frequented by streetwalkers

Strikemaster British BAC 167 ground-attack jet aircraft

Strix Peter Fleming

Stub Toe State Montana

Student of Democracy British Ambassador James Bryce—author of *The American Commonwealth*

Studioland Hollywood, California

Studs Lonigan trilogy comprising *Young Lonigan, Young Manhood of Studs Lonigan, Judgment Day*—James T Farrell's literary portrait of life among lower middleclass Chicago Irish

St Valentine's disease epilepsy

St Vitus' dance epilepsy

Styx Soviet surface-to-surface naval missile

Subbotnik (Russian—Little Saturday)—Red Saturday—annual holiday when people clean up factory sites and neighborhoods as well as parks and other public places

Subcontinent of Asia India

Sublime Porte nickname for the government of the Turkish Empire in the times of the sultans

Submarine Capital Groton, Connecticut

Subtropical Siberia Castro's Cuba

subversive delinquents political prisoners

Successor of Saint Peter the Pope

Sucker State Illinois nickname dating from pioneer days when settlers sucked water from underground springs with long hollow tubes called suckers

Sudanese Port Bur Sudan

Sudanese Sister Cities Khartoum and Omdurman

Suds City Milwaukee, Wisconsin—famous for beer

Sugar Bowl New Orleans, Louisiana

Sugar Country Queensland, Australia

Sugar Islands Leeward Islands of the West Indies

Sugar King Claus Spreckels

sugar of lead lead acetate

Sugar State Louisiana

Suicide Capital of the United States San Francisco

Suicide Capital of the World Budapest

Suliman seal five-pointed pentagrammic star; symbol of Morocco and other Islamic lands

Sulphur King Herman Frasch

Sultan of Swat George Herman (Babe) Ruth

Summerless Southland southernmost New Zealand on its South Island south of Dunedin

Summit City Akron, Ohio

Sun Belt southern United States extending from Florida to Hawaii

Sun Bowl El Paso, Texas

Sun City St Petersburg, Florida; Yuma, Arizona

Sun Flag *Hi-no-maru*—sun flag

of Japan—Land of the Rising Sun—red sun on a white field

Sunflake City Grand Forks, North Dakota

Sunflower(s) Kansan(s)

Sunflower State Kansas

Sun God Adonis (Syrian); Apollo (Roman); Apollon (Greek); Baal (Chaldean); Helios Hyperion (Greek in Homer's time); Horus (symbolized in Upper Egypt by a hawk); Mithras (Persian); Moloch (Canaanite); Osiris (Egyptian); Ra or Re (symbolized in Egypt's Old Kingdom by an obelisk); Sol Invictus (Latin—Sun Invincible)—Romans shortened this to Sol; Surya (Hindu)

Sunken Continent of the Atlantic Atlantis

Sunken Continent of the Pacific Lemuria

Sun King Louis XIV *(Le Roi Soleil)*

Sun of May *El Sol de Mayo*—revolutionary symbol on the great seals of Argentina, Ecuador, and Uruguay; standing for national emergence in the fight for freedom

sunnie(s) sunfish(es)

Sunny Alberta Canada's Province of Alberta

Sunnyside Washington Irving's home near Tarrytown, New York

Sunny South southern United States

Sunrise Poet Sidney Lanier

Sunset Land Arizona

Sunshine Capital of the United States Yuma, Arizona

Sunshine City Saint Petersburg-Tampa, San Diego, Tucson, Yuma and Durban, South Africa—City of Sunshine

Sunshine Coast British Columbia from Lund to Vancouver; Queensland's coast from Brisbane to Noosa

Sunshine Continent Australia

Sunshine Province Alberta, Canada

Sunshine State Florida, New Mexico, and South Dakota; Queensland, Australia; official nickname of Florida

Super Constellation Lockheed transport carrying 99 passengers

Supercop Philadelphia's mayor Frank Rizzo—a former policeman

Superdome City New Orleans,

Louisiana

Super Falcon Hughes air-to-air missile also called Nuclear Falcon

Super Frelon Sud antisubmarine helicopter developed in France

Superman Philosopher Friedrich Wilhelm Nietzsche

Superman of the Prize Ring Joe Louis

Super Mystere Dassault fighter-bomber and jet interceptor built in France

Superpowers U.S.A. and the USSR

Super Sherman U.S. M-4 Sherman tank modernized

Superstition Personified Abessa who sought sanctuary behind convent walls shielding her from truth, according to Spenser's *Faerie Queene*

Supertenor Luciano Pavarotti

Supporter of the Universe Atlas, in the Roman mythology; the ash tree Ygdrasil in Norse mythology

Supreme Genius Leonardo da Vinci

Supreme Genius of Spanish Painting Diego Rodríguez de Silva y Velázquez

Supreme God of the Hindus Brahma

Supreme Governor of the Church of England the King or Queen

Supreme Pontiff of the Universal Church the Pope

Surfburgia California seaside suburban communities such as Malibu, Santa Monica, Seal Beach, Pacific Beach, Imperial Beach

Surgeon of the Rusty Knife Dr José Pedro de Freitas Arigo of Congonhas do Campo, Brazil

Surinam Ports Nieuw Nickerie, Paramaribo, Paranam, Moengo, Albina

Surveyor American program for lunar surface and subsurface exploration

Susan Hayward Edyth Marriner

Susie Susan; Susannah; Suzanne

Sussex Seaport Garden Resort Felixstowe

Suzy Susan; Susanna; Susanne

Svensker(s) Swedish sailor(s)

Sverdlov Soviet class of light cruisers

Sverdlovsk formerly Ekaterin-

burg where the Soviets murdered the last of the czars and his family; terminus of the Trans-Siberian railroad and place where Asia is said to look at Europe

Swamp Fox sobriquet shared by Revolutionary War general Francis Marion as well as by Confederate generals Nathan Bedford Forrest and Philip Dale Roddey

Swamp State South Carolina

Swan of Avon Ben Jonson's name for Shakespeare

Swan City Perth, Western Australia

Swanland southwestern Australia

Swan of Mantua Virgil

Swan of Meander Homer

Swan of Pesaro Gioacchino Antonio Rossini who was born in Pesaro, Italy

Swan River Colony Perth built around the River Swan in Western Australia

Swanside Perth, Western Australia

Swansider(s) inhabitant(s) of Perth on the Swan River estuary of Western Australia

Swatow mainland-China-built fast patrol craft

Swatter Soviet antitank missile

Sweden's Most Popular Sculptor Vilhelm Carl Emil Milles (originally surnamed Anderson)

Sweden's Principal Port Göteborg (Gothenburg)

Swedish Swedish massage (based on Swedish-type physiotherapeutic movements); Swedish mile (10 kilometers); Swedish movements (physiotherapeutic exercises); Swedish putty (spackle + spar varnish waterproofing mixture); Swedish turnip (rutabaga)

Swedish Film Pioneer Ingmar Bergman

Swedish Hanseatic Port City Visby on the island of Gotland

Swedish Nightingale Jenny Lind

Swedish Ports Lysekil, Uddevalla, Göteborg, Varberg, Falkenberg, Halmstad, Hoganas, Viken, Halsingborg, Landskrona, Malmö, Limhamn, Klagshamn, Trelleborg, Ystad, Simrishamn, Ahus, Solvesborg, Karlshamn, Ronnebyhamn, Karlskrona,

Kalmar, Oskarshamn, Slite, Farosund, Visby, Vastervik, Mem, Norrkoping, Oxelosund, Nykoping, Sodertalje, Nynashamn, Stockholm, Vasteras, Oregrund, Skutskar, Kastet, Gavle, Vallvik, Ljusne, Sandarne, Soderhamn, Hudiksvall, Sundsvall, Harnosand, Gustavsvik, Ornskoldsvik, Pitea, Lulea, Haparanda

Swedish West Indies St Barts (St Barthélmy now a French colony)

Sweet Singer of the Sierras Joaquin Miller

Sweyn Forkbeard King Svend of Denmark

Swingfire British BAC airlaunched or ground-launched antitank missile

Swiss Swiss chard (beetlike herb used in stews); Swiss cheese (Emmenthaler cheese characterized by its pale-yellow body and many holes); Swiss lapis (imitation lapis lazuli); Swiss muslin (curtain material); Swiss steak (thin slice of steak doused in flour and vegetables); Swiss watch

Swiss Cheese Capital of the U.S.A. Monroe, Wisconsin

Swiss Day Independence Day (August 1)

Swiss Family of Mathematicians and Scientists the Bernoullis

Swiss Family of Painters the Fuesslis

Swiss Riviera northern shores of Lake Lucerne

Sycamore Bristol four-place helicopter

Sycamore City Terre Haute, Indiana

Sylvia-Ducalis (Latin—'s Hertogenbosch)—also known as Bois le Duc or Sylvia Ducis

symbol of suffering the Christian cross

Synthesizer of Adrenalin Jokichi Takamine

Syrian Ports Latakia and Baniyas

Taco Benders Mexican Americans

Taco Town San Diego, California

Taffie(s) Welsh person(s)

Taffy diminutive of David the tutelar saint of Wales; nickname for a Welshman

Taig Terence

Taiwan (Chinese—Terrace Bay)

—island of Formosa—the Republic of China

Tall Boy Gouverneur Morris—amanuensis of the *Constitution of the United States*

Tall City Midland, Texas

Talleyrand Charles Maurice de Talleyrand-Périgord

Tallinn (Estonian—Dane's Town)—formerly Reval

Talos Bendix long-range surface-to-air missile

Tamaulipas formerly Pánuco

Tammany Boss William M Tweed

Tammies Tamburitzans

Tam(my) Thomas; Tom(my)

Tanana River City Fairbanks, Alaska

Tandjungpriok Djakarta (Jakarta)—formerly Batavia

Tanganyika the mainland of Tanzania

Tangerine Bowl Orlando, Florida

Tanya (Russian nickname—Tatiana, Tatyana)

Tanyu Morinobu Kano (*see* Kano)

Tanzanian Ports Lindi, Dar es Salaam, Tanga, Chake Chake and Zanzibar

Tapatios Mexicans from the state of Jalisco or from Guadalajara—its capital

Target Island Kahoolawe, Hawaii

Tarheeler(s) North Carolinian(s)

Tar Heel State North Carolina

Tartar General Dynamics naval surface-to-air missile

tartar emetic potassium antimony tartrate

Tartu Dorpat

Tasha (Russian nickname—Natasha)

Tasmania modern name for Van Diemen's Land

Tassie(s) Tasmanian(s)

Tassy Tasmania

Tassyland Tasmania (Australian slang)

Tatertown Gleason, Tennessee—shipping point for potatoes grown in the region

Tavern of Europe Paris

Taxachusetts Massachusetts

Taxassee Tallahassee

Tax Day U.S. federal taxes due April 15

Tax-Reform President Ronald Wilson Reagan

Taycheedah Wisconsin Home for Women at Taycheedah

Tay Pay Irish journalist Thomas Power O'Connor

Tchad Chad

T Dorp Schenectady, New York

Teacher of Doctors Sir William Osler

Teacher of Germany Philipp Melanchthon (collaborator of Luther)

Teacher President James Abram Garfield—twentieth President of the United States

Teachers Day September 28 in many Oriental lands where it is also the birthday of Confucius

Teague (nickname for an Irishman); Terence

Teapot Dome U.S. Navy's petroleum reserve near Casper, Wyoming and name of a political scandal during the administration of President Harding

tear gas chloroacetophenone; irritant gas also known as mace (MACE); causes temporary blindness as well as irritation of the mucous membranes and the skin

Tear-jerking Giacomo Giacomo Antonio Domenico Michele Secondo Maria Puccini (1858–1924)

Tebuan Malaysian name for the CL-41 Wasp attack-trainer aircraft

Teche Country Bayou Teche county, Louisiana

Technion Israeli Institute of Technology

Tecumseh Girls' Town correctional facility at Tecumseh, Oklahoma

Ted(dy) Edward; Theodore; Theodosia

Ted Morgan Sanche de Gramont

Teenie Christina

Teflon President Ronald Wilson Reagan

Tegoose Tegucigalpa (Honduras)

Tegusi Tegucigalpa's nickname

Tehachapi Institution California Correctional Institution at Tehachapi

Telemaque Denmark Vesey

Television City Hollywood, California

Teller of Sea Tales Joseph Conrad

Teller of Tall Tales Nathaniel Hawthorne, E T A Hoffmann, Washington Irving, Baron von Munchausen, Edgar Allan

Poe, Aleksander Sergeevich Pushkin, and Mark Twain

Tell Town Altdorf, Switzerland—reputed home of William Tell

Temple of Culture Thomas Jefferson's home, *Monticello*

Temple Mount Jerusalem's sobriquet

Temple of Music New York City's Carnegie Hall, also called the Cathedral of Music

Tennessee Williams Thomas Lanier Williams

Ten Provinces Ten Canadian Provinces (Alberta, British Columbia, Manitoba, New Brunswick, Newfoundland, Nova Scotia, Ontario, Prince Edward Island, Québec, Saskatchewan)

Tenth Muse Sappho, according to Plato, who esteemed the lyric poetess of Mytilene on the island of Lesbos

Tenth State Virginia (*see* First State)

Te Rangi Hiroa Sir Peter Buck

Teri Theresa; Therese

Terrace Bay Taiwan (Formosa)

Terrapin State Maryland

Terre Haute U.S. Penitentiary at Terre Haute, Indiana

The Terrible Ivan IV—Czar of Russia 1547 to 1584

Terrier General Dynamics naval surface-to-air missile

Terry Terence; Teresa; Terrell; Terrill; Theresa; Therese

Tess(ie) Theresa

Texarkana Institution Federal Correctional Institution at Texarkana, Texas

Texas Babe Mildred Didrikson Zaharias

Texas Cow Town Fort Worth

Texas Ports (east to west) Port Arthur, Beaumont, Galveston, Texas City, Houston, Corpus Christi, Brownsville

Texas tea crude oil and natural gas

Thailand's Major Ports Sattiship, Krung Thep (Bangkok)

Thailand's Principal Port Krung Thep (Bangkok)

That Man Franklin Delano Roosevelt

Theda Bara Theodosia Goodman

Thermaic Gulf Gulf of Salonika in the Aegean

Thespian Maids the Nine Muses

Thiefrow nickname for London's Heathrow Airport

thieves of time procrastinators

Third Estate The Commons—the legislature

third-generation money electronically controlled funds

Third International Lenin's organization of communists in Moscow in 1919

Third Reich Nazi Germany (1933–1945)

Third Republic France from 1871 to 1940

Third State New Jersey (*see* First State)

Third Viennese School of Psychotherapy theories proffered by Viktor E Frankl

Third World emerging nations, often former colonies, outside industrialized communist and Western nations; poorer nations

Thirstland waterless country north of Bechuanaland

Thirteen Colonies British North American colonies that became the original thirteen states of the United States

Thirteen States New Hampshire, Massachusetts, Rhode Island, Connecticut, New York, New Jersey, Pennsylvania, Delaware, Maryland, Virginia, North Carolina, South Carolina, Georgia

Thirteenth Apostle Constantine, the Roman emperor who built Constantinople (now Istanbul)

Thirteenth State Rhode Island (*see* First State)

Thirteenth Tribe Khazars converted to Judaism (*see* Twelve Tribes)

Thirtieth State Wisconsin

Thirty-eighth State Colorado

Thirty-fifth State West Virginia

Thirty-first State California

Thirty-fourth State Kansas

Thirty-ninth State North Dakota

Thirty Rock nickname of the National Broadcasting Company (NBC) at Thirty Rockefeller Center in New York City

Thirty-second State Minnesota

Thirty-seventh State Nebraska

Thirty-sixth State Nevada

Thirty-third State Oregon

This Is The Place Salt Lake City, Utah's sobriquet repeating the words of its founder—Brigham Young

thistle symbol of Scotland and the Scots

Thomas Jefferson and Abraham Lincoln of American Music Charles Ives, according to Leonard Bernstein

Thomas Jefferson Snodgrass (pseudonym—Samuel L Clemens)

Thomas Kyd Alfred Bennett Harbage's pseudonym

Thomas of London Thomas à Becket

Three Baltic Duchies Estonia, Latvia, Lithuania

Three Capitals and Five Ports Japanese numerical categories comprising the ancient and modern capitals—Kyoto, Osaka, and Tokyo plus the ports of Hakodate, Kobe, Nagasaki, Niigata, and Yokohama

three-C's Central Criminal Court

Three Kingdoms Denmark, Norway, Sweden

Three Little S's Saba, Sint Eustatius (Statia), Sint Maarten (Dutch Windward Islands—Netherlands Antilles)

Three Poles Mount Everest, North Pole, South Pole

three-R's reading, writing, arithmetic (colloquially: readin', 'ritin', 'rithmetic)

Three Sisters Earth, Mars, Venus

Three Virgins St Croix, St John, St Thomas (United States Virgin Islands)

Throne of Solomon Ethiopia

thruppence threepence

Thunder Bay modern name for the Canadian twin cities of Fort William and Port Arthur on the northwest shore of Lake Superior

Thunderbird British BAC mobile surface-to-air missile

Thunderbolt Republic fighting aircraft F-47; Swedish Viggen jet fighter J-37

Thunderchief Republic single-engine fighter-bomber jet aircraft (F-105)

Thunderjet Republic fighter-bomber F-84

Tib(by) Isabel(la); Ishbel(le)

Tiber River City Rome

Tico Costa Rican; Ticonderoga; USS *Ticonderoga* (attack aircraft carrier)

Ticos Costa Ricans

Tidewater States Maryland, Virginia, North Carolina, South Carolina, Georgia

Tien Percent Tien Suharto, wife of Indonesian President Suharto, so called for her percentage take of business with Indonesia

Tierra del Fuego (Spanish—Land of Fire)—originally the fires of Patagonian Indians but more recently the burning gases from oil rigs in southernmost South America

Tiger II Northrup F-5 twin-jet fighter aircraft

Tigercat Short and Harland surface-to-air missile

Tiger of France Georges Clemenceau (1841–1929)

Tigers of the Sun Sherpas of northern India and Nepal

Tight Little Island Great Britain

Tightrope Walkers Extraordinaire Charles Blondin who crossed Niagara Falls in 1855 on an 1100-foot (336-meter) tightrope suspended 160 feet (48 meters) above the falls and five years later carried his agent across piggyback; in 1974 Philippe Petit crossed between the twin towers of the World Trade Center in New York on a tightwire 1350 feet (412 meters) above the city sidewalk

Tigres River City Baghdad

Tilda Mathilda

Tillie Mathilda

Tilly Mathilda

Time Personified the aged Chronos of the Greeks and Romans—Father Time

Timesqueer New York City's Times Square

Time of the Troubles Russia between 1604 and 1613, from the reign of Boris Godunov to that of Michael Romanov

Timur the Lame Tamerlane

Tina Albertina; Christina; Clementina; Valentina

Tin Islands Indonesia's Banka and Belitung

Tin King Simón Ituri Patiño

Tin Lizzie Model-T Ford's nickname

Tinseltown Hollywood, California

Tintoretto Jacopo Robusti

Tip Thomas Phillip O'Neill, Jr; Timothy

Tip of Canada Ellesmere Land

Tippecanoe William Henry Harrison

Tipton Center State Correctional Center at Tipton, Missouri

Tiradentes (Portuguese—Tooth Puller)—nickname of José Joaquim da Silva Xavier—first Brazilian fighter for independence from Portuguese rule—a dentist

Tire City Akron, Ohio

Tirso de Molina (pseudonym—Gabriel Tellez)

Tish Letitia

Titan two-stage intercontinental ballistic missile (Martin)

Titian Tiziano Vecellio

Tito Josip Broz(ovich)

Titograd formerly Podgorica the capital of Montenegro now called South Yugoslavia

Tito Schipa Raffaele Attilio Amadeo Schipa

Titta Ruffo Ruffo Cafiero Titta

Titulescu James T(homas) Farrell

Tobacco Capital Durham, North Carolina

Tobacco City Winston-Salem, North Carolina

Tobacco Road dilapidated and poverty-striken rural areas (sociological synonym); tobacco-raising areas of the southern United States (generic and economic meaning)

Tobacco State Kentucky

Tóbal Cristóbal

Toby Tobyhanna; Tobias

Togo Port Lomé

To Hell and Back nickname of the Toronto, Buffalo, and Hamilton Railway

Toinette Antoinette

Tokaido Corridor urban strip between Kyoto and Tokyo (Kyoto, Kobe, Osaka, Nara, Nagoya, Hamamatsu, Shizuoka, Yokohama, Tokyo)

Tokyo (Japanese—Eastern)—formerly called Edo or Yedo and now capital of Japan as well as the world's largest city

Tolliver Tagliafiero

Tombs old New York City Prison on the Lower East Side, connected to the Criminal Courts Building by a Bridge of Sighs

Tomb Town Moscow

Tomcat F-14 fighter aircraft

Tom, Dick, and Harry the crowd; ordinary people; the mob; no one in particular

Tom Mix Eugene Blackman's motion-picture name

Tommy nickname for a British soldier; Thomas

Tommy Atkins (nickname for a British Army private)

Tommy the Cork Thomas Corcoran

Tommy gun Thompson submachine gun

Tom o' Bedlam incurable male lunatic

Toms two-dollar bills bearing the portrait of President Thomas Jefferson

tom thumb (Cockney—rum)

Toña Antonia

'Tona Daytona Beach, Florida

Tongan Ports Nukualofa Tongata, Pangai Haapai, Neiafu Vavau

Tongareva Penrhyn Island in the South Pacific

Tongue Troopers Canadian term for government's French-language enforcement squads

Toni Antonia

Tono Tomuelo

Toño Antonio

Tony Anthony; Antoinette Perry Awards (American Theatre Wing)

Tony Curtis Bernie Schwartz

Tony Martin Alvin Morris

Tony Randall Leonard Rosenberg

Tony Sarg Anthony Frederick Sarg

too. time of origin

tooies tuinal (half amobarbital and half secobarbital)

Tooth City Florence, South Carolina

Toothpicks nickname given early settlers of Arkansas who were believed to pick their teeth with bowie knives

Top of Europe northern sections of Finland, Norway, Russia, and Sweden near the Arctic Circle

Top Ten FBI's list of the 10 most wanted fugitives from justice

Top of the World Point Barrow, Alaska

Torchbearer of the Revolution Nathaniel Bacon of Virginia

Tor House home of Robinson Jeffers at Carmel, California

Tornado Alley area between Lawton, Oklahoma and Wichita Falls, Texas

Tortilla Curtain U.S.–Mexican border area between El Paso, Texas and San Diego, California; sometimes extended from Brownsville, Texas

t' other siders the other siders (nickname given east coast

Australians by their west coast counterparts)

Toti dal Monte Antonietta Meneghel

Tough Guy (stock exchange nickname for Texas Gulf Sulphur company)

Tourette's disease convulsive facial tic

Tow Hughes antitank missile designated MGM-71A

Towel Town Kannapolis, North Carolina where Cannon towels are made

The Tower The Tower of London

Town of Floating Gardens Xochimilco, Mexico

Town of Fools Chelm (*see* Chelmer)

Town of Merchants Shanghai

Town on the Water Stockholm

Town of Roses Molde, Norway

Town Too Tough To Die Tombstone, Arizona

Town of the Vikings Manitoba

Toy Bulldog Mickey Walker

Tracker Grumman S-2 antisubmarine search-and-attack aircraft

Tracy Theresa

Trader Horn nickname of Alfred Aloysius Smith

Tragic Patriot Thomas Paine (imprisoned by the Reign of Terror in France and reviled by the clergy in the United States he helped create)

Tragic Queen Marie Antoinette

Tragus Heironymus Bock

tranks nickname for sedative drugs (tranquilizers)

Transcendental Philosopher Ralph Waldo Emerson

Transjordan(ia) Hashemite Kingdom of Jordan better known as Jordan

Traven B Traven (pseudonym used by Berick Traven Torsvan)

Treasure State Montana

Tree of Heaven *Ailanthus* tree

Tree Planter's State Nebraska

Trench Town West Kingston, Jamaica's ghetto

Trent's Town Trenton, New Jersey's capital named for its original settler—William Trent

Tres Marías (Spanish—Three Marys)—María Madre, María Magdalena, María Cleofás—islands off the west coast of Mexico serving as a convict colony

The Tribune Man (pseudonym—Henry Ten Eyck White)

Tribune of the People John Bright

Tricia Patricia

Tri-Cities Florence, Sheffield, and Tuscumbia on Tennessee River near Muscle Shoals in northwestern Alabama; Davenport, Iowa—Moline and Rock Island, Illinois

Tricky Dick President Richard M Nixon

tricolor flag divided into three horizontal or vertical stripes; the Tricolor, initially capitalized, refers to the Tricolor of France consisting of red, white, and blue vertical stripes

Tri-Lingual Land Switzerland where people often speak French, German, and Italian as well as Romansh and English

Trinidad and Tobago Ports Trinidad—Chaguaramas Bay, Port-of-Spain, Pointe a Pierre, San Fernando, La Brea, Brighton, Point Fortin; Tobago—Canaan, Charlotteville, Scarborough

Trinity of Science Experience, Observation, and Reason

Triple Alliance Austria, Germany, and Italy (before outbreak of World War I)

Triple Cities Binghampton, Endicott, Johnson City (also called Tri-cities)

Tripsville Haight-Ashbury district of San Francisco

Trish Patricia; Tricia

Trixie Friganza Delia O'Callahan

Trix(ie)(y) Beatrice; Beatrix

Trojan North American T-28 trainer aircraft

Troldhaugen (Norwegian—Troll's Hill)—Edvard Grieg's home near Bergen

Trondheim modern name of Nidaros

Trooper Turned Physician Thomas Sydenham

Troopship Fokker military version of the 40 to 52-passenger aircraft F-27

Tropical North northern Queensland, Australia

Tropic Metropolis Miami, Florida

Tropics torrid lands and seas between Tropic of Cancer and Tropic of Capricorn

Trotsky Lev Davydovich Bronstein

Trucial States (*see* United Arab Emirates)

Trudy Gertrude

True King of Our Storytellers Jack London, according to Upton Sinclair

Truman Capote Truman Streckfus Persons

Trumpeter of the Last Judgment Gabriel

Trust Buster Theodore Roosevelt—26th President of the United States

The Trust Buster William Howard Taft

Trust Territory Micronesian islands of the Pacific (Carolines, Marianas, Marshalls, Ponape, Truk, Yap, etc.) under American administration

truth serum sodium pentathol

Tsaritsyn czarist name of Volgograd formerly called Stalingrad

Tsarskoe Selo former name of Pushkin near Leningrad

Tsnra Gora (Serbo-Croatian—Black Mountain)—Montenegro now called South Yugoslavia

T-town Tijuana

Tuba Player of the Century Roger Bobo of the Los Angeles Philharmonic

Tula Gertrude; Gertrudis

Tullahoma Vocational Tennessee State Vocational School for Girls at Tullahoma

Tully Marcus Tullius Cicero

Tum-Tum portly Albert Edward, HRH the Prince of Wales who later became King Edward the Seventh

Tung Tree Capital Picayune, Mississippi

Tunisian Ports Susa, Halq al Wadi, Tunis, Banzart, Bizerte, Sfax, and Gabes

tuppenny twopenny

Turkestan Desert includes Kara Kum south of Aral Sea, Kyzyl Kum southeast of Aral Sea, Ust Urt between Aral and Caspian seas

Turkey Capital of the World Berryville, Arkansas and Worthington, Minnesota

Turkey's Principal Port Istanbul (Constantinople)

Turkish Turkish bath (steam bath); Turkish delight (fruit-flavored gelatin candy dusted with confectioner's sugar); Turkish rug; Turkish tobacco (highly aromatic); Turkish towel (water-absorbent long-nap towel)

Turkish Ports Istanbul (Constantinople), Hydarpasa, Izmir (Smyrna), Antalya (Adalia), Mersin, Iskenderun (Alexandretta)

Turkish Towel Actress Brigette Bardot

Turner's syndrome genetic abnormality in females inheriting only forty-five chromosomes, causes retarded sexual development

Turpentine State North Carolina

turquoise hydrargillite (basic hydrated copper aluminum phosphate)

Turtle Bay Bahía de San Blas, Mexico

Tusitala (Samoan—Teller of Tales)—Robert Louis Stevenson's nickname

Tutor Canadair-built jet-trainer aircraft designated CL-41

Tuvalu formerly New Hebrides in the South Pacific between Fiji and New Caledonia

Tver czarist name for Kalinin

Twelfth State North Carolina (*see* First State)

Twelve Apostles twelve Apostle Islands in Lake Superior off northern Wisconsin

Twelve-Tone Technician Arnold Schönberg

Twelve Tribes Twelve Tribes of Israel—Reuben, Simeon, Judah, Zebulun, Issachar, Dan, Gad, Asher, Nephtali, Benjamin, Ephraim, and Mannaseh

Twentieth State Mississippi

Twenty-eighth State Texas

Twenty-fifth State Arkansas

Twenty-first State Illinois

Twenty-fourth State Missouri

Twenty-ninth State Iowa

Twenty-second State Alabama

Twenty-seventh State Florida

Twenty-sixth State Michigan

Twenty-third State Maine

Twiggy Leslie Hornby

Twilight of the Gods Gotterdämmerung (German mythology); Ragnarok (Norse mythology)

Twilight Zone the Mexican Border

Twin Cities Bristol on the Tennessee-Virginia border; Central Falls and Pawtucket, Rhode Island; Champaign and Urbana, Illinois; Minneapolis and St Paul, Minnesota; Texarkana on the Arkansas-Texas border; Winston-Salem, North Carolina

Twin Maples Farm British Columbia facility for treating women alcoholics

Twin Otter DeHavilland light transport (DHC-6)

The Twins Minneapolis and St Paul

Twin Sisters North and South Dakota

Twin States New Hampshire and Vermont

twister dustwhirl, sandspout, tornado, or waterspout wherein ascending and rotating movement of air column is especially apparent

Two Eyes of Greece Athens and Sparta

Two-headed Eagle popular symbol of the Austro-Hungarian Empire, Imperial Russia, and the Holy Roman Empire

Tybalt Theobald

Tybee Savannah Beach, Georgia

Ty Cobb Tyrus Raymond Cobb—idol of baseball fans

type metal antimony-copperlead-tin alloy

Typhoid Mary Mary Mallon (c.1870—1938)

Ubangi Republic Central African Republic

U-boat Führer Gross Admiral Karl Doenitz

Ulan Ude formerly Verkhneudinsk

Ulster Northern Ireland (formerly an ancient province of Ireland; now containing the counties of Antrim, Armagh, Down, Fermanagh, Londonderry, and Tyrone)

Ulster Cradle of U.S. Presidents Northern Ireland—ancestral home of Presidents Arthur, Grant, Jackson, McKinley, Truman, Wilson

Ultima Thule Iceland; Mainland (largest of the Shetland Islands); Norway; or any remote northern place, according to ancient travellers

Ulysses' fifty-dollar bills bearing the portrait of President Ulysses S Grant

Umbrian Historical Painter Pinturicchio (Bernardino di Betto)

Unabashed Atheist Charles Bradlaugh

The Unashamed Accompanist Gerald Moore

UN City Vienna, Austria's International Center (available to the UN cost free)

uncle pawnbroker

Uncle Arthur Arthur Henderson

Uncle Billie General William Tecumseh Sherman, USA

Uncle Dickie affectionate nickname of British military hero Mountbatten of Burma, Admiral of the Fleet and last Viceroy of India

Uncle Gene Eugene Ormandy

Uncle George George Geist

Uncle Ho Ho Chi Minh

Uncle Horace Horace Greeley

Uncle Joe U.S. Representative Joseph Gurney Cannon also known as the Watchdog of the Treasury

Uncle Kwesi Jonathan Kwesi Lamptey

Uncle Remus (pseudonym—Joel Chandler Harris)

Uncle Robert Robert E Lee; Robert L Sheppard

Uncle Sam cartoon symbol and nickname for an American citizen or the United States of America

Uncle Sam's Crib Treasury of the United States

Uncle Sam's Pocket Handkerchief Delaware—second smallest state in the U.S.

Uncle Sugar FBI's nickname

Uncle Tom Josiah Henson (slave immortalized in Harriet Beecher Stowe's *Uncle Tom's Cabin*); submissive African-American

Uncle Whiskers underworld nickname for Uncle Sam

Unconditional Abolitionist William Lloyd Garrison

Unconditional Surrender Grant General Ulysses Simpson Grant, USA

Uncrowned King of Ireland Charles Stewart Parnell

Uncrowned King of the Jews Chaim Weizmann

Underground Railroad Conductor Harriet Tubman who conducted fleeing slaves from the South to the northern United States and Canada

Underworld Statesman and Patriot Meyer (Little Man) Lansky

Union Jack flag flown at forward jackstaff of American vessels—consists of dark-blue rectangular field with 50 five-pointed white stars; flag of United Kingdom symbolizing union of England, Northern Ireland, Scotland, and

Wales—combines crosses of St George (England), St Patrick (Northern Ireland), St Andrew (Scotland)

Union of the Great Islands Comoro Islands

Unitarian Economist David Ricardo

Unitarian Quaker Elias Hicks

United Arab Emirate Ports Ash Shariqah, Dubai, Abu Dhabi (Abu Zaby)

United Arab Emirates Trucial Sheikdoms (Persian Gulf oil-producing country)

United Arab Republic name referring to the former union of Egypt and Syria

United Kingdom's Principal Port London

United Nations Capital New York City

United Provinces colors blue and white displayed in flags of El Salvador, Guatemala, Honduras, and Nicaragua—formerly federated after their liberation from Spain

Universal Genius Leonardo da Vinci (anatomist, architect, cartographer, engineer, inventor, musician, painter, poet, sculptor, zoologist)

University City Cambridge, Massachusetts

Unreconstructed Rebel Senator George Carter Glass, nicknamed by President Franklin D Roosevelt

Unsainted Anthony San Antonio, Texas

Unser Fritz (German—Our Fritz)—Frederick William III of Prussia

Upper Adige the Italian Tyrol also called the Southern Tyrol

Upper Alsace Haut-Rhin department of France

Upper Amazon Amazon River extending from the highlands of Peru to Manaus in northern Brazil

Upper Austria northern Austria bordering Bavaria and Czechoslovakia

Upper Burma inland Burma

Upper California in Spanish-colonial times all of California north of Monterey but today all of California except Baja or Lower California

Upper Canada English-speaking Ontario and the upper St Lawrence region during the 19th century

Upper Egypt Egypt from Cairo

south to the Sudan

Upper Galilee Israel north of the Sea of Galilee

Upper Lakes northernmost Great Lakes—Huron, Michigan, Superior

Upper Michigan the upper peninsula of northern Michigan between Lake Michigan and Lake Superior

Upper Mississippi Mississippi River from Lake Itaska in Minnesota near the Canadian border to Saint Louis, Missouri

Upper Nile Nile River from its headwaters in central East Africa to Khartoum in the Sudan

Upper Palatinate eastern Bavaria

Upper Peninsula northern Michigan between Lake Michigan and Lake Superior

Upper Peru former name for Bolivia

Upper Rhine Rhine River between Basel in Switzerland and Mainz in Germany

uppers nickname for stimulants

Upper Silesia northern Silesia once a Prussian province

Urga former name of Ulan Bator

Urista Uriel da Costa

Ursula Bloom Mrs ACG Robinson's pen name

Ursula Undress Ursula Andress

Uruguayan Ports La Paloma, Maldonado, Montevideo, Puerto Sauce, Nueva Palmira, Fray Bentos, Puerto Concepción, Paysandá, Salto

Uruguay's Largest Port Montevideo

Ushant Ile d'Ouessant—France's most westerly point at the Bay of Biscay entrance to the English Channel

USSR'S Principal Ports Leningrad on the Baltic, Odessa on the Black Sea, Vladivostok on the Pacific

Utilitarian Philosopher Jeremy Bentham; John Stuart Mill

Utopian Author Bacon, Bellamy, Butler, Cabot, Campanella, Fourier, Huxley, More, Morris, Owen, Plato, Proudhon, Rabelais, Rousseau, Saint-Simon, Wells

Vacation City on Casco Bay Portland, Maine

Vacationland Maine

Vacationland of Opportunity

Alaska

Vagen (Norwegian—Bay)—old Bergen and its waterfront along the bay

Valencia Conductor-Pianist José Iturbi (1895–1980)

Valentine State Arizona, admitted on St Valentine's Day —February 14, 1912

Valhalla Hall of the Slain Warriors (Norse or Scandinavian mythology)

Valhalla Girls Women's Correctional Unit at Valhalla, New York

valium diazepam

Valley Between Two Worlds Rio Grande Valley (between Mexico and the United States)

Valley of God's Pleasure Cleveland, Ohio's suburban section around Shaker Heights

The Valley Island Maui, Hawaii

Valley Isle Maui, Hawaii

Valley of Opportunity New York State's Triple Cities area including Binghampton, Endicott, and Johnson City

Valley of Rice Sikkimese name for state in India known as Denjong

Valley of the Sun Arizona's central valley

Valley of Valleys Gudbrandsdalen, Norway

Valley of Wonders Yellowstone National Park (in Idaho, Montana, and Wyoming)

Vampire De Havilland jet fighter-bomber aircraft

Van Cliburn Harvey Lavan Cliburn

Vancoo Vancouver, British Columbia

Van Diemen's Land Tasmania

Vanier Canadian city formerly called Eastview

Vapor City Hot Springs, Arkansas

Varangians (*Russian*—Vikings) —Danes or Norsemen who sailed to America around year 1000

Vatican City State *Stato della Città del Vaticano*

Vautour Sud attack bomber and interceptor made in France

Veecees Vietcongs

Veenees Vietnamese

Veep Vice-President

Vega Alta Industrial School for (criminal) Women at Vega Alta, Puerto Rico

Vegas East Atlantic City, New

Jersey

Vehicle City Flint, Michigan

Velvet Breughel Jan Breughel the Elder

Venerable Nestor of Massachusetts John Quincy Adams —sixth President of the United States

Venereal Disease of the New Morality Herpes Virus type 1—above the waist; Herpes Virus type 2—below the waist

Venetian Venetian ball (glass paperweight containing colorful objects); venetian blind (horizontally slatted sun curtain); Venetian glass (ornamental glassware of the type made in Venice); Venetian red (dark orange red); Venetian window (palladian window)

Venetian Family of Painters the Bellinis and the Tintorettos

Venetian red ferric oxide (Fe_2-O_3)

Venezuelan Poet Laureate Irma De-Sola Ricardo

Venezuelan Ports Maracaibo, Puerto Miranda, Bahía de Amuay, Puerto Cabello, La Guaira, Puerto de Hierro, Puerto Ordaz, Ciudad Bolívar

Venezuela's Principal Port La Guaira

Venom British de Havilland jet fighter aircraft

Vera Zorina Eva Brigitta Hartwig

Vercors pseudonym of Jean Bruller

Vermeer Jan van der Meer van Delft

Vermilionville Lafayette, Louisiana's old name

Vermillion Sea Gulf of California (Mar Bermejo)

Verneur Gouverneur

Vernon Castle Vernon Blythe

Vernon Duke Vladimir Dukelsky

Vernon Lee Violet Paget's pseudonym

Veronese Paolo Cagliari

Veronica Berenice

Veronica Lake Constance Ockleman

Verrocchio Andrea di Michele Cione

Ves Sylvester

Veteran Curmudgeon H(enry) L(ewis) Mencken

Vicar of Christ the Pope

Vichy Government France—1940–1944—which ruled by

collaborationists Marshal Pétain and Pierre Laval who maintained their headquarters in Vichy within unoccupied France

Vicki Victoria

Victim of Religion and Revolt Northern Ireland also called Captive of History

Victor letter V radio code; Handley-Page jet bomber aircraft

Victor Borge Borge Rosenbaum

Victor-Charlie VC; Vietcong

Victoria de los Angeles Victoria Gomez Cima

Victoria Holt Eleanor Burford Hibbert

Victorian Librettist W(illiam) S(chwenck) Gilbert

Victor Seastrom Viktor Sjöström

Victor Serge Victor Lvovich Kibalchich

Victory Personified Nike the Greek goddess or her Roman counterpart Victoria

video (Latin—I see)—picture portion of a tv broadcast

Vienna Vienna brown (bronzetone gold); Vienna green (emerald); Vienna lake (carmine); Vienna lime (Magnesialime polish); Vienna red (vermillion); Vienna sausage (short thin frankfurter)

Vietnamese Ports Cam Pha, Hon Gai, Haiphong, Ben Thuy, Da Nang (Tourane), Cam Ranh Bay, Saigon (Ho Chi Minh City)

Vietnam's Principal Port Ho Chi Minh City (Saigon)

Vieux Carré (French—Old Square)—French Quarter of New Orleans

Viggen Swedish Thunderbolt jet fighter aircraft

Vigilante North American-Rockwell A-5 bomber aircraft

Viki Victoria; Victorine

Viking Capital Oslo, Norway

Viking Genius John Ericsson

Viking Land Norway

Viking Program systematic investigation of Mars from orbit and from the surface with emphasis on the search for life on this planet

Vilhjalmur Stefansson Canadian-born William Stevenson's adopted name

Villa Acuña former name of Ciudad Acuña

The Village Carmel-by-the-Sea

in California; Greenwich Village in New York City; La Jolla's shopping district near San Diego, California

Village of Fools Chelm, Poland

vinegar acetic acid (CH_3COOH)

Vinegar Joe General Joseph Warren Stilwell, USA

Vinland North American coast discovered by Leif Ericsson and Norse sailors in year 1000; collective name for area extending from Labrador to Martha's Vineyard

Vinnie Vincent

Vinny Vincent

Violet Crown Acropolis Hill, Athens

Violet-crowned City Athens

viper's weed marijuana

Virgin Goddesses Artemis, Athena also known as Parthenia, and Hestia

Virginia Ports Alexandria, Newport News, Norfolk, Portsmouth

Virgin Island Ports Charlotte Amalie on St Thomas, Cruz Bay on St John, Frederiksted on St Croix

Virgin Queen Elizabeth I

Virgins American and British Virgin Islands in the West Indies

Virgin Superior St Thomas Island, Virgin Islands

Virtuoso Vladimir pianist Vladimir Horowitz (1903–1989)

Viscount Vickers medium transport aircraft

Viscountess Beaconsfield Mary Anne Disraeli (Mrs Benjamin Disraeli)

Viscount Melbourne William Lamb (former Prime Minister of Great Britain)

Viscount Palmerston Henry John Temple

Viscount Sidmouth Henry Addington

Vissarion Belinsky Vissarion Fondaminsky (1811–1848)

Vistula River Cities Cracow and Warsaw

Vitalis Erik Sjöberg

Vivazza (Italian—Vivacity)— Gioacchino Antonio Rossini's nickname

Viveca Lindfors Elsa Viveca Torstensdotter

Vivien Leigh Vivian Mary Hartley

Vladimir Sirin Vladimir Nabokov's pseudonym

Voice of the American Revolution Patrick Henry

Voice of Australia Dame Edna Everage, also known as Barry Hymphries

Voice of the Century Marian Anderson

Voice of Doom Gabriel Heatter (before and during World War II); Ann Watson (in the uncertain 1970s)

Voice from the Fo'c's'le Richard Henry Dana in *Two Years Before the Mast*; Herman Melville in *Whitejacket*

Voice of Israel Abba Eban

Voice of Liberty Thomas Paine

Voice of Northern Industrialism Daniel Webster

Voice of Polish Nationalism Adam Mickiewicz

Voice of the Revolution Patrick Henry

volatile alkali ammonia

Volcano Island Hawaii

Volcano Land Iceland

Volgograd formerly Stalingrad during Stalin's time and Tsaritsyn during czarist times

Volodya (Russian nickname—Vladimir)

Voltaire assumed name of François-Marie Arouet (*see* Jean Meslier)

Voltaire of the Unitarians Dr Joseph Priestley, according to William Hazlitt

Volunteer(s) Tennessean(a)

Volunteer State Tennessee

Von Economo's disease encephalitis lethargica

von Hofmanns Hugo von Hofmannsthal

von Reuter Israel Beer Josphat (founder of Reuter's news agency)

Voodoo Canadian-built version of the F-101 jet interceptor

Voroshilovgrad formerly Lugansk

Voyager American spacecraft destined for landings on Mars and Venus

V Sirin Vladimir Nabokov

V-spot $5 bill

Vulcan Hawker-Siddeley jet bomber aircraft; US-built six-barrel 20mm cannon

Vulcan (Latin—Hephaistos)— the blacksmith

Waddy Walter

Wade Miller pseudonym of writers Robert Wade and Bill Miller

Waistline of the Western Hemisphere Isthmus of Panama

Walden Henry David Thoreau's

hut on the shores of Walden Pond near Concord, Massachusetts

The Wales The Bank of New South Wales

walking handbag(s) alligator(s); cayman(s); gavial(s); crocodile(s)

Wall of Death drift gill netting

Walleye Martin-Hughes tv-guided glide bomb

Wallows Walla Walla, Washington

Wall Street New York City's financial center

Wall Street of Canada Bay Street in Toronto, Ontario

Walnut City McMinnville, Oregon

Walrussia nickname for Alaska in 1867 when it was purchased from Russia; believed by critics to have nothing but walruses

Walt Disney José Guizao Zamora

Walter Hampden Walter H Dougherty

Walter Huston Walter Houghston

Walter Wanger Walter Feuchtwanger

The Waltz King Johann Strauss, Jr

Wandering Mathematician Johannes Kepler (1571–1630)

War Between the States Civil War; War of the Secession

Warden of the Honour of the North Halifax, Nova Scotia

Warehouse of the East free port of Penang, Malaysia

War Fury Bellona—Roman goddess of war whose Greek counterpart is Enyo

Warhorse of the Confederacy Lieutenant General James Longstreet, CSA

War of Independence American Revolution

Warlord Wotan

Warlord of the First Reich Prince Otto von Bismarck-Schönhausen

Warlord of the Second Reich Kaiser Wilhelm II

Warlord of the Third Reich Führer Adolf Hitler

War of the Pacific Chile vs. Bolivia and Peru (1879–1883)

War of Reparation American Revolution

War of the Secession Southern synonym for Civil War, War of the Rebellion, War be-

tween the States (of the United States)—1861 to 1865

The Wash North Sea inlet between Lincoln and Norwich on east coast of England

washing soda sodium carbonate crystals (Na_2CO_3 + $10H_2O$)

Washington Ports South Bend, Raymond, Aberdeen, Hoquiam, Port Angeles, Port Townsend, Olympia, Tacoma, Seattle, Everett, Anacortes, Bellingham

Washington State Funnypark Washington State Prison near Walla Walla

Washoe early settler's name for Nevada during the Comstock Lode gold-and-silver rush of 1859

Washoe Giant Mark Twain

Wasp Westland naval helicopter built in Britain

Watch City Waltham, Massachusetts

Watchdog of Central Park *New York Times* publisher Adolph S Ochs

Watchdog of the Eastern Pacific Pearl Harbor

Watchdog of the Western Pacific Guam

Waterfront Philosopher Eric Hoffer

Waterfront of the West San Francisco

Watergab Watergate English (Nixon-era federalese)

Watergate Potomac River waterfront of Washington, DC; national scandal first detected at the Watergate office building

Watergate President Richard Milhous Nixon

waterglass sodium silicate (Na_2SiO_3)

Waterland the Netherlands

Waterloo battlefield near Brussels, Belgium, where Napoleon met his final defeat June 18, 1815

Watermelon Capital of the World Hope, Arkansas

Waters William Russell's pseudonym

water turkey anhinga

Water Wonderland Michigan

Wat(ty) Walter

W C Fields William Claude Dukenfield (1880–1946)

Wealth Personified Ploutus (Greek); Plutus (Roman)

Weather Capital of the World a groundhog hole in Punxsutawney, Pennsylvania where

groundhog leaves hibernation every February 2 to forecast winter's end (if he sees his shadow it means six more weeks of ice and snow, and if he doesn't then spring is near)

Webfeet Oregonians

Webster Ford Edgar Lee Master's pseudonym

weed nickname for marijuana; grass, pot

Weegee photographer Arthur Fellig's nom de voir

Wee Willie William Keeler

Weil's disease jaundice

Weimar Republic Germany between 1919 and 1933—the Second Reich

Welcher(s) person(s) of Welsh origin

wellies wellington boots

Wellington Arthur Wellesley (Duke of Wellington); British Hovercraft class of hovercraft

Welsh Welsh cob (horse), Welsh corgi (dog), Welsh dresser (cupboard), Welsh harp, Welsh main (cockfight), Welsh mortgage, Welsh mountain (pony or sheep), Welsh process (smelting), Welsh rabbit (cheese dish also called Welsh rarebit), Welsh runt (cattle), Welsh springer (spaniel), Welsh terrier

Welsh Cradle of U.S. President Wales—ancestral home of President Jefferson

Welsh Landscape Painter Richard Wilson

Welsh Ports Port Dinorwic, Holyhead, Caernarvon, Fishguard, Milford Haven, Llanelly, Swansea, Port Talbot, Barry, Cardiff

Welsh Wizard David Lloyd George

Wessex Westland-built verson of Sikorsky utility helicopter

West Britain Wales

West Country southwestern England—Cornwall, Devonshire, Dorset, Somerset

West End fashionable London

Western Western omelet; movie or novel featuring the Wild West

Western Hemisphere half of the world containing North America, South America, and associated islands

Western Prairie Province Alberta

Western Samoan Port Apia

Western States United States west of the Mississippi River

Western Tip of Florida Pensacola

Western Tip of Texas El Paso

West Indies Greater and Lesser Antilles in the Caribbean Sea

West Irian western half of New Guinea formerly Dutch or Netherlands New Guinea and now part of Indonesia

West Jersey southern and western New Jersey

West Lothian Linlithgow, Scotland

West Malaysia mainland Malaysia plus Singapore before it became independent

Westminster Abbey Collegiate Church of St Peter in Westminster

Westminster Palace Houses of Parliament in London

West North Central States Iowa, Kansas, Minnesota, Missouri, Nebraska, North Dakota, South Dakota

West Point U.S. Military Academy at West Point, New York

West Point of the Air San Antonio, Texas

West Point of Capitalism Harvard Business School

West Point of Law Enforcement FBI National Academy at Quantico, Virginia

Westport Landing pioneer name for Kansas City

West South Central States Arkansas, Louisiana, Oklahoma, and Texas

West Virginie West Virginia

Westway New York City's west side highway extending northward from Battery Park along the Hudson River

Westy Westermoreland; Westmoreland

Wet Tortugas Florida Keys

Whaling City New Bedford, Massachusetts

Whangpoo Hwang Pu

Wharf of North America Nova Scotia

Wheat Energy State North Dakota

Wheatland home of President James Buchanan in Lancaster, Pennsylvania

Wheat Provinces Alberta, Manitoba, Saskatchewan

Wheat State South Australia

Where America's Day Begins Guam

Where the Andes Greet the Caribbean Venezuela

Where It's Springtime All The Time San Diego, California

Where Mexico Meets Uncle Sam nickname of Brownsville, Texas

Where the Turf Meets the Surf Del Mar, California

Whirlwind Westland military helicopter built in Britain

White Africa South Africa

Whitechapel London's East End

White City of the North Helsinki

White City of the South Sucre, Bolivia

White Elephant Thailand ensign bearing a green and red caparisoned white elephant

White Ensign flag of the Royal Navy and the Royal Yacht Club—St George cross on a white ground with the Union Jack in the upper canton corner

white flag symbol of surrender or truce

Whitehall London's street of government offices

White House executive office and residence of the President of the United States in Washington, DC

White Island Ibiza in the Balearics

white lead lead carbonate

white light signal indicating apparatus, craft, or vehicle has power and is illuminated

White Man's Grave equatorial West Africa

White Metropolis Helsinki

White Mountain State New Hampshire

white niggers British racist nickname for communist and socialists

white plague pulmonary tuberculosis

White Russia Byelorussian district around Minsk

whites thick whitish vaginal discharge; synonym for leukorrhea

White Town of Lake Mjosa Gjovik, Norway

white vitriol zinc sulfate

whitewings white-uniformed street cleaners

Whorez trucker's nickname for Ciudad Juarez, Mexico

Why Not Town Minot, North Dakota, nicknamed Why Not Minot?

Widow at Windsor Queen Victoria who was a widow for the last 39 years of her life

Wigwam Tammany Hall

Wild Bill William Joseph (Wild Bill) Donovan; James Butler (Wild Bill) Hickok

Wilderness of Judah western shores of the Dead Sea in Israel

Wilderness Park Nahanni National Park (Canadian Northwest Territories)

Wilderness Trail Blazer Daniel Boone

Wildflower State Western Australia

Wild Man of Borneo orangutan (friendliest of the great apes)

Wild Rose Iowa state flower; Iowa girl's nickname

Wildrose Country Alberta

Wild West western United States

Wild Wonderful State West Virginia

Wilhelm Xylander Wilhelm Holtzmann

Wilkes Land Australian Antarctica

Willa Wilhelmina

Willem De Merode Willem Eduard Keuning's pseudonym

William Ashenden W Somerset Maugham

William B Goodrich Roscoe (Fatty) Arbuckle's pseudonym

William Bolitho William Bolitho Ryall

William the Conqueror William I of Normandy and England

William Haggard Richard Henry Michael Clayton's pseudonym

William Holden William Franklin Beedle

William of Nassau William I—Prince of Orange and Count of Nassau—founder of the Dutch Republic; also called William the Silent

William Penn's Town Philadelphia

William Sharp Fiona Macleod's pseudonym

William the Silent William —Prince of Orange

William Tell's Town Altdorf in Switzerland's Uri Canton

William Trevor William Trevor Cox

Willie William; W(illiam) (Willie) Somerset Maugham

Willie Mays Willie Howard Mays

Willies Good Will Industries

Will Rogers William Penn

Adair Rogers (1879–1935)
Willy Wilhelm; William
Willy Brandt Karl Herbert
Frahm (1913–)
Wilma Wilhelmina
Wilmington Women Correctional Institution for Women at Wilmington, Delaware
Window to Europe Saint Petersburg, Petrograd, Leningrad
Windscale Sellafield, England's nuclear reprocessing plant
Windward Islands Dominica, Grenada, Grenadines, Saint Lucia, Saint Vincent—British West Indies
Windward Passage ocean passage between Cuba and Haiti; connects Atlantic Ocean with Caribbean Sea
Windy City Chicago, Illinois and Wellington on New Zealand's North Island
Windy Wellington Wellington, North Island, New Zealand
Wine-Red Sea the Aegean, according to Homer
Winesburg Sherwood Anderson's name for Clyde, a small town southwest of Sandusky, Ohio and the title of his play—*Winesburg, Ohio*
Winney Winston; Sir Winston Churchill (1874–1965)
Winnie Sir Winston Churchill—British Prime Minister
Win(nie) Winslow; Winston
Wintergarden of the East southern Florida
Wintergarden of the Gulf lower Rio Grande Valley
Wintergarden of the West the Imperial Valley
Winterless Northland northernmost New Zealand
Winter Wonderland British Columbia
Wipers (British slang—Ypres)
Wisconsin Ports Racine, Milwaukee, Port Washington, Sheboygan, Manitowoc, Sturgeon Bay, Green Bay, Marinette, Ashland, Superior
Wisest Man of Greece Socrates
Witchcraft City Salem, Massachusetts
Witch of Wall Street Hetty Green
Wizard of American Drama David Belasco
Wizard from Vienna Franz Anton Mesmer
Wizard of Kinderhook Martin Van Buren—eighth President of the United States

Wizard of Menlo Park Thomas Alva Edison whose research laboratory was in Menlo Park, New Jersey
Wizard of the Saddle Lieutenant General Nathan Bedford Forrest, CSA
Wizard of Scotland Sir Walter Scott
Wizard of Tuskegee George Washington Carver
Wizard of Word Music Edgar Allan Poe
W N P Barbellion Bruce Frederick Cummings (1889–1919)
Wobblies International Workers of the World (so named because Chinese members pronounced IWW as *I Wobbly Wobbly*)
Wolf House Jack London's home in Glen Ellen, California
wolfram iron manganese tungstate
Wolverine nickname for a Michiganite
Wolverine State Michigan
Woman With The Whip Evita Perón
Womb of Nations Scandinavia
Wonder City of the World New York
Wonder State Arkansas
wood alcohol methyl alcohol (CH_3OH)
Wooden Leg Governor Peter Stuyvesant of Nieuw Amsterdam
Woodie Woodmansee; Woodrow
Woodland Capital Boise, Idaho
woodpile xylophone's nickname
Woodstein Bob Woodward and Carl Bernstein (of the *Washington Post*, known for uncovering the Watergate coverup)
Woody Woodrow
Woody Allen Allen Stewart Konigsberg
Woody Herman Woodrow Wilson Herman
Wool and Mohair Capital of the West Del Rio, Texas
Woo Poo cadet's nickname for West Point
Word Jeweler Lafcadio Hearn (1850–1904)
Word King New-Zealand-born British lexicographer Eric Partridge
Workers' Paradise derisive

nickname applied to the communist-controlled USSR
Workshop of the Orient Japan
World Citizen Thomas Paine
World Famous Contralto Ernestine Schumann-Heink (1861–1936)
World Island African-Asian-European land mass
World's Workshop productive nations such as Germany, Great Britain, Japan, and the United States
World War Photographer Edward Steichen
Wreckers Coast England's Cornish coast, Florida's east coast, North Carolina's Outer Banks around Nag's Head, other coasts where wreckers profit from shipwrecks
Wurst City in the World Sheboygan, Wisconsin, where making sausage is a specialty
Wyoming Suffragette Esther Hobart Morris
Xanadu Xamdu (city where Kubla Khan lived and name given the Hearst Castle at San Simeon, California by Orson Welles)
Xanrof Léon Fourneau
Xavante Aermacchi jet-trainer ground-attack aircraft also designated AT-26
Xavier Joseph Xavier Boniface's nom de plume
Xavier Mayne Edward Irenaeus Stevenson
Xenius Eugenio d'Ors
Xenocoj Santo Domingo, Guatemala
Xeres Jerez de la Frontera
Xibaro Jivaro
Xipangu Marco Polo's name for Japan
X-ray Discoverer Wilhelm Konrad Roentgen
Xtet (Swedish—the X)—Sven Erixson
XYY syndrome unusually aggressive male having an extra Y-sex chromosome
Yaacov Agam Yaacov Gipstein
Yafo Tel Aviv's seaport also known as Jaffa or Joppa
Yakumo Koizumi Lafcadio Hearn's Japanese name
Yallerhammer State Alabama
Yankee Athens New Haven, Connecticut
Yankee Clipper Joe Di Maggio
Yankeedom New England; Northeastern United States
Yankee Doodle Dandy George M Cohan (born on July 4)

Yankee State Ohio

Yanko-Spanko Conflict Spanish-American War (so named in 1899 by historian Arthur Bird)

Yaptown Cleveland, Ohio

Yard(s) Montagnard(s)

Yasnaya Polyana (Russian—Clear Glade)—Tolstoy family home near Tula about 177 kilometers (110 miles) south of Moscow

Yedo Tokyo's old name (also written Edo)

Yekké Israeli-born nickname for German Jews

Yellow Belly Youngstown Sheet & Tube's nickname

yellowcake uranium ore (U_3O_8)

Yellow flag yellow signal flown when a vessel requests pratique; letter Q or Quebec in the international code; also called the quarantine flag

Yellowhammer Alabama state bird; symbolic nickname of an Alabaman

Yellowhammer State Alabama

yellowjack quarantine flag; yellow fever; yellow flag

Yellow Silk Journal of Erotic Arts

Yellow Thunder Country around the Dells of the Wisconsin River near Baraboo, Wisconsin

Yemeni Ports Qishn, Al Luhayyah, Kamaran, Al Hudayah, Al Mukalla, Perim Harbour, Aden

Yerba Buena former name of San Francisco

Yerevan Armenia's Erevan or Erivan

Yezo Japanese island of Hokkaido

Yggdrasill tree of the universe, according to Norse mythology

Yogi Lawrence Peter Berra, known as Yogi Berra

Yoknapatawpha William Faulkner's mythical Mississippi county

Yorkshire Queen of Song Susan Sunderland

York State New York State

Young Hickory James K Polk —eleventh President of the United States

Young Napoleon U.S. General George B McClellan (1826–1885)

Young Pretender Bonnie Prince Charlie (Charles Edward Louis Philip Casimir Stuart)—son of the Old Pretender—James Stuart

Youth Personified Juventus (Latin—youth)

You've Got a Friend State Pennsylvania

Yseult Isolde

Yucatan Channel ocean passage between Cuba and Yucatan Peninsula of Mexico; connects Caribbean Sea with Gulf of Mexico

Yucca Country the southwestern United States

Yugoslavia's Principal Seaport Split

Yugoslav Ports Rovinj, Pula, Luka Rijeka, Bakar, Zadar, Sibenik, Split, Gruz, Dubrovnik

Yukio Mishima Kimitake Hiraoka's pseudonym

Yul Brynner Taidje Kahn, Jr

Yu-Lin Betty Yü-Lin Ho

Yuri Bilstin Youry Bildstein

Yves Montand Ivo Livi

Yvonne de Carlo Peggy Yvonne Middleton

Zagreb Agram

Zaire Ports Banana, Boma, Matadi

Zancle old name for Messina

Zane Grey Pearl Grey's pseudonym

zazou (French nickname—zootsuiter)

Zazul Vera Zazulich

Zbig Zbigniew Brzezinski

Zedland English shires of Devon, Dorset, and Somerset

Zeke Ezekiel

zeks (Soviet-Russian slang—prisoners)

Zelda Griselda

Zen Garden of the Atlantic England's Scilly Isles

Zenith imaginary Minnesota city whose leading citizen is George F Babbitt described by Sinclair Lewis in his novel, *Babbitt*

Zenith City of the Unsalted Sea Duluth, Minnesota on Lake Superior leading to the other Great Lakes

Zeppo Marx Herbert Marx

Zero Mostel Sam Mostel

Z-grams Admiral Zumwalt's policy statements

Zilia Clara Josephine Wieck (later Clara Schumann, the wife of the composer-critic Robert Schumann who gave her that pseudonym)

Zilli Cecilia

Zinoviev, Grigori Evseevich Hirsch Apfelbaum

Zipango Marco Polo's name for Japan

Zoé (French nickname—automic pile)

Zola of America Sir Arthur Conan Doyle's nickname for Upton Sinclair

Zonian(s) American(s) of the Panama Canal Zone

Zoological Attic of the World Australia, New Guinea, New Zealand, Tasmania

Zsa Zsa Gabor Sari Gabor

Zsuzsa Zsuzsa Heiligenberg

Zubie Zubin Mehta of the New York Philharmonic

Zuinglius Latinization of Ulrich Zwingli

Zulo Ignacio de Zuloaga

Zungaria Dzungaria or Sungaria region between Mongolia and Russia

Fishing Port Registration Symbols
(Distinguishing Letters)
England

AB	Aberystwith	BK	Berwick-on-Tweed	BR	Briggwater
BD	Bideford	BL	Bristol	BS	Beaumaris
BE	Barnstaple	BM	Brixham	BW	Barrow
BH	Blyth	BN	Boston	CA	Cardigan

CF	Cardiff	LN	Lynn	RX	London	SA	Rye
CH	Chester	LO	London	SA			Swansea
CK	Colchester	LR	Lancaster	SC			Scilly
CL	Carlisle	LT	Lowestoft	SD			Sunderland
CO	Carnarvon	M	Milford	SE			Salcombe
CS	Cowes	MH	Middlesbrough	SH			Scarborough
DH	Dartmouth	MN	Maldon	SM			Shoreham
DR	Dover	MR	Manchester	SN			Shields, North
E	Exeter	MT	Maryport	SS			St Ives
FD	Fleetwood	NE	Newcastle	SSS			Shields, South
FE	Folkstone	NN	Newhaven	ST			Stockton
FH	Falmouth	NT	Newport, Mon.	SU			Southhampton
FY	Fowey	P	Portsmouth	TH			Teignmouth
GE	Goole	PE	Poole	TO			Truro
GR	Gloucester	PH	Plymouth	WA			Whitehaven
GY	Grimsby	PN	Preston	WH			Weymouth
HH	Harwich	PT	Port Talbot	WI			Wisbech
HL	Hartlepool, West	PW	Padstow	WO			Workington
IH	Ipswich	PZ	Penzance	WY			Whitby
LA	Llanelly	R	Ramsgate	YH			Yarmouth (Norfolk)
LI	Littlehampton	RN	Runcorn				
LL	Liverpool	RR	Rochester				

Northern Ireland

B	Belfast	LY	Londonderry
CE	Coleraine	N	Newry

Republic of Ireland

C	Cork	G	Galway	T	Tralee
D	Dublin	L	Limerick	W	Waterford
DA	Drogheda	S	Skibbereen	WD	Wexford
DK	Dundalk	SO	Sligo	WT	Westport

Greek Alphabet

ALPHA	A	α	IOTA	I	ι	RHO	P	ρ
BETA	B	β	KAPPA	K	κ	SIGMA	Σ	σ
GAMMA	Γ	γ	LAMBDA	Λ	λ	TAU	T	τ
DELTA	Δ	δ	MU	M	μ	UPSILON	Υ	υ
EPSILON	E	ε	NU	N	ν	PHI	Φ	φ
ZETA	Z	ζ	XI	Ξ	ξ	CHI	X	χ
ETA	H	η	OMICRON	O	o	PSI	Ψ	ψ
THETA	Θ	θ	PI	Π	π	OMEGA	Ω	ϖ

International Civil Aircraft Markings

AN Nicaragua
AP Pakistan
B Formosa

CB Bolivia
CC Chile
CCCP Soviet Union (USSR)

CF Canada
CR; CS Portugal and colonies
CU Cuba

CX Uruguay
CZ Principality of Monaco
D Western Germany
EC Spain
EI; EJ Ireland
EL Liberia
EP Iran
ET Ethiopia
F France and French Union
G United Kingdom
HA Hungary
HB Switzerland
HC Ecuador
HH Haiti
HI Dominican Republic
HK Colombia
HL Korea
HS Thailand
HZ Saudi Arabia
I Italy
JA Japan
JY Jordan
LN Norway
LV Argentine Republic
LX Luxembourg
LZ Bulgaria

MC Monte Carlo
N United States of America
OB Peru
OD Lebanon
OE Austria
OH Finland
OK Czechoslovakia
OO Belgium
OY Denmark
PH Netherlands
PI Philippine Republic
PJ Curaçao (Netherlands Antilles)
PK Indonesia
PP; PT Brazil
PZ Surinam (Netherlands Guiana)
RX Republic of Panama
SE Sweden
SN Sudan
SP Poland
SU Egypt
SX Greece
TC Turkey
TF Iceland
TG Guatemala

TI Costa Rica
VH Australia
VP; VQ; VR British Colonies and Protectorates
VT India
XA; XB; XC Mexico
XH Honduras
XT China (Nationalist)
XY; XZ Burma
YA Afghanistan
YE Yemen
YI Iraq
YK Syria
YR Romania
YS El Salvador
YU Yugoslavia
YV Venezuela
ZA Albania
ZK; ZL; ZM New Zealand
ZP Paraguay
ZS; ZT; ZU Union of South Africa
$_4$R Ceylon
$_4$X Israel
$_5$A Libya
$_9$G Ghana

International Conversions Simplified

area

a (acres)	x	0.4	= ha (hectares)
cm^2 (square centimeters)	x	0.16	= in.^{2} (square inches)
ft^2 (square feet)	x	0.09	= m^2 (square meters)
ha (hectares)	x	2.5	= a (acres)
in.^{2} (square inches)	x	6.5	= cm^2 (square centimeters)
km^2 (square kilometers)	x	0.4	= mi^2 (square miles)
m^2 (square meters)	x	1.2	= yd^2 (square yards)
mi^2 (square miles)	x	2.6	= km^2 (square kilometers)
yd^2 (square yards)	x	0.8	= m^2 (square meters)

length

cm (centimeters)	x	0.4	= in. (inches)
ft (feet)	x	30.0	= cm (centimeters)
in. (inches)	x	2.54*	= cm (centimeters)	*exactly
km (kilometers)	x	0.6	= mi (miles)
m (meters)	x	3.3	= ft (feet)
m (meters)	x	1.1	= yd (yards)
mi (miles)	x	1.6	= km (kilometers)
mm (millimeters)	x	0.04	= in. (inches)
yd (yards)	x	0.9	= m (meters)

temperature (exact)

C (degrees Celsius or centigrade)   x   9/5 + 32 = F (degrees Fahrenheit)
F (degrees Fahrenheit)   32 x 5/9 = C (degrees Celsius or centigrade)

volume

cups	x	0.24	= l (liters)
fl oz (fluid ounces)	x	30.00	= ml (milliliters)
ft^3 (cubic feet)	x	0.03	= m^3 (cubic meters)

gal (British Imperial gallons)	x	4.6	= 1 (liters)
gal (U.S. gallons)	x	3.8	= 1 (liters)
l (liters)	x	2.1	= pt (pints)
l (liters)	x	1.06	= qt (quarts)
l (liters)	x	0.22	= gal (British Imperial gallons)
l (liters)	x	0.26	= gal (gallons)
m^3 (cubic meters)	x	35.00	= ft^3 (cubic feet)
m^3 (cubic meters)	x	1.3	= yd^3 (cubic yards)
ml (milliliters)	x	0.03	= fl oz (fluid ounces)
pt (pints)	x	0.47	= 1 (liters)
qt (quarts)	x	0.95	= 1 (liters)
tbsp (tablespoons)	x	15.00	= ml (milliliters)
tsp (teaspoons)	x	5.00	= ml (milliliters)
yd^3 (cubic yards)	x	0.76	= m^3 (cubic meters)

weight

g (grams)	x	0.035	= oz (ounces)
kg (kilograms)	x	2.2	= lb (pounds)
lb (pounds)	x	0.45	= kg (kilograms)
oz (ounces)	x	28.00	= g (grams)
st (short tons—2000 pounds)	x	0.9	= t (tonnes)
t (tonnes—1000 kilograms)	x	1.1	= st (short tons)

International Radio Alphabet and Code

A Alpha . _
B Bravo _ . . .
C Charlie _ . _ .
D Delta _ . .
E Echo .
F Foxtrot . . _ .
G Golf _ _ .
H Hotel . . . .
I India . .
J Juliet . _ _ _
K Kilo _ . _
L Lima (leema) . _ . .

M Mike _ _
N November _ .
O Oscar _ _ _
P Papa . _ _ .
Q Quebec (kaybeck) _ _ . _
R Romeo . _ .
S Sierra . . .
T Tango _
U Uniform . . _
V Victor . . . _
W Whiskey . _ _
X Xray _ . . _

Y Yankee _ . _ _
Z Zulu _ _ . .
0 (zee-ro) _ _ _ _ _
1 (wun) . _ _ _ _
2 (too) . . _ _ _
3 (thuh-ree) . . . _ _
4 (fo-wer) . . . . _
5 (fi-yiv) . . . . .
6 (siks) _ . . . .
7 (sev-ven) _ _ . . .
8 (ate) _ _ _ . .
9 (ni-yen) _ _ _ _ .

International Vehicle License Letters

A Austria
ADN Aden (South Yemen, Peoples' Democratic Republic of Yemen)
AFG Afghanistan
AL Albania
AND Andorra
AUS Australia
B Belgium
BD Bangladesh
BDS Barbados
BG Bulgaria
BH Belize (formerly, British Honduras)
BR Brazil
BRN Bahrain
BRU Brunei
BS Bahamas

BT Botswana
BV Bolivia
BUR Burma
C Cuba
CDN Canada
CH Switzerland
CI Ivory Coast
CL Sri Lanka
CO Colombia
CR Costa Rica
CS Czechoslovakia
CU Curaçao
CY Cyprus
D West Germany
DDR East Germany
DK Denmark
DOM Dominican Republic
DY Benin

DZ Algeria
E Spain (España)
EAK Kenya
EAT Tanzania
EAU Uganda
EC Ecuador
ES El Salvador
ET Egypt
F France
FJI Fiji
FL Liechtenstein
FR Faeroe Islands
GB Great Britain
GBA Alderney
GBG Guernsey
GBJ Jersey
GBM Isle of Man
GBZ Gibraltar

GCA Guatemala
GH Ghana
GR Greece
GUY Guyana
HN Honduras
H Hungary
HK Hong Kong
I Italy
IL Israel
IND India
IR Iran
IRL Ireland
IRQ Iraq
IS Iceland
J Japan
JA Jamaica
K Kampuchea
KWT Kuwait
L Luxembourg
LAO Laos
LAR Libya
LB Liberia
LS Lesotho
M Malta
MA Morocco
MAL Malaysia
MC Monaco
MEX Mexico
MS Mauritius
MW Malawi
N Norway
NA Netherlands Antilles
NIC Nicaragua

NL Netherlands
NZ New Zealand
P Portugal
PA Panama
PAK Pakistan
PE Peru
PL Poland
PNG Papua New Guinea
PY Paraguay
R Romania
RA Argentina
RC Taiwan
RCA Central African Republic
RCB Congo
RCH Chile
RH Haiti
RI Indonesia
RIM Mauritania
RL Lebanon
RM Madagascar
RMM Mali
RN Niger
ROK Korea
RP Philippines
RSM San Marino
RSR Zimbabwe
RU Burundi
RWA Rwanda
S Sweden
SD Swaziland
SF Finland
SGP Singapore
SME Suriname

SN Senegal
SP Somalia
SU Soviet Union (USSR)
SY Seychelles
SYR Syria
T Thailand
TC Cameroon
TG Togo
TN Tunisia
TR Turkey
TT Trinidad and Tobago
U Uruguay
USA United States of America
V Vatican
VN Vietnam
WAG Gambia
WAL Sierra Leone
WAN Nigeria
WD Dominica
WG Grenada
WL Saint Lucia
WS Western Samoa
WV Saint Vincent
YMN North Yemen (Yemen Arab Republic)
YU Yugoslavia
YV Venezuela
Z Zambia
ZA South Africa
ZA South-West Africa (Namibia)
ZRE Zaire

International Yacht Racing Union Nationality Codes

Every yacht of an international class recognized by the International Yacht Racing Union must carry on her mainsail, when *racing* in foreign waters, a letter or letters showing her nationality.

A Argentina
AR United Arab Republic
B Belgium
BA Bahamas
BL Brazil
BU Bulgaria
CA Cambodia
CY Ceylon
CZ Czechoslovakia
D Denmark
E Spain
EC Ecuador
F France
G West Germany
GO East Germany
GR Greece
H Holland
HA Netherlands Antilles
I Italy
IR Republic of Ireland

K United Kingdom
KA Australia
KB Bermuda
KC Canada
KG Guyana
KGB Gibraltar
KH Hong Kong
KI India
KJ Jamaica
KK Kenya
KR Zambia, Malawi, Zimbabwe
KS Singapore
KT West Indies
KZ New Zealand
L Finland
LE Lebanon
LX Luxembourg
M Hungary
MA Morocco

MO Monaco
MX Mexico
N Norway
NK Democratic People's Republic of Korea
OE Austria
P Portugal
PH Philippines
PR Puerto Rico
PU Peru
PZ Poland
RC Cuba
RI Indonesia
RM Romania
S Sweden
SA South Africa
SE Senegal
SR Union of Soviet Socialist Republics
T Tunisia

TH Thailand   US United States of America   Y Yugoslavia
TK Turkey   V Venezuela   Z Switzerland
U Uruguay   X Chile

Inventions and Inventors

adding machine Blaise Pascal and William Burroughs
addressograph J S Duncan
air brake George Westinghouse
air conditioning Willis H Carrier
airplane Orville and Wilbur Wright
analog computer Vannevar Bush
autogyro Juan de la Cierva
automatic rifle John Moses Browning
automobile Karl Benz
bakelite Leo H Baekeland
ballpoint pen John Loud
barbed wire Joseph Glidden
barometer Evangelista Torricelli
Bessemer steel Henry Bessemer
bicycle Karl Drais von Sauerbronn
bifocals Benjamin Franklin
Braille printing Louis Braille
Bunsen burner Robert Wilhelm von Bunsen
calculus Wilhelm von Leibniz and Isaac Newton
carbon-filament electric lamp Thomas A Edison
carburetor Gottlieb Daimler
carpet sweeper M R Bissell
cash register James Ritty
cathode-ray tube Sir William Crookes
cellophane Jacques E Brandenberger
celluloid Alexander Parkes
celluloid-film photography Hannibal Goodwin
cement Joseph Aspdin
chronometer John Harrison
coal stove Jordan Mott
color photography Gabriel Lippmann
compound microscope Hans and Zacharias Janssen
condenser steam engine James Watt
cotton gin Eli Whitney
cream separator Gustf de Laval
cyclotron Ernest O Lawrence
cylinder lock Linus Yale
cylinder phonograph Thomas

A Edison
detective story Edgar Allan Poe
Diesel engine Rudolf Diesel
digital computer Howard Aiken
diode-tube radio John Fleming
disk brake Frederick Lanchester
disk phonograph Emile Berliner
dynamite Alfred Nobel
electric battery Alessandro Volta
electric fan Schuyler Wheeler
electric flat iron H W Seeley
electric generators Michael Faraday (disk), Hippolyte Pixii (coil)
electric locomotive Werner von Siemens
electric motors Zenobe Gramme (DC), Nikola Tesla (AC)
electric power plant Thomas A Edison
electric razor Jacob Schick
electric stove W S Hadaway
electric telegraph Samuel F B Morse
electric washing machine Alva Fisher
electric welder Elihu Thompson
electrocardiograph Willem Einthoven
electroencephalograph Hans Berger
electromagnet Joseph Henry, William Sturgeon
electronic television Vladimir Zworykin
electron microscope Ernst Ruska
electroplating George and Henry Elkington
Esperanto Lazarus Ludwig Zamenhof
field ion microscope Erwin W Mueller
flush toilet Joseph Brahmah
foot-operated sewing machine Isaac Singer
fountain pen Lewis Waterman
Franklin stove Benjamin Franklin
frequency modulation Edwin

H Armstrong
frozen food Clarence Birdseye
gas lighting William Murdock
gasoline-powered automobile Karl Benz
gas stove Robert Wilhelm von Bunsen
Geiger counter Hans Geiger
gyrocompass Elmer A Sperry
gyroscope Jean Bernard Léon Foucault
half-tone engraving Frederick Ives
helicopter Igor Sikorsky
holography Dennis Gabor
hot-air balloon Jacques and Joseph Montgolfier
jet aircraft engine Sir Frank Whittle
Kodak camera George Eastman
laser Gordon Gould
lever-fill fountain pen W A Shaeffer
lightning rod Benjamin Franklin
linoleum Frederick Walton
linotype typesetting Ottmar Mergenthaler
liquid-fuel rockets Robert H Goddard, Herman Oberth, Konstantin Tsiolkovsky
lithography Alois Senefelder
long-playing records Peter Goldmark
machine gun Richard J Gatling
magnetic tape recording J A O'Neill
manual washing machine Hamilton Smith
mechanical television John L Baird
metronome Dietrich Nikolaus Winkel
microwave oven Percy L Spencer
milking machine Anna Baldwin
miner's safety lamp Humphry Davy
modern oilwell Edwin Drake
monotype typesetting Tolbert Lanston
motion picture camera Thomas A Edison
motorcycle Gottlieb Daimler
motor scooter Greville Brad-

shaw
movable-type printing Johann Gutenberg
music synthesizer Robert A Moog
neon lamp Georges Claude
nylon Wallace Carothers
offset printing Ira Rubel
oil pipeline Samuel van Syckel
outboard engine Ole Evinrude
parking meter Carlton Magee
pendulum clock Christiaan Huygens
photoelectric cell G R Carey
photoengraving William Fox Talbot
photographic camera Joseph Niepce
photography on metal Joseph Niepce
photography on paper William Fox Talbot
piston steam engine Thomas Newcomen
pneumatic tires John B Dunlop
Polaroid camera Edwin Land
power loom Edmund Cartwright
projected motion pictures Auguste and Louis Lumière
punched cards Herman Hollerith
radar Albert Taylor, Leo Young
radio Guglielmo Marconi
radio telescope Karl Jansky
railway car couplings Eli Jannery
railway sleeping cars George

Pullman
rayon Hilaire Chardonnet
reaping machine Cyrus H McCormick
reflecting telescope Isaac Newton
refracting telescope Galileo Galilei
repeating rifle O F Winchester
revolver Samuel Colt
rigid dirigible airship Ferdinand von Zeppelin
rotary printing Richard Hoe
safety elevator Elisha G Otis
safety pin Walter Hunt
safety razor King C Gillette
science of flight Sir George Cayley
screw propeller John Ericsson
self-propelled torpedo Robert Whitehead
self-starter Charles F Kettering
semaphore telegraph Claude Chappe
sewing machine Elias Howe, Barthélemy Thimonnier
sextant John Hadley, Thomas Godfrey
spectroscope Joseph von Fraunhofer
spinning frame Richard Arkwright
spinning jenny James Hargreaves
stainless steel Harry Brearley
steam locomotive Richard Trevithick

steam railway George Stephenson
steamship Robert Fulton
steam shovel William Otis
steam tractor Nicholas Cugnot
steam turbine Charles Parsons
steel plow John Deere
submarine David Bushnell
swim fins Benjamin Franklin
telescope Hans Lippershey
thermometer Galileo Galilei
thermos bottle James Dewar
transformer Otto Blathy
transistor John Bardeen, Walter Brattain, William Shockley
transparent paper-film photography George Eastman
triode vacuum tube Lee De Forest
tungsten filament Irving Langmuir
tungsten-filament electric lamp William Coolidge
typewriter Carlos Glidden, Christopher Sholes
vacuum cleaner I W McGaffey
vocal radio Valdemer Poulsen
vulcanized rubber Charles Goodyear
waterproof raincoat Charles Macintosh
wire-cable suspension bridge John A Roebling
xerography Chester Carlson
X-ray Wilhelm Roentgen
zipper Whitcomb Judson

Irish Counties Abbreviated

Republic of Ireland

Car County Carlow
Cav County Cavan
Clare County Clare
Cork County Cork
Don County Donegal
Dub County Dublin
Gal County Galway
Ker County Kerry
Kild County Kildare

Kilk County Kilkenny
Leit County Leitrim
Leix County Leix
Lim County Limerick
Long County Longford
Louth County Louth
Mayo County Mayo
Meath County Meath
Monag County Monaghan

Off County Offaly
Ros County Roscommon
Sligo County Sligo
Tipp County Tipperary
Wat County Waterford
Westmeath County Westmeath
Wex County Wexford
Wick County Wicklow

Mexican State Names and Abbreviations

Ags Aguascalientes (inhabitants called Hidrocalidos)
BC Baja California (Baja Californianos)
BC Front Baja California Fron-

teriza (Frontier Baja California)
BC Sur Baja California Sur (Baja California South)
Cam Campeche (Campecha-

nos)
Chih Chihuahua (Chihuahuenses)
Chis Chiapas (Chiapanecos)
Coah Coahuila (Coahuileños or

Coahuilenses)
Col Colima (Colimenses)
DF Distrito Federal (Federal District around Mexico City; Capitolinos)
Dgo Durango (Durangueños or Duranguenses or Durangueses)
Gro Guerrero (Guerreros)
Gto Guanajuato (Guanajuatos)
Hgo Hidalgo (Hidalgos)
Jal Jalisco (Jalisciences)

Méx México (Mexicanos)
Mich Michoacán (Michoacanos)
Mor Morelos (Morelianos)
Nay Nayarit (Nayaritos)
NL Nuevo León (Nuevo Leones)
Oax Oaxaca (Oaxaqueños)
Pue Puebla (Poblanos)
Qro Querétero (Queretanos)
Q Roo Quintana Roo (Quintana Roenses)

Sin Sinaloa (Sinaloenses)
SLP San Luís Potosí (Potiseños)
Son Sonora (Sonorenses)
Tab Tabasco (Tabasqueños)
Tam Tamaulipas (Tamaulipecos)
Tlax Tlaxcala (Tlaxcaltecas)
Ver Veracruz (Veracruzanos)
Yuc Yucatán (Yucatecos)
Zac Zacatecas (Zacatecos)

Musical Superlatives

ABCs of Opera *Aida, Boheme, Carmen*
Adam's Most Popular Ballet *Giselle*
Adieu Chopin's Polonaise in B-flat minor
Aegyptische Die Aegyptische Helena (German—The Egyptian Helen), Richard Strauss's music drama
Aeolian Symphony Rachmaninoff's Symphony No 3 in A minor
Africaine L'Africaine (French—The African Girl), Meyerbeer opera
African-American Arranger-Composer-Conductor Eva Jessye
Age of Anxiety Bernstein's Symphony No 2
Albeniz's Most Popular Piano Piece *Iberia*
Albinoni's Most Popular Piece *Adagio for Strings and Organ*
Al combate corred bayameses (Spanish—Swift in combat, men of Bayamo), Cuban national anthem
Alfven's Most Popular Overture *Midsommervaka* (Swedish—Midsummer-Night Vigil)
All Men are Brothers Symphony No 11 by Alan Hovhaness
Alpine Alpine Symphony (symphonic poem by Richard Strauss—*Eine Alpensinfonie*)
Alpine Waterfall Saint-Saëns Piano Concerto No 3 in E flat major
Alsatian musicians conductor Charles Munch; organists Edvard Nies-Berger and Albert Schweitzer; composer Émile Waldteufel
Amahl Amahl and the Night

Visitors (Menotti opera)
Amelia Amelia Goes to the Ball (Menotti one-act comic opera)
American Dvořák's Quartet in F (opus 96) for two violins, viola, and cello
American Symphony Orchestras of Greatest Distinction Boston Symphony, New York Philharmonic, Philadelphia Orchestra, Chicago Symphony, Cleveland Orchestra, St Louis Symphony, Los Angeles Philharmonic, San Francisco Symphony, Seattle Symphony
An Donau An der schönen blauen Donau (German—On the beautiful blue Danube) —waltz by Johann Strauss Jr
And This Is My Beloved Borodin's Nocturne from his String Quartet No 2
Anniversary Waltz Ivanovici's fanfare titled *Danube Waves*
Antar Rimsky-Korsakov's Symphony No 2
Antheil's Most Popular Work *Ballet Mecarique*
Apotheosis of the Dance Beethoven's Symphony No 7
Appassionata Beethoven's Piano Sonata No 23 in F minor (opus 57); nicknamed for its impassioned mood
Archduke Beethoven's Trio in B minor (opus 97); dedicated to Archduke Rudolph
Arctic Vassilenko's Fourth Symphony
Arensky's Most Popular Ballet *Egyptian Night*
Argentinian Composers Alberto Ginastera, Juan Guiterrez, Roberto Morillo, Alberto Williams
Ariadne Ariadne auf Naxos

(German—Ariadne on Naxos) one-act opera by Richard Strauss
Aristocrat of Orchestras The Boston Symphony Orchestra
Arlésienne L'Arlésienne Suite No 1 and Suite No 2 by Bizet
Arne's Most Popular Oratorio *Judith*
Auber's Most Popular Overture *Masaniello*
Aus der Neuen Welt (German—From the New World) Dvořák's Ninth Symphony
Aus Ital Aus Italien (German—From Italy)—symphonic poem by Richard Strauss
Aus meinem Aus meinem Leben (German—From My Life)—Smetana's autobiographical String Quartet No 1 in E minor
Australian composers Arthur L Benjamin, Peggy Glanville-Hicks, Percy Aldridge Grainger
Australian Duo classical composers Percy Grainger and Arthur Benjamin
Austrian composers Alban Berg, Anton Bruckner, Franz Joseph Haydn, Wolfgang Amadeus Mozart, Arnold Schoenberg, Franz Schubert, the Strauss family (Johann, Johann Jr, the two Josephs)
Austrian Waltz Kings Josef (Franz Karl) Lanner and Johann Strauss, Jr
Babi Yar Symphony No 13 of Shostakovich inspired by poems of Yevtushenko
Bach's Most Popular Chaconne *Partita in D minor for Unaccompanied Violin*
Bach's Most Popular Orchestral Works 6 *Brandenburg*

Concerti
Bach's Most Popular Piece
Air for the G string
Bach's Most Popular Work
Toccata and Fugue in D minor
Balfe's Most Popular Opera
The Bohemian Girl
Ballo Un Ballo in Maschera
(Italian—A Masked Ball),
three-act opera by Verdi
Banditenstreiche (German—
Jolly Robbers) von Suppé
overture
Barber Barber of Seville (Rossini opera)
Barber's Most Popular Symphonic Work *Adagio for Strings*
Barbiere Il Barbiere di Siviglia
(Italian—The Barber of Seville), opera by Rossini
Baritone of the Century
Dietrich Fischer-Dieskau
(1925-)
Bartók's Most Popular Symphonic Work *Concerto for Orchestra*
Bass of the Century Feodor
Chailiapin (1873–1938),
George London (1920–1985)
basset-horn tenor clarinet
Bassoonist of the Century Sol
Schoenbach of the Philadelphia Orchestra
Bayang megiliw (Tagalog—
Land of the Morning) Philippine national anthem
Beethoven 1st, 2nd, 3rd, 4th,
5th, 6th, 7th, 8th, 9th First,
Second, Third *(Eroica),*
Fourth, Fifth, Sixth *(Pastoral),* Seventh, Eight, Ninth
(Choral)—symphonies composed by Beethoven
Beethoven's Most Popular
Overture *Leonore No 3* from
Fidelio
Beethoven's Most Popular Piano Concerto *No 5—Emperor*
Beethoven's Most Popular Piano Sonata *No 14—Moonlight*
Beethoven's Most Popular
Song *Adelaide*
Beethoven's Most Popular
Symphony *No 5 in C minor*
Beggar's The Beggar's Opera
(ballad opera by John Christopher Pepusch with libretto
by John Gay)
Belgian composers Charles
Bériot, César Franck, André
Grétry, Henri Vieuxtemps
Belgian Quartet Belgium's

best-known classical composers, ranked chronologically,
include Gretry, Vieuxtemps,
Franck, and Ysaÿe
Bellini's Most Popular Opera
Norma
Bells (see *The Bells*)
Bells of Zionice Dvořák's
Symphony No 1
Benvenuto Benvenuto Cellini,
Berlioz opera nicknamed *Malvenuto* by critics
Berg's Most Popular Opera
Lulu
Berlin Bach Carl Philipp
Emanuel Bach, also called the
Hamburg Bach
Berlioz's Most Popular Opera
Damnation of Faust
Berlioz's Most Popular Symphonic Work *Symphonie fantastique*
Berlioz symphonies *Symphonie Fantastique* and *Symphonie Funèbre et Triomphale*
Bernstein's Most Popular Ballet *Dybbuk*
Bernstein's Most Popular
Overture *Candide*
best known Russian art song
see *None But The Lonely Heart*
Big Seven America's leading
symphony orchestras—Boston, Chicago, Cleveland, Los
Angeles, New York, Philadelphia, Pittsburgh
Big Six America's leading symphony orchestras—Boston,
Chicago, Cleveland, Los Angeles, New York, Philadelphia
Bird Haydn's String Quartet in
C (opus 33, no 3)
Birds The Birds (Respighi's
symphonic poem—*Gli Uccelli)*
Bizet's Most Popular Opera
Carmen
Bizet's Most Popular Song
Agnus Dei
Bizet's Most Popular Work
L'Arlesienne Suites 1 & 2
Black Conductor of the Century Dean Dixon (1915–
1976)
Black Contralto of the
Century Marian Anderson
(1902-)
Black Key Chopin's Piano
Etude No 5 in G-flat major
Black Orchestra Leader of
the Century Count Basie
(1904–1984)
Black Pianist of the Century
Andre Watts (1946-)
Black Soprano of the Century

Leontyne Price (1927-)
Black Tenor of the Century
Roland Hayes (1887–1976)
Black Trumpeter of the Century Louis (Satchmo) Armstrong (1900–1971)
Blind Composer Joaquin
Rodrigo born near Valencia,
Spain, remembered for his
guitar *Concerto Arunjuez*
Bloch's Most Popular Rhapsody *Schelomo for cello and orchestra*
*Bluebeard's Duke Bluebeard's
Castle* (Bartók's one-act opera)
Boccanegra Simone Boccanegra (three-act Verdi opera)
Bohème La Bohème (Puccini
opera)
Bohemian composers Antonin
Dvořák, Leoš Janacek, Gustav
Mahler, Bedrich Smetana,
Josef Suk
Boieldieu's Most Popular
Overture *Caliph of Bagdad*
Boito's Most Popular Opera
Mefistofele
Boris Boris Godunov (Mussorgsky's opera)
Borodin's Most Popular Opera *Prince Igor*
Brahms' 4 Brahms's four symphonies
Brahms' Most Popular Overture *Academic Festival*
Brahms' Most Popular Symphony *No 1 in C minor*
Brandenburg Bach's Brandenburg Concertos dedicated to
Duke Christian Ludwig of
Brandenburg, Germany
Brazilian composers Antonio
Carlos Gomes, Mozart Camargo Guarnieri, Heitor Villa-Lobos
Brazilian Trio Brazil's three
best-known classical composers—Antonio Carlos Gomes,
Alberto Nepomuceno, and
Heitor Villa-Lobos
Britten's Most Popular Opera
Peter Grimes
Britten's Most Popular Work
*Young Person's Guide to the
Orchestra*
Brooklyn composers Aaron
Copland, George and Ira
Gershwin, Roger Sessions
(see New York composers)
Bruch's Most Popular Violin
Concerto *No 1 in G*
Bruckner Conductor Eugen
Jochum
Bruckner's 10 Bruckner's ten
symphonies comprising *Die*

Nullte (The Zero) and 1 through 9 including *Romantische* (Romantic, No 4)

Bruckner's Most Popular Symphony *No 4—Romantic*

bup-bup-bup-bum Beethovenian kettledrumming

Butterfly Chopin's Piano Etude No 9 in G flat; Puccini's opera *Madame Butterfly*—a Japanese tragedy

Calife Le Calife de Bagdad (French—The Caliph of Bagdad)—one-act opera by Boieldieu

California composers Henry Cowell, Harry Partch

Calm Sea Calm Sea and Prosperous Voyage (Goethe's *Meeresstille und glückliche Fahrt*)—Beethoven's choral work with orchestra or Mendelssohn's concert overture

Campanello Il Campanello di Notte (Italian—The Night Bell)—Donizetti opera

Carpenter's Most Popular Ballet *Skyscrapers*

Cavalleria Cavelleria Rusticana (Mascagni opera)

Cavalleria espanola Massenet's verismo opera *La Navarraise,* also called *Calvélleria espanola* after Emma Calvé

Cav-Pag Cavalleria Rusticana and *I Pagliacci* (Italian operas frequently performed in succession)

Celestial Gate Symphony No 6 by Alan Hovhaness

Cellini Benvenuto Cellini (opera by Berlioz, nicknamed *Malvenuto* by critics)

Cellist of the Century Pablo Casals (1876–1973)

Cellist-Conductor-Composer Pablo Casals

Cellist-Conductors Barbirolli, Casals, Herbert, Kindler, Rostropovitch, Toscanini, Wallenstein

'Cello Symphony Symphony No 2 by Taaffe Zwilic

Cenerentola La Cenerentola (Italian—Cinderella)—Rossini opera

Chabrier's Most Popular Overture *España*

Chaconne Bach's Partita No 2 in D minor for solo violin; or its transcription for the guitar of Segovia by Marc Pincherle; or for piano by Brahms, Busoni, Mendelssohn, Raff, or Schumann; or for orches-

tra by Hubay, Stokowski, or Wilhelmj

Charpentier's Most Popular Opera *Louise*

Chausson's Most Popular Piece *Poème for Violin and Orchestra*

Chavez' Most Popular Symphony *Sinfonia India*

Cherubini's Most Popular Overture *Anacreon*

Chilean composers Pedro Allende, Próspero Biquert, Domingo Santa Cruz

Chilean First Chile's outstanding classical composer—Pedro Umberto Allende

Chopin's Most Popular Piano Concerto *No 1 in E minor*

Choral Beethoven's Symphony No 9 in D minor whose last movement contains Schiller's *Ode to Joy* sung by chorus and soloists with full orchestral support

Choreographic Symphony Ravel's name for his *Daphnis et Chloé* ballet

Christmas sobriquet of Corelli's Concerto Grosso Opus 8 Number 8, Rimsky-Korsakov's *Christmas Eve* opera, Bach's *Christmas Oratorio,* Haydn's *Christmas Symphony* in D minor—No 26 also called *Lamentatione*

Clara Robert Schumann's Symphony No 4 in D minor

Clarinetist of the Century Benny Goodman (1909–1986)

Classical Prokofiev's Symphony No 1

Clock Haydn's Symphony No 101 in D major

Columbia the Gem of the Ocean symphonic poem by Charles Ives

Compliment Beethoven's String Quartet in G major Opus 18 No 2

Composer-Bandmaster Edwin Franko Goldman; Ivan Ivanovici; John Philip Sousa

Composer-Chemist Alexander Borodin

Composer-Conductor Johann Sebastian Bach; Hector Berlioz; Carlos Chávez; Aaron Copland; Edward Elgar; Carlos Gomes; Morton Gould; George Friedrich Handel; Ferde Grofé; Howard Hanson; Aram Khachaturian; Franz Liszt; Gustav Mahler; Felix Mendelssohn; Carl

Nielson; Oscar Straus; Richard Strauss; Franz von Suppé; Peter Ilyich Tchaikovsky; Heitor Villa-Lobos; Carl Maria von Weber; Richard Wagner

Composer-Conductor-Cellist Pablo Casals; Victor Herbert

Composer-Conductor-Critic Hector Berlioz

Composer-Conductor-Educator Leonard Bernstein; Walter Damrosch; Howard Hanson; Edward MacDowell; André Previn

Composer-Conductor-Musicologist Hector Berlioz; Nicholas Slonimsky; Richard Wagner

Composer-Conductor-Organist-Pianist Camille Saint-Saëns, Sir Charles Villiers Stanford, Sir Arthur S Sullivan

Composer-Conductor-Pianist Beethoven, Bernstein, Britten, Damrosch, Dohnanyi, Foss, Gottschalk, Grainger, Liszt, Mendelssohn, Prokofiev, Rachmaninoff, Stravinsky, and Villa-Lobos

Composer-Conductor-Pianist-Statesman Ignacy Jan Paderewski

Composer-Conductor-Pianist-Violinist Georges Enesco; Bedrich Smetana

Composer-Conductor-Pianist-Violinist Teacher Georges Enesco

Composer-Conductor-Violinist Hans Christian Lumbye; Juventino Rosas; Johann Strauss; Johann Strauss Jr; Josef Strauss; Eugéne Ysaye

Composer-Critic Joseph McCabe; Robert Schumann; Carl Shapiro; Virgil Thomson

Composer-Orchestrators Hector Berlioz; Maurice Ravel; Nikolai Rimsky-Korsakov; Richard Strauss; Richard Wagner

Composer-Organist Johann Sebastian Bach; Anton Bruckner; Dietrich Buxtehude; César Franck; Charles Gounod; George Friedrich Handel; Camille Saint-Saëns

Composer-Pianist Ludwig van Beethoven; Johannes Brahms; Frédéric Chopin; George Gershwin; Percy Grainger; Franz Liszt; Edward MacDowell; Wolfgang Amadeus Mozart; Sergei Prokofiev;

Sergei Rachmaninoff; Robert Schumann

Composer-Pianist-Conductor Ludwig van Beethoven; Louis Moreau Gottschalk; Franz Liszt; Sergei Rachmaninoff; Camille Saint-Saëns

Composer-Violinist Kreisler, Paganini, Sarasate, Vieuxtemps, Vivaldi, and Wieniawski

Computer Composer Yannis Xenakis

Conductor-Cellist Barbirolli, Casals, Rostropovich, Toscanini, and Wallenstein

Conductor of the Century Arturo Toscanini (1867–1957), Herbert von Karajan (1908–)

Conductor-Chorus Master Frank Damrosch, Robert Shaw, Roger Wagner

Conductor-Composer George Barati; Pierre Boulez; Stanislaw Skrowaczewski

Conductor-Composer-Pianist Leonard Bernstein; Walter Damrosch; André Previn

Conductor-Double-Bass Serge Koussevitzky, Henry Lewis, Zubin Mehta

Conductor-Educator Leonard Bernstein, Walter Damrosch, André Previn

Conductor-Organist Edouard Nies-Berger; Leopold Stokowski; Walter Teutsch

Conductor-Organist-Pianist Eduard Nies-Berger, Leopold Stokowski, and Walter Teutsch

Conductor-Pianist Ashkenazy, Barenboim, Dello Joio, Foss, Ganz, Hendl, Iturbi, Mitropoulos, Previn, Solti, Szell, von Karajan, Walter, and Zinman

Conductor-Singer Placido Domingo

Conductor-Violinist Boskovsky, Brusilow, Burgin, Giulini; Haitink, Katims; (Daniel) Lewis, Menuhin, Munch, Oisrakh (father and son), Ormandy, Paganini, Piastro, Schneider, Silverstein, Stern, and Zukerman

Connecticut Composer Charles Ives

Conservatory Concerto for the Piano Mendelssohn's *Concerto No 1 in G-minor* opus 25

Conservatory Concerto for the Violin Mendelssohn's

Concerto in E opus 64

Contralto of the Century Marian Anderson (1902–)

Copland's Most Popular Work *Appalachian Spring*

Coq Le Coq d'Or (French—The Golden Cock) Rimsky-Korsakoff opera

Cornerstone of Western Music Beethoven's Ninth Symphony, according to conductor Seiji Ozawa

Coronation Mozart's Mass in C or his Piano Concerto in D major (K 537)

Cosi Cosi Fan Tutti (Italian—Thus Do They All)—opera by Mozart

Cuban composers Alejandro García Caturla, Amadeo Roldán, Eduardo Sanchez de Fuentes y Peláez

Czar Czar und Zimmermann (German—czar and carpenter)—Lortzing opera

Czech Duo Czechoslovakia's best-known classical composers Bedrich Smetana and Antonin Dvořák

d'Albert's Most Popular Opera *Tiefland* (German—lowland)

Dalmatian First Yugoslavia's best-known operetta composer of the last century— Franz von Suppé

Damnación de Fausto (Spanish—Damnation of Faust)— four-part dramatic legend composed by Berlioz

Damnation La Damnation de Faust (French—The Damnation of Faust)—four part legend by Berlioz

Damrosch's Most Popular Opera *Man Without a Country*

Danish Composers Niels Vilhelm Gade, Friedrich Kuhlau, Hans Christian Lumbye, Carl August Nielsen

Danish Quintet Denmark's five best-known classical composers Buxtehude, Kuhlau, Lumbye, Gade, Nielsen

Dannazione di Faust (Italian—Damnation of Faust)— four-part legend by Berlioz

Danse macabre (French—dance of death) Saint-Saëns work

Dante tone poem by Liszt

Danube Waves Ivanovici's fanfare, *Anniversary Waltz*

Daphnis Ravel's ballet *Daphnis et Chloe*

Das Judenthum Das Judenthum in der Musik (German—Jewry in Music)—Richard Wagner essay

Das Lied Das Lied von der Erde (German—The Song of the Earth)—Mahler work

Death and Death and Transfiguration (symphonic poem by Richard Strauss—*Tod und Verklärung*)

Death and Life Gounod's *Mors et Vita* Mass

Death and the Maiden Schubert's Quartet No 14 in D minor

Debussy's Most Popular Opera *Pelléas and Mélisande*

Debussy's Most Popular Piano Piece *Clair de Lune*

Debussy's Most Popular Symphonic Work *La Mer*

Delibes's Most Popular Aria *Bell Song* in *Lakmé*

Delibes's Most Popular Ballet *Coppélia*

Delius's Most Popular Tone Poem *Over the Hills and Far Away*

Dichter und Bauer (German—Poet and Peasant)—von Suppé overture

Die Frau Die Frau Ohne Schatten (German—The Woman Without a Shadow)—opera by Richard Strauss

Die Nullte (German—The Zero)—Bruckner's Symphony No 0 in D minor

Die schöne Galathée (German—The Beautiful Galatea) —von Suppé overture

Dies Irae (Latin—Day of Wrath)—medieval mass for the dead theme used by romantic composers such as Berlioz, Liszt, and Rachmaninov

d'Indy's Most Popular Work *Symphony on a French Mountain Air* for piano and orchestra

Dissonant Mozart's String Quartet in C (K 465)

Divine Poem Scriabin's Symphony No 3

Doàn Quân Vietnam national anthem

Dohnanyi's Most Popular Orchestral Work *Variations on a Nursery Song*

Domestica Symphonia Domestica by Richard Strauss

Donizetti's Most Popular Opera *Lucia di Lammermoor*

Double Bassist of the Century

Gary Karr, Serge Koussevitsky

Dramatic Symphony *Roméo et Juliette* by Hector Berlioz

Dreigroschen Die Dreigroschenoper (German—The Threepenny Opera)—Kurt Weill's modern reworking of the Beggar's Opera

Dreigroschenoper (German— Threepenny Opera)—Kurt Weill work with a libretto by Bertolt Brecht

Drum Roll Haydn's Symphony No 103 in E-flat major

Du Gamla, Du Fria *Du gamla, du fria, du fjällhöga Nord* (Swedish—You ancient, you free, you mountainous North) —Sweden's national anthem

Dukas's Most Popular Orchestral Work *Sorcerer's Apprentice*

Dulce Patria (Spanish—sweet country)—Chile's national anthem

Dumb Girl Dumb Girl of Portici (Auber opera—*La Muette de Portici*)

Dumky Dvořák trio

Dutch composers Cornelis Dopper, Willem Pijper, Jan Pieterszoon Sweelinck, Johan Wagenaar, Bernard Zweers

Dutchman *The Flying Dutchman* (Wagner opera whose German title is *Der Fliegende Holländer*)

Dutch Quartet Wagenaar, Pijper, Badings, and Otterloo

Dvořák's 9 Dvořák's nine symphonies including *Bells of Zlonice* (No 1) and *From the New World* (No 9)

Dvořák's Most Popular Air *Songs My Mother Taught Me*

Dvořák's Most Popular Concerto *Cello Concerto in B minor*

Dvořák's Most Popular Symphony *No 9—New World*

Egyptian Piano Concerto No 5 by Saint-Saëns

Eine Alpensinfonie (German —An Alpine Symphony) —Richard Strauss tone poem

Eine Kleine Nachtmusik (German—A Little Night Music)—Mozart's Serenade for String Orchestra (K 525)

Ein Heldenleben (German—A Hero's Life)—autobiographical symphonic poem by Richard Strauss

Ein Morgen, ein Mittag, ein Abend in Wien (German—

Morning, Noon, and Night in Vienna)—von Suppé overture

Ein Sommernachtstraum (German—A Midsummer Night's Dream)—Mendelssohn overture

Elektra Richard Strauss tragedy

Elgar's Most Popular Orchestral Work *Enigma Variations*

Elgar's Most Popular Song *Salut d'amour*

Elisir *L'Elisir d'Amore* (Italian—The Elixir of Love)— two-act opera by Donizetti

Elvira Madigan motion picture and nickname of Mozart's Piano Concerto No 21 in C major (K 467)

Emperor Beethoven's Piano Concerto No 5 in E flat; Haydn's String Quartet in C (opus 76, no 3)

Emperor Franz-Joseph of the Austro-Hungarian Empire rolls down the Prater in the imperial coach drum-roll and trumpet-punctuated finale of the *Emperor Waltz* by Johann Strauss Jr

Enesco's Most Popular Orchestral Work *Romanian Rhapsody No 1*

English composers Sir Granville Bantock, Sir Arnold Bax, Benjamin Britten, William Byrd, Frederick Delius, Sir Edward Elgar, Gustav Holst, Henry Purcell, Sir Arthur Sullivan, Thomas Tallis, Ralph Vaughan Williams, Sir William Walton

English Hornist Thomas Stacy of the New York Philharmonic

English Nonet England's nine best-known classical composers—Tallis, Byrd, Purcell, Sullivan, Elgar, Delius, Vaughan Williams, Walton, Britten

English Symphonist Ralph Vaughan Williams

Eroica Beethoven's Symphony No 3 in E-flat major *(Sinfonia eroica)*

Eskimo Opera Hakon Axel Einar Børresen's opera about Greenland Eskimos (produced in Copenhagen in 1921 under the title of *Kaddara*)

Eugen *Eugen Onegin* (Tchaikovsky opera

Evgeny Onyegin (Russian— Eugene Onegin)—Tchaikovsky's most popular opera based on a poem by Pushkin

Falla's (de Falla's) Most Popular Work *Nights in the Gardens of Spain*

Fanciulla *La Fanciulla del West* (Italian—The Girl of the Golden West)—Puccini opera whose libretto recalls David Belasco's play

Fantastique *Symphonie Fantastique* (French—Fantastique Symphony)—composed by Berlioz who subtitled it *Épisode de la Vie d'un Artiste* (Episode in the Life of an Artist)

Farewell Beethoven's Piano Sonata No 32 in C minor (opus 111); Haydn's Symphony No 45 in F-sharp minor

Fate Beethoven's Symphony No 5 in C minor (see *Victory*)

Fauré's Most Popular Work *Elégie for Cello and Orchestra*

Faust *Damnation of Faust* by Berlioz; *Faust* opera by Gounod, *Faust Symphony* by Liszt, *Faust Overtures* by Schumann and Wagner

Fausts Verdammnis (German —Damnation of Faust) dramatic legend composed by Berlioz

Favorite American Anthems *America the Beautiful; God Bless America; My Country 'tis of Thee; Star-Spangled Banner*

Favorite Canadian Anthems *O Canada!; The Maple Leaf Forever*

Favorite Christmas Carol *Heilige Nacht* (German—Holy Night)

FBI in War and Peace Hollywood version of Prokofiev's suite entitled "The Love for Three Oranges"

Fiddlers Three Isaac Stern, Itzhak Perlman, Pinchas Zukerman

Fidelio Beethoven opera

Fingal's Höhle (German— Fingal's Hole)—Mendelssohn overture also called *Hebrides* or *Fingal's Cave*

Finnish composers Armas Järnefelt, Selim Palmgren, Jan Sibelius

Finnish First Finland's best known classical composer— Jean Sibelius

Finnish National Composer Jean Sibelius

Fledermaus *Die Fledermaus*

(German—The Bat)—operetta by Johann Strauss, Jr
Flotow's (von Flotow's) Most Popular Overture *Martha*
Flutist of the Century Jean-Pierre Rampal; James Galway
Forellen Quintet (see *Trout*)
Foremost Belgian Operatic Conductor André Cluytens
Foremost Musical Romanticist Berlioz, Liszt, or Schumann
Forza La Forza del Destino (Italian—The Force of Destiny)—Verdi four-act opera
Foss's Most Popular Work *Night Music for Brasses and Orchestra*
Fountains Fountains of Rome (Repi—*Fontane di Roma*)
Four Saints Four Saints in Three Acts, opera by Virgil Thompson with text by Gertrude Stein
Four Seasons Antonio Vivaldi's concerto *Le Quattro Staggioni*
Four Temperaments Hindemith composition for string orchestra; Nielsen's Symphony No 2
Fra Diavolo Michele Pezza—leading character in Auber's opera *Fra Diavolo*
Francesca Francesca da Rimini (Tchaikovsky symphonic fantasia, Zandonai four-act opera)
Franck's Most Popular Work *Symphonic Variations for Piano & Orchestra*
Franck symphony Cesar Franck's Symphony in D
Freischütz Der Freischütz (German—The Freeshooter) von Weber opera
French Baroque Masters Jean Baptiste Lully, Jean Philippe Rameau
French composers Hector Berlioz, Georges Bizet, François Boïeldieu, Pierre Boulez, Emmanuel Chabrier, Ernest Chausson, François Couperin, Achille-Claude Debussy, Vincent d'Indy, Paul Dukas, Gabriel Fauré, Benjamin Godard, Charles Gounod, Jacnegger, Edouard Lalo, Jules Massenet, Giacomo Meyerbeer, Jacques Offenbach, Jean Philippe Rameau, Camille Saint-Saëns, Ambroise Thomas
French Dozen France's twelve best-known classical composers—Lully, Couperin, Rameau, Berlioz, Gounod, Offenbach, Saint-Saëns, Bizet, Massenet, Debussy, Ravel, Milhaud
Frog Haydn's String Quartet in D (opus 50, no 6)
From the Halls From the Halls of Montezuma to the Shores of Tripoli (US Marine Corps anthem)
From My Life Smetana's String Quartet No 1 in E minor
From the New World Dvořák's Symphony No 9 (formerly No 5)
Full Moon and Empty Arms Rachmaninoff's Second Piano Concerto
Funeral March Sonata Piano Sonata in B-flat minor by Chopin
Gaelic Gaelic Symphony by Mrs HHA Beach (first symphonic work by an American woman)
Georgia composer Wallingford Riegger
German composers Johann Sebastian Bach and his sons, Ludwig van Beethovan, Johannes Brahms, Christoph Willibald von Glück, George Friedrich Handel, Paul Hindemith, Felix Mendelssohn, Robert Schumann, Karlheinz Stockhausen, Richard Strauss, Richard Wagner
German Fourteen Germany's fourteen best-known classical composers: Telemann, Handel, Glück, Beethoven, von Weber, Mendelssohn, Schumann, Wagner, Brahms, Bruch, Strauss, Schoenberg, Hindemith, Weill
German-Polish Composer-Pianist-Teacher Franz Xavier Scharwenka
Gershwin's Most Popular Opera *Porgy and Bess*
Gershwin's Most Popular Work *Rhapsody in Blue*
Ghost Beethoven trio in D major
Gianni Gianni Schicchi (Puccini opera)
Gioconda La Gioconda (Ponchielli opera)
Glazunov 6 Alexander Glazunov's six symphonies
Glazunov's Most Popular Ballet *The Seasons*
Gliere's Most Popular Symphony *No 3—Ilya Mourometz*
Glinka's Most Popular Aria *Sussanin's* in *A Life for the*
Tsar Ivan Sussanin
Glinka's Most Popular Work *Russlan & Ludmila Overture*
Glocken von Zlonice (German—*Bells of Zlonice*)— Dvořák's First Symphony
Glück's (von Glück's) Most Popular Opera *Orfeo ed Euridice*
Goldberg Bach's Goldberg Variations; composed for a keyboard pupil named Johann Gottlieb Goldberg
Golden Flutist Georges Barrère
Goldman's Most Popular March *On the Mall*
Goldmark's Most Popular Symphony *Rustic Wedding*
Gomes's Most Popular Overture *Il Guarany*
Gould's Most Popular Orchestral Piece *American Salute*
Gounod's Most Popular Opera *Faust*
Goyescas Enrique Granados opera
Granados' Most Popular Opera *Goyescas*
Grand Canyon Grand Canyon Suite—symphonic work by Ferdé Grofé
Great The Great Symphony No 9 in C major by Schubert (formerly No 7)
Great C-major Schubert's Symphony No 9
Great Organ Mass Haydn's E-flat *Grosse Orgelmesse*
Greek Composers Nicholas Mantzaros, Dimitri Mitropoulos, Iannis Xenakis
Greek First Iannis Xenakis—best-known classical composer of modern Greece
Grieg's Most Popular Orchestral Suite *Peer Gynt*
Grieg's Most Popular Song *Ich liebe Dich* (German—I Love You)
Grieg's Most Popular Work *Concerto in A minor for Piano*
Griffes' Most Popular Orchestral Piece *The White Peacock*
Grimes Peter Grimes (opera by Britten)
Grofé's Most Popular Orchestral Suite *Grand Canyon*
Guatemala feliz (Spanish—happy Guatemala)—Guatemala national anthem
Guitarist of the Century Andrés Segovia (1893–1987)
Haffner Mozart's Serenade

Suite in D or his Symphony No 35 in D major; both honor the Burgomeister of Salzburg—Sigmund Haffner

Halévy's Most Popular Opera *La Juive* (French—The Jewess)

Halka (Polish—Helen)—Moniuszko's most popular opera and the most popular Polish one

Hamlet funeral march by Berlioz; fantasy overture by Tchaikovsky; opera by Thomas

Hammerklavier (German—hammer-keyboard)—pianoforte; Beethoven Piano Sonatas 28 and 29 in B flat (opus 106)

Ham 'n' Eggs musician's nickname for *Cavalleria Rusticana* and *Pagliacci* as these two go well together

Handel's Most Popular Air *Ombra mai fu* (Italian—Shade of My Tree) from *Xerxes*, best known as Handel's *Largo*

Handel's Most Popular Oratorio *Messiah*

Handel's Most Popular Orchestral Suite *Fireworks*

Hänsel Hänsel und Gretel (Humperdinck's Christmastime entertainment and opera about a brother, sister, parents, and an old witch in a gingerbread house)

Hanson's Most Popular Symphony *No 2—Romantic*

Harold Harold en Italie (Italian—Harold in Italy)—Berlioz symphony with viola solo

Harp Beethoven's String Quartet in E-flat major (opus 74); Chopin's Piano Etude in A flat (opus 25, no 1)

Harpist of the Century Alfredo Casella (1883–1947)

Harpsichordist of the Century Wanda Landowska (1877–1959)

Harris's Most Popular Piece *Cimarron* (symphonic overture)

Hatikvah (Hebrew—The Hope)—Israeli anthem

Haydn's 104 Franz Joseph Haydn's 104 symphonies

Haydn's Most Popular Oratorio *The Seasons*

Haydn's Most Popular Symphony *No 104—London*

Hebrew Opera-Oratorio *Samson et Delila* by Saint-Saëns

Hebrides Mendelssohn overture, also called *Fingal's*

Cave Hercules of Music Christoph Willibald Glück

Hindemith's Most Popular Orchestral Work *Mathis der Maler* (German—Mathis the Painter)

Historical Symphony No 6 of Ludwig Spohr

Hobo Composer Harry Partch (inventor of the forty-three microtone to the octave scale)

Holländer Die Fliegende Holländer (German—The Flying Dutchman)—opera by Wagner

Holst's Most Popular Orchestral Work *The Planets*

Holy Mass Haydn's *Heiligesse in B-flat*

Honnegar's Most Popular Orchestral Work *Pacific 231*

Hornist of the Century Dennis Brain (1921–1957)

Horseman Haydn's String Quartet in G minor (opus 74, no 3)

Hovhaness 39 Alan Hovhaness's thirty-nine symphonies

Hugh Hugh the Drover opera by Ralph Vaughan Williams)

Huguenots Les Huguenots (French—The Huguenots)—Meyerbeer

Humperdinck's Most Popular Opera *Hänsel und Gretel*

Hungarian-born Eminent American Conductors Antal Dorati, Eugene Ormandy, Fritz Reiner, Sir Georg Solti, George Szell

Hungarian composers Béla Bartók, Ernst Dohnanyi, Zoltan Kodaly, Franz Liszt

Hungarian Quartet Hungary's four best-known classical composers—Liszt, Dohnanyi, Bartók, Kodaly

Hunting Mozart's String Quartet in B flat (K 458)

Husitská Dvořák's overture honoring Bohemian patriot Jan Huss

Hymn of Praise Mendelssohn's Symphony No 2 in B-flat major (also known as *Lobgesang*)

Ibert's Most Popular Orchestral Suite *Escales* (French—Ports of Call)

Igor Prince Igor (opera by Borodin)

Il Distrato (Italian—The Absent-Minded)—Haydn's Symphony No 60 in C-major

Ilia Mourometz Gliere's Sym-

phony No 3

I'm Always Chasing Rainbows Chopin's Fantasie Impromptu in C-sharp minor (popularized)

Imperial Haydn's Symphony No 99 in E flat

Impresario Mozart opera

In an 18th-century Drawing Room Mozart's Piano Sonata in C

Indian MacDowell's Suite No 2 for Orchestra introducing American Indian themes

Inextinguishable Nielsen's Symphony No 4

Inno de Mameli (Italian—Hymn of Mameli)—Italy's national anthem honoring Goffredo Mameli

Ippolitov-Ivanov's Most Popular Orchestral Work *Caucasian Sketches*

Irish Irish Rhapsody by Victor Herbert; *Irish Symphony* by Sir Charles Villiers Stanford; *Irish Symphony* by Sir Hamilton Harty

Irish composers William Balfe, Sir Hamilton Harty, Sir Charles Villiers Stanford, William Wallace, Charles Wood

Irish First Ireland's best known cellist-composer-conductor was Victor Herbert, remembered for his operettas produced in the United States where he became a naturalized citizen

Israel Ernst Bloch symphony

Israeli National Composer Ernst Bloch

Italian Mendelssohn's Symphony No 4 in A major

Italiana l'Italiana in Algeri (Italian—The Italian Girl in Algiers)—Rossini opera

Italian Baroque Masters Arcangelo Corelli, Claudio Monteverdi, Antonio Vivaldi

Italian composers Tomaso Albinoni, Vincinzo Bellini, Arrigo Boito, Ferruccio Busoni, Alfredo Casella, Mario Caslnuovo-Tedesco, Luigi Cherubini, Domenico Cimarosa, Muzio Clementi, Arcangelo Corelli, Luigi Dallapicola, Andrea and Giovanni Gabrieli, Giuseppe and Tommaso Giordani, Umberto Giordano, Pietro Mascagni, Claudio Monteverdi, Giovanni Palestrina, Giovanni Pergolese, Giacomo Puccini, Ottorino

Respighi, Gioacchino Rossini, Alessandro and Domenico Scarlatti, Francisco Paolo Tosti, Giuseppe Verdi, Antonio Vivaldi, Ermanno Wolf-Ferrari, Riccardo Zandonai

Italian Fourteen Italy's best-known classical composers—Palestrina, Monteverdi, Fescobaldi, Vivaldi, Scarlatti (2), Cherubini, Paganini, Rossini, Donizetti, Bellini, Verdi, Puccini, Respighi

Italian Girl Italian Girl in Algiers (Rossini opera)

Ivan the Terrible Rimsky-Korsakoff opera

Ives 4 four symphonies by Charles Ives

Ives's Most Popular Orchestral Work *The Fourth of July*

Jeremiah Bernstein's Symphony No 1 commemorating the prophet Jeremiah and his prophecies

Jeu des cartes (French—deck of cards)—Stravinsky ballet

Joke Haydn's String Quartet in E flat (opus 33, no 2)

Juive La Juive (French—The Jewess)—Halévy opera

Jupiter (Latin—Zeus)—Mozart's Symphony No 41 in C major—his last

Kabalevsky's Most Popular Overture *Colas Breugnon*

Kaddish Bernstein's Symphony No 3

Kaiser Kaiser-Waltzer (German—Emperor Waltz)—Johann Strauss Jr's opus 437

Kansas City Composer Virgil Thomson

Katerina Katerina Izmaylova (Russian title of Shostakovich's opera *Lady Macbeth of the Mtsensk District*)

Kettledrum Haydn's Kettledrum Mass in C major *(Paukenmesse)*

Khachaturian's Most Popular Ballet *Gayne*

Khovantchina Mussorgsky opera completed after his death by Rimsky-Korsakoff

Knyaz Knyaz Igor (Russian—Prince Igor)—Borodin's uncompleted opera partly orchestrated by Glazunov and Rimsky-Korsakov

Kodaly's Most Popular Suite *Háry János*

Komische Opera (German—comic opera)—Berlin Opera House

Kreutzer Beethoven's Sonata in

A minor (opus 47) for violin and piano; dedicated to his friend the violinist Rudolphe Kreutzer

Kutchka Mogutchaya Kutchka (Russian—Mighty Handful)—Balakirev, Borodin, Cui, Mussorgsky, and Rimsky-Korsakov

La Chasse Haydn's Quartet in B flat (opus 1, no 1); Haydn's Symphony No 73 in D major (The Hunt)

La Damnation de Faust (French—The Damnation of Faust)—four-part dramatic legend composed by Berlioz

Lady Macbeth Lady Macbeth of the Mtsensk District; Lady Macbeth of Mzensk (Shostakovich opera known to Russians as *Katerina Izmaylova*)

Lakmé Delibes opera

Lalo's Most Popular Overture *Le Roi d'Ys* (French—The King of Ys)

Lalo's Most Popular Work *Symphonie espagnole for violin and orchestra*

Lamentatione Haydn's Symphony No 26 in D minor also called the *Christmas Symphony*

Land of Hope and Glory Elgar's *Pomp and Circumstance,* March No 1

La Reine (French—The Queen)—Haydn's Symphony No 85 in B-flat major

Largest Opera House Metropolitan Opera House in New York's Lincoln Center seats 3,800

Lark Haydn's String Quartet in D (opus 64, no 5)

L'Arlésienne Bizet's Suite No 1 and No 2

Last of the American Romantic Composers Edward MacDowell

Last of the Australian Romantic Composers Percy Grainger

Last of the Austrian Romantic Composers Gustav Mahler

Last of the Belgian Romantic Composers César Franck

Last of the Bohemian Romantic Composers Antonín Dvořák

Last of the Brazilian Romantic Composers Heitor Villa-Lobos

Last of the Danish Romantic Composers Carl Nielsen

Last of the Dutch Romantic Composers Johan Wagenaar

Last of the English Romantic Composers Sir Edward Elgar

Last of the Finnish Romantic Composers Jean Sibelius

Last of the French *Opéra Bouffe* **Composers** Jacques Offenbach

Last of the German Romantic Composers Richard Strauss

Last of the Great Romantic Composers Antonín Dvořák, Edward Elgar, Edvard Grieg, Gustav Mahler, Carl Nielsen, Sergei Rachmaninoff, Camille Saint-Saëns, Jean Sibelius, Richard Strauss

Last of the Hungarian Romantic Composers Franz Liszt

Last of the Italian Opera Buffa Composers Rossini

Last of the Italian Romantic Composers Ottorino Respighi

Last of the Norwegian Romantic Composers Christian Sinding

Last of the Polish Romantic Composers Ignace Jan Paderewski

Last of the Romantic Romanian Composers Georges Enesco

Last of the Russian Romantic Composers Sergei Rachmaninoff

Last of the Spanish Romantic Composers Manuel de Falla

Last of the Swedish Romantic Composers Kurt Atterberg

Leading Bel Canto Composer Gioacchino Rossini

Le Divin Poeme (French— The Divine Poem)—Scriabin's Symphony No 3

Lehár's Most Popular Operetta *The Merry Widow*

Leichte Kavallerie (German—Light Cavalry)—von Suppé overture

Lélio, ou Le Retour á la vie (French—Lélio, or the Return to Life)—Berlioz monodrama sequel to his *Symphonie fantastique*

Lenin Shostakovich's Symphony No 12

Leningrad Shostakovich's Symphony No 7

Leoncavallo's Most Popular Opera *I Pagliacci* (Italian—The Players)

Les Adieux Beethoven's Piano

Sonata No 23 in E flat (opus 81a) *Les Adieux, l'absence, et le retour*—the farewell, the absence, and the return

L'Heure *L'Heure Espagnole* (French—The Spanish Hour) —Ravel operatic farce

Liadov's Most Popular Symphonic Poem *Kikimora*

l'Impériale Haydn's Symphony No 53 in D major

Linz Mozart's Symphony No 36 in C major named for the Austrian town of Linz

Liszt's Most Popular Symphonic Poem *Les Preludes*

Liszt's Most Popular Waltz *Mephisto*

Little Schubert's Symphony No 6 in C

Little C-major Schubert's Symphony No 6

Little Russian Tchaikovsky's Symphony No 2 in C minor

Lobgesang (German—Hymn of Praise)—Mendelssohn's Symphony No 2 in B-flat major

Lohengrin Wagner's three-act romantic opera

London Haydn's Trios No 1 and 2 (for two flutes and cello); Haydn's Symphony No 104 in D major; Symphony No 2 by Vaughan Williams—A London Symphony

London Suite *London Again* or *London Every Day* (symphonic suite by Eric Coates)

Lone Ranger gallop music at the end of Rossini's *William Tell* overture

Longest Opera *Die Meistersinger von Nürnberg* by Wagner (performance time: 5 hours 15 minutes)

Longest Symphony Symphony No 3 in D minor by Mahler (performance time: 1 hour 40 minutes)

Loudest Opera *Damnation of Faust* by Berlioz with its Ride to the Abyss

Loudest Oratorio *Requiem* by Berlioz (score calls for a chorus of 210 and an orchestra of 217)

Loudest Symphony Mahler's *Tragic* Symphony No 6 in A minor

Loudest Undersea Songs sounds of humpback whales

Lucia *Lucia di Lammermoor* (three-act opera by Donizetti)

Lully's Most Popular Ballet *Alceste*

MacDowell's Most Popular

Piano Concerto *No 2 in D minor*

Magic Flute Mozart opera

Mahagonny *Aufstieg und Fall der Stadt Mahagonny* (German—Rise and Fall of the City of Mahagonny)—opera by Kurt Weill with text by Bertolt Brecht

Mahler's 10 Mahler's ten symphonies including the *Resurrection* (No 2), the *Symphony of a Thousand* (No 8), the *Unfinished* (No 10)

Mahler's Most Popular Song Cycle *Das Lied von der Erde* (German—The Song of the Earth)

Mahler's Most Popular Work *Symphony No 1 in D major* —the *Titan*

Maine Composer Walter Piston

Malvenuto nickname critics bestowed on the Berlioz opera *Benvenuto Cellini*

Mamelles *Les Mamelles de Tirésias* (French—The Breasts of Tiresias)—comic opera by Poulenc

Manfred *Manfred Overture* (Schumann); *Manfred Symphony* (Tchaikovsky)

Manon Massenet opera

Manon Lescaut Puccini opera

Manzoni Mass Verdi's Requiem

Mascagni's Most Popular Opera *Cavalleria Rusticana* (Italian—Rustic Chivalry)

Masonic Composer Wolfgang Amadeus Mozart who was a freemason and alluded to the ethical laws of Masonry in his opera The Magic Flute *(Il Flauto Magico, La Flûte Enchantée, Die Zauberflöte)*—his Masonic Funeral Music *(Maurerische Trauermusik)* also reveals his affiliation with the Masonic Order

Massachusetts composers Leonard Bernstein, William Billings, Alan Hovhaness, Daniel Gregory Mason, Lowell Mason

Massenet's Most Popular Ballet *Le Cid*

Mathis *Mathis der Mahler* (German—Mathias Grünewald Painter)—symphonic suite by Hindemith

Ma Vlast (Czechoslovakian—My Fatherland)—Smetana's symphonic poem

May Day Shostakovich's Sym-

phony No 3 also called *May First*

May Night overture by Rimsky-Korsakov

Meerestille Mendelssohn's *Calm Sea and Prosperous Voyage* overture more correctly translated as *Becalmed at Sea and Prosperous Voyage*

Mefistofele (Italian—Mephistopheles)—Boito's opera about the Faust legend

Meistersinger *Die Meistersinger von Nürnberg* (Wagner's opera about The Mastersingers of Nuremberg)

Melusine *Die schöne Melusine* (German—lovely Melusina) —Mendelssohn overture

Mendelssohn's 5 Mendelssohn's five symphonies including *Lobgesang* (No 2), *Scottish* (No 3), *Italian* (No 4), and *Reformation* (No 5)

Mendelssohn's Most Popular Oratorio *Elijah*

Mendelssohn's Most Popular Overture *Midsummer Night's Dream*

Mendelssohn's Most Popular Symphony *No 4 in A major*—the *Italian*

Mendelssohn's Most Popular Work *Concerto in E minor for Violin*

Menotti's Most Popular Work *Amahl and the Night Visitors*

Mexican Composer-Conductor Carlos Chávez or Juventino Rosas

Mexican composers Carlos Chavez, Manuel Ponce, Silvestre Revueltas, Juventino Rosas

Mexican Trio Mexico's three best-known classical composers—Manuel Ponce, Carlos Chávez, Silvestre Revueltas

Mezzo-Soprano of the Century Christa Ludwig (1928–)

Midsommarvaka (Swedish—Midsummer Fete)—Hugo Alvén's rhapsody for orchestra

Midsummer Night's Dream Mendelssohn's incidental music to Shakespeare's play

Mighty Five Balakirev, Borodin, Cui, Mussorgsky, and Rimsky-Korsakov

Milhaud's Most Popular Ballet *Le Boeuf sur le toit* (French—The Ox on the Roof)

Militaire Paganini's Violin Caprice (opus 1, no 14)

Military Haydn's Symphony

No 100 in G major
Minute Chopin's Waltz D flat (opus 64, no 1)
Miracle Haydn's Symphony No 96 in D major
Moïse (French—Moses)—Rossini opera
Montreal Composer Henry Dreyfus Brant
Moonlight Beethoven's Piano Sonata No 14 in C-sharp minor (opus 27, no 2) *Sonata quasi una Fantasia*
Moonlight and Roses Tchaikovsky's Andante Cantabile movement, Symphony No 5 in E minor also called *Moon Love*
Mors et Vita Gounod's Death and Life requiem
Moses Moses in Egypt—Rossini's sacred melodrama in four acts *(Mosè in Egitto)*
Most Admired and Most Discussed Anglo-American Conductor Leopold Stokowski
Most Amazing Composer Wolfgang Amadeus Mozart
Most Mispronounced Symphony Tchaikovsky's Symphony No 6—*Pathétique*
Most Popular American Composers of Musicals Gershwin—*Porgy and Bess*, Rodgers and Hammerstein—*Oklahoma!* and *South Pacific*
Most Popular American Folk Opera Gershwin's *Porgy and Bess*
Most Popular American March Sousa's *Stars and Stripes Forever*
Most Popular American Musical Show *Chorus Line*
Most Popular Anglo-American Anthem Britain's *God Save the Queen*, the same tune as America's *My Country 'tis of Thee*
Most Popular Austrian Operas Mozart's *Don Giovanni*, Johann Strauss's *Fledermaus*
Most Popular Beethoven Piano Concerto *Concerto No 5 in E flat*—*Emperor*
Most Popular Belgian Symphony Franck's *Symphony in D minor*
Most Popular Brahms Piano Concerto *Concerto No 1 in D minor*
Most Popular Canadian Anthems *O Canada!* and *The Maple Leaf Forever*
Most Popular Cello Concertos Dvořák *Cello Concerto in*

B minor; Lalo in *D-minor*; Saint-Saëns *No 1 in A minor*
Most Popular Chopin Piano Concerto *Concerto No 2 in F minor*
Most Popular Christmas Carol *Stille Nacht, heilige Nacht* (German—Silent Night, Holy Night)
Most Popular Czech Symphony Dvořák *Symphony No 9 in E minor* from the *New World*
Most Popular English Conductor Sir Thomas Beecham
Most Popular English Operetta *The Mikado* by Gilbert and Sullivan
Most Popular English Orchestral Work Elgar's *Enigma Variations*
Most Popular French Classic Opera von Glück's *Iphigénie en Aulide* (Iphigenia in Aulis)
Most Popular French Opera Buffa *Orpheus in the Underworld* by Offenbach
Most Popular French Operas Bizet's *Carmen*, Gounod's *Faust*
Most Popular French Romantic Opera Bizet's *Carmen*
Most Popular French Symphony Franck's *Symphony in D minor*
Most Popular German Comic Opera Richard Strauss's *Der Rosenkavalier*
Most Popular German Operas Beethoven's *Fidelio*, Richard Strauss's *Rosenkavalier*, Wagner's *Meistersinger*
Most Popular Grand Opera Verdi's *Aida*
Most Popular Grieg Piano Concerto *Concerto in A*
Most Popular Italian Opera Buffa Rossini's *Barber of Seville*
Most Popular Liszt Piano Concerto *Concerto No 1 in E flat*
Most Popular MacDowell Piano Concerto *Concerto No 2 in D minor*
Most Popular Mass Beethoven's *Missa Solemnis*
Most Popular Modern German Opera *Der Rosenkavalier* by Richard Strauss
Most Popular Mozart Piano Concerto *Concerto No 20 in D minor*
Most Popular Norwegian Symphonic Suites Grieg's *Peer Gynt*

Most Popular Opera Bizet's *Carmen*
Most Popular Operas in French *Carmen* by Bizet, *Faust* by Gounod, *Samson and Delilah* by Saint-Saëns, *The Trojans* by Berlioz
Most Popular Operas in German *Fidelio* by Beethoven, *Fledermaus* by Johann Strauss, *Freischütz* by von Weber, *Rosenkavalier* by Richard Strauss
Most Popular Operas in Italian *Aida* by Verdi, *Barber of Seville* by Rossini, *Don Giovanni* by Mozart, *Madama Butterfly* by Puccini
Most Popular Operas in Russian *A Life for the Czar* by Glinka, *Boris Godounov* by Mussorgsky, *Eugen Onegin* by Tchaikovsky, *Prince Igor* by Borodin
Most Popular Overtures Berlioz's *Benvenuto Cellini*; Rossini's *William Tell*; Tchaikovsky's *1812* and *Romeo and Juliet*; Wagner's *Tannhäuser*
Most Popular Piano Concertos Grieg's *Concerto in A minor*; Rachmaninoff's *Concerto No 2 in C*, and *No 3 in D minor*; Tchaikovsky's *No 1 in B flat*
Most Popular Rachmaninoff Piano Concerto *Concerto No 3 in D minor*
Most Popular Requiem Mass Berlioz, Mozart, or Verdi
Most Popular Russian Grand Opera Mussorgsky's *Boris Godonov*
Most Popular Russian Operas Borodin's *Prince Igor*, Mussorgsky's *Boris Godunov*, Tchaikovsky's *Eugen Onegin*
Most Popular Russian Overture Tchaikovsky's *Overture 1812*
Most Popular Russian Symphonic Suite Rimsky-Korsakov's *Scheherazade*
Most Popular Russian Symphony Tchaikovsky's *Symphony No 4 in F minor*
Most Popular Saint-Saëns Piano Concerto *Concerto No 2 in G minor*
Most Popular Spanish Suite for Piano and Orchestra *Rapsodia española* by Albéniz; *Nights in the Gardens of Spain* by de Falla
Most Popular Symphonic

Poem Liszt's *Les Préludes*
Most Popular Symphonies
Beethoven's *Symphony No 5
in C minor (*also known as
the *Victory Symphony)* and
Fifth, Berlioz's *Fantastique,*
Dvořák's *New World,*
Tchaikovsky's *Fourth*
Most Popular Trio Beethoven's *Archduke* for cello, piano, and violin
Most Popular Viola Concerto
Harold in Italy by Berlioz
Most Popular Violin Concertos Beethoven in *D major,*
Brahms *D major,* Mendelssohn *E minor*
Most Popular Waltz *Blue
Danube* by Johann Strauss
(1825–1899)
Most Prolific Composer
Roland de Lassus of the Netherlands, also known as Orlando di Lasso and Orlandus
Lassus
**Most Versatile Musician of
Our Era** Georges Enesco
(Romanian composer, conductor, pianist, teacher, and
violinist)
Mozart's 41 Mozart's forty-one
symphonies including the
Haffner (No 35), the *Linz* (No
36), the *Prague* (No 38), the
Jupiter (No 41)
Mozart's Most Popular Aria
Finch' han dal vino (Italian—Fetch the Wine) from
Don Giovanni
Mozart's Most Popular Opera
Don Giovanni
**Mozart's Most Popular Piano
Concerto** *No 20 in D minor*
**Mozart's Most Popular String
Suite** *Eine Kleine Nachtmusik*
(German—A Little Night Music)
Mozart's Most Popular Symphony *No 41—Jupiter*
Musical Charlotte Russe Tchaikovsky's Andante cantabile
from his Symphony No 5 in E
minor
Musical Dictator of Dalmatia
Franz von Suppé (Francesco
Ezechiale Ermenegildo Cavaliere Suppé Demelli)
Musical Philosopher Alfred
Brendel
Music Capital of America Los
Angeles and New York
Music Capital of Eastern Europe Vienna
Music Capital of Western Europe London
Music City, U.S.A. Nashville,

Tennessee
Music Man Meredith Willson
**Mussorgsky's Most Popular
Opera** *Boris Godunov*
**Mussorgsky's Most Popular
Work** *Pictures at an Exhibition*
Mysterious Mountain Symphony No 2 by Alan Hovhaness
Nabuco *Nabucodonosor* (opera
by Verdi)
Napoleon of the Waltz Johann
Strauss
National Composer of Norway Edward Grieg
Nelson Mass Haydn's *Nelson-messe* in D-minor
Nerone (Italian—Nero)—Boito
opera
New Orleans Composer Louis
Moreau Gottschalk
New World Dvořák's Symphony No 9 in E minor (formerly No 5)
New York composers Elliott
Carter, Norman Dello Joio,
Edward MacDowell, William
Schuman (*see* Brooklyn composers)
Nicolai's Most Popular Overture *Merry Wives of Windsor*
Nielsen's 6 Carl Nielsen's six
symphonies including *Four
Temperaments* (No 2), *Sinfonia Espansiva* (No 3), *Inextinguishable* (No 4), *Sinfonia
Semplice* (No 6)
Nielsen's Most Popular Symphony *No 4—Inextinguishable*
Nigger non-pejorative nickname for Dvořák's *American
Quartet* filled with Negro spiritual themes
Noisiest Opera Massenet's *La
Navarraise* replete with bells,
cannon, castanets, guns, tambourines, and trumpets plus
chorus and orchestra
None But The Lonely Heart
best known Russian art song
by Tchaikovsky
Nordic Hanson's Symphony No
1
Norma Bellini opera
Norwegian composers Edvard
Hagerup Grieg, Christian Sinding, Johan Severin Svendsen
Norwegian First Norway's best-known classical composer—
Edvard Hagerup Grieg
Norwegian National Composer Edvard Grieg
Nozze Le Nozze di Figaro (Italian—The Marriage of Figaro)

—opera by Mozart
Nuits Nuits d'éte (French—
Summer Nights)—song cycle
by Berlioz including Absence,
Villanelle, Le spectre de la
rose, Sur les lagunes, Au
cimetière, L'Île inconnue
Nursery Song *Variations on a
Nursery Song* by Ernst von
Dohnanyi
Oboist of the Century Bruno
Labate of the New York Philharmonic Symphony; John de
Lancie of the Philadelphia Orchestra; Leon Goossens,
Marcel Tabuteau, or Harold
Gomberg
Ocean Symphony No 2 by
Anton Rubinstein
October Revolution Shostakovich's Symphony No 2
Ode to Heavenly Joy Mahler's
Symphony No 4 in G major
Ode to Joy Beethoven's Symphony No 9 in D minor—
whose closing movement is
based on the text of Schiller's
Ode to Joy
Odysseus Symphony No 25 by
Alan Hovhaness
**Offenbach's Most Popular
Opera** *Les Contes d'Hoffmann* (French—The Tales of
Hoffmann)
**Offenbach's Most Popular
Work** *Gaité Parisienne* (French
—Parisian Gaiety)
O Guarani (Portuguese—the
Guarani)—opera by Carlos
Gomes
Oklahoma composer Roy Harris
Oldest American Opera House
Le Petit Opéra Louisianais (The
Little Louisianian Opera House
begun in 1813, later known as
The Old French Opera House
and the St Charles Theatre)
**Oldest American Symphony
Orchestra** New York Philharmonic founded in 1842 and
merged with New York Symphony in 1928
**Oldest Austrian Symphony
Orchestra** *Wiener Philharmonische Konzerte* (German
—Vienna Philharmonic Concerts), 1842
Oldest British Symphony Orchestra London's Royal Philharmonic Orchestra, 1813
**Oldest Canadian Symphony
Orchestra** Philharmonic Society of Montreal, 1848
Oldest Czechoslovakian Symphony Orchestra Prague's

Česká filharmonie (Czech Philharmonic), 1864

Oldest Dutch Symphony Orchestra Amsterdam's *Concertgebouw* (Concert Building), 1883

Oldest European Symphony Orchestra Rome's *Santa Cecilia*, 1566

Oldest French Symphony Orchestra *Société des Concerts du Conservatoire* of Paris, 1828

Oldest German Symphony Orchestra *Leipzig Gewandhaus Konzerte* (Leipzig Cloth Hall Concerts), 1743

Oldest Hungarian Symphony Orchestra *Budapesti Filharmónin Társaság* (Budapest Philharmonic Society), 1853

Oldest Midwestern Symphony Orchestra St Louis Symphony, 1880

Oldest Performing Opera House in America New York's Metropolitan Opera House, 1883

Oldest Performing Opera House in Argentina *El Teatro Colón* in Buenos Aires, 1908

Oldest Performing Opera House in Australia Sydney's Opera House, 1954

Oldest Performing Opera House in France Paris's Opera, 1875

Oldest Performing Opera House in Great Britain London's Covent Garden Theatre, 1732

Oldest Performing Opera House in Italy Milan's La Scala, 1778

Oldest Performing Opera House in Spain Barcelona's *Teatro Liceo* (Lyceum Theater), 1862

Oldest Performing Opera House in the USSR Leningrad's *Kirov*, 1860

Oldest Spanish Symphony Orchestra Barcelona's *Orquesta Pau Casals* (Catalan—Pablo Casals Symphony), 1919

Oldest Swiss Symphony Orchestra Zürich's *Tonhalle* (German—one Hall), 1613

Oldest Symphony Orchestra in Berlin *Berlin Philharmoniker*, 1882

Oldest Symphony Orchestra in California San Francisco Symphony, 1911

Oldest Symphony Orchestra in Eastern Pennsylvania Philadelphia Orchestra, 1900

Oldest Symphony Orchestra in English-speaking Canada Toronto Symphony Orchestra, 1906

Oldest Symphony Orchestra on the Great Lakes Chicago Symphony, 1891

Oldest Symphony Orchestra in Minnesota Minneapolis Symphony Orchestra, 1903

Oldest Symphony Orchestra in New England Boston Symphony Orchestra, 1881

Oldest Symphony Orchestra in Northern Ohio Cleveland Orchestra, 1918

Oldest Symphony Orchestra in Ohio Cincinnati Symphony Orchestra, 1895

Oldest Symphony Orchestra in Pennsylvania Pittsburgh Symphony Orchestra, 1896

Oldest Symphony Orchestra in Southern California Los Angeles Philharmonic, 1919

Old Maid Old Maid and the Thief (opera by Menotti)

Onegin Evgeny Onyegin (Russian—Eugen Onegin)— Tchaikovsky opera based on a poem by Pushkin

Opera of Operas Mozart's *Don Giovanni*

orchestral horses Vortex and Giaour in the *Damnation of Faust* by Berlioz; *Phaeton's* four steeds in Saint-Saëns' tone poem; the nine horses in Wagner's ride of the Valkyries in *Die Walküre*

Orchestral Orgasm nickname of the *Don Juan* tone poem by Richard Strauss

Orfeo opera by Monteverdi; *Orfeo ed Euridice* (Italian— Orpheus and Euridice)— Glück's most popular opera and orchestral suite

Orff's Most Popular Scenic Cantata *Carmina Burana*

Organ Poulenc's Concerto in G for organ, strings, and timpani; Saint-Saëns Symphony No 3 for orchestra and organ

Organist of the Century E Power Biggs (1906–1977)

Orleanskaya Orleanskaya deva (Russian—Maid of Orleans) —Tchaikovsky opera based on Schiller's tale about Joan of Arc

Ory Le Compte Ory (French—The Count Ory)—opera by Rossini

Otello (Italian—Othello)—Rossini opera; Verdi opera

Oxford Haydn's Symphony No 92 in G major

Oy Veh (Yiddish—Oh My God) —Mahler's *Resurrection* Symphony in C minor

Pag I Pagliacci (Italian—The Players)—opera by Leoncavallo

Paganini's Most Popular Work *24 Caprices for Violin*

Paris Mozart's Symphony No 31 in D major

Parisian Composers Bizet, Boulanger, Charpentier, Chausson, Debussy, d'Indy, Dukas, Gounod, Ibert, Poulenc, Rabaud, Saint-Saëns

Paris symphonies Haydn's symphonies 82 through 87, commissioned in Paris, bearing such names as *l'Ours* (The Bear—82), *La Poule* (The Hen—83), *La Reine* (The Queen—85)

Parsifal seven-hour-long music drama by Wagner

Partch's Most Popular Work *Daphne of the Dunes*

Pastoral Beethoven's Piano Sonata No 15 in D (opus 28); Beethoven's Symphony No 6 in F major (opus 68); Dvořák's Symphony No 8 in G major; Symphony No 3 by Vaughan Williams

Pathétique Beethoven's Piano Sonata No 8 in C minor (opus 13); Tchaikovsky's Symphony No 6 in B minor

Patriotic Symphony No 2 in D major by Sibelius

Pêcheurs de Perles (French —The Pearl Fishers)—opera by Bizet

Peer Gynt drama by Ibsen with incidental music by Grieg

Pelléas Pelléas et Melisande (Debussy's opera)

Pennsylvania composers Samuel Barber, Stephen Foster, Peter Mennin

Pepusch's Most Popular Ballad Opera *The Beggar's Opera*

percussion bells, castanets, chimes, clappers, cymbals, drums, glockenspiels, gongs, marimbas, tambourines, triangles, wood blocks, xylophones

Pianist of the Century Artur Rubinstein (1889–1982)

Pictures Pictures at an Exhibition (Mussorgsky's piano

suite frequently presented in the Ravel orchestration)

Pikovaya Pikovaya dama (Russian—La Pique Dame)— Tchaikovsky opera sometimes sung in English under the title *Queen of Spades*

Pinafore HMS Pinafore or *The Lass that Loved a Sailor* (Gilbert and Sullivan operetta)

Pines The Pines of Rome (Respighi's symphonic poem — *Pini di Roma*)

Pique Pique Dame (French —The Queen of Spades) —opera by Tchaikovsky

Pique Dame (French—Queen of Spades)—Tchaikovsky opera; von Suppé overture

Pirates Pirates of Penzance (Gilbert and Sullivan operetta)

Piston's Most Popular Ballet *Incredible Flutist*

Poem of Ecstacy Scriabin's Symphony No 4

Polish Tchaikovsky's Symphony No 3 in D major

Polish composers Frédéric Chopin, Michal Kondracki, Emil Mlynarski, Ignacy Jan Paderewski, Krystof Penderecki, Karol Szymanowski, Alexander Tansman, Henryk Wieniawski

Polish First Poland's best-known classical composer —Frédéric Chopin

Polonia (Polish—Poland)— Wagner overture; Mlynarski symphony

Poulenc's Most Popular Ballet *Les Biches* (French—The Deer Does)

Poulenc's Most Popular Concerto *Concerto for Organ, Strings, and Timpani*

Prague Mozart's Symphony No 38 in D major

Préludes Les Préludes (Liszt symphonic poem)

Prince Prince Igor (Borodin's opera known to Russians as *Knyaz Igor*)

Printer's Symphony Mendelssohn's Symphony No 2 in B-flat major also known as the Hymn of Praise *(Lobgesang)* celebrating the 400th anniversary of the invention of printing

Prisoner's Chorus part of Beethoven's opera, *Fidelio;* the finale of Act I of *Fidelio,* Beethoven's only opera, is an appeal from political prisoners longing for the scent of

open air as they know their prison is a tomb

prisoner's opera Beethoven's *Fidelio* has all three acts set in a Spanish prison run by a tyrant; memorable for the compassion the composer shows political prisoners

prisoner's work songs outstanding collection compiled and edited by Bruce Jackson in *Wake Up Dead Man— Afro-American Worksongs from Texas Prisons,* Harvard University Press, Cambridge, Mass, 1972

prison scenes set to music Beethoven's opera *Fidelio,* the *Damnation of Faust* by Berlioz, Boito's *Mefistofele,* Gounod's *Faust,* and Puccini's *Tosca* present some of the most musically memorable scenes although there are others by Verdi

Prokofiev's 7 Prokoviev's seven symphonies including the *Classical* (No 1)

Prokofiev's Most Popular Ballet *Romeo and Juliet*

Prokofiev's Most Popular Symphony *No 5*

Prokofiev's Most Popular Work *Peter and the Wolf*

Puccini's Most Popular Aria *E lucevan le stelle* (Italian— And the stars shone brightly) in *Tosca*

Puccini's Most Popular Opera *Madama Butterfly*

Purcell's Most Popular Opera *Dido and Aeneas*

Queen The Queen (La Reine) —Haydn's Symphony No 85 in B-flat major

Queen of Spades Tchaikovsky opera; English title of a Tchaikovsky opera called *La Pique Dame* by the French and *Pikovaya dama* by Russians

Queen Symphony Haydn's *La Reine* (No 85)

Quiet Quiet Flows the Don (Dzerzhinsky's opera known to Russians as *Tikhiy Don*)

Quinten Haydn's String Quartet in D (opus 76, no 2) —nickname refers to the fifth form or grade in Austrian schools

Quixote Don Quixote (Fantastic Variations on a Theme of Knightly Character by Cervantes as composed by Richard Strauss)

Rachmaninoff's 3 Rachmaninoff's three symphonies

Rachmaninoff's 4 Rachmaninoff's four piano concertos

Rachmaninoff's Most Popular Symphonic Poem *The Isle of the Dead*

Rachmaninoff's Most Popular Symphony *No 2 in E minor*

Rachmaninoff's Most Popular Work *Concerto No 2 for Piano*

railroad music *Pacific 231* by Arthur Honneger; *Little Train of the Caipira* from *Bachianas Brasileiras No 2* by Heitor Villa-Lobos

Rain Violin and Piano Sonata in G (opus 78) by Brahms

Raindrop Chopin's Piano Prelude No 15 in D-flat major

Rakóczy traditional Hungarian march used by Berlioz in his *Damnation of Faust* and by Liszt in his Hungarian Rhapsody No 15 in A minor

Rameau's Most Popular Opera *Dardanus*

Rape Rape of Lucretia (Britten opera)

Rasumovsky Beethoven's Quartets in F major, E minor, and C major for two violins, viola, and cello (opus 59, nos 1, 2, 3); dedicated to Count Rasumovsky

Ravel's Most Popular Ballet *Daphnis et Chloé*

Ravel's Most Popular Choreographic Poem for Orchestra La Valse (French— The Waltz)

Ravel's Most Popular Opera *L'Heure Espagnole* (French —The Spanish Hour)

Ravel's Most Popular Work *Bolero*

Reformation Mendelssohn's Symphony No 5 in D major

Requiem Mass for the dead; most memorable composed by Berlioz, Brahms, Bruckner, Cherubini, Dvořák, Fauré, Mozart, Palestrina, and Verdi

Respighi's Most Popular Suite *Gli Uccelli* (Italian—The Birds)

Respighi's Most Popular Work *Fountains of Rome*

Resurrection Mahler's Symphony No 2 in C minor

Revolutionary Chopin's Piano Etude No 12 in C minor

Revolutionary Composer Pierre de Geyter best known for the formerly official communist anthem—the *Internationale*

Reznicek's Most Popular Overture *Donna Diana*

Rheingold *Das Rheingold* (Wagner music drama)

Rhenish Schumann's Symphony No 3 in E-flat major

Riegger 4 four symphonies by Wallingford Riegger

Rigoletto Verdi opera

Rimsky-Korsakov's Most Popular Work *Scheherazade*

Ring Cycle The Ring of the Nibelungen *(q.v.)*

Ring of the Nibelungen Wagner's Ring Cycle consisting of *Das Rheingold* (Rhinegold), *Die Walküre* (Valkyries), *Siegfried,* and *Götterdämmerung* (Twilight of the Gods)

Rodrigo's Most Popular Work *Concierto de Aranjuez for Guitar and Orchestra*

rogues' march quickstep played when offenders are drummed out of the army, the marines, the navy, or other military units; at public floggings and executions it was the custom to have a drummer beat out the rhythm of the rogues' march

Romanian First Romania's best-known classical composer-conductor-pianist-violinist—Georges Enesco

Romanian National Composer Georges Enesco

Romantic Bruckner's Symphony No 4; Hanson's Symphony No 2

Roméo *Roméo et Juliette* (Berlioz symphony for chorus, orchestra, and solo voices)

Rosenkavalier *Der Rosenkavalier* (German—The Red Knight)—Richard Strauss's most popular opera

Rose of Venice Haydn's Quartet in D for Strings (opus 20, no 4)

Rossini's Most Popular Opera *Barber of Seville*

Rossini's Most Popular Oratorio *Stabat Mater*

Rossini's Most Popular Overture *William Tell*

Rubinstein's Most Popular Aria *Epithalamium of Vindex* in *Nero*

Ruslan *Russlan and Ludmila* (Glinka's most popular opera)

Russian Haydn's six string quartets—Opus 33; Rachmaninoff's Symphony No 3 in A minor

Russian composers Anton Arensky, Mili Balakirev, Alexander Borodin, Cesar Cui, Alexander Dargomijsky, Alexander Glazunov, Reinhold Gliere, Mikhail Glinka, Alexander Gretchaninov, Dmitri Kabalevsky, Aram Khachaturian, Anatoly Liadov, Modest Mussorgsky, Sergei Prokofiev, Sergei Rachmaninoff, Nikolai Rimsky-Korsakov, Anton Rubinstein, Alexander Scriabin, Dmitri Shostakovich, Igor Stravinsky, Pyotr Ilyich Tchaikovsky

Russian Easter Rimsky-Korsakov's *Russian Easter Festival*—concert overture

Russian Fourteen Russia's fourteen best-known classical composers—Glinka, Borodin, Cui, Balakirev, Mussorgsky, Tchaikovsky, Rimsky-Korsakov, Glazunov, Scriabin, Rachmaninoff, Gliere, Stravinsky, Prokofiev, Shostakovich

Russian National Composer Mikhail Ivanovich Glinka

Russian Symphonist Pyotr Ilyich Tchaikovsky

Russia's Most Russian Composer Tchaikovsky

Russlan *Russlan and Ludmilla* (opera by Glinka)

Russo-American Composer Igor Stravinsky (1882–1971)

Rustic Wedding Karl Goldmark's Symphony in E flat (opus 26)

Sacre *Le Sacre du Printemps* (French—The Rite of Spring)—Stravinsky ballet for orchestra

Sacred operas *Mosè en Egitto* (Moses in Egypt) by Rossini; *Samson et Dalila* (Samson and Delilah) by Saint-Saëns

Saint-Saëns' 5 the five symphonies of Saint-Saëns including his Symphony in A major, the Symphony No 1, Symphony in F major *(Urbs Roma),* the Symphony No 2 in A minor, the Symphony No 3 in C minor *(Organ)* for organ and orchestra

Saint-Saëns' Most Popular Aria *Mon coeur s'ouvre a ta voix* (French—At your voice my heart unfolds) in *Samson et Dalila*

Saint-Saëns' Most Popular Symphony *No 3—Organ*

Saint-Saëns' Most Popular

Work *Danse macabre*

Saint Vartan Symphony No 9 by Alan Hovhaness

Salome music drama by Richard Strauss

Salomon symphonies Haydn's symphonies 93 through 104 bearing such names as *Surprise* (94), *Miracle* (96), *Military* (100), *Clock* (101), *Drum Roll* (103), and *London* (104); series named for the impresario JP Salomon who secured concerts for Haydn in London

Samson *Samson et Dalila* (opera by Saint-Saëns based on the biblical legend of Samson and Delilah)

San Carlo of the Symphony Carlo Maria Giulini

Sarasate's Most Popular Piece *Carmen Fantasy*

Savoy operas Gilbert and Sullivan operettas

Saxophonist of the Century Paul Brodie

Schelomo (Hebrew— Solomon) title of Bloch's composition for 'cello and orchestra

Schoenberg's Most Popular Orchestral Work *Verklärte Nacht* (German—Transfigured Night)

Schubert's 9 nine symphonies of Franz Schubert including *Tragic* (No 4), *Little* (No 6), *Unfinished* (No 8), *The Great* (No 9)

Schubert's Most Popular Symphony *No 8—Unfinished*

Schumann 1st, 2nd, 3rd, 4th First *(Spring),* Second, Third *(Rhenish),* Fourth symphonies composed by Robert Schumann

Schumann's Most Popular Piano Work *Fantasy in C*

Schumann's Most Popular Symphony *No 1—Spring*

Scotch Mendelssohn's Symphony No 3 in A minor

Scottish Mendelssohn's Symphony No 3 in A minor, often called *Scotch Symphony*

Scottish Composers Erik Chisholm, Sir Alexander Campbell Mackenzie, John Blackwood McEwen, Thea Musgrave, Ian Whyte

Scriabin 5 five symphonies by Alexander Scriabin including the *Divine Poem* (No 3), the *Poem of Ecstasy* (No 4), and the *Poem of Fire* (No 5)

Scriabin's Most Popular Orchestral Work *Poem of Ec-*

stasy

Sea Symphony No 1 by Vaughan Williams

Seasons Glazunov's ballet; Haydn's oratorio *Die Jahreszeiten*

Sea Symphony Symphony No 1 by Ralph Vaughan Williams

Semiramide Rossini opera

Shostakovich's 15 Shostakovich's fifteen symphonies including *Leningrad* (No 7), *Year 1905* (No 11), *Lenin* (No 12), *Babi Yar* (No 13)

Shostakovich's Most Popular Ballet *Age of Gold*

Shostakovich's Most Popular Symphony *No 5*

Sibelius' 7 the seven symphonies of Sibelius

Sibelius's Most Popular Symphonic Poem *Finlandia*

Sibelius's Most Popular Symphony *No 2 in D major*

Sibelius's Most Popular Violin Concerto in *D minor*

Siegfried Wagner's music drama

Silver Pilgrimage Symphony No 15 by Alan Hovhaness

Sinfonia Antarctica Symphony No 7 by Vaughan Williams

Sinfonia Concertante Mozart's two are most familiar

Sinfonia Domestica composition reflecting the daily life of Richard Strauss

Sinfonia Espansiva Nielsen's Symphony No 3

Sinfonia India (Spanish—Indian Symphony)—by Carlos Chavez (1899–1978)

Sinfonia Semplice Nielsen's Symphony No 6

Six-Four Time Mass Haydn's *Sechsviertelmesse* in G

Small Organ Mass Haydn's B-flat *Kleine Orgelmesse*

Smetana's Most Popular Opera *The Bartered Bride*

Smetana's Most Popular Piece *Moldau*

Snegurochka (Russian—The Snow Maiden)—Rimsky-Korsakoff opera

Song of the Night Karol Szymanowski's Symphony No 3; Mahler's Symphony No 7 in E minor

Sonnam Sonnambula (Italian—Sleepwalker)—Bellini opera

Sopranos of the Century Elisabeth Schumann (1894–1966), Maria Callas (1923–1977), Joan Sutherland 1926–)

Sousa's Most Popular March *The Stars and Stripes*

South African composers John Joubert, Priaulx Rainier

Soviet Symphonists Serge Prokofiev and Dmitri Shostakovich

Spain's Quartet Spain's four best-known classical composers—de Falla, Albéniz, Granados, Turina

Spanish Caprice Rimsky-Korsakov's *Capriccio espagnol*

Spanish composers Isaac Albéniz, Manuel de Falla, Enrique Granados, Felipe Pedrell, Joaquin Turina

Spanish Dances *Danzas españoles* composed by Granados for the piano

Spanish Hour Ravel's brief but witty opera—*L'Heure espagnole*

Spanish Nights de Falla's *Nights in the Gardens of Spain*

Spanish Overture Glinka's *Jota aragonesa*

Spanish Pieces de Falla's *Piezas españoles* for piano

Spanish Rhapsody Liszt's *Rhapsodie espagnole;* Ravel's *Rapsodie espagnole*

Spanish Song Ravel's *Chanson espagnole* for piano and voice

Spanish Songbook Hugo Wolf's *Spanisches Liederbuch*

Spanish Songs *Cantos de España* composed by Albeniz for the piano

Spanish Suite *Suite Española* by Albéniz

Spanish Symphony Lalo's *Symphonie espagnole* for violin and orchestra

Spirit Beethoven's *Spirit Trio* called *Das Geister Trio* by the Germans

Spirit of Man Prokofiev's name for his Symphony No 5 Opus 100 completed in 1944

Spring Beethoven's Sonata No 5 for Violin and Piano (opus 24); Schumann's Symphony No 1 in B-flat major

Stabat Mater (Latin—standing mother)—liturgical mass set to music by Haydn, Liszt, Palestrina, Rossini, and Verdi

Steppes *In the Steppes of Central Asia* (symphonic sketch by Borodin)

Stokowski silver sizzle the sound of the Philadelphia Orchestra (developed by Leopold Stokowski)

storm-at-sea mus storm-at-sea music, the Sea and Sinbad's Ship section of Rimsky-Korsakoff's *Scherezade*

storm mus storm music (most memorable includes the Thunderstorm movement in Beethoven's *Symphony No 6*— Pastoral, the Royal Hunt and Storm in *Les Troyens* by Berlioz, the Tempesta interlude in Rossini's *Barber of Seville* and the Alpine Storm in his *William Tell*, the Storm sometimes accompanying and often dominating the seduction scene in the *Samson and Delilah* of Saint-Saëns, the Storm movement in the Alpine Symphony of Richard Strauss, the opening incidental music composed by Sir Arthur Sullivan for *The Tempest*, the howling thunder and wind in Act III of Tchaikovsky's *Queen of Spades*)

Strange Music Hollywood adaptation of Grieg's "Wedding Day at Troldhaugen" in the *Song of Norway*

Stranger in Paradise *Kismet* theme adapted from Borodin's *Prince Igor*

Strauss's (Johann Jr) Most Popular Operetta *Die Fledermaus* (German—The Bat)

Strauss's (Johann) Most Popular Waltz *The Blue Danube*

Strauss's (Richard) Most Popular Opera *Der Rosenkavalier* (German—The Knight of the Rose)

Strauss's (Richard) Most Popular Piece for Piano and Orchestra *Burleske*

Strauss's (Richard) Most Popular Symphony *Alpine*

Strauss's (Richard) Most Popular Tone Poem *Also sprach Zarathustra* (German—Thus Spake Zarathustra)

Stravinsky's Most Popular Ballet *Firebird*

Suicide European nickname for Tchaikovsky's Symphony No 6 in B major—the Pathétique

Sullivan's Most Popular Orchestral Work *In Memoriam*

Sunrise Haydn's String Quartet in B flat (opus 76, no 4)

Suor Angelica (Italian—Sister Angelica)—one-act opera by Puccini

Suppe's (von Suppe's) Most Popular Overture *Light Cavalry*

Surprise Haydn's Symphony No 94 in G major

Swan Song Symphony Prokofiev's Symphony No 7 in C-sharp minor

Swedish composers Kurt Atterberg, Franz Adolf Berwald, Hilding Rosenberg, Wilhelm Stenhammar, Dag Wirén

Swedish Quartet Sweden's leading classical composers, Berwald, Rangstrom, Atterberg, and Wiren

Swiss composers Ernest Bloch, Frank Martin, Jean Jacques Rousseau

Swiss Quartet Switzerland's foremost composers of classical music, Raff, Bloch, Martin, and Honegger

Symphonia domestica (German—Domestic Symphony) —autobiographical tone poem by Richard Strauss

Symphonic-Poem composers Franz Liszt, Camille Saint-Saëns, Jean Sibelius, Bedrich Smetana, Richard Strauss

Symphonie Espagnole Edouard Lalo's most popular violin concerto

Symphonie fantastique (French—Fantastic Symphony)— major orchestral work of Berlioz

Symphony of a Thousand Mahler's Symphony No 8 in E-flat major

Symphony of Heavenly Length Schubert's Symphony No 9, according to Schumann

Symphony of Psalms Stravinsky's best-known symphony

Tabarro Il Tabarro (Italian —The Cloak)—opera by Puccini

Tales Tales of Hoffmann (Offenbach opera)

Tann Tannhäuser und der Sängerkrieg auf der Wartburg (German—Tannhäuser and the Singing Contest of the Wartburg)—three-act Wagner opera

Tchaikovsky 1st, 2nd, 3rd, 4th, 5th, 6th First *(Winter Reveries)*, Second *(Little Russian)*, Third *(Polish)*, Fourth, Fifth, Sixth *(Pathétique)* symphonies composed by Tchaikovsky

Tchaikovsky's Most Popular Ballet *Swan Lake*

Tchaikovsky's Most Popular Opera *Eugen Onegin*

Tchaikovsky's Most Popular Overture *1812*

Tchaikovsky's Most Popular Piano Concerto *No 1 in B flat minor*

Tchaikovsky's Most Popular Song *None But The Lonely Heart*

Tchaikovsky's Most Popular Symphony *No 4 in F minor*

Tear-Jerker Composer Giacomo Puccini—opposite of Gioacchino Rossini

Telemann's Most Popular Concerto *Concerto for Trumpet and Strings in D*

Tell Rossini's opera *William Tell*

Tempest Beethoven's Piano Sonata No 17 in D (opus 31, no 2); Tchaikovsky's Symphonic Fantasy—*Tempest*

Tenors of the Century Enrico Caruso (1873–1921), Jussi Björling (1907–1960), Luciano Pavarotti (1935–)

Thaïs Massenet opera

The Bells Rachmaninoff's choral symphony based on Poe's poem *The Bells*

The Five (Russian composers Balakirev, Borodin, Cui, Moussorgsky, Rimsky-Korsakov)

The Great Schubert's Symphony No 9 in C major

The Isle The Isle of the Dead (orchestral work by Rachmaninoff inspired by Arnold Böcklin's painting of this title)

The Lamp Is Low Hollywood version of Ravel's *Pavane pour une infante de'funte*

Thomas's Most Popular Overture *Raymond*

Thousand The Symphony of a Thousand (Mahler's Symphony No 8 in E-flat major)

Three Classic Masses Bach's B minor, Beethoven's *Missa Solemnis*, Bruckner's *Grosse Messe* in F minor

Three Leading Conductors in America 1900–1950 Serge Koussevitzky, Leopold Stokowski, Arturo Toscanini

Three Leading Conductors in America 1950–1990 Leonard Bernstein, Zubin Mehta, Sir Georg Solti

Three Penny Three Penny Opera (composed by Kurt Weill, based on a modernized German version of John Gay's *The Beggar's Opera*)

Tikhiy Tikhiy Don (Russian —Quiet Flows the Don)— Dzerzhinsky's opera

Till Till Eulenspiegels lustige Streiche (German—Till Eulenspiegel's Merry Pranks)— symphonic poem by Richard Strauss

Titan Mahler's Symphony No 1 in D major—he preferred to call it his *Werther* symphony comparing it with Goethe's first novel

Tod und Verklärung (German—Death and Transfiguration)—symphonic poem by Richard Strauss

Tone-Poem Composer Richard Strauss

Tong-hai Moolkwa (Korean— Tong-Hai Sea)—anthem of South Korea

Tonight We Love popular name for Tchaikovsky's Piano Concerto No 1 in B-flat minor

top 25 top 25 concertos featured in many orchestral programs (Bach's concerto for two violins, Beethoven's five piano and one violin concertos, two piano and one violin concertos by Brahms, Bruch's violin concerto, Chopin's two piano concertos, Dvořák's cello concerto, Gershwin's piano concerto, Grieg's piano concerto, Liszt's piano concerto No 1, Mendelssohn's violin concerto, Mozart's piano concerto No 20, Paganini's concerto No 1 for violin, Rachmaninoff's piano concerto No 2, Schumann's piano concerto, the violin concerto of Sibelius, Tchaikovsky's concerto No 1 for piano and his violin concerto)

top 30 top 30 symphonic spectaculars favored on many orchestral programs [Bach's *Toccata and Fugue in D*; Beethoven's *Lenore Overture No 3*; Berlioz's *Symphonie fantastique*; Borodin's *Polovetsian Dances*; Brahms's *Variations on a Theme by Haydn*; Debussy's *La Mer*; Glinka's *Russlan and Ludmilla* Overture; Handel's *Water Music*; Liszt's *Les Preludes*; Mussorgsky's *Night on Bald Mountain*, *Pictures at an Exhibition*; Prokofiev's *Peter and the Wolf*; Rachmaninoff's *Rhapsody on a Theme by Paganini*; Ravel's *Bolero*;

Rimsky-Korsakov's *Scheherazade;* Saint-Saëns's *Carnival of the Animals,* Symphony No 3 (Organ); Sibelius's *Finlandia;* Smetana's *Moldau;* Richard Strauss's tone poems—Don Juan, Don Quixote, Hero's Life *(Heldenleben),* Thus Spake Zarathustra *(Also Sprach Zarathustra),* Till Eulenspiegel; Stravinsky's *Sacre du Printemps;* Tchaikovsky's Overture *1812,* Romeo and Juliet; Wagner's *Flying Dutchman* and *Tannhäuser* overtures, *Tristan* Prelude and Liebestod]

top 40 top 40 symphonies favored on many symphonic programs [Beethoven's 3rd *(Eroica),* 5th, 6th *(Pastoral),* and 9th *(Choral);* the four by Brahms; Bruckner's 4th *(Romantic)* and 9th; Dvořák's 6th and 9th *(New World);* Haydn's 94th *(Surprise),* 100th *(Military),* 101st *(Clock),* 103rd *(Drum Roll),* 104th *(London);* Mahler's 1st, 2nd *(Resurrection),* and 9th; Mendelssohn's 3rd *(Scottish),* 4th *(Italian),* and 5th *(Reformation);* Mozart's 35th *(Haffner)* and 41st *(Jupiter);* Prokofiev's 1st *(Classical)* and 5th; Rachmaninoff's 2nd; Schubert's 8th *(Unfinished)* and 9th; Schumann's 1st *(Spring),* 2nd, 3rd *(Rhenish),* and 4th; Shostakovich's 1st and 5th; Sibelius's 1st; Tchaikovsky's 4th, 5th, and 6th *(Pathétique)]*

Tosca Puccini opera

Totentanz (German—Death Dance)—Liszt's paraphrase on the *Dies Irae* for piano and orchestra

Totentänze (German—Dances of Death)—part of Mahler's Symphony No 9

Toy Toy Symphony usually ascribed to Haydn but now believed to be part of a larger work by Leopold Mozart

Tragic overture by Brahms; Symphony No 6 by Mahler; Symphony No 4 by Schubert

Traviata Verdi opera

Triangle Liszt's Piano Concerto No 1 in E flat

Tristan Tristan und Isolde (German—Tristan and Iseult)— music drama by Wagner

Trittico Il Trittico (The Tryp-

tych)—Puccini's three short operas—*Gianni Schicchi, Suor Angelica,* and *Il Tabarro*

Trojans Les Troyens (French —The Trojans)—opera by Berlioz

Trombonist of the Century Tommy Dorsey

Trout Schubert's Quintet in A major for violin, viola, cello, double bass, and piano

Trov Il Trovatore (Italian—The Troubador)—Verdi opera

Troyens Les Troyens (French —The Trojans)—opera by Berlioz

Trumpeter of the Century Maurice André (1933–)

Tubists of the Century William Bell of the New York Philharmonic, Roger Bobo of the Los Angeles Philharmonic

Turandot Puccini opera

Turina's Most Popular Work *Danzas fantásticas*

Turkish Mozart's Violin Concerto in A major (K 219)

Twentieth-Century Romantics Rachmaninoff, Sibelius, and Richard Strauss

Unfinished Schubert's Symphony No 8 in B minor

Urbs Orba Symphony in F of Saint-Saëns

Valurile Dunării (Romanian— Danube Waves)—popular fanfare also called *Anniversary Waltz*

Variations on an Original Theme *Enigma Variations* of Elgar

Vaughan Williams' Most Popular Fantasia *Greensleeves*

Vaughan Williams' Most Popular Opera *Hugh the Drover*

Vaughan Williams' Most Popular Symphony *No 1—Sea Symphony*

Venezuelan composers Teresa Carreño, Reynaldo Hahn, José Angel Montero, Juan Bautista Plaza, Vicente Emilio Sojo

Venezuelan First Venezuela's composer-conductor Reynaldo Hahn who became music critic of *Le Figaro* and music director of the Paris Opera

Venezuelan Pianist Teresa Carreño

Verdi's Most Popular Mass *Manzoni Requiem*

Verdi's Most Popular Opera *Aida*

verismo (Italian—realism) applied to composers such as Leoncavallo, Mascagni, Puccini

Vespri I Vespri Siciliani (Italian—The Sicilian Vespers)— Verdi opera

Victory nickname for Beethoven's Symphony No 5 in C minor

Vie La Vie Parisienne (French— Parisian Life)—Offenbach opera

Vieuxtemp's Most Popular Concerto *No 5 for Violin*

Villa-Lobos' Most Popular Work *Bachianas Brasileiras No 5 for soprano and 8 celli*

Violinist of the Century Jascha Heifetz (1901–1987); William Primrose (1904–1982)

Violinist-Composer-Conductor Eugène Ysaÿe

Violinist-Conductor Willi Boskovsky; Richard Burgin; Sidney Harth, David Oistrakh; Igor Oistrakh, Joseph Silverstein; Isaac Stern

Violinist-Violist-Conductor Yehudi Menuhin; Pinchas Zukerman

Violin-Maker's Capital Cremona, Italy

Vishnu Symphony No 19 by Alan Hovhaness

Vivaldi's Most Popular Piece *Le Quattro Stagioni* (Italian— The Four Seasons)

Wagner's Most Popular Aria *In fernem Land* (German—In a Far Land) in *Lohengrin*

Wagner's Most Popular Music Drama *Tannhäuser*

Wagner's Most Popular Prelude *Act I—Lohengrin*

Wagner's Most Popular Song *Träume* (German—Dreams)

Waldstein Beethoven's Piano Sonata No 21 in C (opus 53); dedicated to Count von Waldstein

Walküre Die Walküre (German —The Valkyrie)—Wagner music drama

Wallace's Most Popular Overture *Maritana*

Walton's Most Popular Orchestral Work *Belshazzar's Feast*

Waltz King nickname shared by Lanner, Lehar, Lumbye, Kalman, Johann Strauss Sr and Jr, Josef Strauss, and Oskar Straus

Wanderer Schubert's Piano Fantasie in C (opus 15)

Warsaw Warsaw Concerto by Richard Addinsell

Weber's (von Weber's) Most Popular Overture *Oberon*

Welsh composers Alun Hoddinott, Arwel Hughes, Daniel Jones, William Mathias, Grace Williams, David Wynne

Werther Massenet opera

West Point Morton Gould's Symphony No 4 for Band

When the Lights Go On Again popularized version of Beethoven's Minuet in G

William Tell Rossini opera

Winter Reveries Tchaikovsky's Symphony No 1 in G minor (*Rêverie d'Hiver*)

Winter Wind Chopin's Piano Etude No 11 in A minor

Wolf's Most Popular Lieder *Italienisches Liederbuch* (Ger-

man—Italian Lieder Book)

World's Largest Concert Hall Royal Albert Hall, London, with a seating capacity of 10,000

World's Most Musical Country Germany

Wozzeck Alban Berg music drama

Xerxes Handel opera

Year 1905 Shostakovich's Symphony No 11

Year 1917 Shostakovich's Symphony No 12

Youth Kabalevsky's *Concerto No 3 in D major; Youth Symphony* in D minor by Rachmaninoff

Zampa Hérold opera

Zarathustra Thus Spake Zarathustra (symphonic poem by Richard Strauss—*Also sprach Zarathustra*)

Zigeunerbaron Der Zigeuner baron (German—The Gypsy Baron)—operetta by Johann Strauss Jr

Zigeunerweisen (German—Gypsy Melodies)—Pablo de Sarasate work

Zingareska George Antheil's Symphony No 1

Z m Z Z mého Zivota (Czechoslovakian—From my Life) —Smetana's String Quartet No 1

National Capitals

Abidjan Ivory Coast
Abu Dhabi United Arab Emirates
Accra Ghana
Addis Ababa Ethiopia
Algiers Algeria
Amman Jordan
Amsterdam Netherlands
Andorra la Vella Andorra
Ankara Turkey
Antanarivo Madagascar
Apia Western Samoa
Asunción Paraguay
Athens Greece
Baghdad Iraq
Bamako Mali
Bandar Seri Begawan Brunei
Bangkok Thailand
Bangui Central African Republic
Banjui Gambia
Basse-Terre St Kitts and Nevis
Beijing China
Beirut Lebanon
Belgrade Yugoslavia
Belmopan Belize
Berlin Germany
Bern Switzerland
Bissau Guinea-Bissau
Bloemfontein South Africa
Bogota Colombia
Brasilia Brazil
Brazzaville Congo
Bridgetown Barbados
Brussels Belgium
Bucharest Romania
Budapest Hungary
Buenos Aires Argentina
Bujumbura Burundi
Cairo Egypt
Canberra Australia

Cape Town South Africa
Caracas Venezuela
Castries Saint Lucia
Colombo Sri Lanka
Conakry Guinea
Copenhagen Denmark
Dakar Senegal
Damascus Syria
Dar es Salaam Tanzania
Dhaka Bangladesh
Djibouti Djibouti
Doha Qatar
Dublin Ireland
Freetown Sierra Leone
Funafuti Tuvalu
Gaborone Botswana
Georgetown Guyana
Guatemala City Guatemala
Haiti Port-au-Prince
Hanoi Vietnam
Harare Zimbabwe
Havana Cuba
Helsinki Finland
Honiara Solomon Islands
Islamabad Pakistan
Jakarta Indonesia
Jerusalem Israel
Kabul Afghanistan
Kampala Uganda
Kathmandu Nepal
Khartoum Sri Lanka
Kigali Rwanda
Kingston Jamaica
Kingstown St Vincent and Grenadines
Kinshasa Zaire
Kuala Lumpur Malaysia
Kuwait Kuwait
La Paz Bolivia's administrative capital
Libreville Gabon

Lilongwe Malawi
Lima Peru
Lisbon Portugal
Lomé Togo
London United Kingdom
Luanda Angola
Lusaka Zambia
Luxembourg Luxembourg
Madrid Spain
Malabo Equatorial Guinea
Male Maldives
Managua Nicaragua
Manama Bahrain
Manila Philippines
Maputo Mozambique
Maseru Lesotho
Mbabane Swaziland
Mogadishu Somalia
Monaco-Ville Monaco
Monrovia Liberia
Montevideo Uruguay
Moroni Comoros
Moscow USSR (Union of Soviet Socialist Republics)
Muscat Oman
Nairobi Kenya
Nassau Bahamas
N'Djamena Chad
New Delhi India
Niamey Niger
Nicosia Cyprus
Nouakchott Mauritania
Nuku'alofa Tonga
Nuuk Greenland
Oslo Norway
Ottawa Canada
Ouagadougou Burkina Faso
Panama Panama
Paramaribo Suriname
Paris France
Phnom Penh Cambodia

Port-au-Prince Haiti
Port Louis Mauritius
Port Moresby Papua New Guinea
Port-of-Spain Trinidad and Tobago
Porto-Novo Benin
Prague Czechoslovakia
Praia Cape Verde
Pretoria South Africa
Pyongyang North Korea
Quito Ecuador
Rabat Morocco
Rangoon Burma
Reykjavik Iceland
Riyadh Saudi Arabia
Rome Italy
Roseau Dominica
Russia Moscow
St George's Grenada

St John's Antigua and Barbuda
Sanaa Yemen
San José Costa Rica
San Marino San Marino
San Salvador El Salvador
Santiago Chile
Santo Domingo Dominican Republic
São Tomé São Tomé and Principe
Seoul South Korea
Singapore Singapore
Sofia Bulgaria
Stockholm Sweden
Sucre Bolivia's judicial capital
Suva Fiji
Taipei Taiwan
Tarawa Kiribati
Tegucigalpa Honduras
Teheran Iran

Thimphu Bhutan
Tirana Albania
Tokyo Japan
Tripoli Libya
Tunis Tunisia
Ulaanbaatar Mongolia
Vaduz Liechtenstein
Valetta Malta
Victoria Seychelles
Vienna Austria
Vientiane Laos
Vila Vanuatu
Warsaw Poland
Washington DC United States of America
Wellington New Zealand
Yaounde Cameroon
Yaren Nauru
Yerevan Armenia

National Holidays

Afghanistan Independence Day August 19
Albania Independence Day November 28
Algeria Independence Day July 5
Andorra National Day September 8
Angola Independence Day November 11
Antigua and Barbuda Independence Day November 1
Anzac Day April 25 (in Australia, New Zealand, and associated territories)
Arab Emirates Independence Day December 2
Argentina Independence Day July 9
Australia Day January 26
Austria National Day October 26
Bahama Independence Day July 10
Bahrain Independence Day December 16
Bangladesh Independence Day March 26
Barbados Independence Day November 30
Belgium National Days September 9 and 10
Belize Independence Day September 21
Benin Independence Day December 6
Bhutan Independence Day January 4
Birmania Independence Day

January 4
Bolivia Independence Day August 6
Botswana Independence Days September 30 and October 1
Brazil Independence Day September 7
Brunei Independence Day January 1
Bulgaria National Days September 30 and October 1
Burkina Faso Proclamation Day December 11
Burundi Independence Day July 1
Cameroon National Day May 20
Canada Day Dominion Day, July 1
Canada National Day January 1
Cape Verde Independence Day July 5
Central African Republic Independence Day August 13
Central America Day Central American Independence Day, September 15 (in Costa Rica, El Salvador, Guatemala, Honduras, and Nicaragua)
Chile Independence Day September 18
China National Days October 1 and 2
Colombia Independence Day July 20
Commonwealth Day third Monday in May in parts of the British Commonwealth

Comoras Independence Day July 6
Congo National Day August 15
Costa Rica Independence Day September 15
Cuba Liberation Day January 1
Curaçao Day July 26 (celebrated throughout the Netherlands Antilles)
Cyprus Independence Day October 1
Czechoslovakia Liberation Day May 9
Denmark Constitution Day June 5
Día de la Raza (Spanish—Day of the Race)—Columbus Day, October 12
Dieciséis (Spanish—Sixteenth) —September 16 (Mexican Independence Day)
Djibouti Independence Day June 27
Dominica National Days November 2 and 3
Dominican Republic Independence Day February 27
Ecuador Independence Day August 10
Egypt Proclamation Day June 18
El Salvador Independence Day September 15
Empire Day nearest Monday to May 24, celebrated in many parts of the British Commonwealth of Nations
Equatorial Independence Day

October 12
Ethiopia Victory Day April 6
Fawkes Fawkes Day, November 5 (in Great Britain)
Fiji Independence Day October 10
Finland Independence Day December 6
Flag Day June 14 in the United States
France Bastille Day July 14
Gabon Independence Day August 17
Gambia Independence Day February 18
Germany Unity Day June 17
Ghana Independence Day March 6
Greece Independence Day March 25
Grenada Independence Day February 7
Guatemala Independence Day September 15
Guinea Republic Day October 2
Guinea-Bissau National Day September 24
Guyana Republic Day February 23
Haiti Independence Day January 1
Honduras Independence Day September 15
Hostos Eugenio María Hostos Birthday, January 11 (celebrated in Puerto Rico)
Hungary Liberation Day April 4
Iceland National Day June 17
India Independence Day August 15
Indonesia National Day August 17
Iran Day of the Islamic Republic April 1
Iraq Day of the Republic July 14
Ireland's St Patrick's Day March 17
Israel Independence Day May 14
Italy National Day June 2
Ivory Coast Independence Day December 7
Jamaica Independence Day August 6
Japanese New Year December 28 through January 3
Japan National Foundation Day February 11
Jordan Independence Day May 25
Kenya Independence Day December 12
Kiribati Independence Day

July 12
Kuwait Independence Day February 25
Labor Day first Monday in September in the U.S.A.
Laos National Day December 2
Lebanon Independence Day November 22
Lesotho Independence Day October 4
Liberia Independence Day July 26
Libya Revolution Day September 1
Liechtenstein National Day January 1
Luxembourg Day June 23— Grand Duke's Birthday
Madagascar Independence Day June 26
Malawi Republic Day July 6
Malaysia National Day August 31
Maldives Independence Day July 26
Mali Independence Day September 22
Malta Republic Day December 13
Mauritania National Day November 28
Mauritius National Day March 12
Mexico Independence Day September 16
Monaco National Day November 19
Mongolia National Day July 11
Morocco Independence Day November 18
Mozambique Independence Day June 25
Myanmar National Day January 1
Namibia Republic Day May 31
Nauru Independence Day January 31
Nepal National Day February 18
Netherlands Day April 30, Queen's Birthday
New Zealand National Day February 6
Nicaragua Independence Day September 15
Niger Independence Day August 3
Nigeria National Day October 1
North Korea Independence Day September 9
Norway Constitution Day May 17
Oman National Day November 18
Pakistan Independence Day

August 14
Panama Day November 3— Separation from Colombia
Papua New Guinea Independence Day September 16
Paraguay Independence Days May 14 and 15
Peru Independence Day July 28
Philippine Independence Day June 12
Poland Liberation Day July 22
Portugal Day Independence Day, December 1
Portugal Liberty Day April 25
Puerto Rico—American Independence Day July 4
Qatar Independence Day September 3
Remembrance Canada's Remembrance Day, November 11—Armistice Day
Romania National Day August 23
Rwanda Independence Day July 1
St Christopher and Nevis Associate Day February 27
St Lucia Independence Day February 22
St Vincent and the Grenadines Independence Day October 27
San Marino Liberation Day February 5
São Tomé and Principe Independence Day July 12
Saudi Arabia National Day September 23
Senegal Association Day July 14
Seychelles Association Day June 29
Sierra Leone Anniversary Day April 19
Singapore National Day August 9
Solomon Islands Independence Day July 7
Somalia Foundation Day July 1
South Africa Republic Day May 31
South Korea Liberation Day August 15
Spain Hispanic Day October 12
Sri Lanka Independence Day February 4
Stanton's Elizabeth Cady Stanton's Day, November 12
Sudan Independence Day January 1
Suriname Independence Day November 25
Swaziland Independence Day September 6

Sweden Flag Day June 6
Switzerland Confederation Day August 1
Syria National Day November 16
Taiwan Revolution Day October 10
Tanzania Union Day April 26
Thailand King's Birthday December 5
Togo Independence Day April 27
Tonga Emancipation Day June 4
Transkei Independence Day October 26
Trinidad and Tobago Independence Day August 31
Tunisia Independence Day June 1
Turkey Independence Day October 29
Tuvalu Day October 1
Uganda Independence Day October 9
United Kingdom Queen's Birthday April 21
United States of America Independence Day July 4
Uruguay Independence Day August 25
USSR Revolution Day November 7
Venezuela Independence Day July 5
Victoria Day Queen Victoria's Birthday, May 20
Vietnam National Day September 2
Western Samoa Independence Day January 1
Yugoslavia Republic Days November 29 and 30
Zaire Independence Day June 30
Zambia Independence Day October 24
Zimbabwe Independence Day April 18

Numbered Abbreviations

O² both eyes
0-0 zero-zero (no ceiling, no visibility for an aircraft)
000 emergency services (Australia)
007 James Bond (Ian Fleming's international sleuth)
0 deg lat zero degrees latitude (the Equator, encircling widest part of the earth)
0-g zero gravity
0° zero degrees (the Equator)
0°lat zero degrees latitude (the Equator)
0°longitude Greenwich meridian with lines of longitude either east or west of this prime meridian in Greenwich, England
¼ d farthing (fourth of an English penny); a fourthling
¼ h quarter-hard
¼ ly quarterly
¼ ph quarter-phase
¼ rd quarter-round
¼s quarters
½ can narcotics equal to a half can of pipe tobacco
½d halfpenny (half of an English penny); ha'penny
½ gr half-gross
½ h half-hard
½ rd half-round
½ sov half sovereign (10 shillings)
½ sovereign 10 shillings
½ t half title
1 in the beginning; in the year one; No 1 diet (bread and water); Number 1 (urine)
I first violin
1/ a bob (British slang for one

shilling)
1ª primeira (Portuguese—first) —feminine gender; primera (Spanish—first)—feminine gender
I-A available for military service
I-A-O conscientious objector available only for noncombatant military service
1-armed one-armed bandit (gambling device also called a slot machine)
1b first base(man)
1-bagger one-bagger; single
I-BCE first century before the Christian era (Caesar's Century)—Julius Caesar conquered Britain and Egypt before he was assassinated in the Roman senate in the year 44
1¢ one cent
1/c single-conductor
1C member or former member of US armed forces with honorable discharge
I-C first century (Vesuvian Century)—destruction of Pompeii, Herculaneum, and nearby Neapolitan places by the volcano Vesuvius in the year 79 of the Christian era; member of the armed forces, Coast and Geodetic Survey, or Public Health Service
1ce once
1 cent 1 penny (10 mills)
1 Chron The First Book of the Chronicles
1 Cor The First Epistle of Paul the Apostle to the Corinthians

1 crown 5 shillings
1d an English penny
I-D member of reserve component or student taking military training
1 dime 10 cents
1 double eagle $20 (gold)
1/e first edition
1 eagle $10 (gold)
1er(e) premier(e) (French—first)
1ers premières (French—first violins)
1 Esd The First (Apocryphal) Book of Esdras
1 florin 2 shillings
1-fold one-fold (single undivided whole)
1 frogskin 1 bill
1G, 2G, 3G, etc. slang for one, two, three thousand dollars, etc.
I Geigen (German—first violins)
1 guinea 21 shillings
1 half crown 2 shillings, 6 pence
1 half dime 5 cents
1 half dollar 50 cents
1 half eagle $5 (gold)
1 halfpenny 2 farthings
1 Hen IV First part of King Henry IV
1 Hen VI First part of King Henry VI
1 John The First Epistle General of John
1 Kings The First Book of the Kings
1/M First Mate
1ma prima (Italian feminine—first)
1 Macc The First (Apocryphal)

Book of Maccabees
1mo primo (Italian—first)
1 Ne. First Book of Nephi
1° primeiro (Portuguese—first)
—masculine gender; *primero* (Spanish—first)—masculine gender
1° solo
1/O First Officer
I-O conscientious objector available only for civilian work contributing to national health, safety, or interest
1-1-1 one man, one vote, one election
1p one new penny
1-p single pole
1 penny 4 farthings
1 Pet The First Epistle General of Peter
1 ph single-phase
1-piece(r) one-piece bathing suit, coverall, or other one-piece garment
1 pound 20 shillings
1Q first quarter
1Q66 first quarter 1966
1 quarter dollar 25 cents
1 quarter eagle $2.50 (gold)
1s shilling, also called a *bob*
I-S student deferred by statute until end of current school year
1 Sam The First Book of Samuel
1 shilling 12 pence
1 sixpence 6 pence
1 sovereign 1 pound sterling; 20 shillings
1-spot $1 bill
1st first
1st Asst Engr First Assistant Engineer
1st Asst Pur First Assistant Purser
1st cl hon first-class honors (in academic degrees)
1-step one-step (dance or music for such a dance)
1st Lieut First Lieutenant
1st Naval District Boston, Massachusetts
1st Off First Officer
1-striper ensign (USN); third assistant engineer of third mate (merchant marine); private first class (US Army)
1st Sgt First Sergeant
1st State Delaware (first state of the original thirteen states to ratify the *Constitution of the United States*)
1st Vln(s) first violin(s)
1s & 2s mixed 1st- and 2nd-quality lumber
1-suit(er) one-suit (garment bag

or suitcase)
¹/₁₀ net 30 1-percent discount off the face value of an invoice is allowed if invoice is paid in 10 days; otherwise the full amount is payable in 30 days
1 Thess The First Epistle of Paul the Apostle to the Thessalonians
1 threepence 3 pence
1 Tim The First Epistle of Paul the Apostle to Timothy
1-upmanship one step ahead of your adversary or even your best friend
I-W conscientious objector performing civilian work contributing to national health, safety, or interest, or who has completed such work
1-way one-way street; one-way traffic; unilateral
1-wd one-wheel drive
I-Y registrant does not meet present standards; available for military service only in event of war or national emergency
1-y-o one-year-old (child, pet, racehorse, etc.)
1-yTs one-year Treasury securities
1½ striper naval lieutenant, junior grade
2 Number 2 (excrement); soft lead pencil
II in music, second violin
2/ two shillings; also the coin called a florin
2ªf *segunda-feira* (Portuguese —Monday)
II-A registrant deferred because of civilian occupation (except agriculture and activity in study) or an apprentice deferred by statute
II-B registrant deferred because necessary to war production
2b doubles; second base
2-bagger two bagger (double)
II-BCE second century before the Christian era (Roman Century)—Punic wars resulted in destruction of Carthage by the Roman legions—the 100s
2 bits 25 cents
2-B's boiled or bottled waters (essential in many countries)
2¢ two cents
2/c two-conductor
II-C registrant deferred because of agricultural occupation; second century (Aurelian Century)—reign of the Roman emperor-philosopher

Marcus Aurelius—the 100s
2ce twice
2 Chron The Second Book of the Chronicles
2-cycle two-cycle
2d second
2-d two-dimensional
2ᵈᵃ segunda (Italian, Portuguese, or Spanish—second)—feminine gender
2-decker double-decker (sandwich or ship); two-decker
2ᵈᵒ segundo (Italian, Portuguese, or Spanish—second)—masculine gender
2ᵈˢ deuxièmes (French—second violins)
2 Dzerzhinsky Moscow address of the KGB and the Lubyanka prison
2/e second edition
IIᵉ deuxième, second, seconde (French—second)
2 Esd The Second (Apocryphal) Book of Esdras
2ᵉᵗ duet
2-F two-seater fighter aircraft (naval symbol)
2-4-D dichlorophenoxy-acetic acid (weed killer)
2-4-5-T trichlorophenoxy-acetic acid (antiplant agent and defoliant)
2-fer two-fer (two for the price of one)
2-fold double; twofold
2g, 3g, 4g, etc. multiples of acceleration of gravity which at the surface of the earth is 32.2 feet per second
2 Geigen (German—second violins)
2H hard lead pencil
2 Hen IV Second part of *King Henry IV*
2 Hen VI Second part of *King Henry VI*
2 i/c second in command
2 John The Second Epistle of John
2 Kings The Second Book of Kings
2/M Second Mate
2 Macc The Second (Apocryphal) Book of Maccabbees
2n diploid number
2nd Asst Engr Second Assistant Engineer
2nd Asst Pur Second Assistant Purser
2nd Lieut Second Lieutenant
2nd Off Second Officer
2nd State Pennsylvania
2nd Stwd Second Steward
2nd Vln(s) second violin(s)
2 Ne. Second Book of Nephi

2° segundo (Spanish—second)
2/O Second Officer
2p two new pence
2-p double pole
2 pc two-physician certification
2 Pet The Second Epistle General of Peter
2 ph two-phase
2-piece(r) two-piece(r)
2-ply two-ply
2Q second quarter
2Q66 2nd quarter 1966
2s two shillings; also the coin called a florin
2's two-year-old children
II-S registrant deferred because of activity in study
2 Sam The Second Book of Samuel
2-sided two-sided
2/6 two-and-six (two shillings and sixpence); also called half a crown
2-some twosome (two persons or things)
2-spot $2 bill
2-st two-storey
2-step two-step (dance)
2-striper corporal (US Army); lieutenant (USN); second assistant engineer or second mate (merchant marine)
2-suit(er) two-suit (garment bag or suitcase)
2T double throw
2/10-30 2 percent discount if paid in 10 days, net in 30 days
2 Thess The Second Epistle of Paul the Apostle to the Thessalonians
2-13 drug addict
2 Tim Second Epistle of Paul the Apostle to Timothy
2-time(r) two-time(r)
2-tone two-tone(d)
2/2 (music) *breve* time; cut time
2U to you
2U2 to you too
2-way two-way
2-wd 2-wheel drive
2WW Second Weather Wing (Air Force—New York)
2-y-o two-year-old (child, pet, racehorse, etc.)
2½ small glass of milk
2½-striper naval lieutenant commander
III-A registrant with child or children or registrant deferred by reason of extreme hardship to dependents
3ªf *terça-feira* (Portuguese—Tuesday)
3b third base; triples
3-bagger three-bagger (triple)

3-ball three-ball (golf match)
III-BCE third century before the Christian era (Carthaginian Century)—Hannibal crossed the Alps to defeat the Romans—the 200s
3-bravs *bravo* (shout meaning a male performer has done very well); *brava* (for a female performer), *bravi* (for male and female performers)
3-b's bang 'em, blow 'em, bow 'em—musical instruments played by banging them (percussion), blowing them (brasses or woodwinds), or bowing them (strings)
3-Bs Bach, Beethoven, Berlioz; Bach, Beethoven, Brahms; Bach, Beethoven, Bruckner
3/c three-conductor
3C Computer Control Company
III-C third century (the Chinese Century)—Chin dynasty ruled a reunited China—the 200s
3-card three-card (monte)
3ce thrice
3-color three-color (photo, photography, print, or printing process)
3-cord three parallel white cords adorning bibs of British sailors and their allies
3d English threepenny; thruppence; third
3-d dizzy, dopey, and dumb; three dimensional
3de three-day event
3-decker three-decker (sandwich or ship); triple-decker
3-Ds discouragement, disillusionment, disappointment (including frustration and loss)—often leads to suicide, experts insist
3d 10th 40m 3 days 10 hours 40 minutes (Atlantic crossing of SS *United States* in July 1952)
3/e third edition
III° troisième (French—third)
3-e's of the 1990s economy, environment, ethics
3 x 5 3-inch by 5-inch filing card
3-fold threefold; triple
3-gaited three-gaited (horse)
3-Gs *Gegurgel, Gejodel, Geklapper* (German—gurgling, yodelling, clattering)—Wagner's opinion of music composed by Mendelssohn and Meyerbeer
3H very hard lead pencil
3-H Hubert Horatio Humphrey
3-hand(ed)   three-hand(ed)

(card game)
3 Hen VI Third part of *King Henry VI*
3io trio
3-I voters Irish, Israeli, Italian
3 John The Third Epistle of John
3 K's *Kinder, Küche, Kirche* (German—children, kitchen, church)
3-l's latitude, lead, lookout; lead, log, lookout (dead-reckoning essentials)
3M Minnesota Mining and Manufacturing Company
3-M Maintenance and Material Management (USN)
3/M Third Mate
3-m l-l c's three-martini liquid-lunch clubbers (alcoholically befuddled expense-account experts contributing to the higher cost of so many things)
3-m lunch three-martini luncheon
3 mMs three musical Ms (Martinon, Monteux, Munch) —conductor-musicians par excellence
3Ms Macmurdo, Mackintosh, and Morris (British architects)
3-Ms Mozart, Mendelssohn, Mahler
3 Ne. Third Book of Nephi
3° trio
3° *tercero* (Spanish—third)
3/O Third Officer
3-p triple pole
3ph three-phase
3-piece three-piece (garment)
3pl three-point linkage (tractor)
3-p's three phantoms (world depression, world unemployment, world unrest)
3-Ps prosecution, punishment, and persecution (often life-long) faced by every felon
3pt three-point (tractor linkage)
3Q third quarter
3Q66 third quarter 1966
3rd Asst Engr Third Assistant Engineer
3rd degree prolonged interrogation designed to produce a confession of guilt
3rd Naval District New York, New York
3rd Off Third Officer
3rd State New Jersey
3rd Stwd Third Steward
3-ring three-ring circus
3-RRs remedial reading, remedial writing, remedial arithmetic
3-Rs reading, writing, arithmetic (colloquially, readin',

'ritin', 'rithmetic)

3-Rs of productivity recognition, responsibility, rewards (for workers)

3-Rs of war relentless, remorseless, ruthless

3 S's Saba, Sint Eustatius (Statia), Sint Maarten (in the Netherlands Antilles within the Lesser Antilles near the Virgin Islands)

3-st three-storey

3-star admiral or general of three-star rank

3-striper commander (USN); first assistant engineer or first mate (merchant marine); sergeant (US Army)

3T triple throw

3-3 three (school) terms—three courses per term

3-way three-way

3-wheel atv's 3-wheel all-terrain vehicles

3WW Third Weather Wing (Air Force—Nebraska)

4 level 4 (death-dealing dose or injection of breath-stopping barbital or other drug used by executioners)

4a man 38 years or over and deferred from military service by reason of age

IV-A registrant who has completed service or a sole surviving son

4ᵃf *quarta-feira* (Portuguese— Wednesday)

4-As American Association for the Advancement of Atheism

4-AS American Association for the Advancement of Science

IV-B government official deferred by statute

4-bagger four-bagger (home run)

4-banger four-cylinder engine

IV-BCE fourth century before the Christian era (Alexandrian Century)—Alexander the Great of Macedonia defeated the Egyptians, the Persians, and the Indians; encouraged the Greek philosophers and poets—the 300s

4 bits 50 cents

4/c four-conductor

4C Community-Coordinated Child Care Program

IV-C alien; fourth century (Constantinian Century)— Roman emperor Constantine built the city of Constantinople on the site of ancient Byzantium and proclaimed it capital of the Eastern Empire

—the 300s

4-class fourth-class (mail)

4-col p four-color page

4-d meat meat of dead, disabled, diseased, or dying animals

IV-D minister of religion or divinity student

4/e fourth edition

IVᵉ *quatrième* (French—fourth)

IV-E conscientious objector available for, assigned to, or released for work of national importance

4-F find, feel, fornicate, and forget—code of conduct of certain men in search of casual sexual relationships

IV-F registrant not qualified for any military service

4-gls fourth-generation languages

4H very very hard lead pencil

4-H 4-H Club(s)—H standing for head, heart, health, and helping hands

4-hand four-hand(ed)

4 Last Songs ballet based on the *Four Last Songs* of Richard Strauss

4-letter four-letter words (Early English terms such as cock, crap, crud, cunt, dung, fart, fuck, lust, piss, puke, scum, shit, snot, spit, suck)

4-Ls latitude, lead, longitude, lookout

4-Ms Monteverdi, Mozart, Mendelssohn, Mahler

4 Ne. Fourth Book of Nephi

4° quarto (a book about 9 x 12 inches)

4ᵒ cuarto (Spanish—fourth)

4/O Fourth Officer

4 out of 10 4 out of 10 adult Americans are afraid to walk alone at night in their own neighborhood, a recent Gallup Poll revealed

4-p quadruple pole

4-p's product, place, promotion, price

4Q fourth quarter

4Q66 fourth quarter 1966

4R Ceylon aircraft

4-R Act Railroad Revitalization and Regulatory Reform Act

4-Ss shit, shave, shampoo, and shower

4-st four-storey

4-star admiral or general of four-star rank

4-striper captain (merchant marine or USN); chief engineer (merchant marine)

4tet quartet(te)

4th the 4th of July (American holiday celebrating the signing of the *Declaration of Independence* in Philadelphia on July 4, 1776)

4th Asst Engr Fourth Assistant Engineer

4th Naval District Philadelphia, Pennsylvania

4th Off Fourth Officer

4th State Georgia

4U for you

4U2 for you too

4-way four-way

4-wd(s) four-wheel drive vehicle(s)

4-wheel four-wheel (drive)

4-ws (4WS) 4-wheel steering

4WW Fourth Weather Wing (Air Force—Colorado)

4X Israeli aircraft

5 large glass of milk

5A Libyan aircraft

V-A registrant over the age liability for military service

5ᵃf *quinta-feira* (Portuguese— Thursday)

5-and-10 variety store

5b bald man with baywindow, bifocals, bridgework, and bunions (humorous Selective Service rating)

V-BCE fifth century before the Christian era (Athenian Century)—Athenians destroyed Persian fleet at Salamis; completed the Parthenon in Athens—the 400s

5-Bs Bach, Beethoven, Berlioz, Brahms, Bruckner

5-B's Boston baked beans and brownbread

5BX five basic exercises (Royal Canadian Air Force physical fitness program)

5 by 5 radio reception loud and clear (volume and clarity measured on a scale from 1 to 5)

V-C fifth century (Christian Century)—Christianity affirmed as the official faith by two Roman emperors—the 400s

5-Cs 5-Cs of cinematography (camera angles, closeups, composition, continuity, cutting)

5 de cinco de mayo (Mexican holiday celebrating victory over the French at the Battle of Puebla on May 5, 1862)

5-dimensions five dimensions of health: emotional, intellectual, physical, social, spiritual

5 don'ts don't kill, steal, commit adultery, become intoxicated, or lie, advised Buddah

5/e fifth edition

Vᵉ cinquième (French—fifth)

5-er five-dollar bill; fiver; five-pound note

5'er $5 bill; 5-pound note

5/50 five-year, fifty-thousand-mile protection (new car warranty)

5-finger five-finger(ed)

5-fold cinquefold(ed); fivefold(ed)

5 GLs five Great Lakes: Ontario, Erie, Huron, Michigan, Superior

5-HIAA 5-hydroxy indoleacetic acid

5-HT 5-hydroxytryptamine

5-letter woman defined by a five-letter word—bitch

5-lp's five large powers (China, France, Japan, United Kingdom, West Germany)

5 m's man/womanpower, materials, money, machinery, management

5ᵒ quinto (Spanish—fifth)

5½ ballet and *Symphony for Fun*, full title *Symphony 5½*, by Don Gillis

5 o's police

5p five new pence

5-percenter person who for 5 percent arranges introductions leading to valuable orders

5 Préludes Scriabin's last composition for the piano

5-Ps William Oxbery—British player, poet, publican, publisher, and printer; Pootsa Power Publishing Press Publications

5-spot $5 bill

5-star five-star (top quality)

5tet quintet

5th Naval District Norfolk, Virginia

5th State Connecticut

5ᵗᵗᵉ quintet

5 vd's five venereal diseases: chancroid, gonorrhea, granuloma inguinale, lymphogranuloma venereum, syphilis

5 w's the *who, what, when, where,* and *why* reporters attempt to include in writing summary paragraphs

6ᵃf sexta-feira (Portuguese—Friday)

6-banger 6-cylinder engine

VI-BCE sixth century before the Christian era (Babylonian Century)—Babylonians defeated Israelites and made them captive after destroying

the temple of Solomon in Jerusalem—the 500s

6 bits 75 cents

6/c six-conductor

VI-C sixth century (Persian Century)—Khosru Nushirwan made peace with the Byzantine Empire and extended Persian rule throughout the Middle East—the 500s

6d English sixpenny; sixpence

6-dW Six-day War between Arab countries of Egypt, Jordan, Lebanon, and Syria versus Israel; June 5 to 10, 1967

6/e sixth edition

VIᵉ sixième (French—sixth)

6'er leader of a pack of six scouts

6-fold sixfold

6-gun six-chambered revolver; sixshooter

6-hda 6-hydroxydopamine

6-mcd six-month certificate of deposit

6-mo sixmo(s)

6-mTs six-month Treasury securities

6ᵒ sesto; sexto (Spanish—sixth)

6-pack carton containing six of a kind (6 containers of beer, soda, etc.)

6-Rs remedial readin', remedial 'ritin', remedial 'rithmetic

6-shooter revolver holding six cartridges

666 symbol for the anti-Christ

6th Naval District Charleston, South Carolina

6th State Massachusetts

6tt sextet

6WW Sixth Weather Wing (Air Force—Washington, D.C.)

7A Seven Arts Society

7 aa's 7 archangels (Gabriel, Jerahmeel, Michael, Raguel, Raphael, Sariel, and Uriel)

VII-BCE seventh century before the Christian era (Assyrian Century) when Assyria ruled Middle East and conquered Egypt—the 600s

7ber September

7ᵇʳᵉ Septembre (French—September); *septiembre* (Spanish—September)

7/c seven-conductor

VII-C seventh century (Islamic Century)—marked by Mohammed's flight from Mecca to Medina and his death in 632; Islam began expanding throughout the Middle East and North Africa as well as moving toward France and Spain—the 600s

7 cd's 7 chief devils (Aniguel, Anizel, Ariel, Aziel, Barfael, Marbuel, and Mephistopheles, according to the diabolarchy of hell)

7 Dec Pearl Harbor Day (1941)

7ds seven deadly sins—anger, covetousness, envy, gluttony, lechery, pride, sloth

7-d's deliriums, delusions, dementias, depressions, deviations, disorders, dreams

7/e seventh edition

7ᵉ septiembre (Spanish—September)

VIIᵉ septième (French—seventh)

7-fold sevenfold

7 lib ars seven liberal arts: grammar, logic, rhetoric, arithmetic, geometry, astronomy, and music

7 Ms seven (dance) movements: *élancer, étendre, glisser, plier, relever, sauter, tourner*

7ᵒ septimo (Spanish—seventh)

7 Octaves Louis Moreau Gottschalk

7 seas collective nickname of any seven seas; the Aegean, Baltic, Black, Caribbean, Mediterranean, North, and Red seas; the Arafura, Banda, Celebes, Coral, Java, Tasman, and Timor seas; the Aegean, Baltic, Black, Mediterranean, North, Norwegian, and Red seas; the Bering, East and South China seas, the Sea of Okhotsk, the Sea of Japan, the Philippine Sea, and the Yellow Sea; the Aral, Azov or Putrid Sea, Bellinghausen, Caspian, Korean, Marmora, Sulu, White; many mariners define the seven seas as the Antarctic, Arctic, North and South Atlantic, North and South Pacific, and Indian oceans

767–300 Boeing 767 300-passenger airliner

7 smells camphoric (moth repellant), musky (angelica oil), floral (roses), pepperminty (mint-flavored confections), ethereal (dry-cleaning fluids), pungent (vinegar), putrid (rotten eggs)

7th State Maryland

7tt septet

7/24 7 days a week, 24 hours a day

7-Up a carbonated beverage

7WW Seventh Weather Wing

(Air Force—Illinois)

8 numerical symbol for heroin, as H is the eight of the alphabet

8ᵉ **ottava** (Italian—octave)

8-ball behind the eight ball (in a bad position); eight ball (black billiard or pool ball numbered 8)

VIII-BCE eighth century before the Christian era (Chou Century)—eastern Chou dynasty began ruling China for the next five centuries—the 700s

8 bits one dollar

8ᵇʳᵉ **octobre** (French—October); *octubre* (Spanish—October)

VIII-C eighth century (Carolingian Century)—Charlemagne or Charles the Great reigned as King of the Franks and Emperor of the West as well as chief patron of learning—the 700s

8/e eighth edition

8ᵉ octubre (Spanish—October)

VIIIᵉ huitième (French—eighth)

8-fold eightfold

8 great 8 great violin concertos (Bach's for 2 violins, Beethoven, Brahms, Bruch, Mendelssohn, Paganini, Sibelius, Tchaikovsky)

8-h NBC's concert hall studio used by Toscanini and the NBC Symphony of the Air and more recently by Mehta and the New York Philharmonic Symphony

8mm film 8-millimeter film

8N American National 8-thread series

8° octavo (a book about 9¾ inches high)

8° octavo (Spanish—eighth)

8½ x 11 paper size 8½ inches by 11 inches; standard letterhead size

8tet octet

8th Naval District New Orleans, Louisiana

8th State South Carolina

8 to 5 everyday 8 A.M. to 5 P.M. job

8UN Unified 8-thread series

8ᵛᵃ ottava (Italian—octave)

8ᵛᵃ ᵃˡᵗᵒ ottava alto (Italian—octave higher)

8va bass. ottava bassa (Italian—octave lower)

8ᵛᵉ octave

IX-BCE ninth century before the Christian era (Phoenician Century)—Carthage founded by the Phoenicians who

traded in all areas of the Mediterranean—the 800s

9ᵇʳᵉ **novembre** (French—November); *noviembre* (Spanish—November)

IX-C ninth century (Century of Confusion)—Carolingian Empire of Charlemagne disintegrated; European unity dismembered and divided—the 800s

9/e, 10/e, 11/e, 12/e, etc. ninth edition, tenth edition, eleventh edition, twelfth edition, et cetera

9ᵉ **noviembre** (Spanish—November)

IXᵉ neuvième (French—ninth)

9ᵉᵗ nonet

9-fold ninefold

914 9 inches by 14 inches

9° nono; noveno (Spanish—ninth)

9-pins ninepin(s)

9 Ps nine known planets in our solar system according to their distance from the sun: Mercury, Venus, Earth, Mars, Jupiter, Saturn, Uranus, Neptune, Pluto

9th Naval District Great Lakes, Illinois

9th State New Hampshire

9 to 5 everyday 9 A.M. to 5 P.M. job

9-to-5 National Association of Working Women

10 deka (da)

'10 1810 (Bolvarian-type Spanish-American revolutions and wars of liberation, 1810–1826)

10⁻¹ deci (d)

10⁻² centi (c)

10⁻³ milli (m)

10⁻⁶ micro (µ)

10⁻⁹ nano (n)

10⁻¹² pico (p)

10⁻¹⁵ femto (f)

10⁻¹⁸ atto (a)

10² hecto (h)

10³ kilo (k)

10⁶ mega (M)

10⁹ giga (G)

10¹² tera (T)

10 Aug Ecuadorian Independence Day

X-BCE tenth century before the Christian era (Israelian Century)—King Solomon reigned and Israelites defeated all enemies and built the great temple of Jerusalem—the 900s

10ᵇʳᵉ **décembre** (French—December); *diciembre* (Spanish—December)

X-C tenth century (Mayan Century)—great American civilization left monumental ruins strewn from Honduras to Yucatan—the 900s

10 Dec Human Rights Day (Liberia)

10 Downing Street British prime minister's home in west central London

10ᵉ diciembre (Spanish—December)

Xᵉ dixième (French—tenth)

10ᵉᵗ dixet

10-fold tenfold

10-gage 10-gage shotgun

10-gallon hat cowboy hat

10 great 10 great violin concertos (Bach's for 2 violins, Beethoven, Brahms, Bruch, Mendelssohn, Paganini, Saint-Saëns, Sibelius, Tchaikovsky, Vieuxtemps)

10° decimo (Spanish—tenth)

10p 10 new pence

10-pin tenpin (bowling)

10-spot $10 bill

10th Naval District San Juan, Puerto Rico

10th State Virginia

10-V the lowest; the opposite of A-1; the worst

11 once (Spanish—eleven)— stands for aguardiente as this strong alcoholic drink has eleven letters

XI-BCE eleventh century before the Christian era (Century of Saul and David)—King Saul followed by King David as ruler of Israel—the 1000s

XI-C eleventh century (Aztecan and Incan Century)—vast monuments in the highlands of Mexico and Peru stand as mute witnesses to these great American civilizations—the 1000s

11 Downing Street official town residence of the British Chancellor of the Exchequer

11-11-11 eleventh hour, eleventh day, eleventh month of 1918 when Armistice ended World War I

11th-hr eleventh hour (last minute; latest possible time for assistance or a death-sentence reprieve)

11th Naval District San Diego, California

11th State New York

XII-BCE twelfth century before the Christian era (Trojan Century)—Troy fell to the

Greeks after a ten-year siege celebrated in Homer's epic poem, the *Iliad*—the 1100s

XII-C twelfth century (Portuguese Century) when Alfonso I Henriques reigned as king of Portugal, which emerged as a great maritime power—the 1100s

12 Downing Street office of the British Government Whips

12-gage 12-gage shotgun

12N American National 12-thread series

12° twelvemo (a book about 7¾ inches high)

12th Naval District San Francisco, California

12th State North Carolina

12-tone twelve-tone music; twelve-tone row

12UN Unified 12-thread series

13 Gallery 13—prisoner's grave(yard); numerical symbol for marijuana as M is the thirteenth letter of the alphabet; police radio signal call 13 indicates an officer needs help; the boss is nearby; white bread

XIII-BCE thirteenth century before the Christian era (Century of the Exodus)—Moses led the Israelites out of Egypt—the 1200s

XIII-C thirteenth century (Mongol Century)—dominated by the reign of the Mongol emperor Genghiz Khan whose hordes conquered China and Russia —the 1200s

13th Naval District Seattle, Washington

13th State Rhode Island

14 numerical symbol for narcotics as N is the fourteenth letter of the alphabet; special (food) order

XIV-BCE fourteenth century before the Christian era (Century of the Pharaoh Tutankhamen)

XIV-C fourteenth century (Tamerlane's Century)— Mongol emperor Timur (Tamer the Lame) dominated Middle East and western India—the 1300s

14 f 13 Nazi code name for death by gassing

14th Naval District Pearl Harbor, Oahu, Hawaii

14th State Vermont

XV-BCE fifteenth century before the Christian era (Egyptian Century)—Egyptian kingdom extended from the Sahara to beyond the Euphrates —the 1400s

XV-C fifteenth century (Italian Century)—powerful families such as the Borgias and the de Medicis brought about the renewal of art and architecture in Italy—the Italian Renaissance—the 1400s

15th Naval District Balboa, Canal Zone

15th State Kentucky

XVI to XXXII BCE (*see* XXXII-BCE)

XVI-C sixteenth century (Spanish Century) marked by discoveries and colonizations of much of the New World, circumnavigation of the globe, flowering of art and literature —the Golden Age or *Siglo de Oro*—as well as the defeat of the Spanish Armada by the British—the 1500s

16-gage 16-gage shotgun

16mm film 16-millimeter film

16N American National 16-thread series

16° sixteenmo (a book about 6¾ inches high)

16's 16 rpm phonograph records

16th State Tennessee

16UN Unified 16-thread series

XVII-C seventeenth century (Dutch Century) saw the discovery and settlement of what is now New York as well as South Africa and the East Indies by the Dutch who after a war at sea arranged a mutual defense pact with their British rivals—the 1600s (*see* the Elizabethan Age, *Le Grand Siecle, El Siglo de Oro*)

17-D modified yellow-fever virus

17 Tage und 4 Minuten (German—17 days and 4 minutes) —Werner Egk opera

17th Naval District Kodiak, Alaska

17th Parallel line of latitude dividing North and South Vietnam

17th State Ohio

XVIII-C eighteenth century (French Century) of courtesans and kings, poets and playwrights, of great territories acquired and lost, of Louis XVI and Marie Antoinette beheaded by the guillotine only to be replaced by Napoleon—the turbulent 1700s (*see* The Enlightenment)

18-19 Sept Chilean Independence Days

18th State Louisiana

19 banana split

XIX-C nineteenth century (British Century) from Napoleon's defeat by Wellington at Waterloo to the defeat of the Boers in South Africa this century was marked by British advances in invention, in the success of its industrial revolution, in its colonization in all parts of the world, and its maritime supremacy on all the oceans—the 1800s

19°2'40" N lat x 62°17'20" W long 19 degrees, 2 minutes, 40 seconds North latitude by 62 degrees, 17 minutes, 20 seconds West longitude, imaginary location of *Treasure Island* created by Robert Louis Stevenson

19th nineteenth

19th State Indiana

20 twenty

20' twenty-foot-long shipping container

XX-C twentieth century (American Century) characterized by industrial advances, victory in two world wars, as well as the development of inventions, the discovery of the North Pole, the placing of men on the moon, the elevation of living standards, the devotion to democratic ideals —the 1900s

20-gage 20-gage shotgun

'20s 1920–1929

20-spot $20 bill

20th Twentieth Century Limited (New York Central Railroad)

20th-century syndrome hypersensitivity to modern chemical compounds

20th State Mississippi

20 toes 10 toes down and 10 toes up (bedtime frolic)

21 blackjack; limeade

XXI-C twenty-first century (Japanese Century)—providing productivity, standard of living, and other growth factors are not disturbed by large-scale earthquakes and world wars—the 2000s

21st State Illinois

22 customer's check still unpaid

.22 .22-caliber ammunition, pistol, or rifle

22-cal killers assassins using 22-caliber silencer-equipped automatic pistols

22nd State Alabama

22 s-e silencer-equipped 22-caliber revolver

23 get lost; I'm busy; leave me alone; scram

23°27' N lat 23 degrees, 27 minutes North latitude, Tropic of Cancer

23°27' S lat 23 degrees, 27 minutes South latitude, Tropic of Capricorn

23rd twenty-third

23rd State Maine

23½ deg N lat Tropic of Cancer

23½ deg S lat Tropic of Capricorn

24 24 Capricci (Opus 1)—Paganini's Twenty-four Caprices for cadenza-like unaccompanied violin

24° twenty-fourmo (a book about 5¾ inches high)

24-Ps 24 Parganas (Zamindari of Calcutta fiscal divisions in the Ganges delta)

24/7 24 hours a day, 7 days a week

24th twenty-fourth

24th State Missouri

25 LSD as 25 is part of the chemical name—d-lysergic acid diethylamide tartrate 25

.25 .25-caliber ammunition or automatic

25th twenty-fifth

25th State Arkansas

26th twenty-sixth

26th State Michigan

27th twenty-seventh

27th State Florida

28-gage 28-gage shotgun

28th twenty-eighth

28th State Texas

29th twenty-ninth

29th State Iowa

30 finis symbol used by newspaper journalists at end of article or story

30 days, etc. (calendar mnemonic—30 days hath September, April, June, and November; all the rest have 31 save February; 28 are all its score, but in leap year one day more)

'30s 1930–1939

30th State Wisconsin

.30-'06 .30-caliber American cartridge introduced in 1906; used by US Armed Forces in World Wars I and II for rifles and machine guns

31st State California

XXXII-BCE thirty-second century before the Christian era (Dynastic Century) when the first and second of many Egyptian dynasties began a rule lasting for at least seventeen centuries before the power of the pharaohs began to wane—the 3100s

32nd State Minnesota

32° thirty-twomo (a book about 5 inches high)

33rd St New York dialect enriched by longshoremen, taxi-cab drivers, and others who pronounce it *toity-toid street*

33rd State Oregon

33s 33⅓ rpm phonograph records

34th State Kansas

35mm film 35-millimeter film

35th State West Virginia

36 postum

36th State Nevada

37th State Nebraska

.38 .38-caliber ammunition or pistol

38th Parallel line of latitude dividing North and South Korea

38th State Colorado

39th State North Dakota

40 40 acres

40' forty-foot-long shipping container

'40s 1940–1949

40th State South Dakota

40 winks a nap or short sleep

41st State Montana

42nd cousin a distant relative

42nd State Washington

42nd Street New York City's most blatant nightlife area

43rd State Idaho

.44 .44-caliber ammunition or pistol

44th State Wyoming

.45 .45-caliber ammunition, pistol, or submachine gun

45's 45 rpm phonograph records

45th State Utah

46th State Oklahoma

47th State New Mexico

47th Street New York City's diamond center between the Avenue of the Americas and Fifth Avenue on 47th Street

48 48-hour weekend liberty pass; the forty-eight preludes and fugues of Johann Sebastian Bach's collection known as the *Well-tempered Clavier*

48er emigrant who came to America in 1848; participant in German revolution of 1848

48° forty-eightmo (a book about 4 inches high)

48th State Arizona

48½ discharged; fired

49er gold-rush settler who came to California in 1849

49th State Alaska

.50 .50-caliber ammunition or machine gun

50p fifty new pence

50 percenter auto-mechanic causing the need for unnecessary repairs or recommending unnecessary replacements and then splitting the profits with gas-station operators

'50s 1950–1959

50-spot $50 bill

50th State Hawaii

51 one cup of hot chocolate

52 two cups of hot chocolate

54 54 minorities comprising the people of China

55½ small root beer

60 60 minutes (tv program exposing corruption)

'60s 1960–1969

64° sixty-fourmo (a book about 3 inches high)

66 dirty dishes; Phillips Petroleum Company

66 deg 17 min N lat Arctic Circle

66 deg 17 min S lat Antarctic Circle

69 pictorial numerical symbol for oral-genital copulation

70mm film 70-millimeter film

'70s 1970–1979

73 best regards (amateur radio)

75's 75mm cannon

76 Union Oil

'76 1776

78's 78 rpm phonograph records

'80s 1980–1989

81 glass of water

82 two glasses of water

84 naval prison

84 Ave Foch Paris headquarters of the Gestapo during the German occupation of Paris in World War II; its top floor contained cells and its lower floors housed offices and torture chambers

86 don't serve

87½ look at the lovely girl(s) out front

88 Column 88 (neo-Nazi organization based in London); love and kisses (amateur radio); (musician's slang) a pi-

ano

89d 89 days (New York to San Francisco run of American clipper ship *Flying Cloud* in 1854)

89er Oklahoman who settled in 1889 when the territory was opened

90-day wonder officer commissioned after only 90 days of training

90 deg N lat North Pole (zero degrees longitude)

90 deg S lat South Pole (zero degrees longitude)

90° N lat 90 degrees North latitude, North Pole

90° S lat 90 degrees South latitude (South Pole)

'90s 1990–1999

91 a glass of seltzer water

92 St Y 92nd Street Y School of Music

93-score best grade of butter (USDA grade AA)

95 the customer's leaving without paying

'96 1796 (Napoleonic Wars, 1796-1815)

98 assistant manager's nearby —look out

'98 the generation of 1898 (Spain's generation of cultured persons reacting to the illusions and incompetence that led to the loss of Cuba in 1898)—Altamira the historian, Azorín the author, Cossío the art historian and teacher, Ferrer the anarchist teacher, Machado the poet, Ortega y Gasset the essayist, Ramón y Cajal the histologist and surgeon, Unamuno the teacher and writer, etc.

99 manager's nearby—look out

100 Hydrographic Office Publication 100—*Merchant Marine House Flags and Stack Insignia*—U.S. Navy Hydrographic Office; one hundred

100-percenter(s) one-hundred percent patriot(s)

100-proof completely honest; the whole truth

100s hundreds

108-named god of the Hindus Vishnu

110 110 volt(s)

110v 110 volts

110v/60c (60h) 110-volt 60-cycle (60 hertz) electric current

111 emergency services (New Zealand); One-Eleven (British Aircraft Corporation short-take-off-and-landing fan-jet aircraft)

118-island city Venice (Venezia)

150 Publication 150—*World Port Index*—U.S. Naval Oceanographic Office

180° longitude International Date Line where east meets west 180 degrees east or west of the Greenwich or prime meridian

200 200-mile Exclusive Economic Zone (EEZ)

220(v) 220 volt(s)

240 Convair two-engine transport airplane; trotting horse speed—1 mile in 2 minutes and 40 seconds; synonym for high speed

240v 240 volts

280 copper alloy (Muntz metal); yellow metal

291 291 Fifth Avenue (address of An American Place—Alfred Stieglitz's art gallery where much avant-garde lithography, painting, and photography were first shown)

400 the four hundred; the socially elite (originally designated by Ward McAllister, who drew up a list containing the top 400 in New York society)

.410 .410-caliber shotgun

415 PC Section 415 Penal Code—disturbing the peace

486 French formula employed in abortion-inducing pill

500 Festival 500-mile (805-kilometer) auto racing event of the International Automobile Speed Classic at Indianapolis, Indiana

502 drunken driving (police code)

606 arsphenamine compound sold as Salvarsan; 606th compound developed and tested by Paul Ehrlich for treatment of relapsing fevers and syphilis

707 Boeing Stratoliner jet-transport airplane

720 Boeing medium-range jet-transport airplane

727 Boeing jet-transport with three empennage-mounted engines

737 Boeing short-range twin-jet airplane

747 Boeing jumbo jetliner

757 Boeing 136-seat medium-range jetliner

767 Boeing's fuel-saver twin-jet passenger plane

800-C-O-C-A-I-N-E 800–262–

2463 (national cocaine hotline connecting addicts with counselors and psychiatric hospital emergency rooms)

880 Convair 880 jet airplane

911 emergency telephone number

925 Office Workers Union

990 Convair 990 fan-engine jet airplane

999 emergency services (United Kingdom)

1000s Thousand Islands, some 1,500 islets in the St Lawrence River

1011 Lockheed 1011 jumbo jetliner

1080 sodium fluoroacetate

1098 supplies and equipment requisition form (British Army)

1400s fourteen hundreds

1500s fifteen hundreds

1600 Pennsylvania Avenue (Washington, DC, address of the White House)

1600s sixteen hundreds

1700s seventeen hundreds

1800s eighteen hundreds

1812 Tchaikovsky overture for full orchestra and cannon

1814 Leroux opera

1854-1954 century-long life of the French Guiana penal colony popularly known as Cayenne or Devil's Island; world opinion about the inhumanities practiced here did much to stop its existence

1900s nineteen hundreds

"1905" Leon Trotsky's account of the dress-rehearsal Russian revolution of 1905; Dmitri Shostakovich's Symphony No 11—*Year 1905*

1905er Old Bolshevik; participant in the Russian Revolution of 1905; veteran communist

1917 Symphony No 12 by Shostakovich

"1919" novel by John Dos Passos depicting World War I era of American life in series of camera-eye closeups

1941 Prokofiev's symphonic suite containing three movements—*Battle, At Night, Brotherhood of Nations*

1945 Symphony No 9 (E-flat major) by Shostakovich

"1984" *Nineteen eighty-four* (novel of George Orwell describing totalitarian terror in the year 1984; symbol for anti-libertarian trends

2141 2141 islands of Micro-

nesia (greater in area than the continental U.S. but smaller in land mass than Rhode Island)—Trust Territory of the Pacific with its capital on Saipan atop Capitol Hill
2707 Boeing supersonic transport airplane
9653 Convict 9653 (American

Socialist nominee for President—Eugene V Debs when in the U.S. Penitentiary in Atlanta, Georgia)
10,000 Lakes State Minnesota
22445 prisoners' petitions for judicial reviews of their cases
23102a V(ehicle) C(ode) driving under the influence of any

intoxicating liquor or drug
338171 TE　TE　Lawrence (Lawrence of Arabia's number in the British Army; he used this number rather than his name as a final defense against a world he found hostile and unresponsive)

Numeration

Power	Prefix	Abbre-viation	Name
1,000,000,000,000,000,000	10^{18}	eva	e	one quintillion*
1,000,000,000,000,000	10^{15}	peta	p	one quadrillion
1,000,000,000,000	10^{12}	tera	t	one trillion
100,000,000,000	10^{11}			one-hundred billion
10,000,000,000	10^{10}			ten billion
1,000,000,000	10^{9}	giga	g	one billion
100,000,000	10^{8}			one-hundred million
10,000,000	10^{7}			ten million
1,000,000	10^{6}	mega	m	one million
100,000	10^{5}			one-hundred thousand
10,000	10^{4}			ten thousand
1000	10^{3}	kilo	k	one thousand
100	10^{2}			one hundred
10	10^{1}			ten
1	10^{0}			one
0.1	10^{-1}	deci	d	one-tenth
0.01	10^{-2}	centi	c	one-hundredth
0.001	10^{-3}	milli	m	one-thousandth
0.0001	10^{-4}			one ten-thousandth
0.00001	10^{-5}			one hundred-thousandth
0.000001	10^{-6}	micro (μ-mu)		one millionth
0.0000001	10^{-7}			one ten-millionth
0.00000001	10^{-8}			one hundred-millionth
0.000000001	10^{-9}	nano	n	one billionth
0.0000000001	10^{-10}			one ten-billionth
0.00000000001	10^{-11}			one hundred-billionth
0.000000000001	10^{-12}	pico	p	one trillionth
0.0000000000001	10^{-13}			one ten-trillionth
0.00000000000001	10^{-14}			one hundred-trillionth
0.000000000000001	10^{-15}	femto	f	one quadrillionth
0.0000000000000001	10^{-16}			one ten-quadrillionth
0.00000000000000001	10^{-17}			one hundred-quadrillionth
0.000000000000000001	10^{-18}	atto	a	one quintillionth

*Quintillions are followed by sextillions, octillions, nonillions, decillions, undecillions, duodecillions, tredecillions, quattuordecillions, quinquedecillions, sexdecillions, septendecillions, octodecillions, novemdecillions, vigintillions (a thousand novemdecillions).

Oldest North American Settlements

Oldest American Settlement entries refer to the first permanent settlements.

Alabama by the French in the Mobile area around 1702
Alaska Three Saints Bay, Kod-

iak Island, founded in 1784 by Grigori Shelekhov
American Samoa Pago Pago

settled 4,000 years ago on Tutuila Island
American Virgin Islands St

Thomas Island's port of Charlotte Amalie built by the Danish in 1672

Arizona Tombstone, the old mining town booming in the 1880s

Arkansas Arkansas Post established by French settlers in 1686 east-southeast of Pine Bluff on Arkansas River

California San Diego de Alcalá founded in 1769 by Padre Junípero Serra

Colorado Denver, 1858

Connecticut Hartford, founded as a Dutch trading post by Adriaen Block in 1633

Delaware Wilmington, called Fort Christina by the Swedes, in 1638

District of Columbia cornerstone of the capitol's north wing laid by President Washington on September 18, 1793

Florida St Augustine, founded by the Spanish in 1565

Georgia Savannah settled by the English under James Oglethorpe in 1733

Guam Agaña settled by the Spanish in 1668

Hawaii Honolulu, discovered by Captain Cook in 1778

Idaho Mormon settlement at Franklin

Illinois Fort Crève Coeur in 1680 near Peoria

Indiana Vincennes in 1702 on the Wabash River

Iowa Dubuque, 1834

Kansas Fort Leavenworth, 1827

Kentucky Harrodsburg, 1774,

southwest of Lexington

Los Angeles *Nuestra Señora la Reina de los Angeles de Porciuncula* (Our Lady of the Angels of Porciuncula), 1781

Louisiana Natchitoches founded by the French in 1714

Maine Phippsburg, 1607, at the mouth of the Kennebec River

Maryland St Mary's 1634

Massachusetts Plymouth, 1620

Michigan Fort Ponchartrain, 1701, at the site of what is now Detroit

Minnesota Fort Snelling on the site of Minneapolis, 1820

Mississippi Biloxi Bay, 1699

Missouri Sainte Genevieve, 1735

Montana Virginia City, 1850

Nebraska Bellevue, 1823

Nevada Genoa, 1849, near what is now Carson City

New Hampshire Dover, 1624

New Jersey Communipaw, 1620

New Mexico Pueblo San Juan, 1598

New Orleans platted in 1718 by the Sieur de Bienville, made capital of Louisiana in 1722

New York Nieuw Amsterdam, 1664 on the southern tip of Manhattan Island

North America St Augustine, Florida founded by the Spanish in 1565

North Carolina Bath, 1705, after the loss of Roanoke Island's settlers

North Dakota Pembina, 1851

Ohio Marietta, 1788

Oklahoma Salina settled in the early 1800s

Oregon fur-trading post established in 1811 by John Jacob Astoria at the mouth of the Columbia River and the site of Astoria

Pennsylvania Tinicum Island, south of Philadelphia, founded by the Swedes in 1643

Puerto Rico San Juan settled by Ponce de León in 1521

Rhode Island Providence settled by Roger Williams in 1636

San Francisco San Francisco de Asís, now called Mission Dolores, founded in 1776 by the Spanish

Seaport Pensacola, Florida

Seattle began as a lumber town in 1852

South Carolina Charleston 1680

South Dakota Fort Pierre, 1817

Tennessee Jonesboro, 1779 close to the site of Johnson City

Texas Ysleta, 1681, close to El Paso

Utah Salt Lake City, 1847

Vermont Fort Dummer, near Burlington, 1724

Virginia Jamestown, 1607

Washington Fort Okanogan, 1888

West Virginia Romney and Sheperdstown, 1762

Wisconsin Green Bay, 1701

Wyoming Fort Laramie, 1834, a fur-trading post near Cheyenne

Phobias

Opposite of the suffix *phobia*—fear or hatred of—is *phil* or *phile*—admiration or love of—as in *Anglophile* meaning admiration or love of the English, the British Empire, the Commonwealth, or the United Kingdom, whereas *Anglophobia* means fear of or hatred of all things English.

acarophobia delusion one's skin is infested with mites, insects, or worms

acrophobia fear of high places

aelurophobia fear or hatred of cats (*also written* ailurophobia)

aerophobia fear of flying due to fatal accidents, fear of heights, or motion sickness; fear of high winds or violent

storms (*also written* airphobia)

agoraphobia fear of busy streets, crowded churches and stores, stadiums; fear of the marketplace or other open spaces

agyiophobia fear of streets

aichmophobia fear of being touched by a finger or a sharp-pointed instrument or weapon

AIDS-phobia fear of AIDS (ac-

quired immune deficiency syndrome)

ailurophobia fear of cats and other felines (*also written* aelurophobia *and synonym of* felinophobia)

airphobia (*see* aerophobia)

alcoholophobia fear of becoming an alcoholic

algophobia fear of pain or watching anyone in pain

amaxophobia fear of vehicles
Americanophobia fear of Americans or the United States of America
amphidiaphobia fear of amphibians—coecilians, frogs, salamanders, toads
amychophobia fear of being scratched by claws or sharp-pointed nails
androphobia fear of boys and men
anemophobia fear of wind
anginophobia fear of angina pectoris
Anglophobia fear of or hatred of the English, the British Empire, the Commonwealth, the United Kingdom
Angst (German—fear)
anthophobia fear of flowers
anthrophobia fear of people, persons, or society
anthropophobia fear of being eaten alive; fear of cannibalism
antlophobia fear of floods
antrophobia fear of people
anxiety hysteria *see* phobia
anxiety phobias agoraphobia (fear of being alone or in public places); simple or social phobias marked by irrational fear of objects or certain social situations believed to be embarrassing or humiliating
apartheidphobia fear of people who impose apartheid in South Africa
aphephobia fear of being touched by anyone
apiphobia fear of bees or other buzzing insects
apotemnophobia fear of persons with amputations
aquaphobia fear of water
arachnephobia fear of spiders
assassinophobia fear of assassination
asthenophobia fear of weakness
astraphobia fear of lightning and thunder
astrophobia fear of the heavens and the stars
ataxophobia fear of muscular hysteria and incoordination
aulophobia fear of flutes
automysophobia fear of being unclean
autophobia fear of being alone; fear of oneself
bacillophobia fear of bacilli; fear of microbes
bacteriophobia fear of bacteria
ballistophobia fear of bullets;

fear of missiles
basiphobia fear of being unable to walk
basophobia fear of being unable to stand erect (*synonym for* stasiphobia)
bathophobia fear of depths or looking down from high places
batophobia fear of falling objects or high places
batrachiophobia fear of amphibians—coecilians, frogs and toads, newts and salamanders
belonophobia fear of needles, pins, and other sharp-pointed instruments or weapons such as daggers and swords
bibliophobia fear of books and other printed matter
botanophobia fear of forests, jungles, plantlife
boviphobia fear of bulls and other aggressive cattle, such as bison, buffalo, and yaks
brassophobia fear of brass and brassware
bromidrosiphobia fear of body odors
brontophobia fear of thunder
bufonophobia fear of toads
cainotophobia fear of new ideas or new things
cancerophobia fear of cancer
cardiophobia abnormal fear of heart disease or heart failure
cenophobia fear of empty houses or rooms
ceraunophobia fear of thunder
cheloniaphobia fear of terrapins, tortoises, turtles
cherophobia fear of fun and gaiety
chionophobia fear of snow
chirosphobia fear of writer's cramp
chirurphobia fear of surgery
cholerophobia fear of contracting cholera
chrematophobia fear of money
chromophobia fear of certain colors
chronophobia fear of time
cibophobia fear of food
claustrophobia fear of closed spaces
climacophobia fear of climbing; fear of stairs
clinophobia fear of going to bed
clithrophobia fear of being locked in
coitophobia fear of sexual intercourse
coprophobia fear of dung
cosophobia fear of dawn

cremnophobia fear of high or steep places
crocodiliophibia fear of crocodilians—alligators, caymans, crocodiles, gavials
crystallophobia fear of glass
cumacophobia fear of falling downstairs
cumophobia fear of beds
cyclophobia fear of bicycles
cymophobia fear of waves
cynophobia fear of dogs and all their relatives—coyotes, dingos, foxes, hyenas, wolves
cypridophobia fear of venereal disease
cypriphobia fear of venereal disease
decidophobia fear of making decisions
demonophobia fear of demons or devils
dermatopathophobia fear of skin diseases
dextrophobia fear of the right side
dipsophobia fear of drinking
domatophobia fear of being confined in a house
doraphobia fear of touching the hair or fur of any animal
dromophobia fear of crossing streets or wandering
dustophobia fear of dust
Dutchphobia fear of the Dutch
dysmorphophobia fear of deformity
ecophobia fear our planet is increasingly inhospitable
Egyptophobia fear of Egypt or the Egyptians
electrophobia fear of electricity
emetophobia fear of vomiting
entomophobia fear of insects
eosophobia fear of dawn
equinophobia fear of asses, horses, mules, ponies, zebras and other members of the horse family
eratophobia fear of sexual lust
eremophobia fear of being alone, monophobia
ereuthrophobia fear of blushing
ergasiophobia fear of assuming responsibility or doing work of any kind
ergophobia fear of working
erotic pyromania sexual pleasure derived from raging fires
erotophobia fear of sexual lust
erythrophobia fear of blushing or embarrassment; fear of, or aversion to, anything colored red

eurotophobia fear of the female genital organs

felinophobia fear and hatred of all cats—bobcats, cheetahs, jaguars, leopards, lions, lynxes, ocelots, ounces, panthers, pumas, tigers, wildcats; *synonym of* aelurophobia, *also written* ailurophobia

feminophobia fear of females

forgetophobia fear of forgetting

Francophobia fear of France and all things French

Gallophobia *see* Francophobia

gamophobia fear of marriage

gatophobia fear of cats and other felines

genophobia fear of sex

gephyrophobia fear of bodies of water, of crossing bridges over water or traveling on boats or aircraft

Germanophobia fear of the Germans or Germany

gerontophobia fear of old people

graphophobia fear of writing

Grecophobia fear of Greece or the Greeks

gringophobia fear of gringos

gymnophobia fear of nudity

gynophobia fear of women or being in their company

hagiophobia fear of religion and its saints

haphephobia fear of being touched by another person

harpaxophobia fear of robbers

hedonophobia fear of pleasure

heliophobia fear of the sun

helminthophobia fear of worms or delusion of being infested by them

hematophobia fear of blood or the sight of blood

hemophobia fear of bleeding or seeing blood

heresyphobia fear of committing heresy

herpetophobia fear of amphibians (frogs, newts, salamanders, toads) and reptiles (crocodiles, lizards, snakes, turtles)

hierophobia fear of religious objects and people connected with religion

Hispanophobia fear of Spain or Spanish-speaking people

hodophobia fear of travel

homophobia hatred of homosexuals

hydrophobia fear of water; common name for rabies, resulting from a bite by a rabid animal

hydrophobophobia fear of rabies

hyenophobia fear of hyenas

hygrophobia fear of dampness

hylophobia fear of forests

hypnophobia fear of falling asleep

hypsophobia fear of high places, acrophobia, aerophobia

iatrophobia fear of doctors and doctor-induced disorders

ichthyophobia fear of fish

ideophobia fear of mental images

illyngophobia fear of vertigo

incontiphobia fear of becoming incontinent

insectophobia fear of insects

iophobia fear of being poisoned; fear of touching anything rusty

Italophobia fear of the Italians and Italy

jumbophobia fear of bigness

kakorrhaphiophobia fear of failure

keraunophobia fear of thunder and lightning

kleptophobia fear of stealing

koniophobia fear of dust

kopophobia fear of fatigue

lacertiliaphobia fear of lizards

lalophobia fear of speaking for fear of making errors or stammering

leprophobia fear of lepers and leprosy

levophobia fear of objects on the left side of the body

linophobia fear of strings

logizomechanophobia fear of computers, computing machines, and all their works

lyssophobia fear of rabies

maieusophobia fear of pregnancy or childbirth

mechanophobia fear of machines and machinery in operation

metallophobia fear of metallic objects

meteorophobia fear of meteors

microphobia fear of germs; fear of microbes; dread of small objects

monophobia fear of being alone, eremophobia

mysophobia fear of contamination, dirt, disease, and microbes

mythophobia fear of lying and making false statements

necrophobia fear of dead bodies or of death

negrophobia fear of the black

people

neophobia fear of anything new or novel

nephophobia fear of clouds

nictophobia fear of the dark

Nippophobia fear of Japan and the Japanese

noctiphobia fear of night and darkness

noisicophobia fear of noise or electronically amplified music

nomatophobia fear of names of people or place-names

nosophobia fear of a disease or illness

nostophobia fear of returning home

nucleomitiphobia fear of nuclear warfare

nudophobia fear of being naked

numerophobia fear of numbers

nyctophobia fear of darkness or night

ochlophobia fear of populated places and crowds

odontophobia fear of dental surgery or the sight of teeth

ombrophobia fear of black clouds, rain, and storms

ommatophobia fear of eyes

ophidiophobia fear of snakes

optophobia fear of opening your eyes

osmophobia fear of odors

pagophobia fear of eating

panphobia groundless fear of everything

pantophobia fear of everything

paralipophobia fear of forgetting or neglecting to perform some duty

parasitophobia fear of parasites

pathophobia fear of disease

peccatophobia fear of sinning

pediculophobia fear of lice

pediophobia fear of children, dolls, puppets

peniaphobia fear of poverty

pharmacophobia fear of taking medicines

phasmophobia fear of ghosts

phengophobia fear of light

-phobia suffix meaning dread, abnormal fear, or aversion to a subject

phobic characterized by or pertaining to a phobia

phobism(s) affected by phobias

phobist person with a morbid fear of anything

phobophobia fear of acquiring a phobia

phonophobia fear of noise or sound; fear of speaking or

hearing one's own voice
photophobia fear of light
phronemophobia fear of thinking
pnigophobia fear of choking
polyphobia fear of many ideas and things
ponophobia fear of exerting one's self; dread of pain
pornophobia fear of erotic behavior
Portugophobia fear of Portugal and the Portuguese
potamophobia fear of rivers and running water
proctophobia fear of rectal disease
psychoneurophobia fear of emotional and mental disorders
psychophobia fear of the mind; fear of psychological findings
psychrophobia fear of cold
pyrexiophobia fear of fever
pyrolagnia sexual pleasure aroused by raging fires
pyrophobia fear of fire
quizziphobia fear of being quizzed
ranidaphobia fear of frogs
rectophobia fear of rectal disease
reptiliaphobia fear of reptiles—crocodilians, lizards, snakes, tuataras, turtles
rhabdophobia fear of being hit or beaten with a rod
rhypophobia fear of defecation or fecal filth
Romanophobia fear of the Roman Empire or the Romans
Russophobia fear of Russia or Russians
satanophobia fear of the devil
scabiphobia fear of itching; fear of scabies
schoolphobia fear of going to school
scopophobia fear of being seen
scotophobia fear of darkness

sidereaphobia fear of stars
siderophobia fear of touching iron or steel
silverphobia fear of silver and silverware
simple phobia irrational fear of specific objects such as cats, dogs, mice, snakes
Sinophobia fear of China or the Chinese
sitiophobia fear of eating; fear of food (*also written* sitophobia)
social phobia fear of being watched by others
sophophobia fear of going to school; fear of learning
spatiophobia fear of space
spectrophobia fear of mirrors
spermatophobia fear of loss of semen
spinnanphobia fear of spiders
stagiophobia fear of hell
stasiphobia fear of being unable to stand erect, basophobia
stenophobia fear of open places
sudophobia fear of sweat
symbolophobia fear of doing or saying anything that may be interpreted as having symbolic meaning
syphiliphobia fear of syphilis
Syriaphobia fear of Syria and the Syrians
taeniophobia *see* teniophobia
taphephobia fear of being buried alive, tapophobia
tapophobia fear of being buried alive
technophobia fear of technology
teleophobia fear of the existence of a grand design in nature
teniophobia fear of being infested with tapeworms
teratophobia fear of deformed people

terroristophobia fear of terrorists
thallasophobia fear of the sea
thanatophobia fear of death and dying
theophobia fear of the wrath of a god
tocophobia fear of childbirth
tonitorphobia fear of thunder
topophobia fear of performing; fear of a particular place; stage fright
toxicophobia fear of being poisoned
traumatophobia fear of injury
tremophobia fear of trembling
triakaidekaphobia fear of thirteen
trichinophobia fear of developing trichinosis
tricopathophobia women's fear of unwanted hair
tricophobia fear of hair or touching hair
triskaidekaphobia fear of the number 13
tropophobia fear of moving or marine changes
tuberculophobia fear of being infected with tuberculosis
Turkophobia fear of Turkey or the Turks
urbanophobia fear of cities
urophobia fear of urine
ursiphobia fear of bears
vaccinophobia fear of vaccination
verbophobia fear of words
vestiophobia fear of clothing
vomitophobia fear of vomit
vulvaphobia fear of vulval inflammation
Walloonphobia fear of the Walloons
xenophobia fear of foreigners, strangers, or anything foreign or strange
Yankeephobia fear of Yankees
zelophobia fear of jealousy
zoophobia fear of animals

Ports of the World

Ports of the world are listed alphabetically by country or state in Eponyms, Nicknames and Geographical Names. Many also appear among Superlatives and in the main body of the dictionary.

Proofreader's Marks

‖ align; straighten ends of lines	⊔ depress or sink a letter or word	⩔ quotation marks
⩔ apostrophe or single quotation mark	⊓ elevate or raise a letter or word	*rom* set in roman type ;/ semicolon
bf black face or bold face type (run <u>waved line</u> under text matter)	=/ hyphen	*sc* small caps (run double line under material: <u>a.d.</u>)
⊗ broken type; damaged type; imperfect type	*ital* set in *italics* (material to be italicized is <u>underlined</u>)	# space; # # double space; etc.
cap capital letter	*lc* lower case (run / through letter or letters to be set in *l*ower *c*ase)	⑤⑭ spell out (material to be spelled out is encircled: ⊙.⊙.)
≡ capital letters (run triple line under material to be capitalized: <u>G</u>eorge <u>W</u>ashington)	*lead* insert lead spacing between lines	*stet* let stand that which has been deleted; restore crossed out material (indicate by running dots under the letters of the words to be <u>restored</u>)
∧ caret; insertion mark	⊏ move to the left	
◡ close up	⊐ move to the right	
:/ colon	⁋ paragraph	*tr* transpose (indicate in text by ⩑ or ⩗)
⩓ comma	⊙ period	℮ turn letter right side up
d delete or dele; expunge; take out	⊥ push down space which prints as a mark	*wf* wrong font

Railroad Conductor's Cord-Pull Signals Plus Engineer's Whistle Signals

Cord-Pull Signals

1 short cord pull or 1 short whistle toot: apply brakes—stop
1 long whistle toot *when standing:* apply brakes or brakes applied
             when running: approaching grade crossing, junction, or station
2 long cord pulls or 2 long whistle toots: release brakes—proceed
3 short cord pulls or 3 short whistle toots *when standing:* back up
             when running: stop at next passenger station
4 short cord pulls or 4 short whistle toots: call for signals
succession of short cord pulls or short whistle toots: alarm or emergency such as persons or livestock
   on track; stop train until safe to proceed

Engineer's Whistle Signals

1 long toot followed by 3 short toots: flagman protect rear of train
3 short toots followed by 1 long toot: flagman protect front of train
4 long toots: flagman may return from west or south
5 long toots: flagman may return from east or north
1 short toot followed by 1 long toot: flagman inspect train for sticking brakes or leaks
2 long toots followed by a short toot and a long toot *(t o o t   t o o t   toot   t o o t)*: approaching curve, grade crossing, tunnel, or other obscure place; approaching a train standing on an adjacent track

1 long toot followed by a short toot *(t o o t   toot)*: blown when running against the current of traffic approaching curves, grade crossings, junctions, stations, and tunnels or obscure places

In the event of whistle failure, the bell must be rung continuously while the train is enroute. When the train is approaching or leaving a station, the bell is rung to indicate the need for caution and to avoid the noise of the whistle.

Railroads of the World

This listing includes abbreviations, nicknames and reporting marks.

AA Ann Arbor Railroad
AAR Association of American Railroads
A & B Antofagasta and Bolivia
ABB Akron and Barberton Belt Railroad
ABL Alameda Belt Line
AC Algoma Central Railway
ACL Atlantic Coast Line (Seaboard Coast Line Railroad)
ACR Algoma Central Railway
ACY Akron, Canton and Youngstown Railroad
AD Atlantic and Danville Railway
ADN Ashley, Drew and Northern Railway (also AD & N)
AEC Atlantic and East Carolina
AF Alma and Jonquieres Railway
AFE Administracion de los Ferrocarriles del Estado (Spanish—State Railways Administration)—Venezuela
AFL Administracion de los Ferrocarriles del Estado (Spanish—State Railways Administration)—Venezuela
AGS Alabama Great Southern (Southern Railway)
AL Almanor Railroad
ALM Arkansas and Louisiana Missouri Railway (also A & LM)
ALN Albany and Northern Railroad
ALQS Aliquippa and Southern
ALS Alton and Southern Railroad
AL & S Alton and Southern Railroad
Alton Route Gulf, Mobile and Ohio Railroad
AMC Amador Central Railroad
AMR Arcata and Mad River
Amtrak American (railroad) tracks—(government-sponsored program for reviving city-to-city passenger service)
AN Apalachicola Northern Railroad

Ann Arbor Detroit, Toledo and Ironton Railroad
Annie & Mary (nickname— Arcata and Mad River Railroad)—originally the Union Wharf and Plank Walk Company
ANR Angelina and Neches River Railroad; Australian National Railways
APA Apache Railway Company
APD Albany Port District
AR Aberdeen and Rockfish
ARA Arcade and Attica Railroad
ARC Alexander Railroad (Southern)
ARR Alaska Railroad
ART American Refrigerator Transit
ARW Arkansas Western Railway (Kansas City Southern)
A & S Abilene and Southern
ASAB Atlanta and Saint Andrews Bay Railway
ASDA Asbestos and Danville
ASLRA American Short Line Railroad Association
ASR Association of Southeastern Railroads
ATC Arnold Transit Company
ATN Alabama, Tennessee and Northern Railroad
ATSF Atchison, Topeka and Santa Fe Railway (also AT & SF)
ATW Atlantic and Western
AUG Augusta Railroad
AUS Augusta and Summerville
Austrail Railways of Australia
AVL Aroostook Valley Railroad
AW Ahnapee and Western Railway
AWP Atlanta and West Point Rail Road (includes Western Railway of Alabama and Georgia Railroad)—also A & WP
AWW Algers, Winslow and Western Railway
A y B Antofagasta y Bolivia

(Spanish—Antofagasta and Bolivia)—Chilean Railway linking Pacific port with highlands of landlocked Bolivia
AYSS Allegheny and South Side
ba BART (Bay Area Rapid Transit)
B & A Boston and Albany (Penn Central)
B-A-M Baikal-Amur-Magistral (railroad in Pacific Siberia, USSR)
BAP Butte, Anaconda and Pacific Railway (also BA & P)
BAR Bangor and Aroostook Railroad
BARC Baltimore and Annapolis Railroad Company
B & ARR Boston and Albany Railroad
BART Bay Area Rapid Transit (San Francisco Bay Area mass transportation system)
Bay Line Atlanta and Saint Andrews Bay Railway
BB Birmingham Belt Railroad
BCE Route British Columbia Electric Route
BCH British Columbia Hydro and Power Authority
BCK Buffalo Creek Railroad
BCK Bas-Congo au Katanga (French—Lower Congo—Katanga)—railway of Zaire
BCR British Columbia Railway
BCRR Boyne City Railroad
BCYR British Columbia Yukon Railway
BDZ (Cyrillic transliteration— Bulgarian State Railways)
BE Baltimore and Eastern Railroad (Penn Central)
BEDT Brooklyn Eastern District Terminal Railroad
BEEM Beech Mountain Railroad
BEM Beaufort and Morehead Railroad
Bessemer Bessemer and Lake Erie

Railroad
CIW Chicago and Illinois Western Railroad
CIWL Compangie Internationale des Wagon-Lits (French —International Sleeping Car Company)
CKSO Condon, Kinzua and Southern Railroad
CLC Colombia and Cowlitz
CLCO Claremont and Concord
Clinchfield Chinchfield Railroad (Carolina, Clinchfield and Ohio Railway)
CLK Cadillac and Lake City Railway
CLP Clarendon and Pittsford Railroad
CLRR Camp Lejeune Railroad
CMO Chicago, St Paul, Minneapolis and Omaha (Chicago North Western)
C M StP & P Chicago, Milwaukee, St Paul and Pacific
CN Canadian National (includes Canadian National Railways; Central Vermont Railway; Duluth, Winnipeg and Pacific Railway; Grand Trunk Lines in U.S.A.)
C & N Carolina and Northwestern Railway
CNJ Central Railroad of New Jersey
CN & L Columbia, Newberry and Laurens Railroad
CNO & TPR Cincinnati, New Orleans and Texas Pacific Railway
CNR Chiriqui National Railroad (Panama)
CNTP Cincinnati, New Orleans and Texas Pacific
CNW Chicago and North Western Railway (includes Chicago, St Paul, Minneapolis and Omaha; Litchfield and Madison Railway; Minneapolis and St Louis)
C & NW Chicago and North Western Railway
C & O Chesapeake and Ohio (Chessie System)
Coahuila-Zacatecas Railway Ferrocarril Coahuila-Zacatecas—Mexico
Cog Wheel Route Manitou and Pike's Peak Railway
Conrail Consolidated Rail Corporation (Ann Arbor, Central Railroad of New Jersey, Erie Lackawanna, Lehigh and Hudson River, Lehigh Valley, Penn Central, Reading)
COP City of Prineville Railway
COPR Copper Range Railroad

Corn Belt Route St Louis Southwestern Railway
Cotton Belt Cotton Belt Route (St Louis Southwestern Railway—SSW)
CP Canadian Pacific Railway (Dominion Atlantic Railway, Esquimalt and Nanaimo Railway, Grand River Railway, Lake Erie and Northern Railway, Quebec Central Railway, Vancouver and Lulu Island Branch)
CP Companhia des Caminhos de ferro Portuguese (Portuguese—Portuguese Railways)
CPA Coudersport and Port Allegany Railroad
CPF Cotton Plant—Fargo Railway
CP & LT Camino, Placerville and Lake Tahoe Railroad
CPR Canadian Pacific Railroad
CP Rail Canadian Pacific Railroad
CPT Chicago Produce Terminal
CR Commonwealth Railways (Australia and Tasmania); Copper Range Railroad (Michigan, Wisconsin, Illinois)
CRANDIC Route Cedar Rapids and Iowa City Railway
CRC Cameroon Railways Corporation (West Africa); Cumberland Railway Company (Nova Scotia)
CRI Chicago River and Indiana
CR & IC Cedar Rapids and Iowa City Railway
CR & IR Chicago River and Indiana Railroad
CRN Carolina and Northwestern (Southern Railway)
CRP Central Railway of Peru
CRR Clinchfield Railroad
CRRNJ Central Railroad of New Jersey
C & S Colorado and Southern Railway
CSAR Central South African Railways
CSD Cekoslovenske Statni Drahy (Czechoslovakian—Czechoslovak State Railways)
CSL Chicago Short Line Railway
CSP Camas Prairie Railroad
CSS Chicago South Shore and South Bend Railroad
CSS & SBR Chicago South Shore and South Bend Railroad
C St P M & O Chicago, St Paul, Minneapolis and Omaha (Chicago North Western)
CSX Chessie and Seaboard

(railroads consolidated)
C & T Cumbres & Toltec
CTA Chicago Transit Authority (elevated and subway railroads)
CTC Canadian Transport Commission; Cincinnati Transit Company
CTN Canton Railroad
CTS Cleveland Transit System
CUTC Cincinnati Union Terminal Company
CUVA Cuyahoga Valley Railroad
CV Central Vermont Railway
CVRy Cuyahoga Valley Railway
C & W Colorado and Wyoming Railway
C & WC Charleston and Western Carolina Railway (Seaboard Coast Line Railroad)
CWI Chicago and Western Indiana
CWP Chicago, West Pullman and Southern Railroad (also CWP & S)
CWR California Western Railroad
DA Dominion Atlantic Railway (Canadian Pacific)
DB Deutsche Bundesbahn (German—German Railways)
DC Delray Connecting Railroad
DCI Des Moines and Central Iowa
DCR Delray Connecting Railroad (Zug Island Road)
DCT Washington, DC Transit
D & E De Queen and Eastern
Delay Long and Wait nickname for the Delaware, Lackawanna and Western Railroad (derived from the initials DL & W)
D & H Delaware and Hudson
DHR Darjeeling Himalayan Railway
diner dining car
DKS Doniphan, Kensett and Searcy Railway
DL & W Delaware, Lackawanna and Western Railroad (Erie Lackawanna)
D & M Detroit and Mackinac
DM & IRR Duluth, Missabe and Iron Range Railway
DMM Dansville and Mount Morris
DMU Des Moines Union Railway
DMWR Des Moines Western Railway
DNE Duluth and Northeastern Railroad
DO Direct Orient (Orient Ex-

press)
DORR Delaware Otsego Railroad
DQ & ERR De Queen and Eastern Railroad
D & R Dardanelle and Russellville
D & RGW Denver and Rio Grande Western Railroad
DRI Davenport, Rock Island and North Western Railway
DRy Devco Railway
DS Durham and Southern Railway
D & S Durango & Silverton; Durham and Southern Railway
DSB Danske Statsbaner (Danish—Danish State Railways)
DSR Detroit Street Railways
DT Detroit Terminal Railroad
D of T Department of Transportation
DTC Dallas Transit Company
DTI Detroit, Toledo and Ironton Railroad (also DT & I)
D & TS Detroit and Toledo Shore Line Railroad
DVS Delta Valley and Southern Railway
DWP Duluth, Winnipeg and Pacific Railway
E Erie Lackawanna
EAR East African Railways
EARC East African Railways Corporation
EAR & H East African Railways and Harbours
EBR Emu Bay Railway (Tasmania)
EBRy Eastern Bengal Railway (East Pakistan)
EDLR Egyptian Delta Light Railways
EDW El Dorado and Wesson
EEC East Erie Commercial Railroad
EFA Empresa Ferrocarriles Argentinos (Spanish—Argentine Railways Enterprise)
EFE Empresa de los Ferrocarriles del Estado (Spanish—State Railways Enterprise)—Chile
EFEE Empresa de los Ferrocarriles del Estado Ecuatoriano (Spanish—Ecuadorian State Railways Enterprise)
EJ & ERy Elgin, Joliet and Eastern Railway
EJR East Jersey Railroad
El Elevated Railroad
EL Erie Lackawanna Railway (merger of Erie with Delaware, Lackawanna and Western)

ELS Escanaba and Lake Superior Railroad (also E & LSRR)
E & M Edgmoor and Manetta
EN Esquimalt and Nanaimo Railway (Canadian Pacific)
ENF Empresa Nacional de Ferrocarriles (Spanish—National Railways Enterprise)—Bolivia
ER Egyptian Railways
ERBR Eastern Region of British Railways
Erie Erie Railroad (Erie Lackawanna)
ESLJ East St Louis Junction Railroad
ETL Essex Terminal Railway
ET & WNC East Tennessee and Western North Carolina Railroad
Eurailpass European railroad pass (ticket system valid on almost all European railroads)
EW East Washington Railway
EYB Europa Year Book
F & C Frankfort and Cincinnati Railroad
FCAB Ferrocarril Antofagasta-Bolivia (Spanish—Antofagasta and Bolivia Railway)
FC del P Ferrocarril Central del Perú (Spanish—Central Railway of Peru)
FCDN Ferrocarril del Nacozari (Spanish—Nacozari Railroad) —Mexico
FCG Fernwood, Columbia and Gulf Railroad
FCIN Frankfort and Cincinnati
FCM Ferrocarriles Nacionales de México (Spanish—Mexican National Railways)—includes Nacional de México and Nacional de Tehuantepec
FCNM Ferrocarriles Nacionales de México (Spanish—National Railroads of Mexico)
FCP Ferrocarril del Pacífico (Spanish—Pacific Railroad) —links Arizona border with Mazatlan on west coast of Mexico
FCZ Ferrocarril Coahuila-Zacatecas (Spanish—Coahuila-Zacatecas Railway)—Mexico
FDDM Fort Dodge, Des Moines and Southern Railway
F de C Ferrocarriles de Cuba (Spanish—Cuban Railroads) —Unidad Habana (western Cuba) and Unidad Camaguey (eastern Cuba)
F de G a LP Ferrocarril de Guayaquil-La Paz (Spanish—Guayaquil-La Paz Railway)—

Peru
F del N Ferrocarriles del Norte (Spanish—Northern Railways) —Paraguay
F del P Ferrocarril del Pacífico (Spanish—Pacific Railroad) —Mexico
Feather River Route Western Pacific Railroad
FEC Florida East Coast Railway
FEGUA Ferrocarriles de Guatemala (Spanish—Railroads of Guatemala)
FEP Ferrocarril Electrico al Pacífico (Spanish—Pacific Electric Railway)—Costa Rican line linking Pacific port of Puntarenas with mountain capital of San José
FEPASA Federação Paulista Sedada Anomina (Portuguese—Paulist Federation Company)—Brazilian railroad
FER Franco-Ethiopian Railway
FES Ferrocarril de El Salvador (Spanish—El Salvador Railway)
FFAC Federação Ferrocarril Agricola Cotias (Portuguese—Agricultural Cooperative Railway Federation)—Brazil
FICA Ferrocarriles Internacionales de Centro America (Spanish—International Railways of Central America)
FIPC Ferrocarril Industrial del Potosí y Chihuahua (Spanish—Industrial Railroad of Potosi and Chihuahua)—Mexico
FJG Fonda, Johnstown and Gloversville Railroad
FLR Fayum Light Railways (Egypt)
FMS Fort Myers Southern Railroad
FN Ferrocarriles Nacionales (Spanish—National Railways) —Argentina, Chile, Colombia, Cuba, Ecuador, Honduras, Mexico, Panama, Venezuela, etc.)
FNC Ferrocarriles Nacionales de Cuba (National Railroad of Cuba nationalized by Castro government and consisting of Consolidated Railroads of Cuba, The Cuba Railroad, Cuba Northern Railways, Guantanamo and Western Railroad, Guantanamo Railroad, Hershey Cuban Railway, etc.)
FN de H Ferrocarriles Nacionales de Honduras (Spanish—National Railways of Honduras)

FNM Ferrocarriles Nacionales de México (Spanish—National Railways of Mexico)

FOM Ferrocarril Occidental de México (Spanish—Western Railway of Mexico)

FOR Fore River Railroad

FPCAL Ferrocarriles President Carlos Antonio López (Spanish—President Carlos Antonio Lopez Railways)—Paraguay

FPE Fairport, Painesville and Eastern Railroad

FP & ER Fairport, Painesville and Eastern Railway

FPN Ferrocarril del Pacífico de Nicaragua (Spanish—Pacific Railway of Nicaragua)

FR Feather River Railway

FRDN Ferdinand Railroad

Frisco St Louis-San Francisco Railway

FS Ferrovie dello Stato (Italian—State Railway)

FSBC Ferrocarril Sonora-Baja California (Sonora-Baja California Railroad)

FS del P Ferrocarril del Sur del Perú (Spanish—Southern Railway of Peru)

FSVB Fort Smith and Van Buren Railway (Kansas City Southern)

FtD DM & S Fort Dodge, Des Moines and Southern Railway

FUD Ferrocarriles Unidos Dominicanos (Spanish—United Dominican Railways)—Dominican Republic

FUS Ferrocarriles Unidos del Sureste (United Railways of the Southeast)

FUY Ferrocarriles Unidos de Yucatan (Spanish—United Railways of Yucatan)—Mexico

FWB Fort Worth Belt Railway

FW & D Fort Worth and Denver

GA Georgia Railroad

GANO Georgia Northern Railway

GASC Georgia, Ashburn, Sylvester and Camilla Railway

GB & W Green Bay and Western Lines (includes Kewaunee, Green Bay and Western Railroad)

GC Graham County Railroad

GCW Garden City Western Railway

George Washington's Railroad Chesapeake and Ohio

Georgia Georgia Railroad

G & F Georgia and Florida Railway

GFS Grand Falls Central Railway

GH & H Galveston, Houston and Henderson Railroad

GJ Greenwich and Johnsonville Railway

G & J Greenwich and Johnsonville Railway

GM Gainesville Midland Railroad

GM & O Gulf, Mobile and Ohio Railroad

GMRC Green Mountain Railroad Corporation

GN Great Northern Railway

GNA Graysonia, Nashville and Ashdown Railroad

GNW Genessee and Wyoming Railroad

GNWR Genessee and Wyoming Railroad

GO Transit Government of Ontario Transit

G & Q Guayaquil and Quito Railway

Grand Trunk Grand Trunk Railway System (Canadian National) and Grand Trunk Western Railroad

Green Bay Route Green Bay and Western Railroad

GRN Greenville and Northern Railway

GRNR Grand River Railway (Canadian Pacific)

GR & PA Ghana Railway and Port Authority

GRR Georgetown Railroad

GRSS Guyana Railways and Shipping Services

GSF Georgia Southern and Florida (Southern)

GSW Great Southwest Railroad

GTW Grand Trunk Western Railroad (Canadian National)

G&U Grafton and Upton Railroad

GWF Galveston Wharves

GWR Great Western Railway

GWWDR Great Winnipeg Water District Railway

HB Hampton and Branchville

HBLRR Harbor Belt Line Railroad

HBS Hoboken Shore Railroad

HBT Houston Belt and Terminal

HC Heber Creeper (Utah)

HE Hollis and Eastern Railroad

HER Hellenic Electric Railway (Athens-Piraeus subway system linking capital with its seaport)

HH Hamburger Hochbahn (German—Hamburg Elevated Railway)—includes subway system

HI Holton Inter-Urban Railway

HJR Hedjaz Jordan Railway

HLNE Hillsboro and Northeastern

H & M Hudson & Manhattan (Hudson Tubes)

HN Hutchinson and Northern Railway

HNE Harriman and Northeastern (Southern)

hovertrain railroad train supported by an air cushion instead of wheels

HPTD High Point, Thomasville and Denton Railroad

HRT Hartwell Railway

HS Hartford and Slocomb Railroad

HSW Helena Southwestern Railroad

HTW Hoosac Tunnel and Wilmington Railroad

i Illinois Central Gulf Railroad

IAT Iowa Terminal Railroad

IB&TC International Bridge and Terminal Company

IC Illinois Central Gulf (includes Mississippi Central)

ICC Interstate Commerce Commission

ICG Illinois Central Gulf

IGA Indian Government Administration (Railway Board of India)

IHB Indiana Harbor Belt Railroad

IN Illinois Northern Railway

IND Independent (New York subway system)

Indiana Harbor Belt "connects with all Chicago railroads"

Industrial Railway of Potosí and Chihuahua (Ferrocarril Industrial del Potosí y Chihuahua)—Mexico

INT Interstate Railroad

Interstate Interstate Railroad

IPE Indian-Pacific Express [Perth to Sydney—2461 miles (3960 kilometers) in 65 hours]

IR Israel Railways

IRCA International Railways of Central America (El Salvador, Guatemala, and Honduras)

IRN Ironton Railroad

IRRys Iraqi Republic Railways

IRS Iranian State Railway

IRT Interborough Rapid Transit (New York City subway system)

ITC Illinois Terminal Company

ITRC Iowa Transfer Railway Company

IU Indiana Union Railway

JE Jerseyville and Eastern

Jersey Central Lines Central Railroad of New Jersey and

Lehigh and New England

JHSC Johnstown and Stony Creek Railroad

JNR Japanese National Railways (world's fastest)

JRC Jamaica Railway Corporation

JTC Jacksonville Terminal Company

JWR Jane's World Railways

Katy Missouri-Kansas-Texas Railroad (MKT)

KBR Kankakee Belt Route

KCC Kansas City Connecting Railroad

KCMO Kansas City, Mexico and Orient Railway (Ferrocarril Chihuahua al Pacifico)

KCNW Kelley's Creek and Northwestern Railroad

KCPSFO Kansas City Public Service Freight Operation

KCR Kanawha Central Railway

K-C Ry Kowloon-Canton Railway (Hong Kong)

KCS Kansas City Southern Railway (includes Arkansas Western, Fort Smith and Van Buren, Louisiana and Arkansas railways)

KCT Kansas City Terminal Railway

KGB Kewaunee, Green Bay and Western Railroad (Green Bay and Western Lines)—also KGB&W

KIT Kentucky and Indiana Terminal Railroad

K&M Kansas and Missouri Railway and Terminal Company

KMRT Kansas and Missouri Railway and Terminal Company

KNR Klamath Northern Railway; Korean National Railways

KO&G Kansas, Oklahoma and Gulf Railway

KRI Kyle Railway Inc

K&T Kentucky and Tennessee

KTM Keretapi Tanah Malayu (Malayan Railway)

Kyle Kyle Railway

L&A Louisiana and Arkansas Railway (Kansas City Southern)—also LA

LAJ Los Angeles Junction Railway

LA&LR Livonia, Avon and Lakeville Railroad

LAMCO Liberian America Swedish Minerals Company (Liberian Railways)

Land of Evangeline Route Dominion Atlantic Railway

LART Los Angeles Rapid Transit

LAWV Lorain and West Virginia Railway (North and Western)

LBR Lowville and Beaver River Railroad

L&C Lancaster and Chester Railway

LEE Lake Erie and Eastern Railroad

LEF Lake Erie, Franklin and Clarion Railroad

LE&FW Lake Erie and Fort Wayne

LEN Lake Erie and Northern Railway (Canadian Pacific)

LHR Lehigh and Hudson River

LI Long Island Railroad (Metropolitan Transportation Authority)—M

Lickenpurr (Hawaiian nickname—Lahaina-Kaanapal and Pacific Rail Road)—nickname derived from abbreviations—LK & PRR

LIRR Long Island Rail Road

LK & PRR Lahaina-Kaanapal and Pacific Rail Road (Maui, Hawaii)

LM Litchfield and Madison Railway (Chicago North Western)—also L&M

LM Leningrad Metro (Russian—Leningrad subway)

LMC Liberia Mining Company

LMRBR London Midland Region of British Railways

L&N Louisville and Nashville Railroad

LNAC Louisville, New Albany and Corydon Railroad

LNE Lehigh and New England Railway (Central Railroad of New Jersey)

L&NR Ludington and Northern Railway

L&NRY Laona and Northern Railway

L&NW Louisiana and North West Rail Road

LOPG Live Oak, Perry and Gulf (Southern)

LPB Louisiana and Pine Bluff Railway

LPN Longview, Portland and Northern Railway

LRB London Transport Board

lrc (LRC) light, rapid, comfortable (high-speed railroad trains)

LRI Lawndale Transportation Company

LRS Laurinburg and Southern

L&S Laurinburg and Southern

LS&BC La Salle and Bureau County Railroad

LS&I Lake Superior and Ishpeming Railroad

LSO Louisiana Southern Railway (Southern)

LSR Lebanese State Railroads

LST&TRC Lake Superior Terminal and Transfer Railway Company

LT Lake Terminal Railroad (also LTRR)

LV Lehigh Valley Railroad

LW Louisville and Wadley Railway

L&W Louisville and Wadley Railway

LWV Lackawanna and Wyoming Valley Railway

M Metropolitan Transit Authority (New York City's rapid-transit system); Metropolitan Transportation Authority (Long Island Railroad); Monon Railroad

MA Magyan Allamvasutak (Hungarian—Hungarian State Railways)

MACR Minneapolis, Anoka and Guyana Range Railroad

Main Line of Mid-America Illinois Central Railroad

MARR Magma Arizona Railroad

M-A Ry Massawa-Agordad Railway (Ethiopia)

M&B Meridan and Bigbee Railroad

MBI Marianna and Bloustown Railroad

MBT Marianna and Blountstown

MBTA Massachusetts Bay Transportation Authority (Boston's subway system)

MC Michigan Central Railroad (Penn Central)

McR McCloud River Railroad

MCRR Main Central Rail Road; Monongahela Connecting Railroad

MCSA Moscow, Camden and San Augustine Railroad

MD Municipal Docks Railway of the Jacksonville Port Authority

M del P Méxicano del Pacifico (Mexican Pacific Railroad formerly Southern Pacific of Mexico)

MD&W Minnesota, Dakota and Western Railway

M&E Morristown and Erie Railroad

MEC Maine Central Railroad

MER Metropolitan Elevated Railroad

METC Medesto and Empire Traction Company

Metro (French short form—*Chemin de fer Metropolitain*)—Paris subway system

Metropolitano Rome's subway system

Mexican Pacific Railroad Ferrocarril Mexicano del Pacifico—Los Mochis to Camp

MF Middle Fork Railroad

MGA Monongahela Railway

MGU Mobile and Gulf Railroad

MHM Mount Hope Mineral Railroad

M&HMRR Marquette and Huron Mountain Railroad

MI Missouri-Illinois Railroad

MICO Midland Continental Railroad

MID Midway Railroad

MILW Chicago, Milwaukee, St Paul and Pacific Railroad (Milwaukee Road)

MINE Minneapolis Eastern Railway

MIR Minneapolis Industrial Railway

Mitropa *Mitteleuropaische Schlaf und Speiswagen* (German—Middle-European Sleeping Car and Dining Car)

MJ Manufacturers' Junction Railway

MKC McKeesport Connecting Railroad

MKT Missouri-Kansas-Texas Railroad (Katy)

MLD Midland Railway of Manitoba

MLS Manistique and Lake Superior Railroad

MMR Moscow Metro Railway (Moscow's radiating subway system famed for its beautiful stations)

MNCRR Metro-North Commuter Railroad

MNF Morehead North Fork Railroad

MNJ Middletown and New Jersey Railway

MNS Minneapolis, Northfield and Southern Railway

MOB Montreux-Oberland-Bernois (railway)

MON Monon Railroad

Monon Monon Railroad (formerly Chicago, Indianapolis and Louisville Railway)

Mon Rys Mongolian Railways

Montour Montour Railroad (Youngstown and Southern Railway)

MOP Missouri-Pacific Lines

Mo-Pac Missouri-Pacific Lines

MOV Moshassuck Valley Railroad

MOW Montana Western Railway

MP Missouri Pacific Railroad

MPA Maryland and Pennsylvania

MPB Montpelier and Barre Railroad

MPPR Manitou and Pike's Peak Railway

MR McCloud River Railroad (also McRRR)

M of R Ministry of Railways (mainland China)

MRA Malayan Railway Administration

MRL Malawi Railways Limited

MRR Mattagami Railroad (Ontario); Mossi Railroad (Upper Volta)

MRS Manufacturers Railway

MRy Malayan Railway

MSC Mississippi Central (Illinois Central)

MSE Mississippi Export Railroad

M St L Minneapolis and St Louis (Chicago North Western)

M&StL Minneapolis-St Louis (Chicago North Western)

MSTL Minneapolis-St Louis (Chicago North Western)

MSTR Massena Terminal Railroad

MSV Mississippi and Skuna Valley Railroad

MT Ministry of Transport (USSR's administration of twenty-six railway lines including the deluxe Leningrad-Moscow and the transcontinental Trans-Siberian from Moscow with Vladivostok)

MTC Milwaukee Transport Company; Montreal Transportation Commission (subway and surface railways); Mystic Terminal Company (Boston and Maine)

MTFR Minnesota Transfer Railroad

MTH Mount Hood Railway

MTR Montour Railroad

MTW Marinette, Tomahawk and Western Railroad

MTWCR Mt Washington Cog Railway

MWR Muncie and Western Railroad

NAJ Napierville Junction Railway

NAP Narragansett Pier Railroad

NAR Northern Alberta Railways; Northern Australia Railway

National Railroads of Cuba Ferrocarriles Nacionales de Cuba (includes nationalized lines of the Cuba Railroad, Cuba Northern Railways, Guantanamo Railroad, Guantanamo Western, Hershey Cuban Railway, etc.)

National Railways of Mexico Ferrocarriles de México

NB Northampton and Bath Railroad

NC & StL Nashville, Chattanooga and St Louis Railway (L&N)

N de M Nacional de México (National of Mexico)

N de T Nacional de Tehuantepec (Tehuantepec National)

New Haven New York, New Haven and Hartford Railroad

NEZP Nezperce Railroad

NFD Norfolk, Franklin and Danville Railway

NGR Nepalese Government Railway

NH New York, New Haven and Hartford Railroad (Penn Central)

NHIR New Hope and Ivyland Railroad

Nickel Plate New York, Chicago and St Louis Railroad (merged with Norfolk and Western)

NJ Niagara Junction Railway

NJI&I New Jersey, Indiana and Illinois Railroad

NKP Nickel Plate (New York, Chicago and St Louis Railroad)—merged with Norfolk and Western

NLC New Orleans and Lower Coast Railroad

NLG North Louisiana and Gulf Railroad

NM Nagoya Municipality (subway system)

NN Nevada Northern Railway

NNC Northern Navigation Company

NO de M Noroeste de México (Northwestern of Mexico)

NODM Ferrocarril Noroeste de México (Northwest Railway of Mexico—Ferrocarril Chihuahua al Pacifico)

NONE New Orleans and Northeastern Railroad (Southern)

NOPB New Orleans Public Belt Railroad

NOPS New Orleans Public Service

Norf S Norfolk Southern, Norfolk & Western, Southern Railway (merger)

NP Northern Pacific Railway

N&PB Norfolk and Portsmouth Belt Line Railroad

NR Newfoundland Railway (Canadian National); Northern Railway of Costa Rica (from mountain capital of San José to Caribbean seaport of Limón)

NRC Nigerian Railway Corporation

NRPC National Railroad Passenger Corporation (Amtrak)

NRRC National Railroad Company (of Haiti)

NRZ National Railways of Zimbabwe

NS Norfolk Southern Railway

NS Nederlandsche Spoorwagen (Dutch—Netherlands Railway Carriage)—Netherlands Railways

NSB Norges Statsbaner (Norwegian—Norwegian State Railways)

NSL Norwood and St Lawrence Railroad

NSS Newburgh and South Shore Railway

NSWGR New South Wales Government Railways

NUR Natchez, Urania and Ruston Railway

NW Norfolk and Western

N&W Norfolk and Western Railway

NWP Northwestern Pacific Railroad

NWRy North Western Railway (West Pakistan)

NWS Norfolk & Western Southern (merger)

NYC New York Central Railroad (Penn Central)

NYCTA New York City Transit Authority (subway systems include BMT, IRT, INDependent)

NYD New York Dock Railway

NYLB New York and Long Branch Railroad

NYNH&H New York, New Haven and Hartford Railroad

NY O & W New York, Ontario and Western

NYS Nepal Yatayat Samsthan (Nepali—Transport Corporation of Nepal)

NYSW New York, Susquehanna and Western Railroad (NYS&W)

NZGR New Zealand Government Railways

NZR New Zealand Railways

OCE Oregon, California and Eastern Railway

OE Oregon Electric Railway (Spokane, Portland, and Seattle Railway)

OGR Official Guide of the Railways

OKT Oakland Terminal Railway

OL&BR Omaha, Lincoln and Beatrice Railway

OMTB Osaka Metropolitan Transportation Bureau (subway system)

ON Ontario Northland

ONCF Office National des Chemins de Fer (French—National Railways Office)—Morocco

ONRY Ogdensburg and Norwood Railway

ONT Ontario Northland Railway

ONW Oregon and Northwestern

O&NW Oregon and Northwestern

ÖOB Österreichischen Bundesbahnen (German—Austrian State Railways)

OPE Oregon, Pacific and Eastern

ORER Official Railway Equipment Register

OT Oregon Trunk Railway (Spokane, Portland, and Seattle Railway)

OUR & D Ogden Union Railway and Depot

Overland Route Union Pacific Railroad

PA Pittsburgh Authority (rapid transit)

PAA Pennsylvania and Atlantic Railroad

PACC Pacific Coast Railroad

Pacific Railroad Ferrocarril del Pacifico (linking American border at Nogales with Mazatlan on Pacific coast of Mexico)

Pacific Railway of Costa Rica from Pacific port of Puntarenas to San José

Pacific Railways of Nicaragua Ferrocarril del Pacifico de Nicaragua—from Corinto on the Pacific to Granada on Lake Nicaragua

Pac Rail Missouri Pacific, Union Pacific, Western Pacific (merged)

PA&M Pittsburgh, Allegheny and McKees Rocks Railroad

Panama Railroad division of

the Panama Canal linking Cristóbal and Colón on the Atlantic with Balboa and Panama City on the Pacific and running parallel to the Panama Canal

P & AR Pacific and Arctic Railway

PATCO (transportation system linking Camden, New Jersey and Philadelphia, Pennsylvania)

PATH Port Authority Trans-Hudson Corporation (operates Hudson Tubes between New Jersey and New York)

PBNE Philadelphia, Bethlehem and New England Railroad

PBR Patapsco and Back Rivers

PC Penn Central (Pennsylvania New York Central Transportation Company; Pennsylvania Railroad; New York Central Railroad; New York, New Haven, and Hartford Railroad; Baltimore and Eastern Railroad; Canada Southern Railway; Cleveland, Cincinnati, Chicago and St Louis Railway; Michigan Central Railroad; Peoria and Eastern Railway; Waynesburg and Washington Railroad)

PCL Peruvian Corporation Limited

PCN Point Comfort and Northern

PCR Paraguayan Central Railway

PCY Pittsburgh, Chartiers and Youghiogheny Railway

PE Pacific Electric (interurban railway system serving entire Los Angeles area before replacement by smog-producing buses); Pacific Electric Railway of Costa Rica (links Pacific seaport of Puntarenas with mountain capital of San José)—also called *FEP*

P&E Peoria and Eastern Railway (Penn Central)

Pennsy (nickname—Pennsylvania Railroad)—now part of the Penn Central

Peoria Peoria and Pekin Union Railway

P&F Pioneer and Fayette Railroad

PGE Pacific Great Eastern Railway

PH&D Port Huron and Detroit Railroad

P&I Paducah and Illinois Railroad

PIC Pickens Railroad

Pick Pickens Railroad

Pickens Pickens Railroad

PKP *Polskie Koleje Panstwowe* (Polish—Polish State Railways)

P&LE Pittsburgh and Lake Erie Railroad

PLM Paris-Lyon-Mediterranée

P&N Piedmont and Northern Railway

PNKA *Perusahaan Negara Kereta Api* (Indonesian—Indonesian State Railways)

PNR Philippine National Railways

PNW Prescott and Northwestern Railroad

Port St Joe Route Apalachicola Northern Railroad

'Possum Trot Line Reader Railroad

POV Pend Oreille Valley

P&OV Pittsburgh and Ohio Valley

POVA Pend Oreille Valley (railway)

P&PU Peoria and Pekin Union

PR Panama Railroad

P-R Pennsylvania-Reading Seashore Lines

PRC Philippine Railway Company

PRCR Pacific Railway Costa Rica

PRR Pennsylvania Railroad (Penn Central)

PRS Pennsylvania-Reading Seashore Lines

PRTD Portland Railroad and Terminal Division of the Portland Traction Company

PRV Pearl River Valley Railroad

PS Pittsburg and Shawmut Railroad

P&SR Petaluma and Santa Rosa

PTC Peoria Terminal Company; Philadelphia Transportation Company (also called PATCO includes elevated and subway lines of Philadelphia area)

PTM Portland Terminal Company

PTR Parr Terminal Railroad

PTS Port Townsend Railroad

Pullman deluxe railroad cars providing lounging, observation, and sleeping facilities aboard first-class express trains

PVS Pecos Valley Southern

P&WV Pittsburgh and West Virginia Railway (Norfolk and Western)

P y RV *Potosí y Rio Verde* (Spanish—Potosi and Green River Railroad of Chihuahua)

QAP Quanah, Acme and Pacific

QC Quebec Central Railway (Canadian Pacific)

QNS&LRC Quebec North Shore and Labrador Railway Company

QR Queensland Railways

Quanah Route Quanah, Acme and Pacific Railway

QUI Quincy Railroad

RB Rail Box (American box car pool)

RC Railway Corporation (Nigeria)

RCFA-N *Regie du Chemin de Fer Abidjan-Niger* (French—Abidjan-Niger Railway Administration)—Ivory Coast

RD Railway Directorate (Albania)

RDG Reading Company (formerly Philadelphia and Reading Railroad)

REA Railway Express Agency; Reader Railroad

Reading Lines Reading Railway System (formerly Philadelphia and Reading Railroad)

Rebel Route Gulf, Mobile and Ohio Railroad

RENFE *Red Nacional de los Ferrocarriles Españoles* (Spanish—Spanish National Railway System)

RFFSA *Rede Ferroviária Federal SA* (Portuguese—Federal Railway System Corporation)—Brazil

RFP Richmond, Fredericksburg and Potomac Railroad (RF&P)

RF&PRR Richmond, Fredericksburg and Potomac Railroad

RI Chicago, Rock Island and Pacific Railroad; Rail India

Rio Grande Denver and Rio Grande Western

RKG Rockingham Railroad

RM *Rotterdam Metro* (Dutch—Rotterdam Subway)

RNCF *Reseau National des Chemins de Fer* (French—National Railway System)—Madagascar

Rock Island Chicago, Rock Island and Pacific Railroad

RR (abbreviation—Railroad or Rail Road); (reporting mark—Raritan River Rail Road); Rhodesian Railways

RRRR Raritan River Railroad

RRys Rhodesian Railways

RS Roberval and Seguenay Railway

RSP Roscoe, Snyder and Pacific

R-S Pacific Route Roscoe, Snyder and Pacific Railway

RSS Rockdale, Sandow and Southern Railroad

RT River Terminal Railway

RTM Railway Transfer Company of Minneapolis

RV Rahway Valley Railway

Ry Railway

S&A Savannah and Atlanta Railway

SAL Seaboard Airline Railroad (Seaboard Coast Line Railroad is official name adopted to avoid confusion with an airline)

SAN Sandersville Railroad

Santa Fe Atchison, Topeka and Santa Fe Railway

SAR South African Railways; South Australian Railways

SAR&H South African Railways and Harbours

SATS San Antonio Transit System

SAVE Swiss-Alberg-Vienna Express

SB South Buffalo Railway

SBA *Subterraneos de Buenos Aires* (Spanish—Buenos Aires Subways)

SBC Ferrocarril Sonora Baja California (Sonora—Baja California Railway)

SBK South Brooklyn Railway

SC Sumter and Choctaw Railway

SCE Shanghai-Canton Express

SCL Seaboard Coast Line Railroad (Atlantic Coast Line Railroad, Charleston and Western Carolina Railway, Seaboard Air Line Railroad—former name of the Seaboard Coast Line Railroad)

SC&MR Strouds Creek and Muddlety Railroad

SCT Sioux City Terminal Railway

SDAE San Diego and Arizona Eastern Railway

SD & AE San Diego and Arizona Eastearn Railway

SD & IV San Diego & Imperial Valley

SDTS San Diego Transit System

SE Ferrocarril del Sureste (Southeast Railroad)

Seashore Lines Pennsylvania-

Reading Seashore Lines

SE & CR Southeastern and Chatham Railway (nicknamed Seldom Ever Caught Running)

SEMTA Southeastern Michigan Transportation Authority

SERA Sierra Railroad

SFBRR San Francisco Belt Railroad

SFMR San Francisco Municipal Railway (operates the cable cars)

SFSP Santa Fe/Southern Pacific (merger)

SF/SP Santa Fe/Southern Pacific (railroad merger)

SG South Georgia Railway (Southern Railway)

SGR Saudi Government Railroad (Saudi Arabia); Surinam Government Railway (Netherlands Guiana)

SH Steelton and Highspire Railroad

Shawmut The Pittsburg and Shawmut Railroad

SHK Sidirodromi Hellinikou Kratous (Greek—Hellenic State Railways)—Greece

SI Spokane International Railroad

SIR Staten Island Rapid Transit Railway

SIRRI Southern Industrial Railroad Incorporated

SJ Statens Jarnvargar (Swedish—State Railways)

SJB St Joseph Belt Railway

SJL St Johnsbury and Lamoille County Railroad

SJ & LC St Johnsbury and Lamoille County Railroad

SJTR St Joseph Terminal Railroad

SKSL Skaneateles Short Line Railroad

SLC San Luis Central Railroad

SLGW Salt Lake, Garfield and Western Railway

SLR Sierra Leone Railway

SLSF St Louis-San Francisco Railway

SM St Marys Railroad

SMA San Manuel Arizona Railroad

SMR South Manchurian Railway

SMV Santa Maria Valley Railroad

SN Sacramento Northern Railway (also SNRy)

SNCB Société Nationale des Chemins de Fer Belges (French—Belgian National Railways)

SNCF Société Nationale des

Chemins de Fer Français (French—French National Railways)

SNCFA Société Nationale des Chemins de Fer Algeriens (French—Algerian National Railways)

SNY Southern New York Railway

SOE Simplon-Orient Express

SOI Southern Indiana Railway

Sonora—Baja California Railway Ferrocarril Sonora—Baja California—Mexicali to Benjamin Hill

SOO Soo Line Railroad

$oo Line Soo Line Railroad

SOT South Omaha Terminal Railway

Southern Southern Railway System (Alabama Great Southern Railroad; Carolina and Northwestern Railway; Cincinnati, New Orleans and Texas Pacific Railway; Georgia Southern and Florida Railway; Harriman and Northeastern Railroad; Live Oak, Perry and Gulf Railroad; Louisiana Southern Railway; New Orleans and Northeastern Railroad; South Georgia Railway)

Southern Pacific SP

South Shore Line Chicago South Shore and South Bend Railroad

SP Southern Pacific (includes Southern Pacific Lines, Sunset Railway, Texas and Louisiana Lines, Texas and New Orleans, etc.)—in fact many school children once said the United States was bounded on the north by Canada and the Great Lakes, on the east by the Atlantic Ocean, and on the south and southwest by the Southern Pacific

SPGT Springfield Terminal Railway

SPS Spokane, Portland and Seattle Railway (includes Oregon Electric and Oregon Trunk railways)

SR Southern Railway

SRBR Southern Region of British Railways

SRC Salvador Railway Company (El Salvador)

SRN Sabine River and Northern

SRRC Sierra Railroad Company; Strasburg Rail Road Company

SRRCO Sandersville Railroad Company

SRT State Railways of Thailand (Siam)

SSDK Savannah State Docks Railroad

SSLVRR Southern San Luis Valley Railroad

SSRy Sand Springs Railway

SSW St Louis Southwestern Railway (Cotton Belt Route)

STE Stockton Terminal and Eastern Railroad

STRT Stewartstown Railroad

STS Seattle Transit System

SU Stockholm Underground (subway system)

Sub Suburban; Subway

Sud Rys Sudan Railways

SUR Soviet Union Railways (managed by Ministry of Communications and comprising some twenty-six lines including the Trans-Mongolian and the Trans-Siberian as well as the plush Leningrad-Moscow express)

Susquehanna New York, Susquehanna and Western Railroad

Syr Rys Syrian Railways

T symbol for Boston's subways; shortened from MBTA (Massachusetts Bay Transportation Authority)

TAAA Travelers Aid Association of America

TA & G Tennessee, Alabama and Georgia Railway

TAG Route Tennessee, Alabama and Georgia Railway

Tan-Zam Tanzania-Zambia Railroad

TAR Trans-Australian Railways

TAS Tampa Southern Railroad

TASD Terminal Railway Alabama State Docks

TA&W Toledo, Angola and Western Railway

TB Twin Branch Railroad

TBTMG Transportation Bureau of the Tokyo Metropolitan Government (subway)

TC Tennessee Central Railway

TCDD Turkiye Cumhuriyeti Deviet Demiryollari Isletmesi (Turkish—Turkish State Railways)

TCG Tucson, Cornelia and Gila Bend Railroad

TCT Texas City Terminal Railway

TEBRCL The Emu Bay Railway Company Limited

TEE Trans-Europe Express

TENN Tennessee Railroad

TEXC Texas Central Railroad

THB Toronto, Hamilton and

Buffalo Railway

The Q CB&Q (Chicago, Burlington and Quincy)

TM Texas Mexican Railway; Transport Ministry (USSR's administration of twenty-six railway lines)—TM sometimes used on engines

TMR Trans-Mongolian Railway

TN Texas and Northern Railway

T-NM Texas-New Mexico Railway

T & NO Texas and New Orleans (Southern Pacific)—also TNO

TOC Pennsylvania New York Central Transportation Company (Penn Central)

TOE Texas, Oklahoma and Eastern Railroad

TOV Tooele Valley Railway

T&P Texas and Pacific Railway (also TP)

TPMP Texas-Pacific-Missouri Pacific Terminal Railroad of New Orleans

TPT Trenton-Princeton Traction Company

TP & W Toledo, Peoria and Western Railroad

TR Tasmanian Railways

TRA Taiwan Railway Administration

Trans-Sib Trans-Siberian Railway

TRC Tela Railway Company (Honduras); Trona Railway Company (California)

TRRA Terminal Railroad Association of St Louis

TS Tidewater Southern Railway

TS-E Texas South-Eastern

TSR Trans-Siberian Railway

TSU Tulsa-Sapulpa Union Railway

TT Toledo Terminal Railroad

T&T Tijuana and Tecate Railway (freight cars marked TITE)

TTC Toronto Transit Commission (subway and surface railway systems)

Turk-Sib Turkestan-Siberian (railway)

TVG Tavares and Gulf Railroad

TVRy Tooele Valley Railway

Tweetsie (nickname—East Tennessee and Western North Carolina Railroad)—believed to be derived from high-pitched whistles of its engines

T-Z RA Tanzania-Zambia Railway Authority

U Underground (London's subway system)

UBR Ulan Bator Railway

UCR Utah Coal Route

U de Y Unidos de Yucatan (Spanish—United Railways of Yucatan, Mexico)

UFC United Fruit Company (railroads in Costa Rica and Panama)

UMP Upper Merion and Plymouth Railroad

UNF Union Freight Railroad

UNI Unity Railways

UO Union Railroad—Oregon

UP Union Pacific Railroad (includes Oregon Short Line and Oregon-Washington Railroad and Navigation Company)

UR Uganda Railway

URR Union Railroad—Pittsburgh

USSR (Ministry of Railways administers operation of twenty-six railway boards throughout the USSR)

UT Union Terminal Railway

UTA Ulster Transport Authority (railways of six counties in Northern Ireland)

UTAH Utah Railway

Utah Coal Route Utah Railway

UTR Union Transportation Company

V Valtionrautatiet (Finnish—State Railways)

VBR Virginia Blue Ridge Railway

VC Virginia Central Railway

VCS Virginia and Carolina Southern Railroad

VCY Ventura County Railway

VE Visalia Electric Railroad

VGN Virginian Railway (Norfolk and Western)

VIA VIA Rail Canada

Via Rail Canadian National + Canadian Pacific (passenger-carrying consolidation)

Virginian Virginian Railway (Norfolk and Western)

V & LI Vancouver and Lulu Island (branch of Canadian Pacific)

V-MNR Viet-Minh National Railways (North Vietnam)

V-NR Viet-Nam Railways (South Vietnam)

VR Victorian Railways (Australia)

V Ry Verapaz Railway (Guatemala)

VSL Valley and Siletz Railroad

VSO Valdosta Southern Railroad

VSOE Venice-Simplon Orient Express

VTR Vermont Railway

W of A Western Railway of Alabama

WAB Wabash Railroad (Norfolk and Western)

Wabash Wabash Railroad (Norfolk and Western)

WAG Wellsville, Addison and Galeton Railroad

WAGR Western Australian Government Railways

WATC Washington Terminal Company

WAW Waynesburg and Washington Railroad (Penn Central)

WBCRR Wilkes-Barre Connecting Railroad

WBT&SRC Waco, Beaumont, Trinity and Sabine Railway Company

Western Railway of Mexico Ferrocarril Occidental de México—Culiacan to Limoncito

West Point Route Atlanta and West Point Rail Road

Westrain Western Australian Trains

White Pass British Columbia Yukon Railway, British Yukon Railway, Pacific and Arctic Railway

White Pass and Yukon Route British Columbia Yukon Railway, British Yukon Navigation, British Yukon Railway, Pacific and Arctic Railway and Navigation Company

WIM Washington, Idaho and Montana Railway

WL Wagon Lits (French—sleeping cars)

WLO Waterloo Railroad

WM Western Maryland Railway

WMR Wasatch Mountain Railway

WMTA Washington Metropolitan Transit Authority (subway system)

WMWN Weatherford, Mineral Wells and Northwestern Railway

WNF Winfield Railroad

W&NO Wharton and Northern Railroad

WOD Washington and Old Dominion Railroad

W&OV Warren and Ouachita Valley Railway

WP Western Pacific Railroad

WPER West Pittston-Exeter Railroad

WP & Y White Pass and Yukon Railway

WRA Western Railroad Association

WRBR Western Region of British Railways
WRNT Warrenton Railroad
WRWK Warwick Railway
WS Ware Shoals Railroad
WSR Warren and Saline River
WSS Winston-Salem Southbound Railway
WSYP White Sulphur Springs and Yellowstone Park Railway
WTR Wrightsville and Tennille Railroad
WVN West Virginia Northern Railroad
WW Winchester and Western Railroad
WWV Walla Walla Valley Railway
WYS Wyandotte Southern Railroad
WYT Wyandotte Terminal Railroad
X express; transport; transportation (as in many private bulk carriers' names such as GATX—General American Transportation)
Xing crossing (highway or railroad)—also XING
YAN Yancey Railroad
YN Youngstown & Northern (railroad)
Y & N Youngstown and Northern Railroad
YR Yucatan Railways (*Ferro-carriles Unidos del Sureste*—United Railways of the Southeast)—along the Gulf of Mexico from Coatzacoalcos to Merida
YS Youngtown and Southern Railway (Montour)
Y & S Yakutat and Southern Railway
YVT Yakima Valley Transportation Company
YW Yreka Western Railroad
ZJZ *Zajednica Jugoslovenskih Zalesnicca* (Yugoslavian—Community of Yugoslav Railways)
ZR Zambia Railways
Zug Island Road Delray Connecting Railroad (DC)

Roman Numerals

I 1	LV 55	DCCC 800
II 2	LIX 59	CM 900
III 3	LX 60	M 1000
IV 4	LXV 65	MD 1500
V 5	LXIX 69	MDC 1600
VI 6	LXX 70	MDCC 1700
VII 7	LXXV 75	MDCCC 1800
VIII 8	LXXIX 79	MCM or MDCCCC 1900
IX 9	LXXX 80	MCMX 1910
X 10	LXXXV 85	MCMXX 1920
XV 15	LXXXIX 89	MCMXXX 1930
XIX 19	XC 90	MCMXL 1940
XX 20	XCV 95	MCML 1950
XXV 25	XCIX 99	MCMLX 1960
XXIX 29	C 100	MCMLXX 1970
XXX 30	CL 150	MCMLXXX 1980
XXXV 35	CC 200	MCMXC 1990
XXXIX 39	CCC 300	MM 2000
XL 40	CD 400	MMM 3000
XLV 45	D 500	MMMM or M$\overline{V}$ 4000
XLIX 49	DC 600	$\overline{V}$ 5000
L 50	DCC 700	$\overline{M}$ 1,000,000

Rules of the Road—at Sea

Red to red and green to green all is safe to pass abeam.

or

Green to green and red to red perfect safety—go ahead.

If on your starboard red appear it is your duty to keep clear: to act as judgment says is proper— to port, or starboard, back, or stop her.

But when upon your port is seen a steamer's starboard light of green— there's not so much for you to do

for green to port keeps clear of
you.

Both in safety and in doubt always keep   a good lookout;

in danger with no room to
turn ease her, stop her, go
astern.

When Two Ships Meet Head On

When both side lights you see
ahead
port your helm and show your

red.

(Steer to starboard so your red
light will pass the red light of

the approaching vessel, and
thus you'll pass on the left as
people do ashore.)

Russian Alphabet (transliterated)

Russian Capital Letters Sounds	English Capital Letters	Russian Small Letters	English Small Letters	Russian Alphabet Letter Names	Nearest English Equivalent
А	A	а	a	*ah*	*a* as in *a*rch
Б	B	б	b	*beh*	*b* as in *b*it
В	V	в	v	*veh*	*v* as in *v*est
Г	G	г	g	*geh*	*g* as in *g*et
Д	D	д	d	*deh*	*d* as in *d*ay
Е	Ye	е	ye	*yeh*	*y* as in *y*es
Ж	Zh	ж	zh	*zheh*	*zh* sound as in measure
З	Z	з	z	*zeh*	*z* as in *z*ero
И	I	и	i	*ee*	*i* as in p*ee*l
Й	Y	й	y	*ee s krátkoi*	(short *i* after vowels)
К	K	к	k	*kah*	*k* as in *k*ite
Л	L	л	l	*el*	*l* as in woo*l*
М	M	м	m	*em*	*m* as in *m*an
Н	N	н	n	*en*	*n* as in *n*ow
О	O	о	o	*oh*	*o* as in h*o*ax
П	P	п	p	*peh*	*p* as in *p*encil
Р	R	р	r	*err*	*r* as in *r*ye
С	S	с	s	*ess*	*s* as in *s*ay
Т	T	т	t	*teh*	*t* as in *t*ent
У	Oo	у	oo	*ooh*	*oo* as in l*oo*se
Ф	F	ф	f	*eff*	*f* as in *f*ancy
Х	Kh	х	kh	*khan*	*kh* as in lo*ch*
Ц	Ts	ц	ts	*tseh*	*ts* as in ha*ts*
Ч	Ch	ч	ch	*cheh*	*ch* as in *ch*air
Ш	Sh	ш	sh	*shah*	*sh* as in *sh*ave
Щ	Shch	щ	shch	*shchah*	*shch* as in Iri*sh* *ch*uck
Ъ		ъ		*tvyódy znak*	(silent-hard sound)
Ы	Y	ы	y	*yery*	*y* as in h*i*t
Ь		ь		*myakhki znak*	(silent)
Э	Eh	э	eh	*eh oborótnoye*	*eh* sound as in d*e*bt
Ю	Yu	ю	yu	*yoo*	*yu* as in *you*
Я	Ya	я	ya	*yah*	*ya* as in *ya*m

Ship's Bell Time Signals

1 bell —12:30	or 4:30	or 8:30a.m. or p.m.	5 bells— 2:30	6:30	10:30
2 bells— 1:00	5:00	9:00	6 bells— 3:00	7:00	11:00
3 bells— 1:30	5:30	9:30	7 bells— 3:30	7:30	11:30
4 bells— 2:00	6:00	10:00	8 bells— 4:00	8:00	12:00

On many vessels the ship's whistle is blown at noon. On some ships a lightly struck 1 bell announces 15 minutes before the change of watch, usually at 4, 8, and 12 o'clock.

The ship's day starts at noon. The *afternoon watch* is from noon to 4 p.m. The 4 to 8 work period is called the *dogwatch*. From 8 p.m. to midnight is the *first watch*. From midnight to 4 a.m. is the *middle watch*. From 8 a.m. to noon is the *forenoon watch*.

Signs and Symbols Frequently Used

+ add; addition sign; north; plus
& and (ampersand)
&c et cetera (and so forth)
* asterisk
@ at
because
¢ centavo; centime; cent(s)
© copyright
° degree(s)
° ′ ″ degrees, minutes, seconds (used to measure latitude north and south of the Equator and longitude east or west of the Greenwich meridian)
÷ divide; divided by; division sign
$ dollar sign—used universally for monetary units as diverse as Nicaraguan cordobas; Brazilian cruzeiros; Australian, Bahamian, Barbadian, British Honduran, Canadian, Ethiopian, Guyanian, Hong Kongese, Levantine, Liberian, Malaysian, New Zealand, Taiwan, trade, Trinidadian-Tobagonian, U.S., Viet Namese, West Indian, yuan dollars; Portuguese escudos; Honduran lempiras; Brazilian milreis; Chilean, Colombian, Cuban, Dominican, Mexican, Philippine, Uruguayan pesos; Peruvian soles (often with a lower-case dollar sign, $);

Chinese yuans
$A Australian dollar(s)
$b Bolivian peso(s)
$B Bahamian, Barbadian, British dollar(s)
$BH British Honduran dollar(s)
$C Brazilian cruzeiro(s); Canadian dollar(s)
$Col Colombian peso(s)
$E Ethiopian dollar(s)
$Eth Ethiopian dollar(s)
$G Guyanian dollar(s)
$HK Hong Kong dollar(s)
$K $1000 (e.g. $13K $13,000)
$L Levant(ine) dollar(s) —Maria Theresa thaler(s); Liberian dollar(s)
$M Malay(sian) dollar(s)
$Mal Malay(sian) dollar(s)
$Mex Mexican peso(s)
$NT New Taiwan dollar(s)
$NZ New Zealand dollar(s)
$RD Republica Dominicana peso(s)—Dominican Republic monetary unit(s)
$S Singapore dollar(s)
$T Taiwan dollar(s); trade dollar(s); Trinidad(ian) and Tobago(nian) dollar(s)
$TT Trinidad(ian) and Tobago(ian) dollar(s)
$Ur Uruguayan peso(s)
$US United States dollar(s) [also shown as US$, as are other monetary units where

national designations often precede dollar sign: C$—Canadian dollar(s), HK$—Hong Kong dollar(s)]
$VN Viet Namese dollar(s)
$WI West Indian dollar(s); West Indies dollar(s)
$Y yuan dollar(s)
= equality; equals; equal to
♀ female
G Paraguayan guarani(s)
K certified kosher
LC Cyrian pound(s)
LR Rhodesian pound(s)
♂ male
— minus; south; subtract; subtraction sign
× multiplication sign; multiplied by; multiply
≥ equal to or greater than
≤ equal to or less than
> greater than
< less than
>> much greater than
<< much less than
fracture(s) (medical); number(s) or pound(s) (commercial); sharp(s) (musical); space(s) (typographical); tic-tac-toe (game symbol); zinc (alchemical)
₱ Philippine peso(s)
% percent
+ plus; north
± plus or minus

£ pound *(libra)* sign—used universally for monetary units such as the Australian, British, Egyptian, Gambian, Ghanian, Irish, Israeli, Jamaican, Lebanese, Libyan, Malawi, New Zealand, Nigerian, South African, Sudanese, Syrian, Turkish, Western Samoan, Zambian pound

£A pound Australian

£E pound Egyptian (United Arab Republic)

£G pound Gambian; pound Ghanian

£I pound Irish; pound Israeli (also shown as I£)

£J pound Jamaican

£L pound Lebanese; pound Libyan

£M pound Malawi

£N pound Nigerian

£NZ pound New Zealand (also shown as NZ£)

£S pound sterling; pound Sudanese; pound Syrian

£SAf pound South African (also shown as SAf£)

£/s/d pounds, shillings, and pence

£T pound Turkish

£WS pound Western Samoan

£Z pound Zambian

R registered

℞ prescription; receipt; recipe; response; reverse

/ shilling mark; slash; solidus; virgule

T'ai-chi-T'u (Chinese—yin and yang)—ancient symbol for the diagram of the supreme ultimate

∴ therefore

U Union of Orthodox Jewish Congregations of America (symbol for kosher product approved for detergent or dietary use)

XMAS (symbol—commercialized Christmas)

Y Japanese yen

y & y yin and yang—ancient Chinese symbol for the diagram of the supreme ultimate (see *T'ai-chi-T'u*)

States, Nations, and Territories

States, nations, and territories are listed alphabetically in Eponyms, Nicknames, and Geographical Names. Many also appear among Superlatives.

Steamship Lines

A Ahearn Shipping Ltd; Alaska Steamship Company; Alcoa Steamship Company; American Export Isbrandtsen Lines; American Mail Line; American Oil Company; American Steamships; Tidewater Oil (capital A between red wings); etc.

ABC Line Antwerp Bulk Carriers Line

ABRT A/B Rederi Transatlantic (Pacific Australia Direct Line)

AC African Coasters

ACL American Canadian Line; American Cruise Line; Atlantic Container Line

ACS American Coal Shipping

ACSC Australian Coastal Shipping Commission

AD Armement Dieppe

AE African Enterprises

AECL Anglo-European Container Line

AEL Afro Eurasian Line; American Express Line

AFCL Africa Container Lines

AFS American Foreign Steamship

AH Alfred Holt (Blue Funnel Line)

AHB Great Eastern Line

AHL Associated Humber Lines

AJCL Australia-Japan Container Line

AL Admiral Line

Alcoa Alcoa Steamship Company

ALL Anchor Line Limited

All America Cables All America Cables and Radio

AML American Mail Line

AMOCO American Oil Company

AN Anglo Nordic

ANCAP Administracion Nacional de Combustibles Alcohol y Portland (Spanish—National Administration of Flammable Alcohol and Portland Cement)—Uruguay

ANL Australian National Line

ANZECS Australia-New Zealand-Europe Container Service

ANZS Africa-New Zealand Service

AP American Pioneer Lines

AP Atlantska Plovidba (Yugoslavian—Atlantic Line)

APL American President Lines

APT Australian Pacific Traders

ASA Admanthos Shipping Agency

ASC Alcoa Steamship Company

ASCL Australia Straits Container Line

ASFS Alaska State Ferry System

ASN Atlantic Steam Navigation

ASNC Atlantic Steam Navigation Company

ASOK Angfartigas Svenska Östasiatiske Kompaniet (Swedish—Swedish East Asiatic Steamship Company)

AT American Trading

ATLANTIC Atlantic Refining Company

Atlantic Container Line ACL

AUT American Union Transport

AWPL Australia West Pacific Line

B Barber Lines; Booth Line; Branch Lines; Bull Steamship Lines; etc.

BACS Ben Asia Container Service

BAF Belgian African Line

BBS Barber Blue Sea

BCCS British Columbia Coastal Service

BCF British Columbia Ferries

BCL Bermuda Container Line; Bristol City Line

BCSC British Columbia Steamship Company; British and Continental Steamship Company

BDS Bergenske Dampskibsselskab (Norwegian—Bergen Steamship Line)—connecting Norway and United Kingdom ports

Ben Ocean Ben Line, Blue Funnel, and Glen Line

BFL Belgian Fruit Lines; Blue Funnel Lines

BHP Broken Hill Proprietary

BISNC British India Steam Navigation Company

B & I SPC British and Irish Steam Packet Company

BL Bahamas Line; Bank Line; Bergen Line; Bibby Line; Booth Line; etc.

B & L Burns and Laird Lines

BLS Ben Line Steamers

Blue Star Blue Star Line

BM British Methane Limited

BMM Belfast, Mersey and Manchester Steamship Company

BOC Burmah Oil Company

Bore Ro-Ro Bore Roll-on Roll-off Line

BOS British Oil Shipping

BP British Petroleum

BPC British Phosphate Commissioners

BP & Co Burns, Philip and Company

BR British Railways (operates many ferry steamers linking England and Scotland with Belgium, France, Ireland, and Holland)

BSC Baltic Steamship Company

BSL Black Star Line; Blue Sea Line; Blue Star Line; etc.

BSNC Bristol Steam Navigation Company

BSPL Blue Star Port Lines

BTC Bethlehem Transportation Corporation

B&W Brocklebank and Well Lines

C Calmar Line (Bethlehem Steel); Caribbean Steamships Company; Clarke Line; Clyde Line; etc.

"C" Costa Line

CA Carregadores Açoreanos (Portuguese—Azorean Cargo Carriers)

CAROL Caribbean Overseas Lines

CAVN Compañía Anonima Venezolana de Navegacion (Spanish—Venezuelan Navigation Company)—Venezuela Line

CCAL Christensen Canadian African Line

CC Co Commercial Cable Company

CCN Companhia Colonial de Navegacão (Portuguese—Colonial Navigation Company)

CCNI Cía Chilena de Navegación Interoceanica (Spanish—Chilean Interoceanic Navigation Company)

CEA Central Electricity Authority

CF Compagnie de Navigation Fraissinet

CFL Container Fleets Limited

CFPO Compagnie Française des Phosphates de l'Oceanie (French—French Phosphate Company of Oceania)

CGL Canadian Gulf Line

CGM Compagnie Générale Maritime (French Line)

CGS Central Gulf Steamships

CGT Compagnie Générale Transatlantique (Cie G^{le} T^{rans}) (French—General Transatlantic Company)—the French Line

CHEVRON Chevron Shipping (oil tankers)

Chilean Line (see CSAV)

China Merchants Steam Navigation Company CMSNC

CI Catalina Island Steamship Line; Christmas Island Phosphate Commission

Cie G^{le} T^{rans} Compagnie Générale Transatlantique (French—General Transatlantic Company)—the French Line

Cities Service Cities Service Oil Company

CL Ceylon Lines; Coast Lines

Clipper Line Wisconsin and Michigan Steamship Company

CM Compañía Maritima (Spanish—Maritime Company)

CMB Compagnie Maritime Belge (French—Belgian Maritime Company)—Royal Belgian Lloyd

CMSNC China Merchants Steam Navigation Company

CMZ Compagnie Maritime du Zaire

CNC China Navigation Company

CNM Canadian National Marine (steamship line)

CNN Compagnie de Navigation Nationale

CNN Companhia Nacional de Navegacão (Portuguese—National Navigation Company)

CNP Compagnie Navigation Paquet (French—Paquet Navigation Company)—Paquet Line

CNS Canadian National Steamships

Coastal Express see *Hurtigruta*

COLDEMAR Compañía Colombiana de Navegación Maritima (Spanish—Colombian Maritime Navigation Company)

Columbus Line HSDG

COSCO China Ocean Shipping Company

CP Ships Canadian Pacific Steamships (*Empress* vessels)

CPV Corporación Peruana de Vapores (Spanish—Peruvian Steamship Corporation)

Crusader Crusader Line

CSAV Compañía Sud-Americana de Vapores (Spanish—South American Steamship Company)—Chile

CSC Clyde Shipping Company

CSL Canada Steamship Lines

CSO Cities Service Oil

CSS Caribbean Steamship

CSSCo Cunard Steamship Company

CT Cleveland Tankers; Cove Tankers

CT Compania Transmediterranea (Spanish—Transmediterranean Company)

CTE Compañía Transatlantica Española (Spanish—Spanish Transatlantic Line)—The Spanish Line

CTL Coastal Transport Limited

Cunard Cunard Steam-Ship Company, Limited (includes White Star Line)

D Delta Line; Donaldson Line; Red 'D' Line; etc.

'D' Red 'D' Line (merged with Grace Line)

DAL Deutsche-Afrika Linien (German—German Africa Line)

d'Amico d'Amico Line

Day Line Hudson River Day Line

DBK Daiichi Bussan Kaisha

DDSG *Donau-Dampfschiffahrt-Gesellschaft* (German—Danube Steamship Company)—Austria

D-F *Dansk-Franske* (Danish-French Line)

DFDS *Det Forenede Dampskibs-Selskab* (Danish—United Steamship Company)—famous for its ferries

DHX Dependable Hawaiian Express

Djakarta Line DL

DL Djakarta Line; Djakarta Lloyd

DPLC Dundee, Perth and London Shipping Company

DS Dominion Shipping

D-S Ditlev-Simonsen, Halfdan and Company

e El Paso Marine

E American Export Isbrandtsen Lines; Eastern Steamship Line; Exxon Tankers; Hellenic Lines and many Greek lines where the letter E stands for Ellas or Hellas—Greece, or for the last name of an owner as in other lands

E & A Eastern and Australian Steamship Co

EAC East Asiatic Company

E&B Ellerman and Buchnall Steamship Company

EDL Elder Demptser Lines

E&F Elders and Fyffes Ltd

ELMA *Empresa Lineas Maritimas Argentinas* (Spanish—Argentine Maritime Lines)—formerly *FANU* and uses *FANU* house flag

EMC Evergreen Marine Corporation

Empress liners Canadian Pacific ships

ENS Empresa Naviera Santa

ESL Eagle Shipping Ltd

Esso Esso Petroleum Company

EXXON formerly Esso

EY El Yam (bulk carriers)

F Fabre Line; Falcon Tankers; Falkland Islands Trading Company; Farrell Lines; Finnlines; etc.

FAA *Finska Angfartygs Akiebolaget* (Finnish—Finnish Steamship Company)—Finland Line

Falline Federal Atlantic-Lakes Line

FANF *Flota Argentina de Navegación Fluvial* (Spanish—Argentine River Navigation Fleet)

FANU *Flota Argentina de Navegación de Ultramar* (Spanish—Argentine High-Sea Navigation Fleet)

Far East Steamship Company FESCO

FB Franco Belgian Line

FBS Franco-Belgian Services

FCNCo Federal Commerce and Navigation Company

F de P Ferrocarril de Panamá (formerly the Panama Railroad)

Fedpac Federal Pacific Lakes Line

Fedsea Federal South East Asia Line

FESCO Far East Steamship Company

Finald Line (see *FAA*)

FL Ferdinand Laeisz Line; Fesco Pacific Line

FLL Finanglia Line Ltd

FMC Federal Maritime Commission

FMD *Flota Mercante Dominicana* (Spanish—Dominican Merchant Fleet)

FMG *Flota Mercante Grancolombiana* (Spanish—Great Colombian Merchant Fleet)

French Line (see *CGT*)

Frota Frota Oceanica Brasileira

FW Furness, Withy and Company

FWL Furness Warren Line

G Glynafon Shipping; Graig Shipping; Arthur Guiness (the brewer); etc.

GAL German Atlantic Line

GG Guinea Gulf Line

GL Greek Line

GMC Gulf Maritime Company

GO Gulf Oil

GPRL Gulf Puerto Rico Lines

GRACE Grace Line (Prudential-Grace Lines)

Gran Flota Blanca (Spanish—Great White Fleet)—United Fruit Company (fleet of white steamships)—United Brands

GS Galleon Shipping

GSA Gulf and South American Steamship Company

GULF Gulf Oil Corporation

GYSCo Great Yarmouth Shipping Company

H Hansa Line; Heering Line; Horn Line; etc.

HAL Holland Amerika Lijn (NASM—Nederlandsch-Amerikaansche Stoomvaart Maatschappij)—NASM appears on house flag

HANSA Hansa Line

Hanseatic-Vassa Line VL

HAPAG *Hamburg-Amerika Paket Aktiengesselschaft* (German—Hamburg-America

Packet Company)—Hamburg-America Line

Hapag-Lloyd Hamburg-Amerika—North German Lloyd Lines

HB C Hudson's Bay Company

HCL Hamburg-Chicago Line

HFL Hawaii Freight Lines

HH H Hogarth and Sons

HHA HH Andersen Line

HKCL Hong Kong Container Line

HKEL Hong Kong Export Lines

HKIL Hong Kong Islands Line

HKX Hong Kong Express

HL Home Lines

H-L Hapag-Lloyd

HLC Hapag-Lloyd Container (line)

HMM Hyundai Merchant Marine

HMS Her (His) Majesty's Ship (as in HMS *Dreadnought*)

hovercraft marine craft supported by an air cushion instead of a conventional hull

HSAL Hamburg South American Line

HSDG Hamburg-Sudamerikanische Dampfs Gesell (Columbus Line)

HT Hudson Tankers

Hurtigruta (Norwegian—quick way)—coastal steamship lines linking ports from Bergen to Kirkenes

H&W Holm and Wonsild

HWAL Holland West-Afrika Line

I Incres Line; Interocean Steamship Lines; Isthmian Lines (U.S. Steel); Ivaran Lines; etc.

ICI Imperial Chemical Industries

ICSN Indo-China Steam Navigation Company

IFI Inter-Freight International

INSCO Intercontinental Shipping Corporation

Inter-Freight International IFI

IO Ltd Imperial Oil Ltd

IOM SPC Isle of Man Steam Packet Company

IOT Iron Ore Transport

IPL Ital Pacific Line

ISOS International Ship Operating Services

Italia Italian Line

ITI Inagua Transports Incorporated

J Japan Line; John I Jacobs and Company; Johnson Line; etc.

Jadroplov *Jadrarnska Slobodna Plovida* (Yugoslav Great Lakes Line)

JBPS Jamaica Banana Producers' Steamship Co
JL J Lauritzen; Jebsen Line
K Kavolines; Kawasaki Kisen Kaisha; Kerr Lines; Keystone Shipping (Chas Kurz); Kingsport Shipping; Kirkconnel; Klaveness Line; Knutsen Line; etc.
KG Koctug Line
KK Karlander Kangaroo Line
KKL Karlander Kangaroo Line
K Line Kawasaki Kisen Kaisha
KMTC Korea Marine Transport Company
KNC Kingcome Navigation Company
KNSM Koninklijke Nederlandsche Stoomboot Maatschappij (Dutch—Royal Netherlands Steamship Company)
Koctug Line KL
KS Korea Shipping
KSC Korean Shipping Corporation
KSN Karachi Steam Navigation Line
L Lauritzen Line; Luckenbach Line; Lykes Line; etc.
LASH Lighter Aboard Ship Handling
LB Lloyd Brasileiro
L + H Lamport and Holt Line
LL Lauro Line; Link Line
Lloyd's Lloyd's Register of Shipping (LRS)
LPR Lauritzen Peninsula Reefer (line)
LRS Lloyd's Register of Shipping
LT Loyd Triestino (Italian—Trieste Line)
M Maersk Line; Marine Transport Lines; Matson Line; Meyer Line; Montship Lines; Moore-McCormack Lines; Munson Line; etc.
Maersk Maersk Line
MAMENIC Marina Mercante Nicaraguense (Spanish—Nicaraguan Merchant Marine)
MANZ Montreal-Australia-New Zealand (Line)
Maritime Fruit Carriers MFC
MCP Maritime Company of the Philippines
MFC Maritime Fruit Carriers
MILI Micronesia Interocean Line Incorporated
Milwaukee Clipper Wisconsin and Michigan Steamship Company
MISC Malaysian International Shipping Corporation
Mitsui Mitsui OSK Lines
ML Manchester Liners

MLL Manchester Lines Ltd
M. M Messageries Maritimes (French—Maritime Mail, Parcel, and Passenger Service)
MOBIL Mobil Oil
MOLU Mitsui-Osaka Container Line
M/S Motorship
MS Co Melbourne Steamship Company
MSTS Military Sea Transportation Service
MTL Marine Transport Lines
MV Motor Vessel
N Naess Shipping Company; Niarchos Tankers; Nigerian National Line; etc.
NA & G North Atlantic and Gulf Steamship Company
NAL Nigeria America Line
N-A-L Norwegian America Line
NASM (*see* HAL)
NAWAL North American-West African Line
NB Navibel (Belgian Maritime Navigation Company)
NB & C Norfolk, Baltimore and Carolina Line
NCL Norwegian Caribbean Line; Norwegian Cruise Lines
NCP Naviera Chilena del Pacifico (Spanish—Chilean Shipping of the Pacific)
Nedlloyd Nedlloyd and Hoegh Lines
NEE New England Express
NEPU Neptune Orient Container Line
New England Express NEE
NMB Navigation Maritime Bulgare (Bulgarian Maritime Navigation)
NNC Northern Navigation Company
NOL Neptune Orient Line; Norse Oriental Lines
NPCL North Pacific Coast Line
NPL Nauru Pacific Line
NPR Navieras de Puerto Rico
NTGB North Thames Gas Board
NYK Line Nippon Yusen Kaisha
NZL New Zealand Line
NZSCo New Zealand Shipping Company
O Ocean Carriers; Olsen Line; MJ Osorio; etc.
OCL Overseas Container Line; Overseas Containers Limited
Official Steamship Guide International OSGI
OG O Gross and Sons Ltd.
OK Oijekonsumenternas
ØK Østasiatiske Kompagni

(Danish—East Asiatic Company)—EAL
Olympic Olympic Steamship Company
OO Orient Overseas Line
OOCL Orient Overseas Container Line
OOL Odessa Ocean Line; Orient Overseas Line
OS Ocean Steamship
OSGI Official Steamship Guide International
OSK Osaka Syosen Kaisha (Osaka Mercantile Steamship Company)—Mitsui Lines
OW Olof Wallenius Line
P Panama Line (Panama Canal Company); Pocahontas Steamships; Prudential-Grace Lines; Pure Oil; etc.
P-A Pan-Atlantic Steamship Corporation
PACE Pacific America Container Express
Pacific America Container Express PACE
Pacific Australia Direct Line (*see* ABRT)
PAD Pacific Australia Direct (Line)
PAL Pan Asia Line
Petrobras Petroleo Brasileiro (Portuguese—Brazilian Petroleum)
PFEL Pacific Far East Line
PFL Pacific Forum Line; Pacific Freight Line
P-G Prudential Grace Lines
PIL Pacific International Line
PITL Pacific Islands Transport Line (Thor Dahls Hvalfangerselskap)
PL Polynesia Line; Port Line; Poseidon Lines; Prince Line
PLA Port of London Authority
PLL Prince Line Limited
PLO Polskie Linie Oceaniczne (Polish—Polish Line)
PM Petroleos Mexicanos (Spanish—Mexican Petroleum)
PNGL Papua New Guinea Line
PNL Philippine National Lines
P&O Peninsular and Occidental Steamship Company; Peninsular and Oriental Line
POE Pacific Orient Express Line
PPL Philippine President Lines
Princess Line Gothenburg-Frederikshavn Line
PSC Point Shipping Company
PSFL Puget Sound Freight Lines
PSNC Pacific Steam Navigation Company
PT Pope and Talbot

PURE Pure Oil Company

PV Pacific Venture

Q Qatar Petroleum; Quaker Line; Queensland; Quintessence Navigation; etc.

Q&O Quebec and Ontario Transportation

R Rasmussen; Richfield Oil; Ringdal; Robert; etc.

RIL Royal Interocean Lines [*Koninklijke Java-China-Paketvaart Lijnen*—(Dutch— Royal Java-China-Packet Line)]

RL Regent's Line (Grand Union Shipping)

RLR Royal Rotterdam Lloyd

RML Royal Mail Lines

Royal Netherlands Steamship Line (see *KNSM*)

RVL Royal Viking Line

S Saguenay Terminals Ltd; Salen; Seatrain Lines; Sinclair Refining; Socony Mobil Oil; Standard Oil of California; States Marine Lines; States Line (seahorse-shaped red-letter S); Sun Oil; Svea Line; etc.

SA & CL South Atlantic and Caribbean Line

Safmarine South African Marine Corporation

SAL Svenska Amerika Linien (Swedish-America Line)

Santa ships Prudential-Grace Line vessels

SC Submarine Cables Ltd

S & C Star and Crescent

SCC Shipping and Coal Company

SCI Sea Containers Incorporated; Shipping Corporation of India

Scindia Scindia Steam Navigation

SEGB South Eastern Gas Board

Shell Shell Tankers

Shipping Corporation of India SCI

SL Southern Lines

S-L Sea-Land (Line)

SLS Sea-Land Service

SML States Marine Lines

SN Sincere Navigation

SOPONATA Sociedade Portuguesa de Navios Tanques Limitada (Portuguese—Portuguese Tankships Limited)

Sovtorgflot Soviet Merchant Marine Fleet

Spanish Line Compañía Transatlantica Española

SPL Scan Pacific Line

SS Steamship (as in SS *Santa Clara*)

SSS Sea Speed Service (container)

STANVAC Standard-Vacuum Oil Company

STL Seatrain Lines

SUNOCO Sun Oil Company

T Tankers Limited; Texaco (The Texas Company); Thai Mercantile Marine; Thompson Shipping; Thoren Line; Tirrenia; Transatlantic Line; etc.

TCL Transatlantic Carriers Limited

TCR Texas City Refining

Texaco The Texas Company

TFL Trans Freight Line

TH Thorvald Hansen

Thor Dahls Havalfangerselskap Pacific Islands Transport Line

TMM Transportación Maritima Méxicana

TOTE Totem Ocean Trailer Express

Transamerica Trailer Transport TTT

TS Tasmanian Steamers

TSK Tokyo Senpaku Kaisha

TTT Transamerica Trailer Transport

U Union Oil; United Oriental Steamship Company; Universe Tankships; etc.

UA United Africa Company, Ltd

UBC United Baltic Corporation

UBL Union Barge Line

UCMS Union-Castle Mail Steamship

UFC United Fruit Company

UIL Ulster Imperial Line

U.O. Co. Union Oil Company of California

UPL United Philippine Lines

USC Union Steamship Company

USL United States Lines

USMSTS U.S. Military Sea Transport Service

USS United States Ship (as in USS *Constitution*)

USSCo Ulster Steam Ship Company; Union Steam Ship Company

UT United Transports

UYL United Yugoslav Lines

V Vaccaro Line (Standard Fruit); Valentine Chemical Carriers; Vinke Tankers; Von Sydow; Vulcan Shipping; etc.

VA Compañía de Navegación Vasco-Asturiana (Spanish— Basque-Asturian Navigation Company)

VC Victory Carriers

VL Vaasa Line (Hanseatic-Vassa Line)

VLC Valley Line Company

VNGC Van Niervelt, Goudriaan and Company (Rotterdam— South American Line)

VW Volkswagen (auto-carrier ships)

W Waterman Steamship Lines; West Line; Westriver Ore Transports; Weyerhaeuser Line; etc.

W&A Wiel and Amundsen

Wallenius Line OW (Olof Wallenius)

WHMV & NSSA Woods Hole, Martha's Vineyard and Nantucket Steamship Authority

WIL West India Lines

WIT West India Tankers

WL Westfal-Larsen Line

W&L Westcott and Laurance Line (Ellerman's)

WL&Co Westfal-Larsen and Company

W&M SS Co Wisconsin and Michigan Steamship Company (The Clipper Line)

WSFS Washington State Ferry System

WTC Western Transportation Company

X (funnel marking—Chandris America Lines; Southern Cross Steamship Line); Xenophon Navigation Company; etc.

Y Yamashita-Shinnihon Kisen Line; Ybarra Lines; Yukiteru Kaiun; Yung Yang Shipping; etc.

YML Yang Ming Line

YPF Yacimientos Petroliferos Fiscales (Spanish—Fiscal Petroleum Deposits)—Argentine tanker fleet

Y-S Line Yamashita-Shinnihon Line

Z Zacharissen; Zante Navegación; Zillah Shipping; Zim Israel Navigation; Zurga Shipping Company; etc.

Zapata Zapata Bulk Transport

Zim Zim Israel Line

ZPL Zim Passenger Line

ZSC Zeeland Shipping Company; Zeeland Steamship Company

Superlatives

Afghanistan's Highest Peak Tirish Mir—25,263 feet, 7,655 meters—in the Hindu Kush on the Pakistan border
Afghanistan's Largest City Kabul
Africa's Easternmost City Hafun, Somalia
Africa's Easternmost Point Cape Guardafui (Ras Asir), Somalia
Africa's Highest Peak Kilimanjaro—19,340 feet, 5,861 meters—in Tanzania where it is called Kibo
Africa's Largest Black City Ibadan, Nigeria (population more than a million in 1985)
Africa's Largest City Cairo
Africa's Largest Drainage System Congo
Africa's Largest Island Madagascar
Africa's Longest River Nile
Africa's Northernmost City Bizerta, Tunisia
Africa's Northernmost Point Ras el Abiadh (near Bizerta, Tunisia)
Africa's Southernmost City Cape Town, South Africa
Africa's Southernmost Point Cape Agulhas, South Africa
Africa's Westernmost City Dakar, Senegal
Africa's Westernmost Point Cape Almadies, Senegal
Alabama's Deepest Cavern 12-story-deep Cathedral Caverns containing a stalagmite 60 feet (18 meters) high
Alabama's Highest Point Cheaha Mountain—2407 feet, 738 meters
Alabama's Largest City Birmingham
Alabama's Longest River Tombigbee
Alabama's Principal Port Mobile
Alaska's Highest Point Mount McKinley—20,320 feet, 6,187 meters
Alaska's Largest City Anchorage
Alaska's Longest River Yukon
Alaska's Principal Port Valdez Harbor
Alaska's Richest Agricultural Area the Matanuska Valley
Albania's Highest Peak Korab

—9,066 feet, 2,747 meters
Albania's Largest City Tirana
Albania's Principal Port Durazzo
Alberta's Highest Point Mount Columbia—12,294 feet, 3,747 meters
Alberta's Largest City Calgary
Algeria's Highest Peak Lella Khedidja—7,572 feet, 2,295 meters—in the Tell Atlas mountains
Algeria's Largest City Algiers
Algeria's Principal Port Algiers
American Samoa's Highest Point Matafao on Tutuila Island—2,142 feet, 649 meters —and Lata on Tau Island in the Manua Islands—3,056 feet, 926 meters
American Samoa's Largest City Pago Pago
American Samoa's Principal Port Pago Pago
American Virgin Islands' Highest Point Crown Mountain—1,556 feet, 472 meters —on Saint Thomas
American Virgin Islands' Largest Center Charlotte Amalie on Saint Thomas
American Virgin Islands' Principal Ports Charlotte Amalie on Saint Thomas, Cruz Bay on Saint John, Frederiksted on Saint Croix
America's Busiest Airport Chicago's O'Hare International
America's Easternmost City Eastport, Maine
America's Easternmost Point West Quoddy Head, Maine
America's Finest Opera George Gershwin's *Porgy and Bess*
America's First National Park Yellowstone
America's Foremost Modern Composer Aaron Copland
America's Highest Peak Mount McKinley—20,320 feet, 6,158 meters—in Alaska
America's Largest City New York City
America's Largest State Alaska
America's Longest River Mississippi–Missouri

America's Most Beloved Woman The Statue of Liberty
America's Most Used and Abused Drug alcohol
America's Northernmost City Barrow, Alaska
America's Northernmost Point Point Barrow, Alaska
America's Oldest Animal the horseshoe crab *(Limulus polyphemus)*
America's Principal Port New Orleans
America's Smallest State Rhode Island
America's Southernmost City Hilo on the island of Hawaii
America's Southernmost Point Ka Lae (South Cape) on the island of Hawaii
America's Westernmost City Agaña, Guam
America's Westernmost Point Cape Wrangell on Attu Island, Alaska
Andorra's Highest Peak Puig de la Coma Pedrosa—9,665 feet, 2,929 meters—between France and Spain
Andorra's Largest Place Andorra la Vella
Angola's Largest City Luanda
Angola's Principal Port Luanda
Anguilla's Largest Town The Valley
Antarctica's Highest Peak Vinson Massif—16,860 feet, 5,140 meters
Antigua and Barbuda's Principal Port St John's
Antigua and Barbuda's Largest Town St John's
Arctic's Largest Carnivore polar bear
Argentina's Easternmost City Posadas (on the Paraguay border)
Argentina's Highest Peak Aconcagua—22,835 feet, 6,920 meters
Argentina's Largest City Buenos Aires
Argentina's Largest Port Buenos Aires
Argentina's Largest Province Santa Cruz
Argentina's Northernmost City San Salvador (close to the Chilean border)
Argentina's Principal Port

Buenos Aires

Argentina's Smallest Province Tucumán

Argentina's Southernmost City Ushuaia (on Beagle Channel)

Argentina's Westernmost City Mendoza (near the Chilean border)

Arizona's Grandest Canyon Grand Canyon of the Colorado River

Arizona's Highest Point Humphrey's Peak—12,655 feet, 3,835 feet

Arizona's Largest City Phoenix

Arizona's Longest River Colorado

Arkansas' Highest Point Magazine Mountain—2,753 feet, 839 meters

Arkansas' Largest City Little Rock

Arkansas's Largest Spring Blue Spring produces 144 million liters (38 million gallons) of water every day

Arkansas' Longest River Arkansas

Arkansas' Principal Port Little Rock

Asia's Highest Peak Everest —29,028 feet, 8,796 meters— (*see* China's Highest Peak)

Asia's Largest City Tokyo

Asia's Largest Drainage System Ob–Irtysh

Asia's Largest Island Borneo

Asia's Largest Nation USSR, in terms of landmass; China in terms of population

Asia's Longest River Yangtze

Australasia's Largest Island New Guinea

Australia's Easternmost City Brisbane, Queensland

Australia's Easternmost Point Cape Byron, New South Wales

Australia's Highest Peak Mount Kosciusko—7,305 feet, 2,214 meters—in the Australian Alps

Australia's Largest City Sydney

Australia's Largest Port Sydney

Australia's Largest State Western Australia

Australia's Longest River Murray–Darling

Australia's Northernmost City Darwin, Northern Territory

Australia's Northernmost

Point Cape York, Queensland

Australia's Principal Port Sydney

Australia's Southernmost City Hobart, Tasmania

Australia's Southernmost Point South East Cape, Tasmania

Australia's Westernmost City Carnarvon, Western Australia

Australia's Westernmost Point Cape Inscription, Western Australia

Austria's Highest Peak Grossglöckner—12,460 feet, 3,776 meters—on the border with Italy

Austria's Longest River Danube

Azores' Largest Town Ponta Delgada

Bahamas' Largest City Nassau

Bahamas' Principal Port Nassau

Bahrain's Largest City Manama

Bahrain's Principal Port Sitra

Balearic Islands' Largest City Palma de Majorca

Bangladesh's Largest City Dhaka

Bangladesh's Principal Port Chittagong

Barbados' Largest City Bridgetown

Barbados' Principal Port Bridgetown

Belau's Largest Town Koror

Belgium's Largest City Brussels

Belgium's Largest Port Antwerp

Belgium's Longest River Scheldt

Belgium's Principal Port Antwerp

Belize's Highest Peak Victoria —3,681 feet, 1,115 meters— in the Cockscomb Mountains

Belize's Largest City Belize

Belize's Principal Port Belize City

Belle City of the Lakes Racine, Wisconsin

Benin's Largest City Cotonou

Benin's Principal Port Porto–Novo

Bermuda's Largest Town Hamilton

Best of the Bachs Johann Sebastian Bach who fathered Carl Philipp Emanuel, Johann Christina, Johann Christoph Friedrich, Wilhelm Friedmann; known collectively as

die Bäche—the Bachs

Best of the Bonn Boys Ludwig van Beethoven, born in Bonn

Best Butler Hudson (Gordon Jackson on PBS television production of *Upstairs Downstairs*)

Best Sellers in Nazi Germany the *Bible* and *Mein Kampf,* according to Benjamin B Ferencz

Best Title given by Queen Isabel the Catholic appointing Christopher Columbus Admiral of the Ocean Sea, Viceroy and Governor of the Lands You Discover *(Almirante del mar Océano, Virrey y Gobernador de las Tierras que Descubriese)*

Bhutan's Highest Peak Kula Kangri—28,780 feet, 8,721 meters—in the West Assam Himalayas

Bhutan's Largest City Thimphu

Biggest Domestic Dog Saint Bernard, standing up to 25 inches (65 centimeters) and weighing up to 170 pounds (77 kilograms)

Biggest Murder Trial in History Nuremberg Trial of Nazi war criminals

Bolivia's Highest Mountain Ancohuma—21,490 feet, 6,512 meters—in the Andes

Bolivia's Largest City La Paz

Bolivia's Largest Department Santa Cruz

Bolivia's Smallest Department Tarija

Bophuthatswana's Largest City Mmabatho

Botswana's Largest City Gaborone

Brazil's Easternmost City João Pessôa or Recife

Brazil's Easternmost Town Cabedelo (close to João Pesôa)

Brazil's Highest Mountain Pico de Bandeira—9,462 feet, 2,867 meters

Brazil's Largest City São Paulo

Brazil's Largest Port Rio de Janeiro

Brazil's Largest State Amazonas

Brazil's Northernmost City Belém do Pará

Brazil's Northernmost Town Maturuca (close to the Guyana border)

Brazil's Principal Port Rio de

Janeiro

Brazil's Smallest State Sergipe

Brazil's Southernmost City Rio Grande do Sul (close to the Uruguay border)

Brazil's Southernmost Town Jaguarão (on the Uruguay border)

Brazil's Westernmost City Rio Branco (close to the Bolivian border)

Brazil's Westernmost Town Cruziero do Sul (near the Peruvian border)

Britain's Largest City London

British Columbia's Highest Point Mount Fairweather—15,300 feet, 4,663 meters—on the Alaskan border

British Columbia's Largest City Vancouver

British Columbia's Longest River Fraser

British Columbia's Principal Port Vancouver

British superlatives (*see* United Kingdom, Largest British, *and* Smallest British *entries*)

British Virgin Islands' Largest Town Road Town

Brunei's Largest City Seri Begawan

Bulgaria's Highest Peak Musala—9,596 feet, 2,908 meters

Bulgaria's Largest City Sofia

Bulgaria's Largest Seaport Varna (formerly called Stalin)

Bulgaria's Longest River Danube

Bulgaria's Principal Port Varda

Bullfight Capital of the World Madrid

Burma's Highest Peak Mount Victoria—10,016 feet, 3,035 meters

Burma's Largest City Rangoon

Burma's Principal Port Rangoon

Burundi's Highest Peak Bugungu—5,025 feet, 1,523 meters—near the Tanzania border

Burundi's Largest City Bujumbura

California's Highest Point Mount Whitney—14,494 feet, 4,420 meters

California's Largest City Los Angeles

California's Longest River Sacramento

California's Most Spectacular Park Yosemite

California's Principal Port Long Beach in Los Angeles Harbor containing San Pedro, Terminal Island, and Wilmington

Cambodia's Highest Peak Tadet—3,667 feet, 1111 meters—south of Pailin

Cambodia's Largest City Phnom Penh

Cambodia's Principal Port Phnom Penh

Cameroon's Largest City Douala

Cameroon's Principal Port Douala

Canada's Biggest Port on the Atlantic Halifax, Nova Scotia

Canada's Biggest Port on the Great Lakes Sault Sainte Marie, Ontario on a canal connecting Lake Huron and Lake Superior

Canada's Biggest Port on Lake Ontario Toronto, Ontario

Canada's Biggest Port on the Pacific Vancouver, British Columbia

Canada's Biggest Port on the St Lawrence Montreal, Québec

Canada's Easternmost City St John's, Newfoundland

Canada's Easternmost Point Cape Spear, Newfoundland

Canada's Highest Peak Mount Logan—19,850 feet, 6,015 meters—in the southwest Yukon

Canada's Highest Town Lake Louise, Alberta (1,540 meters or 5,051 feet)

Canada's Largest City Montreal

Canada's Largest Province Québec

Canada's Longest River Mackenzie–Peace

Canada's Most Vibrant City Toronto

Canada's Northernmost Deepwater Port Churchill, Manitoba

Canada's Northernmost Point Cape Columbia, Ellesmere Land

Canada's Northernmost Town Inuvik, Northwest Territory

Canada's Oldest City Québec City

Canada's Oldest National Park Banff National Park including lovely Lake Louise

Canada's Principal Port Montreal

Canada's Smallest Province Prince Edward Island

Canada's Southernmost City Kingsville, Ontario

Canada's Southernmost Point Point Pelee, Ontario on Lake Erie

Canada's Tallest Mountain Mount Robson in the Canadian Rockies (12,972 feet or 3,954 meters)

Canada's Westernmost City Dawson, Yukon

Canada's Westernmost Point in Yukon Territory just east of Alaska's Demarcation Point

Canary Islands' Largest City Las Palmas de Gran Canaria

Cape Verde's Largest City Praia

Cape Verde's Principal Port Praia

Cayman Islands' Largest Place Georgetown

Central Africa's Largest City Bangui

Central America's Highest Peak Tajumulco volcano—13,816 feet, 4,605 meters—in Guatemala

Central America's Largest City Guatemala City

Central America's Largest Island Cuba

Chad's Highest Peak *see* Libya's Highest Peak

Chad's Largest City N'Djaména

Channel Islands' Largest City Saint Helier on Jersey

Cheapest Capital Kuala Lumpur, Malaysia

Chile's Easternmost City Santiago

Chile's Highest Peak Aconcagua—22,835 feet, 6,920 meters—shared with Argentina

Chile's Largest City Santiago

Chile's Largest Port Valparaiso

Chile's Largest Region Antofagasta

Chile's Longest River Maipo

Chile's Northernmost City Arica (near Tacna, Peru)

Chile's Principal Port Valparaiso

Chile's Smallest Region Liberator General Bernardo O'Higgins

Chile's Southernmost City Punta Arenas (on the Strait of Magellan)

Chile's Westernmost City Valdivia

China's Highest Peak Everest —29,028 feet, 8,796 meters— called Quomolangma by the Chinese and Sagarmatha by the Nepalese, on the China –Nepal border

China's Largest City Shanghai

China's Largest Port Shanghai

China's Longest River Yangtze

China's Principal Port Shanghai

Cleanest City in the East Singapore

Cleanest City in the West Copenhagen, Oslo, Stockholm

Cleanest Port in the Orient Singapore

Cleanest and Most Decorative Subway System Moscow's

Cleanest Tropical City Singapore

Coldest Spot Vostok, Antarctica

Coldest and Windiest Continent Antarctica

Colombia's Biggest Port on the Caribbean Barranquilla

Colombia's Biggest Port on the Pacific Buenaventura

Colombia's Easternmost City Cúcuta

Colombia's Easternmost Town Puerto Carreño

Colombia's Highest Peak Cristóbal Colón—18,947 feet, 5,742 meters

Colombia's Largest City Bogotá

Colombia's Largest Department Caquetá

Colombia's Largest Port Barranquilla

Colombia's Longest River Magdalena

Colombia's Northernmost City Riohacha

Colombia's Northernmost Town Inosu

Colombia's Smallest Department Quindio

Colombia's Southernmost City Cali

Colombia's Southernmost Town Leticia (on the border of Brazil and Peru)

Colombia's Westernmost City Buenaventura

Colombia's Westernmost Town Tumaco (close to the Ecuador border)

Colorado's Highest Point

Mount Elbert—14,443 feet, 4,398 meters

Colorado's Largest City Denver

Colorado's Largest Flat-topped Mountain Grand Mesa and its National Forest atop the world's largest flat-topped mountain

Colorado's Longest River Colorado

Commonest British Bird blackbird

Comoros' Largest City Moroni on Grande Comore

Comoros' Principal Port Moroni

Congo's Largest City Brazzaville

Connecticut's Biggest Beach Park Hammonasset fronting on Long Island Sound

Connecticut's Highest Point Mount Frissell—2,380 feet, 725 meters

Connecticut's Largest City Bridgeport

Connecticut's Longest River Connecticut

Connecticut's Principal Port New Haven

Cook Islands' Largest Place Avarua on Rarotonga

Costa Rica's Atlantic Port Limón

Costa Rica's Highest Peak Chirripó—12,533 feet, 3798 meters—in the Talamanca Cordillera

Costa Rica's Largest City San José

Costa Rica's Largest Province Puntarenas

Costa Rica's Pacific Port Puntarenas

Costa Rica's Smallest Province Heredia

Cuba's Easternmost Point Cabo Maisi

Cuba's Easternmost and Southernmost City Santiago

Cuba's Highest Peak Pico Turquino—6,560 feet, 1,988 meters—in the Sierra Maestra

Cuba's Largest City Havana

Cuba's Largest Province Camagüey

Cuba's Northernmost City and Point Havana

Cuba's Principal Port Havana

Cuba's Smallest Province Habana

Cuba's Southernmost Point Cabo Cruz

Cuba's Westernmost City Pinar del Rio

Cuba's Westernmost Point Cabo San Antonio

Cultural Capital of the World New York City

Curaçao's Largest Town Willemstad

Curaçao's Principal Port Willemstad

Cyprus' Highest Point Mount Troodos—6,404 feet, 1,941 meters—in the Olympic Mountains

Cyprus' Largest City Nicosia

Cyprus' Principal Port Nicosia

Czechoslovakia's Highest Peak Gerlachovka—8,737 feet, 2,648 meters—in the High Tatras, once named Stalin

Czechoslovakia's Largest City Prague

Czechoslovakia's Longest River Danube

Deepest Gorge Hell's Canyon of the Snake River in Idaho

Deepest Lake Baikal—5,315 feet, 1,620 meters—in southeastern Siberia

Deepest Part of the Arctic Ocean Eurasia Basin between Komsomolets Island and the North Pole (2,980 fathoms or 17,880 feet or 5,450 meters in depth)

Deepest Part of the Atlantic Puerto Rico Trench north of Hispaniola and Puerto Rico (4,729 fathoms or 28,374 feet or 8,648 meters in depth)

Deepest Part of the Caribbean Cayman Trench between the Cayman Islands and Jamaica (3,833 fathoms or 23,000 feet or 7,010 meters deep)

Deepest Part of the Indian Ocean Diamantina Trench south of Western Australia (4,400 fathoms or 26,400 feet or 8,047 meters in depth)

Deepest Part of the Mediterranean off Cape Matapan, Greece (2,406 fathoms or 14,435 feet or 4,400 meters deep)

Deepest Part of the North Sea in the Skaggerak between Denmark and Norway (333 fathoms or 1,998 feet or 605 meters)

Deepest Part of the Ocean Mariana Trench in the Western Pacific east of Saipan (6,033 fathoms or 36,198 feet or 11,034 meters in depth)

Deepest Part of the Pacific

(*see* Deepest Part of the Ocean)

Delaware's Highest Point Centerville—442 feet, 135 meters

Delaware's Largest City Wilmington

Delaware's Longest River Delaware

Delaware's Principal Port Wilmington

Denmark's Highest Peak Mollejoh—561 feet, 170 meters—in East Jutland

Denmark's Largest City Copenhagen

Denmark's Principal Port Copenhagen

District of Columbia's Highest Point Tenleytown—410 feet, 125 meters

Djibouti's Largest City Djibouti

Djibouti's Principal Port Djibouti

Dominica's Largest City Roseau

Dominica's Principal Port Roseau

Dominican Republic's Highest Peak Monte Tina—10,301 feet, 3,434 meters

Dominican Republic's Largest City Santo Domingo

Dominican Republic's Largest Province San Juan

Dominican Republic's Principal Port Santo Domingo

Dominican Republic's Smallest Province Salcedo

Driest Desert Atacama, Chile where at Calama no rain has ever been recorded

Dutch superlatives (*see* Nederlands entries)

Easter Islands' Largest Town Hanga Roa

Easternmost Capital of Africa Mogadishu, Somalia

Easternmost Capital of Asia Tokyo, Japan

Easternmost Capital of Australia Brisbane, Queensland

Easternmost Capital of Europe Moscow, USSR

Easternmost Capital of North America St John's, Newfoundland

Easternmost Capital of South America Brasilia, Brazil

Easternmost Hawaiian Island Hawaii

Easternmost Point of Canada Cape Spear, Newfoundland on Avalon Peninsula near St John's

Easternmost Point of the continental United States West Quoddy Head near Lubec, Maine

Easternmost Point of Mexico Isla de las Mujeres off the Caribbean coast of Quintana Roo

Easternmost Point of the territorial United States East Point on Saint Croix in the Virgin Islands east of Puerto Rico

Ecuador's Easternmost City Quito

Ecuador's Highest Mountain Chimborazo—20,702 feet, 6,275 meters—in the Andes

Ecuador's Largest City Guayaquil

Ecuador's Largest Port Guayaquil

Ecuador's Largest Province Napo

Ecuador's Northernmost City Esmeraldas

Ecuador's Principal Port Guayaquil

Ecuador's Smallest Province Bolívar

Ecuador's Southernmost City Cuenca

Ecuador's Westernmost City Manta

Egypt's Highest Peak Gebel Katherina—8,651 feet, 2,622 meters—in the Sinai

Egypt's Largest City Cairo

Egypt's Principal Port Alexandria

El Salvador's Highest Peak Santa Ana—7,825 feet, 2,371 meters

El Salvador's Largest City San Salvador

El Salvador's Largest Department Usulután

El Salvador's Principal Port La Union

El Salvador's Smallest Department Cuscatlán

England's Extremitude Land's End—westernmost point in Cornwall and nearby Lizard Head the southernmost point

England's Highest Mountain Scafell Pike—3,210 feet, 973 meters—in Cumberland

England's Largest City London

England's Largest Lake Windermere

England's Longest Rivers Severn, Thames

Equatorial Guinea's Largest City Malabo

Equatorial Guinea's Principal Port Malabo

Ethiopia's Highest Peak Ras Dashan—15,158 feet, 4,593 meters

Ethiopia's Largest City Addis Ababa

Ethiopia's Principal Port Masewa

Eurasia's Easternmost Point Mys Dezhneva (East Cape), Siberia

Eurasia's Easternmost Town Uelen, Eastern Siberia (across Bering Strait from Tin City, Alaska)

Eurasia's Northernmost Point Rudolph Island off Franz Josef Land in the Arctic

Eurasia's Northernmost Town Ny Ålesund, Spitsbergen

Eurasia's Southernmost Point Roti Island in the Lesser Sundas of Indonesia

Eurasia's Southernmost Town Kupang on Timor, Indonesia

Eurasia's Westernmost Point Tearaght Island off Ireland's Dingle Peninsula

Eurasia's Westernmost Town Dingle, Ireland

Europe's Easternmost City and Point Gornyatskiy, in the USSR on the western slopes of the Ural Mountains

Europe's Highest Peak Mont Blanc—15,770 feet, 4,779 meters—between France and Italy

Europe's Largest Drainage Basin Rhine

Europe's Largest Island Great Britain

Europe's Largest Nation USSR (Union of Soviet Socialist Republics)

Europe's Longest River Volga

Europe's Northernmost City Hammerfest, Norway

Europe's Northernmost Point Nordkyn, Norway (north of North Cape)

Europe's Southernmost City Nicosia, Cyprus

Europe's Southernmost Point south coast of Cyprus

Europe's Westernmost Place Tearaght Island off Ireland's Dingle Peninsula

Europe's Westernmost Town Dingle, Ireland

Faeroe Islands' Largest Town Tórshavn

Falkland Islands' Largest Town Stanley

Fastest Clipper Ship *Lightning*—designed and built by Donald McKay in his East Boston shipyards—logged 436 nautical miles in 24 hours; McKay launched 16 of his fastest and finest clippers between 1850 and 1853

Fastest Passenger Vessel *United States* of United States Lines averages 35 knots (about 40 miles an hour) —has crossed Atlantic Ocean in less than 4 days

Fastest Railroad *TGV*, Paris-Lyon express operating in one section at 237 miles per hour and at an average speed of more than 130 miles per hour

Fastest Transatlantic Clipper Ship Donald McKay's *James Baines* sailed from Boston to Liverpool in 12 days and 6 hours (more than a 100 years ago)

Fifty Most Expensive Cities in the Late 1900s Tokyo, Kobe, Osaka, Brazzaville, Libreville, Dakar, Douala, Tehran, Abidjan, Geneva, Zurich, Vienna, Copenhagen, Oslo, Helsinki, Munich, Hamburg, Berlin, Paris, Duesseldorf, Frankfurt, Milan, Lyons, Rome, Dublin, Algiers, Amsterdam, Brussels, Taipei, Amman, Stockholm, Luxembourg, Cairo, New York, Tel Aviv, Los Angeles, Chicago, Washington, D.C., San Francisco, Boston, London, Lima, Miami, Houston, Abu Dhabi, Tripoli, Barcelona, Madrid, Port Moresby, Singapore

Fiji's Largest City Suva

Fiji's Principal Port Suva

Finest Harbor in the World Sydney Cove, according to Captain Arthur Phillip when he founded Australia's first settlement in 1788

Finland's Highest Peak Haltia —4,343 feet, 1,316 meters— on the Norwegian border

Finland's Largest City Helsinki

Finland's Longest River Kemi

Finland's Principal Port Helsinki

First Afghan Amir Ahmad Shah

First Albanian King William of Wied

First American Freethinker Mayor Jane Kathryn Conrad of Lochbuie, Colorado

First American Poet Laureate Robert Penn Warren

First American President George Washington

First American Romantic Composer William Mason

First Apartment House in the United States Pontalba Apartments facing Jackson Square in the French Quarter of New Orleans

First Argentinian President Justo José Urquiza

First Australian Prime Minister Edmund Barton

First of the Austrian Romantic Composers Ignaz Pleyel

First Austrian Ruler Otakar

First Austro-Hungarian Empress Maria Theresa

First Bahaman Governor-General Milo Broughton Butler

First Bangladesh President Abu Sayeed Chowdhury

First Barbadan Governor-General John Montague Stow

First Belgian Romantic Composer Henri Vieuxtemps

First Belizian Prime Minister George C Price

First Bohemian Romantic Composer Bedrich Smetana

First Bolivian President Antonio José de Sucre

First Brazilian Emperor Dom Pedro I

First Brazilian President Manuel Deodoro da Fonseca

First Brazilian Romantic Composer Carlos Gomes

First British Prime Minister Sidney Godolphin

First British Sovereign Egbert

First Bulgarian Ruler Prince Alexander of Battenberg

First Burmese President Sao Shwe Thaike

First Canadian Prime Minister John A Macdonald

First Chilean President Bernardo O'Higgins

First Chinese Communist Head of State Mao Tse-tung

First Chinese Nationalist President Chiang Kai-shek

First Chinese Provisional President Sun Yat-sen

First Colombian President Eustorgio Salgar

First Costa Rican President Rafael Iglesias

First Cuban President Tomás Estrada Palma

First Czechoslovakian President Tomás Garrigue Masaryk

First Czech Romantic Composer Antonin Dvořák

First Danish King Gorm the Old

First Danish Romantic Composer Friedrich Kuhlau

First Dominican Republic President Pedro Santana

First Dutch Monarch William I

First Dutch Romantic Composer Johannes Verhulst

First Ecuadorian President Juan José Flores

First Egyptian President Muhammad Naguib

First English Poet Laureate John Dryden

First English Romantic Composer Sir Arthur Sullivan

First English Ruler Egbert

First Ethiopian Emperor in Modern Times Theodore

First Finnish President Kaarlo Juho Stahlberg

First Finnish Romantic Composer Armas Järnefelt

First First Lady Martha Dandridge Custis Washington

First in Flight Orville and Wilbur Wright, Kitty Hawk, North Carolina

First French Fifth Republic President Charles de Gaulle

First French Romantic Composer Hector Berlioz

First German Romantic Composer Ludwig van Beethoven

First German Ruler Charlemagne (Karl the Great)

First Greek King in Modern Times Otto I of Bavaria

First Grenadan Prime Minister Eric Matthew Gairy

First Guatemalan President Rafael Carrera

First Guyanan Prime Minister Lindon Forbes Sampson Burnham

First Haitian Emperor Jean Jacques Dessalines

First Honduran President José M Medina

First Hungarian Regent Admiral Nicholas Horthy de Nagybánya

First Hungarian Romantic Composer Franz Liszt

First Indian Prime Minister Jawaharlal Nehru

First Indonesian President Achmed Sukarno

First Iranian Shah Agha Mohammed Khan

First Irish President William T Cosgrave

First **Irish Prime Minister** Eamon de Valera

First **Irish Romantic Composer** William Wallace

First **Israeli President** Chaim Wiezmann

First **Israeli Prime Minister** David Ben-Gurion

First **Italian King of Sardinia and Italy** Victor Amadeus II

First **Italian Romantic Composer** Gioacchino Rossini

First **Jamaican Prime Minister** Alexander Bustamante

First **Japanese Prime Minister** Hirobumi Ito

First **Jordanian King** Abdullah

First **Lebanese President** Bechara el-Khoury

First **Liberian President** Joseph J Roberts

First **Mexican Emperor** Augustín de Iturbide

First **Mexican Romantic Composer** Manuel Ponce

First **National Park in the United States** Yosemite in California

First **New Zealand Prime Minister** Henry Sewell

First **Nicaraguan President** Frutos Chamorro

First **North Korean Premier** Kim Il Sung

First **Norwegian Romantic Composer** Christian Sinding

First **Norwegian Ruler** Haakon the Good

First **Pakistan Governor-General** Doctor Mohammed Ali Jinnah

First **Panamanian President** Manuel Amador Guerrero

First **Paraguayan President** Cirolo Rivarola

First **Peruvian President** José de San Martín

First **Philippine President** Manuel Roxas

First **Polish President** Josef Pilsudski

First **Polish Romantic Composer** Frédéric Chopin

First **Portuguese Ruler** Afonso I

First **President of Tuskegee Institute** Booker T Washington

First **President of the U.S.** George Washington

First **Puerto Rican Elected Governor** Luis Muñoz Marín

First **Romanian Romantic Composer** Ludwig Wiest

First **Roman Ruler** Octavianus, Emperor Augustus

First **Romanian Ruler** Prince Alexander John I

First **Ruler of England and Great Britain** Egbert

First **Russian Romantic Composer** Mikhail Glinka

First **Russian Ruler** Ivan the Great

First **Salvadoran President** Rafael Zaldivar

First **Saudi Arabian Ruler** Abdul-Aziz

First **Scot to Rule Picts and Scots** Kenneth I MacAlpin

First **South African Prime Minister** Louis Botha

First **South Korean President** Syngman Rhee

First **Soviet Chairman** Nikolai Lenin

First **Spanish Republican Prime Minister** Niceto Alcalá Zamora y Torres

First **Spanish Romantic Composer** Felipe Pedrell

First **Swedish Regent** Earl Birger

First **Swedish Romantic Composer** Ivar Hallström

First **Trinidad and Tobago Prime Minister** Eric Eustace Williams

First **Turkish Sultan** Osman

First **Uruguayan President** José Fructuoso Rivera

First **U.S. Attorney General** Edmund Randolph

First **U.S. Chief Justice** John Jay

First **U.S. Postmaster General** Samuel Osgood

First **U.S. President** George Washington

First **U.S. Secretary of Agriculture** Norman J Colman

First **U.S. Secretary of Commerce** William C Redfield

First **U.S. Secretary of Commerce and Labor** George B Cortelyou

First **U.S. Secretary of Defense** James V Forrestal

First **U.S. Secretary of Education** Shirley Hufstedler

First **U.S. Secretary of Energy** James R Schlesinger

First **U.S. Secretary of Health, Education, and Human Services** Oveta Culp Hobby

First **U.S. Secretary of Health and Human Services** Patricia Roberts Harris

First **U.S. Secretary of Housing and Urban Development** Robert C Weaver

First **U.S. Secretary of the Interior** Thomas Ewing

First **U.S. Secretary of Labor** William B Wilson

First **U.S. Secretary of the Navy** Benjamin Stoddert

First **U.S. Secretary of State** Thomas Jefferson

First **U.S. Secretary of Transportation** Alan S Boyd

First **U.S. Secretary of the Treasury** Alexander Hamilton

First **U.S. Secretary of War** James Knox

First **U.S. Speaker of the House of Representatives** Frederick A C Muhlenberg

First **U.S. Vice President** John Adams

First **Venezuelan President** José Antonio Páez

First **West German Chancellor** Konrad Adenauer

First **Yugoslav President** Josip Broz, better known as Tito

Florida's Highest Point west boundary of Walton County —345 feet, 105 meters

Florida's Largest City Jacksonville

Florida's Largest Subtropical Wilderness Everglades National Park

Florida's Longest River Saint Johns

Florida's Principal Port Tampa

Foremost American Composers Leonard Bernstein; Aaron Copland; Stephen Foster; Louis Moreau Gottschalk; and Edward MacDowell

Foremost Atheist Orator of America Robert G Ingersoll

Foremost Atheist Philosopher of England Bertrand Russell

Foremost Atheist Philosopher of Germany Friedrich Wilhelm Nietzsche

Foremost Austrian Composer Wolfgang Amadeus Mozart

Foremost Belgian Composer César Franck

Foremost Bohemian Composer Antonin Dvořák

Foremost Brazilian Composer Heitor Villa-Lobos

Foremost Classical Composer Johann Sebastian Bach

Foremost Dutch Composer Jan Pieterszoon Sweelinck

Foremost English Composers Sir Edward Elgar, Ralph Vaughan Williams

Foremost Finnish Composer

Jean Sibelius

Foremost French Classical Composer Jean Baptiste Lully

Foremost French Composer Hector Berlioz

Foremost French Modern Composer Maurice Ravel

Foremost German Classical Composer Johann Sebastian Bach

Foremost German Composer Ludwig van Beethoven

Foremost German Modern Composer Richard Strauss

Foremost Hungarian Composer Franz Liszt

Foremost Irish Composer William Vincent Wallace

Foremost Italian Composer Giuseppe Verdi

Foremost Mexican Composer Carlos Chávez

Foremost Norwegian Composer Edvard Hagerup Grieg

Foremost Polish Composer Frédéric François Chopin

Foremost Romanian Composer Georges Enesco

Foremost Romantic Composer Hector Berlioz

Foremost Russian Composer Pyotr Ilyich Tchaikovsky

Foremost Spanish Composer Manuel de Falla

Fourth First Lady Dorothea "Dolley" Payne Todd Madison

France's Easternmost Point Lauterbourg (on the Rhine opposite Karlsruhe)

France's Easternmost Port Menton (at the Italian border on the Mediterranean)

France's Highest Peak Mont Blanc—15,781 feet, 4,782 meters—on the Italian border in the Alps

France's Largest City Paris

France's Largest Metropolitan Region Midi-Pyrénées

France's Longest River Loire

France's Northernmost Point Malo les Bains (on the North Sea at the Belgian border)

France's Northernmost Port Dunkerque (on the Strait of Dover)

France's Principal Port Marseille

France's Smallest Metropolitan Region Alsace (ashore); Corsica (offshore)

France's Southernmost Point Cerbère (opposite Spain's Port Bou on the Mediterra-

nean)

France's Southernmost Port Port Vendres (on the Mediterranean close to the Spanish frontier)

France's Westernmost Point Île d'Ouessant (off the Brittany peninsula)

France's Westernmost Port Brest (on the Brittany peninsula)

French Guiana's Highest Peak Bienvenue—2,723 feet, 825 meters

French Guiana's Largest City Cayenne

French Guiana's Longest River Maroni

French Guiana's Principal Port Cayenne

French Polynesia's Largest City Papeete

French Polynesia's Principal Port Papeete

Gabon's Largest City Libreville

Gabon's Principal Port Banjul

Gabriela Mistral Lucila Godoy Alcayaga

Galápagos Islands' Largest Town Baquerizo Moreno on Chatham; San Cristóbal island

Gambia's Largest City Banjul

Gambia's Principal Port Banjul

Georgia's Highest Point Brasstown Bald—4,784 feet, 1,458 meters

Georgia's Largest City Atlanta

Georgia's Longest River Savannah

Georgia's Most Gorgeous Swamp the Okefenokee it shares with Florida

Georgia's Principal Port Savannah

German Baroque Masters Johann Sebastian Bach, George Frideric Handel, Heinrich Schütz

Germany's Easternmost City Zittau (close to the Czech and Polish borders)

Germany's Highest Peak Zugspitze—9,719 feet, 2,945 meters—in the Bavarian Alps

Germany's Largest City Berlin

Germany's Largest Port Hamburg

Germany's Longest River Elbe

Germany's Northernmost Point List on Sylt (in the North Frisian Islands)

Germany's Principal Ports

Hamburg and Rostack

Germany's Southernmost Region Algäuer Alps (near Oberstdorf on the Austrian border)

Germany's Westernmost City Aachen (Aix-la-Chapelle)— near the Belgian and Netherlands border

Ghana's Largest City Accra

Ghana's Principal Port Accra

Great Britain's Highest Mountain Ben Nevis in Scotland

Great Britain's Highest Peak Ben Nevis—4,406 feet, 1,335 meters—in Scotland

Great Britain's Largest City London

Great Britain's Longest River Severn

Greatest American Composer Charles Ives, according to conductor Leonard Slatkin

Greatest Band in the Land Goldman Band

Greatest Capes of the northern Pacific coast Cape Scott, Cape Cook, Cape Beale, Cape Flattery, Cape Disappointment, Cape Mendocino, Cabo San Lazaro, Cabo San Lucas

Greatest Czechoslovakian Composer Antonin Dvořák

Greatest German Artist Albrecht Dürer

Greatest Guitarist Andrés Segovia

Greatest Island Australia

Greatest Jewish City New York, according to Harry Golden who wrote *Greatest Jewish City in the World*

Greatest Man Who Made a Dictionary Dr Samuel Johnson

Greatest Massacre in Human History 26,300,000 Chinese, according to a Soviet radio broadcast

Greatest Modern Hispanic Poets Ruben Dario, Frederico Garcia Lorca

Greatest Modern Opera Alban Berg's *Wozzeck*

Greatest Opera Bufa Rossini's *Barber of Seville*, according to Verdi

Greatest Opera Ever Written Mozart's *Don Giovanni*

Greatest Political Propagandist in History Thomas Paine

Greatest Predator Man *(Homo sapiens)*

Greatest Primate Gymnasts gibbons (who leap through

the air with the greatest of ease and unaided by any trapeze)

Greatest Roman Catholic Missionary Saint Francis Xavier

Greatest Russian Romantic Composer Pyotr Ilyich Tchaikovsky

Greatest Sculptured Cape of the northern Atlantic coast Cape Cod

Greatest Sculptured Capes of the southern Atlantic coast Cape Hatteras, Cape Lookout, Cape Fear, Cape Canaveral

Greatest Shortcut the Panama Canal

Greatest Show on Earth Barnum & Bailey's three-ring circus (later merged with Ringling Brothers)

Greatest Tropical Killers dysentery, hookworm, malaria, roundworm, sleeping sickness, tapeworm

Greatest Wagnerian Soprano of the 20th Century Kirsten Flagstad

Greatest Woman Poet of Canada Isabella Valancy Crawford

Greatest Woman Poet of Chile Gabriela Mistral

Greatest Woman Poet of Mexico Sor Juana de la Cruz

Greatest Woman Poet in Russian Literature Anna Akhmatova (Anna Andreyevna Gorenko)

Greatest Woman Poet in Venezuelan Literature Irma De Sola de Lovera

Greece's Highest Peak Mount Olympus—9,570 feet, 2,900 meters—legendary abode of the gods

Greece's Largest City Athens

Greece's Principal Port Piraeus

Greenland's Highest Peak Gunnbjorn—12,139 feet, 3,678 meters

Greenland's Largest Town Nuuk (Godthaab)

Greenland's Principal Port Nuuk

Grenada's Highest Point Mount Saint Catherine— 2,749 feet, 833 meters

Grenada's Largest Town St George's

Grenada's Principal Port St George's

Guadeloupe's Largest City Point-à-Pitre

Guadeloupe's Principal Port Basse-Terre

Guam's Best Port Apra on its west coast

Guam's Highest Point Mount Lamlam—1,329 feet, 402 meters

Guam's Largest Town Agaña

Guam's Principal Port Apra

Guatemala's Easternmost Point Puerto Barrios on the Caribbean Sea

Guatemala's Highest Peak Tajumulco—13,816 feet, 4,187 meters—volcano in southeastern sector

Guatemala's Largest City Guatemala City

Guatemala's Largest Department El Petén

Guatemala's Principal Caribbean Port Puerto Barrios

Guatemala's Principal Pacific Port San Jose

Guatemala's Smallest Department Sacatepéquez

Guatemala's Westernmost Point Ocós on the Pacific Ocean near Ciudad Hidalgo

Guinea-Bissau's Largest City Bissau

Guinea-Bissau's Principal Port Bissau

Guinea's Highest Peak Tamgui—4,970 feet, 1,506 meters

Guinea's Largest City Conakry

Guinea's Principal Port Conakry

Guyana's Highest Peak Roraima—9,219 feet, 2,794 meters—mountain shared with Brazil and Venezuela

Guyana's Largest City Georgetown

Guyana's Longest River Essequibo

Guyana's Principal Port Georgetown

Haiti's Highest Peak Massif de la Selle—8,793 feet, 2,665 meters

Haiti's Largest City Port-au-Prince

Haiti's Largest Department l'Artibonite

Haiti's Principal Port Port-au-Prince

Haiti's Smallest Department Nord-Est

Hawaii's Highest Point Mauna Kea—13,796 feet, 4,205 meters—an extinct volcano, on the island of Hawaii

Hawaii's Largest City Honolulu

Hawaii's Principal Port Honolulu

Hawaii's Superlative Volcanoes Kilauea and Mauna Loa in the Hawaii Volcanoes National Park on Hawaii

Heaviest Even-toed Ungulate hippopotamus

Heaviest Living Dog Saint Bernard

Highest American Mountain Mount McKinley

Highest Andean Nation Bolivia

Highest Bridge Royal Gorge Bridge in Colorado

Highest Canadian Mountain Mount Logan

Highest Capital City of Africa Addis Ababa, Ethiopia

Highest Capital City of the Americas La Paz, Bolivia

Highest Capital City of Asia Kathmandu, Nepal

Highest Capital City of Australia Canberra

Highest Capital City of Central America Mexico City

Highest Capital City of Europe Madrid, Spain

Highest Capital City of North America Mexico City

Highest Capital City of South America La Paz, Bolivia

Highest Mountain Ranges and Mountains (in order of height) Himalayas (Everest); Andes (Aconcagua); Rockies (McKinley); Kilimanjaro (Kibo); Caucasus (El'brus); Ellsworth, Antarctica (Vinson Massif); Australian Alps (Kosciusko)

Highest Mountain in the Western Hemisphere Aconcagua —22,835 feet, 6,920 meters —shared by Argentina and Chile

Highest Murder Rate Mexico with 43 registered homicides per 100,000 in 1985

Highest Navigable Lake Titicaca

Highest Peak in the Eastern Hemisphere Everest

Highest Peak in North Africa Djebel Toubkal—13,665 feet, 4,141 meters—in the High Atlas of Morocco

Highest Peak in the Northern Hemisphere Everest

Highest Peak in the Southern Hemisphere Aconcagua

Highest Peak in the USSR *Pik Kommunizma* (Russian—Peak of Communism)—24,590

feet, 7,452 meters—in Central Asia, once named for Stalin
Highest Peak in the Western Hemisphere Aconcagua
Highest Peak in the West Indies Monte Tina—10,301 feet, 3,122 meters—in the Dominican Republic
Highest Point in Africa Kilimanjaro in Tanzania (19,340 feet or 5,963 meters above sea level)
Highest Point in Antarctica Vinson Massif (16,860 feet or 5,140 meters above sea level)
Highest Point in Asia (*see* Highest Point in the World)
Highest Point in Australia Mount Kosciusko in New South Wales (7,328 feet or 2,229 meters above sea level)
Highest Point in Europe Mount Elb'rus in the Caucasus of the USSR (18,567 feet or 5,659 meters above sea level)
Highest Point in North America Mount McKinley in Alaska (20,320 feet or 6,187 meters above sea level)
Highest Point in South America Mount Aconcagua in Argentina (22,835 feet or 6,960 meters above sea level)
Highest Point in the World Mount Everest between Nepal and Tibet in Asia (29,028 feet or 8,848 meters above sea level)
Highest Tides Nova Scotia's Bay of Fundy
Highest Volcanic Island Mountain Mauna Loa (Hawaiian—Long Mountain)—on the island of Hawaii and often ranked as the world's largest active volcano
Highest Waterfalls (in order by height) Angel in Venezuela; Tugela in South Africa; Yosemite in California; Cuquenán in Venezuela; Sutherland in New Zealand; Mardalsfossen in Norway; Ribbon in California; King George VIth in Guyana; Gavarnie in France; Victoria between Zimbabwe and Zambia; Iguazú between Argentina and Brazil; Niagara between Canada and the U.S.
Honduras' Highest Peak *see* Nicaragua's Highest Peak
Honduras' Largest City Tegucigalpa
Honduras' Largest Depart-

ment Gracias a Dios
Honduras' Principal Caribbean Port Puerto Cortés
Honduras' Principal Pacific Port Amapala
Honduras' Smallest Department Islas de la Bahia
Hong Kong's Largest City Victoria
Hottest and Rainiest Continent Africa
Hottest Spot Al'Aziziyah, Libya
Hungary's Highest Point Kekes—3,330 feet, 1,009 meters—in the Matra hills
Hungary's Largest City Budapest
Hungary's Longest River Danube
Iceland's Highest Peak Hvannadalshnjukur—6,952 feet, 2,107 meters
Iceland's Largest City Reykjavik
Iceland's Principal Port Reykjavik
Idaho's Deepest Gorge Hells Canyon within the Grand Canyon of the Snake River with an average depth of 550 feet (1,676 meters) and a maximum of 7,900 feet (2,408 meters) in depth
Idaho's Highest Point Borah Peak—12,662 feet, 3,869 meters
Idaho's Largest City Boise
Illinois' Highest Point Charles Mound—1,235 feet, 377 meters
Illinois' Largest City Chicago
Illinois's Largest Forest Shawnee National Forest extending from the Ohio to the Mississippi rivers in southern Illinois
Illinois' Principal Port Chicago
Indiana's Highest Point Franklin—1,257 feet, 381 meters—in Wayne County
Indiana's Largest Cave Wyandotte Caves complete with five floor levels and an underground mountain
Indiana's Largest City Indianapolis
Indiana's Principal Port Lake Michigan Facility east of Gary
India's Highest Peak Nanda Devi—25,645 feet, 7,771 meters—in the Himalayas
India's Largest City Calcutta
India's Longest River Ganges

India's Principal Port Calcutta
India's Principal East Coast Port Calcutta
India's Principal West Coast Port Bombay
Indo-China's Longest River Mekong
Indonesia's Highest Peak Kinabalu—13,455 feet, 4,077 meters—in North Borneo
Indonesia's Largest City Jakarta
Indonesia's Principal Port Djakarta
Iowa's Highest Point Ocheyedan Mound—1,675 feet, 511 meters—deposited by a melting glacier that contained much sand and gravel
Iowa's Largest City Des Moines
Iran's Highest Peak Damavand—18,602 feet, 5,637 meters
Iran's Largest City Teheran
Iran's Principal Port Khorramshahr
Iraq's Highest Point Kuhe Haji Ebrahim—11,880 feet, 3,600 meters
Iraq's Largest City Baghdad
Iraq's Principal Port Basra
Ireland's Easternmost City Belfast or Dublin
Ireland's Easternmost Point Burr Head or Wicklow Head
Ireland's Highest Mountain Carrauntoohil—3,414 feet, 1,035 meters—*see* Northern Ireland's Highest Peak
Ireland's Largest City Dublin
Ireland's Largest Lake Lough Corrib
Ireland's Longest River Shannon
Ireland's Northernmost City Londonderry
Ireland's Northernmost Point Malin Head
Ireland's Principal Port Dublin
Ireland's Southernmost City Cork
Ireland's Southernmost Point Fastnet Rock
Ireland's Westernmost City Galway
Ireland's Westernmost Point Tearaght Island
Islands of the World (ranked by area) Greenland, New Guinea, Borneo, Madagascar, Baffin Land, Sumatra, Honshu, Great Britain, Victoria, Ellesmere, Sulawesi, New Zealand's South Island,

Java, New Zealand's North Island, Cuba, Newfoundland, Luzon, Iceland, Mindinao, Ireland

Isle of Man's Largest Town Douglas

Israel's Highest Peak Hare Meron—3,963 feet, 1,201 meters—called Jebel Jarmaq by the Arabs

Israel's Largest City Jerusalem

Israel's Principal Port Haifa

Italy's Easternmost Seaport City Trieste (shared with Yugoslavia)

Italy's Easternmost Town Otranto

Italy's Highest Point Monte Bianco (Mont Blanc)—15,781 feet, 4,782 meters—on the French border in the Alps

Italy's Largest City Rome

Italy's Largest Port Genova (Genoa)

Italy's Largest Southern Port Napoli (Naples)

Italy's Longest River Po

Italy's Most Historic Northern Port Venezia (Venice)

Italy's Northernmost Big City Milano (Milan)

Italy's Northernmost Town Brennero (at the Brenner Pass in Upper Adige on the Austrian frontier)

Italy's Principal Port Genoa

Italy's Southernmost Big City Messina (on Sicily)

Italy's Southernmost Town Portopalo (on Sicily)

Italy's Westernmost Big City Torino (Turin)

Italy's Westernmost Town Bardoecchia (on the French frontier)

Ivory Coast's Largest City Abidjan

Ivory Coast's Principal Port Abidjan

Jamaica's Highest Peak Blue Mountain—7,520 feet, 2,279 meters

Jamaica's Largest City Kingston

Jamaica's Largest Parish Saint Elizabeth

Jamaica's Principal Port Kingston

Jamaica's Smallest Parish Kingston

Japan's Easternmost Point Nashappu (on Hokkaido just south of Sakhalin)

Japan's Easternmost Town Habomai (on Hokkaido)

Japan's Highest Peak Fuji—12,388 feet, 3,754 meters—on the island of Honshu, also called Fujisan or Fujiyama

Japan's Largest City Tokyo

Japan's Northernmost Cape and Town Soya (on Hokkaido)

Japan's Principal Port Yokohama

Japan's Southernmost Island and Point Hateruma Shima in the Ryukyu islands

Japan's Westernmost Island Yonaguni Jima (in the Ryukyu islands close to Taiwan)

Japan's Westernmost Town Sonai (on Yonaguni Jima)

Jordan's Highest Peak Jebalal Lawz—8,465 feet, 2,565 meters

Jordan's Largest City Amman

Jordan's Principal Port Aqaba

Kansas' Highest Point Mount Sunflower—4,039 feet, 1,231 meters

Kansas' Largest City Wichita

Kansas' Most Famous Landmark Pawnee Rock on the Santa Fe Trail where Indians and pioneers battled

Kansas' Principal Port Kansas City

Kentucky's Highest Point Black Mountain—4,145 feet, 1,262 meters

Kentucky's Largest City Louisville

Kentucky's Longest River Ohio

Kentucky's Most Scenic Park Mammoth Cave National Park

Kentucky's Principal Port Paducah

Kenya's Highest Peak Mount Kenya—17,040 feet, 5,164 meters—a volcanic cone

Kenya's Largest City Nairobi

Kenya's Principal Port Mombasa

Kiribati's Largest City Bairiki

Kiribati's Principal Port Tarawa

Korea's Highest Peak see North Korea's Highest Peak

Korea's Largest City Seoul

Kuwait's Largest City Kuwait

Kuwait's Principal Port Mina al-Ahmadi

Laos' Highest Peak Phoi Loi—7,425 feet, 2,250 meters

Laos' Largest City Vientiane

Largest Active Volcano Mauna Loa on the island of Hawaii

Largest Afghan City Kabul

Largest African City Cairo

Largest African Nation Sudan in landmass

Largest African Primate gorilla of equatorial Africa

Largest Alabama City Birmingham

Largest Alaska City Anchorage

Largest Albanian City Tirana

Largest Albertan City Edmonton

Largest Algerian City Algiers

Largest American Bird wild turkey (once almost extinct)

Largest American City New York

Largest American City on the Canadian Border Detroit, Michigan

Largest American City on the Mexican Border San Diego, California (opposite Tijuana, Baja California)

Largest American East Coast City New York

Largest American Estuary Chesapeake Bay

Largest American Great Lakes City Chicago

Largest American Gulf Coast City Houston

Largest American Port of Entry San Ysidro, California

Largest American Prison Population California, Texas, New York, Florida, and Ohio

Largest American Samoan City Pago Pago

Largest American Southern City Houston

Largest American State Alaska

Largest American Virgin Island Saint Croix

Largest American West Coast City Los Angeles

Largest Amphibian Chinese giant salamander *(Andreas davidianus)*

Largest Andorran Town Andorra la Vella

Largest Angolan City Luanda

Largest Anteater giant anteater

Largest Anthropoid mountain gorilla

Largest Arboreal Mammals orangutans

Largest Arctic Ocean Island Greenland

Largest Argentine City Buenos Aires

Largest Arizona City Phoenix

Largest Arkansas City Little Rock

Largest Asian City Tokyo

Largest Asian Country China —the mainland People's Republic of China

Largest Asian Nation the USSR

Largest Asian Primate orangutan of Borneo and Sumatra

Largest Atlantic Ocean Island Greenland

Largest Atlantic Port New York (including Brooklyn, New Jersey, and Staten Island ports)

Largest Australian City Sydney

Largest Australian State Western Australia

Largest Austrian City Vienna

Largest Bahaman City Nassau

Largest Bahraini City Manama

Largest Balkan City Athens

Largest Baltic Sea Island Gotland

Largest Bangalee City Dacca (capital of Bangladesh)

Largest Barbadian City Bridgetown (capital of Barbados)

Largest Basotho City Maseru (capital of Lesotho)

Largest Belgian City Brussels

Largest Belizian City Belize

Largest Beninois City Benin

Largest Bermudan City Hamilton

Largest Bhutanese Town Thimphu

Largest Bolivian City La Paz

Largest Botswanan City Gaborone

Largest Brazilian City São Paulo

Largest British Bird of Prey golden eagle

Largest British Carnivore common otter

Largest British City London

Largest British Columbian City Vancouver

Largest British Deer red deer

Largest British Marine Bird the cormorant or the gannet

Largest British Marine Mammal Atlantic seal also called grey seal

Largest British Wading Bird heron

Largest Bulgarian City Sofia

Largest Burman City Rangoon

Largest Burundian City Bujumbura

Largest California City Los Angeles

Largest Cambodian City Phnom Penh (capital of Cambodia or Kampuchea)

Largest Cameroonian City Douala

Largest Canadian Arctic Island Baffin

Largest Canadian City Montreal

Largest Canadian City on the American Border Windsor, Ontario

Largest Canadian Great Lakes City Toronto

Largest Canadian Province Québec

Largest Canadian Provincial Capital Toronto, Ontario

Largest Canadian West Coast City Vancouver

Largest Cape Verde Island Town Praia

Largest Capital of the Eastern World Tokyo

Largest Capital of the Western World Mexico City

Largest Caribbean Sea Islands (by size) Cuba, Hispaniola (Dominican Republic and Haiti), Jamaica, Puerto Rico, Trinidad

Largest Central African Empire City Bangui

Largest Central American Nation Mexico

Largest Central American Primate mantled howler monkey

Largest Central American Republic Nicaragua

Largest Chadian City N'Djamena

Largest Channel Island Jersey

Largest Chilean City Santiago de Chile

Largest Chinese City Shanghai

Largest Cities in North America Mexico City and New York City

Largest City in Africa Cairo, Egypt

Largest City in Alberta Edmonton

Largest City in Asia Tokyo, Japan

Largest City in Australia Sydney

Largest City in Brazil São Paulo

Largest City in the British Isles London

Largest City in California Los Angeles

Largest City in Canada Montreal

Largest City in the Canadian Northwest Territories Yellowknife

Largest City in China Shanghai

Largest City in Europe Paris

Largest City in India Calcutta

Largest City in Indonesia Jakarta

Largest City in Italy Rome

Largest City in Japan Tokyo

Largest City in the Largest State Anchorage, Alaska

Largest City in Latin America Mexico City

Largest City in Manitoba Winnipeg

Largest City in the Middle East Teheran, Iran

Largest City in the Middle West Chicago, Illinois

Largest City in New Brunswick Saint John

Largest City in Newfoundland St John's

Largest City in North America New York

Largest City in Northern Ireland or Ulster Belfast

Largest City in Nova Scotia Halifax

Largest City in Ontario Toronto

Largest City in the Orient Tokyo, Japan

Largest City in the Philippines Manila

Largest City in Prince Edward Island Charlottetown

Largest City in Saskatchewan Regina

Largest City in the South Houston, Texas

Largest City in South America Buenos Aires, Argentina

Largest City in Spain Madrid

Largest City in the USSR Moscow

Largest City in the Yukon Whitehorse

Largest Clipper Ships *Great Republic* built in Boston by Donald McKay and launched in 1853 as the world's largest ship

Largest Coffee Port Santos, Brazil

Largest Colombian City Bogotá

Largest Colorado City Denver

Largest Communist State Peoples' Republic of China

Largest Comoran Town Moroni

Largest Congolese or Zairian City Kinshasa

Largest Connecticut City Hartford
Largest Continent Asia
Largest Costa Rican City San José
Largest Cuban City Habana
Largest Cypriot City Nicosia
Largest Czechoslovakian City Prague
Largest Dahomian City Porto Novo
Largest Danish Arctic Island Greenland
Largest Danish City Copenhagen
Largest Deer moose
Largest Delaware City Wilmington
Largest Djiboutian City Djibouti
Largest Dominican City Santo Domingo
Largest Dutch City Rotterdam
Largest East European Country Soviet Union
Largest Ecuadorean City Guayaquil
Largest Eggs laid by North African ostriches
Largest Egyptian City Cairo
Largest English City London
Largest English East Coast City Manchester
Largest Equatorial Guinean City Bata
Largest Estonian City Tallinn
Largest Ethiopian City Addis Ababa
Largest European City Paris
Largest European Island Great Britain
Largest European Nation Russia
Largest-eyed Primate spectral tarsier of Indonesia and the Philippines
Largest, Fastest, and Most Powerful Present-day Bird ostrich
Largest Fijian City Suva
Largest Filipino City Manila (capital of the Philippines)
Largest Finnish City Helsinki
Largest Fish Market in Tokyo
Largest Flightless Land Bird South African ostrich
Largest Florida Metropolitan Area Miami
Largest Flower rafflesia (found in Indonesia and Malaysia by Sir Stamford Raffles; often 3 feet, or 900 millimeters, in diameter)
Largest Flying Animal extinct pterodactyl *(Quetzocoatlus northropi)*

Largest Flying Bird condor (ranging from the Andes to California and in danger of extinction)
Largest French City Paris
Largest French Guianese City Cayenne
Largest French Polynesian City Papeete, Tahiti
Largest French West Indian City Fort-de-France, Martinique
Largest Freshwater Lake in Africa Victoria between Kenya, Tanzania, and Uganda
Largest Freshwater Lake in Central America Lake Nicaragua between Costa Rica and Nicaragua
Largest Freshwater Lake of Eurasia Baikal in the USSR
Largest Freshwater Lake in Europe Balaton in Hungary
Largest Freshwater Lake in North America Superior between Canada and the U.S.
Largest Freshwater Lake in South America Titicaca between Bolivia and Peru
Largest Freshwater Lake in the World Superior in North America
Largest Gabonese City Libreville
Largest Galápagos Island Albemarle, officially Isabela
Largest Gambian City Banjul
Largest Garden City in the Netherlands Apeldoorn
Largest of the Geese Canada goose
Largest Georgia City Atlanta
Largest German City Berlin
Largest Ghanian City Accra
Largest Gibraltarian City Gibraltar
Largest Gorge Grand Canyon of the Colorado River in Arizona
Largest Greek City Athens
Largest Greenland Settlement Godthaab
Largest Grenadan City St George's
Largest Guadeloupian City Point-à-Pitre
Largest Guam City Agaña
Largest Guatemalan City Guatemala City
Largest Guinea-Bissauan City Bissau
Largest Guinean City Conakry
Largest Guyanese City Georgetown (capital of Guyana)
Largest Haitian City Port-au-

Prince
Largest Hawaii City Honolulu
Largest Hispanic City Mexico City
Largest Honduran City Tegucigalpa
Largest Hospital Ship USS *Mercy*—United States Navy seagoing hospital
Largest Hungarian City Budapest
Largest Icelandic City Reykjavik
Largest Idaho City Boise
Largest Illinois City Chicago
Largest Indiana City Indianapolis
Largest Indian City Calcutta
Largest Indian Ocean Island Madagascar
Largest Indian Ocean Nation Australia
Largest Indian Ocean Port Singapore
Largest Indonesian City Jakarta
Largest Indonesian Island Borneo (Kalimantan)
Largest Industrial City in Mexico Monterrey, Nuevo León
Largest Invertebrate octopus (eight-armed cephalopod sometimes with an armspan of 9 meters or 30 feet)
Largest Iowa City Des Moines
Largest Iranian City Teheran
Largest Iraqi City Baghdad
Largest Irish City Dublin
Largest Island in Australasia New Guinea
Largest Island in Indonesia Borneo (Kalimantan)
Largest Island in the West Indies Cuba
Largest Island in the World Greenland
Largest Island Mountain Mauna Kea (Hawaiian—White Mountain)
Largest Islands (by size) Greenland, New Guinea, Borneo, Madagascar, Baffin, Sumatra, Honshu, Great Britain, Ellesmere, Victoria, Celebes (Sulawesi)
Largest Israeli City Tel Aviv-Yafo
Largest Italian City Rome
Largest Ivoirian City Abidjan (capital of the Ivory Coast)
Largest Jail Complex in the Western democracies Rikers Island, New York
Largest Jamaican City Kingston

Largest Japanese City Tokyo
Largest Japanese Island Honshu
Largest Jordanian City Amman
Largest Kampuchean City Phnom Penh
Largest Kansas City Wichita
Largest Kentucky City Louisville
Largest Kenyan City Nairobi
Largest Korean City Seoul
Largest Kuwaiti City Kuwait City
Largest Lakes (by size) Caspian Sea, Asia; Superior, North America; Victoria, Africa; Aral Sea, USSR; Huron, North America; Tanganyika, Africa; Great Bear, Canada; Baykal, USSR; Nyasa, Africa
Largest Lake in the World Caspian Sea
Largest Land Mammal African bush elephant
Largest Lao City Vientiane
Largest Latin American Republic Brazil
Largest Latvian City Riga
Largest Lebanese City Beirut
Largest Lesothian City Maseru
Largest Liberian City Monrovia
Largest Libyan City Tripoli
Largest Liechtensteiner Town Vaduz
Largest Lithuanian City Vilna or Vilnius
Largest Living Amphibian Japanese giant salamander
Largest Living Antelope north-central Africa's giant eland
Largest Living Bat Kalong fruit bat of Indonesia and Malaysia
Largest Living Bird North African ostrich
Largest Living Canine the endangered timber wolf of the wild (the largest domestic dog is the Saint Bernard while the tallest is the Great Dane or the Irish Wolfhound)
Largest Living Carnivore brown bear, also called the Kodiak bear
Largest Living Clam giant clam of the South Pacific Ocean up to 500 pounds (227 kg) and 5 feet (150 cm) long
Largest Living Crocodilian salt-water crocodile *(Crocodylus porosus)*
Largest Living Crustacean

Japanese spider crab
Largest Living Deer American moose (called elk in Europe)
Largest Living Dog Saint Bernard
Largest Living Elephant African elephant
Largest Living Feline long-furred Manchurian or Siberian tiger
Largest Living Fish whale shark
Largest Living Freshwater Fish Mekong River catfish (the Pla Buk of Laos)
Largest Living Freshwater Terrapin alligator snapping terrapin of the Mississippi River region
Largest Living Frog giant or goliath frog found in the Cameroons of Africa
Largest Living Game Bird peafowl
Largest Living Horse Belgian stallion
Largest Living Land Animal African elephant
Largest Living Land Mammal African elephant
Largest Living Land Tortoise gigantic Galápagos tortoise
Largest Living Lizard Komodo dragon *(Varanus komodoensis)* of Indonesia
Largest Living Mammal blue or sulfur-bottom whale
Largest Living Marine Carnivore the walrus
Largest Living Marsupial Australian red kangaroo
Largest Living Monotreme duck-billed platypus
Largest Living Penguin emperor penguin
Largest Living Primate African mountain gorilla
Largest Living Rabbit Flemish great rabbit
Largest Living Reptile salt-water crocodile
Largest Living Rhinoceros Indian rhinoceros
Largest Living Rodent capybara rat
Largest Living Salamander giant salamander of China and Japan
Largest Living Sea Turtle trunkback *(Dermochelys coriacea)*
Largest Living Shark whale shark
Largest Living Snake regal python *(Python reticulatus)*
Largest Living Species of

Shark 60-foot-long (18-meter) whale shark
Largest Living Starfish Gulf of Mexico's bristling starfish *(Midgardia xandaros)*
Largest Living Terrestrial Carnivore Kodiak Island brown bear
Largest Living Tiger Siberian tiger
Largest Living Tortoise Galapagos tortoise sometimes 4 feet (120 cm) long weighing 500 pounds (225 kg)
Largest Living Turtle leatherback turtle
Largest, Longest Coral Reef Australia's Great Barrier Reef extending 1,200 miles (1,931 km) along the northeast coast
Largest, Longest Live Snake *Colossus*, a reticulated python in the Pittsburgh Zoo; measured 29 feet and weighed 300 pounds (136 kg and 8.8 meters)
Largest Louisiana City New Orleans
Largest of the Low Countries Netherlands
Largest Luxembourger City Luxembourg
Largest Maine City Portland
Largest Malagasy City Tananarive (capital of Madagascar)
Largest Malawian City Blantyre-Limbre
Largest Malaysian City Kuala Lumpur
Largest Maldivian Port Male
Largest Male Prison in Texas Huntsville, north of Houston
Largest Malian City Bamako (capital of Mali)
Largest Maltese City Sliema
Largest Manitoban City Winnipeg
Largest Man-Made Lake Gatun Lake watering the Panama Canal
Largest Marine Mammal blue or sulfur-bottom whale
Largest Marsupial red kangaroo
Largest Maryland City Baltimore
Largest Massachusetts City Boston
Largest Mauritanian City Nouakchott
Largest Mauritanian City Port Louis (capital of Mauritius)
Largest Mediterranean Port Marseille
Largest Mediterranean Sea Island Sicily

Largest Mexican City Mexico City

Largest Mexican City on the American Border Ciudad Juárez, Chihuahua

Largest Mexican City in Jalisco Guadalajara (Mexico's second largest city)

Largest Mexican City on the Mexican Border Ciudad Juárez (opposite El Paso, Texas)

Largest Michigan City Detroit

Largest Middle American Country Mexico

Largest Minnesota Metropolitan Area Minneapolis-St Paul

Largest Mississippi City Jackson

Largest Mississippi River City St Louis

Largest Missouri City Kansas City

Largest Monacan or Monagasque City Monaco

Largest Mongolian City Ulan Bator

Largest Montana City Billings

Largest Moroccan City Casablanca

Largest Mozambican City Maputo

Largest Namibian City Walvis Bay

Largest Nation the USSR

Largest National Areas USSR, Canada, China, U.S.A., Brazil, Australia, India, Argentina, Sudan, Mongolia

Largest Nation in the South Pacific Australia

Largest Nauruan Town Yaren

Largest Nebraska City Omaha

Largest Nepalese City Katmandu

Largest Netherlandic Antillean City Willemstad, Curaçao

Largest Netherlandic City Rotterdam

Largest Nevada City Las Vegas

Largest New Brunswick City Saint John

Largest New England City Boston

Largest Newfoundland City St John's

Largest New Hampshire City Manchester

Largest New Jersey City Newark

Largest New Mexico City Albuquerque

Largest New York State City New York City

Largest New Zealand City Auckland

Largest Nicaraguan City Managua

Largest Nigerian City Lagos

Largest Nigerois City Niamey (capital of Niger)

Largest North American City New York

Largest North American Land Mammal bison, buffalo

Largest North American Nation Canada

Largest North American Turtle alligator snapper

Largest North Atlantic Ocean Nation Canada

Largest North Carolina City Charlotte

Largest North Dakota City Fargo

Largest North Pacific Ocean Nation USSR

Largest North Pacific Ports Los Angeles (including Long Beach, San Pedro, and Wilmington) and Yokohama (including Kawasaki, Tokyo, and Yokosuka)

Largest Northwest Territory Town Yellowknife

Largest Norwegian Arctic Island Svalbard

Largest Norwegian City Oslo

Largest Nova Scotian City Halifax

Largest Oceanic Areas Pacific, Atlantic, Indian, and Arctic followed by the Mediterranean, South China, Bering, and Caribbean seas; the Gulf of Mexico and the Sea of Okhotsk; the East China and Yellow seas; Hudson Bay; the Sea of Japan; the North, Black, Red, and Baltic seas

Largest Oceanic Bird Pacific albatross

Largest of the Oceanic Dolphins killer whale or grampus (*Orcinus orca*)

Largest Oceanic Nation Indonesia

Largest Officially Atheist Nation the USSR

Largest of the Forty-eight Contiguous States Texas

Largest Ohio City Cleveland

Largest Oil Tanker *Seawise Giant* under Liberian registry and 1504 feet (459 meters) overall length

Largest Oklahoma State City Oklahoma City

Largest and Oldest Subway

System New York City

Largest Omani City Matrah

Largest Ontarian City Toronto

Largest Order of Living Birds the perching birds (containing some 6000 species)

Largest Oregon City Portland

Largest Owl fish owl of northernmost Japan

Largest Pacific Ocean Island Papua New Guinea

Largest Pacific Ocean Nation China

Largest Pakistani City Karachi

Largest Panamanian City Panamá City

Largest Papuan City Port Moresby (capital of Papua New Guinea)

Largest Paraguayan City Asunción

Largest Passenger Ship *Norway* under Norwegian registry and 1035 feet (316 meters) overall length

Largest Penguin Emperor

Largest Pennsylvania City Philadelphia

Largest Peruvian City Lima

Largest Philippine City Manila

Largest Planet Jupiter

Largest Polish City Warsaw

Largest Polynesian City Papeete, Tahiti

Largest Polynesian City Outside Polynesia Auckland, New Zealand

Largest Population China (followed by India, the USSR, U.S.A., Pakistan, Indonesia, Japan, Brazil, Germany, UK, Italy, France, Turkey, Spain, Poland)

Largest Port in Southeast Asia Singapore

Largest Portuguese City Lisbon

Largest Primate central Africa's *Gorilla gorilla*

Largest Prince Edward Island City Charlottetown

Largest Producer of Wool Australia's sheep

Largest Province in Canada Québec

Largest Puerto Rican City San Juan

Largest Qatari City Doha

Largest Québecois City Montreal

Largest Rhode Island City Providence

Largest Rhodesian City Salisbury

Largest Rodent capybara

Largest Romanian City Bucharest

Largest Royal Capital of the Eastern World Tokyo

Largest Royal Capital of the Western World London

Largest Russian Arctic Island Novaya Zemlya

Largest Russian City Moscow

Largest Rwandan City Kigali

Largest Saltwater Lake In Asia Aral in the USSR

Largest Saltwater Lake in Australia Torrens in South Australia

Largest Saltwater Lake in the World Caspian Sea in Eurasia

Largest Salvadorean City San Salvador

Largest Samoan City Apia

Largest Sanmarinese Town San Marino

Largest São Tome and Principe Town São Tome

Largest Saskatchewan City Regina

Largest Saudi Arabian City Riyadh

Largest Scottish City Glasgow

Largest Seas (by area) South China, Caribbean, Mediterranean, Bering, Gulf of Mexico, Sea of Okhotsk, Sea of Japan, Hudson Bay, Andaman, Black, Red, North, Baltic

Largest Senegalese City Dakar

Largest Seychelles Port Port Victoria

Largest Sierra Leonean City Freetown

Largest Singaporean City Singapore

Largest Solomon Island Port Honiara

Largest Somali City Mogadishu

Largest South African City Johannesburg

Largest South American City Buenos Aires

Largest South American Country Brazil

Largest South American Nation Brazil

Largest South American Primate muriquis of Brazil

Largest South Atlantic Ocean Nation Brazil

Largest South Carolina City Columbia

Largest South China Sea City Hong Kong

Largest South Dakota City Sioux Falls

Largest Southeast-Asian City Singapore

Largest South Pacific Ocean Nation Australia

Largest South Pacific Port Sydney

Largest South Yemeni City Aden

Largest Soviet City Moscow

Largest Spanish City Madrid

Largest Sri Lankan City Colombo

Largest State East of the Mississippi Georgia

Largest State in the United States Alaska

Largest State West of the Mississippi Alaska

Largest Sudanese City Khartoum

Largest Surinamese City Paramaribo

Largest Swaziland City Mbabane

Largest Swedish City Stockholm

Largest Swiss City Zurich

Largest Syrian City Damascus

Largest Tanzanian City Dar-es-Salaam

Largest Tennessee City Memphis

Largest Texas City Houston

Largest Texas Metropolitan Area Dallas-Ft Worth

Largest Thai City Bangkok

Largest Toad marine toad (Bufo marinus) of South America

Largest Togolese Town Lome (capital of Togo)

Largest Tongan Town Nuku'alofa

Largest Trinidadian City Port-of-Spain

Largest Tunisian City Tunis

Largest Turkish City Istanbul

Largest Turtle leatherback or trunkback sea turtle

Largest Ugandan City Kampala

Largest Ulster City Belfast

Largest United Arab Emirates City Dubai

Largest United Kingdom City London

Largest Upper Voltan City Ouagadougou

Largest Uruguayan City Montevideo

Largest Utah City Salt Lake City

Largest Venezuelan City Caracas

Largest Vermont City Burlington

Largest Vietnamese City Ho Chi Minh City (Saigon)

Largest Village in Europe The Hague

Largest Virginia City Norfolk

Largest Virgin Island City Charlotte Amalie

Largest Volcanos (ranked by height) Aconcagua in Argentina, Lullaillaco in Chile, Chimborazo and Cotopaxi in Ecuador, Kilimanjaro in Tanzania, Antisana in Ecuador, Citlaltepetl in Mexico, Elbruz in the USSR, Demavend in Iran, Popocatapetl in Mexico, Kluchevskaya in the USSR, Karisimbi in Rwanda and Zaire, Wrangell in Alaska, Mauna Loa in Hawaii, Cameroon in Cameroon, Fujiyama in Japan, Erebus in Antarctica, Pico de Teyde in the Canary Islands, Semerou in Indonesia, Nyiragongo in Zaire, Iliamna in Alaska, Etna in Italy, Baker in Washington, Chillan in Chile, Nyamuragira in Zaire, Haleakala in Hawaii, Villarica in Chile, Ruapehu in New Zealand, Paricutin in Mexico

Largest Washington State City Seattle

Largest Welsh City Cardiff

Largest West European Country France

Largest West Indian City Havana, Cuba

Largest West Indian Nation Cuba

Largest West Virginia City Charleston

Largest Whirlpool in the Western Hemisphere off Eastport, Maine

Largest Wisconsin City Milwaukee

Largest Wyoming City Cheyenne

Largest Yemeni City Sana

Largest Yugoslav City Belgrade

Largest Yukon Territory City Whitehorse

Largest Zairian City Kinshasa

Largest Zambian City Lusaka

Largest Zimbabwe-Rhodesian City Salisbury

Largest Zoological Garden in the World San Diego Zoo

Last American Romantic Composer Edward MacDowell

Last Austrian Romantic Composer Anton Bruckner

Last **Belgian Romantic Composer** César Franck
Last **Bohemian Romantic Composer** Gustav Mahler
Last **Brazilian Romantic Composer** Heitor Villa-Lobos
Last **Czech Romantic Composer** Josef Suk
Last **Danish Romantic Composer** Carl Nielsen
Last **Dutch Romantic Composer** Johan Wagenaar
Last **Emperor of Austria–Hungary** Charles I (1916–1918)
Last **Emperor of Brazil** Dom Pedro II (1831–1889)
Last **Emperor of China** Henry Pu-yi (K'ang Ti) 1934–1945
Last **Emperor of Ethiopia** Haile Selassie (1930–1974)
Last **Emperor of France** Napoleon I, Bonaparte (1804–1815)
Last **Emperor of Germany** Wilhelm II (1888–1918)
Last **Emperor of Mexico** Austrian Archduke Ferdinand Maximilian (1864–1867)
Last **English Romantic Composer** Sir Edward Elgar
Last **Finnish Romantic Composer** Jean Sibelius
Last **French Romantic Composer** Camille Saint-Saëns
Last **German Romantic Composer** Richard Strauss
Last of the **Great Romantic Composers**—Carlos Chavez, Mexico; Claude Achille Debussy, France; Antonin Dvorák, Czechoslovakia; Georges Enesco, Romania; Manuel de Falla, Spain; Edvard Hagerup Grieg, Norway; Howard Hanson, United States; Gustav Mahler, Austria; August Nielsen, Denmark; Giacomo Puccini, Italy; Sergei Rachmaninoff, Russia; Richard Strauss, Germany; Ralph Vaughan Williams, Britain; and Heitor Villa-Lobos, Brazil
Last **Hungarian Romantic Composer** Ernst von Dohnanyi
Last **Irish Romantic Composer** Charles Stanford
Last **Italian Romantic Composer** Giacomo Puccini
Last **Mexican Romantic Composer** Carlos Chávez
Last **Norwegian Romantic Composer** Edvard Grieg

Last **Place on Earth** South Pole, Antarctica
Last **Polish Romantic Composer** Stanislaw Moniuszko
Last **Romanian Romantic Composer** Georges Enesco
Last **Russian Romantic Composer** Sergei Vasilievich Rachmaninoff
Last **Spanish Romantic Composer** Manuel de Falla
Last **Swedish Romantic Composer** Kurt Atterberg
Last *Temptation of Christ* novel by Nikos Kazantzakis
Last **Viceroy** Louis Mountbatten, Admiral of the Fleet and 1st Earl Mountbatten of Burma
Least **Populous Canadian Province** Prince Edward Island
Least **Populous State** Alaska
Lebanon's **Highest Peak** Quernet es Sauda—10,131 feet, 3,070 meters
Lebanon's **Largest City** Beirut
Lebanon's **Principal Port** Beirut
Lesotho's **Largest City** Maseru
Liberia's **Highest Peak** Yekepa —5,748 feet, 1,742 meters— in Nimba Mountains
Liberia's **Largest City** Monrovia
Liberia's **Principal Port** Monrovia
Libya's **Highest Peak** Kegueur Terbi—10,335 feet, 3,132 meters
Libya's **Largest City** Tripoli
Libya's **Principal Port** Tripoli
Liechtenstein's **Highest Peak** Naafkopf—8,440 feet, 2,558 meters)
Liechtenstein's **Largest City** Vaduz
Loneliest **Man in the World** Leonardo da Vinci, according to George Santayana
Longest **African River** Nile
Longest **American River** Missouri-Mississippi
Longest **Australian River** Murray-Darling
Longest **Bridge Span** Verrazano–Narrows Bridge, New York
Longest **Canadian River** Mackenzie-Peace
Longest **Chinese River** Yangtze
Longest **Crocodile** saltwater crocodile attaining more than 20 feet (6 meters)
Longest **Eastern Siberian**

River Lena
Longest **English Novel** Richardson's *Clarissa*
Longest **Fjord** Sognefjord (extending 110 miles or 175 kilometers into the heart of Norway)
Longest **Highway Tunnel** Mont Blanc Tunnel between France and Italy
Longest **Indo-Chinese River** Mekong
Longest **Living Vertebrates** Galápagos tortoises
Longest **Lizard** 10-foot (3-meter)-long Komodo dragon lizard of Indonesia
Longest **Natural Beach in the U.S.** Long Beach, Oregon
Longest **North American River** Mississippi-Missouri
Longest **Northeast Asiatic River** Amur
Longest **and Oldest Shopping Street** Copenhagen's Stroget
Longest **Otter** Giant South American Otter attaining 7 feet (213 cm) in length
Longest **Peninsulas (by length)** Scandinavia, Arabia, Kamchatka, Labrador, Baja California, Balkans, Iberia, Korea, Malay, Alaska, Italy, Florida, Yucatan, Nova Scotia, La Guaira
Longest **Railroad in America** Burlington Northern from Vancouver, BC to Mobile, Alabama
Longest **Railroad Tunnel** Simplon Tunnel between Italy and Switzerland
Longest **Rivers** Nile, Amazon, Mississippi-Missouri-Red Rock, Ob-Irtysh, Yangtze, Hwang Ho, Congo, Amur, Lena, Machenzie, Mekong, Niger, Yenisei, Paraná, Plata-Paraguay, Volga, Madeira, St Lawrence, Rio Grande, Orinoco, Yukon, Danube, Euphrates, Murray, Ganges, Irrawaddy, Dneiper, Negro, Don, Orange, Pechora, Marañon, Dneister, Rhine, Donets, Elbe, Gambia, Yellowstone, Vistula, Tagus (Tajo), Oder, Maas (Meuse), Seine, Guadalquivir, Hudson, Thames, Moldau, etc.
Longest **Snake** South American anaconda measuring 25 feet (7.9 meters) or more; *see* Largest, Longest Snake
Longest **South American River** Amazon

Longest Southern African River Congo

Longest Southern South American River Paraná

Longest Train Trip in the World on the Trans-Siberian railway from Moscow in eastern Europe to Nakhodka near Vladivostok on the North Pacific coast

Longest Undersea Tunnel Japan's 33-mile (54-km)-long railway tunnel linking Honshu with Hokkaido

Longest Venemous Snake sea snake of western Australia *(Hydrophis belcheri)*

Longest West African River Niger

Longest Western Russian River Volga

Longest Western Siberian River Ob'-Irtysh

Louisiana's Highest Point Driskill Mountain—535 feet, 163 meters—near Arcadia

Louisiana's Largest City New Orleans

Louisiana's Largest Lake Pontchartrain on the Greater New Orleans Expressway

Louisiana's Longest River Mississippi

Loveliest Fleet of Islands Hawaiian Islands

Lowest Birth Rate Country in the Industrialized World Italy

Lowest Murder Rate Sikkim —an Himalayan protectorate of India

Lowest Place in Africa Lake Assal in Djibouti (512 feet or 156 meters below sea level)

Lowest Place in Antarctica unknown

Lowest Place in Asia (*see* Lowest Place in the World)

Lowest Place in Australia Lake Eyre in South Australia (52 feet or 16 meters below sea level)

Lowest Place in Eastern Europe Caspian Sea between Iran and the USSR (92 feet or 28 meters below sea level)

Lowest Place in North America Badwater in Death Valley between California and Nevada (286 feet or 87.5 meters below sea level)

Lowest Place in South America Valdes Peninsula of Argentina (131 feet or 40 meters below sea level)

Lowest Places in Western Eu-rope coastal areas of the Netherlands (15 feet or 4.6 meters below sea level)

Lowest Place in the World Dead Sea between Israel and Jordan (1,302 feet or 397 meters below sea level)

Luxembourg's Largest City Luxembourg-Ville

Macau's Largest City Macau City

Madagascar's Highest Peak Tsaratanana Massif—9,450 feet, 2,864 meters

Madagascar's Largest City Antananarivo (Tananarive)

Madagascar's Principal Port Tamatave

Madeira's Largest City Funchal

Maine's Highest Point Mount Katahdin—5,268 feet, 1,605 meters

Maine's Largest City Portland

Maine's Longest River Penobscot

Maine's Outermost Islands Matinicus, Monhegan, Ragged, and Seal islands

Maine's Principal Port Portland

Malawi's Largest City Balantyre-Limbe

Malaysia's Highest Peak Gunong Tahan—7,224 feet, 2,189 meters—east of the Cameron Highlands

Malaysia's Largest City Kuala Lumpur

Malaysia's Principal Port George Town

Maldives' Largest City Malé

Maldives' Principal Port Malé Atoll

Mali's Largest City Bamako

Malta's Largest City Valletta

Malta's Principal Port Valletta

Manitoba's Highest Point Mount Baldy—2,729 feet, 832 meters

Manitoba's Largest City Winnipeg

Man's Most Famous Best Friend Nipper, the shorthaired fox terrier adorning the HMV and Victrola record labels

Marshall Islands' Largest Town Majuro

Martinique's Highest Peak Mont Pelée—4,429 feet, 1,342 meters

Martinique's Largest City Fort-de-France

Martinique's Principal Port Fort-de-France

Maryland's Highest Point Backbone Mountain—3,360 feet, 1,024 meters—between Virginia and West Virginia

Maryland's Largest City Baltimore

Maryland's Longest River Potomac

Maryland's Principal Port Baltimore

Maryland's Unspoiled Seashore Assateague Island National Seashore

Massachusetts's Best Known Lake Walden Pond, immortalized by philosopher-naturalist, Henry David Thoreau

Massachusetts' Highest Point Mount Greylock—3,491 feet, 1,064 meters—in the Berkshires

Massachusetts' Largest City Boston

Massachusetts' Longest River Connecticut

Massachusetts' Principal Port Boston

Mauritania's Largest City Nouakchott

Mauritania's Principal Port Nouakchott

Mauritius' Largest City Port-Louis

Mauritius' Principal Port Port-Louis

Mexico's Easternmost City Chetumal, Quintana Roo

Mexico's Easternmost Point Cabo Catoche, Quintana Roo (first Mexican site discovered by the Spaniards)

Mexico's Highest Peak Citlatepetl (Orizaba)—18,697 feet, 5,666 meters

Mexico's Largest City Mexico City

Mexico's Largest State Chihuahua

Mexico's Longest River Rio Bravo, also called Rio Grande

Mexico's Northernmost City Mexicali, Baja California

Mexico's Northernmost Point Los Algodones, Baja California

Mexico's Principal Gulf of Mexico Port Veracruz

Mexico's Principal West Coast Port Mazatlan

Mexico's Smallest State Tlaxcala

Mexico's Southernmost City Tapachula, Chiapas

Mexico's Southernmost Point Ciudad Hidalgo in the state of Chiapas

Mexico's Westernmost City Tijuana, Baja California

Mexico's Westernmost Point Playas de Tijuana, Baja California

Michigan's Highest Point Mount Curwood—1,980 feet, 604 meters—in the Upper Peninsula

Michigan's Largest City Detroit

Michigan's Principal Port Detroit

Michigan's Unspoiled Forest Wilderness Hiawatha National Forest

Micronesia's Largest Place Kolonia on Ponape

Middle America's Largest Country Mexico

Minnesota's Highest Point Eagle Mountain—2,301 feet, 701 meters—in Cook County

Minnesota's Largest City Minneapolis

Minnesota's Longest River Red

Minnesota's Northwesternmost Angle Lake of the Woods

Minnesota's Principal Port Duluth

Mississippi's Highest Point Woodall Mountain—806 feet, 246 meters—in Tishomingo County

Mississippi's Largest City Jackson

Mississippi's Longest River Mississippi

Mississippi's Most Scenic Highway Natchez Trace Parkway linking Alabama, Mississippi, and Tennessee

Mississippi's Principal Port Pascagoula

Missouri's Highest Point Taum Sauk Mountain—1,772 feet, 541 meters

Missouri's Largest City St Louis

Missouri's Longest River Mississippi

Missouri's Principal Port St Louis

Missouri's Superb Riverway Ozark National Scenic Riverways in southeastern Missouri's Ozark Mountains

Monaco's Largest City Monaco-Ville

Monaco's Principal Port Monaco

Mongolia's Highest Peak Tabun Bogdo—15,266 feet, 4,626 meters

Mongolia's Largest City Ulaanbaatar

Montana's Finest Mountain Country Glacier National Park

Montana's Highest Point Granite Peak—12,799 feet, 3,902 meters—in Park County

Montana's Largest City Billings

Montserrat's Largest Place Plymouth

Morocco's Highest Peak see Highest Peak in North Africa

Morocco's Largest City Casablanca

Morocco's Principal Port Tangier

Most Active Volcano in the Continental United States Mount St Helens in southwestern Washington close to Portland, Oregon

Most Admired Fictional Detective Sherlock Holmes

Most Agnostic and Atheistic Century 18th—the Age of Voltaire, the Encyclopedists, and Paine's *Age of Reason*

Most Amazing of All Composers Wolfgang Amadeus Mozart

Most Beautiful College Town in America Princeton, New Jersey

Most Central Continent Africa

Most Dangerous Rattlesnake Eastern North American Diamondback

Most Decent Man in Politics Hubert H Humphrey

Most Densely Populated Nation in Mainland Latin America El Salvador

Most Elegant Salt-Marsh Terrapin commercially cultivated diamondback terrapin of the Carolinas and Georgia

Most Eloquent Englishman Sir Winston Churchill

Most Endangered Species of Sea Turtle Hawksbill or Tortoise-shell Turtle *(Eretmochelys imbricata)*

Most English Town Outside England Christchurch, South Island, New Zealand

Most Exotic West Indian Island Martinique

Most Expensive Cities see Fifty Most Expensive Cities in the Late 1900s

Most Famous Operatic Intermezzi *Cavalleria Rusticana* by Pietro Mascagni

Most Feared Marine Predator shark

Most Glorious Hero of Norwegian Viking Times Olav Tryggvasson

Most Gorgeous Rhetorician Robert G. Ingersoll

Most Mispronounced Geographical and Political Term Caribbean (correctly *Carib-be-an*)

Most Mispronounced Placename in Southern California Jamacha

Most Mysterious Snake the harmless hoop snake is said to put its tail in its mouth whenever it is afraid; thereupon it rolls away like a hoop; but if someone pursues it the snake proceeds to swallow its tail until there is no more snake; people who tell this story usually know the person who saw this but that person lives in another city or country

Most Northern Southern City Tulsa, Oklahoma

Most Popular Guest Symphonic Conductor Danny Kaye

Most Popular Modern Symphony of 20th Century Prokofiev's No 1 in D-minor—*Classical*

Most Popular Piano Concertos of 20th Century Grieg in *A-minor*, Rachmaninoff No 2 in *C-minor*

Most Popular Romantic Symphony of 20th Century Rachmaninoff's No 2 in E-minor

Most Populous American Megalopolis Los Angeles

Most Populous Canadian Province Ontario

Most Populous Nation China

Most Populous State California

Most-Prized Sailing Trophy America's Cup

Most Prolific Composer Wolfgang Amadeus Mozart or Georg Philipp Telemann

Most Prolific Song Writer Cole Porter who wrote a song a day

Most Prolific Symphonist Franz Josef Haydn composer of 104 symphonies

Most Remote Continent Australia

Most Serene Republic of the Sea Venice, Queen of the

Adriatic

Most Sinful City in South America Guayaquil, Ecuador

Most Versatile Character Actor Alec Guinness

Most Versatile Musician of Our Century Georges Enesco (Romanian composer-conductor-pianist-teacher-violinist)

Mountains of the World (by height) Everest, K-2, Kanchenjunga, Makulu, Dhaulagiri, Nanga Parbat, Annapurna, Nanda Devi, and Kemet in the Himalayas; Namcha Barwa and Minya Konka in China; Kommumizma in the Pamirs; Pobedy in the Tian Shan; Aconcagua, Bonete, Ojos del Salado, Huascaran, Lullaillaco, Sajama, and Chimborazo in the Andes, McKinley in Alaska; Logan in the Yukon; Cotopaxi in the Andes; Kilimanjaro in Tanzania; Antisana in the Andes; Ciltlaltepetl in the Sierra Madre; Elbruz in the Caucasus; Mount St Elias in Alaska; Popocatapetl in the Sierra Madre; Foraker in Alaska; Luciana in the Yukon; Tolima in the Andes; Kenya in Kenya; Ararat in Armenia; the Vinson Massif in Antarctica

Mozambique's Highest Peak Serra de Gorongosa—6,112 feet, 1,852 meters

Mozambique's Largest City Maputo

Mozambique's Principal Port Maputo

Namibia's Largest City Windhoek

Namibia's Principal Port Walvis Bay

Nauru's Largest Town Yaren

Nauru's Principal Port Yaren

Nebraska's Highest Point Johnson—5,426 feet, 1,653 meters—in Kimball County near the Colorado and Wyoming

Nebraska's Largest City Omaha

Nebraska's Longest River Missouri

Nebraska's Most Unusual Natural Feature the Sandhills

Nepal's Highest Peak Mount Everest—29,028 feet, 8,796 meters

Nepal's Largest City Kath-

mandu

Netherlands Antilles' Largest City Willemstad on Curaçao

Netherlands' Easternmost Town Nieuwe-Schans

Netherlands' Largest City Amsterdam

Netherlands' Largest Port Rotterdam

Netherlands' Longest River System Rhine–Meuse

Netherlands' Northernmost City Groningen

Netherlands' Northernmost Island Rottumeroog (in the Frisians between the North Sea and the Waddenzee)

Netherlands' Principal Port Rotterdam

Netherlands' Southernmost City Maastricht

Netherlands' Southernmost Point Vaals

Netherlands' Westernmost Port Flushing or Vlissingen

Netherlands' Westernmost Town Sluis

Nevada's Highest Point Boundary Peak—13,140 feet, 4,005 meters

Nevada's Largest City Las Vegas

Nevada's Largest Forest Toiyabe National Forest (also the largest forest in the conterminous U.S.)

New Brunswick's Highest Point Mount Carleton—2,690 feet, 820 meters

New Brunswick's Largest City Saint John

New Brunswick's Principal Port Saint John

New Caledonia's Largest City Noumea

New Caledonia's Principal Port Noumea

New England's Highest Point Mount Washington—6,288 feet, 1,917 meters—in New Hampshire

Newfoundland's Highest Point Mount Cladonia—4,725 feet, 1,441 meters—in Labrador

Newfoundland's Largest City St John's

New Foundland's Principal Port St John's

New Hampshire's Highest Point Mount Washington—6,288 feet, 1,917 meters

New Hampshire's Largest City Manchester

New Hampshire's Largest Lake Winnipesaukee

New Hampshire's Longest River Connecticut

New Hampshire's Principal Port Portsmouth

New Jersey's Highest Point High Point—1,803 feet, 550 meters—in the Kittatinny Mountains

New Jersey's Largest City Newark

New Jersey's Longest River Delaware

New Jersey's Most Spectacular Park Palisades Interstate Park overlooking the Hudson River and much of New York City

New Jersey's Principal Port Newark

New Mexico's Highest Point Wheeler Peak—13,161 feet, 4,011 meters

New Mexico's Largest City Albuquerque

New Mexico's Largest Gypsum Desert the White Sands National Monument, world's largest gypsum deposit

New Mexico's Longest River Rio Grande

New York's Most Spectacular Waterfall Niagara Falls

New York State's Highest Point Mount Marcy—5,344 feet, 1,630 meters—in the Adirondacks

New York State's Largest City New York

New York State's Longest River Hudson

New York State's Principal Port New York City

New Zealand's Easternmost City Gisborne on North Island

New Zealand's Easternmost Point North Island's East Cape

New Zealand's Highest Peak Mount Cook—12,349 feet, 3,742 meters—Aorangi by the Maori, on South Island

New Zealand's Largest City Auckland on North Island

New Zealand's Largest South Island City Christchurch

New Zealand's Northernmost City Whangarei north of Auckland on North Island

New Zealand's Northernmost Point North Island's North Cape

New Zealand's Principal Port Auckland

New Zealand's Southernmost City Invercargill on South Is-

land south of Dunedin
New Zealand's Southernmost Point Southwest Cape on Steward Island south of South Island
New Zealand's Westernmost City Milford Sound on South Island
New Zealand's Westernmost Point Resolution Island west of South Island
Nicaragua's Highest Peak Teotecacinte—above 7,000 feet, 2,121 meters—on the Honduran-Nicaraguan border
Nicaragua's Largest City Managua
Nicaragua's Largest Department Zelaya
Nicaragua's Principal Caribbean Port Bluefields
Nicaragua's Principal Pacific Port Corinto
Nicaragua's Smallest Department Masaya
Nigeria's Highest Peak Jos—5,843 feet, 1,771 meters
Nigeria's Largest City Lagos
Nigeria's Principal Port Port Harcourt
Niger's Highest Peak Iferouane—5,906 feet, 1,790 meters
Niger's Largest City Niamey
Noisiest Opera Massenet's *La Navarraise* with its cannon fire, clapping, castanets, gun shots, tambourines, and bells
North America's Easternmost City St John's, Newfoundland
North America's Easternmost Point Cape Spear, Newfoundland
North America's Largest Cities Mexico, New York
North America's Largest Drainage Basin Mississippi –Missouri
North America's Largest Island Greenland
North America's Largest Nation Canada
North America's Most European City Quebec
North America's Northernmost City Barrow, Alaska
North America's Northernmost Point Canada's Cape Columbia
North America's Southernmost City David, Panamá
North America's Southernmost Point southwesternmost Panamá just northeast of Juradó, Colombia
North America's Western-

most City Seward, Alaska
North America's Westernmost Point Cape Wrangell on Attu Island in the Aleutians
North Atlantic's Largest Nation Canada
North Carolina's Highest Point Mount Michell—6,684 feet, 2,037 meters
North Carolina's Largest City Charlotte
North Carolina's Longest River Catawba
North Carolina's Loveliest Seashore the Outer Banks including the Cape Hatteras National Seashore
North Carolina's Principal Port Morehead City
North Dakota's Highest Point White Butte—3,506 feet, 1,069 meters
North Dakota's Largest City Fargo
North Dakota's Longest River Missouri
North Dakota's Most Exotic Scenery the Badlands along the Little Missouri River within the Theodore Roosevelt National Memorial Park
Northeasternmost Beach of the United States Wilsons Beach, Campobello Island, Maine
Northeasternmost Point of the continental United States West Quoddy Head near Lubec, Maine
Northern Ireland's Highest Mountain Slieve Donard
Northern Ireland's Highest Peak Slieve Downard—2,796 feet, 847 meters
Northern Ireland's Largest City Belfast
Northern Ireland's Largest Lake Lough Neagh
Northernmost Canadian Town Inuvik, Northwest Territories
Northernmost Capital of Africa Tunis, Tunisia
Northernmost Capital of Asia Ulaanbaatar, Mongolia
Northernmost Capital of Australia Darwin, Northern Territory
Northernmost Capital of Europe Reykjavik, Iceland
Northernmost Capital of North America Yellowknife, Northwest Territories
Northernmost Capital of South America Caracas, Venezuela

Northernmost Europe Cape Norkkyn, Norway
Northernmost Metropolis Leningrad (St Petersburg)
Northernmost Point of Canada Cape Columbia, Northwest Territories on the Arctic Ocean at 83°7′ North
Northernmost Point of the continental United States Point Barrow, Alaska
Northernmost Point of Mexico Los Algodones, Baja California
Northernmost Province Québec
Northernmost State Alaska
Northernmost Territories Northwest Territories
Northernmost Town Ny Ålesund in Spitsbergen
North Korea's Highest Peak Kwanmo also called Kambo Ho—8,337 feet, 2,526 meters
North Korea's Largest City Pyongyang
North Korea's Principal Port Chonglin
Northwesternmost Point of the Contiguous United States Natoosh Island off Cape Flattery
Northwesternmost Point of the continental United States Cape Wrangell on Attu Island in the Aleutians
Northwest Territories' Highest Point Sir James MacBrien—9,062 feet, 2,762 meters
Norway's Highest Peak Galdhoppigen—8,097 feet, 2,454 meters—in the Jotunheim mountains
Norway's Largest City Oslo
Norway's Longest River Glomma
Norway's Principal Port Bergen
Nova Scotia's Highest Point North Barren—1,747 feet, 532 meters—on Cape Breton Island
Nova Scotia's Largest City Halifax
Nova Scotia's Principal Port Halifax
Ohio's Greatest Living Museum the Holden Arboretum east of Cleveland
Ohio's Highest Point Campbell Hill—1,550 feet, 472 meters —near Bellefontaine
Ohio's Largest City Cleveland
Ohio's Longest River Ohio
Ohio's Principal Port Cincin-

nati

Oklahoma's Highest Point Black Mesa—4,973 feet, 1,515 meters

Oklahoma's Largest City Oklahoma City

Oklahoma's Largest Gypsum Cave in Alabaster Caverns State Park south of Freedom

Oklahoma's Longest River Red

Oldest American Orchestra New York Philharmonic (founded in 1842)

Oldest Anglican Church site in the Western Hemisphere St Peter's in St George Harbour, Bermuda

Oldest British Colony Newfoundland, discovered in 1497 by John Cabot

Oldest British Columbia Settlement Vancouver, founded by Hudson's Bay Company in 1825

Oldest Canadian Settlement Port Royal (Annapolis, Nova Scotia) founded by Champlain in 1604

Oldest Central American Democracy Costa Rica

Oldest Centralized Nation China

Oldest City in Canada Québec (founded by Champlain in 1608)

Oldest City in Denmark Ribe (whose church was built in the 12th century)

Oldest City in Germany Trier (founded by the Romans in 15 B.C.)

Oldest City in Malaysia Malacca (settled by Malays around 1400)

Oldest City in North America Mexico City (built by the Aztecs in 1325)

Oldest City in the U.S. St Augustine, Florida (founded by the Spaniards in 1565)

Oldest Continually Operating Streetcar Line St Charles Avenue line in New Orleans

Oldest European Settlement in the Far East Macao (leased from China by the Portuguese in 1557)

Oldest European Settlement in the New World Santo Domingo City (founded by the Spaniards in 1496)

Oldest German Ocean Harbor Bremen (created an archbishopric in 845 A.D.)

Oldest Inhabited City Damas-

cus, Syria

Oldest Inhabited Place in the United States Pueblo Acoma near Albuquerque, New Mexico

Oldest Known Canon *Sumer is incumenn in—Ihude sing cuccu*—Reading Rota most likely composed between 1280 and 1310—still sung

Oldest Nova Scotia Settlement Port Royal established in 1605

Oldest Ontario Settlement Fort St Marie founded in 1639

Oldest Popular Song Composer Irving Berlin who died at 101

Oldest President of the United States Ronald Reagan

Oldest Public School in America Boston Latin School

Oldest Publisher Elsevier, also spelled Elzevir, in business since 1581; now with editorial offices in Amsterdam, Oxford, and New York

Oldest Quaintest City in the United States Santa Fé, New Mexico (built in 1621)

Oldest Quebec Settlement city of Québec founded by Champlain in 1608

Oldest Ship in the British Navy HMS *Victory*— launched in 1765 and served as Nelson's flagship at Trafalgar

Oldest Ship in the U.S. Navy USS *Constitution*—launched in 1797 and nicknamed Old Ironsides

Oldest Symphony Orchestra Dresdener Staatskapelle whose performance history dates to 1548

Oldest Synagogue (in continuous use since 1732) in the New World Mikve Israel in Willemstad, Curaçao

Oldest University in the Americas Santo Tómas de Aquino in Santo Domingo, Dominican Republic, founded in 1538

Oman's Principal Port Muscat

Only American Marsupial opossum

Only Great Ape in Asia orangutan

Ontario's Highest Point Ogidaki Mountain—2,183 feet, 665 meters—overlooking Lake Superior's Haviland Bay

Ontario's Largest City

Toronto

Ontario's Longest River St Lawrence

Ontario's Principal Port Toronto

Oregon's Highest Point Mount Hood—11,235 feet, 3,433 meters

Oregon's Largest City Portland

Oregon's Longest River Columbia

Oregon's Most Beautiful and Peaceful Park Crater Lake National Park

Oregon's Principal Port Portland

Pacific Ocean's Largest Nation Australia

Pakistan's Highest Peak Tirich Mir—25,230 feet, 7,645 meters

Pakistan's Largest City Karachi

Pakistan's Principal Port Karachi

Panama Canal's Most Ecologically Protected Island Barro Colorado

Panama's Highest Peak Chiriqui—11,410 feet, 3,458 meters—an inactive volcano near Costa Rica

Panama's Largest City Panama City

Panama's Largest Province Darién

Panama's Principal Caribbean Port Cristobal

Panama's Principal Pacific Port Balboa

Panama's Smallest Province Herrera

Papua New Guinea's Highest Peak Carstensz—16,400 feet, 4,970 meters

Papua New Guinea's Largest City Port Moresby

Paraguay's Largest City Asunción

Paraguay's Largest Department Presidente Hayes·

Paraguay's Longest River Paraguay

Paraguay's Smallest Department Central

Pennsylvania's Highest Point Mount Davis—3,213 feet, 979 meters—in Somerset County

Pennsylvania's Largest City Philadelphia

Pennsylvania's Longest River Delaware

Pennsylvania's Most Pristine Park Allegheny National Forest extending into New York State

Pennsylvania's Principal Port Philadelphia

Peru's Easternmost and Northernmost City Iquitos (on the Amazon)

Peru's Highest Peak Huascaran—22,205 feet, 6,729 meters—in the Andes

Peru's Largest City Lima

Peru's Largest Department Loreto

Peru's Largest Port Callao

Peru's Longest River Marañón

Peru's Principal Port Callao

Peru's Smallest Department Callao

Peru's Southernmost City Tacna (near the Chilean border)

Peru's Westernmost City Talara

Philippines' Highest Peak Apo —9,690 feet, 2,936 meters —volcano on the island of Mindinao

Philippines' Largest City Manila

Philippines' Principal Port Manila

Pinnacle of the Baroque Johann Sebastian Bach

Poland's Highest Peak Rysy— 8,212 feet, 2,488 meters—in the High Tatra near Czechoslovak

Poland's Largest City Warsaw

Poland's Principal Port Gdansk

Poorest Nation in the New World Haiti

Poorest Nations in the World Kampuchea, Laos, Ciskei and Transkei

Poorest-Per-Capita African Country Chad

Poorest-Per-Capita Asiatic Country Laos

Poorest-Per-Capita European Country Albania

Poorest-Per-Capita South American Country Bolivia

Poorest-Per-Capita West Indian Country Haiti

Portugal's Highest Peak Malhão—6,532 feet, 1,979 meters —in the Serra da Estrêla

Portugal's Largest City Lisbon

Portugal's Longest River *Tejo* (Portuguese—Tagus)

Portugal's Principal Port Lisbon

Prince Edward Island's Highest Point between Hartsville

and Stanchel—450 feet, 136 meters

Prince Edward Island's Largest City Charlottetown

Prince Edward Island's Principal Port Charlottetown

Puerto Rico's Best Port San Juan

Puerto Rico's Biggest Port on the Atlantic San Juan

Puerto Rico's Biggest Port on the Caribbean Ponce

Puerto Rico's Easternmost City Fajardo

Puerto Rico's Easternmost Point Culebrita Island east of Culebra and Vieques

Puerto Rico's Highest Peak Cerro de Punta—4,400 feet, 1,467 meters—in the Cordillera Central

Puerto Rico's Largest City San Juan

Puerto Rico's Northernmost Cities Arecibo and San Juan

Puerto Rico's Northernmost Point Punta Jacinto

Puerto Rico's Principal Port San Juan

Puerto Rico's Southernmost City Ponce

Puerto Rico's Southernmost Point Caja de Muertos Island (southeast of Ponce)

Puerto Rico's Westernmost Cities Aguadilla and Mayagüez

Puerto Rico's Westernmost Point Punta Higüero

Qatar's Largest City Doha

Qatar's Principal Port Doha

Québec's Highest Point Mont Jacques Cartier—4,160 feet, 1,269 meters

Québec's Largest City Montreal

Québec's Longest River St Lawrence

Québec's Principal Port Montreal

Rainiest Spot Mount Waialeale, Hawaii

Réunion's Largest City Saint-Denis

Réunion's Principal Port Saint-Denis

Rhode Island's Highest Point Jeremoth Hill—812 feet, 247 meters—near Connecticut

Rhode Island's Largest City Providence

Rhode Island's Most Notable Feature Narragansett Bay

Rhode Island's Principal Port Providence

Richest Arab Country Qatar

Richest Countries Canada, France, Germany, Italy, Japan, United States

Richest Country in the Middle East Kuwait

Richest Hill on Earth Butte, Montana

Richest Island Country Nauru in the western Pacific

Richest Nation in the New World United States of America

Richest Nation in the Old World Qatar

Richest-Per-Capita African Country Algeria

Richest-Per-Capita Asiatic Country Qatar

Richest-Per-Capita Central American Country Panama

Richest-Per-Capita Countries of the World Qatar, Nauru, Sweden, Switzerland, the U.S., Denmark, Norway, Canada, Kuwait, Luxembourg, Japan, Australia

Richest-Per-Capita European Country Sweden

Richest-Per-Capita North American Country the U.S.

Richest-Per-Capita South American Country Venezuela

Richest-Per-Capita West Indian Country Trinidad and Tobago

Rivers of the World (ranked by length) Nile, Amazon, Mississippi-Missouri, Ob'Irtysh, Yangtze, Hwangho, Congo, Amur, Lena, Mackenzie-Peace, Mekong, Niger, Mackenzie, Paraná, Volga, Yenisei, Madeira, Yukon, Arkansas, Colorado, St Lawrence, Rio Grande, Salween, Danube, Euphrates, Indus, Brahmaputra, Zambesi, Murray-Darling

Romania's Highest Peak Negos—8,361 feet, 2,534 meters —in the Transylvanian Alps

Romania's Largest City Bucharest

Romania's Longest River Danube

Romania's Principal Port Constanta

Russia's Greatest Poet Alexander Pushkin

Russia's Longest River Volga

Rwanda's Highest Peak Goma—14,787 feet, 4,481 meters—on the Uganda and Zaire borders

Rwanda's Largest City Kigali

Saba's Highest Peak Gunong Kinabalu—13,510 feet, 4,094 meters—east of Kota Kinabalu
Saba's Largest City Kota Kinabalu
Samoa's Best Port Pago Pago on Tutuila Island
San Marino's Largest Town San Marino
São Tomé and Principe's Largest City São Tomé
São Tomé and Principe's Principal Port São Tomé
Sardinia's Largest City Cagliari
Saskatchewan's Highest Point Cypress Hills—4,567 feet, 1,392 meters
Saskatchewan's Largest City Regina
Saskatchewan's Longest River Nelson–Saskatchewan
Saudi Arabia Mountain Peak *Jebal al Lawz* (Arabic— Lawrence Mountain)—8,465 feet, 2,580 meters—honoring Lawrence of Arabia
Saudi Arabia's Highest Peak in the Asir Highlands— 10,279 feet, 3,133 meters
Saudi Arabia's Largest City Riyadh
Saudi Arabia's Principal Port Jidda
Scotland's Highest Mountain Ben Nevis
Scotland's Highest Peak Ben Nevis—4,406 feet, 1,335 meters
Scotland's Largest City Glasgow
Scotland's Largest Lake Loch Lomond
Scotland's Longest River Tay
Second First Lady Abigail Smith Adams
Senegal's Largest City Dakar
Senegal's Principal Port Dakar
Seychelles' Highest Peak Morne Seychellois—2,993 feet, 907 meters—on Mahé Island
Seychelles' Largest City Port Victoria
Seychelles' Principal Port Port Victoria
Shallowest Sea Baltic (average depth under 190 feet)
Shortest Snake thread snake of Barbados, Martinique, and St Lucia *(Leptophylops bilineata)*
Siberia's Largest City Novosibirsk
Sicily's Largest City Palermo

Sierra Leone's Highest Peak Bintimane Peak—6,390 feet, 1,936 meters
Sierra Leone's Largest City Freetown
Sierra Leone's Principal Port Freetown
Singapore's Largest City Singapore
Singapore's Principal Port Singapore
Smallest African Country The Gambia
Smallest African Nation Comoros
Smallest American Prison Populations North Dakota, Vermont, New Hampshire, Wyoming, and South Dakota
Smallest American State Rhode Island
Smallest American Virgin Island Cockroach
Smallest Asian Country Singapore
Smallest Asian Nation Maldives
Smallest Australian State Tasmania
Smallest British Bird wren
Smallest British Mammal pygmy shrew
Smallest British Mouse harvest mouse
Smallest Canadian Province Prince Edward Island
Smallest Canadian Provincial Capital Charlottetown, Prince Edward Island
Smallest Capital of the Eastern World Yaren, Nauru
Smallest Capital in the U.S. Carson City, Nevada
Smallest Capital of the Western World San Marino, San Marino
Smallest Central American Country El Salvador
Smallest Channel Island Brechau
Smallest Continent Australia
Smallest Domestic Dog Chihuahua (5 inches or 13 cm high and weighing 6 pounds or 2.7 kg)
Smallest East European Country Albania
Smallest European Country San Marino
Smallest European Nation Monaco
Smallest Galápagos Island Duncan, officially Pinzón
Smallest Indian Ocean Nation Seychelles
Smallest Indonesian Island

Krakatoa
Smallest Living Amphibian tree frog *(Hyla ocularis)* of the southeastern United States
Smallest Living Antelope West African royal antelope
Smallest Living Bat Kitt's hog-nosed bat from Thailand
Smallest Living Bird Cuban hummingbird
Smallest Living Cat rusty-spotted cat of India and Sri Lanka
Smallest Living Crocodilian dwarf caiman of the Amazon Basin
Smallest Living Deer Andean dwarf deer called *pudo* in Chile; Ecuadorean pudu
Smallest Living Dog miniature Chihuahua or pygmy Yorkshire terrier
Smallest Living Fish half-inch long (1.3 cm) Philippine goby
Smallest Living Freshwater Turtle musk turtle
Smallest Living Horse Argentina's Falabella breed
Smallest Living Lizard British Virgin Island gecko *(Sphaerodactylus elasmobranchus)*
Smallest Living Mammal pygmy shrew or wood mouse
Smallest Living Marine Mammal sea otter
Smallest Living Marsupial Kimberly marsupial mouse of Western Australia
Smallest Living Primate Madagascar mouse lemur
Smallest Living Reptile pygmy gecko of Israel
Smallest Living Rodent Eurasian harvest mouse
Smallest Living Sea Turtle Kemp's Bastard or Ridley
Smallest Living Snake West Indian thread snake
Smallest Living Terrapin Colombian musk terrapin
Smallest Living Tortoise Egyptian or South African tortoise
Smallest Middle American Country Grenada
Smallest Nation Principality of Monaco—less than 370 acres (150 hectares)
Smallest Nation in the South Pacific Nauru
Smallest Oceanic Country Nauru in the equatorial Pacific
Smallest Officially Atheist Nation Albania
Smallest Pacific Ocean Nation

Nauru

Smallest Penguin Little Blue Penguin of Australia and New Zealand

Smallest Planet Mercury

Smallest Present-Day Bird hummingbird

Smallest Primate Madagascar's mouse lemur *(Microrebus murinus)*

Smallest Royal Capital of the Eastern World Katmandu, Nepal

Smallest Royal Capital of the Western World Monaco-Ville, Monaco

Smallest Sea Baltic

Smallest South American Country French Guiana

Smallest South American Nation Suriname

Smallest South Atlantic Nation Uruguay

Smallest South Pacific Nation Nauru

Smallest Sovereign State Vatican City *(see* World's Smallest Sovereign State)

Smallest State in the United States Rhode Island

Smallest Warm-blooded Creatures hummingbirds

Smallest West European Country San Marino

Smallest West Indian Nation Grenada; St Kitts and Nevis

Solomon Islands' Highest Peak Mount Balbi—10,170 feet, 308 meters—on Bougainville Island

Solomon Islands' Largest City Honiara

Solomon Islands' Principal Port Honiara

Somalia's Largest City Mogadishu

Somalia's Principal Port Mogadishu

South Africa's Highest Peak Thabantshon—11,425 feet, 3,462 meters—in the Drakensberg Range in Basutoland

South Africa's Biggest Black Ghetto Soweto near Johannesburg

South Africa's Largest City Johannesburg

South Africa's Principal Port Durban

South America's Easternmost Cities Brazil's João Pessoa and Recife

South America's Easternmost Points Brazil's Cabo Branco near João Pessoa and Punta de Pedra near Recife (both on

the same longitude—34°37' W)

South America's Largest City Buenos Aires

South America's Largest Drainage Basin Amazon

South America's Largest Island Tierra del Fuego

South America's Largest Nation Brazil

South America's Longest River Amazon

South America's Northernmost City Coro, Venezuela

South America's Northernmost Point Punta Gallinas, Colombia

South America's Southernmost City Punta Arenas, Chile

South America's Southernmost Point Cabo de Hornos (Cape Horn), Chile

South America's Westernmost City Talara, Peru

South America's Westernmost Point Punta Pariñas, Peru

South Atlantic's Largest Nation Brazil

South Carolina's Highest Point Sassafras Mountain —3,560 feet, 1,085 meters

South Carolina's Largest City Columbia

South Carolina's Longest River Savannah

South Carolina's Most Charming City Charleston

South Carolina's Principal Port Charleston

South Dakota's Highest Point Harney Peak—7,242 feet, 2,208 meters

South Dakota's Largest City Sioux Falls

South Dakota's Longest River Missouri

South Dakota's Most Memorable Memorial Mount Rushmore with the carved heads of George Washington, Thomas Jefferson, Abraham Lincoln, and Theodore Roosevelt

Southeast Asia's Longest River Mekong

Southeasternmost Beach of the United States Miami Beach

Southeasternmost Point of the continental United States Cape Florida at the southern tip of Key Biscayne near Miami

Southeasternmost Point of the territorial United States Vagthus Point on Saint Croix in the Virgin Islands east of Puerto Rico

Southernmost American Town Naalehu, Island of Hawaii

Southernmost Beach in the United States South Beach in Key West, Florida

Southernmost Canadian Town Kingsville, Ontario

Southernmost Capital of Africa Cape Town, South Africa

Southernmost Capital of Asia Jakarta, Indonesia

Southernmost Capital of Australia Hobart, Tasmania

Southernmost Capital of Europe Athens, Greece

Southernmost Capital of North America Panamá City, Panamá

Southernmost City of South America Montevideo, Uruguay

Southernmost Europe Crete's south coast

Southernmost Island in the West Indies Grenada

Southernmost Metropolis Melbourne, Australia

Southernmost Point of Canada Middle Island, Ontario on Lake Erie

Southernmost Point of the continental United States South Beach on Key West, Florida

Southernmost Point of the insular United States Ka Lae or South Cape on Hawaii

Southernmost Point of Mexico Barra del Río Suchiate (Mouth of the Suchiate River)

Southernmost Province Ontario

Southernmost State Hawaii

Southernmost Town Puerto Williams, Chile

South Korea's Largest City Seoul

South Korea's Principal Port Pusan

South's Oldest Daily Newspaper Charleston's *The News and Herald*

Southwesternmost Beach in the continental United States Imperial Beach near San Diego, California

Southwesternmost Point of the continental United States Point Loma, California at the entrance to San Diego Bay

Southwesternmost Point of the territorial United States Steps Point on Tutuila Island of American Samoa in the South Pacific

Soviet Union's Largest City Moscow

Spain's Easternmost Point Cabo de Creus (northeast of Barcelona)

Spain's Easternmost Port Barcelona

Spain's Highest Peaks Mulhacén—11,411 feet, 3,458 meters—on the mainland; Pico de Teide—12,200 feet, 3697 meters—in the Canary Islands

Spain's Largest City Madrid

Spain's Largest Province Andalucía

Spain's Longest River Tagus

Spain's Most Precious Jewel Toledo, according to Cervantes

Spain's Northernmost Point Cabo Ortegal

Spain's Northernmost Port El Ferrol del Caudillo

Spain's Principal Port Barcelona

Spain's Smallest Province Vascongadas (Basque Provinces)

Spain's Southernmost Point Punta de Tarifa

Spain's Southernmost Port Algeciras

Spain's Westernmost Point Cabo Finisterre

Spain's Westernmost Port Vigo

Sri Lanka's Highest Peak Pidurutalagala—8,291 feet, 2,512 meters

Sri Lanka's Largest City Colombo

Sri Lanka's Principal Port Colombo

St Christopher and Nevis' Largest Town Basse-Terre

St Christopher (St Kitts) and Nevis' Principal Port Basse-Terre

St Helena's Largest Place Jamestown

St Lucia's Largest City Castries

St Lucia's Principal Port Castries

St Pierre and Miquelon's Largest Town St John's

St Pierre and Miquelon's Principal Port St Pierre

St Vincent's Largest City Kingstown

St Vincent and the Grenadines' Principal Port Kingstown

Strongest Surface Wind Mount Washington, New Hampshire

Sudan's Largest City Khartoum

Sudan's Principal Port Port Sudan

Suriname's Highest Peak Wilhelmina Gebergte—4,200 feet, 1,273 metrs

Suriname's Largest City Paramaribo

Suriname's Longest River Nickerie

Suriname's Principal Port Paramaribo

Swaziland's Largest City Mbabane

Sweden's Highest Peak Galdhoppigen—8,097 feet, 2,454 meters—in the Jotunheim range

Sweden's Largest City Stockholm

Sweden's Longest River Torne

Sweden's Principal Port Gothenburg

Swiftest Mammal cheetah

Switzerland's Highest Peak Dufourspitze—15,203 feet, 4,607 meters—of Monte Rosa

Switzerland's Largest City Zurich

Switzerland's Longest River Rhine

Syria's Highest Peak Mount Hermon—9,232 feet, 2,798 meters—

Syria's Largest City Damascus

Syria's Principal Port Latakia

Tahiti's Highest Peak Mont Orohena—7,618 feet, 2,308 meters

Tahiti's Principal Port Papeete

Taiwan's Highest Peak Mount Morrison—14,000 feet, 4,242 meters

Taiwan's Largest City Taipei

Taiwan's Principal Port Kaosiung

Tallest Domestic Dog Great Dane, standing up to 32 inches (81 cm) and weighing up to 150 pounds (68 kg)

Tallest Living Dog Great Dane or Irish Wolfhound

Tallest Living Mammal giraffe

Tallest North American Bird whooping crane

Tallest TV Tower KTHI-TV in Blanchard, North Dakota—

2,063 feet (629 meters) high

Tanzania's Highest Peak Kibo or Kilimanjaro—19,340 feet, 5,861 meters

Tanzania's Largest City Dar es Salaam

Tanzania's Principal Port Dar es Salaam

Tennessee's Highest Point Clingmans Dome—6,643 feet, 2,025 meters—in the Great Smoky Mountains

Tennessee's Largest City Memphis

Tennessee's Longest River Tennessee

Tennessee's Most Famous Mountain Lookout overlooking the Tennessee Valley and seven states beyond

Tennessee's Principal Port Memphis

Texas' Highest Point Guadalupe Peak—8,751 feet, 2,667 meters—in Culberson County

Texas' Largest City Houston

Texas' Longest River Rio Grande

Texas' Most Spectacular Park the Big Bend National Park

Texas' Principal Port Houston

Thailand's Highest Peak Inthanon—8,452 feet, 2,561 meters

Thailand's Largest City Bangkok

Thailand's Principal Port Bangkok

Third First Lady Martha Wayles Skelton Jefferson

Timpanist of the 20th Century Saul Goodman (1906–) of the New York Philharmonic

Togo's Largest City Lomé

Togo's Principal Port Lomé

Tokyo's Largest Slum Sanya

Tonga's Largest City Nuku'alofa

Tonga's Principal Port Nuku'alofa

Transkei's Largest City Umtata

Trinidad and Tobago's Largest City Port of Spain

Trinidad's Highest Peak Mount Aripo—3,085 feet, 935 meters

Trinidad and Tobago's Principal Port Port of Spain

Tunisia's Largest City Tunis

Turkey's Highest Peak Mount Ararat—16,946 feet, 5,135 meters—in Armenia

Turkey's Largest City Istanbul

Turkey's Principal Port

Istanbul

Turks and Caicos Islands' Largest Town Cockburn Town

Tuvalu's Largest City Fongafale

Tuvalu's Principal Port Funafuti

Uganda's Highest Peak Ruwenzori—16,795 feet, 5,089 meters

Uganda's Largest City Kampala

ultimate marine predator great white shark

United Arab Emirates' Largest City Abu Dhabi

United Arab Emirates' Principal Ports Abu Dhabi, Dubai

United Kingdom's Easternmost Point Lowestoft Ness on Norfolk's east coast

United Kingdom's Easternmost Port Lowestoft, Suffolk

United Kingdom's Northernmost Point Herma Ness and Muckle Flugga Light on Unst in the Shetland Islands

United Kingdom's Northernmost Port Scapa Flow in the Orkney Islands

United Kingdom's Principal Ports London, Liverpool, Cardiff, Glasgow, Belfast

United Kingdom's Southernmost Point Lizard Point

United Kingdom's Southernmost Port Penzance

United Kingdom's Westernmost Point Land's End, Cornwall

United Kingdom's Westernmost Port Penzance, Cornwall

United States superlatives (*see* America's *entries in this section*)

Uruguay's Largest City Montevideo

Uruguay's Largest Department Tucuarembó

Uruguay's Largest Estuary Rio de la Plata combining the Paraná and Uruguay rivers

Uruguay's Principal Port Montevideo

Uruguay's Smallest Department Montevideo

USSR's Easternmost Point Mys Deshneva across the Bering Strait from Alaska

USSR's Highest Peak Pik Kommunizma—24,590 feet, 7,451 meters—in Tadshikistan, once named Stalin

USSR's Largest City Moscow

USSR's Longest River Ob'-Irtysh

USSR's Northernmost Point Mys Chelyuskin, Siberia

USSR's Principal Baltic Port Leningrad

USSR's Principal Black Sea Port Odessa

USSR's Principal Eastern Port Vladivostok

USSR's Principal Northern Port Leningrad

USSR's Principal Pacific Port Vladivostok

USSR's Principal Port Leningrad

USSR's Principal Southern Port Odessa

USSR's Principal Western Port Kaliningrad (formerly Königsberg, East Prussia)

USSR's Southernmost City Kushka, Turkmen (across the frontier from Golran in Afghanistan)

USSR's Westernmost Point and Port Baltiysk (formerly Pillau, Lithuania)

U.S. Virgin Islands' Largest City Charlotte Amalie on St Thomas

Utah's Highest Point King's Peak—13,528 feet, 4,123 meters

Utah's Largest City Salt Lake City

Utah's Saltiest Inland Sea the Great Salt Lake

Uttermost Port of the Earth Patagonia at the southern tip of South America

Uttermost South Cape Horn/Tierra del Fuego region of southern South America south of Patagonia

Uttermost Tip of Africa Cape Alguhas, South Africa

Uttermost Tip of Asia Singapore at the tip of the Malay Penisula

Uttermost Tip of Australia South East Cape, Tasmania

Uttermost Tip of Europe Nordkin, Norway

Uttermost Tip of North America Attu Island, Alaska or Ounta Mariato, Panama

Uttermost Tip of South America Cabo de Hornos (Cape Horn), Chile or Punta Gallinas, Colombia

Vanuatu's Largest Town Vila on Efate Island

Vanuatu's Principal Port Vila

Venezuela's Easternmost City Tucupita (in the delta of the Orinoco)

Venezuela's Easternmost Town La Horqueta (on the Guyana border)

Venezuela's Highest Mountain Pico Bolivar

Venezuela's Highest Peak Bolívar—16,411 feet, 4,973 meters

Venezuela's Highest Town San Rafael de Mucuchies

Venezuela's Largest City Caracas

Venezuela's Largest Port La Guaira

Venezuela's Largest State Bolívar

Venezuela's Longest River Orinoco

Venezuela's Northernmost City Coro

Venezuela's Northernmost Town Pueblo Nuevo (on the Paraguaná Peninsula)

Venezuela's Principal Port Maracaibo

Venezuela's Smallest State Nueva Esparta

Venuzuela's Southernmost City San Cristóbal

Venezuela's Southernmost Town Piedra de Cucuy (at the border of Brazil and Colombia)

Venezuela's Westernmost City Maracaibo

Venezuela's Westernmost Town Barranca (on the Colombian border)

Vermont's Greatest Forest Green Mountain National Forest in western Vermont

Vermont's Highest Point Mount Mansfield—4,393 feet 1,339 meters

Vermont's Largest City Burlington

Vermont's Longest River Connecticut

Vietnam's Highest Peak Fan Si Pan—10,308 feet, 3,124 meters

Vietnam's Largest City Ho Chi Minh City (Saigon)

Vietnam's Principal Port Ho Chi Minh City (Saigon)

Virginia's Highest Point Mount Rogers—5,729 feet, 1,746 meters

Virginia's Largest City Norfolk

Virginia's Longest River Potomac

Virginia's Most Scenic Highway Blue Ridge Parkway through the southern Appalachian Mountains

Virginia's Principal Port Norfolk Harbor

Volcanos—best known Kilimanjaro in Tanzania, Africa; Popocatepétl, Mexico; Rainier, Washington; Pelée, Martinique; Krakatoa, Indonesia; Vesuvius and Vulcano, Italy

Wales's Highest Mountain Snowdon

Wales' Highest Peak Snowdon —3,560 feet, 1,079 meters— in Caernavonshire

Wales' Largest City Cardiff

Wales's Largest Lake Bala Lake (Llyn Tegid)

Wales's Longest River the Towy

Wallis and Fatuna Islands' Largest Town Mata-Utu

Wallis and Fatuna Islands' Principal Port Mata-Utu

Washington's Most Spectacular Region Mount Baker National Forest including volcanic Mounts Adams, Rainier, and Saint Helena

Washington State's Highest Point Mount Rainier, also called Mount Tacoma— 14,410 feet, 4,392 meters

Washington State's Largest City Seattle

Washington State's Longest River Columbia

Washington State's Principal Port Seattle

Waterfalls of the World (ranked by height) Angel in Venezuela, Tugela in South Africa, Yosemite in California, Cuquenán in Venezuela, Sutherland in New Zealand, Mardalsfossen in Norway, Ribbon in California, King George VI in Guyana, Gavarnie in France, Victoria between Zimbabwe and Zambia, Iguazú between Argentina and Brazil, Niagara between Canada and the United States

Western Europe's Largest Prison Fleury Mergois on the outskirts of Paris

Westernmost American Territory Guam

Westernmost American Town Adak, Aleutian Islands, Alaska

Westernmost Canadian Territory Yukon

Westernmost Canadian Town Dawson, Yukon

Westernmost Capital of Africa Dakar, Senegal

Westernmost Capital of Asia Istanbul, Turkey

Westernmost Capital of Australia Perth, Western Australia

Westernmost Capital of Europe Lisbon, Portugal

Westernmost Capital of North America Juneau, Alaska

Westernmost Capital of South America Quito, Ecuador

Westernmost Hawaiian Island Nihau

Westernmost Ireland Tearagt Island off the Dingie Peninsula often called the westernmost peninsula of Europe

Westernmost Point of Canada Mount Saint Elias, Yukon on the Gulf of Alaska

Westernmost Point of the continental United States Cape Wrangell on Attu Island in the Aleutians

Westernmost Point of Mexico Playas de Tijuana in Baja California

Westernmost Point of the territorial United States Orate Point on Guam in the western Pacific

Westernmost Prairie Province Alberta

Westernmost Province British Columbia

Westernmost State Alaska

Western Samoa's Best Port Apia on Upolu Island

Western Samoa's Largest City Apia

Western Samoa's Principal Port Apia

West Virginia's Highest Point Spruce Kbob—4,862 feet, 1,482 meters

West Virginia's Largest City Charleston

West Virginia's Longest River Ohio

West Virginia's Most Historic Park Harper's Ferry National Historical Park originally surveyed by Thomas Jefferson in 1783

West Virginia's Principal Port Huntington

Wickedest City in the World sunken Port Royal beneath Kingston, Jamaica's harbor where pirates once went for recreation

Widest Meteoric Crater in New Québec

Windiest Continent Antarctica

Wisconsin's Largest City Mil-

waukee

Wisconsin's Longest River Mississippi

Wisconsin's Most Scenic Lakeshore Apostle Islands National Lakeshore on southern Lake Superior

Wisconsin's Principal Port Superior

World's Biggest Bauxite Port Weipa, Queensland

World's Biggest Bay Bay of Bengal

World's Biggest Bookend nickname of the Secretariat building of the United Nations

World's Biggest Gamblers Americans, followed by Britishers and Swedes

World's Busiest Airport Chicago's O'Hare International

World's Busiest Border San Diego, California and Tijuana, Baja California

World's Busiest Border Crossing San Ysidro, California

World's Busiest Morgue New York City Medical Examiner's Office at First Avenue and Thirtieth Street in Manhattan

World's Busiest Seaport Rotterdam in the Netherlands

World's Cleanest Cities in New Zealand and Norway; Pago Pago and Singapore

World's Coldest Places Soviet scientists report—158°F (-105.6°C) at Omyakon, Siberia and -194°F (-125.6°C) near Vostok in Antarctica

World's Coldest Seas Arctic and Antarctic oceans (2°C or 28°F)

World's Coolest City Ulan-Bator, Mongolia

World's Crookedest River the Mississippi, according to Mark Twain

World's Deepest Gorge Hells Canyon in Idaho's Snake River (7,900 feet or 2,410 meters)

World's Deepest Lake Baikal in the USSR (1,742 meters or 5,714 feet)

World's Driest City Arica, Chile

World's Driest Place Chile's Atacama Desert

World's Fastest Passenger Liner *United States*

World's First Detective

François Eugène Vidocq
World's First National Park Yellowstone in Idaho, Montana, and Wyoming since 1872
World's First Nuclear-Powered Submarine USS *Nautilus*
World's First Parliament Iceland's Althing founded in 930
World's First Photographer Joseph Nicéphore Niepce
World's First Woman Prime Minister Sirimavo Ratwatte Badaranaike of Ceylon (Sri Lanka)
World's Foremost Debating Society Oxford Union Society
World's Freest and Smallest Jail San Marino's hilltop lockup
World's Greatest Bullshit Factory Hollywood, according to John Dos Passos
World's Greatest Economic Choke Point Strait of Ormuz between Iran and Oman on the Persian Gulf tanker route
World's Greatest Gorge Grand Canyon of the Colorado
World's Greatest Job Presidency of the United States
World's Greatest Library Library of the British Museum
World's Greatest Predator man
World's Greatest Railroad Terminal Grand Central Terminal in New York City
World's Greatest Tides Nova Scotia's Bay of Fundy (53 feet or 16 meters)
World's Greatest Volcanic Eruption in Ancient Times Mount Vesuvius in southwestern Italy killed 16,000 people in Herculaneum and Pompeii in 79
World's Greatest Volcanic Eruption in Modern Times Mount Pelée wiped out nearby city of St Pierre and its 40,000 people on Martinique in 1902
World's Greatest Volcanic Eruption in Recent Times Nevado del Ruiz, Colombia in mid-November 1985 killed more than 25,000 people
World's Highest Capital City La Paz, Bolivia (elevation 11,909 feet or 3630 meters)
World's Highest City Lhasa,

Tibet (3,687 meters or 12,087 feet above sea level)
World's Highest Crime Rate in the U.S. where more than 300 major crimes are committed every hour
World's Highest Lake Titicaca in the Andes between Bolivia and Peru (3,800 meters of 12,500 feet)
World's Highest Large City La Paz, Bolivia (3,632 meters or 11,909 feet above sea level)
World's Highest Mountain Everest in Nepal (29,028 feet or 8,848 meters)
World's Highest Mountains (*see* Mountains of the World)
World's Highest Murder Rate in the U.S. where more than 18,000 known murders occur yearly and about one American in every 10,000 will die at the hands of another
World's Highest Narcotic Addiction Rate in Hong Kong where in the 1980s some 80,000 in its population of 4 million were addicts
World's Highest Navigable Lake Titicaca (Bolivia)
World's Highest Point Mount Everest between Nepal and Tibet
World's Highest Tides at Burntcoat Head in the Bay of Fundy near Noel, Nova Scotia [53 feet (16 meters) and sometimes even 60 feet (18 meters)]
World's Highest Village Aucanquilca, Chile (17,500 feet or 5,334 meters)
World's Highest Waterfall Angel in Venezuela (3,297 feet or 1,005 meters)
World's Highest Waterfalls (*see* Waterfalls of the World)
World's Hottest Place Arizia, Libya (136°F or 58°C)
World's Hottest Sea Persian Gulf (36°C or 97°F)
World's Largest Active Volcano Mauna Loa on the island of Hawaii
World's Largest Archipelago Indonesia's more than 3,000 islands extending from the Indian Ocean to the western Pacific
World's Largest Art Gallery Hermitage and Winter Palace in Leningrad
World's Largest Atoll Kwajalein in the Marshalls

World's Largest Atom Smasher Superconducting Supercollider
World's Largest Bank Bank of America
World's Largest Bell Tsar Kolokol III in Moscow's Kremlin
World's Largest Blimp YEZ-ZA
World's Largest Bookstore Barnes & Noble in New York City
World's Largest Buddhist Nation Japan
World's Largest Carnivore polar bear
World's Largest Cassowary single-wattled cassowary *(Casuarius unappendiculatus)* of northern New Guinea
World's Largest (but unfinished) Cathedral St John the Divine in New York City
World's Largest Cave Big Room in New Mexico's Carlsbad Caverns
World's Largest Church Holy Roman Catholic Church
World's Largest City Greater New York (with Mexico City, Shanghai, and Tokyo not far behind)
World's Largest Cold Current West Wind Drift (circling Antarctica and washing the southernmost shores of Africa, Australia, and South America)
World's Largest Continent Asia
World's Largest Coral Formation Australia's Great Barrier Reef
World's Largest Countries (ranked by size) USSR, Canada, China, U.S.A., Brazil
World's Largest Delta Ganges-Brahmaputra between India and Pakistan
World's Largest Democracy India
World's Largest Desert North Africa's Sahara
World's Largest Dictionary 13-volume *Oxford English Dictionary* plus supplements
World's Largest Electronic Publisher Reuters
World's Largest Exporter of Bananas Ecuador
World's Largest Exporter of Oil Saudi Arabia
World's Largest Fish Market Tokyo
World's Largest Flower

Rafflesia arnoldii found in southwestern Sumatra

World's Largest Flower Auction Aalsmeer near Amsterdam's Schipol Airport

World's Largest Game Reserve Selous Game Reserve (in Tanzania)

World's Largest Gorge Grand Canyon of the Colorado in Arizona

World's Largest Gothic Cathedral St John the Divine, New York City

World's Largest Green-Space City Oslo, Norway followed by Stockholm, Sweden

World's Largest Gulf Gulf of Mexico

World's Largest Hindu Nation India

World's Largest Island Greenland

World's Largest Islands (*see* Islands of the World)

World's Largest Lagoon Truk Island in Micronesia

World's Largest Lakes (ranked by area) Caspian, Superior, Victoria, Aral, Huron, Michigan, Tanganyika, Great Bear, Baikal, Nyasa, Great Slave, Erie, Winnipeg, Ontario, Ladoga, Balkash, Chad, Maracaibo, Onega, Volta, Titicaca, Athabasca, Nicaragua, Eyre, Rudolf, Reindeer, Torrens, Vanern, Albert, Nipigong, Gairdner, Manitoba, Urmia, etc.

World's Largest Land Carnivore polar bear

World's Largest Library Library of Congress, Washington, D.C.

World's Largest Living Structure Australia's Great Barrier Reef

World's Largest Lumber Shipping Port Coos Bay, Oregon

World's Largest Marine Carnivore walrus

World's Largest Mountain Hawaii's Mauna Loa

World's Largest Museum New York City's American Museum of Natural History

World's Largest Nation Union of Soviet Socialist Republics

World's Largest National Park Tsavo (in Kenya)

World's Largest Naval Base Norfolk, Virginia

World's Largest Newspaper *The New York Times*

World's Largest Number of Great Cities (with a million or more people) U.S. followed by China, Latin America, and middle-south Asiatic countries including India

World's Largest Ocean Pacific

World's Largest Open Sewer Mediterranean Sea from the Bosporus and Iskenderun to the Straits of Gibraltar

World's Largest Open-Space Countries the U.S. followed by Canada and Australia

World's Largest Opera House Metropolitan Opera House, Lincoln Center, New York City

World's Largest Passenger Ship *Queen Elizabeth 2* whose deadweight tonnage exceeds the longer *Norway* (formerly the *France*)

World's Largest Peninsula Arabian Peninsula

World's Largest Ports Port of New York and New Jersey in waterfront acreage; Port of Rotterdam in tonnage

World's Largest Postage Stamp 75-cent Marshall Islands postal adhesive measuring some 4 by 6 inches (105 by 150mm)

World's Largest Prison Kharkhov in the Soviet Ukraine where more than 40,000 prisoners have been incarcerated at one time

World's Largest Public Library New York Public Library at Fifth Avenue and 42nd Street plus its more than 80 branches

World's Largest Publisher U.S. Government Printing Office (GPO)

World's Largest River Basin Amazon

World's Largest Roman Catholic Nation Brazil

World's Largest Semiconductor Maker IBM

World's Largest Shark whale shark

World's Largest Shinto Community Japan

World's Largest Ski Village Vancouver, British Columbia

World's Largest Slum Area Mexico City's 500 slums

World's Largest System of Freshwater Lakes Great Lakes of Canada and the United States

World's Largest University

State University of New York

World's Largest Walled Prison Southern Michigan Prison spanning 54 acres (22 hectares)

World's Largest Warm Current Gulf Stream

World's Least Populated Places Greenland followed by French islands in the south Indian Ocean, Svalberg in the Arctic Ocean, the Falklands in the South Atlantic, the once Spanish Sahara claimed by Algeria and Morocco, French Guiana, Namibia, Mongolia, Botswana, and Libya

World's Least Populous Nation Tuvalu

World's Loftiest Metropolis La Paz, Bolivia

World's Loneliest Meeting Place Isla de Pascua (Easter Island) in the South Pacific

World's Longest Railroad Trans-Siberian from Moscow to Vladivostok

World's Longest Rail-and-Road Bridge connects Honshu and Shikoku islands

World's Longest Railway Tunnel 13-mile-long (21-kilometer-long) Dai-shimuzu in Japan

World's Longest Reef Australia's Great Barrier

World's Longest River Nile

World's Longest Rivers (*see* Rivers of the World)

World's Longest Suspension Bridge Verrazano-Narrows Bridge in New York Harbor

World's Lowest City Brawley, California (184 feet or 56 meters below sea level)

World's Lowest Lake Dead Sea between Israel and Jordan (394 meters or 1292 feet below sea level)

World's Lowest Point Dead Sea between Israel and Jordan

World's Lowest Settlement Ein Bobek on the shores of the Dead Sea (396 meters or 1299 feet below sea level)

World's Main Choke Point for the Flow of Oil Strait of Hormuz connecting Persian Gulf countries with the Indian Ocean

World's Most Active Volcano Kilauea on the island of Hawaii

World's Most Densely Populated City Shanghai

World's Most Easterly City

Gisborne, New Zeland
World's Most Exciting City
Hong Kong, London, New
York, Tokyo, and San Francisco
World's Most Isolated City
Perth, Western Australia
World's Most Northerly City
Hammerfest, Norway
World's Most Northerly Settlement Alert, Canada
World's Most Overcrowded City Hong Kong
World's Most Populated Country China
World's Most Populated Islands Barbados, Haiti, Hong Kong, Jamaica, Java, Puerto Rico, Trinidad
World's Most Populated Place Macao
World's Most Population-Exploding Nation India
World's Most Populous Nation China
World's Most Prolific Composer Wolfgang Amadeus Mozart or Georg Philipp Telemann
World's Most Southerly City Punta Arenas, Chile
World's Most Terror-stricken Country Lebanon
World's Most Westerly City Nome, Alaska or Pago Pago, Samoa
World's Most Widely Prevalent Carnivore the red fox
World's Most Widely Spoken Language English
World's Northernmost Point of Land Greenland's Cape Morris Jesup
World's Oldest Capital City Damascus, Syria
World's Oldest Constitutional Democracy the United States of America
World's Oldest Kingdom Denmark
World's Oldest Medical Journal *New England Journal of Medicine*
World's Oldest Monarchy Denmark
World's Oldest Orchestra Dresdener Staatskapelle founded in Dresden in 1548
World's Oldest Parliament Iceland's *Althing* founded in 930
World's Oldest Profession prostitution

World's Only Marine Lizard Galápagos Marine Iguana
World's Only Poisonous Lizards gila monster of the American southwest; Mexican beaded lizard called *escorpion*; Borneo's *Lanthanotus*
World's Rainiest City Monrovia, Liberia
World's Rainiest Place Mount Waileale, Hawaii
World's Richest Countries (per capita income) United Arab Emirates, Qatar, Kuwait, Liechtenstein, Switzerland, Sweden, Monaco, the United States, Canada, Germany, Australia, Denmark, Belgium, Andorra, and Norway
World's Second-Oldest Profession arms making and trading
World's Shortest Poem *I—why?* by Eli Siegal
World's Smallest Bird Cuban hummingbird
World's Smallest Continent Australia
World's Smallest Monkey pygmy marmoset about 6 inches (14 cm) from nose to base of tail
World's Smallest Sovereign State Vatican City within Rome occupies 44 hectares (109 acres)
World's Smoggiest City Mexico City ·
World's Southernmost Town Puerto Williams, Chile
World's Tallest Bird New Zealand's *Diornis maximus*, an extinct flightless ostrich-like bird measuring 13 feet (396 cm) in height
World's Tallest Building 110-story Sears Tower in Chicago (1,454 feet or 444 meters high)
World's Tallest Buildings New York City's World Trade Center (each building 110 stories with the second topped by a television tower)
World's Tallest Self-Supporting Structure 1,821-foot (555-meter) CN Tower, Toronto, Ont
World's Tallest Structure Canadian National Railway's Communication and Observa-

tion Tower in downtown Toronto, Ontario (1,805 feet high—550 meters)
World's Tallest Trees California's redwoods, towering 367 feet (112 meters) high
World's Warmest City Timbuktu, Mali
World's Wettest City Monrovia, Liberia
World's Wettest Place Mount Waialeale on the island of Kauai in Hawaii
World's Windiest Place Adélie coast of Antarctica
World's Worst Shipwreck unsinkable superliner *Titanic* struck an iceberg on April 14, 1912 and sank in the North Atlantic south of Newfoundland with a loss of more than 1500 lives
Wyoming's Highest Point Gannett Peak—13,785 feet, 4201 meters
Wyoming's Largest City Casper
Wyoming's Most Spectacular Park Yellowstone National Park with headquarters at Mammoth Hot Springs
Youngest Central American Democracy Belize (formerly British Honduras)
Youngest Province Newfoundland, Canada including Labrador
Yugoslavia's Highest Peak Triglav—9,395 feet, 2,847 meters—in the Julian Alps
Yugoslavia's Largest City Belgrade
Yugoslavia's Longest River Danube
Yugoslavia's Principal Port Rijeka
Yukon's Highest Point Mount Logan—19,850 feet, 6,050 meters
Yukon's Largest City Whitehorse
Zaire's Highest Peak Mount Stanley—16,795 feet, 5,089 meters
Zaire's Largest City Kinshasa
Zaire's Principal Port Matadi
Zambia's Largest City Lusaka
Zimbabwe's Highest Peak Inyanga—8,514 feet, 2,580 meters
Zimbabwe's Largest City Harare

U.S. Naval Ship Symbols

AD Destroyer Tender
ADG Degaussing Ship
AE Ammunition Ship
AF Store Ship
AFDB Large Auxiliary Floating Dry Dock (non-self-propelled)
AFDL Small Auxiliary Floating Dry Dock (non-self-propelled)
AFDM Medium Auxiliary Floating Dry Dock (non-self-propelled)
AFS Combat Store Ship
AG Miscellaneous
AGDE Escort Research Ship
AGEH Hydrofoil Research Ship
AGER Environmental Research Ship
AGF Miscellaneous Command Ship
AGM Missile Range Instrumentation Ship
AGMR Major Communications Relay Ship
AGOR Oceanographic Research Ship
AGP Patrol Craft Tender
AGR Radar Picket Ship
AGS Surveying Ship
AGSS Auxiliary Submarine
AGTR Technical Research Ship
AH Hospital Ship
AK Cargo Ship
AKD Cargo Ship, Dock
AKL Light Cargo Ship
AKR Vehicle Cargo Ship
AKS Stores Issue Ship
AKV Cargo Ship and Aircraft Ferry
ANL Net Laying Ship
AO Oiler
AOE Fast Combat Support Ship
AOG Gasoline Tanker
AOR Replenishment Oiler
AP Transport
APB Self-propelled Barracks Ship
APL Barracks Craft (non-self-propelled)
AR Repair Ship
ARB Battle Damage Repair Ship
ARC Cable Repairing Ship
ARD Auxiliary Repair Dry Dock (non-self-propelled)
ARDM Medium Auxiliary Repair Dry Dock (non-self-propelled)

ARG Internal Combustion Engine Repair Ship
ARL Landing Craft Repair Ship
ARS Salvage Ship
ARSD Salvage Lifting Ship
ARST Salvage Craft Tender
ARVA Aircraft Repair Ship (aircraft)
ARVE Aircraft Repair Ship (engine)
ARVH Aircraft Repair Ship (helicopter)
AS Submarine Tender
ASPB Assault Support Patrol Boat
ASR Submarine Rescue Ship
ATA Auxiliary Ocean Tug
ATC Armored Troop Carrier
ATF Fleet Ocean Tug
ATS Salvage Tug
ATSS Auxiliary Training Submarine
AV Seaplane Tender
AVM Guided Missile Ship
AVS Aviation Supply Ship
AVT Auxiliary Aircraft Transport
AW Distilling Ship
BB Battleship
CA Heavy Cruiser
CC Command Ship
CCB Command and Control Boat
CG Guided Missile Cruiser
CGN Guided Missile Cruiser (nuclear propulsion)
CL Light Cruiser
CLG Guided Missile Light Cruiser
CVA Attack Aircraft Carrier
CVAN Attack Aircraft Carrier (nuclear propulsion)
CVS ASW Support Aircraft Carrier
CVT Training Aircraft Carrier
DD Destroyer
DDG Guided Missile Destroyer
DE Escort Ship
DEG Guided Missile Escort Ship
DER Radar Picket Escort Ship
DL Frigate
DLG Guided Missile Frigate
DLGN Guided Missile Frigate (nuclear propulsion)
DSRV Deep Submergence Rescue Vessel
DSV Deep Submergence Vehicle

E (Prefix) Experimental Ship
F (Prefix) Ship being built by U.S. for a foreign nation
FDL Fast Deployment Logistics Ship
IX Unclassified Miscellaneous
LCA Landing Craft, Assault
LCC Amphibious Command Ship
LCM Landing Craft, Mechanized
LCPL Landing Craft, Personnel, Large
LCPR Landing Craft, Personnel, Ramped
LCSR Landing Craft Swimmer Reconnaissance
LCU Landing Craft, Utility
LCVP Landing Craft, Vehicle, Personnel
LFR Inshore Fire Support Ship
LFS Amphibious Fire Support Ship
LHA Amphibious Assault Ship (general purpose)
LKA Amphibious Cargo Ship
LPA Amphibious Transport
LPD Amphibious Transport Dock
LPH Amphibious Assault Ship
LPR Amphibious Transport (small)
LPSS Amphibious Transport Submarine
LSD Dock Landing Ship
LSSC Light SEAL Support Craft
LST Tank Landing Ship
LWT Amphibious Warping Tug
MAC MIUW Attack Craft
MCS Mine Countermeasures Ship
MON Monitor
MSB Minesweeping Boat
MSC Minesweeper, Coastal (nonmagnetic)
MSD Minesweeper, Drone
MSF Minesweeper, Fleet (steel hull)
MSI Minesweeper, Inshore
MSL Minesweeping Launch
MSM Minesweeper, River (Converted LCM-6)
MSO Minesweeper, Ocean (nonmagnetic)
MSR Minesweeper, Patrol
MSS Minesweeper, Special (device)
MSSC Medium SEAL Support Craft

NR Submersible Research Vehicle (nuclear propulsion)
PBR River Patrol Boat
PCE Patrol Escort
PCER Patrol Rescue Escort
PCF Patrol Craft, Inshore
PCH Patrol Craft (hydrofoil)
PG Patrol Gunboat
PGH Patrol Gunboat (hydrofoil)
PTF Fast Patrol Craft
QFB Quiet Fast Boat
RUC Riverine Utility Craft
SDV Swimmer Delivery Vehicle
SES Surface-Effect Ship
SS Submarine
SSBN Fleet Ballistic Missile Submarine (nuclear propulsion)
SSG Guided Missile Submarine
SSN Submarine (nuclear propulsion)
SST Target and Training Submarine (self-propelled)
STAB Strike Assault Boat
T (Prefix) Military Sealift Command Ship
W (Prefix) U.S. Coast Guard Ship
X Submersible Craft (self-propelled)
YAG Miscellaneous Auxiliary (self-propelled)
YC Open Lighter (non-self-propelled)
YCF Car Float (non-self-propelled)
YCV Aircraft Transportation Lighter (non-self-propelled)
YD Floating Crane (non-self-propelled)
YDT Diving Tender (non-self-propelled)
YF Covered Lighter (self-propelled)
YFB Ferryboat or Launch (self-propelled)
YFD Yard Floating Dry Dock (non-self-propelled)
YFN Covered Lighter (non-self-propelled)
YFNB Large Covered Lighter (non-self-propelled)
YFND Dry Dock Companion Craft (non-self-propelled)
YFNX Lighter (special purpose) (non-self-propelled)
YFP Floating Power Barge (non-self-propelled)
YFR Refrigerated Covered Lighter (self-propelled)
YFRN Refrigerated Covered Lighter (non-self-propelled)
YFRT Covered Lighter (range-tender) (self-propelled)
YFU Harbor Utility Craft (self-propelled)
YG Garbage Lighter (self-propelled)
YGN Garbage Lighter (non-self-propelled)
YHLC Salvage Lift Craft, Heavy (non-self-propelled)
YLLC Salvage Lift Craft, Light (self-propelled)
YM Dredge (self-propelled)
YMLC Salvage Lift Craft, Medium (non-self-propelled)
YNG Gate Craft (non-self-propelled)
YO Fuel Oil Barge (self-propelled)
YOG Gasoline Barge (self-propelled)
YOGN Gasoline Barge (non-self-propelled)
YON Fuel Oil Barge (non-self-propelled)
YOS Oil Storage Barge (non-self-propelled)
YP Patrol Craft (self-propelled)
YPD Floating Pile Driver (non-self-propelled)
YR Floating Workshop (non-self-propelled)
YRB Repair and Berthing Barge (non-self-propelled)
YRBM Repair, Berthing and Messing Barge (non-self-propelled)
YRDH Floating Dry Dock Workshop (hull) (non-self-propelled)
YRDM Floating Dry Dock Workshop (machine) (non-self-propelled)
YRR Radiological Repair Barge (non-self-propelled)
YRST Salvage Craft Tender (non-self-propelled)
YSD Seaplane Wrecking Derrick (self-propelled)
YSR Sludge Removal Barge (non-self-propelled)
YTB Large Harbor Tug (self-propelled)
YTL Small Harbor Tug (self-propelled)
YTM Medium Harbor Tug (self-propelled)
YW Water Barge (self-propelled)
YWDN Water Distilling Barge (non-self-propelled)
YWN Water Barge (non-self-propelled)

Vehicle Registration Symbols (Index Markers)
British Isles Symbols

This listing includes registration marks for the Republic of Ireland. These marks were introduced before the creation of the Republic of Ireland, which was continued with the same system.

A	London	AJ	Yorkshire (NR)	AU	Nottingham
AA	Hampshire	AK	Bradford	AV	Aberdeenshire
AB	Worcestershire	AL	Nottinghamshire	AW	Salop
AC	Warwickshire	AM	Wilshire	AX	Monmouthshire
AD	Gloucestershire	AN	London	AY	Leicestershire
AE	Bristol	AO	Cumberland	AZ	Belfast
AF	Cornwall	AP	Sussex (East)	B	Lancashire
AG	Ayrshire	AR	Hertfordshire	BA	Salford
AH	Norfolk	AS	Nairnshire	BB	Newcastle upon Type
AI	Meath	AT	Kingston-upon-Hull	BC	Leicester

BD	Northamptonshire	DS	Peeblesshire	GF	London
BE	Lincolnshire (Lindsey)	DT	Doncaster	GG	Glasgow
BF	Staffordshire	DU	Coventry	GH	London
BG	Burkenhead	DV	Devon	GI	London
BH	Buckinghamshire	DW	Newport (Mon)	GK	London
BI	Monaghan	DX	Ipswich	GL	Bath
BJ	Suffolk (East)	DY	Hastings	GM	Motherwell & Wishaw
BK	Portsmouth	DZ	Antrim	GN	London
BL	Berkshire	E	Staffordshire	GO	London
BM	Bedfordshire	EA	West Bromwich	GP	London
BN	Bolton	EB	Cambridge	GR	Sunderland
BO	Cardiff	EC	Westmorland	GS	Perthshire
BP	Sussex (West)	ED	Warrington	GT	London
BR	Sunderland	EE	Grimsby	GU	London
BS	Orkney	EF	West Hartlepool	GV	Suffolk (West)
BT	Yorkshire (ER)	EG	Huntingdon	GW	London
BU	Oldham	EH	Stone-on-Trent	GX	London
BV	Blackburn	EI	Sligo	GY	London
BW	Oxfordshire	EJ	Cardiganshire	GZ	Belfast
BX	Carmarthenshire	EK	Wigan	H	London
BY	London	EL	Bournemouth	HA	Warley
BZ	Down	EM	Bootle	HB	Merthyr Tydfil
C	Yorkshire (WR)	EN	Bury	HC	Eastbourne
CA	Denbighshire	EO	Berrow-in-Furness	HD	Dewsbury
CB	Blackburn	EP	Montgomeryshire	HE	Barnsley
CC	Caernarvonshire	ER	Cambridgeshire	HF	Wallasey
CD	Brighton	ES	Perthshire	HG	Burnley
CE	Cambridgeshire	ET	Rotherham	HH	Carlisle
CF	Suffolk (West)	EU	Breconshire	HI	Tipperary
CG	Hampshire	EV	Essex	HJ	Southend
CH	Derby	EW	Huntingdonshire	HK	Essex
CI	Laoighis	EX	Great Yarmouth	HL	Wakefield
CJ	Herefordshire	EY	Anglesey	HM	London
CK	Preston	EZ	Belfast	HN	Darlington
CL	Norwich	F	Essex	HO	Hampshire
CM	Birkenhead	FA	Burton-on-Trent	HP	Coventry
CN	Gateshead	FB	Bath	HR	Wiltshire
CO	Plymouth	FC	Oxford	HS	Renfrewshire
CP	Halifax	FD	Dudley	HT	Bristol
CR	Southampton	FE	Lincoln	HU	Bristol
CS	Ayshire	FF	Merionethshire	HV	London
CT	Lincolnshire (Kesteven)	FG	Fife	HW	Bristol
CU	South Shields	FH	Gloucester	HX	London
CV	Cornwall	FI	Tipperary (NR)	HY	Bristol
CW	Burnley	FJ	Exeter	HZ	Tyrone
CX	Huddersfield	FK	Worcester	IA	Antrim
CY	Swansea	FL	Huntingdon	IB	Armagh
CZ	Belfast	FM	Chester	IC	Carlow
D	Kent	FN	Canterbury	ID	Cavan
DA	Wolverhampton	FO	Radnorshire	IE	Clare
DB	Stockport	FP	Rutland	IF	Cork (County)
DC	Teesside	FR	Blackpool	IH	Donegal
DD	Gloucestershire	FS	Edinburgh	IJ	Down
DE	Pembrokeshire	FT	Tynemouth	IK	City and County of Dublin
DF	Gloucestershire	FU	Lincolnshire (Lindsey)		
DG	Gloucestershire	FV	Blackpool	IL	Fermanagh
DH	Walsall	FW	Lincolnshire (Lindsey)	IM	Galway
DI	Roscommon	FX	Dorset	IN	Kerry
DJ	St Helens	FY	Southport	IO	Kildare
DK	Rochdale	FZ	Belfast	IP	Kikenny
DL	Isle of Wight	G	Glasgow	IR	Offaly
DM	Flintshire	GA	Glasgow	IT	Leitrim
DN	York	GB	Glasgow	IU	Limerick
DO	Lincolnshire (Holland)	GC	London	IW	Londonderry
DP	Reading	GD	Glasgow	IX	Longford
DR	Plymouth	GE	Glasgow	IY	Louth

IZ	Mayo	LM	London	OA	Birmingham
J	Durham (County)	LN	London	OB	Birmingham
JA	Stockport	LO	London	OC	Birmingham
JB	Berkshire	LP	London	OD	Devon
JC	Caenarvonshire	LR	London	OE	Birmingham
JD	London	LS	Selkirkshire	OF	Birmingham
JE	Cambridge	LT	London	OG	Birmingham
JF	Leicester	LU	London	OH	Birmingham
JG	Canterbury	LV	Liverpool	OI	Belfast
JH	Hertfordshire	LW	London	OJ	Birmingham
JI	Tyrone	LX	London	OK	Birmingham
JJ	London	LY	London	OL	Birmingham
JK	Eastbourne	LZ	Armagh	OM	Birmingham
JL	Lincolnshire (Holland)	M	Cheshire	ON	Birmingham
JM	Westmorland	MA	Cheshire	OO	Essex
JN	Southend	MB	Cheshire	OP	Birmingham
JO	Oxford	MC	London	OR	Hampshire
JP	Wigan	MD	London	OS	Wigtownshire
JR	Northumberland	ME	London	OT	Hampshire
JS	Ross & Cromarty	MF	London	OU	Hampshire
JT	Dorset	MG	London	OV	Birmingham
JU	Leicestershire	MH	London	OW	Southampton
JV	Grimsby	MI	Wexford	OX	Birmingham
JW	Wolverhampton	MJ	Bedfordshire	OY	London
JX	Halifax	MK	London	OZ	Belfast
JY	Plymouth	ML	London	P	Surrey
JZ	Down	MM	London	PA	Surrey
K	Liverpool	MN	Isle of Man	PB	Surrey
KA	Liverpool	MO	Berkshire	PC	Surrey
KB	Liverpool	MP	London	PD	Surrey
KC	Liverpool	MR	Wittshire	PE	Surrey
KD	Liverpool	MS	Stirlingshire	PF	Surrey
KE	Kent	MT	London	PG	Surrey
KF	Liverpool	MU	London	PH	Surrey
KG	Cardiff	MV	London	PI	Cork
KH	Kingston-upon-Hull	MW	Wiltshire	PJ	Surrey
KI	Waterford	MX	London	PK	Surrey
KJ	Kent	MY	London	PL	Surrey
KK	Kent	MZ	Belfast	PM	Sussex (East)
KL	Kent	N	Manchester	PN	Sussex (East)
KM	Kent	NA	Manchester	PO	Sussex (West)
KN	Kent	NB	Manchester	PP	Buckinghamshire
KO	Kent	NC	Manchester	PR	Dorset
KP	Kent	ND	Manchester	PS	Zetland
KR	Kent	NE	Manchester	PT	Durham (County)
KS	Rosburghsh	NF	Manchester	PU	Essex
KT	Kent	NG	Norfolk	PV	Ipswich
KU	Bradford	NH	Northampton	PW	Norfolk
KV	Coventry	NI	Wicklow	PX	Sussex (West)
KW	Bradford	NJ	Sussex (East)	PY	Yorkshire (NR)
KX	Buckinghamshire	NK	Hertfordshire	PZ	Belfast
KY	Bradford	NL	Northumberland		QA QE QJ QN
KZ	Antrim	NM	Bedfordshire		QB QF QK QO
L	Glamorgan	NN	Nottinghamshire		QC QG QL QP
LA	London	NO	Essex		QD QH QM QS
LB	London	NP	Worcestershire		London for vehicles
LC	London	NR	Leicestershire		temporarily imported
LD	London	NS	Sutherland		from abroad
LE	London	NT	Salop	R	Derbyshire
LF	London	NU	Derbyshire	RA	Derbyshire
LG	Cheshire	NV	Northamptonshire	RB	Derbyshire
LH	London	NW	Leeds (B)	RC	Derby
LI	Westmeath	NX	Warwickshire	RD	Reading
LJ	Bournemouth	NY	Glamorgan	RE	Staffordshire
LK	London	NZ	Londonderry	RF	Staffordshire
LL	London	O	Birmingham	RG	Aberdeen

RH	Kingston-upon-Hull		TW	Essex		WK	Coventry
RI	City and County of		TX	Glamorgan		WL	Oxford
	Dublin		TY	Northumberland		WM	Southport
RJ	Salford		TZ	Belfast		WN	Swansea
RK	London		U	Leeds		WO	Monmouthshire
RL	Cornwall		UA	Leeds		WP	Worcestershire
RM	Cumberland		UB	Leeds		WR	Yorkshire (WR)
RN	Preston		UC	London		WS	Edinburgh
RO	Hertfordshire		UD	Oxfordshire		WT	Yorkshire (WR)
RP	Northamptonshire		UE	Warwickshire		WU	Yorkshire (WR)
RR	Nottinghamshire		UF	Brighton		WV	Wiltshire
RS	Aberdeen		UG	Leeds		WW	Yorkshire (WR)
RT	Suffolk (East)		UH	Cardiff		WX	Yorkshire (WR)
RU	Bournemouth		UI	Londonderry		WY	Yorkshire (WR)
RV	Portsmouth		UJ	Salop		WZ	Belfast
RW	Coventry		UK	Wolverhampton		X	Northumberland
RX	Berkshire		UL	London		XA	London/Kirkcaldy
RY	Leicester		UM	Leeds		XB	London/Coatbridge
RZ	Antrim		UN	Denbighshire		XC	London/Solihull
S	Edinburgh		UO	Devon		XD	London/Luton
SA	Aberdeenshire		UP	Durham (County)		XE	LondonLuton
SB	Argyll		UR	Hertfordshire		XF	London/Torbay
SC	Edinburgh		US	Glasgow		XG	Teesside
SD	Ayrshire		UT	Leicestershire		XH	London
SE	Banffshire		UU	London		XI	Belfast
SF	Edinburgh		UV	London		XJ	Manchester
SG	Edinburgh		UW	London		XK	London
SH	Berwickshire		UX	Salop		XL	London
SJ	Bute		UY	Worcestershire		XM	London
SK	Caithness		UZ	Belfast		XN	London
SL	Clarkmannanshire		V	Lanarkshire		XO	London
SM	Dumfriesshire		VA	Lanarkshire		XP	London
SN	Dunbartonshire		VB	London		XR	London
SO	Moray		VC	Coventry		XS	Paisley
SP	Fife		VD	Lanarkshire		XT	London
SR	Angus		VE	Cambridgeshire		XU	London
SS	East Lothian		VF	Norfolk		XV	London
ST	Inverness-shire		VG	Norwich		XW	London
SU	Kincardineshire		VH	Huddersfield		XX	London
SV	Kinross-shire		VJ	Herefordshire		XY	London
SW	Kircudbrightshire		VK	Newcastle upon Tyne		XZ	Armagh
SX	West Lothian		VL	Lincoln		Y	Somerset
SY	Midlothian		VM	Manchester		YA	Somerset
SZ	Down		VN	Yorkshire (NR)		YB	Somerset
T	Devon		VO	Nottinghamshire		YC	Somerset
TA	Devon		VP	Birmingham		YD	Somerset
TB	Lancashire		VR	Manchester		YE	London
TC	Lancashire		VS	Greenock		YF	London
TD	Lancashire		VT	Stoke-on-Trent		YG	Yorkshire (WR)
TE	Lancashire		VU	Manchester		YH	London
TF	Lancashire		VV	Northampton		YI	City and County of
TG	Glamorgan		VW	Essex			Dublin
TH	Carmarthenshire		VX	Essex		YJ	Dundee
TI	Limerick		VY	York		YK	London
TJ	Lancashire		VZ	Tyrone		YL	London
TK	Dorset		W	Sheffield		YM	London
TL	Lincolnshire (Kesteven)		WA	Sheffield		YN	London
TM	Bedfordshire		WB	Sheffield		YO	London
TN	Newcastle upon Tyne		WC	Essex		YP	London
TO	Nottingham		WD	Warwickshire		YR	London
TP	Portsmouth		WE	Sheffield		YS	Glasgow
TR	Southampton		WF	Yorkshire (ER)		YT	London
TS	Dundee		WG	Stirlingshire		YU	London
TT	Devon		WH	Bolton		YV	London
TU	Cheshire		WI	Waterford		YW	London
TV	Nottingham		WJ	Sheffield		YX	London

YY	London	ZF	Cork		Dublin
YZ	Londonderry	ZH	City and County of	ZP	Donegal
Z	City and County of		Dublin	ZR	Wexford
	Dublin	ZI	City and County of	ZT	Cork (County)
ZA	City of County of		Dublin	ZU	City and County of
	Dublin	ZJ	City and County of		Dublin
ZB	Cork (County)		Dublin	ZW	Kildare
ZC	City and County of	ZK	Cork (County)	ZX	Kerry
	Dublin	ZL	City and County of	ZY	Louth
ZD	City and County of		Dublin	ZZ	Dublin for vehicles
	Dublin	ZM	Galway		temporarily imported
ZE	City and County of	ZN	Meath		from abroad
	Dublin	ZO	City and County of		

International Symbols

Albania	BSA	Indonesia	YDNI	Poland	PKNiM		
Algeria	INAPI	Iran	ISIRI	Portugal	IGPAI		
Australia	SAA	Iraq	IOS	Romania	IRS		
Austria	ON	Ireland	IIRS	Saudi Arabia	SASO		
Bangladesh	BDSI	Israel	SII	Singapore	SISIR		
Belgium	IBN	Italy	UNI	South Africa, Rep. of	SABS		
Brazil	ABNT	Jamaica	JBS	Spain	IRA-		
Bulgaria	DKC	Japan	JISC		NOR		
Canada	SCC	Kenya	KEBS	Sri Lanka	BCS		
Chile	INN	Korea, Dem. P. Rep.	CSK	Sudan	SSD		
Colombia	ICON-	of		Sweden	SIS		
	TEC	Korea, Rep. of	KBS	Switzerland	SNV		
Cuba	NC	Lebanon	LIBNOR	Thailand	TISI		
Czechoslovakia	CSN	Malaysia	SIRIM	Turkey	TSE		
Denmark	DS	Mexico	DGN	United Kingdom	BSI		
Egypt, Arab Rep. of	EOS	Morocco	SNIMA	United States of	ANSI		
Ethiopia	ESI	Netherlands	NNI	America			
Finland	SFS	New Zealand	SANZ	Union of Soviet	GOST		
France	AFNOR	Nigeria	NSO	Socialist Republics			
Germany	DIN	Norway	NSF	Venezuela	COVE-		
Ghana	GSB	Pakistan	PSI		NIN		
Greece	NHS	Peru	ITIN-	Yugoslavia	JZS		
Hungary	MSZH		TEC	Zambia	ZSI		
India	ISI	Philippines	PS				

Weather Symbols (Beaufort Scales)

WITH CORRESPONDING SEA STATE CODES

Beaufort number	Wind speed						Estimating wind speed		Hydrographic Office		International	
	knots	mph	meters per second	km per hour	Seaman's term	U.S. Weather Bureau term	Effects observed at sea	Effects observed on land	Term and height of waves, in feet	Code	Term and height of waves, in feet	Code
0	under 1	under 1	0.0–0.2	under 1	Calm		Sea like mirror.	Calm; smoke rises vertically.	Calm, 0	0	Calm, glassy, 0	0
1	1–3	1–3	0.3–1.5	1–5	Light air	Light	Ripples with appearance of scales; no foam crests.	Smoke drift indicates wind direction; vanes do not move.	Smooth, less than 1	1	Rippled, 0–1	1
2	4–6	4–7	1.6–3.3	6–11	Light breeze	Light	Small wavelets; crests of glassy appearance, not breaking.	Wind felt on face; leaves rustle; vanes begin to move.	Slight, 1–3	2	Smooth, 1–2	2
3	7–10	8–12	3.4–5.4	12–19	Gentle breeze	Gentle	Large wavelets; crests begin to break; scattered whitecaps.	Leaves, small twigs in constant motion; light flags extended.	Moderate, 3–5	3	Slight, 2–4	3
4	11–16	13–18	5.5–7.9	20–28	Moderate breeze	Moderate	Small waves, becoming longer; numerous whitecaps.	Dust, leaves, and loose paper raised up; small branches move.	Rough, 5–8	4	Moderate, 4–8	4
5	17–21	19–24	8.0–10.7	29–38	Fresh breeze	Fresh	Moderate waves, taking longer form; many whitecaps; some spray.	Small trees in leaf begin to sway.			Rough, 8–13	5
6	22–27	25–31	10.8–13.8	39–49	Strong breeze	Fresh	Larger waves forming; whitecaps everywhere; more spray.	Larger branches of trees in motion; whistling heard in wires.	Very rough, 8–12	5	Very rough, 13–20	6
7	28–33	32–38	13.9–17.1	50–61	Moderate gale	Strong	Sea heaps up; white foam from breaking waves begins to be blown in streaks.	Whole trees in motion; resistance felt in walking against wind.				
8	34–40	39–46	17.2–20.7	62–74	Fresh gale	Gale	Moderately high waves of greater length; edges of crests begin to break into spindrift; foam is blown in well-marked streaks.	Twigs and small branches broken off trees; progress generally impeded.	High, 12–20	6		
9	41–47	47–54	20.8–24.4	75–88	Strong gale	Gale	High waves; sea begins to roll; dense streaks of foam; spray may reduce visibility.	Slight structural damage occurs; slate blown from roofs.			High, 20–30	7
10	48–55	55–63	24.5–28.4	89–102	Whole gale	Whole gale	Very high waves with overhanging crests; sea takes white appearance as foam is blown in very dense streaks; rolling is heavy and visibility reduced.	Seldom experienced on land; trees broken or uprooted; considerable structural damage occurs.	Very high, 20–40	7		
11	56–63	64–72	28.5–32.6	103–117	Storm	Whole gale	Exceptionally high waves; sea covered with white foam patches; visibility still more reduced.	Very rarely experienced on land; usually accompanied by widespread damage.	Mountainous, 40 and higher	8	Very high, 30–45	8
12	64–71	73–82	32.7–36.9	118–133	Hurricane	Hurricane	Air filled with foam; sea completely white with driving spray; visibility greatly reduced.		Confused	9	Phenomenal, over 45	9
13	72–80	83–92	37.0–41.4	134–149	Hurricane	Hurricane						
14	81–89	93–103	41.5–46.1	150–166	Hurricane	Hurricane						
15	90–99	104–114	46.2–50.9	167–183	Hurricane	Hurricane						
16	100–108	115–125	51.0–56.0	184–201	Hurricane	Hurricane						
17	100–118	126–136	56.1–61.2	202–220	Hurricane	Hurricane						

Note: Since January 1, 1955, weather map symbols have been based upon wind speed in knots, at five-knot intervals, rather than upon Beaufort number.

Wedding Anniversary Symbols

1st - *Paper* (negotiable paper such as bonds, currency, trust certificates, as well as books, napkins, stationery, and towels)

2nd - *Cotton* (bedspreads, curtains, draperies, pillows, sheets, shirts, socks, underwear, etc.)

3rd - *Leather* (belts, handbags, leatherbound books, luggage, shoes, etc.)

4th - *Linen* (bedsheets, napkins, samplers, scarfs, shirts, tablecloths)

5th - *Wood* (furniture as well as boats and bungalows)

6th - *Iron* (hardware, wrought-iron furniture, ornamental ironwork)

7th - *Wool* (blankets, robes, rugs, socks, suits, sweaters, underwear)

8th - *Bronze* (bells, brassware, bronze objects, gongs, statuary)

9th - *Pottery* (kitchenware, planter's pots, pottery ornaments)

10th - *Aluminum* or *in* (kitchenware and ornaments)

11th - *Steel* (automobiles, hardware, recreation vehicles, tools)

12th - *Silk* (casual clothes, scarfs, wraps)

13th - *Lace* (bedspreads, curtains, doilies, tablecloths)

14th - *Ivory* (carvings, desk sets, scrimshaw)

15th - *Crystal* (crystal sculpture and glassware)

20th - *China* (chinaware and porcelain figurines and tableware)

25th - *Silver* (silver coins and silverware)

30th - *Pearl* (jewelry and mother-of-pearl objects)

35th - *Coral* (jewelry and rare collector's items)

40th - *Ruby* (jewelry)

45th - *Sapphire* (jewelry)

50th - *Golden* (gold coins, gold-plated, solid-gold ornaments)

55th - *Emerald* (jewelry)

60th - *Diamond* (jewelry)

65th - *Diamond-and-gold anniversary* (jewelry)

70th - *Diamond-and-emerald anniversary* (jewelry)

75th - *Diamond-emerald-sapphire anniversary* (solid gold dipped in diamond, emerald, and sapphire chips or stones)

80th - (consult your nearest jeweler; contact the media and the police if you have accumulated all the foregoing wedding anniversary gifts; treat yourself to whatever you want)—this is the *time-flies anniversary* and may earn you a place in the *Guinness Book of World Records*

Winds of the World

The wind bloweth where it listeth.
—John 3:8

Afer hot southwest wind in Italy, so called because it comes from Africa; also called Africanus ventus (the African wind), Africino, Africo, Africuo

Antitrades winds blowing above the trade winds but in opposite directions

Apheliotes (Greek—East Wind)

Avalaison steady west wind of western France

Bad-i-sad-o-bist roz (Persian— 120-day wind)—northerly dust-and-salt-laden wind blowing over Seistan province of Iran

from June through September

Baguios hurricane storms characteristic of the Phillippine Islands

Bat Furan (Arabic—Open-Sea Season)—when northeast or winter monsoon wafts over the Arabian Sea with light winds favoring small sailing vessels

Bat Hiddan (Arabic—Closed-Sea Season)—when southwest or summer monsoon agitates the Arabian Sea with high winds

Bergwind foehn wind of South Africa's south coast

Bise cold and dry northerly wind of southern France and Switzerland

Black Roller dust storm common to western United States

Blizzard cold northerly gale occurring during winter months in Canadian provinces and north United States; great Blizzard of 1888 covered much of Canada and northern United States; needlelike ice crystals and fine dry snow make up the blizzard's pattern of penetrating cold

Bohorok foehn wind of Sumatra

Bora cold north wind blowing over the Adriatic and originating in the Dinaric Alps

Boreas (Greek—North Wind)

Brave West Wind westerly winds of the southern hemisphere

Breath of the Sahara the Sirocco

Breva and Tivano afternoon and morning winds blowing over Lake Como—Breva blows from north to south, Tivano blows from south to north

Brickfielder dusty hot wind originating in sandy wastes of central Australia

Buran blizzard of Central Asia

Burster southerly wind of New South Wales

Canterbury Northwester hot dry wind sometimes blowing over New Zealand

Caurus the Northwest Wind

Chemsin (Arabic—Sirocco)

Chergui Moroccan name for the Sirocco

Chichili Algerian name for the Sirocco

Chili Tunisian name for the Sirocco

Chinook foehn wind blowing over the plains east of the Rockies from northern Canada to southern Colorado; warm southwesterly wind characteristic of the lower Columbia River of Oregon and Washington

Choclatero chocolate-colored dusty wind common about

Yucatan

Chubasco rain-filled violent wind threatening west coast of Mexico from May to November

Cordonazo de San Francisco autumnal equinox falling close to St Francis Day—4 October—and often ushered in by a short but violent hurricane; blow struck with a knotted cord or rope like one worn by St Francis; storm felt along west coast of Central America and Mexico around St Francis Day during autumnal equinox

Coromuel southerly land breeze felt from November to May and from sunset to about 9 A.M. around La Paz and nearby entrance to Gulf of California

Cyclones counterclockwise winds of the northern hemisphere—often called tropical cyclones: hurricane in West Indies, typhoon in China Sea, willy-willy off northwestern Australia; clockwise winds of the southern hemisphere—frequently of great force and considerable duration

Doctor sea breeze refreshing inhabitants of African coasts and west coast of Australia

Dust Bowl area suffering from dust storms as in Oklahoma, West Texas, New Mexico, and Arizona

Dust Devil harmless whirls of dust ascending from the desert floor as high as 3000 feet; may be as wide as 10 feet

East Wind rainy wind characteristic of many places such as England, New England, etc.

Eecatl (Aztec—Wind)—derived from the wind god Quetzalcoatl

Etesian Wind northerly summer wind found in the eastern Mediterranean

Euros (Greek—Southeast Wind)

Favonius (Latin—South Wind)—also known as Foehn or Föhn

Foehn warm dry mountainous wind characteristic of the Alps where its downward rush melts snowdrifts rapidly

Fremantle Doctor cool southwest wind coming from the Indian Ocean to the Swan River Valley of Western Australia around Fremantle and Perth

Friagem (Portuguese—Cold

Wave)—sometimes lasts for several days during Brazil's winter season

Furious Fifties storms ranging from west to east in the south fifty latitudes of the southern hemisphere

Gale wind of about 35 miles per hour (56 kilometers per hour)—a high wind

Garmsal hot wind of Turkestan

Ghibli Libyan name for the Sirocco

Greco the Greek wind—easterly wind encountered in the Mediterranean

Gregale northerly wind of south central Mediterranean area—the Greek gale—often a cold northeast wind blowing from eastern Mediterranean

Haboob Sudanese dust storm noted for its many colors and gritty intensity

Harmattan dry dusty desert wind blowing to Atlantic coast of Africa from the Sahara

Helm Wind cold northeasterly wind of northern England

Hubbub Sudanese sandstorm

Huracán (Spanish—Hurricane)

Hurricane devastating rain-filled wind along Atlantic coast of the United States originating in the Caribbean and Gulf of Mexico; hurricane months recalled by these lines: June—too soon; July—stand by; August—look out you must; September—remember, October—all over

Ibe foehn-type wind blowing through Dzungarian Gate in western China near Lake Balkash

Irish Hurricane a flat calm when no wind blows; also called Paddy's hurricane

Jet Stream high-altitude wind

Kaikias (Greek—Northwest Wind)

Karaburan black-dust blizzard of the Gobi Desert

Khamsin Egyptian name for the Sirocco

Lake breeze wind blowing inland from a lake

Land breeze wind blowing seaward from the land

Leste Sirocco in Madeira and nearby North African coastal region

Levanter easterly wind characteristic of southern Spain and Straits of Gibraltar

Leveche hot dry wind found in

southeastern Spain where it comes from North Africa

Libeccio (Italian—Southwest Wind)—Genoese wind blowing inland from the Mediterranean

Lips (Greek—Southwest Wind)

Maestral cold north wind afflicting Genoa and Gulf of Genoa

Maestro northwesterly wind of central Mediterranean area about Italy and Yugoslavia

Mausim (Arabic—Season)—the monsoon, a seasonal wind, is derived from *mausim*

Medina land breeze felt at port of Cadiz in southwestern Spain

Meltemi (Turkish—Etesian Wind)

Mistral (Latin—masterful; masterly)—the Master Wind—cold dry northerly wind blowing down Rhone Valley into Gulf of Lyons—cold north wind characteristic of Marseilles, southern France, and the Rhone Valley

Monsoon Asiatic wind blowing from northeast in winter and southwest in summer

Nevados cold Andean winds found in Ecuador

Nor'easter storm blowing from the northeast characterized by high winds and rain of three days' duration

Norte cold north wind often experienced in Central America and Mexico

Norther cold north wind characteristic of Texas; hot dry foehn-type wind of California

Nor'wester storm blowing from the northwest

Notus (Greek—South Wind)—the Sirocco

Oberwind katabatic wind of the Salzkammergut in Austria

Ora late morning to early afternoon wind blowing over Lake Garda in northern Italy—direction is south to north (*see* Sover or Vento)

Ox's Eye West African sailor's name for the hurricane of the Guinea Coast

Paddy's Hurricane (*see* Irish Hurricane)

Pampero cold south wind blowing offshore in South Atlantic and over adjacent coastal pampas or plains of Argentina and Uruguay—often carries much dust and rain

Papagayo cold north wind of-

ten causing crop damage in Costa Rica

Phyrhenerwind foehn-type wind occurring in the Austrian and Bavarian Alps

Ponente west wind from the western Mediterranean; sea breeze refreshing west coast of Italy and sometimes penetrating as far inland as Rome

Prester waterspout or whirlwind encountered off the Greek Isles

Purga the Siberian blizzard—extremely cold northerly wind filled with cutting needlelike ice crystals and fine dry snow

Quara Bulgarian west wind; also called Karajol

Roaring Forties roaring storms sweeping from west to east around the southern hemisphere in the south forty latitudes

Samiel hot, devilish, and dusty wind of northeast Africa

Samun (Egyptian—Sirocco)

Santa Ana foehn-type hot dry wind of California usually blowing in late spring, summer, and early fall; named for Mexican general who once charged from the north and seemed to take the path of this wind from north to south

Schneefresser (German-Snoweater)—foehn wind warming lower mountainsides and valleys of Switzerland where it melts the snow drifts almost as rapidly as it contacts them

Sea breeze wind blowing inland from the sea

Seistan 120-day wind of Iran in eastern province of Seistan

Shamal northerly wind, like the Seistan, but found in Iraq over the Tigris-Euphrates plains country

Shrieking Sixties shrieking winds coming from the easterly and southerly sections of Antarctica and prevailing in the south sixty latitudes

Simoon name given the Sirocco when it is dirtier and hotter than usual as at this time natives believe it is a poisonous wind; a dry hot wind felt on deserts of Africa and Arabia during spring and summer

Sirocco south wind blowing from Sahara across North Africa, Mediterranean, and southern Europe; in North Africa is dry, dusty, and hot but

after crossing Mediterranean arrives in Europe moist and warm

Skiron (Greek—Northwest Wind)

Snoweater foehn-type Chinook wind blowing down eastern slopes of the Rockies in Canada and the Rocky Mountain states

Solano easterly rainy wind of southern Spain and Straits of Gibraltar—the Levanter

Sou'easter rain-filled southeast wind

South Wind along the Mediterranean this is the Sirocco

Sou'wester rain-filled southwest wind; oilskin hats, coats, and pants are also called sou'westers as they offer protection from rainy winds

Sover or Vento late afternoon winds blowing over Lake Garda in northern Italy—direction is north to south (see Ora)

Squall violent wind of short duration

Suchowej desert wind of the steppes of southern Russia

Sudestadas southeasterly pampero-type gales along coasts of Argentina, Uruguay, and Brazil

Sumatra squall characteristic of Malacca Strait where it occurs during the southwest monsoon season

Taino Haitian hurricane

Tebbad sand-laden hot wind of Turkestan

Tehuantepecer cold north wind often blowing with hurricane force around the Gulf of Tehuantepec and the peninsula of Yucatan

Terral land breeze felt in Valparaiso, Chile

Tornado violent storm best known for its twisting vertical wind responsible for causing great damage with little warning

Tower of Winds octagonal Greek structure near the Acropolis in Athens; each of its eight sides is decorated with a carved-marble allegorical figure representing the principal winds: North, Boreas; Northeast, Kaikias; East, Apheliotes; Southeast, Euros; South, Notos; Southwest, Lips; West, Zephyros; Northwest, Skiron

Trades Trade Winds (Northeast Trades in northern hemisphere blow from northeastern subtropics to the equator; Southeast Trades in southern hemisphere blow from southeastern subtropics to the equator)

Trade Winds northeast in northern hemisphere and southeast in southern hemisphere; the Northeast Trades cool the West Indies and much of the Spanish Main

Tramontana Lake Maggiore's morning wind blowing from the south and followed by the afternoon wind blowing from the north and called the Inverna

Tronada (Spanish—thunderstorm)

Tropical cyclone a hurricane

Twister a vertical spiralling cyclonic wind often called a tornado

Typhoon the hurricane of the western Pacific

Uala-andhi dusty Bay of Bengal squall ushering in the southwest monsoon season (April through June)

Uracano (Spanish American—Hurricane)—originally *huracán*

Vendavales southwesterly winds blowing around eastern Spain and Straits of Gibraltar

Virazon sea breeze cooling Cadiz on southwestern Spanish coast; afternoon sea breeze often reaching gale force at Valparaiso on central coast of Chile; sea breeze felt along coast of Chile and Peru

Westerlies westerly winds

Willie-Willie Indian Ocean hurricane

Williwaw violent squall afflicting mariners attempting passage through the Straits of Magellan

Willyway violent squall characteristic in Straits of Magellan

Willy-Willy Australian cyclone

Wind of One-Hundred-and-Twenty Days (see Bad-i-sado-bist roz)

Xaloch (Catalan—Sirocco)

Xaloque (Spanish—Sirocco)

Yalca Peruvian snowstorm occurring in northern Andean mountain passes

Yellow Wind cold dry wind of eastern Asia depositing loess dust over much of China

Youg hot summer wind of the Mediterranean

Zephyrus (Greek—West Wind)— the balmy Zephyr

Zobaa Egyptian dust whirl or whirlwind

Zonda westerly foehn wind characteristic of Argentina and southern Chile where it descends the eastern slopes of the Andes; enervating hot winds felt in Argentina and Uruguay where the zonda often precedes a cold pampero storm

Zip-Coded Automatic Data-Processing Abbreviations

The abbreviations listed here may be used in addresses on mail. By using the city-state abbreviations, it is possible to enter city, state, and five-digit ZIP Code on the last line of address within a maximum of 28 positions: 13 positions for city, 1 space between city and state abbreviation, 2 positions for state, 2 spaces between state and ZIP Code, and 10 positions for ZIP + 4 code.

Two-Letter State and Possession Abbreviations

Alabama	AL	Kansas	KS	Northern Mariana Islands	MP
Alaska	AK	Kentucky	KY	Ohio	OH
American Samoa	AS	Louisiana	LA	Oklahoma	OK
Arizona	AZ	Maine	ME	Oregon	OR
Arkansas	AR	Marshall Islands	MH	Palau	PW
California	CA	Maryland	MD	Pennsylvania	PA
Colorado	CO	Massachusetts	MA	Puerto Rico	PR
Connecticut	CT	Michigan	MI	Rhode Island	RI
Delaware	DE	Minnesota	MN	South Carolina	SC
District of Columbia	DC	Mississippi	MS	South Dakota	SD
Federated States of	FM	Missouri	MO	Tennessee	TN
Micronesia		Montana	MT	Texas	TX
Florida	FL	Nebraska	NE	Utah	UT
Georgia	GA	Nevada	NV	Vermont	VT
Guam	GU	New Hampshire	NH	Virginia	VA
Hawaii	HI	New Jersey	NJ	Virgin Islands	VI
Idaho	ID	New Mexico	NM	Washington	WA
Illinois	IL	New York	NY	West Virginia	WV
Indiana	IN	North Carolina	NC	Wisconsin	WI
Iowa	IA	North Dakota	ND	Wyoming	WY

Geographic Directional Abbreviations

North	N	West	W	Southwest	SW
East	E	Northeast	NE	Northwest	NW
South	S	Southeast	SE		

Abbreviations for Street Designators (Street Suffixes)

Word	Abbreviation	Word	Abbreviation	Word	Abbreviation
Alley	ALY	Fork	FRK	Pines	PNES
Annex	ANX	Forks	FRKS	Place	PL
Arcade	ARC	Fort	FT	Plain	PLN
Avenue	AVE	Freeway	FWY	Plains	PLNS
Bayou	BYU	Gardens	GDNS	Plaza	PLZ
Beach	BCH	Gateway	GRWY	Point	PT
Bend	BND	Glen	GLN	Port	PRT
Bluff	BLF	Green	GRN	Prairie	PR
Bottom	BTM	Grove	GRV	Radial	RADL
Boulevard	BLVD	Harbor	HBR	Ranch	RNCH
Branch	BR	Haven	HVN	Rapids	RPDS
Bridge	BRG	Heights	HTS	Rest	RST
Brook	BRK	Highway	HWY	Ridge	RDG
Burg	BG	Hill	HL	River	RIV
Bypass	BYP	Hills	HLS	Road	RD
Camp	CP	Hollow	HOLW	Row	ROW
Canyon	CYN	Inlet	INLT	Run	RUN
Cape	CPE	Island	IS	Shoal	SHL
Causeway	CSWY	Islands	ISS	Shoals	SHLS
Center	CTR	Isle	ISLE	Shore	SHR
Circle	CIR	Junction	JCT	Shores	SHRS
Cliffs	CLFS	Key	KY	Spring	SPG
Club	CLB	Knolls	KNLS	Springs	SPGS
Corner	COR	Lake	LK	Spur	SPUR
Corners	CORS	Lakes	LKS	Square	SQ
Course	CRSE	Landing	LNDG	Station	STA
Court	CT	Lane	LN	Stravenue	STRA
Courts	CTS	Light	LGT	Stream	STRM
Cove	CV	Loaf	LF	Street	ST
Creek	CRK	Locks	LCKS	Summit	SMT
Crescent	CRES	Lodge	LDG	Terrace	TER
Crossing	XING	Loop	LOOP	Trace	TRCE
Dale	DL	Mall	MALL	Track	TRAK
Dam	DM	Manor	MNR	Trail	TRL
Divide	DV	Meadows	MDWS	Trailer	TRLR
Drive	DR	Mill	ML	Tunnel	TUNL
Estates	EST	Mills	MLS	Turnpike	TPKE
Expressway	EXPY	Mission	MSN	Union	UN
Extension	EXT	Mount	MT	Valley	VLY
Fall	FALL	Mountain	MTN	Viaduct	VIA
Falls	FLS	Neck	NCK	View	VW
Ferry	FRY	Orchard	ORCH	Village	VLG
Field	FLD	Oval	OVAL	Ville	VL
Fields	FLDS	Park	PARK	Vista	VIS
Flats	FLT	Parkway	PKY	Walk	WALK
Ford	FRD	Pass	PASS	Way	WAY
Forest	FRST	Path	PATH	Wells	WLS
Forge	FRG	Pike	PIKE		

Extended Suffix Table

The table which follows lists some suffix forms which may appear in address files. The corresponding official USPS suffix (as coded in the ZIP + 4 National Directory File) is shown in the adjacent column.

Street Suffix Word or Suffix Abbreviation	Official USPS Street Suffix Abbreviation	Street Suffix Word or Suffix Abbreviation	Official USPS Street Suffix Abbreviation	Street Suffix Word or Suffix Abbreviation	Official USPS Street Suffix Abbreviation
ALLEE	ALY	CANYON	CYN	DALE	DL
ALLEY	ALY	CAPE	CPE	DAM	DM
ALLY	ALY	CAUSEWAY	CSWY	DIV	DV
ALY	ALY	CAUSWAY	CSWY	DIVIDE	DV
ANEX	ANX	CEN	CTR	DL	DL
ANNEX	ANX	CENT	CTR	DM	DM
ANNX	ANX	CENTER	CTR	DR	DR
ANX	ANX	CENTERS	CTR	DRIV	DR
ARC	ARC	CENTR	CTR	DRIVE	DR
ARCADE	ARC	CIR	CIR	DRIVES	DR
AV	AVE	CIRC	CIR	DRV	DR
AVE	AVE	CIRCL	CIR	DV	DV
AVEN	AVE	CIRCLE	CIR	DVD	DV
AVENU	AVE	CIRCLES	CIR	EST	EST
AVENUE	AVE	CLB	CLB	ESTATE	EST
AVN	AVE	CLF	CLFS	ESTATES	EST
AVNUE	AVE	CLFS	CLFS	ESTS	EST
BAYOO	BYU	CLIFF	CLFS	EXP	EXPY
BAYOU	BYU	CLIFFS	CLFS	EXPR	EXPY
BCH	BCH	CLUB	CLB	EXPRESS	EXPY
BEACH	BCH	CMP	CP	EXPW	EXPY
BEND	BND	CNTER	CTR	EXPY	EXPY
BG	BG	CNTR	CTR	EXT	EXT
BLF	BLF	CNYN	CYN	EXTENSION	EXT
BLUF	BLF	COR	COR	EXTN	EXT
BLUFF	BLF	CORNER	COR	EXTNSN	EXT
BLUFFS	BLF	CORNERS	CORS	EXTS	EXT
BLVD	BLVD	CORS	CORS	FALLS	FLS
BND	BND	COURSE	CRSE	FERRY	FRY
BOT	BTM	COURT	CT	FIELD	FLD
BOTTM	BTM	COURTS	CTS	FIELDS	FLDS
BOTTOM	BTM	COVE	CV	FL	FL
BOUL	BLVD	COVES	CV	FLAT	FLT
BOULEVARD	BLVD	CP	CP	FLATS	FLT
BOULV	BLVD	CPE	CPE	FLD	FLD
BR	BR	CRCL	CIR	FLDS	FLDS
BRANCH	BR	CRCLE	CIR	FLS	FLS
BRDGE	BRG	CRECENT	CRES	FLT	FLT
BRG	BRG	CREEK	CRK	FLTS	FLT
BRIDGE	BRG	CRES	CRES	FORD	FRD
BRK	BRK	CRESCENT	CRES	FORDS	FRD
BRNCH	BR	CRESENT	CRES	FOREST	FRST
BROOK	BRK	CRK	CRK	FORESTS	FRST
BROOKS	BRK	CROSSING	XING	FORG	FRG
BTM	BTM	CRSCNT	CRES	FORGE	FRG
BURG	BG	CRSE	CRSE	FORGES	FRG
BURGS	BG	CRSENT	CRES	FORK	FRK
BYP	BYP	CRSNT	CRES	FORKS	FRKS
BYPA	BYP	CRSSNG	XING	FORT	FT
BYPAS	BYP	CSWY	CSWY	FRD	FRD
BYPASS	BYP	CT	CT	FREEWAY	FWY
BYPS	BYP	CTR	CTR	FREEWY	FWY
BYU	BYU	CTS	CTS	FRG	FRG
CAMP	CP	CV	CV	FRK	FRK
CANYN	CYN	CYN	CYN	FRKS	FRKS

Street Suffix Word or Suffix Abbreviation	Official USPS Street Suffix Abbreviation	Street Suffix Word or Suffix Abbreviation	Official USPS Street Suffix Abbreviation	Street Suffix Word or Suffix Abbreviation	Official USPS Street Suffix Abbreviation
FRRY	FRY	ISLES	ISLE	MNTN	MTN
FRST	FRST	ISLND	IS	MNTNS	MTN
FRT	FT	ISLNDS	ISS	MOUNT	MT
FRWAY	FWY	ISS	ISS	MOUNTAIN	MTN
FRWY	FWY	JCT	JCT	MOUNTIN	MTN
FRY	FRY	JCTION	JCT	MSN	MSN
FT	FT	JCTN	JCT	MSSN	MSN
FWY	FWY	JCTNS	JCT	MT	MT
GARDEN	GDNS	JCTS	JCT	MTIN	MTN
GARDENS	GDNS	JUNCTION	JCT	MTN	MTN
GARDN	GDNS	JUNCTN	JCT	NCK	NCK
GATEWAY	GTWY	JUNCTON	JCT	NECK	NCK
GATEWY	GTWY	KEY	KY	ORCH	ORCH
GATWAY	GTWY	KEYS	KY	ORCHARD	ORCH
GDN	GDNS	KNL	KNLS	ORCHRD	ORCH
GDNS	GDNS	KNLS	KNLS	OVAL	OVAL
GLEN	GLN	KNOL	KNLS	OVL	OVAL
GLENS	GLN	KNOLL	KNLS	PARK	PARK
GLN	GLN	KNOLLS	KNLS	PARKS	PARK
GRDEN	GDNS	KY	KY	PARKWAY	PKY
GRDN	GDNS	KYS	KY	PARKWY	PKY
GRDNS	GDNS	LAKE	LK	PASS	PASS
GREEN	GRN	LAKES	LKS	PATH	PATH
GREENS	GRN	LANDING	LNDG	PATHS	PATH
GRN	GRN	LANE	LN	PIKE	PIKE
GROV	GRV	LANES	LN	PIKES	PIKE
GROVE	GRV	LCK	LCKS	PINE	PNES
GROVES	GRV	LCKS	LCKS	PINES	PNES
GRV	GRV	LDG	LDG	PKWAY	PKY
GTWAY	GTWY	LDGE	LDG	PKWY	PKY
GTWY	GTWY	LF	LF	PKWYS	PKY
HARB	HBR	LGT	LGT	PKY	PKY
HARBOR	HBR	LIGHT	LGT	PL	PL
HARBORS	HBR	LIGHTS	LGT	PLACE	PL
HARBR	HBR	LK	LK	PLAIN	PLN
HAVEN	HVN	LKS	LKS	PLAINES	PLNS
HAVN	HVN	LN	LN	PLAZA	PLZ
HBR	HBR	LNDG	LNDG	PLN	PLN
HEIGHT	HTS	LNDNG	LNDG	PLNS	PLNS
HEIGHTS	HTS	LOAF	LF	PLZ	PLZ
HIGHWAY	HWY	LOCK	LCKS	PLZA	PLZ
HIGHWY	HWY	LOCKS	LCKS	PNES	PNES
HILL	HL	LODG	LDG	POINT	PT
HILLS	HLS	LODGE	LDG	POINTS	PT
HIWAY	HWY	LOOP	LOOP	PORT	PRT
HIWY	HWY	LOOPS	LOOP	PORTS	PRT
HL	HL	MALL	MALL	PR	PR
HLLW	HOLW	MANOR	MNR	PRAIRIE	PR
HLS	HLS	MANORS	MNR	PRK	PARK
HOLLOW	HOLW	MDW	MDWS	PRR	PR
HOLW	HOLW	MDWS	MDWS	PRT	PRT
HOLWS	HOLW	MEADOW	MDWS	PRTS	PRT
HRBOR	HBR	MEADOWS	MDWS	PT	PT
HT	HTS	MEDOWS	MDWS	PTS	PT
HTS	HTS	MILL	ML	RAD	RADL
HVN	HVN	MILLS	MLS	RADIAL	RADL
HWAY	HWY	MISSION	MSN	RADIEL	RADL
HWY	HWY	MISSN	MSN	RADL	RADL
INLET	INLT	ML	ML	RANCH	RNCH
INLT	INLT	MLS	MLS	RANCHES	RNCH
IS	IS	MNR	MNR	RAPID	RPDS
ISLAND	IS	MNRS	MNR	RAPIDS	RPDS
ISLANDS	ISS	MNT	MT	RD	RD
ISLE	ISLE	MNTAIN	MTN	RDG	RDG

Street Suffix Word or Suffix Abbreviation	Official USPS Street Suffix Abbreviation	Street Suffix Word or Suffix Abbreviation	Official USPS Street Suffix Abbreviation	Street Suffix Word or Suffix Abbreviation	Official USPS Street Suffix Abbreviation
RDGE	RDG	ST	ST	TUNLS	TUNL
RDGS	RDG	STA	STA	TUNNEL	TUNL
RDS	RD	STATION	STA	TUNNL	TUNL
REST	RST	STATN	STA	TURNPIKE	TPKE
RIDGE	RDG	STN	STA	TURNPK	TPKE
RIDGES	RDG	STR	ST	UN	UN
RIV	RIV	STRA	STRA	UNION	UN
RIVER	RIV	STRAV	STRA	UNIONS	UN
RIVR	RIV	STRAVE	STRA	VALLEY	VLY
RNCH	RNCH	STRAVEN	STRA	VALLEYS	VLY
RNCHS	RNCH	STRAVENUE	STRA	VALLY	VLY
ROAD	RD	STRAVN	STRA	VDCT	VIA
ROADS	RD	STREAM	STRM	VIA	VIA
ROW	ROW	STREET	ST	VIADCT	VIA
RPD	RPDS	STREETS	ST	VIADUCT	VIA
RPDS	RPDS	STREME	STRM	VIEW	VW
RST	RST	STRM	STRM	VIEWS	VW
RUN	RUN	STRT	ST	VILL	VLG
RVR	RIV	STRVN	STRA	VILLAG	VLG
SHL	SHL	STRVNUE	STRA	VILLAGE	VLG
SHLS	SHLS	SUMIT	SMT	VILLE	VL
SHOAL	SHL	SUMITT	SMT	VILLG	VLG
SHOALS	SHLS	SUMMIT	SMT	VILLIAGE	VLG
SHOAR	SHR	TER	TER	VIS	VIS
SHOARS	SHRS	TERR	TER	VIST	VIS
SHORE	SHR	TERRACE	TER	VISTA	VIS
SHORES	SHRS	TPK	TPKE	VL	VL
SHR	SHR	TPKE	TPKE	VLG	VLG
SHRS	SHRS	TRACE	TRCE	VLGS	VLG
SMT	SMT	TRACES	TRCE	VLLY	VLY
SPG	SPG	TRACK	TRAK	VLY	VLY
SPGS	SPGS	TRACKS	TRAK	VLYS	VLY
SPNG	SPG	TRAIL	TRL	VST	VIS
SPNGS	SPGS	TRAILER	TRLR	VSTA	VIS
SPRING	SPG	TRAILS	TRL	VW	VW
SPRINGS	SPGS	TRAK	TRAK	VWS	VW
SPRNG	SPG	TRCE	TRCE	WALK	WALK
SPRNGS	SPGS	TRK	TRAK	WALKS	WALK
SPUR	SPUR	TRKS	TRAK	WAY	WAY
SPURS	SPUR	TRL	TRL	WAYS	WAY
SQ	SQ	TRLR	TRLR	WELL	WLS
SQR	SQ	TRLRS	TRLR	WELLS	WLS
SQRE	SQ	TRLS	TRL	WLS	WLS
SQU	SQ	TRNPK	TPKE	WY	WAY
SQUARE	SQ	TUNEL	TUNL	XING	XING
SQUARES	SQ	TUNL	TUNL		

Zodiacal Signs

♈ : Aries (The Ram), first sign of the zodiac, symbolized by the ram's horns; the sun enters this period on March 21, marking the spring or vernal equinox

♉ : Taurus (The Bull), second sign of the zodiac, symbolized by the bull's head and horns; sun enters this pe-riod April 20

♊ : Gemini (The Twins), third sign of the zodiac, symbolized by wooden statues of Castor and Pollux coupled by horizontal lintels; sun enters this period May 21

♋ : Cancer (The Crab), fourth sign of the zodiac, symbolized by overlapping crab claws; sun enters this period June 22, marking the summer solstice, the longest day of the year

♌ : Leo (The Lion), fifth sign of the zodiac, symbolized by stylized figure representing the lion's tufted tail; sun enters this period on July 23

♍ : Virgo (The Virgin), sixth sign of the zodiac; symbol taken from *par* in *parthenos*, Greek for virgin; sun enters Virgo on August 23

♎ : Libra (The Balance), seventh sign of the zodiac, symbolized by a stylized balance; sun enters this period on September 23, marking the autumnal equinox

♏ : Scorpio (The Scorpion), eighth sign of the zodiac, symbolized by stylized representation of legs and stinger tail of the scorpion; sun enters this period on October 24

♐ : Sagittarius (The Archer), ninth sign of the zodiac; symbolized by archer's bow and arrow; sun enters this period on November 22

♑ : Capricornus (The Goat), tenth sign of the zodiac; symbol taken from *tr* of *tragos*, Greek for goat; sun enters Capricorn on December 22, marking the winter solstice, the shortest day in the year

♒ : Aquarius (The Water Carrier), eleventh sign of the zodiac, symbolized by two parallel water waves; sun enters this period on January 20

♓ : Pisces (The Fishes), twelfth sign of the zodiac; symbolized by two fishes tied by a thong; sun enters this period on February 19